TITLE IX IS LA[...]**BITS SEX DISCRIMI**[...]**GRAMS.** Learn how to use the [...] the average expenditures for male and female athletes in Division I-A schools given information on how much was spent on them together and how much more funding the males received. (See Example 2 from Section 9.2 on page 757.)

ON THE SHORELINE SURROUNDING CERTAIN TYPES OF RECREATIONAL LAKES IN MINNESOTA, THE DEPARTMENT OF NATURAL RESOURCES RESTRICTS THE SIZE OF BUILDINGS. Learn how to use quadratic equations and problem solving to determine how big one can build a structure along the waterfront when area dimensions are imposed. (See Section 3.2 on page 189.)

LEARN HOW FUNCTIONS AND MATRICES CAN EXPLAIN COMPUTER GRAPHICS AND DIGITAL ANIMATION. (See Section 9.6 on page 811.)

3rd edition

PRECALCULUS

with Modeling and Visualization

Gary K. Rockswold

Minnesota State University, Mankato

PEARSON

Addison
Wesley

Boston   San Francisco   New York
London   Toronto   Sydney   Tokyo   Singapore   Madrid
Mexico City   Munich   Paris   Cape Town   Hong Kong   Montreal

Publisher: Greg Tobin
Senior Acquisitions Editor: Anne Kelly
Editorial Assistants: Vanessa Hayes and Toni Moreno
Executive Project Manager: Christine O'Brien
Managing Editor: Karen Wernholm
Senior Production Supervisor: Kathleen A. Manley
Senior Designer: Barbara T. Atkinson
Cover Design: Dardani Gasc Design
Production Coordination and Text Design: Nesbitt Graphics, Inc.
Senior Technical Art Specialist: Joe Vetere
Senior Marketing Manager: Becky Anderson
Marketing Assistant: Maureen McLaughlin
Associate Media Producer: Ceci Fleming
Software Development: Mary Durnwald and Malcolm Litowitz
Senior Manufacturing Buyer: Evelyn Beaton
Rights and Permissions Advisor: Dana Weightman
Compositor: Nesbitt Graphics, Inc.
Illustrations: Techsetters, Inc.
Cover Photo: Allee of Light and Shadow: Sandra Ivany/Picture Arts/Botanica

About the cover: This photo of trees was chosen as a representative application for this course. The height of a tree can be determined using similar triangles if a person who is standing by knows his/her own height and can measure the length of his/her shadow as well as the length of the tree's shadow. This proportion results in a linear equation.

Photo Credits: **p. 1, 2, 46, 69, 140, 206, 363, 475, 504, 512, 594, 616, 769, 941** PhotoDisc; **p. 29, 929** Brand X Pictures; **p. 72** AP Photo/M. Spencer Green; **p. 73** Ramon Santos/Photo Research; **p. 87, 664** Brand X Pictures (RF); **p. 80, 106, 123, 242, 274, 444, 490, 602, 691, 738, 780** Digital Vision; **p. 114, 474, 581, 935** PhotoDisc Blue; **p. 120** NASA GPS; **p. 171** AP Photo/Ron Frehm; **p. 172, 257, 364, 382, 418, 457, 518, 538, 550, 936** Corbis RF; **p. 189** our Brand X Pictures CD; **p. 216, 446** NOAA; **p. 241** Dex Images (RF); **p. 628, 658** ThinkStock; **p. 293** our PhotoDisc V17; **p. 325** Brand X PP; **p. 334, 436, 610** NASA; **p. 399, 653, 845** PhotoDisc (PP); **p. 419, 628, 714, 739** PhotoDisc Red; **p. 433, 582** Digital Vision (PP) **p. 433** USGS; **p. 455, 456, 754,** our PhotoDisc; **p. 772** Rosenfeld Images, Inc/Photos; **p. 797** Rubberball; **p. 811** Kobal Collection; **p. 825, 532** Comstock RF; **p. 753** Sara Anderson; **p. 538, 550, 757** Corbis RF (PP); **p. 840** our PhotoDisc CD2; **p. 844** NASA Hubble; **p. 855** Hank Morgan/Photo Research; **p. 887** AP Photo; **p. 902** our PhotoDisc Vol 24; **p. 920** our PhotoDisc Vol 72; **p. 652** our PhotoDisc BS14; **p. 676** our PhotoDisc Vol. 10; **p. 702** our PhotoDisc BS13

Library of Congress Cataloging-in-Publication Data
Rockswold, Gary K.
    Precalculus with modeling and visualization. — 3rd ed. / Gary K. Rockswold.
        p. cm.
    Includes bibliographical references and index.
    ISBN 0-321-27907-7
        1. Functions. 2. Graph theory. I. Title.
QA331.3.R63 2006
512'.13—dc22   2004054922

1 2 3 4 5 6 7 8 9 10—VH—08070605

To Gaylan

Contents

* Sections marked Optional require the use of a graphing calculator.

v

5   Exponential and Logarithmic Functions   363

6   Trigonometric Functions   474

7   Trigonometric Identities and Equations   581

8   Further Topics in Trigonometry   652

9   Systems of Equations and Inequalities   738

Reference: Basic Concepts from Algebra and Geometry   R-1

Preface

This textbook offers an innovative approach that consistently links mathematical concepts to real-world applications by moving from the concrete to the abstract. It demonstrates the relevance of mathematics and shows why math matters. This text provides a comprehensive curriculum with the balance and flexibility necessary for today's college mathematics courses. The early introduction of functions and graphs allows the instructor to use applications and visualization to present mathematical topics. Real data, graphs, and tables play an important role in the course, giving meaning to the numbers and equations that students encounter. This approach increases students' interest and motivation and their likelihood for success.

Approach

Instructors are free to strike their own balance of skills, rule of four, applications, modeling, and technology. With a flexible approach to the rule of four (verbal, graphical, numerical, and symbolic methods), instructors can easily emphasize one rule more than another to meet their students' needs. This approach also extends to modeling and applications. The use of technology, which helps students visualize mathematical concepts, is *optional* and not a requirement for students to benefit from this approach. Nevertheless, the text still provides a strong option for instructors who wish to implement graphing calculator technology. The text contains numerous applications, including models of real-world data with functions and problem-solving strategies. It is not necessary for an instructor to discuss any particular application; rather, an instructor has the option to choose from a wide variety of topics.

The concept of a function is the unifying theme of the text. It is common for students to examine a type of function and its graph. Then this knowledge is used to solve associated equations and inequalities. For example, students apply their knowledge of quadratic functions and parabolas to solve quadratic equations and inequalities. This visual approach complements the traditional symbolic approach to solving equations and inequalities and allows students to solve a problem by using more than one method. Functions and their graphs are frequently used to solve applications and model real-world data, and students are often asked to interpret their results. Mathematical skills also play an important role in this text. Numerous exercises have been included so that students can practice their skills.

When students arrive in a college-level mathematics class, they often lack a full mastery of intermediate algebra. Rather than reviewing all of this material in the first chapter in hopes that saying it "louder and faster" will help students remember it better, the necessary intermediate algebra skills are integrated seamlessly throughout the early chapters. In addition, review notes and geometry notes appearing in the margins refer students "just in time" to extra help found in the expanded Chapter R: Basic Concepts from Algebra and Geometry.

Content Changes to the Third Edition

The third edition contains several important changes, which are the result of the many comments and suggestions made by instructors, students, and reviewers. This text contains eleven chapters and an expanded review chapter at the end of the text. The breadth and depth of several topics have been increased. Some highlights of this revision include the following:

◆ Approximately 100 new examples and over 2200 new exercises have been added to meet the needs of students with a variety of abilities.

◆ Every exercise set has been revised to ensure that there are sufficient types of exercises for each mathematical concept and that there is a pairing of odd and even exercises. The exercise sets cover a diverse assortment of topics and are carefully graded with several levels of difficulty.

◆ The number of exercises in which students are asked to interpret their answers in the context of the application being solved has been increased.

◆ The real-world data in many applications have been updated.

◆ At the request of reviewers, new sections covering solving linear systems in three variables without matrices (Section 9.3) and mathematical induction (Section 11.5) have been added.

◆ The topics of partial fractions and rotation of axes have been included in Appendices C and D.

◆ Chapter 2 has been reduced from six sections to five, with the midpoint formula now presented in Section 1.2 with the distance formula.

◆ Chapter R: Basic Concepts from Algebra and Geometry has been expanded from six sections to eight. These sections contain review material from intermediate algebra that students can use "just in time." There is extended coverage of exponents, factoring, rational expressions, and radical expressions.

◆ The following topics have been expanded:
  Graphing by hand
  Calculating the average rate of change
  Finding the difference quotient
  Maximizing and minimizing quadratic functions
  Evaluating function notation
  Solving equations reducible to quadratic form
  Graphing rational functions with "holes"
  Solving nonlinear equations and inequalities
  Completing the square
  Writing functions and formulas to solve applications
  Transforming graphs to model data

Features

◆ **Chapter and Section Introductions**
Many college algebra students have little or no understanding of mathematics. Chapter and section introductions motivate and explain some of the reasons for studying mathematics. (See pages 1, 171, 241, and 242.)

NEW! ◆ **Now Try**
This new feature occurs after each example and refers students to a similar exercise they can work to see if they understand the concept presented in the example. (See pages 5 and 76.)

◆ **Applications and Models**
Interesting, relevant applications are a major strength of this textbook, and as a result, students become more effective problem solvers and have a better understanding of how mathematics is used in the real world. Because the applications are intuitive and not overly technical, they can be introduced with a minimum amount of class time. Current data are used to create meaningful mathematical models. A unique feature of this text is that the applications and models are woven into both the discussions and the exercises. It is easier for students to learn how to solve applications if they are discussed within the text. (See pages 29, 181, 239–240, 364, and 378–382.) An Index of Applications is included at the end of the text.

◆ **Sources**
Since there are numerous applications throughout the text, genuine sources and a comprehensive bibliography are given. These sources reinforce for the student the practical applications of mathematics in real life. (See pages 120–121, and 417.)

◆ **Algebra and Geometry Review Notes**

Throughout the text, Algebra and Geometry Review Notes are located in the margins, which direct students "just in time" to Chapter R, where important topics in algebra and geometry are reviewed. Instructors can use this chapter for extra review or refer students to it as needed. This feature *frees* instructors from frequently reviewing material from intermediate algebra and geometry. (See pages 88 and 116.)

◆ **Calculator Help Notes**

The Calculator Help Notes located in the margins direct students "just in time" to Appendix B: Using the Graphing Calculator. This appendix shows students the necessary keystrokes to complete specific examples from the text. This feature *frees* instructors from teaching the specifics of the graphing calculator and gives students a convenient reference written specifically for this text. (See pages 7, 94, and 195.)

◆ **Class Discussion**

This feature is included in most sections and poses a question that can be used for either classroom discussion or homework. (See pages 4, 141, and 248.)

◆ **Making Connections**

This feature occurs throughout the text and helps students see how concepts covered previously are related to new concepts being presented. (See pages 32, 98, and 196.)

◆ **Putting It All Together**

This helpful feature at the end of each section summarizes techniques and reinforces the mathematical concepts presented in the section. This information is given in an easy-to-follow grid. (See pages 9–10, 342–343, and 411–412.)

◆ **Checking Basic Concepts**

This feature includes a small set of exercises provided after every two sections that can be used for review. These exercises require about 15 or 20 minutes to complete and can be used for collaborative learning exercises, if time permits. (See pages 106, 205–206, and 334.)

◆ **Exercise Sets**

The exercise sets are the heart of any mathematics text, and this text includes a large variety of instructive exercises. Each set of exercises involves skill building, mathematical concepts, and applications. Graphical interpretation and tables of data are often used to extend students' understanding of mathematical concepts. The exercise sets are graded carefully and categorized according to topic, which makes it easy for an instructor to select appropriate assignments. (See pages 118–123 and 201–205.)

◆ **Chapter Summaries**

Chapter summaries are presented in an easy-to-read grid. They allow students to quickly review key concepts from the chapter. (See pages 234–236 and 347–352.)

◆ **Chapter Review Exercises**

This exercise set contains both skill-building and applied exercises. These exercises stress different techniques for solving problems and provide students with the review necessary to pass a chapter test. (See pages 66–70 and 352–357.)

◆ **Extended and Discovery Exercises**

Extended and Discovery Exercises occur at the end of selected sections and at the end of every chapter. These exercises are usually more complex and challenging, and often require extension of a topic presented or discovery of a new topic. They can be used for either collaborative learning or extra homework assignments. (See pages 70–71, 239–240, and 256–257.)

NEW! ◆ **Cumulative Review Exercises**

These comprehensive exercise sets occur after every two or three chapters and give students an opportunity to review previously covered material. (See pages 167–170 and 358–362.)

Student Supplements

STUDENT'S SOLUTIONS MANUAL
- By Rockswold/Krieger/Schneider/Block/Purcell
- Provides complete solutions to all odd-numbered text exercises, excluding Extended and Discovery Exercises
  ISBN: 0-321-28074-1

VIDEOTAPE SERIES
- Features an engaging team of lecturers
- Provides comprehensive coverage of each section and topic (with the exception of Ch. R) in the text
  ISBN: 0-321-28075-X

DIGITAL VIDEO TUTOR
- Complete set of digitized videos for student use at home or on campus
- Ideal for distance learning or supplemental instruction
  ISBN: 0-321-28076-8

A REVIEW OF ALGEBRA
- By Heidi Howard, Florida Community College at Jacksonville
- Provides additional support for those students needing further algebra review
  ISBN: 0-201-77347-3

ADDISON-WESLEY MATH TUTOR CENTER
- Provides free tutoring through a registration number that can be packaged with a new textbook or purchased separately
- Staffed by qualified college mathematics instructors
- Accessible via toll-free telephone, toll-free fax, e-mail, and the Internet
- www.aw-bc.com/tutorcenter

Instructor Supplements

ANNOTATED INSTRUCTOR'S EDITION
- Special edition of the text
- Provides answers in the margins to many text exercises, plus a full answer section at the back
  ISBN: 0-321-28625-1

INSTRUCTOR'S SOLUTIONS MANUAL
- Rockswold/Krieger/Schneider/Block/Purcell
- Provides complete solutions to all text exercises
  ISBN: 0-321-28079-2

INSTRUCTOR'S TESTING MANUAL
- By Vincent McGarry, *Austin Community College*
- Provides prepared tests for each chapter of the text, with answers provided
  ISBN: 0-321-28078-4

TESTGEN® WITH QUIZMASTER®
- Enables instructors to build, edit, print, and administer tests
- Features a computerized bank of questions developed to cover all text objectives
- Available on a dual-platform Windows/Macintosh CD-ROM
  ISBN: 0-321-28077-6

NEW! ADJUNCT SUPPORT MANUAL
- Includes resources to help both new and adjunct faculty with course preparation and classroom management
- Provides helpful teaching tips
  ISBN: 0-321-32055-7

POWERPOINT LECTURE PRESENTATION
- Classroom presentation software correlated specifically to this textbook sequence
- Available within MyMathLab or on this book's catalog page within www.aw-bc.com

NEW! ADJUNCT SUPPORT CENTER
- Offers consultation on suggested syllabi, helpful tips on using the textbook support package, assistance with content, and advice on classroom strategies
- Available Sunday–Thursday evenings from 5 P.M. to midnight: telephone: 1-800-435-4084; e-mail: AdjunctSupport@aw.com; fax: 1-877-262-9774

MathXL®

MathXL is an online homework, tutorial, and assessment system that uses algorithmically generated exercises correlated to the objectives in your textbook. Instructors can assign tests and homework provided by Addison-Wesley or create and customize their own tests and homework assignments. Instructors can also track their students' results and tutorial work in an online gradebook. Students can take chapter tests and receive personalized study plans that diagnose weaknesses and link students to areas they need to study and retest. Students can also work unlimited practice exercises that provide tutorial instruction, and they can access animations and video clips directly from selected exercises. **MathXL** can be packaged with new copies of your textbook and is available to qualified adopters. Please contact your local sales representative for details, or visit our Web site at www.mathxl.com.

MyMathLab

MyMathLab is a series of text-specific, easily customizable online courses for Addison-Wesley textbooks in mathematics and statistics. **MyMathLab** is powered by CourseCompass™— Pearson Education's online teaching and learning environment—and by **MathXL**®—our online homework, tutorial, and assessment system. **MyMathLab** gives you the tools you need to deliver all or a portion of your course online, whether your students are in a lab setting or working from home. **MyMathLab** provides a rich and flexible set of course materials, featuring free-response exercises that are algorithmically generated for unlimited practice and mastery. Students can also use online tools, such as video lectures, animations, and a multimedia textbook, to independently improve their understanding and performance. Instructors can use **MyMathLab's** homework and test managers to select and assign online exercises correlated directly to the textbook, and they can import TestGen tests into **MyMathLab** for added flexibility. **MyMathLab's** online gradebook—designed specifically for mathematics and statistics—automatically tracks students' homework and test results and gives the instructor control over how to calculate final grades. MyMathLab is available to qualified adopters. Please contact your local sales representative for more information, or visit our Web site at www.mymathlab.com.

MathXL® Tutorials on CD ISBN 0-321-34823-0

This interactive tutorial CD-ROM provides algorithmically generated practice exercises that are correlated at the objective level to the exercises in the textbook. Every practice exercise is accompanied by an example and a guided solution designed to involve students in the solution process. Selected exercises may also include a video clip to help students visualize concepts. The software provides helpful feedback for incorrect answers and can generate printed summaries of students' progress.

InterAct Math® Tutorial Web site www.interactmath.com

Get practice and tutorial help online! This interactive tutorial Web site provides algorithmically generated practice exercises that correlate directly to the exercises in your textbook. You can retry an exercise as many times as you like with new values each time for unlimited practice and mastery. Every exercise is accompanied by an interactive guided solution that gives you helpful feedback if you enter an incorrect answer, and you can also view a worked-out sample problem that steps you through an exercise similar to the one you're working on.

Acknowledgments

Many individuals contributed to the development of this textbook. I would like to thank the following reviewers, whose comments and suggestions were invaluable in preparing this and previous editions of the text.

Randy Baker	*Fairmont State College*
Victoria Borlaug	*Walters State Community College*
Connie Buller	*Metropolitan Community College*
Bruce S. Burdick	*Roger Williams University*
Rose Cavin	*Chipola Junior College*
Sarah Cook	*Washburn University*
Amy Daniel	*University of New Orleans*
William Fox	*Francis Marion University*
Johanna G. Halsey	*Dutchess Community College*
Libby Higgins	*Greenville Technical College*
Gloria Hitchcock	*Georgia Perimeter College*
Mike Huff	*Austin Community College, Cypress Creek Campus*
Peter Jarvis	*Georgia College and State University*
Tuesday J. Johnson	*New Mexico State University*
Susan Jones	*Nashville State Technical Institute*
Cheryl Kane	*University of Nebraska*
Mike Keller	*St. Johns River Community College*
Dr. Tom Kelly	*Metropolitan State College of Denver*
Mary Kilbride	*Augustana College*
Marlene Kovaly	*Florida Community College, Kent Campus*
Michael Lloyd	*Henderson State University*
Judith Maggiore	*Holyoke Community College*
Mary Martin	*Middle Tennessee State University*
Nancy Mauldin	*College of Charleston*
Joe May	*North Hennepin Community College*
Canda Mueller-Engheta	*University of Kansas*
Rebecca Muller	*Southeastern Louisiana University*
Donna Norman	*Jefferson Community College*
Wing Park	*College of Lake County*
Faith Peters	*Miami Dade Community College*
Nancy Pevey	*Pellissippi State Community College*
David Platt	*Front Range Community College, Larimer*
Joan Raines	*Middle Tennessee State University*
Laura Ralston	*Floyd College at North Metro Tech*
Jolene Rhodes	*Valencia Community College*
Judith D. Smalling	*St. Petersburg Junior College*
Terra Stamps	*Prairie State College*
Christine Wise	*University of Louisiana at Lafayette*

A special thank you is due Terry Krieger for his valuable suggestions, comments, and work with the manuscript, answers, and solutions. I would also like to thank Elina Niemelä and Janis Cimperman at St. Cloud State University for their work with the answers and proofreading. Thank you to Namyong Lee at Minnesota State University, Mankato for his excellent work as an accuracy checker.

Without the excellent cooperation from the professional staff at Addison-Wesley, this project would have been impossible. They are, without a doubt, the best. Thanks go to Greg Tobin for his support of this project. Particular recognition is due Christine O'Brien and Anne Kelly, who gave advice, support, assistance, and encouragement. The outstanding contributions of Kathy Manley, Barbara Atkinson, Becky Anderson, Beth Anderson, Maureen McLaughlin, Toni Moreno, Vanessa Hayes, Ceci Fleming, and Joe Vetere are much appreciated. The work of Janet Nuciforo was instrumental to the success of this project.

Thanks go to Wendy Rockswold and Jessica Rockswold, who proofread the manuscript and gave invaluable encouragement throughout the project.

A special thank you goes to the many students and instructors who used the first and second editions. Their suggestions were insightful. Please feel free to contact me at *gary.rockswold@mnsu.edu* or Department of Mathematics, Minnesota State University, Mankato, MN 56001 with your comments. Your opinion is important.

Gary Rockswold

1

Introduction to Functions and Graphs

Think you can, think you can't; either way you'll be right.

Henry Ford

Perhaps you have injured your foot or ankle. If you have, you are not alone because more than 14.2 million people made visits to the doctor in 2001 as a result of foot, toe, and ankle problems. Some of the reasons for so many visits include the following:

- Walking puts up to 1.5 times your body weight on your foot.
- On average, your feet walk or run about 1000 miles per year.
- Your feet cushion up to 1 million pounds of force during 1 hour of strenuous exercise.

It is common for people to need crutches after they injure their feet or ankles, and it is important to have well-fitted crutches. Various techniques have been developed to determine a proper crutch size. One quick (but not always accurate) way is to choose a pair of crutches that reach 2 to 3 inches below a patient's armpits. Another method uses a *mathematical formula* based on a person's height.

In this chapter we discuss several applications of mathematics, including how to choose a correct crutch length. (See Example 7 in Section 1.3.) Mathematics is rapidly becoming more and more important to our society. It is vital that we know more mathematics so that we can better understand current data and technologies.

Source: American Academy of Orthopaedic Surgeons.

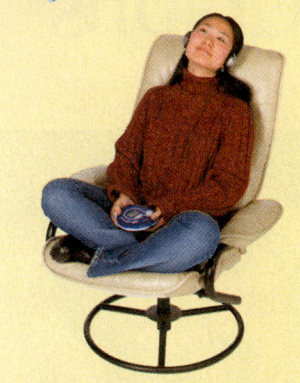

1.1   Numbers, Data, and Problem Solving

Introduction

Throughout history, humans have had a desire to communicate information and knowledge. Both languages and number systems have evolved to satisfy this desire. In order to build roads, design CD players, or forecast the weather, numbers and mathematics are necessary. As our society becomes more complex, the problems that it faces also become more complex. Mathematics plays an increasingly important role in solving these problems. Problem-solving strategies are essential for our society to progress. This section discusses sets of numbers and introduces some basic strategies for solving problems.

- ◆ Recognize common sets of numbers
- ◆ Learn scientific notation and use it in applications
- ◆ Apply problem-solving strategies

Sets of Numbers

One important set of numbers is the set of **natural numbers**. This set comprises the *counting numbers* $N = \{1, 2, 3, 4, \ldots\}$. Natural numbers can be used when data are not presented in fractional parts.

The **integers**

$$I = \{\ldots, -3, -2, -1, 0, 1, 2, 3, \ldots\}$$

are a set of numbers that contains the natural numbers, their additive inverses (negatives), and 0. Historically, negative numbers were not readily accepted. They did not appear to have real meaning. However, today when a person opens a personal checking account for the first time, negative numbers quickly take on meaning. There is a significant difference between a positive and a negative balance.

A **rational number** is any number that can be expressed as the ratio of two integers $\frac{p}{q}$, where $q \neq 0$. Rational numbers include the integers. Some examples of rational numbers are

$$\frac{2}{1}, \frac{1}{3}, -\frac{1}{4}, \frac{-50}{2}, \frac{22}{7}, 0, \sqrt{25}, \text{ and } 1.2.$$

Note that 0 and 1.2 are both rational numbers. They can be represented by the fractions $\frac{0}{1}$ and $\frac{12}{10}$. Because two fractions that look different can be equivalent, rational numbers have more than one form. A rational number can always be expressed in a decimal form that either *repeats* or *terminates*. For example, $\frac{2}{3} = 0.\overline{6}$, a repeating decimal, and $\frac{1}{4} = 0.25$, a terminating decimal. The overbar indicates that $0.\overline{6} = 0.6666666\ldots$.

Real numbers can be represented by decimal numbers. Since every fraction has an equivalent decimal form, real numbers include rational numbers. However, some real numbers cannot be expressed by fractions. These numbers are called **irrational numbers**. The numbers $\sqrt{2}$, $\sqrt{15}$, and π are examples of irrational numbers. They can be represented by nonrepeating, nonterminating decimals. Real numbers include both rational and irrational numbers, and can be *approximated* by a terminating decimal. Examples of real numbers include

$$2, -10, -131.3337, \frac{1}{3} = 0.\overline{3}, -\sqrt{5} \approx -2.2361, \text{ and } \sqrt{11} \approx 3.3166.$$

◆ **CLASS DISCUSSION**

The number 0 was invented well after the natural numbers. Many societies did not have a zero—for example, there is no Roman numeral for 0. Discuss some possible reasons for this. ◆

Note: The symbol "$\approx$" means **approximately equal**. This symbol will be used in place of an equals sign whenever two unequal quantities are close in value. For example, $\frac{1}{4} = 0.25$, whereas $\frac{1}{3} \approx 0.3333$.

EXAMPLE  1   Classifying numbers

Classify each real number as one or more of the following: natural number, integer, rational number, or irrational number.

$$5, -1.2, \frac{13}{7}, -\sqrt{7}, -12, \sqrt{16}$$

SOLUTION

5: Natural number, integer, and rational number

-1.2: Rational number

$\frac{13}{7}$: Rational number

$-\sqrt{7}$: Irrational number

-12: Integer and rational number

$\sqrt{16} = 4$: Natural number, integer, and rational number   *Now Try Exercise 9* ◆

Even though a data set includes only integers, decimals can be used to describe it. One way to analyze data is to find the **average**, or **mean**. To calculate the average (or mean) of a set of n numbers, we add the n numbers and then divide their sum by n. The average of a set of integers need not be an integer—sometimes it is a fraction or decimal.

EXAMPLE 2   Finding the average number of endangered animals

Table 1.1 lists endangered animals in the United States for selected years. Calculate the average number of endangered animals over this 4-year period. Is this average an integer? Is it a rational number?

TABLE 1.1 U.S. Endangered Animals

Year	2000	2001	2002	2003
Number	368	386	387	388

Source: U.S. Fish and Wildlife Services.

SOLUTION To find this average, add the number of endangered animals for each year and then divide by 4.

$$\frac{368 + 386 + 387 + 388}{4} = \frac{1529}{4} = 382.25$$

This average of four numbers is not an integer, but it is a rational number.

Now Try Exercise 91 ◆

Real numbers have applications in many areas such as interest rates, grade point averages, and the Consumer Price Index (CPI). The CPI, often referred to as the "cost of living index," is the numerical scale most commonly used to measure inflation. It tracks the prices of basic consumer goods. More than 40 economists calculate the CPI annually, at a cost of $26 million. The time from 1982 to 1984 is called the *reference period*, where the CPI is defined to be 100.0. If the CPI changes from c_1 to c_2, then the **percent change** equals $\frac{c_2 - c_1}{c_1} \times 100$.

 **Interpreting the Consumer Price Index**

Table 1.2 lists the CPI for selected years from 1960 through 2000. A CPI of 29.6 indicates that a typical sample of consumer goods costing $100 from 1982 to 1984 would have cost $29.60 in 1960.

TABLE 1.2   Consumer Price Index

Year	1960	1965	1970	1975	1980	1985	1990	1995	2000
CPI	29.6	31.5	38.8	53.8	82.4	107.6	130.7	153.5	172.2

Source: Bureau of the Census.

(a) Discuss the change in the CPI over this 40-year period.
(b) Approximate the percent change in prices from 1970 to 1980.
(c) From 1990 to 2000 the average tuition and fees at public colleges and universities rose from $1908 to $3487. Determine if the percent change in tuition and fees was more or less than the percent change in the CPI for the same time period.   (**Source:** The College Board.)

SOLUTION
(a) The CPI increased during every 5-year period between 1960 and 2000. This reflects the fact that prices have consistently risen.

◆ **CLASS DISCUSSION**
If your salary is $10,000 per year and you receive a 300% raise for outstanding work, what is your new salary? ◆

(b) In 1970 the CPI was 38.8 and in 1980 it was 82.4. The percent change was approximately

$$\frac{82.4 - 38.8}{38.8} \times 100 \approx 112\%.$$

(c) In 1990 the CPI was 130.7 and in 2000 it was 172.2. The CPI rose by $\frac{172.2 - 130.7}{130.7} \times 100 \approx 32\%$, while tuition and fees rose by $\frac{3487 - 1908}{1908} \times 100 \approx 83\%$. The rise in tuition and fees outpaced the CPI.   *Now Try Exercise 79* ◆

 **Calculating percent change**

Find the percent change if a quantity changes from price P_1 to price P_2. Round your answer to the nearest tenth of a percent when appropriate.
(a) $P_1 = \$20, P_2 = \$27$ (b) $P_1 = \$1.89, P_2 = \$1.56$

SOLUTION

(a) Percent change $= \frac{P_2 - P_1}{P_1} \times 100 = \frac{27 - 20}{20} \times 100 = 35\%.$

(b) Percent change $= \frac{P_2 - P_1}{P_1} \times 100 = \frac{1.56 - 1.89}{1.89} \times 100 \approx -17.5\%$. The negative sign indicates that the price decreased.   *Now Try Exercise 19* ◆

Calculator Help
To display numbers in scientific notation, see Appendix B (page AP-5).

Scientific Notation

Numbers that are large or small in absolute value occur frequently in applications. For simplicity these numbers are often expressed in scientific notation. Table 1.3 lists examples of numbers in **standard (decimal) form** and in **scientific notation**.

TABLE 1.3

Standard Form	Scientific Notation	Application
93,000,000 mi	9.3×10^7 mi	Distance to the sun
13,060	1.306×10^4	Radio stations in 2001
9,000,000,000	9×10^9	Estimated world population in 2050
0.00000538 sec	5.38×10^{-6} sec	Time for light to travel 1 mile
0.000005 cm	5×10^{-6} cm	Size of a typical virus

To write 0.00000538 in scientific notation, start by moving the decimal point to the right of the first nonzero digit 5 to obtain 5.38. Since the decimal point was moved six places to the *right*, the exponent of 10 is -6. Thus, $0.00000538 = 5.38 \times 10^{-6}$. When the decimal point is moved to the *left*, the exponent of 10 is positive, rather than negative. Here is a formal definition of scientific notation.

SCIENTIFIC NOTATION

A real number r is in **scientific notation** when r is written as $c \times 10^n$, where $1 \le |c| < 10$ and n is an integer.

EXAMPLE 5

Evaluating expressions by hand

Evaluate each expression by hand. Write your result in scientific notation and standard form.

(a) $(3 \times 10^3)(2 \times 10^4)$ **(b)** $(5 \times 10^{-3})(6 \times 10^5)$ **(c)** $\dfrac{4.6 \times 10^{-1}}{2 \times 10^2}$

SOLUTION

(a) $(3 \times 10^3)(2 \times 10^4) = 3 \times 2 \times 10^3 \times 10^4$ Commutative property
$ = 6 \times 10^{3+4}$ Add exponents.
$ = 6 \times 10^7$ Scientific notation
$ = 60,000,000$ Standard form

(b) $(5 \times 10^{-3})(6 \times 10^5) = 5 \times 6 \times 10^{-3} \times 10^5$ Commutative property
$\phantom{(5 \times 10^{-3})(6 \times 10^5)} = 30 \times 10^2$ Add exponents.
$\phantom{(5 \times 10^{-3})(6 \times 10^5)} = 3 \times 10^3$ Scientific notation
$\phantom{(5 \times 10^{-3})(6 \times 10^5)} = 3000$ Standard form

Algebra Review
To review exponents, see Chapter R (page R-13).

(c) $\dfrac{4.6 \times 10^{-1}}{2 \times 10^2} = \dfrac{4.6}{2} \times \dfrac{10^{-1}}{10^2}$ Multiplication of fractions
$\phantom{\dfrac{4.6 \times 10^{-1}}{2 \times 10^2}} = 2.3 \times 10^{-1-2}$ Subtract exponents.
$\phantom{\dfrac{4.6 \times 10^{-1}}{2 \times 10^2}} = 2.3 \times 10^{-3}$ Scientific notation
$\phantom{\dfrac{4.6 \times 10^{-1}}{2 \times 10^2}} = 0.0023$ Standard form

Now Try Exercises 51, 53, and 59

Calculators often use "E" to express powers of 10. For example, 4.2×10^{-3} might be displayed as 4.2E−3. On some calculators, numbers can be entered in scientific notation with the (EE) key, which you can find by pressing (2nd)(,).

 **EXAMPLE 6**   Computing in scientific notation

Calculator Help

To enter numbers in scientific notation, see Appendix B (page AP-5).

Use a calculator to approximate each expression. Write your answer in scientific notation.

(a) $(3.1 \times 10^6)(5.6 \times 10^{-4})$ **(b)** $\left(\dfrac{6 \times 10^3}{4 \times 10^6}\right)(1.2 \times 10^2)$

(c) $\sqrt{4500\pi}\left(\dfrac{103 + 450}{0.233}\right)^3$

SOLUTION

(a) The given expression is entered in two ways in Figure 1.1. Note that in both cases $(3.1 \times 10^6)(5.6 \times 10^{-4}) = 1736 = 1.736 \times 10^3$.

(b) From Figure 1.2, $\left(\dfrac{6 \times 10^3}{4 \times 10^6}\right)(1.2 \times 10^2) = 0.18 = 1.8 \times 10^{-1}$.

(c) Be sure to insert parentheses around 4500π and around the numerator, $103 + 450$, in the ratio. From Figure 1.3 we can see that the result is approximately 1.6×10^{12}.

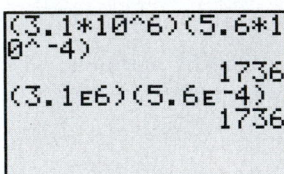

FIGURE 1.1   **FIGURE 1.2**   **FIGURE 1.3**

Now Try Exercises 63 and 65

 **EXAMPLE 7**   Computing with a calculator

Use a calculator to evaluate each expression. Round your answer to the nearest thousandth.

Algebra Review

To review cube roots, see Chapter R (page R-49).

(a) $\sqrt[3]{131}$ **(b)** $\pi^3 + 1.2^2$ **(c)** $\dfrac{1 + \sqrt{2}}{3.7 + 9.8}$ **(d)** $|\sqrt{3} - 6|$

SOLUTION

(a) On some calculators the cube root can be found by using the MATH menu. If your calculator does not have a cube root key, enter $131^{(1/3)}$. From the first two lines in Figure 1.4, we see that $\sqrt[3]{131} \approx 5.079$.

(b) Do *not* use 3.14 for the value of π. Instead, use the built-in key to obtain a more accurate value of π. From the bottom two lines in Figure 1.4, $\pi^3 + 1.2^2 \approx 32.446$.

(c) When evaluating this expression be sure to include parentheses around the numerator and around the denominator. Most calculators have a special square root key that can be used to evaluate $\sqrt{2}$. From the first three lines in Figure 1.5, $\dfrac{1 + \sqrt{2}}{3.7 + 9.8} \approx 0.179$.

(d) The absolute value can be found on some calculators by using the MATH NUM menus. From the bottom two lines in Figure 1.5, $|\sqrt{3} - 6| \approx 4.268$.

To enter expressions such as $\sqrt[3]{131}$, $\sqrt{2}$, π, and $|\sqrt{3}-6|$, see Appendix B (page AP-6).

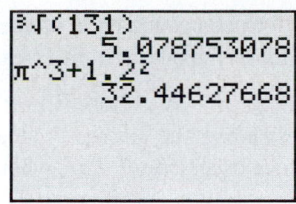

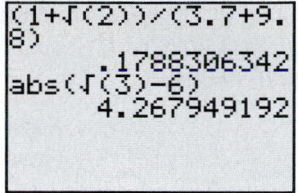

FIGURE 1.4   **FIGURE 1.5**

Now Try Exercises 69, 71, and 75 ◆

 **Applying scientific notation**

A compact disc (CD) is an optically readable device that is capable of storing large amounts of data. Some CDs can store about 680 million bytes of information. (A **byte** is the amount of memory necessary to store a single character, such as one letter or a blank space. The phrase "math test" requires 9 bytes of storage.)   (**Source:** Grolier Electronic Publishing, Inc.)

(a) Express in scientific notation the number of bytes available on this type of CD.
(b) Approximate how many sentences like this one could be stored on a CD. Express your answer in scientific notation.
(c) When playing music on a CD, a sound channel is sampled (or read) at a rate of 44,100 times per second per channel. If two channels are played for 1 minute of stereo music, find the total number of times the two sound channels are sampled. Express your answer in scientific notation.

SOLUTION
(a) Move the decimal point in 680,000,000 to the left 8 places to obtain 6.8. Then 680,000,000 can be written as 6.8×10^8.
(b) The sentence

  "Approximate how many sentences like this one could be stored on a CD."

is composed of 69 characters including the blank spaces and the period.

$$\frac{6.8 \times 10^8}{6.9 \times 10^1} \approx 1 \times 10^7 = 10,000,000$$

The CD could store (save) the sentence approximately 10 million times.
(c) Two channels are being sampled $2 \times 44,100 = 88,200$ times per second. In 60 seconds, they would be sampled $(6.0 \times 10^1)(8.82 \times 10^4) = 52.92 \times 10^5 = 5.292 \times 10^6$ times.

Now Try Exercise 82 ◆

Problem Solving

Many problem-solving strategies are used in algebra. However, in this subsection we focus on two important strategies that are used frequently: making a sketch and applying one or more formulas. These strategies are illustrated in the next examples.

 **Finding the speed of Earth**

Earth travels in an approximately circular orbit around the sun with an average radius of 93 million miles. If Earth takes 1 year, or about 365 days, to complete one orbit, estimate the orbital speed of Earth in miles per hour.

Geometry Review

To find the circumference of a circle, see Chapter R (page R-3).

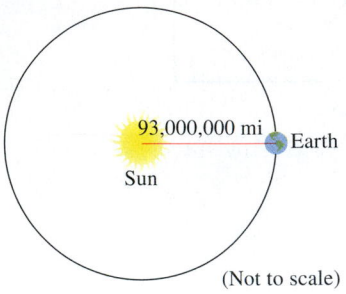

FIGURE 1.6

SOLUTION Speed S equals distance D divided by time T, $S = \frac{D}{T}$. We need to find the number of miles Earth travels in 1 year and then divide it by the number of hours in 1 year.

Distance Traveled   A sketch of Earth orbiting the sun is shown in Figure 1.6. In 1 year Earth travels along the circumference of a circle with a radius of 93 million miles. The circumference of a circle is $2\pi r$, where r is the radius, so the distance D is

$$D = 2\pi r = 2\pi(\mathbf{93{,}000{,}000}) \approx 584{,}300{,}000 \text{ miles.}$$

Hours in 1 Year   The number of hours H in 1 year, or 365 days, equals

$$H = 365 \times 24 = 8760 \text{ hours.}$$

Speed of Earth $S = \frac{D}{H} = \frac{584{,}300{,}000}{8760} \approx 66{,}700$ miles per hour.

Now Try Exercise 83 ◆

Many times in geometry we evaluate formulas to determine quantities, such as perimeter, area, and volume. In the next example we use a formula to determine the number of fluid ounces in a soda can.

 **Finding the volume of a soda can**

Geometry Review

To find the volume of a cylinder, see Chapter R (page R-4).

FIGURE 1.7 A Soda Can

The volume V of the cylindrical soda can in Figure 1.7 is given by $V = \pi r^2 h$, where r is its radius and h is its height.
(a) If $r = 1.4$ inches and $h = 5$ inches, find the volume of the can in cubic inches.
(b) Could this can hold 16 fluid ounces? (*Hint:* 1 cubic inch equals about 0.55 fluid ounces.)

SOLUTION
(a) $V = \pi r^2 h = \pi(\mathbf{1.4})^2(\mathbf{5}) = 9.8\pi \approx 30.8$ cubic inches.
(b) To find the number of fluid ounces, multiply the number of cubic inches by 0.55.

$$30.8 \times 0.55 = 16.94$$

Yes, the can could hold 16 fluid ounces.

Now Try Exercise 93 ◆

Measuring the thickness of a very thin layer of material can be difficult to do by using a direct measurement. For example, it might be difficult to take a ruler and measure the thickness of a sheet of aluminum foil or to measure the thickness of a coat of paint. However, it can be done indirectly using the following formula.

$$\text{Thickness} = \frac{\text{Volume}}{\text{Area}}$$

That is, the thickness of a thin layer equals the volume of the substance divided by the area that it covers. For example, if a volume of 1 cubic inch of paint is spread over an area of 100 square inches, then the thickness of the paint equals $\frac{1}{100}$ inch. This formula is illustrated in the next example.

 **Calculating the thickness of aluminum foil**

A rectangular sheet of aluminum foil is 15 centimeters by 35 centimeters and weighs 5.4 grams. If 1 cubic centimeter of aluminum weighs 2.7 grams, find the thickness of the aluminum foil.   (**Source:** U. Haber-Schaim, *Introductory Physical Science.*)

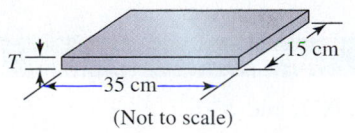

(Not to scale)

FIGURE 1.8

Geometry Review

To find the area of a rectangle and the volume of a box, see Chapter R (page R-3).

SOLUTION Start by making a sketch of a rectangular sheet of aluminum, as shown in Figure 1.8. To complete this problem we need to find the volume V of the aluminum foil and its area A. Then we can determine the thickness T by using the formula $T = \frac{V}{A}$.

Note: For a rectangular box shape, such as the one shown in Figure 1.8,

$$\text{Volume} = \text{Length} \times \text{Width} \times \text{Thickness}.$$

It follows that $\text{Thickness} = \frac{\text{Volume}}{\text{Area}}$.

Volume Because the aluminum foil weighs 5.4 grams and each 2.7 grams equals 1 cubic centimeter, the volume of the aluminum foil is

$$\frac{5.4}{2.7} = 2 \text{ cubic centimeters} \qquad \textit{Divide weight by density.}$$

Area The aluminum foil is rectangular with an area of $15 \times 35 = 525$ square centimeters.

Thickness The thickness of 2 cubic centimeters of aluminum foil with an area of 525 square centimeters is

$$T = \frac{V}{A} = \frac{2}{525} \approx 0.0038 \text{ centimeter.}$$

Now Try Exercise 87 ◆

1.1

Putting it all Together

Data and numbers play central roles in our society. Because of the variety of data, it has been necessary to develop different types of numbers. Without numbers, data could be described qualitatively but not quantitatively. For example, we might say that the day seems hot but would not be able to give an actual number for the temperature. Accurate comparisons with temperatures on other days would be difficult, if not impossible. Problem-solving strategies are used in almost every facet of our lives, providing the procedures needed to systematically complete tasks and perform computations.

The following table summarizes some of the concepts in this section.

Concept	Comments	Examples
Natural numbers	Sometimes referred to as the *counting numbers*	$1, 2, 3, 4, 5, \ldots$
Integers	Includes the natural numbers, their opposites, and 0	$\ldots, -2, -1, 0, 1, 2, \ldots$
Rational numbers	Includes integers; all fractions $\frac{p}{q}$, where p and q are integers with $q \neq 0$; all repeating and all terminating decimals	$\frac{1}{2}, -3, \frac{128}{6}, -0.335, 0,$ $0.25 = \frac{1}{4}, 0.\overline{33} = \frac{1}{3},$ and all fractions

Concept	Comments	Examples
Irrational numbers	Can be written as nonrepeating, nonterminating decimals; cannot be a rational number; a positive integer whose square root is not an integer is an irrational number.	π, $\sqrt{2}$, $-\sqrt{5}$, $\sqrt[3]{7}$, and π^4
Real numbers	Any number that can be expressed in standard, or decimal, form Includes the rational numbers and irrational numbers	π, $\sqrt{7}$, $-\frac{4}{7}$, 0, -10, 1.237 $0.\overline{6} = \frac{2}{3}$, 1000, $\sqrt{15}$, $-\sqrt{5}$
Scientific notation	A number in the form $c \times 10^n$, where $1 \le \lvert c \rvert < 10$ and n is an integer Used to represent numbers that are large or small in absolute value	3.12×10^4 -1.4521×10^{-2} 5×10^9 -3.98715×10^9 1.5987×10^{-6}
Percent change	If a quantity changes from c_1 to c_2, then the percent change equals $\frac{c_2 - c_1}{c_1} \times 100$.	If the price of a gallon of gasoline changes from \$1.00 to \$1.40, then the percent change is $\frac{1.4 - 1}{1} \times 100 = 40\%$.

1.1 — Exercises

Classifying Numbers

Exercises 1–8: Classify the number as one or more of the following: natural number, integer, rational number, or real number.

1. 45,000 (Franchise fee in dollars for a McDonald's restaurant)

2. $\frac{21}{24}$ (Fraction of people in the United States completing at least 4 years of high school)

3. -3 (Annual percent change in the area of tropical rain forests)

4. 18,273 (Average dollar cost of tuition and fees at a private college in 2002)

5. 7.5 (Average number of gallons of water used each minute while taking a shower)

6. 13.1 (Nielsen rating of the TV show *E.R.* during 2002–2003)

7. $90\sqrt{2}$ (Distance in feet from home plate to second base in baseball)

8. -71 (Wind chill when the temperature is $-30°F$ and the wind speed is 40 mph)

Exercises 9–12: Classify each number as one or more of the following: natural number, integer, rational number, or irrational number.

9. π, -3, $\frac{2}{9}$, $\sqrt{9}$, $1.\overline{3}$, $-\sqrt{2}$

10. $\frac{3}{1}$, $-\frac{5}{8}$, $\sqrt{7}$, $0.\overline{45}$, 0, 5.6×10^3

11. $\sqrt{13}$, $\frac{1}{3}$, 5.1×10^{-6}, -2.33, $0.\overline{7}$, $-\sqrt{4}$

12. -103, $\frac{21}{25}$, $\sqrt{100}$, $-\frac{5.7}{10}$, $\frac{2}{9}$, -1.457, $\sqrt{3}$

Exercises 13–18: For the measured quantity, state the set of numbers that is most appropriate to describe it. Choose from the natural numbers, integers, and rational numbers. Explain your answer.

13. Shoe sizes   **14.** Populations of states

15. Gallons of gasoline   **16.** Speed limits

17. Temperatures given in a winter weather forecast in Montana

18. Numbers of compact disc sales

Real Number Computation

Exercises 19–22: **Percent Change** *(Refer to Example 4.) Find the percent change if a quantity changes from P_1 to P_2. Round your answer to the nearest tenth of a percent when appropriate.*

19. $P_1 = \$8$, $P_2 = \$13$ **20.** $P_1 = \$0.90$, $P_2 = \$13.47$

21. $P_1 = 1.4$, $P_2 = 0.85$ **22.** $P_1 = 1256$, $P_2 = 1195$

Exercises 23–36: Write the number in scientific notation.

23. 185,800 (New lung cancer cases reported in 2003)

24. 126,237 (AIDS cases reported in New York City during 2001)

25. 0.03892 (Fraction of U.S. deaths attributed to accidents in 2000)

26. 0.001369 (Fraction of U.S. farmland in New Hampshire in 1998)

27. 2450   **28.** 105.6

29. 0.56   **30.** −0.00456

31. −0.0087   **32.** 1,250,000

33. 206.8   **34.** 0.00007

35. 854,000   **36.** 0.789

Exercises 37–50: Write the number in standard form.

37. 1×10^{-6} (Approximate wavelength in meters of visible light)

38. 9.11×10^{-31} (Weight in kilograms of an electron)

39. 2×10^8 (Number of years for the sun to complete an orbit in our galaxy)

40. 6.75×10^{12} (Federal debt in dollars in 2003)

41. 1.567×10^2 **42.** 0.34×10^4

43. -5.68×10^{-1} **44.** 690×10^{-3}

45. 5×10^5 **46.** 3.5×10^3

47. 0.045×10^5 **48.** -5.4×10^{-5}

49. 67×10^3 **50.** 0.0032×10^{-1}

Exercises 51–62: Evaluate the expression by hand. Write your result in scientific notation and standard form.

51. $(4 \times 10^3)(2 \times 10^5)$ **52.** $(3 \times 10^1)(3 \times 10^4)$

53. $(5 \times 10^2)(7 \times 10^{-4})$ **54.** $(8 \times 10^{-3})(7 \times 10^1)$

55. $(1.2 \times 10^{-3})(1.2 \times 10^{-3})$

56. $(2.5 \times 10^{-2})(2.5 \times 10^{-2})$

57. $\dfrac{4 \times 10^4}{2 \times 10^3}$ **58.** $\dfrac{9 \times 10^7}{3 \times 10^5}$

59. $\dfrac{6.3 \times 10^{-2}}{3 \times 10^1}$ **60.** $\dfrac{8.2 \times 10^2}{2 \times 10^{-2}}$

61. $\dfrac{4 \times 10^{-3}}{8 \times 10^{-1}}$ **62.** $\dfrac{2.4 \times 10^{-5}}{4.8 \times 10^{-7}}$

Exercises 63–68: Use a calculator to approximate the expression. Write your result in scientific notation.

63. $(9.87 \times 10^6)(34 \times 10^{11})$

64. $\dfrac{8.947 \times 10^7}{0.00095}(4.5 \times 10^8)$

65. $\left(\dfrac{101 + 23}{0.42}\right)^2 + \sqrt{3.4 \times 10^{-2}}$

66. $\sqrt[3]{(2.5 \times 10^{-8})} + 10^{-7}$

67. $(8.5 \times 10^{-5})(-9.5 \times 10^7)^2$

68. $\sqrt{\pi(4.56 \times 10^4) + (3.1 \times 10^{-2})}$

Exercises 69–78: Use a calculator to evaluate the expression. Round your result to the nearest thousandth.

69. $\sqrt[3]{192}$ **70.** $\sqrt{(32 + \pi^3)}$

71. $|\pi - 3.2|$ **72.** $\dfrac{1.72 - 5.98}{35.6 + 1.02}$

73. $\dfrac{0.3 + 1.5}{5.5 - 1.2}$ **74.** $3.2(1.1)^2 - 4(1.1) + 2$

75. $\dfrac{1.5^3}{\sqrt{2 + \pi - 5}}$ **76.** $4.3^2 - \dfrac{5}{17}$

77. $15 + \dfrac{4 + \sqrt{3}}{7}$ **78.** $\dfrac{5 + \sqrt{5}}{2}$

Problem Solving

79. Consumer Price Index   (Refer to Example 3.) From 1985 to 2002 the average tuition and fees at private colleges and universities rose from $6,121 to $18,273, while the CPI changed from 107.6 to 179.9. Compare the percent change in tuition and fees to the percent change in the CPI.   (**Sources:** Bureau of the Census, The College Board.)

80. Consumer Price Index   (Refer to Example 3.) Use Table 1.2 to estimate the percent change in prices from 1965 to 1995. In 1965 a gallon of gasoline cost about $0.30. If this price had kept pace with the CPI, what would a gallon of gasoline have cost in 1995?

81. Wages   For some jobs, employees are paid time and a half for every hour worked over 40 hours and double time for all hours on Sunday. Suppose an employee's regular hourly pay is $12. Find the pay for working 47 hours from Monday through Saturday and 8 hours on Sunday.

82. Storage on a CD   (Refer to Example 8.) Estimate the number of times that the word *mathematics* could be saved on a CD with 680 million bytes of memory. Write your answer in scientific notation.

83. Orbital Speed   (Refer to Example 9.) The planet Mars travels in a nearly circular orbit around the sun with a radius of 141 million miles. If it requires 1.88 years for Mars to complete one orbit, estimate the orbital speed of Mars in miles per hour.

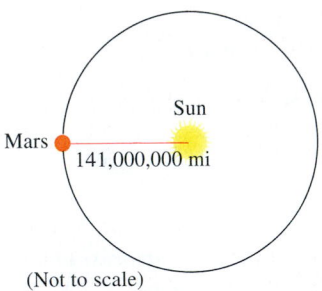

(Not to scale)

84. Size of the Milky Way   The speed of light is about 186,000 miles per second. The Milky Way galaxy has an approximate diameter of 6×10^{17} miles. Estimate the number of years to the nearest thousand it takes for light to travel across the Milky Way. (**Source:** C. Ronan, *The Natural History of the Universe.*)

85. Federal Debt   The amount of federal debt changed dramatically during the 30 years from 1970 to 2000. (**Sources:** Department of the Treasury, Bureau of the Census.)

(a) In 1970 the population of the United States was 203,000,000 and the federal debt was $370 billion. Find the debt per person.

(b) In 2000 the population of the United States was approximately 281,000,000 and the federal debt was $5.54 trillion. Find the debt per person.

(c) If this trend were to continue, estimate the debt per person in the year 2030. (*Hint:* There is more than one way to do this estimate.)

86. Discharge of Water   The Amazon River discharges water into the Atlantic Ocean at an average rate of 4,200,000 cubic feet per second, which is the most of any river in the world. Is this more or less than 1 cubic mile of water per day? Explain your calculations. (*Hint:* 1 cubic mile = 5280^3 cubic feet.)   (**Source:** *The Guinness Book of Records 1993.*)

87. Thickness of an Oil Film   (Refer to Example 11.) A drop of oil measuring 0.12 cubic centimeter is spilled onto a lake. The oil spreads out in a circular shape having a diameter of 23 centimeters. Approximate the thickness of the oil film.

88. Thickness of Gold Foil   (Refer to Example 11.) A flat, rectangular sheet of gold foil measures 20 centimeters by 30 centimeters and has a mass of 23.16 grams. If 1 cubic centimeter of gold has a mass of 19.3 grams, find the thickness of the gold foil.   (**Source:** U. Haber-Schaim, *Introductory Physical Science.*)

89. Analyzing Debt   A 1-inch high stack of $100 bills contains about 250 bills. In 2000 the federal debt was approximately 5.54 trillion dollars.
(a) If the entire federal debt were converted into a stack of $100 bills, how many feet high would it be?

(b) The distance between Los Angeles and New York is approximately 2500 miles. Could this stack of $100 bills reach between these two cities?

90. Military Personnel   The table lists the number of personnel in millions on active duty in the military for selected years.   (**Source:** Department of Defense.)
(a) Discuss the change in the number of active-duty military personnel between 1970 and 2002.

(b) Use the concept of an average or mean to estimate the number of personnel in 1975 and in 2001.

Year	Personnel
1970	3.1
1980	2.1
1985	2.2
1990	2.0
1995	1.5
2000	1.4
2002	1.4

91. *Unemployment Rate*   (Refer to Example 2.) The accompanying table lists the unemployment rate as a percentage over a 5-year period. Calculate the average of these unemployment rates. Is your answer an integer? Is it a rational number?

Year	1998	1999	2000	2001	2002
Unemployment Rate	4.5%	4.2%	4.0%	4.7%	5.8%

Source: Department of Labor.

92. *House Sales*   The accompanying table lists existing one-family homes sold in millions over a 4-year period. Calculate the annual average number of houses sold during this time.

Year	1998	1999	2000	2001
Houses Sold	5.0	5.2	5.2	5.3

Source: National Association of Realtors.

93. *Size of a Soda Can*   (Refer to Example 10.) The volume V of a cylindrical soda can is given by $V = \pi r^2 h$, where r is its radius and h is its height.
   (a) If $r = 1.3$ inches and $h = 4.4$ inches, find the volume of the can in cubic inches.

   (b) Could this can hold 12 fluid ounces? (*Hint:* 1 cubic inch equals about 0.55 fluid ounces.)

94. *Weight of Water*   One cubic centimeter of water weighs about 1 gram. There are about 3.05×10^1 centimeters in 1 foot and there are about 4.54×10^2 grams in 1 pound. How many pounds does 1 cubic foot of water weigh?

95. *Alcohol Consumption*   The table lists the average annual U.S. consumption in gallons of alcohol per person (14 years old or older) from 1940 to 1998.   (**Sources:** Department of Health and Human Services, Bureau of the Census.)

Year	1940	1950	1960	1970
Alcohol (gallons)	1.56	2.04	2.07	2.52

Year	1980	1990	1994	1998
Alcohol (gallons)	2.76	2.46	2.21	2.19

   (a) Discuss the trend in the per capita consumption of alcohol.

   (b) In 1998 there were 225,130,000 people 14 years old or older. Estimate the total gallons of alcohol consumed in the United States in 1998.

   (c) What is the percent change in U.S. consumption of alcohol from 1980 to 1998?

96. *Water in a Lake*   A lake covers 928 acres with an average depth of 17 feet. Estimate the number of cubic feet of water in the lake. If 1 cubic foot equals about 7.48 gallons, approximate the number of gallons of water in the lake. Use scientific notation. (*Hint:* 1 acre equals 43,560 square feet.)

Writing about Mathematics

97. Describe some basic sets of numbers that are used in mathematics.

98. Explain how to calculate percent change.

EXTENDED AND DISCOVERY EXERCISE

1. If you have access to a scale that can weigh in grams, find the thickness of regular and heavy-duty aluminum foil. Is heavy-duty foil worth the price difference? (*Hint:* Each 2.7 grams of aluminum equals 1 cubic centimeter.)

1.2   Visualization of Data

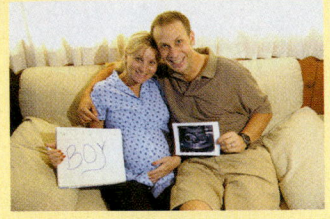

- ◆ Learn to analyze one-variable data
- ◆ Find the domain and range of a relation
- ◆ Graph a relation in the *xy*-plane
- ◆ Calculate the distance between two points
- ◆ Find the midpoint of a line segment
- ◆ Learn to graph equations with a calculator (optional)

Introduction

Computer technology, the Internet, and electronic communication are creating large amounts of data. The challenge is to convert this data into meaningful information that can be used to solve important problems and create new knowledge. Before conclusions can be drawn, data must be analyzed. A powerful tool in this step is visualization. Visual presentations have the capability of communicating vast quantities of information in short periods of time. A typical page of graphics contains one hundred times more information than a page of text. Figure 1.9 shows an MRI scan of a human head and shoulders. Imagine trying to describe this MRI scan with only words. It would be an enormous task, and the final result would not be nearly as understandable as the picture. This section discusses how different types of data can be visualized by using various mathematical techniques.

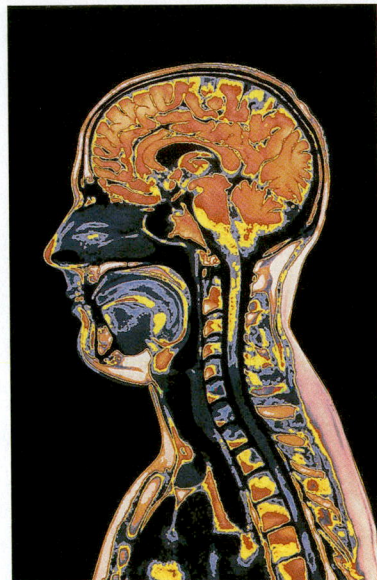

FIGURE 1.9 An MRI Scan

One-Variable Data

Data often occur in the form of a list. A list of test scores without names is an example. The only variable involved is the score. Data of this type are referred to as **one-variable data**. If the values in a list are unique, they can be represented visually on a number line.

EXAMPLE 1   Determining the maximum, minimum, and mean of a list of data

Table 1.4 lists the monthly average temperatures in degrees Fahrenheit at Mould Bay, Canada.   (**Source:** A. Miller and J. Thompson, *Elements of Meteorology.*)

TABLE 1.4   Monthly Average Temperatures at Mould Bay, Canada

Temperature (°F)	−27	−31	−26	−9	12	32	39	36	21	1	−17	−24

(a) Plot these temperatures on a number line.
(b) Find the maximum and minimum temperatures.
(c) Determine the mean of these 12 temperatures.

SOLUTION
(a) In Figure 1.10 the numbers in Table 1.4 are plotted on a number line.

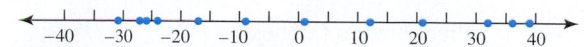

FIGURE 1.10 Monthly Average Temperatures

◆ **CLASS DISCUSSION**
In Example 1(c), the mean of the temperatures is approximately 0.6°F. Interpret this temperature. Explain your reasoning. ◆

(b) The maximum temperature of 39°F is plotted farthest to the right in Figure 1.10. Similarly, the minimum temperature of -31°F is plotted farthest to the left.
(c) The sum of the 12 temperatures in Table 1.4 equals 7. The mean or average of these temperatures is $\frac{7}{12} \approx 0.6$°F.

Now Try Exercise 1 ◆

The **median** of a sorted list is equal to the value that is located in the middle of the list. If there is an odd number of data items, the median is the middle data item. If there is an even number of data items, the median is the average of the two middle items. The **range** of a list of data is the difference between the maximum and minimum values. The range is a measure of the *spread* or *dispersion* in the data. Means, medians, and ranges can be found for one-variable data sets.

EXAMPLE 2   Determining the median and range for a list of data

The monthly average precipitations in inches for Seattle are 5.7, 4.2, 3.7, 2.4, 1.7, 1.4, 0.8, 1.1, 1.9, 3.5, 5.9, and 5.9.   (**Source:** J. Williams.)
(a) Sort these data in increasing order and display them with a table.
(b) Find the median and the range of this data set.
(c) Interpret the median and the range.

SOLUTION
(a) The sorted data are displayed in Table 1.5.

TABLE 1.5   Monthly Average Precipitation for Seattle

Precipitation (inches)	0.8	1.1	1.4	1.7	1.9	2.4	3.5	3.7	4.2	5.7	5.9	5.9

Calculator Help
To calculate the maximum, minimum, mean, and median of the data in Example 2, see Appendix B (page AP-6).

(b) Since there is an even number of data items, the median is the average of the two middle values, 2.4 and 3.5. This is $\frac{2.4 + 3.5}{2} = 2.95$. The largest value is 5.9 and the smallest value is 0.8. The range is equal to their difference, $5.9 - 0.8 = 5.1$.
(c) In Seattle, half the months have an average precipitation of less than 2.95 inches (the median) and half the months have an average precipitation of greater than 2.95 inches. For any two months, the average precipitations vary by at most 5.1 inches (the range).

Now Try Exercise 5 ◆

Two-Variable Data

It may be possible for a relationship to exist between two lists of data. In Example 2, each monthly average precipitation could be associated with a month. By forming a relationship between the month and the precipitation, more information is communicated. Table 1.6 on the next page lists the monthly average precipitation in inches for Portland, Oregon. In this table, 1 corresponds to January, 2 to February, and so on, until 12 represents December.

Showing the relationship between a month and its average precipitation is accomplished by combining the two lists so that corresponding months and precipitations are visually paired.

TABLE 1.6   Average Precipitation for Portland, Oregon

Month	1	2	3	4	5	6	7	8	9	10	11	12
Precipitation (inches)	6.2	3.9	3.6	2.3	2.0	1.5	0.5	1.1	1.6	3.1	5.2	6.4

If x is the month and y is the precipitation, then the **ordered pair** (x, y) represents the average amount of precipitation y during month x. For example, the ordered pair $(5, 2.0)$ indicates that in May the average precipitation is 2.0 inches, whereas the ordered pair $(2, 3.9)$ indicates that the average precipitation in February is 3.9 inches. *Order is important* in an ordered pair.

Since the data in Table 1.6 involve two variables, the month and precipitation, we refer to them as **two-variable data**. It is important to realize that a relation established by two-variable data are between two lists rather than within a single list. January is not related to August and a precipitation of 6.2 inches is not associated with 1.1 inches. Instead January is paired with 6.2 inches, and August is paired with 1.1 inches. We now define the mathematical concept of a relation.

RELATION

A **relation** is a set of ordered pairs.

If we denote the ordered pairs in a relation by (x, y), then the set of all x-values is called the **domain** of the relation and the set of all y-values is called the **range**. The relation shown in Table 1.6 has domain

$$D = \{1, 2, 3, 4, 5, 6, 7, 8, 9, 10, 11, 12\},$$

and range

$$R = \{0.5, 1.1, 1.5, 1.6, 2.0, 2.3, 3.1, 3.6, 3.9, 5.2, 6.2, 6.4\}.$$

EXAMPLE 3   Finding the domain and range of a relation

A physics class measures the time y that it takes for an object to fall x feet, as shown in Table 1.7. The object was dropped twice from each height.

(a) Express the data as a relation S.

(b) Find the domain and range of S.

TABLE 1.7   Falling Object

x (feet)	y (seconds)
20	1.2
20	1.1
40	1.5
40	1.6

SOLUTION

(a) A relation is a set of ordered pairs, so we can write

$$S = \{(20, 1.2), (20, 1.1), (40, 1.5), (40, 1.6)\}.$$

(b) The domain is the set of x-values of the ordered pairs or $D = \{20, 40\}$. The range is the set of y-values of the ordered pairs or $R = \{1.1, 1.2, 1.5, 1.6\}$.

Now Try Exercise 9 ◆

◆ **MAKING CONNECTIONS**

The Different Meanings of Range   Notice that the word *range* has different meanings for one-variable and two-variable data. For one-variable data, the range is a *real number* that describes the spread of the data in a single list. For two-variable data, the range is a *set of numbers* associated with a set of ordered pairs. As is often the case in English, we can determine the correct meaning of *range* from the context in which it is used.

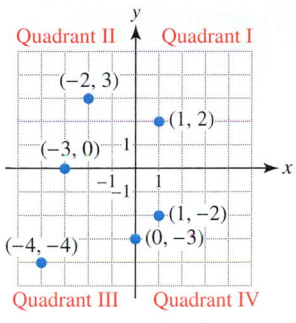

FIGURE 1.11   The *xy*-plane

To visualize a relation, we often use the **Cartesian (rectangular) coordinate plane** or *xy*-**plane**. The horizontal axis is the *x*-**axis** and the vertical axis is the *y*-**axis**. The axes intersect at the **origin** and determine four regions called **quadrants**, numbered I, II, III, and IV, counterclockwise, as shown in Figure 1.11. We can plot the ordered pair (x, y) using the *x*- and *y*-axes. The point $(1, 2)$ is located in quadrant I, $(-2, 3)$ in quadrant II, $(-4, -4)$ in quadrant III, and $(1, -2)$ in quadrant IV. A point lying on a coordinate axis does not belong to any quadrant. The point $(-3, 0)$ is located on the *x*-axis, whereas the point $(0, -3)$ lies on the *y*-axis.

EXAMPLE 4   Graphing a relation

Complete the following for the relation

$$S = \{(5, 10), (5, -5), (-10, 10), (0, 15), (-15, -10)\}.$$

(a) Find the domain and range of the relation.
(b) Determine the maximum and minimum of the *x*-values; of the *y*-values.
(c) Label appropriate scales on the *xy*-axes.
(d) Plot the relation.

SOLUTION
(a) The elements of the domain correspond to the first number in each ordered pair. Thus

$$D = \{-15, -10, 0, 5\}.$$

Similarly, the elements of the range correspond to the second number in each ordered pair. Thus

$$R = \{-10, -5, 10, 15\}.$$

(b) *x*-minimum: -15, *x*-maximum: 5, *y*-minimum: -10, *y*-maximum: 15
(c) An appropriate scale for the *xy*-axes might be -20 to 20 with each tick mark representing a distance of 5. This scale is shown in Figure 1.12.

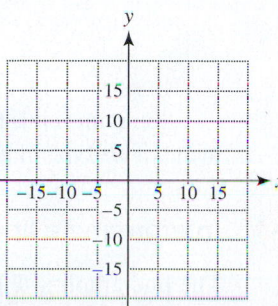

FIGURE 1.12

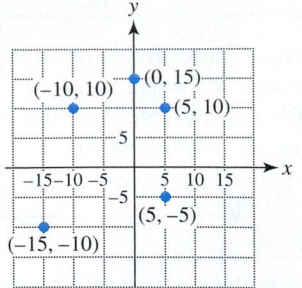

FIGURE 1.13

(d) The points $(5, 10)$, $(5, -5)$, $(-10, 10)$, $(0, 15)$, and $(-15, -10)$ have been plotted in Figure 1.13. Notice that the point $(0, 15)$ lies on the positive y-axis.

Now Try Exercise 61 ◆

The term **scatterplot** is given to a graph in the xy-plane, where distinct data points are plotted. A scatterplot is shown in Figure 1.14. Identify the coordinates of each point.

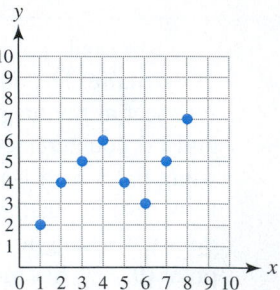

FIGURE 1.14 A Scatterplot

EXAMPLE 5 Making a scatterplot

Use Table 1.6 to make a scatterplot of the monthly average precipitations in Portland, Oregon.

SOLUTION Use the x-axis for the months and the y-axis for the precipitation amounts. To make a scatterplot, simply graph the ordered pairs $(1, 6.2)$, $(2, 3.9)$, $(3, 3.6)$, $(4, 2.3)$, $(5, 2.0)$, $(6, 1.5)$, $(7, 0.5)$, $(8, 1.1)$, $(9, 1.6)$, $(10, 3.1)$, $(11, 5.2)$ and $(12, 6.4)$ in the xy-plane as shown in Figure 1.15.

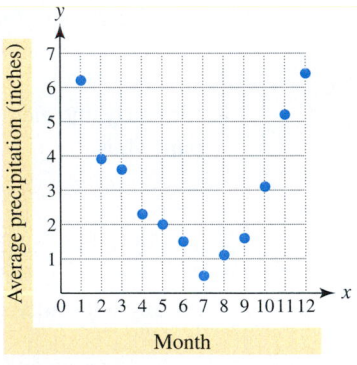

FIGURE 1.15 Monthly Average Precipitation

Now Try Exercise 81 ◆

Sometimes it is helpful to connect the data points in a scatterplot with straight-line segments. This type of graph visually emphasizes changes in the data. It is called a **line graph**.

EXAMPLE 6 Interpreting a line graph

The line graph in Figure 1.16 shows the number of college graduates in thousands at 10-year intervals between 1900 and 2000. (**Sources:** Department of Education, Center for Educational Statistics.)

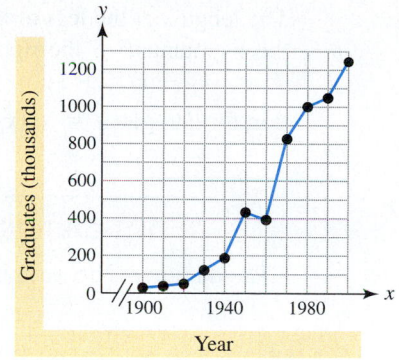

FIGURE 1.16 College Graduates

(a) Did the number of graduates ever decrease during this time period? Explain.
(b) Create a table that contains the same data as the line graph.

SOLUTION
(a) Yes, the number decreased slightly between 1950 and 1960. For this time period, the line segment slopes downward from left to right.
(b) Although we cannot read values from the line graph exactly, we can make reasonable estimates. Table 1.8 lists approximations for the number of college graduates in thousands for the years shown in Figure 1.16. Note that answers often vary slightly when reading a graph.

TABLE 1.8   College Graduates (thousands)

Year	1900	1910	1920	1930	1940	1950	1960	1970	1980	1990	2000
Graduates	20	35	50	110	190	430	390	830	1000	1050	1240

Source: Department of Education.

Now Try Exercise 91 ◆

The Distance Formula

Geometry Review

To review the Pythagorean theorem, see Chapter R (page R-3).

In the *xy*-plane, the length of a line segment with endpoints (x_1, y_1) and (x_2, y_2) can be calculated by using the **Pythagorean theorem**. See Figure 1.17.

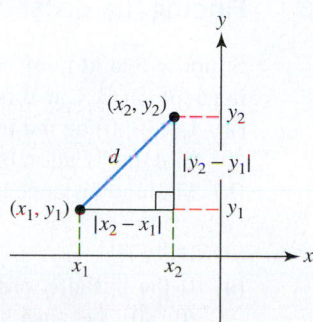

FIGURE 1.17

The lengths of the legs of the right triangle are $|x_2 - x_1|$ and $|y_2 - y_1|$. The distance d is the hypotenuse of the right triangle. Applying the Pythagorean theorem to this triangle, $d^2 = (x_2 - x_1)^2 + (y_2 - y_1)^2$. Because distance is nonnegative, we can solve this equation for d to get $d = \sqrt{(x_2 - x_1)^2 + (y_2 - y_1)^2}$.

DISTANCE FORMULA

The **distance** d between the points (x_1, y_1) and (x_2, y_2) in the xy-plane is

$$d = \sqrt{(x_2 - x_1)^2 + (y_2 - y_1)^2}.$$

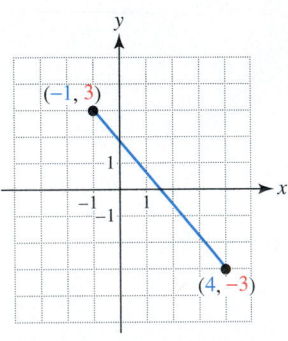

FIGURE 1.18

Figure 1.18 shows a line segment connecting the points $(-1, 3)$ and $(4, -3)$. Its length is

$$\begin{aligned} d &= \sqrt{(4 - (-1))^2 + (-3 - 3)^2} \\ &= \sqrt{61} \\ &\approx 7.81. \end{aligned}$$

EXAMPLE 7 Finding the distance between two points

Find the exact distance between $(3, -4)$ and $(-2, 7)$. Then approximate this distance to the nearest hundredth.

SOLUTION In the distance formula let (x_1, y_1) be $(3, -4)$ and (x_2, y_2) be $(-2, 7)$.

$$\begin{aligned} \sqrt{(x_2 - x_1)^2 + (y_2 - y_1)^2} &= \sqrt{(-2 - 3)^2 + (7 - (-4))^2} && \text{Distance formula} \\ &= \sqrt{(-5)^2 + 11^2} && \text{Subtract.} \\ &= \sqrt{146} && \text{Simplify.} \\ &\approx 12.08 && \text{Approximate.} \end{aligned}$$

The exact distance is $\sqrt{146}$ and the approximate distance rounded to the nearest hundredth is 12.08. Note that we would obtain the same answer if we had let (x_1, y_1) be $(-2, 7)$ and (x_2, y_2) be $(3, -4)$. *Now Try Exercise 21* ◆

In the next example the distance between two moving cars is found.

EXAMPLE 8 Finding the distance between two moving cars

Suppose that at noon car A is traveling south at 20 miles per hour and is located 80 miles north of car B. Car B is traveling east at 40 miles per hour.
(a) Let $(0, 0)$ be the initial coordinates of car B in the xy-plane, where units are in miles. Plot the location of each car at noon and at 1:30 P.M.
(b) Find the distance between the cars at 1:30 P.M.

SOLUTION
(a) If the initial coordinates of car B are $(0, 0)$, then the initial coordinates of car A are $(0, 80)$, because car A is 80 miles north of car B. After 1 hour and 30 minutes or 1.5 hours, car A has traveled $1.5 \times 20 = 30$ miles south, and so it is located 50 miles

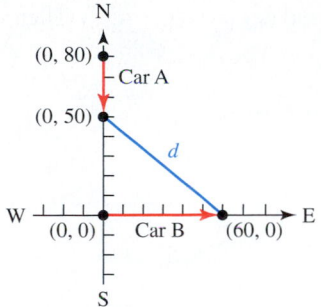

FIGURE 1.19

north of the initial location of car B. Thus its coordinates are (0, 50) at 1:30 PM. Car B traveled $1.5 \times 40 = 60$ miles east, so its coordinates are (60, 0) at 1:30 PM. See Figure 1.19 where these points are plotted.

(b) To find the distance between the cars at 1:30 P.M. we must find the distance d between the points (**0**, **50**) and (**60**, **0**).

$$d = \sqrt{(60 - 0)^2 + (0 - 50)^2} \approx 78.1 \text{ miles}$$

Now Try Exercise 41 ◆

The Midpoint Formula

A common way to make estimations is to average data values. For example, in 1980 the average cost of tuition and fees at public colleges and universities was $800, whereas in 1982 it was $1000. One might estimate the cost of tuition and fees in 1981 to be $900. This averaging is referred to as finding the midpoint. If a line segment is drawn between two data points, then its *midpoint* is the unique point on the line segment that is equidistant from the endpoints.

On a real number line, the midpoint M of two data points x_1 and x_2 is calculated by averaging their coordinates, as shown in Figure 1.20. For example, the midpoint of -3 and 5 is $M = \frac{-3 + 5}{2} = 1$.

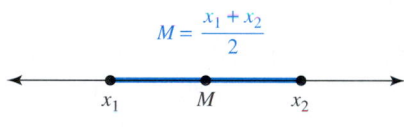

FIGURE 1.20

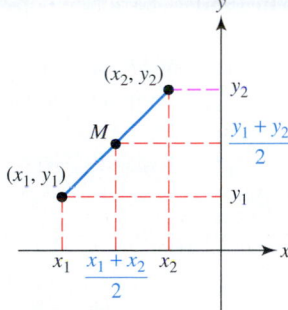

FIGURE 1.21

The midpoint formula in the *xy*-plane is similar to the formula for the real number line, except that both coordinates are averaged. Figure 1.21 shows midpoint M located on the line segment connecting the two data points (x_1, y_1) and (x_2, y_2). The *x*-coordinate of the midpoint M is located halfway between x_1 and x_2 and is $\frac{x_1 + x_2}{2}$. Similarly, the *y*-coordinate of M is the average of y_1 and y_2. For example, if we let (x_1, y_1) be $(-3, 1)$ and (x_2, y_2) be $(-1, 3)$ in Figure 1.21, then the midpoint is computed by

$$\left(\frac{-3 + -1}{2}, \frac{1 + 3}{2} \right) = (-2, 2).$$

The midpoint formula is summarized as follows.

MIDPOINT FORMULA IN THE *xy*-PLANE

The **midpoint** of the line segment with endpoints (x_1, y_1) and (x_2, y_2) in the *xy*-plane is

$$\left(\frac{x_1 + x_2}{2}, \frac{y_1 + y_2}{2} \right).$$

 **9**   **Finding the midpoint**

Find the midpoint of the line segment connecting the points $(6, -7)$ and $(-4, 3)$.

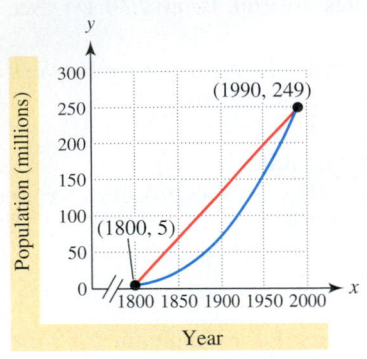

FIGURE 1.22 U.S. Population

SOLUTION In the midpoint formula, let (x_1, y_1) be $(6, -7)$ and (x_2, y_2) be $(-4, 3)$. Then the midpoint M can be found as follows.

$$M = \left(\frac{x_1 + x_2}{2}, \frac{y_1 + y_2}{2} \right) \qquad \text{Midpoint formula}$$

$$= \left(\frac{6 + (-4)}{2}, \frac{-7 + 3}{2} \right) \qquad \text{Substitute.}$$

$$= (1, -2) \qquad \text{Simplify.} \qquad \textit{Now Try Exercise 49} \blacklozenge$$

The population of the United States from 1800 to 1990 is shown by the blue curve in Figure 1.22. The red line segment connecting the points (1800, 5) and (1990, 249) does not accurately describe the population over the time period from 1800 to 1990. However, over a shorter time interval this type of line segment may be more accurate.

EXAMPLE 10 Using the midpoint formula

In 1970 the population of the United States was 203 million and by 1990 it had increased to 249 million. (**Source:** Bureau of the Census.)
(a) Use the midpoint formula to estimate the population in 1980.
(b) Describe this approximation graphically.

SOLUTION
(a) The U.S. populations in 1970 and 1990 is given by the data points (1970, 203) and (1990, 249). The midpoint M of the line segment connecting these points is

$$M = \left(\frac{1970 + 1990}{2}, \frac{203 + 249}{2} \right) = (1980, 226).$$

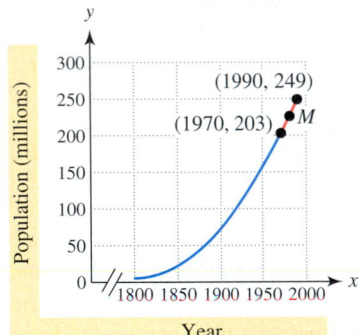

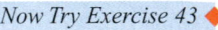

FIGURE 1.23 A Linear Approximation

The midpoint formula estimates a population of 226 million in 1980. (The actual population was 226.5 million.)
(b) Figure 1.23 shows the U.S. population and a line segment with midpoint M connecting the data points (1970, 203) and (1990, 249). Notice that the line segment appears to be an accurate approximation over a relatively short period of 20 years.

Now Try Exercise 43 $\blacklozenge$

Graphing with a Calculator (Optional)

Graphing calculators provide several features beyond those found on scientific calculators. Graphing calculators can be used to create tables, scatterplots, line graphs, and other types of graphs.

The **viewing rectangle**, or **window**, on a graphing calculator is similar to the view finder in a camera. A camera cannot take a picture of an entire scene. It must be centered on a portion of the available scenery. A camera can capture different views of the same scene by zooming in and out. Graphing calculators have similar capabilities. The xy-plane is infinite. The calculator screen can show only a finite, rectangular region of the xy-plane. The viewing rectangle must be specified by setting minimum and maximum values for both the x- and y-axes before a graph can be drawn.

We will use the following terminology regarding the size of a viewing rectangle. **Xmin** is the minimum x-value and **Xmax** is the maximum x-value along the x-axis. Similarly, **Ymin** is the minimum y-value and **Ymax** is the maximum y-value along the y-axis.

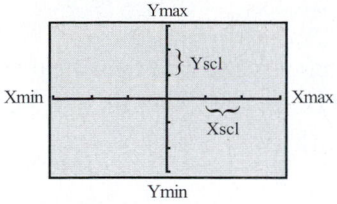

FIGURE 1.24

Most graphs show an *x*-scale and a *y*-scale using tick marks on the respective axes. Sometimes the distance between consecutive tick marks is one unit, and other times it might be five units. The distance represented by consecutive tick marks on the *x*-axis is called **Xscl**, and the distance represented by consecutive tick marks on the *y*-axis is called **Yscl**. See Figure 1.24. This information about the viewing rectangle can be written concisely as **[Xmin, Xmax, Xscl] by [Ymin, Ymax, Yscl]**. For example, [−10, 10, 1] by [−10, 10, 1] means that Xmin = −10, Xmax = 10, Xscl = 1, Ymin = −10, Ymax = 10, and Yscl = 1. This setting is commonly referred to as the **standard viewing rectangle**.

EXAMPLE 11   Setting the viewing rectangle

Show the standard viewing rectangle and the viewing rectangle given by [−30, 40, 10] by [−400, 800, 100] on your calculator.

SOLUTION The window settings and viewing rectangles are displayed in Figures 1.25–1.28. Notice that in Figure 1.26, there are 10 tick marks on the positive *x*-axis, since its length is 10 and the distance between consecutive tick marks is 1.

Calculator Help

To set a viewing rectangle or window, see Appendix B (page AP-7).

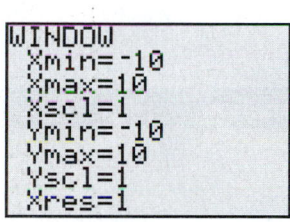

FIGURE 1.25

$[-10, 10, 1]$ by $[-10, 10, 1]$

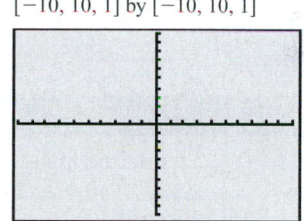

FIGURE 1.26

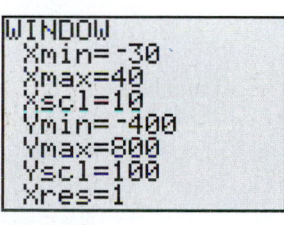

FIGURE 1.27

$[-30, 40, 10]$ by $[-400, 800, 100]$

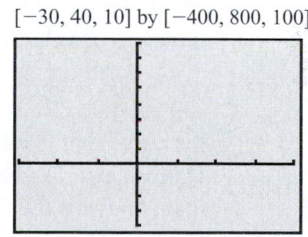

FIGURE 1.28

Now Try Exercise 71 ◆

EXAMPLE 12   Making a scatterplot with a graphing calculator

Plot the points $(-5, -5)$, $(-2, 3)$, $(1, -7)$, and $(4, 8)$ in the standard viewing rectangle.

Calculator Help

To make the scatterplot in Figure 1.29, see Appendix B (page AP-7).

SOLUTION The standard viewing rectangle is given by $[-10, 10, 1]$ by $[-10, 10, 1]$. The points $(-5, -5)$, $(-2, 3)$, $(1, -7)$, and $(4, 8)$ are plotted in Figure 1.29.

$[-10, 10, 1]$ by $[-10, 10, 1]$

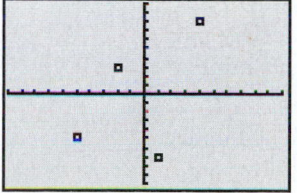

FIGURE 1.29

Now Try Exercise 83 ◆

EXAMPLE 13 Creating a line graph with a graphing calculator

Calculator Help

To make a line graph, see Appendix B (page AP-7).

[1988, 2004, 2] by [0, 100, 20]

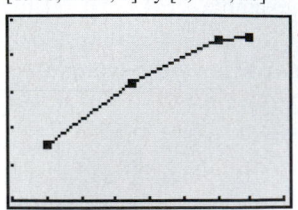

FIGURE 1.30

Table 1.9 lists the percent share of music sales that compact discs (CDs) held from 1990 to 2002. Make a line graph of these sales in the viewing rectangle [1988, 2004, 2] by [0, 100, 20].

TABLE 1.9

Year	1990	1995	2000	2002
CDs (% share)	31	65	89	91

Source: Recording Industry Association of America.

SOLUTION Enter the data points (1990, 31), (1995, 65), (2000, 89), and (2002, 91). A line graph can be created by selecting this option on your graphing calculator. The graph is shown in Figure 1.30.

Now Try Exercise 87 ◆

1.2

Putting it all Together

The following table lists basic concepts in this section.

Concept	Explanation	Examples
Mean or average	To find the mean or average of n numbers, divide their sum by n.	The mean of the four numbers $-3, 5, 6, 9$ is $$\frac{-3 + 5 + 6 + 9}{4} = 4.25.$$
Range (one-variable)	The range of a list of data is the difference between the maximum and the minimum of the numbers or values.	The range of the data $7, 6, -4, 2, 11$ is $$11 - (-4) = 15.$$ (Maximum $-$ Minimum $=$ Range)
Median	The median of a sorted list of numbers equals the value that is located in the middle of the list. Half the data is greater than or equal to the median, and half the data is less than or equal to the median.	The median of $2, 3, 6, 9, 11$ is 6, the middle data item. The median of $2, 3, 6, 9$ is the average of the two middle values: 3 and 6. Therefore the median is $\frac{3 + 6}{2} = 4.5$.
Relation, domain, and range	A relation is a set of ordered pairs (x, y). The set of x-values is called the domain and the set of y-values is called the range.	The relation $S = \{(1, 3), (2, 5), (1, 6)\}$ has domain $D = \{1, 2\}$ and range $R = \{3, 5, 6\}$.
Distance formula	The distance between (x_1, y_1) and (x_2, y_2) is $$d = \sqrt{(x_2 - x_1)^2 + (y_2 - y_1)^2}.$$	The distance between $(2, -1)$ and $(-1, 3)$ is $$d = \sqrt{(-1 - 2)^2 + (3 - (-1))^2} = 5.$$
Midpoint formula	The midpoint of the line segment connecting (x_1, y_1) and (x_2, y_2) is $$M = \left(\frac{x_1 + x_2}{2}, \frac{y_1 + y_2}{2}\right).$$ This formula can be used to make an approximation.	The midpoint of the line segment connecting $(4, 3)$ and $(-2, 5)$ is $$M = \left(\frac{4 + (-2)}{2}, \frac{3 + 5}{2}\right) = (1, 4).$$

The next table summarizes some basic concepts related to one-variable and two-variable data.

Type of Data	Methods of Visualization	Comments
One-variable data	Number line, list, one-column or one-row table	The data items are the same type, and can be described using x-values. Computations of the mean, median, and range are performed on one-variable data.
Two-variable data	Two-column or two-row table, scatterplot, line graph, or other types of graphs in the xy-plane	Two types of data are related, and can be described by using ordered pairs (x, y).

1.2 Exercises

Data Involving One Variable

Exercises 1–4: For the table of data, complete the following.

(a) Plot the numbers on a number line.
(b) Find the maximum and minimum of the data.
(c) Determine the mean of these data.

1. 3   −2   5   0   6   −1

2. 5   −3   4   −2   1   6

3. −10   20   30   −20   0   10

4. 0.5   −1.5   2.0   4.5   −3.5   −1.0

Exercises 5–8: Sort the list of numbers from smallest to largest and display them in a table.

(a) Determine the maximum and minimum values.
(b) Calculate the mean, median, and range. Round each result to the nearest hundredth when appropriate.

5. −10, 25, 15, −30, 55, 61, −30, 45, 5

6. −1.25, 4.75, −3.5, 1.5, 2.5, 4.75, 1.5

7. $\sqrt{15}$, $2^{2.3}$, $\sqrt[3]{69}$, π^2, 2^π, 4.1

8. $\frac{22}{7}$, 3.14, $\sqrt[3]{28}$, $\sqrt{9.4}$, $4^{0.9}$, $3^{1.2}$

Exercises 9–12: For the table of data, complete the following.

(a) Express the data as a relation S.
(b) Find the domain and range of S.

9.

x	−1	2	3	5	9
y	5	2	−1	−4	−5

10.

x	−2	0	2	4	6
y	−4	−2	−1	0	4

11.

x	1	4	5	4	1
y	5	5	6	6	5

12.

x	−1	0	3	−1	−2
y	$\frac{1}{2}$	1	$\frac{3}{4}$	3	$-\frac{5}{6}$

*Exercises 13–16: **Geography** The set of numbers contains data about geographic features of the world.*

(a) Plot the numbers on a number line.
(b) Calculate the mean, median, and range for the set of numbers. Interpret your results.
(c) Try to identify the geographic feature associated with the largest number in the set.

13. {840, 227, 280, 196, 306, 165} (Areas of largest islands in thousands of square miles) (**Source:** National Geographic Society.)

14. {31.7, 22.3, 12.3, 26.8, 24.9, 23.0} (Areas of largest freshwater lakes in thousands of square miles) (**Source:** U.S. National Oceanic and Atmospheric Administration.)

15. {19.3, 18.5, 29.0, 7.31, 16.1, 22.8, 20.3} (Highest elevations of the continents in thousands of feet) (**Source:** *National Geographic Atlas of the World.*)

16. {4145, 3720, 3360, 3740, 3990, 3650, 3590, 3030} (Lengths of longest rivers in miles) (**Source:** U.S. National Oceanic and Atmospheric Administration.)

17. *Basketball Salaries* In the 2003–2004 season, the average salary of the Los Angeles Lakers was $1,500,000 and the minimum salary was $367,000. Suppose that one player earned $26,518,000, while five other players earned the minimum. (**Source:** National Basketball Association.)
  (a) Find the average salary for the six players.

  (b) Find their median salary.

  (c) Discuss why "average salary" can be misleading.

18. *Tennis Earnings* Martina Navratilova had the following earnings for the first six years of her professional tennis career: $173,668, $128,535, $300,317, $450,757, $747,548, and $749,250. (**Source:** J. Leder, *Martina Navratilova.*)

  (a) Determine her mean and median winnings during this time period.

  (b) Find the range of these earnings.

19. *Designing a Data Set* Find a set of three numbers with a mean of 20 and a median of 18. Is your answer unique?

20. *Designing a Data Set* Find a set of five numbers with a mean of 10, a median of 9, and a range of 15. Is your answer unique?

Distance Formula

Exercises 21–38: Find the distance in the xy-plane between the two points. Round approximate results to the nearest hundredth.

21. $(2, -2), (5, 2)$

22. $(0, -3), (12, -8)$

23. $(7, -4), (9, 1)$

24. $(-1, -6), (-8, -5)$

25. $(-6.5, 2.7), (3.6, -2.9)$

26. $(3.6, 5.7), (-2.1, 8.7)$

27. $(-3, 2), (-3, 10)$

28. $(7, 9), (-1, 9)$

29. $\left(\frac{1}{2}, -\frac{1}{2}\right), \left(\frac{3}{4}, \frac{1}{2}\right)$

30. $\left(-\frac{1}{3}, \frac{2}{3}\right), \left(\frac{1}{3}, -\frac{4}{3}\right)$

31. $\left(\frac{2}{5}, \frac{3}{10}\right), \left(-\frac{1}{10}, \frac{4}{5}\right)$

32. $\left(-\frac{1}{2}, \frac{2}{3}\right), \left(\frac{1}{3}, -\frac{5}{2}\right)$

33. $(20, 30), (-30, -90)$

34. $(40, 6), (-20, 17)$

35. $(a, 0), (0, -b)$

36. $(4b, b), (3b, 0)$

37. $(a, b), (a, a)$

38. $(x, y), (1, 2)$

39. *Geometry* An **isosceles triangle** has at least two sides of equal length. Determine whether the triangle with vertices $(0, 0), (3, 4), (7, 1)$ is isosceles.

40. *Geometry* An **equilateral triangle** has sides of equal length. Determine whether the triangle with vertices $(-1, -1), (2, 3), (-4, 3)$ is equilateral.

41. *Distance between Cars* (Refer to Example 8.) At 9:00 A.M. car A is traveling north at 50 miles per hour and is located 50 miles south of car B. Car B is traveling west at 20 miles per hour.
  (a) Let $(0, 0)$ be the initial coordinates of car B in the *xy*-plane, where units are in miles. Plot the locations of each car at 9:00 A.M. and at 11:00 A.M.

  (b) Find the distance *d* between the cars at 11:00 A.M.

42. *Distance between Ships* (Refer to Example 8.) Two ships leave the same harbor at the same time. The first ship heads north at 20 miles per hour and the second ship heads west at 15 miles per hour. Write an expression that gives the distance *d* between the ships after *t* hours.

Midpoint Formula

Exercises 43–48: Use the midpoint formula to complete the following.

43. *U.S. Average Life Expectancy* The average life expectancy of a female born in 1980 was 77.4 years and for a female born in 2000 it was 79.5 years. Estimate the average life expectancy of a female born in 1990. (Actual life expectancy was 78.8.) (**Source:** Bureau of the Census.)

44. *State and Federal Inmates* In 1990 there were 773,919 inmates in state and federal prisons and in 2000 there were 1,391,892. Estimate the number of inmates in 1995. (Actual number was 1,125,874.) (**Source:** Department of Justice.)

45. *State and Federal Inmates* In 1980 there were 329,821 inmates in state and federal prisons and in 1983 there were 436,885. Estimate the number of inmates in 1986. (Actual number was 544,972.) (**Source:** Department of Justice.)

46. *Population Estimates* The population of the United States was 151 million in 1950 and 179 million in 1960. Estimate the population in 1970. (Actual population was 203 million.) (**Source:** Bureau of the Census.)

47. *Olympic Times* In the Olympic Games, the 200-meter dash is run in approximately 20 seconds. Estimate the time to run the 100-meter dash.

48. *Real Numbers* Between any two real numbers a and b there is always another real number. How could such a number be found?

Exercises 49–60: Find the midpoint of the line segment connecting the points.

49. $(1, 2), (5, -3)$ 50. $(-6, 7), (9, -4)$

51. $(-30, 50), (50, -30)$ 52. $(28, -33), (52, 38)$

53. $(1.5, 2.9), (-5.7, -3.6)$ 54. $(9.4, -4.5), (-7.7, 9.5)$

55. $(\sqrt{2}, \sqrt{5}), (\sqrt{2}, -\sqrt{5})$

56. $(\sqrt{7}, 3\sqrt{3}), (-\sqrt{7}, -\sqrt{3})$

57. $(a, b), (-a, 3b)$ 58. $(-a, b), (3a, b)$

59. $(a + b, a - b), (a - b, b - a)$

60. $(3c, c + a), (2c, c - a)$

Data Involving Two Variables

Exercises 61–68: Complete the following.
  (a) *Find the domain and range of the relation.*
  (b) *Determine the maximum and minimum of the x-values; of the y-values.*
  (c) *Label appropriate scales on the xy-axes.*
  (d) *Plot the relation by hand in the xy-plane.*

61. $\{(0, 5), (-3, 4), (-2, -5), (7, -3), (0, 0)\}$

62. $\{(1, 1), (3, 0), (-5, -5), (8, -2), (0, 3)\}$

63. $\{(2, 2), (-3, 1), (-4, -1), (-1, 3), (0, -2)\}$

64. $\{(1, 1), (2, -3), (-1, -1), (-1, 2), (-1, 0)\}$

65. $\{(10, 50), (-35, 45), (0, -55), (75, 25), (-25, -25)\}$

66. $\{(11, 15), (2, -13), (-5, -14), (-7, 19), (-17, -4)\}$

67. $\{(0.1, -0.3), (0.5, 0.4), (-0.7, 0), (0.8, -0.1)\}$

68. $\{(-1.2, 1.5), (1.0, 0.5), (-0.3, 1.1), (-0.8, -1.3)\}$

Exercises 69–76: Show the viewing rectangle on your graphing calculator. Predict the number of tick marks on the positive x-axis and the positive y-axis.

69. Standard viewing rectangle

70. $[-4.7, 4.7, 1]$ by $[-3.1, 3.1, 1]$

71. $[0, 100, 10]$ by $[-50, 50, 10]$

72. $[-30, 30, 5]$ by $[-20, 20, 5]$

73. $[1980, 1995, 1]$ by $[12000, 16000, 1000]$

74. $[1800, 2000, 20]$ by $[5, 20, 5]$

75. $[0, 10, 3]$ by $[-4, 5, 2]$

76. $[-0.5, 0.5, 0.3]$ by $[-1.1, 1.1, 0.2]$

Exercises 77–80: Match the settings for a viewing rectangle with the correct figure (a.–d.).

77. $[-9, 9, 1]$ by $[-6, 6, 1]$

78. $[-6, 6, 1]$ by $[-9, 9, 1]$

79. $[-2, 2, 0.5]$ by $[-4.5, 4.5, 0.5]$

80. $[-4, 8, 1]$ by $[-600, 600, 100]$

a.   b.

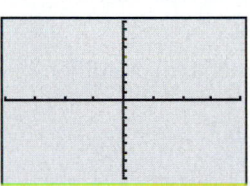 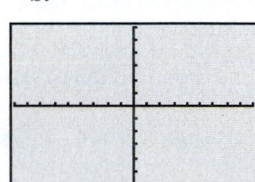

c.   d.

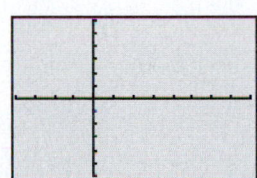 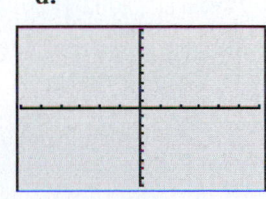

Exercises 81–86: Make a scatterplot of the relation.

81. $\{(1, 3), (-2, 2), (-4, 1), (-2, -4), (0, 2)\}$

82. $\{(6, 8), (-4, -10), (-2, -6), (2, -5)\}$

83. {(10, −20), (−40, 50), (30, 60), (−50, −80), (70, 0), (0, −30)}

84. {(0.01, −0.1), (−0.04, 0.04), (−0.01, 0.09), (0, −1.3)}

85. {(3.1, 6.2), (−5.1, 10.1), (−0.7, −1.4), (1.8, 3.6), (−4.9, −9.8)}

86. {(−1.2, 0.6), (1.0, −0.5), (−0.4, 0.2), (−2.8, 1.4), (2.8, −1.4)}

Exercises 87–90: The table contains real data involving two variables.

    (a) *Determine the maximum and minimum values for each variable in the table.*

    (b) *Use your results from part (a) to determine an appropriate viewing rectangle.*

    (c) *Make a scatterplot of the data.*

    (d) *Make a line graph of the data.*

87. Marijuana use by high school seniors

x (year)	1979	1985	1988	1993	2001
y (percent)	37	25	18	15	22

Source: The Substance Abuse and Mental Health Services Administration.

88. Illicit drug use in past year by people ages 18–25

x (year)	1979	1985	1988	1993	2001
y (percent)	46	38	29	24	28

Source: The Substance Abuse and Mental Health Services Administration.

89. Estimated population of Asian-Americans in millions

x (year)	1996	1998	2000
y (millions)	9.7	10.5	11.2

x (year)	2002	2004
y (millions)	12.0	12.8

Source: U.S. Census Bureau.

90. TV cable subscribers in millions

x (year)	1960	1970	1980	1990	2000
y (millions)	0.6	4	18	55	69

Source: National Cable Television Association.

Exercises 91 and 92: **Graduate Degrees**   *Create a table that contains the same data as the line graph.*

91. Numbers of doctorate degrees in thousands conferred in the United States from 1950 to 2000

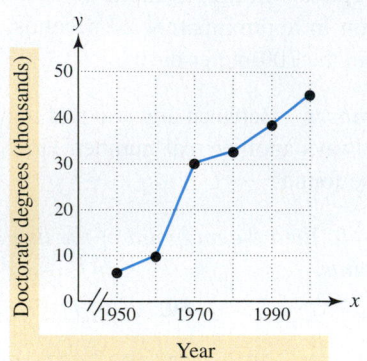

Source: Department of Education.

92. Numbers of master's degrees in thousands conferred in the United States from 1950 to 2000

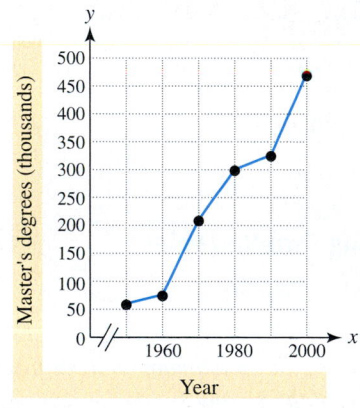

Source: Department of Education.

Writing about Mathematics

93. Describe how one- and two-variable data can be represented. Give one example of each type of data.

94. Give two meanings of the word *range*.

CHECKING BASIC CONCEPTS FOR SECTIONS 1.1 AND 1.2

1. Approximate each expression to the nearest hundredth.

   (a) $\sqrt{4.2(23.1 + 0.5^3)}$ (b) $\dfrac{23 + 44}{85.1 - 32.9}$

2. Write each number using scientific notation.
   (a) 348,500,000   (b) -1237.4

   (c) 0.00198

3. Find the distance in the xy-plane between the points $(-3, 1)$ and $(3, -5)$.

4. Find the midpoint of the line segment connecting the points $(-2, 3)$ and $(4, 2)$.

5. The average depths in feet of the four oceans are 13,215, 12,881, 13,002, and 3953. Calculate the mean, median, and range of these depths.

6. Make a scatterplot and a line graph with the four points $(-5, -4)$, $(-1, 2)$, $(2, -2)$, and $(3, 6)$. State the quadrant in which each point lies.

1.3 # Functions and Their Representations

◆ Learn function notation

◆ Represent a function four different ways

◆ Define a function formally

◆ Identify the domain and range of a function

◆ Use calculators to represent functions (optional)

◆ Identify functions

Introduction

Any concept can be made more or less complicated by how it is represented. Many advances throughout history were the result of new representations of known information. For example, the theory of germs provided a new representation of disease that greatly improved medical science.

Similar events have occurred in mathematics. Historically, fractions were developed much earlier than decimals. The introduction of decimals greatly simplified arithmetic involving fractions. For example, sums like $\frac{1}{2} + \frac{1}{4}$ caused difficulty for people because of unlike denominators. However, their decimal equivalents could be easily combined as the sum $0.50 + 0.25 = 0.75$. Viewed in a new representation, an old problem became easier.

In this section the concept of a function is introduced. Functions are a special type of relation that can provide descriptions of phenomena in our world.   (**Source:** L. Motz and J. Weaver, *The Story of Mathematics.*)

Basic Concepts

Although thunder is caused by lightning, we sometimes see a flash of lightning before we hear the thunder. This is because light travels at 186,000 *miles* per second, whereas sound travels at about 1050 *feet* per second. Since 1 mile equals 5280 feet, sound takes about 5 seconds to travel 1 mile. It follows that the farther away lightning is, the greater the time lapse between seeing the flash of lightning and hearing the thunder. Table 1.10 lists the *approximate* distance y in miles between a person and a bolt of lightning when there is a time lapse of x seconds between seeing the lightning and hearing the thunder. Note that the value of y can be found by dividing the corresponding value of x by 5.

TABLE 1.10   Distance from a Bolt of Lightning

x (seconds)	5	10	15	20	25
y (miles)	1	2	3	4	5

This table establishes a special type of relation between two sets of numbers, where each valid input x in seconds determines *exactly one* output y in miles. We say that Table 1.10

represents a function f, where function f *computes* the distance between an observer and a lightning bolt.

To emphasize that y is a function of x, the notation $y = f(x)$ is used. The expression $f(x)$ does *not* indicate that f and x are multiplied. Rather, the notation $y = f(x)$ is called *function notation*, is read "y equals f of x," and denotes that function f with input x produces output y. That is,

$$f(\text{Input}) = \text{Output},$$

where the input is in seconds and the output is in miles. For example, we write $f(5) = 1$ because if there is a 5-second delay between a lightning bolt and its thunder, then the lightning bolt was about 1 mile away. The expression $f(5)$ represents the output from f when the input is 5. Similarly, $f(10) = 2$ and $f(15) = 3$. Note that a function calculates a set of ordered pairs (x, y), where $y = f(x)$. For example, the ordered pairs $(5, 1)$, $(10, 2)$, and $(15, 3)$ all belong to the relation computed by f. We can think of these ordered pairs as input-output pairs with the form (**input**, **output**). A relation that calculates exactly one output for each valid input is a *function*.

The distance y *depends* on the time x, and so y is called the *dependent variable* and x is called the *independent variable*. For example, if we pick the independent variable x to be 10 seconds then the value of y is dependent on x and equals $\frac{10}{5} = 2$ miles. The y-values depend on the x-values that we choose. The following diagram can help us visualize the computation performed by f.

Input x in seconds $\rightarrow$ Compute $y = f(x)$ by dividing x by 5. $\rightarrow$ **Output y in miles**
(independent variable)   (dependent variable)

This discussion about function notation is summarized as follows.

FUNCTION NOTATION

The notation $y = f(x)$ is called **function notation**. The **input** is x, the **output** is y, and the *name* of the function is f.

$$\text{Name}$$
$$\downarrow$$
$$y = f(x)$$
$$\text{Output} \qquad \text{Input}$$

The variable y is called the **dependent variable** and the variable x is called the **independent variable**. The expression $f(20) = 4$ is read "f of 20 equals 4" and indicates that f outputs 4 when the input is 20. A function computes *exactly one* output for each valid input. The letters f, g, and h are often used to denote names of functions.

The set of valid or meaningful inputs x is called the **domain** of the function and the set of corresponding outputs y is the **range**. For example, suppose that a function f computes the height after x seconds of a ball thrown into the air. Then the domain of f might consist of all times while the ball was in flight, and the range would include all heights attained by the ball.

The following can be computed by functions because they result in *one* output for each valid input.

- Calculating the square of a number x
- Finding the sale price of an item discounted 25% with regular price x
- Naming the biological mother of person x

◆ **MAKING CONNECTIONS** ─────────────────

The Expressions f and $f(x)$ The italic letter f represents the *name* of a function, whereas the expression $f(x)$ represents the function f evaluated for input x. That is, $f(x)$ typically represents a formula for function f that can be used to evaluate f for various values of x. For example, if f represents the name of the squaring function, then $f(x) = x^2$. Thus we speak of the graph of f or the domain of f, but we evaluate $f(3)$ to be $3^2 = 9$.

◆

Representations of Functions

A function can be represented by verbal descriptions, tables, diagrams, symbols, and graphs.

Verbal Representation (Words)   If function f approximates the distance between an observer and a bolt of lightning, then we can verbally describe f with the following sentence: "Divide x seconds by 5 to obtain y miles." We call this a **verbal representation** of f.

Numerical Representation (Table of Values)   A numerical representation for the function f that calculates the distance between a lightning bolt and an observer is given in Table 1.10. A **numerical representation** is a *table of values* that lists input-output pairs for a function. A different numerical representation for f is shown in Table 1.11.

TABLE 1.11   Distance from a Bolt of Lightning

x (seconds)	1	2	3	4	5	6	7
y (miles)	0.2	0.4	0.6	0.8	1.0	1.2	1.4

One difficulty with a numerical representation is that it is often either inconvenient or impossible to list all possible inputs x. For this reason we sometimes refer to a table of this type as a **partial numerical representation** as opposed to a **complete numerical representation**, which would include all elements from the domain of a function. For example, many valid inputs do not appear in Table 1.11, such as $x = 11$ or $x = 0.75$.

Some tables do not represent a function. For example, Table 1.12 represents a relation but *not a function* because input 1 produces two outputs, 3 and 12.

TABLE 1.12   A Relation That Is Not a Function

x	1	2	3	1
y	3	6	9	12

Diagrammatic Representation (Diagram)   Functions are sometimes represented using **diagrammatic representations**, or **diagrams**. Figure 1.31 is a diagram of a function with domain $D = \{5, 10, 15, 20\}$ and range $R = \{1, 2, 3, 4\}$. An arrow is used to show that input x produces output y. For example, input 5 results in output 1 or $f(5) = 1$. Figure 1.32 shows a relation, but not a function, because input 2 results in two different outputs, 5 and 6.

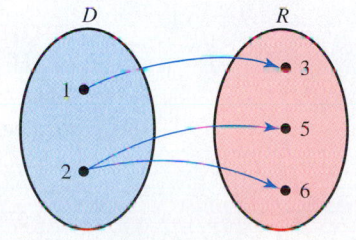

FIGURE 1.31   Function   **FIGURE 1.32**   Not a Function

Symbolic Representation (Formula)   A formula provides a **symbolic representation** of a function. The computation performed by f is expressed by $f(x) = \frac{x}{5}$, where $y = f(x)$. We say that function f is *represented by*, *defined by*, or *given by* $f(x) = \frac{x}{5}$. It follows that $f(\mathbf{6}) = \frac{6}{5} = \mathbf{1.2}$.

Similarly, if a function g computes the square of a number x, then g can be represented by $g(x) = x^2$. A formula is an efficient, but less visual, way to define a function.

Graphical Representation (Graph)   Leonhard Euler (1707–1783) invented the function notation $f(x)$. He also was the first to allow a function to be represented by a graph, rather than only by a formula.

A **graphical representation**, or **graph**, visually pairs an x-input with a y-output. In a graph of a function, the ordered pairs (x, y) are plotted in the xy-plane. The ordered pairs

$$(1, 0.2), (2, 0.4), (3, 0.6), (4, 0.8), (5, 1.0), (6, 1.2), \text{ and } (7, 1.4)$$

from Table 1.11 are plotted in Figure 1.33. This scatterplot suggests a line for the graph of f, as shown in Figure 1.34.

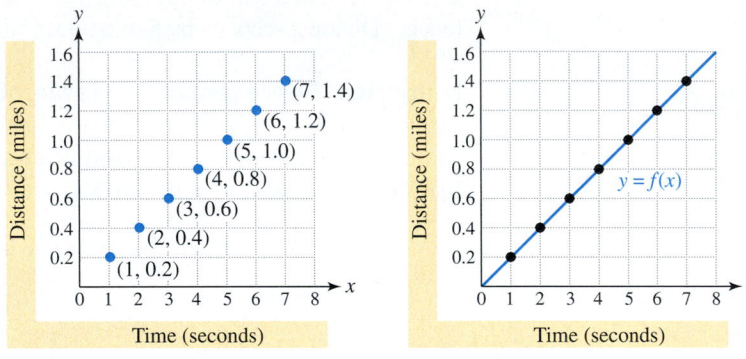

FIGURE 1.33   **FIGURE 1.34**

Note that since $f(5) = 1$, the point $(5, 1)$ lies on the graph of f. Similarly, since $(7, 1.4)$ lies on the graph of f, it follows that $f(7) = 1.4$.

◆ **MAKING CONNECTIONS**

Functions, Points, and Graphs   If $f(a) = b$, then the point (a, b) lies on the graph of f. Conversely, if the point (a, b) lies on the graph of f, then $f(a) = b$. Thus each point on the graph of f can be written in the form $(a, f(a))$.   ◆

 **EXAMPLE 1**   Graphing the absolute value function by hand

Graph $f(x) = |x|$ by hand.

SOLUTION   To graph by hand, it is sometimes helpful to make a table of values first. Start by selecting convenient x-values and then substitute them into $f(x) = |x|$, as shown in Table 1.13. For example, when $x = -2$, then $f(-2) = |-2| = 2$ so the point $(-2, 2)$ is located on the graph of $y = f(x)$.

TABLE 1.13

x	−2	−1	0	1	2
$f(x)$	2	1	0	1	2

Next, we plot the points $(-2, 2)$, $(-1, 1)$, $(0, 0)$, $(1, 1)$, and $(2, 2)$ in the xy-plane, as shown in Figure 1.35. The points appear to be V-shaped, and the graph of f shown in Figure 1.36 results if all possible pairs of the form $(x, |x|)$ are plotted.

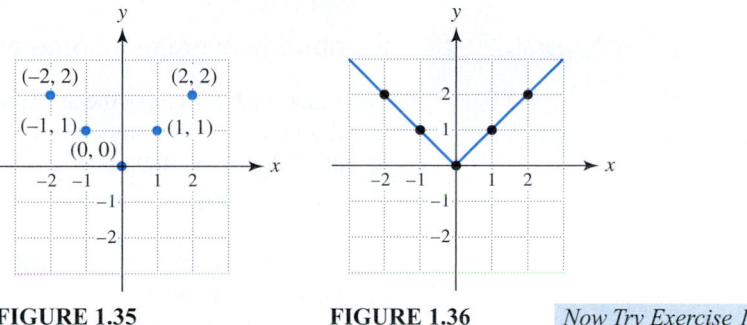

FIGURE 1.35   FIGURE 1.36   *Now Try Exercise 15* ◆

◆ **MAKING CONNECTIONS**

Four Representations of a Function

Symbolic Representation $f(x) = x^2 + 1$

Numerical Representation   *Graphical Representation*

x	y
-2	5
-1	2
0	1
1	2
2	5

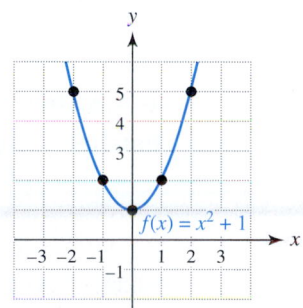

Verbal Representation f squares input x and then adds 1 to produce an output y.

Although there are many ways to represent a function, there is no *best* representation. It depends on both the individual and the application. Sometimes a formula works best—other times a graph or a table is preferable.

Formal Definition of a Function

Because the idea of a function is a fundamental concept in mathematics, it is important that we define a function precisely. This definition should allow for *all* representations of a function. The commonality among representations is the concept of an ordered pair.

A relation is a set of ordered pairs. Would it be sufficient to say that a function is a relation? No, because a function must produce *exactly one output* for each valid input.

FUNCTION

A **function** is a relation in which each element in the domain corresponds to exactly one element in the range.

The set of ordered pairs for a function can be either finite or infinite. The function represented by $f = \{(1, 2), (3, 4), (5, 6)\}$ is a finite set of ordered pairs. In contrast, the function represented by $g(x) = x^2$ with all real numbers as its domain generates an infinite set of ordered pairs, such as $(1, 1)$, $(2, 4)$, and $(2.5, 6.25)$.

EXAMPLE 2 Computing average income as a function of educational attainment

The function f computes the average 1999 individual annual earnings in dollars by educational attainment. This function is defined by $f(N) = 23,400$; $f(H) = 30,400$; $f(B) = 52,200$; $f(M) = 62,300$, where N denotes no diploma, H a high school diploma, B a bachelor's degree, and M a master's degree. (**Sources:** Bureau of the Census, Department of Commerce.)

(a) Write f as a set of ordered pairs.
(b) Give the domain and range of f.
(c) Discuss the relationship between education and income.

SOLUTION
(a) $f = \{(N, 23400), (H, 30400), (B, 52200), (M, 62300)\}$.
(b) The domain D and range R of f are

$$D = \{N, H, B, M\} \quad \text{and} \quad R = \{23400, 30400, 52200, 62300\}.$$

(c) The greater the educational attainment, the greater the annual earnings.

Now Try Exercises 71 and 76 ◆

◆ **MAKING CONNECTIONS**

Relations and Functions   Every function is a relation, whereas every relation is not a function. A function has exactly one output for each valid input. ◆

Unless stated otherwise, the domain of a function f is the set of all real numbers for which its symbolic representation (formula) is defined. The domain can be thought of as the set of all valid inputs that make sense in the expression for $f(x)$. In this case the domain is often referred to as the **implied domain**. Other times the domain of a function must be restricted. For example, if an object falls for 5 seconds, then the distance d that it falls in feet after t seconds is given by $d(t) = 16t^2$ and the domain of d must be restricted to $0 \le t \le 5$.

EXAMPLE 3 Evaluating a function and determining its domain

Let a function f be represented symbolically by $f(x) = \frac{x}{x^2 - 1}$.
(a) Evaluate $f(3)$ and $f(a + 1)$.   **(b)** Find the domain of f.

SOLUTION
(a) To evaluate $f(3)$, substitute 3 for x in the formula: $f(3) = \frac{3}{3^2 - 1} = \frac{3}{8}$.

To evaluate $f(a + 1)$, substitute $(a + 1)$ for x in the formula for $f(x)$.

Algebra Review
To square a binomial, see Chapter R (page R-25).

$$f(a + 1) = \frac{(a + 1)}{(a + 1)^2 - 1} \qquad \text{Let } x = (a + 1).$$

$$= \frac{a + 1}{a^2 + 2a + 1 - 1} \qquad \text{Square the binomial.}$$

$$= \frac{a + 1}{a^2 + 2a} \qquad \text{Simplify.}$$

(b) The expression for $f(x)$ is not defined when the denominator $x^2 - 1 = 0$. Therefore the domain of f is all real numbers except -1 and 1. The domain can be expressed in set-builder notation as $\{x \mid x \neq -1, x \neq 1\}$. *Now Try Exercise 33* ◆

SET-BUILDER NOTATION

The expression $\{x \mid x \neq -1, x \neq 1\}$ is written in **set-builder notation** and represents the set of all real numbers x, such that x does not equal -1 and x does not equal 1. Another example is $\{y \mid 1 < y < 5\}$, which represents the set of all real numbers y, such that y is greater than 1 *and* less than 5.

EXAMPLE 4   Evaluating a function symbolically and graphically

A function g is given by $g(x) = x^2 - 2x$, and its graph is shown in Figure 1.37.
(a) Find the domain of g.
(b) Use $g(x)$ to evaluate $g(-1)$.
(c) Use the graph of g to evaluate $g(-1)$.

SOLUTION

(a) The domain for $g(x) = x^2 - 2x$ includes all real numbers because the formula is defined for all real number inputs x.

(b) To evaluate $g(-1)$, substitute -1 for x in the given formula.

$$g(-1) = (-1)^2 - 2(-1) = 1 + 2 = 3$$

(c) Refer to Figure 1.38. Begin by finding $x = -1$ on the x-axis. Move upward until the graph of g is reached. Then move across to the y-axis. The y-value corresponding to an x-value of -1 is 3. Thus $g(-1) = 3$.

FIGURE 1.37

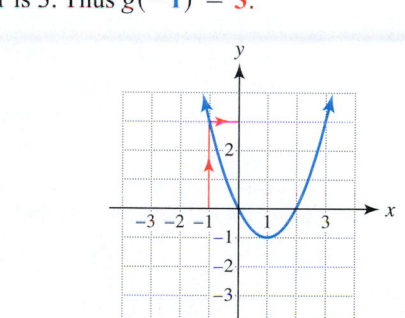

FIGURE 1.38 $g(-1) = 3$ *Now Try Exercise 41* ◆

EXAMPLE 5   Finding the domain and range graphically

A graph of $f(x) = \sqrt{x - 2}$ is shown in Figure 1.39. Find the domain and range of f.

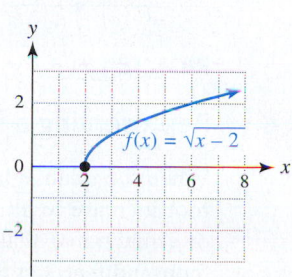

FIGURE 1.39

◆ **CLASS DISCUSSION**

Suppose a golf ball hit into the air reaches a maximum height of 144 feet and strikes the ground after 6 seconds. Sketch an approximate graph of a function f that outputs the height y (in feet) after t seconds. Use your graph to describe the domain and range of f. ◆

SOLUTION The arrow in the graph of f indicates that both the x-values and the y-values increase without reaching a maximum value. In Figure 1.40 the domain and range of f have been labeled by arrows. Note that points appear on the graph for all x greater than or equal to 2. Thus the domain is $D = \{x \mid x \geq 2\}$. The minimum y-value on the graph of f is 0 and it occurs at the point $(2, 0)$. There is no maximum y-value on the graph, so the range is $R = \{y \mid y \geq 0\}$.

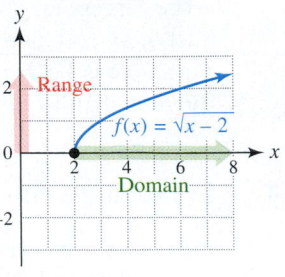

FIGURE 1.40

Now Try Exercise 43 ◆

Graphing Calculators and Functions (Optional)

Graphing calculators can create graphs and tables of a function—usually more efficiently and reliably than pencil-and-paper techniques. However, a graphing calculator uses the same basic method that we might use to draw a graph. For example, one way to sketch a graph of $y = x^2$ is to first make a table of values. See Table 1.14.

TABLE 1.14

x	-3	-2	-1	0	1	2	3
y	9	4	1	0	1	4	9

We can plot these points in the xy-plane, as shown in Figure 1.41. Next we might connect the points with a smooth curve, as shown in Figure 1.42.

A graphing calculator typically follows the same process to draw a graph. It plots numerous points and connects them to make a graph. In Figure 1.43, a graphing calculator has been used to graph $y = x^2$. Note that the graph is not completely smooth, as is the case in Figure 1.42. The reason is that the graphing calculator screen has limited resolution.

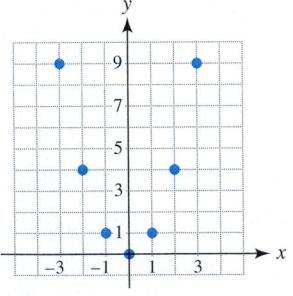

FIGURE 1.41

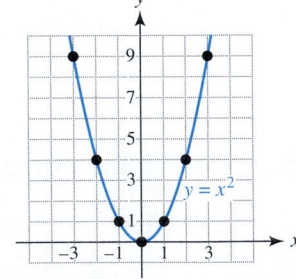

FIGURE 1.42

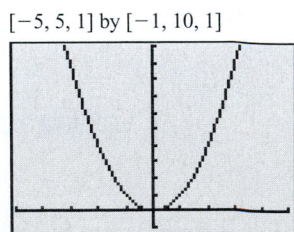

FIGURE 1.43 $y = x^2$

 **EXAMPLE 6** Representing a function

When the relative humidity is less than 100%, air cools at a rate of 5.4°F for every 1000-foot increase in altitude. Give verbal, symbolic, graphical, and numerical representations of a function f that computes this change in temperature for an increase in altitude of x-thousand feet. Let the domain of f be $0 \leq x \leq 6$. (**Source:** L. Battan, *Weather in Your Life.*)

SOLUTION

Verbal Multiply the input x by -5.4 to obtain the change in temperature.

Symbolic Let $f(x) = -5.4x$.

Graphical Since $f(x) = -5.4x$, enter $Y_1 = -5.4X$, as shown in Figure 1.44. Graph y_1 in a viewing rectangle such as $[0, 6, 1]$ by $[-35, 10, 5]$. See Figure 1.45.

Calculator Help

To enter a formula and create a graph, see Appendix B (page AP-8).

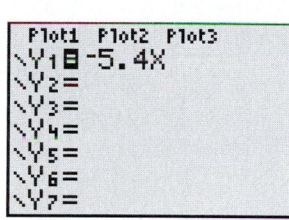

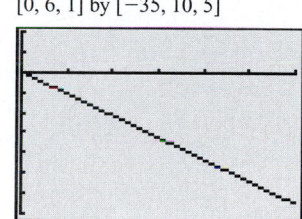
$[0, 6, 1]$ by $[-35, 10, 5]$

FIGURE 1.44   **FIGURE 1.45**

Numerical It is impossible to list all inputs x, since $0 \leq x \leq 6$. However, Figures 1.46 and 1.47 show how to create a table for $Y_1 = -5.4X$ with $x = 0, 1, 2, 3, 4, 5, 6$.

Calculator Help

To create a table similar to Figure 1.47, see Appendix B (page AP-9).

FIGURE 1.46   **FIGURE 1.47**   *Now Try Exercise 105* ◆

 **EXAMPLE 7** ## Evaluating representations of a function

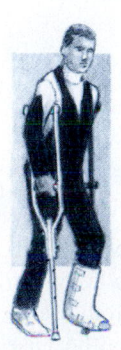

People who sustain leg injuries often require crutches. (Refer to the introduction to this chapter.) It is possible to estimate the proper crutch length without using trial and error. The function given by $f(x) = 0.72x + 2$ can compute the appropriate crutch length in inches for a person with a height of x inches. (**Source:** *Journal of the American Physical Therapy Association.*)

(a) Evaluate $f(65)$ symbolically. Interpret your result.

(b) Evaluate $f(65)$ graphically. Use the window $[60, 90, 10]$ by $[40, 80, 10]$.

(c) Make a table of f starting at $x = 60$, incrementing by 1. Evaluate $f(65)$.

SOLUTION

(a) *Symbolically* Since $f(x) = 0.72x + 2$, $f(65) = 0.72(65) + 2 = 48.8$. This result means that a person 65 inches tall needs about 49-inch crutches.

Calculator Help

To make a graph and evaluate a function graphically, see Appendix B (page AP-9).

(b) *Graphically* Begin by graphing $Y_1 = 0.72X + 2$, as shown in Figure 1.48 on the next page. Two ways to evaluate a function with a graphing calculator are to use either the "trace" or "value" utilities. On some calculators the "value" utility is found under the CALCULATE menu. In Figures 1.49 and 1.50 this utility has been used to evaluate the graph at $x = 65$ to obtain 48.8.

[60, 90, 10] by [40, 80, 10]   [60, 90, 10] by [40, 80, 10]

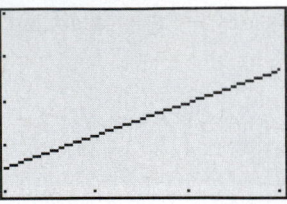

FIGURE 1.48   **FIGURE 1.49**   **FIGURE 1.50**

(c) *Numerically*   To create a table on a graphing calculator, we retain the formula for y_1 and specify both the *starting value* for x (TblStart) and the *increment* between successive x-values (ΔTbl). Here we start the table at $x = 60$ and increment by 1. (There is no viewing rectangle to set.) Figures 1.51 and 1.52 illustrate these steps. When $x = 65$, $y = 48.8$, so $f(65) = 48.8$.

Calculator Help

To create the table in Figure 1.52, see Appendix B (page AP-9).

FIGURE 1.51   **FIGURE 1.52**

Now Try Exercises 63 and 106 ◆

Identifying Functions

By applying the definition of a function, we can determine if a relation is a function.

 **EXAMPLE 8**   Determining if a set of ordered pairs is a function

Determine if each set of ordered pairs represents a function.
(a) $A = \{(-2, 3), (-1, 2), (0, -3), (-2, 4)\}$
(b) $B = \{(1, 4), (2, 5), (-3, -4), (-1, 7), (0, 9)\}$

SOLUTION
(a) Set A does not represent a function because input -2 results in two outputs: 3 and 4.
(b) Set B represents a function because each input (x-value) is paired with exactly one output (y-value).   *Now Try Exercise 87* ◆

Vertical Line Test   To conclude that a graph represents a function, we must be convinced that it is impossible for two distinct points with the same x-coordinate to lie on the graph. For example, the ordered pairs $(4, 2)$ and $(4, -2)$ are distinct points with the same x-coordinate. These two points could not lie on the graph of the same function because input 4 would result in two outputs: 2 and -2. A function has exactly one output for each valid input. When the points $(4, 2)$ and $(4, -2)$ are plotted, they lie on the same vertical line, as shown in Figure 1.53. A graph passing through these points intersects the line twice, as illustrated in Figure 1.54. Therefore, the graph in Figure 1.54 does not represent a function.

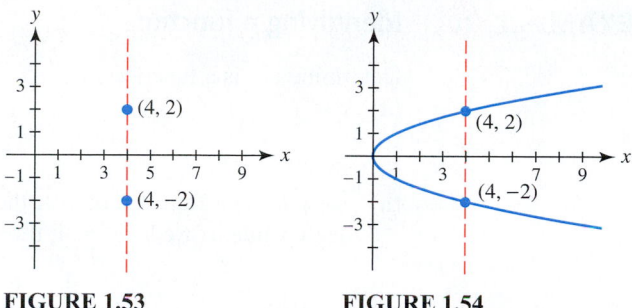

FIGURE 1.53   FIGURE 1.54

To determine if a graph represents a function, simply visualize vertical lines in the *xy*-plane. If every vertical line intersects a graph at no more than one point, then it is a graph of a function. This is called the **vertical line test** for a function.

VERTICAL LINE TEST

If every vertical line intersects a graph at no more than one point, then the graph represents a function.

EXAMPLE 9   Identifying a function graphically

Use the vertical line test to determine if the graph represents a function.

(a)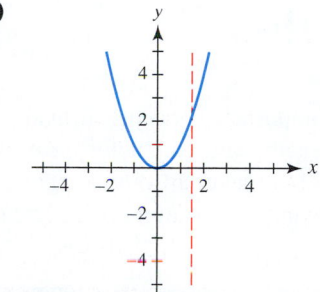

(b)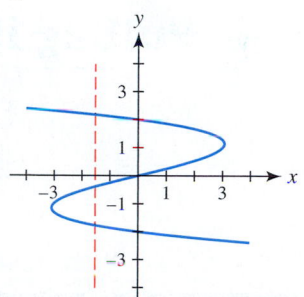

FIGURE 1.55   FIGURE 1.56

SOLUTION
(a) Note in Figure 1.55 that every vertical line that could be visualized would intersect the graph at most once. Therefore the graph represents a function.
(b) The graph in Figure 1.56 does not represent a function because it is possible for a vertical line to intersect the graph more than once.   *Now Try Exercises 77 and 79* ◆

Equations can sometimes define functions. In the next example we determine if an equation defines a function.

EXAMPLE 10   Identifying a function

Determine if y is a function of x.
(a) $x = y^2$ (b) $y = x^2 - 2$

SOLUTION
(a) For y to be a function of x in the equation $x = y^2$, each valid x-value must result in one y-value. If we let $x = 4$, then y could be either -2 or 2 since

$$4 = (-2)^2 \quad \text{and} \quad 4 = (2)^2.$$

Therefore y is not a function of x. A graph of the equation $x = y^2$ is shown in Figure 1.57. Note that this graph fails the vertical line test.
(b) In the equation $y = x^2 - 2$ each x-value determines exactly one y-value, and y is a function of x. A graph of this equation is shown in Figure 1.58. Note that this graph passes the vertical line test.

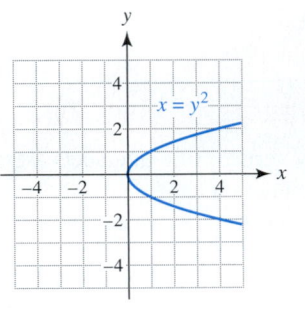

FIGURE 1.57   FIGURE 1.58

Now Try Exercises 93 and 95 ◆

1.3 Putting it all Together

One of the most important concepts in mathematics is that of a function. A function always computes exactly one output for each element in its domain. The domain of a function f consists of the set of all valid x-inputs, and the range is the set of corresponding y-outputs. The following table summarizes some important topics from this section.

Concept	Explanation	Examples
Function	A function is a *relation* in which each valid input results in one output. The *domain* of a function is the set of valid inputs (x-values) and the *range* is the set of resulting outputs (y-values).	$f = \{(1, 3), (2, 6), (3, 9), (4, 12)\}$ The domain is $D = \{1, 2, 3, 4\}$ and the range is $R = \{3, 6, 9, 12\}$.
Domain	When a function is represented by a formula, its domain is the set of all valid inputs (x-values) that are defined or make sense in the formula, unless stated otherwise.	$$f(x) = \sqrt{x + 4}$$ Domain of f: $\{x \mid x \geq -4\}$ $$g(x) = \frac{1}{x - 5}$$ Domain of g: $\{x \mid x \neq 5\}$

Concept	Explanation	Examples
Identifying graphs of functions	The vertical line test can be used to determine if a graph represents a function.	Not a function because a vertical line can intersect the graph more than once
Verbal representation of a function	Words describe precisely what is computed.	A verbal representation of $f(x) = x^2$ is "Square the input x to obtain the output."
Symbolic representation of a function	Mathematical formula	The squaring function is given by $f(x) = x^2$, and the square root function is given by $g(x) = \sqrt{x}$.
Numerical representation of a function	Table of values	A *partial* numerical representation of $f(x) = 3x$ is shown. $\begin{array}{c\|cccc} x & 0 & 1 & 2 & 3 \\ \hline f(x) & 0 & 3 & 6 & 9 \end{array}$
Graphical representation of a function	Graph of ordered pairs (x, y) that satisfy $y = f(x)$	A graph of $f(x) = 2x$ is shown. Each point on the graph satisfies $y = 2x$.

1.3 ▸ Exercises

Evaluating and Representing Functions

1. If $f(-2) = 3$, identify a point on the graph of f.

2. If $f(3) = -9.7$, identify a point on the graph of f.

3. If $(7, 8)$ lies on the graph of f, then $f(\underline{\quad}) = \underline{\quad}$.

4. If $(-3, 2)$ lies on the graph of f, then $f(\underline{\quad}) = \underline{\quad}$.

Exercises 5–22: Graph $y = f(x)$ by hand by first plotting points to determine the shape of the graph.

5. $f(x) = 3$ 6. $f(x) = -2$

7. $f(x) = 2x$ 8. $f(x) = x + 1$

9. $f(x) = 4 - x$ 10. $f(x) = 3 + 2x$

11. $f(x) = -3x + 1$ 12. $f(x) = 2 - 2x$

13. $f(x) = \frac{1}{2}x - 2$ 14. $f(x) = \frac{1}{3}x + 3$

15. $f(x) = |x - 1|$ 16. $f(x) = |0.5x|$

17. $f(x) = |3x|$ 18. $f(x) = |2x - 1|$

19. $f(x) = \frac{1}{2}x^2$ 20. $f(x) = 2x^2$

21. $f(x) = x^2 - 2$ 22. $f(x) = x^2 + 1$

Exercises 23–36: Complete the following.
   (a) Evaluate the given $f(x)$ for the indicated values of x, if possible.
   (b) Find the domain of f.

23. $f(x) = x^3$ for $x = -2, 5$

24. $f(x) = 2x - 1$ for $x = 8, -1$

25. $f(x) = \sqrt{x}$ for $x = -1, a + 1$

26. $f(x) = \sqrt{1 - x}$ for $x = -2, a + 2$

27. $f(x) = \dfrac{1}{x - 1}$ for $x = -1, a + 1$

28. $f(x) = \dfrac{3x - 5}{x + 5}$ for $x = -1, a$

29. $f(x) = -7$ for $x = 6, a - 1$

30. $f(x) = x^2 - x + 1$ for $x = 1, -2$

31. $f(x) = \dfrac{1}{x^2}$ for $x = 4, -7$

32. $f(x) = \sqrt{x - 3}$ for $x = 4, a + 4$

33. $f(x) = \dfrac{1}{x^2 - 9}$ for $x = 4, a - 5$

34. $f(x) = \dfrac{1}{x^2 + 4}$ for $x = -2, a + 4$

35. $f(x) = \dfrac{1}{\sqrt{2 - x}}$ for $x = 1, a + 2$

36. $f(x) = \dfrac{1}{\sqrt{x - 1}}$ for $x = 0, a^2 - a + 1$

Exercises 37–42: (Refer to Example 4.) Use the graph to complete the following.
   (a) Find the domain of g.
   (b) Use the formula for $g(x)$ shown in the graph to evaluate $g(-1)$ and $g(2)$.
   (c) Use the graph of g to evaluate $g(-1)$ and $g(2)$.

37.
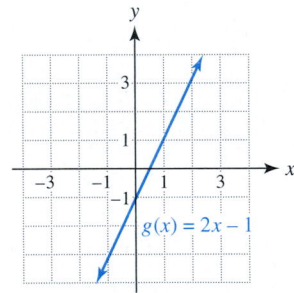
$g(x) = 2x - 1$

38.
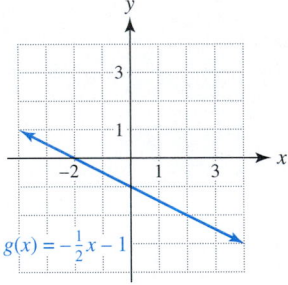
$g(x) = -\frac{1}{2}x - 1$

39.
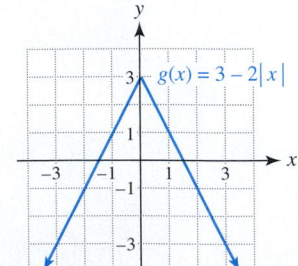
$g(x) = 3 - 2|x|$

40.
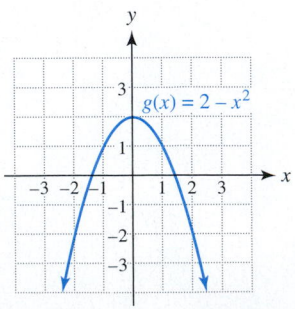
$g(x) = 2 - x^2$

41.

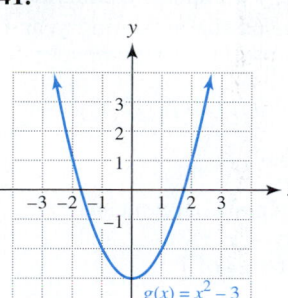

42.

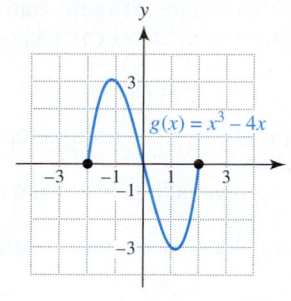

49.

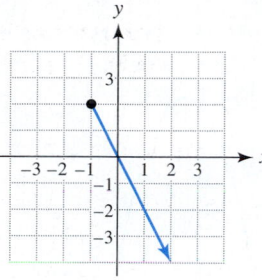

50.

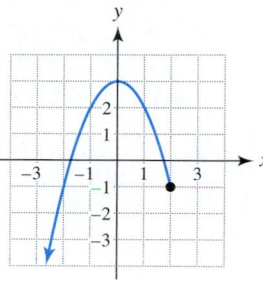

Exercises 43–50: Use the graph of the function f to estimate its domain and range. Evaluate f(0).

Exercises 51 and 52: Use the given diagram to complete the following.

(a) *Evaluate f(2).*
(b) *Write f as a set of ordered pairs.*
(c) *Find the domain and range of f.*

43.

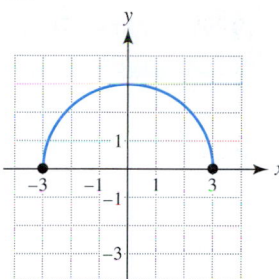

44.

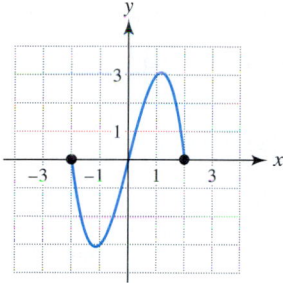

51.

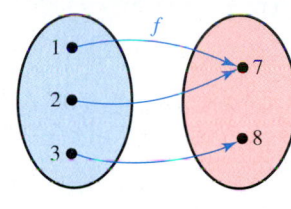

52.

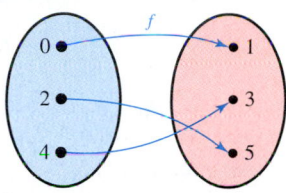

Exercises 53–58: Refer to the graph of f in the figure.

(a) *Evaluate f(0) and f(2).*
(b) *Find all x such that f(x) = 0.*

45.

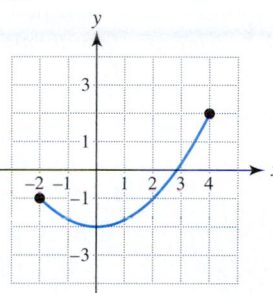

46.

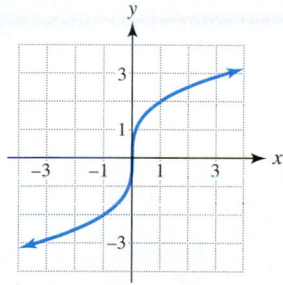

53.

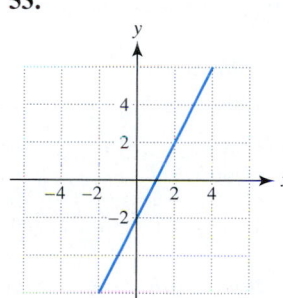

54.

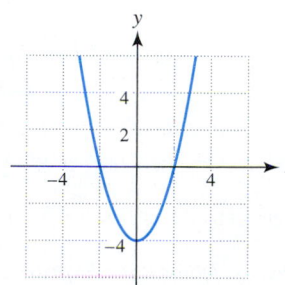

47.

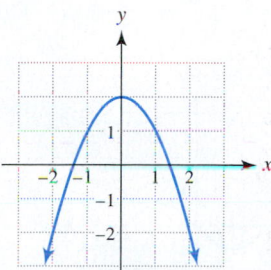

48.

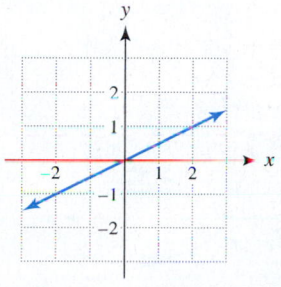

55.

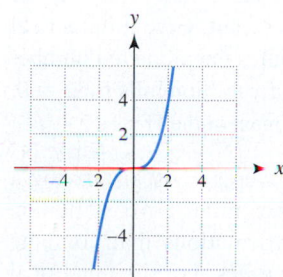

56.

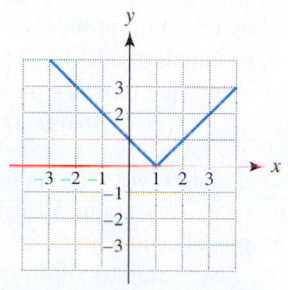

57.

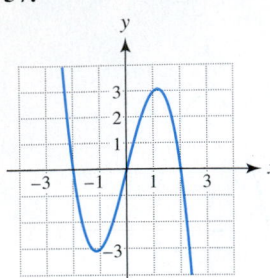

58.

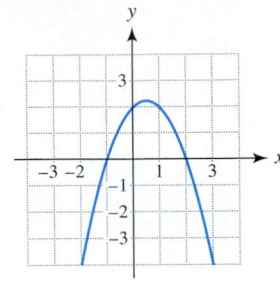

Exercises 59–62: Graph $y = f(x)$ in the viewing rectangle $[-4.7, 4.7, 1]$ by $[-3.1, 3.1, 1]$.

(a) *Use the graph to evaluate $f(2)$.*
(b) *Evaluate $f(2)$ symbolically.*
(c) *Let $x = -3, -2, -1, 0, 1, 2, 3$ and make a table of values for $f(x)$.*

59. $f(x) = 0.25x^2$ **60.** $f(x) = 3 - 1.5x^2$

61. $f(x) = \sqrt{x + 2}$ **62.** $f(x) = |1.6x - 2|$

Exercises 63–70: Use $f(x)$ to determine verbal, graphical, and numerical representations. For the numerical representation use a table with $x = -2, -1, 0, 1, 2$.

63. $f(x) = x^2$ **64.** $f(x) = 2x - 5$

65. $f(x) = |2x + 1|$ **66.** $f(x) = 8$

67. $f(x) = |x|$ **68.** $f(x) = 5 - x$

69. $f(x) = \sqrt{x + 1}$ **70.** $f(x) = x^2 - 1$

Exercises 71 and 72: A function g is defined.

(a) *Write g as a set of ordered pairs.*
(b) *Give the domain and range of g.*

71. $g(-1) = 2, g(0) = 4, g(1) = -3, g(2) = 2$

72. $g(-4) = 5, g(0) = -5, g(4) = 5, g(8) = 0$

Exercises 73 and 74: Express a function f with the specified representation.

73. *Counterfeit Money*   It is estimated that nine out of every one million bills are counterfeit. Give a numerical representation (table) that computes the predicted number of counterfeit bills in a sample of x million bills for $x = 0, 1, 2, \ldots, 6$.   (**Source:** Department of the Treasury.)

74. *Cost of Driving*   In 2004 the average cost of driving a new car was about 50 cents per mile. Give symbolic, graphical, and numerical representations that compute the cost in dollars of driving x miles. For the numerical representation use a table with $x = 1, 2, 3, 4, 5, 6$. (**Source:** Associated Press.)

75. *Radio Stations*   The function f computes the number y in thousands of radio stations on the air during year x. (**Source:** M. Street Corporation.)

$$f = \{(1950, 2.8), (1975, 7.7), (2001, 13.1)\}$$

(a) Draw a diagram to represent f.

(b) Evaluate $f(1975)$ and explain what it means.

(c) Identify the domain and range of f.

76. *Unhealthy Air Quality*   The Environmental Protection Agency (EPA) monitors air quality in U.S. cities. The function f, represented by the table, gives the annual number of days with unhealthy air quality in Los Angeles.

x	1996	1997	1998	1999	2000	2001
$f(x)$	94	60	56	27	48	30

(a) Find $f(1998)$. Interpret the result.

(b) Write f as a set of ordered pairs.

(c) Find the domain and range of f.

Identifying Functions

Exercises 77–82: Determine if the graph represents a function. If it represents a function, determine its domain and range.

77.

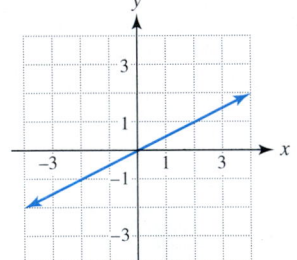

78.

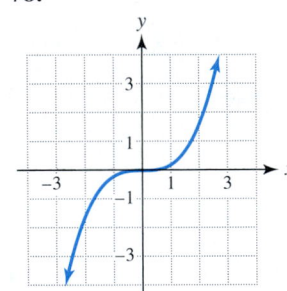

79.

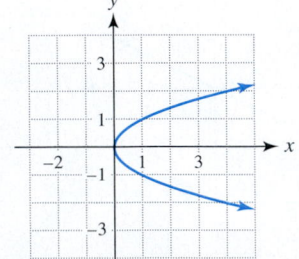

80.

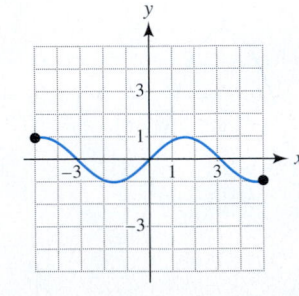

81.   **82.**

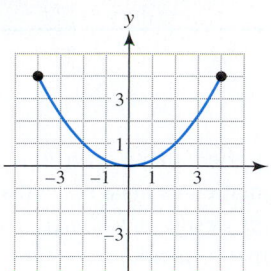   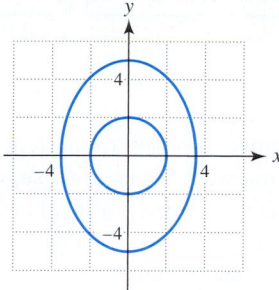

Exercises 83–86: Determine if the following describes a function. Explain your answer.

83. Calculating the cube root of a number

84. Calculating your age

85. Listing the students who passed a given English exam

86. Finding the *x*-values in the domain of a relation

Exercises 87–92: Determine if S is a function.

87. $S = \{(1, 2), (2, 3), (4, 5), (1, 3)\}$

88. $S = \{(-3, 7), (-1, 7), (3, 9), (6, 7), (10, 0)\}$

89. $S = \{(a, 2), (b, 3), (c, 3), (d, 3), (e, 2)\}$

90. $S = \{(a, 2), (a, 3), (b, 5), (-b, 7)\}$

91. *S* is given by the table.

x	1	3	1
y	10.5	2	−0.5

92. *S* is given by the table.

x	1	2	3
y	1	1	1

Exercises 93–96: Determine if y is a function of x.

93. $x = y^4$ **94.** $y^2 = x + 1$

95. $\sqrt{x + 1} = y$ **96.** $x^2 = y - 7$

Exercises 97–102: **Formulas** *Write a symbolic representation (formula) for a function g that calculates the given quantity. Then evaluate g(10) and interpret the result.*

97. The number of inches in *x* feet.

98. The number of quarts in *x* gallons.

99. The number of dollars in *x* quarters.

100. The number of quarters in *x* dollars.

101. The number of seconds in *x* days.

102. The number of feet in *x* miles

Applications

103. *Distance to Lightning* The speed of sound varies with the temperature of the air. Find a formula for a function *f* that computes the distance between an observer and a lightning bolt when the speed of sound is 1150 feet per second. Evaluate *f*(15) and interpret the result.

104. *Distance to Lightning* Give a reasonable domain for the function *f* that you found in Exercise 103. Graph *f* over the domain that you selected. What is the range of your function? (Note that answers may vary.)

105. *Air Temperature* (Refer to Example 6.) When the relative humidity is 100%, air cools 5.8°F for every 1-mile increase in altitude. Give verbal, symbolic, graphical, and numerical representations of a function *f* that computes this change in temperature for an increase of *x* miles. Let the domain of *f* be $0 \le x \le 3$. (**Source:** L. Battan.)

106. *Crutch Length* (Refer to Example 7.) Determine the crutch length for someone 6 feet 3 inches tall. For each 1-inch increase in a person's height, by how much does the recommended crutch length increase?

Writing about Mathematics

107. Explain how you could use a complete numerical representation (table) for a function to determine its domain and range.

108. Explain in your own words what a function is. How is a function different from a relation?

1.4   Types of Functions and Their Rates of Change

♦ Identify and use constant and linear functions

♦ Interpret slope as a rate of change

♦ Identify and use nonlinear functions

♦ Recognize linear and nonlinear data

♦ Use and interpret average rate of change

♦ Calculate the difference quotient

Introduction

A central theme of applied mathematics is describing real-world phenomena with functions. Because applications involving real data are diverse, mathematicians have created a wide assortment of functions. In fact, professional mathematicians design new functions every day for use in business, education, and government. Mathematics is not static—it is dynamic. It requires both ingenuity and creativity to analyze data and make predictions about the future. This section demonstrates how different types of data require a variety of functions to describe them.

Mathematicians often classify problems into one of two general categories: *linear* or *nonlinear*. Functions that are used to solve these problems can also be classified as either linear or nonlinear. Linear functions are usually easier to graph and evaluate than nonlinear functions. A simple example of a linear function is called a *constant* function. For this reason, we begin by discussing constant functions.

Constant Functions

The monthly average wind speeds in miles per hour at Hilo, Hawaii, from May through December are listed in Table 1.15.

TABLE 1.15

Month	May	June	July	Aug	Sept	Oct	Nov	Dec
Wind Speed (mph)	7	7	7	7	7	7	7	7

Source: J. Williams, *The Weather Almanac 1995*.

It is apparent that the monthly average wind speed is constant between May and December. This data can be described by a set f of ordered pairs (x, y), where x is the month and y is the wind speed. The months have been assigned the standard numbers.

$$f = \{(5, 7), (6, 7), (7, 7), (8, 7), (9, 7), (10, 7), (11, 7), (12, 7)\}$$

A scatterplot of f is shown in Figure 1.59. By the vertical line test f is a function.

The function f is given by $f(x) = 7$, where $x = 5, 6, 7, \ldots, 12$. The output of f never changes. We say f is a *constant function*.

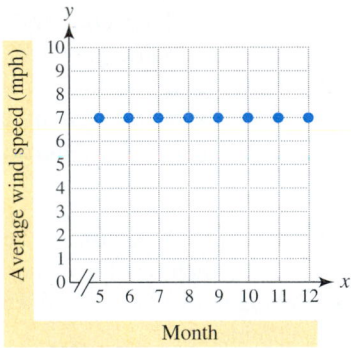

FIGURE 1.59 A Discrete Constant Model

CONSTANT FUNCTION

A function f represented by $f(x) = b$, where b is a constant (fixed number), is a **constant function**.

In the previous discussion $f(x) = 7$, so $b = 7$. The range of f is $R = \{7\}$. The range of a constant function contains one element. The domain of f is $D = \{5, 6, 7, 8, 9, 10, 11, 12\}$. Since f is defined only at individual or discrete values of x, f is called a **discrete function**. The graph of a discrete function suggests a scatterplot.

Sometimes it is more convenient to describe discrete data with a continuous graph. If the domain of $f(x) = 7$ is changed to $D = \{x \mid 5 \le x \le 12\}$, its graph becomes a continuous horizontal line without breaks. See Figure 1.60. The graph of a **continuous function** can be

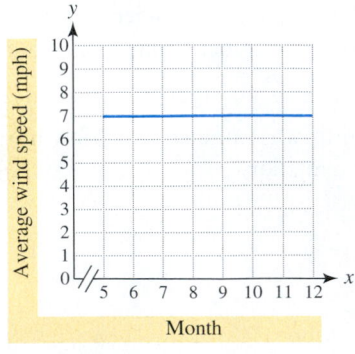

FIGURE 1.60 A Continuous Constant Model

sketched without picking up the pencil. There are no breaks in the graph of a continuous function. The graph of a continuous constant function is a horizontal line.

If the domain includes $x = 6.5$, what does $f(6.5) = 7$ represent? The expression $f(6)$ computes the average wind speed in June, and $f(7)$ gives the average wind speed in July. We might interpret $f(6.5)$ to represent the average wind speed from June 15 to July 15. Other interpretations are possible.

As simple as constant functions might appear, they occur frequently in applications. The following are examples that can be described by constant functions. In both cases the independent variable (input) is time.

• A thermostat computes a constant function regardless of the weather outside by maintaining a set temperature.
• A cruise control in a car computes a constant function by maintaining a fixed speed regardless of the type of road or terrain.

Linear Functions

A car is initially located 30 miles north of the Texas border, traveling north on Interstate 35 at 60 miles per hour. The distances between the automobile and the border are listed in Table 1.16 for various times.

TABLE 1.16

Elapsed Time (hours)	0	1	2	3	4	5
Distance (miles)	30	90	150	210	270	330

It can be seen that the distance increases by 60 miles every hour. These data can be given by a set f of ordered pairs (x, y), where x is the elapsed time and y is the distance from the border.

$$f = \{(0, 30), (1, 90), (2, 150), (3, 210), (4, 270), (5, 330)\}$$

The set f is a function, but not a constant function. A scatterplot of f is shown in Figure 1.61. The scatterplot suggests a line that rises from left to right.

If the car travels for x hours, the distance traveled can be computed by multiplying 60 times x and adding the initial distance of 30 miles. This computation can be expressed as $f(x) = 60x + 30$. The formula is valid for nonnegative values of x. For example, $f(\mathbf{1.5}) = 60(\mathbf{1.5}) + 30 = \mathbf{120}$ means that the car is 120 miles from the border after 1.5 hours. A graph of $f(x) = 60x + 30$ is shown in Figure 1.62. Its graph is a line. We call f a *linear function*.

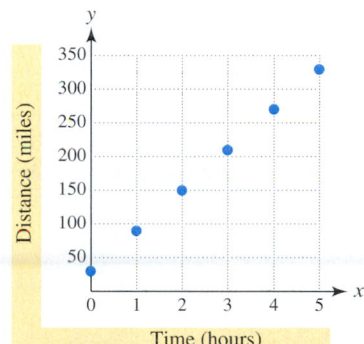

FIGURE 1.61 A Discrete Linear Model

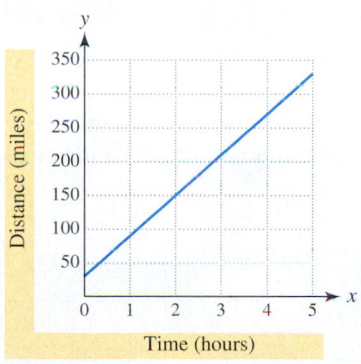

FIGURE 1.62 A Continuous Linear Model

LINEAR FUNCTION

A function f represented by $f(x) = ax + b$, where a and b are constants, is a **linear function**.

If $a = 0$ then $f(x) = b$, which defines a constant function. Thus every constant function is also a linear function. In the example of the moving car, $f(x) = 60x + 30$, so $a = 60$ and $b = 30$. The value of a represents the speed of the car, and b corresponds to the initial distance of the car from the border. Other examples of linear functions include the following. The values of the constants a and b are also given.

$$f(x) = 1.5x - 6 \qquad a = 1.5, \quad b = -6$$
$$g(t) = 8t \qquad\qquad a = 8, \quad\ b = 0$$
$$h(x) = 72 \qquad\qquad a = 0, \quad\ b = 72$$
$$k(t) = 1.9 - 3t \qquad a = -3, \quad b = 1.9$$

A distinguishing feature of a linear function is that each time x increases by one unit, the value of $f(x)$ always changes by an amount equal to a. That is, a linear function f has a **constant rate of change**. (The constant rate of change a is equal to the slope of the graph of f.) The following applications are modeled by linear functions. Try to determine the value of the constant a in each case.

- The wages earned by an individual working x hours at \$6.25 per hour
- The amount of tuition and fees when registering for x credits if each credit costs \$75 and the fees are fixed at \$56
- The distance traveled by light in x seconds if the speed of light is 186,000 miles per second

Slope as a Rate of Change

The graph of a (continuous) linear function is a line. Slope is a real number that measures the "tilt" of a line in the xy-plane. If the input x to a linear function increases by 1 unit, then the output y changes by a constant amount that is equal to the slope of its graph. In Figure 1.63 a line passes through the points (x_1, y_1) and (x_2, y_2). The *change in y* is $y_2 - y_1$, and the *change in x* is $x_2 - x_1$. The ratio of the change in y to the change in x is called the *slope*. We sometimes denote the change in y by Δy (delta y) and the change in x by Δx (delta x). That is, $\Delta y = y_2 - y_1$ and $\Delta x = x_2 - x_1$.

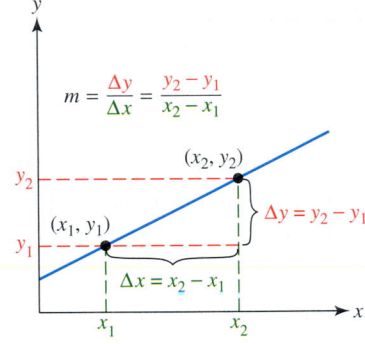

FIGURE 1.63

SLOPE

The **slope** m of the line passing through the points (x_1, y_1) and (x_2, y_2) is

$$m = \frac{\Delta y}{\Delta x} = \frac{y_2 - y_1}{x_2 - x_1},$$

where $x_1 \neq x_2$.

If the slope of a line is positive, the line *rises* from left to right. If the slope is negative, the line *falls* from left to right. Slope 0 indicates that the line is horizontal. When $x_1 = x_2$, the line is vertical and the slope is undefined. Figures 1.64–1.67 illustrate these situations. Slope 2 indicates that a line rises 2 units for every unit increase in x and slope $-\frac{1}{2}$ indicates that the line *falls* $\frac{1}{2}$ unit for every unit increase in x.

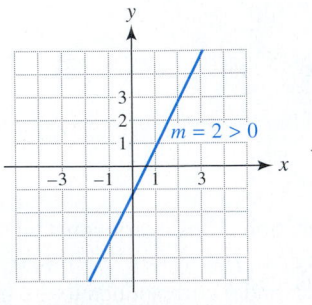

FIGURE 1.64

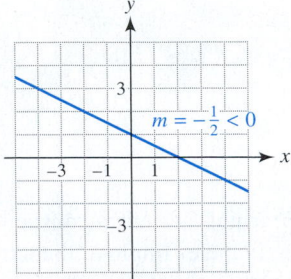

FIGURE 1.65

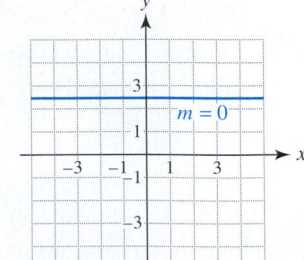

FIGURE 1.66

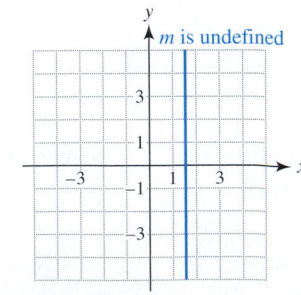

FIGURE 1.67

EXAMPLE 1

Calculating the slope of a line

Find the slope of the line passing through the points $(-2, 3)$ and $(1, -2)$. Plot these points together with the line. Interpret the slope.

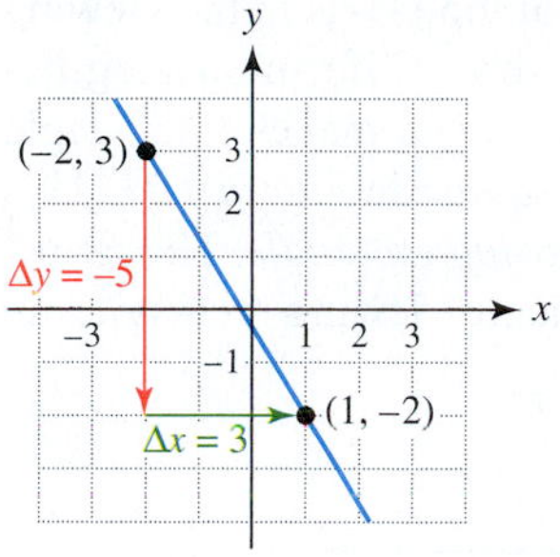

SOLUTION The slope is

$$m = \frac{y_2 - y_1}{x_2 - x_1} = \frac{-2 - 3}{1 - (-2)} = -\frac{5}{3}.$$

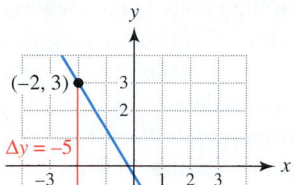

FIGURE 1.68

A graph of the line passing through these two points is shown in Figure 1.68. The change in y is $\Delta y = -5$ and the change in x is $\Delta x = 3$, so $m = \frac{\Delta y}{\Delta x} = -\frac{5}{3}$ indicates that the line falls $\frac{5}{3}$ units for each unit increase in x, or equivalently, the line falls 5 units for each 3-unit increase in x.   *Now Try Exercise 1* ◆

The graph of $f(x) = ax + b$ is a line. Since $f(0) = b$ and $f(1) = a + b$, the graph of f passes through the points $(0, b)$ and $(1, a + b)$. The slope of this line is

$$m = \frac{y_2 - y_1}{x_2 - x_1} = \frac{a + b - b}{1 - 0} = \frac{a}{1} = a.$$

In applications involving linear functions, slope sometimes is interpreted as a (*constant*) *rate of change*. In Example 7 of Section 1.3, the recommended crutch length for a person x inches tall was given by $f(x) = 0.72x + 2$. The slope of its graph is 0.72. One interpretation of this slope is that for each 1-inch increase in the height of a person, the crutch length should be increased by 0.72 inch.

EXAMPLE 2

Interpreting slope as rate of change

Figure 1.69 shows the price of x tons of landscape rock.
(a) Why is it reasonable for the graph to pass through the origin?
(b) Find the slope of the graph.
(c) Interpret the slope as a rate of change.

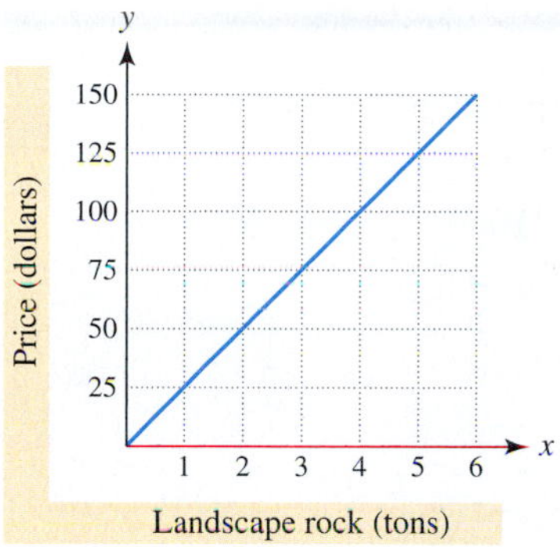

SOLUTION
(a) Because 0 tons of landscape rock cost \$0, the graph passes through the point $(0, 0)$.
(b) The graph passes through the points $(0, 0)$ and $(2, 50)$. The slope of this line is

$$m = \frac{50 - 0}{2 - 0} = 25.$$

(c) The cost of landscape rock is \$25 per ton.   *Now Try Exercise 27* ◆

FIGURE 1.69

◆ **MAKING CONNECTIONS**

Units for Rates of Change   When using a graph, the units for a rate of change can be found by placing the units from the vertical axis over the units from the horizontal axis. For example, in Figure 1.69 the units on the y-axis are *dollars* and the units on the x-axis are *tons*. Thus, the units for the slope or rate of change are *dollars per ton*.

◆

Squaring a Viewing Rectangle (Optional)   The graph of the line $y = x$ in the standard viewing rectangle is shown in Figure 1.70 on the next page. The visual distances between consecutive tick marks on the two axes are not equal. As a result, the line $y = x$ does not appear to have slope 1. Notice that although the box has 5 units on each side, it appears

Calculator Help

To set a square viewing rectangle, see Appendix B (page AP-10).

as a rectangle and not as a square. The visual change in y is less than the visual change in x. The reason for this is that the graphing calculator screen is wider than it is high. If we graph $y = x$ in the window $[-9, 9, 1]$ by $[-6, 6, 1]$, the length of the y-axis in the viewing rectangle is $\frac{2}{3}$ of the length of the x-axis. See Figure 1.71. There is uniform spacing between consecutive tick marks on both axes. In this case, the line $y = x$ makes a 45° angle with the x-axis. The rectangle in Figure 1.70 now appears as a *square* in Figure 1.71. This process of setting the viewing rectangle is called *squaring a viewing rectangle*. The ratio $\frac{2}{3}$ may be different on other calculators. On some calculators a square viewing rectangle can be set automatically by using the ZOOM menu.

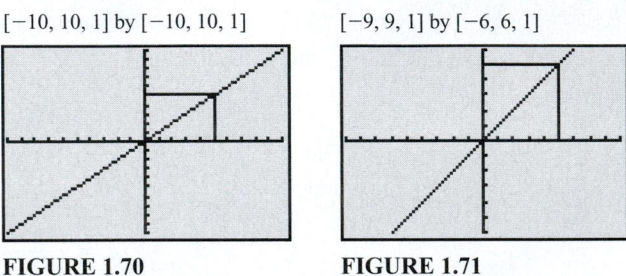

$[-10, 10, 1]$ by $[-10, 10, 1]$ $[-9, 9, 1]$ by $[-6, 6, 1]$

FIGURE 1.70   **FIGURE 1.71**

Nonlinear Functions

Table 1.17 shows the population of the United States at 20-year intervals from 1800 to 2000. Between 1800 and 1820 the population increased by 5 million, between 1820 and 1840 it increased by 7 million, and between 1840 and 1860 the increase was 14 million. If the increases in population for each 20-year period had been equal, or nearly equal, these data could be modeled by a linear function.

TABLE 1.17

Year	Population (millions)	Year	Population (millions)
1800	5	1920	106
1820	10	1940	132
1840	17	1960	179
1860	31	1980	226
1880	50	2000	281
1900	76		

The actual population data are nonlinear. In Figure 1.72 the data together with a nonlinear function f have been plotted. The graph of f is curved rather than straight.

If a function is not linear, then it is called a **nonlinear function**. *The graph of a nonlinear function is not a (straight) line.* With a nonlinear function, it is possible for the input x to increase by 1 unit and the output y to change by different amounts. Nonlinear functions *cannot* be written in the form $f(x) = ax + b$. Examples of nonlinear functions include the following:

$$f(x) = x^2 - 3x + 2, \quad g(x) = \sqrt{x}, \quad \text{and} \quad h(x) = x^3.$$

Real-world phenomena often are modeled by using nonlinear functions. The following are some examples of quantities that can be described by nonlinear functions.

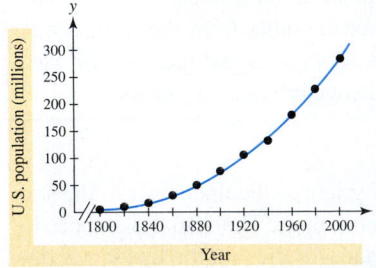

FIGURE 1.72 A Nonlinear Model

◆ **CLASS DISCUSSION**
The time required to drive a distance of 100 miles depends on the average speed. Let the function f compute this time, given the average speed x as input. For example, $f(50) = 2$, since it would take 2 hours to travel 100 miles at an average speed of 50 miles per hour. Make a table of values for f. Is f linear or nonlinear? ◆

- The total number of people who have contracted AIDS beginning in 1980
  (The increase in the number of AIDS cases is not the same each year.)
- The monthly average temperature in Chicago
  (Monthly average temperatures increase and decrease throughout the year.)
- The height of a child between the ages of 2 and 18
  (A child grows faster at certain ages.)

There are many examples of nonlinear functions. You may have encountered some of them in previous mathematics courses. In Figures 1.73–1.76 graphs and formulas are given for four common nonlinear functions. Note that each graph is not a line.

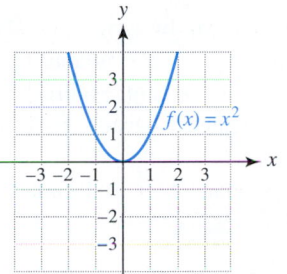

FIGURE 1.73 Square Function

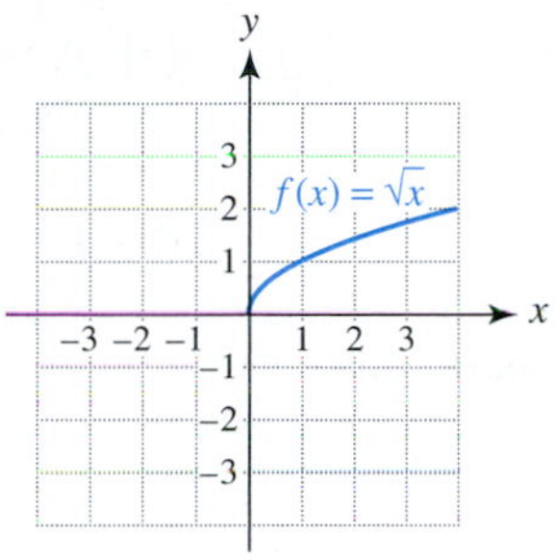

FIGURE 1.74 Square Root Function

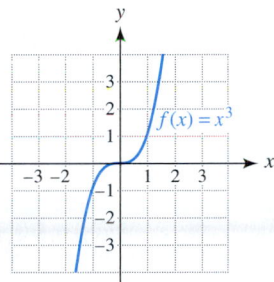

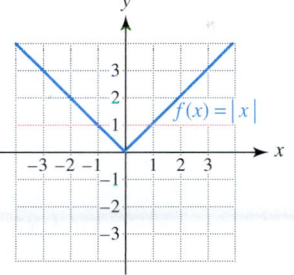

FIGURE 1.75 Cube Function

FIGURE 1.76 Absolute Value Function

 **EXAMPLE 3**   **Recognizing linear and nonlinear data**

For each table decide whether the data are linear or nonlinear. If the data are linear, state the slope of the line that passes through the data points.

(a)

x	0	5	10	15	20
y	−4	−2	0	2	4

(b)

x	−3	0	3	6	9
y	5	7	10	14	19

(c)

x	0	1	2	3	4
y	11	11	11	11	11

(d)

x	0	1	3	6	10
y	3	6	9	12	15

SOLUTION
(a) Between each pair of points, the y-values increase 2 units for each 5-unit increase in x. Therefore the data are linear, but not constant. The slope of the line passing through the data points is $m = \frac{2}{5}$.
(b) The y-values do not increase by a constant amount for each 3-unit increase in x. The data are nonlinear.
(c) These data are linear and constant. The slope of the line passing through these data points is 0.
(d) Although the y-values increase by 3 units between consecutive data points, the data are nonlinear because the corresponding increases in the x-values are not constant. For

example, the slope of the line passing through the points (0, 3) and (1, 6) is 3, whereas the slope of the line passing through the points (1, 6) and (3, 9) is 1.5.

Now Try Exercises 31 and 33 ◆

EXAMPLE 4   ### Recognizing linear and nonlinear functions

Determine if f is a linear or nonlinear function. If f is a linear function, state if f is a constant function. Support your results by graphing f.

(a) $f(x) = 6 - 4x$ **(b)** $f(x) = 3x^2 - 2$ **(c)** $f(x) = 5$

SOLUTION

(a) A linear function can be written as $f(x) = ax + b$, where a and b are real numbers. We can write the given formula as $f(x) = -4x + 6$, so f is linear with $a = -4$ and $b = 6$ but f is not constant. Figure 1.77 shows that the graph of f is a line.

(b) Function f is nonlinear because its formula contains an x^2-term. Its formula cannot be written in the form $f(x) = ax + b$. The graph of f is not a line, as shown in Figure 1.78.

(c) Function f is both linear and constant. Its graph is a horizontal line, as shown in Figure 1.79.

Calculator Help

To set a viewing rectangle and graph a function, see Appendix B (pages AP-7 and AP-9).

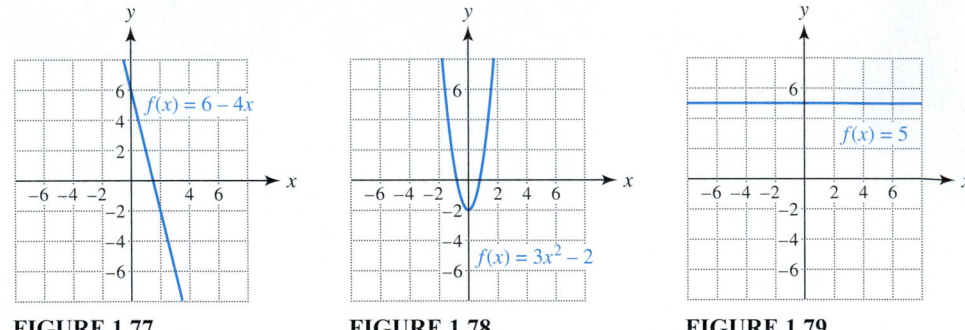

FIGURE 1.77   **FIGURE 1.78**   **FIGURE 1.79**

Now Try Exercises 37, 41, and 45 ◆

Average Rate of Change

The graphs of nonlinear functions are not lines, so there is no notion of a single slope. The slope of the graph of a linear function gives its rate of change. With a nonlinear function we speak of an *average* rate of change. Suppose that the points (x_1, y_1) and (x_2, y_2) lie on the graph of a nonlinear function f. See Figure 1.80. The slope of the line L passing through these two points represents the *average rate of change of f from x_1 to x_2*. The line L is referred to as a **secant line**. If different values for x_1 and x_2 are selected, then a different secant line and a different average rate of change usually result.

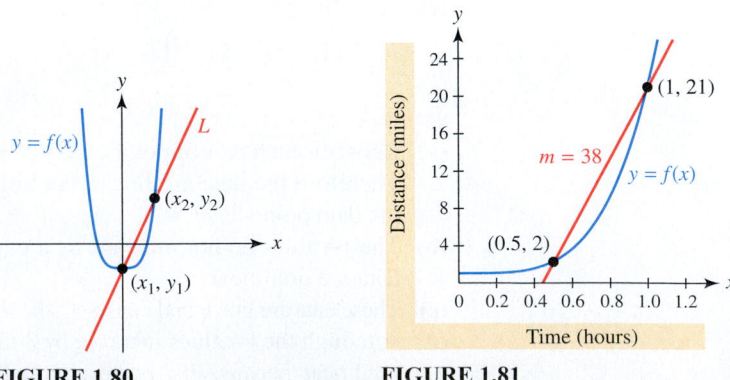

FIGURE 1.80   **FIGURE 1.81**

In applications the average rate of change measures how fast a quantity is changing over an interval of its domain, *on average*. For example, suppose the graph of the function f in Figure 1.81 represents the distance y in miles that a car has traveled on a straight highway (under construction) after x hours. The points $(0.5, 2)$ and $(1, 21)$ lie on this graph. Thus after 0.5 hour the car has traveled 2 miles and after 1 hour the car has traveled 21 miles. The slope of the line passing through these two points is

$$m = \frac{y_2 - y_1}{x_2 - x_1} = \frac{21 - 2}{1 - 0.5} = 38.$$

This means that during the half hour from 0.5 to 1 hour the average rate of change, or average velocity, was 38 miles per hour. These ideas lead to the following definition.

AVERAGE RATE OF CHANGE

Let (x_1, y_1) and (x_2, y_2) be distinct points on the graph of a function f. The **average rate of change of f from x_1 to x_2** is

$$\frac{y_2 - y_1}{x_2 - x_1}.$$

That is, the average rate of change from x_1 to x_2 equals the slope of the line passing through (x_1, y_1) and (x_2, y_2).

Note: Average rate of change equals $\dfrac{f(x_2) - f(x_1)}{x_2 - x_1}$.

If f is a constant function, its average rate of change is zero. For a linear function defined by $f(x) = ax + b$, its average rate of change is equal to a, the slope of its graph. The average rate of change for a nonlinear function varies.

 ### Calculating and interpreting average rates of change

Use Table 1.17 and Figure 1.72 on page 50 to complete the following.
(a) Calculate the average rates of change in the U.S. population from 1800 to 1840 and from 1900 to 1940. Interpret the results.
(b) Illustrate your results from part (a) graphically.

SOLUTION
(a) In 1800 the population was 5 million and in 1840 it was 17 million. Therefore, the average rate of change in the population from 1800 to 1840 was

$$\frac{17 - 5}{1840 - 1800} = 0.3.$$

In 1900 the population was 76 million and in 1940 it was 132 million. Therefore the average rate of change in the population from 1900 to 1940 was

$$\frac{132 - 76}{1940 - 1900} = 1.4.$$

This means that from 1800 to 1840, the U.S. population increased, *on average*, by 0.3 million per year and from 1900 to 1940, the U.S. population increased, *on average*, by 1.4 million per year.

(b) These average rates of change can be illustrated graphically by sketching a line L_1 through the points (1800, 5) and (1840, 17) and another line L_2 through the points (1900, 76) and (1940, 132), as depicted in Figure 1.82. The slope of L_1 is 0.3 and the slope of L_2 is 1.4.

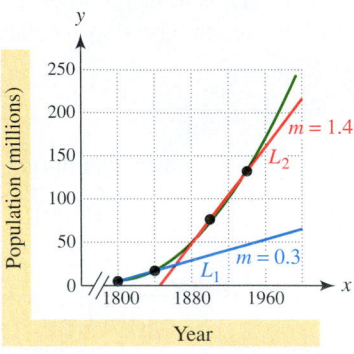

FIGURE 1.82

Now Try Exercise 79 ◆

EXAMPLE 6

Modeling braking distance for a car

On wet, level pavement highway engineers sometimes model the braking distance in feet for a car traveling at x miles per hour by using the formula $f(x) = \frac{1}{9}x^2$. (**Source:** L. Haefner, *Introduction to Transportation Systems.*)

(a) Evaluate $f(30)$ and $f(60)$. Interpret the results.

(b) Calculate the average rate of change of f from 30 to 60. Discuss what the result means.

SOLUTION

(a) $f(\mathbf{30}) = \frac{1}{9}(\mathbf{30})^2 = \mathbf{100}$ and $f(\mathbf{60}) = \frac{1}{9}(\mathbf{60})^2 = \mathbf{400}$. At 30 miles per hour the braking distance is 100 feet and at 60 miles per hour the braking distance is 400 feet.

(b) The average rate of change of f from 30 to 60 is calculated as follows.

$$\frac{\mathbf{400} - \mathbf{100}}{\mathbf{60} - \mathbf{30}} = 10$$

This means that the braking distance increases, on average, by 10 feet for each 1-mile-per-hour increase in speed between 30 and 60 miles per hour.

Now Try Exercise 88 ◆

The Difference Quotient

The difference quotient often occurs in calculus and uses function notation to calculate the average rate of change of a function f in general. Consider the graph of $y = f(x)$ shown in Figure 1.83, and let h be a real number. The points $(x, f(x))$ and $(x + h, f(x + h))$ denote the coordinates of two points on this graph.

The line L passes through these two points and is a *secant line*. The slope m of L is

$$m = \frac{f(x + h) - f(x)}{(x + h) - x} = \frac{f(x + h) - f(x)}{h}.$$

This expression, written in function notation, is called the *difference quotient* and is equal to the average rate of change of f from x to $x + h$.

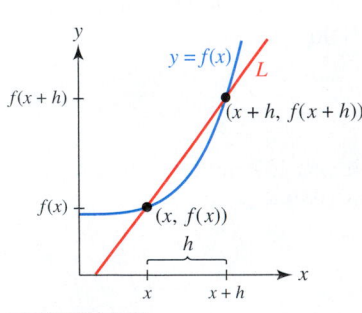

FIGURE 1.83

> **DIFFERENCE QUOTIENT**
>
> The **difference quotient of a function f** is an expression of the form
>
> $$\frac{f(x+h) - f(x)}{h},$$
>
> where $h \neq 0$.

   **Calculating a difference quotient**

Let $f(x) = x^2 - 2x$.
(a) Find $f(x+h)$.
(b) Find the difference quotient of f and simplify the result.

SOLUTION

(a) To calculate $f(x+h)$ substitute $(x+h)$ for x in the expression $x^2 - 2x$.

Algebra Review

To square a binomial, see Chapter R (page R-25).

$$f(\boldsymbol{x+h}) = (\boldsymbol{x+h})^2 - 2(\boldsymbol{x+h}) \qquad f(x) = x^2 - 2x$$
$$= x^2 + 2xh + h^2 - 2x - 2h \qquad \text{Square the binomial; apply the distributive property.}$$

(b) The difference quotient can be calculated as follows.

$$\frac{\boldsymbol{f(x+h)} - f(x)}{h} = \frac{x^2 + 2xh + h^2 - 2x - 2h - (x^2 - 2x)}{h} \qquad \text{Substitute.}$$

$$= \frac{2xh + h^2 - 2h}{h} \qquad \text{Combine like terms.}$$

$$= \frac{\boldsymbol{h}(2x + h - 2)}{\boldsymbol{h}} \qquad \text{Factor out } h.$$

$$= 2x + h - 2 \qquad \text{Simplify.}$$

Now Try Exercise 95 ◆

   **Calculating a difference quotient**

Let the distance d in feet that a racehorse travels after t seconds be given by $d(t) = 2t^2$ for $0 \leq t \leq 10$.
(a) Find $d(t+h)$.
(b) Find the difference quotient of d and simplify the result.
(c) Evaluate the difference quotient for $t = 7$ and $h = 0.1$. Interpret your results.

SOLUTION

(a) To calculate $d(t+h)$ substitute $(t+h)$ for t in the expression $2t^2$.

$$d(\boldsymbol{t+h}) = 2(\boldsymbol{t+h})^2 \qquad \text{Substitute } (t+h) \text{ for } t.$$
$$= 2(t^2 + 2th + h^2) \qquad \text{Square the binomial.}$$
$$= 2t^2 + 4th + 2h^2 \qquad \text{Distributive property}$$

◆ **CLASS DISCUSSION**
How would you estimate the velocity of the racehorse at exactly 7 seconds? ◆

(b) $\dfrac{d(t+h)-d(t)}{h} = \dfrac{2t^2 + 4th + 2h^2 - 2t^2}{h}$ Substitute for $d(t+h)$ and $d(t)$.

$= \dfrac{4th + 2h^2}{h}$ Combine like terms.

$= \dfrac{h(4t + 2h)}{h}$ Factor out h.

$= 4t + 2h$ Simplify.

(c) If $t = 7$ and $h = 0.1$, then the difference quotient becomes

$$4t + 2h = 4(7) + 2(0.1) = 28.2.$$

The average rate of change, or average velocity, of the horse during the time interval from 7 seconds to 7.1 seconds is 28.2 feet per second.   *Now Try Exercise 101* ◆

1.4 Putting it all Together

Functions are either linear or nonlinear. A constant function is a special type of linear function. Slope can be used to describe how a linear function and its graph change. Average rate of change can be used to describe how a nonlinear function and its graph change. The following table summarizes these ideas.

Concept	Formula	Examples
Slope of a line passing through (x_1, y_1) and (x_2, y_2)	$m = \dfrac{\Delta y}{\Delta x} = \dfrac{y_2 - y_1}{x_2 - x_1}$ $\Delta y = y_2 - y_1$ denotes the change in y. $\Delta x = x_2 - x_1$ denotes the change in x.	A line passing through $(-1, 3)$ and $(1, 7)$ has slope $m = \dfrac{7 - 3}{1 - (-1)} = \dfrac{4}{2} = 2$. This slope indicates that the line rises 2 units for each unit increase in x.
Average rate of change of f from x_1 to x_2	If (x_1, y_1) and (x_2, y_2) lie on the graph of f, then the average rate of change from x_1 to x_2 equals $$\dfrac{y_2 - y_1}{x_2 - x_1}.$$	If $f(x) = 3x^2$, then the average rate of change from $x = 1$ to $x = 3$ is given by $$\dfrac{27 - 3}{3 - 1} = 12$$ because $f(3) = 27$ and $f(1) = 3$. This means that, on average, $f(x)$ increases by 12 units for each unit increase in x when $1 \le x \le 3$.
Difference quotient	Calculates average rate of change of f from x to $x + h$. $$\dfrac{f(x+h) - f(x)}{h}, h \ne 0$$	If $f(x) = 2x$, then the difference quotient equals $$\dfrac{2(x+h) - 2x}{h} = \dfrac{2h}{h} = 2.$$

Concept	Formula	Examples
Constant function	$f(x) = b$, where b is a fixed number.	$f(x) = 12$, $g(x) = -2.5$, and $h(x) = 0$. Every constant function is a linear function.
Linear function	$f(x) = ax + b$, where a and b are fixed numbers or constants. The graph of f has slope a.	$f(x) = 3x - 1$, $g(x) = -5$, and $h(x) = \frac{1}{2} - \frac{3}{4}x$. Their graphs have slopes 3, 0, and $-\frac{3}{4}$, respectively.
Nonlinear function	A nonlinear function cannot be expressed in the form $f(x) = ax + b$.	$f(x) = \sqrt{x + 1}$, $g(x) = 4x^3$, and $h(x) = x^{1.01} + 2$

The following table summarizes important concepts related to constant, linear, and nonlinear functions.

Concept	Constant Function	Linear Function	Nonlinear Function
Slope of graph	Always zero	Always constant	No notion of one slope
Average rate of change	Always zero	Always constant	Can vary
Graph	Horizontal line	Nonvertical line	Not a line
Examples	$f(x) = 2$	$f(x) = 2x - 2$	$f(x) = x^2 - 3$
Examples	$f(x) = -3$	$f(x) = -x + 3$	$f(x) = x^3 - 4x$

1.4 — Exercises

Slope

Exercises 1–16: If possible, find the slope of the line passing through each pair of points.

1. $(4, 6), (2, 5)$

2. $(-8, 5), (-3, -7)$

3. $(-1, 4), (5, -2)$

4. $(10, -4), (-15, 7)$

5. $(12, -8), (7, -8)$

6. $(8, -5), (8, 2)$

7. $(0.2, -0.1), (-0.3, 0.4)$

8. $(-0.3, 0.6), (-0.2, 1.1)$

9. $(-0.5, 9.2), (-0.3, 7.6)$

10. $(1.6, 12), (1.6, 5)$

11. $(1997, 5.6), (1994, 7.9)$

12. $(1824, 108), (1900, 380)$

13. $(-5, 6), (-5, 8)$

14. $(17, 7), (19, 7)$

15. $\left(\frac{1}{3}, -\frac{3}{5}\right), \left(-\frac{5}{6}, \frac{7}{10}\right)$

16. $\left(-\frac{13}{15}, -\frac{7}{8}\right), \left(\frac{1}{10}, \frac{3}{16}\right)$

Exercises 17–24: State the slope of the graph of f. Interpret this slope.

17. $f(x) = 2x + 7$

18. $f(x) = 6 - x$

19. $f(x) = 9 - \frac{3}{4}x$

20. $f(x) = \frac{2}{3}x$

21. $f(x) = -5$

22. $f(x) = x + 5$

23. $f(x) = 9 - x$

24. $f(x) = 23$

25. *Velocity of a Car* A driver's distance D in miles from a rest stop on an interstate highway after x hours is given by $D(x) = 75x$.
   (a) How far is the driver from the rest stop after 2 hours?

   (b) Find the slope of the graph of D. Interpret this slope as a rate of change.

26. *Velocity of a Train* A train is initially 150 miles from the station and moving toward the station at 20 miles per hour.
   (a) Write a symbolic representation (formula) for a function D that calculates the distance that the train is from the station after x hours.

   (b) How far is the train from the station after 3 hours?

   (c) Find the slope of the graph of D. Interpret this slope as a rate of change.

27. *Price of Carpet* The graph shows the price of x square yards of carpeting.

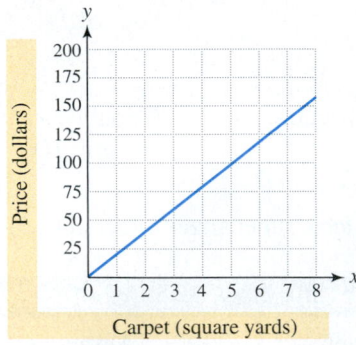

Carpet (square yards)

   (a) Why is it reasonable for the graph to pass through the origin?

   (b) Find the slope of the graph.

   (c) Interpret the slope as a rate of change.

28. *Whirlpool Bath* The graph shows the number of gallons of water in a whirlpool bath x hours past noon.

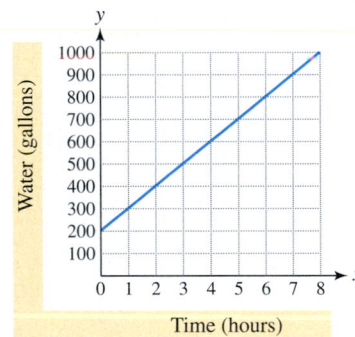

Time (hours)

   (a) The graph does *not* pass through the origin. Interpret this fact.

   (b) Find the slope of the graph.

   (c) Interpret the slope as a rate of change.

29. *Age in the United States* The median age of the U.S. population for each year t between 1970 and 2010 can be approximated by $A(t) = 0.243t - 450.8$. (**Source:** Bureau of the Census.)
   (a) Compute the median ages of the U.S. population in 1980 and 2000.

   (b) What is the slope of the graph of A? Interpret the slope.

30. _Commercial Banks_ From 1987 to 1997 the numbers of federally insured commercial banks could be modeled by $N(t) = -458t + 923,769$, where t is the year. (**Source:** Federal Deposit Insurance Corporation.)

(a) Estimate the number of commercial banks in 1987 and in 1997.

(b) What is the slope of the graph of N? Interpret the slope.

Exercises 31–36: (Refer to Example 3.) Decide whether the data in the table are linear or nonlinear. If the data are linear, state the slope of the line passing through the data points.

31.

x	0	1	2	3	4
y	-1	3	7	11	15

32.

x	-4	-2	0	2	4
y	1.0	-0.5	-2.0	-3.5	-5.0

33.

x	-5	-3	1	3	5
y	-5	-2	1	4	7

34.

x	-4	-2	0	2	4
y	-1	-1	-1	-1	-1

35.

x	-4	0	1	2	5
y	5.0	3.0	2.5	2.0	0.5

36.

x	100	200	250	350	400
y	400	1900	3015	6025	7900

Linear and Nonlinear Functions

Exercises 37–52: (Refer to Example 4.) Determine if f is a linear or nonlinear function. If f is a linear function, state if f is a constant function. Support your answer by graphing f.

37. $f(x) = -2x + 5$

38. $f(x) = -0.01x - 2$

39. $f(x) = 2x^{1.01} - 1$

40. $f(x) = 5.123$

41. $f(x) = 1$

42. $f(x) = \sqrt{(-x)}$

43. $f(x) = 0.2x^3 - 2x + 1$

44. $f(x) = 0.5x^2 - 2x - 3$

45. $f(x) = 5x - x^2$

46. $f(x) = |2x - 1|$

47. $f(x) = |x + 1|$

48. $f(x) = |3|$

49. $f(x) = x^2 - 1$

50. $f(x) = x^3$

51. $f(x) = 2\sqrt{x}$

52. $f(x) = \sqrt{x - 1}$

Exercises 53–56: **Analyzing Real Data** For the given data set complete the following.

(a) Make a line graph of the data. Let this graph represent a function f.

(b) Decide whether f is linear or nonlinear.

53. Interest income after 1 year on an investment earning 7% per year

Investment	$500	$1000	$2000	$3500
Interest	$35	$70	$140	$245

54. World energy consumption in quadrillion Btu (**Note:** 1 quadrillion $= 1 \times 10^{15}$)

Year	1990	1991	1992	1993
Consumption	345	345	345	345

Source: Energy Information Administration, _International Energy Annual._

55. Median incomes of full-time female workers

Year	1970	1980	1990	2000
Income	$5,440	$11,591	$20,591	$32,422

Source: Bureau of the Census, _Current Population Reports._

56. Median incomes of full-time male workers

Year	1970	1980	1990	2000
Income	$9,184	$19,173	$28,979	$37,435

Source: Bureau of the Census, _Current Population Reports._

57. _Average Wind Speed_ The table lists the average wind speed in miles per hour at Myrtle Beach, South Carolina. The months are assigned the standard numbers. (**Source:** J. Williams.)

Month	1	2	3	4	5	6
Wind (mph)	7	8	8	8	7	7

Month	7	8	9	10	11	12
Wind (mph)	7	7	7	6	6	6

(a) Could these data be modeled exactly by a constant function?

(b) Determine a continuous, constant function f that models these data.

(c) Graph f and the data.

58. *Income Tax* The federal income tax rate for single individuals in 2003 was 10% of taxable income between $0 and $7000, and 15% of taxable income between $7000 and $28,400. (**Source:** Internal Revenue Service.)

(a) Find a linear function that computes the tax on taxable incomes x between $0 and $7000. Determine the income tax on a taxable income of $3600.

(b) The linear function given by

$$f(x) = \frac{3}{20}x - 350$$

computes the tax on taxable incomes x between $7000 and $28,400. Determine the income tax on a taxable income of $27,400.

59. Suppose that a car's distance from a service center along a straight highway can be described by a constant function of time. Discuss what can be said about the car's velocity.

60. Suppose that a car's distance in miles from a rest stop along a straight stretch of an interstate highway after x hours can be modeled by $f(x) = ax$. Discuss what can be said about the car's velocity.

Exercises 61–68: Write a symbolic representation (formula) for a function f that computes the following.

61. The number of pounds in x ounces

62. The number of dimes in x dollars

63. The distance traveled by a car moving at 50 miles per hour for x hours

64. The monthly electric bill in dollars for using x kilowatt-hours at 6 cents per kilowatt-hour plus a fixed fee of $6.50

65. The cost of downhill skiing x times with a $500 season pass

66. The total number of hours in day x

67. The distance in miles between a runner and home after x hours if the runner is initially 1 mile from home and jogging *away* from home at 6 miles per hour

68. The speed of a car in feet per second after x minutes if its tires are 2 feet in diameter and rotating 14 times per second

Curve Sketching

*Exercises 69–74: **Critical Thinking** Assume that each function is continuous. Do not use a graphing calculator.*

69. Sketch a graph of a linear function f with $f(3) = -1$ and a positive slope.

70. Sketch a graph of a linear function with a slope of $-\frac{2}{3}$, passing through the point $(-2, 3)$.

71. On the same coordinate axes, sketch the graphs of a constant function f and a nonlinear function g that intersect exactly twice.

72. Sketch a graph of a linear function f that intersects a constant function g exactly once.

73. Sketch a graph of a nonlinear function f that has only positive average rates of change.

74. Sketch a graph of a nonlinear function f that has only negative average rates of change.

Average Rates of Change

Exercises 75–78: Use the graph and formula of f(x) to find the average rates of change of f from −3 to −1 and from 1 to 3.

75. $f(x) = 4$ 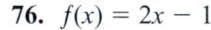 **76.** $f(x) = 2x - 1$

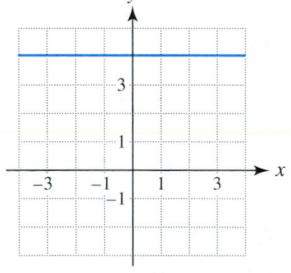 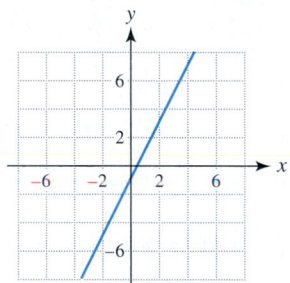

77. $f(x) = -0.3x^2 + 4$ **78.** $f(x) = 0.3x^2 - 4$

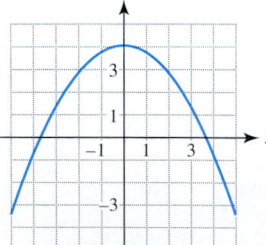 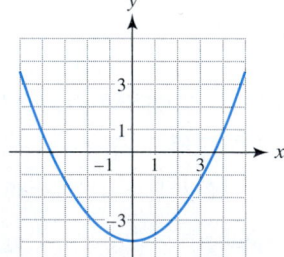

Exercises 79 and 80: (Refer to Example 5.) Use the given f(x) to complete the following.

(a) Calculate the average rate of change of f from $x = 1$ to $x = 2$.

(b) Illustrate your result from part (a) graphically.

79. $f(x) = x^2$ **80.** $f(x) = 4 - x^2$

Exercises 81–86: Compute the average rate of change of f from x_1 to x_2. Round your answer to two decimal places when appropriate. Interpret your result graphically.

81. $f(x) = 7x - 2$, $x_1 = 1$, and $x_2 = 4$

82. $f(x) = -8x + 5$, $x_1 = -2$, and $x_2 = 0$

83. $f(x) = x^3 - 2x$, $x_1 = 2$, and $x_2 = 4$

84. $f(x) = 0.5x^2 - 5$, $x_1 = -1$, and $x_2 = 4$

85. $f(x) = \sqrt{2x - 1}$, $x_1 = 1$, and $x_2 = 3$

86. $f(x) = \sqrt[3]{x + 1}$, $x_1 = 7$, and $x_2 = 26$

87. *Remaining Life Expectancy* The accompanying table lists the average *remaining* life expectancy E in years for females at age x. (**Source:** Department of Health and Human Services.)

x (years)	0	10	20	30	40
E (years)	72.3	69.9	60.1	50.4	40.9

x (years)	50	60	70	80
E (years)	31.6	23.1	15.5	9.2

(a) Make a line graph of the data.

(b) Assume that the graph represents a function f. Compute the average rates of change of f during each 10-year period. Interpret the results.

(c) Determine the life expectancy (not the remaining life expectancy) of a female whose age is 20. What is the life expectancy of a female who is 70 years old? Discuss one reason why these two expectancies are not equal.

88. *Temperature Change* The table gives the outside temperature in degrees Fahrenheit on a summer afternoon.

Time (P.M.)	12:00	1:00	3:15
Temperature (°F)	78	82	88

Time (P.M.)	4:30	6:00
Temperature (°F)	93	91

(a) Calculate the average rate of change between consecutive data points in the table.

(b) Determine the time interval(s) when the thermometer was rising the fastest, on average.

The Difference Quotient

Exercises 89–100: (Refer to Example 7.) Complete the following.

    (a) Find $f(x + h)$.
    (b) Find the difference quotient of f and simplify the result.

89. $f(x) = 3$ **90.** $f(x) = -5$

91. $f(x) = -2x$ **92.** $f(x) = 10x$

93. $f(x) = 2x + 1$ **94.** $f(x) = -3x + 4$

95. $f(x) = 3x^2 + 1$ **96.** $f(x) = x^2 - 2$

97. $f(x) = -x^2 + 2x$ **98.** $f(x) = -4x^2 + 1$

99. $f(x) = 2x^2 - x + 1$ **100.** $f(x) = x^2 + 3x - 2$

101. *Speed of a Car* (Refer to Example 8.) Let the distance in feet that a car travels after t seconds be given by $d(t) = 8t^2$ for $0 \le t \le 6$.
  (a) Find $d(t + h)$.

  (b) Find the difference quotient for d and simplify the result.

  (c) Evaluate the difference quotient when $t = 4$ and $h = 0.05$. Interpret your results.

102. *Draining a Pool* Let the number of gallons G of water in a pool after t hours be given by $G(t) = 4000 - 100t$ for $0 \le t \le 40$.
  (a) Find $G(t + h)$.

  (b) Find the difference quotient for G. Interpret your results.

Writing about Mathematics

103. Describe two methods of determining if a data set can be modeled by a linear function. Create or find a data set that consists of four or more ordered pairs. Determine if your data set can be modeled by a linear function.

104. Suppose you are given a graphical representation of a function f. Explain how you would determine whether f is constant, linear, or nonlinear. How would you determine the type if you were given a numerical or symbolic representation? Give examples.

EXTENDED AND DISCOVERY EXERCISE

1. *Geometry* Suppose that the radius of a circle on a computer monitor is increasing at a constant rate of 1 inch per second.

   (a) Does the circumference of the circle increase at a constant rate? If it does, find this rate in inches per second.

   (b) Does the area of the circle increase at a constant rate? Explain.

CHECKING BASIC CONCEPTS FOR SECTIONS 1.3 AND 1.4

1. Create symbolic, numerical, and graphical representations of a function f that computes the number of feet in x miles. For the numerical representation use a table and let $x = 1, 2, 3, 4, 5$.

2. Sketch the graphs of two relations—one that represents a function and one that does not represent a function. Explain your answer.

3. Let $f(x) = \sqrt{x - 3}$. Identify the domain of f.

4. Let $f(x) = \dfrac{2x}{x - 4}$.

   (a) Find $f(2)$ and $f(a + 4)$.

   (b) Find the domain of f.

5. Graph $f(x) = x^2 - 2$. Identify the domain and range of f.

6. Find the slope of the line passing through the points $(-2, 4)$ and $(4, -5)$. If the graph of $f(x) = ax + b$ passes through these two points, what is the value of a?

7. Identify each function f as constant, linear, or nonlinear. Support your answer graphically.

   (a) $f(x) = -1.4x + 5.1$

   (b) $f(x) = 2x^2 - 5$

   (c) $f(x) = 25$

8. The table shows U.S. advertising expenditures y in billions of dollars for year x.

   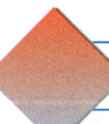 **(a)** Make a line graph of the data. Let this graph represent a function f.

   (b) Compute the average rate of change of f from 1980 to 1990 and from 1990 to 2000. Interpret your results.

x (year)	1980	1990	2000
y ($ billions)	54.8	128.6	243.7

CHAPTER 1 SUMMARY

CONCEPT	EXPLANATION AND EXAMPLES

SECTION 1.1 NUMBERS, DATA, AND PROBLEM SOLVING

SETS OF NUMBERS	*Natural numbers:* $N = \{1, 2, 3, 4, \ldots\}$
	Integers: $I = \{\ldots, -3, -2, -1, 0, 1, 2, 3, \ldots\}$
	Rational numbers: $\frac{p}{q}$, where p and $q \neq 0$ are integers; includes repeating and terminating decimals
	Irrational numbers: Includes nonrepeating, nonterminating decimals
	Real numbers: Any number that can be expressed in decimal form; includes rational and irrational numbers

CONCEPT	EXPLANATION AND EXAMPLES

SECTION 1.1 NUMBERS, DATA, AND PROBLEM SOLVING (CONTINUED)

SCIENTIFIC NOTATION

A real number r is written as $c \times 10^n$, where $1 \le |c| < 10$.

Examples: $1234 = 1.234 \times 10^3$ $0.054 = 5.4 \times 10^{-2}$

PERCENT CHANGE

If a quantity changes from c_1 to c_2, then the percent change equals

$$\frac{c_2 - c_1}{c_1} \times 100.$$

PROBLEM SOLVING

A problem-solving strategy can help find a solution. Two common examples are making a sketch and applying a formula.

SECTION 1.2 VISUALIZATION OF DATA

MEAN (AVERAGE) AND MEDIAN

The mean represents the average of a set of numbers and the median represents the middle of a sorted list.

Example: $4, 6, 9, 13, 15$;   Mean $= \dfrac{4 + 6 + 9 + 13 + 15}{5} = 9.4$;   Median $= 9$

RELATION, DOMAIN, AND RANGE

A relation S is a set of ordered pairs. The domain D is the set of x-values and the range R is the set of y-values.

Example: $S = \{(-1, 2), (4, -5), (5, 9)\}$; $D = \{-1, 4, 5\}$, $R = \{-5, 2, 9\}$

CARTESIAN (RECTANGULAR) COORDINATE SYSTEM OR *xy*-PLANE

The xy-plane has four quadrants and is used to graph ordered pairs.

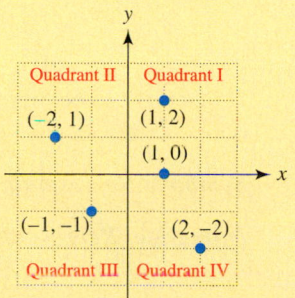

DISTANCE FORMULA

The distance d between the points (x_1, y_1) and (x_2, y_2) is

$$d = \sqrt{(x_2 - x_1)^2 + (y_2 - y_1)^2}.$$

Example: The distance between $(-3, 5)$ and $(2, -7)$ is

$$d = \sqrt{(2 - (-3))^2 + (-7 - 5)^2} = \sqrt{5^2 + (-12)^2} = 13.$$

CONCEPT	EXPLANATION AND EXAMPLES

SECTION 1.2 VISUALIZATION OF DATA (CONTINUED)

MIDPOINT FORMULA

The midpoint M of the line segment with endpoints (x_1, y_1) and (x_2, y_2) is

$$M = \left(\frac{x_1 + x_2}{2}, \frac{y_1 + y_2}{2} \right).$$

Example: The midpoint of the line segment connecting $(1, 2)$ and $(-3, 5)$ is

$$M = \left(\frac{1 + (-3)}{2}, \frac{2 + 5}{2} \right) = \left(-1, \frac{7}{2} \right).$$

SCATTERPLOT AND LINE GRAPH

A scatterplot consists of a set of ordered pairs plotted in the xy-plane. When consecutive points are connected with line segments, a line graph results.

SECTION 1.3 FUNCTIONS AND THEIR REPRESENTATIONS

FUNCTION

A function computes exactly one output for each valid input. The set of valid inputs is called the domain D and the set of outputs is called the range R.

Examples: $f(x) = \sqrt{1 - x}$

$D = \{x \mid x \leq 1\}; R = \{y \mid y \geq 0\}$

$f = \{(-1, 0.5), (0, 4), (2, 4), (6, \pi)\}$

$D = \{-1, 0, 2, 6\}, R = \{0.5, \pi, 4\}$

FUNCTION NOTATION

Example: $f(x) = x^2 - 4$

$f(3) = 3^2 - 4 = 5$

$f(a + 1) = (a + 1)^2 - 4 = a^2 + 2a - 3$

REPRESENTATIONS OF FUNCTIONS

A function can be represented symbolically (formula), graphically (graph), numerically (table of values), and verbally (words).

Symbolic Representation $\quad f(x) = x^2 - 1$

Numerical Representation

Graphical Representation

x	y
-2	3
-1	0
0	-1
1	0
2	3

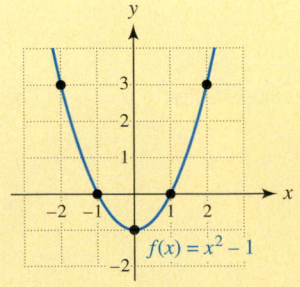

Verbal Representation $\quad f$ computes the square of the input x and then subtracts 1.

VERTICAL LINE TEST

If any vertical line intersects a graph at most once, then the graph represents a function.

CONCEPT	EXPLANATION AND EXAMPLES

SECTION 1.4 TYPES OF FUNCTIONS AND THEIR RATES OF CHANGE

SLOPE

The slope m of the line passing through (x_1, y_1) and (x_2, y_2) is

$$m = \frac{\Delta y}{\Delta x} = \frac{y_2 - y_1}{x_2 - x_1},$$

where $\Delta y = y_2 - y_1$ is the change in y and $\Delta x = x_2 - x_1$ is the change in x.

Example: The slope of the line passing through $(1, -1)$ and $(-2, 3)$ is

$$m = \frac{3 - (-1)}{-2 - 1} = -\frac{4}{3}.$$

CONSTANT FUNCTION

Given by $f(x) = b$, where b is a constant and its graph is a horizontal line

LINEAR FUNCTION

Given by $f(x) = ax + b$ and its graph is a nonvertical line; the slope of its graph is equal to a, which is also equal to its constant rate of change.

Example: The graph of $f(x) = -8x + 100$ has slope -8.
If $G(t) = -8t + 100$ calculates the number of gallons of water in a tank after t minutes, then water is *leaving* the tank at 8 gallons per minute.

NONLINEAR FUNCTION

The graph of a nonlinear function is not a line and *cannot* be written as $f(x) = ax + b$.

Examples: $f(x) = x^2 - 4$; $g(x) = \sqrt[3]{x} - 2$; $h(t) = \frac{1}{t + 1}$

LINEAR AND NONLINEAR DATA

If the slopes of the lines passing through consecutive data points are always equal (or nearly equal), then the data are linear. Otherwise the data are nonlinear.

Example: For each 2-unit increase in x, the y-values increase by 10 units. Consecutive slopes between points are $m = \frac{10}{2} = 5$ so the data are linear.

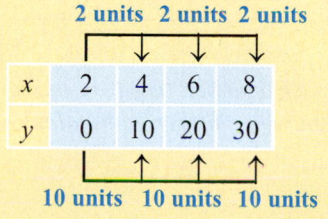

AVERAGE RATE OF CHANGE

If (x_1, y_1) and (x_2, y_2) are distinct points on the graph of f, then the average rate of change from x_1 to x_2 equals the slope of the line passing through these two points and equals

$$\frac{y_2 - y_1}{x_2 - x_1}.$$

Example: $f(x) = x^2$; because $f(2) = 4$ and $f(3) = 9$, the graph of f passes through the points $(2, 4)$ and $(3, 9)$, and the average rate of change from 2 to 3 is $\frac{9 - 4}{3 - 2} = 5$.

CONCEPT	EXPLANATION AND EXAMPLES

SECTION 1.4 TYPES OF FUNCTIONS AND THEIR RATES OF CHANGE (CONTINUED)

DIFFERENCE QUOTIENT

The difference quotient of a function f is an expression of the form

$$\frac{f(x + h) - f(x)}{h},$$

where $h \neq 0$.

Example: Let $f(x) = x^2$. The difference quotient of f is

$$\frac{(x + h)^2 - x^2}{h} = \frac{x^2 + 2xh + h^2 - x^2}{h}$$

$$= 2x + h.$$

REVIEW EXERCISES

Exercises 1 and 2: Classify each number listed as one or more of the following: natural number, integer, rational number, or real number.

1. $-2, \frac{1}{2}, 0, 1.23, \sqrt{7}, \sqrt{16}$

2. $55, 1.5, \frac{104}{17}, 2^3, \sqrt{3}, -1000$

Exercises 3 and 4: Find the percent change when the price of a quantity changes from P_1 to P_2. Round your answer to the nearest tenth of a percent when appropriate.

3. $P_1 = \$1.25, P_2 = \$1.75$

4. $P_1 = \$225, P_2 = \$150$

Exercises 5–8: Write each number in scientific notation.

5. 1,891,000   **6.** $-13,850$

7. 0.0000439   **8.** 0.0001001

Exercises 9 and 10: Write each number in standard form.

9. 1.52×10^4 **10.** -7.2×10^{-3}

11. Evaluate each expression with a calculator. Round answers to the nearest hundredth.

(a) $\sqrt[3]{1.2} + \pi^3$ (b) $\dfrac{3.2 + 5.7}{7.9 - 4.5}$

(c) $\sqrt{5^2 + 2.1}$ (d) $1.2(6.3)^2 + \dfrac{3.2}{\pi - 1}$

12. Evaluate each expression by hand. Write your answer in scientific notation and in standard form.

(a) $(4 \times 10^3)(5 \times 10^{-5})$ (b) $\dfrac{3 \times 10^{-5}}{6 \times 10^{-2}}$

(c) $\dfrac{4.2 \times 10^5}{2 \times 10^8}$

13. Approximate each expression with a calculator. Write the answer in scientific notation.
   (a) $(9.3 \times 10^3)(2.1 \times 10^{-4})$

   (b) $\dfrac{2.46 \times 10^6}{4.1 \times 10^2}$

14. The volume V of a cone is given by $V = \frac{1}{3}\pi r^2 h$, where r is its radius and h is its height. Find V when $r = 4$ inches and $h = 1$ foot. Round your answer to the nearest hundredth.

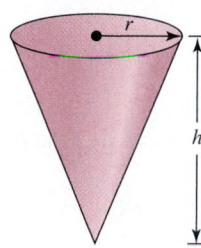

Exercises 15 and 16: Sort each list of numbers from smallest to largest and display them in a table.

   (a) *Determine the maximum and minimum values in the list.*
   (b) *Calculate the mean, median, and range.*

15. $-5, 8, 19, 24, -23$

16. $8.9, -1.2, -3.8, 0.8, 1.7, 1.7$

Exercises 17 and 18: For the given table of data, complete the following.

   (a) *Express the data as a relation S.*
   (b) *Find the domain and range of S.*

17.
x	-15	-10	0	5	20
y	-3	-1	1	3	5

18.
x	-0.6	-0.2	0.1	0.5	1.2
y	10	20	25	30	80

Exercises 19 and 20: Complete the following.

   (a) *Find the domain and range of the relation.*
   (b) *Determine the maximum and minimum of the x-values; of the y-values.*
   (c) *Label appropriate scales on the xy-axes.*
   (d) *Plot the relation by hand in the xy-plane.*

19. $\{(-1, -2), (4, 6), (0, -5), (-5, 3), (1, 0)\}$

20. $\{(10, 20), (-35, -25), (60, 60), (-70, 35), (0, -55)\}$

 Exercises 21 and 22: Make a scatterplot of the relation. Determine if the relation is a function.

21. $\{(10, 13), (-12, 40), (-30, -23), (25, -22), (10, 20)\}$

22. $\{(1.5, 2.5), (0, 2.1), (-2.3, 3.1), (0.5, -0.8), (-1.1, 0)\}$

Exercises 23–26: Find the distance between the points.

23. $(-4, 5), (2, -3)$ 24. $\left(\frac{1}{2}, 0\right), \left(\frac{5}{2}, -\frac{1}{2}\right)$

25. $(-3, -4), (5, -4)$ 26. $(0.2, -4), (0.2, 6)$

Exercises 27 and 28: Find the midpoint of the line segment with the given endpoints.

27. $(24, -16), (-20, 13)$ 28. $\left(\frac{1}{2}, \frac{5}{4}\right), \left(\frac{1}{2}, -\frac{5}{2}\right)$

29. Determine if the triangle with vertices $(1, 2)$, $(-3, 5)$, and $(0, 9)$ is isosceles. (*Hint:* An isosceles triangle has at least two sides with equal measure.)

30. Use the graph to determine the domain and range of each function. Evaluate $f(-2)$.

(a)
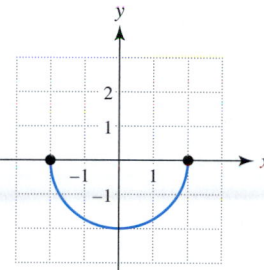

(b)
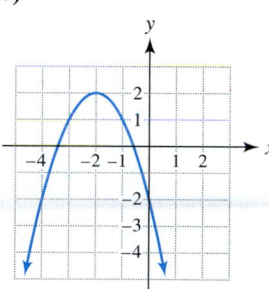

Exercises 31–40: Graph f by first plotting points to determine the shape of the graph.

31. $f(x) = -2$ 32. $f(x) = 3x$

33. $f(x) = -x + 1$ 34. $f(x) = 2x - 3$

35. $f(x) = 4 - 2x^2$ 36. $f(x) = \frac{1}{2}x^2 - 1$

37. $f(x) = |x + 3|$ 38. $f(x) = \left|\frac{1}{3}x\right|$

39. $f(x) = \sqrt{3 - x}$ 40. $f(x) = \sqrt{2x}$

Exercises 41 and 42: Use the verbal representation to express the function f symbolically, graphically, and numerically. Let $y = f(x)$ with $0 \le x \le 100$. For the numerical representation, use a table with $x = 0, 25, 50, 75, 100$.

41. To convert x pounds to y ounces, multiply x by 16.

42. To find the area y of a square, multiply the length x of a side by itself.

43. (a) Use the graph of f to evaluate $f(-3)$ and $f(1)$.

    (b) Find all x such that $f(x) = 0$.

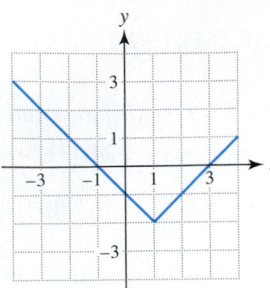

44. *ACT Scores* The function f defined by

$$f(1987) = 20.6, \, f(1990) = 20.6, \, f(1996) = 20.9,$$
$$f(1997) = 21.0, \text{ and } f(2002) = 20.8$$

computes the national average ACT composite score in the year x. (**Source:** The American College Testing Program.)

    (a) Use ordered pairs to describe f.

    (b) Identify the domain and range of f.

    (c) Find all x such that $f(x) = 20.6$.

Exercises 45–52: Complete the following for the function f.

    (a) Evaluate f(x) at the indicated values of x.
    (b) Find the domain of f.

45. $f(x) = \sqrt[3]{x}$ for $x = -8, 1$

46. $f(x) = 3x + 2$ for $x = -2, 5$

47. $f(x) = 5$ for $x = -3, 1.5$

48. $f(x) = 4 - 5x$ for $x = -5, 6$

49. $f(x) = x^2 - 3$ for $x = -10, a + 2$

50. $f(x) = x^3 - 3x$ for $x = -10, a + 1$

51. $f(x) = \dfrac{1}{x^2 - 4}$ for $x = -3, a + 1$

52. $f(x) = \sqrt{x + 3}$ for $x = 1, a - 3$

53. Determine if y is a function of x in the equation $x = y^2 + 5$.

54. Give an example of a relation that involves real data. List some or all of the ordered pairs in the relation.

Exercises 55 and 56: Determine if the graph represents a function.

55.   **56.**

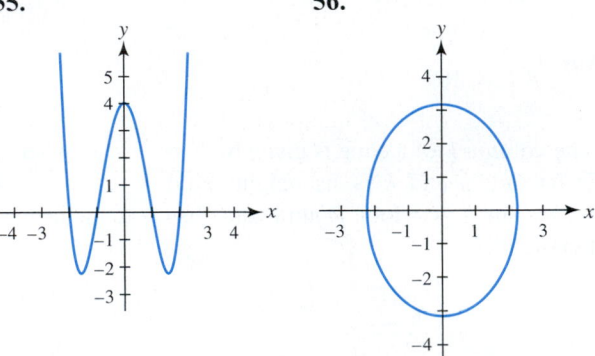

Exercises 57 and 58: Determine if the relation S represents a function.

57. $S = \{(-3, 4), (-1, 2), (3, -5), (4, 2)\}$

58. $S = \{(-1, 3), (0, 2), (-1, 7), (3, -3)\}$

Exercises 59 and 60: Write a symbolic representation (formula) of a function f that computes the following quantity.

59. The number of seconds in t days

60. The tuition for x credits if each credit costs $125

Exercises 61–64: State the slope of the graph of f.

61. $f(x) = 7$ **62.** $f(x) = \frac{1}{3}x - \frac{2}{3}$

63. $f(x) = 4 - 6x$ **64.** $f(x) = x + 10$

Exercises 65–68: If possible, find the slope of the line passing through each pair of points.

65. $(-1, 7), (3, 4)$ **66.** $(1, -4), (2, 10)$

67. $(8, 4), (-2, 4)$ **68.** $\left(-\frac{1}{3}, \frac{2}{3}\right), \left(-\frac{1}{3}, -\frac{5}{6}\right)$

Exercises 69–74: Decide whether the function f is constant, linear, or nonlinear. Support your answer graphically.

69. $f(x) = 8 - 3x$ **70.** $f(x) = 2x^2 - 3x - 8$

71. $f(x) = |x + 2|$ **72.** $f(x) = 2x$

73. $f(x) = 6$ **74.** $f(x) = x^3 - x + 1$

75. Sketch a graph for a 2-hour period showing the distance between two cars meeting on a straight highway, each traveling 60 miles per hour. Assume that the cars are initially 120 miles apart.

76. Sketch a graph of a function f satisfying $f(-2) = 3$, $f(2) = -3$, and $f(4) = 4$. Answers may vary.

77. Determine if the following data are modeled best by a constant, linear, or nonlinear function.

x	−2	0	2	4
y	50	42	34	26

78. Compute the average rate of change of $f(x) = x^2 - x + 1$ from $x_1 = 1$ to $x_2 = 3$.

Exercises 79 and 80: Find the difference quotient for the given $f(x)$.

79. $f(x) = 5x + 1$ **80.** $f(x) = 3x^2 - 2$

Applications

81. *Speed of Light*   The average distance between the planet Mars and the sun is approximately 228 million kilometers. Estimate the time required for sunlight, traveling at 300,000 kilometers per second, to reach Mars. (**Source:** C. Ronan, *The Natural History of the Universe.*)

82. *Flow Rate*   Niagara Falls has an average of 212,000 cubic feet of water passing over it each second. Is this rate more or less than 10 cubic miles per year? Explain your calculations. (*Hint:* 1 cubic mile = 5280^3 cubic feet.)   (**Source:** National Geographic Society.)

83. *Geometry*   Suppose that 0.25 cubic inch of paint is painted onto a circular piece of paper with a diameter of 20 inches. Estimate the thickness of the paint on the paper.

84. *Enclosing a Pool*   A rectangular swimming pool that is 25 feet by 50 feet has a 6-foot wide sidewalk around it.
   (a) How much fencing would be needed to enclose the sidewalk?
   (b) Find the area of the sidewalk.

85. *Distance*   A driver's distance D in miles from a rest stop on an interstate highway after t hours is given by $D(t) = 280 - 70t$.

(a) How far is the driver from the rest stop after 2 hours?

(b) Find the slope of the graph of D. Interpret this slope as a rate of change.

86. *Education*   The line graph shows the number of Catholic secondary schools in thousands for various years.
   (a) Discuss any trends from 1960 to 2000.

   (b) Estimate and interpret the slope of each line segment.

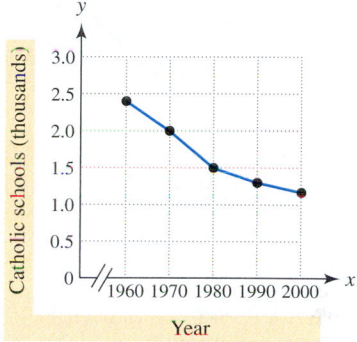

87. *Survival Rates*   The survival rates for song sparrows from 100 eggs are shown in the table. The values listed are the numbers of song sparrows that attain a given age. For example, 6 song sparrows reach an age of 2 years from 100 eggs laid in the wild.   (**Source:** S. Kress, *Bird Life.*)

Age	0	1	2	3	4
Number	100	10	6	3	2

(a) Make a line graph of the data. Interpret the data.

(b) Does this line graph represent a function?

(c) Calculate and interpret the average rate of change for each 1-year period.

88. *Cost of Tuition*   The graph shows the cost of taking *x* credits at a university.

(a) Why is it reasonable for the graph to pass through the origin?

(b) Find the slope of the graph.

(c) Interpret the slope as a rate of change.

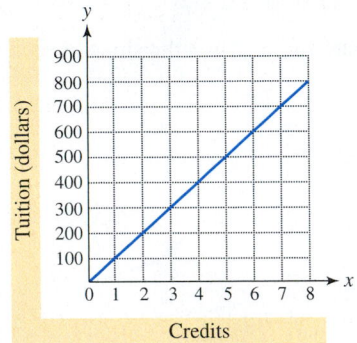

89. *Average Rate of Change*   Let $f(x) = 0.5x^2 + 50$ compute the outside temperature in degrees Fahrenheit at *x* P.M. where $1 \le x \le 5$.

(a) Graph *f*. Is *f* linear or nonlinear?

(b) Calculate the average rate of change of *f* from 1 P.M. to 4 P.M.

(c) Interpret this average rate of change verbally and graphically.

90. *Distance*   At noon car A is traveling north at 30 miles per hour and is located 20 miles north of car B. Car B is traveling west at 50 miles per hour. Approximate the distance between the cars at 12:45 P.M. to the nearest mile.

EXTENDED AND DISCOVERY EXERCISES

Curves designed by engineers for highways and railroads frequently have parabolic, rather than circular, shapes. The reason for this is that a parabolic curve becomes sharper gradually, as shown in the figure at the top of the next column on the left. If railroad tracks changed abruptly from straight to circular, the momentum of the locomotive could cause a derailment. A second figure illustrates straight tracks connecting to a circular curve.   (**Source:** F. Mannering and W. Kilareski, *Principles of Highway Engineering and Traffic Analysis.*)

In order to design a curve and estimate its cost, engineers determine the distance around a curve before it is built. In the next figure the distance along a parabolic curve from *A* to *C* is approximated by two line segments *AB* and *BC*. The distance formula can be used to calculate the length of each

segment. The sum of these two lengths gives a crude estimate of the length of the curve.

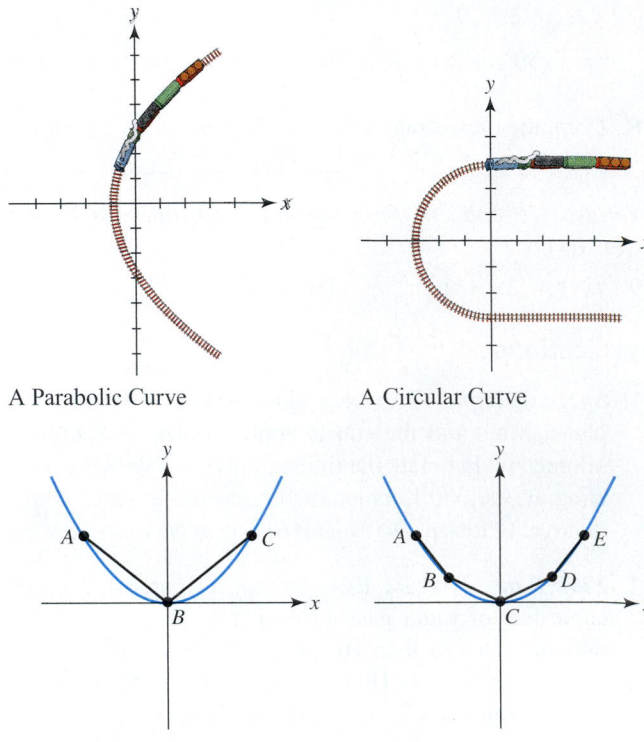

A better estimate using four line segments is shown above on the right. As the number of segments increases, so does the accuracy of the approximation.

1. *Curve Length*   Suppose that a curve designed for railroad tracks is represented by the equation $y = 0.2x^2$, where the units are in kilometers. The points $(-3, 1.8)$, $(-1.5, 0.45)$, $(0, 0)$, $(1.5, 0.45)$, and $(3, 1.8)$ lie on the graph of $y = 0.2x^2$. Approximate the length of the curve from $x = -3$ to $x = 3$ by using line segments connecting these points.

Exercises 2–5: Curve Length   *Use three line segments connecting the four points to estimate the length of the curve on the graph of f from $x = -1$ to $x = 2$. Graph f and a line graph of the four points in the indicated viewing rectangle.*

2. $f(x) = x^2$; $(-1, 1)$, $(0, 0)$, $(1, 1)$, $(2, 4)$; $[-4.5, 4.5, 1]$ by $[-1, 5, 1]$

3. $f(x) = \sqrt[3]{x}$; $(-1, -1)$, $(0, 0)$, $(1, 1)$, $(2, \sqrt[3]{2})$; $[-3, 3, 1]$ by $[-2, 2, 1]$

4. $f(x) = 0.5x^3 + 2$; $(-1, 1.5)$, $(0, 2)$, $(1, 2.5)$, $(2, 6)$; $[-4.5, 4.5, 1]$ by $[0, 6, 1]$

5. $f(x) = 2 - 0.5x^2$; $(-1, 1.5)$, $(0, 2)$, $(1, 1.5)$, $(2, 0)$; $[-3, 3, 1]$ by $[-1, 3, 1]$

6. The distance along the curve of $y = x^2$ from $(0, 0)$ to $(3, 9)$ is approximately 9.747. Use this fact to estimate the distance along the curve of $y = 9 - x^2$ from $(0, 9)$ to $(3, 0)$.

7. Estimate the distance along the curve of $y = \sqrt{x}$ from $(1, 1)$ to $(4, 2)$. (The actual value is approximately 3.168.)

8. *Endangered Species*   The Florida scrub-jay is an endangered species that prefers to live in open landscape with short vegetation. In recent years, NASA has been attempting to create an environment near Kennedy Space Center that supports the habitat for these birds. The following table lists their population for selected years, where $x = 0$ corresponds to 1980, $x = 1$ to 1981, $x = 2$ to 1982, and so on.

x (1980$\longleftrightarrow$0)	0	5	9
y (population)	3697	2512	2176

x (1980$\longleftrightarrow$0)	11	15	19
y (population)	2100	1689	1127

Source: *Mathematics Explorations II*, NASA–AMATYC–NSF.

(a) Make a scatterplot of the data.

(b) Find a linear function f that models the data approximately.

(c) Graph the data and f in the same viewing rectangle.

(d) Estimate the scrub-jay population in 1987 and in 2003.

9. *Global Warming*   If the global climate were to warm significantly, as a result of the greenhouse effect or other climatic change, the Arctic ice cap would start to melt. It is estimated that this ice cap contains the equivalent of 680,000 cubic miles of water. Over 200 million people currently live on soil that is less than 3 feet above sea level. In the United States, several large cities have low average elevations, such as Boston (14 feet), New Orleans (4 feet), and San Diego (13 feet).   (**Sources:** Department of the Interior, Geological Survey.)

(a) Devise a plan to determine how much sea level would rise if the Arctic cap melted. (*Hint:* The radius of Earth is 3960 miles and 71% of its surface is covered by oceans.)

(b) Implement your plan to estimate this rise in sea level.

(c) Discuss the implications of your calculation.

(d) Estimate how much sea level would rise if the 6,300,000 cubic miles of water in the Antarctic ice cap melted.

10. Prove that $\sqrt{2}$ is irrational by assuming that $\sqrt{2}$ is rational and arriving at a contradiction.

2

Linear Functions and Equations

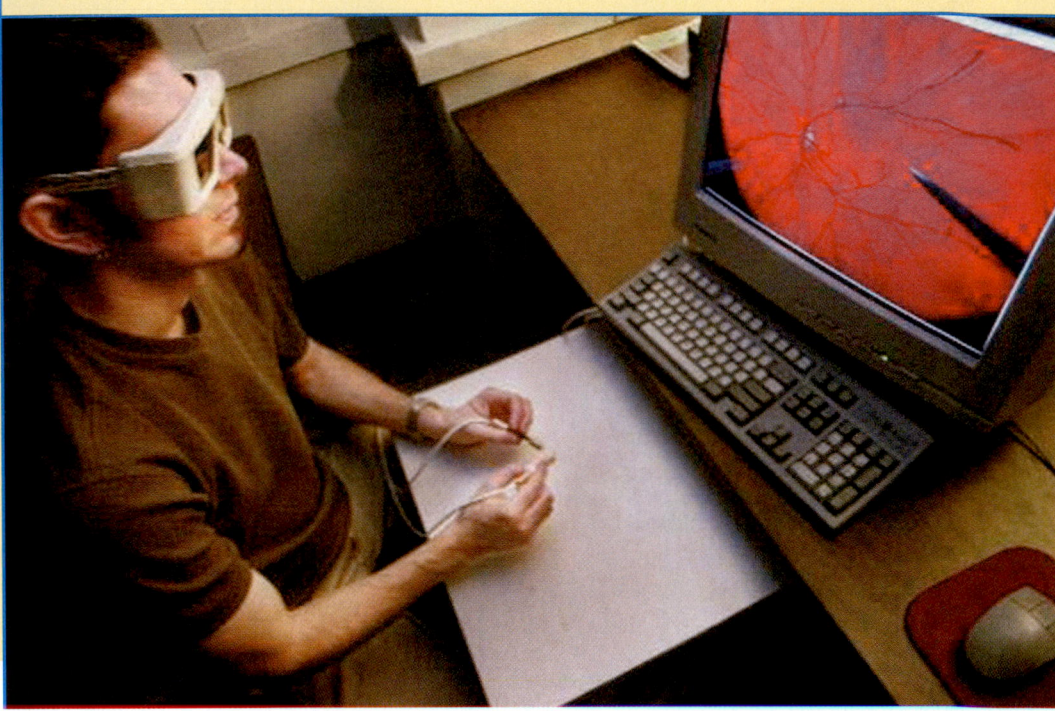

Success is dependent on effort.

Sophocles

The oday a surgical procedure can be simulated in cyberspace before it is performed on the patient. With the help of computers, engineers are able to model the human body. This new technology gives physicians the ability to tinker with a model of the patient's arteries, experimenting with different placements for arterial bypasses. This virtual reality allows doctors to follow hunches, experiment, and even make mistakes before making the first real cut. The software designed to do this task is called ASPIRE— Advanced Surgical Planning Interactive Research Environment. The accompanying figure shows a person's abdominal aorta. Red indicates the greatest blood flow, and blue indicates the least. The areas of blue are of greatest concern to doctors.

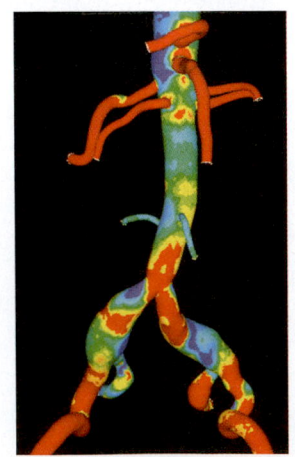

Mathematics played an essential role in the development of this software. Modern technology, such as this, would not be possible without mathematics. In this chapter we discuss how mathematics can be used to model or describe many types of real-life phenomena, such as cellular phones, college tuition, airline travel, computer memory, distant galaxies, and population growth. Mathematics is the *language of technology*.

Source: William Speed. "Downloading Your Body." *Discover,* September 2000. (Photograph reprinted with permission.)

2.1 Linear Functions and Models

- ◆ Recognize exact and approximate models
- ◆ Identify the graph of a linear function
- ◆ Identify a table of values for a linear function
- ◆ Model data with a linear function
- ◆ Use linear regression to model data (optional)

Introduction

Throughout history, people have attempted to explain natural phenomena by creating models. A model is based on observed data. It can be a diagram, a graph, an equation, a verbal expression, or some other form of communication. Models are used in diverse areas such as economics, physics, chemistry, astronomy, psychology, religion, and mathematics. Regardless of where it is used, a model is an *abstraction* that has the following two characteristics:

1. A model is able to explain present phenomena. It should not contradict data and information that are already known to be correct.
2. A model is able to make predictions about data or results. It should be able to use current information to forecast phenomena or create new information.

Mathematical models are used to forecast business trends, design the shapes of cars, estimate ecological trends, control highway traffic, describe epidemics, predict weather, and discover new information when human knowledge is inadequate.

A function can sometimes be a model. For example, in 2002 people spent $25 million on Internet music subscriptions, and in 2003 this amount increased to $50 million. (**Source:** Jupiter Research.) If $t = 0$ corresponding to 2002, $t = 1$ to 2003, and so on, then the linear function defined by $A(t) = 25t + 25$ accurately *models* these *known* dollar amounts (in millions) because

$$A(0) = 25(0) + 25 = 25 \qquad \textit{Amount spent in 2002}$$

and

$$A(1) = 25(1) + 25 = 50. \qquad \textit{Amount spent in 2003}$$

We can use this function A to *predict* the amount spent on Internet music subscriptions in 2006 by evaluating $A(4)$ to obtain

$$A(4) = 25(4) + 25 = 125. \qquad \textit{Amount in year 2006}$$

Thus this model predicts that the amount spent in 2006 on Internet music subscriptions will reach $125 million. Note that this prediction may or may not be correct.

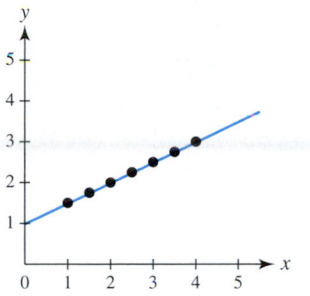

FIGURE 2.1 An Exact Model

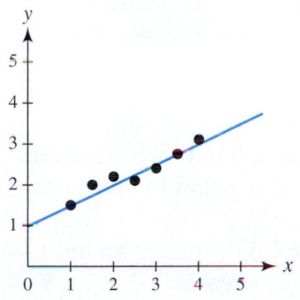

FIGURE 2.2 An Approximate Model

Exact and Approximate Models

Most mathematical models are not exact representations of data. Data might appear to be nearly linear, but not exactly linear. In this case a linear function can be used to provide an *approximate model* of the data. In Figure 2.1 the data are modeled exactly by a linear function, whereas in Figure 2.2 the data are modeled approximately. In the approximate case, the graph follows the general trend in the data. In most real applications, an approximate model is much more likely to occur than an exact model. However, exact models sometimes occur when working with the laws of physics.

EXAMPLE 1   Modeling data with a function

Determine whether $f(t)$ models the data in the table exactly or approximately.
(a) $f(t) = 11.6t + 411$, where t is the number of years after 1995

TABLE 2.1   U.S. Prisoners per 100,000 Residents

Year	1995	2000	2002
Incarceration Rate	411	469	476

Source: Bureau of Justice.

(b) $f(t) = 16t^2$, where t is the number of seconds an object falls

TABLE 2.2   Distance an Object Falls in a Vacuum

Time (seconds)	0	1	2	3
Distance (feet)	0	16	64	144

SOLUTION

(a) Note that $t = 0$ corresponds to 1995, $t = 5$ to 2000, and $t = 7$ to 2002.

$$f(\mathbf{0}) = 11.6(\mathbf{0}) + 411 = \mathbf{411} \qquad \text{\textit{Value agrees with table.}}$$

$$f(\mathbf{5}) = 11.6(\mathbf{5}) + 411 = \mathbf{469} \qquad \text{\textit{Value agrees with table.}}$$

$$f(\mathbf{7}) = 11.6(\mathbf{7}) + 411 = \mathbf{492.2} \neq 476 \qquad \text{\textit{Value does not agree with table.}}$$

Because the values from $f(t)$ do not agree with all the data in Table 2.1, the model is not exact; rather, it is approximate.

(b) Start by evaluating f for $t = 0, 1, 2,$ and 3.

$$f(\mathbf{0}) = 16(\mathbf{0})^2 = \mathbf{0} \qquad \text{\textit{Value agrees with table.}}$$

$$f(\mathbf{1}) = 16(\mathbf{1})^2 = \mathbf{16} \qquad \text{\textit{Value agrees with table.}}$$

$$f(\mathbf{2}) = 16(\mathbf{2})^2 = \mathbf{64} \qquad \text{\textit{Value agrees with table.}}$$

$$f(\mathbf{3}) = 16(\mathbf{3})^2 = \mathbf{144} \qquad \text{\textit{Value agrees with table.}}$$

Because the values from $f(t)$ agree with all the data in Table 2.2, the model is exact.

Now Try Exercises 1 and 3 ◆

Representations of Linear Functions

Any linear function f can be written as $f(x) = ax + b$ and its graph is always a nonvertical line. Before we begin to model data with linear functions, it is helpful to understand how formulas for linear functions relate to their graphs.

Consider the graph of the linear function f shown in Figure 2.3. Its graph is a line that intersects each axis once. From the graph we can see that when y increases by 3 units, x increases by 2 units. Thus the change in y is $\Delta y = 3$, the change in x is $\Delta x = 2$, and the slope is $\frac{\Delta y}{\Delta x} = \frac{3}{2}$. The graph of f intersects the y-axis at the point $(0, 3)$. We say that the **y-intercept** on the graph of f is 3. Note that when we evaluate $f(x) = ax + b$ at $x = 0$ we obtain

$$f(0) = a(0) + b = b.$$

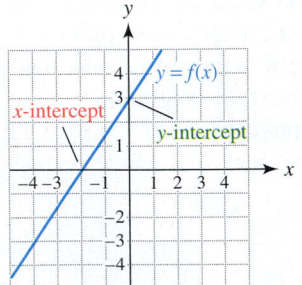

FIGURE 2.3   Linear Function

That is, the point $(0, b)$ lies on the graph of f and the value of b represents the y-intercept. Using this information, we can write a formula or a symbolic representation of f as

$$f(x) = \frac{3}{2}x + 3.$$

slope y-intercept

Note that a function can have at most one y-intercept because $f(0)$ can have at most one output value.

    The graph of f in Figure 2.3 intersects the x-axis at the point $(-2, 0)$. We say that the **x-intercept** on the graph of f is -2. Note that when we substitute $x = -2$ in the formula for f, we obtain

$$f(-2) = \frac{3}{2}(-2) + 3 = 0.$$

An x-intercept corresponds to an input that results in an output of 0. We also say that -2 is a *zero* of the function f, since $f(-2) = 0$. A **zero** of a function f corresponds to an x-intercept on the graph of f. If the slope of the graph of a linear function f is not 0, then the graph of f has exactly one x-intercept.

    A table of values or numerical representation for $f(x) = \frac{3}{2}x + 3$ is shown in Table 2.3. Note that for every 1-unit increase in x, $f(x)$ increases by $\frac{3}{2}$, or 1.5, units. In general, for each 1-unit increase in x, a linear function g, given by $g(x) = ax + b$, changes by a units. This rate of change is an increase when $a > 0$ and a decrease when $a < 0$.

TABLE 2.3
$f(x) = \frac{3}{2}x + 3$

x	$f(x)$
-2	0
-1	1.5
0	3
1	4.5
2	6

(increments of 1.5)

EXAMPLE 2 Finding a formula from a graph

Use the graph of a linear function f in Figure 2.4 to complete the following.
(a) Find the slope, y-intercept, and x-intercept.
(b) Write a formula for f.
(c) Find any zeros of f.

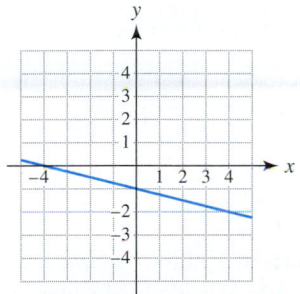

FIGURE 2.4

SOLUTION
(a) The line falls 1 unit each time the x-values increase by 4 units. Therefore the slope is $-\frac{1}{4}$. The graph intersects the y-axis at the point $(0, -1)$ and intersects the x-axis at the point $(-4, 0)$. Therefore the y-intercept is -1, and the x-intercept is -4.
(b) Because the slope is $-\frac{1}{4}$ and the y-intercept is -1, it follows that

$$f(x) = -\frac{1}{4}x - 1.$$

(c) Zeros of f correspond to x-intercepts, so the only zero is -4.   *Now Try Exercise 5* ◆

Modeling with Linear Functions

Linear functions can be used to model data and physical phenomena that change at a constant rate. For example, the distance traveled by a car can be modeled by a linear function *if* the car is traveling at a constant speed.

    When we are modeling data with a linear function f, slope gives us information not only about the "tilt" of the graph of f, but also about how fast a quantity is changing or its *rate of change*. If a quantity changes at a constant rate, then we can model the quantity with $f(x) = ax + b$, where the value of a corresponds to this constant rate of change and the value of b corresponds to the initial amount.

MODELING WITH A LINEAR FUNCTION

To model a quantity that is changing at a constant rate with a linear function f, the following may be used.

$$f(x) = \text{(constant rate of change)} \, x + \text{(initial amount)}$$

Note: The constant rate of change corresponds to the slope of the graph of f and the initial amount corresponds to the y-intercept.

This method is illustrated in the next example.

 **EXAMPLE 3** Modeling ice deposits

A roof has a 0.5-inch layer of ice on it from a previous storm. Another ice storm begins to deposit ice at a rate of 0.25 inch per hour.
(a) Find a formula for a linear function f that models the thickness of the ice on the roof x hours after the second ice storm started.
(b) How thick is the ice after 2.5 hours?

SOLUTION
(a) The roof initially has 0.5 inch of ice and ice is accumulating at a constant rate of 0.25 inch per hour.

$$f(x) = (\textbf{constant rate of change})x + (\textbf{initial amount})$$
$$= \textbf{0.25}x + \textbf{0.5}$$

(b) The thickness of the ice after 2.5 hours is

$$f(2.5) = 0.25(2.5) + 0.5 = 1.125 \text{ inches.} \quad \textit{Now Try Exercise 43} \blacklozenge$$

 **EXAMPLE 4** Finding a symbolic representation

A 100-gallon tank is initially full of water and being drained at a rate of 5 gallons per minute.
(a) Write a formula for a linear function f that models the number of gallons of water in the tank after x minutes.
(b) How much water is in the tank after 4 minutes?
(c) Graph f. Identify the x- and y-intercepts and interpret each.

SOLUTION
(a) The amount of water in the tank is *decreasing* at a constant rate of 5 gallons per minute, so the constant rate of change is -5. The initial amount of water is equal to 100 gallons.

$$f(x) = (\textbf{constant rate of change})x + (\textbf{initial amount})$$
$$= \textbf{-5}x + \textbf{100}$$

(b) After 4 minutes the tank contains $f(4) = -5(4) + 100 = 80$ gallons.
(c) Since $f(x) = -5x + 100$, the graph has y-intercept 100 and slope -5. Its graph is shown in Figure 2.5. The x-intercept is 20, which corresponds to the time in minutes that it takes to empty the tank. The y-intercept corresponds to the gallons of water in the tank initially. Note that the domain of f is $0 \leq x \leq 20$. *Now Try Exercise 41* $\blacklozenge$

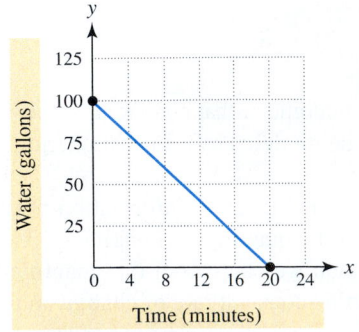

FIGURE 2.5

Writing formulas for functions

Write the formula for a linear function that models each situation. Choose both an appropriate name for the function and an appropriate variable. State what the input variable represents and the domain of the function.

(a) In 2003 the average cost of attending a private college was $25,000 and it is projected to increase, on average, by $1670 per year until 2010.   (**Source:** Cerulli Associates.)

(b) A car's speed is 50 miles per hour, and then it begins to slow down at a constant rate of 10 miles per hour each second.

SOLUTION

(a) Let C be the name of the function and t be the number of years after 2003. Then

$$C(t) = (\text{constant rate of change})t + (\text{initial amount})$$
$$= 1670t + 25{,}000$$

models the cost in dollars of attending a private college t years past 2003. Because this projection is valid only until 2010, or for 7 years past 2003, the domain D of function C is

$$D = \{t \,|\, t = 0, 1, 2, 3, 4, 5, 6, \text{ or } 7\}.$$

Note that t represents a year, so it is probably most appropriate to restrict the domain to integer values for t.

(b) Let S be the name of the function and t be the elapsed time in seconds after the car begins to slow. Then

$$S(t) = (\text{constant rate of change})t + (\text{initial speed})$$
$$= -10t + 50$$

models the speed of the car after an elapsed time of t seconds. Because the car's initial speed is 50 miles per hour and it slows at 10 miles per hour per second, the car can slow down for at most 5 seconds before it comes to a stop. Thus the domain D of function S is

$$D = \{t \,|\, 0 \le t \le 5\}.$$

Note that t represents time in seconds, so t does not need to be restricted to an integer.

Now Try Exercises 45 and 47 ◆

If the slope between consecutive data points is always the same, the data can be modeled exactly by a linear function. If the slope between consecutive data points is nearly the same, then the data can be modeled approximately by a linear function. In the next example we model data approximately with a linear function.

Modeling fuel consumption

TABLE 2.4

x (gallons)	y (miles)
5	84
10	169
15	255
20	338

Table 2.4 shows the distance traveled in miles by a car using x gallons of gasoline.

(a) Make a scatterplot of the data. Could a linear function be used to model this data?

(b) Find values for a and b so that $f(x) = ax + b$ models the distance traveled on x gallons of gasoline. Graph f and the data.

(c) Interpret the slope of the graph of f.

SOLUTION

(a) A scatterplot of the data is shown in Figure 2.6 on the next page. The data appear to be nearly linear so a linear function could be used to model the data.

(b) With 0 gallons of gasoline a car could travel 0 miles. So the data point (0, 0) could be included in the table. Thus the y-intercept is 0, and we can let $b = 0$ in the formula $f(x) = ax + b$.

Next we must determine a. The slopes between consecutive points are

$$\frac{84 - 0}{5 - 0} = \textbf{16.8,} \quad \frac{169 - 84}{10 - 5} = \textbf{17.0,} \quad \frac{255 - 169}{15 - 10} = \textbf{17.2,} \text{ and } \frac{338 - 255}{20 - 15} = \textbf{16.6.}$$

Since they are not exactly equal, f cannot model the data exactly. An initial estimate for a might be about 16.9 or 17. If we let $a = 17$ then $f(x) = 17x + 0$, or $f(x) = 17x$. A graph of f and the data are shown in Figure 2.7. (Values for a and b may vary slightly.)

Calculator Help

To make a scatterplot, see Appendix B (page AP-7). To plot data and graph an equation in the same viewing rectangle, see Appendix B (page AP-11).

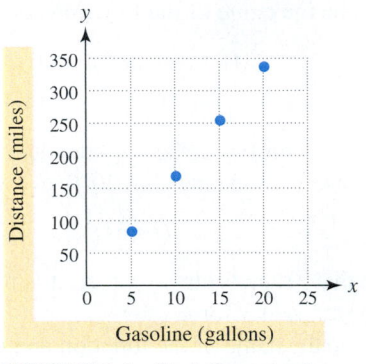

FIGURE 2.6   Fuel Consumption

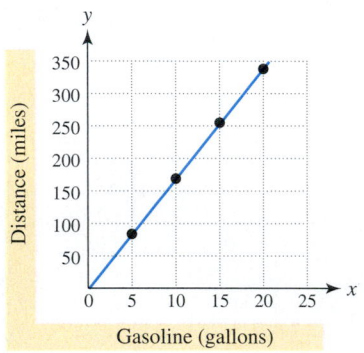

FIGURE 2.7   A Linear Model

(c) A slope of 17 indicates that the mileage for this car is 17 miles per gallon.

Now Try Exercises 49 and 53 ◆

◆ **MAKING CONNECTIONS**

Slope and Approximately Linear Data   When modeling data approximately with $f(x) = ax + b$, one way to obtain an initial value for a is to calculate the slope between the first and last data point in the table. The value for a can then be adjusted visually by graphing f and the data in the same viewing rectangle to obtain a slightly better fit. In Example 6 this would have resulted in

$$a = \frac{338 - 84}{20 - 5} \approx 16.93,$$

which compares favorably with our decision to let $a = 17$. ◆

Linear Regression (Optional)

We have used linear functions to model data involving the variables x and y. Unknown values for y were predicted at given values of x. Problems where one variable is used to predict the behavior of a second variable are called **regression** problems. If a linear function or line is used to approximate the data, then this is referred to as **linear regression**.

We have already solved problems by selecting a line that *visually* fits the data in a scatterplot. However, this technique has some disadvantages. First, it does not produce a unique line. Different people may arrive at different lines to fit the same data. Second, the line is not determined automatically by a calculator or computer. A person must view the data and adjust the line until it "fits." By contrast, a statistical method used to determine a unique linear function or line is based on **least squares**.

Most graphing calculators have the capability to calculate the least-squares regression line automatically after the data points have been entered. When determining the least-squares line, calculators often compute a real number r, called the **correlation coefficient**, where $-1 \leq r \leq 1$. When r is positive and near 1, low x-values correspond to low y-values and high x-values correspond to high y-values. For example, there is a positive correlation between years of education x and income y. More years of education correlate with higher income. When r is near -1, the reverse is true. Low x-values correspond to high y-values and high x-values correspond to low y-values. An example of this is the relation between latitude and average yearly temperature. As latitude increases (moving toward either the north or south poles), the average yearly temperature decreases. Therefore there will be a negative correlation between latitude and average annual temperature. If $r \approx 0$, then there is little or no correlation between the data points. In this case, a linear function does not provide a suitable model. A summary of these concepts regarding correlation is shown in Table 2.5.

TABLE 2.5   Correlation Coefficient r, $(-1 \leq r \leq 1)$

Value of r	Comments	Sample Scatterplot
$r = 1$	There is an exact linear fit. The line passes through all data points and has a positive slope.	
$r = -1$	There is an exact linear fit. The line passes through all data points and has a negative slope.	
$0 < r < 1$	There is a positive correlation. As the x-values increase, so do the y-values. The fit is not exact.	
$-1 < r < 0$	There is a negative correlation. As the x-values increase, the y-values decrease. The fit is not exact.	
$r = 0$	There is no correlation. The data has no tendency toward being linear. A regression line predicts poorly.	

◆ **MAKING CONNECTIONS**

Correlation and Causation When geese begin to fly north, summer is coming and the weather becomes warmer. Geese flying north correlate with warmer weather. However, geese flying north clearly do not *cause* warmer weather. It is important to remember that correlation does not always indicate the cause. ◆

In the next example we use a graphing calculator to find the line of least-squares fit that models three data points.

EXAMPLE 7 Determining a line of least-squares fit

Calculator Help
To find a line of least-squares fit, see Appendix B (page AP-11).

Find the line of least-squares fit for the data points (1, 1), (2, 3), and (3, 4). What is the correlation coefficient? Plot the data and graph the line.

SOLUTION Begin by entering the three data points into the calculator. Refer to Figures 2.8–2.11. Select the LinReg(ax + b) option from the STAT CALC menu. From the home screen we can see that the line (linear function) of least-squares is given by $y = 1.5x - \frac{1}{3}$. The correlation coefficient is $r \approx 0.98$. Since $r \neq 1$, the line does not provide an exact model of the data.

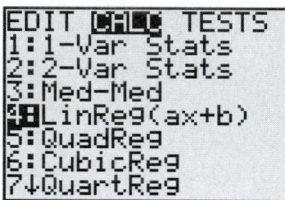

FIGURE 2.8

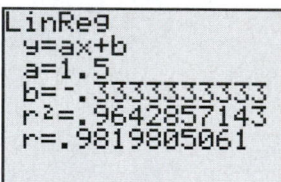

FIGURE 2.9

FIGURE 2.10

[0, 5, 1] by [0, 5, 1]

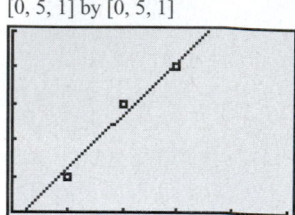

FIGURE 2.11

Now Try Exercise 57 ◆

In the next example we use regression to find a linear function that models numbers of airline passengers at some of our largest airports.

EXAMPLE 8 Predicting airline passengers

Table 2.6 lists the numbers in millions of airline passengers at some of the largest airports in the United States during 1999 and 2002.

TABLE 2.6 Airline Passengers (millions)

Airport	1999	2002
Hartsfield Atlanta	78.1	76.9
Chicago O'Hare	72.6	66.5
Los Angeles (LAX)	64.3	56.2
Dallas/Fort Worth	60.0	52.8
Denver	38.0	35.7

Source: Airports Association Council International.

(a) Make a scatterplot of the data using the 1999 data for x-values and the corresponding 2002 data for y-values. Predict whether the correlation coefficient will be positive or negative.

(b) Use a calculator to find the linear function f based on least-squares regression that models this data. Graph $y = f(x)$ and the data in the same viewing rectangle.

(c) In 1999 Miami International Airport had 33.9 million passengers. Assuming that this airport followed a trend similar to that of the five airports listed in the Table 2.4, use your linear function to estimate the number of passengers at Miami International in 2002. Compare this result to the actual value of 30.0 million passengers.

SOLUTION

(a) A scatterplot of the data is shown in Figure 2.12. Because increasing x-values correspond to increasing y-values, the correlation coefficient is positive.

(b) Because $y = f(x)$, the formula for a linear function f that models this data is given by $f(x) = 0.9811x - 3.7956$, where coefficients have been rounded to four decimal places. See Figure 2.13. A graph of f and the data are shown in Figure 2.14.

(c) We can use $f(x)$ to predict y when $x = 33.9$.

$$y = f(33.9) = 0.9811(33.9) - 3.7956 \approx 29.5 \text{ million}$$

This value is slightly less than the actual value of 30.0 million.

[30, 90, 10] by [30, 90, 10]   [30, 90, 10] by [30, 90, 10]

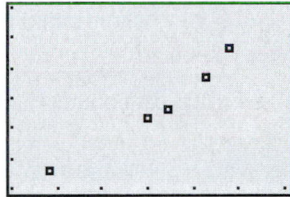

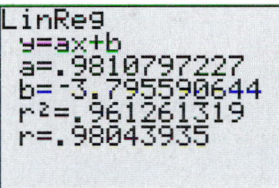

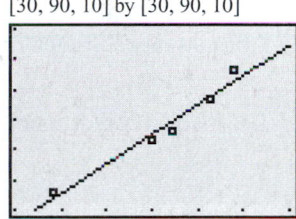

FIGURE 2.12   **FIGURE 2.13**   **FIGURE 2.14**

Now Try Exercise 61 ◆

2.1

Putting it all Together

Any linear function can be written as $f(x) = ax + b$, and its graph is a line with slope a and y-intercept b. If data increase (or decrease) by nearly the same amount for each unit increase in x, then the data can be modeled by a linear function. One method to determine whether a set of data is linear is to calculate the slope between consecutive data points. If the slopes are nearly equal, it may be possible to model the data with a linear function. Another way is to use linear regression.

Concept	Description
Exact and approximate models	An exact model describes data precisely. In real life, models are usually approximate. A good model should describe and explain current data. It should also be able to make predictions and forecast phenomena.

Concept	Description
Graph of a linear function	The graph of a linear function is a line. Important features include its slope, y-intercept, and x-intercept. If $f(x) = ax + b$, then the slope equals a and the y-intercept equals b. The following graph has slope $\frac{1}{2}$, y-intercept 1, and x-intercept -2. The zero of f is the x-intercept, -2. 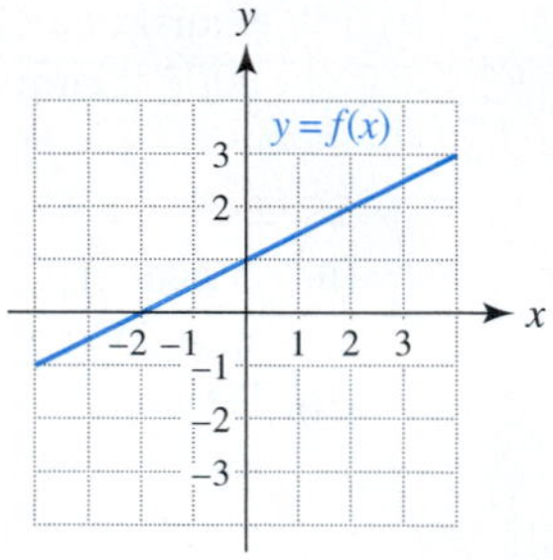
Linear model	If a quantity experiences a constant rate of change, then it can be modeled by a linear function in the form $f(x) = ax + b$. The following can be used to determine a linear modeling function f. $$f(x) = (\textbf{constant rate of change})x + (\textbf{initial value})$$
Correlation coefficient r	The values of r satisfy $-1 \le r \le 1$, where a line fits the data better if r is near -1 or 1. A value near 0 indicates a poor fit.
Least-squares regression line	The line of least-squares fit for the data points $(1, 3)$, $(2, 5)$, and $(3, 6)$ is $$y = \frac{3}{2}x + \frac{5}{3}$$ and $r \approx 0.98$. Try verifying this with a calculator.

2.1 ◆ Exercises

Exact and Approximate Models

Exercises 1–4: (Refer to Example 1.) For each data table a function f is given. Determine whether f models the data exactly or approximately.

1. $f(x) = 5x - 2$

x	1	2	3	4
y	3	8	13	18

2. $f(x) = 1 - 0.2x$

x	5	10	15	20
y	0	-1	-2	-4

3. $f(x) = 3.7 - 1.5x$

x	-6	0	6	12
y	12.7	3.7	-5.4	-14.2

4. $f(x) = 13.3x - 6.1$

x	1	2	5	10
y	7.2	20.5	60.4	126.9

Graphs of Linear Functions

Exercises 5–10: The graph of a linear function f is shown.

    (a) *Identify the slope, y-intercept, and x-intercept.*

    (b) *Write a formula for f.*

    (c) *Estimate the zero of f.*

5.

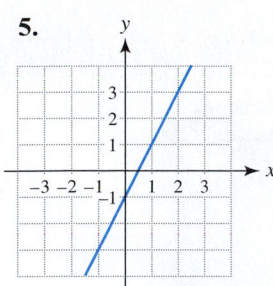

6.

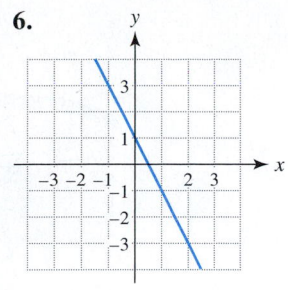

7.

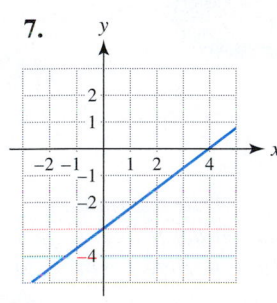

8.

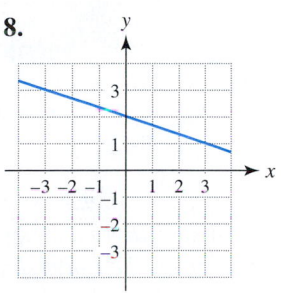

9.

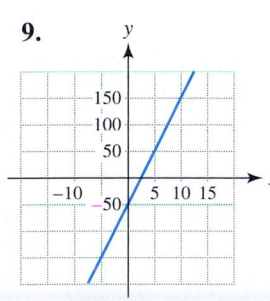

10.

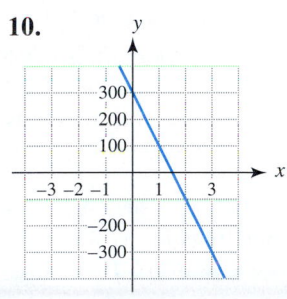

Exercises 11–24: Graph the linear function by hand. Identify the slope and y-intercept.

11. $f(x) = 3x + 2$

12. $f(x) = -\frac{3}{2}x$

13. $f(x) = \frac{1}{2}x - 2$

14. $f(x) = 3 - x$

15. $g(x) = -2$

16. $g(x) = 20 - 10x$

17. $f(x) = 4 - \frac{1}{2}x$

18. $f(x) = 2x - 3$

19. $g(x) = \frac{1}{2}x$

20. $g(x) = 3$

21. $g(x) = 5 - 5x$

22. $g(x) = \frac{3}{4}x - 2$

23. $f(x) = 20x - 10$

24. $f(x) = -30x + 20$

Exercises 25–30: Write a formula for a linear function f whose graph satisfies the conditions.

25. Slope $-\frac{3}{4}$, y-intercept $\frac{1}{3}$

26. Slope -122, y-intercept 805

27. Slope 15, passing through the origin

28. Slope 1.68, passing through $(0, 1.23)$

29. Slope 0.5, passing through $(1, 4.5)$

30. Slope -2, passing through $(-1, 5)$

Exercises 31–36: Find the average rate of change of f from -2 to 2. What is the average rate of change of f from x_1 to x_2, where x_1 and x_2 are distinct real numbers?

31. $f(x) = 10$

32. $f(x) = -5$

33. $f(x) = -\frac{1}{4}x$

34. $f(x) = \frac{5}{3}x$

35. $f(x) = 4 - 3x$

36. $f(x) = 5x + 1$

Modeling with Linear Functions

Exercises 37–40: Match the situation with the graph (a.–d.) that models it best, where x-values represent time.

37. The height of the Empire State Building from 1990 to 2000

38. The average cost of a new car from 1980 to 2000

39. The distance between a runner in a race and the finish line

40. Amount of money earned after x hours when working for an hourly rate of pay

a.   **b.**

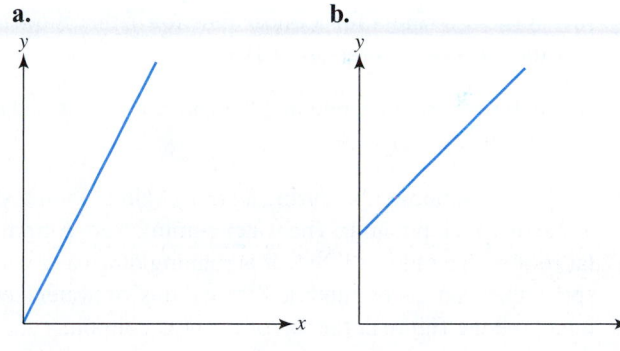

c.   **d.**

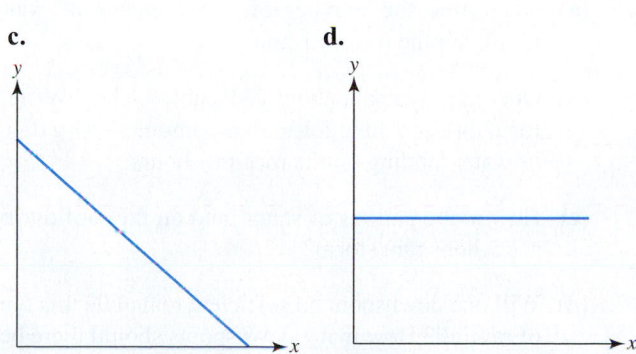

41. *Filling a Tank*   (Refer to Example 4.) A 500-gallon tank initially contains 200 gallons of fuel oil. A pump is filling the tank at a rate of 6 gallons per minute.
   (a) Write a formula for a linear function f that models the number of gallons of fuel oil in the tank after x minutes.

   (b) Graph f. What is an appropriate domain for f?

   (c) Identify the y-intercept and interpret it.

   (d) Does the x-intercept of the graph of f have any physical meaning in this problem? Explain.

42. *HIV Infections*   In 1998 there were 47 million people worldwide who had been infected with HIV. At that time the infection rate was 5.8 million people per year. (**Source:** United Nations AIDS and World Health Organization.)
   (a) Write a formula for a linear function f that models the total number of people in millions who were infected with HIV x years after 1998.

   (b) Estimate the number of people who may have been infected by the year 2006.

43. *Birth Rate*   In 1990 the number of births per 1000 people in the United States was 16.7 and decreasing at 0.26 birth per 1000 people each year.   (**Source:** National Center for Health Statistics.)
   (a) Write a formula for a linear function f that models the birth rate x years after 1990.

   (b) Estimate the birth rate in 2002 and compare it to the actual value of 13.9.

44. *Rainfall*   Suppose that during a storm, rain is falling at a rate of 1 inch per hour. The water coming from a circular roof with a radius of 20 feet is running down a downspout that can accommodate 400 gallons of water per hour. See the figure at the top of the next column.
   (a) Determine the number of cubic inches of water falling on the roof in 1 hour.

   (b) One gallon equals about 231 cubic inches. Write a formula for a function g that computes the gallons of water landing on the roof in x hours.

   (c) How many gallons of water land on the roof during a 2.5-hour rain storm?

   (d) Will one downspout be sufficient to handle this type of rainfall? How many downspouts should there be?

Geometry Review   To review formulas related to circles, see Chapter R (page R-2).

Exercises 45–48: (Refer to Example 5.) Write a formula for a linear function that models the situation described. Choose both an appropriate name and an appropriate variable for the function. State what the input variable represents and the domain of the function.

45. *Velocity of a Falling Object*   A stone is dropped from a water tower and its velocity increases at a rate of 32 feet per second. The stone hits the ground with a velocity of 96 feet per second.

46. *Speed of a Car*   A car's speed is 30 miles per hour, and then it begins to slow down at a constant rate of 6 miles per hour every 4 seconds.

47. *Population Density*   In 1900 the average number of people per square mile in the United States was 21.5 and it increased, on average, by 5.81 people every 10 years until 2000.   (**Source:** Bureau of the Census.)

48. *Injury Rate*   In 1992 the rate of injury cases recorded in private industry per 100 full-time workers was 8.3 and it decreased, on average, by 0.32 injury every year until 2001.   (**Source:** Bureau of Labor Statistics.)

Exercises 49–52: Write a formula for a linear function f that models the data exactly.

49.

x	0	1	2	3	4
$f(x)$	-7	-4	-1	2	5

50.

x	-0.4	-0.2	0	0.2	0.4
$f(x)$	-2.54	-2.82	-3.1	-3.38	-3.66

51.

x	-6	-4	1	3
$f(x)$	15	10	-2.5	-7.5

52.

x	15	20	25	30
$f(x)$	30	32	34	36

53. *Passenger Travel*   The table shows the number of miles (in trillions) traveled by passengers of all types for various years, where $x = 0$ corresponds to 1970, $x = 10$ to 1980, and so on.

Year (1970 $\leftrightarrow$ 0)	0	10	20	30
Miles (trillions)	2.2	2.8	3.7	4.7

Source: Department of Transportation.

(a) Could the data be modeled exactly by a linear function?

(b) Estimate values for a and b so that $f(x) = ax + b$ models the data. (Answers may vary.)

(c) Graph f and the data. Interpret the slope.

(d) Estimate passenger miles in 2003.

54. *School Enrollment*   The table shows the number of students (in millions) attending public school (K–8) for selected years, where $x = 0$ corresponds to 1997, $x = 1$ to 1998, and so on.

Year (1997 $\leftrightarrow$ 0)	0	1	2	3	4
Students (millions)	33.1	33.3	33.5	33.7	33.9

Source: Department of Education.

(a) Find values for a and b so that $f(x) = ax + b$ models the data.

(b) Graph f and the data. Interpret the slope.

(c) Predict the public K–8 enrollment in 2005.

55. *Braking Distance*   The faster a car is traveling, the farther it takes for the car to stop. If a car doubles its speed x, then its braking distance quadruples. Could braking distance be modeled by $f(x) = ax$? Explain your reasoning.

56. *Computer Memory*   Devices for computers often store information in units called megabytes (MB). For example, discs containing more megabytes can hold more pages of text. Is the amount of text that can be stored on a disc a function of the megabytes available? Is the function a linear function? Explain your reasoning.

Linear Regression

Exercises 57–60: Complete the following.

(a) Conjecture whether the correlation coefficient r for the data will be positive, negative, or zero.

(b) Use a calculator to find the equation of the least-squares regression line with the form $y = ax + b$ and the value of r.

(c) Use the regression line to predict y when $x = 2.4$.

57.

x	−3	−2	−1	0
y	−11.1	−8.4	−5.7	−2.6

x	1	2	3
y	1.1	3.9	7.3

58.

x	−4	−2	0	2	4
y	1.2	2.8	5.3	6.7	9.1

59.

x	1	3	5	7	10
y	5.8	−2.4	−10.7	−17.8	−29.3

60.

x	−4	−3	−1	3	5
y	37.2	33.7	27.5	16.4	9.8

61. *Distant Galaxies*   In the late 1920s the famous observational astronomer Edwin P. Hubble (1889–1953) determined both the distance to several galaxies and the velocity at which they were receding from Earth. Four galaxies with their distances in light-years and velocities in miles per second are listed in the table.   (**Sources:** A. Acker and C. Jaschek, *Astronomical Methods and Calculations;* A. Sharov and I. Novikov, *Edwin Hubble: The Discoverer of the Big Bang Universe.*)

Galaxy	Distance	Velocity
Virgo	50	990
Ursa Minor	650	9,300
Corona Borealis	950	15,000
Bootes	1700	25,000

(a) Let x be distance and y be velocity. Plot the data points in $[-100, 1800, 100]$ by $[-1000, 28000, 1000]$.

(b) Find the least-squares regression line that models these data.

(c) If the galaxy Hydra is receding at a speed of 37,000 miles per second, estimate its distance.

62. *Cellular Phones* One of the early problems with cellular phones was the delay involved with placing a call when the system was busy. One study analyzed this delay. The table shows that as the number of calls increased by P percent, the average delay time D to put through a call also increased. (**Source:** A. Mehrotra, *Cellular Radio: Analog and Digital Systems.*)

P (%)	0	20	40	60	80	100
D (minutes)	1	1.6	2.4	3.2	3.8	4.4

(a) Let P correspond to x-values and D to y-values. Find the least-squares regression line that models these data. Plot the data and the regression line.

(b) Estimate the delay for a 50% increase in the number of calls.

63. *Women in Politics* The table lists percentages of women in state legislatures for past years. (**Source:** National Women's Political Caucus.)

Year	1975	1977	1979	1981
Percent	8.0	9.1	10.3	12.1

Year	1983	1985	1987	1989
Percent	13.3	14.8	15.7	17.0

Year	1991	1993	1995	1998
Percent	18.3	20.5	20.7	21.6

(a) Make a scatterplot of the data.

(b) Find the least-squares regression line that models these data.

(c) Interpret the slope.

(d) Assuming that trends continued, use this line to estimate the percentage of women in state legislatures in 2003 and compare it to the actual value of 22.3%.

64. *Average Household Size* The following table lists the average number y of people per household for various years x.

x	1940	1950	1960	1970
y	3.67	3.37	3.33	3.14

x	1980	1990	2000
y	2.76	2.63	2.62

Source: Bureau of the Census.

(a) Make a scatterplot of the data.

(b) Decide whether the correlation coefficient is positive, negative, or zero.

(c) Find the least-squares regression line that models these data. What is the correlation coefficient?

(d) Estimate the number of people per household in 1975 and compare it to the actual value of 2.94.

Writing about Mathematics

65. How can you recognize a symbolic representation (formula) of a linear function? How can you recognize a graph or table of values of a linear function?

66. A student graphs $f(x) = x^2 - x$ in the viewing rectangle [2, 2.1, 0.01] by [1.9, 2.3, 0.1]. Using the graph, the student decides that f is a linear function. How could you convince the student otherwise?

67. Explain how average rate of change relates to a linear function.

68. Find a data set in a newspaper or on the Internet that can be modeled by a linear function. Find the linear modeling function. Is your model exact or approximate? Explain.

EXTENDED AND DISCOVERY EXERCISES

1. *Temperature and Volume*   The following table shows the relationship between the temperature of a sample of helium and its volume.

Temperature (°C)	0	25	50	75	100
Volume (in.³)	30	32.7	35.4	38.1	40.8

(a) Make a scatterplot of the data.

(b) Write a formula for a function f that receives the temperature x as input and outputs the volume y of the helium.

(c) Find the volume of the gas when the temperature is 65°C.

2. *Height and Shoe Size*   In this exercise you will determine if there is a relationship between height and shoe size.

(a) Have classmates write their sex, shoe size, and height in inches on a slip of paper. When you have enough information, complete the following table; one for adult males and one for adult females.

Height (inches)				
Shoe Size				

(b) Make a scatterplot of each table, by letting height correspond to the x-axis and shoe size correspond to the y-axis. Is there any relationship between height and shoe size? Explain.

(c) Try to find a linear function that models each data set.

Exercises 3 and 4: Graph the function f in the standard viewing rectangle.

(a) *Choose any curved portion of the graph of f and repeatedly zoom in. Describe how the graph appears. Repeat this on different portions of the graph.*

(b) *Under what circumstances could a linear function be used to accurately model a nonlinear graph?*

3. $f(x) = 4x - x^3$ 4. $f(x) = x^4 - 5x^2$

2.2 Equations of Lines

- Write the point-slope and slope-intercept forms for a line
- Find the intercepts of a line
- Write equations for horizontal, vertical, parallel, and perpendicular lines
- Model data with lines and linear functions (optional)
- Understand interpolation and extrapolation
- Use direct variation to solve problems

Introduction

Lines are a fundamental geometric concept that have applications in a variety of areas such as computer graphics, business, and science. Any quantity that experiences growth at a constant rate can be modeled by the graph of a linear function, which is a line. In this section we discuss how equations of lines can be determined and some of their applications.

Forms for Equations of Lines

Suppose that a nonvertical line passes through the point (x_1, y_1) with slope m. If (x, y) is any point on this line with $x \neq x_1$, then the change in y is $\Delta y = y - y_1$, the change in x is $\Delta x = x - x_1$, and the slope equals $m = \frac{\Delta y}{\Delta x} = \frac{y - y_1}{x - x_1}$, as illustrated in Figure 2.15 on the next page.

Using this slope formula, the equation of the line can be found.

$$m = \frac{y - y_1}{x - x_1} \qquad \text{Slope formula}$$

$$y - y_1 = m(x - x_1) \qquad \text{Cross multiply.}$$

$$y = m(x - x_1) + y_1 \qquad \text{Add } y_1 \text{ to each side.}$$

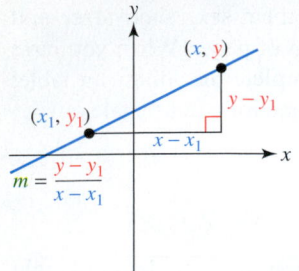

FIGURE 2.15

The equation $y - y_1 = m(x - x_1)$ is traditionally called the *point-slope form* of the equation of a line. Since we think of y as being a function of x, written $y = f(x)$, the equivalent form $y = m(x - x_1) + y_1$ will also be referred to as the point-slope form. The point-slope form is not unique, since any point on the line can be used for (x_1, y_1). However, these point-slope forms are *equivalent*—their graphs are identical.

POINT-SLOPE FORM

The line with slope m passing through the point (x_1, y_1) has an equation

$$y = m(x - x_1) + y_1, \quad \text{or} \quad y - y_1 = m(x - x_1),$$

the **point-slope form** of the equation of a line.

In the next example we find the equation of a line given two points.

EXAMPLE 1 Determining a point-slope form

Find an equation of the line passing through the points $(-2, -3)$ and $(1, 3)$. Plot the points and graph the line by hand.

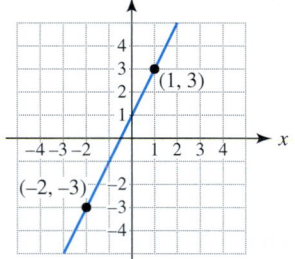

FIGURE 2.16

SOLUTION Begin by finding the slope of the line.

$$m = \frac{3 - (-3)}{1 - (-2)} = \frac{6}{3} = 2$$

Substituting $(x_1, y_1) = (\mathbf{1}, \mathbf{3})$ and $m = \mathbf{2}$ into the point-slope form results in

$$y = \mathbf{2}(x - \mathbf{1}) + \mathbf{3}.$$

If we use the point $(-2, -3)$, the point-slope form is

$$y = 2(x + 2) - 3.$$

A graph of this line passing through the two points is shown in Figure 2.16.

Now Try Exercise 1 ◆

The two point-slope forms found in Example 1 are equivalent, which can be shown as follows.

$y = 2(x - 1) + 3$	First point-slope form
$y = 2x - 2 + 3$	Distributive property
$y = 2x + 1$	Simplify.

Algebra Review

To review the distributive property, see Chapter R (page R-22).

Applying the same steps on the second point-slope form, we obtain the same result.

$y = 2(x + 2) - 3$	Second point-slope form
$y = 2x + 4 - 3$	Distributive property
$y = 2x + 1$	Simplify.

Both point-slope forms simplify to the same equation.

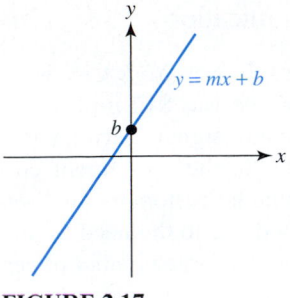

FIGURE 2.17

The form $y = mx + b$ is called the *slope-intercept form* and unlike the point-slope form, it is *unique*. The real number m represents the slope and the real number b represents the y-intercept, as illustrated in Figure 2.17.

SLOPE-INTERCEPT FORM

The line with slope m and y-intercept b is given by

$$y = mx + b,$$

the **slope-intercept form** of the equation of a line.

Note: The slope-intercept form, $y = mx + b$, defines a linear function f, where $y = f(x)$. Thus $f(x) = mx + b$ defines a linear function whose graph has slope m and y-intercept b.

EXAMPLE 2 Finding equations of lines

Find a point-slope form for the line that satisfies the conditions. Then convert this equation into slope-intercept form.

(a) Slope $-\frac{1}{2}$ passing through the point $(-3, -7)$

(b) x-intercept -4, y-intercept 2

SOLUTION

(a) Let $m = -\frac{1}{2}$ and $(x_1, y_1) = (-3, -7)$ in the point-slope form.

$$y = m(x - x_1) + y_1 \qquad \text{Point-slope form}$$

$$y = -\frac{1}{2}(x + 3) - 7 \qquad \text{Substitute.}$$

The slope-intercept form can be found by simplifying.

$$y = -\frac{1}{2}(x + 3) - 7 \qquad \text{Point-slope form}$$

$$y = -\frac{1}{2}x - \frac{3}{2} - 7 \qquad \text{Distributive property}$$

$$y = -\frac{1}{2}x - \frac{17}{2} \qquad \text{Slope-intercept form}$$

(b) The line passes through the points $(-4, 0)$ and $(0, 2)$. Its slope is

$$m = \frac{2 - 0}{0 - (-4)} = \frac{1}{2}.$$

Thus a point-slope form is $y = \frac{1}{2}(x + 4) + 0$, where the point $(-4, 0)$ is used for (x_1, y_1). The slope-intercept form is $y = \frac{1}{2}x + 2$.   *Now Try Exercises 5 and 17* ◆

In the next example we use the point-slope form of a line to model cellular phone growth during the late 1980s.

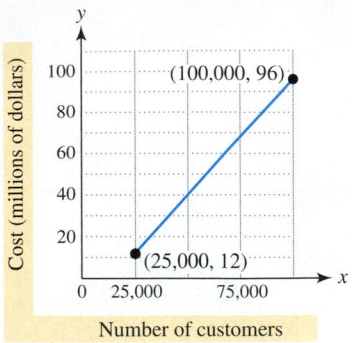

FIGURE 2.18 Cost of a Cellular Site

Estimating growth and investment in cellular communication

Cellular phone use has grown dramatically in the United States. In New York City when there were 25,000 customers, the investment cost per cellular site was $12 million. (A cellular site would include such things as a relay tower to transmit signals between cellular phones.) When the number of customers rose to 100,000, the investment cost rose to $96 million. Although cost usually decreases with additional customers, this was not the case for *early* cellular technology. Instead, cost increased due to the need to purchase expensive real estate and to establish communication among a large number of cellular sites. The relationship between customers and investment costs per site was approximately linear, as shown in Figure 2.18.   (**Source:** M. Paetsch, *Mobile Communications in the U.S. and Europe.*)

(a) Find a point-slope form of the line passing through the points (25000, 12) and (100000, 96). Interpret the slope.

(b) Use this equation to estimate the investment cost per cellular site when there were 70,000 customers in New York City.

(c) Find the slope-intercept form of this line.

SOLUTION

(a) The slope m of the line segment passing through (**25000**, **12**) and (**100000**, **96**) is

$$m = \frac{96 - 12}{100{,}000 - 25{,}000} = 0.00112.$$

A point-slope form of the line passing through (25000, 12) with slope 0.00112 is found as follows.

$$y = m(x - x_1) + y_1 \qquad \text{Point-slope form}$$
$$y = 0.00112(x - 25{,}000) + 12 \qquad \text{Substitute for } (x_1, y_1) \text{ and } m.$$

In this equation y represents the investment cost for each cellular site in millions of dollars when there were x customers. Slope 0.00112 indicates that for each cellular site cellular phone companies spent, on average, $0.00112 million per customer. Moving the decimal 6 places to the right (multiplying by 1,000,000), this values is equivalent to $1120 per customer.

(b) For $x = $ **70,000** customers, the investment cost was

$$y = 0.00112(\mathbf{70{,}000} - 25{,}000) + 12 = 62.4,$$

or $62.4 million per cellular site.

(c) To obtain the slope-intercept form, we apply the distributive property.

$$y = 0.00112(x - 25{,}000) + 12$$
$$y = 0.00112x - 28 + 12 \qquad \text{Distributive property}$$
$$y = 0.00112x - 16 \qquad \text{Simplify.}$$

Now Try Exercise 79 ◆

Determining Intercepts

The point-slope form and the slope-intercept form are not the only forms for the equation of a line. A equation of a line is in **standard form** when it is written as

$$\mathbf{ax + by = c,}$$

where a, b, and c are constants. By using standard form, we can write the equation of any line, including vertical lines which are discussed later in this section. Examples of equations of lines in standard form include

$$2x - 3y = -6, \qquad y = \frac{1}{4}, \qquad x = -3, \qquad \text{and} \qquad -3x + y = \frac{1}{2}.$$

Standard form is a convenient form for finding the x- and y-intercepts of a line. Once the intercepts are found, we can graph the line. For example, to find the x-intercept for the line determined by $3x + 4y = 12$, we let $y = 0$ and solve for x to obtain

$$3x + 4(0) = 12 \quad \text{or} \quad x = 4.$$

The x-intercept is 4. To find the y-intercept, we let $x = 0$ and solve for y to obtain

$$3(0) + 4y = 12 \quad \text{or} \quad y = 3.$$

The y-intercept is 3. Thus the graph of $3x + 4y = 12$ passes through the points $(4, 0)$ and $(0, 3)$. Knowing these two points allows us to graph the line easily. Note that this technique can be used to find intercepts on the graph of any equation, not just lines written in standard form.

FINDING INTERCEPTS

To find any x-intercepts, let $y = 0$ in the equation and solve for x.
To find any y-intercepts, let $x = 0$ in the equation and solve for y.

EXAMPLE 4   Finding intercepts

Locate the x- and y-intercepts on the line whose equation is $4x + 3y = 6$. Use the intercepts to graph the equation.

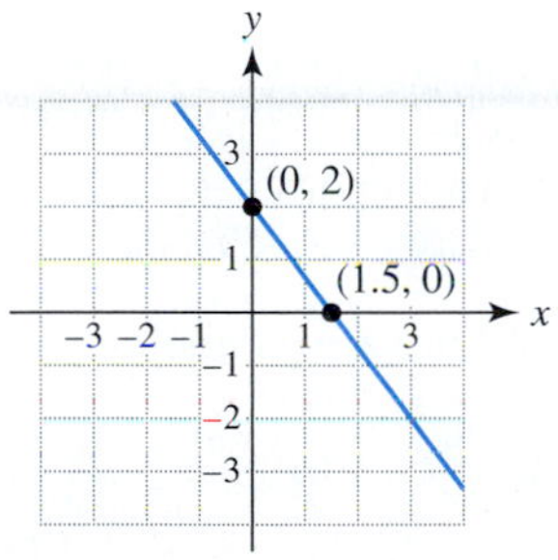

FIGURE 2.19

SOLUTION To locate the x-intercept, let $y = 0$ in the equation.

$$4x + 3(0) = 6 \qquad \text{Let } y = 0.$$
$$x = 1.5 \qquad \text{Divide by 4.}$$

The x-intercept is 1.5. Similarly, to find the y-intercept, substitute $x = 0$ into the equation.

$$4(0) + 3y = 6 \qquad \text{Let } x = 0.$$
$$y = 2 \qquad \text{Divide by 3.}$$

The y-intercept is 2. Therefore the line passes through the points $(1.5, 0)$ and $(0, 2)$, as shown in Figure 2.19.   *Now Try Exercise 59* ◆

◆ **CLASS DISCUSSION**
Find the slope-intercept form of the line in Example 4. What are the slope and y-intercept? Is your answer consistent with Figure 2.19? ◆

Horizontal, Vertical, Parallel, and Perpendicular Lines

The graph of a constant function f, defined by $f(x) = b$, is a horizontal line having slope 0 and y-intercept b.

A vertical line cannot be represented by a function because distinct points on a vertical line have the same x-coordinate. In fact, this is the distinguishing feature about points on a vertical line—they all have the same x-coordinate. The vertical line shown in

Figure 2.20 is $x = 3$. The equation of a vertical line with x-intercept k is given by $x = k$, as shown in Figure 2.21. Horizontal lines have slope 0, and vertical lines have an undefined slope.

◆ **CLASS DISCUSSION**
Why do you think that a vertical line sometimes is said to have "infinite slope"? What are some problems with taking this phrase too literally? ◆

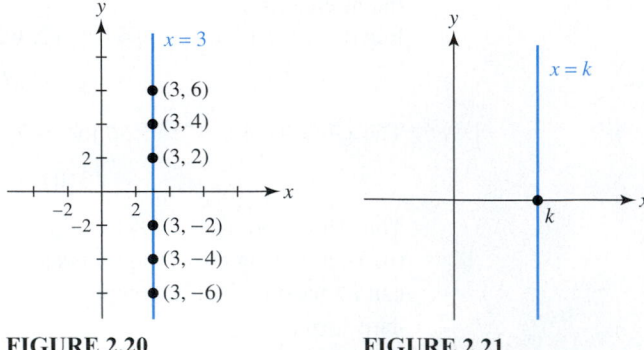

FIGURE 2.20   **FIGURE 2.21**

EQUATIONS OF HORIZONTAL AND VERTICAL LINES

An equation of the horizontal line with y-intercept b is $y = b$. An equation of the vertical line with x-intercept k is $x = k$.

 EXAMPLE 5 Finding equations of horizontal and vertical lines

Find equations of vertical and horizontal lines passing through the point $(8, 5)$.

SOLUTION The x-coordinate of the point $(8, 5)$ is 8. The vertical line $x = 8$ passes through every point in the xy-plane with an x-coordinate of 8, including the point $(8, 5)$. Similarly, the horizontal line $y = 5$ passes through every point with a y-coordinate of 5, including $(8, 5)$.   *Now Try Exercises 39 and 41* ◆

Slope is an important concept when determining whether two lines are parallel or perpendicular. Two nonvertical parallel lines have equal slopes.

PARALLEL LINES

Two lines with slopes m_1 and m_2, neither of which is vertical, are parallel if and only if their slopes are equal, that is $m_1 = m_2$.

Note: The phrase "if and only if" is used when two statements are mathematically equivalent. If two nonvertical lines are parallel, then it is true that $m_1 = m_2$. Conversely, if two nonvertical lines have equal slopes, then they are parallel. Either condition implies the other.

 EXAMPLE 6 Finding parallel lines

Find the slope-intercept form of a line parallel to $y = -2x + 5$, passing through $(-2, 3)$.

SOLUTION The line $y = -2x + 5$ has slope -2, so any parallel line also has slope -2. The line passing through $(-2, 3)$ with slope -2 is determined as follows.

$$y = -2(x + 2) + 3 \qquad \text{Point-slope form}$$
$$y = -2x - 4 + 3 \qquad \text{Distributive property}$$
$$y = -2x - 1 \qquad \text{Slope-intercept form}$$

Now Try Exercise 29 ◆

Cross hairs, found in telescopes, view finders, and video games, are examples of perpendicular lines. In video games, provisions sometimes are made for cross hairs to tilt in order to simulate things such as a craft rolling from side to side. In this situation, the cross hairs remain perpendicular, but they are not always in a horizontal-vertical position. Two lines with nonzero slopes are perpendicular if the product of their slopes is equal to -1.

PERPENDICULAR LINES

Two lines with nonzero slopes m_1 and m_2 are perpendicular if and only if their slopes have product -1, that is $m_1 m_2 = -1$.

For perpendicular lines, m_1 and m_2 are *negative reciprocals*. That is, $m_1 = -\frac{1}{m_2}$ and $m_2 = -\frac{1}{m_1}$. Table 2.7 shows examples for m_1 and m_2 that result in perpendicular lines because $m_1 m_2 = -1$.

TABLE 2.7   Slopes of Perpendicular Lines

m_1	$\frac{1}{2}$	$\frac{6}{5}$	5	-1	$-\frac{2}{3}$
m_2	-2	$-\frac{5}{6}$	$-\frac{1}{5}$	1	$\frac{3}{2}$
$m_1 m_2$	-1	-1	-1	-1	-1

EXAMPLE 7 Finding perpendicular lines

Find the slope-intercept form of the line perpendicular to $y = -\frac{2}{3}x + 2$, passing through the point $(-2, 1)$. Graph the lines.

SOLUTION The line $y = -\frac{2}{3}x + 2$ has slope $-\frac{2}{3}$. The negative reciprocal of $m_1 = -\frac{2}{3}$ is $m_2 = \frac{3}{2}$ because

$$m_1 m_2 = -\frac{2}{3} \cdot \frac{3}{2} = -1.$$

The slope-intercept form of a line having slope $\frac{3}{2}$ and passing through $(-2, 1)$ can be found as follows.

$$y = m(x - x_1) + y_1 \qquad \text{Point-slope form}$$
$$y = \frac{3}{2}(x + 2) + 1 \qquad \text{Let } x_1 = -2 \text{ and } y_1 = 1.$$
$$y = \frac{3}{2}x + 3 + 1 \qquad \text{Distributive property}$$
$$y = \frac{3}{2}x + 4 \qquad \text{Slope-intercept form}$$

Calculator Help

To set a square viewing rectangle, see Appendix B (page AP-10).

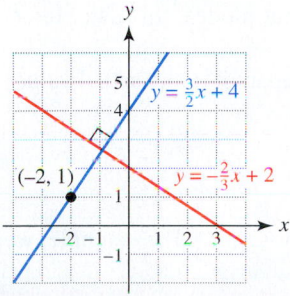

FIGURE 2.22 Perpendicular Lines

Figure 2.22 shows graphs of these perpendicular lines.

Now Try Exercise 31 ◆

Note: If a graphing calculator is used to graph these lines, a square viewing rectangle must be used to have the lines appear perpendicular.

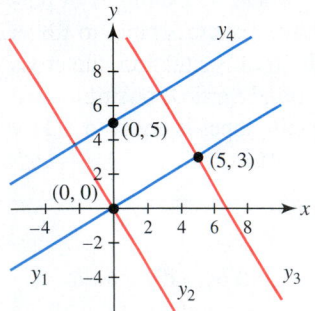

FIGURE 2.23

EXAMPLE 8 Determining a rectangle

In Figure 2.23 a rectangle is outlined by four lines denoted $y_1, y_2, y_3,$ and y_4. Find the equation of each line.

SOLUTION

Line y_1: This line passes through the points $(0, 0)$ and $(5, 3)$, so $m = \frac{3}{5}$ and the *y*-intercept is 0. Its equation is $y_1 = \frac{3}{5}x$.

Line y_2: This line passes through the point $(0, 0)$ and is perpendicular to y_1, so its slope is given by $m = -\frac{5}{3}$ and the *y*-intercept is 0. Its equation is $y_2 = -\frac{5}{3}x$.

Line y_3: This line passes through the point $(5, 3)$ and is parallel to y_2, so its slope is given by $m = -\frac{5}{3}$. In a point-slope form, its equation is $y_3 = -\frac{5}{3}(x - 5) + 3$, which is equivalent to $y_3 = -\frac{5}{3}x + \frac{34}{3}$.

Line y_4: This line passes through the point $(0, 5)$ and is parallel to y_1, so its slope is given by $m = \frac{3}{5}$. Its equation is $y_4 = \frac{3}{5}x + 5$. *Now Try Exercise 95* ◆

◆ **CLASS DISCUSSION**

Check the results from Example 8 by graphing the four equations in the same viewing rectangle. How does your graph compare with Figure 2.23? Why is it important to use a square viewing rectangle? ◆

Modeling Data (Optional)

Point-slope form for the equation of a line can sometimes be valuable when modeling real data. In the next example we model the rise in the cost of tuition of fees at private colleges and universities. This modeling is performed by using a linear function.

EXAMPLE 9 Modeling linear data

Tuition and fees have risen at private colleges. Table 2.8 lists the average tuition and fees for selected years.

TABLE 2.8   Tuition and Fees at Private Colleges

Year	1980	1985	1990	1995	2000
Tuition	$3617	$6121	$9340	$12,432	$16,233

Source: The College Board.

(a) Make a scatterplot of this data. What type of model does the scatterplot suggest?
(b) Find a linear function given by $f(x) = m(x - x_1) + y_1$ that models this data. Interpret the slope m.
(c) Use f to estimate tuition and fees in 1987 and in 1998. Compare it to the actual values of $7048 and $14,508, respectively.

SOLUTION

(a) The scatterplot in Figure 2.24 suggests a linear model.
(b) The data table contains several points that could be used for (x_1, y_1). For example, we could choose the first data point $(1980, 3617)$, then write

$$f(x) = m(x - 1980) + 3617.$$

Calculator Help

To make a scatterplot, see Appendix B (page AP-7).

To plot data and graph an equation in the same viewing rectangle, see Appendix B (page AP-11).

[1978, 2002, 2] by [0, 20000, 5000]

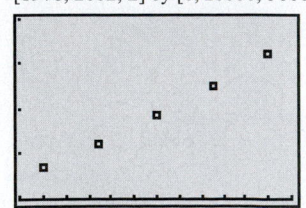

FIGURE 2.24

To estimate a slope m we could choose two points that appear to lie on a line that models the data. For example, if we choose the first data point (1980, 3617) and the third data point (1990, 9340), then the slope m is

$$m = \frac{9340 - 3617}{1990 - 1980} = 572.3.$$

This slope indicates that tuition and fees have risen, on average, $572.30 per year.

Figure 2.25 shows the graph of $f(x) = 572.3(x - 1980) + 3617$ and the data. It is important to realize that answers may vary when modeling real data because if we choose different points, the resulting equation for $f(x)$ would be different.

(c) To estimate the tuition and fees in 1987 and in 1998, evaluate $f(1987)$ and $f(1998)$, respectively.

$$f(\mathbf{1987}) = 572.3(\mathbf{1987} - 1980) + 3617 = \$7623.10$$

$$f(\mathbf{1998}) = 572.3(\mathbf{1998} - 1980) + 3617 = \$13{,}918.40$$

Both values differ from the actual values by less than $600.

Now Try Exercises 83 and 84 ◆

[1978, 2002, 2] by [0, 20000, 5000]

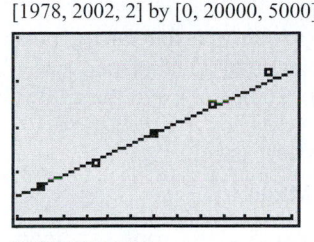

FIGURE 2.25

◆ **MAKING CONNECTIONS** ———————————————

Modeling and Forms of Equations In Example 9 we modeled college tuition and fees by using the formula

$$f(x) = 572.3(x - 1980) + 3617.$$

This point-slope form readily reveals that in 1980 tuition and fees cost $3617 and have risen, on average, by $572.30 per year. In slope-intercept form, this formula becomes

$$f(x) = 572.3x - 1{,}129{,}537.$$

Although the slope is apparent in slope-intercept form, it is less obvious that the actual value of tuition in 1980 was $3617. Which form is more convenient often depends on the problem being solved. ◆

Interpolation and Extrapolation

Total energy production in the Asia-Pacific region of the world increased substantially between 1990 and 2005. Table 2.9 lists this energy production in million metric tons of oil equivalent (Mtoe) for selected years.   (**Source:** R. Andre-Pascal, *Global Energy: The Changing Outlook.*)

TABLE 2.9   Asia-Pacific Energy Production

Year	1990	1995	2000	2005
Energy (Mtoe)	415	530	645	760

If the data in Table 2.9 are used to estimate energy production during 1998, this would be an example of **interpolation** since 1998 lies between 1990 and 2005. If these data are used to estimate energy production in either 1980 or 2010, this would be an example of **extrapolation** because these years do not lie between 1990 and 2005. Interpolation usually provides more reliable results than extrapolation. Extrapolation should be used cautiously.

 **Estimating past and future energy production**

(a) Use Table 2.9 to determine an equation of a line that models the data. Interpret the slope.
(b) Use this line to estimate energy production in 1998 and 1970. Discuss the results.

SOLUTION
(a) The data increases by exactly 115 Mtoe every 5 years, or by $\frac{115}{5} = 23$ Mtoe per year. Thus the line with slope 23 passing through $(1990, 415)$ models these data exactly. Its equation can be written as $y = 23(x - 1990) + 415$. The slope indicates that energy production in the Asia-Pacific region is expected to increase at a rate of 23 Mtoe per year.
(b) To estimate energy production in 1998 and 1970, substitute 1998 and 1970 for x.

$$y = 23(\mathbf{1998} - 1990) + 415 = 599 \text{ Mtoe}$$

$$y = 23(\mathbf{1970} - 1990) + 415 = -45 \text{ Mtoe}$$

The interpolated value for 1998 is reasonable, since it lies between the 1995 and 2000 values of 530 and 645 Mtoe. The extrapolated value for 1970 is not reasonable because it is negative. It might be acceptable to use extrapolation for estimating energy production in 1988 or 2007, since these years lie relatively close to the domain of the data. However, extrapolating 20 years back provides an incorrect result.

 *Now Try Exercise 53* ◆

Direct Variation

When a change in one quantity causes a proportional change in another quantity, the two quantities are said to *vary directly* or to *be directly proportional*. For example, if we work for $8 per hour, our pay is proportional to the number of hours that we work. Doubling the hours doubles the pay, tripling the hours triples the pay, and so on. This is stated more precisely as follows.

DIRECT VARIATION

Let x and y denote two quantities. Then y is **directly proportional** to x, or y **varies directly** with x, if there exists a nonzero number k such that

$$y = kx.$$

The number k is called the **constant of proportionality** or the **constant of variation**.

If a person earns $57.75 working for 7 hours, the constant of proportionality k is the hourly pay rate. If y represents the pay in dollars and x the hours worked, then k is found by substituting values for x and y into the equation $\mathbf{y = kx}$ and solving for k. That is,

$$\mathbf{57.75 = k(7)} \qquad \text{or} \qquad k = \frac{57.75}{7} = 8.25,$$

so the hourly pay rate is $8.25.

Hooke's law states that the distance that an elastic spring stretches beyond its natural length is directly proportional to the amount of weight hung on the spring, as illustrated in Figure 2.26. This law is valid whether the spring is stretched or compressed. The constant of proportionality is called the **spring constant**. Thus if a weight or force F is applied to the spring with spring constant k and the spring stretches a distance x beyond its natural length, then the equation $F = kx$ models this situation, where k is the spring constant.

EXAMPLE 11 Working with Hooke's Law

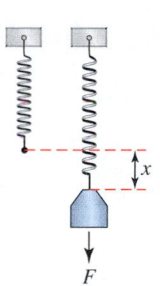

FIGURE 2.26 A Spring Being Stretched

A 12-pound weight is hung on a spring, and it stretches 2 inches.
(a) Find the spring constant.
(b) Determine how far the spring stretches when a 19-pound weight is hung on the spring.

SOLUTION
(a) Let $F = kx$, given that $F = \mathbf{12}$ pounds and $x = \mathbf{2}$ inches. Thus

$$\mathbf{12} = k(\mathbf{2}) \quad \text{or} \quad k = 6,$$

and the spring constant equals 6.

(b) Let $F = 19$ and $k = 6$. Thus $F = kx$ implies that $19 = 6x$, or $x = \frac{19}{6} \approx 3.17$ inches.

Now Try Exercise 107 ◆

Given a set of data points, one method to determine if y is directly proportional to x is to graph the ordered pairs (x, y). If the points lie on a line that passes through the origin, then y varies directly with x and the constant of proportionality k is equal to the slope of the line. A second method is to compute the ratio $\frac{y}{x}$ for each ordered pair. The equation $y = kx$ implies that $k = \frac{y}{x}$, so each ratio $\frac{y}{x}$ will equal k.

EXAMPLE 12 Modeling storage requirements for recording music

Recording music requires large amounts of computer memory. A compact disc (CD) can hold approximately 680 million bytes. One million bytes is commonly referred to as a **megabyte** (MB). (See Example 8, Section 1.1 for an explanation of a byte.) Table 2.10 lists the megabytes x needed to record y seconds of music.

TABLE 2.10

x (MB)	0.129	0.231	0.415	0.491
y (sec)	6.010	10.74	19.27	22.83

x (MB)	0.667	1.030	1.160	1.260
y (sec)	31.00	49.00	55.25	60.18

Source: Gateway 2000 System CD.

(a) Compute the ratio $\frac{y}{x}$ for each musical segment in the table. Interpret these ratios.
(b) Approximate a constant of proportionality k satisfying the equation $y = kx$. Graph the data and the equation together.
(c) Estimate the maximum number of seconds of music that can be placed on a 1.44-megabyte floppy disc.

SOLUTION
(a) The ratios are shown in Table 2.11 on the next page. For example, $\frac{6.010}{0.129} \approx 46.6$. These ratios represent the number of seconds that can be recorded on 1 megabyte. They are approximately equal, which indicates direct variation.

Calculator Help

To plot data and graph an equation in the same viewing rectangle, see Appendix B (page AP-11).

[−0.1, 1.5, 0.25] by [0, 70, 10]

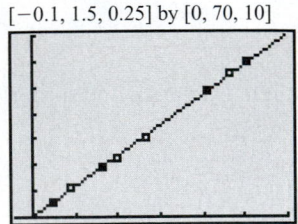

FIGURE 2.27

TABLE 2.11

x (MB)	0.129	0.231	0.415	0.491
y (sec)	6.010	10.74	19.27	22.83
y/x	46.6	46.5	46.4	46.5

x (MB)	0.667	1.030	1.160	1.260
y (sec)	31.00	49.00	55.25	60.18
y/x	46.5	47.6	47.6	47.8

(b) From Table 2.11 it appears that approximately 47 seconds of music can be recorded on 1 megabyte. Therefore let the constant of proportionality be $k = 47$. The data points and the equation $y = 47x$ are graphed in Figure 2.27.

(c) A 1.44-megabyte floppy disc could store at most $47 \cdot 1.44 \approx 68$ seconds of music.

Now Try Exercise 111 ◆

◆ **MAKING CONNECTIONS** ────────

Direct Variation and Linear Functions   If a set of data points (x, y) can be modeled by a linear function f that passes through the origin, then y varies directly with x. In this case the y-intercept is 0 and f can be written as $f(x) = mx$, where m is the constant of variation.

2.2

Putting it all Together

The following table summarizes some important topics.

Concept	Comments	Example
Point-slope form $y = m(x - x_1) + y_1$ or $y - y_1 = m(x - x_1)$	Used to find the equation of a line, given two points or one point and the slope	Given two points (5, 1) and (4, 3), first compute $m = \frac{3 - 1}{4 - 5} = -2$. An equation of this line is $$y = -2(x - 5) + 1.$$
Slope-intercept form $y = mx + b$	A unique equation for a line, determined by the slope m and the y-intercept b	An equation of the line with slope 5 and y-intercept -4 is $y = 5x - 4$.
Interpolation	Estimates values that are between two or more known data values	If the data points (0, 1) and (2, 3) are used to estimate the value of y when $x = 1$, then this would involve interpolation.
Extrapolation	Estimates values that are not between two known data values	If the data points (0, 1) and (2, 3) are used to estimate the value of y when $x = 4$, then this would involve extrapolation.

Concept	Comments	Example
Direct variation	The variable y is directly proportional to x or varies directly with x if $y = kx$ for some nonzero constant k. Constant k is the constant of proportionality or the constant of variation.	The sales tax on a purchase is directly proportional to the amount of the purchase. If the sales tax rate is 7%, sales tax y on a purchase of x dollars is calculated by $y = 0.07x$. The sales tax on a purchase of \$125 is $$y = 0.07(125) = \$8.75.$$

The following table summarizes the important concepts concerning special types of lines.

Concept	Equation(s)	Example
Horizontal line	$y = b$, where b is a constant.	A horizontal line with y-intercept 7 has the equation $y = 7$.
Vertical line	$x = k$, where k is a constant.	A vertical line with x-intercept -8 has the equation $x = -8$.
Parallel lines	$y = m_1x + b_1$ and $y = m_2x + b_2$, where $m_1 = m_2$.	The lines $y = -3x - 1$ and $y = -3x + 5$ are parallel because they both have slope -3.
Perpendicular lines	$y = m_1x + b_1$ and $y = m_2x + b_2$, where $m_1m_2 = -1$.	The lines $y = 2x - 5$ and $y = -\frac{1}{2}x + 2$ are perpendicular because $m_1m_2 = 2\left(-\frac{1}{2}\right) = -1$.

2.2 Exercises

Equations of Lines

Exercises 1–4: (Refer to Example 1.) Find an equation of the line passing through the given points. Use the first point as (x_1, y_1) and write your answer in point-slope form. Plot the points and graph the line by hand.

1. $(1, 2), (3, -2)$ **2.** $(-2, 3), (1, 0)$

3. $(-3, -1), (1, 2)$ **4.** $(-1, 2), (-2, -3)$

Exercises 5–10: Find a point-slope form for the equation of the line satisfying the conditions. Use the first point given for (x_1, y_1). Then convert the equation to slope-intercept form.

5. Slope -2.4, passing through $(4, 5)$

6. Slope 1.7, passing through $(-8, 10)$

7. Passing through $(1, -2)$ and $(-9, 3)$

8. Passing through $(-6, 10)$ and $(5, -12)$

9. Passing through $(1980, 5)$ and $(1990, 25)$

10. Passing through $(1990, -3)$ and $(1996, 19)$

Exercises 11–14: Find the slope-intercept form for the line in the figure.

11.

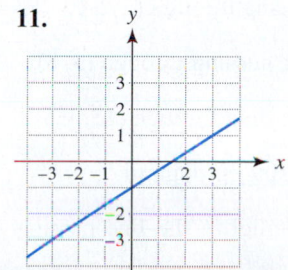

12.

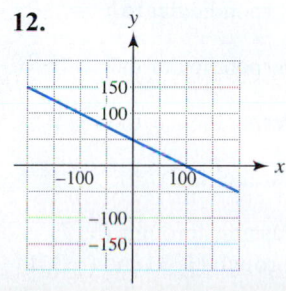

13. 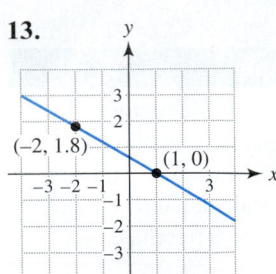   **14.**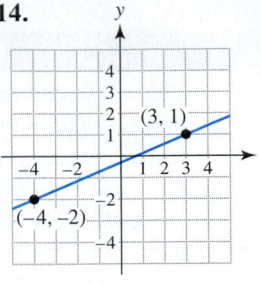

Exercises 15–38: Find the slope-intercept form for the line satisfying the conditions.

15. y-intercept 5, slope -7.8

16. y-intercept -155, slope 5.6

17. y-intercept 45, x-intercept 90

18. x-intercept -6, y-intercept -8

19. Parallel to $y = 4x + 16$, passing through $(-4, -7)$

20. Parallel to the line $y = -\frac{3}{4}(x - 100) - 99$, passing through $(1.5, \sqrt{3})$

21. Perpendicular to the line $y = -\frac{2}{3}(x - 1980) + 5$, passing through $(1980, 10)$

22. Perpendicular to $y = 6x - 1000$, passing through $(15, -7)$

23. Slope -3, passing through $(0, 5)$

24. Slope $\frac{1}{3}$, passing through $\left(\frac{1}{2}, -2\right)$

25. Passing through $(0, -6)$ and $(4, 0)$

26. Passing through $\left(\frac{3}{4}, -\frac{1}{4}\right)$ and $\left(\frac{5}{4}, \frac{7}{4}\right)$

27. Passing through $\left(\frac{1}{2}, \frac{3}{4}\right)$ and $\left(\frac{1}{5}, \frac{2}{3}\right)$

28. Passing through $\left(-\frac{7}{3}, \frac{5}{3}\right)$ and $\left(\frac{5}{6}, -\frac{7}{6}\right)$

29. Parallel to $y = \frac{2}{3}x + 3$, passing through $(0, -2.1)$

30. Parallel to $y = -4x - \frac{1}{4}$, passing through $(2, -5)$

31. Perpendicular to $y = -2x$, passing through $(-2, 5)$

32. Perpendicular to $y = -\frac{6}{7}x + \frac{3}{7}$, passing through $(3, 8)$

33. Perpendicular to $x + y = 4$, passing through $(15, -5)$

34. Parallel to $2x - 3y = -6$, passing through $(4, -9)$

35. Passing through $(5, 7)$ and parallel to the line passing through $(1, 3)$ and $(-3, 1)$

36. Passing through $(1990, 4)$ and parallel to the line passing through $(1980, 3)$ and $(2000, 8)$

37. Passing through $(-2, 4)$ and perpendicular to the line passing through $\left(-5, \frac{1}{2}\right)$ and $\left(-3, \frac{2}{3}\right)$

38. Passing through $\left(\frac{3}{4}, \frac{1}{4}\right)$ and perpendicular to the line passing through $(-3, -5)$ and $(-4, 0)$

Exercises 39–46: Find an equation of the line satisfying the conditions.

39. Vertical, passing through $(-5, 6)$

40. Vertical, passing through $(1.95, 10.7)$

41. Horizontal, passing through $(-5, 6)$

42. Horizontal, passing through $(1.95, 10.7)$

43. Perpendicular to $y = 15$, passing through $(4, -9)$

44. Perpendicular to $x = 15$, passing through $(1.6, -9.5)$

45. Parallel to $x = 4.5$, passing through $(19, 5.5)$

46. Parallel to $y = -2.5$, passing through $(1985, 67)$

Exercises 47–52: Match the equation to its graph (a.–f.).

47. $y = m(x - x_1) + y_1, m > 0$

48. $y = m(x - x_1) + y_1, m < 0$

49. $y = mx, m > 0$

50. $y = mx + b, m < 0$ and $b > 0$

51. $x = k, k > 0$ **52.** $y = b, b < 0$

a.   **b.**

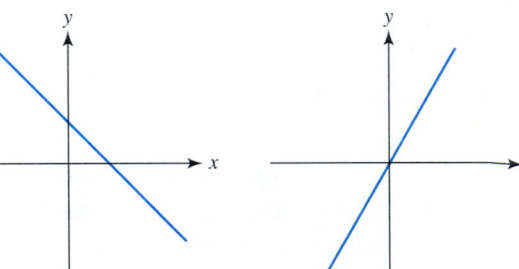

c.   **d.**

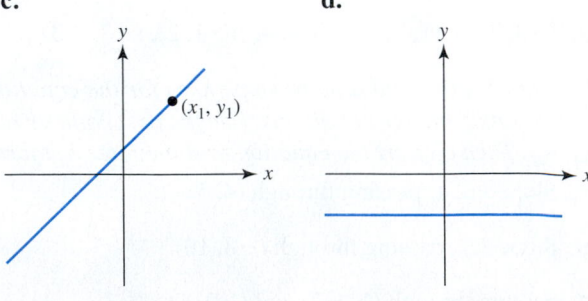

e.   **f.**

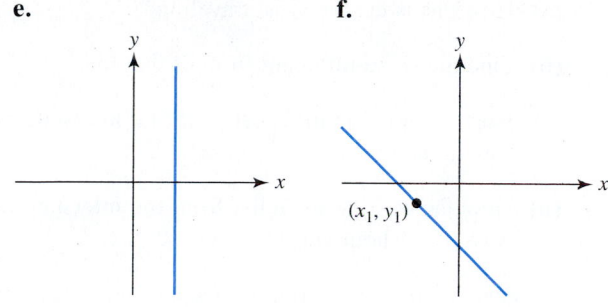

Interpolation and Extrapolation

Exercises 53–56: The table lists data that are exactly linear.

(a) *Find the slope-intercept form of the line that passes through these data points.*

(b) *Predict y when $x = -2.7$ and 6.3. Decide if these calculations involve interpolation or extrapolation.*

53.

x	-3	-2	-1	0	1
y	-7.7	-6.2	-4.7	-3.2	-1.7

54.

x	-2	-1	0	1	2
y	10.2	8.5	6.8	5.1	3.4

55.

x	5	23	32	55	61
y	94.7	56.9	38	-10.3	-22.9

56.

x	-11	-8	-7	-3	2
y	-16.1	-10.4	-8.5	-0.9	8.6

57. *Air Safety Inspectors* The number of air safety inspectors is shown in the accompanying table for selected years.

Year	1998	1999	2000
Inspectors	3305	3185	3089

Source: Federal Aviation Administration.

(a) Find a linear function f that models these data. Is f exact or approximate?

(b) Use f to estimate the number of inspectors in 1996. Compare your answer to the actual value of 2776. Did your estimate involve interpolation or extrapolation?

58. *Deaths on School Grounds* Nationwide deaths on school grounds for school years ending in year x are shown in the table.

x (year)	1998	1999	2000
y (deaths)	43	26	9

Source: FBI.

(a) Find a linear function f that models these data. Is f exact or approximate?

(b) Use f to estimate the number of deaths on school grounds in 1997. Compare your answer to the actual value of 26. Did your estimate involve interpolation or extrapolation?

Determining Intercepts

Exercises 59–72: Determine the x- and y-intercepts on the graph of the equation. Graph the equation.

59. $4x - 5y = 20$ **60.** $-3x - 5y = 15$

61. $x - y = 7$ **62.** $15x - y = 30$

63. $6x - 7y = -42$ **64.** $5x + 2y = -20$

65. $y = 8x - 5$ **66.** $y = -1.5x + 15$

67. $y = 3(x - 2) - 5$ **68.** $y = -2(x + 1) + 7$

69. $y - 3x = 7$ **70.** $4x - 3y = 6$

71. $0.2x + 0.4y = 0.8$ **72.** $\frac{2}{3}y - x = 1$

*Exercises 73–78: The **intercept form of a line** is $\frac{x}{a} + \frac{y}{b} = 1$. Determine the x- and y-intercepts on the graph of the equation. Draw a conclusion about what the constants a and b represent in this form.*

73. $\frac{x}{5} + \frac{y}{7} = 1$ **74.** $\frac{x}{2} + \frac{y}{3} = 1$

75. $\frac{x}{4} + \frac{y}{-3} = 1$ **76.** $\frac{x}{-4} + \frac{y}{2} = 1$

77. $\frac{2x}{3} + \frac{4y}{5} = 1$ **78.** $\frac{5x}{6} - \frac{y}{2} = 1$

Applications

79. *Projected Cost of College*   (Refer to Example 3.) In 2003 the average annual cost of attending a private college or university, including tuition, fees, room, and board, was $25,000. This cost is projected to rise to $37,000 in 2010, as illustrated in the figure.   (**Source:** Cerulli Associates.)

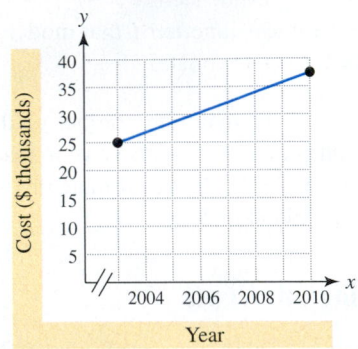

Year

(a) Find a point-slope form of the line passing through the points (2003, 25000) and (2010, 37000). Interpret the slope.

(b) Use the equation to estimate the cost of attending a private college in 2007.

(c) Find the slope-intercept form of this line.

80. *Music on the Internet*   In 2002 sales of premium online music were $1.6 billion. In 2005 this revenue reached $3.6 billion.   (**Source:** Jupiter Research.)
(a) Find a point-slope form of the line passing through the points (2002, 1.6) and (2005, 3.6). Interpret the slope.

(b) Use the equation to estimate projected sales in 2008.

(c) Find the slope-intercept form of this line.

81. *Distance*   A person is riding a bicycle along a straight highway. The accompanying graph shows the rider's distance y in miles from an interstate highway after x hours.

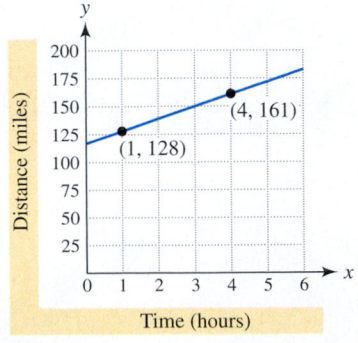

Time (hours)

(a) How fast is the bicyclist traveling?

(b) Find the slope-intercept form of the line.

(c) How far was the bicyclist from the interstate highway initially?

(d) How far was the bicyclist from the interstate highway after 1 hour and 15 minutes?

82. *Water in a Tank*   The graph shows the amount of water y in a 100-gallon tank after x minutes have elapsed.

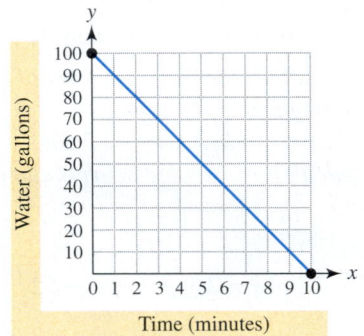

Time (minutes)

(a) Is water entering or leaving the tank? How much water is in the tank after 3 minutes?

(b) Find both the x- and y-intercepts. Interpret each intercept.

(c) Find the slope-intercept form of the equation of the line. Interpret the slope.

(d) Estimate the x-coordinate of the point $(x, 50)$ that lies on the line.

83. *Lots of Spam*   The following table lists the *daily* average number of worldwide spam messages in billions during selected years.

Year	1999	2000	2001	2002	2003	2004
Messages	1.0	2.3	4.0	5.6	7.3	8.8

Source: IDC.

(a) Make a scatterplot of the data.

(b) Find a formula $f(x) = m(x - x_1) + y_1$ so that $f(x)$ models these data. (Answers may vary.) Interpret the slope m.

(c) Use your formula to predict the daily average number of worldwide spam messages during 2007.

84. *Tuition and Fees* (Refer to Example 9.) The following table lists average tuition and fees in dollars at public 4-year colleges for selected years.

Year	1980	1985	1990
Tuitions and Fees	$804	$1318	$1908

Year	1995	2000
Tuitions and Fees	$2811	$3487

Source: The College Board.

(a) Make a scatterplot of the data.

(b) Find $f(x) = m(x - x_1) + y_1$ so that $f(x)$ models the data. Interpret the slope m.

(c) Use $f(x)$ to estimate tuition in 1992. Compare it to the actual value of $2334.

85. *Toyota Vehicles Sold* The following table lists the U.S. sales of Toyota vehicles in millions.

Year	1998	1999	2000	2001	2002
Vehicles	1.4	1.5	1.6	1.7	1.8

Source: Autodata.

(a) Make a scatterplot of the data.

(b) Find $f(x) = m(x - x_1) + y_1$ so that $f(x)$ models these data. Interpret the slope m.

(c) Is $f(x)$ an exact or approximate model for the data listed in the table?

86. *Farm Pollution* In 1988 the number of farm pollution incidents reported in England and Wales was 4000. This number had increased roughly at a rate of 280 per year since 1979. (**Source:** C. Mason, *Biology of Freshwater Pollution.*)

(a) Find an equation of a line $y = m(x - x_1) + y_1$ that models these data, where y represents the number of pollution incidents during the year x.

(b) Estimate the number of incidents in 1975.

87. *Cost of Driving* The cost of driving a car includes both fixed costs and mileage costs. Assume that it costs $350 per month for insurance and car payments and it costs $0.29 per mile for gasoline, oil, and routine maintenance.

(a) Determine a linear function f that computes the annual cost of driving this car x miles.

(b) What does the y-intercept on the graph of f represent?

88. *Average Wages* The average hourly wage (adjusted to 1982 dollars) was $8.03 in 1970 and $7.75 in 1998. (**Source:** Department of Commerce.)

(a) Find an equation of a line that passes through the points (1970, 8.03) and (1998, 7.75).

(b) Interpret the slope.

(c) Approximate the hourly wage in 1990. Compare it to the actual value of $7.52.

Exercises 89 and 90: *Modeling Real Data* *The table contains data that can be modeled by* $f(x) = m(x - x_1) + y_1$.

(a) *Make a scatterplot of the data. (Do not try to plot the undetermined point in the table.)*

(b) *Fit f to the data by approximating values for the constants m, x_1, and y_1. Graph f together with the data on the same coordinate axes.*

(c) *Interpret the slope m.*

(d) *Use f to approximate the undetermined value in the table.*

89. Asian-American population in millions

Year	1996	1997	1998	1999
Population	9.7	10.1	10.5	10.9

Year	2000	2001	2002	2005
Population	11.2	11.6	12.0	?

Source: Bureau of the Census.

90. Population in millions of the western region of the United States

Year	1950	1960	1970
Population	20.2	28.1	34.8

Year	1980	1990	2000
Population	43.2	52.8	?

Source: Bureau of the Census.

Perspectives and Viewing Rectangles

91. Graph $y = \frac{1}{1024}x + 1$ in [0, 3, 1] by [−2, 2, 1].

(a) Is the graph a horizontal line?

(b) Conjecture why the calculator screen appears as it does.

92. Graph the line $y = 1000x + 1000$ in the standard viewing rectangle.
(a) Is the graph a vertical line?

(b) Explain why the calculator screen appears as it does.

93. *Square Viewing Rectangle* Graph the lines $y = 2x$ and $y = -\frac{1}{2}x$ in the standard viewing rectangle.
(a) Do the lines appear to be perpendicular?

(b) Graph the lines in the following viewing rectangles.
   i. $[-15, 15, 1]$ by $[-10, 10, 1]$
   ii. $[-10, 10, 1]$ by $[-3, 3, 1]$
   iii. $[-3, 3, 1]$ by $[-2, 2, 1]$
   Do the lines appear to be perpendicular in any of these viewing rectangles?

(c) Determine the viewing rectangles where perpendicular lines will appear perpendicular. (Answers may vary depending on the model of graphing calculator used.)

94. Continuing with Exercise 93, make a conjecture about which viewing rectangles result in the graph of a circle with radius 5 and center at the origin appearing circular.
   i. $[-9, 9, 1]$ by $[-6, 6, 1]$
   ii. $[-5, 5, 1]$ by $[-10, 10, 1]$
   iii. $[-5, 5, 1]$ by $[-5, 5, 1]$
   iv. $[-18, 18, 1]$ by $[-12, 12, 1]$
   Test your conjecture by graphing this circle in each viewing rectangle. (*Hint:* Graph $y_1 = \sqrt{25 - x^2}$ and $y_2 = -\sqrt{25 - x^2}$ to create the circle.)

Graphing a Rectangle

Exercises 95–98: (Refer to Example 8.) A rectangle is determined by the stated conditions. Find the slope-intercept form of the four lines that outline the rectangle.

95. Vertices $(0, 0)$, $(2, 2)$, and $(1, 3)$

96. Vertices $(1, 1)$, $(5, 1)$, and $(5, 5)$

97. Vertices $(4, 0)$, $(0, 4)$, $(0, -4)$, and $(-4, 0)$

98. Vertices $(1, 1)$ and $(2, 3)$, and the point $(3.5, 1)$ lie on a side of the rectangle

Direct Variation

Exercises 99–102: Find the constant of proportionality k and the undetermined value in the table if y is directly proportional to x. Support your answer by graphing the equation $y = kx$ and the data points.

99.

x	3	5	6	8
y	7.5	12.5	15	?

100.

x	1.2	4.3	5.7	?
y	3.96	14.19	18.81	23.43

101. Sales tax y on a purchase of x dollars

x	$25	$55	?
y	$1.50	$3.30	$5.10

102. Cost y of buying x compact discs having the same price

x	3	4	5
y	$41.97	$55.96	?

103. *Cost of Tuition* The cost of tuition is directly proportional to the number of credits taken. If 11 credits cost $720.50, find the cost of taking 16 credits. What is the constant of proportionality?

104. *Strength of a Beam* The maximum load that a horizontal beam can carry is directly proportional to its width. If a beam 1.5 inches wide can support a load of 250 pounds, find the load that a beam of the same type can support if its width is 3.5 inches.

105. *Antarctic Ozone Layer* Stratospheric ozone occurs in the atmosphere between altitudes of 12 and 18 miles above sea level and is an important filter of ultraviolet light from the sun. Ozone in the stratosphere is frequently measured in Dobson units. The Dobson scale is linear where 300 Dobson units is a midrange value that corresponds to an ozone layer 3 millimeters thick. In 1991 the reported minimum in the Antarctic *ozone hole* was about 110 Dobson units. (**Source:** R. Huffman, *Atmospheric Ultraviolet Remote Sensing.*)
(a) The thickness y of the ozone layer is directly proportional to the Dobson scale x. Find the constant of proportionality k.

(b) How thick was the ozone layer over the Antarctic in 1991?

(c) Was the ozone hole actually a *hole* in the ozone layer?

106. *Weight on Mars* The weight of an object on Earth is directly proportional to the weight of an object on Mars. If a 25-pound object on Earth weighs 10 pounds on Mars, how much would a 195-pound astronaut weigh on Mars?

107. *Hooke's Law* (Refer to Example 11.) Suppose a 15-pound weight stretches a spring 8 inches, as shown in the figure.

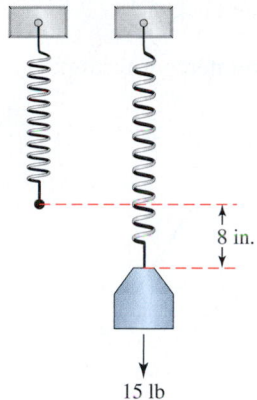

15 lb

  (a) Find the spring constant.

  (b) How far will a 25-pound weight stretch this spring?

108. *Hooke's Law* If an 80-pound force compresses a spring 3 inches, how much force must be applied to copress the spring 7 inches?

109. *Stopping Distance* The approximate stopping distances y in feet for a car traveling x miles per hour are listed in the table. Determine if stopping distance is directly proportional to speed. Interpret your results. (**Source:** L. Haefner, *Introduction to Transportation Systems.*)

x (mph)	0	20	40	60	80
y (ft)	0	118	324	620	1004

110. *Distance Traveled* Suppose a car travels at 60 miles per hour for x hours. Then the distance y that the car travels is directly proportional to x. Find the constant of proportionality. How far does the car travel in 3 hours?

111. *Force of Friction* The table lists the force F needed to push a cargo box weighing x pounds on a smooth wood floor.

x (lb)	150	180	210	320
F (lb)	26	31	36	54

  (a) Compute the ratio $\frac{F}{x}$ for each data pair in the table. Interpret these ratios.

  (b) Approximate a constant of proportionality k satisfying $F = kx$. (The value of k is called the *coefficient of friction*.)

  (c) Graph the data and the equation together.

  (d) Estimate the force needed to push a 275-pound cargo box on the floor.

112. *Electrical Resistance* The electrical resistance of a wire varies directly with its length. If a 255-foot wire has a resistance of 1.2 ohms, find the resistance of 135 feet of the same type of wire. Interpret the constant of proportionality in this situation.

Writing about Mathematics

113. Compare the slope-intercept form with the point-slope form. Give examples of each.

114. Give an example of two quantities in real life that vary directly. Explain your answer. Use an equation to describe the relationship between the two quantities.

EXTENDED AND DISCOVERY EXERCISES

Exercises 1 and 2: ***Estimating Populations*** *Biologists sometimes use direct variation to estimate the number of fish in small lakes. This is done by tagging a small number of fish and then releasing them. Biologists assume that over a period of time, the tagged fish distribute themselves evenly throughout the lake. Later a second sample is collected. The total number of fish in the sample is counted along with the number of tagged fish. To determine the total population of fish in the lake, biologists assume that the proportion of tagged fish in the second sample is equal to the proportion of tagged fish in the entire lake. This technique can also be used to count other types of animals such as birds, when they are not migrating.*

  1. Eighty-five fish are tagged and released into a pond. Later a sample of 94 fish from the pond contains 13 tagged fish. Estimate the number of fish in the pond.

  2. Sixty-three blackbirds are tagged and released. Later it is estimated that out of a sample of 32 blackbirds, only 8 are tagged. Estimate the population of blackbirds in the area.

CHECKING BASIC CONCEPTS FOR SECTIONS 2.1 AND 2.2

1. Graph $f(x) = 4 - 2x$ by hand. Identify the slope, the x-intercept, and the y-intercept.

2. The death rate from heart disease for ages 15 through 24 is 2.7 per 100,000 people.
   (a) Write a function f that models the number of deaths in a population of x million people who are 15 to 24 years old.

   (b) There are about 39 million people in the United States who are 15 to 24 years old. Use f to estimate the number of deaths from heart disease in this age group.

3. A driver of a car is initially 50 miles south of home, driving 60 miles per hour south. Write a function f that models the distance between the driver and home.

4. Find an equation of the line passing through the points $(-3, 4)$ and $(5, -2)$. Give equations of lines that are parallel and perpendicular to this line.

5. Find equations of horizontal and vertical lines that pass through the point $(-4, 7)$.

6. Write the slope-intercept form of the line shown in the figure.

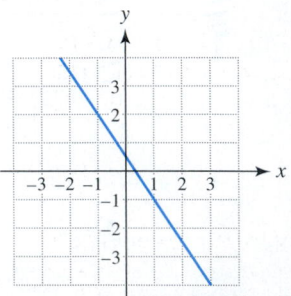

7. Find the x- and y-intercepts on the graph of the equation $-3x + 2y = -18$.

- ◆ Understand basic terminology related to equations
- ◆ Recognize a linear equation
- ◆ Solve linear equations symbolically
- ◆ Solve linear equations graphically and numerically
- ◆ Understand the intermediate value property
- ◆ Solve problems involving percentages
- ◆ Apply problem-solving strategies

2.3   Linear Equations

Introduction

A primary objective of both theoretical and applied mathematics is solving equations. Billions of dollars are spent each year on computers and personnel to solve equations that hold the answers for creating better products. Equations occur in a wide variety of forms. Some equations can be solved easily, whereas others require enormous amounts of resources. Without graphical, numerical, and symbolic techniques for solving equations, our society would not have televisions, CD players, satellites, fiber optics, CAT scans, computers, or accurate weather forecasts.

Chapter 1 introduced the concept of a function and its representations. In this section we see that applications involving functions lead to equations. For example, the function given by $f(x) = 572.3(x - 1980) + 3617$ models the average tuition and fees at private colleges from 1980 to 2000. (See Example 9, Section 2.2.) One way to determine the year x when the average tuition was \$14,000 would be to solve the *equation*

$$572.3(x - 1980) + 3617 = 14,000.$$

Equations

An **equation** is a statement that two mathematical expressions are equal. Equations always contain an equals sign. Some examples of equations include

$$x + 15 = 9x - 1, \qquad x^2 - 2x + 1 = 2x, \qquad z + 5 = 0,$$
$$xy + x^2 = y^3 + x, \qquad \text{and} \qquad 1 + 2 = 3.$$

The first three equations have one variable, the fourth equation has two variables, and the fifth equation contains only constants. For now, our discussion concentrates on equations with one variable.

To **solve** an equation means to find all values for the variable that make the equation a true statement. Such values are called **solutions**. The set of all solutions is the **solution set**. The solutions to the equation $x^2 - 1 = 0$ are 1 or -1, written as $x = \pm 1$. Either value for x **satisfies** the equation. The solution set is $\{-1, 1\}$. Two equations are **equivalent** if they have the same solution set. For example, the equations $x + 2 = 5$ and $x = 3$ are equivalent.

If an equation has no solutions, then its solution set is empty and the equation is called a **contradiction**. The equation $x + 2 = x$ has no solutions and is a contradiction. However, if every (meaningful) value for the variable is a solution, then the equation is an **identity**. The equation $x + x = 2x$ is an identity because every value for x makes the equation true. Any equation that is satisfied by some, but not all, values of the variable is a **conditional equation**. The equation $x^2 - 1 = 0$ is a conditional equation. Only the values -1 and 1 for x make this equation a true statement.

As with functions, equations can be either *linear* or *nonlinear*. A linear equation is one of the simplest types of equations.

LINEAR EQUATION IN ONE VARIABLE

A **linear equation** in one variable is an equation that can be written in the form

$$ax + b = 0,$$

where a and b are real numbers with $a \neq 0$.

If an equation is not linear, then we say that it is a **nonlinear equation**. The following are examples of linear equations. In each case, rules of algebra could be used to write the equation in the form $ax + b = 0$ with $a \neq 0$. A linear equation has *exactly one* solution. Why?

$$x - 12 = 0, \quad 2x - 4 = -x, \quad 2(1 - 4x) = 16, \quad x - 5 + 3(x - 1) = 0$$

Symbolic Solutions

Linear equations can be solved symbolically. One advantage of a symbolic method is that the solution is *always exact*. To solve a linear equation symbolically, we usually apply the *properties of equality* to the given equation and transform it into equivalent equations that are simpler.

PROPERTIES OF EQUALITY

Addition Property of Equality

If a, b, and c are real numbers, then

$$a = b \quad \text{is equivalent to} \quad a + c = b + c.$$

Multiplication Property of Equality

If a, b, and c are real numbers with $c \neq 0$, then

$$a = b \quad \text{is equivalent to} \quad ac = bc.$$

The addition property states that an equivalent equation results if the same number is added to (or subtracted from) each side of an equation. Similarly, the multiplication property states that an equivalent equation results if each side of an equation is multiplied (or divided) by the same nonzero number.

EXAMPLE 1  Solving a linear equation symbolically

Solve the equation $3(x - 4) = 2x - 1$. Check your answer.

SOLUTION

$3(x - 4) = 2x - 1$	Given equation
$3x - 12 = 2x - 1$	Distributive property
$3x - 2x = 12 - 1$	Subtract 2x and add 12.
$x = 11$	Simplify.

The solution is 11.
    We can check our answer as follows.

$3(x - 4) = 2x - 1$	Given equation
$3(\mathbf{11} - 4) \stackrel{?}{=} 2 \cdot \mathbf{11} - 1$	Let x = 11.
$21 = 21$	The answer checks.

Now Try Exercise 17 ◆

EXAMPLE 2 Solving a linear equation symbolically

Solve $3(2x - 5) = 10 - (x + 5)$. Check your answer.

SOLUTION

$3(2x - 5) = 10 - (x + 5)$	Given equation
$6x - 15 = 10 - x - 5$	Distributive property
$6x - 15 = 5 - x$	Simplify.
$7x - 15 = 5$	Add x to each side.
$7x = 20$	Add 15 to each side.
$x = \dfrac{20}{7}$	Divide each side by 7.

The solution is $\frac{20}{7}$.
    To check this answer, let $x = \frac{20}{7}$ in the given equation and simplify.

$3(2x - 5) = 10 - (x + 5)$	Given equation
$3\left(2 \cdot \dfrac{\mathbf{20}}{\mathbf{7}} - 5\right) \stackrel{?}{=} 10 - \left(\dfrac{\mathbf{20}}{\mathbf{7}} + 5\right)$	Let $x = \frac{20}{7}$.
$\dfrac{15}{7} = \dfrac{15}{7}$	The answer checks.

Now Try Exercises 11 and 18 ◆

When fractions or decimals appear in an equation, our work can be made simpler by multiplying each side of the equation by the least common denominator (LCD) (or a common denominator) of all fractions in the equation. This method is illustrated in the next example.

EXAMPLE 3 Eliminating fractions and decimals

Solve each linear equation.

(a) $\dfrac{t-2}{4} - \dfrac{1}{3}t = 5 - \dfrac{1}{12}(3-t)$ (b) $0.03(z-3) - 0.5(2z+1) = 0.23$

SOLUTION
(a) To eliminate fractions, multiply each side (or term in the equation) by the LCD, 12.

$$\dfrac{t-2}{4} - \dfrac{1}{3}t = 5 - \dfrac{1}{12}(3-t) \qquad \text{Given equation}$$

$$\dfrac{12(t-2)}{4} - \dfrac{12}{3}t = 12(5) - \dfrac{12}{12}(3-t) \qquad \text{Multiply each side (term) by 12.}$$

$$3(t-2) - 4t = 60 - (3-t) \qquad \text{Reduce and simplify.}$$

$$3t - 6 - 4t = 60 - 3 + t \qquad \text{Distributive property}$$

$$-t - 6 = 57 + t \qquad \text{Combine like terms on each side.}$$

$$-2t = 63 \qquad \text{Add } -t \text{ and 6 to each side.}$$

$$t = -\dfrac{63}{2} \qquad \text{Divide each side by } -2.$$

The solution is $-\frac{63}{2}$.

(b) To eliminate decimals, multiply each side (or term in the equation) by 100.

$$0.03(z-3) - 0.5(2z+1) = 0.23 \qquad \text{Given equation}$$

$$3(z-3) - 50(2z+1) = 23 \qquad \text{Multiply each side (term) by 100.}$$

$$3z - 9 - 100z - 50 = 23 \qquad \text{Distributive property}$$

$$-97z - 59 = 23 \qquad \text{Combine like terms.}$$

$$-97z = 82 \qquad \text{Add 59 to each side.}$$

$$z = -\dfrac{82}{97} \qquad \text{Divide each side by } -97.$$

The solution is $-\frac{82}{97}$. *Now Try Exercises 25 and 29*

Hand-held personal computers (PCs) are becoming increasingly popular. In the next example we estimate the year when sales of hand-held PCs could reach 19 million.

EXAMPLE 4 Modeling hand-held PCs

In 1998 worldwide sales of hand-held PCs were 4 million, and in 2000 they were 7.7 million. Use a linear function f to estimate the year when sales could reach 19 million. (**Source:** Dataquest.)

SOLUTION The graph of f must pass through the points (**1998**, **4.0**) and (**2000**, **7.7**), and its slope is

$$m = \dfrac{7.7 - 4.0}{2000 - 1998} = 1.85.$$

Thus, $f(x) = \mathbf{1.85}(x - \mathbf{1998}) + \mathbf{4}$ models the data. To determine when sales could reach 19 million, solve the equation $f(x) = 19$.

$$1.85(x - 1998) + 4 = 19 \qquad\qquad f(x) = 19$$

$$1.85(x - 1998) = 15 \qquad\qquad \text{Subtract 4.}$$

$$x - 1998 = \frac{15}{1.85} \qquad\qquad \text{Divide by 1.85.}$$

$$x = 1998 + \frac{15}{1.85} \qquad\qquad \text{Add 1998.}$$

$$x \approx 2006.1 \qquad\qquad \text{Approximate.}$$

This linear model predicts that 19 million hand-held PCs will be sold during 2006.

Now Try Exercise 83 ◆

◆ **CLASS DISCUSSION**

Do you think the prediction in Example 4 is accurate? Explain your answer. ◆

Graphical and Numerical Solutions

The equation $f(x) = g(x)$ results whenever the formulas for two functions f and g are set equal to each other. A solution to this equation corresponds to the x-coordinate of a point where the graphs of f and g intersect. This technique is called the *intersection-of-graphs method*. If the graphs of f and g are lines with different slopes, then their graphs intersect once. For example, if $f(x) = 2x + 1$ and $g(x) = -x + 4$, then the equation $f(x) = g(x)$ becomes $\mathbf{2x + 1 = -x + 4}$. To apply the intersection-of-graphs method, we graph $\mathbf{y_1 = 2x + 1}$ and $\mathbf{y_2 = -x + 4}$, as shown in Figure 2.28.

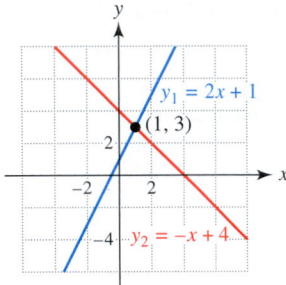

FIGURE 2.28

Their graphs intersect at the point $(1, 3)$. Since the variable in the given equation $2x + 1 = -x + 4$ is x, the solution is 1, the x-coordinate of the point of intersection. When $x = 1$, the functions f and g both assume the value 3, that is, $f(1) = 2(1) + 1 = 3$ and $g(1) = -1 + 4 = 3$. The value 3 is the y-coordinate of the point of intersection $(1, 3)$.

The intersection-of-graphs method is summarized in the following.

INTERSECTION-OF-GRAPHS METHOD

The **intersection-of-graphs method** can be used to solve an equation graphically. To implement this procedure, follow these steps.

STEP 1: Set y_1 equal to the left side of the equation, and set y_2 equal to the right side of the equation.

STEP 2: Graph y_1 and y_2.

STEP 3: Locate any points of intersection. The x-coordinates of these points correspond to solutions to the equation.

EXAMPLE 5   Solving an equation graphically

Solve $2x - 1 = \frac{1}{2}x + 2$ graphically by hand. Check your answer symbolically.

SOLUTION

Graphical Solution   Let $y_1 = 2x - 1$ and $y_2 = \frac{1}{2}x + 2$. The graph of y_1 has slope 2 and y-intercept -1, and the graph of y_2 has slope $\frac{1}{2}$ and y-intercept 2. Their graphs intersect at the point $(2, 3)$, as shown in Figure 2.29 so the solution is 2.

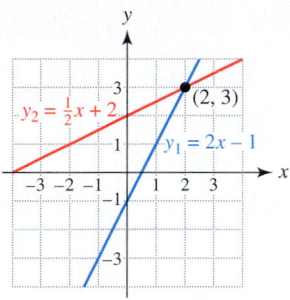

FIGURE 2.29   Intersection-of-Graphs

Symbolic Solution

$2x - 1 = \dfrac{1}{2}x + 2$	*Given equation*
$2x = \dfrac{1}{2}x + 3$	*Add 1 to each side.*
$\dfrac{3}{2}x = 3$	*Subtract $\frac{1}{2}x$ from each side.*
$\dfrac{2}{3} \cdot \dfrac{3}{2}x = \dfrac{2}{3} \cdot 3$	*Multiply each side by $\frac{2}{3}$.*
$x = 2$	*Multiply fractions.*

The solution is 2 and agrees with the graphical solution.   *Now Try Exercise 51* ◆

EXAMPLE 6

Applying the intersection-of-graphs method

The percent share of music sales (in dollars) that compact discs held from 1987 to 1998 could be modeled by $f(x) = 5.91x + 13.7$. During the same time period the percent share of music sales that cassette tapes held could be modeled by $g(x) = -4.71x + 64.7$. In these formulas $x = 0$ corresponds to 1987, $x = 1$ to 1988, and so on. Use the intersection-of-graphs method to estimate the year when the percent share of sales of CDs equaled the percent share of sales of cassettes.   (**Source:** Recording Industry Association of America.)

SOLUTION   We must solve the linear equation $f(x) = g(x)$, or equivalently,

$$5.91x + 13.7 = -4.71x + 64.7.$$

Graph $Y_1 = 5.91X + 13.7$ and $Y_2 = -4.71X + 64.7$, as shown in Figure 2.30. In Figure 2.31 their graphs intersect near the point $(4.8, 42.1)$. Since $x = 0$ corresponds to 1987 and $1987 + 4.8 \approx 1992$, it follows that in 1992 sales of CDs and cassette tapes were approximately equal. Both shared about 42.1% of the sales in 1992.

Calculator Help

To find the point of interesection in Figure 2.31, see Appendix B (page AP-12).

[0, 12, 2] by [0, 100, 10]   [0, 12, 2] by [0, 100, 10]

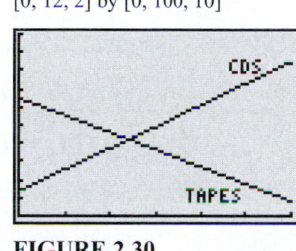

FIGURE 2.30   **FIGURE 2.31**   *Now Try Exercise 111* ◆

In the next example we solve the equation that was presented in the introduction to this section.

EXAMPLE 7

Solving an application symbolically and graphically

The linear function given by $f(x) = 572.3(x - 1980) + 3617$ models average tuition at private colleges from 1980 to 2000. Solve the equation $f(x) = 14{,}000$ symbolically and graphically to determine the year when tuition reached \$14,000.

SOLUTION *Symbolic Solution*

$$f(x) = 14{,}000 \qquad \textcolor{blue}{\text{Given equation}}$$

$$572.3(x - 1980) + 3617 = 14{,}000 \qquad \textcolor{blue}{\text{Substitute for } f(x).}$$

$$572.3(x - 1980) = 10{,}383 \qquad \textcolor{blue}{\text{Subtract 3617.}}$$

$$x - 1980 = \frac{10{,}383}{572.3} \qquad \textcolor{blue}{\text{Divide by 572.3.}}$$

[1980, 2000, 5] by [0, 18000, 1000]

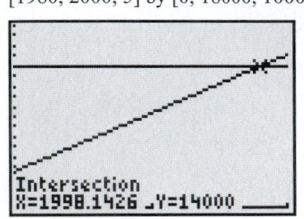

$$x = 1980 + \frac{10{,}383}{572.3} \qquad \textcolor{blue}{\text{Add 1980.}}$$

$$x \approx 1998.14 \qquad \textcolor{blue}{\text{Approximate.}}$$

Tuition at private colleges reached \$14,000 in 1998.

Graphical Solution Graph $Y_1 = 572.3(X - 1980) + 3617$ and $Y_2 = 14{,}000$, as shown in Figure 2.32. Their graphs intersect near the point (1998.14, 14000). Notice that the graphical solution agrees with the symbolic solution. *Now Try Exercise 79* ◆

FIGURE 2.32

EXAMPLE 8 Applying graphical, symbolic and numerical methods

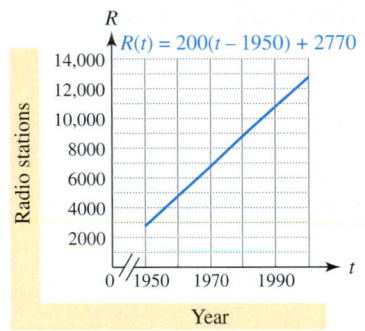

FIGURE 2.33 Radio Stations

Figure 2.33 shows a graph of a linear function R that approximates the number of radio stations on the air in the United States from 1950 to 2000. (**Source:** National Association of Broadcasters.)

(a) Use the graph to estimate the year when 10,000 radio stations were on the air.

(b) Use the formula $R(t) = 200(t - 1950) + 2770$ to estimate the year when 10,000 radio stations were on the air.

(c) Use a table of values to estimate when 10,000 radio stations were on the air.

(d) Compare these graphical, symbolic, and numerical solutions.

SOLUTION

(a) *Graphical Solution* A value of 10,000 on the R-axis appears to correspond to 1986 on the t-axis, as illustrated in Figure 2.34.

(b) *Symbolic Solution* Solve the linear equation $R(t) = 10{,}000$ symbolically.

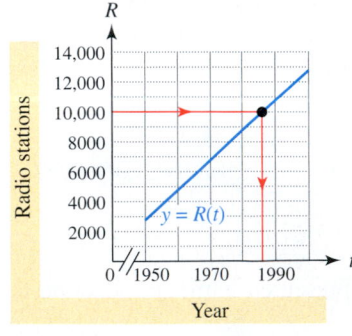

FIGURE 2.34 Radio Stations

$$200(t - 1950) + 2770 = 10{,}000 \qquad \textcolor{blue}{\text{Solve } R(t) = 10{,}000.}$$

$$200(t - 1950) = 7230 \qquad \textcolor{blue}{\text{Subtract 2770 from each side.}}$$

$$t - 1950 = \frac{7230}{200} \qquad \textcolor{blue}{\text{Divide each side by 200.}}$$

$$t = 1950 + \frac{7230}{200} \qquad \textcolor{blue}{\text{Add 1950 to each side.}}$$

$$t = 1986.15 \qquad \textcolor{blue}{\text{Simplify.}}$$

In 1986 there were about 10,000 radio stations.

(c) *Numerical Solution* The equation $R(t) = 10{,}000$ can be solved by letting $Y_1 = 200(X - 1950) + 2770$ and making a table of values. In Figure 2.35 we see that there were approximately 10,000 radio stations in 1986. If desired, we can increment by 0.1 to obtain a more accurate solution, as shown in Figure 2.36. From Figure 2.36 and knowing that y_1 is linear, can you determine the exact solution? Why?

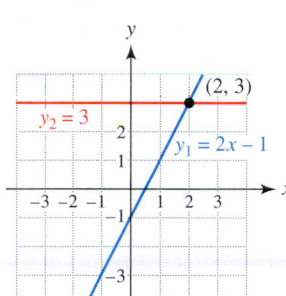

FIGURE 2.35   FIGURE 2.36

(d) The graphical, symbolic, and numerical solutions all give similar results.

Now Try Exercise 80 ◆

Calculator Help
To make a table of values, see Appendix B (page AP-9).

◆ MAKING CONNECTIONS

Symbolic, Graphical, and Numerical Solutions   Linear equations can be solved symbolically, graphically, and numerically. Symbolic solutions to linear equations are *always exact*, whereas graphical and numerical solutions are *sometimes approximate*. The following illustrates how to solve the equation $2x - 1 = 3$ with each method.

Symbolic Solution

$$2x - 1 = 3$$
$$2x = 4$$
$$x = 2$$

Graphical Solution

The solution is 2.

Numerical Solution

x	0	1	**2**	3
$2x - 1$	-1	1	**3**	5

Because $2x - 1$ equals 3 when $x = 2$, the solution to $2x - 1 = 3$ is 2.

In Example 8, the number of radio stations on the air is modeled by a continuous linear function f. In part (c), it is assumed that since $y_1 < 10{,}000$ when $x = 1986$ and $y_1 > 10{,}000$ when $x = 1987$, there is an x-value between 1986 and 1987 where $y_1 = 10{,}000$. This is true because the graph of y_1 is *continuous* with no breaks. This basic concept is referred to as the *intermediate value property*.

INTERMEDIATE VALUE PROPERTY

Let (x_1, y_1) and (x_2, y_2) with $y_1 \neq y_2$ and $x_1 < x_2$ be two points on the graph of a continuous function f. Then, on the interval $x_1 \leq x \leq x_2$, f assumes every value between y_1 and y_2 at least once.

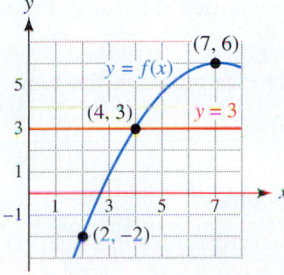

FIGURE 2.37

The intermediate value property is illustrated in Figure 2.37. The points $(2, -2)$ and $(7, 6)$ lie on the graph of a function f. The value 3 lies between the y-values of -2 and 6. Because f is continuous, $f(x)$ must equal 3 for some x on the interval $2 \leq x \leq 7$. This x-value is 4, since the point $(4, 3)$ lies on the graph of f. Loosely speaking, the intermediate value property is saying that we cannot draw a continuous curve that connects the points $(2, -2)$ and $(7, 6)$ without crossing the line $y = 3$ at least once. The only way not to cross this line would be to pick up the pencil. However, this creates a discontinuous graph, rather than a continuous one.

There are many examples of the intermediate value property. Physical motion is usually considered to be continuous. Suppose at one time a car is traveling at 20 miles per hour and at another time it is traveling at 40 miles per hour. It is logical to assume that the car traveled 30 miles per hour at least once between these times. In fact, by the intermediate value property, the car must have assumed all speeds between 20 and 40 miles per hour at least once. Similarly, if a jet airliner takes off and flies at an altitude of 30,000 feet, then by the intermediate value property, we may conclude that the airliner assumed all altitudes between ground level and 30,000 feet at least once.

Percentages   Applications involving percentages often result in linear equations because percentages can be computed by linear functions. Taking P percent of x is performed by $f(x) = Px$, where P is in decimal form. For example, to calculate 35% of x, let $f(x) = 0.35x$. Then 35% of \$150 can be computed by $f(150) = 0.35(150) = 52.5$ or \$52.50.

EXAMPLE 9   **Solving an application involving percentages**

A survey found that 76% of bicycle riders do not wear helmets.   (**Source:** Opinion Research Corporation for Glaxo Wellcome, Inc.)
(a) Find a symbolic representation for a function that computes the number of people who do not wear helmets among x bicycle riders.
(b) There are approximately 38.7 million riders of all ages who do not wear helmets. Write a linear equation whose solution gives the total number of bicycle riders. Solve this equation.

SOLUTION
(a) A linear function f that computes 76% of x is given by $f(x) = 0.76x$.
(b) We must find the x-value for which $f(x) = 38.7$ million. The equation is $0.76x = 38.7$. Solving gives $x = \frac{38.7}{0.76} \approx 50.9$ million bike riders.   *Now Try Exercise 93* ◆

Problem-Solving Strategies

To become more proficient at solving problems, we need to establish a procedure to guide our thinking. The following steps may be helpful in solving application problems.

SOLVING APPLICATION PROBLEMS

STEP 1: Read the problem and make sure you understand it. Assign a variable to what you are being asked to find. If necessary, write other quantities in terms of this variable.

STEP 2: Write an equation that relates the quantities described in the problem. You may need to sketch a diagram and refer to known formulas.

STEP 3: Solve the equation and determine the solution.

STEP 4: Look back and check your solution. Does it seem reasonable?

These steps are applied to the next four examples.

EXAMPLE  **Working together**

A large pump can empty a tank of gasoline in 5 hours, and a smaller pump can empty the same tank in 9 hours. If both pumps are used to empty the tank, how long will it take?

SOLUTION

STEP 1:   We are asked to find the time it takes for both pumps to empty the tank. Let this time be t.

$$t\text{: Time to empty the tank}$$

STEP 2:   In 1 hour the large pump will empty $\frac{1}{5}$ of the tank and the smaller pump will empty $\frac{1}{9}$ of the tank. The fraction of the tank that they will empty together in 1 hour is given by $\frac{1}{5} + \frac{1}{9}$. In 2 hours the large pump will empty $\frac{2}{5}$ of the tank and the smaller pump will empty $\frac{2}{9}$ of the tank. The fraction of the tank that they will empty together in 2 hours is $\frac{2}{5} + \frac{2}{9}$. Similarly, in t hours the fraction of the tank that the two pumps can empty is $\frac{t}{5} + \frac{t}{9}$. Since the tank is empty when this fraction reaches 1, we must solve the following equation.

$$\frac{t}{5} + \frac{t}{9} = 1$$

STEP 3:   Multiply by the LCD, 45, to eliminate fractions.

$$\frac{45t}{5} + \frac{45t}{9} = 45 \qquad \text{Multiply by LCD.}$$

$$9t + 5t = 45 \qquad \text{Simplify.}$$

$$14t = 45 \qquad \text{Add like terms.}$$

$$t = \frac{45}{14} \approx 3.21 \qquad \text{Divide by 14 and approximate.}$$

Working together, the two pumps can empty the tank in about 3.21 hours.

STEP 4:   This sounds reasonable. Working together the two pumps should be able to empty the tank faster than the large pump working alone, but not twice as fast.

Now Try Exercise 97 ◆

EXAMPLE  **Solving an application involving motion**

In 1 hour an athlete travels 10.1 miles by running at 8 miles per hour and then at 11 miles per hour. How long did the athlete run at each speed?

SOLUTION

STEP 1:   We are asked to find the time spent running at each speed. If we let x represent the time in hours running at 8 miles per hour, then $1 - x$ represents the time spent running at 11 miles per hour because the total running time was 1 hour.

$$x\text{: Time spent running at 8 miles per hour}$$

$$1 - x\text{: Time spent running at 11 miles per hour}$$

STEP 2:   Distance d equals rate r times time t; that is, $d = rt$. In this example we have two rates (or speeds) and two times. The total distance must sum to 10.1 miles.

$$d = r_1 t_1 + r_2 t_2 \qquad \text{General equation}$$

$$10.1 = 8x + 11(1 - x) \qquad \text{Substitute.}$$

STEP 3: We can solve this equation symbolically.

$$10.1 = 8x + 11 - 11x \qquad \text{Distributive property}$$
$$10.1 = 11 - 3x \qquad \text{Combine like terms.}$$
$$3x = 0.9 \qquad \text{Add } 3x; \text{ subtract } 10.1.$$
$$x = 0.3 \qquad \text{Divide by } 3.$$

The athlete runs 0.3 hour (18 minutes) at 8 miles per hour and 0.7 hour (42 minutes) at 11 miles per hour.

STEP 4: We can check this solution as follows.

$$8(0.3) + 11(0.7) = 10.1 \qquad \text{It checks.}$$

This sounds reasonable. The runner's average speed was 10.1 miles per hour so the runner must have run longer at 11 miles per hour than at 8 miles per hour.

Now Try Exercise 95 ◆

Similar triangles are often used in applications involving geometry. Similar triangles are used to solve the next application.

EXAMPLE 12 Solving an application involving similar triangles

A person 6 feet tall stands 17 feet from the base of a streetlight, as illustrated in Figure 2.38. If the person's shadow is 8 feet, estimate the height of the street light.

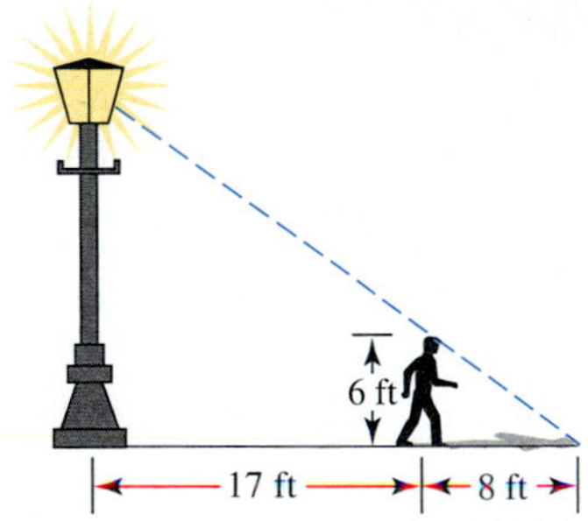

SOLUTION

STEP 1: We are being asked to find the height of the streetlight in Figure 2.38. Let x represent this height.

$$x\text{: Height of the streetlight}$$

FIGURE 2.38

STEP 2: In Figure 2.39, triangle *ACD* is similar to triangle *BCE*. Thus, ratios of corresponding sides are equal.

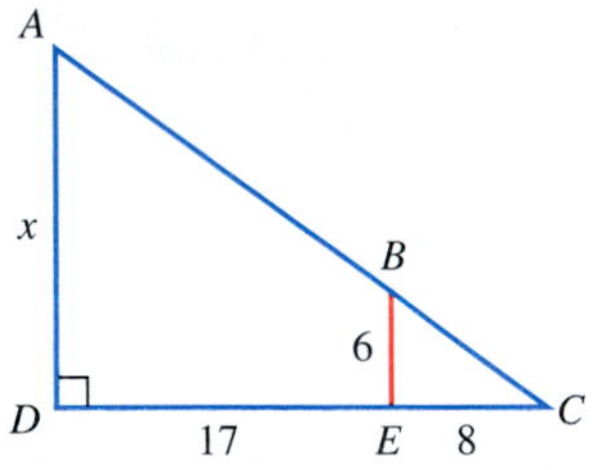

$$\frac{AD}{BE} = \frac{DC}{EC}$$

$$\frac{x}{6} = \frac{17 + 8}{8}$$

FIGURE 2.39

STEP 3: We can solve this equation symbolically.

$$\frac{x}{6} = \frac{25}{8} \qquad \text{Simplify.}$$

$$x = \frac{6 \cdot 25}{8} \qquad \text{Multiply by 6.}$$

$$x = 18.75 \qquad \text{Simplify.}$$

The height of the streetlight is 18.75 feet.

Geometry Review
To review similar triangles, see Chapter R (page R-5).

STEP 4: This sounds reasonable. The streetlight should be taller than the person.

Now Try Exercise 85 ◆

EXAMPLE 13 Mixing acid in chemistry

Pure water is being added to a 30% solution of 153 milliliters of hydrochloric acid. How much water should be added to reduce it to a 13% mixture?

SOLUTION

STEP 1: We are asked to find the amount of water that should be added to 153 milliliters of 30% acid to make it a 13% solution. Let this amount of water be equal to x.

$$x: \text{Amount of pure water to be added}$$

$$x + 153: \text{Final volume of 13\% solution}$$

(beaker diagram labeled: $x + 153$ ml, Acid — 153 ml)

STEP 2: Since only water is added, the total amount of acid in the solution after adding the water must equal the amount of acid before the water is added. The volume of pure acid after the water is added equals 13% of $x + 153$ milliliters, and the volume of pure acid before the water is added equals 30% of 153 milliliters. We must solve the following equation.

$$0.13(x + 153) = 0.30(153)$$

STEP 3: Begin by dividing each side by 0.13.

$$x + 153 = \frac{0.30(153)}{0.13} \qquad \text{Divide by 0.13.}$$

$$x = \frac{0.30(153)}{0.13} - 153 \qquad \text{Subtract 153.}$$

$$x \approx 200.08 \qquad \text{Approximate.}$$

We should add about 200 milliliters of pure water.

STEP 4: This sounds reasonable. If we added 153 milliliters of water, we would have diluted the acid to half its concentration, which would be 15%. It follows that we should add more than 153 milliliters, but not a lot more.

Now Try Exercise 99 ◆

2.3 Putting it all Together

A linear equation can be solved symbolically, graphically, and numerically. Symbolic solutions are exact, whereas graphical and numerical solutions are often approximate. For example, if the exact solution is $\frac{1}{3}$, a graphical or numerical solution might give 0.333333. The following table summarizes some important concepts in this section.

Concept	Description
Linear equation	A linear equation can be written as $ax + b = 0$, $a \neq 0$, and has exactly one solution. To solve an equation symbolically, we simplify the equation until we have $x = k$ for some k. *Example:* $2x - 3 = 5$ $\qquad 2x = 8 \qquad$ Add 3 to each side. $\qquad x = 4 \qquad$ Divide each side by 2.

Concept	Description
Intersection-of-graphs method	This is a graphical method for solving equations. Set y_1 equal to the left side and y_2 equal to the right side. Find the point of intersection. The solution is the x-coordinate of this point. **Example:** $3x = x - 2$ Graph $y_1 = 3x$ and $y_2 = x - 2$. The point of intersection is $(-1, -3)$, so the solution is -1.
Problem solving	Becoming a proficient problem solver takes practice. A general step-by-step procedure for solving applications is found on page 114.

2.3 — Exercises

Identifying Linear and Nonlinear Equations

Exercises 1–6: Determine whether the equation is linear or nonlinear by trying to write it in the form $ax + b = 0$.

1. $3x - 1.5 = 7$ **2.** $100 - 23x = 20x$

3. $2\sqrt{x} + 2 = 1$ **4.** $4x^3 - 7 = 0$

5. $7x - 55 = 3(x - 8) + x$

6. $2(x - 3) = 4 - (5x + 2)$

Solving Linear Equations Symbolically

Exercises 7–30: Solve the equation and check your answer.

7. $2x - 8 = 0$ **8.** $4x - 8 = 0$

9. $-5x + 3 = 23$ **10.** $-9x - 3 = 24$

11. $4(z - 8) = z$ **12.** $-3(2z - 1) = 2z$

13. $-5(3 - 4t) = 65$ **14.** $6(5 - 3t) = 66$

15. $k + 8 = 5k - 4$ **16.** $2k - 3 = k + 3$

17. $2(1 - 3x) + 1 = 3x$ **18.** $5(x - 2) = -2(1 - x)$

19. $-5(3 - 2x) - (1 - x) = 4(x - 3)$

20. $-3(5 - x) - (x - 2) = 7x - 2$

21. $\dfrac{2}{7}n + \dfrac{1}{5} = \dfrac{4}{7}$ **22.** $\dfrac{6}{11} - \dfrac{2}{33}n = \dfrac{5}{11}n$

23. $\dfrac{1}{2}(d - 3) - \dfrac{2}{3}(2d - 5) = \dfrac{5}{12}$

24. $\dfrac{7}{3}(2d - 1) - \dfrac{2}{5}(4 - 3d) = \dfrac{1}{5}d$

25. $\dfrac{x - 5}{3} + \dfrac{3 - 2x}{2} = \dfrac{5}{4}$ **26.** $\dfrac{3x - 1}{5} - 2 = \dfrac{2 - x}{3}$

27. $0.1z - 0.05 = -0.07z$ **28.** $1.1z - 2.5 = 0.3(z - 2)$

29. $0.15t + 0.85(100 - t) = 0.45(100)$

30. $0.35t + 0.65(10 - t) = 0.55(10)$

Exercises 31–46: Complete the following.

   (a) *Solve the equation symbolically.*

   (b) *Classify the equation as a contradiction, an identity, or a conditional equation.*

31. $5x - 1 = 5x + 4$

32. $7 - 9z = 2(3 - 4z) - z$

33. $3(x - 1) = 5$ **34.** $22 = -2(2x + 1.4)$

35. $0.5(x - 2) + 5 = 0.5x + 4$

36. $\frac{1}{2}x - 2(x - 1) = -\frac{3}{2}x + 2$

37. $\dfrac{t + 1}{2} = \dfrac{3t - 2}{6}$ **38.** $\dfrac{2x + 1}{3} = \dfrac{2x - 1}{3}$

39. $\dfrac{1 - 2x}{4} = \dfrac{3x - 1.5}{-6}$ **40.** $3(x - 2) + x = 2(x + 3)$

41. $0.5(x - 1980) + 5 = 10$

42. $5(x - 1995) - 15 = 65$

43. $\frac{1}{2}(x - 3) = \sqrt{2}(4 + x)$

44. $\frac{2}{3}(4 - 3x) + \frac{1}{6}x = \frac{1}{3}x + 10$

45. $\pi(2x - 2.1) = 16 + 2\pi x$

46. $2x^2 + 4x - 6 = x(4 + 2x) - 6$

Solving Linear Equations Graphically

Exercises 47 and 48: A linear equation is solved by using the intersection-of-graphs method. Find the solution by interpreting the graph. Assume that the solution is an integer.

47.   **48.**

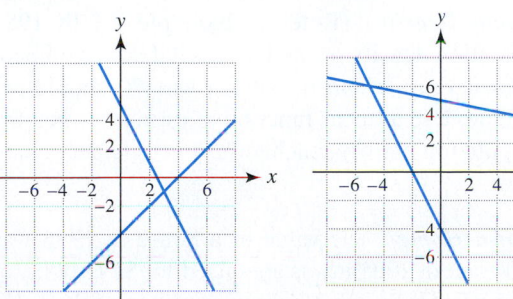

Exercises 49 and 50: Use the graph of $y = f(x)$ to solve each equation.

   (a) $f(x) = -1$ (b) $f(x) = 0$ (c) $f(x) = 2$

49. 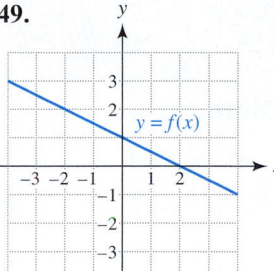   **50.**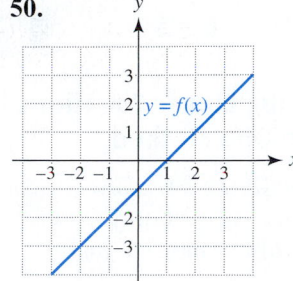

Exercises 51–56: (Refer to Example 5.) Use the intersection-of-graphs method to solve the equation by hand. Check your answer.

51. $x + 4 = 1 - 2x$ **52.** $2x = 3x - 1$

53. $-x + 4 = 3x$ **54.** $1 - 2x = x + 4$

55. $2(x - 1) - 2 = x$ **56.** $-(x + 1) - 2 = 2x$

Exercises 57–64: Solve the linear equation with the intersection-of-graphs method. Approximate the solution to the nearest thousandth whenever appropriate.

57. $5x - 1.5 = 5$ **58.** $8 - 2x = 1.6$

59. $3x - 1.7 = 1 - x$ **60.** $\sqrt{2}x = 4x - 6$

61. $3.1(x - 5) = \frac{1}{5}x - 5$ **62.** $65 = 8(x - 6) - 5.5$

63. $\dfrac{6 - x}{7} = \dfrac{2x - 3}{3}$

64. $\pi(x - \sqrt{2}) = 1.07x - 6.1$

Solving Linear Equations Numerically

Exercises 65 and 66: Let $y_1 = f(x)$ and $y_2 = g(x)$, where f and g are linear functions. Use the table to solve the linear equation $f(x) = g(x)$.

65.   **66.**

X	Y1	Y2
-3	3	7
-2	4	4
-1	5	1
0	6	-2
1	7	-5
2	8	-8
3	9	-11

X= -3

X	Y1	Y2
0	-1	1
1	-.8	.7
2	-.6	.4
3	-.4	.1
4	-.2	-.2
5	0	-.5
6	.2	-.8

X=0

Exercises 67–72: Use tables to solve the equation numerically to the nearest tenth.

67. $2x - 7 = -1$ **68.** $1 - 6x = 7$

69. $2x - 7.2 = 10$ **70.** $5.8x - 8.7 = 0$

71. $5x - 16 = 5 - x$ **72.** $\dfrac{3 - x}{2} + 2x = 96$

Solving Linear Equations by More than One Method

Exercises 73–78: Solve the equation
(a) *symbolically,*
(b) *graphically, and*
(c) *numerically.*

73. $x - 3 = 2x + 1$ **74.** $3(x - 1) = 2x - 1$

75. $6x - 8 = -7x + 18$

76. $5 - 8x = 3(x - 7) + 37$

77. $5.5x - 16 = 2.3x + 8$ **78.** $\dfrac{x + 1}{3} = 3x$

Applications

79. *Income*   The per capita (per person) income from 1970 to 2000 can be modeled by

$$f(x) = 604.8(x - 1970) + 2280,$$

where x is the year. Determine the year when the per capita income was $16,190.   (**Source:** Bureau of the Census.)

80. *Median Age*   In 1970 the median age in the United States was 27.7 years and it increased, on average, by 0.264 year per year.
(a) Write a linear function defined by

$$f(x) = m(x - x_1) + y_1$$

that models this situation.

(b) Determine graphically when the median age reached 35 years.

(c) Solve part (b) numerically.

(d) Solve part (b) symbolically.

81. *Classroom Ventilation*   Ventilation is an effective method for removing indoor air pollutants. According to the American Society of Heating, Refrigerating, and Air-Conditioning Engineers (ASHRAE), a classroom should have a ventilation rate of 900 cubic feet per hour for each person in the classroom.   (**Source:** ASHRAE.)
(a) Find a function V that computes the ventilation rate in cubic feet per hour that is necessary for a classroom containing x people.

(b) Determine the hourly ventilation rate necessary for a classroom containing 50 people.

(c) If a classroom contains 10,000 cubic feet with 50 people, how many times in 1 hour should all of the air in the classroom be replaced?

(d) In areas like bars and lounges that allow smoking, the ventilation rate should be increased to 3000 cubic feet per hour for each person. Compared to classrooms, by what factor should the ventilation rate be increased in smoking areas?

82. *Celestial Orbits*   To escape the gravity of any celestial body, such as Earth or the sun, a spacecraft must attain a certain velocity, called the *escape velocity*, denoted by v_e. In order to go into a circular orbit, a slower velocity v_c is necessary. The relation between v_e and v_c is described by $v_e = \sqrt{2v_c}$.   (**Source:** H. Karttunen, *Fundamental Astronomy*.)

(a) The velocity for a circular orbit around Earth is 17,700 miles per hour. Approximate the escape velocity for Earth.

(b) A spacecraft from Earth requires a velocity of 94,000 miles per hour in order to escape the solar system. What velocity is needed to travel in a circular orbit around the sun?

83. *Population Density*   (Refer to Example 4.) In 1980 the population density of the United States was 64 people per square mile and in 1990 it was 70 people per square mile. Use a linear function to estimate when the U.S. population density reached 72.4 people per square mile.

84. *Value of a Home*   The value of a house in 1995 was $180,000 and in 2005 it was appraised for $245,000.
(a) Find a linear function V that approximates the value of the house during year x.

(b) What does the slope of the graph of V represent?

(c) Use V to estimate the year when the house was worth $219,000.

85. *Shadow Length* (Refer to Example 12.) A person 66 inches tall is standing 15 feet from a streetlight. If the person casts a shadow 84 inches long, how tall is the streetlight?

86. *Height of a Tree* In the accompanying figure, a person 5 feet tall casts a shadow 4 feet long. A nearby tree casts a shadow that is 33 feet long. Find the height of the tree by solving a linear equation.

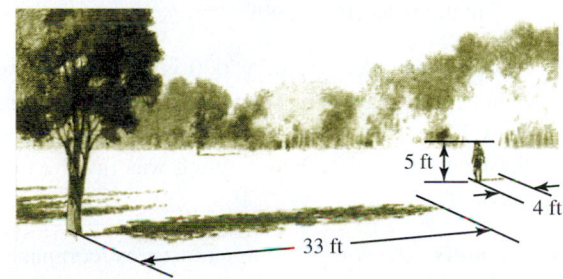

5 ft
4 ft
33 ft

87. *Two-Cycle Engines* Two-cycle engines, used in snowmobiles, chain saws, and outboard motors, require a mixture of gasoline and oil. For certain engines the amount of oil in pints that should be added to x gallons of gasoline is computed by $f(x) = 0.16x$. (**Source:** Johnson Outboard Motor Company.)

(a) Why is it reasonable to expect that f is linear?

(b) Evaluate $f(3)$ and interpret the answer.

(c) How much gasoline should be mixed with 1 quart of oil?

88. *Grades* In order to receive an A in a college course it is necessary to obtain an average of 90% correct on three 1-hour exams of 100 points each and on one final exam of 200 points. If a student scores 82, 88, and 91 on the 1-hour exams, what is the minimum score on the final exam that the person can receive and still earn an A?

89. *Conical Water Tank* A water tank in the shape of an inverted cone has a height of 11 feet and a radius of 3.5 feet, as illustrated in the accompanying figure.

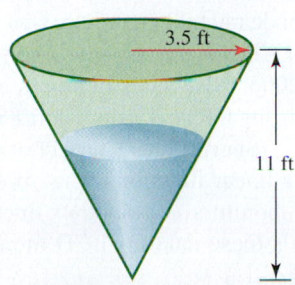

3.5 ft

11 ft

If the volume of the cone is

$$V = \frac{1}{3}\pi r^2 h,$$

find the volume of the water in the tank when the water is 7 feet deep. (*Hint*: Consider using similar triangles.)

Geometry Review To review formulas related to cones, see Chapter R (page R-5).

90. *Dimension of a Cone* (Refer to Exercise 89.) A conical water tank holds 100 cubic feet of water and has a diameter of 6 feet. Estimate its height to the nearest tenth of a foot.

91. *Deaths in the United States* There were 2.31 million deaths in the United States during the year 1997. This number represented 8.6 deaths per 1000 people. Approximate the population of the United States in 1997. (**Source:** Bureau of the Census.)

92. *Skin Cancer* In 2000 the population of the United States was approximately 281 million and there were 54,000 cases of skin cancer reported. (**Sources:** American Cancer Society; Bureau of the Census.)

(a) Determine the rate of skin cancer per 1000 people.

(b) Estimate the population in a region where 15,200 cases of skin cancer were reported.

93. *Sale Price* A store is discounting all regularly priced merchandise by 25%. Find a function f that computes the sale price of an item having a regular price of x. If an item normally costs $56.24, what is its sale price?

94. To continue Exercise 93, use f to find the regular price of an item that costs $19.62 on sale.

95. *Motion* (Refer to Example 11.) A car travels 372 miles in 6 hours, traveling at 55 miles per hour and at 70 miles per hour. How long did the car travel at each speed?

96. *Perimeter* Find the length of the longest side of the rectangle if its perimeter is 25 feet.

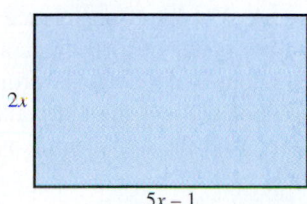

$2x$
$5x - 1$

Geometry Review To review formulas related to rectangles, see Chapter R (page R-1).

97. *Working Together*   (Refer to Example 10.) Suppose that a lawn can be raked by one gardener in 3 hours and it can be raked by a second gardener in 5 hours.
   (a) Mentally estimate how long it will take them to rake the lawn working together.

   (b) Solve part (a) symbolically.

98. *Pumping Water*   Suppose that a large pump can empty a swimming pool in 50 hours and that a small pump can empty the pool in 80 hours. How long does it take to empty the pool if both pumps are used?

99. *Chemistry*   (Refer to Example 13.) Determine how much pure water should be mixed with 5 liters of a 40% solution of sulfuric acid to make a 15% solution of sulfuric acid.

100. *Mixing Antifreeze*   A radiator holds 5 gallons of fluid. If it is full with a 15% solution, how much fluid should be drained and replaced with 65% antifreeze mixture to result in a 40% mixture of antifreeze?

101. *Mixing Candy*   Candy sells for $2.50 per pound and $4.00 per pound. A store clerk is trying to make a 5-pound mixture of candy that is worth $17.60. How much of each type of candy should be added to the mixture?

102. *Running*   At 2:00 P.M. a runner heads north on a highway jogging at 10 miles per hour. At 2:30 P.M. a driver heads north on the same highway to pick up the runner. If the car travels at 55 miles per hour, how long will it take the driver to catch the runner?

103. *Window Dimensions*   A rectangular window has a length that is 18 inches more than its width. If its perimeter is 180 inches, find its dimensions.

104. *Modeling Data*   The following data can be modeled by a linear function. Estimate the value of x when $y = 2.99$.

x	0	1	2	3
y	-5.38	-1.66	2.06	5.78

105. *Geometry*   A 174-foot long fence is being placed around the perimeter of a rectangular swimming pool that has a 3-foot wide sidewalk around it. The actual swimming pool without the sidewalk is twice as long as it is wide. Find the dimensions of the pool without the sidewalk.

106. *Housing Costs*   The following table shows the average costs of a new house in the South for three selected years.

Average Sale Prices of New One-Family Homes

Year	1999	2000	2001
Price ($ thousands)	173	179	186

Source: Bureau of the Census.

   (a) Find the percent increase in sale price from 1999 to 2000 and from 2000 to 2001.

   (b) If a house was worth $200 thousand in 2001, estimate its worth in 1999.

107. *Investments*   A total of $5000 is invested in two accounts. One pays 5% annual interest, and the second pays 7% annual interest. If the interest at the end of the first year is $325, how much was invested in each account?

108. *Business*   A company manufactures compact discs with recorded music. The master disc costs $2000 to produce and copies cost $0.45 each. If a company spends $2990 producing compact discs, how many copies did the company manufacture?

109. *Temperature Scales*   The Celsius and Fahrenheit scales are related by the equation $C = \frac{5}{9}(F - 32)$. These scales have the same temperature reading at a unique value, where $F = C$. Find this temperature.

110. *Indoor Air Pollution*   Formaldehyde is an indoor air pollutant found in plywood, foam insulation, and carpeting. When concentrations in the air reach 33 micrograms per cubic foot ($\mu\text{g/ft}^3$), eye irritation can occur. One square foot of new plywood can emit 140 μg per hour.   (**Source:** A. Hines, *Indoor Air Quality & Control.*)
   (a) A room has 100 square feet of new plywood flooring. Find a linear function f that computes the amount of formaldehyde in micrograms emitted in x hours.

   (b) The room contains 800 cubic feet of air and has no ventilation. Determine how long it takes for concentrations to reach 33 $\mu\text{g/ft}^3$.

111. *Sales of CRT and LCD Screens*   In 2002, 75 million CRT (cathode ray tube) monitors were sold while 29 million flat LCD (liquid crystal display) monitors were sold. In 2006 these sales numbers are expected to be 45 million for CRT monitors and 88 million for LCD monitors.   (**Source:** International Data Corporation.)
   (a) Find a linear function C that models these data for CRT monitors and another linear function L that models these data for LCD monitors. Let x be the year.

(b) Interpret the slope of the graph of C and of L.

(c) Determine graphically the year when sales for these two types of monitors were equal.

(d) Solve part (c) symbolically.

(e) Solve part (c) numerically.

112. *Online Holiday Sales*   In 1999 online holiday sales were \$5 billion and in 2003 they were \$17 billion.
   (a) Find a linear function S that models these data. Let x be the year.

   (b) Interpret the slope of the graph of S.

   (c) Use S to predict when online sales may reach \$26 billion.

Intermediate Value Property

113. The points $(1, -4)$ and $(5, 6)$ lie on the graph of the function $f(x) = 2.5x - 6.5$. Find a point $(x, 3.5)$ that lies on the graph of f. Sketch a graph to illustrate how the intermediate value property applies in this situation.

114. Sketch a graph of a function f that passes through the points $(-2, 3)$ and $(1, -2)$ but never assumes a value of 0. What must be true about the graph of f?

Exercises 115–118: Use the intermediate value property to show that $f(x) = 0$ for some x on the given interval for x.

115. $f(x) = x^2 - 5, 2 \le x \le 3$ (*Hint:* Evaluate $f(2)$ and $f(3)$ and then apply the intermediate value property.)

116. $f(x) = x^3 - x - 1, 1 \le x \le 2$

117. $f(x) = 2x^3 - 1, 0 \le x \le 1$

118. $f(x) = 4x^2 - x - 1, -1 \le x \le 0$

Writing about Mathematics

119. Describe a basic graphical method used to solve a linear equation. Give examples.

120. Describe verbally how to solve the equation $ax + b = 0$. What assumptions have you made about the value of a?

EXTENDED AND DISCOVERY EXERCISES

1. *Geometry*   Suppose that two rectangles are similar and the sides of the first rectangle are twice as long as the corresponding sides of the second rectangle.
   (a) Is the perimeter of the first rectangle twice the perimeter of the second rectangle? Explain.

   (b) Is the area of the first rectangle twice the area of the second rectangle? Explain.

2. *Geometry*   Repeat the previous exercise for an equilateral triangle. Try to make a generalization. (*Hint:* The area of an equilateral triangle is $A = \frac{\sqrt{3}}{4}x^2$, where x is the length of a side.) What will happen to the circumference of a circle if the radius is doubled? What will happen to its area?

2.4 Linear Inequalities

◆ Understand basic terminology related to inequalities
◆ Use interval notation
◆ Solve linear inequalities symbolically
◆ Solve linear inequalities graphically and numerically
◆ Solve compound inequalities

Introduction

If a person who weighs 143 pounds needs to purchase a life preserver for whitewater rafting, it is doubtful that there is one designed exactly for this weight. Life preservers are manufactured to support a range of body weights. A vest that is approved for weights between 120 and 160 pounds would be appropriate. Every airplane has a maximum weight allowance. It is important that this weight limit is accurately determined. However, most people feel more comfortable at takeoff if that maximum has not been reached. This is because any weight that is less than the maximum is also safe and allows a greater margin of error. Both of these situations involve the concept of inequality.

   In mathematics much effort is expended toward solving equations and determining equality. One reason for this is that equality is frequently a boundary between *greater than* and *less than*. The solution to an inequality often can be found by first locating where two expressions are equal. Since equality and inequality are closely related, many of the techniques used to solve equations also can be applied to inequalities.

Inequalities

Inequalities result whenever the equals sign in an equation is replaced with any one of the symbols $<, \leq, >$, or $\geq$. Some examples of inequalities include

$$x + 15 < 9x - 1, \qquad x^2 - 2x + 1 \geq 2x, \qquad z + 5 > 0,$$
$$xy + x^2 \leq y^3 + x, \qquad \text{and} \qquad 2 + 3 > 1.$$

The first three inequalities have one variable, the fourth inequality contains two variables, and the fifth inequality has only constants. As with linear equations, our discussion focuses on inequalities with one variable.

To **solve** an inequality means to find all values for the variable that make the inequality a true statement. Such values are **solutions** and the set of all solutions is the **solution set** to the inequality. Two inequalities are **equivalent** if they have the same solution set. It is common for an inequality to have infinitely many solutions. For instance, the inequality $x - 1 > 0$ has infinitely many solutions because any real number x satisfying $x > 1$ is a solution. The solution set is $\{x \mid x > 1\}$.

Like functions and equations, inequalities in one variable can be classified as *linear* or *nonlinear*.

LINEAR INEQUALITY IN ONE VARIABLE

A **linear inequality** in one variable is an inequality that can be written in the form

$$ax + b > 0,$$

where $a \neq 0$. (The symbol $>$ may be replaced by $\geq$, $<$, or $\leq$.)

Examples of linear inequalities include

$$3x - 4 < 0, \qquad 7x + 5 \geq x, \qquad x + 6 > 23, \qquad \text{and} \qquad 7x + 2 \leq -3x + 6.$$

Using techniques from algebra, each of these inequalities can be transformed into one of the forms $ax + b > 0$, $ax + b \geq 0$, $ax + b < 0$, or $ax + b \leq 0$. For example, by subtracting x from each side of $7x + 5 \geq x$, we obtain the equivalent inequality $6x + 5 \geq 0$. If an inequality is not a linear inequality, it is called a **nonlinear inequality**.

Interval Notation

The solution to a linear inequality is typically an interval on the real number line. For example, the solution set to $x - 2 > 0$ is $\{x \mid x > 2\}$. This set of real numbers is graphed in Figure 2.40. Note that a parenthesis at $x = 2$ indicates that the endpoint is *not* included. The solution set $\{x \mid -1 \leq x \leq 4\}$ is shown in Figure 2.41, and the solution set $\{x \mid -\frac{7}{2} < x < -\frac{1}{2}\}$ is shown in Figure 2.42. Note that brackets, either "[" or "]", are used when endpoints are *included*.

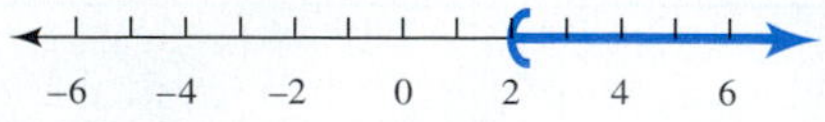

FIGURE 2.40 $x > 2$

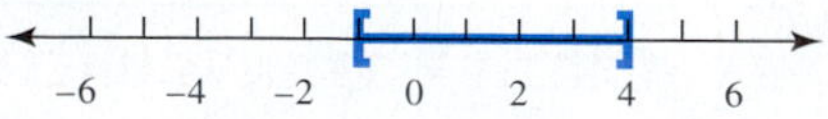

FIGURE 2.41 $-1 \leq x \leq 4$

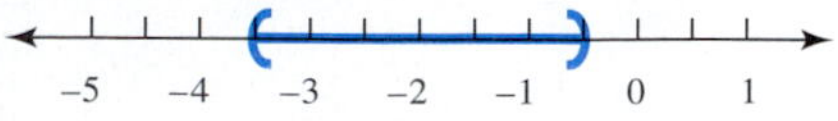

FIGURE 2.42 $-\frac{7}{2} < x < -\frac{1}{2}$

A convenient notation for number line graphs is called **interval notation**. Instead of drawing the entire number line, as in Figure 2.41, the solution set can be expressed as $[-1, 4]$. Because the solution set includes the endpoints -1 and 4, the interval is a **closed interval** and brackets are used. A solution set that includes all real numbers satisfying $-\frac{7}{2} < x < -\frac{1}{2}$ would be expressed as the **open interval** $\left(-\frac{7}{2}, -\frac{1}{2}\right)$. Parentheses indicate that the endpoints are not included in the solution set. An example of a **half-open interval** is $[0, 4)$, which represents the inequality $0 \leq x < 4$.

Table 2.12 provides some examples of interval notation. The symbol ∞ refers to **infinity**. It does not represent a real number. The notation $(1, \infty)$ means $\{x \mid x > 1\}$, or simply $x > 1$. Since this interval has no maximum x-value, ∞ is used in the position of the right endpoint. A similar interpretation holds for the symbol $-\infty$, which represents **negative infinity**.

TABLE 2.12 Interval Notation

Inequality	Interval Notation	Graph
$-2 < x < 2$	$(-2, 2)$ open interval	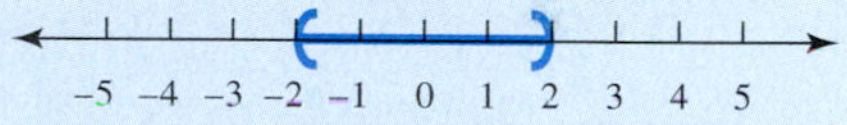
$-1 < x \leq 3$	$(-1, 3]$ half-open interval	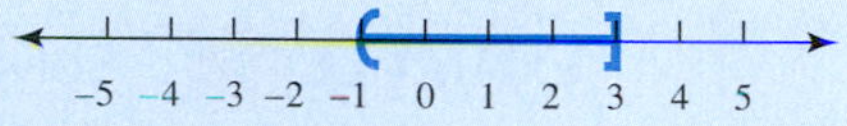
$-3 \leq x \leq 2$	$[-3, 2]$ closed interval	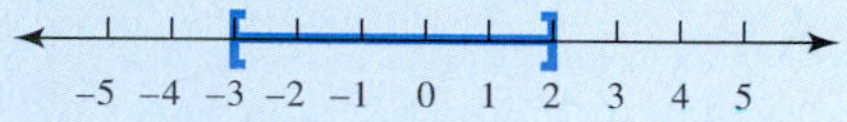
$x > -3$	$(-3, \infty)$ infinite interval	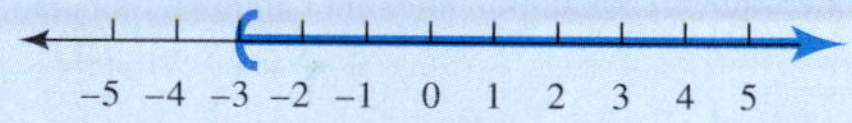
$x \leq 1$	$(-\infty, 1]$ infinite interval	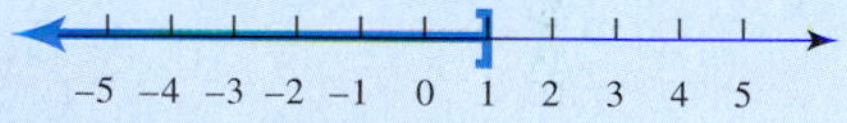
$-\infty < x < \infty$ (entire number line)	$(-\infty, \infty)$ infinite interval	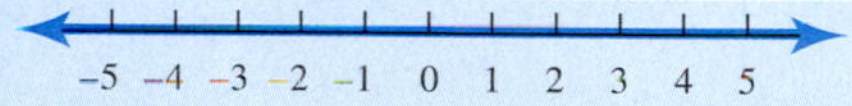

◆ **MAKING CONNECTIONS**

Points and Intervals The expression $(2, 5)$ has two possible meanings. The first is that $(2, 5)$ is an ordered pair, which can be plotted as a point on the xy-plane. The second is that $(2, 5)$ represents the open interval $2 < x < 5$. To alleviate confusion, phrases like "the point $(2, 5)$" or "the interval $(2, 5)$" may be used. ◆

Techniques for Solving Inequalities

Linear inequalities can be solved symbolically, graphically, and numerically. We begin our discussion with symbolic solutions.

Symbolic Solutions To solve linear inequalities symbolically, the following properties of inequalities can be used.

PROPERTIES OF INEQUALITIES

Let a, b, and c be real numbers.

1. $a < b$ and $a + c < b + c$ are equivalent.
   (The same number may be added to or subtracted from each side of an inequality.)
2. If $c > 0$, then $a < b$ and $ac < bc$ are equivalent.
   (Each side of an inequality may be multiplied or divided by the same positive number.)
3. If $c < 0$, then $a < b$ and $ac > bc$ are equivalent.
   (Each side of an inequality may be multiplied or divided by the same negative number provided the inequality symbol is reversed.)

Note: Replacing $<$ with $\leq$ and $>$ with $\geq$ results in similar properties.

The following examples illustrate each property.

Property 1: To solve $x - 5 < 6$, add 5 to each side to obtain $x < 11$.

Property 2: To solve $5x < 10$, divide each side by 5 to obtain $x < 2$.

Property 3: To solve $-5x < 10$, divide each side by -5 to obtain $x > -2$. (Whenever you multiply or divide an inequality by a negative number, reverse the inequality symbol.)

 **Solving linear inequalities symbolically**

Use the properties of inequalities to solve each inequality. Write the solution set in set-builder and interval notations.

(a) $2x - 3 < \dfrac{x + 2}{-3}$ **(b)** $-3(4z - 4) \geq 4 - (z - 1)$

SOLUTION

(a) Use Property 3 by multiplying each side by -3 to clear fractions. Remember to reverse the inequality symbol when multiplying by a negative number.

$$2x - 3 < \frac{x + 2}{-3} \qquad \textit{Given inequality}$$

$$-6x + 9 > x + 2 \qquad \textit{Property 3: Multiply by -3 and reverse the inequality symbol.}$$

$$9 > 7x + 2 \qquad \textit{Property 1: Add 6x.}$$

$$7 > 7x \qquad \textit{Property 1: Add -2 (or subtract 2).}$$

$$1 > x \qquad \textit{Property 2: Divide by 7.}$$

In set-builder notation the solution set is $\{x \mid x < 1\}$ and in interval notation it is written as $(-\infty, 1)$.

(b) Begin by applying the distributive property.

$$-3(4z - 4) \geq 4 - (z - 1) \qquad \textit{Given inequality}$$

$$-12z + 12 \geq 4 - z + 1 \qquad \textit{Distributive property}$$

$$-12z + 12 \geq -z + 5 \qquad \textit{Simplify.}$$

$$-12z + z \geq 5 - 12 \qquad \text{Property 1: Add } z \text{ and } -12.$$
$$-11z \geq -7 \qquad \text{Simplify.}$$
$$z \leq \frac{7}{11} \qquad \begin{array}{l}\text{Property 3: Divide by } -11 \text{ and}\\ \text{reverse inequality symbol.}\end{array}$$

In set-builder notation the solution set is $\{z \mid z \leq \frac{7}{11}\}$ and in interval notation it is written as $\left(-\infty, \frac{7}{11}\right]$.

Now Try Exercises 17 and 19 ◆

Graphical Solutions   The intersection-of-graphs method can be extended to solve inequalities. Figure 2.43 shows the velocity of two cars in miles per hour after x minutes. V_A denotes the velocity of car A, and V_B denotes the velocity of car B. The domains of V_A and V_B are both assumed to be $0 \leq x \leq 6$.

At 3 minutes $V_A = V_B$, and both cars are traveling at 30 miles per hour. To the left of $x = 3$, the graph of V_A is below the graph of V_B so car A is traveling slower than car B. Thus

$$V_A < V_B \quad \text{when} \quad 0 \leq x < 3.$$

To the right of $x = 3$ the graph of V_A is above the graph of V_B, so car A is traveling faster than car B. Thus

$$V_A > V_B \quad \text{when} \quad 3 < x \leq 6.$$

This technique is used in the next example.

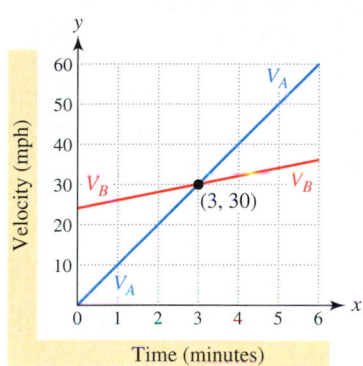

FIGURE 2.43   Velocities of Two Cars

EXAMPLE 2

Solving a linear inequality graphically

Graph $y_1 = \frac{1}{2}x + 2$ and $y_2 = 2x - 1$ by hand. Use the graph to solve the linear inequality $\frac{1}{2}x + 2 > 2x - 1$.

SOLUTION   The graph of $y_1 = \frac{1}{2}x + 2$ is a line with slope $\frac{1}{2}$ and y-intercept 2; the graph of $y_2 = 2x - 1$ is a line with slope 2 and y-intercept -1. Their graphs are shown in Figure 2.44.

Note that the graphs intersect at the point $(2, 3)$. The graph of $y_1 = \frac{1}{2}x + 2$ is above the graph of $y_2 = 2x - 1$ to the left of the point of intersection or when $x < 2$. Thus the solution set to the inequality $\frac{1}{2}x + 2 > 2x - 1$ is $\{x \mid x < 2\}$ or $(-\infty, 2)$.   *Now Try Exercise 41* ◆

In the next example, we use graphical techniques to solve an application from meteorology.

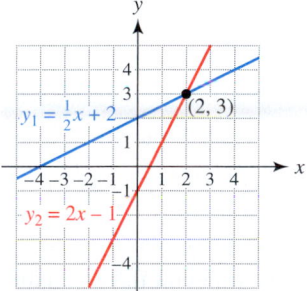

FIGURE 2.44

EXAMPLE 3

Using the intersection-of-graphs method

When the air temperature reaches the dew point, fog may form. This phenomenon also causes clouds to form at higher altitudes. Both the air temperature and the dew point often decrease at a constant rate as the altitude above ground level increases. If the ground-level Fahrenheit temperature and dew point are T_0 and D_0, the air temperature can sometimes be approximated by $T(x) = T_0 - 29x$ and the dew point by $D(x) = D_0 - 5.8x$ at an altitude of x miles.

(a) If $T_0 = 75°F$ and $D_0 = 55°F$, determine the altitudes where clouds will not form. See Figure 2.45.

(b) The slopes of the graphs for the functions T and D are called *lapse rates*. Interpret their meanings. Explain how these two slopes ensure a strong likelihood of clouds forming.   (**Source:** A. Miller and R. Anthes, *Meteorology.*)

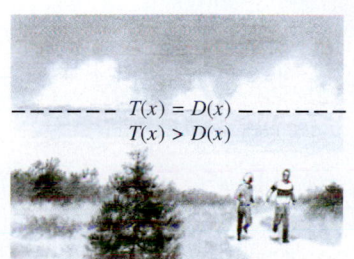

FIGURE 2.45

[0, 2, 1] by [0, 80, 10]

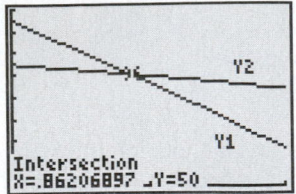

FIGURE 2.46

Calculator Help

To find a point of intersection, see Appendix B (page AP-12).

SOLUTION

(a) Since $T_0 = 75$ and $D_0 = 55$, let $T(x) = 75 - 29x$ and $D(x) = 55 - 5.8x$. Clouds will not form when the air temperature is greater than the dew point. Therefore we must solve the inequality $T(x) > D(x)$. Graph $Y_1 = 75 - 29X$ and $Y_2 = 55 - 5.8X$, as shown in Figure 2.46. The graphs intersect near $(0.86, 50)$. This means that the air temperature and dew point are both 50°F at about 0.86 mile above ground level. Clouds will not form below this altitude or when the graph of y_1 is above the graph of y_2. The solution set is $\{x \mid 0 \le x < 0.86\}$, where the endpoint 0.86 has been approximated.

(b) The slope of the graph of T is -29. This means that for each 1-mile increase in altitude, the air temperature *decreases* by 29°F. Similarly, the slope of the graph of D is -5.8. The dew point *decreases* by 5.8°F for every 1-mile increase in altitude. As the altitude increases, the air temperature decreases at a faster rate than the dew point. As a result, the air temperature typically cools to the dew point at higher altitudes. At or above this altitude clouds may form.   *Now Try Exercise 90* ◆

◆ **CLASS DISCUSSION**

How does the difference between the air temperature and the dew point at ground level affect the altitudes at which clouds may form? Explain. ◆

If a linear inequality can be written in the form $y_1 > 0$, where $>$ may be replaced by $\ge$, $\le$, or $<$, then we can solve this inequality by using the **x-intercept method**. To apply this method for $y_1 > 0$, graph y_1 and find the x-intercept. The solution set includes x-values where the graph of y_1 is above the x-axis. The x-intercept method is applied in the next example.

EXAMPLE 4   Applying the *x*-intercept method

Solve the inequality $1 - x > \frac{1}{2}x - 2$ by using the x-intercept method. Write the solution set in interval notation. Then solve the inequality symbolically.

[−6, 6, 1] by [−4, 4, 1]

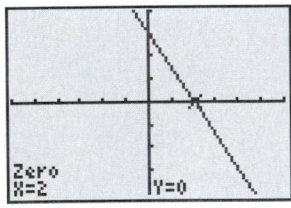

FIGURE 2.47

Calculator Help

To locate a zero or x-intercept on the graph of a function, see Appendix B (page AP-12).

SOLUTION

Graphical Solution   Start by subtracting $\frac{1}{2}x - 2$ from each side to obtain the inequality $1 - x - \left(\frac{1}{2}x - 2\right) > 0$. Then graph $Y_1 = 1 - X - (0.5X - 2)$, as shown in Figure 2.47, where the x-intercept is 2. The graph of y_1 is above the x-axis when $x < 2$. Therefore the solution set to $y_1 > 0$ is $(-\infty, 2)$.

Symbolic Solution

$$1 - x > \frac{1}{2}x - 2 \quad \text{Given inequality}$$
$$-x > \frac{1}{2}x - 3 \quad \text{Subtract 1 from each side.}$$
$$-\frac{3}{2}x > -3 \quad \text{Subtract } \tfrac{1}{2}x \text{ from each side.}$$
$$x < 2 \quad \text{Multiply by } -\tfrac{2}{3}; \text{ reverse inequality.}$$

The solution set is $(-\infty, 2)$.   *Now Try Exercise 53* ◆

Example 4 suggests a general result about linear inequalities. The graph of the equation $y = ax + b$, $a \ne 0$, is a line that intersects the x-axis once and either slopes upward (if $a > 0$) or downward (if $a < 0$). If the x-intercept is k, then the solution set to $ax + b > 0$ either satisfies $x > k$ or $x < k$, as illustrated in Figures 2.48 and 2.49. (Similar remarks hold for linear inequalities that contain the symbols $<$, $\leq$, or $\geq$.) Note that the solution to $ax + b = 0$ is k and the value of k is that boundary between *greater than* and *less than* in either situation.

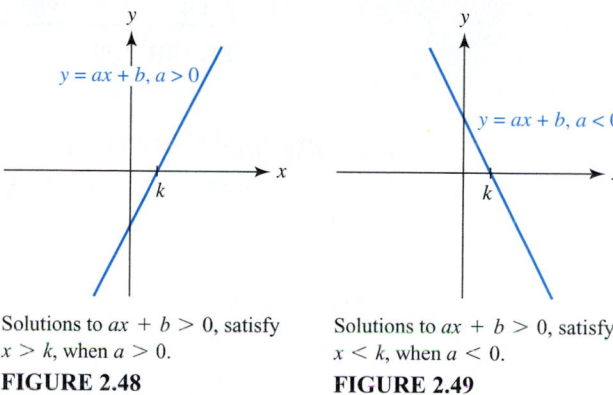

Solutions to $ax + b > 0$, satisfy $x > k$, when $a > 0$.

FIGURE 2.48

Solutions to $ax + b > 0$, satisfy $x < k$, when $a < 0$.

FIGURE 2.49

Numerical Solutions Inequalities can sometimes be solved by using a table of values. The following example helps to explain the mathematical concept behind this method.

Suppose that it costs $5x + 200$ dollars for a company to produce x pairs of headphones and that the company receives $15x$ dollars for selling x pairs of headphones. Then the profit P from selling x pairs of headphones is $P = 15x - (5x + 200) = 10x - 200$. A value of $x = 20$ results in $P = 0$, and so $x = 20$ is called the **boundary number** because it represents the boundary between making money and losing money. To make money, the profit P must be positive, and the inequality

$$10x - 200 > 0$$

must be satisfied. A table of values for $y_1 = 10x - 200$ in Table 2.13 shows the boundary number $x = 20$ along with several **test values**. On one hand, the test values of $x = 17, 18,$ and 19 result in the company losing money. On the other hand, the test values of $x = 21,$ 22, and 23 result in a profit. Therefore the solution set to $10x - 200 > 0$ is $\{x \mid x > 20\}$. These concepts are applied in the next example.

TABLE 2.13

Boundary number

x	17	18	19	**20**	21	22	23
$10x - 200$	-30	-20	-10	**0**	10	20	30

Less than 0   Greater than 0

EXAMPLE 5   Solving a linear inequality with test values

Solve $3(6 - x) + 5 - 2x < 0$ numerically.

SOLUTION Begin by making a table of $Y_1 = 3(6 - X) + 5 - 2X$, as shown in Figure 2.50 on the next page. We can see that the boundary number for this inequality lies between $x = 4$ and $x = 5$. In Figure 2.51 the increment is changed from 1 to 0.1 and the boundary number for the inequality is $x = 4.6$. The test values of $x = 4.7, 4.8,$ and 4.9 indicate that when $x > 4.6$, the inequality $y_1 < 0$ is true. The solution set is $\{x \mid x > 4.6\}$.

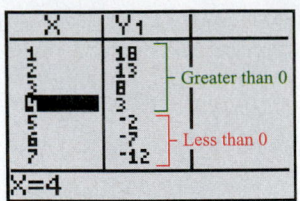 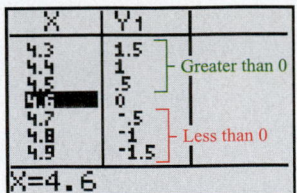

FIGURE 2.50   **FIGURE 2.51**   *Now Try Exercise 73* ◆

◆ **MAKING CONNECTIONS**

Symbolic, Graphical, and Numerical Solutions Linear inequalities can be solved symbolically, graphically, or numerically. Each method is used in the following to solve the inequality $3 - (x + 2) > 0$.

Symbolic Solution

$$3 - (x + 2) > 0$$
$$-(x + 2) > -3$$
$$x + 2 < 3$$
$$x < 1$$

Graphical Solution

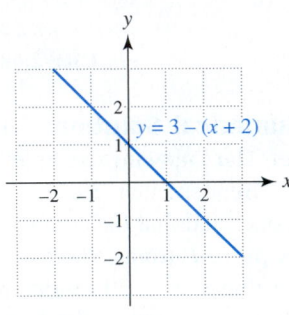

The graph of $y = 3 - (x + 2)$ is above the x-axis when $x < 1$.

Numerical Solution

x	$3 - (x + 2)$
-2	3
-1	2
0	1
1	**0**
2	-1
3	-2

Greater than 0

Less than 0

The values of $3 - (x + 2)$ are greater than 0 when $x < 1$.

Compound Inequalities

Sometimes a variable must satisfy two inequalities. For example, on a freeway there may be a minimum speed limit of 40 miles per hour and a maximum speed limit of 70 miles per hour. If the variable x represents the speed of a vehicle, then x must satisfy the compound inequality

$$x \geq 40 \quad \text{and} \quad x \leq 70.$$

A compound inequality occurs when two inequalities are connected by the word *and* or *or*. When the word *and* connects two inequalities, it can sometimes be written as a **three-part inequality**. For example, the previous compound inequality may be written as the three-part inequality

$$40 \leq x \leq 70.$$

Compound inequalities involving the word *or* are discussed in the next section.

EXAMPLE 6 **Solving a three-part inequality symbolically**

Solve the inequality. Write the solution set in interval notation.

(a) $-4 \leq 5x + 1 < 21$ **(b)** $\dfrac{1}{2} < \dfrac{1 - 2t}{4} < 2$

SOLUTION

(a) Use properties of inequalities to simplify the three-part inequality.

$$-4 \leq 5x + 1 < 21 \qquad \textit{Given inequality}$$
$$-5 \leq 5x < 20 \qquad \textit{Add }-1\textit{ to each part.}$$
$$-1 \leq x < 4 \qquad \textit{Divide each part by 5.}$$

The solution set is $[-1, 4)$.

(b) Begin by multiplying each part by 4 to clear fractions.

$$\frac{1}{2} < \frac{1 - 2t}{4} < 2 \qquad \textit{Given inequality}$$
$$2 < 1 - 2t < 8 \qquad \textit{Multiply by 4.}$$
$$1 < -2t < 7 \qquad \textit{Add }-1\textit{ to each part.}$$
$$-\frac{1}{2} > t > -\frac{7}{2} \qquad \textit{Divide by }-2\textit{; reverse inequalities.}$$
$$-\frac{7}{2} < t < -\frac{1}{2} \qquad \textit{Rewrite the inequality.}$$

The solution set is $\left(-\frac{7}{2}, -\frac{1}{2}\right)$.

Now Try Exercises 27 and 29 ◆

Three-part inequalities occur in applications and can also be solved graphically. This is demonstrated in the next example.

EXAMPLE 7

Solving a compound inequality graphically and symbolically

As the altitude above ground level increases, the air temperature becomes cooler. The function given by $T(x) = T_0 - 29x$ models the Fahrenheit temperature x miles high, where T_0 is the ground temperature. (**Source:** A. Miller.)

(a) Suppose $T_0 = 70°F$. Use the intersection-of-graphs method to determine the altitudes where the air temperature is from 32°F to 50°F.

(b) Solve part (a) symbolically.

SOLUTION

(a) *Graphical Solution*   Let $T(x) = 70 - 29x$. The solution set consists of x-values where the compound inequality $32 \leq T(x) \leq 50$ is true. Graph $Y_1 = 32$, $Y_2 = 70 - 29X$, and $Y_3 = 50$. Their graphs intersect near the points $(0.69, 50)$ and $(1.31, 32)$, as shown in Figures 2.52 and 2.53. The air temperature is from 32°F to 50°F whenever the graph of $y_2 = 70 - 29x$ is between the graphs of $y_1 = 32$ and $y_3 = 50$. Thus the solution set is $\{x \mid 0.69 \leq x \leq 1.31\}$, where the endpoints have been approximated. This means that the air temperature is between 32°F and 50°F from about 0.69 mile to 1.31 miles above ground level.

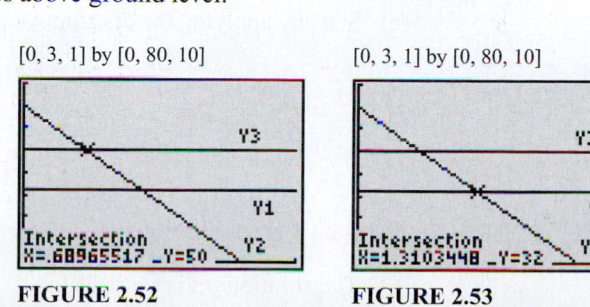

FIGURE 2.52   FIGURE 2.53

(b) *Symbolic Solution*   The air temperature is from 32°F to 50°F for altitudes x satisfying the following.

$$32 \le 70 - 29x \le 50 \qquad \text{Compound inequality}$$

$$-38 \le -29x \le -20 \qquad \text{Subtract 70 from each expression.}$$

$$\frac{-38}{-29} \ge x \ge \frac{-20}{-29} \qquad \text{Divide by } -29. \text{ Reverse each inequality.}$$

$$\frac{38}{29} \ge x \ge \frac{20}{29} \qquad \text{Simplify.}$$

$$\frac{20}{29} \le x \le \frac{38}{29} \qquad \text{Rewrite the inequality.}$$

The solution set is $\left\{x \mid \frac{20}{29} \le x \le \frac{38}{29}\right\}$. Both the graphical and symbolic methods produce similar results since $\frac{20}{29} \approx 0.69$ and $\frac{38}{29} \approx 1.31$. Note that the exact values were found using a symbolic approach rather than a graphical approach.   *Now Try Exercise 89* ◆

EXAMPLE 8   Solving inequalities symbolically

Solve the linear inequalities symbolically. Express the solution set using interval notation.

(a) $-\dfrac{x}{2} + 1 \le 3$ **(b)** $-8 < \dfrac{3x-1}{2} \le 5$ **(c)** $5(x-6) < 2x - 2(1-x)$

SOLUTION

(a) Simplify the inequality as follows.

$$-\frac{x}{2} + 1 \le 3 \qquad \text{Given inequality}$$

$$-\frac{x}{2} \le 2 \qquad \text{Add } -1 \text{ or subtract 1.}$$

$$x \ge -4 \qquad \text{Multiply by } -2. \text{ Reverse the inequality.}$$

In interval notation the solution set is $[-4, \infty)$.

(b) The two parts of this compound inequality can be solved simultaneously.

$$-8 < \frac{3x-1}{2} \le 5 \qquad \text{Given inequality}$$

$$-16 < 3x - 1 \le 10 \qquad \text{Multiply by 2.}$$

$$-15 < 3x \le 11 \qquad \text{Add 1.}$$

$$-5 < x \le \frac{11}{3} \qquad \text{Divide by 3.}$$

The solution set is $\left(-5, \frac{11}{3}\right]$.

(c) Start by applying the distributive property to each side of the inequality.

$$5(x-6) < 2x - 2(1-x) \qquad \text{Given inequality}$$

$$5x - 30 < 2x - 2 + 2x \qquad \text{Distributive property}$$

$$5x - 30 < 4x - 2 \qquad \text{Simplify.}$$

$$x - 30 < -2 \qquad \text{Subtract } 4x.$$

$$x < 28 \qquad \text{Add 30.}$$

The solution set is $(-\infty, 28)$.   *Now Try Exercises 33 and 35* ◆

2.4

Putting it all Together

Applications involving linear functions result in both linear equations and inequalities. Like a linear equation, the solution set for a linear inequality can be found symbolically, graphically, and numerically. One common strategy for solving a linear inequality is first to locate the x-value that results in equality. This boundary number represents the boundary between *greater than* and *less than*. With this value, the solution set for the linear inequality can be found.

The following table includes methods for solving linear inequalities in the form $h(x) > 0$ or $f(x) > g(x)$. Inequalities involving $<$, $\leq$, and $\geq$ are solved in a similar manner.

Concept	Explanation	Example
Interval notation	An efficient notation for writing solutions to inequalities	$x \leq 6$ is equivalent to $(-\infty, 6]$. $x > 3$ is equivalent to $(3, \infty)$. $2 < x \leq 5$ is equivalent to $(2, 5]$.
Compound inequality	Two inequalities connected by the word *and* or *or*	$x \leq 4$ or $x \geq 10$ $x \geq -3$ and $x < 4$ The inequality $x > 5$ and $x \leq 20$ can be written as the three-part inequality $5 < x \leq 20$.
Symbolic method	Use properties of inequalities to simplify $f(x) > g(x)$ to either $x > k$ or $x < k$ for some real number k.	$\dfrac{1}{2}x + 1 > 3 - \dfrac{3}{2}x$ $2x + 1 > 3$ Add $\frac{3}{2}x$. $2x > 2$ Subtract 1. $x > 1$ Divide by 2.
Intersection-of-graphs method	To solve $f(x) > g(x)$, graph $y_1 = f(x)$ and $y_2 = g(x)$. Find the point of intersection. The solution set includes x-values where the graph of y_1 is above the graph of y_2.	$\dfrac{1}{2}x + 1 > 3 - \dfrac{3}{2}x$ Graph $y_1 = \frac{1}{2}x + 1$ and $y_2 = 3 - \frac{3}{2}x$. The solution set for $y_1 > y_2$ is $(1, \infty)$. 

Concept	Explanation	Example
The x-intercept method	Write the inequality as $h(x) > 0$. Graph $y_1 = h(x)$. Solutions occur where the graph is above the x-axis.	$\frac{1}{2}x + 1 > 3 - \frac{3}{2}x$ Graph $y_1 = \frac{1}{2}x + 1 - \left(3 - \frac{3}{2}x\right)$. The solution set for $y_1 > 0$ is $(1, \infty)$.
Numerical method	Write the inequality as $h(x) > 0$. Create a table for $y_1 = h(x)$ and find the boundary number $x = k$ such that $h(k) = 0$. Use the test values in the table to determine if the solution set is $x > k$ or $x < k$.	$\frac{1}{2}x + 1 > 3 - \frac{3}{2}x$ Table $y_1 = \frac{1}{2}x + 1 - \left(3 - \frac{3}{2}x\right)$. The solution set for $y_1 > 0$ is $(1, \infty)$.

Table:

x	-1	0	1	2	3
y_1	-4	-2	0	2	4

Less than 0   Greater than 0

2.4 — Exercises

Interval Notation

Exercises 1–12: Express each of the following in interval notation.

1. $x \geq 5$

2. $x < 100$

3. $4 \leq x < 19$

4. $-4 < x < -1$

5. $x \leq -37$

6. $\{x \mid x \leq -3\}$

7. $\{x \mid -1 \leq x\}$

8. $17 > x \geq -3$

9.
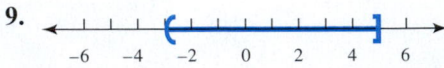

10.

11.

12.

Solving Linear Inequalities Symbolically

Exercises 13–40: Solve the inequality symbolically. Express the solution set in interval notation.

13. $2x + 6 \geq 10$

14. $-4x - 3 < 5$

15. $-2(x - 10) + 1 > 0$ **16.** $3(x + 5) \le 0$

17. $\dfrac{t + 2}{3} \ge 5$ **18.** $\dfrac{2 - t}{6} < 0$

19. $-3(z - 4) \ge 2(1 - 2z)$

20. $-\frac{1}{4}(2z - 6) + z \ge 5$

21. $\dfrac{1 - x}{4} < \dfrac{2x - 2}{3}$ **22.** $\dfrac{3x}{4} < x - \dfrac{x + 2}{2}$

23. $2x - 3 > \frac{1}{2}(x + 1)$ **24.** $5 - (2 - 3x) \le -5x$

25. $-1 \le 2t \le 4$ **26.** $5 < 4t - 1 \le 11$

27. $3 \le 4 - x \le 20$ **28.** $-5 < 1 - 2x < 40$

29. $0 < \dfrac{7x - 5}{3} \le 4$ **30.** $-7 \le \dfrac{1 - 4x}{7} < 12$

31. $5 > 2(x + 4) - 5 > -5$

32. $\frac{8}{3} \ge \frac{4}{3} - (x + 3) \ge \frac{2}{3}$

33. $3 \le \frac{1}{2}x + \frac{3}{4} \le 6$ **34.** $-4 \le 5 - \frac{4}{5}x < 6$

35. $3x - 1 < 2(x - 3) + 1$ **36.** $5x - 2(x + 3) \ge 4 - 3x$

37. $\dfrac{1}{2} \le \dfrac{1 - 2t}{3} < \dfrac{2}{3}$ **38.** $-\dfrac{3}{4} < \dfrac{2 - t}{5} < \dfrac{3}{4}$

39. $\frac{1}{2}z + \frac{2}{3}(3 - z) - \frac{5}{4}z \ge \frac{3}{4}(z - 2) + z$

40. $\dfrac{2}{3}(1 - 2z) - \dfrac{3}{2}z + \dfrac{5}{6}z \ge \dfrac{2z - 1}{3} + 1$

Solving Linear Inequalities Graphically

Exercises 41–48: (Refer to Example 2.) Solve the inequality graphically by hand.

41. $x + 2 \ge 2x$ **42.** $2x - 1 \le x$

43. $\frac{2}{3}x - 2 > -\frac{4}{3}x + 4$ **44.** $-2x \ge -\frac{5}{3}x + 1$

45. $-1 \le 2x - 1 \le 3$ **46.** $-2 < 1 - x < 2$

47. $-3 < x - 2 \le 2$ **48.** $-1 \le 1 - 2x < 5$

Exercises 49–52: Use the graph of $y = ax + b$ to solve each equation and inequality. Write the solution set to each inequality in interval notation.

(a) $ax + b = 0$ (b) $ax + b < 0$ (c) $ax + b \ge 0$

49.
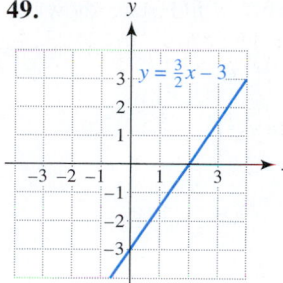

50.
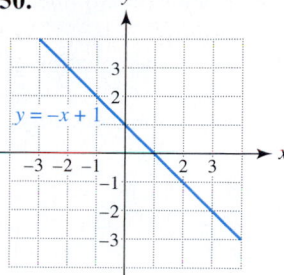

51.
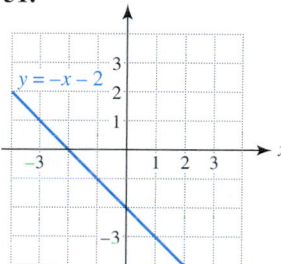

52.
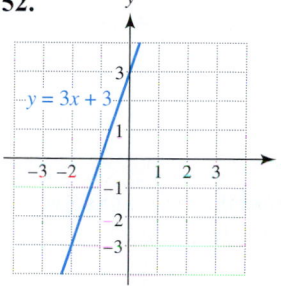

Exercises 53–56: (Refer to Example 4.) Use the x-intercept method to solve the inequality. Write the solution set in interval notation. Then solve the inequality symbolically.

53. $x - 3 \le \frac{1}{2}x - 2$ **54.** $x - 2 \le \frac{1}{3}x$

55. $2 - x < 3x - 2$ **56.** $\frac{1}{2}x + 1 > \frac{3}{2}x - 1$

 Exercises 57–62: Solve the linear inequality graphically. Write the solution set in set-builder notation, and approximate endpoints to the nearest hundredth whenever appropriate.

57. $5x - 4 > 10$ **58.** $-3x + 6 \le 9$

59. $-2(x - 1990) + 55 \ge 60$

60. $\sqrt{2}x > 10.5 - 13.7x$

61. $\sqrt{5}(x - 1.2) - \sqrt{3}x < 5(x + 1.1)$

62. $1.238x + 0.998 \le 1.23(3.987 - 2.1x)$

Exercises 63–68: Solve the compound linear inequality graphically. Write the solution set in interval notation, and approximate endpoints to the nearest tenth whenever appropriate.

63. $3 \le 5x - 17 < 15$ **64.** $-4 < \dfrac{55 - 3.1x}{4} < 17$

65. $1.5 \le 9.1 - 0.5x \le 6.8$ **66.** $0.2x < \dfrac{2x - 5}{3} < 8$

67. $x - 4 < 2x - 5 < 6$ **68.** $-3 \le 1 - x \le 2x$

69. The graphs of two linear functions f and g are shown in the figure.
   (a) Solve the equation $g(x) = f(x)$.

   (b) Solve the inequality $g(x) > f(x)$.

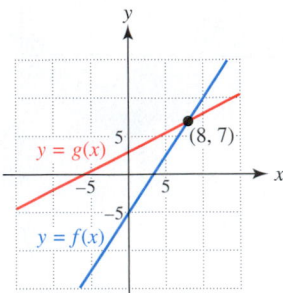

70. Use the figure to solve each equation or inequality.
   (a) $f(x) = g(x)$ **(b)** $g(x) = h(x)$

   (c) $f(x) < g(x) < h(x)$ **(d)** $g(x) > h(x)$

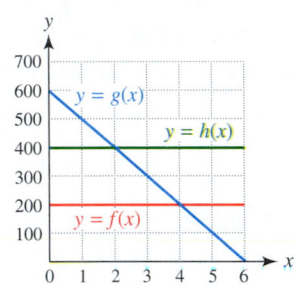

Solving Linear Inequalities Numerically

Exercises 71 and 72: Assume y_1 represents a linear function with the set of real numbers for its domain. Use the table to solve the inequalities. Write the solution set in set-builder notation.

71. $y_1 > 0$, $y_1 \le 0$ **72.** $y_1 < 0$, $y_1 \ge 0$

X	Y1
0	220
1	165
2	110
3	55
4	0
5	-55
6	-110

X=0

X	Y1
-5	-32
-4	-16
-3	0
-2	16
-1	32
0	48
1	64

X=-5

Exercises 73–80: Solve each inequality numerically. Write the solution set in interval notation, and approximate endpoints to the nearest tenth whenever appropriate.

73. $-4x - 6 > 0$ **74.** $1 - 2x \ge 9$

75. $1 \le 3x - 2 \le 10$ **76.** $-5 < 2x - 1 < 15$

77. $-\dfrac{3}{4} < \dfrac{2 - 5x}{3} \le \dfrac{3}{4}$ **78.** $(\sqrt{11} - \pi)x - 5.5 \le 0$

79. $\dfrac{3x - 1}{5} < 15$ **80.** $1.5(x - 0.7) + 1.5x < 1$

You Decide the Method

Exercises 81–84: Solve the inequality. Write the solution set in interval notation. Approximate the endpoints to the nearest thousandth when appropriate.

81. $2x - 8 > 5$

82. $\pi x - 5.12 \le \sqrt{2}x - 5.7(x - 1.1)$

83. $5.1x - \pi \ge \sqrt{3} - 1.7x$

84. $5 < 4x - 2.5$

Applications

85. *Interest* The function f computes the annual interest y on a loan of x dollars with an interest rate of 10%. The graphs of f and the horizontal line $y = 100$ are shown in the figure. Determine the loan amounts that result in the following. Express your answers verbally.
   (a) An annual interest equal to $100

   (b) An annual interest of more than $100

   (c) An annual interest of less than $100

   (d) An annual interest of $100 or more

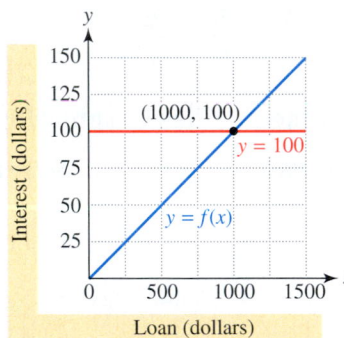

86. *U.S. Population* The function f models the population of the United States from 1970 to 2000. The graphs of f and the horizontal line $y = 226$ are shown in the figure. Use the graphs to determine when each of the following were satisfied. Express your answers verbally.
   (a) A population equal to 226 million

   (b) A population of 226 million or less

   (c) A population of 226 million or more

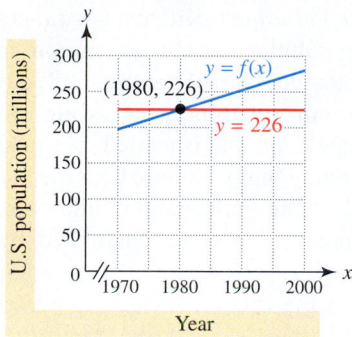

Year

(d) When is the car's distance from Omaha greater than 100 miles?

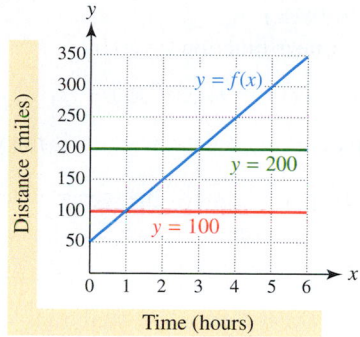

Time (hours)

87. *Distance between Cars*   Cars A and B are both traveling in the same direction. Their distances in miles north of St. Louis after x hours are computed by the functions f_A and f_B, respectively. The graphs of f_A and f_B are shown in the figure for $0 \le x \le 10$.

(a) Which car is traveling faster? Explain.

(b) How many hours elapse before the two cars are the same distance from St. Louis? How far are they from St. Louis when this occurs?

(c) During what time interval is car B farther from St. Louis than car A?

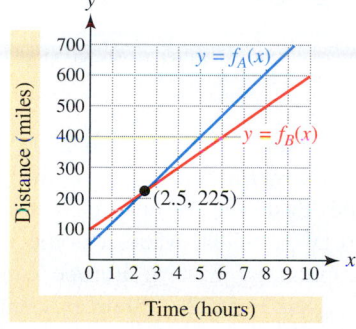

Time (hours)

88. *Distance*   The linear function f computes the distance y in miles between a car and the city of Omaha after x hours, where $0 \le x \le 6$. The graphs of f and the horizontal lines $y = 100$ and $y = 200$ are shown in the accompanying figure at the top of next column. Use the graphs to answer the following.

(a) Is the car moving toward or away from Omaha? Explain.

(b) Determine the times when the car is 100 miles or 200 miles from Omaha.

(c) When is the car from 100 to 200 miles from Omaha?

89. *Temperature and Altitude*   (Refer to Example 7.) Suppose the Fahrenheit temperature x miles above ground level is given by $T(x) = 85 - 29x$.

(a) Use the intersection-of-graphs method to estimate the altitudes where the temperature is below freezing. Assume that the domain of T is $0 \le x \le 6$.

(b) What does the x-intercept on the graph of T represent?

(c) Solve part (a) symbolically.

90. *Clouds and Temperature*   (Refer to Example 3.) Suppose the ground-level temperature is 65°F and the dew point is 50°F.

(a) Use the intersection-of-graphs method to estimate the altitudes where clouds will not form.

(b) Solve part (a) symbolically.

91. *Prices of Homes*   The median prices of a single-family home from 1980 to 1990 can be approximated by $P(x) = 3421x + 61{,}000$, where $x = 0$ corresponds to 1980 and $x = 10$ to 1990.   (**Source:** Department of Commerce.)

(a) Interpret the slope of the graph of P.

(b) Estimate the years when the median price range was from \$71,000 to \$88,000.

92. *Motorcycles*   The number of Harley-Davidson motorcycles manufactured between 1985 and 1995 can be approximated by $N(x) = 6409(x - 1985) + 30{,}300$, where x is the year.   (**Source:** Harley-Davidson.)

(a) Did the demand for Harley-Davidson motorcycles increase or decrease over this time period? Explain your reasoning.

(b) Estimate the years when production was between 56,000 and 75,000.

93. *VISA Debit Cards*   Annual transactions on Visa's debit cards that do not require a PIN code have grown from about $100 billion in 1997 to $400 billion in 2002. (**Source:** CardWeb.)
    (a) Find a linear function given by

$$T(x) = m(x - x_1) + y_1$$

    that models these data, where x is the year.

    (b) Use $T(x)$ to estimate the years when the annual transactions were between $160 billion and $280 billion.

94. *MasterCard Debit Cards*   Annual transactions on MasterCard's debit cards that do not require a PIN code have grown from about $10 billion in 1997 to $110 billion in 2002.   (**Source:** CardWeb.)
    (a) Find a linear function given by

$$T(x) = m(x - x_1) + y_1$$

    that models these data, where x is the year.

    (b) Use $T(x)$ to estimate years when the annual transactions were between $50 billion and $90 billion.

95. *Broadband Internet Connections*   The number of online households using broadband connections, such as cable and DSL, has increased from 6 million in 2000 to 30 million in 2004.   (**Source:** eMarketer.)
    (a) Find a linear function given by

$$B(x) = m(x - x_1) + y_1$$

    that models these data, where x is the year.

    (b) Use $B(x)$ to estimate the years when the number of online households using broadband connections was or will be 24 million or more. Assume that the domain of B is 2000 to 2006.

96. *Online Betting*   Consumer losses on online betting were $4 billion in 2002 and $10 billion in 2005. (**Source:** Christiansen Capital Advisors.)
    (a) Find a linear function given by

$$B(x) = m(x - x_1) + y_1$$

    that models these data, where x is the year.

    (b) Use $B(x)$ to estimate the years when consumer losses on online betting were or will be more than $6 billion. Assume that the domain of B is 2002 to 2007.

97. *Indoor Air Pollution*   Kitchen gas ranges are a source of indoor pollutants such as carbon monoxide and nitrogen dioxide. One of the most effective ways to remove contaminants from the air while cooking is to operate a range hood that is vented. If a range hood is capable of removing x liters of air per second from a kitchen, then the percentage of the contaminants that are also removed may be calculated by the formula $f(x) = 1.06x + 7.18$ where $10 \le x \le 75$. A graph of f is shown in the accompanying figure.   (**Source:** R. L. Rezvan, "Effectiveness of Local Ventilation in Removing Simulated Pollutants from Point Sources.")
    (a) How does increasing the amount of ventilation x affect the percentage of pollutants removed from the kitchen? Interpret the slope of the graph of f.

    (b) Estimate the x-values where 50% to 70% of the pollutants are removed.

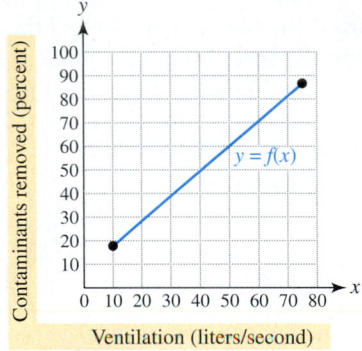

98. *Oil Consumption*   From 1973 to 2005, oil consumption in million metric tons of oil equivalent (Mtoe) by the Middle East can be modeled by the formula $M(x) = 10.1x - 19{,}904$, where x is the year. Similarly, oil consumption by Eastern Europe can be modeled by $G(x) = 2.89x - 5622$.   (**Source:** International Energy Agency, *Global Energy: The Changing Outlook.*)
    (a) Determine the year when the oil consumption by the Middle East and Eastern Europe were approximately equal.

    (b) Determine the years when oil consumption by the Middle East exceeded that of Eastern Europe.

99. *Error Tolerances*   Suppose that an aluminum can is manufactured so that its radius r can vary from 1.99 inches to 2.01 inches. What range of values is possible for the circumference C of the can? Express your answer using a three-part inequality.

100. *Error Tolerances* Suppose that a square picture frame has sides that vary between 9.9 inches and 10.1 inches. What range of values is possible for the perimeter P of the picture frame? Express your answer using a three-part inequality.

101. *Modeling Data* The following data are exactly linear.
(a) Find a linear function f that models the data.

(b) Solve the inequality $f(x) > 2.25$.

x	0	2	4	6
y	−1.5	4.5	10.5	16.5

102. *Modeling Data* The following data are exactly linear.
(a) Find a linear function f that models the data.

(b) Solve the inequality $2 \leq f(x) \leq 8$.

x	1	2	3	4	5
y	0.4	3.5	6.6	9.7	12.8

103. *Modeling Salaries* The following data shows a person's salary for 3 years.
(a) Find a linear function f that models the data.

(b) Solve the inequality $f(x) \leq 52{,}700$. Assume that the domain is $D = \{2002, 2003, 2004, \ldots, 2010\}$. Interpret your answer.

Year	2002	2003	2004
Salary ($)	47,975	48,650	49,325

104. *Modeling Real Data* The following table lists numbers of visitors to national parks.

Year	1979	1985	1990	1999
Visitors (millions)	60	68	79	88

Source: National Park Service

(a) Find a linear function f that models the data. (Answers may vary.)

(b) Solve the inequality $65 \leq f(x) \leq 75$. Interpret your answer.

Writing about Mathematics

105. Suppose the solution to the equation $ax + b = 0$ with $a > 0$ is $x = k$. Discuss how the value of k can be used to help solve the linear inequalities $ax + b > 0$ and $ax + b < 0$. Illustrate this graphically. How would the solution sets change if $a < 0$?

106. Describe how to numerically solve the linear inequality $ax + b \leq 0$. Give an example.

EXTENDED AND DISCOVERY EXERCISES

1. *Arithmetic Mean* The arithmetic mean of two numbers a and b is given by $\frac{a+b}{2}$. Use properties of inequalities to show that if $a < b$, then $a < \frac{a+b}{2} < b$.

2. *Geometric Mean* The geometric mean of two numbers a and b is given by $\sqrt{ab}$. Use properties of inequalities to show that if $0 < a < b$, then $a < \sqrt{ab} < b$.

CHECKING BASIC CONCEPTS FOR SECTIONS 2.3 AND 2.4

1. Solve the linear equation $4(x - 2) = 2(5 - x) - 3$ by using each method. Compare your results.
(a) Graphical (b) Numerical (c) Symbolic

2. Solve the inequality $2(x - 4) > 1 - x$. Express the solution set in set-builder notation.

3. Solve the compound inequality $-2 \leq 1 - 2x \leq 3$. Use interval notation to express the solution set.

4. Use the graph to solve each equation and inequality. Then solve each part symbolically. Use interval notation when possible.
(a) $-3(2 - x) - \frac{1}{2}x - \frac{3}{2} = 0$

(b) $-3(2 - x) - \frac{1}{2}x - \frac{3}{2} > 0$

(c) $-3(2 - x) - \frac{1}{2}x - \frac{3}{2} \leq 0$

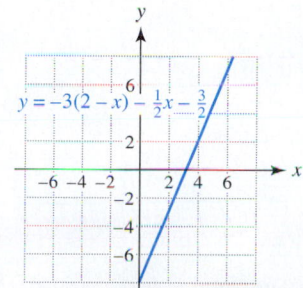

2.5 Piecewise-Defined Functions

◆ Evaluate and graph piecewise-defined functions

◆ Evaluate and graph the greatest integer function

◆ Evaluate and graph the absolute value function

◆ Solve absolute value equations and inequalities

Introduction

When a function f models real data, it is possible that no single formula can conveniently represent f. An example is first-class postage. In 2004 a first-class letter weighing up to 1 ounce cost $0.37 to mail, whereas a letter weighing 1.1 ounces cost $0.60 to mail. It is common to define a postage function in pieces that depend on the weight of the letter. Functions defined in pieces are called **piecewise-defined functions**. They typically use different formulas on various intervals of their domains. Many times the domains of piecewise-defined functions do not include all real numbers; rather, they are restricted to only a portion of the real number line. Our discussion in this section centers on piecewise-defined functions where each piece is linear.

Evaluating and Graphing Piecewise-Defined Functions

In 2004 postage rates in dollars for first-class letters weighing up to 5 ounces could be computed by the piecewise-defined function f.

$$f(x) = \begin{cases} 0.37 & \text{if } 0 < x \le 1 \\ 0.60 & \text{if } 1 < x \le 2 \\ 0.83 & \text{if } 2 < x \le 3 \\ 1.06 & \text{if } 3 < x \le 4 \\ 1.29 & \text{if } 4 < x \le 5 \end{cases}$$

The function f is not a constant function because the cost varies according to the weight x of a letter. However, each piece of f is a constant. Therefore, f is a **piecewise-constant function**. Since each constant piece is linear, f also can be called a **piecewise-linear function**. Notice that the domain of f has been restricted to $0 < x \le 5$ and does not include all real numbers.

The graph of f is composed of horizontal line segments. When $0 < x \le 1$, $f(x) = 0.37$, as shown in Figure 2.54. A small circle occurs at the point $(0, 0.37)$ because $x = 0$ is not in the domain of f. The other pieces of f are graphed similarly. Since there are breaks in the graph, f is not a continuous function. It is **discontinuous** at $x = 1, 2, 3,$ and 4. This type of function is called a **step function**.

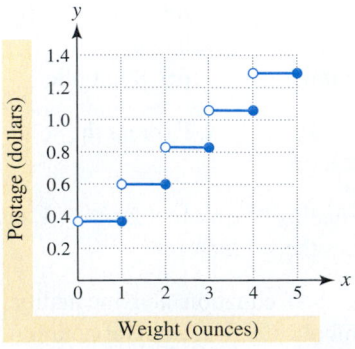

FIGURE 2.54 First-Class Postage in 2004

EXAMPLE 1 Evaluating a graphical representation

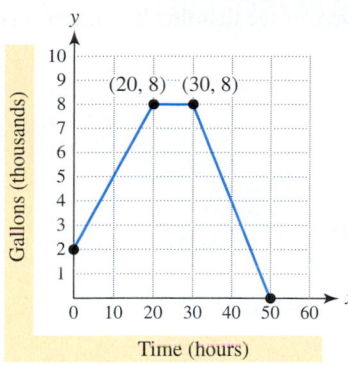

FIGURE 2.55 Water in a Swimming Pool

Figure 2.55 depicts a graph of a piecewise-linear function f. It models the amount of water in thousands of gallons in a swimming pool after x hours have elapsed.
(a) Use the graph to evaluate $f(0)$, $f(25)$, and $f(40)$. Interpret the results.
(b) Discuss how the amount of water in the pool changed. Is f a continuous function on the interval $[0, 50]$?
(c) Interpret the slope of each line segment in the graph.

SOLUTION
(a) From Figure 2.55, when $x = 0$, $y = 2$, so $f(0) = 2$. Initially, there are 2 thousand gallons of water in the pool. The points $(25, 8)$ and $(40, 4)$ lie on the graph of f. Therefore, $f(25) = 8$ and $f(40) = 4$. After 25 hours there were 8 thousand gallons in the pool, and after 40 hours there were 4 thousand gallons.
(b) During the first 20 hours, water in the pool increases at a constant rate from 2 thousand to 8 thousand gallons. For the next 10 hours, the water level is constant. Between 30 and 50 hours, the pool is drained at a constant rate. Since there are no breaks in the graph, f is a continuous function on the interval $[0, 50]$.

♦ **CLASS DISCUSSION**

Suppose that a function g models the amount of water in a swimming pool. Discuss whether g must be continuous. If g were discontinuous at some point in time, what must happen to the amount of water in the pool? ♦

(c) Slope indicates the rate at which water is entering or leaving the pool. The first line segment connects the points $(0, 2)$ and $(20, 8)$. Its slope is $m = \frac{8 - 2}{20 - 0} = 0.3$. Since the units are thousands of gallons and hours, water is entering the pool during this time at a rate of 300 gallons per hour. Between 20 and 30 hours the slope of the line segment is 0. Water is neither entering nor leaving the pool. The slope of the line segment connecting the points $(30, 8)$ and $(50, 0)$ is $m = \frac{0 - 8}{50 - 30} = -0.4$. On this interval, the amount of water is *decreasing* at a rate of 400 gallons per hour.

Now Try Exercise 5 ♦

EXAMPLE 2 Evaluating and graphing a piecewise-defined function

Use $f(x)$ to complete the following.

$$f(x) = \begin{cases} x - 1 & \text{if } -4 \leq x < 2 \\ -2x & \text{if } \;\; 2 \leq x \leq 4 \end{cases}$$

(a) What is the domain of f?   **(b)** Evaluate f at $x = -3, 2, 4,$ and 5.
(c) Sketch a graph of f.   **(d)** Is f a continuous function on its domain?
(e) Find all x-values where $f(x) = -4$.

SOLUTION
(a) Function f is defined for x-values satisfying either $-4 \leq x < 2$ or $2 \leq x \leq 4$. Thus the domain of f is $D = \{x \mid -4 \leq x \leq 4\}$ or in interval notation, $[-4, 4]$.
(b) For x-values satisfying $-4 \leq x < 2$, $f(x) = x - 1$ and so $f(-3) = -3 - 1 = -4$. Similarly, if $2 \leq x \leq 4$, then $f(x) = -2x$. Thus $f(2) = -2 \cdot 2 = -4$ and $f(4) = -2 \cdot 4 = -8$. The expression $f(5)$ is undefined because 5 is not in the domain of f.

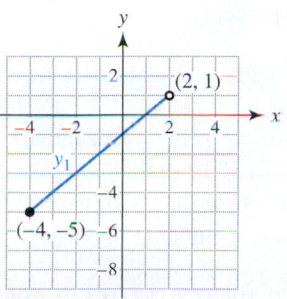

FIGURE 2.56

(c) Let $y_1 = x - 1$. The graph of y_1 is a line with slope 1 and y-intercept -1. However, we must restrict this line to the region where $-4 \leq x < 2$. When $x = -4$, $y = -5$, so place a dot at $(-4, -5)$. Similarly, when $x = 2$, $y_1 = 1$, so place an open circle at $(2, 1)$. Connect the points with a line segment. See Figure 2.56.

Similarly, graph the line $y_2 = -2x$ for x-values satisfying $2 \leq x \leq 4$ by placing dots at $(2, -4)$ and $(4, -8)$, and then connecting them with a line segment. See Figure 2.57 on the next page for the graph of f.

(d) The function f is not continuous because there is a break in its graph at $x = 2$.

(e) To determine all x-values that satisfy the equation $f(x) = -4$, visualize the line $y = -4$ on the graph of f, as shown in Figure 2.58. We can see that this line intersects the graph of f when $x = -3$ or when $x = 2$.

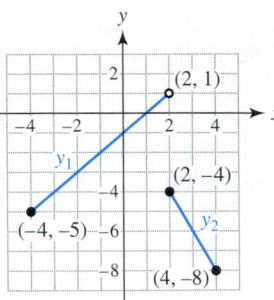

FIGURE 2.57

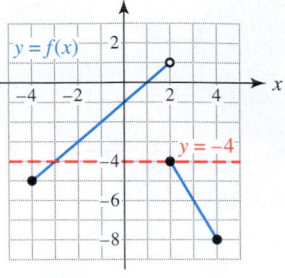

FIGURE 2.58

Now Try Exercise 7

Analyzing pollution and brown trout

Due to acid rain, the percentage of lakes in Scandinavia that lost their population of brown trout increased dramatically between 1940 and 1975. Based on a sample of 2850 lakes, this percentage can be approximated by the piecewise-linear function f. (**Source:** C. Mason, *Biology of Freshwater Pollution*.)

$$f(x) = \begin{cases} \dfrac{11}{20}(x - 1940) + 7 & \text{if } 1940 \le x < 1960 \\[2ex] \dfrac{32}{15}(x - 1960) + 18 & \text{if } 1960 \le x \le 1975 \end{cases}$$

(a) Determine the percentage of lakes that lost brown trout by 1947 and by 1972.

(b) Sketch a graph of f.

(c) Is f a continuous function on its domain?

SOLUTION

(a) Because $1940 \le 1947 < 1960$, $f(1947)$ is calculated using the first formula.

$$f(\mathbf{1947}) = \frac{11}{20}(\mathbf{1947} - 1940) + 7 = 10.85 \text{ (percent)}.$$

Similarly, $f(1972)$ is found using the second formula.

$$f(\mathbf{1972}) = \frac{32}{15}(\mathbf{1972} - 1960) + 18 = 43.6 \text{ (percent)}.$$

Therefore by 1947 approximately 11% of the lakes had lost their population of brown trout. By 1972 this percentage had increased to about 44%.

(b) Because f is a piecewise-linear function, the graph of f consists of two line segments. The first segment is determined by $y = \frac{11}{20}(x - 1940) + 7$. If $x = 1940$ then $y = 7$, and if $x = 1960$ then $y = 18$. Place a dot at $(1940, 7)$ and a small circle at $(1960, 18)$, since the year 1960 is not in the interval $1940 \le x < 1960$. These points are endpoints of the line segment.

　　The second line segment is determined by $y = \frac{32}{15}(x - 1960) + 18$. If $x = 1960$ then $y = 18$, and if $x = 1975$ then $y = 50$. This segment includes $(1960, 18)$ and

(1975, 50), so place a dot at both endpoints. Notice that this fills in the small circle at (1960, 18) from the first line segment. The graph of f is shown in Figure 2.59.

(c) The graph has no breaks, so f is continuous on its domain.

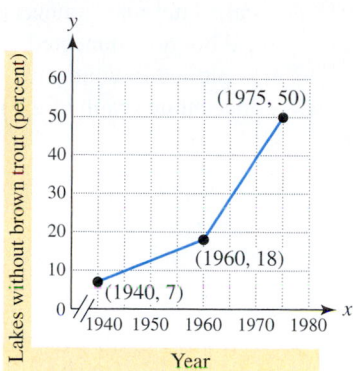

FIGURE 2.59 Lakes Losing Brown Trout

Now Try Exercise 107 ◆

The Greatest Integer Function

A common piecewise-defined function used in mathematics is the greatest integer function, denoted $f(x) = [x]$. The **greatest integer function** is defined as follows.

$[x]$ is the greatest integer less than or equal to x.

Some examples that evaluate $[x]$ include

$$[6.7] = 6, \quad [3] = 3, \quad [-2.3] = -3, \quad [-10] = -10, \quad \text{and} \quad [-\pi] = -4.$$

The graph of $y = [x]$ is shown in Figure 2.60 with its symbolic representation. The greatest integer function is both a piecewise-constant function and a step function.

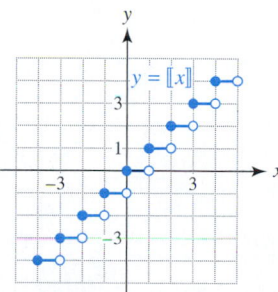

$$[x] = \begin{cases} \vdots \\ -2 & \text{if} & -2 \leq x < -1 \\ -1 & \text{if} & -1 \leq x < 0 \\ 0 & \text{if} & 0 \leq x < 1 \\ 1 & \text{if} & 1 \leq x < 2 \\ 2 & \text{if} & 2 \leq x < 3 \\ \vdots \end{cases}$$

FIGURE 2.60 The Greatest Integer Function

Calculator Help

To access the greatest integer function, see Appendix B (page AP-13).

[0, 10, 1] by [31, 35, 1]

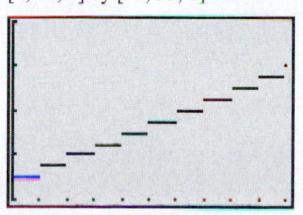

FIGURE 2.61 Dot Mode

In some applications, fractional parts are either not allowed or ignored. Framing lumber for houses is measured in 2-foot multiples, and mileage charges for rental cars may be calculated to the mile.

Suppose a car rental company charges $31.50 per day plus $0.25 for each mile driven, where fractions of a mile are ignored. The function given by $f(x) = 0.25[x] + 31.50$ calculates the cost of driving x miles in one day. For example, the cost of driving 100.4 miles is

$$f(100.4) = 0.25[100.4] + 31.50 = 0.25(100) + 31.50 = \$56.50.$$

On some calculators and computers, the greatest integer function is denoted int(X). A graph $Y_1 = 0.25*\text{int}(X) + 31.5$ is shown in Figure 2.61.

Calculator Help

To set a calculator in dot mode, see Appendix B (page AP-13).

◆ **MAKING CONNECTIONS**

Connected and Dot Modes   It is common for graphing calculators to connect points to make a graph look continuous. However, if a graph has breaks in it, a graphing calculator may connect points where there should be breaks. In dot mode, points are plotted but not connected. Figure 2.62 is the same graph shown in Figure 2.61, except that it is plotted in connected mode. Notice that in this instance connected mode generates an inaccurate graph of the greatest integer function.

[0, 10, 1] by [31, 35, 1]

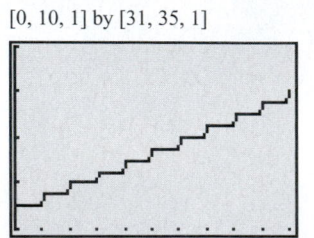

FIGURE 2.62   Connected Mode

The Absolute Value Function

The graph of $y = |x|$ is shown in Figure 2.63. It is V-shaped and cannot be represented by a single linear function. However, it can be represented by the lines $y = x$ (when $x \geq 0$) and $y = -x$ (when $x < 0$). This suggests that the absolute value function can be defined symbolically using a piecewise-linear function.

$$|x| = \begin{cases} -x & \text{if } x < 0 \\ x & \text{if } x \geq 0 \end{cases}$$

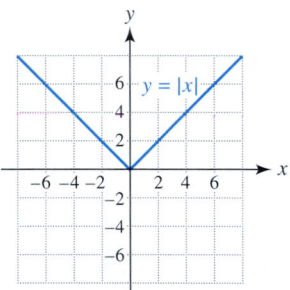

FIGURE 2.63   The Absolute Value Function

There is another formula for the absolute value function. Consider the following examples.

$$\sqrt{3^2} = \sqrt{9} = 3 \quad \text{and} \quad \sqrt{(-3)^2} = \sqrt{9} = 3.$$
$$\sqrt{7^2} = \sqrt{49} = 7 \quad \text{and} \quad \sqrt{(-7)^2} = \sqrt{49} = 7.$$

That is, regardless of whether a real number x is positive or negative, the expression $\sqrt{x^2}$ equals the absolute value of x. This statement is summarized by

$$\sqrt{x^2} = |x| \quad \textbf{for all real numbers } x.$$

 **EXAMPLE 4**

Analyzing the graph of $y = |ax + b|$

For each linear function f, graph $y = f(x)$ and $y = |f(x)|$ separately. Discuss how the absolute value affects the graph of f.
(a) $f(x) = x + 2$ **(b)** $f(x) = -2x + 4$

Calculator Help

To access the absolute value function, see Appendix B (page AP-14).

SOLUTION

(a) The graphs of $y_1 = x + 2$ and $y_2 = |x + 2|$ are shown in Figures 2.64 and 2.65, respectively. The graph of y_1 is a line with x-intercept -2. The graph of y_2 is V-shaped. The graphs are identical to the right of the x-intercept. To the left of the x-intercept, the graph of $y_1 = f(x)$ passes below the x-axis, and the graph of $y_2 = |f(x)|$ is the *reflection* of $y_1 = f(x)$ across the x-axis. The graph of $y_2 = |f(x)|$ does not dip below the x-axis because an absolute value is never negative.

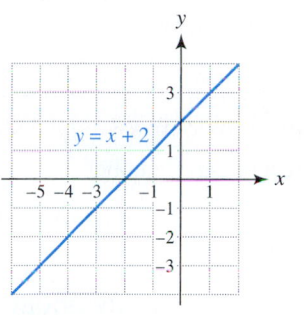

FIGURE 2.64

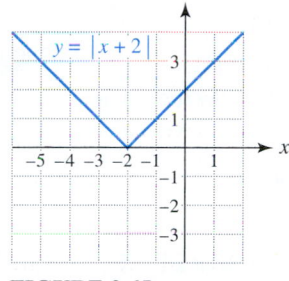

FIGURE 2.65

(b) The graphs of $y_1 = -2x + 4$ and $y_2 = |-2x + 4|$ are shown in Figures 2.66 and 2.67, respectively. Again, the graph of y_2 is V-shaped. The graph of $y_2 = |f(x)|$ is the reflection of f across the x-axis whenever the graph of $y_1 = f(x)$ is below the x-axis.

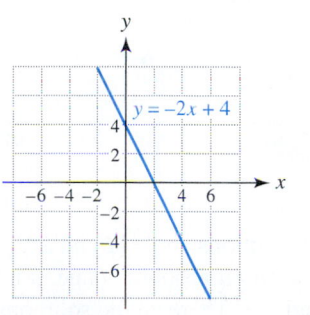

FIGURE 2.66

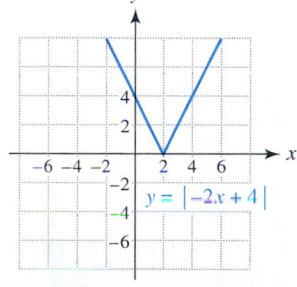

FIGURE 2.67

Now Try Exercises 33 and 37 ◆

Example 4 illustrates the fact that the graph of $y = |ax + b|$ is V-shaped and is never located below the x-axis. The vertex (or point) of the V-shaped graph corresponds to the x-intercept, which can be found by solving the linear equation $ax + b = 0$.

Equations and Inequalities Involving Absolute Values

Absolution Value Equations   The equation $|x| = 5$ has two solutions, ± 5. This fact is shown visually in Figure 2.68 on the next page, where the graph of $y = |x|$ intersects the horizontal line $y = 5$ at the points $(\pm 5, 5)$.

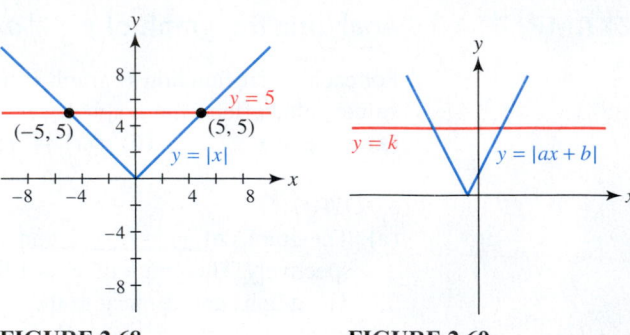

FIGURE 2.68   **FIGURE 2.69**

In general, the graph of $y = |ax + b|$ with $a \neq 0$ is V-shaped. It intersects the horizontal line $y = k$ twice whenever $k > 0$, as illustrated in Figure 2.69. Thus there are two solutions to the equation $|ax + b| = k$.

◆ **CLASS DISCUSSION**

Illustrate graphically how the equation $|ax + b| = k$ can have 0 or 1 solution when $k < 0$ or $k = 0$, respectively. ◆

ABSOLUTE VALUE EQUATIONS

Let k be a positive number. Then

$$|ax + b| = k \text{ is equivalent to } ax + b = \pm k.$$

EXAMPLE 5 Solving absolute value equations

Solve each equation.

(a) $\left|\frac{3}{4}x - 6\right| = 15$ **(b)** $|1 - 2x| = -3$ **(c)** $|3x - 2| - 5 = -2$

SOLUTION

(a) The equation $\left|\frac{3}{4}x - 6\right| = 15$ is satisfied when $\frac{3}{4}x - 6 = \pm 15$.

$$\frac{3}{4}x - 6 = 15 \quad \text{or} \quad \frac{3}{4}x - 6 = -15 \qquad \textit{Equations to solve}$$

$$\frac{3}{4}x = 21 \quad \text{or} \quad \frac{3}{4}x = -9 \qquad \textit{Add 6 to each side.}$$

$$x = 28 \quad \text{or} \quad x = -12 \qquad \textit{Multiply by } \frac{4}{3}.$$

The solutions are -12 and 28.

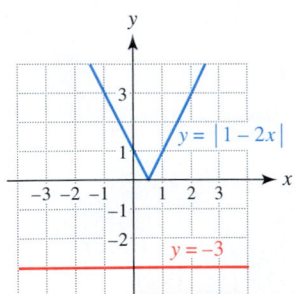

FIGURE 2.70 No Solutions

(b) Because an absolute value is never negative, $|1 - 2x| \geq 0$ for all x and can never equal -3. There are no solutions. This is illustrated graphically in Figure 2.70.

(c) Because the right side of the equation is a negative number, it might appear at first glance that there are no solutions. However, if we add 5 to each side of the equation,

$$|3x - 2| - 5 = -2 \quad \text{becomes} \quad |3x - 2| = 3.$$

This equation is equivalent to $3x - 2 = \pm 3$ and has two solutions.

$$3x - 2 = 3 \quad \text{or} \quad 3x - 2 = -3 \qquad \textit{Equations to solve}$$

$$3x = 5 \quad \text{or} \quad 3x = -1 \qquad \textit{Add 2 to each side.}$$

$$x = \frac{5}{3} \quad \text{or} \quad x = -\frac{1}{3} \qquad \textit{Divide by 3.}$$

The solutions are $-\frac{1}{3}$ and $\frac{5}{3}$.

Now Try Exercises 41, 47 and 49 ◆

In the next example we solve an absolute value equation graphically, numerically, and symbolically.

EXAMPLE 6   ### Solving an equation involving absolute value

Solve the equation $|2x + 5| = 2$ graphically, numerically, and symbolically.

SOLUTION

Graphical Solution   Graph $Y_1 = \text{abs}(2X + 5)$ and $Y_2 = 2$. The V-shaped graph of y_1 intersects the horizontal line at the points $(-3.5, 2)$ and $(-1.5, 2)$, as shown in Figures 2.71 and 2.72. The solutions are -3.5 and -1.5.

Numerical Solution   Table $Y_1 = \text{abs}(2X + 5)$ and $Y_2 = 2$, as in Figure 2.73. The solutions to $y_1 = y_2$ are -3.5 and -1.5.

Calculator Help

To find a point of intersection, see Appendix B (page AP-12).

$[-9, 9, 1]$ by $[-6, 6, 1]$ $[-9, 9, 1]$ by $[-6, 6, 1]$

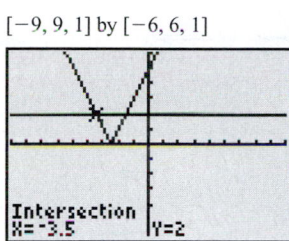   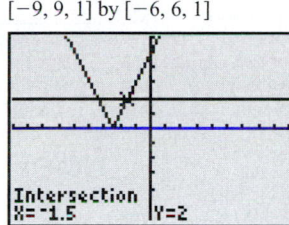   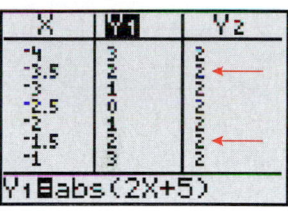

FIGURE 2.71   **FIGURE 2.72**   **FIGURE 2.73**

Symbolic Solution   The equation $|2x + 5| = 2$, is satisfied when $2x + 5 = \pm 2$.

$$2x + 5 = 2 \qquad \text{or} \qquad 2x + 5 = -2$$
$$2x = -3 \qquad \text{or} \qquad 2x = -7$$
$$x = -\frac{3}{2} \qquad \text{or} \qquad x = -\frac{7}{2}$$

Now Try Exercise 61 ◆

Sometimes more than one absolute value sign can occur in an equation. For example, an equation might be in the form

$$|ax + b| = |cx + d|.$$

In this case there are two possibilities: either

$$ax + b = cx + d \qquad \text{or} \qquad ax + b = -(cx + d).$$

This symbolic technique is demonstrated in the next example.

EXAMPLE 7   ### Solving an equation involving two absolute values

Solve the equation $|x - 2| = |1 - 2x|$.

SOLUTION   We must solve both of the following equations.

$$x - 2 = 1 - 2x \qquad \text{or} \qquad x - 2 = -(1 - 2x)$$
$$3x = 3 \qquad \text{or} \qquad x - 2 = -1 + 2x$$
$$x = 1 \qquad \text{or} \qquad -1 = x$$

There are two solutions: -1 and 1.

Now Try Exercise 53 ◆

Absolution Value Inequalities In Figure 2.74 the solutions to $|ax + b| = k$ are labeled s_1 and s_2. The V-shaped graph of $y = |ax + b|$ is below the horizontal line $y = k$ between s_1 and s_2 or when $s_1 < x < s_2$. The solution set for the inequality $|ax + b| < k$ is green on the x-axis. In Figure 2.75 the V-shaped graph is above the horizontal line $y = k$ left of s_1 or right of s_2. That is, when $x < s_1$ or $x > s_2$. The solution set for the inequality $|ax + b| > k$ is green on the x-axis. Note that in both figures, equality (determined by s_1 and s_2) is the boundary between *greater than* and *less than*. For this reason, s_1 and s_2 are called *boundary numbers*.

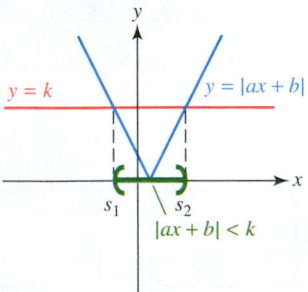

FIGURE 2.74   FIGURE 2.75

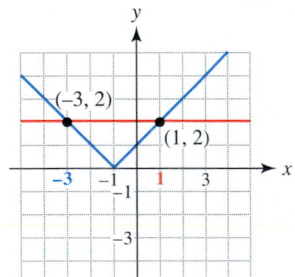

FIGURE 2.76

For example, the graphs of $y = |x + 1|$ and $y = 2$ are shown in Figure 2.76. These graphs intersect at the points $(-3, 2)$ and $(1, 2)$. It follows that the two solutions to

$$|x + 1| = 2$$

are $s_1 = -3$ and $s_2 = 1$. The solutions to $|x + 1| < 2$ lie between $s_1 = -3$ and $s_2 = 1$, which can be written as $-3 < x < 1$. Furthermore, the solutions to $|x + 1| > 2$ lie "outside" $s_1 = -3$ and $s_2 = 1$. This can be written as $x < -3$ or $x > 1$.

These results are generalized.

ABSOLUTE VALUE INEQUALITIES

Let the solutions to $|ax + b| = k$ be s_1 and s_2, where $s_1 < s_2$ and $k > 0$.

1. $|ax + b| < k$ is equivalent to $s_1 < x < s_2$.

2. $|ax + b| > k$ is equivalent to $x < s_1$ or $x > s_2$.

Similar statements can be made for inequalities involving $\leq$ or $\geq$.

Note: The **union symbol** $\cup$ may be used to write $x < s_1$ or $x > s_2$ in interval notation. For example, $x < -2$ or $x > 5$ is written as $(-\infty, -2) \cup (5, \infty)$ in interval notation. This indicates that the solution set includes all real numbers in either $(-\infty, -2)$ or $(5, \infty)$.

 **EXAMPLE 8** Solving inequalities involving absolute values symbolically

Solve the inequality symbolically. Write the solution set in interval notation.
(a) $|2x - 5| \leq 6$ **(b)** $|5 - x| > 3$

SOLUTION

(a) Begin by solving $2x - 5 = \pm 6$.

$$2x - 5 = 6 \qquad \text{or} \qquad 2x - 5 = -6$$
$$2x = 11 \qquad \text{or} \qquad 2x = -1$$
$$x = \frac{11}{2} \qquad \text{or} \qquad x = -\frac{1}{2}$$

The solutions to $|2x - 5| = 6$ are $-\frac{1}{2}$ and $\frac{11}{2}$. The solution set for the inequality $|2x - 5| \leq 6$ includes all real numbers x satisfying $-\frac{1}{2} \leq x \leq \frac{11}{2}$. In interval notation this is written as $\left[-\frac{1}{2}, \frac{11}{2}\right]$.

(b) Begin by solving $5 - x = \pm 3$.

$$5 - x = 3 \qquad \text{or} \qquad 5 - x = -3$$
$$-x = -2 \qquad \text{or} \qquad -x = -8$$
$$x = 2 \qquad \text{or} \qquad x = 8$$

The solutions to $|5 - x| = 3$ are 2 and 8. The solution set for $|5 - x| > 3$ includes all real numbers x left of 2 and right of 8. Thus $|5 - x| > 3$ is equivalent to $x < 2$ or $x > 8$. In interval notation this is written as $(-\infty, 2) \cup (8, \infty)$.

Now Try Exercises 71 and 77 ◆

 **EXAMPLE 9**

Analyzing the temperature range in Santa Fe

The inequality $|T - 49| \leq 20$ describes the range of monthly average temperatures T in degrees Fahrenheit for Santa Fe, New Mexico.   (**Source:** A. Miller and J. Thompson, *Elements of Meteorology*.)

(a) Solve this inequality graphically and symbolically.

(b) The high and low monthly average temperatures satisfy the absolute value equation $|T - 49| = 20$. Use this fact to interpret the results from part (a).

SOLUTION

(a) *Graphical Solution*   Graph $Y_1 = \text{abs}(X - 49)$ and $Y_2 = 20$, as in Figure 2.77. The V-shaped graph of y_1 intersects the horizontal line at the points $(29, 20)$ and $(69, 20)$. See Figures 2.78 and 2.79. The graph of y_1 is below the graph of y_2 between these two points. Thus, the solution set consists of all temperatures T satisfying $29 \leq T \leq 69$.

[20, 80, 5] by [0, 35, 5]   [20, 80, 5] by [0, 35, 5]   [20, 80, 5] by [0, 35, 5]

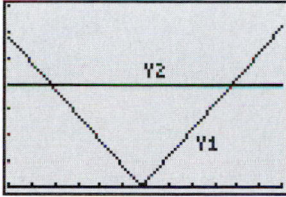   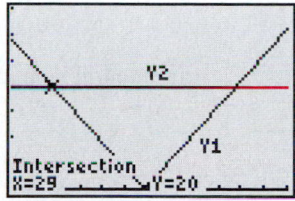   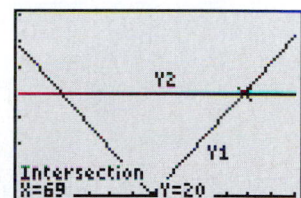

FIGURE 2.77   **FIGURE 2.78**   **FIGURE 2.79**

Symbolic Solution   First solve the related equation $|T - 49| = 20$.

$$T - 49 = -20 \qquad \text{or} \qquad T - 49 = 20$$
$$T = 29 \qquad \text{or} \qquad T = 69$$

Thus by our previous discussion, $|T - 49| \leq 20$ is equivalent to $29 \leq T \leq 69$.

(b) The solutions to $|T - 49| = 20$ are 29 and 69. Therefore, the monthly average temperatures in Sante Fe vary between a low of 29°F (January) and a high of 69°F (July). The monthly averages are always within 20 degrees of 49°F.   *Now Try Exercise 93* ◆

An Alternative Method   There is a second symbolic method that can be used to solve absolute value inequalities. This method is often used in advanced mathematics courses, such as calculus. It is based on the following two properties.

ABSOLUTE VALUE INEQUALITIES (ALTERNATIVE METHOD)

Let k be a positive number.

1. $|ax + b| < k$ is equivalent to $-k < ax + b < k$.
2. $|ax + b| > k$ is equivalent to $ax + b < -k$ or $ax + b > k$.

Similar statements can be made for inequalities involving $\leq$ or $\geq$.

EXAMPLE 10 Using an alternative method

Solve each absolute value inequality. Write your answer in interval notation.
(a) $|4 - 5x| \leq 3$ **(b)** $|-4x - 6| > 2$

SOLUTION

(a) $|4 - 5x| \leq 3$ is equivalent to the following three-part inequality.

$$-3 \leq 4 - 5x \leq 3 \qquad \text{Equivalent inequality}$$

$$-7 \leq -5x \leq -1 \qquad \text{Subtract 4 from each part.}$$

$$\frac{7}{5} \geq x \geq \frac{1}{5} \qquad \text{Divide each part by 5; reverse the inequality.}$$

In interval notation, the solution is $\left[\frac{1}{5}, \frac{7}{5}\right]$.

(b) $|-4x - 6| > 2$ is equivalent to the following compound inequality.

$$-4x - 6 < -2 \quad \text{or} \quad -4x - 6 > 2 \qquad \text{Equivalent compound inequality}$$

$$-4x < 4 \quad \text{or} \quad -4x > 8 \qquad \text{Add 6 to each side.}$$

$$x > -1 \quad \text{or} \quad x < -2 \qquad \text{Divide each by } -4; \text{ reverse the inequality.}$$

In interval notation the solution set is $(-\infty, -2) \cup (-1, \infty)$.

Now Try Exercises 73 and 79 ◆

◆ **CLASS DISCUSSION**

Sketch the graphs of $y = ax + b$, $y = |ax + b|$, $y = -k$, and $y = k$ on one xy-plane. Now use these graphs to explain why the alternative method for solving absolute value inequalities is correct. ◆

2.5 Putting it all Together

Piecewise-defined functions are used when it is not convenient to define a function by using one formula. The greatest integer function and the absolute value function are two examples of piecewise-defined functions.

The following table summarizes some important concepts from this section.

Concept	Explanation	Examples
Piecewise-defined function	A function is piecewise-defined if it has different formulas on different intervals of its domain. Many times the domain is restricted.	$f(x) = \begin{cases} 2x - 3 & \text{if } -3 \le x < 1 \\ x + 5 & \text{if } \ 1 \le x \le 5 \end{cases}$ When $x = 2$ then $f(x) = x + 5$, so $f(2) = 2 + 5 = 7$. The domain of f is $[-3, 5]$.
Absolute value function	$f(x) = \|x\|$: The output from the absolute value function is never negative.	$f(4) = \|4\| = 4$ $f(-2) = \|-2\| = 2$ The graph of $y = \|x\|$ is V-shaped and never goes below the x-axis.
Absolute value equations	$\|ax + b\| = k$ with $k > 0$ is equivalent to $ax + b = \pm k$. If $k < 0$, then the absolute value equation has no solutions.	Solve $\|3x - 5\| = 4$. $3x - 5 = -4 \quad$ or $\quad 3x - 5 = 4$ $3x = 1 \quad$ or $\quad 3x = 9$ $x = \dfrac{1}{3} \quad$ or $\quad x = 3$ $\|4x + 9\| = -2$ has no solutions.
Absolute value inequalities	Let the solutions to $\|ax + b\| = k$, $k > 0$ be s_1 and s_2 with $s_1 < s_2$. The solution set to $\|ax + b\| < k$ is $s_1 < x < s_2$ and the solution set to $\|ax + b\| > k$ is $x < s_1$ or $x > s_2$.	To solve either $\|x - 5\| < 4$ or $\|x - 5\| > 4$, first solve $x - 5 = -4$ and $x - 5 = 4$. The solutions are $x = 1$ and 9. Thus the solution set to $\|x - 5\| < 4$ is $1 < x < 9$, and the solution set to $\|x - 5\| > 4$ is $x < 1$ or $x > 9$.

Concept	Explanation	Examples
Alternative method for solving absolute value inequalities	**1.** $\|ax + b\| < k$ is equivalent to $$-k < ax + b < k.$$ **2.** $\|ax + b\| > k$ is equivalent to $$ax + b < -k \text{ or } ax + b > k.$$	**1.** $\|x - 1\| < 5$ is solved as follows. $$-5 < x - 1 < 5$$ $$-4 < x < 6$$ **2.** $\|x - 1\| > 5$ is solved as follows. $$x - 1 < -5 \quad \text{or} \quad x - 1 > 5$$ $$x < -4 \quad \text{or} \qquad x > 6$$

2.5 Exercises

Evaluating and Graphing Piecewise-Defined Functions

1. **Speed Limits**   The graph of $y = f(x)$ gives the speed limit y along a rural highway after traveling x miles.
  (a) What are the maximum and minimum speed limits along this stretch of highway?

  (b) Estimate the miles of highway with a speed limit of 55 miles per hour.

  (c) Evaluate $f(4)$, $f(12)$, and $f(18)$.

  (d) At what x-values is the graph discontinuous? Interpret each discontinuity.

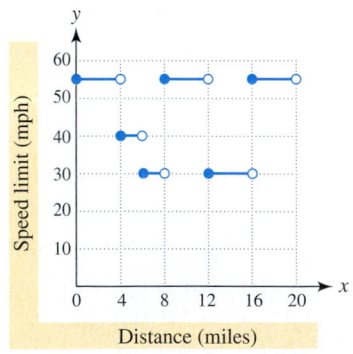

2. **ATM**   The graph of $y = f(x)$ depicts the amount of money y in dollars in an automatic teller machine (ATM) after x minutes.
  (a) Determine the initial and final amounts of money in the ATM.

  (b) Evaluate $f(10)$ and $f(50)$. Is f continuous?

  (c) How many *withdrawals* occurred during this time period?

  (d) When did the largest withdrawal occur? How much was it?

  (e) How much was deposited into the machine?

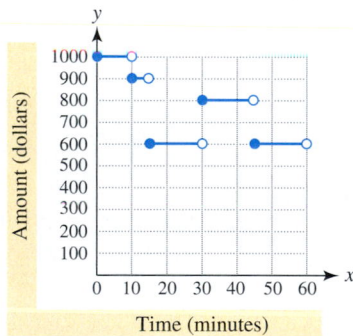

3. **Swimming Pool Levels**   The graph of $y = f(x)$ shows the amount of water in thousands of gallons remaining in a swimming pool after x days.
  (a) Estimate the initial and final amounts of water in the pool.

  (b) When did the amount of water in the pool remain constant?

  (c) Approximate $f(2)$ and $f(4)$.

  (d) At what rate was water being drained from the pool when $1 \le x \le 3$?

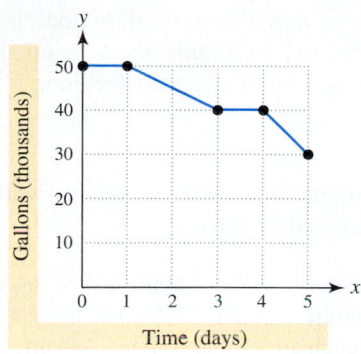

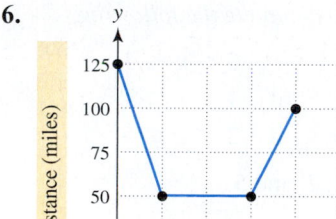

6.

4. *Gasoline Usage* The graph shows the gallons of gasoline y in the gas tank of a car after x hours.
(a) Estimate how much gasoline was in the gas tank when $x = 3$.

(b) Interpret the graph.

(c) During what time interval did the car burn gasoline at the fastest rate?

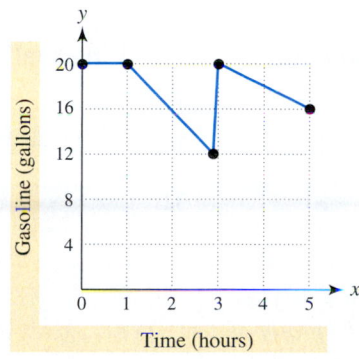

Exercises 5 and 6: An individual is driving a car along a straight road. The graph shows the distance from home that the driver is after x hours.

(a) *Use the graph to evaluate $f(1.5)$ and $f(4)$.*
(b) *Interpret the slope of each line segment.*
(c) *Describe the motion of the car.*

5.

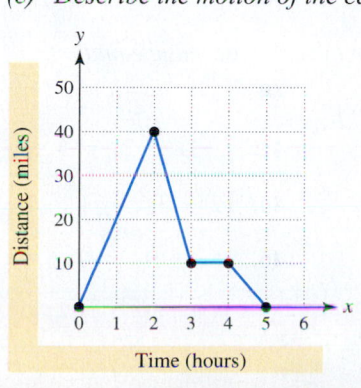

Exercises 7–12: (Refer to Example 2.) Complete the following for $f(x)$.

(a) *Determine the domain of f.*
(b) *Evaluate $f(-2)$, $f(0)$, and $f(3)$.*
(c) *Graph f.*
(d) *Is f continuous on its domain?*

7. $f(x) = \begin{cases} 2 & \text{if } -5 \le x \le -1 \\ x + 3 & \text{if } -1 < x \le 5 \end{cases}$

8. $f(x) = \begin{cases} 2x + 1 & \text{if } -3 \le x < 0 \\ x - 1 & \text{if } 0 \le x \le 3 \end{cases}$

9. $f(x) = \begin{cases} 3x & \text{if } -1 \le x < 1 \\ x + 1 & \text{if } 1 \le x \le 2 \end{cases}$

10. $f(x) = \begin{cases} -2 & \text{if } -6 \le x < -2 \\ 0 & \text{if } -2 \le x < 0 \\ 3x & \text{if } 0 \le x \le 4 \end{cases}$

11. $f(x) = \begin{cases} x & \text{if } -3 \le x \le -1 \\ 1 & \text{if } -1 < x < 1 \\ 2 - x & \text{if } 1 \le x \le 3 \end{cases}$

12. $f(x) = \begin{cases} 3 & \text{if } -4 \le x \le -1 \\ x - 2 & \text{if } -1 < x \le 2 \\ 0.5x & \text{if } 2 < x \le 4 \end{cases}$

Exercises 13 and 14: Graph f.

13. $f(x) = \begin{cases} -\frac{1}{2}x + 1 & -4 \le x \le -2 \\ 1 - 2x & -2 < x \le 1 \\ \frac{2}{3}x + \frac{4}{3} & 1 < x \le 4 \end{cases}$

14. $f(x) = \begin{cases} \frac{3}{2} - \frac{1}{2}x & -3 \le x < -1 \\ -2x & -1 \le x \le 2 \\ \frac{1}{2}x - 5 & 2 < x \le 3 \end{cases}$

Exercises 15–18: Use f(x) to complete the following.

$$f(x) = \begin{cases} 3x - 1 & \text{if } -5 \le x < 1 \\ 4 & \text{if } 1 \le x \le 3 \\ 6 - x & \text{if } 3 < x \le 5 \end{cases}$$

15. Evaluate f at $x = -3, 1, 2$, and 5.

16. On what interval is f constant?

17. Sketch a graph of f. Is f continuous on its domain?

18. Find the x-value(s) where $f(x) = 2$.

Exercises 19–22: Use g(x) to complete the following.

$$g(x) = \begin{cases} -2x - 6 & \text{if } -8 \le x \le -2 \\ x & \text{if } -2 < x < 2 \\ 0.5x + 1 & \text{if } 2 \le x \le 8 \end{cases}$$

19. Evaluate g at $x = -8, -2, 2$, and 8.

20. For what x-values does $g(x)$ have a positive slope?

21. Sketch a graph of g. Is g continuous on its domain?

22. Find the x-value(s) where $g(x) = 0$.

Greatest Integer Function

Exercises 23–26: Complete the following.

　(a) *Use dot mode to graph the function f in the standard viewing rectangle.*
　(b) *Evaluate f(−3.1) and f(1.7).*

23. $f(x) = [\![2x - 1]\!]$　　**24.** $f(x) = [\![x + 1]\!]$

25. $f(x) = 2[\![x]\!] + 1$　　**26.** $f(x) = [\![-x]\!]$

27. *Lumber Costs*　Lumber that is used to frame walls of houses is frequently sold in multiples of 2 feet. If the length of a board is not exactly a multiple of 2 feet, there is often no charge for the additional length. For example, if a board measures at least 8 feet but less than 10 feet, then the consumer is charged for only 8 feet.

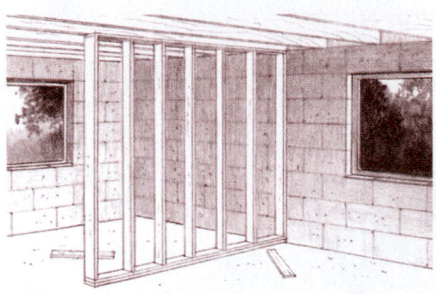

(a) Suppose that the cost of lumber is $0.80 every 2 feet. Find a formula for a function f that computes the cost of a board x feet long for $6 \le x \le 18$.

(b) Graph f.

(c) Determine the costs of boards with lengths of 8.5 feet and 15.2 feet.

28. *Cellular Phone Bills*　Suppose that the charges for a cellular phone call are $0.50 for the first minute and $0.25 for each additional minute. Assume that a fraction of a minute is rounded up.

(a) Determine the cost of a phone call that lasts 3.5 minutes.

(b) Find a formula for a function f that computes the cost of a telephone call that is x minutes long, where $0 < x \le 5$. (*Hint:* Express f as a piecewise-constant function.)

(c) The greatest integer function is not convenient to use in part (b). Explain why.

Absolute Value Equations and Inequalities

Exercises 29–32: Graph by hand.

29. $y = |x + 1|$　　　**30.** $y = |1 - x|$

31. $y = |2x - 3|$　　　**32.** $y = |\frac{1}{2}x + 1|$

Exercises 33–38: (Refer to Example 4.) Do the following.
　(a) *Graph $y = f(x)$.*
　(b) *Use the graph of $y = f(x)$ to sketch a graph of the equation $y = |f(x)|$.*
　(c) *Determine the x-intercept for the graph of the equation $y = |f(x)|$.*

33. $y = 2x$　　　　**34.** $y = 0.5x$

35. $y = 3x - 3$　　　**36.** $y = 2x - 4$

37. $y = 6 - 2x$　　　**38.** $y = 2 - 4x$

Exercises 39–58: Solve the absolute value equation.

39. $|-2x| = 4$　　　**40.** $|3x| = -6$

41. $|5x - 7| = 2$　　　**42.** $|-3x - 2| = 5$

43. $|3 - 4x| = 5$　　　**44.** $|2 - 3x| = 1$

45. $|-6x - 2| = 0$　　　**46.** $|6x - 9| = 0$

47. $|17x - 6| = -3$ **48.** $|-8x - 11| = -7$

49. $|1.2x - 1.7| - 1 = 3$ **50.** $|3 - 3x| - 2 = 2$

51. $|4x - 5| + 3 = 2$ **52.** $|4.5 - 2x| + 1.1 = 9.7$

53. $|2x - 9| = |8 - 3x|$ **54.** $|x - 3| = |8 - x|$

55. $\left|\frac{3}{4}x - \frac{1}{4}\right| = \left|\frac{3}{4} - \frac{1}{4}x\right|$ **56.** $\left|\frac{1}{2}x + \frac{3}{2}\right| = \left|\frac{3}{2}x - \frac{7}{2}\right|$

57. $|15x - 5| = |35 - 5x|$

58. $|20x - 40| = |80 - 20x|$

Exercises 59 and 60: Let $f(x) = |ax + b|$ where $a \neq 0$ and $g(x) = k$, where $k > 0$ is a constant. The graphs of f and g are shown. Solve each equation and inequality.

59. (a) $f(x) = g(x)$ **(b)** $f(x) < g(x)$
  (c) $f(x) > g(x)$

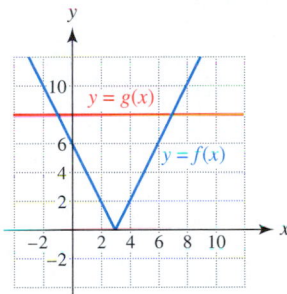

60. (a) $f(x) = g(x)$ **(b)** $f(x) \leq g(x)$
  (c) $f(x) \geq g(x)$

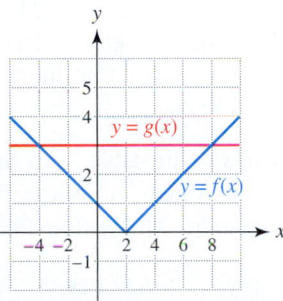

Exercises 61–64: Solve the equation

  (a) graphically,
  (b) numerically, and
  (c) symbolically.
Then solve the related inequality.

61. $|2x - 5| = 10$, $|2x - 5| < 10$

62. $|3x - 4| = 8$, $|3x - 4| \leq 8$

63. $|5 - 3x| = 2$, $|5 - 3x| > 2$

64. $|4x - 7| = 5$, $|4x - 7| \geq 5$

Exercises 65–70: Solve the equation symbolically. Then solve the related inequality.

65. $|2.1x - 0.7| = 2.4$, $|2.1x - 0.7| \geq 2.4$

66. $\left|\frac{1}{2}x - \frac{3}{4}\right| = \frac{7}{4}$, $\left|\frac{1}{2}x - \frac{3}{4}\right| \leq \frac{7}{4}$

67. $|3x| + 5 = 6$, $|3x| + 5 > 6$

68. $|x| - 10 = 25$, $|x| - 10 < 25$

69. $\left|\frac{2}{3}x - \frac{1}{2}\right| = -\frac{1}{4}$, $\left|\frac{2}{3}x - \frac{1}{2}\right| \leq -\frac{1}{4}$

70. $|5x - 0.3| = -4$, $|5x - 0.3| > -4$

You Decide the Method

Exercises 71–82: Solve the inequality. Write the solution in interval notation.

71. $|3x - 1| < 8$ **72.** $|15 - x| < 7$

73. $|7 - 4x| \leq 11$ **74.** $|-3x + 1| \leq 5$

75. $|0.5x - 0.75| < 2$ **76.** $|2.1x - 5| \leq 8$

77. $|2x - 3| > 1$ **78.** $|5x - 7| > 2$

79. $|-3x + 8| \geq 3$ **80.** $|-7x - 3| \geq 5$

81. $|0.25x - 1| > 3$ **82.** $|-0.5x + 5| \geq 4$

Sketching Graphs

83. Sketch a graph that depicts the amount of water in a 100-gallon tank. The tank is initially empty and then filled at a rate of 5 gallons per minute. Immediately after it is full, a pump is used to empty the tank at 2 gallons per minute.

84. Sketch a graph showing the distance from home that a person is after x hours if that individual drives on a straight road at 40 mph to a park 20 miles away, remains at the park for 2 hours, and then returns home at a speed of 20 mph.

85. Sketch a graph showing the amount of money y that a person makes after working x hours in one day, where $0 \leq x \leq 12$. Assume that the worker earns \$8 per hour and receives time-and-a-half pay for all hours over eight.

86. Suppose that two cars, both traveling at a constant speed of 60 miles per hour, approach each other on a straight highway. If they are initially 4 miles apart, sketch a graph of the distance between the cars after x minutes, where $0 \leq x \leq 4$.

Concepts

87. If $f(k) = -6$, what is the value of $|f(k)|$?

88. If $f(k) = 17$, what is the value of $|f(k)|$?

89. If the domain of $f(x)$ is given by $[-2, 4]$, what is the domain of $|f(x)|$?

90. If the domain of $f(x)$ is given by $(-\infty, 0]$, what is the domain of $|f(x)|$?

91. If the range of $f(x)$ is given by $(-\infty, 0]$, what is the range of $|f(x)|$?

92. If the range of $f(x)$ is given by $(-4, 5)$, what is the range of $|f(x)|$?

Applications

Exercises 93–98: Average Temperatures   (Refer to Example 9.) The inequality describes the range of monthly average temperatures T in degrees Fahrenheit at a certain location.

(a) Solve the inequality.
(b) If the high and low monthly average temperatures satisfy equality, interpret the inequality.

93. $|T - 43| \leq 24$, Marquette, Michigan

94. $|T - 62| \leq 19$, Memphis, Tennessee

95. $|T - 50| \leq 22$, Boston, Massachusetts

96. $|T - 10| \leq 36$, Chesterfield, Canada

97. $|T - 61.5| \leq 12.5$, Buenos Aires, Argentina

98. $|T - 43.5| \leq 8.5$, Punta Arenas, Chile

99. *Error in Measurements*   Products are often manufactured to be a given size to within a specified tolerance. For instance, if an aluminum can is supposed to have a diameter of 3 inches, either 2.99 inches or 3.01 inches might be acceptable. If the maximum error in the diameter d of a can is limited to 0.004 inch, then d must satisfy the absolute value inequality

$$|d - 3| \leq 0.004.$$

Solve this inequality and interpret the results.

100. *Error in Measurements*   (Refer to the preceding exercise.) Suppose that a 12-inch ruler must have the correct length L to within 0.0002 inch.
   (a) Write an absolute value inequality for L that describes this requirement.

   (b) Solve this inequality and interpret the results.

101. *Relative Error*   If a quantity is measured to be Q and its exact value is A, then the relative error in Q is $\left|\frac{Q - A}{A}\right|$. If the exact value is $A = 35$ and you want the relative error in Q to be less than or equal to 0.02 (or 2%), what values for Q are possible?

102. *Relative Error*   (Refer to the preceding exercise.) The exact perimeter P of a square is 50 feet. What measured lengths are possible for the side S of the square to have relative error in the perimeter that is less than or equal to 0.04 (or 4%)?

103. *Shoe Size*   Professional basketball player Shaquille O'Neal is 7 feet 1 inch tall and weighs 325 pounds. The table lists his shoe sizes at certain ages.   (**Source:** *USA Today.*)

Age	20	21	22	23
Shoe Size	19	20	21	22

   (a) Write a formula that calculates his shoe size at age $x = 20, 21, 22$, and 23.

   (b) Suppose that after age 23 his shoe size did not change. Sketch a graph of a continuous, piecewise-linear function f that models his shoe size from ages 20 to 26.

104. *ACT Scores*   The table lists the average composite scores on the American College Test (ACT) for selected years.   (**Source:** The American College Testing Program.)

Year	1975	1980	1985	1990	1995	2000
Score	18.6	18.5	18.6	20.6	20.8	21.0

   (a) Make a line graph of the data in the viewing rectangle $[1970, 2005, 5]$ by $[15, 25, 1]$.

   (b) If this graph represents a piecewise-linear function f, find a formula for the piece of f on the interval given by $[1990, 1995]$.

   (c) Evaluate $f(1993)$. (The actual average composite score in 1993 was 20.7.)

105. *Violent Crime* The table lists victims of violent crime per 1000 people by age group using interval notation. (**Source:** Department of Justice.)

Age	[12, 16)	[16, 19)	[19, 24)	[24, 35)
Crime Rate	88	96	68	47

Age	[35, 50)	[50, 65)	[65, 90)
Crime Rate	32	15	4

(a) Sketch the graph of a piecewise-constant function that models the data, where x represents age.

(b) Discuss the impact that age has on the likelihood of being a victim of a violent crime.

106. *Student Loans* The table lists the amounts in millions of dollars of government-guaranteed student loans from 1990 to 1994. (**Source:** USA Today.)

Year	1990	1991	1992	1993	1994
Loans ($ millions)	12.3	13.5	14.7	16.5	18.3

(a) Make a line graph of the data. Can a linear function model the data exactly?

(b) Find values for the constants a, b, c, and d so that $f(x)$ models the data exactly.

$$f(x) = \begin{cases} a(x - 1990) + b & \text{if } 1990 \le x < 1992 \\ c(x - 1992) + d & \text{if } 1992 \le x \le 1994 \end{cases}$$

(c) Evaluate $f(1991)$ and $f(1993)$. Do the results agree with the actual values listed in the table?

107. *Cesarean Births* The percentages of babies delivered by Cesarean birth are listed in the table for selected years. (**Source:** S. Teutsch.)

Year	1970	1975	1980	1985	1990
Cesarean Births (%)	5	11	17	23	23

(a) Make a line graph of the data. Interpret the graph.

(b) Find values for the constants a, b, and c so that $f(x)$ models the data exactly.

$$f(x) = \begin{cases} a(x - 1970) + b & \text{if } 1970 \le x < 1985 \\ c & \text{if } 1985 \le x \le 1990 \end{cases}$$

(c) Evaluate $f(1978)$ and $f(1988)$. Interpret the results.

108. *Flow Rates* A water tank has an inlet pipe with a flow rate of 5 gallons per minute and an outlet pipe with a flow rate of 3 gallons per minute. A pipe can be either closed or completely open. The graph shows the number of gallons of water in the tank after x minutes have elapsed. Use the concept of slope to interpret each piece of this graph.

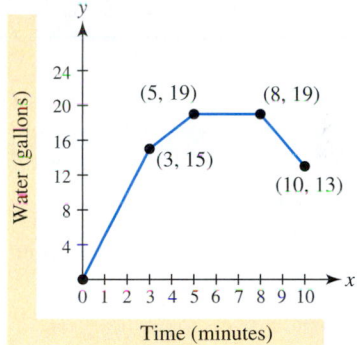

Writing about Mathematics

109. Find an actual piecewise-constant function that describes movie prices for children, adults, and senior citizens. Represent your function graphically and symbolically. Explain your results.

110. Explain why some functions are defined piecewise. Give an example.

EXTENDED AND DISCOVERY EXERCISE

1. *Measurement* The circumference of a circular garden with a radius r is to have a 150-foot fence around it. The length of the fence can vary from 150 feet by at most ± 2 feet. What range of values is possible for r?

CHECKING BASIC CONCEPTS FOR SECTION 2.5

1. Consider the function

$$f(x) = \begin{cases} 2 - x & \text{if } -5 \le x < 0 \\ 2x - 1 & \text{if } \ 0 \le x \le 5. \end{cases}$$

(a) What is the domain of f?
(b) Evaluate $f(-2)$, $f(0)$, and $f(3)$.
(c) Graph f.   (d) Is f continuous on its domain?

2. Evaluate $f(-2.3)$ if $f(x) = [\![x - 2]\!]$.

3. Graph $y = |3x - 2|$ by hand.

4. Solve each equation or inequality. Write the solution set in interval notation for each inequality.
(a) $|2 - 5x| - 4 = -1$ (b) $|3x - 5| \le 4$

(c) $\left|\frac{1}{2}x - 3\right| > 5$

5. (a) Solve the equation $|2x - 1| = 5$.

(b) Use part (a) to solve the absolute value inequalities $|2x - 1| \le 5$ and $|2x - 1| > 5$.

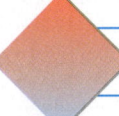

CHAPTER 2 SUMMARY

CONCEPT	EXPLANATION AND EXAMPLES
SECTION 2.1 LINEAR FUNCTIONS AND MODELS	
LINEAR FUNCTION	A linear function can be written as $f(x) = ax + b$. Its graph is a line. **Example:** $f(x) = -2x + 5$; slope $= -2$, y-intercept $= 5$, and x-intercept $= \frac{5}{2}$.
LINEAR MODEL	If a quantity increases or decreases by a constant amount for each unit increase in x, then it can be modeled by a linear function given by $$f(x) = (\text{constant rate of change})x + (\text{initial amount}).$$ **Example:** If water is being emptied from a 100-gallon tank at a rate 7 gallons per minute, then $A(t) = 100 - 7t$ calculates the gallons of water in the tank after t minutes.
LINEAR REGRESSION	One way to determine a linear function or a line that models data is to use the method of least squares. This method determines a unique line and can be found with a calculator. The correlation coefficient r, $(-1 \le r \le 1)$, measures how well a line fits the data. **Example:** The line of least squares modeling the data $(1, 1)$, $(3, 4)$, and $(4, 6)$ is given by $y \approx 1.643x - 0.714$, with $r \approx 0.997$.
SECTION 2.2 EQUATIONS OF LINES	
SLOPE	Given points (x_1, y_1) and (x_2, y_2), then $m = \frac{y_2 - y_1}{x_2 - x_1}$. The change in y is denoted $\Delta y = y_2 - y_1$, the change in x is denoted $\Delta x = x_2 - x_1$, and the slope is given by $m = \frac{\Delta y}{\Delta x}$. **Example:** If $(3, 0)$ and $(2, -4)$ lie on a line, then the slope of the line is $$m = \frac{-4 - 0}{2 - 3} = 4.$$

CONCEPT	EXPLANATION AND EXAMPLES

SECTION 2.2 EQUATIONS OF LINES (CONTINUED)

POINT-SLOPE FORM

If a line passes through (x_1, y_1) with slope m, then

$$y = m(x - x_1) + y_1 \quad \text{or} \quad y - y_1 = m(x - x_1).$$

Example: $y = -\frac{3}{4}(x + 4) + 5$ indicates slope $-\frac{3}{4}$ and the line passes through $(-4, 5)$.

SLOPE-INTERCEPT FORM

If a line has slope m and y-intercept b, then

$$y = mx + b.$$

Example: $y = 3x - 4$ indicates slope 3 and y-intercept -4.

DETERMINING INTERCEPTS

To find the x-intercept(s), let $y = 0$ in the equation and solve for x.
To find the y-intercept(s), let $x = 0$ in the equation and solve for y.

Examples: The x-intercept on the graph of $3x - 4y = 12$ is 4 because $3x - 4(0) = 12$ implies that $x = 4$.

The y-intercept on the graph of $3x - 4y = 12$ is -3 because $3(0) - 4(y) = 12$ implies that $y = -3$.

HORIZONTAL AND VERTICAL LINES

A horizontal line passing through the point (h, k) is given by $y = k$, and a vertical line passing through (h, k) is given by $x = h$.

Examples: The vertical line $x = 4$ passes through $(4, -2)$.

The horizontal line $y = -3$ passes through $(6, -3)$.

PARALLEL AND PERPENDICULAR LINES

Parallel lines have equal slopes satisfying $m_1 = m_2$, and perpendicular lines have slopes satisfying $m_1 m_2 = -1$, provided neither line is vertical.

Examples: The lines $y_1 = 3x - 1$ and $y_2 = 3x + 4$ are parallel.

The lines $y_1 = 3x - 1$ and $y_2 = -\frac{1}{3}x + 4$ are perpendicular.

DIRECT VARIATION

A quantity y is directly proportional to a quantity x, or y varies directly with x, If $y = kx$, where $k \neq 0$. If data vary directly, the ratios $\frac{y}{x}$ are equal to the constant of variation k. These data can be modeled by a line passing through the origin.

Example: If a person works for \$8 per hour, then the person's pay P is directly proportional to or varies directly with the number of hours H that the person works by the equation $P = 8H$, where the constant of variation is $k = 8$.

SECTION 2.3 LINEAR EQUATIONS

PROPERTIES OF EQUALITY

Addition: $a = b$ is equivalent to $a + c = b + c$.

Multiplication: $a = b$ is equivalent to $ac = bc$ provided $c \neq 0$.

Example:

$\frac{1}{2}x - 4 = 3$	Given equation
$\frac{1}{2}x = 7$	Addition property; add 4.
$x = 14$	Multiplication property; multiply by 2.

CONCEPT	EXPLANATION AND EXAMPLES

SECTION 2.3 LINEAR EQUATIONS (CONTINUED)

LINEAR EQUATION

Can be written as $ax + b = 0$ with $a \neq 0$ and has one solution

Example: The solution to $2x - 4 = 0$ is 2 because $2(2) - 4 = 0$.

INTERSECTION-OF-GRAPHS METHOD

Set y_1 equal to the left side of the equation and set y_2 equal to the right side. The x-coordinate of a point of intersection is a solution.

Example: The graphs of $y_1 = 2x$ and $y_2 = x + 1$ intersect at $(1, 2)$, so the solution to the linear equation $2x = x + 1$ is 1.

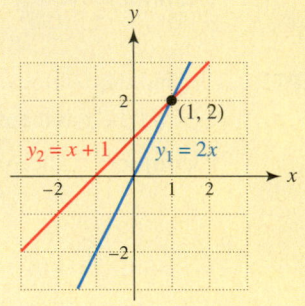

PROBLEM-SOLVING STRATEGIES

STEP 1: Read the problem and make sure you understand it. Assign a variable to what you are being asked to find. If necessary, write other quantities in terms of this variable.

STEP 2: Write an equation that relates the quantities described in the problem. You may need to sketch a diagram and refer to known formulas.

STEP 3: Solve the equation and determine the solution.

STEP 4: Look back and check your solution. Does it seem reasonable?

SECTION 2.4 LINEAR INEQUALITIES

INTERVAL NOTATION

A concise way to express intervals on the number line.

Example: $x < 4$ is expressed as $(-\infty, 4)$.
$-3 \leq x < 1$ is expressed as $[-3, 1)$.
$x \leq 2$ or $x \geq 5$ is expressed as $(-\infty, 2] \cup [5, \infty)$.

LINEAR INEQUALITY

Can be written as $ax + b > 0$ with $a \neq 0$, where $>$ can be replaced by $<$, $\leq$, or $\geq$. If the solution to $ax + b = 0$ is k, then the solution to the linear inequality $ax + b > 0$ is either the interval $(-\infty, k)$ or the interval (k, ∞).

Example: $3x - 1 < 2$ is linear since it can be written as $3x - 3 < 0$. The solution set is $(-\infty, 1)$.

CONCEPT	EXPLANATION AND EXAMPLES

SECTION 2.4 LINEAR INEQUALITIES (CONTINUED)

PROPERTIES OF INEQUALITY

Addition: $a < b$ is equivalent to $a + c < b + c$.

Multiplication: $a < b$ is equivalent to $ac < bc$ when $c > 0$.
$a < b$ is equivalent to $ac > bc$ when $c < 0$.

Example: $-3x - 4 < 14$ Given equation

$-3x < 18$ Addition property; add 4.

$x > -6$ Multiplication property; divide by -3. Reverse the inequality symbol.

COMPOUND INEQUALITY

Example: $x \geq -2$ and $x \leq 4$, or equivalently, $-2 \leq x \leq 4$. This is called a three-part inequality.

SECTION 2.5 PIECEWISE-DEFINED FUNCTIONS

PIECEWISE-DEFINED FUNCTIONS

A function defined by more than one formula on its domain

Example: Step function, greatest integer function, absolute value function, and

$$f(x) = \begin{cases} 4 - x & \text{if } -4 \leq x < 1 \\ 3x & \text{if } 1 \leq x \leq 5. \end{cases}$$

It follows that $f(2) = 6$ because if $1 \leq x \leq 5$, then $f(x) = 3x$. Note that f is continuous on its domain of $[-4, 5]$.

ABSOLUTE VALUE EQUATIONS

$|ax + b| = k$ with $k > 0$ is equivalent to $ax + b = \pm k$.

Example: $|2x - 3| = 4$ is equivalent to $2x - 3 = 4$ or $2x - 3 = -4$.
The solutions are 3.5 and -0.5.

ABSOLUTE VALUE INEQUALITIES

Let the solutions to $|ax + b| = k$ be s_1 and s_2, where $s_1 < s_2$ and $k > 0$.

1. $|ax + b| < k$ is equivalent to $s_1 < x < s_2$.
2. $|ax + b| > k$ is equivalent to $x < s_1$ or $x > s_2$.

Similar statements can be made for inequalities involving $\leq$ or $\geq$.

Example: The solutions to $|2x + 1| = 5$ are given by $x = -3$ and $x = 2$.
The solutions to $|2x + 1| < 5$ are given by $-3 < x < 2$.
The solutions to $|2x + 1| > 5$ are given by $x < -3$ or $x > 2$.

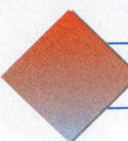

REVIEW EXERCISES

Exercises 1 and 2: The graph of a linear function f is shown.

   (a) *Identify the slope, y-intercept, and x-intercept.*

   (b) *Write a formula for f.*

   (c) *Find any zeros of f.*

1.
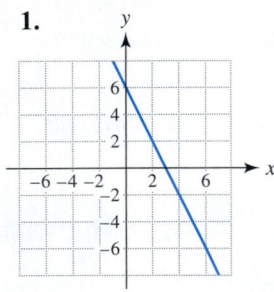

2.
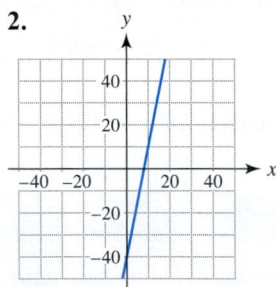

Exercises 3 and 4: Find $f(x) = ax + b$ so that f models the data exactly.

3.

x	1	2	3	4
$f(x)$	2.5	0	-2.5	-5

4.

x	-3	6	15	24	33
$f(x)$	-1.65	-1.2	-0.75	-0.3	0.15

Exercises 5–8: Graph the linear function by hand.

5. $f(x) = -\frac{2}{3}x$ **6.** $f(x) = 3x - 2$

7. $g(x) = 10x - 20$ **8.** $g(x) = 4 - 2x$

9. Write a formula for a linear function whose graph has slope -2 and passes through $(-2, 3)$.

10. Find the average rate of change of $f(x) = -3x + 8$ from -2 to 3.

Exercises 11–16: Find the slope-intercept form of the equation of a line satisfying the conditions.

11. Slope 7, passing through $(-3, 9)$

12. Passing through $(2, -4)$ and $(7, -3)$

13. Passing through $(1, -1)$, parallel to $y = -3x + 1$

14. Passing through the point $(-2, 1)$, perpendicular to the line $y = 2(x + 5) - 22$

15. Parallel to the line segment connecting $(0, 3.1)$ and $(5.7, 0)$, passing through $(1, -7)$

16. Perpendicular to $y = -\frac{5}{7}x$, passing through $\left(\frac{6}{7}, 0\right)$

Exercises 17–22: Find an equation of the line satisfying the conditions.

17. Parallel to the y-axis, passing through $(6, -7)$

18. Parallel to the x-axis, passing through $(-3, 4)$

19. Horizontal, passing through $(1, 3)$

20. Vertical, passing through $(1.5, 1.9)$

21. Horizontal with y-intercept -8

22. Vertical with x-intercept 2.7

Exercises 23–26: Determine the x- and y-intercepts on the graph of the equation. Graph the equation.

23. $5x - 4y = 20$ **24.** $-x + 2y = 3$

25. $y = -2(x + 1) + 4$ **26.** $\dfrac{x}{3} - \dfrac{y}{2} = 1$

Exercises 27–32: Solve the linear equation symbolically or graphically.

27. $5x - 22 = 10$ **28.** $5(4 - 2x) = 16$

29. $-2(3x - 7) + x = 2x - 1$

30. $\pi x + 1 = 6$

31. $5x - \frac{1}{2}(4 - 3x) = \frac{3}{2} - (2x + 3)$

32. $\dfrac{x - 4}{2} = x + \dfrac{1 - 2x}{3}$

Exercises 33 and 34: Use a table to solve each linear equation numerically to the nearest tenth.

33. $3.1x - 0.2 = 2(x - 1.7)$

34. $\sqrt{7} - 3x = 2.1(1 + x)$

35. Solve $x + 1 = 2x - 1$ graphically by hand. Check your answer.

36. Use tables to solve the equation $3x - 1 = -7$ numerically by hand.

Exercises 37–40: Complete the following.

  (a) *Solve the equation symbolically.*

  (b) *Classify the equation as a contradiction, an identity, or a conditional equation.*

37. $4(6 - x) = -4x + 24$

38. $\frac{1}{2}(4x - 3) + 2 = 3x - (1 + x)$

39. $5 - 2(4 - 3x) + x = 4(x - 3)$

40. $\dfrac{x - 3}{4} + \dfrac{3}{4}x - 5(2 - 7x) = 36x - \dfrac{43}{4}$

Exercises 41–46: Express the inequality in interval notation.

41. $x > -3$ **42.** $x \le 4$

43. $-2 \le x < \frac{3}{4}$ **44.** $x \le -2$ or $x > 3$

45. $x < 0$ or $x > 2$ **46.** $-4 < x \le 5$

Exercises 47–52: Solve the linear inequality. Write the solution set in interval notation.

47. $3x - 4 \le 2 + x$ **48.** $-2x + 6 \le -3x$

49. $\dfrac{2x - 5}{2} < \dfrac{5x + 1}{5}$

50. $-5(1 - x) > 3(x - 3) + \dfrac{1}{2}x$

51. $-2 \le 5 - 2x < 7$ **52.** $-1 < \dfrac{3x - 5}{-3} < 3$

Exercises 53–56: Solve the inequality graphically.

53. $2x > x - 1$ **54.** $2 - x \le x + 4$

55. $-1 \le 1 + x \le 2$ **56.** $\frac{1}{2}x - 3 > -1$

57. The graphs of two linear functions f and g are shown in the figure at the top of the next column. Solve each equation or inequality.

  (a) $f(x) = g(x)$

  (b) $f(x) < g(x)$

  (c) $f(x) > g(x)$

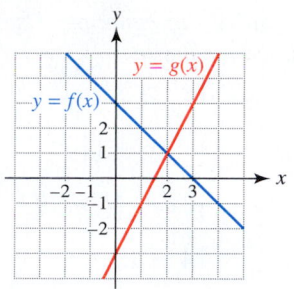

58. The graphs of three linear functions f, g, and h with domains $D = \{x \mid 0 \le x \le 7\}$ are shown in the figure. Solve each equation or inequality.

  (a) $f(x) = g(x)$ (b) $g(x) = h(x)$

  (c) $f(x) < g(x) < h(x)$ (d) $g(x) > h(x)$

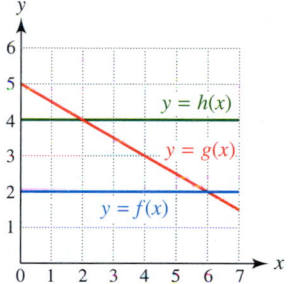

59. Use $f(x)$ to complete the following.

$$f(x) = \begin{cases} 8 + 2x & \text{if } -3 \le x \le -1 \\ 5 - x & \text{if } -1 < x \le 2 \\ x + 1 & \text{if } 2 < x \le 5 \end{cases}$$

  (a) Evaluate f at $x = -2, -1, 2$, and 3.

  (b) Sketch a graph of f. Is f continuous on its domain?

  (c) Determine the x-value(s) where $f(x) = 3$.

60. Graph f. Is f continuous on the interval $[-2, 3]$?

$$f(x) = \begin{cases} 2x - 1 & \text{if } -2 \le x < 0 \\ -1 & \text{if } 0 \le x < 1 \\ -x & \text{if } 1 \le x \le 3 \end{cases}$$

61. Graph $f(x) = [\![x]\!]$ without a graphing calculator on the interval $-4 \le x \le 4$.

62. If $f(x) = [\![2x - 1]\!]$, evaluate $f(-3.1)$ and $f(2.5)$.

Exercises 63–66: Solve the equation.

63. $|2x - 5| - 1 = 8$ **64.** $|3 - 7x| = 10$

65. $|6 - 4x| = -2$ **66.** $|9 + x| = |3 - 2x|$

Exercises 67–70: Solve the equation. Use the solutions to help solve the related inequality.

67. $|x| = 3$, $\qquad$ $|x| > 3$

68. $|-3x + 1| = 2$, $\quad$ $|-3x + 1| < 2$

69. $|3x - 7| = 10$, $\quad$ $|3x - 7| > 10$

70. $|4 - x| = 6$, $\qquad$ $|4 - x| \leq 6$

71. Solve $|x + 1| = |3x + 2|$.

72. Solve $\left|\frac{1}{2}x - 3\right| + 5 > 17$.

Exercises 73–76: Solve the inequality.

73. $|3 - 2x| < 9$ $\qquad$ **74.** $|-2x - 3| > 3$

75. $\left|\frac{1}{3}x - \frac{1}{6}\right| \geq 1$ $\qquad$ **76.** $\left|\frac{1}{2}x\right| - 3 \leq 5$

Applications

77. *Median Income* The median U.S. family income between 1980 and 2000 can be modeled by the formula $f(x) = 1550(x - 1980) + 21{,}000$, where x is the year. (**Source:** Department of Commerce.)

  (a) Solve the equation $f(x) = 36{,}500$ graphically. Interpret the solution.

  (b) Solve part (a) symbolically.

78. *Course Grades* In order to receive a B grade in a college course, it is necessary to have an overall average of 80% correct on two 1-hour exams of 75 points each and one final exam of 150 points. If a person scores 55 and 72 on the 1-hour exams, what is the minimum score on the final exam that the person can receive and still earn a B?

79. *Medicare Costs* Future estimates for Medicare costs in billions of dollars can be modeled by the formula $f(x) = 18x - 35{,}750$, where $1995 \leq x \leq 2007$. Determine when Medicare costs are expected to be from 268 to 358 billion dollars. (**Source:** Office of Management and Budget.)

80. *Temperature Scales* The table shows equivalent temperatures in degrees Celsius and degrees Fahrenheit.

°F	−40	32	59	95	212
°C	−40	0	15	35	100

  (a) Plot the data by having the x-axis correspond to Fahrenheit temperature and the y-axis to Celsius temperature. What type of relation exists between the data?

  (b) Find a function C that receives the Fahrenheit temperature x as input and outputs the corresponding Celsius temperature. Interpret the slope.

  (c) If the temperature is 83°F, what is this temperature in degrees Celsius?

81. *Distance from Home* The graph depicts the distance y that a person driving a car on a straight road is from home after x hours. Interpret the graph. What speeds did the car travel?

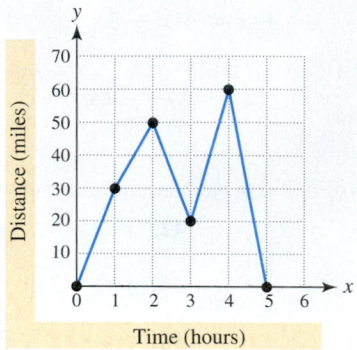

82. *Distance between Bicyclists* The graph shows the distance after x hours between two bicyclists traveling in opposite directions along a straight road.

  (a) At what time did the bicycle riders meet?

  (b) When were they 20 miles apart?

  (c) Find the times when they were less than 20 miles apart.

  (d) Estimate the sum of the speeds of the two bicyclists.

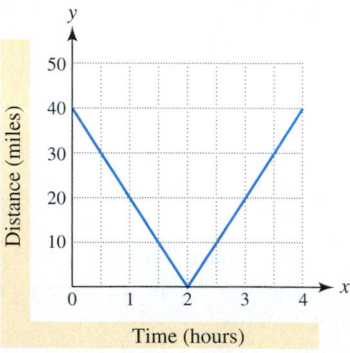

83. *Graphical Model* A 500-gallon water tank is initially full, then emptied at a constant rate of 50 gallons per minute. Then the tank is filled by a pump that outputs 25 gallons of water per minute. Sketch a graph that depicts the amount of water in the tank after x minutes.

84. *ACT Scores* The table lists the average composite ACT scores for selected years. (**Source:** The American College Testing Program.)

Year	1989	1990	1991	1992	1993	1994
Score	20.6	20.6	20.6	20.6	20.7	20.8

(a) Make a line graph of the data.

(b) Let this line graph represent a piecewise-linear function f. Find a symbolic representation for f.

(c) What is the domain of f?

85. *Population Estimates* In 2000 the population of a city was 143,247 and in 2004 it was 167,933. Estimate the population in 2002.

86. *Distance* A driver of a car is initially 455 miles from home, traveling toward home on a straight freeway at 70 miles per hour.
(a) Write a formula for a linear function f that models the distance between the driver and home after x hours.

(b) Graph f. What is an appropriate domain for f?

(c) Identify the x- and y-intercepts. Interpret each.

87. *Working Together* Suppose that one worker can shovel snow from a storefront sidewalk in 50 minutes and another worker can shovel it in 30 minutes. How long will it take if they work together?

88. *Antifreeze* Initially, a tank contains 20 gallons of a 30% antifreeze solution. How many gallons of an 80% antifreeze solution should be added to the tank in order to have the concentration of the antifreeze in the tank increase to 50%?

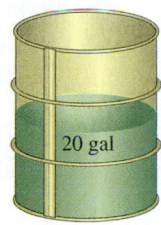

20 gal

89. *Running* An athlete travels 13.5 miles in 1 hour 48 minutes, jogging at 7 miles per hour and then at 8 miles per hour. How long did the runner jog at each speed?

90. *Geometry* A rectangle is twice as long as it is wide and has a perimeter of 78 inches. Find the width and length of this rectangle.

91. *Math SAT Scores* The table lists average math SAT scores for selected years.

Year	1994	1995	1996	1998	2000
Score	504	506	508	512	514

Source: The College Board.

(a) Find a linear function f that models these data. (Answers may vary.)

(b) Use f to approximate the average math SAT score in 1997. Did this estimate involve interpolation or extrapolation?

92. The table lists data that are exactly linear.
(a) Determine the slope-intercept form of the line that passes through these data points.

(b) Predict y when $x = -1.5$ and 3.5. State whether these calculations involve interpolation or extrapolation.

x	-3	-2	-1	0	1	2	3
y	6.6	5.4	4.2	3	1.8	0.6	-0.6

93. *Flow Rates* (Refer to Exercise 108 in Section 2.5.) Suppose the tank is modified so that it has a second inlet pipe, which flows at a rate of 2 gallons per minute. Interpret the graph by determining when each inlet and outlet pipe is open or closed.

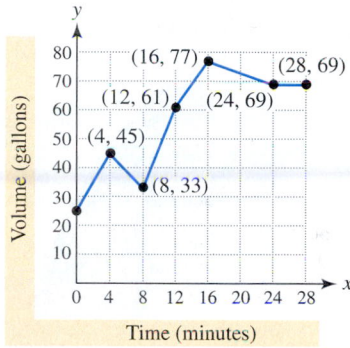

94. *Air Temperature* For altitudes up to 4 kilometers, moist air will cool at a rate of about 6°C per kilometer. If the ground temperature is 25°C, at what altitudes would the air temperature be from 5°C to 15°C? (**Source:** A. Miller and R. Anthes, *Meteorology.*)

95. *Water Pollution* At one time, the Thames River in England supported an abundant community of fish. By 1915, however, pollution destroyed all the fish in a 40-mile stretch near its mouth for a 45-year period. Since then, improved sewage treatment facilities and other ecological steps have resulted in a dramatic increase in the number of different fish present. The number of species present from 1967 to 1978 can be modeled by $f(x) = 6.15x - 12,059$, where x is the year.
(a) Find the year when the number of species first exceeded 70.

(b) The domain of f is $1967 \leq x \leq 1978$. Discuss whether f would continue to give good approximations beyond 1978. What would likely happen to the number of species of fish present after time passes and pollution disappears? (**Source:** C. Mason, *Biology of Freshwater Pollution.*)

96. *Ozone Layer* Ozone is the only gas capable of screening out lethal ultraviolet light. Without ozone, ultraviolet light would kill terrestrial life. Although the ozone layer can vary in thickness, it is typically only 3 millimeters thick. Worldwide, it has been estimated that for each 0.03-millimeter decrease in the thickness of the ozone layer, there would be an additional 100,000 cases of blindness annually, due to cataracts. (**Source:** R. K. Turner, *Environmental Economics.*)

(a) Find a linear function that computes the estimated number of additional cases of blindness each year from cataracts for a decrease of x millimeters in the ozone layer.

(b) Approximate the additional annual cases of blindness that may occur due to a decrease in the ozone layer of 0.05 millimeter.

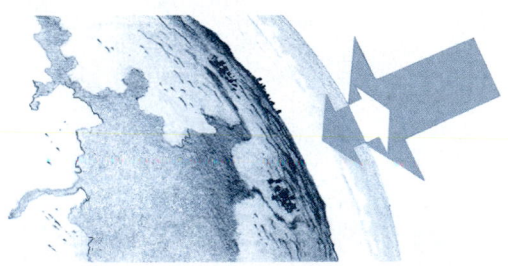

97. *Least-Squares Fit* The accompanying table lists the actual annual cost y to drive a midsize car 15,000 miles per year for selected years x.

x	1960	1970	1980	1990	2000
y	\$1394	\$1763	\$3176	\$5136	\$6880

Source: Runzheimer International.

(a) Predict whether the correlation coefficient is positive, negative, or zero.

(b) Find a least-squares regression line that models these data. What is the correlation coefficient?

(c) Estimate the annual cost of driving a midsize car in 1995.

98. *Relative Error* The actual length of a side of a building is 52.3 feet. How accurately must an appentice carpenter measure this side to have the relative error in the measurement less than 0.003 (0.3%)? (*Hint:* Use $\left| \frac{C - A}{A} \right|$, where C is the carpenter's measurement and A is the actual length.)

EXTENDED AND DISCOVERY EXERCISES

1. *Indoor Air Pollution* Today, indoor air pollution has become more hazardous as people spend 80% to 90% of their time in tightly sealed, energy-efficient buildings that often lack proper ventilation. Many contaminants such as tobacco smoke, formaldehyde, radon, lead, and carbon monoxide often are allowed to increase to unsafe levels. Mathematics plays a central role in risk assessment of pollutants. If x people are exposed to a contaminant, then the resulting number of cancer cases can sometimes be estimated by $f(x) = Cx$. In this formula, C is a constant that represents the excess cancer risk per year. For example, if $C = 0.01$, then an exposed individual has a 1% greater chance of developing cancer during a period of 1 year. (**Source:** A. Hines, *Indoor Air Quality & Control.*)

(a) Environmental tobacco smoke is a health hazard. The EPA has estimated that the annual excess cancer risk is $C = 0.00002$. The population of the United States was 281 million in 2000. Assuming that everyone is exposed to tobacco smoke to some degree, estimate the annual number of cancer cases that are caused by environmental tobacco smoke.

(b) Formaldehyde is a volatile organic compound that has come to be recognized as a highly toxic indoor air pollutant. It is found in building materials such as fiberboard, plywood, foam insulation, and carpeting. Some studies have estimated that $C = 0.0007$ for formaldehyde. In a population of 100,000 people exposed to formaldehyde, how many are likely to develop cancer during a year?

(c) Uranium occurs in small traces in most soils throughout the world. Radon gas occurs naturally in homes and is produced when uranium decays radioactively into lead. Radon enters buildings through basements, water supplies, and natural gas lines. Exposure to radon gas is a known factor in lung cancer risk. According to the EPA, radon causes between 5000 and 20,000 lung cancer cases each year in the United States. Approximate a range for the constant C, if the population of the United States was 250 million when the EPA study was performed.

(d) Discuss why it is reasonable that f is a linear function.

2. *Archeology* It is possible for archeologists to estimate the height of an adult based only on the length of the humerus, a bone located between the elbow and the shoulder. The approximate relationship between the height y of an individual and the length x of the humerus is shown in the table for both males and females. All measurements are recorded in inches. Tables like this are the result of measuring bones from many skeletons. Individuals may vary from these values.

x	8	9	10	11
y (females)	50.4	53.5	56.6	59.7
y (males)	53.0	56.0	59.0	62.0

x	12	13	14
y (females)	62.8	65.9	69.0
y (males)	65.0	68.0	71.0

   (a) Find the estimated height of a female with a 12-inch humerus.

   (b) Plot the ordered pairs (x, y) for both sexes. What type of relation exists between the data?

   (c) For each 1-inch increase in the length of the humerus, what are the corresponding increases in the heights of females and of males?

   (d) Determine linear functions f and g that model these data for females and males, respectively.

   (e) Suppose a humerus is estimated to be between 9.7 and 10.1 inches long. If the sex is not known, use f and g to approximate the range for the height of an individual of each sex.

3. Continuing with Exercise 2, have members of the class measure their heights and the lengths of their humeri (plural of *humerus*) in inches.

   (a) Make a table of the results.

   (b) Find regression lines that fit the data points for males and females.

   (c) Compare your results with the table in the previous exercise.

4. *Comparing Ages* The age of Earth is approximately 4.45 billion years. The earliest evidence of dinosaurs dates back 200 million years, whereas the earliest evidence for *Homo sapiens* dates back 300,000 years. If the age of Earth were condensed into one year, determine the approximate times when dinosaurs and *Homo sapiens* first appeared.

200 million years ago

300,000 years ago

5. *A Puzzle* Three people leave for a city 15 miles away. The first person walks 4 miles per hour and the other two people ride in a car that travels 28 miles per hour. After some time, the second person gets out of the car and walks 4 miles per hour to the city while the driver goes back and picks up the first person. The driver takes the first person to the city. If all three people arrive in the city at the same time, how far did each person walk?

CHAPTERS 1–2 CUMULATIVE REVIEW EXERCISES

1. Find the percent change if the price of gasoline goes from $1.89 to $1.69. Round your answer to the nearest tenth of a percent.

2. Write 123,000 and 0.0051 in scientific notation.

3. Write 6.7×10^6 and 1.45×10^{-4} in standard form.

4. Evaluate $\dfrac{4 + \sqrt{2}}{4 - \sqrt{2}}$. Round you answer to the nearest hundredth.

5. The table represents a relation S.

x	−1	0	1	2	3
y	6	4	3	0	0

   (a) Does S represent a function?

   (b) Determine the domain and range of S.

6. Graph the relation

$$S = \{(-2, 3), (-2, -1), (0, 2), (2, 3)\}.$$

7. Find the exact distance between $(-3, 5)$ and $(2, -3)$.

8. Find the midpoint of the line segment with endpoints $(5, -2)$ and $(-3, 1)$.

9. Find the domain and range of the function shown in the graph. Evaluate $f(-1)$.

   (a)

   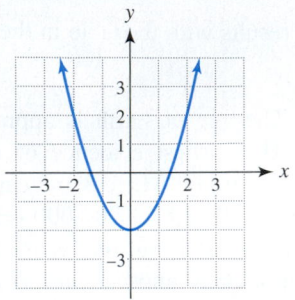

   (b)

   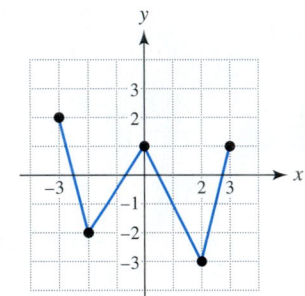

10. Graph f by hand.

    (a) $f(x) = 3 - 2x$ **(b)** $f(x) = |x + 1|$

    (c) $f(x) = x^2 - 3$ **(d)** $f(x) = \sqrt{x + 2}$

Exercises 11–14: Complete the following.

   (a) Evaluate $f(2)$ and $f(a - 1)$.
   (b) Determine the domain of f.

11. $f(x) = 5x - 3$ 12. $f(x) = x^2 - 1$

13. $f(x) = \sqrt{2x - 1}$ 14. $f(x) = \dfrac{1}{3x - 2}$

15. Determine if the graph represents a function. Explain your answer.

   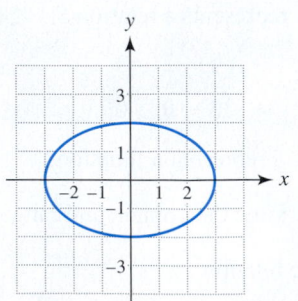

16. Write formula for a function f that computes the cost of taking x credits if credits cost $80 each and fees are fixed at $89.

17. Find the average rate of change of $f(x) = x^2 - 2x + 1$ from $x = 1$ to $x = 2$.

18. Find the difference quotient for $f(x) = 2x^2 - x$.

Exercises 19 and 20: The graph of a linear function f is shown.

   (a) Identify the slope, y-intercept, and x-intercept.
   (b) Write a formula for f.
   (c) Find any zeros of f.

19.   20.

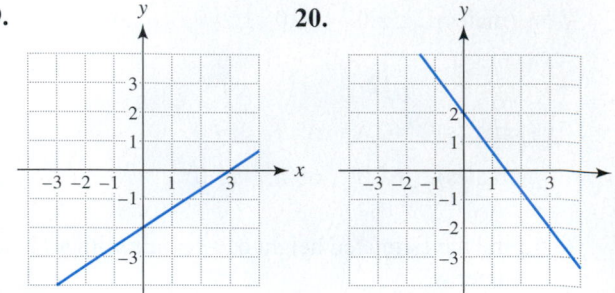

21. Write a formula for a linear function whose graph has slope -3 and passes through $\left(\frac{2}{3}, -\frac{2}{3}\right)$.

22. Write the formula for a constant function whose graph passes through $(-4, 60)$.

23. Find the slope of the line passing through $(2000, 52)$ and $(2004, 26)$.

24. If $G(t) = 200 - 10t$ models the gallons of water in a tank after t minutes, interpret the numbers 200 and -10 in the formula for G.

Exercises 25–30: Write an equation of a line satisfying the given conditions. Use slope-intercept form whenever possible.

25. Passing through $(1, -5)$ and $\left(-3, \frac{1}{2}\right)$

26. Passing through the point $(-3, 2)$ and perpendicular to the line $y = \frac{2}{3}x - 7$

27. Parallel to the y-axis and passing though $(-1, 3)$

28. Slope 30, passing though $(2002, 50)$

29. Passing through $(-3, 5)$ and parallel to the line segment connecting $(2.4, 5.6)$ and $(3.9, 8.6)$

30. Perpendicular to the y-axis and passing through the origin

Exercises 31 and 32: Determine the x- and y-intercepts on the graph of the equation. Graph the equation.

31. $-2x + 3y = 6$ **32.** $x = 2y - 3$

Exercises 33–36: Solve the equation.

33. $4x - 5 = 1 - 2x$ **34.** $\dfrac{2x - 4}{2} = \dfrac{3x}{7} - 1$

35. $\frac{2}{3}(x - 2) - \frac{4}{5}x = \frac{4}{15} + x$

36. $-0.3(1 - x) - 0.1(2x - 3) = 0.4$

37. Solve $x + 1 = 2x - 2$ graphically and numerically.

38. Solve $2x - (5 - x) = \dfrac{1 - 4x}{2} + 5(x - 2)$. Is this equation either an identity or a contradiction?

Exercises 39–42: Express each inequality in interval notation.

39. $x < 5$ **40.** $-2 \le x \le 5$

41. $x < -2$ or $x > 2$ **42.** $x \ge -3$

Exercises 43–46: Solve the inequality. Write the solution set in interval notation.

43. $-5x + 3 < 3 - x$ **44.** $\dfrac{x - 2}{5} \ge \dfrac{2x}{7} - \dfrac{12}{35}$

45. $-3(1 - 2x) + x \le 4 - (x + 2)$

46. $\dfrac{1}{3} \le \dfrac{2 - 3x}{2} < \dfrac{4}{3}$

47. The graphs of two linear functions f and g are shown. Solve each equation or inequality.
   (a) $f(x) = g(x)$ **(b)** $f(x) > g(x)$

   (c) $f(x) \le g(x)$

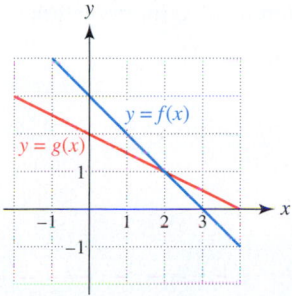

48. Graph f. Is f continuous on the interval $[-4, 4]$?

$$f(x) = \begin{cases} 2 - x & \text{if } -4 \le x < -2 \\ \frac{1}{2}x + 5 & \text{if } -2 \le x < 2 \\ 2x + 1 & \text{if } 2 \le x \le 4 \end{cases}$$

Exercises 49–52: Solve the equation.

49. $|d + 1| = 5$ **50.** $|3 - 2x| = 7$

51. $|2t| - 4 = 10$ **52.** $|11 - 2x| = |3x + 1|$

Exercises 53–56: Solve the inequality. Write the solution set in interval notation.

53. $|2t - 5| \le 5$ **54.** $|5 - 5t| > 7$

55. $|4 - \frac{1}{2}k| \ge 10$ **56.** $|11 - 2x| < 7$

Applications

57. *Volume of a Cylinder* The volume V of a cylinder is given by $V = \pi r^2 h$, where r is the radius of the cylinder and h is its height. If an aluminum can has a volume of 24 cubic inches and a radius of 1.5 inches, find its height to the nearest hundredth of an inch.

58. *Interpreting Slope* The figure shows the weight of a load of gravel in a dump truck. Interpret the slope of each line segment.

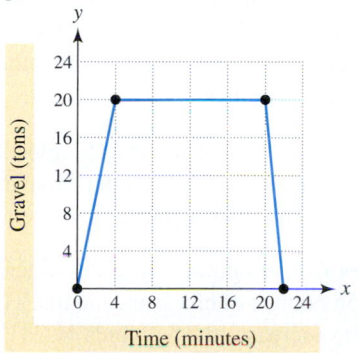

59. *Thickness of an Oil Film* Two cubic inches of oil are spilled into a lake. The resulting oil film spreads into a circular shape that has a diameter of 40 inches. Estimate the thickness of this oil film. (*Hint:* Thickness equals volume divided by area.)

60. *Cost*   A company's cost C in dollars for making x computers is $C(x) = 500x + 20{,}000$.
(a) Evaluate $C(1500)$. Interpret the result.

(b) Find the slope of the graph of C. Interpret the slope.

61. *Distance*   At midnight car A is traveling north at 60 miles per hour and is located 40 miles south of car B. Car B is traveling west at 70 miles per hour. Approximate the distance between the cars at 1:15 A.M. to the nearest tenth of a mile.

62. *Average Rate of Change*   On a warm summer day the Fahrenheit temperature x hours past noon is given by $T(x) = 70 + \frac{3}{2}x^2$.
(a) Find the average rate of change of T from 2:00 P.M. to 4:00 P.M.

(b) Interpret this average rate of change.

63. *Distance from Home*   A driver is initially 270 miles from home, traveling toward home on a straight interstate at 72 miles per hour.
(a) Write a formula for a function D that models the distance between the driver and home after x hours.

(b) What is an appropriate domain for D? Graph D.

(c) Identify the x- and y-intercepts. Interpret each.

64. *Working Together*   Suppose one person can mow a large lawn in 5 hours with a riding mower and it takes another person 12 hours to mow the lawn with a push mower. How long will it take to mow the lawn if the two people work together?

65. *Running*   An athlete travels 15 miles in 1 hour and 45 minutes, running at 8 miles per hour and 10 miles per hour. How long did the athlete run at each speed?

66. *Chicken Consumption*   In 2001 Americans ate, on average, 56 pounds of chicken annually. This amount is expected to increase to 61 pounds in 2012.   (**Source:** Department of Agriculture.)
(a) Determine a formula
$$f(x) = m(x - x_1) + y_1$$
that models these data. Let x be the year.

(b) Estimate the annual chicken consumption in 2007.

67. *Basic Cable*   In 2001 the average basic cable subscription cost $9.30 per month and in 2006 it is expected to rise to $14.25.   (**Source:** Morgan Stanley.)
(a) Determine a formula
$$f(x) = m(x - x_1) + y_1$$
that models these data. Let x be the year.

(b) Estimate the year when the average basic cable subscription was $12.25.

68. *Relative Error*   If the actual value of a quantity is A and its measured value is M, then the relative error in measurement M is $\left| \frac{M - A}{A} \right|$. If $A = 65$, determine the range of values for M to have a relative error that is less than or equal to 0.03 (3%).

69. *Modeling Data*   According to government guidelines, the recommended minimum weight for a person 58 inches tall is 91 pounds, and for a person 64 inches tall it is 111 pounds.
(a) Find an equation of the line that passes through the points $(58, 91)$ and $(64, 111)$.

(b) Use this line to estimate the minimum weight for someone 61 inches tall. Then use the midpoint formula to find this minimum weight. Are the results the same? Why?

70. The following table lists per capita income.

Year	1970	1980	1990	2000
Income	$4095	$10,183	$19,572	$29,760

Source: Bureau of Economic Analysis.

(a) Find the least-squares regression line for the data.

(b) Estimate the per capita income in 1995. Did this calculation involve interpolation or extrapolation?

Quadratic Functions and Equations

The work of mathematicians from even a millennium ago is still routinely used today, and the work of mathematicians from today will be part of the math of the future.

Terry Tao

The last basketball player in the NBA to shoot foul shots underhand was Rick Barry, who retired in 1980. On average, he was able to make about 9 out of 10 shots. Since then, every NBA player has used the overhand style of shooting foul shots—even though this style has often resulted in lower free-throw percentages.

According to Peter Brancazio, a physics professor emeritus from Brooklyn College and author of *Sports Science,* there are good reasons for shooting underhand. An underhand shot obtains a higher arc, and as the ball approaches the hoop, it has a better chance of going through the hoop than does a ball with a flatter arc. If a basketball is tossed at an angle of 32 degrees or less, it will likely hit the back of the rim and bounce

Rick Barry

out. The optimal angle is greater than 45 degrees and depends on the height at which the ball is released. Lower release points require steeper arcs and increase the chances of the ball passing through the hoop. (See the Extended and Discovery Exercise at the end of this chapter to model the arc of a basketball.)

Mathematics plays an important role in analyzing applied problems such as shooting foul shots. Whether NBA players choose to agree with Professor Brancazio is another question, but mathematics tells us that steeper arcs are necessary for accurate foul shooting.

Source: Curtis Rist. "The Physics of Foul Shots." *Discover,* October 2000. (Photograph reprinted with permission.)

3.1 Quadratic Functions and Models

- ◆ Learn basic concepts about quadratic functions and their graphs
- ◆ Complete the square and apply the vertex formula
- ◆ Graph a quadratic function by hand
- ◆ Solve applications and model data
- ◆ Use quadratic regression to model data (optional)

Introduction

Sometimes when data lie on a line or nearly lie on a line, they can be modeled with a linear function and are called linear data. Data that are not linear are called **nonlinear data** and require a nonlinear function to model them. There are many types of nonlinear data, and mathematicians can choose from a wide variety of nonlinear functions to create models. One of the simplest types of nonlinear functions is a quadratic function. Figures 3.1–3.3 illustrate three sets of nonlinear data that can be modeled by a quadratic function.

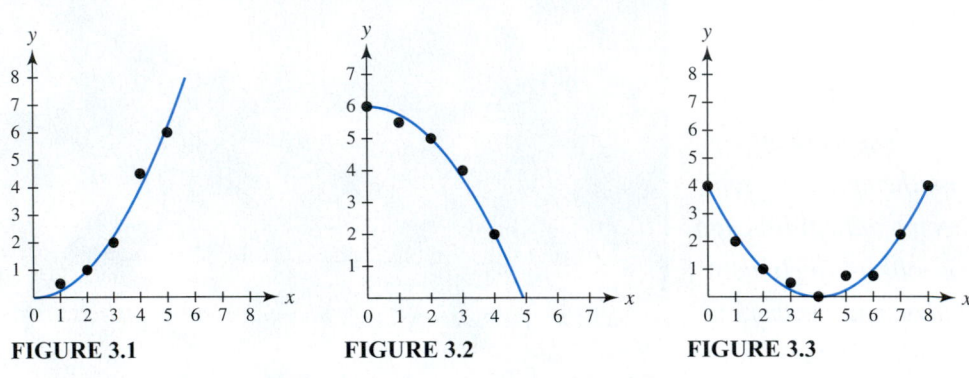

FIGURE 3.1   FIGURE 3.2   FIGURE 3.3

Basic Concepts

Recall that a linear function can be written as $f(x) = ax + b$ (or $f(x) = mx + b$). The formula for a quadratic function is different from that of a linear function because it contains an x^2-term. Examples of quadratic functions include

$$f(x) = 2x^2 - 4x - 1, \qquad g(x) = 4 - x^2, \qquad \text{and} \qquad h(x) = \frac{1}{3}x^2 + \frac{2}{3}x + 1.$$

The following defines a general form for a quadratic function.

QUADRATIC FUNCTION

Let a, b, and c be real numbers with $a \neq 0$. A function represented by

$$f(x) = ax^2 + bx + c$$

is a **quadratic function**.

The domain of a quadratic function includes all real numbers. The leading coefficient of a quadratic function is a. For the previous three functions f, g, and h, the leading coefficients are $a = 2$ for $f(x) = 2x^2 - 4x - 1$, $a = -1$ for $g(x) = 4 - 1x^2$, and $a = \frac{1}{3}$ for $h(x) = \frac{1}{3}x^2 + \frac{2}{3}x + 1$.

The graph of a quadratic function is a **parabola**—a U-shaped graph that opens either upward or downward. Graphs of the three quadratic functions f, g, and h are shown in Figures 3.4–3.6, respectively. A parabola opens upward if a is positive and opens downward if a is negative. For example, since $a = -1$ for $g(x) = 4 - x^2$, the graph of g opens downward. The highest point on a parabola that opens downward or the lowest point on a parabola that opens upward is called the **vertex**. The vertical line passing through the vertex is called the **axis of symmetry**. The vertex and axis of symmetry are shown in each figure.

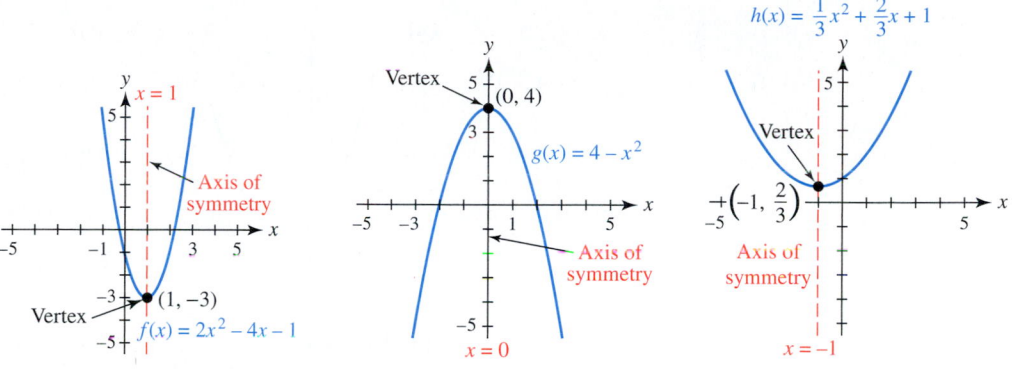

FIGURE 3.4   FIGURE 3.5   FIGURE 3.6

The leading coefficient a of a quadratic function not only determines whether its graph opens upward or downward, but it also controls the width of the parabola. Larger values of $|a|$ result in a narrower parabola, and smaller values of $|a|$ result in a wider parabola. This concept is illustrated in Figures 3.7 and 3.8.

◆ **CLASS DISCUSSION**
What does the graph of a quadratic function resemble if its leading coefficient is nearly 0? ◆

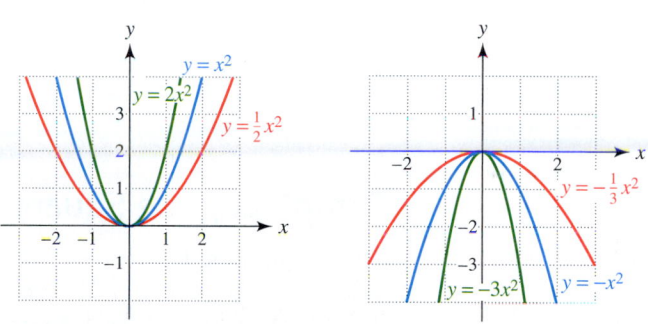

FIGURE 3.7 $y = ax^2, a > 0$ **FIGURE 3.8** $y = ax^2, a < 0$

EXAMPLE 1   Identifying quadratic functions

Identify the function as linear, quadratic, or neither. If it is quadratic, identify the leading coefficient and evaluate the function at $x = 2$.

(a) $f(x) = 3 - 2^2x$ **(b)** $g(x) = 5 + x - 3x^2$ **(c)** $h(x) = \dfrac{3}{x^2 + 1}$

SOLUTION
(a) Because $f(x) = 3 - 2^2x$ can be written as $f(x) = -4x + 3$, f is linear.
(b) Function g can be written as $g(x) = -3x^2 + x + 5$, so g is a quadratic function with $a = -3, b = 1$, and $c = 5$. The leading coefficient is $a = -3$ and

$$g(2) = 5 + 2 - 3(2)^2 = -5.$$

(c) $h(x) = \dfrac{3}{x^2 + 1}$ cannot be written as $h(x) = ax + b$ or $h(x) = ax^2 + bx + c$, so h is neither a linear nor a quadratic function.   *Now Try Exercise 1* ◆

EXAMPLE 2 Analyzing graphs of quadratic functions

Use the graph of each quadratic function to determine the sign of the leading coefficient, its vertex, and the equation of the axis of symmetry.

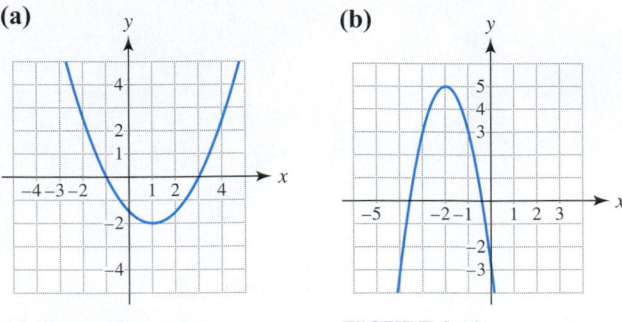

FIGURE 3.9   FIGURE 3.10

SOLUTION
(a) The graph in Figure 3.9 opens upward, so the leading coefficient a is positive. The vertex is the lowest point on the graph and is located at $(1, -2)$. The axis of symmetry is a vertical line passing through the vertex with equation $x = 1$.
(b) The graph in Figure 3.10 opens downward, so the leading coefficient a is negative. The vertex is the highest point on the graph and is located at $(-2, 5)$. The axis of symmetry is a vertical line through the vertex with equation $x = -2$.

Now Try Exercise 9 ◆

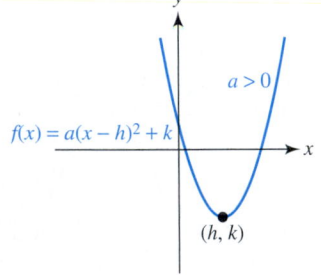

FIGURE 3.11 A Parabola with Vertex (h, k)

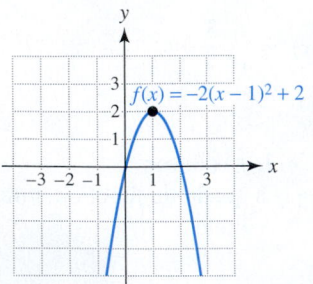

FIGURE 3.12 A Parabola with Vertex $(1, 2)$

Completing the Square and the Vertex Formula

When a quadratic function is expressed as $f(x) = ax^2 + bx + c$, the coordinates of the vertex are not apparent. However, if a quadratic function is written in the form $f(x) = a(x - h)^2 + k$, then the vertex is located at (h, k), as illustrated in Figure 3.11. For example, the graph of $f(x) = -2(x - 1)^2 + 2$ is a parabola opening downward with vertex $(1, 2)$. See Figure 3.12.

To justify that the vertex is indeed (h, k), consider the following. If $a > 0$ in the formula $f(x) = a(x - h)^2 + k$, then the term $a(x - h)^2$ is never negative and the minimum value of $f(x)$ is k. This value occurs when $x = h$ because

$$f(h) = a(h - h)^2 + k = 0 + k = k.$$

Thus the lowest point on the graph of $f(x) = a(x - h)^2 + k$ with $a > 0$ is (h, k), and because this graph is a parabola that opens upward, the vertex must be (h, k). Similarly, if $a < 0$ in the formula $f(x) = a(x - h)^2 + k$, then the term $a(x - h)^2$ is never positive and the maximum value of $f(x)$ is k. Thus the highest point on the graph of $f(x) = a(x - h)^2 + k$ with $a < 0$ is (h, k), and because this parabola opens downward, the vertex must be (h, k).

The formula $f(x) = a(x - h)^2 + k$ is sometimes called the **standard form for a parabola with a vertical axis**. Because the vertex is apparent in this formula, we will simply call it **vertex form**.

> ### VERTEX FORM
>
> The parabolic graph of $f(x) = a(x - h)^2 + k$ with $a \neq 0$ has vertex (h, k). Its graph opens upward when $a > 0$ and opens downward when $a < 0$.

EXAMPLE 3 Converting to $f(x) = ax^2 + bx + c$

Write $f(x) = 2(x - 1)^2 + 4$ in the form $f(x) = ax^2 + bx + c$.

SOLUTION Begin by expanding the expression $(x - 1)^2$.

Algebra Review
To review squaring a binomial, see Chapter R (page R-25).

$$
\begin{aligned}
2(x - 1)^2 + 4 &= 2(x^2 - 2x + 1) + 4 \qquad \text{Square binomial.} \\
&= 2x^2 - 4x + 2 + 4 \qquad \text{Distributive property} \\
&= 2x^2 - 4x + 6 \qquad \text{Add terms.}
\end{aligned}
$$

Thus $f(x) = 2(x - 1)^2 + 4$ is equivalent to $f(x) = 2x^2 - 4x + 6$.

Now Try Exercise 17 ◆

We can convert the general form $f(x) = ax^2 + bx + c$ to vertex form by **completing the square**. If a quadratic expression can be written as $x^2 + kx + \left(\frac{k}{2}\right)^2$, then it is a perfect square trinomial and can be factored as

$$
x^2 + kx + \left(\frac{k}{2}\right)^2 = \left(x + \frac{k}{2}\right)^2.
$$

This technique of converting to vertex form by completing the square is illustrated in the next example.

EXAMPLE 4 Converting to $f(x) = a(x - h)^2 + k$

Write each formula in vertex form by completing the square.

(a) $f(x) = x^2 + 6x - 3$ **(b)** $f(x) = \dfrac{1}{3}x^2 - x + 2$

SOLUTION
(a) Start by letting $y = f(x)$.

Algebra Review
To review perfect square trinomials, see Chapter R (page R-34).

$$
\begin{aligned}
y &= x^2 + 6x - 3 \qquad && \text{Given formula} \\
y + 3 &= x^2 + 6x \qquad && \text{Add 3 to each side.} \\
y + 3 + 9 &= x^2 + 6x + 9 \qquad && \text{Let } k = 6; \text{ add } \left(\frac{k}{2}\right)^2 = \left(\frac{6}{2}\right)^2 = 9. \\
y + 12 &= (x + 3)^2 \qquad && \text{Factor perfect square trinomial.} \\
y &= (x + 3)^2 - 12 \qquad && \text{Subtract 12.}
\end{aligned}
$$

The required form is $f(x) = (x + 3)^2 - 12$.

(b) Start by letting $y = f(x)$.

$$y = \frac{1}{3}x^2 - x + 2 \qquad \text{Given formula}$$

$$3y = x^2 - 3x + 6 \qquad \text{Make leading coefficient 1.}$$

$$3y - 6 = x^2 - 3x \qquad \text{Subtract 6 from each side.}$$

$$3y - 6 + \frac{9}{4} = x^2 - 3x + \frac{9}{4} \qquad \text{Let } k = -3; \text{ add } \left(\frac{k}{2}\right)^2 = \left(\frac{-3}{2}\right)^2 = \frac{9}{4}.$$

$$3y - \frac{15}{4} = \left(x - \frac{3}{2}\right)^2 \qquad \text{Factor perfect square trinomial.}$$

$$3y = \left(x - \frac{3}{2}\right)^2 + \frac{15}{4} \qquad \text{Add } \frac{15}{4} \text{ to each side.}$$

$$y = \frac{1}{3}\left(x - \frac{3}{2}\right)^2 + \frac{5}{4} \qquad \text{Multiply each side by } \frac{1}{3}.$$

The required form is $f(x) = \frac{1}{3}\left(x - \frac{3}{2}\right)^2 + \frac{5}{4}$. 　　　　　*Now Try Exercises 39 and 43* ◆

Derivation of the Vertex Formula The above procedure of completing the square can be done in general to derive a formula to determine the vertex of any parabola.

$$y = ax^2 + bx + c \qquad \text{General equation for a parabola}$$

$$\frac{y}{a} = x^2 + \frac{b}{a}x + \frac{c}{a} \qquad \text{Divide each side by } a \text{ to make leading coefficient 1.}$$

$$\frac{y}{a} - \frac{c}{a} = x^2 + \frac{b}{a}x \qquad \text{Subtract } \frac{c}{a} \text{ from each side.}$$

$$\frac{y}{a} - \frac{c}{a} + \frac{b^2}{4a^2} = x^2 + \frac{b}{a}x + \frac{b^2}{4a^2} \qquad \text{Let } k = \frac{b}{a}; \text{ add } \left(\frac{k}{2}\right)^2 = \left(\frac{b/a}{2}\right)^2 = \frac{b^2}{4a^2}.$$

$$\frac{y}{a} + \frac{b^2 - 4ac}{4a^2} = \left(x + \frac{b}{2a}\right)^2 \qquad \text{Combine left terms; factor perfect square trinomial.}$$

$$\frac{y}{a} = \left(x + \frac{b}{2a}\right)^2 - \frac{b^2 - 4ac}{4a^2} \qquad \text{Isolate } y\text{-term on the left side.}$$

$$y = a\left(x + \frac{b}{2a}\right)^2 - \frac{b^2 - 4ac}{4a} \qquad \text{Multiply by } a.$$

$$y = a\left(x - \left(-\frac{b}{2a}\right)\right)^2 + \frac{4ac - b^2}{4a} \qquad \text{Write } y = a(x - h)^2 + k.$$

$$\underbrace{\qquad}_{h} \qquad \underbrace{\qquad}_{k}$$

Because the coordinates of the vertex are (h, k), the x-coordinate is $-\frac{b}{2a}$. Note that it is not necessary to memorize the expression for k, because the y-coordinate of the vertex can be found by evaluating $y = f(x)$ for $x = -\frac{b}{2a}$. This derivation of the *vertex formula* is now summarized.

VERTEX FORMULA

The *vertex* of the graph of $f(x) = ax^2 + bx + c$ with $a \neq 0$ is the point $\left(-\frac{b}{2a}, f\left(-\frac{b}{2a}\right)\right)$.

 **5** | Using the vertex formula

Find the vertex of the graph of $f(x) = 1.5x^2 - 6x + 4$ symbolically. Support your answer graphically and numerically.

SOLUTION

Symbolic If $f(x) = \mathbf{1.5}x^2 - \mathbf{6}x + 4$, then $a = \mathbf{1.5}$, $b = \mathbf{-6}$, and $c = 4$. The x-coordinate of the vertex is

$$x = -\frac{b}{2a} = -\frac{(\mathbf{-6})}{2(\mathbf{1.5})} = \mathbf{2}.$$

The y-coordinate of the vertex can be found by evaluating $f(2)$.

$$y = f(\mathbf{2}) = 1.5(\mathbf{2})^2 - 6(\mathbf{2}) + 4 = \mathbf{-2}$$

Thus the vertex is $(\mathbf{2}, \mathbf{-2})$.

Graphical This result can be supported by graphing $Y_1 = 1.5X^2 - 6X + 4$, as shown in Figure 3.13. The vertex is the lowest point on the graph.

Calculator Help

To find a minimum point on a graph, see Appendix B (page AP-14).

$[-4.7, 4.7, 1]$ by $[-3.1, 3.1, 1]$

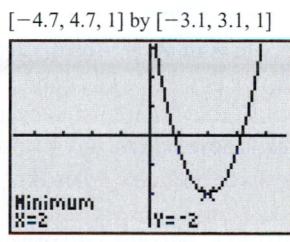

FIGURE 3.13

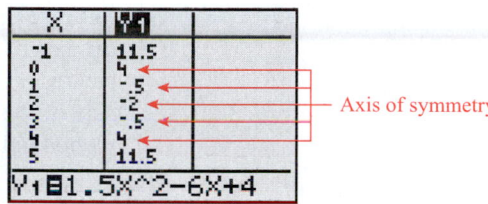

Axis of symmetry

FIGURE 3.14

Numerical Numerical support is shown in Figure 3.14. The minimum y-value of -2 occurs when $x = 2$. Notice that if we move an equal distance left or right of the axis of symmetry, the y-values are equal. For example, when $x = 1$ or $x = 3$, $y_1 = -0.5$. Similarly when $x = 0$ or 4, $y_1 = 4$.

Now Try Exercise 25 ◆

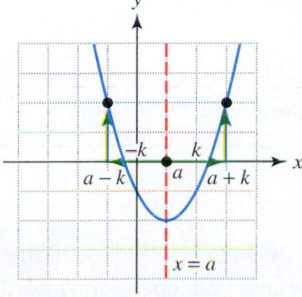

FIGURE 3.15

◆ **MAKING CONNECTIONS**

Graphing and the Axis of Symmetry If the graph of a quadratic function f is folded along the axis of symmetry, the two halves of the parabola match exactly. If the axis of symmetry is $x = a$, then $f(a + k) = f(a - k)$ for every value of k. This is illustrated in Figure 3.15. For example, if the axis of symmetry is $x = 1$, then $f(-1) = f(3)$ because $x = -1$ is two units to the left of the axis of symmetry and $x = 3$ is two units to the right of the axis of symmetry. ◆

The vertex formula can be used to write a quadratic function in vertex form.

EXAMPLE 6   Converting to $f(x) = a(x - h)^2 + k$

Use the vertex formula to write $f(x) = 3x^2 + 12x + 7$ in vertex form.

SOLUTION   Begin by finding the vertex.

$$x = -\frac{b}{2a} \qquad \text{Vertex formula}$$

$$= -\frac{12}{2(3)} \qquad \text{Let } a = 3 \text{ and } b = 12.$$

$$= -2 \qquad \text{Simplify.}$$

Since $f(-2) = 3(-2)^2 + 12(-2) + 7 = -5$, the vertex is $(-2, -5)$. The leading coefficient is $a = 3$, so $f(x)$ can be written as $f(x) = 3(x + 2)^2 - 5$.

Now Try Exercise 47 ◆

The graph of a quadratic function is a parabola. When sketching a parabola, it is important to determine the vertex, the axis of symmetry, and whether the parabola opens upward or downward. In the next example we sketch the graphs of two quadratic functions by hand.

EXAMPLE 7   Graphing quadratic functions by hand

Graph each quadratic function.

(a) $g(x) = 2(x - 1)^2 - 3$ **(b)** $h(x) = -\frac{1}{2}x^2 - x + 2$

SOLUTION

(a) Because the formula is in vertex form, it is apparent that the vertex is $(1, -3)$ and that the axis of symmetry is $x = 1$. The parabola opens upward because the leading coefficient is positive. In Table 3.1 we list the vertex and a few other points located on either side of the vertex. These points and a smooth U-shaped curve that opens upward are plotted in Figure 3.16. Note that when $x = 0$, $y = -1$, and so the y-intercept is -1.

TABLE 3.1

x	y
-1	5
0	-1
1	-3	← Vertex
2	-1
3	5

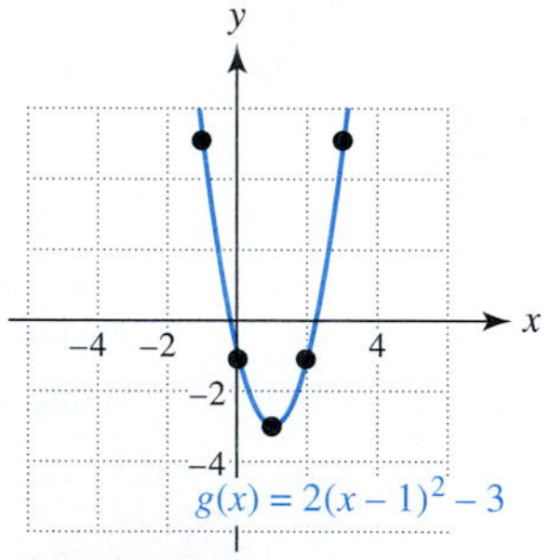

FIGURE 3.16

(b) The given formula is not in vertex form, but we can find the vertex.

$$x = -\frac{b}{2a} = -\frac{-1}{2\left(-\frac{1}{2}\right)} = -1$$

The y-coordinate of the vertex is $h(-1) = -\frac{1}{2}(-1)^2 - (-1) + 2 = \frac{5}{2}$. Thus the vertex is $\left(-1, \frac{5}{2}\right)$, the axis of symmetry is $x = -1$, and the parabola opens downward because the leading coefficient is negative. In Table 3.2 we list the vertex and a few other points located on either side of the vertex. These points and a smooth ∩-shaped curve that opens downward are plotted in Figure 3.17. Note that when $x = 0, y = 2$, and so the y-intercept is 2.

TABLE 3.2

x	y
-4	-2
-3	$\frac{1}{2}$
-2	2
-1	$\frac{5}{2}$	← Vertex
0	2
1	$\frac{1}{2}$
2	-2

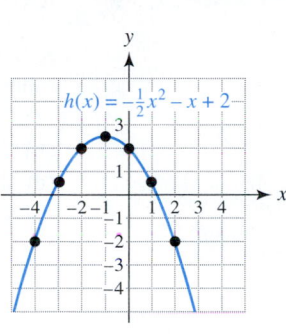

FIGURE 3.17

Now Try Exercises 65 and 75 ◆

Applications and Models

Sometimes when a quadratic function f is used in applications, the vertex provides important information. The reason is that the y-coordinate of the vertex is the minimum value of $f(x)$ when its graph opens upward and is the maximum value of $f(x)$ when its graph opens downward. This concept is applied in the next example.

EXAMPLE 8 Maximizing area

A farmer is fencing a rectangular area for cattle using the straight portion of a river as one side of the rectangle, as illustrated in Figure 3.18. If the farmer has 2400 feet of fence, find the dimensions of the rectangle that give the maximum area for the cattle.

FIGURE 3.18

SOLUTION Let W be the width and L be the length of the rectangle. Because the 2400-foot fence does not go along the river, it follows that

$$W + L + W = 2400 \quad \text{or} \quad L = 2400 - 2W.$$

Area A of a rectangle equals length times width, so

$$A = LW$$
$$= (2400 - 2W)W$$
$$= 2400W - 2W^2.$$

Thus the graph of $A = -2W^2 + 2400W$ is a parabola opening downward, and by the vertex formula, maximum area occurs when

$$W = -\frac{2400}{2(-2)} = 600 \text{ feet.}$$

The corresponding length is $L = 2400 - 2W = 2400 - 2(600) = 1200$ feet. The dimensions that maximize area are 600 feet by 1200 feet.   *Now Try Exercise 79* ◆

Another application of quadratic functions occurs in projectile motion, such as when a baseball is hit up in the air. If air resistance is ignored, then the formula

$$s(t) = -16t^2 + v_0 t + h_0$$

calculates the height s of the object above the ground in feet after t seconds. In this formula,

h_0 represents the initial height of the object in feet and v_0 represents its *vertical* initial velocity in feet per second. If the initial velocity of the projectile is upward, then $v_0 > 0$, and if the initial velocity is downward, then $v_0 < 0$.

EXAMPLE 9 Modeling the flight of a baseball

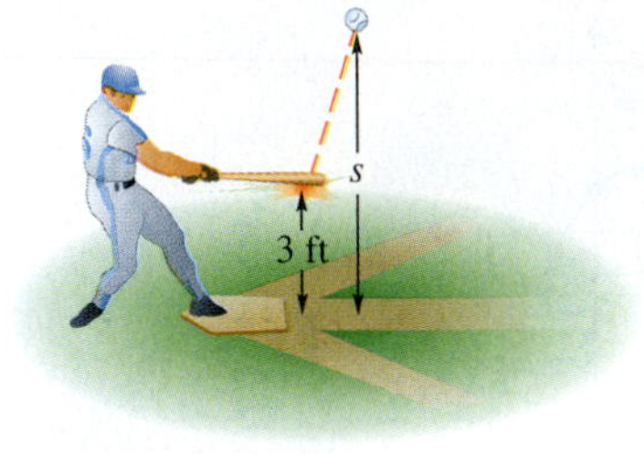

A baseball is hit straight up with an initial velocity of $v_0 = 80$ feet per second (or about 55 miles per hour) and leaves the bat with an initial height of $h_0 = 3$ feet, as shown in Figure 3.19.
(a) Write a formula $s(t)$ that models the height of the baseball after t seconds.
(b) How high is the baseball after 2 seconds?
(c) Find the maximum height of the baseball. Support your answer graphically.

SOLUTION
(a) Because $v_0 = 80$ and the initial height is $h_0 = 3$,

$$s(t) = -16t^2 + v_0t + h_0$$
$$= -16t^2 + 80t + 3.$$

FIGURE 3.19

Calculator Help

To find a maximum point on a graph, see Appendix B (page AP-14).

(b) $s(2) = -16(2)^2 + 80(2) + 3 = 99$, so the baseball is 99 feet high after 2 seconds.
(c) Because $a = -16$, the graph of s is a parabola opening downward. The vertex is the highest point on the graph, with an t-coordinate of

$$t = -\frac{b}{2a} = -\frac{80}{2(-16)} = 2.5.$$

The corresponding y-coordinate of the vertex is

$$s(2.5) = -16(2.5)^2 + 80(2.5) + 3 = 103 \text{ feet}.$$

[0, 5, 1] by [−20, 120, 20]

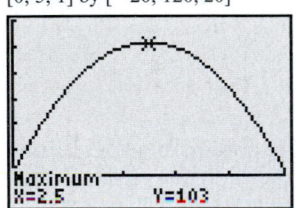

Maximum
X=2.5   Y=103

Thus the vertex is $(2.5, 103)$ and the maximum height of the baseball is 103 feet after 2.5 seconds. Graphical support is shown in Figure 3.20, where $Y_1 = -16X^2 + 80X + 3$ and the vertex is located at $(2.5, 103)$.   *Now Try Exercise 83* ◆

FIGURE 3.20

In the next example we use a quadratic function to model data exactly.

EXAMPLE 10 Modeling quadratic data

TABLE 3.3

x	y
2	10
3	1
4	−2
5	1
6	10

Use $f(x) = a(x - h)^2 + k$ to model the data in Table 3.3 exactly.

SOLUTION Notice that the smallest y-value in Table 3.3 is $y = -2$ when $x = 4$. If x decreases or increases by 1 unit to $x = 3$ or $x = 5$, the y-value increases to 1. If x decreases or increases by 2 units to $x = 2$ or $x = 6$, the y-value increases to 10. This symmetry about $x = 4$ indicates that the axis of symmetry is $x = 4$ and the vertex is $(4, -2)$. It follows that $f(x) = a(x - 4)^2 - 2$. To determine the value of a, we can use any point in the table (other than the vertex) because f models the data exactly. For example, if we note that the graph of f passes through $(5, 1)$, then $f(5) = 1$.

$$f(5) = 1 \qquad \text{The graph passes through } (5, 1).$$
$$a(5 - 4)^2 - 2 = 1 \qquad \text{Let } x = 5 \text{ in } f(x) = a(x - 4)^2 - 2.$$
$$a(5 - 4)^2 = 3 \qquad \text{Add 2.}$$
$$a = 3 \qquad \text{Simplify.}$$

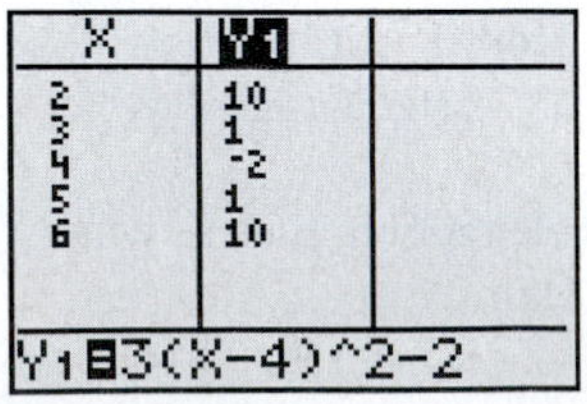

X	Y1
2	10
3	1
4	-2
5	1
6	10

Y1■3(X-4)^2-2

Thus $f(x) = 3(x - 4)^2 - 2$ models the data, which is supported in Figure 3.21.
            Now Try Exercise 93 ◆

FIGURE 3.21

Calculator Help

To create a table similar to Figure 3.21, see Appendix B (page AP-15).

A well-conditioned athlete's heart rate can reach 200 beats per minute (bpm) during strenuous physical activity. Upon stopping an activity, a typical heart rate decreases rapidly at first and then gradually levels off, as shown in Table 3.4.

The data are not linear because for each 2-minute interval the heart rate does not decrease by a fixed amount. A scatterplot of the data is shown in Figure 3.22. In Figure 3.23 the data are modeled with a nonlinear function. Note that the graph of this function resembles the left half of a parabola, which opens upward.

TABLE 3.4   Athlete's Heart Rate

Time (min)	0	2	4	6	8
Heart Rate (bpm)	200	150	110	90	80

Adapted from: V. Thomas, *Science and Sport.*

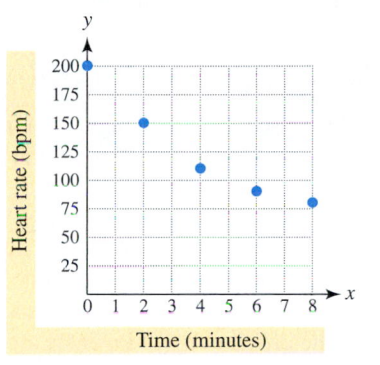

FIGURE 3.22

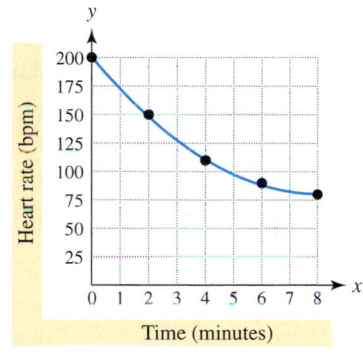

FIGURE 3.23

EXAMPLE 11

Modeling an athlete's heart rate

Find a quadratic function f expressed in vertex form that models the data in Table 3.4. Support your result by graphing f and the data together. What is the domain of your function?

SOLUTION To model the data we use the left half of a parabola. Since the minimum heart rate of 80 beats per minute occurs when $x = 8$, let $(8, 80)$ be the vertex and write

$$f(x) = a(x - 8)^2 + 80.$$

Next, we must determine a value for the leading coefficient a. One possibility is to have the graph of f pass through the first data point $(0, 200)$, or equivalently, let $f(0) = 200$.

$[-0.5, 8.5, 2]$ by $[0, 220, 20]$

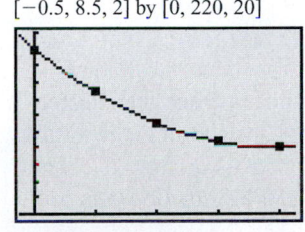

FIGURE 3.24

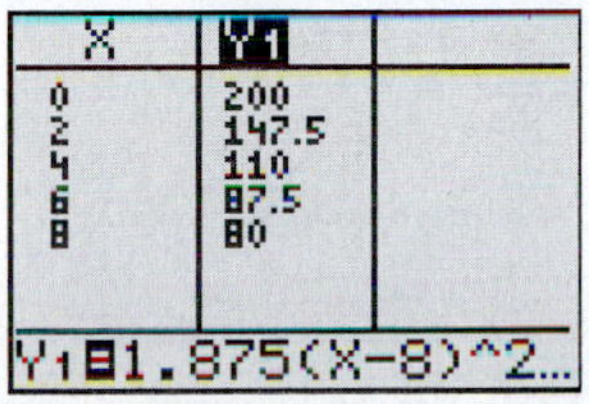

$$
\begin{aligned}
f(\mathbf{0}) &= \mathbf{200} &&\text{\textit{Have the graph pass through (0, 200).}}\\
a(\mathbf{0} - 8)^2 + 80 &= \mathbf{200} &&\text{\textit{Let x = 0 in f(x).}}\\
a(0 - 8)^2 &= 120 &&\text{\textit{Subtract 80.}}\\
a &= \frac{120}{64} &&\text{\textit{Divide by (0 - 8)}}^2 = 64.\\
a &= 1.875 &&\text{\textit{Write as a decimal.}}
\end{aligned}
$$

Thus $f(x) = 1.875(x - 8)^2 + 80$ can be used to model the athlete's heart rate. A graph of f and the data are shown in Figure 3.24, which is similar to Figure 3.23. Figure 3.25 shows a table of $Y_1 = 1.875(X - 8)^2 + 80$. Although the table in Figure 3.25 does not match Table 3.4 exactly, it gives reasonable approximations. (Note that formulas for $f(x)$ may vary. For example, if we had selected the point $(2, 150)$ rather than $(0, 200)$, then $a \approx 1.94$. You may wish to verify this result.)

The domain D of $f(x) = 1.875(x - 8)^2 + 80$ needs to be restricted to $0 \le x \le 8$ because this interval corresponds to the domain of the data in Table 3.4. The vertex on the

X	Y1
0	200
2	147.5
4	110
6	87.5
8	80

Y₁▊1.875(X-8)^2...

FIGURE 3.25

graph of $y = f(x)$ is (8, 80), and so for $x > 8$, the values of $f(x)$ would begin to increase because the graph of $y = f(x)$ is a U-shaped parabola. If we had information that the athlete started exercising again after 8 minutes, then it might be possible for $f(x)$ to model the heart rate beyond the 8-minute period shown in Table 3.4.   *Now Try Exercise 99* ◆

◆ MAKING CONNECTIONS

General Form, Vertex Form, and Modeling   When modeling quadratic data by hand, it is often easier to use the vertex form: $f(x) = a(x - h)^2 + k$, rather than the general form: $f(x) = ax^2 + bx + c$. Because (h, k) corresponds to the vertex of a parabola, it may be appropriate to let (h, k) correspond to either the highest data point or the lowest data point in the scatterplot. Then a value for a can be found by substituting a data point into the formula for $f(x)$. In the next subsection, least-squares regression provides a quadratic modeling function in general form.   ◆

Quadratic Regression (Optional)

In Chapter 2 we discussed how a regression line could be found by the method of least squares. This method can also be applied to quadratic data and is illustrated in the next example.

EXAMPLE 12   Finding a quadratic regression model

In one study the efficiency of photosynthesis in an Antarctic species of grass was investigated. Table 3.5 lists results for various temperatures. The temperature x is in degrees Celsius and the efficiency y is given as a percent. The purpose of the research was to determine the temperature at which photosynthesis is most efficient.   (**Source:** D. Brown and P. Rothery, *Models in Biology: Mathematics, Statistics and Computing.*)

TABLE 3.5

x (°C)	−1.5	0	2.5	5	7	10	12	15	17	20	22	25	27	30
y (%)	33	46	55	80	87	93	95	91	89	77	72	54	46	34

(a) Plot the data. Discuss reasons why a quadratic function might model the data.
(b) Find a least-squares function f given by $f(x) = ax^2 + bx + c$ that models the data.
(c) Graph f and the data in $[-5, 35, 5]$ by $[20, 110, 10]$. Discuss the fit.
(d) Determine the temperature at which f predicts that photosynthesis is most efficient. Compare the results with the findings of the experiment.

SOLUTION
(a) A plot of the data is shown in Figure 3.26. The y-values first increase and then decrease as the temperature x increases. The data suggest a parabolic shape opening downward.
(b) Enter the data into your calculator, and then select quadratic regression from the menu, as shown in Figures 3.27 and 3.28. In Figure 3.29, the modeling function f is given (approximately) by $f(x) = -0.249x^2 + 6.77x + 46.37$.

$[-5, 35, 5]$ by $[20, 110, 10]$

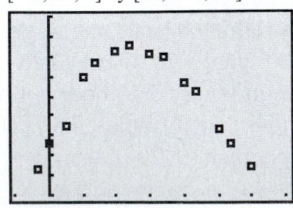

FIGURE 3.26

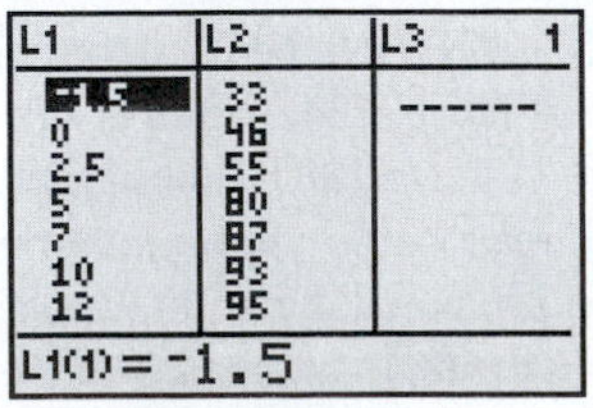

FIGURE 3.27

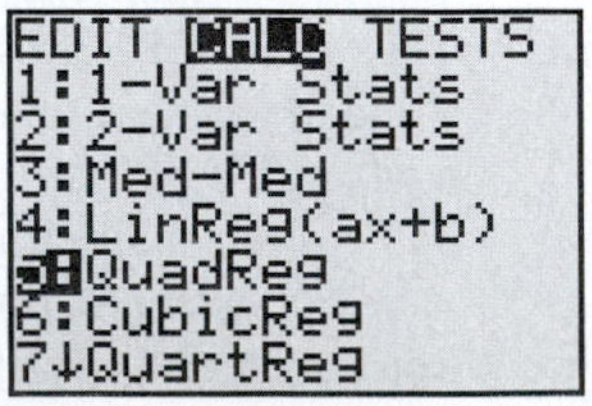

FIGURE 3.28

FIGURE 3.29

Calculator Help

To find an equation of least-squares fit, see Appendix B (page AP-15).

(c) The graph of $Y_1 = -.249X^2 + 6.77X + 46.37$ with the data is shown in Figure 3.30. Although the model is not exact, the parabola describes the general trend in the data. Notice that the parabola is opening downward since $a = -0.249$.

$[-5, 35, 5]$ by $[20, 110, 10]$ $[-5, 35, 5]$ by $[20, 110, 10]$

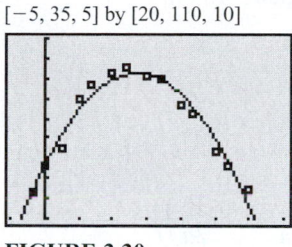

FIGURE 3.30   **FIGURE 3.31**

(d) The vertex is the highest point located on the graph. Its coordinates are approximately (13.6, 92.4), as shown in Figure 3.31. This quadratic model predicts that the highest efficiency is about 92.4%, and it occurs near 13.6°C. Although there are percentages in the table higher than 92.4%, 13.6°C is a reasonable estimate for the optimum temperature. The function f is attempting to model the general trend in the data and predict future results.   *Now Try Exercise 107* ◆

3.1

Putting it all Together

$\mathbf{T}$he following table summarizes some important topics from this section.

Concept	Symbolic Representation	Comments and Examples
Quadratic function	$f(x) = ax^2 + bx + c$, where a, b, and c are constants with $a \neq 0$, (general form).	It models data that are not linear. Its graph is a parabola (U-shape) that opens either upward ($a > 0$) or downward ($a < 0$).
Parabola	The graph of $y = ax^2 + bx + c$, $a \neq 0$, is a parabola.	Vertex: (h, k) Axis of symmetry: $x = h$ Maximum (or minimum) y-value on the graph: k

Concept	Symbolic Representation	Comments and Examples
Completing the square to find vertex form	To complete the square for $x^2 + kx$, add $\left(\frac{k}{2}\right)^2$ to make a perfect square trinomial.	$y = x^2 - 2x + 3$ $y - 3 = x^2 - 2x$ $y - 3 + 1 = x^2 - 2x + 1$ Add $\left(\frac{-2}{2}\right)^2 = 1$. $y - 2 = (x - 1)^2$ $y = (x - 1)^2 + 2$
Vertex formula	The vertex for $f(x) = ax^2 + bx + c$ is the point $$\left(-\frac{b}{2a},\, f\left(-\frac{b}{2a}\right)\right).$$	The x-value of the vertex on the graph of $f(x) = x^2 - 2x + 3$ is $$x = -\frac{-2}{2(1)} = 1.$$ y-value of vertex: $f(1) = 2$ Vertex: $(1, 2)$ Axis of symmetry: $x = 1$
Vertex form (standard form for a parabola with vertical axis)	The vertex form for a quadratic function is $f(x) = a(x - h)^2 + k$, with vertex (h, k).	Let $f(x) = 2(x + 3)^2 - 5$ Parabola opens upward: $a > 0$ Vertex: $(-3, -5)$ Axis of symmetry: $x = -3$ Minimum y-value on graph: -5

3.1 ▷ Exercises

Basics of Quadratic Functions

Exercises 1–8: Identify f as being linear, quadratic, or neither. If f is quadratic, identify the leading coefficient and evaluate f(−2).

1. $f(x) = 1 - 2x + 3x^2$ **2.** $f(x) = -5x + 11$

3. $f(x) = \dfrac{1}{x^2 - 1}$ **4.** $f(x) = (x^2 + 1)^2$

5. $f(x) = \dfrac{1}{2} - \dfrac{3}{10}x$ **6.** $f(x) = \dfrac{1}{5}x^2$

7. $f(x) = -3x^2 + 9$ **8.** $f(x) = 5x^2 - \sqrt{2}x$

Exercises 9–12: Use the graph of the quadratic function to determine the sign of the leading coefficient, its vertex, and the equation of the axis of symmetry.

9.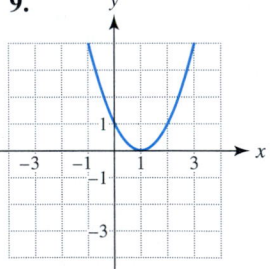

10.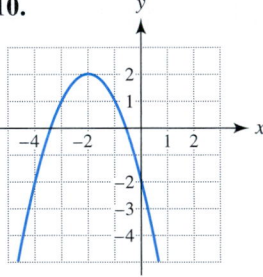

11.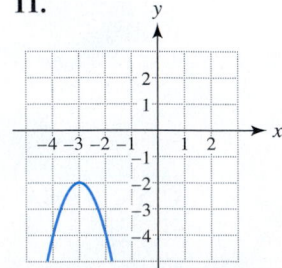

12.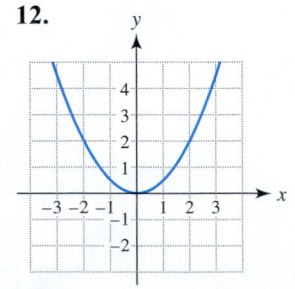

Exercises 13–16: The formulas for $f(x)$ and $g(x)$ are identical except for their leading coefficients a. Compare the graphs of f and g. You may want to support your answers by graphing f and g together.

13. $f(x) = x^2,$ $\qquad g(x) = 2x^2$

14. $f(x) = \frac{1}{2}x^2,$ $\qquad g(x) = -\frac{1}{2}x^2$

15. $f(x) = 2x^2 + 1,$ $\qquad g(x) = -\frac{1}{3}x^2 + 1$

16. $f(x) = x^2 + x,$ $\qquad g(x) = \frac{1}{4}x^2 + x$

Vertex Formula

Exercises 17–24: Identify the vertex and leading coefficient. Then write the expression as $f(x) = ax^2 + bx + c$.

17. $f(x) = -3(x - 1)^2 + 2$ **18.** $f(x) = 5(x + 2)^2 - 5$

19. $f(x) = 5 - 2(x - 4)^2$ **20.** $f(x) = \frac{1}{2}(x + 3)^2 - 5$

21. $f(x) = \frac{3}{4}(x + 5)^2 - \frac{7}{4}$ **22.** $f(x) = -5(x - 4)^2$

23. $f(x) = \frac{1}{2}\left(x + \frac{3}{4}\right)^2$ **24.** $f(x) = \frac{7}{5} - \frac{2}{3}\left(x - \frac{1}{3}\right)^2$

Exercises 25–38: Use the vertex formula to determine the vertex of the graph of f. Support your results graphically or numerically.

25. $f(x) = 6 - x^2$ **26.** $f(x) = 2x^2 - 2x + 1$

27. $f(x) = x^2 - 6x$ **28.** $f(x) = -2x^2 + 4x + 5$

29. $f(x) = 2x^2 - 4x + 1$ **30.** $f(x) = -3x^2 + x - 2$

31. $f(x) = \frac{1}{2}x^2 + 10$ **32.** $f(x) = \frac{9}{10}x^2 - 12$

33. $f(x) = -\frac{3}{4}x^2 + \frac{1}{2}x - 3$ **34.** $f(x) = -\frac{4}{5}x^2 - \frac{1}{5}x + 1$

35. $f(x) = x^2 - 3.8x - 2$ **36.** $f(x) = -4x^2 + 16x$

37. $f(x) = 1.5 - 3x - 6x^2$ **38.** $f(x) = 0.25x^2 - 1.5x + 1$

Exercises 39–50: Write the given expression in the form $f(x) = a(x - h)^2 + k$. Identify the vertex.

39. $f(x) = x^2 + 4x - 5$ **40.** $f(x) = x^2 + 10x + 7$

41. $f(x) = x^2 - 3x$ **42.** $f(x) = x^2 - 7x + 5$

43. $f(x) = 2x^2 - 5x + 3$ **44.** $f(x) = 3x^2 + 6x + 2$

45. $f(x) = -\frac{1}{2}x^2 - \frac{3}{2}x + 1$ **46.** $f(x) = \frac{1}{3}x^2 + x + 1$

47. $f(x) = 2x^2 - 8x - 1$ **48.** $f(x) = -\frac{1}{2}x^2 - x$

49. $f(x) = 2 - 9x - 3x^2$ **50.** $f(x) = 6 + 5x - 10x^2$

Exercises 51–58: Use the graph of the quadratic function f to write its formula as $f(x) = a(x - h)^2 + k$.

51.
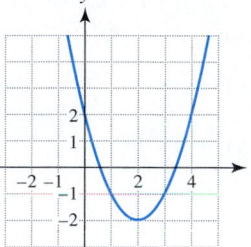

52.
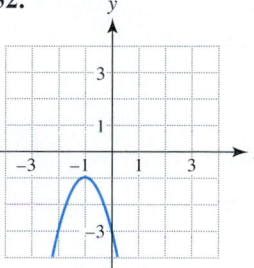

53.
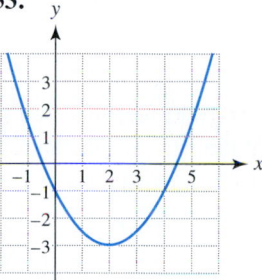

54.
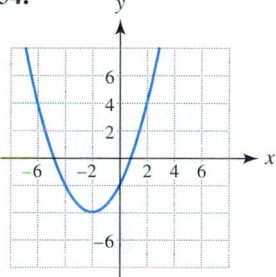

55.
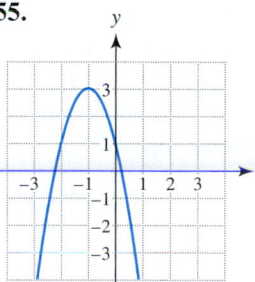

56.
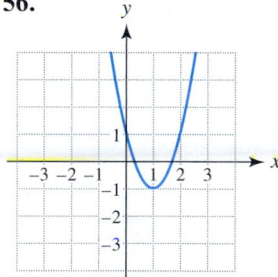

57.
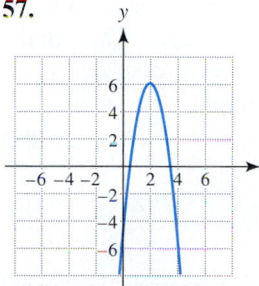

58.
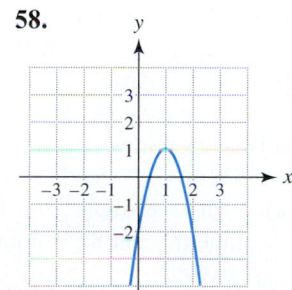

Exercises 59–78: Sketch a graph of f.

59. $f(x) = x^2$ **60.** $f(x) = -2x^2$

61. $f(x) = -\frac{1}{2}x^2$ **62.** $f(x) = 4 - x^2$

63. $f(x) = x^2 - 3$ **64.** $f(x) = x^2 + 2$

65. $f(x) = (x - 2)^2 + 1$ **66.** $f(x) = (x + 1)^2 - 2$

67. $f(x) = -3(x + 1)^2 + 3$

68. $f(x) = -2(x - 1)^2 + 1$

69. $f(x) = x^2 - 2x - 2$ **70.** $f(x) = x^2 - 4x$

71. $f(x) = -x^2 + 4x - 2$ **72.** $f(x) = -x^2 + 2x + 1$

73. $f(x) = 2x^2 - 4x - 1$ **74.** $f(x) = 3x^2 + 6x$

75. $f(x) = -3x^2 - 6x + 1$

76. $f(x) = -2x^2 + 4x - 1$

77. $f(x) = \frac{1}{2}x^2 - 2x + 2$ **78.** $f(x) = -\frac{1}{2}x^2 + x + 1$

Applications and Models

79. *Maximizing Area*   A farmer has 1000 feet of fence to enclose a rectangular area. What dimensions for the rectangle result in the maximum area enclosed by the fence?

80. *Maximizing Area*   A homeowner has 80 feet of fence to enclose a rectangular garden. What dimensions for the garden give a maximum area?

81. *Maximizing Revenue*   Suppose the revenue R in thousands of dollars that a company receives from producing x thousand compact disc players is given by the formula $R(x) = x(40 - 2x)$.
   (a) Evaluate $R(2)$ and interpret the result.

   (b) How many CD players should the company produce to maximize its revenue?

   (c) What is the maximum revenue?

82. *Maximizing Revenue*   A large hotel is considering the following group discount on room rates. The regular price for a room is $120, but for each room rented the price decreases by $2. For example, one room costs $118, two rooms cost $116 × 2 = $232, three rooms cost $114 × 3 = $342, and so on.
   (a) Write a formula for a function R that gives the revenue for renting x rooms.

   (b) Sketch a graph of R. What is a reasonable domain for R?

   (c) Determine the maximum revenue and the number of rooms that should be rented.

83. *Hitting a Baseball*   A baseball is hit so that its height in feet after t seconds is $s(t) = -16t^2 + 44t + 4$.
   (a) How high is the baseball after 1 second?

   (b) Find the maximum height of the baseball. Support your answer graphically.

84. *Throwing a Stone*   A stone is thrown *downward* with a velocity of 66 feet per second (45 miles per hour) from a bridge that is 120 feet above a river, as illustrated in the accompanying figure.
   (a) Write a formula $s(t)$ that models the height of the stone after t seconds.

   (b) Does the stone hit the water within the first 2 seconds? Explain.

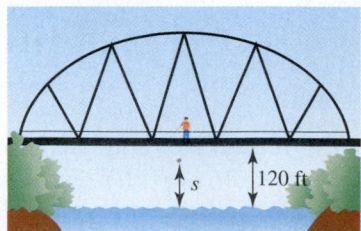

Exercises 85–88: Quadratic Models   Match the physical situation with the graph of the quadratic function (a.–d.) that models it best.

85. The height y of a stone thrown from ground level after x seconds

86. The number of people attending a popular movie x weeks after its opening

87. The temperature after x hours in a house where the furnace quits and a repairperson fixes it

88. The cumulative number of reported AIDS cases from 1980 to year x

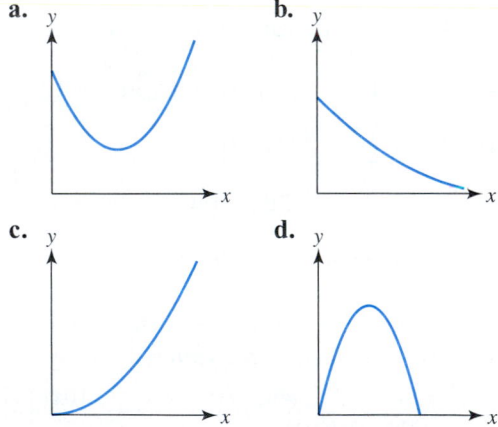

Exercises 89–92: Maximizing Altitude   If air resistance is ignored, the height h of a projectile above the ground after x seconds is given by $h(x) = -\frac{1}{2}gx^2 + v_0x + h_0$. In this formula g is the acceleration due to gravity. The leading coefficient is negative since gravity pulls downward. This formula is valid not only for Earth, but also for other celestial bodies. A ball is thrown straight up at 88 feet per second from a height of 25 feet.

   (a) *For the given value of g, graphically estimate both the maximum height and the time when this occurs.*

   (b) *Solve part (a) symbolically.*

89. $g = 32$ (Earth)

90. $g = 5.1$ (Moon)

91. $g = 13$ (Mars)

92. $g = 88$ (Jupiter)

Exercises 93–96: Find $f(x) = a(x - h)^2 + k$ so that f models the data exactly.

93.

x	-1	0	1	2	3
y	5	-1	-3	-1	5

94.

x	-4	-3	-2	-1	0
y	1	-0.5	-1	-0.5	1

95.

x	-2	-1	0	1	2
y	2	4	2	-4	-14

96.

x	-4	-3	-2	-1	0
y	8	9	8	5	0

Exercises 97 and 98: **Modeling Real Data** *Each table contains real data that can be modeled approximately by $f(x) = a(x - h)^2 + k$.*

(a) *Make a scatterplot of the data. (Do not try to plot the undetermined point(s) listed in the table.)*

(b) *Find values for a, h, and k. Graph f together with the data in the same viewing rectangle.*

(c) *Approximate the undetermined value(s) in the table.*

97. Total cumulative AIDS cases in the United States

Year	1982	1984	1986	1988
Cases	1586	10,927	41,910	106,304

Year	1990	1992	1994	1998
Cases	196,576	329,205	441,528	?

Source: Department of Health and Human Services.

98. U.S. population in millions

Year	1800	1820	1840	1860
Population	5	10	17	31

Year	1870	1880	1900	1920
Population	?	50	76	106

Year	1940	1960	1980	2000
Population	132	179	226	?

Source: Bureau of the Census.

99. *Heart Rate* (Refer to Example 11.) In the accompanying table an athlete's heart rate is shown upon stopping a moderate activity.

Time (min)	0	1	2	3	4
Heart Rate (bpm)	122	108	98	92	90

(a) Model the data with $H(t) = a(t - h)^2 + k$. What is the domain of H?

(b) Estimate the athlete's heart rate when $t = 1.5$ minutes.

100. *Maximizing Area* (Refer to Example 8.) A farmer is fencing a rectangular area for cattle using the straight portion of a river as one side of the rectangle. If the farmer has P feet of fence, find the dimensions of the rectangle that give the maximum area for the cattle.

101. *Women in the Work Force* The number of women in millions that were gainfully employed in the work force for selected years is shown in the table.

Year	1900	1910	1920	1930	1940
Work Force	5.3	7.4	8.6	10.8	12.8

Year	1950	1960	1970	1980	1990	2000
Work Force	18.4	23.2	31.5	45.5	56.6	65.6

Source: Department of Labor.

(a) Find a quadratic function f so that the formula $f(x) = a(x - h)^2 + k$ models the data. Support your result graphically.

(b) Predict the number of women in the labor force in 2010.

102. *Suspension Bridge* The cables that support a suspension bridge, such as the Golden Gate Bridge, can be modeled by parabolas. Suppose that a 300-foot long suspension bridge has towers at its ends that are 120 feet tall, as illustrated in the accompanying figure. If the cable comes within 20 feet of the road in the center of the bridge, find a quadratic function that models the height of the cable above the road a distance of x feet from the center of the bridge.

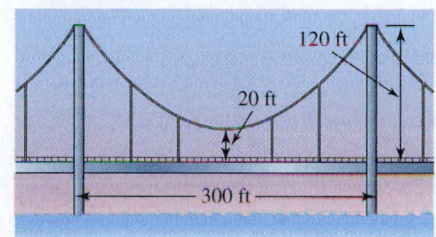

Quadratic Regression

*Exercises 103–106: **Quadratic Models** Use least-squares regression to find a quadratic function f that models the data given in the table. Then estimate f(3.5) to the nearest hundredth.*

103.

x	0	2	4	6
$f(x)$	−1	16	57	124

104.

x	10	20	30	40
$f(x)$	4.2	24.3	84.1	184

105.

x	1	2	3	4
$f(x)$	4.14	10.12	18.92	30.56

106.

x	−2	−1	0	1	2
$f(x)$	−2.88	1.95	1.5	−4.25	−15.28

107. *Household Size* The following table lists the average number of people aged 18 or older in a U.S. household for selected years.

Year	1940	1950	1960	1970
Household Size	2.53	2.31	2.12	2.05

Year	1980	1990	2000
Household Size	1.97	1.93	1.93

Source: Bureau of the Census.

(a) Find a quadratic function that models the data. Support your result graphically.

(b) Estimate the average number of people aged 18 or older in a household during 1975. Compare it to the actual value of 2.01.

108. *Consumer Price Index* The following table lists the consumer price index (CPI) for selected years.

Year	1950	1960	1970	1980	1990	2000
CPI	24.1	29.6	38.8	82.4	130.7	172.2

Source: Bureau of the Census.

(a) Find a quadratic function that models the data. Support your result graphically.

(b) Estimate the CPI in 1995 and compare it to the actual value of 152.4.

Writing about Mathematics

109. How do the values of a, h, and k affect the graph of $f(x) = a(x - h)^2 + k$?

110. Explain why the vertex is important when you are trying to find either the maximum y-value or the minimum y-value on the graph of a quadratic function.

EXTENDED AND DISCOVERY EXERCISES

*Exercises 1–4: **Linear and Quadratic Data** Suppose we are given a table that lists ordered pairs (x, y), where the x-values increase by a fixed amount between consecutive data points. If the data are linear, then the differences between consecutive y-values are always the same. For example, the following table contains linear data generated by y = 3x + 1. The increases in consecutive y-values are 7 − 4 = 3, 10 − 7 = 3, 13 − 10 = 3, and 16 − 13 = 3. The increases always equal 3 for each unit of increase in x. Thus when x = 6, y = 16 + 3 = 19, and this agrees with the formula y = 3x + 1.*

x	1	2	3	4	5
y	4	7	10	13	16

Quadratic data can be determined in a similar manner. The following table contains quadratic data generated by y = x² + 2. The increases in consecutive y-values are 6 − 3 = 3, 11 − 6 = 5, 18 − 11 = 7, and 27 − 18 = 9. The sequence of numbers 3, 5, 7, 9 is determined by adding 2 to the previous number. According to this pattern, the next increase should be 9 + 2 = 11 so when x = 6, y = 27 + 11 = 38. This agrees with the formula y = x² + 2 when x = 6.

x	1	2	3	4	5
y	3	6	11	18	27

Use this method to determine if the data in the table are linear, quadratic, or neither. If the data are either linear or quadratic, find the next y-value in the table.

1.

x	1	2	3	4	5	6
y	−6	−9	−16	−27	−42	?

2.

x	5	10	15	20	25	30
y	9	8	7	6	5	?

3.

x	−2	0	2	4	6	8
y	−15	−7	1	18	43	?

4.

x	−4	−2	0	2	4	6
y	−9	−7	−1	9	23	?

*Exercises 5–8: **Difference Quotient** Complete the following for the given $f(x)$.*

(a) *Evaluate $f(x)$ for each x-value in the table.*

x	1	2	3	4	5
$f(x)$					

(b) *Calculate the average rate of change of f between consecutive data points in the table.*

(c) *Find the difference quotient for $f(x)$. Then let $h = 1$ in the difference quotient.*

(d) *Evaluate this difference quotient for $x = 1, 2, 3,$ and 4. How do these results compare to your results in part (b)?*

5. $f(x) = x^2 - 3$ **6.** $f(x) = 2x - x^2$

7. $f(x) = -2x^2 + 3x - 1$ **8.** $f(x) = 3x^2 + x + 2$

3.2 Quadratic Equations and Problem Solving

- ◆ Understand basic concepts about quadratic equations
- ◆ Use factoring, the square root property, completing the square, and the quadratic formula to solve quadratic equations
- ◆ Understand the discriminant
- ◆ Solve problems involving quadratic equations

Introduction

In Chapter 2, applications involving linear functions led to linear equations. In a similar manner, applications involving nonlinear functions often result in nonlinear equations. Linear equations can always be solved symbolically. This is not the case with some types of nonlinear equations. Their solutions must be approximated using numerical or graphical techniques.

Quadratic equations are one of the simplest types of nonlinear equations. They can be solved symbolically, as well as numerically and graphically.

Basic Concepts

The scatterplot in Figure 3.32 shows cumulative numbers of AIDS deaths in the United States from 1984 to 1994. In Figure 3.33, the data are modeled by a quadratic function D, given by

$$D(x) = 2375x^2 + 5134x + 5020,$$

where $x = 0$ corresponds to 1984, $x = 1$ to 1985, and so on until $x = 10$ corresponds to 1994. (Note that function D was found using quadratic regression.)

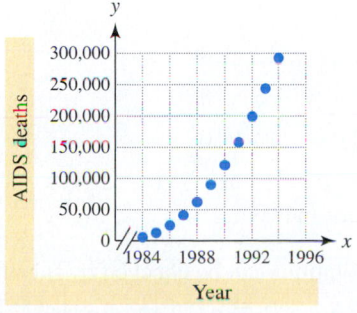

FIGURE 3.32 AIDS Deaths

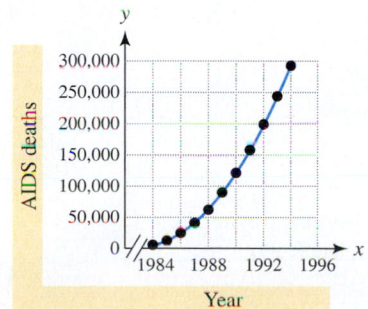

FIGURE 3.33 Modeling AIDS Deaths

Once the function D has been found, it is natural to use D to obtain information about AIDS. For example, to estimate the year when the number of AIDS deaths reached 500,000, we could solve the following equation.

$$D(x) = 500,000$$

$$2375x^2 + 5134x + 5020 = 500,000 \qquad \text{\color{teal}Substitute for D(x).}$$

$$2375x^2 + 5134x - 494,980 = 0 \qquad \text{\color{teal}Subtract 500,000 from both sides.}$$

The equation $2375x^2 + 5134x - 494,980 = 0$ is a *quadratic equation*. We solve this equation in Example 8.

QUADRATIC EQUATION

A **quadratic equation** in one variable is an equation that can be written in the form
$$ax^2 + bx + c = 0,$$
where a, b, and c are real numbers with $a \neq 0$.

Solving Quadratic Equations

Quadratic equations can always be solved symbolically. Four basic symbolic strategies include factoring, the square root property, completing the square, and the quadratic formula. We begin by discussing factoring.

Factoring   Factoring is a common technique used to solve equations. It is based on the **zero-product property**, which states that if $ab = 0$, then $a = 0$ or $b = 0$ or both. It is important to remember that this property works only for 0. For example, if $ab = 1$, then this equation does *not* imply that either $a = 1$ or $b = 1$. For example, $a = \frac{1}{2}$ and $b = 2$ also satisfies $ab = 1$.

 **Solving a quadratic equation with factoring**

Solve the quadratic equation. Check your results.
(a) $2x^2 + 2x - 11 = 1$ **(b)** $12t^2 = t + 1$

SOLUTION
(a) Start by writing the equation in the form $ax^2 + bx + c = 0$.

$$2x^2 + 2x - 11 = 1 \qquad \text{\color{teal}Given equation}$$

$$2x^2 + 2x - 12 = 0 \qquad \text{\color{teal}Subtract 1 from each side.}$$

$$x^2 + x - 6 = 0 \qquad \text{\color{teal}Divide each side by 2. (Optional step)}$$

$$(x + 3)(x - 2) = 0 \qquad \text{\color{teal}Factor.}$$

$$x + 3 = 0 \quad \text{or} \quad x - 2 = 0 \qquad \text{\color{teal}Zero-product property}$$

$$x = -3 \quad \text{or} \quad x = 2 \qquad \text{\color{teal}Solve.}$$

Algebra Review

To review factoring trinomials, see Chapter R (page R-29).

These solutions can be checked by substituting them into the given equation.

$$2(-3)^2 + 2(-3) - 11 \overset{?}{=} 1 \qquad\qquad 2(2)^2 + 2(2) - 11 \overset{?}{=} 1$$

$$1 = 1 \qquad\qquad\qquad\qquad\qquad 1 = 1$$

(b) Write the equation in the form $at^2 + bt + c = 0$.

$$12t^2 = t + 1 \qquad \text{Given equation}$$
$$12t^2 - t - 1 = 0 \qquad \text{Subtract } t \text{ and } 1.$$
$$(3t - 1)(4t + 1) = 0 \qquad \text{Factor.}$$
$$3t - 1 = 0 \quad \text{or} \quad 4t + 1 = 0 \qquad \text{Zero-product property}$$
$$t = \frac{1}{3} \quad \text{or} \quad t = -\frac{1}{4} \qquad \text{Solve.}$$

To check these solutions, substitute them into the given equation.

$$12\left(\frac{1}{3}\right)^2 = \frac{1}{3} + 1 \qquad\qquad 12\left(-\frac{1}{4}\right)^2 = -\frac{1}{4} + 1$$
$$\frac{4}{3} = \frac{4}{3} \qquad\qquad\qquad \frac{3}{4} = \frac{3}{4}$$

Now Try Exercises 1 and 5 ◆

We can also use factoring to find the *x*-intercepts of the graph of a quadratic equation.

EXAMPLE 2 Finding *x*-intercepts

Find the exact values for the *x*-intercepts shown in Figure 3.34.

SOLUTION From the graph it is difficult to determine the *exact x*-intercepts. However, they can be determined symbolically.

$$24x^2 + 7x - 6 = 0 \qquad \text{Set expression equal to 0.}$$
$$(3x + 2)(8x - 3) = 0 \qquad \text{Factor.}$$
$$3x + 2 = 0 \quad \text{or} \quad 8x - 3 = 0 \qquad \text{Zero-product property}$$
$$x = -\frac{2}{3} \quad \text{or} \quad x = \frac{3}{8} \qquad \text{Solve.}$$

$y = 24x^2 + 7x - 6$

FIGURE 3.34

The *x*-intercepts are $-\frac{2}{3}$ and $\frac{3}{8}$.

Now Try Exercise 25 ◆

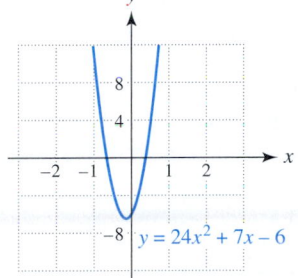

EXAMPLE 3 Solving a quadratic equation symbolically, graphically, and numerically

On the shoreline surrounding certain types of recreational lakes in Minnesota, the Department of Natural Resources (DNR) restricts the sizes of buildings. Storage buildings built within 100 feet of the shoreline are sometimes limited to a maximum area of 120 square feet. Suppose a person is building a rectangular shed that has an area of 120 square feet and is 7 feet longer than it is wide. See Figure 3.35. Determine its dimensions symbolically, graphically, and numerically.

SOLUTION
Symbolic Solution Let *x* be the width of the shed. Then $x + 7$ represents the length. Since area is equal to width times length, we solve the following equation.

FIGURE 3.35

$$x(x + 7) = 120 \qquad \text{Area is equal to 120 square feet.}$$
$$x^2 + 7x = 120 \qquad \text{Distributive property}$$
$$x^2 + 7x - 120 = 0 \qquad \text{Subtract 120 from each side.}$$
$$(x + 15)(x - 8) = 0 \qquad \text{Factor.}$$
$$x + 15 = 0 \quad \text{or} \quad x - 8 = 0 \qquad \text{Zero-product property}$$
$$x = -15 \quad \text{or} \quad x = 8 \qquad \text{Solve.}$$

Since dimensions cannot be negative, the solution that has meaning is $x = 8$. The length is 7 feet longer than the width, so the dimensions of the shed are 8 feet by 15 feet.

Graphical Solution The graphs of $Y_1 = X(X + 7)$ and $Y_2 = 120$ intersect at the points $(-15, 120)$ and $(8, 120)$, as shown in Figures 3.36 and 3.37. The solutions to the equation are -15 and 8.

Calculator Help

To find a point of intersection, see Appendix B (page AP-12).

Numerical Solution Table y_1 and y_2. The solution of 8 is shown in Figure 3.38, and the solution of -15 may be found by scrolling through the table. The equation $y_1 = y_2$ is satisfied when $x = -15$ or 8. The symbolic, graphical, and numerical solutions agree.

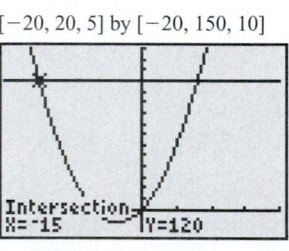
$[-20, 20, 5]$ by $[-20, 150, 10]$

Intersection
X=-15   Y=120

FIGURE 3.36

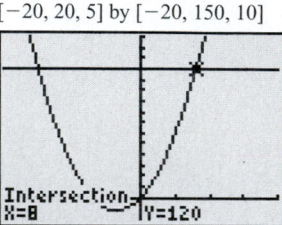
$[-20, 20, 5]$ by $[-20, 150, 10]$

Intersection
X=8   Y=120

FIGURE 3.37

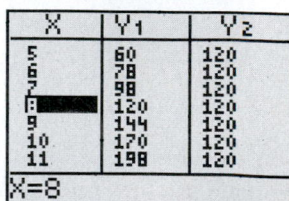

X	Y₁	Y₂
5	60	120
6	78	120
7	98	120
8	120	120
9	144	120
10	170	120
11	198	120

X=8

FIGURE 3.38

Note: If the solutions are fractions or irrational numbers, it may be difficult or even impossible to determine them with a table of values.   *Now Try Exercise 107* ◆

The Square Root Property Some types of quadratic equations can be written as $x^2 = k$, where k is a nonnegative number. The solutions to this equation are $\pm\sqrt{k}$. (Recall that the symbol $\pm$ represents *plus or minus*.) For example, $x^2 = 16$ has two solutions, ± 4. We refer to this as the **square root property**.

Algebra Review

To review square roots, see Chapter R (page R-48).

> ## SQUARE ROOT PROPERTY
>
> Let k be a nonnegative number. Then the solutions to the equation
> $$x^2 = k$$
> are given by $x = \pm\sqrt{k}$.

EXAMPLE 4   Using the square root property

If a metal ball is dropped 100 feet from a water tower, its height h in feet above the ground after t seconds is given by $h(t) = 100 - 16t^2$. See Figure 3.39. Determine how long it takes the ball to hit the ground.

SOLUTION The ball strikes the ground when the equation $100 - 16t^2 = 0$ is satisfied.

$$100 - 16t^2 = 0$$
$$100 = 16t^2 \qquad \text{Add } 16t^2 \text{ to each side.}$$
$$\frac{100}{16} = t^2 \qquad \text{Divide each side by 16.}$$
$$\pm\sqrt{\frac{100}{16}} = t \qquad \text{Square root property}$$

FIGURE 3.39

$$\pm\frac{10}{4} = t \qquad \text{Simplify.}$$

$$t = \pm 2.5 \qquad \text{Rewrite.}$$

In this example only positive values for time are valid, so after 2.5 seconds the ball strikes the ground.

Now Try Exercise 97 ◆

Functions can be defined by formulas, graphs, tables, and diagrams. Functions can also be defined by equations. In the next example we solve equations for y, and then determine if y is a function of x, where $y = f(x)$.

EXAMPLE 5 Determining if equations represent functions

Solve each equation for y. Determine if y is a function of x.

(a) $x^2 + (y-1)^2 = 4$ **(b)** $2y = \frac{x+y}{2}$

SOLUTION

(a) Start by subtracting x^2 from each side of the equation.

$$(y-1)^2 = 4 - x^2 \qquad \text{Subtract } x^2 \text{ from each side.}$$

$$y - 1 = \pm\sqrt{4-x^2} \qquad \text{Square root property}$$

$$y = 1 \pm \sqrt{4-x^2} \qquad \text{Add 1 to each side.}$$

There are two formulas: $y = 1 + \sqrt{4-x^2}$ and $y = 1 - \sqrt{4-x^2}$, which indicates that y is not a function of x. That is, one x-input can produce two y-outputs.

(b) Clear fractions by multiplying each side by 2.

$$4y = x + y \qquad \text{Multiply each side by 2.}$$

$$3y = x \qquad \text{Subtract } y \text{ from each side.}$$

$$y = \frac{x}{3} \qquad \text{Divide each side by 3.}$$

Algebra Review
To clear fractions, see Chapter R (page R-46).

The equation $y = \frac{x}{3}$ defines a linear function, so y is a function of x.

Now Try Exercises 85 and 87 ◆

Completing the Square Another technique that can be used to solve a quadratic equation is *completing the square*. If a quadratic equation can be written in the form $x^2 + kx = d$, where k and d are constants, then the equation can be solved using

$$x^2 + kx + \left(\frac{k}{2}\right)^2 = \left(x + \frac{k}{2}\right)^2.$$

For example, $k = 6$ in $x^2 + 6x = 7$, so add $\left(\frac{k}{2}\right)^2 = \left(\frac{6}{2}\right)^2 = 9$ to each side.

$$x^2 + 6x = 7 \qquad \text{Given equation}$$

$$x^2 + 6x + 9 = 7 + 9 \qquad \text{Add 9 to each side.}$$

$$(x+3)^2 = 16 \qquad \text{Factor the perfect square.}$$

$$x + 3 = \pm 4 \qquad \text{Square root property}$$

$$x = -3 \pm 4 \qquad \text{Subtract 3 from each side.}$$

$$x = 1 \quad \text{or} \quad x = -7 \qquad \text{Simplify.}$$

Algebra Review
To review factoring perfect square trinomials, see Chapter R (page R-35).

Completing the square is useful when solving quadratic equations that do not factor easily.

EXAMPLE 6   Completing the square

Solve each equation.
(a) $x^2 - 8x + 9 = 0$ (b) $2x^2 - 8x = 7$

SOLUTION
(a) Start by writing the equation in the form $x^2 + kx = d$.

$$x^2 - 8x + 9 = 0 \qquad \text{Given equation}$$
$$x^2 - 8x = -9 \qquad \text{Subtract 9 from each side.}$$
$$x^2 - 8x + \mathbf{16} = -9 + \mathbf{16} \qquad \text{Add } \left(\tfrac{k}{2}\right)^2 = \left(\tfrac{-8}{2}\right)^2 = 16.$$
$$(x - 4)^2 = 7 \qquad \text{Factor the perfect square.}$$
$$x - 4 = \pm\sqrt{7} \qquad \text{Square root property}$$
$$x = 4 \pm \sqrt{7} \qquad \text{Add 4 to each side.}$$

(b) Divide each side by 2 to obtain a 1 for the leading coefficient.

$$2x^2 - 8x = 7 \qquad \text{Given equation}$$
$$x^2 - 4x = \tfrac{7}{2} \qquad \text{Divide by 2.}$$
$$x^2 - 4x + \mathbf{4} = \tfrac{7}{2} + \mathbf{4} \qquad \text{Add } \left(\tfrac{-4}{2}\right)^2 = 4 \text{ to each side.}$$
$$(x - 2)^2 = \tfrac{15}{2} \qquad \text{Factor the perfect square.}$$
$$x - 2 = \pm\sqrt{\tfrac{15}{2}} \qquad \text{Square root property}$$
$$x = 2 \pm \sqrt{\tfrac{15}{2}} \qquad \text{Add 2 to each side.}$$

Now Try Exercises 67 and 69 ◆

The Quadratic Formula   The quadratic formula can be used to find the solutions to any quadratic equation. Its derivation is based on completing the square and left as an exercise (see Exercise 114.)

QUADRATIC FORMULA

The solutions to the quadratic equation $ax^2 + bx + c = 0$, where $a \neq 0$, are given by

$$x = \frac{-b \pm \sqrt{b^2 - 4ac}}{2a}.$$

EXAMPLE 7   Using the quadratic formula

Solve the equation $3x^2 - 6x + 2 = 0$.

SOLUTION   Let $a = 3, b = -6$, and $c = 2$.

$$x = \frac{-b \pm \sqrt{b^2 - 4ac}}{2a} \qquad \text{Quadratic formula}$$

$$x = \frac{-(-6) \pm \sqrt{(-6)^2 - 4(3)(2)}}{2(3)}$$ Substitute for a, b, and c.

$$x = \frac{6 \pm \sqrt{12}}{6}$$ Simplify.

$$x = 1 \pm \frac{1}{6}\sqrt{12}$$ Divide.

Algebra Review

To review simplifying square roots, see Chapter R (page R-56).

Note: Because $\sqrt{12} = \sqrt{4} \cdot \sqrt{3} = 2\sqrt{3}$, we can write $1 \pm \frac{1}{6}\sqrt{12}$ as $1 \pm \frac{1}{3}\sqrt{3}$ if desired.

Now Try Exercise 19 ◆

◆ **CLASS DISCUSSION**

Use the results of Example 7 to evaluate each expression mentally.

$$3\left(1 + \tfrac{1}{6}\sqrt{12}\right)^2 - 6\left(1 + \tfrac{1}{6}\sqrt{12}\right) + 2, \qquad 3\left(1 - \tfrac{1}{6}\sqrt{12}\right)^2 - 6\left(1 - \tfrac{1}{6}\sqrt{12}\right) + 2 \;\blacklozenge$$

EXAMPLE 8

Estimating AIDS deaths

Solve the quadratic equation $2375x^2 + 5134x - 494{,}980 = 0$ (as discussed earlier in this section on page 190) to estimate when the number of AIDS deaths reached 500,000. Use symbolic and graphical methods.

SOLUTION

Symbolic Solution Apply the quadratic formula with $a = 2375$, $b = 5134$, and $c = -494{,}980$.

$$x = \frac{-b \pm \sqrt{b^2 - 4ac}}{2a}$$

$$x = \frac{-5134 \pm \sqrt{5134^2 - 4(2375)(-494{,}980)}}{2(2375)}$$

$$x = \frac{-5134 \pm \sqrt{4{,}728{,}667{,}956}}{4750}$$

$$x \approx 13.4 \qquad \text{or} \qquad x \approx -15.6$$

Calculator Help

To find a zero, or x-intercept, see Appendix B (page AP-12).

Since $x = 0$ corresponds to 1984, $x = -15.6$ represents $1984 - 15.6 = 1968.4$, and $x = 13.4$ represents $1984 + 13.4 = 1997.4$. AIDS was unknown in 1968, so the model estimates that about 500,000 AIDS deaths were reported by 1997.

Graphical Solution Graph of $Y_1 = 2375X^2 + 5134X - 494980$. Two x-intercepts, or zeros, are located at $x \approx -15.6$ and $x \approx 13.4$, as shown in Figures 3.40 and 3.41. These graphs support our symbolic solutions.

◆ **CLASS DISCUSSION**

The number of reported AIDS deaths through 1997 was actually 390,242. Discuss a reason why the number of deaths was not 500,000, as was predicted by function D in Example 8. What are some problems with making predictions with modeling functions? ◆

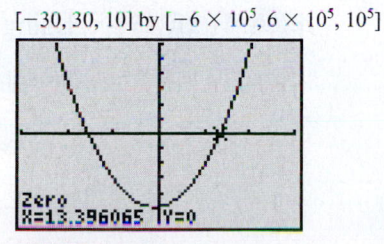

$[-30, 30, 10]$ by $[-6 \times 10^5, 6 \times 10^5, 10^5]$ $[-30, 30, 10]$ by $[-6 \times 10^5, 6 \times 10^5, 10^5]$

FIGURE 3.40   **FIGURE 3.41**

Now Try Exercise 99 ◆

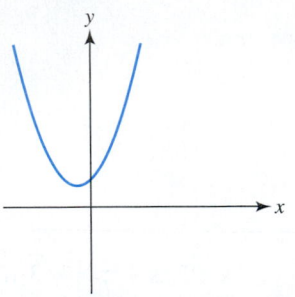

FIGURE 3.42 No Real Solutions

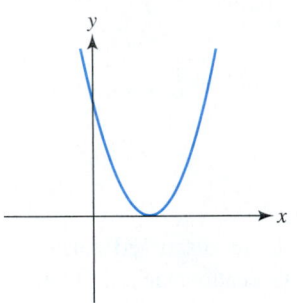

FIGURE 3.43 One Real Solution

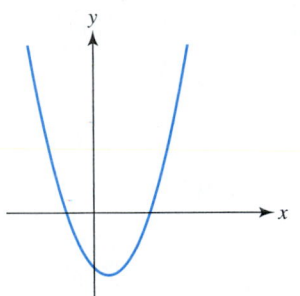

FIGURE 3.44 Two Real Solutions

◆ **MAKING CONNECTIONS**

Quadratic Equations and Graphical Solutions   When solving a quadratic equation graphically, such as $4x^2 - 2x = 3$, we can graph $y_1 = 4x^2 - 2x$ and $y_2 = 3$. The x-coordinate of each point of intersection represents a solution. This *intersection-of-graphs method* was used in Example 3. A second approach is to write the equation as $4x^2 - 2x - 3 = 0$ and graph $y_1 = 4x^2 - 2x - 3$. Each x-intercept, or zero, represents a solution. This *x-intercept method* was used in Example 8. You can use either graphical method to solve a quadratic equation.

The Discriminant   One important difference between linear and nonlinear equations is that nonlinear equations can have more than one solution. If the quadratic equation $ax^2 + bx + c = 0$ is solved graphically, the parabola $y = ax^2 + bx + c$ can intersect the x-axis 0, 1, or 2 times, as illustrated in Figures 3.42–3.44, respectively. Each x-intercept is a real solution to the quadratic equation $ax^2 + bx + c = 0$.

   The quantity $b^2 - 4ac$ in the quadratic formula is called the **discriminant**. It provides information about the number of real solutions to a quadratic equation.

QUADRATIC EQUATIONS AND THE DISCRIMINANT

To determine the number of real solutions to $ax^2 + bx + c = 0$, with $a \neq 0$, evaluate the discriminant $b^2 - 4ac$.

1. If $b^2 - 4ac > 0$, there are two real solutions.
2. If $b^2 - 4ac = 0$, there is one real solution.
3. If $b^2 - 4ac < 0$, there are no real solutions.

Note: When $b^2 - 4ac < 0$, the solutions to a quadratic equation may be expressed as complex numbers. Complex numbers are discussed in Section 4.4. If a, b, and c are *integers* and the discriminant is a perfect square, then the solutions to the quadratic equation are rational numbers.

   In Example 7 the discriminant is

$$b^2 - 4ac = (-6)^2 - 4(3)(2) = 12 > 0.$$

Because the discriminant is positive, there are two real solutions.

EXAMPLE 9   Using the discriminant

Use the discriminant to determine the number of solutions to the quadratic equation $9x^2 - 12.6x + 4.41 = 0$. Then solve the equation by using the quadratic formula. Support your result graphically.

SOLUTION
Symbolic Solution   Let $a = 9$, $b = -12.6$, and $c = 4.41$. The discriminant is given by

$$b^2 - 4ac = (-12.6)^2 - 4(9)(4.41) = 0.$$

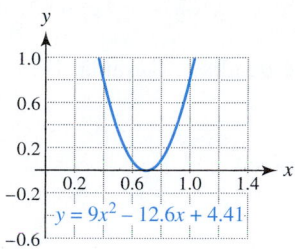

FIGURE 3.45

Since the discriminant is 0, there is one solution.

$$x = \frac{-b \pm \sqrt{b^2 - 4ac}}{2a} \qquad \text{Quadratic formula}$$

$$x = \frac{-(-12.6) \pm \sqrt{0}}{18} \qquad \text{Substitute.}$$

$$x = 0.7 \qquad \text{Simplify.}$$

The only solution is 0.7.

Graphical Solution   A graph of $y = 9x^2 - 12.6x + 4.41$ is shown in Figure 3.45. Notice that the graph suggests that there is one x-intercept corresponding to 0.7.

Now Try Exercise 47 ◆

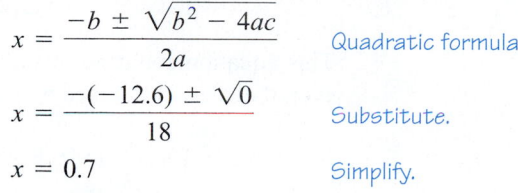

◆ **CLASS DISCUSSION**
Explain how you could determine that the quadratic equation $ax^2 + bx + c = 0$ has no real solutions (a) symbolically and (b) graphically. ◆

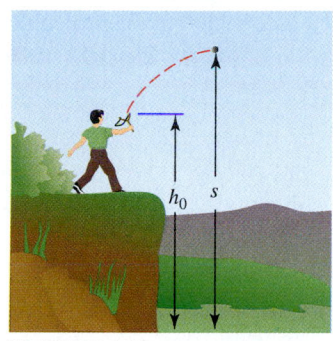

FIGURE 3.46

In Section 3.1 we learned that the height s of an object propelled into the air is modeled by $s(t) = -16t^2 + v_0 t + h_0$, where v_0 represents the object's (vertical) initial velocity in feet per second and h_0 represents the object's initial height in feet, as illustrated in Figure 3.46. In the next example we model the position of a projectile.

EXAMPLE 10 Modeling projectile motion

Table 3.6 shows the height of a projectile fired into the air.

Table 3.6   Height of a Projectile

t (seconds)	0	2	4	6	8
$s(t)$ (feet)	96	400	576	624	544

(a) Use $s(t) = -16t^2 + v_0 t + h_0$ to model the data.
(b) After how many seconds did the projectile strike the ground?

SOLUTION
(a) If $t = 0$, then $s(0) = 96$ so the initial height is $h_0 = 96$, and

$$s(t) = -16t^2 + v_0 t + 96.$$

The value of v_0 can be determined by noting that when $t = 2$, $s(2) = \mathbf{400}$. (Other values for t and $s(t)$ could be used.) Substituting gives the following result.

$$-16(\mathbf{2})^2 + v_0(\mathbf{2}) + 96 = \mathbf{400} \qquad s(2) = 400$$

$$2v_0 = 368 \qquad \text{Subtract 32 from each side.}$$

$$v_0 = 184 \qquad \text{Divide each side by 2.}$$

Thus $s(t) = -16t^2 + 184t + 96$ models the height of the projectile.

(b) The projectile strikes the ground when $s(t) = 0$, or when

$$-16t^2 + 184t + 96 = 0.$$

This equation could be solved graphically or by using the quadratic formula. However, it also can be solved by factoring.

$$-16t^2 + 184t + 96 = 0 \qquad \text{\textit{Equation to be solved.}}$$
$$2t^2 - 23t - 12 = 0 \qquad \text{\textit{Divide each term by }} -8.$$
$$(2t + 1)(t - 12) = 0 \qquad \text{\textit{Factor the trinomial.}}$$
$$t = -\frac{1}{2} \quad \text{or} \quad t = 12 \qquad \text{\textit{Solve the equation.}}$$

Thus, the projectile strikes the ground after 12 seconds. The solution of $-\frac{1}{2}$ has no physical meaning in this problem because it corresponds to a time before the projectile is fired into the air.   *Now Try Exercise 105* ◆

Problem Solving

Many types of applications involve quadratic equations. To solve these problems, we use the steps for "Solving Application Problems" from Section 2.3 on page 114. In the next example the dimensions of a box are found by solving a quadratic equation.

 **Solving a construction problem**

A box is being constructed by cutting 2-inch squares from the corners of a rectangular piece of cardboard that is 6 inches longer than it is wide, as illustrated in Figure 3.47. If the box is to have a volume of 224 cubic inches, find the dimensions of the piece of cardboard.

Geometry Review
To review formulas related to boxes, see Chapter R (page R-3).

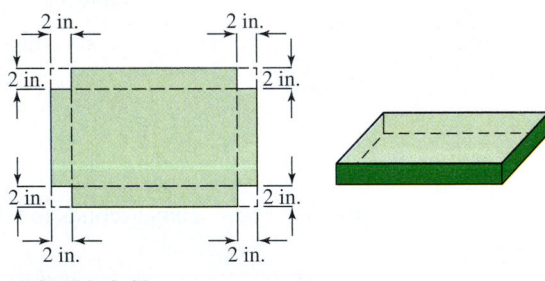

FIGURE 3.47

SOLUTION

STEP 1: The rectangular piece of cardboard is 6 inches longer than it is wide. Let x be its width and $x + 6$ be its length.

x: Width of the cardboard in inches

$x + 6$: Length of the cardboard in inches

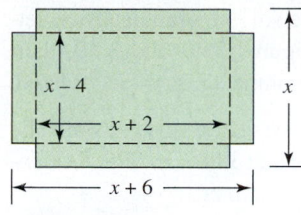

FIGURE 3.48

STEP 2: First make a drawing of the box with the appropriate labeling, as shown in Figure 3.48. The width of the bottom of the box is $x - 4$ inches, because two square corners, with sides of 2 inches have been removed. Similarly, the length of the bottom of the box is $x + 2$ inches. Because the height times the width times the length must equal the volume, or 224 cubic inches, it follows that

$$2(x - 4)(x + 2) = 224$$

or

$$(x - 4)(x + 2) = 112.$$

STEP 3: Write the quadratic equation in the form $ax^2 + bx + c = 0$ and factor.

$x^2 - 2x - 8 = 112$	*Equation to be solved*
$x^2 - 2x - 120 = 0$	*Subtract 112.*
$(x - 12)(x + 10) = 0$	*Factor.*
$x = 12$ or $x = -10$	*Zero-product property*

Since the dimensions cannot be negative, the width of the cardboard is 12 inches and the length is 6 inches more, or 18 inches.

STEP 4: After the 2-inch square corners are cut out, the dimensions of the bottom of the box are $12 - 4 = 8$ inches by $18 - 4 = 14$ inches. The volume of the box is then $2 \cdot 8 \cdot 14 = 224$ cubic inches, which checks.   *Now Try Exercise 109* ◆

When compact discs (CDs) are manufactured, a discount is sometimes given to a customer who makes a large order. Discounts affect the revenue that a company receives. We discuss this situation in the next example.

EXAMPLE 12    ### Determining revenue

A company charges \$5 to burn (make) one CD, but it reduces this cost by \$0.05 per CD for each additional CD ordered, up to a maximum of 60 CDs. For example, the price for one CD is \$5, the price for two CDs is 2(\$4.95) = \$9.90, the price for 3 CDs is 3(\$4.90) = \$14.70, and so on. If the total price is \$95, how many CDs were ordered?

SOLUTION

STEP 1: We are asked to find the number of CDs that results in an order of \$95. Let this number be x.

$$x: \text{Number of CDs ordered}$$

STEP 2: Revenue equals the number of CDs sold times the price of each CD. If x CDs are sold, then the price in dollars of each CD is $5 - 0.05(x - 1)$. (Note that when $x = 1$ the price is $5 - 0.05(1 - 1) = \$5$.) The revenue R is given by

$$R(x) = x(5 - 0.05(x - 1))$$

and we must solve the equation

$$x(5 - 0.05(x - 1)) = 95 \quad \text{or} \quad -0.05x^2 + 5.05x - 95 = 0.$$

STEP 3: Although this equation could be solved symbolically, we solve it graphically by letting $Y_1 = X(5 - 0.05(X - 1))$ and $Y_2 = 95$. In Figures 3.49 and 3.50, their graphs intersect at $(25, 95)$ and $(76, 95)$. Either 25 or 76 compact discs were ordered.

STEP 4: If 25 compact discs are ordered, then the cost of each disc would be $5 - 0.05(24) = \$3.80$ and the total revenue would be $25(3.80) = \$95$. The solution for 76 compact discs can be checked in a similar manner.

[0, 100, 10] by [0, 150, 50]

[0, 100, 10] by [0, 150, 50]

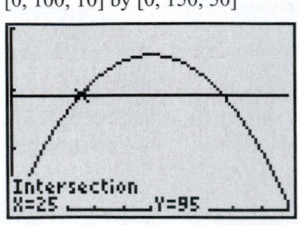

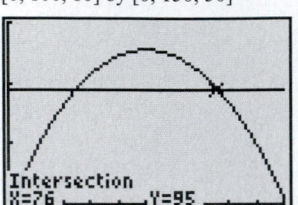

FIGURE 3.49   **FIGURE 3.50**   *Now Try Exercise 112* ◆

3.2   Putting it all Together

The following table summarizes important topics related to quadratic equations.

Concept	Explanation	Examples
Quadratic equation	$ax^2 + bx + c = 0$, where a, b, and c are constants with $a \neq 0$.	A quadratic equation can have zero, one, or two real solutions. $$x^2 = -5 \qquad \text{(No real solutions)}$$ $$(x - 2)^2 = 0 \qquad \text{(One real solution)}$$ $$x^2 - 4 = 0 \qquad \text{(Two real solutions)}$$
Factoring	A symbolic technique for solving equations; based on the zero-product property: if $ab = 0$, then either $a = 0$ or $b = 0$.	$$x^2 - 3x = -2$$ $$x^2 - 3x + 2 = 0$$ $$(x - 1)(x - 2) = 0$$ $$x - 1 = 0 \quad \text{or} \quad x - 2 = 0$$ $$x = 1 \quad \text{or} \quad x = 2$$
Square root property	The solutions to $x^2 = k$ are $x = \pm \sqrt{k}$, where $k \geq 0$.	$x^2 = 9$ is equivalent to $x = \pm 3$. $x^2 = 11$ is equivalent to $x = \pm \sqrt{11}$.
Completing the square	To solve $x^2 + kx = d$ symbolically, add $\left(\frac{k}{2}\right)^2$ to each side to obtain a perfect trinomial square. Use the square root property to solve.	$$x^2 - 6x = 1$$ $$x^2 - 6x + 9 = 1 + 9 \qquad \left(\frac{-6}{2}\right)^2 = 9$$ $$(x - 3)^2 = 10$$ $$x - 3 = \pm \sqrt{10}$$ $$x = 3 \pm \sqrt{10}$$

Concept	Explanation	Examples
Quadratic formula	The solutions to $ax^2 + bx + c = 0$ are given by $$x = \frac{-b \pm \sqrt{b^2 - 4ac}}{2a}.$$ Always gives the *exact* solutions	To solve $2x^2 - x - 4 = 0$, let $a = 2$, $b = -1$, and $c = -4$. $$x = \frac{-(-1) \pm \sqrt{(-1)^2 - 4(2)(-4)}}{2(2)}$$ $$= \frac{1 \pm \sqrt{33}}{4} \approx 1.69 \text{ or } -1.19$$
Graphical solution	To solve $ax^2 + bx + c = 0$, let y_1 equal the left side of the equation and graph y_1. The solutions correspond to the x-intercepts.	To solve $x^2 + x - 2 = 0$, graph $y = x^2 + x - 2$. The x-intercepts are -2 and 1. $y = x^2 + x - 2$
Numerical solution	To solve $ax^2 + bx + c = 0$, let y_1 equal the left side of the equation and create a table for y_1. The zeros of y_1 are the solutions. May not be a good method when solutions are fractions or irrational numbers	To solve $2x^2 + x - 1 = 0$, make a table for $Y_1 = 2X^2 + X - 1$. The solutions are $x = -1$ or $x = 0.5$.

3.2 Exercises

Quadratic Equations

Exercises 1–24: Solve the quadratic equation.

1. $x^2 + x - 11 = 1$

2. $x^2 - 9x + 10 = -8$

3. $t^2 = 2t$

4. $t^2 - 7t = 0$

5. $2z^2 + 13z + 15 = 0$

6. $4z^2 + 29z + 7 = 0$

7. $x(3x + 14) = 5$

8. $x(5x + 19) = 4$

9. $6x^2 + \frac{5}{2} = 8x$

10. $8x^2 + 63 = -46x$

11. $(t + 3)^2 = 5$

12. $(t - 2)^2 = 11$

13. $4x^2 - 13 = 0$

14. $9x^2 - 11 = 0$

15. $2(x - 1)^2 + 4 = 0$

16. $-3(x + 5)^2 - 6 = 0$

17. $\frac{1}{2}x^2 - 3x + \frac{1}{2} = 0$

18. $\frac{3}{4}x^2 + \frac{1}{2}x - \frac{1}{2} = 0$

19. $-3z^2 - 2z + 4 = 0$ **20.** $-4z^2 + z + 1 = 0$

21. $25k^2 + 1 = 10k$ **22.** $49k^2 + 4 = -28k$

23. $-0.3x^2 + 0.1x = -0.02$ **24.** $-0.1x^2 + 1 = 0.5x$

Exercises 25–30: Find the exact values of any x-intercepts shown in the graph.

25.
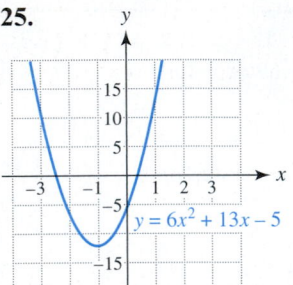
$y = 6x^2 + 13x - 5$

26.
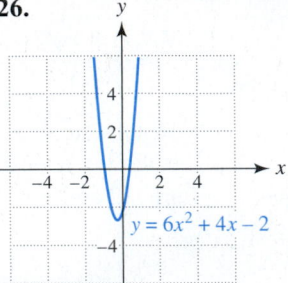
$y = 6x^2 + 4x - 2$

27.
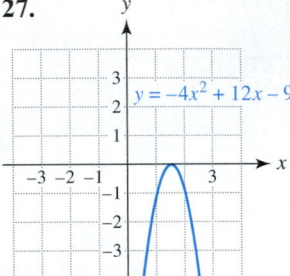
$y = -4x^2 + 12x - 9$

28.
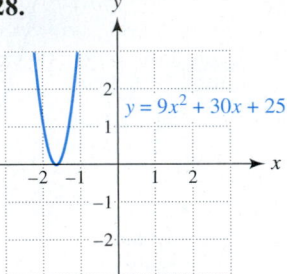
$y = 9x^2 + 30x + 25$

29.
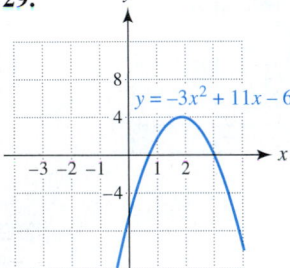
$y = -3x^2 + 11x - 6$

30.
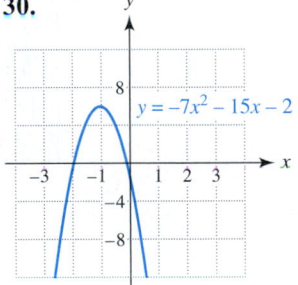
$y = -7x^2 - 15x - 2$

Exercises 31–38: Solve each quadratic equation

(a) *graphically,*
(b) *numerically, and*
(c) *symbolically.*

Express graphical and numerical solutions to the nearest tenth when appropriate.

31. $x^2 + 2x = 0$ **32.** $x^2 - 4 = 0$

33. $x^2 - x - 6 = 0$ **34.** $2x^2 + 5x - 3 = 0$

 35. $2x^2 = 6$ **36.** $x^2 - 225 = 0$

37. $4x^2 - 12x = -9$ **38.** $-4x(x - 1) = 1$

Exercises 39–44: Solve the quadratic equation graphically.

39. $20x^2 + 11x = 3$ **40.** $x^2 - 3.1x = 0.32$

41. $2.5x^2 = 4.75x - 2.1$ **42.** $-2x^2 + 4x = 1.595$

43. $x(x + 24) = 6912$

44. $x(31.4 - 2x) = -348$

Exercises 45–60: Complete the following.

(a) *Write the equation as $ax^2 + bx + c = 0$.*
(b) *Calculate the discriminant $b^2 - 4ac$ and determine the number of real solutions.*
(c) *Solve the equation.*

45. $3x^2 = 12$ **46.** $8x^2 - 2 = 14$

47. $x^2 - 2x = -1$ **48.** $6x^2 = 4x$

49. $4x - 2 = x^2$ **50.** $2x^2 - 3x - 3 = 0$

51. $x^2 + 1 = x$ **52.** $2x^2 + x = 2$

53. $2x^2 + 3x = 12 - 2x$ **54.** $3x^2 - 2 = 5x$

55. $\frac{1}{4}x^2 + 3x = x - 4$ **56.** $9x(x - 4) = -36$

57. $x\left(\frac{1}{2}x + 1\right) = -\frac{13}{2}$ **58.** $4x = 6 + x^2$

59. $3x^2 = 1 - x$ **60.** $x(5x - 3) = 1$

Exercises 61–64: The graph of $f(x) = ax^2 + bx + c$ is shown in the figure.

(a) *State whether $a > 0$ or $a < 0$.*
(b) *Solve the equation $ax^2 + bx + c = 0$.*
(c) *Determine if the discriminant is positive, negative, or zero.*

61.
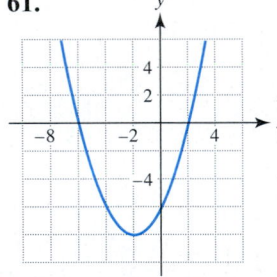

62.
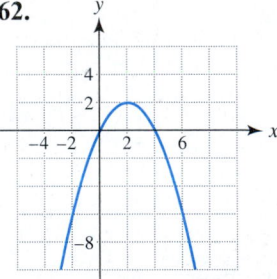

63.
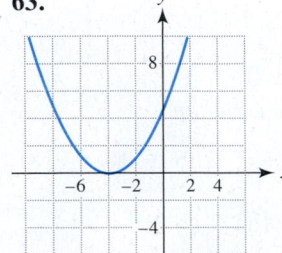

64.
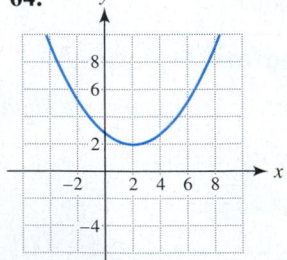

Exercises 65–80: Solve the equation by completing the square.

65. $x^2 + 4x = 6$ **66.** $x^2 - 10x = 1$

67. $x^2 + 5x - 4 = 0$ **68.** $2x^2 + 4x - 5 = 0$

69. $3x^2 - 6x - 2 = 0$ **70.** $2x^2 - 3x + 1 = 0$

71. $x^2 - 8x = 10$ **72.** $5x^2 - 10x = 2$

73. $\frac{1}{2}t^2 - \frac{3}{2}t = 1$ **74.** $\frac{1}{3}t^2 + \frac{1}{2}t = 2$

75. $-2z^2 + 3z + 1 = 0$ **76.** $-3z^2 - 5z + 3 = 0$

77. $-5x^2 + 7x + 2 = 0$ **78.** $-4x^2 + 5x + 2 = 0$

79. $-\frac{3}{2}z^2 - \frac{1}{4}z + 1 = 0$ **80.** $-\frac{1}{5}z^2 - \frac{1}{2}z + 2 = 0$

Exercises 81–84: Find the domain of the function. Write your answer in set-builder notation.

81. $f(x) = \dfrac{1}{x^2 - 5}$ **82.** $f(x) = \dfrac{4x}{7 - x^2}$

83. $g(t) = \dfrac{5 - t}{t^2 - t - 2}$ **84.** $g(t) = \dfrac{t + 1}{2t^2 - 11t - 21}$

Exercises 85–90: (Refer to Example 5.) Solve the equation for y. Determine if y is a function of x.

85. $4x^2 + 3y = \dfrac{y + 1}{3}$ **86.** $\dfrac{x^2 + y}{2} = y - 2$

87. $3x^2 + 4y^2 = 12$ **88.** $2x^2 + 9y^2 = 18$

89. $x - 25y^2 = 50$ **90.** $4x - 2y^2 = 9$

Exercises 91–96: Solve the equation for the specified variable.

91. $V = \frac{1}{3}\pi r^2 h$ for r **92.** $V = \frac{1}{2}gt^2 + h$ for t

93. $K = \frac{1}{2}mv^2$ for v **94.** $W = I^2R$ for I

95. $s = -16t^2 + 100t$ for t

96. $T^2 - kT - k^2 = 0$ for T

Applications and Models

97. *Height of Baseball* (Refer to Example 4.) A baseball is dropped from a stadium seat that is 75 feet above the ground. Its height s in feet after t seconds is given by $s(t) = 75 - 16t^2$. Estimate to the nearest tenth of a second how long it takes for the baseball to strike the ground.

98. *Height of Baseball* A baseball is thrown *downward* with an initial velocity of 30 feet per second from a stadium seat that is 80 feet above the ground. Estimate

to the nearest tenth of a second how long it takes for the baseball to strike the ground.

99. *U.S. AIDS Deaths* (Refer to Example 8.) The equation

$$D(x) = 2375x^2 + 5134x + 5020$$

models the cumulative number of AIDS deaths x years after 1984. Estimate the year when there were 90,000 deaths.

100. *U.S. AIDS Cases* From 1984 to 1994 the cumulative number of AIDS cases can be modeled by the equation

$$C(x) = 3034x^2 + 14{,}018x + 6400,$$

where x represents years after 1984. Estimate the year when 200,000 AIDS cases had been diagnosed.

101. *Safe Runway Speed* The road (or taxiway) used by aircraft to exit from the runway should not contain sharp curves. The safe radius for any curve depends on the taxiing speed of an airplane. The table lists the recommended minimum radius R of these exit curves, where the taxiing speed of the airplane is x miles per hour. (**Source:** Federal Aviation Administration.)

x (mph)	10	20	30	40	50	60
R (ft)	50	200	450	800	1250	1800

(a) If the taxiing speed x of the plane doubles, what happens to the minimum radius R of the curve?

(b) The Federal Aviation Administration (FAA) used $R(x) = ax^2$ to compute the values in the table. Determine the constant a.

(c) If $R = 500$, find x. Interpret your results.

102. *Falling Object* The table lists the velocity and distance traveled by a falling object.

Elapsed Time (sec)	0	1	2	3	4	5
Velocity (ft/sec)	0	32	64	96	128	160
Distance (ft)	0	16	64	144	256	400

(a) Make a scatterplot of the ordered pairs determined by (time, velocity) and (time, distance) in the same viewing rectangle $[-1, 6, 1]$ by $[-10, 450, 50]$.

(b) Find a function v that models the velocity.

(c) The distance is modeled by $d(x) = ax^2$. Find the constant a.

(d) Use the table to mentally estimate the elapsed time

when the distance traveled by the falling object is 200 feet. Determine this time symbolically using d. Find the velocity at this time.

103. *Wal-Mart Employees*   The following table lists actual and projected numbers of Wal-Mart employees E in millions x years after 1987. (**Source:** Wal-Mart.)

x	0	5	10	15	20
E	0.20	0.38	0.68	1.4	2.2

   (a) Evaluate $E(15)$ and interpret the result.

   (b) Find a quadratic function E that models these data.

   (c) Graph the data and function E in the same xy-plane.

   (d) Use E to estimate the year when the number of employees may reach 3 million.

104. *Biology*   Some types of worms have a remarkable capacity to live without moisture. The table shows the number of worms y surviving after x days in one study. (**Source:** D. Brown and P. Rothery, *Models in Biology: Mathematics, Statistics and Computing*.)

x (days)	0	20	40	80	120	160
y (worms)	50	48	45	36	20	3

   (a) Find a quadratic function in the form $f(x) = a(x - h)^2 + k$ that models these data.

   (b) Solve the quadratic equation $f(x) = 0$. Do both solutions have real meaning? Explain.

105. *Projectile Motion*   (Refer to Example 10.) The accompanying table shows the height of a projectile that is fired into the air.

t (seconds)	0	1	2	3	4
s (feet)	32	176	288	368	416

   (a) Find $s(t) = -16t^2 + v_0 t + h_0$ so that s models the data.

   (b) After how many seconds did the projectile strike the ground?

106. *Braking Distance*   Braking distance for cars can be approximated by $D(x) = \frac{x^2}{30k}$. The input x is the car's velocity in miles per hour and the output $D(x)$ is the braking distance in feet. The positive constant k is a measure of the traction of the tires. Small values of k indicate a

slippery road.   (**Source:** L. Haefner, *Introduction to Transportation Systems.*)

   (a) Let $k = 0.3$. Evaluate $D(60)$ and interpret the result.

   (b) If $k = 0.25$, find the velocity x that corresponds to a braking distance of 300 feet.

107. *Screen Dimensions*   The width of a rectangular computer screen is 2.5 inches more than its height. If the area of the screen is 93.5 square inches, determine its dimensions.

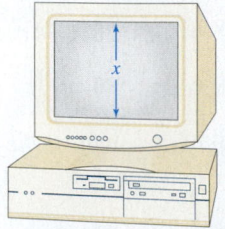

108. *Maximizing Area*   A rectangular pen for a pet is under construction using 100 feet of fence.
   (a) Determine the dimensions that result in an area of 576 square feet.

   (b) Find the dimensions that give maximum area.

109. *Construction*   (Refer to Example 11.) A box is being constructed by cutting 4-inch squares from the corners of a rectangular sheet of metal that is 10 inches longer than it is wide. If the box is to have a volume of 476 cubic inches, find the dimensions of the metal sheet.

110. *Construction*   A cylindrical aluminum can is being constructed to have a height h of 4 inches, as illustrated in the accompanying figure. If the can is to have a volume of 28 cubic inches, approximate its radius r.

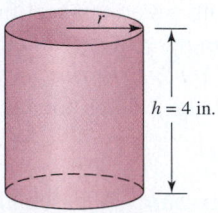

Geometry Review   To review formulas related to cylinders, see Chapter R (page R-4).

111. *Picture Frame* A frame for a picture is 2 inches wide. The picture inside the frame is 4 inches longer than it is wide. See the accompanying figure. If the area of the picture is 320 square inches, find the outside dimensions of the picture frame.

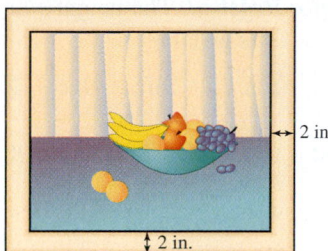

2 in

2 in.

112. *Ticket Prices* One airline ticket costs $250. For each additional airline ticket sold to a group, the price of all the tickets is reduced by $2. For example, 2 tickets cost $2 \cdot 248 = \$496$ and 3 tickets cost $3 \cdot 246 = \$738$.
   (a) Write a quadratic function that gives the total cost of buying x tickets.

   (b) What is the cost of 5 tickets?

   (c) How many tickets were sold if the total cost was $5200?

   (d) What number of tickets sold results in the greatest cost?

113. *Window Dimensions* A window comprises a square with sides of length x and a semicircle with diameter x, as shown in the accompanying figure. If the total area

of the window is 463 square inches, estimate the value of x to the nearest hundredth of an inch.

Geometry Review To review formulas related to circles, see Chapter R (page R-2).

114. *Quadratic Formula* Prove the quadratic formula by completing the following.
   (a) Write $ax^2 + bx + c = 0$ as $x^2 + \frac{b}{a}x = -\frac{c}{a}$.

   (b) Complete the square to obtain
   $$\left(x + \frac{b}{2a}\right)^2 = \frac{b^2 - 4ac}{4a^2}.$$

   (c) Use the square root property and solve for x.

Writing about Mathematics

115. Discuss three symbolic methods for solving a quadratic equation. Make up a quadratic equation and use each method to find the solution set.

116. Explain how to solve a quadratic equation graphically.

CHECKING BASIC CONCEPTS FOR SECTIONS 3.1 AND 3.2

1. Graph $f(x) = (x - 1)^2 - 4$. Identify the vertex, axis of symmetry, and x-intercepts.

2. A graph of $y = ax^2 + bx + c$ is shown to the right.
   (a) Which of the following is true: $a > 0, a < 0$, or $a = 0$?

   (b) Find the vertex and axis of symmetry.

   (c) Solve $ax^2 + bx + c = 0$.

   (d) Is the discriminant positive, negative, or zero?

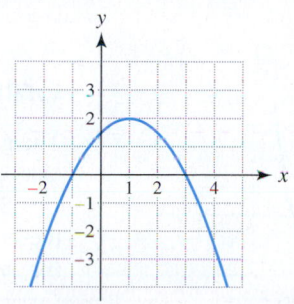

3. Use $f(x) = a(x - h)^2 + k$ to model the data in the table exactly.

x	-3	-2	-1	0	1
$f(x)$	11	5	3	5	11

4. Find the vertex on the graph of $y = 3x^2 - 9x - 2$.

5. Write $f(x) = x^2 + 4x - 3$ as $f(x) = a(x - h)^2 + k$. What are the coordinates of the vertex? What is the minimum y-value on the graph of f?

6. Solve the quadratic equations.
   (a) $16x^2 = 81$

   (b) $2x^2 + 3x = 2$

   (c) $x^2 = x - 3$

   (d) $2x^2 = 3x + 4$

7. *Dimensions of a Rectangle* A rectangle is 4 inches longer than it is wide and has an area of 165 square inches. Find its dimensions.

8. *Height of a Baseball* The height s of a baseball in feet after t seconds is given by

$$s(t) = -16t^2 + 96t + 2.$$

   (a) What is the height of the baseball after 1 second?

   (b) After how many seconds is the baseball 142 feet high?

   (c) Find the maximum height of the baseball.

   (d) How long is the baseball in the air?

3.3 Quadratic Inequalities

♦ Solve quadratic inequalities graphically

♦ Solve quadratic inequalities symbolically

Introduction

Highway engineers often use quadratic functions to model safe stopping distances for cars. For example, $f(x) = \frac{1}{12}x^2 + \frac{11}{5}x$ is sometimes used to model the stopping distance for a car traveling at x miles per hour on dry, level pavement. If a driver can see only **200** feet ahead on a highway with a sharp curve, then safe driving speeds x satisfy the quadratic inequality

$$\frac{1}{12}x^2 + \frac{11}{5}x \le 200$$

or equivalently,

$$\frac{1}{12}x^2 + \frac{11}{5}x - 200 \le 0.$$

A quadratic equation can be written as $ax^2 + bx + c = 0$. If the equals sign is replaced by $>$, $\ge$, $<$, or $\le$, a **quadratic inequality** results. Since equality is (usually) the boundary between *greater than* and *less than*, a first step in solving a quadratic inequality is to determine the x-values where equality occurs. These x-values are the *boundary numbers*. We begin by discussing graphical solutions of quadratic inequalities.

Graphical Solutions

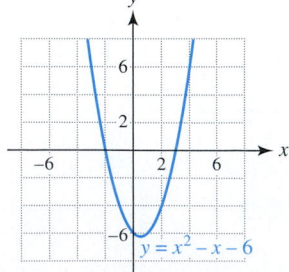

FIGURE 3.51

The graph of a quadratic function is a parabola that opens either upward or downward. For example, Figure 3.51 shows the parabola $y = x^2 - x - 6$.

Since $a = 1$, this parabola opens upward. It has x-intercepts -2 and 3, which satisfy the equation $x^2 - x - 6 = 0$. The parabola lies below the x-axis between the intercepts (or boundary numbers), so the solution set to the inequality $x^2 - x - 6 < 0$ is $\{x \mid -2 < x < 3\}$ or $(-2, 3)$ in interval notation. Similarly, the solutions to $x^2 - x - 6 > 0$ include x-values either left of $x = -2$ or right of $x = 3$, where the parabola is above the x-axis. This solution set is $\{x \mid x < -2 \text{ or } x > 3\}$ or $(-\infty, -2) \cup (3, \infty)$ in interval notation.

◆ MAKING CONNECTIONS

Graphs and Inequalities   Suppose that $y = ax^2 + bx + c$. Then the solution set to $y > 0$ includes all x-values where the graph of y is *above* the x-axis. Similarly, the solution set to $y < 0$ includes all x-values where the graph of y is *below* the x-axis. This is because the x-axis corresponds to $y = 0$. This discussion is illustrated in Figures 3.52 and 3.53.

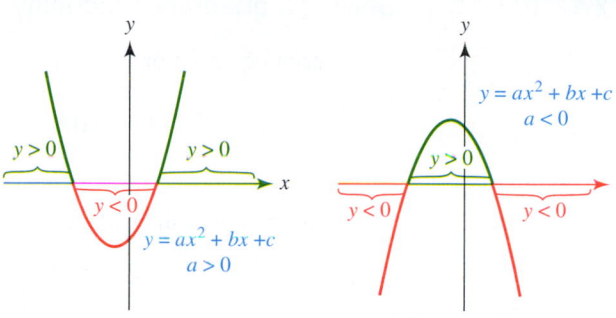

FIGURE 3.52   FIGURE 3.53

EXAMPLE 1   Solving quadratic inequalities graphically

Use the graphs of $y = ax^2 + bx + c$ in Figures 3.54–3.57 to solve the inequalities:
i. $ax^2 + bx + c \leq 0$ and **ii.** $ax^2 + bx + c > 0$.

(a) 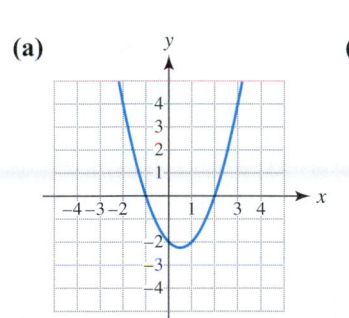   (b) 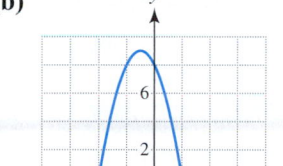   (c) 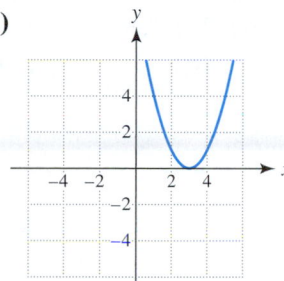   (d)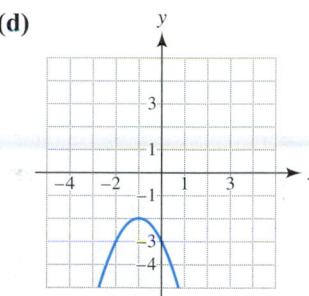

FIGURE 3.54   FIGURE 3.55   FIGURE 3.56   FIGURE 3.57

SOLUTION

(a) **i.** $ax^2 + bx + c \leq 0$ The x-intercepts or boundary numbers are -1 and 2. Between these x-values, the graph is below the x-axis, and the y-values are negative. The solution set is $\{x \mid -1 \leq x \leq 2\}$.
ii. $ax^2 + bx + c > 0$ The graph is above the x-axis, and the y-values are positive, either left of $x = -1$ or right of $x = 2$. The solution set is $\{x \mid x < -1 \text{ or } x > 2\}$.

(b) **i.** $ax^2 + bx + c \leq 0$ The parabola opens downward with x-intercepts or boundary numbers of -4 and 2. The graph is below the x-axis, and the y-values are negative, either left of $x = -4$ or right of $x = 2$. The solution set is $\{x \mid x \leq -4 \text{ or } x \geq 2\}$.
ii. $ax^2 + bx + c > 0$ Between $x = -4$ and $x = 2$, the graph is above the x-axis and the y-values are positive. The solution set is $\{x \mid -4 < x < 2\}$.

(c) **i.** $ax^2 + bx + c \leq 0$ The graph never dips below the x-axis so the y-values are never negative. There is one x-intercept at $x = 3$, where $y = 0$. The solution set is $\{x \mid x = 3\}$.
ii. $ax^2 + bx + c > 0$ The graph is always above the x-axis except at $x = 3$. Thus every real number except $x = 3$ is a solution. The solution set is $\{x \mid x \neq 3\}$.

(d) i. $ax^2 + bx + c \leq 0$ The graph always lies below the x-axis so the y-values are negative and always satisfy the given inequality. The solution set includes all real numbers or $\{x \mid -\infty < x < \infty\}$.

ii. $ax^2 + bx + c > 0$ The graph is never above the x-axis so the solution set is empty.   *Now Try Exercises 1, 3, and 5* ◆

EXAMPLE 2   Solving a quadratic inequality

Solve each equation or inequality. Write the solution set for each inequality in interval notation.

(a) $2x^2 - 3x - 2 = 0$ **(b)** $2x^2 - 3x - 2 < 0$ **(c)** $2x^2 - 3x - 2 > 0$

SOLUTION

(a) This equation can be solved by factoring.

$$(2x + 1)(x - 2) = 0 \qquad \text{\color{blue}Factor trinomial.}$$

$$x = -\frac{1}{2} \quad \text{or} \quad x = 2 \qquad \text{\color{blue}Zero-product property}$$

The solutions are $-\frac{1}{2}$ and 2.

(b) The graph of $y = 2x^2 - 3x - 2$ is a parabola opening upward. Its x-intercepts are $-\frac{1}{2}$ and 2. See Figure 3.58. This parabola is below the x-axis ($y < 0$) for x-values between $-\frac{1}{2}$ and 2, so the solution set is $\left(-\frac{1}{2}, 2\right)$.

(c) In Figure 3.59, the graph of $y = 2x^2 - 3x - 2$ is above the x-axis ($y > 0$) for x-values less than $-\frac{1}{2}$ or greater than 2, so the solution set is $\left(-\infty, -\frac{1}{2}\right) \cup \left(2, \infty\right)$.

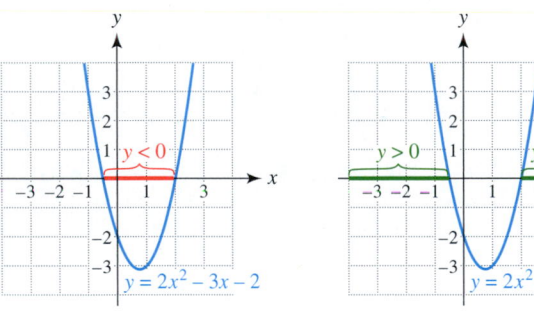

FIGURE 3.58   **FIGURE 3.59**

***Note:** It is not necessary to graph the parabola.* Instead you can simply *visualize* the parabola opening upward with x-intercepts $-\frac{1}{2}$ and 2.   *Now Try Exercise 11* ◆

◆ **CLASS DISCUSSION**

Sketch a graph of the equation $y = ax^2 + bx + c$ if the quadratic inequality $ax^2 + bx + c < 0$ satisfies the following conditions.

(a) $a > 0$, solution set: $\{x \mid -1 < x < 3\}$

(b) $a < 0$, solution set: $\{x \mid x \neq 1\}$

(c) $a < 0$, solution set: $\{x \mid x < -2 \text{ or } x > 2\}$ ◆

 **Determining safe speeds**

In the introduction to this section the quadratic inequality

$$\frac{1}{12}x^2 + \frac{11}{5}x \le 200$$

was explained. (Recall that $f(x) = \frac{1}{12}x^2 + \frac{11}{5}x$ computes the stopping distance in feet for a car traveling x miles per hour on dry, level pavement.) Solve this inequality to determine safe speeds on a curve where a driver can see the road ahead for at most 200 feet. What might be a safe speed limit to drive on this curve?

SOLUTION We can solve this inequality by graphing $Y_1 = X^2/12 + 11X/5$ and $Y_2 = 200$, as shown in Figure 3.60. Since we are interested in *positive* speeds, we need to locate only the point of intersection where x is positive. This occurs when $x \approx 37.5$. For positive x-values to the left of $x \approx 37.5$, $y_1 < 200$. Thus safe speeds are less than 37.5 miles per hour. A reasonable speed limit might be 35 miles per hour. *Now Try Exercise 63* ◆

[−100, 100, 50] by [−300, 300, 100]

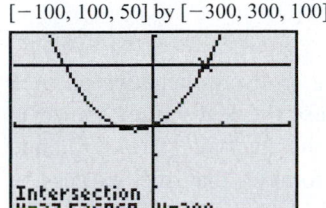

FIGURE 3.60

Calculator Help

To find a point of intersection, see Appendix B (page AP-12).

Symbolic Solutions

Although it is usually easier to solve a quadratic inequality graphically by visualizing a parabola and its x-intercepts, we can also solve a quadratic inequality symbolically without the aid of a graph. This symbolic method involves a table and is often used to solve more complicated inequalities. (See Section 4.6.)

SOLVING QUADRATIC INEQUALITIES

STEP 1: If necessary, write the inequality as $ax^2 + bx + c < 0$, where $<$ may be replaced by $>$, $\le$, or $\ge$

STEP 2: Solve the equation $ax^2 + bx + c = 0$ either graphically or symbolically. The solutions are called boundary numbers.

STEP 3: Use the boundary numbers to separate the number line into disjoint intervals. Note that on each interval, $y = ax^2 + bx + c$ is either always positive or always negative.

STEP 4: To solve the inequality, use either a graph or a table. For example, the solution set to $ax^2 + bx + c < 0$ corresponds to intervals where the graph of $y = ax^2 + bx + c$ is below the x-axis or to intervals where test values result in negative values.

For example, to solve

$$2x^2 - 5x - 12 < 0,$$

symbolically with a table, replace $<$ with $=$ and solve the equation by factoring.

$2x^2 - 5x - 12 = 0$	Quadratic equation
$(2x + 3)(x - 4) = 0$	Factor.
$2x + 3 = 0$ or $x - 4 = 0$	Zero-product property
$x = -\frac{3}{2}$ or $x = 4$	Solve.

Algebra Review

To review factoring see Chapter R (page R-29).

The boundary numbers $-\frac{3}{2}$ and 4 separate the number line into three disjoint intervals:

$$\left(-\infty, -\frac{3}{2}\right), \left(-\frac{3}{2}, 4\right), \text{ and } (4, \infty),$$

as illustrated in Figure 3.61.

The expression $2x^2 - 5x - 12$ is either always positive or always negative on each interval. To determine where $2x^2 - 5x - 12 < 0$ we can use the **test values** shown in Table 3.7. For example, since the test value -2 lies in the interval $\left(-\infty, -\frac{3}{2}\right)$, and $2x^2 - 5x - 12$ evaluated at $x = -2$ equals $6 > 0$, it follows that the expression $2x^2 - 5x - 12$ is always positive for $\left(-\infty, -\frac{3}{2}\right)$. This interval has + signs on the number line in Figure 3.62.

FIGURE 3.61   FIGURE 3.62

From Table 3.7 we can see that the expression $2x^2 - 5x - 12$ is negative when $x = 0$, so it is always negative between the boundary numbers of $-\frac{3}{2}$ and 4. Negative signs are shown on the number line in Figure 3.62 in the interval $\left(-\frac{3}{2}, 4\right)$. Finally, when $x = 6$, $2x^2 - 5x - 12 = 30 > 0$ and + signs are placed along the x-axis in Figure 3.62 when $x > 4$. Therefore, the solution set for $2x^2 - 5x - 12 < 0$ is $\left(-\frac{3}{2}, 4\right)$. Note that it is important to choose one test value less than $-\frac{3}{2}$, one test value between $-\frac{3}{2}$ and 4, and one test value greater than 4.

TABLE 3.7

Interval	Test Value x	$2x^2 - 5x - 12$	Positive or Negative?
$\left(-\infty, -\frac{3}{2}\right)$	-2	6	Positive
$\left(-\frac{3}{2}, 4\right)$	0	-12	Negative
$(4, \infty)$	6	30	Positive

 **EXAMPLE 4**   Solving a quadratic inequality

Solve $x^2 \geq 2 - x$ symbolically. Write the solution set in interval notation.

SOLUTION

STEP 1: Rewrite the inequality as $x^2 + x - 2 \geq 0$.

STEP 2: Solve $x^2 + x - 2 = 0$.

$$(x + 2)(x - 1) = 0 \qquad \textcolor{teal}{\text{Factor.}}$$

$$x = -2 \text{ or } x = 1 \qquad \textcolor{teal}{\text{Zero-product property}}$$

STEP 3: These two boundary numbers separate the number line into three disjoint intervals:

$$(-\infty, -2), \quad (-2, 1), \quad \text{and} \quad (1, \infty).$$

STEP 4: We choose the test values $x = -3, x = 0$, and $x = 2$. From Table 3.8, the expression $x^2 + x - 2$ is positive when $x = -3$ and $x = 2$.

TABLE 3.8

Interval	Test Value x	$x^2 + x - 2$	Positive or Negative?
$(-\infty, -2)$	-3	4	**Positive**
$(-2, 1)$	0	-2	**Negative**
$(1, \infty)$	2	4	**Positive**

Therefore the expression $x^2 + x - 2$ is positive when $x < -2$ or $x > 1$. Thus the solution set is $(-\infty, -2] \cup [1, \infty)$. The boundary numbers, -2 and 1, are included because the inequality involves $\geq$ rather than $>$.   *Now Try Exercise 55* ◆

An Application   In the next example we model an athlete's heart rate by using a quadratic function.

EXAMPLE  **5**   Modeling heart rate

Suppose that a person's heart rate, x minutes after vigorous exercise has stopped, can be modeled by $f(x) = \frac{4}{5}(x - 10)^2 + 80$. The output is in beats per minute, where the domain of f is $0 \leq x \leq 10$.
(a) Evaluate $f(0)$ and $f(2)$. Interpret the result.
(b) Estimate the times when the person's heart rate was between 100 and 120 beats per minute, inclusively.

SOLUTION
(a) $f(0) = \frac{4}{5}(0 - 10)^2 + 80 = 160$ and $f(2) = \frac{4}{5}(2 - 10)^2 + 80 = 131.2$. Initially when the person stops exercising the heart rate is 160 beats per minute, and after 2 minutes the heart rate has dropped to about 131 beats per minute.
(b) Graph $Y_1 = 100$, $Y_2 = (.8)(X - 10)^2 + 80$, and $Y_3 = 120$. Then find their points of intersection, as shown in Figures 3.63 and 3.64. The person's heart rate is between 100 and 120, when the graph of y_2 is between the graphs of y_1 and y_3. This occurs between 2.9 minutes (approximately) and 5 minutes after the person stops exercising. Numerical support is shown in Figure 3.65, where $y_1 \approx 120$ when $x = 3$, and $y_1 = 100$ when $x = 5$.

[0, 10, 1] by [60, 180, 20]   [0, 10, 1] by [60, 180, 20]

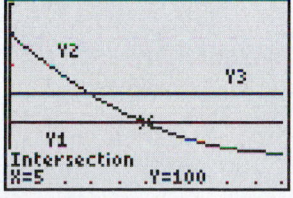   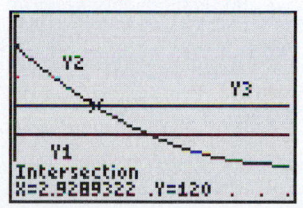

FIGURE 3.63   **FIGURE 3.64**   **FIGURE 3.65**

Now Try Exercise 61 ◆

3.3

Putting it all Together

In this section we solved different types of quadratic inequalities both graphically and symbolically. The following table summarizes important concepts related to solving quadratic inequalities.

Concept	Description
Quadratic inequality	Can be written as $ax^2 + bx + c < 0$, where $<$ may be replaced by $>$, $\leq$, or $\geq$. **Example:** $-x^2 + x < -2$ is a quadratic inequality because it can be written as $-x^2 + x + 2 < 0$.
Graphical solution	Write the inequality in the form $ax^2 + bx + c < 0$, where $<$ may be replaced by $>$, $\leq$, or $\geq$. Graph $y = ax^2 + bx + c$, and locate the x-intercepts, or boundary numbers. Use these numbers to determine x-values where the graph is below (above) the x-axis. In the accompanying figure, the inequality $-x^2 + x + 2 < 0$ is satisfied when either $x < -1$ or $x > 2$. The solution set is $(-\infty, -1) \cup (2, \infty)$.
Symbolic solution	Write the inequality as $ax^2 + bx + c < 0$, where $<$ may be replaced by $>$, $\leq$, or $\geq$. Solve the equation $ax^2 + bx + c = 0$. To determine where $y = ax^2 + bx + c$ is positive or negative, use a table of test values. **Example:** $-x^2 + x + 2 < 0$ Solving $-x^2 + x + 2 = 0$ results in $x = -1$ or $x = 2$. From the table the solution set is $(-\infty, -1) \cup (2, \infty)$.

Interval	Test Value x	$-x^2 + x + 2$	Positive or Negative?
$(-\infty, -1)$	-2	-4	**Negative**
$(-1, 2)$	0	2	**Positive**
$(2, \infty)$	3	-4	**Negative**

3.3 Exercises

Quadratic Inequalities

Exercises 1–6: The graph of $f(x) = ax^2 + bx + c$ is shown in the figure. Solve each inequality.

1. (a) $f(x) < 0$ **2.** (a) $f(x) > 0$

  (b) $f(x) \geq 0$ (b) $f(x) < 0$

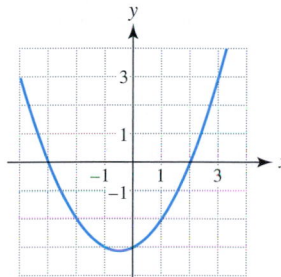 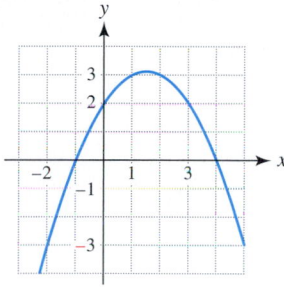

3. (a) $f(x) \leq 0$ **4.** (a) $f(x) \geq 0$

  (b) $f(x) > 0$ (b) $f(x) \leq 0$

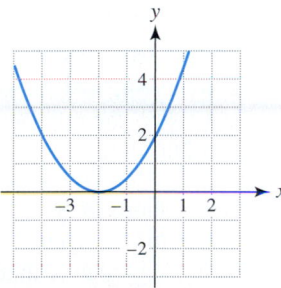 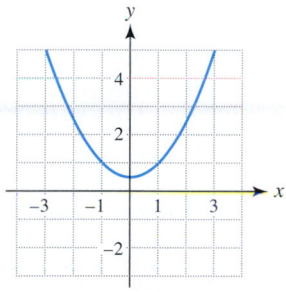

5. (a) $f(x) > 0$ **6.** (a) $f(x) \geq 0$

  (b) $f(x) < 0$ (b) $f(x) < 0$

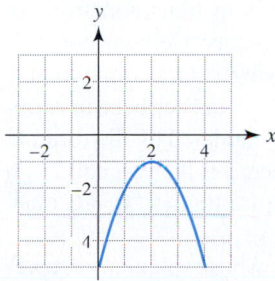 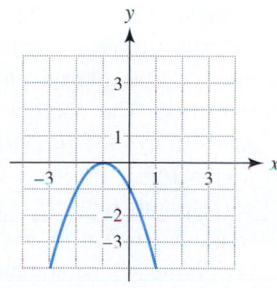

Exercises 7–10: Use the graph of $y = f(x)$ to solve each equation or inequality. Write the solution set for each inequality in interval notation.

  (a) $f(x) = 0$ (b) $f(x) < 0$ (c) $f(x) > 0$

7. 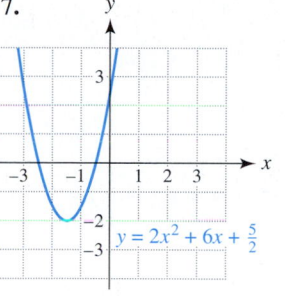   **8.**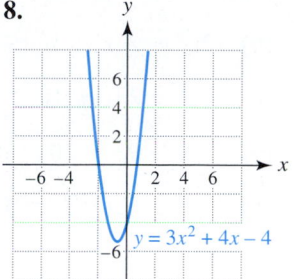

9. 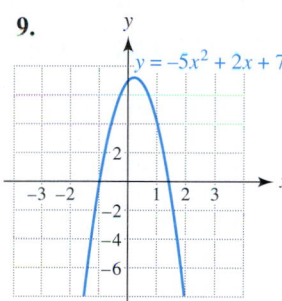   **10.**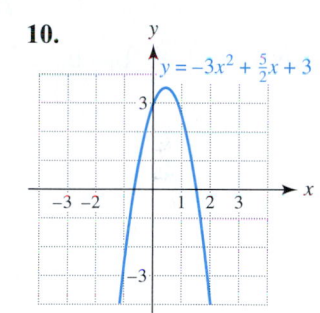

Exercises 11–22: Solve each equation and inequality. Write the solution set for each inequality in interval notation.

11. (a) $x^2 - x - 12 = 0$ **12.** (a) $x^2 - 8x + 12 = 0$

  (b) $x^2 - x - 12 < 0$ (b) $x^2 - 8x + 12 < 0$

  (c) $x^2 - x - 12 > 0$ (c) $x^2 - 8x + 12 > 0$

13. (a) $k^2 - 5 = 0$ **14.** (a) $n^2 - 17 = 0$

  (b) $k^2 - 5 \leq 0$ (b) $n^2 - 17 \leq 0$

  (c) $k^2 - 5 \geq 0$ (c) $n^2 - 17 \geq 0$

15. (a) $3x^2 + 8x = 0$ **16.** (a) $7x^2 - 4x = 0$

  (b) $3x^2 + 8x \leq 0$ (b) $7x^2 - 4x \leq 0$

  (c) $3x^2 + 8x \geq 0$ (c) $7x^2 - 4x \geq 0$

17. (a) $-4x^2 + 12x - 9 = 0$

  (b) $-4x^2 + 12x - 9 < 0$

  (c) $-4x^2 + 12x - 9 > 0$

18. (a) $x^2 + 2x + 1 = 0$

    (b) $x^2 + 2x + 1 < 0$

    (c) $x^2 + 2x + 1 > 0$

19. (a) $12z^2 - 23z + 10 = 0$

    (b) $12z^2 - 23z + 10 \leq 0$

    (c) $12z^2 - 23z + 10 \geq 0$

20. (a) $18z^2 + 9z - 20 = 0$

    (b) $18z^2 + 9z - 20 \leq 0$

    (c) $18z^2 + 9z - 20 \geq 0$

21. (a) $x^2 + 2x - 1 = 0$ **22. (a)** $2x^2 + 4x - 3 = 0$

    (b) $x^2 + 2x - 1 < 0$ **(b)** $2x^2 + 4x - 3 < 0$

    (c) $x^2 + 2x - 1 > 0$ **(c)** $2x^2 + 4x - 3 > 0$

Exercises 23–26: The accompanying table contains test values for a quadratic function $f(x) = ax^2 + bx + c$. *Use the table to determine the solution to each inequality.*

    (a) $f(x) > 0$ **(b)** $f(x) \leq 0$

23.

x	-2	-1	0	1	2
$f(x)$	3	0	-1	0	3

24.

x	-6	-4	1	5	8
$f(x)$	-22	0	20	0	-36

25.

x	-6	-4	-2	0	2
$f(x)$	0	4	0	-12	-32

26.

x	-4	-2	0	2	4
$f(x)$	16	0	-8	-8	0

Exercises 27–52: Solve the inequality.

27. $x^2 - 3x - 4 < 0$ **28.** $2x^2 + 5x + 2 \leq 0$

29. $x^2 + x > 6$ **30.** $-3x \geq 9 - 12x^2$

31. $x^2 \leq 4$ **32.** $2x^2 > 16$

33. $x(x - 4) \geq -4$ **34.** $x^2 - 3x - 10 < 0$

35. $-x^2 + x + 6 \leq 0$ **36.** $-x^2 - 2x + 8 > 0$

37. $6x^2 - x < 1$ **38.** $5x^2 \leq 10 - 5x$

39. $(x + 4)(x - 10) \leq 0$ **40.** $(x - 3.1)(x + 2.7) > 0$

41. $x^2 + 4x + 3 < 0$ **42.** $2x^2 + x + 4 < 0$

43. $x^2 + 2x \geq 35$ **44.** $9x^2 + 4 > 12x$

45. $x^2 \geq x$ **46.** $x^2 \geq -3$

47. $x(x - 1) \geq 6$ **48.** $x^2 - 9 < 0$

49. $x^2 - 5 \leq 0$ **50.** $0.5x^2 - 3.2x > -0.9$

51. $7x^2 + 515.2 \geq 179.8x$ **52.** $-10 < 3x - x^2$

Exercises 53–60: (Refer to Example 4.) Use a table to solve each inequality.

53. $x^2 - 9x + 14 \leq 0$ **54.** $x^2 + 10x + 21 > 0$

55. $x^2 \geq 3x + 10$ **56.** $x^2 < 3x + 4$

57. $x^2 - \frac{1}{2}x - 5 < 0$ **58.** $\frac{1}{8}x^2 + x + 2 \geq 0$

59. $x^2 > 3 - 4x$ **60.** $2x^2 \leq 1 - 4x$

Applications

61. *AIDS Deaths*   Let $f(x) = 2375x^2 + 5134x + 5020$ estimate the number of AIDS deaths from 1984 to 1994, where $x = 0$ corresponds to 1984, $x = 1$ to 1985, and so on. Estimate the years when the number of AIDS deaths was from 90,000 to 200,000.

62. *Safe Driving Speeds*   The stopping distance d in feet for a car traveling at x miles per hour is given by $d(x) = \frac{1}{12}x^2 + \frac{11}{9}x$. Determine the driving speeds that correspond to stopping distances between 300 and 500 feet, inclusively. Round speeds to the nearest mile per hour.

63. *Cellular Phones*   Our society is in transition from an industrial to an informational society. Cellular communication has played an increasingly large role in this transition. The number of cellular subscribers in thousands from 1985 to 1991 can be modeled by the formula $f(x) = 163x^2 - 146x + 205$, where x is the year and $x = 0$ corresponds to 1985. (**Source:** M. Paetsch, *Mobile Communication in the U.S. and Europe.*)

    (a) Write a quadratic inequality whose solution set represents the years when there were more than 2 million subscribers.

    (b) Solve this inequality.

64. **Air Density** As the altitude increases, air becomes thinner or less dense. An approximation of the density of air at an altitude of x meters above sea level is given by

$$d(x) = (3.32 \times 10^{-9})x^2 - (1.14 \times 10^{-4})x + 1.22.$$

The output is the density of air in kilograms per cubic meter. The domain of d is $0 \le x \le 10,000$. (**Source:** A. Miller and J. Thompson, *Elements of Meteorology*.)

(a) Denver is sometimes referred to as the mile-high city. Compare the density of air at sea level and in Denver. (*Hint:* 1 ft $\approx$ 0.305 m.)

(b) Determine the altitudes where the density is greater than 1 kilogram per cubic meter.

65. **Modeling Water Flow** A cylindrical container measuring 16 centimeters high and 12.7 centimeters in diameter with a 0.5-centimeter hole at the bottom was completely filled with water. As water leaked out, the height of the water level inside the container was measured every 15 seconds. The results of the experiment appear in the accompanying table.

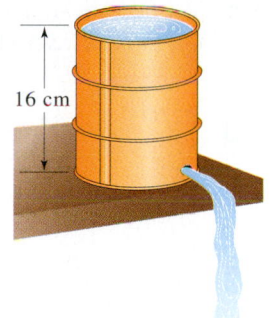

16 cm

Time (sec)	0	15	30	45	60
Height (cm)	16	13.8	11.6	9.8	8.1

Time (sec)	75	90	105	120	135
Height (cm)	6.6	5.3	4.1	3.1	2.3

Time (sec)	150	165	180
Height (cm)	1.4	0.8	0.5

(a) Explain why the data cannot be modeled by a linear function.

(b) Use the table to estimate when the height of the water was from 5 to 10 centimeters.

(c) The data can be modeled by

$$f(x) = 0.0003636x^2 - 0.1511x + 15.92,$$

where x represents the time. Graph f and the data in the same viewing rectangle.

(d) Solve part (b) using $f(x)$.

66. **Heart Rate** The accompanying table shows a person's heart rate after exercise has stopped.

Time (min)	0	2	4
Heart Rate (bpm)	154	106	90

(a) Find values for the constants a, h, and k so that the formula $f(x) = a(x - h)^2 + k$ models the data, where x represents time $0 \le x \le 4$.

(b) Evaluate $f(1)$ and interpret the result.

(c) Estimate the times when the heart rate was from 115 to 125 beats per minute.

Writing about Mathematics

67. Explain how a table of values can be used to help solve a quadratic inequality provided that the boundary numbers are listed in the table.

68. Explain how to determine the solution set for the inequality $ax^2 + bx + c < 0$, where $a > 0$. How would the solution set change if $a < 0$?

3.4 Transformations of Graphs

- Graph functions using vertical and horizontal translations
- Graph function using stretching and shrinking
- Graph function using reflections
- Combine transformations
- Model data with transformations (optional)

Introduction

Graphs are often used to model different types of phenomena. For example, when a cold front moves across the United States, we might use a circular arc on a weather map to describe its shape. (See Exercise 7 in the Extended and Discovery Exercises at the end of this chapter.) If the front does not change its shape significantly, we could model the movement of the front on a television weather map by translating the circular arc. Before we can portray a cold front on a weather map, we need to discuss how to transform graphs of functions. (**Sources:** S. Hoggar, *Mathematics for Computer Graphics;* A. Watt, *3D Computer Graphics.*)

Vertical and Horizontal Translations

Graphs of two functions defined by $f(x) = x^2$, and $g(x) = \sqrt{x}$ will be used to demonstrate translations in the xy-plane. Symbolic, numerical, and graphical representations of these functions are shown in Figures 3.66 and 3.67, respectively. Points listed in the table are plotted on the graph.

x	-2	-1	0	1	2
y	4	1	0	1	4

x	0	1	4
y	0	1	2

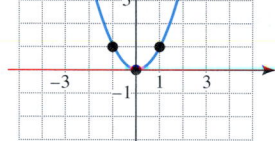

FIGURE 3.66 **FIGURE 3.67**

Vertical Shifts If 2 is added to the formula of each function, the original graphs are shifted upward 2 units. The graphs of $y = x^2 + 2$ and $y = \sqrt{x} + 2$ are shown in Figures 3.68 and 3.69. A table of points is included, together with the graph of the original function. Notice that the y-values in both the graphical and numerical representations increase by 2 units.

x	-2	-1	0	1	2
y	6	3	2	3	6

x	0	1	4
y	2	3	4

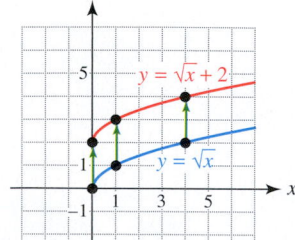

FIGURE 3.68 **FIGURE 3.69**

If 2 is subtracted from each of the original formulas, the graphs would translate downward 2 units. Verify this by graphing $y = x^2 - 2$ and $y = \sqrt{x} - 2$. Translations of this type are called **vertical shifts**, or **vertical translations**. They do not alter the shape of the graph—only its position. A vertical shift is an example of an **isometry** because the distance between any two points on the graph does not change. The original and shifted graphs are congruent.

Horizontal Shifts   If the variable x is replaced by $(x - 2)$ in each of the formulas for f and g, a different type of shift results. Figures 3.70 and 3.71 show the graphs and tables of $y = (x - 2)^2$ and $y = \sqrt{(x - 2)}$ together with the graphs of the original functions.

x	0	1	2	3	4
y	4	1	0	1	4

x	2	3	6
y	0	1	2

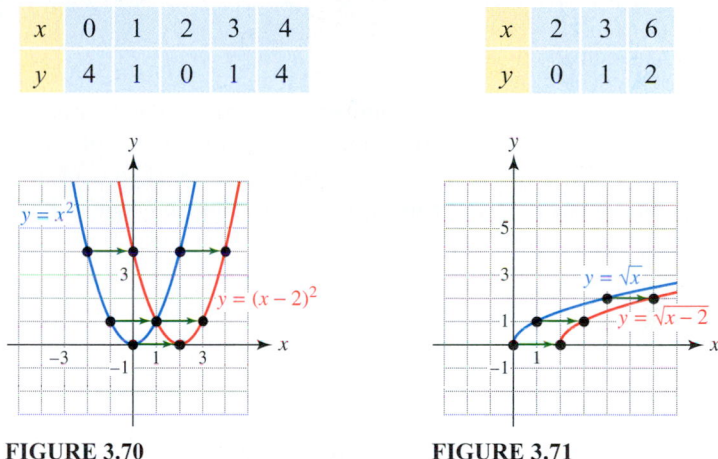

FIGURE 3.70   **FIGURE 3.71**

Each new graph suggests a shift of the original graph to the *right* by 2 units. Notice that a table for a graph shifted right 2 units can be obtained from the original table by *adding* 2 to each x-value.

If the variable x is replaced by $(x + 3)$ in each equation, the original graphs are translated to the *left* 3 units. The graphs of $y = (x + 3)^2$ and $y = \sqrt{(x + 3)}$ and their tables are shown in Figures 3.72 and 3.73. This type of translation is a **horizontal shift**, or **horizontal translation**. Horizontal shifts are also examples of isometries. The table for a graph shifted left 3 units is obtained from the original table by *subtracting* 3 from each x-value.

x	-5	-4	-3	-2	-1
y	4	1	0	1	4

x	-3	-2	1
y	0	1	2

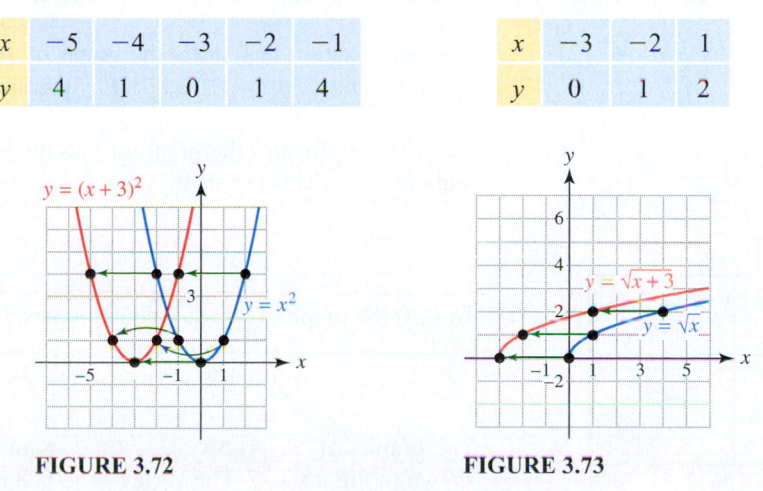

FIGURE 3.72   **FIGURE 3.73**

These ideas are summarized in the following box on the next page.

VERTICAL AND HORIZONTAL SHIFTS

Let f be a function, and let k be a positive number.

To graph	Shift the graph of $y = f(x)$ by k units
$y = f(x) + k$	upward
$y = f(x) - k$	downward
$y = f(x - k)$	right
$y = f(x + k)$	left

Shifts can be combined to translate a graph of $y = f(x)$ both vertically and horizontally. For example, to shift the graph of $y = f(x)$ to the right 2 units and downward 4 units, we graph $y = f(x - 2) - 4$. If $y = |x|$, then $y = f(x - 2) - 4 = |x - 2| - 4$, as shown in Figures 3.74–3.76. This technique is illustrated in the next example.

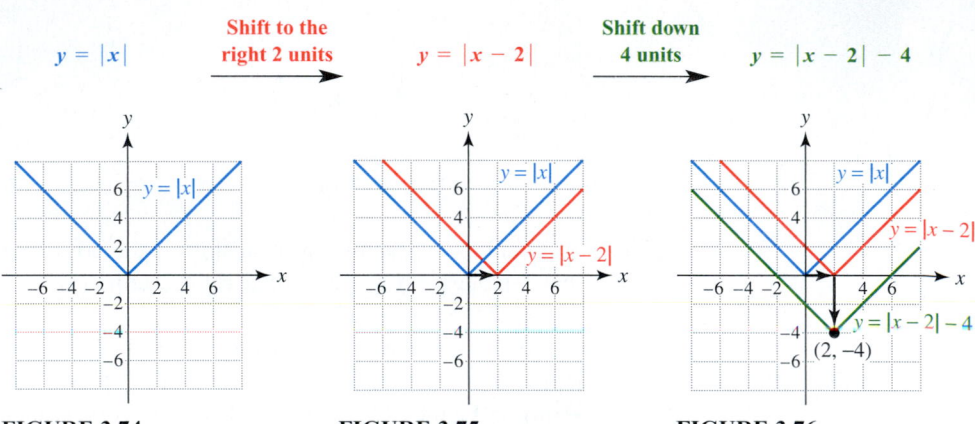

FIGURE 3.74   **FIGURE 3.75**   **FIGURE 3.76**

EXAMPLE 1 Combining vertical and horizontal shifts

Find an equation that shifts the graph of $f(x) = \frac{1}{2}x^2 - 4x + 6$ to the left 8 units and upward 4 units. Support this result by graphing.

SOLUTION To shift the graph of f to the left 8 units, replace x with $(x + 8)$ in the formula for $f(x)$. This results in

$$y = f(x + 8) = \frac{1}{2}(x + 8)^2 - 4(x + 8) + 6.$$

To shift the graph of this new equation upward 4 units, add 4 to the formula to obtain

$$y = f(x + 8) + 4 = \frac{1}{2}(x + 8)^2 - 4(x + 8) + (6 + 4).$$

$[-9, 9, 2]$ by $[-6, 6, 2]$

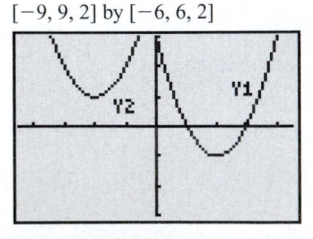

FIGURE 3.77

The graphs of $Y_1 = .5X^2 - 4X + 6$ and $Y_2 = .5(X + 8)^2 - 4(X + 8) + 10$ are shown in Figure 3.77. The vertex of y_2 is 8 units to the left and 4 units upward compared to the vertex of y_1.

Now Try Exercise 29 ◆

EXAMPLE 2 **Writing formulas**

Write a formula for a function g whose graph is similar to $f(x) = 4x^2 - 2x + 1$ but is shifted right 1980 units and upward 50 units. Do not simplify the formula.

SOLUTION Replace x with $(x - 1950)$ in the formula for $f(x)$ and then add 50.

$$g(x) = f(x - 1980) + 50$$

$$= 4(x - 1980)^2 - 2(x - 1980) + 1 + 50$$

$$= 4(x - 1980)^2 - 2(x - 1980) + 51 \qquad \text{\textit{Now Try Exercise 33}} \blacklozenge$$

Stretching and Shrinking

The graph of $f(x) = \sqrt{x}$ in Figure 3.78 can be stretched or shrunk *vertically*. On one hand, the graph of $y = 2f(x)$, or $y = 2\sqrt{x}$, in Figure 3.79 represents a vertical stretching of the graph of $y = \sqrt{x}$. On the other hand, the graph of $y = \frac{1}{2}f(x)$, or equivalently $y = \frac{1}{2}\sqrt{x}$, in Figure 3.80 represents a vertical shrinking of the graph of $y = \sqrt{x}$.

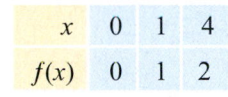

x	0	1	4
$f(x)$	0	1	2

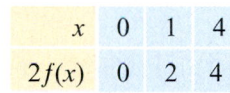

x	0	1	4
$2f(x)$	0	2	4

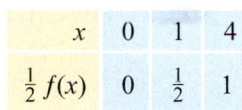

x	0	1	4
$\frac{1}{2}f(x)$	0	$\frac{1}{2}$	1

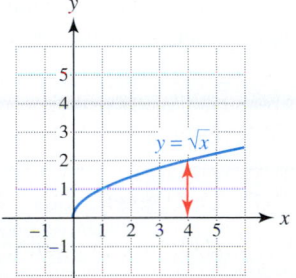

FIGURE 3.78 Given Function

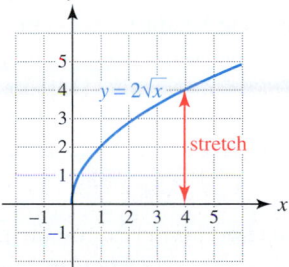

FIGURE 3.79 Vertical Stretching

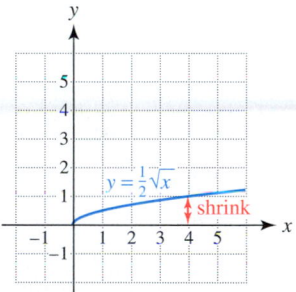

FIGURE 3.80 Vertical Shrinking

These results can be generalized for any function f.

VERTICAL STRETCHING AND SHRINKING

If the point (x, y) lies on the graph of $y = f(x)$, then the point (x, cy) lies on the graph of $y = cf(x)$. If $c > 1$, the graph of $y = cf(x)$ is a vertical stretching of the graph of $y = f(x)$, whereas if $0 < c < 1$, the graph of $y = cf(x)$ is a vertical shrinking of the graph of $y = f(x)$.

For example, if the point $(4, 2)$ is on the graph of $y = f(x)$, then the point $(4, 4)$ is on the graph $y = 2f(x)$ and the point $(4, 1)$ is on the graph of $y = \frac{1}{2}f(x)$.

The line graph in Figure 3.81 can be stretched or shrunk *horizontally*. On one hand, if the line graph represents the graph of a function f, then the graph of $y = f\left(\frac{1}{2}x\right)$ in Figure 3.82 is a horizontal stretching of the graph of $y = f(x)$. On the other hand, the graph of $y = f(2x)$ in Figure 3.83 represents a horizontal shrinking of the graph of $y = f(x)$.

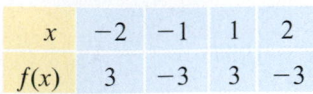

x	-2	-1	1	2
$f(x)$	3	-3	3	-3

x	-4	-2	2	4
$f\left(\frac{1}{2}x\right)$	3	-3	3	-3

x	-1	$-\frac{1}{2}$	$\frac{1}{2}$	1
$f(2x)$	3	-3	3	-3

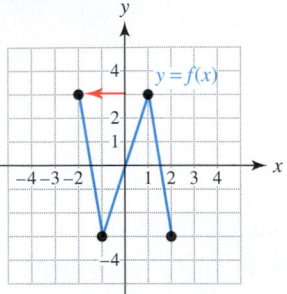

FIGURE 3.81 Given Function

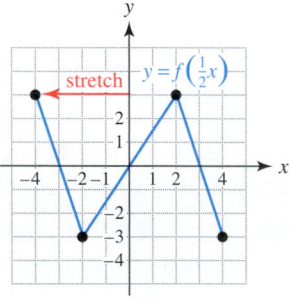

FIGURE 3.82 Horizontal Stretching

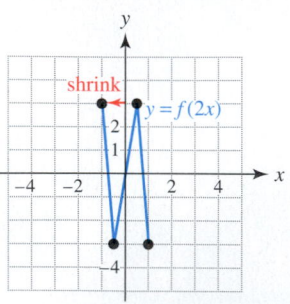

FIGURE 3.83 Horizontal Shrinking

If one were to imagine the graph of $y = f(x)$ as a flexible wire, then a horizontal stretching would happen if the wire were pulled on each end, and a horizontal shrinking would happen if the wire were compressed. Note that horizontal stretching and shrinking does not change the height (maximum or minimum y-values) of the graph, nor does it change the y-intercept. Horizontal stretching and shrinking can be generalized for any function f.

HORIZONTAL STRETCHING AND SHRINKING

If the point (x, y) lies on the graph of $y = f(x)$, then the point $\left(\frac{x}{c}, y\right)$ lies on the graph of $y = f(cx)$. If $c > 1$, the graph of $y = f(cx)$ is a horizontal shrinking of the graph of $y = f(x)$, whereas if $0 < c < 1$, the graph of $y = f(cx)$ is a horizontal stretching of the graph of $y = f(x)$.

For example, if the point $(-2, 4)$ is on the graph of $y = f(x)$, then the point $(-1, 4)$ is on the graph of $y = f(2x)$, and the point $(-4, 4)$ is on the graph of $y = f\left(\frac{1}{2}x\right)$.
The next example illustrates the previous discussion.

 **Stretching and shrinking of a graph**

Use the graph of $y = f(x)$ in Figure 3.84 to sketch a graph of each equation.

(a) $y = 3f(x)$ **(b)** $y = f\left(\frac{1}{2}x\right)$

SOLUTION

(a) The graph of $y = 3f(x)$ is a vertical stretching of the graph of $y = f(x)$ shown in Figure 3.84 and can be obtained by multiplying each y-coordinate on the graph of

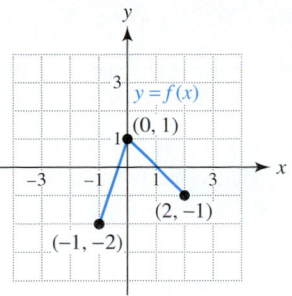

FIGURE 3.84

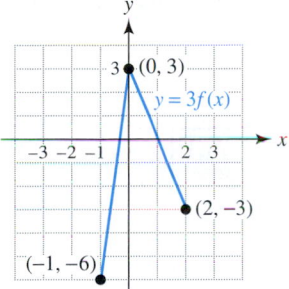

FIGURE 3.85 Vertical Stretching

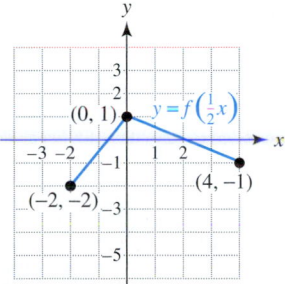

FIGURE 3.86 Horizontal Stretching

$y = f(x)$ by 3. When the y-coordinates of the points $(-1, -2)$, $(0, 1)$, and $(2, -1)$ are multiplied by 3, the following results.

$$(-1, \mathbf{3} \cdot -2) = (-1, -6)$$
$$(0, \mathbf{3} \cdot 1) = (0, 3)$$
$$(2, \mathbf{3} \cdot -1) = (2, -3)$$

Next we sketch a line graph with the points $(-1, -6)$, $(0, 3)$, and $(2, -3)$, as shown in Figure 3.85.

(b) The graph of $y = f\left(\frac{1}{2}x\right)$ is a horizontal stretching of the graph of $y = f(x)$ and can be obtained by dividing each x-coordinate on the graph of $y = f(x)$ by $\frac{1}{2} = 0.5$. For the points $(-1, -2)$, $(0, 1)$, and $(2, -1)$, we obtain the following.

$$\left(\frac{-1}{\mathbf{0.5}}, -2\right) = (-2, -2)$$

$$\left(\frac{0}{\mathbf{0.5}}, 1\right) = (0, 1)$$

$$\left(\frac{2}{\mathbf{0.5}}, -1\right) = (4, -1)$$

Next we sketch a line graph with the points $(-2, -2)$, $(0, 1)$, and $(4, -1)$, as shown in Figure 3.86.

Now Try Exercise 17 ◆

Reflection of Graphs

Another type of translation is called a **reflection**. For example, the reflection of the blue graph of $f(x) = x^2 - x + 2$ across the x-axis is shown in Figure 3.87 as a red curve.

If (x, y) is a point on the graph of f, then $(x, -y)$ lies on the graph of its reflection across the x-axis, as shown in Figure 3.87. Thus a reflection of $y = f(x)$ is given by $-y = f(x)$, or equivalently, $y = -f(x)$. For example, a reflection of the graph of $f(x) = x^2 - x + 2$ is obtained by graphing $y = -f(x) = -(x^2 - x + 2)$.

If a point (x, y) lies on the graph of a function f, then the point $(-x, y)$ lies on the graph of its reflection across the y-axis, as shown in Figure 3.88. Thus a reflection of $y = f(x)$ across the y-axis is given by $y = f(-x)$. For example, a reflection of the graph of $f(x) = (x - 2)^2$ across the y-axis is shown in Figure 3.89 as a red curve. Its reflection is given by

$$f(-x) = (-x - 2)^2.$$

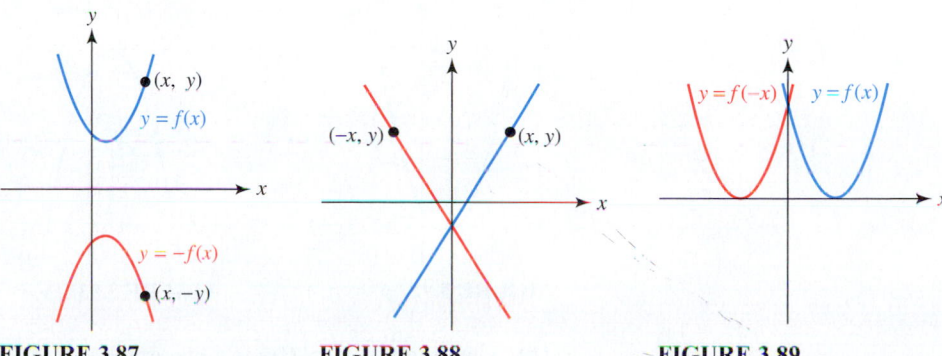

FIGURE 3.87   **FIGURE 3.88**   **FIGURE 3.89**

These results are summarized.

REFLECTIONS OF GRAPHS ACROSS THE x- AND y-AXES

1. The graph of $y = -f(x)$ is a reflection of the graph of $y = f(x)$ across the x-axis.
2. The graph of $y = f(-x)$ is a reflection of the graph of $y = f(x)$ across the y-axis.

If a graphing calculator is capable of using function notation, equations for reflections of a function f can be entered easily. For example, if $f(x) = (x-4)^2$, then let $Y_1 = (X-4)\text{\textasciicircum}2$, $Y_2 = -Y_1$, and $Y_3 = Y_1(-X)$. The graph of y_2 is the reflection of f across the x-axis, and y_3 is the reflection of y_1 across the y-axis. See Figures 3.90 and 3.91. However, it is not necessary to have this feature to graph reflections.

Calculator Help

To access the variable Y_1, as shown in Figure 3.90, see Appendix B (page AP-16).

$[-10, 10, 1]$ by $[-10, 10, 1]$

FIGURE 3.90

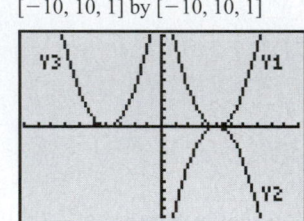

FIGURE 3.91

 **EXAMPLE 4**   Reflecting graphs of functions

For each representation of a function f, graph its reflection across the x-axis and across the y-axis.

(a) $f(x) = x^2 + 2x - 3$
(b) The graph of f is a line graph determined by Table 3.9.

TABLE 3.9

x	$f(x)$
-2	1
-1	-3
0	-1
3	2

SOLUTION

(a) The graph of $f(x) = x^2 + 2x - 3$ is a parabola with vertex $(-1, -4)$ and x-intercepts -3 and 1, as shown in Figure 3.92. To obtain its reflection across the x-axis, graph $y = -f(x)$, or $y = -(x^2 + 2x - 3)$, as shown in Figure 3.93. The vertex is now $(-1, 4)$, and the x-intercepts have not changed. To obtain the reflection of f across the y-axis, let $y = f(-x)$, or $y = (-x)^2 + 2(-x) - 3$ and graph, as shown in Figure 3.94. Note that the vertex is now $(1, -4)$, and the x-intercepts have changed to -1 and 3.

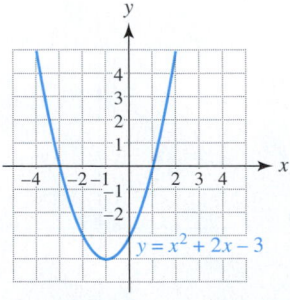

FIGURE 3.92 $y = f(x)$

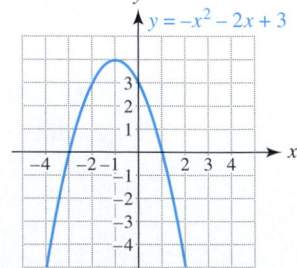

FIGURE 3.93 $y = -f(x)$

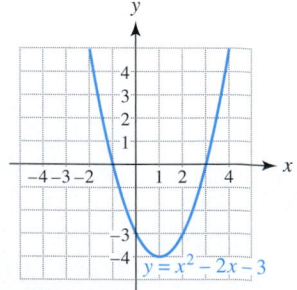

FIGURE 3.94 $y = f(-x)$

(b) The graph of $y = f(x)$ is a line graph, which is shown in Figure 3.95. To graph the reflection of f across the x-axis, start by making a table of values for $y = -f(x)$.

TABLE 3.10

x	$-f(x)$
-2	-1
-1	3
0	1
3	-2

TABLE 3.11

x	$f(-x)$
2	1
1	-3
0	-1
-3	2

See Table 3.10, where each point (x, y) in Table 3.9 has been changed to $(x, -y)$. Plot the points $(-2, -1)$, $(-1, 3)$, $(0, 1)$, and $(3, -2)$. Note that each point in Figure 3.95 has been reflected across the x-axis in Figure 3.96.

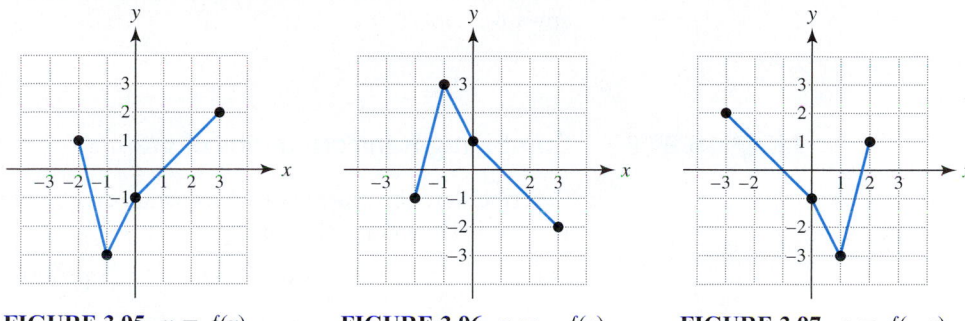

FIGURE 3.95 $y = f(x)$ **FIGURE 3.96** $y = -f(x)$ **FIGURE 3.97** $y = f(-x)$

To graph the reflection of f across the y-axis, make a table of values for $y = f(-x)$. To do this, change each point (x, y) in Table 3.9 to $(-x, y)$, as shown in Table 3.11. Plot the points $(2, 1)$, $(1, -3)$, $(0, -1)$, and $(-3, 2)$. Note that each point in Figure 3.95 has been reflected across the y-axis in Figure 3.97.

Now Try Exercises 77 and 83 ◆

Combining Transformations

Transformation of graphs can be combined to create new graphs. For example, the graph of $y = -2(x - 1)^2 + 3$ can be obtained by performing four transformations on the graph of $y = x^2$.

1. Shift the graph of $y = x^2$ to the right 1 unit: $y = (x - 1)^2$.
2. Vertically stretch the graph of $y = (x - 1)^2$ by a factor of 2: $y = 2(x - 1)^2$.
3. Reflect the graph of $y = 2(x - 1)^2$ across the x-axis: $y = -2(x - 1)^2$.
4. Shift the graph of $y = -2(x - 1)^2$ upward 3 units: $y = -2(x - 1)^2 + 3$.

These steps are summarized by the following.

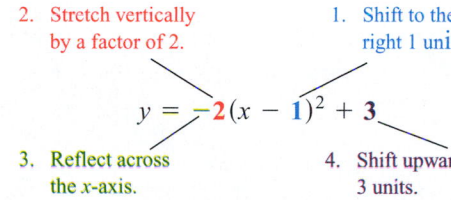

2. Stretch vertically by a factor of 2.   **1.** Shift to the right 1 unit.

$$y = -2(x - 1)^2 + 3$$

3. Reflect across the x-axis.   **4.** Shift upward 3 units.

The resulting sequence of graphs is shown in Figures 3.98–3.101.

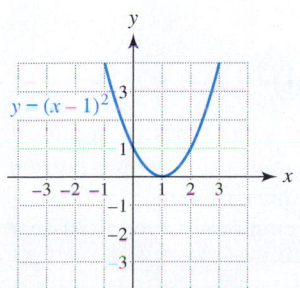

FIGURE 3.98 Shift Right

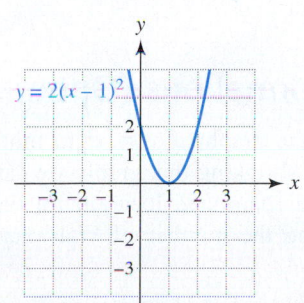

FIGURE 3.99 Vertical Stretch

FIGURE 3.100 Reflect Across x-axis

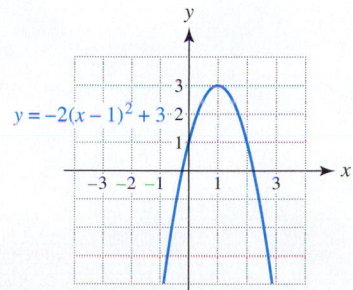

FIGURE 3.101 Shift Upward

◆ **CLASS DISCUSSION**

Does the order in which transformations are performed affect the resulting graph? For example, if a graph is translated upward 2 units and then reflected across the *x*-axis, do we obtain the same graph as when the graph is reflected across the *x*-axis and then translated upward 2 units? ◆

EXAMPLE 5   Combining transformations of graphs

Describe how the graph of each equation can be obtained by transforming the graph of $y = \sqrt{x}$. Then graph the equation.

(a) $y = -\dfrac{1}{2}\sqrt{x}$ **(b)** $y = \sqrt{-(x + 2)} - 1$

SOLUTION

(a) Vertically shrink the graph of $y = \sqrt{x}$ by a factor of $\frac{1}{2}$ and then reflect it across the *x*-axis. The graph of $y = \sqrt{x}$ and the resulting graph of $y = -\frac{1}{2}\sqrt{x}$ are shown in Figure 3.102.

(b) The following transformations can be used to obtain the graph of the equation $y = \sqrt{-(x + 2)} - 1$ from $y = \sqrt{x}$.

1. Reflect the graph of $y = \sqrt{x}$ across the *y*-axis: $y = \sqrt{-x}$.
2. Shift the graph of $y = \sqrt{-x}$ left 2 units: $y = \sqrt{-(x + 2)}$.
3. Shift the graph of $y = \sqrt{-(x + 2)}$ down 1 unit: $y = \sqrt{-(x + 2)} - 1$.

The graph of $y = \sqrt{x}$ and the resulting graph are shown in Figure 3.103.

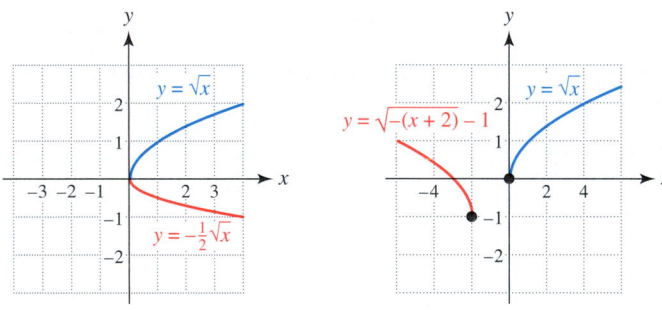

FIGURE 3.102   **FIGURE 3.103**

Now Try Exercises 19 and 63 ◆

Modeling with Transformations (Optional)

We can use transformations of the graph of $y = x^2$ to model some types of nonlinear data. By shifting, stretching, and shrinking this graph, we can transform it into a *portion of a parabola* that has a desired shape and location. In the next example we demonstrate this technique by modeling the number of Wal-Mart employees over a 20-year period.

EXAMPLE 6 Modeling data with a quadratic function

TABLE 3.12
Wal-Mart Employees
(in millions)

Year	Employees
1987	0.20
1992	0.37
1997	0.68
2002	1.4
2007	2.2

Source: Wal-Mart.

Table 3.12 lists actual and projected numbers of Wal-Mart employees in millions for selected years.
(a) Make a scatterplot of the data.
(b) Use transformations of graphs to determine $f(x) = a(x - h)^2 + k$ so that $f(x)$ models this data. Graph $y = f(x)$ together with the data.
(c) Use $f(x)$ to predict the number of Wal-Mart employees in 2010.

SOLUTION
(a) A scatterplot of the data is shown in Figure 3.104. This plot suggests that the data could be modeled by the right half of a parabola that opens upward.
(b) Because the parabola opens upward, it follows that $a > 0$ and the vertex is the lowest point on the parabola. The minimum number of employees is 0.20 million in 1987. One possibility for the vertex (h, k) is (1987, 0.20). Translate the graph of $y = x^2$ right 1987 units and upward 0.20 units. Thus $f(x) = a(x - 1987)^2 + 0.20$. To determine a value for a, graph the data and $y = f(x)$ for different values of a. Figure 3.105 shows $y = f(x)$ with values of $a = 0.001$ and $a = 0.01$. From this graph it can be seen that the desired value of a is between 0.001 and 0.01. With a little experimentation, a reasonable value for a near 0.005 is found. A scatterplot of the data and $f(x) = 0.005(x - 1987)^2 + 0.2$ is shown in Figure 3.106. Note that answers may vary.

[1985, 2010, 5] by [0, 3, 0.5]

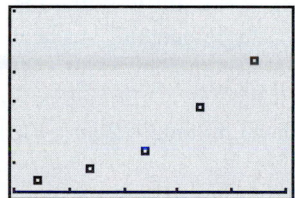

FIGURE 3.104

[1985, 2010, 5] by [0, 3, 0.5]

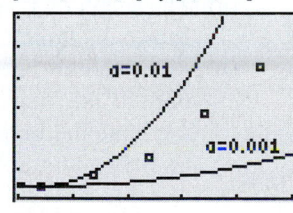

FIGURE 3.105

[1985, 2010, 5] by [0, 3, 0.5]

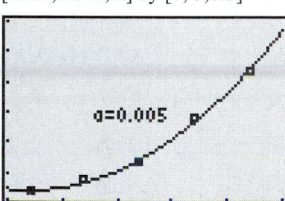

FIGURE 3.106

(c) To predict the number of employees in 2010, evaluate $f(2010)$.

$$f(\mathbf{2010}) = 0.005(\mathbf{2010} - 1987)^2 + 0.2 = 2.845$$

This model estimates there will be about 2.8 million Wal-Mart employees in 2010.

Now Try Exercise 93 ◆

Translations of graphs and figures play an important role in computer graphics. In early motion pictures, it was common to have the background move to create the appearance that the actors and actresses were moving. This same technique is used in two-dimensional graphics.

In older video games, the background often is translated to give the illusion that the player in the game is moving. A simple scene of a mountain and an airplane is shown in

Figure 3.107. In order to make it appear as though the airplane is flying to the right, the image of the mountain could be translated horizontally to the left, as shown in Figure 3.108. (**Source:** C. Pokorny and C. Gerald, *Computer Graphics.*)

FIGURE 3.107   **FIGURE 3.108**

 **Using translations to model movement**

Suppose the mountain in Figure 3.107 can be described by the quadratic function represented by $f(x) = -0.4x^2 + 4$ and that the airplane is located at the point $(1, 5)$.

(a) Graph f in $[-4, 4, 1]$ by $[0, 6, 1]$, where the units are in kilometers. Plot a point to mark the position of the airplane.

(b) Assume that the airplane is moving horizontally to the right at 0.4 kilometer per second. To give a video player the illusion that the airplane is moving, graph the image of the mountain and the position of the airplane after 5 seconds and then after 10 seconds.

SOLUTION

(a) The graph of $y = f(x) = -0.4x^2 + 4$ and the position of the airplane at $(1, 5)$ are shown in Figure 3.109. The "mountain" has been shaded to emphasize its position.

(b) Five seconds later, the airplane has moved $5(0.4) = 2$ kilometers right. In 10 seconds it has moved $10(0.4) = 4$ kilometers right. To graph these new positions, translate the graph of the mountain 2 and 4 kilometers to the left. Replace x with $(x + 2)$ and graph

$$y = f(x + 2) = -0.4(x + 2)^2 + 4$$

together with the point $(1, 5)$. In a similar manner graph

$$y = f(x + 4) = -0.4(x + 4)^2 + 4.$$

The results are shown in Figures 3.110 and 3.111. The position of the airplane at $(1, 5)$ has not changed. However, it has the appearance that it has flown to the right.

Calculator Help

To shade below a parabola, see Appendix B (page AP-16).

[−4, 4, 1] by [0, 6, 1]   [−4, 4, 1] by [0, 6, 1]   [−4, 4, 1] by [0, 6, 1]

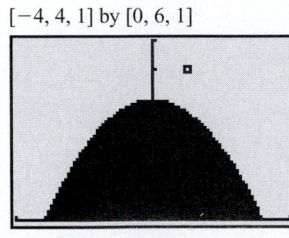 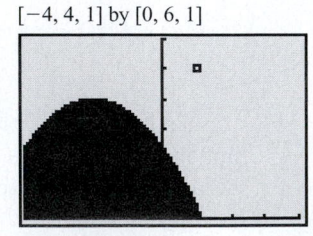 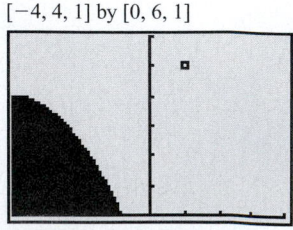

FIGURE 3.109   **FIGURE 3.110**   **FIGURE 3.111**

Now Try Exercise 99 ◆

◆ **CLASS DISCUSSION**

Discuss how one might create the illusion of the airplane moving to the left and *gaining altitude* as it passes over the mountain. ◆

3.4

Putting it all Together

In this section we discussed several transformations of graphs. The following table summarizes how these transformations affect the graph of $y = f(x)$.

Equation	Effect on Graph of $y = f(x)$
Let $k > 0$.	
$y = f(x) + k$	Graph is shifted upward k units.
$y = f(x) - k$	Graph is shifted downward k units.
$y = f(x + k)$	Graph is shifted to the left k units.
$y = f(x - k)$	Graph is shifted to the right k units.
	Examples: (a) $y = f(x) - 1$ (b) $y = f(x + 1)$
Let $c > 0$. $y = cf(x)$	If (x, y) lies on the graph of $y = f(x)$, then (x, cy) lies on the graph of $y = cf(x)$. The graph is vertically stretched if $c > 1$ and vertically shrunk if $0 < c < 1$.
	Examples: (a) $y = 2f(x)$ (b) $y = \dfrac{1}{2}f(x)$

Equation	Effect on Graph of $y = f(x)$
Let $c > 0$. $y = f(cx)$	If (x, y) lies on the graph of $y = f(x)$, then $\left(\frac{x}{c}, y\right)$ lies on the graph of $y = f(cx)$. The graph is horizontally shrunk if $c > 1$ and horizontally stretched if $0 < c < 1$. **Examples:** **(a)** $y = f(2x)$ **(b)** $y = f\left(\frac{1}{2}x\right)$
$y = -f(x)$ $y = f(-x)$	Graph is reflected across the x-axis. Graph is reflected across the y-axis. **Examples:** **(a)** $y = -f(x)$ **(b)** $y = f(-x)$

3.4 Exercises

Vertical and Horizontal Translations

Exercises 1–10: Write the equation of the graph. (Note: The given graph represents a translation of the graph of one of the following equations: $y = x^2$, $y = \sqrt{x}$, or $y = |x|$.)

1.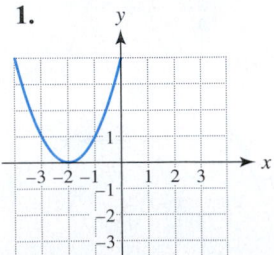

2.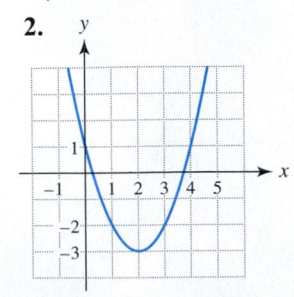

3.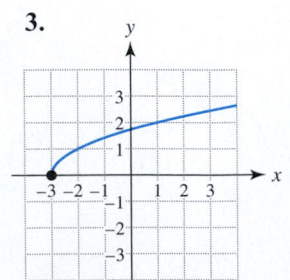

4.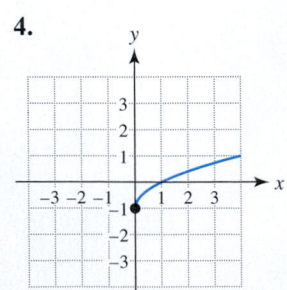

5.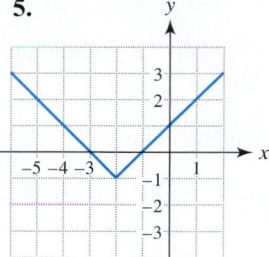

6.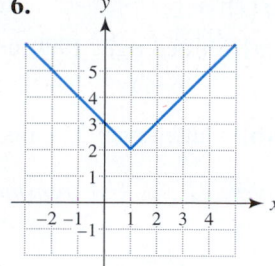

7.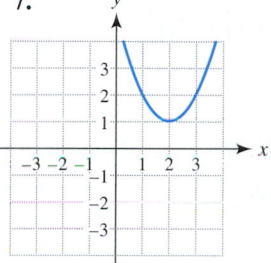

8.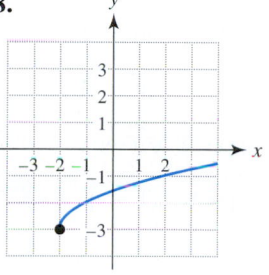

9.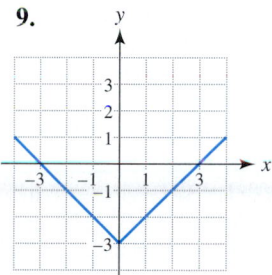

10.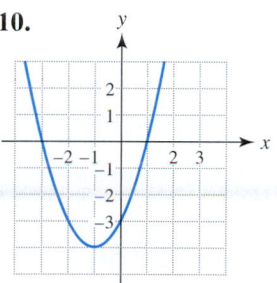

Transforming Graphical Representations

Exercises 11–18: Use the accompanying graph of $y = f(x)$ to sketch a graph of each equation.

11. (a) $y = f(x) + 2$

    (b) $y = f(x - 2) - 1$

    (c) $y = -f(x)$

12. (a) $y = f(x + 1)$

    (b) $y = -f(x)$

    (c) $y = 2f(x)$

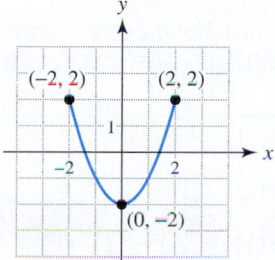

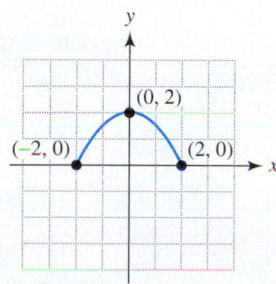

13. (a) $y = f(x + 3) - 2$

    (b) $y = f(-x)$

    (c) $y = \frac{1}{2}f(x)$

14. (a) $y = f(x - 1) - 2$

    (b) $y = -f(x) + 1$

    (c) $y = f\left(\frac{1}{2}x\right)$

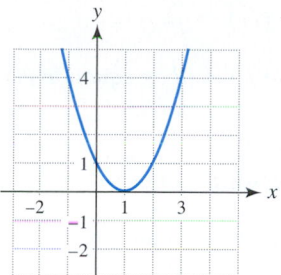

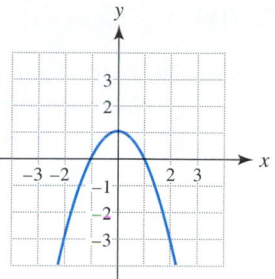

15. (a) $y = f(x) + 1$

    (b) $y = -f(x) - 1$

    (c) $y = 2f\left(\frac{1}{2}x\right)$

16. (a) $y = f(x) - 2$

    (b) $y = f(x - 1) + 2$

    (c) $y = 2f(x)$

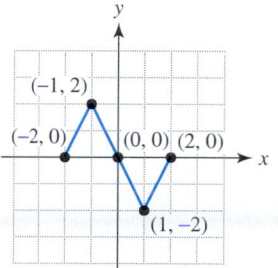

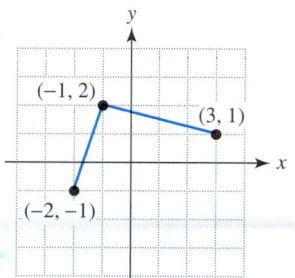

17. (a) $y = f(2x) + 1$

    (b) $y = 2f\left(\frac{1}{2}x\right) + 1$

    (c) $y = \frac{1}{2}f(x - 2)$

18. (a) $y = f(2x)$

    (b) $y = f\left(\frac{1}{2}x\right) - 1$

    (c) $y = 2f(x) - 1$

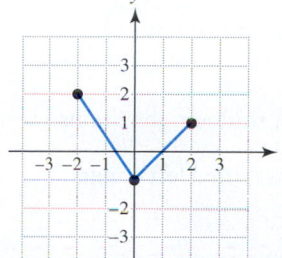

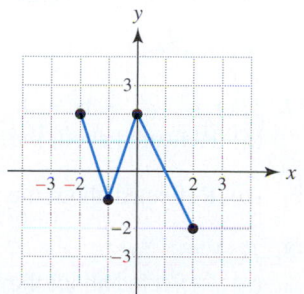

Exercises 19–28: Use transformations to explain how the graph of f can be found by using the graph of $y = x^2$, $y = \sqrt{x}$, or $y = |x|$. You do not need to graph $y = f(x)$.

19. $f(x) = (x - 3)^2 + 1$ **20.** $f(x) = (x + 2)^2 - 3$

21. $f(x) = \frac{1}{4}(x + 1)^2$ **22.** $f(x) = 2(x - 4)^2$

23. $f(x) = -\sqrt{x + 5}$ **24.** $f(x) = -\sqrt{x} - 3$

25. $f(x) = 2\sqrt{-x}$ **26.** $f(x) = \sqrt{-\frac{1}{2}x}$

27. $f(x) = |-(x + 1)|$ **28.** $f(x) = |4 - x|$

Exercises 29–32: (Refer to Example 1.) Find an equation that shifts the graph of f by the desired amounts. Graph f and the shifted graph in the same xy-plane.

29. $f(x) = x^2$;   right 2 units, downward 3 units

30. $f(x) = 3x - 4$;   left 3 units, upward 1 unit

31. $f(x) = x^2 - 4x + 1$;   left 6 units, upward 4 units

32. $f(x) = 5 - 3x - \frac{1}{2}x^2$;   right 5 units, downward 8 units

Exercises 33–40: (Refer to Example 2.) Write a formula for a function g whose graph is similar to f(x) but satisfies the given conditions. Do not simplify the formula.

33. $f(x) = 3x^2 + 2x - 5$
   (a) Shifted left 3 units

   (b) Shifted downward 4 units

34. $f(x) = 2x^2 - 3x + 2$
   (a) Shifted right 8 units

   (b) Shifted upward 2 units

35. $f(x) = -x^3 + 2x^2 - 4x + 1$
   (a) Shifted right 2 units and upward 4 units

   (b) Shifted left 8 units and downward 5 units

36. $f(x) = -3x^4 + 5x^2 - 3$
   (a) Shifted left 10 units and downward 6 units

   (b) Shifted right 1 unit and upward 10 units

37. $f(x) = 2x^2 - 3x + 1$
   (a) Reflected about the x-axis, shifted right 2 units

   (b) Reflected about the y-axis, shifted left 4 units

38. $f(x) = -5x^2 + x + 5$
   (a) Reflected about the x-axis, shifted left 10 units

   (b) Reflected about the y-axis, shifted right 3 units

39. $f(x) = x^2$
   (a) Shifted right 2000 units, upward 80 units, and vertically stretched by a factor of 10

   (b) Shifted left 20 units and reflected about the y-axis

40. $f(x) = |x|$
   (a) Shifted right 1990 units, downward 200 units, and vertically stretched by a factor of 5

   (b) Shifted right 25 units and reflected about the x-axis

Exercises 41–68: Use transformations of graphs to sketch a graph of f.

41. $f(x) = x^2 - 3$ **42.** $f(x) = -x^2$

43. $f(x) = (x + 4)^2$ **44.** $f(x) = (x - 5)^2 + 3$

45. $f(x) = -\sqrt{x}$ **46.** $f(x) = 2(x - 1)^2 + 1$

47. $f(x) = -x^2 + 4$ **48.** $f(x) = \sqrt{-x}$

49. $f(x) = \sqrt{x + 1}$ **50.** $f(x) = \sqrt{x} + 1$

51. $f(x) = |x| - 4$ **52.** $f(x) = |x - 4|$

53. $f(x) = \sqrt{x - 3} + 2$ **54.** $f(x) = |x + 2| - 3$

55. $f(x) = \sqrt{2x}$ **56.** $f(x) = \frac{1}{2}(x + 2)^2$

57. $f(x) = |2x|$ **58.** $f(x) = \frac{1}{2}|x|$

59. $f(x) = 1 - \sqrt{x}$ **60.** $f(x) = 2\sqrt{(x - 2)} - 1$

61. $f(x) = -\sqrt{1 - x}$ **62.** $f(x) = \sqrt{-x} - 1$

63. $f(x) = \sqrt{-(x + 1)}$ **64.** $f(x) = 2 + \sqrt{-(x - 3)}$

65. $f(x) = (x - 1)^3$ **66.** $f(x) = (x + 2)^3$

67. $f(x) = -x^3$ **68.** $f(x) = (-x)^3 + 1$

Exercises 69–76: The points $(-12, 6)$, $(0, 8)$, and $(8, -4)$ lie on the graph of $y = f(x)$. Determine three points that lie on the graph of $y = g(x)$.

69. $g(x) = f(x) + 2$ **70.** $g(x) = f(x) - 3$

71. $g(x) = f(x - 2) + 1$ **72.** $g(x) = f(x + 1) - 1$

73. $g(x) = -\frac{1}{2}f(x)$ **74.** $g(x) = -2f(x)$

75. $g(x) = f(-2x)$ **76.** $g(x) = f\left(-\frac{1}{2}x\right)$

Exercises 77–84: (Refer to Example 4.) For the given representation of a function f, graph its reflection across the x-axis and across the y-axis.

77. $f(x) = x^2 - 2x - 3$ **78.** $f(x) = 4 - 7x - 2x^2$

79. $f(x) = 2x^2 + x - 1$ **80.** $f(x) = 1 - 2x$

81. $f(x) = |x + 1| - 1$ **82.** $f(x) = \frac{1}{2}|x - 2| + 2$

83. Line graph determined by the table

x	−3	−1	1	2
$f(x)$	2	3	−1	−2

84. Line graph determined by the table

x	−4	−2	0	1
$f(x)$	−1	−4	2	2

Transforming Numerical Representations

Exercises 85–92: Two functions f and g are related by the given equation. Use the numerical representation of f to make a numerical representation of g.

85. $g(x) = f(x) + 7$

x	1	2	3	4	5
$f(x)$	5	1	6	2	7

86. $g(x) = f(x) - 10$

x	0	5	10	15	20
$f(x)$	−5	11	21	32	47

87. $g(x) = f(x - 2)$

x	−4	−2	0	2	4
$f(x)$	5	2	−3	−5	−9

88. $g(x) = f(x + 50)$

x	−100	−50	0	50	100
$f(x)$	25	80	120	150	100

89. $g(x) = f(x + 1) - 2$

x	1	2	3	4	5
$f(x)$	2	4	3	7	8

90. $g(x) = f(x - 3) + 5$

x	−3	0	3	6	9
$f(x)$	3	8	15	27	31

91. $g(x) = f(-x) + 1$

x	−2	−1	0	1	2
$f(x)$	11	8	5	2	−1

92. $g(x) = -f(x + 2)$

x	−4	−2	0	2	4
$f(x)$	5	8	10	8	5

Applications

Exercises 93 and 94: (Refer to Example 6.) Use transformations of graphs to model the table of data with the formula $f(x) = a(x - h)^2 + k$. (Answers may vary.)

93. Percentage of malpractice awards above $1 million

Year	1997	1998	1999	2000	2001
Percent	3.8	4.3	4.9	5.9	7.9

Source: Physician Insurers Association of America.

94. Number of titles released for DVD rentals

Year	1998	1999	2000	2001	2002
Titles	2049	4787	8723	14,321	21,260

Source: DVD Release Report.

95. *AIDS Deaths* The function D defined by

$$D(x) = 2375x^2 + 5134x + 5020$$

models AIDS deaths where x is the year and $x = 0$ corresponds to 1984. Write a formula $g(x)$ that computes AIDS deaths during year x, where x is the actual year.

96. *Public College Tuition* The function f, represented by $f(x) = 122.8x + 786.2$, models the average tuition at public four-year colleges from 1981 to 1994, where x is the year and $x = 1$ corresponds to 1981. (**Source:** The College Board.) Use transformations to determine a function g that computes the average tuition at four-year colleges during the year x, where x is the actual year.

97. *Daylight Hours* The accompanying figure on the next page shows a partial graph of the number of daylight hours at latitude 60°N. The variable x is measured in months where $x = 0$ corresponds to December 21, the day with the least amount of daylight. The domain is $-6 \leq x \leq 6$. For example, $x = 2$ represents February 21 and $x = -3$ represents September 21. (**Source:** J. Williams, *The Weather Almanac 1995*.)

(a) Estimate the number of daylight hours on February 21.

(b) Make a conjecture about the number of daylight hours on October 21 when $x = -2$.

(c) Sketch the left side of the graph for $-6 \leq x \leq 0$.

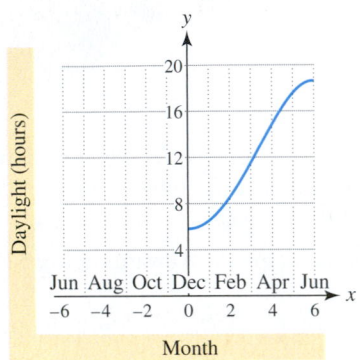

Month

98. *Daylight Hours* (Continuation of Exercise 97.) Sketch a graph that shows the daylight hours at a latitude of 60°S, which is in the Southern Hemisphere rather than in the Northern Hemisphere.

Using Transformations to Model Motion

99. *Computer Graphics* (Refer to Example 7.) Suppose that the airplane in Figure 3.107 is flying at 0.2 kilometer per second to the left, rather than to the right. If the position of the airplane is fixed at $(-1, 5)$, graph the image of the mountain and the position of the airplane after 15 seconds.

100. *Computer Graphics* (Refer to Example 7.) Suppose that the airplane in Figure 3.107 is traveling to the right at 0.1 kilometer per second and gaining altitude at 0.05 kilometer per second. If the airplane's position is fixed at $(-1, 5)$, graph the image of the mountain and the position of the airplane after 20 seconds.

101. *Modeling a Weather Front* Suppose a cold front is passing through the United States at noon with a shape described by the function $y = \frac{1}{20}x^2$. Each unit represents 100 miles. Des Moines, Iowa, is located at $(0, 0)$, and the positive y-axis points north. See the accompanying figure.

(a) If the cold front is moving south at 40 miles per hour and retains its present shape, graph its location at 4 P.M.

(b) Suppose that by midnight the vertex of the front has moved 250 miles south and 210 miles east of Des Moines, maintaining the same shape. Columbus, Ohio, is located approximately 550 miles east and 80 miles south of Des Moines. Plot the locations of Des Moines and Columbus together with the new position of the cold front. Determine whether the cold front has reached Columbus by midnight.

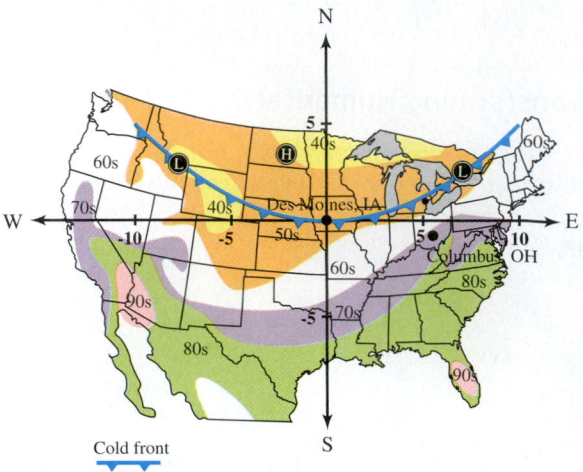

Cold front

102. *Modeling Motion* The first figure contains a picture that is composed of lines and curves. In this exercise we will model only the semicircle that outlines the top of the silo. In order to make it appear that the person is walking to the right, the background must be translated horizontally to the left as shown in the second figure.

The semicircle at the top of the silo in the first figure is described by $f(x) = \sqrt{9 - x^2} + 12$.

(a) Graph f in the (square) viewing rectangle $[-12, 12, 1]$ by $[0, 16, 1]$.

(b) To give the illusion that the person is walking to the right at 2 units per second, graph the top of the silo after 1 second, and after 4 seconds.

Writing about Mathematics

103. Explain how to graph the reflection of $y = f(x)$ across the x-axis. Give an example.

104. Let k be a positive number. Explain how to shift the graph of $y = f(x)$ upward, downward, left, or right k units. Give examples.

105. If the graph of $y = f(x)$ undergoes a vertical stretch or shrink to become the graph of $y = g(x)$, do these two graphs have the same x-intercepts? y-intercepts? Explain your answers.

106. If the graph of $y = f(x)$ undergoes a horizontal stretch or shrink to become the graph of $y = g(x)$, do these two graphs have the same x-intercepts? y-intercepts? Explain your answers.

EXTENDED AND DISCOVERY EXERCISES

Exercises 1–4: ***Commutative Property*** *In the following exercises you will determine if transformations of a graph are commutative. That is, does the order in which transformations of graphs are applied affect the final graph? To answer this question, start by determining if the two sequences of transformations are equivalent for all possible graphs. You may want to try some examples to help decide.*

1. Reflect across the x-axis, shift upward 1 unit.
Shift upward 1 unit, reflect across the x-axis.

2. Shift upward 2 units, shift left 3 units.
Shift left 3 units, shift upward 2 units.

3. Reflect across the x-axis, reflect across the y-axis.
Reflect across the y-axis, reflect across the x-axis.

4. Stretch in the vertical direction, shift downward 2 units.
Shift downward 2 units, stretch in the vertical direction.

CHECKING BASIC CONCEPTS FOR SECTIONS 3.3 AND 3.4

1. A graph of $y = f(x)$ is shown in the figures. Solve the inequalities $f(x) \le 0$ and $f(x) > 0$ for each figure. Write your answer in interval notation.

(a)
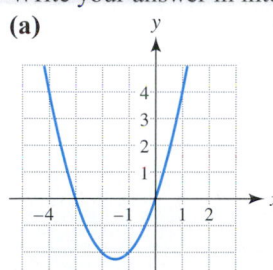

(b)
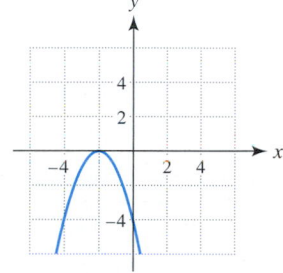

2. Solve each equation and inequality. Write the solution set for each inequality in interval notation.
(a) $2x^2 + 7x - 4 = 0$ (b) $2x^2 + 7x - 4 < 0$

(c) $2x^2 + 7x - 4 > 0$

3. Solve each inequality.
(a) $x^2 - 5 \ge 0$ (b) $4x^2 + 9 > 9x$

(c) $2x(x - 1) \le 2$

4. Predict how the graph of each equation will appear compared to the graph of $f(x) = x^2$. Test your prediction by graphing f and the equation in the same viewing rectangle.

(a) $y = (x + 4)^2$ (b) $y = x^2 - 3$

(c) $y = (x - 5)^2 + 3$

5. A graph of $y = f(x)$ is shown in the figure. Sketch a graph of each equation.
(a) $y = -2f(x)$ (b) $y = f\left(-\frac{1}{2}x\right)$

(c) $y = f(x - 1) + 1$

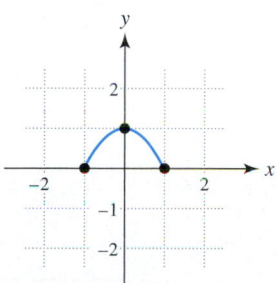

6. *Safe Driving Speeds* The stopping distance d in feet for a car traveling x miles per hour on wet, level pavement can be estimated by $d(x) = \frac{1}{9}x^2 + \frac{11}{3}x$. Determine the driving speeds that correspond to stopping distances between 80 and 180 feet, inclusively.

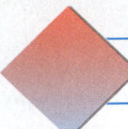

CHAPTER 3 SUMMARY

CONCEPT	EXPLANATION AND EXAMPLES

SECTION 3.1 QUADRATIC FUNCTIONS AND MODELS

QUADRATIC FUNCTION

General form: $f(x) = ax^2 + bx + c, a \neq 0$

Examples: $f(x) = x^2; f(x) = -3x^2 + x + 5$

PARABOLA

The graph of a quadratic function is a parabola.

Vertex form: $f(x) = a(x - h)^2 + k$ (standard form for a parabola with a vertical axis)

Leading coefficient: a

Vertex: (h, k)

Axis of symmetry: $x = h$

Example:

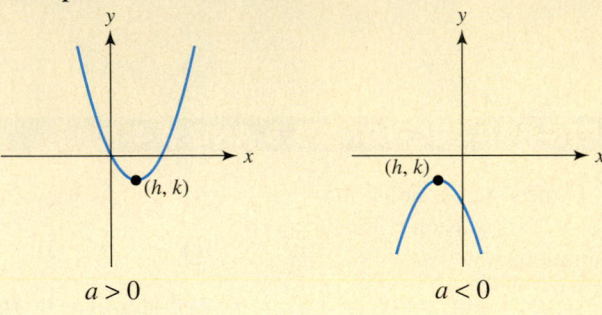

$a > 0$ $a < 0$

COMPLETING THE SQUARE TO FIND THE VERTEX

The vertex of a parabola can be found by completing the square.

Example: $y = x^2 - 4x + 1$

$y - 1 = x^2 - 4x$ Subtract 1.

$y - 1 + 4 = x^2 - 4x + 4$ Add $\left(\frac{-4}{2}\right)^2 = 4.$

$y + 3 = (x - 2)^2$ Perfect square trinomial

$y = (x - 2)^2 - 3$ Subtract 3.

The vertex is $(2, -3)$.

VERTEX FORMULA

x-coordinate: $x = -\dfrac{b}{2a}$

y-coordinate: $f\left(-\dfrac{b}{2a}\right)$

Example: $f(x) = 2x^2 + 4x - 4$

$x = -\dfrac{4}{2(2)} = -1, y = 2(-1)^2 + 4(-1) - 4 = -6$

Vertex: $(-1, -6)$

CONCEPT	EXPLANATION AND EXAMPLES

SECTION 3.2 QUADRATIC EQUATIONS AND PROBLEM SOLVING

QUADRATIC EQUATION

Can be written as $ax^2 + bx + c = 0$, $a \neq 0$

A quadratic equation can have zero, one, or two real solutions.

Examples: $x^2 + 1 = 0$, $x^2 + 2x + 1 = 0$, $x(x - 1) = 20$

FACTORING

Write an equation in the form $ab = 0$ and apply the zero-product property.

Example:
$$x^2 - 3x = -2$$
$$x^2 - 3x + 2 = 0$$
$$(x - 1)(x - 2) = 0$$
$$x = 1 \quad \text{or} \quad x = 2$$

SQUARE ROOT PROPERTY

If $x^2 = k$ and $k \geq 0$, then $x = \pm\sqrt{k}$.

Example: $x^2 = 16$ implies $x = \pm 4$.

COMPLETING THE SQUARE

If $x^2 + kx = d$, then add $\left(\frac{k}{2}\right)^2$ to each side.

Example:
$$x^2 - 4x = 2 \qquad\qquad k = -4$$
$$x^2 - 4x + 4 = 2 + 4 \qquad \text{Add } \left(\frac{-4}{2}\right)^2 = 4.$$
$$(x - 2)^2 = 6 \qquad\qquad \text{Perfect square trinomial}$$
$$x - 2 = \pm\sqrt{6} \qquad\qquad \text{Square root property}$$
$$x = 2 \pm \sqrt{6} \qquad\qquad \text{Add 2.}$$

QUADRATIC FORMULA

$$x = \frac{-b \pm \sqrt{b^2 - 4ac}}{2a} \qquad \text{Always works}$$

Example: $2x^2 - 5x - 3 = 0$

$$x = \frac{-(-5) \pm \sqrt{(-5)^2 - 4(2)(-3)}}{2(2)} = \frac{5 \pm 7}{4} = 3, -\frac{1}{2}$$

DISCRIMINANT

The number of real solutions to $ax^2 + bx + c = 0$, with $a \neq 0$, can be found by evaluating the discriminant, $b^2 - 4ac$.

1. If $b^2 - 4ac > 0$, there are two real solutions.
2. If $b^2 - 4ac = 0$, there is one real solution.
3. If $b^2 - 4ac < 0$, there are no real solutions.

SECTION 3.3 QUADRATIC INEQUALITIES

QUADRATIC INEQUALITY

$ax^2 + bx + c < 0$, where $<$ may be replaced by $\leq$, $>$, or $\geq$.

Example: $3x^2 - x + 1 \leq 0$

CONCEPT	EXPLANATION AND EXAMPLES

SECTION 3.3 QUADRATIC INEQUALITIES (CONTINUED)

GRAPHICAL SOLUTION

Graph $y = ax^2 + bx + c$ and locate the x-intercepts; then determine x-values where the inequality is satisfied.

Example: $-x^2 - x + 2 > 0$.

The x-intercepts are -2 and 1.

Solution set is $\{x \mid -2 < x < 1\}$ or $(-2, 1)$.

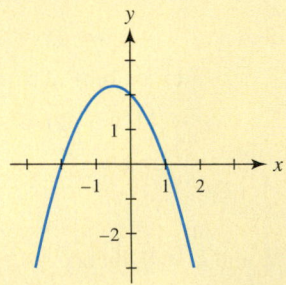

SYMBOLIC SOLUTION

Solve $ax^2 + bx + c = 0$ and use a table of values to determine the intervals where the inequality is satisfied.

Example: $x^2 - 4 \geq 0$.

$x^2 - 4 = 0$ implies $x = \pm 2$.

Solution set is $\{x \mid x \leq -2 \text{ or } x \geq 2\}$, or $(-\infty, -2] \cup [2, \infty)$

Interval	Test Value x	$x^2 - 4$	Positive or Negative?
$(-\infty, -2)$	-3	5	**Positive**
$(-2, 2)$	0	-4	**Negative**
$(2, \infty)$	3	5	**Positive**

SECTION 3.4 TRANSFORMATIONS OF GRAPHS

VERTICAL SHIFTS WITH $k > 0$

$y = f(x) + k$ shifts the graph of $y = f(x)$ upward k units.
$y = f(x) - k$ shifts the graph of $y = f(x)$ downward k units.

HORIZONTAL SHIFTS WITH $k > 0$

$y = f(x - k)$ shifts the graph of $y = f(x)$ to the right k units.
$y = f(x + k)$ shifts the graph of $y = f(x)$ to the left k units.

VERTICAL STRETCHING AND SHRINKING

$y = cf(x)$ vertically stretches the graph of $y = f(x)$ when $c > 1$ and shrinks the graph when $0 < c < 1$.

HORIZONTAL STRETCHING AND SHRINKING

$y = f(cx)$ horizontally shrinks the graph of $y = f(x)$ when $c > 1$ and stretches the graph when $0 < c < 1$.

REFLECTIONS

$y = -f(x)$ is a reflection of $y = f(x)$ across the x-axis.
$y = f(-x)$ is a reflection of $y = f(x)$ across the y-axis.

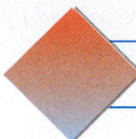

REVIEW EXERCISES

Exercises 1 and 2: Use the graph of the quadratic function to determine the sign of the leading coefficient, its vertex, and the equation of the axis of symmetry.

1.

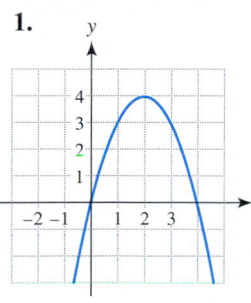

2.

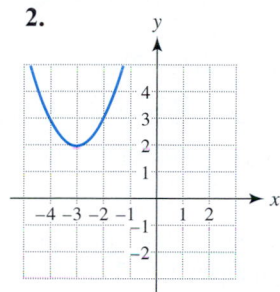

Exercises 3 and 4: Write $f(x)$ in the form $f(x) = ax^2 + bx + c$, and identify the leading coefficient.

3. $f(x) = -2(x - 5)^2 + 1$ **4.** $f(x) = \frac{1}{3}(x + 1)^2 - 2$

Exercises 5 and 6: Write $f(x)$ in the form $f(x) = a(x - h)^2 + k$, and identify the vertex.

5. $f(x) = x^2 + 6x - 1$ **6.** $f(x) = 2x^2 + 4x - 5$

Exercises 7 and 8: Use the vertex formula to determine the vertex on the graph of f.

7. $f(x) = -3x^2 + 2x - 4$ **8.** $f(x) = x^2 + 8x - 5$

Exercises 9 and 10: Use the graph of the quadratic function f to write it as $f(x) = a(x - h)^2 + k$.

9.

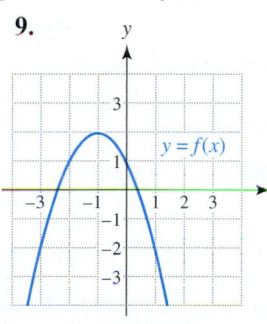

10.

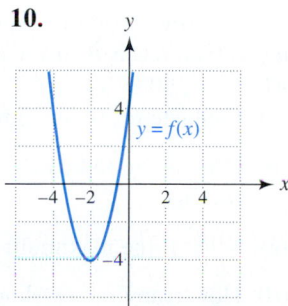

Exercises 11–22: Sketch a graph of the function.

11. $f(x) = -3x^2 + 3$ **12.** $f(x) = \frac{1}{2}x^2$

13. $g(x) = -\frac{1}{2}(x + 2)^2 + 1$ **14.** $g(x) = 2(x - 1)^2 - 3$

15. $h(t) = t^2 - 4t - 1$ **16.** $h(t) = -\frac{1}{4}t^2 + t + 2$

17. $f(x) = -|x + 3|$ **18.** $f(x) = |x - 1| - 2$

19. $g(t) = t^2 - 2t$ **20.** $g(t) = 3t - t^2$

21. $f(x) = -\sqrt{x + 1}$ **22.** $f(x) = \sqrt{2 - x}$

Exercises 23 and 24: The graph of $f(x) = ax^2 + bx + c$ is given.

  (a) *State whether $a > 0$ or $a < 0$.*
  (b) *Estimate the real solutions to $ax^2 + bx + c = 0$.*
  (c) *Determine if the discriminant is positive, negative, or zero.*

23.

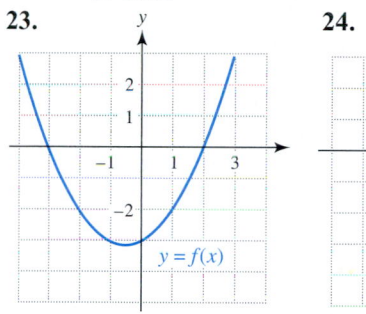

24.

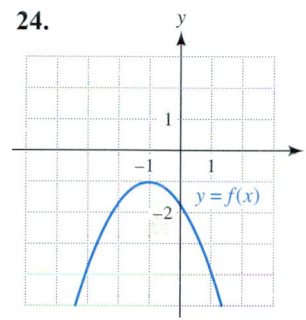

Exercises 25–36: Solve the quadratic equation.

25. $x^2 - x - 20 = 0$ **26.** $3x^2 + 4x = -1$

27. $x^2 = 4x$ **28.** $-5x^2 - 3x = 0$

29. $4z^2 - 7 = 0$ **30.** $25z^2 = 9$

31. $-2t^2 - 3t + 14 = 0$ **32.** $\frac{1}{2}t^2 + \frac{3}{4}t + \frac{1}{4} = 0$

33. $0.1x^2 - 0.3x = 1$ **34.** $x(6 - x) = -16$

35. $(k - 1)^2 = \frac{9}{4}$ **36.** $(k + 2)^2 = 7$

Exercises 37–42: Solve by completing the square.

37. $x^2 + 2x = 5$ **38.** $x^2 - 3x = 3$

39. $2z^2 - 6z - 1 = 0$ **40.** $-3z^2 - 2z + 2 = 0$

41. $\frac{1}{2}x^2 - 4x + 1 = 0$ **42.** $-\frac{1}{4}x^2 - \frac{1}{2}x + 1 = 0$

43. Solve the equation $2x^2 - 3y^2 = 6$ for y. Is y a function of x?

44. Solve $h = -\frac{1}{2}gt^2 + 100$ for t.

45. Solve the equation or inequality.
  (a) $x^2 - 3x + 2 = 0$ (b) $x^2 - 3x + 2 < 0$
  (c) $x^2 - 3x + 2 > 0$

46. Solve $2x^2 + 1.3x \le 0.4$ graphically.

Exercises 47–52: Solve the inequality.

47. $x^2 - 3x + 2 \le 0$ **48.** $x^2 - 2x \ge 0$

49. $10x^2 + 19x + 5 < 0$ **50.** $9x^2 - 4 > 0$

51. $n(n - 2) \ge 15$ **52.** $n^2 + 4 \le 6n$

53. Use the graph of $y = f(x)$ to solve the inequality.
  (a) $f(x) > 0$ **(b)** $f(x) \le 0$

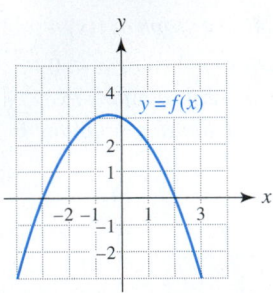

54. The accompanying table contains test values for a quadratic function $f(x)$. Use the table to determine the solution set to each inequality.

x	-6	-5	0	3	4
$f(x)$	9	0	-15	0	9

  (a) $f(x) < 0$ **(b)** $f(x) \ge 0$

55. If $f(x) = 2x^2 - 3x + 1$, use transformations to graph $y = -f(x)$ and $y = f(-x)$.

56. Use the given graph of $y = f(x)$ to sketch a graph of each expression.
  (a) $y = f(x + 1) - 2$ **(b)** $y = -2f(x)$

  (c) $y = f(2x)$

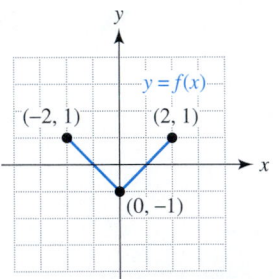

Exercises 57–64: Use transformations of graphs to sketch a graph of f.

57. $f(x) = x^2 - 4$ **58.** $f(x) = 3|x - 2| + 1$

59. $f(x) = (x + 4)^2 - 2$ **60.** $f(x) = \sqrt{x + 4}$

61. $f(x) = \sqrt{x + 2}$ **62.** $f(x) = -4\sqrt{-x}$

63. $f(x) = -2(x - 2)^2 + 3$ **64.** $f(x) = -|x - 3|$

Exercises 65 and 66: Two functions f and g are related by the given equation. Use the numerical representation of f to construct a numerical representation of g.

65. $g(x) = f(x) + 2$

x	1	2	3	4
$f(x)$	-3	3	4	7

66. $g(x) = f(x - 5)$

x	0	5	10	15
$f(x)$	-5	11	21	32

67. Sketch a graph of a quadratic function f with axis of symmetry $x = 2$ and that passes through the points $(2, 3)$ and $(4, -1)$. Find a formula for the function.

68. Find a function f given by $f(x) = a(x - h)^2 + k$ that models the data exactly.

x	2	3	4	5
y	4	7	16	31

Applications

69. *Maximizing Area* A homeowner has 44 feet of fence to enclose a rectangular garden. One side of the garden needs no fencing because it is along the wall of the house. What dimensions for the garden will maximize its area?

70. *Maximizing Revenue* The revenue R in dollars that an individual receives from selling x radios is given by $R(x) = x(90 - x)$.
  (a) Evaluate $R(20)$ and interpret the result.

  (b) How many radios should be sold to maximize the revenue?

  (c) What is the maximum revenue?

  (d) How many radios should be sold for revenue to be $2000 or more?

71. *Projectile* A slingshot is used to propel a stone upward so that its height h in feet after t seconds is given by $h(t) = -16t^2 + 88t + 5$.
  (a) Evaluate $h(0)$ and interpret the result.

  (b) How high was the stone after 2 seconds?

  (c) Find the maximum height of the stone.

  (d) At what time(s) was the stone 117 feet high?

72. *Modeling Water Flow*   Water is leaking out of a hole in the bottom of a 100-gallon tank. The accompanying table lists the volume V of the water after t minutes. Decide whether $f(t) = 0.4t^2 - 9.9t + 100$ or $g(t) = 100 - 9.5t$ models the data better.

t (min)	0	1	2	3	4
V (gal)	100	90.5	81.9	74.1	67.0

73. *World Population*   The function given by the formula $f(x) = 0.000478x^2 - 1.813x + 1720.1$ models the world population in billions from 1950 to 2000, where x is the year.
(a) Evaluate $f(1985)$ and interpret the result.

(b) Use $f(x)$ to estimate world population during the year 2000.

(c) When does this model predict that world population may reach 7 billion?

74. *Construction*   A box is being constructed by cutting 3-inch squares from the corners of a rectangular sheet of metal that is 4 inches longer than it is wide. If the box is to have a volume of 135 cubic inches, find the dimensions of the metal sheet.

75. *Room Prices*   Room prices are regularly $100, but for each additional room rented by a group, the price is reduced by $3 for each room. For example, 1 room costs $100, 2 rooms cost $2 \times \$97 = \$194$, and 3 rooms cost $3 \times \$94 = \$282$.
(a) Write a quadratic function C that gives the total cost of renting x rooms.

(b) What is the total cost of renting 6 rooms?

(c) How many rooms are rented if the total cost is $730?

(d) What number of rooms rented results in the greatest total cost?

76. *Irrigation and Yield*   The table shows how irrigation of rice crops affects yield, where x represents the percent of total area that is irrigated and y is the rice yield in tons per hectare. (1 hectare $\approx$ 2.47 acres.)   (**Source:** D. Grigg, *The World Food Problem.*)

x	0	20	40	60	80	100
y	1.6	1.8	2.2	3.0	4.5	6.1

(a) Use least-squares regression to find a quadratic function that models the data.

(b) Solve the equation $f(x) = 3.7$. Interpret the results.

77. *Credit Card Debt*   The table lists the outstanding balances on Visa and MasterCard credit cards in billions of dollars.   (**Sources:** Bankcard Holders of America, Federal Reserve Board.)

Year	1980	1984	1988	1992	1996
Debt	82	108	172	254	444

(a) Explain why a linear function does not model the data.

(b) Find a quadratic function f that models these data. Solve the equation $f(x) = 212$. Interpret the results.

78. *Internet*   Downloading files from the Internet can be both slow and unreliable when standard phone lines are being used. Phone modems can transmit 56,000 bits per second, whereas cable TV modems are capable of transmitting 2–10 million bits per second. Demand for this new technology increased from 1996 to 2000. The table shows this increase in millions of accounts.   (**Source:** *New York Times Service, Pioneer Press Research.*)

Year	1996	1997	1998	1999	2000
Accounts	0.5	1	2	4	7

(a) Let $f(x) = a(x - 1996)^2 + 0.5$. Approximate a so that f models the data.

(b) Solve the equation $f(x) = 10$ and interpret the result.

EXTENDED AND DISCOVERY EXERCISES

1. *Shooting a Foul Shot*   (Refer to the introduction of this chapter.) To make a foul shot in basketball, the ball must follow a parabolic arc. This arc depends on both the angle and velocity with which the basketball is released. If a person shoots the basketball overhand from a position 8 feet above the floor, then the path can sometimes be modeled by the parabola

$$y = \frac{-16x^2}{0.434v^2} + 1.15x + 8,$$

where v is the velocity of the ball in feet per second, as illustrated in the accompanying figure on the next page. (**Source:** C. Rist, "The physics of foul shots.")

(a) If the basketball hoop is 10 feet high and located 15 feet away, what initial velocity v should the basketball have?

(b) Check your answer from part (a) graphically. Plot the point $(0, 8)$ where the ball is released and the point $(15, 10)$ where the basketball hoop is. Does your graph pass through both points?

(c) What is the maximum height of the basketball?

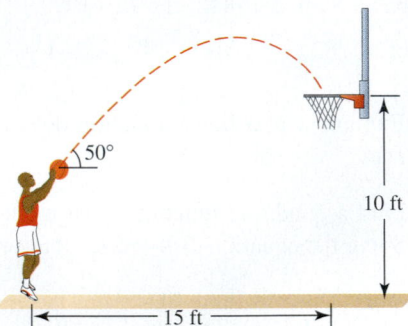

2. (Continuation of Exercise 1) If a person releases a basketball underhand from a position 3 feet above the floor, it often has a steeper arc than if it is released overhand and sometimes may be modeled by

$$y = \frac{-16x^2}{0.117v^2} + 2.75x + 3.$$

See the accompanying figure. Complete parts (a), (b), and (c) for Exercise 1. Then compare the paths for an overhand shot and an underhand shot.

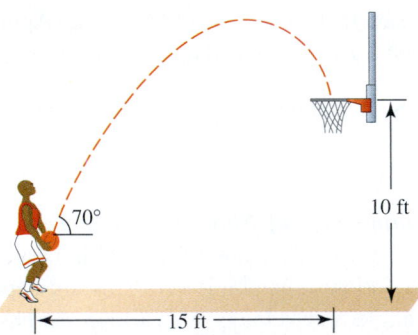

Exercises 3–6: **Reflecting Functions** *Computer graphics frequently use reflections. Reflections can speed up the generation of a picture or create a figure that appears perfectly symmetrical.* (**Source:** S. Hoggar, *Mathematics for Computer Graphics.*)

(a) *For the given $f(x)$, constant k, and viewing rectangle, graph $x = k$, $y = f(x)$, and $y = f(2k - x)$.*

(b) *Generalize how the graph of $y = f(2k - x)$ compares to the graph of $y = f(x)$.*

3. $f(x) = \sqrt{x}, k = 2, [-1, 8, 1]$ by $[-4, 4, 1]$

4. $f(x) = x^2, k = -3, [-12, 6, 1]$ by $[-6, 6, 1]$

5. $f(x) = x^4 - 2x^2 + 1, k = -6, [-15, 3, 1]$ by $[-3, 9, 1]$

6. $f(x) = 4x - x^3, k = 5, [-6, 18, 1]$ by $[-8, 8, 1]$

7. **Modeling a Cold Front** A weather map of the United States on April 22, 1996, is shown in the figure. There was a cold front roughly in the shape of a circular arc passing north of Dallas and west of Detroit. The center of the arc was located near Pierre, South Dakota, with a radius of about 750 miles. If Pierre has the coordinates $(0, 0)$ and the positive y-axis points north, then the equation of the front can be modeled by

$$f(x) = -\sqrt{750^2 - x^2},$$

where $0 \le x \le 750$. (**Source:** AccuWeather, Inc.)

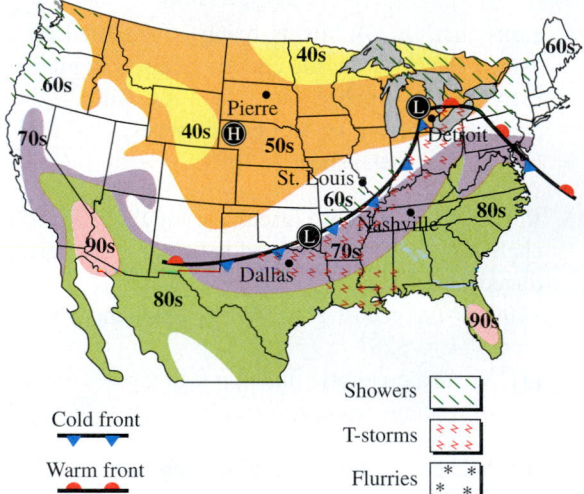

(a) St. Louis is located at $(535, -400)$ and Nashville is at $(730, -570)$ where units are in miles. Plot these points and graph f in the window $[0, 1200, 100]$ by $[-800, 0, 100]$. Did the cold front reach these cities?

(b) During the next 12 hours, the center of the front moved approximately 110 miles south and 160 miles east. Assuming the cold front did not change shape, use transformations of graphs to determine an equation that models its new location.

(c) Use graphing to determine visually if the cold front reached both cities.

4 Nonlinear Functions and Equations

The wonderful things you learn in schools are the work of many generations, produced by enthusiastic effort and infinite labor in every country.

Albert Einstein

Mathematics can be both abstract and applied. Abstract mathematics is focused on axioms, theorems, and proofs. It can be derived independently of empirical evidence. Theorems that were proved centuries ago are still valid today. In this sense, abstract mathematics transcends time. Yet, even though mathematics can be developed in an abstract setting—separate from science and all measured data—it also has countless applications.

There is a common misconception that theoretical mathematics is unimportant, yet many of the ideas that eventually had great practical importance were first born in the abstract. For example, in 1854 George Boole published *Laws of Thought*, which outlined the basis for Boolean algebra. This was 85 years before the invention of the first digital computer. However, Boolean algebra became the basis by which modern computer hardware operates.

Much like Boolean algebra, the purpose of complex numbers was at first theoretical. However, today complex numbers are used in the design of electrical circuits, ships, and airplanes. Basic quantum physics and certain features of the theory of relativity would not have been developed without complex numbers. Even the *fractal image* shown in the figure below would not have been discovered without complex numbers.

In this chapter we discuss some important topics in algebra that have had an impact on society. We are privileged to read in a few hours what took people centuries to discover. To ignore either the abstract beauty or the profound applicability of mathematics is like seeing a rose but never smelling one.

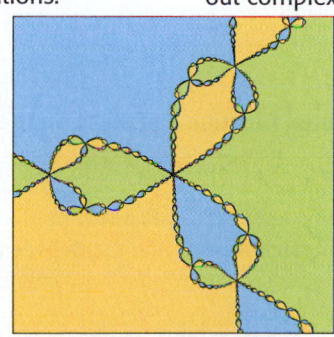

The Cube Roots of Unity

Source: From *Numerical Analysis, Mathematics of Scientific Computing* 1st edition by KINCAID/CHENEY. © 1991.

241

4.1 Nonlinear Functions and Their Graphs

◆ Learn terminology about polynomial functions

◆ Identify intervals where a function is increasing or decreasing

◆ Find extrema of a function

◆ Identify symmetry in a graph of a function

◆ Determine if a function is odd, even, or neither

Introduction

Monthly average high temperatures at Daytona Beach are shown in Table 4.1.

TABLE 4.1 Monthly Average High Temperatures at Daytona Beach

Month	1	2	3	4	5	6	7	8	9	10	11	12
Temperature (°F)	69	70	75	80	85	88	90	89	87	81	76	70

Source: J. Williams, *The USA Weather Almanac, 1995.*

Figure 4.1 shows a scatterplot of the data. A linear function would not model this data because the data do not lie on a line. One possibility is to model the data with a quadratic function, as shown in Figure 4.2. However, a better fit can be obtained with the nonlinear function f whose graph is shown in Figure 4.3 and is given by

$$f(x) = 0.0145x^4 - 0.426x^3 + 3.53x^2 - 6.23x + 72.0,$$

where $x = 1$ corresponds to January, $x = 2$ to February, and so on. (Least-squares regression was used to determine f.) Function f is a *polynomial function* with degree 4.

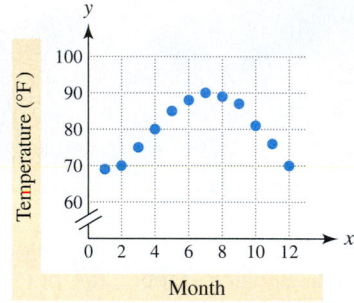

FIGURE 4.1 Temperature Data

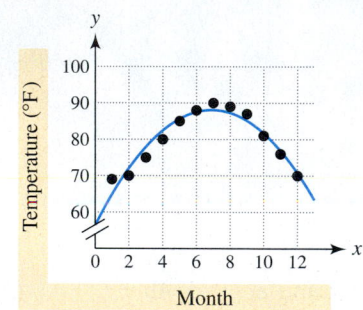

FIGURE 4.2 Quadratic Model

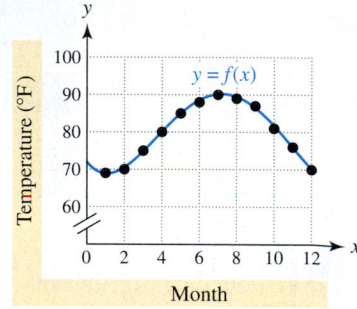

FIGURE 4.3 New Nonlinear Model

Polynomial Functions

Polynomial functions are frequently used to approximate data. The domain of a polynomial function is all real numbers, and its graph is continuous without breaks or sharp edges.

POLYNOMIAL FUNCTION

A **polynomial function** f **of degree** n **in the variable** x can be represented by

$$f(x) = a_n x^n + a_{n-1} x^{n-1} + \cdots + a_2 x^2 + a_1 x + a_0$$

where each coefficient a_k is a real number, $a_n \neq 0$, and n is a nonnegative integer. The **leading coefficient** is a_n and the **degree** is n.

Formulas, degrees, and leading coefficients of some polynomial functions include the following.

Algebra Review
To review polynomials, see
Chapter R (page R-20).

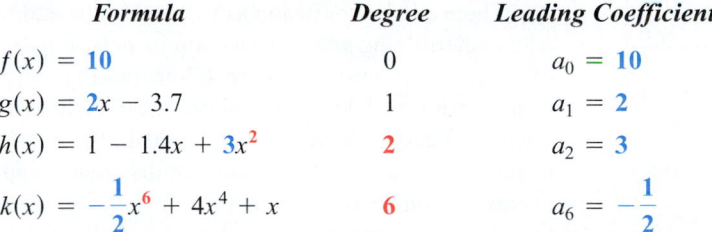

Formula	Degree	Leading Coefficient
$f(x) = 10$	0	$a_0 = 10$
$g(x) = 2x - 3.7$	1	$a_1 = 2$
$h(x) = 1 - 1.4x + 3x^2$	2	$a_2 = 3$
$k(x) = -\dfrac{1}{2}x^6 + 4x^4 + x$	6	$a_6 = -\dfrac{1}{2}$

A polynomial function of degree 2 or higher is a nonlinear function. Functions f and g are linear, whereas functions h and k are nonlinear. As a result, polynomial functions are used to model both linear and nonlinear data.

Note: Quadratic functions are examples of nonlinear functions, which we discussed in Chapter 3. This chapter introduces more nonlinear functions.

Functions that contain radicals, ratios, or absolute values of variables are not polynomials. For example, $f(x) = 2\sqrt{x}$, $g(x) = \frac{1}{x-1}$, and $h(x) = |2x + 5|$ are *not* polynomials.

Increasing and Decreasing Functions

The concepts of increasing and decreasing relate to whether the graph of a function rises or falls. Intuitively, if we could walk *from left to right* along the graph of an increasing function, it would be uphill. For a decreasing function, we would walk downhill. We speak of a function f increasing or decreasing *over an interval of its domain*. For example, in Figure 4.4 the function is decreasing (the graph falls) when $-2 \le x \le 0$ and increasing (the graph rises) when $0 \le x \le 2$. In interval notation the graph decreases on $[-2, 0]$ and increases on $[0, 2]$.

Increasing and decreasing functions are defined as follows.

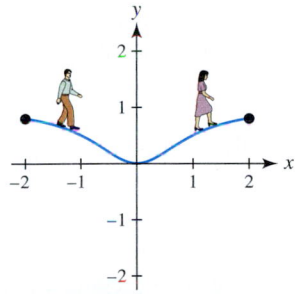

FIGURE 4.4

INCREASING AND DECREASING FUNCTIONS

Suppose that a function f is defined over an interval I on the number line. If x_1 and x_2 are in I,

(a) f **increases** on I if, whenever $x_1 < x_2$, $f(x_1) < f(x_2)$;
(b) f **decreases** on I if, whenever $x_1 < x_2$, $f(x_1) > f(x_2)$.

Figures 4.5 and 4.6 illustrate these concepts.

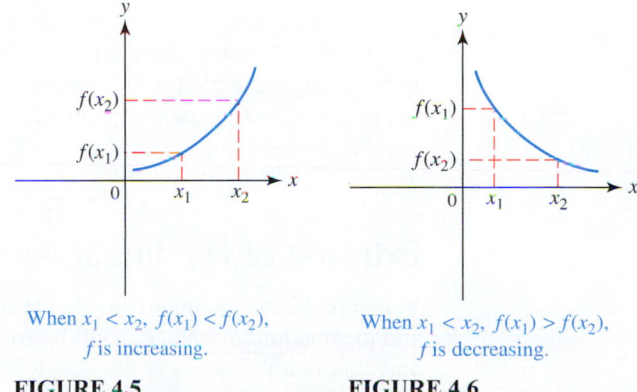

When $x_1 < x_2$, $f(x_1) < f(x_2)$,   When $x_1 < x_2$, $f(x_1) > f(x_2)$,
f is increasing. f is decreasing.

FIGURE 4.5   **FIGURE 4.6**

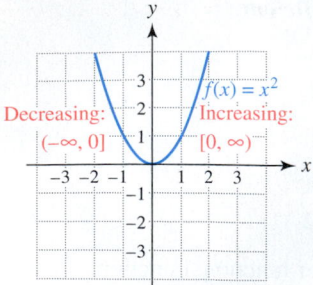

FIGURE 4.7

There can be confusion as to whether the endpoints of an interval should be included when determining where a function is increasing or decreasing. For example, is the graph of $f(x) = x^2$, shown in Figure 4.7, increasing on $[0, \infty)$ or just on $(0, \infty)$? The definition of increasing and decreasing allows us to include 0 as part of the interval I where f is increasing, because if we let $x_1 = 0$, then $f(0) < f(x_2)$ whenever $0 < x_2$. Thus $f(x) = x^2$ is increasing on $[0, \infty)$. A similar discussion can be used to show that $f(x) = x^2$ is decreasing on $(-\infty, 0]$ by letting $x_2 = 0$. Do *not* confuse these concepts by saying that function f both increases and decreases at the point $(0, 0)$. The concepts of increasing and decreasing apply only to intervals of the real number line and not to individual points.

Note that it is *not* incorrect to say that $f(x) = x^2$ is increasing on $(0, \infty)$ because $f(x) = x^2$ also increases on $[1, \infty)$, $(10, \infty)$, and $[100, 200]$. However, we generally give the largest interval possible, $[0, \infty)$, when determining where a function is increasing or decreasing. (**Reference:** J. Stewart, *Calculus*, Fourth edition, p. 21.)

EXAMPLE 1 Determining where a function is increasing or decreasing

A graph of a function f shown in Figure 4.8 gives the tides at Clearwater Beach, Florida, x hours after midnight on August 28, 1998. (**Source:** D. Pentcheff, *WWW Tide and Current Predictor*.) Use the graph to determine when water levels were increasing and when they were decreasing for $0 \le x \le 27$.

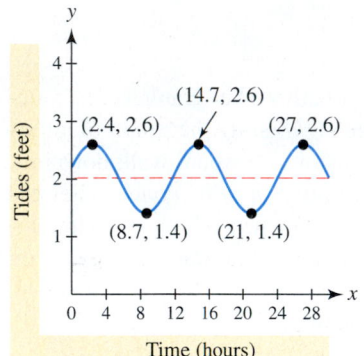

FIGURE 4.8 Tides at Clearwater Beach

SOLUTION Water levels are rising when the graph of f increases from left to right. This occurs during the following time intervals:

$$[0, 2.4], \quad [8.7, 14.7], \quad \text{and} \quad [21, 27].$$

For example, the tides rose from 1.4 feet to 2.6 feet between 8.7 and 14.7 hours after midnight. Similarly, the tides were falling during the following time intervals:

$$[2.4, 8.7] \quad \text{and} \quad [14.7, 21]. \qquad \textit{Now Try Exercise 115(a)} \blacklozenge$$

Note: When determining where a function is increasing and where it is decreasing, it is important to give x-intervals and not y-intervals.

EXAMPLE 2 Determining where a function is increasing or decreasing

Use the graph of $f(x) = 4x - \frac{1}{3}x^3$ (shown in Figure 4.9) and interval notation to identify where f is increasing or decreasing.

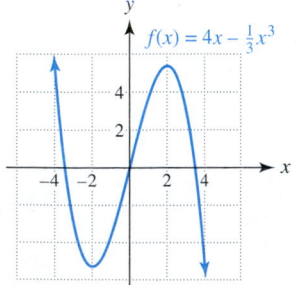

FIGURE 4.9

SOLUTION Moving from left to right on the graph of f, the y-values decrease until $x = -2$, increase until $x = 2$, and decrease thereafter. Thus $f(x) = 4x - \frac{1}{3}x^3$ is decreasing on $(-\infty, -2]$, increasing on $[-2, 2]$, and decreasing again on $[2, \infty)$. In interval notation f is decreasing on $(-\infty, -2] \cup [2, \infty)$. $\qquad \textit{Now Try Exercise 13} \blacklozenge$

Extrema of Nonlinear Functions

In Figure 4.3 the minimum monthly average temperature of 69°F occurs in January ($x = 1$) and the maximum average monthly temperature of 90°F occurs in July ($x = 7$). Minimum and maximum y-values on the graph of a function often represent important data. Graphs of polynomial functions often have "hills" or "valleys." For example, consider the two polynomial functions f and g given by

$$f(x) = \frac{1}{2}x^2 - 2x - 4 \qquad \text{and} \qquad g(x) = -\frac{1}{4}x^4 + \frac{2}{3}x^3 + \frac{5}{2}x^2 - 6x.$$

Their graphs are shown in Figures 4.10 and 4.11. (Some values have been rounded.)

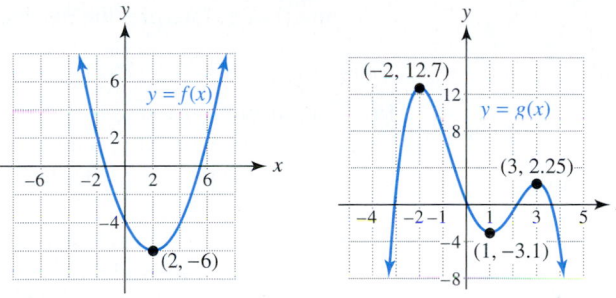

FIGURE 4.10   FIGURE 4.11

Figure 4.10 shows the graph of a parabola. If one traces along this graph from left to right, the y-values decrease until the vertex $(2, -6)$ is reached. To the right of the vertex, the y-values increase. The minimum y-value on the graph of f is -6. It is called the *absolute minimum* of f. The function f has no *absolute maximum* because there is no largest y-value on a parabola opening upward.

In Figure 4.11 the highest "hill" on the graph of g is located at $(-2, 12.7)$. Therefore the absolute maximum of g is **12.7**. There is a smaller peak located at the point $(3, 2.25)$. In a small interval near $x = 3$, the y-value of 2.25 is locally the largest. We say that g has a *local maximum* of **2.25**. Similarly, a "valley" occurs on the graph of g. The lowest point is $(1, -3.1)$. The value -3.1 is not the smallest y-value on the entire graph of g. Therefore it is not an absolute minimum. Rather, -3.1 is a *local minimum*.

Maximum and minimum values that are either absolute or local are called **extrema** (plural of extremum). A function may have several local extrema, but at most one absolute maximum and one absolute minimum. However, it is possible for a function to assume an absolute extremum at two values of x. In Figure 4.12 the absolute maximum is 11. It occurs at $x = \pm 2$. Note that 11 is also a local maximum, because near $x = -2$ and $x = 2$ it is the largest y-value.

Sometimes an absolute maximum (minimum) is called a *global maximum* (*minimum*). Similarly, sometimes a local maximum (minimum) is called a *relative maximum* (*minimum*).

This discussion is summarized in the following.

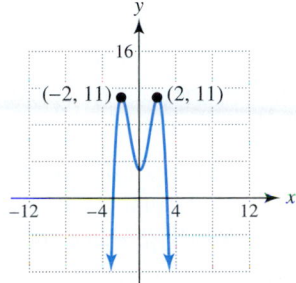

FIGURE 4.12

ABSOLUTE AND LOCAL EXTREMA

Let c be in the domain of f.

$f(c)$ is an **absolute (global) maximum** if $f(c) \geq f(x)$ *for all x in the domain of f.*

$f(c)$ is an **absolute (global) minimum** if $f(c) \leq f(x)$ *for all x in the domain of f.*

$f(c)$ is a **local (relative) maximum** if $f(c) \geq f(x)$ *when x is near c.*

$f(c)$ is a **local (relative) minimum** if $f(c) \leq f(x)$ *when x is near c.*

Note: The expression "near c" means that there is an open interval in the domain of f containing c, where $f(c)$ satisfies the inequality.

Identifying and interpreting extrema

Figure 4.13 shows the graph of a function f that models the volume of air in a person's lungs measured in liters after x seconds. (**Source:** Adapted from V. Thomas, *Science and Sport.*)

(a) Find the absolute maximum and the absolute minimum of f. Interpret the results.

(b) Identify two local maxima (plural of maximum) and two local minima (plural of minimum) of f. Interpret the results.

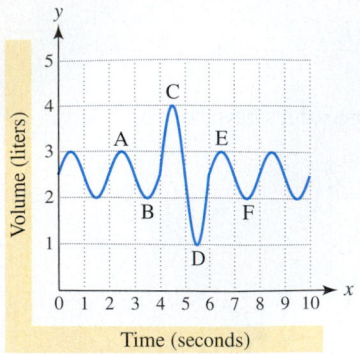

FIGURE 4.13 Volume of Air in a Person's Lungs

SOLUTION

(a) The absolute maximum is 4 liters and occurs at C. The absolute minimum is 1 liter and occurs at D. At C a deep breath has been taken and the lungs are more inflated. After C, the person exhales beyond normal breathing until the lungs contain only 1 liter of air at D.

(b) One local maximum is 3 liters. It occurs at A and E and represents the amount of air in a person's lungs after inhaling normally. One local minimum is 2 liters. It occurs at B and F and represents the amount of air after exhaling normally. Another local maximum is 4 liters, which was also the absolute maximum. Similarly, 1 liter is a local minimum and also the absolute minimum. *Now Try Exercise 115(b)* ◆

Identifying extrema

Use the graph of f in Figure 4.14 to estimate any local and absolute extrema.

SOLUTION

Local Extrema The points $(-2, 8)$ and $(1, -19)$ on the graph of f correspond to the lowest point in a "valley." Thus there are local minimums of 8 and -19. The point $(-1, 13)$ corresponds to the highest point on a "hill." Thus there is a local maximum of 13.

Absolute Extrema Because the arrows point upward, there is no maximum y-value on the graph. Thus there is no absolute maximum. However, the minimum y-value on the graph of f occurs at the point $(1, -19)$. The absolute minimum is -19. *Now Try Exercise 41* ◆

FIGURE 4.14

Modeling ocean temperatures

The monthly average ocean temperature in degrees Fahrenheit at Bermuda can be modeled by $f(x) = 0.0215x^4 - 0.648x^3 + 6.03x^2 - 17.1x + 76.4$, where $x = 1$ corresponds to January and $x = 12$ to December. The domain of f is $D = \{x \mid 1 \le x \le 12\}$. (**Source:** J. Williams, *The Weather Almanac 1995.*)

(a) Graph f in $[1, 12, 1]$ by $[50, 90, 10]$.

(b) Estimate the absolute extrema. Interpret the results.

SOLUTION

(a) The graph of $Y_1 = .0215X^4 - .648X^3 + 6.03X^2 - 17.1X + 76.4$ is shown in Figure 4.15.

(b) Many graphing calculators have the capability to find maximum and minimum y-values. The points associated with absolute extrema are shown in Figures 4.15 and 4.16. An absolute minimum of about 61.5 corresponds to the point $(2.01, 61.5)$. This means that the monthly average ocean temperature is coldest during the month of February ($x \approx 2$) when it reaches a minimum of about 61.5°F.

An absolute maximum of approximately 82 corresponds to the point (7.61, 82.0). The warmest average ocean temperature occurs during August ($x \approx 8$) when it reaches a maximum of 82°F. We might also say that this maximum occurs in late July, since $x \approx 7.61$.

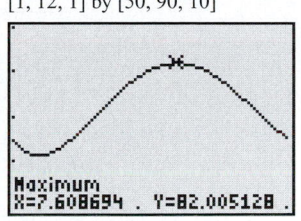

[1, 12, 1] by [50, 90, 10]   [1, 12, 1] by [50, 90, 10]

Minimum   Maximum
X=2.0128359 Y=61.479564   X=7.608694 . Y=82.005128

FIGURE 4.15   FIGURE 4.16

Now Try Exercise 119 ◆

Calculator Help
To find a minimum or maximum point on a graph, see Appendix B (page AP-14).

Symmetry

Symmetry is used frequently in art, mathematics, and computer graphics. Many objects are symmetric along a vertical line so that the left and right sides are mirror images. If an automobile is viewed from the front, the left side is typically a mirror image of the right side. Similarly, animals and people usually have an approximate left-right symmetry. Graphs of functions may also exhibit this type of symmetry, as shown in Figures 4.17–4.19.

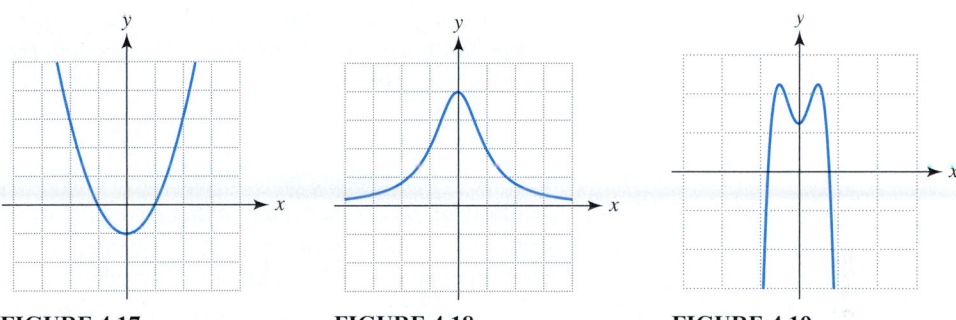

FIGURE 4.17   FIGURE 4.18   FIGURE 4.19

If each graph were folded along the y-axis, the left and right halves would match. These graphs are **symmetric with respect to the y-axis**. A function whose graph satisfies this characteristic is called an *even function.*

Figure 4.20 shows a graph of an even function f. Since the graph is symmetric with respect to the y-axis, the points (x, y) and $(-x, y)$ both lie on the graph of f. Thus $f(x) = y$ and $f(-x) = y$, and so $f(x) = f(-x)$ for an even function. This means that if we change the sign of the input, the output does not change. For example, if $g(x) = x^2$, then $g(2) = g(-2) = 4$. Since this is true for *every input*, g is an even function.

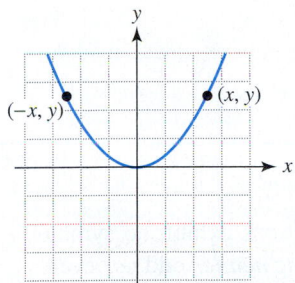

FIGURE 4.20 An Even Function

EVEN FUNCTION

A function f is an **even function** if $f(-x) = f(x)$ for every x in its domain. The graph of an even function is symmetric with respect to the y-axis.

A second type of symmetry is shown in Figures 4.21–4.23 on the next page. If we could spin or rotate the graph about the origin, the original graph would reappear after half a turn. These graphs are **symmetric with respect to the origin** and represent *odd functions.*

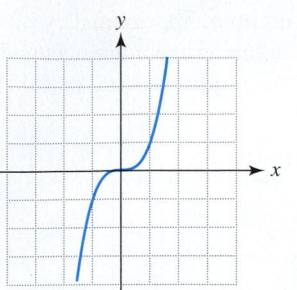

FIGURE 4.21

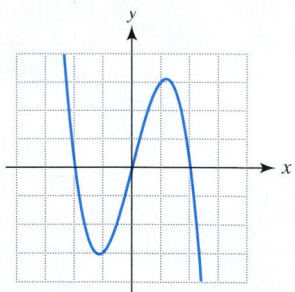

FIGURE 4.22

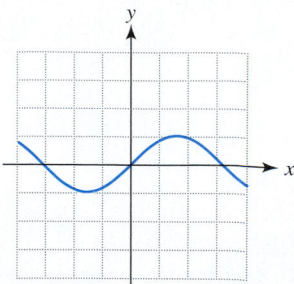

FIGURE 4.23

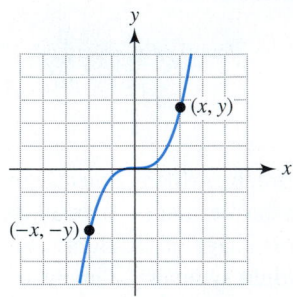

FIGURE 4.24 An Odd Function

In Figure 4.24 the point (x, y) lies on the graph of an odd function f. If this point spins half a turn or 180° around the origin, its new location is $(-x, -y)$. Thus $f(x) = y$ and $f(-x) = -y$. It follows that $f(-x) = -y = -f(x)$ for any odd function f. Changing the sign of the input only changes the sign of the output. For example, if $g(x) = x^3$ then $g(3) = 27$ and $g(-3) = -27$. Since this is true for *every input*, g is an odd function.

> **ODD FUNCTION**
>
> A function f is an **odd function** if $f(-x) = -f(x)$ for every x in its domain. The graph of an odd function is symmetric with respect to the origin.

The terms *odd* and *even* have special meaning when they are applied to a polynomial function f. If $f(x)$ contains terms that have only odd powers of x, then f is an odd function. Similarly, if $f(x)$ contains terms that have only even powers of x (and possibly a constant term), then f is an even function. For example, $f(x) = x^6 - 4x^4 - 2x^2 + 5$ is an even function, whereas $g(x) = x^5 + 4x^3$ is an odd function. This can be shown symbolically.

$$f(-x) = (-x)^6 - 4(-x)^4 - 2(-x)^2 + 5 \qquad \text{Substitute } -x \text{ for } x.$$
$$= x^6 - 4x^4 - 2x^2 + 5 \qquad \text{Simplify.}$$
$$= f(x) \qquad f \text{ is an even function.}$$
$$g(-x) = (-x)^5 + 4(-x)^3 \qquad \text{Substitute } -x \text{ for } x.$$
$$= -x^5 - 4x^3 \qquad \text{Simplify.}$$
$$= -g(x) \qquad g \text{ is an odd function.}$$

It is important to remember that the graphs of many functions exhibit no symmetry with respect to either the y-axis or the origin. These functions are *neither* odd *nor* even.

◆ **CLASS DISCUSSION**

If 0 is in the domain of an odd function f, what point must lie on its graph? Explain your reasoning.
◆

 **Identifying odd and even functions**

For each representation of a function f, identify whether f is odd, even, or neither.
(a)   TABLE 4.2

x	−3	−2	−1	0	1	2	3
$f(x)$	10.5	2	−0.5	−2	−0.5	2	10.5

(b)

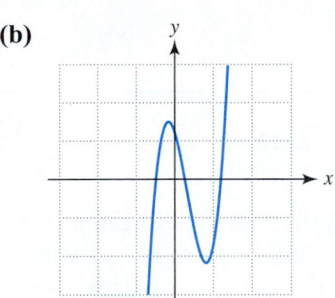

FIGURE 4.25

(c) $f(x) = x^3 - 5x$ **(d)** f is the cube root function.

◆ **CLASS DISCUSSION**

Discuss the possibility of the graph of a function being symmetric with respect to the x-axis. ◆

SOLUTION

(a) The function defined by Table 4.2 has domain $D = \{-3, -2, -1, 0, 1, 2, 3\}$. Notice that $f(-3) = 10.5 = f(3)$ and $f(-2) = 2 = f(2)$. The function f satisfies the statement $f(-x) = f(x)$ for every x in D. Thus f is an even function.

(b) If we fold the graph in Figure 4.25 on the y-axis, the two halves do not match, so f is not an even function. Similarly, f is not an odd function since spinning its graph half a turn about the origin does not result in the same graph. The function f is neither odd nor even.

(c) Since f is a polynomial containing only odd powers of x, it is an odd function. This also can be shown symbolically as follows.

$$
\begin{aligned}
f(-x) &= (-x)^3 - 5(-x) && \text{Substitute } -x \text{ for } x. \\
&= -x^3 + 5x && \text{Simplify.} \\
&= -(x^3 - 5x) && \text{Distributive property} \\
&= -f(x) && f \text{ is an odd function.}
\end{aligned}
$$

This result is supported graphically in Figure 4.26. Spinning the graph $180°$ about the origin results in the same graph.

Calculator Help

On some calculators, the cube root can be found under the MATH menu. See Appendix B (page AP-6).

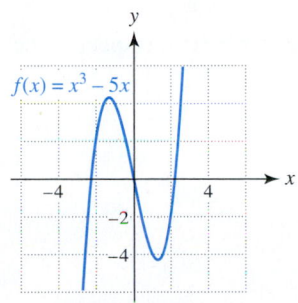

FIGURE 4.26

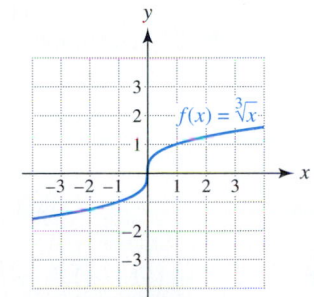

FIGURE 4.27

Algebra Review

To review cube roots, see Chapter R (page R-49).

(d) Note that $\sqrt[3]{-8} = -2$ and that $\sqrt[3]{8} = 2$. In general $\sqrt[3]{-x} = -\sqrt[3]{x}$, which indicates that $f(-x) = -f(x)$, where $f(x) = \sqrt[3]{x}$. Thus f is an odd function. This fact can also be seen by graphing $f(x) = \sqrt[3]{x}$, as shown in Figure 4.27. Spinning the graph of $f(x) = \sqrt[3]{x}$ a half a turn about the origin results in the same graph.

Now Try Exercises 69, 79, 83, and 91 ◆

4.1 Putting it all Together

Nonlinear functions have graphs that can increase or decrease over their domains. Extrema and symmetry are used to describe nonlinear graphs. Polynomial functions of degree 2 or higher are examples of nonlinear functions. The following table summarizes some important concepts related to the graphs of nonlinear functions.

Concept	Explanation	Graphical Example
Increasing	f increases on an interval, if whenever $x_1 < x_2$, then $f(x_1) < f(x_2)$.	f is increasing on $(-\infty, -1] \cup [0, 1]$.
Decreasing	f decreases on an interval, if whenever $x_1 < x_2$, then $f(x_1) > f(x_2)$.	f is decreasing on $[-2, 2]$.
Even function	$f(-x) = f(x)$ The graph is symmetric with respect to the y-axis.	
Odd function	$f(-x) = -f(x)$ The graph is symmetric with respect to the origin.	

Concept	Explanation	Graphical Example
Absolute, or global, maximum (minimum)	The maximum (minimum) y-value on the graph $y = f(x)$	
Local, or relative, maximum (minimum)	A maximum (minimum) y-value on the graph $y = f(x)$ in an open interval of the domain of f	

4.1 Exercises

Polynomials

Exercises 1–10: Determine if the given function is a polynomial function. If it is, state its degree and leading coefficient.

1. $f(x) = 2x^3 - x + 5$ **2.** $f(x) = -x^4 + 1$

3. $f(x) = \sqrt{x}$ **4.** $f(x) = 2x^3 - \sqrt[3]{x}$

5. $f(x) = 1 - 4x - 5x^4$ **6.** $f(x) = \dfrac{1}{1-x}$

7. $g(t) = \dfrac{1}{t^2 + 3t - 1}$ **8.** $g(t) = 5 - 4t$

9. $g(t) = 22$ **10.** $g(t) = |2t|$

Increasing and Decreasing Functions

Exercises 11–16: Use the graph of f to determine intervals where f is increasing and where f is decreasing. Use interval notation.

11.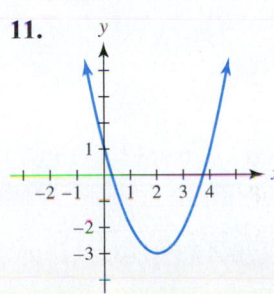

12.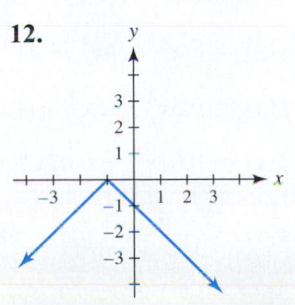

13.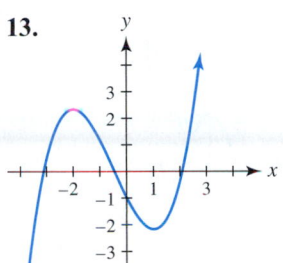

14.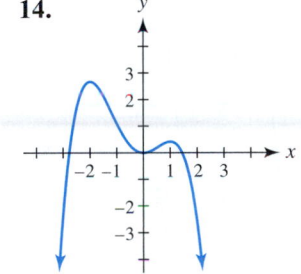

15.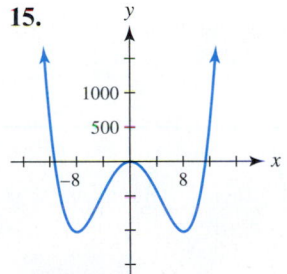

16.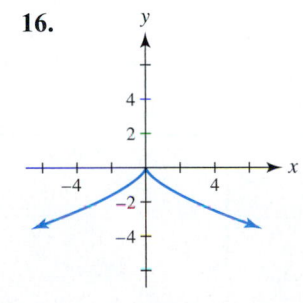

Exercises 17–38: Use interval notation to identify where f is increasing and where f is decreasing. (Hint: Consider the graph $y = f(x)$.)

17. $f(x) = -3$ **18.** $f(x) = 5$

19. $f(x) = 2x - 1$ **20.** $f(x) = 4 - x$

21. $f(x) = x^2 - 2$ **22.** $f(x) = -\frac{1}{2}x^2$

23. $f(x) = 2x - x^2$ **24.** $f(x) = x^2 - 4x$

25. $f(x) = 3(x + 1)^2 - 5$ **26.** $f(x) = -2(x - 1)^2 + 3$

27. $f(x) = \sqrt{x - 1}$ **28.** $f(x) = -\sqrt{x + 1}$

29. $f(x) = |x + 3|$ **30.** $f(x) = |x - 1|$

31. $f(x) = x^3$ **32.** $f(x) = \sqrt[3]{x}$

 33. $f(x) = \frac{1}{3}x^3 - 4x$ **34.** $f(x) = x^3 - 3x$

35. $f(x) = 2x^3 + 3x^2 - 12x$

36. $f(x) = -x^3 + 3x^2 + 9x - 10$

37. $f(x) = -\frac{1}{4}x^4 + \frac{1}{3}x^3 + x^2$

38. $f(x) = \frac{1}{4}x^4 - 2x^2$

Exercises 39–48: Use the graph of f to estimate the

  (a) *local extrema, and*

  (b) *absolute extrema.*

(Hint: *Local extrema cannot occur at endpoints, but absolute extrema can.*)

39. 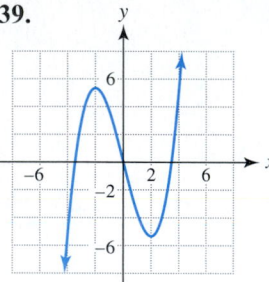 **40.**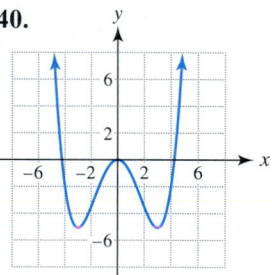

41. 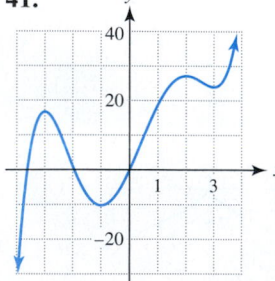 **42.**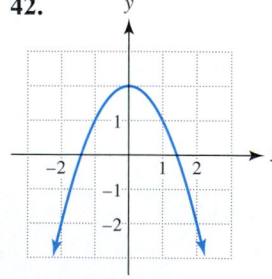

43. 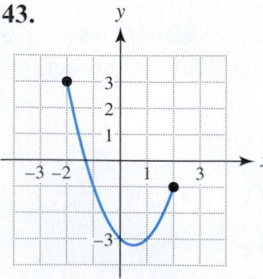 **44.**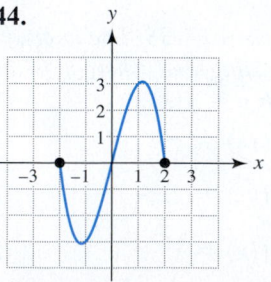

45. 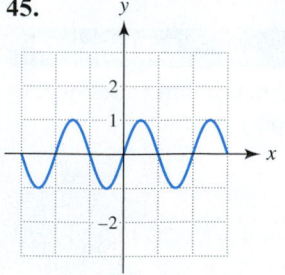 **46.**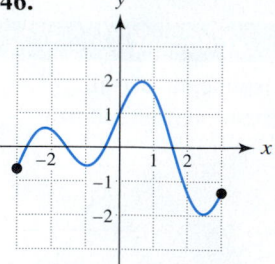

47. 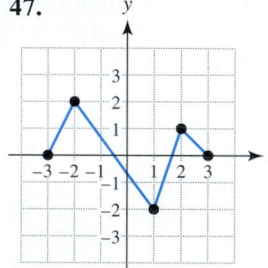 **48.**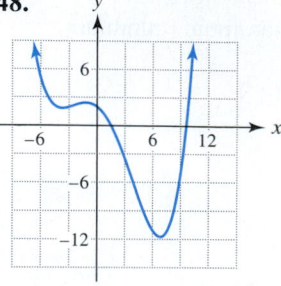

Exercises 49–62: Determine any

  (a) *local extrema, and*

  (b) *absolute extrema.*

(Hint: *Consider the graph* $y = g(x)$.)

49. $g(x) = 1 - 3x$ **50.** $g(x) = \frac{1}{4}x$

51. $g(x) = x^2 + 1$ **52.** $g(x) = 1 - x^2$

53. $g(x) = -2(x + 3)^2 + 4$

54. $g(x) = \frac{1}{3}(x - 1)^2 - 2$

55. $g(x) = 2x^2 - 3x + 1$ **56.** $g(x) = -3x^2 + 4x - 1$

57. $g(x) = |x + 3|$ **58.** $g(x) = -|x| + 2$

59. $g(x) = \sqrt[3]{x}$ **60.** $g(x) = -x^3$

61. $g(x) = 3x - x^3$ **62.** $g(x) = \dfrac{1}{1 + |x|}$

Exercises 63–68: Determine graphically any

  (a) *local extrema, and*

  (b) *absolute extrema.*

63. $f(x) = -3x^4 + 8x^3 + 6x^2 - 24x$

64. $f(x) = -x^4 + 4x^3 - 4x^2$

65. $f(x) = 0.5x^4 - 5x^2 + 4.5$

66. $f(x) = 0.01x^5 + 0.02x^4 - 0.35x^3 - 0.36x^2 + 1.8x$

67. $f(x) = \dfrac{8}{1 + x^2}$ **68.** $f(x) = \dfrac{6}{x^2 + 2x + 2}$

Symmetry

Exercises 69 and 70: Use the graph to determine whether the function f is odd or even. Then use the point shown to find a second point on the graph.

69.

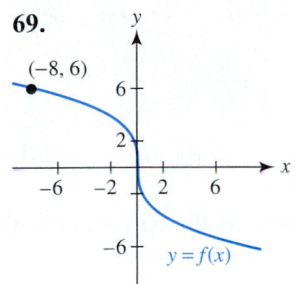

70.

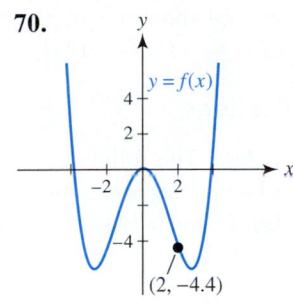

Exercises 71–90: (Refer to Example 6.) Determine if f is odd, even, or neither.

71. $f(x) = 5x$

72. $f(x) = -3x$

73. $f(x) = x + 3$

74. $f(x) = 2x - 1$

75. $f(x) = x^2 - 10$

76. $f(x) = 8 - 2x^2$

77. $f(x) = x^4 - 6x^2 + 2$

78. $f(x) = -x^6 + 5x^2$

79. $f(x) = x^3 - 2x$

80. $f(x) = -x^5$

81. $f(x) = x^2 - x^3$

82. $f(x) = 3x^3 - 1$

83. $f(x) = \sqrt[3]{x^2}$

84. $f(x) = \sqrt{-x}$

85. $f(x) = \sqrt{1 - x^2}$

86. $f(x) = \sqrt{x^2}$

87. $f(x) = \dfrac{1}{1 + x^2}$

88. $f(x) = \dfrac{1}{x}$

89. $f(x) = |x + 2|$

90. $f(x) = \dfrac{1}{x + 1}$

91. The table is a complete representation of f. Decide if f is even, odd, or neither.

x	-100	-10	-1	0	1	10	100
$f(x)$	56	-23	5	0	-5	23	-56

92. The table is a complete representation of f. Decide if f is even, odd, or neither.

x	-5	-3	-1	1	2	3
$f(x)$	-4	-2	1	1	-2	-4

93. Complete the table if f is an even function.

x	-3	-2	-1	0	1	2	3
$f(x)$	21		-25			-12	

94. Complete the table if f is an odd function.

x	-5	-3	-2	0	2	3	5
$f(x)$	13		-5			-1	

95. Complete the table so the function f is neither odd nor even. Answers may vary.

x	-2	-1	0	1	2
$f(x)$	5			3	

96. Complete the table if the function f is odd.

x	-2	0	2
$f(x)$	5		

97. A partial graph of an odd function f with domain given by $D = \{x \mid -2 \le x \le 2\}$ is shown in the figure. Make a sketch of the complete graph.

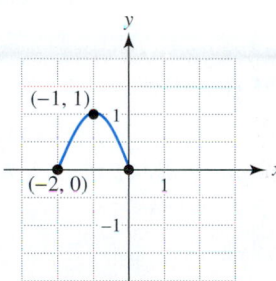

98. A partial graph of an even function f with domain given by $D = \{x \mid -10 \le x \le 10\}$ is shown in the figure. Make a sketch of the complete graph.

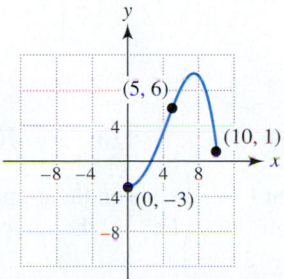

Concepts

99. Sketch a graph of a linear function that is an odd function.

100. Sketch a graph of a linear function that is an even function.

101. Does there exist a continuous odd function that is always increasing and whose graph passes through the points $(-3, -4)$ and $(2, 5)$? Explain.

102. Is there an even function whose domain is all real numbers and is always decreasing? Explain.

103. Sketch a graph of a continuous function with an absolute minimum of -3 at $x = -2$ and a local minimum of -1 at $x = 2$.

104. Sketch a graph of a continuous function with no absolute extrema but with a local minimum of -2 at $x = -1$ and a local maximum of 2 at $x = 1$.

105. Sketch a graph of a continuous function that is increasing on $(-\infty, 2]$ and decreasing on $[2, \infty)$. Could this function be quadratic?

106. Sketch a graph of a continuous function with a local maximum of 2 at $x = -1$ and a local maximum of 0 at $x = 1$.

Translations of Graphs

Exercises 107–110: Use the graph of $f(x) = 4x - \frac{1}{3}x^3$ and translations of graphs to sketch the graph of the equation.

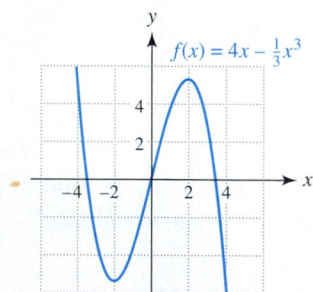

107. $y = f(x + 1)$ **108.** $y = f(x) - 2$

109. $y = 2f(x)$ **110.** $y = f\left(\frac{1}{2}x\right)$

111. If the point $(-2, 3)$ lies on the graph of $y = f(x)$, then what point must lie on the graph of the equation $y = f(x + 2) - 1$?

112. If the point $(-2, 3)$ lies on the graph of $y = f(x)$, then what point must lie on the graph of $y = -f(-x)$?

113. If the graph of $y = f(x)$ is increasing on $[1, 4]$, then where is the graph of $y = f(x + 1) - 2$ increasing? Where is the graph of $y = -f(x - 2)$ decreasing?

114. If the graph of f is decreasing on $[0, \infty)$, then what can be said about the graph of $y = f(-x) + 1$? the graph of $y = -f(x) - 1$?

Applications

115. *Tides* The following figure shows tide levels in feet x hours after midnight.
 (a) Estimate when the tides were rising on the interval $0 \le x \le 26.8$.

 (b) Identify all local extrema and interpret each.

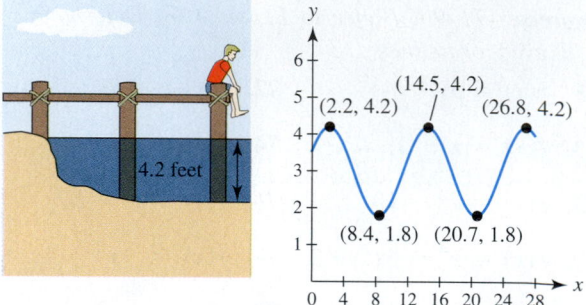

116. *Wiretaps* The line graph shows the approximate number of state and federal wiretaps approved from 1968 to 1992. If the line graph represents a function f, find the local extrema. (*Hint:* Local extrema do not occur at endpoints.) (**Source:** Administrative Office of the U.S. Courts.)

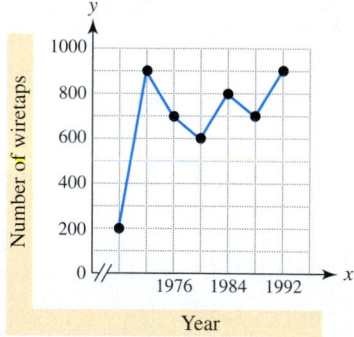

117. *Energy* The U.S. consumption of energy by residential and commercial users from 1950 to 1980 can be modeled by

$$f(x) = -0.00113x^3 + 0.0408x^2 - 0.0432x + 7.66,$$

where $x = 0$ corresponds to 1950 and $x = 30$ to 1980. Consumption is measured in quadrillion Btu. (**Source:** Department of Energy.)

(a) Evaluate $f(5)$ and interpret the result.

(b) Graph f in [0, 30, 5] by [6, 16, 1]. Describe the energy usage during this time period.

(c) Approximate the local maximum and interpret the result.

118. Natural Gas The U.S. consumption of natural gas from 1965 to 1980 can be modeled by

$$f(x) = 0.0001234x^4 - 0.005689x^3 + 0.08792x^2 - 0.5145x + 1.514,$$

where $x = 6$ corresponds to 1966 and $x = 20$ to 1980. Consumption is measured in trillion cubic feet. (**Source:** Department of Energy.)

(a) Evaluate $f(10)$ and interpret the result.

(b) Graph f in [6, 20, 5] by [0.4, 0.8, 0.1]. Describe the energy usage during this time period.

(c) Determine the local extrema and interpret the results.

119. Heating Costs In colder climates of the United States the cost for natural gas to heat homes can vary dramatically from one month to the next. The polynomial function given by

$$f(x) = -0.1213x^4 + 3.462x^3 - 29.22x^2 + 64.68x + 97.69$$

models the monthly cost in dollars of heating a typical home. The input x represents the month, where $x = 1$ corresponds to January and $x = 12$ to December. (**Source:** Minnegasco, A NORAM Energy Company.)

(a) Where might the absolute extrema occur for $1 \le x \le 12$?

(b) Graph f in [1, 12, 1] by [0, 150, 10]. Identify the absolute extrema in this graph and interpret the results.

120. Daylight Hours The accompanying graph shows the daylight hours at 60°N latitude, where $x = 1$ corresponds to January 1, $x = 2$ to February 1, and so on.
(a) Estimate when daylight hours are increasing for $1 \le x \le 12$.

(b) Identify and interpret the local maximum.

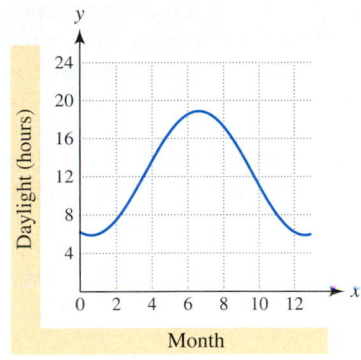

121. Average Temperature The accompanying graph approximates the monthly average temperatures in degrees Fahrenheit in Austin, Texas. In this graph x represents the month, where $x = 0$ corresponds to July.
(a) Is this a graph of an odd or even function?

(b) June corresponds to $x = -1$ and August to $x = 1$. The average temperature in June is 83°F. What is the average temperature in August?

(c) March corresponds to $x = -4$ and November to $x = 4$. According to the graph, how do their average temperatures compare?

(d) Interpret what this type of symmetry implies about average temperatures in Austin.

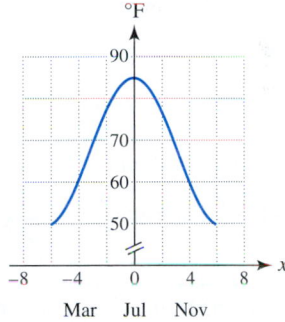

122. Average Temperature The accompanying graph on the next page shows the deviation of the monthly average temperatures from the yearly average temperature in Anchorage, Alaska. In this graph x represents the month, where $x = 0$ corresponds to October. For example, the average temperature in October is equal to the average yearly temperature, so its deviation is 0°F. The deviation in July ($x = -3$) is 22°F above the yearly average.
(a) October and what other month have a temperature deviation of 0°F from the yearly average?

(b) Is this a graph of an odd or even function?

(c) September corresponds to $x = -1$ and November to $x = 1$. The average temperature in November is 11°F below the average yearly temperature. What does this indicate about the average temperature in September?

(d) August corresponds to $x = -2$ and December to $x = 2$. How do their temperatures compare to the average yearly temperature?

(e) Interpret what this type of symmetry implies about the average monthly temperatures in Anchorage.

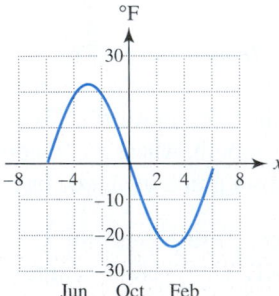

123. *Height of a Projectile* If a projectile is shot into the air, it attains a maximum height and then falls back to the ground. Suppose that $x = 0$ corresponds to the time when the projectile's height is maximum. If air resistance is ignored, its height h above the ground at any time x may be modeled by $h(x) = -16x^2 + h_{max}$, where h_{max} is the projectile's maximum height above the ground. Height is measured in feet and time in seconds. Let $h_{max} = 400$ feet.

(a) Evaluate $h(-2)$ and $h(2)$. Interpret these results.

(b) Evaluate $h(-5)$ and $h(5)$. Interpret these results.

(c) Graph h for $-5 \le x \le 5$. Is h an even or odd function?

(d) How do $h(x)$ and $h(-x)$ compare when $-5 \le x \le 5$? What does this indicate about the flight of a projectile?

124. *Velocity of a Projectile* (Refer to Exercise 123.) The velocity in feet per second of a projectile shot into the air is given by $v(t) = -32t$, where $-5 \le t \le 5$ and $t = 0$ corresponds to the time when the maximum height is attained. A positive velocity indicates that the projectile is traveling up, whereas a negative velocity indicates that it is falling.

(a) Evaluate $v(-2)$ and $v(2)$. Interpret each result.

(b) Graph v for $-5 \le t \le 5$. Is v an even or odd function?

(c) How do $v(t)$ and $v(-t)$ compare when $-5 \le t \le 5$? What does this mean about the velocity of a projectile?

Writing about Mathematics

125. Explain the difference between a local maximum and an absolute maximum. Are extrema x-values, y-values, or points (x, y)?

126. Describe ways to determine if a polynomial function is odd, even, or neither. Give examples.

EXTENDED AND DISCOVERY EXERCISES

1. Find the dimensions of a rectangle of maximum area that can be inscribed in a semicircle with radius 3. Assume that the rectangle is positioned as shown in the accompanying figure.

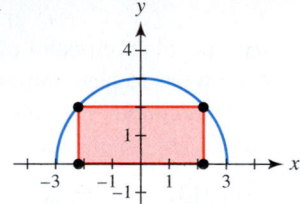

2. *Minimizing Area* A piece of wire 20 inches long is cut into two pieces. One piece is bent into a square and the other is bent into an equilateral triangle, as illustrated in the figure.

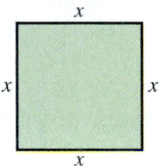 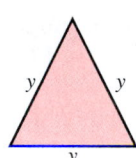

(a) Write a formula that gives the area A of the two shapes in terms of x.

(b) Find the length of wire to the nearest tenth of an inch that should be used for the square if the combined area of the two shapes is a minimum?

3. *Minimizing Time* A person is in a rowboat 3 miles from the closest point on a straight shoreline, as illustrated in the figure. The person would like to reach a cabin that is 8 miles down the shoreline. The person can row at 4 miles per hour and jog at 7 miles per hour.

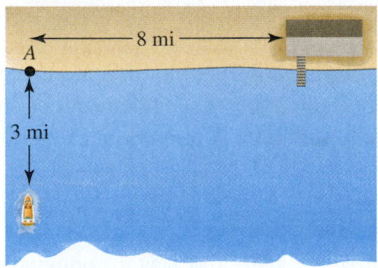

(a) How long does it take to reach the cabin if the person rows straight toward shore at point A and then jogs to the cabin?

(b) How long does it take to reach the cabin if the person rows straight to the cabin and does no jogging?

(c) Find the minimum time for the person to reach the cabin.

4.2 Polynomial Functions and Models

◆ Understand the graphs of polynomial functions

◆ Evaluate and graph piecewise-defined functions

◆ Use polynomial regression to model data (optional)

Introduction

The study of higher degree polynomial equations dates back to Old Babylonian civilization in about 1800–1600 B.C. Gottfried Leibniz (1646–1716) was the first mathematician to generalize polynomial functions of degree n. Many eighteenth-century mathematicians devoted their lives to studying polynomial equations. Today, polynomial functions are used to model a wide variety of real data. (**References:** *Historical Topics for the Mathematics Classroom, Thirty-first Yearbook,* NCTM; L. Motz and J. H. Weaver, *The Story of Mathematics.*)

The consumption of natural gas by the United States has varied throughout past decades. Table 4.3 lists energy consumption (in quadrillion Btu) for selected years, which increased, decreased, and then increased again. A scatterplot of the data is shown in Figure 4.28, and one possibility for a polynomial modeling function f is shown in Figure 4.29. Notice that f is neither linear nor quadratic. What degree of polynomial might we use to model this data? We learn the answer to this question in this section.

TABLE 4.3   Natural Gas Consumption

Year	1960	1970	1980	1990	2000
Consumption	12.4	21.8	20.4	19.3	24.0

Source: Department of Energy.

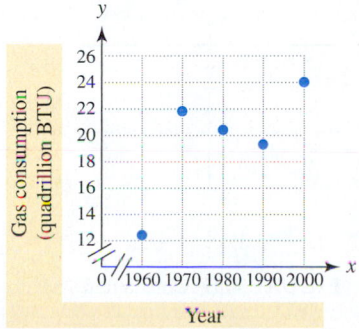

FIGURE 4.28   A Scatterplot

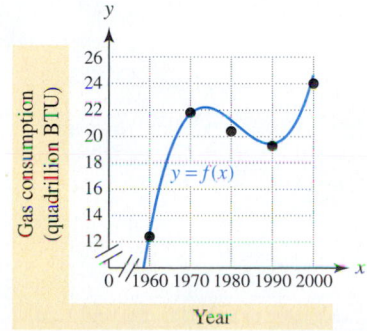

FIGURE 4.29   A Modeling Function

Graphs of Polynomial Functions

In Section 4.1 polynomial functions were defined. Polynomial functions have been used in several applications. Their graphs are continuous; they have no breaks or sharp edges. The domain of a polynomial function is all real numbers. A polynomial function f of degree n can be expressed as

$$f(x) = a_n x^n + \cdots + a_2 x^2 + a_1 x + a_0,$$

where each coefficient a_k is a real number, $a_n \neq 0$, and n is a nonnegative integer. The *leading coefficient* is a_n where n is the largest exponent of x.

The expression $a_n x^n + \cdots + a_2 x^2 + a_1 x + a_0$ is a **polynomial of degree n** and the equation $a_n x^n + \cdots + a_2 x^2 + a_1 x + a_0 = 0$ is a **polynomial equation of degree n**. Thus, f is a polynomial function, $f(x)$ is a polynomial, and $f(x) = 0$ is a polynomial equation. For example, the function f, given by $f(x) = x^3 - 3x^2 + x - 5$, is a polynomial function, $x^3 - 3x^2 + x - 5$ is a polynomial, and $x^3 - 3x^2 + x - 5 = 0$ is a polynomial equation.

A **turning point** occurs whenever the graph of a polynomial function changes from increasing to decreasing or from decreasing to increasing. Turning points are associated with "hills" or "valleys" on a graph. The y-value at a turning point is either a local maximum or local minimum of the function. In Figure 4.30 the graph has two turning points, $(-2, 8)$ and $(2, -8)$. A local maximum is 8 and a local minimum is -8.

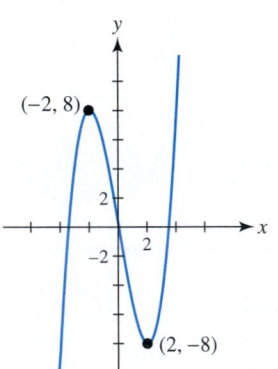

FIGURE 4.30

◆ **MAKING CONNECTIONS**

Turning Points and Local Extrema   A turning point (x, y) is a point on the graph of a function that is located where the graph changes from increasing to decreasing or from decreasing to increasing. A local extrema is a y-value, not a point, and often corresponds to the y-value of a turning point.

We discuss the graphs of polynomial functions, starting with degree 0 and continuing to degree 5.

Constant Polynomial Functions   If $f(x) = a$ and $a \neq 0$, then f is both a constant function and a polynomial function of degree 0. (If $a = 0$ then f has an undefined degree.) Its graph is a horizontal line that does not coincide with the x-axis. Graphs of $f(x) = 4$ and $g(x) = -3$ are shown in Figures 4.31 and 4.32. A graph of a polynomial function of degree 0 has no x-intercepts or turning points.

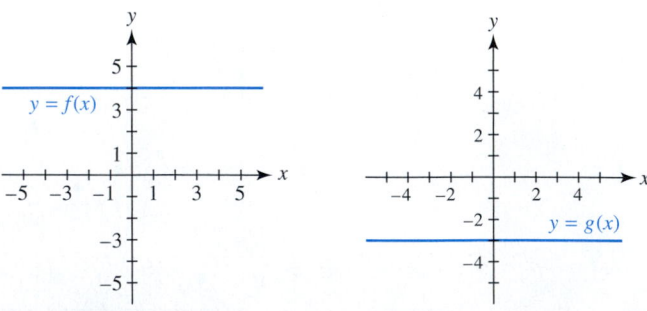

FIGURE 4.31 Constant, $a > 0$ FIGURE 4.32 Constant, $a < 0$

Linear Polynomial Functions   If $f(x) = ax + b$ and $a \neq 0$, then f is both a linear function and a polynomial function of degree 1. Its graph is a line that is neither horizontal nor vertical. The graphs of $f(x) = 2x - 3$ and $g(x) = -1.6x + 5$ are shown in Figures 4.33 and 4.34. A polynomial function of degree 1 has one x-intercept and no turning points.

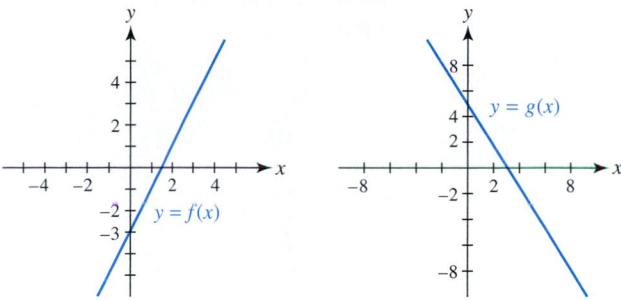

FIGURE 4.33 Linear, $a > 0$ **FIGURE 4.34** Linear, $a < 0$

The graph of $f(x) = ax + b$ with $a > 0$ is a line sloping upward from left to right. As one traces from left to right, the y-values become larger without a maximum. We say that the **end behavior** of the graph tends to $-\infty$ on the left and ∞ on the right. (Strictly speaking, the graph of a polynomial has infinite length and does not have an end.) More formally we say that $\boldsymbol{f(x) \to -\infty}$ **as** $\boldsymbol{x \to -\infty}$ and $\boldsymbol{f(x) \to \infty}$ **as** $\boldsymbol{x \to \infty}$.

If $a < 0$, then the end behavior is switched. The line slopes downward from left to right. The y-values on the left side of the graph become large positive values without a maximum and the y-values on the right side become negative without a minimum. The end behavior tends to ∞ on the left and $-\infty$ on the right, or $\boldsymbol{f(x) \to \infty}$ **as** $\boldsymbol{x \to -\infty}$ and $\boldsymbol{f(x) \to -\infty}$ **as** $\boldsymbol{x \to \infty}$.

Quadratic Polynomial Functions   If $f(x) = ax^2 + bx + c$ and $a \neq 0$, then f is both a quadratic function and a polynomial function of degree 2. Its graph is a parabola that either opens upward ($a > 0$) or downward ($a < 0$). The graphs of $f(x) = 0.5x^2 + 2$, $g(x) = x^2 + 4x + 4$, and $h(x) = -x^2 + 3x + 4$ are shown in Figures 4.35–4.37, respectively. Quadratic functions can have zero, one, or two x-intercepts. A parabola has exactly one turning point, which is also the vertex.

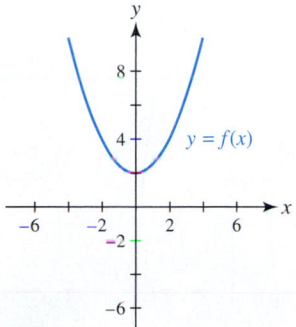 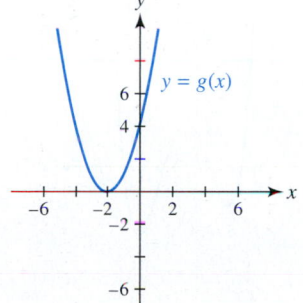 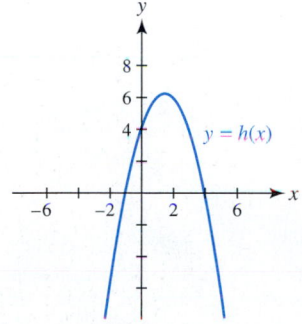

FIGURE 4.35 Quadratic, $a > 0$ **FIGURE 4.36** Quadratic, $a > 0$ **FIGURE 4.37** Quadratic, $a < 0$

If $a > 0$, as in Figure 4.35, then both sides of the graph go up. The end behavior tends to ∞ on both sides, or $\boldsymbol{f(x) \to \infty}$ **as** $\boldsymbol{x \to \pm\infty}$. If $a < 0$, as in Figure 4.37, then the end behavior is switched and tends to $-\infty$ on both sides, or $\boldsymbol{f(x) \to -\infty}$ **as** $\boldsymbol{x \to \pm\infty}$.

Cubic Polynomial Functions   If $f(x) = ax^3 + bx^2 + cx + d$ and $a \neq 0$, then f is both a **cubic function** and a polynomial function of degree 3. The graph of a cubic function can have zero or two turning points. The graph of $f(x) = -x^3 + 4x + 1$ in Figure 4.38 has two turning points, whereas the graph of $g(x) = \frac{1}{4}x^3$ in Figure 4.39 has no turning points.

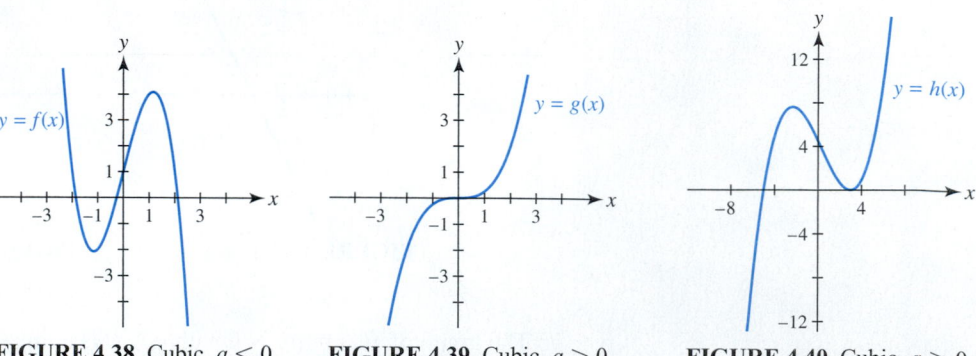

FIGURE 4.38 Cubic, $a < 0$ FIGURE 4.39 Cubic, $a > 0$ FIGURE 4.40 Cubic, $a > 0$

If $a > 0$, the graph of a cubic function falls to the left and rises to the right, as in Figure 4.39. If $a < 0$, its graph rises to the left and falls to the right, as in Figure 4.38. The end behavior of a cubic function is similar to that of a linear function, and tends to ∞ on one side and $-\infty$ on the other. Therefore its graph must cross the x-axis at least once. A cubic function can have up to three x-intercepts. The graph of $h(x) = 0.1x^3 - 0.1x^2 - 2.1x + 4.5$ in Figure 4.40 has two x-intercepts.

Quartic Polynomial Functions   If $f(x) = ax^4 + bx^3 + cx^2 + dx + e$ and $a \neq 0$, then f is both a **quartic function** and a polynomial function of degree 4. The graph of a quartic function can have up to four x-intercepts and three turning points, and the graph of $f(x) = 0.1x^4 - 0.3x^3 - 1.2x^2 + 1.8x + 2$ in Figure 4.41 is an example. The graph of $g(x) = -x^4 + 2x^3$ in Figure 4.42 has one turning point and two x-intercepts, and the graph of $h(x) = -x^4 + x^3 + 3x^2 - x - 2$ in Figure 4.43 has three turning points and three x-intercepts.

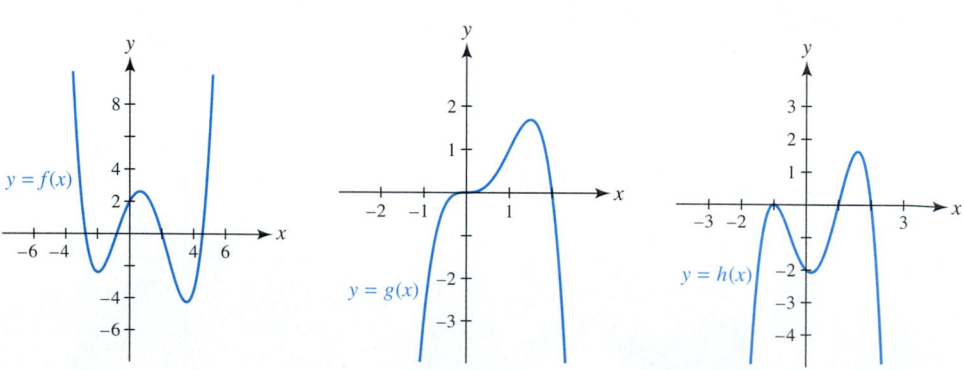

FIGURE 4.41 Quartic, $a > 0$ FIGURE 4.42 Quartic, $a < 0$ FIGURE 4.43 Quartic, $a < 0$

◆ **CLASS DISCUSSION**

Can a quartic function have both an absolute maximum and an absolute minimum? Explain. ◆

If $a > 0$, then both ends of the graph of a quartic function go up, as in Figure 4.41. If $a < 0$, then both ends of its graph go down, as in Figures 4.42 and 4.43. The end behaviors of quartic and quadratic functions are similar.

Quintic Polynomial Functions   If $f(x) = ax^5 + bx^4 + cx^3 + dx^2 + ex + k$ and $a \neq 0$, then f is both a **quintic function** and a polynomial function of degree 5. The graph of a quintic function may have up to five x-intercepts and four turning points. An example is shown in Figure 4.44, given by

$$f(x) = 0.01x^5 + 0.03x^4 - 0.63x^3 - 0.67x^2 + 8.46x - 7.2.$$

Other quintic functions are shown in Figures 4.45 and 4.46. They are defined by

$$g(x) = \frac{1}{5}x^5 - 3 \quad \text{and}$$

$$h(x) = -0.02x^5 - 0.14x^4 + 0.04x^3 + 0.92x^2 - 1.3x + 0.5.$$

The function g has one x-intercept and no turning points. The graph of h appears to have two x-intercepts and two turning points. Notice that the end behavior of a quintic function is similar to linear and cubic functions.

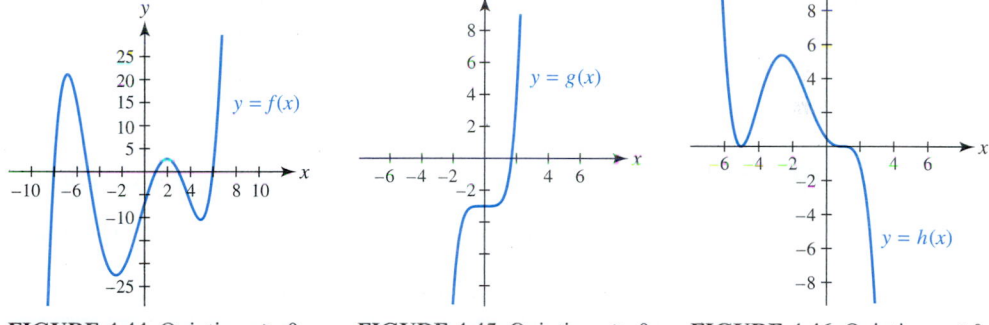

FIGURE 4.44 Quintic, $a > 0$ **FIGURE 4.45** Quintic, $a > 0$ **FIGURE 4.46** Quintic, $a < 0$

The end behavior of polynomial functions is summarized in the following.

◆ **CLASS DISCUSSION**
Can you sketch the graph of a quadratic function with no turning points, a cubic function with one turning point, or a quartic function with two turning points? Explain.
◆

END BEHAVIOR OF POLYNOMIAL FUNCTIONS

Let f be a polynomial function with leading coefficient a and degree n.

1. $n \geq 2$ is even.
   $a > 0$ implies the graph of f rises both to the left and to the right. That is, $f(x) \to \infty$ as $x \to \pm\infty$.
   $a < 0$ implies the graph of f falls both to the left and to the right. That is, $f(x) \to -\infty$ as $x \to \pm\infty$.
2. $n \geq 1$ is odd.
   $a > 0$ implies the graph of f falls to the left and rises to the right. That is, $f(x) \to -\infty$ as $x \to -\infty$ and $f(x) \to \infty$ as $x \to \infty$.
   $a < 0$ implies the graph of f rises to the left and falls to the right. That is, $f(x) \to \infty$ as $x \to -\infty$ and $f(x) \to -\infty$ as $x \to \infty$.

The maximum number of x-intercepts and turning points on the graph of a polynomial function of degree n are summarized on the next page.

DEGREE, x-INTERCEPTS, AND TURNING POINTS

The graph of a polynomial function of degree $n \geq 1$ has at most n x-intercepts and at most $n - 1$ turning points.

EXAMPLE 1   Analyzing the graph of a polynomial function

Figure 4.47 shows the graph of a polynomial function f.
(a) How many turning points and x-intercepts are there?
(b) Is the leading coefficient a positive or negative? Is the degree odd or even?
(c) Determine the minimum possible degree of f.

SOLUTION
(a) There are four turning points corresponding to the two "hills" and two "valleys." There appear to be four x-intercepts.
(b) The left side of the graph rises and the right side falls. Therefore, $a < 0$ and the polynomial function has odd degree.
(c) The graph has four turning points. A polynomial of degree n can have at most $n - 1$ turning points. Therefore, f must be at least degree 5.   *Now Try Exercise 7* ◆

FIGURE 4.47

EXAMPLE 2   Analyzing the graph of a polynomial function

Graph $f(x) = x^3 - 2x^2 - 5x + 6$, and then complete the following.
(a) Identify the x-intercepts.
(b) Approximate the coordinates of any turning points to the nearest hundredth.
(c) Use the turning points to approximate any local extrema.

SOLUTION
(a) A graph of f, shown in Figure 4.48, appears to intersect the x-axis at the points $(-2, 0)$, $(1, 0)$, and $(3, 0)$. Therefore the x-intercepts are -2, 1, and 3.
(b) There are two turning points. From Figures 4.49 and 4.50 their coordinates are approximately $(-0.79, 8.21)$ and $(2.12, -4.06)$.
(c) There is a local maximum of about 8.21 and a local minimum of about -4.06.

Calculator Help

To find a minimum or a maximum point on a graph, see Appendix B (page AP-14).

$[-10, 10, 1]$ by $[-10, 10, 1]$ $[-10, 10, 1]$ by $[-10, 10, 1]$ $[-10, 10, 1]$ by $[-10, 10, 1]$

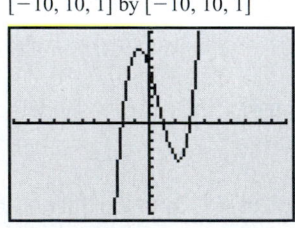   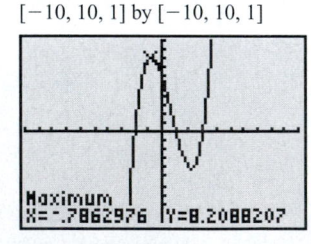   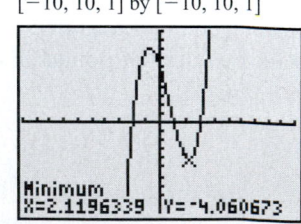

FIGURE 4.48   **FIGURE 4.49**   **FIGURE 4.50**

Now Try Exercise 37 ◆

EXAMPLE 3   Analyzing the end behavior of a graph

Let $f(x) = 2 + 3x - 3x^2 - 2x^3$.
(a) Give the degree and leading coefficient.
(b) State the end behavior of the graph of f.

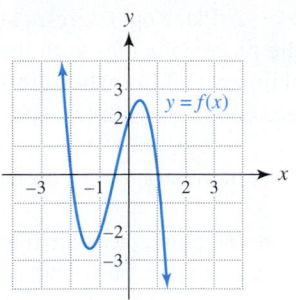

FIGURE 4.51 Cubic with $a < 0$

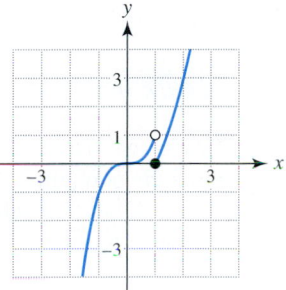

FIGURE 4.52

SOLUTION

(a) Rewriting gives $f(x) = -2x^3 - 3x^2 + 3x + 2$. The term with highest degree is $-2x^3$ so the degree is **3** and the leading coefficient is **−2**.

(b) The degree of $f(x)$ is odd and the leading coefficient is negative. Therefore the graph of f rises to the left and falls to the right. More formally,

$$f(x) \to \infty \text{ as } x \to -\infty \text{ and } f(x) \to -\infty \text{ as } x \to \infty.$$

This conclusion is supported by Figure 4.51.   *Now Try Exercise 25* ◆

Piecewise-Defined Polynomial Functions

In Section 2.5 piecewise-defined functions were discussed. If each piece is a polynomial, then the function is a **piecewise-defined polynomial function** or **piecewise-polynomial function**. An example is given by $f(x)$.

$$f(x) = \begin{cases} x^3 & \text{if } x < 1 \\ x^2 - 1 & \text{if } x \geq 1 \end{cases}$$

The graph of f can be determined by graphing $y = x^3$ when $x < 1$ and graphing $y = x^2 - 1$ when $x \geq 1$. At $x = 1$ there is a break in the graph, where the graph of f is discontinuous. See Figure 4.52.

EXAMPLE 4   Evaluating a piecewise-defined polynomial function

Evaluate $f(x)$ at $x = -3, -2, 1,$ and 2.

$$f(x) = \begin{cases} x^2 - x & \text{if } -5 \leq x < -2 \\ -x^3 & \text{if } -2 \leq x < 2 \\ 4 - 4x & \text{if } 2 \leq x \leq 5 \end{cases}$$

SOLUTION   To evaluate $f(-3)$ we use the formula $f(x) = x^2 - x$ because -3 is in the interval $-5 \leq x < -2$.

$$f(-3) = (-3)^2 - (-3) = 12$$

To evaluate $f(-2)$ we use formula $f(x) = -x^3$ because -2 is in the interval $-2 \leq x < 2$.

$$f(-2) = -(-2)^3 = -(-8) = 8$$

Similarly, $f(1) = -1^3 = -1$ and $f(2) = 4 - 4(2) = -4$.   *Now Try Exercise 71* ◆

EXAMPLE 5   Graphing a piecewise-defined function

Complete the following.

(a) Sketch a graph of f.

(b) Determine if f is continuous on its domain.

(c) Solve the equation $f(x) = 1$.

$$f(x) = \begin{cases} \frac{1}{2}x^2 - 2 & \text{if } -4 \leq x \leq 0 \\ 2x - 2 & \text{if } 0 < x < 2 \\ 2 & \text{if } 2 \leq x \leq 4 \end{cases}$$

SOLUTION

(a) For the first piece, graph the parabola determined by $y = \frac{1}{2}x^2 - 2$ on the interval $-4 \leq x \leq 0$. Place dots at the endpoints, which are $(-4, 6)$ and $(0, -2)$. See Figure 4.53.

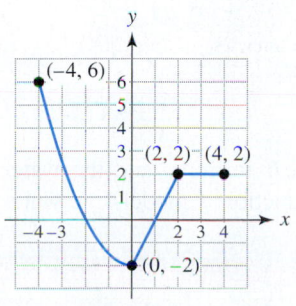

FIGURE 4.53

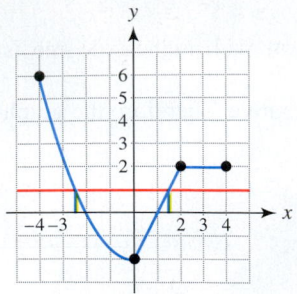

FIGURE 4.54

For the second piece, graph the line determined by $y = 2x - 2$. Place open circles at $(0, -2)$ and $(2, 2)$. Note that the left endpoint of the middle piece coincides with the right endpoint of the first piece. Finally graph the horizontal line $y = 2$ from the points $(2, 2)$ to $(4, 2)$. Note that the left endpoint of the third piece coincides with the open circle on the right for the middle piece.

(b) The domain of f is $-4 \le x \le 4$. Because there are no breaks in the graph of f on its domain, the graph of f is continuous.

(c) The horizontal line $y = 1$ intersects the graph of $y = f(x)$ at two points, as shown in Figure 4.54. The x-coordinates of these two points of intersection can be found by solving the equations

$$\frac{1}{2}x^2 - 2 = 1 \quad \text{and} \quad 2x - 2 = 1.$$

The solutions are $-\sqrt{6} \approx -2.45$ and $\frac{3}{2}$.   *Now Try Exercise 75* ◆

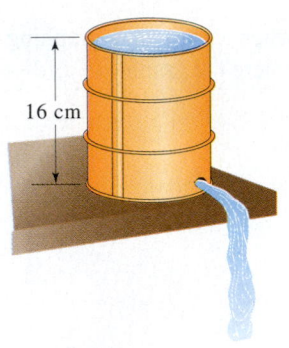

FIGURE 4.55

A cylindrical container has a height of 16 centimeters. Water entered the container at a constant rate until it was completely filled. Then, water was allowed to leak out through a small hole in the bottom. See Figure 4.55. The height of the water in the container was recorded every half minute in Table 4.4 over a 5-minute period.

TABLE 4.4

Time (min)	0	0.5	1.0	1.5	2.0	2.5	3.0	3.5	4.0	4.5	5.0
Height (cm)	0	4	8	12	16	11.6	8.1	5.3	3.1	1.4	0.5

During the first 2 minutes, the water level rose at a constant rate of 4 centimeters every half minute or 8 centimeters per minute. After 2 minutes, water was drained at a nonconstant rate. A scatterplot of the data in Table 4.4 is shown in Figure 4.56.

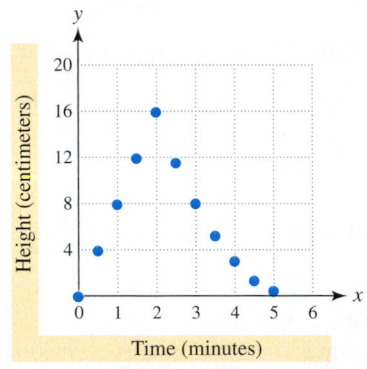

FIGURE 4.56

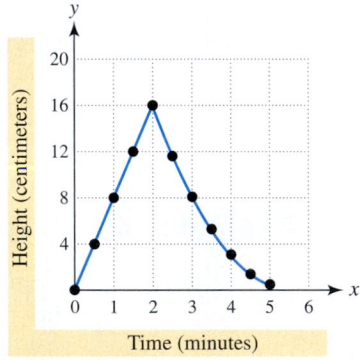

FIGURE 4.57

Water entering the container can be modeled with a linear formula, and the water leaking out can be modeled with a nonlinear formula. The function f models the water level in this example.

$$f(x) = \begin{cases} 8x & \text{if } 0 \le x \le 2 \\ 1.32x^2 - 14.4x + 39.39 & \text{if } 2 < x \le 5 \end{cases}$$

The graph of f and a scatterplot of the data are shown in Figure 4.57. By defining f in pieces, f is capable of modeling water flowing into and out of the container. (Quadratic regression was used to find the formula for $f(x)$ when $2 < x \leq 5$.)

EXAMPLE 6 Analyzing a piecewise-defined nonlinear function

Use the preceding function f to complete the following.
(a) Approximate the water level in the container after 1.25 and after 3.2 minutes.
(b) Estimate the time when the water level was 5 centimeters.

SOLUTION
(a) When $x = 1.25$, water is flowing into the tank and $f(x) = 8x$.

$$f(\mathbf{1.25}) = 8(\mathbf{1.25}) = 10 \text{ cm}$$

When $x = 3.2$ minutes, water is flowing out of the tank and we use the second formula.

$$f(\mathbf{3.2}) = 1.32(\mathbf{3.2})^2 - 14.4(\mathbf{3.2}) + 39.39 = 6.8268 \approx 7 \text{ cm}$$

(b) The water level in Figure 4.57 equals 5 centimeters twice—once when water is entering the container and once when it is leaking out. To find these times, solve the equations $8x = 5$ and $1.32x^2 - 14.4x + 39.39 = 5$. Thus, after $\frac{5}{8}$ minute or 37.5 seconds, the water level was 5 centimeters. The equation $1.32x^2 - 14.4x + 39.39 = 5$ is solved graphically in Figure 4.58. However, the quadratic formula could be used to solve this equation instead. The solution satisfying $2 < x \leq 5$ is $x \approx 3.53$ minutes.

[2, 5, 1] by [0, 20, 5]

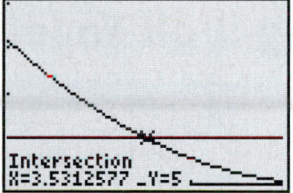

FIGURE 4.58 *Now Try Exercise 90* ◆

Polynomial Regression (Optional)

We now have the mathematical understanding to model the data presented in the introduction to this section. The polynomial modeling function f shown in Figure 4.29 on page 257 falls to the left and rises to the right, so it has odd degree and the leading coefficient is positive. Since the graph of f has two turning points, it must be at least degree 3. A cubic polynomial $f(x)$ is a possible choice, where

$$f(x) = ax^3 + bx^2 + cx + d.$$

Trial and error would be a difficult method to find values for a, b, c, and d. Instead, we can use least-squares regression, which was also discussed in Sections 2.1 and 3.1, for linear and quadratic functions. The next example illustrates cubic regression.

EXAMPLE 7 Determining a cubic modeling function

The data in Table 4.3 (page 257) list natural gas consumption in the United States.
(a) Find a polynomial function of degree 3 that models the data.
(b) Graph f and the data together.
(c) Estimate the natural gas consumption in 1982.

SOLUTION

(a) Enter the five data points (1960, 12.4), (1970, 21.8), (1980, 20.4), (1990, 19.3), and (2000, 24.0) into your calculator. Then select cubic regression, as shown in Figure 4.59. The equation for $f(x)$ is shown in Figure 4.60.

(b) A graph of f and a scatterplot of the data are shown in Figure 4.61.

(c) To estimate natural gas consumption in 1982, evaluate $f(1982)$ to obtain about 20.3 quadrillion Btu.

Calculator Help

To find an equation of least-squares fit, see Appendix B (page AP-15). To copy a regression equation into Y_1, see (page AP-17.)

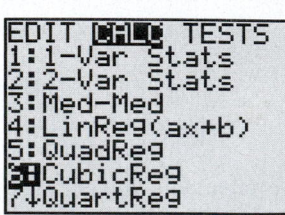

FIGURE 4.59

FIGURE 4.60

[1955, 2005, 5] by [10, 25, 5]

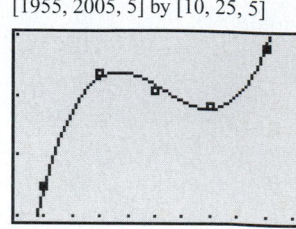

FIGURE 4.61

Now Try Exercise 87 ◆

4.2 Putting it all Together

Higher degree polynomials generally have more complicated graphs. Each additional degree allows the graph to have possibly one more turning point and one more x-intercept. The graph of a polynomial function is continuous and smooth; it has no breaks or sharp edges. Its domain includes all real numbers. The end behavior of a polynomial always tends to either ∞ or $-\infty$. End behavior describes what happens to the y-values as $|x|$ becomes large.

Piecewise-defined functions occur when a function is defined by using two or more formulas for different intervals of the domain. If each formula is a polynomial, it is called a piecewise-polynomial function. Piecewise-polynomial functions are valuable when describing data that changes its character over time.

The following summary shows important concepts regarding graphs of polynomial functions.

Function Type	Characteristics	Example Graphs
Constant (degree 0)	No x-intercepts and no turning points $f(x) = a, a \neq 0$	

Function Type	Characteristics	Example Graphs
Linear (degree 1)	One x-intercept and no turning points $f(x) = ax + b, a \neq 0$	
Quadratic (degree 2)	At most two x-intercepts and exactly one turning point $f(x) = ax^2 + bx + c, a \neq 0$	
Cubic (degree 3)	At most three x-intercepts and up to two turning points $f(x) = ax^3 + bx^2 + cx + d, a \neq 0$	
Quartic (degree 4)	At most four x-intercepts and up to three turning points $f(x) = ax^4 + bx^3 + cx^2 + dx + e, a \neq 0$	
Quintic (degree 5)	At most five x-intercepts and up to four turning points $f(x) = ax^5 + bx^4 + cx^3 + dx^2 + ex + k, a \neq 0$	

4.2 — Exercises

Graphs of Polynomial Functions

1. A runner is working out on a straight track. The graph shows the runner's distance y in hundreds of feet from the starting line after t minutes.
   (a) Estimate the turning points.

   (b) Interpret each turning point.

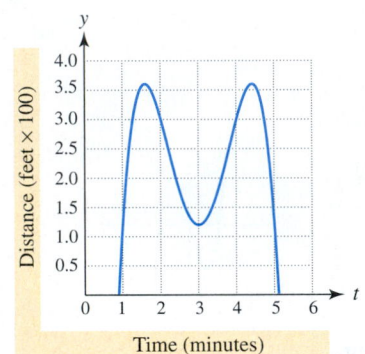

2. A stone is thrown into the air. Its height y in feet after t seconds is shown in the graph.
   (a) Estimate the turning point.

   (b) Interpret this point.

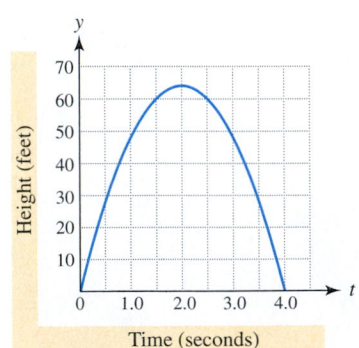

Exercises 3–12: Use the graph of the polynomial function f to complete the following.

(a) Determine the number of turning points and estimate any x-intercepts.
(b) State whether the leading coefficient is positive or negative.
(c) Determine the minimum degree of f.

3.
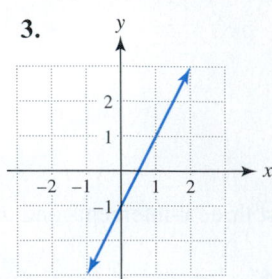

4.
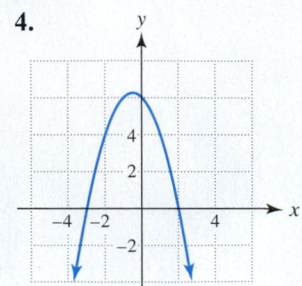

5.
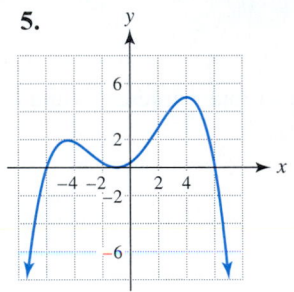

6.
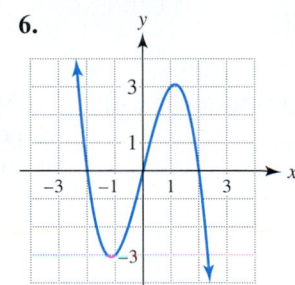

7.
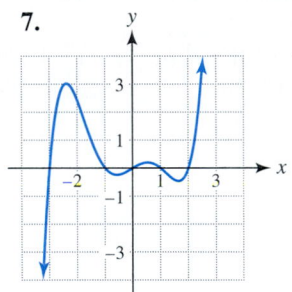

8.
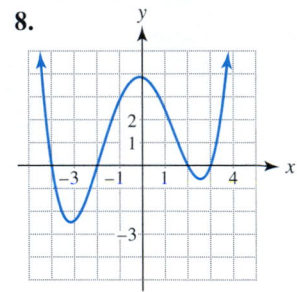

9.
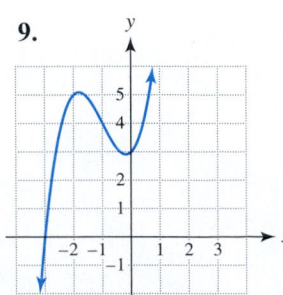

10.
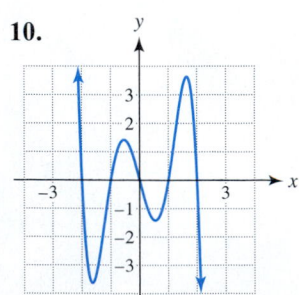

11.

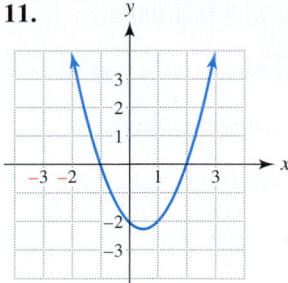

12.

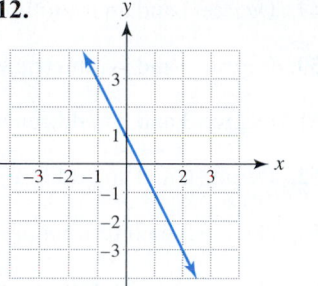

e.

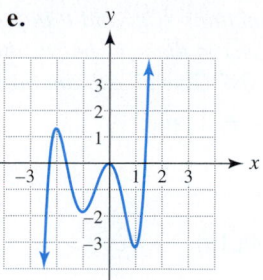

f.

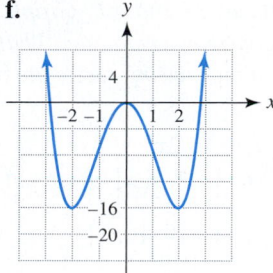

Exercises 13–18: Complete the following without a calculator.

   (a) *Match the equation with its graph (a.–f.).*
   (b) *Identify the turning points.*
   (c) *Estimate the x-intercepts.*
   (d) *Estimate any local extrema.*
   (e) *Estimate any absolute extrema.*

13. $f(x) = 1 - 2x + x^2$

14. $f(x) = 3x - x^3$

15. $f(x) = x^3 + 3x^2 - 9x$

16. $f(x) = x^4 - 8x^2$

17. $f(x) = 8x^2 - x^4$

18. $f(x) = x^5 + \frac{5}{2}x^4 - \frac{5}{3}x^3 - 5x^2$

a.

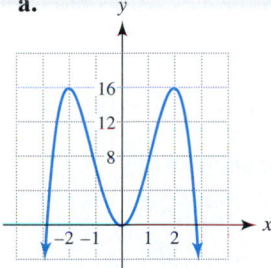

b.

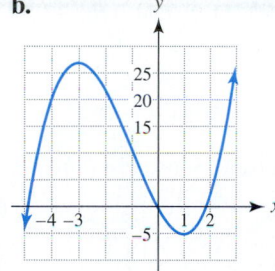

c.

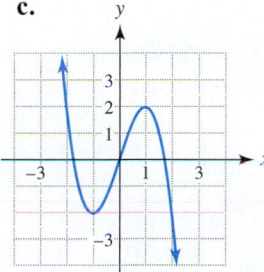

d.

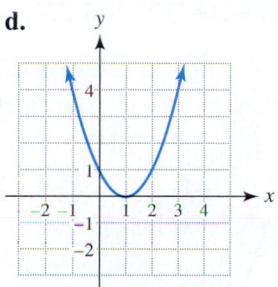

Exercises 19–30: For each polynomial function f complete the following.

   (a) *State the degree and leading coefficient.*
   (b) *Predict the end behavior of the graph of f.*

19. $f(x) = -2x + 3$ **20.** $f(x) = \frac{2}{3}x - 2$

21. $f(x) = x^2 + 4x$ **22.** $f(x) = 5 - \frac{1}{2}x^2$

23. $f(x) = -2x^3$ **24.** $f(x) = 4x - \frac{1}{3}x^3$

25. $f(x) = x^2 - x^3 - 4$

26. $f(x) = x^4 - 4x^3 + 3x^2 - 3$

27. $f(x) = 0.1x^5 - 2x^2 - 3x + 4$

28. $f(x) = 3x^3 - 2 - x^4$

29. $f(x) = 4 + 2x - \frac{1}{2}x^2$

30. $f(x) = -0.2x^5 + 4x^2 - 3$

Exercises 31–38: Complete the following.

   (a) *Graph $y = f(x)$ in the standard viewing rectangle.*
   (b) *Approximate the coordinates of each turning point.*
   (c) *Use the turning points to estimate any local extrema.*

31. $f(x) = \frac{1}{9}x^3 - 3x$ **32.** $f(x) = x^2 - 4x - 3$

33. $f(x) = 0.025x^4 - 0.45x^2 - 5$

34. $f(x) = -\frac{1}{8}x^4 + \frac{1}{3}x^3 + \frac{5}{4}x^2 - 3x + 3$

35. $f(x) = 1 - 2x + 3x^2$ **36.** $f(x) = 4x - \frac{1}{3}x^3$

37. $f(x) = \frac{1}{3}x^3 + \frac{1}{2}x^2 - 2x$

38. $f(x) = \frac{1}{4}x^4 + \frac{2}{3}x^3 - \frac{1}{2}x^2 - 2x + 1$

Exercises 39 and 40: Graph the functions f, g, and h in the same viewing rectangle. What happens to their graphs as the size of the viewing rectangle increases? Give an explanation.

39. $f(x) = 2x^4$, $g(x) = 2x^4 - 5x^2 + 1$, and
$h(x) = 2x^4 + 3x^2 - x - 2$
  i. $[-4, 4, 1]$ by $[-4, 4, 1]$

  ii. $[-10, 10, 1]$ by $[-100, 100, 10]$

  iii. $[-100, 100, 10]$ by $[-10^6, 10^6, 10^5]$

40. $f(x) = -x^3$, $g(x) = -x^3 + x^2 + 2$, and
$h(x) = -x^3 - 2x^2 + x - 1$
  i. $[-4, 4, 1]$ by $[-4, 4, 1]$

  ii. $[-10, 10, 1]$ by $[-100, 100, 10]$

  iii. $[-100, 100, 10]$ by $[-10^5, 10^5, 10^4]$

Exercises 41–44: The data table has been generated by a linear, quadratic, or cubic function f. All zeros of f are real numbers located in the interval $[-3, 3]$.

  (a) Make a line graph of the data.
  (b) Conjecture the degree of f.

41.

x	-3	-2	-1	0	1	2	3
$f(x)$	11	9	7	5	3	1	-1

42.

x	-3	-2	-1	0	1	2	3
$f(x)$	3	-8	-7	0	7	8	-3

43.

x	-3	-2	-1	0	1	2	3
$f(x)$	14	7	2	-1	-2	-1	2

44.

x	-3	-2	-1	0	1	2	3
$f(x)$	-13	-6	-1	2	3	2	-1

Sketching Graphs of Polynomials

Exercises 45–56: If possible, sketch a graph of a polynomial that satisfies the conditions.

45. Degree 3 with three real zeros and a positive leading coefficient

46. Degree 4 with four real zeros and a negative leading coefficient

47. Linear with a negative leading coefficient

48. Cubic with one real zero and a positive leading coefficient

49. Degree 4 and an even function with four turning points

50. Degree 5 and symmetric with respect to the y-axis

51. Degree 3 and an odd function with no x-intercepts

52. Degree 6 and an odd function with five turning points

53. Degree 3 with turning points $(-1, 2)$ and $\left(1, \frac{2}{3}\right)$

54. Degree 4 with turning points $(-1, -1)$, $(0, 0)$, and $(1, -1)$

55. Degree 2 with turning point $(-1, 2)$, passing through $(-3, 4)$ and $(1, 4)$

56. Degree 5 and an odd function with five x-intercepts and a negative leading coefficient

Average Rates of Change

57. Compare the average rates of change from 0 to $\frac{1}{2}$ for $f(x) = x$, $g(x) = x^2$, and $h(x) = x^3$.

58. Compare the average rates of change from 1 to $\frac{3}{2}$ for $f(x) = x$, $g(x) = x^2$, and $h(x) = x^3$.

Exercises 59–62: Calculate the average rate of change of f on each interval. What happens to this average rate of change as the interval decreases in length?

  (a) $[1.9, 2.1]$ (b) $[1.99, 2.01]$ (c) $[1.999, 2.001]$

59. $f(x) = x^3$ **60.** $f(x) = 4x - \frac{1}{3}x^3$

61. $f(x) = \frac{1}{4}x^4 - \frac{1}{3}x^3$ **62.** $f(x) = 4x^2 - \frac{1}{2}x^4$

Exercises 63–66: Find the difference quotient of g.

63. $g(x) = 3x^3$ **64.** $g(x) = -2x^3$

65. $g(x) = 1 + x - x^3$ **66.** $g(x) = \frac{1}{2}x^3 - 2x$

Piecewise-Defined Functions

Exercises 67–74: Evaluate $f(x)$ at the given values of x.

67. $x = -2$ and 1   **68.** $x = -1$, 0, and 3

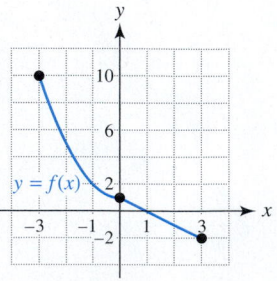

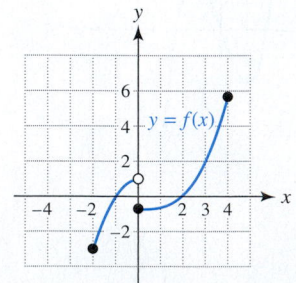

69. $x = -1, 1,$ and 2   **70.** $x = -2, 0,$ and 2

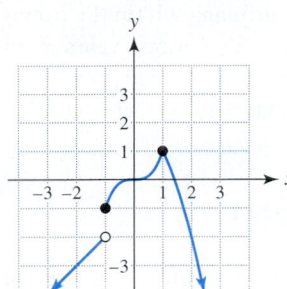   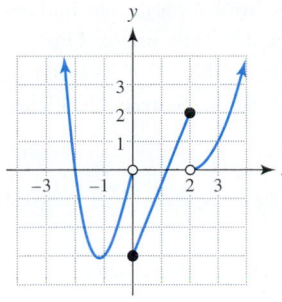

71. $x = -3, 1,$ and 4

$$f(x) = \begin{cases} x^3 - 4x^2 & \text{if } x \leq -3 \\ 3x^2 & \text{if } -3 < x < 4 \\ x^3 - 54 & \text{if } x \geq 4 \end{cases}$$

72. $x = -4, 0,$ and 4

$$f(x) = \begin{cases} -4x & \text{if } x \leq -4 \\ x^3 + 2 & \text{if } -4 < x \leq 2 \\ 4 - x^2 & \text{if } x > 2 \end{cases}$$

73. $x = -2, 1,$ and 2

$$f(x) = \begin{cases} x^2 + 2x + 6 & \text{if } -5 \leq x < 0 \\ x + 6 & \text{if } 0 \leq x < 2 \\ x^3 + 1 & \text{if } 2 \leq x \leq 5 \end{cases}$$

74. $x = 1975, 1980,$ and 1998

$$f(x) = \begin{cases} 0.2(x - 1970)^3 + 60 & \text{if } 1970 \leq x < 1980 \\ 190 - (x - 1980)^2 & \text{if } 1980 \leq x < 1990 \\ 2(x - 1990) + 100 & \text{if } 1990 \leq x \leq 2000 \end{cases}$$

Exercises 75–80: Complete the following.

(a) Sketch a graph of f.
(b) Determine if f is continuous on its domain.
(c) Solve $f(x) = 0$.

75. $f(x) = \begin{cases} 4 - x^2 & \text{if } -3 \leq x \leq 0 \\ x^2 - 4 & \text{if } 0 < x \leq 3 \end{cases}$

76. $f(x) = \begin{cases} x^2 & \text{if } -2 \leq x < 0 \\ x + 1 & \text{if } 0 \leq x \leq 2 \end{cases}$

77. $f(x) = \begin{cases} 2x & \text{if } -5 \leq x < -1 \\ -2 & \text{if } -1 \leq x < 0 \\ x^2 - 2 & \text{if } 0 \leq x \leq 2 \end{cases}$

78. $f(x) = \begin{cases} 0.5x^2 & \text{if } -4 \leq x \leq -2 \\ x & \text{if } -2 < x < 2 \\ x^2 - 4 & \text{if } 2 \leq x \leq 4 \end{cases}$

79. $f(x) = \begin{cases} x^3 + 3 & \text{if } -2 \leq x \leq 0 \\ x + 3 & \text{if } 0 < x < 1 \\ 4 + x - x^2 & \text{if } 1 \leq x \leq 3 \end{cases}$

80. $f(x) = \begin{cases} -2x & \text{if } -3 \leq x < -1 \\ x^2 + 1 & \text{if } -1 \leq x \leq 2 \\ \frac{1}{2}x^3 + 1 & \text{if } 2 < x \leq 3 \end{cases}$

Applications

81. *Electronics*   The **Heaviside function** H is used in the study of electrical circuits and is defined by

$$H(t) = \begin{cases} 0 & \text{if } t < 0. \\ 1 & \text{if } t \geq 0. \end{cases}$$

(a) Evaluate $H(-2)$, $H(0)$, and $H(3.5)$.

(b) Graph $y = H(t)$.

82. *A Strange Graph*   The following definition is discussed in advanced mathematics courses.

$$f(x) = \begin{cases} 0 & \text{if } x \text{ is a rational number.} \\ 1 & \text{if } x \text{ is an irrational number.} \end{cases}$$

(a) Evaluate $f\left(-\frac{3}{4}\right)$, $f(-\sqrt{2})$, and $f(\pi)$.

(b) Is f a function? Explain.

(c) Discuss the difficulty with graphing $y = f(x)$.

83. *Modeling Temperature*   In the accompanying figure the monthly average temperature in degrees Fahrenheit from January to December in Minneapolis is modeled by a polynomial function f, where $x = 1$ corresponds to January and $x = 12$ to December.   (**Source:** A. Miller and J. Thompson, *Elements of Meteorology*.)

(a) Estimate the turning points.

(b) Interpret each turning point.

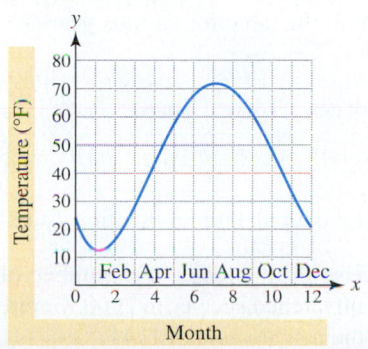

84. *Natural Gas Consumption* Refer to Figure 4.29 and Table 4.3 in the introduction to this section.

(a) Solve the equation $f(x) = 20$ graphically. Interpret the solution set.

(b) Calculate the average rate of change in natural gas consumption from 1970 to 1980. Interpret the result.

85. *Highway Design* In order to allow enough distance for cars to pass on two-lane highways, engineers calculate minimum sight distances between curves and hills. See the accompanying figure. The table shows the minimum sight distance y in feet for a car traveling at x miles per hour. (**Source:** L. Haefner, *Introduction to Transportation Systems*.)

x (mph)	20	30	40	50
y (ft)	810	1090	1480	1840

x (mph)	60	65	70
y (ft)	2140	2310	2490

(a) Make a scatterplot of the data. As x increases, describe how y changes.

(b) What is the minimum degree of a polynomial that might model these data? Find such a polynomial, $D(x)$.

(c) If a car is traveling at 43 miles per hour, estimate the minimum sight distance.

86. *Endangered Species* The total number y of endangered and threatened species in the United States are given in the table for various years x. (**Source:** Fish and Wildlife Services.)

x (yr)	1980	1985	1990	1995
y (total)	786	941	1181	1599

(a) Find a polynomial function f that models the data.

(b) Use f to predict the number of endangered and threatened species in 2000. Compare your answer to the actual value of 1741.

87. *Marijuana Use* The table lists the percentage y of high school seniors that had used marijuana within the previous month in the United States for various years x. In this table $x = 0$ corresponds to 1975 and $x = 20$ to 1995. (**Source:** Health and Human Services Department.)

x (yr)	0	3	5	10	15	20
y (%)	27	37	33	25	14	21

(a) Use least-squares regression to find a polynomial function f that models the data.

(b) Use f to estimate marijuana use in 1997. Compare your answer to the actual value of 23%.

88. *Aging of America* The table lists the number in thousands of Americans over 100 years old for selected years. (**Source:** Bureau of the Census.)

Year	Number (thousands)
1994	50
1996	56
1998	65
2000	75
2002	94
2004	110

(a) Use least-squares regression to find a function f that models the data. Let $x = 0$ correspond to 1994.

(b) Estimate the number of Americans over 100 in the year 2008.

89. *Modeling* An object is lifted rapidly into the air at a constant speed and then dropped. Its height after x seconds is listed in the table.

Time (sec)	0	1	2	3
Height (ft)	0	36	72	108

Time (sec)	4	5	6	7
Height (ft)	144	128	80	0

(a) Make a line graph of the data. At what time does it appear that the object was dropped?

(b) Identify the time interval when the height could be modeled by a linear function. When could it be modeled by a nonlinear function?

(c) Determine values for the constants m, a, and b so that f models the data.

$$f(x) = \begin{cases} mx & \text{if } 0 \le x \le 4 \\ a(x-4)^2 + b & \text{if } 4 < x \le 7 \end{cases}$$

(d) Solve the equation $f(x) = 100$ and interpret your answer.

90. *Modeling* (Refer to Example 6.) A tank of water is filled with a hose and then drained. The accompanying table shows the number of gallons y in the tank after t minutes.

t (min)	0	1	2	3	4	5	6	7
y (gal)	0	9	18	27	36	16	4	0

The following piecewise-polynomial function f models the data in the accompanying table.

$$f(t) = \begin{cases} 9t & \text{if } 0 \le t \le 4 \\ 4t^2 - 56t + 196 & \text{if } 4 < t \le 7 \end{cases}$$

Solve the equation $f(t) = 12$ and interpret the results.

Writing about Mathematics

Exercises 91–94: Discuss possible local or absolute extrema on the graph of f. Assume that $a > 0$.

91. $f(x) = ax + b$ 92. $f(x) = ax^2 + bx + c$

93. $f(x) = ax^3 + bx^2 + cx + d$

94. $f(x) = a|x|$

CHECKING BASIC CONCEPTS FOR SECTIONS 4.1 AND 4.2

1. Use the graph of f to complete the following.

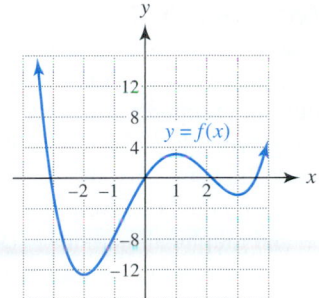

(a) Determine where f is increasing or decreasing. Use interval notation.

(b) Identify any local extrema.

(c) Identify any absolute extrema.

(d) Approximate the x-intercepts and zeros of f. Then solve $f(x) = 0$. How are the x-intercepts, zeros, and solutions to $f(x) = 0$ related?

2. Use the graph to complete the following.

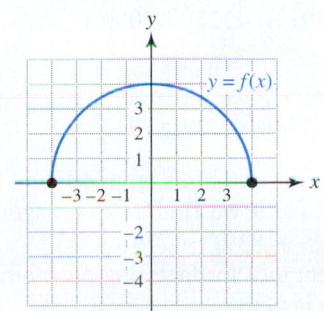

(a) Evaluate $f(-4)$, $f(0)$, and $f(4)$.

(b) What type of symmetry does the graph of f exhibit?

(c) Is f an odd function or an even function? Why?

(d) Find the domain and range of f. Use interval notation.

3. If possible, sketch a graph of a cubic polynomial with a negative leading coefficient that satisfies each of the following.
   (a) Zero x-intercepts   (b) One x-intercept
   (c) Two x-intercepts   (d) Four x-intercepts

4. Plot the data in the accompanying table.

x	-3.2	-2	0	2	3.2
y	-11	15	-10	15	-11

(a) What is the minimum degree of a polynomial function f that would be needed to model these data? Explain.

(b) Should function f be odd, even, or neither? Explain.

(c) Should the leading coefficient of f be positive or negative? Explain.

5. Use least-squares regression to find a polynomial that models the data in Exercise 4.

6. *Torricelli's Law*   A cylindrical tank contains 500 gallons of water. A plug is pulled from the bottom of the tank and it takes 10 minutes to drain the tank. The amount A of water in gallons remaining in the tank after t minutes is approximated by

$$A(t) = 500\left(1 - \frac{t}{10}\right)^2.$$

(a) What is a reasonable domain for A?

(b) Evaluate $A(1)$ and interpret the result.

(c) What is the degree of $A(t)$? What is the leading coefficient?

(d) Is half the water drained from the tank after 5 minutes? Does this agree with your intuition? Explain.

4.3 Real Zeros of Polynomial Functions

- ◆ Divide polynomials
- ◆ Understand the division algorithm, remainder theorem, and factor theorem
- ◆ Factor higher degree polynomials
- ◆ Analyze polynomials with multiple zeros
- ◆ Find rational zeros
- ◆ Solve polynomial equations

Introduction

In Section 3.2 the quadratic formula was used to solve $ax^2 + bx + c = 0$. Are there similar formulas for higher degree polynomial equations? One of the most spectacular mathematical achievements during the sixteenth century was the discovery of formulas for solving cubic and quartic equations. This was accomplished by the Italian mathematicians Tartaglia, Cardano, Fior, del Ferro, and Ferrari between 1515 and 1545. These formulas are quite complicated and typically used only in computer software. Between 1750 and 1780 both Euler and Lagrange failed at finding symbolic solutions to the quintic equation $ax^5 + bx^4 + cx^3 + dx^2 + ex + k = 0$. Later, in about 1805, the Italian physician Ruffini proved that formulas for quintic or higher degree equations were impossible. His results make it necessary for us to rely on numerical and graphical methods.   (**Source:** H. Eves, *An Introduction to the History of Mathematics.*)

Division of Polynomials

Some species of birds, such as robins, have two nesting periods each summer. Because the survival rate for young birds is low, bird populations can fluctuate greatly during the summer months.   (**Source:** S. Kress, *Bird Life.*)

   The graph of $f(x) = x^3 - 61x^2 + 839x + 4221$ shown in Figure 4.62 models a population of birds in a small county, where $x = 1$ corresponds to June 1, $x = 2$ to June 2, $x = 3$ to June 3, and so on.

   If we want to determine the dates when the population was 5000, we can solve the equation

$$x^3 - 61x^2 + 839x + 4221 = 5000,$$

which can also be written as

$$x^3 - 61x^2 + 839x - 779 = 0.$$

From the graph of f it appears that there were 5000 birds around June 1 ($x = 1$), June 20 ($x = 20$), and July 10 ($x = 40$). In Example 9 we find an exact symbolic solution by factoring the polynomial $g(x) = x^3 - 61x^2 + 839x - 779$.

   Division of polynomials is an important tool for factoring polynomials. We begin by reviewing how to divide a monomial into a polynomial.

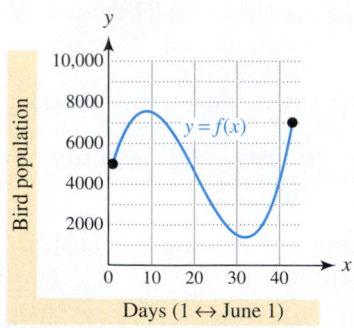

FIGURE 4.62   A Summer Bird Population

 **Dividing by a monomial**

Divide $6x^3 - 3x^2 + 2$ by $2x^2$.

SOLUTION Write the problem as $\frac{6x^3 - 3x^2 + 2}{2x^2}$. Then divide $2x^2$ into *every* term in the numerator.

$$\frac{6x^3 - 3x^2 + 2}{2x^2} = \frac{6x^3}{2x^2} - \frac{3x^2}{2x^2} + \frac{2}{2x^2}$$

$$= 3x - \frac{3}{2} + \frac{1}{x^2}$$

Now Try Exercise 3 ◆

Before dividing a polynomial by a binomial, we briefly review division of natural numbers.

$$
\begin{array}{r}
\text{quotient} \rightarrow 58 \\
\text{divisor} \rightarrow 3\overline{)175} \leftarrow \text{dividend} \\
15 \\
\hline
25 \\
24 \\
\hline
1 \leftarrow \text{remainder}
\end{array}
$$

This result is checked as follows: $3 \cdot 58 + 1 = 175$. The quotient and remainder also can be expressed as $58\frac{1}{3}$. Since 3 does not divide into 175 evenly, 3 is not a factor of 175. When the remainder is 0, the divisor is a factor of the dividend. Division of polynomials is similar to division of natural numbers.

 **Dividing polynomials**

Divide $2x^3 - 3x^2 - 11x + 7$ by $x - 3$. Check the result.

SOLUTION Begin by dividing x into $2x^3$.

$$
\begin{array}{r}
2x^2 \\
x - 3\overline{)2x^3 - 3x^2 - 11x + 7} \\
2x^3 - 6x^2 \\
\hline
3x^2 - 11x
\end{array}
$$

$\frac{2x^3}{x} = 2x^2$

$2x^2(x - 3) = 2x^3 - 6x^2$
Subtract. Bring down $-11x$.

In the next step, divide x into $3x^2$.

$$
\begin{array}{r}
2x^2 + 3x \\
x - 3\overline{)2x^3 - 3x^2 - 11x + 7} \\
2x^3 - 6x^2 \\
\hline
3x^2 - 11x \\
3x^2 - 9x \\
\hline
-2x + 7
\end{array}
$$

$\frac{3x^2}{x} = 3x$

$3x(x - 3) = 3x^2 - 9x$
Subtract. Bring down 7.

Now divide x into $-2x$.

$$
\begin{array}{r}
2x^2 + 3x - 2 \\
x - 3 \overline{)\, 2x^3 - 3x^2 - 11x + 7} \\
\underline{2x^3 - 6x^2} \\
3x^2 - 11x \\
\underline{3x^2 - 9x} \\
-2x + 7 \\
\underline{-2x + 6} \\
1
\end{array}
$$

$\dfrac{-2x}{x} = -2$

$-2(x - 3) = -2x + 6$

Subtract. Remainder is 1.

The quotient is $2x^2 + 3x - 2$ and the remainder is 1. This result can be written as $2x^2 + 3x - 2 + \dfrac{1}{x - 3}$ in the same manner as $175 \div 3$ is expressed as $58\frac{1}{3}$. Polynomial division is checked by multiplying the divisor with the quotient and then adding the remainder. That is, the equation

$$(\textbf{Divisor})(\textbf{Quotient}) + \textbf{Remainder} = \textbf{Dividend}$$

can be used to check the result.

Algebra Review

To review multiplication of polynomials, see Chapter R (page R-23).

$$
\begin{aligned}
(x - 3)(2x^2 + 3x - 2) + 1 &= x(2x^2 + 3x - 2) - 3(2x^2 + 3x - 2) + 1 \\
&= 2x^3 + 3x^2 - 2x - 6x^2 - 9x + 6 + 1 \\
&= 2x^3 - 3x^2 - 11x + 7
\end{aligned}
$$

Now Try Exercise 17 ◆

EXAMPLE 3 Dividing polynomials

Divide each expression. Check your answer.

(a) $\dfrac{6x^2 + 5x - 10}{2x + 3}$ **(b)** $(5x^3 - 4x^2 + 7x - 2) \div (x^2 + 1)$.

SOLUTION

(a) Begin by dividing $2x$ into $6x^2$.

$$
\begin{array}{r}
3x \\
2x + 3 \overline{)\, 6x^2 + 5x - 10} \\
\underline{6x^2 + 9x} \\
-4x - 10
\end{array}
$$

$\dfrac{6x^2}{2x} = 3x$

$3x(2x + 3) = 6x^2 + 9x$

Subtract: $5x - 9x = -4x$.

Bring down the -10.

In the next step, divide $2x$ into $-4x$

$$
\begin{array}{r}
\textbf{3x - 2} \\
\textbf{2x + 3} \overline{)\, 6x^2 + 5x - 10} \\
\underline{6x^2 + 9x} \\
-4x - 10 \\
\underline{-4x - 6} \\
\textbf{-4}
\end{array}
$$

$\dfrac{-4x}{2x} = -2$

$-2(2x + 3) = -4x - 6$

Subtract: $-10 - (-6) = -4$.

The quotient is $3x - 2$ with remainder -4. This result can also be written as follows.

$$\textbf{3x - 2} + \dfrac{\textbf{-4}}{\textbf{2x + 3}} \qquad \text{Quotient} + \dfrac{\text{Remainder}}{\text{Divisor}}$$

To check this result use the equation

$$(\text{Divisor})(\text{Quotient}) + \text{Remainder} = \text{Dividend}.$$

This result can be checked.

$$(2x + 3)(3x - 2) + (-4) = 6x^2 + 5x - 6 - 4$$

(Divisor)(Quotient) +
Remainder = Dividend.

$$= 6x^2 + 5x - 10$$

The result checks.

(b) Begin by writing $x^2 + 1$ as $x^2 + 0x + 1$.

$$
\begin{array}{r}
5x - 4 \\
x^2 + 0x + 1 \overline{\smash{)}\, 5x^3 - 4x^2 + 7x - 2} \\
\underline{5x^3 + 0x^2 + 5x} \\
-4x^2 + 2x - 2 \\
\underline{-4x^2 + 0x - 4} \\
2x + 2
\end{array}
$$

The quotient is $5x - 4$ with remainder is $2x + 2$. This result can also be written as

$$5x - 4 + \frac{2x + 2}{x^2 + 1}.$$

This result can be checked.

$$(x^2 + 1)(5x - 4) + 2x + 2 = 5x^3 - 4x^2 + 5x - 4 + 2x + 2$$

$$= 5x^3 - 4x^2 + 7x - 2$$

The result checks.

Now Try Exercises 21 and 25 ◆

Synthetic Division   A shortcut called **synthetic division** can be used to divide $x - k$ into a polynomial. For example, to divide $x - 2$ into $3x^4 - 7x^3 - 4x + 5$, we perform the following steps. The equivalent steps involving long division are shown to the right.

$$
\begin{array}{r|rrrrr}
2 & 3 & -7 & 0 & -4 & 5 \\
  &   & 6 & -2 & -4 & -16 \\
\hline
  & 3 & -1 & -2 & -8 & -11
\end{array}
$$

$$
\begin{array}{r}
3x^3 - x^2 - 2x - 8 \\
x - 2 \overline{\smash{)}\, 3x^4 - 7x^3 + 0x^2 - 4x + 5} \\
\underline{3x^4 - 6x^3} \\
-1x^3 + 0x^2 \\
\underline{-1x^3 + 2x^2} \\
-2x^2 - 4x \\
\underline{-2x^2 + 4x} \\
-8x + 5 \\
\underline{-8x + 16} \\
-11
\end{array}
$$

Notice how the highlighted numbers in the expression for long division correspond to the third row in synthetic division. The remainder is -11, which is the last number in the third row. The quotient, $3x^3 - x^2 - 2x - 8$, is one degree less than $f(x)$. Its coefficients are 3, -1, -2, and -8 and are found in the third row. The steps to divide a polynomial $f(x)$ by $x - k$ using synthetic division are summarized.

1. Write k to the left and the coefficients of $f(x)$ to the right in the top row. If any power of x does not appear in $f(x)$, include a 0 for that term. In this example an x^2-term did not appear, so a 0 is included in the first row.

2. Copy the leading coefficient of $f(x)$ into the third row and multiply it by k. Write the result below the next coefficient of $f(x)$ in the second row. Add the numbers in the second column and place the result in the third row. Repeat the process. In this example, the leading coefficient is 3 and $k = 2$. Since $3 \cdot 2 = 6$, 6 is placed below -7. Then add to obtain $-7 + 6 = -1$. Multiply -1 by 2 and repeat.

3. The last number in the third row is the remainder. If the remainder is 0, then the binomial $x - k$ is a factor of $f(x)$. The other numbers in the third row are the coefficients of the quotient in descending powers.

EXAMPLE 4   Performing synthetic division

Use synthetic division to divide $2x^3 + 4x^2 - x + 5$ by $x + 2$.

SOLUTION   Let $k = -2$ and perform the following.

$$
\begin{array}{r|rrrr}
-2 & 2 & 4 & -1 & 5 \\
   &   & -4 & 0 & 2 \\
\hline
   & 2 & 0 & -1 & 7
\end{array}
$$

The remainder is 7 and the quotient is $2x^2 + 0x - 1 = 2x^2 - 1$. This result is expressed by the equation

$$
\frac{2x^3 + 4x^2 - x + 5}{x + 2} = 2x^2 - 1 + \frac{7}{x + 2}.
$$

Now Try Exercise 15 ◆

This process of dividing two polynomials is summarized by the following *division algorithm for polynomials*.

DIVISION ALGORITHM FOR POLYNOMIALS

Let $f(x)$ and $d(x)$ be two polynomials with the degree of $d(x)$ greater than zero and less than the degree of $f(x)$. Then there exist unique polynomials $q(x)$ and $r(x)$ such that

$$
f(x) = d(x) \cdot q(x) + r(x),
$$

Dividend = Divisor · Quotient + Remainder

where either $r(x) = 0$ or the degree of $r(x)$ is less than the degree of $d(x)$. The polynomial $r(x)$ is called the remainder.

◆ **CLASS DISCUSSION**
Two polynomials are divided and the remainder is 0. The divisor has degree $m > 0$, the quotient has degree n, and the dividend has degree $p > m$. What is the relationship among m, n, and p? ◆

If the divisor $d(x)$ is $x - k$, then the division algorithm for polynomials simplifies to

$$
f(x) = (x - k)q(x) + r,
$$

where r is a constant. If we let $x = k$ in this equation, then

$$
f(k) = (k - k)q(k) + r = r.
$$

Thus $f(k)$ is equal to the remainder obtained in synthetic division. In Example 4 when $f(x) = 2x^3 + 4x^2 - x + 5$ is divided by $x + 2$, the remainder is 7. It follows that $f(-2) = 7$. This result is summarized by the *remainder theorem*.

REMAINDER THEOREM

If a polynomial $f(x)$ is divided by $x - k$, the remainder is $f(k)$.

Factoring Polynomials

The polynomial $f(x) = x^2 - 3x + 2$ can be factored as $f(x) = (x - 1)(x - 2)$. Note that $f(1) = 0$ and $(x - 1)$ is a factor of $f(x)$. Similarly, $f(2) = 0$ and $(x - 2)$ is a factor.

This discussion can be generalized. By the remainder theorem we know that

$$f(x) = (x - k)q(x) + r,$$

where r is the remainder. If $r = 0$, then $f(x) = (x - k)q(x)$ and $(x - k)$ is a factor of $f(x)$. Similarly, if $(x - k)$ is a factor of $f(x)$, then $r = 0$. That is,

$$f(x) = (x - k)q(x)$$

and $f(k) = (k - k)q(x) = 0 \cdot q(x) = 0$. This discussion justifies the *factor theorem*.

FACTOR THEOREM

A polynomial $f(x)$ has a factor $x - k$ if and only if $f(k) = 0$.

EXAMPLE 5 Applying the factor theorem

Use the graph of $f(x) = x^3 - 2x^2 - 5x + 6$ in Figure 4.63 and the factor theorem to list the factors of $f(x)$.

SOLUTION Figure 4.63 shows that the zeros (or x-intercepts) of f are -2, 1, and 3. Since $f(-2) = 0$, the factor theorem states that $(x + 2)$ is a factor of $f(x)$. Similarly, $f(1) = 0$ implies that $(x - 1)$ is a factor, and $f(3) = 0$ implies that $(x - 3)$ is a factor. Thus the factors of $x^3 - 2x^2 - 5x + 6$ are $(x + 2)$, $(x - 1)$, and $(x - 3)$. *Now Try Exercise 35* ◆

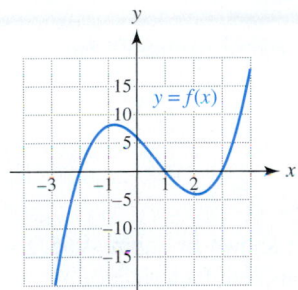

FIGURE 4.63

◆ **MAKING CONNECTIONS**

***x*-Intercepts, Zeros, and Factors**   Let $f(x)$ be a polynomial with degree 1 or more. The following statements are equivalent.

1. The graph of $y = f(x)$ has x-intercept k.
2. A zero of $f(x)$ is k. That is, $f(k) = 0$.
3. A factor of $f(x)$ is $(x - k)$.

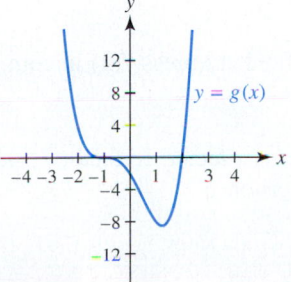

FIGURE 4.64

The polynomial $f(x) = x^2 + 4x + 4$ can be written as $f(x) = (x + 2)^2$. Since the factor $(x + 2)$ occurs twice in $f(x)$, the zero -2 is called a **zero of multiplicity 2**. The polynomial $g(x) = (x + 1)^3(x - 2)$ has zeros -1 and 2 with *multiplicities* 3 and 1, respectively. A graph of g is shown in Figure 4.64, where the x-intercepts coincide with the zeros of g. *Counting multiplicities*, a polynomial of degree n has at most n real zeros. For $g(x)$, the sum of the multiplicities is $3 + 1 = 4$, which equals its degree.

These concepts together with the factor theorem can be used to find the *complete factored form* of a polynomial.

COMPLETE FACTORED FORM

Suppose a polynomial

$$f(x) = a_n x^n + \cdots + a_2 x^2 + a_1 x + a_0$$

has n real zeros $c_1, c_2, c_3, \ldots, c_n$, where distinct zeros are listed as many times as their multiplicities. Then $f(x)$ can be written in **complete factored form** as

$$f(x) = a_n(x - c_1)(x - c_2)(x - c_3) \cdots (x - c_n).$$

EXAMPLE 6 Finding a complete factorization

Write the complete factorization for each polynomial with the given zeros.

(a) $f(x) = 13x^2 - \frac{91}{2}x + 39$; zeros: $\frac{3}{2}$ and 2

(b) $f(x) = 7x^3 - 21x^2 - 7x + 21$; zeros: -1, 1, and 3

SOLUTION

(a) The leading coefficient is **13** and its zeros are $\frac{3}{2}$ and **2**. By the factor theorem, $\left(x - \frac{3}{2}\right)$ and $(x - 2)$ are factors. The complete factorization is

$$f(x) = \mathbf{13}\left(x - \frac{\mathbf{3}}{\mathbf{2}}\right)(x - \mathbf{2}).$$

(b) The leading coefficient is 7 and its zeros are -1, 1, and 3. The complete factorization is

$$f(x) = 7(x + 1)(x - 1)(x - 3).$$

Now Try Exercises 39 and 41 ◆

EXAMPLE 7 Factoring a polynomial graphically

$[-5, 5, 1]$ by $[-25, 25, 5]$

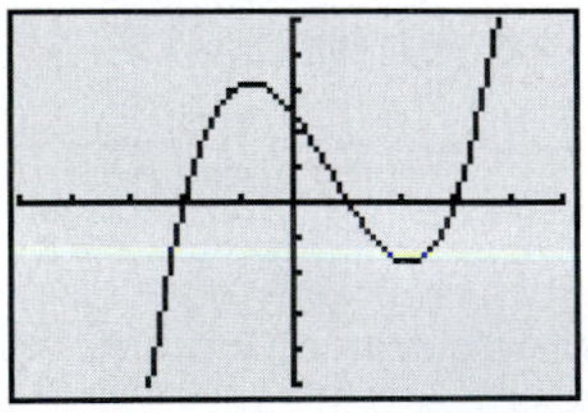

FIGURE 4.65

Use graphing to factor $f(x) = 2x^3 - 4x^2 - 10x + 12$.

SOLUTION A graph of $Y_1 = 2X^3 - 4X^2 - 10X + 12$ is shown in Figure 4.65. Its zeros are -2, 1, and 3. Since the leading coefficient is 2, the complete factorization is

$$f(x) = 2(x + 2)(x - 1)(x - 3).$$

Now Try Exercise 59 ◆

EXAMPLE 8 Factoring a polynomial symbolically

The polynomial $f(x) = 2x^3 - 2x^2 - 34x - 30$ has a zero of -1. Express $f(x)$ in complete factored form.

SOLUTION If -1 is a zero, then by the factor theorem $(x + 1)$ is a factor. To factor $f(x)$, divide $x + 1$ into $2x^3 - 2x^2 - 34x - 30$ by using synthetic division.

$$
\begin{array}{r|rrrr}
-1 & 2 & -2 & -34 & -30 \\
   &   & -2 &   4 & 30 \\
\hline
   & 2 & -4 & -30 &   0 \\
\end{array}
$$

The remainder is 0, so $x + 1$ divides evenly into the dividend. By the division algorithm,

$$2x^3 - 2x^2 - 34x - 30 = (x + 1)(2x^2 - 4x - 30).$$

The quotient $2x^2 - 4x - 30$ can be factored further.

$$2x^2 - 4x - 30 = 2(x^2 - 2x - 15)$$
$$= 2(x + 3)(x - 5)$$

Algebra Review
To review factoring trinomials, see Chapter R (page R-29).

Thus the complete factored form is $f(x) = 2(x + 1)(x + 3)(x - 5)$.

Now Try Exercise 63 ◆

In the next example we factor the polynomial $g(x) = x^3 - 61x^2 + 839x - 779$, which was presented earlier, and then use the complete factorization to determine the days when the bird population was 5000.

EXAMPLE 9   ## Factoring a polynomial

Factor $g(x) = x^3 - 61x^2 + 839x - 779$. Use the zeros of $g(x)$ to determine when the bird population was 5000.

SOLUTION   From Figure 4.62 on page 274, it appears that the bird population was 5000 when $x = 1$. If we substitute $x = 1$ in this polynomial, the result is 0.

$$g(1) = 1^3 - 61(1)^2 + 839(1) - 779 = 0$$

By the factor theorem, $(x - 1)$ is a factor of $g(x)$. We can use synthetic division to divide $x^3 - 61x^2 + 839x - 779$ by $x - 1$.

$$\begin{array}{r|rrrr} 1 & 1 & -61 & 839 & -779 \\ & & 1 & -60 & 779 \\ \hline & 1 & -60 & 779 & 0 \end{array}$$

By the division algorithm,

$$x^3 - 61x^2 + 839x - 779 = (x - 1)(x^2 - 60x + 779).$$

Since it is not obvious how to factor $x^2 - 60x + 779$, we can use the quadratic formula to find its zeros.

$$x = \frac{-b \pm \sqrt{b^2 - 4ac}}{2a}$$

$$= \frac{-(-60) \pm \sqrt{(-60)^2 - 4(1)(779)}}{2(1)}$$

$$= \frac{60 \pm 22}{2}$$

$$= 41 \text{ or } 19$$

The zeros of $g(x) = x^3 - 61x^2 + 839x - 779$ are 1, 19, and 41, and its leading coefficient is 1. The complete factorization is $g(x) = (x - 1)(x - 19)(x - 41)$. The bird population equals 5000 on June 1 ($x = 1$), June 19 ($x = 19$), and July 11 ($x = 41$).

Now Try Exercise 137 ◆

Graphs and Multiple Zeros

The polynomial $f(x) = 0.02(x + 3)^3(x - 3)^2$ has zeros -3 and 3 with multiplicities **3** and **2**, respectively. At the zero of even multiplicity the graph does not cross the x-axis, whereas the graph does cross the x-axis at the zero of odd multiplicity. See Figure 4.66.

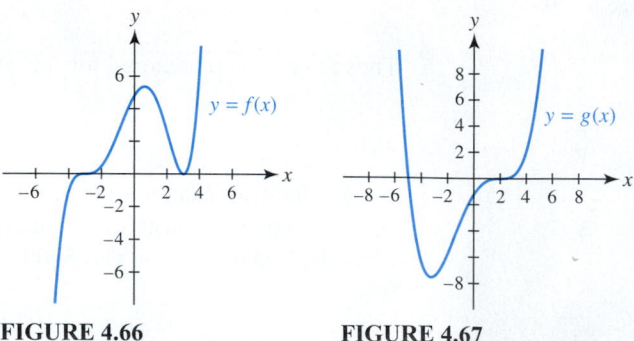

FIGURE 4.66   **FIGURE 4.67**

The graph of $g(x) = 0.03(x + 5)(x - 2)^3$ is shown in Figure 4.67. The zeros of g are -5 and 2 with multiplicities of 1 and 3, respectively. Since both zeros have odd multiplicity, the graph crosses the x-axis at -5 and 2. Notice that the zero 2 has a higher multiplicity and the graph levels off more near $x = 2$ than it does near $x = -5$. The higher the multiplicity of a zero, the more the graph of a polynomial levels off near the zero.

◆ **MAKING CONNECTIONS**

Zeros and Multiplicity   If a zero of a polynomial $f(x)$ has odd multiplicity, then its graph crosses the x-axis at the zero. If a zero has even multiplicity, then its graph intersects, but does not cross, the x-axis at the zero. The higher the multiplicity of a zero, the more the graph levels off at the zero. These concepts are illustrated in the following figures.

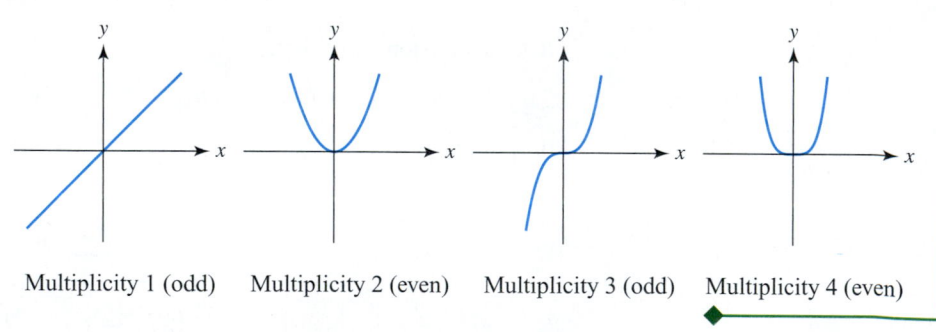

Multiplicity 1 (odd)   Multiplicity 2 (even)   Multiplicity 3 (odd)   Multiplicity 4 (even)

 EXAMPLE 10   Finding multiplicities graphically

Figure 4.68 shows the graph of a sixth-degree polynomial $f(x)$ with leading coefficient 1. All zeros are integers. Write $f(x)$ in complete factored form.

SOLUTION   The x-intercepts or zeros of f are -2, 0, and 4. Since the graph crosses the x-axis at -2 and 4, these zeros have odd multiplicity. The graph of f levels off more at $x = 4$ than at $x = -2$, so 4 has a higher multiplicity than -2. At $x = 0$ the graph of f does not cross the x-axis. Thus, 0 has even multiplicity. If -2 has multiplicity 1, 0 has multiplicity 2,

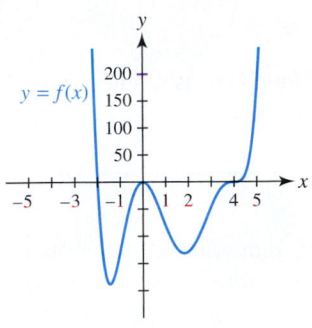

FIGURE 4.68

and 4 has multiplicity 3, then the sum of the multiplicities is $1 + 2 + 3 = 6$, which equals the degree of $f(x)$. List the zeros as

$$-2, \quad 0, \quad 0, \quad 4, \quad 4, \quad \text{and} \quad 4.$$

The leading coefficient is 1, so the complete factorization of $f(x)$ is

$$f(x) = 1(x + 2)(x - 0)(x - 0)(x - 4)(x - 4)(x - 4) \quad \text{or}$$
$$f(x) = x^2(x + 2)(x - 4)^3.$$

Now Try Exercise 77 ◆

Multiple zeros sometimes have physical significance in an application. The next example shows how a multiple zero represents the boundary between an object's floating or sinking.

EXAMPLE 11 Interpreting a multiple zero

The polynomial $f(x) = \frac{\pi}{3}x^3 - 5\pi x^2 + \frac{500\pi d}{3}$ can be used to find the depth that a ball, 10 centimeters in diameter, sinks in water. The constant d is the density of the ball, where the density of water is 1. The smallest *positive* zero of $f(x)$ equals the depth that the sphere sinks. Approximate this depth for each material and interpret the results.
(a) A wood ball with $d = 0.8$
(b) A solid aluminum sphere with $d = 2.7$
(c) A water balloon with $d = 1$

SOLUTION
(a) Let $d = 0.8$ and graph $Y_1 = (\pi/3)X^3 - 5\pi X^2 + 500\pi(0.8)/3$. In Figure 4.69 the smallest positive zero is near 7.13. This means that the 10-centimeter wood ball sinks about 7.13 centimeters into the water.
(b) Let $d = 2.7$ and graph $Y_1 = (\pi/3)X^3 - 5\pi X^2 + 500\pi(2.7)/3$. In Figure 4.70 there is no smallest positive zero. The aluminum sphere is more dense than water and sinks.
(c) Let $d = 1$ and graph $Y_1 = (\pi/3)X^3 - 5\pi X^2 + 500\pi/3$. The graph in Figure 4.71 has one positive zero of 10 with multiplicity 2. The water balloon has the same density as water and "floats" even with the surface. The value of $d = 1$ represents the boundary between sinking and floating. If the ball floats, $f(x)$ has two positive zeros; if it sinks, $f(x)$ has no positive zeros. With the water balloon there is one positive zero with multiplicity 2 that represents a transition point between floating and sinking.

◆ **CLASS DISCUSSION**
Make a conjecture about the depth that a ball with a 10-centimeter diameter will sink in water if $d = 0.5$. Test your conjecture graphically. ◆

$[-20, 20, 5]$ by $[-300, 500, 100]$

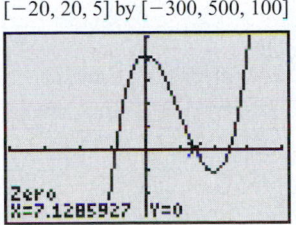

FIGURE 4.69

$[-20, 20, 5]$ by $[-500, 2000, 500]$

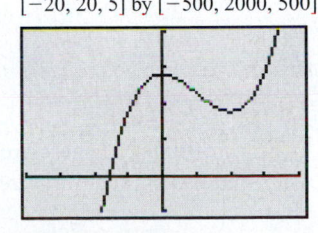

FIGURE 4.70

$[-20, 20, 5]$ by $[-300, 600, 100]$

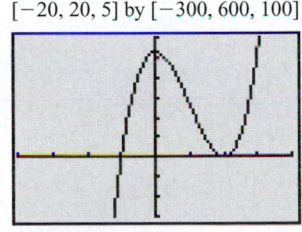

FIGURE 4.71

Now Try Exercise 133 ◆

Rational Zeros

If a polynomial has a zero that is a rational number, it can be found by using the *rational zero test*.

RATIONAL ZERO TEST

Let $f(x) = a_n x^n + \cdots + a_2 x^2 + a_1 x + a_0$, where $a_n \neq 0$, represent a polynomial function f with integer coefficients. If $\frac{p}{q}$ is a rational number written in lowest terms and if $\frac{p}{q}$ is a zero of f, then p is a factor of the constant term a_0 and q is a factor of the leading coefficient a_n.

The following example illustrates how to find rational zeros by using this test.

 **Finding rational zeros of a polynomial**

Find all rational zeros of $f(x) = 6x^3 - 5x^2 - 7x + 4$. Write $f(x)$ in complete factored form.

SOLUTION   If $\frac{p}{q}$ is a rational zero in lowest terms, then p is a factor of the constant term 4 and q is a factor of the leading coefficient 6. The possible values for p and q are as follows.

$$p: \quad \pm 1, \quad \pm 2, \quad \pm 4$$
$$q: \quad \pm 1, \quad \pm 2, \quad \pm 3, \quad \pm 6$$

As a result, any rational zero of $f(x)$ in the form $\frac{p}{q}$ must occur in the list

$$\pm\frac{1}{6}, \quad \pm\frac{1}{3}, \quad \pm\frac{1}{2}, \quad \pm\frac{2}{3}, \quad \pm\frac{1}{1}, \quad \pm\frac{4}{3}, \quad \pm\frac{2}{1}, \quad \text{or} \quad \pm\frac{4}{1}.$$

Evaluate $f(x)$ at each value in the list. See Table 4.5.

TABLE 4.5

x	$f(x)$	x	$f(x)$	x	$f(x)$	x	$f(x)$
$\frac{1}{6}$	$\frac{49}{18}$	$\frac{1}{2}$	0	1	-2	2	18
$-\frac{1}{6}$	5	$-\frac{1}{2}$	$\frac{11}{2}$	-1	0	-2	-50
$\frac{1}{3}$	$\frac{4}{3}$	$\frac{2}{3}$	$-\frac{10}{9}$	$\frac{4}{3}$	0	4	280
$-\frac{1}{3}$	$\frac{50}{9}$	$-\frac{2}{3}$	$\frac{14}{3}$	$-\frac{4}{3}$	$-\frac{88}{9}$	-4	-432

From Table 4.5 there are three rational zeros of -1, $\frac{1}{2}$, and $\frac{4}{3}$. Since a third degree polynomial has at most three zeros, the complete factored form of $f(x)$ is

$$f(x) = 6(x + 1)\left(x - \frac{1}{2}\right)\left(x - \frac{4}{3}\right).$$

Now Try Exercise 89 ◆

Although $f(x)$ in Example 12 had only rational zeros, it is important to realize that many polynomials have irrational zeros. Irrational zeros cannot be found using the rational zero test.

Polynomial Equations

In Section 3.2, factoring was used to solve quadratic equations. Factoring also can be used to solve polynomial equations with degrees greater than 2. This technique is illustrated in the next two examples.

EXAMPLE 13 Solving a cubic equation

Solve $x^3 + 3x^2 - 4x = 0$ symbolically. Support your answer graphically and numerically.

SOLUTION
Symbolic Solution

$[-5, 5, 1]$ by $[-15, 15, 5]$

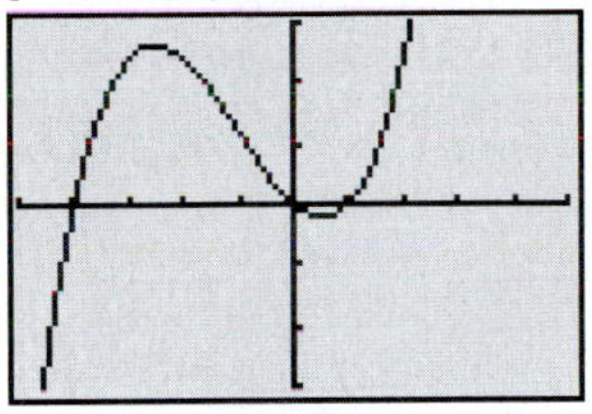

FIGURE 4.72

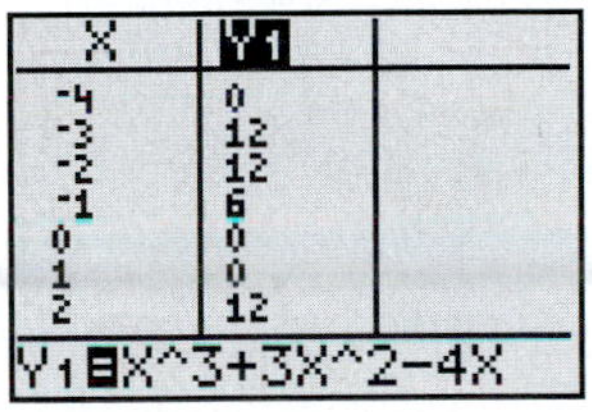

X	Y1	
-4	0	
-3	12	
-2	12	
-1	6	
0	0	
1	0	
2	12	

Y1◼X^3+3X^2-4X

FIGURE 4.73

$$x^3 + 3x^2 - 4x = 0 \qquad \text{Given equation}$$
$$x(x^2 + 3x - 4) = 0 \qquad \text{Factor out } x.$$
$$x(x + 4)(x - 1) = 0 \qquad \text{Factor the quadratic expression.}$$
$$x = 0, x + 4 = 0, \text{ or } x - 1 = 0 \qquad \text{Zero-product property}$$
$$x = 0, -4, \text{ or } 1 \qquad \text{Solve.}$$

Graphical Solution Graph $Y_1 = X^3 + 3X^2 - 4X$ as in Figure 4.72. The x-intercepts are $-4, 0,$ and 1, which correspond to the solutions.

Numerical Solution Table $Y_1 = X^3 + 3X^2 - 4X$ as in Figure 4.73. The zeros of Y_1 occur at $x = -4, 0,$ and 1. *Now Try Exercise 97* ◆

◆ **MAKING CONNECTIONS**

Functions and Equations Each time a new type of function is defined, we can define a new type of equation. For example, cubic functions given by $f(x) = ax^3 + bx^2 + cx + d$ can be used to define cubic equations of the form $ax^3 + bx^2 + cx + d = 0$. This concept will be used to define other types of equations. ◆

EXAMPLE 14 Solving a polynomial equation

Find all real solutions to each equation symbolically.
(a) $4x^4 - 5x^2 - 9 = 0$ **(b)** $2x^3 + 12 = 3x^2 + 8x$

SOLUTION
(a) The expression $4x^4 - 5x^2 - 9$ can be factored in a manner similar to the way quadratic expressions are factored.

$$4x^4 - 5x^2 - 9 = 0 \qquad \text{Given equation}$$
$$(4x^2 - 9)(x^2 + 1) = 0 \qquad \text{Factor.}$$
$$4x^2 - 9 = 0 \quad \text{or} \quad x^2 + 1 = 0 \qquad \text{Zero-product property}$$
$$4x^2 = 9 \quad \text{or} \quad x^2 = -1 \qquad \text{Add 9 or subtract 1.}$$
$$x^2 = \frac{9}{4} \quad \text{or} \quad x^2 = -1 \qquad \text{Divide by 4.}$$
$$x = \pm\frac{3}{2} \quad \text{or} \quad x^2 = -1 \qquad \text{Square root property}$$

Since $x^2 \geq 0$ for all x, the equation $x^2 = -1$ has no real solutions. Thus the only solutions are $\pm\frac{3}{2}$.

(b) First transpose each term on the right side of the equation to the left side of the equation. Then use grouping to factor the polynomial.

Algebra Review
To review factoring by grouping, see Chapter R (page R-28).

$$2x^3 + 12 = 3x^2 + 8x \qquad \text{Given equation}$$

$$2x^3 - 3x^2 - 8x + 12 = 0 \qquad \text{Rewrite the equation.}$$

$$(2x^3 - 3x^2) + (-8x + 12) = 0 \qquad \text{Associative property}$$

$$\mathbf{x^2(2x - 3) - 4(2x - 3)} = 0 \qquad \text{Factor.}$$

$$\mathbf{(x^2 - 4)(2x - 3)} = 0 \qquad \text{Factor out } 2x - 3.$$

$$x^2 - 4 = 0 \quad \text{or} \quad 2x - 3 = 0 \qquad \text{Zero-product property}$$

$$x = \pm 2 \quad \text{or} \quad x = \frac{3}{2} \qquad \text{Solve each equation.}$$

The solutions are ± 2 and $\frac{3}{2}$.   *Now Try Exercises 105 and 115* ◆

Some types of polynomial equations cannot be solved easily by factoring. The next example illustrates how we can obtain an approximate solution graphically.

EXAMPLE 15   Finding a solution graphically

Solve the equation $\frac{1}{2}x^3 - 2x - 4 = 0$ graphically. Round any solutions to the nearest hundredth.

SOLUTION   A graph of $Y_1 = .5X^3 - 2X - 4$ is shown in Figure 4.74. Since there is only one x-intercept, the equation has one real solution of $x \approx 2.65$.

Calculator Help
To find a zero of a function, see Appendix B (page AP-12).

$[-9, 9, 1]$ by $[-6, 6, 1]$

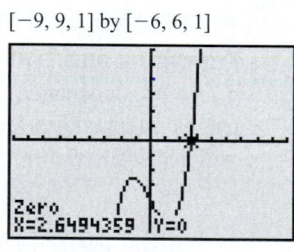

FIGURE 4.74   *Now Try Exercise 123* ◆

4.3   Putting it all Together

The following table lists some important concepts related to a polynomial $f(x)$.

Concept	Explanation	Example
Division by a monomial	Be sure to divide the denominator into *every term* in the numerator.	$\dfrac{5a^3 - 10a^2}{5a^2} = \dfrac{5a^3}{5a^2} - \dfrac{10a^2}{5a^2} = a - 2$

Concept	Explanation	Example
Division by a polynomial	Division by a polynomial can be done in a manner similar to long division of natural numbers. See Examples 2 and 3.	When $6x^3 + 5x^2 - 8x + 4$ is divided by $2x - 1$, the quotient is $3x^2 + 4x - 2$ with remainder 2 and can be written as $$\frac{6x^3 + 5x^2 - 8x + 4}{2x - 1} = 3x^2 + 4x - 2 + \frac{2}{2x - 1}.$$
Remainder theorem	If a polynomial $f(x)$ is divided by $x - k$, the remainder is $f(k)$.	If $f(x) = 3x^2 - 2x + 6$ is divided by $x - 2$ the remainder is $$f(2) = 3(2)^2 - 2(2) + 6 = 14.$$
Factor theorem	$(x - k)$ is a factor of $f(x)$ if and only if $f(k) = 0$.	$f(x) = x^2 + 3x - 4$ and $f(1) = 0$ implies that $(x - 1)$ is a factor of $f(x)$. That is, $f(x) = (x - 1)(x + 4)$.
Complete factored form	$f(x) = a_n(x - c_1) \cdots (x - c_n)$, where the c_k are zeros of f, listed as many times as their multiplicities.	$f(x) = 3(x - 5)(x + 3)(x + 3)$ $= 3(x - 5)(x + 3)^2$ $a_n = 3, c_1 = 5, \quad c_2 = -3, \quad c_3 = -3$
Division algorithm	Dividend = Divisor · Quotient + Remainder If the divisor is $(x - k)$, then $f(x) = (x - k)q(x) + r$, where r is the remainder. See p. 278.	$f(x) = (x - 1)(x^2 - x - 2) + 3$ $q(x) = x^2 - x - 2$ and $r = 3$
Zero with odd multiplicity	The graph of $y = f(x)$ crosses the x-axis at a zero of odd multiplicity.	$f(x) = (x + 1)^3(x - 3)$ Both zeros of -1 and 3 have odd multiplicity.
Zero with even multiplicity	The graph of $y = f(x)$ intersects but does not cross the x-axis at a zero of even multiplicity.	$f(x) = (x + 1)^2(x - 3)^4$ Both zeros of -1 and 3 have even multiplicity.
Factoring a polynomial graphically (Only real zeros)	Graph $y = f(x)$ and locate all the zeros, or x-intercepts. If leading coefficient is a, and the zeros are c_1, c_2, and c_3, then $f(x) = a(x - c_1)(x - c_2)(x - c_3)$.	$f(x) = 2x^3 + 4x^2 - 2x - 4$ has zeros $-2, -1$, and 1 with leading coefficient 2. Thus $$f(x) = 2(x + 2)(x + 1)(x - 1).$$

4.3 — Exercises

Division of Polynomials

Exercises 1–6: Divide the expression.

1. $\dfrac{5x^4 - 15}{10x}$

2. $\dfrac{x^2 - 5x}{5x}$

3. $\dfrac{3x^4 - 2x^2 - 1}{3x^3}$

4. $\dfrac{5x^3 - 10x^2 + 5x}{15x^2}$

5. $\dfrac{x^3 - 4}{4x^3}$

6. $\dfrac{2x^4 - 3x^2 + 4x - 7}{-4x}$

Exercises 7–12: Divide the first polynomial by the second. State the quotient and remainder.

7. $x^3 - 2x^2 - 5x + 6$ $x - 3$

8. $3x^3 - 10x^2 - 27x + 10$ $x + 2$

9. $2x^4 - 7x^3 - 5x^2 - 19x + 17$ $x + 1$

10. $x^4 - x^3 - 4x + 1$ $x - 2$

11. $3x^3 - 7x + 10$ $x - 1$

12. $x^4 - 16x^2 + 1$ $x + 4$

Exercises 13 and 14: Use the division algorithm to complete the following.

13. $\dfrac{x^3 - 8x^2 + 15x - 6}{x - 2} = x^2 - 6x + 3$ implies

    $(x - 2)(x^2 - 6x + 3) = \underline{\quad ? \quad}$.

14. $\dfrac{x^4 - 15}{x + 2} = x^3 - 2x^2 + 4x - 8 + \dfrac{1}{x + 2}$

    implies $x^4 - 15 = (x + 2) \times \underline{\quad ? \quad} + \underline{\quad ? \quad}$.

Exercises 15–30: Divide the expression.

15. $\dfrac{x^4 - 3x^3 - x + 3}{x - 3}$

16. $\dfrac{x^3 - 2x^2 - x + 3}{x + 1}$

17. $\dfrac{4x^3 - x^2 - 5x + 6}{x - 1}$

18. $\dfrac{x^4 + 3x^3 - 4x + 1}{x + 2}$

19. $\dfrac{x^3 + 1}{x + 1}$

20. $\dfrac{x^5 + 3x^4 - x - 3}{x + 3}$

21. $\dfrac{6x^3 + 5x^2 - 8x + 4}{2x - 1}$

22. $\dfrac{12x^3 - 14x^2 + 7x - 7}{3x - 2}$

23. $\dfrac{3x^4 - 7x^3 + 6x - 16}{3x - 7}$

24. $\dfrac{20x^4 + 6x^3 - 2x^2 + 15x - 2}{5x - 1}$

25. $\dfrac{5x^4 - 2x^2 + 6}{x^2 + 2}$

26. $\dfrac{x^3 - x^2 + 2x - 3}{x^2 + 3}$

27. $\dfrac{8x^3 + 10x^2 - 12x - 15}{2x^2 - 3}$

28. $\dfrac{3x^4 - 2x^2 - 5}{3x^2 - 5}$

29. $\dfrac{2x^4 - x^3 + 4x^2 + 8x + 7}{2x^2 + 3x + 2}$

30. $\dfrac{3x^4 + 2x^3 - x^2 + 4x - 3}{x^2 + x - 1}$

Remainder Theorem

Exercises 31–34: Use the remainder theorem to find the remainder when $f(x)$ is divided by the given $x - k$.

31. $f(x) = 5x^2 - 3x + 1$ $x - 1$

32. $f(x) = -4x^2 + 6x - 7$ $x + 4$

33. $f(x) = 4x^3 - x^2 + 4x + 2$ $x + 2$

34. $f(x) = -x^4 + 4x^3 - x + 3$ $x - 3$

Factor Theorem

Exercises 35–38: Use the factor theorem to decide if $x - k$ is a factor of $f(x)$ for the given k.

35. $f(x) = x^3 - 6x^2 + 11x - 6$, $k = 2$

36. $f(x) = x^3 + x^2 - 14x - 24$, $k = -3$

37. $f(x) = x^4 - 2x^3 - 13x^2 - 10x$, $k = 3$

38. $f(x) = 2x^4 - 11x^3 + 9x^2 + 14x$, $k = \frac{1}{2}$

Factoring Polynomials

Exercises 39–44: Use the given zeros to write the complete factored form of $f(x)$.

39. $f(x) = 2x^2 - 25x + 77$; zeros: $\frac{11}{2}$ and 7

40. $f(x) = 6x^2 + 21x - 90$; zeros: -6 and $\frac{5}{2}$

41. $f(x) = x^3 - 2x^2 - 5x + 6$; zeros: -2, 1, and 3

42. $f(x) = x^3 + 6x^2 + 11x + 6$; zeros: -3, -2, and -1

43. $f(x) = -2x^3 + 3x^2 + 59x - 30$; zeros: -5, $\frac{1}{2}$, and 6

44. $f(x) = 3x^4 - 8x^3 - 67x^2 + 112x + 240$;
zeros: -4, $-\frac{4}{3}$, 3, and 5

45. Let $f(x)$ be a quadratic polynomial with leading coefficient 7. Suppose that $f(-3) = 0$ and $f(2) = 0$. Write the complete factored form of $f(x)$.

46. Let $g(x)$ be a cubic polynomial with leading coefficient -4. Suppose that $g(-2) = 0$, $g(1) = 0$, and $g(4) = 0$. Write the complete factored form of $g(x)$.

Exercises 47–50: The graph of a polynomial $f(x)$ with leading coefficient ± 1 and integer zeros is shown in the figure. Write its complete factored form.

47.

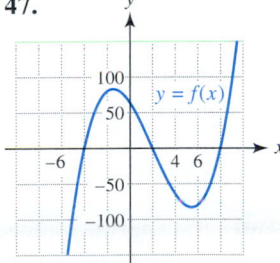

48.

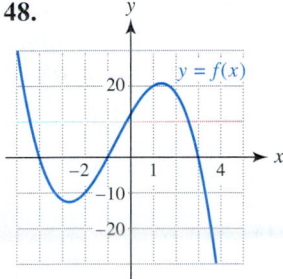

49.

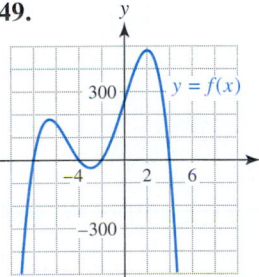

50.

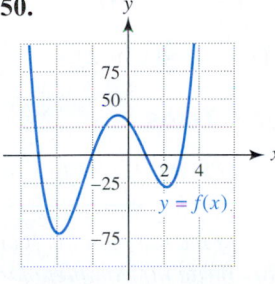

51. Let $f(x)$ be a cubic polynomial with zeros -1, 2, and 3. If the graph of f passes through the point $(0, 3)$, write the complete factored form of $f(x)$.

52. Let $g(x)$ be a quartic polynomial with zeros -2, -1, 1, and 2. If the graph of g passes through the point $(0, 8)$, write the complete factored form of $g(x)$.

Exercises 53–56: The graph of a polynomial $f(x)$ with integer zeros is shown in the figure. Write its complete factored form. Note that the leading coefficient of $f(x)$ is not ± 1.

53.

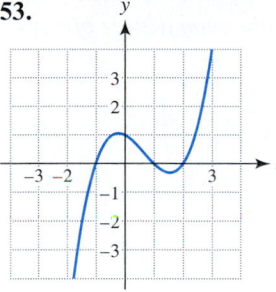

54.

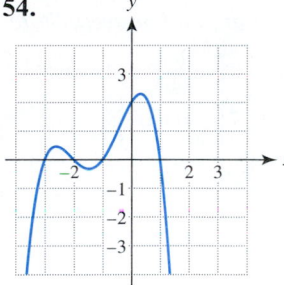

55.

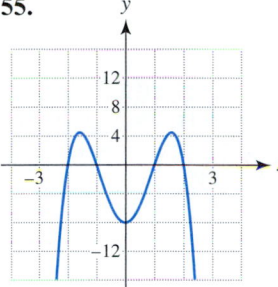

56.

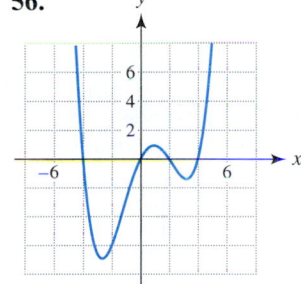

Exercises 57–62: (Refer to Example 7.) Use graphing to factor $f(x)$.

57. $f(x) = 10x^2 + 17x - 6$

58. $f(x) = 2x^3 + 7x^2 + 2x - 3$

59. $f(x) = -3x^3 - 3x^2 + 18x$

60. $f(x) = \frac{1}{2}x^3 + \frac{5}{2}x^2 + x - 4$

61. $f(x) = x^4 + \frac{5}{2}x^3 - 3x^2 - \frac{9}{2}x$

62. $f(x) = 10x^4 + 7x^3 - 27x^2 + 2x + 8$

Exercises 63–68: (Refer to Example 8.) Write the complete factored form of the polynomial $f(x)$, given that k is a zero.

63. $f(x) = x^3 - 9x^2 + 23x - 15$, $\qquad k = 1$

64. $f(x) = 2x^3 + x^2 - 11x - 10$, $\qquad k = -2$

65. $f(x) = -4x^3 - x^2 + 51x - 36$, $\qquad k = -4$

66. $f(x) = 3x^3 - 11x^2 - 35x + 75$, $\qquad k = 5$

67. $f(x) = 2x^4 - x^3 - 13x^2 - 6x$, $\qquad k = -2$

68. $f(x) = 35x^4 + 48x^3 - 41x^2 + 6x$, $\qquad k = \frac{3}{7}$

Graphs and Multiple Zeros

Exercises 69–72: The graph of a polynomial $f(x)$ is shown in the figure. Estimate the zeros and state whether their multiplicities are odd or even. State the minimum degree of $f(x)$.

69.

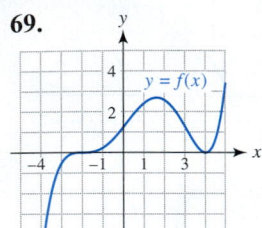

70.

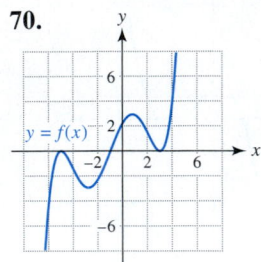

71.

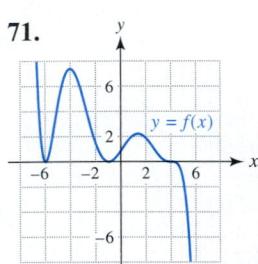

72.

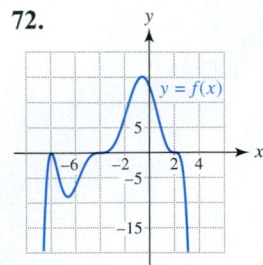

Exercises 73–76: Write a polynomial $f(x)$ in complete factored form that satisfies the conditions. Let the leading coefficient be 1.

73. Degree 3; zeros: -1 with multiplicity 2, and 6 with multiplicity 1

74. Degree 4; zeros: 5 and 7, both with multiplicity 2

75. Degree 4; zeros: 2 with multiplicity 3, and 6 with multiplicity 1

76. Degree 5; zeros: -2 with multiplicity 2, and 4 with multiplicity 3

Exercises 77–82: The graph of either a cubic, quartic, or quintic polynomial $f(x)$ with integer zeros is shown. Write the complete factored form of $f(x)$. (Hint: In Exercises 81 and 82 the leading coefficient is not ± 1.)

77.

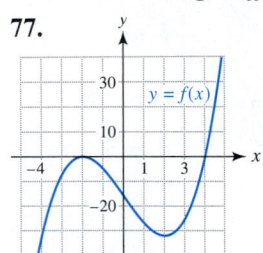

78.

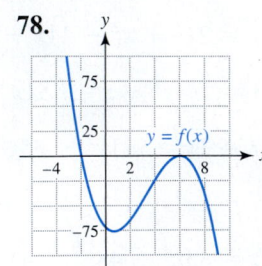

79.

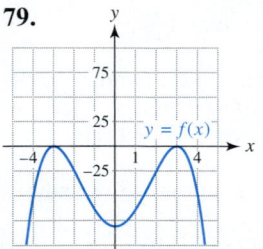

80.

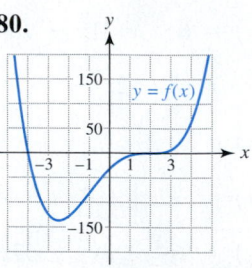

81.

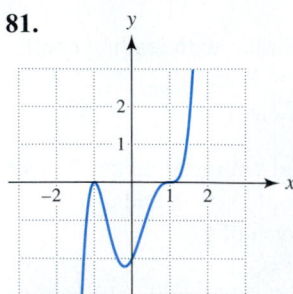

82.

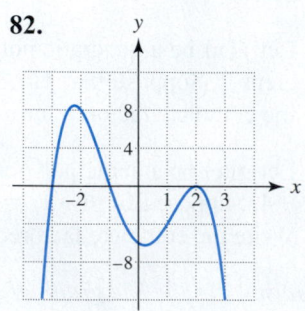

Exercises 83–88: Complete the following.

(a) Find the x- and y-intercepts.
(b) Determine the multiplicity of each zero of f.
(c) Sketch a graph of $y = f(x)$ by hand.

83. $f(x) = 2(x + 2)(x + 1)^2$

84. $f(x) = -3(x - 1)^3$

85. $f(x) = -(x + 1)(x - 1)(x - 2)$

86. $f(x) = x^2(x + 2)(x - 2)$

87. $f(x) = \frac{1}{3}(x + 3)^2(x + 1)^2$

88. $f(x) = -\frac{1}{2}(x + 2)^2(x - 1)^3$

Rational Zeros

Exercises 89–96: (Refer to Example 12.)

(a) Use the rational zero test to find any rational zeros of the polynomial $f(x)$.
(b) Write the complete factored form of $f(x)$.

89. $f(x) = 2x^3 + 3x^2 - 8x + 3$

90. $f(x) = x^3 - 7x + 6$

91. $f(x) = 2x^4 + x^3 - 8x^2 - x + 6$

92. $f(x) = 2x^4 + x^3 - 19x^2 - 9x + 9$

93. $f(x) = 3x^3 - 16x^2 + 17x - 4$

94. $f(x) = x^3 + 2x^2 - 3x - 6$

95. $f(x) = x^3 - x^2 - 7x + 7$

96. $f(x) = 2x^3 - 5x^2 - 4x + 10$

Polynomial Equations

Exercises 97–102: Solve the equation

   (a) *symbolically,*

   (b) *graphically, and*

   (c) *numerically.*

97. $x^3 + x^2 - 6x = 0$ **98.** $2x^2 - 8x + 6 = 0$

99. $x^4 - 1 = 0$ **100.** $x^4 - 5x^2 + 4 = 0$

101. $-x^3 + 4x = 0$ **102.** $6 - 4x - 2x^2 = 0$

Exercises 103–122: Solve the equation.

103. $x^3 - 25x = 0$ **104.** $x^4 - x^3 - 6x^2 = 0$

105. $x^4 - x^2 = 2x^2 + 4$ **106.** $x^4 + 5 = 6x^2$

107. $x^3 - 3x^2 - 18x = 0$ **108.** $x^4 - x^2 = 0$

109. $2x^3 = 4x^2 - 2x$ **110.** $x^3 = x$

111. $12x^3 = 17x^2 + 5x$ **112.** $3x^3 + 3x = 10x^2$

113. $9x^4 + 4 = 13x^2$ **114.** $4x^4 + 7x^2 - 2 = 0$

115. $4x^3 + 4x^2 - 3x - 3 = 0$

116. $9x^3 + 27x^2 - 2x - 6 = 0$

117. $2x^3 + 4 = x(x + 8)$ **118.** $3x^3 + 18 = x(2x + 27)$

119. $8x^4 = 30x^2 - 27$ **120.** $4x^4 - 21x^2 + 20 = 0$

121. $x^6 - 19x^3 = 216$ **122.** $x^6 = 7x^3 + 8$

Exercises 123–128: (Refer to Example 15.) Solve the equation graphically. Round your answers to the nearest hundredth.

123. $x^3 - 1.1x^2 - 5.9x + 0.7 = 0$

124. $x^3 + x^2 - 18x + 13 = 0$

125. $-0.7x^3 - 2x^2 + 4x + 2.5 = 0$

126. $3x^3 - 46x^2 + 180x - 99 = 0$

127. $2x^4 - 1.5x^3 + 13 = 24x^2 + 10x$

128. $-x^4 + 2x^3 + 20x^2 = 22x + 41$

Applications

129. *Area of a Rectangle*   Use the figure to find the length L of the rectangle from its width and area A. Determine L when $x = 10$ feet.

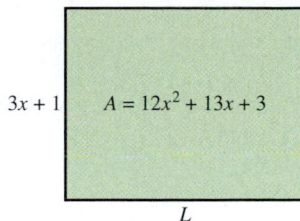

$3x + 1$ $A = 12x^2 + 13x + 3$

L

130. *Area of a Rectangle*   Use the figure to find the width W of the rectangle from its length and area A. Determine W when $x = 5$ inches.

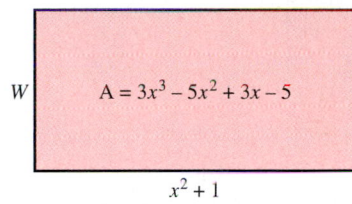

W $A = 3x^3 - 5x^2 + 3x - 5$

$x^2 + 1$

131. *Water Pollution*   In one study, freshwater mussels were used to monitor copper discharge into a river from an electroplating works. Copper in high doses can be lethal to aquatic life. The table lists copper concentrations in mussels after 45 days at various distances downstream from the plant. The concentration C is measured in micrograms of copper per gram of mussel x kilometers downstream. (**Sources:** R. Foster and J. Bates, "Use of mussels to monitor point source industrial discharges"; C. Mason, *Biology of Freshwater Pollution.*)

x	5	21	37	53	59
C	20	13	9	6	5

(a) Describe the relationship between x and C.

(b) These data are modeled by

$$C(x) = -0.000068x^3 + 0.0099x^2 - 0.653x + 23.$$

Graph C and the data.

(c) Concentrations above 10 are lethal to mussels. Locate this region in the river.

132. *Dog Years*   There is an old saying that every year of a dog's life is equal to 7 years for a human. A more accurate approximation is given by the graph of f. Given a dog's age x, where $x \geq 1$, $f(x)$ models the equivalent age in human years. According to the Bureau of the Census, middle age for people begins at age 45. (**Source:** J. Brearley and A. Nicholas, *This Is the Bichon Frise.*)

(**a**) Use the graph of f to estimate the equivalent age for dogs.

(**b**) Use the formula

$$f(x) = -0.001183x^4 + 0.05495x^3 - 0.8523x^2 + 9.054x + 6.748$$

to solve part (a) graphically or numerically.

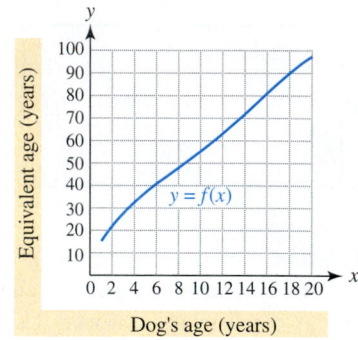

Dog's age (years)

133. *Floating Ball*   (Refer to Example 11.) If a ball has a 20-centimeter diameter, then

$$f(x) = \frac{\pi}{3}x^3 - 10\pi x^2 + \frac{4000\pi d}{3}$$

can be used to determine the depth that it sinks in water. Find the depth that this size ball sinks when $d = 0.6$.

134. *Floating Ball*   (Refer to Example 11.) Determine the depth that a pine ball with a 10-centimeter diameter sinks in water, if $d = 0.55$.

135. *Modeling Temperature*   Complete the following.
(**a**) Approximate the complete factored form of $f(x) = -0.184x^3 + 1.45x^2 + 10.7x - 27.9$.

(**b**) The cubic polynomial $f(x)$ models monthly average temperature at Trout Lake, Canada, in degrees Fahrenheit, where $x = 1$ corresponds to January and $x = 12$ represents December. Interpret the zeros of f.

136. *Average High Temperatures*   The monthly average high temperatures in degrees Fahrenheit at Daytona Beach can be modeled by

$$f(x) = 0.0151x^4 - 0.438x^3 + 3.60x^2 - 6.49x + 72.5,$$

where $x = 1$ corresponds to January and $x = 12$ represents December.

(**a**) Find the average high temperature during March and July.

(**b**) Graph f in [0.5, 12.5, 1] by [60, 100, 10]. Interpret the graph.

(**c**) Estimate graphically and numerically when the average high temperature is 80°F.

137. *Bird Populations*   (Refer to Example 9.) A bird population can be modeled by

$$f(x) = x^3 - 66x^2 + 1052x + 652,$$

where $x = 1$ corresponds to June 1, $x = 2$ to June 2, and so on. Find the days when f estimates that there were 2500 birds.

138. *Geometry*   A rectangular box has sides with lengths x, $x + 1$, and $x + 2$. If the volume of the box is 504 cubic inches, find the dimensions of the box.

Writing about Mathematics

139. Suppose that $f(x)$ is a quintic polynomial with distinct real zeros. Assuming you have access to technology, explain how to factor $f(x)$ approximately. Have you used the factor theorem? Explain.

140. Explain how to determine graphically whether a zero of a polynomial is a multiple zero. Sketch examples.

4.4   The Fundamental Theorem of Algebra

- ◆ Perform arithmetic operations on complex numbers
- ◆ Solve quadratic equations having complex solutions
- ◆ Apply the fundamental theorem of algebra
- ◆ Factor polynomials having complex zeros
- ◆ Solve polynomial equations having complex solutions

Introduction

Throughout history, people have invented new numbers to solve equations and describe data. Often these new numbers were met with resistance and were regarded as being imaginary or unreal. The number 0 was not invented at the same time as the natural numbers. There was no Roman numeral for 0. No doubt there were skeptics who wondered why a number was needed to represent nothing. Negative numbers also met strong resistance. After all, how could one possibly have -6 apples? The same was true for complex numbers. However, complex numbers are no more imaginary than any other number created by mathematicians.

Complex Numbers

The graph of $y = x^2 + 1$, has no x-intercepts, as shown in Figure 4.75. Therefore the equation $x^2 + 1 = 0$ has no real solutions.

If we attempt to solve $x^2 + 1 = 0$ symbolically, it results in $x^2 = -1$. Since $x^2 \geq 0$ for any real number x, there are no real number solutions. However, we can invent a solution.

$$x^2 = -1$$
$$x = \pm\sqrt{-1}$$

We now define a new number called the *imaginary unit*, denoted i.

PROPERTIES OF THE IMAGINARY UNIT i

$$i = \sqrt{-1}, \qquad i^2 = -1$$

By inventing the number i, the solutions to the equation $x^2 + 1 = 0$ are i and $-i$. Using the real numbers and the imaginary unit i, complex numbers can be defined. A **complex number** can be written in **standard form** as $a + bi$, where a and b are real numbers. The **real part** is a and the **imaginary part** is b. Every real number a is also a complex number because it can be written as $a + 0i$. A complex number $a + bi$ with $b \neq 0$ is an **imaginary number**. Table 4.6 lists several complex numbers with their real and imaginary parts.

Using the imaginary unit i, square roots of negative numbers can be written as complex numbers. For example, $\sqrt{-3} = i\sqrt{3}$, and $\sqrt{-16} = i\sqrt{16} = 4i$. This can be summarized as follows.

THE EXPRESSION $\sqrt{-a}$

If $a > 0$, then $\sqrt{-a} = i\sqrt{a}$.

Arithmetic operations are also defined for complex numbers.

Addition and Subtraction   To add the complex numbers $(-2 + 3i)$ and $(4 - 6i)$, simply combine the real and imaginary parts.

$$(-2 + 3i) + (4 - 6i) = -2 + 4 + 3i - 6i$$
$$= 2 - 3i$$

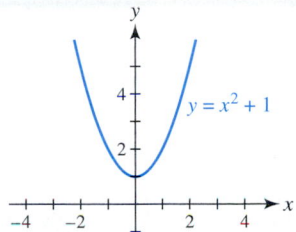

FIGURE 4.75

TABLE 4.6

$a + bi$	a	b
$-3 + 2i$	-3	2
5	5	0
$-3i$	0	-3
$-1 + 7i$	-1	7
$-5 - 2i$	-5	-2
$4 + 6i$	4	6

This same process works for subtraction.

$$(5 - 7i) - (8 + 3i) = 5 - 8 - 7i - 3i$$
$$= -3 - 10i$$

Multiplication   Two complex numbers can be multiplied. The property $i^2 = -1$ is applied when appropriate.

Algebra Review
Before multiplying complex numbers, you may want to review multiplication of binomials, see Chapter R (page R-23).

$$(-5 + i)(7 - 9i) = -5(7) + -5(-9i) + (i)(7) + (i)(-9i)$$
$$= -35 + 45i + 7i - 9i^2$$
$$= -35 + 52i - 9(-1)$$
$$= -26 + 52i$$

Note: When performing arithmetic with complex numbers, express the result in standard form $a + bi$.

Division   The **conjugate** of $a + bi$ is $a - bi$. To find the conjugate, change the sign of the imaginary part b. Table 4.7 lists examples of complex numbers and their conjugates.

TABLE 4.7

$a + bi$	$2 + 5i$	$6 - 3i$	$-2 + 7i$	$-1 - i$	5	$-4i$
$a - bi$	$2 - 5i$	$6 + 3i$	$-2 - 7i$	$-1 + i$	5	$4i$

Calculator Help
To perform arithmetic on complex numbers, see Appendix B (page AP-17).

To simplify the quotient $\frac{3 + 2i}{5 - i}$, first multiply both the numerator and the denominator by the conjugate of the denominator.

$$\frac{3 + 2i}{5 - i} = \frac{(3 + 2i)(5 + i)}{(5 - i)(5 + i)} \qquad \text{Multiply by conjugate.}$$

$$= \frac{3(5) + (3)(i) + (2i)(5) + (2i)(i)}{(5)(5) + (5)(i) + (-i)(5) + (-i)(i)} \qquad \text{Expand.}$$

$$= \frac{15 + 3i + 10i + 2i^2}{25 + 5i - 5i - i^2} \qquad \text{Simplify.}$$

$$= \frac{15 + 13i + 2(-1)}{25 - (-1)} \qquad i^2 = -1$$

$$= \frac{13 + 13i}{26} \qquad \text{Simplify.}$$

$$= \frac{1}{2} + \frac{1}{2}i \qquad \frac{a + bi}{c} = \frac{a}{c} + \frac{b}{c}i$$

The last step expresses the quotient as a complex number in standard form.

Evaluating Complex Arithmetic with Technology   Some graphing calculators perform complex arithmetic. The evaluation of the previous examples are shown in Figures 4.76 and 4.77.

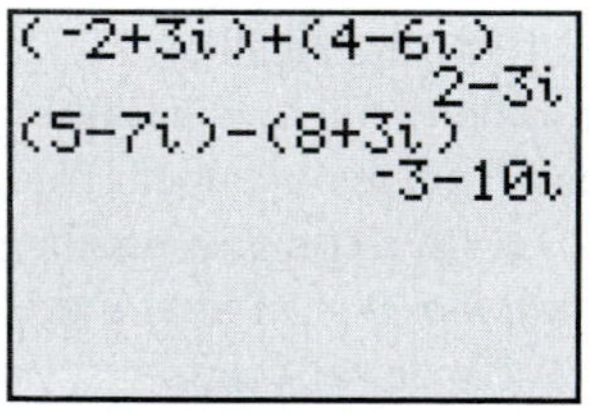

```
(-2+3i)+(4-6i)
             2-3i
(5-7i)-(8+3i)
            -3-10i
```

FIGURE 4.76

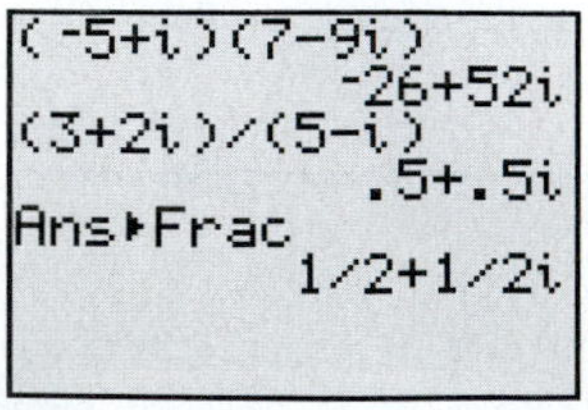

```
(-5+i)(7-9i)
          -26+52i
(3+2i)/(5-i)
            .5+.5i
Ans▶Frac
         1/2+1/2i
```

FIGURE 4.77

EXAMPLE 1 Performing complex arithmetic

Write each expression in standard form. Support your results using a calculator.
(a) $(-3 + 4i) + (5 - i)$ **(b)** $(-7i) - (6 - 5i)$
(c) $(-3 + 2i)^2$ **(d)** $\dfrac{17}{4 + i}$

```
(-3+4i)+(5-i)
              2+3i
(-7i)-(6-5i)
             -6-2i
```

FIGURE 4.78

```
(-3+2i)²
          5-12i
17/(4+i)
          4-i
```

FIGURE 4.79

SOLUTION

(a) $(-3 + 4i) + (5 - i) = -3 + 5 + 4i - i = 2 + 3i$

(b) $(-7i) - (6 - 5i) = -6 - 7i + 5i = -6 - 2i$

(c) $(-3 + 2i)^2 = (-3 + 2i)(-3 + 2i)$
$$= 9 - 6i - 6i + 4i^2$$
$$= 9 - 12i + 4(-1)$$
$$= 5 - 12i$$

(d) $\dfrac{17}{4 + i} = \dfrac{17}{4 + i} \cdot \dfrac{4 - i}{4 - i}$
$$= \frac{68 - 17i}{16 - i^2}$$
$$= \frac{68 - 17i}{17}$$
$$= 4 - i$$

Standard forms can be found using a calculator. See Figures 4.78 and 4.79.

Now Try Exercises 11, 13, 23, and 27 ◆

◆ **MAKING CONNECTIONS**

Complex, Real, and Imaginary Numbers   The following diagram illustrates the relationship among complex, real, and imaginary numbers, where a and b are real numbers. Note that complex numbers comprise two disjoint sets of numbers: the real numbers and the imaginary numbers.

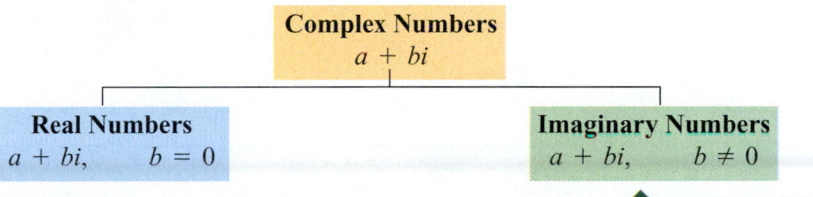

Quadratic Equations with Complex Solutions

We can use the quadratic formula to solve the quadratic equation $ax^2 + bx + c = 0$. If the discriminant, $b^2 - 4ac$, is negative, then there are no real solutions, and the graph of $y = ax^2 + bx + c$ does not intersect the x-axis. However, there are solutions that can be expressed as imaginary numbers. This is illustrated in the next example.

EXAMPLE 2   Solving quadratic equations with imaginary solutions

Solve the quadratic equation

(a) $\dfrac{1}{2}x^2 + 17 = 5x$ **(b)** $x^2 + 3x + 5 = 0$ **(c)** $2x^2 + 3 = 0$

SOLUTION

(a) Rewrite the equation as $\frac{1}{2}x^2 - 5x + 17 = 0$, and let $a = \frac{1}{2}$, $b = -5$, and $c = 17$ in the quadratic formula.

$$x = \frac{-b \pm \sqrt{b^2 - 4ac}}{2a}$$

$$= \frac{5 \pm \sqrt{(-5)^2 - 4(0.5)(17)}}{2(0.5)}$$

$$= 5 \pm \sqrt{-9}$$

$$= 5 \pm 3i$$

Algebra Review

To review the quadratic formula, see Section 3.2.

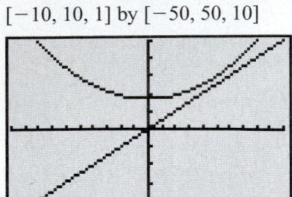

$[-10, 10, 1]$ by $[-50, 50, 10]$

FIGURE 4.80

Note that if we graph $Y_1 = .5X^2 + 17$ and $Y_2 = 5X$, as shown in Figure 4.80, their graphs do not intersect. This indicates that there are no real solutions. However, there are two complex solutions of $5 + 3i$ and $5 - 3i$.

(b) Let $a = 1$, $b = 3$, and $c = 5$ and apply the quadratic formula.

$$x = \frac{-b \pm \sqrt{b^2 - 4ac}}{2a}$$

$$= \frac{-3 \pm \sqrt{3^2 - 4(1)(5)}}{2(1)}$$

$$= \frac{-3 \pm \sqrt{-11}}{2}$$

$$= \frac{-3 \pm i\sqrt{11}}{2}$$

$$= -\frac{3}{2} \pm \frac{i\sqrt{11}}{2}$$

(c) Rather than use the quadratic formula for this equation, we apply the square root property because the equation contains no x-term.

$$2x^2 + 3 = 0 \qquad \text{Given equation}$$

$$2x^2 = -3 \qquad \text{Subtract 3.}$$

$$x^2 = -\frac{3}{2} \qquad \text{Divide by 2.}$$

$$x = \pm\sqrt{-\frac{3}{2}} \qquad \text{Square root property}$$

$$x = \pm i\sqrt{\frac{3}{2}} \qquad \sqrt{-a} = i\sqrt{a}$$

Now Try Exercises 45, 49, and 51 ◆

◆ **CLASS DISCUSSION**

What is the result if each expression is evaluated? (See Example 2(b).)

$$\left(-\frac{3}{2} + \frac{i\sqrt{11}}{2}\right)^2 + 3\left(-\frac{3}{2} + \frac{i\sqrt{11}}{2}\right) + 5$$

$$\left(-\frac{3}{2} - \frac{i\sqrt{11}}{2}\right)^2 + 3\left(-\frac{3}{2} - \frac{i\sqrt{11}}{2}\right) + 5 \ ◆$$

Fundamental Theorem of Algebra

One of the most brilliant mathematicians of all time, Carl Friedrich Gauss, at age 20 proved the fundamental theorem of algebra as part of his doctoral thesis. Although his theorem and proof were completed in 1797, they are still valid today.

FUNDAMENTAL THEOREM OF ALGEBRA

A polynomial $f(x)$ of degree $n \geq 1$ has at least one complex zero.

The fundamental theorem of algebra guarantees that every polynomial has a complete factorization, if we are allowed to use complex numbers. (A complex number can be written as $a + bi$. If $b = 0$, then the complex number $a + bi$ is also a real number.)

If $f(x)$ is a polynomial of degree 1 or higher, then by the fundamental theorem of algebra there is a zero c_1 such that $f(c_1) = 0$. By the factor theorem $(x - c_1)$ is a factor of $f(x)$ and

$$f(x) = (x - c_1)q_1(x)$$

for some polynomial $q_1(x)$. If $q_1(x)$ has positive degree, then by the fundamental theorem of algebra there exists a zero c_2 of $q_1(x)$. By the factor theorem $q_1(x)$ can be written as

$$q_1(x) = (x - c_2)q_2(x).$$

Then,

$$f(x) = (x - c_1)q_1(x)$$
$$= (x - c_1)(x - c_2)q_2(x).$$

If $f(x)$ has degree n, this process can be continued until $f(x)$ is written in the complete factored form

$$f(x) = a_n(x - c_1)(x - c_2) \cdots (x - c_n),$$

where a_n is the leading coefficient and the c_k are complex zeros of $f(x)$. If each c_k is distinct, then $f(x)$ has n zeros. However, in general the c_k may not be distinct since multiple zeros are possible.

NUMBER OF ZEROS THEOREM

A polynomial of degree n has at most n distinct zeros.

EXAMPLE 3 Classifying zeros

All zeros for the given polynomials are distinct. Use Figures 4.81–4.83 to determine graphically the number of real zeros and the number of imaginary zeros.

(a) $f(x) = 3x^3 - 3x^2 - 3x - 5$ **(b)** $g(x) = 2x^2 + x + 1$

(c) $h(x) = -x^4 + 4x^2 + 4$

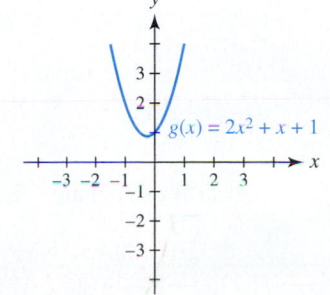

$f(x) = 3x^3 - 3x^2 - 3x - 5$

$g(x) = 2x^2 + x + 1$

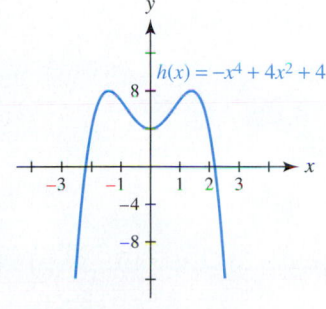

$h(x) = -x^4 + 4x^2 + 4$

FIGURE 4.81   **FIGURE 4.82**   **FIGURE 4.83**

SOLUTION

(a) The graph of $f(x)$ in Figure 4.81 crosses the x-axis once so there is one real zero. Since f is degree 3 and all zeros are distinct, there are two imaginary zeros.

(b) The graph of $g(x)$ in Figure 4.82 never crosses the x-axis. Since g is degree 2, there are no real zeros and two imaginary zeros.

(c) The graph of $h(x)$ is shown in Figure 4.83. Since h is degree 4, there are two real zeros and the remaining two zeros are imaginary.

Now Try Exercises 69, 71, and 73 ◆

EXAMPLE 4 Constructing a polynomial with prescribed zeros

Represent a polynomial $f(x)$ of degree 4 with a leading coefficient of 2 and zeros of $-3, 5$, i, and $-i$ in

(a) complete factored form, and   (b) expanded form.

SOLUTION

(a) Let $a_n = 2$, $c_1 = -3$, $c_2 = 5$, $c_3 = i$, and $c_4 = -i$. Then,

$$f(x) = 2(x + 3)(x - 5)(x - i)(x + i).$$

(b) To expand this expression for $f(x)$, perform the following.

$$2(x + 3)(x - 5)(x - i)(x + i) = 2(x + 3)(x - 5)(x^2 + 1)$$
$$= 2(x + 3)(x^3 - 5x^2 + x - 5)$$
$$= 2(x^4 - 2x^3 - 14x^2 - 2x - 15)$$
$$= 2x^4 - 4x^3 - 28x^2 - 4x - 30$$

Thus $f(x) = 2x^4 - 4x^3 - 28x^2 - 4x - 30$.

Now Try Exercise 81 ◆

EXAMPLE 5 Factoring a cubic polynomial with imaginary zeros

Determine the complete factored form for $f(x) = x^3 + 2x^2 + 4x + 8$.

Algebra Review

To review factoring by grouping, see Chapter R (page R-28).

SOLUTION We can use factoring by grouping to determine the complete factored form.

$$x^3 + 2x^2 + 4x + 8 = (x^3 + 2x^2) + (4x + 8) \qquad \text{Associative property}$$
$$= x^2(x + 2) + 4(x + 2) \qquad \text{Distributive property}$$
$$= (x^2 + 4)(x + 2) \qquad \text{Factor out } x + 2.$$

To factor $x^2 + 4$, first find its zeros.

$$x^2 + 4 = 0$$
$$x^2 = -4$$
$$x = \pm\sqrt{-4}$$
$$x = \pm 2i$$

The zeros of $f(x)$ are $-2, 2i$, and $-2i$. Its complete factored form is

$$f(x) = (x + 2)(x - 2i)(x + 2i).$$

Now Try Exercise 93 ◆

Notice that in Example 5 both $2i$ and $-2i$ were zeros of $f(x)$. The numbers $2i$ and $-2i$ are conjugates. This result can be generalized for any polynomial with real coefficients.

CONJUGATE ZEROS THEOREM

If a polynomial $f(x)$ has only real coefficients and if $a + bi$ is a zero of $f(x)$, then the conjugate $a - bi$ is also a zero of $f(x)$.

EXAMPLE 6 Constructing a polynomial with prescribed zeros

Find a cubic polynomial $f(x)$ with real coefficients, a leading coefficient of 2, and zeros of 3 and $5i$. Express f in
(a) complete factored form, and **(b)** expanded form.

SOLUTION
(a) Since $f(x)$ has real coefficients, it must also have a third zero of $-5i$, the conjugate of $5i$. Let $c_1 = 3$, $c_2 = 5i$, $c_3 = -5i$, and $a_n = 2$. The complete factored form is

$$f(x) = 2(x - 3)(x - 5i)(x + 5i).$$

(b) To expand $f(x)$ perform the following steps.

$$
\begin{aligned}
2(x - 3)(x - 5i)(x + 5i) &= 2(x - 3)(x^2 + 25) \\
&= 2(x^3 - 3x^2 + 25x - 75) \\
&= 2x^3 - 6x^2 + 50x - 150
\end{aligned}
$$

Now Try Exercise 83 ◆

EXAMPLE 7 Finding imaginary zeros of a polynomial

Find the zeros of $f(x) = x^4 + x^3 + 2x^2 + x + 1$, given that one zero is $-i$.

SOLUTION By the conjugate zeros theorem it follows that i must also be a zero of $f(x)$. Therefore $(x - i)$ and $(x + i)$ are factors of $f(x)$. Because $(x - i)(x + i) = x^2 + 1$, we can use long division to find another quadratic factor of $f(x)$.

$$
\begin{array}{r}
x^2 + x + 1 \\
x^2 + 0x + 1 \overline{\smash{\big)}\, x^4 + x^3 + 2x^2 + x + 1} \\
\underline{x^4 + 0x^3 + x^2 } \\
x^3 + x^2 + x \\
\underline{x^3 + 0x^2 + x } \\
x^2 + 0x + 1 \\
\underline{x^2 + 0x + 1} \\
0
\end{array}
$$

The quotient is $x^2 + x + 1$ with remainder 0. By the division algorithm,

$$x^4 + x^3 + 2x^2 + x + 1 = (x^2 + 1)(x^2 + x + 1).$$

We can use the quadratic formula to find the zeros of $x^2 + x + 1$.

$$x = \frac{-b \pm \sqrt{b^2 - 4ac}}{2a}$$

$$= \frac{-1 \pm \sqrt{1^2 - 4(1)(1)}}{2(1)}$$

$$= -\frac{1}{2} \pm i\frac{\sqrt{3}}{2}$$

The four zeros of $f(x)$ are $\pm i$ and $-\frac{1}{2} \pm i\frac{\sqrt{3}}{2}$.

Now Try Exercise 89 ◆

Polynomial Equations with Complex Solutions

Every polynomial equation of degree n can be written in the form

$$a_n x^n + \cdots + a_2 x^2 + a_1 x + a_0 = 0.$$

If we let $f(x) = a_n x^n + \cdots + a_2 x^2 + a_1 x + a_0$ and write $f(x)$ in complete factored form as

$$f(x) = a_n(x - c_1)(x - c_2) \cdots (x - c_n),$$

then the solutions to the polynomial equation are the zeros $c_1, c_2, \ldots, c_n$ of $f(x)$. Solving a polynomial equation with this technique is illustrated in the next two examples.

EXAMPLE 8   Solving a polynomial equation

Solve $x^3 = 3x^2 - 7x + 21$.

$[-5, 5, 1]$ by $[-30, 30, 10]$

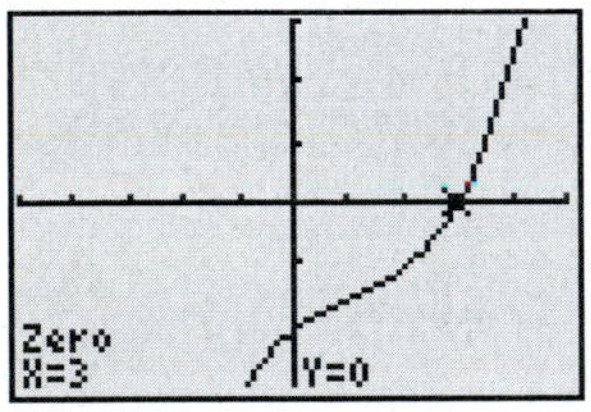

Zero
X=3   Y=0

FIGURE 4.84

SOLUTION Write the equation as $f(x) = 0$, where $f(x) = x^3 - 3x^2 + 7x - 21$. Although we could use factoring by grouping, as is done in Example 5, we use graphing instead to find one real zero of $f(x)$. Figure 4.84 shows that 3 is a zero of $f(x)$. By the factor theorem, $x - 3$ is a factor of $f(x)$. Using synthetic division, we divide $x - 3$ into $f(x)$.

$$
\begin{array}{r|rrrr}
3 & 1 & -3 & 7 & -21 \\
  &   & 3 & 0 & 21 \\
\hline
  & 1 & 0 & 7 & 0
\end{array}
$$

Thus, $x^3 - 3x^2 + 7x - 21 = (x - 3)(x^2 + 7)$ and we can solve as follows.

$x^3 - 3x^2 + 7x - 21 = 0$		$f(x) = 0$
$(x - 3)(x^2 + 7) = 0$		Factor.
$x - 3 = 0$ or $x^2 + 7 = 0$		Zero-product property
$x = 3$ or $x^2 = -7$		Solve.
$x = 3$ or $x = \pm i\sqrt{7}$		Property of i

The solutions are 3 and $\pm i\sqrt{7}$.

Now Try Exercise 101 ◆

EXAMPLE 9   Solving a polynomial equation

Solve $x^4 + x^2 = x^3$.

SOLUTION Write the equation as $f(x) = 0$, where $f(x) = x^4 - x^3 + x^2$.

$x^4 - x^3 + x^2 = 0$		$f(x) = 0$
$x^2(x^2 - x + 1) = 0$		Factor out x^2.
$x^2 = 0$ or $x^2 - x + 1 = 0$		Zero-product property

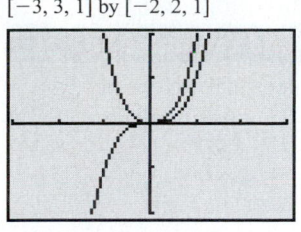

[−3, 3, 1] by [−2, 2, 1]

FIGURE 4.85

The only solution to $x^2 = 0$ is 0. To solve $x^2 - x + 1 = 0$, use the quadratic formula, as in Example 7. The solutions are 0 and $\frac{1}{2} \pm i\frac{\sqrt{3}}{2}$.

The graphs of $Y_1 = X^4 + X^2$ and $Y_2 = X^3$ are shown in Figure 4.85. Notice that they appear to intersect only at the origin. This indicates that the only real solution is 0.

Now Try Exercise 105 ◆

4.4   Putting it all Together

Some of the important topics in this section are summarized in the following table.

Concept	Explanation	Comments and Examples
Imaginary unit	$i = \sqrt{-1}, i^2 = 1$	The imaginary unit i allows us to define a new set of numbers called the complex numbers.
The expression $\sqrt{-a}$ with $a > 0$	$\sqrt{-a} = i\sqrt{a}$	$\sqrt{-4} = 2i$ $\sqrt{-5} = i\sqrt{5}$
Complex number	$a + bi$, where a and b are real numbers	Every real number is a complex number. $5 - 4i, 5, 2 + i$, and $-9i$ are examples of complex numbers.
Standard form of a complex number	$a + bi$, where a and b are real numbers	Converting to standard form: $$\frac{3 \pm 4i}{2} = \frac{3}{2} + 2i \quad \text{or} \quad \frac{3}{2} - 2i$$
Conjugates	The conjugate of $a + bi$ is $a - bi$.	*Number* *Conjugate* $5 - 6i$ $5 + 6i$ $12i$ $-12i$
Arithmetic operations on complex numbers	Complex numbers may be added, subtracted, multiplied, or divided.	$(2 + 3i) + (-3 - i) = -1 + 2i$ $(5 + i) - (3 - i) = 2 + 2i$ $(1 + i)(5 - i) = 5 - i + 5i - i^2$ $\qquad\qquad\qquad = 6 + 4i$ $\dfrac{3 + i}{1 - i} = \dfrac{(3 + i)(1 + i)}{(1 - i)(1 + i)} = 1 + 2i$
Number of zeros theorem	A polynomial of degree n has at most n distinct zeros.	The cubic polynomial, $$ax^3 + bx^2 + cx + d,$$ has *at most* 3 distinct zeros.

Concept	Explanation	Comments and Examples
Fundamental theorem of algebra	A polynomial of degree $n \geq 1$ has at least one complex zero.	This theorem guarantees that we can always factor a polynomial $f(x)$ into complete factored form: $$f(x) = a_n(x - c_1) \cdots (x - c_n),$$ where the c_k are complex numbers.
Conjugate zeros theorem	If a polynomial has real coefficients and $a + bi$ is a zero, then $a - bi$ is also a zero.	Since $\frac{1}{2} + \frac{1}{2}i$ is a zero of $2x^2 - 2x + 1$, it follows that $\frac{1}{2} - \frac{1}{2}i$ is also a zero.

4.4 — Exercises

Complex Numbers

Exercises 1–8: Simplify the expression using the imaginary unit i.

1. $\sqrt{-4}$

2. $\sqrt{-16}$

3. $\sqrt{-100}$

4. $\sqrt{-49}$

5. $\sqrt{-23}$

6. $\sqrt{-11}$

7. $\sqrt{-12}$

8. $\sqrt{-32}$

Exercises 9–36: Write the expression in standard form.

9. $3i + 5i$

10. $-7i + 5i$

11. $(3 + i) + (-5 - 2i)$

12. $(-4 + 2i) + (7 + 35i)$

13. $(12 - 7i) - (-1 + 9i)$

14. $2i - (-5 + 23i)$

15. $3 - (4 - 6i)$

16. $(7 + i) - (-8 + 5i)$

17. $(2)(2 + 4i)$

18. $(-5)(-7 + 3i)$

19. $(1 + i)(2 - 3i)$

20. $(-2 + i)(1 - 2i)$

21. $(-3 + 2i)(-2 + i)$

22. $(2 - 3i)(1 + 4i)$

23. $(-2 + 3i)^2$

24. $(2 - 3i)^2$

25. $2i(1 - i)^2$

26. $-i(5 - 2i)^2$

27. $\dfrac{1}{1 + i}$

28. $\dfrac{1 - i}{2 + 3i}$

29. $\dfrac{4 + i}{5 - i}$

30. $\dfrac{10}{1 - 4i}$

31. $\dfrac{2i}{10 - 5i}$

32. $\dfrac{3 - 2i}{1 + 2i}$

33. $\dfrac{3}{-i}$

34. $\dfrac{4 - 2i}{i}$

35. $\dfrac{-2 + i}{(1 + i)^2}$

36. $\dfrac{3}{(2 - i)^2}$

Exercises 37–42: Evaluate the expression with a calculator.

37. $(23 - 5.6i) + (-41.5 + 93i)$

38. $(-8.05 - 4.67i) + (3.5 + 5.37i)$

39. $(17.1 - 6i) - (8.4 + 0.7i)$

40. $\left(\frac{3}{4} - \frac{1}{10}i\right) - \left(-\frac{1}{8} + \frac{4}{25}i\right)$

41. $(-12.6 - 5.7i)(5.1 - 9.3i)$

42. $(7.8 + 23i)(-1.04 + 2.09i)$

Exercises 43 and 44: Perform the complex division using a calculator. Express your answer as a + bi, where a and b are rounded to the nearest thousandth.

43. $\dfrac{17 - 135i}{18 + 142i}$

44. $\dfrac{141 + 52i}{102 - 31i}$

Quadratic Equations with Complex Solutions

Exercises 45–62: Solve the quadratic equation. Write complex solutions in standard form.

45. $x^2 + 5 = 0$

46. $4x^2 + 3 = 0$

47. $5x^2 + 1 = 3x^2$

48. $x(3x + 1) = -1$

49. $3x = 5x^2 + 1$

50. $4x^2 = x - 1$

51. $x^2 - 4x + 5 = 0$

52. $2x^2 + x + 1 = 0$

53. $x^2 = 3x - 5$

54. $3x - x^2 = 5$

55. $6 = x^2 + 2x + 10$

56. $x(x - 4) = -8$

57. $3x^2 - 4x = x^2 - 3$

58. $2x^2 + 3 = 1 - x$

59. $2x(x - 2) = x - 4$

60. $3x^2 + x = x(5 - x) - 2$

61. $3x(3 - x) - 8 = x(x - 2)$

62. $-x(7 - 2x) = -6 - (3 - x)$

Zeros of Quadratic Polynomials

Exercises 63–68: The graph and the formula for a quadratic polynomial $f(x)$ are given.

    (a) Use the graph to predict the number of real zeros and the number of imaginary zeros.

    (b) Find these zeros using the quadratic formula.

63. $f(x) = 2x^2 - x - 3$

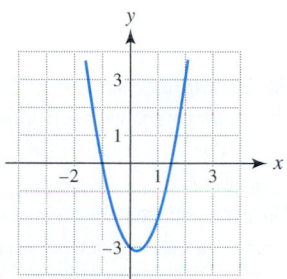

64. $f(x) = -x^2 + 4.6x - 5.29$

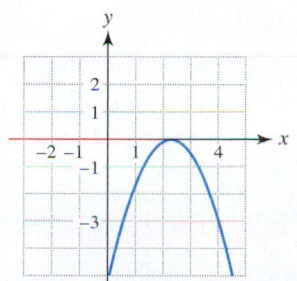

65. $f(x) = x^2 + x + 2$

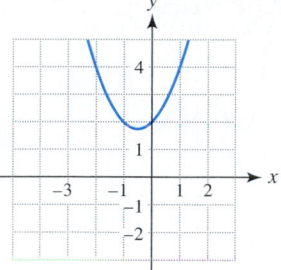

66. $f(x) = -2x^2 + 2x - 3$

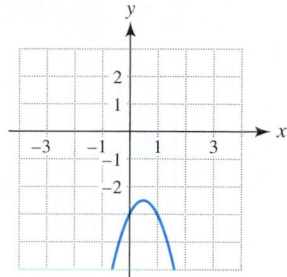

67. $f(x) = 5x^2 + 4x + 1$

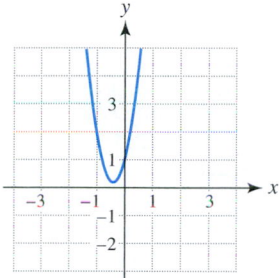

68. $f(x) = 9x^2 - 12x + 4$

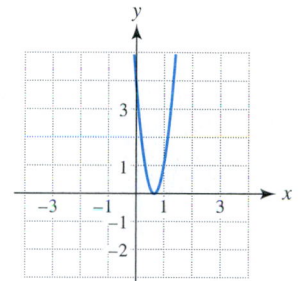

Zeros of Polynomials

Exercises 69–76: The graph and degree of a polynomial $f(x)$ are given. Determine the number of real zeros and the number of imaginary zeros. Assume that all zeros of $f(x)$ are distinct.

69. Degree 2

70. Degree 2

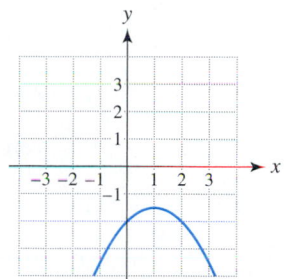

71. Degree 3

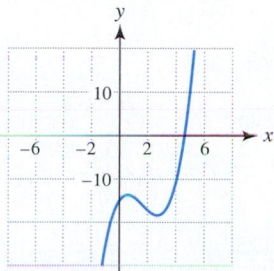

72. Degree 3

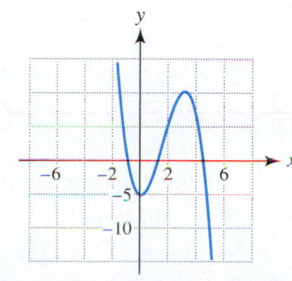

73. Degree 4

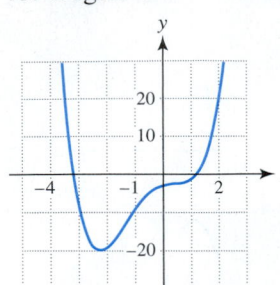

74. Degree 4

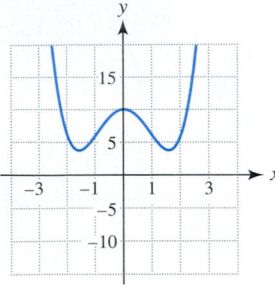

75. Degree 5

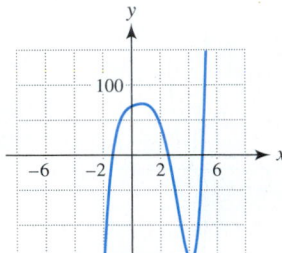

76. Degree 5

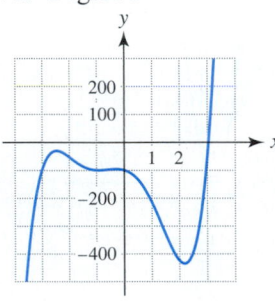

Exercises 77–86: Let a_n be the leading coefficient.

(a) *Find the complete factored form of a polynomial with real coefficients $f(x)$ that satisfies the given conditions.*

(b) *Express $f(x)$ in expanded form.*

77. Degree 2; $a_n = 1$;   zeros $6i$ and $-6i$

78. Degree 3; $a_n = 5$;   zeros 2, i, and $-i$

79. Degree 3; $a_n = -1$;   zeros -1, $2i$, and $-2i$

80. Degree 4; $a_n = 3$;   zeros -2, 4, i, and $-i$

81. Degree 4; $a_n = 10$;   zeros 1, -1, $3i$, and $-3i$

82. Degree 2; $a_n = -5$;   zeros $1 + i$ and $1 - i$

83. Degree 4; $a_n = \frac{1}{2}$;   zeros $-i$ and $2i$

84. Degree 3; $a_n = -\frac{3}{4}$;   zeros $-3i$ and $\frac{2}{5}$.

85. Degree 3; $a_n = -2$;   zeros $1 - i$ and 3.

86. Degree 4; $a_n = 7$;   zeros $2i$ and $3i$.

Exercises 87–90: (Refer to Example 7.) Find the zeros of $f(x)$ given that one zero is k.

87. $f(x) = 3x^3 - 5x^2 + 75x - 125$ $k = \frac{5}{3}$

88. $f(x) = x^4 + 2x^3 + 8x^2 + 8x + 16$ $k = 2i$

89. $f(x) = 2x^4 - x^3 + 19x^2 - 9x + 9$ $k = -3i$

90. $f(x) = 7x^3 + 5x^2 + 12x - 4$ $k = \frac{2}{7}$

Exercises 91–98: Complete the following.

(a) *Find all zeros of $f(x)$.*

(b) *Write the complete factored form of $f(x)$.*

91. $f(x) = x^2 + 25$ **92.** $f(x) = x^2 + 11$

93. $f(x) = 3x^3 + 3x$ **94.** $f(x) = 2x^3 + 10x$

95. $f(x) = x^4 + 5x^2 + 4$ **96.** $f(x) = x^4 + 4x^2$

97. $f(x) = x^4 + 2x^3 + x^2 + 8x - 12$

98. $f(x) = x^3 + 2x^2 + 16x + 32$

Exercises 99–110: Solve the polynomial equation.

99. $x^3 + x = 0$ **100.** $2x^3 - x + 1 = 0$

101. $x^3 = 2x^2 - 7x + 14$ **102.** $x^2 + x + 2 = x^3$

103. $x^4 + 5x^2 = 0$ **104.** $x^4 - 2x^3 + x^2 - 2x = 0$

105. $x^4 = x^3 - 4x^2$ **106.** $x^5 + 9x^3 = x^4 + 9x^2$

107. $x^4 + x^3 = 16 - 8x - 6x^2$

108. $x^4 + 2x^2 = x^3$

109. $3x^3 + 4x^2 + 6 = x$

110. $2x^3 + 5x^2 + x + 12 = 0$

Applications

*Exercises 111–116: **Electricity**   Complex numbers are used in the study of electrical circuits, such as the current found in a household outlet. Impedance Z (or the opposition to the flow of electricity), voltage V, and current I can all be represented by complex numbers. They are related by the equation $Z = \frac{V}{I}$. Use this equation to find the value of the missing variable.*

111. $V = 50 + 98i$ $I = 8 + 5i$

112. $V = 30 + 60i$ $I = 8 + 6i$

113. $I = 1 + 2i$ $Z = 3 - 4i$

114. $I = \frac{1}{2} + \frac{1}{4}i$ $Z = 8 - 9i$

115. $Z = 22 - 5i$ $V = 27 + 17i$

116. $Z = 10 + 5i$ $V = 10 + 8i$

Writing about Mathematics

117. Could a cubic function have only imaginary zeros? Explain.

118. Give an example of a polynomial function that has only imaginary zeros and a polynomial function that has only real zeros. Explain how to determine graphically if a function has only imaginary zeros.

EXTENDED AND DISCOVERY EXERCISE

1. The properties of the imaginary unit are $i = \sqrt{-1}$ and $i^2 = -1$.

(a) Begin simplifying the expressions $i, i^2, i^3, i^4, i^5, \ldots$, until a pattern is discovered. For example, $i^3 = i \cdot i^2 = i \cdot (-1) = -i$.

(b) Summarize your findings by describing how to simplify i^n for any natural number n.

CHECKING BASIC CONCEPTS FOR SECTIONS 4.3 AND 4.4

1. Simplify the expression $\dfrac{5x^4 - 10x^3 + 5x^2}{5x^2}$.

2. Divide the expression.

(a) $\dfrac{x^3 - x^2 + 4x - 4}{x - 1}$ **(b)** $\dfrac{2x^3 - 3x^2 + 4x + 4}{2x + 1}$

(c) $\dfrac{x^4 - 3x^3 + 6x^2 - 13x + 9}{x^2 + 4}$

3. Use the graph of the cubic polynomial $f(x)$ to determine its complete factored form. State the multiplicity of each zero. Assume that all zeros are integers and that the leading coefficient is *not* ± 1.

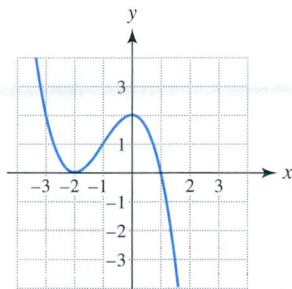

4. Solve $x^3 - 2x^2 - 15x = 0$.

5. Determine graphically the zeros of

$$f(x) = x^4 - x^3 - 18x^2 + 16x + 32.$$

Write $f(x)$ in complete factored form.

6. Find a quadratic polynomial $f(x)$ with zeros $\pm 4i$ and leading coefficient 3. Write $f(x)$ in complete factored form and expanded form.

7. Sketch a graph of a quartic function (degree 4) with a negative leading coefficient, two real zeros, and two imaginary zeros.

8. Write $x^3 - x^2 + 4x - 4$ in complete factored form.

9. Solve each equation.

(a) $2x^3 + 45 = 5x^2 - 18x$ **(b)** $3x^2 - x + 2 = 0$

(c) $x^4 + 5x^2 = 36$

4.5   Rational Functions and Models

- ♦ Identify a rational function and state its domain
- ♦ Find and interpret vertical asymptotes
- ♦ Find and interpret horizontal asymptotes
- ♦ Solve rational equations
- ♦ Solve applications involving rational equations
- ♦ Solve applications involving variation

Introduction

Rational functions are nonlinear functions that frequently occur in applications. For example, rational functions are used to design curves for railroad tracks, determine stopping distances on hills, and calculate the average number of people waiting in a line.

Rational Functions

A rational number can be expressed as a ratio $\frac{p}{q}$, where p and q are integers with $q \neq 0$. A rational function is defined similarly by using the concept of a polynomial.

Algebra Review
To review rational expressions, see Chapter R (page R-38).

RATIONAL FUNCTION

A function f represented by $f(x) = \frac{p(x)}{q(x)}$, where $p(x)$ and $q(x)$ are polynomials and $q(x) \neq 0$, is a **rational function**.

The domain of a rational function includes all real numbers *except* the zeros of the denominator $q(x)$. The graph of a rational function is continuous except at x-values where $q(x) = 0$.

EXAMPLE 1 Identifying rational functions

Determine if the function is rational and state its domain.

(a) $f(x) = \dfrac{2x - 1}{x^2 + 1}$ **(b)** $g(x) = \dfrac{1}{\sqrt{x}}$ **(c)** $h(x) = \dfrac{x^3 - 2x^2 + 1}{x^2 - 3x + 2}$

SOLUTION

(a) Both the numerator, $2x - 1$, and the denominator, $x^2 + 1$, are polynomials so f is a rational function. The domain of f includes all real numbers because $x^2 + 1 \neq 0$ for any real number x.

(b) Since $\sqrt{x}$ is not a polynomial, g is not a rational function. The domain of g is $\{x \mid x > 0\}$.

(c) Both the numerator and the denominator are polynomials, so h is a rational function. Because

$$x^2 - 3x + 2 = (x - 1)(x - 2) = 0$$

when $x = 1$ or $x = 2$, the domain of h is $\{x \mid x \neq 1, x \neq 2\}$.

Now Try Exercises 1 and 9 ◆

◆ **CLASS DISCUSSION**

Is an integer a rational number? Is a polynomial function a rational function? ◆

Vertical Asymptotes

If cars leave a parking ramp randomly and stop to pay the parking attendant on the way out, then the average length of the line depends on two factors: the average traffic rate at which cars are exiting the ramp and the average rate at which the parking attendant can wait on cars. For instance, if the average traffic rate is three cars per minute and the parking attendant can serve four cars per minute, then at times a line may form if cars arrive in a random manner. The **traffic intensity** x is the ratio of the average traffic rate to the average working rate of the attendant. In this example, $x = \frac{3}{4}$. (**Source:** F. Mannering and W. Kilareski, *Principles of Highway Engineering and Traffic Control.*)

EXAMPLE 2 Estimating the length of parking ramp lines

If the traffic intensity is x, then the average number of cars waiting in line to exit a parking ramp can be computed by $N(x) = \frac{x^2}{2 - 2x}$, where $0 \leq x < 1$.

(a) Evaluate $N(0.5)$ and $N(0.9)$. Interpret the results.

(b) Use the graph of $y = N(x)$ shown in Figure 4.86 to explain what happens to the length of the line as the traffic intensity x approaches 1 (or $x \to 1$.)

SOLUTION

(a) $N(\mathbf{0.5}) = \frac{\mathbf{0.5}^2}{2 - 2(\mathbf{0.5})} = 0.25$ and $N(\mathbf{0.9}) = \frac{\mathbf{0.9}^2}{2 - 2(\mathbf{0.9})} = 4.05$. This means that if the traffic intensity is 0.5, there is little waiting in line. As the traffic intensity increases to 0.9, the average line has more than four cars.

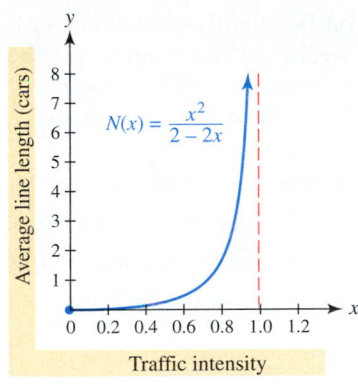

$N(x) = \dfrac{x^2}{2 - 2x}$

Average line length (cars)

Traffic intensity

FIGURE 4.86 Parking Ramp Lines

(b) As the traffic intensity x approaches 1 from the left, the graph of f increases rapidly without bound. Numerical support is given in Table 4.8. With a traffic intensity slightly less than 1, the attendant has difficulty keeping up. If cars occasionally arrive in groups, long lines will form. At $x = 1$ the denominator, $2 - 2x$, equals 0 and $N(x)$ is undefined.

TABLE 4.8

x	0.94	0.95	0.96	0.97	0.98	0.99	1
$\dfrac{x^2}{2 - 2x}$	7.36	9.03	11.52	15.68	24.01	49.01	—

Now Try Exercise 101

In Figure 4.86, the red vertical line $x = 1$ is a *vertical asymptote* of the graph of f. A graph of a different rational function f is shown in Figure 4.87. The $f(x)$-values *decrease without bound* as x approaches 2 from the left. This is denoted $f(x) \to -\infty$ **as** $x \to 2^-$. Similarly, the $f(x)$-values *increase without bound* as x approaches 2 from the right. This is expressed as $f(x) \to \infty$ **as** $x \to 2^+$. The line $x = 2$ is a vertical asymptote of the graph of f.

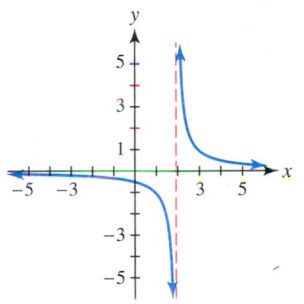

FIGURE 4.87

VERTICAL ASYMPTOTE

The line $x = k$ is a **vertical asymptote** of the graph of f if $f(x) \to \infty$ or $f(x) \to -\infty$, as x approaches k from either the left or the right.

If $x = k$ is a vertical asymptote of the graph of f, then k is not in the domain of f. Furthermore, the graph of a rational function f does *not* cross a vertical asymptote.

Horizontal Asymptotes

Probability sometimes can be described with rational functions. Suppose that a container holds ten identical balls numbered 1 through 10, where one ball has the winning number. Then the probability or likelihood of randomly drawing the winning ball is 1 in 10. This probability can be expressed as the ratio $\frac{1}{10}$. Probabilities vary between 0 and 1, where 1 represents an event that is certain to occur.

If there are x balls, the probability of randomly drawing the winning ball is given by $P(x) = \frac{1}{x}$, where x is a natural number. A graph of P is shown in Figure 4.88. As the number of balls increases, the probability of drawing the winning ball decreases toward 0, without becoming 0. Even if there are one million balls in the container, there still is a slight chance of drawing the winning ball. As a result, the graph of P comes closer and closer to the x-axis ($y = 0$) without ever actually touching it. The x-axis is a *horizontal asymptote* of the graph of P.

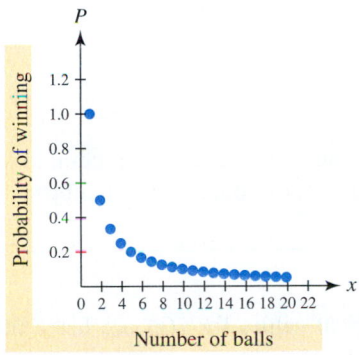

Probability of winning

Number of balls

FIGURE 4.88

HORIZONTAL ASYMPTOTE

The line $y = b$ is a **horizontal asymptote** of the graph of f, if $f(x) \to b$ as x approaches either ∞ or $-\infty$.

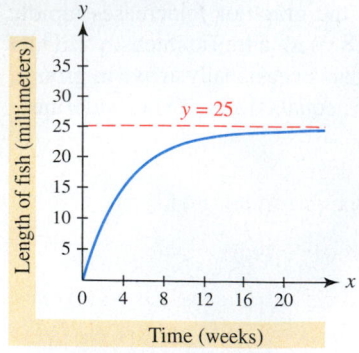

FIGURE 4.89 Size of a Small Fish

The graph of f in Figure 4.89 is an example of a von Bertalanffy growth curve. It models the length in millimeters of a small fish after x weeks. After several weeks the length of the fish begins to level off near 25 millimeters. Thus $y = 25$ is a horizontal asymptote of the graph of f. This is denoted by $f(x) \to 25$ as $x \to \infty$. (**Source:** D. Brown and P. Rothery, *Models in Biology: Mathematics, Statistics and Computing*.)

Another example of a function whose graph has a horizontal asymptote is given by $f(x) = \frac{x^2}{x^2 + 1}$. The horizontal asymptote is $y = 1$ because as $x \to \infty, y \to 1$. This is shown numerically in Table 4.9. Note that numerator in the ratio $\frac{x^2}{x^2 + 1}$ is always 1 less than the denominator and that as x becomes large, the ratio approaches 1. Similarly, it can also be shown that as $x \to -\infty, y \to 1$.

TABLE 4.9

x	1	5	10	50	100	500
$\frac{x^2}{x^2 + 1}$	$\frac{1}{2}$	$\frac{25}{26}$	$\frac{100}{101}$	$\frac{2500}{2501}$	$\frac{10,000}{10,001}$	$\frac{250,000}{250,001}$

A graph of $y = \frac{x^2}{x^2 + 1}$ is shown in Figure 4.90 with the horizontal asymptote $y = 1$ shown in red. Note that an asymptote is not part of the graph of a function; rather, it is an aid that can be used to sketch a graph of a function.

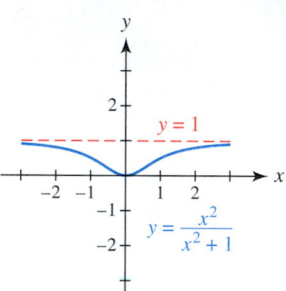

FIGURE 4.90

Identifying Asymptotes

The graph in Figure 4.88 limits the domain of P to natural numbers because a fraction of a ball is not allowed. However, the domain of $f(x) = \frac{1}{x}$ is the set of nonzero real numbers.

EXAMPLE 3

Analyzing the graph of $f(x) = \frac{1}{x}$

The graph of $f(x) = \frac{1}{x}$ is shown in Figure 4.91. Numerical values for $f(x)$ are listed in Tables 4.10 and 4.11. Relate these values to the graph of f.

TABLE 4.10

x	0	0.001	0.01	0.1	1	10	100	1000
$\frac{1}{x}$	—	1000	100	10	1	0.1	0.01	0.001

TABLE 4.11

x	-1000	-100	-10	-1	-0.1	-0.01	-0.001	0
$\frac{1}{x}$	-0.001	-0.01	-0.1	-1	-10	-100	-1000	—

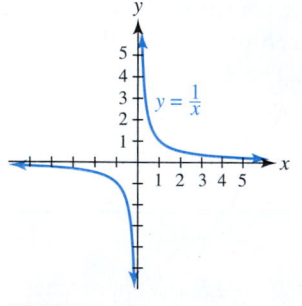

FIGURE 4.91

SOLUTION Table 4.10 shows positive values of x. As x approaches 0 from the right, the $f(x)$-values increase without bound. This is denote $\boldsymbol{f(x) \to \infty}$ **as** $\boldsymbol{x \to 0^+}$. Slightly right of the y-axis there are large y-values on the graph of $f(x) = \frac{1}{x}$. The y-axis $(x = 0)$ is a vertical asymptote.

As x assumes large positive values in Table 4.10, the $f(x)$-values decrease and tend toward 0. This is expressed as $\boldsymbol{f(x) \to 0}$ **as** $\boldsymbol{x \to \infty}$. As a result, the graph of f levels off above the x-axis. The x-axis $(y = 0)$ is a horizontal asymptote.

Table 4.11 shows negative values of x. As x approaches 0 from the left, the $f(x)$-values decrease without bound. This is denoted by $\boldsymbol{f(x) \to -\infty}$ **as** $\boldsymbol{x \to 0^-}$. The graph of $f(x) = \frac{1}{x}$ falls rapidly just left of the y-axis, indicating that the y-axis $(x = 0)$ is a vertical asymptote.

When x decreases in Table 4.11, the $f(x)$-values are negative and tend toward 0. This is denoted by $\boldsymbol{f(x) \to 0}$ **as** $\boldsymbol{x \to -\infty}$. The graph of f levels off below the x-axis $(y = 0)$, which is a horizontal asymptote.   *Now Try Exercise 13* ◆

Transformations of graphs can be used to graph some types of rational functions by hand. For example, we can graph $g(x) = \frac{1}{x-1} + 2$ by translating the graph of $f(x) = \frac{1}{x}$ right 1 unit and upward 2 units. That is, $g(x)$ can be written in terms of $f(x)$ by using the formula $g(x) = f(x-1) + 2$. Because the graph of f in Figure 4.91 has vertical asymptote $x = 0$ and horizontal asymptote $y = 0$, the graph of g in Figure 4.92 has vertical asymptote $x = 1$ and horizontal asymptote $y = 2$.

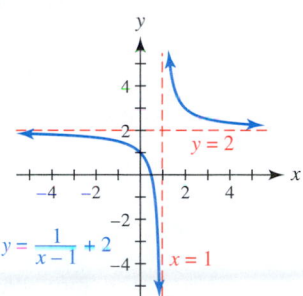

FIGURE 4.92

EXAMPLE 4

Graphing a rational function

Use the graph of $f(x) = \frac{1}{x^2}$ shown in Figure 4.93 to sketch a graph of $g(x) = -\frac{1}{(x+2)^2}$. Include all asymptotes in your graph. Write $g(x)$ in terms of $f(x)$.

SOLUTION Note that the graph of $y = \frac{1}{x^2}$ has vertical asymptote $x = 0$ and horizontal asymptote $y = 0$. The graph of $g(x) = -\frac{1}{(x+2)^2}$ is a translation of the graph of $f(x) = \frac{1}{x^2}$ left 2 units and then a reflection across the x-axis. The vertical asymptote for $y = g(x)$ is $x = -2$ and the horizontal asymptote is $y = 0$, as shown in Figure 4.94. We can write $g(x)$ in terms of $f(x)$ as $g(x) = -f(x+2)$.

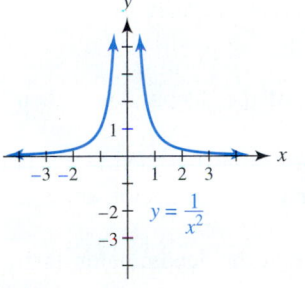

FIGURE 4.93

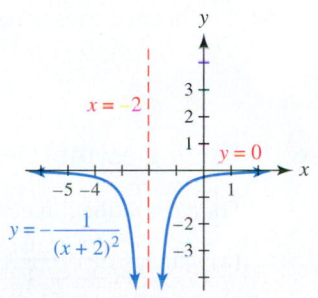

FIGURE 4.94   *Now Try Exercise 45* ◆

EXAMPLE 5 Determining horizontal and vertical asymptotes visually

Use the graph of each rational function to determine any vertical or horizontal asymptotes.

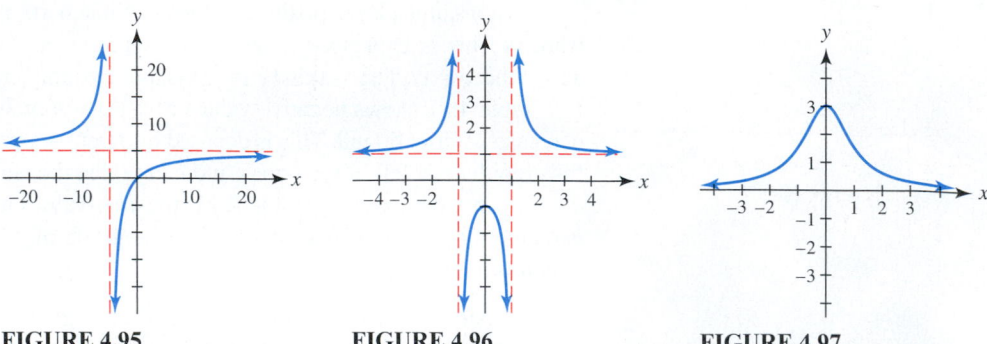

FIGURE 4.95   FIGURE 4.96   FIGURE 4.97

SOLUTION In Figure 4.95, $y = 5$ is a horizontal asymptote and $x = -5$ is a vertical asymptote. In Figure 4.96, $x = \pm 1$ are vertical asymptotes and $y = 1$ is a horizontal asymptote. In Figure 4.97 there are no vertical asymptotes. The x-axis ($y = 0$) is a horizontal asymptote.

Now Try Exercises 15, 17, and 19 ◆

The following can be used to find vertical and horizontal asymptotes.

FINDING VERTICAL AND HORIZONTAL ASYMPTOTES

Let f be a rational function given by $f(x) = \frac{p(x)}{q(x)}$ written in lowest terms.

Vertical Asymptote

To find a vertical asymptote, set the denominator, $q(x)$, equal to 0 and solve. If k is a zero of $q(x)$, then $x = k$ is a vertical asymptote. *Caution:* If k is a zero of both $q(x)$ *and* $p(x)$, then $f(x)$ is *not* written in lowest terms, and $x - k$ is a common factor.

Horizontal Asymptote

(a) If the degree of the numerator is less than the degree of the denominator, then $y = 0$ (the x-axis) is a horizontal asymptote.

(b) If the degree of the numerator equals the degree of the denominator, then $y = \frac{a}{b}$ is a horizontal asymptote, where a is the leading coefficient of the numerator, and b is the leading coefficient of the denominator.

(c) If the degree of the numerator is greater than the degree of the denominator, then there are no horizontal asymptotes.

EXAMPLE 6 Finding asymptotes

For each rational function, determine any horizontal or vertical asymptotes.

(a) $f(x) = \dfrac{6x - 1}{3x + 3}$ **(b)** $g(x) = \dfrac{x + 1}{x^2 - 4}$ **(c)** $h(x) = \dfrac{x^2 - 1}{x + 1}$

SOLUTION

(a) The degrees of the numerator and the denominator are both 1. Since the ratio of the leading coefficients is $\frac{6}{3} = 2$, the graph of f has a horizontal asymptote of $y = 2$. This is supported numerically in Figures 4.98 and 4.99, where the y-values approach 2 as the x-values increase or decrease.

Calculator Help

To make a table of values, see Appendix B (page AP-9).

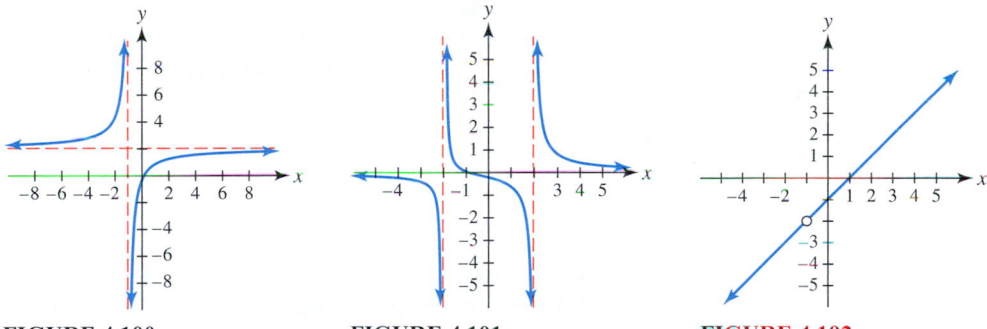

FIGURE 4.98 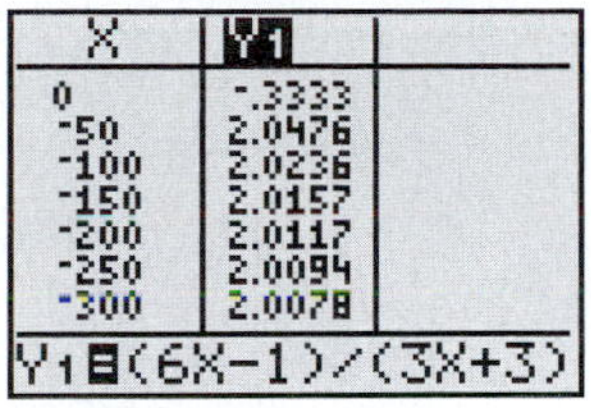   **FIGURE 4.99**

When $x = -1$ the denominator, $3x + 3$, equals 0 and the numerator, $6x + 1$, does not equal 0. Thus, $x = -1$ is a vertical asymptote. A graph of f is shown in Figure 4.100.

FIGURE 4.100   **FIGURE 4.101**   **FIGURE 4.102**

(b) The degree of the numerator is one less than the degree of the denominator, so the x-axis, or $y = 0$, is a horizontal asymptote. The denominator, $x^2 - 4$, equals 0 and the numerator, $x + 1$, does not equal 0 when $x = \pm 2$. Thus, $x = \pm 2$ are vertical asymptotes. See Figure 4.101. Note that the graph crosses the horizontal asymptote.

(c) The degree of the numerator is greater than the degree of the denominator so there are no horizontal asymptotes. When $x = -1$, both numerator and denominator equal 0. We can simplify $h(x)$ as follows.

Algebra Review

To review simplifying rational expressions, see Chapter R (page R-38).

$$h(x) = \frac{x^2 - 1}{x + 1} = \frac{(x + 1)(x - 1)}{x + 1} = x - 1, \qquad x \neq -1$$

The graph of $h(x)$ is the line $y = x - 1$ with the point $(-1, -2)$ missing. There are no vertical asymptotes. See Figure 4.102.   *Now Try Exercises 21, 23, and 31* ◆

EXAMPLE 7   Graphing a rational function having a "hole"

Graph $f(x) = \frac{2x^2 - 5x + 2}{x^2 - 3x + 2}$ by hand.

SOLUTION

First factor the numerator and the denominator.

$$\frac{2x^2 - 5x + 2}{x^2 - 3x + 2} = \frac{(2x - 1)(x - 2)}{(x - 1)(x - 2)} = \frac{2x - 1}{x - 1}, \qquad x \neq 2$$

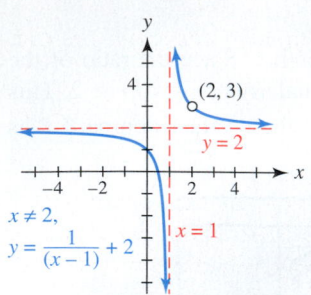

FIGURE 4.103

Thus $f(x) = \frac{2x - 1}{x - 1}$ provided $x \neq 2$, and there is a "hole," or point missing, in the graph of f when $x = 2$. The horizontal asymptote is $y = 2$ and the vertical asymptote is $x = 1$. Long division can be used to rewrite $f(x) = \frac{2x - 1}{x - 1}$.

$$
\begin{array}{r}
2 \\
x - 1 \overline{\smash{)}2x - 1} \\
\underline{2x - 2} \\
1
\end{array}
$$

$\frac{2x}{x} = 2$

$2(x - 1) = 2x - 2$

Subtract: $-1 - (-2) = 1$

Thus $f(x) = 2 + \frac{1}{x - 1}$ and its graph is similar to Figure 4.92 except that the point $(2, 3)$ is missing and an open circle appears in its place. See Figure 4.103.

Now Try Exercise 53 ◆

Calculator Help

To set dot mode, see Appendix B (page AP-13); to set a decimal window, see Appendix B (page AP-18).

Note: Calculators typically graph in connected or dot mode. (See Example 8.) If connected mode is used when graphing a rational function, it may appear as though the calculator is graphing vertical asymptotes automatically. However, in most instances the calculator is connecting points inappropriately.

Sometimes rational functions can be graphed in connected mode using a *decimal* or *friendly* viewing rectangle. A decimal window is used in Example 9.

EXAMPLE 8   **Analyzing a rational function with technology**

Let $f(x) = \frac{2x^2 + 1}{x^2 - 4}$.

(a) Use a calculator to graph f. Find the domain of f.
(b) Identify any vertical or horizontal asymptotes.
(c) Sketch a graph of f that includes the asymptotes.

$[-6, 6, 1]$ by $[-6, 6, 1]$

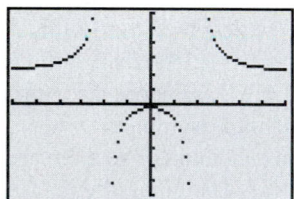

FIGURE 4.104

SOLUTION

(a) A calculator graph of f is shown in Figure 4.104 using dot mode. The function is undefined when $x^2 - 4 = 0$, or when $x = \pm 2$. The domain of f is given by $D = \{x \mid x \neq 2, x \neq -2\}$.
(b) When $x = \pm 2$ the denominator, $x^2 - 4$, equals 0 and the numerator, $2x^2 + 1$, does not equal 0. Therefore, $x = \pm 2$ are vertical asymptotes. The degree of the numerator equals the degree of the denominator, and the ratio of the leading coefficients is $\frac{2}{1} = 2$. A horizontal asymptote of the graph of f is $y = 2$.
(c) A second graph of f and its asymptotes is shown in Figure 4.105.

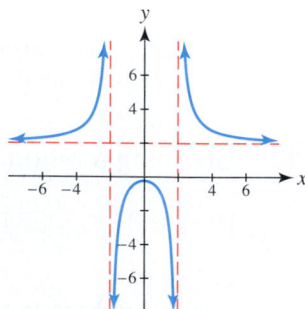

FIGURE 4.105

Now Try Exercise 59 ◆

A third type of asymptote, which is neither vertical nor horizontal, occurs when the numerator of a rational function has degree *one more* than the degree of the denominator.

For example, let $f(x) = \frac{x^2 + 2}{x - 1}$. If $x - 1$ is divided into $x^2 + 2$, the quotient is $x + 1$ with remainder 3. Thus

$$f(x) = x + 1 + \frac{3}{x - 1}$$

is an equivalent representation of f. For large values of $|x|$ the ratio $\frac{3}{x - 1}$ approaches 0 and the graph of f approaches $y = x + 1$. The line $y = x + 1$ is called a **slant asymptote** or **oblique asymptote** of the graph of f. A graph of f with vertical asymptote $x = 1$ and slant asymptote $y = x + 1$ is shown in Figure 4.106.

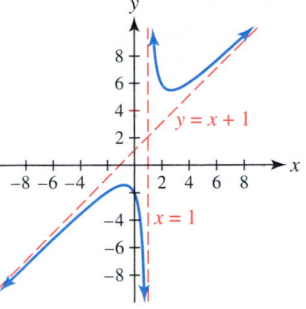

FIGURE 4.106

Rational Equations

A rational equation can be written in the form $f(x) = 0$, where f is a rational function. Rational equations often contain rational expressions. Examples of rational equations include

$$\frac{x^2 - 1}{x^2 + x + 3} = 0, \qquad \frac{3x}{x^3 + x} = 5, \qquad \text{and} \qquad \frac{1}{x - 1} + \frac{1}{x} = 5.$$

Rational equations can be solved symbolically, graphically, and numerically.

EXAMPLE 9 Solving a rational equation

Solve $\frac{4x}{x - 1} = 6$ symbolically, graphically, and numerically.

SOLUTION

Symbolic Solution

$\dfrac{4x}{x - 1} = 6$	Given equation
$4x = 6(x - 1)$	Cross multiply.
$4x = 6x - 6$	Distributive property
$-2x = -6$	Subtract 6x.
$x = 3$	Divide by −2.

Graphical Solution Graph $Y_1 = 4X/(X - 1)$ and $Y_2 = 6$. Their graphs intersect at $(3, 6)$, so the solution is **3**. See Figure 4.107.

$[-9.4, 9.4, 1]$ by $[-9.4, 9.4, 1]$

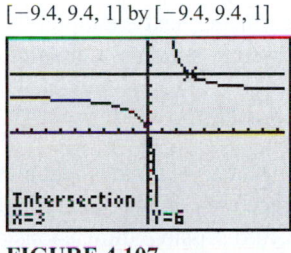

FIGURE 4.107   **FIGURE 4.108**

Numerical Solution In Figure 4.108, $y_1 = y_2$ when $x = 3$.   *Now Try Exercise 79* ◆

A common approach to solve rational equations symbolically is to multiply each side of the equation by a common denominator. This technique clears fractions from an equation and is used in Examples 10 and 11.

 **Solving a rational equation**

Solve $\frac{6}{x^2} - \frac{5}{x} = 1$ symbolically.

SOLUTION

Algebra Review
To review clearing fractions,
see Chapter R (page R-44).

$$\frac{6}{x^2} - \frac{5}{x} = 1 \qquad \text{Given equation}$$

$$\frac{6}{x^2} \cdot x^2 - \frac{5}{x} \cdot x^2 = 1 \cdot x^2 \qquad \text{Multiply each term by } x^2.$$

$$6 - 5x = x^2 \qquad \text{Simplify.}$$

$$0 = x^2 + 5x - 6 \qquad \text{Add 5x and subtract 6.}$$

$$0 = (x + 6)(x - 1) \qquad \text{Factor.}$$

$$x + 6 = 0 \quad \text{or} \quad x - 1 = 0 \qquad \text{Zero-product property}$$

$$x = -6 \quad \text{or} \quad x = 1 \qquad \text{Solve.} \qquad \textit{Now Try Exercise 95} \blacklozenge$$

The next example illustrates the importance of checking possible solutions when solving rational equations.

 **Solving a rational equation**

Solve $\frac{1}{x + 3} + \frac{1}{x - 3} = \frac{6}{x^2 - 9}$ symbolically. Check the result.

SOLUTION

$$\frac{1}{x + 3} + \frac{1}{x - 3} = \frac{6}{x^2 - 9} \qquad \text{Given equation}$$

$$\frac{(x + 3)(x - 3)}{x + 3} + \frac{(x + 3)(x - 3)}{x - 3} = \frac{6(x + 3)(x - 3)}{x^2 - 9} \qquad \begin{array}{l}\text{Multiply by} \\ (x + 3)(x - 3).\end{array}$$

$$(x - 3) + (x + 3) = 6 \qquad \text{Reduce.}$$

$$2x = 6 \qquad \text{Combine terms.}$$

$$x = 3 \qquad \text{Divide by 2.}$$

When 3 is substituted for x, two expressions in the given equation are undefined. There are no solutions. $\qquad \textit{Now Try Exercise 99} \blacklozenge$

Rational equations are used in real-world applications such as in construction, which is discussed in the next example. Steps for solving application problems (see page 114) have been used to structure our solution.

 **Designing a box**

A box with rectangular sides and a top is being designed to hold 324 cubic inches and to have a surface area of 342 square inches. If the length of the box is four times longer than the height, find possible dimensions of the box.

SOLUTION

STEP 1: We are asked to find the dimensions of a box. If x is the height of the box, and y is the width, then the length of the box is four times longer than the height, or $4x$.

Geometry Review
To review formulas related to box shapes, see Chapter R (page R-3).

x: Height of the box

y: Width of the box

$4x$: Length of the box

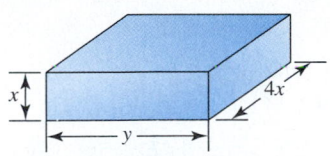

FIGURE 4.109

STEP 2: To relate these variables to an equation, sketch a box as illustrated in Figure 4.109. The volume V of the box is height times width times length.

$$V = xy(4x)$$
$$= 4x^2y$$

The surface area A of this box is determined by finding the area of the 6 rectangular sides: left and right sides, front and back, and top and bottom.

$$A = 2(4x \cdot x) + 2(x \cdot y) + 2(4x \cdot y)$$
$$= 8x^2 + 10xy.$$

If we solve $V = 4x^2y$ for y, and let $V = 324$, we obtain

$$y = \frac{V}{4x^2} = \frac{324}{4x^2} = \frac{81}{x^2}.$$

Substituting $y = \frac{81}{x^2}$ into the formula for A eliminates the y variable.

$$A = 8x^2 + 10xy \qquad \textcolor{blue}{\text{Area formula}}$$

$$= 8x^2 + 10x \cdot \frac{81}{x^2} \qquad \textcolor{blue}{\text{Let } y = \frac{81}{x^2}.}$$

$$= 8x^2 + \frac{810}{x} \qquad \textcolor{blue}{\text{Simplify.}}$$

Since the surface area is $A = 342$ square inches, the height x can be determined by solving the rational equation

$$8x^2 + \frac{810}{x} = 342.$$

STEP 3: Figures 4.110 and 4.111 show the graphs of $Y_1 = 8X^2 + 810/X$ and $Y_2 = 342$. There are two solutions, $x = 3$ and $x = 4.5$.

Note: This equation can be written as $8x^3 - 342x + 810 = 0$ and the rational zeros test or factoring could be used to find the solutions, 3 and $\frac{9}{2}$, if technology is not available.

Calculator Help

To find a point of intersection, see Appendix B (page AP-12).

[0, 6, 1] by [300, 400, 20]   [0, 6, 1] by [300, 400, 20]

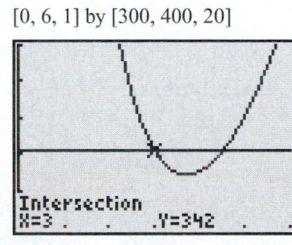

```
Intersection
X=3        Y=342
```

FIGURE 4.110

```
Intersection
X=4.5      Y=342
```

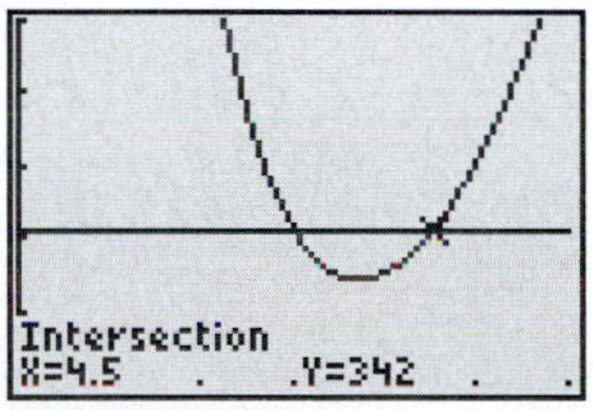

FIGURE 4.111

If the height is $x = 3$ inches, then the length is $4 \cdot 3 = 12$ inches, and the width is $y = \frac{81}{3^2} = 9$ inches. (Note that $y = \frac{81}{x^2}$.) If the height is 4.5 inches, then the length is $4 \cdot 4.5 = 18$ inches, and the width is $y = \frac{81}{4.5^2} = 4$ inches. Thus the dimensions of the box in inches can be either $3 \times 9 \times 12$ or $4.5 \times 4 \times 18$.

STEP 4: We can check our results directly. If the dimensions are $3 \times 9 \times 12$, then

$$V = 3 \cdot 9 \cdot 12 = 324 \qquad \text{and}$$
$$S = 2(3 \cdot 9) + 2(3 \cdot 12) + 2(9 \cdot 12) = 342.$$

If the dimensions are 4.5 × 4 × 18, then

$$V = 4.5 \cdot 4 \cdot 18 = 324 \qquad \text{and}$$
$$S = 2(4.5 \cdot 4) + 2(4.5 \cdot 18) + 2(4 \cdot 18) = 342.$$

In both cases our results check.

Now Try Exercise 107 ◆

Variation

In Section 2.2 direct variation was discussed. Sometimes a quantity y varies directly as a power of a variable. For example, the area A of a circle varies directly as the second power of the radius r. That is, $A = \pi r^2$.

DIRECT VARIATION AS THE *n*th POWER

Let x and y denote two quantities and n be a positive number. Then y is **directly proportional to the *n*th power** of x, or y **varies directly as the *n*th power** of x, if there exists a nonzero number k such that

$$y = kx^n.$$

The number k is called the *constant of variation* or the *constant of proportionality*. In the formula $A = \pi r^2$, $k = \pi$.

Modeling a pendulum

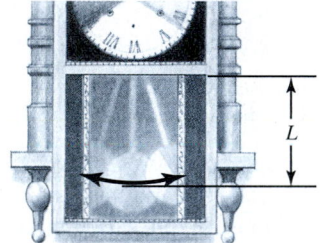

The time T required for a pendulum to swing back and forth once is called its *period*. The length L of a pendulum is directly proportional to the *n*th power of T, where n is a positive integer. Table 4.12 lists the period T for various lengths L.
(a) Find the constant of proportionality k and the value of n.
(b) Predict T for a pendulum having a length of 5 feet.

TABLE 4.12

L (ft)	1.0	1.5	2.0	2.5	3.0	3.5	4.0
T (sec)	1.11	1.36	1.57	1.76	1.92	2.08	2.22

SOLUTION

(a) The equation $L = kT^n$ models the pendulum. It follows that $k = \frac{L}{T^n}$. The ratio $\frac{L}{T^n}$ equals the constant k for some n. To find k and n see Table 4.13. The ratio is constant when $n = 2$. In this case $k \approx 0.81$ and $L = 0.81T^2$.

TABLE 4.13

L	T	L/T	L/T^2	L/T^3
1.0	1.11	0.90	**0.81**	0.73
1.5	1.36	1.10	**0.81**	0.60
2.0	1.57	1.27	**0.81**	0.52
2.5	1.76	1.42	**0.81**	0.46
3.0	1.92	1.56	**0.81**	0.42
3.5	2.08	1.68	**0.81**	0.39
4.0	2.22	1.80	**0.81**	0.37

(b) If $L = 5$, then $5 = 0.81T^2$. It follows that $T = \sqrt{5/0.81} \approx 2.48$ seconds.

Now Try Exercise 129 ◆

When two quantities vary inversely, an increase in one quantity results in a decrease in the second quantity. For example, it takes 4 hours to travel 100 miles at 25 miles per hour and 2 hours to travel 100 miles at 50 miles per hour. Greater speed results in less travel time. If s represents the average speed of a car and t is the time to travel 100 miles, then $s \cdot t = 100$ or $t = \frac{100}{s}$. Doubling the speed cuts the time in half, tripling the speed reduces the time by one-third. The quantities t and s are said to *vary inversely*. The constant of variation is 100.

INVERSE VARIATION AS THE *n*th POWER

Let x and y denote two quantities and n be a positive number. Then y is **inversely proportional to the *n*th power** of x, or y **varies inversely as the *n*th power** of x, if there exists a nonzero number k such that

$$y = \frac{k}{x^n}.$$

If $y = \frac{k}{x}$, then y is **inversely proportional** to x or y **varies inversely** as x.

Inverse variation occurs when measuring the intensity of light. If we increase our distance from a lightbulb, the intensity of the light decreases. Intensity I is inversely proportional to the second power of the distance d. The equation $I = \frac{k}{d^2}$ models this phenomenon.

 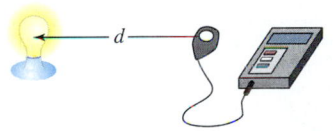

EXAMPLE 14 Modeling the intensity of light

At a distance of 3 meters, a 100-watt bulb produces an intensity of 0.88 watt per square meter. (**Source:** R. Weidner and R. Sells, *Elementary Classical Physics*, Volume 2.)
(a) Find the constant of proportionality k.
(b) Determine the intensity at a distance of 2 meters.

SOLUTION
(a) Substitute $d = 3$ and $I = 0.88$ into the equation $I = \frac{k}{d^2}$. Solve for k.

$$0.88 = \frac{k}{3^2} \quad \text{or} \quad k = 7.92$$

(b) Let $I = \frac{7.92}{d^2}$ and $d = 2$. Then $I = \frac{7.92}{2^2} = 1.98$. The intensity at 2 meters is 1.98 watts per square meter.

Now Try Exercise 137 ◆

4.5

Putting it all Together

$\mathbf{T}$he following table summarizes some important concepts about rational functions and equations.

Concept	Explanation	Examples
Rational function	$f(x) = \frac{p(x)}{q(x)}$, where $p(x)$ and $q(x)$ are polynomials with $q(x) \neq 0$.	$f(x) = \dfrac{x - 1}{x^2 + 2x + 1}$ $g(x) = 1 + \dfrac{1}{x}$ $\qquad \left(Note: 1 + \dfrac{1}{x} = \dfrac{x + 1}{x}.\right)$
Graph of a rational function	The graph of a rational function is continuous, except at x-values where the denominator equals zero.	The graph of $f(x) = \frac{3x^2 + 1}{x^2 - 4}$ is discontinuous at $x = \pm 2$. It has vertical asymptotes of $x = \pm 2$ and a horizontal asymptote of $y = 3$. (graph: vertical asymptotes $x = -2$, $x = 2$; horizontal asymptote $y = 3$)
Vertical asymptote	If k is a zero of the denominator, but not of the numerator, then $x = k$ is a vertical asymptote.	The graph of $f(x) = \frac{2x + 1}{x - 2}$ has vertical asymptotes at $x = 2$ because 2 is a zero of $x - 2$, but not a zero of $2x + 1$.
Horizontal asymptote	A horizontal asymptote occurs when the degree of the numerator is less than or equal to the degree of the denominator.	$f(x) = \dfrac{-4x^2 + 1}{3x^2 - x}$ Horizontal asymptote: $y = -\frac{4}{3}$. $f(x) = \dfrac{x}{4x^2 + 2x}$ Horizontal asymptote: $y = 0$.
Rational equation	A rational equation contains rational expressions. Be sure to check your solutions.	To solve the rational equation $\dfrac{4}{2x - 1} = 8,$ multiply each side by $2x - 1$. $4 = 8(2x - 1)$ $12 = 16x$ $x = \dfrac{3}{4}$

The following table summarizes some concepts about variation.

Concept	Equation	Examples
Varies directly as the nth power of x	$y = kx^n$	Let y vary directly as the third power of x. If the constant of variation is $k = 5$, then $y = 5x^3$.
Varies inversely as the nth power of x	$y = \dfrac{k}{x^n}$	Let y vary inversely as the square of x. If the constant of variation is $k = 3$, then $y = \dfrac{3}{x^2}$.

4.5 Exercises

Rational Functions

Exercises 1–12: Determine whether f is a rational function and state its domain.

1. $f(x) = \dfrac{x^3 - 5x + 1}{4x - 5}$

2. $f(x) = \dfrac{6}{x^2}$

3. $f(x) = x^2 - x - 2$

4. $f(x) = \dfrac{x^2 + 1}{\sqrt{x - 8}}$

5. $f(x) = \dfrac{|x - 1|}{x + 1}$

6. $f(x) = \dfrac{4}{x} + 1$

7. $f(x) = \dfrac{3x}{x^2 + 1}$

8. $f(x) = \dfrac{|x + 1|}{x + 1}$

9. $f(x) = \dfrac{3 - \sqrt{x}}{x^2 + x}$

10. $f(x) = \dfrac{x^3 - 3x + 1}{x^2 - 5}$

11. $f(x) = 4 - \dfrac{3}{x + 1}$

12. $f(x) = 5x^3 - 4x$

13. (Refer to Example 3.) The graph of $f(x) = \frac{1}{x^2}$ is shown in Figure 4.93. Numerical values for $f(x)$ are listed in the table. Relate these values to the graph of f.

x	0	± 0.01	± 0.1	± 1	± 10	± 100
$\dfrac{1}{x^2}$	—	10,000	100	1	0.01	0.0001

14. Is $f(x) = \frac{1}{x}$ an odd or even function? Is $g(x) = \frac{1}{x^2}$ an odd or even function?

Asymptotes

Exercises 15–20: Identify any horizontal or vertical asymptotes in the graph. State the domain of f.

15.
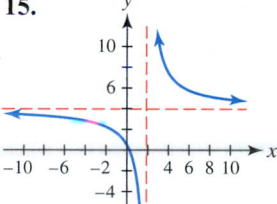

16.
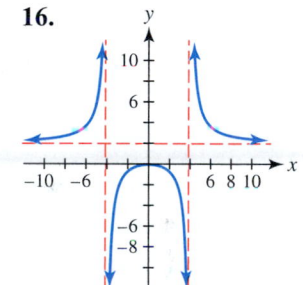

17.
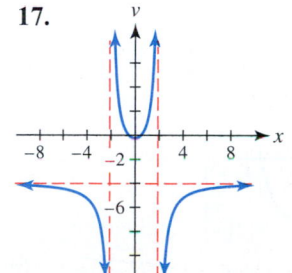

18.
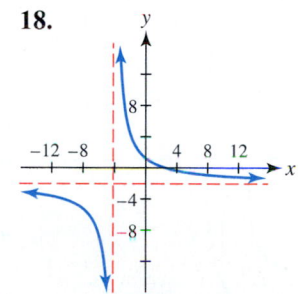

19.
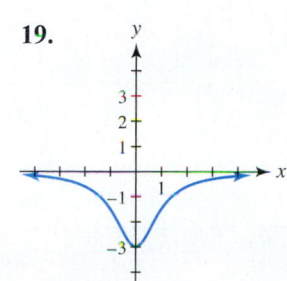

20.
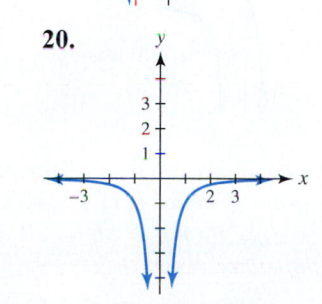

Exercises 21–32: Find any vertical or horizontal asymptotes.

21. $f(x) = \dfrac{4x + 1}{2x - 6}$

22. $f(x) = \dfrac{x + 6}{5 - 2x}$

23. $f(x) = \dfrac{3}{x^2 - 5}$

24. $f(x) = \dfrac{3x^2}{x^2 - 9}$

25. $f(x) = \dfrac{x^4 + 1}{x^2 + 3x - 10}$

26. $f(x) = \dfrac{4x^3 - 2}{x + 2}$

27. $f(x) = \dfrac{x^2 + 2x + 1}{2x^2 - 3x - 5}$

28. $f(x) = \dfrac{6x^2 - x - 2}{2x^2 + x - 6}$

29. $f(x) = \dfrac{3x(x + 2)}{(x + 2)(x - 1)}$

30. $f(x) = \dfrac{x}{x^3 - x}$

31. $f(x) = \dfrac{x^2 - 9}{x + 3}$

32. $f(x) = \dfrac{2x^2 - 3x + 1}{2x - 1}$

Exercises 33–36: Let a be a positive constant. Match $f(x)$ with its graph (a.–d.) without using a calculator.

33. $f(x) = \dfrac{a}{x - 1}$

34. $f(x) = \dfrac{2x + a}{x - 1}$

35. $f(x) = \dfrac{x - a}{x + 2}$

36. $f(x) = \dfrac{-2x}{x^2 - a}$

a.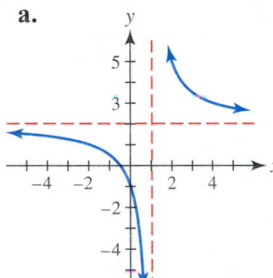

b.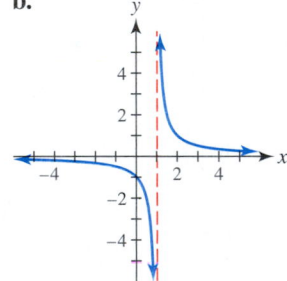

c.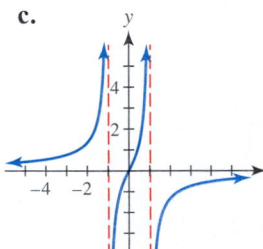

d.

Exercises 37–46: (Refer to Example 4.) Use transformations of the graph of either $f(x) = \frac{1}{x}$ or $h(x) = \frac{1}{x^2}$ to sketch a graph of $y = g(x)$ by hand. Show all asymptotes. Write $g(x)$ in terms of either $f(x)$ or $h(x)$.

37. $g(x) = \dfrac{1}{x - 3}$

38. $g(x) = \dfrac{1}{x + 2}$

39. $g(x) = \dfrac{1}{x} + 2$

40. $g(x) = 1 - \dfrac{2}{x}$

41. $g(x) = \dfrac{1}{x + 1} - 2$

42. $g(x) = \dfrac{1}{x - 2} + 1$

43. $g(x) = -\dfrac{2}{(x - 1)^2}$

44. $g(x) = \dfrac{1}{x^2} - 1$

45. $g(x) = \dfrac{1}{(x + 1)^2} - 2$

46. $g(x) = 1 - \dfrac{1}{(x - 2)^2}$

Exercises 47–56: (Refer to Example 7.) Graph $y = f(x)$ by hand. You may want to use division, factoring, or transformations of the graph of $y = \frac{1}{x}$ as an aid. Show all asymptotes and "holes."

47. $f(x) = \dfrac{x^2 - 2x + 1}{x - 1}$

48. $f(x) = \dfrac{4x^2 + 4x + 1}{2x + 1}$

49. $f(x) = \dfrac{x + 2}{x + 1}$

50. $f(x) = \dfrac{2x + 3}{x + 1}$

51. $g(x) = \dfrac{(2x + 1)(x - 2)}{(x - 2)^2}$

52. $g(x) = \dfrac{(x - 2)(x + 1)}{(x - 3)(x + 1)}$

53. $f(x) = \dfrac{2x^2 + 9x + 9}{2x^2 + 7x + 6}$

54. $f(x) = \dfrac{x^2 - 4}{x^2 - x - 6}$

55. $f(x) = \dfrac{-2x^2 + 11x - 14}{x^2 - 5x + 6}$

56. $f(x) = \dfrac{2x^2 - 3x - 14}{x^2 - 2x - 8}$

Exercises 57–64: Complete the following.

(a) Find the domain of f.

 (b) Graph f in an appropriate viewing rectangle.

(c) Find any vertical or horizontal asymptotes.

(d) Sketch a graph of f that includes any asymptotes.

57. $f(x) = \dfrac{x + 3}{x - 2}$

58. $f(x) = \dfrac{6 - 2x}{x + 3}$

59. $f(x) = \dfrac{4x + 1}{x^2 - 4}$

60. $f(x) = \dfrac{0.5x^2 + 1}{x^2 - 9}$

61. $f(x) = \dfrac{4}{1 - 0.25x^2}$ **62.** $f(x) = \dfrac{x^2}{1 + 0.25x^2}$

63. $f(x) = \dfrac{x^2 - 4}{x - 2}$ **64.** $f(x) = \dfrac{4(x - 1)}{x^2 - x - 6}$

Exercises 65 and 66: In the accompanying table, Y_1 represents a rational function. Give a possible equation for a horizontal asymptote of the graph of Y_1.

65.

X	Y1
50	2.8654
100	2.9314
150	2.9539
200	2.9653
250	2.9722
300	2.9768
350	2.9801

X=50

66.

X	Y1
-10	4.8922
-20	4.9726
-30	4.9878
-40	4.9931
-50	4.9956
-60	4.9969
-70	4.9978

X=-10

Exercises 67–70: Write a symbolic representation of a rational function f that satisfies the conditions.

67. Vertical asymptote $x = -3$,
horizontal asymptote $y = 1$

68. Vertical asymptote $x = 4$,
horizontal asymptote $y = -3$

69. Vertical asymptotes $x = \pm 3$,
horizontal asymptote $y = 0$

70. Vertical asymptotes $x = -2$ and $x = 4$,
horizontal asymptote $y = 5$

Exercises 71–78: Complete the following.

  (a) Find any vertical or slant asymptotes.
  (b) Graph $y = f(x)$. Show all asymptotes.

71. $f(x) = \dfrac{x^2 + 1}{x + 1}$ **72.** $f(x) = \dfrac{2x^2 - 5x - 2}{x - 2}$

73. $f(x) = \dfrac{0.5x^2 - 2x + 2}{x + 2}$ **74.** $f(x) = \dfrac{0.5x^2 - 5}{x - 3}$

75. $f(x) = \dfrac{x^2 + 2x + 1}{x - 1}$ **76.** $f(x) = \dfrac{2x^2 + 3x + 1}{x - 2}$

77. $f(x) = \dfrac{4x^2}{2x - 1}$ **78.** $f(x) = \dfrac{4x^2 + x - 2}{4x - 3}$

Rational Equations

Exercises 79–84: Solve the rational equation

  (a) symbolically,
  (b) graphically, and
  (c) numerically.

79. $\dfrac{2x}{x + 2} = 6$ **80.** $\dfrac{3x}{2x - 1} = 3$

81. $2 - \dfrac{5}{x} + \dfrac{2}{x^2} = 0$ **82.** $\dfrac{1}{x^2} + \dfrac{1}{x} = 2$

83. $\dfrac{1}{x + 1} + \dfrac{1}{x - 1} = \dfrac{1}{x^2 - 1}$ **84.** $\dfrac{4}{x - 2} = \dfrac{3}{x - 1}$

Exercises 85–100: Find all real solutions to the equation. Check your results.

85. $\dfrac{x + 1}{x - 5} = 0$ **86.** $\dfrac{x - 2}{x + 3} = 1$

87. $\dfrac{6(1 - 2x)}{x - 5} = 4$ **88.** $\dfrac{2}{5(2x + 5)} + 3 = -1$

89. $\dfrac{1}{x + 2} + \dfrac{1}{x} = 1$ **90.** $\dfrac{2x}{x - 1} = 5 + \dfrac{2}{x - 1}$

91. $\dfrac{1}{x} - \dfrac{2}{x^2} = 5$ **92.** $\dfrac{1}{x^2 - 2} = \dfrac{1}{x}$

93. $\dfrac{x^3 - 4x}{x^2 + 1} = 0$

94. $\dfrac{1}{x + 2} + \dfrac{1}{x + 3} = \dfrac{2}{x^2 + 5x + 6}$

95. $\dfrac{35}{x^2} = \dfrac{4}{x} + 15$ **96.** $6 - \dfrac{35}{x} + \dfrac{36}{x^2} = 0$

97. $\dfrac{x + 5}{x + 2} = \dfrac{x - 4}{x - 10}$ **98.** $\dfrac{x - 1}{x + 1} = \dfrac{x + 3}{x - 4}$

99. $\dfrac{1}{x - 2} - \dfrac{2}{x - 3} = \dfrac{-1}{x^2 - 5x + 6}$

100. $\dfrac{1}{x - 1} + \dfrac{3}{x + 1} = \dfrac{4}{x^2 - 1}$

Applications

101. ***Time Spent in Line*** If two parking attendants can wait on 8 vehicles per minute and vehicles are leaving the parking lot randomly at an average rate of x vehicles per minute, then the average time T in minutes spent waiting in line *and* paying the attendant is given by the formula $T(x) = -\dfrac{1}{x - 8}$, where $0 \le x < 8$. A graph of T is shown in the figure on the next page.

(a) Evaluate $T(4)$ and $T(7.5)$. Interpret the results.

(b) Explain what happens to the wait as vehicles start arriving at an average rate that approaches 8 cars per minute.

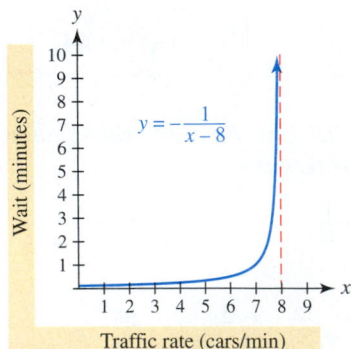

102. **Time Spent in Line** (Refer to Exercise 101.) If the parking attendants can wait on 5 vehicles per minute, the average time T in minutes spent waiting in line *and* paying the attendant becomes $T(x) = -\frac{1}{x - 5}$.

(a) What is a reasonable domain for T?

(b) Graph $y = T(x)$. Include any vertical asymptotes.

(c) Explain what happens to $T(x)$ as $x \to 5^-$.

103. **Interpreting an Asymptote** Suppose that an insect population in millions is modeled by $f(x) = \frac{10x + 1}{x + 1}$, where $x \geq 0$ is in months.

(a) Graph f in $[0, 14, 1]$ by $[0, 14, 1]$. Find the equation of the horizontal asymptote.

(b) Determine the initial insect population.

(c) What happens to the population after several months?

(d) Interpret the horizontal asymptote.

104. **Interpreting an Asymptote** Suppose that the population of a species of fish in thousands is modeled by $f(x) = \frac{x + 10}{0.5x^2 + 1}$, where $x \geq 0$ is in years.

(a) Graph f in $[0, 12, 1]$ by $[0, 12, 1]$. What is the horizontal asymptote?

(b) Determine the initial population.

(c) What happens to the population of this fish after many years?

(d) Interpret the horizontal asymptote.

105. **Time Spent in Line** Suppose the average number of vehicles arriving at the main gate of an amusement park is equal to 10 per minute, while the average number of vehicles being admitted through the gate per minute is equal to x. Then the average waiting time in minutes for each vehicle at the gate can be computed by $f(x) = \frac{x - 5}{x^2 - 10x}$, where $x > 10$. (**Source:** F. Mannering.)

(a) Estimate the admittance rate x that results in an average wait of 15 seconds.

(b) If one attendant can serve 5 vehicles per minute, how many attendants are needed to keep the average wait to 15 seconds or less?

106. **Length of Lines** (Refer to Example 2.) Determine the traffic intensity x when the average number of vehicles in line equals 3.

107. **Construction** (Refer to Example 12.) Find possible dimensions for a box with a volume of 196 cubic inches, a surface area of 280 square inches, and a length that is twice the width.

108. **Minimizing Surface Area** An aluminum can is being designed to hold a volume of 100π cubic centimeters. See the accompanying figure.

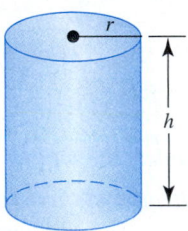

Geometry Review To review formulas for cylinders, see Chapter R (page R-4).

(a) Write a formula for the volume V of the can in terms of r and h.

(b) Write a formula for a function S that calculates the outside surface area of the can in terms of only r. Evaluate $S(2)$ and interpret the result.

(c) Find the dimensions of the can that result in the least amount of aluminum used in the construction of the can.

109. **Minimizing Cost** A cardboard box with no top is being constructed with a square base and must have a volume of 108 cubic inches. Let x be the length of a side of its base in inches.

(a) Write a formula for a function A in terms of x that calculates the outside surface area in square feet of the box.

(b) If cardboard costs $0.10 per square foot, write a formula for a function C in terms of x that gives the cost in dollars of the cardboard in the box.

(c) Find the dimensions of the box that would minimize the cost of the cardboard.

110. *Cost-Benefit*   A cost-benefit function C computes the cost in millions of dollars of implementing a city recycling project when x percent of the citizens participate, where $C(x) = \frac{1.2x}{100 - x}$.

(a) Graph C in $[0, 100, 10]$ by $[0, 10, 1]$. Interpret the graph as x approaches 100.

(b) If 75% participation is expected, determine the cost for the city.

(c) The city plans to spend $5 million on this recycling project. Estimate the percentage of participation that can be expected.

111. *Train Curves*   When curves are designed for trains, sometimes the outer rail is elevated or banked, so that a locomotive can safely negotiate the curve at a higher speed. See the accompanying figure. Suppose a circular curve is being designed for 60 miles per hour. The rational function given by $f(x) = \frac{2540}{x}$ computes the elevation y in inches of the outer track for a curve with a radius of x feet, where $y = f(x)$.   (**Source:** L. Haefner, *Introduction to Transportation Systems.*)

(a) Evaluate $f(400)$ and interpret the result.

(b) Graph f in $[0, 600, 100]$ by $[0, 50, 5]$. Discuss how the elevation of the outer rail changes with the radius x.

(c) Interpret the horizontal asymptote.

(d) What radius is associated with an elevation of 12.7 inches?

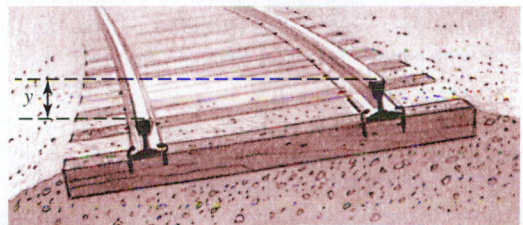

112. *Probability*   A container holds x balls numbered 1 through x. Only one ball has the winning number.

(a) Determine a rational function f that computes the probability or likelihood of *not* drawing the winning ball.

(b) What is the domain of f?

(c) What happens to the probability of *not* drawing the winning ball as the number of balls increases?

(d) Interpret the horizontal asymptote of the graph of f.

113. *Braking Distance*   The **grade** x of a hill is a measure of its steepness. For example, if a road rises 10 feet for every 100 feet of horizontal distance, then it has an uphill grade of $x = \frac{10}{100}$ or 10%. See the accompanying figure. Grades are typically kept quite small—usually less than 10%. The braking distance D for a car traveling at 50 miles per hour on a wet, uphill grade is given by $D(x) = \frac{2500}{30(0.3 + x)}$.   (**Source:** L. Haefner.)

(a) Evaluate $D(0.05)$ and interpret the result.

(b) Describe what happens to the braking distance as the hill becomes steeper. Does this agree with your driving experience?

(c) Estimate the grade associated with a braking distance of 220 feet.

114. *Braking Distance*   (Refer to the previous exercise.) If a car is traveling 50 miles per hour downhill, then its braking distance on a wet pavement is given by $D(x) = \frac{2500}{30(0.3 + x)}$, where $x < 0$ for a downhill grade.

(a) Evaluate $D(-0.1)$ and interpret the result.

(b) What happens to the braking distance as the downhill grade becomes steeper? Does this agree with your driving experience?

(c) The graph of D has a vertical asymptote at $x = -0.3$. Give the physical significance of this asymptote.

(d) Estimate the grade associated with a braking distance of 350 feet.

115. *Slippery Roads*   If a car is moving at 50 miles per hour on a level highway, then its braking distance depends on the road conditions. This distance in feet can be computed by $D(x) = \frac{2500}{30x}$, where x is the coefficient of friction between the tires and the road and $0 < x \le 1$. A smaller value of x indicates that the road is more slippery.
(a) Identify and interpret the vertical asymptote.

(b) Estimate the coefficient of friction associated with a braking distance of 340 feet.

116. *Concentration of a Drug*   The concentration of a drug in a medical patient's bloodstream is given by the formula $f(t) = \frac{5}{t^2 + 1}$, where the input $t \ge 0$ is in hours and the output is in milligrams per liter.
(a) Does the concentration of the drug increase or decrease? Explain.

(b) The patient should not take a second dose until the concentration is below 1.5 milligrams per liter. How long should the patient wait before taking a second dose?

Variation

Exercises 117–120: Find the constant of proportionality k for the given conditions.

117. $y = \frac{k}{x}$, and $y = 2$ when $x = 3$

118. $y = \frac{k}{x^2}$, and $y = \frac{1}{4}$ when $x = 8$

119. $y = kx^3$, and $y = 64$ when $x = 2$

120. $y = kx^{3/2}$, and $y = 96$ when $x = 16$

Exercises 121–124: Solve the variation problem.

121. Suppose T varies directly as the $\frac{3}{2}$ power of x. When $x = 4$, $T = 20$. Find T when $x = 16$.

122. Suppose y varies directly as the second power of x. When $x = 3$, $y = 10.8$. Find y when $x = 1.5$.

123. Let y be inversely proportional to x. When $x = 6$, $y = 5$. Find y when $x = 15$.

124. Let z be inversely proportional to the third power of t. When $t = 5$, $z = 0.08$. Find z when $t = 2$.

Exercises 125–128: Assume that the constant of proportionality is positive.

125. Let y be inversely proportional to x. If x doubles, what happens to y?

126. Let y vary inversely as the second power of x. If x doubles, what happens to y?

127. Suppose y varies directly as the third power of x. If x triples, what happens to y?

128. Suppose y is directly proportional to the second power of x. If x is halved, what happens to y?

Exercises 129 and 130: The data in the table satisfy the equation $y = kx^n$, where n is a positive integer. Determine k and n.

129.

x	2	3	4	5
y	2	4.5	8	12.5

130.

x	3	5	7	9
y	32.4	150	411.6	874.8

Exercises 131 and 132: The data in the table satisfy the equation $y = \frac{k}{x^n}$, where n is a positive integer. Determine k and n.

131.

x	2	3	4	5
y	1.5	1	0.75	0.6

132.

x	2	4	6	8
y	9	2.25	1	0.5625

133. *Allometric Growth*   The weight y of a fiddler crab is directly proportional to the 1.25 power of the weight x of its claws. A crab with a body weight of 1.9 grams has claws weighing 1.1 grams. Estimate the weight of a fiddler crab with claws weighing 0.75 gram. (**Source:** D. Brown.)

134. *Gravity*   The weight of an object varies inversely as the second power of the distance from the *center* of Earth. The radius of Earth is approximately 4000 miles. If a person weighs 160 pounds on Earth's surface, what would this individual weigh 8000 miles above the surface of Earth?

135. *Hubble Telescope*   The brightness or intensity of starlight varies inversely as the square of its distance from Earth. The Hubble Telescope can see stars whose intensities are $\frac{1}{50}$ of the faintest star now seen by ground-based telescopes. Determine how much farther the Hubble Telescope can see into space than ground-based telescopes. (**Source:** National Aeronautics and Space Administration.)

136. *Volume*   The volume V of a cylinder with a fixed height is directly proportional to the square of its radius r. If a cylinder with a radius of 10 inches has a volume of 200 cubic inches, what is the volume of a cylinder with the same height and a radius of 5 inches?

137. *Electrical Resistance*   The electrical resistance R of a wire varies inversely as the square of its diameter d. If a 25-foot wire with a diameter of 2 millimeters has a resistance of 0.5 ohm, find the resistance of a wire having the same length and a diameter of 3 millimeters.

138. *Strength of a Beam*   The strength of a rectangular wood beam varies directly as the square of the depth of its cross section. If a beam with a depth of 3.5 inches can support 1000 pounds, how much weight can the same type of beam hold if its depth is 10.5 inches?

Exercises 139 and 140: ***Violin String***   *The frequency F of a vibrating string is directly proportional to the square root of the tension T on the string and inversely proportional to the length L of the string.*

139. If both the tension and length are doubled, what happens to F?

140. Give two ways to double the frequency F.

Writing about Mathematics

141. Suppose that the formula for a rational function f is given.
   (a) Explain how to find any vertical or horizontal asymptotes of the graph of f.
   (b) Discuss what a horizontal asymptote represents.

142. Discuss how to find the domain of a rational function symbolically and graphically.

◆ Solve polynomial inequalities
◆ Solve rational inequalities

4.6   Polynomial and Rational Inequalities

Introduction

Waiting in line has become part of almost everyone's life. When people arrive randomly at a line, rational functions can be used to estimate the average number of people standing in line. For example, if an attendant at a ticket booth can wait on 30 customers per hour, and if customers arrive at an average rate of x per hour, then the average number of customers waiting in line is computed by

$$f(x) = \frac{x^2}{900 - 30x},$$

where $0 \le x < 30$. For example, $f(28) \approx 13$ indicates that if customers arrive, on average, at 28 per hour, then the average number of people in line is 13. If a line length of 8 customers or fewer is acceptable, then $f(x)$ can be used to estimate customer arrival rates x that one attendant can accommodate by solving the rational inequality

$$\frac{x^2}{900 - 30x} \le 8.$$

In this section we discuss solving polynomial inequalities, and then use similar techniques to solve rational inequalities.   (**Source:** N. Garber and L. Hoel, *Traffic and Highway Engineering.*)

Polynomial Inequalities

In Section 3.3 a strategy for solving quadratic inequalities was presented. This strategy involves first finding boundary numbers (x-values) where equality holds. Once the boundary numbers are known, a graph or a table of test values can be used to determine the intervals where inequality holds. This strategy can be applied to other types of inequalities.

For example, consider the graph of $p(x) = -x^4 + 5x^2 - 4$ with boundary numbers -2, -1, 1, and 2 shown in Figure 4.112. The solution set to the inequality $p(x) > 0$ can be written in interval notation as $(-2, -1) \cup (1, 2)$ and corresponds to where the graph is above the x-axis. (The symbol $\cup$ means union and indicates that x can be in either interval.) The intervals $(-2, -1)$ and $(1, 2)$ are shaded on the x-axis in Figure 4.113. Similarly, the solution set to the inequality $p(x) < 0$ is $(-\infty, -2) \cup (-1, 1) \cup (2, \infty)$ and corresponds to where the graph of $y = p(x)$ is below the x-axis. These intervals are shaded on the x-axis in Figure 4.114.

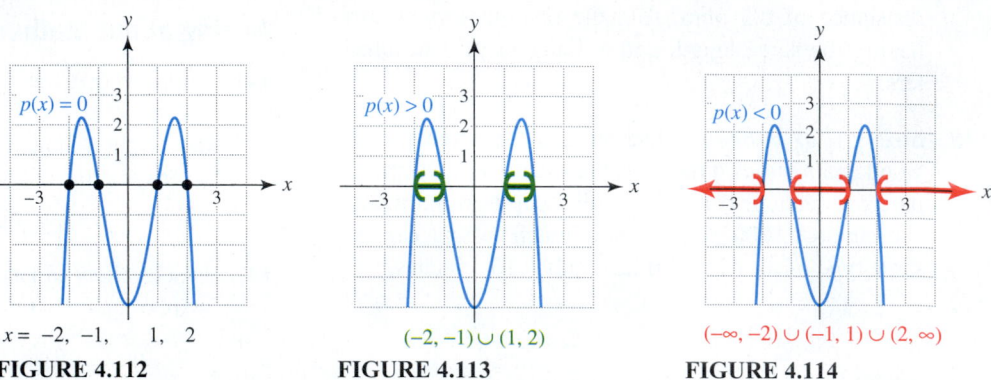

$x = -2, -1, \quad 1, \quad 2$ $\qquad$ $(-2, -1) \cup (1, 2)$ $\qquad$ $(-\infty, -2) \cup (-1, 1) \cup (2, \infty)$

FIGURE 4.112 $\qquad\qquad$ **FIGURE 4.113** $\qquad\qquad$ **FIGURE 4.114**

◆ **MAKING CONNECTIONS**

Visualization and Inequalities A precise graph of p is not necessary to solve the polynomial inequality $p(x) > 0$ or $p(x) < 0$. Once the x-intercepts are determined, we can use our knowledge about graphs of quartic polynomials (see Section 4.2) to visualize the M-shaped graph of p shown in Figure 4.112. We need to make only a rough sketch and then determine where the graph of p is above the x-axis and where it is below the x-axis.

Polynomial inequalities can also be solved symbolically. For example, to solve the inequality $-x^4 + 5x^2 - 4 > 0$, begin by finding the boundary numbers.

$$-x^4 + 5x^2 - 4 = 0 \qquad \text{Replace > with =.}$$
$$x^4 - 5x^2 + 4 = 0 \qquad \text{Multiply by } -1.$$
$$(x^2 - 4)(x^2 - 1) = 0 \qquad \text{Factor.}$$
$$x^2 - 4 = 0 \quad \text{or} \quad x^2 - 1 = 0 \qquad \text{Zero-product property}$$
$$x = \pm 2 \quad \text{or} \quad x = \pm 1 \qquad \text{Square root property}$$

Algebra Review

To review interval notation, see Section 2.4.

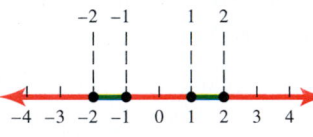

FIGURE 4.115

The boundary numbers -2, -1, 1, and 2 separate the number line into five intervals:

$$(-\infty, -2), (-2, -1), (-1, 1), (1, 2), \text{ and } (2, \infty),$$

as illustrated in Figure 4.115.

The polynomial $p(x) = -x^4 + 5x^2 - 4$ is either always positive or always negative on each of these intervals. To determine which is the case, we can choose one test value (x-value) from each interval and substitute this test value into $p(x)$. From Table 4.14 we see that the test value $x = -1.5$ results in

$$p(-1.5) = -(1.5)^4 + 5(1.5)^2 - 4 = 2.1875 > 0.$$

Since $x = -1.5$ is in the interval $(-2, -1)$, it follows that $p(x)$ is positive on this interval. The sign of $p(x)$ on the other intervals is determined similarly. Thus the solution set to $p(x) = -x^4 + 5x^2 - 4 > 0$ is $(-2, -1) \cup (1, 2)$, which is consistent with the graphical solution in Figure 4.113.

TABLE 4.14

Interval	Test Value x	$-x^4 + 5x^2 - 4$	Positive or Negative?
$(-\infty, -2)$	-3	-40	Negative
$(-2, -1)$	-1.5	2.1875	Positive
$(-1, 1)$	0	-4	Negative
$(1, 2)$	1.5	2.1875	Positive
$(2, \infty)$	3	-40	Negative

These symbolic and graphical procedures are now summarized verbally.

SOLVING POLYNOMIAL INEQUALITIES

STEP 1: If necessary, write the inequality as $p(x) < 0$, where $p(x)$ is a polynomial and $<$ may be replaced by $>$, $\leq$, or $\geq$.

STEP 2: Solve $p(x) = 0$ either symbolically or graphically. The solutions are called boundary numbers.

STEP 3: Use the boundary numbers to separate the number line into disjoint intervals. On each interval, $p(x)$ is either always positive or always negative.

STEP 4: To solve the inequality, either make a table of test values for $p(x)$ or use a graph of $p(x)$. For example, the solution set for $p(x) < 0$ corresponds to intervals where test values result in negative outputs or to intervals where the graph of $p(x)$ is below the x-axis.

EXAMPLE **1**   Solving a polynomial inequality

Solve $x^3 \geq 2x^2 + 3x$ symbolically and graphically.

SOLUTION

Symbolic Solution

STEP 1: Begin by writing the inequality as $x^3 - 2x^2 - 3x \geq 0$.

STEP 2: Replace the $\geq$ symbol with an equals sign and solve the resulting equation.

$$x^3 - 2x^2 - 3x = 0 \qquad \text{Replace } \geq \text{ with } =.$$
$$x(x^2 - 2x - 3) = 0 \qquad \text{Factor out } x.$$
$$x(x + 1)(x - 3) = 0 \qquad \text{Factor the trinomial.}$$
$$x = 0 \quad \text{or} \quad x = -1 \quad \text{or} \quad x = 3 \qquad \text{Zero-product property}$$

Algebra Review

To review factoring, see Chapter R, (page R-27).

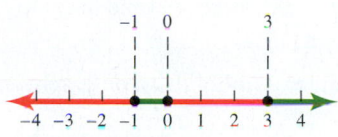

FIGURE 4.116

The boundary numbers are -1, 0, and 3.

STEP 3: The boundary numbers separate the number line into four disjoint intervals:

$$(-\infty, -1), (-1, 0), (0, 3), \text{ and } (3, \infty),$$

which is illustrated in Figure 4.116.

STEP 4: In Table 4.15 on the next page the expression $x^3 - 2x^2 - 3x$ is evaluated at test values from each interval. The solution set is $[-1, 0] \cup [3, \infty)$. In Figure 4.117 a graphing calculator has been used to evaluate the same test values. (*Note:* The boundary numbers are included in the solution set because the inequality involves $\geq$ rather than $>$.)

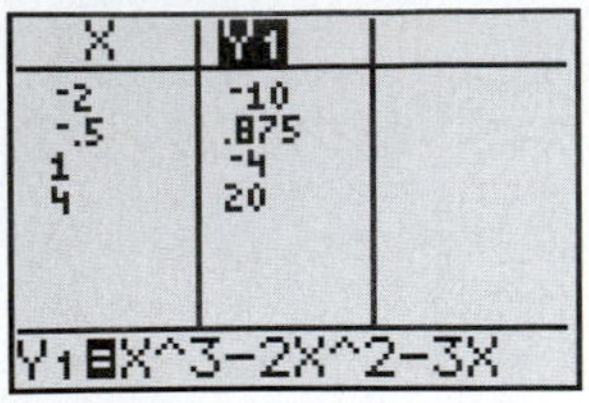

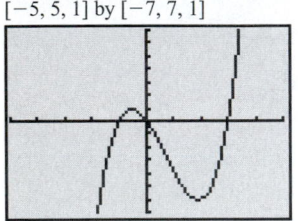

FIGURE 4.117

$[-5, 5, 1]$ by $[-7, 7, 1]$

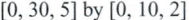

FIGURE 4.118

TABLE 4.15

Interval	Test Value x	$x^3 - 2x^2 - 3x$	Positive or Negative?
$(-\infty, -1)$	-2	-10	**Negative**
$(-1, 0)$	-0.5	0.875	**Positive**
$(0, 3)$	1	-4	**Negative**
$(3, \infty)$	4	20	**Positive**

Graphical Solution Graph $Y_1 = X^3 - 2X^2 - 3X$, as shown in Figure 4.118. The zeros or x-intercepts are located at -1, 0, and 3. The graph of y_1 is positive (or above the x-axis) when $-1 < x < 0$ or when $3 < x < \infty$. If we include the boundary numbers, this result agrees with the symbolic solution. *Now Try Exercise 15* ◆

Rational Inequalities

In the introduction we discussed how a rational inequality can be used to estimate the number of people standing in line at a ticket booth. In the next example, this inequality is solved graphically.

EXAMPLE 2

Modeling customers in a line

A ticket booth attendant can wait on 30 customers per hour. To keep the time waiting in line reasonable, the line length should not exceed 8 customers. Solve the inequality $\frac{x^2}{900 - 30x} \le 8$ to determine the rates x at which customers can arrive before a second attendant is needed. Note that the x-values are limited to $0 \le x < 30$.

$[0, 30, 5]$ by $[0, 10, 2]$

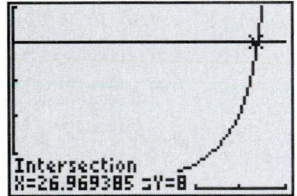

FIGURE 4.119

SOLUTION Graph $Y_1 = X^2/(900 - 30X)$ and $Y_2 = 8$ for $0 \le x \le 30$, as shown in Figure 4.119. The only point of intersection on this interval is near $(26.97, 8)$. The graph of y_1 is below the graph of y_2 for x-values to the left of this point. We conclude that if the arrival rate is about 27 customers per hour or less, then the line length does not exceed 8 customers. If the arrival rate is more than 27 customers per hour, a second ticket booth attendant is needed. *Now Try Exercise 47* ◆

To solve rational inequalities, we can use the same basic techniques that we used to solve polynomial inequalities, with one important modification: boundary numbers also occur at x-values where the denominator of any rational expression in the equation equals 0. The following steps can be used to solve a rational inequality.

SOLVING RATIONAL INEQUALITIES

STEP 1: If necessary, write the inequality in the form $\frac{p(x)}{q(x)} > 0$, where $p(x)$ and $q(x)$ are polynomials. Note that $>$ may be replaced by $<$, $\le$, or $\ge$.

STEP 2: Solve the equations $p(x) = 0$ and $q(x) = 0$ either symbolically or graphically. The solutions are boundary numbers.

STEP 3: Use the boundary numbers to separate the number line into disjoint intervals. On each interval, $\frac{p(x)}{q(x)}$ is either always positive or always negative.

STEP 4: Use a table of test values or a graph to solve the inequality in **STEP 1**.

EXAMPLE 3

Solving a rational inequality

Solve $\frac{2-x}{2x} > 0$ symbolically. Support your answer graphically.

SOLUTION

Symbolic Solution   Since the inequality is already written the form $\frac{p(x)}{q(x)} > 0$, **STEP 1** is unnecessary.

STEP 2: Set the numerator and the denominator equal to 0 and solve.

Numerator	Denominator
$2 - x = 0$	$2x = 0$
$x = 2$	$x = 0$

STEP 3: The boundary numbers are 0 and 2, which separate the number line into three disjoint intervals: $(-\infty, 0)$, $(0, 2)$, and $(2, \infty)$.

STEP 4: Table 4.16 shows a table of test values. We can see that the expression is positive between the two boundary numbers or when $0 < x < 2$. In interval notation the solution set is $(0, 2)$.

TABLE 4.16

Interval	Test Value x	$(2 - x)/(2x)$	Positive or Negative?
$(-\infty, 0)$	-1	-1.5	**Negative**
$(0, 2)$	1	0.5	**Positive**
$(2, \infty)$	4	-0.25	**Negative**

$[-4.7, 4.7, 1]$ by $[-3.1, 3.1, 1]$

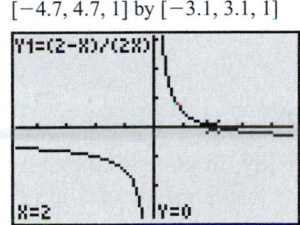

FIGURE 4.120

Graphical Solution   Graph $Y_1 = (2 - X)/(2X)$, as shown in Figure 4.120. The graph has a vertical asymptote at $x = 0$ and an x-intercept at $x = 2$. Between these boundary numbers the graph of y_1 is positive (or above the x-axis.) This agrees with our symbolic solution.

Now Try Exercise 33 ◆

EXAMPLE 4

Solving a rational inequality symbolically

Solve $\frac{1}{x} \le \frac{2}{x+1}$.

SOLUTION

STEP 1: Begin by writing the inequality in the form $\frac{p(x)}{q(x)} \le 0$.

Algebra Review

To review subtraction of rational expressions, see Chapter R (page R-42).

$$\frac{1}{x} - \frac{2}{x+1} \le 0 \qquad \text{Subtract } \tfrac{2}{x+1}.$$

$$\frac{1}{x} \cdot \frac{(x+1)}{(x+1)} - \frac{2}{x+1} \cdot \frac{x}{x} \le 0 \qquad \text{Common denominator is } x(x+1).$$

$$\frac{x+1}{x(x+1)} - \frac{2x}{x(x+1)} \le 0 \qquad \text{Multiply.}$$

$$\frac{1-x}{x(x+1)} \le 0 \qquad \text{Subtract and combine terms.}$$

STEP 2: Find the zeros of the numerator and the denominator.

Numerator	*Denominator*
$1 - x = 0$	$x(x + 1) = 0$
$x = 1$	$x = 0 \quad \text{or} \quad x = -1$

STEP 3: The boundary numbers are $-1, 0,$ and 1, which separate the number line into four disjoint intervals: $(-\infty, -1), (-1, 0), (0, 1),$ and $(1, \infty)$.

STEP 4: Table 4.17 can be used to solve the inequality $\frac{1-x}{x(x+1)} \leq 0$. The solution set is $(-1, 0) \cup [1, \infty)$. (*Note:* The boundary numbers -1 and 0 are not included in the solution set because the given inequality is undefined when $x = -1$ or $x = 0$.)

TABLE 4.17

Interval	Test Value x	$(1 - x)/(x(x + 1))$	Positive or Negative?
$(-\infty, -1)$	-2	1.5	**Positive**
$(-1, 0)$	-0.5	-6	**Negative**
$(0, 1)$	0.5	$0.\overline{6}$	**Positive**
$(1, \infty)$	2	$-0.1\overline{6}$	**Negative**

Now Try Exercise 41 ◆

When solving a rational inequality, it is essential *not* to multiply or divide each side of the inequality by the LCD (least common denominator) if the LCD contains a *variable*. This technique often leads to an incorrect solution set.

For example, if each side of the rational inequality $\frac{1}{x} < 2$ is multiplied by x to clear fractions, the inequality becomes $1 < 2x$ or $x > \frac{1}{2}$. However, this solution set is clearly incomplete because $x = -1$ is also a solution to the given inequality. In general, the variable x can be either negative or positive. If $x < 0$, the inequality symbol should be reversed, whereas if $x > 0$, the inequality symbol should not be reversed. Because we have no way of knowing ahead of time which is the case, this technique of multiplying by a variable should be avoided.

4.6

Putting it all Together

The following table outlines basic concepts for solving polynomial and rational inequalities.

Concept	Description
Polynomial inequality	Can be written as $p(x) < 0$, where $p(x)$ is a polynomial and $<$ may be replaced by $>, \leq,$ or $\geq$. **Examples:** $x^3 - x \leq 0$ $2x^4 - 3x^2 \geq 5x + 1$

Concept	Description
Solving a polynomial inequality	Follow the steps to solve a polynomial inequality presented in this section. (See page 327.) Either graphical or symbolic methods can be used. **Example:** A graph of $y = x^3 - 2x^2 - 5x + 6$ is shown. The boundary numbers are -2, 1, and 3. The solution set to $x^3 - 2x^2 - 5x + 6 > 0$ is $(-2, 1) \cup (3, \infty)$. $y = x^3 - 2x^2 - 5x + 6$
Rational inequality	Can be written as $\frac{p(x)}{q(x)} < 0$, where $p(x)$ and $q(x) \neq 0$ are polynomials and $<$ may be replaced by $>$, $\leq$, or $\geq$. **Examples:** $\dfrac{x - 3}{x + 2} \geq 0$ $2x - \dfrac{2}{x^2 - 1} > 5$
Solving a rational inequality	Follow the steps to solve a rational inequality presented in this section. (See page 328.) Either graphical or symbolic methods can be used. **Example:** A graph of $y = \frac{x - 1}{2x + 1}$ is shown. Boundary numbers occur where either the numerator or denominator equals zero: $x = 1$ or $x = -\frac{1}{2}$. The solution set to $\frac{x - 1}{2x + 1} < 0$ is $\left(-\frac{1}{2}, 1\right)$. $y = \dfrac{x - 1}{2x + 1}$

◆ **4.6** ◆ **Exercises**

Graphical Solutions

Exercises 1–6: Use the graph of f to solve the equation and inequalities.

    (a) $f(x) = 0$
    (b) $f(x) > 0$
    (c) $f(x) < 0$

1.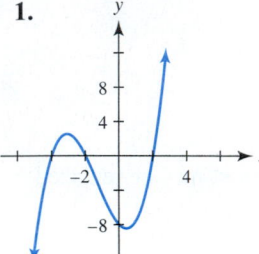

2.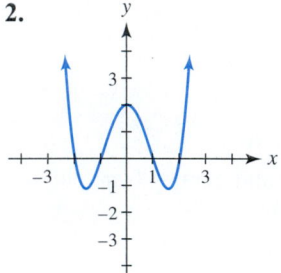

3.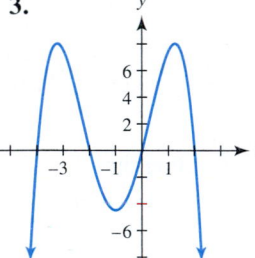

4.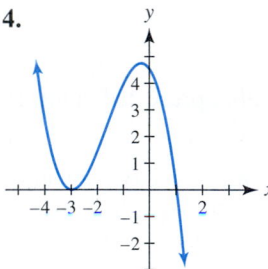

5.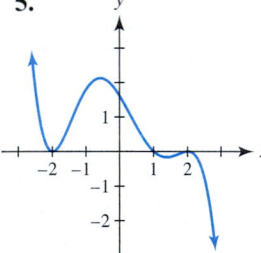

6.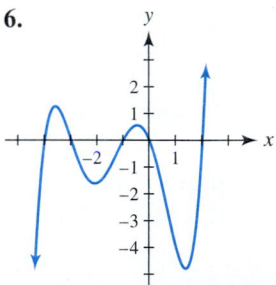

Exercises 7–12: Use the graph of the rational function f to complete the following.

    (a) *Identify x-values where either f(x) is undefined or f(x) = 0.*
    (b) *Solve f(x) > 0.*
    (c) *Solve f(x) < 0.*

7.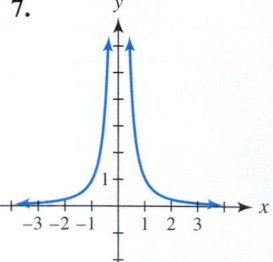

8.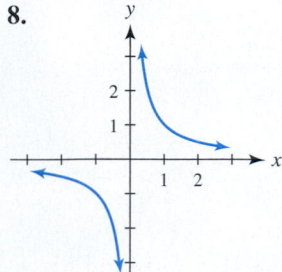

9.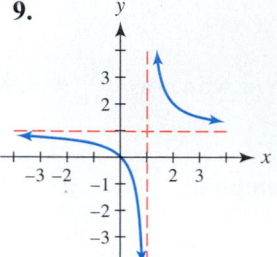

10.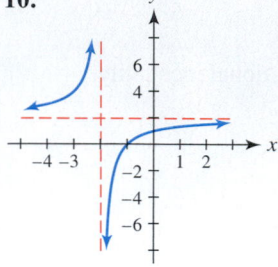

11.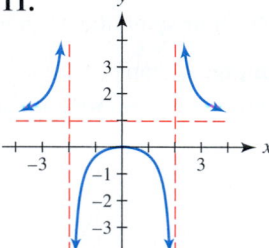

12.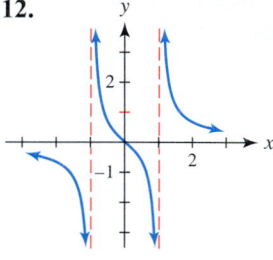

Polynomial Inequalities

Exercises 13–18: Solve the polynomial inequality

    (a) *symbolically, and*
    (b) *graphically.*

13. $x^3 - x > 0$ **14.** $8x^3 < 27$

15. $x^3 + x^2 \geq 2x$ **16.** $2x^3 \leq 3x^2 + 5x$

17. $x^4 - 13x^2 + 36 < 0$ **18.** $4x^4 - 5x^2 - 9 \geq 0$

Exercises 19–24: Solve the polynomial inequality.

19. $7x^4 \geq 14x^2$ **20.** $3x^4 - 4x^2 < 7$

21. $(x - 1)(x - 2)(x + 2) \geq 0$

22. $-(x + 1)^2(x - 2) \geq 0$

23. $2x^4 + 2x^3 \leq 12x^2$ 24. $x^3 + 6x^2 + 9x > 0$

Exercises 25–28: Solve the polynomial inequality graphically.

25. $x^3 - 7x^2 + 14x \leq 8$ 26. $2x^3 + 3x^2 - 3x < 2$

27. $3x^4 - 7x^3 - 2x^2 + 8x > 0$

28. $x^4 - 5x^3 \leq 5x^2 + 45x + 36$

Rational Inequalities

Exercises 29–34: Solve the rational inequality

(a) *symbolically, and*
(b) *graphically.*

29. $\dfrac{1}{x} < 0$ 30. $\dfrac{1}{x^2} > 0$

31. $\dfrac{4}{x + 3} \geq 0$ 32. $\dfrac{x - 1}{x + 1} < 0$

33. $\dfrac{5}{x^2 - 4} < 0$ 34. $\dfrac{x}{x^2 - 1} \geq 0$

Exercises 35–46: Solve the rational inequality.

35. $\dfrac{(x + 1)^2}{x - 2} \leq 0$ 36. $\dfrac{2x}{(x - 2)^2} > 0$

37. $\dfrac{3 - 2x}{1 + x} < 0$ 38. $\dfrac{x + 1}{4 - 2x} \geq 1$

39. $\dfrac{(x + 1)(x - 2)}{(x + 3)} < 0$ 40. $\dfrac{x(x - 3)}{x + 2} \geq 0$

41. $\dfrac{2x - 5}{x^2 - 1} \geq 0$ 42. $\dfrac{5 - x}{x^2 - x - 2} < 0$

43. $\dfrac{1}{x - 3} \leq \dfrac{5}{x - 3}$ 44. $\dfrac{3}{2 - x} > \dfrac{x}{2 + x}$

45. $2 - \dfrac{5}{x} + \dfrac{2}{x^2} \geq 0$ 46. $\dfrac{1}{x - 1} + \dfrac{1}{x + 1} > \dfrac{3}{4}$

Applications

47. *Waiting in Line* (Refer to Example 2.) A parking lot attendant can wait on 40 cars per hour. If cars arrive randomly at a rate of x cars per hour, then the average line length to enter the ramp is given by $f(x) = \dfrac{x^2}{1600 - 40x}$, where $0 \leq x < 40$.

(a) Solve the inequality $f(x) \leq 8$.

(b) Interpret your answer from part (a).

48. *Time Spent in Line* If a parking ramp attendant can wait on 3 vehicles per minute, and vehicles are leaving the ramp at x vehicles per minute, then the average wait in minutes for a car trying to exit is given by $f(x) = \dfrac{1}{3 - x}$.

(a) Solve the three-part inequality $5 \leq \dfrac{1}{3 - x} \leq 10$.

(b) Interpret your result from part (a).

49. *Slippery Roads* The coefficient of friction x measures the friction between the tires of a car and the road, where $0 < x \leq 1$. A smaller value of x indicates that the road is more slippery. If a car is traveling at 60 miles per hour, then the braking distance D in feet is given by $D(x) = \dfrac{120}{x}$.

(a) What happens to the braking distance as the coefficient of friction becomes smaller?

(b) Find values for the coefficient of friction x that correspond to a braking distance of 400 feet or more.

50. *Average Temperature* The monthly average high temperature in degrees Fahrenheit at Daytona Beach, Florida, can be approximated by

$$f(x) = 0.0145x^4 - 0.426x^3 + 3.53x^2 - 6.22x + 72.0,$$

where $x = 1$ corresponds to January, $x = 2$ to February, and so on. Estimate graphically when the monthly average high temperature is 75°F or more.

51. *Geometry* A cubical box is being manufactured to hold 213 cubic inches. If this measurement can vary between 212.8 cubic inches and 213.2 cubic inches, inclusively, by how much can the length x of a side of the cube vary?

52. *Construction* A cylindrical aluminum can is being manufactured so that its height h is 8 centimeters more than its radius r. Estimate values for the radius to the nearest hundredth that result in the can having a volume between 1000 and 1500 cubic centimeters inclusively.

Writing about Mathematics

53. Describe the steps to graphically solve a polynomial inequality in the form $p(x) > 0$.

54. Describe the steps to symbolically solve a rational inequality in the form $f(x) > 0$.

CHECKING BASIC CONCEPTS FOR SECTIONS 4.5 AND 4.6

1. Let $f(x) = \frac{1}{x-1} + 2$.
   (a) Find the domain of f.

   (b) Identify any vertical or horizontal asymptotes.

   (c) Sketch a graph of f that includes all asymptotes.

2. Find any horizontal or vertical asymptotes for the graph of $f(x) = \frac{4x^2}{x^2-4}$. State the domain of f.

3. Solve each equation. Check your result.
   (a) $\frac{3x-1}{1-x} = 1$ (b) $3 + \frac{8}{x} = \frac{35}{x^2}$

   (c) $\frac{x-3}{5-x} = \frac{x+4}{1-x}$

   (d) $\frac{1}{x-1} - \frac{1}{3(x+2)} = \frac{1}{x^2+x-2}$

4. Solve $2x^3 + x^2 - 6x < 0$. Write the solution set in interval notation.

5. Solve $\frac{x^2-1}{x+2} \geq 0$. Write the solution set in interval notation.

6. *Time Spent in Line*   Suppose a parking ramp attendant can wait on 4 vehicles per minute and vehicles are leaving the ramp randomly at an average rate of x vehicles per minute. Then the average time T in minutes spent waiting in line and paying the attendant is given by

$$T(x) = \frac{1}{4-x},$$

   where $0 \leq x < 4$.   (**Source:** N. Garber and L. Hoel, *Traffic and Highway Engineering.*)
   (a) Evaluate $T(2)$ and interpret the result.

   (b) Graph T for $0 \leq x < 4$.

   (c) What happens to the waiting time as x increases from 0 to (nearly) 4?

   (d) Find x if the waiting time is 5 minutes.

4.7 Power Functions and Radical Equations

- ◆ **Learn properties of rational exponents**
- ◆ **Learn radical notation**
- ◆ **Understand properties and graphs of power functions**
- ◆ **Use power functions to model data**
- ◆ **Solve equations involving rational exponents**
- ◆ **Solve equations involving radical expressions**
- ◆ **Use power regression to model data (optional)**

Introduction

Johannes Kepler (1571–1630) was the first to recognize that the orbits of planets are elliptical, rather than circular. He also found that a power function models the relationship between a planet's distance from the sun and its period of revolution. Table 4.18 lists the average distance x from the sun and the time y in years for several planets to orbit the sun. The distance x has been normalized so that Earth is one unit away from the sun. For example, Jupiter is 5.2 times farther from the sun than Earth and requires 11.9 years to orbit the sun.   (**Source:** C. Ronan, *The Natural History of the Universe.*)

TABLE 4.18

Planet	x (distance)	y (period)
Mercury	0.387	0.241
Venus	0.723	0.615
Earth	1.00	1.00
Mars	1.52	1.88
Jupiter	5.20	11.9
Saturn	9.54	29.5

A scatterplot of the data in Table 4.18 is shown in Figure 4.121. To model this data we might try a polynomial, such as $f(x) = x$ or $g(x) = x^2$. Figure 4.122 shows that $f(x) = x$ increases too slowly, and $g(x) = x^2$ increases too fast. To model this data, a new type of function is required. That is, we need a function in the form $h(x) = x^b$, where $1 < b < 2$. Polynomials only allow the exponent b to be a nonnegative integer, whereas *power functions* allow b to be any real number. See Figure 4.123.

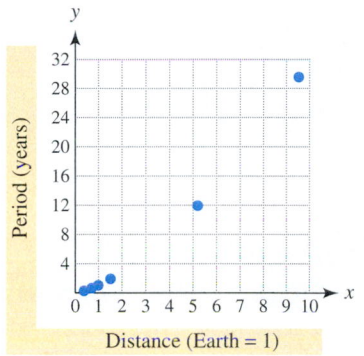

FIGURE 4.121 Orbital Data

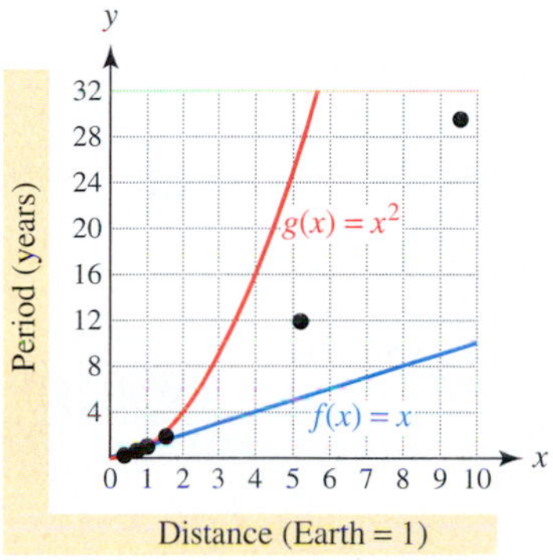

FIGURE 4.122 Polynomial Model

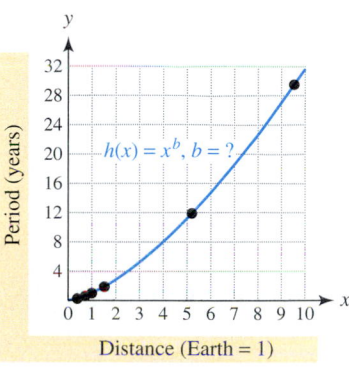

FIGURE 4.123 Power Model

In this section we discuss power functions and use a power function in Example 5 to model Kepler's data. Because power functions use rational numbers as exponents, we begin by reviewing properties of exponents and radical notation.

Rational Exponents and Radical Notation

The following properties can be used to simplify expressions with rational exponents.

Algebra Review
To review integer exponents, see Chapter R (page R-13). To review radical notation and rational exponents, see Chapter R (page R-50).

PROPERTIES OF RATIONAL EXPONENTS

Let m and n be positive integers with $\dfrac{m}{n}$ in lowest terms and $n \geq 2$. Let r and p be rational numbers. Assume that b is a nonzero real number and that each expression is a real number.

Property	*Example*
1. $b^{m/n} = (b^m)^{1/n} = (b^{1/n})^m$	$4^{3/2} = (4^3)^{1/2} = (4^{1/2})^3 = 2^3 = 8$
2. $b^{m/n} = \sqrt[n]{b^m} = (\sqrt[n]{b})^m$	$8^{2/3} = \sqrt[3]{8^2} = (\sqrt[3]{8})^2 = 2^2 = 4$
3. $(b^r)^p = b^{rp}$	$(2^{3/2})^4 = 2^6 = 64$
4. $b^{-r} = \dfrac{1}{b^r}$	$4^{-1/2} = \dfrac{1}{4^{1/2}} = \dfrac{1}{2}$
5. $b^r b^p = b^{r+p}$	$3^{5/2} \cdot 3^{3/2} = 3^{(5/2)+(3/2)} = 3^4 = 81$
6. $\dfrac{b^r}{b^p} = b^{r-p}$	$\dfrac{5^{5/4}}{5^{3/4}} = 5^{(5/4)-(3/4)} = 5^{1/2}$

EXAMPLE 1   Applying properties of exponents

Simplify each expression by hand.

(a) $16^{3/4}$ (b) $\dfrac{4^{1/3}}{4^{5/6}}$ (c) $27^{-2/3} \cdot 27^{1/3}$ (d) $(5^{3/4})^{2/3}$ (e) $(-125)^{-4/3}$

SOLUTION

(a) $16^{3/4} = (\sqrt[4]{16})^3 = (2)^3 = 8$

(b) $\dfrac{4^{1/3}}{4^{5/6}} = 4^{(1/3)-(5/6)} = 4^{-1/2} = \dfrac{1}{\sqrt{4}} = \dfrac{1}{2}$

(c) $27^{-2/3} \cdot 27^{1/3} = 27^{(-2/3)+(1/3)} = 27^{-1/3} = \dfrac{1}{\sqrt[3]{27}} = \dfrac{1}{3}$

(d) $(5^{3/4})^{2/3} = 5^{(3/4)(2/3)} = 5^{1/2}$ or $\sqrt{5}$

(e) $(-125)^{-4/3} = \dfrac{1}{(\sqrt[3]{-125})^4} = \dfrac{1}{(-5)^4} = \dfrac{1}{625}$

Now Try Exercises 3, 9, 11, 13, and 15 ◆

EXAMPLE 2   Writing radicals with rational exponents

Use positive rational exponents to write each expression.

(a) $\sqrt{x}$ (b) $\sqrt[3]{x^2}$ (c) $(\sqrt[4]{z})^{-5}$ (d) $\sqrt{\sqrt[3]{y} \cdot \sqrt[4]{y}}$

SOLUTION

(a) $\sqrt{x} = x^{1/2}$ (b) $\sqrt[3]{x^2} = (x^2)^{1/3} = x^{2/3}$

(c) $(\sqrt[4]{z})^{-5} = (z^{1/4})^{-5} = z^{-5/4} = \dfrac{1}{z^{5/4}}$

(d) $\sqrt{\sqrt[3]{y} \cdot \sqrt[4]{y}} = \left(y^{1/3} \cdot y^{1/4}\right)^{1/2} = \left(y^{(1/3)+(1/4)}\right)^{1/2} = \left(y^{7/12}\right)^{1/2} = y^{7/24}$

Now Try Exercises 25, 27, and 31 ◆

◆ **MAKING CONNECTIONS**

Rational Exponents and Radical Notation Any expression that is written with (noninteger) rational exponents can be written in radical notation, and any expression written in radical notation can be written with rational exponents. Normally we choose the form that is most convenient for a given situation. ◆

Power Functions and Models

Functions with rational exponents are often used to model physical characteristics of living organisms. For example, taller animals tend to be heavier than shorter animals. There is a relationship between an animal's height and weight. (See also Example 4.) This area of study in biology is called *allometry*, which attempts to model the relative sizes of different characteristics of an organism.

Power functions typically have rational exponents, and a special type of power function is a root function. These functions are defined in the following.

POWER FUNCTION

A function f given by $f(x) = x^b$, where b is a constant, is a **power function**. If $b = \frac{1}{n}$ for some integer $n \geq 2$, then f is a **root function** given by $f(x) = x^{1/n}$, or equivalently, $f(x) = \sqrt[n]{x}$.

Examples of power functions include

$$f_1(x) = x^2, \qquad f_2(x) = x^{3/4}, \qquad f_3(x) = x^{0.4}, \qquad \text{and} \qquad f_4(x) = \sqrt[3]{x^2}.$$

Frequently, the domain of a power function f is restricted to nonnegative numbers. Suppose the rational number $\frac{p}{q}$ is written in lowest terms. Then the domain of $f(x) = x^{p/q}$ is all real numbers whenever q is odd and all nonnegative real numbers whenever q is even. (If b is an irrational number, the domain of $f(x) = x^b$ is all nonnegative real numbers.) For example, the domain of $f(x) = x^{1/3} (f(x) = \sqrt[3]{x})$ is all real numbers, whereas the domain of $g(x) = x^{1/2} (g(x) = \sqrt{x})$ is all nonnegative numbers.

Graphs of three common power functions are shown in Figures 4.124–4.126.

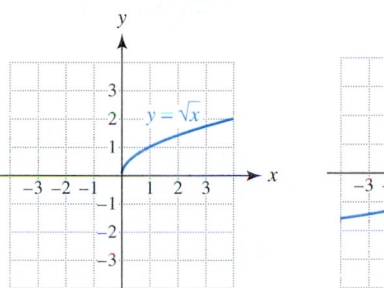

FIGURE 4.124 Square Root   **FIGURE 4.125** Cube Root

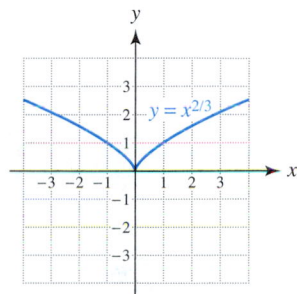

FIGURE 4.126 Cube Root Squared

EXAMPLE 3

Graphing power functions

Graph $f(x) = x^b$, $b = 0.3$, 1, and 1.7, for $x \geq 0$. Discuss the effect that b has on the graph of f.

SOLUTION The graphs of $y = x^{0.3}$, $y = x^1$, and $y = x^{1.7}$ are shown in Figure 4.127. Larger values of b cause the graph of f to increase faster. Note that each graph passes through the point $(1, 1)$. Why?   *Now Try Exercise 37* ◆

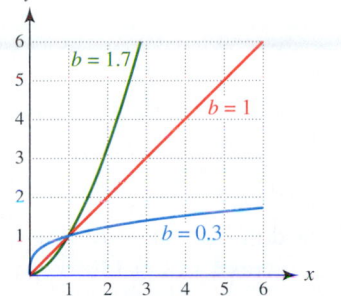

FIGURE 4.127 $f(x) = x^b$, $x \geq 0$

Note: Graphs of power functions with negative exponents are investigated in the Extended and Discovery Exercises from this section.

In the next example we model the size of the wings of birds.

EXAMPLE 4

Modeling wing size of a bird

Heavier birds have larger wings with more surface areas than do lighter birds. For some species of birds, this relationship can be modeled by $S(w) = 0.2w^{2/3}$, where w is the weight of the bird in kilograms and S is the surface area of the wings in square meters. (**Source:** C. Pennycuick, *Newton Rules Biology.*)

(a) Approximate $S(0.5)$ and interpret the result.

(b) What weight corresponds to a surface area of 0.25 square meter?

SOLUTION

(a) $S(0.5) = 0.2(0.5)^{2/3} \approx 0.126$. The wings of a bird that weighs 0.5 kilogram have a surface area of about 0.126 square meter.

(b) To answer this question we must solve the equation $0.2w^{2/3} = 0.25$.

$$0.2w^{2/3} = 0.25$$

$$w^{2/3} = \frac{0.25}{0.2} \qquad \text{Divide by 0.2.}$$

$$(w^{2/3})^3 = \left(\frac{0.25}{0.2}\right)^3 \qquad \text{Cube each side.}$$

$$w^2 = \left(\frac{0.25}{0.2}\right)^3 \qquad \text{Simplify.}$$

$$w = \pm\sqrt{\left(\frac{0.25}{0.2}\right)^3} \qquad \text{Square root property}$$

$$w \approx \pm 1.4 \qquad \text{Approximate.}$$

Since w must be positive, this means that the wings of a 1.4-kilogram bird have a surface area of about 0.25 square meter.

Now Try Exercise 83 ◆

In the next example we use a power function to model the planetary data presented in the introduction to this section.

EXAMPLE 5 Modeling the period of planetary orbits

Use the data in Table 4.18 on page 334 to complete the following.
(a) Make a scatterplot of the data. Estimate graphically a value for b so that $f(x) = x^b$ models the data.
(b) Numerically check the accuracy of f.
(c) The average distances of Uranus, Neptune, and Pluto from the sun are 19.2, 30.1, and 39.5, respectively. Use f to estimate the periods of revolution for these planets. Compare these answers to the actual values of 84.0, 164.8, and 248.5 years.

SOLUTION
(a) Make a scatterplot of the data and then graph $y = x^b$ for different values of b. By viewing the graphs of $y = x^{1.4}$, $y = x^{1.5}$, and $y = x^{1.6}$ in Figures 4.128–4.130, it can be seen that $b \approx 1.5$.

Calculator Help

To make a scatterplot, see Appendix B (page AP-7). To make a table like Figure 4.131, see Appendix B (page AP-15).

[0, 10, 1] by [0, 30, 10]   [0, 10, 1] by [0, 30, 10]   [0, 10, 1] by [0, 30, 10]

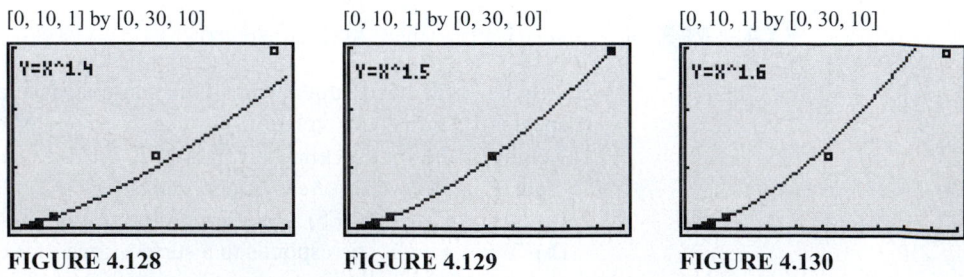

FIGURE 4.128   **FIGURE 4.129**   **FIGURE 4.130**

(b) Let $f(x) = x^{1.5}$ and table $Y_1 = X^{1.5}$. The values shown in Figure 4.131 model the data in Table 4.18 remarkably well.

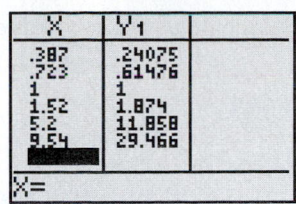

FIGURE 4.131

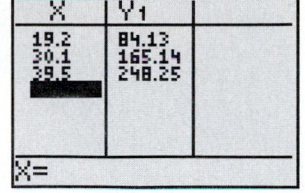

FIGURE 4.132

◆ **CLASS DISCUSSION**

Suppose that in Example 5 we let $f(x) = x^b$ and require that the equation $f(9.54) = 29.5$ be true. Can we symbolically solve the resulting equation, $9.54^b = 29.5$? Explain. ◆

(c) To approximate the number of years for Uranus, Neptune, and Pluto to orbit the sun, evaluate $f(x) = x^{1.5}$ at $x = 19.2, 30.1,$ and $39.5,$ as shown in Figure 4.132. These values are close to the actual values.   *Now Try Exercise 87* ◆

Equations Involving Rational Exponents

Equations sometimes have rational exponents. In the next example we demonstrate a basic technique used to solve these types of equation.

EXAMPLE 6   Solving an equation with rational exponents

Solve $2x^{5/2} - 7 = 23$. Approximate the answer to the nearest hundredth, and give graphical support.

SOLUTION

Symbolic Solution   Start by adding 7 to each side.

$$2x^{5/2} = 30 \qquad \text{Add 7 to each side.}$$

$$x^{5/2} = 15 \qquad \text{Divide by 2.}$$

$$(x^{5/2})^2 = 15^2 \qquad \text{Square each side.}$$

$$x^5 = 225 \qquad \text{Properties of exponents}$$

$$x = 225^{1/5} \qquad \text{Take the fifth root.}$$

$$x \approx 2.95 \qquad \text{Approximate.}$$

Graphical Solution   Graphical support is shown in Figure 4.133, where the graphs of $Y_1 = 2X^{\wedge}(5/2) - 7$ and $Y_2 = 23$ intersect near $(2.95, 23)$.   *Now Try Exercise 53* ◆

Equations that have rational exponents are sometimes reducible to quadratic form.

EXAMPLE 7   Solving an equation having negative exponents

Solve $15n^{-2} - 19n^{-1} + 6 = 0$.

SOLUTION Two methods to solve this equation are presented.
Method 1:   Use the substitution $u = n^{-1} = \frac{1}{n}$ and $u^2 = n^{-2} = \frac{1}{n^2}$.

$$15n^{-2} - 19n^{-1} + 6 = 0 \qquad \text{Given equation}$$

$$15u^2 - 19u + 6 = 0 \qquad \text{Let } u = n^{-1} \text{ and } u^2 = n^{-2}.$$

$$(3u - 2)(5u - 3) = 0 \qquad \text{Factor.}$$

$$u = \frac{2}{3} \text{ or } u = \frac{3}{5} \qquad \text{Zero-product property}$$

Because $u = \frac{1}{n}$, it follows that $n = \frac{1}{u}$. Thus $n = \frac{3}{2}$ or $n = \frac{5}{3}$.

Calculator Help

When entering $X^{\wedge}(5/2)$, be sure to put parentheses around the fraction 5/2.

$[-5, 5, 1]$ by $[-40, 40, 10]$

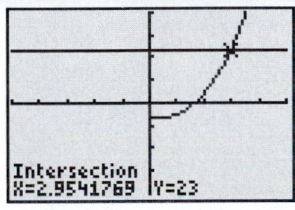

FIGURE 4.133

Method 2: Another way to solve this equation is to multiply each side by n^2 to clear fractions.

$$15n^{-2} - 19n^{-1} + 6 = 0 \qquad \text{\textit{Given equation}}$$
$$n^2(15n^{-2} - 19n^{-1} + 6) = n^2(0) \qquad \text{\textit{Multiply each side by } } n^2.$$
$$15n^2n^{-2} - 19n^2n^{-1} + 6n^2 = 0 \qquad \text{\textit{Distributive property}}$$
$$15 - 19n + 6n^2 = 0 \qquad \text{\textit{Properties of exponents}}$$
$$6n^2 - 19n + 15 = 0 \qquad \text{\textit{Rewrite the equation.}}$$
$$(2n - 3)(3n - 5) = 0 \qquad \text{\textit{Factor.}}$$
$$n = \frac{3}{2} \text{ or } n = \frac{5}{3} \qquad \text{\textit{Zero-product property}}$$

Now Try Exercise 57 ◆

In the next example we solve an equation with fractional exponents that can be written in quadratic form by using substitution.

EXAMPLE 8 Solving an equation having fractions for exponents

Solve $2x^{2/3} + 5x^{1/3} - 3 = 0$.

SOLUTION To solve this equation, use the substitutions $u = x^{1/3}$.

$$2x^{2/3} + 5x^{1/3} - 3 = 0 \qquad \text{\textit{Given equation}}$$
$$2(x^{1/3})^2 + 5(x^{1/3}) - 3 = 0 \qquad \text{\textit{Properties of exponents}}$$
$$2u^2 + 5u - 3 = 0 \qquad \text{\textit{Let } } u = x^{1/3}.$$
$$(2u - 1)(u + 3) = 0 \qquad \text{\textit{Factor.}}$$
$$u = \frac{1}{2} \text{ or } u = -3 \qquad \text{\textit{Zero-product property}}$$

Because $u = x^{1/3}$, it follows that $x = u^3$. Thus $x = \left(\frac{1}{2}\right)^3 = \frac{1}{8}$ or $x = (-3)^3 = -27$.

Now Try Exercise 61 ◆

Equations Involving Radicals

When solving equations that contain square roots, it is common to square each side of an equation. This is done in the next example.

EXAMPLE 9 Solving an equation containing a square root

Solve $x = \sqrt{15 - 2x}$. Check your solutions.

SOLUTION Begin by squaring each side of the equation.

$$x = \sqrt{15 - 2x} \qquad \text{\textit{Given equation}}$$
$$x^2 = (\sqrt{15 - 2x})^2 \qquad \text{\textit{Square each side.}}$$
$$x^2 = 15 - 2x \qquad \text{\textit{Simplify.}}$$
$$x^2 + 2x - 15 = 0 \qquad \text{\textit{Add 2x and subtract 15.}}$$
$$(x + 5)(x - 3) = 0 \qquad \text{\textit{Factor.}}$$
$$x = -5 \quad \text{or} \quad x = 3 \qquad \text{\textit{Solve.}}$$

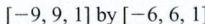

$[-9, 9, 1]$ by $[-6, 6, 1]$

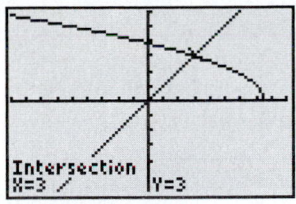

FIGURE 4.134

Now substitute these values in the given equation $x = \sqrt{15 - 2x}$.

$$-5 \neq \sqrt{15 - 2(-5)} = 5, \qquad 3 = \sqrt{15 - 2(3)}$$

Thus 3 is the only solution. This result is supported graphically in Figure 4.134, where $Y_1 = X$ and $Y_2 = \sqrt{(15 - 2X)}$. Notice that *no* point of intersection occurs when $x = -5$.

Now Try Exercise 69 ◆

The value -5 in Example 9 is called an **extraneous solution** because it does not satisfy the given equation. It is important to check results whenever *squaring* has been used to solve an equation.

Sometimes it is necessary to cube both sides of an equation, as demonstrated in the next example.

EXAMPLE 10   Solving an equation containing a cube root

Solve $\sqrt[3]{2x + 5} - 2 = 1$.

SOLUTION   Start by adding 2 to each side. Then cube each side.

$$\sqrt[3]{2x + 5} = 3 \qquad \text{Add 2 to each side.}$$
$$(\sqrt[3]{2x + 5})^3 = 3^3 \qquad \text{Cube each side.}$$
$$2x + 5 = 27 \qquad \text{Simplify.}$$
$$2x = 22 \qquad \text{Subtract 5 from each side.}$$
$$x = 11 \qquad \text{Divide by 2.}$$

The only solution is 11.

Now Try Exercise 77 ◆

Power Regression (Optional)

Rather than visually fit a curve to data, as was done in Example 5, we can use least-squares regression to fit the data. Least-squares regression was introduced in Section 2.1. In the next example we apply this technique to data from biology.

EXAMPLE 11   Modeling the length of a bird's wing

Table 4.19 lists the weight W and the wingspan L for birds of a particular species.

TABLE 4.19   Weights and Wingspans

W (kilograms)	0.5	1.5	2.0	2.5	3.0
L (meters)	0.77	1.10	1.22	1.31	1.40

Source: C. Pennycuick.

(a) Use power regression to model the data with $L = aW^b$. Graph the data and the equation.
(b) Approximate the wingspan for a bird weighing 3.2 kilograms.

SOLUTION
(a) Let x be the weight W and y be the length L. Enter the data, and then select power regression (PwrReg), as shown in Figures 4.135 and 4.136 on the next page. The results are shown in Figure 4.137. Let

$$y = 0.9674x^{0.3326} \qquad \text{or} \qquad L = 0.9674W^{0.3326}.$$

Calculator Help
To find an equation of least-squares
fit, see Appendix B (page AP-15).

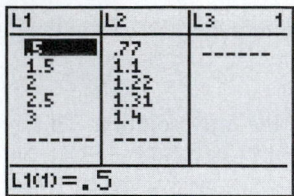

FIGURE 4.135

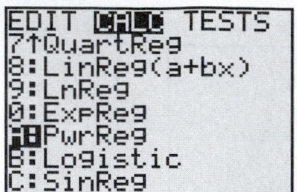

FIGURE 4.136

$[0, 4, 1]$ by $[0.5, 1.5, 0.5]$

FIGURE 4.137

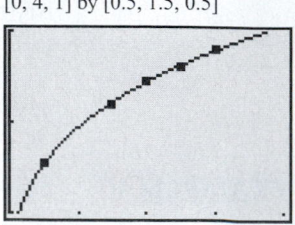

FIGURE 4.138

The data and equation are graphed in Figure 4.138.

(b) If a bird weighs 3.2 kilograms, this model predicts the wingspan to be

$$L = 0.9674(3.2)^{0.3326} \approx 1.42 \text{ meters.}$$ *Now Try Exercise 93* ◆

4.7

Putting it all Together

The following table outlines important concepts in this section.

Concept	Symbolic Representation	Examples
Rational exponents	$x^{m/n} = (x^m)^{1/n}$ $= (x^{1/n})^m$	$9^{3/2} = (9^3)^{1/2} = (729)^{1/2} = 27$ $9^{3/2} = (9^{1/2})^3 = (3)^3 = 27$
Radical notation	$x^{1/2} = \sqrt{x}$ $x^{1/3} = \sqrt[3]{x}$ $x^{m/n} = \sqrt[n]{x^m}$ $= (\sqrt[n]{x})^m$	$25^{1/2} = \sqrt{25} = 5$ $27^{1/3} = \sqrt[3]{27} = 3$ $8^{2/3} = \sqrt[3]{8^2} = \sqrt[3]{64} = 4$ $\sqrt{4^3} = (\sqrt{4})^3 = (2)^3 = 8$
Power function	$f(x) = x^b$, where b is a constant.	$f(x) = x^{5/4}$ $g(x) = x^{-3.14}$
Root function	$f(x) = x^{1/n}$, where $n \geq 2$ is an integer.	$f(x) = x^{1/2}$ or $f(x) = \sqrt{x}$ $g(x) = x^{1/5}$ or $g(x) = \sqrt[5]{x}$

In this section we studied power functions and root functions. In the first four chapters, we studied several types of functions. The following summary of functions lists several types of functions. This summary may be used as a reference for future work. Unless specified otherwise, each tick mark represents 1 unit.

Type of Function	Examples	Graphs
Linear function $f(x) = ax + b$	$f(x) = 0.5x - 1$ $g(x) = -3x + 2$ $h(x) = 2$	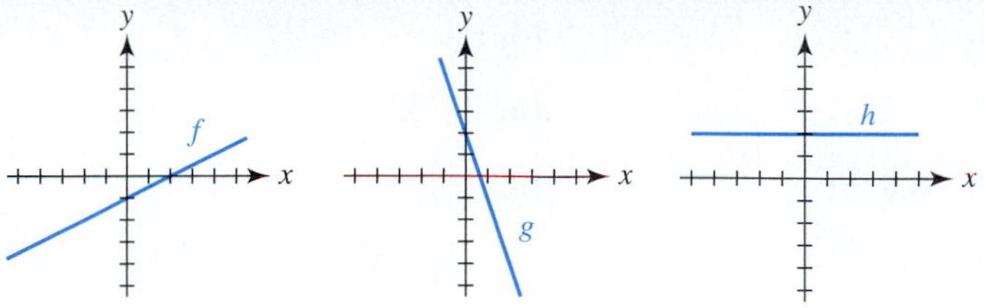
Polynomial function $f(x) = a_n x^n + \cdots + a_2 x^2 + a_1 x + a_0$	$f(x) = x^2 - 1$ $g(x) = x^3 - 4x - 1$ $h(x) = -x^4 + 4x^2 - 2$	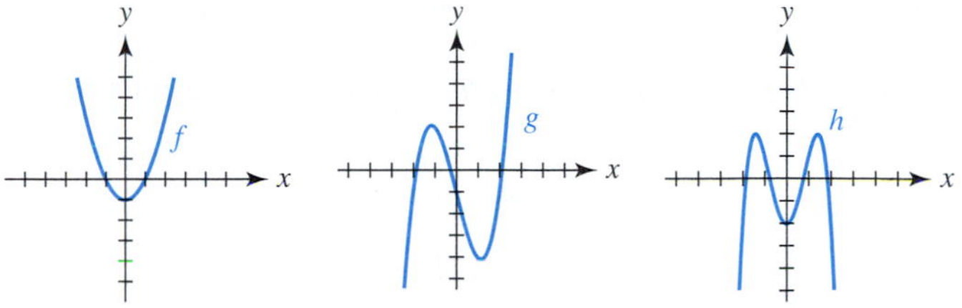
Rational function $f(x) = \frac{p(x)}{q(x)}$, where $p(x)$ and $q(x)$ are polynomials	$f(x) = \dfrac{1}{x}$ $g(x) = \dfrac{2x-1}{x+2}$ $h(x) = \dfrac{1}{x^2-1}$	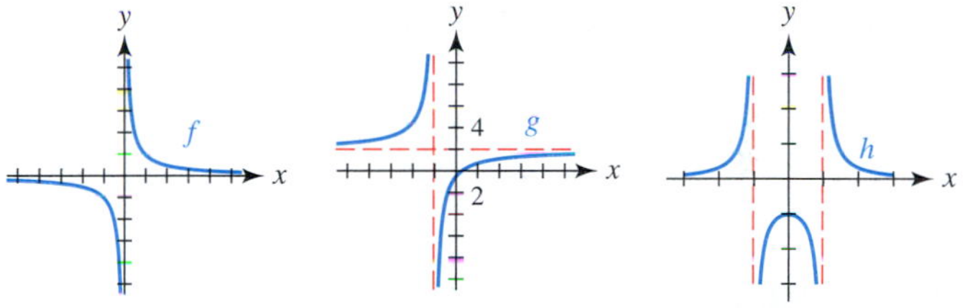
Root function $f(x) = x^{1/n}$, where $n \geq 2$ is an integer	$f(x) = x^{1/2} = \sqrt{x}$ $g(x) = x^{1/3} = \sqrt[3]{x}$ $h(x) = x^{1/4} = \sqrt[4]{x}$	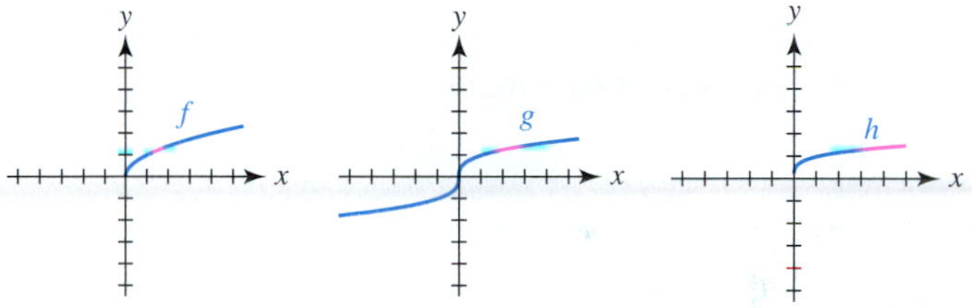
Power function $f(x) = x^b$, where b is a constant	$f(x) = x^{2/3}$ $g(x) = x^{1.41}$ $h(x) = x^3$	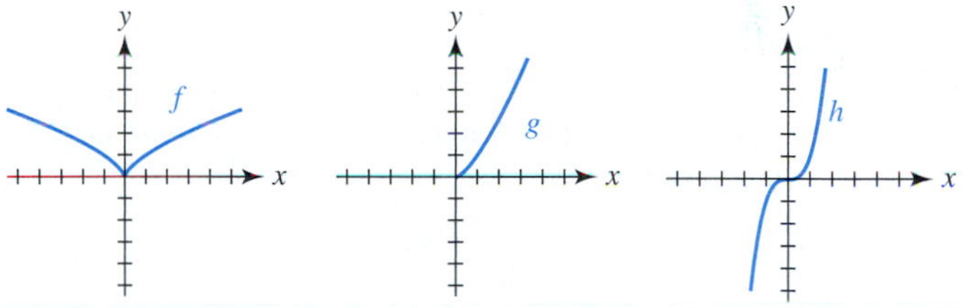

4.7 Exercises

Properties of Exponents

Exercises 1–22: Evaluate the expression by hand. Check your result with a calculator if you are uncertain about your answer.

1. $8^{2/3}$

2. $-16^{3/2}$

3. $16^{-3/4}$

4. $25^{-3/2}$

5. $-81^{0.5}$

6. $32^{1/5}$

7. $64^{1/6}$

8. $16^{-0.25}$

9. $(-9^{3/4})^2$

10. $(4^{-1/2})^{-4}$

11. $\dfrac{8^{5/6}}{8^{1/2}}$

12. $\dfrac{4^{-1/2}}{4^{3/2}}$

13. $27^{5/6} \cdot 27^{-1/6}$

14. $16^{2/3} \cdot 16^{-1/6}$

15. $(-27)^{-5/3}$

16. $(-32)^{-3/5}$

17. $(0.5^{-2})^2$

18. $\left(\tfrac{1}{4}\right)^{-3/2}$

19. $\left(\tfrac{2}{3}\right)^{-2}$

20. $\left(\tfrac{1}{9}\right)^{-1/2} + \left(\tfrac{1}{16}\right)^{-1/2}$

21. $1^{-1} + 2^{-1} + 3^{-1}$

22. $(8^{-1/3} + 27^{-1/3})^2$

Exercises 23–32: Use positive rational exponents to rewrite the expression.

23. $\sqrt[3]{2x}$

24. $\sqrt{x+1}$

25. $\sqrt[3]{z^5}$

26. $\sqrt[5]{x^2}$

27. $(\sqrt[4]{y})^{-3}$

28. $(\sqrt[3]{y^2})^{-5}$

29. $\sqrt{x} \cdot \sqrt[3]{x}$

30. $(\sqrt[5]{z})^{-3}$

31. $\sqrt{y} \cdot \sqrt[4]{y}$

32. $\dfrac{\sqrt[3]{x}}{\sqrt{x}}$

Power Functions

Exercises 33–36: Evaluate $f(x)$ at the given x. Approximate each result to the nearest hundredth.

33. $f(x) = x^{1.62}$, $\qquad x = 1.2$

34. $f(x) = x^{-0.71}$, $\qquad x = 3.8$

35. $f(x) = x^{3/2} - x^{1/2}$, $\qquad x = 50$

36. $f(x) = x^{5/4} - x^{-3/4}$, $\qquad x = 7$

Exercises 37 and 38: Match $f(x)$ with its graph. Assume that a and b are constants with $0 < a < 1 < b$.

37. $f(x) = x^a$

38. $f(x) = x^b$

a.

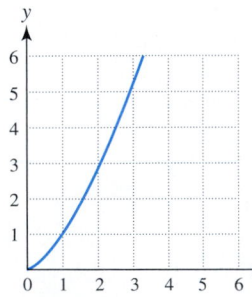

b.

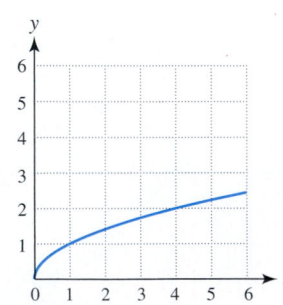

Exercises 39–44: Use translations of graphs to help sketch a graph of f.

39. $f(x) = \sqrt{x} + 1$

40. $f(x) = \sqrt[3]{x} - 1$

41. $f(x) = x^{2/3} - 1$

42. $f(x) = \sqrt{x-1}$

43. $f(x) = \sqrt{x+2} - 1$

44. $f(x) = (x-1)^{2/3}$

Equations Involving Rational Exponents

Exercises 45–64: Solve the equation. Check your answers.

45. $x^3 = 8$

46. $x^4 = \tfrac{1}{81}$

47. $x^{1/4} = 3$

48. $x^{1/3} = \tfrac{1}{5}$

49. $x^{2/5} = 4$

50. $x^{2/3} = 16$

51. $2(x^{1/5} - 2) = 0$

52. $x^{1/2} + x^{1/2} = 8$

53. $2x^{1/3} - 5 = 1$

54. $4x^{3/2} + 5 = 21$

55. $n^{-2} + 3n^{-1} + 2 = 0$

56. $2n^{-2} - n^{-1} = 3$

57. $5n^{-2} + 13n^{-1} = 28$

58. $3n^{-2} - 19n^{-1} + 20 = 0$

59. $x^{2/3} - x^{1/3} - 6 = 0$

60. $x^{2/3} + 9x^{1/3} + 14 = 0$

61. $6x^{2/3} - 11x^{1/3} + 4 = 0$

62. $10x^{2/3} + 29x^{1/3} + 10 = 0$

63. $x^{3/4} - x^{1/2} - x^{1/4} + 1 = 0$

64. $x^{3/4} - 2x^{1/2} - 4x^{1/4} + 8 = 0$

Equations Involving Radicals

Exercises 65–82: Solve the equation. Check your answers.

65. $\sqrt{x+2} = x - 4$

66. $\sqrt{2x+1} = 13$

67. $\sqrt{3x+7} = 3x + 5$

68. $\sqrt{1-x} = x + 5$

69. $\sqrt{5x-6} = x$

70. $x - 5 = \sqrt{5x-1}$

71. $\sqrt{x+5} + 1 = x$

72. $\sqrt{4-3x} = x + 8$

73. $\sqrt{x+1} + 3 = \sqrt{3x+4}$ (*Hint:* You will need to square the equation twice.)

74. $\sqrt{x} = \sqrt{x-5} + 1$

75. $\sqrt{2x+3} - \sqrt{x+1} = 1$

76. $\sqrt{2x-4} + 2 = \sqrt{3x+4}$

77. $\sqrt[3]{z+1} = -3$

78. $\sqrt[3]{z} + 5 = 4$

79. $\sqrt[3]{x+1} = \sqrt[3]{2x-1}$

80. $\sqrt[3]{2x^2+1} = \sqrt[3]{1-x}$

81. $\sqrt[4]{x-2} + 4 = 20$

82. $\sqrt[4]{2x+3} = \sqrt{x+1}$

Applications

83. *Modeling Wing Size* Suppose that the surface area S of a bird's wings in square feet can be modeled by $S(w) = 1.27w^{2/3}$, where w is the weight of the bird in pounds. Estimate the weight of a bird with wings having a surface area of 3 square feet.

84. *Modeling Wingspan* The wingspan L in feet of a bird weighing W pounds is given by $L = 2.43W^{0.3326}$. Estimate the wingspan of a bird that weighs 5.2 pounds.

85. *Modeling Planetary Orbits* The formula $f(x) = x^{1.5}$ calculates the number of years it would take for a planet to orbit the sun if its average distance from the sun is x times farther than Earth. If there were a planet located 15 times farther from the sun than Earth, how many years would it take for the planet to orbit the sun?

86. *Modeling Planetary Orbits* (Refer to Exercise 85.) If there were a planet that took 200 years to orbit the sun, what would be its average distance x from the sun compared to Earth?

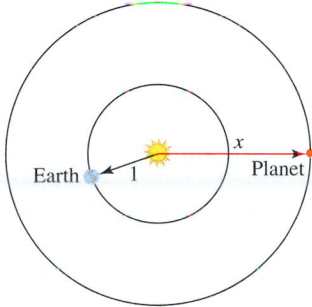

87. *Trout and Pollution* Rainbow trout are sensitive to zinc ions in the water. High concentrations are lethal. The average survival times x in minutes for trout in various concentrations of zinc ions y in milligrams per liter (mg/L) are listed in the table. (**Source:** C. Mason, *Biology of Freshwater Pollution.*)

x (min)	0.5	1	2	3
y (mg/L)	4500	1960	850	525

(a) These data can be modeled by $f(x) = ax^b$, where a and b are constants. Determine an appropriate value for a. (*Hint:* Let $f(1) = 1960$.)

(b) Estimate b.

(c) Evaluate $f(4)$ and interpret the result.

88. *Asbestos and Cancer* Insulation workers who were exposed to asbestos and employed before 1960 experienced an increased likelihood of lung cancer. If a group of insulation workers have a cumulative total of 100,000 years of work experience with their first date of employment x years ago, then the number of lung cancer cases occurring within the group can be modeled by $N(x) = 0.00437x^{3.2}$. (**Source:** A. Walker, *Observation and Inference.*)

(a) Calculate $N(x)$ when $x = 5$, 10, and 20. What happens to the likelihood of cancer as x increases?

(b) If x doubles, does the number of cancer cases also double?

(c) Solve the equation $N(x) = 100$, and interpret your answer.

89. *Fiddler Crab Size* Allometric relations often can be modeled by $f(x) = ax^b$, where a and b are constants. One study of the male fiddler crab showed a connection between the weight of the large claws and its total body weight. For a crab weighing over 0.75 gram, the weight of its claws can be estimated by $f(x) = 0.445x^{1.25}$. The input x is the weight of the crab in grams and the output $f(x)$ is the weight of the claws in grams. (**Sources:** J. Huxley, *Problems of Relative Growth;* D. Brown and P. Rothery, *Models in Biology: Mathematics, Statistics and Computing.*)

(a) Predict the weight of the claws for a 2-gram crab.

(b) Approximate graphically the weight of a crab that has 0.5-gram claws.

(c) Solve part (b) symbolically.

90. *Weight and Height* Allometry can be applied to height-weight relationships for humans. The average weight in pounds for men and women can be estimated by $f(x) = ax^{1.7}$, where x represents a person's height in inches and a is a constant that depends on the sex of the individual.

(a) If an average weight for a 68-inch tall man is 152 pounds, approximate a. Use f to estimate the average weight of a 66-inch tall man.

(b) If an average weight for a 68-inch tall woman is 137 pounds, approximate a. Use f to estimate the average weight of a 70-inch tall woman.

Exercises 91 and 92: Power Regression The table contains data that can be modeled by a function of the form $f(x) = ax^b$. Use regression to find the constants a and b to the nearest hundredth. Graph f and the data.

91.

x	2	4	6	8
$f(x)$	3.7	4.2	4.6	4.9

92.

x	3	6	9	12
$f(x)$	23.8	58.5	99.2	144

93. *Pulse Rate and Weight* According to one model, the rate at which an animal's heart beats varies with its weight. Smaller animals tend to have faster pulses, whereas larger animals tend to have slower pulses. The accompanying table lists average pulse rates in beats per minute (bpm) for animals with various weights in pounds (lb). Use regression (or some other method) to find values for a and b so that $f(x) = ax^b$ models these data.

Weight (lb)	40	150	400	1000	2000
Pulse (bpm)	140	72	44	28	20

Source: C. Pennycuick.

94. *Pulse Rate and Weight* (Continuation of Exercise 93) Use the results in the previous exercise to calculate the pulse rates for a 60-pound dog and a 2-ton whale.

Writing about Mathematics

95. Can a function be both a polynomial function and a power function? Explain.

96. Explain the basic steps needed to solve equations that contain square roots of variables.

EXTENDED AND DISCOVERY EXERCISES

1. *Odd Root Functions* Graph $y = \sqrt[n]{x}$ for $n = 3, 5,$ and 7. State some generalizations about a graph of an odd root function.

2. *Even Root Functions* Graph $y = \sqrt[n]{x}$ for $n = 2, 4,$ and 6. State some generalizations about a graph of an even root function.

3. *Power Functions* Graph $y = x^b$ for $b = -1, -3,$ and -5. State some generalizations about a graph of a power function, where b is a negative odd integer.

4. *Power Functions* Graph $y = x^b$ for $b = -2, -4,$ and -6. State some generalizations about a graph of a power function, where b is a negative even integer.

5. Find the difference quotient for $f(x) = \sqrt{x}$.

6. Find the difference quotient for $f(x) = \frac{1}{x}$.

CHECKING BASIC CONCEPTS FOR SECTION 4.7

1. Simplify each expression without a calculator.
(a) $-4^{3/2}$ (b) $(8^{-2})^{1/3}$ (c) $\sqrt[3]{27^2}$

2. Solve the equation $4x^{3/2} - 3 = 29$.

3. Solve the equation $\sqrt{5x - 4} = x - 2$.

4. Solve each equation.
(a) $n^{-2} + 6n^{-1} = 16$

(b) $2x^{2/3} + 5x^{1/3} - 12 = 0$

5. Find constants a and b so that $f(x) = ax^b$ models the data.

x	1	2	3	4
$f(x)$	2	2.83	3.46	4

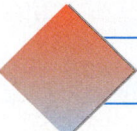

CHAPTER 4 SUMMARY

CONCEPT	EXPLANATION AND EXAMPLES

SECTION 4.1 NONLINEAR FUNCTIONS AND THEIR GRAPHS

POLYNOMIAL FUNCTION

Can be represented by $f(x) = a_n x^n + \cdots + a_2 x^2 + a_1 x + a_0$
The leading coefficient is $a_n \neq 0$ and the degree is n.

Example: $f(x) = -4x^3 - 2x^2 + 6x + \frac{1}{2}$
Leading coefficient: -4; degree: 3

INCREASING/DECREASING

f increases on interval I if, whenever $x_1 < x_2$, $f(x_1) < f(x_2)$.
f decreases on interval I if, whenever $x_1 < x_2$, $f(x_1) > f(x_2)$.

Example: $f(x) = x^2$ increases on $[0, \infty)$ and decreases on $(-\infty, 0]$.

ABSOLUTE AND LOCAL EXTREMA

The accompanying graph has the following extrema.
Absolute maximum: none   Absolute minimum: -14.8
Local maximum: 1   Local minimums: $-4.3, -14.8$

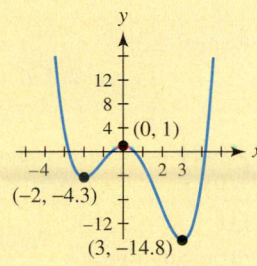

EVEN FUNCTION

$f(-x) = f(x)$; the graph is symmetric with respect to the y-axis.

Example: $f(x) = x^4 - 3x^2$

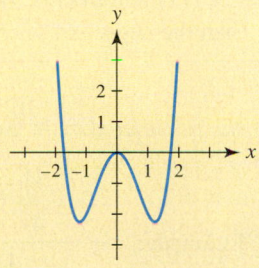

CONCEPT	EXPLANATION AND EXAMPLES

SECTION 4.1 NONLINEAR FUNCTIONS AND THEIR GRAPHS (CONTINUED)

ODD FUNCTION

$f(-x) = -f(x)$; the graph is symmetric with respect to the origin.

Example: $f(x) = x - x^3$

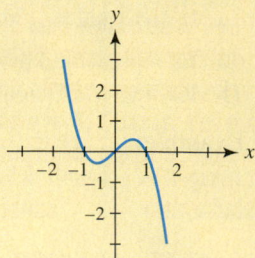

SECTION 4.2 POLYNOMIAL FUNCTIONS AND MODELS

GRAPHS OF POLYNOMIAL FUNCTIONS

Their graphs are continuous with no breaks and their domains include all real numbers. The graph of a polynomial function of degree $n \geq 1$ has at most n x-intercepts and at most $n - 1$ turning points.

Examples: See Putting It All Together for Section 4.2 on pages 266–267.

PIECEWISE-POLYNOMIAL FUNCTION

Example:

$$f(x) = \begin{cases} x^2 - 2 & \text{if } x < 0 \\ 1 - 2x & \text{if } x \geq 0 \end{cases}$$

$f(-2) = (-2)^2 - 2 = 2$
$f(0) = 1 - 2(0) = 1$
$f(2) = 1 - 2(2) = -3$

f is discontinuous at $x = 0$.

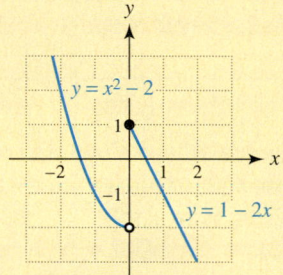

SECTION 4.3 REAL ZEROS OF POLYNOMIAL FUNCTIONS

DIVISION ALGORITHM

Let $f(x)$ and $d(x)$ be two polynomials with the degree of $d(x)$ greater than zero and less than the degree of $f(x)$. Then there exist unique polynomials $q(x)$ and $r(x)$ such that

$$f(x) = d(x) \cdot q(x) + r(x),$$

where either $r(x) = 0$ or the degree of $r(x)$ is less than the degree of $d(x)$. That is,

Dividend = (Divisor) · (Quotient) + Remainder.

Example: $\dfrac{x^2 - 4x + 5}{x - 1} = x - 3 + \dfrac{2}{x - 1}$. That is,

$$x^2 - 4x + 5 = (x - 1)(x - 3) + 2.$$

REMAINDER THEOREM

If a polynomial $f(x)$ is divided by $x - k$, the remainder is $f(k)$.

Example: If $x^2 - 4x + 5$ is divided by $x - 1$ the remainder is

$$f(1) = 1^2 - 4(1) + 5 = 2.$$

CONCEPT	EXPLANATION AND EXAMPLES

SECTION 4.3   REAL ZEROS OF POLYNOMIAL FUNCTIONS (CONTINUED)

FACTOR THEOREM

A polynomial $f(x)$ has a factor $x - k$ if and only if $f(k) = 0$.

Example: $f(x) = x^2 - 3x + 2$;
$f(1) = 0$ implies $(x - 1)$ is a factor of $x^2 - 3x + 2$.

COMPLETE FACTORED FORM

$f(x) = a_n(x - c_1)(x - c_2) \cdots (x - c_n)$, where a_n is the leading coefficient of the polynomial $f(x)$ and the c_k are its zeros.

Example: $f(x) = -2x^3 + 8x$ has zeros of $-2, 0, 2$.
$f(x) = -2(x + 2)(x - 0)(x - 2)$

MULTIPLE ZEROS

Odd multiplicity: graph crosses the x-axis.
Even multiplicity: graph touches but does not cross the x-axis.

Example: Let $f(x) = -2(x + 1)^3(x - 4)^2$; $f(x)$ has a zero of -1 with odd multiplicity 3 and a zero of 4 with even multiplicity 2.

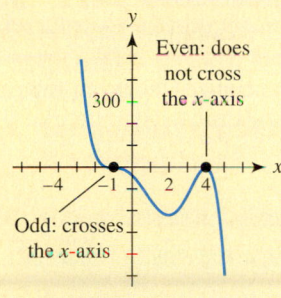

SECTION 4.4   THE FUNDAMENTAL THEOREM OF ALGEBRA

IMAGINARY UNIT

$i = \sqrt{-1}, \qquad i^2 = -1$

Examples: $\sqrt{-4} = 2i, \ \sqrt{-7} = i\sqrt{7}$

COMPLEX NUMBER

$a + bi$, where a and b are real numbers (standard form)
Complex numbers include all real numbers. We can add, subtract, multiply, and divide complex numbers.

Example: $(2 - 3i) + (1 + 5i) = (2 + 1) + (-3 + 5)i = 3 + 2i$

COMPLEX SOLUTIONS

The quadratic formula can be used to solve quadratic equations with complex solutions.

Example: The solutions to $x^2 - x + 2 = 0$ are

$$x = \frac{1 \pm \sqrt{(-1)^2 - 4(1)(2)}}{2(1)} = \frac{1}{2} \pm i\frac{\sqrt{7}}{2}.$$

CONCEPT	EXPLANATION AND EXAMPLES

SECTION 4.4 THE FUNDAMENTAL THEOREM OF ALGEBRA (CONTINUED)

FUNDAMENTAL THEOREM OF ALGEBRA

A polynomial $f(x)$ of degree $n \geq 1$ has at least one complex zero.

Explanation: With complex numbers, any polynomial can be written in complete factored form.

Examples:
$$x^2 + 1 = (x + i)(x - i)$$
$$3x^2 - 3x - 6 = 3(x + 1)(x - 2)$$

NUMBER OF ZEROS THEOREM

A polynomial of degree n has at most n distinct zeros.

Example: A cubic polynomial has at most 3 distinct zeros.

SECTION 4.5 RATIONAL FUNCTIONS AND MODELS

RATIONAL FUNCTION

$f(x) = \dfrac{p(x)}{q(x)}$, where $p(x)$ and $q(x) \neq 0$ are polynomials.

Example: $f(x) = \dfrac{2x - 3}{x - 1}$

Horizontal asymptote: $y = 2$
Vertical asymptote: $x = 1$

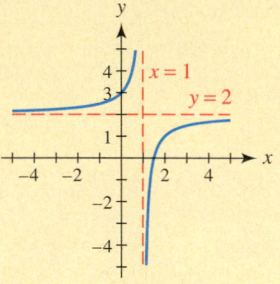

SOLVING RATIONAL EQUATIONS

Multiply each side of the equation by a common denominator.

Example: $\dfrac{-24}{x - 3} - 4 = x + 3$

$-24 - 4(x - 3) = (x + 3)(x - 3)$ Multiply by $x - 3$.

$-24 - 4x + 12 = x^2 - 9$

$0 = x^2 + 4x + 3$

$0 = (x + 3)(x + 1)$

$x = -3$ or $x = -1$ Both solutions check.

DIRECT VARIATION

Let x and y denote two quantities and n be a positive number. Then y is *directly proportional to the nth power of x*, or y *varies directly as the nth power of x*, if there exists a nonzero number k such that $y = kx^n$.

Example: Because $V = \frac{4}{3}\pi r^3$, the volume of a sphere varies directly as the 3rd power of the radius.

CONCEPT	EXPLANATION AND EXAMPLES

SECTION 4.5 RATIONAL FUNCTIONS AND MODELS (CONTINUED)

INVERSE VARIATION

Let x and y denote two quantities and n be a positive number. Then y is *inversely proportional to the nth power of x*, or *y varies inversely as the nth power of x*, if there exists a nonzero number k such that $y = \frac{k}{x^n}$.

Example: Because $I = \frac{k}{D^2}$, the intensity of a light source varies inversely as the square of the distance from the light source.

SECTION 4.6 POLYNOMIAL AND RATIONAL INEQUALITIES

POLYNOMIAL INEQUALITY

Write the inequality as $p(x) > 0$, where $>$ may be replaced by $\geq$, $<$, or $\leq$. Replace the inequality sign with an equals sign, and solve this equation. The solutions are called boundary numbers. Then use a graph or table to find the solution set to the given inequality.

Example: $4x - x^3 > 0$; Boundary numbers $-2, 0, 2$
Solution set: $(-\infty, -2) \cup (0, 2)$

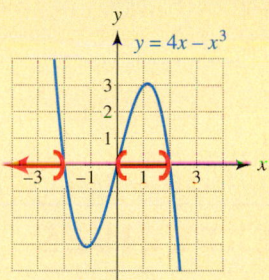

RATIONAL INEQUALITY

As with polynomial inequalities, find the boundary numbers, including x-values where any expressions are undefined.

Example: $\frac{(x + 2)(x - 3)}{x} \geq 0$; Boundary numbers: $-2, 0, 3$
Solution set: $[-2, 0) \cup [3, \infty)$

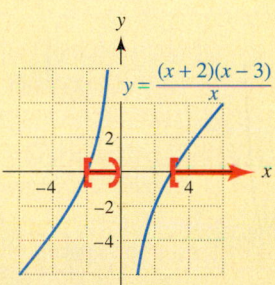

SECTION 4.7 POWER FUNCTIONS AND RADICAL EQUATIONS

RATIONAL EXPONENTS

$x^{m/n} = \sqrt[n]{x^m} = (\sqrt[n]{x})^m$

Example: $25^{3/2} = \sqrt{25^3} = (\sqrt{25})^3 = 5^3 = 125$

POWER FUNCTION

$f(x) = x^b$,   where b is a constant.

Examples: $f(x) = x^{4/3}$, $g(x) = x^{1.72}$

CONCEPT	EXPLANATION AND EXAMPLES

SECTION 4.7 POWER FUNCTIONS AND RADICAL EQUATIONS (CONTINUED)

SOLVING RADICAL EQUATIONS

When an equation contains a square root, isolate the square root and then square each side. *Be sure to check your answers.*

Example: $x + \sqrt{3x - 3} = 1$

$$\sqrt{3x - 3} = 1 - x \qquad \text{Subtract } x.$$

$$3x - 3 = (1 - x)^2 \qquad \text{Square each side.}$$

$$3x - 3 = 1 - 2x + x^2 \qquad \text{Expand.}$$

$$0 = x^2 - 5x + 4 \qquad \text{Combine terms.}$$

$$0 = (x - 1)(x - 4) \qquad \text{Factor.}$$

$$x = 1 \quad \text{or} \quad x = 4 \qquad \text{Solve.}$$

Check: 1 is a solution, but 4 is not.

$$1 + \sqrt{3(1) - 3} = 1 \qquad 4 + \sqrt{3(4) - 3} \neq 1$$

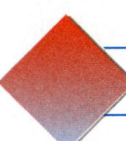

REVIEW EXERCISES

1. State the degree and leading coefficient of the polynomial $f(x) = 4 + x - 2x^2 - 7x^3$.

2. Use the graph of f to estimate where f is increasing or decreasing. Use interval notation.

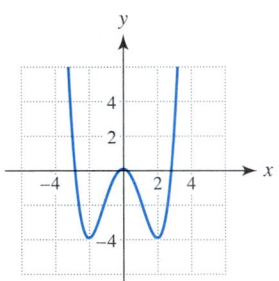

Exercises 3 and 4: Complete the following.
   (a) Graph f.
   (b) Estimate where f is increasing or decreasing. Use interval notation.
3. $f(x) = x^2 - x$ 4. $f(x) = x^3 - 4x$

Exercises 5 and 6: Use the graph of f to estimate any
   (a) local extrema, and
   (b) absolute extrema.

5.
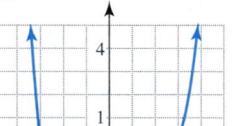

6.
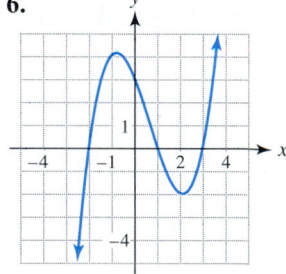

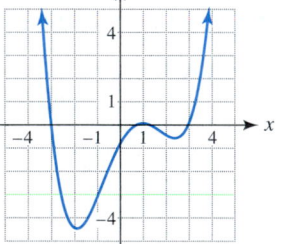

7. Graph $f(x) = -0.25x^4 + 0.67x^3 + 9.5x^2 - 20x - 50$.
   (a) Approximate any local extrema.

   (b) Approximate any absolute extrema.

   (c) Determine where f is increasing or decreasing. Use interval notation.

8. Find any absolute extrema on the graph of the polynomial $g(x) = -3(x - 2)^2 + 4$.

9. Graph each quartic (degree 4) polynomial. Count the local maxima, local minima, and x-intercepts.
   (a) $f(x) = x^4 + 2x^3 - 9x^2 - 2x + 20$

(b) $g(x) = -0.4x^4 + 3.6x^2$

(c) $h(x) = x^4 + 2x^3 - 21x^2 - 22x + 40$

10. Graph each quintic (degree 5) polynomial. Estimate any turning points.

(a) $f(x) = 0.03x^5 - 0.21x^4 + 0.21x^3 + 0.57x^2$
$- 0.48x - 0.6$

(b) $g(x) = -0.02x^5 + 1$

(c) $h(x) = 0.01x^5 - 0.09x^4 - 0.22x^3 + 2.72x^2$
$- 0.96x - 12.8$

Exercises 11–14: Determine if f is even, odd, or neither.

11. $f(x) = 2x^6 - 5x^4 - x^2$ 12. $f(x) = -5x^3 - 18$

13. $f(x) = 7x^5 + 3x^3 - x$ 14. $f(x) = \dfrac{1}{1 + x^2}$

Exercises 15 and 16: The table is a complete representation of f. Decide if f is even, odd, or neither.

15.

x	-4	-2	0	2	4
$f(x)$	13	7	0	-7	-13

16.

x	-5	-3	-1	1	3	5
$f(x)$	-6	2	7	7	2	-6

17. If $f(2) = 4$, then give a point that lies on the graph of $y = -f(x + 1) - 2$.

18. Is there an odd function f such that $f(-2) = 4$ and $f(2) = 4$? Explain.

Exercises 19–22: Sketch a graph of a polynomial function that satisfies the given conditions.

19. Cubic, two x-intercepts, and a positive leading coefficient

20. Quartic (degree 4) with no x-intercepts

21. Even with no absolute minimum, increasing on $(-\infty, 0]$, and decreasing on $[0, \infty)$

22. Degree 4 with a positive leading coefficient, three turning points, and one x-intercept

Exercises 23 and 24: Use the graph of the polynomial function to complete the following.

(a) Determine the number of turning points and estimate any x-intercepts.

(b) State whether the leading coefficient is positive or negative.

(c) Determine the minimum degree of f.

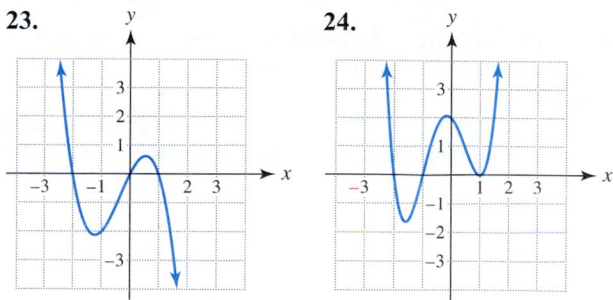

23.

24.

Exercises 25 and 26: Predict the end behavior of the graph of f.

25. $f(x) = -2x^3 + 4x - 2$

26. $f(x) = 1 - 2x - x^4$

27. Find the average rate of change of $f(x) = x^3 + 1$ from $x = -2$ to $x = -1$.

28. Find the difference quotient for $g(x) = 4x^3$.

29. Let $f(x)$ be given by

$$f(x) = \begin{cases} 2x & \text{if } 0 \le x < 2 \\ 8 - x^2 & \text{if } 2 \le x \le 4. \end{cases}$$

(a) Sketch a graph of f. Is f continuous on its domain?

(b) Evaluate $f(1)$ and $f(3)$.

(c) Solve the equation $f(x) = 2$.

30. Determine the type of symmetry that the graph of $g(x) = x^5 - 4x^3$ exhibits.

Exercises 31–34: Divide the expression.

31. $\dfrac{14x^3 - 21x^2 - 7x}{7x}$ 32. $\dfrac{2x^3 - x^2 - 4x + 1}{x + 2}$

33. $\dfrac{4x^3 - 7x + 4}{2x + 3}$ 34. $\dfrac{3x^3 - 5x^2 + 13x - 18}{x^2 + 4}$

35. The polynomial given by $f(x) = \frac{1}{2}x^3 - 3x^2 + \frac{11}{2}x - 3$ has zeros 1, 2, and 3. Write its complete factored form.

36. Find the complete factored form of the polynomial $f(x) = x^4 + 2x^3 - 13x^2 - 14x + 24$ given that 1 and 3 are zeros.

37. Find the complete factored form of the polynomial $f(x) = 2x^3 + 3x^2 - 18x + 8$ graphically.

38. Write a complete factored form of a quintic (degree 5) polynomial $f(x)$ that has zeros -2 and 2 with multiplicities 2 and 3, respectively.

39. Find the complete factored form of the polynomial $f(x) = x^3 - 2x^2 - 5x + 6$, given 3 is a zero.

40. Use the graph of $y = f(x)$ to write its complete factored form. (Do not assume that the leading coefficient is ± 1.)

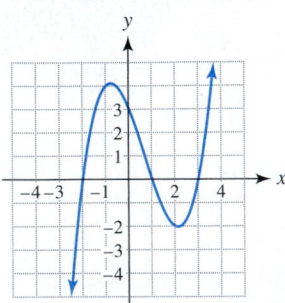

41. Use the graph of the fourth-degree polynomial f to estimate its zeros. Write its complete factored form if its leading coefficient is ± 1.

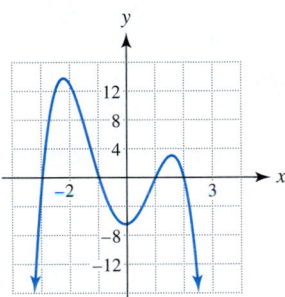

42. Write the complete factored form of a polynomial of degree six with leading coefficient 5 and three zeros: -2 with multiplicity one, 1 with multiplicity two, and 4 with multiplicity three.

43. What is the maximum number of times that a horizontal line can intersect the graph of each type of polynomial?
(a) linear (degree 1)   **(b)** quadratic

(c) cubic

44. (a) Give the maximum number of times that a horizontal line can intersect the graph of a polynomial of degree $n \geq 1$.

(b) State the maximum number of real solutions to the polynomial equation $a_n x^n + \cdots + a_2 x^2 + a_1 x + a_0 = k$, where k is a constant and $n \geq 1$.

Exercises 45 and 46: Use the rational zero test to determine any rational zeros of $f(x)$.

45. $f(x) = 2x^3 + x^2 - 13x + 6$

46. $f(x) = x^3 + x^2 - 11x - 11$

Exercises 47–54: Solve the equation.

47. $9x = 3x^3$ **48.** $x^3 = 4x$

49. $x^3 - x^2 - 6x = 0$ **50.** $x^3 - 2x^2 - 4x = 0$

51. $x^4 - 3x^2 + 2 = 0$ **52.** $4x^4 + 3 = 7x^2$

53. $2x^3 + x^2 = 6x + 3$

54. $3x^3 + 50 = -x(2x + 75)$

Exercises 55 and 56: Solve the equation graphically. Round your answers to the nearest hundredth.

55. $x^3 - 3x + 1 = 0$ **56.** $x^4 - 2x = 2$

Exercises 57–60: Write the expression in standard form.

57. $(2 - 3i) + (-3 + 3i)$ **58.** $(-5 + 3i) - (-3 - 5i)$

59. $(3 + 2i)(-4 - i)$ **60.** $\dfrac{3 + 2i}{2 - i}$

Exercises 61–64: Find all real and imaginary solutions.

61. $4x^2 + 9 = 0$ **62.** $2x^2 + 3 = 2x$

63. $x^3 + x = 0$ **64.** $x^4 + 3x^2 + 2 = 0$

65. Determine graphically the number of real zeros and the number of imaginary zeros of $f(x) = x^3 - 3x^2 + 3x - 9$.

66. Write a polynomial $f(x)$ in complete factored form that has degree 3, leading coefficient 4, and zeros 1, $3i$, and $-3i$. Then write $f(x)$ in expanded form.

67. Find the complete factorization of $f(x) = 2x^2 + 4$.

68. Use the graph of $f(x) = 2x^4 - x^2 - 1$ to predict the number of real zeros and the number of imaginary zeros of f. Find these zeros symbolically.

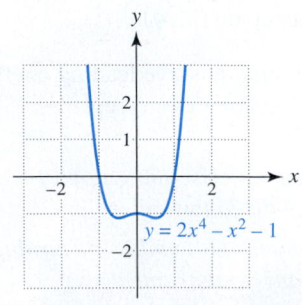

Exercises 69 and 70: Find the complete factored form of the polynomial f(x) given that one zero is k.

69. $f(x) = x^4 + x^3 + 2x^2 + x + 1$ $k = -i$

70. $f(x) = 2x^3 - 5x^2 + 9x - 9$ $k = \frac{3}{2}$

71. State the domain of $f(x) = \frac{3x - 2}{5x + 4}$. Identify any horizontal or vertical asymptotes in the graph of f.

72. Find any horizontal or vertical asymptotes in the graph of $f(x) = \frac{2x^2 + x - 15}{3x^2 + 8x - 3}$.

73. Identify any horizontal or vertical asymptotes in the graph of f. State the domain of f.

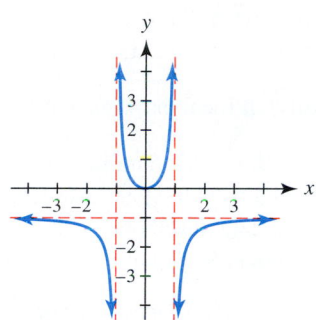

74. Let $f(x) = \frac{2x^2}{x^2 - 4}$.

  (a) Find the domain of f.

  (b) Identify any vertical or horizontal asymptotes.

  (c) Graph f with a graphing calculator.

  (d) Sketch a graph of f that includes all asymptotes.

Exercises 75–78: Graph y = g(x) by hand.

75. $g(x) = \frac{1}{x + 1} - 2$ **76.** $g(x) = \frac{x}{x - 1}$

77. $g(x) = \frac{x^2 - 1}{x^2 + 2x + 1}$ **78.** $g(x) = \frac{2x - 3}{2x^2 + x - 6}$

79. Sketch a graph of a function f with vertical asymptote $x = -2$ and horizontal asymptote $y = 2$.

80. Solve the equation $\frac{3x}{x - 2} = 2$ symbolically, graphically, and numerically.

Exercises 81–88: Solve the equation. Check your results.

81. $\frac{5x + 1}{x + 3} = 3$ **82.** $\frac{1}{x} - \frac{1}{x^2} + 2 = 0$

83. $\frac{1}{x + 2} + \frac{1}{x - 2} = \frac{1}{x^2 - 4}$

84. $x - \frac{1}{x} = 4$

85. $\frac{x + 5}{x - 2} = \frac{x - 1}{x + 1}$ **86.** $\frac{2x - 5}{3x + 1} = 5$

87. $\frac{1}{x - 3} - \frac{2}{x + 3} = \frac{4}{x^2 - 9}$ **88.** $2 - \frac{3}{x} = \frac{9}{x^2}$

Exercises 89–90: Use the graph of f to solve each inequality.

    (a) $f(x) > 0$ *(b)* $f(x) < 0$

89. 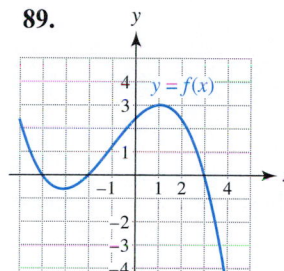   **90.**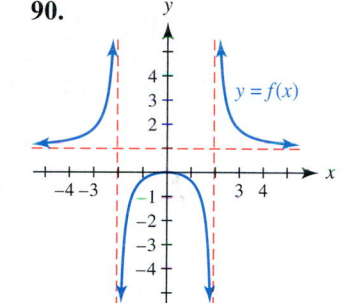

Exercises 91–96: Solve the inequality either graphically or symbolically.

91. $x^3 + x^2 - 6x > 0$ **92.** $2x^3 + x^2 \le 10x$

93. $x^4 + 4 < 5x^2$ **94.** $\frac{1 - x}{x} < 0$

95. $\frac{2x - 1}{x + 2} > 0$ **96.** $\frac{1}{x} + \frac{1}{x + 2} \le \frac{4}{3}$

Exercises 97–100: Evaluate the expression by hand.

97. $(36^{3/4})^2$ **98.** $(9^{-3/2})^{-2}$

99. $(2^{-3/2} \cdot 2^{1/2})^{-3}$ **100.** $\left(\frac{4}{9}\right)^{-3/2}$

Exercises 101–104: Write the expression using positive exponents.

101. $\sqrt[3]{x^4}$ **102.** $(\sqrt[4]{z})^{-1/2}$

103. $\sqrt[3]{y} \cdot \sqrt{y}$ **104.** $\sqrt{x} \cdot \sqrt[3]{x^2} \cdot \sqrt[4]{x^3}$

Exercises 105 and 106: Give the domain of the power function. Approximate f(3) to the nearest hundredth.

105. $f(x) = x^{5/2}$ **106.** $f(x) = x^{-2/3}$

Exercises 107–120: Solve the equation. Check your results.

107. $x^5 = 1024$ **108.** $x^{1/3} = 4$

109. $\sqrt{x - 2} = x - 4$ **110.** $x^{3/2} = 27$

111. $2x^{1/4} + 3 = 6$ **112.** $\sqrt{x - 2} = 14 - x$

113. $\sqrt[3]{2x - 3} + 1 = 4$ **114.** $x^{1/3} + 3x^{1/3} = -2$

115. $2n^{-2} - 5n^{-1} = 3$

116. $m^{-3} + 2m^{-2} + m^{-1} = 0$

117. $k^{2/3} - 4k^{1/3} - 5 = 0$

118. $x^{3/4} - 16x^{1/4} = 0$

119. $\sqrt{x + 1} + 1 = \sqrt{2x}$

120. $\sqrt{x - 2} = 5 - \sqrt{x + 3}$

Applications

121. *Allometry*   One of the earliest studies in allometry was performed by Bryan Robinson during the eighteenth century. He found that the pulse rate of an animal could be approximated by $f(x) = 1607x^{-0.75}$. The input x is the length of the animal in inches, and the output $f(x)$ is the approximate number of heartbeats per minute.   (**Source:** H. Lancaster, *Quantitative Methods in Biology and Medical Sciences.*)
   (a) Use f to estimate the pulse rates of a 2-foot dog and a 5.5-foot person.

   (b) What length corresponds to a pulse rate of 400 beats per minute?

122. *Electrical Circuits*   The figure shows the voltage y in a household electrical circuit over a time interval of $\frac{1}{60}$ second.
   (a) What is the maximum voltage?

   (b) Determine the times when the voltage is decreasing.

   (c) Use your results from part (b) to determine when the voltage is increasing. Let $0 \le x \le \frac{1}{60}$.

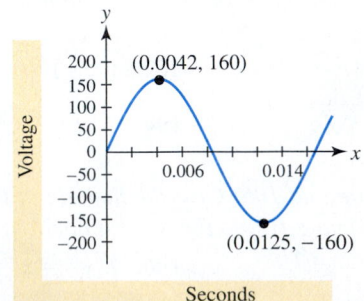

123. *Wind Speed*   The figure shows a graph of the wind speed x hours past noon.

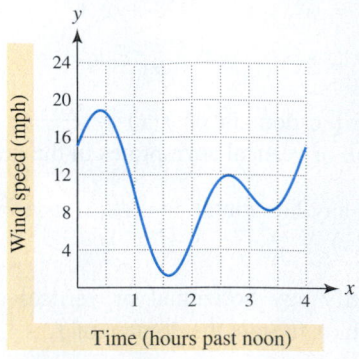

   (a) Estimate when the wind speeds were increasing between noon and 4 P.M.

   (b) Identify all local extrema and interpret each.

124. *Corporate Mergers*   The number of corporate mergers changed during the past two decades. The table lists the approximate numbers of mergers from 1987 to 1995.   (**Source:** Securities Data Company.)

Year	1987	1988	1989	1990	1991
Mergers	3100	3800	5300	5600	5100

Year	1992	1993	1994	1995
Mergers	5300	6200	7500	8800

   (a) Make a line graph of the data.

   (b) Suggest a possible degree for a polynomial that models these data. Explain your reasoning.

   (c) Use least-squares regression to find a polynomial that models these data.

125. *Modeling Ocean Temperatures*   The formula

$$T(m) = -0.064m^3 + 0.56m^2 + 2.9m + 61$$

approximates the ocean temperature in degrees Fahrenheit at Naples, Florida. In this formula m is the month with $m = 1$ corresponding to January.
   (a) What is the average ocean temperature in May?

   (b) Estimate the absolute maximum of T on the interval $[1, 12]$ and interpret the result.

126. *Waiting in Line*   If a parking attendant can wait on 70 cars per hour and the vehicles are arriving randomly at a rate of x vehicles per hour, the average time T in hours spent waiting in line and paying the attendant is given by $T(x) = \frac{1}{70 - x}$, where $0 \le x < 70$.

(a) Evaluate $T(0)$ and $T(60)$. Interpret your results.

(b) Find rate x if the wait is 3 minutes.

📟 **127.** *Minimizing Surface Area* Find possible dimensions that minimize the surface area of a box with no top that has a volume of 96 cubic inches and a length that is three times the width.

128. *Falling Object* If an object is dropped from a height h, then the time t for the object to strike the ground is directly proportional to the square root of h. If it requires 1 second for an object to fall 16 feet, how long does it take for an object to fall 256 feet?

129. *Animals and Trotting Speeds* Taller animals tend to take longer, but fewer, steps per second than shorter animals. The relationship between the shoulder height h in meters of an animal and an animal's stepping frequency F in steps per second, while *trotting*, is shown in the table. (**Source:** C. Pennycuick, *Newton Rules Biology.*)

h	0.5	1.0	1.5	2.0	2.5
F	2.6	1.8	1.5	1.3	1.2

📟 **(a)** Find values for constants a and b so that $f(x) = ax^b$ models the data.

(b) Estimate the stepping frequency for an elephant with a 3-meter shoulder height.

EXTENDED AND DISCOVERY EXERCISES

Exercises 1 and 2: **Velocity** *Suppose that a person is riding a bicycle on a straight road, and that $f(t)$ computes the total distance in feet that the rider has traveled after t seconds. To calculate the person's average velocity between time t_1 and time t_2, we can evaluate the difference quotient*

$$\frac{f(t_2) - f(t_1)}{t_2 - t_1}.$$

For example, if $f(t) = t^2$ then $f(10) = 100$ indicates that the person has traveled 100 feet after 10 seconds, and $f(15) = 225$ indicates that the person has traveled 225 feet after 15 seconds. Therefore the average velocity between 10 seconds and 15 seconds is

$$\frac{225 - 100}{15 - 10} = 25 \text{ feet per second.}$$

(a) *For the given $f(t)$ and the indicated values of t_1 and t_2, calculate the average velocity of the bike rider. Make a table to organize your work.*

(b) *Make a conjecture about the velocity of the bike rider precisely at time t_1.*

1. $f(t) = t^2$
   (i) $t_1 = 10, t_2 = 11$
   (ii) $t_1 = 10, t_2 = 10.1$
   (iii) $t_1 = 10, t_2 = 10.01$
   (iv) $t_1 = 10, t_2 = 10.001$

2. $f(t) = \sqrt{t}$
   (i) $t_1 = 4, t_2 = 5$
   (ii) $t_1 = 4, t_2 = 4.1$
   (iii) $t_1 = 4, t_2 = 4.01$
   (iv) $t_1 = 4, t_2 = 4.001$

Exercises 3–8: **Instantaneous Velocity** *In Exercises 1 and 2 we calculated the average velocity of a bike rider over smaller and smaller time intervals. If we calculate the average velocity as the time interval approaches 0, we say that we are calculating the **instantaneous velocity**. To find the instantaneous velocity, we can calculate the average velocity between times $t_1 = a$ and $t_2 = a + h$. The difference quotient becomes*

$$\frac{f(t_2) - f(t_1)}{t_2 - t_1} = \frac{f(a + h) - f(a)}{h}.$$

For example, if $f(t) = t^2$, the difference quotient can be simplified as follows.

$$\frac{f(a + h) - f(a)}{h} = \frac{(a + h)^2 - a^2}{h}$$

$$= \frac{a^2 + 2ah + h^2 - a^2}{h}$$

$$= \frac{2ah + h^2}{h}$$

$$= 2a + h$$

If we let h approach 0 ($h \to 0$), then the instantaneous velocity becomes $2a + 0 = 2a$ at time $t_1 = a$. For example, if $a = 10$, then $v = 2(10) = 20$. That is, after 10 seconds the bike rider is traveling at 20 feet per second. Does this result agree with your findings from Exercise 1?

(a) *For the given $f(t)$, calculate the difference quotient $\frac{f(a + h) - f(a)}{h}$. Be sure to simplify completely.*

(b) *Let h approach 0 and determine a formula for the instantaneous velocity at time $t_1 = a$.*

(c) *Calculate the instantaneous velocity at times $a = 5$, 10, and 15.*

3. $f(t) = 5t$ **4.** $f(t) = t^2 + 1$

5. $f(t) = t^2 + 2t$ **6.** $f(t) = \dfrac{200}{t}$

7. $f(t) = t^3$

8. $f(t) = \sqrt{t}$ (*Hint*: To simplify the ratio, multiply the numerator and denominator by $\sqrt{a + h} + \sqrt{a}$. This procedure is called *rationalizing the numerator*.)

Algebra Review   To review rationalizing the denominator, see Chapter R (page R-60).

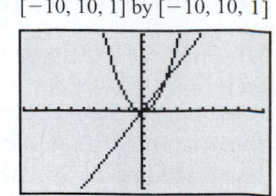

[−10, 10, 1] by [−10, 10, 1]

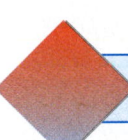 *Exercises 9–12:* ***Average Rates of Change***   *These exercises investigate the relationship between polynomial functions and their average rates of change. For example, the average rate of change of $f(x) = x^2$ from x to $x + 0.001$ for any x can be calculated and graphed as shown in the accompanying figures. The graph of f is a parabola, and the graph of its average rate of change is a line. There is an interesting relationship between a graph of a polynomial function and a graph of its average rate of change. Try to discover what it is by completing the following.*

(a) *Graph each function and its average rate of change from x to $x + 0.001$.*

(b) *Compare the graph of each function to the graph of its average rate of change. How are turning points on the graph of a function related to its average rate of change?*

(c) *Make a generalization about these results. Test your generalization.*

9. *Linear Functions*
$f_1(x) = 3x + 1$ $f_2(x) = -2x + 6$
$f_3(x) = 1.5x - 5$ $f_4(x) = -4x - 2.5$

10. *Quadratic Functions*
$f_1(x) = 2x^2 - 3x + 1$ $f_2(x) = -0.5x^2 + 2x + 2$
$f_3(x) = x^2 + x - 2$ $f_4(x) = -1.5x^2 - 4x + 6$

11. *Cubic Functions*
$f_1(x) = 0.5x^3 - x^2 - 2x + 1$
$f_2(x) = -x^3 + x^2 + 3x - 5$
$f_3(x) = 2x^3 - 5x^2 + x - 3$
$f_4(x) = -x^3 + 3x - 4$

12. *Quartic Functions*
$f_1(x) = 0.05x^4 + 0.2x^3 - x^2 - 2.4x$
$f_2(x) = -0.1x^4 + 0.1x^3 + 1.3x^2 - 0.1x - 1.2$
$f_3(x) = 0.1x^4 + 0.4x^3 - 0.2x^2 - 2.4x - 2.4$

CHAPTERS 1–4 CUMULATIVE REVIEW EXERCISES

1. Find the percent change if a quantity Q changes from 45 to 54.

2. Write 0.065 in scientific notation and 7.88×10^5 in standard notation.

3. Evaluate $\dfrac{1.2 - 5.6}{0.4 + 9^2}$. Round your answer to three decimal places.

4. Let $S = \{(-3, 4), (-1, -2), (0, 4), (1, -2), (-1, 5)\}$.
   (a) Find the domain and range of S.

   (b) Is S a function?

   (c) Graph S.

5. Find the exact distance between $(-1, 4)$ and $(3, -9)$.

6. Use the graph in the next column to express the domain and range of f in interval notation. Then evaluate $f(0)$.

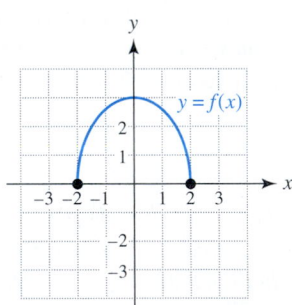

7. Graph $y = g(x)$ by hand.
   (a) $g(x) = 2 - 3x$ (b) $g(x) = |2x - 1|$

   (c) $g(x) = \frac{1}{2}(x - 2)^2 + 2$ (d) $g(x) = x^3 - 1$

   (e) $g(x) = \sqrt{-x}$ (f) $g(x) = \sqrt[3]{x}$

   (g) $g(x) = \dfrac{1}{x - 4} + 2$ (h) $g(x) = x^2 - x$

8. Does the graph of a parabola with a vertical axis of symmetry represent a function? Explain your answer.

Exercises 9 and 10: Complete the following.

   (a) Determine the domain of f.

   (b) Evaluate $f(2)$.

9. $f(x) = \sqrt{x^2 - 4}$ **10.** $\dfrac{2x - 3}{3x^2 + 11x - 4}$

11. The monthly cost of driving a car is $200 for maintenance plus $0.25 a mile. Write a formula for a function C that calculates the monthly cost of driving a car x miles. Evaluate $C(2000)$ and interpret the result.

12. The graph of a linear function f is shown.

   (a) Identify the slope, y-intercept, and x-intercept.

   (b) Write a formula for $f(x)$.

   (c) Evaluate $f(-3)$ symbolically and graphically.

   (d) Find any zeros of f.

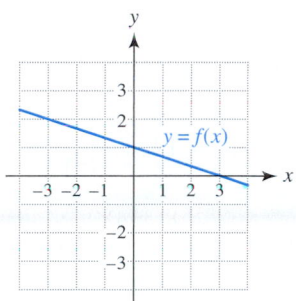

13. Find the average rate of change of $f(x) = x^3 - x$ from $x = -3$ to $x = -2$.

14. Find the difference quotient for $f(x) = x^2 + 6x$.

15. Write the slope-intercept form for a line that passes through $(-2, 5)$ and $(3, -4)$.

16. Write the slope-intercept form for a line that passes through $(-1, 4)$ and is perpendicular to the line $3x - 4y = 12$.

17. Write an equation of a line that is parallel to the x-axis and passes through $(4, -5)$.

18. Determine the x- and y-intercepts on the graph of $5x - 4y = 10$. Graph the equation.

19. If $C(x) = 15x + 2000$ calculates the cost in dollars of producing x radios, interpret the numbers 15 and 2000 in the formula for $C(x)$.

20. Solve $-2.4x - 2.1 = \sqrt{3x} + 1.7$ graphically and numerically. Round your answer to the nearest tenth.

Exercises 21–32: Solve the equation.

21. $-3(2 - 3x) - (-x - 1) = 1$

22. $\dfrac{5 - 3x}{6} = \dfrac{x - (3 - 4x)}{2}$

23. $|3x - 4| + 1 = 5$ **24.** $x^3 + 5 = 5x^2 + x$

25. $7x^2 + 9x = 10$ **26.** $2x^2 + x + 2 = 0$

27. $2x^3 + 4x^2 = 6x$ **28.** $x^4 + 9 = 10x^2$

29. $3x^{2/3} + 5x^{1/3} - 2 = 0$ **30.** $\sqrt{5 + 2x} + 4 = x + 5$

31. $\dfrac{2x - 3}{5 - x} = \dfrac{4x - 3}{1 - 2x}$ **32.** $\sqrt[3]{x - 4} - 1 = 3$

33. Solve $\frac{1}{2}x - (4 - x) + 1 = \frac{3}{2}x - 5$. Is this equation either an identity or a contradiction?

34. Graph f. Is f continuous on its domain? Evaluate $f(1)$.

$$f(x) = \begin{cases} x^2 - 1 & \text{if } -3 \le x \le -1 \\ x + 1 & \text{if } -1 < x < 1 \\ 1 - x^2 & \text{if } 1 \le x \le 3 \end{cases}$$

Exercises 35–40: Solve the inequality. Write the solution set in interval notation.

35. $-\frac{1}{3}x - (1 + x) > \frac{2}{3}x$ **36.** $-4 \le 4x - 6 < \frac{5}{2}$

37. $|5x - 7| \ge 3$ **38.** $5x^2 + 13x - 6 < 0$

39. $x^3 - 9x \le 0$ **40.** $\dfrac{4x - 3}{x + 2} > 0$

41. The graph of a nonlinear function f is shown. Solve each equation or inequality. Write the solution set for each inequality in interval notation.

   (a) $f(x) = 0$ **(b)** $f(x) > 0$ **(c)** $f(x) \le 0$

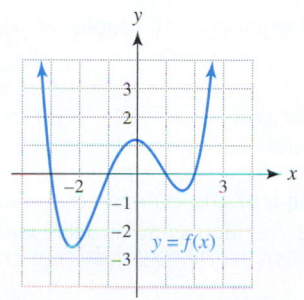

42. Write the quadratic polynomial $f(x) = 2x^2 - 4x + 1$ in the form $f(x) = a(x - h)^2 + k$.

43. Find the vertex on the graph of $f(x) = -\frac{1}{2}x^2 + 3x - 2$.

44. A graph of a quadratic function $f(x) = ax^2 + bx + c$ is shown.
  (a) State whether $a > 0$ or $a < 0$.

  (b) Estimate the real solutions to $ax^2 + bx + c = 0$.

  (c) Determine if the discriminant $b^2 - 4ac$ is positive, negative, or zero.

  (d) Write the complete factored form of $f(x)$.

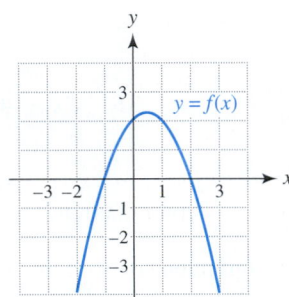

45. Solve $x^2 - 3x = 1$ by completing the square.

46. Use the given graph of $y = f(x)$ to sketch a graph of each expression.
  (a) $y = f(x + 2) - 1$ **(b)** $y = -2f(x)$

  (c) $y = f(-x) + 1$ **(d)** $y = f(\frac{1}{2}x)$

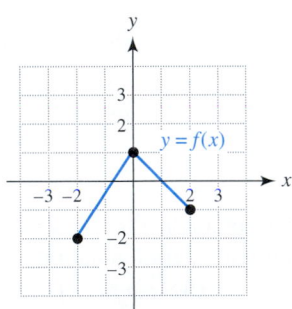

47. Use transformations of graphs to sketch a graph of $y = 2\sqrt{x} + 1$.

48. Where is the graph of $f(x) = -\frac{1}{2}(x - 2)^2 + 3$ increasing? Use interval notation.

49. Use the graph of f to estimate each of the following.
  (a) Where f is increasing or decreasing

(b) The zeros of f

(c) The coordinates of any turning points

(d) Any local extrema

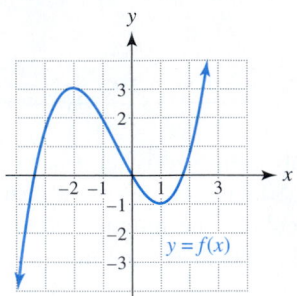

50. Determine whether the polynomial functions defined by $f(x) = x^4 - 5x^3 - 7$ and $g(x) = \sqrt{9 - x^2}$ are even, odd, or neither.

51. Sketch a graph of a quartic (degree 4) function with a negative leading coefficient, three x-intercepts, and three turning points.

52. Predict the end behavior of $f(x) = 4 + 3x - x^3$.

53. Divide each expression.
  (a) $\dfrac{4a^3 - 8a^2 + 12}{4a^2}$ **(b)** $\dfrac{2x^3 - 4x + 1}{x - 1}$

  (c) $\dfrac{x^4 + 2x^3 - x^2 + 5x - 2}{x^2 + 2}$

54. A cubic function f has zeros $-1, 0$, and 2 with leading coefficient -5. Write the complete factored form of $f(x)$.

55. A degree 6 function f has zeros $-3, 1$, and 4 with multiplicities 1, 2, and 3, respectively. If the leading coefficient is 4, write the complete factored form of $f(x)$.

56. A quintic (degree 5) function f with real coefficients has leading coefficient $\frac{1}{2}$ and zeros $-2, i$, and $-2i$. Write $f(x)$ in complete factored form and expanded form.

57. Write the complete factored form of the polynomial given by $f(x) = 2x^3 - x^2 - 6x + 3$.

58. Find all zeros, real or imaginary, of
$$f(x) = x^4 + x^3 + 12x^2 + 9x + 27$$
given that one zero is $-3i$.

59. Use the graph to write the complete factored form of the cubic polynomial $f(x)$.

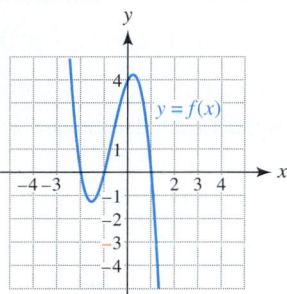

60. Use the graph to write the complete factored form of the degree 5 polynomial $f(x)$.

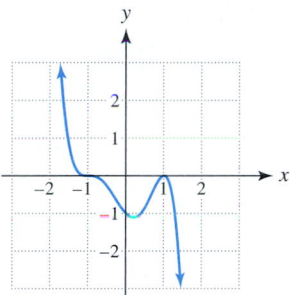

61. Write $\frac{3 + 4i}{1 - i}$ in standard form.

62. Find all solutions, real or imaginary, to $x^4 - 25 = 0$.

63. State the domain of $f(x) = \frac{2x - 5}{x^2 - 3x - 4}$. Find any vertical or horizontal asymptotes.

64. Write $\sqrt[3]{x^5}$ using rational exponents. Evaluate the expression for $x = 8$.

Applications

65. *Dimensions of a Drinking Cup* The volume V of a cone is given by $V = \frac{1}{3}\pi r^2 h$, where r is the radius and h is the height. If a paper cup in the shape of a cone has a volume of 6 cubic inches and a radius of $\frac{5}{4}$ inches, find the height of the cone to the nearest hundredth of an inch.

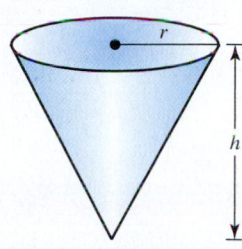

66. *Water in a Pool* The graph shows the amount of water in a swimming pool x hours past noon. Find the slope of each line segment and interpret each slope.

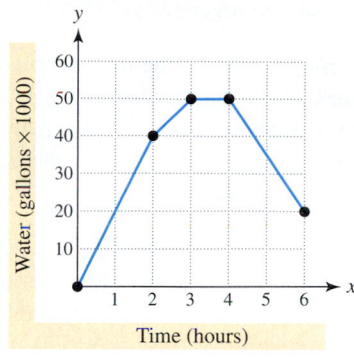

67. *Distance* At noon one runner heads south at 8 miles per hour and is located 2 miles north of a second runner, who is heading west at 7 miles per hour. Approximate the distance between the runners to the nearest tenth of a mile at 12:30 P.M.

68. *Average Rate of Change* The total distance D in feet traveled by a car after t seconds is given by $D(t) = 4t^2$ for $0 \le t \le 7$.
(a) Find the average rate of change of D from 0 to 2 and 2 to 4.

(b) Interpret these average rates of change.

69. *Working Together* Suppose one person can paint a room in 10 hours and another person can paint the same room in 8 hours. How long will it take to paint the room if they work together?

70. *Mixing Acid* Two liters of a 35% sulfuric acid solution need to be diluted to a 20% solution. How many liters of a 12% sulfuric acid solution should be mixed with the 2-liter solution?

71. *U.S. Population* In 1990 the population of the United States was about 250 million and in 2000 it was about 280 million. (**Source:** Bureau of the Census.)
(a) Determine a formula in the form

$$P(t) = m(t - t_1) + y_1$$

that models these data. Let P be in millions and t be the year.

(b) Estimate the year when the population of the United States could reach 300 million.

72. *Maximizing Revenue*   The revenue R in dollars from selling x thousand toy figures is given by the formula $R(x) = x(800 - x)$. How many toy figures should be sold to maximize revenue?

73. *Construction*   A box is being constructed by cutting 2-inch squares from the corners of a rectangular sheet of metal that is 6 inches longer than it is wide. If the box is to have a volume of 270 cubic inches, find the dimensions of the metal sheet.

74. *Group Rates*   Round trip airline tickets to Hawaii are regularly $800, but for each additional ticket purchased, the price is reduced by $5. For example, 1 ticket costs $800, 2 tickets cost $2(795) = \$1590$, and 3 tickets cost $3(790) = \$2370$.
    (a) Write a quadratic function C that gives the total cost of purchasing t tickets.

    (b) Solve $C(t) = 17{,}000$ and interpret the result.

    (c) Find the absolute maximum for C and interpret your result. Assume that t must be an integer.

75. *Modeling Data*   Find a quadratic function in the form $f(x) = a(x - h)^2 + k$ that models the data in the table. Graph $y = f(x)$ and the data if a graphing calculator is available.

x	4	6	8	10
y	6	15	37	80

76. *Health Care Cost*   Average health care cost per employee has risen dramatically since 1998. The table lists the annual percent change from 1998 to 2002.

Year	1998	1999	2000	2001	2002
Increase (%)	6.1	7.3	8.1	11.2	14.7

Source: Mercer U.S. Health Care Survey, 2002.

    (a) Find a polynomial function that models these data. (Answers may vary.)

    (b) Estimate the annual percent change in 2004.

77. *Americans and Weight*   The percentage of Americans that are 30 or more pounds above a healthy weight has increased during the past 40 years. The table lists this percentage for various years.

Year	1961	1978	1991	2000
Percentage	13	15	23	31

Source: National Health and Nutrition Examination Survey.

    (a) Find a polynomial function that models these data. (Answers may vary.)

    (b) Estimate the year when this percentage may reach 40%. Compare your prediction with the experts' prediction of 2009.

78. *Minimizing Surface Area*   A cylindrical can is being constructed to have a volume of 10π cubic inches. Find the dimensions of the can that result in the least amount of aluminum used in the construction of the can.

5 Exponential and Logarithmic Functions

The important thing is not to stop questioning.

Albert Einstein

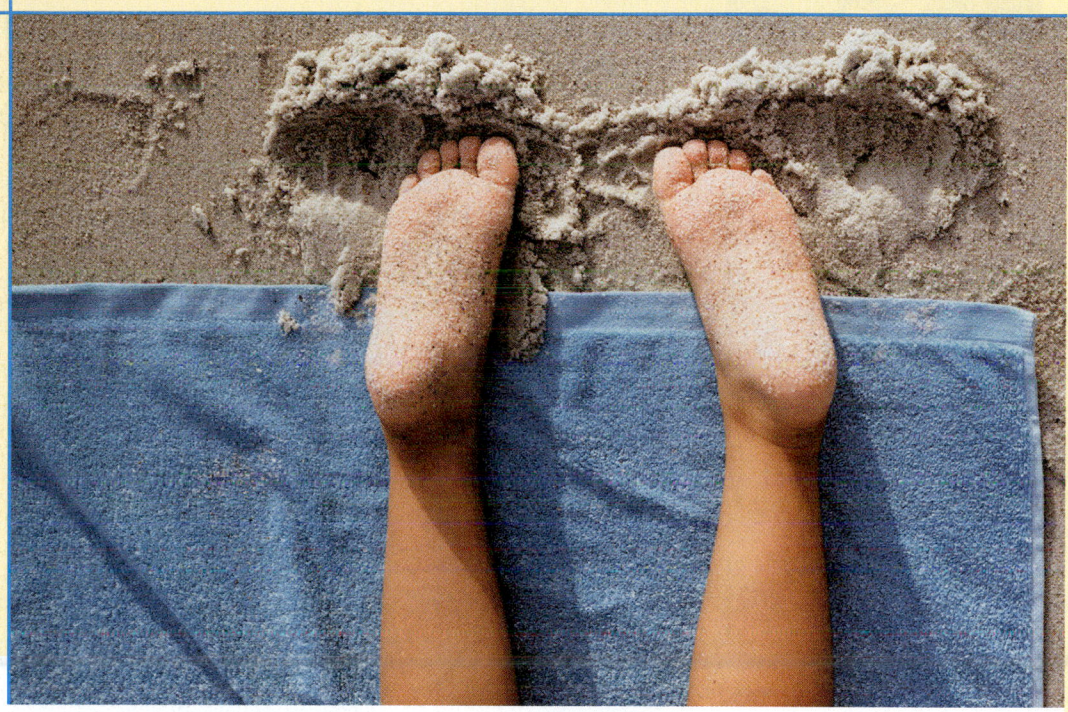

In 1900 the population of the world was approximately 1.6 billion. At that time the Swedish scientist Svante Arrhenius first predicted a greenhouse effect resulting from emissions of carbon dioxide by the industrialized countries. His classic calculation made use of logarithms and predicted that a doubling of the carbon dioxide concentration in the atmosphere would raise the average global temperature by 7°F to 11°F. Since then, the world population has increased dramatically to approximately 6 billion. With this increase in population, there has been a corresponding increase in emissions of greenhouse gases such as carbon dioxide, methane, and chlorofluorocarbons. These emissions have created the potential to alter the earth's climate and destroy portions of the ozone layer.

When quantities such as population and pollution increase rapidly, nonlinear functions and equations are used to model their growth. In this chapter, exponential and logarithmic functions are introduced. They occur in a wide variety of applications. Examples include acid rain, the decline of the bluefin tuna, air pollution, global warming, salinity of the ocean, demand for organ transplants, diversity of bird species, the relationship between caloric intake and land ownership in developing countries, hurricanes, earthquakes, and increased skin cancer due to a decrease in the ozone layer. Understanding these issues and discovering solutions will require creativity, innovation, and mathematics. This chapter presents some of the mathematical tools necessary for modeling these trends.

Source: M. Kraljic, *The Greenhouse Effect.*

5.1

Combining Functions

♦ Perform arithmetic operations on functions

♦ Perform composition of functions

Introduction

Addition, subtraction, multiplication, and division can be performed on numbers and variables. Arithmetic operations also can be used to combine functions. For example, to model the stopping distance of a car, we compute two quantities. The first quantity is the *reaction distance*, which is the distance that a car travels between the time when a driver first recognizes a hazard and the time when the brakes are applied. The second quantity is *braking distance*, which is the distance that a car travels after the brakes have been applied. *Stopping distance* is equal to the sum of the reaction distance and the braking distance. One way to determine stopping distance is to find one function that models the reaction distance and a second function that calculates the braking distance. The stopping distance is then found by adding the two functions.

Arithmetic Operations on Functions

Highway engineers frequently assume that drivers have a reaction time of 2.5 seconds or less. During this time, a car continues to travel at a constant speed, until a driver is able to move his or her foot from the accelerator to the brake pedal. If a car travels at x miles per hour, then $r(x) = \left(\frac{11}{3}\right)x$ computes the distance in feet that a car travels in 2.5 seconds. For example, a driver traveling at **55** miles per hour might have a reaction distance of $r(\mathbf{55}) = \left(\frac{11}{3}\right)(\mathbf{55}) \approx 201.7$ feet.

The braking distance in feet for a car traveling on wet, level pavement at x miles per hour can be approximated by $b(x) = x^2/9$. For instance, a car traveling at 55 miles per hour would need $b(\mathbf{55}) = \mathbf{55}^2/9 \approx 336.1$ feet to stop after the brakes have been applied. The estimated stopping distance s at 55 miles per hour is the sum of these distances: $201.7 + 336.1 = 537.8$ feet. See the accompanying figure.   (**Source:** L. Haefner, *Introduction to Transportation Systems.*)

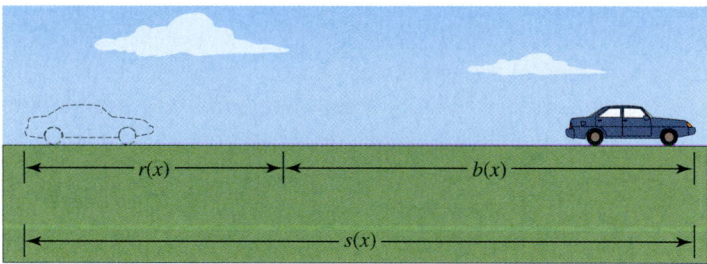

The concept of finding the sum of two functions can be represented symbolically, graphically, and numerically, as illustrated in the next example.

 **EXAMPLE 1**

Representing stopping distance symbolically, graphically, and numerically

A driver traveling at 60 miles per hour with a reaction time of 2.5 seconds attempts to stop on wet, level pavement in order to avoid an accident.
(a) Write a symbolic representation for a function s using r and b that computes the stopping distance for a car traveling at x miles per hour. Evaluate $s(60)$.
(b) Graph r, b, and s in [45, 70, 5] by [0, 800, 100]. Interpret the graph.
(c) Illustrate the relationship among r, b, and s numerically.

[45, 70, 5] by [0, 800, 100]

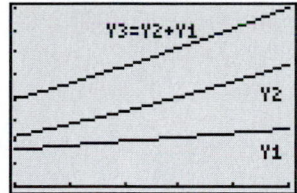

FIGURE 5.1

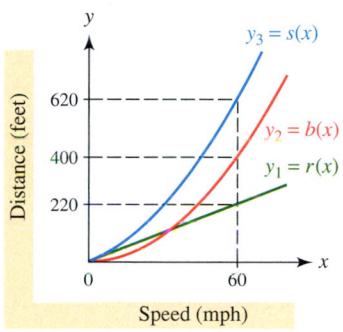

FIGURE 5.2

SOLUTION

(a) *Symbolic Representation*   Let $s(x)$ be the sum of the two formulas for r and b.

$$s(x) = r(x) + b(x) = \frac{11}{3}x + \frac{1}{9}x^2$$

The stopping distance for a car traveling at 60 miles per hour is

$$s(60) = \frac{11}{3}(60) + \frac{1}{9}(60)^2 = 620 \text{ feet.}$$

(b) *Graphical Representation*   Let $Y_1 = (11/3)X$, $Y_2 = (X^2)/9$, and $Y_3 = (11/3)X + (X^2)/9$ or $Y_3 = Y_1 + Y_2$, as shown in Figure 5.1. For any x-value in the graph, the sum of y_1 and y_2 equals y_3. Figure 5.2 illustrates that

$$s(60) = r(60) + b(60) = 220 + 400 = 620.$$

(c) *Numerical Representation*   Table 5.1 shows numerical representations of $r(x)$, $b(x)$, and $s(x) = r(x) + b(x)$. Notice that values for $s(x)$ can be found by adding $r(x)$ and $b(x)$. For example, when $x = 60$, $r(60) = 220$, $b(60) = 400$, and $s(60) = 220 + 400 = 620$.

TABLE 5.1

x	0	12	24	36	48	60
$r(x)$	0	44	88	132	176	220
$b(x)$	0	16	64	144	256	400
$s(x)$	0	60	152	276	432	620

Now Try Exercise 109 ◆

We now formally define arithmetic operations on functions.

OPERATIONS ON FUNCTIONS

If $f(x)$ and $g(x)$ both exist, the sum, difference, product, and quotient of two functions f and g are defined by

$$(f + g)(x) = f(x) + g(x),$$
$$(f - g)(x) = f(x) - g(x),$$
$$(fg)(x) = f(x) \cdot g(x), \quad \text{and}$$
$$\left(\frac{f}{g}\right)(x) = \frac{f(x)}{g(x)}, \quad \text{where } g(x) \neq 0.$$

The domains of the sum, difference, and product of f and g include x-values that are in *both* the domain of f and the domain of g. The domain of the quotient f/g includes all x-values in both the domain of f and the domain of g, where $g(x) \neq 0$.

EXAMPLE 2 ▸   Performing arithmetic operations on functions symbolically

Let $f(x) = 2 + \sqrt{x - 1}$ and $g(x) = x^2 - 4$.
(a) Find the domain of $f(x)$ and $g(x)$. Then find the domains of $(f + g)(x)$, $(f - g)(x)$, $(fg)(x)$, and $(f/g)(x)$.
(b) If possible, evaluate $(f + g)(5)$, $(f - g)(1)$, $(fg)(0)$, and $(f/g)(3)$.
(c) Write expressions for $(f + g)(x)$, $(f - g)(x)$, $(fg)(x)$, and $(f/g)(x)$.

SOLUTION

(a) Whenever $x \geq 1$, $f(x) = 2 + \sqrt{x - 1}$ is defined. Therefore the domain of f is $\{x \mid x \geq 1\}$. The domain of $g(x) = x^2 - 4$ is all real numbers. The domains of $f + g$, $f - g$, and fg include all x-values in *both* the domain of f and the domain of g. Thus their domains are $\{x \mid x \geq 1\}$.

To determine the domain of f/g, we must also exclude x-values for which $g(x) = x^2 - 4 = 0$. This occurs when $x = \pm 2$. Thus the domain of f/g is $\{x \mid x \geq 1, x \neq 2\}$. (Note that $x \neq -2$ is satisfied if $x \geq 1$.)

(b) The expressions can be evaluated as follows.

$$(f + g)(5) = f(5) + g(5) = (2 + \sqrt{5 - 1}) + (5^2 - 4) = 4 + 21 = 25$$

$$(f - g)(1) = f(1) - g(1) = (2 + \sqrt{1 - 1}) - (1^2 - 4) = 2 - (-3) = 5$$

$(fg)(0)$ is undefined since 0 is not in the domain of $f(x)$.

$$\left(\frac{f}{g}\right)(3) = \frac{f(3)}{g(3)} = \frac{2 + \sqrt{3 - 1}}{3^2 - 4} = \frac{2 + \sqrt{2}}{5}$$

(c) The sum, difference, product, and quotient of f and g are calculated as follows.

$$(f + g)(x) = f(x) + g(x) = (2 + \sqrt{x - 1}) + (x^2 - 4) = \sqrt{x - 1} + x^2 - 2$$

$$(f - g)(x) = f(x) - g(x) = (2 + \sqrt{x - 1}) - (x^2 - 4) = \sqrt{x - 1} - x^2 + 6$$

$$(fg)(x) = f(x) \cdot g(x) = (2 + \sqrt{x - 1})(x^2 - 4)$$

$$\left(\frac{f}{g}\right)(x) = \frac{f(x)}{g(x)} = \frac{2 + \sqrt{x - 1}}{x^2 - 4}$$

Now Try Exercises 11 and 17 ◆

EXAMPLE 3 Evaluating combinations of functions

If possible, use the given representations of the functions f and g to evaluate $(f + g)(4)$, $(f - g)(-2)$, $(fg)(1)$, and $(f/g)(0)$.

(a)

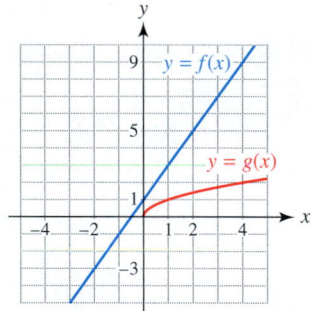

FIGURE 5.3

(b) **TABLE 5.2**

x	-2	0	1	4
$f(x)$	-3	1	3	9
$g(x)$	—	0	1	2

(c) $f(x) = 2x + 1$, $g(x) = \sqrt{x}$

SOLUTION

(a) *Graphical Evaluation* In Figure 5.3, $f(4) = 9$ and $g(4) = 2$. Thus

$$(f + g)(4) = f(4) + g(4) = 9 + 2 = 11.$$

Although $f(-2) = -3$, $g(-2)$ is undefined because -2 is not in the domain of g. Thus $(f - g)(-2)$ is undefined. The domains of f and g include 1, and it follows $(fg)(1) = f(1)g(1) = 3(1) = 3$. The graph of g intersects the origin, so $g(0) = 0$. Thus $(f/g)(0)$ is undefined.

(b) *Numerical Evaluation*   Numerical evaluation of f and g is performed using Table 5.2. From the table $f(4) = 9$ and $g(4) = 2$. As in part (a),

$$(f + g)(4) = f(4) + g(4) = 9 + 2 = 11.$$

A dash in the table indicates that $g(-2)$ is undefined, so $(f - g)(-2)$ is also undefined. The calculations of $(fg)(1)$ and $(f/g)(0)$ are done in a similar manner.

(c) *Symbolic Evaluation*   Use the formulas $f(x) = 2x + 1$ and $g(x) = \sqrt{x}$.

$$(f + g)(4) = f(4) + g(4) = (2 \cdot 4 + 1) + \sqrt{4} = 9 + 2 = 11$$

$$(f - g)(-2) = f(-2) - g(-2) = (2 \cdot (-2) + 1) - \sqrt{-2} \text{ is undefined.}$$

$$(fg)(1) = f(1)g(1) = (2 \cdot 1 + 1)\sqrt{1} = 3(1) = 3$$

$$\left(\frac{f}{g}\right)(0) = \frac{f(0)}{g(0)} \text{ is undefined, since } g(0) = 0. \quad \textit{Now Try Exercises 1, 5, and 9} \blacklozenge$$

The next example involves an application from business where the difference between two functions occurs.

EXAMPLE 4   Finding the difference of two functions

Once the sound track for a compact disc (CD) has been recorded, the cost of producing the master disc can be significant. After this disc has been made, additional discs can be produced inexpensively. A typical cost for producing the master disc is $2000, while additional discs cost approximately $2 each.   (**Source:** Windcrest Productions.)

(a) Assuming no other expenses, find a function C that outputs the cost of producing the master disc plus x additional compact discs. Find the cost of making the master disc and 1500 additional compact discs.

(b) Suppose that each CD is sold for $12. Find a function R that computes the revenue received from selling x compact discs. Find the revenue from selling 1500 compact discs.

(c) Assuming that the master CD is not sold, determine a function P that outputs the profit from selling x compact discs. How much profit is there from selling 1500 discs?

SOLUTION

(a) The cost of producing the master disc for $2000 plus x additional discs at $2 each is given by $C(x) = 2x + 2000$. The $2000 cost is sometimes called a *fixed cost*. The cost of manufacturing the master disc and 1500 additional compact discs is

$$C(1500) = 2(1500) + 2000 = \$5000.$$

(b) The revenue from x discs at $12 each is computed by $R(x) = 12x$. The revenue from selling 1500 compact discs is $R(1500) = 12(1500) = \$18,000.$

(c) Profit P is equal to revenue minus cost. This can be written using function notation.

Algebra Review

To review arithmetic operations on polynomials, see Chapter R (page R-20).

$$P(x) = R(x) - C(x)$$

$$= 12x - (2x + 2000)$$

$$= 12x - 2x - 2000$$

$$= 10x - 2000$$

If $x = 1500$, $P(1500) = 10(1500) - 2000 = \$13,000$.   *Now Try Exercise 99* $\blacklozenge$

Review of Function Notation   In the next example we review how to evaluate function notation before we discuss composition of functions.

EXAMPLE 5   Evaluating function notation

Let $g(x) = 3x^2 - 6x + 2$. Evaluate each expression.
(a) $g(2)$ **(b)** $g(k)$ **(c)** $g(x^2)$ **(d)** $g(x + 2)$

SOLUTION

Algebra Review
To review squaring a binomial, see Chapter R (page R-25).

(a) $g(2) = 3(2)^2 - 6(2) + 2 = 12 - 12 + 2 = 2$
(b) $g(k) = 3k^2 - 6k + 2$
(c) $g(x^2) = 3(x^2)^2 - 6(x^2) + 2 = 3x^4 - 6x^2 + 2$
(d) $g(x + 2) = 3(x + 2)^2 - 6(x + 2) + 2$
$$= 3(x^2 + 4x + 4) - 6(x + 2) + 2$$
$$= 3x^2 + 12x + 12 - 6x - 12 + 2$$
$$= 3x^2 + 6x + 2$$

Now Try Exercise 43 ◆

Composition of Functions

Many tasks in life are performed in sequence. For example, to go to a movie we might get into a car, drive to the movie theater, and get out of the car. The order in which these tasks are performed is important.

A similar situation occurs with functions. For example, to convert miles to inches we might first convert miles to feet and then feet to inches. Since there are 5280 feet in a mile, $f(x) = 5280x$ converts x miles to an equivalent number of feet. Then $g(x) = 12x$ changes feet to inches. To convert x miles to inches, we combine the functions f and g in sequence. Figure 5.4 illustrates how to convert 5 miles to inches. First, $f(5) = 5280(5) = 26,400$. Then the output of 26,400 feet from f is used as input for g. The number of inches in 26,400 feet is $g(26,400) = 12(26,400) = 316,800$. This computation is called the *composition* of g and f.

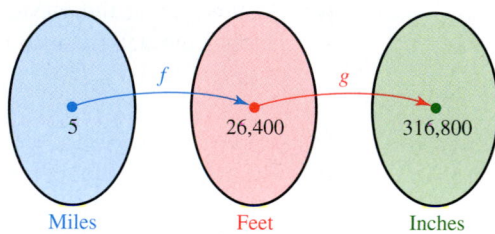

Miles   Feet   Inches

FIGURE 5.4

The composition of g and f shown in Figure 5.4 can be expressed symbolically. The symbol ∘ is used to denote composition of two functions.

$$(g \circ f)(5) = g(f(5)) \qquad \text{First compute } f(5).$$
$$= g(5280 \cdot 5) \qquad f(x) = 5280x$$
$$= g(26,400) \qquad \text{Simplify.}$$
$$= 12(26,400) \qquad g(x) = 12x$$
$$= 316,800 \qquad \text{Simplify.}$$

A distance of 5 miles is equivalent to 316,800 inches. The concept of composition of two functions is now defined formally.

COMPOSITION OF FUNCTIONS

If f and g are functions, then the **composite function** $g \circ f$, or **composition** of g and f is defined by

$$(g \circ f)(x) = g(f(x)).$$

Note: We read $g(f(x))$ as "g of f of x."

The domain of $g \circ f$ is all x in the domain of f such that $f(x)$ is in the domain of g.

 **6**   Finding a symbolic representation of a composite function

Find a symbolic representation for the composite function $g \circ f$ that converts x miles into inches.

SOLUTION Let $f(x) = 5280x$ and $g(x) = 12x$.

$$(g \circ f)(x) = g(\mathbf{f(x)}) \qquad \text{Definition of composition.}$$
$$= g(\mathbf{5280x}) \qquad f(x) = 5280x \text{ is the input for } g.$$
$$= 12(5280x) \qquad g \text{ multiplies the input by 12.}$$
$$= 63{,}360x \qquad \text{Simplify.}$$

Thus $(g \circ f)(x) = 63{,}360x$ converts x miles to inches.   *Now Try Exercise 101* ◆

◆ **MAKING CONNECTIONS**

Product and Composition of Two Functions Computing a product of two functions is fundamentally different from computing a composition of two functions. With the product $(fg)(x) = f(x) \cdot g(x)$, both f and g receive the *same* input x. Then their outputs are multiplied. However, with the composition $(f \circ g)(x) = f(g(x))$, the output $g(x)$ provides the input for f. These concepts are illustrated in Figure 5.5.

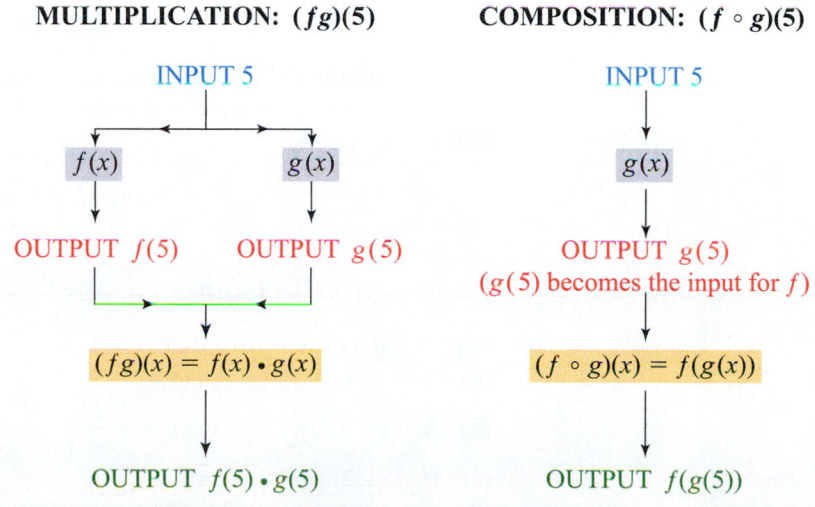

FIGURE 5.5 Comparison of Multiplication and Composition

EXAMPLE 7 Evaluating a composite function symbolically

Let $f(x) = x^2 + 3x + 2$ and $g(x) = \frac{1}{x}$.
(a) Evaluate $(f \circ g)(2)$ and $(g \circ f)(2)$. How do they compare?
(b) Find symbolic expressions for the composite functions defined by $(f \circ g)(x)$ and $(g \circ f)(x)$. Are they equivalent expressions?
(c) Find the domains of $(f \circ g)(x)$ and $(g \circ f)(x)$.

SOLUTION
(a) $(f \circ g)(2) = f(g(2)) = f\left(\frac{1}{2}\right) = \left(\frac{1}{2}\right)^2 + 3\left(\frac{1}{2}\right) + 2 = \frac{15}{4} = 3.75$.
 $(g \circ f)(2) = g(f(2)) = g(2^2 + 3 \cdot 2 + 2) = g(12) = \frac{1}{12} \approx 0.0833$.
 The results are not equal.
(b) Symbolic representations for $(f \circ g)(x)$ and $(g \circ f)(x)$ also can be found.

$$(f \circ g)(x) = f(g(x)) = f\left(\frac{1}{x}\right) = \left(\frac{1}{x}\right)^2 + 3\left(\frac{1}{x}\right) + 2 = \frac{1}{x^2} + \frac{3}{x} + 2$$

$$(g \circ f)(x) = g(f(x)) = g(x^2 + 3x + 2) = \frac{1}{x^2 + 3x + 2}$$

The expressions for $(f \circ g)(x)$ and $(g \circ f)(x)$ are not equivalent.
(c) The domain of f is all real numbers and the domain of g is $\{x \mid x \neq 0\}$. The domain of $(f \circ g)(x) = f(g(x))$ consists of all x in the domain of g such that $g(x)$ is in the domain of f. Thus the domain of $(f \circ g)(x) = \frac{1}{x^2} + \frac{3}{x} + 2$ is $\{x \mid x \neq 0\}$.
 The domain of $(g \circ f)(x) = g(f(x))$ consists of all x in the domain of f such that $f(x)$ is in the domain of g. Since $x^2 + 3x + 2 = 0$ when $x = -1$ or $x = -2$, the domain of $(g \circ f)(x) = \frac{1}{x^2 + 3x + 2}$ is $\{x \mid x \neq -1, x \neq -2\}$.

Now Try Exercises 61 and 65 ◆

◆ **MAKING CONNECTIONS**

Composition and Domains To find the domain of a composition of two functions, it is sometimes helpful not to immediately simplify the resulting expression. For example, if $f(x) = x^2$ and $g(x) = \sqrt{x - 1}$, then

$$(f \circ g)(x) = (\sqrt{x - 1})^2.$$

From this unsimplified expression, we can see that the domain (input) of $f \circ g$ must be restricted to $x \geq 1$ for the output to be a real number. As a result, we can write the simplified expression

$$(f \circ g)(x) = x - 1, x \geq 1.$$

EXAMPLE 8 Finding symbolic representations for composite functions

Find $(f \circ g)(x)$ and $(g \circ f)(x)$.
(a) $f(x) = x + 2, \qquad g(x) = x^3 - 2x^2 - 1$
(b) $f(x) = \sqrt{2x}, \qquad g(x) = \frac{1}{x + 1}$
(c) $f(x) = 2x - 3, \qquad g(x) = x^2 + 5$

SOLUTION

(a) Begin by writing $(f \circ g)(x) = f(g(x)) = f(x^3 - 2x^2 - 1)$. Function f adds 2 to the input. That is, $f(\text{input}) = (\text{input}) + 2$ because $f(x) = x + 2$. Thus

$$f(x^3 - 2x^2 - 1) = (x^3 - 2x^2 - 1) + 2 = x^3 - 2x^2 + 1.$$

To find $(g \circ f)(x)$, begin by writing $(g \circ f)(x) = g(f(x)) = g(x + 2)$. Function g does the following to its input: $g(\text{input}) = (\text{input})^3 - 2(\text{input})^2 - 1$. Thus

$$g(x + 2) = (x + 2)^3 - 2(x + 2)^2 - 1.$$

(b) $(f \circ g)(x) = f(g(x)) = f\left(\dfrac{1}{x + 1}\right) = \sqrt{2 \cdot \dfrac{1}{x + 1}} = \sqrt{\dfrac{2}{x + 1}}$

$(g \circ f)(x) = g(f(x)) = g(\sqrt{2x}) = \dfrac{1}{\sqrt{2x} + 1}$

(c) $(f \circ g)(x) = f(g(x)) = f(x^2 + 5) = 2(x^2 + 5) - 3 = 2x^2 + 7$

$(g \circ f)(x) = g(f(x)) = g(2x - 3) = (2x - 3)^2 + 5 = 4x^2 - 12x + 14$

Now Try Exercises 65, 67, and 69 ◆

The next example shows how to evaluate a composition of functions graphically.

EXAMPLE 9   Evaluating a composite function graphically

Use the graphs of f and g shown in Figure 5.6 to evaluate each expression.
(a) $(f \circ g)(2)$ **(b)** $(g \circ f)(-3)$

SOLUTION

(a) Because $(f \circ g)(2) = f(g(2))$, first evaluate $g(2)$. From Figure 5.7, $g(2) = 1$ and

$$(f \circ g)(2) = f(g(2)) = f(1).$$

To complete the evaluation of $(f \circ g)(2)$, use Figure 5.8 to determine that $f(1) = 3$. Thus $(f \circ g)(2) = 3$.

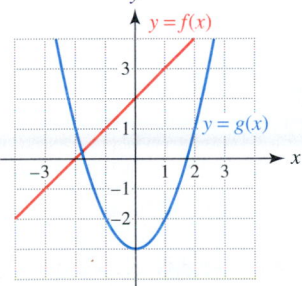

FIGURE 5.6

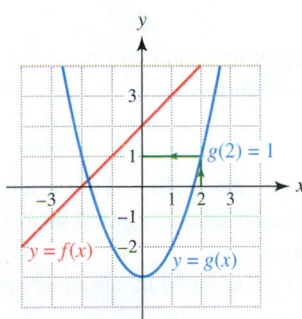

FIGURE 5.7

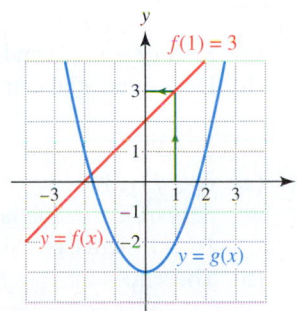

FIGURE 5.8

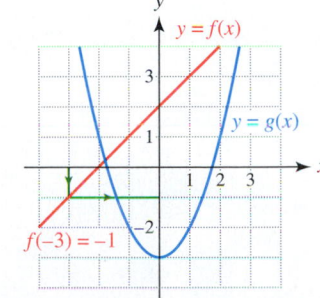

FIGURE 5.9

(b) Because $(g \circ f)(-3) = g(f(-3))$, first evaluate the expression $f(-3)$. From Figure 5.9, $f(-3) = -1$ and

$$(g \circ f)(-3) = g(f(-3)) = g(-1).$$

From Figure 5.10 on the next page, $g(-1) = -2$. Thus $(g \circ f)(-3) = -2$.

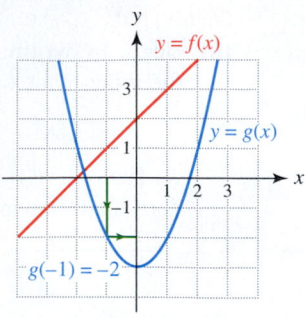

FIGURE 5.10

Now Try Exercise 57 ◆

 **10**   **Evaluating a composite function numerically**

Ozone in the stratosphere filters out approximately 90% of the harmful ultraviolet (UV) rays from the sun. Depletion of the ozone layer has caused an increase in the amount of UV radiation reaching the surface of the earth. An increase in UV radiation is associated with skin cancer. In Table 5.3 the function f computes the approximate percent *increase* in UV radiation resulting from an x percent *decrease* in the thickness of the ozone layer. The function g shown in Table 5.4 computes the expected percent increase in cases of skin cancer resulting from an x percent increase in UV radiation.   (**Source:** R. Turner, D. Pearce, and I. Bateman, *Environmental Economics.*)

TABLE 5.3   Percent Increase in UV Radiation

x	0	1	2	3	4	5	6
$f(x)$	0	1.5	3.0	4.5	6.0	7.5	9.0

TABLE 5.4   Percent Increase in Skin Cancer

x	0	1.5	3.0	4.5	6.0	7.5	9.0
$g(x)$	0	5.25	10.5	15.75	21.0	26.25	31.5

(a) Find $(g \circ f)(2)$ and interpret this calculation.
(b) Determine a tabular representation for $g \circ f$. Describe what $(g \circ f)(x)$ computes.

SOLUTION
(a) $(g \circ f)(2) = g(f(2)) = g(3.0) = 10.5$. This means that a 2% decrease in the thickness of the ozone layer results in a 3% increase in UV radiation. This could cause a 10.5% increase in skin cancer.
(b) The values for $(g \circ f)(x)$ can be found in a manner similar to part (a). See Table 5.5.

TABLE 5.5

x	0	1	2	3	4	5	6
$(g \circ f)(x)$	0	5.25	10.5	15.75	21.0	26.25	31.5

The composition $(g \circ f)(x)$ computes the percent increase in cases of skin cancer resulting from an x percent decrease in the ozone layer.

Now Try Exercise 103 ◆

EXAMPLE 11   Modeling the urban heat island phenomenon

Cities are made up of large amounts of concrete and asphalt that heat up in the daytime from sunlight but do not cool off completely at night. As a result, urban areas tend to be warmer than the surrounding rural areas. This effect is called the *urban heat island* and has been documented in cities throughout the world. In Figure 5.11 function f computes the average increase in nighttime summer temperatures in degrees Celsius at Sky Harbor Airport in Phoenix from 1948 to 1990. In this graph 1948 is the base year with a zero temperature increase. This rise in urban temperature increased peak demand for electricity. In Figure 5.12 function g computes the percent increase in electrical demand for an average nighttime temperature increase of x degrees Celsius.   (**Source:** W. Cotton and R. Pielke, *Human Impacts on Weather and Climate.*)

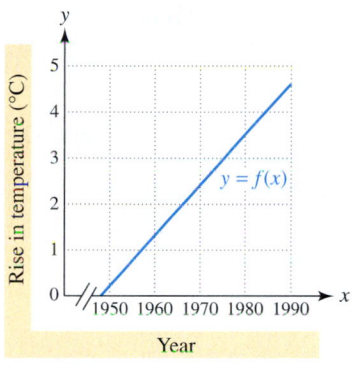

FIGURE 5.11 Nighttime Temperature Increase

FIGURE 5.12 Percent Increase in Electrical Demand

(**a**) Evaluate $(g \circ f)(1975)$ graphically.   (**b**) Interpret $(g \circ f)(x)$.

SOLUTION

(**a**) To evaluate $(g \circ f)(1975) = g(f(1975))$ graphically, first find $f(1975)$. In Figure 5.13, $f(1975) \approx 3$. Next let 3 be input for $g(x)$. In Figure 5.14, $g(3) \approx 4.5$. It follows that $(g \circ f)(1975) = g(f(1975)) \approx g(3) \approx 4.5$. In 1975 the average nighttime temperature had risen about 3°C since 1948. This resulted in a 4.5% increase in peak demand for electricity.

(**b**) $(g \circ f)(x)$ computes the percent increase in peak electrical demand during year x.

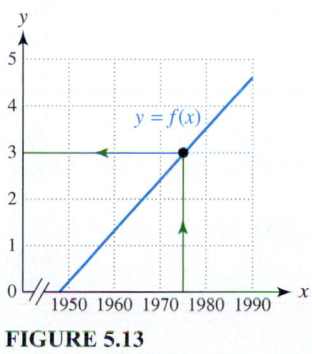 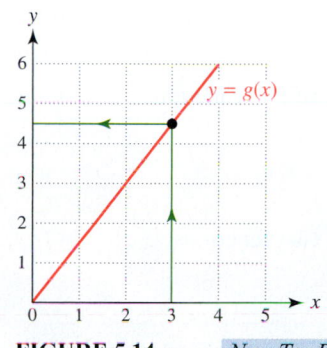

FIGURE 5.13   **FIGURE 5.14**   *Now Try Exercise 105* ◆

When you are solving problems, it is sometimes helpful to recognize a function as the composition of two simpler functions. For example, $h(x) = \sqrt[3]{x^2}$ can be thought of as the composition of the cube root function, $g(x) = \sqrt[3]{x}$, and the squaring function, $f(x) = x^2$. Then function h can be written as $h(x) = g(f(x)) = \sqrt[3]{x^2}$. This concept is demonstrated in the next example. Note that answers may vary.

EXAMPLE 12   Writing a function as a composition of two functions

Find functions f and g so that $h(x) = (g \circ f)(x)$.

(a) $h(x) = (x + 3)^2$ **(b)** $h(x) = \sqrt{2x - 7}$ **(c)** $h(x) = \dfrac{1}{x^2 + 2x}$

SOLUTION

(a) Let $f(x) = x + 3$ and $g(x) = x^2$. Then,

$$(g \circ f)(x) = g(f(x)) = g(x + 3) = (x + 3)^2.$$

(b) Let $f(x) = 2x - 7$ and $g(x) = \sqrt{x}$. Then,

$$(g \circ f)(x) = g(f(x)) = g(2x - 7) = \sqrt{2x - 7}.$$

(c) Let $f(x) = x^2 + 2x$ and $g(x) = \frac{1}{x}$. Then,

$$(g \circ f)(x) = g(f(x)) = g(x^2 + 2x) = \frac{1}{x^2 + 2x}.$$

Now Try Exercises 85, 87, and 95 ◆

5.1

Putting it all Together

Addition, subtraction, multiplication, and division can be used to combine functions. Composition of functions is another way to combine functions, and it is fundamentally different from arithmetic operations on functions. In the composition $(g \circ f)(x)$, the *output* $f(x)$ is used as *input* for g.

The following table summarizes some concepts involved with combining functions.

Concept	Notation	Examples
Sum of two functions	$(f + g)(x) = f(x) + g(x)$	$f(x) = x^2, g(x) = 2x + 1$ $(f + g)(3) = f(3) + g(3) = 9 + 7 = 16$ $(f + g)(x) = f(x) + g(x) = x^2 + 2x + 1$
Difference of two functions	$(f - g)(x) = f(x) - g(x)$	$f(x) = 3x, g(x) = 2x + 1$ $(f - g)(1) = f(1) - g(1) = 3 - 3 = 0$ $(f - g)(x) = f(x) - g(x) = 3x - (2x + 1)$ $\qquad\qquad\qquad = x - 1$
Product of two functions	$(fg)(x) = f(x) \cdot g(x)$	$f(x) = x^3, g(x) = 1 - 3x$ $(fg)(-2) = f(-2) \cdot g(-2) = (-8)(7) = -56$ $(fg)(x) = f(x) \cdot g(x) = x^3(1 - 3x) = x^3 - 3x^4$
Quotient of two functions	$\left(\dfrac{f}{g}\right)(x) = \dfrac{f(x)}{g(x)}, g(x) \neq 0$	$f(x) = x^2 - 1, g(x) = x + 2$ $\left(\dfrac{f}{g}\right)(2) = \dfrac{f(2)}{g(2)} = \dfrac{3}{4}$ $\left(\dfrac{f}{g}\right)(x) = \dfrac{f(x)}{g(x)} = \dfrac{x^2 - 1}{x + 2}, x \neq -2$

Concept	Notation	Examples
Composition of two functions	$(g \circ f)(x) = g(f(x))$	$f(x) = x^3, g(x) = x^2 - 2x + 1$ $(g \circ f)(2) = g(f(2)) = g(8)$ $\qquad = 64 - 16 + 1 = 49$ $(g \circ f)(x) = g(f(x)) = g(x^3)$ $\qquad = (x^3)^2 - 2(x^3) + 1$ $\qquad = x^6 - 2x^3 + 1$

5.1 Exercises

Arithmetic Operations on Functions

Exercises 1 and 2: Use the table to evaluate each expression, if possible.

   (a) $(f + g)(2)$ (b) $(f - g)(4)$
   (c) $(fg)(-2)$ (d) $(f/g)(0)$

1.

x	-2	0	2	4
$f(x)$	0	5	7	10
$g(x)$	6	0	-2	5

2.

x	-2	0	2	4
$f(x)$	-4	8	5	0
$g(x)$	2	-1	4	0

3. Use the table in Exercise 1 to complete the following table.

x	-2	0	2	4
$(f + g)(x)$				
$(f - g)(x)$				
$(fg)(x)$				
$(f/g)(x)$				

4. Use the table in Exercise 2 to complete the table in Exercise 3.

Exercises 5–8: Use the graph to evaluate each expression.

5. (a) $(f + g)(2)$

   (b) $(f - g)(1)$

   (c) $(fg)(0)$

   (d) $(f/g)(1)$

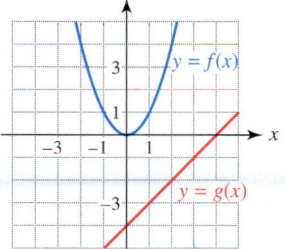

6. (a) $(f + g)(1)$

   (b) $(f - g)(0)$

   (c) $(fg)(-1)$

   (d) $(f/g)(1)$

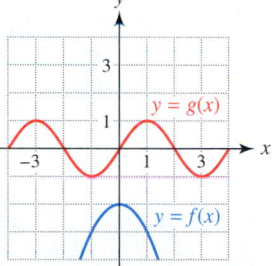

7. (a) $(f + g)(0)$

   (b) $(f - g)(-1)$

   (c) $(fg)(1)$

   (d) $(f/g)(2)$

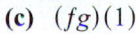

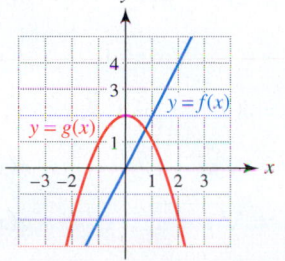

8. (a) $(f + g)(-1)$

(b) $(f - g)(-2)$

(c) $(fg)(0)$

(d) $(f/g)(2)$

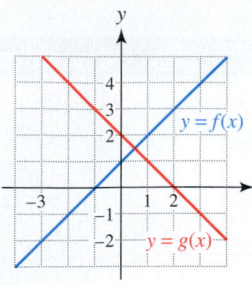

Exercises 9–12: Use f(x) and g(x) to evaluate each expression symbolically.

9. $f(x) = 2x - 3, g(x) = 1 - x^2$

   (a) $(f + g)(3)$ **(b)** $(f - g)(-1)$

   (c) $(fg)(0)$ **(d)** $(f/g)(2)$

10. $f(x) = 4x - x^3, g(x) = x + 3$

   (a) $(g + g)(-2)$ **(b)** $(f - g)(0)$

   (c) $(gf)(1)$ **(d)** $(g/f)(-3)$

11. $f(x) = 2x + 1, g(x) = \dfrac{1}{x}$

   (a) $(f + g)(2)$ **(b)** $(f - g)\left(\tfrac{1}{2}\right)$

   (c) $(fg)(4)$ **(d)** $(f/g)(0)$

12. $f(x) = \sqrt[3]{x^2}, g(x) = |x - 3|$

   (a) $(f + g)(-8)$ **(b)** $(f - g)(-1)$

   (c) $(fg)(0)$ **(d)** $(f/g)(27)$

Exercises 13–36: Use f(x) and g(x) to find each expression symbolically. Identify its domain.

   (a) $(f + g)(x)$ *(b)* $(f - g)(x)$

   (c) $(fg)(x)$ *(d)* $(f/g)(x)$

13. $f(x) = 2x,$ $g(x) = x^2$

14. $f(x) = 1 - 4x,$ $g(x) = 3x + 1$

15. $f(x) = x^2 - 1,$ $g(x) = x^2 + 1$

16. $f(x) = 4x^3 - 8x^2,$ $g(x) = 4x^2$

17. $f(x) = x - \sqrt{x - 1},$ $g(x) = x + \sqrt{x - 1}$

18. $f(x) = 3 + \sqrt{2x + 9},$ $g(x) = 3 - \sqrt{2x + 9}$

19. $f(x) = \sqrt{x} - 1,$ $g(x) = \sqrt{x} + 1$

20. $f(x) = \sqrt{1 - x},$ $g(x) = x^3$

21. $f(x) = \dfrac{1}{x + 1},$ $g(x) = \dfrac{3}{x + 1}$

22. $f(x) = 1 - x^2,$ $g(x) = 3x^2 + 5$

23. $f(x) = 2x,$ $g(x) = 3 - 4x$

24. $f(x) = x^{1/2},$ $g(x) = 3$

25. $f(x) = \dfrac{1}{2x - 4},$ $g(x) = \dfrac{x}{2x - 4}$

26. $f(x) = \dfrac{1}{x},$ $g(x) = x^3$

27. $f(x) = x + 2,$ $g(x) = x^2 + 2x$

28. $f(x) = 6 - x,$ $g(x) = \sqrt{x - 4}$

29. $f(x) = x^2 - 1,$ $g(x) = |x + 1|$

30. $f(x) = |2x - 1|,$ $g(x) = |2x + 1|$

31. $f(x) = \dfrac{x^2 - 3x + 2}{x + 1},$ $g(x) = \dfrac{x^2 - 1}{x - 2}$

32. $f(x) = \dfrac{4x - 2}{x + 2},$ $g(x) = \dfrac{2x - 1}{3x + 6}$

33. $f(x) = \dfrac{2}{x^2 - 1},$ $g(x) = \dfrac{x + 1}{x^2 - 2x + 1}$

34. $f(x) = \dfrac{1}{x + 2},$ $g(x) = x^2 + x - 2$

35. $f(x) = x^{5/2} - x^{3/2},$ $g(x) = x^{1/2}$

36. $f(x) = x^{2/3} - 2x^{1/3} + 1,$ $g(x) = x^{1/3} - 1$

37. Let $f(x) = \sqrt{x}$ and $g(x) = x + 1$.

   (a) Graph $f(x)$, $g(x)$, and $(f + g)(x)$ in the viewing rectangle $[0, 9, 1]$ by $[0, 15, 1]$.

   (b) Explain how the graph of $(f + g)(x)$ can be found using the graphs of $f(x)$ and $g(x)$.

38. Let $f(x) = 0.2x + 5$ and $g(x) = 0.1x + 1$.

   (a) Graph $f(x)$, $g(x)$, and $(f - g)(x)$ in the viewing rectangle $[0, 10, 1]$ by $[0, 10, 1]$.

   (b) Explain how the graph of $(f - g)(x)$ can be found using the graphs of $f(x)$ and $g(x)$.

Review of Function Notation

Exercises 39–52: For the given g(x), evaluate each expression.

(a) $g(-3)$ (b) $g(b)$ (c) $g(x^3)$ (d) $g(2x - 3)$

39. $g(x) = 2x + 1$ **40.** $g(x) = 5 - \frac{1}{2}x$

41. $g(x) = 2(x + 3)^2 - 4$ **42.** $g(x) = -(x - 1)^2$

43. $g(x) = \frac{1}{2}x^2 + 3x - 1$ **44.** $g(x) = 2x^2 - x - 9$

45. $g(x) = \sqrt{x + 4}$ **46.** $g(x) = \sqrt{2 - x}$

47. $g(x) = |3x - 1| + 4$ **48.** $g(x) = 2|1 - x| - 7$

49. $g(x) = \dfrac{4x}{x + 3}$ **50.** $g(x) = \dfrac{x + 3}{2}$

51. $g(x) = 4x^{2/3}$ **52.** $g(x) = \sqrt[3]{\frac{1}{2}x}$

Composition of Functions

Exercises 53 and 54: Numerical representations for the functions f and g are given. Evaluate the expression, if possible.

(a) $(g \circ f)(1)$ (b) $(f \circ g)(4)$ (c) $(f \circ f)(3)$

53.

x	1	2	3	4
$f(x)$	4	3	1	2

x	1	2	3	4
$g(x)$	2	3	4	5

54.

x	1	3	4	6
$f(x)$	2	6	5	7

x	2	3	5	7
$g(x)$	4	2	6	0

55. Use the tables for $f(x)$ and $g(x)$ in Exercise 53 to complete the composition shown in the diagram.

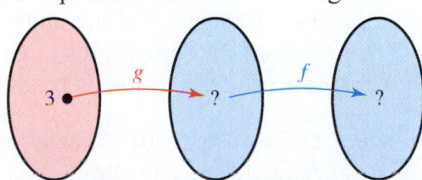

56. Use the tables for $f(x)$ and $g(x)$ in Exercise 54 to complete the composition shown in the diagram.

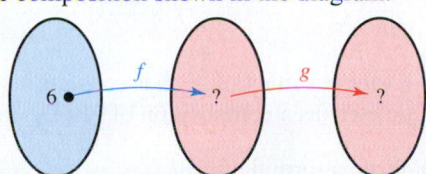

Exercises 57–60: Use the graph to evaluate each expression.

57. (a) $(f \circ g)(4)$

(b) $(g \circ f)(3)$

(c) $(f \circ f)(2)$

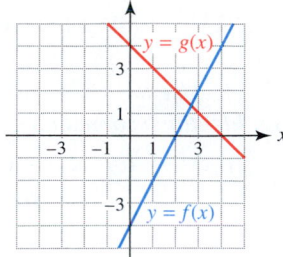

58. (a) $(f \circ g)(2)$

(b) $(g \circ g)(0)$

(c) $(g \circ f)(4)$

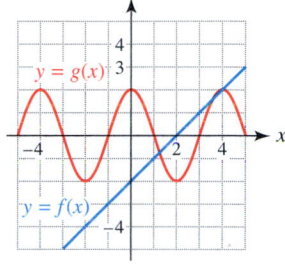

59. (a) $(f \circ g)(1)$

(b) $(g \circ f)(-2)$

(c) $(g \circ g)(-2)$

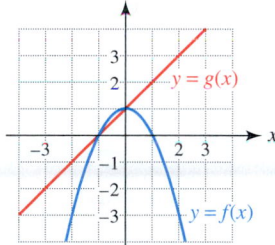

60. (a) $(f \circ g)(-2)$

(b) $(g \circ f)(1)$

(c) $(f \circ f)(0)$

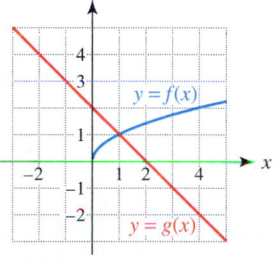

Exercises 61–64: Use the given f(x) and g(x) to evaluate each expression.

61. $f(x) = \sqrt{x + 5}, g(x) = x^2$
(a) $(f \circ g)(2)$ (b) $(g \circ f)(-1)$

62. $f(x) = |x^2 - 4|, g(x) = 2x^2 + x + 1$
(a) $(f \circ g)(1)$ (b) $(g \circ f)(-3)$

63. $f(x) = 5x - 2, g(x) = |x|$
(a) $(f \circ g)(-4)$ (b) $(g \circ f)(5)$

64. $f(x) = \dfrac{1}{x-4}, g(x) = 5$

(a) $(f \circ g)(3)$ (b) $(g \circ f)(8)$

Exercises 65–84: Use the given f(x) and g(x) to find each of the following. Identify its domain.

(a) $(f \circ g)(x)$ (b) $(g \circ f)(x)$ (c) $(f \circ f)(x)$

65. $f(x) = x^3,$ $g(x) = x^2 + 3x - 1$

66. $f(x) = 2 - x,$ $g(x) = \dfrac{1}{x^2}$

67. $f(x) = x^2,$ $g(x) = \sqrt{1-x}$

68. $f(x) = x + 2,$ $g(x) = x^4 + x^2 - 3x - 4$

69. $f(x) = 2 - 3x,$ $g(x) = x^3$

70. $f(x) = \sqrt{x},$ $g(x) = 1 - x^2$

71. $f(x) = x - 4,$ $g(x) = 5 - x^2$

72. $f(x) = \sqrt{4-x},$ $g(x) = x^2$

73. $f(x) = \dfrac{1}{x+1},$ $g(x) = 5x$

74. $f(x) = \dfrac{1}{3x},$ $g(x) = \dfrac{2}{x-1}$

75. $f(x) = x + 4,$ $g(x) = \sqrt{4-x^2}$

76. $f(x) = 2x + 1,$ $g(x) = 4x^3 - 5x^2$

77. $f(x) = \sqrt{x-1},$ $g(x) = 3x$

78. $f(x) = x^2 - 1,$ $g(x) = 2 - 3x^2$

79. $f(x) = \dfrac{x-3}{2}$ $g(x) = 2x + 3$

80. $f(x) = 1 - 5x$ $g(x) = \dfrac{1-x}{5}$

81. $f(x) = \sqrt[3]{x-1}$ $g(x) = x^3 + 1$

82. $f(x) = \dfrac{1}{kx}, k > 0$ $g(x) = \dfrac{1}{kx}, k > 0$

83. $f(x) = \dfrac{ax+b}{3}, a > 0$ $g(x) = \dfrac{x}{a} - b, a > 0$

84. $f(x) = ax^2, a > 0$ $g(x) = \sqrt{ax}, a > 0$

Exercises 85–98: (Refer to Example 12.) Find functions f and g so that h(x) = (g \circ f)(x).

85. $h(x) = \sqrt{x-2}$ **86.** $h(x) = (x+2)^4$

87. $h(x) = 5(x+2)^2 - 4$ **88.** $h(x) = \dfrac{1}{x+2}$

89. $h(x) = 4(2x+1)^3$ **90.** $h(x) = \sqrt[3]{x^2+1}$

91. $h(x) = (x^3-1)^2$ **92.** $h(x) = 4(x-5)^{-2}$

93. $h(x) = -4|x+2| - 3$

94. $h(x) = 5\sqrt{x-1}$

95. $h(x) = \dfrac{1}{(x-1)^2}$ **96.** $h(x) = \dfrac{2}{x^2-x+1}$

97. $h(x) = x^{3/4} - x^{1/4}$ **98.** $h(x) = x^{2/3} - 5x^{1/3} + 4$

Applications

99. *Revenue, Cost, and Profit* Suppose that it costs $150,000 to produce a master disc for a music video and $1.50 to produce each copy.

(a) Write a cost function C that outputs the cost of producing the master disc and x copies.

(b) If the music video is sold for $6.50 each, find a function R that outputs the revenue received from selling x music videos. What is the revenue from selling 8000 videos?

(c) Assuming that the master disc is not sold, find a function P that outputs the profit from selling x music videos. What is the profit from selling 40,000 videos?

(d) How many videos must be sold to break even? That is, how many videos must be sold for the revenue to equal the cost?

100. *Profit* (Refer to Example 4.) Determine a profit function P that results if the compact discs are sold for $15 each. Find the profit from selling 3000 compact discs.

101. *Converting Units* There are 36 inches in a yard and about 2.54 centimeters in an inch.

(a) Write a function I that converts x yards to inches.

(b) Write a function C that converts x inches to centimeters.

(c) Express a function F that converts x yards to centimeters as a composition of two functions.

(d) Write a formula for F.

102. **Converting Units** There are 4 quarts in 1 gallon, 4 cups in 1 quart, and 16 tablespoons in 1 cup.
    (a) Write a function Q that converts x gallons to quarts.

    (b) Write a function C that converts x quarts to cups.

    (c) Write a function T that converts x cups to tablespoons.

    (d) Express a function F that converts x gallons to tablespoons as a composition of *three* functions.

    (e) Write a formula for F.

103. **Skin Cancer** (Refer to Example 10 and Tables 5.3 and 5.4.) If possible, calculate the composition and interpret the result.
    (a) $(g \circ f)(1)$ (b) $(f \circ g)(21)$

104. **Skin Cancer** In Example 10 the functions f and g are both linear.
    (a) Find symbolic representations for f and g.

    (b) Determine $(g \circ f)(x)$.

    (c) Evaluate $(g \circ f)(3.5)$ and interpret the result.

105. **Urban Heat Island** (Refer to Example 11 and Figures 5.11 and 5.12.) If possible, calculate the composition and interpret the result.
    (a) $(g \circ f)(1980)$ (b) $(f \circ g)(3)$

106. **Urban Heat Island** (Refer to Example 11.) The functions f and g are given by
    $$f(x) = 0.11(x - 1948) \quad \text{and} \quad g(x) = 1.5x.$$
    (a) Evaluate $(g \circ f)(1960)$.

    (b) Find $(g \circ f)(x)$.

    (c) What type of functions are f, g, and $g \circ f$?

107. **Swimming Pools** In the accompanying figures at the top of the next column, function f computes the cubic feet of water in a pool after x days, and g converts cubic feet to gallons.
    (a) Estimate the gallons of water in the pool when $x = 2$ days.

    (b) Interpret $(g \circ f)(x)$.

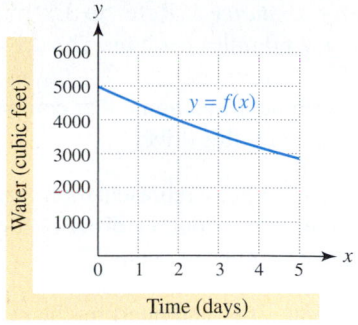

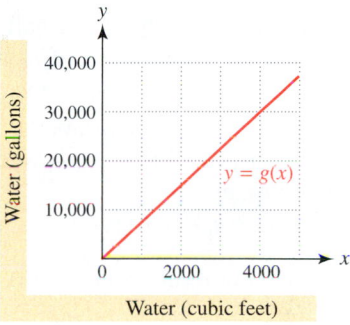

108. **Temperature** The figures show graphs of f and g. The function f computes the temperature on a summer day after x hours, and g converts Fahrenheit temperature to Celsius temperature.
    (a) Evaluate $(g \circ f)(2)$.

    (b) Interpret $(g \circ f)(x)$.

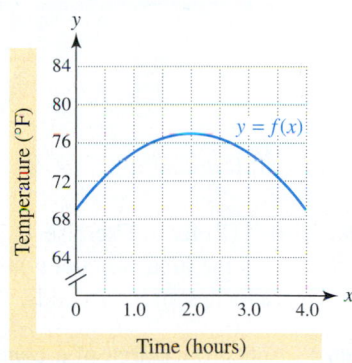

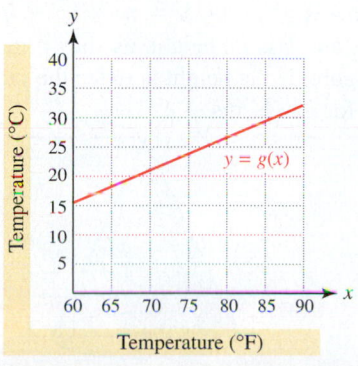

109. *Stopping Distance* (Refer to Example 1.) A car is traveling at 60 miles per hour. The driver has a reaction time of 1.25 seconds.

(a) Determine a function r that computes the reaction distance for this driver.

(b) Find a symbolic representation for a function s that computes the stopping distance for this driver traveling at x miles per hour.

(c) Evaluate $s(60)$ and interpret the result.

110. *Energy of a Falling Object* A ball with mass m is dropped from an initial height of h_0 and lands with a final velocity of v_f. The kinetic energy of the ball is $K(v) = \frac{1}{2}mv^2$, where v is its velocity, and the potential energy of the ball is $P(h) = mgh$, where h is its height and g is a constant.

(a) Show that $P(h_0) = K(v_f)$. (*Hint:* $v_f = \sqrt{2gh_0}$.)

(b) Interpret your result from part (a).

111. *Circular Wave* A marble is dropped into a lake and it results in a circular wave, whose radius increases at a rate of 6 inches per second. Write a formula for C that gives the circumference of the circular wave in inches after t seconds.

112. *Circular Wave* (Refer to Exercise 111.) Write a function A that gives the area contained inside the circular wave in square inches after t seconds.

113. *Geometry* The surface area of a cone (excluding the bottom) is given by $S = \pi r\sqrt{r^2 + h^2}$, where r is its radius and h is its height, as shown in the accompanying figure. If the height is twice the radius, write a formula for S in terms of r.

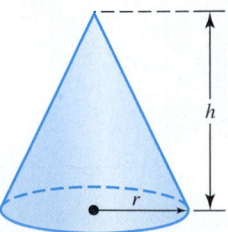

114. *Sphere* The volume V of a sphere with radius r is given by $V = \frac{4}{3}\pi r^3$, and the surface area S is given by $S = 4\pi r^2$. Show that $V = \frac{4}{3}\pi\left(\frac{S}{4\pi}\right)^{3/2}$.

115. *Acid Rain* A common air pollutant responsible for acid rain is sulfur dioxide (SO_2). Emissions of SO_2 from burning coal during year x are computed by $C(x)$ in the table. Emissions of SO_2 from burning oil are computed by $O(x)$. Amounts are given in millions of tons. (**Source:** B. Freedman.)

x	1860	1900	1940	1970	2000
$C(x)$	2.4	12.6	24.2	32.4	55.0
$O(x)$	0.0	0.2	2.3	17.6	23.0

(a) Evaluate $(C + O)(1970)$.

(b) Interpret $(C + O)(x)$.

(c) Make a table for $(C + O)(x)$.

116. *Acid Rain* (Refer to Exercise 115.) Make a table for a function h defined by $h(x) = (C/O)(x)$. Round values for $h(x)$ to the nearest hundredth. What information does h give regarding the use of coal and oil in the United States?

117. *Methane Emissions* Methane is a greenhouse gas. It lets sunlight into the atmosphere but blocks heat from escaping the earth's atmosphere. Methane is a by-product of burning fossil fuels. In the table, f models the predicted methane emissions in millions of tons produced by *developed* countries. The function g models the same emissions for *developing* countries. (**Source:** A. Nilsson, *Greenhouse Earth.*)

x	1990	2000	2010	2020	2030
$f(x)$	27	28	29	30	31
$g(x)$	5	7.5	10	12.5	15

(a) Make a table for a function h that models the total predicted methane emissions for developed *and* developing countries.

(b) Write an equation that relates $f(x)$, $g(x)$, and $h(x)$.

118. *Methane Emissions* (Refer to the previous exercise.) The accompanying figure shows graphs of the functions *f* and *g* that model methane emissions. Use their graphs to sketch a graph of the function *h*.

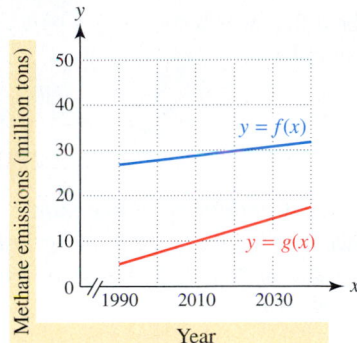

119. *Methane Emissions* (Refer to Exercises 117 and 118.) Symbolic representations for *f* and *g* are $f(x) = 0.1x - 172$ and $g(x) = 0.25x - 492.5$, where *x* is the year. Find a symbolic representation for *h*.

120. *Petroleum Spillage* Large amounts of petroleum products enter the oceans each year. In the table *f* computes the total amount of petroleum that entered the world's oceans. The function *g* computes petroleum spillage into the oceans caused by oil tankers. Amounts are given in thousands of tons. (**Source:** B. Freedman, *Environmental Ecology*.)

Year	1973	1979	1981	1983	1989
$f(x)$	6110	4670	3570	3200	570
$g(x)$	1380	900	1050	1100	—

(a) Define a function *h* by $h(x) = f(x) - g(x)$. Make a table for $h(x)$. What is the domain of *h*?

(b) Interpret what *h* computes.

121. *China's Energy Production* Predicted energy production in China is shown in the figure at the top of the next column. The function *f* computes total coal production, and the function *g* computes total coal *and* oil production. Energy units are in million metric tons of oil equivalent (Mtoe). Let the function *h* compute China's oil production. (**Source:** A. Pascal, *Global Energy: The Changing Outlook*.)
(a) Write an equation that relates $f(x)$, $g(x)$, and $h(x)$.

(b) Evaluate $h(1995)$ and $h(2000)$.

(c) Determine a symbolic representation for *h*.

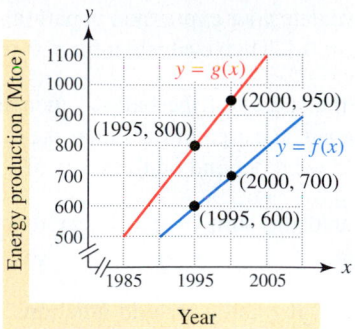

122. *AIDS* The accompanying table lists both the cumulative numbers of AIDS cases and cumulative deaths reported for selected years *x*. The numbers of AIDS cases are modeled by $f(x) = 3200(x - 1982)^2 + 1586$ and the numbers of deaths are modeled by the formula $g(x) = 1900(x - 1982)^2 + 619$. (**Source:** Department of Health and Human Services.)

Year	Cases	Deaths
1982	1586	619
1984	10,927	5605
1986	41,910	24,593
1988	106,304	61,911
1990	196,576	120,811
1992	329,205	196,283
1994	441,528	270,533

(a) Graph $h(x) = \frac{g(x)}{f(x)}$ in [1982, 1994, 2] by [0, 1, 0.1]. Interpret the graph.

(b) Compute the ratio $\frac{\text{Deaths}}{\text{Cases}}$ for each year. Compare these ratios with the results from part (a).

123. *Equilateral Triangle* The area of an equilateral triangle with sides of length *s* is given by

$$A(s) = \frac{\sqrt{3}}{4}s^2.$$

(a) Find $A(4s)$ and interpret the result.

(b) Find $A(s + 2)$ and interpret the result.

124. *Surface Area of a Balloon* The surface area *A* of a balloon with radius *r* is given by $A(r) = 4\pi r^2$. Suppose that the radius of the balloon increases from *r* to $r + h$, where *h* is a small, positive number.
(a) Find $A(r + h) - A(r)$. What does this expression represent?

(b) Evaluate your expression in part (a) when $r = 3$ and $h = 0.1$. Then evaluate it for $r = 6$ and $h = 0.1$.

(c) If the radius of the balloon increases by 0.1, does the surface area always increase by a fixed amount or does it depend on the value of r?

125. Show that the sum of two linear functions is a linear function.

126. Show that if f and g are odd functions, then the composition $g \circ f$ is also an odd function.

127. Let $f(x) = k$ and $g(x) = ax + b$, where k, a, and b are constants.
(a) Find $(f \circ g)(x)$. What type of function is $f \circ g$?

(b) Find $(g \circ f)(x)$. What type of function is $g \circ f$?

128. Show that if $f(x) = ax + b$ and $g(x) = cx + d$, then $(g \circ f)(x)$ also represents a linear function. Find the slope of the graph of $(g \circ f)(x)$.

Writing about Mathematics

129. Describe differences between $(fg)(x)$ and $(f \circ g)(x)$. Give examples.

130. Describe differences between $(f \circ g)(x)$ and $(g \circ f)(x)$. Give examples.

EXTENDED AND DISCOVERY EXERCISES

Exercises 1–4: ***Combining Polynomials*** *Suppose that $f(x)$ is a polynomial with degree n and $g(x)$ is a polynomial with degree m and $m > n \geq 1$. What can be said about the degree D of each expression?*

1. $(gf)(x)$

2. $(g/f)(x)$ (Assume that $f(x)$ divides into $g(x)$ with no remainder.)

3. $(g \circ f)(x)$ **4.** $(f + g)(x)$

5.2   Inverse Functions and Their Representations

◆ Calculate inverse operations

◆ Identify one-to-one functions

◆ Find inverse functions symbolically

◆ Use other representations to find inverse functions

Introduction

Many actions are reversible. A closed door can be opened—an open door can be closed. One hundred dollars can be withdrawn from and deposited into a savings account. These actions undo or cancel each other. But not all actions are reversible. Explosions and weather are two examples. In mathematics this concept of reversing a calculation and arriving at the original result is associated with an inverse.

To introduce the basic concept of an inverse function, refer to Table 5.6, which can be used to convert gallons to pints. It is apparent that there are 8 pints in a gallon, and $f(x) = 8x$ converts x gallons to an equivalent number of pints. However, if we want to reverse the computation and convert pints to gallons, a different function g is required. This conversion is calculated by $g(x) = \frac{x}{8}$, where x is the number of pints. Function g performs the inverse operation of f. We say that f and g are inverse functions and write this as $g(x) = f^{-1}(x)$. We read f^{-1} as "f inverse."

TABLE 5.6   Converting Gallons to Pints

x (gallons)	1	2	3	4	5
$f(x)$ (pints)	8	16	24	32	40

Inverse Operations

Throughout time, codes have been used for secrecy. Every code requires a consistent method to decode it. A simple code that was used during the time of the Roman Empire is shown in Table 5.7.

TABLE 5.7

Letter	A	B	C	D	E	F	G	H	I	J	K	L	M
Code	C	D	E	F	G	H	I	J	K	L	M	N	O

Letter	N	O	P	Q	R	S	T	U	V	W	X	Y	Z
Code	P	Q	R	S	T	U	V	W	X	Y	Z	A	B

Source: A. Sinkov, *Elementary Cryptanalysis: A Mathematical Approach.*

Table 5.7 can be used to code the word HELP as JGNR and decode it back to HELP. Coding and decoding are inverse actions or operations. It is essential that both the coding and decoding procedures give exactly one output for each input.

Actions and their inverses occur in everyday life. Suppose a person opens a car door, gets in, and starts the engine. What are the inverse actions? They are turn off the engine, get out, and close the car door. Notice that the order must be reversed as well as applying the inverse operation at each step.

In mathematics there are basic operations that can be considered inverse operations of each other. For example, if we begin with 10 and add 5, the result is 15. To undo this operation, subtract 5 from 15 to obtain 10. Addition and subtraction are inverse operations. The same is true for multiplication and division. If we multiply a number by 2 and then divide by 2, the original number is obtained. Division and multiplication are inverse operations.

 **EXAMPLE 1**

Finding inverse actions and operations

For each of the following, state the inverse actions or operations.
(a) Put on a coat and go outside.
(b) Subtract 7 from x and divide the result by 2.

SOLUTION
(a) To find the inverse actions, reverse the order and apply the inverse action at each step. The inverse actions would be to come inside and take off the coat.
(b) We must reverse the order and apply the inverse operation at each step. The inverse operations would be to multiply x by 2 and add 7. The original operations could be expressed as $\frac{x-7}{2}$, and the inverse operations could be written as $2x + 7$.

Now Try Exercises 1 and 5 ◆

Inverse operations can be described by functions. If $f(x) = x + 5$, then the *inverse function* of f is given by $f^{-1}(x) = x - 5$. For example, $f(\mathbf{5}) = \mathbf{10}$, and $f^{-1}(\mathbf{10}) = \mathbf{5}$. If input x produces output y with function f, input y produces output x with function f^{-1}. This can be seen numerically in Table 5.8.

TABLE 5.8

x	0	5	10	15
$f(x)$	5	10	15	20

x	5	10	15	20
$f^{-1}(x)$	0	5	10	15

From Table 5.8, note that $f(0) = 5$ and $f^{-1}(5) = 0$, and that $f(15) = 20$ and $f^{-1}(20) = 15$. In general, if $f(a) = b$ then $f^{-1}(b) = a$. That is, if f outputs b with input a, then f^{-1} must output a with input b. *Outputs and inputs are interchanged for inverse functions.* This statement is illustrated in Figure 5.15.

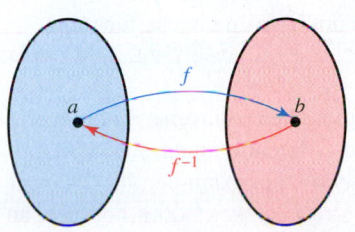

FIGURE 5.15 $f(a) = b; f^{-1}(b) = a$

When $f(x) = x + 5$ and $f^{-1}(x) = x - 5$ are applied in sequence, the output of f is used as input for f^{-1}. This is composition of functions.

$$
\begin{aligned}
(f^{-1} \circ f)(x) &= f^{-1}(\mathbf{f(x)}) && \text{Definition of composition} \\
&= f^{-1}(\mathbf{x + 5}) && f(x) = x + 5 \\
&= (x + 5) - 5 && f^{-1} \text{ subtracts 5 from its input.} \\
&= x && \text{Simplify.}
\end{aligned}
$$

The composition $f^{-1} \circ f$ with input x produces output x. The same action occurs for $f \circ f^{-1}$.

$$
\begin{aligned}
(f \circ f^{-1})(x) &= f(\mathbf{f^{-1}(x)}) && \text{Definition of composition} \\
&= f(\mathbf{x - 5}) && f^{-1}(x) = x - 5 \\
&= (x - 5) + 5 && f \text{ adds 5 to its input.} \\
&= x && \text{Simplify.}
\end{aligned}
$$

◆ **MAKING CONNECTIONS**

The Notation f^{-1} and Negative Exponents If a represents a real number, then $a^{-1} = \frac{1}{a}$. For example, $4^{-1} = \frac{1}{4}$. On the other hand, if f represents a function, then $f^{-1}(x) \neq \frac{1}{f(x)}$. Instead, $f^{-1}(x)$ represents the inverse function of f. For instance, if $f(x) = 5x$, then $f^{-1}(x) = \frac{x}{5} \neq \frac{1}{5x}$.

Algebra Review
To review negative exponents, see Chapter R (page R-14).

Another pair of inverse functions is given by $f(x) = x^3$ and $f^{-1}(x) = \sqrt[3]{x}$. Before a formal definition of inverse functions is given, we must discuss the concept of one-to-one functions.

One-to-One Functions

Does every function have an inverse function? The next example answers this question.

 **EXAMPLE 2** Determining if a function has an inverse function

Table 5.9 represents a function C that computes the percentage of the time that the sky is cloudy in Augusta, Georgia, where x corresponds to the standard numbers for the months. Determine if C has an inverse function.

TABLE 5.9   Cloudy Skies in Augusta

x (month)	1	2	3	4	5	6	7	8	9	10	11	12
$C(x)$ (%)	43	40	39	29	28	26	27	25	30	26	31	39

Source: J. Williams, *The Weather Almanac 1995*.

SOLUTION For each input, C computes exactly one output. For example, $C(3) = 39$ means that during March the sky is cloudy 39% of the time. If C has an inverse function, the inverse must receive 39 as input and produce exactly one output. Both March and December have cloudy skies 39% of the time. Given an input of 39, it is impossible for an inverse *function* to output both 3 and 12. Therefore C does not have an inverse function.

Now Try Exercise 19 ◆

If different inputs of a function f produce the same output, then an inverse function of f does not exist. However, if different inputs always produce different outputs, f is a *one-to-one function. Every one-to-one function has an inverse function*. For example, $f(x) = x^2$ is not one-to-one because $f(-2) = 4$ and $f(2) = 4$. Therefore $f(x) = x^2$ does not have an inverse function because an inverse *function* cannot receive input 4 and produce both -2 and 2 as outputs.

ONE-TO-ONE FUNCTION

A function f is a **one-to-one function** if, for elements c and d in the domain of f,

$$c \neq d \quad \text{implies} \quad f(c) \neq f(d).$$

That is, different inputs always result in different outputs.

EXAMPLE 3 **Determining if a function is one-to-one graphically**

Use the graph of f to determine if f is a one-to-one function in Figures 5.16 and 5.17.

(a) 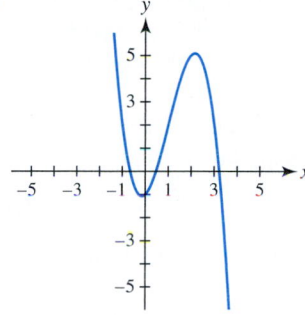   **(b)**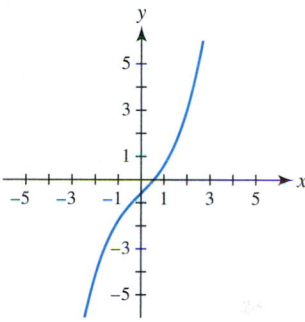

FIGURE 5.16   **FIGURE 5.17**

SOLUTION

(a) To decide if the graph represents a one-to-one function, determine if different inputs (x-values) always correspond to different outputs (y-values). In Figure 5.18 the horizontal line $y = 2$ intersects the graph of f at $(-1, 2)$, $(1, 2)$, and $(3, 2)$. This means that $f(-1) = f(1) = f(3) = 2$. Three distinct inputs, -1, 1, and 3, produce the same output, 2. Therefore f is not one-to-one and does not have an inverse function.

(b) Any horizontal line will intersect the graph of f at most once. For example, if the line $y = 3$ is graphed, it intersects the graph of f at one point, $(2, 3)$. See Figure 5.19. Only input 2 results in output 3. This is true in general. Therefore f is one-to-one and has an inverse function.

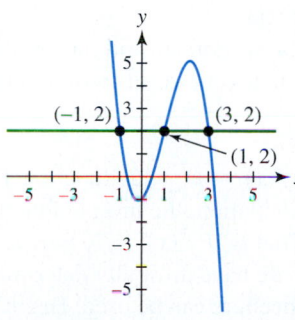   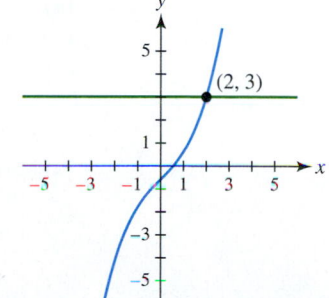

FIGURE 5.18   **FIGURE 5.19**

Now Try Exercises 13 and 15 ◆

Note: To show that f is not one-to-one, it is not necessary to find the actual points of intersection—we have to show only that a horizontal line can intersect the graph of f more than once.

The technique of visualizing horizontal lines to determine if a graph represents a one-to-one function is called the *horizontal line test*.

HORIZONTAL LINE TEST

If every horizontal line intersects the graph of a function f at most once, then f is a one-to-one function.

If a function f is increasing on its domain, then a horizontal line will intersect the graph of f at most once. By the horizontal line test, f is a one-to-one function. For example, the function shown in Figure 5.17 is always increasing on its domain and so it is one-to-one. Similarly, if a function g is decreasing on its domain, then g is a one-to-one function.

Symbolic Representations of Inverse Functions

If a function f is one-to-one, then an inverse function f^{-1} exists. Therefore $f(a) = b$ implies $f^{-1}(b) = a$ for every a in the domain of f. That is,

$$(f^{-1} \circ f)(a) = f^{-1}(f(a)) = f^{-1}(b) = a.$$

Similarly, $f^{-1}(b) = a$ implies $f(a) = b$ for every b in the domain of f^{-1} and so

$$(f \circ f^{-1})(b) = f(f^{-1}(b)) = f(a) = b.$$

These two properties can be used to define an inverse function.

INVERSE FUNCTION

Let f be a one-to-one function. Then f^{-1} is the **inverse function** of f, if

$(f^{-1} \circ f)(x) = f^{-1}(f(x)) = x$ for every x in the domain of f, and

$(f \circ f^{-1})(x) = f(f^{-1}(x)) = x$ for every x in the domain of f^{-1}.

In the next example, we find an inverse function and verify that it is correct.

EXAMPLE 4   Finding and verifying an inverse function

Let f be the one-to-one function given by $f(x) = x^3 - 2$.
(a) Find $f^{-1}(x)$.
(b) Identify the domain and range of f^{-1}.
(c) Verify that your result from part (a) is correct.

SOLUTION
(a) Since $f(x) = x^3 - 2$, function f cubes the input x and then subtracts 2. To reverse this calculation, the inverse function must add 2 to the input x and then take the cube root. That is, $f^{-1}(x) = \sqrt[3]{x + 2}$.

If we have difficulty determining the formula for an inverse mentally, the following procedure can be used. Begin by solving $y = f(x)$ for x.

$$y = x^3 - 2 \qquad \textcolor{teal}{y = f(x)}$$
$$y + 2 = x^3 \qquad \textcolor{teal}{\text{Add 2.}}$$
$$\sqrt[3]{y + 2} = x \qquad \textcolor{teal}{\text{Take the cube root.}}$$

Interchange x and y to obtain $y = \sqrt[3]{x + 2}$. This gives us the formula for $f^{-1}(x)$.
(b) Both the domain and range of the cube root function include all real numbers. The graph of $f^{-1}(x) = \sqrt[3]{x + 2}$ is the graph of the cube root function shifted left 2 units. Therefore the domain and range of f^{-1} also include all real numbers.

(c) To verify that $f^{-1}(x) = \sqrt[3]{x + 2}$ is indeed the inverse of $f(x) = x^3 - 2$, we must show that $f^{-1}(f(x)) = x$ and that $f(f^{-1}(x)) = x$.

$$
\begin{aligned}
f^{-1}(f(x)) &= f^{-1}(x^3 - 2) & f(x) = x^3 - 2 \\
&= \sqrt[3]{(x^3 - 2) + 2} & f^{-1}(x) = \sqrt[3]{x + 2} \\
&= \sqrt[3]{x^3} & \text{Combine terms.} \\
&= x & \text{Simplify.} \\
f(f^{-1}(x)) &= f(\sqrt[3]{x + 2}) & f^{-1}(x) = \sqrt[3]{x + 2} \\
&= (\sqrt[3]{x + 2})^3 - 2 & f(x) = x^3 - 2 \\
&= (x + 2) - 2 & \text{Cube the expression.} \\
&= x & \text{Combine terms.}
\end{aligned}
$$

Algebra Review
To review rational exponents and radical notation, see Chapter R (page R-50).

These calculations verify that our result is correct.   *Now Try Exercise 73* ◆

This symbolic technique is now summarized verbally.

FINDING A SYMBOLIC REPRESENTATION FOR f^{-1}

To find a formula for f^{-1}, perform the following steps.

STEP 1: Verify that f is a one-to-one function.

STEP 2: Solve the equation $y = f(x)$ for x, resulting in the equation $x = f^{-1}(y)$.

STEP 3: Interchange x and y to obtain $y = f^{-1}(x)$.

To verify $f^{-1}(x)$, show that $(f^{-1} \circ f)(x) = x$ and $(f \circ f^{-1})(x) = x$.

EXAMPLE 5   ## Finding and verifying an inverse function

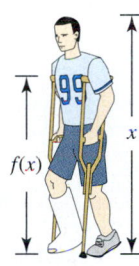

Suppose that a person x inches tall sprains an ankle. An approximation for the length of a crutch (in inches) that this person might need is calculated by $f(x) = \frac{18}{25}x + 2$.
(a) Explain why f is a one-to-one function.
(b) Find a symbolic representation for f^{-1}.
(c) Verify that this is the inverse function of f.
(d) Interpret the meaning of $f^{-1}(56)$.

SOLUTION
(a) Since f is a linear function, its graph is a line with a nonzero slope of $\frac{18}{25}$. Every horizontal line intersects it at most once. By the horizontal line test, f is one-to-one.
(b) To find $f^{-1}(x)$, solve the equation $y = f(x)$ for x.

$$
y = \frac{18}{25}x + 2 \qquad y = f(x)
$$

$$
y - 2 = \frac{18}{25}x \qquad \text{Subtract 2.}
$$

$$
\frac{25}{18}(y - 2) = x \qquad \text{Multiply by } \tfrac{25}{18}, \text{ the reciprocal of } \tfrac{18}{25}.
$$

Now interchange x and y to obtain $y = \frac{25}{18}(x - 2)$. The formula for the inverse is, therefore,

$$
f^{-1}(x) = \frac{25}{18}(x - 2).
$$

(c) We must show that $(f^{-1} \circ f)(x) = x$ and $(f \circ f^{-1})(x) = x$.

$$(f^{-1} \circ f)(x) = f^{-1}(f(x)) \qquad \text{Definition of composition}$$

$$= f^{-1}\left(\frac{18}{25}x + 2\right) \qquad f(x) = \frac{18}{25}x + 2$$

$$= \frac{25}{18}\left(\left(\frac{18}{25}x + 2\right) - 2\right) \qquad f^{-1}(x) = \frac{25}{18}(x - 2)$$

$$= \frac{25}{18}\left(\frac{18}{25}x\right) \qquad \text{Simplify.}$$

$$= x \qquad \text{It checks.}$$

Similarly,

$$(f \circ f^{-1})(x) = f(f^{-1}(x)) \qquad \text{Definition of composition}$$

$$= f\left(\frac{25}{18}(x - 2)\right) \qquad f^{-1}(x) = \frac{25}{18}(x - 2)$$

$$= \frac{18}{25}\left(\frac{25}{18}(x - 2)\right) + 2 \qquad f(x) = \frac{18}{25}x + 2$$

$$= (x - 2) + 2 \qquad \text{Simplify.}$$

$$= x \qquad \text{It checks.}$$

(d) The expression $f(x)$ calculates the proper crutch length for a person x inches tall, and $f^{-1}(x)$ computes the height of a person requiring a crutch x inches long. Therefore $f^{-1}(56) = \frac{25}{18}(56 - 2) = 75$ means that a 56-inch crutch would be appropriate for a person 75 inches tall. See Figure 5.20.   *Now Try Exercise 127* ◆

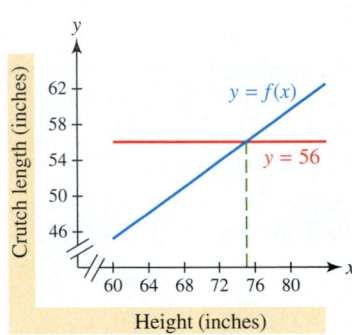

FIGURE 5.20

EXAMPLE 6   Restricting the domain of a function

Let $f(x) = (x - 1)^2$.
(a) Does f have an inverse function? Explain.
(b) Restrict the domain of f so that f^{-1} exists.
(c) Find $f^{-1}(x)$ for the restricted domain.

SOLUTION
(a) The graph of $f(x) = (x - 1)^2$, shown in Figure 5.21, does not pass the horizontal line test. Therefore f is not one-to-one and does not have an inverse function.

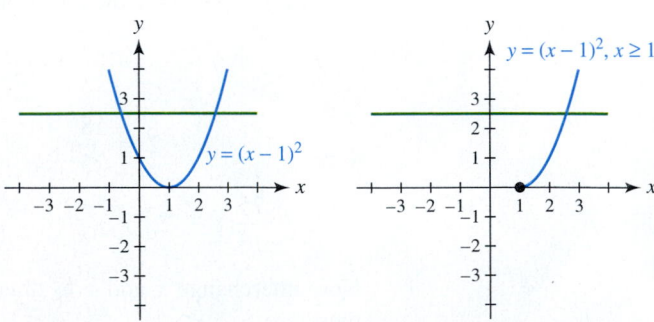

FIGURE 5.21 Not One-to-One

FIGURE 5.22 Restricting the Domain

(b) If we restrict the domain of f to $D = \{x \mid x \geq 1\}$, then f becomes a one-to-one function. To illustrate this, the graph of $y = (x - 1)^2$ for $x \geq 1$ is shown in Figure 5.22. This graph passes the horizontal line test and now f^{-1} exists.

(c) Assume that $x \geq 1$ and solve the equation $y = f(x)$ for x.

$$y = (x - 1)^2 \qquad y = f(x)$$

Take the positive square root.

$$\sqrt{y} = x - 1$$

Note: $x \geq 1$ implies that $x - 1 \geq 0$.

$$\sqrt{y} + 1 = x \qquad \text{Add 1.}$$

Thus $f^{-1}(x) = \sqrt{x} + 1$.

Now Try Exercise 63 ◆

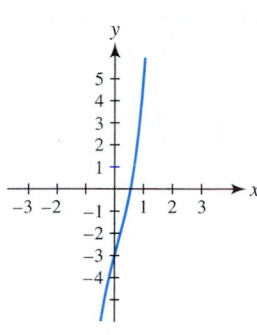

FIGURE 5.23

The graph of $f(x) = x^5 + 3x^3 - x^2 + 5x - 3$ is shown in Figure 5.23. By the horizontal line test, f is a one-to-one function and f^{-1} exists. However, it would be difficult to obtain a formula for f because we would need to solve $y = x^5 + 3x^3 - x^2 + 5x - 3$ for x. It is important to realize that many functions have inverses, but their inverses can be either difficult or impossible to find symbolically.

Other Representations of Inverse Functions

Given a symbolic representation of a one-to-one function f, we can often find its inverse f^{-1} by using the techniques discussed in the previous subsection. Numerical and graphical representations of a one-to-one function can also be used to find its inverse.

Numerical Representations   In Table 5.10, f has domain $D = \{1940, 1970, 2000\}$ and it computes the percentage of the U.S. population with 4 or more years of college in year x.

Function f is one-to-one because different inputs produce different outputs. Therefore f^{-1} exists. Since $f(1940) = 5$, it follows that $f^{-1}(5) = 1940$. Similarly, $f^{-1}(11) = 1970$ and $f^{-1}(27) = 2000$. Table 5.11 shows a numerical representation of f^{-1}.

TABLE 5.10

x	$f(x)$
1940	5
1970	11
2000	27

TABLE 5.11

x	$f^{-1}(x)$
5	1940
11	1970
27	2000

The domain of f is $\{1940, 1970, 2000\}$ and the range is $\{5, 11, 27\}$. The domain of f^{-1} is $\{5, 11, 27\}$ and the range of f^{-1} is $\{1940, 1970, 2000\}$. The functions f and f^{-1} interchange domains and ranges. This result is true in general for inverse functions.

The diagrams in Figures 5.24 and 5.25 demonstrate this property. To obtain f^{-1} from f, the arrows for f are simply reversed. This reversal causes the domains and ranges to be interchanged.

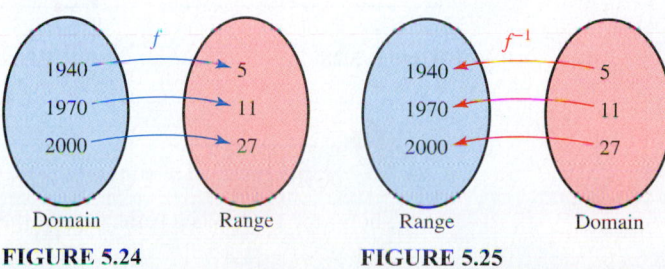

FIGURE 5.24   **FIGURE 5.25**

Figure 5.26 shows a function f that is not one-to-one. In Figure 5.27 the arrows defining f have been reversed. This is a relation. However, this relation does not represent the inverse *function* because input 4 produces two outputs, 1 and 2.

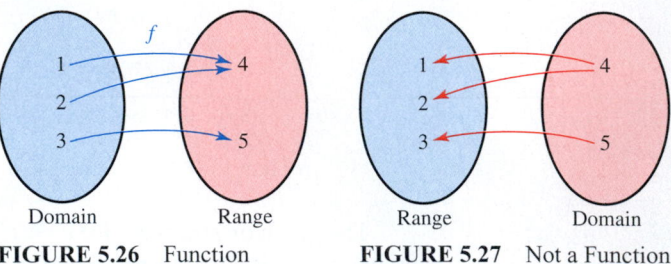

FIGURE 5.26   Function   FIGURE 5.27   Not a Function

The relationship between domains and ranges is summarized in the following.

DOMAINS AND RANGES OF INVERSE FUNCTIONS
The domain of f equals the range of f^{-1}.
The range of f equals the domain of f^{-1}.

Graphical Representations   If the point $(2, 5)$ lies on the graph of f, then $f(2) = 5$ and $f^{-1}(5) = 2$. Therefore the point $(5, 2)$ must lie on the graph of f^{-1}. In general, if the point (a, b) lies on the graph of f, then the point (b, a) lies on the graph of f^{-1}. Refer to Figure 5.28. If a line segment is drawn between the points (a, b) and (b, a), the line $y = x$ is a perpendicular bisector of this line segment. Figure 5.29 shows pairs of points in the form (a, b) and (b, a). Figure 5.30 shows continuous graphs of f and f^{-1} passing through these points. The graph of f^{-1} is a *reflection* of the graph of f across the line $y = x$.

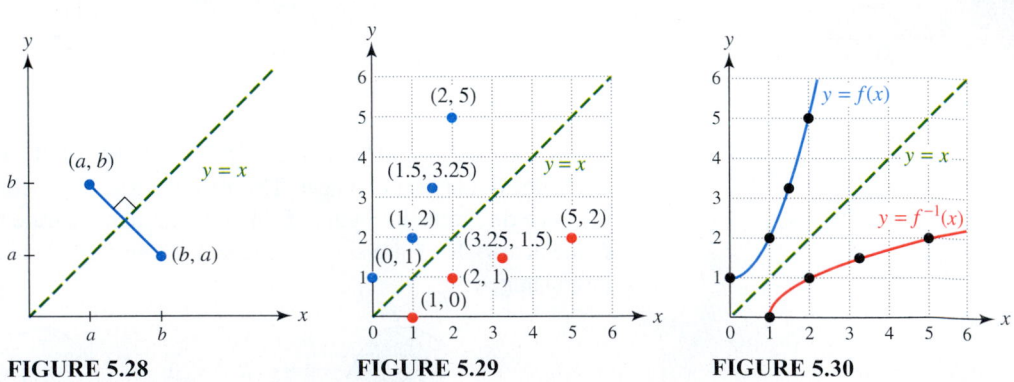

FIGURE 5.28   FIGURE 5.29   FIGURE 5.30

GRAPHS OF FUNCTIONS AND THEIR INVERSES
The graph of f^{-1} is a reflection of the graph of f across the line $y = x$.

EXAMPLE 7 Representing an inverse function graphically

Let $f(x) = x^3 + 2$. Graph f. Then sketch a graph of f^{-1}.

SOLUTION Figure 5.31 shows a graph of f. To sketch a graph of f^{-1}, reflect the graph of f across the line $y = x$. The graph of f^{-1} appears as though it were the image of the graph of f in a mirror located along $y = x$. See Figure 5.32.

Calculator Help

To graph an inverse function, see Appendix B, (page AP-18).

$[-5, 5, 1]$ by $[-5, 5, 1]$

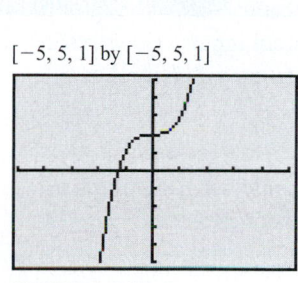

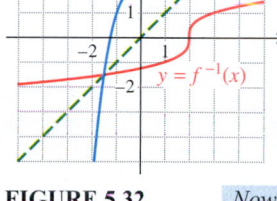

FIGURE 5.31   **FIGURE 5.32**   *Now Try Exercise 113* ◆

EXAMPLE 8 Graphically evaluating f and f^{-1}

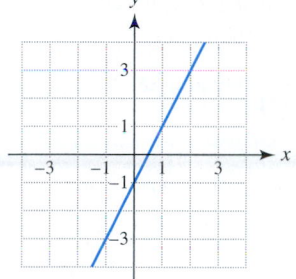

FIGURE 5.33

Use the graph of f in Figure 5.33 to evaluate each expression.
(a) $f(2)$ **(b)** $f^{-1}(3)$ **(c)** $f^{-1}(-3)$

SOLUTION
(a) To evaluate $f(2)$, find 2 on the x-axis, move upward to the graph of f, and then move across to the y-axis to obtain $f(2) = 3$, as shown in Figure 5.34.
(b) Evaluating $f^{-1}(3)$ is a slightly different process because we are given a graph of f, not f^{-1}. We must reverse the method used in part (a). Start by finding 3 on the y-axis, move across to the graph of f, and then downward to the x-axis to obtain $f^{-1}(3) = 2$, as shown in Figure 5.35. Notice from part (a) that $f(2) = 3$ and $f^{-1}(3) = 2$ in part (b).
(c) Find -3 on the y-axis, move left to the graph of f, and then upward to the x-axis. We can see from Figure 5.36 that $f^{-1}(-3) = -1$.

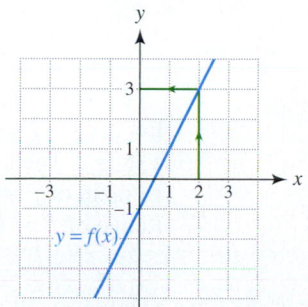

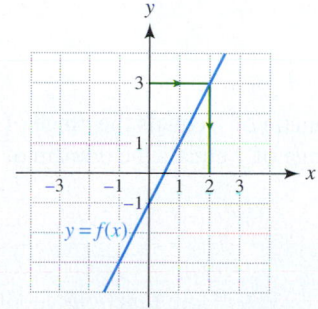

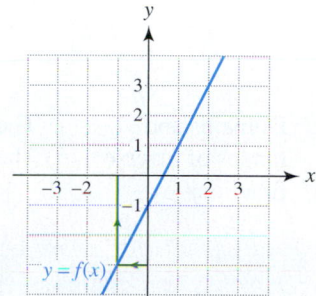

FIGURE 5.34 $f(2) = 3$ **FIGURE 5.35** $f^{-1}(3) = 2$ **FIGURE 5.36** $f^{-1}(-3) = -1$

Now Try Exercise 103 ◆

◆ **CLASS DISCUSSION**

Does an inverse function for $f(x) = |2x - 1|$ exist? Explain. What difficulties would you encounter if you tried to evaluate $f^{-1}(3)$ graphically? ◆

5.2

Putting it all Together

$\mathbf{T}$he following table summarizes some important concepts about inverse functions.

Concept	Comments	Examples
One-to-one function	f is one-to-one if different inputs always result in different outputs. That is, $a \neq b$ implies $f(a) \neq f(b)$.	$f(x) = x^2 - 4x$ is not one-to-one because $f(0) = 0$ *and* $f(4) = 0$. With this function, *different* inputs can result in the *same* output.
Horizontal line test	If every horizontal line intersects the graph of f at most once, then f is one-to-one.	 f is not one-to-one because a horizontal line can intersect it more than once.
Inverse function	If a function is one-to-one, it has an inverse function f^{-1} that satisfies both $$(f^{-1} \circ f)(x) = f^{-1}(f(x)) = x$$ and $$(f \circ f^{-1})(x) = f(f^{-1}(x)) = x.$$	$f(x) = 3x - 1$ is one-to-one and has inverse function $f^{-1}(x) = \frac{x+1}{3}$. $$f^{-1}(f(x)) = f^{-1}(3x - 1)$$ $$= \frac{(3x - 1) + 1}{3}$$ $$= x$$ $$f(f^{-1}(x)) = f\left(\frac{x+1}{3}\right)$$ $$= 3\left(\frac{x+1}{3}\right) - 1$$ $$= x$$
Domains and ranges of inverse functions	The domain of f equals the range of f^{-1}. The range of f equals the domain of f^{-1}.	Let $f(x) = (x + 2)^2$ with domain $x \geq -2$. Then, $f^{-1}(x) = \sqrt{x} - 2$. Domain of $f = \{x \mid x \geq -2\} = $ range of f^{-1}. Range of $f = \{y \mid y \geq 0\} = $ domain of f^{-1}.

Both functions and their inverses have verbal, symbolic, numerical, and graphical representations. The following table summarizes these representations.

Concept	Examples
Verbal representations	Given a verbal representation of f, apply the inverse operations in reverse order to find f^{-1}. *Example:* The function f multiplies 2 times x and then adds 25. The function f^{-1} subtracts 25 from x and divides the result by 2.

Concept	Examples
Symbolic representations	If possible, solve the equation $y = f(x)$ for x. For example, let $f(x) = 3x - 5$. $$y = 3x - 5 \quad \text{is equivalent to} \quad \frac{y + 5}{3} = x.$$ Interchange x and y to obtain $f^{-1}(x) = \dfrac{x + 5}{3}$.
Numerical representations	<table><tr><td>x</td><td>1</td><td>2</td><td>3</td></tr><tr><td>$f(x)$</td><td>0</td><td>5</td><td>7</td></tr></table> <table><tr><td>x</td><td>0</td><td>5</td><td>7</td></tr><tr><td>$f^{-1}(x)$</td><td>1</td><td>2</td><td>3</td></tr></table>
Graphical representations	The graph of f^{-1} can be obtained by reflecting the graph of f across the line $y = x$.

5.2 Exercises

Inverse Operations

Exercises 1–4: State the inverse action or actions.

1. Opening a window

2. Climbing up a ladder

3. Walking into a classroom, sitting down, and opening a book

4. Opening the door and turning on the lights

Exercises 5–12: Describe verbally the inverse of the statement. Then, express both the statement and its inverse symbolically.

5. Add 2 to x.

6. Multiply x by 5.

7. Subtract 2 from x and multiply the result by 3.

8. Divide x by 20 and add 10.

9. Take the cube root of x and add 1.

10. Multiply x by -2 and add 3.

11. Take the reciprocal of a nonzero number x.

12. Take the square root of a positive number x.

One-to-One Functions

Exercises 13–18: Use the graph of f to determine if f is one-to-one.

13.

14.

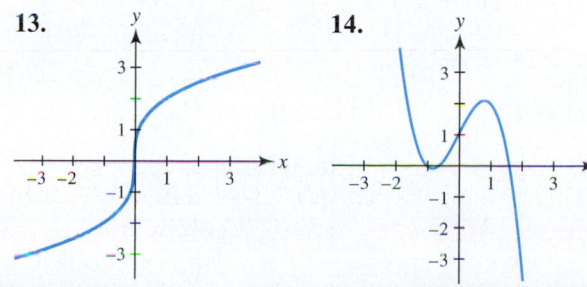

15.

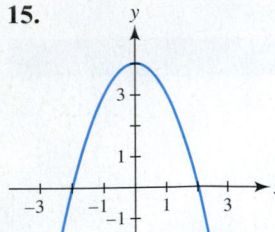

16.

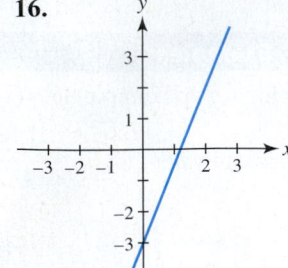

17.

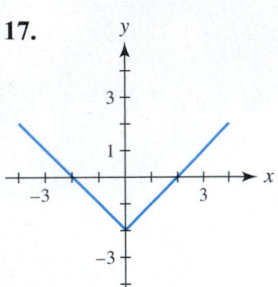

18.

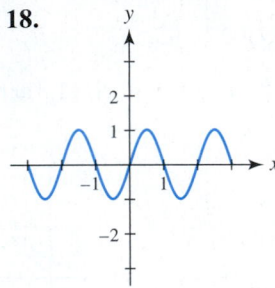

Exercises 19–22: The table is a complete representation of f. Use the table to determine if f is one-to-one and has an inverse.

19.

x	1	2	3	4
$f(x)$	4	3	3	5

20.

x	-2	0	2	4
$f(x)$	4	2	0	-2

21.

x	0	2	4	6	8
$f(x)$	-1	0	4	1	-3

22.

x	-2	-1	0	1	2
$f(x)$	4	1	0	1	4

Exercises 23–36: Determine if f is one-to-one. You may want to graph $y = f(x)$ and apply the horizontal line test.

23. $f(x) = 2x - 7$ **24.** $f(x) = x^2 - 1$

25. $f(x) = -2x^2 + x + 1$

26. $f(x) = 4 - \frac{3}{4}x$

27. $f(x) = x^4$ **28.** $f(x) = |2x - 5|$

29. $f(x) = |x - 1|$ **30.** $f(x) = x^3$

31. $f(x) = \dfrac{1}{1 + x^2}$ **32.** $f(x) = \dfrac{1}{x}$

33. $f(x) = 3x - x^3$ **34.** $f(x) = x^{2/3}$

35. $f(x) = x^{1/2}$ **36.** $f(x) = x^3 - 4x$

Exercises 37–40: Decide if the situation could be modeled by a one-to-one function.

37. The distance between the ground and a person who is riding a Ferris wheel after x seconds

38. The cumulative numbers of AIDS cases from 1980 to present

39. The population of the United States from 1980 to 2000

40. The height y of a stone thrown upward after x seconds

Symbolic Representations of Inverse Functions

Exercises 41–62: Find a symbolic representation for $f^{-1}(x)$.

41. $f(x) = \sqrt[3]{x}$ **42.** $f(x) = 2x$

43. $f(x) = -2x + 10$ **44.** $f(x) = x^3 + 2$

45. $f(x) = 3x - 1$ **46.** $f(x) = \dfrac{x - 1}{2}$

47. $f(x) = 2x^3 - 5$ **48.** $f(x) = 1 - \frac{1}{2}x^3$

49. $f(x) = x^2 - 1, \quad x \geq 0$

50. $f(x) = (x + 2)^2, \quad x \leq -2$

51. $f(x) = \dfrac{1}{2x}$ **52.** $f(x) = \dfrac{2}{\sqrt{x}}$

53. $f(x) = \frac{1}{2}(4 - 5x) + 1$ **54.** $f(x) = 6 - \frac{3}{4}(2x - 4)$

55. $f(x) = \dfrac{x}{x + 2}$ **56.** $f(x) = \dfrac{3x}{x - 1}$

57. $f(x) = \dfrac{2x + 1}{x - 1}$ **58.** $f(x) = \dfrac{1 - x}{3x + 1}$

59. $f(x) = \dfrac{1}{x} - 3$ **60.** $f(x) = \dfrac{1}{x + 5} + 2$

61. $f(x) = \dfrac{1}{x^3 - 1}$ **62.** $f(x) = \dfrac{2}{2 - x^3}$

Exercises 63–70: Restrict the domain of $f(x)$ so that f is one-to-one. Then find $f^{-1}(x)$.

63. $f(x) = 4 - x^2$ **64.** $f(x) = 2(x + 3)^2$

65. $f(x) = (x - 2)^2 + 4$ **66.** $f(x) = x^4 - 1$

67. $f(x) = x^{2/3} + 1$ **68.** $f(x) = 2(x + 3)^2$

69. $f(x) = \sqrt{9 - 2x^2}$ **70.** $f(x) = \sqrt{25 - x^2}$

Exercises 71–84: Find a symbolic representation for $f^{-1}(x)$. Identify the domain and range of f^{-1}. Verify that f and f^{-1} are inverses.

71. $f(x) = 5x - 15$

72. $f(x) = (x + 3)^2, x \geq -3$

73. $f(x) = \sqrt[3]{x - 5}$ **74.** $f(x) = 6 - 7x$

75. $f(x) = \dfrac{x - 5}{4}$ **76.** $f(x) = \dfrac{x + 2}{9}$

77. $f(x) = \sqrt{x - 5}, x \geq 5$

78. $f(x) = \sqrt{5 - 2x}, x \leq \frac{5}{2}$

79. $f(x) = \dfrac{1}{x + 3}$ **80.** $f(x) = \dfrac{2}{x - 1}$

81. $f(x) = 2x^3$ **82.** $f(x) = 1 - 4x^3$

83. $f(x) = x^2, x \geq 0$ **84.** $f(x) = \sqrt[3]{1 - x}$

Numerical Representations of Inverse Functions

Exercises 85–88: Use the table for $f(x)$ to find a table for $f^{-1}(x)$. Identify the domains and ranges of f and f^{-1}.

85.

x	1	2	3
$f(x)$	5	7	9

86.

x	1	10	100
$f(x)$	0	1	2

87.

x	0	1	2	3	4
$f(x)$	0	1	4	9	16

88.

x	−2	−1	0	1	2
$f(x)$	$\frac{1}{4}$	$\frac{1}{2}$	1	2	4

Exercises 89–92: Use $f(x)$ to complete the table for $f^{-1}(x)$.

89. $f(x) = x + 5$

x	−3	0	3	6
$f^{-1}(x)$				

90. $f(x) = 4x$

x	0	2	4	6
$f^{-1}(x)$				

91. $f(x) = x^3$

x	−8	−1	8	27
$f^{-1}(x)$				

92. $f(x) = \dfrac{1}{x}$

x	−3	−1	4	5
$f^{-1}(x)$				

Exercises 93–100: Use the tables to evaluate the following.

x	0	1	2	3	4
$f(x)$	1	3	5	4	2

x	−1	1	2	3	4
$g(x)$	0	2	1	4	5

93. $f^{-1}(3)$ **94.** $f^{-1}(5)$

95. $g^{-1}(4)$ **96.** $g^{-1}(0)$

97. $(f \circ g^{-1})(1)$ **98.** $(g^{-1} \circ g^{-1})(2)$

99. $(g \circ f^{-1})(5)$ **100.** $(f^{-1} \circ g)(4)$

Graphs and Inverse Functions

101. *Interpreting an Inverse*   The graph of f computes the number of dollars in a savings account after x years. Estimate each of the following. Interpret what $f^{-1}(x)$ computes.

(a) $f(1)$ **(b)** $f^{-1}(110)$ **(c)** $f^{-1}(160)$

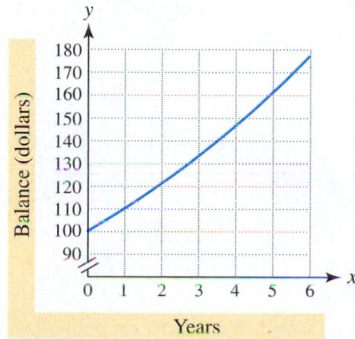

102. *Interpreting an Inverse* The graph of f computes the Celsius temperature of a pan of water after x minutes. Estimate each of the following. Interpret what $f^{-1}(x)$ computes.

(a) $f(4)$ (b) $f^{-1}(90)$ (c) $f^{-1}(80)$

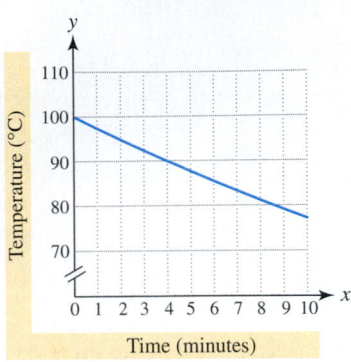

Time (minutes)

Exercises 103 and 104: Use the graph of f to evaluate each expression.

103. (a) $f(-1)$

(b) $f^{-1}(-2)$

(c) $f^{-1}(0)$

(d) $(f^{-1} \circ f)(3)$

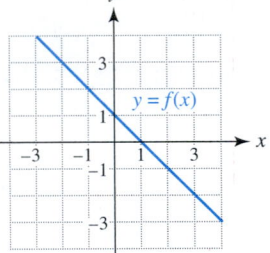

104. (a) $f(1)$

(b) $f^{-1}(1)$

(c) $f^{-1}(4)$

(d) $(f \circ f^{-1})(2.5)$

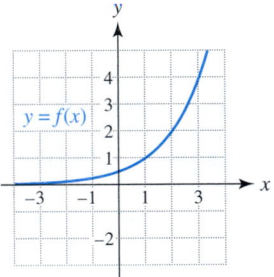

Exercises 105–110: Use the graph of f to sketch a graph of f^{-1}.

105.

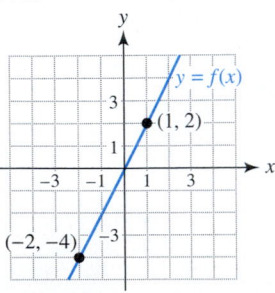

106.

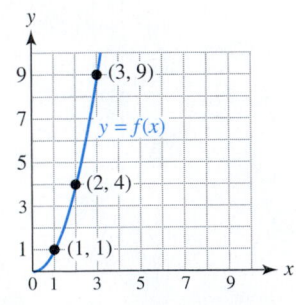

107.

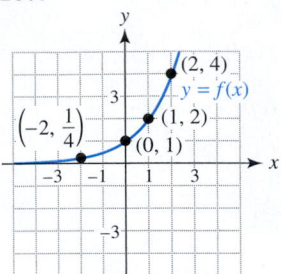

108.

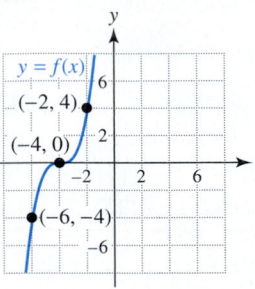

109.

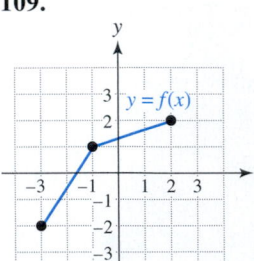

110.

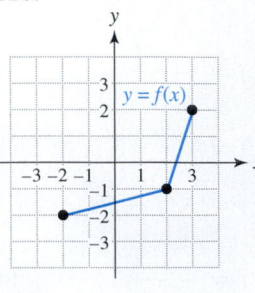

Exercises 111–116: Graph $y = f(x)$ and $y = x$. Then graph $y = f^{-1}(x)$.

111. $f(x) = 2x - 1$ **112.** $f(x) = -\frac{1}{2}x + 1$

113. $f(x) = x^3 - 1$ **114.** $f(x) = \sqrt[3]{x - 1}$

115. $f(x) = (x + 1)^2, x \geq -1$

116. $f(x) = \sqrt{x + 1}$

Exercises 117–120: Graph $y = f(x)$, $y = f^{-1}(x)$, and $y = x$ in a square viewing rectangle such as $[-4.7, 4.7, 1]$ by $[-3.1, 3.1, 1]$.

117. $f(x) = 3x - 1$ **118.** $f(x) = \dfrac{3 - x}{2}$

119. $f(x) = \frac{1}{3}x^3 - 1$ **120.** $f(x) = \sqrt[3]{x - 1}$

Calculator Help To learn about a square viewing rectangle, see Appendix B (page AP-10).

Applications

121. *Advertising Costs* The line graph at the top of the next page represents a function C that computes the cost of a 30-second commercial during a Super Bowl telecast. Perform each calculation and interpret the results. (**Source:** *USA Today.*)

(a) Evaluate $C(1995)$.

(b) Solve $C(x) = 1$ for x.

(c) Evaluate $C^{-1}(1)$.

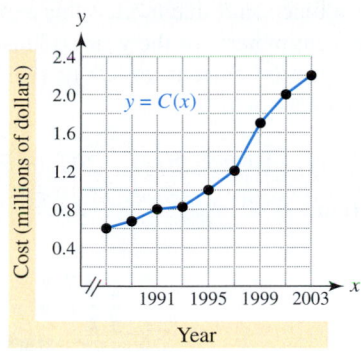

Year

122. *Social Security*   The line graph represents a function N that models the number of Social Security recipients from 1990 to 2030.   (**Source:** Social Security Administration.)

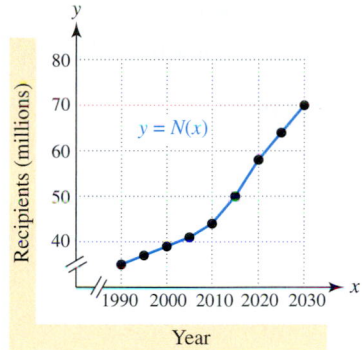

Year

(a) Solve the equation $N(x) = 50$ for x.

(b) Evaluate $N^{-1}(50)$.

123. *Volume*   The volume V of a sphere with radius r is given by $V = \frac{4}{3}\pi r^3$.
(a) Does V represent a one-to-one function?

(b) What does the inverse of V compute?

(c) Find a formula for the inverse.

(d) Normally we interchange x and y to find the inverse of a function. Does it make sense to interchange V and r in part (c) of this exercise? Explain.

124. *Temperature*   The formula $F = \frac{9}{5}C + 32$ converts a Celsius temperature to an equivalent Fahrenheit temperature.
(a) Find a formula for the inverse.

(b) Normally we interchange x and y to find the inverse function. Does it makes sense to interchange F and C in part (a) of this exercise? Explain.

(c) What Celsius temperature is equivalent to 68°F?

125. *Codes*   Use the code in Table 5.7 on page 383 to code MATH, then decode HWPEVKQPU. If a function f is used to code a word x, why is it essential that f is one-to-one?

126. *Codes*   Consider the following method to code a word. First let A = 1, B = 2, C = 3, . . . , Z = 26. For a given word, add the numbers associated with each letter in the word. The sum represents the code for the word. For example, since M = 13, A = 1, T = 20, and H = 8, the word MATH is coded as 13 + 1 + 20 + 8 = 42. Is this a good way to code words? Explain.

127. *Height and Weight*   The formula $W = \frac{25}{7}h - \frac{800}{7}$ approximates the recommended minimum weight for a person h inches tall, where $h \geq 62$.
(a) What is the recommended minimum weight for someone 70 inches tall?

(b) Does W represent a one-to-one function?

(c) Find a formula for the inverse.

(d) Evaluate the inverse for 150 pounds and interpret the result.

(e) What does the inverse compute?

128. *Planetary Orbits*   The formula $T(x) = x^{3/2}$ calculates the time in years that it takes a planet to orbit the sun if the planet is x times farther from the sun than Earth.
(a) Find the inverse of T.

(b) What does the inverse of T calculate?

129. *Converting Units*   The accompanying tables represent a function F that converts yards to feet and a function Y that converts miles to yards. Evaluate each expression and interpret the results.

x (yd)	1760	3520	5280	7040	8800
$F(x)$ (ft)	5280	10,560	15,840	21,120	26,400

x (mi)	1	2	3	4	5
$Y(x)$ (yd)	1760	3520	5280	7040	8800

(a) $(F \circ Y)(2)$ (b) $F^{-1}(26,400)$

(c) $(Y^{-1} \circ F^{-1})(21,120)$

130. *Converting Units* (Refer to the previous exercise.)
(a) Find symbolic representations for $F(x)$, $Y(x)$, and $(F \circ Y)(x)$.

(b) Express $(Y^{-1} \circ F^{-1})(x)$ symbolically. What does this function compute?

131. *Converting Units* The accompanying tables represent a function C that converts tablespoons to cups and a function Q that converts cups to quarts. Evaluate each expression and interpret the results.

x (tbsp)	32	64	96	128
$C(x)$ (c)	2	4	6	8

x (c)	2	4	6	8
$Q(x)$ (qt)	0.5	1	1.5	2

(a) $(Q \circ C)(96)$

(b) $Q^{-1}(2)$

(c) $(C^{-1} \circ Q^{-1})(1.5)$

132. *Converting Units* (Refer to the previous exercise.)
(a) Find symbolic representations for $C(x)$, $Q(x)$, and $(Q \circ C)(x)$.

(b) Express $(C^{-1} \circ Q^{-1})(x)$ symbolically. What does this function compute?

133. *Air Pollution* Tiny particles suspended in the air are necessary for clouds to form. Because of this fact, experts believe that air pollutants may cause an increase in cloud cover. From 1930 to 1980 the percentage of cloud cover over the world's oceans was monitored. The linear function given by $f(x) = 0.06(x - 1930) + 62.5$, where $1930 \le x \le 1980$, approximates this percentage. (**Source:** W. Cotton and R. Pielke, *Human Impacts on Weather and Climate.*)
(a) Evaluate the expressions $f(1930)$ and $f(1980)$. How did the amount of cloud cover over the oceans change during this 50-year period?

(b) What does $f^{-1}(x)$ compute?

(c) Use your results from part (a) to evaluate $f^{-1}(62.5)$ and $f^{-1}(65.5)$.

(d) Find $f^{-1}(x)$ symbolically.

134. *Rise in Sea Level* Due to the greenhouse effect, the global sea level could rise through thermal expansion and partial melting of the polar ice caps. The table rep-

resents a function R that models this expected rise in sea level in centimeters for the year t. (This model assumes no changes in current trends.) (**Source:** A. Nilsson, *Greenhouse Earth.*)

t (yr)	1990	2000	2030	2070	2100
$R(t)$ (cm)	0	1	18	44	66

(a) Is R a one-to-one function? Explain.

(b) Use $R(t)$ to find a table for $R^{-1}(t)$. Interpret R^{-1}.

Writing about Mathematics

135. Explain how to find verbal, numerical, graphical, and symbolic representations of an inverse function. Give examples.

136. Can a one-to-one function have more than one x-intercept or more than one y-intercept? Explain.

137. If the graphs of $y = f(x)$ and $y = f^{-1}(x)$ intersect at a point (x, y), what can be said about this point? Explain your reasoning.

138. If $f(x) = ax^2 + bx + c$ with $a \ne 0$, does $f^{-1}(x)$ exist? Explain.

EXTENDED AND DISCOVERY EXERCISES

1. *Interpreting an Inverse* Let $f(x)$ compute the distance traveled in miles after x hours by a car with a velocity of 60 miles per hour.
(a) Explain what $f^{-1}(x)$ computes.

(b) Interpret the solution to the equation $f(x) = 200$.

(c) Explain how to solve the equation in part (b) using $f^{-1}(x)$.

2. *Interpreting an Inverse* Let $f(x)$ compute the height in feet of a rocket after x seconds of upward flight.
(a) Explain what $f^{-1}(x)$ computes.

(b) Interpret the solution to the equation $f(x) = 5000$.

(c) Explain how to solve the equation in part (b) using $f^{-1}(x)$.

3. If the graph of f lies entirely in quadrants I and II, in which quadrant(s) does the graph of f^{-1} lie?

CHECKING BASIC CONCEPTS FOR SECTIONS 5.1 AND 5.2

1. Use the table to evaluate each expression, if possible.

x	-2	-1	0	1	2
$f(x)$	0	1	-2	-1	2
$g(x)$	1	-2	-1	2	0

(a) $(f + g)(1)$ **(b)** $(f - g)(-1)$ **(c)** $(fg)(0)$

(d) $(f/g)(2)$ **(e)** $(f \circ g)(2)$ **(f)** $(g \circ f)(-2)$

2. Use the graph to evaluate each expression at the top of the next column, if possible.

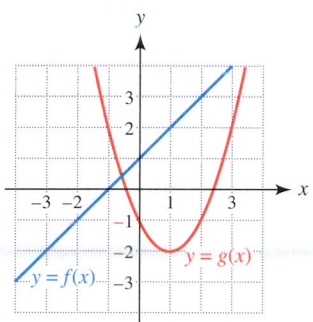

(a) $(f + g)(1)$ **(b)** $(g - f)(0)$

(c) $(fg)(2)$ **(d)** $(g/f)(-1)$

(e) $(f \circ g)(2)$ **(f)** $(g \circ f)(1)$

3. Let $f(x) = x^2 + 3x - 2$ and $g(x) = 3x - 1$. Find each expression.

(a) $(f + g)(x)$ **(b)** $(f/g)(x)$ **(c)** $(f \circ g)(x)$

4. If $f(x) = 5 - 2x$, find $f^{-1}(x)$.

5. Use the graph in Exercise 2 to answer the following.

(a) Is f a one-to-one function? Does f^{-1} exist? If so, find it.

(b) Is g a one-to-one function? Does g^{-1} exist? If so, find it.

6. Graph $f(x) = \sqrt[3]{x}$ and $y = x$. Then graph $y = f^{-1}(x)$.

5.3 Exponential Functions and Models

Introduction

Even though modern exponential notation was not developed until 1637 by the great French mathematician René Descartes, some of the earliest applications involving exponential functions occurred in the calculation of interest. The custom of charging interest dates back to at least 2000 B.C. in ancient Babylon, where interest rates ran as high as 33%. Today, exponential functions are used not only to calculate interest, but also to model a wide variety of phenomena in business, biology, medicine, engineering, and education. This section discusses exponential functions and their representations. (**Source:** *Historical Topics for the Mathematics Classroom, Thirty-first Yearbook*, NCTM.)

◆ Distinguish between linear and exponential growth
◆ Model data with exponential functions
◆ Calculate compound interest
◆ Use the natural exponential function in applications

Linear and Exponential Growth

TABLE 5.12
A Linear Function

x	$y = g(x)$
0	3
1	5
2	7
3	9
4	11
5	13

A linear function g can be written as $g(x) = mx + b$, where m represents the rate of change in $g(x)$ for each unit increase in x. For example, Table 5.12 shows a numerical representation of a linear function g. Each time x increases by 1 unit, $g(x)$ increases by 2 units. We can write a formula $g(x) = \mathbf{2}x + \mathbf{3}$ because the rate of change is $\mathbf{2}$ and $g(0) = \mathbf{3}$.

An exponential function is fundamentally different from a linear function. Rather than *adding* a fixed amount to the previous y-value for each unit increase in x, an exponential function *multiplies* the previous y-value by a fixed amount for each unit increase in x. Table 5.13 shows an exponential function f. Note that in Table 5.13 consecutive y-values are found by multiplying the previous y-value by 2.

TABLE 5.13   An Exponential Function

x	0	1	2	3	4	5
$y = f(x)$	3	6	12	24	48	96

Compare the following patterns for calculating the linear function g and the exponential function f.

Linear Growth

$g(0) = 3$

$g(1) = \underset{g(0)}{\underline{3}} + \mathbf{2} = 3 + 2 \cdot 1 = 5$

$g(2) = \underset{g(1)}{\underline{3 + 2}} + \mathbf{2} = 3 + 2 \cdot 2 = 7$

$g(3) = \underset{g(2)}{\underline{3 + 2 + 2}} + \mathbf{2} = 3 + 2 \cdot 3 = 9$

$g(4) = \underset{g(3)}{\underline{3 + 2 + 2 + 2}} + \mathbf{2} = 3 + 2 \cdot 4 = 11$

$g(5) = \underset{g(4)}{\underline{3 + 2 + 2 + 2 + 2}} + \mathbf{2} = 3 + 2 \cdot 5 = 13$

Exponential Growth

$f(0) = 3$

$f(1) = \underset{f(0)}{\underline{3}} \cdot \mathbf{2} = 3 \cdot 2^1 = 6$

$f(2) = \underset{f(1)}{\underline{3 \cdot 2}} \cdot \mathbf{2} = 3 \cdot 2^2 = 12$

$f(3) = \underset{f(2)}{\underline{3 \cdot 2 \cdot 2}} \cdot \mathbf{2} = 3 \cdot 2^3 = 24$

$f(4) = \underset{f(3)}{\underline{3 \cdot 2 \cdot 2 \cdot 2}} \cdot \mathbf{2} = 3 \cdot 2^4 = 48$

$f(5) = \underset{f(4)}{\underline{3 \cdot 2 \cdot 2 \cdot 2 \cdot 2}} \cdot \mathbf{2} = 3 \cdot 2^5 = 96$

Notice that if x is a positive integer then

$$g(x) = 3 + \underset{x \text{ terms}}{\underline{(2 + 2 + \cdots + 2)}} \quad \text{and} \quad f(x) = 3 \cdot \underset{x \text{ factors}}{\underline{(2 \cdot 2 \cdot \cdots \cdot 2)}}.$$

Using these patterns we can write formulas for $g(x)$ and $f(x)$ as follows.

$$g(x) = 3 + 2x \quad \text{and} \quad f(x) = 3 \cdot 2^x$$

This discussion gives motivation for the following definition.

EXPONENTIAL FUNCTION

A function f represented by

$$f(x) = Ca^x, \quad a > 0, \quad a \neq 1, \quad \text{and} \quad C > 0,$$

is an **exponential function with base a and coefficient C.**

Note: Some definitions for an exponential function require that $C = 1$.

Examples of exponential functions include

$$f(x) = 3^x, \qquad g(x) = 5(1.7)^x, \quad \text{and} \quad h(x) = 4\left(\frac{1}{2}\right)^x.$$

Their bases are **3**, **1.7**, and $\frac{1}{2}$, respectively. When evaluating exponential functions, remember that any nonzero number raised to the 0 power equals 1. Thus, $f(\mathbf{0}) = 3^{\mathbf{0}} = 1$, $g(\mathbf{0}) = 5(1.7)^{\mathbf{0}} = 5$, and $h(\mathbf{0}) = 4\left(\frac{1}{2}\right)^{\mathbf{0}} = 4$.

Note: If $f(x) = Ca^x$, then $f(0) = Ca^0 = C(1) = C$. That is, C equals the value of the function at $x = 0$. If x represents time, then C often equals the *initial value* of the quantity being modeled by f.

For large values of x, an exponential function with $a > 1$ grows faster than any linear function. Figures 5.37 and 5.38 show graphs of $f(x) = 2^x$ and $g(x) = 2x$, respectively. The points shown in the table have also been plotted. Notice that the graph of the exponential function f increases more rapidly than the graph of the linear function g for large values of x.

x	-1	0	1	2	3	4
2^x	$\frac{1}{2}$	1	2	4	8	16

x	-1	0	1	2	3	4
$2x$	-2	0	2	4	6	8

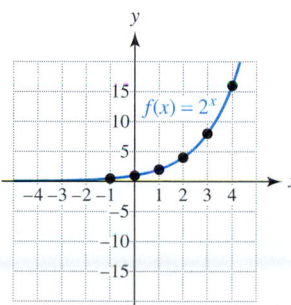

FIGURE 5.37 An Exponential Function

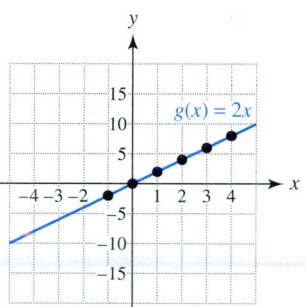

FIGURE 5.38 A Linear Function

EXAMPLE 1

Recognizing linear and exponential data

For each table of data, find either a linear or exponential function that models the data.

(a)

x	0	1	2	3
y	-3	-1.5	0	1.5

(b)

x	0	1	2	3	4
y	16	4	1	$\frac{1}{4}$	$\frac{1}{16}$

(c)

x	0	1	2	3
y	3	4.5	6.75	10.125

(d)

x	0	1	2	3	4
y	16	12	8	4	0

SOLUTION

(a) For each unit increase in x, the y-values increase by 1.5, so the data are linear. Because $y = -3$ when $x = 0$, it follows that the data can be modeled by $f(x) = 1.5x - 3$.

(b) For each unit increase in x, the y-values are multiplied by $\frac{1}{4}$. This is an exponential function with $C = 16$ and $a = \frac{1}{4}$, so $f(x) = 16\left(\frac{1}{4}\right)^x$.

(c) Since the data do not change by a fixed amount for each unit increase in x, the data are not linear. To determine if the data are exponential, calculate ratios of consecutive y-values.

$$\frac{4.5}{3} = 1.5, \qquad \frac{6.75}{4.5} = 1.5, \qquad \frac{10.125}{6.75} = 1.5$$

For each unit increase in x, the next y-value in the table can be found by multiplying the previous y-value by 1.5, so let $a = 1.5$. Since $y = 3$ when $x = 0$, let $C = 3$. Thus $f(x) = 3(1.5)^x$.

(d) For each unit increase in x, the next y-value is found by adding -4 to the previous y-value, so the data can be modeled by $f(x) = -4x + 16$.

Now Try Exercises 19 and 21 ◆

EXAMPLE 2   Finding linear and exponential functions

A college graduate signs a contract for an annual salary of $50,000 and can choose either Option A: a fixed annual raise of $6000 each year, or Option B: a 10% raise each year. However, the graduate must stay with the same option.

(a) Write a formula that gives the annual salary during the nth year for each option.
(b) Discuss which option is better.

SOLUTION

(a) Option A: The salary for the first year is $50,000, for the second year is $56,000, for the third year is $62,000, and so on. Next year's salary is found by adding $6000 to this year's salary. These salaries can be modeled by a linear function g with $m = 6000$, passing through the point $(1, 50000)$. Using the point-slope form for a line, we obtain

$$g(n) = 6000(n - 1) + 50{,}000 \quad \text{or} \quad g(n) = 6000n + 44{,}000.$$

Algebra Review

To review point-slope form of a line, see Section 2.2

Option B: The salary for the first year is $50,000. Since 10% of $50,000 is $5000, the salary for the second year is $50{,}000 + \$5000 = \$55{,}000$. This is equivalent to multiplying $50,000 by $1 + 0.10 = 1.10$. In general, to calculate next year's salary we can multiply this year's salary by 1.10.

$$f(1) = 50{,}000 = 50{,}000(1.10)^0 = \$50{,}000$$

$$f(2) = 50{,}000(1.1) = 50{,}000(1.10)^1 = \$55{,}000$$

$$f(3) = 50{,}000(1.1)(1.1) = 50{,}000(1.10)^2 = \$60{,}500$$

$$f(4) = 50{,}000(1.1)(1.1)(1.1) = 50{,}000(1.10)^3 = \$66{,}550$$

From the pattern above, we can write an exponential function f that models this situation

$$f(n) = 50{,}000(1.10)^{n-1}.$$

Note: The exponent for function f must be $n - 1$ rather than n, because during the first year ($n = 1$) the salary is $50,000, not $55,000.

(b) Table 5.14 shows the salaries under each option. During the 5th year the two options have comparable salaries. If the graduate is planning on changing jobs after a few years, Option A is better. However, if the graduate plans to stay for a long time, Option B is better.

TABLE 5.14   Salaries with Linear and Exponential Growth

x (year)	1	3	5	7	9	11	13
Option A	$50,000	$62,000	$74,000	$86,000	$98,000	$110,000	$122,000
Option B	$50,000	$60,500	$73,205	$88,578	$107,179	$129,687	$156,921

Now Try Exercises 25 and 26 ◆

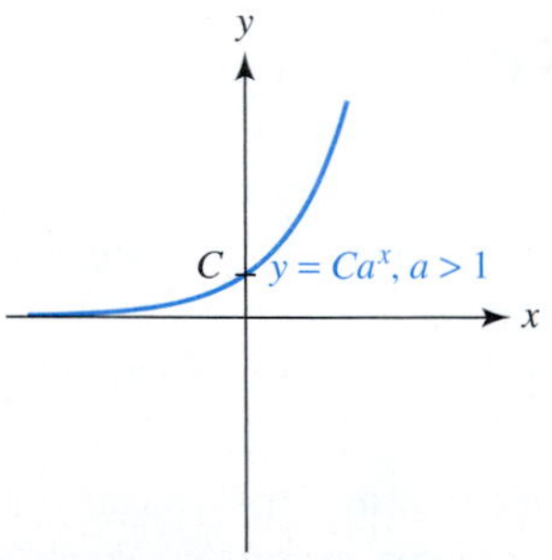

FIGURE 5.39 Exponential Growth

If an exponential function is written as $f(x) = Ca^x$ with $a > 1$, then $f(x)$ experiences **exponential growth**, as illustrated in Figure 5.39. In Example 2, the salary described

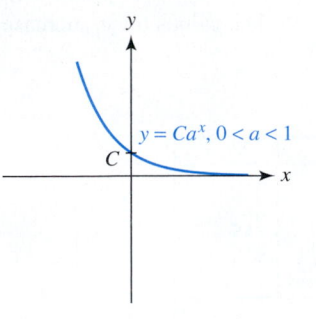

FIGURE 5.40 Exponential Decay

by Option B experienced exponential growth, with a **growth factor** of $a = 1.10$. If $0 < a < 1$, then the y-values decrease by a factor of a for each unit increase in x. In this case $f(x) = Ca^x$ experiences **exponential decay** with **decay factor** a. See Figure 5.40. Exponential decay was modeled in Example 1(b), where the decay factor was $a = \frac{1}{4}$.

An *exponential function* f, defined by $f(x) = Ca^x$ with $a > 0$, $a \neq 1$, and $C > 0$ has the following properties.

1. The domain of f is $(-\infty, \infty)$ and the range of f is $(0, \infty)$.
2. The graph of f is continuous with no breaks. The x-axis is a horizontal asymptote. There are no x-intercepts and the y-intercept is C.
3. If $a > 1$, f is increasing on its domain and if $0 < a < 1$, f is decreasing on its domain.
4. f is one-to-one.

Algebra Review

To review properties of exponents, see Chapter R (page R-13) and Section 4.7.

To investigate further the effect that a has on the graph of an exponential function, we can let $C = 1$ in $f(x) = Ca^x$, and graph $y = 1^x$, $y = 1.3^x$, $y = 1.7^x$, and $y = 2.5^x$, as shown in Figure 5.41. (Note that $f(x) = 1^x$ does *not* represent an exponential function. However, the graph of $y = 1^x$ represents the boundary between exponential growth and decay.) As a increases, the graph of $y = a^x$ increases at a faster rate. On the other hand, the graphs of the equations $y = 0.7^x$, $y = 0.5^x$, and $y = 0.15^x$ decrease faster as a decreases. See Figure 5.42. The graph of $y = a^x$ is increasing when $a > 1$ and decreasing when $0 < a < 1$. Note that the graph of $y = a^x$, $a > 0$, always passes through the point $(0, 1)$.

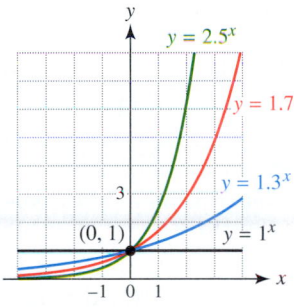

FIGURE 5.41

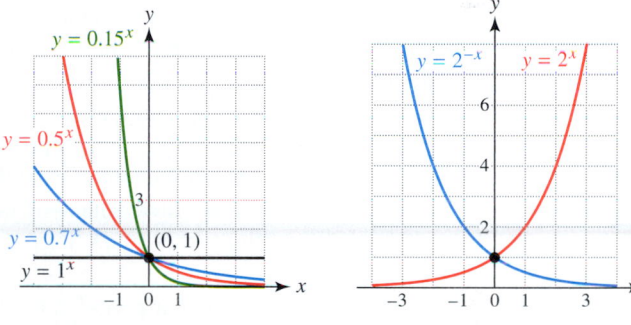

FIGURE 5.42   **FIGURE 5.43**

In Section 3.4 we learned that the graph of $y = f(-x)$ is a reflection of the graph of $y = f(x)$ across the y-axis. As a result, the graph of $y = a^{-x}$ is a reflection of $y = a^x$ across the y-axis. For example, the graphs of $y = 2^x$ and $y = 2^{-x}$ are reflections across the y-axis. See Figure 5.43. Note that by properties of exponents, $2^{-x} = \left(\frac{1}{2}\right)^x$.

◆ **MAKING CONNECTIONS**

Exponential Functions and Polynomial Functions   An exponential function has a *variable* for an exponent, whereas a polynomial function has a *constant* exponent. For example, $f(x) = 3^x$ represents an exponential function, and $g(x) = x^3$ represents a polynomial function.

 **EXAMPLE 3**   Comparing exponential and polynomial functions

Compare $f(x) = 3^x$ and $g(x) = x^3$ graphically and numerically for $x \geq 0$.

SOLUTION

Graphical Comparison   The graphs $Y_1 = 3\text{^}X$ and $Y_2 = X\text{^}3$ are shown in Figure 5.44 on the next page. For $x \geq 6$, the graph of the exponential function y_1 increases significantly faster than the graph of the polynomial function y_2.

Numerical Comparison   Table y_1 and y_2, as shown in Figure 5.45. The values for y_1 increase faster than the values for y_2.

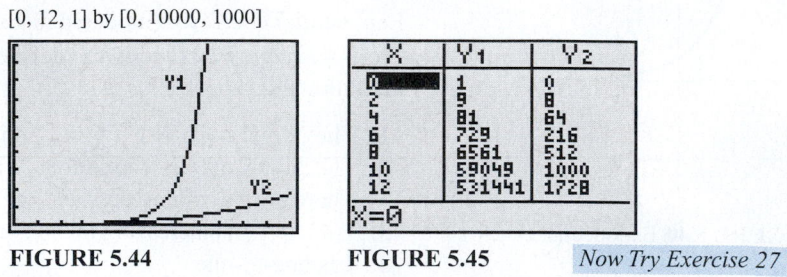

[0, 12, 1] by [0, 10000, 1000]

FIGURE 5.44   FIGURE 5.45   *Now Try Exercise 27* ◆

The results of Example 3 are true in general. For large enough inputs, exponential functions with $a > 1$ eventually become greater than any polynomial function.

Exponential Models

Radioactivity is an application of exponential decay. When an element such as uranium undergoes radioactive decay, atoms change from one element to another. The time it takes for half of the atoms to decay into a different element is called the **half-life**, and this time varies for different elements. For example, radioactive carbon-14, which is found in all living things, has a half-life of about 5700 years and can be used to date fossils. While animals are alive, they breathe both carbon dioxide and oxygen. Because a small portion of normal atmospheric carbon dioxide is made up of radioactive carbon-14, a fixed percentage of their bodies is composed of carbon-14. When an animal dies, it quits breathing and the carbon-14 continues to disintegrate without being replaced. To determine the time since an animal died, scientists measure the percentage of carbon-14 remaining in its bones. If initially there was 1 gram of carbon-14 present, then after 5700 years there would be $\frac{1}{2}$ gram, after $2 \cdot 5700 = 11{,}400$ years there would be $\frac{1}{4}$ gram, and in general after each 5700-year period, the amount of carbon-14 would be reduced by half. Because the amount of carbon-14 is being reduced by a constant factor of $\frac{1}{2}$ every 5700 years, we can model the amount of carbon-14 with an exponential function.

If the initial amount of carbon-14 equals C grams, then $f(x) = Ca^x$ models the amount of carbon-14 present after x years, where a needs to be determined for carbon-14. After 5700 years, there will be $\frac{1}{2}C$ grams, so $f(5700) = \frac{1}{2}C$ implies that

$$Ca^{5700} = \frac{1}{2}C.$$

To solve this equation for a, begin by dividing each side by C.

$$a^{5700} = \frac{1}{2} \qquad \text{Divide each side by } C.$$

$$(a^{5700})^{1/5700} = \left(\frac{1}{2}\right)^{1/5700} \qquad \text{Raise to the } \tfrac{1}{5700}\text{th power.}$$

$$a = \left(\frac{1}{2}\right)^{1/5700} \qquad \text{Properties of exponents}$$

Using properties of exponents, we write $f(x) = Ca^x$ as

$$f(x) = C\left(\left(\frac{1}{2}\right)^{1/5700}\right)^x = C\left(\frac{1}{2}\right)^{x/5700}.$$

This discussion is summarized in the following.

MODELING RADIOACTIVE DECAY

If a radioactive sample containing C units has a half-life of k years, then the amount A remaining after x years is given by

$$A(x) = C\left(\frac{1}{2}\right)^{x/k}.$$

For example, the half-life of radium-226 is about **1600** years. After **9600** years a **2**-gram sample decays to

$$A(\textbf{9600}) = \textbf{2}\left(\frac{1}{2}\right)^{\textbf{9600}/\textbf{1600}} = 0.03125 \text{ gram}.$$

EXAMPLE 4 Finding the age of a fossil

A fossil contains 5% of the amount of the carbon-14 that it contained when it was alive. Graphically estimate its age. This situation is illustrated in Figure 5.46.

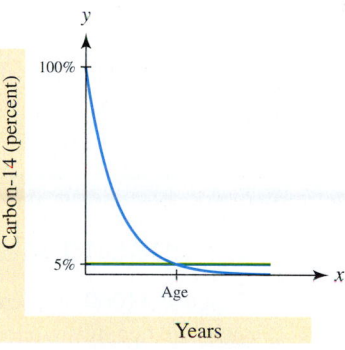

FIGURE 5.46

SOLUTION The initial amount of carbon-14 is 100% or 1, the final amount is 5% or 0.05, and the half-life is 5700 years, so $A = 0.05$, $C = 1$, and $k = 5700$ in $A(x) = C\left(\frac{1}{2}\right)^{x/k}$. To determine the age of the fossil, solve

$$0.05 = 1\left(\frac{1}{2}\right)^{x/5700}$$

for x. Graph $Y_1 = 0.05$ and $Y_2 = 0.5^\wedge(X/5700)$, as shown in Figure 5.47. Their graphs intersect near (24635, 0.05), so the fossil is about 24,635 years old.   *Now Try Exercise 89*

[0, 50000, 10000] by [0, 0.1, 0.01]

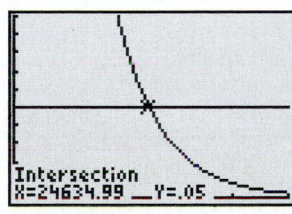

FIGURE 5.47

EXAMPLE 5 Modeling atmospheric CO_2 concentrations

Predicted concentrations of atmospheric carbon dioxide (CO_2) in parts per million (ppm) are shown in Table 5.15 on the next page. (These concentrations assume that current trends continue.) The CO_2 levels in the year 2000 were greater than they have been in any time in the previous 160,000 years. The increase in concentrations of CO_2 has been accelerated by the burning of fossil fuels and by deforestation.   (**Source:** R. Turner, *Environmental Economics.*)

TABLE 5.15 **Concentrations of Atmospheric CO_2**

Year	2000	2050	2100	2150	2200
CO_2 (ppm)	364	467	600	769	987

(a) Let $x = 0$ correspond to 2000 and $x = 200$ to 2200. Find values for C and a so that $f(x) = Ca^x$ models these data.

(b) Estimate CO_2 concentrations for the year 2025.

SOLUTION

(a) The concentration is 364 when $x = 0$, so $C = 364$. This gives $f(x) = 364a^x$. Next, we estimate a value for a. One possibility for determining a is to require that the graph of f pass through the point $(2200, 987)$. It then follows that $f(200) = 987$.

$$364a^{200} = 987 \qquad \textcolor{blue}{f(200) = 987}$$

$$a^{200} = \frac{987}{364} \qquad \textcolor{blue}{\text{Divide by 364.}}$$

$$(a^{200})^{1/200} = \left(\frac{987}{364}\right)^{1/200} \qquad \textcolor{blue}{\text{Take the } \tfrac{1}{200}\text{th power.}}$$

$$a = \left(\frac{987}{364}\right)^{1/200} \qquad \textcolor{blue}{\text{Properties of exponents}}$$

$$a \approx 1.005 \qquad \textcolor{blue}{\text{Approximate.}}$$

Thus $f(x) = 364(1.005)^x$. Answers for $f(x)$ may vary slightly.

(b) Since 2025 corresponds to $x = 25$, evaluate $f(25)$.

$$f(\textbf{25}) = 364(1.005)^{\textbf{25}} \approx 412.$$

Concentration of carbon dioxide could reach 412 ppm by 2025.

Now Try Exercise 93 ◆

Compound Interest

Suppose \$1000 are deposited in an account paying 10% annual interest. At the end of 1 year, the account will contain \$1000 plus 10% of \$1000, or \$100, in interest. Let P represent the **principal** or initial amount deposited. Then, the amount A_1 in the account after 1 year can be computed as follows.

$$A_1 = P + (0.10)P \qquad \textcolor{blue}{\text{Principal plus interest}}$$

$$= P(1 + 0.10) \qquad \textcolor{blue}{\text{Factor.}}$$

$$= 1000(1.10) \qquad \textcolor{blue}{P = \$1000}$$

$$= 1100 \qquad \textcolor{blue}{\text{Simplify.}}$$

The sum of the principal and interest is \$1100.

During the second year, the account earns interest on \$1100. The amount A_2 after the second year will be equal to A_1 plus 10% of A_1.

$$A_{\textbf{2}} = A_1 + (0.10)A_1$$

$$= \textcolor{red}{A_1}(1 + 0.10) \qquad \textcolor{blue}{\text{Factor.}}$$

$$= \textcolor{red}{P(1 + 0.10)}(1 + 0.10) \qquad \textcolor{blue}{A_1 = P(1 + 0.10)}$$

$$= P(1 + 0.10)^2$$

$$= 1000(1.10)^2$$

$$= 1210$$

After 2 years there will be \$1210.

We would like to determine a general formula for A, the amount in the account after t years. To do this, compute A_3 and observe a pattern.

$$
\begin{aligned}
A_3 &= A_2 + (0.10)A_2 \\
&= A_2(1 + 0.10) \\
&= P(1 + 0.10)^2(1 + 0.10) \qquad A_2 = P(1 + 0.10)^2 \\
&= P(1 + 0.10)^3 \\
&= 1000(1.10)^3 \\
&= 1331
\end{aligned}
$$

The amount is \$1331. We can see that in general, $A = P(1 + 0.10)^t$. This type of interest is said to be *compounded annually*, because it is paid once a year.

If the interest rate had been r, *expressed in decimal form*, then $A = P(1 + r)^t$. Notice that in the expression $P(1 + r)^t$, the variable t occurs as an exponent.

 EXAMPLE 6   Calculating an account balance

If the principal is \$2000 and the interest rate is 8% compounded annually, calculate the account balance after 4 years.

SOLUTION   The principal is $P = 2000$, the interest rate is $r = 0.08$, and the number of years is $t = 4$.

$$
A = P(1 + r)^t = 2000(1 + 0.08)^4 \approx 2720.98
$$

After 4 years the account contains \$2720.98.   *Now Try Exercise 69* ◆

In most savings accounts, interest is paid more often than once a year. In this case a smaller amount of interest is paid more frequently. For example, suppose \$1000 are deposited in an account paying 10% annual interest, compounded quarterly. After 3 months the interest would amount to one-fourth of 10% or 2.5% of \$1000. The account balance would be $1000(1 + 0.025) = \$1025$. After the next 3-month period, interest would be paid on the \$1025. In a manner similar to annual compounding, the balance would be equal to $\$1000(1 + 0.025)^2 \approx \$1050.63$ after 6 months, $\$1000(1 + 0.025)^3 \approx \$1076.89$ after 9 months, and $\$1000(1 + 0.025)^4 \approx \$1103.81$ after a year. With annual compounding the amount is \$1100. The difference of \$3.81 is due to compounding quarterly. Although this amount is small after 1 year, compounding more frequently can have a dramatic effect over a long time.

COMPOUND INTEREST

If a principal of P dollars is deposited in an account paying an annual rate of interest r (expressed in decimal form) compounded (paid) n times per year, then after t years the account will contain A dollars, where

$$
A = P\left(1 + \frac{r}{n}\right)^{nt}.
$$

 **EXAMPLE 7**   Comparing compound interest

Suppose \$1000 are deposited by a 20-year-old worker in an Individual Retirement Account (IRA) that pays an annual interest rate of 12%. Describe the effect on the balance at age 65, if interest were compounded annually and quarterly.

◆ CLASS DISCUSSION
In Example 7, make a conjecture about the effect on the IRA balance after 45 years if the interest rate were 6% instead of 12%. Test your conjecture. ◆

SOLUTION
Compounded Annually   Let $P = 1000$, $r = 0.12$, $n = 1$, and $t = 65 - 20 = 45$.

$$A = P\left(1 + \frac{r}{n}\right)^{nt} = 1000(1 + 0.12)^{45} \approx \$163{,}987.60$$

Compounded Quarterly   Let $P = 1000$, $r = 0.12$, $n = 4$, and $t = 45$.

$$A = 1000\left(1 + \frac{0.12}{4}\right)^{4(45)} = 1000(1 + 0.03)^{180} \approx \$204{,}503.36$$

Quarterly compounding results in an increase of \$40,515.76!   *Now Try Exercise 77* ◆

The Natural Exponential Function

In Example 7, compounding interest quarterly rather than annually made a significant difference in the balance after 45 years. What would happen if interest were compounded daily or even hourly? Would there be a limit to the amount of interest that could be earned? To answer these questions, suppose \$1 was deposited in an account at the very high interest rate of 100%. Table 5.16 shows the amount of money after 1 year, compounding n times during the year. The first column represents the time interval between compound interest payments. In Table 5.16, the formula $A = P\left(1 + \frac{r}{n}\right)^{nt}$ reduces to $A = \left(1 + \frac{1}{n}\right)^{n}$, since $P = t = r = 1$.

TABLE 5.16

Time	n	A
Year	1	2.000000
Month	12	2.613035
Day	365	2.714567
Hour	8760	2.718127
Minute	525,600	2.718279
Second	31,536,000	2.718282

Notice that A levels off near \$2.72. If compounding were done more frequently, by letting n become large without bound, it would be called **continuous compounding**. The exponential expression $\left(1 + \frac{1}{n}\right)^{n}$ reaches a limit of approximately 2.718281828 as $n \to \infty$. This value is so important in mathematics that it has been given its own symbol, e, sometimes called **Euler's number**. The number e has many of the same characteristics as π. Its decimal expansion never terminates or repeats in a pattern. It is an irrational number.

VALUE OF e

To nine decimal places, $e \approx 2.718281828$.

Continuous compounding can be applied to population growth. Compounding annually would mean that all births and deaths occur on December 31. Similarly, compounding quarterly would mean that births and deaths occur at the end of March, June, September, and December. In large populations, births and deaths occur *continuously* throughout the year. Compounding continuously is a *natural* way to model large populations.

> ### THE NATURAL EXPONENTIAL FUNCTION
>
> The function f, represented by $f(x) = e^x$, is the **natural exponential function**.

EXAMPLE 8

Evaluating the natural exponential function

Use a calculator to approximate to four decimal places $f(x) = e^x$ when $x = 1, 0.5,$ and -2.56.

Calculator Help

When evaluating e^x, be sure to use the built-in key for e^x, rather than use an approximation for e, such as 2.72.

SOLUTION $f(1) = e^1 \approx 2.7183$, $f(0.5) = e^{0.5} \approx 1.6487$, and $f(-2.56) = e^{-2.56} \approx 0.0773$, as shown in Figure 5.48.

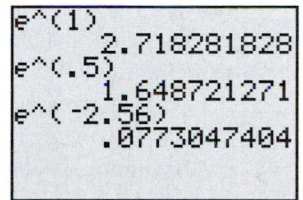

FIGURE 5.48

Now Try Exercise 63 ◆

◆ **CLASS DISCUSSION**

Graph $y = 2^x$ and $y = 3^x$ on the same coordinate axes, by using the viewing rectangle $[-3, 3, 1]$ by $[0, 4, 1]$. Make a conjecture about how the graph of $y = e^x$ will appear. Test your conjecture.

When interest is compounded continuously, the following can be used to compute the amount of money in an account.

> ### CONTINUOUSLY COMPOUNDED INTEREST
>
> If a principal of P dollars is deposited in an account paying an annual rate of interest r (expressed in decimal form), compounded continuously, then after t years the account will contain A dollars, where
>
> $$A = Pe^{rt}.$$

EXAMPLE 9

Calculating continuously compounded interest

Suppose $1000 are deposited into an IRA with an interest rate of 12%, compounded continuously.
(a) How much money will there be after 45 years?
(b) Determine graphically and numerically the 10-year period when the account balance increases the most.

SOLUTION
(a) Let $P = \mathbf{1000}$, $r = \mathbf{0.12}$, and $t = \mathbf{45}$. Then $A = \mathbf{1000}e^{(0.12)45} \approx \$221,406.42$. This is more than the $204,503.36 that resulted from compounding quarterly in Example 7.
(b) Graph $Y_1 = 1000e^{\wedge}(.12X)$, as shown in Figure 5.49. From the graph we can see that the account balance increases the most during the last 10 years. Numerical support for this conclusion is shown in Figure 5.50.

Calculator Help

To make a table like Figure 5.50, see Appendix B (page AP-15).

[0, 45, 5] by [0, 250000, 100000]

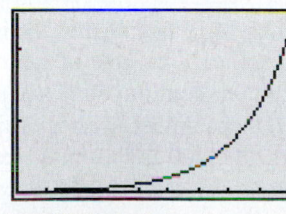

FIGURE 5.49

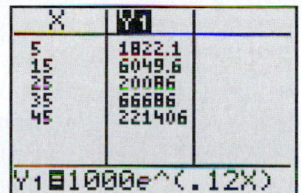

FIGURE 5.50

Now Try Exercise 97 ◆

Modeling traffic flow

At an intersection cars arrive randomly with an average rate of 30 cars per hour. Highway engineers estimate the likelihood or probability that at least one car will enter the intersection within a period of x minutes with $f(x) = 1 - e^{-0.5x}$. (**Source:** F. Mannering and W. Kilareski, *Principles of Highway Engineering and Traffic Analysis.*)

(a) Evaluate $f(2)$ and interpret the answer.

(b) Graph f for $0 \le x \le 60$. What happens to the likelihood that at least one car enters the intersection during a 60-minute period?

$[0, 60, 10]$ by $[0, 1.2, 0.2]$

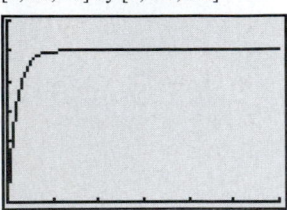

FIGURE 5.51

SOLUTION

(a) $f(2) = 1 - e^{-0.5(2)} = 1 - e^{-1} \approx 0.63$. There is a 63% chance that at least one car will enter the intersection during a 2-minute period.

(b) Graph $Y_1 = 1 - e^{\wedge}(-0.5X)$, as shown in Figure 5.51. As time progresses, the probability increases and begins to approach 1. That is, it is almost certain that at least one car will enter the intersection during a 60-minute period. (Note that a horizontal asymptote is $y = 1$.)   *Now Try Exercise 95* ◆

The natural exponential function is often used to model growth of a quantity. If A_0 is the initial amount of a quantity A at time $t = 0$ and if k is a *positive* constant, then exponential growth of A can be modeled by

$$A(t) = A_0 e^{kt}. \quad \textit{Growth}$$

Figure 5.52 illustrates this type of growth graphically. Similarly,

$$A(t) = A_0 e^{-kt} \quad \textit{Decay}$$

can be used to model exponential decay provided $k > 0$. Figure 5.53 illustrates this type of decay graphically. A larger value of k causes the graph of A to increase or decrease more rapidly. That is, a larger value of k causes the rate of change in A to be greater in absolute value.

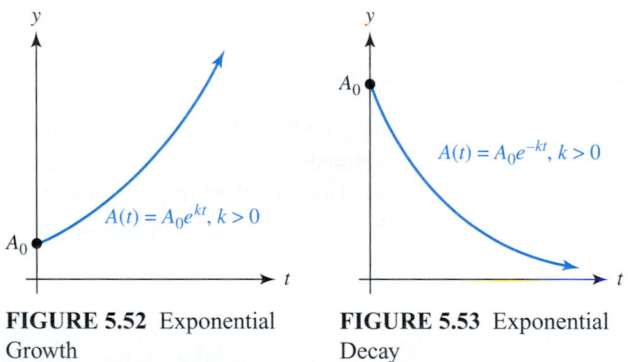

FIGURE 5.52 Exponential Growth

FIGURE 5.53 Exponential Decay

Sometimes variables, such as N or Q, are used instead of the variable A in these two formulas. In the next example, we model the number of bacteria N in a sample that undergoes exponential growth.

Modeling the growth of *E. coli* bacteria

A type of bacteria that inhabits the intestines of animals is named *E. coli* (*Escherichia coli*). These bacteria are capable of rapid growth and can be dangerous to humans—especially children. In one study, *E. coli* bacteria were capable of doubling in number about every 49.5 minutes. Their number N after t minutes could be modeled by $N(t) = N_0 e^{0.014t}$. Suppose that $N_0 = 500{,}000$ is the initial number of bacteria per milliliter. (**Source:** G. S. Stent, *Molecular Biology of Bacterial Viruses.*)

(a) Make a conjecture about the number of bacteria per milliliter after 99 minutes. Verify your conjecture.

(b) Determine graphically the elapsed time when there were 25 million bacteria per milliliter.

SOLUTION

(a) Since the bacteria double every 49.5 minutes, there would be 1,000,000 per milliliter after 49.5 minutes and 2,000,000 after 99 minutes. This is verified by evaluating

$$N(99) = 500,000e^{0.014(99)} \approx 2,000,000.$$

(b) *Graphical Solution* Solve $N(t) = 25,000,000$ by graphing $Y_1 = 500000e^{\wedge}(0.014X)$ and $Y_2 = 25000000$. Their graphs intersect near $(279.4, 25,000,000)$, as shown in Figure 5.54. Thus in a 1 milliliter sample, half a million *E. coli* bacteria could increase to 25 million in approximately 279 minutes, or 4 hours and 39 minutes.

Now Try Exercise 101 ◆

$[0, 400, 100]$ by $[0, 3 \times 10^7, 1 \times 10^7]$

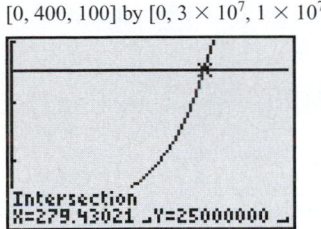

FIGURE 5.54

 5.3

Putting it all Together

The following table summarizes some important concepts about exponential functions and types of growth and decay.

Concept	Comments	Examples
Exponential function $f(x) = Ca^x,$ where $a > 0, a \neq 1,$ and $C > 0$.	Exponential growth occurs when $a > 1$, and exponential decay occurs when $0 < a < 1$. C often represents the initial amount present.	$f(x) = 5(0.8)^x$ *Decay* $g(x) = 3^x$ *Growth* Decay: $0 < a < 1$ Growth: $a > 1$
Linear growth $y = mx + b$	If data increase by a fixed amount m for each unit increase in x, they can be modeled by a linear function.	The following data can be modeled by $f(x) = 3x + 2$ because the data increase 3 units for each unit increase in x, and because $y = 2$ when $x = 0$. x: 0, 1, 2, 3 y: 2, 5, 8, 11 $+3$ $+3$ $+3$

Concept	Comments	Examples
Exponential growth $y = Ca^x$	If data increase by a constant factor a for each unit increase in x, they can be modeled by an exponential function.	The following data can be modeled by $f(x) = 2(3)^x$ because the data increase by a factor of 3 for each unit increase in x and because $y = 2$ when $x = 0$. <table><tr><td>x</td><td>0</td><td>1</td><td>2</td><td>3</td></tr><tr><td>y</td><td>2</td><td>6</td><td>18</td><td>54</td></tr></table> $\times 3 \quad \times 3 \quad \times 3$
Radioactive decay	If a radioactive sample has a half-life of k years and contains C units, then the amount A remaining after x years is given by $$A(x) = C\left(\frac{1}{2}\right)^{x/k}.$$	A 5-gram sample of radioactive material with a half-life of 300 years is modeled by $$A(x) = 5\left(\frac{1}{2}\right)^{x/300}.$$
Interest compounded n times per year $A = P\left(1 + \dfrac{r}{n}\right)^{nt}$	P is the principal, r is the interest rate (expressed in decimal form), n is the number of times interest is paid each year, t is the number of years, and A is the amount after t years.	$\$500$ at 8%, compounded monthly for 3 years yields: $$500\left(1 + \frac{0.08}{12}\right)^{12(3)}$$ $$\approx \$635.12.$$
The number e	The number e is an irrational number and important in mathematics, much like π.	$e \approx 2.718282$
Natural exponential function	This function is an exponential function with base e.	$f(x) = e^x$
Interest compounded continuously $A = Pe^{rt}$	P is the principal, r is the interest rate (expressed in decimal form), t is the number of years, and A is the amount after t years.	$\$500$ at 8% compounded continuously for 3 years yields: $500e^{0.08(3)} \approx \$635.62.$

5.3 Exercises

Exponents

Exercises 1–16: Simplify the expression without a calculator.

1. 2^{-3}

2. $(-3)^{-2}$

3. $3(4)^{1/2}$

4. $5\left(\frac{1}{2}\right)^{-3}$

5. $-2(27)^{2/3}$

6. $-4(8)^{-2/3}$

7. $\left(\frac{1}{8}\right)^{-1}$

8. $a^{0.5}a^{1.5}$

9. $4^{1/6}4^{1/3}$

10. $\frac{9^{5/6}}{9^{1/3}}$

11. $e^x e^x$

12. $16^{-1/2}27^{1/3}$

13. 3^0

14. $5\left(\frac{3}{4}\right)^0$

15. $(5^{101})^{1/101}$

16. $(8^{27})^{1/27}$

Exercises 17 and 18: Use a calculator to approximate the expression to the nearest hundredth.

17. (a) $3^{5/2}$ (b) 2^{π} (c) $-e^{\sqrt{3}}$

18. (a) π^2 (b) e^{-1} (c) $2^{0.145}$

Linear and Exponential Growth

Exercises 19–24: (Refer to Example 1.) Find either a linear or an exponential function that models the data in the table.

19.

x	0	1	2	3	4
y	2	0.75	-0.50	-1.75	-3

20.

x	0	1	2	3	4
y	2	8	32	128	512

21.

x	-3	-2	-1	0	1
y	64	32	16	8	4

22.

x	-2	-1	0	1	2
y	3	5.5	8	10.5	13

23.

x	-4	-2	0	2	4
y	0.3125	1.25	5	20	80

24.

x	-15	-5	5	15	25
y	22	24	26	28	30

25. *Job Offer* A company offers a college graduate $40,000 per year with a guaranteed 8% raise each year. Is this an example of linear or exponential growth? Find a function f that computes the salary during the nth year.

26. *Job Offer* A new employee is offered $35,000 per year with a guaranteed $5000 raise each year. Is this an example of linear or exponential growth? Find a function f that computes the salary during the nth year.

27. *Comparing Growth* Which function becomes larger for $0 \le x \le 10$: $f(x) = 2^x$ or $g(x) = x^2$?

28. *Salaries* If you were offered 1¢ for the first week of work, 3¢ for the second week, 5¢ for the third week, 7¢ for the fourth week, and so on for a year, would you accept the offer? Would you accept an offer that pays 1¢ for the first week of work, 2¢ for the second week, 4¢ for the third week, 8¢ for the fourth week, and so on for a year? Explain your answers.

Exponential Functions

Exercises 29–36: Find C and a so that $f(x) = Ca^x$ satisfies the given conditions.

29. $f(0) = 5$ and for each unit increase in x, the output is multiplied by 1.5.

30. $f(1) = 3$ and for each unit increase in x, the output is multiplied by $\frac{3}{4}$.

31. $f(0) = 10$ and $f(1) = 20$

32. $f(0) = 7$ and $f(-1) = 1$

33. $f(1) = 9$ and $f(2) = 27$

34. $f(-1) = \frac{1}{4}$ and $f(1) = 4$

35. $f(-2) = \frac{9}{2}$ and $f(2) = \frac{1}{18}$

36. $f(-2) = \frac{3}{4}$ and $f(2) = 12$

Exercises 37–40: Find C and a so that $f(x) = Ca^x$ models the situation described. State what the variable x represents in your formula. (Answers may vary.)

37. There are initially 5000 bacteria and this sample doubles in size every hour.

38. Fifteen hundred dollars are deposited in an account that triples in value every decade.

39. In 2000 a house was worth $200,000 and its value *decreases* by 5% each year thereafter.

40. A fish population is initially 6000 and decreases by half each year.

41. *Tire Pressure* The pressure in a tire with a leak is initially 30 pounds per square inch and can be modeled by $f(x) = 30(0.9)^x$ after x minutes. What is the tire's pressure after 9.5 minutes?

42. *Population* The population of California in 1998 was about 33 million and increased by a factor of 1.012 each year. Estimate the population of California in 2000.

Exercises 43–52: Sketch a graph of $y = f(x)$.

43. $f(x) = 2^x$

44. $f(x) = 4^x$

45. $f(x) = 3^{-x}$

46. $f(x) = 3(2^{-x})$

47. $f(x) = 2\left(\frac{1}{3}\right)^x$

48. $f(x) = 2(3^x)$

49. $f(x) = \left(\frac{1}{2}\right)^x$

50. $f(x) = \left(\frac{1}{4}\right)^x$

51. $f(x) = 8(2)^{-x}$

52. $f(x) = 9\left(\frac{1}{3}\right)^x$

Exercises 53–56: Use the graph of $y = Ca^x$ to determine values for C and a.

53.

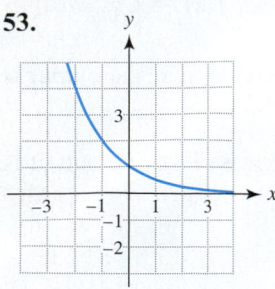

54.

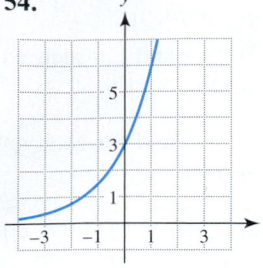

55.

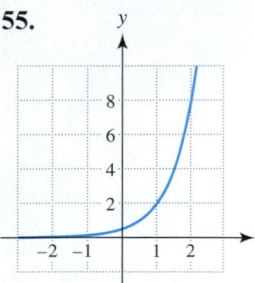

56.

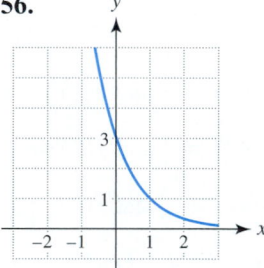

57. Let $f(x) = 7\left(\frac{1}{8}\right)^x$.

(a) What are the domain and range of f?

(b) Is f either increasing or decreasing on its domain?

(c) Find any asymptotes on the graph of f.

(d) Find any x- or y-intercepts on the graph of f.

(e) Is f a one-to-one function? Does f have an inverse?

58. Repeat Exercise 57 with $f(x) = e^x$.

59. Match the symbolic representation of f with its graphical representation (a.–d.). Do not use a calculator.

   (i) $f(x) = e^x$ (ii) $f(x) = 3^{-x}$

   (iii) $f(x) = 1.5^x$ (iv) $f(x) = 0.99^x$

a.

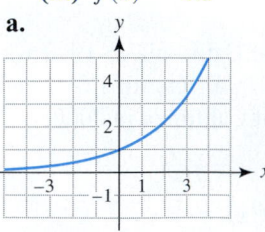

b.

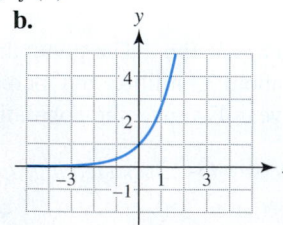

c.

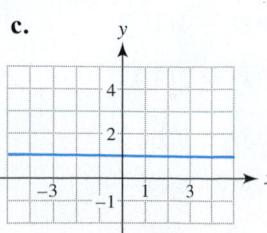

d.

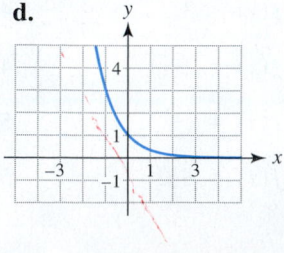

60. *Modeling Phenomena* Match the situation with the graph (a.–d.) that models it best.

  (i) Balance of an account after x years earning 10% interest compounded continuously

  (ii) Balance of an account after x years earning 5% interest compounded annually

  (iii) Air pressure in a car tire with a large hole in it after x minutes

  (iv) Air pressure in a car tire with a pinhole in it after x minutes

a.

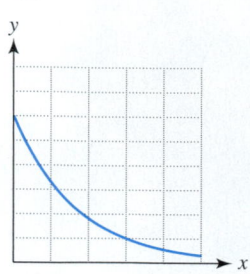

b.

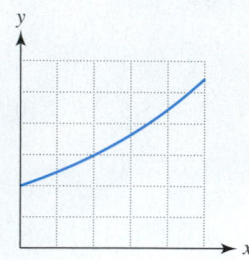

c.

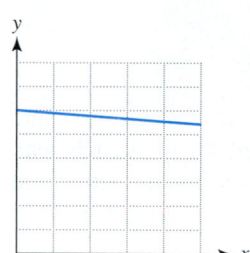

d.

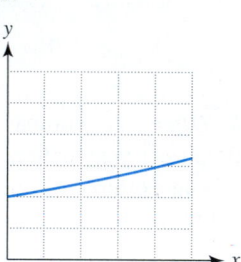

Exercises 61 and 62: The graph of $y = f(x)$ is shown in the accompanying figure. Sketch a graph of each equation using translations of graphs and reflections. Do not use a graphing calculator.

61. $f(x) = 2^x$

(a) $y = 2^x - 2$

(b) $y = 2^{x-1}$

(c) $y = 2^{-x}$

(d) $y = -2^x$

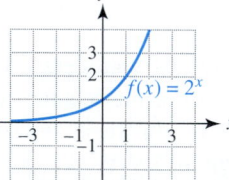

62. $f(x) = e^{-0.5x}$

(a) $y = -e^{-0.5x}$

(b) $y = e^{-0.5x} - 3$

(c) $y = e^{-0.5(x-2)}$

(d) $y = e^{0.5x}$

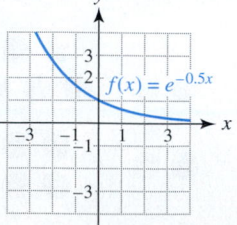

Exercises 63–68: Approximate f(x) to four decimal places.

63. $f(x) = e^x$, $\qquad x = 3.1$

64. $f(x) = e^{2x}$, $\qquad x = -0.43$

65. $f(x) = 4e^{-1.2x}$, $\qquad x = -2.4$

66. $f(x) = -2.1e^{-0.71x}$, $\qquad x = 1.9$

67. $f(x) = \dfrac{e^x - e^{-x}}{2}$, $\qquad x = -0.7$

68. $f(x) = 4(e^{-0.3x} - e^{-0.6x})$, $\qquad x = 1.6$

Compound Interest

Exercises 69–76: Use the compound interest formula to determine the final value of each amount.

69. $600 at 7% compounded annually for 5 years

70. $2300 at 11% compounded semiannually for 10 years

71. $950 at 3% compounded daily for 20 years (*Hint:* Let $n = 365$.)

72. $3300 at 8% compounded quarterly for 2 years

73. $2000 at 10% compounded continuously for 8 years

74. $100 at 19% compounded continuously for 50 years

75. $1600 at 10.4% compounded monthly for 2.5 years

76. $2000 at 8.7% compounded annually for 5 years

77. *Investments* Compare investing $2000 at 10% compounded monthly for 20 years with investing $2000 at 13% compounded monthly for 20 years.

78. *Lake Property* In some states, lake shore property is increasing in value by 15% per year. Determine the value of a $90,000 lake lot after 5 years.

79. *College Tuition* If college tuition is currently $8000 per year, inflating at 6% per year, what will be the cost of tuition in 10 years?

80. *Doubling Time* How long does it take for an investment to double its value if the interest is 12% compounded annually? 6% compounded annually?

81. *Interest* A principal of $100 is deposited into an account paying 5% interest compounded monthly. In a similar account, $200 are deposited. Make a conjecture about how the amounts in each account will compare after 10 years. Explain your reasoning.

82. *Interest* A principal of $2000 is deposited into two different accounts. One account pays 10% interest compounded annually, and the other pays 5% interest compounded annually.
(a) Make a conjecture whether the 10% account will accrue twice the interest as the 5% account after 5 years.
(b) Test your conjecture by computing the actual amounts.

83. *Federal Debt* In fiscal year 2003 the federal budget deficit was about $300 billion. At the same time, 30-year treasury bonds were paying 4.95% interest. Suppose the American taxpayer loaned $300 billion to the federal government at 4.95% compounded annually. If the federal government waited 30 years to pay the entire amount back, including the interest, how much would this be? (**Source:** Department of the Treasury.)

84. *Federal Debt* In the previous exercise suppose that interest rates were 2% higher. How much would the federal government owe after 30 years? Is the national debt sensitive to interest rates?

85. *Annuity* If x dollars are deposited every 2 weeks (26 times per year) into an account paying an annual interest rate r, expressed in decimal form, then the amount A in the account after t years can be approximated by the formula

$$A = x\left[\frac{(1 + r/26)^{26t} - 1}{(r/26)}\right].$$

If $50 is deposited every 2 weeks into an account paying 8% interest, approximate the amount in the account after 10 years.

86. *Annuity* (Refer to the previous exercise.) Suppose a retirement account pays 10% annual interest. Determine how much a 20-year-old worker should deposit in this account every 2 weeks in order to have 1 million dollars at age 65.

Applications

87. *Radioactive Radium-226* The half-life of radium-226 is about 1600 years. After 3000 years what percentage P of a sample of radium remains?

88. *Radioactive Strontium-90* Radioactive strontium-90 has a half-life of about 28 years and sometimes contaminates the soil. After 50 years, what percentage of a sample of radioactive strontium would remain?

89. *Radioactive Carbon-14* (Refer to Example 4.) A fossil contains 10% of the carbon-14 it contained when it was alive. Graphically estimate its age.

90. *Radioactive Carbon-14*   If a fossil contained 0.01 milligram of carbon-14 when it was alive and now contains 0.002 milligram of carbon-14, estimate its age.

91. *Radioactive Cesium-137*   Radioactive cesium-137 was emitted in large amounts in the Chernobyl nuclear power station accident in Russia on April 26, 1986. The amount of cesium remaining after x years in an initial sample of 100 milligrams can be described by the formula $A(x) = 100e^{-0.02295x}$. (**Source:** C. Mason, *Biology of Freshwater Pollution.*)

   (a) How much is remaining after 50 years? Is the half-life of cesium more or less than 50 years?

   (b) Estimate graphically the half-life of cesium-137.

92. *Radioactive Polonium-210*   The accompanying table shows the amount y of polonium in milligrams remaining after x days from an initial sample of 2 milligrams.

x (days)	0	100	200	300
y (milligrams)	2	1.22	0.743	0.453

   (a) Use the table to determine mentally if the half-life of polonium is more or less than 200 days.

   (b) Give a general formula that can be used to approximate the amount A of polonium in the table after x days.

   (c) Estimate graphically the half-life of polonium.

93. *Greenhouse Gases*   (Refer to Example 5.) Chlorofluorocarbons (CFCs) are gases created by people that increase the greenhouse effect and damage the ozone layer. CFC-12 is one type of chlorofluorocarbon used in refrigeration, air conditioning, and foam insulation. The following table lists future concentrations of CFC-12 in parts per billion (ppb), if current trends continue.

Year	2000	2005	2010	2015	2020
CFC-12 (ppb)	0.72	0.88	1.07	1.31	1.60

   (a) Let $x = 0$ correspond to 2000 and $x = 20$ to 2020. Find values for C and a so that $f(x) = Ca^x$ models these data.

   (b) Estimate the CFC-12 concentration in 2013.

94. *Swimming Pool Maintenance*   Chlorine is frequently used to disinfect swimming pools. The chlorine concentration should remain between 1.5 and 2.5 parts per million. On warm, sunny days with many swimmers agitating the water, 30% of the chlorine can dissipate into the air or combine with other chemicals. (**Source:** D. Thomas, *Swimming Pool Operators Handbook.*)

   (a) Let $C(x) = 2.5(0.7)^x$ model the amount of chlorine in a pool after x days in parts per million. What is the initial concentration of chlorine in the pool?

   (b) If no chlorine is added, estimate graphically or numerically the number of days before chlorine should be added.

95. *Modeling Traffic Flow*   At an intersection cars arrive randomly with an average rate of 50 cars per hour. The likelihood or probability that at least one car will enter the intersection within a period of x minutes can be estimated by $P(x) = 1 - e^{-5x/6}$.

   (a) Find the likelihood that at least one car enters the intersection during a 3-minute period.

   (b) Graphically determine the value of x that gives a 50–50 chance of at least one car entering the intersection during an interval of x minutes.

96. *Filters*   Impurities in water are frequently removed using filters. Suppose that a 1-inch filter allows 10% of the impurities to pass through it. The other 90% is trapped in the filter.

   (a) Find a formula in the form $f(x) = 100a^x$ that calculates the percentage of impurities passing through x inches of this type of filter.

   (b) Use $f(x)$ to estimate the percentage of impurities passing through 2.3 inches of the filter.

97. *Saving for Retirement*   Suppose $1500 are deposited into an IRA with an interest rate of 6%, compounded continuously.

   (a) How much money will there be in the account after 30 years?

   (b) Determine graphically or numerically the 10-year period when the account balance increases the most.

98. *Population Growth*   The population of Phoenix, Arizona, in 2000 was 1.3 million and growing continuously at a 3% rate.

   (a) Assuming this trend continues, estimate the population of Phoenix in 2010.

   (b) Determine graphically or numerically when this population might reach 2 million.

99. *Thickness of Runways*   Heavier aircraft require runways with thicker pavement for landings and takeoffs. A pavement 6 inches thick can accommodate an aircraft weighing 80,000 pounds, whereas a 12-inch pavement is necessary for a 350,000-pound plane. The relation between pavement thickness t in inches and gross weight W in thousands of pounds can be modeled by $W = 18.29(1.279)^t$. Complete the table. Round values to the nearest thousand pounds. (**Source:** Federal Aviation Administration.)

t (inches)	6	7.5	9	10.5	12
W (lb × 1000)	80				350

100. *Trains and Horsepower*   The faster a locomotive travels, the more horsepower is needed. The function given by $H(x) = 0.157(1.033)^x$ calculates this horsepower for a level track. The input x is in miles per hour and the output $H(x)$ is the horsepower required per ton of cargo. (**Source:** L. Haefner, *Introduction to Transportation Systems.*)

(a) Graph H in $[0, 100, 10]$ by $[0, 5, 1]$. Is H increasing or decreasing?

(b) Evaluate $H(30)$ and interpret the result.

(c) Determine the horsepower needed to move a 5000-ton train 30 miles per hour.

(d) Some types of locomotives are rated for 1350 horsepower. How many locomotives of this type would be needed in part (c)?

101. *E. Coli Growth*   (Refer to Example 11.)
(a) Approximate the number of *E. coli* after 3 hours.

(b) Estimate graphically the elapsed time before there are 10 million bacteria per milliliter.

102. *Drug Concentrations*   Sometimes after a patient takes a drug, the amount of medication A in the bloodstream can be modeled by $A = A_0 e^{-rt}$, where A_0 is the initial concentration in milligrams per liter, r is the hourly percentage decrease (in decimal form) of the drug in the bloodstream, and t is the elapsed time in hours. Suppose that a drug's concentration is initially 2 milligrams per liter and that $r = 0.2$. Find the drug concentration after 3.5 hours.

103. *Tree Density*   Ecologists studied the spacing between individual trees in a forest in British Columbia. This lodgepole pine forest was 40 to 50 years old and had approximately 1600 trees per acre that were randomly spaced. The probability or likelihood that there is at least one tree located in a circle with a radius of x feet is estimated by $P(x) = 1 - e^{-0.1144x}$. For example, $P(7) \approx 0.55$ means that if a person picks a point at random in the forest, there is a 55% chance that at least one tree will be located within 7 feet. See the accompa-

nying figure.   (**Source:** E. Pielou, *Populations and Community Ecology.*)

(a) Evaluate $P(2)$ and $P(20)$, and interpret the results.

(b) Graph P. Explain verbally why it is logical for P to be an increasing function. Does the graph have a horizontal asymptote?

(c) Solve $P(x) = 0.5$ and interpret the result.

104. *Computer Growth*   Between 1971 and 1995 the numbers of transistors that could be placed on a single chip are shown in the table and modeled by the formula $N(t) = 0.00168 e^{(t-1970)/3}$. The input t is the year and the output $N(t)$ is the number of transistors in millions. (**Source:** Intel.)

Year	Chip	Transistors (millions)
1971	4004	0.0023
1986	386DX	0.275
1989	486DX	1.2
1993	Pentium	3.3
1995	P6	5.5

(a) Graph N and the data pairs (year, transistors) in the window $[1968, 1998, 2]$ by $[-1, 6, 1]$.

(b) Use $N(t)$ to estimate an average annual percentage increase in the number of transistors between 1971 and 1995.

105. *Incubation Time of AIDS*   With some diseases, such as AIDS, there is a delay between the initial infection and the time when symptoms begin. This delay time can vary greatly from one individual to the next, but it appears to have an overall pattern. In one study the formula $N(x) = N_0 e^{-0.1x^2/2}$ models the delay time for AIDS patients who received infected blood transfusions between 1978 and 1986. The formula $N(x)$ computes the number of people who have not developed AIDS x years after their initial infection. (**Source:** G. Medley, "Incubation period of AIDS in patients infected via blood transfusions.")
(a) Evaluate $N(0)$ and interpret N_0.

(b) Let $N_0 = 100$. Graphically approximate the number of years before 50% of the patients have developed symptoms of AIDS.

106. *Survival of Reindeer* For all types of animals, the percentage that survive into the next year decreases. In one study, the survival rate of a sample of reindeer was modeled by $S(t) = 100(0.999993)^{t^5}$. The function S outputs the percentage of reindeer that survive t years. (**Source:** D. Brown.)

(a) Evaluate $S(4)$ and $S(15)$. Interpret the results.

(b) Graph S in $[0, 15, 5]$ by $[0, 110, 10]$. Interpret the graph. Does the graph have a horizontal asymptote?

Writing about Mathematics

107. Explain how a linear function and an exponential function differ.

108. Discuss the domain and range of an exponential function f. Is f one-to-one? Explain.

EXTENDED AND DISCOVERY EXERCISES

Exercises 1–4: **Present Value** *In the compound interest formula $A = P(1 + r/n)^{nt}$, we can think of P as the present value of an investment and A as the future value of an investment after t years. For example, if we are saving for college and need a future value of A dollars, then P would represent the amount needed in an account today to reach our goal in t years at an interest rate of r, compounded n times per year. If we solve the equation for P, it results in*

$$P = A(1 + r/n)^{-nt}.$$

1. Verify that the two formulas are equivalent by transforming the first equation into the second.

2. What should the present value of a savings account be to cover \$30,000 of college expenses in 12.5 years, if the account pays 7.5% interest compounded quarterly?

3. If you need \$15,000 to buy a car in 3 years, what should the present value of a savings account be to reach this goal, if the account pays 5% compounded monthly?

4. A parent expects college costs to reach \$40,000 in 6 years. How much should the parent deposit in an account that pays 6% interest compounded *continuously* to cover the \$40,000 in future expenses?

Exercises 5–8: **Average Rate of Change of e^x** *Complete the following. Round your answers to two decimal places.*

 (a) Find the average rate of change of $f(x) = e^x$ from x to $x + 0.001$ for the given x.
 (b) Approximate $f(x) = e^x$ for the given x.
 (c) Compare your answers in parts (a) and (b).

5. $x = 0$ **6.** $x = -2$

7. $x = -0.5$ **8.** $x = 1.5$

9. *Average Rate of Change* What is the pattern in the results from Exercises 5–8? You may want to test your conjecture by trying different values of x.

10. *Average Rate of Change* (Refer to Exercises 5–8.) For any real number k, what is a good approximation for the average rate of change of $f(x) = e^x$ on a small interval near $x = k$? Explain how your answer relates to the graph of $f(x) = e^x$.

5.4 Logarithmic Functions and Models

◆ Evaluate the common logarithmic function

◆ Solve basic exponential and logarithmic equations

◆ Evaluate logarithms with other bases

◆ Solve general exponential and logarithmic equations

Introduction

In Chapter 1 numbers were introduced as a way to measure and quantify data. Later, functions were developed to model and describe data. Applications involving functions frequently result in equations. Although we have solved many equations, one equation that we have not solved *symbolically* is $a^x = k$, where the variable x is an exponent. This exponential equation occurs frequently in applications. In this section we discuss logarithmic functions and how they can be used to solve exponential equations.

The Common Logarithmic Function

Exponential functions are capable of modeling data that exhibit rapid growth. However, a new type of nonlinear function is needed to model data that grows much slower. Table 5.17 lists the growth of a bacteria colony at 1-day intervals and illustrates exponential growth: each time x increases by 1 day, the number of bacteria y increases by a factor of 10. This growth is modeled by $f(x) = 10^x$, which calculates the number of bacteria on day x.

TABLE 5.17   Growth of Bacteria

x (day)	0	1	2	3	4
y (bacteria in thousands)	1	10	100	1000	10,000

Now consider Table 5.18, where the roles of x and y have been interchanged. Table 5.18 can be used to calculate the day when there were x bacteria. Notice that the function represented by this table grows very slowly: each time x increases by a factor of 10, y increases by 1. This function is called the *common (or base-10) logarithmic function*, and is written as $g(x) = \log x$.

TABLE 5.18   Growth of a Bacteria Colony

x (bacteria in thousands)	1	10	100	1000	10,000
y (day)	0	1	2	3	4

The common exponential function and the common logarithmic function are represented numerically in Tables 5.19 and 5.20.

TABLE 5.19   Common Exponential Function

x	-4	-3	-2	-1	0	0.5	1	2	π
10^x	10^{-4}	10^{-3}	10^{-2}	10^{-1}	10^0	$10^{0.5}$	10^1	10^2	10^π

TABLE 5.20   Common Logarithmic Function

x	10^{-4}	10^{-3}	10^{-2}	10^{-1}	10^0	$10^{0.5}$	10^1	10^2	10^π
$\log x$	-4	-3	-2	-1	0	0.5	1	2	π

Note: The common (or base-10) logarithmic function denoted by $\log x$ outputs the *exponent* k if the input x can be written as $x = 10^k$ for some real number k. That is, $\log 10^k = k$ for all real numbers k. This concept is illustrated visually on the next page.

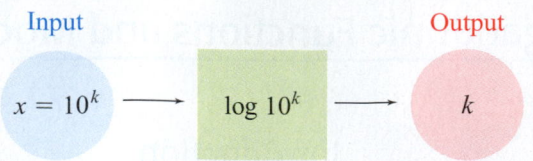

Common Logarithm

The points $(10^{-1}, -1)$, $(10^0, 0)$, $(10^{0.5}, 0.5)$, and $(10^1, 1)$ are located on the graph of $y = \log x$. They are plotted in Figure 5.55. Any *positive* real number x can be expressed as $x = 10^k$ for some real number k. In this case $\log x = \log 10^k = k$. The graph of $y = \log x$ is a continuous curve, as shown in Figure 5.56. The y-axis is a vertical asymptote. The common logarithmic function is one-to-one and always increasing. Its domain is all positive real numbers, and its range is all real numbers.

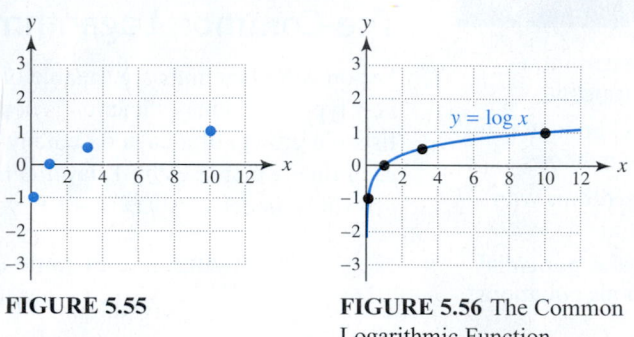

FIGURE 5.55

FIGURE 5.56 The Common
Logarithmic Function

We have shown *verbal*, *numerical*, *graphical*, and *symbolic* representations of the common logarithmic function. A formal definition of the common logarithm is now given.

COMMON LOGARITHM

The **common logarithm of a positive number** x, denoted $\log x$, is defined by

$$\log x = k \quad \text{if and only if} \quad x = 10^k,$$

where k is a real number. The function given by

$$f(x) = \log x$$

is called the **common logarithmic function**.

Note: The common logarithmic function outputs an *exponent* k, which may be positive, negative, or zero. However, a valid input must be positive. Thus its range is $(-\infty, \infty)$ and its domain is $(0, \infty)$.

 **EXAMPLE 1** Evaluating common logarithms

Simplify each logarithm by hand.

(a) $\log 1$ **(b)** $\log \dfrac{1}{1000}$ **(c)** $\log \sqrt{10}$ **(d)** $\log(-2)$

Algebra Review

To review properties of exponents, see Section 4.7, and Chapter R (page R-13).

SOLUTION

(a) Since $1 = 10^0$, $\log 1 = \log 10^0 = 0$.

(b) $\log \dfrac{1}{1000} = \log 10^{-3} = -3$.

(c) $\log \sqrt{10} = \log(10^{1/2}) = \dfrac{1}{2}$.

(d) The domain of $f(x) = \log x$ is $(0, \infty)$, so $\log(-2)$ is undefined because the input is negative.

Now Try Exercise 3 ◆

FIGURE 5.57

Much like the square root function, the common logarithmic function does not have an easy-to-evaluate formula. For example, $\sqrt{4} = 2$ and $\sqrt{100} = 10$ can be calculated mentally, but for $\sqrt{2}$ we usually rely on a calculator. Similarly, $\log 100 = 2$ can be found mentally since $100 = 10^2$, whereas $\log 12$ can be approximated using a calculator. See Figure 5.57. To check that $\log 12 \approx 1.0791812$, evaluate $10^{1.0791812} \approx 12$. Another similarity between the square root function and the common logarithmic function is that their domains do not include negative numbers. If outputs are restricted to real numbers, both $\sqrt{-3}$ and $\log(-3)$ are undefined expressions.

◆ **MAKING CONNECTIONS**

Logarithms and Exponents *A logarithm is an exponent.* For example, $\log 1000$ represents the exponent k such that $10^k = 1000$. Thus $\log 1000 = 3$ because $10^3 = 1000$.

◆

The human ear is extremely sensitive and is able to detect pressures on the eardrum that range from 10^{-16} w/cm^2 (watts per square centimeter) to 10^{-4} w/cm^2. Sound with an intensity of 10^{-4} w/cm^2 is painful to the human eardrum. Because of this wide range of intensities, scientists use logarithms to measure sound in decibels.

EXAMPLE 2 Applying common logarithms to sound

Sound levels in decibels (db) can be computed by $D(x) = 10 \log(10^{16}x)$, where x is the intensity of the sound in watts per square centimeter. (**Source:** R. Weidner and R. Sells, *Elementary Classical Physics.*)

(a) At what decibel level does the threshold for pain occur?
(b) Make a table of D for the intensities $x = 10^{-13}, 10^{-12}, 10^{-11}, \ldots, 10^{-4}$.
(c) If the intensity x increases by a factor of 10, what is the corresponding increase in decibels?

SOLUTION

(a) The threshold for pain occurs when $x = 10^{-4}$.

$$
\begin{aligned}
D(10^{-4}) &= 10 \log(10^{16} \cdot 10^{-4}) && \text{Substitute } x = 10^{-4}. \\
&= 10 \log(10^{12}) && \text{Add exponents when multiplying like bases.} \\
&= 10(12) && \text{Evaluate the logarithm.} \\
&= 120 && \text{Multiply.}
\end{aligned}
$$

The human eardrum begins to hurt at 120 decibels.

(b) Other values for x in Table 5.21 are computed in a similar manner.

◆ **CLASS DISCUSSION**

If the sound level increases by 60 db, by what factor does the intensity increase? ◆

TABLE 5.21

x (w/cm^2)	10^{-13}	10^{-12}	10^{-11}	10^{-10}	10^{-9}	10^{-8}	10^{-7}	10^{-6}	10^{-5}	10^{-4}
$D(x)$ (db)	30	40	50	60	70	80	90	100	110	120

(c) From Table 5.21, a tenfold increase in x results in an increase of 10 decibels.

Now Try Exercise 111 ◆

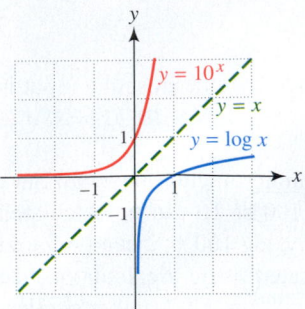

FIGURE 5.58

Basic Equations

The graphs of $y = 10^x$ and $y = \log x$ are shown in Figure 5.58. Notice that they are reflections of each other across the line $y = x$. Both functions are one-to-one.

If $f(x) = 10^k$, then its inverse function is given by $f^{-1}(x) = \log k$. To see this numerically, consider Table 5.22. Notice that $\log 10^k = k$ for each input k. In a similar manner, $10^{\log k} = k$ for any positive k. For example, $\log 10^2 = 2$ and $10^{\log 100} = 100$.

TABLE 5.22

x	-2	-1	0	1	2
$f(x) = 10^x$	10^{-2}	10^{-1}	10^0	10^1	10^2

x	10^{-2}	10^{-1}	10^0	10^1	10^2
$f^{-1}(x) = \log x$	-2	-1	0	1	2

Inverse properties are summarized in the following.

> ### INVERSE PROPERTIES OF THE COMMON LOGARITHM
> The following inverse properties hold for the common logarithm.
>
> $$\log 10^x = x \qquad \text{for any real number } x, \text{ and}$$
> $$10^{\log x} = x \qquad \text{for any positive number } x.$$

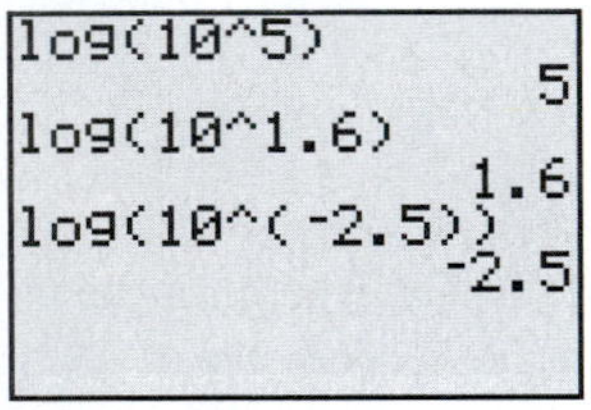

```
log(10^5)
              5
log(10^1.6)
            1.6
log(10^(-2.5))
           -2.5
```

FIGURE 5.59

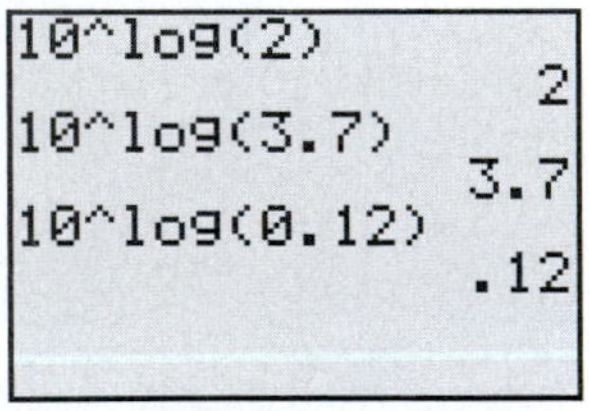

A graphing calculator has been used to illustrate these properties in Figures 5.59 and 5.60.

An exponential equation has a variable that occurs as an exponent in an expression. To solve the exponential equation $10^x = k$, we take the common logarithm of each side and then apply the inverse property $\log 10^x = x$ for any real number x. The following example illustrates this method.

$$10^x = 5 \qquad \textit{Given exponential equation}$$
$$\log 10^x = \log 5 \qquad \textit{Take the common logarithm.}$$
$$x = \log 5 \qquad \textit{Inverse property: } \log 10^k = k$$

```
10^log(2)
              2
10^log(3.7)
            3.7
10^log(0.12)
            .12
```

FIGURE 5.60

Converting $10^x = 5$ to the equivalent equation $x = \log 5$ is sometimes referred to as changing **exponential form** to **logarithmic form**. In the second step, the common logarithm is taken of each side of the equation. We include this second step to emphasize the fact that inverse properties are being used to solve these equations.

EXAMPLE 3 Solving equations of the form $10^x = k$

Solve each equation, if possible.
(a) $10^x = 0.001$ (b) $10^x = 55$ (c) $10^x = -1$

SOLUTION
(a) Take the common logarithm of each side of the equation $10^x = 0.001$. Then,

$$\log 10^x = \log 0.001 \qquad \text{or} \qquad x = \log 10^{-3} = -3.$$

(b) In a similar manner, $10^x = 55$ is equivalent to $x = \log 55 \approx 1.7404$.
(c) The equation $10^x = -1$ has no real solution, because -1 is not in the range of 10^x. Figure 5.61 shows that the graphs of $y_1 = 10^x$ and $y_2 = -1$ do not intersect. Note that $\log(-1)$ is undefined.

Now Try Exercise 13 ◆

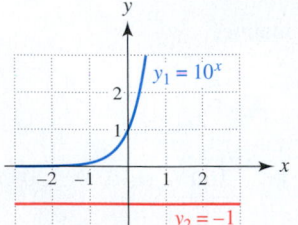

FIGURE 5.61

EXAMPLE 4 Solving an exponential equation

Solve $4(10^{3x}) = 244$.

SOLUTION Begin by dividing each side by 4.

$$4(10^{3x}) = 244 \qquad \text{Given equation}$$

$$10^{3x} = 61 \qquad \text{Divide by 4.}$$

$$\log 10^{3x} = \log 61 \qquad \text{Take the common logarithm.}$$

$$3x = \log 61 \qquad \text{Inverse property: } \log 10^k = k$$

$$x = \frac{\log 61}{3} \qquad \text{Divide by 3.}$$

$$x \approx 0.595 \qquad \text{Approximate.} \qquad \textit{Now Try Exercise 59}\; \blacklozenge$$

A logarithmic equation contains logarithms. To solve logarithmic equations we *exponentiate* each side of the equation and then apply the inverse property $10^{\log x} = x$.

$$\log x = 2.5 \qquad \text{Given logarithmic equation}$$

$$\mathbf{10}^{\log x} = \mathbf{10}^{2.5} \qquad \text{Exponentiate each side; base 10.}$$

$$x = 10^{2.5} \qquad \text{Inverse property: } \log 10^k = k$$

$$x \approx 316.23 \qquad \text{Approximate.}$$

Converting the equation $\log x = k$ to the equivalent equation $x = 10^k$ is called changing *logarithmic form* to *exponential form*.

EXAMPLE 5 Solving equations of the form $\log x = k$

Solve each equation.
(a) $\log x = 3$ **(b)** $\log x = -2$ **(c)** $\log x = 2.7$

SOLUTION
(a)

$$\log x = 3 \qquad \text{Given equation}$$

$$\mathbf{10}^{\log x} = \mathbf{10}^3 \qquad \text{Exponentiate each side; base 10.}$$

$$x = 10^3 \qquad \text{Inverse property: } 10^{\log k} = k$$

$$x = 1000 \qquad \text{Simplify.}$$

(b) Similarly, $\log x = -2$ is equivalent to $x = 10^{-2} = 0.01$.
(c) $\log x = 2.7$ is equivalent to $x = 10^{2.7} \approx 501.2$. $\qquad$ *Now Try Exercises 69 and 75* $\blacklozenge$

EXAMPLE 6 Solving a logarithmic equation

Solve $5 \log 2x = 16$.

SOLUTION Begin by dividing each side by 5.

$$5 \log 2x = 16 \qquad \text{Given equation}$$

$$\log 2x = 3.2 \qquad \text{Divide each side by 5.}$$

$$10^{\log 2x} = 10^{3.2} \qquad \text{Exponentiate each side; base 10.}$$

$$2x = 10^{3.2} \qquad \text{Inverse property: } 10^{\log k} = k$$

$$x = \frac{10^{3.2}}{2} \qquad \text{Divide each side by 2.}$$

$$x \approx 792.4 \qquad \text{Approximate.} \qquad \textit{Now Try Exercise 77}\; \blacklozenge$$

Some types of data grow slowly and can be modeled by $f(x) = a + b \log x$. For example, a larger area of land tends to have a wider variety of birds. However, if the land area doubles, the number of species of birds does not double. In actuality, the land area has to more than double before the number of species doubles.

EXAMPLE 7   **Modeling data with logarithms**

Table 5.23 lists the number of species of birds on islands of different sizes near New Guinea.

(a) Find values for a and b so that $f(x) = a + b \log x$ models the data.

(b) Determine symbolically and graphically the island size that might have 50 different species of birds.

TABLE 5.23

Area (mi²)	0.1	1	10	100	1000
Species of Birds	26	39	52	65	78

Source: B. Freedman, *Environmental Ecology.*

SOLUTION

(a) Begin by substituting $x = 1$ into the formula to determine a.

$$a + b\,\mathbf{\log 1} = 39 \qquad f(1) = 39$$

$$a + b(\mathbf{0}) = 39 \qquad \log 1 = \log 10^0 = 0$$

$$a = 39 \qquad \text{Simplify.}$$

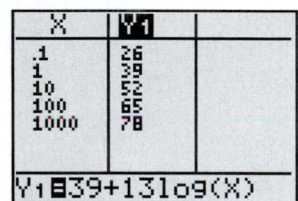

FIGURE 5.62

Thus $f(x) = 39 + b \log x$. Although we can use any data point in Table 5.23 to find b, we let $x = 10$.

$$39 + b \log 10 = 52 \qquad f(10) = 52$$

$$b \log 10 = 13 \qquad \text{Subtract 39.}$$

$$b = 13 \qquad \log 10^1 = 1$$

The data is modeled by $f(x) = 39 + 13 \log x$ which is verified in Figure 5.62.

(b) *Symbolic Solution*   We must solve the equation $f(x) = 50$.

$$39 + 13 \log x = 50 \qquad f(x) = 50$$

$$13 \log x = 11 \qquad \text{Subtract 39.}$$

$$\log x = \frac{11}{13} \qquad \text{Divide by 13.}$$

$$10^{\log x} = 10^{11/13} \qquad \text{Exponentiate each side; base 10.}$$

$$x = 10^{11/13} \qquad \text{Inverse property: } 10^{\log k} = k$$

$$x \approx 7 \qquad \text{Approximate.}$$

$[0, 15, 5]$ by $[0, 70, 10]$

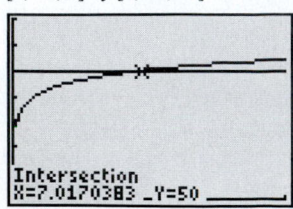

FIGURE 5.63

To have 50 species of birds, an island should be about 7 square miles.

Graphical Solution   The graphs of $Y_1 = 39 + 13 \log (X)$ and $Y_2 = 50$ in Figure 5.63 intersect near the point $(7.02, 50)$. This result agrees with our symbolic solution.

Now Try Exercise 113 ◆

Air pollutants frequently cause acid rain. A measure of the acidity is pH, which ranges between 1 and 14. Pure water is neutral and has a pH of 7. Acidic solutions have a pH less than 7, whereas alkaline solutions have a pH greater than 7. A pH value measures the concentration of the hydrogen ions in a solution. It can be computed by $f(x) = -\log x$, where x represents the hydrogen ion concentration in moles per liter. Pure water exposed to normal carbon dioxide in the atmosphere has a pH of about 5.6. If the pH of a lake drops below this level, it is indicative of an *acid lake.*

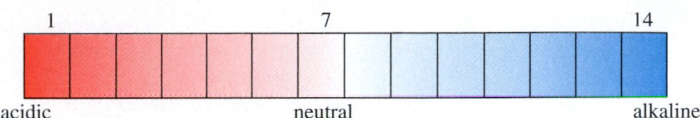

acidic   neutral   alkaline

EXAMPLE 8 Analyzing acid rain

In rural areas of Europe, rainwater typically has a hydrogen ion concentration of about $x = 10^{-4.7}$.   (**Source:** G. Howells, *Acid Rain and Acid Water.*)
(a) Find its pH. What effect might this have on a lake with a pH of 5.6?
(b) Seawater has a pH of 8.2. Compared to seawater, how many times greater is the hydrogen ion concentration in rainwater from rural Europe?

SOLUTION
(a) $f(10^{-4.7}) = -\log 10^{-4.7} = -(-4.7) = 4.7$. The pH is 4.7, which could cause the lake to become more acidic.
(b) To find the hydrogen ion concentration in seawater, solve $f(x) = 8.2$ for x.

$$-\log x = 8.2 \qquad \textcolor{blue}{f(x) = 8.2}$$
$$\log x = -8.2 \qquad \textcolor{blue}{\text{Multiply by } -1.}$$
$$10^{\log x} = 10^{-8.2} \qquad \textcolor{blue}{\text{Exponentiate each side; base 10.}}$$
$$x = 10^{-8.2} \qquad \textcolor{blue}{\text{Inverse property}}$$

Since $\frac{10^{-4.7}}{10^{-8.2}} = 10^{3.5}$, the hydrogen ion concentration in the rainwater is $10^{3.5} \approx 3162$ times greater than it is in seawater.
Now Try Exercise 119 ◆

Logarithms with Other Bases

The development of base-a logarithms can be done with any positive base $a \neq 1$. For example, in computer science base-2 logarithms are frequently used. A numerical representation of the base-2 logarithmic function, denoted $f(x) = \log_2 x$, is shown in Table 5.24. If x can be expressed in the form $x = 2^k$ for some k, then $\log_2 x = k$. Thus $\log_2 x$ is an exponent. A graph of $y = \log_2 x$ is shown in Figure 5.64.

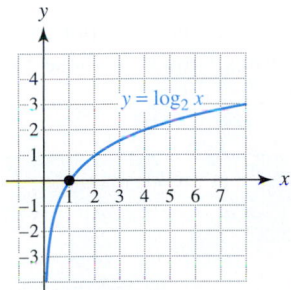

FIGURE 5.64

TABLE 5.24 Base-2 Logarithm

x	$2^{-3.1}$	2^{-2}	$2^{-0.5}$	2^0	$2^{0.5}$	2^2	$2^{3.1}$
$\log_2 x$	-3.1	-2	-0.5	0	0.5	2	3.1

In a similar manner, a numerical representation for the base-e logarithm is shown in Table 5.25 on the next page. The base-e logarithm is referred to as the **natural logarithm** and denoted either $\log_e x$ or $\ln x$. Natural logarithms are used in mathematics, science, economics, and technology. A graph of $y = \ln x$ is shown in Figure 5.65.

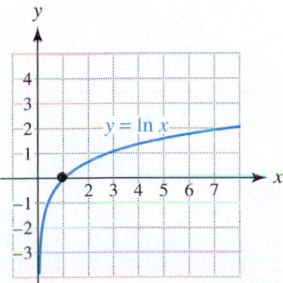

FIGURE 5.65

◆ **CLASS DISCUSSION**
Make a numerical representation for a base-4 logarithm. Evaluate $\log_4 16$. ◆

TABLE 5.25   Natural Logarithm

x	$e^{-3.1}$	e^{-2}	$e^{-0.5}$	e^0	$e^{0.5}$	e^2	$e^{3.1}$
$\ln x$	-3.1	-2	-0.5	0	0.5	2	3.1

A base-a logarithm is now defined. Its domain is $(0, \infty)$ and its range is $(-\infty, \infty)$.

LOGARITHM

The **logarithm with base a of a positive number x**, denoted by $\log_a x$, is defined by

$$\log_a x = k \quad \text{if and only if} \quad x = a^k,$$

where $a > 0$, $a \neq 1$, and k is a real number. The function, given by

$$f(x) = \log_a x,$$

is called the **logarithmic function with base a**.

EXAMPLE 9 Finding the domain of a logarithmic function

Write the domain of f in interval notation.
(a) $f(x) = \log_2 (x - 4)$ **(b)** $f(x) = \ln (10^x)$

SOLUTION
(a) *The input to a logarithmic function must be positive.* Thus any element of the domain of f must satisfy $x - 4 > 0$, or equivalently, $x > 4$. Thus $D = (4, \infty)$.
(b) The expression 10^x is positive for all real numbers x. (See Figure 5.58 where the graph of $y = 10^x$ is above the x-axis for all values of x.) Thus $D = (-\infty, \infty)$.

Now Try Exercises 91 and 95 ◆

Remember that *a logarithm is an exponent.* The expression $\log_a x$ is the exponent k such that $a^k = x$. Logarithms with base a satisfy inverse properties similar to those for common logarithms.

INVERSE PROPERTIES

The following inverse properties hold for logarithms with base a.

$$\log_a a^x = x \quad \text{for any real number } x, \text{ and}$$
$$a^{\log_a x} = x \quad \text{for any positive number } x.$$

EXAMPLE 10 Applying inverse properties

Use inverse properties to evaluate each expression.
(a) $\log_6 6^{-1.3}$ **(b)** $5^{\log_5 (x+8)}$ **(c)** $\log_{1/2} \left(\frac{1}{2}\right)^{45}$

SOLUTION
(a) $\log_a a^x = x$, so $\log_6 6^{-1.3} = -1.3$.
(b) $a^{\log_a x} = x$, so $5^{\log_5 (x+8)} = x + 8$, provided $x > -8$.
(c) $\log_a a^x = x$, so $\log_{1/2}\left(\frac{1}{2}\right)^{45} = 45$. Note that the base of logarithmic function can be a positive fraction less than 1.

Now Try Exercises 37, 39, and 44 ◆

The inverse function of $f(x) = a^x$ is $f^{-1}(x) = \log_a x$. Therefore the graph of $y = \log_a x$ is a reflection of the graph of $y = a^x$ across the line $y = x$. Figure 5.66 shows graphs of $f(x) = 2^x$ and $f^{-1}(x) = \log_2 x$. The graph of f^{-1} is a reflection of f in the line $y = x$. The rapid growth in f is called *exponential growth*. On the other hand, the graph of f^{-1} begins to level off. This slower rate of growth is called **logarithmic growth**. Since the range of f is all positive numbers, the domain of f^{-1} is all positive numbers. Notice that when $0 < x < 1$, $f^{-1}(x) = \log_2 x$ outputs negative numbers, and when $x > 1$, $f^{-1}(x)$ outputs positive numbers.

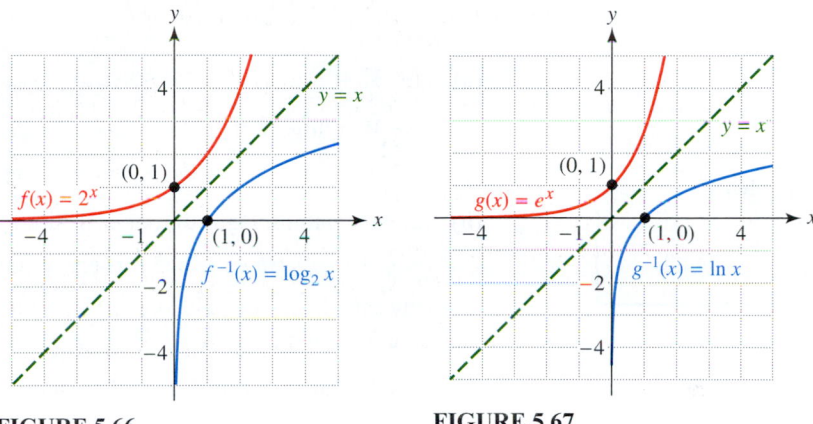

FIGURE 5.66   **FIGURE 5.67**

Graphs of $g(x) = e^x$ and $g^{-1}(x) = \ln x$ are shown in Figure 5.67. Since $e \approx 2.7183 > 2$, the graph of g increases faster than $f(x) = 2^x$, and the graph of g^{-1} levels off faster than the graph of $f^{-1}(x) = \log_2 x$.

◆ **MAKING CONNECTIONS**

Exponential and Logarithmic Functions   The inverse of an exponential function is a logarithmic function, and the inverse of a logarithmic function is an exponential function. For example,

$$\text{if } f(x) = 10^x \quad \text{then} \quad f^{-1}(x) = \log x,$$
$$\text{if } g(x) = \ln x \quad \text{then} \quad g^{-1}(x) = e^x, \quad \text{and}$$
$$\text{if } h(x) = 2^x \quad \text{then} \quad h^{-1}(x) = \log_2 x.$$

EXAMPLE 11   Evaluating logarithms

Evaluate each logarithm.

(a) $\log_2 8$ **(b)** $\log_5 \dfrac{1}{25}$ **(c)** $\log_7 49$ **(d)** $\ln e^{-7}$

SOLUTION

(a) To determine $\log_2 8$, express 8 as 2^k for some k. Since $8 = 2^3$, $\log_2 8 = \log_2 2^3 = \mathbf{3}$.

(b) $\log_5 \dfrac{1}{25} = \log_5 \dfrac{1}{5^2} = \log_5 5^{-2} = \mathbf{-2}$ **(c)** $\log_7 49 = \log_7 7^2 = 2$

(d) $\ln e^{-7} = \log_e e^{-7} = -7$ *Now Try Exercises 19, 23, and 25* ◆

General Logarithmic Equations

To solve the equation $a^x = k$, take the base-a logarithm of each side.

EXAMPLE 12 Solving equations of the form $a^x = k$

Solve each equation.

(a) $3^x = \dfrac{1}{27}$ **(b)** $e^x = 5$

SOLUTION

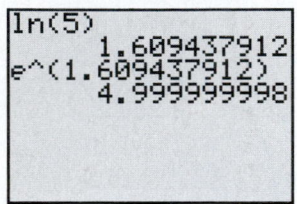

FIGURE 5.68

(a) $\log_3 3^x = \log_3 \dfrac{1}{27}$ *Take the base-3 logarithm of each side.*

$\log_3 3^x = \log_3 3^{-3}$ *Properties of exponents*

$x = -3$ *Inverse property:* $\log_a a^k = k$

(b) Take the natural logarithm of each side. Then,

$$\ln e^x = \ln 5 \quad \text{is equivalent to} \quad x = \ln 5 \approx 1.6094.$$

Many calculators are able to compute natural logarithms. The evaluation of $\ln 5$ is shown in Figure 5.68. Notice that $e^{1.609437912} \approx 5$.   *Now Try Exercises 49 and 53* ◆

To solve the equation $\log_a x = k$, exponentiate each side of the equation by using base a. This is illustrated in the next example.

EXAMPLE 13 Solving equations of the form $\log_a x = k$

Solve each equation.
(a) $\log_2 x = 5$ **(b)** $\log_5 x = -2$ **(c)** $\ln x = 4.3$

SOLUTION

(a) $\log_2 x = 5$ *Given equation*

$2^{\log_2 x} = 2^5$ *Exponentiate each side; base 2.*

$x = 2^5$ *Inverse property:* $a^{\log_a k} = k$

$x = 32$ *Simplify.*

(b) In a similar manner, $\log_5 x = -2$ is equivalent to $x = 5^{-2} = \dfrac{1}{25}$.
(c) $\ln x = 4.3$ is equivalent to $x = e^{4.3} \approx 73.7$.   *Now Try Exercises 71 and 73* ◆

EXAMPLE 14 Solving exponential and logarithmic equations

Solve each equation symbolically. Support your results graphically.
(a) $5e^x - 8 = 37$ **(b)** $5 \ln 2x + 3 = 10$

SOLUTION

(a) *Symbolic Solution*   Begin by adding 8 to each side.

$5e^x = 45$ *Add 8 to each side.*

$e^x = 9$ *Divide each side by 5.*

$\ln e^x = \ln 9$ *Take the natural logarithm.*

$x = \ln 9$ *Inverse property:* $\log_a a^k = k$

$x \approx 2.197$ *Approximate.*

[0, 5, 1] by [0, 50, 10]

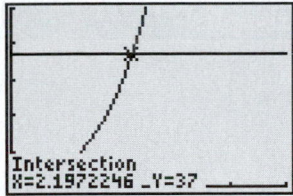

FIGURE 5.69

[−5, 5, 1] by [−20, 20, 5]

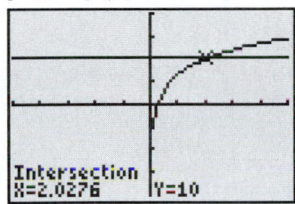

FIGURE 5.70

Graphical Solution   The graphs of $Y_1 = 5*e^{\wedge}(X) - 8$ and $Y_2 = 37$ intersect near the point (2.197, 37), as shown in Figure 5.69.

(b) *Symbolic Solution*   Start by subtracting 3 from each side.

$$5 \ln 2x + 3 = 10 \qquad \text{Given equation}$$

$$5 \ln 2x = 7 \qquad \text{Subtract 3.}$$

$$\ln 2x = \frac{7}{5} \qquad \text{Divide by 5.}$$

$$e^{\ln 2x} = e^{7/5} \qquad \text{Exponentiate each side; base } e.$$

$$2x = e^{7/5} \qquad \text{Inverse property: } a^{\log_a k} = k$$

$$x = \frac{e^{7/5}}{2} \qquad \text{Divide by 2.}$$

$$x \approx 2.028 \qquad \text{Approximate.}$$

Graphical Solution   The graphs of $Y_1 = 5 \ln(2X) + 3$ and $Y_2 = 10$ in Figure 5.70 intersect near the point (2.028, 10). This result agrees with the symbolic solution.

Now Try Exercises 65 and 85 ◆

5.4

Putting it all Together

The following table summarizes some important concepts about base-a logarithms. Common and natural logarithms satisfy the same properties.

Concept	Explanation	Examples
Base-a logarithm	The base-a logarithm of a positive number x is $$\log_a x = k$$ if and only if $$x = a^k.$$ That is, a logarithm is an exponent k.	$\log 100 = \log 10^2 = 2$ $\log_2 8 = \log_2 2^3 = 3$ $\log_3 \sqrt[3]{3} = \log_3 3^{1/3} = \frac{1}{3}$ $\ln 5 \approx 1.609$ (using a calculator)
Graph of $y = \log_a x$	The graph of the base-a logarithm *always* passes through the point (1, 0) because $\log_a 1 = \log_a a^0 = 0$. The y-axis is a vertical asymptote.	*y* graph: $y = \log_a x, a > 1$, passing through $(1, 0)$

Concept	Explanation	Examples
Inverse properties	a^x and $\log_a x$ calculate inverse operations. That is, $$\log_a a^x = x$$ and $$a^{\log_a x} = x \text{ for } x > 0.$$	$\log 10^{-3} = -3$ $\log_4 4^6 = 6$ $10^{\log x} = x, x > 0$ $4^{\log_4 2x} = 2x, x > 0$ $e^{\ln 5} = 5$
Inverse functions	The inverse function of $f(x) = a^x$ is $f^{-1}(x) = \log_a x$	$f(x) = 10^x \qquad f^{-1}(x) = \log x$ $g(x) = e^x \qquad g^{-1}(x) = \ln x$ $h(x) = \log_2 x \qquad h^{-1}(x) = 2^x$
Exponential equations	To solve $a^x = k$, take the base-a logarithm of each side.	$10^x = 15 \qquad\qquad e^x = 20$ $\log 10^x = \log 15 \qquad \ln e^x = \ln 20$ $x = \log 15 \qquad\qquad x = \ln 20$
Logarithmic equations	To solve $\log_a x = k$, exponentiate each side; base a.	$\log x = 3 \qquad\qquad \ln x = 5$ $10^{\log x} = 10^3 \qquad e^{\ln x} = e^5$ $x = 1000 \qquad\qquad x = e^5$

5.4 Exercises

Common Logarithms

Exercises 1 and 2: Complete the table for the common logarithm.

1.

x	10^0	10^4	10^8	$10^{1.2}$
$\log x$		4		

2.

x	10^{-20}	$10^{-\pi}$	10^{55}	$10^{7.89}$
$\log x$			55	

Exercises 3–8: Evaluate the following expressions by hand, if possible.

3. (a) $\log(-3)$ (b) $\log \frac{1}{100}$

    (c) $\log \sqrt{0.1}$ (d) $\log 5^0$

4. (a) $\log 10{,}000$ (b) $\log(-\pi)$

    (c) $\log \sqrt{0.001}$ (d) $\log 8^0$

5. (a) $\log 10$ (b) $\log 10{,}000$

    (c) $20 \log 0.1$ (d) $\log 10 + \log 0.001$

6. (a) $\log 100$ (b) $\log 1{,}000{,}000$

    (c) $5 \log 0.01$ (d) $\log 0.1 - \log 1000$

7. (a) $2 \log 0.1 + 4$ (b) $\log 10^{1/2}$

    (c) $3 \log 100 - \log 1000$ (d) $\log(-10)$

8. (a) $\log(-4)$ (b) $\log 1$

    (c) $\log 0$ (d) $-6 \log 100$

Exercises 9 and 10: Determine mentally an integer n so that the logarithm is between n and $n + 1$. Check your result with a calculator.

9. (a) $\log 79$ (b) $\log 500$

    (c) $\log 5$ (d) $\log 0.5$

10. (a) $\log 63$ **(b)** $\log 5000$

(c) $\log 9$ **(d)** $\log 0.04$

Exercises 11–16: Solve each equation.

11. (a) $10^x = 1000$ **(b)** $10^x = 1{,}000{,}000$

(c) $10^x = 0.01$ **(d)** $10^x = 0.0001$

12. (a) $10^x = 100$ **(b)** $10^x = \sqrt{10}$

(c) $10^x = \frac{1}{100}$ **(d)** $10^x = 10^6$

13. (a) $10^x = 1$ **(b)** $10^x = 50$

(c) $10^x = \frac{1}{10{,}000}$ **(d)** $10^x = -3$

14. (a) $10^x = 7^0$ **(b)** $10^x = 25$

(c) $10^x = \frac{1}{100{,}000}$ **(d)** $10^x = -7$

15. (a) $10^x = 300$ **(b)** $10^x = 5$ **(c)** $10^x = 0.2$

16. (a) $10^x = 250$ **(b)** $10^x = 4$ **(c)** $10^x = 0.5$

Exercises 17 and 18: Find the exact value of each expression.

17. (a) $\log \sqrt{1000}$ **(b)** $\log \sqrt[3]{10}$

(c) $\log \sqrt{10} + \log \sqrt[5]{0.1}$

(d) $\log \sqrt{0.01} - \log \sqrt{1}$

18. (a) $\log \sqrt{100{,}000}$ **(b)** $\log \sqrt[3]{100}$

(c) $2 \log \sqrt{0.1}$ **(d)** $10 \log \sqrt[5]{10}$

General Logarithms

Exercises 19–44: Simplify the expression.

19. $\log_2 64$ **20.** $\log_2 \frac{1}{4}$

21. $\log_4 2$ **22.** $\log_3 9$

23. $\ln 1$ **24.** $\ln e$

25. $\ln \sqrt[3]{e}$ **26.** $\log_7 49$

27. $\log_{1/2}\left(\frac{1}{4}\right)$ **28.** $\log_{1/3}\left(\frac{1}{27}\right)$

29. $\log_{1/6} 36$ **30.** $\log_{1/4} 64$

31. $\log_a \dfrac{1}{a}$ **32.** $\log_a (a^2 \cdot a^3)$

33. $\log_5 5^0$ **34.** $\ln \sqrt{e}$

35. $\log_2 \frac{1}{16}$ **36.** $\log_8 8^k$

37. $2^{\log_2 k}$ **38.** $\ln e^4 + \ln e^{-4}$

39. $\log_5 5^\pi$ **40.** $\log_6 6^9$

41. $3^{\log_3(x-1)}$ **42.** $8^{\log_8(\pi+1)}$

43. $7^{\log_7 (x^2+2)}$ **44.** $\log_{1/3}\left(\frac{1}{3}\right)^{-6} + \log_4 4^2$

Exercises 45–48: Complete the table without a calculator.

45. $f(x) = \log_2 x$

x	$\frac{1}{16}$	1	8	32
$f(x)$				

46. $f(x) = \log_4 x$

x	$\frac{1}{4}$	4	16	64
$f(x)$				

47. $f(x) = 2 \log_2 (x - 5)$ **48.** $f(x) = 2 \log_3 (2x)$

x	6	7	21
$f(x)$			

x	$\frac{1}{18}$	$\frac{3}{2}$	$\frac{9}{2}$
$f(x)$			

Solving Equations

Exercises 49–68: Solve the equation.

49. $2^x = 72$ **50.** $3^x = 101$

51. $5^x = 0.25$ **52.** $4^x = 11$

53. $e^x = 25$ **54.** $e^x = 0.23$

55. $e^{-x} = 3$ **56.** $e^{-x} = \frac{1}{2}$

57. $10^x - 5 = 95$ **58.** $2 \cdot 10^x = 66$

59. $10^{3x} = 100$ **60.** $4 \cdot 10^{2x} + 1 = 21$

61. $e^x + 1 = 24$ **62.** $1 - 2e^x = -5$

63. $2^x + 1 = 15$ **64.** $3 \cdot 5^x = 125$

65. $5e^x + 2 = 20$ **66.** $6 - 2e^{3x} = -10$

67. $8 - 3(2)^{0.5x} = -40$ **68.** $2(3)^{-2x} + 5 = 167$

Exercises 69–88: Solve the equation. Approximate answers to four decimal places when appropriate.

69. $\log x = 2.3$ **70.** $\log_2 x = -3$

71. $\log_2 x = 1.2$ **72.** $\log_4 x = 3.7$

73. $\ln x = -2$ **74.** $\ln x = 2$

75. $2 \log x = 6$

76. $\log 4x = 2$

77. $2 \log 5x = 4$

78. $6 - \log x = 3$

79. $4 \ln x = 3$

80. $\ln 5x = 8$

81. $5 \ln x - 1 = 6$

82. $2 \ln 3x = 8$

83. $4 \log_2 x = 16$

84. $\log_3 5x = 10$

85. $5 \ln (2x) + 6 = 12$

86. $16 - 4 \ln 3x = 2$

87. $9 - 3 \log_4 2x = 3$

88. $7 \log_6 (4x) + 5 = -2$

Exercises 89 and 90: Find values for a and b so that f(x) models the data exactly.

89. $f(x) = a + b \log x$

x	1	10	100	1000
y	5	7	9	11

90. $f(x) = a + b \log_2 x$

x	1	2	4	8
y	3.1	6	8.9	11.8

Domains of Logarithmic Functions

Exercises 91–98: Find the domain of f and write it in interval notation.

91. $f(x) = \log (x + 3)$

92. $f(x) = \ln (2x - 4)$

93. $f(x) = \log_2 (x^2 - 1)$

94. $f(x) = \log_4 (4 - x^2)$

95. $f(x) = \log_3 (3^x)$

96. $f(x) = \log_5 (5^x - 25)$

97. $f(x) = \ln (\sqrt{3 - x} - 1)$

98. $f(x) = \log(4 - \sqrt{2 - x})$

Graphs of Logarithmic Equations

Exercises 99 and 100: Use the graph of f to sketch a graph of f^{-1}. Give a symbolic representation of f^{-1}.

99. $f(x) = e^x$

100. $f(x) = \log_4 x$

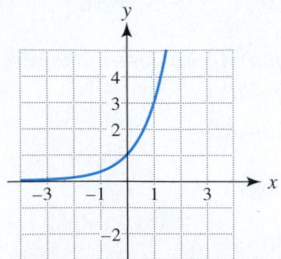

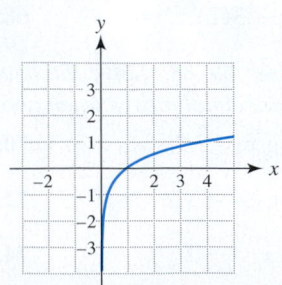

Exercises 101–104: Match f with its graph (a.–d.).

101. $f(x) = \log_2 x$

102. $f(x) = \log x$

103. $f(x) = \log (x + 2)$

104. $f(x) = 2 + \log_2 x$

a.

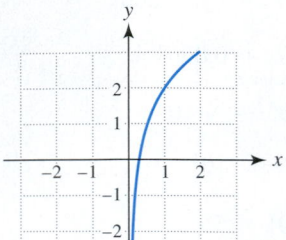

b.

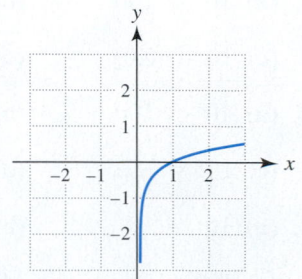

c.

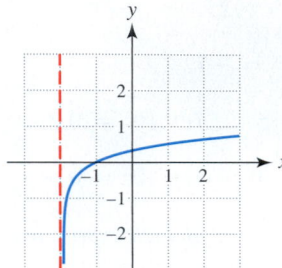

d.

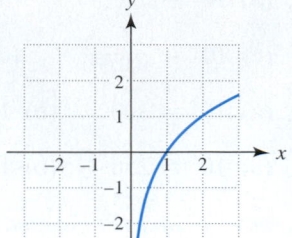

Exercises 105–108: Graph f and state its domain.

105. $f(x) = \log (x + 1)$

106. $f(x) = \log (x - 3)$

107. $f(x) = \ln (-x)$

108. $f(x) = \ln (x^2 + 1)$

Exercises 109 and 110: Graph $y = f(x)$. Is f increasing or decreasing on its domain?

109. $f(x) = \log_{1/2} x$

110. $f(x) = \log_{1/3} x$

Applications

111. *Decibels* (Refer to Example 2.) Use the formula $D(x) = 10 \log (10^{16}x)$ to determine the decibels when the intensity of a sound is $x = 10^{-11/2}$ watts per square centimeter.

112. *Decibels* (Refer to Example 2.) If the decibels of a sound increase by 15, by what factor did the intensity of the sound increase?

113. *Diversity of Birds* (Refer to Example 7.) The accompanying table lists the number of species of birds on islands of various sizes. Find values for a and b so that $f(x) = a + b \log x$ models these data. Then use f to estimate the size of an island that might have 16 species of birds.

Area (km²)	0.1	1	10	100	1000
Species of Birds	3	7	11	15	19

114. *Diversity of Insects*   The accompanying table lists the number of types of insects found in wooded regions with various acreages. Find values for a and b so that $f(x) = a + b \log x$ models these data. Then use f to estimate an acreage that might have 1200 types of insects.

Area (acres)	10	100	1000	10,000	100,000
Types of Insects	500	800	1100	1400	1700

115. *Growth of Bacteria*   The accompanying table lists the growth of a colony of bacteria in millions after an elapsed time of x days.

x (days)	0	1	2	3	4
y (millions of bacteria)	3	6	12	24	48

   (a) Find values for C and a so that $f(x) = Ca^x$ models the data.

   (b) Estimate when there were 16 million bacteria in the colony.

116. *Growth of an Investment*   The growth of an investment is shown in the accompanying table.

x (years)	0	5	10	15	20
y (dollars)	100	300	900	2700	8100

   (a) Find values for C and a so that $f(x) = Ca^x$ models the data.

   (b) Estimate when the account contained $2000.

117. *Runway Length*   There is a mathematical relation between an airplane's weight x and the runway length L required at takeoff. For some airplanes the minimum runway length L in thousands of feet may be modeled by $L(x) = 3 \log x$, where x is measured in thousands of pounds. (**Source:** L. Haefner, *Introduction to Transportation Systems*.)

(a) Graph L for $0 < x \le 50$. Interpret the graph.

(b) If the weight of an airplane increases tenfold from 10,000 to 100,000 pounds, does the length of the required runway also increase by a factor of 10? Explain.

(c) Generalize your answer from part (b).

118. *Runway Length*   (Refer to the previous exercise.) Estimate the maximum weight of a plane that can take off from a runway that is 5 thousand feet long.

119. *Acid Rain*   (Refer to Example 8.) Find the hydrogen ion concentration for the following pH levels of acid rain.   (**Source:** G. Howells.)

   (a) 4.92 (pH of rain at Amsterdam Islands in the Indian Ocean)

   (b) 3.9 (pH of rain at some locations in the eastern United States)

120. *Growth in Salary*   Suppose that a person's salary is initially $30,000 and is modeled by $f(x)$, where x represents the number of years of experience. Use $f(x)$ to approximate the years of experience when the salary exceeds $60,000.
   (a) $f(x) = 30,000(1.1)^x$

   (b) $f(x) = 30,000 \log (10 + x)$

Would most people prefer that their salaries increase exponentially or logarithmically?

121. *Earthquakes*   The Richter scale is used to measure the intensity of earthquakes. Intensity corresponds to the amount of energy released by an earthquake. If an earthquake has an intensity of x then its *magnitude*, as computed by the Richter scale, is given by the formula $R(x) = \log \frac{x}{I_0}$, where I_0 is the intensity of a small, measurable earthquake.

(a) On July 26, 1963, an earthquake in Yugoslavia had a magnitude of 6.0 on the Richter scale, and on

August 19, 1977, an earthquake in Indonesia measured 8.0. Find the intensity x for each of these earthquakes if $I_0 = 1$.

(b) How many times more intense was the Indonesian earthquake than the Yugoslavian earthquake?

122. *Modeling Algae Growth*   When sewage was accidentally dumped into Lake Tahoe, the concentration of the algae *Selenastrum* increased from 1000 cells per milliliter to approximately 1,000,000 cells per milliliter within 6 days. Let f model the algae concentration after x days have elapsed.   (**Source:** A. Payne, "Responses of the three test algae of the algal assay procedure: bottle test.")

(a) Compute the average rate of change in f from 0 to 6 days.

(b) The specific growth rate r is defined by the formula $r = \dfrac{\log N_2 - \log N_1}{x_2 - x_1}$, where N_1 is the algae concentration at time x_1 and N_2 is the algae concentration at time x_2. Compute r in this example.

(c) Discuss why environmental scientists might use the specific growth rate r, rather than the average rate of change, to describe algae growth.

123. *Hurricanes*   Hurricanes are some of the largest storms on earth. They are very low pressure areas with diameters of over 500 miles. The barometric air pressure in inches of mercury at a distance of x miles from the eye of a severe hurricane is modeled by the formula $f(x) = 0.48 \ln(x + 1) + 27$.   (**Source:** A. Miller and R. Anthes, *Meteorology.*)

(a) Evaluate $f(0)$ and $f(100)$. Interpret the results.

(b) Graph f in $[0, 250, 50]$ by $[25, 30, 1]$. Describe how air pressure changes as one moves away from the eye of the hurricane.

(c) At what distance from the eye of the hurricane is the air pressure 28 inches of mercury?

124. *Predicting Wind Speed*   Wind speed typically varies in the first 20 meters above the ground. Close to the ground, wind speed is often less than it is at 20 meters above the ground. For this reason, the National Weather Service usually measures wind speeds at heights between 5 and 10 meters. For a particular day, let the formula $f(x) = 1.2 \ln x + 2.3$ compute the wind speed in meters per second at a height x meters above the ground for $x \geq 1$.   (**Source:** A. Miller.)

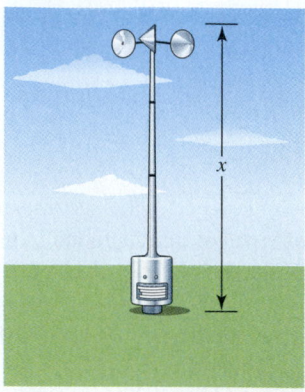

(a) Find the wind speed at a height of 5 meters.

(b) Graph f in the window $[0, 20, 5]$ by $[0, 7, 1]$. Interpret the graph.

(c) Estimate the height where the wind speed is 5 meters per second.

125. *Cooling an Object*   A pot of boiling water with a temperature of 100°C is set in a room with a temperature of 20°C. The temperature T of the water after x hours is given by $T(x) = 20 + 80e^{-x}$.

(a) Estimate the temperature of the water after 1 hour.

(b) How long did it take the water to cool to 60°C?

126. *Warming an Object*   A can of soda with a temperature of 5°C is set in a room with a temperature of 20°C. The temperature T of the soda after x minutes is given by $T(x) = 20 - 15(10)^{-0.05x}$.

(a) Estimate the temperature of the soda after 5 minutes.

(b) After how many minutes was the temperature of the soda 15°C?

127. *Modeling Traffic Flow*   (Refer to Example 10, Section 5.3.) At an intersection cars arrive randomly with an average rate of 20 cars per hour. The likelihood or probability that no car enters the intersection within a period of x minutes can be estimated by $f(x) = e^{-x/3}$.

(a) What is the probability that no car enters the intersection during a 5-minute period?

(b) Determine the value of x that gives a 30% chance that no car enters the intersection during an interval of x minutes.

128. *Population Growth*   The population of Tennessee in millions is given by $P(x) = 4.88e^{0.0133x}$, where $x = 0$ corresponds to 1990.   (**Source:** Bureau of the Census.)

(a) Determine symbolically the year when the population of Tennessee was 5.4 million.

(b) Solve part (a) graphically.

Writing about Mathematics

129. Describe the relationship between an exponential function and a logarithmic function with the same base. Explain why logarithms are needed to solve exponential equations.

130. Give verbal, numerical, graphical, and symbolic representations of a base-5 logarithmic function.

EXTENDED AND DISCOVERY EXERCISES

1. *Average Rate of Change of* ln *x*   Find the average rate of change of $f(x) = \ln x$ from x to $x + 0.001$ for each value of x. Round your answers to two decimal places.
   (a) $x = 1$ **(b)** $x = 2$

   (c) $x = 3$ **(d)** $x = 4$

2. *Average Rate of Change of* ln *x*   (Refer to Exercise 1.) Compare each average rate of change of ln x to x. What is the pattern? Make a generalization.

3. *Greenhouse Effect*   According to one model, the future increases in average global temperatures (due to carbon dioxide levels exceeding 280 parts per million) can be estimated using $T = 6.5 \ln (C/280)$, where C is the concentration of atmospheric carbon dioxide in parts per million (ppm) and T is in degrees Fahrenheit. Let future amounts of carbon dioxide be modeled by the formula $C(x) = 364(1.005)^x$, where $x = 0$ corresponds to the year 2000 and $x = 100$ to 2100, and so on.   (**Source:** W. Clime, *The Economics of Global Warming*.)

   (a) Use composition of functions to write T as a function of x. Evaluate T when $x = 100$ and interpret the result.

   (b) Graph $C(x)$ in [0, 200, 50] by [0, 1000, 100] and $T(x)$ in [0, 200, 50] by [0, 10, 1]. Describe each graph.

   (c) How does an exponential growth in carbon dioxide concentrations affect the rise in global temperature?

CHECKING BASIC CONCEPTS FOR SECTIONS 5.3 AND 5.4

1. If the principal is $1200 and the interest rate is 9.5% compounded monthly, calculate the account balance after 4 years. Determine the balance if the interest is compounded continuously.

2. Find values for C and a so that $f(x) = Ca^x$ models the data in the table.

x	0	1	2	3
y	4	2	1	0.5

3. Explain verbally what $\log_2 15$ represents. Is it equal to an integer? (Do not use a calculator.)

4. Evaluate each of the following logarithms by hand.

   (a) $\log_6 36$ **(b)** $\log \sqrt{10} + \log 0.01$ **(c)** $\ln \dfrac{1}{e^2}$

5. Solve each equation.
   (a) $e^x = 5$ **(b)** $10^x = 25$ **(c)** $\log x = 1.5$

6. Solve each equation.
   (a) $2e^x + 1 = 25$ **(b)** $\log 2x = 2.3$

   (c) $\log x^2 = 1$

7. *Population*   In July 1994 the population of New York state in millions was modeled by $f(x) = 18.2e^{0.001x}$, and the population of Florida in millions was modeled by $g(x) = 14e^{0.0168x}$. In both formulas x is the year, where $x = 0$ corresponds to July 1994.   (**Source:** Bureau of the Census.)

   (a) Find the population of New York and Florida in July 1994.

   (b) Assuming these trends continue, estimate the year when the population of Florida will equal the population of New York. What will their populations be at this time?

5.5 Properties of Logarithms

- Apply basic properties of logarithms
- Use the change of base formula

Introduction

The discovery of logarithms by John Napier (1550–1617) played an important role in the history of science. Logarithms were instrumental for Johannes Kepler (1571–1630) to calculate the positions of the planet Mars, which led to his discovery of the laws of planetary motion. Kepler's laws were used by Isaac Newton (1642–1727) to discover the universal laws of gravitation. Although calculators and computers have made tables of logarithms obsolete, applications involving logarithms still play an important role in modern-day computation. One reason for this is that logarithms possess several important properties. For example, the loudness of a sound can be measured in decibels by the formula $f(x) = 10 \log (10^{16}x)$, where x is the intensity of the sound in watts per square centimeter. In Example 4, we use properties of logarithms to simplify this formula.

Basic Properties of Logarithms

Logarithms possess several important properties. One property of logarithms states that the sum of the logarithms of two numbers equals the logarithm of their product. For example, we see in Figure 5.71 that

$$\log 5 + \log 2 = \log 10 \qquad 5 \cdot 2 = 10$$

and in Figure 5.72 that

$$\log 4 + \log 25 = \log 100. \qquad 4 \cdot 25 = 100$$

These calculations illustrate a basic property of logarithms: $\log_a m + \log_a n = \log_a (mn)$.
Four properties of logarithms are as follows.

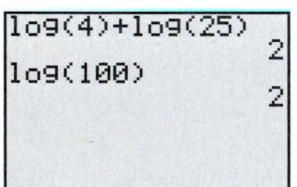

FIGURE 5.71

FIGURE 5.72

PROPERTIES OF LOGARITHMS

For positive numbers m, n, and $a \neq 1$, and any real number r:

1. $\log_a 1 = 0$ and $\log_a a = 1$
2. $\log_a m + \log_a n = \log_a (mn)$
3. $\log_a m - \log_a n = \log_a \left(\dfrac{m}{n} \right)$
4. $\log_a (m^r) = r \log_a m$

Algebra Review

To review properties of exponents, see Section 4.7, and Chapter R (page R-13).

The properties of logarithms are a direct result of the properties of exponents and the inverse property $\log_a a^k = k$, as shown in the following.

Property 1:   This property is a direct result of the inverse property: $\log_a a^x = x$.

$$\log_a 1 = \log_a a^0 = 0 \quad \text{and} \quad \log_a a = \log_a a^1 = 1$$

Examples: $\log 1 = 0$ and $\ln e = 1$.

Property 2:   If m and n are positive numbers, then we can write $m = a^c$ and $n = a^d$ for some real numbers c and d.

$$\log_a m + \log_a n = \log_a a^c + \log_a a^d = c + d$$

$$\log_a (mn) = \log_a (a^c a^d) = \log_a (a^{c+d}) = c + d$$

Thus $\log_a m + \log_a n = \log_a (mn)$.

Example: Let $m = 100$ and $n = 1000$.

$$\log m + \log n = \log 100 + \log 1000 = \log 10^2 + \log 10^3 = 2 + 3 = 5$$

$$\log (mn) = \log (100 \cdot 1000) = \log 100{,}000 = \log 10^5 = 5$$

Property 3:   Let $m = a^c$ and $n = a^d$ for some real numbers c and d.

$$\log_a m - \log_a n = \log_a a^c - \log_a a^d = c - d$$

$$\log_a \left(\frac{m}{n} \right) = \log_a \left(\frac{a^c}{a^d} \right) = \log_a (a^{c-d}) = c - d$$

Thus $\log_a m - \log_a n = \log_a \left(\frac{m}{n} \right)$.

Example: Let $m = 100$ and $n = 1000$.

$$\log m - \log n = \log 100 - \log 1000 = \log 10^2 - \log 10^3 = 2 - 3 = -1$$

$$\log \left(\frac{m}{n} \right) = \log \left(\frac{100}{1000} \right) = \log \left(\frac{1}{10} \right) = \log (10^{-1}) = -1$$

Property 4:   Let $m = a^c$ and r be any real number.

$$\log_a m^r = \log_a (a^c)^r = \log_a (a^{cr}) = cr$$

$$r \log_a m = r \log_a a^c = rc$$

Thus $\log_a (m^r) = r \log_a m$.

Example: Let $m = 100$ and $r = 3$.

$$\log m^r = \log 100^3 = \log 1{,}000{,}000 = \log 10^6 = 6$$

$$r \log m = 3 \log 100 = 3 \log 10^2 = 3 \cdot 2 = 6$$

Caution: $\log_a (m + n) \neq \log_a m + \log_a n;$ $\log_a (m - n) \neq \log_a m - \log_a n$

EXAMPLE 1

Recognizing properties of logarithms

Use a calculator to evaluate each pair of expressions. Then state which property of logarithms this calculation illustrates.

(a) $\ln 5 + \ln 4$, $\ln 20$ **(b)** $\log 10 - \log 5$, $\log 2$ **(c)** $\log 5^2$, $2 \log 5$

```
ln(5)+ln(4)
        2.995732274
ln(20)
        2.995732274
```

FIGURE 5.73

SOLUTION

(a) From Figure 5.73, we see that the two expressions are equal. These calculations illustrate Property 2 because $\ln 5 + \ln 4 = \ln (5 \cdot 4) = \ln 20$.

(b) The two expressions are equal in Figure 5.74 on the next page, and these calculations illustrate Property 3 because

$$\log 10 - \log 5 = \log \frac{10}{5} = \log 2.$$

(c) The two expressions are equal in Figure 5.75, and these calculations illustrate Property 4 because

$$\log 5^{\mathbf{2}} = \mathbf{2} \log 5.$$

FIGURE 5.74   FIGURE 5.75

Now Try Exercises 1, 3, and 5 ◆

EXAMPLE 2   **Expanding logarithmic expressions**

Use properties of logarithms to expand each expression. Write your answers without exponents.

(a) $\log xy$ **(b)** $\ln \dfrac{6}{z}$ **(c)** $\log_4 \dfrac{\sqrt[3]{x}}{\sqrt{k}}$

SOLUTION

(a) By Property 2, $\log xy = \log x + \log y$.
(b) By Property 3, $\ln \frac{6}{z} = \ln 6 - \ln z$.
(c) Begin by using Property 3.

$$\log_4 \frac{\sqrt[3]{x}}{\sqrt{k}} = \log_4 \sqrt[3]{x} - \log_4 \sqrt{k} \qquad \text{Property 3}$$

$$= \log_4 x^{1/3} - \log_4 k^{1/2} \qquad \text{Properties of exponents}$$

$$= \frac{1}{3} \log_4 x - \frac{1}{2} \log_4 k \qquad \text{Property 4}$$

Now Try Exercises 7, 11, and 25 ◆

EXAMPLE 3   **Applying properties of logarithms**

Expand each expression. Write your answers without exponents.

(a) $\log 2x^4$ **(b)** $\ln \dfrac{7x^3}{k}$ **(c)** $\log \dfrac{\sqrt{x+1}}{(x-2)^3}$

SOLUTION

(a)
$$\log 2x^4 = \log 2 + \log x^4 \qquad \text{Property 2}$$
$$= \log 2 + 4 \log x \qquad \text{Property 4}$$

(b)
$$\ln \frac{7x^3}{k} = \ln 7x^3 - \ln k \qquad \text{Property 3}$$
$$= \ln 7 + \ln x^3 - \ln k \qquad \text{Property 2}$$
$$= \ln 7 + 3 \ln x - \ln k \qquad \text{Property 4}$$

(c)
$$\log \frac{\sqrt{x+1}}{(x-2)^3} = \log \sqrt{x+1} - \log (x-2)^3 \qquad \text{Property 3}$$
$$= \log (x+1)^{1/2} - \log (x-2)^3 \qquad \text{Properties of exponents}$$
$$= \frac{1}{2} \log (x+1) - 3 \log (x-2) \qquad \text{Property 4}$$

Now Try Exercises 9, 15, and 27 ◆

Sometimes properties of logarithms are used in applications to simplify a formula. This is illustrated in the next example.

EXAMPLE 4   ### Analyzing sound with decibels

Sound levels in decibels (db) can be computed by $D(x) = 10 \log (10^{16}x)$.
(a) Use properties of logarithms to simplify the formula for D.
(b) Ordinary conversation has an intensity of $x = 10^{-10}$ w/cm^2. Find the decibel level.

SOLUTION
(a) To simplify the formula use Property 2.

$$
\begin{aligned}
D(x) &= 10 \log (10^{16}x) \\
&= 10(\log 10^{16} + \log x) && \text{Property 2} \\
&= 10(16 + \log x) && \text{Evaluate the first logarithm.} \\
&= 160 + 10 \log x && \text{Distributive property}
\end{aligned}
$$

(b) $D(10^{-10}) = 160 + 10 \log (10^{-10}) = 160 + 10(-10) = 160 - 100 = 60$
Ordinary conversation occurs at about 60 decibels.   *Now Try Exercise 81* ◆

The next two examples demonstrate how properties of logarithms can be used to combine logarithmic expressions.

EXAMPLE 5   ### Applying properties of logarithms

Write each expression as the logarithm of a single expression.

(a) $\ln 2e + \ln \dfrac{1}{e}$ **(b)** $\log_2 27 + \log_2 x^3$ **(c)** $\log x^3 - \log x^2$

SOLUTION
(a) By Property 2, $\ln 2e + \ln \frac{1}{e} = \ln \left(2e \cdot \frac{1}{e}\right) = \ln 2$.
(b) By Property 2, $\log_2 27 + \log_2 x^3 = \log_2 (27x^3)$.
(c) By Property 3, $\log x^3 - \log x^2 = \log \frac{x^3}{x^2} = \log x$.   *Now Try Exercises 39, 41, and 45* ◆

EXAMPLE 6   ### Collecting terms in logarithmic expressions

Write each expression as the logarithm of a single expression.

(a) $\log 5 + \log 15 - \log 3$ **(b)** $2 \ln x - \frac{1}{2} \ln y - 3 \ln z$
(c) $5 \log_3 x + \log_3 2x - \log_3 y$

Algebra Review
To review rational exponents and radical notation, see Chapter R (page R-50).

SOLUTION
(a)
$$
\begin{aligned}
\log 5 + \log 15 - \log 3 &= \log (5 \cdot 15) - \log 3 && \text{Property 2} \\
&= \log \left(\frac{5 \cdot 15}{3}\right) && \text{Property 3} \\
&= \log 25 && \text{Simplify.}
\end{aligned}
$$

(b) $2 \ln x - \dfrac{1}{2} \ln y - 3 \ln z = \ln x^2 - \ln y^{1/2} - \ln z^3$ *Property 4*

$$= \ln \left(\dfrac{x^2}{y^{1/2}} \right) - \ln z^3 \qquad \textit{Property 3}$$

$$= \ln \dfrac{x^2}{y^{1/2} z^3} \qquad \textit{Property 3}$$

$$= \ln \dfrac{x^2}{z^3 \sqrt{y}} \qquad \textit{Properties of exponents}$$

(c) $5 \log_3 x + \log_3 2x - \log_3 y = \log_3 x^5 + \log_3 2x - \log_3 y$ *Property 4*

$$= \log_3 (x^5 \cdot 2x) - \log_3 y \qquad \textit{Property 2}$$

$$= \log_3 \dfrac{2x^6}{y} \qquad \textit{Property 3}$$

Now Try Exercises 37, 47, and 51

Change of Base Formula

Occasionally it is necessary to evaluate a logarithm with a base other than 10 or e. This computation can be accomplished by using a change of base formula.

CHANGE OF BASE FORMULA

Let x, $a \neq 1$, and $b \neq 1$ be positive real numbers. Then,

$$\log_a x = \dfrac{\log_b x}{\log_b a}.$$

The change of base formula can be derived as follows.

$$y = \log_a x$$

$$a^y = a^{\log_a x} \qquad \textit{Exponentiate each side; base a.}$$

$$a^y = x \qquad \textit{Inverse property}$$

$$\log_b a^y = \log_b x \qquad \textit{Take base-b logarithm of each side.}$$

$$y \log_b a = \log_b x \qquad \textit{Property 4}$$

$$y = \dfrac{\log_b x}{\log_b a} \qquad \textit{Divide by } \log_b a.$$

$$\log_a x = \dfrac{\log_b x}{\log_b a} \qquad \textit{Substitute } \log_a x \textit{ for y.}$$

To calculate $\log_2 5$, evaluate $\dfrac{\log 5}{\log 2} \approx 2.322$. The change of base formula was used with $x = 5$, $a = 2$, and $b = 10$. We could have also evaluated $\dfrac{\ln 5}{\ln 2} \approx 2.322$.

 **Applying the change of base formula**

Use a calculator to approximate each expression to the nearest thousandth.
(a) $\log_4 20$ **(b)** $\log_2 125 + \log_7 39$

SOLUTION

(a) Using the change of base formula, we have $\log_4 20 = \dfrac{\log 20}{\log 4} \approx 2.161$. We could also evaluate $\dfrac{\ln 20}{\ln 4}$ to obtain the same result, as shown in Figure 5.76.

(b) $\log_2 125 + \log_7 39 = \dfrac{\log 125}{\log 2} + \dfrac{\log 39}{\log 7} \approx 8.848$. See Figure 5.77.

FIGURE 5.76   **FIGURE 5.77**

Now Try Exercises 65 and 69 ◆

The change of base formula can be used to graph logarithmic functions with bases other than 10 or *e*.

EXAMPLE 8   Using the change of base formula

Estimate graphically the solution to the equation $\log_2 (x^3 + x - 1) = 5$.

SOLUTION

Graph $Y_1 = \log(X\^3 + X - 1)/\log(2)$ and $Y_2 = 5$. See Figure 5.78. Their graphs intersect near the point (3.104, 5). The solution is given by $x \approx 3.104$.

Now Try Exercise 75 ◆

[−10, 10, 1] by [−10, 10, 1]

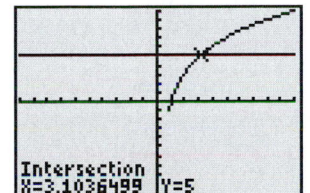

FIGURE 5.78

5.5

Putting it all Together

The following table summarizes some properties of logarithms.

Concept	Explanation	Examples
Properties of logarithms	1. $\log_a 1 = 0$ and $\log_a a = 1$ 2. $\log_a m + \log_a n = \log_a (mn)$ 3. $\log_a m - \log_a n = \log_a \left(\dfrac{m}{n}\right)$ 4. $\log_a (m^r) = r \log_a m$	$\ln 1 = 0$ and $\log_2 2 = 1$ $\log 3 + \log 6 = \log(3 \cdot 6) = \log 18$ $\log_3 8 - \log_3 2 = \log_3 \dfrac{8}{2} = \log_3 4$ $\log 6^7 = 7 \log 6$
Change of base formula	Let x, $a \neq 1$, and $b \neq 1$ be positive real numbers. Then, $$\log_a x = \dfrac{\log_b x}{\log_b a}.$$	$\log_3 6 = \dfrac{\log 6}{\log 3} = \dfrac{\ln 6}{\ln 3} \approx 1.631$
Graphing logarithmic functions	Use the change of base formula to graph $y = \log_a x$, whenever $a \neq 10$ or $a \neq e$.	To graph $y = \log_2 x$, let $Y_1 = \log(X)/\log(2)$ or $Y_1 = \ln(X)/\ln(2)$.

5.5 — Exercises

Properties of Logarithms

Exercises 1–6: (Refer to Example 1.) Use a calculator to evaluate each pair of expressions. Then state which property of logarithms this calculation illustrates.

1. $\log 4 + \log 7$, $\log 28$ **2.** $\ln 12 + \ln 5$, $\ln 60$

3. $\ln 72 - \ln 8$, $\ln 9$ **4.** $3 \log 4$, $\log 4^3$

5. $10 \log 2$, $\log 1024$

6. $\log_2 100 - \log_2 20$, $\log_2 5$

Exercises 7–30: (Refer to Examples 2 and 3.) Expand the expression. If possible, write your answer without exponents.

7. $\log_2 ab$ **8.** $\ln 3x$

9. $\ln 7a^4$ **10.** $\log \dfrac{a^3}{3}$

11. $\log \dfrac{6}{z}$ **12.** $\ln \dfrac{xy}{z}$

13. $\log \dfrac{x^2}{3}$ **14.** $\log 3x^6$

15. $\ln \dfrac{2x^7}{3k}$ **16.** $\ln \dfrac{kx^3}{5}$

17. $\log_2 4k^2x^3$ **18.** $\log \dfrac{5kx^2}{11}$

19. $\log_5 \dfrac{25x^3}{y^4}$ **20.** $\log_2 \dfrac{32}{xy^2}$

21. $\log_4 (0.25(x + 2)^3)$ **22.** $\log (0.001(a - b)^{-3})$

23. $\log_5 \dfrac{x^3}{(x - 4)^4}$ **24.** $\log_8 \dfrac{(3x - 2)^2}{x^2 + 1}$

25. $\log_2 \dfrac{\sqrt{x}}{z^2}$ **26.** $\log \sqrt{\dfrac{xy^2}{z}}$

27. $\ln \sqrt[3]{\dfrac{2x + 6}{(x + 1)^5}}$ **28.** $\log \dfrac{\sqrt{x^2 + 4}}{\sqrt[3]{x - 1}}$

29. $\log_2 \dfrac{\sqrt[3]{x^2 - 1}}{\sqrt{1 + x^2}}$ **30.** $\log_8 \sqrt[3]{\dfrac{x + y^2}{2z + 1}}$

Exercises 31–52: (Refer to Examples 5 and 6.) Write the expression as a logarithm of a single expression.

31. $\log 2 + \log 3$ **32.** $\log \sqrt{2} + \log \sqrt[3]{2}$

33. $\ln \sqrt{5} - \ln 25$ **34.** $\ln 33 - \ln 11$

35. $\log 20 + \log \frac{1}{10}$ **36.** $\log_2 24 + \log_2 \frac{1}{48}$

37. $\log 4 + \log 3 - \log 2$

38. $\log_3 5 - \log_3 10 - \log_3 \frac{1}{2}$

39. $\log_7 5 + \log_7 k^2$ **40.** $\log_6 45 + \log_6 b^3$

41. $\ln x^6 - \ln x^3$ **42.** $\log 10x^5 - \log 5x$

43. $\log \sqrt{x} + \log x^2 - \log x$

44. $\log \sqrt[4]{x} + \log x^4 - \log x^2$

45. $\ln \dfrac{1}{e^2} + \ln 2e$ **46.** $\ln 4e^3 - \ln 2e^2$

47. $2 \ln x - 4 \ln y + \frac{1}{2} \ln z$

48. $\frac{1}{3} \log_5 (x + 1) + \frac{1}{3} \log_5 (x - 1)$

49. $\log 4 - \log x + 7 \log \sqrt{x}$

50. $\ln 3e - \ln \dfrac{1}{4e}$

51. $2 \log (x^2 - 1) + 4 \log (x - 2) - \frac{1}{2} \log y$

52. $\log_3 x + \log_3 \sqrt{x + 3} - \frac{1}{3} \log_3 (x - 4)$

Exercises 53–60: Complete the following.

   (a) *Make a table of $f(x)$ and $g(x)$ starting at $x = 1$, incrementing by 1. Determine whether $f(x) = g(x)$.*
   (b) *If possible, use properties of logarithms to show that $f(x) = g(x)$.*

53. $f(x) = \log 3x + \log 2x$, $g(x) = \log 6x^2$

54. $f(x) = \log 2 + \log x$, $g(x) = \log 2x$

55. $f(x) = \ln 3x - \ln 2x$, $g(x) = \ln x$

56. $f(x) = \ln 2x^2 - \ln x$, $g(x) = \ln 2x$

57. $f(x) = \log x^4$, $g(x) = 4 \log x$

58. $f(x) = \log x^2 + \log x^3$, $g(x) = 5 \log x$

59. $f(x) = \ln x^4 - \ln x^2$, $g(x) = 2 \ln x$

60. $f(x) = (\ln x)^2$, $g(x) = 2 \ln x$

Exercises 61–64: Sketch a graph of f.

61. $f(x) = \log_2 x$ **62.** $f(x) = \log_2 x^2$

63. $f(x) = \log_3 |x|$ **64.** $f(x) = \log_4 2x$

Change of Base Formula

Exercises 65–74: Use the change of base formula to approximate the logarithm to the nearest thousandth.

65. $\log_2 25$ **66.** $\log_3 67$

67. $\log_5 130$ **68.** $\log_6 0.77$

69. $\log_2 5 + \log_2 7$ **70.** $\log_9 85 + \log_7 17$

71. $\sqrt{\log_4 46}$ **72.** $2 \log_5 15 + \sqrt[3]{\log_3 67}$

73. $\dfrac{\log_2 12}{\log_2 3}$ **74.** $\dfrac{\log_7 125}{\log_7 25}$

Exercises 75–78: Solve the equation graphically. Express any solutions to the nearest thousandth.

75. $\log_2 (x^3 + x^2 + 1) = 7$

76. $\log_3 (1 + x^2 + 2x^4) = 4$

77. $\log_2 (x^2 + 1) = 5 - \log_3 (x^4 + 1)$

78. $\ln (x^2 + 2) = \log_2 (10 - x^2)$

Applications

79. *Runway Length*   (Refer to Exercise 117, Section 5.4.) Use a natural logarithm (instead of a common logarithm) to write the formula $L(x) = 3 \log x$. Evaluate $L(50)$ for each formula. Do your answers agree?

80. *Allometry*   The equation $y = bx^a$ is used in applications involving allometry. Another form of this equation is $\log y = \log b + a \log x$. Use properties of logarithms to obtain this second equation from the first. (**Source:** H. Lancaster, *Quantitative Methods in Biological and Medical Sciences.*)

81. *Decibels*   If the intensity x of a sound increases by a factor of 10, then the decibel level increases by how much?

82. *Decibels*   (Refer to Example 4.) Use a natural logarithm to write the formula

$$f(x) = 160 + 10 \log x.$$

Evaluate $f(5 \times 10^{-8})$ for each formula. Do your answers agree?

83. *Light Absorption*   When sunlight passes through lake water, its initial intensity I_0 decreases to a weaker intensity I at a depth of x feet according to the formula

$$\ln I - \ln I_0 = -kx,$$

where k is a positive constant. Solve this equation for I.

84. *Dissolving Salt*   If C grams of salt are added to a sample of water, the amount A of undissolved salt is modeled by $A = Ca^x$, where x is time. Solve the equation for x.

85. Solve $A = Pe^{rt}$ for t.

86. Solve $P = P_0 e^{r(t - t_0)} + 5$ for t.

87. Write the sum

$$\log 1 + 2 \log 2 + 3 \log 3 + 4 \log 4 + 5 \log 5$$

as a logarithm of a single expression.

88. Show that

$$\log_2 (x + \sqrt{x^2 - 4}) + \log_2 (x - \sqrt{x^2 - 4}) = 2$$

is an identity. What is the domain of the expression on the left side of the equation? Use interval notation.

Writing about Mathematics

89. A student insists that

$$\log (x + y) \quad \text{and} \quad \log x + \log y$$

are equal. How could you convince the student otherwise?

90. A student insists that

$$\log \left(\dfrac{x}{y} \right) \quad \text{and} \quad \dfrac{\log x}{\log y}$$

are equal. How could you convince the student otherwise?

5.6 Exponential and Logarithmic Equations

◆ Solve exponential equations
◆ Solve logarithmic equations

Introduction

The population of the world has grown rapidly during the past century. As a result, heavy demands have been made on the world's resources. Exponential functions and equations are often used to model this rapid growth, whereas logarithms are used to model slower growth.

Exponential Equations

The population P of the world was 3 billion in 1960, 6 billion in 1999, and can be modeled by $P(x) = 3(1.018)^{x-1960}$, where x is the year. We can use P to estimate the year when world population reached **5** billion by solving the exponential equation

$$3(1.018)^{x-1960} = 5.$$

An equation where one or more variables occur in the exponent of an expression is called an **exponential equation**. In the next example we use Property 4 of logarithms, given by $\log_a (m)^r = r \log_a m$, to solve this equation.

 **EXAMPLE 1**

Modeling world population

World population in billions during year x can be modeled by $P(x) = 3(1.018)^{x-1960}$ and is shown in Figure 5.79. Solve the equation $3(1.018)^{x-1960} = 5$ symbolically to estimate the year when world population reached 5 billion.

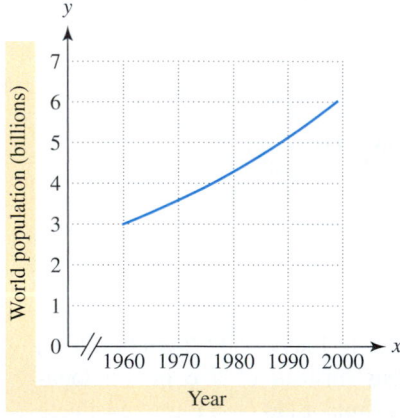

FIGURE 5.79

SOLUTION First, divide each side by 3, and then take the common logarithm of each side. (The natural logarithm could also be used.)

$$3(1.018)^{x-1960} = 5 \qquad \text{Given equation}$$

$$(1.018)^{x-1960} = \frac{5}{3} \qquad \text{Divide by 3.}$$

$$\log (1.018)^{x-1960} = \log \frac{5}{3} \qquad \text{Take the common logarithm.}$$

$$(x - 1960) \log (1.018) = \log \frac{5}{3} \qquad \text{Property 4: } \log (m)^r = r \log m$$

$$x - 1960 = \frac{\log (5/3)}{\log (1.018)} \qquad \text{Divide by log (1.018).}$$

$$x = 1960 + \frac{\log (5/3)}{\log (1.018)} \qquad \text{Add 1960.}$$

$$x \approx 1988.6 \qquad \text{Approximate.}$$

This model predicts world population reached 5 billion during 1988.

Now Try Exercise 67 ◆

Exponential equations occur in a variety of applications and can be solved symbolically, graphically, and numerically. In the next example these techniques are used to solve an exponential equation that describes the minimum thickness of the pavement for an airport runway.

EXAMPLE 2 Calculating the thickness of a runway

Heavier aircraft require runways with thicker pavement for landings and takeoffs. The relation between the thickness of the pavement t in inches and gross weight W in thousands of pounds can be approximated by

$$W(t) = 18.29e^{0.246t}.$$

(a) Determine the required thickness of the runway for a 130,000-pound plane.
(b) Solve part (a) graphically and numerically.

SOLUTION
(a) *Symbolic Solution* Because the unit for W is thousands of pounds, we solve the equation $W(t) = 130$.

$$18.29e^{0.246t} = 130 \qquad W(t) = 130.$$

$$e^{0.246t} = \frac{130}{18.29} \qquad \text{Divide by 18.29.}$$

$$\ln e^{0.246t} = \ln \frac{130}{18.29} \qquad \text{Take the natural logarithm of each side.}$$

$$0.246t = \ln \frac{130}{18.29} \qquad \text{Inverse property: } \ln e^k = k.$$

$$t = \frac{\ln(130/18.29)}{0.246} \qquad \text{Divide by 0.246.}$$

$$t \approx 7.97 \qquad \text{Approximate.}$$

The runway should be about 8 inches thick.

(b) *Graphical Solution* Let $Y_1 = 18.29e^{\wedge}(.246X)$ and $Y_2 = 130$. In Figure 5.80 their graphs intersect near (7.97, 130).

Numerical Solution Numerical support is shown in Figure 5.81, where $y_1 \approx y_2$ when $x = 8$.

Now Try Exercise 69 ◆

To solve an exponential equation with a base other than 10 or e, Property 4 of logarithms can be used. This technique was demonstrated in Example 1 and is also used in the next example.

Calculator Help

To find a point of intersection, see Appendix B (page AP-12).

[0, 10, 1] by [0, 200, 50]

Intersection
X=7.9722764 Y=130

FIGURE 5.80

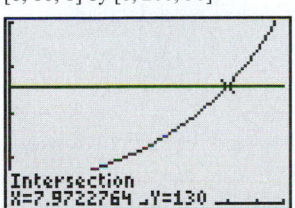

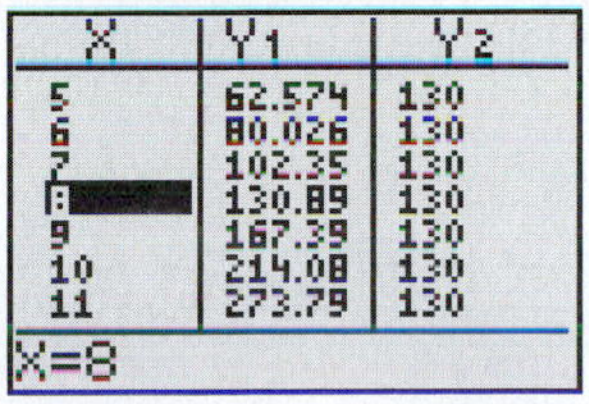

X	Y₁	Y₂
5	62.574	130
6	80.026	130
7	102.35	130
8	130.89	130
9	167.39	130
10	214.08	130
11	273.79	130

X=8

FIGURE 5.81

 **Modeling the decline of bluefin tuna**

Bluefin tuna are large fish that can weigh 1500 pounds and swim at speeds of 55 miles per hour. Because they are used for sushi, a prime fish can be worth over $30,000. As a result, the western Atlantic bluefin tuna have been exploited, and their numbers have declined exponentially. Their numbers in thousands from 1974 to 1991 can be modeled by the formula $f(x) = 230(0.881)^x$, where x is the year and $x = 0$ corresponds to 1974. (**Source:** B. Freedman, *Environmental Ecology*.)

(a) Estimate the number of bluefin tuna in 1974 and 1991.

(b) Determine symbolically the year when their numbers were 50 thousand.

SOLUTION

(a) To determine their numbers in 1974 and 1991, evaluate $f(0)$ and $f(17)$.

$$f(\mathbf{0}) = 230(0.881)^{\mathbf{0}} = 230(1) = 230$$

$$f(\mathbf{17}) = 230(0.881)^{\mathbf{17}} \approx 26.7$$

Bluefin tuna decreased from 230 thousand in 1974 to fewer than 27 thousand in 1991.

(b) Solve the equation $f(x) = 50$.

$$230(0.881)^x = 50 \qquad \textcolor{blue}{f(x) = 50}$$

$$0.881^x = \frac{5}{23} \qquad \textcolor{blue}{\text{Divide by 230.}}$$

$$\textbf{ln}\, 0.881^x = \textbf{ln}\, \frac{5}{23} \qquad \textcolor{blue}{\begin{array}{l}\text{Take the natural logarithm of}\\ \text{each side. (The common}\\ \text{logarithm could also be used.)}\end{array}}$$

$$x\, \ln 0.881 = \ln \frac{5}{23} \qquad \textcolor{blue}{\text{Property 4: } \ln m^r = r \ln m}$$

$$x = \frac{\ln(5/23)}{\ln 0.881} \qquad \textcolor{blue}{\text{Divide by ln 0.881.}}$$

$$x \approx 12.04 \qquad \textcolor{blue}{\text{Approximate } x.}$$

The population of the bluefin tuna was about 50 thousand in $1974 + 12.04 \approx 1986$.

Now Try Exercise 71 ◆

In real applications, a modeling function is seldom provided with the data. However, it is not uncommon to be given the general form of a function that describes a data set.

 **Finding a modeling function**

The gap between available organs for transplants and people who need them has widened. In 1999 about 60,000 people were waiting for organ transplants and in 2002 this number increased to 80,000. This trend can be modeled by $T(x) = Ca^{x-1999}$, where C and a are constants. (**Source:** Scientific Registry of Transplant Recipients.)

(a) Approximate C and a if the unit for T is thousands of people.

(b) Use $T(x)$ to estimate the number of people that could be waiting for an organ transplant in 2006.

SOLUTION

(a) In 1999 there were 60,000 people waiting, so $T(1999) = 60$. This equation can be used to find C.

$$T(1999) = Ca^{1999-1999} = Ca^0 = C(1) = 60.$$

Thus $C = 60$ and $T(x) = 60a^{x-1999}$. In 2002 the number waiting was 80,000, so we can solve the equation $T(2002) = 80$ to find the value of a.

$$60a^{2002-1999} = 80 \qquad \textcolor{blue}{T(2002) = 80}$$

$$a^3 = \frac{4}{3} \qquad \textcolor{blue}{\text{Divide by 60; simplify.}}$$

$$a = \sqrt[3]{\frac{4}{3}} \qquad \textcolor{blue}{\text{Take the cube root of each side.}}$$

$$a \approx 1.10 \qquad \textcolor{blue}{\text{Approximate if desired.}}$$

Thus T is given by $T(x) = 60(1.1)^{x-1999}$.

(b) In 2006 there may be

$$T(2006) = 60(1.1)^{2006-1999} \approx 117$$

thousand people waiting for an organ transplant. *Now Try Exercise 73* ◆

Exponential equations can occur in many forms. Although some types of exponential equations cannot be solved symbolically, Example 5 shows four equations that can.

EXAMPLE 5 Solving exponential equations symbolically

Solve each equation.

(a) $10^{x+2} = 10^{3x}$ **(b)** $5(1.2)^x + 1 = 26$ **(c)** $\left(\frac{1}{4}\right)^{x-1} = \frac{1}{10}$

(d) $5^{x-3} = e^{2x}$

SOLUTION

(a) Start by taking the common logarithm of each side.

$$10^{x+2} = 10^{3x} \qquad \textcolor{blue}{\text{Given equation}}$$

$$\log 10^{x+2} = \log 10^{3x} \qquad \textcolor{blue}{\text{Take the common logarithm.}}$$

$$x + 2 = 3x \qquad \textcolor{blue}{\text{Inverse property: } \log 10^k = k}$$

$$2 = 2x \qquad \textcolor{blue}{\text{Subtract } x.}$$

$$x = 1 \qquad \textcolor{blue}{\text{Divide by 2; rewrite.}}$$

(b) Begin by subtracting 1 from each side of the equation.

$$5(1.2)^x = 25 \qquad \textcolor{blue}{\text{Subtract 1.}}$$

$$(1.2)^x = 5 \qquad \textcolor{blue}{\text{Divide by 5.}}$$

$$\log (1.2)^x = \log 5 \qquad \textcolor{blue}{\text{Take the common logarithm.}}$$

$$x \log (1.2) = \log 5 \qquad \textcolor{blue}{\text{Property 4: } \log m^r = r \log m}$$

$$x = \frac{\log 5}{\log (1.2)} \qquad \textcolor{blue}{\text{Divide by } \log (1.2).}$$

$$x \approx 8.827 \qquad \textcolor{blue}{\text{Approximate.}}$$

(c) Begin by taking the common logarithm of each side.

$$\left(\frac{1}{4}\right)^{x-1} = \frac{1}{10} \qquad \textcolor{blue}{\text{Given equation}}$$

$$\log \left(\frac{1}{4}\right)^{x-1} = \log \frac{1}{10} \qquad \textcolor{blue}{\text{Take the common logarithm.}}$$

$$(x - 1) \log \left(\frac{1}{4} \right) = \log \frac{1}{10} \qquad \textit{Property 4: } \log m^r = r \log m$$

$$(x - 1) = \frac{\log (1/10)}{\log (1/4)} \qquad \textit{Divide by } \log \frac{1}{4}.$$

$$x = 1 + \frac{\log (1/10)}{\log (1/4)} \approx 2.661 \qquad \textit{Add 1 and approximate.}$$

This also could have been solved by taking the natural logarithm of each side.

(d) Begin by taking the natural logarithm of each side. (The common logarithm could also be used.)

$$5^{x-3} = e^{2x} \qquad \textit{Given equation}$$

$$\ln 5^{x-3} = \ln e^{2x} \qquad \textit{Take natural logarithm.}$$

$$(x - 3) \ln 5 = 2x \qquad \textit{Property 4; inverse property}$$

$$x \ln 5 - 3 \ln 5 = 2x \qquad \textit{Distributive property}$$

$$x \ln 5 - 2x = 3 \ln 5 \qquad \textit{Subtract 2x; add 3 ln 5.}$$

$$x(\ln 5 - 2) = 3 \ln 5 \qquad \textit{Factor out x.}$$

$$x = \frac{3 \ln 5}{\ln 5 - 2} \qquad \textit{Divide by ln 5 −2.}$$

$$x \approx -12.36 \qquad \textit{Approximate.}$$

Now Try Exercises 7, 13, 17, and 23 ◆

If a hot object is put in a room with temperature T_0, then according to **Newton's law of cooling**, the temperature of the object after time t is modeled by

$$T(t) = T_0 + Da^t,$$

where $0 < a < 1$ and D is the initial temperature *difference* between the object and the room. This phenomenon is discussed in the next example.

EXAMPLE 6 Modeling coffee cooling

A pot of coffee with a temperature of 100°C is set down in a room with a temperature of 20°C. The coffee cools to 60°C after 1 hour.
(a) Find values for T_0, D, and a so that $T(t) = T_0 + Da^t$ models the data.
(b) Find the temperature of the coffee after half an hour.
(c) How long did it take for the coffee to reach 50°C? Support your result graphically.

SOLUTION
(a) The room has temperature $T_0 = 20°C$ and the initial temperature difference between the coffee and the room is $D = 100 - 20 = 80°C$. It follows that $T(t) = 20 + 80a^t$. We can use the fact that the temperature of the coffee after 1 hour was 60°C to determine a.

$$T(1) = 60$$

$$20 + 80a^1 = 60 \qquad \textit{Let t = 1 in T(t) = 20 + 80a}^t.$$

$$80a = 40 \qquad \textit{Subtract 20.}$$

$$a = \frac{1}{2} \qquad \textit{Divide by 80.}$$

Thus $T(t) = 20 + 80\left(\frac{1}{2}\right)^t$.

(b) After half an hour the temperature is

$$T\left(\frac{1}{2}\right) = 20 + 80\left(\frac{1}{2}\right)^{1/2} \approx 76.6°C.$$

(c) *Symbolic Solution*   To determine when the coffee reached 50°C, solve $T(t) = 50$.

$$20 + 80\left(\frac{1}{2}\right)^t = 50 \qquad\qquad \textcolor{blue}{T(t) = 50}$$

$$80\left(\frac{1}{2}\right)^t = 30 \qquad\qquad \textcolor{blue}{\text{Subtract 20.}}$$

$$\left(\frac{1}{2}\right)^t = \frac{3}{8} \qquad\qquad \textcolor{blue}{\text{Divide by 80.}}$$

$$\log\left(\frac{1}{2}\right)^t = \log\frac{3}{8} \qquad\qquad \textcolor{blue}{\text{Take the common logarithm.}}$$

$$t\,\log\left(\frac{1}{2}\right) = \log\frac{3}{8} \qquad\qquad \textcolor{blue}{\text{Property 4}}$$

$$t = \frac{\log\,(3/8)}{\log\,(1/2)} \qquad\qquad \textcolor{blue}{\text{Divide by } \log\tfrac{1}{2}.}$$

$$t \approx 1.415 \qquad\qquad \textcolor{blue}{\text{Approximate.}}$$

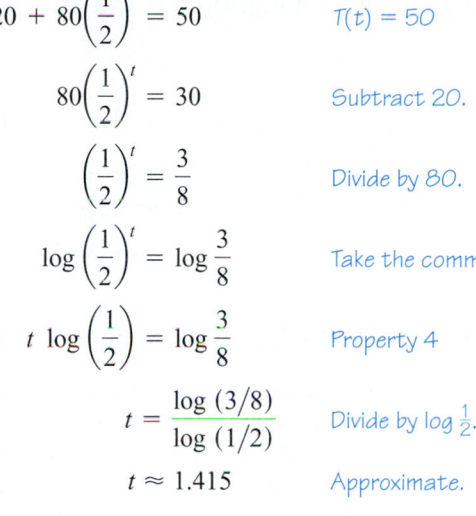

[0, 3, 1] by [0, 100, 10]

Intersection
X=1.4150375 Y=50

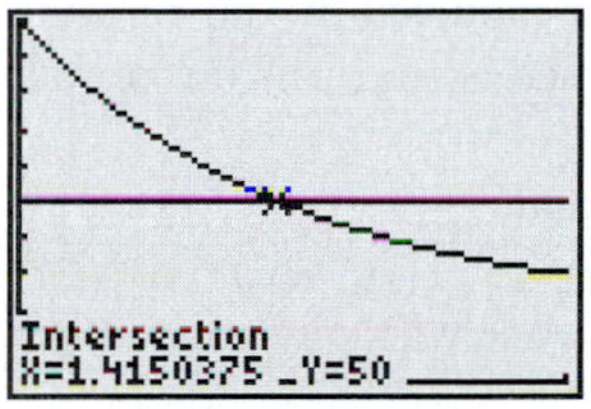

FIGURE 5.82

The temperature of the coffee reaches 50°C after about 1.415 hours or after about 1 hour and 25 minutes.

Graphical Solution   The graphs of $Y_1 = 20 + 80(1/2)^{\wedge}X$ and $Y_2 = 50$ intersect near (1.415, 50), as shown in Figure 5.82. This result agrees with the symbolic solution.

Now Try Exercise 83 ◆

Note: The formula for Newton's law of cooling can also be used to model the temperature of a cold object that is brought into a warm room. In this case, the temperature difference D is negative.

Some exponential equations cannot be solved symbolically, but they can be solved graphically. This is demonstrated in the next example.

EXAMPLE 7 Solving an exponential equation graphically

Solve $e^{-x} + 2x = 3$ graphically. Approximate all solutions to the nearest hundredth.

SOLUTION The graphs of $Y_1 = e^{\wedge}(-X) + 2X$ and $Y_2 = 3$ intersect near the points $(-1.92, 3)$ and $(1.37, 3)$, as shown in Figures 5.83 and 5.84. Thus the solutions are approximately -1.92 and 1.37.

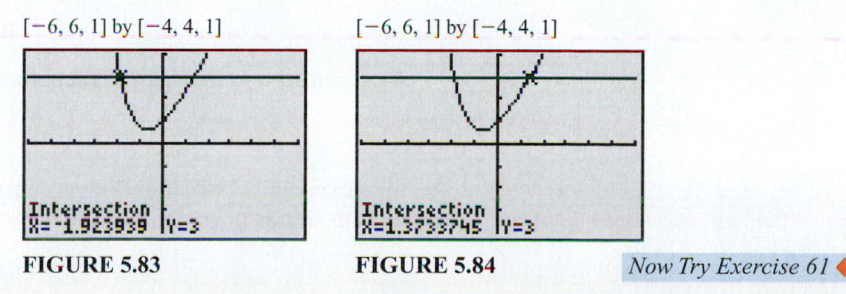

[−6, 6, 1] by [−4, 4, 1]

Intersection
X=-1.923939 Y=3

FIGURE 5.83

[−6, 6, 1] by [−4, 4, 1]

Intersection
X=1.3733745 Y=3

FIGURE 5.84

Now Try Exercise 61 ◆

Logarithmic Equations

Logarithmic equations contain logarithms. Like exponential equations, logarithmic equations also occur in applications. To solve a logarithmic equation, we use the inverse property $a^{\log_a x} = x$. This technique is illustrated in the next example.

EXAMPLE 8   Solving a logarithmic equation

Solve $3 \log_3 x = 12$.

SOLUTION   Begin by dividing each side by 3.

$$\log_3 x = 4 \qquad \text{Divide by 3.}$$

$$3^{\log_3 x} = 3^4 \qquad \text{Exponentiate each side; base 3.}$$

$$x = 81 \qquad \text{Inverse property: } a^{\log_a k} = k. \qquad \textit{Now Try Exercise 47} \blacklozenge$$

EXAMPLE 9   Solving a logarithmic equation symbolically

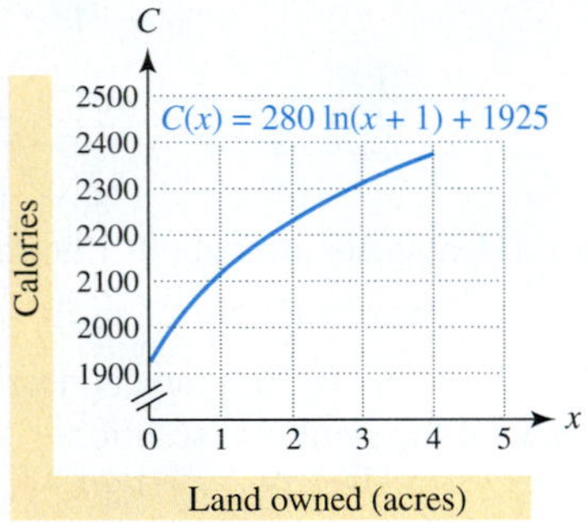

In developing countries there is a relationship between the amount of land a person owns and the average daily calories consumed. This relationship is modeled by the formula $C(x) = 280 \ln (x + 1) + 1925$, where x is the amount of land owned in acres and $0 \le x \le 4$. (**Source:** D. Grigg, *The World Food Problem.*)

(a) Find the average caloric intake for a person who owns no land.

(b) A graph of C is shown in Figure 5.85. Interpret the graph.

(c) Determine symbolically the number of acres owned by someone whose average intake is 2000 calories per day.

FIGURE 5.85

SOLUTION

(a) Since $C(0) = 280 \ln (0 + 1) + 1925 = 1925$, a person without land consumes an average of 1925 calories per day.

(b) As the amount of land x increases, the caloric intake y also increases. However, it begins to level off. This would be expected because there is a limit to the number of calories an average person would eat, regardless of his or her economic status.

(c) Solve the equation $C(x) = 2000$.

$$280 \ln (x + 1) + 1925 = 2000 \qquad C(x) = 2000$$

$$280 \ln (x + 1) = 75 \qquad \text{Subtract 1925.}$$

$$\ln (x + 1) = \frac{75}{280} \qquad \text{Divide by 280.}$$

$$e^{\ln (x+1)} = e^{75/280} \qquad \text{Exponentiate each side; base } e.$$

$$x + 1 = e^{75/280} \qquad \text{Inverse property: } e^{\ln k} = k$$

$$x = e^{75/280} - 1 \qquad \text{Subtract 1.}$$

$$x \approx 0.307 \qquad \text{Approximate.}$$

A person who owns about 0.3 acre has an average intake of 2000 calories per day.

Now Try Exercise 93 $\blacklozenge$

Like exponential equations, logarithmic equations can occur in many forms. The next example illustrates three equations that can be solved symbolically.

EXAMPLE 10 Solving logarithmic equations symbolically

Solve each equation.
(a) $\log(2x + 1) = 2$ (b) $\log_2 4x = 2 - \log_2 x$
(c) $2\ln(x + 1) = \ln(1 - 2x)$

SOLUTION
(a) To solve the equation, exponentiate each side of the equation using base 10.

$\log(2x + 1) = 2$	*Given equation*
$10^{\log(2x+1)} = 10^2$	*Exponentiate each side; base 10.*
$2x + 1 = 100$	*Inverse property: $10^{\log k} = k$*
$x = 49.5$	*Solve for x.*

(b) To solve this equation, we apply properties of logarithms.

$\log_2 4x = 2 - \log_2 x$	*Given equation*
$\log_2 4x + \log_2 x = 2$	*Add $\log_2 x$.*
$\log_2 4x^2 = 2$	*Property 2: $\log_a m + \log_a n = \log_a(mn)$*
$2^{\log_2 4x^2} = 2^2$	*Exponentiate each side; base 2.*
$4x^2 = 4$	*Inverse property: $a^{\log_a k} = k$*
$x = \pm 1$	*Solve for x.*

However, -1 is not a solution since $\log_2 x$ is undefined for negative values of x. Thus the only solution is 1.

(c) Start by applying Property 4 to the given equation.

$2\ln(x + 1) = \ln(1 - 2x)$	*Given equation*
$\ln(x + 1)^2 = \ln(1 - 2x)$	*Property 4*
$e^{\ln(x+1)^2} = e^{\ln(1-2x)}$	*Exponentiate; base e.*
$(x + 1)^2 = 1 - 2x$	*Inverse property: $a^{\log_a k} = k$*
$x^2 + 2x + 1 = 1 - 2x$	*Expand binomial.*
$x^2 + 4x = 0$	*Combine terms.*
$x(x + 4) = 0$	*Factor.*
$x = 0$ or $x = -4$	*Zero-product property*

Substituting $x = 0$ and $x = -4$ into the given equation shows that 0 is a solution, but -4 is not a solution.   *Now Try Exercises 51, 53, and 57*◆

◆ **CLASS DISCUSSION**
Compare the first step for solving each pair of equations.

i. $10^x = 5$ and $\log x = 5$
ii. $e^x = 5$ and $\ln x = 5$
iii. $4^x = 5$ and $\log_4 x = 5$ ◆

EXAMPLE 11 Modeling the life span of a robin

The life spans of 129 robins were monitored over a 4-year period in one study. The equation $y = \frac{2 - \log(100 - x)}{0.42}$ can be used to calculate the number of years y for x percent of the robin population to die. For example, to find the time when 40% of the robins had died, substitute $x = 40$ into the equation. The result is $y \approx 0.53$, or about half a year. (**Source:** D. Lack, *The Life of a Robin.*)
(a) Graph y in $[0, 100, 10]$ by $[0, 5, 1]$. Interpret the graph.
(b) Estimate graphically the percentage of the robins that died after 2 years.

[0, 100, 10] by [0, 5, 1]

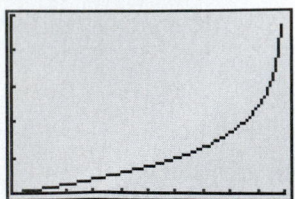

FIGURE 5.86

[0, 100, 10] by [0, 5, 1]

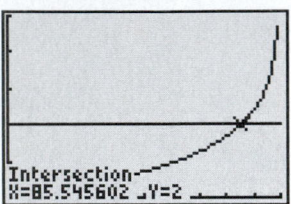

FIGURE 5.87

SOLUTION

(a) The graph of $Y_1 = (2 - \log(100 - X))/.42$ is shown in Figure 5.86. At first, the graph increases slowly between 0% and 60%. Since years are listed on the y-axis and percentage on the x-axis, this slow increase in f indicates that not many years pass by before a large percentage of the robins die. When $x \geq 60$, the graph begins to increase dramatically. This indicates that a small percentage has a comparatively long life span.

(b) To determine the percentage that died after 2 years, graph y_1 and $y_2 = 2$. Their graphs intersect near (85.5, 2), as shown in Figure 5.87. After 2 years, approximately 85.5% of the robins had died—a surprisingly high percentage.   *Now Try Exercise 91* ◆

◆ **MAKING CONNECTIONS** ─────────────

Solving Exponential and Logarithmic Equations   At some point in the process of solving an exponential equation, we often take a logarithm of each side of the equation. Similarly, when solving a logarithmic equation, we often exponentiate each side of the equation.   ◆

5.6

Putting it all Together

$\mathbf{T}$he following table summarizes techniques that can be used to solve some types of exponential and logarithmic equations symbolically.

Concept	Explanation	Example
Exponential equations	Typical form: $Ca^x = k$ Solve for a^x. Then take a base-a logarithm of each side. Use the inverse property: $$\log_a a^x = x$$	$4e^x = 24$ $e^x = 6$ $\ln e^x = \ln 6$ $x = \ln 6 \approx 1.79$
Logarithmic equations	*Equation*: $C \log_a x = k$ Solve for $\log_a x$. Then exponentiate each side with base a. Use the inverse property: $$a^{\log_a x} = x$$	$4 \log x = 10$ $\log x = 2.5$ $10^{\log x} = 10^{2.5}$ $x = 10^{2.5} \approx 316$
	Equation: $\log_a bx \pm \log_a cx = k$ When more than one logarithm with the same base occurs, use properties of logarithms to combine logarithms. Be sure to check any solutions.	$\log x + \log 4x = 2$ $\log 4x^2 = 2$ $4x^2 = 10^2$ $x^2 = 25$ $x = \pm 5$ The only solution is 5.

5.6 Exercises

Solving Exponential Equations

Exercises 1–4: The graphical and symbolic representations of f and g are shown.

(a) Use the graph to solve $f(x) = g(x)$.
(b) Solve $f(x) = g(x)$ symbolically.

1.

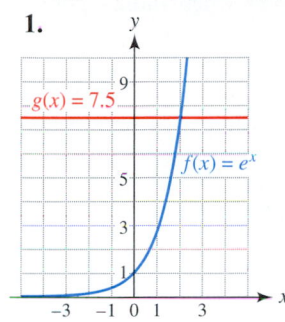

2.

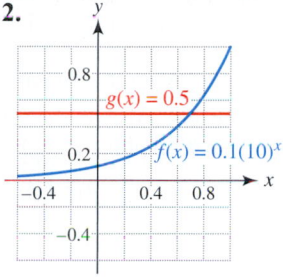

3.

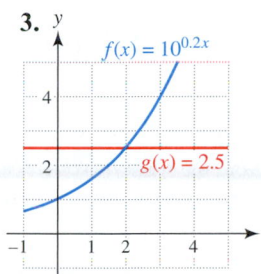

4.

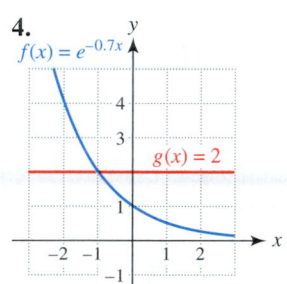

Exercises 5–32: Solve the exponential equation.

5. $4e^x = 5$

6. $2e^{-x} = 8$

7. $2(10)^x + 5 = 45$

8. $100 - 5(10)^x = 7$

9. $2.5e^{-1.2x} = 1$

10. $9.5e^{0.005x} = 19$

11. $1.2(0.9)^x = 0.6$

12. $0.05(1.15)^x = 5$

13. $4(1.1)^{x-1} = 16$

14. $3(2)^{x-2} = 99$

15. $5(1.2)^{3x-2} + 94 = 100$ **16.** $1.4(2)^{x+3} = 2.8$

17. $5^{3x} = 5^{1-2x}$

18. $7^{x^2} = 7^{4x-3}$

19. $10^{(x^2)} = 10^{3x-2}$

20. $e^{2x} = e^{5x-3}$

21. $\left(\frac{1}{5}\right)^x = -5$

22. $2^x = -4$

23. $4^{x-1} = 3^{2x}$

24. $3^{1-2x} = e^{0.5x}$

25. $e^{x-3} = 2^{3x}$

26. $6^{x+1} = 4^{2x-1}$

27. $3(1.4)^x - 4 = 60$

28. $2(1.05)^x + 3 = 10$

29. $5(1.015)^{x-1980} = 8$

30. $30 - 3(0.75)^{x-1} = 29$

31. $4\left(\frac{3}{4}\right)^{x+1} = \frac{1}{81}\left(\frac{3}{2}\right)^{5+x}$

32. $5\left(\frac{2}{5}\right)^{x+1} = \frac{1}{125}\left(\frac{4}{5}\right)^{x-3}$

Solving Logarithmic Equations

Exercises 33–60: Solve the logarithmic equation.

33. $3 \log x = 2$

34. $5 \ln x = 10$

35. $\ln 2x = 5$

36. $\ln 4x = 1.5$

37. $\log 2x^2 = 2$

38. $\log (2 - x) = 0.5$

39. $\log_2 (3x - 2) = 4$

40. $\log_3 (1 - x) = 1$

41. $\log_5 (8 - 3x) = 3$

42. $\log_6 (2x + 4) = 2$

43. $160 + 10 \log x = 50$

44. $160 + 10 \log x = 120$

45. $\ln x + \ln x^2 = 3$

46. $\log x^5 = 4 + 3 \log x$

47. $2 \log_2 x = 4.2$

48. $3 \log_2 3x = 1$

49. $\log x + \log 2x = 2$

50. $\ln 2x + \ln 3x = \ln 6$

51. $\ln (x - 1) = 1$

52. $\log x + \log (2x + 5) = \log 7$

53. $2 \ln x = \ln (2x + 1)$

54. $\log (x^2 + 3) = 2 \log (x + 1)$

55. $\log (x + 1) + \log (x - 1) = \log 3$

56. $\ln (x^2 - 4) - \ln (x + 2) = \ln (3 - x)$

57. $\log_2 2x = 4 - \log_2 (x + 2)$

58. $\log_3 x + \log_3 (x + 2) = \log_3 24$

59. $\log_5 (x + 1) + \log_5 (x - 1) = \log_5 15$

60. $\log_7 4x - \log_7 (x + 3) = \log_7 x$

Solving Equations Graphically

Exercises 61–66: The following equations cannot be solved symbolically. Solve these equations graphically and round your answers to the nearest hundredth.

61. $2x + e^x = 2$

62. $xe^x - 1 = 0$

63. $x^2 - x \ln x = 2$

64. $x \ln|x| = -2$

65. $xe^{-x} + \ln x = 1$

66. $2^{x-2} = \log x^4$

Applications

67. *Population Growth* World population P in billions during year x can be modeled by

$$P(x) = 3(1.018)^{x-1960},$$

where $1960 \le x \le 2000$. Estimate the year when world population reached 4 billion.

68. *Population of Arizona* The population P of Arizona has been increasing at an annual rate of 3.5%. In 1990 the population of Arizona was 3.7 million.
(a) Write a formula for $P(x)$, where x represents the year.

(b) Use $P(x)$ to estimate the population of Arizona in 2005.

69. *Credit Cards* From 1987 to 1996 the number of Visa cards and MasterCards was up 80% to 376 million. The function given by $f(x) = 36.2e^{0.14x}$ models the amount of credit card spending from Thanksgiving to Christmas in billions of dollars. In this formula $x = 0$ corresponds to 1987 and $x = 9$ to 1996. (**Source:** National Credit Counseling Services.)
(a) Determine symbolically the year when this amount reached $55 billion.

(b) Solve part (a) graphically or numerically.

70. *Gambling* U.S. gambling revenues in billions of dollars from 1991 to 1995 can be modeled by $f(x) = 26.6e^{0.131x}$, where x is the year and $x = 0$ corresponds to 1991. Determine symbolically the year when revenues reached $45 billion. Support your answer graphically.

71. *Bluefin Tuna* The number of Atlantic bluefin tuna in thousands can be modeled by $f(x) = 230(0.881)^x$, where x is the year and $x = 0$ corresponds to 1974. (See Example 3.) Estimate the year when the number of bluefin tuna reached 95 thousand.

72. *Organ Transplants* (Refer to Example 4.) Determine when the number of individuals waiting for organ transplants might reach 140,000.

73. *Population Growth* In 2000 the population of India reached 1 billion and in 2025 it is projected to be 1.4 billion. (**Source:** Bureau of the Census.)
(a) Find values for C and a so that $P(x) = Ca^{x-2000}$ models the population of India in year x.

(b) Estimate India's population in 2010.

(c) Use P to determine the year when India's population might reach 1.5 billion.

74. *Midair Near Collisions* The table shows the number of airliner near collisions y in year x. (**Source:** Federal Aviation Administration.)

x	1989	1991	1993	1995
y	131	78	44	34

(a) Approximate constants C and a so that the formula $f(x) = Ca^{(x-1989)}$ models the data.

(b) Support your answer by graphing f and the data.

*Exercises 75 and 76: **Continuous Compounding** Suppose that P dollars are deposited in a savings account paying 9% interest compounded continuously. After t years the account will contain $A(t) = Pe^{0.09t}$ dollars.*
(a) Solve the equation $A(t) = k$ for the given values of P and k.
(b) Interpret your results.

75. $P = 500$ and $k = 750$

76. $P = 1000$ and $k = 2000$

77. *Doubling Time* How long does it take for $1000 to double in value if the interest rate is 8.5% compounded quarterly?

78. *Doubling Time* How long does it take for $1000 to double in value if the interest rate is 12% compounded continuously?

79. *Radioactive Carbon-14* The percentage P of radioactive carbon-14 remaining in a fossil after t years is given by $P = 100\left(\frac{1}{2}\right)^{t/5700}$. Suppose a fossil contains 35% of the carbon-14 it contained when it was alive. Estimate the age of the fossil.

80. *Radioactive Radium-226*   The amount A of radium in milligrams remaining in a sample after t years is given by $A(t) = 0.02\left(\frac{1}{2}\right)^{t/1600}$. How many years will it take for the sample to decay to 0.004 milligrams?

81. *Modeling Traffic Flow*   Cars arrive randomly at an intersection with an average traffic volume of 1 car per minute. The likelihood or probability that at least one car enters the intersection during a period of x minutes can be estimated by $f(x) = 1 - e^{-x}$.
(a) What is the probability that at least one car enters the intersection during a 5-minute period?

(b) Determine the value of x that gives a 40% chance that at least one car enters the intersection during an interval of x minutes.

82. *Drug Concentrations*   The concentration of a drug in a patient's bloodstream after t hours is modeled by the formula $C(t) = 11(0.72)^t$, where C is measured in milligrams per liter.
(a) What is the initial concentration of the drug?

(b) How long does it take for the concentration to decrease to 50% of its initial level?

83. *Newton's Law of Cooling*   A pan of boiling water with a temperature of 212°F is set in a bin of ice with a temperature of 32°F. The pan cools to 70°F in 30 minutes.
(a) Find values for T_0, D, and a so that the formula $T(t) = T_0 + Da^t$ models the data where t is in hours.

(b) Find the temperature of the pan after 10 minutes.

(c) How long did it take the pan to reach 40°F? Support your result graphically.

84. *Warming an Object*   (Refer to the previous exercise.) A pan of cold water with a temperature of 35°F is brought into a room with a temperature of 75°F. After 1 hour the temperature of the pan of water is 45°F.
(a) Find values for T_0, D, and a so that the formula $T(t) = T_0 + Da^t$ models the data.

(b) Find the temperature of the water after 3 hours.

(c) How long would it take the water to reach 60°F?

85. *Warming a Soda Can*   Suppose that a can of soda, initially at 5°C, warms to 18°C after 2 hours in a room that has a temperature of 20°C.
(a) Find the temperature of the soda can after 1.5 hours.

(b) How long did it take for the soda can to warm to 15°C?

86. *Cooling a Soda Can*   A soda can at 80°F is put into a cooler containing ice at 32°F. The temperature of the soda can after t minutes is modeled by the formula $T(t) = 32 + 48(0.9)^t$.
(a) Evaluate $T(30)$ and interpret your results.

(b) How long did it take for the soda can to cool to 50°F?

87. *Modeling Bacteria Growth*   Suppose that the concentration of a bacteria sample is 100,000 bacteria per milliliter. If the concentration of bacteria doubles every 2 hours, how long will it take for the concentration to reach 350,000 bacteria per milliliter?

88. *Modeling Compound Interest*   Suppose that $2000 are deposited in an account and the balance increases to $2300 after 4 years. How long will it take for the account to grow to $3200?

89. *Modeling Radioactive Decay*   Suppose that a 0.05-gram sample of a radioactive substance decays to 0.04 gram in 20 days. How long will it take for the sample to decay to 0.025 gram?

90. *Atmospheric Pressure*   As altitude increases, air pressure decreases. The atmospheric pressure P in millibars (mb) at a given altitude x in meters is listed in the table. (**Source:** A. Miller and J. Thompson, *Elements of Meteorology.*)

Altitude (m)	0	5000	10,000	15,000
Pressure (mb)	1013	541	265	121

Altitude (m)	20,000	25,000	30,000
Pressure (mb)	55	26	12

(a) The data can be modeled by $P(x) = P_0 e^{kx}$. Approximate the constant k and P_0.

(b) Graph P and the data in the same viewing rectangle. Comment on the fit.

(c) Estimate the atmospheric pressure at an altitude of 23,000 feet.

91. *Life Span of Robins*   Solve Example 11, part (b) symbolically.

92. *Salinity*   The salinity of the oceans changes with latitude and with depth. In the tropics, the salinity increases on the surface of the ocean due to rapid evaporation. In the higher latitudes, there is less evaporation and rainfall causes the salinity to be less on the surface than at lower depths. The function given by

$$S(x) = 31.5 + 1.1 \log (x + 1)$$

models salinity to depths of 1000 meters at a latitude of 57.5°N. The input x is the depth in meters and the output $S(x)$ is in grams of salt per kilogram of seawater.   (**Source:** D. Hartman, *Global Physical Climatology.*)

(a)   Make a table for S starting at $x = 0$, incrementing by 100. Discuss any trends.

(b)   Numerically approximate the depth where the salinity equals 33.

(c)   Solve part (b) symbolically.

93. *Caloric Intake*   The formula

$$C(x) = 280 \ln (x + 1) + 1925$$

models the number of calories consumed daily by a person owning x acres of land in a developing country. (See Example 9.) Estimate the number of acres owned for the average intake to be 2300 calories per day.

94. *Modeling Data*   Use the data in the table to complete the following.

x	1	2	5	10
y	2.5	2.1	1.6	1.2

(a)   Find values for a and b so that the formula $f(x) = a + b \log x$ models the data.

(b)   Solve the equation $f(x) = 1.8$.

95. *Reducing Carbon Emissions*   When fossil fuels are burned, carbon is released into the atmosphere. Governments could reduce carbon emissions by placing a tax on fossil fuels. For example, a higher tax might lead people to use less gasoline. The **cost-benefit** equation

$$\ln(1 - P) = -0.0034 - 0.0053x$$

estimates the relationship between a tax of x dollars per ton of carbon and the percent P reduction in emissions

of carbon, where P is in decimal form. Determine P when $x = 60$. Interpret the result.   (**Source:** W. Clime, *The Economics of Global Warming.*)

96.   The cost-benefit function from the previous exercise can be represented as $P(x) = 1 - e^{-0.0034 - 0.0053x}$.

(a)   Graph P in [0, 1000, 100] by [0, 1, 0.1]. Discuss the benefit of continuing to raise taxes on fossil fuels.

(b)   According to this model, what tax would result in a 50% reduction in carbon emissions?

(c)   Verify that

$$\ln (1 - P) = -0.0034 - 0.0053x$$

and

$$P = 1 - e^{-0.0034 - 0.0053x}$$

are equivalent by solving the first equation for P.

97. *Investments*   The formula $A = P\left(1 + \frac{r}{n}\right)^{nt}$ can be used to calculate the future value of an investment. Solve the equation for t.

98. *Decibels*   The formula $D = 160 + 10 \log x$ can be used to calculate loudness of a sound. Solve the equation for x.

Writing about Mathematics

99.   Explain how to solve the equation $Ca^x = k$ symbolically, where C and k are constants. Demonstrate your method.

100.   Explain how to solve the equation $b \log_a x = k$ symbolically, where b and k are constants. Demonstrate your method.

CHECKING BASIC CONCEPTS FOR SECTIONS 5.5 AND 5.6

1. Use properties of logarithms to expand $\log \frac{x^2 y^3}{\sqrt[3]{z}}$. Write your answer without exponents.

2. Use properties of logarithms to write the expression $\frac{1}{2} \ln x - 3 \ln y + \ln z$ as a logarithm of a single expression.

3. Solve each equation.
   (a) $5(1.4)^x - 4 = 25$ (b) $4^{2-x} = 4^{2x+1}$

4. Solve each equation.
   (a) $5 \log_2 2x = 25$

   (b) $\ln (x + 1) + \ln (x - 1) = \ln 3$

5. The temperature T of a cooling object in degrees Fahrenheit after x minutes is given by
$$T = 80 + 120(0.9)^x.$$
   (a) What happens to the temperature of the object after a long time?

   (b) After how long is the object's temperature 100°F?

5.7 Constructing Nonlinear Models

- Find an exponential model
- Find a logarithmic model
- Find a logistic model

Introduction

If data change at a constant rate, then they can be modeled with a linear function. However, real-life data usually change at a nonconstant rate. For example, a tree grows slowly when it is small and then gradually grows faster as it becomes larger. Finally, when the tree is mature, its height begins to level off. This type of growth is nonlinear.

Three types of nonlinear data are shown in Figures 5.88–5.90, where t represents time. In Figure 5.88 the data increase rapidly, and an exponential function might be an appropriate modeling function. In Figure 5.89 we see data that are growing, but at a slower rate than in Figure 5.88. These data could be modeled by a logarithmic function. Finally in Figure 5.90 the data increase slowly, then faster, and finally level off. These data might represent the height of a tree over a 50-year period. To model these data, we need a new type of function called a *logistic function*.

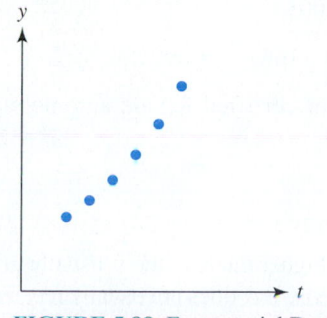

FIGURE 5.88 Exponential Data

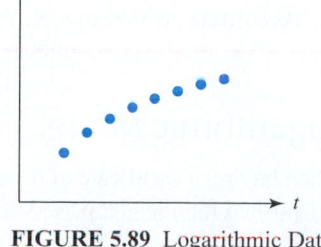

FIGURE 5.89 Logarithmic Data

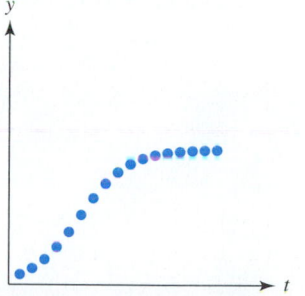

FIGURE 5.90 Logistic Data

Exponential Model

The numbers of certified female automotive technicians changed dramatically during the early 1990s. In the next example we find an exponential function that models these numbers.

 **Using exponential regression**

The National Institute for Automotive Service Excellence (ASE) reported that the number of females working in automotive repair is increasing. Table 5.26 shows the number of female ASE-certified technicians for selected years.

TABLE 5.26

Year	1988	1989	1990	1991	1992	1993	1994	1995
Total	556	614	654	737	849	1086	1329	1592

Source: National Institute for Automotive Service Excellence.

(a) What type of function might model these data?

(b) Use least-squares regression to find an exponential function given by $f(x) = ab^x$ that models the data.

(c) Use f to estimate the number of certified female technicians in 2005. Round the result to the nearest hundred.

SOLUTION

(a) Let y be the number of female technicians and x be the year, where $x = 0$ corresponds to 1988, $x = 1$ to 1989, and so on, until $x = 7$ corresponds to 1995. A scatterplot of the data is shown in Figure 5.91. The data is increasing rapidly, and an exponential function might model these data.

Calculator Help

To find an equation of least-squares fit, see Appendix B (page AP-15).

(b) The formula $f(x) = 507.1(1.166)^x$ is found by using least-squares regression, as shown in Figures 5.92 and 5.93. A scatterplot of the data, together with a graph of f, is shown in Figure 5.94.

$[-1, 8, 1]$ by $[400, 1700, 100]$

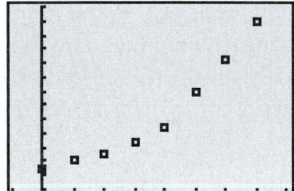

FIGURE 5.91

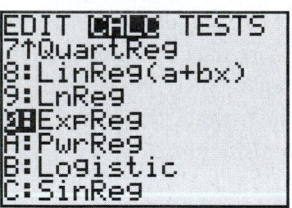

FIGURE 5.92

FIGURE 5.93

$[-1, 8, 1]$ by $[400, 1700, 100]$

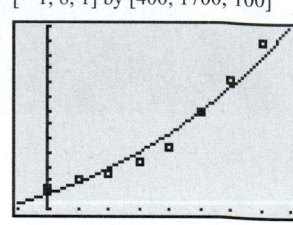

FIGURE 5.94

(c) Since $x = 17$ corresponds to the year 2005,

$$f(17) = 507.1(1.166)^{17} \approx 6902.$$

According to this model, the number of certified female automotive technicians in 2005 could be about 6900.   *Now Try Exercise 11* ◆

Logarithmic Model

When buying a certificate of deposit (CD), a higher interest rate is usually given if the money is deposited for a longer period of time. However, one does not usually receive twice the interest rate for a 2-year CD than for a 1-year CD. Instead the rate of interest gradually increases with time. In the next example we model interest rates with a logarithmic function.

 **2** Modeling interest rates

Table 5.27 lists the interest rates for certificates of deposit during January 1997. Use the data to complete the following.

TABLE 5.27   Yield on Certificates of Deposit

Time	6 mo	1 yr	2.5 yr	5 yr
Yield	4.75%	5.03%	5.25%	5.54%

Source: *USA Today.*

(a) Make a scatterplot of the data. What type of function might model these data?
(b) Use least-squares regression to obtain a formula, $f(x) = a + b \ln x$, that models these data.
(c) Graph f and the data in the same viewing rectangle.

SOLUTION

(a) Enter the data points (0.5, 4.75), (1, 5.03), (2.5, 5.25), and (5, 5.54) into your calculator. A scatterplot of the data is shown in Figure 5.95. The data increase but are gradually leveling off. A logarithmic modeling function may be appropriate.
(b) In Figures 5.96 and 5.97 least-square regression has been used to find a logarithmic function f given (approximately) by $f(x) = 5 + 0.33 \ln x$.
(c) A graph of f and the data are shown in Figure 5.98.

[0, 6, 1] by [4, 6, 1]

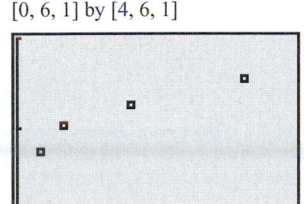

FIGURE 5.95

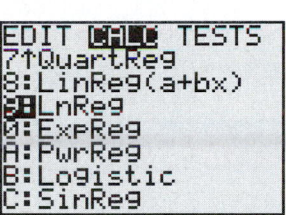

FIGURE 5.96

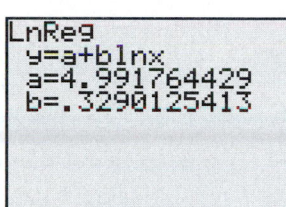

FIGURE 5.97

[0, 6, 1] by [4, 6, 1]

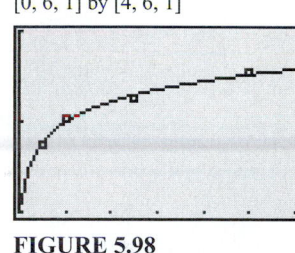

FIGURE 5.98

Now Try Exercise 10 ◆

Logistic Model

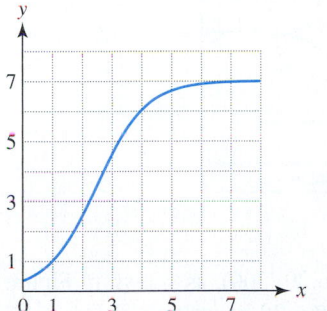

FIGURE 5.99 A Sigmoidal Curve

In real life, populations of bacteria, insects, and animals do not continue to grow indefinitely. Initially, population growth may be slow. Then, as their numbers increase, so does the rate of growth. After a region has become heavily populated or saturated, the population usually levels off because of limited resources.

This type of growth may be modeled by a **logistic function** represented by $f(x) = \dfrac{c}{1 + ae^{-bx}}$, where a, b, and c are positive constants. A typical graph of a logistic function f is shown in Figure 5.99. The graph of f is referred to as a **sigmoidal curve**. The next example demonstrates how a logistic function can be used to describe the growth of a yeast culture.

 **3** Modeling logistic growth

One of the earliest studies about population growth was done using yeast plants in 1913. A small amount of yeast was placed in a container with a fixed amount of nourishment. The units of yeast were recorded every 2 hours. The data are listed in Table 5.28 on the next page.   (**Source:** T. Carlson, *Biochem.*; D. Brown, *Models in Biology.*)

TABLE 5.28

Time	0	2	4	6	8	10	12	14	16	18
Yeast	9.6	29.0	71.1	174.6	350.7	513.3	594.8	640.8	655.9	661.8

(a) Make a scatterplot of the data in Table 5.28. Describe the growth.

(b) Use least-squares regression to find a logistic function f that models the data.

(c) Graph f and the data in the same viewing rectangle.

(d) Approximate graphically the time when the amount of yeast was 200 units.

SOLUTION

(a) A scatterplot of the data is shown in Figure 5.100. The yeast increase slowly at first. Then they grow more rapidly, until the amount of yeast gradually levels off. The limited amount of nourishment causes this leveling off.

$[-2, 20, 1]$ by $[-100, 800, 100]$

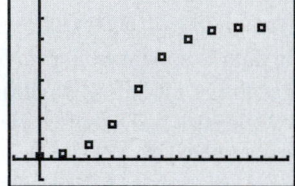

FIGURE 5.100

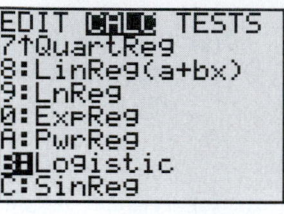

FIGURE 5.101

FIGURE 5.102

(b) In Figures 5.101 and 5.102 we see least-squares regression being used to find a logistics function f given by

$$f(x) = \frac{661.8}{1 + 74.46e^{-0.552x}}.$$

(c) In Figure 5.103 the data and f are graphed in the same viewing rectangle. The fit for this real data is remarkably good.

$[-2, 20, 1]$ by $[-100, 800, 100]$ $[-2, 20, 1]$ by $[-100, 800, 100]$

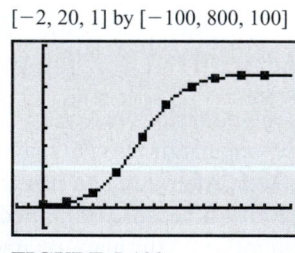

FIGURE 5.103

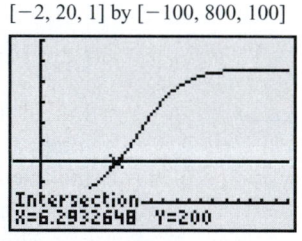

FIGURE 5.104

(d) The graphs of $Y_1 = f(x)$ and $Y_2 = 200$ intersect near $(6.29, 200)$, as shown in Figure 5.104. The amount of yeast reached 200 units after about 6.29 hours.

Now Try Exercises 8 and 15 ◆

◆ **CLASS DISCUSSION**

In Example 3, suppose that after 18 hours the experiment had been extended and more nourishment were provided for the yeast plants. Sketch a possible graph of the amount of yeast. ◆

◆ **MAKING CONNECTIONS** ───────

Logistic Functions and Horizontal Asymptotes   If a logistic function is given by $f(x) = \frac{c}{1 + ae^{-bx}}$, where a, b, and c are positive constants, then the graph of f has a horizontal asymptote of $y = c$. (Try to explain why this is true.) In Example 3, the value of c was 661.8. This means that the amount of yeast leveled off at 661.8 units.

5.7

Putting it all Together

$\mathbf{T}$he following table summarizes the basics of exponential, logarithmic, and logistic models. Least-squares regression can be used to determine the constants a, b, c, and C.

Concept	Explanation	Examples
Exponential model	$f(x) = Ca^x,$ $f(x) = ab^x,$ or $A(t) = A_0e^{kt}$	Exponential functions can be used to model data that increase or decrease rapidly over time.
Logarithmic model	$f(x) = a + b \log x$ or $f(x) = a + b \ln x$	Logarithmic functions can be used to model data that increase gradually over time.
Logistic model	$f(x) = \dfrac{c}{1 + ae^{-bx}}$	Logistic functions can be used to model data that at first increase slowly, then increase rapidly, and finally level off.

5.7   Exercises

Selecting a Model

Exercises 1–4: Select an appropriate type of modeling function for the data shown in the graph. Choose from the following.

   i. Exponential
   ii. Logarithmic
   iii. Logistic

1.   **2.**

3.   **4.**

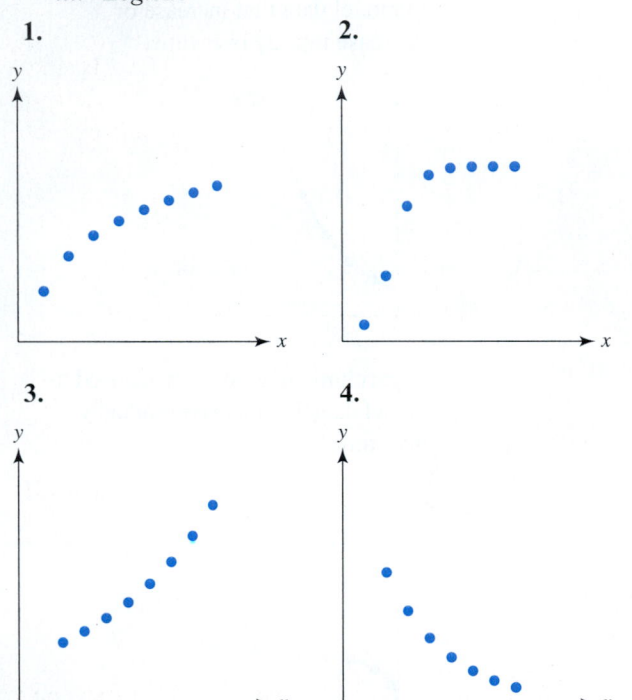

Exercises 5–8: Make a scatterplot of the data. Then find an exponential, logarithmic, or logistic function f that best models the data.

5.

x	1	2	3	4
y	2.04	3.47	5.90	10.02

6.

x	1	2	3	4	5
y	1.98	2.35	2.55	2.69	2.80

7.

x	1	2	3	4	5
y	1.1	3.1	4.3	5.2	5.8

8.

x	1	2	3	4	5	6
y	1	2	4	7	9	10

Applications

9. *Heart Disease Death Rates* The accompanying table contains heart disease death rates per 100,000 people in 1996 for selected ages.

Age	30	40	50	60	70
Death Rate	30.5	108.2	315	776	2010

Source: Department of Health and Human Services.

(a) Make a scatterplot of the data in the viewing rectangle [25, 75, 5] by [−100, 2100, 200].

(b) Find a function f that models the data.

(c) Estimate the heart disease death rate for people who are 80 years old.

10. *Hurricanes* The accompanying table shows the air pressure y in inches of mercury x miles from the eye of a hurricane.

x (mi)	2	4	8
y (in. of mercury)	27.3	27.7	28.04

x (mi)	15	30	100
y (in. of mercury)	28.3	28.7	29.3

Source: A. Miller and R. Anthes, *Meteorology.*

(a) Make a scatterplot of the data in the viewing rectangle [0, 110, 10] by [27, 30, 1].

(b) Find a function f that models the data.

(c) Estimate the air pressure at 50 miles.

11. *Atmospheric Density* The table lists the atmospheric density y in kilograms per cubic meter (kg/m^3) at an altitude of x meters.   (**Source:** A. Miller.)

x (m)	0	5000	10,000	15,000
y (kg/m^3)	1.2250	0.7364	0.4140	0.1948

x (m)	20,000	25,000	30,000
y (kg/m^3)	0.0889	0.0401	0.0184

(a) Use exponential regression to estimate the constants a and b so that $f(x) = ab^x$ models the data.

(b) Predict the density at 7000 meters. (The actual value is 0.59 kg/m^3.)

12. *Bird Populations*   Near New Guinea there is a relationship between the number of bird species found on an island and the size of the island. The accompanying table lists numbers of species of birds y found on an island with an area of x square kilometers.   (**Source:** B. Freedman, *Environmental Ecology.*)

x (km^2)	0.1	1	10	100	1000
y (species)	10	15	20	25	30

(a) Find a function f that models the data.

(b) Predict the number of bird species on an island of 5000 square kilometers.

13. *Fertilizer Usage*   Between 1950 and 1980 the use of chemical fertilizers increased. The table lists worldwide average usage y in kilograms per hectare of cropland during year x, where $x = 0$ corresponds to 1950. (*Note:* 1 hectare ≈ 2.47 acres.)   (**Source:** D. Grigg.)

x	0	13	22	29
y	12.4	27.9	54.3	77.1

(a) Graph the data. Are the data linear?

(b) Find a function f that models the data.

(c) Predict the chemical fertilizer usage in 1989. The actual value was 98.7 kilograms per hectare. What does this indicate about usage of fertilizer during the 1980s?

14. *Social Security*   If major reform occurs in the Social Security system, individuals may be able to invest some of their contributions into individual accounts. Many of these accounts would be managed by financial firms that charge fees. The following table lists the amount in billions of dollars that may be collected if fees are 0.93% of the assets each year.   (**Source:** Social Security Advisory Council.)

Year	2005	2010	2015	2020
Fees ($ billions)	20	41	80	136

(a) Use exponential regression to find a and b so that $f(x) = ab^x$ models the data. Let $x = 5$ correspond to 2005, $x = 10$ to 2010, and so on.

(b) Graph f and the data.

(c) Estimate the fees in 2013.

15. *Insect Population*   The accompanying table shows the density y of a species of insect measured in thousands per acre during a period of x days.

x (days)	2	4	6	8
y ($\times$ 1000/acre)	0.38	1.24	2.86	4.22

x (days)	10	12	14
y ($\times$ 1000/acre)	4.78	4.94	4.98

(a) Find a function f that models the data.

(b) Use f to estimate the insect density after a long time.

16. *Telecommuting*   Some workers use technology such as fax machines, e-mail, computers, and multiple phone lines to work at home, rather than in the office. However, because of the need for teamwork and collaboration in the workplace, fewer employees are telecommuting than expected. The table lists the expected telecommuters in millions from 1997 to 2006.   (**Source:** *USA Today.*)

Year	1997	1998	1999	2000
Telecommuters	9.2	9.6	10.0	10.4

Year	2001	2002	2003	2004
Telecommuters	10.6	11.0	11.1	11.2

Year	2005	2006
Telecommuters	11.3	11.4

Find values for a and b so that $f(x) = a + b \ln x$ models the data, where $x = 1$ corresponds to 1997, $x = 2$ to 1998, and so on.

17. *Heart Disease*   As age increases, so does the likelihood of coronary heart disease (CHD). The fraction of people x years old with some CHD is modeled by the formula $f(x) = \dfrac{0.9}{1 + 271e^{-0.122x}}$.   (**Source:** D. Hosmer and S. Lemeshow, *Applied Logistic Regression.*)

(a) Evaluate $f(25)$ and $f(65)$. Interpret the results.

(b) At what age does this likelihood equal 50%?

18. *Modeling Tree Growth*   The height of a tree in feet after x years is modeled by $f(x) = \dfrac{50}{1 + 47.5e^{-0.22x}}$.

(a) Make a table of f starting at $x = 10$, incrementing by 10. What appears to be the maximum height of the tree?

(b) Graph f and identify the horizontal asymptote. Explain its significance.

(c) After how long was the tree 30 feet tall?

Writing about Mathematics

19. How can you distinguish between data that illustrate exponential growth and data that illustrate logarithmic growth?

20. Give an example of data that could be modeled by a logistic function and explain why.

EXTENDED AND DISCOVERY EXERCISE

1. For medical reasons, dyes may be injected into the bloodstream to determine the health of internal organs. In one study involving animals, the dye BSP was injected to assess the blood flow in the liver. The results are listed in the accompanying table, where x represents the elapsed time in minutes and y is the concentration of the dye in the bloodstream in milligrams per milliliter (mg/ml).   (**Sources:** F. Harrison, "The measurement of liver blood flow in conscious calves"; D. Brown and P. Rothery, *Models in Biology: Mathematics, Statistics and Computing.*)

x (min)	1	2	3	4
y (mg/ml)	0.102	0.077	0.057	0.045

x (min)	5	7	9	13
y (mg/ml)	0.036	0.023	0.015	0.008

x (min)	16	19	22
y (mg/ml)	0.005	0.004	0.003

(a) Find a function that models the data.

(b) Estimate the elapsed time when the concentration of the dye reaches 30% of its initial concentration of 0.133 mg/ml.

CHECKING BASIC CONCEPTS FOR SECTION 5.7

Exercises 1–3: Find a function that models the data in the table. Choose from exponential, logarithmic, or logistic functions.

1.

x	2	3	4	5	6	7
y	0.72	0.86	1.04	1.24	1.49	1.79

2.

x	2	3	4	5	6	7
y	0.08	1.30	2.16	2.83	3.38	3.84

3.

x	2	3	4	5	6	7
y	0.25	0.86	2.19	3.57	4.23	4.43

CHAPTER 5 SUMMARY

CONCEPT	EXPLANATION AND EXAMPLES

SECTION 5.1   COMBINING FUNCTIONS

ARITHMETIC OPERATIONS ON FUNCTIONS

Addition: $(f + g)(x) = f(x) + g(x)$
Subtraction: $(f - g)(x) = f(x) - g(x)$
Multiplication: $(fg)(x) = f(x) \cdot g(x)$

Division: $(f/g)(x) = \dfrac{f(x)}{g(x)}, \; g(x) \neq 0$

Examples: Let $f(x) = x^2 - 5$, $g(x) = x^2 - 4$.
$(f + g)(x) = (x^2 - 5) + (x^2 - 4) = 2x^2 - 9$
$(f - g)(x) = (x^2 - 5) - (x^2 - 4) = -1$
$(fg)(x) = (x^2 - 5)(x^2 - 4) = x^4 - 9x^2 + 20$
$(f/g)(x) = \dfrac{x^2 - 5}{x^2 - 4}, \; x \neq \pm 2$

CONCEPT	EXPLANATION AND EXAMPLES

SECTION 5.1 COMBINING FUNCTIONS (CONTINUED)

COMPOSITION OF FUNCTIONS

Composition: $(g \circ f)(x) = g(f(x))$
$(f \circ g)(x) = f(g(x))$

Examples: Let $f(x) = 3x + 2$, $g(x) = 2x^2 - 4x + 1$.
$g(f(x)) = g(3x + 2)$
$= 2(3x + 2)^2 - 4(3x + 2) + 1$
$f(g(x)) = f(2x^2 - 4x + 1)$
$= 3(2x^2 - 4x + 1) + 2$
$= 6x^2 - 12x + 5$

SECTION 5.2 INVERSE FUNCTIONS AND THEIR REPRESENTATIONS

INVERSE FUNCTION

The inverse function of f is f^{-1} if
$f^{-1}(f(x)) = x$ for every x in the domain of f, and
$f(f^{-1}(x)) = x$ for every x in the domain of f^{-1}.
Note: If $f(a) = b$, then $f^{-1}(b) = a$.

Example: Find the inverse function of $f(x) = 4x - 5$.

$y = 4x - 5$ is equivalent to $\frac{y + 5}{4} = x$. (Solve for x.)

Thus $f^{-1}(x) = \frac{x + 5}{4}$.

Check: $f^{-1}(f(x)) = f^{-1}(4x - 5)$

$= \frac{(4x - 5) + 5}{4}$

$= x$

$f(f^{-1}(x)) = f\left(\frac{x + 5}{4}\right)$

$= 4\left(\frac{x + 5}{4}\right) - 5$

$= x$ It checks.

ONE-TO-ONE FUNCTION

If different inputs always result in different outputs, then f is one-to-one. That is, $a \neq b$ implies $f(a) \neq f(b)$.
Note: If f is one-to-one, then f has an inverse denoted f^{-1}.

Example: $f(x) = x^2 + 1$ is not one-to-one because $f(2) = f(-2) = 5$.

HORIZONTAL LINE TEST

If every horizontal line intersects the graph of a function f at most once, then f is a one-to-one function.

SECTION 5.3 EXPONENTIAL FUNCTIONS AND MODELS

EXPONENTIAL FUNCTION

$f(x) = Ca^x$, $a > 0$, $a \neq 1$, and $C > 0$.
Exponential growth: $a > 1$; Exponential decay: $0 < a < 1$

Examples: $f(x) = 3(2)^x$ (growth); $f(x) = 1.2(0.5)^x$ (decay)

CONCEPT	EXPLANATION AND EXAMPLES

SECTION 5.3 EXPONENTIAL FUNCTIONS AND MODELS (CONTINUED)

NATURAL EXPONENTIAL FUNCTION

$f(x) = e^x$, where $e \approx 2.718282$.

COMPOUND INTEREST

$A = P\left(1 + \dfrac{r}{n}\right)^{nt}$, where P is the principal, r is the interest expressed as a decimal, n is the number of times interest is paid each year, and t is the number of years.

Example: $A = 2000\left(1 + \dfrac{0.10}{12}\right)^{12(4)} \approx \$2978.71$

calculates the future value of \$2000 invested at 10% compounded monthly for 4 years.

CONTINUOUSLY COMPOUNDED INTEREST

$A = Pe^{rt}$, where P is the principal, r is the interest expressed as a decimal, t is the number of years.

Example: $A = 2000e^{0.10(4)} \approx \$2983.65$

calculates the future value of \$2000 invested at 10% compounded continuously for 4 years.

EXPONENTIAL DATA

For each unit increase in x, the y-values increase (or decrease) by a constant factor a.

Example: The data in the table are modeled by $y = 5(2)^x$.

x	0	1	2	3
y	5	10	20	40

SECTION 5.4 LOGARITHMIC FUNCTIONS AND MODELS

COMMON LOGARITHM

$\log x = k$ if and only if $x = 10^k$

NATURAL LOGARITHM

$\ln x = k$ if and only if $x = e^k$

GENERAL LOGARITHM

$\log_a x = k$ if and only if $x = a^k$

Examples: $\log 100 = 2$ because $100 = 10^2$.

$\ln \sqrt{e} = \dfrac{1}{2}$ because $\sqrt{e} = e^{1/2}$.

$\log_2 \dfrac{1}{8} = -3$ because $\dfrac{1}{8} = 2^{-3}$.

INVERSE PROPERTIES

$\log 10^k = k, \qquad 10^{\log k} = k, k > 0$

$\ln e^k = k, \qquad e^{\ln k} = k, k > 0$

$\log_a a^k = k, \qquad a^{\log_a k} = k, k > 0$

Examples: $10^{\log 100} = 100$

$e^{\ln 23} = 23$

$\log_4 64 = \log_4 4^3 = 3$

CONCEPT	EXPLANATION AND EXAMPLES

SECTION 5.4 LOGARITHMIC FUNCTIONS AND MODELS (CONTINUED)

INVERSE FUNCTIONS

The inverse function of $f(x) = a^x$ is $f^{-1}(x) = \log_a x$.

Examples: If $f(x) = 10^x$, then $f^{-1}(x) = \log x$.
If $f(x) = \ln x$, then $f^{-1}(x) = e^x$.
If $f(x) = \log_5 x$, then $f^{-1}(x) = 5^x$.

SECTION 5.5 PROPERTIES OF LOGARITHMS

**PROPERTIES OF
LOGARITHMS**

1. $\log_a 1 = 0$ and $\log_a a = 1$
2. $\log_a m + \log_a n = \log_a (mn)$

3. $\log_a m - \log_a n = \log_a \left(\dfrac{m}{n} \right)$

4. $\log_a (m^r) = r \log_a m$

Examples: 1. $\log_4 1 = 0$ and $\log_4 4 = 1$
2. $\log 2 + \log 5 = \log (2 \cdot 5) = \log 10 = 1$
3. $\log 500 - \log 5 = \log (500/5) = \log 100 = 2$
4. $3 \log_2 2 = \log_2 2^3 = 3$

**CHANGE OF BASE
FORMULA**

$$\log_a x = \frac{\log_b x}{\log_b a}$$

Example: $\log_3 23 = \dfrac{\log 23}{\log 3} \approx 2.854$

SECTION 5.6 EXPONENTIAL AND LOGARITHMIC EQUATIONS

**SOLVING EXPONENTIAL
EQUATIONS**

To solve an exponential equation we typically take a logarithm of each side.

Example:	$4e^x = 48$	Given equation
	$e^x = 12$	Divide by 4.
	$\ln e^x = \ln 12$	Take the natural logarithm.
	$x = \ln 12$	Inverse property

**SOLVING LOGARITHMIC
EQUATIONS**

To solve a logarithmic equation we typically need to exponentiate each side.

Example:	$5 \log_3 x = 10$	Given equation
	$\log_3 x = 2$	Divide by 5.
	$3^{\log_3 x} = 3^2$	Exponentiate; base 3.
	$x = 9$	Inverse property

SECTION 5.7 CONSTRUCTING NONLINEAR MODELS

EXPONENTIAL MODEL

$f(x) = Ca^x$, $f(x) = ab^x$, or $A(t) = A_0 e^{kt}$
Models data that can increase or decrease rapidly

LOGARITHMIC MODEL

$f(x) = a + b \ln x$ or $f(x) = a + b \log x$
Models data that increase slowly

CONCEPT	EXPLANATION AND EXAMPLES

SECTION 5.7 CONSTRUCTING NONLINEAR MODELS (CONTINUED)

LOGISTIC MODEL

$f(x) = \dfrac{c}{1 + ae^{-bx}}$, a, b, and c are positive constants.

Models data that increase slowly at first, then increase more rapidly, and finally level off near the value of c.

REVIEW EXERCISES

1. Use the table to evaluate each expression, if possible.

x	-1	0	1	3
$f(x)$	3	5	7	9
$g(x)$	-2	0	1	9

   (a) $(f + g)(1)$ **(b)** $(f - g)(3)$

   (c) $(fg)(-1)$ **(d)** $(f/g)(0)$

2. Use the graph to evaluate each expression.
   (a) $(f - g)(2)$ **(b)** $(fg)(0)$

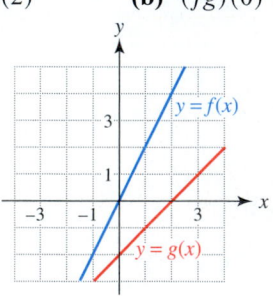

3. Use $f(x) = x^2$ and $g(x) = 1 - x$ to evaluate each expression.
   (a) $(f + g)(3)$ **(b)** $(f - g)(-2)$

   (c) $(fg)(1)$ **(d)** $(f/g)(3)$

4. Use $f(x) = x^2 + 3x$ and $g(x) = x^2 - 1$ to find each expression. Identify its domain.
   (a) $(f + g)(x)$ **(b)** $(f - g)(x)$

   (c) $(fg)(x)$ **(d)** $(f/g)(x)$

5. Numerical representations for f and g are given by the tables in the next column. Evaluate the expressions.

   (a) $(g \circ f)(-2)$ **(b)** $(f \circ g)(3)$

x	-2	0	2	4
$f(x)$	1	4	3	2

x	1	2	3	4
$g(x)$	2	4	-2	0

6. Use the graph to evaluate each expression.
   (a) $(f \circ g)(2)$ **(b)** $(g \circ f)(0)$

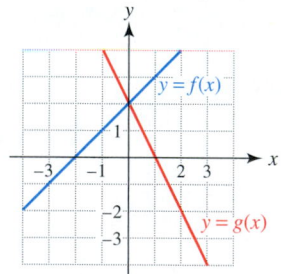

7. Use $f(x) = \sqrt{x}$ and $g(x) = x^2 + x$ to evaluate each expression.
   (a) $(f \circ g)(2)$ **(b)** $(g \circ f)(9)$

8. Use $f(x) = x^2 + 1$ and $g(x) = x^3 - x^2 + 2x + 1$ to find the following.
   (a) $(f \circ g)(x)$ **(b)** $(g \circ f)(x)$

Exercises 9–12: Find $(f \circ g)(x)$ and identify its domain.

9. $f(x) = x^3 - x^2 + 3x - 2$ $g(x) = \dfrac{1}{x}$

10. $f(x) = \sqrt{x + 3}$ $g(x) = 1 - x^2$

11. $f(x) = \sqrt[3]{2x - 1}$ $g(x) = \frac{1}{2}x^3 + \frac{1}{2}$

12. $f(x) = \dfrac{2}{x - 5}$ $g(x) = \dfrac{1}{x + 1}$

Exercises 13 and 14: Find f and g so that $h(x) = (g \circ f)(x)$.

13. $h(x) = \sqrt{x^2 + 3}$ **14.** $h(x) = \dfrac{1}{(2x + 1)^2}$

Exercises 15 and 16: Describe the inverse operations of the given statement. Then, express both the statement and its inverse symbolically.

15. Divide x by 10 and add 6.

16. Subtract 5 from x and take the cube root.

Exercises 17 and 18: Determine if f is one-to-one.

17. $f(x) = 3x - 1$ **18.** $f(x) = 3x^2 - 2x + 1$

19. The table is a complete representation of f. Use the table to determine if f is one-to-one and if f has an inverse.

x	−1	−2	0	4
$f(x)$	−5	4	3	4

20. The table is a complete representation of f. Use the table of f to determine a table for f^{-1}. Identify the domains and ranges of f and f^{-1}.

x	−1	0	4	6
$f(x)$	6	4	3	1

Exercises 21 and 22: Use the graph of f to determine if f is one-to-one.

21.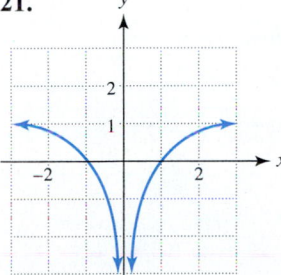

22.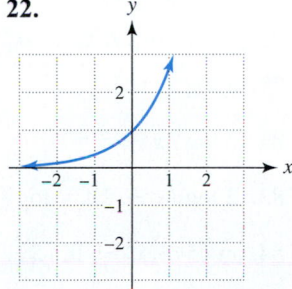

23. The function f computes the dollars in a savings account after x years. Use the graph at the top of the next column to evaluate the following. Interpret f^{-1}.
(a) $f(1)$ (b) $f^{-1}(1200)$

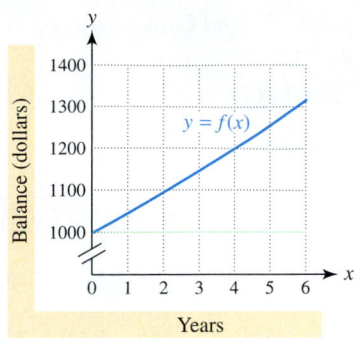

24. Use the graph of f to sketch a graph of f^{-1}.

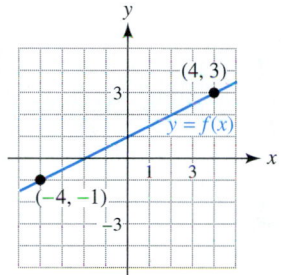

Exercises 25 and 26: Find $f^{-1}(x)$.

25. $f(x) = 3x - 5$ **26.** $f(x) = \sqrt[3]{x} + 1$

27. If $f(x) = \dfrac{3x}{x + 7}$, find $f^{-1}(x)$.

28. Verify that $f(x) = 2x - 1$ and $f^{-1}(x) = \dfrac{x + 1}{2}$ are inverses.

29. Restrict the domain of $f(x) = 2(x - 4)^2 + 3$ so that f is one-to-one. Then find $f^{-1}(x)$.

30. Find $f^{-1}(x)$ if $f(x) = \sqrt{x + 1}, x \geq -1$. Identify the domain and range of f and of f^{-1}.

Exercises 31 and 32: Use the tables to evaluate each expression.

x	0	1	2	3	4
$f(x)$	4	3	2	1	0

x	0	1	2	3	4
$g(x)$	0	2	3	4	6

31. $(f \circ g^{-1})(4)$ **32.** $(g^{-1} \circ f^{-1})(1)$

Exercises 33 and 34: Simplify the expression.

33. $e^x e^{-2x}$ **34.** $2(27)^{2/3} - 4$

Exercises 35 and 36: Find C and a so that $f(x) = Ca^x$ satisfies the given conditions.

35. $f(0) = 3$ and $f(3) = 24$

36. $f(-1) = 8$ and $f(1) = 2$

Exercises 37–40: Sketch a graph of $y = f(x)$. Identify the domain of f.

37. $f(x) = 4(2)^{-x}$ **38.** $f(x) = 3^{x-1}$

39. $f(x) = \log_4 x$ **40.** $f(x) = \log(x + 1)$

Exercises 41 and 42: Use the graph of $y = Ca^x$ to determine values for C and a.

41.   **42.**

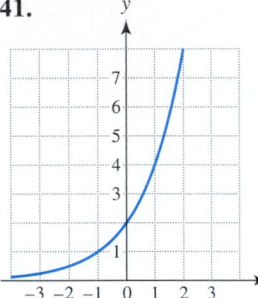   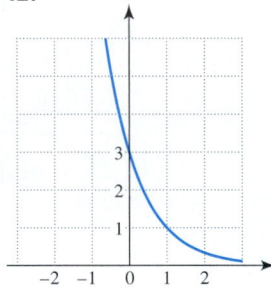

43. Determine the final value of $1200 invested at 9% compounded semiannually for 3 years.

44. Determine the final value of $500 invested at 6.5% compounded continuously for 8 years.

Exercises 45 and 46: Evaluate $f(x)$ for the given x. Approximate the result to the nearest thousandth.

45. $f(x) = e^{-2x}, x = -1.2$

46. $f(x) = 4e^{0.23x}, x = 5.7$

47. Solve $e^x = 19$ symbolically. Support your result graphically or numerically.

 48. Solve $2^x - x^2 = x$ graphically. Round each solution to the nearest thousandth.

Exercises 49–52: Evaluate the expression without a calculator.

49. $\log 1000$ **50.** $\log 0.001$

51. $10 \log 0.01 + \log \frac{1}{10}$ **52.** $\log 100 + \log \sqrt[3]{10}$

Exercises 53–56: Evaluate the logarithm without a calculator.

53. $\log_3 9$ **54.** $\log_5 \frac{1}{25}$

55. $\ln e$ **56.** $\log_2 32$

Exercises 57 and 58: Approximate the logarithm to the nearest thousandth.

57. $\log_3 18$ **58.** $\log_2 173$

Exercises 59–66: Solve the equation.

59. $10^x = 125$ **60.** $1.5^x = 55$

61. $e^{0.1x} = 5.2$ **62.** $4e^{2x} - 5 = 3$

63. $5^{-x} = 10$ **64.** $3(10)^{-x} = 6$

65. $50 - 3(0.78)^{x-10} = 21$ **66.** $5(1.3)^x + 4 = 104$

Exercises 67 and 68: Find either a linear or an exponential function that models the data in the table.

67.

x	0	1	2	3	4
y	1.5	3	6	12	24

68.

x	0	1	2	3	4
y	3	4.5	6	7.5	9

Exercises 69–74: Solve the equation.

69. $\log x = 1.5$ **70.** $\log_3 x = 4$

71. $\ln x = 3.4$ **72.** $\log x = 2.2$

73. $2 \log_4 (x + 2) + 5 = 12$

74. $4 - \ln (5 - x) = \frac{5}{2}$

Exercises 75 and 76: Use properties of logarithms to write the expression as a logarithm of a single expression.

75. $\log 6 + \log 5x$ **76.** $\log \sqrt{3} - \log \sqrt[3]{3}$

77. Expand $\ln \frac{4}{x^2}$.   **78.** Expand $\log \frac{4x^3}{k}$.

Exercises 79–84: Solve the logarithmic equation.

79. $8 \log x = 2$ **80.** $\ln 2x = 2$

81. $2 \log 3x + 5 = 15$ **82.** $5 \log_2 x = 25$

83. $2 \log (x + 2) = \log (x + 8)$

84. $\ln (5 - x) - \ln (5 + x) = -\ln 9$

85. Suppose that b is the y-intercept on the graph of a one-to-one function f. What is the x-intercept on the graph of f^{-1}? Explain your reasoning.

86. Let f be a linear function given by $f(x) = ax + b$ with $a \neq 0$.

(a) Show that f^{-1} is also a linear function by finding $f^{-1}(x)$.

(b) How is the slope of the graph of f related to the slope of the graph of f^{-1}?

Applications

87. *Bacteria Growth*   There are initially 4000 bacteria per milliliter in a sample and after 1 hour their concentration increases to 6000 bacteria per milliliter.
   (a) How many bacteria are there after 2.5 hours?

   (b) After how long are there 8500 bacteria per milliliter?

88. *Value of a House*   A house is purchased for $180,000 and the house appreciates 4% per year. Estimate the time it will take for the house to have a value of $300,000.

89. *Newton's Law of Cooling*   A pan of boiling water with a temperature of 100°C is set in a room with a temperature of 20°C. The water cools to 50°C in 40 minutes.
   (a) Find values for $T_0, D,$ and a so that the formula $T(t) = T_0 + Da^t$ models the data, where t is in hours.

   (b) Find the temperature of the water after 90 minutes.

   (c) How long did it take the water to reach 30°C?

90. *Combining Functions*   The total number of gallons of water passing through a pipe after x seconds is computed by $f(x) = 10x$. Another pipe delivers $g(x) = 5x$ gallons after x seconds. Find a function h that computes the total volume of water passing through both pipes in x seconds.

91. *Converting Units*   The accompanying figures show graphs of a function f that converts fluid ounces to pints and a function g that converts pints to quarts. Evaluate each expression. Interpret the results.
   (a) $(g \circ f)(32)$ (b) $f^{-1}(1)$

   (c) $(f^{-1} \circ g^{-1})(1)$

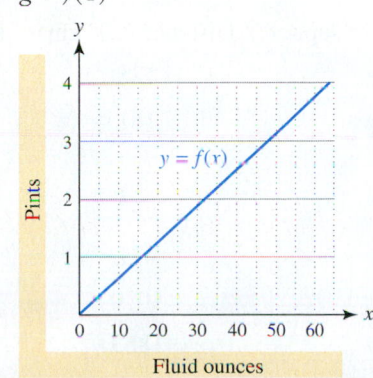

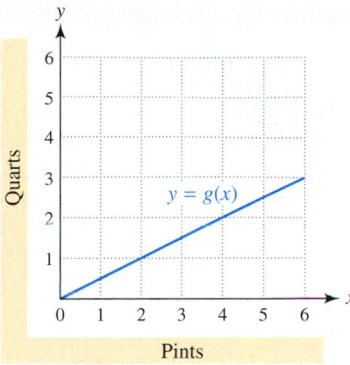

92. *Population and Tax Revenue*   The accompanying graph of f shows the population in thousands of a city from 1980 until 2000. The graph of g computes the expected tax revenue in millions of dollars that can be raised from a population of x thousand. Evaluate and interpret $(g \circ f)(1988)$.

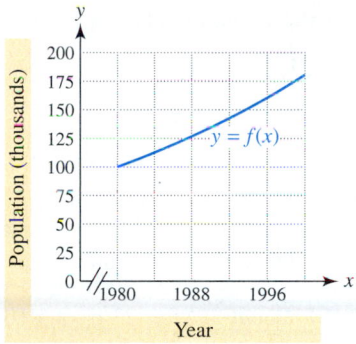

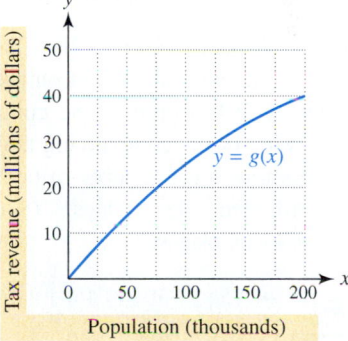

93. *Modeling Growth* The function given by

$$W(x) = 175.6(1 - 0.66e^{-0.24x})^3$$

models the weight in milligrams of a small fish called the *Lebistes reticulatus* after x weeks, where $0 \le x \le 14$. (**Source:** D. Brown and P. Rothery, *Models in Biology: Mathematics, Statistics and Computing.*)

(a) Evaluate $W(1)$.

(b) Solve the equation $W(x) = 50$ graphically. Interpret the result.

(c) Solve part (b) symbolically.

94. *Test Scores* Let scores on a standardized test be modeled by $f(x) = 36e^{-(x-20)^2/49}$. The function f computes the number in thousands of people that received score x.

(a) Graph f in [0, 40, 10] by [0, 40, 10]. What score did the largest number of people receive?

(b) Solve the equation $f(x) = 30$ symbolically. Interpret your result.

(c) Solve part (b) graphically.

95. *Tire Pressure* A car tire has a small leak, and the tire pressure in pounds per square inch after t minutes is given by $P(t) = 32e^{-0.2t}$. After how many minutes was the pressure 15 pounds per square inch?

96. *Radioactive Decay* After 23 days a 10-milligram sample of a radioactive material decays to 5 milligrams. After how many days will there be 1 milligram of the material?

97. *Modeling Epidemics* In 1666 the village of Eyam, located in England, experienced an outbreak of the Great Plague. Out of 261 people in the community, only 83 people survived. The accompanying table shows a function f that computes the number of people who had not (yet) been infected after x days. (**Source:** G. Raggett, "Modeling the Eyam plague.")

x	0	15	30	45
$f(x)$	254	240	204	150

x	60	75	90	125
$f(x)$	125	103	97	83

(a) Use a table to represent a function g that computes the number of people in Eyam that were infected after x days.

(b) Write an equation that shows the relationship between $f(x)$ and $g(x)$.

(c) Use graphing to decide which equation represents $g(x)$ better,

$$y_1 = \frac{171}{1 + 18.6e^{-0.0747x}} \quad \text{or} \quad y_2 = 18.3(1.024)^x.$$

(d) Use your results from parts (b) and (c) to find a formula for $f(x)$.

98. *Greenhouse Gases* Methane is a greenhouse gas that is produced when fossil fuels are burned. In 1600 methane had an atmospheric concentration of 700 parts per billion (ppb), whereas in 2000 its concentration was about 1700 ppb. (**Source:** D. Wuebbles and J. Edmonds, *Primer on Greenhouse Gases.*)

(a) Let $x = 0$ correspond to 1600 and $x = 400$ to 2000. Then, $f(x) = 700(10^{0.000964x})$ models the methane concentrations. Explain why f is one-to-one.

(b) Find $f^{-1}(x)$. Evaluate $f^{-1}(1000)$ and interpret the result.

99. *Exponential Regression* The data in the table can be modeled by $f(x) = ab^x$. Use regression to estimate the constants a and b. Graph f and the data.

x	1	2	3	4
y	2.59	1.92	1.42	1.05

100. *Logarithmic Regression* The data in the table can be modeled by $f(x) = a + b \ln x$. Use regression to estimate the constants a and b. Graph f and the data.

x	2	3	4	5
y	2.93	3.42	3.76	4.03

EXTENDED AND DISCOVERY EXERCISES

1. *Lunar Orbits* The table lists the orbital distances and periods of several moons of Jupiter. Let x represent the

Moons of Jupiter	Distance (10^3 km)	Period (days)
Metis	128	0.29
Almathea	181	0.50
Thebe	222	0.67
Io	422	1.77
Europa	671	3.55
Ganymede	1070	7.16
Callisto	1883	16.69

distance and y the period. Find a function f that models these data. Try more than one type of function such as linear, quadratic, power, and exponential. Graph each function and the data. Discuss which function models the data best.   (**Source:** M. Zeilik, *Introductory Astronomy and Astrophysics*.)

2. *Modeling Data*   There is a procedure to determine whether data can be modeled by $y = ax^b$, where a and b are constants. That is, the following procedure can be used to model data with a power function. Start by taking the natural logarithm of each side of this equation.

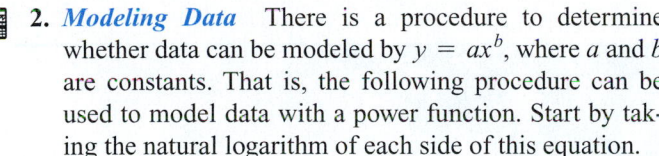

$$\ln y = \ln (ax^b)$$

$$\ln y = \ln a + \ln x^b \qquad \text{Property 2:}$$
$$\qquad\qquad\qquad\qquad\quad \ln (mn) = \ln m + \ln n$$

$$\ln y = \ln a + b \ln x \qquad \text{Property 4:}$$
$$\qquad\qquad\qquad\qquad\quad \ln (m^r) = r \ln m$$

If we let $z = \ln y$, $d = \ln a$, and $w = \ln x$, then the equation $\ln y = \ln a + b \ln x$ becomes $z = d + bw$. Thus the data points $(w, z) = (\ln x, \ln y)$ lie on the line having a slope of b and y-intercept $d = \ln a$. The following steps provide a procedure for finding the constants a and b.

MODELING DATA WITH THE EQUATION $y = ax^b$

If a data set (x, y) can be modeled by the equation $y = ax^b$, then the following procedure can be applied to determine the constants a and b.

STEP 1: Let $w = \ln x$ and $z = \ln y$ for each data point. Graph the points (w, z). If these data are not linear, then do *not* use this procedure.

STEP 2: Find an equation of a line in the form given by $z = bw + d$ that models the data points (w, z). (Linear regression may be used.)

STEP 3: The slope of the line equals the constant b. The value of a is given by $a = e^d$.

Apply this procedure to the data in Exercise 1.

3. *Global Warming*   Greenhouse gases such as carbon dioxide trap heat from the sun. Presently, the net incoming solar radiation reaching Earth's surface is approximately 240 watts per square meter (w/m^2). Any portion of this amount that is due to greenhouse gases is called *radiative forcing*. The accompanying table lists the estimated increase in radiative forcing R over the levels in 1750.   (**Source:** A. Nilsson, *Greenhouse Earth*.)

x (year)	1800	1850	1900	1950	2000
$R(x)(w/m^2)$	0.2	0.4	0.6	1.2	2.4

(a) Estimate constants C and k so that $R(x) = Ce^{kx}$ models the data. Let $x = 0$ correspond to 1800.

(b) Estimate the year when the additional radiative forcing could reach 3 w/m^2.

4. *Global Warming*   (Refer to the previous exercise.) The relationship between radiative forcing R and the increase in average global temperature T in degrees Fahrenheit can be modeled by $T(R) = 1.03R$. For example, $T(2) = 1.03(2) = 2.06$ means that if the earth's atmosphere traps an additional 2 watts per square meter, then the average global temperature may increase by 2.06°F, *if* all other factors remained constant.   (**Source:** W. Clime, *The Economics of Global Warming*.)

(a) Use $R(x)$ from the previous exercise to express $(T \circ R)(x)$ symbolically, where x is the year and $x = 0$ corresponds to the year 1800.

(b) Evaluate $(T \circ R)(100)$ and interpret its meaning.

Exercises 5 and 6: Evaluate the expression as accurately as possible. Try to decide if it is precisely an integer. (Hint: Use computer software capable of calculating a large number of decimal places.)

5. $\left(\dfrac{1}{\pi} \ln (640{,}320^3 + 744) \right)^2$

(**Source:** I. J. Good, "What is the most amazing approximate integer in the universe?")

6. $e^{\pi \sqrt{163}}$ (**Source:** W. Cheney and D. Kincaid, *Numerical Mathematics and Computing*.)

6 Trigonometric Functions

*Education is not the
filling of a pail, but
the lighting of a fire.*

William Butler Yeats

Trigonometry has been used for millennia to solve problems involving astronomy, surveying, and construction. The word *trigonometry* is derived from two Greek words which mean triangle measurement. In fact, the Greek astronomer Hipparchus is usually given credit for first studying the trigonometric properties of angles. Trigonometry originated when people tried to correlate shadow lengths with the time of day. Prior to the fifteenth century, astronomy had the greatest influence on the development of trigonometry, and it was not until the thirteenth century that astronomy and trigonometry could be regarded as separate entities.

In modern times trigonometric relationships have been viewed more in terms of functions and graphs. One reason for this is the explosion of new technologies in our society. Trigonometric functions often are used to model phenomena that involve periodic motion or rotation. Some examples of where trigonometry is used include average monthly temperature, daylight hours, tides, tidal currents, the Global Positioning System (GPS), movement of robotic arms, electricity, highway design, orbits of satellites, the phases of the moon, and music. Trigonometric functions have been essential to the development of our modern society. They are similar to other types of functions that we have studied and have graphical, numerical, and symbolic representations.

Sources: H. Freebury, *A History of Mathematics; Historical Topics for the Mathematics Classroom, Thirty-first Yearbook,* NCTM.

6.1 Angles and Their Measure

◆ Learn basic concepts about angles

◆ Apply degree measure to problems

◆ Apply radian measure to problems

◆ Calculate arc length

◆ Calculate the area of a sector

Introduction

The concept of an angle dates back thousands of years. Degree measure began in Babylonia (5000–4000 B.C.), where angles were frequently used in astronomy. Astronomy was important to society because of its connections to the calendar, the seasons, and planting times. Around 1873 a second unit to measure angles, called a *radian,* was developed independently by Thomas Muir and James Thomson, a mathematician and a physicist. Radian measure is used because it often results in simpler formulas.   (**Source:** *Historical Topics for the Mathematics Classroom, Thirty-first Yearbook,* NCTM.)

Angles

An **angle** is formed by rotating a ray about its endpoint. The starting position of the ray is called the **initial side** and the final position of the ray is the **terminal side**. If the rotation of the ray is counterclockwise, the angle has *positive measure,* whereas if the rotation is clockwise, the angle has *negative measure.* For simplicity we will refer to an angle as being **positive** or **negative**. The endpoint of the ray is called the **vertex** of the angle. See Figures 6.1 and 6.2, where the Greek letter θ (theta) has been used to denote an angle.

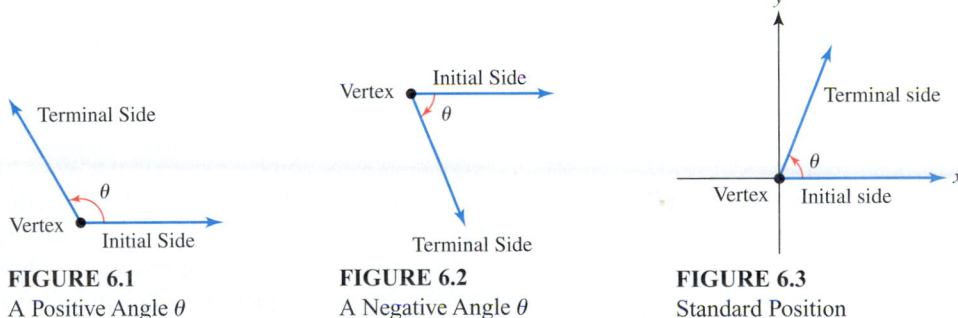

FIGURE 6.1
A Positive Angle θ

FIGURE 6.2
A Negative Angle θ

FIGURE 6.3
Standard Position

If the vertex is positioned so it corresponds to the origin in the *xy*-plane and the initial side coincides with the positive *x*-axis, then the angle is in **standard position**, as shown in Figure 6.3.

There are two common systems to measure the size of an angle: *degree measure* and *radian measure.*

Degree Measure

In **degree measure**, one complete rotation of a ray about its endpoint contains 360 degrees. One degree, denoted 1°, represents $\frac{1}{360}$ of a complete rotation. Figure 6.4 shows some examples of angles with their degree measure.

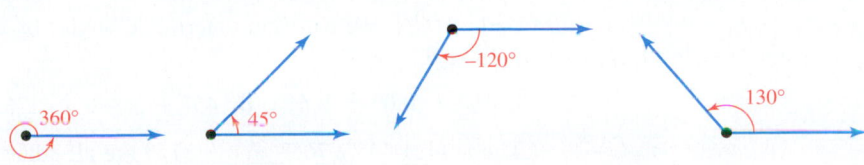

FIGURE 6.4 Degree Measure

A **right angle** has measure 90°, and a **straight angle** has measure 180°. The measure of an **acute angle** is greater than 0° but less than 90°, whereas the measure of an **obtuse angle** is greater than 90° but less than 180°. Examples are shown in Figure 6.5.

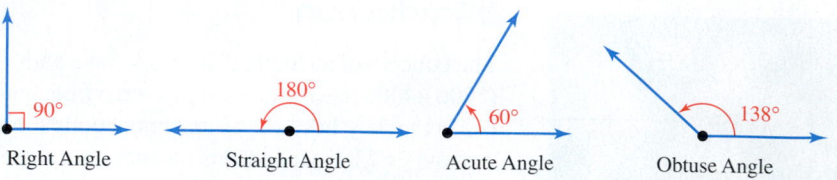

FIGURE 6.5 Types of Angles

We will use the Greek letters α (alpha), β (beta), γ (gamma), and θ (theta) to denote angles. For simplicity, we sometimes refer to an angle θ having measure 45° as a 45° angle, or an angle of 45°. This may be expressed as $\theta = 45°$. Two angles with the same initial and terminal sides are **coterminal angles**. Examples of coterminal angles are shown in Figure 6.6.

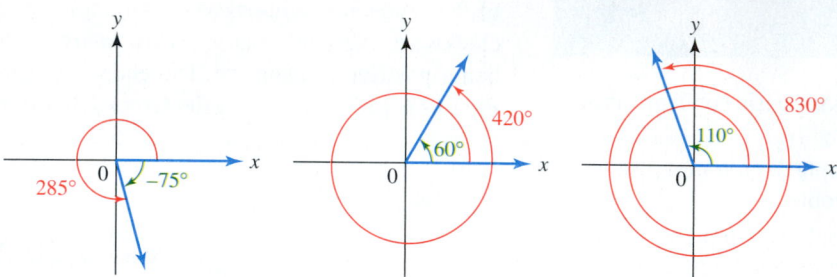

FIGURE 6.6 Coterminal Angles

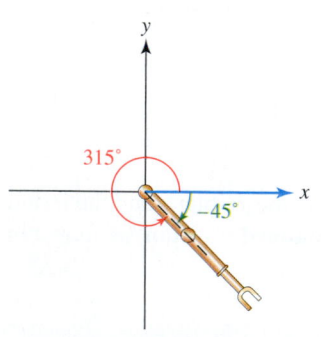

FIGURE 6.7 A Robotic Arm

Use of robotic arms has increased dramatically throughout society. Robots assemble products, paint vehicles, prepare fast foods, perform surgeries, and even link space stations. Coterminal angles have applications in robotics. Suppose a robotic arm, called a *simple polar manipulator,* picks up a bolt at an initial position corresponding to the positive x-axis and inserts the bolt at a final position, as illustrated in Figure 6.7. There are different angles through which the shoulder of the robotic arm could rotate to accomplish this task. For example, the arm could either rotate 45° clockwise or 315° counterclockwise. These rotations are represented by coterminal angles of −45° or 315°. Either angle accomplishes the task. However, an angle of −45° is usually preferable, because it involves less movement for the robotic arm and saves time.   (**Source:** J. Craig, *Introduction to Robotics: Mechanics and Control.*)

EXAMPLE 1 Finding coterminal angles

Find three angles coterminal to $\theta = 45°$, where θ is in standard position. Sketch these angles in standard position.

SOLUTION We can find coterminal angles by either adding or subtracting multiples of 360° to θ.

i. $45° + 360° = 405°$ **ii.** $45° + 2(360°) = 765°$ **iii.** $45° - 360° = -315°$

The angles 405°, 765°, and −315° are all coterminal to a 45° angle. These three angles and θ are sketched in Figure 6.8.

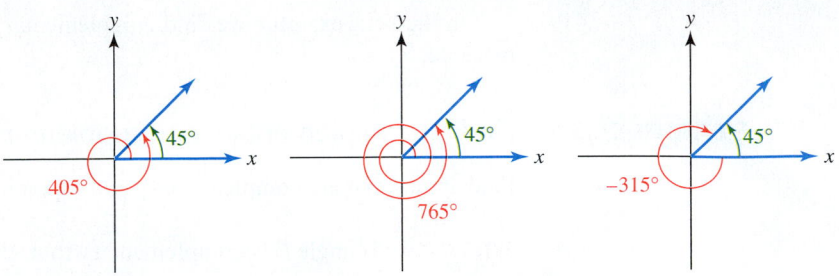

FIGURE 6.8

Now Try Exercise 13 ◆

Two positive angles are **complementary angles** if their sum equals 90° and are **supplementary angles** if their sum is 180°. For example, $\alpha = 35°$ and $\beta = 55°$ are complementary angles, whereas $\alpha = 60°$ and $\beta = 120°$ are supplementary angles. See Figures 6.9 and 6.10.

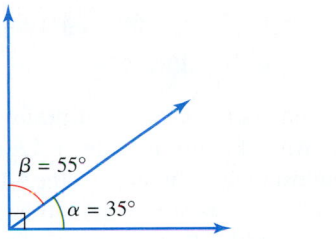

FIGURE 6.9 Complementary Angles

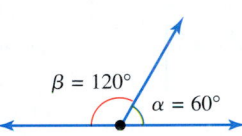

FIGURE 6.10 Supplementary Angles

Fractions of a degree may be measured using **minutes** and **seconds**. One minute, written $1'$, equals $\frac{1}{60}$ of a degree and one second, written $1''$, equals $\frac{1}{60}$ of a minute or $\frac{1}{3600}$ of a degree. The measurement $25°45'30''$ represents 25 degrees, 45 minutes, 30 seconds. Expressed in decimal degrees, this measurement is

$$24°45'30'' = 24° + \left(\frac{45}{60}\right)^° + \left(\frac{30}{3600}\right)^° = 24.758\overline{3}°.$$

We can also convert decimal degrees to degrees, minutes, and seconds as illustrated in the next example.

 **Converting to degrees, minutes, and seconds**

Convert 34.41° to degrees, minutes, and seconds.

SOLUTION

$$
\begin{aligned}
34.41° &= 34° + 0.41° && \text{Rewrite the expression.}\\
&= 34° + 0.41(60') && 1° = 60'\\
&= 34° + 24.6' && \text{Multiply.}\\
&= 34° + 24' + 0.6' && \text{Rewrite the expression.}\\
&= 34° + 24' + 0.6(60'') && 1' = 60''\\
&= 34° + 24' + 36'' && \text{Multiply.}\\
&= 34°24'36'' && \text{Rewrite.}
\end{aligned}
$$

Now Try Exercise 31 ◆

In the next example we find complementary and supplementary angles using degree measure.

EXAMPLE 3   Finding complementary and supplementary angles

Find angles that are complementary and supplementary to $\alpha = 34°19'42''$.

SOLUTION   If angle β is complementary to α, then $\beta = 90° - \alpha$.

$$\beta = 90° - 34°19'42''$$
$$= 89°59'60'' - 34°19'42''   \qquad 90° = 89°59'60''$$
$$= 55°40'18''$$

A supplementary angle to α is given by $\gamma = 180° - \alpha$.

$$\gamma = 180° - 34°19'42''$$
$$= 179°59'60'' - 34°19'42''   \qquad 180° = 179°59'60''$$
$$= 145°40'18''$$

Now Try Exercise 23 ◆

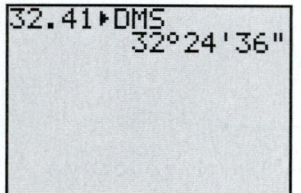

FIGURE 6.11 Degree Mode

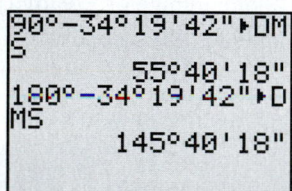

FIGURE 6.12 Degree Mode

Some calculators are capable of performing arithmetic using degrees, minutes, and seconds, as shown in Figures 6.11 and 6.12. (See the ANGLE menu.)

A new and exciting technology is the **Global Positioning System** (GPS). This system involves 24 satellites in nearly circular orbits that can be used to determine locations and velocities on Earth with a high degree of accuracy. See Figure 6.13. Private individuals can purchase handheld GPS devices that determine coordinates within meters and velocities within 2 meters per second.   (**Source:** J. Van Sickle, *GPS for Land Surveyors.*)

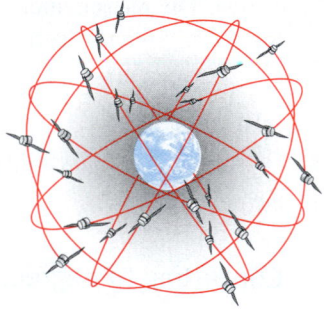

FIGURE 6.13 GPS Satellite Coverage

EXAMPLE 4   Determining the accuracy of GPS

Some GPS receivers display the latitude of a location to within $\pm \frac{1}{60,000}^{\circ}$. Convert this error to seconds.   (**Source:** G. West, "Differential GPS—how accurate is it?")

SOLUTION   Since there are 3600 seconds in one degree, it follows that

$$\pm \frac{1}{60,000}^{\circ} \times 3600 = \pm \frac{3''}{50} = \pm 0.06''$$

Thus this GPS device can display a latitude measurement to within $\pm 0.06''$.

Now Try Exercise 77 ◆

Radian Measure

A second unit of angle measure is *radians*. Radian measure is a common unit of measurement in many technical fields, including calculus. Radian measure often results in formulas being simpler and easier to use. Two examples of this are arc length and area of a sector, which are introduced later in this section.

Angle θ in Figure 6.14 has a measure of one *radian*. The vertex of θ is located at the center of the circle, and its initial and terminal sides intercept an arc whose length is equal to the radius of the circle.

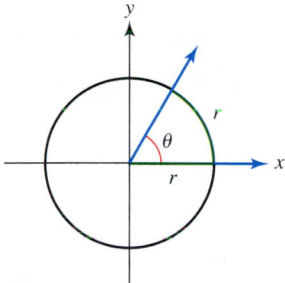

FIGURE 6.14   One Radian

RADIAN MEASURE

An angle that has its vertex at the center of a circle and intercepts an arc on the circle equal in length to the radius of the circle has a measure of **one radian**.

The circumference of a circle is $C = 2\pi r$. If we mark off distances of r along the circumference of a circle, it will appear as in Figure 6.15, where $2\pi \approx 6.28$ distances of r are shown. Therefore one rotation contains $2\pi \approx 6.28$ radians, and so $360°$ is equivalent to 2π radians. Radian measure can be compared to degree measure using proportions. Since $180°$ is equivalent to π radians, it follows that

$$\frac{\text{radian measure}}{\text{degree measure}} = \frac{\pi}{180°}.$$

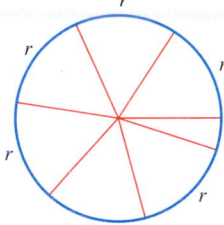

FIGURE 6.15   One Revolution Contains 2π Radians

Solving for radian measure results in

$$\textbf{radian measure} = \textbf{degree measure} \times \frac{\pi}{180°},$$

and solving for degree measure results in

$$\textbf{degree measure} = \textbf{radian measure} \times \frac{180°}{\pi}.$$

The following statements verbally summarize the preceding discussion.

CONVERTING BETWEEN DEGREES AND RADIANS

To convert *degrees to radians*, multiply a degree measure by $\frac{\pi}{180°}$.

To convert *radians to degrees*, multiply a radian measure by $\frac{180°}{\pi}$.

EXAMPLE 5 **Converting degrees to radians**

Convert each degree measure to radian measure.
(a) 90°   **(b)** 225°

SOLUTION

(a) To convert degrees to radians, multiply by $\frac{\pi}{180°}$.

$$90° \times \frac{\pi}{180°} = \frac{\pi}{2} \text{ radians}$$

Thus 90° are equivalent to $\frac{\pi}{2}$ radians.

(b) $225° \times \frac{\pi}{180°} = \frac{5\pi}{4}$ radians.

Now Try Exercise 39 ◆

Table 6.1 shows some equivalent measures in degrees and radians.

TABLE 6.1

Degrees	0°	30°	45°	60°	90°	180°	360°
Radians	0	$\frac{\pi}{6}$	$\frac{\pi}{4}$	$\frac{\pi}{3}$	$\frac{\pi}{2}$	π	2π

EXAMPLE 6 **Converting radians to degrees**

Convert each radian measure to degree measure.
(a) $\frac{4\pi}{3}$ **(b)** $\frac{5\pi}{6}$

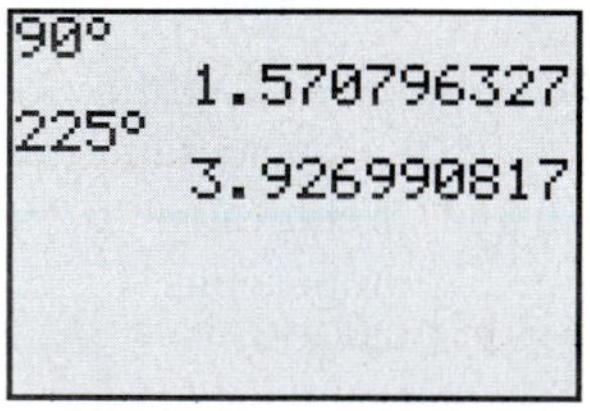

SOLUTION

(a) To convert radians to degrees, multiply by $\frac{180°}{\pi}$.

$$\frac{4\pi}{3} \times \frac{180°}{\pi} = 240°$$

Thus $\frac{4\pi}{3}$ radians are equivalent to 240°.

(b) $\frac{5\pi}{6} \times \frac{180°}{\pi} = 150°$.

Now Try Exercise 43 ◆

Some calculators can convert degrees to radians and radians to degrees, as shown in Figures 6.16 and 6.17, respectively. When converting from degrees to radians, many calculators give only decimal approximations rather than exact values. For example, $\frac{\pi}{2}$ may be expressed as 1.570796327.

FIGURE 6.16 Radian Mode

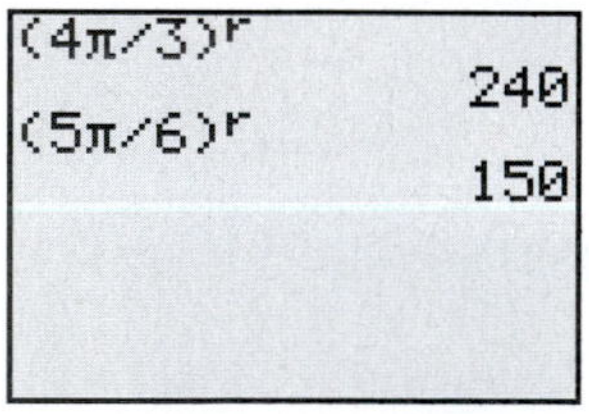

FIGURE 6.17 Degree Mode

Arc Length

From geometry we know that the arc length s on a circle is proportional to the measure of the central angle θ. See Figure 6.18. A central angle of 2π radians corresponds to an arc length that equals the circumference $C = 2\pi r$. Using proportions,

$$\frac{s}{\theta} = \frac{2\pi r}{2\pi},$$

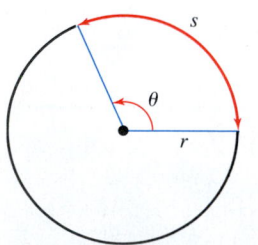

FIGURE 6.18 Arc Length s

which simplifies to $s = r\theta$.

ARC LENGTH

The **arc length** s intercepted on a circle of radius r by a central angle of θ *radians* is given by

$$s = r\theta.$$

Note: Angle θ *must* be in radian measure when using the arc length formula $s = r\theta$.

EXAMPLE 7   Finding arc length

A circle has a radius of 25 inches. Find the length of an arc intercepted by a central angle of 45°.

SOLUTION   First convert 45° to radian measure.

$$45° \times \frac{\pi}{180°} = \frac{\pi}{4}$$

The arc length s is given by

$$s = r\theta$$
$$= 25\left(\frac{\pi}{4}\right)$$
$$= 6.25\pi \text{ inches.}$$

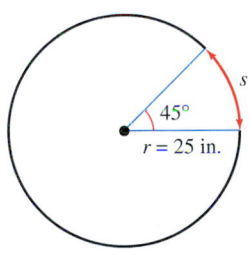

FIGURE 6.19

The arc length s shown in Figure 6.19 is $6.25\pi \approx 19.6$ inches.   *Now Try Exercise 47* ◆

EXAMPLE 8   Finding distance between cities

Albuquerque, New Mexico, and Glasgow, Montana, have the same longitude of 106°37′ W. The latitude of Albuquerque is 35°03′ N and the latitude of Glasgow is 48°13′ N. If the radius of Earth is approximately 3955 miles, estimate the distance between Albuquerque and Glasgow. See Figure 6.20.   (**Source:** J. Williams, *The Weather Almanac 1995.*)

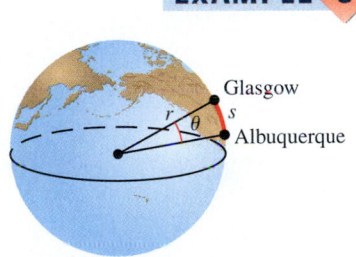

FIGURE 6.20   Distance between Two Cities (Not to scale)

SOLUTION   This distance can be estimated using the arc length formula. Start by converting $\theta = 48°13′ - 35°03′ = 13°10′$ to radian measure.

$$\left[13° + \left(\frac{10}{60}\right)^{\circ}\right] \times \frac{\pi}{180°} \approx 0.2298 \text{ radian}$$

The distance between Albuquerque and Glasgow is approximated by

$$s = r\theta$$
$$\approx 3955(0.2298)$$
$$\approx 909 \text{ miles.}$$ *Now Try Exercise 81* ◆

The human joint that can be flexed the fastest is the wrist, which can rotate through 90°, or $\frac{\pi}{2}$ radians, in 0.045 second while holding a tennis racket. See Figure 6.21 on the next page. **Angular speed** ω (omega) measures the speed of rotation and is defined by

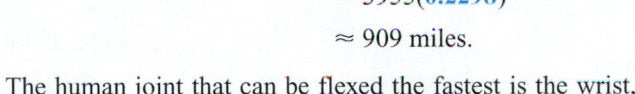

$$\omega = \frac{\theta}{t},$$

FIGURE 6.21 Flexing the Wrist in Tennis

where θ is the angle of rotation and t is time. The angular speed of a human wrist holding a tennis racket is

$$\frac{\pi/2}{0.045} \approx 34.9 \text{ rad/sec,}$$

or about 35 radians per second.

The **linear speed** v at which the tip of the racket travels as a result of flexing the wrist is given by $v = r\omega$, where r is the radius (distance) from the tip of the racket to the wrist joint. If $r = 2$ feet, then the speed at the tip of the racket is

$$v = r\omega \approx (2)(35) = 70 \text{ ft/sec,}$$

or about 48 miles per hour. In a tennis serve the arm rotates at the shoulder, so the final speed of the racket is considerably faster. (**Source:** J. Cooper and R. Glassow, *Kinesiology.*)

◆ **CLASS DISCUSSION**

A human shoulder can rotate at about 25 radians per second. Estimate how much this rotation increases the speed of a racket. ◆

EXAMPLE 9 Finding the speed of a GPS satellite

Each of the 24 satellites used in the GPS is located 16,526 miles from the center of Earth and has a nearly circular orbit with a period of 12 hours. (**Source:** Y. Zhao, *Vehicle Location and Navigation Systems.*)

(a) Find the angular speed of a satellite.

(b) Estimate the linear speed of a satellite using the formula $v = r\omega$.

SOLUTION

(a) A GPS satellite circles Earth once every 12 hours. Its angular speed is

$$\frac{2\pi}{12} = \frac{\pi}{6} \approx 0.5236 \text{ rad/hr.}$$

(b) Its linear speed is $v = r\omega = (16{,}526)(0.5236) \approx 8653$ miles per hour.

Now Try Exercise 79 ◆

Area of a Sector

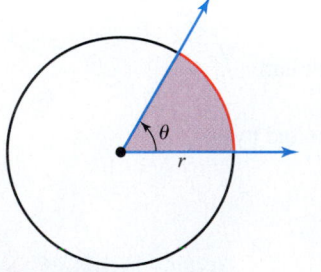

FIGURE 6.22 Sector of a Circle

The **sector of a circle** is the portion of the interior of a circle intercepted by a central angle. The shaded region in Figure 6.22 shows a sector of a circle with radius r and central angle θ.

The area of a sector is proportional to the measure of the central angle. If the central angle is 2π radians, then the area of the sector is the entire interior of the circle, which has an area of πr^2. Using proportions,

$$\frac{\text{area of a sector}}{\theta} = \frac{\pi r^2}{2\pi}.$$

Algebra Review

To review formulas related to circles, see Chapter R (page R-2).

Solving the equation for the area of the sector results in

$$\text{area of a sector} = \frac{1}{2}r^2\theta.$$

AREA OF A SECTOR

The **area of a sector** A of a circle of radius r and central angle θ in *radians* is given by

$$A = \frac{1}{2}r^2\theta.$$

Note: Angle θ *must* be in radian measure when using the area formula $A = \frac{1}{2}r^2\theta$.

EXAMPLE 10 Finding the area of a sector

A circle has a radius of 6 inches. Find the area of the sector if its central angle is $60°$.

SOLUTION Since $60°$ is equivalent to $\frac{\pi}{3}$ radians, the area of the sector is given by

$$A = \frac{1}{2}r^2\theta$$
$$= \frac{1}{2}(6)^2\left(\frac{\pi}{3}\right)$$
$$= 6\pi \text{ square inches.}$$

This region of $6\pi \approx 18.8$ square inches is illustrated in Figure 6.23.

Now Try Exercise 65 ◆

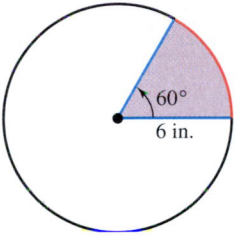

FIGURE 6.23

Consider the robotic arm shown in Figure 6.24. The *work space* of the robotic arm is the shaded region and corresponds to the places that the arm can reach either by rotating or by changing its length. (**Source:** W. Stadler, *Analytical Robotics and Mechatronics.*)

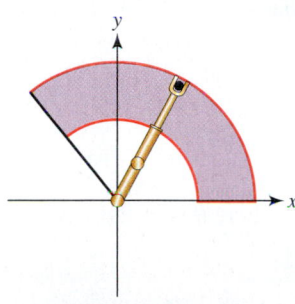

FIGURE 6.24 Work Space of a Robotic Arm

EXAMPLE 11 Finding the area of the work space for a robotic arm

Suppose that a robotic arm similar to the one in Figure 6.24 can rotate between $\theta = 10°$ and $\theta = 130°$. If the length of the robotic arm can vary between 5 inches and 20 inches, find the area of its work space.

SOLUTION The work space can be thought of as a large sector having radius $r_1 = 20$ inches with a small sector of radius $r_2 = 5$ inches removed. The arm can rotate

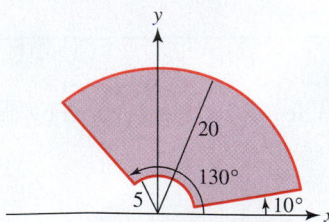

FIGURE 6.25

through $130° - 10° = 120°$ or $\frac{2\pi}{3}$ radians. See Figure 6.25. The area A of the work space is as follows.

$$A = \frac{1}{2}r_1^2\theta - \frac{1}{2}r_2^2\theta$$

$$= \frac{1}{2}\theta(r_1^2 - r_2^2)$$

$$= \frac{1}{2}\left(\frac{2\pi}{3}\right)(20^2 - 5^2)$$

$$= 125\pi$$

The work space is $125\pi \approx 392.7$ square inches.

Now Try Exercise 73 ◆

6.1   Putting it all Together

Some concepts involving angles are summarized in the following table.

Concept	Explanation or Formula	Examples
Degree measure	One complete rotation contains 360°.	A right angle contains 90°. A straight angle contains 180°. An acute angle α satisfies $0° < \alpha < 90°$. An obtuse angle β satisfies $90° < \beta < 180°$.
Radian measure	One complete rotation contains 2π radians.	2π radians are equivalent to 360°. π radians are equivalent to 180°. $\frac{\pi}{2}$ radians are equivalent to 90°. $\frac{\pi}{3}$ radians are equivalent to 60°. $\frac{\pi}{4}$ radian is equivalent to 45°. $\frac{\pi}{6}$ radian is equivalent to 30°.
Arc length	$s = r\theta$, where θ is in radians.	If $r = 12$ feet and $\theta = 90°$, then $s = (12)\frac{\pi}{2} = 6\pi \approx 18.8$ feet
Area of a sector	$A = \frac{1}{2}r^2\theta$, where θ is in radians.	If $r = 6$ inches and $\theta = 45°$, then $A = \frac{1}{2}(6)^2\left(\frac{\pi}{4}\right) = 4.5\pi \approx 14.1$ square inches.
Angular speed	$\omega = \frac{\theta}{t}$, where θ is the angle of rotation and t is time.	If $\theta = 5$ radians and $t = 0.1$ second then $\omega = \frac{5}{0.1} = 50$ radians per second
Linear speed of a rotating object	$v = r\omega$, where r is the radius and ω is the angular speed.	If $r = 3$ feet and $\omega = 5$ radians per second, then $v = (3)(5) = 15$ feet per second.

Angles

Exercises 1 and 2: Sketch the following angles in standard position.

1. **(a)** $45°$ **(b)** $-150°$

    (c) $\dfrac{\pi}{3}$ **(d)** $-\dfrac{3\pi}{4}$

2. **(a)** $-90°$ **(b)** $225°$

    (c) $-\dfrac{2\pi}{3}$ **(d)** $\dfrac{\pi}{6}$

Exercises 3–10: Sketch an angle θ in standard position that satisfies the conditions.

3. Acute

4. Obtuse

5. A positive straight angle

6. Complementary to $60°$

7. Positive and the terminal side lies in quadrant III

8. Negative and the terminal side lies in quadrant IV

9. Negative and coterminal to $\alpha = 90°$ if α is in standard position

10. Positive and coterminal to $\alpha = -135°$ if α is in standard position

11. What fraction of a complete revolution is each of the following angles?
    (a) $90°$ **(b)** $30°$

    (c) $\dfrac{\pi}{3}$ **(d)** $\dfrac{\pi}{4}$

12. What angle is its own complement? What angle is its own supplement?

Exercises 13–20: Find a positive angle and a negative angle that are coterminal to the given angle.

13. $150°$ 14. $65°$

15. $-72°$ 16. $-330°$

17. $\dfrac{\pi}{2}$ 18. $\dfrac{5\pi}{6}$

19. $-\dfrac{\pi}{5}$ 20. $-\dfrac{2\pi}{3}$

Exercises 21–26: Find the complementary angle α and the supplementary angle β to θ.

21. $\theta = 55.9°$ 22. $\theta = 71.5°$

23. $\theta = 85°23'45''$ 24. $\theta = 5°45'30''$

25. $\theta = 23°40'35''$ 26. $\theta = 67°25'10''$

Exercises 27–30: Express the angle in decimal degrees.

27. $125°15'$ 28. $15°30'$

29. $108°45'36''$ 30. $256°06'12''$

Exercises 31–34: Convert the given angle to degrees, minutes, and seconds.

31. $125.3°$ 32. $15.25°$

33. $51.36°$ 34. $22.46°$

Exercises 35–38: Use the figure to determine the radian measure of angle θ. Then approximate the degree measure of θ to the nearest tenth of a degree.

35.
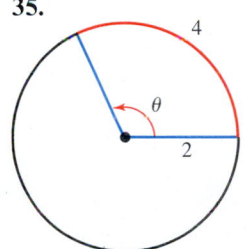

36.
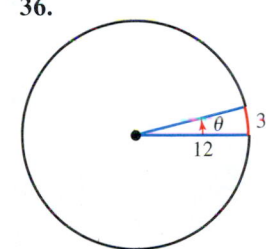

37.
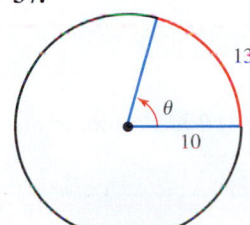

38.
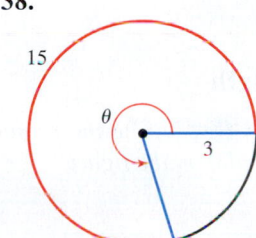

Exercises 39–42: Convert each angle from degree measure to radian measure. Round to the nearest hundredth of a radian when appropriate.

39. (a) 45° (b) 135°

   (c) −120° (d) −210°

40. (a) 105° (b) 245°

   (c) −255° (d) −80°

41. (a) 37° (b) 123.4°

   (c) −92°25′ (d) 230°17′

42. (a) 56° (b) 88.7°

   (c) 122°15′ (d) −7°48′

Exercises 43–46: Convert each angle from radian measure to degree measure. Round to the nearest hundredth of a degree when appropriate.

43. (a) $\dfrac{\pi}{6}$ (b) $\dfrac{\pi}{15}$

   (c) $-\dfrac{5\pi}{3}$ (d) $-\dfrac{7\pi}{6}$

44. (a) $-\dfrac{\pi}{12}$ (b) $-\dfrac{5\pi}{2}$

   (c) $\dfrac{17\pi}{15}$ (d) $\dfrac{5\pi}{6}$

45. (a) $\dfrac{\pi}{4}$ (b) $\dfrac{\pi}{7}$

   (c) 3.1 (d) $-\dfrac{5}{2}$

46. (a) $\dfrac{7\pi}{2}$ (b) $\dfrac{2\pi}{5}$

   (c) −4.1 (d) $-\dfrac{2}{3}$

Arc Length

Exercises 47–52: Use the formula $s = r\theta$ to determine the missing value in the figure.

47. 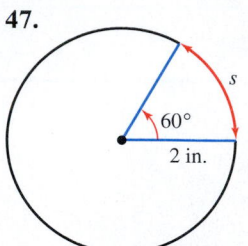 **48.**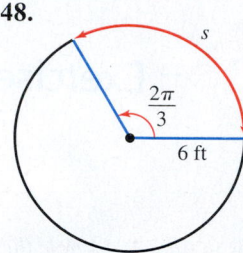

49. 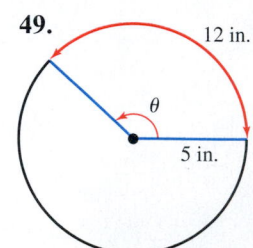 **50.**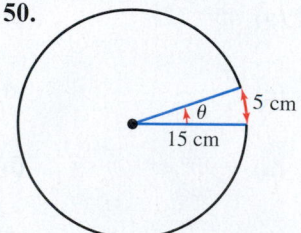

51. 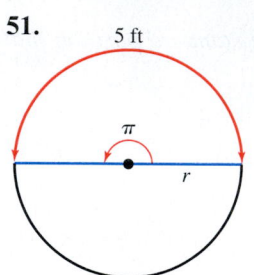 **52.**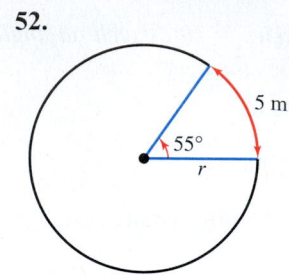

Exercises 53–58: Find the length of the arc intercepted by a central angle θ in a circle of radius r.

53. $r = 3$ m, $\theta = \dfrac{\pi}{12}$ **54.** $r = 7.3$ mm, $\theta = \dfrac{7\pi}{4}$

55. $r = 12$ ft, $\theta = 15°$ **56.** $r = 5$ cm, $\theta = 240°$

57. $r = 2$ mi, $\theta = 1°45′$ **58.** $r = 3$ mi, $\theta = 4°15′09″$

Exercises 59–62: A minute hand on a clock is 4 inches long. Determine how far the tip of the minute hand travels between the given times. Find the linear speed of the tip.

59. 10:15 A.M., 10:30 A.M. **60.** 1:00 P.M., 1:40 P.M.

61. 3:00 P.M., 4:15 P.M. **62.** 11:00 A.M., 1:25 P.M.

63. A bicycle has a tire 26 inches in diameter that is rotating at 15 radians per second. Approximate the speed of the bicycle in feet per second and in miles per hour.

64. The wheels on a skateboard have a diameter of 2.25 inches. If a skateboarder is traveling downhill at 15 miles per hour, determine the angular velocity of the wheels in radians per second.

Area of a Sector

Exercises 65–68: Find the area of the shaded sector.

65.

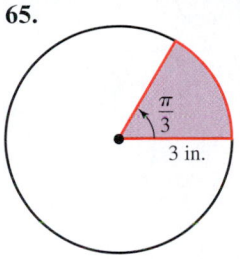

$\frac{\pi}{3}$

3 in.

66.

$\frac{5\pi}{6}$

2 ft

67.

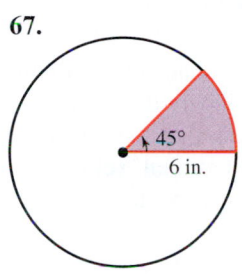

45°

6 in.

68.

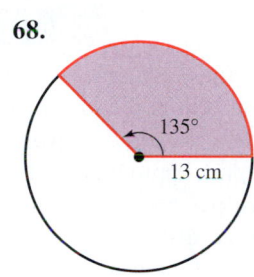

135°

13 cm

Exercises 69–72: Find the area of the sector of a circle having radius r and central angle θ.

69. $r = 13.1$ cm, $\theta = \dfrac{\pi}{15}$ **70.** $r = 7.3$ m, $\theta = \dfrac{5\pi}{4}$

71. $r = 1.5$ ft, $\theta = 30°$ **72.** $r = 5.5$ in., $\theta = 225°$

Exercises 73–76: **Robotics** *(Refer to Example 11.) Find the area of the work space for a robotic arm that can rotate between angles θ_1 and θ_2 and can change its length from r_1 to r_2. See the accompanying figure.*

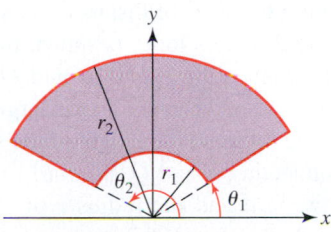

73. $\theta_1 = -45°, \theta_2 = 90°, r_1 = 6$ in., $r_2 = 26$ in.

74. $\theta_1 = -60°, \theta_2 = 60°, r_1 = 0.5$ ft, $r_2 = 2.5$ ft

75. $\theta_1 = 15°, \theta_2 = 195°, r_1 = 21$ cm, $r_2 = 95$ cm

76. $\theta_1 = 43°, \theta_2 = 178°, r_1 = 0.4$ m, $r_2 = 1.8$ m

Applications

77. *Global Positioning System* (Refer to Example 4.) Suppose that a GPS receiver displays the latitude of a location to within $\pm\frac{1}{100,000}$ of a degree. Convert this error to seconds of a degree.

78. *Ferris Wheel* A large Ferris wheel has a diameter of 140 feet. It completes 1 revolution every 420 seconds.
 (a) Find the angular velocity of the Ferris wheel in radians per second.

 (b) What is the linear speed of a person who is riding this Ferris wheel?

79. *Fan Speed* The blades of a fan have a 30-inch diameter and rotate at 500 revolutions per minute.
 (a) Find the angular velocity of a fan blade.

 (b) Estimate the linear speed at the tip of a fan blade.

80. *Earth's Rotation* Earth rotates 1 complete revolution every 24 hours and has a radius of about 3955 miles.
 (a) Find the angular velocity of a person standing at the equator in radians per hour.

 (b) Estimate the linear speed in miles per hour at the equator due to Earth's rotation.

81. *Distance between Cities* (Refer to Example 8.) Daytona Beach, Florida, and Akron, Ohio, have nearly the same longitude of 81° W. The latitude of Daytona Beach is 29°11′ and the latitude of Akron is 40°55′. Approximate the distance between these two cities if the average radius of Earth is 3955 miles. (**Source:** J. Williams.)

82. *Nautical Miles* Nautical miles are used by ships and airplanes. They are different from statute miles, which equal 5280 feet. A nautical mile is defined to be the arc length along the equator intercepted by a central angle *AOB* of 1 minute, as illustrated in the following figure. If the equatorial radius of Earth is 3963 miles, use the arc length formula to approximate the number of statute miles in 1 nautical mile. Round your answer to two decimal places.

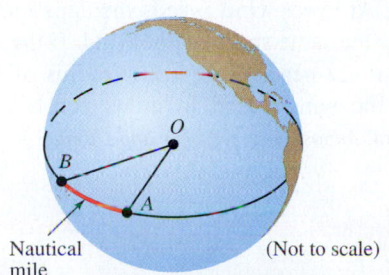

O

B

A

Nautical mile

(Not to scale)

83. *Pulleys* Approximate how many inches the weight in the following figure will rise if $r = 11$ inches and the pulley is rotated through an angle of $75.3°$.

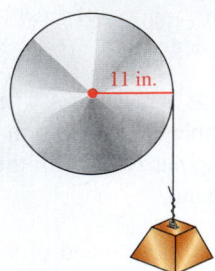

11 in.

84. *Pulleys* Use the figure in the previous exercise to estimate the angle θ through which the pulley should be rotated to raise the weight 5 inches.

85. *Bicycle Chain Drive* The figure shows the chain drive of a bicycle. The radius of the sprocket wheel that the pedals are attached to is 3.75 inches and the radius of the other sprocket wheel is 1.5 inches.
   (a) If the pedals are rotated one revolution, determine the number of revolutions that the bicycle tire rotates.

   (b) If the bicycle has a tire with a 26-inch diameter, determine how fast the bicycle travels in feet per second if the pedals turn through two revolutions per second.

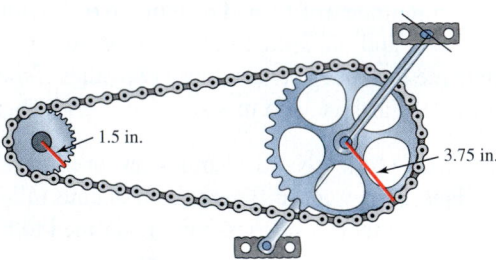

1.5 in.   3.75 in.

86. *Wind Speed* One of the most common ways to measure wind speed is with a *three-cup anemometer,* as shown in the figure at the top of the next column. The cups catch the wind and cause the vertical shaft to rotate. At lower wind speeds the cups move at approximately the same speed as the wind. If the cups are rotating 5 times per second with a radius of 6 inches, estimate the wind speed in miles per hour. (**Source:** J. Navarra, *Atmosphere, Weather and Climate.*)

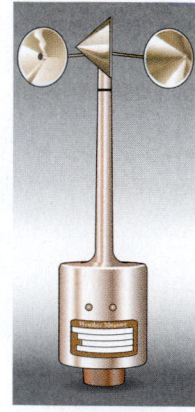

87. *Velocity of Planets* The average distance D in millions of miles from the sun and the orbital period P in years of various planets are given. Assuming that the orbits are circular, approximate the average orbital velocity in miles per hour for each planet. Discuss the effect that average distance from the sun has on orbital velocity. (**Source:** C. Ronan, *The Natural History of the Universe.*)
   (a) Venus: $D = 67.2$, $P = 0.615$

   (b) Earth: $D = 92.9$, $P = 1$

   (c) Jupiter: $D = 483.6$, $P = 11.86$

   (d) Neptune: $D = 2794$, $P = 164.8$

88. *Speed of a Propeller* A 90-horsepower outboard motor at full-throttle rotates its propeller 5000 revolutions per minute. Find the angular velocity of the propeller in radians per second. What is the linear speed in inches per second of a point at the tip of the propeller if its diameter is 10 inches?

89. *Surveying* The *subtense bar method* is a technique used in surveying to measure distances. A subtense bar, which is usually 2 meters long, is shown in the accompanying figure connecting points P and Q. If the distance d from the surveyor to the bar is large, then there is little difference between the length of the subtense bar and the arc connecting P and Q. Similarly, there is little difference between d and the radius r of the arc intercepted by the subtense bar. If θ is measured to be $0.835°$, approximate d using the arc length formula. (**Source:** I. Mueller and K. Ramsayer, *Introduction to Surveying.*)

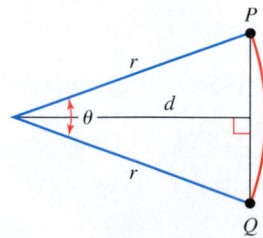

P

r

d

θ

r

Q

90. *Diameter of the Moon* (Refer to the previous exercise.) The distance to the moon is approximately 238,900 miles. Use the arc length formula to estimate the diameter d of the moon if angle θ in the accompanying figure is measured to be 0.517°.

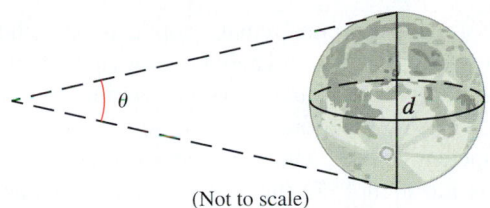

(Not to scale)

91. *Global Positioning System* A GPS location device can consistently determine latitude to within ±0.001°. If the average radius of Earth is 3955 miles, approximate how accurately the north-south position of an object can be found. (**Source:** Y. Zhao.)

92. *Doppler Radar* Radar is used to identify severe weather. If Doppler radar can detect weather within a 240-mile radius and creates a new image every 48 seconds, find the area scanned by the radar in 1 second.

93. *Arc Length Formula* Modify the arc length formula $s = r\theta$ so that angle θ can be given in degrees rather than radians. Which of the two formulas is simpler?

94. *Area of a Sector Formula* Modify the area formula $A = \frac{1}{2}r^2\theta$ so that angle θ can be given in degrees rather than radians. Which of the two formulas is simpler?

95. *Solar Power Plant* A 150-megawatt solar power plant requires approximately 475,000 square meters of land to collect the required amount of energy from sunlight. (**Source:** C. Winter, *Solar Power Plants.*)
(a) If this land area is circular, approximate its radius.
(b) If this land area is a sector of a circle with $\theta = 70°$, approximate its radius.

96. *Location of the North Star* Presently the North Star, Polaris, is located near the true North Pole. However, because Earth is inclined 23.5°, the moon's gravitational pull on Earth is uneven. As a result, Earth precesses like a spinning top and the direction of the celestial North Pole traces out a circular path once every 26,000 years, as shown in the figure at the top of the next column. For example, in the year 14,000 the star Vega and not Polaris will be located at the celestial North Pole. As viewed from the center C of this circular path, calculate the angle in seconds that the celestial North Pole moves each year. (**Source:** M. Zeilik et al., *Introductory Astronomy and Astrophysics.*)

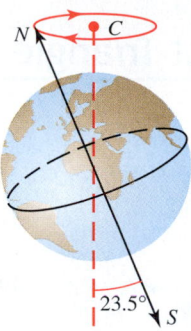

97. *Measuring the Circumference of Earth* The first accurate estimate of the distance around Earth was done by the Greek astronomer Eratosthenes (276–195 B.C.), who noted that the noontime position of the sun at the summer solstice differed by 7°12′ from the city of Syene to the city of Alexander. See the accompanying figure. The distance between these two cities is 496 miles. Use the arc length formula to estimate the radius of Earth. Then find the circumference of Earth. (**Source:** M. Zeilik.)

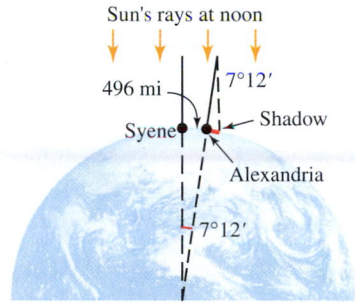

98. *Club Speed in Golf* The shoulder joint can rotate at about 25 radians per second. Assuming that a golfer's arm is straight and the distance from the shoulder to the club head is 5 feet, estimate the linear speed of the club head from shoulder rotation. (**Source:** J. Cooper and R. Glassow, *Kinesiology.*)

Writing about Mathematics

99. Give definitions for 1 degree and 1 radian. Compare these two units of angle measure. Which unit of measure do you prefer? Explain why.

100. Suppose a central angle θ of a circle remains fixed. Describe what happens to the arc length intercepted by θ and the area of the corresponding sector as the radius r doubles and triples.

6.2 Right Triangle Trigonometry

◆ Learn basic concepts about trigonometric functions

◆ Apply right triangle trigonometry

◆ Understand complementary angles and cofunctions

Introduction

A right triangle is a basic geometric shape that occurs in many applications such as astronomy, surveying, construction, highway design, GPS, and aerial photography. Trigonometric functions are used to *solve* triangles. Solving a triangle involves finding the measure of each side and angle in the triangle. The sine function is one of the earliest trigonometric functions and it dates back to the ancient Greeks, who invented a similar function called the *chord function*. However, it was not until 1550 that the sine function was formally defined in terms of right triangles by Georg Rhaeticus. The six trigonometric functions are similar to other functions that we have encountered previously in that they have symbolic, graphical, and numerical representations and can be used to model data and a variety of physical phenomena. (**Source:** *Historical Topics for the Mathematics Classroom, Thirty-first Yearbook,* NCTM.)

Basic Concepts of Trigonometric Functions

The following **standard labeling** is used to designate vertices, angles, and sides of triangle *ABC*, as shown in Figure 6.26. The vertices are denoted *A*, *B*, and *C*, the angles α, β, and γ, and the lengths of the sides opposite these angles *a*, *b*, and *c*. If triangle *ABC* is a right triangle, then we let $\gamma = 90°$.

Many important ideas in trigonometry depend on the properties of similar triangles. **Similar triangles** have congruent corresponding angles, but similar triangles are not necessarily the same size. An example of similar triangles is shown in Figure 6.27. Corresponding sides of similar triangles are proportional.

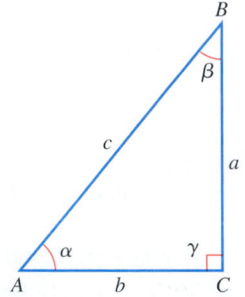

FIGURE 6.26 Standard Labeling

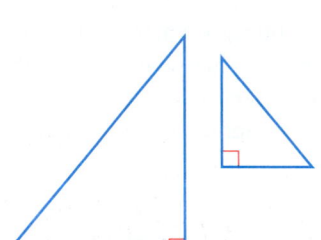

FIGURE 6.27 Similar Triangles

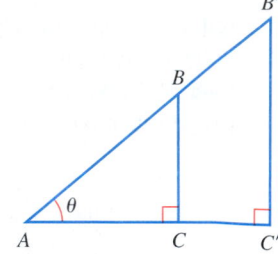

FIGURE 6.28

The right triangles *ABC* and *AB′C′* shown in Figure 6.28 are similar triangles. Using the properties of similar triangles the following ratios are equal.

$$\frac{BC}{AB} = \frac{B'C'}{AB'}$$

Geometry Review

To review similar triangles see Chapter R (page R-5).

That is, the ratio of the side opposite angle θ to the hypotenuse is constant for a given angle θ and does not depend on the size of the right triangle. If the measure of θ changes, then the ratio of the side opposite to the hypotenuse also changes. This concept can be

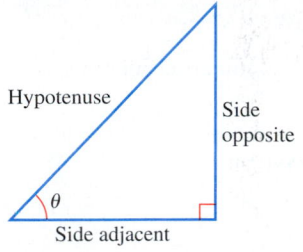

FIGURE 6.29

used to define a new function called the *sine function*. That is, if θ is an acute angle in a right triangle, as shown in Figure 6.29, then we define the sine of θ as

$$\sin\theta = \frac{\text{side opposite}}{\text{hypotenuse}},$$

where $\sin\theta$ denotes the sine function with input θ.

◆ **CLASS DISCUSSION**

Is it possible that $\sin\theta > 1$ for some angle θ? Explain your reasoning. ◆

EXAMPLE 1

Evaluating the sine function

Find $\sin 30°$. Support your answer by using a calculator.

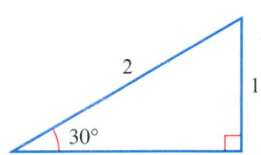

FIGURE 6.30

SOLUTION   Since the sine function depends only on the measure of θ, we can choose any size right triangle to evaluate $\sin\theta$. For convenience let the length of the hypotenuse equal 2, as shown in Figure 6.30. From geometry we know that the length of the shortest leg in a $30°-60°$ right triangle is half the hypotenuse. Thus the side opposite equals 1 and

$$\sin 30° = \frac{\text{side opposite}}{\text{hypotenuse}} = \frac{1}{2}.$$

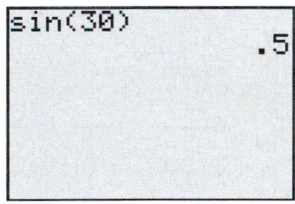

FIGURE 6.31 Degree Mode

This result is supported in Figure 6.31, where the sine function has been evaluated at $30°$ using a calculator set in degree mode.   *Now Try Exercise 5* ◆

Using Figures 6.28 and 6.29, we can define other trigonometric functions. Since

$$\frac{AC}{AB} = \frac{AC'}{AB'},$$

the ratio of the side adjacent to the hypotenuse is constant for a fixed angle θ and does not depend on the size of the right triangle. We define the *cosine function* to be

$$\cos\theta = \frac{\text{side adjacent}}{\text{hypotenuse}}.$$

The six trigonometric functions of angle θ are called **sine**, **cosine**, **tangent**, **cosecant**, **secant**, and **cotangent**. We use the customary abbreviations for each trigonometric function.

RIGHT TRIANGLE-BASED DEFINITIONS OF TRIGONOMETRIC FUNCTIONS

Let θ be an acute angle in a right triangle. Then the six trigonometric functions of θ may be evaluated as follows.

$$\sin\theta = \frac{\text{side opposite}}{\text{hypotenuse}} \qquad \cos\theta = \frac{\text{side adjacent}}{\text{hypotenuse}} \qquad \tan\theta = \frac{\text{side opposite}}{\text{side adjacent}}$$

$$\csc\theta = \frac{\text{hypotenuse}}{\text{side opposite}} \qquad \sec\theta = \frac{\text{hypotenuse}}{\text{side adjacent}} \qquad \cot\theta = \frac{\text{side adjacent}}{\text{side opposite}}$$

The next example illustrates how to evaluate the trigonometric functions.

Evaluating trigonometric functions

Consider the right triangle shown in Figure 6.32. Find the six trigonometric functions of θ.

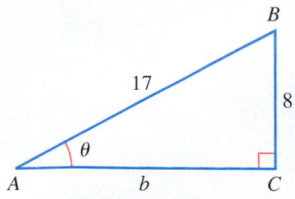

FIGURE 6.32

Algebra Review
To review the Pythagorean theorem, see Chapter R (page R-3).

SOLUTION In triangle ABC the side opposite angle θ is $a = 8$ and the hypotenuse is $c = 17$. To find the adjacent side b we apply the Pythagorean theorem.

$$c^2 = a^2 + b^2 \qquad \text{Pythagorean theorem}$$
$$b^2 = c^2 - a^2 \qquad \text{Solve for } b^2.$$
$$b^2 = 17^2 - 8^2 \qquad \text{Let } c = 17 \text{ and } a = 8.$$
$$b^2 = 225 \qquad \text{Simplify.}$$
$$b = 15 \qquad \text{Solve for } b, \text{ where } b > 0.$$

Thus the six trigonometric functions of θ are as follows.

$$\sin\theta = \frac{\text{side opposite}}{\text{hypotenuse}} = \frac{8}{17} \qquad \csc\theta = \frac{\text{hypotenuse}}{\text{side opposite}} = \frac{17}{8}$$

$$\cos\theta = \frac{\text{side adjacent}}{\text{hypotenuse}} = \frac{15}{17} \qquad \sec\theta = \frac{\text{hypotenuse}}{\text{side adjacent}} = \frac{17}{15}$$

$$\tan\theta = \frac{\text{side opposite}}{\text{side adjacent}} = \frac{8}{15} \qquad \cot\theta = \frac{\text{side adjacent}}{\text{side opposite}} = \frac{15}{8}$$

Now Try Exercise 17 ◆

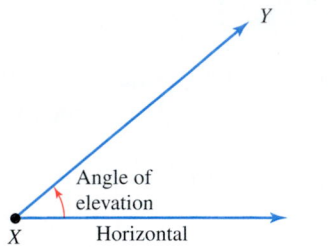

FIGURE 6.33 Angle of Elevation

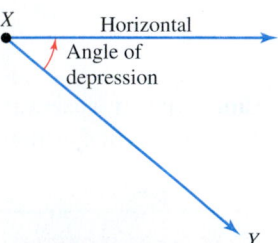

FIGURE 6.34 Angle of Depression

If an object is located above the horizontal, then the angle between the horizontal and the line of sight XY is called the **angle of elevation**. See Figure 6.33. If an object is located below the horizontal, then the angle between the horizontal and the line of sight XY is called the **angle of depression**. See Figure 6.34.

Trigonometry allows people to determine distances and heights without measuring them directly. For example, the altitude of the cloud base is particularly important at airports. (The cloud base is where the lowest layer of clouds begins to form.) Although it is not practical to measure the altitude of the cloud base directly, trigonometry can indirectly determine this height at nighttime. In Figure 6.35 a bright spotlight is directed vertically upward. It creates a bright spot on the cloud base. From a known horizontal distance d from the spotlight, the angle of elevation θ is measured. The side adjacent to θ is d and the side opposite is h, where h represents the height of the cloud base. (**Source:** F. Cole, *Introduction to Meteorology.*) It follows that

$$\tan\theta = \frac{\text{side opposite}}{\text{side adjacent}} = \frac{h}{d}.$$

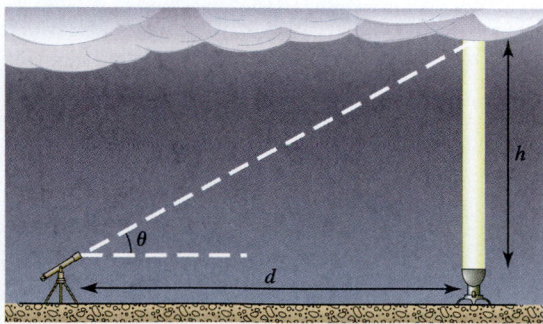

FIGURE 6.35

EXAMPLE 3

Determining the cloud base

Suppose that in Figure 6.35 $\theta = 55°$ and $d = 1150$ feet. Estimate the height of the cloud base. (Neglect the height of the telescope in Figure 6.35.)

SOLUTION Solve the equation $\tan \theta = \frac{h}{d}$ for h and then substitute values for θ and d.

$$\tan \theta = \frac{h}{d}$$

$$h = \mathbf{d} \tan \theta \qquad \text{Cross multiply.}$$

$$= \mathbf{1150} \tan \mathbf{55°} \qquad \text{Substitute for } d \text{ and } \theta.$$

$$\approx 1150 \, (1.4281) \qquad \text{Approximate } \tan 55°.$$

$$\approx 1642 \text{ feet} \qquad \text{Multiply.}$$

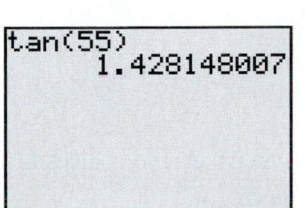

```
tan(55)
        1.428148007
```

FIGURE 6.36 Degree Mode

Thus the cloud base is about 1642 feet. A calculator was used to approximate $\tan 55°$, as shown in Figure 6.36.   *Now Try Exercise 57* ◆

We can use trigonometric functions to find unknown sides of a right triangle. This process is sometimes referred to as *solving a triangle* and is demonstrated in the next example.

EXAMPLE 4

Solving a triangle

Find the lengths of the unknown sides a and c for the right triangle shown in Figure 6.37. Round each value to the nearest hundredth.

SOLUTION We are given angle $\theta = 40°$ and side $b = 35$, which is adjacent to angle θ. Side a is opposite angle θ. Because the tangent function involves the opposite and adjacent sides, we use it to find side a.

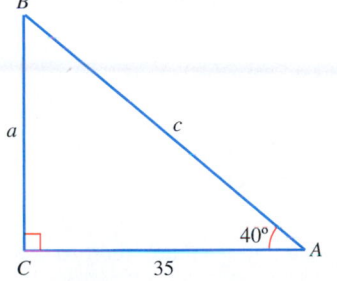

FIGURE 6.37

$$\tan 40° = \frac{a}{35} \qquad \tan \theta = \frac{\text{side opposite}}{\text{side adjacent}}$$

$$35 \tan 40° = a \qquad \text{Multiply by 35.}$$

$$a \approx 29.37 \qquad \text{Rewrite; approximate (if desired).}$$

To find the length of hypotenuse c we can use $\cos \theta$, which involves the adjacent side and the hypotenuse of the right triangle.

$$\cos 40° = \frac{35}{c} \qquad \cos \theta = \frac{\text{side adjacent}}{\text{hypotenuse}}$$

$$c \cos 40° = 35 \qquad \text{Multiply by } c.$$

$$c = \frac{35}{\cos 40°} \qquad \text{Divide by } \cos 40°.$$

$$c \approx 45.69 \qquad \text{Rewrite; approximate (if desired).}$$

Now Try Exercise 21 ◆

In most applications, calculators are used to approximate values of the trigonometric functions. However, with the aid of geometry we can determine exact values for the trigonometric functions of some special angles such as 30°, 45°, and 60°.

Evaluating trigonometric functions

Evaluate the six trigonometric functions of $\theta = 45°$.

SOLUTION Begin by drawing a right triangle with a 45° angle, as shown in Figure 6.38. The lengths of the legs in this triangle are equal. Since the size of the right triangle does not affect the values of the trigonometric functions, let the lengths of both legs equal 1. Using the Pythagorean theorem, we can find the length of the hypotenuse as follows.

$$c^2 = a^2 + b^2 \qquad \textcolor{blue}{\textit{Pythagorean theorem}}$$
$$c^2 = 1^2 + 1^2 \qquad \textcolor{blue}{a = b = 1}$$
$$c^2 = 2 \qquad \textcolor{blue}{\textit{Simplify.}}$$
$$c = \sqrt{2} \qquad \textcolor{blue}{\textit{Solve for c, where c > 0.}}$$

The hypotenuse has length $\sqrt{2}$. Evaluating the six trigonometric functions gives the following.

$$\sin 45° = \frac{\text{side opposite}}{\text{hypotenuse}} = \frac{1}{\sqrt{2}} \qquad \csc 45° = \frac{\text{hypotenuse}}{\text{side opposite}} = \frac{\sqrt{2}}{1} = \sqrt{2}$$

$$\cos 45° = \frac{\text{side adjacent}}{\text{hypotenuse}} = \frac{1}{\sqrt{2}} \qquad \sec 45° = \frac{\text{hypotenuse}}{\text{side adjacent}} = \frac{\sqrt{2}}{1} = \sqrt{2}$$

$$\tan 45° = \frac{\text{side opposite}}{\text{side adjacent}} = \frac{1}{1} = 1 \qquad \cot 45° = \frac{\text{side adjacent}}{\text{side opposite}} = \frac{1}{1} = 1$$

Now Try Exercise 39 ◆

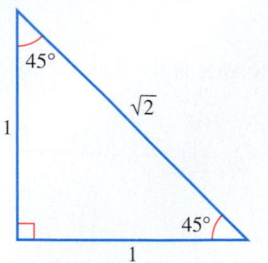

FIGURE 6.38 A 45°–45° Right Triangle

In Example 5, $\sin 45° = \frac{1}{\sqrt{2}}$ and $\csc \theta = \frac{\sqrt{2}}{1}$. Since

$$\sin \theta = \frac{\text{side opposite}}{\text{hypotenuse}} \quad \text{and} \quad \csc \theta = \frac{\text{hypotenuse}}{\text{side opposite}},$$

it follows that $\csc \theta = \frac{1}{\sin \theta}$ in general. That is, the values of $\csc \theta$ and $\sin \theta$ are *reciprocals*. In a similar manner,

$$\sec \theta = \frac{1}{\cos \theta} \quad \text{and} \quad \cot \theta = \frac{1}{\tan \theta}.$$

Most calculators have keys to evaluate the sine, cosine, and tangent functions, but do not have keys to evaluate the cosecant, secant, and cotangent functions. These three functions may be evaluated by using the following *reciprocal identities*.

$$\csc \theta = \frac{1}{\sin \theta}, \qquad \sec \theta = \frac{1}{\cos \theta}, \qquad \cot \theta = \frac{1}{\tan \theta}$$

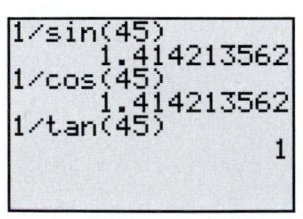

FIGURE 6.39 Degree Mode

Some of the results from Example 5 are supported in Figure 6.39, where these reciprocal identities have been applied to evaluate $\csc 45°$, $\sec 45°$, and $\cot 45°$ with a calculator. Note that $\sqrt{2} \approx 1.414213562$.

Caution: Do not use the $\sin^{-1}$, $\cos^{-1}$, and $\tan^{-1}$ keys on your calculator to evaluate reciprocals because they represent inverse functions, which will be discussed in Section 6.6. You may want to use the x^{-1} key to evaluate reciprocals instead.

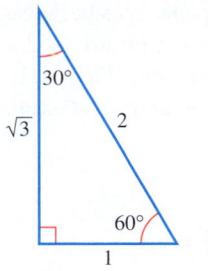

FIGURE 6.40 A 30°–60°
Right Triangle

Using the right triangles in Figures 6.38 and 6.40, we can evaluate the six trigonometric functions at 30°, 45°, and 60° without the aid of a calculator. Table 6.2 lists the values of the six trigonometric functions for selected angles.

TABLE 6.2

θ	$\sin\theta$	$\cos\theta$	$\tan\theta$	$\csc\theta$	$\sec\theta$	$\cot\theta$
30°	$\frac{1}{2}$	$\frac{\sqrt{3}}{2}$	$\frac{1}{\sqrt{3}}$	2	$\frac{2}{\sqrt{3}}$	$\sqrt{3}$
45°	$\frac{1}{\sqrt{2}}$	$\frac{1}{\sqrt{2}}$	1	$\sqrt{2}$	$\sqrt{2}$	1
60°	$\frac{\sqrt{3}}{2}$	$\frac{1}{2}$	$\sqrt{3}$	$\frac{2}{\sqrt{3}}$	2	$\frac{1}{\sqrt{3}}$

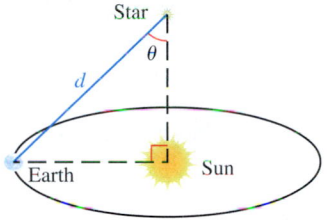

FIGURE 6.41 Parallax of a Star
(Not to scale)

Applications of Right Triangle Trigonometry

For centuries astronomers wanted to know how far it was to the stars. Not until 1838 did the astronomer Friedrich Bessel determine the distance to a star called 61 Cygni. He used a *parallax* method that relied on the measurement of very small angles. See Figure 6.41. As Earth revolves around the sun, the observed parallax of 61 Cygni is $\theta \approx 0.0000811°$. Because stars are so distant, parallax angles are very small. (**Sources:** H. Freebury, *A History of Mathematics;* M. Zeilik et al., *Introductory Astronomy and Astrophysics.*)

EXAMPLE 6   Calculating the distance to a star

One of the nearest stars is Alpha Centauri, which has a parallax of $\theta \approx 0.000212°$. (**Source:** M. Zeilik et al.)

(a) Calculate the distance to Alpha Centauri if the Earth–Sun distance is 93,000,000 miles.

(b) A light-year is defined to be the distance that light travels in 1 year and equals about 5.9 trillion miles. Find the distance to Alpha Centauri in light-years.

SOLUTION

(a) Let d represent the distance between Earth and Alpha Centauri. From Figure 6.41, it can be seen that

$$\sin\theta = \frac{93,000,000}{d} \quad \text{or} \quad d = \frac{93,000,000}{\sin\theta}.$$

Substituting for θ gives the following result.

$$d = \frac{93,000,000}{\sin 0.000212°} \approx 2.51 \times 10^{13} \text{ miles}$$

(b) This distance equals $\dfrac{2.51 \times 10^{13}}{5.9 \times 10^{12}} \approx 4.3$ light-years.

Now Try Exercise 63 ◆

Water is often an obstacle to surveyors in the field when measuring distances between two points. For example, to measure the distance between points P and Q in Figure 6.42 a baseline PR, perpendicular to PQ, is determined. Angle PRQ is then measured. Then right triangle trigonometry can be used to determine the length of PQ. (**Source:** P. Kissam, *Surveying Practice.*)

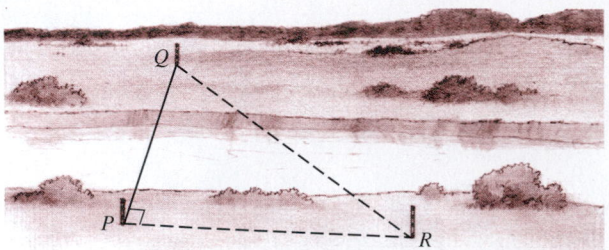

FIGURE 6.42

EXAMPLE 7 Finding distance

Suppose in Figure 6.42 the length of PR is 94.75 feet and angle PRQ has measure 41.6°. Estimate the distance between points P and Q.

SOLUTION Let angle PRQ be θ. Since $\tan\theta = \frac{PQ}{PR}$, it follows that

$$PQ = PR\tan\theta = 94.75\tan(41.6°) \approx 84.12 \text{ feet.}$$

Now Try Exercise 71 ◆

In the next example we derive a formula that is used in the design of highways.

EXAMPLE 8 Deriving a formula for the design of highway curves

One common type of highway curve is a *simple horizontal curve*. It consists of two straight segments of highway connected by a circular arc with radius r, as shown in Figure 6.43. The distance d is called the *external distance*. (**Source:** F. Mannering and W. Kilareski, *Principles of Highway Engineering and Traffic Analysis.*)
(a) Derive a formula for d that involves r and θ.
(b) Find d for a curve with a 750-foot radius and $\theta = 36°$.

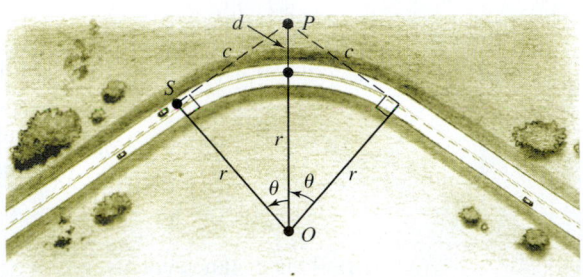

FIGURE 6.43 A Simple Horizontal Curve

SOLUTION

(a)

$$\cos\theta = \frac{r}{r+d} \qquad \text{Use triangle OSP.}$$

$$(r+d)\cos\theta = r \qquad \text{Multiply by } r+d.$$

$$r + d = \frac{r}{\cos\theta} \qquad \text{Divide by } \cos\theta.$$

$$d = \frac{r}{\cos\theta} - r \qquad \text{Subtract } r.$$

$$d = r\left(\frac{1}{\cos\theta} - 1\right) \qquad \text{Factor out } r.$$

(b) $d = 750\left(\dfrac{1}{\cos 36°} - 1\right) \approx 177$ feet.

Now Try Exercise 77 ◆

Complementary Angles and Cofunctions

In Figure 6.44, α and β are complementary angles since their measures sum to 90°. The six trigonometric functions for α and β can be expressed as follows.

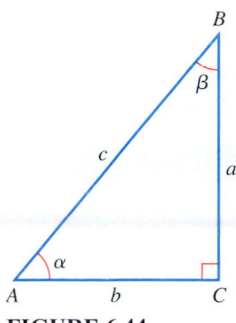

FIGURE 6.44

$$\sin\alpha = \frac{a}{c} = \cos\beta \qquad \cos\alpha = \frac{b}{c} = \sin\beta$$

$$\tan\alpha = \frac{a}{b} = \cot\beta \qquad \cot\alpha = \frac{b}{a} = \tan\beta$$

$$\sec\alpha = \frac{c}{b} = \csc\beta \qquad \csc\alpha = \frac{c}{a} = \sec\beta$$

Notice that the value of a trigonometric function of α equals the value of the trigonometric cofunction for β. For example, $\sin\alpha = \cos\beta$ and $\sin\beta = \cos\alpha$. This is how cofunctions were named. In 1620 Edmund Gunter combined the words "complement" and "sine" to obtain *co*sine. In a similar manner the cosecant and cotangent functions are the "complementary functions" of the secant and tangent functions, respectively, and their names were shortened to *co*secant and *co*tangent.

COFUNCTION FORMULAS

$$\sin\theta = \cos(90° - \theta) \qquad \cos\theta = \sin(90° - \theta)$$

$$\tan\theta = \cot(90° - \theta) \qquad \cot\theta = \tan(90° - \theta)$$

$$\sec\theta = \csc(90° - \theta) \qquad \csc\theta = \sec(90° - \theta)$$

EXAMPLE 9 **Evaluating trigonometric functions using complementary angles**

Write an equivalent expression using a cofunction. Then evaluate the expression using a calculator.

(a) $\cot 23°$ **(b)** $\sec 70°$ **(c)** $\cos 12°$

SOLUTION

(a) The complementary angle of $23°$ is $90° - 23° = 67°$. Thus

$$\cot 23° = \tan 67° \approx 2.3559.$$

(b) $\sec \mathbf{70°} = \csc (90° - \mathbf{70°}) = \csc 20° = \dfrac{1}{\sin 20°} \approx 2.9238$

(c) $\cos \mathbf{12°} = \sin (90° - \mathbf{12°}) = \sin 78° \approx 0.9781$

Now Try Exercise 53 ◆

6.2

Putting it all Together

The following table summarizes some properties of right triangle trigonometry.

Concept	Formulas and Figures
Trigonometric functions	Let θ be an acute angle in a right triangle ABC. $\sin\theta = \dfrac{\text{side opposite}}{\text{hypotenuse}}$ $\cos\theta = \dfrac{\text{side adjacent}}{\text{hypotenuse}}$ $\tan\theta = \dfrac{\text{side opposite}}{\text{side adjacent}}$ $\csc\theta = \dfrac{\text{hypotenuse}}{\text{side opposite}}$ $\sec\theta = \dfrac{\text{hypotenuse}}{\text{side adjacent}}$ $\cot\theta = \dfrac{\text{side adjacent}}{\text{side opposite}}$
Cofunction formulas	Let α and β be complementary angles. $\sin\alpha = \cos(90° - \alpha) = \cos\beta$ $\tan\alpha = \cot(90° - \alpha) = \cot\beta$ $\sec\alpha = \csc(90° - \alpha) = \csc\beta$

6.2 Exercises

Basic Concepts

Exercises 1–4: Sketch a right triangle with the following properties. Label the measure of each angle and side.

1. Acute angles of 30° and 60°, and a hypotenuse with length 2

2. Acute angle of 45° and a leg with length 1

3. Isosceles and a hypotenuse with length 4

4. Acute angle of 60° and the shorter leg with length 3

Exercises 5–10: (Refer to Example 1.) Use a 30°–60° right triangle to find the exact value of the trigonometric expression.

5. sin 60°   6. tan 30°

7. cos 30°   8. cot 30°

9. sec 60°   10. csc 60°

Exercises 11–16: Use the 45°–45° right triangle in Figure 6.38 to find the exact value of the trigonometric expression.

11. tan 45°   12. sec 45°

13. cot 45°   14. csc 45°

15. sin 45°   16. cos 45°

Exercises 17–20: Find the six trigonometric functions of θ.

17.
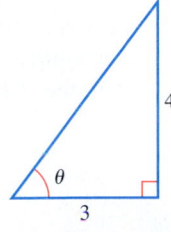

18.
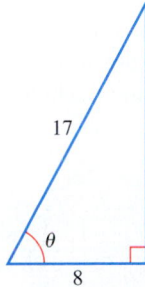

19.
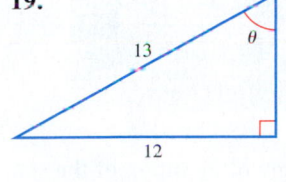

20.
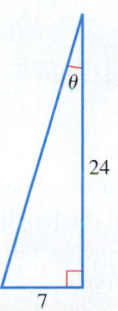

Exercises 21–28: Find the lengths of the unknown sides in the right triangle. Round values to the nearest hundredth.

21.
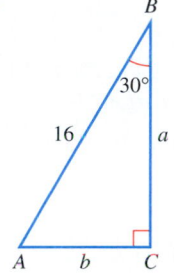

22.
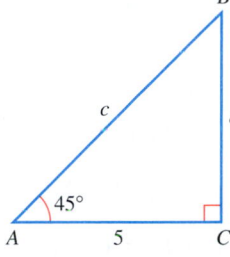

23.
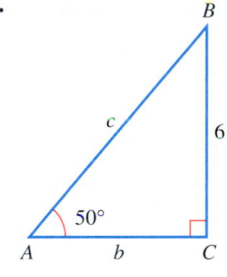

24.
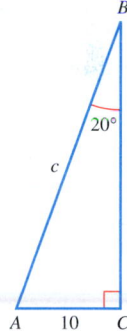

25.
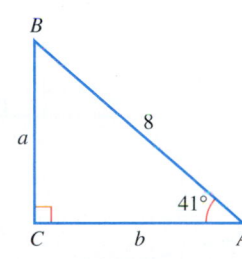

26.
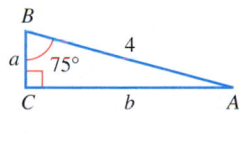

27.
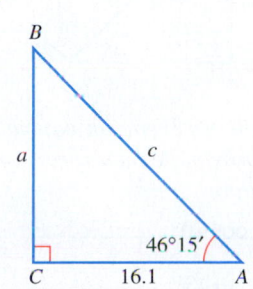

28.
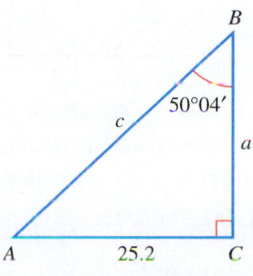

Exercises 29–32: Let right triangle ABC have the standard labeling shown in Figure 6.26 and complete the following. Approximate the answer to the nearest hundredth.

29. Find a if $b = 12$ and $\alpha = 60°$.

30. Find b if $c = 23$ and $\beta = 45°$.

31. Find c if $a = 100$ and $\beta = 53°43'$.

32. Find a if $b = 64$ and $\alpha = 78°15'$.

Exercises 33–38: Let θ be an acute angle. Find the unknown trigonometric value using the given information.

33. $\sec \theta$ if $\cos \theta = \frac{1}{3}$ **34.** $\cot \theta$ if $\tan \theta = 5$

35. $\csc \theta$ if $\sin \theta = \frac{12}{13}$ **36.** $\sin \theta$ if $\csc \theta = \frac{5}{4}$

37. $\tan \theta$ if $\cot \theta = \frac{7}{24}$ **38.** $\cos \theta$ if $\sec \theta = \frac{7}{5}$

Exercises 39–48: Find the six trigonometric functions of the given angle. Approximate to three decimal places when appropriate.

39. $60°$ **40.** $45°$

41. $25°$ **42.** $30°$

43. $5°35'$ **44.** $85°35'33''$

45. $13°45'30''$ **46.** $45°44'$

47. $1.05°$ **48.** $0.161°$

Exercises 49–52: Find the exact length of each side labeled with a variable in the figure.

49.
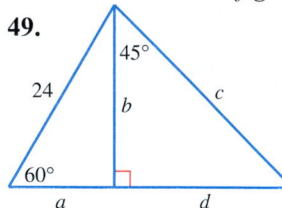

50.
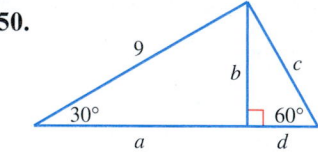

51.
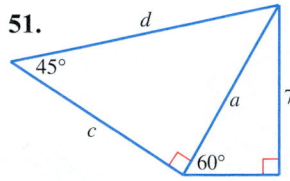

52.
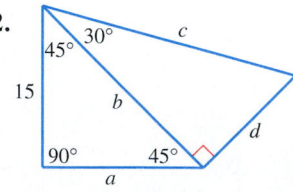

Exercises 53–56: (Refer to Example 9.) Write an equivalent expression using a cofunction. Approximate the expression to four decimal places using a calculator.

53. **(a)** $\sin 70°$ **(b)** $\cos 40°$

54. **(a)** $\cot 23°$ **(b)** $\tan 48°$

55. **(a)** $\csc 49°$ **(b)** $\sec 63°$

56. **(a)** $\cot 87°$ **(b)** $\sec 72°$

Applications

57. *Height of the Cloud Base* (Refer to Example 3 and Figure 6.35.) From a distance of 1500 feet from the spotlight, the angle of elevation θ equals $37°30'$. Find the height of the cloud base. (**Source:** F. Cole.)

58. *Weather Tower* A 410-foot weather tower used to measure wind speed has a guy wire attached to it 175 feet above the ground. The angle between the wire and the vertical tower is $57°$, as shown in the figure. Approximate the length of the guy wire. (**Source:** Brookhaven National Laboratory.)

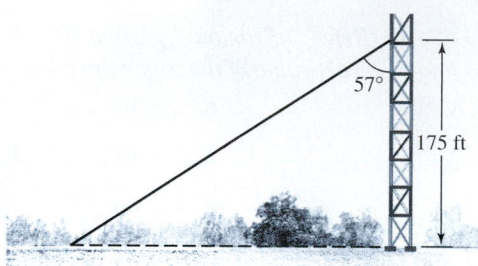

59. *Height of a Tree* One hundred feet from the trunk of a tree on level ground, the angle of elevation of the top of the tree is $35°$. Estimate the height of the tree to the nearest foot.

60. *Height of a Building* From a window 30 feet above the street the angle of elevation to the top of the building across the street is $50°$ and the angle of depression to the base of this building is $20°$. See the accompanying figure. Find the height of the building across the street.

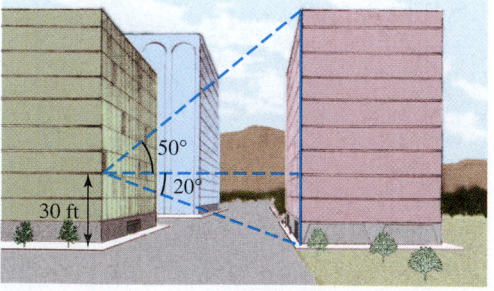

61. *Length of a Shadow* The angle of elevation of the sun is $34°$. Find the length of a shadow cast by a person who is 5 feet 3 inches tall. Round your answer to the nearest tenth of a foot.

62. *Angle of Depression* An airplane is flying near a football stadium at 12,000 feet above level ground. The angle of depression from the airplane to the stadium is 13°. How far horizontally must the airplane fly to be directly over the football stadium? Round your answer to the nearest thousand feet.

63. *Distance to Nearby Stars* (Refer to Example 6 and Figure 6.41.) The table lists the parallax θ in degrees for some nearby stars. Approximate the distance in miles from Earth to each star. Estimate this distance in light-years.

Star	θ (degrees)
Barnard's Star	1.52×10^{-4}
Sirius	1.05×10^{-4}
61 Cygni	8.11×10^{-5}
Procyon	7.97×10^{-5}

Source: M. Zeilik et al.

64. *Parallax and Distance* (Refer to the previous exercise.) When the parallax θ is equal in measure to 1 second, a star is said to have a distance from Earth of 1 parsec. If the distance between Earth and the sun is 93,000,000 miles, approximate the number of miles in 1 parsec. How many light-years is this? (**Source:** M. Zeilik.)

65. *Observing Mercury* The planet Mercury is closer to the sun than Earth. For this reason it can only be observed low in the horizon around sunset or sunrise. See the accompanying figure, where angle θ is called the *elongation*. Because Mercury's orbit is not circular, the elongation varies between 18° and 28°. Approximate the minimum and maximum distances between Mercury and the sun. (**Source:** M. Zeilik.)

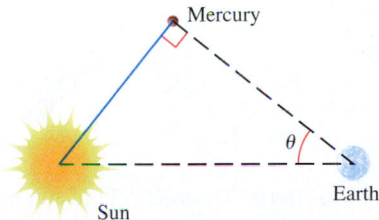

66. *Observing Venus* (Refer to the previous exercise.) The orbit of Venus is nearly circular with an elongation of 48°. Estimate the distance between Venus and the sun.

67. *Orbital Height of a GPS Satellite* The accompanying figure illustrates a satellite in the Global Positioning System (GPS) orbiting over the equator, where $r = 3963$ miles and $\theta = 76.1°$. Use $d = r\left(\frac{1}{\cos\theta} - 1\right)$ from Example 8 to determine the altitude of the GPS satellite above Earth's surface. (**Source:** Y. Zhao, *Vehicle Location and Navigation Systems.*)

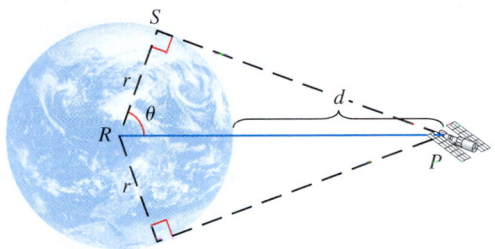

68. *Height of a Mountain* From a point A the angle of elevation of Mount Kilimanjaro in Africa is 13.7° and from a point B directly behind A, the angle of elevation is 10.4°. See the accompanying figure. If the distance between A and B is 5 miles, approximate the height of Mount Kilimanjaro to the nearest hundred feet.

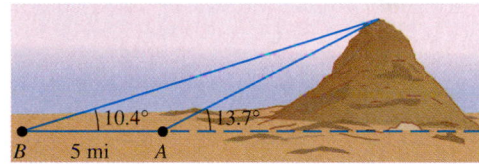

69. *Heights of Lunar Mountains* The lunar mountain peak Huygens has a height of 21,000 feet. The shadow of Huygens on a photograph was 2.8 mm, and the nearby mountain Bradley had a shadow of 1.8 mm on the same photograph. Use similar triangles to calculate the height of Bradley to the nearest hundred feet. (**Source:** T. Webb, *Celestial Objects for Common Telescopes.*)

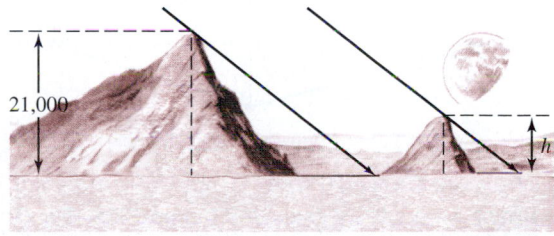

70. *Depths of Moon Craters* (Refer to the previous exercise.) The depths of unknown craters on the moon can be found by comparing their shadows to shadows of nearby craters with known depths. The crater Aristillus is 11,000 feet deep and its shadow was measured as 1.5 mm on a photograph. Its companion crater, Autolycus, had a shadow of 1.3 mm on the same photograph. Estimate the depth of the crater Autolycus. (**Source:** T. Webb.)

71. *Surveying*   (Refer to Example 7 and Figure 6.42.) Find the distance from P to Q if PR is 85.62 feet and angle PQR is 23.76°.

72. *Aerial Photography*   An aerial photograph is taken directly above a building. The length of the building's shadow is 48 feet when the angle of elevation of the sun is 35.3°. Estimate the height of the building.

73. *Surveying*   The *subtense bar method* is used by surveyors to find a distance d between two points P and Q. A subtense bar 2 meters long is centered at Q, perpendicular to the line of sight from P to Q, as shown in the figure. If angle θ is measured, then d can be found using trigonometry.   (**Source:** I. Mueller and K. Ramsayer, *Introduction to Surveying.*)

   (a) Find a formula for d involving a trigonometric function of θ.

   (b) Find d if $\theta = 1°45'15''$.

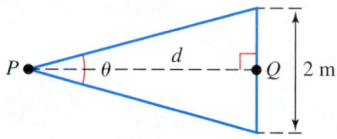

74. *Surveying*   (Refer to the previous exercise.) A variation of the subtense bar method that is used to determine larger distances between two points P and Q is shown in the figure. A subtense bar with length b is placed between the points P and Q so that the line of sight connecting P and Q is a perpendicular bisector. If $\alpha = 0.63°$, $\beta = 0.78°$, and $b = 2$ meters, find the distance from P to Q. (**Source:** I. Mueller.)

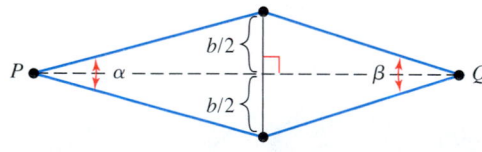

75. *Highway Curve Design*   Highway curves are sometimes banked so that the outside of the curve is slightly elevated or inclined above the inside of the curve, as shown in the figure at the top of the next column. This inclination is called the *superelevation*. Both the curve's radius and superelevation must be correct for a given speed limit. The relationship between a car's velocity v in feet per second, the safe radius r of the curve in feet, and the superelevation θ in degrees is given by $r = \dfrac{v^2}{4.5 + 32.2\tan\theta}$. (**Source:** F. Mannering and W. Kilareski.)

(a) A curve has a speed limit of 66 feet per second (45 mph) and a superelevation of $\theta = 3°$. Approximate the safe radius r.

(b) Find r if $\theta = 5°$ and $v = 66$.

(c) Make a conjecture about how increasing θ affects the safe radius r. Verify your conjecture by tabling r starting at $\theta = 0$ and incrementing by 1. Let $v = 66$.

76. *Highway Design*   (Refer to the previous exercise.) A highway curve has a radius $r = 1150$ feet and has a superelevation of $\theta = 2.1°$. What should be the speed limit (in miles per hour) for this curve?

77. *Highway Design*   (Refer to Example 8 and Figure 6.43.) Find the external distance d for a highway curve with $r = 625$ feet and $\theta = 54°$.

78. *Highway Design*   (Refer to Example 8.) A simple horizontal curve is shown in the accompanying figure. The points P and S mark the beginning and end of the curve. Let Q be the point of intersection where the two straight sections of highway leading into the curve would meet if extended. The radius of the curve is r and the angle θ denotes how many degrees the curve turns. If $r = 765$ feet and $\theta = 83°$, find the distance between P and Q. (**Source:** F. Mannering and W. Kilareski.)

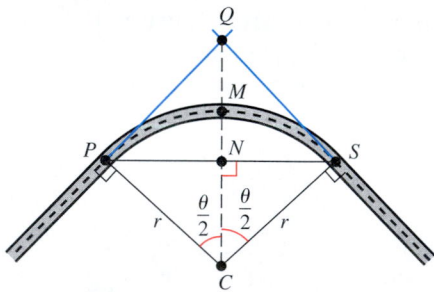

79. *Area of an Equilateral Triangle*   Find the area of the equilateral triangle shown in the figure in terms of s.

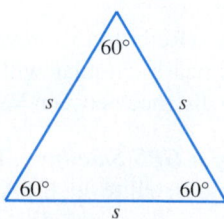

80. *Area of a Hexagon* Write the area of the hexagon shown in the figure in terms of x. Assume that the six triangles that comprise the hexagon are equilateral and congruent.

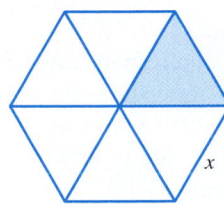

Writing about Mathematics

81. Most calculators have built-in keys to compute the sine, cosine, and tangent functions, but not the secant, cosecant, and cotangent functions. Is it possible to evaluate all of the trigonometric functions with this type of calculator? Explain and include examples.

82. The sine function is defined in terms of right triangles as $\sin\theta = \frac{\text{side opposite}}{\text{hypotenuse}}$. Suppose that a fixed angle θ occurs in two right triangles. If the length of the hypotenuse in the first triangle has twice the length of the hypotenuse in the second triangle, what can be said about the sides opposite in each triangle? How does the value $\sin\theta$ compare in each triangle? Explain.

EXTENDED AND DISCOVERY EXERCISES

1. (a) Use the right triangle in the accompanying figure to show symbolically that

$$\sin^2\theta + \cos^2\theta = 1,$$

where θ is an acute angle in a right triangle. Note that $\sin^2\theta = (\sin\theta)^2$ and $\cos^2\theta = (\cos\theta)^2$.

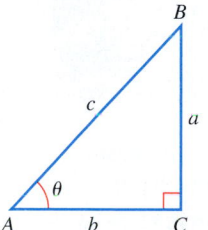

(b) Support your result numerically by tabling $Y_1 = (\sin(X))^\wedge 2 + (\cos(X))^\wedge 2$ starting at $x = 0$ and incrementing by 10. Use degree mode.

2. Repeat the previous exercise and verify the identity $\sec^2\theta - \tan^2\theta = 1$.

CHECKING BASIC CONCEPTS FOR SECTIONS 6.1 AND 6.2

1. Find the radian measure of each angle.
   (a) $45°$ (b) $75°$

2. Find the degree measure of each angle.
   (a) $\dfrac{\pi}{6}$ (b) $\dfrac{5\pi}{4}$

3. Find the arc length intercepted by a central angle of $\theta = 30°$ in a circle with radius $r = 12$ inches. Calculate the area of the sector determined by r and θ.

4. Use the right triangle in the figure to the right to find the six trigonometric functions of θ.

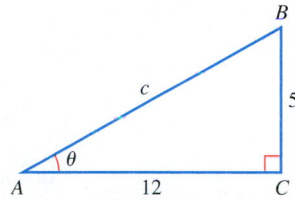

5. Evaluate the six trigonometric functions of $\theta = 60°$ by hand. Support your results by using a calculator.

6. If $\alpha = 63°$ and $a = 9$ in right triangle ABC, approximate the length of the hypotenuse to the nearest tenth.

6.3

The Sine and Cosine Functions and Their Graphs

* Define the sine and cosine functions for any angle
* Use the unit circle to define trigonometric functions of any real number
* Represent the sine and cosine functions
* Use the sine and cosine functions in applications
* Model with the sine function (optional)

Introduction

The sine and cosine functions are used not only in applications involving right triangles, they are also used to model phenomena involving rotation and periodic motion. To accomplish this, we must extend the domains of the trigonometric functions from acute angles to angles of any measure. This will allow us to model a wide variety of phenomena such as biorhythms, weather, tides, electricity, and robotic arms.

Definitions

Robotics is a rapidly growing field that requires extensive mathematics. One basic problem that occurs when designing a robotic arm is determining the location of the robot's hand. Suppose we have a robotic arm that rotates at the shoulder and is controlled by changing the angle θ and the length of the arm r, as illustrated in Figure 6.45. We would like to find a relation between the xy-coordinates of the hand and the values for r and θ. (**Source:** W. Stadler, *Analytical Robotics and Mechatronics.*)

Notice that for a fixed angle θ, if the length of the arm is changed from r_1 to r_2, triangle ABC in Figure 6.46 and triangle DEF in Figure 6.47 are similar triangles. Thus the following ratios are equal and depend only on the measure of θ.

$$\frac{x_1}{r_1} = \frac{x_2}{r_2} \quad \text{and} \quad \frac{y_1}{r_1} = \frac{y_2}{r_2}$$

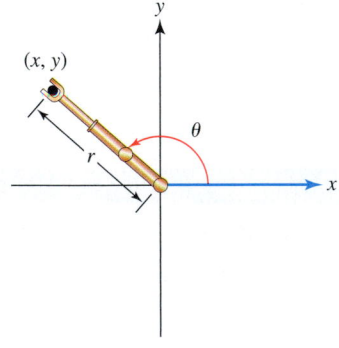

FIGURE 6.45

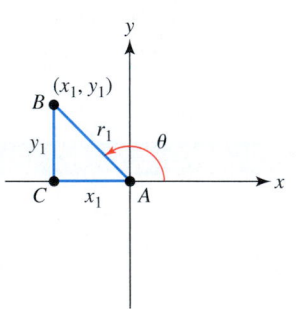

FIGURE 6.46

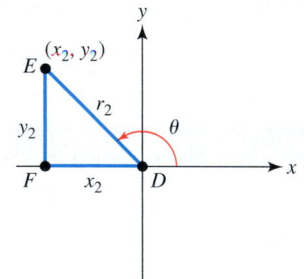

FIGURE 6.47

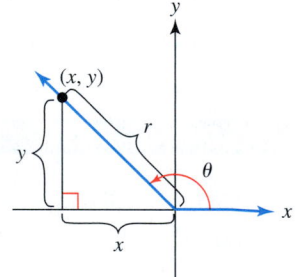

FIGURE 6.48

In Figure 6.48 the Pythagorean theorem gives $r^2 = x^2 + y^2$. Since $r > 0$, it follows that $r = \sqrt{x^2 + y^2}$. The ratios $\frac{y}{r}$ and $\frac{x}{r}$ can be used to define the *sine* and *cosine* functions of any angle θ.

THE SINE AND COSINE FUNCTIONS OF ANY ANGLE θ

Let angle θ be in standard position with the point (x, y) lying on the angle's terminal side. If $r = \sqrt{x^2 + y^2}$, then

$$\sin\theta = \frac{y}{r} \quad \text{and} \quad \cos\theta = \frac{x}{r} \quad (r \neq 0).$$

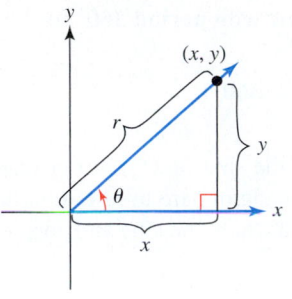

FIGURE 6.49

Although the terminal side of θ was shown in the second quadrant, these definitions are valid for any angle θ.

◆ **MAKING CONNECTIONS**

Right Triangle Trigonometry   If $0° < \theta < 90°$, then x corresponds to the length of the adjacent side, y corresponds to the length of the opposite side, and r corresponds to the length of the hypotenuse. See Figure 6.49. These new definitions for sine and cosine are consistent with the definitions presented in Section 6.2 when θ is an acute angle.

◆

EXAMPLE 1   Evaluating sine and cosine for coterminal angles

Suppose a robotic hand is located at the point $(15, -8)$, where all units are in inches.
(a) Find the length of the arm.
(b) Let α satisfy $0° \le \alpha < 360°$ and represent the angle between the positive x-axis and the robotic arm. Find $\sin \alpha$ and $\cos \alpha$.
(c) Let β satisfy $-360° \le \beta < 0°$ and represent the angle between the positive x-axis and the robotic arm. Find $\sin \beta$ and $\cos \beta$. How do the values for $\sin \beta$ and $\cos \beta$ compare with the values for $\sin \alpha$ and $\cos \alpha$?

SOLUTION
(a) Since $r = \sqrt{15^2 + (-8)^2} = 17$, the length of the arm is 17 inches. See Figure 6.50.
(b) Let $x = 15$, $y = -8$, and $r = 17$. Then,

$$\sin \alpha = \frac{y}{r} = -\frac{8}{17} \qquad \text{and} \qquad \cos \alpha = \frac{15}{17}.$$

(c) In Figure 6.51, β satisfies $-360° \le \beta < 0°$. Since the values of x, y, and r do not change, the trigonometric values of β are the same as those for α in part (b).

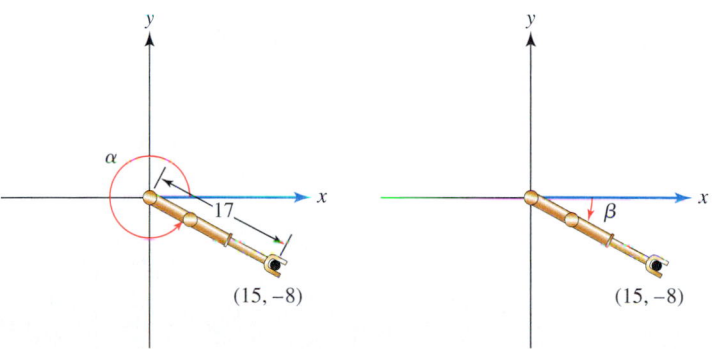

FIGURE 6.50   **FIGURE 6.51**

Now Try Exercise 1 ◆

The results of Example 1 can be generalized. If α and β are coterminal angles, then

$$\sin \alpha = \sin \beta \qquad \text{and} \qquad \cos \alpha = \cos \beta.$$

The angles θ and $\theta + 360°$ are coterminal for any θ. Their terminal sides pass through the same point (x, y), as shown in Figure 6.52. Therefore for all θ, $\sin \theta = \sin (\theta + 360°)$ and $\cos \theta = \cos (\theta + 360°)$. In general, if n is any integer, then

$$\sin \theta = \sin (\theta + n \cdot 360°) \qquad \text{and} \qquad \cos \theta = \cos (\theta + n \cdot 360°).$$

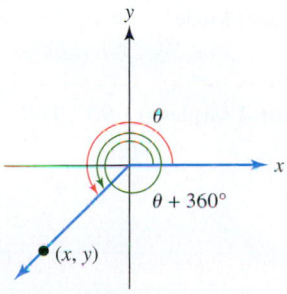

FIGURE 6.52

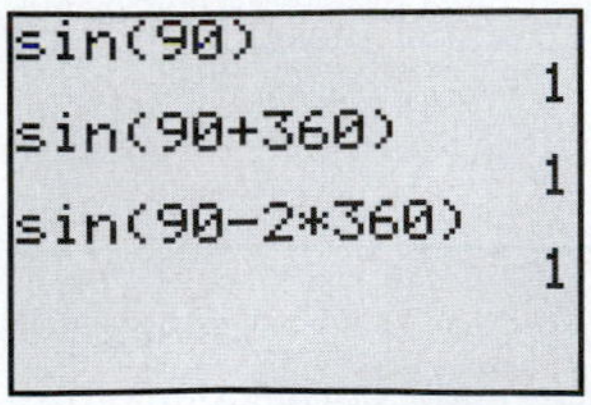

```
sin(90)
                    1
sin(90+360)
                    1
sin(90-2*360)
                    1
```

FIGURE 6.53 Degree Mode

As a result, we say that the sine and cosine functions are **periodic** with **period 360°** (or 2π radians.) For example,

$$\sin 90° = \sin (90° + \mathbf{360°}) = \sin (90° - 2 \cdot \mathbf{360°}).$$

See Figure 6.53.

If an angle is a multiple of 30° or 45°, exact evaluation of the sine function or cosine function is possible by hand. However, in most applications calculators are used to obtain numerical approximations. The next example illustrates an angle where the sine and cosine functions can be evaluated exactly by hand.

EXAMPLE 2 Finding values of $\sin \theta$ and $\cos \theta$

Find $\sin 120°$ and $\cos 120°$. Support your answer using a calculator.

SOLUTION When evaluating the sine or cosine function by hand, we can select any positive value for r and not change the resulting values for $\sin \theta$ and $\cos \theta$. For convenience we let $r = 2$, as shown in Figure 6.54. The length of the shortest leg in a 30°–60° right triangle is half the hypotenuse. Since the terminal side of θ is in the second quadrant, $x < 0$ and so $x = -1$. Next we find y.

$$x^2 + y^2 = r^2 \qquad \text{Pythagorean theorem}$$
$$y^2 = r^2 - x^2 \qquad \text{Subtract } x^2.$$
$$y^2 = 2^2 - (-1)^2 \qquad \text{Let } r = 2 \text{ and } x = -1.$$
$$y = \pm\sqrt{3} \qquad \text{Square root property}$$

Since the terminal side of θ is in the second quadrant, $y > 0$ and so $y = \sqrt{3}$. Thus

$$\sin 120° = \frac{y}{r} = \frac{\sqrt{3}}{2} \approx 0.8660, \qquad \text{and}$$

$$\cos 120° = \frac{x}{r} = -\frac{1}{2} = -0.5.$$

These results are supported in Figure 6.55.

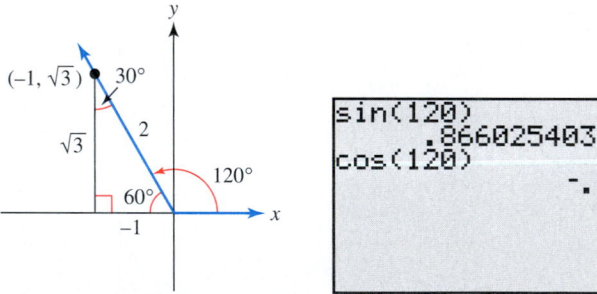

FIGURE 6.54

FIGURE 6.55 Degree Mode

Now Try Exercise 11 ◆

The values of the sine and cosine functions for the **quadrantal angles** 0°, 90°, 180°, and 270° are shown in Table 6.3. Try to verify these values.

TABLE 6.3

θ	0°	90°	180°	270°
$\sin \theta$	0	1	0	-1
$\cos \theta$	1	0	-1	0

Note: If the terminal side of an angle θ in standard position lies on the y-axis, $\sin \theta = \pm 1$ and $\cos \theta = 0$. If the terminal side of an angle θ in standard position lies on the x-axis, $\sin \theta = 0$ and $\cos \theta = \pm 1$.

Trigonometric functions can also be evaluated using radian mode. For example, since $90°$ is equivalent to $\frac{\pi}{2}$ radians, it follows that

$$\sin \frac{\pi}{2} = 1 \qquad \text{and} \qquad \cos \frac{\pi}{2} = 0.$$

In Figure 6.56 a calculator has been used to perform these evaluations in *radian mode*.

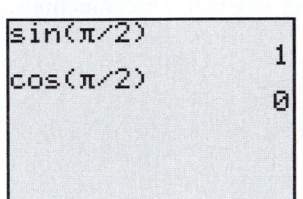

FIGURE 6.56 Radian Mode

Grade or slope is a measure of steepness and indicates whether a highway is uphill or downhill. A 5% grade indicates that a road is increasing 5 vertical feet for each 100-foot increase in horizontal distance. *Grade resistance R* is the gravitational force acting on a vehicle and is given by

$$R = W \sin\theta,$$

where W is the weight of the vehicle and θ is the angle associated with the grade. See Figure 6.57. For an uphill grade $\theta > 0$ and for a downhill grade $\theta < 0$.   (**Source:** F. Mannering and W. Kilareski, *Principles of Highway Engineering and Traffic Analysis.*)

FIGURE 6.57

EXAMPLE 3

Calculating the grade resistance

A downhill highway grade is modeled by the line $y = -0.06x$ in the fourth quadrant.
(a) Find the grade of the road.
(b) Determine the grade resistance for a 3000-pound car. Interpret the result.

SOLUTION

(a) The slope of the line is -0.06, so when x *increases* by 100 feet, y *decreases* by 6 feet. See Figure 6.58. Thus this road has a grade of -6%.

(b) First we must find $\sin \theta$. From Figure 6.58 we see that the point $(\mathbf{100}, \mathbf{-6})$ lies on the terminal side of θ. Since

$$r = \sqrt{\mathbf{100}^2 + (\mathbf{-6})^2} = \sqrt{\mathbf{10{,}036}},$$

it follows that

$$\sin\theta = \frac{y}{r} = \frac{\mathbf{-6}}{\sqrt{\mathbf{10{,}036}}}.$$

The grade resistance is

$$R = W \sin\theta$$
$$= 3000 \left(\frac{-6}{\sqrt{10{,}036}} \right)$$
$$\approx -179.7 \text{ lb.}$$

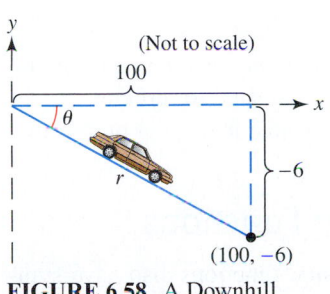

FIGURE 6.58 A Downhill Grade of -6%

On this stretch of highway, gravity would pull a 3000-pound vehicle *downhill* with a force of about 180 pounds. Note that a downhill grade results in a negative grade resistance.

Now Try Exercise 67

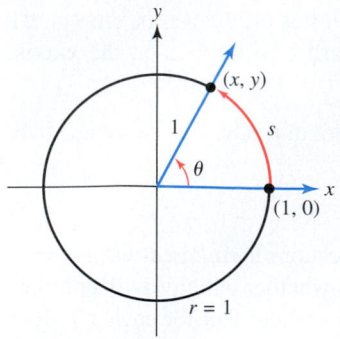

FIGURE 6.59

The Unit Circle

Trigonometric functions were first associated with angles. As applications became more diverse, real numbers were included in the domains of the six trigonometric functions. Real numbers were needed to represent quantities such as time and distance. To extend the domains of the sine and cosine functions to include all real numbers we will introduce the unit circle.

The **unit circle** has radius 1 and equation $x^2 + y^2 = 1$. Let (x, y) be a point on the unit circle and let s be the arc length along the unit circle from the point $(1, 0)$ to the point (x, y) determined by a counterclockwise rotation. See Figure 6.59. Since $r = 1$, the arc length formula $s = r\theta$ reduces to $s = \theta$. That is, the real number s representing arc length is numerically equal to the radian measure of θ. This discussion gives motivation for the following.

TRIGONOMETRIC FUNCTIONS OF REAL NUMBERS

The value of a trigonometric function at the real number s is equal to its value at s radians.

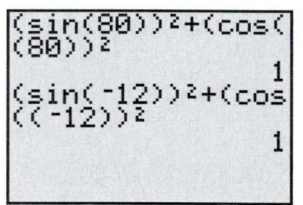

FIGURE 6.60 Degree Mode

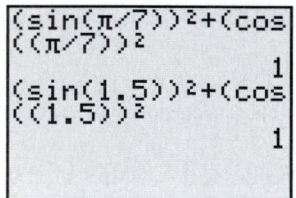

FIGURE 6.61 Radian Mode

Note: To evaluate a trigonometric function of a real number s with a calculator, use *radian* mode.

If (x, y) is a point on the unit circle, lying on the terminal side of an angle θ in standard position, then

$$\sin\theta = \frac{y}{r} = y \qquad \text{and} \qquad \cos\theta = \frac{x}{r} = x$$

because $r = 1$. Points (x, y) on the unit circle can be expressed as $(\cos\theta, \sin\theta)$. Because trigonometric functions can be defined using the unit circle, they are referred to as **circular functions**.

The unit circle is described by $x^2 + y^2 = 1$. Since $x = \cos\theta$ and $y = \sin\theta$, it follows that

$$(\cos\theta)^2 + (\sin\theta)^2 = 1, \qquad \text{or equivalently,}$$

$$\sin^2\theta + \cos^2\theta = 1.$$

This equation is an example of a trigonometric identity. An identity is true for all meaningful values of the variable. In Figures 6.60 and 6.61 this identity is evaluated for different values of θ. In every case the result is 1, regardless of whether θ is measured in degrees or radians.

Representations of the Sine and Cosine Functions

Like other functions that we have studied, the sine and cosine functions also have symbolic, numerical, and graphical representations. Both functions are nonlinear.

The Sine Function   A *symbolic representation* of the sine function is $f(t) = \sin t$. The domain of the sine function is all real numbers. There is no simple formula that can be used to evaluate the sine function. Instead we generally rely on other methods for its evaluation.

A *numerical representation* of $f(t) = \sin t$ is shown in Table 6.4. (These values can be found by hand as in Example 2 or with a calculator.) Since outputs from the sine function correspond to a y-coordinate on the unit circle, the range of the sine function is $-1 \le y \le 1$.

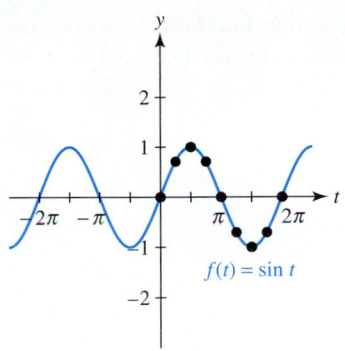

FIGURE 6.62 The Sine Function

TABLE 6.4

t	0	$\frac{\pi}{4}$	$\frac{\pi}{2}$	$\frac{3\pi}{4}$	π	$\frac{5\pi}{4}$	$\frac{3\pi}{2}$	$\frac{7\pi}{4}$	2π
$\sin t$	0	$\frac{1}{\sqrt{2}}$	1	$\frac{1}{\sqrt{2}}$	0	$-\frac{1}{\sqrt{2}}$	-1	$-\frac{1}{\sqrt{2}}$	0

A graphical representation of $f(t) = \sin t$ is shown in Figure 6.62, where the points in Table 6.4 are also plotted. The graph of the sine function increases from 0 to 1 for $0 \le t \le \frac{\pi}{2}$, decreases from 1 to -1 for $\frac{\pi}{2} \le t \le \frac{3\pi}{2}$, and then increases from -1 to 0 for $\frac{3\pi}{2} \le t \le 2\pi$. The graph repeats itself both to the left and the right every 2π units on the x-axis since the sine function has period 2π or $360°$.

◆ **MAKING CONNECTIONS**

The Unit Circle and the Sine Graph   The graph of the sine function may be understood better if we use a unit circle. If a point P on the unit circle starts at $(1, 0)$ and rotates counterclockwise, then the y-coordinate of P corresponds to $\sin \theta$. Refer to Figure 6.59. As a result, $\sin \theta$ first increases from 0 to 1, then decreases from 1 to -1, and finally increases from -1 to 0, where P completes one rotation and returns to the point $(1, 0)$. If P continues to rotate, it passes through the same points a second time and so the sine function has period 2π or $360°$.

EXAMPLE 4

Evaluating the sine function

Evaluate $f(t) = \sin t$ at $t = \frac{3\pi}{2}$ by hand.

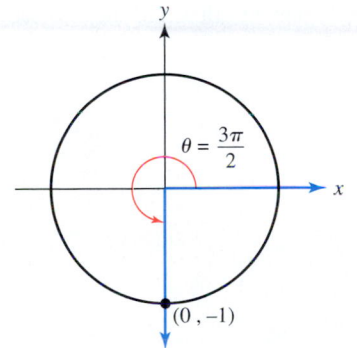

FIGURE 6.63

SOLUTION   An angle θ of $\frac{3\pi}{2}$ radians in standard position has a terminal side that intersects the unit circle at the point $(0, -1)$. See Figure 6.63. Therefore $\sin\left(\frac{3\pi}{2}\right) = -1$.

Now Try Exercise 39 ◆

The Cosine Function   The cosine function, *represented symbolically by* $f(t) = \cos t$, has numerical and graphical representations similar to those of the sine function. Its domain is all real numbers. Like the sine function, the cosine function does not have a simple formula.

A *numerical representation* of $f(t) = \cos t$ is shown in Table 6.5. Since an output from the cosine function corresponds to an x-coordinate on the unit circle, the range of the cosine function is from -1 to 1. Like the sine function, the cosine function also has period 2π or $360°$.

TABLE 6.5

t	0	$\frac{\pi}{4}$	$\frac{\pi}{2}$	$\frac{3\pi}{4}$	π	$\frac{5\pi}{4}$	$\frac{3\pi}{2}$	$\frac{7\pi}{4}$	2π
$\cos t$	1	$\frac{1}{\sqrt{2}}$	0	$-\frac{1}{\sqrt{2}}$	-1	$-\frac{1}{\sqrt{2}}$	0	$\frac{1}{\sqrt{2}}$	1

A graphical representation of $f(t) = \cos t$ is shown in Figure 6.64 on the next page, where the points in Table 6.5 are also plotted. Notice that the overall shape of the graph of the cosine function is similar to that of the graph of the sine function. However, the cosine function decreases from 1 to -1 for $0 \le t \le \pi$ and increases from -1 to 1 for $\pi \le t \le 2\pi$.

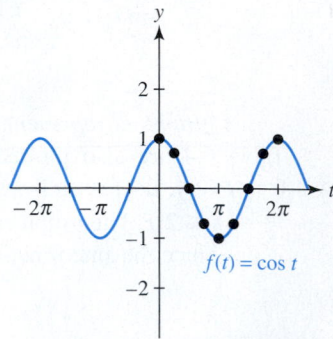

FIGURE 6.64 The Cosine Function

◆ **CLASS DISCUSSION**

If a point P on the unit circle starts at $(1, 0)$ and rotates counterclockwise in Figure 6.59, then the x-coordinate of P corresponds to $\cos \theta$. Explain how this relates to the graph of the cosine function. ◆

EXAMPLE 5 Evaluating the cosine function

Evaluate $f(t) = \cos t$ at $t = \frac{5\pi}{6}$ by hand.

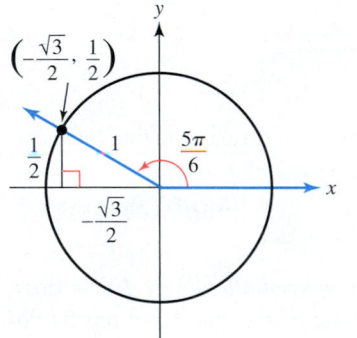

FIGURE 6.65

SOLUTION An angle of $\frac{5\pi}{6}$ radians in standard position has a terminal side that intersects the unit circle in the second quadrant. See Figure 6.65. To find the point of intersection (x, y) notice that the hypotenuse of the 30°–60° right triangle has length 1 and the shorter leg has length $y = \frac{1}{2}$. We can determine x symbolically.

$$x^2 + y^2 = 1 \qquad \text{Equation of the unit circle}$$

$$x^2 + \left(\frac{1}{2}\right)^2 = 1 \qquad y = \frac{1}{2}$$

$$x^2 = \frac{3}{4} \qquad \text{Solve for } x^2.$$

$$x = \pm\frac{\sqrt{3}}{2} \qquad \text{Square root property}$$

Since the point (x, y) is located in the second quadrant, choose $x = -\frac{\sqrt{3}}{2}$. Because $\cos t = x$, it follows that $\cos \frac{5\pi}{6} = -\frac{\sqrt{3}}{2} \approx -0.8660$. *Now Try Exercise 41* ◆

Applications of the Sine and Cosine Functions

Periodic graphs that are similar in shape to the graphs of the sine and cosine functions are **sinusoidal**. Of all the periodic graphs, sinusoidal graphs are the most important in applications because they occur in nearly every aspect of physical science.

Because the moon orbits Earth, we observe different phases of the moon during the period of a month. In Figure 6.66 angle θ is called the *phase angle*. The *phase F* of the moon is computed by

$$F(\theta) = \frac{1}{2}(1 - \cos\theta),$$

and gives the fraction of the moon's face that is illuminated by the sun.   (**Source:** P. Duffet-Smith, *Practical Astronomy with Your Calculator.*)

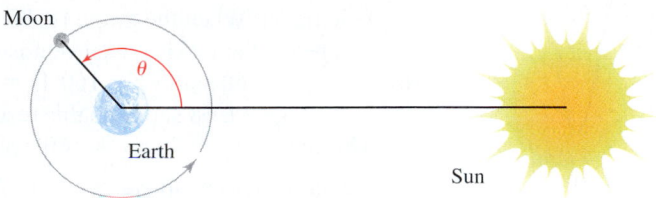

FIGURE 6.66   Phase Angle θ (Not to scale)

EXAMPLE 6   Modeling the phases of the moon

Let $F(\theta) = \frac{1}{2}(1 - \cos\theta)$.
(a) A graph of F is shown in Figure 6.67. Discuss how the graph relates to the phases of the moon.
(b) Evaluate $F(0)$, $F\left(\frac{\pi}{2}\right)$, $F(\pi)$, and $F\left(\frac{3\pi}{2}\right)$. Interpret each result.

SOLUTION
(a) The phases of the moon are periodic. Each peak represents a full moon and each valley represents a new moon.
(b) $F(0) = \frac{1}{2}(1 - \cos 0) = \frac{1}{2}(1 - 1) = 0$. When $\theta = 0$ the moon is located between Earth and the sun. Since $F = 0$, the face of the moon is not visible, which corresponds to a *new moon*.
$F\left(\frac{\pi}{2}\right) = \frac{1}{2}\left(1 - \cos\frac{\pi}{2}\right) = \frac{1}{2}(1 - 0) = \frac{1}{2}$. When $\theta = \frac{\pi}{2}, F = \frac{1}{2}$. Thus half the face of the moon is visible. This phase is called the *first quarter.*
$F(\pi) = \frac{1}{2}(1 - \cos\pi) = \frac{1}{2}(1 - (-1)) = 1$. When $\theta = \pi$, Earth is between the moon and the sun. Since $F = 1$, the face of the moon is completely visible, which corresponds to a *full moon.*
$F\left(\frac{3\pi}{2}\right) = \frac{1}{2}\left(1 - \cos\frac{3\pi}{2}\right) = \frac{1}{2}(1 - 0) = \frac{1}{2}$. When $\theta = \frac{3\pi}{2}, F = \frac{1}{2}$. Thus half the face of the moon is visible. This phase is called the *last quarter.*   *Now Try Exercise 72* ◆

Sinusoidal curves occur when modeling the voltage in a common electrical outlet. An electrical circuit can sometimes be compared to water flowing through a hose. *Voltage* in an electrical circuit corresponds to the water pressure. Higher water pressure causes more water to flow through a hose in the same way that higher voltage causes more current to flow through a wire. Electrical current (or *amperage*) corresponds to the amount of water flowing through a hose. More electrons moving through a wire per second result in greater amperage. Electrical *resistance* can be compared to the diameter of a hose. A smaller diameter reduces the flow of water in the same way that increasing electrical resistance reduces current in a wire.

Common household current is called *alternating current* (AC) because the current changes direction in a wire 120 times per second.

EXAMPLE 7   Analyzing household current

The voltage V in a household outlet can be modeled by $V(t) = 160\sin(120\pi t)$, where t represents time in seconds.
(a) A graph of $y = V(t)$ is shown in Figure 6.68 on the next page. Describe the voltage.
(b) Evaluate $V\left(\frac{1}{240}\right)$ and interpret the result.
(c) One way to estimate the "average" voltage in a circuit is to use the **root mean square** voltage, which equals the maximum voltage divided by $\sqrt{2}$. Find this "average" voltage.

FIGURE 6.67

F

Phase

2.0

1.5

1.0

0.5

$F(\theta) = \frac{1}{2}(1 - \cos\theta)$

θ

π 2π 3π 4π 5π 6π

Phase angle (radians)

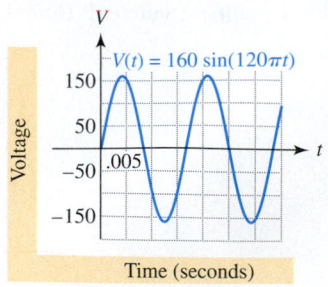

$V(t) = 160 \sin(120\pi t)$

FIGURE 6.68

SOLUTION

(a) The graph of V in Figure 6.68 is sinusoidal and varies between -160 volts and 160 volts. This corresponds to the fact that the current is changing direction in a household outlet. When the graph is above the x-axis the current is one direction, and when it is below the x-axis the current is in the opposite direction.

(b) $V\left(\frac{1}{240}\right) = 160 \sin\left(\frac{\pi}{2}\right) = 160(1) = 160$. After $\frac{1}{240} \approx 0.004$ second, the voltage is 160 volts. Figure 6.68 supports this result.

(c) The maximum voltage is 160 volts. The "average" voltage is $\left(\frac{160}{\sqrt{2}}\right) \approx 113$ volts. Common household electricity is often rated at 110–120 volts. *Now Try Exercise 69* ◆

Modeling with the Sine Function (Optional)

The study of *biological clocks* is a fascinating field. Many living organisms undergo regular biological rhythms. A simple example is a flower that opens during daylight and closes at nighttime. One amazing result is that flowers often continue to open and close even when they are placed in continual darkness. (**Source:** F. Brown et al., *The Biological Clock.*)

EXAMPLE 8

Modeling biological rhythms

Some types of water plants are *luminescent*—they radiate a type of "cold" light that is similar to the light emitted from a firefly. A sample of a luminescent plant (*Gonyaulax polyedra*) was put into continual dim light. The luminescence was measured at 6-hour intervals and the data in Table 6.6 summarize the results. Noon corresponds to $t = 0$ and the y-units of luminescence are arbitrary. (**Source:** E. Bünning, *The Physiological Clock.*)

TABLE 6.6

t (hour)	0	6	12	18	24	30	36	42	48
y (luminescence)	1	4	7	4	1	4	7	4	1

(a) Make a scatterplot of the data in $[-6, 54, 6]$ by $[0, 8, 1]$. Interpret the data.

(b) Graph the data and $f(t) = 3 \sin(0.27t - 1.7) + 4$. Comment on how well f models the data.

(c) Estimate the luminescence when $t = 33$.

SOLUTION

(a) A scatterplot of the data is shown in Figure 6.69. Luminescence appears to be periodic, increasing and decreasing at regular intervals even though the lighting was always dim. The plant is the most luminescent during times that correspond to midnight: $t = 12$ and 36. It was least luminescent at times corresponding to noon: $t = 0, 24$, and 48.

(b) Graph $Y_1 = 3 \sin(.27X - 1.7) + 4$ and the data, as shown in Figure 6.70. The graph of f models the periodic data quite well.

Calculator Help

To make a scatterplot and graph an equation, see Appendix B (pages AP-7 and AP-11).

$[-6, 54, 6]$ by $[0, 8, 1]$

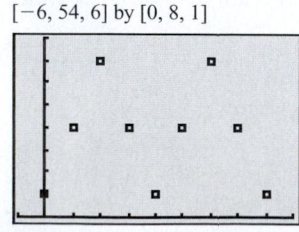

FIGURE 6.69

$[-6, 54, 6]$ by $[0, 8, 1]$

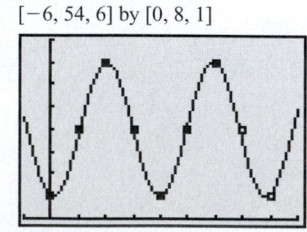

FIGURE 6.70

(c) Evaluate $f(t)$ at $t = 33$ using radian mode.

$$f(\mathbf{33}) = 3\sin(0.27(\mathbf{33}) - 1.7) + 4 \approx 6.4$$

The luminescence was about 6.4 after 33 hours or at 9 P.M.

Now Try Exercise 71 ◆

6.3 Putting it all Together

$\mathbf{I}$n this section we extended the domains of the sine and cosine functions to include any angle θ and any real number t. Both of these functions are nonlinear functions and their ranges include values satisfying $-1 \le y \le 1$. Some concepts about the sine and cosine functions are summarized in the following table.

Concept	Formulas and Figures
Trigonometric functions of angles	$\sin\theta = \dfrac{y}{r}$ and $\cos\theta = \dfrac{x}{r}$, where $r = \sqrt{x^2 + y^2}$
The unit circle	$x^2 + y^2 = 1 \quad (r = 1)$ $\sin t = y$ and $\cos t = x$ The period of the sine and cosine functions is 2π or $360°$. Graphs of the sine and cosine functions are shown in Figures 6.62 and 6.64.

6.3 Exercises

Basic Concepts

Exercises 1–4: The xy-coordinates of the hand for a robotic arm are shown in the figure.

  (a) *Find the length of the arm.*
  (b) *Find the sine and cosine functions for the angle θ.*

1.
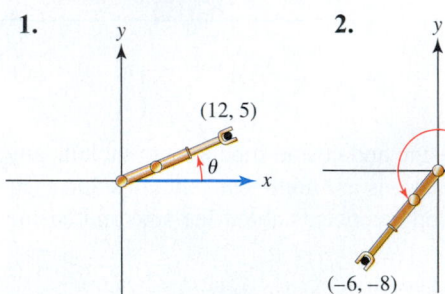

(12, 5)

2.

(−6, −8)

3.
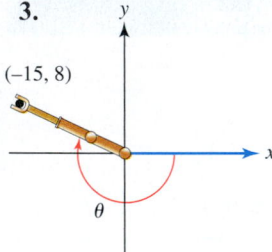

(−15, 8)

4.
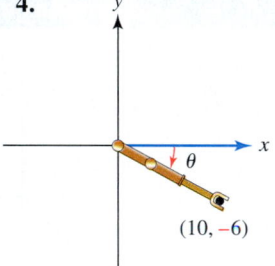

(10, −6)

Exercises 5–10: Find sin θ and cos θ.

5.
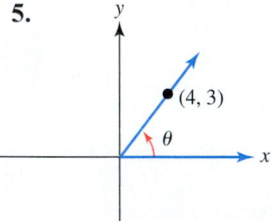

(4, 3)

6.
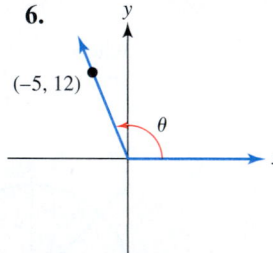

(−5, 12)

7.
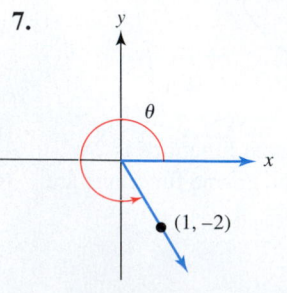

(1, −2)

8.
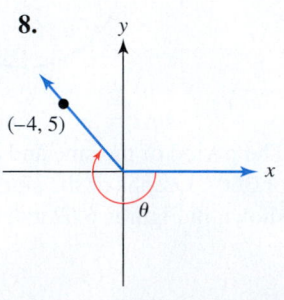

(−4, 5)

9.
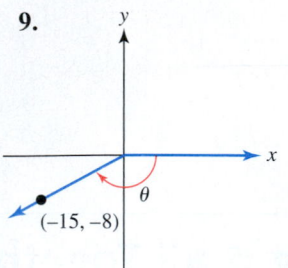

(−15, −8)

10.
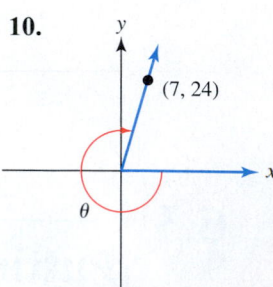

(7, 24)

Exercises 11–20: (Refer to Example 2.) Find the sine and cosine functions by hand for the given angle. Then support your answer using a calculator.

11. $45°$ **12.** $150°$

13. $-30°$ **14.** $-180°$

15. $\dfrac{\pi}{3}$ **16.** $\dfrac{5\pi}{4}$

17. $-\dfrac{\pi}{2}$ **18.** -2π

19. $\dfrac{7\pi}{6}$ **20.** $\dfrac{4\pi}{3}$

Exercises 21–26: The terminal side of an angle θ in standard position lies on the line in the given quadrant. Find the sine and cosine functions of θ. (Hint: Find a point (x, y) lying on the terminal side of θ.)

21. $y = 2x$,   Quadrant I

22. $y = -\tfrac{1}{2}x$,   Quadrant II

23. $y = -3x$,   Quadrant IV

24. $y = \tfrac{3}{4}x$,   Quadrant III

25. $y = -\tfrac{2}{3}x$,   Quadrant II

26. $y = \tfrac{5}{3}x$,   Quadrant I

Exercises 27–34: Approximate the sine and cosine of each angle to four decimal places.

27. $93.2°$ **28.** $-43°$

29. $123°50'$ **30.** $12°40'45''$

31. −4

32. 1.56

33. $\dfrac{11\pi}{7}$

34. $-\dfrac{7\pi}{5}$

Exercises 35–38: The figure shows angle θ in standard position with its terminal side intersecting the unit circle. Evaluate sin θ and cos θ.

35.

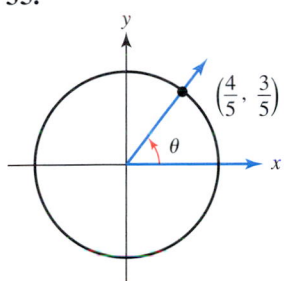

$\left(\dfrac{4}{5}, \dfrac{3}{5}\right)$

36.

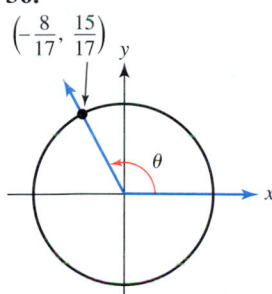

$\left(-\dfrac{8}{17}, \dfrac{15}{17}\right)$

37.

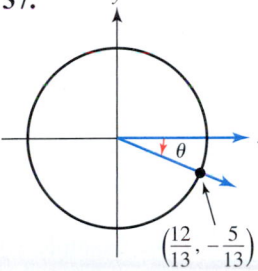

$\left(\dfrac{12}{13}, -\dfrac{5}{13}\right)$

38.

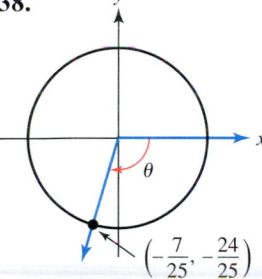

$\left(-\dfrac{7}{25}, -\dfrac{24}{25}\right)$

Exercises 39–48: (Refer to Examples 4 and 5.) Use a unit circle to evaluate sin t and cos t by hand at the given values of t.

39. $t = \dfrac{\pi}{2}$

40. $t = \pi$

41. $t = \dfrac{7\pi}{6}$

42. $t = -\dfrac{\pi}{4}$

43. $t = -\dfrac{3\pi}{4}$

44. $t = \dfrac{5\pi}{3}$

45. $t = \dfrac{5\pi}{2}$

46. $t = \dfrac{11\pi}{6}$

47. $t = -\dfrac{\pi}{3}$

48. $t = -\dfrac{3\pi}{2}$

Representations of Functions

Exercises 49–54: Use the graph of $f(t) = \sin t$ at the top of the next column to complete the following.

    (a) Evaluate $f(t)$ graphically for the given t.
    (b) Support your result by evaluating $f(t)$ with a calculator.

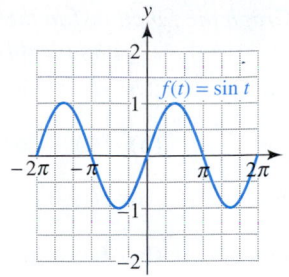

49. $t = 0$

50. $t = \pi$

51. $t = \dfrac{\pi}{2}$

52. $t = \dfrac{3\pi}{2}$

53. $t = -\dfrac{\pi}{6}$

54. $t = -\dfrac{\pi}{2}$

Exercises 55–60: Use the graph of $f(t) = \cos t$ to complete the following.

    (a) Evaluate $f(t)$ graphically for the given t.
    (b) Support your result by evaluating $f(t)$ with a calculator.

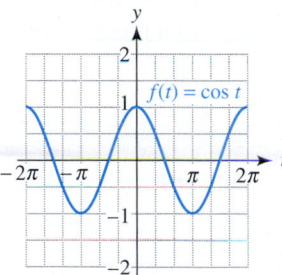

55. $t = 0$

56. $t = -\dfrac{\pi}{2}$

57. $t = \dfrac{\pi}{3}$

58. $t = \dfrac{\pi}{2}$

59. $t = -\dfrac{3\pi}{2}$

60. $t = \pi$

61. Give graphical and numerical representations of the functions $f(t) = 2 \sin t$ and $g(t) = \sin (2t)$. For the numerical representation, start at $t = 0$, incrementing by $\frac{\pi}{4}$. Are f and g the same functions? Explain.

62. Give graphical and numerical representations of the functions $f(t) = 3 \cos t$ and $g(t) = \cos (3t)$. Are f and g the same functions? Explain.

Exercises 63–66: Graph the function f in the viewing rectangle $[-2\pi, 2\pi, \pi/2]$ *by* $[-4, 4, 1]$ *and identify the range of f.*

63. (a) $f(t) = 3 \sin t$ (b) $f(t) = \sin (3t)$

64. (a) $f(t) = 2 \cos t$ (b) $f(t) = \cos (2t)$

65. (a) $f(t) = 2 \cos (t) + 1$ (b) $f(t) = \cos (2t) - 1$

66. (a) $f(t) = 3 \sin (t) + 1$ (b) $f(t) = \sin (3t) - 1$

Applications

67. **Highway Grade** (Refer to Example 3.) Suppose an uphill grade of a highway can be modeled by the line $y = 0.03x$ in the first quadrant.
    (a) Find the grade of the hill.

    (b) Determine the grade resistance for a gravel truck weighing 25,000 pounds.

68. **Highway Design** When an automobile travels along a circular curve with radius r, trees and buildings situated on the inside of the curve can obstruct a driver's vision. See the accompanying figure. To ensure a safe stopping distance S, the *minimum* distance d that should be cleared on the inside of a highway curve is given by the equation $d = r\left(1 - \cos \frac{\beta}{2}\right)$, where β in radians is determined by $\beta = \frac{S}{r}$. (**Source:** F. Mannering.)

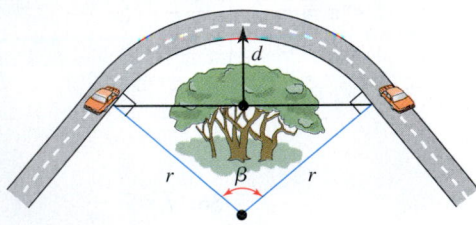

    (a) At 45 miles per hour, $S = 390$ feet. If $r = 600$ feet, approximate d.

    (b) At 60 miles per hour, $S = 620$ feet. Approximate d for the same curve.

    (c) Discuss how the speed limit affects the amount of land that should be cleared on the inside of the curve.

69. **Voltage** (Refer to Example 7.) Electric ranges and ovens often use a higher voltage than that found in normal household outlets. This voltage can be modeled by
$$V(t) = 310 \sin (120\pi t),$$
where t represents time in seconds.

(a) A graph of $y = V(t)$ is shown in the figure. Describe the voltage in the circuit.

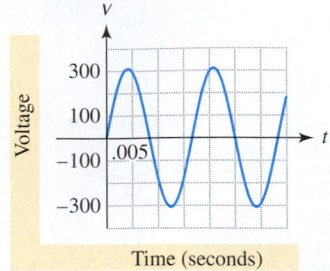

(b) Evaluate $V\left(\frac{1}{120}\right)$ and interpret the result.

(c) Approximate the root mean square voltage.

70. **Dead Reckoning** Suppose an airplane flies the path shown in the figure, where distance is measured in hundreds of miles. Approximate the final coordinates of the airplane if its initial coordinates are $(0, 0)$.

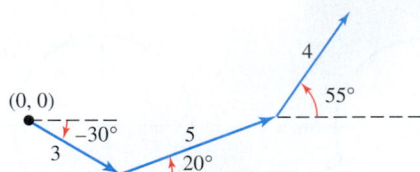

71. **Modeling Biological Rhythms** (Refer to Example 8.) The accompanying table shows the time y when a typical flying squirrel becomes active in the afternoon during each month x. (**Source:** J. Harker, *The Physiology of Diurnal Rhythms.*)

x (month)	1	2	3	4
y (P.M.)	4:30	5:15	5:45	6:30

x (month)	5	6	7	8
y (P.M.)	7:00	7:30	7:45	7:30

x (month)	9	10	11	12
y (P.M.)	6:45	6:15	5:00	4:15

(a) Make a scatterplot of the data. (*Hint:* Represent 4:30 P.M. by 4.5.)

(b) The function given by
$$f(x) = 1.9 \sin (0.42x - 1.2) + 5.7$$
models the time of sunset, where x represents the month. Graph f and the data together. Interpret the graph.

72. *Phases of the Moon* (Refer to Example 6.) Find all phase angles θ that correspond to a full moon and all phase angles that correspond to a new moon. Assume that θ can be any angle.

73. *Braking Distance and Grade* The braking distance D in feet for a typical automobile to change its velocity from V_1 to V_2 on dry pavement can be estimated using the equation

$$D = \frac{1.05\left(V_1^2 - V_2^2\right)}{27 + 64.4 \sin \theta}.$$

In this equation θ represents the angle of the grade of the highway and velocity is measured in feet per second. See Figure 6.57. (**Source:** F. Mannering.)

(a) Compute the number of feet required to slow a car from 88 feet per second (60 mph) to 44 feet per second (30 mph), while traveling uphill with $\theta = 3°$.

(b) Repeat part (a) with $\theta = -3°$.

(c) How is braking distance affected by θ? Does this agree with your driving experience?

74. *Braking Distance* (Refer to the previous exercise.) An automobile is traveling at 88 feet per second (60 mph) on a highway with $\theta = -4°$. The driver sees a stalled truck in the road and applies the brakes 250 feet from the vehicle. Assuming that a collision cannot be avoided, how fast is the car traveling when it collides with the truck?

75. *Music and the Sine Function* A *pure tone* can be modeled by a sine wave. Pure tones typically sound dull and uninteresting. An example of a pure tone is the sound heard when a tuning fork is lightly struck. The pure tone of the first A-note above middle C can be modeled by $f(t) = \sin(880\pi t)$. (**Source:** J. Pierce, *The Science of Musical Sound.*)

(a) In [0, 1/100, 1/880] by [−1.5, 1.5, 0.5], graph the function f.

(b) Find the period P of this tone graphically.

(c) Frequency gives the number of vibrations or cycles per second in a sinusoidal graph. The human ear can hear frequencies from 16.4 to 16,000 cycles per second. Frequency F may be determined using the equation $F = \frac{1}{P}$, where P is the period. Find the frequency of this A-note.

76. *Music* (Continuation of the previous exercise) Middle C has a frequency of 261.6 cycles per second and can be modeled by $g(t) = \sin(523.2\pi t)$. (**Source:** J. Pierce.)

(a) Estimate graphically the period of middle C.

(b) Graph f from the previous exercise and g in the window [0, 1/100, 1/880] by [−1.5, 1.5, 0.5]. Compare their graphs.

Writing about Mathematics

77. Discuss whether the sine and cosine functions are linear or nonlinear functions. Use graphical and numerical representations to justify your reasoning.

78. Describe two ways to define the sine and cosine functions. Compare and contrast these definitions. Give examples.

EXTENDED AND DISCOVERY EXERCISES

1. Graph $f(t) = \sin t$ and $g(t) = \cos t$ in the same viewing rectangle. Then translate the graph of f to the left $\frac{\pi}{2}$ units by graphing $y = f\left(t + \frac{\pi}{2}\right)$. How does this translated graph compare to the graph of g?

2. (Continuation of the previous exercise) Translate the graph of $g(t) = \cos t$ to the right $\frac{3\pi}{2}$ units by graphing $y = g\left(t - \frac{3\pi}{2}\right)$. How does this translated graph compare to the graph of $f(t) = \sin t$?

6.4

Other Trigonometric Functions and Their Graphs

- ◆ Learn definitions and basic identities
- ◆ Represent other trigonometric functions
- ◆ Solve applications

Introduction

Like the sine and cosine functions, we can extend the domains of the tangent, cotangent, secant, and cosecant functions. This will be accomplished using techniques similar to those presented in the previous section.

The origins of the tangent and cotangent functions were not with astronomy, as was the case with the sine function. Rather, they were developed as part of surveying land, finding heights of objects, and determining time. The cotangent function was computed as early as 1500 B.C. in Egypt, where sundials were used to determine time. Depending on the position of the sun in the sky, a vertical stick casts shadows of different lengths. See Figure 6.71. If θ represents the angle of elevation of the sun and the stick is 1 unit long, then the length of the shadow equals $\cot\theta$. This is a simple device for evaluating the cotangent function. (**Source:** *Historical Topics for the Mathematics Classroom, Thirty-first Yearbook.* NCTM.)

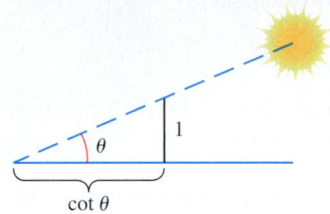

FIGURE 6.71   Modeling Shadow Length

Definitions and Basic Identities

Using Figure 6.72, the other trigonometric functions may be defined for any angle θ in a manner similar to the way sine and cosine functions were defined.

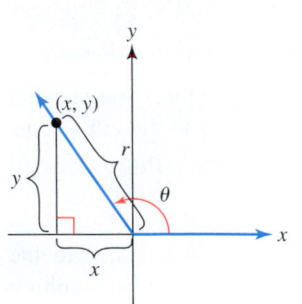

FIGURE 6.72

TRIGONOMETRIC FUNCTIONS OF ANY ANGLE θ

Let (x, y) be a point other than the origin on the terminal side of an angle θ in standard position. If $r = \sqrt{x^2 + y^2}$, then the six trigonometric functions are as follows.

$$\sin\theta = \frac{y}{r} \qquad\qquad \csc\theta = \frac{r}{y} \ (y \neq 0)$$

$$\cos\theta = \frac{x}{r} \qquad\qquad \sec\theta = \frac{r}{x} \ (x \neq 0)$$

$$\tan\theta = \frac{y}{x} \ (x \neq 0) \qquad \cot\theta = \frac{x}{y} \ (y \neq 0)$$

The domains of both the sine and cosine functions include all angles. However, the cosecant and cotangent functions are undefined when $y = 0$, which corresponds to angles whose terminal sides lie on the x-axis. Examples include $0°$, $\pm180°$, and $\pm360°$. The domains of the cosecant and cotangent functions are as follows.

$$D = \{\theta \mid \theta \neq 180° \cdot n\}, \quad n \text{ an integer}$$

Similarly, the tangent and secant functions are undefined whenever $x = 0$. This corresponds to angles whose terminal sides lie on the y-axis. Examples include $\pm 90°$, $\pm 270°$, and $\pm 450°$. The domains of the tangent and secant functions are as follows.

$$D = \{\theta \mid \theta \neq 90° + 180° \cdot n\}, \quad n \text{ an integer}$$

In some situations we can evaluate the six trigonometric functions without a calculator. This is illustrated in the next example.

EXAMPLE 1 Finding values of trigonometric functions

The point $(5, -12)$ is located on the terminal side of an angle θ in standard position. Find the six trigonometric functions of θ.

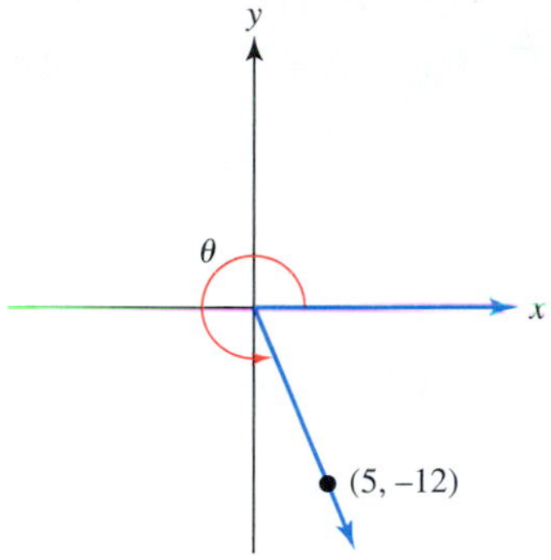

SOLUTION There are many coterminal angles that have the point $(5, -12)$ located on their terminal sides. However, the trigonometric values for coterminal angles are equal. In Figure 6.73 one possibility for θ is shown. Begin by calculating r.

$$\begin{aligned} r &= \sqrt{x^2 + y^2} \\ &= \sqrt{5^2 + (-12)^2} \\ &= 13 \end{aligned}$$

FIGURE 6.73

Since $x = 5$, $y = -12$, and $r = 13$, the values of the six trigonometric functions are as follows.

$$\sin \theta = \frac{y}{r} = -\frac{12}{13} \qquad \csc \theta = \frac{r}{y} = -\frac{13}{12}$$

$$\cos \theta = \frac{x}{r} = \frac{5}{13} \qquad \sec \theta = \frac{r}{x} = \frac{13}{5}$$

$$\tan \theta = \frac{y}{x} = -\frac{12}{5} \qquad \cot \theta = \frac{x}{y} = -\frac{5}{12}$$

Now Try Exercise 1 ◆

Since $\sin \theta = \frac{y}{r}$ and $\csc \theta = \frac{r}{y}$, it follows that $\sin \theta$ and $\csc \theta$ are reciprocals. This may be expressed using either of the *reciprocal identities,*

$$\sin \theta = \frac{1}{\csc \theta} \qquad \text{or} \qquad \csc \theta = \frac{1}{\sin \theta}.$$

In Example 1 it was shown that $\sin \theta = -\frac{12}{13}$. It follows that

$$\csc \theta = \frac{1}{-12/13} = -\frac{13}{12}.$$

Identities are equations that are true for all meaningful values of a variable. Reciprocal identities also hold for $\tan \theta$ and $\cot \theta$ as well as $\cos \theta$ and $\sec \theta$.

RECIPROCAL IDENTITIES

$$\sin \theta = \frac{1}{\csc \theta} \qquad \cos \theta = \frac{1}{\sec \theta} \qquad \tan \theta = \frac{1}{\cot \theta}$$

$$\csc \theta = \frac{1}{\sin \theta} \qquad \sec \theta = \frac{1}{\cos \theta} \qquad \cot \theta = \frac{1}{\tan \theta}$$

The expressions $\cot\theta$ and $\tan\theta$ can be written in terms of the $\sin\theta$ and $\cos\theta$. For example,

$$\cot\theta = \frac{x}{y} = \frac{x/r}{y/r} = \frac{\cos\theta}{\sin\theta}.$$

A *quotient identity* can be obtained for $\tan\theta$ similarly.

QUOTIENT IDENTITIES

$$\tan\theta = \frac{\sin\theta}{\cos\theta} \qquad \cot\theta = \frac{\cos\theta}{\sin\theta}$$

 **EXAMPLE 2**   Using identities

If $\sin\theta = \frac{3}{5}$ and $\cos\theta = -\frac{4}{5}$, find the other four trigonometric functions of θ.

SOLUTION   To find the other four trigonometric functions, apply the quotient and reciprocal identities.

$$\tan\theta = \frac{\sin\theta}{\cos\theta} = \frac{3/5}{-4/5} = -\frac{3}{4}$$

$$\cot\theta = \frac{\cos\theta}{\sin\theta} = \frac{-4/5}{3/5} = -\frac{4}{3}$$

$$\sec\theta = \frac{1}{\cos\theta} = \frac{1}{-4/5} = -\frac{5}{4}$$

$$\csc\theta = \frac{1}{\sin\theta} = \frac{1}{3/5} = \frac{5}{3}$$

Now Try Exercise 7 ◆

In Section 6.3 we showed that $\sin^2\theta + \cos^2\theta = 1$ for all real numbers (or angles) θ. In the next example we apply this identity to find the values of the trigonometric functions.

EXAMPLE 3   Applying the identity $\sin^2\theta + \cos^2\theta = 1$

If $\sin\theta = \frac{4}{5}$ and $\cos\theta < 0$, find the values of the other five trigonometric functions.

SOLUTION   The identity $\sin^2\theta + \cos^2\theta = 1$ can be used to determine $\cos\theta$.

$$\sin^2\theta + \cos^2\theta = 1 \qquad \text{Identity}$$

$$\cos^2\theta = 1 - \sin^2\theta \qquad \text{Subtract } \sin^2\theta.$$

$$\cos\theta = \pm\sqrt{1 - \sin^2\theta} \qquad \text{Square root property}$$

$$\cos\theta = \pm\sqrt{1 - \left(\frac{4}{5}\right)^2} \qquad \text{Let } \sin\theta = \tfrac{4}{5}.$$

$$\cos\theta = \pm\sqrt{\frac{9}{25}} \qquad \text{Simplify.}$$

$$\cos\theta = -\frac{3}{5} \qquad \cos\theta < 0$$

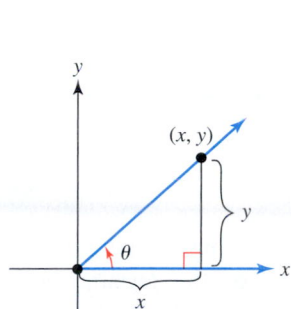

FIGURE 6.74

Now we can use $\sin\theta = \frac{4}{5}$ and $\cos\theta = -\frac{3}{5}$ to determine the other four trigonometric functions.

$$\tan\theta = \frac{\sin\theta}{\cos\theta} = \frac{4/5}{-3/5} = -\frac{4}{3}$$

$$\cot\theta = \frac{1}{\tan\theta} = \frac{1}{-4/3} = -\frac{3}{4}$$

$$\sec\theta = \frac{1}{\cos\theta} = \frac{1}{-3/5} = -\frac{5}{3}$$

$$\csc\theta = \frac{1}{\sin\theta} = \frac{1}{4/5} = \frac{5}{4}$$

Now Try Exercise 13 ◆

Suppose the terminal side of an angle θ in standard position intersects the unit circle at the point (x, y). See Figure 6.74. In this case $r = 1$ and the definitions of the six trigonometric functions become as follows.

$$\sin\theta = y \qquad \cos\theta = x \qquad \tan\theta = \frac{y}{x}$$

$$\csc\theta = \frac{1}{y} \qquad \sec\theta = \frac{1}{x} \qquad \cot\theta = \frac{x}{y}$$

Like the sine and cosine functions, evaluating the other trigonometric functions at a real number t is equivalent to evaluating these functions at t radians.

Representations of Other Trigonometric Functions

Like $\sin t$ and $\cos t$, there are no simple formulas to evaluate the other four trigonometric functions. However, these functions do have numerical and graphical representations. We begin by discussing the tangent function.

The Tangent Function   Figure 6.75 shows angle θ in standard position with its terminal side passing through the point (x, y). The slope of this terminal side is

$$m = \frac{y - 0}{x - 0} = \frac{y}{x} = \tan\theta.$$

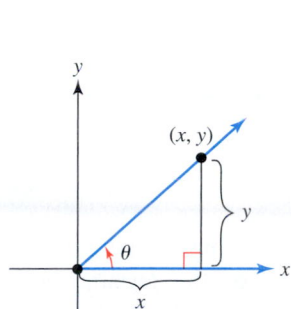

It can be shown that the slope of the terminal side of θ equals $\tan\theta$, regardless of the quadrant containing θ. Using this fact, we will analyze the tangent function by letting θ vary from $-\frac{\pi}{2}$ to $\frac{\pi}{2}$, or equivalently, $-90°$ to $90°$.

When $\theta = -\frac{\pi}{2}$ the slope of the terminal side of θ is undefined, and so, $\tan\left(-\frac{\pi}{2}\right)$ is undefined. Refer to Figure 6.76. For values of θ slightly greater than $-\frac{\pi}{2}$, the

FIGURE 6.75

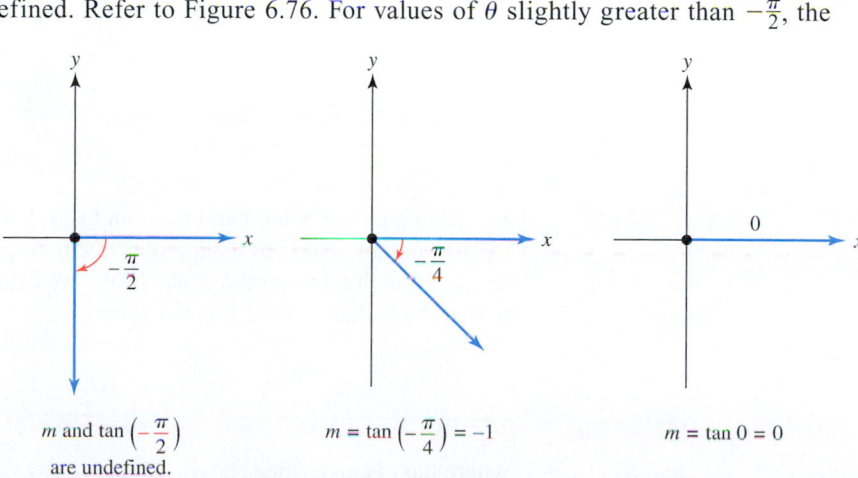

m and $\tan\left(-\frac{\pi}{2}\right)$ are undefined.

$m = \tan\left(-\frac{\pi}{4}\right) = -1$

$m = \tan 0 = 0$

FIGURE 6.76

slope of the terminal side of θ is both negative and large in absolute value. At $\theta = -\frac{\pi}{4}$, the slope of the terminal side is $m = -1$ and $\tan\left(-\frac{\pi}{4}\right) = -1$. The slope is negative and increasing for $-\frac{\pi}{2} < \theta < 0$. When $\theta = 0$ then $m = 0$ and $\tan 0 = 0$.

As θ increases from 0 to $\frac{\pi}{2}$, the slope m is increasing and positive. Refer to Figure 6.77. At $\theta = \frac{\pi}{4}$ the slope is $m = 1$ and $\tan\left(\frac{\pi}{4}\right) = 1$. As θ increases toward $\frac{\pi}{2}$ the slope of the terminal side increases without bound and becomes undefined when $\theta = \frac{\pi}{2}$. Thus $\tan\left(\frac{\pi}{2}\right)$ is undefined.

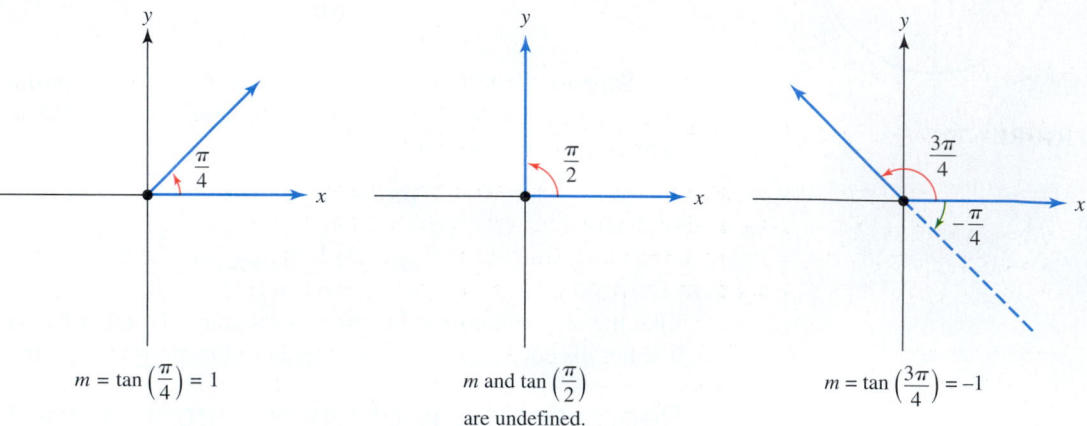

$m = \tan\left(\dfrac{\pi}{4}\right) = 1$ m and $\tan\left(\dfrac{\pi}{2}\right)$ are undefined. $m = \tan\left(\dfrac{3\pi}{4}\right) = -1$

FIGURE 6.77   FIGURE 6.78

As θ becomes greater than $\frac{\pi}{2}$ the values of $\tan \theta$ repeat. For example, when $\theta = \frac{3\pi}{4}$, the slope of the terminal side of θ is equal to the slope of the terminal side of $\theta = -\frac{\pi}{4}$. See Figure 6.78. As a result, the tangent function has period π or 180°.

A *numerical representation* of $f(t) = \tan t$ for selected values of t satisfying $-\frac{\pi}{2} \le t \le \frac{\pi}{2}$ is shown in Table 6.7. (A dash indicates that $\tan t$ is undefined.)

TABLE 6.7

t	$-\dfrac{\pi}{2}$	$-\dfrac{\pi}{3}$	$-\dfrac{\pi}{4}$	$-\dfrac{\pi}{6}$	0	$\dfrac{\pi}{6}$	$\dfrac{\pi}{4}$	$\dfrac{\pi}{3}$	$\dfrac{\pi}{2}$
$\tan t$	—	$-\sqrt{3}$	-1	$-\dfrac{1}{\sqrt{3}}$	0	$\dfrac{1}{\sqrt{3}}$	1	$\sqrt{3}$	—

A graph of $y = \tan t$ and the data in Table 6.7 appear in Figure 6.79, where t is measured in radians. Vertical asymptotes occur when $\tan t$ is undefined at $t = \pm\frac{\pi}{2}$. Since the tangent function has period π (or 180°), we can extend the graph of $y = \tan t$, as in Figure 6.80. Notice that vertical asymptotes occur at

$$t = \pm\frac{\pi}{2}, \pm\frac{3\pi}{2}, \pm\frac{5\pi}{2}, \ldots$$

where $\tan t$ is undefined.

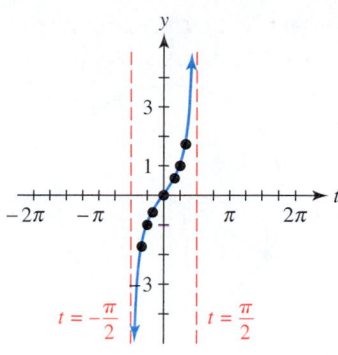

FIGURE 6.79

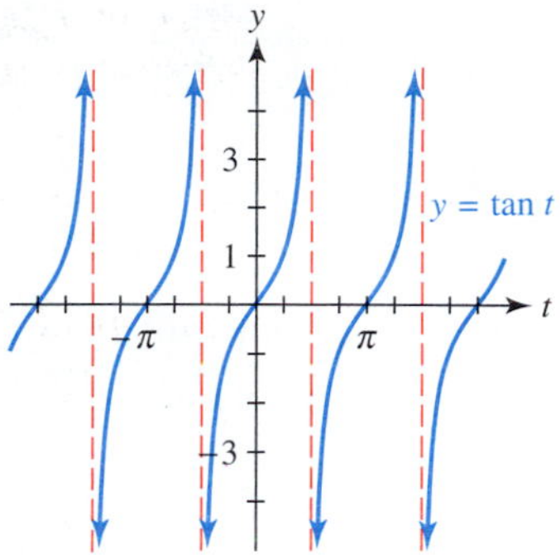

FIGURE 6.80 The Tangent Function

EXAMPLE 4

Evaluating the tangent function

Evaluate $f(t) = \tan t$ at $t = -\frac{\pi}{4}$ by hand.

SOLUTION If angle $-\frac{\pi}{4}$ is in standard position, the terminal side intersects the unit circle at the point $\left(\frac{1}{\sqrt{2}}, -\frac{1}{\sqrt{2}}\right)$. See Figure 6.81. Therefore

$$\tan\left(-\frac{\pi}{4}\right) = \frac{y}{x} = \frac{-1/\sqrt{2}}{1/\sqrt{2}} = -1.$$

Note that to determine the length of the sides of the triangle in Figure 6.81, the 45°–45° triangle in Figure 6.82 can be used as and aid. Also note that the slope of the terminal side of the angle is -1 and equals $\tan\left(-\frac{\pi}{4}\right)$.

Now Try Exercise 17 ◆

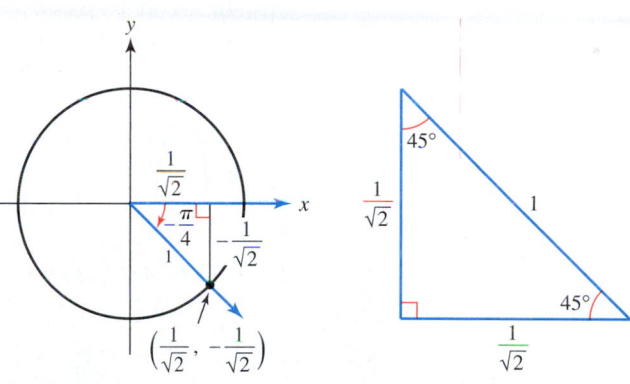

FIGURE 6.81

FIGURE 6.82

The Cosecant Function The cosecant function can also be evaluated by hand. For example, if $t = \frac{7\pi}{6}$, then the terminal side of $\theta = \frac{7\pi}{6}$ intersects the unit circle at the point $\left(\frac{-\sqrt{3}}{2}, -\frac{1}{2}\right)$. See Figure 6.83 on the next page. It follows that

$$\csc\frac{7\pi}{6} = \frac{1}{y} = \frac{1}{-1/2} = -2.$$

Note that to determine the length of the sides of the triangle in Figure 6.83 on the next page, the 30°–60° triangle in Figure 6.84 can be used as an aid.

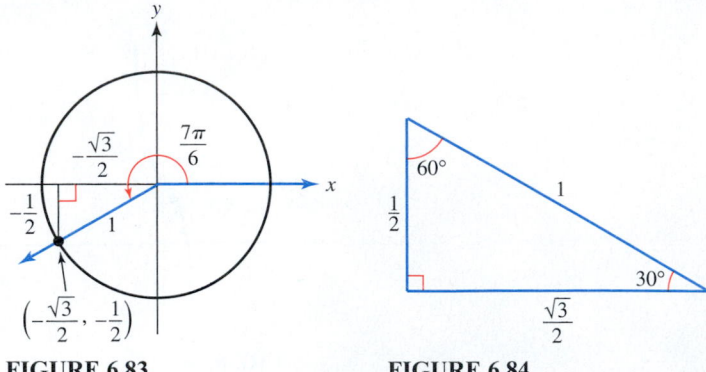

FIGURE 6.83   FIGURE 6.84

A *numerical representation* of $f(t) = \csc t$ is shown in Table 6.8 for selected values of t satisfying $-\pi \le t \le \pi$. Note that $\csc t = \frac{1}{\sin t}$. (A dash indicates that $\csc t$ is undefined.)

TABLE 6.8

t	$-\pi$	$-\frac{3\pi}{4}$	$-\frac{\pi}{2}$	$-\frac{\pi}{4}$	0	$\frac{\pi}{4}$	$\frac{\pi}{2}$	$\frac{3\pi}{4}$	π
$\csc t$	—	$-\sqrt{2}$	-1	$-\sqrt{2}$	—	$\sqrt{2}$	1	$\sqrt{2}$	—

A graph of $\csc t$ is given in Figure 6.85. The domain of the cosecant function is $D = \{t \mid t \ne \pi n\}$, where n is an integer and vertical asymptotes occur whenever $t = \pi n$. The period of the cosecant function equals 2π.

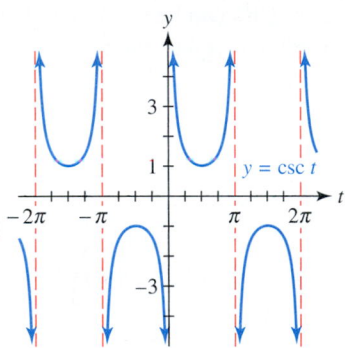   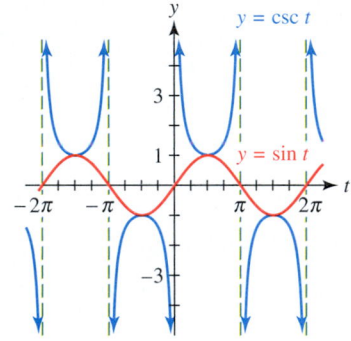

FIGURE 6.85 The Cosecant Function   **FIGURE 6.86** The Sine and Cosecant Functions

Since $\csc t = \frac{1}{\sin t}$, we can sketch the graph of $y = \csc t$ by using the graph of $y = \sin t$ as an aid. See Figure 6.86. When $\sin t = 0$, $\csc t$ is undefined and a vertical asymptote occurs on the graph of $\csc t$. When $\sin t = \pm 1$, $\csc t = \pm 1$. Whenever $\sin t$ increases, $\csc t$ decreases and whenever $\sin t$ decreases, $\csc t$ increases. Because $|\sin t| \le 1$ for all t, it follows that $|\csc t| \ge 1$ for all $t \ne \pi n$.

The Cotangent Function   A graph of $y = \cot t$ is shown in Figure 6.87. Its period is π and its domain is $D = \{t \mid t \ne \pi n\}$, where n is an integer. Vertical asymptotes occur at $t = \pi n$.

Trigonometric functions may be represented graphically in either radian or degree mode. If a friendly window is selected using degree mode, the graph of a trigonometric function may be accurately graphed in connected mode. Graphs of the cosecant and cotangent functions are shown in Figures 6.88 and 6.89.

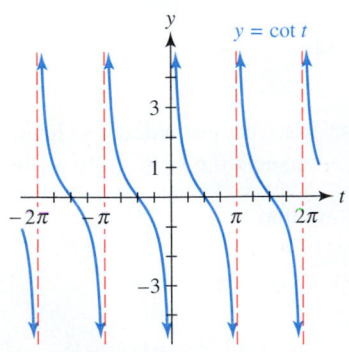

FIGURE 6.87 The Cotangent Function

The Secant Function   A graph of $y = \sec t$ is shown in Figure 6.90. Its period is 2π and its domain is $D = \{t \mid t \ne \frac{\pi}{2} + \pi n\}$, where n is an integer. Vertical asymptotes occur at

$[-352.5°, 352.5°, 90°]$ by $[-4, 4, 1]$

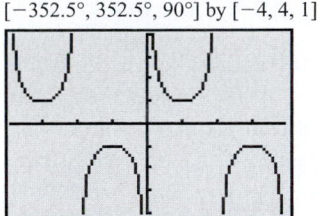

FIGURE 6.88 The Cosecant Function

$[-352.5°, 352.5°, 90°]$ by $[-4, 4, 1]$

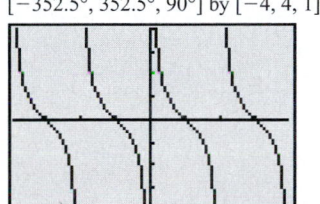

FIGURE 6.89 The Cotangent Function

$t = \frac{\pi}{2} + \pi n$. The graph of the cosine function may be used as an aid to graph the secant function, since the zeros of $y = \cos t$ correspond to the vertical asymptotes on the graph of $y = \sec t$. See Figure 6.91.

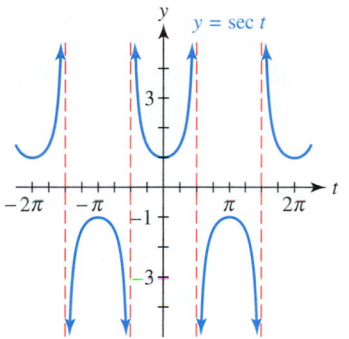

FIGURE 6.90 The Secant Function

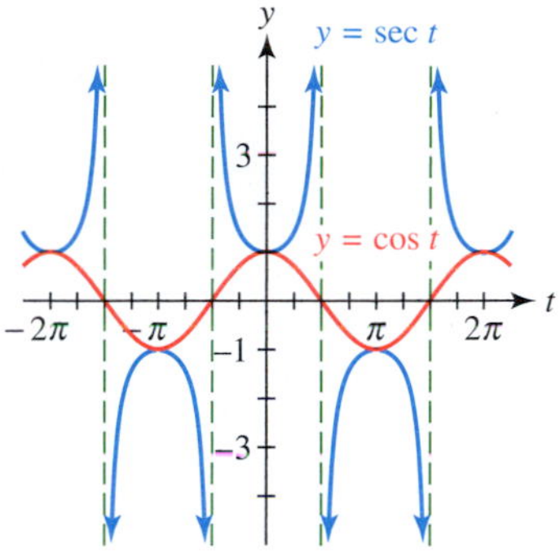

FIGURE 6.91 The Cosine and Secant Functions

The next two examples illustrate how to evaluate the six trigonometric functions by hand and with a calculator.

EXAMPLE 5 Evaluating the trigonometric functions

If $\theta = \frac{3\pi}{2}$, find the six trigonometric functions of θ.

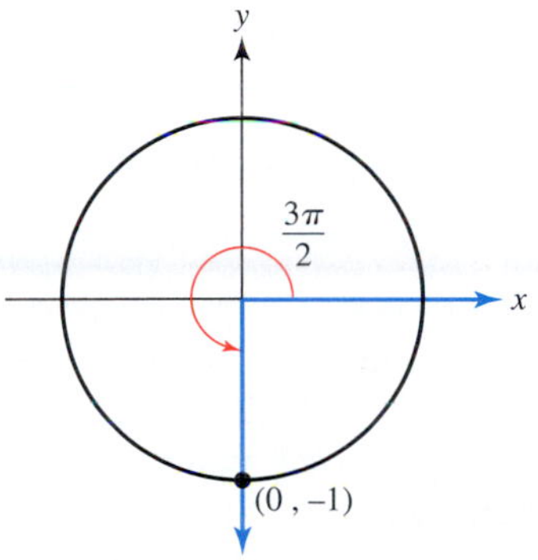

SOLUTION If θ is in standard position, then its terminal side intersects the unit circle at the point $(0, -1)$. See Figure 6.92. Thus $x = 0$ and $y = -1$. The values of the six trigonometric functions are as follows.

$$\sin \frac{3\pi}{2} = y = -1 \qquad\qquad \csc \frac{3\pi}{2} = \frac{1}{y} = \frac{1}{-1} = -1$$

$$\cos \frac{3\pi}{2} = x = 0 \qquad\qquad \sec \frac{3\pi}{2} = \frac{1}{x} \text{ is undefined, since } x = 0$$

$$\tan \frac{3\pi}{2} = \frac{y}{x} \text{ is undefined, since } x = 0 \qquad \cot \frac{3\pi}{2} = \frac{x}{y} = \frac{0}{-1} = 0$$

FIGURE 6.92

Now Try Exercise 27

EXAMPLE 6 Using a calculator to evaluate trigonometric functions

Find values of the six trigonometric functions of θ.
(a) $\theta = 102.6°$ **(b)** $\theta = 2.56$

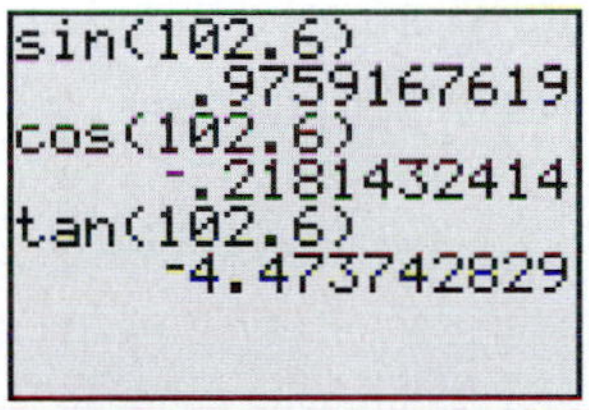

```
sin(102.6)
      .9759167619
cos(102.6)
      -.2181432414
tan(102.6)
      -4.473742829
```

FIGURE 6.93 Degree Mode

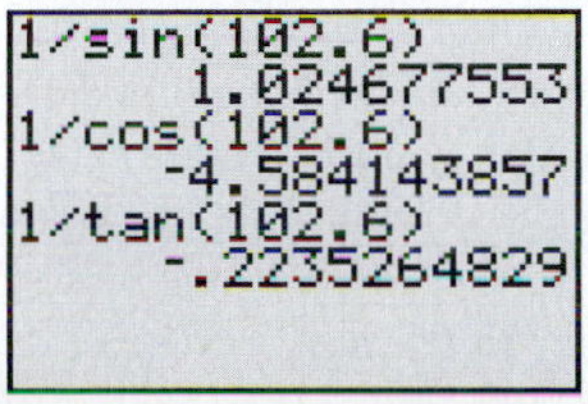

SOLUTION
(a) Most calculators do not have special keys for secant, cosecant, and cotangent. To evaluate these functions, we use the reciprocal identities. For example, to find $\sec(102.6°)$, evaluate $1/\cos(102.6°)$. The values of the trigonometric functions are shown in Figures 6.93 and 6.94.

(b) Since there is no degree symbol, use radian mode. With the aid of a calculator, the values of the six trigonometric functions are approximated as follows.

```
1/sin(102.6)
      1.024677553
1/cos(102.6)
      -4.584143857
1/tan(102.6)
      -.2235264829
```

FIGURE 6.94 Degree Mode

$$\sin 2.56 \approx 0.5494 \qquad \cos 2.56 \approx -0.8356 \qquad \tan 2.56 \approx -0.6574$$
$$\csc 2.56 \approx 1.8203 \qquad \sec 2.56 \approx -1.1968 \qquad \cot 2.56 \approx -1.5210$$

Now Try Exercises 45 and 49

Applications of Trigonometric Functions

When a stick is partially put into a glass of water, it appears to bend at the surface of the water. A similar phenomenon occurs when starlight enters Earth's atmosphere. To an observer on the ground, a star's apparent position α is different from its true position. This is referred to as *atmospheric refraction*. See Figure 6.95. The amount that starlight is bent is given by angle θ measured in seconds, where $\theta = 57.3 \tan \alpha$ with $0° \leq \alpha \leq 45°$. (**Source:** W. Schlosser, *Challenges of Astronomy*.)

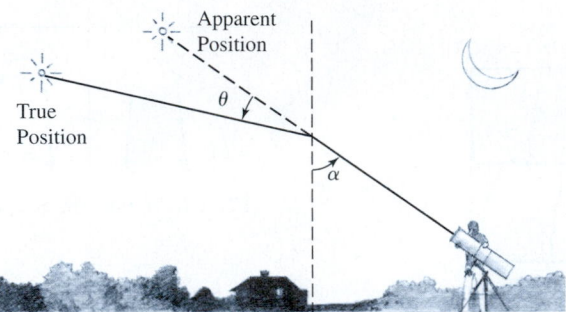

FIGURE 6.95   Atmospheric Refraction

EXAMPLE 7   Calculating refraction of starlight

Refer to Figure 6.95 and the formula $\theta = 57.3 \tan \alpha$.
(a) Calculate θ when $\alpha = 32°35'21''$.
(b) Is starlight bent more or less for stars that appear lower in the horizon? Explain.

SOLUTION
(a) $\theta = 57.3 \tan (32°35'21'') \approx 36.6''$. The star is actually 36.6'' lower in the sky than it appears.
(b) As α increases from 0° to 45° the star appears lower in the horizon, and the graph of $y = \tan \alpha$ increases. Thus $\theta = 57.3 \tan \alpha$ also increases and starlight is bent *more* for stars that appear lower in the horizon. This is supported numerically in Figure 6.96, where $Y_1 = 57.3 \tan (X)$ increases as x increases.

Calculator Help

To review how to make a table, see Appendix B (page AP-9).

X	Y1
0	0
9	9.0754
18	18.618
27	29.196
36	41.631
45	57.3
54	78.867

Y1=57.3tan(X)

FIGURE 6.96   Degree Mode   *Now Try Exercise 75* ◆

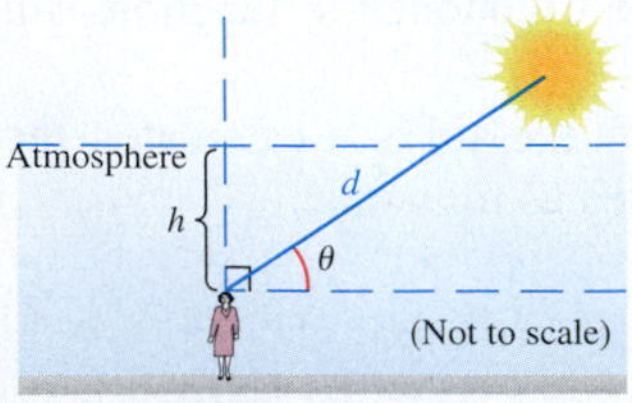

FIGURE 6.97

The shortest path through Earth's atmosphere for the sun's rays occurs when the sun is directly overhead. As the sun moves lower in the horizon, sunlight travels through more atmosphere. See Figure 6.97, where $d = h \csc \theta$. If the angle of elevation of the sun is θ, then the path length d of sunlight through the atmosphere increases by a factor of $\csc \theta$ from its noontime value of h. (Due to the curvature of Earth, this model is not accurate when $\theta < 20°$ and the sun is positioned near the horizon. See Exercise 78.) When sunlight passes through more atmosphere, less ultraviolet light reaches Earth's surface. This is one reason why some experts recommend sun tanning either earlier or later in the day.   (**Source:** C. Winter, *Solar Power Plants*.)

Measuring the intensity of the sun

Assuming that the sun is directly overhead at noon, calculate the percent increase in atmospheric distance that sunlight must pass through at 10 A.M. compared to noon.

SOLUTION   From Figure 6.97, $\theta = 90°$ at noon. The sun moves $360°$ in the sky in 24 hours or $15°$ per hour. Thus 2 hours earlier $\theta = 60°$. Since $\csc 90° = 1$ and $\csc 60° = 1/\sin 60° \approx 1.15$, this means that sunlight travels through about 15% more atmosphere at 10 A.M. than at noon.

Now Try Exercise 77 ◆

6.4   Putting it all Together

There are six trigonometric functions. The following table summarizes the domain, range, and period of each trigonometric function. In this table n is an integer and t is a real number.

Function	Domain	Range	Period
$\sin t$	$-\infty < t < \infty$	$-1 \le \sin t \le 1$	2π
$\cos t$	$-\infty < t < \infty$	$-1 \le \cos t \le 1$	2π
$\tan t$	$t \ne \dfrac{\pi}{2} + \pi n$	$-\infty < \tan t < \infty$	π
$\cot t$	$t \ne \pi n$	$-\infty < \cot t < \infty$	π
$\sec t$	$t \ne \dfrac{\pi}{2} + \pi n$	$|\sec t| \ge 1$	2π
$\csc t$	$t \ne \pi n$	$|\csc t| \ge 1$	2π

Graphs of the six trigonometric functions, including any asymptotes, are as follows.

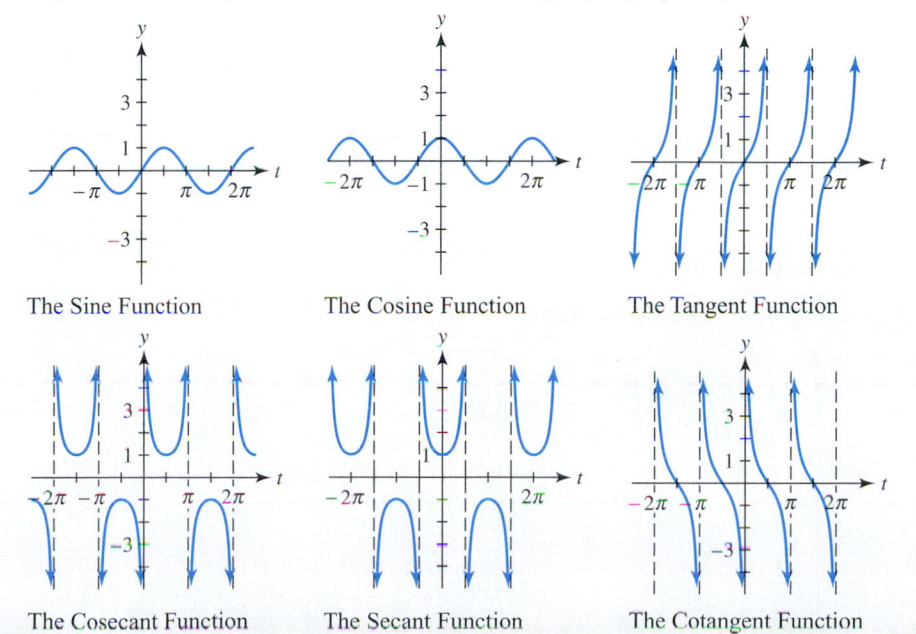

The Sine Function   The Cosine Function   The Tangent Function

The Cosecant Function   The Secant Function   The Cotangent Function

The following figure can be used to evaluate the trigonometric functions for certain angles. For example, if $\theta = 120°$ or $\frac{2\pi}{3}$ radians is in standard position, then its terminal side intersects the unit circle at the point $\left(-\frac{1}{2}, \frac{\sqrt{3}}{2}\right)$. It follows that $\cos 120° = -\frac{1}{2}$ and $\sin 120° = \frac{\sqrt{3}}{2}$. Other trigonometric values can be found using these values. Note that $\frac{\sqrt{2}}{2} = \frac{1}{\sqrt{2}}$.

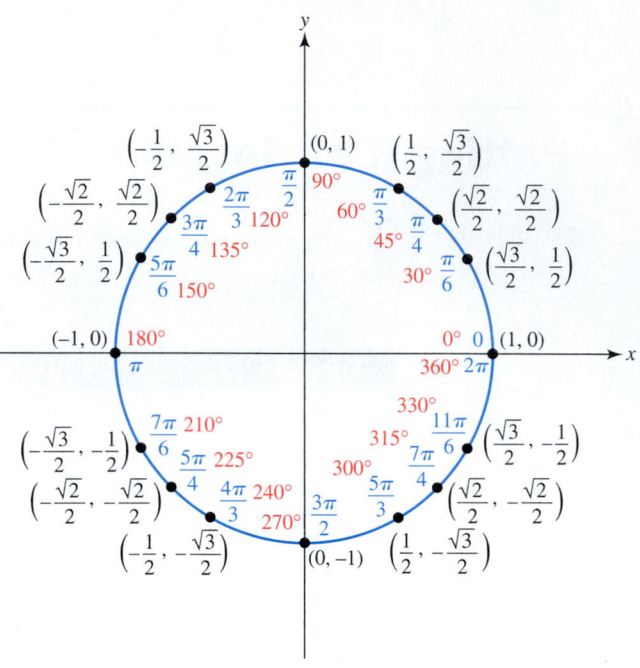

The Unit Circle and Special Angles

6.4 — Exercises

Basic Concepts

Exercises 1–6: Find the six trigonometric functions of θ.

1.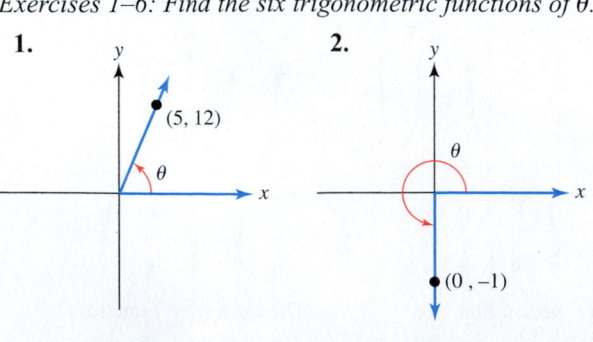

2.

3.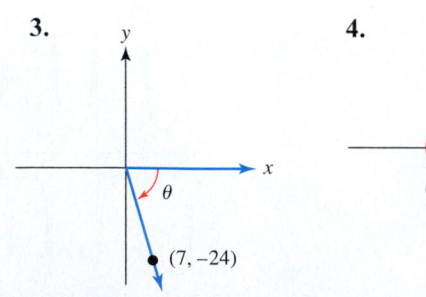

4.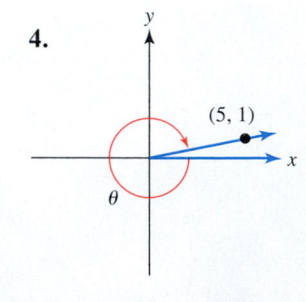

5.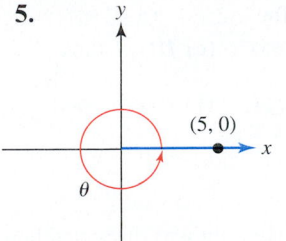

6.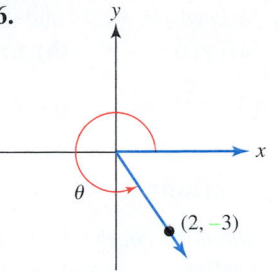

23.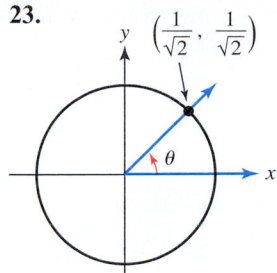

24.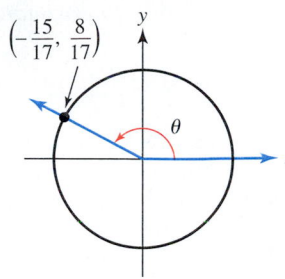

Exercises 7–16: Determine the values of the trigonometric functions of θ by using the given information.

7. $\sin\theta = \frac{3}{5}$ and $\cos\theta = \frac{4}{5}$

8. $\sin\theta = -\frac{7}{25}$ and $\cos\theta = \frac{24}{25}$

9. $\csc\theta = -\frac{17}{15}$ and $\sec\theta = -\frac{17}{8}$

10. $\csc\theta = 2$ and $\sec\theta = -\frac{2}{\sqrt{3}}$

11. $\tan\theta = \frac{5}{12}$ and $\cos\theta = \frac{12}{13}$
(*Hint:* $\sin\theta = \tan\theta\cos\theta$.)

12. $\sin\theta = \frac{3}{5}$ and $\cot\theta = -\frac{4}{3}$

13. $\sin\theta = -\frac{3}{5}$ and $\cos\theta > 0$

14. $\sin\theta = -\frac{12}{13}$ and $\cos\theta < 0$

15. $\cos\theta = -\frac{4}{5}$ and $\sin\theta < 0$

16. $\cos\theta = \frac{7}{25}$ and $\sin\theta > 0$

17. Find $\tan\frac{3\pi}{4}$ by hand.

18. Find $\csc\left(-\frac{\pi}{4}\right)$ by hand.

Exercises 19–22: (Refer to Figure 6.72 on page 518.) Let angle θ be in standard position.

   (a) *Locate a point (x, y) on its terminal side.*
   (b) *Calculate the distance r that this point is from the origin.*
   (c) *Find* $\tan\theta$, $\cot\theta$, $\sec\theta$, *and* $\csc\theta$.

19. $\theta = \frac{\pi}{2}$ **20.** $\theta = -\frac{5\pi}{3}$

21. $\theta = -135°$ **22.** $\theta = 330°$

Exercises 23–26: The figure at the top of the next column shows angle θ in standard position with its terminal side intersecting the unit circle. Evaluate the six trigonometric functions of θ.

25.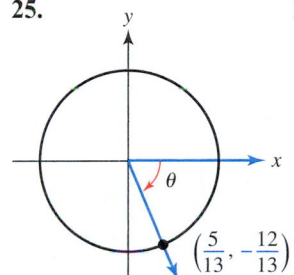

26.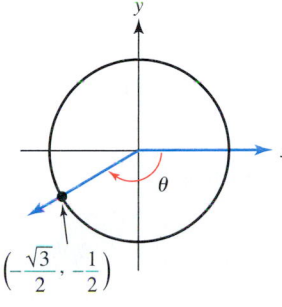

Exercises 27–40: Find the six trigonometric functions of the given angle by hand. Support your answer using a calculator.

27. $90°$ **28.** $135°$

29. $-45°$ **30.** $-180°$

31. π **32.** $\frac{3\pi}{4}$

33. $-\frac{\pi}{3}$ **34.** $-\frac{5\pi}{6}$

35. $-\frac{\pi}{2}$ **36.** $\frac{3\pi}{2}$

37. $360°$ **38.** $60°$

39. $\frac{\pi}{6}$ **40.** $-\frac{2\pi}{3}$

Exercises 41–44: The terminal side of an angle θ lies on the line in the given quadrant. Find the six trigonometric functions of θ. How does the slope of the line compare to $\tan\theta$?

41. $y = -4x$,   Quadrant II

42. $y = \frac{1}{2}x$,   Quadrant I

43. $y = 6x$,   Quadrant III

44. $y = -\frac{2}{3}x$,   Quadrant IV

Exercises 45–52: Approximate the following to four decimal places.

45. (a) $\sin 93.2°$

(b) $\csc 93.2°$

46. (a) $\cos (-43°)$

(b) $\sec (-43°)$

47. (a) $\tan 234°33'$

(b) $\cot 234°33'$

48. (a) $\sec 123°44'25''$

(b) $\cos 123°44'25''$

49. (a) $\cot (-4)$

(b) $\tan (-4)$

50. (a) $\csc 1.56$

(b) $\sin 1.56$

51. (a) $\cos \left(\frac{11\pi}{7}\right)$

(b) $\sec \left(\frac{11\pi}{7}\right)$

52. (a) $\tan \left(\frac{7\pi}{5}\right)$

(b) $\cot \left(\frac{7\pi}{5}\right)$

Exercises 53–58: Graph $y = f(t)$. Discuss the symmetry of the graph.

53. $f(t) = \sin t$

54. $f(t) = \cos t$

55. $f(t) = \tan t$

56. $f(t) = \cot t$

57. $f(t) = \sec t$

58. $f(t) = \csc t$

Exercises 59–64: Use the graph of f to identify its domain, range, and period.

59. $f(t) = \sin t$

60. $f(t) = \cos t$

61. $f(t) = \tan t$

62. $f(t) = \cot t$

63. $f(t) = \sec t$

64. $f(t) = \csc t$

Exercises 65–68: Graph f using degree mode in a friendly window such as $[-352.5°, 352.5°, 90°]$ by $[-4, 4, 1]$. Then use the trace feature to evaluate $f(t)$ at $t = -90°, -45°, 0°, 45°, 90°$, if possible.

65. $f(t) = \tan t$

66. $f(t) = \cot t$

67. $f(t) = \csc t$

68. $f(t) = \sec t$

Exercises 69 and 70: Graph f and g for $-2\pi \le t \le 2\pi$. Explain how the graph of f is related to the graph of g.

69. $f(t) = \sin t, \qquad g(t) = \csc t$

70. $f(t) = \cos t, \qquad g(t) = \sec t$

71. Determine where $f(t) = 0$ for $-2\pi < t \le 2\pi$.

(a) $f(t) = \sin t$ (b) $f(t) = \cos t$ (c) $f(t) = \tan t$

(d) $f(t) = \csc t$ (e) $f(t) = \sec t$ (f) $f(t) = \cot t$

72. Find where $f(t)$ is undefined for $-2\pi < t \le 2\pi$.

(a) $f(t) = \sin t$ (b) $f(t) = \cos t$ (c) $f(t) = \tan t$

(d) $f(t) = \csc t$ (e) $f(t) = \sec t$ (f) $f(t) = \cot t$

Applications

73. *Shadow Length* In the introduction it was discussed how shadows can be used to compute the cotangent function.

(a) Graph $f(t) = \cot t$ for $0 \le t \le \pi$.

(b) Let $t = 0$ correspond to sunrise, $t = \frac{\pi}{2}$ to noon, and $t = \pi$ to sunset. Explain how the graph of f models the length of a shadow cast by a vertical stick with length 1.

74. *Shadow Length* (Refer to Figure 6.71 on page 518.) Calculate the shadow length of a 2-foot stick when the angle of elevation of the sun is $\theta = 27°31'$.

75. *Refraction* (Refer to Example 7.) Use the formula $\theta = 57.3 \tan \alpha$ to calculate the refraction θ in seconds when $\alpha = 17°23'43''$.

76. *GPS Satellite Communication* Artificial satellites that orbit Earth often use VHF signals to communicate with the ground. Because VHF signals travel in straight lines, a satellite orbiting Earth can only communicate with a fixed location on the ground during certain times. The height h in miles of an orbit with communication time T is given by

$$h = 3955 (\sec (\pi T/P) - 1),$$

where P is the period for the satellite to orbit Earth. Suppose a GPS satellite orbit has a period of $P = 12$ hours and can communicate with a person at the North Pole for $T = 5.08$ hours during each orbit. Approximate the height h of its orbit. (**Sources:** W. Schlosser, *Challenges of Astronomy*; Y. Zhao.)

77. *Intensity of the Sun* (Refer to Example 8.) If the sun is directly overhead at noon, calculate the percent increase in atmospheric distance that sunlight must pass through at 3:00 P.M. compared to noon.

78. *Intensity of the Sun* (Refer to the previous exercise.) The formula

$$y_1 = \csc \theta = \frac{1}{\sin \theta}$$

presented in Example 8 to calculate the path length of sunlight through the atmosphere relative to noon is not accurate when the sun's elevation is less than $20°$.

A more accurate formula for small values of θ is given by

$$y_2 = \frac{1}{\sin\theta + 0.5(6° + \theta)^{-1.64}},$$

where θ is measured in degrees. Make a table of y_1 and y_2 starting at $\theta = 2°$ and incrementing by 1°. How do the values of y_1 and y_2 compare as θ increases?   (**Source:** C. Winter.)

*Exercises 79 and 80: **Projectile Flight**   If a projectile is fired with an initial velocity of v feet per second at an angle θ with the horizontal, it will follow a parabolic path described by $y = \dfrac{-16x^2}{v^2(\cos\theta)^2} + x\tan\theta$. See the accompanying figure.*

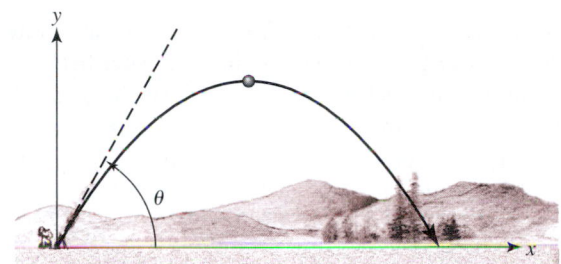

79. If $v = 750$ and $\theta = 30°$, graph the path of the projectile.
(**a**) Determine the coordinates when the maximum height occurs.

(**b**) Assuming the ground is flat, find the total distance traveled by the projectile graphically.

80. If $v = 500$ and $\theta = 45°$, graph the path of the projectile.
(**a**) Determine the maximum height graphically.

(**b**) Assume that the projectile is fired toward a hill that rises with a constant slope of $\frac{1}{4}$. Approximate graphically the total horizontal distance traveled by the projectile.

Writing about Mathematics

81. Discuss whether any of the six trigonometric functions are linear functions. Justify your reasoning.

82. If α and β are coterminal angles, what can be said about the six trigonometric functions of α and β? Explain your answer using an example.

EXTENDED AND DISCOVERY EXERCISE

1. The accompanying figure shows the unit circle and an acute angle θ in standard position. Explain why the trigonometric functions $\sin\theta$, $\cos\theta$, $\tan\theta$, and $\sec\theta$ are equal to the lengths of the line segments shown.

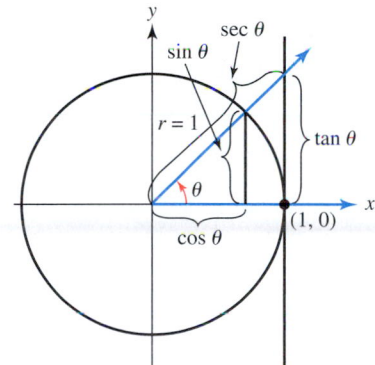

CHECKING BASIC CONCEPTS FOR SECTIONS 6.3 AND 6.4

1. Find the six trigonometric functions of θ, if θ is in standard position and its terminal side passes through the point $(-7, 6)$.

2. Evaluate $\sin 45°$ by hand. Check your results with a calculator.

3. Sketch a graph of the sine, cosine, and tangent functions.

4. Evaluate the six trigonometric functions by hand at the given real number. Support your results using a calculator.

(**a**) $-\pi$ (**b**) $\dfrac{3\pi}{4}$ (**c**) $\dfrac{7\pi}{6}$

6.5 Graphing Trigonometric Functions

Introduction

Trigonometric graphs can be used to model periodic data. For example, monthly average temperatures are usually periodic. They can vary dramatically between January and December, but tend to be periodic from one year to the next. Tides are also periodic. By performing basic transformations of graphs we can model a variety of phenomena.

- ◆ Learn basis transformations of trigonometric graphs
- ◆ Graph trigonometric functions by hand
- ◆ Understand simple harmonic motion
- ◆ Model real data with trigonometric functions (optional)

Transformations of Trigonometric Graphs

Before modeling real data, we will discuss how the constants a and b affect the graphs of $y = a \sin bx$ and $y = a \cos bx$. The constant a controls the *amplitude* of a sinusoidal wave. If $y = 3 \sin x$ then the amplitude is $|a| = 3$. The graph of $f(x) = \sin x$ oscillates between -1 and 1, whereas the graph of $g(x) = 3 \sin x$ oscillates between -3 and 3. See Figure 6.98.

When the constant a is negative, the graphs are reflected across the x-axis. For example, Figure 6.99 shows the graphs of $f(x) = 2 \cos x$ and $g(x) = -2 \cos x$. The graph of g is a reflection of the graph of f across the x-axis.

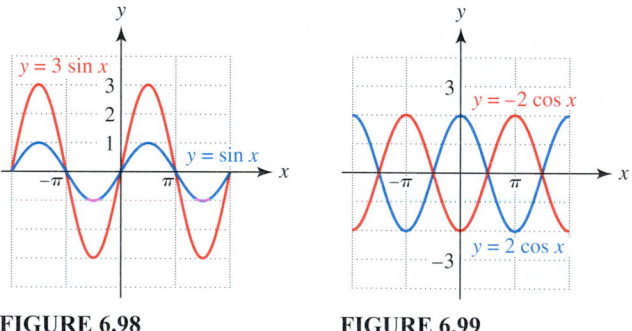

FIGURE 6.98   **FIGURE 6.99**

Algebra Review

To review transformations of graphs, see Section 3.4.

◆ **CLASS DISCUSSION**

Predict how the graph of the given equation will compare with the graph of $f(x) = \cos x$. Check your prediction by graphing both f and the equation.

i. $y = 2 \cos x$

ii. $y = -\frac{1}{2} \cos x$

iii. $y = -3 \cos x$ ◆

The constant b controls the number of oscillations in each interval of length 2π. For example, if $b = 2$ then there are two complete oscillations in every interval of length 2π. As a result, the graph repeats every π units and the period of both $y = \sin 2x$ and $y = \cos 2x$ is π. The *period P* of a sinusoidal graph can be computed by using the formula $P = \frac{2\pi}{b}$, where $b > 0$.

For example, Figure 6.100 shows graphs of $f(x) = \sin x$ and $g(x) = \sin 2x$. Note that the graph of g oscillates "twice as fast" as the graph of f. On the other hand, if $b = \frac{1}{2}$, the graph of $g(x) = \sin \frac{1}{2} x$ oscillates "half as fast" as the graph of $f(x) = \sin x$ and has period 4π. That is, the graph of g only completes half an oscillation every interval of length 2π, or equivalently, the graph of g completes one oscillation every interval of length 4π. See Figure 6.101.

This discussion can also be understood in terms of stretching and shrinking graphs. When $b > 1$ the graph of $y = \sin bx$ is horizontally *shrunk* compared to the graph of $y = \sin x$. When $0 < b < 1$ the graph of $y = \sin bx$ is horizontally *stretched* compared to the graph of $y = \sin x$.

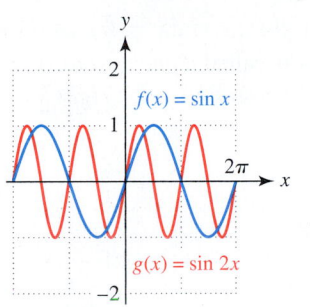

FIGURE 6.100

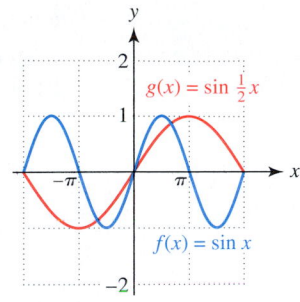

FIGURE 6.101

EXAMPLE 1 Sketching graphs of trigonometric functions

Sketch a graph of each equation. Identify the period and amplitude.

(a) $y = 3 \cos \frac{1}{2}x$ **(b)** $y = -2 \sin 3x$

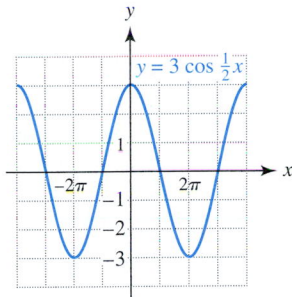

FIGURE 6.102

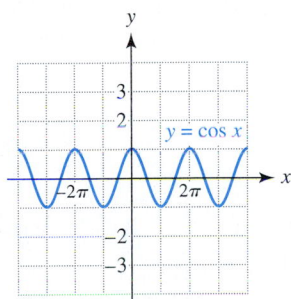

FIGURE 6.103

SOLUTION

(a) The graph of $y = 3 \cos \frac{1}{2}x$ is similar to the graph of $y = \cos x$, except that its period P is

$$P = \frac{2\pi}{b} = \frac{2\pi}{1/2} = 4\pi$$

rather than 2π and its amplitude is 3 rather than 1. A graph of $y = 3 \cos \frac{1}{2}x$ is shown in Figure 6.102, and a graph $y = \cos x$ is shown in Figure 6.103 as a comparison.

(b) First consider the graph of $y = \sin 3x$, which is shown in Figure 6.104 and has period

$$P = \frac{2\pi}{b} = \frac{2\pi}{3}.$$

It oscillates three times every interval of length 2π, whereas the graph of $y = \sin x$ shown in Figure 6.105 oscillates once every interval of length 2π. The graph of $y = -2 \sin 3x$ is similar to the graph of $y = \sin 3x$, except its amplitude is 2 rather than 1 and its graph is reflected across the x-axis. Figure 6.106 shows the required graph of $y = -2 \sin 3x$.

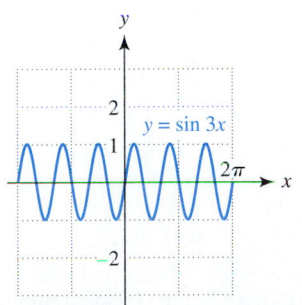

FIGURE 6.104

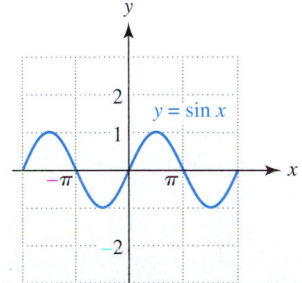

FIGURE 6.105

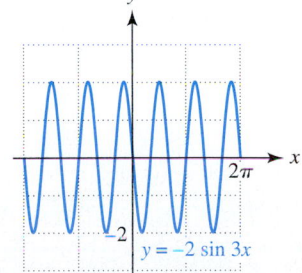

FIGURE 6.106

Now Try Exercises 1 and 3 ◆

Trigonometric graphs can be shifted vertically or horizontally. For example, to shift the graph of $f(x) = \sin x$ upward 1 unit, graph $g(x) = \sin(x) + 1$, as shown in Figure 6.107 on the next page. Similarly, to shift the graph of $f(x) = \sin x$ right $\frac{\pi}{2}$ units, graph the equation $g(x) = \sin\left(x - \frac{\pi}{2}\right)$, as shown in Figure 6.108. A horizontal shift of a trigonometric graph is called a *phase shift*. The phase shift for $g(x) = \sin\left(x - \frac{\pi}{2}\right)$, is $\frac{\pi}{2}$ because the graph of g is shifted to the *right* $\frac{\pi}{2}$ units.

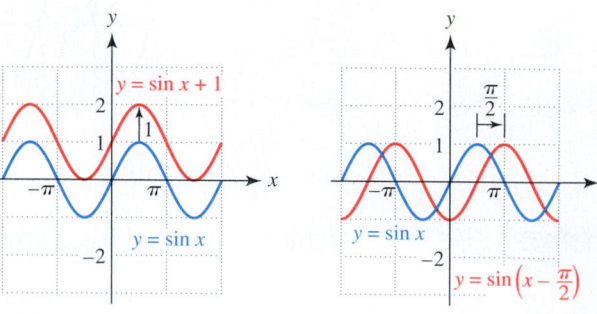

FIGURE 6.107   **FIGURE 6.108**

These concepts are applied in the next example.

 **EXAMPLE 2**

Transforming sinusoidal graphs

Explain how the graph of f can be obtained from the graph of either the sine or cosine function. Then graph f.

(a) $f(x) = 3\cos(2x) + 1$ **(b)** $f(x) = -2\sin\left(x - \frac{\pi}{2}\right)$

SOLUTION

(a) The graph of f can be obtained from the graph of the cosine function by performing the following steps. Shorten the period to π, increase the amplitude to 3, and shift the graph upward 1 unit. These three steps are shown in Figures 6.109–6.111, where the equations $y_1 = \cos 2x$, $y_2 = 3\cos 2x$, and $y_3 = 3\cos(2x) + 1$ are graphed, respectively.

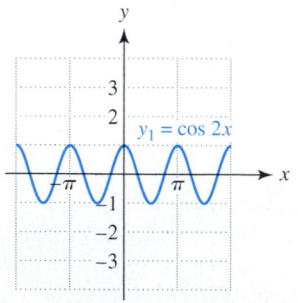 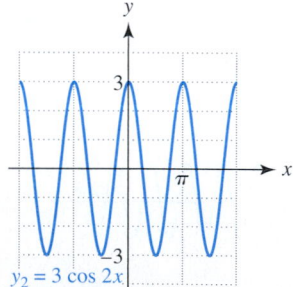 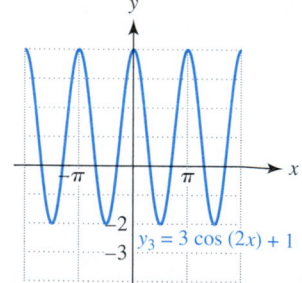

FIGURE 6.109 Reduce Period   **FIGURE 6.110** Increase Amplitude   **FIGURE 6.111** Shift Upward

(b) The graph of f can be obtained from the graph of the sine function by shifting the graph right $\frac{\pi}{2}$ units and increasing the amplitude to 2. The negative sign will cause the graph to be reflected across the x-axis. These steps are shown in Figures 6.112–6.114, where $y_1 = \sin\left(x - \frac{\pi}{2}\right)$, $y_2 = 2\sin\left(x - \frac{\pi}{2}\right)$, and $y_3 = -2\sin\left(x - \frac{\pi}{2}\right)$ are graphed, respectively.

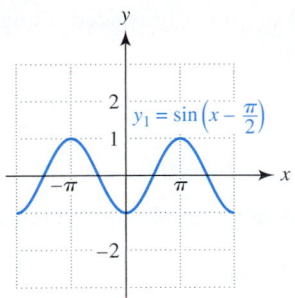

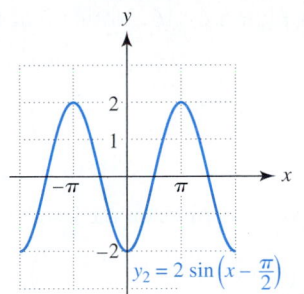

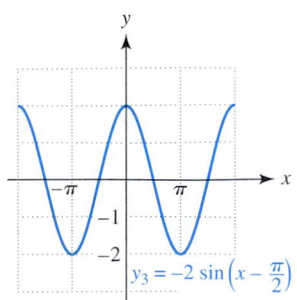

FIGURE 6.112 Shift Right   **FIGURE 6.113** Increase Amplitude   **FIGURE 6.114** Reflect Graph

Now Try Exercises 7 and 9 ◆

Our discussion is summarized by the following.

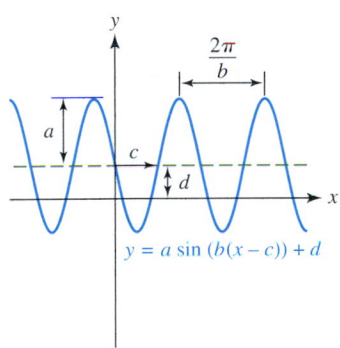

FIGURE 6.115 Transforming the Sine Graph

AMPLITUDE, PERIOD, PHASE SHIFT, AND VERTICAL SHIFT

The **amplitude**, **period**, **phase shift**, and **vertical shift** for the graphs of

$$y = a \sin(b(x - c)) + d \quad \text{and} \quad y = a \cos(b(x - c)) + d$$

with $b > 0$ may be determined as follows.

$$\text{Amplitude} = |a|, \quad \text{Period} = \frac{2\pi}{b}, \quad \text{Phase shift} = c, \quad \text{Vertical shift} = d$$

A vertical shift of $|d|$ units upward occurs when $d > 0$ and $|d|$ units downward when $d < 0$. The graph of $y = a \sin(b(x - c)) + d$ in Figure 6.115 illustrates the role of each constant, assuming that all constants are positive.

EXAMPLE 3   Identifying amplitude, period, phase shift, and vertical shift

For the graph of each equation, identify the amplitude, period, phase shift, and vertical shift.

(a) $y = 7 \sin\left(4\left(x + \frac{\pi}{3}\right)\right) + 2$ **(b)** $y = -3 \cos(5x - 3) - 6$

SOLUTION

(a) The amplitude is 7, and the period is

$$P = \frac{2\pi}{b} = \frac{2\pi}{4} = \frac{\pi}{2}.$$

The phase shift is $-\frac{\pi}{3}$ and the vertical shift is 2.

(b) Start by applying the distributive property to rewrite the equation as

$$y = -3 \cos\left(5\left(x - \frac{3}{5}\right)\right) - 6.$$

The amplitude is 3, and the period is $P = \frac{2\pi}{b} = \frac{2\pi}{5}$. The phase shift is $\frac{3}{5}$ and the vertical shift is -6.

Now Try Exercises 13 and 15 ◆

Graphing Trigonometric Functions by Hand

If desired, graphs of trigonometric functions can be sketched by hand. To sketch the graph $y = \sin x$ by hand, it is helpful to first locate five **key points**. They are labeled in Figure 6.116. Notice that they are equally spaced along the interval $[0, 2\pi]$ on the x-axis.

$(0, 0)$ $\left(\dfrac{\pi}{2}, 1\right)$ $(\pi, 0)$ $\left(\dfrac{3\pi}{2}, -1\right)$ $(2\pi, 0)$

x-intercept   Maximum x-intercept   Minimum x-intercept

Five key points on the graph of $y = \cos x$ are shown in Figure 6.117.

$(0, 1)$ $\left(\dfrac{\pi}{2}, 0\right)$ $(\pi, -1)$ $\left(\dfrac{3\pi}{2}, 0\right)$ $(2\pi, 1)$

Maximum x-intercept   Minimum x-intercept   Maximum

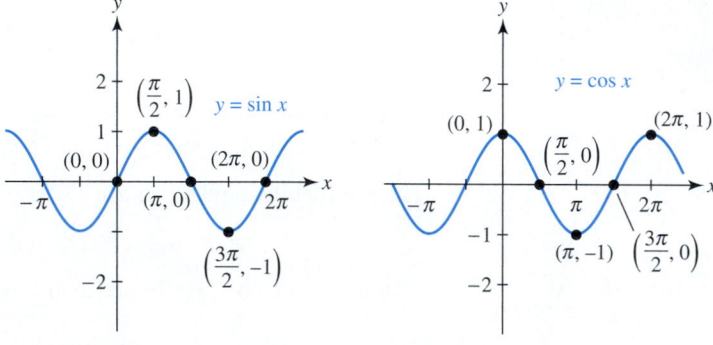

FIGURE 6.116 Key Points on the Sine Graph

FIGURE 6.117 Key Points on the Cosine Graph

One way to sketch a sinusoidal graph by hand is to find the transformed key points. You can use these points and the fact that the graph is periodic to make a sketch. This is illustrated for the sine function in the next example.

EXAMPLE 4   ### Sketching a sinusoidal graph by hand

Find the amplitude, period, and phase shift of the graph of $f(x) = 3 \sin 2x$. Sketch a graph of f by hand on the interval $[-2\pi, 2\pi]$.

SOLUTION If $f(x) = 3 \sin 2x$, then the amplitude is 3 and the period is π. There is no horizontal or vertical shifting. The graph of f passes through the point $(0, 0)$ and completes one oscillation in π units. Thus the graph passes through the point $(\pi, 0)$. An amplitude of 3 will change the maximum and minimum y-values to 3 and -3, respectively. The transformed key points are equally spaced on the graph of f.

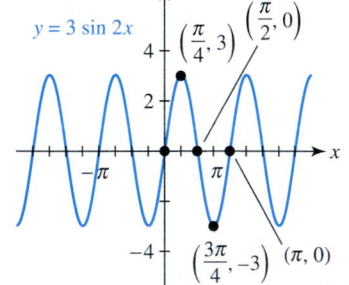

FIGURE 6.118

$(0, 0)$ $\left(\dfrac{\pi}{4}, 3\right)$ $\left(\dfrac{\pi}{2}, 0\right)$ $\left(\dfrac{3\pi}{4}, -3\right)$ $(\pi, 0)$

x-intercept   Maximum x-intercept   Minimum x-intercept

These points and the graph of f are plotted in Figure 6.118.   *Now Try Exercise 39* ◆

EXAMPLE 5   ### Sketching a graph by hand

Graph $y = 2 \cos (4x + \pi)$ by hand.

SOLUTION First write the equation in the form $y = a \cos (b(x - c))$ by factoring out 4.

$$y = 2 \cos \left(4 \left(x + \dfrac{\pi}{4} \right) \right)$$

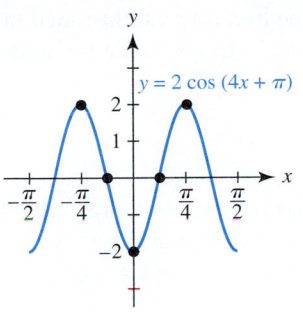

$y = 2 \cos(4x + \pi)$

FIGURE 6.119

◆ **CLASS DISCUSSION**

How would you modify the graph in Figure 6.119 to obtain the graph of

$$y = 2 \cos(4x + \pi) - 1?$$

What are the key points for this graph? ◆

The amplitude is 2 and the period is $\frac{2\pi}{4} = \frac{\pi}{2}$. The graph is translated $\frac{\pi}{4}$ unit to the left compared to the graph of $y = 2 \cos 4x$. Because the graph is translated $\frac{\pi}{4}$ unit to the left, start the x-values in the key points at $0 - \frac{\pi}{4} = -\frac{\pi}{4}$. The first period ends at $-\frac{\pi}{4} + \frac{\pi}{2} = \frac{\pi}{4}$. The key points for this cosine graph are as follows.

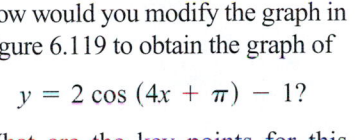

$\left(-\dfrac{\pi}{4}, 2\right)$, $\left(-\dfrac{\pi}{8}, 0\right)$, $(0, -2)$, $\left(\dfrac{\pi}{8}, 0\right)$, $\left(\dfrac{\pi}{4}, 2\right)$

Maximum   x-intercept   Minimum   x-intercept   Maximum

Plot these key points to complete one period of the graph. The graph can be extended to show two periods, ranging from $-\frac{\pi}{2}$ to $\frac{\pi}{2}$, as shown in Figure 6.119.   *Now Try Exercise 51* ◆

The graphs of the other four trigonometric functions can also be transformed. The period of the tangent function and the cotangent function is π. As a result, the graphs of $y = \tan(b(x - c))$ and $y = \cot(b(x - c))$ have a period of $P = \frac{\pi}{b}$ and a phase shift of c. Since the ranges of the tangent and cotangent functions include all real numbers, their graphs do not have an amplitude. The next example illustrates tangent and cotangent graphs. Graphs of the secant and cosecant functions are done in Exercises 73–78.

EXAMPLE 6   Graphing other trigonometric functions

Find the period and phase shift for the graph of f. Graph f on the interval $[-2\pi, 2\pi]$. Identify where asymptotes occur in the graph.

(a) $f(x) = \tan \dfrac{1}{2}x$ **(b)** $f(x) = \cot\left(x + \dfrac{\pi}{2}\right) + 1$

SOLUTION

(a) If $f(x) = \tan \frac{1}{2}x$, then $b = \frac{1}{2}$ and $c = 0$. The period is $P = \frac{\pi}{b} = 2\pi$ and there is no phase shift. Graph $y = \tan\frac{1}{2}x$, as shown in Figure 6.120. On the interval $[-2\pi, 2\pi]$, vertical asymptotes occur at $x = \pm \pi$. The graph of f is horizontally stretched compared to the graph of the tangent function.

(b) If $f(x) = \cot\left(x + \frac{\pi}{2}\right) + 1$, then $b = 1$ and $c = -\frac{\pi}{2}$. The period is π, the phase shift is $-\frac{\pi}{2}$, and the graph is shifted upward 1 unit. Vertical asymptotes occur at $x = \pm \frac{\pi}{2}$ and $x = \pm \frac{3\pi}{2}$. The graph of $y = \cot\left(x + \frac{\pi}{2}\right) + 1$ is shown in Figure 6.121.

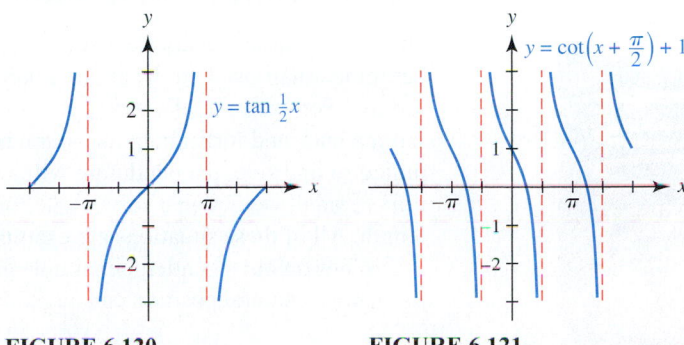

$y = \tan \frac{1}{2}x$

$y = \cot\left(x + \frac{\pi}{2}\right) + 1$

FIGURE 6.120   **FIGURE 6.121**

Now Try Exercises 65 and 71 ◆

An Application   A *sag curve* occurs when a highway goes downhill and then uphill. Improperly designed sag curves can be dangerous at night because a vehicle's headlights point downward and may not illuminate the uphill portion of the highway, as illustrated in Figure 6.122. The minimum safe length L for a typical sag curve with a 40-mile-per-hour speed limit is computed by the equation

$$L = \frac{2700}{h + 3\tan\alpha}.$$

The variable h represents the height of the headlights above the road surface and α represents a small angle associated with the vertical alignment of the headlight shown in Figure 6.122. (**Source:** F. Mannering and W. Kilareski, *Principles of Highway Engineering and Traffic Analysis.*)

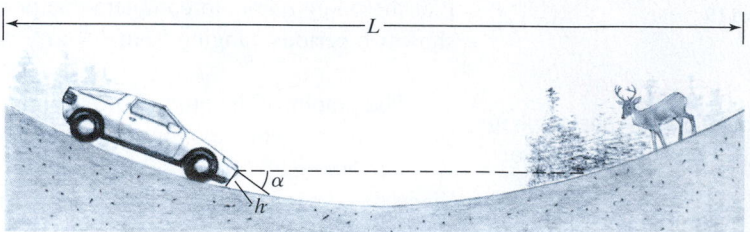

FIGURE 6.122   Designing a Safe Sag Curve (Not to scale)

EXAMPLE 7   Designing a sag curve

Calculate L for a car with headlights 2.5 feet above the ground and $\alpha = 1°$. Repeat the calculations for a truck with headlights 4 feet above the ground and $\alpha = 2°$.

SOLUTION   For the car, let $h = 2.5$ and $\alpha = 1°$.

$$L = \frac{2700}{2.5 + 3\tan 1°}$$
$$\approx 1058 \text{ feet}$$

For the truck, let $h = 4$ and $\alpha = 2°$.

$$L = \frac{2700}{4 + 3\tan 2°}$$
$$\approx 658 \text{ feet}$$

Since both cars and trucks use the same highways, engineers typically use the larger of the two distances to design a safe sag curve.   *Now Try Exercise 91* ◆

Simple Harmonic Motion

Harmonic motion occurs frequently in nature and physical science. A particle or object undergoing small oscillations about a point of stable equilibrium executes simple harmonic motion. For example, if a string on a guitar is plucked gently, any point on the string will vibrate back and forth about its natural position of equilibrium. If a pendulum on a clock is pulled to one side, the pendulum will swing back and forth about its stable, vertical position. A small weight on a spring will bounce up and down when displaced from its natural length. All of these situations are examples of simple harmonic motion.

When an object undergoes simple harmonic motion, its distance or displacement from its stable or natural position can be modeled by either

$$s(t) = a\sin bt \quad \text{or} \quad s(t) = a\cos bt,$$

where $s(t)$ represents the displacement being experienced at time t. To better understand this, consider the spring and weight shown in Figure 6.123. If $s = 0$ corresponds to the

spring's natural length, then $s = -2$ inches indicates that the spring is stretched 2 units beyond its natural length and $s = 2$ indicates that the spring is compressed 2 units. When a spring is stretched and let go, it will oscillate up and down. If the displacement s is plotted after t seconds, a sinusoidal graph results, as shown in Figure 6.124.

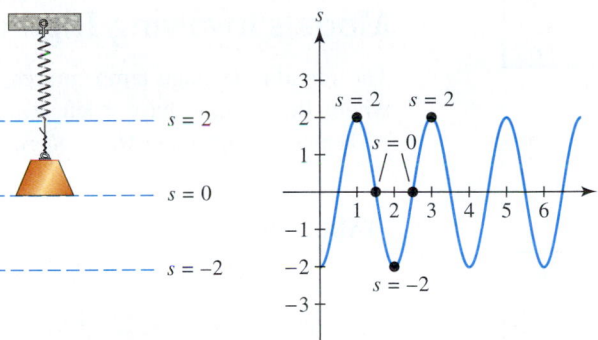

FIGURE 6.123   **FIGURE 6.124** Modeling
Displacement of a Spring   Oscillations of a Spring

The amplitude of each oscillation equals $|a|$ and the period P is given by $\frac{2\pi}{b}$. The number of oscillations per unit time is the **frequency**. The frequency F equals the reciprocal of the period P. That is, $F = \frac{1}{P}$. Since $P = \frac{2\pi}{b}$, it follows that $F = \frac{b}{2\pi}$, or equivalently, $b = 2\pi F$. Substituting for b in the formulas for $s(t)$ results in

$$s(t) = a \sin(2\pi F t) \quad \text{or} \quad s(t) = a \cos(2\pi F t).$$

EXAMPLE 8   Modeling simple harmonic motion

Suppose that the weight and spring in Figure 6.123 have a period of 0.4 second. Initially the weight is lifted 3 inches above its natural length and then let go.
(a) Find an equation in the form $s(t) = a \cos bt$ that models the displacement s of the weight.
(b) Estimate s after 0.92 second. Determine if the weight is moving upward or downward at this time. Support your results graphically.

SOLUTION
(a) Let $s(t) = a \cos(2\pi F t)$. The spring is initially displaced 3 inches, so the amplitude is 3. Let $a = 3$. The period is $P = 0.4$, so the frequency is

$$F = \frac{1}{0.4} = 2.5.$$

This indicates that the weight and spring oscillate up and down 2.5 times per second. It follows that $b = 2\pi F = 5\pi$ and $s(t) = 3 \cos(5\pi t)$. Note that the initial position is $s(0) = 3$ inches.
(b) After 0.92 second, the displacement is

$$s(0.92) = 3 \cos(5\pi(0.92)) \approx -0.927.$$

The weight is about 0.93 inch *below* its natural position after 0.92 second.

The weight initially moves downward and oscillates with a period of 0.4 second. During the time intervals $(0, 0.2)$, $(0.4, 0.6)$, and $(0.8, 1.0)$ the weight is moving downward, and during the intervals $(0.2, 0.4)$, $(0.6, 0.8)$, and $(1.0, 1.2)$ the weight is moving upward. Thus the weight is moving downward when $t = 0.92$.

[0, 1.6, 0.2] by [−4, 4, 1]

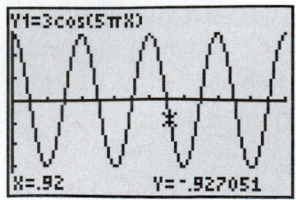

FIGURE 6.125

[0, 25, 2] by [−5, 70, 10]

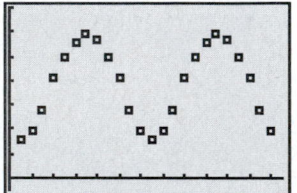

FIGURE 6.126

[0, 25, 2] by [−5, 70, 10]

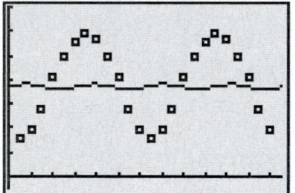

FIGURE 6.127

[0, 25, 2] by [−5, 70, 10]

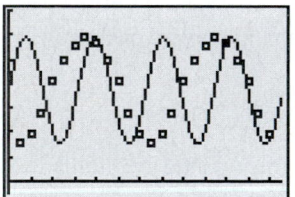

FIGURE 6.128

[0, 25, 2] by [−5, 70, 10]

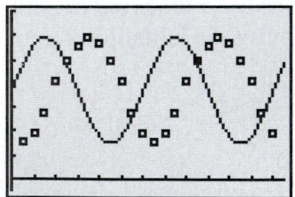

FIGURE 6.129

By tracing the graph of $s(t) = 3 \cos(5\pi t)$, we see that $s(0.92) \approx -0.927$. See Figure 6.125. Since the graph is decreasing in this region the weight is moving downward.

Now Try Exercise 105 ◆

Models Involving Trigonometric Functions (Optional)

The monthly average temperatures for Prince George, Canada, are shown in Table 6.9, where the months have been assigned the standard numbers. (**Source:** A. Miller and J. Thompson, *Elements of Meteorology.*)

TABLE 6.9

Month	1	2	3	4	5	6	7	8	9	10	11	12
Temperature (°F)	15	19	28	41	50	55	59	57	50	41	28	19

Since the data are periodic, they are plotted over a 2-year period in Figure 6.126. For example, both $x = 1$ and $x = 13$ correspond to January. Notice that the scatterplot suggests that a sinusoidal graph might model these data.

Data of the type shown in Figure 6.126 sometimes can be modeled by

$$f(x) = a \sin(b(x - c)) + d,$$

where a, b, c, and d are constants. To estimate values for these constants, we will perform transformations on the graph of $y = \sin x$.

The addition of a constant d causes a vertical shift to the graph of f. The monthly average temperatures at Prince George vary from 15°F to 59°F. The average of these two temperatures is 37°F. If we let $d = 37$ and graph $y = \sin(x) + 37$, the resulting graph is shown in Figure 6.127. Compared to the graph of the sine function, this graph is shifted upward 37 units. (Be sure to use radian mode.)

From Figure 6.127 it is apparent that the amplitude of the oscillations in the graph of $y = \sin(x) + 37$ are not large enough to model the data. The monthly average temperatures have a range of 59°F − 15°F = 44°F. If we let $a = \frac{44}{2} = 22$, the peaks and valleys on the graph of $y = 22 \sin(x) + 37$ will model the data better. See Figure 6.128.

In Figure 6.128 the oscillations are too frequent, which indicates that the period is too small. Since the temperature cycles every 12 months, the period is $P = \frac{2\pi}{b} = 12$. Thus

$$b = \frac{2\pi}{12} = \frac{\pi}{6} \approx 0.524.$$

The graph of $y = 22 \sin\left(\frac{\pi}{6}x\right) + 37$ is shown in Figure 6.129.

We can obtain a reasonable fit by shifting the graph horizontally to the right. The maximum of $y = 22 \sin\left(\frac{\pi}{6}x\right) + 37$ is at $x = 3$, which corresponds to March. We would like this maximum to occur in July ($x = 7$), so we translate the graph 4 units to the right by replacing x with $x - 4$ to obtain

$$y = 22 \sin\left(\frac{\pi}{6}(x - 4)\right) + 37.$$

This equation is graphed in Figure 6.130. The phase shift is $c = 4$.

$$[0, 25, 2] \text{ by } [-5, 70, 10]$$

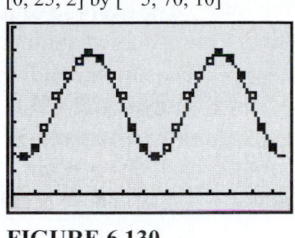

FIGURE 6.130

◆ CLASS DISCUSSION

Is the value of $c = 4$ the only phase shift that can be used to model the temperature data? Explain your reasoning. ◆

Calculator Help

To plot data and graph an equation in the same viewing rectangle, see Appendix B (page AP-11).

Note: Linear and nonlinear regression were introduced in previous chapters. **Sine regression** can also be performed using a sinusoidal function of the form

$$f(x) = a \sin(bx + c) + d.$$

Sine regression may be performed on a 2-year interval of the temperature data in Table 6.9. See Figures 6.131 and 6.132. This function is similar to the one that we found except that the variable c in the regression equation does *not* represent the phase shift.

Calculator Help

To find an equation of least-squares fit, see Appendix B (page AP-15).

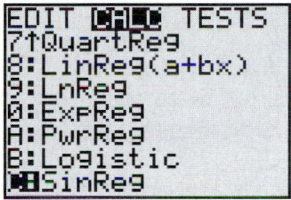

FIGURE 6.131   **FIGURE 6.132**

Tides represent the largest collective motion of water on Earth and cause water levels in oceans to vary during the course of a day. Tides usually occur once or twice a day and change with the phases of the moon. In deep oceans water levels can vary little, whereas near coastlines they can vary as much as 24 feet. The largest tides occur when the moon is either full or new.

EXAMPLE 9   ### Modeling tides

Figure 6.133 shows a function f that models the tides in feet at Clearwater Beach, Florida, x hours after midnight starting on August 26, 1998.   (**Source:** D. Pentcheff, *WWW Tide and Current Predictor.*)

(a) Find the time between high tides.
(b) What is the difference in water levels between high tide and low tide?
(c) Determine a, b, c, and d so that $f(x) = a\cos(b(x - c)) + d$ models the data. Graph f and the data in the same viewing rectangle.

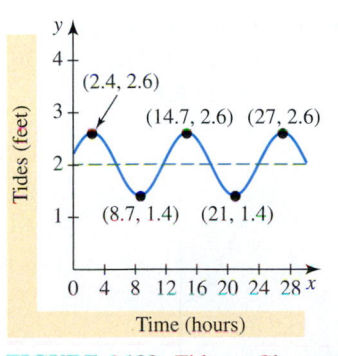

FIGURE 6.133 Tides at Clearwater Beach

SOLUTION

(a) A high tide corresponds to a peak on the graph. The time between peaks is 12.3 hours, since $14.7 - 2.4 = 12.3$ and $27 - 14.7 = 12.3$.
(b) High tides were 2.6 feet and low tides were 1.4 feet. Their difference is 1.2 feet.
(c) The amplitude of f is given by

$$a = \frac{2.6 - 1.4}{2} = 0.6.$$

Since the period is 12.3 hours, the value of b is

$$b = \frac{2\pi}{12.3} \approx 0.511.$$

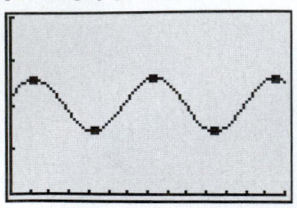

[0, 28, 2] by [0, 4, 1]

FIGURE 6.134

The average of high tide and low tide is $\frac{2.6 + 1.4}{2} = 2$, so we let $d = 2$. Finally, a peak occurs at about 2.4 hours after midnight. Since midnight corresponds to $x = 0$ and the cosine function has a peak at $x = 0$, we translate the graph of f right **2.4** units by letting $c = 2.4$. Thus

$$f(x) = \mathbf{0.6} \cos(\mathbf{0.511}(x - \mathbf{2.4})) + \mathbf{2}.$$

A graph of f and the data are shown in Figure 6.134.   *Now Try Exercise 97* ◆

In the next example we use the cosine function to model daylight hours at San Antonio, Texas.

EXAMPLE 10 Modeling daylight hours

San Antonio, Texas, has a latitude of 29.5°N. Table 6.10 lists the number of daylight hours on the first day of each month at San Antonio. (**Source:** J. Williams, *The Weather Almanac 1995*.)

TABLE 6.10

Month	1	2	3	4	5	6	7	8	9	10	11	12
Daylight (hr)	10.2	10.7	11.5	12.5	13.3	13.9	14.1	13.6	12.7	11.9	11.0	10.4

(a) Plot the data over a 2-year period.
(b) Model these data using $f(x) = a \cos(b(x - c)) + d$.
(c) Estimate the daylight hours on February 15.

SOLUTION

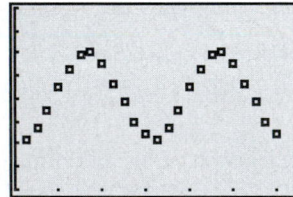

[0, 25, 4] by [8, 16, 1]

FIGURE 6.135

(a) A scatterplot of the data is shown in Figure 6.135. The graph suggests that a sinusoidal graph might model these data.

(b) The maximum number of daylight hours is 14.1 and the minimum is 10.2. The average of these values is $\frac{14.1 + 10.2}{2} = 12.15$ and their difference is $14.1 - 10.2 = 3.9$. Let $d = 12.15$ and $a = \frac{3.9}{2} = 1.95$. Since daylight hours cycle every 12 months, let $b = \frac{2\pi}{12} = \frac{\pi}{6}$. The maximum of the cosine graph occurs at $x = 0$. The maximum in the table occurs on July 1 ($x = 7$), so let $c = 7$. The graph of the function given by $f(x) = 1.95 \cos\left(\frac{\pi}{6}(x - 7)\right) + 12.15$ together with the data is shown in Figure 6.136. A slightly better fit is obtained if we let $c = 6.6$, as shown in Figure 6.137.

◆ **CLASS DISCUSSION**

A phase shift of $c = 7$ corresponds to the maximum number of daylight hours occurring on July 1. Conjecture why $c = 6.6$ fits the data slightly better. ◆

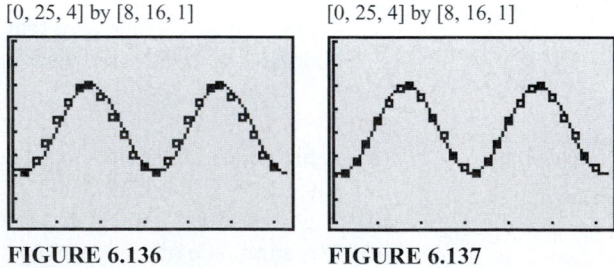

[0, 25, 4] by [8, 16, 1]   [0, 25, 4] by [8, 16, 1]

FIGURE 6.136   **FIGURE 6.137**

(c) February 15 corresponds to $x \approx 2.5$.

$$f(2.5) = 1.95 \cos\left(\frac{\pi}{6}(2.5 - 6.6)\right) + 12.15 \approx 11.1 \text{ hours.}$$

Now Try Exercise 89 ◆

6.5 Putting it all Together

Some concepts about trigonometric models are summarized in the following table.

Concept	Explanation	Examples
Sinusoidal Model $s(x) = a \sin(b(x - c)) + d$ $s(x) = a \cos(b(x - c)) + d$ with $b > 0$	Amplitude $= \lvert a \rvert$. Period $= \dfrac{2\pi}{b}$. Phase shift $= c$. Vertical shift $= d$, upward if $d > 0$ and downward if $d < 0$. If $b = 2\pi F$, then F represents the frequency.	Let $s(x) = 3 \sin(2(x - \pi)) - 1$. Amplitude $= 3$. Period $= \dfrac{2\pi}{2} = \pi$. Phase shift $= \pi$. Vertical shift downward 1 unit. The frequency is $F = \dfrac{b}{2\pi} = \dfrac{1}{\pi}$, which is approximately 0.32 oscillation per unit of time.
		  $y = s(x)$
Simple Harmonic Motion $s(t) = a \sin bt$ or $s(t) = a \cos bt$	An object that oscillates about a stable equilibrium point undergoes simple harmonic motion.	A pendulum on a clock A weight on a spring

6.5 Exercises

Graphs of Trigonometric Functions

Exercises 1–6: Sketch a graph of the equation on the interval $[-4\pi, 4\pi]$. Identify the period and amplitude.

1. $y = 3 \sin \frac{1}{2}x$
2. $y = 2 \cos \frac{1}{3}x$

3. $y = -2 \cos 3x$
4. $y = -3 \sin 2x$

5. $y = \frac{1}{2} \sin \pi x$
6. $y = -\frac{3}{2} \cos \frac{\pi}{2}x$

Exercises 7–12: (Refer to Example 2.) Explain how the graph of f can be obtained from the graph of either the sine or cosine function. Then sketch a graph of f on the interval $[-2\pi, 2\pi]$.

7. $y = 3 \sin (2x) - 1$
8. $y = -2 \cos (3x) + 1$

9. $y = -2 \cos \left(x + \frac{\pi}{2}\right)$
10. $y = -\frac{1}{2} \sin(x + \pi)$

11. $y = \frac{1}{2} \cos (\pi x - 1)$
12. $y = \sin \left(\frac{2}{3}x\right) + 2$

Exercises 13–18: For the graph of the equation, identify the amplitude, period, phase shift, and vertical shift. Do not graph the equation.

13. $y = 3 \sin \left(4\left(x - \frac{\pi}{4}\right)\right) - 4$

14. $y = -5 \sin \left(\frac{1}{2}(x - \pi)\right) + 7$

15. $y = -4 \cos \left(\frac{\pi}{2}(x - 1)\right) + 6$

16. $y = \frac{4}{5} \cos \left(\pi x + \frac{\pi}{3}\right) - \frac{2}{3}$

17. $y = \frac{2}{3} \sin (6x + 3\pi) - \frac{5}{2}$

18. $y = 20 \cos \left(\frac{2}{3}x + \pi\right) + 2$

Exercises 19–22: A graph of the equation $y = a \sin bx$ is shown, where b is a positive constant. Estimate the values for a and b.

19.

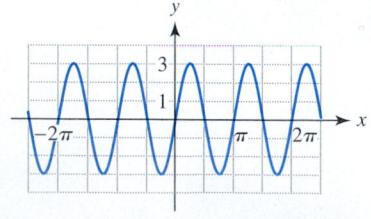

20.

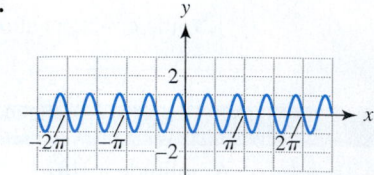

21.

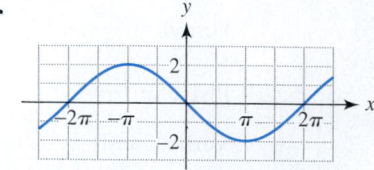

22.

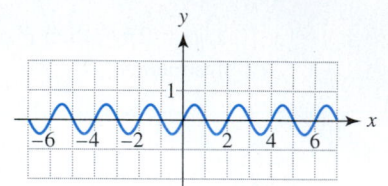

Exercises 23–26: The graph of a trigonometric function f represented by $f(x) = a \sin (b(x - c))$ is shown, where a, b, and c are nonnegative. State the amplitude, period, and phase shift.

23.

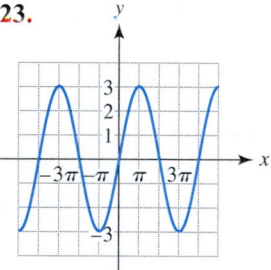

24.

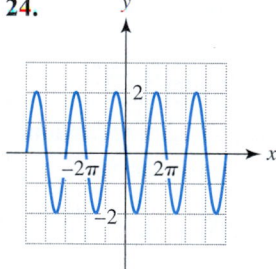

25.

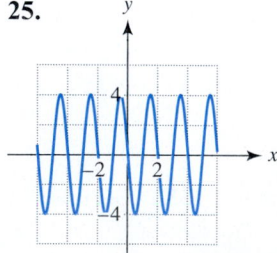

26.

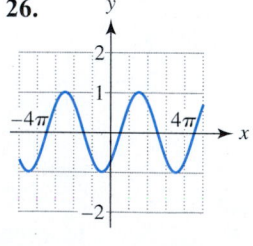

Exercises 27–32: Match the function f with its graph (a.–f.). Do not use a calculator.

27. $f(t) = 2 \sin\left(\frac{1}{2}t\right)$　　**28.** $f(t) = -\sin(2t)$

29. $f(t) = 3 \cos(\pi t)$　　**30.** $f(t) = 2 \sin\left(t - \frac{\pi}{4}\right)$

31. $f(t) = \cos\left(t + \frac{\pi}{2}\right)$　　**32.** $f(t) = -3 \cos t$

a.

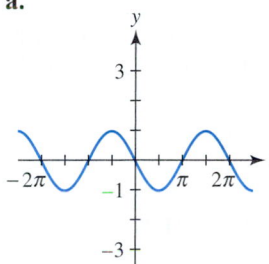

b.

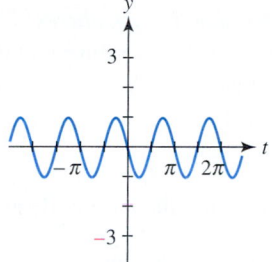

c.

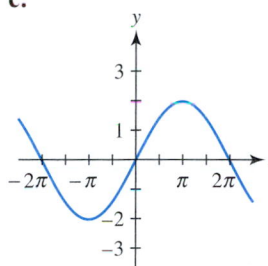

d.

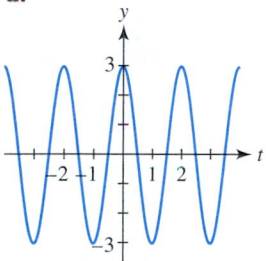

e.

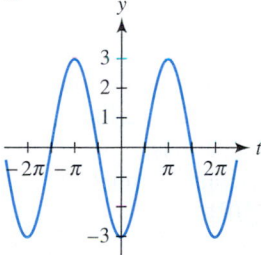

f.

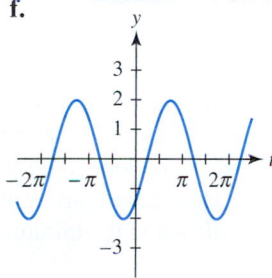

Exercises 33–36: Determine an equation $y = a \sin(b(x - c))$ for the graph shown in the exercise. Assume that a, b, and c are nonnegative.

33. Exercise 23　　**34.** Exercise 24

35. Exercise 25　　**36.** Exercise 26

Exercises 37–54: Find the amplitude, period, and phase shift of f. Then sketch a graph of f by hand on the interval $[-2\pi, 2\pi]$.

37. $f(t) = 2 \sin t$　　**38.** $f(x) = -3 \sin x$

39. $f(x) = \sin \frac{1}{2}x$　　**40.** $f(t) = \cos 2t$

41. $f(x) = 1 + \sin x$　　**42.** $f(x) = -2 + 2 \cos x$

43. $f(t) = \cos(\pi t) + 2$

44. $f(t) = 2 \sin\left(t - \frac{\pi}{2}\right)$　　**45.** $f(t) = -\sin(2(t + \pi))$

46. $f(t) = -3 \cos \frac{1}{2}t$　　**47.** $f(x) = -\cos\left(x - \frac{\pi}{2}\right)$

48. $f(x) = -2 \cos(x + \pi)$　　**49.** $f(x) = -\frac{1}{2}\sin 2x$

50. $f(x) = 3 \sin 4x$

51. $f(x) = 2 \cos\left(2x + \frac{\pi}{2}\right) - 1$

52. $f(x) = -\frac{1}{2}\sin(3x + \pi) + 2$

53. $f(t) = \cos\left(2\left(t - \frac{\pi}{2}\right)\right)$　　**54.** $f(t) = 3 \sin\left(\frac{1}{2}(t - \pi)\right)$

Exercises 55–64: Graph f in $[-2\pi, 2\pi, \pi/2]$ by $[-4, 4, 1]$. State the amplitude, period, and phase shift.

55. $f(t) = 2 \sin(2t)$　　**56.** $f(t) = -3 \sin(t - \pi)$

57. $f(t) = \frac{1}{2}\cos\left(3\left(t + \frac{\pi}{3}\right)\right)$

58. $f(t) = 1.5 \cos\left(\frac{1}{2}\left(t + \frac{\pi}{2}\right)\right)$　　**59.** $f(t) = -\sin(4t)$

60. $f(t) = -2.5 \cos\left(2t + \frac{\pi}{2}\right)$　　**61.** $f(t) = -\cos(2\pi t) + 1$

62. $f(t) = -2.5 \sin\left(\pi t + \frac{\pi}{2}\right) - 1$

63. $f(x) = -2 \cos\left(2\pi x + \frac{\pi}{4}\right) + 1$

64. $f(x) = \frac{3}{4}\sin(\pi x + \pi) - 2$

Exercises 65–72: (Refer to Example 6.) Find the period and phase shift for the graph of f. Graph f on the interval $[-2\pi, 2\pi]$. Identify where asymptotes occur in the graph.

65. $f(t) = \tan 2t$　　**66.** $f(t) = \tan \frac{1}{2}t$

67. $f(t) = \tan\left(t - \frac{\pi}{2}\right)$　　**68.** $f(t) = \cot\left(\frac{1}{3}\left(t - \frac{\pi}{2}\right)\right)$

69. $f(t) = -\cot 2t$　　**70.** $f(t) = -\cot\left(t + \frac{\pi}{2}\right)$

71. $f(x) = \cot\left(2\left(x - \frac{\pi}{4}\right)\right) - 1$

72. $f(x) = \tan\left(\frac{1}{2}x - \frac{\pi}{2}\right)$

Exercises 73–78: Use the directions for Exercises 65–72 to graph f.

73. $f(t) = \sec \frac{1}{2}t$

74. $f(t) = \sec\left(2\left(t - \frac{\pi}{2}\right)\right)$

75. $f(t) = \csc(t - \pi)$

76. $f(t) = -\csc 2t$

77. $f(x) = \sec\left(\frac{1}{3}\left(t - \frac{\pi}{6}\right)\right)$

78. $f(x) = \csc(\pi(x - 1))$

Applications

Exercises 79–82: Match the physical situation with the graph (a.–d.) that models it best.

79. The height y in feet that a person is above the ground while riding a Ferris wheel after t seconds, where $t = 0$ corresponds to when the person began the ride

80. The number of hours y of darkness at 30°N latitude during month t, where $t = 1$ corresponds to January

81. The length y of a shadow cast by a horizontal stick of length 1 on a vertical wall between sunrise and noon, where angle t is shown in the figure

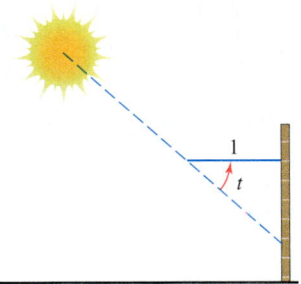

82. The length y of a shadow cast by a vertical stick of length 1 between sunrise and noon, where t is the angle of elevation of the sun

a.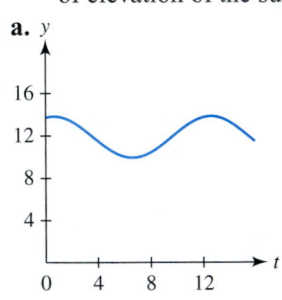

b.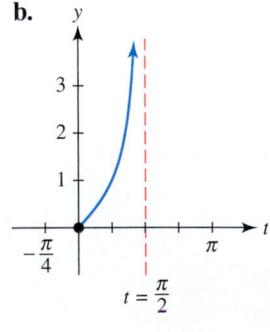

c.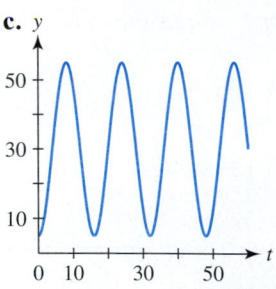

d.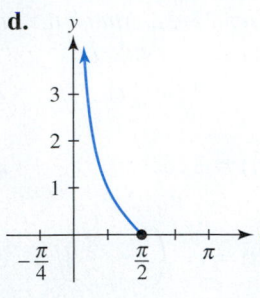

83. *Average Temperatures* The graph models the monthly average temperature y in degrees Fahrenheit for a city in Canada, where x is the month.
  (a) Find the maximum and minimum monthly average temperatures.

  (b) Find the amplitude and period. Interpret the results.

  (c) Explain what the x-intercepts represent.

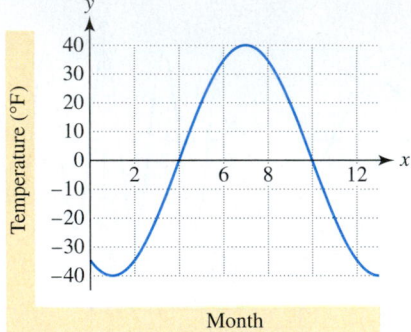

84. *Average Temperatures* The graph in the previous exercise is given by $y = 40 \cos\left(\frac{\pi}{6}(x - 7)\right)$. Modify this equation to model the following situations.
  (a) The maximum monthly average temperature is 50°F and the minimum is −50°F.

  (b) The maximum monthly average temperature is 60°F and the minimum is −20°F.

  (c) The maximum monthly average temperature occurs in August and the minimum occurs in February.

85. *Average Temperatures* The monthly average temperatures in degrees Fahrenheit at Mould Bay, Canada, may be modeled by the equation

$$f(x) = 34 \sin\left(\frac{\pi}{6}(x - 4.3)\right),$$

where x is the month and $x = 1$ corresponds to January. (**Source:** A. Miller and J. Thompson.)

(a) Graph f over the 2-year interval $1 \leq x \leq 25$. Find the amplitude, period, and phase shift.

(b) Approximate the average temperature during May and December.

(c) Estimate the *yearly* average temperature at Mould Bay.

86. *Average Temperatures*   The monthly average temperatures in degrees Fahrenheit at Austin, Texas, are given by

$$f(x) = 17.5 \sin\left(\frac{\pi}{6}(x - 4)\right) + 67.5,$$

where x is the month and $x = 1$ corresponds to January. (**Source:** A. Miller and J. Thompson.)

(a) Graph f over the 2-year interval $1 \leq x \leq 25$. Determine the amplitude, period, phase shift, and vertical shift.

(b) Determine the maximum and minimum monthly average temperature and the months when they occur.

(c) Make a conjecture as to how the *yearly* average temperature might be related to $f(x)$.

87. *Modeling Temperatures*   The monthly average temperatures in Vancouver, Canada, are shown in the table.

Month	1	2	3	4	5	6
Temperature (°F)	36	39	43	48	55	59

Month	7	8	9	10	11	12
Temperature (°F)	64	63	57	50	43	39

Source: A. Miller and J. Thompson.

(a) Plot the average monthly temperature over a 24-month period by letting $x = 1$ and $x = 13$ correspond to January.

(b) Find the constants a, b, c, and d so that the function $f(x) = a \sin(b(x - c)) + d$ models the data.

(c) Graph f together with the data.

88. *Modeling Temperatures*   The monthly average temperatures in Chicago, Illinois, are shown in the table.

Month	1	2	3	4	5	6
Temperature (°F)	25	28	36	48	61	72

Month	7	8	9	10	11	12
Temperature (°F)	74	75	66	55	39	28

Source: A. Miller and J. Thompson.

(a) Plot the monthly average temperature over a 24-month period by letting $x = 1$ and $x = 13$ correspond to January.

(b) Determine the constants a, b, c, and d so that the function $f(x) = a \sin(b(x - c)) + d$ models the data.

(c) Graph f and the data together.

89. *Modeling Temperatures*   The monthly average high temperatures in Augusta, Georgia, are shown in the table.

Month	1	2	3	4	5	6
Temperature (°F)	58	60	68	77	82	90

Month	7	8	9	10	11	12
Temperature (°F)	92	91	83	77	68	60

Source: J. Williams.

(a) Model these data using a function of the form

$$f(x) = a \cos(b(x - c)) + d.$$

(b) Are different values for c possible? Explain.

90. *Modeling Temperatures*   The maximum monthly average temperature in Anchorage, Alaska, is 57°F and the minimum is 12°F.

(a) Using only these two temperatures, determine $f(x) = a \cos(b(x - c)) + d$ so that $f(x)$ models the monthly average temperatures in Anchorage.

(b) Graph f and the actual data in the table over a 2-year period.

Month	1	2	3	4	5	6
Temperature (°F)	12	18	23	36	46	55

Month	7	8	9	10	11	12
Temperature (°F)	57	55	48	36	23	16

Source: A. Miller and J. Thompson.

91. *Highway Design*   (Refer to Example 7.) Calculate the minimum length L of a typical sag curve on a highway with a 40-mile-per-hour speed limit for a car with headlights 2 feet above the ground and alignment set at $\alpha = 1.5°$.

92. *Highway Design*   Repeat the previous exercise for a truck with headlights 3.5 feet above the ground and alignment set at $\alpha = 2.5°$.

93. *Daylight Hours* The graph models the daylight hours at 60°N latitude, where $x = 1$ corresponds to January 1, $x = 2$ to February 1, and so on.
  (a) Estimate the maximum number of daylight hours. When does this occur?

  (b) Estimate the minimum number of daylight hours. When does this occur?

  (c) Interpret the amplitude and period.

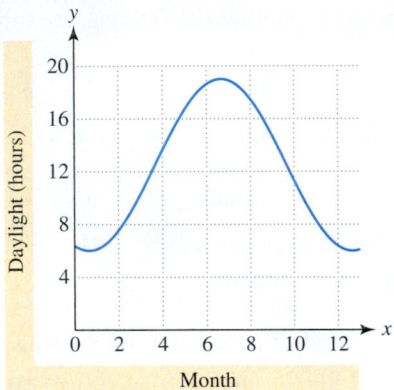

94. *Daylight Hours* The graph in the previous exercise is given by $y = 6.5 \sin\left(\frac{\pi}{6}(x - 3.65)\right) + 12.4$. Modify this equation to model the following situations.
  (a) At 50°N latitude the maximum daylight is about 16.3 hours and the minimum is about 8.3 hours.

  (b) The daylight hours at 60°S latitude

  (c) The daylight hours at the equator

95. *Average Precipitation* The graph models the monthly average precipitation in inches at Mount Adams, Washington, over a 3-year period, where x is the month.

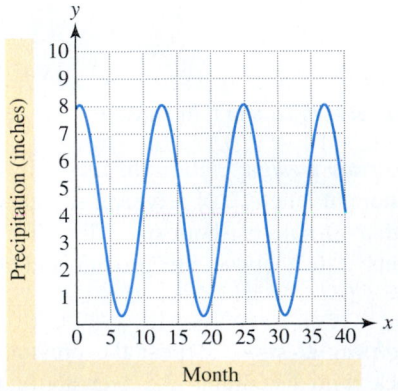

  (a) Find the maximum and minimum monthly average precipitation.

  (b) Find the amplitude and interpret the result.

  (c) Could a graph that models precipitation have an x-intercept? If it could, what would this indicate?

  (d) Estimate the yearly average precipitation.

96. *Average Precipitation* Suppose that the monthly average precipitation at a particular location varies sinusoidally between a maximum of 6 inches in January to a minimum of 2 inches in July. Let $t = 1$ correspond to January and $t = 12$ correspond to December.
  (a) Find values for a, b, c, and d so that
  $$f(t) = a\cos(b(t - c)) + d$$
  models these conditions.

  (b) Support your result in part (a) by evaluating $f(1)$ and $f(7)$.

97. *Modeling Tidal Currents* Tides cause ocean currents to flow into and out of harbors and canals. The table shows the speed of the ocean current at Cape Cod Canal in bogo-knots (bk) x hours after midnight on August 26, 1998. (**Source:** D. Pentcheff, *WWW Tide and Current Predictor.*)

Time (hr)	3.7	6.75	9.8
Current (bk)	−18	0	18

Time (hr)	13.0	16.1	22.2
Current (bk)	0	−18	18

  (a) Find constants a, b, c, and d so that
  $$f(x) = a\cos(b(x - c)) + d$$
  models the data in the table.

  (b) Graph f and the data in $[0, 24, 4]$ by $[-20, 20, 5]$. Interpret the graph.

98. *Tides and Periodic Functions* The accompanying figure shows the tides at Santa Monica Bay, California, on August 28, 1997. This type of tide is *semi-diurnal*, since high tides occur twice a day. (**Source:** Zihua Software.)
  (a) Estimate the time between the two high tides shown in the graph.

  (b) Explain why $f(x) = a\sin(b(x - c)) + d$ cannot model these tide levels accurately.

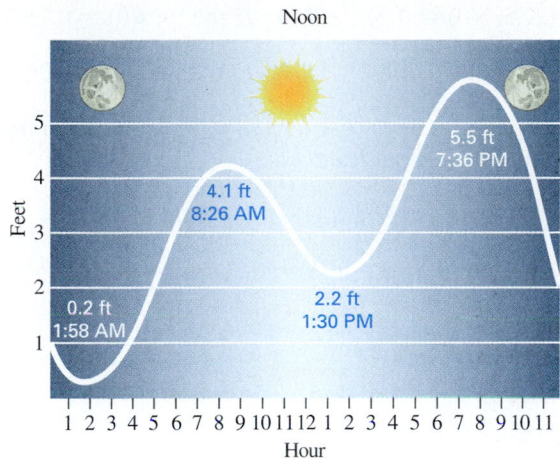

Noon

99. **Ocean Temperatures** The graph models the Gulf of Mexico water temperatures in degrees Fahrenheit at St. Petersburg, Florida. (**Source:** J. Williams.)
   (a) Estimate the maximum and minimum water temperatures. When do they occur?

   (b) What would happen to the amplitude of the graph if the minimum water temperature decreased to 50°F? Make a sketch of this situation.

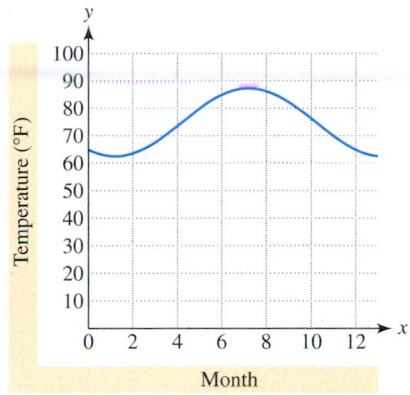

100. **Ocean Temperatures** The graph in the previous exercise is described by the equation

$$y = 12.4 \sin\left(\frac{\pi}{6}(x - 4.2)\right) + 75.$$

Modify this equation to model the following situations.
   (a) The monthly average water temperatures vary between 60°F and 90°F.

   (b) The monthly average water temperatures vary between 50°F and 70°F.

101. **Modeling Ocean Temperature** The following table lists the monthly average ocean temperatures in degrees Fahrenheit at Veracruz, Mexico.

Month	1	2	3	4	5	6
Temperature (°F)	72	73	74	78	81	83

Month	7	8	9	10	11	12
Temperature (°F)	84	85	84	82	78	74

Source: J. Williams.

(a) Make a scatterplot of the data over a 2-year period.

(b) Find the constants a, b, c, and d so that
$$f(x) = a \sin(b(x - c)) + d$$
models the data.

102. **Interpreting a Model** Graph
$$y = 20 + 15 \sin\frac{\pi t}{12}$$
for $0 \le t \le 12$. Let y represent the outdoor temperature in degrees Celsius at time t in hours, where $t = 0$ corresponds to 9 A.M. Interpret the graph.

103. **Carbon Dioxide Levels in Hawaii** At Mauna Loa, Hawaii, atmospheric carbon dioxide levels in parts per million (ppm) have been measured regularly since 1958. The equation

$$L(x) = 0.022x^2 + 0.55x + 316 + 3.5\sin(2\pi x)$$

may be used to model these levels, where x is the year and $x = 0$ corresponds to 1960.
   (a) Graph L in [20, 35, 5] by [320, 370, 10] and interpret the graph.

   (b) The function L is represented by the sum of a quadratic function and a sine function. How does each function affect the shape of the graph? Discuss reasons for each function. (**Source:** A. Nilsson, *Greenhouse Earth.*)

104. **Carbon Dioxide Levels in Alaska** (Refer to the previous exercise.) The carbon dioxide content in the atmosphere at Barrow, Alaska, in parts per million (ppm) can be modeled using the equation
$$C(x) = 0.04x^2 + 0.6x + 330 + 7.5\sin(2\pi x),$$
where $x = 0$ corresponds to 1970. (**Source:** M. Zeilik, *Introductory Astronomy and Astrophysics.*)
   (a) Graph C in [10, 25, 5] by [320, 380, 10]. Compare it with the graph for L in the previous exercise.

   (b) Discuss possible reasons for differences between the two graphs.

Simple Harmonic Motion

Exercises 105–108: **Springs** *(Refer to Example 8.) Suppose that a weight on a spring has an initial position of s(0) and a period of P.*

 (a) *Find a function s given by $s(t) = a \cos(2\pi Ft)$ that models the displacement of the weight.*
 (b) *Evaluate s(1). Is the weight moving upward, downward, or neither when t = 1?*

105. $s(0) = 2$ inches, $P = 0.5$ second

106. $s(0) = 5$ inches, $P = 1.5$ seconds

107. $s(0) = -3$ inches, $P = 0.8$ second

108. $s(0) = -4$ inches, $P = 1.2$ seconds

Exercises 109–112: **Music** *A note on the piano has the given frequency F. Suppose the maximum displacement at the center of the piano wire is given by s(0). Find constants a and b so that $s(t) = a \cos bt$ models this displacement. Graph f in $[0, 0.05, 0.01]$ by $[-0.3, 0.3, 0.1]$.*

109. $F = 27.5$, $s(0) = 0.21$ **110.** $F = 110$, $s(0) = 0.11$

111. $F = 55$, $s(0) = 0.14$ **112.** $F = 220$, $s(0) = 0.06$

Exercises 113–116: **Sine Regression** *Use regression to find constants a, b, c, and d so that*

$$f(x) = a \sin(bx + c) + d,$$

models the real data given in the exercise. Graph the data and f together.

113. Exercise 87   **114.** Exercise 88

115. Exercise 89   **116.** Exercise 90

Writing about Mathematics

117. Discuss how the constants a, b, c, and d affect the graph of $y = a \sin(b(x - c)) + d$. Give an example.

118. Discuss some types of real data that could be modeled by $y = a \cos(b(x - c)) + d$. Give an example.

6.6 Inverse Trigonometric Functions

Introduction

In construction it is sometimes necessary to determine angles. For example, the pitch or slope of a roof frequently is expressed as the ratio $\frac{k}{12}$, where k represents a k-foot rise for every 12 feet of run in horizontal distance. See Figure 6.138. A typical roof pitch for homes is $\frac{6}{12}$. To correctly cut the rafters, a carpenter needs to know the measure of angle θ. This problem can be solved easily using inverse trigonometric functions. Before introducing inverse trigonometric functions, we briefly review inverse functions.

◆ Review inverse functions
◆ Define and use the inverse sine function
◆ Define and use the inverse cosine function
◆ Define and use the inverse tangent function
◆ Solve triangles and equations

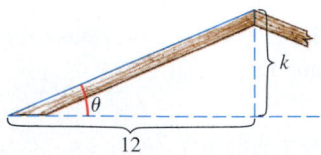

FIGURE 6.138

Review of Inverses

The inverse function f^{-1} will "undo" the computation performed by f. If $f(a) = b$, then $f^{-1}(b) = a$, and as a result, f and f^{-1} interchange domains and ranges. The inverse function of $f(x) = 3x$ is $f^{-1}(x) = \frac{x}{3}$, since dividing by 3 is the inverse operation of multiplying

Algebra Review
To review inverse functions,
see Section 5.2.

by 3. A function f must be one-to-one for f^{-1} to exist. A function is one-to-one if different inputs *always* produce different outputs. The horizontal line test can be used to determine if a function is one-to-one. Inverse functions can be represented verbally, symbolically, numerically, and graphically, as illustrated in the next example.

EXAMPLE 1 **Finding representations of an inverse function**

The function given by $f(x) = -5.5x + 80$ computes the Fahrenheit temperature at an altitude of x-thousand feet when the ground level temperature is 80°F. Find verbal, symbolic, numerical, and graphical representations of f and f^{-1}. Interpret f^{-1}. (**Source:** A. Miller and R. Anthes, *Meteorology.*)

SOLUTION
Verbal Representation   The function f multiplies the input x by -5.5 and then adds 80 to the result. To find the inverse function, apply the inverse operations in reverse order. That is, f^{-1} subtracts 80 from the input x and divides the result by -5.5.

Symbolic Representation   Symbolic representations of f^{-1} include either
$$f^{-1}(x) = \frac{x - 80}{-5.5} \quad \text{or} \quad f^{-1}(x) = \frac{80 - x}{5.5}.$$

Numerical Representation   A numerical representation of f is shown in Table 6.11.

TABLE 6.11

x	0	2	4	6	8
$f(x)$	80	69	58	47	36

Since $f(a) = b$ implies $f^{-1}(b) = a$, it follows that a numerical representation of $f^{-1}(x)$ is shown in Table 6.12.

TABLE 6.12

x	36	47	58	69	80
$f^{-1}(x)$	8	6	4	2	0

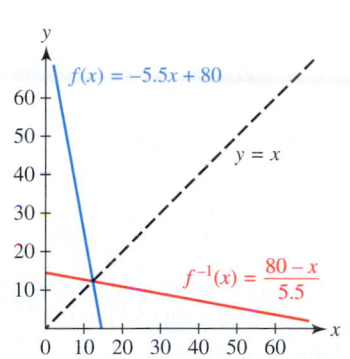

FIGURE 6.139

Graphical Representation   The graph of f^{-1} can be found by reflecting the graph of f across the line $y = x$. See Figure 6.139.

Since $f(x)$ computes the Fahrenheit temperature at x-thousand feet, $f^{-1}(x)$ computes the altitude in thousands of feet at which the temperature is x degrees Fahrenheit.

Now Try Exercise 13 ◆

The Inverse Sine Function

A numerical representation of the sine function is shown in Table 6.13.

TABLE 6.13

x	0	$\frac{\pi}{6}$	$\frac{\pi}{4}$	$\frac{\pi}{3}$	$\frac{\pi}{2}$	$\frac{2\pi}{3}$	$\frac{3\pi}{4}$	$\frac{5\pi}{6}$	π
$f(x)$	0	$\frac{1}{2}$	$\frac{1}{\sqrt{2}}$	$\frac{\sqrt{3}}{2}$	1	$\frac{\sqrt{3}}{2}$	$\frac{1}{\sqrt{2}}$	$\frac{1}{2}$	0

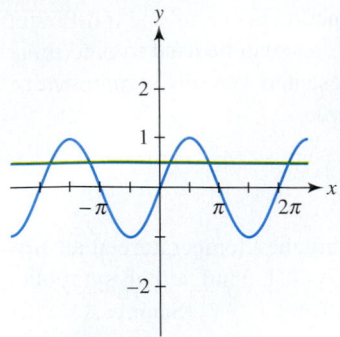

FIGURE 6.140 The Sine Function

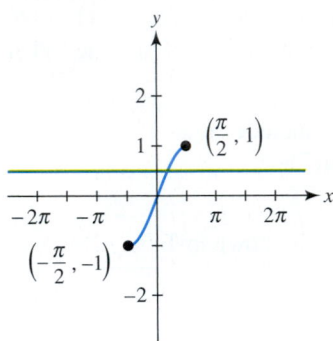

FIGURE 6.141 Restricting the Domain

Notice that different inputs do not always result in different outputs. Therefore the sine function is not one-to-one and so an inverse function does not exist. For example, $f(0) = 0$ and $f(\pi) = 0$, so $f^{-1}(0)$ cannot be defined as a single output, which is necessary for an inverse *function* to exist. By the horizontal line test we also can see that the sine function is not one-to-one. See Figure 6.140, where it is possible for a horizontal line to intersect the sine graph infinitely many times.

If we restrict the domain of $f(x) = \sin x$ to $-\frac{\pi}{2} \leq x \leq \frac{\pi}{2}$, as shown in Figure 6.141, the graph of $y = \sin x$ is one-to-one since a horizontal line intersects this graph at most once. On this restricted domain the sine function has a unique inverse called the *inverse sine function*.

INVERSE SINE FUNCTION

The **inverse sine function**, denoted $\sin^{-1} x$ or $\arcsin x$, is defined by the following: $y = \sin^{-1} x$ or $y = \arcsin x$ means $x = \sin y$ for $-1 \leq x \leq 1$ and y in the interval $\left[-\frac{\pi}{2}, \frac{\pi}{2}\right]$.

Note: When evaluating the inverse sine function, it may be helpful to think of $\sin^{-1} x$ as an *angle* θ, where $\sin \theta = x$ and θ satisfies $-\frac{\pi}{2} \leq \theta \leq \frac{\pi}{2}$.

The next example illustrates how to evaluate the inverse sine function.

EXAMPLE 2   Evaluating the inverse sine function

Evaluate each of the following by hand and then support your results with a calculator.

(a) $\sin^{-1} 1$ (b) $\arcsin\left(-\dfrac{1}{2}\right)$

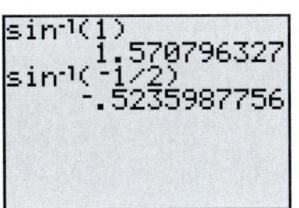

FIGURE 6.142 Radian Mode

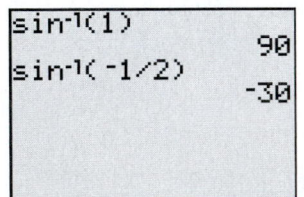

FIGURE 6.143 Degree Mode

SOLUTION

(a) The expression $\sin^{-1} 1$ represents the angle (or real number) θ whose sine equals 1 and satisfies $-\frac{\pi}{2} \leq \theta \leq \frac{\pi}{2}$. Thus $\theta = \sin^{-1}(1) = \frac{\pi}{2} \approx 1.57$. In degrees, $\sin^{-1}(1) = 90°$.

(b) The expression $\arcsin\left(-\frac{1}{2}\right)$ represents the angle (or real number) θ whose sine equals $-\frac{1}{2}$ and satisfies $-\frac{\pi}{2} \leq \theta \leq \frac{\pi}{2}$. Thus $\theta = \arcsin\left(-\frac{1}{2}\right) = -\frac{\pi}{6} \approx -0.52$. In degrees, $\sin^{-1}\left(-\frac{1}{2}\right) = -30°$. Figures 6.142 and 6.143 support these results both in radian mode and degree mode.   *Now Try Exercise 25* ◆

Functions and their inverses interchange domains and ranges. The range of the sine function is $-1 \leq y \leq 1$. Therefore the domain of the inverse sine function is $-1 \leq x \leq 1$. Since the domain of the sine function has been restricted to $-\frac{\pi}{2} \leq x \leq \frac{\pi}{2}$, it follows that the range of the inverse sine function is $-\frac{\pi}{2} \leq y \leq \frac{\pi}{2}$. That is, $\sin^{-1} x$ outputs angles only in the interval $\left[-\frac{\pi}{2}, \frac{\pi}{2}\right]$. The following are properties of the inverse sine function.

$$\sin^{-1}(\sin x) = x \quad \text{for} \quad -\frac{\pi}{2} \leq x \leq \frac{\pi}{2}$$

$$\sin(\sin^{-1} x) = x \quad \text{for} \quad -1 \leq x \leq 1$$

◆ **MAKING CONNECTIONS**

Notation and Inverse Functions   When inverse functions were introduced in Section 5.2, it was discussed that $f^{-1}(x) \neq \frac{1}{f(x)}$. The same is true for the inverse sine function: $\sin^{-1} x \neq \frac{1}{\sin x}$. For example, $\sin^{-1} 1 = \frac{\pi}{2} \approx 1.57$ and $\frac{1}{\sin 1} \approx 1.19$.

EXAMPLE 3

Finding representations of the inverse sine function

Represent the inverse sine function verbally, numerically, graphically, and symbolically.

SOLUTION

Verbal Representation   To compute $\sin^{-1} x$ for $-1 \leq x \leq 1$, determine the angle (or real number) θ such that $\sin \theta = x$ and $-\frac{\pi}{2} \leq \theta \leq \frac{\pi}{2}$.

Numerical Representation   Table 6.14 shows a numerical representation of $\sin x$ on the interval $\left[-\frac{\pi}{2}, \frac{\pi}{2}\right]$. It follows that a numerical representation of $\sin^{-1} x$ is shown in Table 6.15. Notice that if $\sin a = b$, then $\sin^{-1} b = a$, provided $-\frac{\pi}{2} \leq a \leq \frac{\pi}{2}$.

TABLE 6.14

x	$-\frac{\pi}{2}$	$-\frac{\pi}{3}$	$-\frac{\pi}{4}$	$-\frac{\pi}{6}$	0	$\frac{\pi}{6}$	$\frac{\pi}{4}$	$\frac{\pi}{3}$	$\frac{\pi}{2}$
$\sin x$	-1	$-\frac{\sqrt{3}}{2}$	$-\frac{1}{\sqrt{2}}$	$-\frac{1}{2}$	0	$\frac{1}{2}$	$\frac{1}{\sqrt{2}}$	$\frac{\sqrt{3}}{2}$	1

TABLE 6.15

x	-1	$-\frac{\sqrt{3}}{2}$	$-\frac{1}{\sqrt{2}}$	$-\frac{1}{2}$	0	$\frac{1}{2}$	$\frac{1}{\sqrt{2}}$	$\frac{\sqrt{3}}{2}$	1
$\sin^{-1} x$	$-\frac{\pi}{2}$	$-\frac{\pi}{3}$	$-\frac{\pi}{4}$	$-\frac{\pi}{6}$	0	$\frac{\pi}{6}$	$\frac{\pi}{4}$	$\frac{\pi}{3}$	$\frac{\pi}{2}$

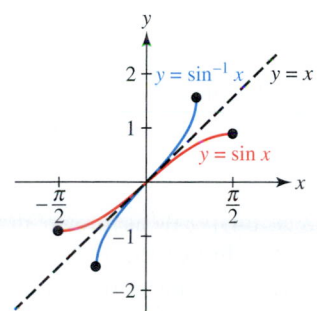

FIGURE 6.144   The Inverse Sine Function

Graphical Representation   The graph of $y = \sin^{-1} x$ can be found by reflecting the graph of $y = \sin x$ for $-\frac{\pi}{2} \leq x \leq \frac{\pi}{2}$ across the line $y = x$, as shown in Figure 6.144.

Symbolic Representation   A symbolic representation of the inverse sine function can be written as either $f(x) = \sin^{-1} x$ or $f(x) = \arcsin x$. There is no simple formula to evaluate $\sin^{-1} x$.

Now Try Exercise 49 ◆

EXAMPLE 4

Evaluating the inverse sine function

Approximate each of the following with a calculator if possible. Express your result in both radians and degrees.
(a) $\sin^{-1} 0.7$ **(b)** $\arcsin 2.1$

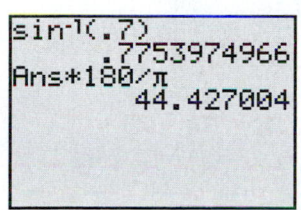

FIGURE 6.145   Radian Mode

SOLUTION

(a) In Figure 6.145, $\sin^{-1} 0.7$ is evaluated in radian mode. We see that $\sin^{-1} 0.7 \approx 0.775$ radians, or equivalently, about $44.4°$.

(b) To evaluate $\arcsin 2.1$, we must find an angle θ such that $\sin \theta = 2.1$. Since $-1 \leq \sin \theta \leq 1$, $\theta = \arcsin 2.1$ is undefined.   *Now Try Exercises 31(a) and 32(b)* ◆

In track and field, when an athlete throws the shot, the distance that the shot travels depends on the angle θ that the initial direction of the shot makes with the horizontal.

Angle θ in Figure 6.146 is called the *projection angle*. The optimum projection angle θ that results in maximum distance for the shot may be calculated by

$$\theta = \sin^{-1}\sqrt{\frac{v^2}{2v^2 + 64.4h}},$$

where v is the initial speed in feet per second of the shot and h is the height in feet of the shot when it is released. (**Source:** J. Cooper and R. Glassow, *Kinesiology.*)

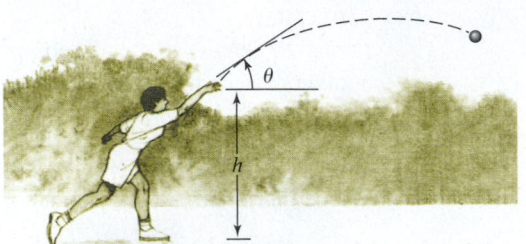

FIGURE 6.146 Projection Angle θ

 **EXAMPLE 5** Finding the optimum projection angle for a shot-putter

Suppose that an athlete releases a shot 8 feet above the ground with velocity v. Give graphical and numerical representations of the optimum projection angle θ. Interpret the results.

SOLUTION Graph and table $Y_1 = \sin^{-1} (\sqrt{(X^2/(2X^2 + 64.4*8))}$, as shown in Figures 6.147 and 6.148. We can see that the faster a person throws the shot, the greater the optimal projection angle θ becomes. For example, if a shot is thrown at 25 feet per second, then $\theta \approx 36.5°$, whereas if the shot is thrown at 50 feet per second then $\theta \approx 42.3°$.

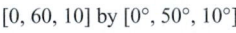
$[0, 60, 10]$ by $[0°, 50°, 10°]$

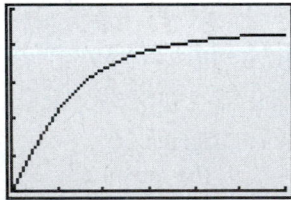

FIGURE 6.147 Degree Mode

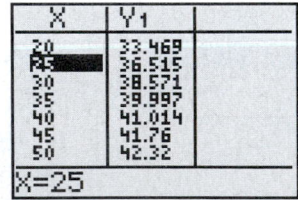

FIGURE 6.148 Degree Mode

Now Try Exercise 89 ◆

◆ **CLASS DISCUSSION**
If a cannonball is shot from ground level, what is the optimum projection angle? ◆

The Inverse Cosine Function

By the horizontal line test, the cosine function is not one-to-one, as illustrated in Figure 6.149. If we restrict the domain of $f(x) = \cos x$ to $0 \leq x \leq \pi$, then the resulting function is one-to-one and has an inverse function. See Figure 6.150. This inverse function is called the *inverse cosine function*.

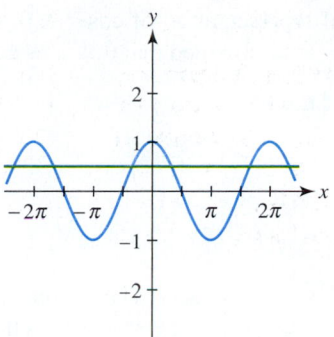

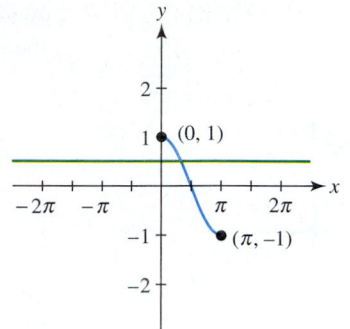

FIGURE 6.149   The Cosine Function

FIGURE 6.150   Restricting the Domain

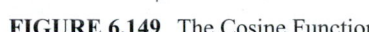

INVERSE COSINE FUNCTION

The **inverse cosine function,** denoted $\cos^{-1} x$ or arccos x, is defined by the following:
$y = \cos^{-1} x$ or $y = \arccos x$ means $x = \cos y$ for $-1 \le x \le 1$ and y in the interval $[0, \pi]$.

EXAMPLE 6

Evaluating the inverse cosine function

Evaluate each of the following.
(a) $\cos^{-1} 1$ **(b)** $\arccos(-0.75)$

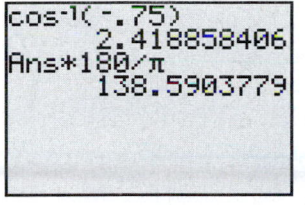

FIGURE 6.151   Radian Mode

SOLUTION

(a) The expression $\cos^{-1} 1$ represents the angle (or real number) θ whose cosine equals 1 and satisfies $0 \le \theta \le \pi$. Thus $\theta = \cos^{-1}(1) = 0$.

(b) A calculator is often necessary to evaluate inverse trigonometric functions. In Figure 6.151, $\cos^{-1}(-0.75) \approx 2.42$ radians or about $138.6°$.

Now Try Exercises 27 and 31(c) ◆

The following are properties of the inverse cosine function.

$$\cos^{-1}(\cos x) = x \quad \text{for } 0 \le x \le \pi$$

$$\cos(\cos^{-1} x) = x \quad \text{for } -1 \le x \le 1$$

EXAMPLE 7

Finding representations of the inverse cosine function

Represent the inverse cosine function verbally, numerically, graphically, and symbolically.

SOLUTION

Verbal Representation   To compute $\cos^{-1} x$ for $-1 \le x \le 1$, determine the angle θ (or real number) such that $\cos \theta = x$ and $0 \le \theta \le \pi$.

Numerical Representation   A numerical representation of $\cos x$ on the interval $[0, \pi]$ is shown in Table 6.16.

TABLE 6.16

x	0	$\frac{\pi}{6}$	$\frac{\pi}{4}$	$\frac{\pi}{3}$	$\frac{\pi}{2}$	$\frac{2\pi}{3}$	$\frac{3\pi}{4}$	$\frac{5\pi}{6}$	π
$\cos x$	1	$\frac{\sqrt{3}}{2}$	$\frac{1}{\sqrt{2}}$	$\frac{1}{2}$	0	$-\frac{1}{2}$	$-\frac{1}{\sqrt{2}}$	$-\frac{\sqrt{3}}{2}$	-1

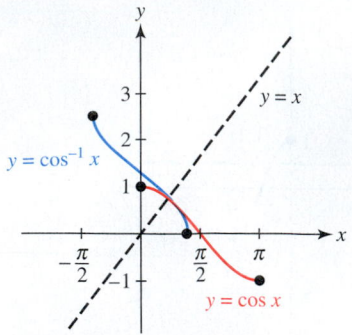

FIGURE 6.152 The Inverse Cosine Function

 CLASS DISCUSSION

Explain the results in Figure 6.153. The calculator was in degree mode.

FIGURE 6.153 Degree Mode

A numerical representation of $\cos^{-1} x$ is shown in Table 6.17. Notice that if $\cos \boldsymbol{a} = \boldsymbol{b}$, then $\cos^{-1} \boldsymbol{b} = \boldsymbol{a}$, provided that $0 \le a \le \pi$.

TABLE 6.17

x	-1	$-\frac{\sqrt{3}}{2}$	$-\frac{1}{\sqrt{2}}$	$-\frac{1}{2}$	0	$\frac{1}{2}$	$\frac{1}{\sqrt{2}}$	$\frac{\sqrt{3}}{2}$	1
$\cos^{-1} x$	π	$\frac{5\pi}{6}$	$\frac{3\pi}{4}$	$\frac{2\pi}{3}$	$\frac{\pi}{2}$	$\frac{\pi}{3}$	$\frac{\pi}{4}$	$\frac{\pi}{6}$	0

Graphical Representation The graph of $y = \cos^{-1} x$ can be found by reflecting the graph of $y = \cos x$ for $0 \le x \le \pi$ across the line $y = x$ as shown in Figure 6.152.

Symbolic Representation A symbolic representation of the inverse cosine function can be written as either $f(x) = \cos^{-1} x$ or $f(x) = \arccos x$. There is no simple formula to evaluate $\cos^{-1} x$.

Now Try Exercise 50 ◆

The Inverse Tangent Function

By the horizontal line test, the tangent function is not one-to-one, as shown in Figure 6.154. If we restrict the domain of $f(x) = \tan x$ to $-\frac{\pi}{2} < x < \frac{\pi}{2}$, then the resulting function is one-to-one and has an inverse function. See Figure 6.155. This inverse function is called the *inverse tangent function*.

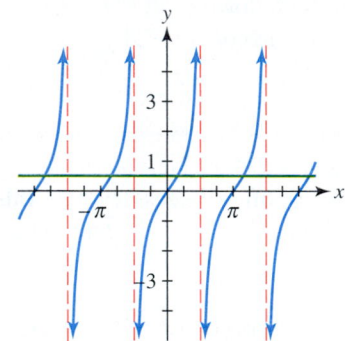

FIGURE 6.154 The Tangent Function

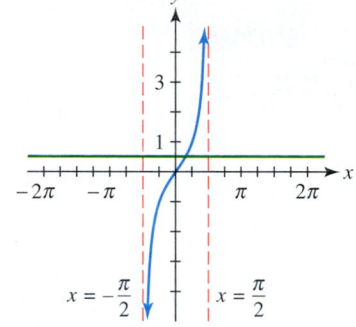

FIGURE 6.155 Restricting the Domain

INVERSE TANGENT FUNCTION

The **inverse tangent function,** denoted $\tan^{-1} x$ or $\arctan x$, is defined by the following: $y = \tan^{-1} x$ or $y = \arctan x$ means $x = \tan y$ for y in the interval $\left(-\frac{\pi}{2}, \frac{\pi}{2}\right)$.

The following are properties of the inverse tangent function.

$$\tan^{-1}(\tan x) = x \quad \text{for } -\frac{\pi}{2} < x < \frac{\pi}{2}$$

$$\tan(\tan^{-1} x) = x \quad \text{for all real numbers } x$$

 **8** Evaluating the inverse tangent function

Evaluate each of the following. Support your answer using a calculator.

(a) $\tan^{-1} 1$ **(b)** $\arctan\left(-\sqrt{3}\right)$

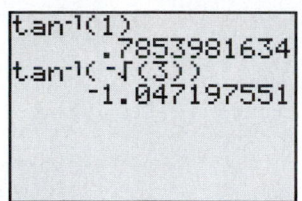

FIGURE 6.156 Radian Mode

SOLUTION

(a) The expression $\tan^{-1} 1$ represents the angle (or real number) θ whose tangent equals 1 and satisfies $-\frac{\pi}{2} < \theta < \frac{\pi}{2}$. Thus $\theta = \tan^{-1}(1) = \frac{\pi}{4} \approx 0.7854$. In degrees, $\tan^{-1}(1) = 45°$.

(b) The expression $\arctan(-\sqrt{3})$ represents the angle (or real number) θ whose tangent equals $-\sqrt{3}$ and satisfies $-\frac{\pi}{2} < \theta < \frac{\pi}{2}$. It follows that $\theta = \arctan(-\sqrt{3}) = -\frac{\pi}{3} \approx -1.047$. Numerical support is shown in Figure 6.156. In degrees, $\arctan(-\sqrt{3}) = -60°$.

Now Try Exercise 29 ◆

◆ **CLASS DISCUSSION**
Give verbal, numerical, and graphical representations of $y = \tan^{-1} x$. ◆

The next example illustrates the use of the inverse tangent function in robotics.

EXAMPLE 9   Using robots to spray paint

In industry it is common to use robots to spray paint. The robotic arm in the accompanying figure is being used to paint a flat surface. Because the spray gun must move at a constant speed v, parallel to the surface being painted, the angle of the arm θ_1 and the angle of the spray gun θ_2 must be continually adjusted. Using Figure 6.157, it can be shown that

$$\theta_1 = \arctan \frac{h}{vt} \quad \text{and}$$

$$\theta_2 = 90° - \theta_1,$$

where $t > 0$ is time in seconds.   (**Source:** W. Stadler, *Analytical Robotics and Mechatronics.*)

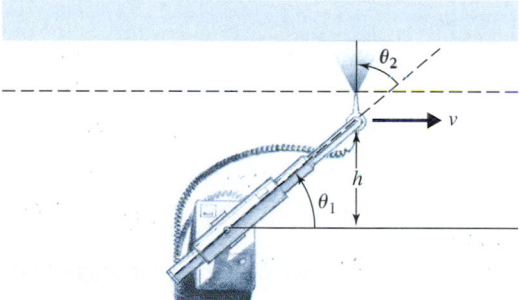

FIGURE 6.157

(a) Let $v = 3$ inches per second and $h = 24$ inches. Graph θ_1 in $[0, 25, 5]$ by $[0, 100, 10]$ using *degree mode*. Describe how θ_1 changes over this 25-second interval.

(b) Determine the degree measure of θ_1 and θ_2 after 10 seconds.

[0, 25, 5] by [0, 100, 10]

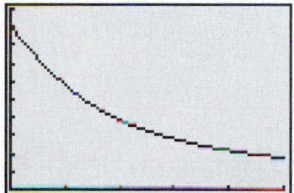

FIGURE 6.158 Degree Mode

◆ **CLASS DISCUSSION**
In Example 9 it was given that $\theta_1 = \arctan \frac{h}{vt}$ and $\theta_2 = 90° - \theta_1$. Use Figure 6.157 to verify this. ◆

SOLUTION

(a) Substitute $h = 24$ and $v = 3$. Graph $Y_1 = \tan^{-1}(24/(3X))$, as shown in Figure 6.158. Initially, the robotic arm is vertical and $\theta_1 = 90°$. As the spray gun moves to the right, angle θ_1 decreases—faster at first and then more slowly.

(b) When $t = 10$, $\theta_1 = \arctan\left(\frac{24}{3(10)}\right) \approx 38.7°$ and $\theta_2 = 90° - \theta_1 \approx 51.3°$.

Now Try Exercise 94 ◆

Solving Triangles and Equations

In Figure 6.159 *standard labeling* is used to denote the vertices, sides, and angles of a right triangle. Finding the measures of the angles and sides in a triangle is called *solving a triangle*. The next example illustrates this process.

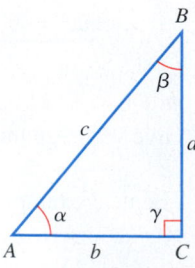

FIGURE 6.159

EXAMPLE 10 Solving a right triangle

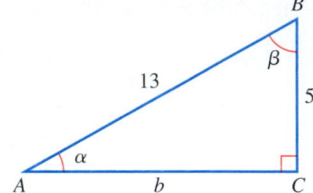

FIGURE 6.160

Solve triangle ABC if $a = 5$ and $c = 13$. See Figure 6.160. Round values to the nearest tenth.

SOLUTION We are given $a = 5$, $c = 13$, and $\gamma = 90°$. We must find b, α, and β. We begin by finding b using the Pythagorean theorem.

$$a^2 + b^2 = c^2$$
$$b^2 = c^2 - a^2$$
$$b^2 = 13^2 - 5^2$$
$$b^2 = 144$$

Thus $b = 12$. We can find angle α as follows.

$$\sin\alpha = \frac{5}{13}$$
$$\alpha = \sin^{-1}\frac{5}{13}$$
$$\alpha \approx 22.6°$$

Since β is complementary to α, $\beta \approx 90° - 22.6° = 67.4°$. *Now Try Exercise 45* ◆

Note: There is more than one way to solve the triangle in Example 10. We could have let $\alpha = \tan^{-1}\frac{5}{12} \approx 22.6°$ and $\beta = \cos^{-1}\frac{5}{13} \approx 67.4°$. There are other possibilities.

EXAMPLE 11 Finding angles in a triangle

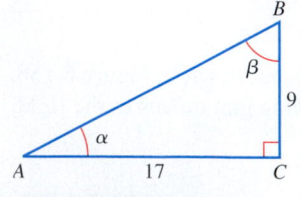

FIGURE 6.161

Approximate the degree measure of the angles α and β shown in Figure 6.161.

SOLUTION From Figure 6.161, we see that $\tan\alpha = \frac{9}{17}$. Thus $\alpha = \tan^{-1}\frac{9}{17} \approx 27.9°$. Since α and β are complementary angles, $\beta \approx 90° - 27.9° = 62.1°$. *Now Try Exercise 43* ◆

Grade resistance is the force F that causes a car to roll down a hill. It can be calculated by $F = W\sin\theta$, where θ represents the angle of the grade and W represents the weight of the the vehicle. (Refer to Figure 6.57 in Section 6.3 on page 507.)

EXAMPLE 12

Calculating highway grade

Find the angle θ for which a 3000-pound car has grade resistance of 500 pounds.

SOLUTION Solve the equation for θ.

$$F = W \sin\theta \qquad \text{Given equation}$$

$$500 = 3000 \sin\theta \qquad \text{Let } W = 3000 \text{ and } F = 500.$$

$$\sin\theta = \frac{1}{6} \qquad \text{Solve for } \sin\theta.$$

$$\theta = \sin^{-1}\frac{1}{6} \qquad \text{Property of inverse sine}$$

$$\theta \approx 9.6° \qquad \text{Approximate.}$$

Thus if a road is inclined at approximately 9.6°, a 3000-pound car would experience a force of 500 pounds pulling downhill.   *Now Try Exercise 84* ◆

During the course of a month, the moon goes through different phases, as illustrated in Figure 6.162. (See also Example 6 in Section 6.3.) Angle θ is called the *phase angle* of the moon and varies from 0° to 360° as the moon completes one orbit of Earth. The *phase F* of the moon may be determined by the equation

$$F = \frac{1}{2}(1 - \cos\theta).$$

The value of F corresponds to the fraction of the face of the moon that is illuminated. For example, if $\theta = 180°$ then $F = \frac{1}{2}(1 - \cos 180°) = 1$ and 100% of the moon is illuminated. This corresponds to a full moon.   (**Source:** P. Duffett-Smith, *Practical Astronomy with Your Calculator.*)

FIGURE 6.162 Phase Angle θ (Not to scale)

EXAMPLE 13

Determining the phase angle

Find the phase angle θ associated with $F = \frac{1}{2}$, assuming that $0° \le \theta \le 180°$. Interpret your result.

SOLUTION

$$F = \frac{1}{2}(1 - \cos\theta) \qquad \text{Given equation}$$

$$\frac{1}{2} = \frac{1}{2}(1 - \cos\theta) \qquad F = \frac{1}{2}$$

$$1 = 1 - \cos\theta \qquad \text{Multiply by 2.}$$

$$\cos\theta = 0 \qquad \text{Solve for } \cos\theta.$$

$$\theta = \cos^{-1}0 \qquad \text{Property of inverses}$$

$$\theta = 90° \qquad \text{Evaluate.}$$

◆ **CLASS DISCUSSION**

Refer to the introduction of this section. If the roof pitch is $\frac{6}{12}$, at what angle θ should the ends of the rafters be cut? ◆

When $\frac{1}{2}$ or 50% of the face of the moon is illuminated (first quarter), the phase angle is 90°.

Now Try Exercise 87 ◆

 **EXAMPLE 14**   **Solving a trigonometric equation**

Solve $9\cos^2\theta = 4$, where θ is an acute angle.

SOLUTION   Begin by dividing each side of the equation by 9.

$$9\cos^2\theta = 4 \qquad \textcolor{blue}{\textit{Given equation}}$$

$$\cos^2\theta = \frac{4}{9} \qquad \textcolor{blue}{\textit{Divide by 9.}}$$

$$\cos\theta = \pm\frac{2}{3} \qquad \textcolor{blue}{\textit{Square root property}}$$

Because θ is an acute angle, $\cos\theta$ must be positive. Thus $\cos\theta = \frac{2}{3}$, and the solution to the equation is $\theta = \cos^{-1}\frac{2}{3} \approx 48.2°$.   *Now Try Exercise 79* ◆

6.6

Putting it all Together

$\mathbf{T}$he six trigonometric functions are not one-to-one functions. However, by restricting their domains to appropriate intervals, inverse trigonometric functions can be defined. In this section we defined inverse functions for the sine, cosine, and tangent functions. Inverse functions can be used to solve triangles and equations.

The following table summarizes some properties of inverse trigonometric functions.

Function	Explanation	Examples
Inverse sine	*Description* $f(x) = \sin^{-1}x$ or $f(x) = \arcsin x$ computes the angle or number in $\left[-\frac{\pi}{2}, \frac{\pi}{2}\right]$ whose sine equals x, where $-1 \le x \le 1$. *Domain* $\{x \mid -1 \le x \le 1\}$ *Range* $\left\{y \mid -\frac{\pi}{2} \le y \le \frac{\pi}{2}\right\}$ *Inverse Properties* $$\sin^{-1}(\sin x) = x \quad \text{for } -\frac{\pi}{2} \le x \le \frac{\pi}{2}$$ $$\sin(\sin^{-1}x) = x \quad \text{for } -1 \le x \le 1$$ *Graph*	$\sin^{-1}1 = 90°$ or $\dfrac{\pi}{2}$ $\sin^{-1}\left(-\dfrac{1}{2}\right) = -30°$ or $-\dfrac{\pi}{6}$ $\sin^{-1}0 = 0°$ or 0 $\sin^{-1}2$ is undefined. $\sin^{-1}0.3 \approx 17.5°$ or 0.305 $\arcsin\left(\dfrac{1}{2}\right) = 30°$ or $\dfrac{\pi}{6}$

Function	Explanation	Examples
Inverse cosine	*Description* $f(x) = \cos^{-1}x$ or $f(x) = \arccos x$ computes the angle or number in $[0, \pi]$ whose cosine equals x, where $-1 \leq x \leq 1$. *Domain* $\{x \mid -1 \leq x \leq 1\}$ *Range* $\{y \mid 0 \leq y \leq \pi\}$ *Inverse Properties* $\quad \cos^{-1}(\cos x) = x \quad$ for $0 \leq x \leq \pi$ $\quad \cos(\cos^{-1}x) = x \quad$ for $-1 \leq x \leq 1$ *Graph* 	$\cos^{-1} 1 = 0°$ or 0 $\cos^{-1}\left(-\dfrac{1}{2}\right) = 120°$ or $\dfrac{2\pi}{3}$ $\cos^{-1} 0 = 90°$ or $\dfrac{\pi}{2}$ $\cos^{-1}(-5)$ is undefined. $\cos^{-1} 0.8 \approx 36.9°$ or 0.644 $\arccos\left(\dfrac{1}{2}\right) = 60°$ or $\dfrac{\pi}{3}$
Inverse tangent	*Description* $f(x) = \tan^{-1}x$ or $f(x) = \arctan x$ computes the angle or number in $\left(-\dfrac{\pi}{2}, \dfrac{\pi}{2}\right)$ whose tangent equals x, where x is any real number. *Domain* $\{x \mid -\infty < x < \infty\}$ (all real numbers) *Range* $\left\{y \mid -\dfrac{\pi}{2} < y < \dfrac{\pi}{2}\right\}$ *Inverse Properties* $\quad \tan^{-1}(\tan x) = x \quad$ for $-\dfrac{\pi}{2} < x < \dfrac{\pi}{2}$ $\quad \tan(\tan^{-1}x) = x \quad$ for all real numbers x *Graph* 	$\tan^{-1} 1 = 45°$ or $\dfrac{\pi}{4}$ $\tan^{-1}\left(-\sqrt{3}\right) = -60°$ or $-\dfrac{\pi}{3}$ $\tan^{-1} 0 = 0°$ or 0 $\tan^{-1} 8 \approx 82.9°$ or 1.446 $\arctan(-1) = -45°$ or $-\dfrac{\pi}{4}$

6.6 Exercises

Review of Inverses

1. For a function f to have an inverse, f must be _____.

2. A function is one-to-one if different inputs always result in _____ outputs.

3. If $f(\pi) = -1$, then $f^{-1}(-1) = $ _____.

4. If $f(c) = d$, then $f^{-1}(d) = $ _____.

5. If $f^{-1}(0) = 1$, then $f(1) = $ _____.

6. If $f^{-1}(b) = a$, then $f(a) = $ _____.

Exercises 7–12: Find a formula for $f^{-1}(x)$.

7. $f(x) = 3x$

8. $f(x) = 5x - 4$

9. $f(x) = \sqrt[3]{x}$

10. $f(x) = x^2, x \geq 0$

11. $f(x) = (x - 1)^2, x \geq 1$

12. $f(x) = x^3 - 1$

Exercises 13–16: (Refer to Example 1.) Give verbal, numerical, graphical, and symbolic representations of $f^{-1}(x)$.

13. $f(x) = x + 6$

14. $f(x) = \dfrac{x}{3}$

15. $f(x) = 2x - 1$

16. $f(x) = \sqrt[3]{x - 1}$

Exercises 17 and 18: A graph of a function f that is not one-to-one is given. Estimate the largest interval $[a, b]$ where f is one-to-one, assuming $a < 0 < b$.

17.
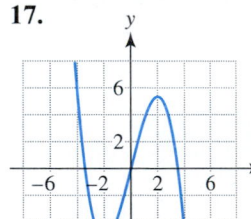

18.
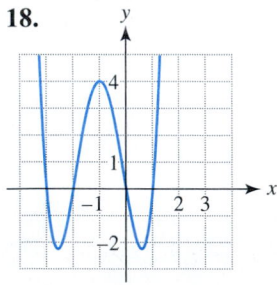

Inverse Trigonometric Functions

19. Since $\sin \frac{\pi}{2} = 1$ and $\frac{\pi}{2}$ is in the interval $\left[-\frac{\pi}{2}, \frac{\pi}{2}\right]$, $\sin^{-1} 1 = $ _____.

20. Since $\cos \frac{\pi}{3} = \frac{1}{2}$ and $\frac{\pi}{3}$ is in the interval $[0, \pi]$, $\cos^{-1} \frac{1}{2} = $ _____.

21. Since $\tan\left(-\frac{\pi}{4}\right) = -1$ and $-\frac{\pi}{4}$ is in the interval $\left(-\frac{\pi}{2}, \frac{\pi}{2}\right)$, $\tan^{-1}(-1) = $ _____.

22. Since $\sin\left(-\frac{\pi}{6}\right) = -\frac{1}{2}$ and $-\frac{\pi}{6}$ is in the interval $\left[-\frac{\pi}{2}, \frac{\pi}{2}\right]$, $\sin^{-1}\left(-\frac{1}{2}\right) = $ _____.

23. Since $\cos\left(\frac{2\pi}{3}\right) = -\frac{1}{2}$ and $\frac{2\pi}{3}$ is in the interval $[0, \pi]$, $\cos^{-1}\left(-\frac{1}{2}\right) = $ _____.

24. Since $\tan\left(\frac{\pi}{3}\right) = \sqrt{3}$ and $\frac{\pi}{3}$ is in the interval $\left(-\frac{\pi}{2}, \frac{\pi}{2}\right)$, $\tan^{-1}\sqrt{3} = $ _____.

Exercises 25–30: Evaluate each of the following, if possible. Give results in both radians and degrees.

25. (a) $\sin^{-1} 1$ (b) $\arcsin 0$
    (c) $\arcsin\left(-\frac{\sqrt{3}}{2}\right)$

26. (a) $\arcsin \frac{1}{2}$ (b) $\sin^{-1}(-2)$
    (c) $\sin^{-1}(-1)$

27. (a) $\cos^{-1} 0$ (b) $\arccos(-1)$
    (c) $\cos^{-1}\frac{1}{2}$

28. (a) $\arccos \frac{\sqrt{3}}{2}$ (b) $\cos^{-1}\left(-\frac{1}{2}\right)$
    (c) $\arccos 1$

29. (a) $\tan^{-1} 1$ (b) $\arctan(-1)$
    (c) $\tan^{-1}\sqrt{3}$

30. (a) $\arctan(-\sqrt{3})$ (b) $\tan^{-1} 0$
    (c) $\tan^{-1}\left(-\frac{1}{\sqrt{3}}\right)$

Exercises 31 and 32: Approximate the following to a hundredth of a radian and a tenth of a degree.

31. (a) $\sin^{-1} 1.5$ (b) $\tan^{-1} 10$
    (c) $\arccos(-0.75)$

32. (a) $\cos^{-1}\left(-\frac{1}{3}\right)$ (b) $\arcsin(-0.54)$
    (c) $\arctan(-2.5)$

Exercises 33 and 34: Evaluate each expression using the figure to obtain either α or β.

33. (a) $\tan^{-1}\frac{4}{3}$

(b) $\sin^{-1}\frac{3}{5}$

(c) $\arccos\frac{3}{5}$

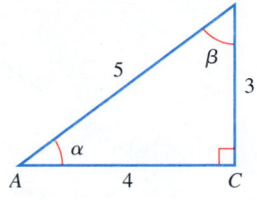

34. (a) $\arcsin\frac{12}{13}$

(b) $\cos^{-1}\frac{5}{13}$

(c) $\tan^{-1}\frac{5}{12}$

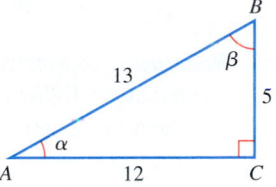

Exercises 35–40: Evaluate each expression.

35. $\sin(\sin^{-1} 1)$

36. $\sin^{-1}\left(\sin\frac{\pi}{4}\right)$

37. $\cos^{-1}\left(\cos\frac{5\pi}{4}\right)$

38. $\cos(\cos^{-1}(-3))$

39. $\tan(\tan^{-1}(-3))$

40. $\tan^{-1}\left(\tan\frac{\pi}{5}\right)$

Exercises 41 and 42: Evaluate the expression in degree mode. Make a generalization about the result and then test your conjecture.

41. (a) $\sin^{-1}\frac{3}{5} + \cos^{-1}\frac{3}{5}$ **(b)** $\sin^{-1}\frac{1}{3} + \cos^{-1}\frac{1}{3}$

(c) $\sin^{-1}\frac{2}{7} + \cos^{-1}\frac{2}{7}$

42. (a) $\tan^{-1}\frac{3}{4} + \tan^{-1}\frac{4}{3}$ **(b)** $\tan^{-1}\frac{5}{12} + \tan^{-1}\frac{12}{5}$

(c) $\tan^{-1}\frac{1}{4} + \tan^{-1} 4$

Exercises 43–48: (Refer to Example 10.) Solve the right triangle.

43.

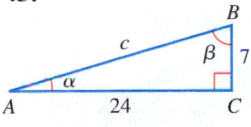

44.

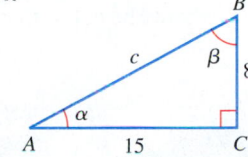

45.

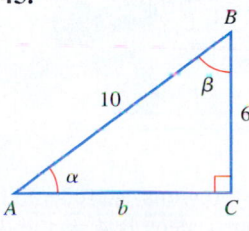

46.

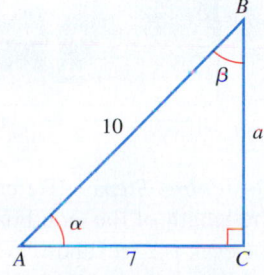

47.

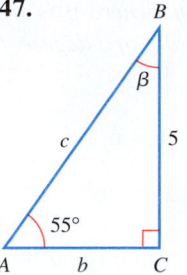

48.

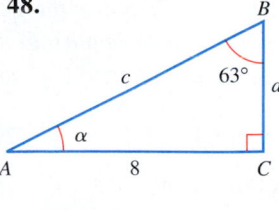

49. Represent $f(x) = \sin^{-1} 2x$ verbally, numerically, and graphically.

50. Represent $f(x) = \cos^{-1} 2x$ verbally, numerically, and graphically.

Exercises 51–54: Let θ be an acute angle. Evaluate the indicated trigonometric function of θ, where θ is an acute angle determined by an inverse trigonometric function. (Hint: Make a sketch of a right triangle containing angle θ.)

51. $\tan\theta$, if $\theta = \sin^{-1} x$

52. $\sin\theta$, if $\theta = \tan^{-1}\frac{x}{\sqrt{1 - x^2}}$

53. $\cos\theta$, if $\theta = \sin^{-1}\frac{x}{\sqrt{1 + x^2}}$

54. $\tan\theta$, if $\theta = \cos^{-1}\frac{1}{x}$

Exercises 55–60: Write the expression as an algebraic expression of u if $0 < u < 1$.

55. $\sin(\cos^{-1} u)$ *(Hint: Let $\theta = \cos^{-1} u$. Sketch a right triangle having an acute angle θ and then label each side of the right triangle.)*

56. $\cos(\sin^{-1} u)$

57. $\tan(\cos^{-1} u)$ **58.** $\sin(\tan^{-1} u)$

59. $\cot\left(\tan^{-1}\frac{1}{u}\right)$ **60.** $\sec(\sin^{-1} u)$

Solving Trigonometric Equations

Exercises 61–66: Solve the trigonometric equation for θ, where $0° \le \theta \le 90°$.

61. $\sin\theta = 1$ **62.** $\cos\theta = \frac{1}{2}$

63. $\tan\theta = 1$ **64.** $\sin\theta = \frac{\sqrt{3}}{2}$

65. $\cos\theta = 0$ **66.** $\tan\theta = \frac{1}{\sqrt{3}}$

Exercises 67–72: Solve the equation for θ, where θ is an acute angle. Approximate θ to the nearest tenth of a degree.

67. $2 \cos \theta = \frac{1}{4}$

68. $3 \sin \theta = \frac{4}{5}$

69. $\tan \theta - 1 = 5$

70. $4 \cos \theta + 1 = 6$

71. $\sin^2 \theta = 0.87$

72. $\tan^3 \theta - 2 = 1.65$

Exercises 73–80: Solve the equation for t, where t is a real number in the given interval. Approximate t to three decimal places.

73. $\tan t = -\frac{1}{5}, \left(-\frac{\pi}{2}, \frac{\pi}{2}\right)$

74. $\sin t = -\frac{1}{3}, \left[-\frac{\pi}{2}, \frac{\pi}{2}\right]$

75. $\cos t = 0.452, [0, \pi]$

76. $\tan t = 5.67, \left(-\frac{\pi}{2}, \frac{\pi}{2}\right)$

77. $2 \sin t = -0.557, \left[-\frac{\pi}{2}, \frac{\pi}{2}\right]$

78. $3 \cos t + 1 = 0.333, [0, \pi]$

79. $\cos^2 t = \frac{1}{25}, [0, \pi]$

80. $\sin^2 t = \frac{1}{16}, \left[-\frac{\pi}{2}, \frac{\pi}{2}\right]$

Applications

81. *Angle of Elevation*   Find the angle of elevation θ of the top of a 50-foot tree at a distance of 85 feet. See the accompanying figure.

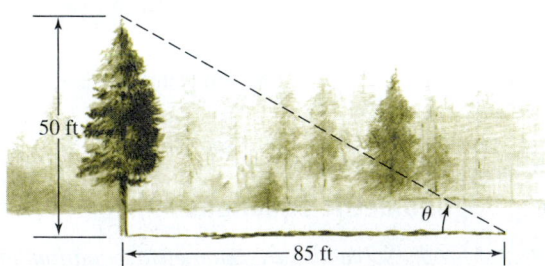

82. *Angle of Elevation*   A 28-foot building casts a 40-foot shadow on level ground. Estimate the angle of elevation θ of the sun to the nearest tenth of a degree. See the figure at the top of the next column.

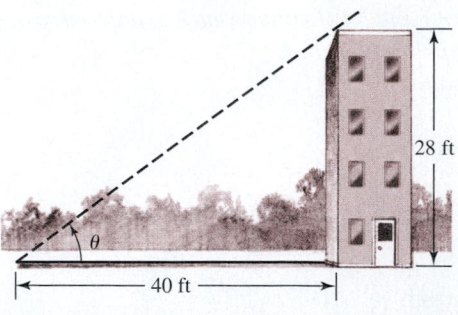

83. *Robotics*   Approximate the angle θ if the robotic hand is located at the following points, where $-90° < \theta < 90°$. See the accompanying figure.
   (a) $(5, 11)$ **(b)** $(1, -3)$

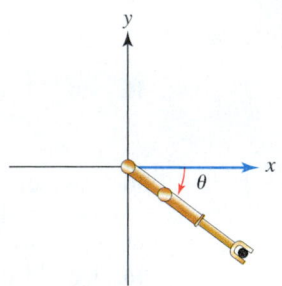

84. *Grade Resistance*   (Refer to Example 12.) Approximate θ to the nearest tenth of a degree for the given grade resistance F and vehicle weight W.
   (a) $F = 400$ lb, $W = 5000$ lb

   (b) $F = 130$ lb, $W = 3500$ lb

   (c) $F = -200$ lb, $W = 4000$ lb

85. *Designing Steps*   Steps are being attached to a deck as shown in the figure. The bottom of the steps should land 10 feet from the deck. If the deck is 4 feet above the ground, estimate the angle θ at which the side boards of the steps should be cut so that the ends lie flat on level ground.

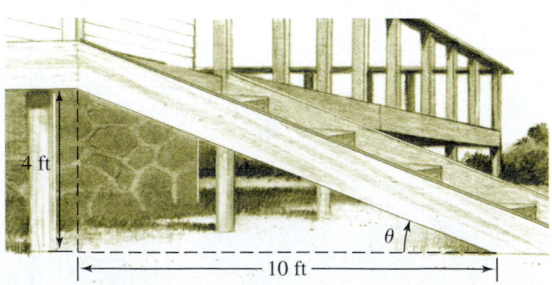

86. *Designing Steps*   (Refer to the previous exercise.) If the length of the side boards for the steps is 4 feet and the deck is 2 feet above the ground, find angle θ.

87. *Phases of the Moon*   (Refer to Example 13.) Find the phase angle θ for the given phase F. Assume that $0° \le \theta \le 180°$.
    (a) $F = \frac{1}{4}$
    (b) $F = \frac{3}{4}$

88. *Roof Pitch*   The pitch or slope of a roof may be expressed in the form $k/12$, where k represents a k-foot rise for every 12 feet of run in horizontal distance. Determine angle θ in Figure 6.138 for each pitch.
    (a) $\frac{3}{12}$ (b) $\frac{4}{12}$ (c) $\frac{6}{12}$ (d) $\frac{12}{12}$

89. *Shot Put*   (Refer to Example 5.) Suppose that a shot is released 7 feet above the ground with a velocity of 43 feet per second. Find the optimum projection angle.

90. *Shot-Putting on the Moon*   Repeat the previous example with $v = 50$ feet per second if the shot is thrown on the moon and the optimal projection angle is given by
$$\theta = \sin^{-1}\sqrt{\frac{v^2}{2v^2 + 10.2h}}.$$

91. *Calculating Daylight Hours*   The ability to calculate the number of daylight hours H at any location is important for estimating the potential solar energy production. The value of H on the longest day can be calculated using the formula
$$\cos(0.1309H) = -0.4336 \tan L,$$
where L is the latitude. Using *radian* mode, calculate the greatest number of daylight hours H during the year for the various cities and their latitudes.   (**Source:** C. Winter, *Solar Power Plants.*)
    (a) Akron, Ohio; $L = 40°55'$

    (b) Corpus Christi, Texas; $L = 27°46'$

    (c) Richmond, Virginia; $L = 37°30'$

92. *Shortest Day*   (Refer to the previous exercise.) The value of H on the shortest day can be calculated using the formula
$$\cos(0.1309H) = 0.4336 \tan L.$$
Find the least number of daylight hours at the following locations.
    (a) Anchorage, Alaska; $L = 61°10'$

    (b) Atlantic City, New Jersey; $L = 39°27'$

    (c) Honolulu, Hawaii; $L = 21°20'$

93. *Snell's Law*   When a ray of light enters water, it is bent. This is because light travels slower in water than in air. This change in direction can be calculated using Snell's law. See the accompanying figure. The angles θ_1 and θ_2 are related by the equation
$$n_1 \sin \theta_1 = n_2 \sin \theta_2,$$
where n_1 and n_2 are constants called *indexes of refraction*. For air $n_1 = 1$ and for water $n_2 = 1.33$. If a ray of light enters the water with $\theta_1 = 40°$, estimate θ_2.   (**Source:** R. Weidner and R. Sells, *Elementary Classical Physics*, Vol. 2.)

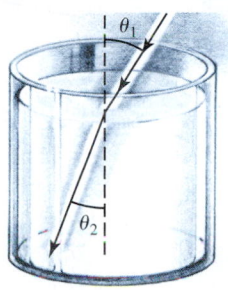

94. *Robotics*   (Refer to Example 9.)
    (a) Let $v = 5$ inches per second and $h = 18$. Graph $\theta_1(t) = \tan^{-1}\left(\frac{h}{vt}\right)$ in the window [0, 15, 5] by [0, 100, 10] using *degree mode*. Describe how θ_1 changes over this 15-second interval.

    (b) Determine the degree measure of θ_1 and θ_2 after 5 seconds.

95. *Landscaping Formula*   A shrub is planted in a 100-foot-wide space between buildings measuring 75 feet and 150 feet tall. The location of the shrub determines how much sun it receives each day. Show that if θ is the angle in the figure and x is the distance of the shrub from the taller building, then the value of θ (in radians) is given by
$$\theta = \pi - \arctan\left(\frac{75}{100 - x}\right) - \arctan\left(\frac{150}{x}\right).$$

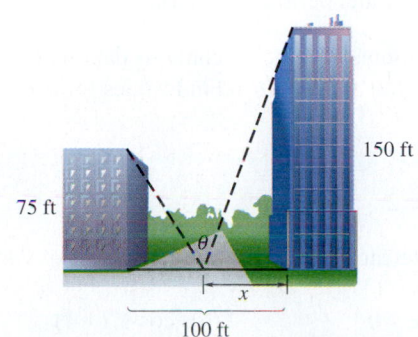

96. *Communications Satellite Coverage* The figure shows a stationary communications satellite positioned 20,000 miles above the equator. What percent of the equator can be seen from the satellite? The diameter of Earth is 7927 miles at the equator.

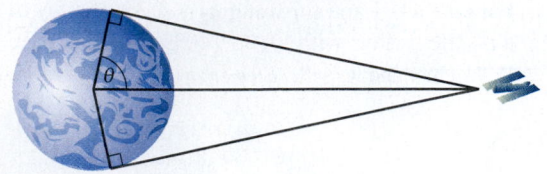

Writing about Mathematics

97. Explain verbally what each expression computes. Give examples.
   (a) $\sin^{-1} x$ **(b)** $\cos^{-1} x$ **(c)** $\tan^{-1} x$

98. Explain why

$$\sin^{-1}\left(\sin\frac{\pi}{2}\right) = \frac{\pi}{2},$$

but

$$\sin^{-1}\left(\sin\frac{5\pi}{2}\right) \neq \frac{5\pi}{2}.$$

Give a similar example using $\cos x$ and $\cos^{-1} x$.

EXTENDED AND DISCOVERY EXERCISE

 1. *Movie Screen* A 10-foot-high movie screen is mounted on a vertical wall so that the bottom of the screen is 6 feet above a horizontal floor. A person sits on a level floor x feet from the screen. If eye level is 3 feet above the floor, then angle θ in the accompanying figure can be expressed as

$$\theta = \tan^{-1}\left(\frac{10x}{x^2 + 39}\right).$$

Graph θ in [0, 50, 10] by [0, 50, 10] using degree mode. Determine where a person should sit to maximize θ.

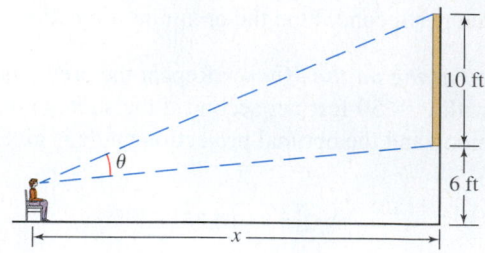

CHECKING BASIC CONCEPTS FOR SECTIONS 6.5 AND 6.6

1. Graph $f(t) = 3 \sin\left(2\left(t - \frac{\pi}{4}\right)\right)$ for $-\pi \le t \le \pi$. State the amplitude, period, and phase shift.

2. The accompanying table contains data that can be modeled by $f(t) = a \cos(bt)$. Find values for a and b.

t	0	0.5	1.0	1.5	2.0	2.5	3.0
$f(t)$	2	0	-2	0	2	0	-2

3. Evaluate each of the following, expressing your answer in degrees.
   (a) $\sin^{-1} 0$ **(b)** $\cos^{-1}(-1)$
   (c) $\tan^{-1}(-1)$ **(d)** $\sin^{-1}\frac{1}{2}$
   (e) $\tan^{-1}\sqrt{3}$

4. Solve the right triangle if $a = 30$ and $b = 40$.

5. Use a calculator to solve each equation, where t is in the indicated interval.
   (a) $\sin t = 0.55, \left[-\frac{\pi}{2}, \frac{\pi}{2}\right]$
   (b) $\cos t = -0.35, [0, \pi]$
   (c) $\tan t = -2.9, \left(-\frac{\pi}{2}, \frac{\pi}{2}\right)$

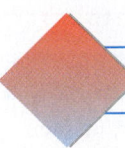

CHAPTER 6 SUMMARY

CONCEPT	EXPLANATION AND EXAMPLES

SECTION 6.1 ANGLES AND THEIR MEASURE

ANGLE MEASURE

Degree measure: $360° =$ one revolution
$1° = 60'$ (minutes)
$1' = 60''$ (seconds)

Radian measure: 2π radians $=$ one revolution

$$\text{radian measure} \times \frac{180°}{\pi} = \text{degree measure}$$

$$\text{degree measure} \times \frac{\pi}{180°} = \text{radian measure}$$

ARC LENGTH

$s = r\theta$, where θ is in radians

Example: The arc length intercepted by a central angle of $120°$ with a radius of 5 inches is

$$s = 5\left(\frac{2\pi}{3}\right) = \frac{10\pi}{3} \approx 10.47 \text{ inches.}$$

AREA OF SECTOR

$A = \frac{1}{2}r^2\theta$, where θ is in radians

Example: The area of the sector determined by a central angle of $45°$ with a radius of 10 inches is

$$A = \frac{1}{2}(10^2)\left(\frac{\pi}{4}\right) = \frac{25\pi}{2} \approx 39.27 \text{ square inches.}$$

SECTION 6.2 RIGHT TRIANGLE TRIGONOMETRY

TRIGONOMETRIC FUNCTIONS (RIGHT TRIANGLES)

$$\sin\theta = \frac{\text{side opposite}}{\text{hypotenuse}} \qquad \csc\theta = \frac{\text{hypotenuse}}{\text{side opposite}}$$

$$\cos\theta = \frac{\text{side adjacent}}{\text{hypotenuse}} \qquad \sec\theta = \frac{\text{hypotenuse}}{\text{side adjacent}}$$

$$\tan\theta = \frac{\text{side opposite}}{\text{side adjacent}} \qquad \cot\theta = \frac{\text{side adjacent}}{\text{side opposite}}$$

Example: The six trigonometric functions of θ in the accompanying figure are as follows.

$$\sin\theta = \frac{11}{61} \qquad \csc\theta = \frac{61}{11}$$

$$\cos\theta = \frac{60}{61} \qquad \sec\theta = \frac{61}{60}$$

$$\tan\theta = \frac{11}{60} \qquad \cot\theta = \frac{60}{11}$$

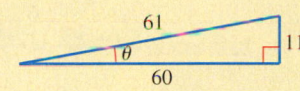

CONCEPT	EXPLANATION AND EXAMPLES

SECTION 6.3 THE SINE AND COSINE FUNCTIONS AND THEIR GRAPHS

SINE AND COSINE

If angle θ is in standard position and its terminal side passes through the point (x, y), then

$$\sin \theta = \frac{y}{r} \quad \text{and} \quad \cos \theta = \frac{x}{r},$$

where $r = \sqrt{x^2 + y^2}$. In the accompanying figure $x = 3$, $y = -4$, and $r = \sqrt{3^2 + (-4)^2} = 5$.

$$\sin \theta = -\frac{4}{5} \qquad \cos \theta = \frac{3}{5}$$

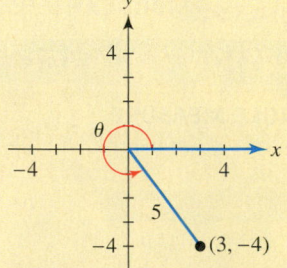

UNIT CIRCLE

If the terminal side of an angle t intersects the unit circle at the point (x, y), then $\sin t = y$ and $\cos t = x$. The domains of the sine and cosine functions include all real numbers and their ranges include all real numbers y such that $-1 \le y \le 1$. In the accompanying figure,

$$x = -\frac{1}{2} \text{ and } y = \frac{\sqrt{3}}{2}.$$

$$\sin t = \frac{\sqrt{3}}{2} \qquad \cos t = -\frac{1}{2}$$

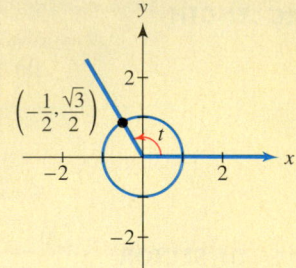

SECTION 6.4 OTHER TRIGONOMETRIC FUNCTIONS AND THEIR GRAPHS

TRIGONOMETRIC FUNCTIONS

The domains, ranges, periods, and graphs of the six trigonometric functions are discussed in Putting it all Together in Section 6.4.

If angle θ is in standard position and its terminal side passes through the point (x, y), then

$$\tan \theta = \frac{y}{x}, \quad \cot \theta = \frac{x}{y}, \quad \sec \theta = \frac{r}{x} \quad \text{and} \quad \csc \theta = \frac{r}{y},$$

where $r = \sqrt{x^2 + y^2}$. In the accompanying figure $x = -2$, $y = 3$, and $r = \sqrt{(-2)^2 + 3^2} = \sqrt{13}$.

$$\tan \theta = -\frac{3}{2}$$

$$\cot \theta = -\frac{2}{3}$$

$$\sec \theta = -\frac{\sqrt{13}}{2}$$

$$\csc \theta = \frac{\sqrt{13}}{3}$$

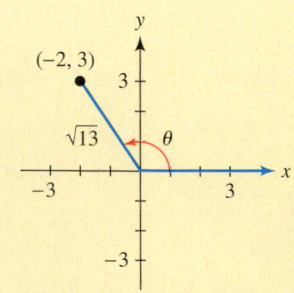

CONCEPT	EXPLANATION AND EXAMPLES

SECTION 6.4 OTHER TRIGONOMETRIC FUNCTIONS AND THEIR GRAPHS (CONTINUED)

UNIT CIRCLE

If the terminal side of an angle t intersects the unit circle at the point (x, y), then

$$\tan t = \frac{y}{x}, \quad \cot t = \frac{x}{y}, \quad \sec t = \frac{1}{x}, \quad \text{and} \quad \csc t = \frac{1}{y}.$$

In the accompanying figure, $x = 0$ and $y = 1$.

$\tan t$ is undefined.

$\cot t = 0$

$\sec t$ is undefined.

$\csc t = 1$

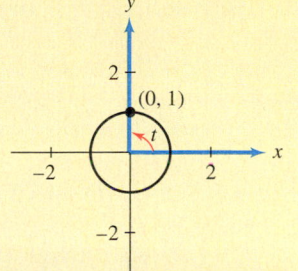

SECTION 6.5 GRAPHING TRIGONOMETRIC FUNCTIONS

MODELING WITH SINE AND COSINE FUNCTIONS

$$f(x) = a \sin (b(x - c)) + d \quad \text{or} \quad f(x) = a \cos (b(x - c)) + d$$

Amplitude $= |a|$ Period $= \dfrac{2\pi}{b}, b > 0$

Phase shift $= c$ Vertical shift $= d$

If $b = 2\pi F$, then F represents the frequency of the sinusoidal graph.

SIMPLE HARMONIC MOTION

Can be modeled by $s(t) = a \sin bt$ or $s(t) = a \cos bt$

Example: If a spring is initially compressed 3 inches and oscillates 4 times per second, then $a = 3, b = 2\pi F = 8\pi$, and its motion can be modeled by $s(t) = 3 \cos (8\pi t)$.

SECTION 6.6 INVERSE TRIGONOMETRIC FUNCTIONS

INVERSE SINE FUNCTION

$\theta = \sin^{-1} x$ implies that $\sin \theta = x$ and either $-\frac{\pi}{2} \leq \theta \leq \frac{\pi}{2}$ or $-90° \leq \theta \leq 90°$.

$\sin^{-1} x$ is also denoted $\arcsin x$.

Examples: $\sin^{-1} 1 = \dfrac{\pi}{2}, \quad \sin^{-1}\left(-\dfrac{1}{2}\right) = -30°, \quad \arcsin 0 = 0°$

INVERSE COSINE FUNCTION

$\theta = \cos^{-1} x$ implies that $\cos \theta = x$ and either $0 \leq \theta \leq \pi$ or $0° \leq \theta \leq 180°$.

$\cos^{-1} x$ is also denoted $\arccos x$.

Examples: $\cos^{-1} 1 = 0, \quad \cos^{-1}(-1) = 180°, \quad \arccos \dfrac{1}{2} = 60°$

INVERSE TANGENT FUNCTION

$\theta = \tan^{-1} x$ implies that $\tan \theta = x$ and either $-\frac{\pi}{2} < \theta < \frac{\pi}{2}$ or $-90° < \theta < 90°$.

$\tan^{-1} x$ is also denoted $\arctan x$.

Examples: $\tan^{-1} 1 = \dfrac{\pi}{4}, \quad \tan^{-1}(-1) = -45°, \quad \arctan \sqrt{3} = 60°$

CONCEPT	EXPLANATION AND EXAMPLES

SECTION 6.6 INVERSE TRIGONOMETRIC FUNCTIONS (CONTINUED)

SOLVING TRIANGLES

Inverse trigonometric functions can be used to solve equations and find angles in triangles.

Example: In triangle ABC inverse trigonometric functions can be used to find α and β.

$$\alpha = \sin^{-1}\frac{3}{5} \approx 36.9°$$

$$\alpha = \tan^{-1}\frac{3}{4} \approx 36.9°$$

$$\beta = \cos^{-1}\frac{3}{5} \approx 53.1°$$

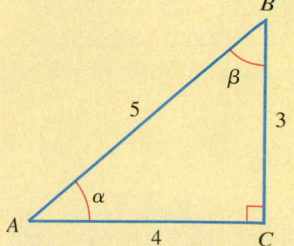

REVIEW EXERCISES

1. Sketch the following angles in standard position.
   (a) $60°$ (b) $-120°$

   (c) $\dfrac{3\pi}{2}$ (d) $-\dfrac{5\pi}{6}$

2. Find the complementary angle and the supplementary angle to $\theta = 61°40'$.

3. Convert each angle from radian measure to degree measure.
   (a) $\dfrac{\pi}{3}$ (b) $\dfrac{\pi}{36}$ (c) $-\dfrac{5\pi}{6}$ (d) $-\dfrac{7\pi}{4}$

4. Convert each angle from degree measure to radian measure.
   (a) $30°$ (b) $165°$ (c) $-90°$ (d) $-105°$

5. Find the length of the arc intercepted by a central angle $\theta = 60°$ and a radius $r = 6$ feet.

6. Find the area of the sector of a circle having a radius $r = 5$ inches and a central angle $\theta = 150°$.

Exercises 7–12: Use either a 30°–60° right triangle or a 45°–45° right triangle to find the exact value of each trigonometric expression.

7. $\sin 30°$ 8. $\tan 45°$

9. $\cot 60°$ 10. $\cos 60°$

11. $\sec 45°$ 12. $\csc 30°$

13. Find the six trigonometric functions of θ.

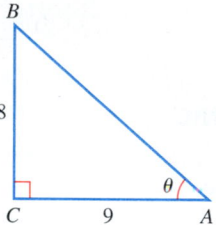

14. Solve triangle ABC.

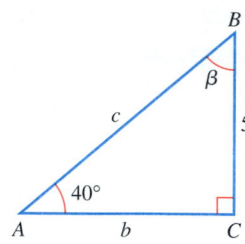

15. Find $\csc\theta$ if $\sin\theta = \frac{1}{3}$.

16. Find $\cot\theta$ if $\sin\theta = \frac{5}{13}$ and $\cos\theta = -\frac{12}{13}$.

Exercises 17 and 18: Approximate the six trigonometric functions of θ to three decimal places.

17. $\theta = 25°$ 18. $\theta = -\dfrac{6\pi}{7}$

Exercises 19 and 20: Find the six trigonometric functions of angle θ.

19.

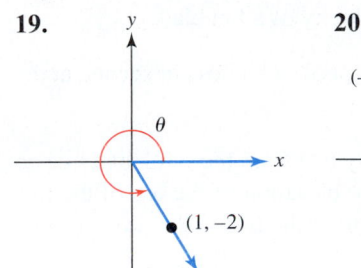

20.

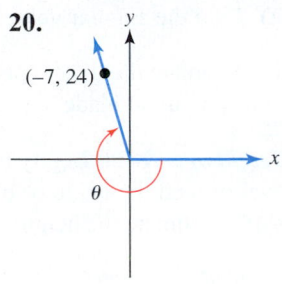

Exercises 21 and 22: Find the six trigonometric functions of the given angle by hand. Support your answer using a calculator.

21. $-45°$

22. $\dfrac{5\pi}{6}$

Exercises 23 and 24: The accompanying figure shows angle θ in standard position with its terminal side intersecting the unit circle. Find the six trigonometric functions of θ.

23.

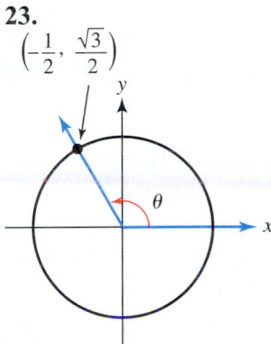

24.

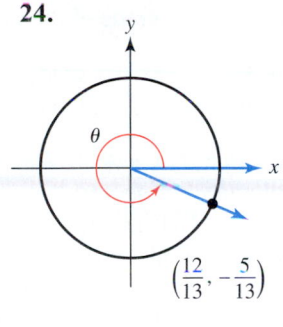

Exercises 25–28: Evaluate the function f at the given value of t.

25. $f(t) = \sin t,\ t = -\dfrac{\pi}{2}$ **26.** $f(t) = \cos t,\ t = \pi$

27. $f(t) = \tan t,\ t = -3\pi$ **28.** $f(t) = \csc t,\ t = \dfrac{\pi}{2}$

29. Find the other trigonometric functions of θ if $\sin\theta = -\dfrac{4}{5}$ and $\cos\theta = \dfrac{3}{5}$.

30. Convert $65°45'36''$ to decimal degrees.

Exercises 31 and 32: Determine which of the six trigonometric functions is represented by the table.

31.

x	$-\dfrac{\pi}{2}$	$-\dfrac{\pi}{6}$	0	$\dfrac{\pi}{6}$	$\dfrac{\pi}{2}$
y	-1	$-\dfrac{1}{2}$	0	$\dfrac{1}{2}$	1

32.

x	$-\dfrac{\pi}{3}$	$-\dfrac{\pi}{4}$	0	$\dfrac{\pi}{4}$	$\dfrac{\pi}{3}$
y	$-\sqrt{3}$	-1	0	1	$\sqrt{3}$

Exercises 33–38: Graph f for $-2\pi \le x \le 2\pi$. State the amplitude, period, and phase shift.

33. $f(t) = 3\cos(2t)$ **34.** $f(t) = -2\sin(2t + \pi)$

35. $f(t) = \sin\left(3\left(t - \dfrac{\pi}{3}\right)\right)$

36. $f(t) = 1.5\cos\left(\dfrac{1}{2}\left(t - \dfrac{\pi}{2}\right)\right)$

37. $f(x) = -3\sin\left(\dfrac{1}{2}(x - \pi)\right) + 1$

38. $f(x) = \dfrac{1}{2}\cos\left(\dfrac{\pi}{2}(x - 1)\right) - 2$

Exercises 39 and 40: A graph of $y = a\cos bx$ is shown, where b is a positive constant. Estimate the values for a and b.

39.

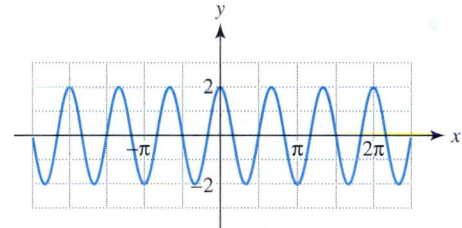

40.

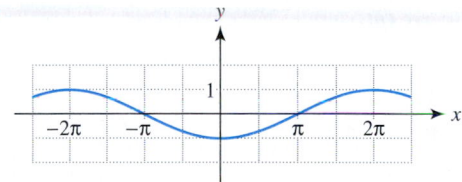

Exercises 41–44: Graph f for $-2\pi \le x \le 2\pi$. State the period and phase shift.

41. $f(t) = \cot 2t$ **42.** $f(t) = \tan\left(t - \dfrac{\pi}{2}\right)$

43. $f(t) = \csc(t + \pi)$ **44.** $f(t) = \sec 2t$

Exercises 45 and 46: If possible, evaluate each of the following in both radians and degrees.

45. **(a)** $\sin^{-1}(-1)$ **(b)** $\arccos\dfrac{1}{2}$ **(c)** $\tan^{-1} 1$

46. **(a)** $\arcsin 3$ **(b)** $\cos^{-1} 0$ **(c)** $\arctan\left(-\sqrt{3}\right)$

47. Approximate the following to a hundredth of a radian and a tenth of a degree.

(a) $\sin^{-1}(-0.6)$ **(b)** $\tan^{-1} 5$ **(c)** $\arccos(0.12)$

48. Evaluate each expression.

(a) $\sin(\sin^{-1} 0.5)$ (b) $\tan^{-1}(\tan 45°)$

(c) $\cos^{-1}\left(\cos\dfrac{3\pi}{2}\right)$

Exercises 49 and 50: Solve the right triangle ABC.

49.

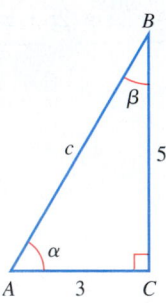

50.

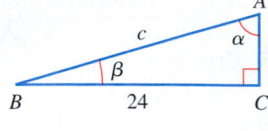

Exercises 51–54: Solve the equation for θ, where 0° ≤ θ ≤ 90°.

51. $\sin\theta = \frac{1}{2}$

52. $\cos\theta = 1$

53. $\tan\theta = \dfrac{1}{\sqrt{3}}$

54. $\sin\theta = 0$

Exercises 55–58: Solve the equation for θ, where θ is an acute angle. Approximate θ to the nearest tenth of a degree.

55. $\cos\theta = \frac{1}{5}$

56. $3\sin\theta = \frac{15}{13}$

57. $2\tan\theta - 1 = 5$

58. $2\cos\theta = \frac{4}{7}$

Exercises 59–62: Solve the equation for t, where t is a real number located in the indicated interval. Approximate t to four decimal places.

59. $\tan t = -\dfrac{3}{4}, \left(-\dfrac{\pi}{2}, \dfrac{\pi}{2}\right)$

60. $\sin t = -\dfrac{3}{5}, \left[-\dfrac{\pi}{2}, \dfrac{\pi}{2}\right]$

61. $2\cos t = 1.8, [0, \pi]$

62. $3\tan t + 2 = 4.7, \left(-\dfrac{\pi}{2}, \dfrac{\pi}{2}\right)$

Applications

63. *Ferris Wheel*   A Ferris wheel has a diameter of 50 feet and completes 1 revolution every 50 seconds.

(a) Find the angular velocity of the Ferris wheel in radians per second.

(b) What is the linear speed in feet per second of a person who is riding this Ferris wheel?

64. *Fan Speed*   The blades of a fan have a 25-inch diameter and rotate at 400 revolutions per minute.

(a) Find the angular velocity of a fan blade.

(b) Estimate the linear speed in inches per second at the tip of a fan blade.

65. *Height of a Tree*   Eighty feet from the trunk of a tree on level ground the angle of elevation of the top of the tree is 48°. Estimate the height of the tree to the nearest foot.

66. *Angle of Elevation*   Find the angle of elevation of the top of a 35-foot building at a horizontal distance of 52 feet.

67. *Grade Resistance*   Approximate θ to the nearest tenth of a degree for the given grade resistance F and vehicle weight W, where $F = W\sin\theta$.

(a) $F = 350$ lb, $W = 6000$ lb

(b) $F = 160$ lb, $W = 4500$ lb

68. *Highway Grade*   (Refer to the previous exercise.) Suppose an uphill grade of a highway can be modeled by the line $y = 0.05x$.

(a) Find the grade of the hill.

(b) Determine the grade resistance for a gravel truck that weighs 30,000 pounds.

69. *Distance between Cities*   Cheyenne, Wyoming, and Colorado Springs, Colorado, have nearly the same longitude of 104°45′ W. The latitude of Cheyenne is 41°09′ and the latitude of Colorado Springs is 38°49′. Approximate the distance between these two cities if the average radius of Earth is 3955 miles.   (**Source:** J. Williams, *The Weather Almanac 1995.*)

70. *Safe Distance for a Tree*   From a distance of 45 feet from the base of a tree, the angle of elevation to the top of a tree is 57°, as shown in the figure. A building is located 52 feet from the base of the tree. Determine if the tree could fall in a storm and damage the building.

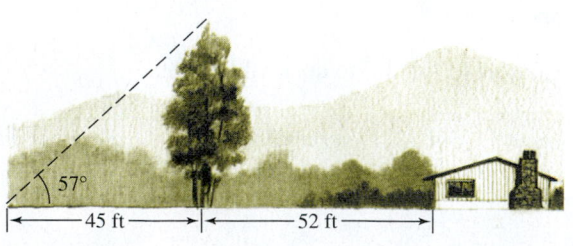

71. *Modeling Temperatures* The monthly average low temperatures in Green Bay, Wisconsin, are shown in the table.

Month	1	2	3	4	5	6
Temperature (°F)	6	10	22	35	45	52

Month	7	8	9	10	11	12
Temperature (°F)	58	56	48	38	26	11

Source: A. Miller and J. Thompson, *Elements of Meteorology.*

(a) Plot the average monthly temperature over a 24-month period by letting $x = 1$ and $x = 13$ correspond to January.

(b) Find the constants a, b, c, and d so that

$$f(x) = a \cos(b(x - c)) + d$$

models the data.

(c) Graph f together with the data.

72. *Light and Sinusoidal Waves* Light frequently is modeled using a sinusoidal wave. Light waves have very high frequencies compared to sound waves. Each color of light has a different frequency. The function given by $f(t) = \sin(2\pi F t)$ can be used to model various colors of light, where F is the frequency in cycles per second. Graph f in $[0, 10^{-14}, 10^{-15}]$ by $[-1.5, 1.5, 0.5]$. Find the period P for each color of light, where $P = \frac{1}{F}$. Interpret the results. (**Source:** R. Weidner and R. Sells, *Elementary Classical Physics,* Vol. 2.)

(a) Violet: $F = 7.5 \times 10^{14}$

(b) Green: $F = 6 \times 10^{14}$

(c) Red: $F = 4 \times 10^{14}$

73. *Vehicle Navigation Systems* Vehicle location and navigation systems for cars are becoming increasingly popular. Devices that depend on Earth's magnetic field are not sufficiently accurate because they can be affected by going through a car wash, having the rear defroster turned on, or even traveling near a large metal truck. The accompanying graph approximates typical errors when using these devices. In this graph 0° corresponds to north, 90° to east, 180° to south, and 270° to west. For example, if a car is traveling in a direction corresponding to 180°, the error in a magnetic compass may be −4°.

Thus the car is actually traveling in a direction of 184°. Find the equation of the graph using $y = a \sin(x - c)$, where $0° \le x \le 360°$. (**Source:** Y. Zhao, *Vehicle Location and Navigation Systems.*)

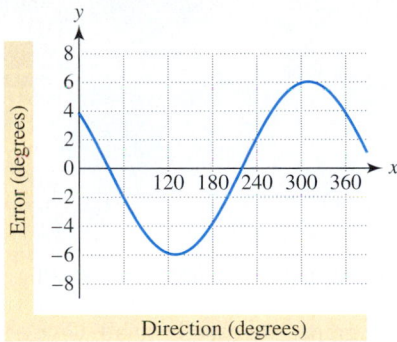

Direction (degrees)

74. *Phases of the Moon* If the phase angle of the moon is θ, then the phase F of the moon is given by

$$F = \frac{1}{2}(1 - \cos\theta).$$

Solve this equation for θ.

EXTENDED AND DISCOVERY EXERCISES

1. *Surveying* The first fundamental problem of surveying is to determine the coordinates of a point Q given the coordinates of a point P, the distance between P and Q, and the bearing θ from P to Q. See the accompanying figure. (**Source:** I. Mueller and K. Ramsayer, *Introduction to Surveying.*)

(a) Find a formula for the coordinates (x_Q, y_Q) of the point Q given θ, the coordinates (x_P, y_P) of P, and the distance d between P and Q.

(b) Use your formula to find (x_Q, y_Q) if $(x_P, y_P) = (152, 186)$, $\theta = 23.2°$, and $d = 208$ feet.

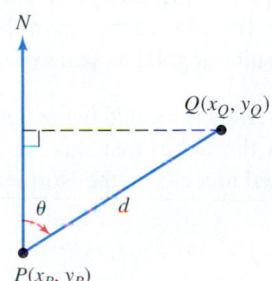

2. *Highway Grade* (Refer to Example 3, Section 6.3.) Complete the table for the trigonometric values of θ to five decimal places.

θ	$\sin\theta$	$\tan\theta$
0°		
1°		
2°		
3°		
4°		

(a) How do $\sin\theta$ and $\tan\theta$ compare for small values of θ?

(b) Highway grades are usually small. Give an approximation to the grade resistance given by $F = W\sin\theta$ that uses the tangent function instead of the sine function.

(c) A stretch of highway has a 4-foot vertical rise for every 100 feet of horizontal run. Use your approximation to estimate the grade resistance for a 3000-pound car.

(d) Compare your result to the exact answer using $F = W\sin\theta$.

(**Source:** F. Mannering and W. Kilareski, *Principles of Highway Engineering and Traffic Analysis.*)

3. *Average Temperature* The maximum average monthly temperature in Buenos Aires, Argentina, is 74°F and the minimum average monthly temperature is 49°F.

(a) Using these two temperatures, find values for a, b, c, and d, so that

$$f(x) = a\cos(b(x - c)) + d$$

models the monthly average temperature.

(b) On the same coordinate axes, graph f for a 2-year period together with the actual data values found in the table at the top of the next column. Are your results as good as you expected? Explain.

(c) Buenos Aires is located in the Southern Hemisphere. Discuss the effect that this has on the graph of f compared to a city in the Northern Hemisphere.

Month	1	2	3	4	5	6
Temperature (°F)	74	73	69	61	55	50

Month	7	8	9	10	11	12
Temperature (°F)	49	51	55	60	66	71

Source: A. Miller and J. Thompson, *Elements of Meteorology.*

4. (Refer to the previous exercise.) The maximum average monthly temperature in Melbourne, Australia, is 68°F and the minimum is 49°F.

(a) Determine $f(x) = a\cos(b(x - c)) + d$ so that f models the monthly average temperature in Melbourne.

(b) Graph f together with the actual data shown in the accompanying table over a 2-year period.

Month	1	2	3	4	5	6
Temperature (°F)	68	68	65	60	54	51

Month	7	8	9	10	11	12
Temperature (°F)	49	51	54	58	61	65

Source: A. Miller and J. Thompson.

5. Suppose two cities, one in the Northern Hemisphere and one in the Southern Hemisphere, both have average monthly temperatures that can be modeled by

$$f(x) = a\cos(b(x - c)) + d,$$

where x is the month. Suppose also that for both cities the warmest monthly average temperature is 80°F and the coldest is 40°F.

(a) Find $f(x)$ for each city. (Assume that the warmest month for the northern city is July and the coldest month is January, and the opposite is true for the southern city.)

(b) Explain how $f(x)$ for one city could be used to obtain $f(x)$ for the other city.

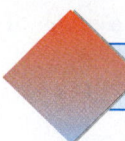

CHAPTERS 1–6 CUMULATIVE REVIEW EXERCISES

1. Find the percent change if the price of gold changes from \$350 an ounce to \$400 an ounce.

2. Write 125,000 in scientific notation and 4.67×10^{-3} in standard notation.

3. Evaluate $\dfrac{2 - \sqrt{2}}{2 + \sqrt{2}}$. Round your answer to two decimal places.

4. Find the midpoint of the line segment connecting $(-3, 2)$ and $(-1, 6)$.

5. Express the domain and range of f in interval notation. Then evaluate $f(-0.5)$.

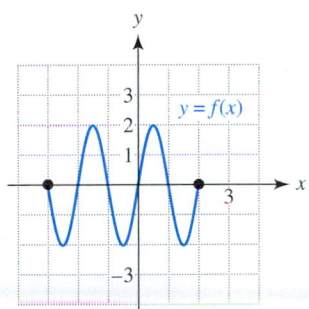

6. Graph $y = g(x)$ by hand.
   (a) $g(x) = 2x - 3$ (b) $g(x) = |x + 2|$

   (c) $g(x) = -2(x + 1)^2 - 3$

   (d) $g(x) = 2^x$

   (e) $g(x) = \ln x$ (f) $g(x) = \sqrt{x - 2}$

   (g) $g(x) = \dfrac{1}{x + 1}$ (h) $g(x) = x^3$

Exercises 7 and 8: Complete the following.
   (a) Determine the domain of f.
   (b) Evaluate $f(-1)$ and $f(2a)$.

7. $f(x) = \sqrt{4 - x}$ 8. $\dfrac{x - 2}{4x^2 - 16}$

9. The graph of a linear function f is shown.
   (a) Identify the slope, y-intercept, and x-intercept.

   (b) Write a formula for $f(x)$.

(c) Evaluate $f(-2)$ symbolically and graphically.

(d) Find any zeros of f.

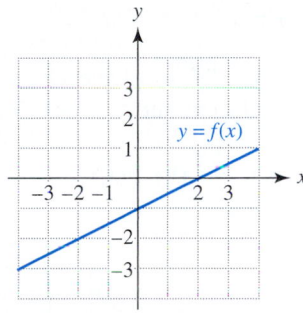

10. Find the average rate of change of $f(x) = 10^x$ from $x = 1$ to $x = 2$.

11. Find the difference quotient for $f(x) = 3x^2$.

12. Write the slope-intercept form of a line that passes through $(2, -3)$ and is parallel to the line $2x + 3y = 6$.

13. Determine the x- and y-intercepts on the graph of the equation $-2x + 5y = 20$.

14. If $R(x) = \frac{1}{2}x + 2$ calculates the amount of rainfall in inches x hours past midnight, interpret the numbers $\frac{1}{2}$ and 2 in the formula for $R(x)$.

15. *Solve each equation.*
    (a) $2(1 - 2x) = 5 - (4 - x)$ (b) $2e^x - 1 = 27$

    (c) $|4 - 5x| = 8$ (d) $2x^2 + x = 1$

    (e) $x^3 - 3x^2 + 2x = 0$ (f) $x^4 + 8 = 6x^2$

    (g) $x^2 - x - 2 = 0$

    (h) $\sqrt{2x - 1} = x - 2$

    (i) $\dfrac{x}{x - 2} = \dfrac{2x - 1}{x + 1}$

    (j) $\log_2(x + 1) = 16$

16. Graph f. Is f continuous on its domain? Evaluate $f(1)$.

$$f(x) = \begin{cases} 1 - x & \text{if } -4 \le x \le -1 \\ -2x & \text{if } -1 < x < 2 \\ \frac{1}{2}x^2 & \text{if } 2 \le x \le 4 \end{cases}$$

17. Solve the inequality. Write the solution set in interval notation.

   (a) $-3(2 - x) < 4 - (2x + 1)$

   (b) $-3 \le 4 - 3x < 6$

   (c) $|4x - 3| \ge 9$ **(d)** $x^2 - 5x + 4 \le 0$

   (e) $t^3 - t > 0$ **(f)** $\dfrac{1}{t + 2} - 3 \ge 0$

18. Use the graph of f to solve each equation or inequality. Write the solution set for each inequality in interval notation.

   (a) $f(x) = 0$ **(b)** $f(x) > 0$ **(c)** $f(x) \le 0$

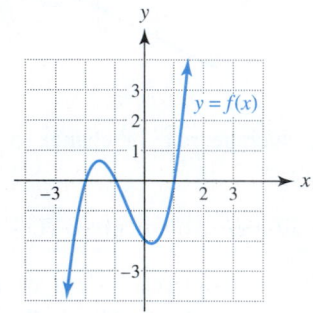

19. Write $f(x) = -2x^2 + 6x - 1$ in the form given by $f(x) = a(x - h)^2 + k$.

20. Find the vertex on the graph of $f(x) = 3x^2 - 4x + 1$.

21. A graph of a quadratic function $f(x) = ax^2 + bx + c$ is shown.

   (a) State whether $a > 0$ or $a < 0$.

   (b) Estimate the real solutions to $ax^2 + bx + c = 0$.

   (c) Determine if the discriminant, $b^2 - 4ac$, is positive, negative, or zero.

   (d) Write the complete factored form of $f(x)$.

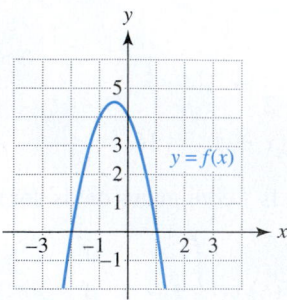

22. Solve $2x^2 + 4x = 1$ by completing the square.

23. Use the given graph of $y = f(x)$ to sketch a graph of each expression.

   (a) $y = f(x - 1) + 2$ **(b)** $y = \frac{1}{2}f(x)$

   (c) $y = -f(-x)$ **(d)** $y = f(2x)$

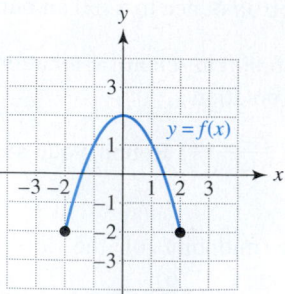

24. Use the graph of f to estimate each of the following.

   (a) Where f is increasing or decreasing

   (b) The zeros of f

   (c) The coordinates of any turning points

   (d) Any local extrema

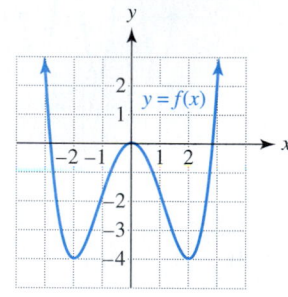

25. Sketch a graph of a cubic function with a negative leading coefficient, two x-intercepts, and two turning points.

26. Predict the end behavior of $f(x) = -x^4 + 2x + 5$.

27. Divide each expression.

   (a) $\dfrac{5a^4 - 2a^2 + 4}{2a^2}$ **(b)** $\dfrac{x^4 + 2x^3 - 2x}{x - 1}$

   (c) $\dfrac{x^3 - 3x^2 + x + 1}{x^2 + 1}$

28. Write the complete factored form for the polynomial given by $f(x) = 2x^3 + x^2 - 8x - 4$.

29. A degree 5 function f has zeros -2, -1, and 3 with multiplicities 1, 2, and 2, respectively. If the leading coefficient is -2, write the complete factored form of $f(x)$.

30. Find all zeros, real or imaginary, of

$$f(x) = x^3 - x^2 + 4x - 4$$

given that one zero is $2i$.

31. Use the graph to write the complete factored form of the quartic (degree 4) polynomial $f(x)$.

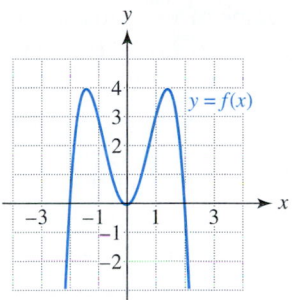

32. Write $(2 + i)^2 + (2 - 3i)$ in standard form.

33. State the domain of $f(x) = \frac{2x + 5}{3x - 7}$. Find any vertical or horizontal asymptotes.

34. Write $x^{2/3}$ in radical notation. Evaluate the expression for $x = 27$.

35. Use the tables for f and g to evaluate each expression, if possible.

x	0	1	2	3	4
$f(x)$	4	3	2	1	0

x	0	1	2	3	4
$g(x)$	0	4	3	2	1

(a) $(f + g)(2)$ **(b)** $(g/f)(4)$

(c) $(f \circ g)(3)$ **(d)** $(f^{-1} \circ g)(1)$

36. Use the graphs of f and g to complete the following.

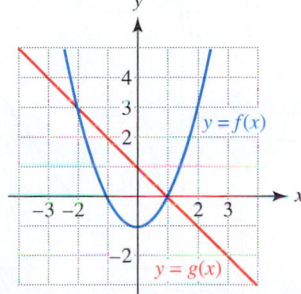

(a) $(f - g)(-1)$ **(b)** $(fg)(2)$

(c) $(g \circ f)(0)$ **(d)** $(g^{-1} \circ f)(2)$

37. Let $f(x) = x^2 + 3x - 2$ and $g(x) = x - 2$. Find the following.
(a) $(f + g)(2)$ **(b)** $(g \circ f)(1)$

(c) $(f - g)(x)$ **(d)** $(f \circ g)(x)$

38. Find $f^{-1}(x)$ if $f(x) = 2\sqrt[3]{x} + 1$.

39. Find the domains of $(f + g)(x)$ and $(f \circ g)(x)$ if $f(x) = \frac{1}{x}$ and $g(x) = \sqrt{2 - x}$.

40. Sketch a graph of the equations $f(x) = 3x - 2, y = x$, and $y = f^{-1}(x)$.

41. Find either a linear or an exponential function f that models the data in the table.

x	0	1	2	3	4
$f(x)$	9	6	4	$\frac{8}{3}$	$\frac{16}{9}$

42. There are initially 2000 bacteria in a sample and this number doubles every 3 hours. Find C and a so that $f(x) = Ca^x$ models the number of bacteria after x hours.

43. Use the graph of $y = Ca^x$ to determine values for C and a.

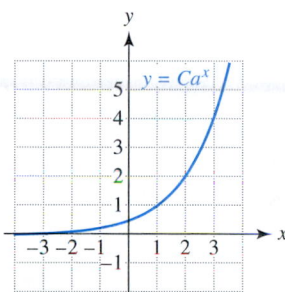

44. Five hundred dollars are deposited in an account that pays 5% annual interest compounded monthly. Find the amount in the account after 10 years.

45. Simplify each logarithm by hand.
(a) $\log 100$ **(b)** $\log_2 16$

(c) $\ln \frac{1}{e^2}$ **(d)** $\log_4 \sqrt[3]{4}$

(e) $\log 4 + \log 25$ **(f)** $\log_6 24 - \log_6 4$

46. Find the domain and range of each function f.
(a) $f(x) = e^x$ **(b)** $f(x) = \log_2 x$

(c) $f(x) = \log(-x)$ **(d)** $f(x) = 2^{-x} + 1$

47. Expand the expression $\log_2 \dfrac{\sqrt[3]{x^2 - 4}}{\sqrt{x^2 + 4}}$.

48. Write $3 \log x - 4 \log y + \frac{1}{2} \log z$ as a logarithm of a single expression.

49. Approximate $\log_3 125$ to three decimal places.

50. Solve the equation.
   (a) $3(2)^{-2x} + 4 = 100$ **(b)** $2 \log_3(3x) = 4$

   (c) $\log_3 4x + \log_3\left(x + \frac{7}{4}\right) = \log_3 2$

51. Find the complementary angle α and the supplementary angle β to $10°34'$.

52. Convert $152.24°$ to degrees, minutes, and seconds.

53. Convert $150°$ to radians.

54. Convert $\frac{5\pi}{4}$ radians to degrees.

55. Find the length of the arc intercepted by a central angle of $15°$ in a circle with a radius of 3 feet.

56. Find the six trigonometric functions of θ.

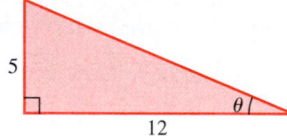

57. Find the length of a in the figure.

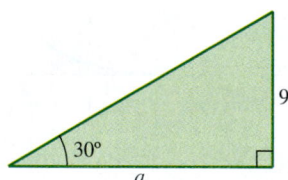

58. Approximate sec (1.24) to the nearest hundredth.

59. Find the six trigonometric functions of θ.

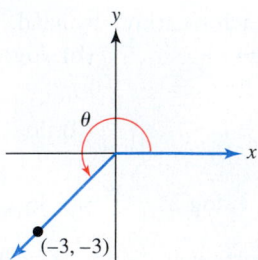

60. Find the exact values of the six trigonometric functions of $\theta = \frac{7\pi}{6}$.

61. Find the exact values of the six trigonometric functions of θ if $\cos \theta = \frac{5}{13}$ and $\sin \theta < 0$.

62. State the amplitude, period, phase shift, and vertical shift for the graph of

$$f(x) = 5 \sin\left(\frac{\pi}{2}(x - 1)\right) - 2.$$

63. Sketch a graph of f on the interval $[-2\pi, 2\pi]$.
   (a) $f(x) = 2 \sin 3x$ **(b)** $f(x) = \sec x$

   (c) $f(x) = \tan \frac{1}{2}x$

   (d) $f(x) = -2 \cos\left(2x + \frac{\pi}{2}\right) + 1$

64. Evaluate $\sin^{-1}\left(-\frac{1}{2}\right)$.

65. Solve the right triangle.

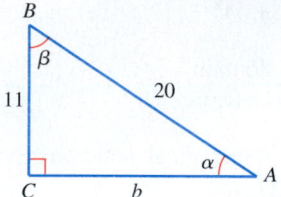

66. Solve $2 \cos \theta = -1$ if $0 \leq \theta \leq 180°$.

Applications

67. *Dimensions of an Aluminum Can*   The volume V of an aluminum can is given by $V = \pi r^2 h$, where r is the radius and h is the height. If an aluminum can has a volume of 12 cubic inches and a diameter of 2 inches, find the height of the can to the nearest hundredth of an inch.

68. *Distance from Home*   The graph shows the distance that the driver of a car on a straight highway is from home. Find the slope of each line segment and interpret each slope.

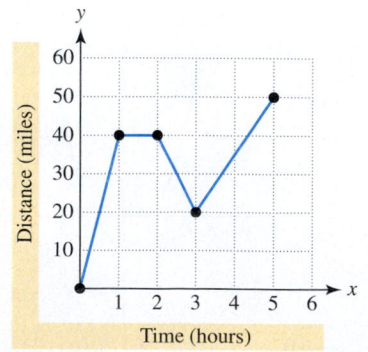

69. *Average Rate of Change*   The total distance D in feet that an object falls after t seconds is given by $D(t) = 16t^2$ for $0 \leq t \leq 3$.
(a) Find the average rate of change of D from 0 to 1 and 1 to 2.

(b) Interpret these average rates of change.

(c) Find the difference quotient of D.

70. *Working Together*   Suppose one person can mow a large lawn in 4 hours and another person can mow the same lawn in 6 hours. How long will it take to mow the lawn if they work together?

71. *Inverse Variation*   The force of gravity F varies inversely as the square of the distance d from the *center* of Earth. If a person weighs 150 pounds on the surface of Earth ($d = 4000$ miles), how much would this person weigh at a distance of 10,000 miles from the center of Earth?

72. *Linear Programming*   Find the minimum value of $C = 3x + y$ subject to the following constraints.

$$x + y \geq 1$$
$$2x + 3y \leq 6$$
$$x \geq 0, y \geq 0$$

73. *College Tuition*   In 1980 the annual cost of tuition and fees at public colleges and universities was about \$800 and in 2000 it was about \$3500. (**Source:** The College Board.)
(a) Determine a formula in the form

$$C(t) = m(t - t_1) + y_1$$

that models these data. Let C be in dollars and t be the year.

(b) Estimate the year when tuition and fees reached \$3000 and compare it to the actual year of 1996.

74. *Hotel Rooms*   Rooms at a hotel are regularly \$110, but for each additional room rented, the cost is reduced by \$2.
(a) Write a quadratic function C that gives the total cost of renting x rooms.

(b) Solve $C(x) = 1470$ and interpret the result.

(c) Find the absolute maximum for C and interpret your result.

75. *Volume of a Balloon*   The radius r in inches of a spherical balloon after t seconds is given by $r = \sqrt{t}$.
(a) Is the radius increasing or decreasing?

(b) Write a formula for a function V that calculates the volume of the sphere after t seconds.

(c) Evaluate $V(4)$ and interpret the result.

76. *Inverse Function*   The function given by the formula $f(x) = \frac{5}{9}(x - 32)$ converts degrees Fahrenheit to degrees Celsius.
(a) Find $f^{-1}(x)$.

(b) What does f^{-1} compute?

77. *Bacteria Growth*   There are initially 200,000 bacteria per milliliter in a sample. The number of bacteria grows exponentially and reaches 300,000 per milliliter after 3 hours.
(a) Use the formula $N(t) = N_0 e^{kt}$ to model the concentration of bacteria after t hours.

(b) Evaluate $N(5)$ and interpret the result.

(c) After how long did the concentration reach 500,000 per milliliter?

78. *Specific Growth Rate*   If the concentration of algae in a sample changes from N_1 to N_2 from time t_1 to time t_2, then the *specific growth rate r* is defined to be

$$r = \frac{\log N_2 - \log N_1}{t_2 - t_1}.$$

(a) Calculate r if $N_1 = 2000$, $N_2 = 8000$, $t_1 = 1$, and $t_2 = 2$.

(b) Find r if $N_1 = 500$, $N_2 = 4500$, $t_1 = 2$, and $t_2 = 4$.

(c) Interpret what 10^r represents.

79. *Interest Formula*   Solve $A = P(1 + \frac{r}{n})^{nt}$ for t.

80. *Bicycle*   A bicycle has a tire that is 24 inches in diameter and is rotating at 20 radians per second. Approximate the speed of the bicycle in feet per second.

81. *Length of a Shadow*   If the angle of elevation of the sun is $43°$, find the length of the shadow cast by a person who is 6 feet tall. Round your answer to the nearest tenth of a foot.

82. *Projectile Motion*   If a projectile is fired with an initial velocity of v feet per second at an angle of elevation θ, then it will follow a parabolic path described by

$$y = \frac{-16x^2}{v^2(\cos\theta)^2} + x\tan\theta.$$

(a) Let $v = 200$ and $\theta = 40°$. Graph the path of the projectile.

(b) Find the maximum height of the projectile.

(c) Determine the horizontal distance that the projectile traveled.

83. *Modeling Temperature*   The monthly average high temperatures in New Orleans are shown in the table. Model these data by using a function of the form

$$f(x) = a\sin(b(x - c)) + d.$$

Month	1	2	3	4	5	6
Temperature (°F)	61	65	70	78	85	89

Month	7	8	9	10	11	12
Temperature (°F)	90	89	88	80	70	63

Source: J. Williams, *The Weather Almanac.*

84. *Angle of Elevation*   A 52-foot building cast a 63-foot shadow. Estimate the angle of elevation of the sun to the nearest tenth of a degree.

7 Trigonometric Identities and Equations

The struggle is what teaches us.
Sue Grafton

Music is both art and science. During the Greek and Roman eras, music played an important role in philosophy and science. Although Pythagoras is usually associated with the Pythagorean theorem, in 500 B.C. he also discovered the mathematical relationships between lengths of strings and musical intervals. This discovery of the mathematical ratios that governed pitch and motion was the beginning of the science of musical sound. In the Middle Ages, music was studied with arithmetic, geometry, and astronomy as part of the liberal arts curriculum. Later in 1862 the psychologist and scientist Hermann von Helmholtz published his classic work that opened a new direction for music using mathematics and technology. Then in 1957, Max Mathews created complex musical sounds with a computer. Mathematics has had an essential role in the development and reproduction of music.

The communications industry also uses mathematics. For example, each number on a touch-tone phone has a unique sound that is a combination of two different tones. These unique tones are easily distinguished when transmitted over phone lines. As a result, consumers can communicate with banks and other businesses using touch-tone phones. Trigonometric functions play an important role in clear, reliable communication systems.

Many applications of mathematics began as theoretical mathematics. This is why both pure and applied mathematics are important to society. In this chapter we introduce the concept of verifying an identity. Identities are important because they allow us to write trigonometric expressions in simpler and more convenient forms. Verifying identities requires both effort and concentration. However, by learning to manipulate trigonometric expressions, we will be able to solve complex problems and equations.

Source: J. Pierce, *The Science of Musical Sound.*

7.1 Fundamental Identities

◆ Learn and apply the reciprocal and quotient identities

◆ Learn and apply the Pythagorean identities

◆ Learn and apply the negative-angle identities

Introduction

Trigonometric expressions can often be written in more than one way. For example, in Section 6.4 we learned that $\cot\theta$ is equivalent to $\frac{\cos\theta}{\sin\theta}$. The equation

$$\cot\theta = \frac{\cos\theta}{\sin\theta}$$

is a trigonometric identity. This identity is true for every value of θ, provided $\sin\theta \neq 0$. Trigonometric identities are used to help solve equations and model physical phenomena in music, science, and electricity. They are also used in calculus. We begin our discussion with the reciprocal and quotient identities.

Reciprocal and Quotient Identities

In Section 6.4 the following definitions were presented for the trigonometric functions of any angle θ. See Figure 7.1.

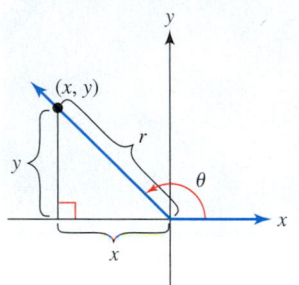

FIGURE 7.1

TRIGONOMETRIC FUNCTIONS OF ANY ANGLE θ

Let (x, y) be a point other than the origin on the terminal side of an angle θ in standard position. If $r = \sqrt{x^2 + y^2}$, then the six trigonometric functions are as follows.

$$\sin\theta = \frac{y}{r} \qquad\qquad \csc\theta = \frac{r}{y}\ (y \neq 0)$$

$$\cos\theta = \frac{x}{r} \qquad\qquad \sec\theta = \frac{r}{x}\ (x \neq 0)$$

$$\tan\theta = \frac{y}{x}\ (x \neq 0) \qquad \cot\theta = \frac{x}{y}\ (y \neq 0)$$

These definitions allow us to write several identities. For example, since

$$\cos\theta = \frac{x}{r} \qquad \text{and} \qquad \sec\theta = \frac{r}{x},$$

it follows that

$$\sec\theta = \frac{1}{x/r} = \frac{1}{\cos\theta} \qquad \text{and} \qquad \cos\theta = \frac{1}{r/x} = \frac{1}{\sec\theta}.$$

These identities are examples of *reciprocal identities*.

RECIPROCAL IDENTITIES

$$\sin\theta = \frac{1}{\csc\theta} \qquad \cos\theta = \frac{1}{\sec\theta} \qquad \tan\theta = \frac{1}{\cot\theta}$$

$$\csc\theta = \frac{1}{\sin\theta} \qquad \sec\theta = \frac{1}{\cos\theta} \qquad \cot\theta = \frac{1}{\tan\theta}$$

 **Applying a reciprocal identity**

In Example 8, Section 6.2 the equation

$$d = r\left(\frac{1}{\cos\theta} - 1\right)$$

was used to calculate the external distance for a highway curve. Use a reciprocal identity to rewrite this equation.

SOLUTION Since $\sec\theta = \frac{1}{\cos\theta}$, we can express the equation as

$$d = r(\sec\theta - 1).$$ *Now Try Exercise 79* ◆

Because $\sin\theta = \frac{y}{r}$ and $\cos\theta = \frac{x}{r}$, it is possible to write the other four trigonometric functions in terms of $\sin\theta$ and $\cos\theta$. For example,

$$\tan\theta = \frac{y}{x} = \frac{y/r}{x/r} = \frac{\sin\theta}{\cos\theta}.$$

This identity is an example of a *quotient identity*. The quotient identities are given in the following.

QUOTIENT IDENTITIES

$$\tan\theta = \frac{\sin\theta}{\cos\theta} \qquad \cot\theta = \frac{\cos\theta}{\sin\theta}$$

If $\sin\theta$ and $\cos\theta$ are known, the reciprocal and quotient identities can be used to find the other four trigonometric functions of θ, as illustrated in the next example.

 **Using reciprocal and quotient identities**

If $\sin\theta = \frac{7}{25}$ and $\cos\theta = -\frac{24}{25}$, find the other four trigonometric functions of θ.

SOLUTION We can use identities to find $\tan\theta$, $\cot\theta$, $\sec\theta$, and $\csc\theta$.

$$\tan\theta = \frac{\sin\theta}{\cos\theta} = \frac{7/25}{-24/25} = -\frac{7}{24}$$

$$\cot\theta = \frac{\cos\theta}{\sin\theta} = \frac{-24/25}{7/25} = -\frac{24}{7}$$

$$\sec\theta = \frac{1}{\cos\theta} = \frac{1}{-24/25} = -\frac{25}{24}$$

$$\csc\theta = \frac{1}{\sin\theta} = \frac{1}{7/25} = \frac{25}{7}$$ *Now Try Exercise 7* ◆

 **Using identities to find trigonometric values**

If $\tan\theta = -\frac{8}{15}$ and $\cos\theta = -\frac{15}{17}$, find the other four trigonometric functions of θ.

SOLUTION Using the reciprocal identities, we can find $\cot\theta$ and $\sec\theta$.

$$\cot\theta = \frac{1}{\tan\theta} = \frac{1}{-8/15} = -\frac{15}{8}$$

$$\sec\theta = \frac{1}{\cos\theta} = \frac{1}{-15/17} = -\frac{17}{15}$$

To find $\sin\theta$, consider the following.

$$\tan\theta\cos\theta = \frac{\sin\theta}{\cos\theta} \cdot \cos\theta = \sin\theta$$

Thus

$$\sin\theta = \tan\theta\cos\theta = \left(-\frac{8}{15}\right)\left(-\frac{15}{17}\right) = \frac{8}{17}$$

and, using a reciprocal identity, we find

$$\csc\theta = \frac{1}{\sin\theta} = \frac{1}{8/17} = \frac{17}{8}.$$

Now Try Exercise 9 ◆

Pythagorean Identities

In Section 6.3 the unit circle, shown in Figure 7.2, was used to define the sine and cosine functions.

$$\sin\theta = y \qquad \text{and} \qquad \cos\theta = x$$

An equation for the unit circle is given by $x^2 + y^2 = 1$. By substitution it follows that

$$(\cos\theta)^2 + (\sin\theta)^2 = 1$$

or equivalently,

$$\cos^2\theta + \sin^2\theta = 1.$$

This identity can be supported graphically and numerically by letting

$$Y_1 = (\cos(X))^\wedge 2 + (\sin(X))^\wedge 2,$$

as shown in Figures 7.3 and 7.4. The graph of Y_1 is the horizontal line $y = 1$. Either radian or degree mode may be used. Notice that regardless of the value of x, $Y_1 = 1$.

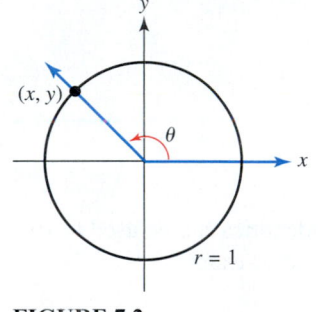

FIGURE 7.2

$[-352.5, 352.5, 90]$ by $[-2, 2, 1]$

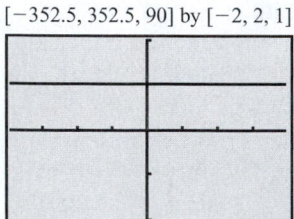

FIGURE 7.3 Degree Mode

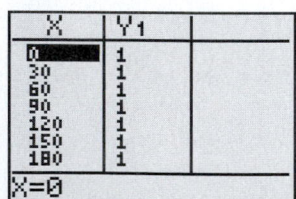

FIGURE 7.4 Degree Mode

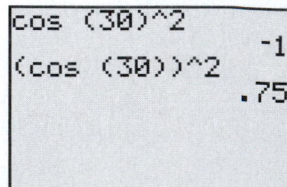

FIGURE 7.5 Degree Mode

Note: On some graphing calculators $\cos^2 x$ must be entered as $(\cos(X))^\wedge 2$, rather than $\cos(X)^\wedge 2$, since $\cos(X)^\wedge 2$ may be interpreted as $\cos(x^2)$. See Figure 7.5.

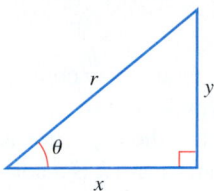

FIGURE 7.6

Using the right triangle shown in Figure 7.6, we also can verify the identity $\cos^2 \theta + \sin^2 \theta = 1$ geometrically when θ is acute.

$$x^2 + y^2 = r^2 \qquad \text{Pythagorean theorem}$$

$$\frac{x^2}{r^2} + \frac{y^2}{r^2} = \frac{r^2}{r^2} \qquad \text{Divide each side by } r^2.$$

$$\left(\frac{x}{r}\right)^2 + \left(\frac{y}{r}\right)^2 = 1 \qquad \text{Properties of exponents}$$

$$\cos^2 \theta + \sin^2 \theta = 1 \qquad \cos \theta = \tfrac{x}{r}, \sin \theta = \tfrac{y}{r}$$

For this reason, $\sin^2 \theta + \cos^2 \theta = 1$ is an example of a Pythagorean identity. Two other Pythagorean identities can be derived as follows.

$$\sin^2 \theta + \cos^2 \theta = 1$$

$$\frac{\sin^2 \theta}{\cos^2 \theta} + \frac{\cos^2 \theta}{\cos^2 \theta} = \frac{1}{\cos^2 \theta} \qquad \text{Divide by } \cos^2 \theta.$$

$$\tan^2 \theta + 1 = \sec^2 \theta \qquad \tan \theta = \tfrac{\sin \theta}{\cos \theta}, \sec \theta = \tfrac{1}{\cos \theta}$$

◆ **CLASS DISCUSSION**
Use the triangle in Figure 7.6 to justify the Pythagorean identities

$$1 + \tan^2 \theta = \sec^2 \theta$$

and

$$1 + \cot^2 \theta = \csc^2 \theta,$$

when θ is acute. ◆

In a similar manner, a third Pythagorean identity can be found.

$$\sin^2 \theta + \cos^2 \theta = 1$$

$$\frac{\sin^2 \theta}{\sin^2 \theta} + \frac{\cos^2 \theta}{\sin^2 \theta} = \frac{1}{\sin^2 \theta} \qquad \text{Divide by } \sin^2 \theta.$$

$$1 + \cot^2 \theta = \csc^2 \theta \qquad \cot \theta = \tfrac{\cos \theta}{\sin \theta}, \csc \theta = \tfrac{1}{\sin \theta}$$

This discussion is summarized in the following.

PYTHAGOREAN IDENTITIES

$$\sin^2 \theta + \cos^2 \theta = 1 \qquad 1 + \tan^2 \theta = \sec^2 \theta \qquad 1 + \cot^2 \theta = \csc^2 \theta$$

In the next example we use a reciprocal, quotient, and Pythagorean identity to simplify a trigonometric expression.

EXAMPLE 4 Applying identities to an expression

Simplify the expression $1 + \sin^2 \theta \sec^2 \theta$.

SOLUTION Begin by applying a reciprocal identity.

$$1 + \sin^2 \theta \sec^2 \theta = 1 + \sin^2 \theta \cdot \frac{1}{\cos^2 \theta} \qquad \text{Reciprocal identity: } \sec \theta = \tfrac{1}{\cos \theta}$$

$$= 1 + \frac{\sin^2 \theta}{\cos^2 \theta} \qquad \text{Multiply.}$$

$$= 1 + \left(\frac{\sin \theta}{\cos \theta}\right)^2 \qquad \text{Properties of exponents}$$

$$= 1 + \tan^2 \theta \qquad \text{Quotient identity: } \tan \theta = \tfrac{\sin \theta}{\cos \theta}$$

$$= \sec^2 \theta \qquad \text{Pythagorean identity: } 1 + \tan^2 \theta = \sec^2 \theta$$

Now Try Exercise 15 ◆

The next example illustrates use of identities in electronics.

FIGURE 7.7 An Inductor and a Capacitor

EXAMPLE 5 Applying a Pythagorean identity to radios

Tuners in radios select a radio station by adjusting the frequency. These tuners may contain an inductor L and a capacitor C, as illustrated in Figure 7.7. The energy stored in the inductor at time t is given by $L(t) = k \sin^2(2\pi Ft)$ and the energy stored in the capacitor is given by $C(t) = k \cos^2(2\pi Ft)$, where F is the frequency of the radio station and k is a constant. The total energy E in the circuit is given by $E(t) = L(t) + C(t)$. Show that E is a constant function. (**Source:** R. Weidner and R. Sells, *Elementary Classical Physics,* Vol. 2.)

SOLUTION

$$E(t) = L(t) + C(t) \qquad \text{Given equation}$$
$$= k \sin^2(2\pi Ft) + k \cos^2(2\pi Ft) \qquad \text{Substitute.}$$
$$= k(\sin^2(2\pi Ft) + \cos^2(2\pi Ft)) \qquad \text{Factor.}$$
$$= k(1) \qquad \sin^2\theta + \cos^2\theta = 1\ (\theta = 2\pi Ft)$$
$$= k \qquad k \text{ is constant.} \qquad \textit{Now Try Exercise 83} \blacklozenge$$

If an angle θ is in standard position and its terminal side lies in quadrant II, as shown in Figure 7.8, then θ is *contained in quadrant II* or θ is a *second quadrant angle.* Similar statements can be made for angles whose terminal sides lie in other quadrants.

Let the point (x, y) lie on the terminal side of θ with $r = \sqrt{x^2 + y^2} > 0$. Then $\sin\theta = \frac{y}{r}$ is positive when $y > 0$ and negative when $y < 0$. As a result, $\sin\theta$ is positive for first and second quadrant angles and negative for third and fourth quadrant angles. Similarly, a point (x, y) has a positive x-coordinate in either quadrant I or IV and a negative x-coordinate in either quadrant II or III. Thus $\cos\theta = \frac{x}{r}$ is positive in quadrants I and IV and negative in quadrants II and III. In Figure 7.9 the signs of the six trigonometric functions for angles in each quadrant are listed. Notice that the six trigonometric functions are all positive for angles in the first quadrant.

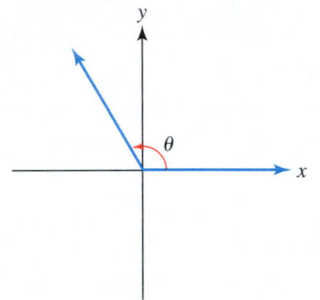

FIGURE 7.8 Angle θ in Quadrant II

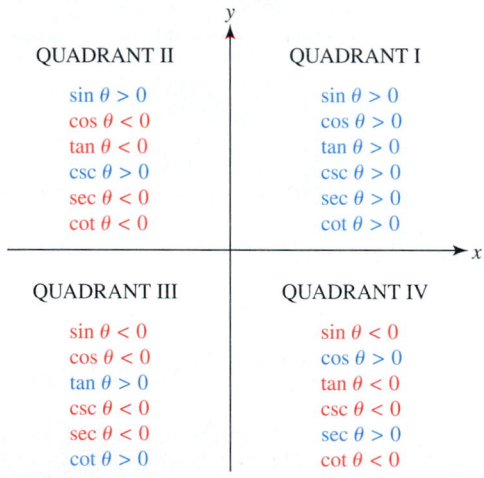

FIGURE 7.9

EXAMPLE 6 Finding the quadrant containing an angle

If $\sin\theta > 0$ and $\cos\theta < 0$, find the quadrant that contains θ. Support your results graphically and numerically.

SOLUTION If $\sin\theta > 0$, $\cos\theta < 0$ then any point (x, y) on the terminal side of θ must satisfy $y > 0$ and $x < 0$. Thus θ is contained in quadrant II.

Graphical Support   In Figure 7.10 the graphs of $Y_1 = \sin(X)$ and $Y_2 = \cos(X)$ are shown. Notice that when $\theta = 135°$ (a second quadrant angle) the graph of $\sin\theta$ is above the x-axis and the graph of $\cos\theta$ is below the x-axis.

Numerical Support   In Figure 7.11 numerical support is given, where angles in quadrant II have positive sine values and negative cosine values.

$[0, 352.5, 90]$ by $[-2, 2, 1]$

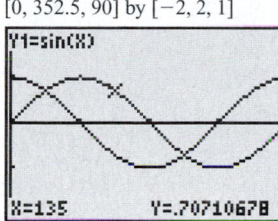

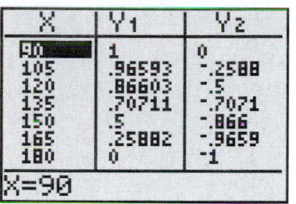

FIGURE 7.10 Degree Mode   **FIGURE 7.11** Degree Mode

Now Try Exercise 35 ◆

If a trigonometric function of θ is known and the quadrant containing θ is also known, then we can find the other five trigonometric functions of θ, as illustrated in the next example.

EXAMPLE 7

Using identities to find trigonometric values

If $\sin\theta = -\frac{3}{5}$ and θ is a third quadrant angle, find the values of the other trigonometric functions.

SOLUTION   Start by finding $\cos\theta$.

$$\sin^2\theta + \cos^2\theta = 1 \qquad \textit{Pythagorean identity}$$
$$\cos^2\theta = 1 - \sin^2\theta \qquad \textit{Subtract } \sin^2\theta.$$
$$\cos\theta = \pm\sqrt{1 - \sin^2\theta} \qquad \textit{Square root property}$$
$$\cos\theta = \pm\sqrt{1 - (-3/5)^2} \qquad \textit{Substitute for } \sin\theta.$$
$$\cos\theta = \pm\frac{4}{5} \qquad \textit{Simplify.}$$

In quadrant III, $x < 0$ and so $\cos\theta = -\frac{4}{5}$. The other four trigonometric functions of θ can be found as follows.

$$\tan\theta = \frac{\sin\theta}{\cos\theta} = \frac{-3/5}{-4/5} = \frac{3}{4}$$

$$\cot\theta = \frac{\cos\theta}{\sin\theta} = \frac{-4/5}{-3/5} = \frac{4}{3}$$

$$\sec\theta = \frac{1}{\cos\theta} = \frac{1}{-4/5} = -\frac{5}{4}$$

$$\csc\theta = \frac{1}{\sin\theta} = \frac{1}{-3/5} = -\frac{5}{3}$$

Now Try Exercise 49 ◆

EXAMPLE 8

Using identities to find trigonometric values

If $\tan \theta = -\frac{7}{3}$ and $\sin \theta > 0$, find the values of the other trigonometric functions.

SOLUTION Since $\tan \theta = -\frac{7}{3}$, it follows that $\cot \theta = -\frac{3}{7}$. Next we find $\sec \theta$.

$$\sec^2 \theta = 1 + \tan^2 \theta \qquad \text{Pythagorean identity}$$

$$\sec \theta = \pm \sqrt{1 + \tan^2 \theta} \qquad \text{Square root property}$$

$$= \pm \sqrt{1 + (-7/3)^2} \qquad \text{Substitute for } \tan \theta.$$

$$= \pm \frac{\sqrt{58}}{3} \qquad \text{Simplify.}$$

Since $\tan \theta < 0$ and $\sin \theta > 0$, θ is a second quadrant angle. It follows that

$$\sec \theta = -\frac{\sqrt{58}}{3} \quad \text{and} \quad \cos \theta = -\frac{3}{\sqrt{58}}.$$

Since $\sin \theta = \tan \theta \cos \theta$,

$$\sin \theta = \left(-\frac{7}{3}\right)\left(-\frac{3}{\sqrt{58}}\right) = \frac{7}{\sqrt{58}}.$$

Using a reciprocal identity

$$\csc \theta = \frac{1}{\sin \theta} = \frac{\sqrt{58}}{7}. \qquad \textit{Now Try Exercise 43} \blacklozenge$$

We can also express any trigonometric function in terms of any other trigonometric function. This fact is demonstrated in the next example.

EXAMPLE 9

Expressing one function in terms of another

Write $\cos x$ and $\cot x$ in terms of $\sin x$, if $\sec x < 0$.

SOLUTION We can write $\cos x$ in terms of $\sin x$ by applying a Pythagorean identity. Note that $\sec x < 0$ implies that $\cos x < 0$ because $\cos x = \frac{1}{\sec x}$ by a reciprocal identity.

◆ **CLASS DISCUSSION**

Write $\sec x$ and $\tan x$ in terms of $\sin x$ if $\cos x > 0$. ◆

$$\sin^2 x + \cos^2 x = 1 \qquad \text{Pythagorean identity}$$

$$\cos^2 x = 1 - \sin^2 x \qquad \text{Subtract } \sin^2 x.$$

$$\cos x = \pm \sqrt{1 - \sin^2 x} \qquad \text{Square root property}$$

Because $\cos x < 0$, we let $\cos x = -\sqrt{1 - \sin^2 x}$. Now we can write $\cot x$ in terms of $\sin x$ by using a quotient identity.

$$\cot x = \frac{\cos x}{\sin x} \qquad \text{Quotient identity}$$

$$= -\frac{\sqrt{1 - \sin^2 x}}{\sin x} \qquad \text{Substitute for } \cos x.$$

$$\textit{Now Try Exercise 53} \blacklozenge$$

EXAMPLE 10 **Using fundamental identities to find trigonometric expressions**

If $\sin\theta = x$ and θ is a quadrant IV angle, find an expression for $\sec\theta$ in terms of x. Approximate $\sec\theta$ if $\sin\theta = -0.7813$.

SOLUTION We begin by writing $\sec\theta$ in terms of $\sin\theta$.

$$\sin^2\theta + \cos^2\theta = 1 \qquad \text{Pythagorean identity}$$

$$\cos^2\theta = 1 - \sin^2\theta \qquad \text{Subtract } \sin^2\theta.$$

$$\cos\theta = \pm\sqrt{1 - \sin^2\theta} \qquad \text{Square root property}$$

$$\sec\theta = \pm\frac{1}{\sqrt{1 - \sin^2\theta}} \qquad \sec\theta = \frac{1}{\cos\theta}$$

$$\sec\theta = \pm\frac{1}{\sqrt{1 - x^2}} \qquad \sin\theta = x$$

Angle θ is in quadrant IV. Thus $\sec\theta > 0$ and

$$\sec\theta = \frac{1}{\sqrt{1 - x^2}}.$$

Since $\sin\theta = x$, let $x = -0.7813$. Then,

$$\sec\theta = \frac{1}{\sqrt{1 - (-0.7813)^2}} \approx 1.602. \qquad \textit{Now Try Exercise 59} \blacklozenge$$

Negative-Angle Identities

In Section 4.1 we discussed odd and even functions. The graphs of odd functions are symmetric with respect to the origin and the graphs of even functions are symmetric with respect to the y-axis. Symbolically, we say that an odd function f satisfies

$$f(-x) = -f(x)$$

for all x in the domain of f. Similarly, an even function f satisfies

$$f(-x) = f(x)$$

for all x in its domain. The graphs of all six trigonometric functions have symmetry, as shown in Figures 7.12–7.14 and in Figures 7.15–7.17 on the next page.

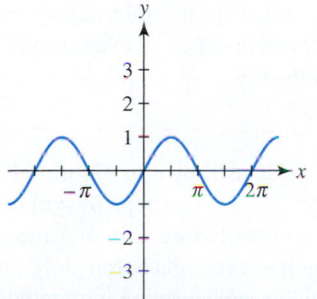

FIGURE 7.12 The Sine Function (An Odd Function)

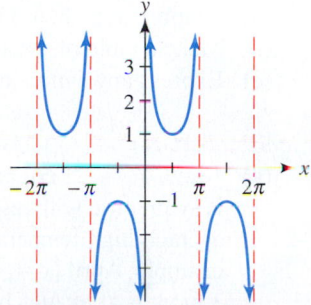

FIGURE 7.13 The Cosecant Function (An Odd Function)

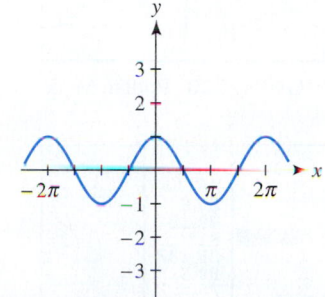

FIGURE 7.14 The Cosine Function (An Even Function)

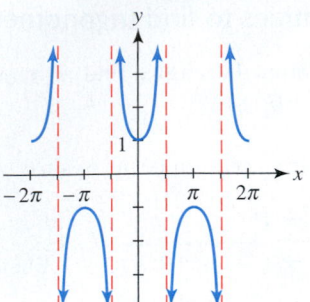

FIGURE 7.15 The Secant Function (An Even Function)

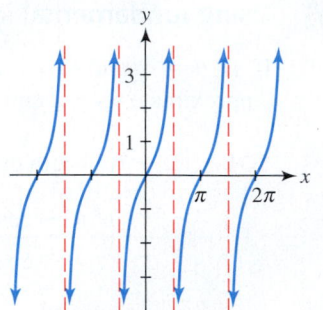

FIGURE 7.16 The Tangent Function (An Odd Function)

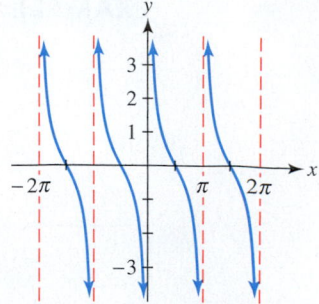

FIGURE 7.17 The Cotangent Function (An Odd Function)

The cosine and secant functions are even functions, and the other four trigonometric functions are odd functions. For even functions the sign of the input does not affect the output, whereas for an odd function changing the sign of the input only changes the sign of the output. These results can be expressed using the negative-angle identities.

NEGATIVE-ANGLE IDENTITIES

$$\sin(-\theta) = -\sin\theta \qquad \cos(-\theta) = \cos\theta \qquad \tan(-\theta) = -\tan\theta$$

$$\csc(-\theta) = -\csc\theta \qquad \sec(-\theta) = \sec\theta \qquad \cot(-\theta) = -\cot\theta$$

EXAMPLE 11 Modeling temperature

The monthly average high temperatures in degrees Fahrenheit at Chattanooga, Tennessee, can be modeled by

$$f(x) = 21\cos\left(\frac{\pi x}{6}\right) + 70,$$

where x is the month with $x = -6$ corresponding to January, $x = 0$ to July, and $x = 5$ to December.   (**Source:** J. Williams, *The Weather Almanac.*)
(a) Graph f in $[-6, 6, 1]$ by $[40, 100, 10]$. Interpret any symmetry in the graph.
(b) Make a table of f and discuss whether f is an even or odd function.
(c) Express any symmetry symbolically.

$[-6, 6, 1]$ by $[40, 100, 10]$

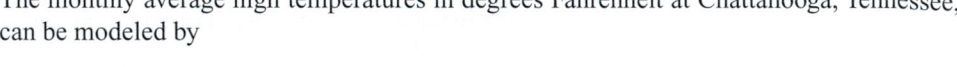

FIGURE 7.18 Radian Mode

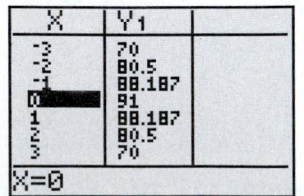

FIGURE 7.19 Radian Mode

SOLUTION
(a) Graph $Y_1 = 21\cos(\pi X/6) + 70$, as shown in Figure 7.18 in radian mode. The graph is symmetric with respect to the y-axis. This type of symmetry implies that the monthly average high temperatures x months before July or x months after July are equal. For example, April ($x = -3$), which is 3 months before July, and October ($x = 3$), which is 3 months after July, both have the same average high temperature of 70°.
(b) A table of Y_1 is shown in Figure 7.19. Notice that f is an even function since the sign of the input does not affect the output.
(c) Symbolically this symmetry can be expressed as $f(-x) = f(x)$.   *Now Try Exercise 85* ◆

7.1 Putting it all Together

In this section we discussed the reciprocal identities, the quotient identities, the Pythagorean identities, and the negative-angle identities. Collectively, they are called fundamental identities. If $\sin\theta$ and $\cos\theta$ are known, then the reciprocal and quotient identities can be used to find the remaining four trigonometric functions of θ. The Pythagorean identities can be used to find trigonometric values when only one trigonometric value is known. The negative-angle identities can be used to determine if a trigonometric function is an odd function or an even function. Cosine and secant are even functions, and sine, cosecant, tangent, and cotangent are odd functions.

Identity	General Form		
Reciprocal	$\sin\theta = \dfrac{1}{\csc\theta}$	$\cos\theta = \dfrac{1}{\sec\theta}$	$\tan\theta = \dfrac{1}{\cot\theta}$
	$\csc\theta = \dfrac{1}{\sin\theta}$	$\sec\theta = \dfrac{1}{\cos\theta}$	$\cot\theta = \dfrac{1}{\tan\theta}$
Quotient	$\tan\theta = \dfrac{\sin\theta}{\cos\theta}$	$\cot\theta = \dfrac{\cos\theta}{\sin\theta}$	
Pythagorean	$\sin^2\theta + \cos^2\theta = 1$	$1 + \tan^2\theta = \sec^2\theta$	$1 + \cot^2\theta = \csc^2\theta$
Negative-angle	$\sin(-\theta) = -\sin\theta$	$\cos(-\theta) = \cos\theta$	$\tan(-\theta) = -\tan\theta$
	$\csc(-\theta) = -\csc\theta$	$\sec(-\theta) = \sec\theta$	$\cot(-\theta) = -\cot\theta$

7.1 Exercises

Fundamental Identities

Exercises 1–6: Use a reciprocal identity to find the indicated trigonometric function of θ.

1. $\cot\theta$ if $\tan\theta = \frac{1}{2}$

2. $\csc\theta$ if $\sin\theta = -\frac{5}{6}$

3. $\sec\theta$ if $\cos\theta = \frac{2}{7}$

4. $\tan\theta$ if $\cot\theta = -\frac{3}{7}$

5. $\cos\theta$ if $\sec\theta = -4$

6. $\sin\theta$ if $\csc\theta = 7$

Exercises 7–14: Find the other trigonometric functions of θ.

7. $\sin\theta = \frac{3}{5}$ and $\cos\theta = -\frac{4}{5}$

8. $\tan\theta = -\frac{12}{5}$ and $\cos\theta = \frac{5}{13}$

9. $\cot\theta = \frac{7}{24}$ and $\sin\theta = -\frac{24}{25}$

10. $\sin\theta = \frac{12}{13}$ and $\cos\theta = \frac{5}{13}$

11. $\sin\theta = -\frac{60}{61}$ and $\cos\theta = -\frac{11}{61}$

12. $\sin\theta = \frac{2}{3}$ and $\cos\theta = -\frac{\sqrt{5}}{3}$

13. $\csc\theta = \sqrt{2}$ and $\sec\theta = -\sqrt{2}$

14. $\csc\theta = -\frac{13}{12}$ and $\sec\theta = -\frac{13}{5}$

Exercises 15–34: (Refer to Example 4.) Simplify each expression using fundamental identities.

15. $\sec\theta\cos\theta$

16. $\tan\theta\cot\theta$

17. $\sin\theta\csc\theta$

18. $\tan\theta\cos\theta$

19. $(\sin^2\theta + \cos^2\theta)^3$

20. $(1 + \tan^2\theta)\cos^2\theta$

21. $1 - \sin^2\theta$

22. $1 - \cos^2(-\theta)$

23. $\sec^2\theta - 1$

24. $1 + \dfrac{\cos^2\theta}{\sin^2\theta}$

25. $\dfrac{\sin(-\theta)}{\cos(-\theta)}$

26. $\sin\theta(\csc\theta + \sec\theta)$

27. $\dfrac{\sin^2\theta + \cos^2\theta}{\cos\theta}$

28. $\dfrac{1 + \tan^2\theta}{\sec^2\theta}$

29. $\dfrac{\sec^2(-\theta)}{\csc^2\theta}$

30. $\dfrac{1 - \cos^2\theta}{\sin^2\theta + \cos^2\theta}$

31. $\dfrac{\cot x}{\csc x}$

32. $\dfrac{\tan x}{\sec x}$

33. $(\sin^2 x)(1 + \cot^2 x)$

34. $\dfrac{\cos x}{\sin x\cot x}$

Exercises 35–40: (Refer to Example 6.) Determine the quadrant that contains θ. Support your result either graphically or numerically, if you have a graphing calculator.

35. $\sin\theta < 0$ and $\cos\theta > 0$

36. $\tan\theta > 0$ and $\cos\theta < 0$

37. $\sec\theta < 0$ and $\sin\theta < 0$

38. $\csc\theta > 0$ and $\tan\theta > 0$

39. $\cot\theta < 0$ and $\sin\theta > 0$

40. $\cos\theta > 0$ and $\cot\theta < 0$

Exercises 41–52: Find the other trigonometric functions of θ.

41. $\cos\theta = \frac{1}{2}$ and $\sin\theta < 0$

42. $\csc\theta = \sqrt{3}$ and $\cos\theta < 0$

43. $\tan\theta = -\frac{11}{60}$ and $\csc\theta < 0$

44. $\sec\theta = -\frac{5}{4}$ and $\sin\theta > 0$

45. $\sin\theta = \frac{7}{25}$ and $\cos\theta > 0$

46. $\cot\theta = \frac{12}{5}$ and $\sec\theta < 0$

47. $\sin\theta = -\frac{1}{3}$ and $\sec\theta < 0$

48. $\tan\theta = -\frac{1}{2}$ and $\sin\theta > 0$

49. $\sec\theta = 3$ and θ in quadrant IV

50. $\tan\theta = \frac{3}{4}$ and θ in quadrant I

51. $\csc\theta = \frac{7}{3}$ and θ in quadrant II

52. $\cos\theta = -\frac{3}{5}$ and θ in quadrant III

53. Write $\sin x$ and $\tan x$ in terms of $\cos x$, if $\csc x > 0$.

54. Write $\cos x$ and $\sec x$ in terms of $\sin x$, if $\cos x < 0$.

55. Write $\sin x$ and $\sec x$ in terms of $\tan x$, if $\cos x < 0$.

56. Write $\cos x$ and $\csc x$ in terms of $\cot x$, if $\sin x > 0$.

57. Write $\cot x$ and $\cos x$ in terms of $\csc x$, if $\cot x < 0$.

58. Write $\tan x$ and $\sin x$ in terms of $\sec x$, if $\tan x > 0$.

Exercises 59–62: (Refer to Example 10.) Write the trigonometric function in terms of x. Then evaluate this trigonometric function if $x = 0.5126$.

59. $\cos\theta$, if $\sin\theta = x$ and θ is acute

60. $\sec\theta$, if $\cos\theta = x$

61. $\sin\theta$, if $\cot\theta = x$ and θ is in quadrant III

62. $\tan\theta$, if $\cos\theta = x$ and θ is in quadrant IV

63. Write $\tan\theta$ in terms of x if $\sin\theta = x$ and θ is acute.

64. Write $\sin\theta$ in terms of x if $\sec\theta = x$ and θ is acute.

Exercises 65–70: Use a negative-angle identity to write an equivalent trigonometric expression involving a positive angle.

65. $\sin(-13°)$

66. $\cos\left(-\dfrac{\pi}{7}\right)$

67. $\tan\left(-\dfrac{\pi}{11}\right)$

68. $\cot(-75°)$

69. $\sec\left(-\dfrac{2\pi}{5}\right)$

70. $\csc(-160°)$

Exercises 71–76: Determine if the equation represents an identity. If you have a graphing calculator, support your answer by making a table of the left and right sides of the equation.

71. $\sin^2\theta + \cos^2\theta + \tan^2\theta = \sec^2\theta$

72. $\sec\theta\cot\theta = \csc\theta$

73. $(\sin\theta + \cos\theta)^2 = 1$

74. $\cot^2 \theta - \csc^2 \theta = 1$ **75.** $1 + \tan^2 \theta = \dfrac{1}{\cos^2 \theta}$

76. $1 + \cot \theta = \csc \theta$

77. A student writes "$\cos^2 + \sin^2 = 1$." Comment on the correctness of this expression.

78. Since $\sec^2 \theta = 1 + \tan^2 \theta$ is an identity, does it follow that $\sec \theta = 1 + \tan \theta$? Explain your reasoning.

Applications

79. *Distance to the Stars*   In Example 6, Section 6.2 the distance d to a star was found by using

$$d = \frac{93{,}000{,}000}{\sin \theta},$$

where θ is the parallax of the star. Use a reciprocal identity to rewrite this formula.

80. *Height of a Building*   If the angle of elevation of the sun is θ, then a building 40 feet high will cast a shadow x feet long, where $x = \frac{40}{\tan \theta}$. Use a reciprocal identity to rewrite this formula.

81. *Oscillating Spring*   The distance or displacement y of a weight attached to an oscillating spring from its natural position is modeled by $y = 4 \cos(2\pi t)$, where t is in seconds. See the accompanying figure. Potential energy is the energy of position and is given by $P = ky^2$, where k is a constant. The weight has the greatest potential energy when the spring is stretched or compressed the most.   (**Source:** R. Weidner and R. Sells, *Elementary Classical Physics*, Vol. 1.)

  (a) Write an expression for P that involves the cosine function.

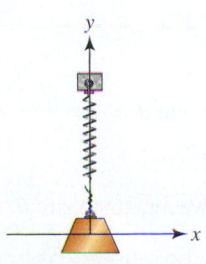 **(b)** Let $k = 2$ and graph P in the viewing rectangle $[0, 2, 0.5]$ by $[-1, 40, 8]$. For $0 \le t \le 2$, at what times is P maximum and at what times is P minimum? Interpret your result.

  (c) Use a fundamental identity to write P in terms of the sine function.

82. *Energy in an Oscillating Spring*   (Refer to the previous exercise.) Two types of mechanical energy are kinetic energy and potential energy. Kinetic energy is the energy of motion and potential energy is the energy of position. A stretched spring has potential energy, which is converted to kinetic energy when it is released. If the potential energy of a weight attached to a spring is $P(t) = k \cos^2(4\pi t)$, where k is a constant and t is in seconds, then its kinetic energy is given by $K(t) = k \sin^2(4\pi t)$. The total mechanical energy E is given by the equation $E(t) = P(t) + K(t)$.

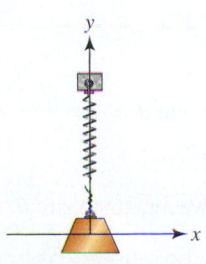 **(a)** If $k = 2$ graph P, K, and E in $[0, 0.5, 0.25]$ by $[-1, 3, 1]$. Interpret the graph.

 (b) Make a table of K, P, and E starting at $t = 0$, incrementing by 0.05. Interpret the results.

  (c) Use a fundamental identity to derive a simplified expression for $E(t)$.

83. *Radio Tuners*   (Refer to Example 5.) Let the energy stored in the inductor be given by the formula $L(t) = 3 \cos^2(6{,}000{,}000t)$ and the energy in the capacitor be given by $C(t) = 3 \sin^2(6{,}000{,}000t)$, where t is time in seconds. The total energy E in the circuit is given by $E(t) = L(t) + C(t)$.

 (a) Graph L, C, and E in the viewing rectangle $[0, 10^{-6}, 10^{-7}]$ by $[-1, 4, 1]$. Interpret the graph.

 (b) Make a table of L, C, and E starting at $t = 0$, incrementing by 10^{-7}. Interpret your results.

  (c) Use a fundamental identity to derive a simplified expression for $E(t)$.

84. *Intensity of a Lamp*   According to Lambert's law, the intensity of light from a single source on a flat surface at point P is given by $I = k \cos^2 \theta$, where k is a constant. See the accompanying figure.   (**Source:** C. Winter, *Solar Power Plants*.)

 (a) Let $k = 1$ and use degree mode to graph I in $[-90, 90, 45]$ by $[-1, 2, 1]$. For what value of θ is I maximum?

  (b) Write I in terms of the sine function.

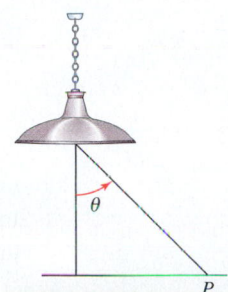

Exercises 85 and 86: *Modeling Temperature (Refer to Example 11.)* *Suppose that the monthly high temperature at a location can be modeled by* $f(x)$, *where* x *is the month with* $x = -6$ *corresponding to January,* $x = 0$ *to July, and* $x = 5$ *to December.*

  (a) *Graph* f *in* $[-6, 6, 1]$ *by* $[0, 100, 10]$. *Interpret any symmetry in the graph.*

  (b) *Make a table of* f *and discuss whether* f *is an even or odd function.*

  (c) *Express any symmetry symbolically.*

85. $f(x) = 40 \cos\left(\dfrac{\pi x}{6}\right) + 50$

86. $f(x) = -15 \cos\left(\dfrac{\pi x}{6}\right) + 60$

Writing about Mathematics

87. Explain in your own words what a trigonometric identity is. Give two examples.

88. Answer each of the following.

  (a) Give two characteristics of an even function. Which of the trigonometric functions are even?

  (b) Give two characteristics of an odd function. Which of the trigonometric functions are odd?

7.2   Verifying Identities

◆ Simplify trigonometric identities

◆ Learn how to verify identities

Introduction

In Example 5 of the previous section we used the identity $\sin^2\theta + \cos^2\theta = 1$ to show that the energy stored in a particular type of electrical circuit is constant. Trigonometric identities are used in both applications and calculus. Before we can use an identity, we must verify that it is correct. Although the equality of two trigonometric expressions can be *supported* graphically and numerically, symbolic *verification* is necessary to be certain that an equation is indeed an identity.

Simplifying Trigonometric Expressions

Many of the algebraic skills that you have already learned can be used to simplify trigonometric expressions. For example, suppose we would like to multiply the following expression.

$$(1 - \cos\theta)(1 + \cos\theta)$$

In algebra we learned that

$$(1 - x)(1 + x) = 1 + x - x - x^2$$
$$= 1 - x^2.$$

If we substitute $\cos\theta$ for x, then

$$(1 - \cos\theta)(1 + \cos\theta) = 1 + \cos\theta - \cos\theta - \cos^2\theta$$
$$= 1 - \cos^2\theta.$$

In algebra we do not simplify $1 - x^2$ further. However, since $\sin^2\theta + \cos^2\theta = 1$, it follows that $\sin^2\theta = 1 - \cos^2\theta$. As a result,

$$(1 - \cos\theta)(1 + \cos\theta) = \sin^2\theta.$$

◆ **MAKING CONNECTIONS**

Algebraic and Trigonometric Expressions   Many of the techniques we use to simplify algebraic expressions can also be used to simplify trigonometric expressions. Here are some examples.

$\tan^2 \theta - 4$

$\quad = (\tan \theta - 2)(\tan \theta + 2)$ is similar to $x^2 - 4 = (x - 2)(x + 2).$

$\cos \theta (\sin \theta + \cos \theta)$

$\quad = \cos \theta \sin \theta + \cos^2 \theta$ is similar to $x(y + x) = xy + x^2.$

$\dfrac{\sin \theta}{\cos \theta} + \dfrac{1}{\cos \theta} = \dfrac{\sin \theta + 1}{\cos \theta}$ is similar to $\dfrac{y}{x} + \dfrac{1}{x} = \dfrac{y + 1}{x}.$

The next example illustrates the addition of two trigonometric expressions.

EXAMPLE 1    **Adding two trigonometric expressions**

Write $\tan t + \cot t$ as a product of two trigonometric functions.

SOLUTION   We begin by writing $\tan t$ and $\cot t$ as ratios involving $\sin t$ and $\cos t.$

$$\tan t + \cot t = \frac{\sin t}{\cos t} + \frac{\cos t}{\sin t}$$

In algebra we combine $\frac{y}{x} + \frac{x}{y}$ by using the common denominator xy as follows.

Algebra Review

To review addition of rational expressions, see Chapter R (page R-42).

$$\frac{y}{x} + \frac{x}{y} = \frac{y}{x} \cdot \frac{y}{y} + \frac{x}{y} \cdot \frac{x}{x} \qquad \textcolor{blue}{\text{Multiply each ratio by 1.}}$$

$$= \frac{y^2}{xy} + \frac{x^2}{xy} \qquad \textcolor{blue}{\text{Simplify.}}$$

$$= \frac{y^2 + x^2}{xy} \qquad \textcolor{blue}{\text{Add.}}$$

Now substitute $\cos t$ for x and $\sin t$ for $y.$

$$\tan t + \cot t = \frac{\sin t}{\cos t} + \frac{\cos t}{\sin t} \qquad \textcolor{blue}{\text{Quotient identities}}$$

$$= \frac{\sin t}{\cos t} \cdot \frac{\mathbf{\sin t}}{\mathbf{\sin t}} + \frac{\cos t}{\sin t} \cdot \frac{\mathbf{\cos t}}{\mathbf{\cos t}} \qquad \textcolor{blue}{\text{Multiply each ratio by 1.}}$$

$$= \frac{\sin^2 t}{\cos t \sin t} + \frac{\cos^2 t}{\cos t \sin t} \qquad \textcolor{blue}{\text{Simplify.}}$$

$$= \frac{\mathbf{\sin^2 t + \cos^2 t}}{\cos t \sin t} \qquad \textcolor{blue}{\text{Add.}}$$

$$= \frac{\mathbf{1}}{\cos t \sin t} \qquad \textcolor{blue}{\sin^2 t + \cos^2 t = 1}$$

$$= \sec t \csc t \qquad \textcolor{blue}{\sec t = \tfrac{1}{\cos t}; \csc t = \tfrac{1}{\sin t}}$$

Thus $\tan t + \cot t$ is equivalent to $\sec t \csc t.$ *Now Try Exercise 11* ◆

Note: To simplify a trigonometric expression, it is *not* necessary to first write the equation using x and y, as was done in Example 1. However, sometimes you may find this technique helpful.

EXAMPLE 2 Factoring a trigonometric expression

Factor each expression.
(a) $\sec^2 \theta - 1$ (b) $2 \sin^2 t + \sin t - 1$

SOLUTION

Algebra Review
To review factoring polynomials,
see Chapter R (page R-27).

(a) In algebra we factor $x^2 - 1$ as $(x - 1)(x + 1)$. This can be applied to the given expression.

$$\sec^2 \theta - 1 = (\sec \theta - 1)(\sec \theta + 1)$$

(b) Since $2y^2 + y - 1$ can be factored as $(2y - 1)(y + 1)$, it follows that

$$2 \sin^2 t + \sin t - 1 = (2 \sin t - 1)(\sin t + 1).$$

Now Try Exercises 25 and 27 ◆

Trigonometric expressions are used in applications involving electricity. This is illustrated in the next example.

EXAMPLE 3 Analyzing electromagnets

Electromagnets are used in a variety of situations, such as lifting scrap metal, ringing door bells, and opening door locks in apartments. Let the wattage W consumed by an electromagnet at t seconds be

$$W(t) = 100 \sin^2 (120 \pi t),$$

and the voltage V in the circuit be

$$V(t) = 160 \cos (120 \pi t).$$

(**Source:** A. Howatson, *Electrical Circuits and Systems*.)
(a) Express $W(t)$ in terms of the cosine function. When V is maximum or minimum, what is the value of W? Explain.
(b) Support your answer in part (a) by graphing W and V in the viewing rectangle $[0, 1/30, 1/60]$ by $[-180, 180, 20]$.

SOLUTION

(a) Let $\theta = 120 \pi t$. Since $\sin^2 \theta = 1 - \cos^2 \theta$, W can be expressed as

$$W(t) = 100(1 - \cos^2 (120 \pi t)).$$

Since $V(t) = 160 \cos (120 \pi t)$, the voltage is maximum (160) or minimum (-160), whenever $\cos (120 \pi t) = \pm 1$. When $\cos (120 \pi t) = \pm 1$, it follows that the value of W is given by $W(t) = 100(1 - (\pm 1)^2) = 0$.

(b) Graph $Y_1 = 100(\sin (120 \pi X))^\wedge 2$ and $Y_2 = 160 \cos (120 \pi X)$, as shown in Figure 7.20. The wattage Y_1 is 0 whenever the voltage Y_2 is ± 160.

Now Try Exercise 75 ◆

[0, 1/30, 1/60] by [−180, 180, 20]

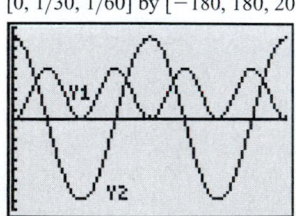

FIGURE 7.20

Verification of Identities

When verifying that an equation is an identity, we usually begin with one side of the equation and write a sequence of equivalent expressions until it is transformed into the other side. This is illustrated in the following examples.

EXAMPLE 4 Verifying an identity

Verify that $\frac{\sin t}{1 - \cos t} = \frac{1 + \cos t}{\sin t}$ is an identity.

SOLUTION

$$\frac{\sin t}{1 - \cos t} = \frac{\sin t}{1 - \cos t} \cdot \frac{1 + \cos t}{1 + \cos t} \qquad \text{Multiply the ratio by 1.}$$

$$= \frac{\sin t (1 + \cos t)}{1 - \cos^2 t} \qquad \text{Simplify.}$$

$$= \frac{\sin t (1 + \cos t)}{\sin^2 t} \qquad \sin^2 t = 1 - \cos^2 t$$

$$= \frac{1 + \cos t}{\sin t} \qquad \text{Simplify.} \qquad \textit{Now Try Exercise 59} \blacklozenge$$

EXAMPLE 5 Verifying an identity

Verify that $(\cos \theta + \sin \theta)^2 = 1 + 2 \sin \theta \cos \theta$.

SOLUTION In algebra the expression $(x + y)^2$ can be expanded as follows.

$$(x + y)^2 = (x + y)(x + y)$$

$$= x^2 + xy + yx + y^2$$

$$= x^2 + 2xy + y^2$$

Algebra Review
To review multiplication of polynomials, see Chapter R (page R-23).

We can perform similar steps with the left side of the trigonometric equation.

$$(\cos \theta + \sin \theta)^2 = (\cos \theta + \sin \theta)(\cos \theta + \sin \theta)$$

$$= \cos^2 \theta + \cos \theta \sin \theta + \sin \theta \cos \theta + \sin^2 \theta$$

$$= \cos^2 \theta + 2 \sin \theta \cos \theta + \sin^2 \theta$$

$$= 1 + 2 \sin \theta \cos \theta$$

The last step is true since $\sin^2 \theta + \cos^2 \theta = 1$.   *Now Try Exercise 33* $\blacklozenge$

EXAMPLE 6 Verifying identities

Verify each identity.

(a) $\dfrac{\sin^2 \theta}{1 + \cos \theta} = 1 - \cos \theta$ **(b)** $\dfrac{\csc t}{\cot t} - \dfrac{\cot t}{\csc t} = \tan t \sin t$

SOLUTION
(a) Begin by applying a Pythagorean identity.

$$\frac{\sin^2 \theta}{1 + \cos \theta} = \frac{1 - \cos^2 \theta}{1 + \cos \theta} \qquad \sin^2 \theta + \cos^2 \theta = 1$$

$$= \frac{(1 - \cos \theta)(1 + \cos \theta)}{1 + \cos \theta} \qquad \text{Factor.}$$

$$= 1 - \cos \theta \qquad \text{Simplify.}$$

(b) Begin by finding a common denominator for the expressions on the left side of the equation.

$$\frac{\csc t}{\cot t} - \frac{\cot t}{\csc t} = \frac{\csc t}{\cot t} \cdot \frac{\mathbf{\csc t}}{\mathbf{\csc t}} - \frac{\cot t}{\csc t} \cdot \frac{\mathbf{\cot t}}{\mathbf{\cot t}} \qquad \text{Find a common denominator.}$$

$$= \frac{\csc^2 t}{\cot t \csc t} - \frac{\cot^2 t}{\csc t \cot t} \qquad \text{Multiply.}$$

$$= \frac{\mathbf{\csc^2 t} - \cot^2 t}{\cot t \csc t} \qquad \text{Subtract the expressions.}$$

$$= \frac{\mathbf{1 + \cot^2 t} - \cot^2 t}{\cot t \csc t} \qquad \csc^2 t = 1 + \cot^2 t$$

$$= \frac{1}{\cot t \csc t} \qquad \text{Simplify.}$$

$$= \frac{1}{\cot t} \cdot \frac{1}{\csc t} \qquad \text{Rewrite the expression.}$$

$$= \tan t \, \sin t \qquad \text{Reciprocal identities}$$

Now Try Exercises 41 and 47 ◆

EXAMPLE 7   Verifying an identity

Verify that $\dfrac{\sec x + \tan x}{\sec x - \tan x} = \dfrac{1 + \sin x}{1 - \sin x}$.

SOLUTION

Because $\sec x = \frac{1}{\cos x}$ and $\tan x = \frac{\sin x}{\cos x}$, we start to simplify the expression by multiplying the numerator and denominator by $\cos x$.

$$\frac{\sec x + \tan x}{\sec x - \tan x} = \frac{(\sec x + \tan x)\mathbf{\cos x}}{(\sec x - \tan x)\mathbf{\cos x}} \qquad \text{Multiply the ratio by 1.}$$

$$= \frac{\sec x \cos x + \tan x \cos x}{\sec x \cos x - \tan x \cos x} \qquad \text{Distributive property}$$

$$= \frac{\frac{1}{\cos x} \cdot \cos x + \frac{\sin x}{\cos x} \cdot \cos x}{\frac{1}{\cos x} \cdot \cos x - \frac{\sin x}{\cos x} \cdot \cos x} \qquad \text{Reciprocal and quotient identities}$$

$$= \frac{1 + \sin x}{1 - \sin x} \qquad \text{Simplify.} \qquad \textit{Now Try Exercise 55} ◆$$

In the last example we show how to verify an identity and then give graphical and numerical support.

EXAMPLE 8   Verifying an identity

Verify that $\frac{\tan \theta}{\sec \theta} = \sin \theta$ symbolically. Give graphical and numerical support.

SOLUTION

Symbolic Verification   We will start with the more complicated expression $\frac{\tan \theta}{\sec \theta}$ and simplify it to $\sin \theta$. Begin by writing $\tan \theta$ and $\sec \theta$ in terms of $\sin \theta$ and $\cos \theta$.

$$\frac{\tan \theta}{\sec \theta} = \frac{\mathbf{\sin \theta / \cos \theta}}{\mathbf{1 / \cos \theta}} \qquad \text{Quotient and reciprocal identities}$$

$$= \frac{\mathbf{\sin \theta}}{\mathbf{\cos \theta}} \cdot \frac{\mathbf{\cos \theta}}{\mathbf{1}} \qquad \text{Invert and multiply.}$$

$$= \sin \theta \qquad \text{Simplify.}$$

These steps verify that $\frac{\tan \theta}{\sec \theta} = \sin \theta$ is an identity.

Graphical Support Graph $Y_1 = \tan(X)/(1/\cos(X))$ and $Y_2 = \sin(X)$, as shown in Figures 7.21 and 7.22. Their graphs appear to be identical.

Numerical Support See Figure 7.23. Notice that when $\theta = \frac{\pi}{2} \approx 1.5708$, the ratio $\frac{\tan\theta}{\sec\theta}$ is undefined, whereas $\sin\theta = 1$. However, the equation $\frac{\tan\theta}{\sec\theta} = \sin\theta$ is nonetheless an identity because $y_1 = y_2$ whenever both expressions are defined.

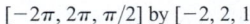
$[-2\pi, 2\pi, \pi/2]$ by $[-2, 2, 1]$ $[-2\pi, 2\pi, \pi/2]$ by $[-2, 2, 1]$

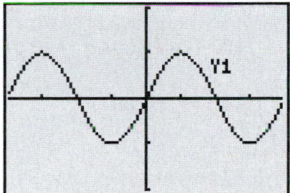
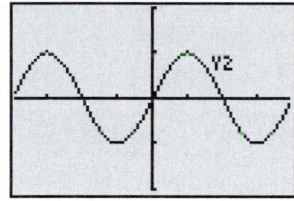

FIGURE 7.21   **FIGURE 7.22**   **FIGURE 7.23** Radian Mode

Now Try Exercise 69 ◆

7.2 Putting it all Together

Becoming proficient at verifying identities requires practice. Many of the skills learned in algebra can be used to help verify identities. The following table lists some suggestions that may be helpful.

Suggestions for Verifying Identities

1. Become familiar with the fundamental identities found in Section 7.1.
2. Use your knowledge of how to simplify algebraic expressions to help guide you. This is particularly true when factoring or combining ratios.
3. When verifying an identity, start by simplifying the more complicated side of the equation. Otherwise, choose a side of the equation that you can transform into a different expression.
4. If you are simplifying the left side of the equation, then work toward making the left side appear more like the right. For example, if the left side contains an addition sign but the right side does not, then add the terms on the left side.
5. If you are uncertain how to proceed, one strategy is to write each trigonometric function in terms of sine and cosine and then simplify.
6. If a ratio contains $1 + \sin\theta$, it is sometimes helpful to multiply the numerator and denominator by $1 - \sin\theta$. Then

$$(1 + \sin\theta)(1 - \sin\theta) = 1 - \sin^2\theta = \cos^2\theta.$$

Similar statements can be made for $1 - \sin\theta$, $1 + \cos\theta$, and $1 - \cos\theta$. See Example 4.

7.2 Exercises

Simplifying Expressions

Exercises 1–4: Multiply the algebraic expression. Then multiply the corresponding trigonometric expression. If possible, simplify the resulting trigonometric expression.

1. (a) $(1 + x)(1 - x)$ **(b)** $(1 + \sin \theta)(1 - \sin \theta)$

2. (a) $(x - 1)(x + 1)$ **(b)** $(\csc \theta - 1)(\csc \theta + 1)$

3. (a) $x(x - 1)$ **(b)** $\sec \theta(\sec \theta - 1)$

4. (a) $(x + 1)(2x - 1)$ **(b)** $(\tan \theta + 1)(2 \tan \theta - 1)$

Exercises 5–8: Factor the algebraic expression. Then factor the corresponding trigonometric expression.

5. (a) $x^2 + 2x + 1$ **(b)** $\cos^2 \theta + 2 \cos \theta + 1$

6. (a) $2x^2 - 3x + 1$ **(b)** $2 \sin^2 t - 3 \sin t + 1$

7. (a) $x^2 - 2x$ **(b)** $\sec^2 t - 2 \sec t$

8. (a) $3x - 9x^2$ **(b)** $3 \tan \theta - 9 \tan^2 \theta$

Exercises 9–16: Simplify the algebraic expression. Then simplify the corresponding trigonometric expression. If possible, simplify the resulting trigonometric expression.

9. (a) $\dfrac{1}{1 - x} + \dfrac{1}{1 + x}$ **(b)** $\dfrac{1}{1 - \cos \theta} + \dfrac{1}{1 + \cos \theta}$

10. (a) $x + \dfrac{1}{x}$ **(b)** $\tan t + \dfrac{1}{\tan t}$

11. (a) $\dfrac{x}{y} + \dfrac{y}{x}$ **(b)** $\dfrac{\cos t}{\sin t} + \dfrac{\sin t}{\cos t}$

12. (a) $\dfrac{1}{y} - \dfrac{x^2}{y}$ **(b)** $\dfrac{1}{\sin \theta} - \dfrac{\cos^2 \theta}{\sin \theta}$

13. (a) $\dfrac{1}{1/y^2} + \dfrac{1}{1/x^2}$ **(b)** $\dfrac{1}{\csc^2 t} + \dfrac{1}{\sec^2 t}$

14. (a) $\left(\dfrac{1}{x} + x\right)^2$ **(b)** $(\cot \theta + \tan \theta)^2$

15. (a) $\dfrac{x/y}{1/y}$ **(b)** $\dfrac{\cot \theta}{\csc \theta}$

16. (a) $\dfrac{1 - x^2}{1 + x}$ **(b)** $\dfrac{1 - \cos^2 \theta}{1 + \cos \theta}$

Exercises 17–24: Perform the indicated operations and simplify.

17. $\cos \theta \tan \theta$ **18.** $\sin^2 \theta \csc \theta$

19. $\tan \theta (\cos \theta - \csc \theta)$ **20.** $(\sin \theta - \cos \theta)^2$

21. $(1 + \tan t)^2$ **22.** $(\sin t - 1)(\sin t + 1)$

23. $\dfrac{\csc^2 \theta - 1}{\csc^2 \theta}$ **24.** $\sin^2 t (1 + \cot^2 t)$

Exercises 25–30: Factor the trigonometric expression and simplify, if possible.

25. $1 - \tan^2 \theta$ **26.** $\sin^2 t - \cos^2 t$

27. $\sec^2 t - \sec t - 6$ **28.** $\cos \theta \sin^2 \theta + \cos^3 \theta$

29. $\tan^4 \theta + 3 \tan^2 \theta + 2$ **30.** $\sin^4 t - \cos^4 t$

Verifying Identities

Exercises 31–66: Verify the identity.

31. $\csc^2 \theta - \cot^2 \theta = 1$ **32.** $\dfrac{\tan^2 \theta + 1}{\sec \theta} = \sec \theta$

33. $(1 - \sin t)^2 = 1 - 2 \sin t + \sin^2 t$

34. $\dfrac{\sin^2 t}{\cos t} = \sec t - \cos t$ **35.** $\dfrac{\sin t + \cos t}{\sin t} = 1 + \cot t$

36. $\sec^4 \theta - \sec^2 \theta = \tan^4 \theta + \tan^2 \theta$

37. $\sec^2 \theta - 1 = \tan^2 \theta$ **38.** $\dfrac{\csc^2 \theta}{\cot \theta} = \csc \theta \sec \theta$

39. $\dfrac{\tan^2 t}{\sec t} = \sec t - \cos t$ **40.** $\dfrac{\sec^2 \theta - 1}{\sec^2 \theta} = \sin^2 \theta$

41. $\dfrac{\sec t}{1 + \sec t} = \dfrac{1}{\cos t + 1}$

42. $\sec^2 t + \csc^2 t = \sec^2 t \csc^2 t$

43. $(\sec t - 1)(\sec t + 1) = \tan^2 t$

44. $\csc^4 \theta - \cot^4 \theta = \csc^2 \theta + \cot^2 \theta$

45. $\dfrac{1 - \sin^2 \theta}{\cos \theta} = \cos \theta$ **46.** $\dfrac{\tan^2 t - 1}{1 + \tan^2 t} = 1 - 2 \cos^2 t$

47. $\dfrac{\sec t}{\tan t} - \dfrac{\tan t}{\sec t} = \cos t \cot t$

48. $\dfrac{\sin^4 t - \cos^4 t}{\sin^2 t - \cos^2 t} = 1$ **49.** $\dfrac{\cot^2 t}{\csc t + 1} = \csc t - 1$

50. $\sec \theta - \cos \theta = \tan \theta \sin \theta$ **51.** $\dfrac{\cot t}{\cot t + 1} = \dfrac{1}{1 + \tan t}$

52. $\cos^4 t - \sin^4 t = 2 \cos^2 t - 1$

53. $\dfrac{1}{1 - \sin t} + \dfrac{1}{1 + \sin t} = 2 \sec^2 t$

54. $\cot \theta + \tan \theta = \csc \theta \sec \theta$

55. $\dfrac{\csc t + \cot t}{\csc t - \cot t} = (\csc t + \cot t)^2$

56. $\dfrac{\csc t}{1 + \csc t} - \dfrac{\csc t}{1 - \csc t} = 2 \sec^2 t$

57. $\dfrac{\cos^2 t}{1 - \sin t} = 1 + \sin t$ **58.** $\csc t + \dfrac{\sec t}{\tan t} = \dfrac{2}{\sin t}$

59. $\dfrac{1}{1 + \sin \theta} = \dfrac{1 - \sin \theta}{\cos^2 \theta}$

60. $\dfrac{2 \sin^2 t + 3 \sin t - 2}{\sin t + 2} = 2 \sin t - 1$

61. $\sqrt{1 - \sin^2 \theta} = \cos \theta$, where θ is acute

62. $\sqrt{\sec^2 \theta - 1} = \tan \theta$, where θ is acute

63. $\dfrac{1 + 2 \sin x + \sin^2 x}{\cos^2 x} = \dfrac{1 + \sin x}{1 - \sin x}$

64. $\dfrac{\tan t - \cot t}{\sin t \cos t} = \sec^2 t - \csc^2 t$

65. $(1 - \cos^2 x)(1 + \cos^2 x) = 2 \sin^2 x - \sin^4 x$

66. $\sin^4 x - \cos^4 x = 2 \sin^2 x - 1$

Exercises 67–74: Verify the identity. Give graphical or numerical support if you have a graphing calculator.

67. $\cot \theta \sin \theta = \cos \theta$ **68.** $\tan \theta \cos \theta = \sin \theta$

69. $(1 - \cos^2 \theta)(1 + \tan^2 \theta) = \tan^2 \theta$

70. $\cos^2 \theta (1 + \cot^2 \theta) = \cot^2 \theta$

71. $\cos t (\tan t - \sec t) = \sin t - 1$

72. $\dfrac{\cos \theta}{1 - \sin \theta} = \sec \theta + \tan \theta$ **73.** $\dfrac{\tan (-\theta)}{\sin (-\theta)} = \sec \theta$

74. $\tan^2 t - \sin^2 t = \tan^2 t \sin^2 t$

Application

75. *Electromagnets*   (Refer to Example 3.) Let the wattage consumed by an electromagnet be given by the formula $W(t) = 5 \cos^2 (120 \pi t)$ and the voltage be given by the formula $V(t) = 25 \sin (120 \pi t)$, where t is in seconds.
  (a) Express $W(t)$ in terms of the sine function. When V is maximum or minimum, what is the value of W? Explain.

  (b) Support your answer in part (a) by graphing W and V in $[0, 1/15, 1/60]$ by $[-30, 30, 10]$.

Writing about Mathematics

76. Create a trigonometric identity of your own. Verify the identity symbolically and then give graphical and numerical support.

77. Explain how to show that an equation is not an identity. Give an example.

CHECKING BASIC CONCEPTS FOR SECTIONS 7.1 AND 7.2

1. Determine the quadrant containing θ if $\cot \theta > 0$ and $\sin \theta < 0$.

2. Find the other trigonometric functions of θ using the given information.
  (a) $\sin \theta = \dfrac{5}{13}$ and $\cos \theta = -\dfrac{12}{13}$

  (b) $\sec \theta = \dfrac{5}{4}$ and $\sin \theta < 0$

  (c) $\tan \theta = -\dfrac{1}{2}$ and $\cos \theta = \dfrac{2}{\sqrt{5}}$

3. Simplify each expression.
  (a) $(1 - \sin \theta)(1 + \sin \theta)$ **(b)** $\tan^2 t \csc^2 t - 1$

4. Factor the trigonometric expression.
  (a) $\tan^2 t - 1$ **(b)** $3 \sin^2 t + \sin t - 2$

5. Verify each identity.
  (a) $(1 - \sin^2 \theta)(1 + \cot^2 \theta) = \cot^2 \theta$

  (b) $\dfrac{\cot^2 t}{\csc t} = \csc t - \sin t$

7.3 Trigonometric Equations

◆ Find and use reference angles

◆ Solve trigonometric equations

◆ Solve inverse trigonometric equations

Introduction

In previous chapters we learned how applications that involve functions result in the need to solve equations. In a similar manner, applications that involve trigonometric functions result in the need to solve trigonometric equations. In Example 10, Section 6.5 we learned that the number of daylight hours near 30°N latitude can be modeled by

$$f(x) = 1.95 \cos\left(\frac{\pi}{6}(x - 6.6)\right) + 12.15,$$

where $x = 1$ corresponds to January 1, $x = 2$ to February 1, and so on. To estimate when there are 11 hours of daylight we can solve the trigonometric equation $f(x) = 11$ or

$$1.95 \cos\left(\frac{\pi}{6}(x - 6.6)\right) + 12.15 = 11.$$

Like other types of equations, trigonometric equations can be solved graphically, numerically, and symbolically. We begin by discussing reference angles, which will be used to solve trigonometric equations.

Reference Angles

A **reference angle** for an angle θ, written θ_R, is the acute angle made by the terminal side of θ and the x-axis. It is assumed that θ is in standard position and its terminal side does not lie on either the x- or y-axis. Examples of reference angles in the four quadrants are shown in Figures 7.24–7.27.

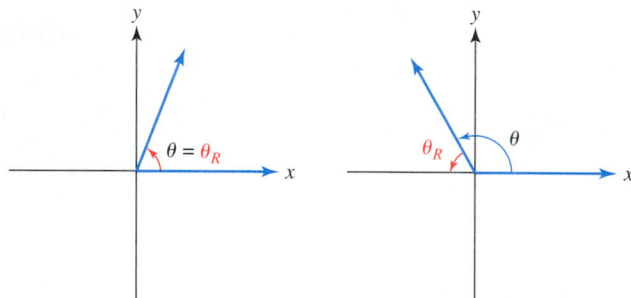

FIGURE 7.24 Quadrant I   **FIGURE 7.25** Quadrant II

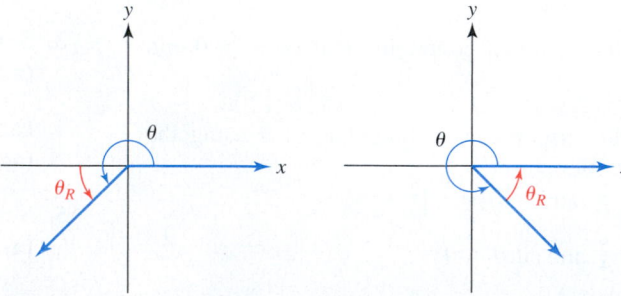

FIGURE 7.26 Quadrant III   **FIGURE 7.27** Quadrant IV

EXAMPLE 1 Finding reference angles

Find the reference angle for θ.

(a) $\theta = 43°$ (b) $\theta = \dfrac{2\pi}{3}$ (c) $\theta = -55°$ (d) $\theta = -\dfrac{3\pi}{4}$

SOLUTION

(a) Since θ is in quadrant I, θ and θ_R are equal. Thus $\theta_R = 43°$.

(b) The terminal side of $\theta = \frac{2\pi}{3}$ lies in quadrant II. This is similar to Figure 7.25. In this case, the acute angle between the terminal side of θ and the x-axis is given by

$$\theta_R = \pi - \theta = \pi - \frac{2\pi}{3} = \frac{\pi}{3}.$$

(c) The terminal side of $\theta = -55°$ lies in quadrant IV and the acute angle between it and the x-axis is $\theta_R = 55°$. See Figure 7.28.

(d) The terminal side of $\theta = -\frac{3\pi}{4}$ (or $-135°$) lies in quadrant III and the acute angle between it and the x-axis is $\theta_R = \frac{\pi}{4}$. See Figure 7.29.

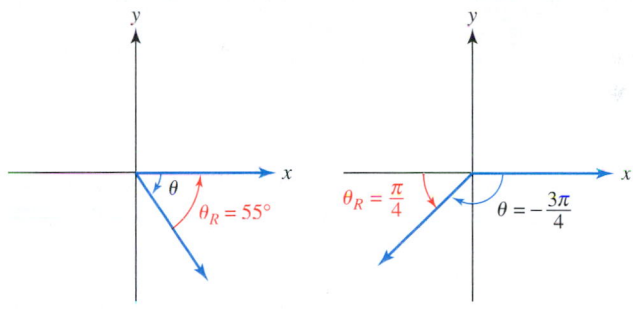

FIGURE 7.28   **FIGURE 7.29**

Now Try Exercises 1 and 9 ◆

The reference angle is important because it can help determine trigonometric values.

REFERENCE ANGLES AND TRIGONOMETRIC FUNCTIONS

Let θ be an angle in standard position with reference angle θ_R. Then

$$|\sin\theta| = \sin\theta_R \qquad |\cos\theta| = \cos\theta_R \qquad |\tan\theta| = \tan\theta_R$$
$$|\csc\theta| = \csc\theta_R \qquad |\sec\theta| = \sec\theta_R \qquad |\cot\theta| = \cot\theta_R.$$

The signs of the trigonometric functions of θ are determined by the quadrant that contains θ. (See Figure 7.9.)

EXAMPLE 2 Solving trigonometric equations using reference angles

Solve the following equations.

(a) $\sin\theta = -\frac{1}{2}$ for θ in $[0°, 360°)$. (b) $\tan\theta = 1$ for θ in $[0, 2\pi)$.

SOLUTION

(a) We start by solving the equation $\sin\theta_R = \frac{1}{2}$. The solution to this equation is $\theta_R = \sin^{-1}\frac{1}{2} = 30°$. The sine function is negative in quadrants III and IV. See Figure 7.9.

Therefore the solution to $\sin\theta = -\frac{1}{2}$ is an angle θ located in either quadrant III or quadrant IV that has a reference angle of 30°. There are two such angles: 210° and 330°. See Figures 7.30 and 7.31.

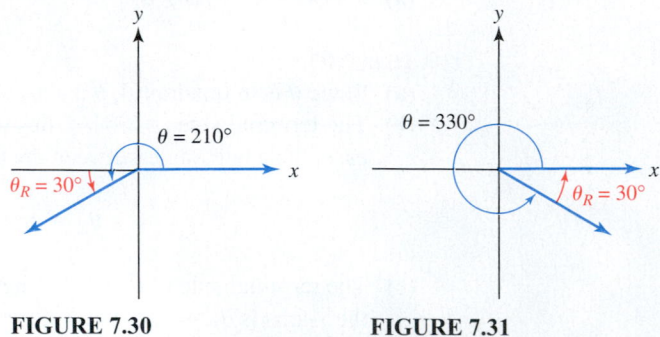

FIGURE 7.30   FIGURE 7.31

◆ **CLASS DISCUSSION**

Let $0 < \theta < 2\pi$. Find expressions for θ_R given the quadrant containing θ. ◆

(b) The solution to $\tan\theta_R = 1$ is $\theta_R = \tan^{-1} 1 = \frac{\pi}{4}$. The tangent function is positive in quadrants I and III. See Figure 7.9. Thus $\frac{\pi}{4}$ and $\frac{5\pi}{4}$ are solutions. See Figures 7.32 and 7.33.

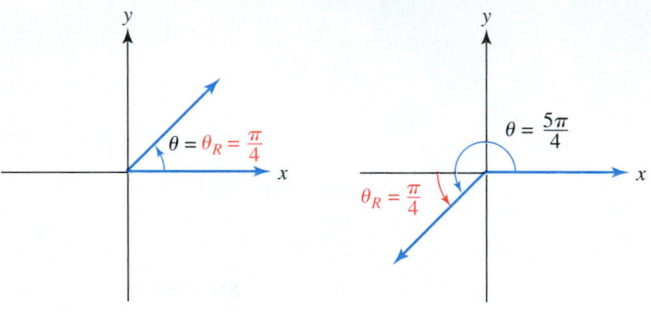

FIGURE 7.32   FIGURE 7.33

Now Try Exercises 15 and 17 ◆

Solving Trigonometric Equations

In the previous section we verified trigonometric *identities*. Identities are equations that are true for all meaningful values of the variable. In this section we discuss trigonometric equations that are *conditional*. Conditional equations are satisfied by some but not all values of the variable. For example,

$$\cos\theta = 1$$

is a conditional trigonometric equation since the cosine function equals 1 only for certain values of θ, such as $\theta = 0°$ or $\theta = 360°$.

 **EXAMPLE 3**   Solving trigonometric equations

Solve $2\sin\theta - 1 = 1$ on the interval $[0°, 360°)$ symbolically and graphically.

SOLUTION

Symbolic Solution   Begin by solving the given equation for $\sin\theta$.

$$2\sin\theta - 1 = 1 \qquad \textcolor{blue}{\text{Given equation}}$$
$$2\sin\theta = 2 \qquad \textcolor{blue}{\text{Add 1.}}$$
$$\sin\theta = 1 \qquad \textcolor{blue}{\text{Divide by 2.}}$$

The only solution to $\sin\theta = 1$ on the interval $[0°, 360°)$ is $\theta = \sin^{-1} 1 = 90°$.

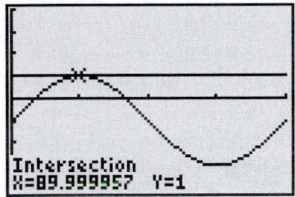

[0, 360, 90] by [−4, 4, 1]

FIGURE 7.34 Degree Mode

Graphical Solution   Graph $Y_1 = 2 \sin(X) - 1$ and $Y_2 = 1$, as in Figure 7.34. Their graphs intersect at $x = 90°$.

Now Try Exercise 23 ◆

In the next example we find all the phase angles associated with a particular phase F of the moon. The fraction of the moon that appears illuminated is called the *phase F*. The *phase angle* θ is shown in Figure 7.35. Refer to Example 13, Section 6.6.   (**Source:** M. Zeilik et al., *Introductory Astronomy and Astrophysics*.)

FIGURE 7.35 Phase Angle θ

Finding phase angles for the moon

The phase F associated with a phase angle θ is given by

$$F = \frac{1}{2}(1 - \cos\theta).$$

Find all phase angles θ in degrees when $F = 0.75$. (Note that $F = 0.5$ corresponds to a quarter moon and $F = 1$ corresponds to a full moon.)

SOLUTION   Let $F = 0.75$ and solve the given equation.

$$0.75 = \frac{1}{2}(1 - \cos\theta) \qquad \text{Let } F = 0.75.$$

$$1.5 = 1 - \cos\theta \qquad \text{Multiply by 2.}$$

$$\cos\theta = -0.5 \qquad \text{Solve for } \cos\theta.$$

Start by solving the equation $\cos\theta_R = 0.5$. The solution is $\theta_R = \cos^{-1} 0.5 = 60°$. The cosine function is negative in quadrants II and III. Angles in these quadrants with a 60° reference angle are 120° and 240°. Since the cosine function has period 360°, all solutions can be written in the form

$$\theta = 120° + 360° \cdot n \qquad \text{or} \qquad \theta = 240° + 360° \cdot n,$$

where n is an integer. For example, 120°, 120° ± 360°, and 120° ± 720° are solutions, as well as 240°, 240° ± 360°, and 240° ± 720°.

Now Try Exercise 93 ◆

Many of the techniques used to solve polynomial equations can be applied to trigonometric equations, as illustrated in the next example.

EXAMPLE 5

Solving trigonometric equations

Find all solutions to each equation. Express your results in radians.
(a) $2 \cot t + 1 = -1$ **(b)** $2 \sin^2 t - 5 \sin t + 2 = 0$

SOLUTION

(a) In algebra, the equation $2x + 1 = -1$ implies $x = -1$. In a similar manner,

$$2 \cot t + 1 = -1 \quad \text{implies} \quad \cot t = -1.$$

If $\cot t = -1$, then $\tan t = -1$ and t has a reference angle of $\tan^{-1} 1 = \frac{\pi}{4}$. The cotangent is negative in quadrants II and IV, so the solutions to $\cot t = -1$ in $[0, 2\pi)$ are $t = \frac{3\pi}{4}$ and $t = \frac{7\pi}{4}$. Note that these angles both have a reference angle of $\frac{\pi}{4}$. Since cotangent has a period of π, all solutions can be expressed in the form

$$t = \frac{3\pi}{4} + \pi n \qquad \text{or} \qquad t = \frac{7\pi}{4} + \pi n,$$

where n is an integer. These solutions are equivalent to just $t = \frac{3\pi}{4} + \pi n$ since the difference between $\frac{7\pi}{4}$ and $\frac{3\pi}{4}$ is π.

(b) In algebra, the equation $2x^2 - 5x + 2 = 0$ can be solved by factoring.

Algebra Review
To review factoring of trinomials, see Chapter R (page R-27).

$$2x^2 - 5x + 2 = (2x - 1)(x - 2) = 0$$

The solutions are $x = \frac{1}{2}$ and 2. In a similar manner, we can factor a trigonometric expression.

$$2 \sin^2 t - 5 \sin t + 2 = (2 \sin t - 1)(\sin t - 2) = 0$$

We must solve the equations $\sin t = \frac{1}{2}$ and $\sin t = 2$. If $\sin t = \frac{1}{2}$, the reference angle is $\sin^{-1} \frac{1}{2} = \frac{\pi}{6}$. The sine function is positive in quadrants I and II, so the solutions in $[0, 2\pi)$ are $t = \frac{\pi}{6}$ and $t = \frac{5\pi}{6}$. Since $-1 \leq \sin t \leq 1$ for all t, the equation $\sin t = 2$ has no solutions. The sine function has period 2π, and all solutions to the given equation can be expressed as

$$t = \frac{\pi}{6} + 2\pi n \qquad \text{or} \qquad t = \frac{5\pi}{6} + 2\pi n.$$

Now Try Exercises 43 and 45 ◆

◆ **MAKING CONNECTIONS**

Polynomial and Trigonometric Equations A polynomial equation of degree n has at most n solutions. However, a trigonometric equation typically has an infinite number of solutions, as demonstrated in Examples 4 and 5.

In the next example we solve an equation for all real numbers t, where the argument of the trigonometric function is $4t$.

 **EXAMPLE 6** Solving a trigonometric equation

Solve $-0.6 \sin 4t = 0.3$, where t is any real number.

SOLUTION First we let $\theta = 4t$. Then the given equation becomes

$$-0.6 \sin \theta = 0.3.$$

Next we solve this modified equation for all real numbers θ.

$$-0.6 \sin \theta = 0.3 \qquad \text{Let } \theta = 4t.$$

$$\sin \theta = -\frac{1}{2} \qquad \text{Divide by } -0.6.$$

From Example 2(a) the solutions to $\sin \theta = -\frac{1}{2}$ on $[0°, 360°)$ are $210°$ and $330°$. In radian measure these solutions are $\theta = \frac{7\pi}{6}$ and $\frac{11\pi}{6}$. Thus all real-number solutions to $-0.6 \sin \theta = 0.3$ are

$$\theta = \frac{7\pi}{6} + 2\pi n \qquad \text{or} \qquad \theta = \frac{11\pi}{6} + 2\pi n,$$

where n is an integer. Because $\theta = 4t$, we can determine t by substituting $4t$ for θ.

$$4t = \frac{7\pi}{6} + 2\pi n \qquad \text{or} \qquad 4t = \frac{11\pi}{6} + 2\pi n.$$

Finally we divide each equation by 4 to obtain

$$t = \frac{7\pi}{24} + \frac{\pi n}{2} \qquad \text{or} \qquad t = \frac{11\pi}{24} + \frac{\pi n}{2}. \qquad \textit{Now Try Exercise 55} \blacklozenge$$

Solar power plants are interested in the number of daylight hours during different times of the year and at different latitudes. In the next example, we solve the trigonometric equation presented in the introduction to determine when there are 11 hours of daylight at 30°N latitude.

 **EXAMPLE 7** ### Analyzing daylight hours

The number of daylight hours at 30°N latitude can be modeled by

$$f(x) = 1.95 \cos\left(\frac{\pi}{6}(x - 6.6)\right) + 12.15,$$

where $x = 1$ corresponds to January 1, $x = 2$ to February 1, and so on. Estimate graphically and numerically when there are 11 hours of daylight.

SOLUTION

Graphical Solution   Graph $Y_1 = 1.95 \cos(\pi/6(X - 6.6)) + 12.15$ and $Y_2 = 11$ in radian mode. Their graphs intersect near $x = 2.4$ and $x = 10.8$. See Figures 7.36 and 7.37. These values correspond to about February 11 and October 25. (Note that four tenths of February is $0.4 \times 28 \approx 11$ days and eight tenths of October is $0.8 \times 31 \approx 25$ days.)

Numerical Solution   Make a table of Y_1, starting at $x = 2$, incrementing by 0.1. The table in Figure 7.38 shows that $Y_1 \approx 11$ when $x = 2.4$. By scrolling down the table, $Y_1 \approx 11$ when $x = 10.8$.

[0, 13, 1] by [8, 16, 1]   [0, 13, 1] by [8, 16, 1]

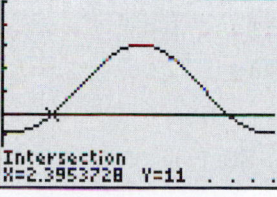

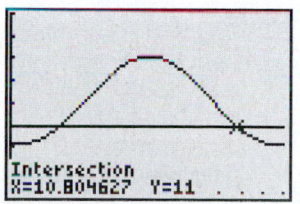

X	Y₁	Y₂
2	10.701	11
2.1	10.771	11
2.2	10.845	11
2.3	10.923	11
2.4	11.004	11
2.5	11.088	11
2.6	11.175	11

X=2.4

FIGURE 7.36   **FIGURE 7.37**   **FIGURE 7.38** Radian Mode

Now Try Exercise 95 $\blacklozenge$

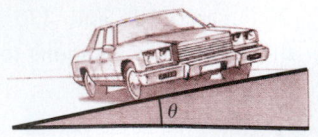

FIGURE 7.39

Highway curves are sometimes banked so that the outside of the curve is slightly elevated or inclined, as shown in Figure 7.39. This inclination is called the *superelevation*. The relationship between a car's velocity v in feet per second, the safe radius r of the curve in feet, and the superelevation θ in degrees is given by

$$r = \frac{v^2}{4.5 + 32.2 \tan \theta}.$$

(**Source:** F. Mannering and W. Kilareski, *Principles of Highway Engineering and Traffic Analysis.*)

EXAMPLE 8 Determining superelevation for a highway curve

A highway curve with a radius of 700 feet and a speed limit of 88 feet per second (60 mph) is being designed. Find the appropriate superelevation for the curve.

SOLUTION Let $r = 700$ and $v = 88$ and then solve the equation for θ.

$$700 = \frac{88^2}{4.5 + 32.2 \tan \theta} \qquad \text{Let } r = 700 \text{ and } v = 88.$$

$$4.5 + 32.2 \tan \theta = \frac{88^2}{700} \qquad \text{Properties of ratios}$$

$$32.2 \tan \theta = \frac{88^2}{700} - 4.5 \qquad \text{Subtract 4.5.}$$

$$\tan \theta = \frac{88^2/700 - 4.5}{32.2} \qquad \text{Divide by 32.2.}$$

$$\tan \theta \approx 0.2038 \qquad \text{Approximate.}$$

$$\theta \approx \tan^{-1} 0.2038 \approx 11.5° \qquad \text{Apply the inverse tangent.}$$

The superelevation should be about 11.5°. *Now Try Exercise 94* ◆

Some trigonometric equations contain more than one type of trigonometric function. In these situations it is sometimes helpful to use trigonometric identities to rewrite the equation in terms of one trigonometric function. This is illustrated in the next example.

EXAMPLE 9 Solving a trigonometric equation

Solve $2 \tan \theta = \sec^2 \theta$ symbolically on the interval $[0, 2\pi)$.

SOLUTION This equation contains two different trigonometric functions. We begin by applying the identity $1 + \tan^2 \theta = \sec^2 \theta$ to rewrite the equation only in terms of $\tan \theta$.

$$2 \tan \theta = \mathbf{sec^2\, \theta} \qquad \text{Given equation}$$

$$2 \tan \theta = \mathbf{1 + tan^2\, \theta} \qquad sec^2\, \theta = 1 + \tan^2 \theta$$

$$\tan^2 \theta - 2 \tan \theta + 1 = 0 \qquad \text{Rewrite the equation.}$$

$$(\tan \theta - 1)(\tan \theta - 1) = 0 \qquad \text{Factor.}$$

$$\tan \theta = 1 \qquad \text{Solve for } \tan \theta.$$

The solutions are $\frac{\pi}{4}$ and $\frac{5\pi}{4}$. See Example 2(b). *Now Try Exercise 37* ◆

In the next example we solve a trigonometric equation by squaring each side. When squaring each side of an equation, it is important to check the answers.

   **Solving a trigonometric equation by squaring**

Solve $\sec t = 1 + \tan t$ for $0 \le t < 2\pi$.

SOLUTION   Begin by squaring each side of the equation.

$$\sec t = 1 + \tan t \qquad \textcolor{blue}{\textit{Given equation}}$$
$$\sec^2 t = (1 + \tan t)^2 \qquad \textcolor{blue}{\textit{Square each side.}}$$
$$\sec^2 t = 1 + 2\tan t + \tan^2 t \qquad \textcolor{blue}{\textit{Square the expression.}}$$
$$\sec^2 t = \sec^2 t + 2\tan t \qquad \textcolor{blue}{1 + \tan^2 t = \sec^2 t.}$$
$$0 = 2\tan t \qquad \textcolor{blue}{\textit{Subtract } \sec^2 t \textit{ from each side.}}$$
$$\tan t = 0 \qquad \textcolor{blue}{\textit{Divide by 2 and rewrite.}}$$
$$t = 0 \quad \text{or} \quad t = \pi \qquad \textcolor{blue}{\textit{Solve for t when } 0 \le t < 2\pi.}$$

Algebra Review

To review squaring a binomial, see Chapter R (page R-25).

Since we squared each side of the equation, we must check $t = 0$ and $t = \pi$ in the given equation.

$$\sec 0 \stackrel{?}{=} 1 + \tan 0 \qquad \sec \pi \stackrel{?}{=} 1 + \tan \pi$$
$$1 = 1 \qquad\qquad -1 \ne 1$$

The only solution is 0. The value of π is called an *extraneous solution*.

Now Try Exercise 53 ◆

One step in the process used by astronomers to calculate the position of a planet as it orbits the sun involves finding the solution of Kepler's equation. Kepler's equation is a trigonometric equation that *cannot* be solved symbolically. It must be solved either graphically or numerically. In real applications, it is not uncommon to encounter equations that cannot be solved symbolically.

   **Solving the equation of Kepler**

One example of Kepler's equation is $\theta = 0.087 + 0.093 \sin \theta$. Solve this equation graphically. The following source may be used to learn more about Kepler's equation.   (**Source:** J. Meeus, *Astronomical Algorithms.*)

SOLUTION   The given equation is equivalent to

$$\theta - 0.087 - 0.093 \sin \theta = 0.$$

To solve this equation graphically, let $Y_1 = X - 0.087 - 0.093 \sin X$ and graph. (Be sure to use radian mode.) The *x*-intercept is near 0.096, as shown in Figure 7.40.

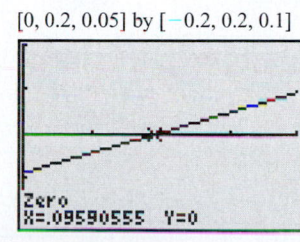

$[0, 0.2, 0.05]$ by $[-0.2, 0.2, 0.1]$

FIGURE 7.40

Now Try Exercise 101 ◆

Solving Inverse Trigonometric Equations

Some types of equations contain inverse trigonometric functions. To solve these equations we often make use of the following inverse properties.

$$\sin\,(\sin^{-1}x) = x \qquad \text{for} \qquad -1 \le x \le 1$$
$$\cos\,(\cos^{-1}x) = x \qquad \text{for} \qquad -1 \le x \le 1$$
$$\tan\,(\tan^{-1}x) = x \qquad \text{for} \qquad -\infty < x < \infty$$

These types of equations are solved in the next example.

EXAMPLE 12   Solving inverse trigonometric equations

Solve each equation symbolically.

(a) $\cos^{-1} x = \pi$ **(b)** $\dfrac{\pi}{2} - \tan^{-1} x = \dfrac{\pi}{4}$ **(c)** $\sin^{-1} 2x = \dfrac{\pi}{3}$

SOLUTION

(a) Begin by taking the cosine of each side of the equation.

$$\cos^{-1} x = \pi \qquad \text{\textit{Given equation}}$$
$$\cos\,(\cos^{-1} x) = \cos \pi \qquad \text{\textit{Take the cosine of each side.}}$$
$$x = -1 \qquad \text{\textit{Simplify.}}$$

(b) Begin solving the equation for $\tan^{-1} x$.

$$\frac{\pi}{2} - \tan^{-1} x = \frac{\pi}{4} \qquad \text{\textit{Given equation}}$$
$$\frac{\pi}{4} = \tan^{-1} x \qquad \text{\textit{Subtract $\frac{\pi}{4}$; add $\tan^{-1}x$.}}$$
$$\tan \frac{\pi}{4} = \tan\,(\tan^{-1} x) \qquad \text{\textit{Take the tangent of each side.}}$$
$$1 = x \qquad \text{\textit{Simplify.}}$$

(c) Begin by taking the sine of each side of the equation.

$$\sin^{-1} 2x = \frac{\pi}{3} \qquad \text{\textit{Given equation}}$$
$$\sin\,(\sin^{-1} 2x) = \sin \frac{\pi}{3} \qquad \text{\textit{Take the sine of each side.}}$$
$$2x = \frac{\sqrt{3}}{2} \qquad \text{\textit{Simplify.}}$$
$$x = \frac{\sqrt{3}}{4} \qquad \text{\textit{Divide by 2.}}$$

Now Try Exercises 83, 89, and 90 ◆

7.3

Putting it all Together

A reference angle θ_R for an angle θ in standard position is the acute angle made by the terminal side of θ and the x-axis. A reference angle can be used to help solve trigonometric equations.

Algebraic skills such as factoring and solving equations can also be used to solve trigonometric equations. Unlike polynomial equations, which always have a finite number of solutions, trigonometric equations frequently have infinitely many solutions.

The following table gives examples of the concepts presented in this section.

Concept	Comments	Examples
Reference angle	If an angle θ is in standard position, then its reference angle is the acute angle made by the terminal side of θ and the x-axis.	**Example:** If $\theta = 155°$ then $\theta_R = 25°$. **Example:** If $\theta = \frac{11\pi}{6}$ then $\theta_R = \frac{\pi}{6}$.
Trigonometric equation	First, use techniques from algebra to isolate any trigonometric functions. Then solve these simpler equations for the given variable.	**Example:** $\sqrt{3}\tan\theta = -1$ $$\tan\theta = -\frac{1}{\sqrt{3}}$$ The reference angle is $\theta_R = \tan^{-1}\left(\frac{1}{\sqrt{3}}\right) = 30°$. Since $\tan\theta$ is negative in quadrants II and IV, the solutions in $[0°, 360°)$ are $150°$ and $330°$. **Example:** $\cos^2 t - 3\cos t + 2 = 0$ $(\cos t - 2)(\cos t - 1) = 0$ $\cos t = 2$ or $\cos t = 1$ $\cos t = 2$ has no solutions. The only solution to $\cos t = 1$ on $[0, 2\pi)$ is $t = 0$. All solutions can be written as $t = 2\pi n$, where n is an integer.

Concept	Comments	Examples
Inverse trigonometric equations	Equations that involve inverse trigonometric functions can sometimes be solved by using the following properties. $\sin(\sin^{-1} x) = x,\ -1 \le x \le 1$ $\cos(\cos^{-1} x) = x,\ -1 \le x \le 1$ $\tan(\tan^{-1} x) = x,\ -\infty < x < \infty$	**Example:** $\tan^{-1} x = \dfrac{\pi}{3}$ $\tan(\tan^{-1} x) = \tan\dfrac{\pi}{3}$ $x = \sqrt{3}$ **Example:** $3\sin^{-1}(x-1) = \dfrac{\pi}{2}$ $\sin^{-1}(x-1) = \dfrac{\pi}{6}$ $\sin(\sin^{-1}(x-1)) = \sin\dfrac{\pi}{6}$ $x - 1 = \dfrac{1}{2}$ $x = \dfrac{3}{2}$

7.3 Exercises

Reference Angles

Exercises 1–10: Find the reference angle for θ.

1. $\theta = 120°$ **2.** $\theta = 230°$

3. $\theta = 85°$ **4.** $\theta = -130°$

5. $\theta = -65°$ **6.** $\theta = 340°$

7. $\theta = \dfrac{5\pi}{6}$ **8.** $\theta = \dfrac{7\pi}{4}$

9. $\theta = -\dfrac{2\pi}{3}$ **10.** $\theta = -\dfrac{5\pi}{4}$

Solving Trigonometric Equations

Exercises 11–14: Use the graph to estimate any solutions to the given equation for $0 \le t < 2\pi$.

11. $\sin t = \cos t$ **12.** $\csc t = \sec t$

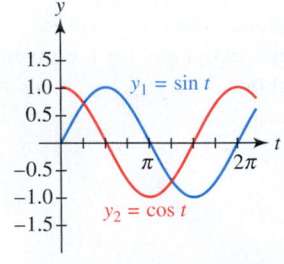

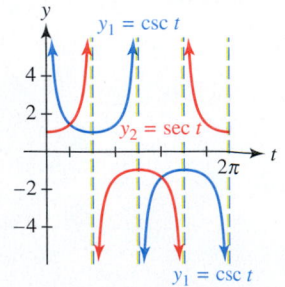

13. $3 \cot t = 2 \sin t$ **14.** $2 \cos^2 t = 1 - \cos t$

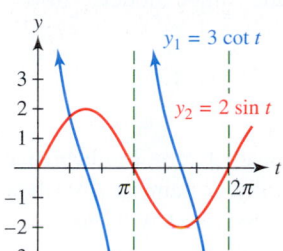

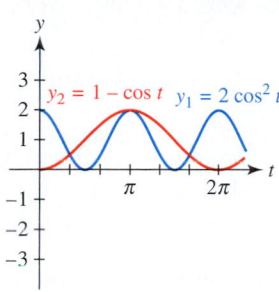

Exercises 15–22: Solve the equation for θ in [0°, 360°) and in [0, 2π).

15. (a) $\sin \theta = 1$ **(b)** $\sin \theta = -1$

16. (a) $\cos \theta = \frac{1}{2}$ **(b)** $\cos \theta = -\frac{1}{2}$

17. (a) $\tan \theta = \sqrt{3}$ **(b)** $\tan \theta = -\sqrt{3}$

18. (a) $\cot \theta = 1$ **(b)** $\cot \theta = -1$

19. (a) $\sec \theta = 2$ **(b)** $\sec \theta = -2$

20. (a) $\csc \theta = \sqrt{2}$ **(b)** $\csc \theta = -\sqrt{2}$

21. (a) $\sin \theta = 3$ **(b)** $\sin \theta = -3$

22. (a) $\cos \theta = \frac{\sqrt{3}}{2}$ **(b)** $\cos \theta = -\frac{\sqrt{3}}{2}$

Exercises 23–30: Solve the algebraic equation for x. Then solve the trigonometric equation for 0° ≤ θ < 360°.

23. (a) $2x - 1 = 0$ **(b)** $2 \sin \theta - 1 = 0$

24. (a) $x - 1 = 0$ **(b)** $\cot \theta - 1 = 0$

25. (a) $x^2 = x$ **(b)** $\sin^2 \theta = \sin \theta$

26. (a) $x^2 - x = 0$ **(b)** $\cos^2 \theta - \cos \theta = 0$

27. (a) $x^2 + 1 = 2$ **(b)** $\tan^2 \theta + 1 = 2$

28. (a) $(x - 1)(x + 1) = 0$ **(b)** $(\sin \theta - 1)(\sin \theta + 1) = 0$

29. (a) $x^2 + x = 2$ **(b)** $\cos^2 \theta + \cos \theta = 2$

30. (a) $2x^2 + 3x = -1$ **(b)** $2 \sin^2 \theta + 3 \sin \theta = -1$

Exercises 31–42: Solve the equation for t on the interval [0, 2π).

31. $\tan^2 t - 3 = 0$ **32.** $2 \sin t = \sqrt{3}$

33. $3 \cos t + 4 = 0$ **34.** $\cos^2 t + \cos t - 6 = 0$

35. $\sin t \cos t = \cos t$ **36.** $\cos^2 t - \sin^2 t = 0$

37. $\csc^2 t = 2 \cot t$ **38.** $2 \sin^2 t - 3 \cos t = 3$

39. $\sin^2 t = \frac{1}{4}$ **40.** $\cos^2 t = -\frac{1}{2}$

41. $\sin t \cos t = 0$ **42.** $\tan t - \cot t = 0$

Exercises 43–54: Find all solutions to the equation. Express your results in radians.

43. $\tan^2 t - 1 = 0$ **44.** $2 \cos t = -1$

45. $\sin^2 t + \sin t - 20 = 0$ **46.** $3 \cos t - 5 = 0$

47. $\cos t \sin t = \sin t$ **48.** $2 \cos^2 t - 1 = 0$

49. $\sec^2 t = 2 \tan t$ **50.** $\cos^2 t - 2 \sin t - 1 = 0$

51. $\sin^2 t \cos^2 t = 0$ **52.** $2 \cot^2 t \sin t - \cot^2 t = 0$

53. $\sin t + \cos t = 1$ (*Hint:* Square each side.)

54. $\sin t - \cos t = 1$

Exercises 55–72: (Refer to Example 6.) Solve the equation, where t is any real number. Approximate t to three decimal places when appropriate.

55. $\sin 3t = \frac{1}{2}$ **56.** $\cos 2t = -\frac{1}{2}$

57. $\cos 4t = -\frac{\sqrt{3}}{2}$ **58.** $\sin 4t = \frac{1}{\sqrt{2}}$

59. $\tan 5t = 1$ **60.** $\cot 3t = -\sqrt{3}$

61. $2 \sin 4t = -1$ **62.** $5 \cos 6t = 2.5$

63. $-\sec 4t = \sqrt{2}$ **64.** $\sqrt{3} \csc 3t = -2$

65. $2 \sin 8t - 3 = -1$ **66.** $3 \cos 8t - 4 = -4$

67. $\cot 4t + 5 = 6$ **68.** $\sqrt{3} \tan 2t + 2 = 3$

69. $5 \cos 3t = 1$ **70.** $7 \cos 5t = -2$

71. $\sin 2t = \frac{1}{3}$ **72.** $\frac{1}{7} \sin 7t = \frac{1}{20}$

Exercises 73–78: The following equations cannot be solved symbolically. Approximate to two decimal places any solutions on [0, 2π) graphically or numerically.

73. $\tan x = x$ **74.** $x - \cos x = 0$

75. $\sin x = (x - 1)^2$ **76.** $\sin^2 x - \ln x = 0$

77. $2x \cos(x + 1) = \sin(\cos x)$ **78.** $e^{-0.1x} \cos x = x \sin x$

Exercises 79–82: Use the accompanying table to find the solutions to the given equation on [0°, 360°). Then write all solutions to the equation.

79. $1 - 4\cos^2\theta = 0$ **80.** $\tan\theta - \sin\theta = 0$

X	Y1
0	-3
60	0
120	0
180	-3
240	0
300	0
360	-3

Y1☐1-4(cos(X))²

X	Y1
0	0
60	.86603
120	-2.598
180	0
240	2.5981
300	-.866
360	0

Y1☐tan(X)-sin(X)

81. $\tan\theta - \dfrac{1}{\sqrt{3}} = 0$ **82.** $2\sin^2\theta - 1 = 0$

X	Y1
30	0
90	ERR:
150	-1.155
210	0
270	ERR:
330	-1.155
390	0

Y1☐tan(X)-1/√(3)

X	Y1
45	0
90	1
135	0
180	-1
225	0
270	1
315	0

Y1☐2(sin(X))²-1

Solving Inverse Trigonometric Equations

Exercises 83–92: Solve the equation.

83. $\sin^{-1}x = \dfrac{\pi}{2}$ **84.** $\sin^{-1}2x = -\dfrac{\pi}{4}$

85. $2\cos^{-1}x = \dfrac{5\pi}{3}$ **86.** $\cos^{-1}x = 0$

87. $\pi + \tan^{-1}x = \dfrac{3\pi}{4}$ **88.** $\tan^{-1}x = -\dfrac{\pi}{3}$

89. $\tan^{-1}(3x + 1) = \dfrac{\pi}{4}$ **90.** $\dfrac{\pi}{4} + \sin^{-1}(x + 1) = \dfrac{\pi}{2}$

91. $\cos^{-1}x + 3\cos^{-1}x = \pi$

92. $\dfrac{\pi}{6} + \sin^{-1}4x = \dfrac{\pi}{3}$

Applications

93. *Phases of the Moon* (Refer to Example 4.) Find all phase angles in degrees where $F = 1$ or $F = 0.25$.

94. *Designing Highway Curves* (Refer to Example 8.) A highway curve with a radius of 800 feet and a speed limit of 66 feet per second (45 mph) is being designed. Find the appropriate superelevation for the curve.

95. *Daylight Hours* (Refer to Example 7.) The number of daylight hours y at 60°N latitude can be modeled by

$$y = 6.5\sin\left(\dfrac{\pi}{6}(x - 3.65)\right) + 12.4,$$

where $x = 1$ corresponds to January 1, $x = 2$ to February 1, and so on. Estimate graphically or numerically when there are 9 hours of daylight. (**Source:** J. Williams.)

96. *Average Temperatures* The monthly average high temperature y in degrees Fahrenheit at Phoenix, Arizona, can be modeled by

$$y = 20.3\sin(0.53x - 2.18) + 83.8,$$

where $x = 1$ corresponds to January, $x = 2$ to February, and so on. Estimate graphically or numerically when the average monthly high temperature is 93°F. (**Source:** J. Williams.)

97. *Daylight Hours* Solve Exercise 95 symbolically.

98. *Average Temperatures* Solve Exercise 96 symbolically.

99. *Maximum Monthly Sunshine* The maximum number of hours of sunshine each month is listed in the table for 50°N latitude.

Month	1	2	3	4	5	6
Hours of Sunshine	261	279	363	407	471	482

Month	7	8	9	10	11	12
Hours of Sunshine	486	442	374	329	267	246

Source: C. Winter, *Solar Power Plants.*

(a) The function f given by

$$f(x) = 122.3\sin(0.524x - 1.7) + 367$$

models these data. Graph f and the data.

(b) Estimate graphically any solutions to the inequality $f(x) \geq 350$ on the interval $[1, 12]$. Interpret the result.

100. *Maximum Yearly Sunshine* The maximum number of hours of sunshine in 1 year is not constant at each latitude. This is due to the eccentricity of Earth's orbit and the tilt of Earth's axis. The accompanying table lists the total hours of daylight in 1 year for selected latitudes. (Note that a negative latitude indicates a location in the Southern Hemisphere.)

Latitude	$-90°$	$-80°$	$-60°$	$-40°$
Hours of Sunshine	4290	4297	4347	4367

Latitude	$-20°$	$0°$	$20°$	$40°$
Hours of Sunshine	4376	4383	4390	4399

Latitude	$60°$	$80°$	$90°$
Hours of Sunshine	4419	4469	4476

Source: C. Winter, *Solar Power Plants.*

  (a) Make a scatterplot of the data.

  (b) Predict whether $y = a \sin(b(x - c)) + d$ could be used to model these data.

  (c) Estimate the latitude where the annual hours of sunshine equals 4320 hours.

Exercises 101 and 102: **Equation of Kepler** *(Refer to Example 11.) Solve Kepler's equation to within two decimal places graphically or numerically.*

101. $\theta = 0.26 + 0.017 \sin\theta$ **102.** $\theta = 0.18 + 0.249 \sin\theta$

103. *Music and Pure Tones* A pure tone can be described by a sinusoidal graph. The graph of $P = 0.004 \sin(100\pi t)$ shown in the accompanying figure represents the pressure of a pure tone on an eardrum in pounds per square foot at time t in seconds. (**Source:** J. Roederer, *Introduction to the Physics and Psychophysics of Music.*)

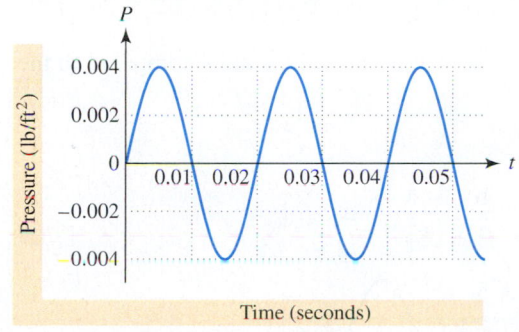

  (a) Estimate all solutions to the equation $P = 0.004$ on the interval $[0, 0.05]$.

  (b) Interpret these solutions.

104. *Music and Pure Tones* The following graphs show the pressure wave resulting from a musical tone, where t is time in seconds and P is pressure in pounds per square foot. (**Source:** J. Roederer.)

  (a) For each graph describe what a person might hear.

  (b) Count the number of solutions to the equation $P = 0.003$ in each graph for $0 \le t \le 0.2$.

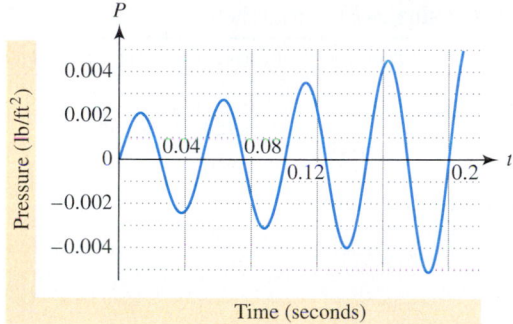

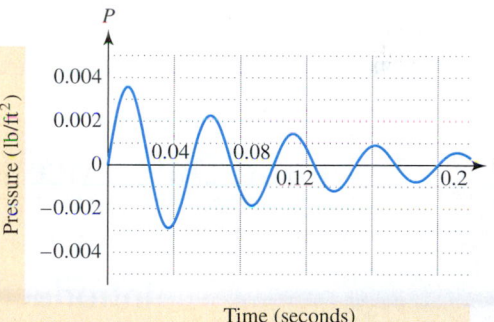

105. *Sums of Pure Tones* If two loudspeakers located at different positions produce the same pure tone, the human ear may hear one sound that is equal to the sum of the individual tones. Since the sources are at different locations, their sinusoidal waves will have different phase angles. (**Source:** N. Fletcher and T. Rossing, *The Physics of Musical Instruments.*)

  (a) Let two musical tones be given by

$$P_1 = 0.003 \sin(880\pi t - 0.7) \quad \text{and}$$
$$P_2 = 0.002 \sin(880\pi t + 0.6).$$

Graph P_1, P_2, and their sum $P = P_1 + P_2$ separately in the viewing rectangle given by $[0, 0.01, 0.005]$ by $[-0.005, 0.005, 0.001]$.

  (b) Determine the maximum pressure for P.

  (c) Is the maximum pressure for P equal to the sum of the maximums of P_1 and P_2? Explain.

106. *Sound*   (Refer to the previous exercise.) Suppose that two loudspeakers located at different positions produce pure tones given by

$$P_1 = A_1 \sin(2\pi Ft + \alpha) \quad \text{and}$$
$$P_2 = A_2 \sin(2\pi Ft + \beta),$$

where F is their common frequency. Then the resulting tone heard by a listener may be written as $P = A \sin(2\pi Ft + \theta)$, where

$$A = \sqrt{(A_1\cos\alpha + A_2\cos\beta)^2 + (A_1\sin\alpha + A_2\sin\beta)^2}$$

and

$$\theta = \arctan\left[\frac{A_1 \sin\alpha + A_2 \sin\beta}{A_1 \cos\alpha + A_2 \cos\beta}\right].$$

(**Source:** N. Fletcher.)

(a) Find A and θ for P if $F = 440$, $A_1 = 0.003$, $\alpha = -0.7$, $A_2 = 0.002$, and $\beta = 0.6$.

(b) Graph $P = A \sin(2\pi Ft + \theta)$ and $y = P_1 + P_2$ in $[0, 0.01, 0.005]$ by $[-0.005, 0.005, 0.001]$. Do the graphs appear to be identical?

Writing about Mathematics

107. Explain the difference between a conditional equation and an identity. Give one example of each.

108. Explain why knowledge of algebra is important when solving trigonometric equations. Give one example of how knowledge of algebra can be applied to solving a trigonometric equation.

7.4

Sum and Difference Identities

◆ Apply the sum and difference identities for cosine

◆ Apply other sum and difference identities

Introduction

Music is made up of vibrations that create pressure on our eardrums. Musical tones can sometimes be modeled with sinusoidal graphs. When more than one tone is played, the resulting pressure is equal to the sum of the individual pressures. Sum and difference identities are sometimes helpful in the analysis of music. In this section we are introduced to several trigonometric identities and some of their applications.

Sum and Difference Identities for Cosine

The graph of $y = \cos\left(t - \frac{\pi}{2}\right)$ is translated to the right $\frac{\pi}{2}$ units compared to the graph of $y = \cos t$, as shown in Figure 7.41.

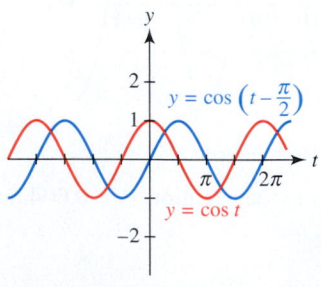

FIGURE 7.41

If the graph of $y = \cos t$ is translated right $\frac{\pi}{2}$ units, it coincides with the graph of $y = \sin t$, which is shown in Figure 7.42. This discussion suggests that

$$\cos\left(t - \frac{\pi}{2}\right) = \sin t.$$

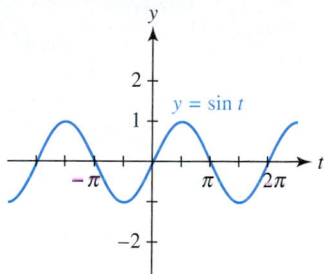

FIGURE 7.42

It is important to notice that

$$\cos\left(t - \frac{\pi}{2}\right) \neq \cos t - \cos\frac{\pi}{2}$$
$$= \cos t - 0$$
$$= \cos t.$$

To verify this symbolically, a new identity is needed. Suppose that α and β represent any two angles. Then the following identity can be used to calculate the cosine of their difference. (Its proof is given at the end of this section.)

$$\cos(\alpha - \beta) = \cos\alpha\cos\beta + \sin\alpha\sin\beta$$

The next example demonstrates how to apply this identity.

EXAMPLE 1 Using the cosine difference identity

Verify the identity $\cos\left(t - \frac{\pi}{2}\right) = \sin t$.

SOLUTION Start by letting $\alpha = t$ and $\beta = \frac{\pi}{2}$ in the cosine difference identity

$$\cos(\alpha - \beta) = \cos\alpha\cos\beta + \sin\alpha\sin\beta.$$

Then,

$$\cos\left(t - \frac{\pi}{2}\right) = \cos t\cos\frac{\pi}{2} + \sin t\sin\frac{\pi}{2}$$
$$= \cos t\,(0) + \sin t\,(1)$$
$$= \sin t. \qquad\qquad\qquad \textit{Now Try Exercise 29} \blacklozenge$$

In the next example we use a difference identity to find the exact value of $\cos 15°$.

EXAMPLE 2   Applying the cosine difference identity

Find the exact value of $\cos 15°$. Use a calculator to support your result.

SOLUTION   Since $45° - 30° = 15°$ and the exact trigonometric values for $45°$ and $30°$ are known, we proceed as follows.

$$\cos 15° = \cos(45° - 30°)$$
$$= \cos 45° \cos 30° + \sin 45° \sin 30° \qquad \textcolor{blue}{\textit{Difference identity for cosine}}$$
$$= \frac{\sqrt{2}}{2} \cdot \frac{\sqrt{3}}{2} + \frac{\sqrt{2}}{2} \cdot \frac{1}{2} \qquad \textcolor{blue}{\textit{Evaluate each function.}}$$
$$= \frac{\sqrt{6} + \sqrt{2}}{4} \qquad \textcolor{blue}{\textit{Simplify the exact value.}}$$

FIGURE 7.43   Degree Mode

In Figure 7.43 we see that the value of $\cos 15°$ agrees with the symbolic result.

Now Try Exercise 5 ◆

With the aid of the difference identity for cosine we can derive a sum identity for cosine.

$$\cos(\alpha + \beta) = \cos(\alpha - (-\beta))$$
$$= \cos \alpha \cos(-\beta) + \sin \alpha \sin(-\beta) \qquad \textcolor{blue}{\textit{Difference identity for cosine}}$$
$$= \cos \alpha \cos \beta - \sin \alpha \sin \beta \qquad \textcolor{blue}{\cos(-\beta) = \cos\beta;}$$
$$\textcolor{blue}{\sin(-\beta) = -\sin(\beta)}$$

Sum and difference identities for cosine are as follows.

> ### COSINE OF SUM OR DIFFERENCE
>
> $$\cos(\alpha + \beta) = \cos \alpha \cos \beta - \sin \alpha \sin \beta$$
> $$\cos(\alpha - \beta) = \cos \alpha \cos \beta + \sin \alpha \sin \beta$$

In Section 6.2 we introduced the cofunction identities for an acute angle θ. These identities are true for any real number t. We verify one of these identities.

EXAMPLE 3   Verifying a cofunction identity

Verify that $\cos\left(\frac{\pi}{2} - t\right) = \sin t$.

SOLUTION   Let $\alpha = \frac{\pi}{2}$ and $\beta = t$ in the cosine difference identity.

$$\cos\left(\frac{\pi}{2} - t\right) = \textcolor{blue}{\cos}\frac{\pi}{2}\cos t + \textcolor{blue}{\sin}\frac{\pi}{2}\sin t$$
$$= (\textcolor{blue}{0})\cos t + (\textcolor{blue}{1})\sin t$$
$$= \sin t$$

Now Try Exercise 19 ◆

The following cofunction identities are valid for any real number t.

COFUNCTION IDENTITIES FOR ANY REAL NUMBER t

$$\cos\left(\frac{\pi}{2} - t\right) = \sin t \qquad \sin\left(\frac{\pi}{2} - t\right) = \cos t$$

$$\cot\left(\frac{\pi}{2} - t\right) = \tan t \qquad \tan\left(\frac{\pi}{2} - t\right) = \cot t$$

$$\csc\left(\frac{\pi}{2} - t\right) = \sec t \qquad \sec\left(\frac{\pi}{2} - t\right) = \csc t$$

Other Sum and Difference Identities

There are also sum and difference identities for sine.

$$\sin(\alpha + \beta) = \cos\left(\frac{\pi}{2} - (\alpha + \beta)\right) \qquad \text{Cofunction identity}$$

$$= \cos\left(\left(\frac{\pi}{2} - \alpha\right) - \beta\right) \qquad \text{Associative property}$$

$$= \mathbf{cos}\left(\frac{\pi}{2} - \alpha\right)\cos\beta + \mathbf{sin}\left(\frac{\pi}{2} - \alpha\right)\sin\beta \qquad \begin{array}{l}\text{Difference identity}\\\text{for cosine}\end{array}$$

$$= \mathbf{sin}\,\alpha\,\cos\beta + \mathbf{cos}\,\alpha\,\sin\beta \qquad \text{Cofunction identities}$$

In a similar manner the difference identity for sine can be derived. We now give the sum and difference identities for sine.

SINE OF SUM OR DIFFERENCE

$$\sin(\alpha + \beta) = \sin\alpha\,\cos\beta + \cos\alpha\,\sin\beta$$

$$\sin(\alpha - \beta) = \sin\alpha\,\cos\beta - \cos\alpha\,\sin\beta$$

 **Analyzing an identity graphically, verbally, and symbolically**

Give graphical and verbal support for $\sin(\theta + \pi) = -\sin\theta$. Then verify the identity symbolically.

SOLUTION

Graphical Support Graph $y = \sin(\theta + \pi)$ and $y = -\sin\theta$, as shown in Figures 7.44 and 7.45 on the next page. Their graphs appear to be identical.

Verbal Support The graph of $y = \sin(\theta + \pi)$ is similar to the graph of $y = \sin\theta$ except that it is translated left π units. The graph of $y = -\sin\theta$ is similar to the graph of $y = \sin\theta$ except that it is reflected across the x-axis. Translating the sine graph left π units or reflecting the sine graph across the x-axis results in the same graph. Therefore we suspect that $\sin(\theta + \pi) = -\sin\theta$ is an identity.

Symbolic Verification Let $\alpha = \theta$ and $\beta = \pi$ in the sum identity for sine.

$$\begin{aligned}\sin(\theta + \pi) &= \sin\theta\cos\pi + \cos\theta\sin\pi \\ &= \sin\theta(-1) + \cos\theta(0) \\ &= -\sin\theta\end{aligned}$$

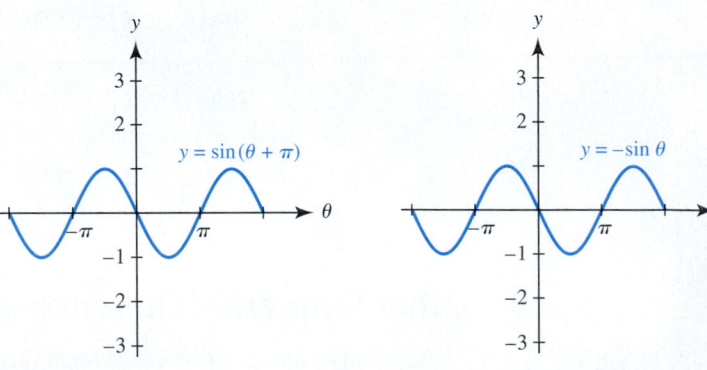

FIGURE 7.44

FIGURE 7.45

Now Try Exercise 11

Sum and difference identities can be used to find exact trigonometric values.

EXAMPLE 5 Applying sum identities for sine and cosine

Let $\sin\alpha = \frac{4}{5}$ and $\cos\beta = \frac{3}{5}$. If α is in quadrant II and β is in quadrant IV, find each of the following.

(a) $\sin(\alpha + \beta)$ **(b)** $\cos(\alpha + \beta)$ **(c)** $\tan(\alpha + \beta)$
(d) The quadrant containing $\alpha + \beta$

FIGURE 7.46

FIGURE 7.47

SOLUTION

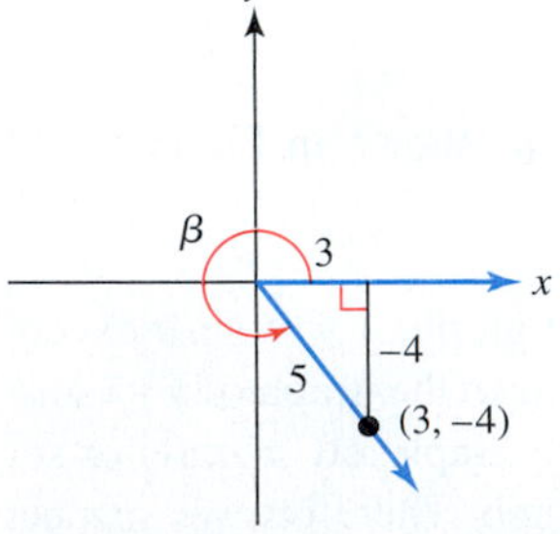

(a) First sketch possible angles for α and for β, as shown in Figures 7.46 and 7.47. From these figures we can see that $\cos\alpha = -\frac{3}{5}$ and $\sin\beta = -\frac{4}{5}$. Now apply the sum identities for sine.

$$\begin{aligned}\sin(\alpha + \beta) &= \sin\alpha\cos\beta + \cos\alpha\sin\beta \\ &= \left(\frac{4}{5}\right)\left(\frac{3}{5}\right) + \left(-\frac{3}{5}\right)\left(-\frac{4}{5}\right) \\ &= \frac{24}{25}\end{aligned}$$

(b) To find $\cos(\alpha + \beta)$, apply the sum identity for cosine.

$$\begin{aligned}\cos(\alpha + \beta) &= \cos\alpha\cos\beta - \sin\alpha\sin\beta \\ &= \left(-\frac{3}{5}\right)\left(\frac{3}{5}\right) - \left(\frac{4}{5}\right)\left(-\frac{4}{5}\right) \\ &= \frac{7}{25}\end{aligned}$$

(c) $\tan(\alpha + \beta) = \dfrac{\sin(\alpha + \beta)}{\cos(\alpha + \beta)} = \dfrac{24/25}{7/25} = \dfrac{24}{7}$

(d) Since both $\sin(\alpha + \beta)$ and $\cos(\alpha + \beta)$ are positive, $\alpha + \beta$ is in quadrant I.

Now Try Exercise 21

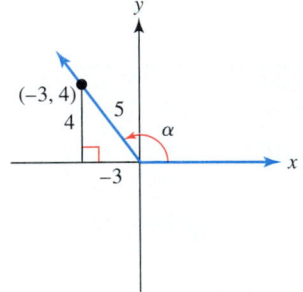

Verifying an identity

Verify the identity $\dfrac{\sin(\alpha - \beta)}{\sin\alpha\sin\beta} = \cot\beta - \cot\alpha$.

SOLUTION Begin by expanding the expression $\sin(\alpha - \beta)$.

$$\frac{\sin(\alpha - \beta)}{\sin\alpha\sin\beta} = \frac{\sin\alpha\cos\beta - \cos\alpha\sin\beta}{\sin\alpha\sin\beta} \qquad \text{Difference identity}$$

$$= \frac{\sin\alpha\cos\beta}{\sin\alpha\sin\beta} - \frac{\cos\alpha\sin\beta}{\sin\alpha\sin\beta} \qquad \frac{a-b}{c} = \frac{a}{c} - \frac{b}{c}$$

$$= \frac{\cos\beta}{\sin\beta} - \frac{\cos\alpha}{\sin\alpha} \qquad \text{Simplify each ratio.}$$

$$= \cot\beta - \cot\alpha \qquad \text{Quotient identity}$$

Now Try Exercise 45 ◆

Because human joints both bend and rotate, trigonometry frequently is applied to human physiology. The next example shows how to calculate the force exerted by a person's back muscles and gives a rather amazing result.

Analyzing stress on a person's back

If a person with weight W bends at the waist with a straight back, then the force F exerted by the lower back muscles may be approximated using $F = 2.89W\sin\left(\theta + \frac{\pi}{2}\right)$, where θ is the angle between a person's torso and the horizontal. See Figure 7.48.   (**Source:** H. Metcalf, *Topics in Classical Biophysics.*)

(a) Let $W = 155$ pounds. The graph of $F = 2.89W\sin\left(\theta + \frac{\pi}{2}\right)$ is shown in Figure 7.49. Interpret the graph.

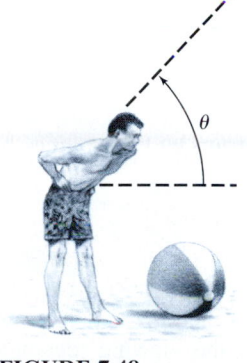

FIGURE 7.48

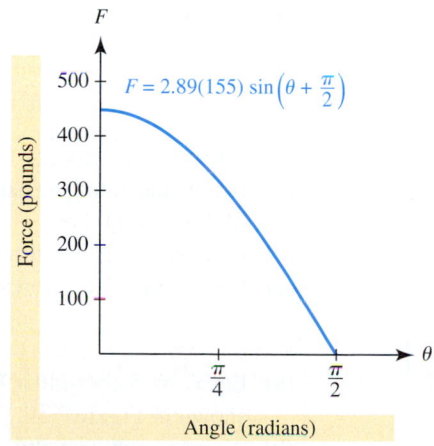

FIGURE 7.49

(b) Show that $F = 2.89W\cos\theta$.

(c) For what value of θ does F equal 400 pounds?

SOLUTION

(a) When $\theta = 0$, the person's back is parallel to the ground and force F exerted by the back muscles has a maximum of about 450 pounds. This is nearly three times the person's weight! As the person straightens up, θ increases, while F decreases. When $\theta = \frac{\pi}{2}$, the person is standing straight up and $F = 0$.

(b) Apply a sum identity for sine.

$$F = 2.89W \sin\left(\theta + \frac{\pi}{2}\right)$$

$$= 2.89W\left(\sin\theta \cos\frac{\pi}{2} + \cos\theta \sin\frac{\pi}{2}\right)$$

$$= 2.89W\left(\sin\theta\,(0) + \cos\theta\,(1)\right)$$

$$= 2.89W \cos\theta$$

(c) Let $F = 400$, $W = 155$, and solve for θ.

$$400 = 2.89(155)\cos\theta \qquad\qquad F = 400$$

$$\cos\theta = \frac{400}{2.89(155)} \qquad\qquad \text{Solve for } \cos\theta.$$

$$\theta = \cos^{-1}\left(\frac{400}{2.89(155)}\right) \qquad \text{Solve for } \theta.$$

$$\theta \approx 0.467 \text{ or } 26.8° \qquad\qquad \text{Approximate } \theta. \quad \textit{Now Try Exercise 53} \blacklozenge$$

Music is composed of tones with various frequencies. Pressure exerted on the eardrum by a pure tone may be modeled by either $P(t) = a \cos bt$ or $P(t) = a \sin bt$, where a and b are constants and t represents time. When two tuning forks produce the same pure tone, the human ear hears only one sound that is equal to the sum of the individual tones. Trigonometry can be used to model this situation. (**Source:** N. Fletcher and T. Rossing, *The Physics of Musical Instruments.*)

EXAMPLE 8 Modeling musical tones

Let the pressure P in grams per square meter exerted on the eardrum by two sources be modeled by

$$P_1(t) = 5 \cos(440\pi t) \quad \text{and}$$

$$P_2(t) = 3 \sin(440\pi t),$$

where t is time in seconds.

(a) Graph the total pressure, $P = P_1 + P_2$, on the eardrum in the viewing rectangle $[0, 0.01, 0.001]$ by $[-8, 8, 1]$.

(b) Use the graph to estimate values for a and k such that $P = a \sin(440\pi t + k)$.

(c) Use a sum or difference identity for sine to verify that $P \approx P_1 + P_2$.

SOLUTION

(a) Let $Y_1 = 5 \cos(440\pi X)$, $Y_2 = 3 \sin(440\pi X)$, and $Y_3 = Y_1 + Y_2$. The graph of y_3 is shown in Figure 7.50.

(b) A maximum y-value occurs near $(0.00039116, 5.831)$. (Note that the x- and y-values shown in Figure 7.50 may vary slightly.) Thus let $a = 5.831$. Since $\sin\theta$ is maximum when $\theta = \frac{\pi}{2}$, we let $t = 0.00039116$ and solve the following equation for k.

$$440\pi(0.00039116) + k = \frac{\pi}{2}$$

$$k = \frac{\pi}{2} - 440\pi(0.00039116)$$

$$\approx 1.0301$$

Thus let $P \approx 5.831 \sin(440\pi t + 1.0301)$. (Other values for k are possible.)

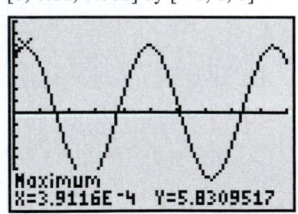

$[0, 0.01, 0.001]$ by $[-8, 8, 1]$

Maximum
X=3.9116E⁻4 Y=5.8309517

FIGURE 7.50 Radian Mode

Calculator Help

To find a maximum point on a graph, see Appendix B (page AP-14).

(c) Apply the sum identity for sine.

$$P \approx 5.831 \sin(440\pi t + 1.0301)$$

$$= 5.831(\sin(440\pi t)\cos(1.0301) + \cos(440\pi t)\sin(1.0301))$$

$$\approx 5.831(\sin(440\pi t)(0.5147) + \cos(440\pi t)(0.8574))$$

$$\approx 3.00\sin(440\pi t) + 5.00\cos(440\pi t))$$

$$= P_1 + P_2 \qquad \qquad \textit{Now Try Exercise 55} \blacklozenge$$

Sum and difference identities can also be found for the tangent function.

TANGENT OF SUM OR DIFFERENCE

$$\tan(\alpha + \beta) = \frac{\tan\alpha + \tan\beta}{1 - \tan\alpha\tan\beta}$$

$$\tan(\alpha - \beta) = \frac{\tan\alpha - \tan\beta}{1 + \tan\alpha\tan\beta}$$

These identities are a result of the sum and difference identities for sine and cosine. For example, the difference identity for tangent can be verified as follows.

$$\tan(\alpha - \beta) = \frac{\sin(\alpha - \beta)}{\cos(\alpha - \beta)} \qquad \textit{Use } \tan\theta = \frac{\sin\theta}{\cos\theta} \textit{ with } \theta = \alpha - \beta.$$

$$= \frac{\sin\alpha\cos\beta - \cos\alpha\sin\beta}{\cos\alpha\cos\beta + \sin\alpha\sin\beta} \qquad \textit{Apply difference identities.}$$

$$= \frac{\dfrac{\sin\alpha\cos\beta}{\cos\alpha\cos\beta} - \dfrac{\cos\alpha\sin\beta}{\cos\alpha\cos\beta}}{\dfrac{\cos\alpha\cos\beta}{\cos\alpha\cos\beta} + \dfrac{\sin\alpha\sin\beta}{\cos\alpha\cos\beta}} \qquad \textit{Divide each term by } \cos\alpha\cos\beta.$$

$$= \frac{\tan\alpha - \tan\beta}{1 + \tan\alpha\tan\beta} \qquad \textit{Simplify.}$$

EXAMPLE 9 **Using the tangent difference identity**

Use Figure 7.51 to find $\tan\gamma$ if $\tan\alpha = \frac{4}{3}$ and $\tan\beta = \frac{3}{4}$.

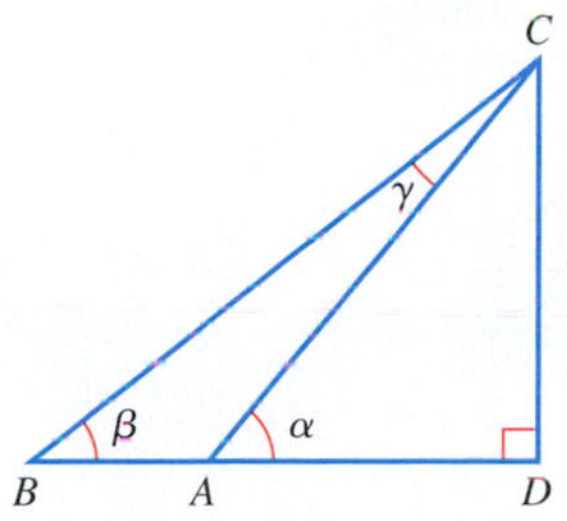

SOLUTION From Figure 7.51, α and $\beta + \gamma$ are both supplements of angle BAC. Thus $\alpha = \beta + \gamma$ or $\gamma = \alpha - \beta$.

$$\tan\gamma = \tan(\alpha - \beta) \qquad \gamma = \alpha - \beta$$

$$= \frac{\tan\alpha - \tan\beta}{1 + \tan\alpha\tan\beta} \qquad \textit{Tangent difference identity}$$

$$= \frac{(4/3) - (3/4)}{1 + (4/3)(3/4)} \qquad \tan\alpha = \frac{4}{3}; \tan\beta = \frac{3}{4}$$

$$= \frac{7}{24} \qquad \textit{Simplify.} \qquad \textit{Now Try Exercise 47} \blacklozenge$$

FIGURE 7.51

Derivation of an Identity

We conclude this section by deriving the difference identity for cosine. Begin by considering the angles α and β in standard position and the unit circle, as shown in Figure 7.52. The terminal side of α intersects the unit circle at $(\cos \alpha, \sin \alpha)$ and the terminal side of β intersects the unit circle at the point $(\cos \beta, \sin \beta)$. The angle formed between the terminal sides of α and β equals $\alpha - \beta$. Now consider angle $\alpha - \beta$ in standard position. Its terminal side intersects the unit circle at the point $(\cos (\alpha - \beta), \sin (\alpha - \beta))$. This is shown in Figure 7.53.

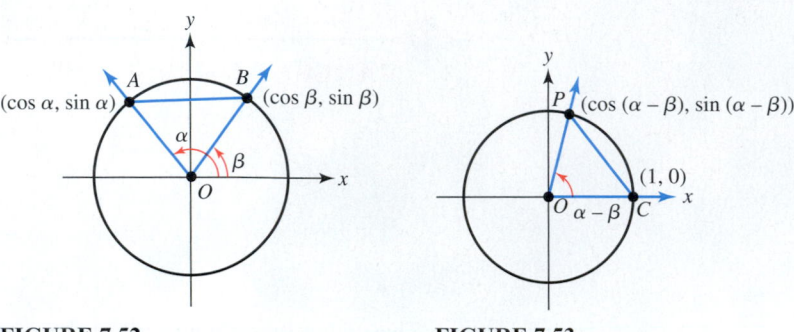

FIGURE 7.52   **FIGURE 7.53**

Since triangles ABO and PCO are congruent, the distance from A to B in Figure 7.52 equals the distance from P to C in Figure 7.53.

$$\sqrt{(\cos \alpha - \cos \beta)^2 + (\sin \alpha - \sin \beta)^2} =$$
$$\sqrt{(\cos (\alpha - \beta) - 1)^2 + (\sin (\alpha - \beta) - 0)^2}$$

Squaring each side and clearing parentheses produces

$$\cos^2 \alpha - 2 \cos \alpha \cos \beta + \cos^2 \beta + \sin^2 \alpha - 2 \sin \alpha \sin \beta + \sin^2 \beta =$$
$$\cos^2 (\alpha - \beta) - 2 \cos (\alpha - \beta) + 1 + \sin^2 (\alpha - \beta).$$

Since $\sin^2 \theta + \cos^2 \theta = 1$ for any θ, the above equation simplifies to

$$2 - 2 \cos \alpha \cos \beta - 2 \sin \alpha \sin \beta = 2 - 2 \cos (\alpha - \beta).$$

Solving this equation for $\cos (\alpha - \beta)$ gives the cosine difference identity

$$\cos (\alpha - \beta) = \cos \alpha \cos \beta + \sin \alpha \sin \beta.$$

7.4 Putting it all Together

Sum and difference identities occur in a variety of applications such as music, physiology, and electricity. The sum and difference identities are also used in calculus. The following table lists the important identities in this section.

Identities	General Form
Cosine sum and difference	$\cos(\alpha + \beta) = \cos\alpha\cos\beta - \sin\alpha\sin\beta$ $\cos(\alpha - \beta) = \cos\alpha\cos\beta + \sin\alpha\sin\beta$
Sine sum and difference	$\sin(\alpha + \beta) = \sin\alpha\cos\beta + \cos\alpha\sin\beta$ $\sin(\alpha - \beta) = \sin\alpha\cos\beta - \cos\alpha\sin\beta$
Tangent sum and difference	$\tan(\alpha + \beta) = \dfrac{\tan\alpha + \tan\beta}{1 - \tan\alpha\tan\beta}$ $\tan(\alpha - \beta) = \dfrac{\tan\alpha - \tan\beta}{1 + \tan\alpha\tan\beta}$
Cofunction	$\cos\left(\dfrac{\pi}{2} - t\right) = \sin t \qquad \sin\left(\dfrac{\pi}{2} - t\right) = \cos t$ $\cot\left(\dfrac{\pi}{2} - t\right) = \tan t \qquad \tan\left(\dfrac{\pi}{2} - t\right) = \cot t$ $\csc\left(\dfrac{\pi}{2} - t\right) = \sec t \qquad \sec\left(\dfrac{\pi}{2} - t\right) = \csc t$

7.4 Exercises

Sum and Difference Identities

Exercises 1–10: Find the exact value for each expression. Use a calculator to support your result numerically.

1. $\sin 15°$

2. $\sin 105°$ *(Hint: $105° = 45° + 60°$)*

3. $\tan 15°$ **4.** $\sin 75°$

5. $\cos 75°$ **6.** $\cos 105°$

7. $\sin\dfrac{\pi}{12}$ *$\left(Hint: \dfrac{\pi}{3} - \dfrac{\pi}{4} = \dfrac{\pi}{12}\right)$*

8. $\cos\dfrac{5\pi}{12}$

9. $\sin\dfrac{5\pi}{12}$ **10.** $\tan\dfrac{\pi}{12}$

Exercises 11–16: Complete the following for the identity.

    (a) Give graphical and verbal support for the identity.
    (b) Verify the identity symbolically.

11. $\sin\left(t + \dfrac{\pi}{2}\right) = \cos t$ **12.** $\sin\left(t + \dfrac{3\pi}{2}\right) = -\cos t$

13. $\cos(t + \pi) = -\cos t$

14. $\cos\left(t + \dfrac{3\pi}{2}\right) = \sin t$

15. $\sec\left(t - \dfrac{\pi}{2}\right) = \csc t$

16. $\tan\left(t + \dfrac{\pi}{2}\right) = -\cot t$

Exercises 17–20: Use a difference identity to verify the co-function identity.

17. $\sin\left(\dfrac{\pi}{2} - t\right) = \cos t$

18. $\tan\left(\dfrac{\pi}{2} - t\right) = \cot t$

19. $\sec\left(\dfrac{\pi}{2} - t\right) = \csc t$

20. $\csc\left(\dfrac{\pi}{2} - t\right) = \sec t$

Exercises 21–28: (Refer to Example 5.) Find the following.

   (a) $\sin(\alpha + \beta)$ (b) $\cos(\alpha + \beta)$

   (c) $\tan(\alpha + \beta)$ (d) *The quadrant containing* $\alpha + \beta$

21. $\sin\alpha = \frac{3}{5}$ and $\sin\beta = \frac{5}{13}$, α and β in quadrant I

22. $\cos\alpha = -\frac{12}{13}$ and $\cos\beta = -\frac{5}{13}$, α and β in quadrant II

23. $\sin\alpha = -\frac{8}{17}$ and $\cos\beta = \frac{11}{61}$, α in quadrant III and β in quadrant I

24. $\cos\alpha = -\frac{24}{25}$ and $\sin\beta = \frac{4}{5}$, α in quadrant II and β in quadrant I

25. $\cos\alpha = -\frac{3}{5}$ and $\cos\beta = \frac{12}{13}$, α in quadrant III and β in quadrant IV

26. $\tan\alpha = \frac{3}{4}$ and $\cos\beta = -\frac{4}{5}$, α in quadrant I and β in quadrant III

27. $\tan\alpha = -\frac{5}{12}$ and $\sec\beta = -\frac{61}{11}$, α and β in quadrant II

28. $\cot\alpha = \frac{3}{4}$ and $\csc\beta = \frac{25}{24}$, α and β in quadrant I

Exercises 29–46: Verify the identity.

29. $\cos\left(t - \dfrac{\pi}{4}\right) = \dfrac{\sqrt{2}}{2}(\cos t + \sin t)$

30. $\sin\left(t + \dfrac{\pi}{4}\right) = \dfrac{\sqrt{2}}{2}(\cos t + \sin t)$

31. $\tan\left(t + \dfrac{\pi}{4}\right) = \dfrac{1 + \tan t}{1 - \tan t}$

32. $\tan(45° - \theta) = \dfrac{1 - \tan\theta}{1 + \tan\theta}$

33. $\dfrac{\cos(x - y)}{\cos(x + y)} = \dfrac{1 + \tan x \tan y}{1 - \tan x \tan y}$

34. $\dfrac{\sin(x - y)}{\sin(x + y)} = \dfrac{\tan x - \tan y}{\tan x + \tan y}$

35. $\dfrac{\cos(\alpha - \beta)}{\cos\alpha\sin\beta} = \tan\alpha + \cot\beta$

36. $\cos(\theta + \theta) = 1 - 2\sin^2\theta$

37. $\sin 2t = 2\sin t \cos t$

38. $\cos 2t = \cos^2 t - \sin^2 t$

39. $\sin(\alpha + \beta) + \sin(\alpha - \beta) = 2\sin\alpha\cos\beta$

40. $\cos(\alpha + \beta) + \cos(\alpha - \beta) = 2\cos\alpha\cos\beta$

41. $\tan(\pi - \theta) = -\tan\theta$

42. $\tan(\theta + \pi) = \tan\theta$

43. $\dfrac{\sin(x - y)}{\sin y} + \dfrac{\cos(x - y)}{\cos y} = \dfrac{\sin x}{\sin y \cos y}$

44. $\tan(x - y) - \tan(y - x) = \dfrac{2(\tan x - \tan y)}{1 + \tan x \tan y}$

45. $\dfrac{\sin(x + y)}{\cos x \cos y} = \tan x + \tan y$

46. $\dfrac{\tan(x + y) - \tan y}{1 + \tan(x + y)\tan y} = \tan x$

Exercises 47 and 48: Solve Example 9 using the given information.

47. $\tan\alpha = \frac{6}{7}$ and $\tan\beta = \frac{5}{7}$ **48.** $\cot\alpha = \frac{8}{13}$ and $\cot\beta = \frac{11}{13}$

Lines and Slopes

Exercises 49–52: Suppose two lines, l_1 and l_2, intersect the x-axis making angles α and β, as shown in the accompanying figure. Then the slopes of l_1 and l_2 satisfy $m_1 = \tan\alpha$ and $m_2 = \tan\beta$, respectively. If l_1 and l_2 intersect with angle θ as shown, then it follows that $\beta = \alpha + \theta$, or equivalently, $\theta = \beta - \alpha$.

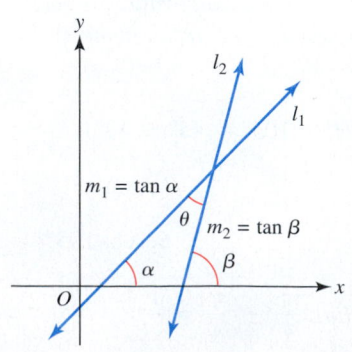

49. Use a difference identity for tangent to show that $\tan\theta = \frac{m_2 - m_1}{1 + m_1 m_2}$.

50. Is the formula in the previous exercise valid if β is an obtuse angle? Explain.

51. Find θ for the two intersecting lines given by $y = 2x - 3$ and $y = \frac{3}{5}x + 1$.

52. Find θ for the two intersecting lines given by $y = \frac{1}{2}x + 1$ and $y = 3 - x$.

Applications

53. *Back Stress* (Refer to Example 7.)
   (a) Suppose a 200-pound person bends at the waist so that $\theta = \frac{\pi}{4}$. Estimate the force exerted by the person's back muscles.

   (b) Approximate the value of θ that results in the back muscles exerting a force of 400 pounds.

54. *Sound Waves* Sound is a result of waves applying pressure to a person's eardrum. For a particular sound wave radiating outward, the trigonometric function $P(r) = \frac{a}{r}\cos(\pi r - 1000t)$ can be used to express the pressure at a radius of r feet from the source after t seconds. In this formula, a is the maximum sound pressure at the source measured in pounds per square foot. (**Source:** L. Beranek, *Noise and Vibration Control.*)
   (a) Let $a = 0.4$, $t = 1$, and graph the sound pressure for $0 \le r \le 20$. What happens to the pressure P as the radius r increases?

   (b) Use a difference identity to simplify the expression for $P(r)$ when r is an even integer.

55. *Modeling Musical Tones* (Refer to Example 8.) Let the pressure exerted by two sound waves in grams per square meter be given by
$$P_1(t) = 4\cos(220\pi t) \quad \text{and} \quad P_2(t) = 3\sin(220\pi t),$$
where t is in seconds.
   (a) Graph the total pressure $P = P_1 + P_2$ in the window $[0, 0.02, 0.001]$ by $[-6, 6, 1]$.

   (b) Use the graph to estimate values for a and k such that $P = a\sin(220\pi t + k)$.

   (c) Use a sum or difference identity for sine to verify that
$$a\sin(220\pi t + k) \approx 3\sin(220\pi t) + 4\cos(220\pi t).$$

56. *Electricity* When voltages $V_1 = 50\sin(120\pi t)$ and $V_2 = 120\cos(120\pi t)$ are applied to the same circuit, the resulting voltage V is equal to their sum. (**Source:** D. Bell, *Fundamentals of Electric Circuits.*)

   (a) Graph $V = V_1 + V_2$ in the viewing rectangle $[0, 0.05, 0.01]$ by $[-160, 160, 40]$.

   (b) Use the graph to estimate values for a and k so that $V = a\sin(120\pi t + k)$.

   (c) Use a sum or difference identity for sine to verify part (b).

Writing about Mathematics

57. Are $\sin(45° + 30°)$ and $\sin 45° + \sin 30°$ equivalent expressions? Explain your answer.

58. Are the expressions $\cos(\alpha - \beta)$ and $\cos\alpha - \cos\beta$ equal? Give an example to justify your answer.

EXTENDED AND DISCOVERY EXERCISES

Modeling Musical Beats Musicians sometimes tune instruments by playing the same tone on two instruments and listening for a phenomenon known as *beats*. Beats occur when two musical tones vary slightly in frequency. When the two instruments are in tune, the beats will disappear. The human ear hears beats because the sound pressure slowly rises and falls as a result of this slight variation in the frequency. The pressure P on an eardrum can be modeled by $P = a\sin(2\pi Ft)$, where F is the frequency of the tone, t is time in seconds, and P is in pounds per square foot. This phenomenon can be modeled with a graphing calculator. (**Source:** J. Pierce, *The Science of Musical Sound.*)

1. Consider two tones with similar frequencies of 440 and 443 cycles per second and pressures
$$P_1 = 0.006\sin(880\pi t) \quad \text{and} \quad P_2 = 0.004\sin(886\pi t),$$
respectively.
   (a) Graph $P = P_1 + P_2$ in $[0.15, 1.15, 0.05]$ by $[-0.01, 0.01, 0.001]$, where P is the total pressure exerted by the tones on an eardrum. How many beats are there in this 1-second interval?

   (b) Repeat part (a) with frequencies of 220 and 224.

   (c) Determine a way to find the number of beats per second if the frequencies of the tones are F_1 and F_2.

Exercises 2 and 3: Music and Beats (Refer to the previous exercise.) Given two musical tones P_1 and P_2, graph their sum in $[0.2, 1.2, 0.05]$ by $[-0.01, 0.01, 0.001]$. Count the number of beats in 1 second.

2. $P_1 = 0.007\sin(450\pi t)$, $P_2 = 0.005\sin(454\pi t)$

3. $P_1 = 0.004\cos(830\pi t)$, $P_2 = 0.005\sin(836\pi t)$

CHECKING BASIC CONCEPTS FOR SECTIONS 7.3 AND 7.4

1. Find the reference angle of each angle.

    (a) $225°$ (b) $\dfrac{5\pi}{6}$
        $45°$

2. Solve each equation for θ in $[0°, 360°)$ symbolically. Give graphical or numerical support.

    (a) $\cos\theta = \dfrac{1}{2}$ (b) $\sin\theta = -\dfrac{\sqrt{3}}{2}$

3. Find all solutions to the given equation where t is a real number.

    (a) $\sin t = -\cos t$ (b) $2\sin^2 t = 1 - \cos t$

4. Use a sum or difference identity to find $\cos\dfrac{\pi}{12}$.

5. Verify the identity $\sin(t - \pi) = -\sin t$ symbolically. Give graphical or numerical support.

7.5 Multiple-Angle Identities

- Learn and use the double-angle identities
- Learn and use the half-angle formulas
- Solve equations
- Learn and use product-to-sum and sum-to-product identities

Introduction

In 1831 Michael Faraday discovered that when a wire is passed near a magnet, a small electric current is produced in the wire. This property is used to generate electric current for homes, schools, and businesses throughout the world. By rotating thousands of wires near large electromagnets, massive amounts of electricity can be produced. In 1 year, utilities in the United States generate enough electricity to power a 100-watt light bulb for over 3 billion years!

Voltage, amperage, and wattage are quantities that can be modeled by sinusoidal graphs and functions. To model electricity and other phenomena, trigonometric functions and identities are used. In this section we are introduced to several important multiple-angle identities. (**Sources:** R. Weidner and R. Sells, *Elementary Classical Physics*, Vol. 2; J. Wright, *The Universal Almanac 1997.*)

Double-Angle Identities

The sum identities can be used to derive double-angle identities for sine, cosine, and tangent.

Sine Double-Angle Identity

$$\sin 2\theta = \sin(\theta + \theta)$$
$$= \sin\theta\cos\theta + \cos\theta\sin\theta$$
$$= 2\sin\theta\cos\theta$$

Cosine Double-Angle Identities

$$\cos 2\theta = \cos(\theta + \theta)$$
$$= \cos\theta\cos\theta - \sin\theta\sin\theta$$
$$= \cos^2\theta - \sin^2\theta$$

Applying the identity $\sin^2\theta + \cos^2\theta = 1$, the expression $\cos^2\theta - \sin^2\theta$ can be written as

$$\cos^2\theta - \mathbf{\sin^2\theta} = \cos^2\theta - (\mathbf{1 - \cos^2\theta})$$
$$= 2\cos^2\theta - 1$$

or as

$$\cos^2 \theta - \sin^2 \theta = (1 - \sin^2 \theta) - \sin^2 \theta$$
$$= 1 - 2 \sin^2 \theta.$$

Tangent Double-Angle Identity

$$\tan 2\theta = \tan (\theta + \theta)$$
$$= \frac{\tan \theta + \tan \theta}{1 - \tan \theta \tan \theta}$$
$$= \frac{2 \tan \theta}{1 - \tan^2 \theta}$$

A summary of these double-angle identities is given.

DOUBLE-ANGLE IDENTITIES

$$\sin 2\theta = 2 \sin \theta \cos \theta$$
$$\cos 2\theta = \cos^2 \theta - \sin^2 \theta = 2 \cos^2 \theta - 1 = 1 - 2 \sin^2 \theta$$
$$\tan 2\theta = \frac{2 \tan \theta}{1 - \tan^2 \theta}$$

The next example illustrates that $\sin 2\theta \neq 2 \sin \theta$.

 EXAMPLE 1   Using double-angle identities

Verify graphically and symbolically that the expressions $\sin 2\theta$ and $2 \sin \theta$ are *not* equivalent.

SOLUTION

Graphical Verification   Graph $y = \sin 2\theta$ and $y = 2 \sin \theta$, as shown in Figures 7.54 and 7.55. Notice that the graphs are different.

Symbolic Verification   The expressions $\sin 2\theta$ and $2 \sin \theta$ are not equivalent, rather

$$\sin 2\theta = 2 \sin \theta \cos \theta \neq 2 \sin \theta.$$

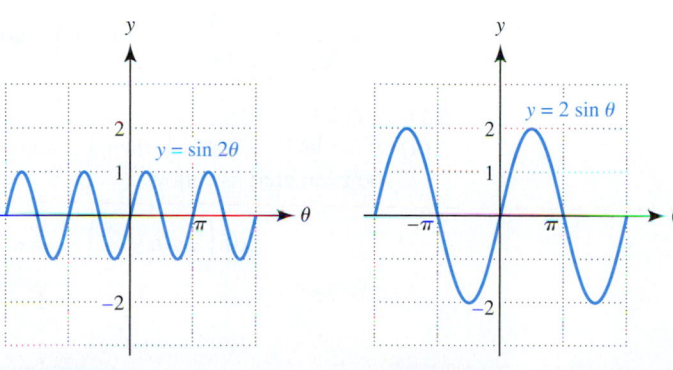

FIGURE 7.54   **FIGURE 7.55**

Now Try Exercise 7 ◆

◆ **CLASS DISCUSSION**
Verify both graphically and symbolically that $\cos 2\theta$ and $2\cos\theta$ are not equivalent expressions. ◆

EXAMPLE 2 Using double-angle identities

Given $\cos\theta = -\frac{12}{13}$ and $\sin\theta > 0$, find $\sin 2\theta$, $\cos 2\theta$, and $\tan 2\theta$. Use a calculator to support your result.

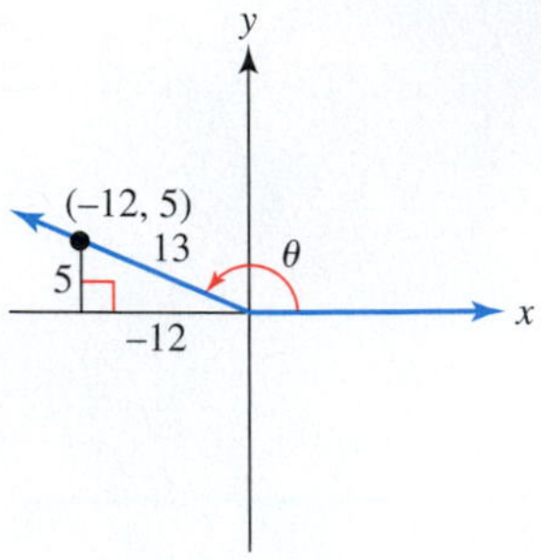

SOLUTION

Symbolic Solution Since $\cos\theta < 0$ and $\sin\theta > 0$, θ is contained in quadrant II. One possibility for θ is shown in Figure 7.56. We see that $\sin\theta = \frac{5}{13}$. Using double-angle identities, we obtain the following results.

$$\sin 2\theta = 2\sin\theta\cos\theta = 2\cdot\frac{5}{13}\cdot\left(-\frac{12}{13}\right) = -\frac{120}{169}$$

$$\cos 2\theta = \cos^2\theta - \sin^2\theta = \left(-\frac{12}{13}\right)^2 - \left(\frac{5}{13}\right)^2 = \frac{119}{169}$$

$$\tan 2\theta = \frac{\sin 2\theta}{\cos 2\theta} = \frac{-120/169}{119/169} = -\frac{120}{119}$$

Calculator Support Since θ is in quadrant II, we can support these results by letting $\theta = \arccos\left(-\frac{12}{13}\right) \approx 157.38°$ and performing the calculations shown in Figures 7.57 and 7.58.

FIGURE 7.56

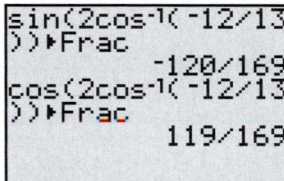

```
sin(2cos⁻¹(-12/13
))▶Frac
              -120/169
cos(2cos⁻¹(-12/13
))▶Frac
               119/169
```

```
tan(2cos⁻¹(-12/13
))▶Frac
              -120/119
```

FIGURE 7.57 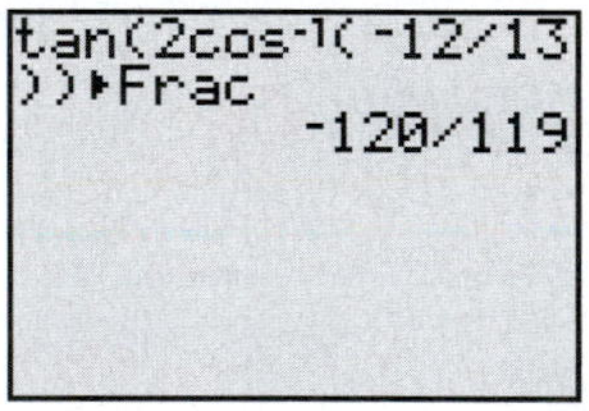   **FIGURE 7.58**

Now Try Exercise 11 ◆

EXAMPLE 3 Evaluating expressions with double-angle identities

Use a double-angle identity to evaluate each expression.

(a) $\cos\left(2\sin^{-1}\frac{1}{3}\right)$ **(b)** $\sin\left(2\cos^{-1}\left(-\frac{3}{5}\right)\right)$ **(c)** $\sin(2\tan^{-1}x)$, $x > 0$

SOLUTION

(a) If we let $\theta = \sin^{-1}\frac{1}{3}$, then it follows that $\sin\theta = \frac{1}{3}$. The expression $\cos\left(2\sin^{-1}\frac{1}{3}\right)$ can be evaluated as follows.

$$\cos\left(2\sin^{-1}\frac{1}{3}\right) = \cos(2\theta) \qquad \color{blue}{\theta = \sin^{-1}\frac{1}{3}}$$

$$= 1 - 2\sin^2\theta \qquad \color{blue}{\text{Double-angle identity}}$$

$$= 1 - 2\left(\frac{1}{3}\right)^2 \qquad \color{blue}{\sin\theta = \frac{1}{3}}$$

$$= \frac{7}{9} \qquad \color{blue}{\text{Simplify.}}$$

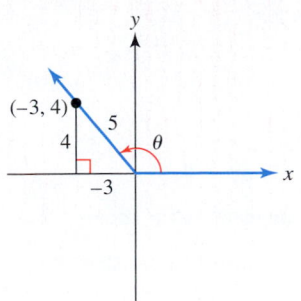

FIGURE 7.59

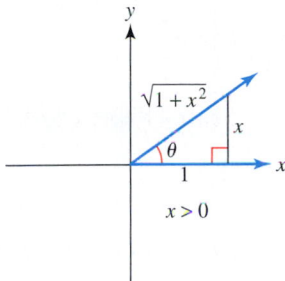

FIGURE 7.60

(b) Begin by letting $\theta = \cos^{-1}\left(-\frac{3}{5}\right)$ and sketching angle θ in standard position, as shown in Figure 7.59. Notice that $\cos\theta = -\frac{3}{5}$ and that θ is a second quadrant angle. From Figure 7.59 it follows that $\sin\theta = \frac{4}{5}$. The expression $\sin\left(2\cos^{-1}-\frac{3}{5}\right)$ can be evaluated as follows.

$$\sin\left(2\cos^{-1}-\frac{3}{5}\right) = \sin(2\theta) \qquad \textcolor{blue}{\theta = \cos^{-1}\left(-\frac{3}{5}\right)}$$

$$= 2\sin\theta\cos\theta \qquad \textcolor{blue}{\text{Double-angle identity}}$$

$$= 2\left(\frac{4}{5}\right)\left(-\frac{3}{5}\right) \qquad \textcolor{blue}{\sin\theta = \frac{4}{5},\ \cos\theta = -\frac{3}{5}}$$

$$= -\frac{24}{25} \qquad \textcolor{blue}{\text{Simplify.}}$$

(c) Begin by letting $\theta = \tan^{-1}x$ and sketching angle θ in standard position, as shown in Figure 7.60. Since $x > 0$, we have drawn a first quadrant angle θ whose tangent function equals $\frac{x}{1}$. That is, $\tan\theta = x$. From Figure 7.60 it follows that $\sin\theta = \frac{x}{\sqrt{1+x^2}}$ and $\cos\theta = \frac{1}{\sqrt{1+x^2}}$. The expression $\sin(2\tan^{-1}x)$ can be evaluated as follows.

$$\sin(2\tan^{-1}x) = \sin(2\theta) \qquad \textcolor{blue}{\theta = \tan^{-1}x}$$

$$= 2\sin\theta\cos\theta \qquad \textcolor{blue}{\text{Double-angle identity}}$$

$$= 2\left(\frac{x}{\sqrt{1+x^2}}\right)\left(\frac{1}{\sqrt{1+x^2}}\right) \qquad \textcolor{blue}{\text{Substitute.}}$$

$$= \frac{2x}{1+x^2} \qquad \textcolor{blue}{\text{Simplify.}}$$

Now Try Exercises 17, 19, and 24 ◆

In the next example we verify an identity by using a double-angle identity.

 **Verifying an identity**

Verify the identity $\dfrac{\sec^2\theta}{1 - \tan^2\theta} = \sec 2\theta$.

SOLUTION

Begin by applying a reciprocal identity.

$$\frac{\sec^2\theta}{1 - \tan^2\theta} = \frac{1}{\cos^2\theta(1 - \tan^2\theta)} \qquad \textcolor{blue}{\text{Reciprocal identity}}$$

$$= \frac{1}{\cos^2\theta\left(1 - \dfrac{\sin^2\theta}{\cos^2\theta}\right)} \qquad \textcolor{blue}{\text{Quotient identity}}$$

$$= \frac{1}{\cos^2\theta - \sin^2\theta} \qquad \textcolor{blue}{\text{Distributive property}}$$

$$= \frac{1}{\cos 2\theta} \qquad \textcolor{blue}{\text{Double-angle identity}}$$

$$= \sec 2\theta \qquad \textcolor{blue}{\text{Reciprocal identity}}$$

Now Try Exercise 59 ◆

EXAMPLE 5 Deriving a triple-angle identity

Write $\cos 3\theta$ in terms of $\cos \theta$.

SOLUTION

$$\cos 3\theta = \cos(2\theta + \theta) \qquad\qquad 3\theta = 2\theta + \theta$$

$$= \mathbf{\cos 2\theta} \cos \theta - \mathbf{\sin 2\theta} \sin \theta \qquad \text{Sum identity for cosine}$$

$$= (\mathbf{2\cos^2 \theta - 1}) \cos \theta - (\mathbf{2\sin\theta\cos\theta}) \sin \theta \qquad \text{Double-angle identities}$$

$$= 2\cos^3 \theta - \cos \theta - 2\mathbf{\sin^2 \theta} \cos \theta \qquad \text{Multiply.}$$

$$= 2\cos^3 \theta - \cos \theta - 2(\mathbf{1 - \cos^2 \theta}) \cos \theta \qquad \text{Apply } \sin^2 \theta + \cos^2 \theta = 1.$$

$$= 2\cos^3 \theta - \cos \theta - 2\cos \theta + 2\cos^3 \theta \qquad \text{Distributive property}$$

$$= 4\cos^3 \theta - 3\cos \theta \qquad\qquad \text{Combine like terms.}$$

Now Try Exercise 63.

Half-Angle Formulas

Power-reducing identities for sine, cosine, and tangent can be derived using the double-angle identities.

Sine Power-Reducing Identity Since

$$\cos 2\theta = 1 - 2\sin^2 \theta,$$

we can solve for $\sin^2 \theta$ to obtain

$$\sin^2 \theta = \frac{1 - \cos 2\theta}{2}.$$

Cosine Power-Reducing Identity Solving the identity

$$\cos 2\theta = 2\cos^2 \theta - 1$$

for $\cos^2 \theta$ gives

$$\cos^2 \theta = \frac{1 + \cos 2\theta}{2}.$$

Tangent Power-Reducing Identity We can use the power-reducing identities

$$\sin^2 \theta = \frac{1 - \cos 2\theta}{2} \quad \text{and} \quad \cos^2 \theta = \frac{1 + \cos 2\theta}{2}$$

to derive

$$\tan^2 \theta = \frac{\sin^2 \theta}{\cos^2 \theta} = \frac{1 - \cos 2\theta}{1 + \cos 2\theta}.$$

A summary of these identities is now given.

POWER-REDUCING IDENTITIES

$$\sin^2 \theta = \frac{1 - \cos 2\theta}{2} \qquad \cos^2 \theta = \frac{1 + \cos 2\theta}{2} \qquad \tan^2 \theta = \frac{1 - \cos 2\theta}{1 + \cos 2\theta}$$

EXAMPLE 6 Using a power-reducing identity

Find the exact value of $\sin^2(22.5°)$. Use a calculator to support your results.

SOLUTION Let $\theta = 22.5°$ and $2\theta = 45°$ and apply the sine power-reducing identity.

$$\sin^2 22.5° = \frac{1 - \cos 45°}{2}$$

$$= \frac{1 - \sqrt{2}/2}{2}$$

$$= \frac{2 - \sqrt{2}}{4}$$

FIGURE 7.61 Degree Mode

Support for this result is shown in Figure 7.61.   *Now Try Exercise 37* ◆

In the next example we apply a power-reducing identity to electrical circuits.

EXAMPLE 7 Using a power-reducing identity to analyze wattage

Amperage I is a measure of the amount of electricity passing through a wire and voltage V is a measure of the force "pushing" the electricity. The wattage W consumed by an electrical device can be calculated using the equation $W = VI$.   (**Source:** G. Wilcox and C. Hesselberth, *Electricity for Engineering Technology.*)

(a) Voltage in a household circuit is given by $V = 160 \sin(120\pi t)$, where t is in seconds. Suppose that the amperage flowing through a toaster is given by $I = 12 \sin(120\pi t)$. Graph the wattage W consumed by the toaster in $[0, 0.04, 0.01]$ by $[-200, 2200, 200]$.
(b) Write the wattage as $W = a \cos(k\pi t) + d$, where a, k, and d are constants.
(c) Compare the periods of the voltage, amperage, and wattage.
(d) The wattage of this toaster equals half the maximum of W. Find the wattage.

SOLUTION

(a) Since $W = VI$, graph the equation $Y_3 = Y_1 * Y_2$, where $Y_1 = 160 \sin(120\pi X)$ and $Y_2 = 12 \sin(120\pi X)$, as shown in Figure 7.62 where radian mode is used.

$[0, 0.04, 0.01]$ by $[-200, 2200, 200]$

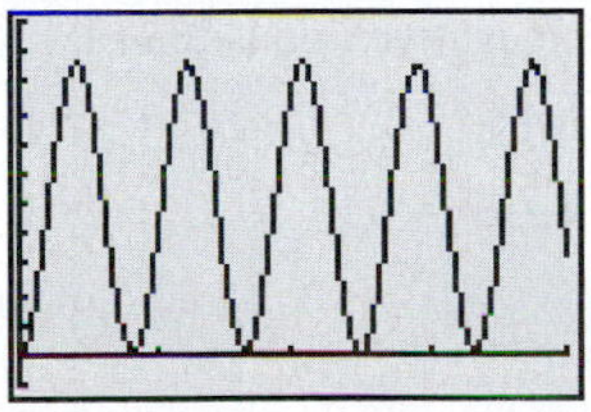

(b)
$$W = VI$$

$$= 160 \sin(120\pi t)\, 12 \sin(120\pi t) \qquad \text{Substitute for } V \text{ and } I.$$

$$= 1920\, \sin^2(120\pi t) \qquad \text{Multiply.}$$

$$= 1920 \cdot \frac{1 - \cos(240\pi t)}{2} \qquad \text{Power-reducing identity}$$

$$= 960 - 960 \cos(240\pi t) \qquad \text{Simplify.}$$

FIGURE 7.62 Wattage W

Thus let $a = -960$, $k = 240$, and $d = 960$. Then the wattage can be written as

$$W = -960 \cos(240\pi t) + 960.$$

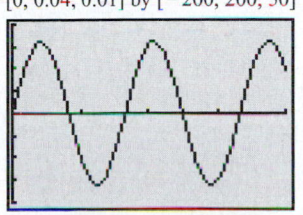

$[0, 0.04, 0.01]$ by $[-200, 200, 50]$

(c) The period for both V and I is $\frac{2\pi}{120\pi} = \frac{1}{60}$ second, and the period for W is $\frac{2\pi}{240\pi} = \frac{1}{120}$ second. This result is supported in Figure 7.63, where the graph of V requires twice as much time as W to complete one oscillation.
(d) The maximum wattage is 1920 watts. Half this amount is 960 watts, which is the wattage rating for the toaster.   *Now Try Exercise 109* ◆

FIGURE 7.63 Voltage V

We can obtain half-angle formulas by using the power-reducing identities.

$$\sin^2 x = \frac{1 - \cos 2x}{2} \qquad \text{Power-reducing identity}$$

$$\sin x = \pm\sqrt{\frac{1 - \cos 2x}{2}} \qquad \text{Square root property}$$

$$\sin \frac{\theta}{2} = \pm\sqrt{\frac{1 - \cos \theta}{2}} \qquad \text{Let } x = \frac{\theta}{2} \text{ and } 2x = \theta.$$

The following gives some half-angle formulas. (Verification of the second and third half-angle formulas for tangent is done in Exercises 75 and 76.)

HALF-ANGLE FORMULAS

$$\sin \frac{\theta}{2} = \pm\sqrt{\frac{1 - \cos \theta}{2}} \qquad \cos \frac{\theta}{2} = \pm\sqrt{\frac{1 + \cos \theta}{2}}$$

$$\tan \frac{\theta}{2} = \pm\sqrt{\frac{1 - \cos \theta}{1 + \cos \theta}} \qquad \tan \frac{\theta}{2} = \frac{1 - \cos \theta}{\sin \theta} \qquad \tan \frac{\theta}{2} = \frac{\sin \theta}{1 + \cos \theta}$$

To decide whether a positive or negative sign should be used in a half-angle formula, we must determine the quadrant containing $\frac{\theta}{2}$. This is illustrated in the next example.

EXAMPLE 8 Using a half-angle formula to find an exact value

Find the exact value of $\sin(-15°)$.

SOLUTION We can use the fact that $\cos(-30°) = \frac{\sqrt{3}}{2}$ to find $\sin(-15°)$. We choose the negative sign in the half-angle formula since the sine function is negative in quadrant IV.

◆ **CLASS DISCUSSION**
Find the exact value of $\tan(-15°)$. Use a calculator to support your result. ◆

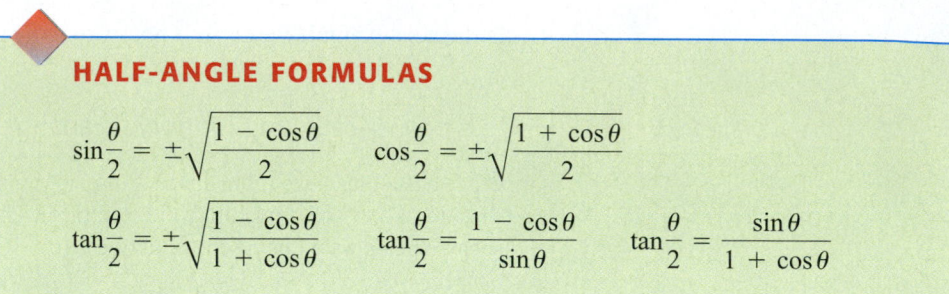

$$\sin(-15°) = \sin\left(\frac{-30°}{2}\right) \qquad \text{Let } \frac{\theta}{2} = -15° \text{ and } \theta = -30°.$$

$$= -\sqrt{\frac{1 - \cos(-30°)}{2}} \qquad \text{Half-angle formula}$$

$$= -\sqrt{\frac{1 - \sqrt{3}/2}{2}} \qquad \cos(-30°) = \frac{\sqrt{3}}{2}$$

$$= -\sqrt{\frac{2 - \sqrt{3}}{4}} \qquad \text{Multiply numerator and denominator by 2.}$$

$$= -\frac{\sqrt{2 - \sqrt{3}}}{2} \qquad \sqrt{4} = 2 \qquad \qquad \textit{Now Try Exercise 41} ◆$$

EXAMPLE 9   ### Using half-angle formulas to find exact values

If $\cos\theta = -\frac{3}{5}$ and $90° \leq \theta \leq 180°$, find $\sin\frac{\theta}{2}$, $\cos\frac{\theta}{2}$, and $\tan\frac{\theta}{2}$.

SOLUTION   Since $90° \leq \theta \leq 180°$, it follows that $45° \leq \frac{\theta}{2} \leq 90°$. In quadrant I the values for $\sin\frac{\theta}{2}$, $\cos\frac{\theta}{2}$, and $\tan\frac{\theta}{2}$ are all positive.

$$\sin\frac{\theta}{2} = \sqrt{\frac{1 - \cos\theta}{2}} = \sqrt{\frac{1 + 3/5}{2}} = \sqrt{\frac{4}{5}} \quad \text{or} \quad \frac{2\sqrt{5}}{5}$$

$$\cos\frac{\theta}{2} = \sqrt{\frac{1 + \cos\theta}{2}} = \sqrt{\frac{1 - 3/5}{2}} = \sqrt{\frac{1}{5}} \quad \text{or} \quad \frac{\sqrt{5}}{5}$$

$$\tan\frac{\theta}{2} = \sqrt{\frac{1 - \cos\theta}{1 + \cos\theta}} = \sqrt{\frac{1 + 3/5}{1 - 3/5}} = 2 \qquad \textit{Now Try Exercise 51} \blacklozenge$$

Solving Equations

When trigonometric functions are used in modeling, it is common to use these functions to make predictions. This often results in trigonometric equations.

EXAMPLE 10   ### Solving a trigonometric equation

Solve the trigonometric equation $\cos\theta - \sin 2\theta = 0$ symbolically, graphically, and numerically for $0° \leq \theta < 360°$.

SOLUTION
Symbolic Solution   We begin by applying the double-angle identity for sine.

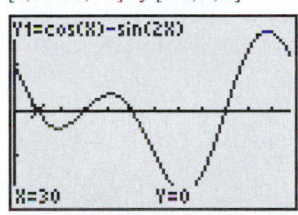

[0, 352.5, 30] by [−2, 2, 1]

FIGURE 7.64   Degree Mode

$\cos\theta - \sin 2\theta = 0$		*Given equation*
$\cos\theta - 2\sin\theta\cos\theta = 0$		*Double-angle identity*
$\cos\theta(1 - 2\sin\theta) = 0$		*Factor out $\cos\theta$.*
$\cos\theta = 0$ or	$1 - 2\sin\theta = 0$	*Zero-product property*
$\cos\theta = 0$ or	$\sin\theta = \dfrac{1}{2}$	*Solve for $\sin\theta$.*
$\theta = 90°, 270°$ or	$\theta = 30°, 150°$	*Solve for θ.*

On the interval $[0°, 360°)$, the solutions are $30°$, $90°$, $150°$, and $270°$.

Graphical Solution   In Figure 7.64 graphical support is given, where the equation $Y_1 = \cos(X) - \sin(2X)$ is shown. Note that four *x*-intercepts correspond to the four symbolic solutions.

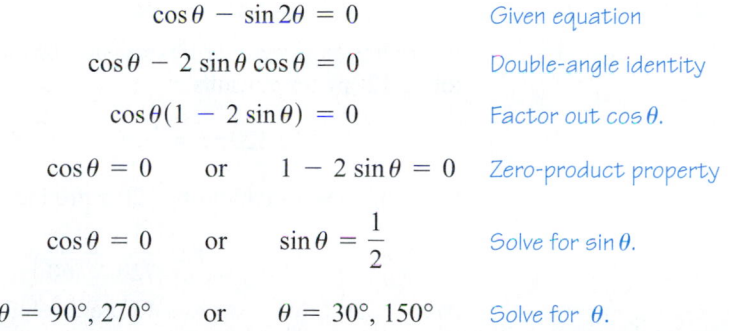

X	Y₁
30	0
90	0
150	0
210	-1.732
270	0
330	1.7321
390	0

X=30

FIGURE 7.65   Degree Mode

Numerical Solution   In Figure 7.65 numerical support is shown.

Now Try Exercise 85 $\blacklozenge$

EXAMPLE 11   Solving a trigonometric equation

Solve the equation $4\cos\frac{x}{2} - 2 = 0$ for $0° \leq x < 360°$.

SOLUTION   Begin by solving for $\cos\frac{x}{2}$.

$$4\cos\frac{x}{2} - 2 = 0 \qquad \textcolor{blue}{\textit{Given equation}}$$

$$\cos\frac{x}{2} = \frac{1}{2} \qquad \textcolor{blue}{\textit{Solve for } \cos\frac{x}{2}.}$$

$$\frac{x}{2} = 60° \quad \text{or} \quad \frac{x}{2} = 300° \qquad \textcolor{blue}{\textit{Solve for } \frac{x}{2}.}$$

$$x = 120° \quad \text{or} \quad x = 600° \qquad \textcolor{blue}{\textit{Multiply by 2.}}$$

The only solution on $[0°, 360°)$ is $120°$.   *Now Try Exercise 95* ◆

EXAMPLE 12   Solving a trigonometric equation from electronics

In Example 7 the voltage in a household electrical outlet was modeled by

$$V(t) = 160\sin(120\pi t),$$

where t is time in seconds. Find the times when the voltage equals 80 volts.

SOLUTION   We must solve the equation $V(t) = 80$ for t.

$$160\sin(120\pi t) = 80 \qquad \textcolor{blue}{V(t) = 80}$$

$$\sin(120\pi t) = \frac{1}{2} \qquad \textcolor{blue}{\textit{Divide by 160.}}$$

Let $\theta = 120\pi t$. The equation $\sin\theta = \frac{1}{2}$ is satisfied whenever

$$\theta = \frac{\pi}{6} + 2\pi n \qquad \text{or} \qquad \theta = \frac{5\pi}{6} + 2\pi n,$$

where n is an integer. The expression $2\pi n$ is included because $\sin\theta$ has period 2π. Substituting $120\pi t$ for θ results in

$$120\pi t = \frac{\pi}{6} + 2\pi n \qquad \text{or} \qquad 120\pi t = \frac{5\pi}{6} + 2\pi n.$$

Dividing these equations by 120π produces the following solutions.

$$t = \frac{1}{720} + \frac{n}{60} \qquad \text{or} \qquad t = \frac{5}{720} + \frac{n}{60}$$

Note: Each value of n gives different solutions. For example, if $n = 1$ then

$$t = \frac{1}{720} + \frac{1}{60} = \frac{13}{720} \qquad \text{and} \qquad t = \frac{5}{720} + \frac{1}{60} = \frac{17}{720}.$$

Two times when the voltage equals 80 volts are after $\frac{13}{720}$ second and after $\frac{17}{720}$ second.
                                                        Now Try Exercise 111 ◆

EXAMPLE 13   Solving trigonometric equations

Find all solutions expressed in radians.
(a) $\cos 2t + 2\cos^2 t = 0$ **(b)** $2\sin 2t = \sqrt{3}$

SOLUTION

(a) Begin by using a double-angle identity.

$$\cos 2t + 2\cos^2 t = 0 \qquad \text{Given equation}$$

$$2\cos^2 t - 1 + 2\cos^2 t = 0 \qquad \text{Double-angle identity}$$

$$4\cos^2 t - 1 = 0 \qquad \text{Combine terms.}$$

$$\cos^2 t = \frac{1}{4} \qquad \text{Solve for } \cos^2 t.$$

$$\cos t = \pm\frac{1}{2} \qquad \text{Square root property}$$

On the interval $[0, 2\pi)$, there are four angles whose cosines equal either $\frac{1}{2}$ or $-\frac{1}{2}$. They are $\frac{\pi}{3}, \frac{2\pi}{3}, \frac{4\pi}{3}$, and $\frac{5\pi}{3}$. Since $\cos t$ has period 2π, all solutions can be written as

$$\frac{\pi}{3} + 2\pi n, \quad \frac{2\pi}{3} + 2\pi n, \quad \frac{4\pi}{3} + 2\pi n, \quad \text{or} \quad \frac{5\pi}{3} + 2\pi n,$$

where n is an integer. Note that since $\frac{4\pi}{3} - \frac{\pi}{3} = \pi$, the solutions $\frac{\pi}{3} + 2\pi n$ and $\frac{4\pi}{3} + 2\pi n$ can be combined and written as $\frac{\pi}{3} + \pi n$. Similarly, the solutions $\frac{2\pi}{3} + 2\pi n$ and $\frac{5\pi}{3} + 2\pi n$ can be written as $\frac{2\pi}{3} + \pi n$. As a result, all solutions can also be written as

$$\frac{\pi}{3} + \pi n \qquad \text{or} \qquad \frac{2\pi}{3} + \pi n.$$

(b) Begin by dividing each side by 2.

$$2\sin 2t = \sqrt{3} \qquad \text{Given equation}$$

$$\sin 2t = \frac{\sqrt{3}}{2} \qquad \text{Divide by 2.}$$

Let $\theta = 2t$ and find all values of θ on the interval $[0, 2\pi)$, where $\sin\theta = \frac{\sqrt{3}}{2}$. Because the reference angle for θ is $\theta_R = \sin^{-1}\frac{\sqrt{3}}{2} = \frac{\pi}{3}$ and the sine function is positive in quadrants I and II, it follows that $\theta = \frac{\pi}{3}$ or $\theta = \frac{2\pi}{3}$. Thus all possible solutions in terms of θ can be written as

$$\theta = \frac{\pi}{3} + 2\pi n \qquad \text{or} \qquad \theta = \frac{2\pi}{3} + 2\pi n.$$

Since $\theta = 2t$, we can write the solutions to the given equation in terms of t.

$$2t = \frac{\pi}{3} + 2\pi n \qquad \text{or} \qquad 2t = \frac{2\pi}{3} + 2\pi n$$

$$t = \frac{\pi}{6} + \pi n \qquad \text{or} \qquad t = \frac{\pi}{3} + \pi n$$

Now Try Exercises 93 and 99. ◆

Product-to-Sum and Sum-to-Product Identities

The sum and difference identities for sine and cosine can be used to derive several identities that make it possible to rewrite a product as a sum. For example, adding the identities for $\cos(\alpha + \beta)$ and $\cos(\alpha - \beta)$ results in the following.

$$\cos(\alpha + \beta) = \cos\alpha\cos\beta - \sin\alpha\sin\beta$$

$$\cos(\alpha - \beta) = \cos\alpha\cos\beta + \sin\alpha\sin\beta$$

$$\overline{\cos(\alpha + \beta) + \cos(\alpha - \beta) = 2\cos\alpha\cos\beta}$$

Rewriting gives

$$\cos\alpha\cos\beta = \frac{1}{2}(\cos(\alpha+\beta) + \cos(\alpha-\beta)).$$

Four product-to-sum identities are given. The other three identities are derived in a similar manner.

PRODUCT-TO-SUM IDENTITIES

$$\cos\alpha\cos\beta = \frac{1}{2}(\cos(\alpha+\beta) + \cos(\alpha-\beta))$$

$$\sin\alpha\sin\beta = \frac{1}{2}(\cos(\alpha-\beta) - \cos(\alpha+\beta))$$

$$\sin\alpha\cos\beta = \frac{1}{2}(\sin(\alpha+\beta) + \sin(\alpha-\beta))$$

$$\cos\alpha\sin\beta = \frac{1}{2}(\sin(\alpha+\beta) - \sin(\alpha-\beta))$$

 **Using a product-to-sum identity**

Write $\cos 5\theta \cos 3\theta$ as a sum.

SOLUTION We begin by applying the first product-to-sum identity with the substitution $\alpha = 5\theta$ and $\beta = 3\theta$.

$$\cos 5\theta \cos 3\theta = \frac{1}{2}(\cos(5\theta + 3\theta) + \cos(5\theta - 3\theta))$$

$$= \frac{1}{2}(\cos 8\theta + \cos 2\theta) \qquad \textit{Now Try Exercise 77} \blacklozenge$$

By rewriting the product-to-sum identities, we can derive four sum-to-product identities. Sum-to-product identities have applications in areas such as musical sounds and touch-tone phones. If we let $a = \alpha + \beta$ and $b = \alpha - \beta$, then the identity

$$\cos\alpha\cos\beta = \frac{1}{2}(\cos(\alpha+\beta) + \cos(\alpha-\beta))$$

reduces to

$$2\cos\frac{a+b}{2}\cos\frac{a-b}{2} = \cos a + \cos b.$$

Note that

$$\frac{a+b}{2} = \frac{\alpha+\beta+\alpha-\beta}{2} = \alpha \quad \text{and}$$

$$\frac{a-b}{2} = \frac{(\alpha+\beta)-(\alpha-\beta)}{2} = \beta.$$

The other three sum-to-product identities can be derived in a similar manner.

SUM-TO-PRODUCT IDENTITIES

$$\cos a + \cos b = 2 \cos \frac{a+b}{2} \cos \frac{a-b}{2}$$

$$\cos a - \cos b = -2 \sin \frac{a+b}{2} \sin \frac{a-b}{2}$$

$$\sin a + \sin b = 2 \sin \frac{a+b}{2} \cos \frac{a-b}{2}$$

$$\sin a - \sin b = 2 \cos \frac{a+b}{2} \sin \frac{a-b}{2}$$

Touch-tone phones are often used to register for classes or make business transactions. One reason this can be done is that each number has a unique pair of frequencies. For example, when 1 is pressed, two frequencies of 697 hertz and 1209 hertz are simultaneously transmitted through the phone line. A *hertz* (Hz) is equal to one cycle per second. When 2 is pressed, the pair of frequencies transmitted is 697 hertz and 1336 hertz. As a result, 2 has a different tone from 1. Table 7.1 shows the frequency pairs for the numbers 0 through 9.

TABLE 7.1   Frequencies Used in Touch-Tone Phones

Number	0	1	2	3	4	5	6	7	8	9
Frequency 1 (Hz)	941	697	697	697	770	770	770	852	852	852
Frequency 2 (Hz)	1336	1209	1336	1477	1209	1336	1477	1209	1336	1477

A tone with frequencies F_1 and F_2 can be modeled by

$$a_1 \cos (2\pi F_1 t) + a_2 \cos (2\pi F_2 t).$$

If both tones have the same intensity, then we can let $a_1 = a_2 = 1$.

 EXAMPLE 15   Analyzing touch-tone phones

(a) Write an expression that models the sound of a 5 on a touch-tone phone.
(b) Rewrite the expression in part (a) as a product of trigonometric expressions.
(c) Give graphical and numerical support for your answer in part (b).

SOLUTION

(a) From Table 7.1 we can see that for number 5, $F_1 = 770$ and $F_2 = 1336$. Thus

$$y = \cos (1540\pi t) + \cos (2672\pi t).$$

(b) Let $a = 1540\pi t$ and $b = 2672\pi t$ in the first sum-to-product identity.

$$\cos 1540\pi t + \cos 2672\pi t$$

$$= 2 \cos \frac{1540\pi t + 2672\pi t}{2} \cos \frac{1540\pi t - 2672\pi t}{2}$$

$$= 2 \cos (2106\pi t) \cos (-566\pi t)$$

$$= 2 \cos (2106\pi t) \cos (566\pi t) \qquad\qquad \cos(-\theta) = \cos\theta$$

(c) This result can be supported by graphing and making a table of

$$Y_1 = \cos(1540\pi X) + \cos(2672\pi X) \qquad \text{and}$$

$$Y_2 = 2\cos(2106\pi X)\cos(566\pi X),$$

as shown in Figures 7.66–7.68. Notice that the graphs of Y_1 and Y_2 appear to be identical.

[0, 0.005, 0.001] by [−2, 2, 1]

[0, 0.005, 0.001] by [−2, 2, 1]

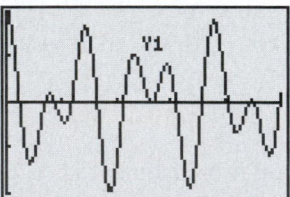

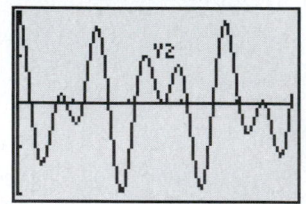

X	Y₁	Y₂
0	2	2
.5236	⁻.5087	⁻.5087
1.0472	.39189	.39189
1.5708	⁻1.861	⁻1.861
2.0944	.36143	.36143
2.618	⁻.024	⁻.024
3.1416	1.4802	1.4802

X=0

FIGURE 7.66 Radian Mode   **FIGURE 7.67** Radian Mode   **FIGURE 7.68** Radian Mode

Now Try Exercise 117 ◆

7.5

Putting it all Together

In this section we introduced several multiple-angle identities. If the trigonometric functions of θ are known, these identities can be used to find the trigonometric functions of $\frac{\theta}{2}$ and 2θ. Some multiple-angle identities or formulas are summarized in the following table.

Identity or Formula	General Form
Double-angle	$\sin 2\theta = 2\sin\theta\cos\theta$
	$\cos 2\theta = \cos^2\theta - \sin^2\theta = 2\cos^2\theta - 1 = 1 - 2\sin^2\theta$
	$\tan 2\theta = \dfrac{2\tan\theta}{1 - \tan^2\theta}$
Power-reducing	$\sin^2\theta = \dfrac{1 - \cos 2\theta}{2} \qquad \cos^2\theta = \dfrac{1 + \cos 2\theta}{2} \qquad \tan^2\theta = \dfrac{1 - \cos 2\theta}{1 + \cos 2\theta}$
Half-angle	$\sin\dfrac{\theta}{2} = \pm\sqrt{\dfrac{1 - \cos\theta}{2}} \qquad \cos\dfrac{\theta}{2} = \pm\sqrt{\dfrac{1 + \cos\theta}{2}} \qquad \tan\dfrac{\theta}{2} = \pm\sqrt{\dfrac{1 - \cos\theta}{1 + \cos\theta}}$
	$\tan\dfrac{\theta}{2} = \dfrac{1 - \cos\theta}{\sin\theta} \qquad \tan\dfrac{\theta}{2} = \dfrac{\sin\theta}{1 + \cos\theta}$

7.5 Exercises

Double-Angle Identities

Exercises 1–6: If possible, evaluate each expression and compare their values.

1. (a) $\sin 30° + \sin 30°$
   (b) $\sin 60°$

2. (a) $\sin 45° + \sin 45°$
   (b) $\sin 90°$

3. (a) $\cos 60° + \cos 60°$
   (b) $\cos 120°$

4. (a) $\cos 90° + \cos 90°$
   (b) $\cos 180°$

5. (a) $\tan 45° + \tan 45°$
   (b) $\tan 90°$

6. (a) $\tan 30° + \tan 30°$
   (b) $\tan 60°$

Exercises 7 and 8: (Refer to Example 1.) Verify graphically and symbolically that the two expressions are not equivalent.

7. $\tan 2\theta,\ 2\tan\theta$

8. $\cos 3\theta,\ 3\cos\theta$

Exercises 9–16: (Refer to Example 2.) Complete each of the following.

   (a) *Find* $\sin 2\theta,\ \cos 2\theta,$ *and* $\tan 2\theta.$
   (b) *Use a calculator to support your results.*

9. $\cos\theta = \frac{4}{5}$ and $\sin\theta = \frac{3}{5}$

10. $\sin\theta = \frac{12}{13}$ and $\cos\theta = \frac{5}{13}$

11. $\sin\theta = -\frac{24}{25}$ and $\cos\theta > 0$

12. $\cos\theta = -\frac{7}{25}$ and $\tan\theta > 0$

13. $\sin\theta = -\frac{11}{61}$ and $\sec\theta > 0$

14. $\csc\theta = -2$ and $\sec\theta > 0$

15. $\tan\theta = \frac{7}{24}$ and $\cos\theta < 0$

16. $\cot\theta = \frac{5}{12}$ and $\sin\theta > 0$

Exercises 17–26: (Refer to Example 3.) Use an identity to evaluate the expression.

17. $\sin\left(2\cos^{-1} 1\right)$

18. $\cos\left(2\sin^{-1}\frac{1}{2}\right)$

19. $\cos\left(2\sin^{-1}\frac{7}{25}\right)$

20. $\sin\left(2\tan^{-1}\frac{3}{4}\right)$

21. $\cos\left(3\sin^{-1}\frac{5}{13}\right)$ (*Hint:* Use the results from Example 5.)

22. $\tan\left(2\tan^{-1}\frac{1}{2}\right)$

23. $\cos\left(2\sin^{-1} x\right),\ x > 0$

24. $\sin\left(2\tan^{-1} x\right),\ x < 0$

25. $\cos\left(\sin^{-1}\frac{3}{5} - \sin^{-1}\frac{4}{5}\right)$ (*Hint:* Let $\alpha = \sin^{-1}\frac{3}{5}$, $\beta = \sin^{-1}\frac{4}{5}$, and apply a difference identity from Section 7.4.)

26. $\sin\left(\tan^{-1}\frac{12}{5} + \cos^{-1}\frac{12}{13}\right)$

Exercises 27–32: Rewrite the expression using a double-angle identity.

27. $2\cos\theta\sin\theta$

28. $2\sin 2\theta\cos 2\theta$

29. $\sin\theta\cos\theta$

30. $(\sin\theta - \cos\theta)(\sin\theta + \cos\theta)$

31. $2\cos^2 2\theta - 1$

32. $1 - 2\sin^2 3\theta$

Exercises 33–36: Use a fundamental identity to write the expression as one term.

33. $\sin^2 3\theta + \cos^2 3\theta$

34. $1 + \tan^2 2\theta$

35. $\csc^2 5x - 1$

36. $\sin^2 8x + \cos^2 8x$

Power-Reducing Identities and Half-Angle Formulas

Exercises 37–40: Use a power-reducing identity to find the exact value of the expression. Use a calculator to support your result.

37. $\cos^2 (22.5°)$

38. $\sin^2 (15°)$

39. $\tan^2 (75°)$

40. $\csc^2 (105°)$

Exercises 41–44: Use a half-angle formula to find the exact value of the expression. Use a calculator to support your result.

41. (a) $\cos 15°$
    (b) $\tan(-15°)$

42. (a) $\sin 67.5°$
    (b) $\cos(-67.5°)$

43. (a) $\tan\dfrac{\pi}{8}$
    (b) $\sin\left(-\dfrac{\pi}{8}\right)$

44. (a) $\cos\dfrac{5\pi}{12}$
    (b) $\cos\left(-\dfrac{5\pi}{12}\right)$

Exercises 45–50: Use a half-angle formula to simplify the expression. Use a calculator to support your result.

45. $\sqrt{\dfrac{1 - \cos 60°}{2}}$ **46.** $\sqrt{\dfrac{1 + \cos 60°}{2}}$

47. $\sqrt{\dfrac{1 + \cos 50°}{2}}$ **48.** $\sqrt{\dfrac{1 - \cos 50°}{2}}$

49. $\sqrt{\dfrac{1 - \cos 40°}{1 + \cos 40°}}$ **50.** $\sqrt{\dfrac{1 + \cos 26°}{1 - \cos 26°}}$

Exercises 51–56: (Refer to Example 9.) Find $\sin \frac{\theta}{2}$, $\cos \frac{\theta}{2}$, and $\tan \frac{\theta}{2}$.

51. $\cos \theta = \frac{4}{5}$ and $0° < \theta < 90°$

52. $\cos \theta = \frac{1}{3}$ and $0° < \theta < 90°$

53. $\tan \theta = -\frac{5}{12}$ and $-90° < \theta < 0°$

54. $\sec \theta = -2$ and $90° < \theta < 180°$

55. $\csc \theta = \frac{25}{24}$ and $90° < \theta < 180°$

56. $\sin \theta = \frac{4}{5}$ and $0° < \theta < 90°$

Verifying Identities

Exercises 57–66: Verify the identity. Give graphical or numerical support.

57. $4 \sin 2x = 8 \sin x \cos x$ **58.** $\cos 4\theta = 1 - 2 \sin^2 2\theta$

59. $\dfrac{2 - \sec^2 x}{\sec^2 x} = \cos 2x$

60. $(\sin x + \cos x)^2 = \sin 2x + 1$

61. $\sec 2x = \dfrac{1}{1 - 2 \sin^2 x}$ **62.** $2 \csc 2t = \csc t \sec t$

63. $\sin 3\theta = 3 \sin \theta - 4 \sin^3 \theta$

64. $\dfrac{2 \tan x}{1 + \tan^2 x} = \sin 2x$

65. $\sin 4\theta = 4 \sin \theta \cos \theta \cos 2\theta$

66. $\cos 4t = 8 \cos^4 t - 8 \cos^2 t + 1$

Exercises 67–76: Verify the identity.

67. $\dfrac{\sin 2\theta}{\sin \theta} = 2 \cos \theta$

68. $2 \sin^2 4\theta = 1 - \cos 8\theta$ **69.** $2 \cos^2 \left(\dfrac{\theta}{2}\right) = 1 + \cos \theta$

70. $\dfrac{\sin^2 2\theta}{1 + \cos 2\theta} = 2 \sin^2 \theta$ **71.** $\cos^4 \theta - \sin^4 \theta = \cos 2\theta$

72. $\dfrac{1 - \tan^2 x}{1 + \tan^2 x} = \cos 2x$ **73.** $\csc 2t = \dfrac{\csc t}{2 \cos t}$

74. $\tan \theta + \cot \theta = \dfrac{2}{\sin 2\theta}$

75. $\tan \dfrac{x}{2} = \dfrac{\sin x}{1 + \cos x}$ $\left(Hint: \text{Let } \tan \dfrac{x}{2} = \dfrac{\sin (x/2)}{\cos (x/2)}.\right)$

76. $\tan \dfrac{x}{2} = \dfrac{1 - \cos x}{\sin x}$ (*Hint:* Use the identity in Exercise 75.)

Product-to-Sum and Sum-to-Product Identities

Exercises 77–80: Write each expression as a sum or difference of trigonometric functions.

77. **(a)** $\cos 50° \sin 20°$ **(b)** $\cos 40° \cos 20°$

78. **(a)** $2 \sin 74° \sin 24°$ **(b)** $8 \sin 144° \cos 104°$

79. **(a)** $\sin 7\theta \cos 3\theta$ **(b)** $\sin 8x \sin 4x$

80. **(a)** $2 \cos 5x \cos 7x$ **(b)** $4 \cos 9\theta \sin 2\theta$

Exercises 81–84: Write each expression as a product of trigonometric functions.

81. **(a)** $\sin 40° + \sin 30°$ **(b)** $\cos 45° + \cos 35°$

82. **(a)** $\cos 104° - \cos 24°$ **(b)** $\sin 32° - \sin 64°$

83. **(a)** $\cos 6\theta + \cos 4\theta$ **(b)** $\sin 7x + \sin 4x$

84. **(a)** $\sin 3x - \sin 5x$ **(b)** $\cos 3\theta - \cos \theta$

Solving Equations

Exercises 85–92: Find the solutions to the equation in the interval $[0°, 360°)$

   (a) symbolically,
   (b) graphically, and
   (c) numerically.

85. $\cos 2\theta = 1$ **86.** $\sin 2\theta = \frac{1}{2}$

87. $\sin 2\theta = 0$ **88.** $\cos \dfrac{\theta}{2} = -\dfrac{\sqrt{3}}{2}$

89. $\sin \dfrac{\theta}{2} = 1$ **90.** $\cos 2\theta + \cos \theta = 0$

91. $\sqrt{2} \sin \dfrac{\theta}{2} - 1 = 0$ **92.** $2 \cos \dfrac{\theta}{2} + 1 = 0$

Exercises 93–106: *Find all solutions expressed in radians.*

93. $\sin 2t + \sin t = 0$ **94.** $\sin t - \cos 2t = 0$

95. $2 \sin \dfrac{t}{2} - 1 = 0$ **96.** $\sin 2t = 2 \cos^2 t$

97. $\cos 2t = \sin t$ **98.** $\cos 2t - \cos t = 0$

99. $\tan 2t = 1$ **100.** $\cot 2t = \sqrt{3}$

101. $2 \cos \dfrac{t}{2} = 1$ **102.** $\tan \dfrac{t}{2} = 1$

103. $\cos 2t = 2 \sin t \cos t$ **104.** $2 \cos^2 2t = 1 - \cos 2t$

105. $2 \sin^2 2t + \sin 2t - 1 = 0$

106. $2 \cos \dfrac{t}{2} + 1 = 0$

Exercises 107 and 108: *Approximate to the nearest thousandth all solutions on* $[0, 2\pi)$.

107. $\sin t + \sin 2t = \cos t$

108. $\sin 3t + \sin 2t = 2 \cos t$

Applications

109. *Electricity*   (Refer to Example 7.) Suppose the voltage in a 220-volt electrical circuit is modeled by $V(t) = 310 \sin (120\pi t)$ and the amperage flowing through a heater is $I = 7 \sin (120\pi t)$.   (**Source:** G. Wilcox and C. Hesselberth.)

  (a) Graph the wattage $W = VI$ consumed by the heater in $[0, 0.04, 0.01]$ by $[-500, 2500, 500]$.

  (b) Find values for the constants a, k, and d so that $W = a \cos (k\pi t) + d$.

110. *Electricity*   If a toaster is plugged into a common household outlet, the wattage W used varies according to the equation $W = \dfrac{V^2}{R}$, where V is the voltage and R is a constant that measures the resistance of the toaster in ohms.   (**Source:** D. Bell, *Fundamentals of Electric Circuits.*)

  (a) Graph W if $R = 15$ and $V = 163 \sin (120\pi t)$ in $[0, 0.05, 0.01]$ by $[-500, 2000, 500]$.

  (b) Approximate the maximum wattage consumed by the toaster.

  (c) Use a power-reducing identity to express the wattage as $W = a \cos (240\pi t) + d$, where a and d are constants.

Exercises 111 and 112: *Electricity*   (Refer to Example 12.) Let $V(t) = 320 \sin (120\pi t)$ denote the voltage in an electrical outlet at time t in seconds. Find the times when V equals the following values.

111. 160 volts   **112.** $160\sqrt{3}$ volts

113. *Highway Curves*   When an automobile travels along a circular curve, objects like trees and buildings situated on the inside of the curve can obstruct a driver's vision. If the cars in the accompanying figure are a safe stopping distance apart, then the distance d that should be cleared on the inside of the curve is $d = r \left(1 - \cos \dfrac{\beta}{2} \right)$, where r is the radius of the curve and β is the central angle between the cars.   (**Source:** F. Mannering and W. Kilareski, *Principles of Highway Engineering and Traffic Analysis.*)

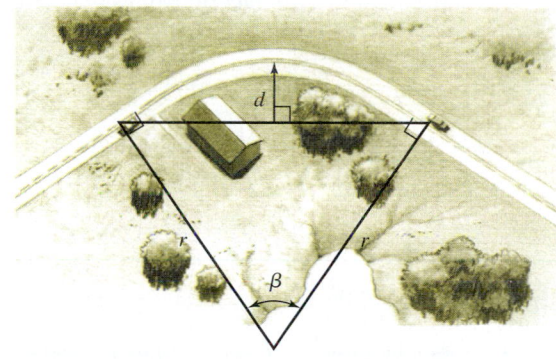

  (a) Find d if $\beta = 80°$ and $r = 600$ feet.

  (b) Use the figure to justify this formula.

  (c) Is the given formula equivalent to the formula $d = r\left(1 - \frac{1}{2} \cos \beta\right)$? Explain.

114. *Highway Curves*   The figure below represents a circular curve with radius r and central angle θ. The tangent length T is an important distance used by surveyors. (**Source:** F. Mannering.)

  (a) Show that $T = r \tan \dfrac{\theta}{2}$.

  (b) Find T for a curve with 1500-foot radius and $\theta = 80°$.

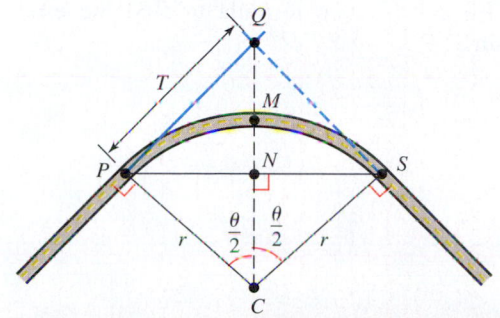

115. *Silver Box* (Refer to Example 15.) A *silver box* contains the standard touch-tone phone keys along with special keys for A, B, C, and D. It has 16 different tones rather than the typical 12 tones. These extra tones are sometimes used to enter security codes. Their frequency pairs in hertz are listed in the table.

Letter	A	B	C	D
Frequency 1 (Hz)	697	770	852	941
Frequency 2 (Hz)	1633	1633	1633	1633

(a) Write a function f expressed as the *sum* of two trigonometric expressions that models the tone generated by A.

(b) Write f using a *product* of two trigonometric functions.

(c) Graph f in $[0, 0.005, 0.001]$ by $[-2, 2, 1]$.

116. *Touch-Tone Phones* (Refer to the previous exercise.) The sound generated by the letter D can be modeled by

$$f(t) = \cos(1882\pi t) + \cos(3266\pi t).$$

(a) Graph f in $[0, 0.005, 0.001]$ by $[-2, 2, 1]$.

(b) Does the graph of f appear to be periodic?

(c) Conjecture whether f is periodic.

117. *Touch-Tone Phones* (Refer to Example 15.) For the numbers 3 and 4 on a touch-tone phone complete the following.

(a) Write an expression that is the *sum* of two trigonometric functions that models the tone generated by each number.

(b) Write the expression for each tone written as a *product* of two trigonometric functions.

(c) Graph the expression for each number in the window $[0, 0.005, 0.001]$ by $[-2, 2, 1]$. Do the two tones for 3 and 4 appear to be different? Explain why they should be different.

118. *Musical Tones and Beats* If two musical tones with nearly the same frequency are played simultaneously, a phenomenon called *beats* occurs. Let $P_1 = 0.04\cos(110\pi t)$ and $P_2 = 0.04\cos(116\pi t)$ represent these tones.

(a) Graph $P = P_1 + P_2$ in the viewing rectangle $[0.2, 1.2, 0.2]$ by $[-0.08, 0.08, 0.01]$. How many beats are there?

(b) Use a sum-to-product identity to write $P_1 + P_2$ as a product of trigonometric expressions.

(c) Give numerical support for your result in part (b).

Writing about Mathematics

119. Suppose a student believes that an equation is an identity but cannot verify it symbolically. Discuss techniques that the student could use to support this belief.

120. Does the equation $\sin^2 3\theta + \cos^2 2\theta = 1$ represent an identity? Explain your reasoning.

CHECKING BASIC CONCEPTS FOR SECTION 7.5

1. Find values for $\sin 2\theta$ and $\cos 2\theta$ if $\cos\theta = -\frac{7}{25}$ and $\sin\theta = \frac{24}{25}$.

2. Use a half-angle formula to find the exact value of $\sin 22.5°$.

3. Find $\sin\frac{\theta}{2}$ and $\cos\frac{\theta}{2}$ if $\sin\theta = \frac{4}{5}$ and θ is acute.

4. Verify that $\dfrac{\cos 2\theta}{\cos^2\theta} = 2 - \sec^2\theta$.

5. Solve $\sin 2\theta = 2\cos\theta$ for $0° \le \theta < 360°$.

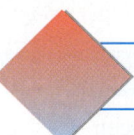

CHAPTER 7 SUMMARY

CONCEPT	EXPLANATION AND EXAMPLES

SECTION 7.1 FUNDAMENTAL IDENTITIES

RECIPROCAL IDENTITIES

$$\sin\theta = \frac{1}{\csc\theta} \qquad \cos\theta = \frac{1}{\sec\theta} \qquad \tan\theta = \frac{1}{\cot\theta}$$

$$\csc\theta = \frac{1}{\sin\theta} \qquad \sec\theta = \frac{1}{\cos\theta} \qquad \cot\theta = \frac{1}{\tan\theta}$$

Example: If $\sin\theta = \dfrac{3}{5}$, then $\csc\theta = \dfrac{5}{3}$.

QUOTIENT IDENTITIES

$$\tan\theta = \frac{\sin\theta}{\cos\theta} \qquad \cot\theta = \frac{\cos\theta}{\sin\theta}$$

Example: If $\sin\theta = -\dfrac{3}{5}$ and $\cos\theta = \dfrac{4}{5}$, then $\tan\theta = \dfrac{-3/5}{4/5} = -\dfrac{3}{4}$.

PYTHAGOREAN IDENTITIES

$$\sin^2\theta + \cos^2\theta = 1 \qquad 1 + \tan^2\theta = \sec^2\theta \qquad 1 + \cot^2\theta = \csc^2\theta$$

Examples: $\sin^2 30° + \cos^2 30° = 1$

$$1 + \tan^2\frac{\pi}{4} = \sec^2\frac{\pi}{4}$$

NEGATIVE-ANGLE IDENTITIES

$$\sin(-\theta) = -\sin\theta \qquad \cos(-\theta) = \cos\theta \qquad \tan(-\theta) = -\tan\theta$$

$$\csc(-\theta) = -\csc\theta \qquad \sec(-\theta) = \sec\theta \qquad \cot(-\theta) = -\cot\theta$$

Note: Cosine and secant are even functions, having graphs that are symmetric with respect to the *y*-axis. Sine, cosecant, tangent, and cotangent are odd functions, having graphs that are symmetric with respect to the origin.

Examples: $\cos(-45°) = \cos 45°, \qquad \sin(-60°) = -\sin(60°)$

SECTION 7.2 VERIFYING IDENTITIES

VERIFYING IDENTITIES

To verify an identity, simplify one side of the equation until it equals the other side of the equation. Make use of the fundamental identities from Section 7.1.

To give either graphical or numerical support for an identity, let Y_1 be the left side of the equation and Y_2 be the right side of the equation. The graphs and tables for Y_1 and Y_2 should be identical wherever Y_1 and Y_2 are defined.

Example: Verify that $\dfrac{1 - \sin^2\theta}{\cos\theta} = \cos\theta$.

$$\frac{1 - \sin^2\theta}{\cos\theta} = \frac{\cos^2\theta}{\cos\theta} \qquad \textcolor{blue}{\sin^2\theta + \cos^2\theta = 1}$$

$$= \cos\theta \qquad \textcolor{blue}{\text{Simplify.}}$$

CONCEPT	EXPLANATION AND EXAMPLES

SECTION 7.3 TRIGONOMETRIC EQUATIONS

REFERENCE ANGLE

The reference angle for an angle θ in standard position is the acute angle between the terminal side of θ and the x-axis.

Example: The reference angle for $\theta = 120°$ is $\theta_R = 60°$.

TRIGONOMETRY EQUATIONS

Unlike polynomial equations, which have a finite number of solutions, trigonometric equations can have infinitely many solutions.

Example: Find all solutions to $2 \sin \theta - 1 = 0$.

$$2 \sin \theta - 1 = 0$$
$$2 \sin \theta = 1$$
$$\sin \theta = \frac{1}{2}$$

Since $\theta_R = \sin^{-1} \frac{1}{2} = 30°$ and the sine function is positive in quadrants I and II, it follows that the solutions to $\sin \theta = \frac{1}{2}$ for $0° \leq \theta < 360°$ are $30°$ and $150°$. Since $\sin \theta$ has period $360°$, all solutions to the equation can be written as

$$\theta = 30° + 360° \cdot n \quad \text{or} \quad \theta = 150° + 360° \cdot n.$$

SECTION 7.4 SUM AND DIFFERENCE IDENTITIES

COSINE SUM AND DIFFERENCE

$$\cos(\alpha + \beta) = \cos \alpha \cos \beta - \sin \alpha \sin \beta$$
$$\cos(\alpha - \beta) = \cos \alpha \cos \beta + \sin \alpha \sin \beta$$

Example: $\cos(60° - 45°) = \cos 60° \cos 45° + \sin 60° \sin 45°$

SINE SUM AND DIFFERENCE

$$\sin(\alpha + \beta) = \sin \alpha \cos \beta + \cos \alpha \sin \beta$$
$$\sin(\alpha - \beta) = \sin \alpha \cos \beta - \cos \alpha \sin \beta$$

Example: $\sin(60° - 45°) = \sin 60° \cos 45° - \cos 60° \sin 45°$

TANGENT SUM AND DIFFERENCE

$$\tan(\alpha + \beta) = \frac{\tan \alpha + \tan \beta}{1 - \tan \alpha \tan \beta}$$

$$\tan(\alpha - \beta) = \frac{\tan \alpha - \tan \beta}{1 + \tan \alpha \tan \beta}$$

Example: $\tan(60° - 45°) = \dfrac{\tan 60° - \tan 45°}{1 + \tan 60° \tan 45°}$

COFUNCTION IDENTITIES

$$\cos\left(\frac{\pi}{2} - t\right) = \sin t \qquad \sin\left(\frac{\pi}{2} - t\right) = \cos t$$

$$\cot\left(\frac{\pi}{2} - t\right) = \tan t \qquad \tan\left(\frac{\pi}{2} - t\right) = \cot t$$

$$\csc\left(\frac{\pi}{2} - t\right) = \sec t \qquad \sec\left(\frac{\pi}{2} - t\right) = \csc t$$

Example: $\tan \dfrac{\pi}{3} = \cot \dfrac{\pi}{6}$

CONCEPT	EXPLANATION AND EXAMPLES

SECTION 7.5 MULTIPLE-ANGLE IDENTITIES

DOUBLE-ANGLE IDENTITIES

$\sin 2\theta = 2 \sin \theta \cos \theta$

$\cos 2\theta = \cos^2 \theta - \sin^2 \theta = 2 \cos^2 \theta - 1 = 1 - 2 \sin^2 \theta$

$\tan 2\theta = \dfrac{2 \tan \theta}{1 - \tan^2 \theta}$

Example: $\sin 120° = 2 \sin 60° \cos 60°$

POWER-REDUCING IDENTITIES

$\sin^2 \theta = \dfrac{1 - \cos 2\theta}{2} \qquad \cos^2 \theta = \dfrac{1 + \cos 2\theta}{2} \qquad \tan^2 \theta = \dfrac{1 - \cos 2\theta}{1 + \cos 2\theta}$

Example: $\sin^2 30° = \dfrac{1 - \cos 60°}{2}$

HALF-ANGLE FORMULAS

$\sin \dfrac{\theta}{2} = \pm \sqrt{\dfrac{1 - \cos \theta}{2}} \qquad \cos \dfrac{\theta}{2} = \pm \sqrt{\dfrac{1 + \cos \theta}{2}}$

$\tan \dfrac{\theta}{2} = \pm \sqrt{\dfrac{1 - \cos \theta}{1 + \cos \theta}} \qquad \tan \dfrac{\theta}{2} = \dfrac{1 - \cos \theta}{\sin \theta} \qquad \tan \dfrac{\theta}{2} = \dfrac{\sin \theta}{1 + \cos \theta}$

Example: $\cos 15° = \sqrt{\dfrac{1 + \cos 30°}{2}}$

PRODUCT-TO-SUM IDENTITIES

$\cos \alpha \cos \beta = \dfrac{1}{2} (\cos (\alpha + \beta) + \cos (\alpha - \beta))$

$\sin \alpha \sin \beta = \dfrac{1}{2} (\cos (\alpha - \beta) - \cos (\alpha + \beta))$

$\sin \alpha \cos \beta = \dfrac{1}{2} (\sin (\alpha + \beta) + \sin (\alpha - \beta))$

$\cos \alpha \sin \beta = \dfrac{1}{2} (\sin (\alpha + \beta) - \sin (\alpha - \beta))$

Example: $\cos 45° \cos 60° = \dfrac{1}{2} (\cos (45° + 60°) + \cos (45° - 60°))$

SUM-TO-PRODUCT IDENTITIES

$\cos a + \cos b = 2 \cos \dfrac{a + b}{2} \cos \dfrac{a - b}{2}$

$\cos a - \cos b = -2 \sin \dfrac{a + b}{2} \sin \dfrac{a - b}{2}$

$\sin a + \sin b = 2 \sin \dfrac{a + b}{2} \cos \dfrac{a - b}{2}$

$\sin a - \sin b = 2 \cos \dfrac{a + b}{2} \sin \dfrac{a - b}{2}$

Example: $\cos 60° + \cos 30° = 2 \cos \dfrac{60° + 30°}{2} \cos \dfrac{60° - 30°}{2}$

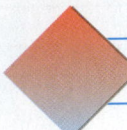

REVIEW EXERCISES

Exercises 1 and 2: Determine the quadrant that contains θ.

1. $\sec\theta < 0$ and $\sin\theta > 0$

2. $\cot\theta > 0$ and $\cos\theta < 0$

Exercises 3–6: Use the given information to find the other trigonometric functions of θ.

3. $\sin\theta = \frac{3}{5}$ and $\cos\theta = -\frac{4}{5}$

4. $\sec\theta = -\frac{13}{12}$ and $\csc\theta = -\frac{13}{5}$

5. $\tan\theta = -\frac{7}{24}$ and $\cos\theta = \frac{24}{25}$

6. $\cot\theta = -\frac{1}{2}$ and $\sin\theta > 0$

Exercises 7–10: Use a negative-angle identity to write an equivalent trigonometric expression involving a positive angle.

7. $\sin(-13°)$

8. $\cos(-106°)$

9. $\sec\left(-\dfrac{3\pi}{7}\right)$

10. $\tan\left(-\dfrac{5\pi}{11}\right)$

11. Explain the difference between an identity and a conditional equation. Give an example of each.

12. Discuss one application where a trigonometric equation or identity occurs.

13. Is there a quadrant where one of the trigonometric functions is positive and the other five are negative? Explain your reasoning.

14. Is it possible for $\sin\theta < 0$, $\cos\theta > 0$, and $\tan\theta > 0$ for some value of θ? Explain.

Exercises 15–20: Simplify each expression.

15. $\sec\theta\cot\theta\sin\theta$

16. $\sin\theta\csc\theta$

17. $(\sec^2 t - 1)(\csc^2 t - 1)$

18. $\dfrac{\sec\theta}{\csc\theta} + \dfrac{\sin\theta}{\cos\theta}$

19. $\dfrac{\csc\theta\sin\theta}{\sec\theta}$

20. $\dfrac{\cos^2\theta}{1-\sin\theta}$

Exercises 21–24: Use a calculator to approximate the other trigonometric functions of θ to four decimal places.

21. $\tan\theta = 1.2367$ and θ is acute

22. $\sin\theta = -0.3434$ and θ is in quadrant IV

23. $\cos\theta = -0.4544$ and θ is in quadrant II

24. $\tan\theta = -0.8595$ and θ is in quadrant IV

Exercises 25–28: Factor the trigonometric expression.

25. $\sin^2\theta + 2\sin\theta + 1$

26. $2\cos^2 t - 3\cos t + 1$

27. $\tan^2\theta - 9$

28. $2\sec^2\theta - 3\sec\theta - 5$

Exercises 29–42: Verify the identity.

29. $(\sec\theta - 1)(\sec\theta + 1) = \tan^2\theta$

30. $(\cos\theta + \sin\theta)^2 + (\cos\theta - \sin\theta)^2 = 2$

31. $(1 + \tan t)^2 = \sec^2 t + 2\tan t$

32. $(1 - \cos^2 t)(1 + \tan^2 t) = \tan^2 t$

33. $\sin(x - \pi) = -\sin x$

34. $\cos(\pi + x) = -\cos x$

35. $\sin 8x = 2\sin 4x\cos 4x$

36. $\cos^4 x - \sin^4 x = \cos 2x$

37. $\sec 2x = \dfrac{1}{2\cos^2 x - 1}$

38. $\dfrac{1 + \tan^2 x}{\sin^2 x + \cos^2 x} = \sec^2 x$

39. $\cos^4 x\sin^3 x = (\cos^4 x - \cos^6 x)\sin x$

40. $\sin^4 x = \frac{3}{8} - \frac{1}{2}\cos 2x + \frac{1}{8}\cos 4x$

41. $\sec^4\theta - \tan^4\theta = 1 + 2\tan^2\theta$

42. $\dfrac{1 + \cos\theta}{\sin\theta} + \dfrac{\sin\theta}{1 + \cos\theta} = 2\csc\theta$

Exercises 43–46: Find the reference angle for θ.

43. $\theta = 240°$

44. $\theta = 320°$

45. $\theta = \dfrac{9\pi}{7}$

46. $\theta = -\dfrac{7\pi}{6}$

Exercises 47 and 48: Use the graph to estimate any solutions to the trigonometric equation for $[0, 2\pi)$. Then solve the equation symbolically.

47. $2 \sin t \cos t - \cos t = 0$

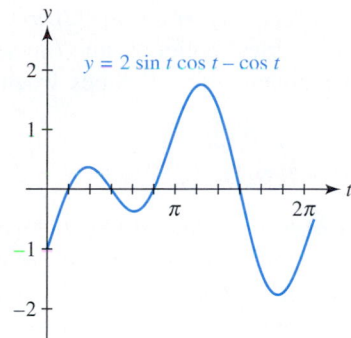

$y = 2 \sin t \cos t - \cos t$

48. $\cos^2 \theta - 2 \cos \theta = 0$

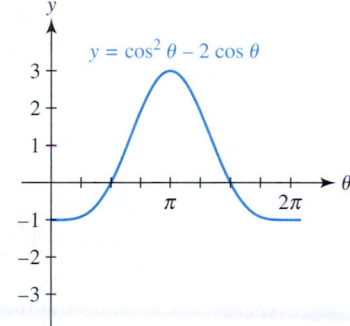

$y = \cos^2 \theta - 2 \cos \theta$

Exercises 49 and 50: Solve the given trigonometric equation for $0° \leq \theta < 360°$.

49. (a) $\tan \theta = \sqrt{3}$ **50.** (a) $\sin \theta = 1$

   (b) $\cot \theta = -\sqrt{3}$ (b) $\cos \theta = -1$

Exercises 51–54: Solve the equation for $0° \leq \theta < 360°$.

51. $2 \cos \theta - 1 = 0$ **52.** $\cot^2 \theta = 1$

53. $2 \sin^2 \theta + \sin \theta - 3 = 0$

54. $\sin^2 \theta + 2 \cos \theta = 1$

Exercises 55 and 56: Solve the equation for t on the interval $[0, 2\pi)$.

55. $\tan^2 t - 2 \tan t + 1 = 0$ **56.** $2 \sin t = \sqrt{3}$

Exercises 57–60: Find all solutions to the equation. Express your results in both degrees and radians.

57. $3 \tan^2 t - 1 = 0$ **58.** $2 \sin^2 t - \sin t - 1 = 0$

59. $\sin 2t + 3 \cos t = 0$ **60.** $\cos 2t = 1$

Exercises 61 and 62: Use a half-angle formula to find the exact value for the expression. Use a calculator to support your result.

61. $\cos 105°$ **62.** $\sin \frac{\pi}{12}$

Exercises 63 and 64: Estimate any solutions in the interval $[-2\pi, 2\pi]$ graphically to two decimal places.

63. $\tan x = x + 1$ **64.** $\sin(\cos x) = \tan x$

Exercises 65 and 66: Complete the following for the identity.

   (a) *Justify the identity verbally using transformations of graphs.*
   (b) *Verify the identity symbolically.*
   (c) *Give graphical support.*

65. $\sin\left(t - \frac{\pi}{2}\right) = -\cos t$ **66.** $\cos\left(t - \frac{\pi}{2}\right) = \sin t$

Exercises 67 and 68: Find the following.

   (a) $\sin(\alpha + \beta)$ (b) $\cos(\alpha + \beta)$
   (c) $\tan(\alpha + \beta)$ (d) *The quadrant containing* $\alpha + \beta$

67. $\cos \alpha = \frac{3}{5}$ and $\cos \beta = \frac{12}{13}$, α and β are in quadrant I

68. $\sin \alpha = -\frac{12}{13}$ and $\tan \beta = -\frac{4}{3}$, α and β are in quadrant IV

Exercises 69 and 70: Complete the following.

   (a) *Find* $\sin 2\theta$, $\cos 2\theta$, *and* $\tan 2\theta$.
   (b) *Use a calculator to support your results.*

69. $\sin \theta = \frac{4}{5}$ and $\tan \theta = -\frac{4}{3}$

70. $\sin \theta = -\frac{12}{13}$ and $\cos \theta = -\frac{5}{13}$

Exercises 71 and 72: Complete the following.

   (a) *Find* $\sin \frac{\theta}{2}$, $\cos \frac{\theta}{2}$, *and* $\tan \frac{\theta}{2}$.
   (b) *Use a calculator to support your results.*

71. $\cos \theta = \frac{1}{4}$ and $0° < \theta < 90°$

72. $\tan \theta = -\frac{8}{15}$ and $-90° < \theta < 0°$

Exercises 73 and 74: Evaluate the expression.

73. $\cos\left(2 \tan^{-1} \frac{11}{60}\right)$ **74.** $\sin(2 \sin^{-1} x), x > 0$

Applications

75. *Daylight Hours*   The number of daylight hours y at 20° S latitude can be modeled by

$$y = 1.2 \cos\left(\frac{\pi}{6}(x - 0.7)\right) + 12.1,$$

where $x = 1$ corresponds to January 1, $x = 2$ to February 1, and so on. Estimate graphically or numerically when the number of daylight hours equals 11.5 hours. (**Source:** J. Williams.)

76. *Music and Pure Tones*   A pure tone is modeled by the graph of $P(t) = 0.006 \cos(50\pi t)$, where P represents the pressure on an eardrum in pounds per square foot at time t in seconds.   (**Source:** J. Roederer, *Introduction to the Physics and Psychophysics of Music.*)
(a) Graph P in $[0, 0.1, 0.01]$ by $[-0.008, 0.008, 0.001]$.

(b) Estimate all solutions to the equation $P = 0$ on the interval $0 \le t \le 0.1$.

77. *Modeling Musical Tones*   Let the pressure exerted on the eardrum by two sound waves in pounds per square foot be given by

$$P_1(t) = 0.006 \cos(100\pi t) \quad \text{and}$$
$$P_2(t) = 0.008 \sin(100\pi t).$$

(a) Graph the pressure on the eardrum $P = P_1 + P_2$ in the window $[0, 0.06, 0.01]$ by $[-0.012, 0.012, 0.002]$.

(b) Use the graph to find values for a and k such that $P = a\sin(100\pi t + k)$.

(c) Use a sum or difference identity for sine to verify your result in part (b).

78. *Electricity*   Suppose the voltage in an electrical circuit is given by $V(t) = 17\sin(120\pi t)$ and the amperage flowing through a heater is given by $I(t) = 2\sin(120\pi t)$ where t is in seconds.   (**Source:** G. Wilcox and C. Hesselberth, *Electricity for Engineering Technology.*)
(a) Graph the wattage $W = VI$ consumed by the heater in $[0, 0.04, 0.01]$ by $[-40, 40, 10]$.

(b) Use a power-reducing identity to express the wattage in the form $W = a\cos(k\pi t) + d$, where a, k, and d are constants.

79. *Electromagnets*   Let the wattage consumed by an electromagnet be given by $W(t) = 7\cos^2(240\pi t)$ and the voltage be given by $V(t) = 50\sin(240\pi t)$, where t is in seconds. Express $W(t)$ in terms of the sine function. When V is maximum or minimum, what is the value of W? Explain.

80. *Average Temperatures*   Let the monthly average high temperature y in degrees Fahrenheit at a location be given by

$$y = 15\sin\left(\frac{\pi}{6}(x - 4)\right) + 60,$$

where x is the month and $x = 1$ corresponds to January. Estimate when the monthly average high temperature is $60°$ F.

81. *Back Stress*   The force exerted by the back muscles of a 100-pound person can be estimated by $F = 289\cos\theta$. (See Figure 7.48 on page 621.) Find θ in degrees and radians when $F = 250$ pounds.

82. *Electricity*   Let $V(t) = 80\sin(120\pi t)$ denote the voltage in an electrical outlet at time t in seconds, where t is a real number. Find all times when the voltage is 40 volts.

EXTENDED AND DISCOVERY EXERCISES

1. *Piano Strings*   A piano string vibrates at more than one frequency when it is struck. It produces a complex wave that can be modeled by a sum of several pure tones. If a piano key with a frequency of f_1 is played, then the corresponding string will not only vibrate at f_1 but it will also vibrate at the higher frequencies of $2f_1$, $3f_1$, $4f_1$, and so on. The *fundamental frequency* of the string is f_1 and the higher frequencies are called the *upper harmonics*. The human ear will hear the sum of these frequencies as one complex tone.   (**Source:** J. Roederer, *Introduction to the Physics and Psychophysics of Music.*)
(a) If the A note above middle C is played, its fundamental frequency is $f_1 = 440$ hertz. (One hertz equals one cycle per second.) The piano string also vibrates at frequencies of $f_2 = 2(440) = 880$, $f_3 = 3(440) = 1320$, $f_4 = 4(440) = 1760$, and so on. The pressure for each frequency in pounds per square foot is modeled by

$$P_1 = 0.002\sin(2\pi(440)t),$$
$$P_2 = \frac{0.002}{2}\sin(2\pi(880)t),$$
$$P_3 = \frac{0.002}{3}\sin(2\pi(1320)t),$$
$$P_4 = \frac{0.002}{4}\sin(2\pi(1760)t), \quad \text{and}$$
$$P_5 = \frac{0.002}{5}\sin(2\pi(2200)t),$$

where t is in seconds. Graph each of the following expressions for P in the viewing rectangle $[0, 0.01, 0.002]$ by $[-0.005, 0.005, 0.001]$.
i. $P = P_1$
ii. $P = P_1 + P_2$
iii. $P = P_1 + P_2 + P_3$
iv. $P = P_1 + P_2 + P_3 + P_4$
v. $P = P_1 + P_2 + P_3 + P_4 + P_5$

(b) The final graph of P models what the human ear hears. Describe this graph.

(c) Estimate the maximum pressure of
$$P = P_1 + P_2 + P_3 + P_4 + P_5.$$

(d) A pure tone with a frequency of 440 hertz is modeled by $P = P_1$, whereas a piano generates the graph of $P = P_1 + P_2 + P_3 + P_4 + P_5$. Compare and contrast these two graphs.

2. *Plucking a String* (Refer to the previous exercise.) If a string with a fundamental frequency of 110 hertz is *plucked in the middle*, it will vibrate at the odd harmonics or frequencies of 110, 330, and 550, but not at the even harmonics of 220, 440, and 660. The resulting pressure P caused by this sound wave may be modeled by

$$P = 0.002 \sin(220\pi t) + \frac{0.002}{3} \sin(660\pi t)$$
$$+ \frac{0.002}{5} \sin(1100\pi t) + \frac{0.002}{7} \sin(1540\pi t).$$

(a) Graph P in $[0, 0.03, 0.01]$ by $[-0.004, 0.004, 0.001]$.

(b) Describe the graph of P.

(c) At lower frequencies, the inner ear hears a tone only when the eardrum is moving outward or when $P < 0$. Estimate the times in the interval $0 \leq t \leq 0.03$ when this occurs. (**Source:** A. Benade, *Fundamentals of Musical Acoustics*.)

3. *Low Tones and Small Speakers* Small speakers found in older radios and telephones often cannot vibrate at frequencies that are less than 200 hertz. Thirty-five keys on a piano have frequencies less than 200 hertz. Nonetheless these notes can still be heard on these speakers. When a piano string creates a tone of 110 hertz, it also creates tones at 220, 330, 440, 550, and 660 hertz. A small speaker cannot reproduce the 110 hertz vibration but it can reproduce the higher frequencies, which are called the upper harmonics. The low tones can still be heard because the speaker produces *difference tones* of the upper harmonics. The difference between consecutive frequencies is 110 hertz and this difference

tone will be heard on a small speaker even though the speaker cannot vibrate at 110 hertz. This phenomenon can be visualized with a graphing calculator. (**Source:** A. Benade.)

(a) Graph the upper harmonics represented by the pressure wave

$$P = \frac{1}{2} \sin(2\pi(220)t) + \frac{1}{3} \sin(2\pi(330)t)$$
$$+ \frac{1}{4} \sin(2\pi(440)t)$$

in $[0, 0.03, 0.01]$ by $[-1.2, 1.2, 0.5]$.

(b) Estimate the t-values on the interval $0 \leq t \leq 0.03$ where P is maximum.

(c) Approximate the frequency of these maximum values. What does a person hear in addition to the frequencies of 220, 330, and 440 hertz?

(d) Discuss the advantage of having large speakers instead of smaller ones. (*Hint:* Try graphing the pressure produced by a speaker that can vibrate both at 110 hertz and at the upper harmonics.)

4. *Piano Strings* When a string is set into vibration by striking it, the amplitude A of the vibrations decreases over time, while the frequency of the vibration remains constant. This phenomenon is called *exponential decay* and can be modeled by $A = A_0 e^{-kt} \sin(2\pi F t)$, where F is the frequency, t is time in seconds, and k and A_0 are positive constants. (**Source:** J. Roederer.)

(a) Graph A when $A_0 = 0.1$, $F = 15$, and $k = 1.2$ in the window $[0, 1, 0.1]$ by $[-0.15, 0.15, 0.05]$.

(b) Now graph the equations $y_1 = -0.1e^{-1.2t}$ and $y_2 = 0.1e^{-1.2t}$ with A. Describe how the graphs of y_1 and y_2 relate to the graph of A.

(c) The *decay half-time* is the time it takes for the maximum amplitude of A_0 to decrease to $\frac{1}{2}A_0$. Estimate this time graphically. (The decay half-time for a typical piano string is about 0.4 second.)

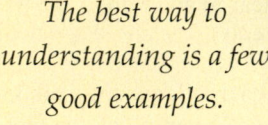

8 Further Topics in Trigonometry

The best way to understanding is a few good examples.

—Isaac Newton

Trigonometry plays an important role in our society. Almost any phenomenon that involves angles, triangles, or circular motion requires trigonometric functions to model it. For example, aerial photography has become essential for predicting weather, surveying land, promoting national security, and even making archaeological discoveries. Aerial photography began in 1858 when Gaspard Tournachon, a French photographer, took pictures of Paris from a hot-air balloon that had a makeshift darkroom. The first archaeological aerial photographs were taken of Stonehenge in 1906. By searching these photographs for unusual soil markings caused by structures lying below the ground, Stonehenge Avenue was discovered. Today, trigonometry is used extensively in aerial photography, and hot-air balloons have been replaced by airplanes and satellites.

In this chapter we use trigonometry in a variety of applications, which include aerial photography, computer graphics, robotics, navigation, GPS, highway design, physics, weather, solar energy, art, electronics, construction, astronomy, surveying, and simulating motion.

Sources: R. Brooks and J. Dieter, *Phytoarchaeology;* F. Moffitt, *Photogrammetry.*

8.1 Law of Sines

Introduction

In Chapter 6 we solved right triangles. Solving a triangle involves finding the length of each side and the measure of each angle in the triangle. In many areas of study, triangles without right angles often occur. To solve these triangles, we will use the law of sines and the law of cosines. In this section the law of sines is derived and used to solve several problems.

Oblique Triangles

◆ Learn about oblique triangles

◆ Solve triangles

◆ Solve the ambiguous case

If a triangle is not a right triangle, then it is called an **oblique triangle**. There are four different situations or cases that can occur when attempting to solve an oblique triangle.

1. All three sides are given. This situation determines a unique triangle and is referred to as SSS. See Figure 8.1. Note that the length of any one side must be less than the sum of the lengths of the other two sides.
2. Two sides and the angle included are given. This situation determines a unique triangle and is referred to as SAS. See Figure 8.2.
3. One side and two angles are given. This situation determines a unique triangle and is referred to as AAS or ASA. See Figure 8.3. Note that whenever two angles of a triangle are known, then the third angle can be found by using the fact that the sum of the measures of the angles equals 180°.
4. Two sides and an angle opposite one of the sides are given. This situation does *not always* determine a unique triangle and is referred to as SSA. There may be 0, 1, or 2 triangles that can satisfy SSA. As a result, we call SSA the **ambiguous case**. See Figures 8.4–8.6.

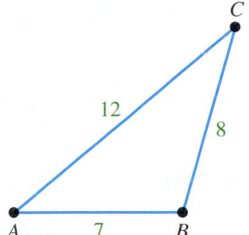

FIGURE 8.1 Case 1: **SSS**

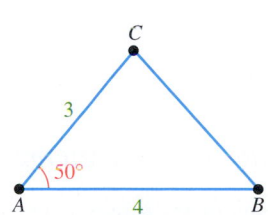

FIGURE 8.2 Case 2: **SAS**

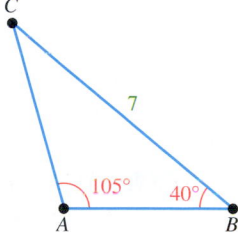

FIGURE 8.3 Case 3: **AAS** or **ASA**

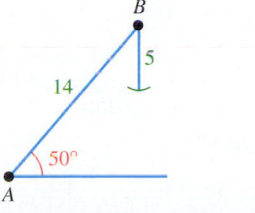

FIGURE 8.4 Case 4: **SSA** (No Triangles)

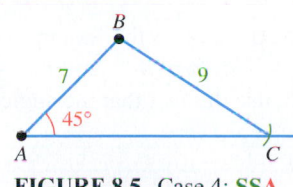

FIGURE 8.5 Case 4: **SSA** (One Triangle)

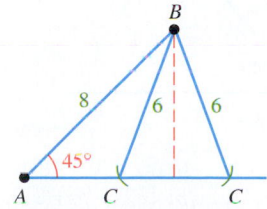

FIGURE 8.6 Case 4: **SSA** (Two Triangles)

Cases 1 and 2 are solved in the next section using the *law of cosines*, and Cases 3 and 4 are solved using the *law of sines*.

Solving Triangles

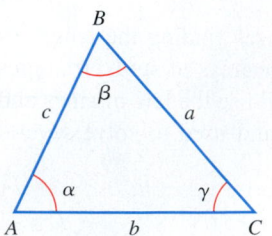

FIGURE 8.7 Standard Labeling

We will label triangles, as shown in Figure 8.7, and refer to this labeling as the **standard labeling**. For example, angle α is located at vertex A and side a is opposite angle α. Note that angle γ need not be a right angle.

The law of sines can be derived using the oblique triangle shown in Figure 8.8.

$$\sin \alpha = \frac{h}{c} \quad \text{or} \quad h = c \sin \alpha$$

$$\sin \gamma = \frac{h}{a} \quad \text{or} \quad h = a \sin \gamma$$

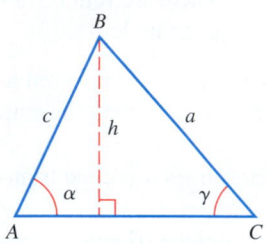

FIGURE 8.8

Since $h = c \sin \alpha$ and $h = a \sin \gamma$, it follows that

$$c \sin \alpha = a \sin \gamma,$$

or dividing each side by ac gives

$$\frac{\sin \alpha}{a} = \frac{\sin \gamma}{c}.$$

In a similar manner it can be shown that

$$\frac{\sin \alpha}{a} = \frac{\sin \beta}{b}.$$

This discussion supports the following result.

LAW OF SINES

Any triangle with standard labeling satisfies

$$\frac{\sin \alpha}{a} = \frac{\sin \beta}{b} = \frac{\sin \gamma}{c}, \quad \text{or equivalently,} \quad \frac{a}{\sin \alpha} = \frac{b}{\sin \beta} = \frac{c}{\sin \gamma}.$$

The next example illustrates how the law of sines is used in aerial photography.

EXAMPLE 1

Using aerial photography to find distances (ASA)

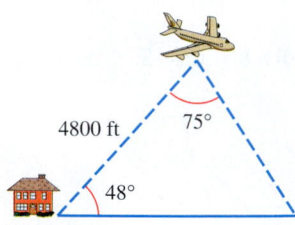

FIGURE 8.9

Trigonometry is used extensively in aerial photography. In Figure 8.9 a camera lens has an angular coverage of 75°. As a picture is taken over level ground, the airplane's distance is 4800 feet from a house located on the edge of the photograph and the angle of elevation of the airplane from the house is 48°. Find the ground distance shown in the photograph. (**Source:** F. Moffitt, *Photogrammetry*.)

SOLUTION In this example we are given two angles and the side included (ASA), so let $\alpha = 75°$, $\beta = 48°$, and $c = 4800$, as shown in Figure 8.10. To find the third angle γ we can use the fact that the angles sum to 180°.

$$\gamma = 180° - \alpha - \beta$$
$$= 180° - 75° - 48°$$
$$= 57°$$

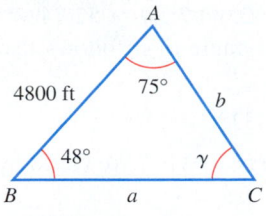

FIGURE 8.10

Side a corresponds to the ground distance shown in the photograph and can be found by using the law of sines.

$$\frac{a}{\sin \alpha} = \frac{c}{\sin \gamma}$$ Law of sines

$$\frac{a}{\sin 75°} = \frac{4800}{\sin 57°}$$ Substitute.

$$a = \frac{4800 \sin 75°}{\sin 57°}$$ Solve for a.

$$a \approx 5528$$ Approximate a.

The picture will show about 5528 feet of ground distance from one edge of the photograph to the other.

Now Try Exercises 11 and 37 ◆

EXAMPLE 2   **Solving a triangle (AAS)**

If $\beta = 85°$, $\gamma = 40°$, and $b = 26$, solve triangle ABC.

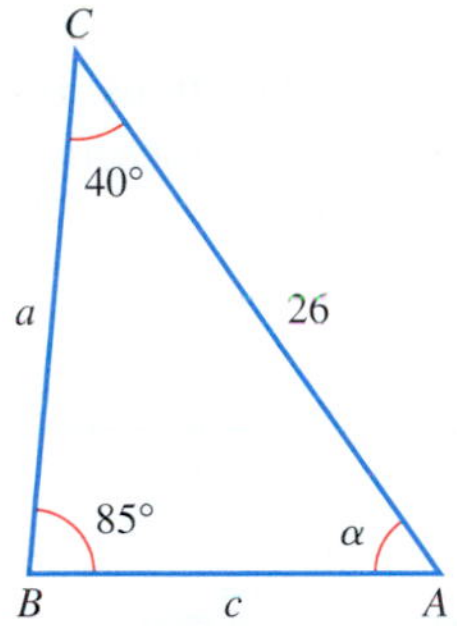

SOLUTION Sketch triangle ABC, as shown in Figure 8.11.

$$\alpha = 180° - \beta - \gamma$$
$$= 180° - 85° - 40°$$
$$= 55°$$

Side a can be found by using the law of sines.

$$\frac{a}{\sin \alpha} = \frac{b}{\sin \beta}$$ Law of sines

$$\frac{a}{\sin 55°} = \frac{26}{\sin 85°}$$ Substitute.

$$a = \frac{26 \sin 55°}{\sin 85°}$$ Solve for a.

$$a \approx 21.4$$ Approximate a.

Side c can be found in a similar manner.

$$\frac{c}{\sin \gamma} = \frac{b}{\sin \beta}$$ Law of sines

$$\frac{c}{\sin 40°} = \frac{26}{\sin 85°}$$ Substitute.

$$c = \frac{26 \sin 40°}{\sin 85°}$$ Solve for c.

$$c \approx 16.8$$ Approximate c.   *Now Try Exercise 19* ◆

FIGURE 8.11

EXAMPLE 3   **Estimating the distance to the moon**

Since the moon is a relatively close celestial object, its distance can be measured directly using trigonometry. To find this distance, two photographs of the moon were taken at precisely the same time from two locations. On April 29, 1976, at 11:35 A.M. the lunar angles of elevation during a partial solar eclipse at Bochum in upper Germany and at Donaueschingen in lower Germany were measured as $\alpha = 52.6997°$ and $\theta = 52.7430°$, respectively. See Figure 8.12 on the next page. If the two cities are 398.02 kilometers apart, approximate the distance to the moon. Disregard the curvature of Earth in this calculation. (**Source:** W. Schlosser, *Challenges of Astronomy*.)

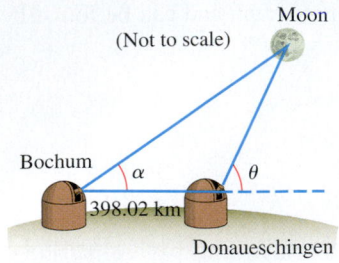

FIGURE 8.12

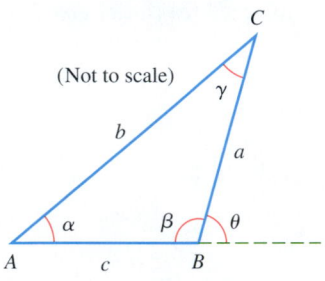

FIGURE 8.13

SOLUTION Consider triangle ABC in Figure 8.13, where $\alpha = 52.6997°$, $\theta = 52.7430°$, and $c = 398.02$. Because $\alpha + \gamma$ and θ are both supplements of angle β, it follows that $\alpha + \gamma = \theta$ and so

$$\gamma = \theta - \alpha = 52.7430° - 52.6997° = 0.0433°.$$

The distance to the moon can be approximated by finding either a or b. Why? We find a by applying the law of sines.

$$\frac{a}{\sin \alpha} = \frac{c}{\sin \gamma} \qquad \text{Law of sines}$$

$$\frac{a}{\sin 52.6997°} = \frac{398.02}{\sin 0.0433°} \qquad \text{Substitute.}$$

$$a = \frac{398.02 \sin 52.6997°}{\sin 0.0433°} \qquad \text{Solve for } a.$$

$$a \approx 419{,}000 \text{ km} \qquad \text{Approximate } a.$$

The distance to the moon on that day was about 419,000 kilometers.

Now Try Exercise 38 ◆

Bearings are used in both surveying and aerial navigation to determine directions. If a single angle is used for a **bearing**, then it is understood that the bearing is measured *clockwise* from due north. Some examples of bearings are shown in Figure 8.14.

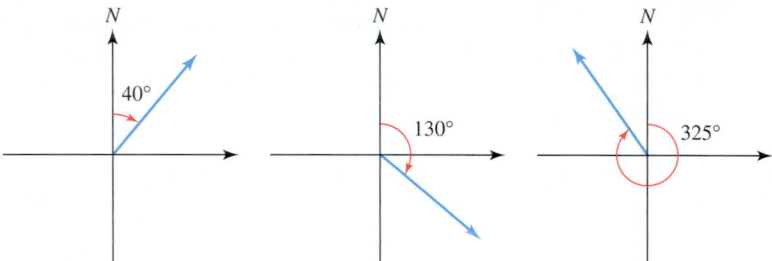

FIGURE 8.14 Examples of Bearings

In the next example we locate a fire by using bearings. Locating fires precisely is particularly important for firefighters in rugged areas with limited access. (**Source:** I. Mueller and K. Ramsayer, *Introduction to Surveying.*)

 **EXAMPLE 4** Determining the location of a forest fire

A fire is spotted from two ranger stations that are 4 miles apart, as illustrated in Figure 8.15. From station A the bearing of the fire is 35° and from station B the bearing of the fire is 335°. Find the distance between the fire and each ranger station if station A lies directly west of station B. Approximate the results to the nearest hundredth of a mile.

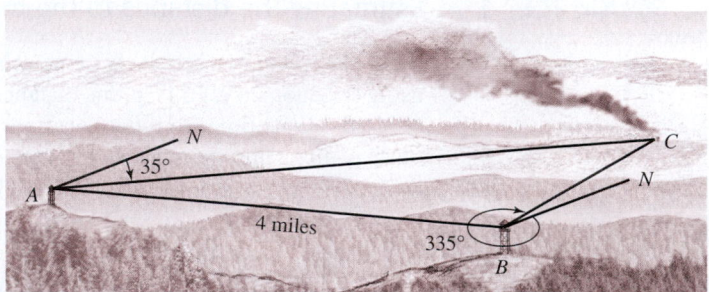

FIGURE 8.15

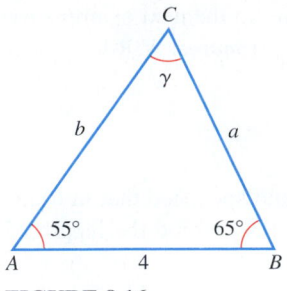

FIGURE 8.16

SOLUTION   Consider triangle ABC in Figure 8.16, where $\alpha = 55°$, $\beta = 335° - 270° = 65°$, and $c = 4$. Thus

$$\gamma = 180° - 55° - 65° = 60°.$$

Using the law of sines, we can find a.

$$\frac{a}{\sin \alpha} = \frac{c}{\sin \gamma} \qquad \textit{Law of sines}$$

$$\frac{a}{\sin 55°} = \frac{4}{\sin 60°} \qquad \textit{Substitute.}$$

$$a = \frac{4 \sin 55°}{\sin 60°} \qquad \textit{Solve for a.}$$

$$a \approx 3.78 \text{ mi} \qquad \textit{Approximate a.}$$

In a similar manner, we can find b.

$$\frac{b}{\sin \beta} = \frac{c}{\sin \gamma} \qquad \textit{Law of sines}$$

$$\frac{b}{\sin 65°} = \frac{4}{\sin 60°} \qquad \textit{Substitute.}$$

$$b = \frac{4 \sin 65°}{\sin 60°} \qquad \textit{Solve for b.}$$

$$\approx 4.19 \text{ mi} \qquad \textit{Approximate b.}$$

The fire is 4.19 miles from station A and 3.78 miles from station B.

Now Try Exercise 39 ◆

The Ambiguous Case

If we are given two sides and an angle opposite one of the sides (SSA), there may be 0, 1, or 2 triangles that satisfy these conditions. Refer to Figures 8.4–8.6. For this reason SSA is called the *ambiguous case*. We will discuss how the law of sines can be used to solve each of these situations.

EXAMPLE 5   Solving the ambiguous case (no solutions)

Let $\alpha = 62°$, $a = 6$, and $b = 10$. If possible, solve the triangle.

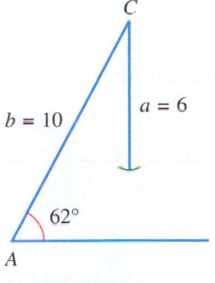

FIGURE 8.17

SOLUTION   We begin by attempting to find β by using the law of sines.

$$\frac{\sin \beta}{b} = \frac{\sin \alpha}{a} \qquad \textit{Law of sines}$$

$$\frac{\sin \beta}{10} = \frac{\sin 62°}{6} \qquad \textit{Substitute.}$$

$$\sin \beta = \frac{10 \sin 62°}{6} \qquad \textit{Solve for sin } \beta.$$

$$\sin \beta \approx 1.47 > 1 \qquad \textit{Approximate sin } \beta.$$

Since the sine function is never greater than 1, there are no solutions for β. No such triangle exists. See Figure 8.17.

Now Try Exercise 29 ◆

Trusses are used in construction to support roofs, radio towers, bridges, and aircraft frames. A truss is composed of straight segments joined together at their ends. Many times the smaller shapes within a truss are triangles. See Figure 8.18 on the next page. If trusses are designed properly, they can be relatively light and very strong. In the next example we determine that a particular truss design results in a unique truss.   (**Source:** W. Riley, *Statics and Mechanics of Materials.*)

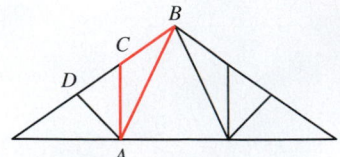

FIGURE 8.18   A Bridge Truss

EXAMPLE 6   Solving the ambiguous case (one solution)

Suppose that an engineer has designed the truss in Figure 8.18 and specified that in triangle *ABC*, *BC* = 17 feet, *AC* = 22 feet, and angle *ABC* = 32°. Determine the length of *AB* to the nearest tenth of a foot. Is this value for *AB* unique?

SOLUTION   Using standard labeling, let $a = 17$, $b = 22$, and $\beta = 32°$. Start by finding angle *CAB* or α.

$$\frac{\sin \alpha}{a} = \frac{\sin \beta}{b} \qquad \text{Law of sines}$$

$$\frac{\sin \alpha}{17} = \frac{\sin 32°}{22} \qquad \text{Substitute.}$$

$$\sin \alpha = \frac{17 \sin 32°}{22} \qquad \text{Solve for } \sin \alpha.$$

$$\sin \alpha \approx 0.4095 \qquad \text{Approximate } \sin \alpha.$$

There are two values possible for angle α if $\sin \alpha \approx 0.4095$. Angle α lies in quadrants I or II with reference angle $\alpha_R \approx \sin^{-1}(0.4095) \approx 24.2°$. Thus

$$\alpha \approx 24.2° \quad \text{or} \quad \alpha \approx 180° - 24.2° = 155.8°.$$

However, if $\alpha \approx 155.8°$ then $\alpha + \beta = 155.8° + 32° > 180°$, which is impossible in a triangle. Therefore $\alpha \approx 24.2°$ is the only possibility and

$$\gamma \approx 180° - 24.2° - 32° = 123.8°.$$

The law of sines allows us to find *AB*, or side *c*.

$$\frac{c}{\sin \gamma} = \frac{b}{\sin \beta} \qquad \text{Law of sines}$$

$$\frac{c}{\sin 123.8°} \approx \frac{22}{\sin 32°} \qquad \text{Substitute.}$$

$$c \approx \frac{22 \sin 123.8°}{\sin 32°} \qquad \text{Solve for } c.$$

$$c \approx 34.5 \text{ ft} \qquad \text{Approximate } c.$$

♦ **CLASS DISCUSSION**

Suppose we are given *a*, *b*, and α, and calculate $\sin \beta = 1$. Discuss the number of solutions for triangle *ABC*. ♦

Thus *AB* is 34.5 feet long and this value is unique.   *Now Try Exercise 23* ♦

EXAMPLE 7   Solving the ambiguous case (two solutions)

Let $\beta = 55°$, $a = 8.5$, and $b = 7.3$. Solve the triangle. Round to the nearest tenth.

SOLUTION   Begin by finding α.

$$\frac{\sin \alpha}{a} = \frac{\sin \beta}{b} \qquad \text{Law of sines}$$

$$\frac{\sin \alpha}{8.5} = \frac{\sin 55°}{7.3} \qquad \text{Substitute.}$$

$$\sin \alpha = \frac{8.5 \sin 55°}{7.3} \qquad \text{Solve for } \sin \alpha.$$

$$\sin \alpha \approx 0.9538 \qquad \text{Approximate } \sin \alpha.$$

Two angles that satisfy $\sin \alpha \approx 0.9538$ in quadrants I and II are

$$\alpha \approx \sin^{-1}(0.9538) \approx 72.5° \quad \text{and} \quad \alpha \approx 180° - 72.5° = 107.5°.$$

Both of these values for α are valid since they do not result in the sum of the angles exceeding $180°$. There are two solutions.

Solution 1   Let $\alpha \approx 72.5°$. Then,

$$\gamma \approx 180° - 72.5° - 55° = 52.5°.$$

Side c can then be found.

$$\frac{c}{\sin \gamma} = \frac{b}{\sin \beta} \qquad \text{Law of sines}$$

$$c \approx \frac{7.3 \sin 52.5°}{\sin 55°} \qquad \text{Substitute and solve for } c.$$

$$c \approx 7.1 \qquad \text{Approximate } c.$$

A sketch of triangle ABC is shown in Figure 8.19.

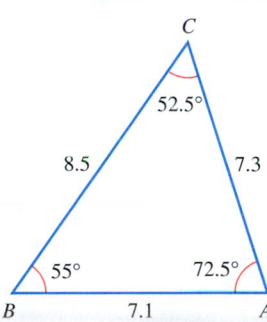

FIGURE 8.19 Solution 1

Solution 2   Let $\alpha \approx 107.5°$. Then,

$$\gamma \approx 180° - 107.5° - 55° = 17.5°.$$

Then side c can be found.

$$\frac{c}{\sin \gamma} = \frac{b}{\sin \beta} \qquad \text{Law of sines}$$

$$c \approx \frac{7.3 \sin 17.5°}{\sin 55°} \qquad \text{Substitute and solve for } c.$$

$$c \approx 2.7 \qquad \text{Approximate } c.$$

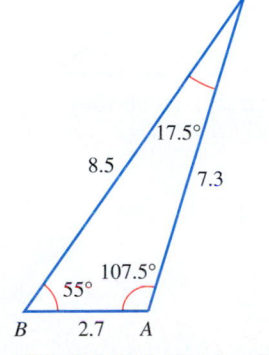

FIGURE 8.20 Solution 2

A sketch of triangle ABC is shown in Figure 8.20.   *Now Try Exercise 21* ◆

◆ **CLASS DISCUSSION**

Suppose you are given a, b, and α. If $a > b$, what can be said about the number of solutions? Explain your reasoning. ◆

8.1

Putting it all Together

The law of sines can be expressed as either

$$\frac{\sin \alpha}{a} = \frac{\sin \beta}{b} = \frac{\sin \gamma}{c} \qquad \text{or} \qquad \frac{a}{\sin \alpha} = \frac{b}{\sin \beta} = \frac{c}{\sin \gamma}.$$

When using the law of sines, a good strategy is to select an equation with the unknown variable in the numerator. The law of sines can be used to solve triangles when we are given ASA, AAS, or SSA. The case where we are given SSA is called the ambiguous case, since there may be 0, 1, or 2 triangles that satisfy the conditions. Several of these situations are shown in the accompanying table.

The Ambiguous Case	Examples
No solutions for triangle *ABC*	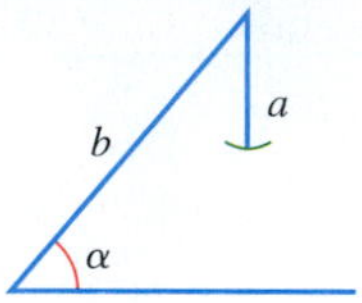 $\sin \beta > 1$ (α acute) 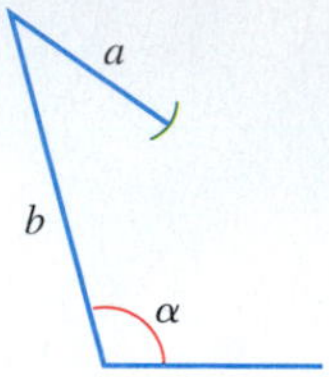 $\sin \beta > 1$ (α obtuse)
One solution for triangle *ABC*	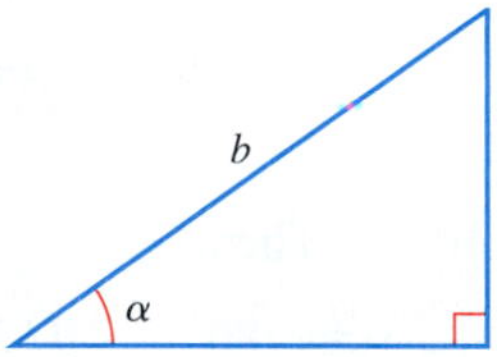 $\sin \beta = 1$ (α acute) 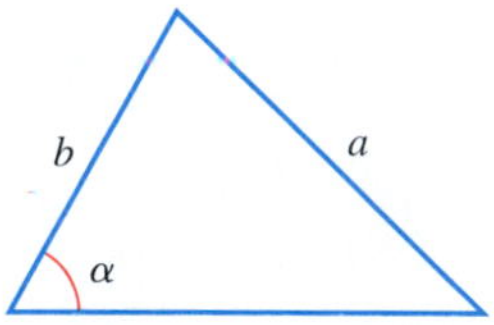 $a > b$ (α acute) 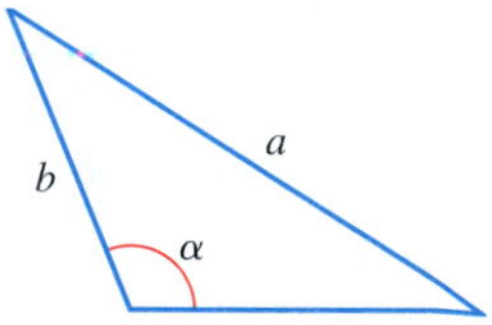 $a > b$ (α obtuse)
Two solutions for triangle *ABC*	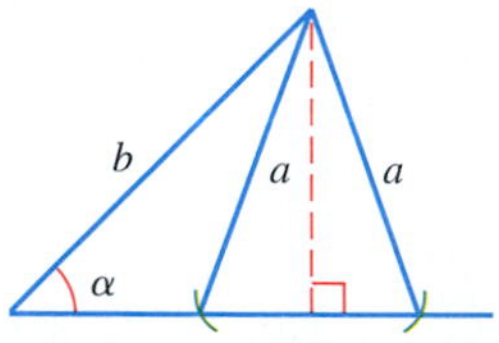 $\sin \beta < 1$ and $a < b$ (α acute)

8.1 — Exercises

Recognizing the Ambiguous Case

Exercises 1–8: Let triangle ABC have standard labeling. Given the following angles and sides, decide if solving the triangle results in the ambiguous case.

1. α, β, and a
2. α, γ, and c
3. a, b, and c
4. α, a, and b
5. β, b, and c
6. α, b, and c
7. γ, a, and c
8. β, b, and α

Solving Triangles

Exercises 9–18: Solve the triangle, if possible. If the triangle represents the ambiguous case, solve for all possible triangles. Approximate values to the nearest tenth.

9.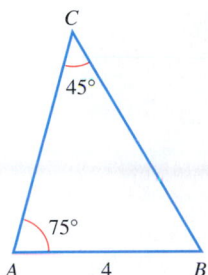

10.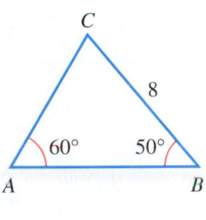

11.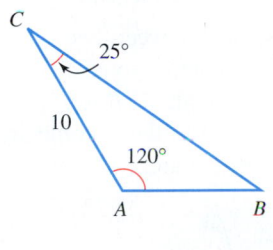

12.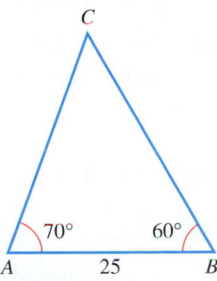

13.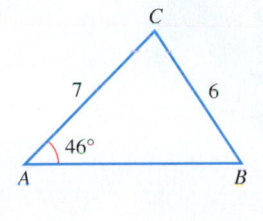

14.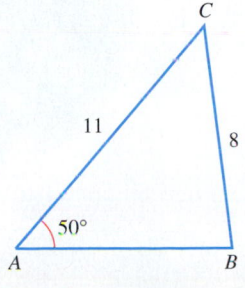

15.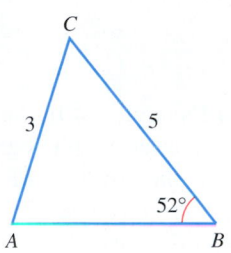

16.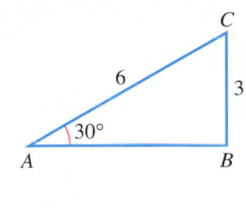

17.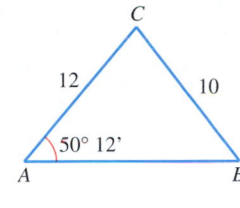

18.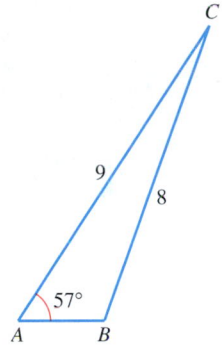

Exercises 19–36: Solve the triangle if possible. Approximate values to the nearest tenth when appropriate.

19. $\alpha = 32°$, $\beta = 55°$, $b = 12$

20. $\beta = 20°$, $\gamma = 67°$, $c = 9$

21. $\alpha = 20°$, $b = 9$, $a = 7$

22. $\alpha = 20°$, $b = 7$, $a = 9$

23. $b = 10$, $\beta = 30°$, $c = 20$

24. $a = 13.5$, $\alpha = 46°$, $c = 27.8$

25. $\gamma = 102°$, $c = 51.6$, $a = 42.1$

26. $\beta = 43°$, $b = 22.1$, $c = 30.7$

27. $\alpha = 55.2°$, $\gamma = 114.8°$, $b = 19.5$

28. $c = 225$, $\alpha = 103.2°$, $\beta = 62.5°$

29. $b = 6.2$, $c = 7.4$, $\beta = 73°$

30. $\alpha = 45°$, $a = 5$, $b = 5\sqrt{2}$

31. $\alpha = 35°15'$, $a = 5$, $b = 12$

32. $\gamma = 71°35'$, $c = 6$, $b = 9$

33. $\beta = 46°45'$, $a = 6$, $b = 5$

34. $\alpha = 54°12'$, $c = 12$, $a = 10$

35. $\alpha = 56°30'$, $\beta = 23°45'$, $c = 100$

36. $\beta = 56°48'$, $\gamma = 10°12'$, $a = 55$

Applications

37. *Aerial Photography* (Refer to Example 1.) In the accompanying figure, a plane takes an aerial photograph with a camera lens that has an angular coverage of 70°. The ground below is inclined at 7°. If the angle of elevation of the plane at B is 52° and distance BC is 3500 feet, estimate the ground distance AB to the nearest foot that appears in the picture. (**Source:** F. Moffit.)

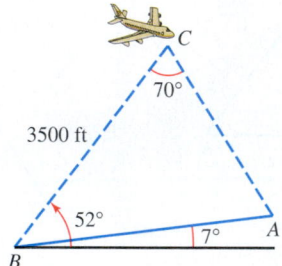

38. *Distance to the Moon* (Refer to Example 3.) Suppose that the lunar angle at Bochum in upper Germany has been measured as $\alpha = 52.6901°$ instead of 52.6997°. Determine the effect that this would have on the estimation of the distance to the moon. Interpret the result.

39. *Locating a Ship* The accompanying figure shows the bearings of a ship on Lake Ontario from two observation points located on a straight shoreline. The bearing from the first observation point is 54.3° and the bearing from the second observation point is 325.2°. If the distance between these points is 15 miles, how far is it from the ship to shore? Assume that the first observation point is directly west of the second observation point.

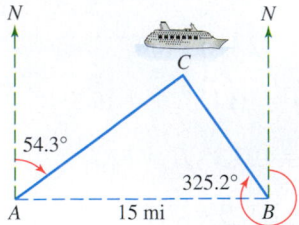

40. *Airplane Navigation* An airplane takes off with a bearing of 55° and flies 480 miles, after which it changes its course to a bearing of 285°. Finally the air-

plane flies back to its starting point with a bearing of 180°. Find the total mileage flown by the airplane.

41. *Distance Across a River* To find the distance AB across a river, a distance $BC = 354$ meters is measured off on one side of the river. See the accompanying figure. It is found that angle $ABC = 112°10'$ and that angle $BCA = 15°20'$. Find the distance AB.

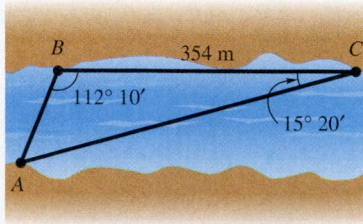

42. *Distance Across a Canyon* To find the distance RS across a canyon, a distance $TR = 582$ yards is measured off one side of the canyon. See the accompanying figure. It is found that angle $TRS = 102°20'$ and that angle $RTS = 32°50'$. Find the distance RS.

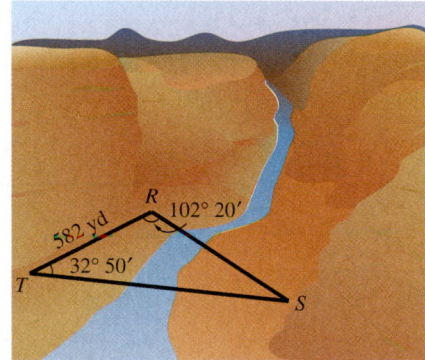

43. *Surveying* To find the distance between two points A and B on the opposite sides of a small pond, a surveyor determines that AC is 97.3 feet, angle ACB is 55.1°, and angle CAB is 75.7°, as illustrated in the accompanying figure. Find the distance between A and B.

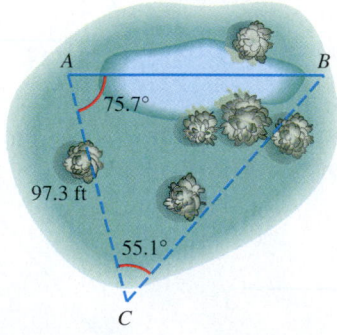

44. *Trigonometric Leveling*   In surveying it is often necessary to determine the height of an inaccessible point P, as illustrated in the accompanying figure. Points A, B, and C lie on level ground.   (**Source:** P. Kissam, *Surveying Practice.*)

(a) If angle ABP is $50°$, angle PAB is $53.3°$, and AB is 102 feet, find PB.

(b) If angle PBC is $47°$, find PC.

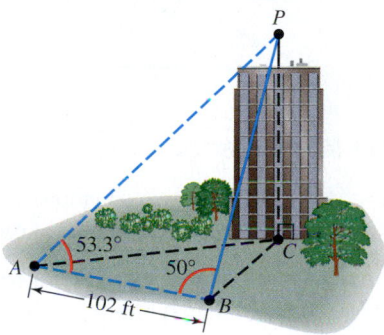

45. *Distance*   An ore ship is traveling west toward Duluth on Lake Superior at 18 miles per hour. The bearing of Split Rock Lighthouse is $285°$. After 1 hour the bearing of the lighthouse is $340°$. Find the distance between the ship and the lighthouse when the second bearing was determined.

46. *Locating a Ship*   From two observation points A and B, a sinking ship is spotted at point C. Angle CAB and angle ABC are measured as $28°$ and $60°$ and the distance AB is 4.12 miles, as illustrated in the figure.

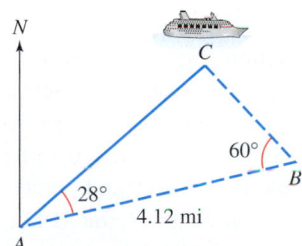

(a) How far is the ship from point A to the nearest hundredth of a mile?

(b) If the coordinates of A are $(0, 0)$ and the coordinates of B are $(4, 1)$, find the bearing of the ship from point A to the nearest degree.

47. *Truss Construction*   In the figure at the top of the next column a truss is shown, where AB is 24.2 feet, angle ABD is $118°$, and angle BDF is $28°$. Find the length of BD.   (**Source:** W. Riley.)

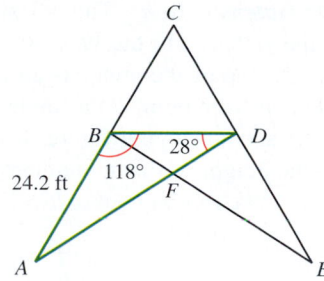

48. *Truss Construction*   Use the results of Example 6 to solve triangle ACD in Figure 8.18 if angle BAD is $90°$.

49. *Highway Construction*   A *reverse curve* or S-curve uses two circular curves to connect two straight portions of highway that are offset, as illustrated in the accompanying figure. Angles α and β will not be equal if the two straight portions of highway have different directions. Typically the same radius r is used for both portions of the reverse curve. If $r = 480$ feet, $\alpha = 38°$, $\beta = 15°$, and $\theta = 75°$, find the distance between A and B to the nearest foot.   (**Source:** P. Kissam.)

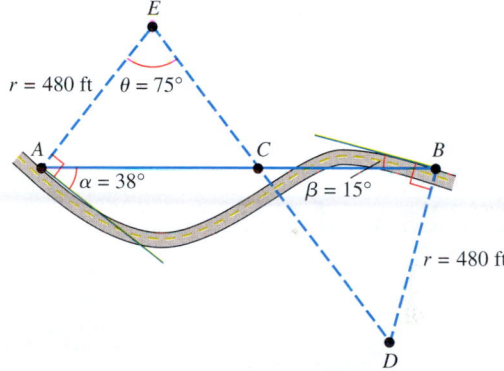

50. *Height of a Hot-Air Balloon*   Two observation points A and B are 1500 feet apart. From these points the angles of elevation of a hot-air balloon are $43°$ and $47°$, as illustrated in the accompanying figure. Find the height of the balloon to the nearest foot.

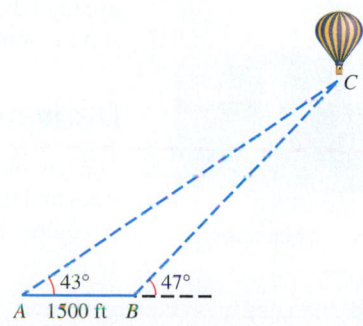

51. *Height of the Gateway Arch* The tallest monument in the world is the Gateway to the West Arch in St. Louis. From point *A* the top of the arch has an angle of elevation of 64.91° and from point *B* the angle of elevation is 60.81°. See the accompanying figure. If distance *AB* is 57 feet, find the height of this monument to the nearest foot. (**Source:** *The Guinness Book of Records*, 1995.)

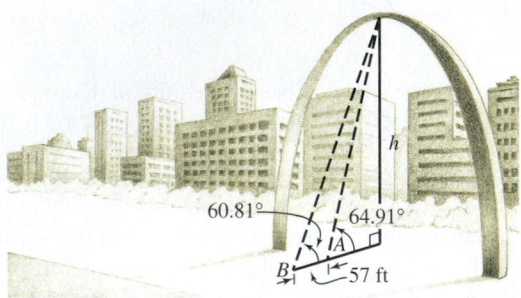

52. *Height of a Tower* A vertical tower supporting a cable for gondolas to transport skiers up a mountain is located on a ski-slope inclined at 28°, as illustrated in the figure in the next column. If the length of the tower's shadow is 21 feet along the mountain side when the angle of the sun is 57° with respect to the ski-slope, calculate the height of the tower.

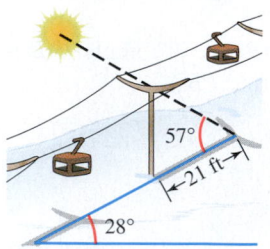

Writing about Mathematics

53. In your own words describe two situations where the law of sines can be applied. Give an example of each situation.

54. Suppose that you are given α, a, and b for triangle *ABC*. If α is obtuse, what is the maximum number of triangles that could satisfy these conditions? What is the maximum number if α is acute? Explain your reasoning and give examples.

8.2 Law of Cosines

♦ Derive the law of cosines
♦ Solve triangles
♦ Find areas of triangles

Introduction

Surveying has been used for centuries in construction and in the determination of boundaries. Today the Global Positioning System (GPS) is being used to determine distances on Earth. The signal from a GPS satellite contains information necessary for hand-held receivers to calculate both the position of a GPS satellite and its distance from the receiver. This information can be used to accurately calculate distances and angles between points on the ground. The law of cosines is a generalization of the Pythagorean theorem and is used in GPS calculations. In this section we introduce the law of cosines and use it to solve several applications. (**Source:** J. Van Sickle, *GPS for Land Surveyors*.)

Derivation of the Law of Cosines

The law of cosines can be used to solve a triangle given either all three sides (SSS) or two sides and the angle included (SAS). In both cases a unique triangle is formed, as illustrated in Figures 8.21 and 8.22.

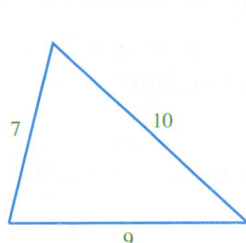

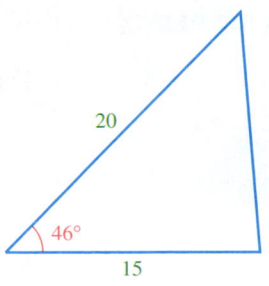

FIGURE 8.21 Given **SSS**   **FIGURE 8.22** Given **SAS**

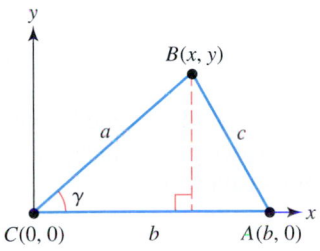

FIGURE 8.23

Consider triangle *ABC* shown in Figure 8.23 with point *B* having coordinates (x, y). Using definitions of sine and cosine, we find

$$\cos \gamma = \frac{x}{a} \quad \text{and} \quad \sin \gamma = \frac{y}{a},$$

or equivalently,

$$x = a \cos \gamma \quad \text{and} \quad y = a \sin \gamma.$$

As a result, the coordinates of *B* are $(a \cos \gamma, a \sin \gamma)$. Since the coordinates of point *A* are $(b, 0)$, the distance *c* between points *A* and *B* can be found.

$c = \sqrt{(a \cos \gamma - b)^2 + (a \sin \gamma - 0)^2}$	Distance formula
$c^2 = (a \cos \gamma - b)^2 + (a \sin \gamma - 0)^2$	Square each side.
$c^2 = a^2 \cos^2 \gamma - 2ab \cos \gamma + b^2 + a^2 \sin^2 \gamma$	Expand each expression.
$c^2 = a^2(\cos^2 \gamma + \sin^2 \gamma) - 2ab \cos \gamma + b^2$	Distributive property
$c^2 = a^2 + b^2 - 2ab \cos \gamma$	$\cos^2 \gamma + \sin^2 \gamma = 1$

This result is valid for any triangle *ABC* and is known as the *law of cosines*. Since the vertices in Figure 8.23 could be rearranged, three possible equations are associated with the law of cosines.

Algebra Review
To review squaring a binomial, see Chapter R (page R-25).

◆ **CLASS DISCUSSION**
Let $\gamma = 90°$ in the formula

$$c^2 = a^2 + b^2 - 2ab \cos \gamma$$

and simplify. Discuss the relationship between the law of cosines and the Pythagorean theorem. ◆

LAW OF COSINES

Any triangle with standard labeling satisfies

$$a^2 = b^2 + c^2 - 2bc \cos \alpha$$
$$b^2 = a^2 + c^2 - 2ac \cos \beta$$
$$c^2 = a^2 + b^2 - 2ab \cos \gamma.$$

Solving Triangles

A common problem in surveying is to find the distance between two points *A* and *B* situated on opposite sides of a building, as illustrated in Figure 8.24 on the next page. This distance can be found by applying the law of cosines.   (**Source:** P. Kissam, *Surveying Practice*.)

Finding the distance between two points (SAS)

In Figure 8.24 a surveyor determines that CA is 75 feet, CB is 58 feet, and angle ACB is 83°. Find distance AB to the nearest foot.

FIGURE 8.24

SOLUTION Note that $CA = b$, $CB = a$, and angle $ACB = \gamma$. So let $b = 75$, $a = 58$, and $\gamma = 83°$. To find c, apply the law of cosines.

$$
\begin{aligned}
c^2 &= a^2 + b^2 - 2ab \cos \gamma && \text{Law of cosines} \\
&= (58)^2 + (75)^2 - 2(58)(75) \cos (83°) && \text{Substitute.} \\
&\approx 7929 && \text{Approximate.} \\
c &\approx 89.04 && \text{Take the square root.}
\end{aligned}
$$

The points A and B are about 89 feet apart. *Now Try Exercises 9 and 49* ◆

In the next example the distance between two GPS receivers is found. Finding the distance between two points might be important to surveyors or to search parties looking for lost hikers. The distance between two GPS receivers is sometimes called the *baseline*.

Using GPS to find a baseline distance (SAS)

A search party and an injured hiker both have hand-held GPS receivers, as illustrated in Figure 8.25. The distance from the satellite to the search party is $b = 20{,}231.15$ kilometers, and the distance from the satellite to the hiker is $c = 20{,}231.57$ kilometers. If it is determined that $\alpha = 0.01456°$, estimate the baseline a between the search party and the hiker to the nearest hundredth of a kilometer.

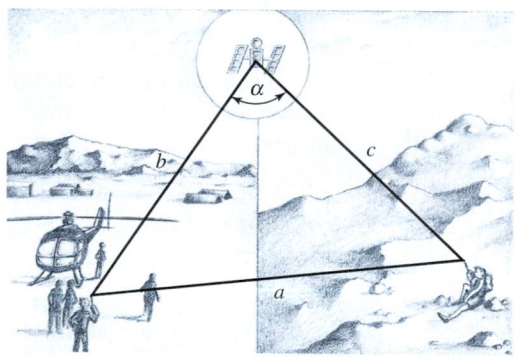

FIGURE 8.25

SOLUTION We can use the law of cosines to find a.

$$
\begin{aligned}
a^2 &= b^2 + c^2 - 2bc \cos \alpha \\
&= (20{,}231.15)^2 + (20{,}231.57)^2 - 2(20{,}231.15)(20{,}231.57) \cos (0.01456°) \\
&\approx 26.61 \\
a &\approx 5.16
\end{aligned}
$$

The distance between the search party and the hiker is about 5.16 kilometers.

Now Try Exercises 19 and 53 ◆

Trusses are frequently used to support roofs on buildings, as illustrated in Figure 8.26. The simplest type of roof truss is a triangle, as shown in Figure 8.27. One basic task when constructing a roof truss is to cut the ends of the rafters so that the roof has the correct slope.   (**Source:** W. Riley, *Statics and Mechanics of Materials*.)

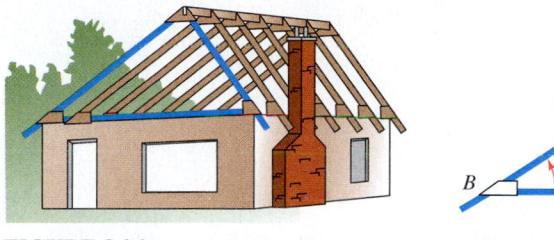

FIGURE 8.26   **FIGURE 8.27** Roof Truss

EXAMPLE 3   Designing a roof truss (SSS)

Find β for the truss shown in Figure 8.27.

SOLUTION Begin by letting $a = 11$, $b = 6$, and $c = 9$, and then use the law of cosines to find β to the nearest degree.

$$b^2 = a^2 + c^2 - 2ac\cos\beta \qquad \text{Law of cosines}$$

$$2ac\cos\beta = a^2 + c^2 - b^2 \qquad \text{Transpose terms.}$$

$$\cos\beta = \frac{a^2 + c^2 - b^2}{2ac} \qquad \text{Divide by 2ac.}$$

$$\cos\beta = \frac{11^2 + 9^2 - 6^2}{2(11)(9)} \qquad \text{Substitute.}$$

$$\cos\beta \approx 0.8384 \qquad \text{Approximate.}$$

Thus $\beta \approx \cos^{-1}(0.8384) \approx 33°$.   *Now Try Exercise 63* ◆

EXAMPLE 4   Solving a triangle (SSS)

Find α, β, and γ in triangle ABC to the nearest tenth of a degree if $a = 5$, $b = 6$, and $c = 9$.

SOLUTION The largest angle, γ, is opposite the largest side, c. We start by finding γ.

$$c^2 = a^2 + b^2 - 2ab\cos\gamma \qquad \text{Law of cosines}$$

$$2ab\cos\gamma = a^2 + b^2 - c^2 \qquad \text{Transpose terms.}$$

$$\cos\gamma = \frac{a^2 + b^2 - c^2}{2ab} \qquad \text{Divide by 2ab.}$$

$$\cos\gamma = \frac{5^2 + 6^2 - 9^2}{2(5)(6)} \qquad \text{Substitute.}$$

$$\cos\gamma = -\frac{1}{3} \qquad \text{Simplify.}$$

Thus $\gamma \approx \cos^{-1}\left(-\frac{1}{3}\right) \approx 109.5°$. The law of cosines could be used again to find either α or β. However, we use the law of sines to find α.

$$\frac{\sin \alpha}{a} = \frac{\sin \gamma}{c} \qquad \text{Law of sines}$$

$$\sin \alpha = \frac{a \sin \gamma}{c} \qquad \text{Multiply by } a.$$

$$\sin \alpha = \frac{5 \sin 109.5°}{9} \qquad \text{Substitute.}$$

$$\sin \alpha \approx 0.5237 \qquad \text{Simplify.}$$

Because γ is the largest angle, it follows that α must be an acute angle and that $\alpha \approx \sin^{-1}(0.5237) \approx 31.6°$. To find β we use the fact that the measures of the angles sum to $180°$ in a triangle.

$$\beta \approx 180° - 109.5° - 31.6° = 38.9°$$

Now Try Exercise 17 ◆

Area Formulas

One task that is frequently performed by surveyors is to find the acreage of a lot using a technique called triangulation. *Triangulation* divides a parcel of land into triangles. The area of the lot equals the sum of the areas of the triangles. We begin our discussion by developing some area formulas for triangles.

The area K of any triangle is given by $K = \frac{1}{2}bh$, where b is its base and h is its height. Using trigonometry, we can find a formula for the area of the triangle shown in Figure 8.28.

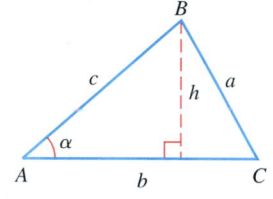

FIGURE 8.28

$$\sin \alpha = \frac{h}{c} \qquad \text{or} \qquad h = c \sin \alpha$$

Thus the area equals

$$K = \frac{1}{2}b\mathbf{h}$$

$$= \frac{1}{2}bc\,\mathbf{\sin\,\alpha}.$$

Since the labels for the vertices in triangle ABC could be rearranged, three area formulas can be written as follows. Notice that these formulas can be applied when we are given SAS.

AREA OF A TRIANGLE

For any triangle with standard labeling, the area K is given by

$$K = \frac{1}{2}ab \sin \gamma, \qquad K = \frac{1}{2}ac \sin \beta, \qquad \text{or} \qquad K = \frac{1}{2}bc \sin \alpha.$$

EXAMPLE 5

Finding the area of a triangle (SAS)

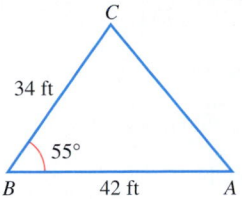

FIGURE 8.29

Find the area of triangle ABC in Figure 8.29 to the nearest square foot.

SOLUTION We are given $\beta = 55°$, $a = 34$ feet, and $c = 42$ feet. Thus the area K is given by the following.

$$K = \frac{1}{2}ac \sin\beta \qquad \text{Area formula}$$

$$= \frac{1}{2}(34)(42) \sin 55° \qquad \text{Substitute.}$$

$$\approx 585 \text{ square feet} \qquad \text{Approximate.}$$

Now Try Exercise 33 ◆

The next formula can be used to find the area of a triangle when the lengths of three sides are known. It is named after the Greek mathematician Heron.

> **HERON'S FORMULA**
>
> If a triangle has sides with lengths a, b, and c, then its area K is given by
> $$K = \sqrt{s(s-a)(s-b)(s-c)},$$
> where $s = \frac{1}{2}(a+b+c)$ and s is called the **semiperimeter**.

EXAMPLE 6

Finding the area of a triangle (SSS)

Approximate the area of triangle ABC with sides $a = 4$, $b = 5$, and $c = 7$.

SOLUTION Begin by calculating s.

$$s = \frac{1}{2}(4 + 5 + 7) = 8$$

Then the area is

$$K = \sqrt{8(8-4)(8-5)(8-7)} = \sqrt{96} \approx 9.8$$

Now Try Exercise 35 ◆

One method for finding the area of a lot is called the *distance method*. This method can be used to find the area of an irregular lot, as illustrated in the next example. The distance method does not measure angles, rather it measures only distances between points. Triangulation and Heron's formula can be used to find the area of the lot. (**Source:** I. Mueller and K. Ramsayer, *Introduction to Surveying*.)

EXAMPLE 7

Applying the distance method to find the area of a lot

Find the area of the parcel of land determined by $ABCDE$ in Figure 8.30 on the next page.

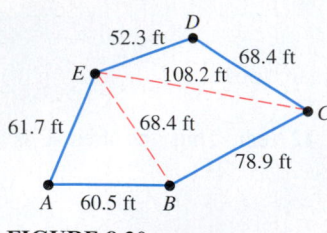

FIGURE 8.30

SOLUTION In triangle ABE, $s = \frac{1}{2}(60.5 + 68.4 + 61.7) = 95.3$ and its area is

$$K_1 = \sqrt{95.3(95.3 - 60.5)(95.3 - 68.4)(95.3 - 61.7)} \approx 1731.$$

In triangle BCE, $s = \frac{1}{2}(78.9 + 108.2 + 68.4) = 127.75$ and its area is

$$K_2 = \sqrt{127.75(127.75 - 78.9)(127.75 - 108.2)(127.75 - 68.4)} \approx 2691.$$

In triangle CDE, $s = \frac{1}{2}(68.4 + 52.3 + 108.2) = 114.45$ and its area is

$$K_3 = \sqrt{114.45(114.45 - 68.4)(114.45 - 52.3)(114.45 - 108.2)} \approx 1431.$$

The area of the lot is

$$K_1 + K_2 + K_3 \approx 1731 + 2691 + 1431 = 5853 \text{ square feet.}$$

Now Try Exercise 67 ◆

8.2 Putting it all Together

The law of cosines can be used to solve triangles when given either three sides (SSS) or two sides and the angle included (SAS). Heron's formula can be used to find the area of a triangle when the lengths of all three sides are known. Important formulas found in this section are given in the table, where it is assumed that triangle ABC has standard labeling.

Concept	Explanation	Examples
Law of cosines	$a^2 = b^2 + c^2 - 2bc \cos \alpha$ $b^2 = a^2 + c^2 - 2ac \cos \beta$ $c^2 = a^2 + b^2 - 2ab \cos \gamma$ **Can be used to solve a triangle given either SSS or SAS**	If $b = 3$, $c = 4$, and $\alpha = 60°$, then $a^2 = 3^2 + 4^2 - 2(3)(4) \cos 60°$ $= 13$ and $a = \sqrt{13}$.
Area formulas	$K = \dfrac{1}{2}ab \sin \gamma$ $K = \dfrac{1}{2}ac \sin \beta$ $K = \dfrac{1}{2}bc \sin \alpha$ **Can be used to find the area of a triangle given SAS**	If $a = 2$ feet, $b = 3$ feet, and $\gamma = 30°$, then the area of the triangle is $K = \dfrac{1}{2}(2)(3) \sin 30° = 1.5 \text{ ft}^2.$

Concept	Explanation	Examples
Heron's formula	$K = \sqrt{s(s-a)(s-b)(s-c)}$, where $s = \dfrac{1}{2}(a+b+c)$. Can be used to find the area of a triangle given SSS	If $a = 3$ feet, $b = 5$ feet, and $c = 4$ feet, then $s = \dfrac{1}{2}(3+5+4) = 6$ and the area of the triangle is $K = \sqrt{6(6-3)(6-5)(6-4)}$ $= 6 \text{ ft}^2.$

8.2 Exercises

Determining a Method to Solve a Triangle

Exercises 1–8: Assume triangle ABC has standard labeling and complete the following.

(a) Determine if AAS, ASA, SSA, SAS, or SSS is given.
(b) Decide if the law of sines or the law of cosines should be used to solve the triangle.

1. a, b, and γ

2. α, γ, and c

3. a, b, and α

4. a, b, and c

5. α, β, and c

6. a, c, and α

7. β, a, and γ

8. b, c, and α

Solving Triangles

Exercises 9–14: Solve the triangle. Approximate values to the nearest tenth.

9.

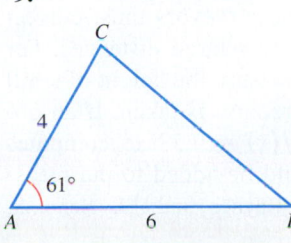

10.

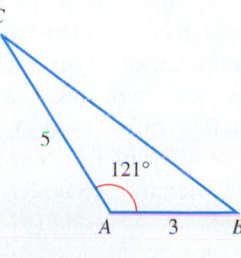

11.

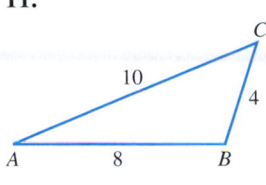

12.

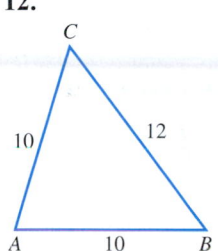

13.

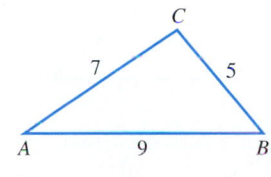

14.

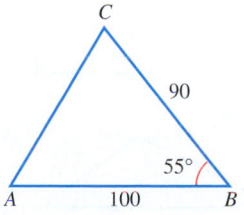

Exercises 15–26: Solve the triangle. Round values to the nearest tenth when appropriate.

15. $a = 45, \gamma = 35°, b = 24$

16. $c = 7.9, \beta = 52°, a = 9.6$

17. $a = 2.4, b = 1.7, c = 1.4$

18. $a = 43, b = 41, c = 34$

19. $\alpha = 10°30', b = 24.1, c = 15.8$

20. $a = 12.8, b = 15.8, \gamma = 36°$

21. $a = 10.6, b = 25.8, c = 20.6$

22. $a = 104, b = 121, c = 111$

23. $\beta = 122°10', a = 20, c = 15$

24. $b = 9.1, \alpha = 43°30', c = 12.5$

25. $a = 5.3, b = 6.7, c = 7.1$

26. $a = 4.2, b = 5.1, c = 3.7$

Does This Triangle Exist?

Exercises 27–32: Decide if a triangle exists that satisfies the conditions. Justify your answer.

27. $a = 10, b = 12, c = 25$

28. $a = 10, \beta = 51°, c = 5$

29. $\alpha = 89°, b = 63, \gamma = 112°$

30. $a = 2, b = 10, \alpha = 50°$

31. $\gamma = 54°, b = 63, \alpha = 63°$

32. $a = 5, b = 6, c = 8$

Area of Triangles

Exercises 33–36: Approximate the area of the triangle to the nearest tenth.

33.
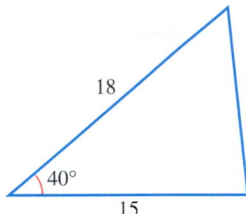

34.
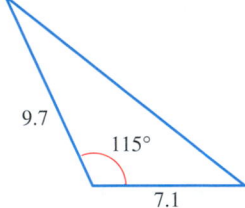

35.
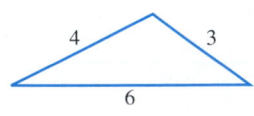

36.
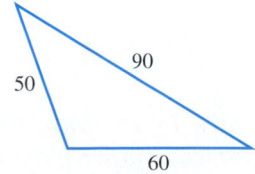

Exercises 37–48: Approximate the area of the triangle to the nearest tenth.

37. $a = 10, b = 12, \gamma = 58°$

38. $\alpha = 40°, b = 5.8, c = 8.8$

39. $\beta = 78°, a = 5.5, c = 6.8$

40. $\alpha = 23°, \gamma = 47°, b = 53$

41. $\beta = 31°, \alpha = 54°, a = 2.6$

42. $a = 7, b = 8, c = 9$

43. $a = 5.5, b = 6.7, c = 9.2$

44. $a = 104, b = 98, c = 112$

45. $a = 11, b = 13, c = 20$

46. $a = 13, b = 14, c = 15$

47. $a = 21, \alpha = 42°, c = 16$

48. $b = 35, c = 38, \gamma = 50°48'$

Applications

49. *Obstructed View* (Refer to Example 1.) In the accompanying figure, a surveyor is attempting to find the distance between points A and B. A grove of trees is obstructing the view so the surveyor determines that AC is 143 feet, BC is 123 feet, and angle ACB is 78°35'. Find the distance between A and B to the nearest foot.

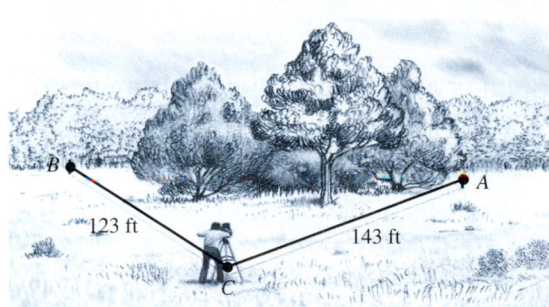

50. *Surveyor* A surveyor measures the sides of a triangular lot to be $a = 145.2$, $b = 136.8$, and $c = 95.3$, where measurements are in feet.
  (a) Approximate angles α, β, and γ.

  (b) What is the area of the lot to the nearest square foot?

51. *Curvature of the Earth* In times past when sailing ships came into port, people would see the top of the sails before they saw the entire ship. This is one reason why people knew the earth was not flat. Because of the curvature of the earth, surveyors must correct height measurements made over large distances. For example, a surveyor might measure the height of a hill to be too small unless a correction is taken. If an object is x miles away, then $f(x) = 0.585x^2$ computes the number of feet that should be added to the measured height. (**Source:** W. Rayner and M. Schmidt, *Elementary Surveying.*)

(a) In the accompanying figure, a surveyor measures a small hill 3 miles away to be 96 feet high. Find the actual height of the hill.

(b) Approximate the correction angle θ to the nearest thousandth of a degree in the figure by assuming that triangle ABC is isosceles.

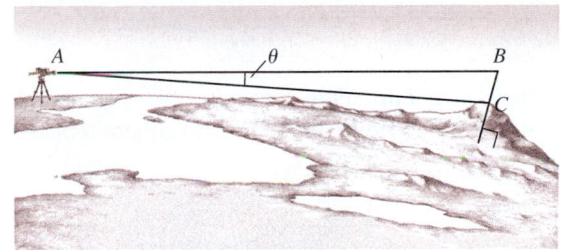

52. *Curvature and Surveying* Suppose that the hill in the previous exercise were 5 miles away. Find the correction angle θ and compare it with the value found in the previous exercise.

53. *Ship Navigation* Two ships set sail with bearings of $52°$ and $121°$, traveling at 20 miles per hour and 14 miles per hour, respectively. Approximate the distance between the ships after 1.5 hours.

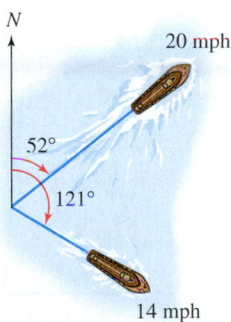

54. *Air Navigation* An airplane flies in a triangular course shown in the figure. Find the bearings of the plane to the nearest degree, while traveling from A to B and from B to C.

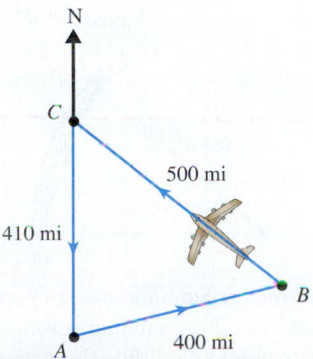

55. *Diagonals of a Parallelogram* One side of a parallelogram is 3.5 feet and another side is 5.2 feet. The angle between these two sides is $56°$. Find the lengths of the diagonals of the parallelogram to the nearest tenth of a foot.

56. *Angle in a Parallelogram* One side of a parallelogram is 6.4 yards and another side is 5.3 yards. The shorter diagonal is 3.5 yards. Find the angle opposite the shorter diagonal to the nearest tenth of a degree.

57. *Distance between Airports* Airports A and B are 515 miles apart, and airport A is directly west of airport B. Airport C is located in a northeasterly direction from airport A and is 357 miles from airport B. See the accompanying figure. If the bearing from airport C to airport B is $125°$, find the distance between airports A and C to the nearest mile.

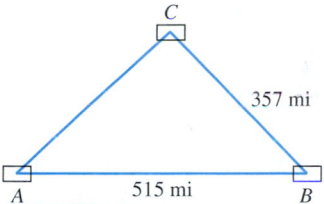

58. *Navigation* A ship is sailing east. At one point the bearing of a submerged rock is $38°45'$. After the ship sails 20.4 miles, the bearing of the rock is $291°15'$. Find the distance between the ship and the rock to the nearest tenth of a mile when the second bearing is taken.

59. *Painting* A painter needs to cover a triangular region with sides of 25 feet and 15 feet. The angle between these two sides is $128°$. Find the area of the region to the nearest tenth of a square foot.

60. *Painting* A painter needs to cover a triangular region with sides of 30 feet, 40 feet, and 38 feet. If each can of paint covers 125 square feet, how many cans of paint are needed?

61. *Area of Regular Polygons* If a regular polygon has n sides of equal length L, then its area A is computed by

$$A = \frac{nL^2}{4} \cot\left(\frac{\pi}{n}\right).$$

(**Source:** M. Mortenson, *Computer Graphics.*)
(a) Using this formula, find the area of an equilateral triangle with sides of 6 inches.

(b) Using Heron's formula, find the area of this triangle. Compare answers.

62. *Area of Regular Polygons* (Refer to the previous exercise.) The measure of an interior angle in a regular polygon with n sides is given by $180°\left(1 - \frac{2}{n}\right)$. For example, a square is a regular polygon with $n = 4$ and each interior angle equals $180°\left(1 - \frac{2}{4}\right) = 90°$.

(a) Find the area of a regular pentagon with sides of length 8 inches using the formula given in the previous exercise. See the accompanying figure.

(b) Using triangulation and Heron's formula, find the area of this regular pentagon.

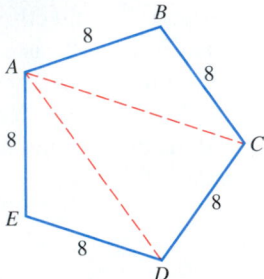

63. *Truss Construction* (Refer to Example 3.) A triangular truss is shown in the figure. Find angle θ.

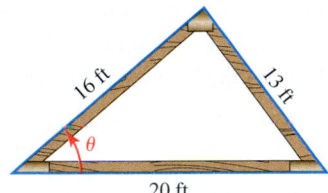

64. *Robotics* The figure illustrates the MIT Scheinman robotic arm. Suppose the length of the upper arm is 20 centimeters and the combined length of the forearm and hand is 30 centimeters. If the arm is positioned so that $\theta = 126°$, find the distance between the hand at point A and the shoulder joint at point B. (**Sources:** G. Beni and S. Hackwood, *Recent Advances in Robotics.*)

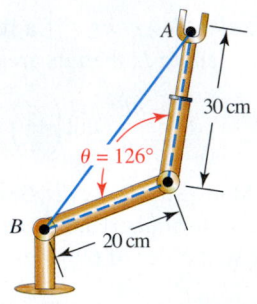

65. *Area of a Lot* Find the area of the lot in the figure to the nearest square foot.

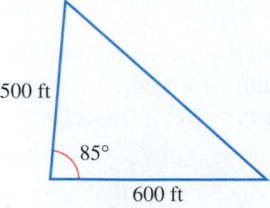

66. *Area of a Lot* Find the area of the quadrangular lot shown in the figure to the nearest square foot.

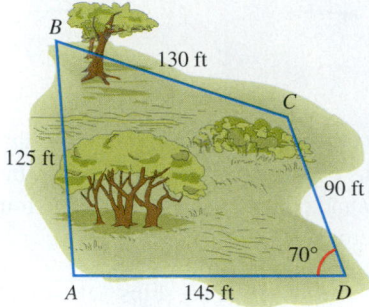

67. *Area of a Lot* Apply the distance method discussed in Example 7 to find the area of the lot to the nearest square foot.

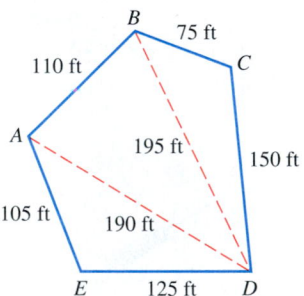

68. *Highway Curve* The most common highway curve consists of a circular arc connecting two sections of straight road, as illustrated in the figure. Find the straight-line distance between PC (point of curve) and PT (point of tangency). (**Source:** P. Kissam.)

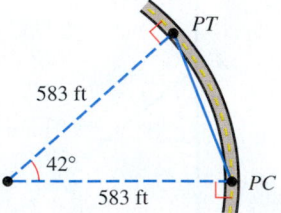

69. *Distance between a Satellite and a Tracking Station* A satellite traveling in a circular orbit 1600 kilometers above Earth is due to pass directly over a tracking station at noon. (See the figure.) Assume that the satellite takes 2 hours to make an orbit and that the radius of Earth is 6400 kilometers. Find the distance between the satellite and the tracking station at 12:03 P.M. (**Source:** *Space Mathematics* by B. Kastner, Ph.D. Copyright © 1972 by the National Aeronautics and Space Administration. Courtesy of NASA.)

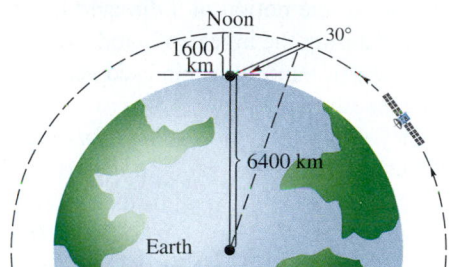

70. ***Distance between a Ship and a Submarine*** From an airplane flying over the ocean, the angle of depression to a submarine lying just under the surface is 24°10′. At the same moment, the angle of depression from the airplane to a battleship is 17°30′. See the figure. The distance from the airplane to the battleship is 5120 feet. Find the distance between the battleship and the submarine. (Assume the airplane, submarine, and battleship are in a vertical plane.)

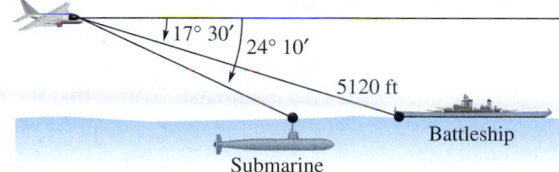

71. ***Computer Graphics*** When shading a triangle on a computer screen, three pixels are used to determine the vertices of a triangle. Many computer screens are 1024 pixels by 800 pixels and each pixel represents 0.015 inch by 0.015 inch, as illustrated in the figure, where pixel (7, 3) is shown. If three pixels $A(100, 300)$,

$B(500, 200)$, and $C(320, 600)$ represent vertices of a triangle, approximate the area of triangle ABC on the computer screen to the nearest tenth of a square inch. (**Source:** J. Foley, *Introduction to Computer Graphics.*)

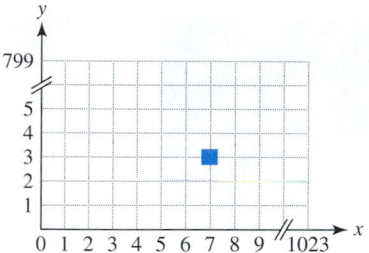

72. ***Computer Graphics*** Scaling factors are used in computer graphics to change the size of an object. For example, if a triangle has vertices $A(1, 2)$, $B(4, 0)$, and $C(2, 4)$ with scaling factors of s_x and s_y, then multiplying the x-coordinates of the vertices by s_x and the y-coordinates by s_y changes the size of the triangle. Scaling factors of $s_x = 2$ and $s_y = 3$ result in a new triangle with vertices $D(2, 6)$, $E(8, 0)$, and $F(4, 12)$. (**Source:** J. Foley.)

(a) Calculate the area of triangles ABC and DEF. How do they compare?

(b) If triangle ABC is scaled using $s_x = 2$ and $s_y = \frac{1}{2}$, find its area.

(c) Predict how the area A of a triangle changes for scaling factors of s_x and s_y.

Writing about Mathematics

73. Describe two different situations where the law of cosines can be applied. Give an example of each situation.

74. Describe two methods to find the area of a triangle. What information do you need to apply each method? Give an example of each situation.

CHECKING BASIC CONCEPTS FOR SECTIONS 8.1 AND 8.2

1. Solve triangle ABC if $\alpha = 44°$, $\gamma = 62°$, and $a = 12$.

2. Solve triangle ABC if $\alpha = 32°$, $a = 6$, and $b = 8$. How many solutions are there?

3. Use the law of cosines to solve the triangles.

(a)
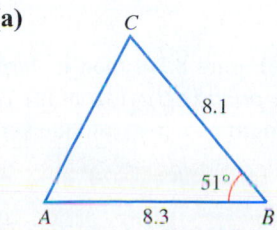

(b)
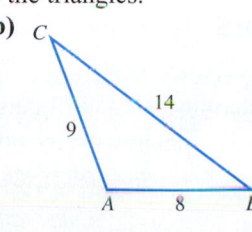

4. Find the area of triangle ABC to the nearest hundredth.
(a) $a = 4.5, b = 5.2, \gamma = 55°$

(b) $a = 6, b = 7, c = 9$

8.3 Vectors

• Learn basic concepts about vectors
• Perform operations on vectors
• Learn and apply the dot product
• Use vectors to calculate work

Introduction

The beginnings of vectors go back centuries to the notion of a directed line segment, but the formal development of vectors occurred during the nineteenth and twentieth centuries, after the invention of complex numbers. It was not until Einstein used vectors in his theory of relativity that their importance became readily accepted.

Vectors are a profound invention. They provide a simple model for science and technology to visualize difficult concepts such as force, velocity, and electric fields. Vectors are essential in creating today's amazing computer graphics. In this section we discuss some of the important properties and applications of vectors. (**Sources:** *Historical Topics for the Mathematical Classroom, Thirty-first Yearbook, NCTM; M. Mortenson, Computer Graphics.*)

Basic Concepts

Many quantities in mathematics can be described using real numbers or **scalars**. Examples include a person's weight, the cost of a CD player, and the gas mileage of a car. Other quantities must be represented using vector quantities. A **vector quantity** involves both *magnitude* and *direction*. Magnitude can be interpreted as size or length. For example, if a car is traveling north at 50 miles per hour, then the direction *north* coupled with a *speed* of 50 miles per hour represents a vector quantity called *velocity*. In science a distinction is made between speed and velocity—speed is the magnitude of velocity.

A vector quantity can be represented by a directed line segment called a **vector**. A vector **v** representing a velocity of a car traveling 50 miles per hour north is shown in Figure 8.31, and the vector **u** represents a velocity of 25 miles per hour east. Notice that the length of **u** is half the length of **v**. Vectors do *not* have position, rather they have magnitude and direction. A vector can be translated, provided its direction and magnitude (length) do not change. Two vectors are **equal** if they have the same magnitude and direction. In Figure 8.32 each directed line segment represents the same vector **v**.

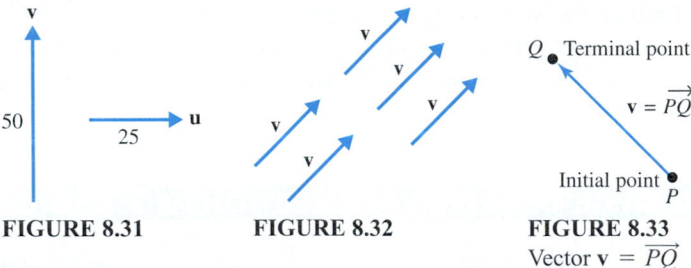

FIGURE 8.31   FIGURE 8.32   FIGURE 8.33
Vector $\mathbf{v} = \overrightarrow{PQ}$

A vector is usually represented symbolically by a letter printed in boldface type, such as **a**, **b**, **v**, or **F**. A second way to denote a vector is to use two points. If the **initial point** of a vector **v** is P and its **terminal point** is Q, then $\mathbf{v} = \overrightarrow{PQ}$, as illustrated in Figure 8.33.

Operations on Vectors

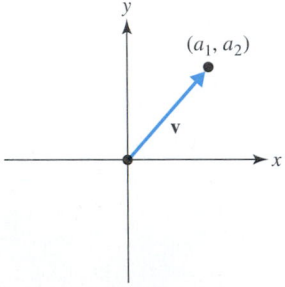

FIGURE 8.34
Vector $\mathbf{v} = \langle a_1, a_2 \rangle$

If we place the initial point of vector **v** at the origin, as in Figure 8.34, then its terminal point (a_1, a_2) can be used to determine **v**. To distinguish the *point* (a_1, a_2) from the *vector* **v**, we use the notation $\mathbf{v} = \langle a_1, a_2 \rangle$. The **horizontal component** of **v** is a_1 and the **vertical component** of **v** is a_2.

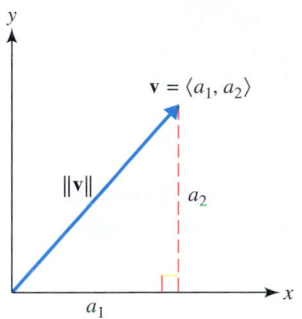

FIGURE 8.35

The length of a vector equals its magnitude. If $\mathbf{v} = \langle a_1, a_2 \rangle$, then the *magnitude* of $\mathbf{v}$ is denoted $\|\mathbf{v}\|$. By applying the Pythagorean theorem to Figure 8.35, $\|\mathbf{v}\| = \sqrt{a_1^2 + a_2^2}$.

MAGNITUDE OF A VECTOR

If $\mathbf{v} = \langle a_1, a_2 \rangle$, then the **magnitude** (or length) of $\mathbf{v}$ is given by

$$\|\mathbf{v}\| = \sqrt{a_1^2 + a_2^2}.$$

If $\|\mathbf{v}\| = 1$, then $\mathbf{v}$ is a **unit vector**.

If a vector has initial point P with coordinates (a_1, b_1) and terminal point Q with coordinates (a_2, b_2), then vector $\overrightarrow{PQ}$ is given by $\overrightarrow{PQ} = \langle a_2 - a_1, b_2 - b_1 \rangle$. This is illustrated in the next example. Note that $\overrightarrow{QP} = \langle a_1 - a_2, b_1 - b_2 \rangle$.

Finding a vector graphically and symbolically

Let P have coordinates $(-1, 2)$ and Q have coordinates $(3, 4)$. Find vector $\overrightarrow{PQ}$ graphically and symbolically. Calculate the magnitude of $\overrightarrow{PQ}$.

SOLUTION   To graph $\overrightarrow{PQ}$, plot the points P and Q. Then sketch a directed line segment from P to Q, as shown in Figure 8.36. We can see that the horizontal component is 4 and the vertical component is 2. A symbolic representation of $\overrightarrow{PQ}$ is given by

$$\overrightarrow{PQ} = \langle 3 - (-1), 4 - 2 \rangle = \langle 4, 2 \rangle.$$

The magnitude or length of $\overrightarrow{PQ}$ is

$$\|\overrightarrow{PQ}\| = \sqrt{4^2 + 2^2} = \sqrt{20} \approx 4.47. \qquad \textit{Now Try Exercise 21} \blacklozenge$$

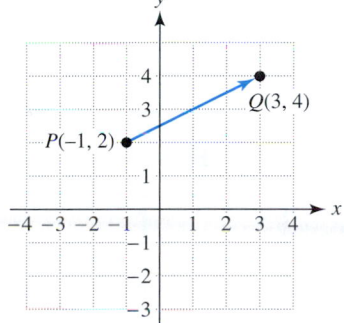

FIGURE 8.36

Vector Addition   Suppose that a swimmer heads directly across a river at 3 miles per hour. If the current is 4 miles per hour, then the person will be carried a distance downstream before reaching the other side, as illustrated in Figure 8.37. We can use vectors to visually find the direction and speed that the swimmer will travel across the river.

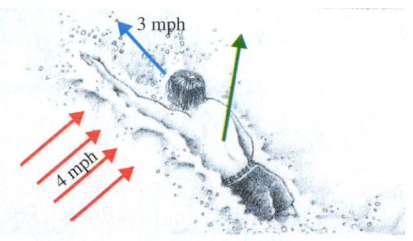

FIGURE 8.37   A Swimmer in a Current

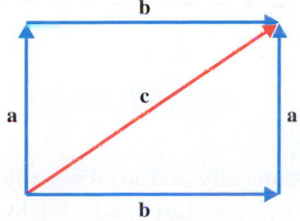

FIGURE 8.38   Vector Addition

Let vector $\mathbf{a}$ represent the speed and direction of the swimmer with no current, vector $\mathbf{b}$ represent the direction and speed of the current, and vector $\mathbf{c}$ represent the final direction and speed of the swimmer. We can find the length and direction of $\mathbf{c}$ by applying the **parallelogram rule**, as shown in Figure 8.38. The speed and direction of the swimmer are represented by the diagonal $\mathbf{c}$ of the parallelogram (rectangle), which is determined by $\mathbf{a}$ and $\mathbf{b}$. Vector $\mathbf{c}$ is called the **sum** or **resultant** of vectors $\mathbf{a}$ and $\mathbf{b}$.

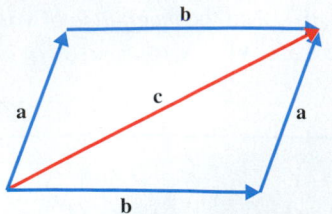

FIGURE 8.39 The Parallelogram Rule: $\mathbf{c} = \mathbf{a} + \mathbf{b}$

Symbolically, we can represent the velocity of the swimmer with no current by $\mathbf{a} = \langle 0, 3 \rangle$, the velocity of the current by $\mathbf{b} = \langle 4, 0 \rangle$, and the velocity of the swimmer in the current by $\mathbf{c} = \langle 4, 3 \rangle$. Vector $\mathbf{c}$ is the sum of vectors $\mathbf{a}$ and $\mathbf{b}$ and can be found as follows.

$$\mathbf{a} + \mathbf{b} = \langle 0, 3 \rangle + \langle 4, 0 \rangle = \langle 0 + 4, 3 + 0 \rangle = \langle 4, 3 \rangle = \mathbf{c}$$

Since $\|\mathbf{c}\| = \sqrt{4^2 + 3^2} = 5$, the swimmer moves 5 miles per hour in the direction of $\mathbf{c}$.

Figure 8.39 illustrates graphically how to find $\mathbf{c} = \mathbf{a} + \mathbf{b}$ in general by using the parallelogram rule. The following defines vector addition symbolically.

VECTOR ADDITION

If $\mathbf{a} = \langle a_1, a_2 \rangle$ and $\mathbf{b} = \langle b_1, b_2 \rangle$, then the **sum** of $\mathbf{a}$ and $\mathbf{b}$ is given by

$$\mathbf{a} + \mathbf{b} = \langle a_1, a_2 \rangle + \langle b_1, b_2 \rangle = \langle a_1 + b_1, a_2 + b_2 \rangle.$$

Suppose that vector $\mathbf{a}$ represents a force of 80 pounds pulling on a water-ski towrope and $\mathbf{b}$ represents a force of 60 pounds pulling on a second towrope. See Figure 8.40. The resultant force $\mathbf{c} = \mathbf{a} + \mathbf{b}$ is given by the diagonal of the parallelogram shown in Figure 8.41. Vector $\mathbf{c}$ represents the net force exerted by the two water skiers.

FIGURE 8.40

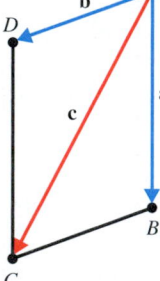

FIGURE 8.41

EXAMPLE 2 Applying the parallelogram rule

Find the magnitude of the resultant force on the ski boat in the preceding discussion if the angle between the towropes is 25°.

SOLUTION The magnitude of the force equals the length of the diagonal AC in Figure 8.41. Since angle ABC is $180° - 25° = 155°$, we find AC by applying the law of cosines.

$$AC^2 = 60^2 + 80^2 - 2(60)(80) \cos 155°$$

$$\approx 18{,}701$$

$$AC \approx 136.8 \text{ pounds}$$

Now Try Exercise 78 ◆

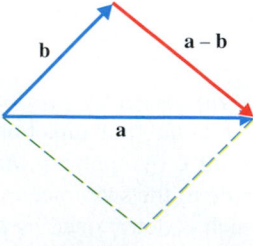

FIGURE 8.42

Vector Subtraction Subtraction can be defined both graphically and symbolically. The difference $\mathbf{a} - \mathbf{b}$ is shown graphically in Figure 8.42. Notice that by the parallelogram rule $\mathbf{b} + (\mathbf{a} - \mathbf{b}) = \mathbf{a}$.

Vector subtraction can be defined symbolically as follows.

VECTOR SUBTRACTION

If $\mathbf{a} = \langle a_1, a_2 \rangle$ and $\mathbf{b} = \langle b_1, b_2 \rangle$, then the **difference** of $\mathbf{a}$ and $\mathbf{b}$ is given by

$$\mathbf{a} - \mathbf{b} = \langle a_1, a_2 \rangle - \langle b_1, b_2 \rangle = \langle a_1 - b_1, a_2 - b_2 \rangle.$$

EXAMPLE 3

Adding and subtracting vectors

Let $\mathbf{a} = \langle -3, 4 \rangle$ and $\mathbf{b} = \langle 5, -6 \rangle$. Find $\mathbf{a} + \mathbf{b}$ and $\mathbf{a} - \mathbf{b}$.

SOLUTION To add two vectors, we add corresponding components.

$$\begin{aligned} \mathbf{a} + \mathbf{b} &= \langle -3, 4 \rangle + \langle 5, -6 \rangle \\ &= \langle -3 + 5, 4 + (-6) \rangle \\ &= \langle 2, -2 \rangle \end{aligned}$$

To subtract two vectors, we subtract corresponding components.

$$\begin{aligned} \mathbf{a} - \mathbf{b} &= \langle -3, 4 \rangle - \langle 5, -6 \rangle \\ &= \langle -3 - 5, 4 - (-6) \rangle \\ &= \langle -8, 10 \rangle \end{aligned}$$

Now Try Exercise 31 ◆

Scalar Multiplication Scalar multiplication results when a vector $\mathbf{v}$ is multiplied by a real number or *scalar* k to form $k\mathbf{v}$. Vectors $\mathbf{v}$ and $k\mathbf{v}$ are parallel if $k \neq 0$. Vector $k\mathbf{v}$ points in the *same* direction as $\mathbf{v}$ if $k > 0$, and $k\mathbf{v}$ points in the *opposite* direction of $\mathbf{v}$ if $k < 0$. The magnitude of $k\mathbf{v}$ is $|k|$ times the magnitude of $\mathbf{v}$. For example, suppose that a 10-mile-per-hour wind is blowing from the north. Since the wind is blowing toward the south, the wind may be represented by $\mathbf{v} = \langle 0, -10 \rangle$. If the wind speed doubles, but does not change direction, then the *scalar product* $2\mathbf{v}$ models this situation.

$$2\mathbf{v} = 2\langle 0, -10 \rangle = \langle 2 \cdot 0, 2 \cdot (-10) \rangle = \langle 0, -20 \rangle$$

The wind is now blowing at 20 miles per hour toward the south. See Figure 8.43.

Next, suppose that a wind from the southwest is modeled by $\mathbf{v} = \langle 4, 4 \rangle$. Then

$$-\tfrac{1}{2}\mathbf{v} = -\tfrac{1}{2}\langle 4, 4 \rangle = \langle -\tfrac{1}{2} \cdot 4, -\tfrac{1}{2} \cdot 4 \rangle = \langle -2, -2 \rangle$$

represents a wind from the northeast in the opposite direction of $\mathbf{v}$ with half the speed. See Figure 8.44. This discussion motivates the following definition of scalar multiplication.

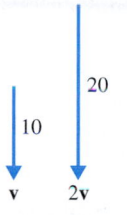

FIGURE 8.43
North Winds

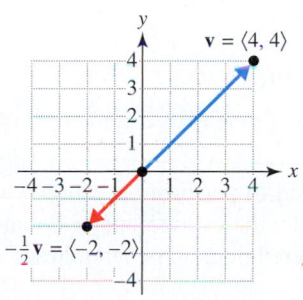

FIGURE 8.44

SCALAR MULTIPLICATION

If $\mathbf{v} = \langle v_1, v_2 \rangle$ and k is a real number, then the **scalar product** $k\mathbf{v}$ is given by

$$k\mathbf{v} = k\langle v_1, v_2 \rangle = \langle kv_1, kv_2 \rangle.$$

Sums, differences, and scalar products can be calculated graphically and symbolically, as illustrated in the next example.

EXAMPLE 4   **Performing operations on vectors**

Find each of the following expressions graphically and symbolically if $\mathbf{a} = \langle -3, 4 \rangle$ and $\mathbf{b} = \langle -1, -2 \rangle$.

(a) $\|\mathbf{a}\|$ (b) $-2\mathbf{b}$ (c) $\mathbf{a} + 2\mathbf{b}$

SOLUTION

(a) Graph $\mathbf{a} = \langle -3, 4 \rangle$, as shown in Figure 8.45. The length of $\mathbf{a}$ appears to be about 5. This can be verified symbolically.

$$\|\mathbf{a}\| = \sqrt{(-3)^2 + (4)^2} = 5$$

(b) Graph $\mathbf{b} = \langle -1, -2 \rangle$. The scalar product $-2\mathbf{b}$ points in the opposite direction of $\mathbf{b}$ with twice the length. See Figure 8.46, where $-2\mathbf{b} = \langle 2, 4 \rangle$. Symbolically this is given by

$$-2\mathbf{b} = -2\langle -1, -2 \rangle = \langle -2 \cdot (-1), -2 \cdot (-2) \rangle = \langle 2, 4 \rangle.$$

(c) Graph $\mathbf{a} = \langle -3, 4 \rangle$ and $2\mathbf{b} = \langle -2, -4 \rangle$, as shown in Figure 8.47. By the parallelogram rule, the diagonal represents $\mathbf{a} + 2\mathbf{b} = \langle -5, 0 \rangle$. This can be verified symbolically.

$$
\begin{aligned}
\mathbf{a} + 2\mathbf{b} &= \langle -3, 4 \rangle + 2\langle -1, -2 \rangle \\
&= \langle -3, 4 \rangle + \langle -2, -4 \rangle \\
&= \langle -5, 0 \rangle
\end{aligned}
$$

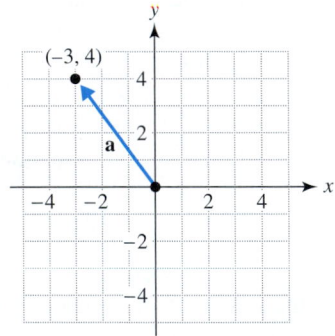

FIGURE 8.45

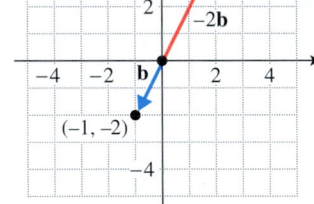

FIGURE 8.46

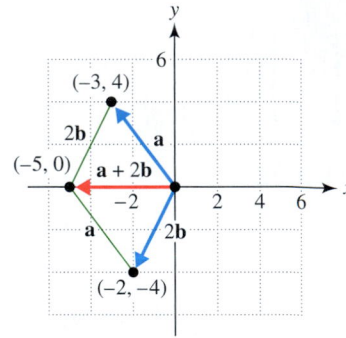

FIGURE 8.47

Now Try Exercise 41 ◆

FIGURE 8.48

On some graphing calculators the list feature can be used to perform operations on vectors. In Figure 8.48 this calculator has been used to evaluate the expressions in Example 4. On other calculators, vectors can be represented by ordered pairs with parentheses.

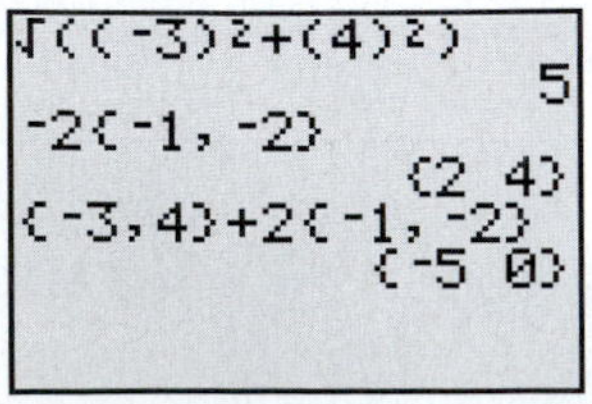

An Application in Robotics   Robotic arms are sometimes modeled using vectors. Consider the *planar two-arm manipulator* in Figure 8.49. If $\overrightarrow{AB} = \mathbf{a}$ and $\overrightarrow{BC} = \mathbf{b}$, then the position of the hand is given by $\overrightarrow{AC} = \mathbf{c}$. Since $\mathbf{c} = \mathbf{a} + \mathbf{b}$, we can easily locate the position of the hand if $\mathbf{a}$ and $\mathbf{b}$ are known.   (**Source:** J. Craig, *Introduction to Robotics*.)

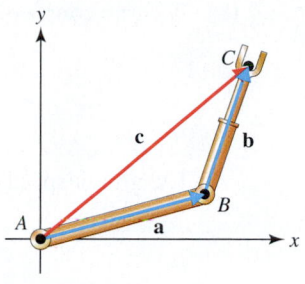

FIGURE 8.49

EXAMPLE 5 Using vectors to locate a robotic arm

Let $\mathbf{a} = \langle 3.1, 1.5 \rangle$ and $\mathbf{b} = \langle 1.4, 2.4 \rangle$ in Figure 8.49.
(a) Find the position of the robotic hand.
(b) Suppose the upper arm represented by $\mathbf{a}$ doubles its length and the forearm represented by $\mathbf{b}$ reduces its length by half. Find the new position of the hand.

SOLUTION
(a) To find the position of the hand, evaluate $\mathbf{a} + \mathbf{b}$.

$$\mathbf{a} + \mathbf{b} = \langle 3.1, 1.5 \rangle + \langle 1.4, 2.4 \rangle = \langle 4.5, 3.9 \rangle$$

The hand is located at the point (4.5, 3.9).
(b) The new position is represented by

$$2\mathbf{a} + \frac{1}{2}\mathbf{b} = 2\langle 3.1, 1.5 \rangle + \frac{1}{2}\langle 1.4, 2.4 \rangle$$

$$= \langle 6.2, 3.0 \rangle + \langle 0.7, 1.2 \rangle$$

$$= \langle 6.9, 4.2 \rangle.$$

The new coordinates of the robotic hand are (6.9, 4.2). *Now Try Exercise 81* ◆

EXAMPLE 6 Using vectors in navigation

An airplane is flying with an airspeed of 300 miles per hour and a bearing of 40° in a 30-mile-per-hour west wind.
(a) Find vectors $\mathbf{v}$ and $\mathbf{u}$ that model the velocity of the airplane and the velocity of the wind, respectively.
(b) Use vectors to determine the groundspeed of the plane.
(c) Find the final bearing of the plane in the wind.

SOLUTION
(a) Consider Figure 8.50, which shows vectors $\mathbf{v}$ and $\mathbf{u}$ graphically. Since $\mathbf{u}$ models a west wind, it points to the right with length 30 and can be represented symbolically by $\mathbf{u} = \langle 30, 0 \rangle$. Let a_1 be the horizontal component and a_2 be the vertical component of $\mathbf{v}$. Since $\|\mathbf{v}\| = 300$,

$$\cos 50° = \frac{a_1}{300} \quad \text{and} \quad \sin 50° = \frac{a_2}{300}.$$

It follows that $a_1 = 300 \cos 50°$ and $a_2 = 300 \sin 50°$. Thus

$$\mathbf{v} = \langle 300 \cos 50°, 300 \sin 50° \rangle \approx \langle 192.8, 229.8 \rangle.$$

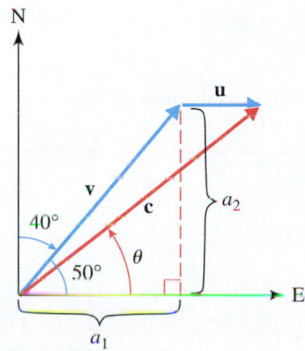

FIGURE 8.50

(b) The true course of the plane is given by $\mathbf{c} = \mathbf{v} + \mathbf{u}$

$$\mathbf{c} = \mathbf{v} + \mathbf{u}$$
$$= \langle 300 \cos 50°, 300 \sin 50° \rangle + \langle 30, 0 \rangle$$
$$= \langle 300 \cos 50° + 30, 300 \sin 50° \rangle$$

The groundspeed of the plane equals $\|\mathbf{c}\|$.

$$\|\mathbf{c}\| = \sqrt{(300 \cos 50° + 30)^2 + (300 \sin 50°)^2}$$
$$\approx 320.1$$

The groundspeed of the airplane is approximately 320 miles per hour.

(c) Since $\mathbf{c} = \langle 300 \cos 50° + 30, 300 \sin 50° \rangle$, angle θ is determined by the vector $\mathbf{c}$ and the positive x-axis (East). Thus $\tan \theta = \frac{300 \sin 50°}{300 \cos 50° + 30} \approx 1.0313$ and $\theta \approx \tan^{-1} 1.0313 \approx 45.9°$. The final bearing of the plane in the wind equals $90° - 45.9° = 44.1°$.   *Now Try Exercise 73* ◆

Vector Notation Using i and j   A second type of vector notation involves the vectors $\mathbf{i} = \langle 1, 0 \rangle$ and $\mathbf{j} = \langle 0, 1 \rangle$. Given a vector $\mathbf{a} = \langle a_1, a_2 \rangle$, it can be expressed as

$$\mathbf{a} = \langle a_1, a_2 \rangle = a_1 \langle 1, 0 \rangle + a_2 \langle 0, 1 \rangle = a_1 \mathbf{i} + a_2 \mathbf{j}.$$

For example, $\langle 3, -4 \rangle$ and $3\mathbf{i} - 4\mathbf{j}$ represent the same vector.

◆ **MAKING CONNECTIONS**

Imaginary unit *i* and unit vector i   In Section 4.4 we discussed the *imaginary unit i*, where $i = \sqrt{-1}$ and $i^2 = -1$. In this section we introduced the *vector* $\mathbf{i} = \langle 1, 0 \rangle$. Each represents a different mathematical concept. ◆

The Dot Product

Thus far we have discussed addition, subtraction, and scalar multiplication of vectors. Another operation on vectors is called the *dot product*, which is important because it can be used to find angles between vectors. The dot product has applications in computer graphics, solar energy, and physics. We begin by defining the dot product.

> **DOT PRODUCT**
>
> Let $\mathbf{a} = \langle a_1, a_2 \rangle$ and $\mathbf{b} = \langle b_1, b_2 \rangle$. The **dot product** of $\mathbf{a}$ and $\mathbf{b}$, denoted $\mathbf{a} \cdot \mathbf{b}$, is a *real number* given by
>
> $$\mathbf{a} \cdot \mathbf{b} = a_1 b_1 + a_2 b_2.$$

In the next example we calculate dot products. Notice that the dot product of two vectors is a real number, rather than a vector.

EXAMPLE **7**   Calculating dot products

Calculate $\mathbf{a} \cdot \mathbf{b}$.
(a) $\mathbf{a} = \langle 4, -3 \rangle$, $\mathbf{b} = \langle -1, 2 \rangle$ **(b)** $\mathbf{a} = 2\mathbf{i} + 5\mathbf{j}$, $\mathbf{b} = -3\mathbf{i} + 2\mathbf{j}$

SOLUTION
(a) $\mathbf{a} \cdot \mathbf{b} = \langle 4, -3 \rangle \cdot \langle -1, 2 \rangle = (4)(-1) + (-3)(2) = -10$
(b) $\mathbf{a} \cdot \mathbf{b} = (2\mathbf{i} + 5\mathbf{j}) \cdot (-3\mathbf{i} + 2\mathbf{j}) = (2)(-3) + (5)(2) = 4$ *Now Try Exercise 51(a)* ◆

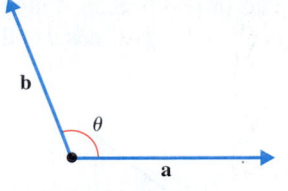

FIGURE 8.51

In Figure 8.51 the *angle between vectors* **a** *and* **b** is θ, where $0° \leq \theta \leq 180°$. If $\theta = 90°$ the vectors are **perpendicular**, and if $\theta = 0°$ or $180°$ the vectors are **parallel**. If $\theta = 0°$ the vectors point in the *same* direction, and if $\theta = 180°$ they point in *opposite* directions.

It is shown in Exercise 6 of the Extended Exercises at the end of the chapter that for any two nonzero vectors **a** and **b**,

$$\mathbf{a} \cdot \mathbf{b} = \|\mathbf{a}\| \|\mathbf{b}\| \cos \theta.$$

This result can be used to find the angle θ between **a** and **b**.

ANGLE BETWEEN TWO VECTORS

If **a** and **b** are nonzero vectors, then the **angle θ between a and b** is given by

$$\theta = \cos^{-1}\left(\frac{\mathbf{a} \cdot \mathbf{b}}{\|\mathbf{a}\| \|\mathbf{b}\|}\right).$$

Vectors **a** and **b** are perpendicular if and only if $\mathbf{a} \cdot \mathbf{b} = 0$.

EXAMPLE 8

Finding the angle between two vectors

Sketch the vectors **a** and **b**. Then find the angle θ between **a** and **b**.
(a) $\mathbf{a} = 2\mathbf{i} - 3\mathbf{j}, \mathbf{b} = 3\mathbf{i} + 2\mathbf{j}$ **(b)** $\mathbf{a} = \langle -4, 3 \rangle, \mathbf{b} = \langle 1, -2 \rangle$

SOLUTION
(a) Vectors **a** and **b** appear to be perpendicular in Figure 8.52. Since

$$\mathbf{a} \cdot \mathbf{b} = (2)(3) + (-3)(2) = 0,$$

the vectors are perpendicular and $\theta = 90°$.
(b) A sketch of the vectors is shown in Figure 8.53. They are neither perpendicular nor parallel. Since

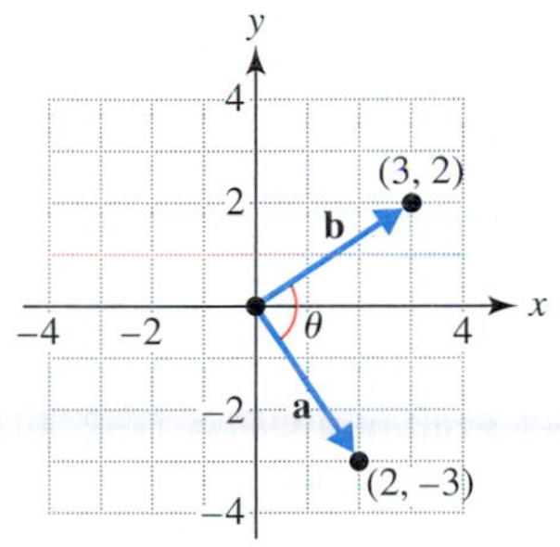

$$\|\mathbf{a}\| = \sqrt{(-4)^2 + (3)^2} = 5 \quad \text{and} \quad \|\mathbf{b}\| = \sqrt{(1)^2 + (-2)^2} = \sqrt{5},$$

it follows that

$$\theta = \cos^{-1}\left(\frac{(-4)(1) + (3)(-2)}{5\sqrt{5}}\right) \approx 153.4°.$$

Now Try Exercise 51(b) and (c) ◆

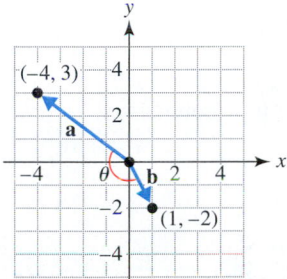

FIGURE 8.52

FIGURE 8.53

Work

In science a force does work only when an object moves. For example, a person pushing against a brick wall does no work, whereas a person lifting a 20-pound weight does work. Work equals force times distance, *provided* the force is in the same direction as the movement of the object. If a 150-pound person climbs up a 20-foot rope, then the work W done is

$$W = 150 \times 20 = 3000 \text{ foot-pounds.}$$

A **foot-pound** equals the work required to lift 1 pound a vertical distance of 1 foot. If the force is not in the same direction as the movement, then we must use trigonometry to determine work.

Consider a person pulling a wagon, as shown in Figure 8.54 on the next page, where **F** represents the force on the handle and **D** represents the distance and direction that the wagon is pulled. The force **F** can be expressed as the sum of a horizontal vector in the direction of **D** and a vertical vector perpendicular to **D**, as illustrated in Figure 8.55. Using the right triangle in Figure 8.56, the horizontal component of **F** is given by $\|\mathbf{F}\| \cos \theta$ and the vertical component of **F** is given by $\|\mathbf{F}\| \sin \theta$.

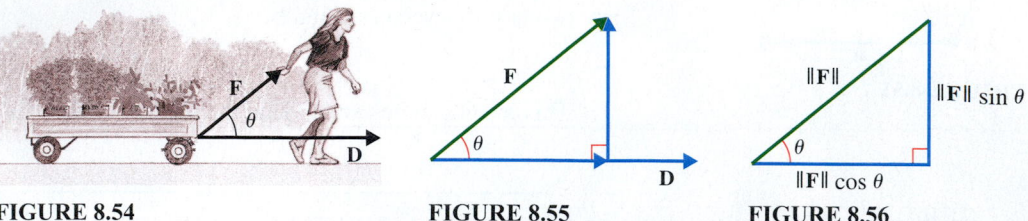

| FIGURE 8.54 | FIGURE 8.55 | FIGURE 8.56 |

The work done pulling the wagon is equal to the horizontal component $\|\mathbf{F}\| \cos \theta$ times the distance, which is given by $\|\mathbf{D}\|$. That is,

$$W = \|\mathbf{F}\| \, \|\mathbf{D}\| \cos \theta = \mathbf{F} \cdot \mathbf{D}.$$

Work equals the dot product of the force vector **F** and the displacement vector **D**.

WORK

If a constant force **F** is applied to an object that moves along a vector **D**, then the work W done is

$$W = \mathbf{F} \cdot \mathbf{D}.$$

EXAMPLE 9 Calculating work

Find the work done when a force $\mathbf{F} = \langle 3, -2 \rangle$ moves an object from point $P = (-2, 1)$ to point $Q = (3, -1)$, where force is measured in pounds and distance in feet.

SOLUTION First we must find the displacement vector $\mathbf{D} = \overrightarrow{PQ}$, where

$$\overrightarrow{PQ} = \langle 3 - (-2), -1 - 1 \rangle = \langle 5, -2 \rangle.$$

The work W done can be calculated as follows.

$$W = \mathbf{F} \cdot \mathbf{D} = \langle 3, -2 \rangle \cdot \langle 5, -2 \rangle = 15 + 4 = 19 \text{ foot-pounds}$$

Now Try Exercise 67 ◆

EXAMPLE 10 Calculating work

A 150-pound person walks 500 feet up a hiking trail that is inclined at 20°. Use vectors to compute the work done by the person, as illustrated in Figure 8.57.

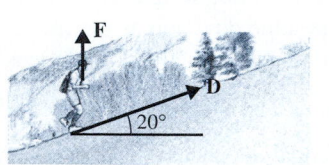

FIGURE 8.57

SOLUTION Vector **D** is given by $D = \langle 500 \cos 20°, 500 \sin 20° \rangle$. Since gravity pulls downward, the force exerted by the person against gravity is $\mathbf{F} = \langle 0, 150 \rangle$. The work done is

$$W = \mathbf{F} \cdot \mathbf{D} = (0)(500 \cos 20°) + (150)(500 \sin 20°) \approx 25{,}650 \text{ foot-pounds.}$$

Now Try Exercise 85 ◆

◆ **CLASS DISCUSSION**

Suppose a force vector **F** is perpendicular to **D**. How much work is done? Interpret your answer. ◆

8.3

Putting it all Together

Vectors are an important invention of the nineteenth and twentieth centuries that have enabled science and technology to model a wide variety of phenomena. The following table summarizes some basic concepts regarding vectors.

Concept	Explanation	Examples
Vectors	Vector quantities denote both magnitude and direction. A vector $\mathbf{a}$ can be expressed as either $\mathbf{a} = \langle a_1, a_2 \rangle$ or $\mathbf{a} = a_1\mathbf{i} + a_2\mathbf{j}$. Its magnitude is given by $\|\mathbf{a}\| = \sqrt{a_1^2 + a_2^2}$. The numbers a_1 and a_2 are called the horizontal and vertical components of $\mathbf{a}$, respectively.	$\mathbf{a} = \langle 1, 2 \rangle$ and $\mathbf{a} = \mathbf{i} + 2\mathbf{j}$ represent the same vector. The magnitude of $\mathbf{a}$ is given by $\|\mathbf{a}\| = \sqrt{1^2 + 2^2} = \sqrt{5}$. The horizontal component is 1, and the vertical component is 2.
Operations on vectors	Let $\mathbf{a} = \langle a_1, a_2 \rangle$, $\mathbf{b} = \langle b_1, b_2 \rangle$, and k be a real number. $$\mathbf{a} + \mathbf{b} = \langle a_1, a_2 \rangle + \langle b_1, b_2 \rangle$$ $$= \langle a_1 + b_1, a_2 + b_2 \rangle \quad \text{Sum}$$ $$\mathbf{a} - \mathbf{b} = \langle a_1, a_2 \rangle - \langle b_1, b_2 \rangle$$ $$= \langle a_1 - b_1, a_2 - b_2 \rangle \quad \text{Difference}$$ $$k\mathbf{a} = k\langle a_1, a_2 \rangle = \langle ka_1, ka_2 \rangle \quad \text{Scalar product}$$	Let $\mathbf{a} = \langle 4, 1 \rangle$, $\mathbf{b} = \langle 3, 2 \rangle$. $$\mathbf{a} + \mathbf{b} = \langle 4, 1 \rangle + \langle 3, 2 \rangle$$ $$= \langle 7, 3 \rangle$$ $$\mathbf{a} - \mathbf{b} = \langle 4, 1 \rangle - \langle 3, 2 \rangle$$ $$= \langle 1, -1 \rangle$$ $$3\mathbf{a} = 3\langle 4, 1 \rangle = \langle 12, 3 \rangle$$
Dot product	If $\mathbf{a} = \langle a_1, a_2 \rangle$ and $\mathbf{b} = \langle b_1, b_2 \rangle$, then $\mathbf{a} \cdot \mathbf{b} = a_1 b_1 + a_2 b_2$.	Let $\mathbf{a} = \langle 2, -2 \rangle$, $\mathbf{b} = \langle 3, 1 \rangle$. $$\mathbf{a} \cdot \mathbf{b} = (2)(3) + (-2)(1) = 4$$
Angle θ between two vectors	If $\mathbf{a} = \langle a_1, a_2 \rangle$ and $\mathbf{b} = \langle b_1, b_2 \rangle$, then $$\theta = \cos^{-1}\left(\frac{\mathbf{a} \cdot \mathbf{b}}{\|\mathbf{a}\| \, \|\mathbf{b}\|}\right).$$ Vectors $\mathbf{a}$ and $\mathbf{b}$ are perpendicular ($\theta = 90°$) if and only if $\mathbf{a} \cdot \mathbf{b} = 0$.	$\mathbf{a} = \langle 1, 0 \rangle$, $\mathbf{b} = \langle 3, 4 \rangle$ $$\theta = \cos^{-1}\left(\frac{\langle 1, 0 \rangle \cdot \langle 3, 4 \rangle}{\|\langle 1, 0 \rangle\| \, \|\langle 3, 4 \rangle\|}\right)$$ $$= \cos^{-1}\frac{3}{(1)(5)}$$ $$\approx 53.1°$$ $\mathbf{a} = \langle 4, -3 \rangle$, $\mathbf{b} = \langle 3, 4 \rangle$ $$\mathbf{a} \cdot \mathbf{b} = (4)(3) + (-3)(4) = 0$$ Thus $\mathbf{a}$ and $\mathbf{b}$ are perpendicular.

Concept	Explanation	Examples
Work	If a constant force **F** is applied to an object that moves along a vector **D**, then the work done is $W = \mathbf{F} \cdot \mathbf{D}$.	If force $\mathbf{F} = 3\mathbf{i} - 4\mathbf{j}$ moves an object along a path described by $\mathbf{D} = 10\mathbf{i} - 20\mathbf{j}$, then the work W done is $$W = \mathbf{F} \cdot \mathbf{D}$$ $$= (3)(10) + (-4)(-20)$$ $$= 110 \text{ foot-pounds},$$ where units are in feet and pounds.

8.3 — Exercises

Representing Vectors and Their Magnitudes

*Exercises 1–4: Use the graphical representation of **v** to complete the following.*

   (a) Estimate integer values for a_1 and a_2 so that $\mathbf{v} = \langle a_1, a_2 \rangle$.

   (b) Calculate $\|\mathbf{v}\|$.

1.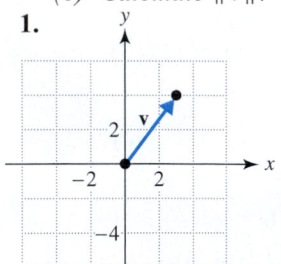

2.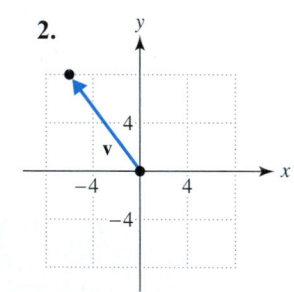

3.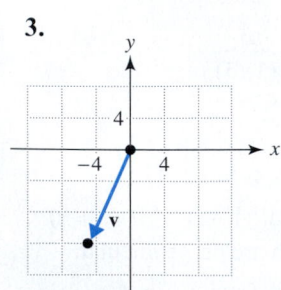

4.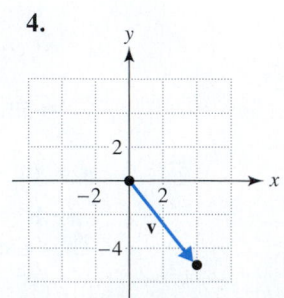

Exercises 5–10: Complete the following.

   *(a) Sketch a vector **v** that models the situation.*

   *(b) Express **v** as $\langle a_1, a_2 \rangle$.*

   (c) Find $2\mathbf{v}$ and $-\frac{1}{2}\mathbf{v}$. Interpret each result.

5. A 20-mile-per-hour north wind

6. A 10-mile-per-hour west wind

7. A 5-mile-per-hour northwest wind

8. A 7-mile-per-hour southeast wind

9. A 30-pound force upward

10. A 15-pound force pulling at a 45° angle in standard position

*Exercises 11–18: Complete the following for vector **v**.*

   (a) Find the horizontal and vertical components.

   *(b) Calculate $\|\mathbf{v}\|$ and decide if **v** is a unit vector.*

   *(c) Graph **v** and interpret $\|\mathbf{v}\|$.*

11. $\mathbf{v} = \langle 1, 1 \rangle$ **12.** $\mathbf{v} = \langle -1, 0 \rangle$

13. $\mathbf{v} = \langle 3, -4 \rangle$ **14.** $\mathbf{v} = \langle -2, -2 \rangle$

15. $\mathbf{v} = \mathbf{i}$ **16.** $\mathbf{v} = -3\mathbf{j}$

17. $\mathbf{v} = 5\mathbf{i} + 12\mathbf{j}$ **18.** $\mathbf{v} = -\frac{3}{5}\mathbf{i} - \frac{4}{5}\mathbf{j}$

*Exercises 19–24: A vector **v** has initial point P and terminal point Q.*
(a) *Graph $\overrightarrow{PQ}$.*
(b) *Write $\overrightarrow{PQ}$ as **v** = $\langle a_1, a_2 \rangle$.*
(c) *Find the magnitude of $\overrightarrow{PQ}$.*

19. $P = (0, 0)$, $Q = (-1, 2)$

20. $P = (0, 0)$, $Q = (4, -6)$

21. $P = (1, 2)$, $Q = (3, 6)$

22. $P = (-1, -2)$, $Q = (4, 4)$

23. $P = (-2, 4)$, $Q = (3, -2)$

24. $P = (0, -4)$, $Q = (1, 3)$

Operations on Vectors

Exercises 25–30: Use the figure to evaluate each of the following.
(a) **a** + **b**
(b) **a** − **b**
(c) −**a**

25.
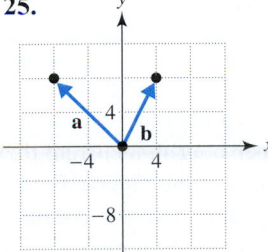

26.
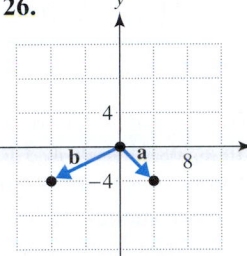

27.
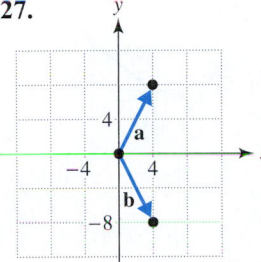

28.
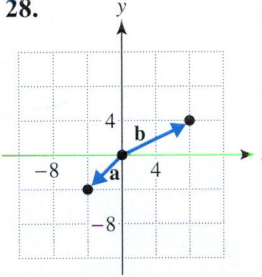

29.
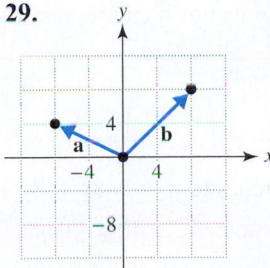

30.
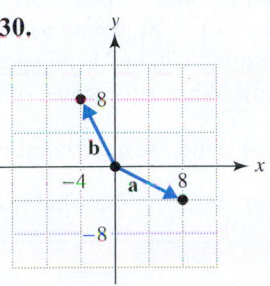

Exercises 31–38: Evaluate each of the following.
(a) **a** + **b**
(b) **a** − **b**

31. **a** = $\langle 0, 2 \rangle$,   **b** = $\langle 3, 0 \rangle$

32. **a** = $\langle 1, 1 \rangle$,   **b** = $\langle -2, 3 \rangle$

33. **a** = 2**i** + **j**,   **b** = **i** − 2**j**

34. **a** = **i** + 2**j**,   **b** = −2**i** + 3**j**

35. **a** = $\left\langle -\sqrt{2}, \frac{1}{2} \right\rangle$, **b** = $\left\langle \sqrt{2}, -\frac{3}{4} \right\rangle$

36. **a** = $\left\langle \frac{4}{5}, -\frac{5}{6} \right\rangle$,   **b** = $\left\langle \frac{3}{10}, \frac{2}{3} \right\rangle$

37. **a** = $\left(\cos \frac{\pi}{4} \right)$**i** + $\left(\sin \frac{\pi}{4} \right)$**j**,

 b = $\left(\cos \frac{\pi}{2} \right)$**i** + $\left(\sin \frac{\pi}{2} \right)$**j**

38. **a** = $\left(\cos \frac{3\pi}{2} \right)$**i** + $\left(\sin \frac{3\pi}{2} \right)$**j**,   **b** = $(\cos \pi)$**i** + $(\sin \pi)$**j**

Exercises 39–42: Evaluate each of the following graphically and symbolically.
(a) $\| \mathbf{a} \|$
(b) 2**a**
(c) 2**a** + 3**b**

39. **a** = 2**i**,   **b** = **i** + **j**

40. **a** = −**i** + **j**,   **b** = **i** − **j**

41. **a** = $\langle -1, 2 \rangle$,   **b** = $\langle 3, 0 \rangle$

42. **a** = $\langle -2, -1 \rangle$,   **b** = $\langle -3, 2 \rangle$

*Exercises 43–46: Approximate the horizontal and vertical components of **v** shown in the figure to the nearest hundredth.*

43.
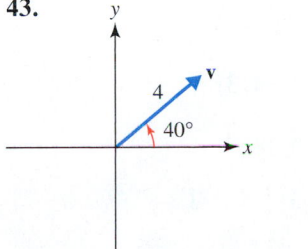

44.
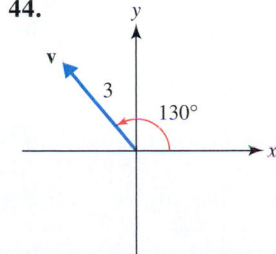

45.

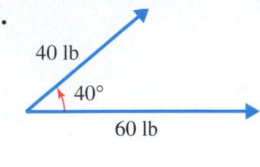

46.

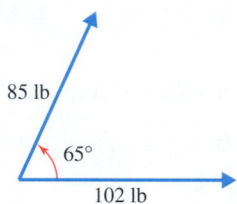

Exercises 47–50: Use the parallelogram rule to find the magnitude of the resultant force for the two forces shown in the figure to the nearest tenth of a pound.

47.

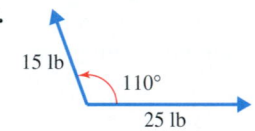

48.

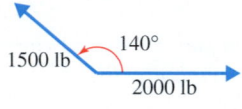

49.

50.

Dot Product and Work

*Exercises 51–58: Complete the following for vectors **a** and **b**.*
(a) *Find **a** · **b**.*
(b) *Approximate the angle θ between **a** and **b** to within a tenth of a degree.*
(c) *State if vectors **a** and **b** are perpendicular, parallel, or neither. If **a** and **b** are parallel, state whether they point in the same direction or in opposite directions.*

51. $\mathbf{a} = \langle 1, -2 \rangle$, $\mathbf{b} = \langle 3, 1 \rangle$

52. $\mathbf{a} = \langle 4, -5 \rangle$, $\mathbf{b} = \langle 2, -2 \rangle$

53. $\mathbf{a} = \langle 6, 8 \rangle$, $\mathbf{b} = \langle -4, 3 \rangle$

54. $\mathbf{a} = \langle 1, -2 \rangle$, $\mathbf{b} = \langle -2, 4 \rangle$

55. $\mathbf{a} = 5\mathbf{i} + 6\mathbf{j}$, $\mathbf{b} = 10\mathbf{i} + 12\mathbf{j}$

56. $\mathbf{a} = -2\mathbf{i} + 6\mathbf{j}$, $\mathbf{b} = 3\mathbf{i} + \mathbf{j}$

57. $\mathbf{a} = \mathbf{i} + 3\mathbf{j}$, $\mathbf{b} = 0.5\mathbf{i} - 1.5\mathbf{j}$

58. $\mathbf{a} = -12\mathbf{i} + 16\mathbf{j}$, $\mathbf{b} = -3\mathbf{i} + 4\mathbf{j}$

Exercises 59–62: Find the work done in each situation.

59. Lifting a 30-pound weight 5 feet into the air

60. Lifting a 15-pound bucket 8 feet into the air

61. Pushing a stalled car on level ground with a force of 100 pounds for 1000 feet

62. A 150-pound person running up 5 flights of steps with 10 feet between floors

*Exercises 63–66: Find the work done when a constant force **F** is applied to an object that moves along the vector **D**, where units are in pounds and feet. Find the magnitude of **F**.*

63. $\mathbf{F} = \langle 10, 20 \rangle$, $\mathbf{D} = \langle 15, 22 \rangle$

64. $\mathbf{F} = \langle 64, 36 \rangle$, $\mathbf{D} = \langle 22, -33 \rangle$

65. $\mathbf{F} = 5\mathbf{i} - 3\mathbf{j}$, $\mathbf{D} = 3\mathbf{i} - 4\mathbf{j}$

66. $\mathbf{F} = 7\mathbf{i} - 24\mathbf{j}$, $\mathbf{D} = -2\mathbf{i} - 5\mathbf{j}$

*Exercises 67–70: Calculate the work done when the force **F** = 5**i** + 3**j** moves an object from P to Q.*

67. $P = (-2, 3)$, $Q = (1, 6)$

68. $P = (-2, -1)$, $Q = (1, 3)$

69. $P = (2, -3)$, $Q = (4, -5)$

70. $P = (1, 1)$, $Q = (-1, 6)$

Applications

71. *Swimming in a Current* A swimmer heads directly north across a river at 3 miles per hour in a current that is 2 miles per hour, as illustrated in the figure. Find a vector that models the resulting direction and speed of the swimmer. With what speed is the swimmer moving in the river?

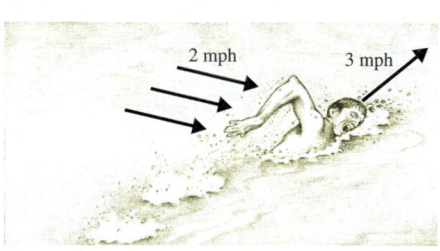

72. *Wind and Vectors* A wind can be described by $\mathbf{v} = 6\mathbf{i} + 8\mathbf{j}$, where vector **j** points north and represents a south wind of 1 mile per hour.
(a) What is the speed of the wind?
(b) Find 3**v**. Interpret the result.
(c) Interpret the wind if it switches to $\mathbf{u} = -8\mathbf{i} + 8\mathbf{j}$.

73. *Air Navigation* (Refer to Example 6.) An airplane heads west at 400 miles per hour in a 50-mile-per-hour northwest wind. Find a vector that models the resulting direction and speed of the airplane. Find the ground-speed and bearing of the airplane in the wind.

74. *Air Navigation* A plane with an airspeed of 240 miles per hour is headed on a bearing of 110°. A north wind is blowing (from the north) at 18 miles per hour. Find the groundspeed and the final bearing of the plane in the wind.

75. *Course and Groundspeed* A plane flies 450 miles per hour on a bearing of 160°. A 20-mile-per-hour wind is blowing from the south. Find the groundspeed and the final bearing of the plane in the wind.

76. *Course and Groundspeed* A plane flies on a bearing of 230° at 350 miles per hour. A wind is blowing from the west at 30 miles per hour. Find the groundspeed and the final bearing of the plane in the wind.

77. *Airspeed and Groundspeed* A pilot wants to fly on a course of 75°. By flying due east, the pilot finds that a 40-mile-per-hour wind, blowing from the south, puts the plane on course. Find the airspeed and the groundspeed.

78. *Force and Water-Ski Towropes* (Refer to Example 2.) Forces of 65 pounds and 110 pounds are exerted by two water-ski towropes. If the angle between the towropes is 19°, find the magnitude of the resultant force.

79. *Measuring Rainfall* Suppose that vector **R** models the amount of rainfall in inches and the direction it falls, and vector **A** models the area in square inches and orientation of the opening of a rain gauge, as illustrated in the figure at the top of the next column. The total volume V of water collected in the rain gauge is given by $V = |\mathbf{R} \cdot \mathbf{A}|$. This formula calculates the volume of water collected even if the wind is blowing the rain in a slanted direction or the rain gauge is not exactly vertical. Let $\mathbf{R} = \mathbf{i} - 2\mathbf{j}$ and $\mathbf{A} = 0.5\mathbf{i} + \mathbf{j}$.
   (a) Find $\|\mathbf{R}\|$ and $\|\mathbf{A}\|$. Interpret your results.

   (b) Calculate V and interpret this result.

   (c) For the rain gauge to collect the maximum amount of water, what must be true about vectors **R** and **A**?

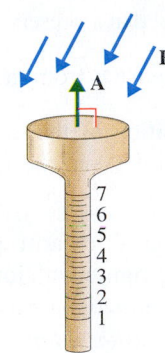

80. *Solar Panels* Suppose that the sun's intensity (in watts per square centimeter) and direction are given by vector **I**, and a solar panel's area (in square centimeters) and orientation are given by vector **A**, as illustrated in the figure. Then the total number of watts W that are collected by the solar panel is given by $W = |\mathbf{I} \cdot \mathbf{A}|$. Let $\mathbf{I} = 0.01\mathbf{i} - 0.02\mathbf{j}$ and $\mathbf{A} = 400\mathbf{i} + 300\mathbf{j}$.

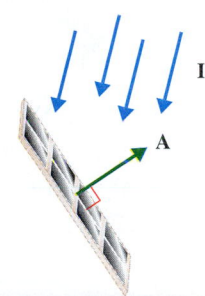

   (a) Find $\|\mathbf{I}\|$ and $\|\mathbf{A}\|$. Interpret your results.

   (b) Calculate W and interpret this result.

   (c) For the solar panel to absorb maximum wattage, what must be true about vectors **I** and **A**?

81. *Robotics* (Refer to Example 5.) Consider the planar two-arm manipulator shown in the accompanying figure. Let the upper arm be modeled by $\mathbf{a} = \langle 3, 2 \rangle$ and the fore-arm be modeled by $\mathbf{b} = \langle -2, 2 \rangle$, where units are in feet. (**Source:** J. Craig.)

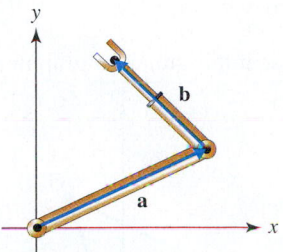

(a) Find a vector **c** that represents the position of the hand.

(b) How far is the hand from the origin?

(c) Find the position of the hand if the length of the upper arm triples and the length of the forearm is reduced by half.

82. *Robotics* A planar three-arm manipulator is shown in the accompanying figure with joint angles measured relative to a positive horizontal axis. (**Source:** R. Murray, *A Mathematical Introduction to Robotic Manipulation*.)
(a) Find vectors **a**, **b**, and **c** that represent each part of the robotic arm. (*Hint:* Find the horizontal and vertical components for each part.)

(b) Find a vector **d** that represents the position of the hand. How far is the hand from the origin?

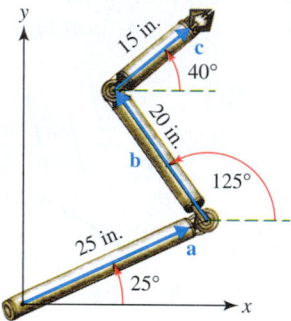

83. *Translations in Computer Graphics* Vectors are used in computer graphics to compute translations of points. For example, suppose we would like to translate the point $(-1, 2)$ by $\mathbf{v} = \langle 2, 1 \rangle$, as illustrated in the accompanying figure, where the *point* $(-1, 2)$ has been represented by the *vector* $\mathbf{a} = \langle -1, 2 \rangle$. The new location of $(-1, 2)$ is modeled by

$$\mathbf{b} = \mathbf{a} + \mathbf{v} = \langle -1, 2 \rangle + \langle 2, 1 \rangle = \langle 1, 3 \rangle.$$

Thus $(-1, 2)$ has been translated by **v** to $(1, 3)$. (**Source:** J. Foley, *Introduction to Computer Graphics*.)
(a) Find the new coordinates of $(-2, 4)$ if it is translated by $\mathbf{v} = \langle 4, -2 \rangle$.

(b) Represent this situation graphically.

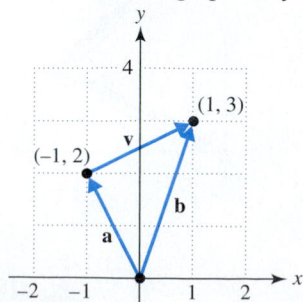

84. *Translations in Computer Graphics* (Refer to the previous exercise.) Let triangle ABC have vertices $(1, 1)$, $(3, 0)$, and $(4, 3)$.
(a) Find the new vertices if the triangle ABC is translated by $\mathbf{v} = \langle -2, 1 \rangle$.

(b) Describe the change in triangle ABC if it were translated by $-2\mathbf{v}$.

85. *Work* (Refer to Example 10.) A 145-pound person walks 1.5 miles up a hiking trail inclined at 15°. Use a dot product to calculate the work W done in foot-pounds.

86. *Work* A wagon is pulled 500 feet using a force of 10 pounds applied to the handle, which makes a 40° angle with the horizontal. See Figure 8.54. Use a dot product to calculate the work W done.

87. *Air Navigation* A pilot would like to fly to a city that is 200 miles away and has a bearing of 135°. The wind is blowing from the north at 30 miles per hour and the trip is to take 1 hour. Find the direction and speed that the pilot should head the plane to accomplish this.

88. *Work* Calculate the work required to push an 1800-pound car up a 7° incline for 0.1 mile.

Writing about Mathematics

89. State the basic properties of a vector. Does a vector have position? Explain. Give two examples of how to write a vector.

90. State one application of vectors. Give a specific example of a vector for this application and explain how the vector models that application.

EXTENDED AND DISCOVERY EXERCISES

1. *Computer Graphics* Vectors frequently are used to determine the color of pixels on computer screens. For example, suppose that we would like to color the right side of the screen blue and the left side yellow, where the boundary is vector $\overrightarrow{PQ}$, as illustrated in the figure below.

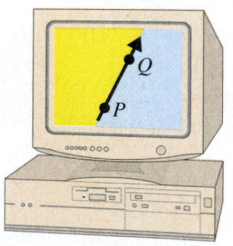

First, we find $\overrightarrow{PR}$ perpendicular to $\overrightarrow{PQ}$. Then to determine if a pixel at point S should be blue or yellow, consider the angle θ between vectors $\overrightarrow{PS}$ and $\overrightarrow{PR}$. If θ is acute, then S must be on the same side of $\overrightarrow{PQ}$ as R and is colored blue, as illustrated in the left figure below. If angle θ is obtuse, then S is on the opposite side of $\overrightarrow{PQ}$ as R and is colored yellow, as shown in the right figure below.

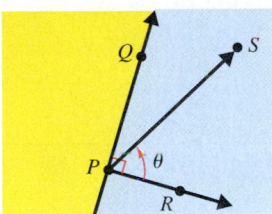 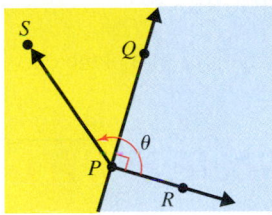

We can determine whether θ is acute or obtuse by calculating the dot product $\overrightarrow{PS} \cdot \overrightarrow{PR}$. If the dot product is positive, then

$$\theta = \cos^{-1}\left(\frac{\overrightarrow{PS} \cdot \overrightarrow{PR}}{\|\overrightarrow{PS}\| \, \|\overrightarrow{PR}\|}\right)$$

is acute and S should be blue. If the dot product is negative, then θ is obtuse and S should be yellow. (**Source:** J. Foley, *Introduction to Computer Graphics.*)

Let P be $(-1, 2)$, S be $(-2, -5)$, $\overrightarrow{PQ} = \mathbf{i} + 5\mathbf{j}$, and $\overrightarrow{PR} = 5\mathbf{i} - \mathbf{j}$. Determine the appropriate color at point S. See the figure below.

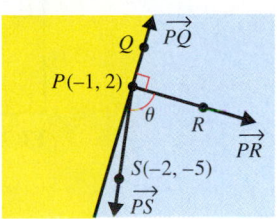

2. ***Dot Products in Computer Graphics*** A computer screen is gray to the left of vector $\overrightarrow{PQ} = 2\mathbf{i} - 3\mathbf{j}$ and blue to the right, where vector $\overrightarrow{PR} = 3\mathbf{i} + 2\mathbf{j}$ is perpendicular to $\overrightarrow{PQ}$. Let point P be $(2, 1)$. Determine the color of a pixel located at S.
   (a) $S = (3, -2)$ (b) $S = (2, 2)$

3. ***Dot Products in Computer Graphics*** A computer screen is to be blue above vector $\overrightarrow{PQ} = 5\mathbf{i} - \mathbf{j}$ and white below, where point P is $(2, 1)$. Determine the color of a pixel located at S.
   (a) $S = (100, -10)$ (b) $S = (-500, 50)$

8.4 Parametric Equations

◆ Learn basic concepts about parametric equations

◆ Use parametric equations to solve applications

Introduction

We have used functions to model curves in the xy-plane. Sometimes a curve cannot be modeled by a function. For example, a circle cannot be described by a single function because a circle fails the vertical line test. Parametric equations represent a different approach to describing curves in the xy-plane. They are used in industry to draw complicated curves and surfaces, such as the hood of an automobile. Parametric equations are also used in computer graphics, engineering, and physics. (**Source:** F. Hill, *Computer Graphics.*)

Basic Concepts

In previous chapters we graphed curves in the xy-plane determined by $y = f(x)$. Some curves cannot be represented by $y = f(x)$, but they can be represented by parametric equations. Some sample graphs of parametric equations are shown in Figures 8.58–8.60 on the next page.

$[-6, 6, 1]$ by $[-4, 4, 1]$

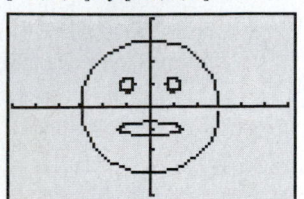

FIGURE 8.58

$[-6, 6, 1]$ by $[-4, 4, 1]$

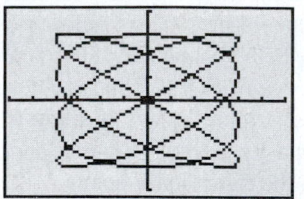

FIGURE 8.59

$[-6, 6, 1]$ by $[-4, 4, 1]$

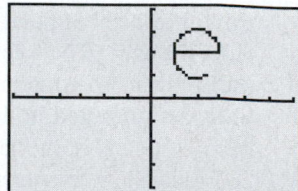

FIGURE 8.60

We now define parametric equations of a plane curve.

PARAMETRIC EQUATIONS OF A PLANE CURVE

A **plane curve** is a set of points (x, y) such that $x = f(t)$ and $y = g(t)$, where f and g are continuous functions on an interval $a \leq t \leq b$. The equations $x = f(t)$ and $y = g(t)$ are **parametric equations** with **parameter** t.

Parametric equations can be represented symbolically, numerically, and graphically and described verbally. This is illustrated in the next example.

EXAMPLE 1 Representing parametric equations

Let $x = t + 3$ and $y = t^2$ for $-3 \leq t \leq 3$.
(a) Make a table of values for x and y with $t = -3, -2, -1, \ldots, 3$.
(b) Plot the points in the table and graph the curve. Add arrows to show how the curve is traced out.
(c) Describe the curve.

SOLUTION
(a) *Numerical Representation* A numerical representation of the parametric equations is shown in Table 8.1. For example, if $t = 2$, then $x = 2 + 3 = 5$ and $y = 2^2 = 4$.

TABLE 8.1

t	-3	-2	-1	0	1	2	3
x	0	1	2	3	4	5	6
y	9	4	1	0	1	4	9

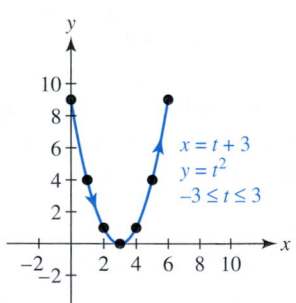

FIGURE 8.61

(b) *Graphical Representation* In Figure 8.61 each ordered pair (x, y) in Table 8.1 is plotted, and then the points are connected to obtain the graph of the curve. Notice that the curve starts at $(0, 9)$ and ends at $(6, 9)$. The curve is traced out from left to right as t increases from -3 to 3.
(c) *Verbal Representation* The curve in Figure 8.61 appears to be the lower portion of a parabola opening upward with vertex $(3, 0)$.   *Now Try Exercise 1* ◆

Graphing calculators are capable of using parametric equations to make tables and graphs. In addition to setting values for the viewing rectangle, the interval for t must also be specified. A window setting, table, and graph for the parametric equations in Example 1 are shown in Figures 8.62–8.64. The variable Tstep represents the increment in the parameter t and has a value of 0.1 in this case.

Calculator Help

To create the graph shown in Figure 8.64 with parametric equations, see Appendix B (page AP-25).

$[-2, 10, 1]$ by $[-2, 10, 1]$

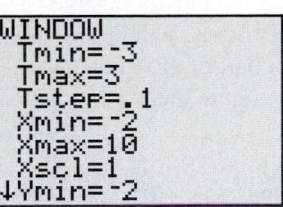

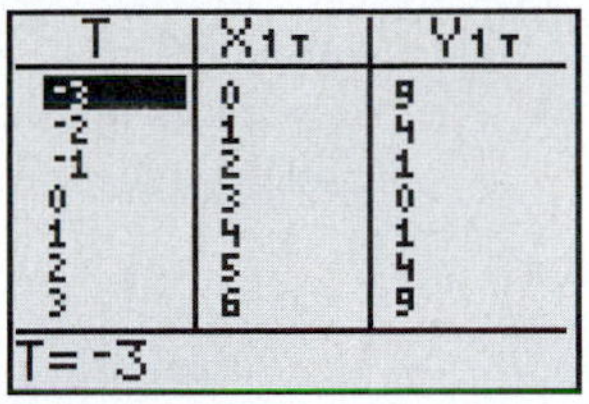

FIGURE 8.62

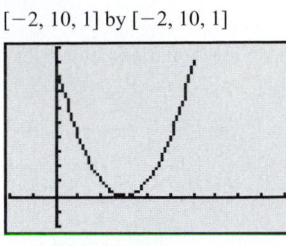

FIGURE 8.63

FIGURE 8.64

We can verify symbolically that the curve in Figure 8.61 is indeed a portion of a parabola, as demonstrated in the next example.

EXAMPLE 2 Finding an equivalent rectangular equation

Find an equivalent rectangular equation for $x = t + 3$ and $y = t^2$, where $-3 \leq t \leq 3$. Note that these parametric equations were discussed in Example 1.

SOLUTION Begin by solving $x = t + 3$ for t to obtain $t = \textbf{\textit{x}} - \textbf{3}$. Substituting for t in $y = t^2$ results in

$$y = (\textbf{\textit{x}} - \textbf{3})^2,$$

which represents a parabola with vertex $(3, 0)$. When $t = -3$ then $x = 0$, and when $t = 3$ then $x = 6$. Thus the domain is restricted to $0 \leq x \leq 6$. *Now Try Exercise 15* ◆

EXAMPLE 3 Finding an equivalent rectangular equation

Find an equivalent rectangular equation for each pair of parametric equations. Use the rectangular equation to help graph the parametric equation. Add arrows to show how the parametric curve is traced out.

(a) $x = 4t$, $y = t - 3$; t is any real number
(b) $x = \sqrt{4 - t^2}$, $y = t$; $-2 \leq t \leq 2$

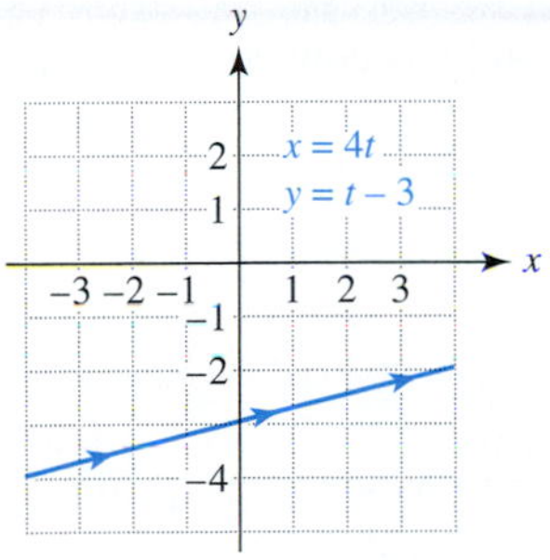

SOLUTION

(a) Start by solving $x = 4t$ for t to obtain $t = \frac{1}{4}x$. Substitute for t in the given parametric equation for y.

$$y = \textbf{\textit{t}} - 3 \qquad \textit{Given parametric equation}$$

$$y = \frac{1}{4}x - 3 \qquad \textit{Let } t = \tfrac{1}{4}x.$$

Because $y = \frac{1}{4}x - 3$, these parametric equations trace out a line with slope $\frac{1}{4}$ and y-intercept -3. As t increases, x also increases so this line is traced out from left to right, as illustrated in Figure 8.65.

(b) Because $y = t$, it follows that a rectangular equation is $x = \sqrt{4 - y^2}$. To determine the graph of this equation, square each side.

$$x = \sqrt{4 - y^2} \qquad \textit{Rectangular equation}$$

$$x^2 = 4 - y^2 \qquad \textit{Square each side.}$$

$$x^2 + y^2 = 4 \qquad \textit{Add } y^2 \textit{ to each side.}$$

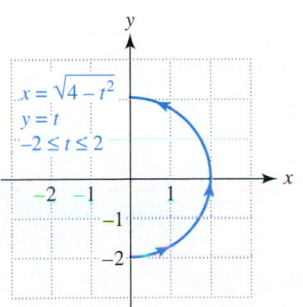

FIGURE 8.65

FIGURE 8.66

Algebra Review

To review equations of circles, see Chapter R (page R-10).

The equation $x^2 + y^2 = 4$ is a circle with circle with center (0, 0) and radius 2. Because $\sqrt{4 - y^2}$ is never negative, it follows that $x \geq 0$. Thus the parametric equation only traces out the right half of this circle. See Figure 8.66. Because $y = t$, this semicircle is traced from bottom to the top as y increases from –2 to 2.

Now Try Exercises 9 and 13 ◆

Parametric equations can model a circle, as shown in the next example.

EXAMPLE 4

Graphing a circle with parametric equations

Graph $x = 2 \cos t$ and $y = 2 \sin t$ for $0 \leq t \leq 2\pi$. Find an equivalent equation by using rectangular coordinates.

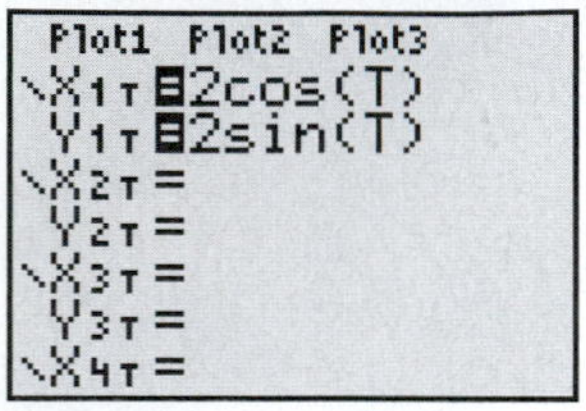

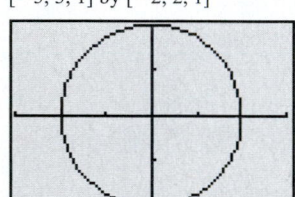

FIGURE 8.67

[−3, 3, 1] by [−2, 2, 1]

FIGURE 8.68

SOLUTION Let $X_1 = 2 \cos (T)$ and $Y_1 = 2 \sin (T)$ and graph these parametric equations as shown in Figures 8.67 and 8.68. Be sure to have the mode of the calculator set for parametric equations. A square viewing rectangle is necessary for the curve to appear circular rather than elliptical.

To verify that this is a circle, consider the following.

$$
\begin{aligned}
x^2 + y^2 &= (2 \cos t)^2 + (2 \sin t)^2 && x = 2 \cos t,\ y = 2 \sin t \\
&= 4 \cos^2 t + 4 \sin^2 t && \text{Properties of exponents} \\
&= 4(\cos^2 t + \sin^2 t) && \text{Distributive property} \\
&= 4 && \cos^2 t + \sin^2 t = 1
\end{aligned}
$$

The parametric equations are equivalent to $x^2 + y^2 = 4$, which is a circle with center (0, 0) and radius 2.

Now Try Exercise 19 ◆

In the next two examples an equation written in terms of x and y is converted to parametric equations.

EXAMPLE 5

Converting to parametric equations

Convert $x = y^2 - 4y + 4$ to parametric equations.

SOLUTION There is more than one way to convert this equation to parametric equations. One simple way is to let $y = t$ and then write the parametric equations as

$$
x = t^2 - 4t + 4, \quad y = t,
$$

where t is any real number. To write a different pair of parametric equations, note that

$$
x = y^2 - 4y + 4 = (y - 2)^2.
$$

Let $t = y - 2$ and then another pair of parametric equations is

$$
x = t^2, \quad y = t + 2.
$$

Now Try Exercise 49 ◆

EXAMPLE 6

Converting to parametric equations

Given the equation $x^2 + y^2 = 1$ complete the following.
(a) Find parametric equations for this equation.
(b) What portion of the graph appears for $0 \leq t \leq \pi$?

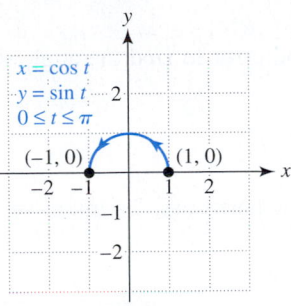

FIGURE 8.69

SOLUTION

(a) The graph of $x^2 + y^2 = 1$ is the unit circle. From trigonometry we know that on the unit circle $x = \cos t$ and $y = \sin t$. Since $\cos^2 t + \sin^2 t = 1$ for all t, we have the following result.

$$x^2 + y^2 = \cos^2 t + \sin^2 t = 1$$

Thus parametric equations for the unit circle are

$$x = \cos t, \qquad y = \sin t; \qquad 0 \le t \le 2\pi.$$

(b) When t increases from 0 to π, the upper half of the circle is graphed, moving from the point $(1, 0)$ to the point $(-1, 0)$. See Figure 8.69. *Now Try Exercise 45* ◆

Applications of Parametric Equations

Parametric equations are used to simulate motion. If a ball is thrown with a velocity of v feet per second at an angle θ with the horizontal, its flight can be modeled by the parametric equations

$$x = (v \cos \theta)t \qquad \text{and} \qquad y = (v \sin \theta)t - 16t^2 + h,$$

where t is in seconds and h is the ball's initial height above the ground. The term $-16t^2$ occurs because gravity is pulling downward. See Figure 8.70. These equations ignore air resistance.

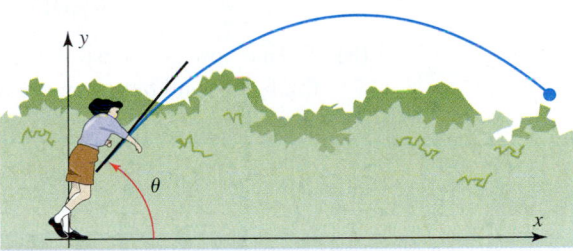

FIGURE 8.70

EXAMPLE 7 Simulating motion with parametric equations

Three golf balls are hit simultaneously into the air at 132 feet per second (90 miles per hour) making angles of 30°, 50°, and 70° with the horizontal.

(a) Assuming the ground is level, determine graphically which ball travels the farthest. Estimate this distance.

(b) Which ball reaches the greatest height? Estimate this height.

SOLUTION

(a) The three sets of parametric equations determined by the three golf balls are as follows since $h = 0$.

$$X_1 = 132 \cos(30)T, \qquad Y_1 = 132 \sin(30)T - 16T\text{^}2$$
$$X_2 = 132 \cos(50)T, \qquad Y_2 = 132 \sin(50)T - 16T\text{^}2$$
$$X_3 = 132 \cos(70)T, \qquad Y_3 = 132 \sin(70)T - 16T\text{^}2$$

The graphs of the three sets of parametric equations are shown in Figures 8.71 and 8.72, where $0 \le t \le 9$. A graphing calculator in simultaneous mode has been used so that we can view all three balls in flight at the same time. From the second graph we can see that the ball hit at 50° travels the farthest distance. Using the trace feature, we estimate this distance to be about 540 feet.

(b) The ball hit at 70° reaches the greatest height of about 240 feet. *Now Try Exercise 67* ◆

[0, 600, 50] by [0, 400, 50]

FIGURE 8.71

[0, 600, 50] by [0, 400, 50]

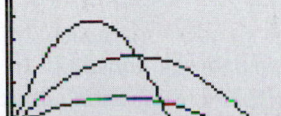

FIGURE 8.72

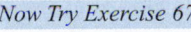

◆ **CLASS DISCUSSION**
If a golf ball is hit at 88 feet per second (60 mph), use trial and error to find the angle θ that results in a maximum distance for the ball. ◆

EXAMPLE 8

Modeling the flight of a baseball

A baseball is hit from a height of 4 feet at a 30° angle above the horizontal. Its initial velocity is 128 feet per second.
(a) Write parametric equations that model the flight of the baseball.
(b) Determine the horizontal distance that the ball travels in the air, assuming that the ground is level.
(c) What is the maximum height of the baseball?
(d) Would the ball clear a 4-foot-high fence that is 400 feet from the batter?

SOLUTION
(a) Let $v = 128$, $\theta = 30°$, and $h = 4$. Then the parametric equations become

$$x = (128 \cos 30°)t \quad \text{and} \quad y = (128 \sin 30°)t - 16t^2 + 4.$$

Since $\cos 30° = \frac{\sqrt{3}}{2}$ and $\sin 30° = \frac{1}{2}$, these equations can be rewritten as

$$x = (64\sqrt{3})t \quad \text{and} \quad y = 64t - 16t^2 + 4.$$

(b) To find how far the ball travels, we first determine the length of time that the ball is in flight. The ball hits the ground when $y = 0$.

$$64t - 16t^2 + 4 = 0 \qquad \text{Substitute for } y.$$
$$16t^2 - 64t - 4 = 0 \qquad \text{Rewrite quadratic equation.}$$
$$t = \frac{64 \pm \sqrt{(-64)^2 - 4(16)(-4)}}{2(16)} \qquad \text{Quadratic formula}$$
$$t \approx 4.0616 \quad \text{or} \quad t \approx -0.0616 \qquad \text{Approximate.}$$

After 4.0616 seconds, the ball traveled *horizontally* $x = 64\sqrt{3}(4.0616) \approx 450.2$ feet.

(c) The graph of $y = 64t - 16t^2 + 4$ is a parabola that opens downward. Using the vertex formula, the maximum height of the ball occurs after

$$t = -\frac{b}{2a} = -\frac{64}{2(-16)} = 2 \text{ seconds.}$$

The maximum height is $y = 64(2) - 16(2)^2 + 4 = 68$ feet.

(d) To determine how long it takes the ball to reach the fence, solve the equation $x = 400$ for t.

$$(64\sqrt{3})t = 400 \quad \text{or} \quad t = \frac{400}{64\sqrt{3}} \approx 3.61 \text{ seconds}$$

After 3.61 seconds the ball has traveled horizontally 400 feet and is

$$y = 64(3.61) - 16(3.61)^2 + 4 \approx 27 \text{ feet}$$

high. The baseball easily clears the 4-foot fence.   *Now Try Exercise 71* ◆

Parametric equations are used frequently in computer graphics to design a variety of figures and letters. Computer fonts are sometimes designed using parametric equations. In the next example, we use parametric equations to design a "smiley" face consisting of a head, two eyes, and a mouth. (**Source:** F. Hill.)

 **EXAMPLE 9** Creating drawings with parametric equations

Graph a "smiley" face using parametric equations.

SOLUTION

Head We can use a circle centered at the origin for the head. If the radius is 2, then let $x = 2\cos t$ and $y = 2\sin t$ for $0 \le t \le 2\pi$. This is graphed in Figure 8.73.

$[-3, 3, 1]$ by $[-2, 2, 1]$

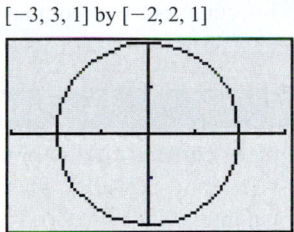

FIGURE 8.73

Eyes For the eyes we can use two small circles. The eye in the first quadrant can be modeled by $x = 1 + 0.3\cos t$ and $y = 1 + 0.3\sin t$ for $0 \le t \le 2\pi$. This represents a circle centered at $(1, 1)$ with radius 0.3. The eye in the second quadrant can be modeled by $x = -1 + 0.3\cos t$ and $y = 1 + 0.3\sin t$ for $0 \le t \le 2\pi$, which is a circle centered at $(-1, 1)$ with radius 0.3. These equations are shown in Figure 8.74.

Mouth For the smile, we can use the lower half of a circle. Using trial and error, we might arrive at $x = 0.5\cos\frac{1}{2}t$ and $y = -0.5 - 0.5\sin\frac{1}{2}t$. This is a semicircle centered at $(0, -0.5)$ with radius 0.5. Since we are letting $0 \le t \le 2\pi$, the term $\frac{1}{2}t$ ensures that only half a circle (semicircle) is drawn. The minus sign before $0.5\sin\frac{1}{2}t$ in the y-equation results in the lower half of the semicircle being drawn rather than the upper half. The final result is shown in Figure 8.75. The pupils have been added by plotting the points $(1, 1)$ and $(-1, 1)$, and the coordinate axes have been turned off.

$[-3, 3, 1]$ by $[-2, 2, 1]$ $[-3, 3, 1]$ by $[-2, 2, 1]$

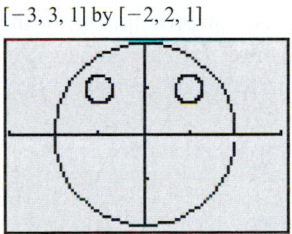 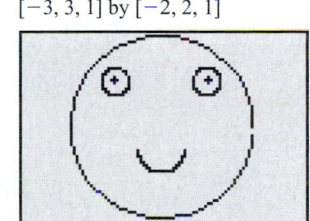

FIGURE 8.74   **FIGURE 8.75**

Now Try Exercise 65 ◆

◆ **CLASS DISCUSSION**

Modify the face in Example 9 so that it is frowning. Try to find a way to make the right eye shut rather than open. ◆

8.4 Putting it all Together

$\mathbf{P}$arametric equations can be used to model a wide variety of curves in the xy-plane that cannot be represented by a single function. Some important concepts about parametric equations are summarized in the following table.

Concept	Explanation	Examples
Plane curve and parametric equations	A plane curve is a set of points (x, y) such that $x = f(t)$ and $y = g(t)$, where f and g are continuous on an interval $a \le t \le b$. The equations $x = f(t)$ and $y = g(t)$ are parametric equations with parameter t.	If $x = 2t$ and $y = t^2$ for $-1 \le t \le 2$, then the resulting graph is a portion of a parabola.
Writing parametric equations in terms of x and y	Solve one of the parametric equations for t and substitute into the second equation.	If $x = t^3$ and $y = t^2 - 2$, solve $x = t^3$ for t to obtain $t = x^{1/3}$. Substituting gives $y = x^{2/3} - 2$.
Converting to parametric equations	If possible, solve the equation for one of the variables. Let the other variable equal t. Now write the first variable in terms of t. Answers may vary.	If $4x = y^2$, solve the equation for x to obtain $x = \frac{1}{4}y^2$. Let $y = t$. The resulting parametric equations are $$x = \tfrac{1}{4}t^2 \quad \text{and} \quad y = t.$$ Another possibility is $$x = t^2 \quad \text{and} \quad y = 2t.$$

Exercises

Graphs of Parametric Equations

Exercises 1–8: Use the parametric equations to complete the following.

 (a) *Make a table of values for $t = 0, 1, 2, 3$.*

 (b) *Plot the points from the table and graph the curve for $0 \leq t \leq 3$. Add arrows to show how the curve is traced out.*

 (c) *Describe the curve.*

1. $x = t - 1, \qquad y = 2t$

2. $x = t + 1, \qquad y = t - 2$

3. $x = t + 2, \qquad y = (t - 2)^2$

4. $x = \frac{1}{3}t^2, \qquad y = t - 1$

5. $x = \sqrt{9 - t^2}, \qquad y = t$

6. $x = t^2, \qquad y = 2t + 1$

7. $x = t, \qquad y = \sqrt{9 - t^2}$

8. $x = 3t, \qquad y = t^2 + 2$

Exercises 9–20: Find a rectangular equation for each curve and describe the curve. Support your result by graphing the parametric equations.

9. $x = 3t, \qquad y = t - 1; \qquad -\infty < t < \infty$

10. $x = t + 3, \qquad y = 2t; \qquad -\infty < t < \infty$

11. $x = 3t^2, \qquad y = t + 1; \qquad -\infty < t < \infty$

12. $x = t^2 - 2t + 1, \; y = t - 1; \qquad -\infty < t < \infty$

13. $x = \sqrt{1 - t^2}, \qquad y = t; \qquad -1 \leq t \leq 1$

14. $x = \frac{1}{2}t, \qquad y = \sqrt{4 - t^2}; \; -2 \leq t \leq 2$

15. $x = t, \qquad y = \frac{1}{2}t^2; \qquad -2 \leq t \leq 2$

16. $x = \sqrt[3]{t}, \qquad y = t; \qquad -2 \leq t \leq 2$

17. $x = t - 2, \qquad y = t^2 + 1; \qquad -1 \leq t \leq 2$

18. $x = 2t, \qquad y = t^2 + 1; \qquad -1 \leq t \leq 2$

19. $x = 3 \sin t, \qquad y = 3 \cos t; \qquad -\pi \leq t \leq \pi$

20. $x = 2 \cos^2 t, \qquad y = 2 \sin^2 t; \qquad 0 \leq t \leq \pi/2$

Exercises 21–38: Graph the parametric equations.

21. $x = \frac{1}{3}t, \qquad y = \frac{2}{3}t + 1; \qquad -\infty < t < \infty$

22. $x = t + 3, \qquad y = 2t - 1; \qquad -\infty < t < \infty$

23. $x = t^2, \qquad y = 2t; \qquad -\infty < t < \infty$

24. $x = \frac{1}{2}(t + 2)^2, \quad y = t + 2; \qquad -\infty < t < \infty$

25. $x = \cos t, \qquad y = \sin t; \qquad 0 \leq t \leq \pi$

26. $x = 2 \sin t, \qquad y = 2 \cos t; \qquad -\pi \leq t \leq 0$

27. $x = t^3, \qquad y = t^2; \qquad -2 \leq t \leq 2$

28. $x = e^t, \qquad y = t - 1; \qquad -2 \leq t \leq 2$

29. $x = t^2, \qquad y = \ln t; \qquad 0 < t \leq 2$

30. $x = t^3 - t, \qquad y = e^t; \qquad -1.5 \leq t \leq 1.5$

31. $x = t - \sin t, \qquad y = 1 - \cos t; \qquad 0 \leq t \leq 6\pi$

32. $x = t^3 + 3t, \qquad y = 2 \cos t; \qquad -1 \leq t \leq 1$

33. $x = 2 + \cos t, \qquad y = \sin t - 1; \qquad 0 \leq t \leq 2\pi$

34. $x = -2 + \cos t, \; y = \sin t + 1; \qquad 0 \leq t \leq 2\pi$

35. $x = \cos^3 t, \qquad y = \sin^3 t; \qquad 0 \leq t \leq 2\pi$

36. $x = \cos^5 t, \qquad y = \sin^5 t; \qquad 0 \leq t \leq 2\pi$

37. $x = |3 \sin t|, \qquad y = |3 \cos t|; \qquad 0 \leq t \leq \pi$

38. $x = 3 \sin 2t, \qquad y = 3 \cos t; \qquad 0 \leq t \leq 2\pi$

Exercises 39–50: Convert the given equation to parametric equations. Answers may vary.

39. $2x + y = 4$

40. $5x - 4y = 20$

41. $y = 4 - x^2$

42. $x = y^2 - 2$

43. $x = y^2 + y - 3$

44. $5x = y^3 + 1$

45. $x^2 + y^2 = 4$

46. $x^2 + y^2 = 9$

47. $\ln y = 0.1x^2$

48. $e^x = |1 - y|$

49. $x = y^2 - 2y + 1$

50. $x = 4y^2 + 4y + 1$

Exercises 51–56: Graph each pair of parametric equations for $0 \leq t \leq 2\pi$. Describe any differences in the two graphs.

51. (a) $x = 3 \cos t$, $y = 3 \sin t$
   (b) $x = 3 \cos 2t$, $y = 3 \sin 2t$

52. (a) $x = 2 \cos t$, $y = 2 \sin t$
   (b) $x = 2 \cos t$, $y = -2 \sin t$

53. (a) $x = 3 \cos t$, $y = 3 \sin t$
   (b) $x = 3 \sin t$, $y = 3 \cos t$

54. (a) $x = t$, $y = t^2$
   (b) $x = t^2$, $y = t$

55. (a) $x = -1 + \cos t$, $y = 2 + \sin t$
   (b) $x = 1 + \cos t$, $y = 2 + \sin t$

56. (a) $x = 2 \cos \frac{1}{2}t$, $y = 2 \sin \frac{1}{2}t$
   (b) $x = 2 \cos t$, $y = 2 \sin t$

Designing Shapes and Figures

Exercises 57–60: Graph the following set of parametric equations for $0 \leq t \leq 2\pi$ in the viewing rectangle $[0, 6, 1]$ by $[0, 4, 1]$. Identify the letter of the alphabet that is being graphed.

57. $x_1 = 1$, $y_1 = 1 + t/\pi$
   $x_2 = 1 + t/(3\pi)$, $y_2 = 2$
   $x_3 = 1 + t/(2\pi)$, $y_3 = 3$

58. $x_1 = 1$, $y_1 = 1 + t/\pi$
   $x_2 = 1 + t/(3\pi)$, $y_2 = 2$
   $x_3 = 1 + t/(2\pi)$, $y_3 = 3$
   $x_4 = 1 + t/(2\pi)$, $y_4 = 1$

59. $x_1 = 1$, $y_1 = 1 + t/\pi$
   $x_2 = 1 + 1.3 \sin (0.5t)$, $y_2 = 2 + \cos (0.5t)$

60. $x_1 = 2 + 0.8 \cos (0.85t)$, $y_1 = 2 + \sin (0.85t)$
   $x_2 = 1.2 + t/(1.3\pi)$, $y_2 = 2$

*Exercises 61–64: **Designing Letters.** Find a set of parametric equations that results in a letter similar to the one shown in the figure. Use the viewing rectangle given by $[-4.7, 4.7, 1]$ by $[-3.1, 3.1, 1]$ and turn off the coordinate axes. Answers may vary.*

61.   **62.**

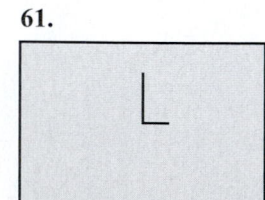   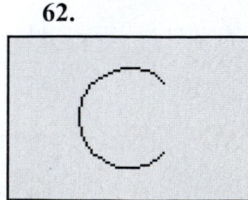

63.   **64.**

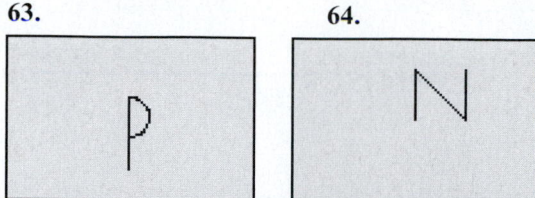

65. *Designing a Face*   (Refer to Example 9.) Use parametric equations to create your own "smiley" face. This face should have a head, a mouth, and eyes.

66. *Designing a Face*   Add a nose to the face that you designed in the previous example.

Applications

67. *Flight of a Golf Ball*   (Refer to Example 7.) Two golf balls are hit into the air at 66 feet per second (45 mph) making angles of 35° and 50° with the horizontal. If the ground is level, estimate the horizontal distance traveled by each golf ball.

68. *Flight of a Golf Ball*   Solve the previous exercise if, instead of being level, the ground is inclined with a slope of $m = 0.1$.

69. *Flight of a Golf Ball*   If a golf ball is hit at 88 feet per second (60 mph) making an angle of 45° with the horizontal, will it go over a fence 10 feet high that is 200 feet away on level ground?

70. *Simulating Gravity on the Moon*   (Refer to Example 7.) If an object is thrown on the moon, then the parametric equations of flight are

$$x = (v \cos \theta)t \quad \text{and} \quad y = (v \sin \theta)t - 2.66t^2 + h.$$

Estimate the horizontal distance that a golf ball hit 88 feet per second (60 mph) at an angle of 45° with the horizontal travels on the moon if the moon's surface is level.

71. *Flight of a Baseball*   (Refer to Example 8.) A baseball is hit with an angle of elevation of 45°, from the top of a ridge that is 50 feet above an area of level ground. The initial velocity of the ball is 88 feet per second or 60 miles per hour. Find the horizontal distance traveled by the ball in the air.

72. *Flight of a Baseball*   A baseball is hit from a height of 3 feet at a 60° angle above the horizontal. Its initial velocity is 64 feet per second.
   (a) Write parametric equations that model the flight of the baseball.
   (b) Determine the horizontal distance traveled by the ball in the air. Assume that the ground is level.

(c) What is the maximum height of the baseball? At that time, how far has the ball traveled horizontally?

(d) Would the ball clear a 5-foot high fence that is 100 feet from the batter?

*Exercises 73–76: **Lissajous Figures** Lissajous figures occur in electronics and may be used to find the frequency of an unknown voltage. (See Extended and Discovery Exercises 1–4 at the end of this section.) Graph the lissajous figure for $0 \leq t \leq 6.5$ in the viewing rectangle $[-6, 6, 1]$ by $[-4, 4, 1]$.*

73. $x = 2 \cos t, \qquad y = 3 \sin 2t$

74. $x = 3 \cos 2t, \qquad y = 3 \sin 3t$

75. $x = 3 \sin 4t, \qquad y = 3 \cos 3t$

76. $x = 4 \sin 4t, \qquad y = 3 \sin 5t$

Writing about Mathematics

77. Describe the basic form of parametric equations. Give an example. Explain how graphs of parametric equations can differ from graphs of functions.

78. Suppose that a function is defined by $y = f(x)$, where the domain of f is $a \leq x \leq b$. Explain how we could represent f with parametric equations. Apply your method to $f(x) = x^2 + 1$, where $-2 \leq x \leq 2$.

EXTENDED AND DISCOVERY EXERCISES

*Exercises 1–4: **Electronic Technology** Parametric equations have applications in electricity. If two sinusoidal voltages, denoted by $x = V_1(t)$ and $y = V_2(t)$, are applied to an oscilloscope, a stationary pattern called a lissajous figure may appear, as shown in the figure in the next column. If the frequency F_1 of V_1 is known and the frequency F_2 of V_2 is unknown, then a lissajous figure may be used to find F_2. The ratio $\frac{F_1}{F_2}$ is equal to the ratio of the corresponding number of tangents to the enclosing rectangle. The number of tangents along a vertical side of the rectangle corresponds to F_1 and the number of tangents along a horizontal side corresponds to F_2. In this figure $\frac{F_1}{F_2} = \frac{3}{2}$. Therefore $F_2 = \frac{2}{3}F_1$. Determine F_2 given the lissajous figure and the frequency F_1 in cycles per second.* (**Source:** R. Smith and R. Dorf, Circuits, Devices, and Systems.)

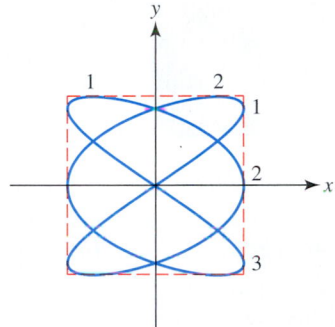

1. $F_1 = 150$

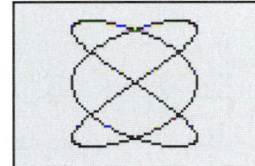

2. $F_1 = 60$

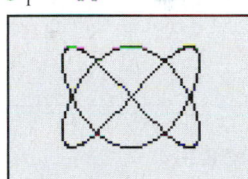

3. $F_1 = 400$

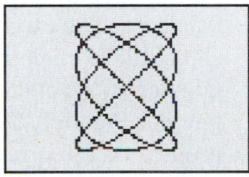

4. $F_1 = 1200$

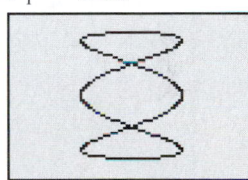

CHECKING BASIC CONCEPTS FOR SECTIONS 8.3 AND 8.4

1. Let the point P be $(-1, 3)$ and the point Q be $(3, 7)$. Find the following.
   (a) $\mathbf{v} = \overrightarrow{PQ}$ **(b)** $\|\mathbf{v}\|$ **(c)** $\overrightarrow{PQ} + \overrightarrow{QP}$

2. Let $\mathbf{v} = 2\mathbf{i} - \mathbf{j}$ and $\mathbf{u} = -3\mathbf{i} + 2\mathbf{j}$. Find the following graphically and symbolically.
   (a) $2\mathbf{v} + \mathbf{u}$ **(b)** $2\mathbf{v}$ **(c)** $\mathbf{v} - 3\mathbf{u}$

3. Let $\mathbf{a} = \langle 3, -2 \rangle$ and $\mathbf{b} = \langle -1, 3 \rangle$. Find the following.
   (a) $\mathbf{a} \cdot \mathbf{b}$

   (b) The angle θ between $\mathbf{a}$ and $\mathbf{b}$ rounded to the nearest tenth of a degree

4. Graph the parametric equations given by $x = t + 1$ and $y = (t - 1)^2$ for $-1 \leq t \leq 5$. Write these parametric equations in terms of x and y.

8.5 Polar Equations

- ◆ Learn the polar coordinate system
- ◆ Graph polar equations
- ◆ Graph polar equations with graphing calculators (optional)
- ◆ Solve polar equations

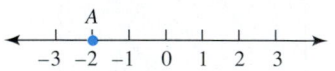

FIGURE 8.76

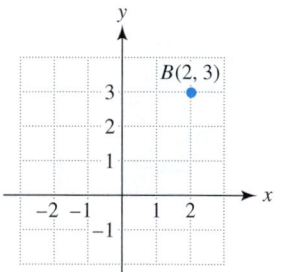

FIGURE 8.77

Introduction

Many times a change in a frame of reference can have a profound effect on the solution of a problem. Thus far we have graphed functions only in the xy-plane. Many interesting curves cannot be represented by a function since these curves fail the vertical line test. By creating a new coordinate system, some types of equations become simpler. For example, the equation $x^2 + y^2 = 1$ describes the unit circle in the rectangular coordinate system. Every point lying on the unit circle is 1 unit from the origin. If we specify a new variable r that represents the radius of the circle, then $r = 1$ also describes the unit circle. A change of variable has resulted in a simpler equation.

In this section we introduce a new coordinate system called the *polar coordinate system* that originated with Jakob Bernoulli (1654–1705). This new system allows us to express many curves with relatively simple equations.   (**Source:** H. Eves, *An Introduction to the History of Mathematics.*)

The Polar Coordinate System

In Figure 8.76 point A is located at $x = -2$. One number completely determines the location of a point on a one-dimensional number line. In two dimensions, two numbers are necessary to locate a point. For example, two numbers are needed to locate point B at $(2, 3)$ in Figure 8.77. In the xy-plane we are accustomed to identifying points using (x, y), where x and y are real numbers. However, the xy-plane is not the only way to locate a point in a plane.

The **polar coordinate system** uses r and θ instead of x and y to locate a point P, as shown in Figure 8.78. The distance OP between P and the origin is represented by $|r|$. Angle θ can be expressed in either degrees or radians and is assumed to be in standard position. If $r > 0$, point P lies on the terminal side of θ and if $r < 0$, point P lies on the ray pointing in the opposite direction of the terminal side of θ, a distance of $|r|$ from the origin. See Figure 8.79. In the polar coordinate system, the origin is called the **pole** and the positive x-axis corresponds to the **polar axis**.

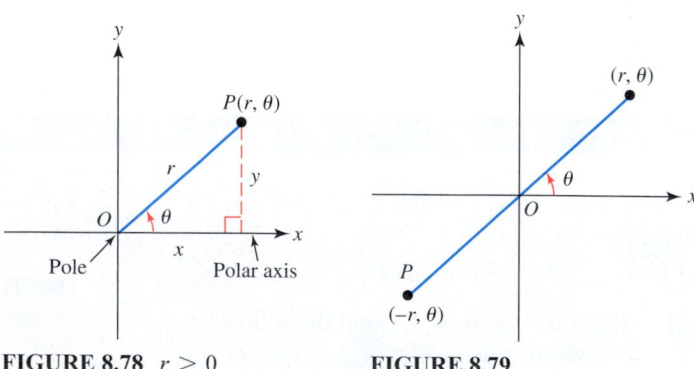

FIGURE 8.78 $r > 0$ **FIGURE 8.79**

Using Figure 8.78 and trigonometry, we can establish the following relationships between rectangular and polar coordinates.

RECTANGULAR AND POLAR COORDINATES

If a point has rectangular coordinates (x, y) and polar coordinates (r, θ), then these coordinates are related as follows.

$$x = r \cos \theta, \qquad y = r \sin \theta$$

$$r^2 = x^2 + y^2, \qquad \tan \theta = \frac{y}{x} \ (x \neq 0)$$

EXAMPLE **1**   Plotting points in polar coordinates

Plot the points (r, θ) on a polar grid.

(a) $(2, 45°)$ (b) $(-3, 150°)$ (c) $\left(3.5, -\dfrac{\pi}{3}\right)$

SOLUTION

(a) Let $r = 2$ and $\theta = 45°$. Plot a point 2 units from the pole on the terminal side of $\theta = 45°$, as shown in Figure 8.80.

(b) Since $r = -3 < 0$ and $\theta = 150°$, begin by locating the terminal side of θ in the second quadrant. Next plot a point 3 units from the pole in the opposite direction of the terminal side of θ, as shown in Figure 8.81.

(c) Since $r = 3.5$ and $\theta = -\frac{\pi}{3}$ (radians), the point is located in the fourth quadrant. See Figure 8.82.

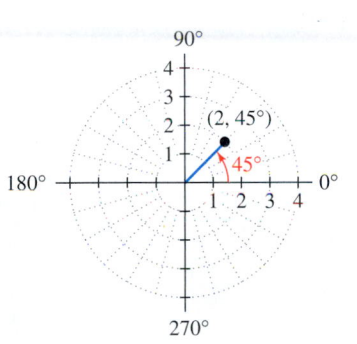

FIGURE 8.80

FIGURE 8.81

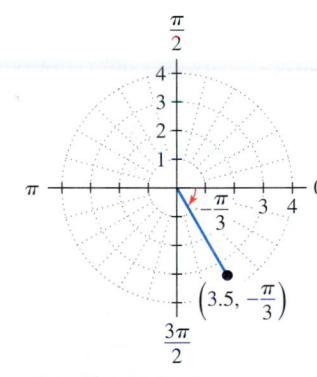

FIGURE 8.82

Now Try Exercises 1 and 3 ◆

◆ **CLASS DISCUSSION**

What can be said about a point (r, θ) if $r = 0$? ◆

It is important to note that unlike the xy-coordinate system, polar coordinates are *not* unique. For example, the $r\theta$-coordinates of $(2, 0°)$, $(2, 360°)$, $(2, -360°)$, and $(-2, 180°)$ all represent the same point.

In the next example, we convert polar coordinates to rectangular coordinates.

EXAMPLE  **2**   Converting to rectangular coordinates

Convert the coordinates of each point from polar coordinates to rectangular coordinates.

(a) $(5, 180°)$ (b) $\left(3, -\dfrac{\pi}{3}\right)$

SOLUTION

(a) To convert $(5, 180°)$ to rectangular coordinates, use the equations $x = r\cos\theta$ and $y = r\sin\theta$ with $r = \textbf{5}$ and $\theta = \textbf{180°}$.

$$x = \textbf{5}\cos(\textbf{180°}) = -5$$
$$y = \textbf{5}\sin(\textbf{180°}) = 0$$

The corresponding rectangular coordinates are $(-5, 0)$.

(b) To convert $\left(3, -\frac{\pi}{3}\right)$ to rectangular coordinates, let $r = \textbf{3}$ and $\theta = -\frac{\pi}{\textbf{3}}$, where θ is in radians.

$$x = \textbf{3}\cos\left(-\frac{\pi}{\textbf{3}}\right) = \frac{3}{2}$$

$$y = \textbf{3}\sin\left(-\frac{\pi}{\textbf{3}}\right) = -\frac{3\sqrt{3}}{2} \approx -2.6$$

The corresponding rectangular coordinates, are $\left(\frac{3}{2}, -\frac{3\sqrt{3}}{2}\right)$.

Now Try Exercises 11 and 15 ◆

In the next example, points expressed in rectangular coordinates are converted to polar coordinates. However, this conversion is not unique because, unlike rectangular coordinates, polar coordinates are not unique.

EXAMPLE 3

Expressing points in polar coordinates

Given the point $(1, \sqrt{3})$ in rectangular coordinates, find polar coordinates (r, θ) that satisfy each condition.
(a) $r > 0,$ $0° \le \theta < 360°$
(b) $r > 0,$ $-360° \le \theta < 0°$
(c) $r < 0,$ $0° \le \theta < 360°$

SOLUTION

(a) Because $r^2 = x^2 + y^2$, let $r = \sqrt{1 + 3} = 2$. The point $(1, \sqrt{3})$ is located in the first quadrant of the xy-plane. Therefore $\tan\theta = \frac{y}{x} = \frac{\sqrt{3}}{1}$ and $\theta = \tan^{-1}\sqrt{3} = 60°$. Let $(r, \theta) = (2, 60°)$. See Figure 8.83.

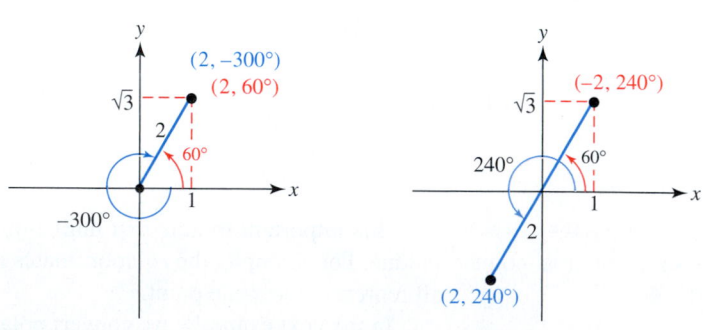

FIGURE 8.83　　　　　　　**FIGURE 8.84**

(b) Rather than let $\theta = 60°$, we can use $\theta = -300°$, so $(r, \theta) = (2, -300°)$. See Figure 8.83. Notice that $60°$ and $-300°$ are coterminal angles.

(c) We can let $r = -2$, but then we need to let angle $\theta = 60° + 180° = 240°$. Thus $(r, \theta) = (-2, 240°)$, as illustrated in Figure 8.84.

Now Try Exercise 23 ◆

Graphs of Polar Equations

When we graph $y = f(x)$ in the xy-coordinate system, we are graphing a function. A vertical line can intersect the graph of a function at most once. As a result, many shapes such as circles, hearts, and leaves cannot be represented by a function. Like parametric equations, polar equations can be valuable when representing curves.

EXAMPLE 4   Representing polar equations

Make a table of values and graph each curve. Then describe the curve.

(a) $\theta = \dfrac{\pi}{4}$ (b) $r = 3$ (c) $r = 2 + 2 \cos \theta$

SOLUTION

(a) *Numerical Representation*   Since $\theta = \frac{\pi}{4}$, every point lying on this graph is of the form $\left(r, \frac{\pi}{4}\right)$, where r can be any real number. In Table 8.2 five of these points are listed.

Graphical Representation   A graph of $\theta = \frac{\pi}{4}$ and the points in Table 8.2 are shown in Figure 8.85.

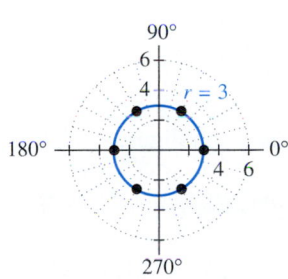

FIGURE 8.85   A Line

TABLE 8.2

θ	$\dfrac{\pi}{4}$	$\dfrac{\pi}{4}$	$\dfrac{\pi}{4}$	$\dfrac{\pi}{4}$	$\dfrac{\pi}{4}$
r	-2	-1	0	1	2

Verbal Representation   The graph is a line with a slope 1 passing through the pole.

(b) *Numerical Representation*   A table of values is shown in Table 8.3.

TABLE 8.3

θ	0°	60°	120°	180°	240°	300°
r	3	3	3	3	3	3

Graphical Representation   A graph and these points are shown in Figure 8.86.

Verbal Representation   The polar equation represents a circle with radius 3.

(c) *Numerical Representation*   A table of values for $r = 2 + 2 \cos \theta$ is shown in Table 8.4. For example, when $\theta = 60°$, then

$$r = 2 + 2 \cos 60° = 2 + 2(0.5) = 3.$$

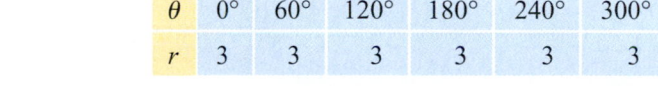

TABLE 8.4

θ	0°	60°	120°	180°	240°	300°	360°
r	4	3	1	0	1	3	4

Graphical Representation   To help graph the equation we can plot the points in Table 8.4. See Figure 8.87. Notice that since the cosine function has period 360°, the graph repeats after 360°.

Verbal Representation   The polar equation represents a heart-shaped graph that is called a **cardioid**.

Now Try Exercises 31, 33, and 37 ◆

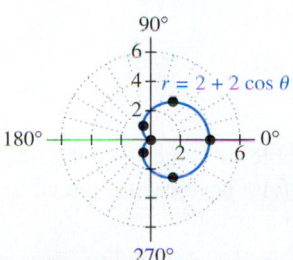

FIGURE 8.86   A Circle

FIGURE 8.87   A Cardioid

 Graphing in polar coordinates

Graph each polar equation.
(a) $r = 2 \sin \theta$ **(b)** $r = 2 + \cos \theta$

SOLUTION

Algebra Review
To review equations of circles and how to find their centers, see Chapter R (page R-10).

(a) The polar equation $r = 2 \sin \theta$ can be converted to rectangular coordinates by first multiplying each side of the equation by r.

$r = 2 \sin \theta$	*Given polar equation*
$r^2 = 2r \sin \theta$	*Multiply by r.*
$x^2 + y^2 = 2y$	*Convert to xy-coordinates.*
$x^2 + y^2 - 2y = 0$	*Subtract 2y.*
$x^2 + y^2 - 2y + \mathbf{1} = \mathbf{1}$	*Complete the square by adding 1.*
$x^2 + (y - 1)^2 = 1$	*Perfect square trinomial*

This final equation represents a circle with center $(0, 1)$ and radius 1. The graph of $r = 2 \sin \theta$ is shown in Figure 8.88.

(b) To graph this polar equation, start by making a table of values like the one shown in Table 8.5. Values are rounded to the nearest tenth. (Degree measure could also be used.)

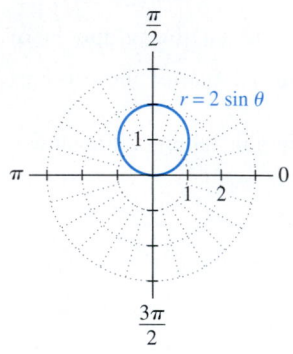

FIGURE 8.88

TABLE 8.5

θ	0	$\dfrac{\pi}{4}$	$\dfrac{\pi}{2}$	$\dfrac{3\pi}{4}$	π	$\dfrac{5\pi}{4}$	$\dfrac{3\pi}{2}$	$\dfrac{7\pi}{4}$
$r = 2 + \cos \theta$	3	2.7	2	1.3	1	1.3	2	2.7

The points in Table 8.5 and the graph of $r = 2 + \cos \theta$ are shown in Figure 8.89. It is called a *limaçon* without an inner loop.

Now Try Exercises 47 and 51 ◆

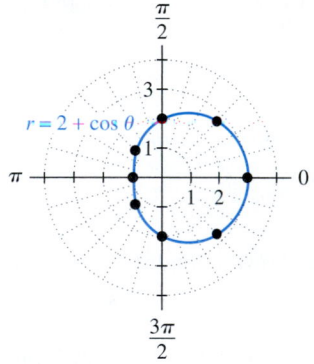

FIGURE 8.89

Polar equations in the form $r = a \sin n\theta$ or $r = a \cos n\theta$ result in graphs of rose curves. It can be shown that when n is odd there are n leaves and when n is even there are $2n$ leaves. In the next example, we graph this type of equation.

 Graphing a four-leaved rose

Graph $r = 3 \cos 2\theta$.

SOLUTION To graph this polar equation, start by making a table of values like the one shown in Table 8.6. Degree measure has been used, and values for r have been rounded to the nearest tenth. Notice that the values repeat themselves starting at 180°.

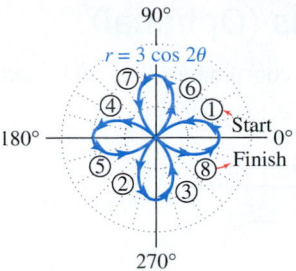

FIGURE 8.90

TABLE 8.6

θ	0°	15°	30°	45°	60°	75°	90°
$r = 3 \cos 2\theta$	3	2.6	1.5	0	−1.5	−2.6	−3

θ	105°	120°	135°	150°	165°	180°
$r = 3 \cos 2\theta$	−2.6	−1.5	0	1.5	2.6	3

Plotting these points in order gives the graph, called a **four-leaved rose**. Note in Figure 8.90 that the graph is developed with a continuous curve, beginning with the upper half of the right horizontal leaf and ending with the lower half of that leaf. As the graph is traced, the curve passes through the pole four times. Each leaf has length 3.

Now Try Exercise 55 ◆

EXAMPLE 7 Writing polar equations in rectangular form

Write the polar equation in terms of x and y. Describe its graph.

(a) $r = 3 \csc \theta$ (b) $r = \dfrac{2}{4 \cos \theta - 3 \sin \theta}$

SOLUTION

(a) Begin by applying a reciprocal identity.

$$r = 3 \csc \theta \qquad \text{Given equation}$$

$$r = \frac{3}{\sin \theta} \qquad \text{Reciprocal identity}$$

$$r \sin \theta = 3 \qquad \text{Multiply by } \sin \theta.$$

$$y = 3 \qquad y = r \sin \theta$$

Its graph is a horizontal line.

(b) Start by cross-multiplying.

$$r = \frac{2}{4 \cos \theta - 3 \sin \theta} \qquad \text{Given equation}$$

$$4r \cos \theta - 3r \sin \theta = 2 \qquad \text{Cross-multiply.}$$

$$4x - 3y = 2 \qquad \text{Substitute.}$$

$$y = \frac{4}{3}x - \frac{2}{3} \qquad \text{Solve for } y.$$

Its graph is a line with slope $\frac{4}{3}$ and y-intercept $-\frac{2}{3}$.

Now Try Exercises 69 and 71 ◆

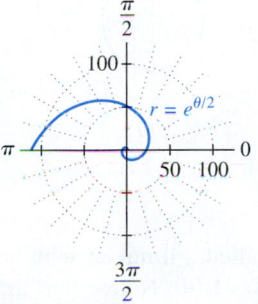

FIGURE 8.91 A Logarithmic Spiral

In 1638 René Descartes described a *logarithmic spiral* using the complicated equation $y = x \tan (\ln (x^2 + y^2))$. This curve, shown in Figure 8.91, cannot be represented by a function. Using polar coordinates, this rectangular equation reduces to the much simpler equation $r = e^{\theta/2}$. Johann Bernoulli was so entranced by this remarkable curve that he ordered it carved on his tombstone. (**Source:** H. Resnikoff and R. Wells, *Mathematics in Civilization.*)

Graphing Calculators and Polar Equations (Optional)

Technology can be used to make tables and graphs in polar coordinates. The table and graph in Example 4(c) are shown in Figures 8.92 and 8.93.

Calculator Help

To create the graph shown in Figure 8.93 with a polar equation, see Appendix B (page AP-26).

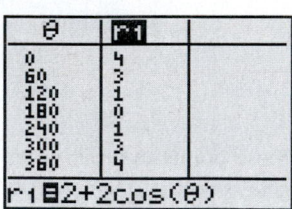

FIGURE 8.92 Degree Mode

$[-6, 6, 1]$ by $[-4, 4, 1]$

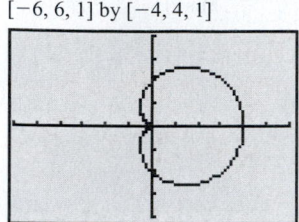

FIGURE 8.93 A Cardioid

As is the case with rectangular and parametric equations, a viewing rectangle must be selected before graphing a polar equation. First choose whether the graph should be plotted in radian or degree mode. Next determine an interval for θ. Many times the interval $0° \leq \theta \leq 360°$ is sufficient to have the entire graph generated. Then select a square viewing rectangle if possible.

EXAMPLE 8 Representing polar equations

Graph each curve. Then describe the curve.
(a) $r = 3 \cos 2\theta$ **(b)** $r = 1 - 2 \sin \theta$

SOLUTION

(a) This equation is graphed by hand in Example 6. In Figure 8.94 a graphing calculator has been used to create a graph of $r = 3 \cos 2\theta$. Turn on the polar grid and try tracing the graph. This shows how each leaf of the rose is generated by the polar equation. Notice, for example, the location of the point $(-1.5, 60°)$ in Figure 8.94.

$[-6, 6, 1]$ by $[-4, 4, 1]$

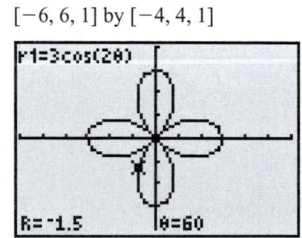

FIGURE 8.94 A Four-Leaved Rose

$[-6, 6, 1]$ by $[-4, 4, 1]$

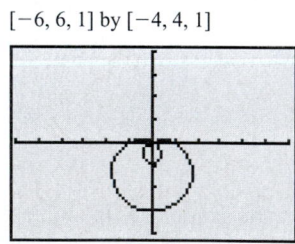

FIGURE 8.95 Degree Mode

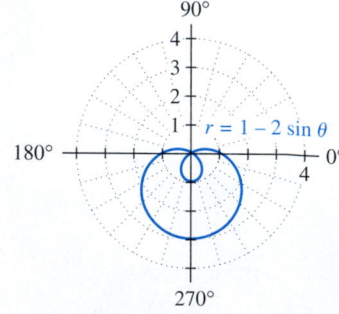

FIGURE 8.96 A Limaçon

(b) A graph of $r = 1 - 2 \sin \theta$ is shown in Figure 8.95. It is called a **limaçon** with an inner loop. A graph with better resolution is shown in Figure 8.96. Notice that the inner loop occurs when $30° \leq \theta \leq 150°$ and $r \leq 0$.

Now Try Exercises 79 and 83 ◆

Polar Equations of Conics The polar equation $r = \dfrac{a(1 - e^2)}{1 + e \cos \theta}$ can be used to model the orbits of planets and comets, where a is the average distance of the celestial body from the sun in astronomical units and e is a constant called the *eccentricity*. (Smaller values of e indicate that an orbit is more circular, whereas larger values indicate a more elliptical orbit. Note that values for e vary between 0 and 1. One astronomical unit equals 93 million miles.) The sun is located at the pole. Table 8.7 lists a and e for the outer planets. (**Source:** H. Karttunen et al., *Fundamental Astronomy.*)

TABLE 8.7 Distances and Eccentricities of the Outer Planets

Planet	a	e
Jupiter	5.20	0.048
Saturn	9.54	0.056
Uranus	19.2	0.047
Neptune	30.1	0.009
Pluto	39.4	0.249

 **EXAMPLE 9** Determining the farthest planet from the sun

Use graphing to determine if Pluto is always the farthest planet from the sun.

SOLUTION We will compare the orbits of Pluto and Neptune since Neptune is closest to Pluto. Their orbital equations are given and then graphed in Figure 8.97.

$$\text{Neptune: } r_1 = \frac{30.1(1 - 0.009^2)}{1 + 0.009 \cos \theta}$$

$$\text{Pluto: } r_2 = \frac{39.4(1 - 0.249^2)}{1 + 0.249 \cos \theta}$$

The graph shows that their orbits pass near each other. By zooming in we can determine that the orbit of Pluto actually passes inside the orbit of Neptune. See Figure 8.98. Therefore there are times when Neptune—not Pluto—is the farthest planet from the sun. However, Pluto's average distance from the sun is considerably greater than Neptune's average distance. Neptune was the farthest planet from the sun for a 20-year period that ended in 1999.

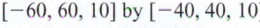

$[-60, 60, 10]$ by $[-40, 40, 10]$

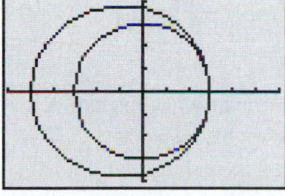

FIGURE 8.97

$[27, 33, 1]$ by $[-2, 2, 1]$

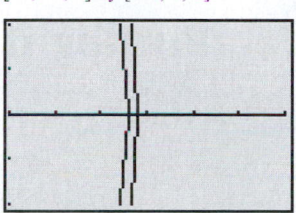

FIGURE 8.98

Now Try Exercise 95 ◆

Solving Polar Equations

We can solve polar equations symbolically, graphically, and numerically.

EXAMPLE 10 Solving a polar equation

Find values for θ where the circle $r = 3$ intersects the cardioid $r = 2 + 2 \cos \theta$. Let $0° \leq \theta \leq 360°$.

SOLUTION

Symbolic Solution   Begin by setting the two equations equal.

$$3 = 2 + 2 \cos \theta \qquad \text{Set equations equal.}$$

$$\cos \theta = \frac{1}{2} \qquad \text{Solve for } \cos \theta.$$

$$\theta = 60° \text{ or } 300° \qquad \text{Solve for } \theta.$$

(Be sure to check symbolic solutions.)

Graphical Solution   Using the intersection-of-graphs method, let $r_1 = 3$ and let $r_2 = 2 + 2 \cos (\theta)$. Their graphs intersect when $\theta = 60°$ and $300°$, as shown in Figures 8.99 and 8.100.

Numerical Solution   Numerical support is shown in Figure 8.101, where $r_1 = r_2 = 3$ when $\theta = 60°$ and $300°$.

Now Try Exercise 91 ◆

[−6, 6, 1] by [−4, 4, 1]   [−6, 6, 1] by [−4, 4, 1]

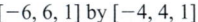 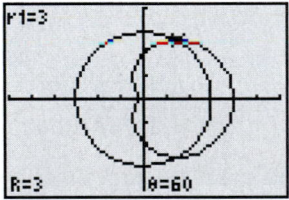 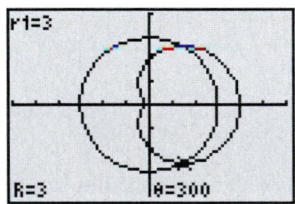

FIGURE 8.99 Degree Mode   **FIGURE 8.100** Degree Mode   **FIGURE 8.101** Degree Mode

8.5 Putting it all Together

Polar equations and graphs can be used to describe curves that are not easily represented by equations involving x and y. The polar coordinates (r, θ) are not unique for a given point, whereas in rectangular coordinates a point is determined uniquely by x and y. Polar equations can be solved graphically, numerically, and symbolically. Some basic concepts regarding polar coordinates and graphs of polar equations are summarized in the following table.

Concept	Examples and Explanations
Polar coordinates	A point is determined by r and θ. To convert between rectangular and polar coordinates use $x = r \cos \theta$, $y = r \sin \theta$, $\tan \theta = \dfrac{y}{x}$, and $r^2 = x^2 + y^2$.

Concept	Examples and Explanations
Graphs in polar coordinates	**Circle** $r = a$ & **Cardioid** $r = a \pm a \sin \theta$ or $r = a \pm a \cos \theta$

Rose Curve

$$r = a \sin n\theta \quad \text{or} \quad r = a \cos n\theta$$

The number of leaves equals n when n is odd and is twice n when n is even.

$n = 3$

$n = 4$

Limaçon $(b \neq a)$

$$r = a \pm b \sin \theta \quad \text{or} \quad r = a \pm b \cos \theta$$

$a > b$; no inner loop $a < b$; one inner loop

8.5 — Exercises

Polar Coordinates

Exercises 1–4: Plot the points (r, θ).

1. (a) $(2, 0°)$

   (b) $(3, 120°)$

   (c) $(-1, 135°)$

3. (a) $\left(2, \frac{\pi}{3}\right)$

   (b) $\left(-3, -\frac{\pi}{6}\right)$

   (c) $\left(0, \frac{3\pi}{4}\right)$

2. (a) $(-2, 60°)$

   (b) $(1, 120°)$

   (c) $(2, 270°)$

4. (a) $(4, \pi)$

   (b) $\left(1, \frac{\pi}{4}\right)$

   (c) $\left(-3, -\frac{3\pi}{2}\right)$

Exercises 5–10: Determine if the pair of polar coordinates represents the same point.

5. $(2, 180°), (2, -180°)$

6. $(1, 90°), (-1, -90°)$

7. $(3, 45°), (3, -45°)$

8. $(-2, 135°), (2, -135°)$

9. $(0, 40°), (0, 50°)$

10. $(-3, 30°), (3, 210°)$

Exercises 11–20: Change the polar coordinates (r, θ) to rectangular coordinates (x, y).

11. $(3, 45°)$

12. $(-4, 225°)$

13. $(10, 90°)$

14. $\left(-1, \frac{\pi}{3}\right)$

15. $(5, 2\pi)$

16. $\left(-3, -\frac{\pi}{2}\right)$

17. $(-3, 60°)$

18. $(4, \pi)$

19. $\left(-2, -\frac{3\pi}{2}\right)$

20. $\left(10, \frac{2\pi}{3}\right)$

Exercises 21–26: For the point given in rectangular coordinates, find equivalent polar coordinates (r, θ) that satisfy the conditions.

   (a) $r > 0,$ $0° \leq \theta < 360°$
   (b) $r < 0,$ $-180° < \theta \leq 180°$

21. $(0, 3)$

22. $(-3, 0)$

23. $(-1, -\sqrt{3})$

24. $(\sqrt{3}, -1)$

25. $(3, -3)$

26. $(2, 2)$

Exercises 27–30: For the point given in rectangular coordinates, find equivalent polar coordinates (r, θ) that satisfy $r > 0$ and $0 \leq \theta < 2\pi$. Approximate θ to the nearest hundredth of a radian.

27. $(7, 24)$

28. $(3, -4)$

29. $(-5, 12)$

30. $(11, -60)$

Graphs of Polar Equations

Exercises 31 and 32: Graph the equation.

31. $\theta = 60°$

32. $\theta = -135°$

Exercises 33–40: Use the polar equation to complete the following.

   (a) Make a table of values with $\theta = 0°, 90°, 180°, 270°$.
   (b) Plot the points from the table and graph the curve for $0° \leq \theta \leq 360°$.

33. $r = 2$

34. $r = \cos \theta$

35. $r = 3 \sin \theta$

36. $r = 2 + 3 \sin \theta$

37. $r = 2 + 2 \sin \theta$

38. $r = 3 - 2 \cos \theta$

39. $r = 2 - \cos \theta$

40. $r = 2 + \sin \theta$

Exercises 41–58: Graph the polar equation by hand.

41. $\theta = 60°$

42. $\theta = -\frac{\pi}{4}$

43. $r = 3$

44. $r = 1$

45. $r \sin \theta = 3$

46. $r \cos \theta = -2$

47. $r = 2 \cos \theta$

48. $r = -2 \sin \theta$

49. $r = \sin 2\theta$

50. $r = \cos 2\theta$

51. $r = 3 + \cos \theta$

52. $r = 3 - \sin \theta$

53. $r = 1 - 2 \sin \theta$

54. $r = 1 + 2 \cos \theta$

55. $r = 3 \sin 2\theta$

56. $r = 2 \cos 2\theta$

57. $r = 2 \cos 3\theta$

58. $r = 4 \sin 3\theta$

Exercises 59–66: Write the equation in polar form.

59. $y = 3$

60. $x = -5$

61. $y = x$

62. $y = -\sqrt{3}$

63. $x^2 + y^2 = 9$

64. $x^2 + y^2 = 36$

65. $x^2 + y^2 = 2x$

66. $x^2 + y^2 = -4y$

Exercises 67–74: (Refer to Example 7.) Write the polar equation in terms of x and y.

67. $r = 3$

68. $r = 5$

69. $r = 2 \sec \theta$

70. $r = 2 \csc \theta$

71. $r = \dfrac{3}{2 \cos \theta + 4 \sin \theta}$

72. $r = \dfrac{2}{5 \cos \theta - \sin \theta}$

73. $r = \cos \theta$ (*Hint: Multiply both sides by r.*)

74. $r = 2 \sin \theta$

Exercises 75–88: Graph the curve.

75. $r = 3 + 3 \cos \theta$ (cardioid)

76. $r = 2 - 2 \sin \theta$ (cardioid)

77. $r = 3 - 2 \sin \theta$ (limaçon)

78. $r = 4 + \cos \theta$ (limaçon)

79. $r = 2 - 4 \cos \theta$ (limaçon with a loop)

80. $r = 1 + 2 \sin \theta$ (limaçon with a loop)

81. $r = 4 \sin \theta$ (circle)

82. $r = 2 \cos 3\theta$ (three-leaved rose)

83. $r = 2 \cos 5\theta$ (five-leaved rose)

84. $r = 3 \sin 4\theta$ (eight-leaved rose)

85. $r = \dfrac{\theta}{2}$ (spiral)

86. $r = e^{\theta/4}$ (logarithmic spiral)

87. $r^2 = 2 \sin 2\theta$ (lemniscate)

88. $r^2 = 4 \cos 2\theta$ (lemniscate)

Solving Equations in Polar Coordinates

Exercises 89 and 90: Find values for θ that satisfy the equation $r_1 = r_2$, where $0° \le \theta \le 360°$. Check any solutions.

89. $r_1 = 3, r_2 = 2 + 2 \sin \theta$

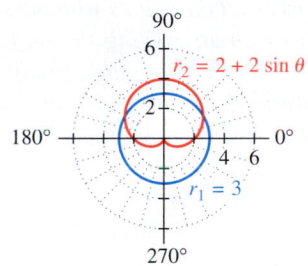

90. $r_1 = 1, r_2 = 2 \cos \theta$

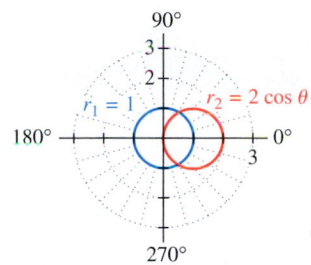

Exercises 91–94: Solve the polar equation $r_1 = r_2$, where $0° \le \theta \le 360°$,

    (a) symbolically, (b) graphically, and (c) numerically.

91. $r_1 = 3, r_2 = 2 - 2 \sin \theta$

92. $r_1 = 3, r_2 = 2 - \sin \theta$ **93.** $r_1 = 1, r_2 = 2 \sin \theta$

94. $r_1 = 2 - \sin \theta, r_2 = 2 + \cos \theta$

Applications

*Exercises 95 and 96: **Planetary Orbits**   (Refer to Example 9 and Table 8.7.) Graph the planetary orbits in a square viewing rectangle using polar coordinates.*

95. Saturn and Uranus   **96.** Jupiter and Neptune

*Exercises 97 and 98: **Planetary Orbits**   (Refer to Example 9.)*

Planet	a	e
Mercury	0.39	0.206
Venus	0.78	0.007
Earth	1.00	0.017
Mars	1.52	0.093

Source: H. Karttunen.

97. Graph the orbits of the four inner planets in the same square viewing rectangle.

98. NASA is planning future missions to Mars. Estimate graphically the closest distance possible between Earth and Mars.

Exercises 99 and 100: ***Broadcasting Patterns*** *Many times radio stations do not broadcast in all directions with the same intensity. To avoid interference with an existing station to the north, a new station may be licensed to broadcast only east and west. To create an east-west signal, two radio towers are sometimes used, as illustrated in the accompanying figure. Locations where the radio signal is received correspond to the interior of the curve defined by* $r^2 = 40{,}000 \cos 2\theta$, *where the polar axis (or positive x-axis) points east.* (**Source:** R. Weidner and R. Sells, *Elementary Classical Physics, Vol. 2.*)

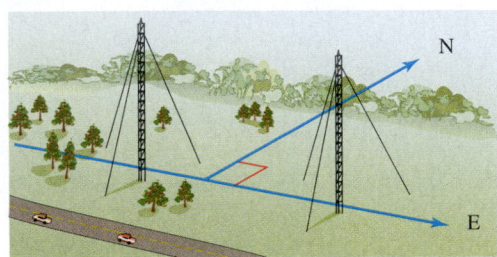

99. Graph $r^2 = 40{,}000 \cos 2\theta$ for $0° \le \theta \le 360°$, where units are in miles. Assuming the radio towers are located near the origin, use the graph to describe the regions where the signal can be received and where the signal cannot be received.

100. (Refer to the previous exercise.) Suppose a radio signal pattern is given by $r^2 = 22{,}500 \sin 2\theta$. Graph this pattern and interpret the results.

Writing about Mathematics

101. Explain why (r, θ) and $(-r, \theta + 180°)$ represent the same points in polar coordinates. Give two examples.

102. Give an example of a curve other than a circle that is more convenient to express in polar coordinates than in rectangular coordinates. Give an example of a curve that is more convenient to express using rectangular coordinates. Explain your reasoning.

EXTENDED AND DISCOVERY EXERCISES

1. ***Logarithmic Spiral*** Figure 8.91 shows a logarithmic spiral that can be described in rectangular coordinates by $y = x \tan (\ln (x^2 + y^2))$ and can be described in polar coordinates by the simpler equation $r = e^{\theta/2}$. Show that the first equation reduces to the second equation by assuming that $-\frac{\pi}{2} < \theta < \frac{\pi}{2}$. (*Hint:* Let $\tan \theta = \frac{y}{x}$ and $r^2 = x^2 + y^2$.)

2. ***Polar Graphs*** Consider the graphs of $r = a \cos n\theta$ and $r = a \sin n\theta$, where n is a positive integer and θ is given by $0° \le \theta < 360°$.
(**a**) How many times are these graphs traced over when n is even?

(**b**) How many times are these graphs traced over when n is odd?

8.6   Trigonometric Form and Roots of Complex Numbers

◆ Learn trigonometric form
◆ Find products and quotients of complex numbers
◆ Apply De Moivre's theorem
◆ Find roots of complex numbers

Introduction

One of the earliest encounters with the square root of a negative number was in A.D. 50 when Heron of Alexandria derived the expression $\sqrt{81 - 144}$. Square roots of negative numbers resulted in the invention of complex numbers. As late as the sixteenth and seventeenth centuries, mathematicians felt uneasy about negative numbers and square roots of negative numbers. The famous mathematician René Descartes rejected complex numbers and coined the term "imaginary" numbers.

The historical development of our present-day number system was often met with resistance to the introduction of new numbers. Today, complex numbers are readily accepted and play an important role in the design of airplanes, ships, electrical circuits, and fractals. (**Sources:** M. Kline, *Mathematics: The Loss of Certainty; Historical Topics for the Mathematics Classroom, Thirty-first Yearbook*, NCTM.)

Trigonometric Form

Complex numbers were introduced in Section 4.4. Any complex number can be expressed in *standard form* as $a + bi$, where a and b are real numbers. The *real part* is a and the *imaginary part* is b.

Real numbers can be plotted on a number line. Since complex numbers are determined by both the real part a and the imaginary part b, we use the **complex plane** to plot complex numbers. The horizontal axis is the **real axis** and the vertical axis is the **imaginary axis**. For example, $2 + 3i$ can be plotted in the complex plane as the point $(2, 3)$ and $3 - 4i$ can be plotted as the point $(3, -4)$. See Figure 8.102.

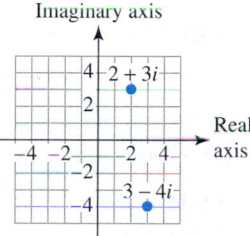

FIGURE 8.102
The Complex Plane

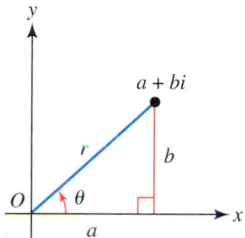

FIGURE 8.103
Trigonometric Form

A second form for complex numbers is called *trigonometric form,* which uses the variables r and θ to locate a complex number, as illustrated in Figure 8.103. We can see that

$$\cos \theta = \frac{a}{r} \qquad \text{and} \qquad \sin \theta = \frac{b}{r}.$$

Solving for a and b gives

$$a = r \cos \theta \qquad \text{and} \qquad b = r \sin \theta.$$

As a result, we can write the complex number $a + bi$ as follows.

$$a + bi = r \cos \theta + (r \sin \theta)i$$
$$= r(\cos \theta + i \sin \theta)$$

By the Pythagorean theorem, $r = \sqrt{a^2 + b^2}$. It also follows that

$$\tan \theta = \frac{b}{a} \quad (a \neq 0).$$

TRIGONOMETRIC FORM OF A COMPLEX NUMBER

The expression

$$r(\cos \theta + i \sin \theta)$$

is called a **trigonometric form** of the complex number $a + bi$, where $a = r \cos \theta$ and $b = r \sin \theta$. The number $r = \sqrt{a^2 + b^2}$ is the **modulus** of $a + bi$ and θ is the **argument** of $a + bi$.

Note: The expression $\cos \theta + i \sin \theta$ can be written as cis θ. The expression $|a + bi|$ is sometimes used to denote the *modulus* of the complex number $a + bi$.

EXAMPLE 1 Converting standard form to trigonometric form

Find the trigonometric form for each complex number, where $0° \leq \theta \leq 360°$.
(a) $1 + i$ **(b)** $-1 - i\sqrt{3}$

SOLUTION

(a) Plot $1 + i$ in the complex plane, as shown in Figure 8.104. The modulus r is

$$r = \sqrt{1^2 + 1^2} = \sqrt{2}.$$

We can see that $\tan \theta = \frac{b}{a} = \frac{1}{1}$. Therefore $\theta = \tan^{-1} 1 = 45°$. The trigonometric form is

$$\sqrt{2}(\cos 45° + i \sin 45°).$$

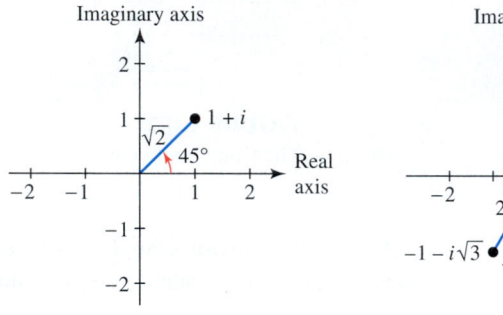

FIGURE 8.104   **FIGURE 8.105**

(b) Plot $-1 - i\sqrt{3}$, as shown in Figure 8.105. The modulus r is

$$r = \sqrt{(-1)^2 + (-\sqrt{3})^2} = 2.$$

The argument θ is in quadrant III and satisfies $\tan = \frac{-\sqrt{3}}{-1} = \sqrt{3}$. The reference angle for θ is $\theta_R = \tan^{-1}(\sqrt{3}) = 60°$. Thus $\theta = 240°$ and the trigonometric form is $2(\cos 240° + i \sin 240°)$. *Now Try Exercises 25 and 27* ◆

Some calculators have the capability to convert the complex number $a + bi$ to trigonometric or **polar form**. This is illustrated in Figures 8.106 and 8.107, where the first computation gives the modulus and the second gives the argument. Notice that angles of $240°$ and $-120°$ are coterminal angles. The value of θ is *not unique* in a trigonometric form. If θ_1 and θ_2 are coterminal angles, then

$$r(\cos \theta_1 + i \sin \theta_1) \quad \text{and} \quad r(\cos \theta_2 + i \sin \theta_2)$$

represent equivalent trigonometric forms.

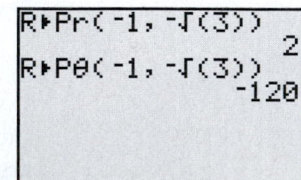

FIGURE 8.106 Degree Mode   **FIGURE 8.107** Degree Mode

EXAMPLE 2    Converting trigonometric form to standard form

Write the complex number as $a + bi$, where a and b are real numbers.

(a) $4\left(\cos \dfrac{\pi}{2} + i \sin \dfrac{\pi}{2}\right)$ **(b)** $\sqrt{3}(\cos 150° + i \sin 150°)$

SOLUTION

(a) $4\left(\cos \dfrac{\pi}{2} + i \sin \dfrac{\pi}{2}\right) = 4(0 + i(1)) = 4i$

(b) $\sqrt{3}(\cos 150° + i \sin 150°) = \sqrt{3}\left(-\dfrac{\sqrt{3}}{2} + \dfrac{1}{2}i\right) = -\dfrac{3}{2} + \dfrac{\sqrt{3}}{2}i$

Now Try Exercises 5 and 9 ◆

Products and Quotients of Complex Numbers

If two complex numbers, z_1 and z_2, are expressed in trigonometric form, it is straightforward to find either their product or quotient. Let

$$z_1 = r_1(\cos \theta_1 + i \sin \theta_1) \qquad \text{and} \qquad z_2 = r_2(\cos \theta_2 + i \sin \theta_2).$$

Then,

$$z_1 z_2 = r_1(\cos \theta_1 + i \sin \theta_1) \cdot r_2(\cos \theta_2 + i \sin \theta_2)$$
$$= r_1 r_2(\cos \theta_1 \cos \theta_2 + i \cos \theta_1 \sin \theta_2 + i \sin \theta_1 \cos \theta_2 + i^2 \sin \theta_1 \sin \theta_2)$$
$$= r_1 r_2((\cos \theta_1 \cos \theta_2 - \sin \theta_1 \sin \theta_2) + i(\cos \theta_1 \sin \theta_2 + \sin \theta_1 \cos \theta_2)).$$

Using the sum identities for cosine and sine, the last expression reduces to

$$z_1 z_2 = r_1 r_2(\cos (\theta_1 + \theta_2) + i \sin (\theta_1 + \theta_2)).$$

Using similar reasoning, it can be shown that

$$\frac{z_1}{z_2} = \frac{r_1}{r_2} (\cos (\theta_1 - \theta_2) + i \sin (\theta_1 - \theta_2)).$$

These results are summarized in the following.

PRODUCTS AND QUOTIENTS OF COMPLEX NUMBERS

Let $z_1 = r_1(\cos \theta_1 + i \sin \theta_1)$ and $z_2 = r_2(\cos \theta_2 + i \sin \theta_2)$. Then

$$z_1 z_2 = r_1 r_2(\cos (\theta_1 + \theta_2) + i \sin (\theta_1 + \theta_2))$$

$$\frac{z_1}{z_2} = \frac{r_1}{r_2} (\cos (\theta_1 - \theta_2) + i \sin (\theta_1 - \theta_2)), \ r_2 \neq 0.$$

EXAMPLE 3    Finding products and quotients

Find the product and quotient of

$$z_1 = 4(\cos 45° + i \sin 45°) \qquad \text{and} \qquad z_2 = 2(\cos 135° + i \sin 135°).$$

Express the answer in standard form.

SOLUTION

$$z_1 z_2 = 4(\cos 45° + i \sin 45°) \cdot 2(\cos 135° + i \sin 135°)$$
$$= (4 \cdot 2)(\cos (45° + 135°) + i \sin (45° + 135°))$$
$$= 8(\cos 180° + i \sin 180°)$$
$$= 8(-1 + 0)$$
$$= -8$$

$$\frac{z_1}{z_2} = \frac{4(\cos 45° + i \sin 45°)}{2(\cos 135° + i \sin 135°)}$$
$$= \frac{4}{2}(\cos (45° - 135°) + i \sin (45° - 135°))$$
$$= 2(\cos (-90°) + i \sin (-90°))$$
$$= 2(0 + -1i)$$
$$= -2i$$

Now Try Exercise 35 ◆

De Moivre's Theorem

If a complex number z is expressed in trigonometric form, then z^n for any positive integer n can be computed easily. Let $z = r(\cos \theta + i \sin \theta)$ and consider the following.

$$z^2 = r(\cos \theta + i \sin \theta) \cdot r(\cos \theta + i \sin \theta)$$
$$= r^2(\cos (\theta + \theta) + i \sin (\theta + \theta))$$
$$= r^2(\cos 2\theta + i \sin 2\theta)$$
$$z^3 = zz^2$$
$$= r(\cos \theta + i \sin \theta) \cdot r^2(\cos 2\theta + i \sin 2\theta)$$
$$= r^3(\cos 3\theta + i \sin 3\theta)$$

In general it can be shown that

$$z^n = r^n(\cos n\theta + i \sin n\theta).$$

This result is summarized in the following theorem, which is due to Abraham De Moivre (1667–1754), a French Huguenot, who was a close friend of Isaac Newton. This theorem has become a keystone of analytic trigonometry. (**Source:** H. Eves, *An Introduction to the History of Mathematics.*)

DE MOIVRE'S THEOREM

Let $z = r(\cos \theta + i \sin \theta)$ and n be a positive integer. Then

$$z^n = r^n(\cos n\theta + i \sin n\theta).$$

EXAMPLE 4 **Finding a power of a complex number**

Use De Moivre's theorem to evaluate $(1 + i)^8$ and express the result in standard form.

SOLUTION From Example 1(a), the trigonometric form of $z = 1 + i$ is

$$z = \sqrt{2}(\cos 45° + i \sin 45°).$$

By De Moivre's theorem

$$z^8 = (\sqrt{2})^8 (\cos (\mathbf{8} \cdot 45°) + i \sin (\mathbf{8} \cdot 45°))$$

$$= 16(\cos 360° + i \sin 360°)$$

$$= 16(1 + 0i)$$

$$= 16.$$

Now Try Exercise 47 ◆

During the past 20 years, computer graphics and complex numbers have made it possible to produce many beautiful fractals. In 1977 Benoit B. Mandelbrot first used the term *fractal*. Largely because of his efforts, fractal geometry has become a new field of study. A fractal is an enchanting geometric figure with an endless self-similarity property that repeats itself infinitely with ever decreasing dimensions. If you look at smaller and smaller portions of the figure, you will continue to see the whole—much like looking into two parallel mirrors that are facing each other. Fractals not only have aesthetic appeal, they also have applications in science. An example of a fractal is the *Mandelbrot set* shown in Figure 8.108. (**Source:** From "The Fractal Geometry of Nature" © by Benoit Mandelbrot 198213.)

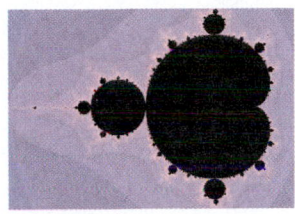

FIGURE 8.108

 **EXAMPLE 5**   Analyzing the Mandelbrot set

The fractal called the Mandelbrot set is shown in Figure 8.108. To determine if a complex number $z = a + bi$ is in the Mandelbrot set, we can perform the following sequence of calculations. Let

$$z_0 = z$$

$$z_1 = z_0^2 + z_0$$

$$z_2 = z_1^2 + z_0$$

$$z_3 = z_2^2 + z_0$$

and so on. If the modulus of any z_k ever exceeds 2, then z is not in the Mandelbrot set, otherwise z is in the Mandelbrot set. Determine if the complex number belongs to the Mandelbrot set.   (**Source:** F. Hill.)
(a) $z = 1 + i$ (b) $z = 0.5i$

SOLUTION
(a) Let $z_0 = 1 + i$. Then,

$$z_1 = (1 + i)^2 + (1 + i) = 1 + 3i.$$

Since the modulus of z_1 is

$$|1 + 3i| = \sqrt{1^2 + 3^2} = \sqrt{10} > 2,$$

the complex number $1 + i$ is not in the Mandelbrot set.
(b) Let $z_0 = 0.5i$. Then,

$$z_1 = (0.5i)^2 + 0.5i = -0.25 + 0.5i$$

$$z_2 = (-0.25 + 0.5i)^2 + 0.5i = -0.1875 + 0.25i$$

$$z_3 = (-0.1875 + 0.25i)^2 + 0.5i \approx -0.0273 + 0.406i$$

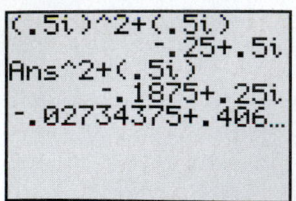

FIGURE 8.109

The modulus of each consecutive z_k never exceeds 2. Thus $0.5i$ is in the Mandelbrot set. You may find it helpful to use a calculator to perform these calculations. See Figure 8.109.

Now Try Exercises 63 and 65 ◆

Roots of Complex Numbers

A number w is the **nth root** of a number z if $w^n = z$. De Moivre's theorem can be used to find roots of complex numbers. To see this, let the trigonometric forms of w and z be

$$w = s(\cos \alpha + i \sin \alpha) \quad \text{and} \quad z = r(\cos \theta + i \sin \theta).$$

Then by De Moivre's theorem $w^n = z$ implies that

$$s^n(\cos n\alpha + i \sin n\alpha) = r(\cos \theta + i \sin \theta).$$

It follows that $s^n = r$ or $s = \sqrt[n]{r}$. Furthermore the following two equations must be satisfied.

$$\cos n\alpha = \cos \theta$$

$$\sin n\alpha = \sin \theta$$

Since the cosine and sine functions have period 360°,

$$n\alpha = \theta + 360° \cdot k$$

for some integer k, or

$$\alpha = \frac{\theta + 360° \cdot k}{n}.$$

Substituting these results in the trigonometric form for w gives

$$w_k = \sqrt[n]{r}\left(\cos \frac{\theta + 360° \cdot k}{n} + i \sin \frac{\theta + 360° \cdot k}{n}\right).$$

We obtain a unique value of w_k for $k = 0, 1, 2, \ldots, n - 1$. This discussion is summarized in the following.

ROOTS OF A COMPLEX NUMBER

Let $z = r(\cos \theta + i \sin \theta)$ be a nonzero complex number and n be any positive integer. Then z has exactly n distinct nth roots given by

$$w_k = \sqrt[n]{r}\left(\cos \frac{\theta + 360° \cdot k}{n} + i \sin \frac{\theta + 360° \cdot k}{n}\right),$$

where $k = 0, 1, 2, \ldots, n - 1$. If radian measure is used then let

$$w_k = \sqrt[n]{r}\left(\cos \frac{\theta + 2\pi k}{n} + i \sin \frac{\theta + 2\pi k}{n}\right).$$

EXAMPLE 6 Finding cube roots of a complex number

Find the three cube roots of $8i$. Check your results with a calculator.

SOLUTION First write the complex number $8i$ in trigonometric form.

$$8i = 8(\cos 90° + i \sin 90°)$$

The three cube roots of $8i$ can be found by letting $n = 3$, $r = 8$, $\theta = 90°$, and $k = 0, 1, 2$.

$$w_0 = \sqrt[3]{8}\left(\cos\frac{90° + 360° \cdot 0}{3} + i \sin\frac{90° + 360° \cdot 0}{3}\right)$$

$$= 2(\cos 30° + i \sin 30°)$$

$$= 2\left(\frac{\sqrt{3}}{2} + \frac{1}{2}i\right)$$

$$= \sqrt{3} + i$$

$$w_1 = \sqrt[3]{8}\left(\cos\frac{90° + 360° \cdot 1}{3} + i \sin\frac{90° + 360° \cdot 1}{3}\right)$$

$$= 2(\cos 150° + i \sin 150°)$$

$$= 2\left(-\frac{\sqrt{3}}{2} + \frac{1}{2}i\right)$$

$$= -\sqrt{3} + i$$

$$w_2 = \sqrt[3]{8}\left(\cos\frac{90° + 360° \cdot 2}{3} + i \sin\frac{90° + 360° \cdot 2}{3}\right)$$

$$= 2(\cos 270° + i \sin 270°)$$

$$= 2(0 - i)$$

$$= -2i$$

The three cube roots of $8i$ are $\sqrt{3} + i$, $-\sqrt{3} + i$, and $-2i$. These can be checked using a calculator, as shown in Figure 8.110.

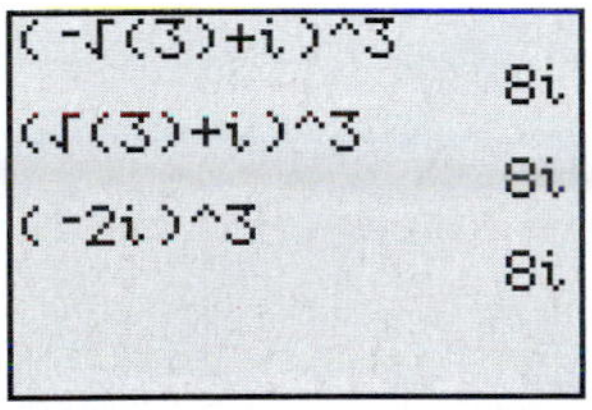

Now Try Exercise 57 ◆

FIGURE 8.110

◆ **CLASS DISCUSSION**

One cube root of the complex number $z = -1$ is $w = -1$. Find the other two cube roots of z graphically. ◆

If the three cube roots of $8i$ are plotted in the complex plane, they lie on a circle of radius 2, equally spaced 120° apart, as shown in Figure 8.111. In general, the nth roots of a complex number $z = r(\cos \theta + i \sin \theta)$ will lie equally spaced on a circle of radius $\sqrt[n]{r}$.

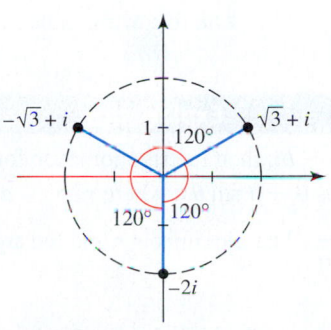

FIGURE 8.111
The Cube Roots of $8i$

EXAMPLE 7 Finding square roots of a complex number

Find the two square roots of $1 + i\sqrt{3}$.

SOLUTION First write the complex number $1 + i\sqrt{3}$ in trigonometric form.

$$1 + i\sqrt{3} = 2\left(\cos\frac{\pi}{3} + i\sin\frac{\pi}{3}\right)$$

The two square roots of $1 + i\sqrt{3}$ can be found by letting $n = 2, r = 2, \theta = \frac{\pi}{3}$, and $k = 0, 1$.

$$w_0 = \sqrt{2}\left(\cos\frac{\frac{\pi}{3} + 2\pi \cdot 0}{2} + i\sin\frac{\frac{\pi}{3} + 2\pi \cdot 0}{2}\right)$$

$$= \sqrt{2}\left(\cos\frac{\pi}{6} + i\sin\frac{\pi}{6}\right)$$

$$= \sqrt{2}\left(\frac{\sqrt{3}}{2} + \frac{1}{2}i\right)$$

$$= \frac{\sqrt{6}}{2} + \frac{\sqrt{2}}{2}i$$

$$w_1 = \sqrt{2}\left(\cos\frac{\frac{\pi}{3} + 2\pi \cdot 1}{2} + i\sin\frac{\frac{\pi}{3} + 2\pi \cdot 1}{2}\right)$$

$$= \sqrt{2}\left(\cos\frac{7\pi}{6} + i\sin\frac{7\pi}{6}\right)$$

$$= \sqrt{2}\left(-\frac{\sqrt{3}}{2} - \frac{1}{2}i\right)$$

$$= -\frac{\sqrt{6}}{2} - \frac{\sqrt{2}}{2}i$$

Thus the two square roots of $1 + i\sqrt{3}$ are $\frac{\sqrt{6}}{2} + \frac{\sqrt{2}}{2}i$ and $-\frac{\sqrt{6}}{2} - \frac{\sqrt{2}}{2}i$.

Now Try Exercise 55 ◆

8.6 Putting it all Together

The following table summarizes some of the important topics in this section.

Concept	Explanation	Examples
Trigonometric form	If $z = a + bi$, then its trigonometric form is $z = r(\cos\theta + i\sin\theta)$, where $r = \sqrt{a^2 + b^2}$ and $\tan\theta = \dfrac{b}{a}$. The modulus is r and the argument is θ.	If $z = 1 + i\sqrt{3}$, then $$r = \sqrt{1^2 + (\sqrt{3})^2} = 2$$ and $$\theta = \tan^{-1}\frac{\sqrt{3}}{1} = 60°.$$ Thus $z = 2(\cos 60° + i\sin 60°)$.

Concept	Explanation	Examples
Products and quotients	Let $z_1 = r_1(\cos\theta_1 + i\sin\theta_1)$ and $z_2 = r_2(\cos\theta_2 + i\sin\theta_2)$. Then $$z_1 z_2 = r_1 r_2 (\cos(\theta_1 + \theta_2) + i\sin(\theta_1 + \theta_2))$$ and $$\frac{z_1}{z_2} = \frac{r_1}{r_2}(\cos(\theta_1 - \theta_2) + i\sin(\theta_1 - \theta_2)),$$ $r_2 \neq 0$.	If $z_1 = 3(\cos 66° + i\sin 66°)$ and $z_2 = 2(\cos 22° + i\sin 22°)$, then it follows that $$z_1 z_2 = 6(\cos 88° + i\sin 88°)$$ and $$\frac{z_1}{z_2} = \frac{3}{2}(\cos 44° + i\sin 44°).$$
De Moivre's theorem	Let $z = r(\cos\theta + i\sin\theta)$. Then $z^n = r^n(\cos n\theta + i\sin n\theta)$.	If $z = 2(\cos 7° + i\sin 7°)$, then $$z^3 = 2^3(\cos 21° + i\sin 21°).$$
Roots of complex numbers	Let $z = r(\cos\theta + i\sin\theta)$ and n be any positive integer. Then the nth roots of z are given by $$w_k = \sqrt[n]{r}\left(\cos\frac{\theta + 360° \cdot k}{n} + i\sin\frac{\theta + 360° \cdot k}{n}\right),$$ where $k = 0, 1, 2, \ldots, n-1$.	The three cube roots of $$8 = 8(\cos 0° + i\sin 0°)$$ are as follows: $$w_0 = \sqrt[3]{8}\left(\cos\frac{0° + 360° \cdot 0}{3} + i\sin\frac{0° + 360° \cdot 0}{3}\right)$$ $$= 2$$ $$w_1 = \sqrt[3]{8}\left(\cos\frac{0° + 360° \cdot 1}{3} + i\sin\frac{0° + 360° \cdot 1}{3}\right)$$ $$= 2(\cos 120° + i\sin 120°)$$ $$= -1 + i\sqrt{3}$$ $$w_2 = \sqrt[3]{8}\left(\cos\frac{0° + 360° \cdot 2}{3} + i\sin\frac{0° + 360° \cdot 2}{3}\right)$$ $$= 2(\cos 240° + i\sin 240°)$$ $$= -1 - i\sqrt{3}$$

8.6 — Exercises

The Complex Plane

Exercises 1–4: Plot the numbers in the complex plane.

1. (a) $3 + 2i$

   (b) $-1 + i$

   (c) $3i$

2. (a) $-2i$

   (b) $2 + 2i$

   (c) $2 - 2i$

3. (a) -3

   (b) $4 - 2i$

   (c) $-1 - 3i$

4. (a) $-1 - i$

   (b) $4 + 3i$

   (c) 4

Trigonometric Form

Exercises 5–12: Write the number in standard form.

5. $5(\cos 180° + i \sin 180°)$ **6.** $3(\cos 90° + i \sin 90°)$

7. $2(\cos 45° + i \sin 45°)$ **8.** $\cos 150° + i \sin 150°$

9. $4\left(\cos \frac{3\pi}{2} + i \sin \frac{3\pi}{2}\right)$ **10.** $2\left(\cos \frac{\pi}{6} + i \sin \frac{\pi}{6}\right)$

11. $3(\cos 2\pi + i \sin 2\pi)$ **12.** $5\left(\cos \frac{3\pi}{4} + i \sin \frac{3\pi}{4}\right)$

Exercises 13–20: Find the modulus of the number.

13. $1 + i$ **14.** $3 - 4i$

15. $12 - 5i$ **16.** $-24 + 7i$

17. -6 **18.** $15i$

19. $2 - 3i$ **20.** $11 - 60i$

Exercises 21–30: Write the number in trigonometric form. Let $0° \le \theta < 360°$.

21. $1 + i$ **22.** $1 - i$

23. 5 **24.** -3

25. $4i$ **26.** $-i$

27. $-1 + i\sqrt{3}$ **28.** $-\sqrt{2} - i\sqrt{2}$

29. $\sqrt{3} + i$ **30.** $-\frac{\sqrt{3}}{2} + \frac{1}{2}i$

Exercises 31–34: Write the number in trigonometric form. Let $0 \le \theta < 2\pi$.

31. -2 **32.** $4i$

33. $-2 + 2i$ **34.** $1 + i\sqrt{3}$

Exercises 35–40: Find $z_1 z_2$ and $\frac{z_1}{z_2}$. Express your answer in standard form.

35. $z_1 = 9(\cos 45° + i \sin 45°)$,
$z_2 = 3(\cos 15° + i \sin 15°)$

36. $z_1 = 5(\cos 90° + i \sin 90°)$,
$z_2 = 2(\cos 30° + i \sin 30°)$

37. $z_1 = 6\left(\cos \frac{3\pi}{4} + i \sin \frac{3\pi}{4}\right)$,
$z_2 = \cos \frac{\pi}{4} + i \sin \frac{\pi}{4}$

38. $z_1 = 4(\cos 300° + i \sin 300°)$,
$z_2 = 2(\cos 60° + i \sin 60°)$

39. $z_1 = \cos 15° + i \sin 15°$,
$z_2 = \cos\left(-\frac{\pi}{4}\right) + i \sin\left(-\frac{\pi}{4}\right)$

40. $z_1 = 11\left(\cos \frac{2\pi}{3} + i \sin \frac{2\pi}{3}\right)$,
$z_2 = 22(\cos 30° + i \sin 30°)$

Powers of Complex Numbers

Exercises 41–46: Use De Moivre's theorem to evaluate the expression. Write the result in standard form.

41. $(2(\cos 30° + i \sin 30°))^3$

42. $(3(\cos 45° + i \sin 45°))^4$

43. $(\cos 10° + i \sin 10°)^{36}$

44. $(\cos 1° + i \sin 1°)^{90}$

45. $(5(\cos 60° + i \sin 60°))^2$

46. $(2(\cos 90° + i \sin 90°))^5$

Exercises 47–50: Use De Moivre's theorem to evaluate the expression. Write the result in standard form and check it using a calculator.

47. $(1 + i)^3$ **48.** $(3i)^4$

49. $(\sqrt{3} + i)^5$ **50.** $(2 - 2i)^6$

Roots of Complex Numbers

Exercises 51–62: Find the following roots and express them in standard form. Check your results with a calculator.

51. The square roots of $4(\cos 120° + i \sin 120°)$

52. The cube roots of $27(\cos 180° + i \sin 180°)$

53. The cube roots of $\cos 180° + i \sin 180°$

54. The fourth roots of $16(\cos 240° + i \sin 240°)$

55. The square roots of i **56.** The cube roots of 1

57. The cube roots of -8 **58.** The square roots of $-4i$

59. The cube roots of $64i$ **60.** The fourth roots of -1

61. The fourth roots of 81

62. The square roots of $-1 + i\sqrt{3}$

Fractals

Exercises 63–66: Mandelbrot Set (Refer to Example 5.)
Determine if the complex number belongs to the Mandelbrot set.

63. $-0.4i$

64. $0.5 + i$

65. $1 + i$

66. $-0.2 + 0.2i$

Applications

67. Electrical Circuits *Impedance* is a measure of the opposition to the flow of current in an electrical circuit. It consists of two parts called the *resistance* and *reactance*. Light bulbs add resistance to an electrical circuit, and reactance occurs when electricity passes through coils of wire like those found in electric motors. Impedance Z in ohms (Ω) may be expressed as a complex number, where the real part represents the resistance and the imaginary part represents the reactance. For example, if the resistive part is 3 ohms and the reactive part is 4 ohms, then the impedance could be described by the complex number $Z = 3 + 4i$. The modulus of Z gives the total impedance in ohms. In a series circuit like the one shown in the figure, the total impedance is the sum of the individual impedances. (**Source:** R. Smith and R. Dorf, *Circuits, Devices and Systems.*)

(a) The circuit contains two light bulbs and two electric motors. If it is assumed that the light bulbs represent resistance and the motors represent reactance, express impedance as $Z = a + bi$.

(b) Find total impedance in ohms by calculating the modulus of Z.

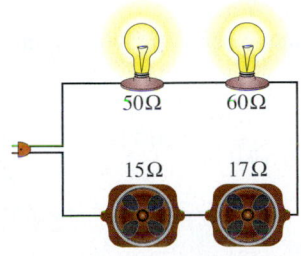

68. Electrical Circuits (Continuation of the previous exercise.) In the parallel electrical circuit shown in the figure, impedance Z is given by

$$Z = \frac{1}{\dfrac{1}{Z_1} + \dfrac{1}{Z_2}},$$

where Z_1 and Z_2 represent impedances for each branch of the circuit. (**Source:** G. Wilcox and C. Hesselberth, *Electricity for Engineering Technology.*)

(a) Find Z.

(b) Find total impedance to the nearest tenth of an ohm by calculating the modulus of Z.

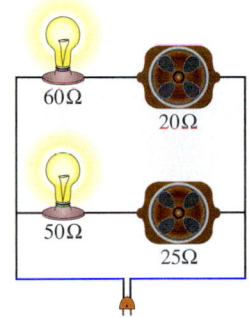

Writing about Mathematics

69. Explain how to find a trigonometric form of a complex number $a + bi$. Give an example. Is trigonometric form unique for a given complex number z? Explain.

70. Suppose that one fourth root w of a complex number z is known. Explain how to graphically find the other fourth roots of z.

CHECKING BASIC CONCEPTS FOR SECTIONS 8.5 AND 8.6

1. Plot the following points (r, θ) on a polar grid.
  (a) $(2, 30°)$ **(b)** $(3, -60°)$ **(c)** $(-4, 120°)$

2. Graph the equation.
  (a) $r = 2$
  (b) $\theta = -\frac{\pi}{4}$
  (c) $r = 3 + 3\cos\theta$
  (d) $r = 3\cos 2\theta$

3. Plot the numbers in the complex plane.
  (a) $-3 + 2i$ **(b)** $-4 - 3i$

4. Find the trigonometric form of $1 + i\sqrt{3}$.

5. Find the three cube roots of i.

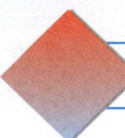

CHAPTER 8 SUMMARY

CONCEPT **EXPLANATION AND EXAMPLES**

SECTION 8.1 LAW OF SINES

LAW OF SINES

$$\frac{\sin \alpha}{a} = \frac{\sin \beta}{b} = \frac{\sin \gamma}{c} \quad \text{or} \quad \frac{a}{\sin \alpha} = \frac{b}{\sin \beta} = \frac{c}{\sin \gamma}$$

The law of sines can be used to solve triangles given ASA, AAS, or SSA. SSA is called the ambiguous case and can have 0, 1, or 2 solutions.

Example: Given $\beta = 32°$, $\gamma = 46°$, and $c = 10$, find b.
We are given AAS, as illustrated in the accompanying figure.

$$\frac{b}{\sin \beta} = \frac{c}{\sin \gamma} \quad \text{implies that} \quad b = \frac{10 \sin 32°}{\sin 46°} \approx 7.37.$$

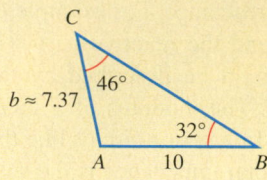

SECTION 8.2 LAW OF COSINES

LAW OF COSINES

$$a^2 = b^2 + c^2 - 2bc \cos \alpha$$
$$b^2 = a^2 + c^2 - 2ac \cos \beta$$
$$c^2 = a^2 + b^2 - 2ab \cos \gamma$$

The law of cosines can be used to solve triangles given SAS or SSS. Each situation results in a unique solution.

Example: Given $a = 5$, $b = 6$, and $c = 7$, find α.
We are given SSS, as illustrated in the accompanying figure.
$a^2 = b^2 + c^2 - 2bc \cos \alpha$ implies that

$$\cos \alpha = \frac{b^2 + c^2 - a^2}{2bc} = \frac{6^2 + 7^2 - 5^2}{2(6)(7)} \approx 0.714.$$

Thus $\alpha \approx \cos^{-1}(0.714) \approx 44.4°.$

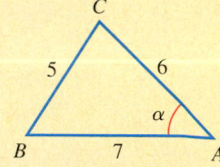

SECTION 8.3 VECTORS

VECTORS

A vector is a directed line segment that has both magnitude (length) and direction. Three different representations for a vector are

$$\mathbf{v} = \langle a_1, a_2 \rangle, \quad \mathbf{v} = a_1 \mathbf{i} + a_2 \mathbf{j}, \quad \text{and} \quad \mathbf{v} = \overrightarrow{PQ}.$$

Horizontal component $= a_1$; vertical component $= a_2$
Magnitude: $\|\mathbf{v}\| = \sqrt{a_1^2 + a_2^2}$

CONCEPT	EXPLANATION AND EXAMPLES

SECTION 8.3 VECTORS (CONTINUED)

CALCULATIONS INVOLVING VECTORS

Let $\mathbf{a} = \langle a_1, a_2 \rangle$ and $\mathbf{b} = \langle b_1, b_2 \rangle$.
Sum: $\mathbf{a} + \mathbf{b} = \langle a_1 + b_1, a_2 + b_2 \rangle$
Difference: $\mathbf{a} - \mathbf{b} = \langle a_1 - b_1, a_2 - b_2 \rangle$
Scalar Product: $k\mathbf{a} = \langle ka_1, ka_2 \rangle$
Dot Product: $\mathbf{a} \cdot \mathbf{b} = a_1 b_1 + a_2 b_2$

Angle θ between $\mathbf{a}$ and $\mathbf{b}$: $\theta = \cos^{-1}\left(\dfrac{\mathbf{a} \cdot \mathbf{b}}{\|\mathbf{a}\| \, \|\mathbf{b}\|} \right)$

WORK

If a constant force $\mathbf{F}$ is applied to an object that moves along a vector $\mathbf{D}$, then the work done is $W = \mathbf{F} \cdot \mathbf{D}$.

SECTION 8.4 PARAMETRIC EQUATIONS

PLANE CURVE AND PARAMETRIC EQUATIONS

Curve defined by the parametric equations $x = f(t)$ and $y = g(t)$, where f and g are continuous and t is the parameter

Example: $x = \cos t, y = \sin t; 0 \le t \le 2\pi$

Since $x^2 + y^2 = \cos^2 t + \sin^2 t = 1$, this curve describes the unit circle. See the accompanying figure.

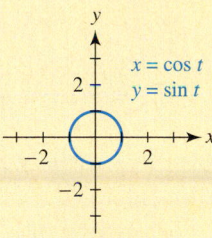

SECTION 8.5 POLAR EQUATIONS

POLAR COORDINATES AND EQUATIONS

Points are identified using r and θ instead of x and y. Polar equations are plotted in the polar plane, where the pole corresponds to the origin and the polar axis corresponds to the positive x-axis. Polar coordinates are not unique.

Example: $r = \cos 5\theta$ (rose curve with 5 leaves)

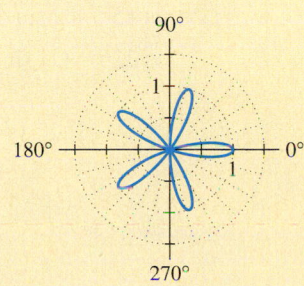

CONCEPT	EXPLANATION AND EXAMPLES

SECTION 8.6 TRIGONOMETRIC FORM AND ROOTS OF COMPLEX NUMBERS

TRIGONOMETRIC FORM AND COMPLEX NUMBERS

The expression $r(\cos \theta + i \sin \theta)$ is the trigonometric form of $a + bi$, where $a = r \cos \theta$ and $b = r \sin \theta$.

Modulus: $|z| = r = \sqrt{a^2 + b^2}$; argument: θ

Example: $z = 2(\cos 30° + i \sin 30°) = \sqrt{3} + i$

Modulus: $r = 2$, argument: $\theta = 30°$

OPERATIONS ON COMPLEX NUMBERS AND TRIGONOMETRIC FORM

Let $z_1 = r_1(\cos \theta_1 + i \sin \theta_1)$ and $z_2 = r_2(\cos \theta_2 + i \sin \theta_2)$.

$$z_1 z_2 = r_1 r_2 (\cos (\theta_1 + \theta_2) + i \sin (\theta_1 + \theta_2))$$

$$\frac{z_1}{z_2} = \frac{r_1}{r_2} (\cos (\theta_1 - \theta_2) + i \sin (\theta_1 - \theta_2))$$

$$z_1^n = r_1^n (\cos (n\theta_1) + i \sin (n\theta_1)) \qquad \text{De Moivre's theorem}$$

Example: Let $z_1 = 6(\cos 120° + i \sin 120°)$ and $z_2 = 2(\cos 80° + i \sin 80°)$.

$$z_1 z_2 = 12(\cos 200° + i \sin 200°)$$

$$\frac{z_1}{z_2} = 3(\cos 40° + i \sin 40°)$$

$$z_1^4 = 6^4 (\cos (4 \cdot 120°) + i \sin (4 \cdot 120°))$$

ROOTS OF COMPLEX NUMBERS

If $z = r(\cos \theta + i \sin \theta)$, then the nth roots of z are given by

$$w_k = \sqrt[n]{r} \left(\cos \frac{\theta + 360° \cdot k}{n} + i \sin \frac{\theta + 360° \cdot k}{n} \right),$$

where $k = 0, 1, 2, \ldots, n - 1$.

Example: Let $z = 4i = 4(\cos 90° + i \sin 90°)$.
The square roots of z are as follows.

$$w_k = \sqrt{4} \left(\cos \frac{90° + 360° \cdot k}{2} + i \sin \frac{90° + 360° \cdot k}{2} \right)$$

for $k = 0$ and 1. Simplifying gives

$$w_0 = 2(\cos 45° + i \sin 45°) = \sqrt{2} + i\sqrt{2} \quad \text{and}$$
$$w_1 = 2(\cos 225° + i \sin 225°) = -\sqrt{2} - i\sqrt{2}.$$

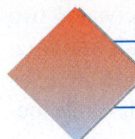

REVIEW EXERCISES

Exercises 1–4: Solve the triangle. Approximate values to the nearest tenth.

1.

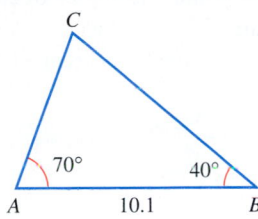

2.

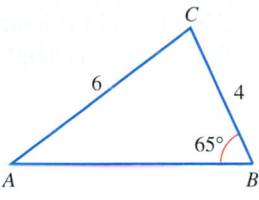

3.

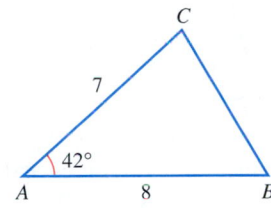

4.

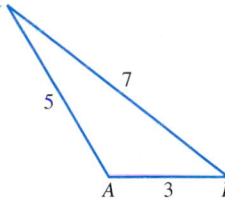

Exercises 5–10: Solve the triangle. Approximate values to the nearest tenth.

5. $\alpha = 19°, \beta = 46°, b = 13$

6. $\alpha = 30°, b = 10, a = 8$

7. $\gamma = 20°, b = 8, c = 11$

8. $\alpha = 70°, b = 17, a = 5$

9. $b = 23, \gamma = 35°, a = 18$

10. $a = 65, b = 45, c = 32$

Exercises 11–14: Approximate the area of each triangle to the nearest tenth.

11. $a = 12.3, b = 13.7, \gamma = 39°$

12. $\alpha = 40°, \beta = 55°, c = 67$

13. $a = 34, b = 67, c = 53$

14. $a = 2.1, b = 1.7, c = 2.2$

Exercises 15 and 16: Complete the following for vector v.

   (a) Give the horizontal and vertical components.
   (b) Find $\|\mathbf{v}\|$.
   (c) Graph v and interpret $\|\mathbf{v}\|$.

15. $\mathbf{v} = \langle 3, 4 \rangle$ **16.** $\mathbf{v} = -5\mathbf{i} + 12\mathbf{j}$

*Exercises 17 and 18: A vector **v** has initial point P and terminal point Q.*

   (a) Graph $\overrightarrow{PQ}$.
   (b) Write $\overrightarrow{PQ}$ as $\mathbf{v} = a_1\mathbf{i} + a_2\mathbf{j}$.
   (c) Find $\|\overrightarrow{PQ}\|$.

17. $P = (0, 0), Q = (-2, -4)$

18. $P = (3, 2), Q = (-3, -1)$

Exercises 19–22: Find each of the following.

   (a) $2\mathbf{a}$ *(b) $\mathbf{a} - 3\mathbf{b}$*   *(c) $\mathbf{a} \cdot \mathbf{b}$*
   *(d) The angle θ between **a** and **b** rounded to a tenth of a degree*

19. $\mathbf{a} = \langle 3, -2 \rangle, \mathbf{b} = \langle 1, 1 \rangle$

20. $\mathbf{a} = \langle 3, 2 \rangle, \mathbf{b} = \langle -2, -3 \rangle$

21. $\mathbf{a} = 2\mathbf{i} + 2\mathbf{j}, \mathbf{b} = \mathbf{i} + \mathbf{j}$

22. $\mathbf{a} = \mathbf{i} - 2\mathbf{j}, \mathbf{b} = 2\mathbf{i} + \mathbf{j}$

23. *Resultant Force*   Use the parallelogram rule to find the magnitude of the resultant force of the two forces shown in the figure.

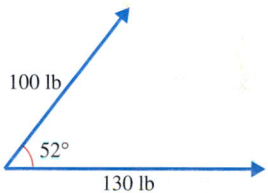

24. *Work*   Find the work done when a constant force $\mathbf{F} = 300\mathbf{i} + 400\mathbf{j}$ is applied to an object that moves along the vector $\mathbf{D} = 10\mathbf{i} - 2\mathbf{j}$, where units are in pounds and feet. Approximate the magnitude of $\mathbf{F}$ and interpret the result.

Exercises 25–28: Graph the parametric equations.

25. $x = t + 2, \quad y = t^2 - 3; \quad -2 \le t \le 2$

26. $x = t^3 - 4, \quad y = t - 1; \quad 0 \le t \le 2$

27. $x = 2 \cos t, \quad y = -2 \sin t; \quad 0 \le t \le 2\pi$

28. $x = 3 \sin t, \quad y = 2 \cos t; \quad -\pi \le t \le \pi$

29. *Designing Letters* Find parametric equations whose graphs resemble the given letter.
    (a) O
    (b) H

30. Change the polar coordinates (r, θ) to rectangular coordinates (x, y).
    (a) $(2, 135°)$
    (b) $(-1, 60°)$

Exercises 31 and 32: Use the polar equation to complete the following.

    (a) *Make a table of values with $\theta = 0°, 90°, 180°, 270°$.*
    (b) *Plot the points from the table and graph the curve.*

31. $r = 1 + \cos \theta$
32. $r = \sin \theta$

Exercises 33–36: Graph each polar equation.

33. $r = 3 \sin 3\theta$
34. $r = 2 - \cos \theta$

35. $r = 3 + 3 \sin \theta$
36. $r = 1 - 2 \sin \theta$

37. Plot the numbers in the complex plane.
    (a) $4 - i$
    (b) $-2 + 2i$

    (c) $-2i$
    (d) -4

38. Write the complex number in trigonometric form. Let θ satisfy $0° \le \theta < 360°$.
    (a) $-2 + 2i$
    (b) $\sqrt{3} + i$

    (c) $5i$
    (d) -6

39. Find $z_1 z_2$ and $\dfrac{z_1}{z_2}$, if

$$z_1 = 4(\cos 150° + i \sin 150°), \text{ and}$$
$$z_2 = 2(\cos 30° + i \sin 30°).$$

Write the results in standard form.

40. Use De Moivre's theorem to evaluate z^4 if the trigonometric form of z is $z = 2(\cos 45° + i \sin 45°)$. Write the result in standard form.

Exercises 41 and 42: Find the following roots. Check your results with a calculator.

41. The square roots of $4(\cos 60° + i \sin 60°)$

42. The cube roots of $27i$

Applications

43. *Airplane Navigation* An airplane takes off with a bearing of 130° and flies 350 miles. Then it changes its course to a bearing of 60° and flies for 500 miles. Determine how far the plane is from its takeoff point.

44. *Obstructed View* To find the distance between two points A and B on the opposite sides of a small building, a surveyor measures AC as 63.15 feet, angle ACB as 43.56°, and CB as 103.53 feet. Find the distance between A and B to the nearest tenth of a foot.

45. *Height of an Airplane* Two observation points A and B are 950 feet apart. From these points the angles of elevation of an airplane are 52° and 57°, as illustrated in the figure. Find the height of the airplane to the nearest foot.

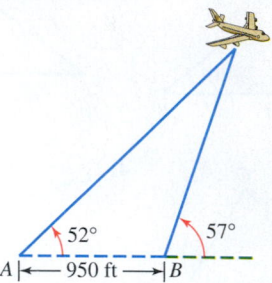

46. *Area of a Lot* A surveyor measures two sides and the included angle of a triangular lot as $a = 93.6$ feet, $b = 110.6$ feet, and $\gamma = 51.8°$. Find the area of the lot.

47. *Area of a Lot* Find the area of the quadrangular lot shown in the figure to the nearest square foot.

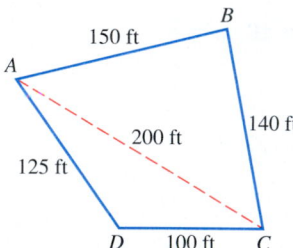

48. *Interpreting a Vector* A boat is heading west at 20 miles per hour in a current that is flowing at 6 miles per hour toward the south. Find a vector $\mathbf{v}$ that models the direction and speed of the boat. What does $\|\mathbf{v}\|$ represent?

49. *Solar Panels* Suppose that the sun's intensity in watts per square meter and direction are given by vector $\mathbf{I}$, and a solar panel's area in square meters and orientation are given by vector $\mathbf{S}$. Then the total number of watts W that can be absorbed by the panel is given by $W = |\mathbf{I} \cdot \mathbf{S}|$. Suppose $\mathbf{I} = 40\mathbf{i} - 180\mathbf{j}$ and $\mathbf{S} = 2\mathbf{i} + 5\mathbf{j}$.
    (a) Find $\|\mathbf{I}\|$ and $\|\mathbf{S}\|$. Interpret your results.

    (b) Calculate W and interpret the results.

50. *Robotics*   Consider the planar two-arm manipulator shown in the accompanying figure, where units are in centimeters. Let the upper arm be modeled by the vector **a** = 40**i** − 20**j** and the forearm be modeled by the vector **b** = 20**i** + 30**j**. (**Source:** J. Craig, *Introduction to Robotics*.)

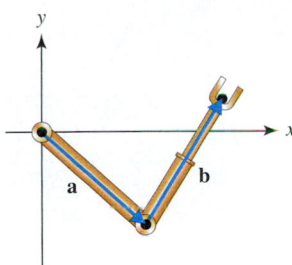

(a) Find a vector **c** that represents the position of the hand.

(b) How far is the hand from the origin?

(c) Find the position of the hand if the length of the forearm doubles.

51. *Work*   A 200-pound person walks 0.75 mile up a hiking trail inclined at 15°. Use a dot product to compute the work done in foot-pounds.

52. *Flight of a Golf Ball*   A golf ball is hit at 50 feet per second making an angle of 45° with the horizontal as it leaves the club. If the ground is level, estimate the horizontal distance traveled by the golf ball in the air.

53. *Aerial Photography*   The distance covered by an aerial photograph is determined by both the focal length of the camera lens and the tilt of the camera from the perpendicular to the ground. Although the tilt is usually small, archaeological and Canadian aerial photographs often use larger tilts. A camera lens with a 12-inch focal length will have an angular coverage of 60°. If an aerial photograph is taken with this camera tilted $\theta = 35°$ at an altitude of 5000 feet, calculate the ground distance d that will be shown in this photograph to the nearest foot. See the accompanying figure. (**Sources:** R. Brooks and J. Dieter, *Phytoarchaeology*; F. Moffitt, *Photogrammetry*.)

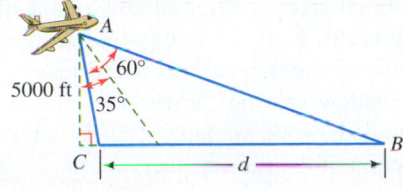

54. *Aerial Photography*   A camera lens with a 6-inch focal length has an angular coverage of 86°. Suppose an aerial photograph is taken vertically with no tilt at an altitude of 3500 feet over ground with an increasing slope of 5°, as shown in the figure. Calculate the ground distance CB that would appear in the resulting photograph to the nearest foot. (**Source:** F. Moffitt.)

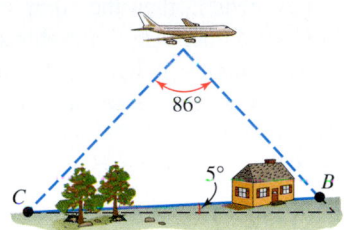

EXTENDED AND DISCOVERY EXERCISES

1. *Velocity of a Star*   The velocity vector **v** of a star relative to the sun can be expressed as the resultant vector of two perpendicular vectors—the radial velocity $\mathbf{v}_r$ and the tangential velocity $\mathbf{v}_t$, where $\mathbf{v} = \mathbf{v}_r + \mathbf{v}_t$, as illustrated in the figure. If a star is located near the sun and its velocity is large, then its motion across the sky will also be large. Barnard's Star is relatively close to the sun with a distance of 35 trillion miles. It moves across the sky through an angle of 10.34″ per year, which is the largest of any known star. Its radial velocity is $\mathbf{v}_r = 67$ miles per second toward the sun. (**Sources:** A. Acker and C. Jaschek, *Astronomical Methods and Calculations*; M. Zeilik, *Introductory Astronomy and Astrophysics*.)

(a) Approximate $\|\mathbf{v}_t\|$ for Barnard's Star in miles per second. (*Hint:* Use $s = r\theta$.)

(b) Compute $\|\mathbf{v}\|$.

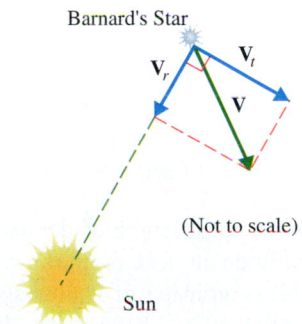

2. *Fractals* The fractal called the *Julia set* is shown in the figure. To determine if a complex number $z = a + bi$ belongs to this set, repeatedly compute the sequence of values

$$z_1 = z^2 - 1, \quad z_2 = z_1^2 - 1, \quad z_3 = z_2^2 - 1,$$

and so on. If the modulus of any of the resulting complex numbers exceeds 2, then the complex number z is not in the Julia set. Otherwise z is in this set. Determine if the complex numbers belong to the Julia set. (**Source:** R. Crownover, *Introduction to Fractals and Chaos. Reprinted with permission.*)

(a) $z = 0 + 0i$ **(b)** $z = 1 + i$

(c) $z = -0.2i$

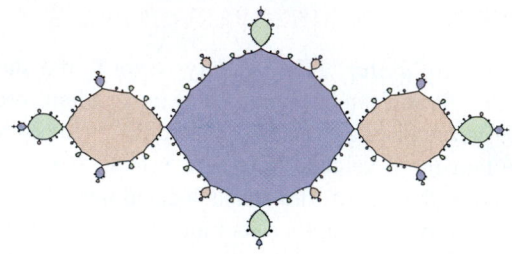

3. *Aerial Photography* Aerial photography from satellites and planes has become important to many applications such as map-making, national security, and surveying. If a photograph is taken from a plane so that the camera is tilted at an angle θ, then trigonometry can be used to find the ground coordinates of an object, as illustrated in the figure in the next column. If an object's photographic coordinates in inches are (x, y), then its ground coordinates (X, Y) in feet can be computed using the formulas

$$X = \frac{ax}{f \sec \theta - y \sin \theta},$$

$$Y = \frac{ay \cos \theta}{f \sec \theta - y \sin \theta},$$

where f is the focal length of the camera in inches and a is the altitude in feet of the airplane. Suppose the photographic coordinates of a house and nearby forest fire are $(x_H, y_H) = (0.9, 3.5)$ and $(x_F, y_F) = (2.1, -2.4)$, respectively. (**Source:** F. Moffitt, *Photogrammetry.*)

(a) Find the distance between the house and the fire on the photograph to the nearest hundredth of an inch.

(b) If the photograph was taken at 7400 feet by a camera with a focal length of 6 inches and a tilt of

$\theta = 4.1°$, find the ground distance in feet between the house and the fire.

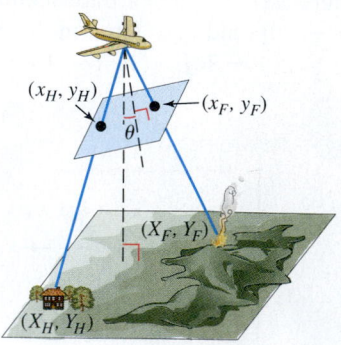

4. *Aerial Photographs and Surveying* To locate important points in aerial photographs, surveyors use basic control monuments established by both the U.S. Coast and Geodetic Survey and the U.S. Geological Survey. These monuments have published *xy*-coordinates called *state plane coordinates*. Using these monuments, coordinates of other important points can be determined. Two basic control monuments A and B have *xy*-coordinates of $x_A = 2,101,345.1$, $y_A = 998,764.3$ and $x_B = 2,131,667.8$, $y_B = 923,541.7$, respectively, where units are in feet. The coordinates of an unknown point P are needed. If angles PAB and PBA are measured to be $37°41'37''$ and $57°52'04''$, respectively, determine the state plane coordinates of P. See the accompanying figure. (**Source:** F. Moffitt.)

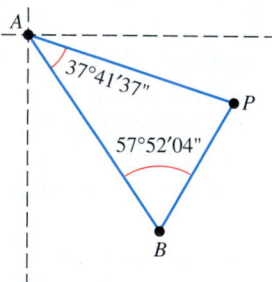

5. *Shadows in Computer Graphics* Vectors are used frequently in computer graphics to simulate realistic shadows. For example, suppose an airplane is taking off from a runway, as illustrated in the figure on the next page. Let the length and direction of an airplane at takeoff be given by vector $\mathbf{L}$. If the sunlight is assumed to be perpendicular to the runway, then the length of the airplane's shadow cast on the runway equals $\|\mathbf{L}\| \cos \theta$. From previous work, we know that if vector $\mathbf{R}$ points in the direction of the runway, then

$$\mathbf{L} \cdot \mathbf{R} = \|\mathbf{L}\| \, \|\mathbf{R}\| \cos \theta.$$

Solving for $\|\mathbf{L}\|\cos\theta$ results in

$$\|\mathbf{L}\|\cos\theta = \frac{\mathbf{L}\cdot\mathbf{R}}{\|\mathbf{R}\|}.$$

The expression $\dfrac{\mathbf{L}\cdot\mathbf{R}}{\|\mathbf{R}\|}$ represents the **component of L in the direction of R**. Find the length of the shadow on the runway for each **L** and **R**. Assume units are in feet. (**Source:** C. Pokorny and C. Gerald, *Computer Graphics*.)

(a) $\mathbf{L} = 40\mathbf{i} + 10\mathbf{j}$, $\mathbf{R} = \mathbf{i}$

(b) $\mathbf{L} = 35\mathbf{i} + 5\mathbf{j}$, $\mathbf{R} = 10\mathbf{i} + \mathbf{j}$

(c) $\mathbf{L} = 100\mathbf{i} + 8\mathbf{j}$, $\mathbf{R} = 30\mathbf{i} + 2\mathbf{j}$

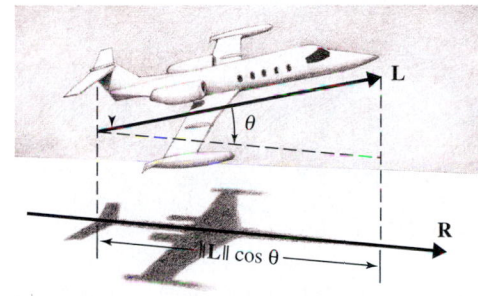

6. *The Dot Product*   In the accompanying figure

$$\mathbf{a} = \langle a_1, a_2\rangle, \qquad \mathbf{b} = \langle b_1, b_2\rangle, \qquad \text{and}$$
$$\mathbf{a} - \mathbf{b} = \langle a_1 - b_1, a_2 - b_2\rangle.$$

Apply the law of cosines to the triangle and derive the equation

$$\mathbf{a}\cdot\mathbf{b} = \|\mathbf{a}\|\,\|\mathbf{b}\|\cos\theta.$$

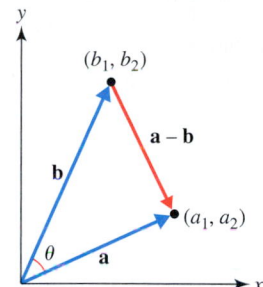

CHAPTERS 1–8 CUMULATIVE REVIEW EXERCISES

1. Write 91,200 in scientific notation and 6.734×10^{-3} in standard notation.

2. Evaluate $\sqrt{\sin^2(1.2) + 5}$. Round your answer to three decimal places.

3. Find the exact distance between $(3, -2)$ and $(7, -9)$.

4. Graph $y = g(x)$ by hand.

   (a) $g(x) = \frac{1}{2}x - 1$ (b) $g(x) = |x + 2|$

   (c) $g(x) = \frac{1}{2}(x + 1)^2 - 2$

   (d) $g(x) = \sqrt{x + 1}$ (e) $g(x) = \dfrac{1}{x + 1}$

   (f) $g(x) = x^3 - 4x$ (g) $g(x) = \sin x$

   (h) $g(x) = -3\cos\left(2\left(x - \frac{\pi}{2}\right)\right)$

   (i) $g(x) = \tan\frac{1}{2}x$ (j) $g(x) = \sec x$

5. Determine the domain of $f(x) = \sqrt{4 - x}$ and evaluate $f(-5)$.

6. The graph of a quadratic function f is shown.
   (a) Identify the vertex and axis of symmetry.

   (b) Identify the x-intercepts.

   (c) Find a formula for $f(x)$.

   (d) Is the leading coefficient positive or negative?

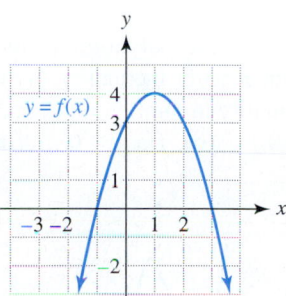

7. Find the average rate of change of $f(x) = 3x^2 - 2x$ from $x = 1$ to $x = 3$.

8. Find the difference quotient for $f(x) = 4x^2$.

9. Write the slope-intercept form for a line that passes through $(-1, 4)$ and $(1, -3)$.

10. Determine the x- and y-intercepts on the graph of $4x + 3y = -12$. Graph the equation.

11. If $G(t) = 300 - 10t$ calculates the gallons of water in a tank after t seconds, interpret the numbers 300 and -10 in the formula for $G(t)$.

12. Solve each equation.
    (a) $-2(1 - 2x) = -(5 - x) + 1$

    (b) $|2x - 5| = 6$ (c) $6x^2 + 22x = 8$

    (d) $x^3 = x$ (e) $x^4 - 2x^2 - 3 = 0$

    (f) $\dfrac{1}{2x} - \dfrac{2}{x^2} = -7$ (g) $3x^{2/3} = 12$

    (h) $2e^{3x} - 1 = 50$ (i) $\ln x - \ln(x - 1) = \ln 2$

    (j) $\sin t = \frac{1}{2}, 0 \le t < 2\pi$

    (k) $2\cos^2 t + \cos t = 1$

    (l) $\tan 2t = -\sqrt{3}$

13. Evaluate $f(-1)$ and graph $y = f(x)$. Is f continuous on its domain?
$$f(x) = \begin{cases} x + 2 & \text{if } -3 \le x \le -1 \\ -2x - 1 & \text{if } -1 < x < 1 \\ x^2 - 4 & \text{if } 1 \le x \le 3 \end{cases}$$

14. Solve each inequality. Use interval notation.
    (a) $4(x - 3) > 1 - x$ (b) $|2x - 1| \le 3$

    (c) $x^2 - 2x - 3 > 0$ (d) $x^3 - 4x > 0$

    (e) $\dfrac{x}{x - 1} \le 0$ (f) $-4 \le 4 - 3x \le 12$

15. The graph of a cubic polynomial function f is shown. Solve each equation or inequality. Write the solution set for each inequality in interval notation.
    (a) $f(x) = 0$ (b) $f(x) > 0$ (c) $f(x) \le 0$

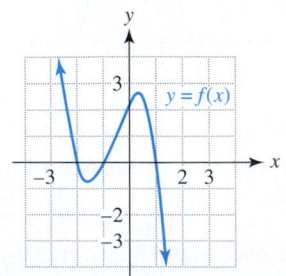

16. Write $f(x) = \frac{1}{2}x^2 + 2x - 1$ in the form given by $f(x) = a(x - h)^2 + k$.

17. Find the vertex on the graph of $f(x) = 2x^2 - 4x + 1$.

18. A graph of a quadratic function $f(x) = ax^2 + bx + c$ is shown.
    (a) State whether $a > 0$ or $a < 0$.

    (b) Estimate the real solutions to $ax^2 + bx + c = 0$.

    (c) Determine if the discriminant, $b^2 - 4ac$, is positive, negative, or zero.

    (d) Write the complete factored form of $f(x)$.

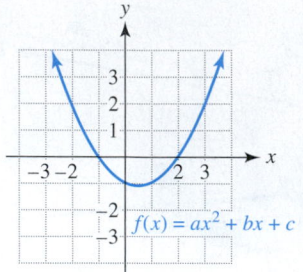

19. Let $f(x) = x^2 + 2x - 3$. If the graph of $y = g(x)$ is similar to the graph of $y = f(x)$, except that it is shifted 2 units left and 3 units down, write a formula for $g(x)$. Simplify your formula.

20. Where is the graph of $f(x) = -2(x + 3)^2 - 4$ decreasing? Use interval notation.

21. Use the graph of f to estimate each of the following.
    (a) Where f is increasing or decreasing

    (b) The zeros of f

    (c) The coordinates of any turning points

    (d) Any local extrema

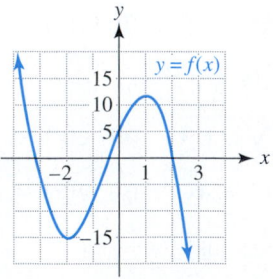

22. Determine if the function defined by the formula $f(x) = x^3 - 2x + 1$ is even, odd, or neither.

23. Predict the end behavior of $f(x) = 4 + 3x + x^2 - 4x^4$.

24. Divide each expression.

(a) $\dfrac{5a^3 - 10a^2 + 20}{2a}$ **(b)** $\dfrac{3x^3 - x + 2}{x + 2}$

(c) $\dfrac{2x^3 - 3x^2 + x - 1}{2x - 1}$

25. A cubic function f has zeros $-2, 3$, and 5 with leading coefficient 4. Write the complete factored form of $f(x)$.

26. A degree 7 function f has zeros $-2, 2$, and 6 with multiplicities $2, 3$, and 2, respectively. If the leading coefficient is -3, write the complete factored form of $f(x)$.

27. A quartic (degree 4) function f with real coefficients has leading coefficient 6 and zeros $1, -1$, and i. Write $f(x)$ in complete factored form and expanded form.

28. Use the graph to write the complete factored form of the cubic polynomial $f(x)$.

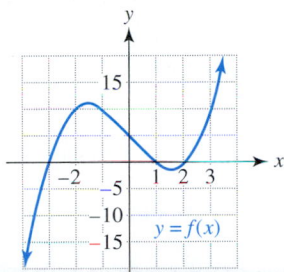

29. Write $(2 + 3i)(4 - 2i)$ in standard form.

30. Find all solutions, real or imaginary, to $x^4 - 4 = 0$.

31. State the domain of $f(x) = \dfrac{3x + 4}{2 - 3x}$. Find any vertical or horizontal asymptotes on the graph of $y = f(x)$.

32. Use the tables for f and g to evaluate each expression.

x	0	1	2	3	4
$f(x)$	1	2	3	4	5

x	0	1	2	3	4
$g(x)$	4	1	2	3	0

(a) $(f - g)(2)$ **(b)** $(gf)(4)$

(c) $(g \circ f)(3)$ **(d)** $(g \circ f^{-1})(1)$

33. Let $f(x) = \dfrac{1}{x^2 - 3}$ and $g(x) = 2x + 1$. Find each of the following.

(a) $(f + g)(2)$ **(b)** $(g \circ f)(2)$

(c) $(g/f)(x)$ **(d)** $(f \circ g)(x)$

34. Find $f^{-1}(x)$ if $f(x) = 3x - 2$.

35. Find an exponential function given by $f(x) = Ca^x$ that models the data in the table.

x	0	1	2	3
$f(x)$	2	6	18	54

36. There are initially 5000 bacteria in a sample and this number doubles in size every 1.5 hours. Find C and a so that $f(t) = Ca^t$ models the number of bacteria after t hours.

37. One thousand dollars are deposited in an account that pays 7% annual interest compounded quarterly. Find the amount in the account after 8 years.

38. Simplify each logarithm by hand.
(a) $\log 1000$ **(b)** $\log_3 \dfrac{1}{27}$

(c) $\ln \dfrac{1}{e^3}$ **(d)** $\log \sqrt[3]{10}$

(e) $\log 4 + \log 25$ **(f)** $\log_4 32 - \log_4 \dfrac{1}{2}$

39. Expand the expression $\ln \sqrt[3]{\dfrac{x^3 y}{z^2}}$.

40. Write $2 \log_2 x + 3 \log_2 y + \dfrac{1}{2}\log_2 z$ as a logarithm of a single expression.

41. Find the complementary angle α and the supplementary angle β to $22°55'$.

42. Covert $75.25°$ to degrees and minutes.

43. Covert $225°$ to radians.

44. Convert $\dfrac{11\pi}{6}$ radians to degrees.

45. Find the six trigonometric functions of θ.

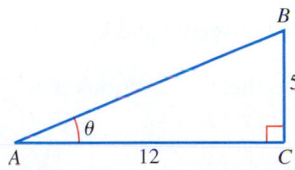

46. Approximate $\cos \dfrac{\pi}{7}$ to the nearest hundredth.

47. Find the values of the six trigonometric functions of an angle θ in standard position with its terminal side passing through the point $(11, -60)$.

48. Find the exact values of the six trigonometric functions of $\theta = \dfrac{2\pi}{3}$.

49. Find the values of the six trigonometric functions of θ if $\sin \theta = -\frac{7}{25}$ and $\sec \theta < 0$.

50. Evaluate $\tan^{-1}(\sqrt{3})$.

51. Solve the right triangle shown in Exercise 45.

52. Simplify $(1 - \cos t)(1 + \cos t)$.

53. Factor $\cot^2 \theta - 2 \cot \theta + 1$.

54. Verify each identity.
(a) $(1 - \cos^2 \theta)(1 + \tan^2 \theta) = \tan^2 \theta$

(b) $\dfrac{\sin(\alpha + \beta)}{\cos \alpha \cos \beta} = \tan \alpha + \tan \beta$

(c) $\dfrac{\csc^2 \theta}{1 - \cot^2 \theta} = -\sec 2\theta$

55. Find $\cos(\alpha - \beta)$ if $\sin \alpha = \frac{5}{13}$, $\cos \beta = -\frac{4}{5}$, and α and β are in quadrant II.

56. Solve $3 \tan \frac{t}{2} = \sin 2t$ graphically to the nearest hundredth on the interval $[0, 2\pi)$.

57. Solve triangle ABC. Approximate values to the nearest tenth.
(a) $\alpha = 31°, \gamma = 53°, b = 15$

(b) $\alpha = 31°, a = 6, b = 5$

(c) $\beta = 56°, a = 6, c = 8$

(d) $a = 6, b = 7, c = 8$

58. Find the area of a triangle with sides of length 7, 10, and 15 feet.

59. Let $\mathbf{a} = \langle -5, 12 \rangle$ and $\mathbf{b} = \langle 7, -24 \rangle$. Find the following.
(a) $\|\mathbf{b}\|$ (b) $2\mathbf{a} - 3\mathbf{b}$ (c) $\mathbf{a} \cdot \mathbf{b}$

(d) The angle between $\mathbf{a}$ and $\mathbf{b}$

60. Graph the parametric equations $x = \frac{1}{2}t, y = (t - 1)^2$ for any real number t.

61. Graph the polar equation $r = 3 - 2 \sin \theta$ for θ satisfying $0° \leq \theta < 360°$.

62. Find the three cube roots of $27i$.

Applications

63. *Distance*   The graph shows the distance in miles that a jogger on a straight highway is from home after x hours. Find the slope of each line segment and interpret each slope.

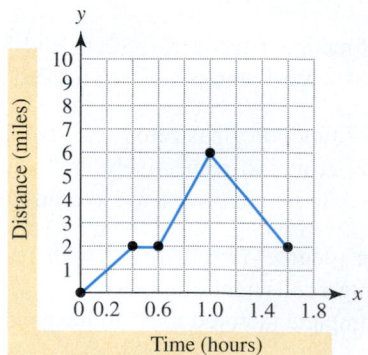

64. *Average Rate of Change*   The total distance D in feet traveled by a race car after t seconds is given by $D(t) = 12t^2$ for $0 \leq t \leq 7$.
(a) Find the average rate of change of D from 1 to 3 and 4 to 6.

(b) Interpret these average rates of change.

65. *Mixing Candy*   A 5-pound mixture of two types of candy costs $21.25. The cost of each type of candy is $3.50 per pound and $4.50 per pound. How much of each type of candy is in the mixture?

66. *Construction*   A box is being constructed by cutting 3-inch squares from the corners of a rectangular sheet of metal that is 4 inches longer than it is wide. If the box is to have a volume of 351 cubic inches, find the dimensions of the metal sheet.

67. *Modeling Data*   Find a quadratic function in the form $f(x) = a(x - h)^2 + k$ that models the data in the table *exactly*.

x	-1	0	1	2	3	4
y	-26	-11	-2	1	-2	-11

68. *Designing a Box*   A box with rectangular sides and a top is being designed to hold 288 cubic inches and to have a surface area of 288 square inches. If the width is half the length, find possible dimensions for the box.

69. *Dimensions of an Aluminum Can*   The volume V of an aluminum can is given by $V = \pi r^2 h$, where r is the radius and h is the height. If an aluminum can has a volume of 20 cubic inches and a diameter of 3.44 inches, find the height of the can to the nearest hundredth of an inch.

70. *Inverse Variation*   The force of gravity F varies inversely as the square of the distance r from the *center* of the moon. If a rock weighs 50 pounds on the surface of the moon ($r = 1750$ kilometers), how much would this rock weigh at a distance of 7000 kilometers from the center of the moon?

71. *College Tuition*   In 1980 the annual cost of tuition and fees at private colleges and universities was about $3600 and in 2000 it was about $16,200. (**Source:** The College Board.)

(a) Determine a formula in the form

$$C(t) = m(t - t_1) + y_1$$

that models these data. Let C be in dollars and t be the year.

(b) Estimate the year when tuition and fees reached $11,000 and compare it to the actual year of 1993.

72. *Inverse Function*   The function given by $f(x) = 36x$ converts x yards to inches.
(a) Find $f^{-1}(x)$.

(b) What does f^{-1} compute?

73. *Bicycle*   A bicycle tire is 2 feet in diameter and rotating at 25 radians per second. Approximate the speed of the bicycle in feet per second.

74. *Length of a Shadow*   The angle of elevation of the sun is 63°. Find the length of the shadow cast by a person who is 5 feet tall. Round your answer to the nearest tenth of a foot.

75. *Modeling Temperature*   The monthly average high temperatures for a location are shown in the table at the top of the next column. Model these data by using a function of the form

$$f(x) = a \sin (b(x - c)) + d.$$

Month	1	2	3	4	5	6
Temperature (°F)	25	28	37	50	63	72

Month	7	8	9	10	11	12
Temperature (°F)	75	72	62	50	38	28

76. *Angle of Elevation*   An 85-foot tree casts a 57-foot shadow. Estimate the angle of elevation of the sun to the nearest tenth of a degree.

77. *Average Temperatures*   The monthly average temperatures in degrees Fahrenheit at Austin, Texas, are given by

$$T(x) = 17.5 \sin\left(\frac{\pi}{6}(x - 4)\right) + 67.5,$$

where x is the month and $x = 1$ corresponds to January. Estimate when the monthly average temperature is 76°F.

78. *Distance*   An ore ship is traveling east at 20 miles per hour. The bearing of a submerged rock is 75°. After 2 hours the bearing of the rock is 305°. Find the distance between the ship and the rock when the second bearing is determined.

79. *Surveyor*   A surveyor measures two sides of a triangular lot to be $a = 242$ feet and $b = 165$ feet. The angle between these sides is $\gamma = 72°$.
(a) Find the length of the third side c.

(b) Estimate the area of the lot.

80. *Work*   A wagon is pulled 1000 feet using a force of 15 pounds applied to the handle, which makes a 35° angle with the horizontal. Use a dot product to calculate the work done.

81. *Flight of a Golf Ball*   A golf ball is hit into the air at 96 feet per second making an angle of 60° with the horizontal. Use parametric equations to estimate the horizontal distance traveled by the golf ball before it strikes the ground.

9 Systems of Equations and Inequalities

The future belongs to those who believe in the beauty of their dreams.

Eleanor Roosevelt

One of the most important inventions of the twentieth century was the electronic digital computer. In 1940 John Atanasoff, a physicist at Iowa State University, needed to find the solution to a system of 29 equations. Since this task was too difficult to solve by hand, it led him to invent the first fully electronic digital computer. Today modern supercomputers compute trillions of arithmetic operations in a single second and solve over 600,000 equations simultaneously. The need to solve a *mathematical problem* resulted in one of the most profound inventions of our time.

A recent breakthrough in visualizing our world is the invention of digital photography. By the year 2006, TV stations may be required by the Federal Communications Commission (FCC) to transmit digital signals. Digital pictures are represented by numbers rather than film. They are crystal clear without interference. One of the first digital pictures was done by James Blinn of NASA in 1981. It was a simulation of the Voyager-Saturn flyby and required substantial mathematics to create.

Mathematics plays a central role in new technology. In this chapter mathematics is used to solve systems of equations, represent digital pictures, compute movement in computer graphics, and even represent color on computer monitors. Throughout history, many important discoveries and inventions have been based on the creative insights of a few people. These insights have had a profound impact on our society.

Sources: A. Tucker, *Fundamentals of Computing I; USA Today; NASA/JPL.*

9.1 Functions and Equations in Two Variables

- ◆ Evaluate functions of two variables
- ◆ Understand basic concepts about systems of equations in two variables
- ◆ Apply the method of substitution
- ◆ Apply graphical and numerical methods to systems of equations
- ◆ Solve problems involving joint variation

Introduction

Many quantities in everyday life depend on more than one variable. Calculating the area of a rectangular room requires both its length and width. The heat index is a function of temperature and humidity, and windchill is determined using temperature and wind speed. Information about grades and credit hours is necessary to compute a grade point average.

In earlier chapters we discussed functions of one input or one variable. Quantities determined by more than one variable often are computed by a function of more than one variable. The mathematical concepts concerning functions of one input also apply to functions of more than one input. One unifying concept about all functions is that each produces *at most one output* each time it is evaluated.

Functions of Two Variables

The arithmetic operations of addition, subtraction, multiplication, and division are computed by functions of two inputs. In order to perform addition, two numbers must be provided. The addition of x and y results in one output z. The addition function f can be represented symbolically by $f(x, y) = x + y$, where $z = f(x, y)$. For example, the addition of 3 and 4 can be written as $z = f(3, 4) = 3 + 4 = 7$. In this case, $f(x, y)$ is a **function of two inputs** or a **function of two variables**. The **independent variables** are x and y, and z is the **dependent variable**. The output z depends on the inputs x and y. In a similar manner, a division function can be defined by $g(x, y) = \frac{x}{y}$, where $z = g(x, y)$.

◆ **MAKING CONNECTIONS** ─────────────────────

Independent and Dependent Variables In Chapters 1 through 5 the input for a function f was usually represented by x and the output by y. This was expressed as $y = f(x)$. For functions of two inputs, it is common to use x and y as inputs and z as the output. This is expressed as $z = f(x, y)$.

EXAMPLE 1 Evaluating functions of more than one input

For each function, evaluate the expression and interpret the result.
(a) $f(3, -4)$, where $f(x, y) = xy$ represents the multiplication function
(b) $M(120, 5)$, where $M(m, g) = \frac{m}{g}$ computes the gas mileage when traveling m miles on g gallons of gasoline
(c) $V(0.5, 2)$, where $V(r, h) = \pi r^2 h$ calculates the volume of a cylindrical barrel with a radius r feet and height h feet. (See Figure 9.1.)

FIGURE 9.1

SOLUTION
(a) $f(3, -4) = (3)(-4) = -12$. The product of 3 and -4 is -12.
(b) $M(120, 5) = \frac{120}{5} = 24$. If a car travels 120 miles on 5 gallons of gasoline, its mileage is 24 miles per gallon.
(c) $V(0.5, 2) = \pi(0.5)^2(2) = 0.5\pi \approx 1.57$. If a barrel has a radius of 0.5 foot and a height of 2 feet, it holds about 1.57 cubic feet of liquid. *Now Try Exercise 1* ◆

The surface area of the skin covering the human body is a function of more than one variable. A taller person tends to have a larger surface area, as does a heavier person. Both height and weight influence the surface area of a person's body. The following example demonstrates a function that can be used to estimate this surface area.

EXAMPLE 2 Estimating the surface area of the human body

A formula to determine the surface area of a person's body in square meters is computed by $S(w, h) = 0.007184(w)^{0.425}(h)^{0.725}$, where w is weight in kilograms and h is height in centimeters. Use S to estimate the surface area of a human body that is 65 inches tall and weighs 154 pounds. (**Source:** H. Lancaster, *Quantitative Methods in Biological and Medical Sciences.*)

SOLUTION First, convert 65 inches to centimeters. There are approximately 2.54 centimeters in 1 inch, so this person's height is $65 \cdot 2.54 = 165.1$ centimeters. Next, convert pounds to kilograms. There are about 2.2 pounds in 1 kilogram, so the individual weighs $\frac{154}{2.2} = 70$ kilograms.

$$S(\mathbf{70}, \mathbf{165.1}) = 0.007184(\mathbf{70})^{0.425}(\mathbf{165.1})^{0.725} \approx 1.77$$

Thus a person who weighs 70 kilograms (154 pounds) and is 165.1 centimeters (65 inches) tall has a surface area of approximately 1.77 square meters. *Now Try Exercise 81* ◆

We often use formulas with more than one variable. Sometimes it is necessary to solve an equation for a variable. For example, to find the radius r of a circle that has an area A of 50 square inches, we might first solve the equation $A = \pi r^2$ for r.

$$A = \pi r^2 \qquad \text{Area of a circle}$$

$$\frac{A}{\pi} = r^2 \qquad \text{Divide by } \pi.$$

$$\pm\sqrt{\frac{A}{\pi}} = r \qquad \text{Square root property}$$

$$r = \sqrt{\frac{A}{\pi}} \qquad \text{For this problem, } r > 0.$$

If a circle has an area of 50 square inches, then its radius is

$$r = \sqrt{\frac{50}{\pi}} \approx 3.99 \text{ inches.}$$

EXAMPLE 3 Solving for a variable

The equation $P = 2L + 2W$ calculates the perimeter of a rectangle with length L and width W, as illustrated in Figure 9.2.
(a) Solve $P = 2L + 2W$ for L.
(b) Find L for a rectangle with $P = 21$ feet and $W = 3.5$ feet.

SOLUTION

(a)
$$P = 2\mathbf{L} + 2W \qquad \text{Given formula}$$
$$P - 2W = 2\mathbf{L} \qquad \text{Subtract } 2W.$$
$$\frac{P - 2W}{2} = L \qquad \text{Divide by 2.}$$

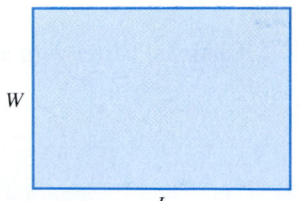

W

L

FIGURE 9.2

Given the perimeter P and the width W, the length L can be computed with this formula.

(b) $L = \dfrac{P - 2W}{2} = \dfrac{21 - 2(3.5)}{2} = 7$ feet.   *Now Try Exercises 13 and 19* ◆

Systems of Equations

A **linear equation in two variables** can be written in the form

$$ax + by = k,$$

where a, b, and k are constants, and a and b are not equal to 0. Examples of linear equations in two variables include

$$2x - 3y = 4, \qquad -x - 5y = 0, \qquad \text{and} \qquad 5x - y = 10.$$

The average of two numbers, x and y, can be found with the formula $\frac{x+y}{2}$. To find two numbers whose average is 10, we solve the linear equation in two variables

$$\frac{x + y}{2} = 10.$$

(Note that $\frac{x+y}{2} = 10$ is a linear equation because it can be written as $\frac{1}{2}x + \frac{1}{2}y = 10$.) The numbers 5 and 15 average to 10, as do the numbers 0 and 20. In fact, any pair of numbers whose sum is 20 will satisfy this equation. There are *infinitely many solutions* to this linear equation. For a unique solution, another restriction must be placed on the variables x and y.

Many situations involving two variables result in the need to determine values for x and y that satisfy two equations. For example, suppose that we would like to find a pair of numbers whose average is 10 and whose difference is 2. The function $f(x, y) = \frac{x+y}{2}$ calculates the average of two numbers and $g(x, y) = x - y$ computes their difference. The solution could be found by solving two linear equations.

$$f(x, y) = 10$$
$$g(x, y) = 2$$

These equations can be written as follows.

$$\frac{x + y}{2} = 10$$
$$x - y = 2$$

This pair of equations is called a **system of linear equations** because we are solving more than one linear equation at once. A **solution** to a system of equations consists of an x-value and a y-value that satisfy *both* equations simultaneously. The set of all solutions is called the **solution set**. Using trial and error, we see that $x = 11$ and $y = 9$ satisfy both equations. This is the only solution and can be expressed as the *ordered pair* $(11, 9)$.

 **EXAMPLE 4** Determining a system of equations

In 2003 gross sales totaled $60.9 billion at McDonald's and Burger King. McDonald's sales exceeded Burger King's by $22.1 billion.   (**Source:** Technomic.)

(a) Write a system of equations involving two variables, whose solution gives the individual sales of each company.

(b) Is the resulting system of equations linear or nonlinear?

SOLUTION

(a) When setting up a system of equations, it is essential that we identify what each variable represents. Let x represent the gross sales at McDonald's and y the gross sales at

Burger King. Their combined sales are $60.9 billion, so $x + y = 60.9$. McDonald's sales exceed Burger King's by $22.1 billion, so $x - y = 22.1$. Thus the system of equations is as follows.

$$x + y = 60.9$$
$$x - y = 22.1$$

(b) *Both* equations are written in the form $ax + by = k$. Therefore it is a system of linear equations.

Now Try Exercise 85(a) ◆

Systems of equations that are not linear are called **systems of nonlinear equations**. The following are two examples of systems of nonlinear equations.

$$x^2 - 5y = 8 \qquad\qquad x^2 + y^2 = 5$$
$$2x + 3y = 3 \qquad\qquad \sqrt{x} - y = -2$$

The Method of Substitution

The next example describes the **method of substitution**. This procedure is often used to solve systems of equations involving two variables.

 **EXAMPLE 5   Solving a linear system of equations**

Use the method of substitution to solve the system of equations in Example 4. Interpret the result, and check the solution.

SOLUTION   In this system x represents 2003 gross sales at McDonald's in billions of dollars, and y represents the corresponding sales at Burger King.

$$x + y = 60.9$$
$$x - y = 22.1$$

Begin by solving an equation for a convenient variable. Then substitute the result into the other equation. For example, if we solve the first equation for y, the result is $y = 60.9 - x$. Then, substitute $(\mathbf{60.9 - x})$ for y into the second equation.

$$x - (\mathbf{60.9 - x}) = 22.1$$

Algebra Review

To review how to solve a linear equation, see Section 2.3.

This equation can be solved for x.

$$x - (60.9 - x) = 22.1$$
$$x - 60.9 + x = 22.1 \qquad \text{Distributive property}$$
$$2x = 83 \qquad \text{Combine x-terms and add 60.9.}$$
$$x = 41.5 \qquad \text{Divide by 2.}$$

Thus McDonald's gross sales reached $41.5 billion in 2003. Since $y = 60.9 - x$, it follows that $y = 60.9 - 41.5 = 19.4$. Burger King's share of the market was $19.4 billion.

To check if $(41.5, 19.4)$ is a solution, substitute $x = 41.5$ and $y = 19.4$ into both equations.

$$41.5 + 19.4 \stackrel{?}{=} 60.9 \qquad \text{True}$$
$$41.5 - 19.4 \stackrel{?}{=} 22.1 \qquad \text{True}$$

Since *both* equations are satisfied, $(41.5, 19.4)$ is a solution.

Now Try Exercise 85(b) ◆

The method of substitution is summarized verbally in the following.

THE METHOD OF SUBSTITUTION

To use the method of substitution to solve a system of two equations in two variables, perform the following steps.

STEP 1: Choose a variable in one of the two equations. Solve the equation for that variable.

STEP 2: Substitute the result from **STEP 1** into the other equation and solve for the remaining variable.

STEP 3: Use the value of the variable from **STEP 2** to determine the value of the other variable. To do this, you may want to use the equation you found in **STEP 1**.

Note: To check your answer, substitute the value of each variable into the *given* equations. These values should satisfy *both* equations.

EXAMPLE 6   Using the method of substitution

Solve the system symbolically. Check your answer.

$$5x - 2y = -16$$
$$x + 4y = \;\;-1$$

SOLUTION

STEP 1: Begin by solving one of the equations for one of the variables. One possibility is to solve the second equation for x

$$x + 4y = -1 \qquad \text{Second equation}$$
$$x = -4y - 1 \quad \text{Subtract 4y from each side.}$$

STEP 2: Next, substitute $(-4y - 1)$ for x in the first equation and solve the resulting equation for y.

$$5x - 2y = -16 \qquad \text{First equation}$$
$$5(-4y - 1) - 2y = -16 \qquad \text{Let x = -4y - 1.}$$
$$-20y - 5 - 2y = -16 \qquad \text{Distributive property}$$
$$-5 - 22y = -16 \qquad \text{Combine like terms.}$$
$$-22y = -11 \qquad \text{Add 5 to each side.}$$
$$y = \frac{1}{2} \qquad \text{Divide each side by -22.}$$

STEP 3: Now find the value of x by using the equation $x = -4y - 1$ from **STEP 1**. Since $y = \frac{1}{2}$, it follows that $x = -4\left(\frac{1}{2}\right) - 1 = -3$. The solution can be written as an ordered pair: $\left(-3, \frac{1}{2}\right)$.

To check this answer substitute $x = -3$ and $y = \frac{1}{2}$ in both equations.

$$5(-3) - 2\left(\frac{1}{2}\right) \overset{?}{=} -16 \qquad \text{True}$$

$$-3 + 4\left(\frac{1}{2}\right) \overset{?}{=} -1 \qquad \text{True}$$

Both equations are satisfied, so the solution is $\left(-3, \frac{1}{2}\right)$. *Now Try Exercise 31* ◆

The method of substitution can also be used to solve systems of nonlinear equations. In the next example we solve a system of nonlinear equations with two solutions. In general, a system of nonlinear equations can have *any number of solutions*.

EXAMPLE 7 Solving a system of nonlinear equations

Solve the system symbolically.

$$6x + 2y = 10$$
$$2x^2 - 3y = 11$$

SOLUTION

STEP 1: Begin by solving one of the equations for one of the variables. One possibility is to solve the first equation for y.

$6x + 2y = 10$	First equation
$2y = 10 - 6x$	Subtract 6x from each side.
$y = 5 - 3x$	Divide each side by 2.

STEP 2: Next, substitute $(5 - 3x)$ for y in the second equation and solve the resulting quadratic equation for x.

$2x^2 - 3y = 11$	Second equation
$2x^2 - 3(5 - 3x) = 11$	Let y = 5 − 3x.
$2x^2 - 15 + 9x = 11$	Distributive property
$2x^2 + 9x - 26 = 0$	Subtract 11 and simplify.
$(2x + 13)(x - 2) = 0$	Factor.
$x = -\dfrac{13}{2}$ or $x = 2$	Zero-product property

Algebra Review
To review factoring, see
Chapter R (page R-29).

STEP 3: Now find the corresponding y-values for each x-value. From **STEP 1**, we know that $y = 5 - 3x$, so it follows that $y = 5 - 3\left(-\frac{13}{2}\right) = \frac{49}{2}$ or $y = 5 - 3(2) = -1$. Thus the solutions are $\left(-\frac{13}{2}, \frac{49}{2}\right)$ and $(2, -1)$.

Now Try Exercise 51 ◆

EXAMPLE 8 Solving a system of nonlinear equations

A circle with radius r, centered at the origin, has an equation $x^2 + y^2 = r^2$. Use the method of substitution to determine the points where the graph of $y = 2x$ intersects this circle when $r = \sqrt{5}$. Sketch a graph that illustrates the solutions.

SOLUTION Since $r^2 = 5$, solve the following system.

$$x^2 + y^2 = 5$$
$$y = 2x$$

Substitute $(2x)$ for y in the first equation.

Geometry Review
To review equations of circles,
see Chapter R (page R-2).

$x^2 + y^2 = 5$	First equation
$x^2 + (2x)^2 = 5$	y = 2x
$x^2 + 4x^2 = 5$	Square the expression.
$5x^2 = 5$	Add like terms.
$x^2 = 1$	Divide by 5.
$x = \pm 1$	Square root property

Since $y = 2x$ we see that when $x = 1$, $y = 2$ and when $x = -1$, $y = -2$. The graphs of $x^2 + y^2 = 5$ and $y = 2x$ intersect at the points $(1, 2)$ and $(-1, -2)$. This nonlinear system has two solutions, which are shown in Figure 9.3.

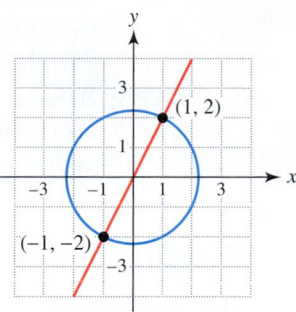

FIGURE 9.3

Now Try Exercise 67

EXAMPLE 9   Solving a system with zero or infinitely many solutions

If possible, solve each system of equations.

(a) $\begin{aligned} x^2 + y &= 1 \\ x^2 - y &= -2 \end{aligned}$ (b) $\begin{aligned} 2x - 4y &= 5 \\ -x + 2y &= -\tfrac{5}{2} \end{aligned}$

SOLUTION

(a) **STEP 1:** Solve the second equation for y, which gives $y = x^2 + 2$.

STEP 2: Substitute $(x^2 + 2)$ for y in the first equation and then solve for x, if possible.

$$x^2 + y = 1 \qquad \text{First equation}$$
$$x^2 + (x^2 + 2) = 1 \qquad \text{Let } y = x^2 + 2.$$
$$2x^2 + 2 = 1 \qquad \text{Combine like terms.}$$
$$2x^2 = -1 \qquad \text{Subtract 2.}$$

Because $2x^2 \geq 0$, it follows that there are no real solutions.

FIGURE 9.4

STEP 3: This step is not necessary because there are no real solutions for x. In Figure 9.4 the graphs of the two equations, which are parabolas, do not intersect. (To graph the given equations by hand, rewrite them as $y = 1 - x^2$ and $y = x^2 + 2$.)

(b) **STEP 1:** First solve the second equation for x to obtain $x = 2y + \tfrac{5}{2}$.

STEP 2: Substitute $\left(2y + \tfrac{5}{2}\right)$ for x in the first equation and then solve for y.

$$2x - 4y = 5 \qquad \text{First equation}$$
$$2\left(2y + \frac{5}{2}\right) - 4y = 5 \qquad \text{Let } x = 2y + \tfrac{5}{2}.$$
$$4y + 5 - 4y = 5 \qquad \text{Distributive property.}$$
$$5 = 5 \qquad \text{Combine like terms.}$$

The equation $5 = 5$ is an identity that is always true and indicates that there are infinitely many solutions. Note that we can multiply each side of the second equation by -2 to obtain the first equation.

$$-2(-x + 2y) = -2\left(-\frac{5}{2}\right) \qquad \text{Multiply second equation by } -2.$$
$$2x - 4y = 5 \qquad \text{Distributive property}$$

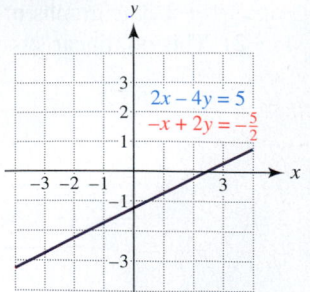

FIGURE 9.5

This fact indicates that the two equations are equivalent, and in Figure 9.5 their graphs appear to be identical. The solution set is $\{(x, y) \mid 2x - 4y = 5\}$. That is, any ordered pair (x, y) that satisfies the first equation also satisfies the second equation. For example $\left(0, -\frac{5}{4}\right)$, $\left(1, -\frac{3}{4}\right)$, and $\left(2, -\frac{1}{4}\right)$ are all solutions because they all satisfy the equation $2x - 4y = 5$. *Note:* Any point on the line in Figure 9.5 represents a solution.

STEP 3: Because there are infinitely many solutions, we (obviously) do not find the coordinates for each point. Rather, the solution set is simply written in set-builder notation. *Now Try Exercises 39 and 59* ◆

Graphical and Numerical Methods

The next example illustrates how a system with two variables can be solved graphically and numerically.

EXAMPLE 10   Modeling roof trusses

Linear systems occur in the design of roof trusses for homes and buildings. See Figure 9.6. One of the simplest types of roof trusses is an equilateral triangle. If a 200-pound force is applied to the peak of a truss, as shown in Figure 9.7, then the weights W_1 and W_2 exerted on each rafter of the truss are determined by the following system of linear equations. (**Source:** R. Hibbeler, *Structural Analysis*.)

$$W_1 - W_2 = 0$$

$$\frac{\sqrt{3}}{2}(W_1 + W_2) = 200$$

Estimate the solution graphically and numerically.

FIGURE 9.6   **FIGURE 9.7**

SOLUTION

Graphical Solution   Begin by solving each equation for the variable W_2.

$$W_2 = W_1$$

$$W_2 = \frac{400}{\sqrt{3}} - W_1$$

Graph the equations $Y_1 = X$ and $Y_2 = (400/\sqrt{3}) - X$. Their graphs intersect near the point $(115.47, 115.47)$, as shown in Figure 9.8. This means that each rafter supports a weight of approximately 115 pounds. (*Remark:* W_1 and W_2 represent the forces parallel to the rafters. Most of the weight is pushing downward, while a smaller portion is attempting to spread the two rafters apart.)

Numerical Solution Numerical support is shown in Figure 9.9, where $y_1 \approx y_2$ when $x = 115$.

[0, 200, 50] by [0, 200, 50]

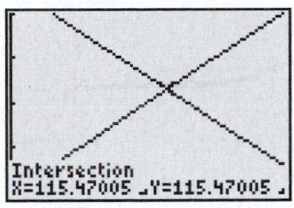

FIGURE 9.8   **FIGURE 9.9**   *Now Try Exercise 93* ◆

EXAMPLE 11 Determining the dimensions of a cylinder

The volume V of a cylindrical container with a radius r and height h is computed by $V(r, h) = \pi r^2 h$. See Figure 9.10. The lateral surface area S of the container, excluding the circular top and bottom, is computed by $S(r, h) = 2\pi r h$.

Geometry Review

To review formulas related to cylinders, see Chapter R (page R-4).

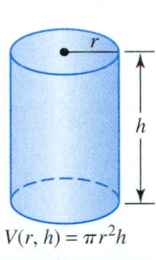

$V(r, h) = \pi r^2 h$

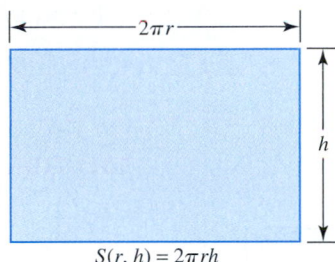

$S(r, h) = 2\pi r h$

FIGURE 9.10

(a) Write a system of equations whose solution is the dimensions for a cylinder with a volume of 38 cubic inches and a lateral surface area of 63 square inches.

(b) Solve the system of equations graphically and symbolically.

SOLUTION

(a) The equations $V(r, h) = 38$ and $S(r, h) = 63$ must be satisfied. This results in the following system of nonlinear equations.

$$\pi r^2 h = 38$$

$$2\pi r h = 63$$

(b) *Graphical Solution* To find the solution graphically, we can solve each equation for h, and then apply the intersection-of-graphs method.

[0, 4, 1] by [0, 20, 5]

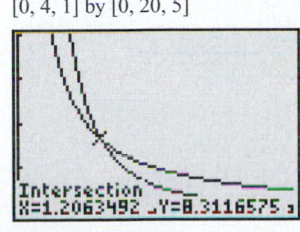

FIGURE 9.11

$$h = \frac{38}{\pi r^2}$$

$$h = \frac{63}{2\pi r}$$

Let r correspond to x and h to y. Graph $Y_1 = 38/(\pi X^{\wedge}2)$ and $Y_2 = 63/(2\pi X)$. Their graphs intersect near the point $(1.206, 8.312)$, as shown in Figure 9.11. Therefore a cylinder with a radius of $r \approx 1.206$ inches and height of $h \approx 8.312$ inches has a volume of 38 cubic inches and lateral surface area of 63 square inches.

Symbolic Solution   Because $h = \frac{38}{\pi r^2}$ and $h = \frac{63}{2\pi r}$, we can determine r by solving the following equation.

$$\frac{38}{\pi r^2} = \frac{63}{2\pi r} \qquad \text{Equation to be solved}$$

$$2\pi r^2\left(\frac{38}{\pi r^2}\right) = 2\pi r^2\left(\frac{63}{2\pi r}\right) \qquad \text{Multiply by the LCD, } 2\pi r^2.$$

$$76 = 63r \qquad \text{Simplify.}$$

$$\frac{76}{63} = r \qquad \text{Divide by 63.}$$

Because $r = \frac{76}{63} \approx 1.206$, $h = \frac{63}{2\pi r} = \frac{63}{2\pi(76/63)} \approx 8.312$, the symbolic result verifies our graphical result.   *Now Try Exercise 89*

Sometimes it is either difficult or impossible to solve a system of nonlinear equations symbolically. However, it might be possible to solve such a system graphically.

EXAMPLE 12   Solving a system of nonlinear equations graphically

Solve the system graphically to the nearest thousandth.

$$2x^3 - y = 2$$
$$\ln x^2 - 3y = -1$$

SOLUTION   Begin by solving both equations for y. The first equation becomes $y = 2x^3 - 2$. Solving the second equation for y gives the following results.

$$\ln x^2 - 3y = -1 \qquad \text{Second equation}$$

$$\ln x^2 + 1 = 3y \qquad \text{Add } 3y \text{ and 1 to each side.}$$

$$\frac{\ln x^2 + 1}{3} = y \qquad \text{Divide each side by 3.}$$

$[-6, 6, 1]$ by $[-4, 4, 1]$

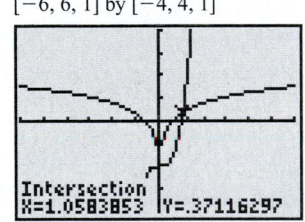

FIGURE 9.12

The graphs of $Y_1 = 2X^3 - 2$ and $Y_2 = (\ln (X^2) + 1)/3$ in Figure 9.12 intersect at one point. To the nearest thousandth, the solution is $(1.058, 0.371)$.   *Now Try Exercise 73*

Joint Variation

A quantity may depend on more than one variable. For example, the volume V of a cylinder is given by $V = \pi r^2 h$. We say that V *varies jointly* as h and the square of r. The *constant of variation* is π.

> ### JOINT VARIATION
>
> Let m and n be real numbers. Then z **varies jointly** as the mth power of x and the nth power of y if a nonzero real number k exists such that
>
> $$z = kx^m y^n.$$

EXAMPLE 13   Modeling the amount of wood in a tree

In forestry it is common to estimate the volume of timber in a given area of forest. To do this, formulas are developed to find the amount of wood contained in a tree with height h in feet and diameter d in inches. See Figure 9.13. One study concluded that the volume V

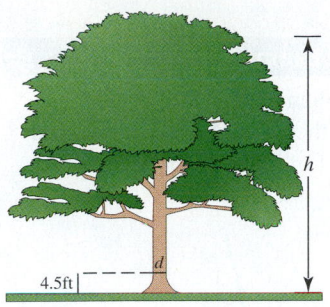

of wood in a tree varies jointly as the 1.12 power of h and the 1.98 power of d. (The diameter is measured 4.5 feet above the ground.)   (**Source:** B. Ryan, B. Joiner, and T. Ryan, *Minitab Handbook*.)

(a) Write an equation that relates V, h, and d.

(b) A tree with a 13.8-inch diameter and a 64-foot height has a volume of 25.14 cubic feet. Estimate the constant of variation k.

(c) Estimate the volume of wood in a tree with $d = 11$ inches and $h = 47$ feet.

SOLUTION

(a) $V = kh^{1.12}d^{1.98}$, where k is the constant of variation.

(b) Substitute $d = 13.8$, $h = 64$, and $V = 25.14$ into the equation and solve for k.

$$25.14 = k(64)^{1.12}(13.8)^{1.98}$$

$$k = \frac{25.14}{(64)^{1.12}(13.8)^{1.98}} \approx 0.00132$$

Thus let $V = 0.00132h^{1.12}d^{1.98}$.

(c) $V = 0.00132(47)^{1.12}(11)^{1.98} \approx 11.4$ cubic feet.

Now Try Exercise 105 ◆

FIGURE 9.13

9.1 Putting it all Together

Many mathematical concepts that were presented in earlier chapters can be applied to functions and equations of two variables. Functions of two variables often lead to equations involving two variables. Systems of equations in two variables can be solved symbolically, graphically, or numerically.

The following table summarizes some mathematical concepts involved with functions and equations in two variables.

Concept	Comments	Example
Function of two inputs or variables	$z = f(x, y)$, where x and y are inputs and z is the output.	$f(x, y) = x^2 + 5y$ $f(2, 3) = 2^2 + 5(3) = 19$
System of two linear equations	Equations can be written as $ax + by = k$. A solution is an ordered pair (x, y) that satisfies both equations.	$2x - 3y = 6$ $5x + 4y = -8$ Solution: $(0, -2)$
System of two nonlinear equations	A system of equations that is not linear is nonlinear. A solution is an ordered pair (x, y) that satisfies both equations.	$5x^2 - 4xy = -3$ $\dfrac{5}{x} - 2y = 1$ Solutions: $(1, 2)$, $\left(-\frac{7}{5}, -\frac{16}{7}\right)$
Method of substitution	Solve one equation for a variable and substitute the result in the second equation and solve.	$x - y = 1$ $x + y = 5$ If $x - y = 1$, then $x = 1 + y$. Substitute this in the second equation, $(1 + y) + y = 5$. This results in $y = 2$ and $x = 3$. The solution is $(3, 2)$.

Concept	Comments	Example
Graphical method for two equations	Solve both equations for the same variable. Then apply the intersection-of-graphs method.	If $x + y = 3$ then $y = 3 - x$. If $4x - y = 2$ then $y = 4x - 2$. Graph and locate the point of intersection at $(1, 2)$.

9.1 Exercises

Functions of More than One Input

Exercises 1 and 2: Evaluate the function for the indicated inputs and interpret the result.

1. $A(5, 8)$, where $A(b, h) = \frac{1}{2}bh$ (A computes the area of a triangle with base b and height h.)

2. $A(20, 35)$, where $A(w, l) = wl$ (A computes the area of a rectangle with width w and length l.)

Exercises 3–8: Evaluate the expression for the given $f(x, y)$.

3. $f(2, -3)$ if $f(x, y) = x^2 + y^2$

4. $f(-1, 3)$ if $f(x, y) = 2x^2 - y^2$

5. $f(-2, 3)$ if $f(x, y) = 3x - 4y$

6. $f(5, -2)$ if $f(x, y) = 6y - \frac{1}{2}x$

7. $f\left(\frac{1}{2}, -\frac{7}{4}\right)$ if $f(x, y) = \dfrac{2x}{y + 3}$

8. $f(0.2, 0.5)$ if $f(x, y) = \dfrac{5x}{2y + 1}$

Exercises 9–12: Write a symbolic representation for $f(x, y)$, if the function f computes the following quantity.

9. The sum of y and twice x

10. The product of x^2 and y^2

11. The product of x and y divided by $1 + x$

12. The square root of the sum of x and y

Exercises 13–18: Solve the equation for x and then solve it for y.

13. $3x - 4y = 7$

14. $-x - 5y = 4$

15. $x - y^2 = 5$

16. $2x^2 + y = 4$

17. $\dfrac{2x - y}{3y} = 1$

18. $\dfrac{x + y}{x - y} = 2$

Exercises 19–22: Solve the given equation for the indicated variable.

19. $A = \frac{1}{2}bh$
   (a) b (b) h

20. $V = \frac{1}{3}\pi r^2 h$
   (a) h (b) r

21. $z = \dfrac{2x^2}{y + 1}$
   (a) x (b) y

22. $z = 2x + 4y$
   (a) x (b) y

Systems of Equations

Exercises 23–26: Determine which ordered pairs are solutions to the given system of equations.

23. $(2, 1), (-2, 1), (1, 0)$
   $2x + y = 5$
   $x + y = 3$

24. $(3, 2), (3, -4), (5, 0)$
   $x - y = 5$
   $2x + y = 10$

25. $(4, -3), (0, 5), (4, 3)$
   $x^2 + y^2 = 25$
   $2x + 3y = -1$

26. $(4, 8), (8, 4), (-4, -8)$
   $xy = 32$
   $x + y = 12$

Exercises 27–30: The figure shows the graph of a system of two linear equations. Use the graph to estimate the solution to this system of equations. Then solve the system symbolically.

27.

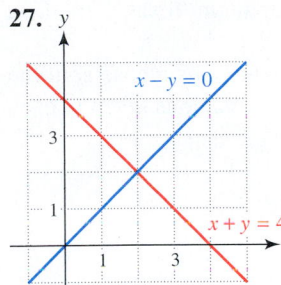

28.

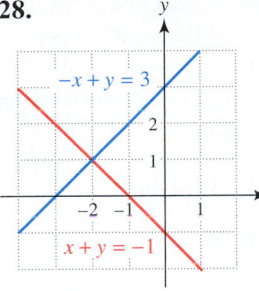

29.

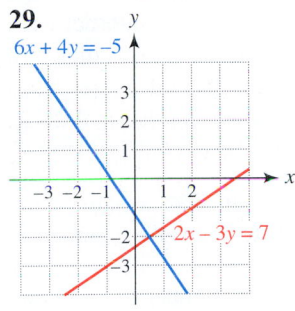

30.

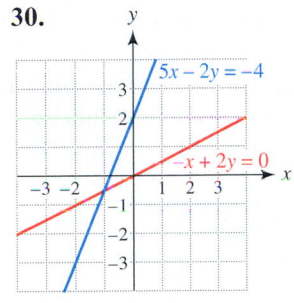

Exercises 31–46: If possible, solve the system of linear equations and check your answer.

31. $x + 2y = 0$
$3x + 7y = 1$

32. $-2x - y = -2$
$3x + 4y = -7$

33. $3x - 5y = -38$
$-4x + y = 28$

34. $7x - 2y = 5$
$x + 9y = 10$

35. $2x - 9y = -17$
$8x + 5y = 14$

36. $3x + 6y = 0$
$4x - 2y = -5$

37. $\frac{1}{2}x - y = -5$
$x + \frac{1}{2}y = 10$

38. $-x - \frac{1}{3}y = -4$
$\frac{1}{3}x + 2y = 7$

39. $3x - 2y = 5$
$-6x + 4y = -10$

40. $\frac{1}{2}x - \frac{3}{4}y = \frac{1}{2}$
$\frac{1}{5}x - \frac{3}{10}y = \frac{1}{5}$

41. $2x - 7y = 8$
$-3x + \frac{21}{2}y = 5$

42. $0.6x - 0.2y = 2$
$-1.2x + 0.4y = 3$

43. $0.2x - 0.1y = 0.5$
$0.4x + 0.3y = 2.5$

44. $0.3x + 0.2y = 9$
$0.5x - 0.1y = 2$

45. $20x - 10y = 30$
$-5x - 20y = -30$

46. $100x + 200y = 300$
$200x + 100y = 0$

Exercises 47–50: Solve the system of linear equations
  (a) graphically,   (b) numerically, and
  (c) symbolically.

47. $2x + y = 1$
$x - 2y = 3$

48. $3x + 2y = -2$
$2x - y = -6$

49. $-2x + y = 0$
$7x - 2y = 3$

50. $x - 4y = 15$
$3x - 2y = 15$

Exercises 51–66: If possible, solve the system of nonlinear equations.

51. $x^2 - y = 0$
$2x + y = 0$

52. $x^2 - y = 3$
$x + y = 3$

53. $xy = 8$
$x + y = 6$

54. $2x - y = 0$
$2xy = 4$

55. $x^2 + y^2 = 20$
$y = 2x$

56. $x^2 + y^2 = 9$
$x + y = 3$

57. $x^2 + y^2 = 6$
$\sqrt{x} - y = 0$

58. $x^2 + y^2 = 2$
$2x - y = 1$

59. $\sqrt{x} - 2y = 0$
$x - y = -2$

60. $x^2 + y^2 = 4$
$2x^2 + y = -3$

61. $2x^2 - y = 5$
$-4x^2 + 2y = -10$

62. $-6\sqrt{x} + 2y = -3$
$2\sqrt{x} - \frac{2}{3}y = 1$

63. $x^2 - y = 4$
$x^2 + y = 4$

64. $x^2 + x = y$
$2x^2 - y = 2$

65. $x^3 - x = 3y$
$x - y = 0$

66. $x^4 + y = 4$
$3x^2 - y = 0$

Exercises 67–70: Solve the system of nonlinear equations
  (a) symbolically, and   (b) graphically.

67. $x^2 + y^2 = 16$
$x - y = 0$

68. $x^2 - y = 1$
$3x + y = -1$

69. $xy = 12$
$x - y = 4$

70. $x^2 + y^2 = 2$
$x^2 - y = 0$

Exercises 71 and 72: The area of a rectangle with length l and width w is computed by $A(l, w) = lw$, and its perimeter is calculated by $P(l, w) = 2l + 2w$. Assume that $l > w$ and use the method of substitution to solve the system of equations for l and w. Interpret the solution.

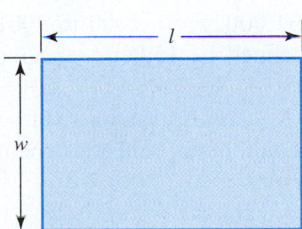

71. $A(l, w) = 35$
$P(l, w) = 24$

72. $A(l, w) = 300$
$P(l, w) = 70$

Exercises 73–78: Approximate, to the nearest thousandth, any solutions to the system of nonlinear equations graphically.

73. $x^3 - 3x + y = 1$
$\quad\quad x^2 + 2y = 3$

74. $x^2 + y = 5$
$\quad\quad x + y^2 = 6$

75. $2x^3 - x^2 = 5y$
$\quad\quad 2^{-x} - y = 0$

76. $x^4 - 3x^3 = y$
$\quad\quad \log x^2 - y = 0$

77. $e^{2x} + y = 4$
$\quad\quad \ln x - 2y = 0$

78. $3x^2 + y = 3$
$\quad\quad (0.3)^x + 4y = 1$

Applications

Exercises 79–82: **Skin and the Human Body** *(Refer to Example 2.) Estimate, to the nearest tenth, the surface area of a human body with weight w and height h.*

79. $w = 86$ kilograms, $h = 185$ centimeters

80. $w = 68$ kilograms, $h = 165$ centimeters

81. $w = 132$ pounds, $h = 62$ inches

82. $w = 220$ pounds, $h = 75$ inches

Exercises 83 and 84: **Life Expectancy** *The table lists life expectancy by birth year and sex. Let this table be a numerical representation of a function E, where $E(x, y)$ computes the life expectancy at birth of someone born in the year x whose sex is y.* (**Source:** Department of Health and Human Services.)

Year	Male	Female
1920	53.6	54.6
1940	60.8	65.2
1960	66.6	73.1
1980	70.0	77.4
2000	74.1	79.5

83. Evaluate $E(1960, \text{Male})$. Interpret the result.

84. Evaluate $E(2000, \text{Female}) - E(2000, \text{Male})$. Interpret the result.

85. **Robberies** The total number of robberies in 2000 and 2001 was 831,000. From 2000 to 2001 the number of robberies declined by 15,000. (**Source:** Federal Bureau of Investigation.)

(a) Write a system of equations whose solution represents the number of robberies committed in each of these years.

(b) Solve the system symbolically.

(c) Solve the system graphically.

86. **Smallpox Vaccinations** During the first four months of 2004, a total of 6767 people were vaccinated for smallpox in Florida and Texas. There were 273 more people vaccinated in Florida than in Texas. (**Source:** Federal Bureau of Investigation.)

(a) Set up a system of linear equations whose solution represents the number of vaccinations given in each state.

(b) Solve the system symbolically.

(c) Solve the system graphically.

87. **Time on the Internet** From 2001 to 2002 the average number of hours that a user spent on the Internet each week increased by 13%. This percent increase amounted to 1.3 hours. Find the average number of hours that a user spent on the Internet each week in 2001 and 2002. (**Source:** UCLA Center for Communications Policy.)

88. **Card Catalogs** Libraries have moved toward using computer catalogs. From 1968 to 1996 the number of 3×5-inch cards sold to libraries by the Library of Congress declined by 78.19 million. The number of cards sold in 1996 was only 0.72% of the 1968 number. How many cards were sold in 1968 and in 1996? Round your answers to the nearest hundredth of a million. (**Source:** Library of Congress.)

89. **Geometry** (Refer to Example 11.) Approximate the radius and height of a cylindrical container with a volume of 50 cubic inches and a lateral surface area of 65 square inches.

90. **Geometry** (Refer to Example 11.) Determine if it is possible to construct a cylindrical container, *including* the top and bottom, with a volume of 38 cubic inches and a surface area of 38 square inches.

91. **Heart Rate** In one study the maximum heart rates of conditioned athletes were examined. A group of athletes was exercised to exhaustion. Let x represent an athlete's heart rate 5 seconds after stopping exercise and y this rate after 10 seconds. It was found that the maximum heart rate H for these athletes satisfied the following two equations.

$$H = 0.491x + 0.468y + 11.2$$
$$H = -0.981x + 1.872y + 26.4$$

If an athlete had a maximum heart rate of $H = 180$, determine x and y graphically. Interpret your answer. (**Source:** V. Thomas, *Science and Sport.*)

92. **Heart Rate** Repeat the previous exercise for an athlete with a maximum heart rate of 195.

93. *Roof Truss* (Refer to Example 10.) The forces or weights W_1 and W_2 exerted on each rafter for the roof truss shown in the figure are determined by the system of linear equations. Solve the system.

$$W_1 + \sqrt{2}W_2 = 300$$
$$\sqrt{3}W_1 - \sqrt{2}W_2 = 0$$

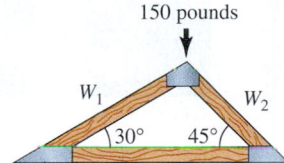

150 pounds

W_1 W_2

30°   45°

94. *Height and Weight* The relationship between a professional basketball player's height h in inches and weight w in pounds was modeled using two samples of players. The resulting equations that modeled each sample were $w = 7.46h - 374$ and $w = 7.93h - 405$. Assume that $65 \leq h \leq 85$.

(a) Use each equation to predict the weight of a professional basketball player who is $6'11''$.

(b) Determine graphically the height where the two models give the same weight.

(c) For each model, what change in weight is associated with a 1-inch increase in height?

95. *Dimensions of a Box* A box has an open top, rectangular sides, and a square base. The volume of the box is 576 cubic inches, and the surface area of the outside of the box is 336 square inches. Find the dimensions of the box.

96. *Dimensions of a Box* A box has rectangular sides and a rectangular top and base that are twice as long as they are wide. The volume of the box is 588 cubic inches, and the surface area of the outside of the box is 448 square inches. Find the dimensions of the box.

97. *Investments* A student invests $5000 at two annual interest rates, 5% and 7%. After 1 year the student receives a total of $325 in interest. How much did the student invest at each interest rate?

98. *Plane Speed* A plane flies 1500 miles against the wind in 3 hours and 45 minutes. The return trip with the wind takes 3 hours. Assume that the wind speed stays constant. Find the speed of the wind and the speed of the airplane with no wind.

Joint Variation

Exercises 99 and 100: Approximate the constant of variation to the nearest hundredth.

99. The variable z varies jointly as the second power of x and the third power of y. When $x = 2$ and $y = 2.5$, $z = 31.9$.

100. The variable z varies jointly as the 1.5 power of x and the 2.1 power of y. When $x = 4$ and $y = 3.5$, $z = 397$.

101. The variable z varies jointly as the square root of x and the cube root of y. If $z = 10.8$ when $x = 4$ and $y = 8$, find z when $x = 16$ and $y = 27$.

102. The variable z varies jointly as the third powers of x and y. If $z = 2160$ when $x = 3$ and $y = 4$, find z when $x = 2$ and $y = 5$.

103. *Wind Power* The electrical power generated by a windmill varies jointly as the square of the diameter of the area swept out by the blades and the cube of the wind velocity. If a windmill with an 8-foot diameter and a 10-mile-per-hour wind generates 2405 watts, how much power would be generated if the blades swept out an area 6 feet in diameter and the wind was 20 miles per hour?

104. *Strength of a Beam* The strength of a rectangular beam varies jointly as its width and the square of its thickness. If a beam 5.5 inches wide and 2.5 inches thick supports 600 pounds, how much can a similar beam that is 4 inches wide and 1.5 inches thick support?

105. *Volume of Wood* (Refer to Example 13.) One cord of wood contains 128 cubic feet. Estimate the number of cords in a tree that is 105 feet tall and has a diameter of 38 inches.

106. *Carpeting* The cost of carpet for a rectangular room varies jointly as its width and length. If a room 10 feet wide and 12 feet long costs $1560 to carpet, find the cost to carpet a room that is 11 feet by 23 feet. Interpret the constant of variation.

107. *Surface Area* Use the results of Example 2 to find a formula for $S(w, h)$ that calculates the surface area of a person's body if w is given in pounds and h is given in inches.

108. *Surface Area* Use the results of the previous exercise to solve Exercises 81 and 82.

Writing about Mathematics

109. Give an example of a quantity occurring in everyday life that can be computed by a function of more than one input. Identify the inputs and the output.

110. Give an example of a system of linear equations with two variables. Explain how to solve the system graphically and symbolically.

EXTENDED AND DISCOVERY EXERCISES

Exercises 1–6: Colors on Computer Monitors Although it might not seem possible to describe color using mathematics, it is done every day on computer screens. Colored light can be broken down into three basic colors: red, green, and blue. Some computer monitors are capable of creating 256 intensities for each of these three colors, which may be numbered from 0 to 255. The number 0 indicates the absence of a color, while 255 represents the brightest intensity of a color. The color displayed on a screen can be computed by $f(r, g, b)$, where r represents the brightness of red light, g of green light, and b of blue light. The expression $f(255, 0, 0)$ indicates the brightest intensity of red with no green or blue

light. Therefore, $f(255, 0, 0) = red$. In a similar manner, $f(0, 255, 0) = green$, and $f(0, 0, 255) = blue$. By mixing intensities of these three basic colors, thousands of colors can be generated. For example, $f(255, 255, 0) = yellow$, since red and green light combine to make yellow light. (**Source:** I. Kerlow, *The Art of 3-D Computer Animation and Imaging.*)

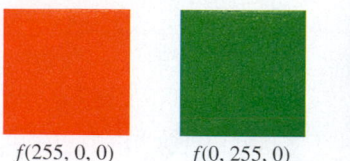 

$f(255, 0, 0)$ $f(0, 255, 0)$ $f(0, 0, 255)$

1. Make a conjecture about what color $f(255, 0, 255)$ represents.

2. Make a conjecture about what inputs create white light.

3. Turquoise is a greenish-blue color. Make a conjecture about inputs that create turquoise light.

4. Make a conjecture about whether $f(141, 43, 17)$ represents rust or cream color.

5. If there are 256 intensities for each basic color, how many colors can be created on a computer screen by mixing different intensities of these colors?

6. Some computers are capable of generating 65,536 intensities for each of the three basic colors. Approximate the number of colors that could be generated by this computer.

9.2   Systems of Equations and Inequalities in Two Variables

♦ Recognize different types of linear systems

♦ Apply the elimination method

♦ Solve systems of linear and nonlinear inequalities

♦ Learn basic concepts about linear programming

Introduction

Large systems of equations and inequalities involving thousands of variables are solved by economists, scientists, engineers, and mathematicians every day. Computers and sophisticated numerical methods are essential for finding their solutions. In this section we focus on systems involving two variables. Graphical, numerical, and symbolic techniques can be applied to equations in two variables. Many of the concepts presented in this section apply to larger systems of linear equations.

Types of Linear Systems in Two Variables

Section 9.1 demonstrated that a system of equations in two variables can have zero, one, two, or infinitely many solutions. In fact, a system of *nonlinear* equations can have any number of solutions. However if the equations are all *linear*, then only zero, one, or infinitely many solutions are possible.

Any system of linear equations in two variables can be written in the form

$$a_1 x + b_1 y = c_1$$
$$a_2 x + b_2 y = c_2,$$

where a_1, b_1, c_1, a_2, b_2, and c_2 are constants. The graph of this system consists of *two* lines in the xy-plane. The lines can intersect at a point, coincide, or be parallel, as illustrated in Figures 9.14–9.16. **Coincident lines** are identical lines and indicate that the two equations are equivalent and have the same graph.

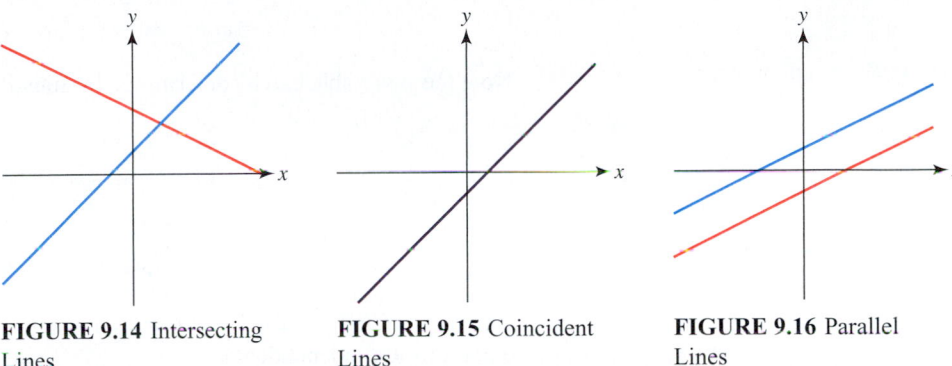

FIGURE 9.14 Intersecting Lines

FIGURE 9.15 Coincident Lines

FIGURE 9.16 Parallel Lines

◆ **CLASS DISCUSSION**
Explain why a system of linear equations in two variables cannot have two or three solutions.

A **consistent** system of linear equations has either one solution or infinitely many solutions. If the system has one solution, then the equations are **independent**, and they can be represented by lines that intersect at one point, as shown in Figure 9.14. If the system has infinitely many solutions, then the equations are **dependent**, and they can be represented by coincident lines, as illustrated in Figure 9.15. An **inconsistent** system has no solutions and can be represented by parallel lines, as shown in Figure 9.16.

This discussion is summarized by the following.

POSSIBLE GRAPHS OF A SYSTEM OF TWO LINEAR EQUATIONS IN TWO VARIABLES

1. The graphs of the two equations are distinct lines that intersect at one point. The system is *consistent*. There is one solution, which is given by the coordinates of the point of intersection. In this case the equations are *independent*.
2. The graphs of the two equations are the same line. The system is *consistent*. There are infinitely many solutions, and the equations are *dependent*.
3. The graphs of the two equations are distinct parallel lines. The system is *inconsistent*. There are no solutions.

The Elimination Method

The substitution method, presented in Section 9.1, is a symbolic method for solving small systems of equations. A second symbolic method is the **elimination method**. The next example demonstrates this technique.

 **EXAMPLE 1** Using elimination to solve a system

Use elimination to solve each system of equations, if possible. Identify the system as consistent or inconsistent. If the system is consistent, state whether the equations are dependent or independent. Support your results graphically.

(a) $2x - y = -4$
$3x + y = -1$

(b) $4x - y = 10$
$-4x + y = -10$

(c) $x - y = 6$
$x - y = 3$

SOLUTION

(a) *Symbolic Solution*   We can eliminate the *y*-variable by adding the equations.

$$
\begin{array}{r}
2x - y = -4 \\
\underline{3x + y = -1} \\
5x \quad\;\; = -5 \quad \text{or} \quad x = -1
\end{array}
$$

Now the *y*-variable can be determined by substituting $x = -1$ in either equation.

$$
\begin{aligned}
2x - y &= -4 && \text{First equation} \\
2(-1) - y &= -4 && \text{Let x} = -1. \\
-y &= -2 && \text{Add 2.} \\
y &= 2 && \text{Multiply by } -1.
\end{aligned}
$$

The solution is $(-1, 2)$. There is a unique solution so the system is consistent and the equations are independent.

Graphical Solution   For a graphical solution, start by solving each equation for *y*.

$$
\begin{aligned}
2x - y = -4 &\quad \text{is equivalent to} \quad y = \;\;\;2x + 4. \\
3x + y = -1 &\quad \text{is equivalent to} \quad y = -3x - 1.
\end{aligned}
$$

The graphs of $y = 2x + 4$ and $y = -3x - 1$ intersect at the point $(-1, 2)$, as shown in Figure 9.17.

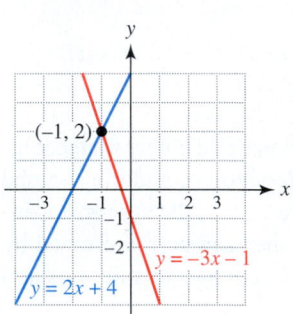

FIGURE 9.17

(b) *Symbolic Solution*   If we add the equations, we obtain the following result.

$$
\begin{array}{r}
4x - y = \;\;\;10 \\
\underline{-4x + y = -10} \\
0 = 0
\end{array}
$$

The equation $0 = 0$ is an identity that is always true. The two equations are equivalent: if we multiply the first equation by -1, we obtain the second equation. Thus there are infinitely many solutions, and we can write the solution set in set-builder notation.

$$
\{(x, y) \,|\, 4x - y = 10\}
$$

Some examples of solutions are $(3, 2)$, $(4, 6)$, and $(1, -6)$. The system is consistent and the equations are dependent.

Graphical Solution   For a graphical solution, start by solving each equation for *y*.

$$
\begin{aligned}
4x - y = 10 &\quad \text{is equivalent to} \quad y = 4x - 10. \\
-4x + y = -10 &\quad \text{is equivalent to} \quad y = 4x - 10.
\end{aligned}
$$

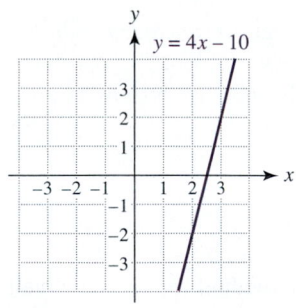

FIGURE 9.18

Clearly, the graphs of $y = 4x - 10$ and $y = 4x - 10$, shown in Figure 9.18, are identical. Their graphs are coincident. Any point on the line represents a solution.

(c) *Symbolic Solution*   If we subtract the second equation from the first, we obtain the following result.

$$
\begin{array}{r}
x - y = 6 \\
\underline{x - y = 3} \\
0 = 3
\end{array}
$$

FIGURE 9.19

The equation $0 = 3$ is a contradiction that is never true. Therefore there are no solutions, and the system is inconsistent.

Graphical Solution For a graphical solution start by solving each equation for y.

$$x - y = 6 \quad \text{is equivalent to} \quad y = x - 6.$$

$$x - y = 3 \quad \text{is equivalent to} \quad y = x - 3.$$

The graphs of $y = x - 6$ and $y = x - 3$, shown in Figure 9.19, are parallel lines. These lines never intersect, so there are no solutions.

Now Try Exercises 5, 9, and 13 ◆

The next example illustrates how a linear system can occur in an application involving real data. We solve this system symbolically, graphically, and numerically.

 **Using the elimination method**

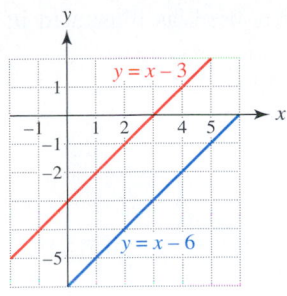

Title IX is landmark legislation that prohibits sex discrimination in sports programs. In 1997 the national average spent on two varsity athletes, one female and one male, was $6050 per athlete for Division I-A schools. However, average expenditures for a male athlete exceeded those for a female athlete by $3900. Determine how much was spent per varsity athlete for each gender. Give graphical and numerical support. (**Source:** *USA Today.*)

SOLUTION

Symbolic Solution Let x represent average expenditures per male athlete and y average expenditures per female athlete in 1997. Since the average amount spent on one female and one male athlete was $6050 per athlete, $(x + y)/2 = 6050$ must be satisfied. The expenditures for a male athlete exceeded those for a female athlete by $3900. Thus $x - y = 3900$.

$$\frac{x + y}{2} = 6050$$

$$x - y = 3900$$

Mutiply the first equation by 2 and then add the resulting equations.

$$
\begin{array}{ll}
x + y = 12{,}100 & \text{Multiply by 2.} \\
\underline{x - y = 3900} & \text{Second equation} \\
2x = 16{,}000 & \text{Add.}
\end{array}
$$

By adding the equations, we *eliminate* the y-variable. The solution to $2x = 16{,}000$ is given by $x = 8000$. Thus the average expenditure per male athlete in 1997 was $8000. To determine y, we can substitute $x = 8000$ in either equation.

$$8000 - y = 3900 \quad \text{implies} \quad y = 4100.$$

The average amount spent on each female athlete was $4100.

Graphical Solution Solve each equation for y and then graph $Y_1 = 12100 - X$ and $Y_2 = X - 3900$. Their graphs intersect at $(8000, 4100)$. See Figure 9.20. This agrees with the symbolic solution.

Numerical Solution Make a table of y_1 and y_2. When $x = 8000$, $y_1 = y_2 = 4100$. See Figure 9.21.

Now Try Exercise 77 ◆

[0, 15000, 5000] by [0, 15000, 5000]

FIGURE 9.20

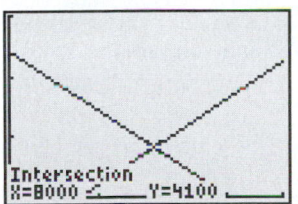

FIGURE 9.21

Sometimes multiplication is performed before elimination is used, as illustrated in Example 2 and the next example.

EXAMPLE 3   Multiplying before using elimination

Solve each system of equations by using elimination.

(a) $2x - 3y = 18$ **(b)** $5x + 10y = 10$
         $5x + 2y = 7$ $x + 2y = 2$

SOLUTION

(a) If we multiply the first equation by 2 and the second equation by 3, then the y-coefficients become -6 and 6. Addition eliminates the y-variable.

$$\begin{aligned}
4x - 6y &= 36 && \text{Multiply by 2.}\\
\underline{15x + 6y} &= \underline{21} && \text{Multiply by 3.}\\
19x &= 57 \quad \text{or} \quad x = 3 && \text{Add.}
\end{aligned}$$

Substituting $x = 3$ in $2x - 3y = 18$ results in

$$2(3) - 3y = 18 \quad \text{or} \quad y = -4.$$

The solution is $(3, -4)$.

(b) If the second equation is multiplied by 5, subtraction eliminates both variables.

$$\begin{aligned}
5x + 10y &= 10\\
\underline{5x + 10y} &= \underline{10} && \text{Multiply by 5.}\\
0 &= 0 && \text{Subtract.}
\end{aligned}$$

The statement $0 = 0$ is an identity that is always true. This means that the equations are dependent. There are infinitely many solutions. The solution set is $\{(x, y) \mid x + 2y = 2\}$.

Now Try Exercises 17 and 21 ◆

EXAMPLE 4   Using elimination to solve a nonlinear system

Solve the system of equations.

$$\begin{aligned}
x^2 + y^2 &= 4\\
2x^2 - y &= 7
\end{aligned}$$

SOLUTION If we multiply each side of the first equation by 2 and subtract the second equation from the first equation, the variable x is eliminated.

$$\begin{aligned}
2x^2 + 2y^2 &= 8 && \text{Multiply by 2.}\\
\underline{2x^2 - y} &= \underline{7}\\
2y^2 + y &= 1 && \text{Subtract.}
\end{aligned}$$

Next we solve $2y^2 + y - 1 = 0$ for y.

$$\begin{aligned}
(y + 1)(2y - 1) &= 0 && \text{Factor.}\\
y + 1 = 0 \quad \text{or} \quad 2y - 1 &= 0 && \text{Zero-product property}\\
y = -1 \quad \text{or} \quad y &= \frac{1}{2} && \text{Solve each equation.}
\end{aligned}$$

Solving the first equation for x results in $x = \pm\sqrt{4 - y^2}$. If $y = \frac{1}{2}$, then $x = \pm\sqrt{\frac{15}{4}}$, which can be written as $\pm\dfrac{\sqrt{15}}{2}$. If $y = -1$, then $x = \pm\sqrt{3}$. Thus there are four solutions: $\left(\pm\frac{\sqrt{15}}{2}, \frac{1}{2}\right)$ and $(\pm\sqrt{3}, -1)$.

A graph of the system of equations is shown in Figure 9.22. The four points of intersection correspond to the four solutions. In Figure 9.23 the four points of intersection are labeled.

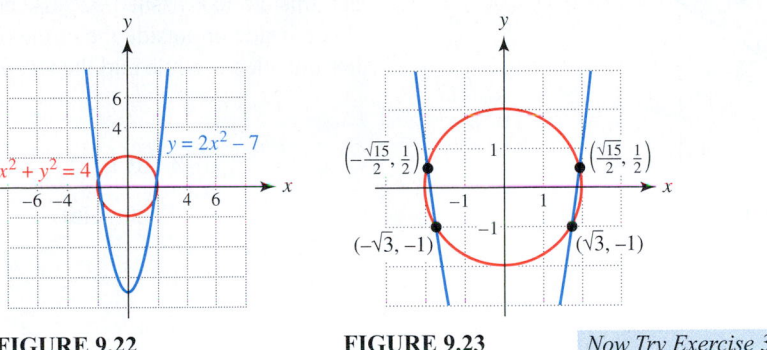

FIGURE 9.22   FIGURE 9.23   *Now Try Exercise 33* ◆

Systems of Linear and Nonlinear Inequalities

A linear inequality in two variables can be written as

$$ax + by \leq c,$$

where a, b, and c are constants with a and b not equal to zero. (The symbol $\leq$ can be replaced by $\geq$, $<$, or $>$.) If an ordered pair (x, y) makes the inequality a true statement, then (x, y) is a solution. The set of all solutions is called the *solution set*. The graph of an inequality includes all points (x, y) in the solution set.

The graph of a linear inequality is a **half-plane**, which may include the boundary. For example, to graph the linear inequality $3x - 2y \leq 6$, solve the inequality for y. This results in $y \geq \frac{3}{2}x - 3$. Then graph the line $y = \frac{3}{2}x - 3$, which is the boundary. The solution set to the inequality includes the line and the half-plane above the line, as shown in Figure 9.24.

A second approach is to write the equation $3x - 2y = 6$ as $y = \frac{3}{2}x - 3$ and graph. To determine which half-plane to shade, select a **test point**. For example, the point $(0, 0)$ satisfies the inequality $3x - 2y \leq 6$, so shade the region containing $(0, 0)$.

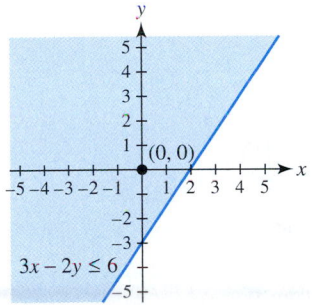

FIGURE 9.24

EXAMPLE 5

Graphing inequalities

Graph the solution set to each inequality.
(a) $2x - 3y \leq -6$ **(b)** $x^2 + y^2 < 9$

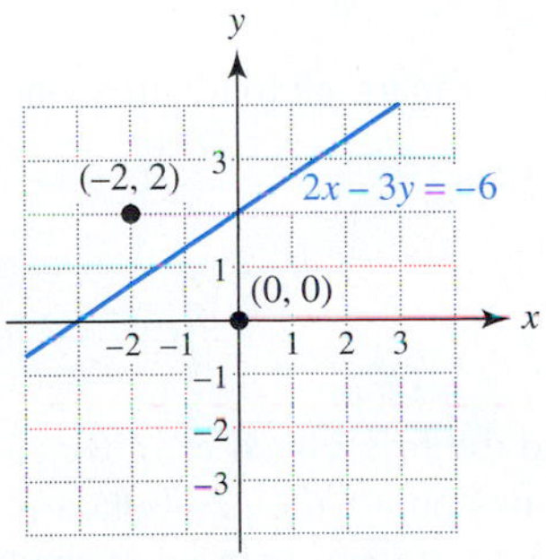

SOLUTION

(a) Start by graphing the line determined by $2x - 3y = -6$, as shown in Figure 9.25. Note that this line is solid because equality is included. We can determine which side of the line to shade by using test points. For example, the test point $(-2, 2)$ lies above the line and the test point $(0, 0)$ lies below the line. See Table 9.1.

FIGURE 9.25

TABLE 9.1

Test Point	$2x - 3y \leq -6$	True or False?
$(-2, 2)$	$2(-2) - 3(2) \overset{?}{\leq} -6$	True
$(0, 0)$	$2(0) - 3(0) \overset{?}{\leq} -6$	False

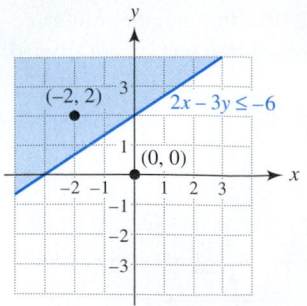

FIGURE 9.26

The test point $(-2, 2)$ satisfies the given inequality, so shade the region above the line that contains the point $(-2, 2)$, but not the point $(0, 0)$. See Figure 9.26.

(b) Start by graphing the circle determined by $x^2 + y^2 = 9$, as shown in Figure 9.27. Note that this circle is dashed because equality is *not* included. We can determine whether to shade inside or outside the circle by using test points. For example, the test point $(3, 3)$ lies outside the circle and the test point $(0, 0)$ lies inside the circle. See Table 9.2.

TABLE 9.2

Test Point	$x^2 + y^2 < 9$	True or False?
$(3, 3)$	$3^2 + 3^2 \overset{?}{<} 9$	False
$(0, 0)$	$0^2 + 0^2 \overset{?}{<} 9$	True

The test point $(0, 0)$ satisfies the given inequality, so shade the region inside the circle. Note that the actual circle is not part of the solutions set. See Figure 9.28.

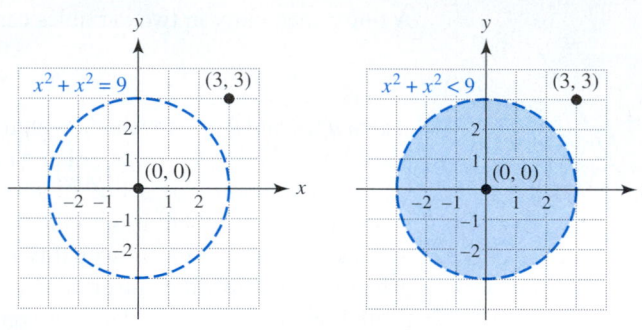

FIGURE 9.27   **FIGURE 9.28**

Now Try Exercises 47 and 49 ◆

In Section 9.1, we learned that systems of equations could be linear or nonlinear. In an analogous manner, systems of inequalities can be linear or nonlinear. The next example illustrates a system of each type. Both are solved graphically.

EXAMPLE 6   Solving systems of inequalities graphically

Solve each system of inequalities by shading the solution set. Use the graph to identify one solution.

(a) $y > x^2$ **(b)** $x + 3y \leq 9$
$\quad\;\; x + y < 4$ $\quad\;\; 2x - y \leq -1$

SOLUTION

(a) This is a nonlinear system. Graph the parabola $y = x^2$ and the line $y = 4 - x$. Since $y > x^2$ and $y < 4 - x$, the region satisfying the system lies above the parabola and below the line. It does not include the boundaries, which are shown using a dashed line and curve. See Figure 9.29.

Any point in the shaded region represents a solution. For example, the test point $(0, 2)$ lies in the shaded region and is a solution since $x = 0$ and $y = 2$ satisfy *both* inequalities.

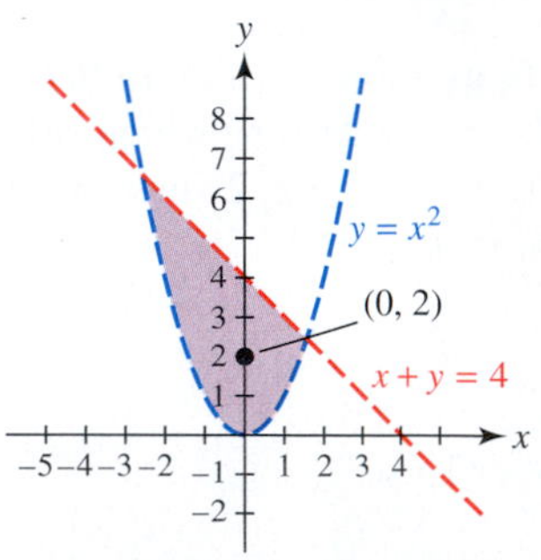

FIGURE 9.29

(b) Begin by solving each linear inequality for y.

$$y \le \frac{9 - x}{3}$$

$$y \ge 2x + 1$$

Graph $y = \frac{9 - x}{3}$ $\left(\text{or } y = -\frac{1}{3}x + 3\right)$ and $y = 2x + 1$. The region satisfying the system is below the line $x + 3y = 9$ and above the line $2x - y = -1$. Because equality is included, the boundaries are part of the region, which are shown as solid lines in Figure 9.30. The test point $(-3, 0)$ is a solution since it lies in the shaded region and satisfies both inequalities.

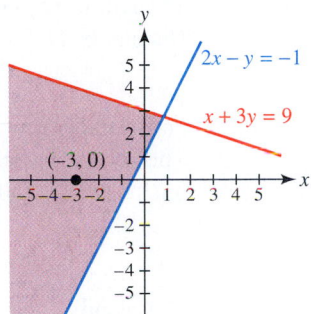

FIGURE 9.30   *Now Try Exercises 57 and 63* ◆

Graphing calculators can be used to shade regions in the xy-plane. See Figure 9.31. The solution set shown in Figure 9.29 is also shown in Figure 9.32, where a graphing calculator has been used. However, the boundary is not dashed.

Figures 9.33 and 9.34 show a different method of shading a solution set; we have shaded the area below $Y_1 = (9 - X)/3$ and above $Y_2 = 2X + 1$. The solution set is the region shaded with both vertical *and* horizontal lines and corresponds to the shaded region in Figure 9.30.

FIGURE 9.31

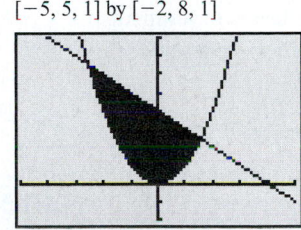

$[-5, 5, 1]$ by $[-2, 8, 1]$

FIGURE 9.32

Calculator Help

To shade a graph, see Appendix B (page AP-18).

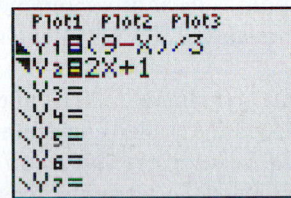

FIGURE 9.33

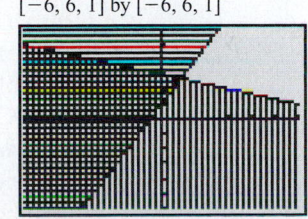

$[-6, 6, 1]$ by $[-6, 6, 1]$

FIGURE 9.34

Modeling plant growth

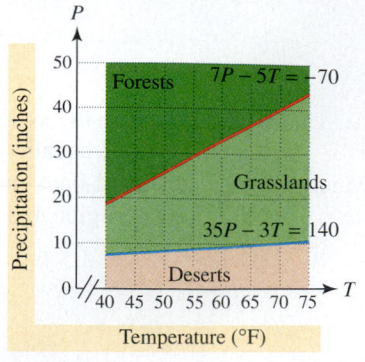

FIGURE 9.35 Effect of Temperature and Precipitation on Plant Growth

Two factors that have a critical effect on plant growth are temperature and precipitation. If a region has too little precipitation, it will be a desert. Forests tend to grow in regions where trees can exist at relatively low temperatures and there is sufficient rainfall. At other levels of precipitation and temperature, grasslands may prevail. Figure 9.35 illustrates the relationship among forests, grasslands, and deserts, as suggested by annual average temperature T in degrees Fahrenheit and precipitation P in inches.   (**Source:** A. Miller and J. Thompson, *Elements of Meteorology.*)

(a) Determine a system of linear inequalities that describes where grasslands are likely to occur.

(b) Bismarck, North Dakota, has annual average temperature of 40°F and precipitation of 15 inches. According to the graph, what type of plant growth would you expect near Bismarck? Do these values satisfy the system of inequalities from part (a)?

SOLUTION

(a) Grasslands occur for ordered pairs (T, P) lying between the two lines in Figure 9.35. The boundary between deserts and grassland is determined by the equation $35P - 3T = 140$. Solving for P results in

$$P = \frac{3}{35}T + \frac{140}{35}.$$

Grasslands grow where values of P are above the line. This region is described by $P > \frac{3}{35}T + \frac{140}{35}$, or equivalently, $35P - 3T > 140$. In a similar manner, the region below the boundary between grasslands and forests is represented by $7P - 5T < -70$. Thus grasslands satisfy the following system of inequalities.

$$35P - 3T > 140$$
$$7P - 5T < -70$$

(b) For Bismarck, $T = 40$ and $P = 15$. From Figure 9.35, it appears that the point $(40, 15)$ lies between the two lines, so the graph predicts that grasslands will exist around Bismarck. Substituting these values for T and P into the system of inequalities results in the following true statements.

$$35(15) - 3(40) = 405 > 140$$
$$7(15) - 5(40) = -95 < -70$$

◆ **CLASS DISCUSSION**

Use Example 7 to find an inequality that describes levels of temperature and precipitation in deserts. ◆

The temperature and precipitation values for Bismarck satisfy the system of inequalities for grasslands.
Now Try Exercise 83 ◆

Linear Programming

An important application of mathematics used in business and social sciences is linear programming. **Linear programming** is a procedure used to optimize quantities such as cost and profit. It was developed during World War II as a method of efficiently allocating supplies. Linear programming applications frequently contain thousands of variables and are solved by computers. However here we focus on problems involving two variables.

  A linear programming problem consists of a linear **objective function** and a system of linear inequalities called **constraints**. The solution set for the system of linear inequalities is called the set of **feasible solutions**. The objective function describes a quantity that is to be optimized. For example, linear programming is often used to maximize profit or minimize cost. The following example illustrates these concepts.

EXAMPLE 8   Finding maximum profit

Suppose a small company manufacturers two products—radios and CD players. Each radio results in a profit of $15 and each CD player provides a profit of $35. Due to demand, the company must produce at least 5 and not more than 25 radios per day. The number of radios cannot exceed the number of CD players, and the number of CD players cannot exceed 30. How many of each should the company manufacture to obtain maximum profit?

SOLUTION   Let x be the number of radios produced daily and y be the number of CD players produced daily. Since the profit from x radios is $15x$ dollars and the profit from y CD players is $35y$ dollars, the total daily profit P is given by

$$P = 15x + 35y.$$

The company produces from 5 to 25 radios per day, so the inequalities

$$x \geq 5 \quad \text{and} \quad x \leq 25$$

must be satisfied. The requirements that the number of radios cannot exceed the number of CD players and the number of CD players cannot exceed 30 indicate that

$$x \leq y \quad \text{and} \quad y \leq 30.$$

Since the number of radios and CD players cannot be negative, we have

$$x \geq 0 \quad \text{and} \quad y \geq 0.$$

Listing all the constraints on production gives

$$x \geq 5, \quad x \leq 25, \quad y \leq 30, \quad x \leq y, \quad x \geq 0, \quad \text{and} \quad y \geq 0.$$

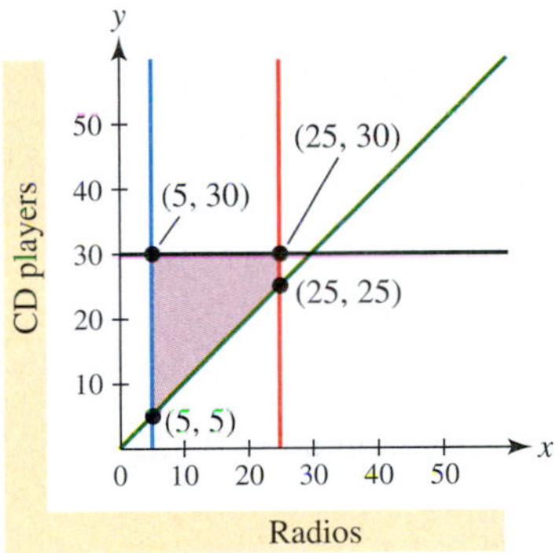

Graphing these constraints results in the shaded region shown in Figure 9.36. This shaded region is the set of feasible solutions. The vertices (or corners) of this region are (5, 5), (25, 25), (25, 30), and (5, 30).

It can be shown that maximum profit occurs at a vertex of the region of feasible solutions. Thus we evaluate P at each vertex, as shown in Table 9.3.

TABLE 9.3

Vertex	$P = 15x + 35y$	
(5, 5)	$15(5) + 35(5) = 250$	
(25, 25)	$15(25) + 35(25) = 1250$	
(25, 30)	$15(25) + 35(30) = 1425$	⟵ **Maximum Profit**
(5, 30)	$15(5) + 35(30) = 1125$	

The maximum value of P is 1425 at vertex (25, 30). Thus the maximum profit is $1425 and it occurs when 25 radios and 30 CD players are manufactured.

Now Try Exercise 103

The following holds for linear programming problems.

FUNDAMENTAL THEOREM OF LINEAR PROGRAMMING

If the optimal value for a linear programming problem exists, then it occurs at a vertex of the region of feasible solutions.

FIGURE 9.36

(graph with x-axis labeled "Radios" from 0 to 50, y-axis labeled "CD players" from 10 to 50; points labeled (5, 30), (25, 30), (25, 25), (5, 5))

EXAMPLE 9 Finding the minimum of an objective function

Find the minimum value of $C = 2x + 3y$ subject to the following constraints.

$$x + y \geq 4$$
$$2x + y \leq 8$$
$$x \geq 0, \quad y \geq 0$$

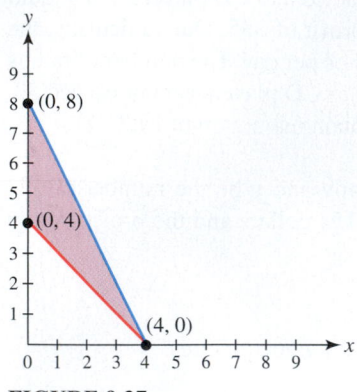

FIGURE 9.37

SOLUTION Sketch the region determined by the constraints and find all vertices, as shown in Figure 9.37.

Evaluate the objective function C at each vertex, as shown in Table 9.4.

TABLE 9.4

Vertex	$C = 2x + 3y$	
$(4, 0)$	$2(4) + 3(0) = 8$	⟵ **Minimum Value**
$(0, 8)$	$2(0) + 3(8) = 24$	
$(0, 4)$	$2(0) + 3(4) = 12$	

The minimum value for C is 8 and it occurs at vertex $(4, 0)$, or when $x = 4$ and $y = 0$.

Now Try Exercise 99

The following procedure describes how to solve a linear programming problem.

SOLVING A LINEAR PROGRAMMING PROBLEM

STEP 1: Read the problem carefully. Consider making a table to display the information given.

STEP 2: Use the table to write the objective function and all the constraints.

STEP 3: Sketch a graph of the region of feasible solutions. Identify all vertices or corner points.

STEP 4: Evaluate the objective function at each vertex. A maximum (or a minimum) occurs at a vertex. *Note:* If the region is unbounded, a maximum (or minimum) may not exist.

 **Minimizing cost**

A breeder is buying two brands of food, A and B, for her animals. Each serving is a mixture of the two foods and should contain at least 40 grams of protein and 30 grams of fat. Brand A costs 90 cents per unit and Brand B costs 60 cents per unit. Each unit of Brand A contains 20 grams of protein and 10 grams of fat, whereas each unit of Brand B contains 10 grams of protein and 10 grams of fat. Determine how much of each brand should be bought to obtain a minimum cost per serving.

SOLUTION

STEP 1: After reading the problem carefully, begin by listing the information, as illustrated in Table 9.5.

TABLE 9.5

Brand	Units	Protein	Fat	Cost
A	x	20	10	90¢
B	y	10	10	60¢
Minimum		40	30	

STEP 2: If x units of Brand A are purchased at 90¢ per unit and if y units of Brand B are purchased at 60¢ per unit, then the cost C is given by $C = 90x + 60y$. Each serving requires at least 40 grams of protein. If x units of Brand A are bought, each containing 20 grams of protein, if y units of Brand B are bought, each containing 10 grams of protein, and if each serving requires at least 40 grams of protein, then we can write $20x + 10y \geq 40$. Similarly, since each serving requires at least 30 grams of fat, we can write $10x + 10y \geq 30$. The linear programming problem can be written as follows.

$$\text{Minimize:} \quad C = 90x + 60y \qquad \textit{Cost}$$
$$\text{Subject to:} \quad 20x + 10y \geq 40 \qquad \textit{Protein}$$
$$10x + 10y \geq 30 \qquad \textit{Fat}$$
$$x \geq 0, \quad y \geq 0$$

STEP 3: The region that contains the feasible solutions is shown in Figure 9.38. The vertices for this region are $(0, 4)$, $(1, 2)$, and $(3, 0)$.

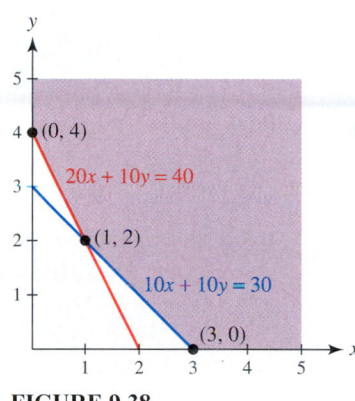

FIGURE 9.38

STEP 4: Evaluate the objective function C at each vertex, as shown in Table 9.6.

TABLE 9.6

Vertex	$C = 90x + 60y$	
$(0, 4)$	240	
$(1, 2)$	210	⟵ **Minimum Cost**
$(3, 0)$	270	

The minimum cost occurs when 1 unit of Brand A and 2 units of Brand B are mixed, at a cost of $2.10 per serving.   *Now Try Exercise 105* ◆

9.2

Putting it all Together

Systems of linear equations can have no solutions, one solution, or infinitely many solutions. The substitution and elimination methods may be used to solve systems of equations. Graphical and numerical methods can also be applied to systems of equations and inequalities in two variables.

The following table summarizes some important mathematical concepts from this section.

Concept	Comments	Example
Consistent system of linear equations in two variables.	A consistent linear system has either one or infinitely many solutions. Its graph is either distinct, intersecting lines or identical lines.	$\begin{aligned} x + y &= 10 \\ x - y &= 4 \\ \hline 2x &= 14 \end{aligned}$ Solution is given by $x = 7$, and $y = 3$. Since the equations have exactly one solution, they are independent.
Dependent system of linear equations in two variables	A dependent linear system has infinitely many solutions. The graph consists of two identical lines.	$2x + 2y = 2$ and $x + y = 1$ are equivalent (dependent) equations. The solution set is $\{(x, y) \mid x + y = 1\}$.
Inconsistent system of linear equations in two variables	An inconsistent linear system has no solutions. The graph is two parallel lines.	$\begin{aligned} x + y &= 1 \\ x + y &= 2 \quad \text{Subtract} \\ \hline 0 &= -1 \quad \text{Always false} \end{aligned}$ The solution set is empty.
Elimination method	By performing arithmetic operations on a system, a variable is eliminated.	$\begin{aligned} 2x + y &= 5 \\ x - y &= 1 \quad \text{Add.} \\ \hline 3x &= 6 \quad \text{so} \quad x = 2 \end{aligned}$ and $y = 1$.
Linear inequality in two variables	$ax + by \leq c$ ($\leq$ may be replaced by $<$, $>$, or $\geq$.) The solution set is typically a shaded region in the xy-plane.	$2x - 3y \leq 12$ $-2x + y \leq 4$

Concept	Comments	Example
Linear programming	In a linear programming problem, the maximum or minimum of an objective function is found, subject to constraints. If a solution exists, it occurs at a vertex in the region of feasible solutions.	Maximize the objective function $$P = 2x + 3y,$$ subject to the following constraints. $$2x + \ y \le 6$$ $$x + 2y \le 6$$ $$x \ge 0, y \ge 0$$

The maximum of $P = 10$ occurs at vertex $(2, 2)$.

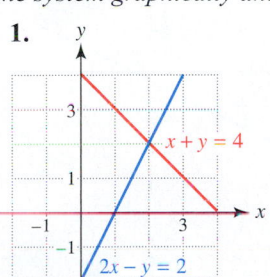

9.2 — Exercises

Consistent and Inconsistent Linear Systems

Exercises 1–4: The figure represents a system of linear equations. Classify the system as consistent or inconsistent. Solve the system graphically and symbolically, if possible.

1.

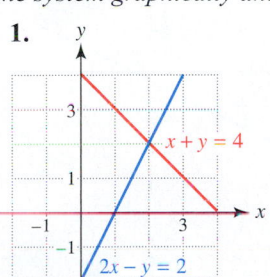

2.

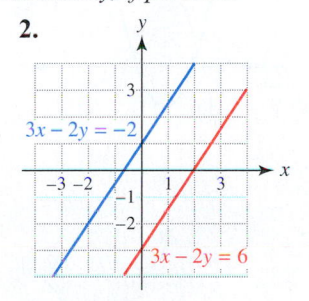

3.

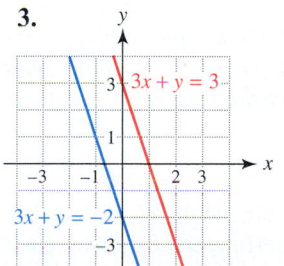

4.

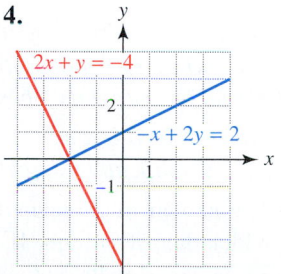

The Elimination Method

Exercises 5–16: Use elimination to solve the system of equations, if possible. Identify the system as consistent or inconsistent. If the system is consistent, state whether the equations are dependent or independent. Support your results graphically or numerically.

5. $x + y = 20$
$\quad\ x - y = \ 8$

6. $2x + y = 15$
$\quad\quad x - y = \ 0$

7. $x + 3y = 10$
$\quad x - 2y = -5$

8. $4x + 2y = 10$
$\quad -2x - y = 10$

9. $x + y = 500$
$\quad \frac{1}{20}x + \frac{1}{20}y = 25$

10. $2x + 3y = 4$
$\quad x - 2y = -5$

11. $5x + y = -5$
$\quad 7x - 3y = -29$

12. $2x + 3y = 5$
$\quad 5x - 2y = 3$

13. $2x + 4y = 7$
$\quad -x - 2y = 5$

14. $4x - 3y = 5$
$\quad 3x + 4y = 2$

15. $2x + 3y = 2$
$\quad x - 2y = -5$

16. $x - 3y = 1$
$\quad 2x - 6y = 2$

Exercises 17–28: Solve the system, if possible.

17. $\frac{1}{2}x - y = 5$
$\quad x - \frac{1}{2}y = 4$

18. $\frac{1}{2}x - \frac{1}{3}y = 1$
$\quad \frac{1}{3}x - \frac{1}{2}y = 1$

19. $2x - 3y = \frac{5}{3}$
$\quad 6x - 2y = \frac{8}{3}$

20. $4x + 3y = 8$
$\quad -2x + 6y = 1$

21. $7x - 3y = -17$
$\quad -21x + 9y = 51$

22. $-\frac{1}{3}x + \frac{1}{6}y = -1$
$\quad 2x - y = 6$

23. $\frac{2}{3}x + \frac{4}{3}y = \frac{1}{3}$
$\quad -2x - 4y = 5$

24. $5x - 2y = 7$
$\quad 10x - 4y = 6$

25. $-5x + 2y = 15$
$\quad x - 6y = -17$

26. $2x - 3y = 1$
$\quad 3x - 2y = 2$

27. $0.2x + 0.3y = 8$
$\quad -0.4x + 0.2y = 0$

28. $\frac{1}{2}x - \frac{3}{2}y = 9$
$\quad -\frac{3}{2}x + \frac{5}{2}y = -17$

Exercises 29–32: Write a system of linear equations with two variables whose solution satisfies the problem. State what each variable represents. Then solve the system

 (a) graphically, and
(b) symbolically using elimination.

29. The screen of a rectangular television set is 2 inches wider than it is high. If the perimeter of the screen is 38 inches, find its dimensions.

30. The sum of two numbers is 300 and their difference is 8. Find the two numbers.

31. Admission prices to a movie are $4 for children and $7 for adults. If 75 tickets were sold for $456, how many of each type of ticket were sold?

32. A sample of 16 dimes and quarters has a value of $2.65. How many of each type of coin are there?

Exercises 33–40: Use elimination to solve the system of non-linear equations.

33. $x^2 + y = 12$
$\quad x^2 - y = 6$

34. $x^2 + 2y = 15$
$\quad 2x^2 - y = 10$

35. $x^2 + y^2 = 25$
$\quad x^2 + 7y = 37$

36. $x^2 + y^2 = 36$
$\quad x^2 - 6y = 36$

37. $x^2 + y^2 = 4$
$\quad 2x^2 + y^2 = 8$

38. $x^2 + y^2 = 4$
$\quad x^2 - y^2 = 4$

39. $\sqrt{x} - y = 0$
$\quad x + 2y = 3$

40. $x + y^2 = 6$
$\quad 3x - y^2 = 4$

Inequalities

Exercises 41–52: Graph the solution set to the inequality.

41. $x \ge y$

42. $y > -3$

43. $x < 1$

44. $y > 2x$

45. $x + y \le 2$

46. $x + y > -3$

47. $2x + y > 4$

48. $2x + 3y \le 6$

49. $x^2 + y^2 > 4$

50. $x^2 + y^2 \le 1$

51. $x^2 + y \le 2$

52. $2x^2 - y < 1$

Exercises 53–56: Match the system of inequalities with the appropriate graph (a. – d.). Use the graph to identify one solution.

53. $x + y \ge 2$
$\quad x - y \le 1$

54. $2x - y > 0$
$\quad x - 2y \le 1$

55. $\frac{1}{2}x^3 - y > 0$
$\quad 2x - y \le 1$

56. $x^2 + y \le 4$
$\quad x^2 - y \le 2$

a.

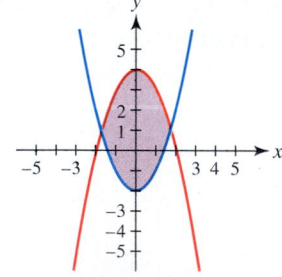

b.

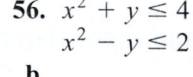
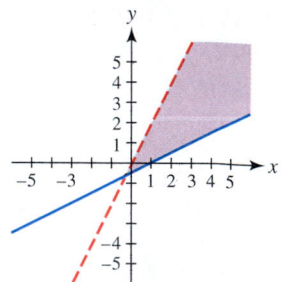

c.

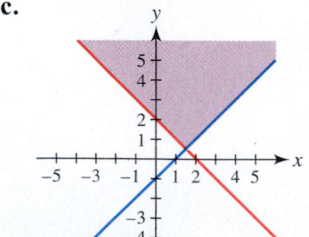

d.

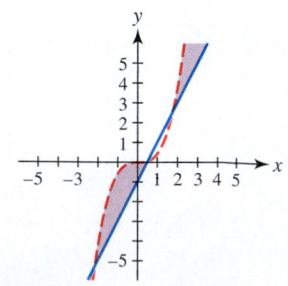

Exercises 57–62: Graph the solution set to the system of inequalities. Use the graph to identify one solution.

57. $y \geq x^2$
        $x + y \leq 6$

58. $y \leq \sqrt{x}$
        $y \geq 1$

59. $x + 2y > -2$
        $x + 2y < 5$

60. $x - y \leq 3$
        $x + y \leq 3$

61. $x^2 + y^2 \leq 16$
        $x + y < 2$

62. $x^2 + y \leq 4$
        $x^2 - y \leq 3$

Exercises 63–70: Graph the solution set to the system of inequalities.

63. $x + 2y \leq 4$
        $2x - y \geq 6$

64. $3x - y \leq 3$
        $x + 2y \leq 2$

65. $3x + 2y < 6$
        $x + 3y \leq 6$

66. $4x + 3y \geq 12$
        $2x + 6y \geq 4$

67. $x - 2y \geq 0$
        $x - 3y \leq 3$

68. $2x - 4y \geq 4$
        $x + y \leq 0$

69. $x^2 + y^2 \leq 4$
        $y \geq 1$

70. $x^2 - y \leq 0$
        $x^2 + y^2 \leq 6$

Applications

71. *Student Loans* A student takes out two loans totaling $3000 to help pay for college expenses. One loan is at 8% interest, and the other is at 10%. Interest for both loans is compounded annually.
   (a) If the first-year interest is $264, determine a system of linear equations whose solution is the amount of each loan.

   (b) Find the amount of each loan.

72. *Student Loans* (Refer to the previous exercise.) Suppose that both loans have an interest rate of 10% and the total first-year interest is $300. If possible, determine the amount of each loan. Interpret your results.

73. *Student Loans* (Refer to the previous two exercises.) Suppose that both loans are at 10% and the total annual interest is $264. If possible, determine the amount of each loan. Interpret your results.

74. *Maximizing Area* Suppose a rectangular pen for a pet is to be made using 40 feet of fence. Let l represent its length and w its width with $l \geq w$.
   (a) Find the dimensions that result in an area of 91 square feet.

   (b) Write a formula for the area A in terms of w.

   (c) What is the maximum area possible for the pen? Interpret this result.

75. *Television Sets* In 1999 there were 32 million television sets sold. For every 10 television sets sold with stereo sound, 19 television sets were sold without stereo sound. (**Source:** Consumer Electronics.)
   (a) Write a linear system whose solution gives the number of television sets sold with stereo sound and the number of television sets sold without stereo sound.

   (b) Solve the system and interpret the result.

76. *The Toll of War* American battlefield deaths in World War I and II totaled about 345,000. There were about 5.5 times as many deaths in World War II as World War I. Find the number of American battlefield deaths in each war. Round your answers to the nearest whole number. (**Source:** Defense Department.)

77. *Cigarettes Sold Overseas* The total number of cigarettes sold overseas in 1990 and 2000 was 10.9 trillion cigarettes. The number sold in 2000 was 100 billion more than the number sold in 1990. How many cigarettes were sold each year? (**Source:** Department of Agriculture.)

78. *Online Shopping* Americans spent $40.4 billion on the Internet in 2002 for travel and computer hardware. The amount spent for travel was triple the amount spent for computer hardware. Find the amount spent on travel and the amount spent on computer hardware. (**Source:** Comscore Networks.)

79. *Air Speed* A jet airliner travels 1680 miles in 3 hours with a tail wind. The return trip, into the wind, takes 3.5 hours. Find both the air speed of the jet and the wind speed. (*Hint:* First find the ground speed of the airplane in each direction.)

80. *River Current* A tugboat can pull a barge 60 miles upstream in 15 hours. The same tugboat and barge can make the return trip downstream in 6 hours. Determine the speed of the current in the river.

81. **Traffic Control** The figure shows two intersections labeled A and B that involve one-way streets. The numbers and variables represent the average traffic flow rates measured in vehicles per hour. For example, an average of 500 vehicles per hour enter intersection A from the west, whereas 150 vehicles per hour enter this intersection from the north. A stoplight will control the unknown traffic flow denoted by the variables x and y. Use the fact that the number of vehicles entering an intersection must equal the number leaving to determine x and y.

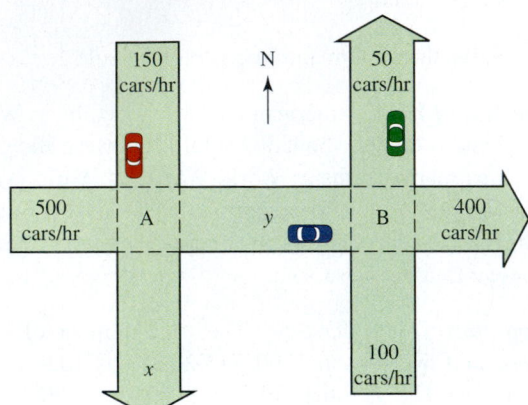

82. **Traffic Control** (Refer to the previous exercise and accompanying figure.) Suppose that the number of vehicles entering intersection A from the west varies between 400 and 600. If all other traffic flows remain the same as in the figure, what effect does this have on the ranges of the values for x and y?

*Exercises 83–86: **Weight and Height*** *The following graph shows a weight and height chart. The weight w is listed in pounds and the height h in inches. The shaded area is a recommended region.* (**Source:** Department of Agriculture.)

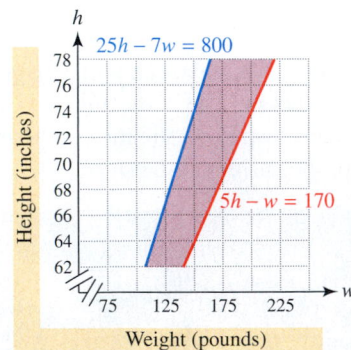

83. What does this chart indicate about an individual who weighs 125 pounds and is 70 inches tall?

84. If a person is 74 inches tall, use the graph to estimate the recommended weight range.

85. Use the graph to find a system of linear inequalities that describes the recommended region.

86. Explain why inequalities, rather than equalities, are more appropriate to describe recommended weight and height combinations.

Linear Programming

Exercises 87–90: Shade the region of feasible solutions for the following constraints.

87. $x + y \le 4$
    $x + y \ge 1$
    $x \ge 0, y \ge 0$

88. $x + 2y \le 8$
    $2x + y \ge 2$
    $x \ge 0, y \ge 0$

89. $3x + 2y \le 12$
    $2x + 3y \le 12$
    $x \ge 0, y \ge 0$

90. $x + y \le 4$
    $x + 4y \ge 4$
    $x \ge 0, y \ge 0$

Exercises 91 and 92: The graph shows a region of feasible solutions. Find the maximum and minimum values of P over this region.

91. $P = 3x + 5y$

92. $P = 6x + y$

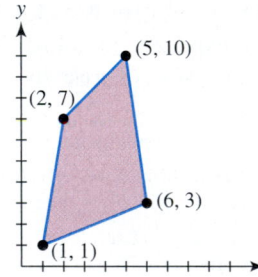
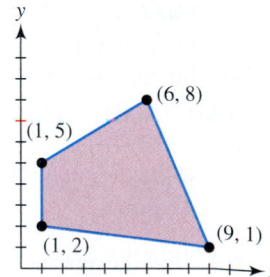

Exercises 93–96: The graph shows a region of feasible solutions. Find the maximum and minimum of C over this region.

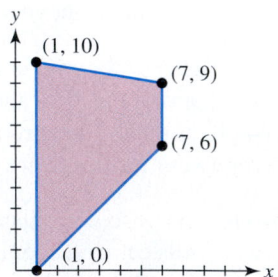

93. $C = 3x + 5y$

94. $C = 5x + 5y$

95. $C = 10y$

96. $C = 3x - y$

Exercises 97 and 98: Write a system of linear inequalities that describes the shaded region.

97.   **98.**

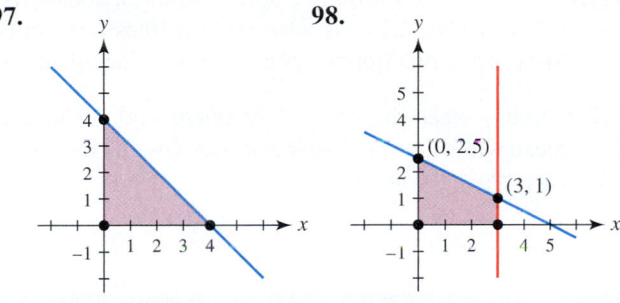

99. Find the minimum value of $C = 4x + 2y$ subject to the following constraints.

$$x + y \geq 3$$
$$2x + 3y \leq 12$$
$$x \geq 0, y \geq 0$$

100. Find the maximum value of $P = 3x + 5y$ subject to the following constraints.

$$3x + y \leq 8$$
$$x + 3y \leq 8$$
$$x \geq 0, y \geq 0$$

Exercises 101 and 102: If possible, maximize and minimize z subject to the given constraints.

101. $z = 7x + 6y$

$$x + y \leq 8$$
$$x + y \geq 4$$
$$x \geq 0, y \geq 0$$

102. $z = 8x + 3y$

$$4x + y \geq 12$$
$$x + 2y \geq 6$$
$$x \geq 0, y \geq 0$$

103. *Maximizing Profit* Rework Example 8 if the profit from each radio is $20 and the profit from each CD player is $15.

104. *Maximizing Revenue* A refinery produces both gasoline and fuel oil, and sells gasoline for $1.00 per gallon and fuel oil for $0.90 per gallon. The refinery can produce at most 600,000 gallons a day, but must produce at least 2 gallons of fuel oil for every gallon of gasoline. Furthermore, at least 150,000 gallons of fuel oil must be produced each day for the coming winter. Determine how much of each type of fuel should be produced to maximize revenue.

105. *Minimizing Cost* (Refer to Example 10.) A breeder is mixing Brand A and Brand B. Each serving should contain at least 60 grams of protein and 30 grams of fat. Brand A costs 80 cents per unit and Brand B costs 50 cents per unit. Each unit of Brand A contains 15 grams of protein and 10 grams of fat, whereas each unit of Brand B contains 20 grams of protein and 5 grams of fat. Determine how much of each food should be bought to achieve a minimum cost per serving.

106. *Pet Food Cost* A pet owner is buying two brands of food, X and Y, for his animals. Each serving of the mixture of the two foods should contain at least 60 grams of protein and 40 grams of fat. Brand X costs 75 cents per unit and Brand Y costs 50 cents per unit. Each unit of Brand X contains 20 grams of protein and 10 grams of fat, whereas each unit of Brand Y contains 10 grams of protein and 10 grams of fat. Determine how much of each brand should be bought to obtain a minimum cost per serving.

107. *Raising Animals* A breeder can raise no more than 50 hamsters and mice but no more than 20 hamsters. If she sells the hamsters for $15 each and the mice for $10 each, find the maximum revenue produced.

108. *Maximizing Storage* An office manager wants to buy filing cabinets. Cabinet X costs $100, requires 6 square feet of floor space, and holds 8 cubic feet. Cabinet Y costs $200, requires 8 square feet of floor space, and holds 12 cubic feet. No more than $1400 can be spent, and the office has room for no more than 72 square feet of cabinets. The office manager wants the maximum storage capacity within the limits imposed by funds and space. How many of each type of cabinet should be bought?

109. *Maximizing Profit* A business manufactures two parts, X and Y. Machines A and B are needed to make each part. To make part X, machine A is needed 4 hours and machine B is needed 2 hours. To make part Y, machine A is needed 1 hour and machine B is needed 3 hours. Machine A is available for 40 hours each week and machine B is available for 30 hours. The profit from part X is $500 and the profit from part Y is $600. How many parts of each type should be made to maximize weekly profit?

110. *Minimizing Cost*   Two substances, X and Y, are found in pet food. Each substance contains the ingredients A and B. Substance X is 20% ingredient A and 50% ingredient B. Substance Y is 50% ingredient A and 30% ingredient B. The cost of substance X is $2 per pound and the cost of substance Y is $3 per pound. The pet store needs at least 251 pounds of ingredient A and at least 200 pounds of ingredient B. If cost is to be minimal, how many pounds of each substance should be ordered? Find the minimum cost.

Writing about Mathematics

111. Give the general form of a system of linear equations in two variables. Discuss what distinguishes a system of linear equations from a system of nonlinear equations.

112. Discuss what the words *consistent* and *inconsistent* mean with regard to linear systems. Give an example of each type of system.

CHECKING BASIC CONCEPTS FOR SECTIONS 9.1 AND 9.2

1. Evaluate $d(13, 18)$ if
$$d(x, y) = \sqrt{(x - 1)^2 + (y - 2)^2}.$$
(The function d computes the distance from the point (x, y) to the point $(1, 2)$.)

2. Solve the system of nonlinear equations using the method of substitution.
$$2x^2 - y = 0$$
$$3x + 2y = 7$$

3. Solve $z = x^2 + y^2$ for y.

4. Solve the system of linear equations by using elimination.
$$3x - 2y = 4$$
$$-x + 6y = 8$$

5. Graph the solution set to $3x - 2y \leq 6$.

6. Graph the solution set to the system of inequalities. Use the graph to identify one solution.
$$x^2 - y < 3$$
$$x - y \geq 1$$

7. In 1996 the United States and the former Soviet Union had a combined total of 13,820 strategic nuclear warheads. The United States had 480 more warheads than the former Soviet Union.   (**Sources:** Arms Control Association, Natural Resources Defense Council.) Find a system of linear equations whose solution gives the number of nuclear warheads for each country. Solve the system.

9.3 Systems of Linear Equations in Three Variables

Introduction

In Sections 9.1 and 9.2 we discussed how to solve systems of linear equations in two variables. Systems of linear equations can have any number of variables. For example, in the design of modern aircraft, engineers routinely solve systems of linear equations containing 100,000 variables. Computers are necessary to solve these systems efficiently. However, in this section we discuss solving systems of linear equations containing three variables by hand.

Basic Concepts

- Learn basic concepts about systems in three variables
- Solve systems using elimination and substitution
- Solve systems with no solutions
- Solve systems with infinitely many solutions

When writing systems of linear equations in three variables it is common, but not necessary, to use the variables x, y, and z. For example,
$$2x - 3y + 4z = 4$$
$$-y + 2z = 0$$
$$x + 5y - 6z = 7$$
represents a system of linear equations in three variables. The solution to this system is given by $x = 3$, $y = 2$, and $z = 1$ because each equation is satisfied when these values for the variables are substituted in the system of linear equations.

$$2(\mathbf{3}) - 3(\mathbf{2}) + 4(\mathbf{1}) \overset{?}{=} 4 \qquad \text{True}$$
$$-(\mathbf{2}) + 2(\mathbf{1}) \overset{?}{=} 0 \qquad \text{True}$$
$$(\mathbf{3}) + 5(\mathbf{2}) - 6(\mathbf{1}) \overset{?}{=} 7 \qquad \text{True}$$

The solution to this system can be written as the **ordered triple, $(\mathbf{3}, \mathbf{2}, \mathbf{1})$**. This system of linear equations has exactly one solution. In general, systems of linear equations can have zero, one, or infinitely many solutions.

EXAMPLE 1   Checking for solutions

Determine whether $(-1, -3, 2)$ or $(1, -10, -13)$ is the solution to the system of linear equations.

$$x - 4y + 2z = 15$$
$$4x - y + z = 1$$
$$6x - 2y - 3z = -6$$

SOLUTION   First, substitute $x = -\mathbf{1}, y = -\mathbf{3}$, and $z = \mathbf{2}$ in the system of linear equations, and then substitute $x = \mathbf{1}, y = -\mathbf{10}$, and $z = -\mathbf{13}$.

$(-\mathbf{1}) - 4(-\mathbf{3}) + 2(\mathbf{2}) \overset{?}{=} 15$ True $(\mathbf{1}) - 4(-\mathbf{10}) + 2(-\mathbf{13}) \overset{?}{=} 15$ True

$4(-\mathbf{1}) - (-\mathbf{3}) + (\mathbf{2}) \overset{?}{=} 1$ True $4(\mathbf{1}) - (-\mathbf{10}) + (-\mathbf{13}) \overset{?}{=} 1$ True

$6(-\mathbf{1}) - 2(-\mathbf{3}) - 3(\mathbf{2}) \overset{?}{=} -6$ True $6(\mathbf{1}) - 2(-\mathbf{10}) - 3(-\mathbf{13}) \overset{?}{=} -6$ False

The ordered triple $(-1, -3, 2)$ satisfies all three equations, so it is the solution to the system of equations. The ordered triple $(1, -10, -13)$ is not a solution to the system of equations because it satisfies only two of the three equations.   *Now Try Exercise 5* ◆

Solving with Elimination and Substitution

We can solve systems of linear equations in three variables by hand. The following procedure uses substitution and elimination and assumes that the variables are $x, y,$ and z.

> **SOLVING A SYSTEM OF LINEAR EQUATIONS IN THREE VARIABLES**
>
> **STEP 1:** Eliminate one variable, such as x, from two of the equations.
>
> **STEP 2:** Apply the techniques discussed in Sections 9.1 and 9.2 to solve the two resulting equations in two variables from **STEP 1**. If x is eliminated, then solve these equations to find y and z.
>
> *Note:* If there are no solutions for y and z, then the given system also has no solutions. If there are infinitely many solutions for y and z, then write y in terms of z and proceed to **STEP 3**.
>
> **STEP 3:** Substitute the values for y and z in one of the given equations to find x. The solution is (x, y, z).

EXAMPLE 2   Solving a linear system in three variables

Solve the following system.

$$x - y + 2z = 6$$
$$2x + y - 2z = -3$$
$$-x - 2y + 3z = 7$$

SOLUTION

STEP 1: We begin by eliminating the variable x from the second and third equations. To eliminate x from the second equation we multiply the first equation by -2 and then add it to the second equation. To eliminate x from the third equation we add the first and third equations.

$$
\begin{array}{ll}
-2x + 2y - 4z = -12 & \text{First equation times } -2 \\
\underline{2x + y - 2z = -3} & \text{Second equation} \\
3y - 6z = -15 & \text{Add.}
\end{array}
\qquad
\begin{array}{ll}
x - y + 2z = 6 & \text{First equation} \\
\underline{-x - 2y + 3z = 7} & \text{Third equation} \\
-3y + 5z = 13 & \text{Add.}
\end{array}
$$

STEP 2: Take the two resulting equations from **STEP 1** and eliminate either variable. Here we add the two equations to eliminate the variable y.

$$
\begin{array}{ll}
3y - 6z = -15 & \\
\underline{-3y + 5z = 13} & \\
-z = -2 & \text{Add the equations.} \\
z = \mathbf{2} & \text{Multiply by } -1.
\end{array}
$$

Now we can use substitution to find the value of y. We let $z = 2$ in either equation used in **STEP 2** to find y.

$$
\begin{array}{ll}
3y - 6z = -15 & \text{Equation from Step 2.} \\
3y - 6(\mathbf{2}) = -15 & \text{Substitute } z = 2. \\
3y - 12 = -15 & \text{Multiply.} \\
3y = -3 & \text{Add 12.} \\
y = -\mathbf{1} & \text{Divide by 3.}
\end{array}
$$

STEP 3: Finally, we substitute $y = -1$ and $z = 2$ in any of the given equations to find x.

$$
\begin{array}{ll}
x - \mathbf{y} + 2\mathbf{z} = 6 & \text{First given equation} \\
x - (\mathbf{-1}) + 2(\mathbf{2}) = 6 & \text{Let } y = -1 \text{ and } z = 2. \\
x + 1 + 4 = 6 & \text{Simplify.} \\
x = \mathbf{1} & \text{Subtract 5.}
\end{array}
$$

The solution is $(\mathbf{1}, -\mathbf{1}, \mathbf{2})$. Check this solution. *Now Try Exercise 9* ◆

In the next example we determine numbers of tickets sold at a play.

EXAMPLE 3 Finding numbers of tickets sold

One thousand tickets were sold for a play, which generated $3800 in revenue. The prices of the tickets were $3 for children, $4 for students, and $5 for adults. There were 100 fewer student tickets sold than adult tickets. Find the number of each type of ticket sold.

SOLUTION Let x be the number of tickets sold to children, y be the number of tickets sold to students, and z be the number of tickets sold to adults. The total number of tickets sold was 1000, so

$$x + y + z = 1000.$$

Each child's ticket costs $3, so the revenue generated from selling x tickets would be $3x$. Similarly, the revenue generated from students would be $4y$, and the revenue from adults would be $5z$. Total ticket sales were $3800, so

$$3x + 4y + 5z = 3800.$$

The equation $z - y = 100$ or $y - z = -100$ must also be satisfied, as 100 fewer tickets were sold to students than adults.

To find the price of a ticket we need to solve the following system of linear equations.

$$x + y + z = 1000$$
$$3x + 4y + 5z = 3800$$
$$y - z = -100$$

STEP 1: We begin by eliminating the variable x from the first equation. To do so, we multiply the first equation by 3 and subtract the second equation.

$$3x + 3y + 3z = 3000 \qquad \textit{First given equation times 3}$$
$$\underline{3x + 4y + 5z = 3800} \qquad \textit{Second equation}$$
$$-y - 2z = -800 \qquad \textit{Subtract.}$$

STEP 2: We then use the resulting equation from **STEP 1** and the third equation to eliminate y.

$$-y - 2z = -800 \qquad \textit{Equation from Step 1}$$
$$\underline{y - z = -100} \qquad \textit{Third given equation}$$
$$-3z = -900 \qquad \textit{Add the equations.}$$

Thus $z = 300$. To find y we can substitute $z = 300$ in the third equation.

$$y - z = -100 \qquad \textit{Third given equation}$$
$$y - 300 = -100 \qquad \textit{Let z = 300.}$$
$$y = 200 \qquad \textit{Add 300.}$$

STEP 3: Finally, we substitute $y = 200$ and $z = 300$ in the first equation.

$$x + y + z = 1000 \qquad \textit{First given equation}$$
$$x + 200 + 300 = 1000 \qquad \textit{Let y = 200 and z = 300.}$$
$$x = 500 \qquad \textit{Subtract 500.}$$

Thus 500 tickets were sold to children, 200 to students, and 300 to adults.

Now Try Exercise 31 ◆

Systems with No Solutions

Regardless of the number of variables, a system of linear equations can have zero, one, or infinitely many solutions. In the next example a system of linear equations has no solutions.

Solving a system with no solutions

Three students buy lunch in the cafeteria. One student buys 2 hamburgers, 1 order of fries, and 1 soda for $9. Another student buys 1 hamburger, 2 orders of fries, and 1 soda for $8. The third student buys 3 hamburgers, 3 fries, and 2 sodas for $18. If possible, find the cost of each item. Interpret the results.

SOLUTION Let x be the cost of a hamburger, y be the cost of an order of fries, and z be the cost of a soda. Then the purchases of the three students can be expressed as a system of linear equations.

$$2x + y + z = 9$$
$$x + 2y + z = 8$$
$$3x + 3y + 2z = 18$$

STEP 1: We can eliminate z in the first equation by subtracting the second equation from the first equation. We can eliminate z in the third equation by subtracting twice the second equation from the third equation.

$2x + y + z = 9$	First equation		$3x + 3y + 2z = 18$	Third equation
$x + 2y + z = 8$	Second equation		$2x + 4y + 2z = 16$	Twice second equation
$x - y \quad\;\; = 1$	Subtract.		$x - y \quad\;\; = 2$	Subtract.

STEP 2: The equations $x - y = 1$ and $x - y = 2$ are *inconsistent* because the difference between two numbers cannot be both 1 and 2. **STEP 3** is not necessary—the system of equations has no solutions.

In this problem the third student bought the same amount of food as the first and second students bought together. Therefore the third student should have paid $\$9 + \$8 = \$17$ rather than $\$18$. *Inconsistent pricing* led to an *inconsistent system* of linear equations.

Now Try Exercise 33 ◆

Systems with Infinitely Many Solutions

Some systems of linear equations have infinitely many solutions. In this case, we say that the system of linear equations is consistent, but the equations are dependent. A system of dependent equations is solved in the next example.

 **EXAMPLE 5**   **Solving a system with infinitely many solutions**

Solve the following system of linear equations.

$$\begin{aligned} x + y - z &= -2 \\ x + 2y - 2z &= -3 \\ y - z &= -1 \end{aligned}$$

SOLUTION

STEP 1: Because x does not appear in the third equation, we begin by eliminating x from the first equation. To do this, subtract the second equation from the first equation.

$x + y - z = -2$	First equation
$x + 2y - 2z = -3$	Second equation
$-y + z = 1$	Subtract.

STEP 2: Adding the resulting equation from **STEP 1** and the third given equation gives the equation $0 = 0$, which indicates that there are infinitely many solutions.

$-y + z = 1$
$y - z = -1$
$0 = 0$ Add.

The variable y can be written in terms of z as $y = z - 1$.

STEP 3: To find x, substitute the results from **STEP 2** in the first given equation.

$x + y - z = -2$	First given equation
$x + (z - 1) - z = -2$	Let $y = z - 1$.
$x = -1$	Solve for x.

Solutions to the given system are of the form $(-1, z - 1, z)$, where z is any real number. For example, if $z = 2$, then $(-1, 1, 2)$ is one possible solution.

Now Try Exercise 15 ◆

◆ **CLASS DISCUSSION**

Three students buy lunch in the cafeteria. One student buys 1 hamburger, 1 order of fries, and 1 soda for $5. Another student buys 2 hamburgers, 2 orders of fries, and 2 sodas for $10. The third student buys 3 hamburgers, 3 orders of fries, and 3 sodas for $15. Can you find the cost of each item? Interpret your answer. ◆

9.3

Putting it all Together

In this section we discussed how to solve a system of three linear equations in three variables by hand. Systems of linear equations can have no solutions, one solution, or infinitely many solutions. The following table summarizes some of the important concepts presented in this section.

Concept	Explanation
System of linear equations in three variables	The following is a system of three linear equations in three variables. $$x - 2y + z = 0$$ $$-x + y + z = 4$$ $$-y + 4z = 10$$
Solution to a linear system in three variables	The solution to a linear system in three variables is an ordered triple, expressed as (x, y, z). The solution to the preceding system is $(1, 2, 3)$ because substituting $x = 1, y = 2$, and $z = 3$ in each equation results in a true statement. $$(1) - 2(2) + (3) = 0 \quad \text{True}$$ $$-(1) + (2) + (3) = 4 \quad \text{True}$$ $$-(2) + 4(3) = 10 \quad \text{True}$$
Solving a linear system with substitution and elimination	Refer to Example 2. **STEP 1:** Eliminate one variable, such as x, from two of the equations. **STEP 2:** Apply the techniques discussed in Sections 9.1 and 9.2 to solve the two resulting equations in two variables from **STEP 1**. If x is eliminated, then solve these equations to find y and z. **STEP 3:** Substitute the values for y and z in one of the given equations to find x. The solution is (x, y, z).

9.3 Exercises

1. Can a system of linear equations have exactly three solutions?

2. Does the ordered triple $(1, 2, 3)$ satisfy the equation $3x + 2y + z = 10$?

3. To solve a system of linear equations in two variables, how many equations do you usually need?

4. To solve a system of linear equations in three variables, how many equations do you usually need?

Exercises 5–8: Determine whether each ordered triple is a solution to the system of linear equations.

5. $(0, 2, -2), (-1, 3, -2)$
$$x + y - z = 4$$
$$-x + y + z = 2$$
$$x + y + z = 0$$

6. $(5, 2, 2), (2, -1, 1)$
$$2x - 3y + 3z = 10$$
$$x - 2y - 3z = 1$$
$$4x - y + z = 10$$

7. $\left(-\frac{5}{11}, \frac{20}{11}, -2\right), (1, 2, -1)$
$$x + 3y - 2z = 9$$
$$-3x + 2y + 4z = -3$$
$$-2x + 5y + 2z = 6$$

8. $(1, 2, 3), (11, 16, -3)$
$$4x - 2y + 2z = 6$$
$$2x - 4y - 6z = -24$$
$$-3x + 3y + 2z = 9$$

Exercises 9–30: If possible, solve the system of linear equations.

9. $x + y + z = 6$
$-x + 2y + z = 6$
$y + z = 5$

10. $x - y + z = -2$
$x - 2y + z = 0$
$y - z = 1$

11. $x + 2y + 3z = 4$
$2x + y + 3z = 5$
$x - y + z = 2$

12. $x - y + z = 2$
$3x - 2y + z = -1$
$x + y = -3$

13. $3x + y + z = 0$
$4x + 2y + z = 1$
$2x - 2y - z = 2$

14. $-x - 5y + 2z = 2$
$x + y + 2z = 2$
$3x + y - 4z = -10$

15. $x + 3y + z = 6$
$3x + y - z = 6$
$x - y - z = 0$

16. $2x - y + 2z = 6$
$-x + y + z = 0$
$-x - 3z = -6$

17. $x - 4y + 2z = -2$
$x + 2y - 2z = -3$
$x - y = 4$

18. $2x + y + 3z = 4$
$-3x - y - 4z = 5$
$x + y + 2z = 0$

19. $4a - b + 2c = 0$
$2a + b - c = -11$
$2a - 2b + c = 3$

20. $a - 4b + 3c = 2$
$-a - 2b + 5c = 9$
$a + 2b + c = 6$

21. $a + b + c = 0$
$a - b - c = 3$
$a + 3b + 3c = 5$

22. $a - 2b + c = -1$
$a + 5b = -3$
$2a + 3b + c = -2$

23. $3x + 2y + z = -1$
$3x + 4y - z = 1$
$x + 2y + z = 0$

24. $x - 2y + z = 1$
$x + y + 2z = 2$
$2x + 3y + z = 6$

25. $-x + 3y + z = 3$
$2x + 7y + 4z = 13$
$4x + y + 2z = 7$

26. $x + 2y + z = 0$
$3x + 2y - z = 4$
$-x + 2y + 3z = -4$

27. $-x + 2z = -9$
$y + 4z = -13$
$3x + y = 13$

28. $x + y + z = -1$
$2x + z = -6$
$2y + 3z = 0$

29. $\frac{1}{2}x - y + \frac{1}{2}z = -4$
$x + 2y - 3z = 20$
$-\frac{1}{2}x + 3y + 2z = 0$

30. $\frac{3}{4}x + y + \frac{1}{2}z = -3$
$x + y - z = -8$
$\frac{1}{4}x - 2y + z = -4$

Applications

31. *Tickets Sold* Five hundred tickets were sold for a play, which generated \$3560 in revenue. The prices of the tickets were \$5 for children, \$7 for students, and \$10 for adults. There were 180 more student tickets sold than adult tickets. Find the number of each type of ticket sold.

32. *Tickets Sold* One thousand tickets were sold for a baseball game. There were one hundred more adult tickets sold than student tickets, and there were four times as many tickets sold to students as to children. How many of each type of ticket were sold?

33. *Buying Lunch* Three students buy lunch in the cafeteria. One student buys 2 hamburgers, 2 orders of fries, and 1 soda for \$9. Another student buys 1 hamburger, 1 order of fries, and 1 soda for \$5. The third student buys 1 hamburger and 1 order of fries for \$5. If possible, find the cost of each item. Interpret the results.

34. *Cost of CDs* The accompanying table at the top of the next page shows the total cost of purchasing various combinations of differently priced CDs. The types of CDs are labeled A, B, and C.

A	B	C	Total Cost
2	1	1	$48
3	2	1	$71
1	1	2	$53

(a) Let a be the cost of a CD of type A, b be the cost of a CD of type B, and c be the cost of a CD of type C. Write a system of three linear equations whose solution gives the cost of each type of CD.

(b) Solve the system of equations and check your answer.

35. *Geometry* The largest angle in a triangle is 25° more than the smallest angle. The sum of the measures of the two smaller angles is 30° more than the measure of the largest angle.
(a) Let x, y, and z be the measures of the three angles from largest to smallest. Write a system of three linear equations whose solution gives the measure of each angle.

(b) Solve the system of equations and check your answer.

36. *Geometry* The perimeter of a triangle is 105 inches. The longest side is 22 inches longer than the shortest side. The sum of the lengths of the two shorter sides is 15 inches more than the length of the longest side. Find the lengths of the sides of the triangle.

37. *Investment Mixture* A sum of $20,000 is invested in three mutual funds. In one year the first fund grew by 5%, the second by 7%, and the third by 10%. Total earnings for the year were $1650. The amount invested in the third fund was 4 times the amount invested in the first fund. Find the amount invested in each fund.

38. *Predicting Home Prices* Selling prices of homes can depend on several factors such as size and age. The table at the top of the next column shows the selling price for three homes. In this table, price P is given in thousands of dollars, age A in years, and home size S in thousands of square feet. These data may be modeled by the equation $P = a + bA + cS$.

Price (P)	Age (A)	Size (S)
190	20	2
320	5	3
50	40	1

(a) Write a system of linear equations whose solution gives a, b, and c.

(b) Solve this system of linear equations.

(c) Predict the price of a home that is 10 years old and has 2500 square feet.

39. *Mixture Problem* One type of lawn fertilizer consists of a mixture of nitrogen, N, phosphorus, P, and potassium, K. An 80-pound sample contains 8 more pounds of nitrogen and phosphorus than potassium. There is 9 times as much potassium as phosphorus.
(a) Write a system of three equations whose solution gives the amount of nitrogen, phosphorus, and potassium in this sample.

(b) Solve the system of equations.

40. *Business Production* A business has three machines that manufacture containers. Together they make 100 containers per day, whereas the two fastest machines can make 80 containers per day. The fastest machine makes 34 more containers per day than the slowest machine.
(a) Let x, y, and z be the number of containers that the machines make from fastest to slowest. Write a system of three equations whose solution gives the number of containers each machine can make.

(b) Solve the system of equations.

Writing about Mathematics

41. When using elimination and substitution, explain how to recognize a system of linear equations that has no solutions.

42. When using elimination and substitution, explain how to recognize a system of linear equations that has infinitely many solutions.

9.4 Solutions to Linear Systems Using Matrices

- ◆ Represent systems of linear equations with matrices
- ◆ Learn row-echelon form
- ◆ Perform Gaussian elimination
- ◆ Solve systems of linear equations with technology (optional)

Introduction

The methods of elimination and substitution can be applied to linear systems that involve more than three variables. As the number of variables increases, these methods can become cumbersome and inefficient. However, by properly combining these two methods, they provide a state-of-the-art numerical method, capable of solving systems of linear equations that contain thousands of variables. This numerical method is called *Gaussian elimination with backward substitution*. Even though this method dates back to Carl Friedrich Gauss (1777–1855), it continues to be one of the most efficient methods for solving systems of linear equations.

Representing Systems of Linear Equations with Matrices

Arrays of numbers occur frequently in many different situations. Spreadsheets often make use of arrays, where data are displayed in a tabular format. A **matrix** is a rectangular array of elements. The following are examples of matrices whose elements are real numbers.

$$\begin{bmatrix} 4 & -7 \\ -2 & 9 \end{bmatrix} \quad \begin{bmatrix} -1 & -5 & 3 \\ 1.2 & 0 & -1.3 \\ 4.1 & 5 & 7 \end{bmatrix} \quad \begin{bmatrix} -3 & -6 & 9 & 5 \\ \sqrt{2} & -8 & -8 & 0 \\ 3 & 0 & 19 & -7 \\ -11 & -3 & 7 & 8 \end{bmatrix} \quad \begin{bmatrix} 5 & -2 \\ -2 & \pi \\ 1 & -1 \end{bmatrix} \quad \begin{bmatrix} 1 & -0.5 & 9 \\ 5 & 0.4 & -3 \end{bmatrix}$$

$$2 \times 2 \qquad\qquad 3 \times 3 \qquad\qquad\qquad 4 \times 4 \qquad\qquad\qquad 3 \times 2 \qquad\qquad 2 \times 3$$

The dimension of a matrix is given much like the dimensions of a rectangular room. We might say a room is m feet long and n feet wide. The **dimension** of a matrix is $m \times n$ (m by n), if it has m rows and n columns. For example, the last matrix has a dimension of 2×3, because it has 2 rows and 3 columns. If the number of rows and columns are equal, the matrix is a **square matrix**. The first three matrices are square matrices.

Matrices are frequently used to represent systems of linear equations. A linear system with three equations and three variables can be written as follows.

$$a_1x + b_1y + c_1z = d_1$$
$$a_2x + b_2y + c_2z = d_2$$
$$a_3x + b_3y + c_3z = d_3$$

The a_k, b_k, c_k, and d_k are constants and x, y, and z are variables. The coefficients of the variables can be represented using a 3×3 matrix. This matrix is called the **coefficient matrix** of the linear system.

$$\begin{bmatrix} a_1 & b_1 & c_1 \\ a_2 & b_2 & c_2 \\ a_3 & b_3 & c_3 \end{bmatrix}$$

◆ **CLASS DISCUSSION**
Give a general form of a system of linear equations with four equations and four variables. Write its augmented matrix. ◆

If we enlarge the matrix to include the constants d_k, the system may be written as

$$\left[\begin{array}{ccc|c} a_1 & b_1 & c_1 & d_1 \\ a_2 & b_2 & c_2 & d_2 \\ a_3 & b_3 & c_3 & d_3 \end{array}\right].$$

Because this matrix is an enlargement of the coefficient matrix, it is commonly called an **augmented matrix**. The vertical line between the third and fourth columns corresponds to where the equals sign occurs in each equation.

EXAMPLE 1 Representing a linear system with an augmented matrix

Express each linear system with an augmented matrix. State the dimension of the matrix.

(a) $\begin{aligned} 3x - 4y &= 6 \\ -5x + y &= -5 \end{aligned}$ (b) $\begin{aligned} 2x - 5y + 6z &= -3 \\ 3x + 7y - 3z &= 8 \\ x + 7y &= 5 \end{aligned}$

SOLUTION

(a) This system has two equations with two variables. It can be represented by an augmented matrix having dimension 2×3.

$$\left[\begin{array}{rr|r} 3 & -4 & 6 \\ -5 & 1 & -5 \end{array}\right]$$

(b) This system has three equations with three variables. Note that variable z does not appear in the third equation. A value of 0 is inserted for its coefficient.

$$\left[\begin{array}{rrr|r} 2 & -5 & 6 & -3 \\ 3 & 7 & -3 & 8 \\ 1 & 7 & 0 & 5 \end{array}\right]$$

Since this matrix has 3 rows and 4 columns, its dimension is 3×4.

Now Try Exercises 3 and 5 ◆

EXAMPLE 2 Converting an augmented matrix into a linear system

Write the linear system represented by the augmented matrix. Let the variables be x, y, and z.

(a) $\left[\begin{array}{rrr|r} 1 & 0 & 2 & -3 \\ 2 & 2 & 10 & 3 \\ -1 & 2 & 3 & 5 \end{array}\right]$ (b) $\left[\begin{array}{rrr|r} 1 & 2 & 3 & -4 \\ 0 & 1 & -6 & 7 \\ 0 & 0 & 1 & 8 \end{array}\right]$

SOLUTION

(a) The first column corresponds to x, the second to y, and the third to z. When a 0 appears, the variable for that column does not appear in the equation. The vertical line corresponds to the location of the equals sign. The last column represents the constant terms.

$$\begin{aligned} x \qquad + 2z &= -3 \\ 2x + 2y + 10z &= 3 \\ -x + 2y + 3z &= 5 \end{aligned}$$

(b) The augmented matrix represents the following linear system.

$$\begin{aligned} x + 2y + 3z &= -4 \\ y - 6z &= 7 \\ z &= 8 \end{aligned}$$

Now Try Exercises 7 and 9 ◆

Row-Echelon Form

To solve linear systems with augmented matrices, a convenient form is row-echelon form. The following matrices are in row-echelon form.

$$\begin{bmatrix} 1 & 3 & 0 & -1 \\ 0 & 1 & -6 & 1 \\ 0 & 0 & 1 & -2 \end{bmatrix} \quad \begin{bmatrix} 1 & 2 & 0 \\ 0 & 1 & 4 \end{bmatrix} \quad \begin{bmatrix} 1 & 3 & -1 & 5 \\ 0 & 1 & -1 & 3 \\ 0 & 0 & 1 & 0 \end{bmatrix} \quad \begin{bmatrix} 1 & 3 & -1 & 5 \\ 0 & 0 & 1 & 3 \\ 0 & 0 & 0 & 0 \end{bmatrix} \quad \begin{bmatrix} 1 & 3 & 5 \\ 0 & 0 & 1 \end{bmatrix}$$

The elements of the **main diagonal** are blue in each matrix. Scanning down the main diagonal of a matrix in row-echelon form, we see that this diagonal first contains only 1's, and then possibly 0's. The first nonzero element in each nonzero row is 1. This 1 is called a **leading 1**. If two rows contain a leading 1, then the row with the left-most leading 1 is listed first in the matrix. Rows containing only 0's occur last in the matrix. All elements below the main diagonal are 0.

The next example demonstrates a technique called **backward substitution**. It can be used to solve linear systems represented by an augmented matrix in row-echelon form.

EXAMPLE 3 Solving a linear system with backward substitution

Solve the system of linear equations.

(a) $\begin{bmatrix} 1 & 1 & 3 & | & 12 \\ 0 & 1 & -2 & | & -4 \\ 0 & 0 & 1 & | & 3 \end{bmatrix}$

(b) $\begin{bmatrix} 1 & -1 & 5 & | & 5 \\ 0 & 1 & 3 & | & 3 \\ 0 & 0 & 0 & | & 0 \end{bmatrix}$

SOLUTION

(a) The linear system is represented by

$$\begin{aligned} x + y + 3z &= 12 \\ y - 2z &= -4 \\ z &= 3. \end{aligned}$$

Since $z = 3$, substitute this value in the second equation to find y.

$$y - 2(3) = -4 \quad \text{or} \quad y = 2$$

Then $y = 2$ and $z = 3$ can be substituted in the first equation to determine x.

$$x + 2 + 3(3) = 12 \quad \text{or} \quad x = 1$$

The solution is given by $x = 1, y = 2$, and $z = 3$, and can be expressed as the *ordered triple* $(1, 2, 3)$.

(b) The linear system is represented by

$$\begin{aligned} x - y + 5z &= 5 \\ y + 3z &= 3 \\ 0 &= 0. \end{aligned}$$

The last equation $0 = 0$ is an identity and is always true. Its presence usually indicates infinitely many solutions. Use the second equation to write y in terms of z.

$$y = 3 - 3z$$

Next, substitute $(3 - 3z)$ for y in the first equation and write x in terms of z.

$$x - (3 - 3z) + 5z = 5$$
$$x - 3 + 3z + 5z = 5 \qquad \text{Distributive property}$$
$$x = 8 - 8z \qquad \text{Solve for } x.$$

All solutions can be written as the ordered triple $(8 - 8z, 3 - 3z, z)$, where z is any real number. There are infinitely many solutions. Sometimes we say that all solutions can be written in terms of the **parameter z**, where z is any real number. For example, if we let $z = 1$, then $y = 3 - 3(1) = 0$ and $x = 8 - 8(1) = 0$. Thus one solution is $(0, 0, 1)$.

Now Try Exercises 17 and 19 ◆

Gaussian Elimination

If an augmented matrix is not in row-echelon form, it can be transformed into row-echelon form using *Gaussian elimination*. This method uses three basic matrix row transformations.

MATRIX ROW TRANSFORMATIONS

For any augmented matrix representing a system of linear equations, the following row transformations result in an equivalent system of linear equations.

1. Any two rows may be interchanged.
2. The elements of any row may be multiplied by a nonzero constant.
3. Any row may be changed by adding to (or subtracting from) its elements a multiple of the corresponding elements of another row.

When we transform a matrix into row-echelon form, we also are transforming a system of linear equations. The next two examples illustrate how Gaussian elimination with backward substitution is performed.

EXAMPLE 4 Transforming a matrix into row-echelon form

Use Gaussian elimination with backward substitution to solve the linear system of equations.

$$x + y + z = 1$$
$$-x + y + z = 5$$
$$y + 2z = 5$$

SOLUTION The linear system is written to the right and illustrates how each row transformation affects the corresponding system of linear equations. Note that it is *not* necessary to write the system of equations to the right of the augmented matrix.

Augmented Matrix

$$\begin{bmatrix} 1 & 1 & 1 & | & 1 \\ -1 & 1 & 1 & | & 5 \\ 0 & 1 & 2 & | & 5 \end{bmatrix}$$

Linear Systems

$$x + y + z = 1$$
$$-x + y + z = 5$$
$$y + 2z = 5$$

The first step is to add the first equation to the second equation to obtain a 0 where the coefficient of x in the second row is highlighted. This row operation is denoted $R_2 + R_1$, and the result becomes the new row 2. It is important to write down each row operation so that we can check our work easily. The row that is changing is written first.

$$R_2 + R_1 \rightarrow \quad \begin{bmatrix} 1 & 1 & 1 & | & 1 \\ 0 & 2 & 2 & | & 6 \\ 0 & 1 & 2 & | & 5 \end{bmatrix} \qquad \begin{aligned} x + y + z &= 1 \\ 2y + 2z &= 6 \\ y + 2z &= 5 \end{aligned}$$

To have the matrix in row-echelon form, we need the highlighted 2 in the second row to be a 1. Multiply each element in row 2 by $\frac{1}{2}$ and denote it $\frac{1}{2}R_2$.

$$\frac{1}{2}R_2 \rightarrow \quad \begin{bmatrix} 1 & 1 & 1 & | & 1 \\ 0 & 1 & 1 & | & 3 \\ 0 & 1 & 2 & | & 5 \end{bmatrix} \qquad \begin{aligned} x + y + z &= 1 \\ y + z &= 3 \\ y + 2z &= 5 \end{aligned}$$

Finally, for the matrix to be in row-echelon form, we need a 0 where the 1 is highlighted in row 3. Subtract row 2 from row 3 and denote it $R_3 - R_2$.

$$R_3 - R_2 \rightarrow \quad \begin{bmatrix} 1 & 1 & 1 & | & 1 \\ 0 & 1 & 1 & | & 3 \\ 0 & 0 & 1 & | & 2 \end{bmatrix} \qquad \begin{aligned} x + y + z &= 1 \\ y + z &= 3 \\ z &= 2 \end{aligned}$$

Because we have a 1 in the highlighted box, the matrix is now in row-echelon form, and we see that $z = 2$. Backward substitution may be applied to find the solution. Substituting $z = 2$ in the second equation gives

$$y + 2 = 3 \quad \text{or} \quad y = 1.$$

Finally, let $y = 1$ and $z = 2$ in the first equation to determine x.

$$x + 1 + 2 = 1 \quad \text{or} \quad x = -2$$

The solution of the system is given by $x = -2$, $y = 1$, and $z = 2$, or $(-2, 1, 2)$.

Now Try Exercise 27 ◆

EXAMPLE 5 Transforming a matrix into row-echelon form

Use Gaussian elimination with backward substitution to solve the linear system of equations.

$$\begin{aligned} 2x + 4y + 4z &= 4 \\ x + 3y + z &= 4 \\ -x + 3y + 2z &= -1 \end{aligned}$$

SOLUTION The initial linear system and augmented matrix are written first.

Augmented Matrix	*Linear System*

$$\begin{bmatrix} 2 & 4 & 4 & | & 4 \\ 1 & 3 & 1 & | & 4 \\ -1 & 3 & 2 & | & -1 \end{bmatrix} \qquad \begin{aligned} 2x + 4y + 4z &= 4 \\ x + 3y + z &= 4 \\ -x + 3y + 2z &= -1 \end{aligned}$$

First we obtain a 1 where the coefficient of x in the first row is highlighted. This can be accomplished by either multiplying the first equation by $\frac{1}{2}$, or interchanging rows 1 and 2. We multiply row 1 by $\frac{1}{2}$. This operation is denoted $\frac{1}{2}R_1$, to indicate that row 1 in the previous augmented matrix is multiplied by $\frac{1}{2}$.

$$\frac{1}{2}R_1 \rightarrow \quad \begin{bmatrix} 1 & 2 & 2 & | & 2 \\ 1 & 3 & 1 & | & 4 \\ -1 & 3 & 2 & | & -1 \end{bmatrix} \qquad \begin{aligned} x + 2y + 2z &= 2 \\ x + 3y + z &= 4 \\ -x + 3y + 2z &= -1 \end{aligned}$$

The next step is to eliminate the x-variable in rows 2 and 3 by obtaining zeros in the highlighted positions. To do this, subtract row 1 from row 2, and add row 1 to row 3.

$$
\begin{array}{l} \\ R_2 - R_1 \to \\ R_3 + R_1 \to \end{array}
\left[\begin{array}{ccc|c} 1 & 2 & 2 & 2 \\ 0 & 1 & -1 & 2 \\ 0 & 5 & 4 & 1 \end{array}\right]
\qquad
\begin{aligned} x + 2y + 2z &= 2 \\ y - z &= 2 \\ 5y + 4z &= 1 \end{aligned}
$$

Since we have a 1 for the coefficient of y in the second row, the next step is to eliminate the y-variable in row 3, and obtain a zero where the y-coefficient of 5 is highlighted. Multiply row 2 by 5, and subtract the result from row 3.

$$
\begin{array}{l} \\ \\ R_3 - 5R_2 \to \end{array}
\left[\begin{array}{ccc|c} 1 & 2 & 2 & 2 \\ 0 & 1 & -1 & 2 \\ 0 & 0 & 9 & -9 \end{array}\right]
\qquad
\begin{aligned} x + 2y + 2z &= 2 \\ y - z &= 2 \\ 9z &= -9 \end{aligned}
$$

Finally, make the coefficient of z in the third row equal 1 by multiplying row 3 by $\frac{1}{9}$.

$$
\begin{array}{l} \\ \\ \frac{1}{9}R_3 \to \end{array}
\left[\begin{array}{ccc|c} 1 & 2 & 2 & 2 \\ 0 & 1 & -1 & 2 \\ 0 & 0 & 1 & -1 \end{array}\right]
\qquad
\begin{aligned} x + 2y + 2z &= 2 \\ y - z &= 2 \\ z &= -1 \end{aligned}
$$

The final matrix is in row-echelon form. Backward substitution may be applied to find the solution. Substituting $z = -1$ in the second equation gives

$$y - (-1) = 2 \quad \text{or} \quad y = 1.$$

Next, substitute $y = 1$ and $z = -1$ in the first equation to determine x.

$$x + 2(1) + 2(-1) = 2 \quad \text{or} \quad x = 2$$

The solution of the system is $(2, 1, -1)$.

Now Try Exercise 31 ◆

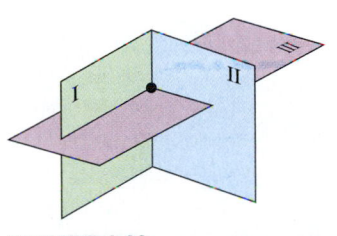

FIGURE 9.39

◆ **MAKING CONNECTIONS** —————————————

A Geometric Interpretation of Systems of Linear Equations   Solving a linear equation in one variable is reduced to finding the x-value on the number line that satisfies the equation. Solving two linear equations in two variables is often equivalent to finding the xy-coordinates where two lines intersect.

Solving linear equations in three variables also has a geometric interpretation. The graph of a linear equation in the three variables x, y, and z is a flat plane. Finding a unique solution is equivalent to locating a point where three planes in space intersect, as illustrated in Figure 9.39.

◆

Sometimes it is convenient to express a matrix in reduced row-echelon form. A matrix in row-echelon form is in **reduced row-echelon form** if every element above and below a leading 1 in a column is 0. The following matrices are examples of reduced row-echelon form.

$$
\begin{bmatrix} 1 & 0 \\ 0 & 1 \end{bmatrix}
\quad
\begin{bmatrix} 1 & 0 \\ 0 & 0 \end{bmatrix}
\quad
\begin{bmatrix} 1 & 0 & 0 \\ 0 & 1 & 0 \\ 0 & 0 & 1 \end{bmatrix}
\quad
\begin{bmatrix} 1 & 0 & 3 \\ 0 & 1 & -2 \end{bmatrix}
\quad
\begin{bmatrix} 1 & 0 & 0 & 3 \\ 0 & 1 & 0 & 1 \\ 0 & 0 & 1 & -1 \end{bmatrix}
\quad
\begin{bmatrix} 1 & 0 & 4 & 8 \\ 0 & 1 & -1 & 2 \\ 0 & 0 & 0 & 0 \end{bmatrix}
$$

If an augmented matrix is in reduced row-echelon form, solving the system of linear equations is often straightforward.

EXAMPLE 6 Determining a solution from a matrix in reduced row-echelon form

Each matrix represents a system of linear equations. Find the solution.

(a) $\begin{bmatrix} 1 & 0 & | & 6 \\ 0 & 1 & | & -5 \end{bmatrix}$
(b) $\begin{bmatrix} 1 & 0 & 0 & | & 3 \\ 0 & 1 & 0 & | & -1 \\ 0 & 0 & 1 & | & 2 \end{bmatrix}$

(c) $\begin{bmatrix} 1 & 0 & 0 & | & 4 \\ 0 & 1 & 0 & | & 3 \\ 0 & 0 & 0 & | & 2 \end{bmatrix}$
(d) $\begin{bmatrix} 1 & 0 & -2 & | & -3 \\ 0 & 1 & 2 & | & 1 \\ 0 & 0 & 0 & | & 0 \end{bmatrix}$

SOLUTION

(a) The system involves two equations and two unknowns. The solution is $(6, -5)$.
(b) The top row represents $1x + 0y + 0z = 3$ or $x = 3$. Using similar reasoning for the second and third rows: $y = -1$ and $z = 2$. The solution is $(3, -1, 2)$.
(c) The last row represents $0x + 0y + 0z = 2$, which has no solutions because $0 \neq 2$. Therefore there are no solutions to the system of equations.
(d) The last row simplifies to $0 = 0$, which is an identity and is always true. The second row gives $y + 2z = 1$, or $y = -2z + 1$. The first row represents $x - 2z = -3$, or $x = 2z - 3$. Thus, this system of linear equations has infinitely many solutions. All solutions can be written as an ordered triple in the form $(2z - 3, -2z + 1, z)$, where z can be any real number. *Now Try Exercises 49, 51, 53, and 55* ◆

Gaussian elimination may be used to transform an augmented matrix into reduced row-echelon form. It requires more effort than transforming a matrix into row-echelon form, but often eliminates the need for backward substitution.

EXAMPLE 7 Transforming a matrix into reduced row-echelon form

Use Gaussian elimination to transform the augmented matrix of the linear system into reduced row-echelon form. State the solution.

$$2x + y + 2z = 10$$
$$x \qquad + 2z = 5$$
$$x - 2y + 2z = 1$$

SOLUTION The linear system has been written to the right for illustrative purposes.

Augmented Matrix	*Linear System*

$\begin{bmatrix} 2 & 1 & 2 & | & 10 \\ 1 & 0 & 2 & | & 5 \\ 1 & -2 & 2 & | & 1 \end{bmatrix}$
$\quad$ $2x + y + 2z = 10$
$x \qquad + 2z = 5$
$x - 2y + 2z = 1$

Start by obtaining a leading 1 in the first row by interchanging rows 1 and 2.

$\begin{matrix} R_2 \to \\ R_1 \to \\ {} \end{matrix}$ $\begin{bmatrix} 1 & 0 & 2 & | & 5 \\ 2 & 1 & 2 & | & 10 \\ 1 & -2 & 2 & | & 1 \end{bmatrix}$
$\quad$ $x \qquad + 2z = 5$
$2x + y + 2z = 10$
$x - 2y + 2z = 1$

Next subtract 2 times row 1 from row 2. Then, subtract row 1 from row 3. This eliminates the x-variable from the second and third equations.

$$\begin{array}{c} \\ R_2 - 2R_1 \rightarrow \\ R_3 - R_1 \rightarrow \end{array} \left[\begin{array}{ccc|c} 1 & 0 & 2 & 5 \\ 0 & 1 & -2 & 0 \\ 0 & -2 & 0 & -4 \end{array}\right] \qquad \begin{array}{rcr} x + 2z &=& 5 \\ y - 2z &=& 0 \\ -2y &=& -4 \end{array}$$

To eliminate the y-variable in row 3, add 2 times row 2 to row 3.

$$\begin{array}{c} \\ \\ R_3 + 2R_2 \rightarrow \end{array} \left[\begin{array}{ccc|c} 1 & 0 & 2 & 5 \\ 0 & 1 & -2 & 0 \\ 0 & 0 & -4 & -4 \end{array}\right] \qquad \begin{array}{rcr} x + 2z &=& 5 \\ y - 2z &=& 0 \\ -4z &=& -4 \end{array}$$

To obtain a leading 1 in row 3, multiply by $-\frac{1}{4}$.

$$\begin{array}{c} \\ \\ -\dfrac{1}{4}R_3 \rightarrow \end{array} \left[\begin{array}{ccc|c} 1 & 0 & 2 & 5 \\ 0 & 1 & -2 & 0 \\ 0 & 0 & 1 & 1 \end{array}\right] \qquad \begin{array}{rcr} x + 2z &=& 5 \\ y - 2z &=& 0 \\ z &=& 1 \end{array}$$

The matrix is in row-echelon form. It can be transformed into reduced row-echelon form by subtracting 2 times row 3 from row 1, and adding 2 times row 3 to row 2.

$$\begin{array}{c} R_1 - 2R_3 \rightarrow \\ R_2 + 2R_3 \rightarrow \\ \end{array} \left[\begin{array}{ccc|c} 1 & 0 & 0 & 3 \\ 0 & 1 & 0 & 2 \\ 0 & 0 & 1 & 1 \end{array}\right] \qquad \begin{array}{rcr} x &=& 3 \\ y &=& 2 \\ z &=& 1 \end{array}$$

This matrix is in reduced row-echelon form. The solution is $(3, 2, 1)$.

Now Try Exercise 59 ◆

EXAMPLE 8   Transforming a system that has no solutions

Use Gaussian elimination to solve the system of linear equations, if possible.

$$\begin{aligned} x - 2y + 3z &= 2 \\ 2x + 3y + 2z &= 7 \\ 4x - y + 8z &= 8 \end{aligned}$$

SOLUTION Because it is not necessary to write the linear system next to the matrix, we omit those steps in this example and write the matrices in a horizontal format.

$$\left[\begin{array}{ccc|c} 1 & -2 & 3 & 2 \\ 2 & 3 & 2 & 7 \\ 4 & -1 & 8 & 8 \end{array}\right] \begin{array}{c} \\ R_2 - 2R_1 \rightarrow \\ R_3 - 4R_1 \rightarrow \end{array} \left[\begin{array}{ccc|c} 1 & -2 & 3 & 2 \\ 0 & 7 & -4 & 3 \\ 0 & 7 & -4 & 0 \end{array}\right] \begin{array}{c} \\ \\ R_3 - R_2 \rightarrow \end{array} \left[\begin{array}{ccc|c} 1 & -2 & 3 & 2 \\ 0 & 7 & -4 & 3 \\ 0 & 0 & 0 & -3 \end{array}\right]$$

The last row of the last matrix represents $0x + 0y + 0z = -3$, which has no solutions because $0 \neq -3$. Therefore there are no solutions to the system of equations.

Now Try Exercise 63 ◆

Solving Systems of Linear Equations with Technology (Optional)

Gaussian elimination is a numerical method. If the arithmetic at each step is done exactly, then it may be thought of as an exact symbolic procedure. However, in real applications the augmented matrix is usually quite large and its elements are not all integers. As a result, calculators and computers often are used to solve systems of equations. Their solutions usually are approximate. The next three examples use a graphing calculator to solve systems of linear equations.

 **EXAMPLE 9** Solving a system of equations using technology

Use a graphing calculator to solve the system of linear equations in Example 7.

SOLUTION To solve the system

$$2x + y + 2z = 10$$
$$x + 2z = 5$$
$$x - 2y + 2z = 1$$

enter the augmented matrix

$$A = \begin{bmatrix} 2 & 1 & 2 & | & 10 \\ 1 & 0 & 2 & | & 5 \\ 1 & -2 & 2 & | & 1 \end{bmatrix},$$

Calculator Help

To enter the elements of a matrix, see Appendix B (page AP-19).

as shown in Figures 9.40–9.42.

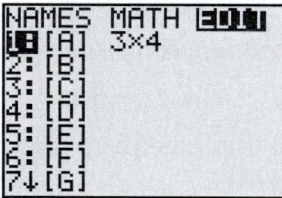

FIGURE 9.40

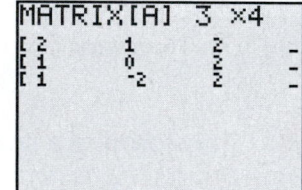

FIGURE 9.41

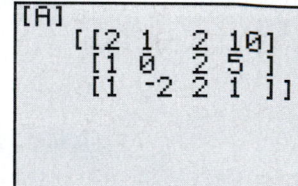

FIGURE 9.42

Now we can use a graphing calculator to transform matrix A into reduced row-echelon form, as illustrated in Figures 9.43 and 9.44. Notice that the reduced row-echelon form obtained from the graphing calculator agrees with our results from Example 7. The solution is (3, 2, 1).

Calculator Help

To transform a matrix into reduced row-echelon form, see Appendix B (page AP-20).

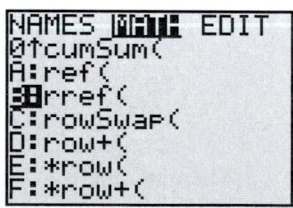

FIGURE 9.43

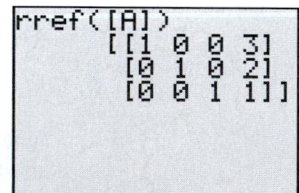

FIGURE 9.44   *Now Try Exercise 65* ◆

 **EXAMPLE 10** Transforming a matrix into reduced row-echelon form

Three food shelters are operated by a charitable organization. Three different quantities are computed, which include monthly food costs F in dollars, number of people served per month N, and monthly charitable receipts R in dollars. The data are shown in Table 9.7. (**Source:** D. Sanders, *Statistics: A First Course.*)

TABLE 9.7

Food Costs (F)	Number Served (N)	Charitable Receipts (R)
3000	2400	8000
4000	2600	10,000
8000	5900	14,000

(a) Model these data by using the equation $F = aN + bR + c$, where a, b, and c are constants.

(b) Predict the food costs for a shelter that serves 4000 people and receives charitable receipts of \$12,000. Round your answer to the nearest hundred dollars.

SOLUTION

(a) Since $F = aN + bR + c$, the constants a, b, and c satisfy the following equations.

$$3000 = a(2400) + b(8000) + c$$

$$4000 = a(2600) + b(10{,}000) + c$$

$$8000 = a(5900) + b(14{,}000) + c$$

This system can be rewritten as

$$2400a + 8000b + c = 3000$$

$$2600a + 10{,}000b + c = 4000$$

$$5900a + 14{,}000b + c = 8000.$$

The resulting augmented matrix is

$$A = \begin{bmatrix} 2400 & 8000 & 1 & 3000 \\ 2600 & 10{,}000 & 1 & 4000 \\ 5900 & 14{,}000 & 1 & 8000 \end{bmatrix}.$$

Figure 9.45 shows the matrix A. The fourth column of A may be viewed by using the arrow keys. In Figure 9.46, A has been transformed into reduced row-echelon form. From Figure 9.46, we see that $a \approx 0.6897$, $b \approx 0.4310$, and $c \approx -2103$. Thus let $F = 0.6897N + 0.431R - 2103$.

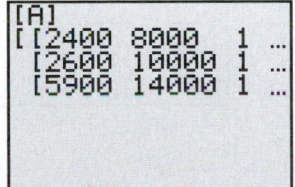

FIGURE 9.45

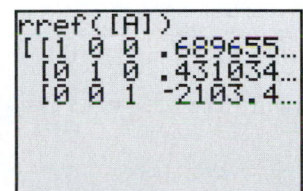

FIGURE 9.46

(b) To predict the food costs for a shelter that serves 4000 people and receives charitable receipts of \$12,000, let $N = 4000$ and $R = 12{,}000$ and evaluate F.

$$F = 0.6897(4000) + 0.431(12{,}000) - 2103 = 5827.8.$$

This model predicts monthly food costs of approximately \$5800.

Now Try Exercise 71 ◆

The next example shows how a system of linear equations can be used to determine a quadratic function.

 **EXAMPLE 11**   Determining a quadratic function

More than half of private-sector employees cannot carry vacation days into a new year. The average number y of paid days off for full-time workers at medium to large companies after x years is listed in Table 9.8.   (**Source:** Bureau of Labor Statistics.)

TABLE 9.8

x (years)	1	15	30
y (days)	9.4	18.8	21.9

(a) Determine the coefficients for $f(x) = ax^2 + bx + c$ so that f models these data.
(b) Graph f with the data in $[-4, 32, 5]$ by $[8, 23, 2]$.
(c) Estimate the number of paid days off after 3 years of experience. Compare it to the actual value of 11.2 days.

SOLUTION
(a) For f to model the data, the equations $f(1) = 9.4$, $f(15) = 18.8$, and $f(30) = 21.9$ must be satisfied.

$$f(1) = a(1)^2 + b(1) + c = 9.4$$
$$f(15) = a(15)^2 + b(15) + c = 18.8$$
$$f(30) = a(30)^2 + b(30) + c = 21.9$$

The associated augmented matrix is

$$\begin{bmatrix} 1^2 & 1 & 1 & 9.4 \\ 15^2 & 15 & 1 & 18.8 \\ 30^2 & 30 & 1 & 21.9 \end{bmatrix}.$$

Figure 9.47 shows a portion of the matrix represented in reduced row-echelon form.

Calculator Help

To plot data and to graph an equation, see Appendix B (page AP-11).

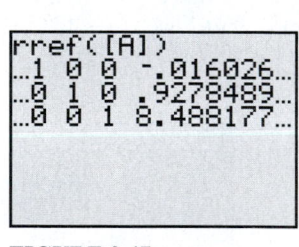

FIGURE 9.47

$[-4, 32, 5]$ by $[8, 23, 2]$

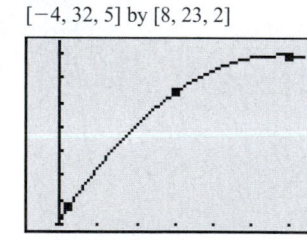

FIGURE 9.48

The solution is $a \approx -0.016026$, $b \approx 0.92785$, and $c \approx 8.4882$.
(b) Graph $Y_1 = -0.016026X^2 + 0.92785X + 8.4882$ along with the points $(1, 9.4)$, $(15, 18.8)$, and $(30, 21.9)$. The graph of f passes through the points as expected. See Figure 9.48.
(c) To estimate the number of paid days off after 3 years, evaluate $f(3)$.

$$f(3) = -0.016026(3)^2 + 0.92785(3) + 8.4882 \approx 11.1$$

This is quite close to the actual value of 11.2 days.   *Now Try Exercise 81* ◆

9.4

Putting it all Together

A matrix is a rectangular array of elements. The elements are often real numbers. An augmented matrix may be used to represent a system of linear equations. One of the most common methods for solving a system of linear equations is Gaussian elimination with backward substitution. Through a sequence of matrix row operations, an augmented matrix is transformed into row-echelon form or reduced row-echelon form. Backward substitution is frequently used to find the solution when a matrix is in row-echelon form. Technology can be used to solve systems of linear equations.

The following table summarizes some mathematical concepts involved with solving systems of linear equations in three variables.

Augmented Matrix

A linear system can be represented by an augmented matrix.

$$\left[\begin{array}{ccc|c} 2 & 0 & -3 & 2 \\ -1 & 2 & -2 & -5 \\ 1 & -2 & -1 & 7 \end{array}\right] \qquad \begin{array}{r} 2x - 3z = 2 \\ -x + 2y - 2z = -5 \\ x - 2y - z = 7 \end{array}$$

Row-Echelon Form

The following matrices are in row-echelon form. They represent three possible situations: no solutions, one solution, and infinitely many solutions.

$$\left[\begin{array}{ccc|c} 1 & -2 & 1 & 0 \\ 0 & 1 & 2 & 3 \\ 0 & 0 & 0 & 1 \end{array}\right] \qquad \left[\begin{array}{ccc|c} 1 & -2 & 1 & 0 \\ 0 & 1 & 2 & 3 \\ 0 & 0 & 1 & 1 \end{array}\right] \qquad \left[\begin{array}{ccc|c} 1 & -2 & 1 & 0 \\ 0 & 1 & 2 & 3 \\ 0 & 0 & 0 & 0 \end{array}\right]$$

No solutions   One solution   Infinitely many solutions
                           $(1, 1, 1)$ $(6 - 5z, 3 - 2z, z)$

Backward Substitution

Backward substitution can be used to solve a system of linear equations represented by an augmented matrix in row-echelon form.

$$\left[\begin{array}{ccc|c} 1 & -2 & 1 & 3 \\ 0 & 1 & -2 & -3 \\ 0 & 0 & 1 & 2 \end{array}\right]$$

From the last row, $z = 2$.
Substitute $z = 2$ in the second row, $y - 2(2) = -3$ or $y = 1$.
Let $z = 2$ and $y = 1$ in the first row, $x - 2(1) + 2 = 3$ or $x = 3$.
The solution is $(3, 1, 2)$.

9.4 — Exercises

Dimensions of Matrices and Augmented Matrices

Exercises 1 and 2: State the dimension of each matrix.

1. (a) $\begin{bmatrix} 1 \\ 2 \\ 3 \end{bmatrix}$ (b) $\begin{bmatrix} a & b & c \\ d & e & b \end{bmatrix}$ (c) $\begin{bmatrix} 3 & 0 \\ 1 & -4 \end{bmatrix}$

2. (a) $\begin{bmatrix} -1 & 1 \end{bmatrix}$ (b) $\begin{bmatrix} 1 & -1 \\ 7 & 5 \\ -4 & 0 \end{bmatrix}$

(c) $\begin{bmatrix} 1 & 3 & 8 & -3 \\ 1 & -1 & 1 & -2 \\ 4 & 5 & 0 & -1 \end{bmatrix}$

Exercises 3–6: Represent the linear system by an augmented matrix, and state the dimension of the matrix.

3. $\begin{aligned} 5x - 2y &= 3 \\ -x + 3y &= -1 \end{aligned}$ **4.** $\begin{aligned} 3x + y &= 4 \\ -x + 4y &= 5 \end{aligned}$

5. $\begin{aligned} -3x + 2y + z &= -4 \\ 5x \quad\quad - z &= 9 \\ x - 3y - 6z &= -9 \end{aligned}$ **6.** $\begin{aligned} x + 2y - z &= 2 \\ -2x + y - 2z &= -3 \\ 7x + y - z &= 7 \end{aligned}$

Exercises 7–10: Write the system of linear equations that the augmented matrix represents.

7. $\left[\begin{array}{cc|c} 3 & 2 & 4 \\ 0 & 1 & 5 \end{array}\right]$ **8.** $\left[\begin{array}{cc|c} -2 & 1 & 5 \\ 7 & 9 & 2 \end{array}\right]$

9. $\left[\begin{array}{ccc|c} 3 & 1 & 4 & 0 \\ 0 & 5 & 8 & -1 \\ 0 & 0 & -7 & 1 \end{array}\right]$ **10.** $\left[\begin{array}{ccc|c} 1 & -1 & 3 & 2 \\ -2 & 1 & 1 & -2 \\ -1 & 0 & -2 & 1 \end{array}\right]$

Row-Echelon Form

Exercises 11 and 12: Determine if the matrix is in row-echelon form.

11. (a) $\left[\begin{array}{cc|c} 1 & 3 & 2 \\ 0 & 1 & -1 \end{array}\right]$ (b) $\left[\begin{array}{ccc|c} 1 & 4 & -1 & 0 \\ 0 & -1 & 1 & 3 \\ 0 & 2 & 1 & 7 \end{array}\right]$

(c) $\left[\begin{array}{ccc|c} 1 & 6 & -8 & 5 \\ 0 & 1 & 7 & 9 \\ 0 & 0 & 1 & 11 \end{array}\right]$

12. (a) $\left[\begin{array}{cc|c} 1 & 3 & 2 \\ 0 & -1 & -1 \end{array}\right]$ (b) $\left[\begin{array}{ccc|c} 1 & 3 & -1 & 8 \\ 0 & 1 & 5 & 3 \\ 0 & 0 & 0 & 0 \end{array}\right]$

(c) $\left[\begin{array}{ccc|c} 0 & 0 & 1 & 1 \\ 0 & 1 & 7 & 9 \\ 1 & 2 & -1 & 11 \end{array}\right]$

Exercises 13–22: The augmented matrix is in row-echelon form and represents a linear system. Solve the system by using backward substitution, if possible. Write the solution as either an ordered pair or an ordered triple.

13. $\left[\begin{array}{cc|c} 1 & 2 & 3 \\ 0 & 1 & -1 \end{array}\right]$ **14.** $\left[\begin{array}{cc|c} 1 & -5 & 6 \\ 0 & 0 & 1 \end{array}\right]$

15. $\left[\begin{array}{cc|c} 1 & -1 & 2 \\ 0 & 1 & 0 \end{array}\right]$ **16.** $\left[\begin{array}{cc|c} 1 & 4 & -2 \\ 0 & 1 & 3 \end{array}\right]$

17. $\left[\begin{array}{ccc|c} 1 & 1 & -1 & 4 \\ 0 & 1 & -1 & 2 \\ 0 & 0 & 1 & 1 \end{array}\right]$ **18.** $\left[\begin{array}{ccc|c} 1 & -2 & -1 & 0 \\ 0 & 1 & -3 & 1 \\ 0 & 0 & 1 & 2 \end{array}\right]$

19. $\left[\begin{array}{ccc|c} 1 & 2 & -1 & 5 \\ 0 & 1 & -2 & 1 \\ 0 & 0 & 0 & 0 \end{array}\right]$ **20.** $\left[\begin{array}{ccc|c} 1 & -1 & 2 & 8 \\ 0 & 1 & -4 & 2 \\ 0 & 0 & 1 & -1 \end{array}\right]$

21. $\left[\begin{array}{ccc|c} 1 & 2 & 1 & -3 \\ 0 & 1 & -3 & \frac{1}{2} \\ 0 & 0 & 0 & 4 \end{array}\right]$ **22.** $\left[\begin{array}{ccc|c} 1 & 0 & -4 & \frac{3}{4} \\ 0 & 1 & 2 & 1 \\ 0 & 0 & 0 & -3 \end{array}\right]$

Solving Systems with Gaussian Elimination

Exercises 23–26: Perform each row operation on the given matrix by completing the matrix at the right.

23. $\left[\begin{array}{ccc|c} 2 & -4 & 6 & 10 \\ -3 & 5 & 3 & 2 \\ 4 & 8 & 4 & -8 \end{array}\right]$ $\begin{array}{l}\frac{1}{2}R_1 \to \\[18pt] \frac{1}{4}R_3 \to\end{array}$ $\left[\begin{array}{ccc|c} 1 & & & \\ -3 & 5 & 3 & 2 \\ & & 1 & \end{array}\right]$

24. $\left[\begin{array}{ccc|c} 1 & -2 & 1 & 3 \\ 1 & 4 & 0 & -1 \\ 2 & 0 & 1 & 5 \end{array}\right]$ $\begin{array}{l} \\ R_2 - R_1 \to \\ R_3 - 2R_1 \to\end{array}$ $\left[\begin{array}{ccc|c} 1 & -2 & 1 & 3 \\ & 6 & & \\ & & & -1 \end{array}\right]$

25. $\begin{bmatrix} 1 & -1 & 1 & | & 2 \\ -1 & 2 & -2 & | & 0 \\ 1 & 7 & 0 & | & 5 \end{bmatrix}$ $\begin{matrix} \\ R_2 + R_1 \to \\ R_3 - R_1 \to \end{matrix}$ $\begin{bmatrix} 1 & -1 & 1 & | & 2 \\ & & & | & \\ & & & | & \end{bmatrix}$

26. $\begin{bmatrix} 1 & -2 & 3 & | & 6 \\ 2 & 1 & 4 & | & 5 \\ -3 & 5 & 3 & | & 2 \end{bmatrix}$ $\begin{matrix} \\ R_2 - 2R_1 \to \\ R_3 + 3R_1 \to \end{matrix}$ $\begin{bmatrix} 1 & -2 & 3 & | & 6 \\ & & & | & \\ & & & | & \end{bmatrix}$

Exercises 27–36: Use Gaussian elimination with backward substitution to solve the system of linear equations. Write the solution as an ordered pair or an ordered triple, whenever possible.

27.
$x + 2y = 3$
$-x - y = 7$

28.
$2x + 4y = 10$
$x - 2y = -3$

29.
$x + 2y + z = 3$
$x + y - z = 3$
$-x - 2y + z = -5$

30.
$x + y + z = 6$
$2x + 3y - z = 3$
$x + y + 2z = 10$

31.
$3x + y + 3z = 14$
$x + y + z = 6$
$-2x - 2y + 3z = -7$

32.
$x + 3y - 2z = 3$
$-x - 2y + z = -2$
$2x - 7y + z = 1$

33.
$2x + 5y + z = 8$
$x + 2y - z = 2$
$3x + 7y = 5$

34.
$x + y + z = 3$
$x + y + 2z = 4$
$2x + 2y + 3z = 7$

35.
$-x + 2y + 4z = 10$
$3x - 2y - 2z = -12$
$x + 2y + 6z = 8$

36.
$4x - 2y + 4z = 8$
$3x - 7y + 6z = 4$
$-x - 5y + 2z = 7$

Exercises 37–48: Solve the system, if possible.

37.
$x - y + z = 1$
$x + 2y - z = 2$
$y - z = 0$

38.
$x - y - 2z = -11$
$x - 2y - z = -11$
$-x + y + 3z = 14$

39.
$2x - 4y + 2z = 11$
$x + 3y - 2z = -9$
$4x - 2y + z = 7$

40.
$x - 4y + z = 9$
$3y - 2z = -7$
$-x + z = 0$

41.
$3x - 2y + 2z = -18$
$-x + 2y - 4z = 16$
$4x - 3y - 2z = -21$

42.
$2x - y - z = 0$
$x - y - z = -2$
$3x - 2y - 2z = -2$

43.
$x - 4y + 3z = 26$
$-x + 3y - 2z = -19$
$-y + z = 10$

44.
$4x - y - z = 0$
$4x - 2y = 0$
$2x + z = 1$

45.
$5x + 4z = 7$
$2x - 4y = 6$
$3y + 3z = 3$

46.
$y + 2z = -5$
$3x - 2z = -6$
$-x - 4y = 11$

47.
$5x - 2y + z = 5$
$x + y - 2z = -2$
$4x - 3y + 3z = 7$

48.
$2x - 4y - z = 2$
$x + y - 3z = 10$
$-x - 7y + 8z = 2$

Exercises 49–56: (Refer to Example 6.) The augmented matrix is in reduced row-echelon form and represents a system of linear equations. If possible, solve the system.

49. $\begin{bmatrix} 1 & 0 & | & 12 \\ 0 & 1 & | & 3 \end{bmatrix}$

50. $\begin{bmatrix} 1 & -1 & | & 1 \\ 0 & 0 & | & 0 \end{bmatrix}$

51. $\begin{bmatrix} 1 & 0 & 0 & | & -2 \\ 0 & 1 & 0 & | & 4 \\ 0 & 0 & 1 & | & \frac{1}{2} \end{bmatrix}$

52. $\begin{bmatrix} 1 & 0 & 0 & | & 7 \\ 0 & 1 & 0 & | & -9 \\ 0 & 0 & 1 & | & 3 \end{bmatrix}$

53. $\begin{bmatrix} 1 & 0 & 2 & | & 4 \\ 0 & 1 & -1 & | & -3 \\ 0 & 0 & 0 & | & 0 \end{bmatrix}$

54. $\begin{bmatrix} 1 & 0 & 1 & | & -2 \\ 0 & 1 & 3 & | & 5 \\ 0 & 0 & 0 & | & 0 \end{bmatrix}$

55. $\begin{bmatrix} 1 & 0 & 0 & | & \frac{3}{4} \\ 0 & 1 & 0 & | & -1 \\ 0 & 0 & 0 & | & \frac{2}{3} \end{bmatrix}$

56. $\begin{bmatrix} 1 & 0 & 0 & | & 10 \\ 0 & 1 & 0 & | & 21 \\ 0 & 0 & 0 & | & -2 \end{bmatrix}$

Exercises 57–64: **Reduced Row-Echelon Form** *Find the solution by transforming the augmented matrix of the linear system into reduced row-echelon form.*

57.
$x - y = 1$
$x + y = 5$

58.
$2x + 3y = 1$
$x - 2y = -3$

59.
$x + 2y + z = 3$
$y - z = -2$
$-x - 2y + 2z = 6$

60.
$x + z = 2$
$x - y - z = 0$
$-2x + y = -2$

61.
$x - y + 2z = 7$
$2x + y - 4z = -27$
$-x + y - z = 0$

62.
$2x - 4y - 6z = 2$
$x - 3y + z = 12$
$2x + y + 3z = 5$

63.
$2x + y - z = 2$
$x - 2y + z = 0$
$x + 3y - 2z = 4$

64.
$-2x - y + z = 3$
$x + y - 3z = 1$
$x - 2y - 4z = 2$

Exercises 65–70: **Technology** *Use technology to find the solution. Approximate values to the nearest thousandth.*

65.
$5x - 7y + 9z = 40$
$-7x + 3y - 7z = 20$
$5x - 8y - 5z = 15$

66.
$12x - 4y - 7z = 8$
$-8x - 6y + 9z = 7$
$34x + 6y - 2z = 5$

67.
$2.1x + 0.5y + 1.7z = 4.9$
$-2x + 1.5y - 1.7z = 3.1$
$5.8x - 4.6y + 0.8z = 9.3$

68. $53x + 95y + 12z = 108$
$\quad\;\; 81x - 57y - 24z = -92$
$\quad\; -9x + 11y - 78z = \;\;\; 21$

69. $0.1x + 0.3y + 1.7z = 0.6$
$\quad\;\, 0.6x + 0.1y - 3.1z = 6.2$
$\quad\quad\quad\;\; 2.4y + 0.9z = 3.5$

70. $\quad 103x - 886y + 431z = 1200$
$\quad -55x + 981y \quad\quad\quad\; = 1108$
$\quad -327x + 421y + 337z = \quad 99$

Applications

71. *Food Shelters* (Refer to Example 10.) Three food shelters have monthly food costs F in dollars, number of
people served per month N, and monthly charitable receipts R in dollars, as shown in the table.

Food Costs (F)	Number Served (N)	Charitable Receipts (R)
1300	1800	5000
5300	3200	12,000
6500	4500	13,000

(a) Model these data using $F = aN + bR + c$, where
a, b, and c are constants.

(b) Predict the food costs for a shelter that serves 3500
people and receives charitable receipts of $12,500.
Round your answer to the nearest hundred dollars.

72. *Computing Time* When computers are programmed to
solve large linear systems involved in applications, such
as designing aircraft or large electrical circuits, they frequently use Gaussian elimination with backward substitution. Solving a linear system with n equations and n
variables requires a computer to perform a total of
$T(n) = \frac{2}{3}n^3 + \frac{3}{2}n^2 - \frac{7}{6}n$ arithmetic operations (addi-
tions, subtractions, multiplications, and divisions).
(**Source:** R. Burden and J. Faires, *Numerical Analysis.*)
(a) John Atanasoff, the inventor of the modern digital
computer, needed to solve a system of 29 linear
equations. Evaluate $T(29)$ to find the number of
arithmetic operations this would require. Would it
be too many to do by hand?

(b) Compute T for $n = 10, 100, 1000, 10,000$, and
100,000. List the results in a table.

(c) If the number of equations and variables increases
by a factor of 10, does the number of arithmetic operations also increase by a factor of 10? Explain.

(d) Discuss why very fast computers are needed to
solve large systems of linear equations.

73. *Pumping Water* Three pumps are being used to empty a
small swimming pool. The first pump is twice as fast as
the second pump. The first two pumps can empty the pool
in 8 hours, while all three pumps can empty it in 6 hours.
How long would it take each pump to empty the pool indi-
vidually? (*Hint:* Let x represent the fraction of the pool that
the first pump can empty in 1 hour. Let y and z represent
this fraction for the second and third pumps, respectively.)

74. *Pumping Water* Suppose in the previous exercise the
first pump is three times as fast as the third pump, the
first and second pumps can empty the pool in 6 hours,
and all three pumps can empty the pool in 8 hours.
(a) Are these data realistic? Explain your reasoning.

(b) Make a conjecture about a mathematical solution to
these data.

(c) Test your conjecture by solving the problem.

75. *Investment* A sum of $5000 is invested in three mutual
funds that pay 8%, 11%, and 14% annual interest rates. The
amount of money invested in the fund paying 14% equals
the total amount of money invested in the other two funds,
and the total annual interest from all three funds is $595.
(a) Write a system of equations whose solution gives
the amount invested in each mutual fund. Be sure to
state what each variable represents.

(b) Solve the system of equations.

76. *Investment* A sum of $10,000 is invested in three ac-
counts that pay 6%, 8%, and 10% interest. Twice as
much money is invested in the account paying 10% as
the account paying 6%, and the total annual interest
from all three accounts is $842.
(a) Write a system of equations whose solution gives
the amount invested in each account. Be sure to
state what each variable represents.

(b) Solve the system of equations.

77. *Electricity* In the study of electrical circuits, the application of Kirchoff's rules frequently results in systems

of linear equations. To determine the current (in amperes) in each branch of the circuit shown in the figure, solve the system of linear equations. Round values to the nearest hundredth.

$$I_1 = I_2 + I_3$$
$$15 + 4I_3 = 14I_2$$
$$10 + 4I_3 = 5I_1$$

78. *Electricity*   (Refer to the previous exercise.) Find the current (in amperes) in each branch of the circuit shown in the figure, by solving the system of linear equations. Round values to the nearest hundredth.

$$I_1 = I_2 + I_3$$
$$20 = 4I_1 + 7I_3$$
$$10 + 7I_3 = 6I_2$$

*Exercises 79 and 80: **Traffic Flow**   The accompanying figure shows three one-way streets with intersections A, B, and C. Numbers indicate the average traffic flow in vehicles per minute. The variables x, y, and z denote unknown traffic flows that need to be determined for timing of stoplights.*

(a) *If the number of vehicles per minute entering an intersection must equal the number exiting an intersection, verify that the accompanying system of linear equations describes the traffic flow.*

(b) *Rewrite the system and solve.*

(c) *Interpret your solution.*

79. A: $x + 5 = y + 7$
B: $z + 6 = x + 3$
C: $y + 3 = z + 4$

80. A: $x + 7 = y + 4$
B: $4 + 5 = x + z$
C: $y + 8 = 9 + 4$

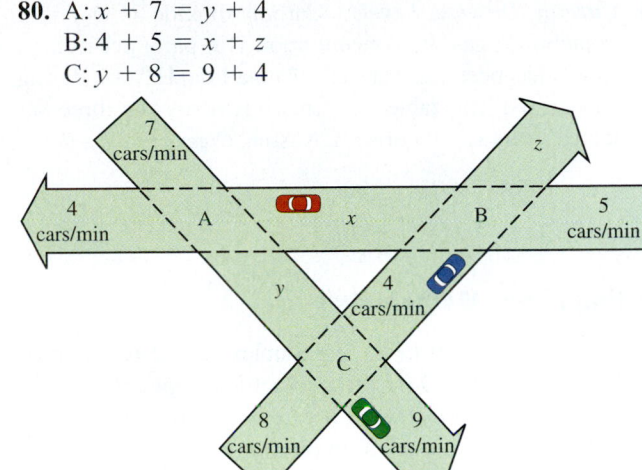

Exercises 81–84: Each set of data can be modeled by $f(x) = ax^2 + bx + c$, where x represents the year.

(a) *Find a linear system whose solution represents values of a, b, and c.*

(b) *Use technology to find the solution.*

(c) *Graph f and the data in the same viewing rectangle.*

(d) *Make your own prediction using f.*

81. *Chronic Health Care*   A large percentage of the U.S. population will require chronic health care in the coming decades. The average caregiving age is 50–64, while the typical person needing chronic care is 85 or older. The ratio y of potential caregivers to those needing chronic health care will shrink in the coming years x, as shown in the table.   (**Source:** Robert Wood Johnson Foundation, *Chronic Care in America: A 21st Century Challenge.*)

x	1990	2010	2030
y	11	10	6

82. *Home Health Care*   The table shows the cost of Medicare home health care y in billions of dollars during the year x. In this table, $x = 0$ corresponds to 1990, and $x = 6$ corresponds to 1996.   (**Source:** Health Care Financing Administration.)

x	0	3	6
y	3.9	10.5	18.1

83. *Women in the Military*   The table shows the percentage y of the enlisted people in the military who are women. In this table $x = 3$ corresponds to 1973 and $x = 26$ to 1996.   (**Source:** Department of Defense.)

x	3	18	26
y	2.2	10.4	12.8

84. *Carbon Dioxide Levels* Carbon dioxide (CO_2) is a greenhouse gas. Its concentration y in parts per million (ppm) has been measured at Mauna Loa, Hawaii, during past years. The table lists measurements for three selected years x.   (**Source:** A. Nilsson, *Greenhouse Earth*.)

x	1958	1973	2003
y	315	325	376

Writing about Mathematics

85. A linear equation in three variables can be represented by a flat plane. Describe geometrically possible situations that can occur when a system of three linear equations has either no solution or an infinite number of solutions.

86. Give an example of an augmented matrix in row-echelon form that represents a system of linear equations that has no solution. Explain your reasoning.

EXTENDED AND DISCOVERY EXERCISE

1. *Weight of a Bear* The accompanying table shows the weight W, neck size N, overall length L, and chest size C for four bears.   (**Sources:** M. Triola, *Elementary Statistics*; Minitab, Inc.)

W (pounds)	N (inches)	L (inches)	C (inches)
125	19	57.5	32
316	26	65	42
436	30	72	48
514	30.5	75	54

(a) We can model these data with the equation $W = a + bN + cL + dC$, where a, b, c, and d are constants. To do so, represent a system of linear equations by a 4×5 augmented matrix whose solution gives values for a, b, c, and d.

(b) Solve the system. Round each value to the nearest thousandth.

(c) Predict the weight of a bear with $N = 24$, $L = 63$, and $C = 39$. Interpret the result.

CHECKING BASIC CONCEPTS FOR SECTIONS 9.3 AND 9.4

1. If possible, solve the system of linear equations.

(a)
$$x - 2y + z = -2$$
$$x + y + 2z = 3$$
$$2x - y - z = 5$$

(b)
$$x - 2y + z = -2$$
$$x + y + 2z = 3$$
$$2x - y + 3z = 1$$

(c)
$$x - 2y + z = -2$$
$$x + y + 2z = 3$$
$$2x - y + 3z = 5$$

2. *Tickets Sold* Two thousand tickets were sold for a play, which generated $19,700. The prices of the tickets were $5 for children, $10 for students, and $12 for adults. There were 100 more adult tickets sold than student tickets. Find the number of each type of ticket sold.

3. Solve the system of linear equations using Gaussian elimination and backward substitution.

$$x \qquad + z = 2$$
$$x + y - z = 1$$
$$-x - 2y - z = 0$$

4. Use technology to solve the system of linear equations in the previous exercise.

Properties and Applications of Matrices

9.5

- ◆ Learn matrix notation
- ◆ Find sums, differences, and scalar multiples of matrices
- ◆ Find matrix products
- ◆ Use technology (optional)

Introduction

Matrices occur in many fields of study and have a wide variety of applications. New technologies, such as digital photography and computer graphics, frequently utilize matrices to achieve their goals. Many of these technologies have become specialized fields that require substantial interdisciplinary skills, including mathematics. In this section we discuss properties of matrices and some of their applications.

Matrix Notation

The following notation is used to denote elements in a matrix.

$$
\begin{bmatrix} a_{11} & a_{12} \\ a_{21} & a_{22} \end{bmatrix}
\quad
\begin{bmatrix} a_{11} & a_{12} & a_{13} \\ a_{21} & a_{22} & a_{23} \\ a_{31} & a_{32} & a_{33} \end{bmatrix}
\quad
\begin{bmatrix} a_{11} & a_{12} & a_{13} & a_{14} \\ a_{21} & a_{22} & a_{23} & a_{24} \\ a_{31} & a_{32} & a_{33} & a_{34} \\ a_{41} & a_{42} & a_{43} & a_{44} \end{bmatrix}
\quad
\begin{bmatrix} a_{11} & a_{12} \\ a_{21} & a_{22} \\ a_{31} & a_{32} \end{bmatrix}
\quad
\begin{bmatrix} a_{11} & a_{12} & a_{13} \\ a_{21} & a_{22} & a_{23} \end{bmatrix}
$$

A general element of the matrix A is denoted by a_{ij}. This refers to the element in the ith row, jth column. For example, a_{23} would be the element of A located in the second row, third column. Two m by n matrices A and B are **equal** if corresponding elements are equal. If A and B have different dimensions, they cannot be equal. For example,

$$
\begin{bmatrix} 3 & -3 & 7 \\ 2 & 6 & -2 \\ 4 & 2 & 5 \end{bmatrix}
=
\begin{bmatrix} 3 & -3 & 7 \\ 2 & 6 & -2 \\ 4 & 2 & 5 \end{bmatrix}
$$

because all corresponding elements are equal. However,

$$
\begin{bmatrix} 1 & 4 \\ -3 & 2 \\ 4 & -7 \end{bmatrix}
\neq
\begin{bmatrix} 1 & 4 \\ -3 & 2 \\ 5 & -7 \end{bmatrix}
$$

because $4 \neq 5$ in row 3 and column 1, and

$$
\begin{bmatrix} 1 & 2 & 3 \\ 4 & 5 & 6 \end{bmatrix}
\neq
\begin{bmatrix} 1 & 2 \\ 4 & 5 \end{bmatrix}
$$

because the matrices have different dimensions.

 **EXAMPLE 1**   ## Determining matrix elements

Let a_{ij} denote a general element in A and b_{ij} a general element in B, where

$$
A = \begin{bmatrix} 3 & -3 & 7 \\ 1 & 6 & -2 \\ 4 & 2 & 5 \end{bmatrix}
\quad \text{and} \quad
B = \begin{bmatrix} 3 & x & 7 \\ 1 & 6 & -2 \\ 4 & 5 & 2 \end{bmatrix}.
$$

(a) Identify a_{12}, b_{32}, and a_{13}.
(b) Compute $a_{31}b_{13} + a_{32}b_{23} + a_{33}b_{33}$.
(c) Is there a value for x that will make the statement $A = B$ true?

SOLUTION

(a) The element a_{12} is located in the first row, second column of A. Thus, $a_{12} = -3$. In a similar manner, $b_{32} = 5$ and $a_{13} = 7$.

(b) $a_{31}b_{13} + a_{32}b_{23} + a_{33}b_{33} = (4)(7) + (2)(-2) + (5)(2) = 34$

(c) No, since $a_{32} = 2 \neq 5 = b_{32}$ and $a_{33} = 5 \neq 2 = b_{33}$. Even if we let $x = -3$, there are other corresponding elements in A and B that are not equal. *Now Try Exercise 1* ◆

Sums, Differences, and Scalar Multiples of Matrices

The FCC has mandated all television stations to transmit digital signals by the year 2006. Digital photography is a new technology in which matrices play an important role. Figure 9.49 shows a black-and-white photograph. In Figure 9.50 a grid has been laid over this photograph. Each square in the grid represents a pixel in a digitized photograph. A gray scale is illustrated in Figure 9.51. In this scale 0 corresponds to white, and 11 represents black. As the scale increases from 0 to 11, the shade of gray darkens.

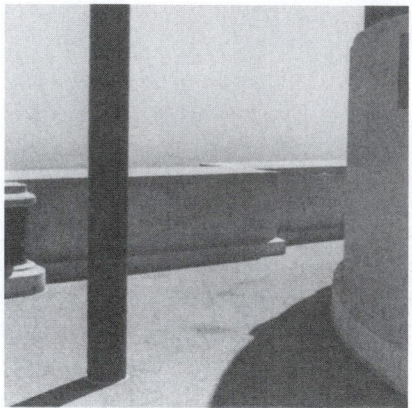

FIGURE 9.49 Photograph to Be Digitized

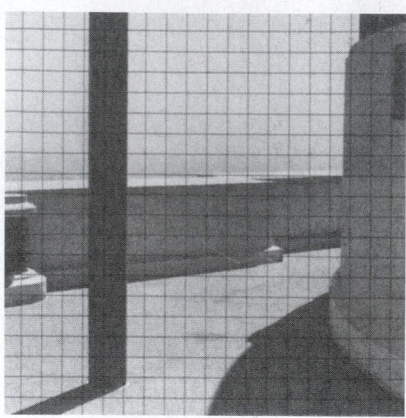

FIGURE 9.50 Placing a Grid over the Photograph

In Figure 9.52 the shades of gray are stored as numbers in a matrix with dimension 20×20. Notice how the dark pole in the picture is represented by the fifth and sixth columns in the matrix. Both these columns contain the numbers 10 and 11, which indicates a dark portion in the photograph. The matrix is a numerical or digital representation of the photograph, which can be stored and transmitted over the Internet. In real applications, low-resolution, digital photographs use a grid with 512 rows and 512 columns. The gray scale typically contains 64 levels instead of the 12 in our example. High-resolution photography uses 2048 rows and columns in the grid with 256 levels of gray.

0	1	2	3	4	5	6	7	8	9	10	11

FIGURE 9.51 Gray Levels

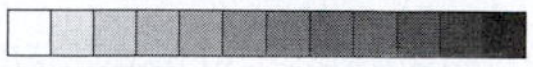

1	1	1	1	10	11	1	1	1	1	1	1	1	1	1	1	1	1	9	8
1	1	1	1	10	11	1	1	1	1	1	1	1	1	1	1	1	2	8	9
1	1	1	1	10	11	1	1	1	1	1	1	1	1	1	1	1	4	8	10
1	1	1	1	10	11	1	1	1	1	1	1	1	1	1	1	1	4	8	10
1	1	1	1	10	11	1	1	1	1	1	1	1	1	1	1	1	5	8	10
1	1	1	1	10	11	1	1	1	1	1	1	1	1	1	1	1	5	8	8
1	0	0	0	10	11	0	0	0	0	0	0	1	1	0	0	0	5	8	8
0	2	2	2	10	11	2	3	3	4	4	5	5	7	7	7	7	6	8	8
3	5	9	8	10	11	7	7	7	7	8	8	8	7	6	6	6	6	8	8
11	9	9	6	10	11	7	7	8	8	8	8	8	6	5	5	5	6	8	8
10	9	9	6	10	11	7	8	8	8	8	8	8	3	5	6	4	6	8	8
10	7	6	6	10	11	6	7	7	6	6	6	6	4	3	2	2	5	8	8
4	3	8	7	10	11	5	4	3	2	2	2	2	2	2	2	4	8	8	8
6	4	2	2	10	11	2	2	2	2	2	3	2	2	2	7	9	8	8	8
2	2	2	2	10	11	2	2	2	2	2	2	2	5	7	9	9	7	8	8
2	2	2	2	10	11	2	2	2	2	4	2	8	9	9	10	9	8	7	8
2	2	2	2	10	11	2	2	2	2	2	5	7	9	10	11	9	10	9	7
2	2	2	2	10	11	2	2	2	2	2	7	9	9	9	10	11	10	9	7
2	2	5	7	9	8	2	2	2	2	7	9	9	9	10	11	11	11	10	7
7	9	7	5	2	2	2	2	2	2	8	9	9	10	11	11	11	11	11	10

FIGURE 9.52 Digitized Photograph

Source (Figures 9.49–9.52): *Visualization* by Friedhoff and Benzon © 1991 by W. H. Freeman and Company. Used with permission.

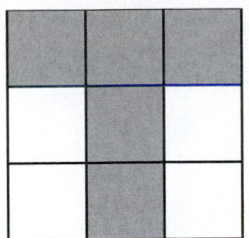

FIGURE 9.53

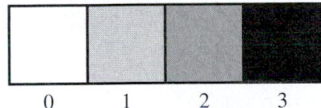

0   1   2   3

FIGURE 9.54

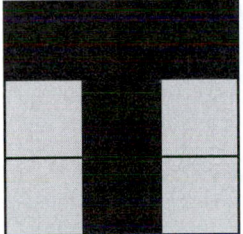

FIGURE 9.55

Matrix Addition and Subtraction   To simplify the concept of digital photography, we reduce the grid to 3 × 3 and have four gray levels, where 0 represents white, 1 light gray, 2 dark gray, and 3 black. Suppose that we would like to digitize the letter T shown in Figure 9.53. The gray levels are shown in Figure 9.54.

Since the T is dark gray and the background is white, Figure 9.53 can be represented by

$$A = \begin{bmatrix} 2 & 2 & 2 \\ 0 & 2 & 0 \\ 0 & 2 & 0 \end{bmatrix}.$$

Suppose that we want to make the entire picture darker. If we change every element in A to 3, the entire picture would be black. A more acceptable solution would be to darken each pixel by one gray level. This corresponds to adding 1 to each element in the matrix A and can be accomplished efficiently using matrix notation.

$$\begin{bmatrix} 2 & 2 & 2 \\ 0 & 2 & 0 \\ 0 & 2 & 0 \end{bmatrix} + \begin{bmatrix} 1 & 1 & 1 \\ 1 & 1 & 1 \\ 1 & 1 & 1 \end{bmatrix} = \begin{bmatrix} 2+1 & 2+1 & 2+1 \\ 0+1 & 2+1 & 0+1 \\ 0+1 & 2+1 & 0+1 \end{bmatrix} = \begin{bmatrix} 3 & 3 & 3 \\ 1 & 3 & 1 \\ 1 & 3 & 1 \end{bmatrix}$$

To add two matrices of equal dimension, add corresponding elements. The result is shown as a picture in Figure 9.55. Notice that the background is now light gray and the T is black. The entire picture is darker.

To lighten the picture in Figure 9.55, subtract 1 from each element. To subtract two matrices of equal dimension, subtract corresponding elements.

$$\begin{bmatrix} 3 & 3 & 3 \\ 1 & 3 & 1 \\ 1 & 3 & 1 \end{bmatrix} - \begin{bmatrix} 1 & 1 & 1 \\ 1 & 1 & 1 \\ 1 & 1 & 1 \end{bmatrix} = \begin{bmatrix} 3-1 & 3-1 & 3-1 \\ 1-1 & 3-1 & 1-1 \\ 1-1 & 3-1 & 1-1 \end{bmatrix} = \begin{bmatrix} 2 & 2 & 2 \\ 0 & 2 & 0 \\ 0 & 2 & 0 \end{bmatrix}$$

Increasing the contrast in a picture causes a light area to become lighter and a dark area to become darker. As a result, there are fewer pixels with intermediate gray levels. Changing contrast is different from making the entire picture lighter or darker.

The digital picture in Figure 9.56 was taken by *Voyager 1* and faintly shows the rings of Saturn. By increasing the contrast in the picture, as shown in Figure 9.57, the rings are much clearer. Notice that light gray features have become lighter, while dark gray areas have become darker. With digital photography, contrast enhancement could be performed on Earth after *Voyager 1* sent back the original picture.

◆ CLASS DISCUSSION
Discuss how the negative image of a digital picture might be represented. ◆

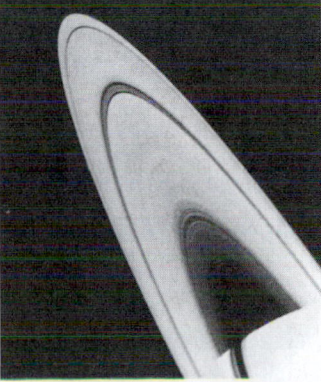

FIGURE 9.56 Rings of Saturn
Source: NASA.

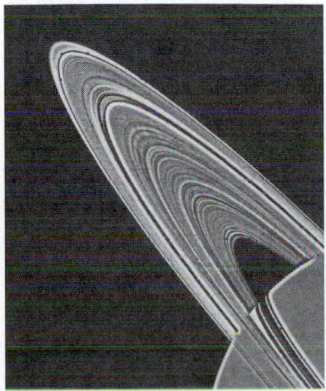

FIGURE 9.57 Enhanced Photograph

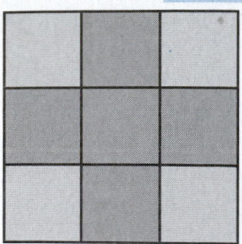

FIGURE 9.58

FIGURE 9.59

EXAMPLE 2   Applying addition of matrices to digital photography

Increase the contrast of the $+$ sign in Figure 9.58 by changing light gray to white and dark gray to black. Use matrices to represent this computation.

SOLUTION   Figure 9.58 can be represented by the matrix A.

$$A = \begin{bmatrix} 1 & 2 & 1 \\ 2 & 2 & 2 \\ 1 & 2 & 1 \end{bmatrix}$$

To change the contrast, we can reduce each 1 in the matrix A to 0 and increase each 2 to 3. The addition of the matrix B can accomplish this task.

$$A + B = \begin{bmatrix} 1 & 2 & 1 \\ 2 & 2 & 2 \\ 1 & 2 & 1 \end{bmatrix} + \begin{bmatrix} -1 & 1 & -1 \\ 1 & 1 & 1 \\ -1 & 1 & -1 \end{bmatrix} = \begin{bmatrix} 0 & 3 & 0 \\ 3 & 3 & 3 \\ 0 & 3 & 0 \end{bmatrix}$$

The picture corresponding to $A + B$ is shown in Figure 9.59.

Now Try Exercises 25 and 27 ◆

EXAMPLE 3   Adding and subtracting matrices

If $A = \begin{bmatrix} 7 & 8 & -1 \\ 0 & -1 & 6 \end{bmatrix}$ and $B = \begin{bmatrix} 5 & -2 & 10 \\ -3 & 2 & 4 \end{bmatrix}$, find the following.

(a) $A + B$ **(b)** $B + A$ **(c)** $A - B$

SOLUTION

(a) $A + B = \begin{bmatrix} 7 & 8 & -1 \\ 0 & -1 & 6 \end{bmatrix} + \begin{bmatrix} 5 & -2 & 10 \\ -3 & 2 & 4 \end{bmatrix}$

$= \begin{bmatrix} 7 + 5 & 8 + (-2) & -1 + 10 \\ 0 + (-3) & -1 + 2 & 6 + 4 \end{bmatrix}$

$= \begin{bmatrix} 12 & 6 & 9 \\ -3 & 1 & 10 \end{bmatrix}$

(b) $B + A = \begin{bmatrix} 5 & -2 & 10 \\ -3 & 2 & 4 \end{bmatrix} + \begin{bmatrix} 7 & 8 & -1 \\ 0 & -1 & 6 \end{bmatrix}$

$= \begin{bmatrix} 5 + 7 & -2 + 8 & 10 + (-1) \\ -3 + 0 & 2 + (-1) & 4 + 6 \end{bmatrix}$

$= \begin{bmatrix} 12 & 6 & 9 \\ -3 & 1 & 10 \end{bmatrix}$

Notice that $A + B = B + A$. The commutative property for matrix addition holds in general, provided that A and B have the same dimension.

(c) $A - B = \begin{bmatrix} 7 & 8 & -1 \\ 0 & -1 & 6 \end{bmatrix} - \begin{bmatrix} 5 & -2 & 10 \\ -3 & 2 & 4 \end{bmatrix}$

$= \begin{bmatrix} 7 - 5 & 8 - (-2) & -1 - 10 \\ 0 - (-3) & -1 - 2 & 6 - 4 \end{bmatrix}$

$= \begin{bmatrix} 2 & 10 & -11 \\ 3 & -3 & 2 \end{bmatrix}$

◆ **CLASS DISCUSSION**

If matrices A and B have the same dimension, does $A - B = B - A$?
◆

Now Try Exercise 11 ◆

Multiplication of a Matrix by a Scalar   The matrix

$$B = \begin{bmatrix} 1 & 1 & 1 \\ 1 & 1 & 1 \\ 1 & 1 & 1 \end{bmatrix}$$

can be used to darken a digital picture. Suppose that a photograph is represented by a matrix A with gray levels 0 through 11. Every time the matrix B is added to A, the picture becomes slightly darker. For example, if

$$A = \begin{bmatrix} 0 & 5 & 0 \\ 5 & 5 & 5 \\ 0 & 5 & 0 \end{bmatrix}$$

then the addition of $A + B + B$ would darken the picture by two gray levels, and could be computed by

$$A + B + B = \begin{bmatrix} 0 & 5 & 0 \\ 5 & 5 & 5 \\ 0 & 5 & 0 \end{bmatrix} + \begin{bmatrix} 1 & 1 & 1 \\ 1 & 1 & 1 \\ 1 & 1 & 1 \end{bmatrix} + \begin{bmatrix} 1 & 1 & 1 \\ 1 & 1 & 1 \\ 1 & 1 & 1 \end{bmatrix} = \begin{bmatrix} 2 & 7 & 2 \\ 7 & 7 & 7 \\ 2 & 7 & 2 \end{bmatrix}.$$

A simpler way to write the expression $A + B + B$ is $A + 2B$. Multiplying B by 2 to obtain $2B$ is called **scalar multiplication**.

$$2B = 2\begin{bmatrix} 1 & 1 & 1 \\ 1 & 1 & 1 \\ 1 & 1 & 1 \end{bmatrix} = \begin{bmatrix} 2(1) & 2(1) & 2(1) \\ 2(1) & 2(1) & 2(1) \\ 2(1) & 2(1) & 2(1) \end{bmatrix} = \begin{bmatrix} 2 & 2 & 2 \\ 2 & 2 & 2 \\ 2 & 2 & 2 \end{bmatrix}$$

Each element of B is multiplied by the real number 2.

Sometimes a matrix B is denoted $B = [b_{ij}]$, where b_{ij} represents the element in the ith row, jth column. In this way, we could write $2B$ as $2[b_{ij}] = [2b_{ij}]$. This indicates that to calculate $2B$, multiply each b_{ij} by 2. In a similar manner, a matrix A is sometimes denoted by $[a_{ij}]$.

Some operations on matrices are now summarized.

OPERATIONS ON MATRICES

Matrix Addition

The sum of two $m \times n$ matrices A and B is the $m \times n$ matrix $A + B$, in which each element is the sum of the corresponding elements of A and B. This is written as $A + B = [a_{ij}] + [b_{ij}] = [a_{ij} + b_{ij}]$. If A and B have different dimensions, then $A + B$ is undefined.

Matrix Subtraction

The difference of two $m \times n$ matrices A and B is the $m \times n$ matrix $A - B$, in which each element is the difference of the corresponding elements of A and B. This is written as $A - B = [a_{ij}] - [b_{ij}] = [a_{ij} - b_{ij}]$. If A and B have different dimensions, then $A - B$ is undefined.

Multiplication of a Matrix by a Scalar

The product of a scalar (real number) k and an $m \times n$ matrix A is the $m \times n$ matrix kA, in which each element is k times the corresponding element of A. This is written as $kA = k[a_{ij}] = [ka_{ij}]$.

EXAMPLE 4 Performing scalar multiplication

If $A = \begin{bmatrix} 2 & 7 & 11 \\ -1 & 3 & -5 \\ 0 & 9 & -12 \end{bmatrix}$, find $-4A$.

SOLUTION

$$-4A = -4 \begin{bmatrix} 2 & 7 & 11 \\ -1 & 3 & -5 \\ 0 & 9 & -12 \end{bmatrix} = \begin{bmatrix} -4(2) & -4(7) & -4(11) \\ -4(-1) & -4(3) & -4(-5) \\ -4(0) & -4(9) & -4(-12) \end{bmatrix} = \begin{bmatrix} -8 & -28 & -44 \\ 4 & -12 & 20 \\ 0 & -36 & 48 \end{bmatrix}$$

Now Try Exercise 13(b) ◆

EXAMPLE 5 Performing operations on matrices

If possible, perform the indicated operations using

$$A = \begin{bmatrix} 4 & -2 \\ 3 & 5 \end{bmatrix}, B = \begin{bmatrix} 0 & 1 \\ -2 & 3 \end{bmatrix}, C = \begin{bmatrix} 1 & -1 \\ 0 & 7 \\ -4 & 2 \end{bmatrix}, \text{ and } D = \begin{bmatrix} -1 & -3 \\ 9 & -7 \\ 1 & 8 \end{bmatrix}.$$

(a) $A + 3B$ **(b)** $A - C$ **(c)** $-2C - 3D$

SOLUTION

(a) $A + 3B = \begin{bmatrix} 4 & -2 \\ 3 & 5 \end{bmatrix} + 3 \begin{bmatrix} 0 & 1 \\ -2 & 3 \end{bmatrix} = \begin{bmatrix} 4 & -2 \\ 3 & 5 \end{bmatrix} + \begin{bmatrix} 0 & 3 \\ -6 & 9 \end{bmatrix} = \begin{bmatrix} 4 & 1 \\ -3 & 14 \end{bmatrix}$

(b) $A - C$ is undefined because the dimension of A is 2×2 and unequal to the dimension of C, which is 3×2.

(c) $-2C - 3D = -2 \begin{bmatrix} 1 & -1 \\ 0 & 7 \\ -4 & 2 \end{bmatrix} - 3 \begin{bmatrix} -1 & -3 \\ 9 & -7 \\ 1 & 8 \end{bmatrix}$

$$= \begin{bmatrix} -2 & 2 \\ 0 & -14 \\ 8 & -4 \end{bmatrix} - \begin{bmatrix} -3 & -9 \\ 27 & -21 \\ 3 & 24 \end{bmatrix} = \begin{bmatrix} 1 & 11 \\ -27 & 7 \\ 5 & -28 \end{bmatrix}$$

Now Try Exercises 15 and 17 ◆

Matrix Products

Addition, subtraction, and multiplication can be performed on numbers, variables, and functions. The same operations apply to matrices. Matrix multiplication is different from scalar multiplication.

Suppose two students are taking day classes at one college and night classes at another, in order to graduate on time. Tables 9.9 and 9.10 list the number of credits taken by the students and the cost per credit at each college.

TABLE 9.9 Credits

	College A	College B
Student 1	10	7
Student 2	11	4

TABLE 9.10

	Cost per Credit
College A	$60
College B	$80

The cost of tuition is computed by multiplying the number of credits times the cost of each credit. Student 1 is taking 10 credits at \$60 each and 7 credits at \$80 each. The total tuition for Student 1 is $10(\$60) + 7(\$80) = \$1160$. In a similar manner, the tuition for Student 2 is given by $11(\$60) + 4(\$80) = \$980$.

The information in these tables can be represented by matrices. Let A represent Table 9.9 and B represent Table 9.10.

$$A = \begin{bmatrix} 10 & 7 \\ 11 & 4 \end{bmatrix} \quad \text{and} \quad B = \begin{bmatrix} 60 \\ 80 \end{bmatrix}$$

The matrix product AB calculates total tuition for each student.

$$AB = \begin{bmatrix} 10 & 7 \\ 11 & 4 \end{bmatrix}\begin{bmatrix} 60 \\ 80 \end{bmatrix} = \begin{bmatrix} 10(60) + 7(80) \\ 11(60) + 4(80) \end{bmatrix} = \begin{bmatrix} 1160 \\ 980 \end{bmatrix}$$

Generalizing from this example provides the following definition of matrix multiplication.

MATRIX MULTIPLICATION

The **product** of an $m \times n$ matrix A and an $n \times k$ matrix B is the $m \times k$ matrix AB, which is computed as follows. To find the element of AB in the ith row and jth column, multiply each element in the ith row of A by the corresponding element in the jth column of B. The sum of these products will give the element of row i, column j in AB.

Note: In order to compute the product of two matrices, the number of columns in the first matrix must equal the number of rows in the second matrix, as illustrated in Figure 9.60.

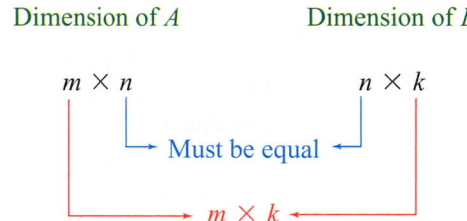

FIGURE 9.60

EXAMPLE 6   Multiplying matrices

If possible, compute each product using

$$A = \begin{bmatrix} 1 & -1 \\ 0 & 3 \\ 4 & -2 \end{bmatrix}, B = \begin{bmatrix} -1 \\ -2 \end{bmatrix}, C = \begin{bmatrix} 1 & 2 & 3 \\ 4 & 5 & 6 \end{bmatrix}, \text{and } D = \begin{bmatrix} 1 & -1 & 2 \\ 0 & 3 & -2 \\ -3 & 4 & 5 \end{bmatrix}.$$

(a) AB **(b)** CA **(c)** DC **(d)** CD

SOLUTION

(a) The dimension of A is 3×2 and the dimension of B is 2×1. The dimension of AB is 3×1 and can be found as follows.

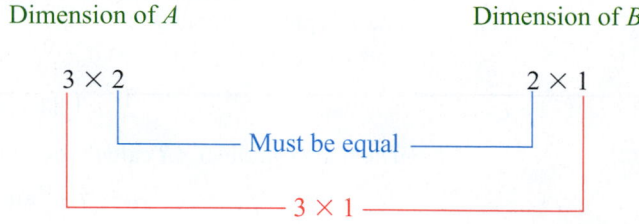

Dimension of AB

$$AB = \begin{bmatrix} \mathbf{1} & -\mathbf{1} \\ \mathbf{0} & 3 \\ \mathbf{4} & -2 \end{bmatrix} \begin{bmatrix} -\mathbf{1} \\ -\mathbf{2} \end{bmatrix} = \begin{bmatrix} (\mathbf{1})(-\mathbf{1}) + (-\mathbf{1})(-\mathbf{2}) \\ (\mathbf{0})(-\mathbf{1}) + (3)(-\mathbf{2}) \\ (\mathbf{4})(-\mathbf{1}) + (-\mathbf{2})(-\mathbf{2}) \end{bmatrix} = \begin{bmatrix} 1 \\ -6 \\ 0 \end{bmatrix}$$

(b) The dimension of C is 2×3, and the dimension of A is 3×2. Thus CA has dimension 2×2.

$$CA = \begin{bmatrix} 1 & 2 & 3 \\ 4 & 5 & 6 \end{bmatrix} \begin{bmatrix} 1 & -1 \\ 0 & 3 \\ 4 & -2 \end{bmatrix}$$

$$= \begin{bmatrix} 1(1) + 2(0) + 3(4) & 1(-1) + 2(3) + 3(-2) \\ 4(1) + 5(0) + 6(4) & 4(-1) + 5(3) + 6(-2) \end{bmatrix}$$

$$= \begin{bmatrix} 13 & -1 \\ 28 & -1 \end{bmatrix}$$

(c) The dimension of D is 3×3 and the dimension of C is 2×3. Therefore DC is undefined. Note that D has 3 columns and C has only 2 rows.

(d) The dimension of C is 2×3 and the dimension of D is 3×3. Thus CD has dimension 2×3.

$$CD = \begin{bmatrix} 1 & 2 & 3 \\ 4 & 5 & 6 \end{bmatrix} \begin{bmatrix} 1 & -1 & 2 \\ 0 & 3 & -2 \\ -3 & 4 & 5 \end{bmatrix}$$

$$= \begin{bmatrix} 1(1) + 2(0) + 3(-3) & 1(-1) + 2(3) + 3(4) & 1(2) + 2(-2) + 3(5) \\ 4(1) + 5(0) + 6(-3) & 4(-1) + 5(3) + 6(4) & 4(2) + 5(-2) + 6(5) \end{bmatrix}$$

$$= \begin{bmatrix} -8 & 17 & 13 \\ -14 & 35 & 28 \end{bmatrix}$$

Now Try Exercises 41, 45, 47, and 57 ◆

◆ **MAKING CONNECTIONS**

The Commutative Property and Matrix Multiplication   In Example 6 it is shown that $CD \neq DC$. Unlike multiplication of numbers, variables, and functions, matrix multiplication is *not* commutative. Instead, matrix multiplication is similar to function composition, where for a general pair of functions $f \circ g \neq g \circ f$. ◆

Square matrices have the same number of rows as columns and have dimension $n \times n$ for some natural number n. When we multiply two square matrices, both having dimension $n \times n$, the resulting matrix also has dimension $n \times n$, as illustrated in the next example.

 EXAMPLE 7 Multiplying square matrices

If $A = \begin{bmatrix} 1 & 0 & 7 \\ 3 & 2 & -1 \\ -5 & -2 & 5 \end{bmatrix}$ and $B = \begin{bmatrix} 4 & -6 & 7 \\ 8 & 9 & 10 \\ 0 & 1 & -3 \end{bmatrix}$, find AB.

SOLUTION

$$AB = \begin{bmatrix} 1 & 0 & 7 \\ 3 & 2 & -1 \\ -5 & -2 & 5 \end{bmatrix} \begin{bmatrix} 4 & -6 & 7 \\ 8 & 9 & 10 \\ 0 & 1 & -3 \end{bmatrix}$$

$$= \begin{bmatrix} 1(4) + 0(8) + 7(0) & 1(-6) + 0(9) + 7(1) & 1(7) + 0(10) + 7(-3) \\ 3(4) + 2(8) - 1(0) & 3(-6) + 2(9) - 1(1) & 3(7) + 2(10) - 1(-3) \\ -5(4) - 2(8) + 5(0) & -5(-6) - 2(9) + 5(1) & -5(7) - 2(10) + 5(-3) \end{bmatrix}$$

$$= \begin{bmatrix} 4 & 1 & -14 \\ 28 & -1 & 44 \\ -36 & 17 & -70 \end{bmatrix}$$

Now Try Exercise 51 ◆

Real numbers satisfy the commutative, associative, and distributive properties for various arithmetic operations. Matrices also satisfy some of these properties, provided that their dimensions are valid so that the resulting expressions are defined. The following box summarizes these basic properties.

◆ **PROPERTIES OF MATRICES**

Let A, B, and C be matrices. Assume that each matrix operation is defined.

1. $A + B = B + A$ Commutative property for matrix addition (No commutative property for matrix multiplication)

2. $(A + B) + C = A + (B + C)$ Associative property for matrix addition

3. $(AB)C = A(BC)$ Associative property for matrix multiplication

4. $A(B + C) = AB + BC$ Distributive property

Technology and Matrices (Optional)

Computing arithmetic operations on large matrices by hand can be a difficult task, prone to errors. Many graphing calculators have the capability to perform addition, subtraction, multiplication, and scalar multiplication with matrices, as the next two examples demonstrate.

 **Multiplying matrices with technology**

Use a graphing calculator to find the product AB from Example 7.

SOLUTION First enter the matrices A and B into your calculator, as illustrated in Figures 9.61 and 9.62. Then find their product on the home screen, as shown in Figure 9.63. Notice that the answer agrees with our results from Example 7.

Calculator Help

To enter the elements of a matrix, see Appendix B (page AP-19). To multiply two matrices, see Appendix B (page AP-20).

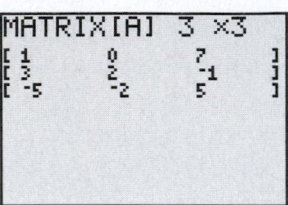

FIGURE 9.61

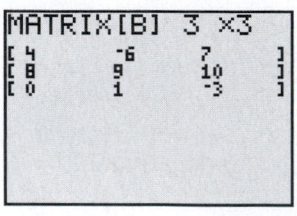

FIGURE 9.62

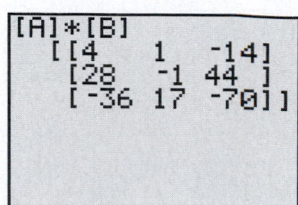

FIGURE 9.63

Now Try Exercise 61 ◆

 **Using technology to evaluate a matrix expression**

Evaluate the expression $2A + 3B^3$, where

$$A = \begin{bmatrix} 3 & -1 & 2 \\ -1 & 6 & -1 \\ 2 & -1 & 9 \end{bmatrix} \quad \text{and} \quad B = \begin{bmatrix} 1 & -2 & 5 \\ 3 & 1 & -1 \\ 5 & 2 & 1 \end{bmatrix}.$$

SOLUTION In the expression $2A + 3B^3$, B^3 is equal to BBB. Enter each matrix into a calculator and evaluate the expression. Figure 9.64 shows the result of this computation.

Now Try Exercise 63 ◆

FIGURE 9.64

 **9.5**

Putting it all Together

Addition, subtraction, and multiplication can be performed on numbers, variables, and functions. In this section we learned how these operations also apply to matrices. The following table provides examples of these operations.

Matrix Addition

$$\begin{bmatrix} 1 & 2 & 3 \\ 5 & 6 & 7 \end{bmatrix} + \begin{bmatrix} -1 & 0 & 8 \\ 9 & -2 & 10 \end{bmatrix} = \begin{bmatrix} 1 + (-1) & 2 + 0 & 3 + 8 \\ 5 + 9 & 6 + (-2) & 7 + 10 \end{bmatrix} = \begin{bmatrix} 0 & 2 & 11 \\ 14 & 4 & 17 \end{bmatrix}$$

Both matrices must have the same dimension for their sum to be defined.

Matrix Subtraction

$$\begin{bmatrix} 1 & -4 \\ -3 & 4 \\ 2 & 7 \end{bmatrix} - \begin{bmatrix} 5 & 1 \\ 3 & 6 \\ 8 & -9 \end{bmatrix} = \begin{bmatrix} 1 - 5 & -4 - 1 \\ -3 - 3 & 4 - 6 \\ 2 - 8 & 7 - (-9) \end{bmatrix} = \begin{bmatrix} -4 & -5 \\ -6 & -2 \\ -6 & 16 \end{bmatrix}$$

Both matrices must have the same dimension for their difference to be defined.

Scalar Multiplication

$$3\begin{bmatrix} 3 & -2 \\ 0 & 1 \end{bmatrix} = \begin{bmatrix} 3(3) & 3(-2) \\ 3(0) & 3(1) \end{bmatrix} = \begin{bmatrix} 9 & -6 \\ 0 & 3 \end{bmatrix}$$

Matrix Multiplication

$$\begin{bmatrix} 0 & 1 \\ 2 & -3 \end{bmatrix}\begin{bmatrix} 3 & -5 \\ 4 & 6 \end{bmatrix} = \begin{bmatrix} 0(3) + 1(4) & 0(-5) + 1(6) \\ 2(3) + (-3)(4) & 2(-5) + (-3)(6) \end{bmatrix} = \begin{bmatrix} 4 & 6 \\ -6 & -28 \end{bmatrix}$$

For a matrix product to be defined, the number of columns in the first matrix must equal the number of rows in the second matrix. Matrix multiplication is not commutative. That is, $AB \neq BA$ in general.

9.5 Exercises

Elements of Matrices

Exercises 1–4: Determine each of the following for the given matrix A, if possible.

(a) $a_{12}, a_{21},$ and a_{32} (b) $a_{11} a_{22} + 3a_{23}$

1. $\begin{bmatrix} 1 & 2 & 3 \\ 4 & 5 & 6 \end{bmatrix}$

2. $\begin{bmatrix} 1 & 2 & 3 & 4 \\ 5 & 6 & 7 & 8 \\ 9 & 10 & 11 & 12 \end{bmatrix}$

3. $\begin{bmatrix} 1 & -1 & 4 \\ 3 & -2 & 5 \\ 7 & 0 & -6 \end{bmatrix}$

4. $\begin{bmatrix} 1 & -2 \\ -4 & 5 \end{bmatrix}$

Exercises 5–8: If possible, find values for x and y so that the matrices A and B are equal.

5. $A = \begin{bmatrix} x & 2 \\ -2 & 1 \end{bmatrix}$, $B = \begin{bmatrix} 1 & 2 \\ -2 & y \end{bmatrix}$

6. $A = \begin{bmatrix} 1 & x+y & 3 \\ 4 & -1 & 6 \\ 3 & 7 & -2 \end{bmatrix}$, $B = \begin{bmatrix} 1 & 2 & 3 \\ 4 & -1 & 6 \\ 3 & y & -2 \end{bmatrix}$

7. $A = \begin{bmatrix} x & 3 \\ 6 & -2 \end{bmatrix}$, $B = \begin{bmatrix} 1 & y & 0 \\ 6 & -2 & 0 \\ 0 & 0 & 0 \end{bmatrix}$

8. $A = \begin{bmatrix} 4 & -2 \\ 3 & -4 \\ x & y \end{bmatrix}$, $B = \begin{bmatrix} 4 & -2 & -2 \\ 3 & -4 & -4 \\ 7 & 8 & 8 \end{bmatrix}$

Addition, Subtraction, and Scalar Multiples

Exercises 9–12: For the given matrices A and B find each of the following.

(a) $A + B$ (b) $B + A$ (c) $A - B$

9. $A = \begin{bmatrix} 4 & -1 \\ -1 & 4 \end{bmatrix}$, $B = \begin{bmatrix} -1 & 4 \\ 4 & -1 \end{bmatrix}$

10. $A = \begin{bmatrix} 2 & -4 \\ -1 & \frac{1}{2} \\ 3 & -2 \end{bmatrix}$, $B = \begin{bmatrix} 5 & 0 \\ 3 & \frac{1}{2} \\ -1 & 1 \end{bmatrix}$

11. $A = \begin{bmatrix} 3 & 4 & -1 \\ 0 & -3 & 2 \\ -2 & 5 & 10 \end{bmatrix}$, $B = \begin{bmatrix} 11 & 5 & -2 \\ 4 & -7 & 12 \\ 6 & 6 & 6 \end{bmatrix}$

12. $A = \begin{bmatrix} 1 & 6 & 1 & -2 \\ 0 & 1 & 3 & 5 \\ 0 & 0 & 1 & -2 \end{bmatrix}$, $B = \begin{bmatrix} 1 & 0 & 0 & 9 \\ 3 & 1 & 0 & 3 \\ -1 & 4 & 1 & -2 \end{bmatrix}$

Exercises 13–18: If possible, find each of the following.

(a) $A + B$ (b) $3A$ (c) $2A - 3B$

13. $A = \begin{bmatrix} 2 & -6 \\ 3 & 1 \end{bmatrix}$, $B = \begin{bmatrix} -1 & 0 \\ -2 & 3 \end{bmatrix}$

14. $A = \begin{bmatrix} 1 & -2 & 5 \\ 3 & -4 & -1 \end{bmatrix}$, $B = \begin{bmatrix} 0 & -1 & -5 \\ -3 & 1 & 2 \end{bmatrix}$

15. $A = \begin{bmatrix} 1 & -1 & 0 \\ 1 & 5 & 9 \\ -4 & 8 & -5 \end{bmatrix}$, $B = \begin{bmatrix} 2 & 8 & -1 \\ 6 & -1 & 3 \end{bmatrix}$

16. $A = \begin{bmatrix} 6 & 2 & 9 \\ 3 & -2 & 0 \\ -1 & 4 & 8 \end{bmatrix}$, $B = \begin{bmatrix} 1 & 0 & -1 \\ 3 & 0 & 7 \\ 0 & -2 & -5 \end{bmatrix}$

17. $A = \begin{bmatrix} -2 & -1 \\ -5 & 1 \\ 2 & -3 \end{bmatrix}$, $B = \begin{bmatrix} 2 & -1 \\ 3 & 1 \\ 7 & -5 \end{bmatrix}$

18. $A = \begin{bmatrix} 0 & 1 \\ 3 & 2 \\ 4 & -9 \end{bmatrix}$, $B = \begin{bmatrix} 5 & 2 & -7 \\ 8 & -2 & 0 \end{bmatrix}$

Exercises 19–24: Evaluate the matrix expression.

19. $2\begin{bmatrix} 2 & -1 \\ 5 & 1 \\ 0 & 3 \end{bmatrix} + \begin{bmatrix} 5 & 0 \\ 7 & -3 \\ 1 & 1 \end{bmatrix} - \begin{bmatrix} 9 & -4 \\ 4 & 4 \\ 1 & 6 \end{bmatrix}$

20. $-3\begin{bmatrix} 3 & 8 \\ -1 & -9 \end{bmatrix} + 5\begin{bmatrix} 4 & -8 \\ 1 & 6 \end{bmatrix}$

21. $\begin{bmatrix} 4 & 6 \\ 3 & -7 \end{bmatrix} - 2\begin{bmatrix} 1 & 0 \\ -4 & 1 \end{bmatrix}$

22. $\begin{bmatrix} 5 & -1 & 6 \\ -2 & 10 & 12 \\ 5 & 2 & 9 \end{bmatrix} - \begin{bmatrix} -1 & 2 & 2 \\ 2 & -1 & 2 \\ 2 & 2 & -1 \end{bmatrix}$

23. $2\begin{bmatrix} 2 & -1 & -1 \\ -1 & 2 & -1 \\ -1 & -1 & 2 \end{bmatrix} + 3\begin{bmatrix} 1 & 2 & 3 \\ 2 & 1 & 3 \\ 2 & 3 & 1 \end{bmatrix}$

24. $3\begin{bmatrix} 1 & 0 & 3 & -1 \\ 0 & 1 & 2 & -1 \\ 1 & 0 & -3 & 1 \end{bmatrix} - 4\begin{bmatrix} -1 & 0 & 0 & 4 \\ 0 & -1 & 3 & 2 \\ 2 & 0 & 1 & -1 \end{bmatrix}$

Matrices and Digital Photography

Exercises 25–28: ***Digital Photography*** *(Refer to the discussion of digital photography in this section.) Consider the following simplified digital photograph that has a 3 × 3 grid with four gray levels numbered from 0 to 3. It shows the number 1 in dark gray on a light gray background. Let A be the 3 × 3 matrix that represents this figure digitally.*

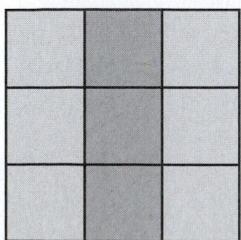

25. Find the matrix A.

26. Find a matrix B such that $A + B$ represents the entire picture becoming one gray level darker. Evaluate the expression $A + B$.

27. (Refer to Example 2.) Find a matrix B such that $A + B$ will result in a contrast enhancement of A by one gray level. Evaluate $A + B$.

28. Find a matrix B such that $A - B$ represents the entire picture becoming lighter by one gray level. Evaluate the expression $A - B$.

29. ***Negative Image*** The negative image of a picture interchanges black and white. The number 1 is represented by the matrix A. Determine a matrix B such that $B - A$ represents the negative image of the picture represented by A. Evaluate $B - A$.

$$A = \begin{bmatrix} 0 & 3 & 0 \\ 0 & 3 & 0 \\ 0 & 3 & 0 \end{bmatrix}$$

30. ***Negative Image*** (Refer to the previous exercise.) Consider the matrix A representing a digital photograph. Find a matrix B that represents the negative image of this picture.

$$A = \begin{bmatrix} 0 & 3 & 0 \\ 1 & 3 & 1 \\ 2 & 3 & 2 \end{bmatrix}$$

Exercises 31 and 32: ***Digital Photography*** *The accompanying digital photograph represents the letter* F *using 20 pixels in a 5 × 4 grid. Assume that there are four gray levels from 0 to 3.*

31. Find a matrix A that represents a digital photograph of this letter F.

32. (Continuation of the previous exercise)
   (a) Find a matrix B such that $B - A$ represents the negative image of A.

   (b) Find a matrix C where $A + C$ represents a decrease in the contrast of A by one gray level.

Exercises 33–36: **Digitizing Letters**   *Complete the following.*

   (a) *Design a matrix A with dimension* 4×4 *that represents a digital photograph of the given letter. Assume that there are four gray levels from 0 to 3.*
   (b) *Find a matrix B such that* $B - A$ *represents the negative image of the picture represented by the matrix A from part (a).*

33. Z   **34.** N

35. L   **36.** O

Matrix Multiplication

Exercises 37–40: **Tuition Costs**   *(Refer to the discussion before Example 6.)*

   (a) *Find a matrix A and a column matrix B that describe the following tables involving credits and tuition costs.*
   (b) *Find the matrix product AB, and interpret the result.*

37.

	College A	College B
Student 1	12	4
Student 2	8	7

	Cost per Credit
College A	$55
College B	$70

38.

	College A	College B
Student 1	15	2
Student 2	12	4

	Cost per Credit
College A	$90
College B	$75

39.

	College A	College B
Student 1	10	5
Student 2	9	8
Student 3	11	3

	Cost per Credit
College A	$60
College B	$70

40.

	College A	College B	College C
Student 1	6	0	3
Student 2	11	3	0
Student 3	0	12	3

	Cost per Credit
College A	$50
College B	$65
College C	$60

Exercises 41–58: If possible, determine the matrix products AB and BA.

41. $A = \begin{bmatrix} 1 & -1 \\ 2 & 0 \end{bmatrix},$ $B = \begin{bmatrix} -2 & 3 \\ 1 & 2 \end{bmatrix}$

42. $A = \begin{bmatrix} -3 & 5 \\ 2 & 7 \end{bmatrix},$ $B = \begin{bmatrix} -1 & 2 \\ 0 & 7 \end{bmatrix}$

43. $A = \begin{bmatrix} 5 & -7 & 2 \\ 0 & 1 & 5 \end{bmatrix},$ $B = \begin{bmatrix} 9 & 8 & 7 \\ 1 & -1 & -2 \end{bmatrix}$

44. $A = \begin{bmatrix} 2 & 1 & -1 \\ 0 & 2 & 1 \\ 3 & 2 & -1 \end{bmatrix},$ $B = \begin{bmatrix} 1 & 0 \\ 2 & -1 \\ 3 & 1 \end{bmatrix}$

45. $A = \begin{bmatrix} 3 & -1 \\ 1 & 0 \\ -2 & -4 \end{bmatrix},$ $B = \begin{bmatrix} -2 & 5 & -3 \\ 9 & -7 & 0 \end{bmatrix}$

46. $A = \begin{bmatrix} -1 & 0 & -2 \\ 4 & -2 & 1 \end{bmatrix},$ $B = \begin{bmatrix} 2 & -2 \\ 5 & -1 \\ 0 & 1 \end{bmatrix}$

47. $A = \begin{bmatrix} 1 & -1 & 0 \\ 2 & -1 & 5 \\ 6 & 1 & -4 \end{bmatrix},$ $B = \begin{bmatrix} -1 & 3 & -1 \\ 7 & -7 & 1 \end{bmatrix}$

48. $A = \begin{bmatrix} 2 & -1 & -5 \\ 4 & -1 & 6 \\ -2 & 0 & 9 \end{bmatrix},$ $B = \begin{bmatrix} 1 & 2 \\ -1 & -1 \\ 2 & 0 \end{bmatrix}$

49. $A = \begin{bmatrix} 2 & -3 \\ 5 & 3 \end{bmatrix},$ $B = \begin{bmatrix} -3 \\ 4 \\ 1 \end{bmatrix}$

50. $A = \begin{bmatrix} 3 & -1 \\ 2 & -2 \\ 0 & 4 \end{bmatrix},$ $B = \begin{bmatrix} 1 & -4 & 0 \\ -1 & 3 & 2 \end{bmatrix}$

51. $A = \begin{bmatrix} 2 & -1 & 3 \\ 0 & 1 & 0 \\ 2 & -2 & 3 \end{bmatrix}$, $B = \begin{bmatrix} 1 & 5 & -1 \\ 0 & 1 & 3 \\ -1 & 2 & 1 \end{bmatrix}$

52. $A = \begin{bmatrix} 1 & -2 & 5 \\ 1 & 0 & -2 \\ 1 & 3 & 2 \end{bmatrix}$, $B = \begin{bmatrix} -1 & 4 & 2 \\ -3 & 0 & 1 \\ 5 & 1 & 0 \end{bmatrix}$

53. $A = \begin{bmatrix} 2 & -1 \\ 3 & 1 \end{bmatrix}$, $B = \begin{bmatrix} 1 \\ 3 \end{bmatrix}$

54. $A = \begin{bmatrix} 5 & -3 \end{bmatrix}$, $B = \begin{bmatrix} 1 \\ 3 \end{bmatrix}$

55. $A = \begin{bmatrix} -3 & 1 \\ 2 & -4 \end{bmatrix}$, $B = \begin{bmatrix} 1 & 0 & -2 \\ -4 & 8 & 1 \end{bmatrix}$

56. $A = \begin{bmatrix} 6 & 1 & 0 \\ -2 & 5 & 1 \\ 4 & -7 & 10 \end{bmatrix}$, $B = \begin{bmatrix} 10 \\ 20 \\ 30 \end{bmatrix}$

57. $A = \begin{bmatrix} 1 & 0 & -2 \\ 3 & -4 & 1 \\ 2 & 0 & 5 \end{bmatrix}$, $B = \begin{bmatrix} 1 \\ -1 \\ 3 \end{bmatrix}$

58. $A = \begin{bmatrix} 1 & -1 & 3 & -2 \\ 1 & 0 & 3 & 4 \\ 2 & -2 & 0 & 8 \end{bmatrix}$, $B = \begin{bmatrix} 1 & -1 \\ 0 & 5 \\ 2 & 3 \\ -5 & 4 \end{bmatrix}$

59. *Auto Parts* A store owner makes two separate orders for three types of auto parts: I, II, and III. The number of parts ordered is represented by the matrix A.

$$A = \begin{bmatrix} 3 & 4 & 8 \\ 5 & 6 & 2 \end{bmatrix} \begin{matrix} \text{Order 1} \\ \text{Order 2} \end{matrix}$$

For example, in Order 1 there were 4 parts of Type II ordered. The cost in dollars of each part can be represented by the matrix B.

$$B = \begin{bmatrix} 10 \\ 20 \\ 30 \end{bmatrix} \begin{matrix} \text{Part I} \\ \text{Part II} \\ \text{Part III} \end{matrix}$$

Find AB and interpret the result.

60. *Car Sales* Two car dealers buy four different makes of cars: I, II, III, and IV. The number and make of automobiles bought by each dealer is represented by the matrix A.

$$A = \begin{bmatrix} 1 & 3 & 8 & 4 \\ 3 & 5 & 7 & 0 \end{bmatrix} \begin{matrix} \text{Dealer 1} \\ \text{Dealer 2} \end{matrix}$$

For example, Dealer 2 buys 7 cars of Type III. The cost in thousands of dollars of each type of car can be represented by the matrix B.

$$B = \begin{bmatrix} 15 \\ 21 \\ 28 \\ 38 \end{bmatrix} \begin{matrix} \text{Make I} \\ \text{Make II} \\ \text{Make III} \\ \text{Make IV} \end{matrix}$$

Find AB and interpret the result.

Exercises 61–64: **Properties of Matrices** *Use a graphing calculator to evaluate the expression with the given matrices A, B, and C. Compare your answers for parts (a) and (b). Then, interpret the results.*

$$A = \begin{bmatrix} 2 & -1 & 3 \\ 1 & 3 & -5 \\ 0 & -2 & 1 \end{bmatrix}, B = \begin{bmatrix} 6 & 2 & 7 \\ 3 & -4 & -5 \\ 7 & 1 & 0 \end{bmatrix},$$
$$C = \begin{bmatrix} 1 & 4 & -3 \\ 8 & 1 & -1 \\ 4 & 6 & -2 \end{bmatrix}$$

61. (a) $A(B + C)$ **(b)** $AB + AC$

62. (a) $(A - B)C$ **(b)** $AC - BC$

63. (a) $(A - B)^2$ **(b)** $A^2 - AB - BA + B^2$

64. (a) $(AB)C$ **(b)** $A(BC)$

Writing about Mathematics

65. Discuss whether matrix multiplication is more like multiplication of functions or composition of functions. Explain your reasoning.

66. Describe one application of matrices.

EXTENDED AND DISCOVERY EXERCISES

Exercises 1–4: **Representing Colors** *Colors for computer monitors are often described using ordered triples. One model, called the RGB system, uses red, green, and blue to generate all colors. The accompanying figure describes the relationships of these colors in this system. For example, red is (1, 0, 0), green is (0, 1, 0), and blue is (0, 0, 1). Since equal amounts of red and green combine to form yellow, yellow is represented by (1, 1, 0). Similarly, magenta (a deep* reddish purple) is a mixture of blue and red and is represented by $(1, 0, 1)$. Cyan is $(0, 1, 1)$, since it is a mixture of blue and green.

Another color model uses cyan, magenta, and yellow. It is referred to as the CMY model and is used in the four-color printing process for textbooks like this one. In this system, cyan is $(1, 0, 0)$, magenta is $(0, 1, 0)$, and yellow is $(0, 0, 1)$. In the CMY model, red is created by mixing magenta and yellow. Thus, red is $(0, 1, 1)$ in this system. To convert ordered triples in the RGB model to ordered triples in the CMY model, we can use the following matrix equation. In both of these systems, color intensities vary between 0 and 1. (**Sources:** I. Kerlow, The Art of 3-D Computer Animation and Imaging; R. Wolff.)

$$\begin{bmatrix} C \\ M \\ Y \end{bmatrix} = \begin{bmatrix} 1 \\ 1 \\ 1 \end{bmatrix} - \begin{bmatrix} R \\ G \\ B \end{bmatrix}$$

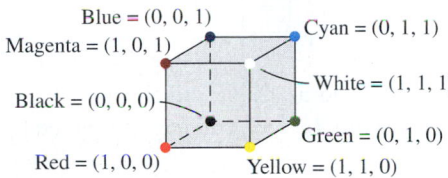

Blue = $(0, 0, 1)$
Magenta = $(1, 0, 1)$
Cyan = $(0, 1, 1)$
Black = $(0, 0, 0)$
White = $(1, 1, 1)$
Red = $(1, 0, 0)$
Green = $(0, 1, 0)$
Yellow = $(1, 1, 0)$

1. In the RGB model, aquamarine is $(0.631, 1, 0.933)$. Use the matrix equation to determine the mixture of cyan, magenta, and yellow that makes aquamarine in the CMY model

2. In the RGB model, rust is $(0.552, 0.168, 0.066)$. Use the matrix equation to determine the mixture of cyan, magenta, and yellow that makes rust in the CMY model.

3. Use the given matrix equation to find a matrix equation that changes colors represented by ordered triples in the CMY model into ordered triples in the RGB model.

4. In the CMY model, $(0.012, 0, 0.597)$ is a cream color. Use the matrix equation from the previous exercise to determine the mixture of red, green, and blue that makes a cream color in the RGB model.

9.6 Inverses of Matrices

◆ Understand matrix inverses
◆ Find inverses symbolically
◆ Represent linear systems with matrix equations
◆ Solve linear systems with matrix inverses

Introduction

In Section 5.2 we discussed how the inverse function f^{-1} will undo or cancel the computation performed by the function f. Like functions, some matrices have inverses. The inverse of a matrix A will undo or cancel the computation performed by A. For example, matrices play an important role in computer graphics. If a matrix A is capable of rotating a figure on a screen 90° clockwise, then the inverse matrix would cause the figure to rotate 90° counterclockwise. This section discusses matrix inverses and some of their applications.

Matrix Inverses

In computer graphics the matrix

$$A = \begin{bmatrix} 1 & 0 & h \\ 0 & 1 & k \\ 0 & 0 & 1 \end{bmatrix}$$

is used to translate a point (x, y), horizontally h units and vertically k units. The translation is to the right if $h > 0$ and to the left if $h < 0$. Similarly, the translation is upward if $k > 0$ and downward if $k < 0$. A point (x, y) is represented by the 3×1 **column matrix**

$$X = \begin{bmatrix} x \\ y \\ 1 \end{bmatrix}.$$

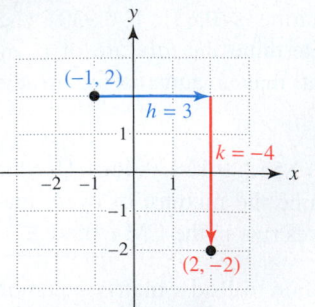

FIGURE 9.65 Translation of a Point

The third element in X is always equal to 1. For example, the point $(-1, 2)$ could be translated 3 units right and 4 units downward by computing the matrix product

$$AX = \begin{bmatrix} 1 & 0 & 3 \\ 0 & 1 & -4 \\ 0 & 0 & 1 \end{bmatrix} \begin{bmatrix} -1 \\ 2 \\ 1 \end{bmatrix} = \begin{bmatrix} 2 \\ -2 \\ 1 \end{bmatrix} = Y.$$

Its new location is $(2, -2)$. In the matrix A, $h = 3$ and $k = -4$. See Figure 9.65.   (**Source:** C. Pokorny and C. Gerald, *Computer Graphics.*)

If A translates a point 3 units right and 4 units downward, then the inverse matrix translates a point 3 units left and 4 units upward. This would return a point to its original position after being translated by A. Therefore the inverse matrix of A, denoted A^{-1}, is given by

$$A^{-1} = \begin{bmatrix} 1 & 0 & -3 \\ 0 & 1 & 4 \\ 0 & 0 & 1 \end{bmatrix}.$$

In A^{-1}, $h = -3$ and $k = 4$. The matrix product $A^{-1}Y$ results in

$$A^{-1}Y = \begin{bmatrix} 1 & 0 & -3 \\ 0 & 1 & 4 \\ 0 & 0 & 1 \end{bmatrix} \begin{bmatrix} 2 \\ -2 \\ 1 \end{bmatrix} = \begin{bmatrix} -1 \\ 2 \\ 1 \end{bmatrix} = X.$$

The matrix A^{-1} translates the point located at $(2, -2)$ to its original coordinates of $(-1, 2)$. The two translations acting on the point $(-1, 2)$ can be represented by the following computation.

$$A^{-1}AX = \begin{bmatrix} 1 & 0 & -3 \\ 0 & 1 & 4 \\ 0 & 0 & 1 \end{bmatrix} \begin{bmatrix} 1 & 0 & 3 \\ 0 & 1 & -4 \\ 0 & 0 & 1 \end{bmatrix} \begin{bmatrix} -1 \\ 2 \\ 1 \end{bmatrix}$$

$$= \begin{bmatrix} 1 & 0 & 0 \\ 0 & 1 & 0 \\ 0 & 0 & 1 \end{bmatrix} \begin{bmatrix} -1 \\ 2 \\ 1 \end{bmatrix}$$

$$= \begin{bmatrix} -1 \\ 2 \\ 1 \end{bmatrix} = X$$

That is, the action of A followed by A^{-1} on the point $(-1, 2)$ results in $(-1, 2)$. In a similar manner, if we reverse the order of A^{-1} and A to compute $AA^{-1}X$, the result is again X.

$$AA^{-1}X = \begin{bmatrix} 1 & 0 & 3 \\ 0 & 1 & -4 \\ 0 & 0 & 1 \end{bmatrix} \begin{bmatrix} 1 & 0 & -3 \\ 0 & 1 & 4 \\ 0 & 0 & 1 \end{bmatrix} \begin{bmatrix} -1 \\ 2 \\ 1 \end{bmatrix}$$

$$= \begin{bmatrix} 1 & 0 & 0 \\ 0 & 1 & 0 \\ 0 & 0 & 1 \end{bmatrix} \begin{bmatrix} -1 \\ 2 \\ 1 \end{bmatrix}$$

$$= \begin{bmatrix} -1 \\ 2 \\ 1 \end{bmatrix} = X$$

Notice that both matrix products $A^{-1}A$ and AA^{-1} resulted in a matrix with 1's on its main diagonal and 0's elsewhere. This matrix is called the **identity matrix**.

THE $n \times n$ IDENTITY MATRIX

The $n \times n$ **identity matrix**, denoted I_n, has only 1's on its main diagonal and 0's elsewhere.

Some examples of identity matrices are

$$I_2 = \begin{bmatrix} 1 & 0 \\ 0 & 1 \end{bmatrix}, \qquad I_3 = \begin{bmatrix} 1 & 0 & 0 \\ 0 & 1 & 0 \\ 0 & 0 & 1 \end{bmatrix}, \qquad \text{and} \qquad I_4 = \begin{bmatrix} 1 & 0 & 0 & 0 \\ 0 & 1 & 0 & 0 \\ 0 & 0 & 1 & 0 \\ 0 & 0 & 0 & 1 \end{bmatrix}.$$

If A is any $n \times n$ matrix, then $I_n A = A$ and $A I_n = A$. For instance, if

$$A = \begin{bmatrix} 2 & 3 \\ 4 & 5 \end{bmatrix}$$

then

$$I_2 A = \begin{bmatrix} 1 & 0 \\ 0 & 1 \end{bmatrix}\begin{bmatrix} 2 & 3 \\ 4 & 5 \end{bmatrix} = \begin{bmatrix} 2 & 3 \\ 4 & 5 \end{bmatrix} = A, \qquad \text{and}$$

$$A I_2 = \begin{bmatrix} 2 & 3 \\ 4 & 5 \end{bmatrix}\begin{bmatrix} 1 & 0 \\ 0 & 1 \end{bmatrix} = \begin{bmatrix} 2 & 3 \\ 4 & 5 \end{bmatrix} = A.$$

Next we formally define the inverse of an $n \times n$ matrix A, whenever it exists.

INVERSE OF A SQUARE MATRIX

Let A be an $n \times n$ matrix. If there exists an $n \times n$ matrix, denoted A^{-1}, that satisfies

$$A^{-1}A = I_n \qquad \text{and} \qquad AA^{-1} = I_n,$$

then A^{-1} is the **inverse** of A.

If A^{-1} exists, then A is **invertible** or **nonsingular**. On the other hand, if a matrix A is not invertible then it is **singular**. Not every matrix has an inverse. For example, the **zero matrix** with dimension 3×3 is given by

$$O_3 = \begin{bmatrix} 0 & 0 & 0 \\ 0 & 0 & 0 \\ 0 & 0 & 0 \end{bmatrix}.$$

The matrix O_3 does not have an inverse. The product of O_3 with any 3×3 matrix B would again be O_3, rather than the identity matrix I_3.

 **Verifying an inverse**

Determine if B is the inverse of A, where

$$A = \begin{bmatrix} 5 & 3 \\ -3 & -2 \end{bmatrix} \qquad \text{and} \qquad B = \begin{bmatrix} 2 & 3 \\ -3 & -5 \end{bmatrix}.$$

SOLUTION For B to be the inverse of A, it must satisfy the equations $AB = I_2$ and $BA = I_2$.

$$AB = \begin{bmatrix} 5 & 3 \\ -3 & -2 \end{bmatrix} \begin{bmatrix} 2 & 3 \\ -3 & -5 \end{bmatrix} = \begin{bmatrix} 1 & 0 \\ 0 & 1 \end{bmatrix} = I_2$$

$$BA = \begin{bmatrix} 2 & 3 \\ -3 & -5 \end{bmatrix} \begin{bmatrix} 5 & 3 \\ -3 & -2 \end{bmatrix} = \begin{bmatrix} 1 & 0 \\ 0 & 1 \end{bmatrix} = I_2$$

Thus B is the inverse of A. That is, $B = A^{-1}$.   *Now Try Exercise 1* ◆

The next example discusses the significance of an inverse matrix in computer graphics.

EXAMPLE 2   Interpreting an inverse matrix

The matrix A can be used to rotate a point 90° clockwise about the origin, where

$$A = \begin{bmatrix} 0 & 1 & 0 \\ -1 & 0 & 0 \\ 0 & 0 & 1 \end{bmatrix} \quad \text{and} \quad A^{-1} = \begin{bmatrix} 0 & -1 & 0 \\ 1 & 0 & 0 \\ 0 & 0 & 1 \end{bmatrix}.$$

(a) Use A to rotate the point $(-2, 0)$ clockwise 90° about the origin.
(b) Make a conjecture about the effect of A^{-1} on the resulting point.
(c) Test this conjecture.

SOLUTION
(a) First, let the point $(-2, 0)$ be represented by the column matrix

$$X = \begin{bmatrix} -2 \\ 0 \\ 1 \end{bmatrix}.$$

Then compute

$$AX = \begin{bmatrix} 0 & 1 & 0 \\ -1 & 0 & 0 \\ 0 & 0 & 1 \end{bmatrix} \begin{bmatrix} -2 \\ 0 \\ 1 \end{bmatrix} = \begin{bmatrix} 0 \\ 2 \\ 1 \end{bmatrix} = Y.$$

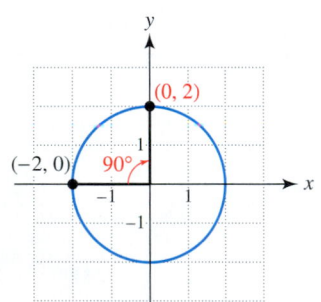

FIGURE 9.66 Rotating a Point about the Origin

If the point $(-2, 0)$ is rotated 90° clockwise about the origin, its new location is $(0, 2)$. See Figure 9.66.
(b) Since A^{-1} represents the inverse operation of A, A^{-1} will rotate the point located at $(0, 2)$ counterclockwise 90°, back to $(-2, 0)$.
(c) This conjecture is correct since

$$A^{-1}Y = \begin{bmatrix} 0 & -1 & 0 \\ 1 & 0 & 0 \\ 0 & 0 & 1 \end{bmatrix} \begin{bmatrix} 0 \\ 2 \\ 1 \end{bmatrix} = \begin{bmatrix} -2 \\ 0 \\ 1 \end{bmatrix} = X.$$

Now Try Exercise 17 ◆

◆ **CLASS DISCUSSION**
What will the results be of the computations AAX and $A^{-1}A^{-1}X$? ◆

Finding Inverses Symbolically

The inverse matrix can be found symbolically by first forming the augmented matrix $[A \mid I_n]$, and then performing matrix row operations, until the left side of the augmented matrix becomes the identity matrix. The resulting augmented matrix can be written as $[I_n \mid A^{-1}]$, where the right side of the matrix is A^{-1}.

In Example 2 we were given the matrix A^{-1}. However, in this example we find A^{-1} for ourselves.

EXAMPLE 3 Finding an inverse symbolically

Find A^{-1} if

$$A = \begin{bmatrix} 0 & 1 & 0 \\ -1 & 0 & 0 \\ 0 & 0 & 1 \end{bmatrix}.$$

SOLUTION Begin by forming the following 3×6 augmented matrix with the 3×3 identity matrix on the right half.

$$\left[\begin{array}{ccc|ccc} 0 & 1 & 0 & 1 & 0 & 0 \\ -1 & 0 & 0 & 0 & 1 & 0 \\ 0 & 0 & 1 & 0 & 0 & 1 \end{array}\right]$$

Next we use row transformations to obtain the 3×3 identity on the left side. Negate the elements in row 2 and then interchange row 1 and row 2. The same row transformations are also applied to the right side of the augmented matrix.

$$\left[\begin{array}{ccc|ccc} 0 & 1 & 0 & 1 & 0 & 0 \\ -1 & 0 & 0 & 0 & 1 & 0 \\ 0 & 0 & 1 & 0 & 0 & 1 \end{array}\right] \begin{array}{c} -R_2 \to \\ R_1 \to \\ R_3 \to \end{array} \left[\begin{array}{ccc|ccc} 1 & 0 & 0 & 0 & -1 & 0 \\ 0 & 1 & 0 & 1 & 0 & 0 \\ 0 & 0 & 1 & 0 & 0 & 1 \end{array}\right]$$

Because the left side of the augmented matrix is now the 3×3 identity, we stop. The right side of the augmented matrix is A^{-1}. Thus

$$A^{-1} = \begin{bmatrix} 0 & -1 & 0 \\ 1 & 0 & 0 \\ 0 & 0 & 1 \end{bmatrix},$$

and our result agrees with the information in Example 2. *Now Try Exercise 27* ◆

Many times finding inverses requires several steps of row transformations. In the next two examples we find the inverse of a 2×2 matrix and a 3×3 matrix.

EXAMPLE 4 Finding the inverse of a 2×2 matrix symbolically

Find A^{-1} if

$$A = \begin{bmatrix} 1 & 4 \\ 2 & 9 \end{bmatrix}.$$

SOLUTION Begin by forming a 2×4 augmented matrix. Perform matrix row operations to obtain the identity matrix on the left side, and perform the same operation on the right side of this matrix.

$$\left[\begin{array}{cc|cc} 1 & 4 & 1 & 0 \\ 2 & 9 & 0 & 1 \end{array}\right] \begin{array}{c} \\ R_2 - 2R_1 \to \end{array} \left[\begin{array}{cc|cc} 1 & 4 & 1 & 0 \\ 0 & 1 & -2 & 1 \end{array}\right] \begin{array}{c} R_1 - 4R_2 \to \\ \\ \end{array} \left[\begin{array}{cc|cc} 1 & 0 & 9 & -4 \\ 0 & 1 & -2 & 1 \end{array}\right]$$

Since the 2×2 identity matrix appears on the left side, it follows that the right side equals A^{-1}. That is,

$$A^{-1} = \begin{bmatrix} 9 & -4 \\ -2 & 1 \end{bmatrix}.$$

Furthermore, it can be verified that $A^{-1}A = I_2 = AA^{-1}$.

Now Try Exercise 21 ◆

EXAMPLE 5  **Finding the inverse of a 3×3 matrix symbolically**

Find A^{-1} if

$$A = \begin{bmatrix} 1 & 0 & 1 \\ 2 & 1 & 3 \\ -1 & 1 & 1 \end{bmatrix}.$$

SOLUTION Begin by forming the following 3×6 augmented matrix. Perform matrix row operations to obtain the identity matrix on the left side, and perform the same operation on the right side of this matrix.

$$\begin{bmatrix} 1 & 0 & 1 & | & 1 & 0 & 0 \\ 2 & 1 & 3 & | & 0 & 1 & 0 \\ -1 & 1 & 1 & | & 0 & 0 & 1 \end{bmatrix} \quad \begin{matrix} R_2 - 2R_1 \rightarrow \\ R_3 + R_1 \rightarrow \end{matrix} \quad \begin{bmatrix} 1 & 0 & 1 & | & 1 & 0 & 0 \\ 0 & 1 & 1 & | & -2 & 1 & 0 \\ 0 & 1 & 2 & | & 1 & 0 & 1 \end{bmatrix}$$

$$\begin{matrix} \\ \\ R_3 - R_2 \rightarrow \end{matrix} \begin{bmatrix} 1 & 0 & 1 & | & 1 & 0 & 0 \\ 0 & 1 & 1 & | & -2 & 1 & 0 \\ 0 & 0 & 1 & | & 3 & -1 & 1 \end{bmatrix} \quad \begin{matrix} R_1 - R_3 \rightarrow \\ R_2 - R_3 \rightarrow \\ \\ \end{matrix} \begin{bmatrix} 1 & 0 & 0 & | & -2 & 1 & -1 \\ 0 & 1 & 0 & | & -5 & 2 & -1 \\ 0 & 0 & 1 & | & 3 & -1 & 1 \end{bmatrix}$$

The right side is equal to A^{-1}. That is,

$$A^{-1} = \begin{bmatrix} -2 & 1 & -1 \\ -5 & 2 & -1 \\ 3 & -1 & 1 \end{bmatrix}.$$

It can be verified that $A^{-1}A = I_3 = AA^{-1}$.

Now Try Exercise 31 ◆

Note: If it is not possible to obtain the identity matrix on the left side of the augmented matrix by using matrix row operations, then A^{-1} does not exist.

Representing Linear Systems with Matrix Equations

In Section 9.4 linear systems were solved using Gaussian elimination with backward substitution. This method used an augmented matrix to represent a system of linear equations. A system of linear equations can also be represented by a matrix equation.

$$3x - 2y + 4z = 5$$
$$2x + y + 3z = 9$$
$$-x + 5y - 2z = 5$$

Let A, X, and B be matrices defined as

| *Coefficient Matrix* | *Variable Matrix* | | *Constant Matrix* |

$$A = \begin{bmatrix} 3 & -2 & 4 \\ 2 & 1 & 3 \\ -1 & 5 & -2 \end{bmatrix}, \quad X = \begin{bmatrix} x \\ y \\ z \end{bmatrix}, \quad \text{and} \quad B = \begin{bmatrix} 5 \\ 9 \\ 5 \end{bmatrix}.$$

The matrix product AX is given by

$$AX = \begin{bmatrix} 3 & -2 & 4 \\ 2 & 1 & 3 \\ -1 & 5 & -2 \end{bmatrix}\begin{bmatrix} x \\ y \\ z \end{bmatrix} = \begin{bmatrix} 3x + (-2)y + 4z \\ 2x + 1y + 3z \\ (-1)x + 5y + (-2)z \end{bmatrix} = \begin{bmatrix} 3x - 2y + 4z \\ 2x + y + 3z \\ -x + 5y - 2z \end{bmatrix}.$$

Thus the matrix equation $AX = B$ simplifies to

$$\begin{bmatrix} 3x - 2y + 4z \\ 2x + y + 3z \\ -x + 5y - 2z \end{bmatrix} = \begin{bmatrix} 5 \\ 9 \\ 5 \end{bmatrix}.$$

This matrix equation $AX = B$ is equivalent to the original system of linear equations. Any system of linear equations can be represented by a matrix equation in the form $AX = B$.

EXAMPLE 6 Representing linear systems with matrix equations

Represent each system of linear equations in the form $AX = B$.

(a) $\begin{aligned} 3x - 4y &= 7 \\ -x + 6y &= -3 \end{aligned}$

(b) $\begin{aligned} x - 5y &= 2 \\ -3x + 2y + z &= -7 \\ 4x + 5y + 6z &= 10 \end{aligned}$

SOLUTION

(a) This linear system comprises two equations and two variables. The equivalent matrix equation is

$$AX = \begin{bmatrix} 3 & -4 \\ -1 & 6 \end{bmatrix}\begin{bmatrix} x \\ y \end{bmatrix} = \begin{bmatrix} 7 \\ -3 \end{bmatrix} = B.$$

(b) The equivalent matrix equation is

$$AX = \begin{bmatrix} 1 & -5 & 0 \\ -3 & 2 & 1 \\ 4 & 5 & 6 \end{bmatrix}\begin{bmatrix} x \\ y \\ z \end{bmatrix} = \begin{bmatrix} 2 \\ -7 \\ 10 \end{bmatrix} = B.$$

Now Try Exercises 45 and 49 ◆

Solving Linear Systems with Inverses

The matrix equation $AX = B$ can be solved by using A^{-1}, if it exists.

$$\begin{aligned} AX &= B && \text{Linear system} \\ A^{-1}AX &= A^{-1}B && \text{Multiply each side by } A^{-1}. \\ I_n X &= A^{-1}B && A^{-1}A = I_n \\ X &= A^{-1}B && I_n X = X \text{ for any } n \times 1 \text{ matrix } X. \end{aligned}$$

To solve a linear system, multiply each side of the matrix equation $AX = B$ by A^{-1}, if it exists. The solution to the system is unique and can be written as $X = A^{-1}B$.

Note: Since matrix multiplication is not commutative, it is essential to multiply each side of the equation on the *left* by A^{-1}. That is, $X = A^{-1}B \neq BA^{-1}$ in general.

EXAMPLE 7 Solving a linear system using the inverse of a 2 × 2 matrix

Write the linear system as a matrix equation in the form $AX = B$. Find A^{-1} and solve for X.

$$x + 4y = 3$$
$$2x + 9y = 5$$

SOLUTION The linear system can be written as

$$AX = \begin{bmatrix} 1 & 4 \\ 2 & 9 \end{bmatrix} \begin{bmatrix} x \\ y \end{bmatrix} = \begin{bmatrix} 3 \\ 5 \end{bmatrix} = B.$$

The matrix A^{-1} was found in Example 4. Thus we can solve for X without having to find A^{-1} first.

$$X = A^{-1}B = \begin{bmatrix} 9 & -4 \\ -2 & 1 \end{bmatrix} \begin{bmatrix} 3 \\ 5 \end{bmatrix} = \begin{bmatrix} 7 \\ -1 \end{bmatrix}$$

The solution to the system is $(7, -1)$. Check this.   *Now Try Exercise 53* ◆

In the next two examples, we use technology to solve the system of linear equations. Technology is especially helpful when finding A^{-1}.

EXAMPLE 8 Solving a linear system using the inverse of a 3 × 3 matrix

Write the linear system as a matrix equation in the form $AX = B$. Find A^{-1} and solve for X.

$$x + 3y - z = 6$$
$$-2y + z = -2$$
$$-x + y - 3z = 4$$

SOLUTION The linear system can be written as

$$AX = \begin{bmatrix} 1 & 3 & -1 \\ 0 & -2 & 1 \\ -1 & 1 & -3 \end{bmatrix} \begin{bmatrix} x \\ y \\ z \end{bmatrix} = \begin{bmatrix} 6 \\ -2 \\ 4 \end{bmatrix} = B.$$

The matrix A^{-1} can be found by hand or with a graphing calculator, as shown in Figure 9.67. The solution to the system is given by $x = 4.5$, $y = -0.5$, and $z = -3$. See Figure 9.68.

Calculator Help

To find the inverse of a matrix, see Appendix B (page AP-21). To solve a linear system with a matrix inverse, see Appendix B (page AP-22).

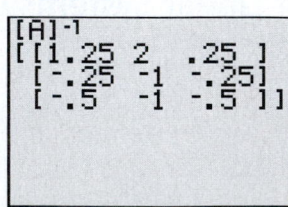

FIGURE 9.67

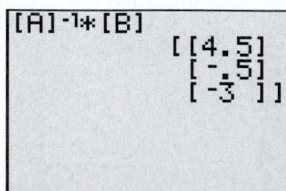

FIGURE 9.68   *Now Try Exercise 65* ◆

EXAMPLE 9   Modeling blood pressure

In one study of adult males, it was believed that systolic blood pressure P was affected by both age A in years and weight W in pounds. This was modeled by $P(A, W) = a + bA + cW$, where a, b, and c are constants. Table 9.11 lists three individuals with representative blood pressures for the group.   (**Source:** C. H. Brase and C. P. Brase, *Understandable Statistics*.)

(a) Use Table 9.11 to approximate values for the constants a, b, and c.

(b) Estimate a typical systolic blood pressure for an individual who is 55 years old and weighs 175 pounds.

TABLE 9.11

P	A	W
113	39	142
138	53	181
152	65	191

SOLUTION

(a) Determine the constants a, b, and c in $P(A, W) = a + bA + cW$ by solving the following three equations.

$$P(39, 142) = a + b(39) + c(142) = 113$$
$$P(53, 181) = a + b(53) + c(181) = 138$$
$$P(65, 191) = a + b(65) + c(191) = 152$$

These three equations can be rewritten as follows.

$$a + 39b + 142c = 113$$
$$a + 53b + 181c = 138$$
$$a + 65b + 191c = 152$$

This system can be represented by the matrix equation $AX = B$.

$$AX = \begin{bmatrix} 1 & 39 & 142 \\ 1 & 53 & 181 \\ 1 & 65 & 191 \end{bmatrix} \begin{bmatrix} a \\ b \\ c \end{bmatrix} = \begin{bmatrix} 113 \\ 138 \\ 152 \end{bmatrix} = B$$

The solution, $X = A^{-1}B$, is shown in Figure 9.69. The values for the constants are $a \approx 32.78$, $b \approx 0.9024$, and $c \approx 0.3171$. Thus P is given by the equation $P(A, W) = 32.78 + 0.9024A + 0.3171W$.

```
[A]⁻¹*[B]
   [[32.7804878 ]
    [.9024390244]
    [.3170731707]]
```

FIGURE 9.69

(b) Evaluate $P(55, 175) = 32.78 + 0.9024(55) + 0.3171(175) \approx 137.9$. This model predicts that a typical (male) individual 55 years old, weighing 175 pounds, has a systolic blood pressure of approximately 138. Clearly, this could vary greatly among individuals.

Now Try Exercise 73 ◆

9.6   Putting it all Together

$\mathbf{T}$he inverse of a matrix A is denoted A^{-1}, if it exists. The inverse matrix will undo or cancel the operations performed by A. Inverse matrices frequently are used in computer graphics. They also can be used to solve systems of linear equations.

The following table summarizes some of the mathematical concepts presented in this section.

Concept	Comments	Examples
Identity matrix	The $n \times n$ identity matrix I_n has only 1's on the main diagonal and 0's elsewhere. When it is multiplied by any $n \times n$ matrix A, the result is A.	$\begin{bmatrix} 1 & 0 \\ 0 & 1 \end{bmatrix} \begin{bmatrix} 2 & 3 \\ 4 & 5 \end{bmatrix} = \begin{bmatrix} 2 & 3 \\ 4 & 5 \end{bmatrix}$ and $\begin{bmatrix} 2 & 3 \\ 4 & 5 \end{bmatrix} \begin{bmatrix} 1 & 0 \\ 0 & 1 \end{bmatrix} = \begin{bmatrix} 2 & 3 \\ 4 & 5 \end{bmatrix}$, where $I_2 = \begin{bmatrix} 1 & 0 \\ 0 & 1 \end{bmatrix}$.
Matrix inverse	If an $n \times n$ matrix A has an inverse it is unique, is denoted A^{-1}, and satisfies the equations $AA^{-1} = I_n$ and $A^{-1}A = I_n$. Matrix inverses can be found by using technology. They can also be found with pencil and paper by performing matrix row operations on the augmented matrix $[A\,|\,I_n]$ until it is transformed to $[I_n\,|\,A^{-1}]$.	If $A = \begin{bmatrix} 2 & 3 \\ 3 & 5 \end{bmatrix}$, then $A^{-1} = \begin{bmatrix} 5 & -3 \\ -3 & 2 \end{bmatrix}$ because $AA^{-1} = \begin{bmatrix} 2 & 3 \\ 3 & 5 \end{bmatrix} \begin{bmatrix} 5 & -3 \\ -3 & 2 \end{bmatrix} = \begin{bmatrix} 1 & 0 \\ 0 & 1 \end{bmatrix}$ and $A^{-1}A = \begin{bmatrix} 5 & -3 \\ -3 & 2 \end{bmatrix} \begin{bmatrix} 2 & 3 \\ 3 & 5 \end{bmatrix} = \begin{bmatrix} 1 & 0 \\ 0 & 1 \end{bmatrix}$.
Systems of linear equations	Systems of linear equations can be written by using the matrix equation $AX = B$. If A is invertible, then there will be a unique solution given by $X = A^{-1}B$. If A is not invertible, then there could be either no solution or infinitely many solutions. In the latter case, Gaussian elimination should be applied.	The linear system $\begin{aligned} 2x - y &= 3 \\ x + 2y &= 4 \end{aligned}$ can be written as $AX = B$, where $A = \begin{bmatrix} 2 & -1 \\ 1 & 2 \end{bmatrix}$, $X = \begin{bmatrix} x \\ y \end{bmatrix}$, and $B = \begin{bmatrix} 3 \\ 4 \end{bmatrix}$. The solution to the system is given by $X = A^{-1}B = \begin{bmatrix} 0.4 & 0.2 \\ -0.2 & 0.4 \end{bmatrix} \begin{bmatrix} 3 \\ 4 \end{bmatrix} = \begin{bmatrix} 2 \\ 1 \end{bmatrix}$. The solution is $(2, 1)$.

9.6 — Exercises

Identifying Inverse Matrices

Exercises 1–4: Determine if B is the inverse matrix of A by calculating AB and BA.

1. $A = \begin{bmatrix} 4 & 3 \\ 5 & 4 \end{bmatrix}$, $\qquad B = \begin{bmatrix} 4 & -3 \\ -5 & 4 \end{bmatrix}$

2. $A = \begin{bmatrix} -1 & 2 \\ -3 & 8 \end{bmatrix}$, $\qquad B = \begin{bmatrix} -4 & 1 \\ -2 & 0.5 \end{bmatrix}$

3. $A = \begin{bmatrix} 1 & -1 & 2 \\ 0 & 1 & -1 \\ 1 & 0 & 2 \end{bmatrix}$, $B = \begin{bmatrix} 2 & 2 & -1 \\ -1 & 0 & 1 \\ -1 & -1 & 1 \end{bmatrix}$

4. $A = \begin{bmatrix} 2 & 1 & 1 \\ -1 & 0 & -1 \\ 0 & 2 & -1 \end{bmatrix}$, $B = \begin{bmatrix} 2 & 3 & -1 \\ -1 & -2 & 1 \\ -2 & -4 & 1 \end{bmatrix}$

Exercises 5–8: Determine the value of the constant k in the matrix B so that $B = A^{-1}$.

5. $A = \begin{bmatrix} 1 & 1 \\ 1 & 2 \end{bmatrix}$, $\qquad B = \begin{bmatrix} 2 & -1 \\ -1 & k \end{bmatrix}$

6. $A = \begin{bmatrix} -2 & 2 \\ 1 & -2 \end{bmatrix}$, $B = \begin{bmatrix} -1 & k \\ -0.5 & -1 \end{bmatrix}$

7. $A = \begin{bmatrix} 1 & 3 \\ -1 & -5 \end{bmatrix}$, $B = \begin{bmatrix} k & 1.5 \\ -0.5 & -0.5 \end{bmatrix}$

8. $A = \begin{bmatrix} -2 & 5 \\ -3 & 4 \end{bmatrix}$, $B = \begin{bmatrix} \frac{4}{7} & -\frac{5}{7} \\ k & -\frac{2}{7} \end{bmatrix}$

Exercises 9–12: Predict the results of $I_n A$ and $A I_n$. Then, verify your prediction.

9. $I_2 = \begin{bmatrix} 1 & 0 \\ 0 & 1 \end{bmatrix}$, $\qquad A = \begin{bmatrix} 1 & -2 \\ 4 & 3 \end{bmatrix}$

10. $I_3 = \begin{bmatrix} 1 & 0 & 0 \\ 0 & 1 & 0 \\ 0 & 0 & 1 \end{bmatrix}$, $A = \begin{bmatrix} 1 & -4 & 3 \\ 1 & 9 & 5 \\ 3 & -5 & 0 \end{bmatrix}$

11. $I_3 = \begin{bmatrix} 1 & 0 & 0 \\ 0 & 1 & 0 \\ 0 & 0 & 1 \end{bmatrix}$, $A = \begin{bmatrix} 0 & 0 & 0 \\ 0 & 0 & 0 \\ 0 & 0 & 0 \end{bmatrix}$

12. $I_4 = \begin{bmatrix} 1 & 0 & 0 & 0 \\ 0 & 1 & 0 & 0 \\ 0 & 0 & 1 & 0 \\ 0 & 0 & 0 & 1 \end{bmatrix}$, $A = \begin{bmatrix} 5 & -2 & 6 & -3 \\ 0 & 1 & 4 & -1 \\ -5 & 7 & 9 & 8 \\ 0 & 0 & 3 & 1 \end{bmatrix}$

Interpreting Inverses

Exercises 13 and 14: **Translations** *(Refer to the discussion in this section about translating a point.) The matrix product AX performs a translation on the point (x, y), where*

$$A = \begin{bmatrix} 1 & 0 & h \\ 0 & 1 & k \\ 0 & 0 & 1 \end{bmatrix} \quad \text{and} \quad X = \begin{bmatrix} x \\ y \\ 1 \end{bmatrix}.$$

(a) *Predict the new location of the point (x, y), when it is translated by A. Compute $Y = AX$ to verify your prediction.*

(b) *Make a conjecture as to what $A^{-1}Y$ represents. Find A^{-1} and calculate $A^{-1}Y$ to test your conjecture.*

(c) *What will AA^{-1} and $A^{-1}A$ equal?*

13. $A = \begin{bmatrix} 1 & 0 & 2 \\ 0 & 1 & 3 \\ 0 & 0 & 1 \end{bmatrix}$, $(x, y) = (0, 1)$, and $X = \begin{bmatrix} 0 \\ 1 \\ 1 \end{bmatrix}$

14. $A = \begin{bmatrix} 1 & 0 & -4 \\ 0 & 1 & 5 \\ 0 & 0 & 1 \end{bmatrix}$, $(x, y) = (4, 2)$, and $X = \begin{bmatrix} 4 \\ 2 \\ 1 \end{bmatrix}$

Exercises 15 and 16: **Translations** *(Refer to the discussion in this section about translating a point.) Find a 3 × 3 matrix A that performs the following translation of a point (x, y) represented by X. Find A^{-1} and describe what it computes.*

15. 3 units to the left and 5 units downward

16. 6 units to the right and 1 unit upward

17. **Rotation** *(Refer to Example 2.) The matrix B rotates the point (x, y) clockwise about the origin 45°, where*

$$B = \begin{bmatrix} \frac{1}{\sqrt{2}} & \frac{1}{\sqrt{2}} & 0 \\ -\frac{1}{\sqrt{2}} & \frac{1}{\sqrt{2}} & 0 \\ 0 & 0 & 1 \end{bmatrix} \quad \text{and} \quad B^{-1} = \begin{bmatrix} \frac{1}{\sqrt{2}} & -\frac{1}{\sqrt{2}} & 0 \\ \frac{1}{\sqrt{2}} & \frac{1}{\sqrt{2}} & 0 \\ 0 & 0 & 1 \end{bmatrix}.$$

(a) Let X represent the point $(-\sqrt{2}, -\sqrt{2})$. Compute $Y = BX$.

(b) Find $B^{-1}Y$. Interpret the computation performed by B^{-1}.

18. *Rotation* (Continuation of the previous exercise.) Predict the result of the computations $BB^{-1}X$ and $B^{-1}BX$ for any point (x, y) represented by X. Explain this result geometrically.

19. *Translations* The matrix A translates a point to the right 4 units and downward 2 units, and the matrix B translates a point to the left 3 units and upward 3 units, where

$$A = \begin{bmatrix} 1 & 0 & 4 \\ 0 & 1 & -2 \\ 0 & 0 & 1 \end{bmatrix} \quad \text{and} \quad B = \begin{bmatrix} 1 & 0 & -3 \\ 0 & 1 & 3 \\ 0 & 0 & 1 \end{bmatrix}.$$

(a) Let X represent the point $(1, 1)$. Predict the result of $Y = ABX$. Check your prediction.

(b) Predict the form of the matrix product AB, and then compute AB.

(c) In this exercise, would you expect $AB = BA$? Verify your answer.

(d) Find $(AB)^{-1}$ mentally. Explain your reasoning.

20. *Rotation* (Refer to Exercises 13 and 17 for matrices A and B.)
(a) Let X represent the point $(0, \sqrt{2})$. If this point is rotated about the origin 45° clockwise, and then translated 2 units to the right and 3 units upward, determine its new coordinates geometrically.

(b) Compute the matrix product $Y = ABX$, and explain the result.

(c) Is your answer in part (b) equal to BAX? Interpret your answer.

(d) Find a matrix that translates Y back to X. Test your matrix.

Calculating Inverses

Exercises 21–34: (Refer to Examples 3–5.) Let A be the given matrix. Find A^{-1} without a calculator.

21. $\begin{bmatrix} 1 & 2 \\ 1 & 3 \end{bmatrix}$

22. $\begin{bmatrix} 1 & 0 \\ 1 & -1 \end{bmatrix}$

23. $\begin{bmatrix} -1 & 2 \\ 3 & -5 \end{bmatrix}$

24. $\begin{bmatrix} 1 & 3 \\ 2 & 5 \end{bmatrix}$

25. $\begin{bmatrix} 8 & 5 \\ 2 & 1 \end{bmatrix}$

26. $\begin{bmatrix} -2 & 4 \\ -5 & 9 \end{bmatrix}$

27. $\begin{bmatrix} 0 & 0 & 1 \\ 1 & 0 & 0 \\ 0 & 1 & 0 \end{bmatrix}$

28. $\begin{bmatrix} 1 & 0 & 0 \\ 1 & 1 & 0 \\ 0 & 1 & 1 \end{bmatrix}$

29. $\begin{bmatrix} 1 & 0 & 1 \\ 2 & 1 & 3 \\ -1 & 1 & 1 \end{bmatrix}$

30. $\begin{bmatrix} -2 & 1 & 0 \\ 1 & 0 & 1 \\ -1 & 1 & 0 \end{bmatrix}$

31. $\begin{bmatrix} 1 & 2 & -1 \\ 2 & 5 & 0 \\ -1 & -1 & 2 \end{bmatrix}$

32. $\begin{bmatrix} 2 & -2 & 1 \\ 1 & 3 & 2 \\ 4 & -2 & 4 \end{bmatrix}$

33. $\begin{bmatrix} -2 & 1 & -3 \\ 0 & 1 & 2 \\ 1 & -2 & 1 \end{bmatrix}$

34. $\begin{bmatrix} 1 & -1 & 1 \\ -1 & 2 & 1 \\ 0 & 2 & 1 \end{bmatrix}$

Exercises 35–44: Let A be the given matrix. Find A^{-1}.

35. $\begin{bmatrix} 0.5 & -1.5 \\ 0.2 & -0.5 \end{bmatrix}$

36. $\begin{bmatrix} -0.5 & 0.5 \\ 3 & 2 \end{bmatrix}$

37. $\begin{bmatrix} 1 & 2 & 0 \\ -1 & 4 & -1 \\ 2 & -1 & 0 \end{bmatrix}$

38. $\begin{bmatrix} -2 & 0 & 1 \\ 5 & -4 & 1 \\ 1 & -2 & 0 \end{bmatrix}$

39. $\begin{bmatrix} 2 & -2 & 1 \\ 0 & 5 & 8 \\ 0 & 0 & -1 \end{bmatrix}$

40. $\begin{bmatrix} 2 & 0 & 2 \\ 1 & 5 & 0 \\ -1 & 0 & 2 \end{bmatrix}$

41. $\begin{bmatrix} 3 & -1 & -1 \\ -1 & 3 & -1 \\ -1 & -1 & 3 \end{bmatrix}$

42. $\begin{bmatrix} 2 & -3 & 1 \\ 5 & -6 & 3 \\ 3 & 2 & 0 \end{bmatrix}$

43. $\begin{bmatrix} 1 & -1 & 0 & 0 \\ -1 & 5 & -1 & 0 \\ 0 & -1 & 5 & -1 \\ 0 & 0 & -1 & 1 \end{bmatrix}$

44. $\begin{bmatrix} 3 & 1 & 0 & 0 \\ 1 & 3 & 1 & 0 \\ 0 & 1 & 3 & 1 \\ 0 & 0 & 1 & 3 \end{bmatrix}$

Exercises 45–52: Represent the system of linear equations in the form $AX = B$.

45. $\begin{aligned} 2x - 3y &= 7 \\ -3x - 4y &= 9 \end{aligned}$

46. $\begin{aligned} -x + 3y &= 10 \\ 2x - 6y &= -1 \end{aligned}$

47. $\begin{aligned} \tfrac{1}{2}x - \tfrac{3}{2}y &= \tfrac{1}{4} \\ -x + 2y &= 5 \end{aligned}$

48. $\begin{aligned} -1.1x + 3.2y &= -2.7 \\ 5.6x - 3.8y &= -3.0 \end{aligned}$

49. $\begin{aligned} x - 2y + z &= 5 \\ 3y - z &= 6 \\ 5x - 4y - 7z &= 0 \end{aligned}$

50. $\begin{aligned} 4x - 3y + 2z &= 8 \\ -x + 4y + 3z &= 2 \\ -2x - 5z &= 2 \end{aligned}$

51.
$$4x - y + 3z = -2$$
$$x + 2y + 5z = 11$$
$$2x - 3y = -1$$

52.
$$x - 2y + z = 12$$
$$4y + 3z = 13$$
$$-2x + 7y = -2$$

Exercises 53–60: Complete the following.

(a) *Write the system in the form $AX = B$.*
(b) *Solve the system by finding A^{-1} and then using the equation $X = A^{-1}B$. (Hint: Some of your answers from Exercises 21–34 may be helpful.)*

53.
$$x + 2y = 3$$
$$x + 3y = 6$$

54.
$$2x + y = 4$$
$$-x + 2y = -1$$

55.
$$-x + 2y = 5$$
$$3x - 5y = -2$$

56.
$$x + 3y = -3$$
$$2x + 5y = -2$$

57.
$$x + z = -7$$
$$2x + y + 3z = -13$$
$$-x + y + z = -4$$

58.
$$-2x + y = -5$$
$$x + z = -5$$
$$-x + y = -4$$

59.
$$x + 2y - z = 2$$
$$2x + 5y = -1$$
$$-x - y + 2z = 0$$

60.
$$2x - 2y + z = 1$$
$$x + 3y + 2z = 3$$
$$4x - 2y + 4z = 4$$

Solving Linear Systems

Exercises 61–68: Complete the following for the given system of linear equations.

(a) *Write the system in the form $AX = B$.*
(b) *Solve the linear system by computing $X = A^{-1}B$ with a calculator. Approximate the solution to the nearest hundredth when appropriate.*

61.
$$1.5x + 3.7y = 0.32$$
$$-0.4x - 2.1y = 0.36$$

62.
$$31x + 18y = 64.1$$
$$5x - 23y = -59.6$$

63.
$$0.08x - 0.7y = -0.504$$
$$1.1x - 0.05y = 0.73$$

64.
$$-231x + 178y = -439$$
$$525x - 329y = 2282$$

65.
$$3.1x + 1.9y - z = 1.99$$
$$6.3x - 9.9z = -3.78$$
$$-x + 1.5y + 7z = 5.3$$

66.
$$17x - 22y - 19z = -25.2$$
$$3x + 13y - 9z = 105.9$$
$$x - 2y + 6.1z = -23.55$$

67.
$$3x - y + z = 4.9$$
$$5.8x - 2.1y = -3.8$$
$$-x + 2.9z = 3.8$$

68.
$$1.2x - 0.3y - 0.7z = -0.5$$
$$-0.4x + 1.3y + 0.4z = 0.9$$
$$1.7x + 0.6y + 1.1z = 1.3$$

Applications

69. **Cost of CDs** A music store has compact discs that sell for three prices marked A, B, and C. The last column in the table shows the total cost of a purchase. Use this information to determine the cost of one CD of each type by setting up a matrix equation and solving it with an inverse.

A	B	C	Total
2	3	4	$120.91
1	4	0	$62.95
2	1	3	$79.94

70. **Determining Prices** A group of students bought 3 soft drinks and 2 boxes of popcorn at a movie for $8.50. A second group bought 4 soft drinks and 3 boxes of popcorn for $12.
(a) Find a matrix equation $AX = B$ whose solution gives the individual prices of a soft drink and a box of popcorn. Solve this matrix equation using A^{-1}.

(b) Could these prices be determined if both groups had bought 3 soft drinks and 2 boxes of popcorn for $8.50? Try to calculate A^{-1} and explain your results.

71. **Traffic Flow** (Refer to Exercises 79 and 80 in Section 9.4.) The accompanying figure shows four one-way streets with intersections A, B, C, and D. Numbers indicate the average traffic flow in vehicles per minute. The variables x_1, x_2, x_3, and x_4 denote unknown traffic flows.
(a) The number of vehicles per minute entering an intersection equals the number exiting an intersection. Verify that the given system of linear equations describes the traffic flow.

A: $x_1 + 5 = 4 + 6$ B: $x_2 + 6 = x_1 + 3$
C: $x_3 + 4 = x_2 + 7$ D: $6 + 5 = x_3 + x_4$

(b) Write the system as $AX = B$ and solve using A^{-1}.

(c) Interpret your results.

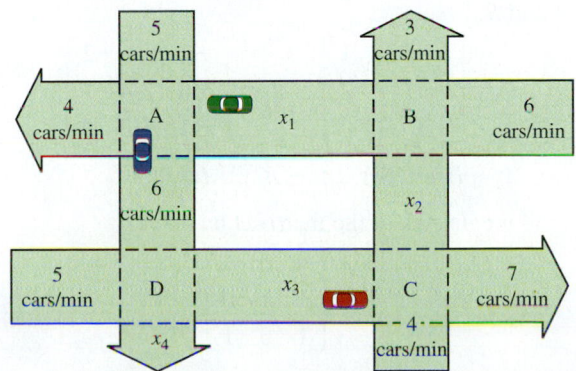

72. *Tire Sales* In one study the relationship among annual tire sales T in thousands, automobile registrations A in millions, and personal disposable income I in millions of dollars was investigated. Representative data for three different years are shown in the table. (**Source:** J. Jarrett, *Business Forecasting Methods.*)

T	A	I
10,170	113	308
15,305	133	622
21,289	155	1937

The data were modeled by $T = aA + bI + c$, where a, b, and c are constants.

(a) Use the data to write a system of linear equations, whose solution gives a, b, and c.

(b) Solve this linear system. Write a formula for T.

(c) If $A = 118$ and $I = 311$, predict T. (The actual value for T was 11,314.)

73. *Home Prices* The selling price of a home can depend on several factors. Real-estate companies sometimes study how the selling price of a home is related to its size and condition. The accompanying table contains representative data for sales of three homes. Price P is measured in thousands of dollars, home size S in square feet, and condition C is rated on a scale from 1 to 10, where 10 represents excellent condition. The variables were found to be related by the linear equation $P = a + bS + cC$.

P	S	C
122	1500	8
130	2000	5
158	2200	10

(a) Use the table to write a system of linear equations whose solution gives a, b, and c.

(b) Estimate the selling price of a home with 1800 square feet and a condition of 7.

74. *Plate Glass Sales* The amount of plate glass sales G can be affected by the number of new building contracts B issued and automobiles A produced, since plate glass is used in buildings and cars. A plate glass company in California wanted to forecast sales. The table contains sales data for three consecutive years. All units are in millions. (**Source:** S. Makridakis and S. Wheelwright, *Forecasting Methods for Management.*)

G	A	B
603	5.54	37.1
657	6.93	41.3
779	7.64	45.6

The data were modeled by $G = aA + bB + c$, where a, b, and c are constants.

(a) Write a system of linear equations whose solution gives a, b, and c.

(b) Solve this linear system. Write a formula for G.

(c) For the following year, it was estimated that $A = 7.75$ and $B = 47.4$. Predict G. (The actual value for G was $878 million.)

Writing about Mathematics

75. Discuss how to solve the matrix equation $AX = B$ if A^{-1} exists.

76. Give an example of a 2×2 matrix A without any 0 elements that does not have an inverse. Explain what happens if one attempts to find the inverse of A symbolically.

CHECKING BASIC CONCEPTS FOR SECTIONS 9.5 AND 9.6

1. Perform the following operations on the given matrices A and B.
$$A = \begin{bmatrix} 1 & 0 & 1 \\ -1 & 1 & 2 \\ 1 & 3 & 0 \end{bmatrix}, \quad B = \begin{bmatrix} -1 & 1 & 2 \\ 0 & 4 & 1 \\ 1 & -2 & 0 \end{bmatrix}$$
(a) $A + B$ (b) $2A - B$ (c) AB

2. Find the inverse of the matrix A by hand.
$$A = \begin{bmatrix} 0 & 0 & 1 \\ 1 & 1 & 0 \\ 1 & 0 & 1 \end{bmatrix}$$

3. Write the following system of linear equations as a matrix equation $AX = B$. Solve the system utilizing A^{-1}.
(a) $x - 2y = 13$
$\quad 2x + 3y = 5$
(b) $x - y + z = 2$
$\quad -x + y + z = 4$
$\quad y - z = -1$
(c) $3.1x - 5.3y = -2.682$
$\quad -0.1x + 1.8y = 0.787$

4. Find A^{-1} if $A = \begin{bmatrix} 2 & -3 & 5 \\ 4 & -3 & 2 \\ 1 & 5 & -4 \end{bmatrix}$.

9.7 Determinants

- ◆ Define and calculate determinants
- ◆ Use determinants to find areas of regions
- ◆ Apply Cramer's rule
- ◆ Learn about limitations on the method of cofactors and Cramer's rule

Introduction

Determinants are used in mathematics for theoretical purposes. However, they also are used to test if a matrix is invertible and to find the area of certain geometric figures. A *determinant* is a real number associated with a square matrix. We begin our discussion by defining a determinant for a 2 × 2 matrix.

Definition and Calculation of Determinants

The determinant of a matrix with dimension 2 × 2 is a straightforward arithmetic calculation.

DETERMINANT OF A 2 × 2 MATRIX

The **determinant** of

$$A = \begin{bmatrix} a & b \\ c & d \end{bmatrix}$$

is a real number defined by

$$\det A = ad - cb.$$

Later we define determinants for any $n \times n$ matrix. The following can be used to determine if a matrix is invertible.

INVERTIBLE MATRIX

A square matrix A is invertible if and only if $\det A \neq 0$.

EXAMPLE 1 Determining if a 2 × 2 matrix is invertible

Determine if A^{-1} exists by computing the determinant of the matrix A.

(a) $A = \begin{bmatrix} 3 & -4 \\ -5 & 9 \end{bmatrix}$ (b) $A = \begin{bmatrix} 52 & -32 \\ 65 & -40 \end{bmatrix}$

SOLUTION

(a) The determinant of the 2 × 2 matrix A is calculated as follows.

$$\det A = \det \begin{bmatrix} 3 & -4 \\ -5 & 9 \end{bmatrix} = (3)(9) - (-5)(-4) = 7$$

Since $\det A = 7 \neq 0$, the matrix A is invertible and A^{-1} exists.

(b) In a similar manner,

$$\det A = \det \begin{bmatrix} 52 & -32 \\ 65 & -40 \end{bmatrix} = (52)(-40) - (65)(-32) = 0$$

Since $\det A = 0$, A^{-1} does not exist. Try finding A^{-1}. What happens?

Now Try Exercises 1 and 3 ◆

We can use determinants of 2×2 matrices to find determinants of larger square matrices. In order to do this, we first define the concepts of a *minor* and a *cofactor*.

MINORS AND COFACTORS

The **minor**, denoted by M_{ij}, for element a_{ij} in the square matrix A is the real number computed by performing the following steps.

STEP 1: Delete the ith row and jth column from the matrix A.

STEP 2: M_{ij} is equal to the determinant of the resulting matrix.

The **cofactor**, denoted A_{ij}, for a_{ij} is defined by $A_{ij} = (-1)^{i+j} M_{ij}$.

EXAMPLE 2 Calculating minors and cofactors

Find the following minors and cofactors for the matrix A.

$$A = \begin{bmatrix} 2 & -3 & 1 \\ -2 & 1 & 0 \\ 0 & -1 & 4 \end{bmatrix}$$

(a) M_{11} and M_{21} **(b)** A_{11} and A_{21}

SOLUTION

(a) To obtain the minor M_{11}, begin by crossing out the first row and first column of A.

$$A = \begin{bmatrix} 2 & 3 & 1 \\ -2 & 1 & 0 \\ 0 & -1 & 4 \end{bmatrix}$$

The remaining elements form the 2×2 matrix

$$B = \begin{bmatrix} 1 & 0 \\ -1 & 4 \end{bmatrix}.$$

The minor M_{11} is equal to $\det B = (1)(4) - (-1)(0) = 4$.

M_{21} is found by crossing out the second row and first column of A.

$$A = \begin{bmatrix} 2 & -3 & 1 \\ -2 & 1 & 0 \\ 0 & -1 & 4 \end{bmatrix}$$

The resulting matrix is

$$B = \begin{bmatrix} -3 & 1 \\ -1 & 4 \end{bmatrix}.$$

Thus $M_{21} = \det B = (-3)(4) - (-1)(1) = -11$.

(b) Since $A_{ij} = (-1)^{i+j} M_{ij}$, A_{11} and A_{21} can be computed as follows.

$$A_{11} = (-1)^{1+1} M_{11} = (-1)^2(4) = 4$$

$$A_{21} = (-1)^{2+1} M_{21} = (-1)^3(-11) = 11 \qquad \textit{Now Try Exercise 5} \blacklozenge$$

Using the concept of a cofactor, we can calculate the determinant of any square matrix.

DETERMINANT OF A MATRIX USING THE METHOD OF COFACTORS

Multiply each element in any row or column of the matrix by its cofactor. The sum of the products is equal to the determinant.

To compute the determinant of a 3×3 matrix A, begin by selecting a row or column.

$$A = \begin{bmatrix} a_{11} & a_{12} & a_{13} \\ a_{21} & a_{22} & a_{23} \\ a_{31} & a_{32} & a_{33} \end{bmatrix}$$

For example, if the second row of A is selected, the elements are a_{21}, a_{22}, and a_{23}. Then,

$$\det A = a_{21}A_{21} + a_{22}A_{22} + a_{23}A_{23}.$$

On the other hand, utilizing the elements of a_{11}, a_{21}, and a_{31} in the first column gives

$$\det A = a_{11}A_{11} + a_{21}A_{21} + a_{31}A_{31}.$$

Regardless of the row or column selected, the value of $\det A$ is the same. The calculation is easier if some elements in the selected row or column equal 0.

EXAMPLE 3   Evaluating the determinant of a 3×3 matrix

Find $\det A$, if

$$A = \begin{bmatrix} 2 & -3 & 1 \\ -2 & 1 & 0 \\ 0 & -1 & 4 \end{bmatrix}.$$

SOLUTION   To find the determinant of A, we can select any row or column. If we begin expanding about the first column of A, then

$$\det A = a_{11}A_{11} + a_{21}A_{21} + a_{31}A_{31}.$$

In the first column, $a_{11} = 2$, $a_{21} = -2$, $a_{31} = 0$. In Example 2, the cofactors A_{11} and A_{21} were computed as 4 and 11, respectively. Since A_{31} is multiplied by $a_{31} = 0$, its value is unimportant. Thus

$$\begin{aligned} \det &= a_{11}A_{11} + a_{21}A_{21} + a_{31}A_{31} \\ &= 2(4) + (-2)(11) + (0)A_{31} \\ &= -14. \end{aligned}$$

We could have also expanded about the second row.

$$\begin{aligned} \det A &= a_{21}A_{21} + a_{22}A_{22} + a_{23}A_{23} \\ &= (-2)A_{21} + (1)A_{22} + (0)A_{23} \end{aligned}$$

◆ **CLASS DISCUSSION**
If a row or column in matrix A contains only zeros, what is det A? ◆

To complete this computation we need to determine only A_{22}, since A_{21} is known to be 11 and A_{23} is multiplied by 0. To compute A_{22}, delete the second row and column of A to obtain M_{22}.

$$M_{22} = \det \begin{bmatrix} 2 & 1 \\ 0 & 4 \end{bmatrix} = 8 \quad \text{and} \quad A_{22} = (-1)^{2+2}(8) = 8$$

Thus det $A = (-2)(11) + (1)(8) + (0)A_{23} = -14$. The same value for det A is obtained in both calculations.

Now Try Exercise 17 ◆

Instead of calculating $(-1)^{i+j}$ for each cofactor, the following sign matrix can be utilized to find determinants of 3×3 matrices. The checkerboard pattern can be expanded to include larger matrices.

$$\begin{bmatrix} + & - & + \\ - & + & - \\ + & - & + \end{bmatrix}$$

For example, if

$$A = \begin{bmatrix} 2 & 3 & 7 \\ -3 & -2 & -1 \\ 4 & 0 & 2 \end{bmatrix},$$

we can compute det A by expanding about the second column to take advantage of the 0. The second column contains $-$, $+$, and $-$ signs. Therefore

$$\det A = -(\mathbf{3}) \det \begin{bmatrix} -3 & -1 \\ 4 & 2 \end{bmatrix} + (-\mathbf{2}) \det \begin{bmatrix} 2 & 7 \\ 4 & 2 \end{bmatrix} - (\mathbf{0}) \det \begin{bmatrix} 2 & 7 \\ -3 & -1 \end{bmatrix}$$
$$= -3(-2) + (-2)(-24) - (0)(19)$$
$$= 54.$$

Many graphing calculators can evaluate the determinant of a matrix, as illustrated in the next example.

EXAMPLE 4 Using technology to find a determinant

Find the determinant of A.

(a) $A = \begin{bmatrix} 2 & -3 & 1 \\ -2 & 1 & 0 \\ 0 & -1 & 4 \end{bmatrix}$ (b) $A = \begin{bmatrix} 2 & -3 & 1 & 5 \\ 7 & 1 & -8 & 0 \\ 5 & 4 & 9 & 7 \\ -2 & 3 & 3 & 0 \end{bmatrix}$

SOLUTION
(a) The determinant of this matrix was calculated in Example 3 by hand. To use technology, enter the matrix and evaluate its determinant, as shown in Figure 9.70. The result is det $A = -14$, which agrees with our earlier calculation.

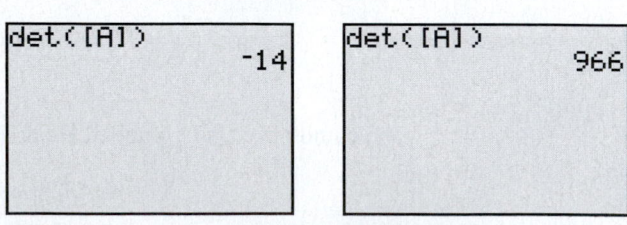

FIGURE 9.70 **FIGURE 9.71**

Calculator Help
To calculate a determinant, see Appendix B (page AP-22).

(b) The determinant of a 4×4 matrix can be computed using cofactors. However, it is considerably easier to use technology. From Figure 9.71 we see that det $A = 966$.

Now Try Exercises 23 and 24 ◆

Area of Regions

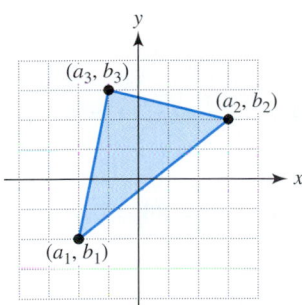

FIGURE 9.72

Determinants may be used to find the area of a triangle. If a triangle has vertices (a_1, b_1), (a_2, b_2), and (a_3, b_3), as shown in Figure 9.72, then its area is equal to the absolute value of D, where

$$D = \frac{1}{2} \det \begin{bmatrix} a_1 & a_2 & a_3 \\ b_1 & b_2 & b_3 \\ 1 & 1 & 1 \end{bmatrix}.$$

If the vertices are entered into the columns of D in a *counterclockwise* direction, then D will be positive.   (**Source:** W. Taylor, *The Geometry of Computer Graphics.*)

EXAMPLE 5   Computing area of a parallelogram

Use determinants to calculate the area of the parallelogram in Figure 9.73.

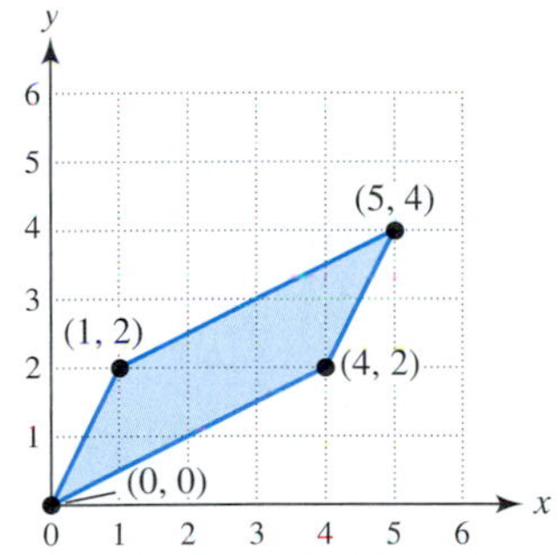

FIGURE 9.73

SOLUTION To find the area of the parallelogram, we view the parallelogram as comprising two triangles. One possible triangle has vertices at (**0**, **0**), (**4**, **2**), and (**1**, **2**), and the other triangle has vertices at (4, 2), (5, 4), and (1, 2). The area of the parallelogram is equal to the sum of the areas of the two triangles. Since these triangles are congruent, we can calculate the area of one triangle and double it. The area of one triangle is equal to D, where

$$D = \frac{1}{2} \det \begin{bmatrix} \mathbf{0} & \mathbf{4} & \mathbf{1} \\ \mathbf{0} & \mathbf{2} & \mathbf{2} \\ 1 & 1 & 1 \end{bmatrix} = 3.$$

◆ **CLASS DISCUSSION**

Suppose we are given three distinct vertices and $D = 0$. What must be true about the three points? ◆

Since the vertices were entered in a counterclockwise direction, D is positive. The area of one triangle is equal to 3. Therefore the area of the parallelogram is twice this value or 6.

Now Try Exercise 27 ◆

Cramer's Rule

Determinants can be used to solve linear systems by employing a method called **Cramer's rule**. Cramer's rule for linear systems in three variables is discussed in the Extended and Discovery Exercise in this section.

CRAMER'S RULE FOR LINEAR SYSTEMS IN TWO VARIABLES

The solution to the linear system

$$a_1 x + b_1 y = c_1$$
$$a_2 x + b_2 y = c_2$$

is given by $x = \frac{E}{D}$ and $y = \frac{F}{D}$, where

$$E = \det \begin{bmatrix} c_1 & b_1 \\ c_2 & b_2 \end{bmatrix}, \quad F = \det \begin{bmatrix} a_1 & c_1 \\ a_2 & c_2 \end{bmatrix}, \quad \text{and} \quad D = \det \begin{bmatrix} a_1 & b_1 \\ a_2 & b_2 \end{bmatrix} \neq 0.$$

EXAMPLE 6   Using Cramer's rule to solve a linear system in two variables

Use Cramer's rule to solve the linear system

$$4x + y = 146$$
$$9x + y = 66.$$

SOLUTION   In this system $a_1 = 4$, $b_1 = 1$, $c_1 = 146$, $a_2 = 9$, $b_2 = 1$, and $c_2 = 66$. By Cramer's rule, the solution can be found as follows.

$$E = \det \begin{bmatrix} c_1 & b_1 \\ c_2 & b_2 \end{bmatrix} = \det \begin{bmatrix} 146 & 1 \\ 66 & 1 \end{bmatrix} = (146)(1) - (66)(1) = \mathbf{80}$$

$$F = \det \begin{bmatrix} a_1 & c_1 \\ a_2 & c_2 \end{bmatrix} = \det \begin{bmatrix} 4 & 146 \\ 9 & 66 \end{bmatrix} = (4)(66) - (9)(146) = \mathbf{-1050}$$

$$D = \det \begin{bmatrix} a_1 & b_1 \\ a_2 & b_2 \end{bmatrix} = \det \begin{bmatrix} 4 & 1 \\ 9 & 1 \end{bmatrix} = (4)(1) - (9)(1) = \mathbf{-5}$$

The solution is

$$x = \frac{E}{D} = \frac{\mathbf{80}}{\mathbf{-5}} = -16 \quad \text{and} \quad y = \frac{F}{D} = \frac{\mathbf{-1050}}{\mathbf{-5}} = 210.$$

Now Try Exercise 29 ◆

Limitations on the Method of Cofactors and Cramer's Rule

Systems of linear equations involving more than two variables can be solved with Cramer's rule. (Cramer's rule for linear systems in three variables is discussed in the Extended and Discovery Exercises at the end of this section.) If a linear system has n equations, then Cramer's rule requires the computation of $n + 1$ determinants with dimension $n \times n$. Cramer's rule is seldom employed in real applications because of the substantial number of arithmetic operations needed to compute determinants of large matrices.

It can be shown that the cofactor method of calculating the determinant of an $n \times n$ matrix, $n > 2$, generally involves more than $n!$ multiplication operations. For example, a 4×4 determinant requires over $4! = 4 \cdot 3 \cdot 2 \cdot 1 = 24$ multiplication operations, and a 10×10 determinant would involve over

$$10! = 10 \cdot 9 \cdot 8 \cdot 7 \cdot 6 \cdot 5 \cdot 4 \cdot 3 \cdot 2 \cdot 1 = 3{,}628{,}800$$

multiplication operations. Factorials grow rapidly and are discussed in Chapter 8.

In real-life applications, it is not uncommon to solve linear systems that involve thousands of equations. Suppose that we were to solve a modest linear system that involved 20 equations with Cramer's rule, and expand each determinant by the method of cofactors. Then, the calculation of one 20×20 determinant would require over $20!$ multiplication operations. Supercomputers can perform 1 trillion (10^{12}) multiplication operations per second. On this supercomputer, just one 20×20 determinant would require over

$$\frac{20!}{10^{12}} \approx 2{,}432{,}902 \text{ seconds} \approx 28.2 \text{ days}$$

to compute. We would need to compute 21 of these determinants. This would require approximately $21 \times 28.2 = 592.2$ days, which is the best part of 2 years. Also, it is not uncommon for the cost of supercomputer time to exceed $2000 per hour. In addition to the time element, these computations could cost over $28 million to perform.

It becomes obvious why this method is not implemented even on a modest linear system, regardless of the technology available. Modern software packages are more efficient than Cramer's rule on linear systems that involve as few as three variables. (**Source:** Minnesota Supercomputer Institute.)

9.7 Putting it all Together

$\mathbf{T}$he determinant of a square matrix A is a real number, denoted det A. If det $A \neq 0$, then the matrix A is invertible. The computation of a determinant frequently involves a large number of arithmetic operations. Graphing calculators can be a valuable aid in computing determinants. Cramer's rule is a method for solving systems of linear equations that involves determinants.

The following table summarizes the calculation of 2×2 and 3×3 determinants by hand.

Determinants of 2 × 2 Matrices

The determinant of a 2×2 matrix A is given by

$$\det A = \det \begin{bmatrix} a & b \\ c & d \end{bmatrix} = ad - cb.$$

Example: $\det \begin{bmatrix} 6 & -2 \\ 3 & 7 \end{bmatrix} = (6)(7) - (3)(-2) = 48$

Determinants of 3 × 3 Matrices

The determinant of a 3×3 matrix A can be reduced to calculating the determinants of three 2×2 matrices. This calculation can be performed using cofactors.

$$\det A = \det \begin{bmatrix} a_1 & b_1 & c_1 \\ a_2 & b_2 & c_2 \\ a_3 & b_3 & c_3 \end{bmatrix}$$

$$= a_1 \det \begin{bmatrix} b_2 & c_2 \\ b_3 & c_3 \end{bmatrix} - a_2 \det \begin{bmatrix} b_1 & c_1 \\ b_3 & c_3 \end{bmatrix} + a_3 \det \begin{bmatrix} b_1 & c_1 \\ b_2 & c_2 \end{bmatrix}$$

$$= a_1(b_2c_3 - b_3c_2) - a_2(b_1c_3 - b_3c_1) + a_3(b_1c_2 - b_2c_1)$$

Example: $\det \begin{bmatrix} 1 & -2 & 3 \\ 4 & 5 & -1 \\ -3 & 7 & 8 \end{bmatrix} = (1) \det \begin{bmatrix} 5 & -1 \\ 7 & 8 \end{bmatrix} - (4) \det \begin{bmatrix} -2 & 3 \\ 7 & 8 \end{bmatrix} + (-3) \det \begin{bmatrix} -2 & 3 \\ 5 & -1 \end{bmatrix}$

$$= (1)(47) - 4(-37) - 3(-13) = 234$$

9.7 — Exercises

Calculating Determinants

Exercises 1–4: Determine if the matrix A is invertible by calculating det A.

1. $A = \begin{bmatrix} 4 & 3 \\ 5 & 4 \end{bmatrix}$

2. $A = \begin{bmatrix} 1 & -3 \\ 2 & 6 \end{bmatrix}$

3. $A = \begin{bmatrix} -4 & 6 \\ -8 & 12 \end{bmatrix}$

4. $A = \begin{bmatrix} 10 & -20 \\ -5 & 10 \end{bmatrix}$

Exercises 5–8: Find the specified minor and cofactor for the matrix A.

5. M_{12} and A_{12} if $A = \begin{bmatrix} 1 & -1 & 3 \\ 2 & 3 & -2 \\ 0 & 1 & 5 \end{bmatrix}$

6. M_{23} and A_{23} if $A = \begin{bmatrix} 1 & 2 & -1 \\ 4 & 6 & -3 \\ 2 & 3 & 9 \end{bmatrix}$

7. M_{22} and A_{22} if $A = \begin{bmatrix} 7 & -8 & 1 \\ 3 & -5 & 2 \\ 1 & 0 & -2 \end{bmatrix}$

8. M_{31} and A_{31} if $A = \begin{bmatrix} 0 & 0 & -1 \\ 6 & -7 & 1 \\ 8 & -9 & -1 \end{bmatrix}$

Exercises 9–12: Let A be the given matrix. Find det A by expanding about the first column. State whether A^{-1} exists.

9. $\begin{bmatrix} 1 & 4 & -7 \\ 0 & 2 & -3 \\ 0 & -1 & 3 \end{bmatrix}$

10. $\begin{bmatrix} 0 & 2 & 8 \\ -1 & 3 & 5 \\ 0 & 4 & 1 \end{bmatrix}$

11. $\begin{bmatrix} 5 & 1 & 6 \\ 0 & -2 & 0 \\ 0 & 4 & 0 \end{bmatrix}$

12. $\begin{bmatrix} 3 & 2 & 3 \\ 2 & 2 & 2 \\ 1 & 3 & 1 \end{bmatrix}$

Exercises 13–20: Let A be the given matrix. Find det A by using the method of cofactors.

13. $\begin{bmatrix} 2 & 0 & 0 \\ 0 & 3 & 0 \\ 0 & 0 & 5 \end{bmatrix}$

14. $\begin{bmatrix} 0 & 0 & 2 \\ 0 & 3 & 0 \\ 5 & 0 & 0 \end{bmatrix}$

15. $\begin{bmatrix} 0 & 0 & 0 \\ -8 & 3 & -9 \\ 15 & 5 & 9 \end{bmatrix}$

16. $\begin{bmatrix} 1 & 1 & 5 \\ -3 & -3 & 0 \\ 7 & 0 & 0 \end{bmatrix}$

17. $\begin{bmatrix} 3 & -1 & 2 \\ 0 & 5 & 7 \\ 1 & 0 & -1 \end{bmatrix}$

18. $\begin{bmatrix} 3 & 0 & -1 \\ 2 & 3 & -4 \\ 6 & -5 & 1 \end{bmatrix}$

19. $\begin{bmatrix} 1 & -5 & 2 \\ -7 & 1 & 3 \\ 0 & 4 & -2 \end{bmatrix}$

20. $\begin{bmatrix} 1 & -1 & 2 \\ -2 & 0 & 1 \\ 1 & 1 & -1 \end{bmatrix}$

Exercises 21–24: Let A be the given matrix. Use technology to calculate det A.

21. $\begin{bmatrix} 11 & -32 \\ 1.2 & 55 \end{bmatrix}$

22. $\begin{bmatrix} 17 & -4 & 3 \\ 11 & 5 & -15 \\ 7 & -9 & 23 \end{bmatrix}$

23. $\begin{bmatrix} 2.3 & 5.1 & 2.8 \\ 1.2 & 4.5 & 8.8 \\ -0.4 & -0.8 & -1.2 \end{bmatrix}$

24. $\begin{bmatrix} 1 & -1 & 3 & 7 \\ 9 & 2 & -7 & -4 \\ 5 & -7 & 1 & -9 \\ 7 & 1 & 3 & 6 \end{bmatrix}$

Calculating Area

Exercises 25–28: Use a determinant to find the area of the figure.

25.
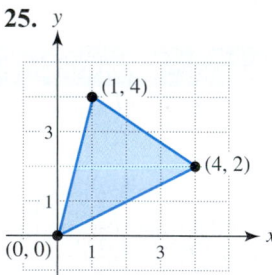

26.
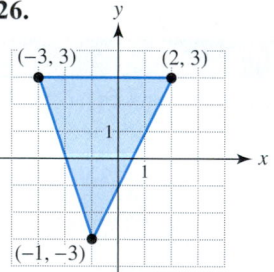

27.
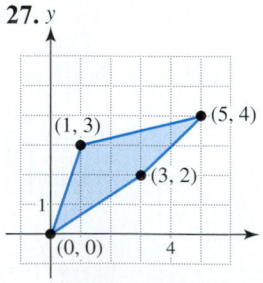

28.
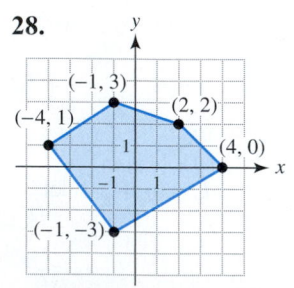

Cramer's Rule

Exercises 29–36: Use Cramer's rule to solve the system of linear equations.

29. $-x + 2y = 5$
$3x + 3y = 1$

30. $2x + y = -3$
$-4x - 6y = -7$

31. $-2x + 3y = 8$
$4x - 5y = 3$

32. $5x - 3y = 4$
$-3x - 7y = 5$

33. $7x + 4y = 23$
$11x - 5y = 70$

34. $-7x + 5y = 8.2$
$6x + 4y = -0.4$

35. $1.7x - 2.5y = -0.91$
$-0.4x + 0.9y = 0.423$

36. $-2.7x + 1.5y = -1.53$
$1.8x - 5.5y = -1.68$

Applying a Concept

Exercises 37–40: Use the concept of the area of a triangle to determine if the three points are collinear.

37. $(1, 3), (-3, 11), (2, 1)$

38. $(3, 6), (-1, -6), (5, 11)$

39. $(-2, -5), (4, 4), (2, 3)$

40. $(4, -5), (-2, 10), (6, -10)$

Writing about Mathematics

41. Choose two matrices A and B with dimension 2×2. Calculate det A, det B, and det (AB). Repeat this process until you are able to discover how these three determinants are related. Summarize your results.

42. Calculate det A and det A^{-1} for different matrices. Compare the determinants. Try to generalize your results.

EXTENDED AND DISCOVERY EXERCISES

Exercises 1–6: **Cramer's Rule** *Cramer's rule can be applied to systems of three linear equations in three variables. For the system of equations*

$$a_1x + b_1y + c_1z = d_1$$
$$a_2x + b_2y + c_2z = d_2$$
$$a_3x + b_3y + c_3z = d_3,$$

the solution can be written as follows.

$$D = \det\begin{bmatrix} a_1 & b_1 & c_1 \\ a_2 & b_2 & c_2 \\ a_3 & b_3 & c_3 \end{bmatrix}, \quad E = \det\begin{bmatrix} d_1 & b_1 & c_1 \\ d_2 & b_2 & c_2 \\ d_3 & b_3 & c_3 \end{bmatrix}$$

$$F = \det\begin{bmatrix} a_1 & d_1 & c_1 \\ a_2 & d_2 & c_2 \\ a_3 & d_3 & c_3 \end{bmatrix}, \quad G = \det\begin{bmatrix} a_1 & b_1 & d_1 \\ a_2 & b_2 & d_2 \\ a_3 & b_3 & d_3 \end{bmatrix}$$

If $D \neq 0$, a unique solution exists and is given by

$$x = \frac{E}{D}, \quad y = \frac{F}{D}, \quad z = \frac{G}{D}.$$

Use Cramer's rule to solve the equations.

1. $x + y + z = 6$
$2x + y + 2z = 9$
$y + 3z = 9$

2. $y + z = 1$
$2x - y - z = -1$
$x + y - z = 3$

3. $x + z = 2$
$x + y = 0$
$y + 2z = 1$

4. $x + y + 2z = 1$
$-x - 2y - 3z = -2$
$y - 3z = 5$

5. $x + 2z = 7$
$-x + y + z = 5$
$2x - y + 2z = 6$

6. $x + 2y + 3z = -1$
$2x - 3y - z = 12$
$x + 4y - 2z = -12$

CHECKING BASIC CONCEPTS FOR SECTION 9.7

1. Find the determinant of the matrix A by using the method of cofactors. Is A invertible?

$$A = \begin{bmatrix} 1 & -1 & 2 \\ 2 & 3 & 1 \\ 0 & -2 & 5 \end{bmatrix}$$

2. Use Cramer's rule to solve the system of linear equations.

$$3x - 4y = 7$$
$$-4x + 3y = 5$$

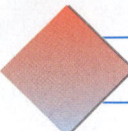

CHAPTER 9 SUMMARY

CONCEPT	EXPLANATION AND EXAMPLES

SECTION 9.1 FUNCTIONS AND EQUATIONS IN TWO VARIABLES

FUNCTIONS OF TWO VARIABLES

$z = f(x, y)$, where x and y are inputs to f

Example: $f(x, y) = 2x - 3y$
$f(4, -1) = 2(4) - 3(-1) = 11$

SYSTEM OF LINEAR EQUATIONS IN TWO VARIABLES

General form
$$a_1 x + b_1 y = c_1$$
$$a_2 x + b_2 y = c_2$$

A linear system can have zero, one, or infinitely many solutions. A solution can be written as an ordered pair. A linear system may be solved symbolically, graphically, or numerically.

Example: $x - y = 2$
$2x + y = 7$ Solution: $(3, 1)$

METHOD OF SUBSTITUTION FOR TWO EQUATIONS

May be used to solve systems of linear or nonlinear equations

Example: $x - y = -3$
$x + 4y = 17$

Solve the first equation for x to obtain $x = y - 3$. Substitute this result in the second equation and solve for y.

$$(y - 3) + 4y = 17 \quad \text{implies that} \quad y = 4.$$

Then $x = 4 - 3 = 1$ and the solution is $(1, 4)$.

JOINT VARIATION

Let m and n be real numbers. Then z *varies jointly* as the mth power of x and the nth power of y if a nonzero real number k exists such that $z = kx^m y^n$.

Example: The area of a triangle varies jointly as the base b and the height h because $A = \frac{1}{2}bh$. Note that $k = \frac{1}{2}$, $m = 1$, and $n = 1$ in this example.

SECTION 9.2 SYSTEMS OF EQUATIONS AND INEQUALITIES IN TWO VARIABLES

TYPES OF LINEAR SYSTEMS WITH TWO VARIABLES

Consistent system: Has either one solution (independent equations) or infinitely many solutions (dependent equations)

Inconsistent system: Has no solutions

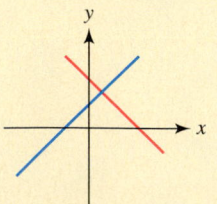

One Solution

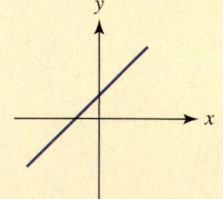

Infinitely Many Solutions

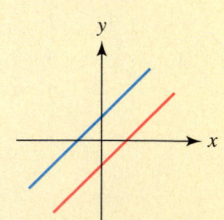

No Solutions

CONCEPT	EXPLANATION AND EXAMPLES

SECTION 9.2 SYSTEMS OF EQUATIONS AND INEQUALITIES IN TWO VARIABLES (CONTINUED)

METHOD OF ELIMINATION

Can be used to solve systems of equations.

Example:
$$2x - 3y = 4$$
$$\underline{x + 3y = 11}$$
$$3x \quad\quad = 15 \quad \text{or} \quad x = 5 \quad \text{Add.}$$

Substituting $x = 5$ in the first equation gives $y = 2$. The solution is $(5, 2)$.

SYSTEM OF INEQUALITIES IN TWO VARIABLES

The solution set is often a shaded region in the xy-plane.

Example:
$$x + y \le 4$$
$$y \ge 0$$
$$x \ge 0$$

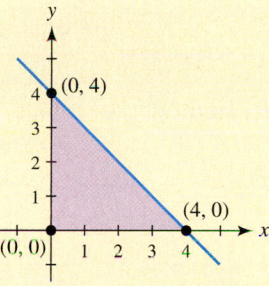

LINEAR PROGRAMMING

Method for maximizing (or minimizing) an objective function subject to a set of constraints

Example: Maximize $P = 2x + 4y$, subject to

$$x + y \le 4, x \ge 0, y \ge 0.$$

The maximum of $P = 16$ occurs at the vertex $(0, 4)$, in the region of feasible solutions. See figure above.

SECTION 9.3 SYSTEMS OF LINEAR EQUATIONS IN THREE VARIABLES

SOLUTION TO A SYSTEM OF LINEAR EQUATIONS IN THREE VARIABLES

An ordered triple (x, y, z) that satisfies *every* equation

Example:
$$x - 2y + 3z = 6$$
$$-x + 3y + 4z = 17$$
$$3x + 4y - 5z = -4$$

The solution is $(1, 2, 3)$ because $x = 1$, $y = 2$, and $z = 3$ satisfies all three equations. (Check this fact.)

ELIMINATION AND SUBSTITUTION

Systems of linear equations in three variables can be solved by using elimination and substitution. The following three steps outline this process.

STEP 1: Eliminate one variable, such as x, from two of the equations.

STEP 2: Apply the techniques discussed in Sections 9.1 and 9.2 to solve the two resulting equations in two variables from **STEP 1**. If x is eliminated, then solve these equations to find y and z.

CONCEPT	EXPLANATION AND EXAMPLES

SECTION 9.3 SYSTEMS OF LINEAR EQUATIONS IN THREE VARIABLES (CONTINUED)

Note: If there are no solutions for y and z, then the given system has no solutions. If there are infinitely many solutions for y and z, then write y in terms of z and go to **STEP 3**.

STEP 3: Substitute the values for y and z in one of the given equations to find x. The solution is (x, y, z).

SECTION 9.4 SOLUTIONS TO LINEAR SYSTEMS USING MATRICES

MATRICES AND SYSTEMS OF LINEAR EQUATIONS

An augmented matrix can be used to represent a system of linear equations.

Augmented Matrix

Example:

$$\begin{aligned} x - 2y + z &= 0 \\ -x + 4y - z &= 4 \\ 2x + y - 3z &= -5 \end{aligned} \qquad \left[\begin{array}{rrr|r} 1 & -2 & 1 & 0 \\ -1 & 4 & -1 & 4 \\ 2 & 1 & -3 & -5 \end{array}\right]$$

ROW-ECHELON FORM

Examples:

$$\begin{bmatrix} 1 & 2 & -1 \\ 0 & 1 & 2 \end{bmatrix} \qquad \left[\begin{array}{rrr|r} 1 & 3 & -2 & 7 \\ 0 & 1 & 4 & 5 \\ 0 & 0 & 1 & -3 \end{array}\right]$$

GAUSSIAN ELIMINATION WITH BACKWARD SUBSTITUTION

Gaussian elimination can be used to transform a system of linear equations into row-echelon form. Then backward substitution can be used to solve the resulting system of linear equations. Graphing calculators can also be used to solve systems of equations.

SECTION 9.5 PROPERTIES AND APPLICATIONS OF MATRICES

OPERATIONS ON MATRICES

Matrices can be added, subtracted, and multiplied, but there is no division of matrices.

ADDITION

$$\begin{bmatrix} 2 & 4 \\ 5 & 6 \end{bmatrix} + \begin{bmatrix} -2 & 1 \\ 7 & 3 \end{bmatrix} = \begin{bmatrix} 0 & 5 \\ 12 & 9 \end{bmatrix}$$

SUBTRACTION

$$\begin{bmatrix} -3 & 0 \\ 4 & -4 \end{bmatrix} - \begin{bmatrix} 1 & 2 \\ 6 & -7 \end{bmatrix} = \begin{bmatrix} -4 & -2 \\ -2 & 3 \end{bmatrix}$$

SCALAR MULTIPLICATION

$$3\begin{bmatrix} 5 & 1 & 6 & -1 \\ 0 & -2 & 3 & 2 \end{bmatrix} = \begin{bmatrix} 15 & 3 & 18 & -3 \\ 0 & -6 & 9 & 6 \end{bmatrix}$$

MULTIPLICATION

$$\begin{bmatrix} 2 & -1 \\ 0 & 3 \\ -7 & 1 \end{bmatrix} \cdot \begin{bmatrix} 1 & -1 & 0 \\ 3 & -5 & -4 \end{bmatrix} = \begin{bmatrix} -1 & 3 & 4 \\ 9 & -15 & -12 \\ -4 & 2 & -4 \end{bmatrix}$$

CONCEPT	EXPLANATION AND EXAMPLES

SECTION 9.6 INVERSES OF MATRICES

MATRIX INVERSES

The inverse of an $n \times n$ matrix A, denoted A^{-1}, satisfies $A^{-1}A = I_n$ and $AA^{-1} = I_n$, where I_n is the $n \times n$ identity matrix. The inverse of a matrix can be found by hand or with technology.

Example: $A = \begin{bmatrix} 5 & 2 \\ 2 & 1 \end{bmatrix}$ and $A^{-1} = \begin{bmatrix} 1 & -2 \\ -2 & 5 \end{bmatrix}$

$$\begin{bmatrix} 5 & 2 \\ 2 & 1 \end{bmatrix}\begin{bmatrix} 1 & -2 \\ -2 & 5 \end{bmatrix} = \begin{bmatrix} 1 & 0 \\ 0 & 1 \end{bmatrix} = I_2$$

$$\begin{bmatrix} 1 & -2 \\ -2 & 5 \end{bmatrix}\begin{bmatrix} 5 & 2 \\ 2 & 1 \end{bmatrix} = \begin{bmatrix} 1 & 0 \\ 0 & 1 \end{bmatrix} = I_2$$

MATRIX EQUATIONS

A system of linear equations can be written as a matrix equation.

Example: $2x - 2y = 3$
$-3x + 4y = 2$

$$AX = \begin{bmatrix} 2 & -2 \\ -3 & 4 \end{bmatrix}\begin{bmatrix} x \\ y \end{bmatrix} = \begin{bmatrix} 3 \\ 2 \end{bmatrix} = B$$

The solution can be found as follows.

$$X = A^{-1}B = \begin{bmatrix} 2 & 1 \\ 1.5 & 1 \end{bmatrix}\begin{bmatrix} 3 \\ 2 \end{bmatrix} = \begin{bmatrix} 8 \\ 6.5 \end{bmatrix}$$

SECTION 9.7 DETERMINANTS

DETERMINANT OF A 2 × 2 MATRIX

$$\det A = \begin{bmatrix} a & b \\ c & d \end{bmatrix} = ad - cb$$

Example: $\det \begin{bmatrix} 1 & 4 \\ 3 & 5 \end{bmatrix} = (1)(5) - (3)(4) = -7$

DETERMINANT OF A 3 × 3 MATRIX

A 3×3 determinant can be reduced to calculating the determinants of 2×2 matrices by using cofactors. If $\det A \neq 0$, then A^{-1} exists.

Example: $\det \begin{bmatrix} 3 & 1 & -1 \\ 2 & 2 & 0 \\ 0 & 1 & -3 \end{bmatrix}$

$$= 3\begin{bmatrix} 2 & 0 \\ 1 & -3 \end{bmatrix} - 2\begin{bmatrix} 1 & -1 \\ 1 & -3 \end{bmatrix} + 0\begin{bmatrix} 1 & -1 \\ 2 & 0 \end{bmatrix}$$

$$= 3(-6) - 2(-2) + 0(2)$$

$$= -14$$

CRAMER'S RULE

Cramer's rule makes use of determinants to solve systems of linear equations. However, Gaussian elimination with backward substitution is usually more efficient.

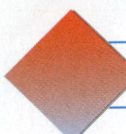

REVIEW EXERCISES

Exercises 1 and 2: Evaluate the function for the indicated inputs.

1. $A(3, 6)$, where $A(b, h) = \frac{1}{2}bh$.

2. $V(2, 5)$, where $V(r, h) = \pi r^2 h$

3. Solve the equation $3x - 4y = -2$ for y.

4. Solve the equation $V = \pi r^2 h$ for r if $r > 0$.

Exercises 5–8: Solve the system of equations

   (a) graphically, and

   (b) symbolically.

5. $\begin{aligned} 3x + y &= 1 \\ 2x - 3y &= 8 \end{aligned}$

6. $\begin{aligned} x^2 - y &= 1 \\ x + y &= 1 \end{aligned}$

7. $\begin{aligned} x^2 + y^2 &= 25 \\ 4x - 3y &= 0 \end{aligned}$

8. $\begin{aligned} \frac{1}{2}x - \frac{3}{4}y &= \frac{5}{4} \\ \frac{3}{4}x - \frac{1}{2}y &= -\frac{5}{4} \end{aligned}$

Exercises 9–12: Use the elimination method to solve each system of linear equations, if possible. Identify the system as consistent or inconsistent.

9. $\begin{aligned} 2x + y &= 7 \\ x - 2y &= -4 \end{aligned}$

10. $\begin{aligned} 3x + 3y &= 15 \\ -x - y &= -4 \end{aligned}$

11. $\begin{aligned} 6x - 15y &= 12 \\ -4x + 10y &= -8 \end{aligned}$

12. $\begin{aligned} 3x - 4y &= -10 \\ 4x + 3y &= -30 \end{aligned}$

Exercises 13 and 14: Use elimination to solve the system of nonlinear equations.

13. $\begin{aligned} x^2 - 3y &= 3 \\ x^2 + 2y^2 &= 5 \end{aligned}$

14. $\begin{aligned} 2x - 3y &= 1 \\ 2x^2 + y &= 1 \end{aligned}$

Exercises 15–18: Graph the solution set to the inequality.

15. $y \geq -1$

16. $x < 2$

17. $2x - y < 4$

18. $3x + 5y \geq -15$

Exercises 19–22: Graph the solution set to the system of inequalities. Use the graph to find one solution.

19. $\begin{aligned} x^2 + y^2 &< 9 \\ x + y &> 3 \end{aligned}$

20. $\begin{aligned} y &\leq 4 - x^2 \\ y &\geq 2 - x \end{aligned}$

21. $\begin{aligned} x + 3y &\geq 3 \\ x + y &\leq 4 \end{aligned}$

22. $\begin{aligned} x - 2y &> -4 \\ 2x + y &< 2 \end{aligned}$

Exercises 23–26: If possible, solve the system of linear equations.

23. $\begin{aligned} x - y + z &= -2 \\ x + 2y - z &= 2 \\ 2y + 3z &= 7 \end{aligned}$

24. $\begin{aligned} x - 3y + 2z &= -10 \\ 2x - y + 3z &= -9 \\ -x - y + z &= -1 \end{aligned}$

25. $\begin{aligned} -x + 2y + 2z &= 9 \\ x + y - 3z &= 6 \\ 3y - z &= 8 \end{aligned}$

26. $\begin{aligned} 2x - y - 3z &= -9 \\ x - 8z &= -23 \\ -3x + 2y - 2z &= -5 \end{aligned}$

Exercises 27–30: The augmented matrix represents a system of linear equations. Solve the system.

27. $\begin{bmatrix} 1 & 5 & | & 6 \\ 0 & 1 & | & 3 \end{bmatrix}$

28. $\begin{bmatrix} 1 & 2 & -2 & | & 8 \\ 0 & 1 & 1 & | & 5 \\ 0 & 0 & 0 & | & 0 \end{bmatrix}$

29. $\begin{bmatrix} 1 & 0 & 0 & | & -2 \\ 0 & 1 & 0 & | & 3 \\ 0 & 0 & 1 & | & 0 \end{bmatrix}$

30. $\begin{bmatrix} 1 & 0 & 0 & | & -5 \\ 0 & 1 & -4 & | & 1 \\ 0 & 0 & 0 & | & 5 \end{bmatrix}$

Exercises 31–34: Use Gaussian elimination with backward substitution to solve the system of linear equations.

31. $\begin{aligned} x + 3y &= 8 \\ -x + y &= 4 \end{aligned}$

32. $\begin{aligned} x + z &= 4 \\ x + y - 2z &= -3 \\ -x + y + z &= 4 \end{aligned}$

33. $\begin{aligned} 2x - y + 2z &= 10 \\ x - 2y + z &= 8 \\ 3x - y + 2z &= 11 \end{aligned}$

34. $\begin{aligned} x - 2y + z &= 1 \\ 2x - 5y + 3z &= 4 \\ 2x - 3y + z &= 0 \end{aligned}$

Exercises 35 and 36: Let a_{ij} denote the general term for the matrix A. Find each of the following.

   (a) $a_{12} + a_{22}$ *(b) $a_{11} - 2a_{23}$*

35. $A = \begin{bmatrix} -2 & 3 & -1 \\ 5 & 2 & 4 \end{bmatrix}$

36. $\begin{bmatrix} -1 & 2 & 5 \\ 1 & -3 & 7 \\ 0 & 7 & -2 \end{bmatrix}$

Exercises 37 and 38: Evaluate the following.

   (a) $A + 2B$ *(b) $A - B$*   *(c) $-4A$*

37. $A = \begin{bmatrix} 1 & -3 \\ 2 & -1 \end{bmatrix}$,

$B = \begin{bmatrix} 3 & 2 \\ -5 & 1 \end{bmatrix}$

38. $A = \begin{bmatrix} 4 & 0 & 1 \\ -2 & 8 & 9 \end{bmatrix}$,

$B = \begin{bmatrix} -5 & 3 & 2 \\ -4 & 0 & 7 \end{bmatrix}$

Exercises 39–42: If possible, find AB and BA.

39. $A = \begin{bmatrix} 2 & 0 \\ -5 & 3 \end{bmatrix}$, $B = \begin{bmatrix} -1 & -2 \\ 4 & 7 \end{bmatrix}$

40. $A = \begin{bmatrix} 1 & -2 \\ 2 & 3 \end{bmatrix}$, $B = \begin{bmatrix} 1 & 0 & 2 \\ -1 & 3 & 4 \end{bmatrix}$

41. $A = \begin{bmatrix} 2 & -1 & 3 \\ 2 & 4 & 0 \end{bmatrix}$, $B = \begin{bmatrix} 1 & 0 \\ -1 & 2 \\ 0 & 3 \end{bmatrix}$

42. $A = \begin{bmatrix} 1 & -1 & 2 \\ 0 & 3 & 4 \\ 1 & 0 & 2 \end{bmatrix}$, $B = \begin{bmatrix} -1 & 0 & 0 \\ 2 & 0 & -1 \\ 1 & 4 & 2 \end{bmatrix}$

Exercises 43 and 44: Determine if B is the inverse matrix of A by evaluating AB and BA.

43. $A = \begin{bmatrix} 8 & 5 \\ 6 & 4 \end{bmatrix}$, $B = \begin{bmatrix} 2 & -2.5 \\ -3 & 4 \end{bmatrix}$

44. $A = \begin{bmatrix} -1 & 1 & 2 \\ 1 & 0 & -1 \\ 0 & 1 & 2 \end{bmatrix}$, $B = \begin{bmatrix} -1 & 0 & 1 \\ 2 & 2 & -1 \\ -1 & -1 & -1 \end{bmatrix}$

Exercises 45–48: Complete the following.
(a) *Write the system of linear equations in the form AX = B.*
(b) *Solve the linear system by computing $X = A^{-1}B$.*

45. $x - 3y = 4$
$2x - y = 3$

46. $x - 2y + z = 0$
$2x + y + 2z = 10$
$y + z = 3$

47. $11x + 31y = -27.6$
$37x - 19y = 240$

48. $12x + 7y - 3z = 14.6$
$8x - 11y + 13z = -60.4$
$-23x + 9z = -14.6$

49. Graph the solution set to the system of inequalities.

$$x + y \le 6$$
$$3x + 2y \ge 6$$
$$x \ge 0, y \ge 0$$

50. If possible, graphically approximate the solution of each system of equations to the nearest thousandth. Identify each system as consistent or inconsistent. If the system is consistent, determine if the equations are dependent or independent.
(a) $3.1x + 4.2y = 6.4$
$1.7x - 9.1y = 1.6$

(b) $6.3x - 5.1y = 9.3$
$4.2x - 3.4y = 6.2$

(c) $0.32x - 0.64y = 0.96$
$-0.08x + 0.16y = -0.72$

Exercises 51 and 52: Let A be the given matrix. Find A^{-1}.

51. $\begin{bmatrix} 1 & -2 \\ -1 & 1 \end{bmatrix}$ **52.** $\begin{bmatrix} 1 & 0 & 1 \\ 1 & 1 & 1 \\ 0 & 1 & -1 \end{bmatrix}$

Exercises 53 and 54: Let A be the given matrix. Find det A by using the method of cofactors.

53. $\begin{bmatrix} 2 & 1 & 3 \\ 0 & 3 & 4 \\ 1 & 0 & 5 \end{bmatrix}$ **54.** $\begin{bmatrix} 3 & 0 & 2 \\ 1 & 3 & 5 \\ -5 & 2 & 0 \end{bmatrix}$

Exercises 55 and 56: Let A be the given matrix. Use technology to find det A. State whether A is invertible.

55. $\begin{bmatrix} 13 & 22 \\ 55 & -57 \end{bmatrix}$ **56.** $\begin{bmatrix} 6 & -7 & -1 \\ -7 & 3 & -4 \\ 23 & 54 & 77 \end{bmatrix}$

Applications

57. *Ultraviolet Light*   There are two types of ultraviolet light: UV-A and UV-B. UV-B is responsible for both skin and eye damage. The table lists the maximum doses of UV-B in the northern hemisphere for selected latitudes L and dates D of the year.   (**Sources:** Orbital Sciences Corporation; J. Williams, *The Weather Almanac 1995*.)

L/D	Mar 21	June 21	Sept 21	Dec 21
0°	325	254	325	272
10°	311	275	280	220
20°	249	292	256	143
30°	179	248	182	80
40°	99	199	127	34
50°	57	143	75	13

(a) A student living in Topeka, Kansas, (39°N) is planning spring break in Honolulu, Hawaii, (21°N) during March 19–23. Approximate how many times stronger the UV-B light is in Hawaii than in Topeka.

(b) Compare the sun's UV-B strength in Glasgow, Montana, (48°N) on June 21 with its strength in Honolulu on December 21.

58. *Wave Height*   The table lists the wave heights produced on the ocean for various wind speeds and durations. (**Source:** J. Navarra, *Atmosphere, Weather and Climate.*)

Wind Speed (mph)	Duration of the Wind			
	10 hr	20 hr	30 hr	40 hr
11.5	2 ft	2 ft	2 ft	2 ft
17.3	4 ft	5 ft	5 ft	5 ft
23.0	7 ft	8 ft	9 ft	9 ft
34.5	13 ft	17 ft	18 ft	19 ft
46.0	21 ft	28 ft	31 ft	33 ft
57.5	29 ft	40 ft	45 ft	48 ft

(a) What is the expected wave height if a 46 mile-per-hour wind blows for 30 hours?

(b) If the wave height is 5 feet, can the speed of the wind and its duration be determined?

(c) Describe the relationship among wave height, wind speed, and duration of the wind.

(d) If the duration of the wind doubles, does the wave height double?

(e) Estimate the wave height if a 46 mile-per-hour wind blows for 15 hours.

59. *Area and Perimeter*   Let l represent the length of a rectangle and w its width, where $l \geq w$. Then, its area can be computed by $A(l, w) = lw$ and its perimeter by $P(l, w) = 2l + 2w$. Solve the system of equations determined by $A(l, w) = 77$ and $P(l, w) = 36$.

60. *Cylinder*   Approximate the radius r and height h of a cylindrical container with a volume V of 30 cubic inches and a lateral surface area S of 45 square inches.

61. *Student Loans*   A student takes out two loans totaling $2000 to help pay for college expenses. One loan is at 7% interest, and the other is at 9%. Interest for both loans is compounded annually.

(a) If the combined total interest for both loans the first year is $156, find the amount of each loan symbolically.

(b) Determine the amount of each loan graphically or numerically

62. *Dimensions of a Screen*   The screen of a rectangular television set is 3 inches wider than it is high. If the perimeter of the screen is 42 inches, find its dimensions by writing a system of linear equations and solving.

63. *CD Prices*   A music store sells compact discs at two prices marked A and B. Each row in the table represents a purchase. Determine the cost of each type of CD by using a matrix inverse.

A	B	Total
1	2	$37.47
2	3	$61.95

64. *Digital Photography*   Design a 3×3 matrix A that represents a digital photograph of the letter T in black on a white background. Find a matrix B, such that $A + B$ darkens only the white background by one gray level.

65. *Area*   Find the area of the triangle whose vertices are (0, 0), (5, 2), and (2, 5).

66. *Voter Turnout*   The table shows the percent y of voter turnout in the United States for three presidential elections in year x, where $x = 0$ corresponds to 1900. Find a quadratic function defined by $f(x) = ax^2 + bx + c$ that models these data. Graph f together with the data. (**Source:** Committee for the Study of the American Electorate.)

x	24	60	96
y	48.9	62.8	48.8

67. *Geometry*   Complete the following.
(a) Write a system of inequalities that describes possible dimensions for a cylinder with a volume V greater than or equal to 30 cubic inches, and a lateral surface area S less than or equal to 45 square inches.

(b) Graph and identify the region of solutions.

(c) Use the graph to estimate one solution to the system of inequalities.

68. *Snowmobile Fatalities*   Through February during the 1996–1997 winter, snowmobile fatalities in Michigan exceeded the average number for an entire winter by 5 fatalities. This was a 19.2% increase over the yearly average, with 2 months still left in the winter. Approximate the average number of snowmobile fatalities in Michigan annually, and the number of fatalities through February during the winter of 1996–1997.   (**Source:** *USA Today.*)

69. *Joint Variation*   Suppose P varies jointly as the square of x and the cube of y. If $P = 432$ when $x = 2$ and $y = 3$, find P when $x = 3$ and $y = 5$.

70. *Linear Programming*   Find the maximum value of $P = 3x + 4y$ subject to the following constraints.

$$x + 3y \le 12$$
$$3x + y \le 12$$
$$x \ge 0, y \ge 0$$

EXTENDED AND DISCOVERY EXERCISES

1. To form the **transpose** of a matrix A, denoted A^T, let the first row of A be the first column of A^T, the second row of A be the second column of A^T, and so on, for each row of A. The following are examples of A and A^T. If A has dimension $m \times n$, then A^T has dimension $n \times m$.

$$A = \begin{bmatrix} 3 & -3 & 7 \\ 1 & 6 & -2 \\ 4 & 2 & 5 \end{bmatrix} \quad A^T = \begin{bmatrix} 3 & 1 & 4 \\ -3 & 6 & 2 \\ 7 & -2 & 5 \end{bmatrix}$$

$$A = \begin{bmatrix} 1 & 2 \\ 3 & 4 \\ 5 & 6 \end{bmatrix} \quad A^T = \begin{bmatrix} 1 & 3 & 5 \\ 2 & 4 & 6 \end{bmatrix}$$

Find the transpose of each matrix A.

(a) $A = \begin{bmatrix} 3 & -3 \\ 2 & 6 \\ 4 & 2 \end{bmatrix}$ **(b)** $A = \begin{bmatrix} 0 & 1 & -2 \\ 2 & 5 & 4 \\ -4 & 3 & 9 \end{bmatrix}$

(c) $A = \begin{bmatrix} 5 & 7 \\ 1 & -7 \\ 6 & 3 \\ -9 & 2 \end{bmatrix}$

Least-Squares Fit and Matrices

The table shows the average cost of tuition and fees y in dollars at 4-year public colleges. In this table $x = 0$ represents 1980 and $x = 15$ corresponds to 1995. (**Source:** The College Board.)

x	0	5	10	15
y	804	1318	1908	2860

The data are modeled using a line in the accompanying figure.

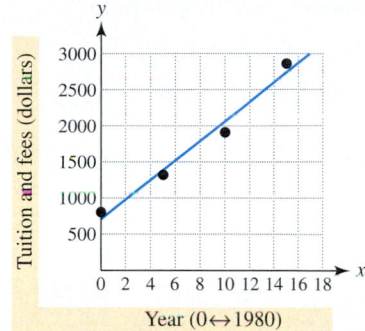

Year ($0 \leftrightarrow 1980$)

The equation of this line can be found using linear regression based on least squares. (Refer to Section 2.1.) If possible, we would like $f(x) = ax + b$ to satisfy the following four equations.

$$f(0) = a(0) + b = 804$$
$$f(5) = a(5) + b = 1318$$
$$f(10) = a(10) + b = 1908$$
$$f(15) = a(15) + b = 2860$$

Since the data points are not collinear, it is impossible for the graph of f to pass through all four points. These four equations can be written as

$$AX = \begin{bmatrix} 0 & 1 \\ 5 & 1 \\ 10 & 1 \\ 15 & 1 \end{bmatrix} \begin{bmatrix} a \\ b \end{bmatrix} = \begin{bmatrix} 804 \\ 1318 \\ 1908 \\ 2860 \end{bmatrix} = B.$$

The least-squares solution is found by solving the **normal equations**

$$A^T A X = A^T B$$

for X. The solution is $X = (A^T A)^{-1} A^T B$. Using technology, $a = 135.16$ and $b = 708.8$. Thus, f is given by the formula $f(x) = 135.16x + 708.8$. The function f and the data can be graphed. See the accompanying figures.

$[-1, 17, 1]$ by $[400, 3200, 200]$

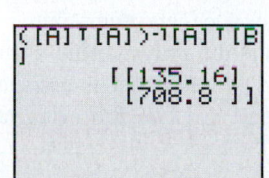
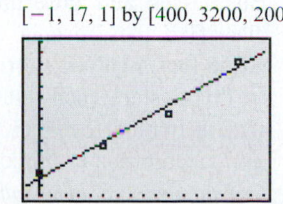

Exercises 2–4: Least-Square Models Solve the normal equations to model the data with $f(x) = ax + b$. Plot the data and f in the same viewing rectangle.

2. Tuition and Fees The table shows average cost of tuition and fees y in dollars at private 4-year colleges. In this table $x = 0$ corresponds to 1980 and $x = 15$ to 1995. (**Source:** The College Board.)

x	0	5	10	15
y	3617	6121	9340	12,432

3. Satellite TV The table lists the number of satellite television subscribers y in millions. In this table $x = 0$ corresponds to 1995 and $x = 5$ to the year 2000. (**Source:** USA Today.)

x	0	1	2	3	4	5
y	2.2	4.5	7.9	10.5	13	15

4. Federal Deficit The table lists the annual federal deficit y in billions of dollars from 1992 to 1996, where $x = 2$ corresponds to 1992 and $x = 6$ to 1996. (**Source:** Department of the Treasury.)

x	2	3	4	5	6
y	290	255	203	164	107

Exercises 5 and 6: Nonlinear Least-Squares Normal equations can be used to solve problems involving nonlinear regression. Solve the normal equations so that f models the given data.

5. $f(x) = \dfrac{a}{x} + \dfrac{b}{x^2}$

x	1	2	3	4
y	6.30	1.88	0.967	0.619

6. $f(x) = a \log x + bx + c$ (*Hint:* The matrix A has dimension 5×3.)

x	2	3	4	5	6
y	2.45	3.12	3.70	4.25	4.77

Cryptography

7. Businesses and government agencies frequently send classified messages in code. One cryptographic technique that involves matrices is the **polygraphic system**. In this system each letter in the alphabet is associated with a number between 1 and 26. The following table gives a common example. (**Source:** A. Sinkov, *Elementary Cryptanalysis: A Mathematical Approach.*)

A	B	C	D	E	F	G	H	I	J
1	2	3	4	5	6	7	8	9	10

K	L	M	N	O	P	Q	R	S	T
11	12	13	14	15	16	17	18	19	20

U	V	W	X	Y	Z
21	22	23	24	25	26

For example, the word MATH is coded as 13 1 20 8. Enter these numbers in a matrix B.

$$B = \begin{bmatrix} 13 & 20 \\ 1 & 8 \end{bmatrix}$$

To code these letters a 2×2 matrix, such as

$$A = \begin{bmatrix} 2 & 1 \\ -5 & -2 \end{bmatrix},$$

is multiplied times B to form the product AB.

$$AB = \begin{bmatrix} 2 & 1 \\ -5 & -2 \end{bmatrix} \begin{bmatrix} 13 & 20 \\ 1 & 8 \end{bmatrix} = \begin{bmatrix} 27 & 48 \\ -67 & -116 \end{bmatrix}$$

Since the resulting elements of AB are less than 1 or greater than 26, they may be scaled between 1 and 26 by adding or subtracting multiples of 26.

$$27 - 1(26) = \mathbf{1} \qquad 48 - 1(26) = \mathbf{22}$$
$$-67 + 3(26) = \mathbf{11} \qquad -116 + 5(26) = \mathbf{14}$$

Thus the word MATH is coded as **1 11 22 14** or AKVN. The advantage of this coding technique is that a particular letter is not always coded the same each time. Use the matrix A to code the following words.

(a) HELP

(b) LETTER (*Hint:* B has dimension 2×3.)

8. (Refer to the previous exercise.) To decode a message, A^{-1} is used. For example, to decode AKVN, write the sequence **1 11 22 14**. This is stored as

$$C = \begin{bmatrix} 1 & 22 \\ 11 & 14 \end{bmatrix}.$$

To decode the message, evaluate $A^{-1}C$, where

$$A^{-1} = \begin{bmatrix} -2 & -1 \\ 5 & 2 \end{bmatrix}.$$

$$A^{-1}C = \begin{bmatrix} -2 & -1 \\ 5 & 2 \end{bmatrix} \begin{bmatrix} 1 & 22 \\ 11 & 14 \end{bmatrix} = \begin{bmatrix} -13 & -58 \\ 27 & 138 \end{bmatrix}$$

Scaling the matrix elements of $A^{-1}C$ between 1 and 26 results in the following.

$$-13 + (1)26 = \mathbf{13} \qquad -58 + 3(26) = \mathbf{20}$$
$$27 - 1(26) = \mathbf{1} \qquad 138 - 5(26) = \mathbf{8}$$

The number **13 1 20 8** represents MATH. Use A^{-1} to decode each of the following.

(a) UBNL   **(b)** QNABMV

10 Conic Sections

Enthusiasm is the most important thing in life.

Tennessee Williams

Throughout history, people have been fascinated by the universe around them and compelled to understand it. Conic sections have played an important role in gaining this understanding. Although conic sections were described and named by the Greek astronomer Apollonius in 200 B.C., it was not until much later that they were used to model motion in the universe. In the sixteenth century Tycho Brahe, the greatest observational astronomer of the age, recorded precise data on planetary movement in the sky. Using Brahe's data in 1619, Johannes Kepler determined that planets move in elliptical orbits around the sun. In 1686 Newton used Kepler's work to show that elliptical orbits are the result of his famous theory of gravitation. We now know that all celestial objects—including planets, comets, asteroids, and satellites—travel in paths described by conic sections. Today scientists search the sky for information about the universe with enormous radio telescopes in the shape of parabolic dishes.

The search to understand the universe has been directed toward both the infinite and the infinitesimal. In 1911 Ernest Rutherford determined the basic structure of the atom. Small atomic particles are capable of traveling in trajectories described by conic sections.

Parabolas, ellipses, and hyperbolas have had a profound influence on our understanding of ourselves and the cosmos around us. In this chapter we learn about these age-old curves.

Source: *Historical Topics for the Mathematics Classroom, Thirty-first Yearbook*, NCTM.

10.1 Parabolas

◆ Find equations of parabolas
◆ Graph parabolas
◆ Learn the reflective property of parabolas
◆ Translate parabolas

Introduction

Conic sections are named after the different ways that a plane can intersect a cone. See Figure 10.1. The three basic conic sections are parabolas, ellipses, and hyperbolas. A circle is an example of an ellipse.

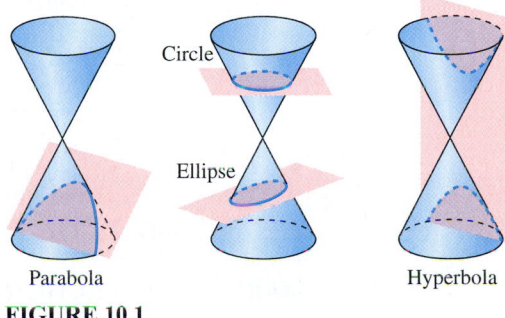

FIGURE 10.1

Figures 10.2–10.4 show examples of conic sections in the xy-plane.

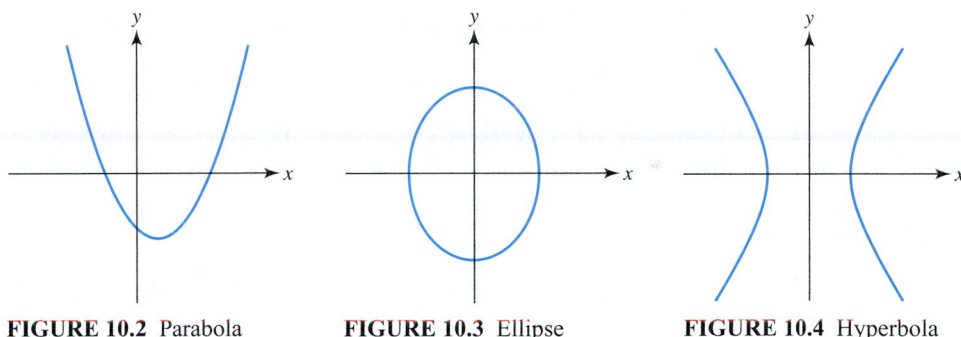

FIGURE 10.2 Parabola   **FIGURE 10.3** Ellipse   **FIGURE 10.4** Hyperbola

In this section we discuss parabolas.

Equations and Graphs of Parabolas

In Chapter 3 we learned that a parabola with vertex $(0, 0)$ can be represented symbolically by the equation $y = ax^2$. With this representation, a parabola can open either upward when $a > 0$ or downward when $a < 0$. The following definition of a parabola allows it to open in any direction.

PARABOLA

A **parabola** is the set of points in a plane that are equidistant from a fixed point and a fixed line. The fixed point is called the **focus** and the fixed line is called the **directrix** of the parabola.

Figures 10.5 and 10.6 show two parabolas with vertex V and focus F. The first parabola has a vertical axis (of symmetry) and directrix $y = -p$, and the second parabola has a horizontal axis (of symmetry) and directrix $x = -p$. For any point P located at (x, y) on the parabola, the distance d_1 from F to P is equal to the perpendicular distance d_2 from P to the directrix. By the vertical line test, the parabola in Figure 10.6 cannot be represented by a function.

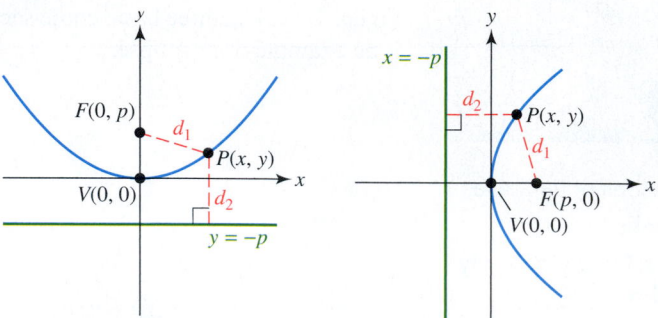

FIGURE 10.5 Vertical Axis   **FIGURE 10.6** Horizontal Axis

◆ MAKING CONNECTIONS

Functions and Points   In Figures 10.5 and 10.6, the point P is labeled $P(x, y)$. This resembles function notation involving two inputs, since the point P is determined by x and y.

We can derive an equation of the parabola shown in Figure 10.5. Since $d_1 = d_2$, the distance formula can be used to express the variables x, y, and p in an equation.

$$d_1 = d_2$$

$\sqrt{(x - 0)^2 + (y - p)^2} = \sqrt{(x - x)^2 + (y - (-p))^2}$	Distance formula
$x^2 + (y - p)^2 = 0^2 + (y + p)^2$	Square each side.
$x^2 + y^2 - 2py + p^2 = y^2 + 2py + p^2$	Expand binomials.
$x^2 - 2py = 2py$	Subtract y^2 and p^2.
$x^2 = 4py$	Add $2py$.

Algebra Review
To review squaring binomials, see Chapter R (page R-25).

If the value of p is known, then the equation of a parabola with vertex $(0, 0)$ can be found using one of the following equations.

EQUATION OF A PARABOLA WITH VERTEX (0, 0)

Vertical Axis

The parabola with a focus at $(0, p)$ and directrix $y = -p$ has equation

$$x^2 = 4py.$$

The parabola opens upward if $p > 0$ and downward if $p < 0$.

Horizontal Axis

The parabola with a focus at $(p, 0)$ and directrix $x = -p$ has equation

$$y^2 = 4px.$$

The parabola opens to the right if $p > 0$ and to the left if $p < 0$.

EXAMPLE 1 Sketching graphs of parabolas

Sketch a graph of each parabola. Label the vertex, focus, and directrix.
(a) $x^2 = 8y$ **(b)** $y^2 = -2x$

SOLUTION
(a) The equation $x^2 = 8y$ is in the form $x^2 = 4py$, where $8 = 4p$. Therefore the parabola has a vertical axis with $p = 2$. Since $p > 0$, the parabola opens upward. The focus is located at $(0, p)$ or $(0, 2)$, and the directrix is $y = -p$ or $y = -2$. The graph of the parabola is shown in Figure 10.7.

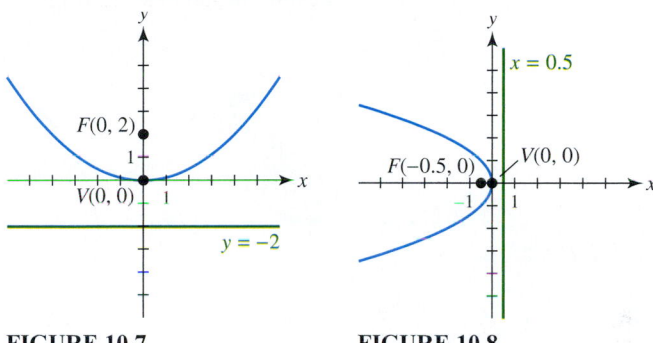

FIGURE 10.7   **FIGURE 10.8**

(b) The equation $y^2 = -2x$ has the form $y^2 = 4px$, where $-2 = 4p$. Therefore the parabola has a horizontal axis with $p = -0.5$. Since $p < 0$, the parabola opens to the left. The focus is located at $(-0.5, 0)$, and the directrix is $x = 0.5$. See Figure 10.8.

Now Try Exercises 15 and 17 ◆

In the next example we find the equation of a parabola, given its focus and directrix.

EXAMPLE 2 Finding the equation of a parabola

Find the equation of the parabola with focus $(-1.5, 0)$ and directrix $x = 1.5$, as shown in Figure 10.9. Sketch a graph of the parabola.

SOLUTION A parabola always opens toward the focus and away from the directrix. From Figure 10.9 we see that the parabola should open to the left. It follows that $p < 0$ in the equation $y^2 = 4px$. The distance between the focus at $(-1.5, 0)$ and the vertex at $(0, 0)$ is 1.5, and so $p = -1.5 < 0$. (Note that the vertex of the parabola is $(0, 0)$ because the vertex always lies midway between the focus and directrix.) The equation of the parabola is $y^2 = 4(-1.5)x$ or $y^2 = -6x$, and a graph of the parabola is shown in Figure 10.10.

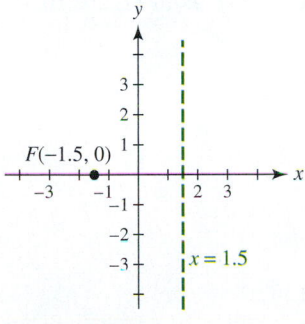

FIGURE 10.9

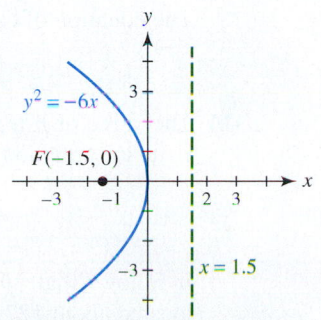

FIGURE 10.10

Now Try Exercise 29 ◆

Reflective Property of Parabolas

When a parabola is rotated about its axis, it sweeps out a shape called a **paraboloid**, as shown in Figure 10.11. Paraboloids have a special reflective property. When incoming, parallel rays of light from the sun or distant stars strike the surface of a paraboloid, each ray is reflected toward the focus. See Figure 10.12. If the rays are sunlight, intense heat is produced, which can be used to generate solar heat. Radio signals from distant space also concentrate at the focus. Scientists can analyze these signals by placing a receiver at the focus. This property of a paraboloid can also be used in reverse. If a light source is placed at the focus, then the light is reflected straight ahead, as shown in Figure 10.13. Searchlights, flashlights, and car headlights make use of this property.

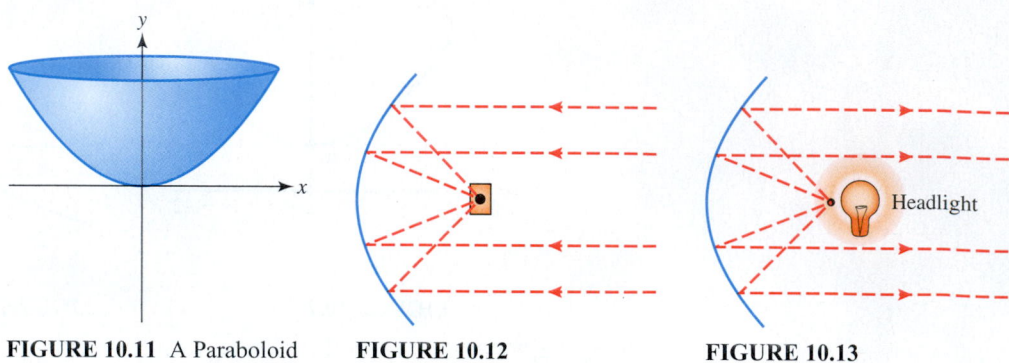

FIGURE 10.11 A Paraboloid   **FIGURE 10.12**   **FIGURE 10.13**

The next example illustrates this reflective property for radio telescopes.

EXAMPLE 3   Locating the receiver for a radio telescope

The U.S. Naval Research Laboratory designed a giant radio telescope weighing 3450 tons. Its parabolic dish has a diameter of 300 feet and a depth of 44 feet. See Figure 10.14. (**Source:** J. Mar, *Structure Technology for Large Radio and Radar Telescope Systems.*)
(a) Find an equation in the form $y = ax^2$ that describes a cross section of this dish.
(b) If the receiver is located at the focus, how far should it be from the vertex?

FIGURE 10.14

SOLUTION
(a) Locate a parabola that passes through $(-150, 44)$ and $(\mathbf{150}, \mathbf{44})$, as in Figure 10.15. Substitute either point into $y = ax^2$.

$$y = ax^2$$
$$\mathbf{44} = a(\mathbf{150})^2$$
$$a = \frac{44}{150^2} = \frac{11}{5625}$$

The equation of the parabola is

$$y = \frac{11}{5625}x^2, \quad \text{where } -150 \le x \le 150.$$

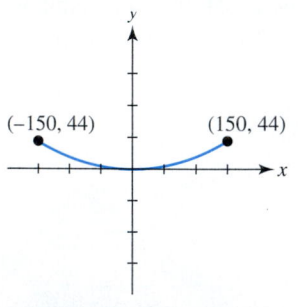

FIGURE 10.15

(b) The value of p represents the distance from the vertex to the focus. To determine p, write the equation in the form $x^2 = 4py$. Then,

$$y = \frac{11}{5625}x^2 \quad \text{is equivalent to} \quad x^2 = \frac{5625}{11}y.$$

It follows that $4p = \frac{5625}{11}$ or $p = \frac{5625}{44} \approx 127.84$. Therefore the receiver should be located about 127.84 feet from the vertex.

Now Try Exercise 85 ◆

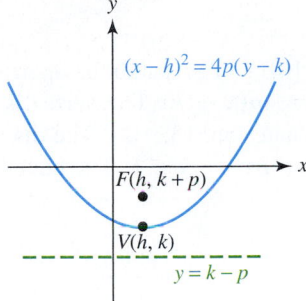

FIGURE 10.16 Vertex (h, k); Vertical Axis

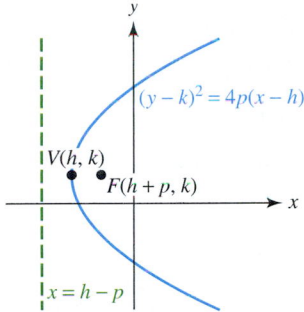

FIGURE 10.17 Vertex (h, k); Horizontal Axis

Translations of Parabolas

If the equation of a parabola is given by either $x^2 = 4py$ or $y^2 = 4px$, then its vertex is $(0, 0)$. We can use translations of graphs to find the equation of a parabola with vertex (h, k) rather than $(0, 0)$. This translation can be obtained by replacing x with $(x - h)$ and y with $(y - k)$.

$$(x - h)^2 = 4p(y - k) \qquad \text{Vertex } (h, k); \text{ vertical axis}$$

$$(y - k)^2 = 4p(x - h) \qquad \text{Vertex } (h, k); \text{ horizontal axis}$$

These two parabolas with $p > 0$ are shown in Figures 10.16 and 10.17, respectively. These results are summarized as follows.

EQUATION OF A PARABOLA WITH VERTEX (h, k)

$(x - h)^2 = 4p(y - k)$ **Vertical axis; vertex: (h, k)**
$p > 0$: opens upward; $p < 0$: opens downward
Focus: $(h, k + p)$; directrix: $y = k - p$
$(y - k)^2 = 4p(x - h)$ **Horizontal axis; vertex: (h, k)**
$p > 0$: opens to the right; $p < 0$: opens to the left
Focus: $(h + p, k)$; directrix: $x = h - p$

In the next example we graph a parabola whose vertex is not $(0, 0)$.

EXAMPLE 4

Graphing a parabola with vertex (h, k)

Graph the parabola given by the equation $x = -\frac{1}{8}(y + 3)^2 + 2$. Label the vertex, focus, and directrix.

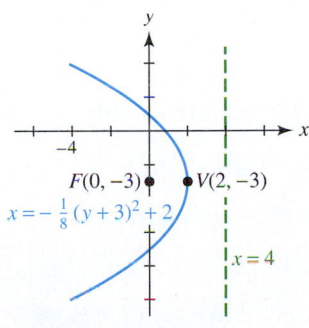

FIGURE 10.18

SOLUTION Rewrite the equation in the form $(y - k)^2 = 4p(x - h)$.

$$x = -\frac{1}{8}(y + 3)^2 + 2 \qquad \textit{Given equation}$$

$$x - 2 = -\frac{1}{8}(y + 3)^2 \qquad \textit{Subtract 2.}$$

$$-8(x - 2) = (y + 3)^2 \qquad \textit{Multiply by } -8.$$

$$(y + 3)^2 = -8(x - 2) \qquad \textit{Rewrite equation.}$$

It follows that the vertex is $(2, -3)$, $4p = -8$ or $p = -2$, and the parabola opens to the left. The focus is located 2 units to the left of the vertex, and the directrix is located 2 units to the right of the vertex. Therefore the focus is $(0, -3)$, and the directrix is $x = 4$. See Figure 10.18.

Now Try Exercise 55 ◆

EXAMPLE 5 Finding the equation of a parabola with vertex (h, k)

Find the equation of the parabola with focus $(3, -4)$ and directrix $y = 2$. Sketch a graph of the parabola. Label the focus, directrix, and vertex.

SOLUTION The focus and directrix are shown in Figure 10.19. The parabola opens downward $(p < 0)$, and its equation has the form $(x - h)^2 = 4p(y - k)$. The vertex is located midway between the focus and directrix, so its coordinates are $(3, -1)$. The distance between the focus $(3, -4)$ and the vertex $(3, -1)$ is 3, so $p = -3 < 0$. The equation of the parabola is

$$(x - 3)^2 = -12(y + 1),$$

and its graph is shown in Figure 10.20.

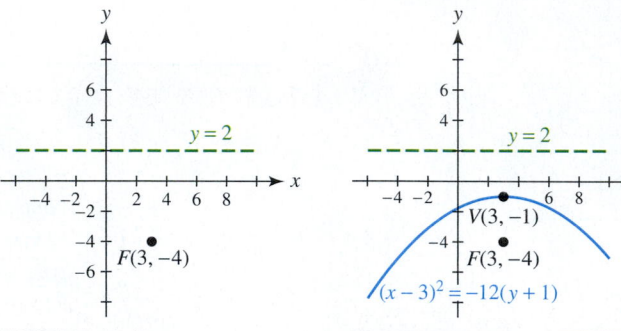

FIGURE 10.19   **FIGURE 10.20**   *Now Try Exercise 61* ◆

EXAMPLE 6 Finding the equation of a parabola

Write $2x = y^2 + 4y + 12$ in the form $(y - k)^2 = a(x - h)$.

Algebra Review
To review completing the square, see Section 3.1. To review perfect square trinomials, see Chapter R (page R-34).

SOLUTION We can write the given equation in the required form by completing the square.

$$2x = y^2 + 4y + 12 \qquad \text{Given equation}$$
$$2x - 12 = y^2 + 4y \qquad \text{Subtract 12 from each side.}$$

To complete the square on the right side of the equation, add $\left(\frac{4}{2}\right)^2 = 4$ to each side.

$$2x - 12 + 4 = y^2 + 4y + 4 \qquad \text{Add 4 to each side.}$$
$$2x - 8 = (y + 2)^2 \qquad \text{Perfect square trinomial}$$
$$2(x - 4) = (y + 2)^2 \qquad \text{Distributive property}$$

The given equation is equivalent to $(y + 2)^2 = 2(x - 4)$.   *Now Try Exercise 67* ◆

◆ **CLASS DISCUSSION**
Sketch a graph of the parabola in Example 6. Identify the vertex, focus, and directrix. ◆

Graphing calculators can graph parabolas with horizontal axes, as illustrated in the next example.

EXAMPLE 7   Graphing a parabola with technology

Graph the equation $(y - 1)^2 = -0.5(x - 2)$ with a graphing calculator.

[−4.7, 4.7, 1] by [−3.1, 3.1, 1]

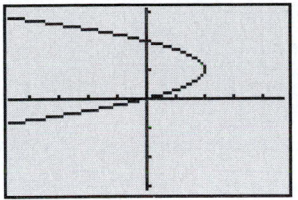

FIGURE 10.21

SOLUTION   Begin by solving the equation for y.

$$(y - 1)^2 = -0.5(x - 2) \qquad \text{Given equation}$$
$$y - 1 = \pm \sqrt{-0.5(x - 2)} \qquad \text{Square root property}$$
$$y = 1 \pm \sqrt{-0.5(x - 2)} \qquad \text{Add 1.}$$

Let $Y_1 = 1 + \sqrt{(-0.5(X - 2))}$ and $Y_2 = 1 - \sqrt{(-0.5(X - 2))}$. The graph of y_1 creates the upper portion of the parabola, and the graph of y_2 creates the lower portion of the parabola, as shown in Figure 10.21.

Now Try Exercise 77 ◆

10.1

Putting it all Together

$\mathbf{T}$he following table summarizes some important concepts about parabolas.

Concept	Equation	Example
Parabola with vertex $(0, 0)$ and vertical axis	$x^2 = 4py$ $p > 0$: opens upward $p < 0$: opens downward Focus: $(0, p)$ Directrix: $y = -p$	$x^2 = -2y$ has $4p = -2$ or $p = -\frac{1}{2}$. The parabola opens downward with vertex $(0, 0)$, focus $\left(0, -\frac{1}{2}\right)$, and directrix $y = \frac{1}{2}$.
Parabola with vertex $(0, 0)$ and horizontal axis	$y^2 = 4px$ $p > 0$: opens to the right $p < 0$: opens to the left Focus: $(p, 0)$ Directrix: $x = -p$	$y^2 = 4x$ has $4p = 4$ or $p = 1$. The parabola opens to the right with vertex $(0, 0)$, focus $(1, 0)$, and directrix $x = -1$.

Concept	Equation	Example
Parabola with vertex (h, k) and vertical axis	$(x - h)^2 = 4p(y - k)$ $p > 0$: opens upward $p < 0$: opens downward Focus: $(h, k + p)$ Directrix: $y = k - p$	$(x - 1)^2 = 8(y - 3)$ has $p = 2$. The parabola opens upward with vertex $(1, 3)$, focus $(1, 5)$, and directrix $y = 1$ 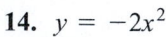
Parabola with vertex (h, k) and horizontal axis	$(y - k)^2 = 4p(x - h)$ $p > 0$: opens to the right $p < 0$: opens to the left Focus: $(h + p, k)$ Directrix: $x = h - p$	$(y + 1)^2 = -2(x + 2)$ has $p = -\frac{1}{2}$. The parabola opens to the left with vertex $(-2, -1)$, focus $\left(-\frac{5}{2}, -1\right)$, and directrix $x = -\frac{3}{2}$ 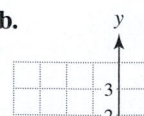

 Exercises

Parabolas with Vertex (0, 0)

Exercises 1–8: Sketch a graph of the parabola.

1. $x^2 = y$ **2.** $x^2 = -y$

3. $y^2 = -x$ **4.** $y^2 = x$

5. $y^2 = -4x$ **6.** $x^2 = 4y$

7. $y^2 = -\frac{1}{2}x$ **8.** $8x = y^2$

Exercises 9–14: Match the equation of the parabola with its graph (a.–f.).

9. $x^2 = 2y$ **10.** $x^2 = -2y$

11. $y^2 = -8x$ **12.** $y^2 = 4x$

13. $x = \frac{1}{2}y^2$ **14.** $y = -2x^2$

a.

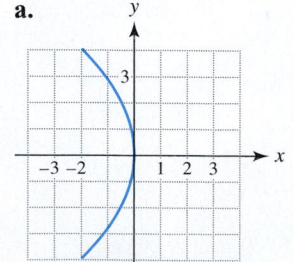

b.

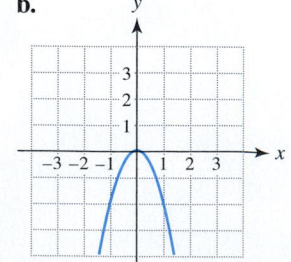

c.

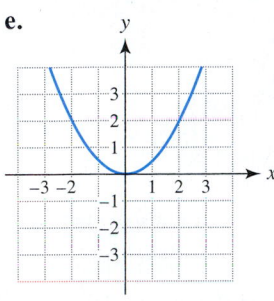

d.

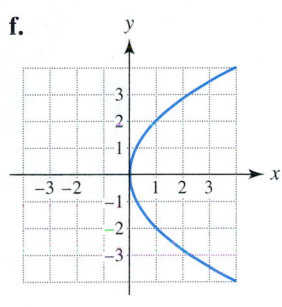

e.

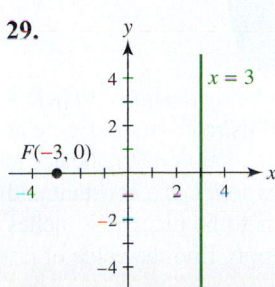

f.

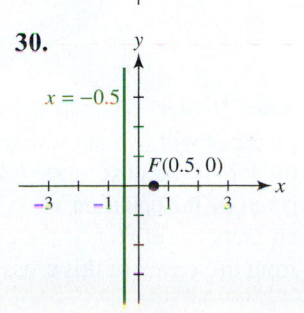

Exercises 15–26: Graph the parabola. Label the vertex, focus, and directrix.

15. $16y = x^2$

16. $y = -2x^2$

17. $x = \frac{1}{8}y^2$

18. $-y^2 = 6x$

19. $-4x = y^2$

20. $\frac{1}{2}y^2 = 3x$

21. $x^2 = -8y$

22. $-\frac{1}{8}x^2 = y$

23. $2y^2 = 6x$

24. $-\frac{1}{2}x^2 = 3y$

25. $2y^2 = -8x$

26. $-3x = \frac{1}{4}y^2$

Exercises 27–30: Sketch a graph of a parabola with focus and directrix as shown in the figure. Find an equation of the parabola.

27.

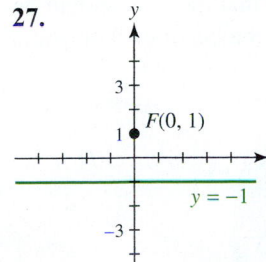

28.

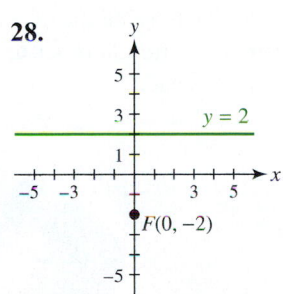

29.

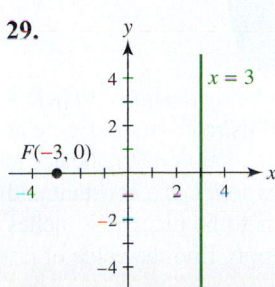

30.

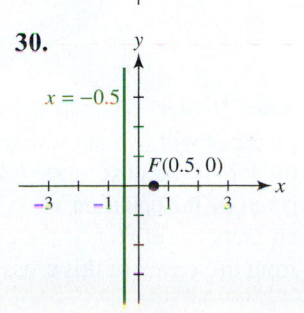

Exercises 31–40: Find an equation of the parabola with vertex $(0, 0)$ that satisfies the given conditions. Sketch its graph.

31. Focus $\left(0, \frac{3}{4}\right)$

32. Directrix $y = 2$

33. Directrix $x = 2$

34. Focus $(-1, 0)$

35. Focus $(1, 0)$

36. Focus $\left(0, -\frac{1}{2}\right)$

37. Directrix $x = \frac{1}{4}$

38. Directrix $y = -1$

39. Horizontal axis, passing through $(1, -2)$

40. Vertical axis, passing through $(-2, 3)$

Exercises 41–44: Find an equation of a parabola that satisfies the given conditions.

41. Focus $(0, -3)$ and directrix $y = 3$

42. Focus $(0, 2)$ and directrix $y = -2$

43. Focus $(-1, 0)$ and directrix $x = 1$

44. Focus $(3, 0)$ and directrix $x = -3$

Parabolas with Vertex (*h*, *k*)

Exercises 45–48: Sketch a graph of the parabola.

45. $(x - 1)^2 = (y - 2)$ **46.** $(x - 2)^2 = -(y + 1)$

47. $(y - 1)^2 = -(x + 1)$ **48.** $(y + 2)^2 = 2x$

Exercises 49–52: Match the equation of the parabola with its graph (a.–d.).

49. $(x - 1)^2 = 4(y - 1)$ **50.** $(x + 1)^2 = -4(y - 2)$

51. $(y - 2)^2 = -8x$ **52.** $(y + 1)^2 = 8(x + 3)$

a.

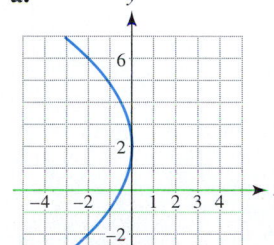

b.

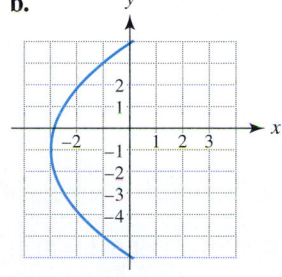

c.

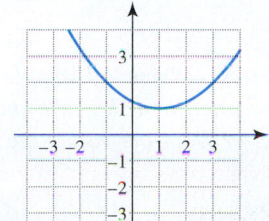

d.

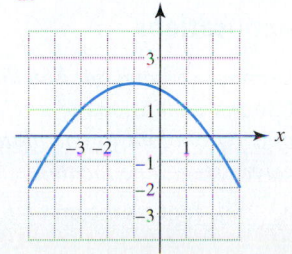

Exercises 53–58: Graph the parabola. Label the vertex, focus, and directrix.

53. $(x - 2)^2 = 8(y + 2)$ **54.** $\frac{1}{16}(x + 4)^2 = -(y - 4)$

55. $x = -\frac{1}{4}(y + 3)^2 + 2$ **56.** $x = 2(y - 2)^2 - 1$

57. $y = -\frac{1}{4}(x + 2)^2$ **58.** $-2(y + 1) = (x + 3)^2$

Exercises 59–66: Find an equation of a parabola that satisfies the given conditions.

59. Focus $(0, 2)$ and vertex $(0, 1)$

60. Focus $(-1, 2)$ and vertex $(3, 2)$

61. Focus $(0, 0)$ and directrix $x = -2$

62. Focus $(2, 1)$ and directrix $x = -1$

63. Focus $(-1, 3)$ and directrix $y = 7$

64. Focus $(1, 2)$ and directrix $y = 4$

65. Horizontal axis, vertex $(-2, 3)$, passing through $(-4, 0)$

66. Horizontal axis, vertex $(-1, 2)$, passing through $(2, 3)$

Exercises 67–74: (Refer to Example 6.) Write the equation either in the form $(y - k)^2 = a(x - h)$ or in the form $(x - h)^2 = a(y - k)$.

67. $-2x = y^2 + 6x + 10$ **68.** $y^2 + 8x - 8 = 4x$

69. $x = 2y^2 + 4y - 1$ **70.** $x = 3y^2 - 6y - 2$

71. $x^2 - 3x + 4 = 2y$ **72.** $-3y = -x^2 + 4x - 6$

73. $4y^2 + 4y - 5 = 5x$ **74.** $-2y^2 + 5y + 1 = -x$

Graphing Parabolas with Technology

Exercises 75–82: Graph the parabola.

75. $0.5y^2 = x$ **76.** $y^2 = -1.2x$

77. $(y + 0.75)^2 = -3x$ **78.** $(y - 3)^2 = \frac{1}{7}x$

79. $(y - 0.5)^2 = 3.1(x + 1.3)$

80. $1.4(y - 1.5)^2 = 0.5(x + 2.1)$

81. $x = 2.3(y + 1)^2$ **82.** $(y - 2.5)^2 = 4.1(x + 1)$

Applications

*Exercises 83 and 84: **Satellite Dishes** (Refer to Example 3.) Use the dimensions of a television satellite dish in the shape of a paraboloid to calculate how far from the vertex the receiver should be located.*

83. Six-foot diameter, nine inches deep

84. Nine-inch radius, two inches deep

85. *Radio Telescope* (Refer to Example 3.) The Parkes radio telescope has the shape of a parabolic dish with a diameter of 210 feet and a depth of 32 feet. The dish is shown in the figure.   (**Source:** J. Mar.)

(a) Determine an equation of the form $y = ax^2$ describing a cross section of the dish.

(b) The receiver is placed at the focus. How far from the vertex is the receiver located?

86. *Comets* A comet sometimes travels along a parabolic path as it passes the sun. In this situation the sun is located at the focus of the parabola and the comet passes the sun once and does not orbit the sun. Suppose the path of a comet is given by $y^2 = 100x$, where units are in millions of miles.

(a) Find the coordinates of the sun.

(b) Find the minimum distance between the sun and the comet.

87. *Headlight* A headlight is being constructed in the shape of a paraboloid with a depth of 4 inches and a diameter of 5 inches, as illustrated in the accompanying figure. Determine the distance d that the bulb should be from the vertex in order to have the beam of light shine straight ahead.

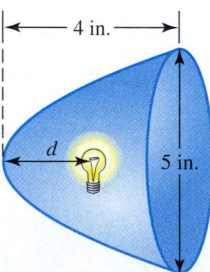

88. *Solar Heater* A solar heater is being designed to heat a pipe that will contain water, as illustrated in the figure at the top of the next page. A cross section of the heater is given by the equation $x^2 = ky$, where k is a constant and all units are in feet. If the pipe is to be placed 18 inches from the vertex of this cross section, find the value of k.

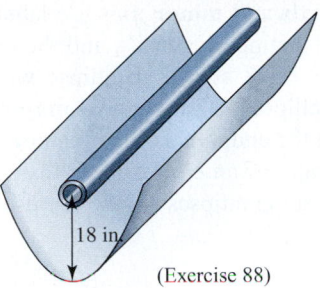

18 in.

(Exercise 88)

Writing about Mathematics

89. Explain how the distance between the focus and the vertex of a parabola affects the shape of the parabola.

90. Explain how to determine the direction that a parabola opens, given the focus and the directrix.

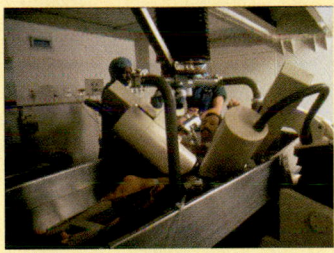

10.2 Ellipses

- ◆ Find equations of ellipses
- ◆ Graph ellipses
- ◆ Learn the reflective property of ellipses
- ◆ Translate ellipses
- ◆ Find the center and radius of a circle
- ◆ Solve systems of nonlinear equations and inequalities

Introduction

When planets travel around the sun, they usually do not travel in circular orbits. Instead they travel in elliptical orbits, which look more like ovals than circles. This discovery by Johannes Kepler made it possible for astronomers to determine the precise position of all types of celestial objects, such as asteroids, comets, and moons. The predictions of both solar and lunar eclipses are made easier because of Kepler's discovery. Ellipses are also used in construction and medicine. In this section we learn about the basic properties of ellipses.   (**Source:** *Historical Topics for the Mathematics Classroom, Thirty-first Yearbook*, NCTM.)

Equations and Graphs of Ellipses

One method for sketching an ellipse is to tie a string to two nails driven into a flat board. If a pencil is placed inside the loop formed by the string, the resulting curve shown in Figure 10.22 is an ellipse. The sum of the distances d_1 and d_2 between the pencil and each of the nails is always fixed by the string. The locations of the nails correspond to the *foci* of the ellipse. If the two nails coincide, the ellipse becomes a circle. As the nails spread farther apart, the ellipse becomes more elongated, or eccentric.

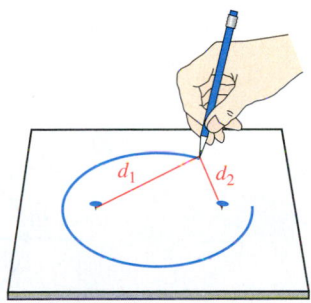

d_1 d_2

FIGURE 10.22

This method of sketching an ellipse suggests the following definition.

ELLIPSE

An **ellipse** is the set of points in a plane, the sum of whose distances from two fixed points is constant. Each fixed point is called a **focus** (plural **foci**) of the ellipse.

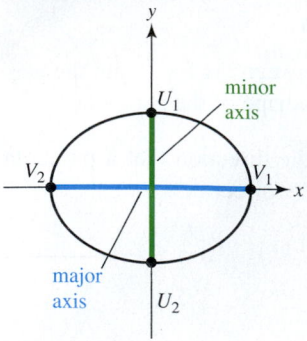

FIGURE 10.23

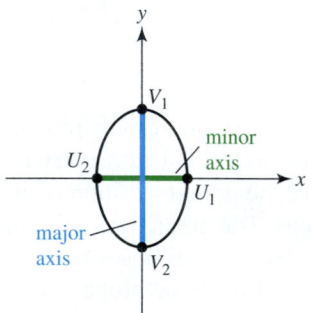

FIGURE 10.24

In Figures 10.23 and 10.24 the **major axis** and **minor axis** are labeled for each ellipse. The major axis is the line segment connecting V_1 and V_2, and the minor axis is the line segment connecting U_1 and U_2. Figure 10.23 shows an ellipse with a **horizontal major axis**, and Figure 10.24 illustrates an ellipse with a **vertical major axis**. The **vertices**, V_1 and V_2, of each ellipse are located at the endpoints of the major axis.

Since a vertical line can intersect the graph of an ellipse more than once, an ellipse cannot be described by a function. However, some ellipses can be represented by the following equations.

STANDARD EQUATIONS FOR ELLIPSES CENTERED AT (0, 0)

The ellipse with center at the origin, *horizontal* major axis, and equation

$$\frac{x^2}{a^2} + \frac{y^2}{b^2} = 1 \qquad (a > b > 0)$$

has vertices $(\pm a, 0)$, endpoints of the minor axis $(0, \pm b)$, and foci $(\pm c, 0)$, where $c^2 = a^2 - b^2$ and $c \geq 0$.

The ellipse with center at the origin, *vertical* major axis, and equation

$$\frac{x^2}{b^2} + \frac{y^2}{a^2} = 1 \qquad (a > b > 0)$$

has vertices $(0, \pm a)$, endpoints of the minor axis $(\pm b, 0)$, and foci $(0, \pm c)$, where $c^2 = a^2 - b^2$ and $c \geq 0$.

Note: If $a = b$, then the ellipse is a circle with radius $r = a$ and center $(0, 0)$.

Figures 10.25 and 10.26 show two ellipses. The first has a horizontal major axis and the second has a vertical major axis. The coordinates of the vertices V_1 and V_2, foci F_1 and F_2, and endpoints of the minor axis U_1 and U_2 are labeled. In each figure $a > b > 0$.

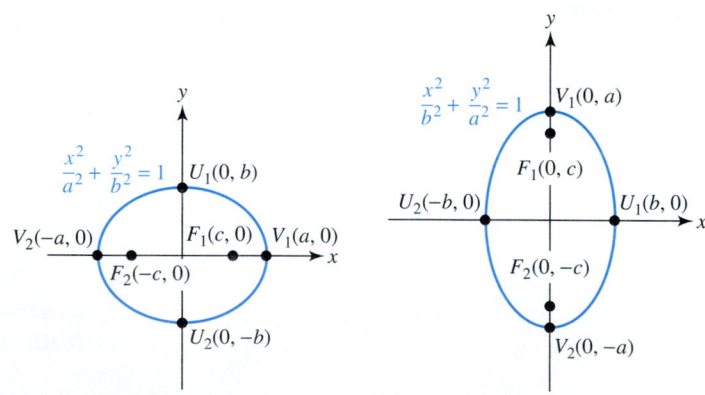

FIGURE 10.25   **FIGURE 10.26**

EXAMPLE 1   Sketching graphs of ellipses

Sketch a graph of each ellipse. Label the vertices, foci, and endpoints of the minor axes.

(a) $\dfrac{x^2}{9} + \dfrac{y^2}{4} = 1$ **(b)** $25x^2 + 16y^2 = 400$

SOLUTION

(a) The equation $\frac{x^2}{9} + \frac{y^2}{4} = 1$ describes an ellipse with $a = 3$ and $b = 2$. The ellipse has a horizontal major axis with vertices $(\pm 3, 0)$. The endpoints of the minor axis are $(0, \pm 2)$. To locate the foci, find c.

$$c^2 = a^2 - b^2 = 9 - 4 = 5 \quad \text{or}$$
$$c = \sqrt{5} \approx 2.24.$$

The foci are located on the major axis with coordinates $(\pm \sqrt{5}, 0)$. See Figure 10.27.

(b) The equation $25x^2 + 16y^2 = 400$ can be written in standard form by dividing each side by 400.

$$25x^2 + 16y^2 = 400 \qquad \textit{Given equation}$$

$$\frac{25x^2}{400} + \frac{16y^2}{400} = \frac{400}{400} \qquad \textit{Divide by 400.}$$

$$\frac{x^2}{16} + \frac{y^2}{25} = 1 \qquad \textit{Reduce each fraction.}$$

This ellipse has a vertical major axis with $a = 5$ and $b = 4$. The value of c is given by

$$c^2 = 5^2 - 4^2 = 9 \qquad \text{or} \qquad c = 3.$$

The ellipse has foci $(0, \pm 3)$, vertices $(0, \pm 5)$, and endpoints of the minor axis located at $(\pm 4, 0)$. See Figure 10.28.

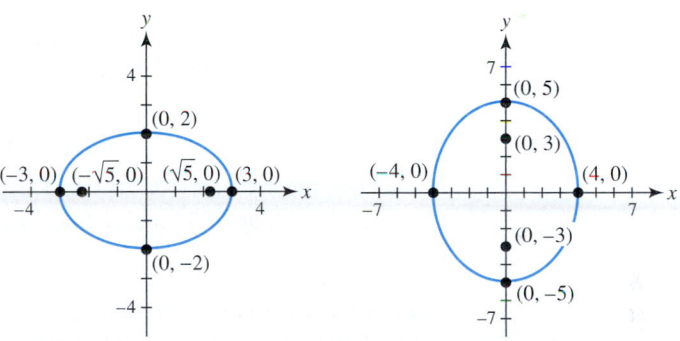

FIGURE 10.27   **FIGURE 10.28**   *Now Try Exercises 1 and 3* ◆

In the next two examples we find standard equations for ellipses.

EXAMPLE 2 Finding the equation of an ellipse

Find the standard equation of the ellipse shown in Figure 10.29. Identify the coordinates of the vertices and the foci.

SOLUTION The ellipse is centered at $(0, 0)$ and has a horizontal major axis. Its standard equation has the form

$$\frac{x^2}{a^2} + \frac{y^2}{b^2} = 1.$$

The endpoints of the major axis are $(\pm 4, 0)$, and the endpoints of the minor axis are $(0, \pm 2)$. It follows that $a = 4$ and $b = 2$, and the standard equation is

$$\frac{x^2}{16} + \frac{y^2}{4} = 1.$$

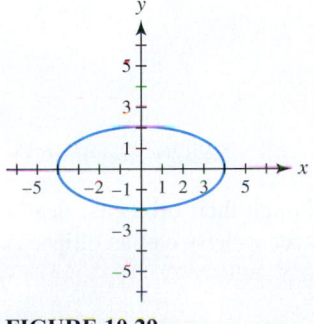

FIGURE 10.29

The foci lie on the horizontal major axis and can be determined as follows.

$$c^2 = a^2 - b^2 = 16 - 4 = 12$$

Thus $c = \sqrt{12} \approx 3.46$, and the coordinates of the foci are $(\pm\sqrt{12}, 0)$. A graph of the ellipse with the vertices and foci plotted is shown in Figure 10.30.

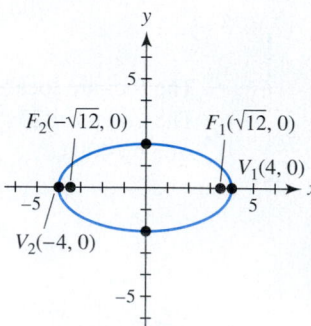

FIGURE 10.30

Now Try Exercises 9 and 11

EXAMPLE 3 Finding the equation of an ellipse

Find the standard equation of the ellipse with foci $(0, \pm 1)$ and vertices $(0, \pm 3)$. Sketch its graph.

SOLUTION Since the foci and vertices lie on the y-axis, the ellipse has a vertical major axis. Its standard equation has the form

$$\frac{x^2}{b^2} + \frac{y^2}{a^2} = 1.$$

Because the foci are $(0, \pm 1)$ and the vertices are $(0, \pm 3)$, it follows that $c = 1$ and $a = 3$. The value of b^2 can be found by using the equation $c^2 = a^2 - b^2$.

$$b^2 = a^2 - c^2 = 9 - 1 = 8$$

Thus the equation of the ellipse is $\frac{x^2}{8} + \frac{y^2}{9} = 1$. Its graph is shown in Figure 10.31. To graph the ellipse, it is helpful to note that the endpoints of the minor axis are $(\pm b, 0)$ or $(\pm\sqrt{8}, 0)$ where $\sqrt{8} \approx 2.83$.

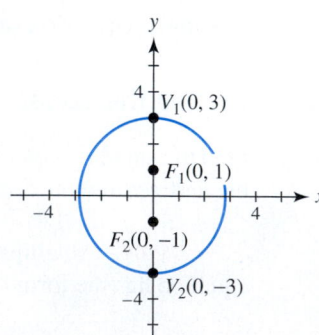

FIGURE 10.31

Now Try Exercise 17

The planets travel around the sun in elliptical orbits. Although their orbits are nearly circular, many planets have a slight eccentricity to them. The **eccentricity** e of an ellipse is defined by

$$e = \frac{\sqrt{a^2 - b^2}}{a} = \frac{c}{a}.$$

Since the foci of an ellipse lie inside the ellipse, $0 \le c < a$ and $0 \le \frac{c}{a} < 1$. Therefore the eccentricity e of an ellipse satisfies $0 \le e < 1$. If $e = 0$, then $a = b$ and the ellipse is a circle. See Figure 10.32. As e increases, the foci spread apart and the ellipse becomes more elongated. See Figures 10.33 and 10.34.

◆ **MAKING CONNECTIONS**

The Number e and the Variable e   Do not confuse the *variable e*, which is used to denote the eccentricity of an ellipse, with the irrational *number* $e \approx 2.72$, which is the base of the natural exponential function $f(x) = e^x$ and of the natural logarithmic function $g(x) = \ln x$. ◆

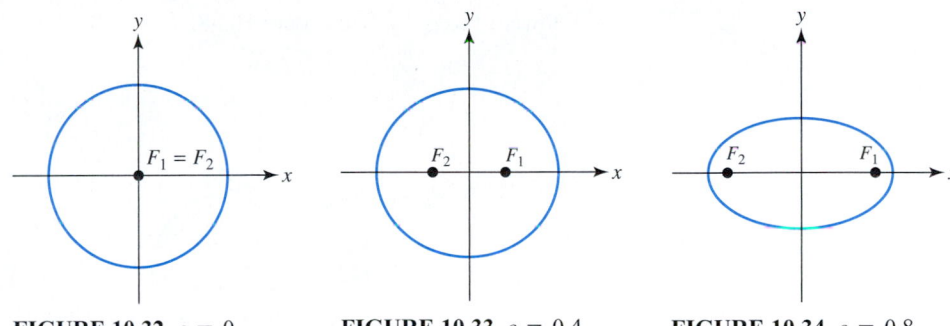

FIGURE 10.32 $e = 0$ FIGURE 10.33 $e = 0.4$ FIGURE 10.34 $e = 0.8$

Astronomers have measured values of a and e for each planet. With this information and the fact that the sun is located at one focus of the ellipse, the equation of a planet's orbit can be found.

EXAMPLE 4   **Finding the orbital equation of the planet Pluto**

The planet Pluto has $a = 39.44$ and $e = 0.249$, the greatest eccentricity of any planet. (For Earth, $a = 1$.) Graph the orbit of Pluto and the position of the sun in $[-60, 60, 10]$ by $[-40, 40, 10]$.   (**Source:** M. Zeilik, *Introductory Astronomy and Astrophysics*.)

Calculator Help

To learn how to access the variable Y_1, see Appendix B (page AP-16).

SOLUTION   Let the orbit of Pluto be given by $\frac{x^2}{a^2} + \frac{y^2}{b^2} = 1$. Then,

$$e = \frac{c}{a} = 0.249 \qquad \text{implies} \qquad c = 0.249a = 0.249(39.44) \approx 9.821.$$

To find b, solve the equation $c^2 = a^2 - b^2$ for b.

$$b = \sqrt{a^2 - c^2}$$
$$= \sqrt{39.44^2 - 9.821^2} \approx 38.20$$

Pluto's orbit is modeled by $\frac{x^2}{39.44^2} + \frac{y^2}{38.20^2} = 1$. Since $c \approx 9.821$, the foci are $(\pm 9.821, 0)$. The sun could be located at either focus. We locate the sun at $(9.821, 0)$.

To graph this ellipse on a graphing calculator, the equation must be solved for y. This results in two equations to graph. For example,

$$\frac{x^2}{39.44^2} + \frac{y^2}{38.20^2} = 1$$

$$\frac{y^2}{38.20^2} = 1 - \frac{x^2}{39.44^2}$$

$$\frac{y}{38.20} = \pm\sqrt{1 - \frac{x^2}{39.44^2}}$$

$$y = \pm 38.20\sqrt{1 - \frac{x^2}{39.44^2}}.$$

FIGURE 10.35

$[-60, 60, 10]$ by $[-40, 40, 10]$

FIGURE 10.36

See Figures 10.35 and 10.36.

Now Try Exercise 91 ◆

Reflective Property of Ellipses

Like parabolas, ellipses also have an important reflective property. If an ellipse is rotated about the x-axis, an **ellipsoid** is formed, which resembles the shell of an egg, as illustrated in Figure 10.37. If a light source is placed at focus F_1, then every beam of light emanating from the light source, regardless of its direction, is reflected at the surface of the ellipsoid toward focus F_2. See Figure 10.38.

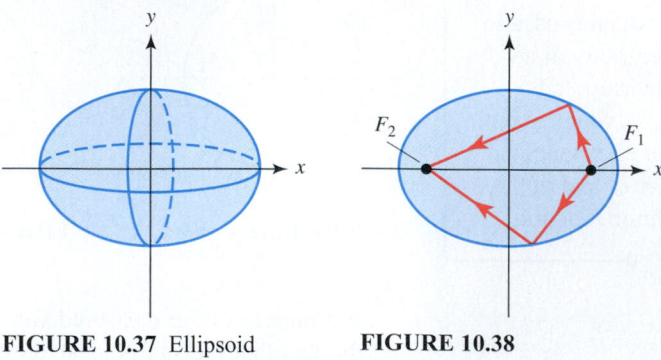

FIGURE 10.37 Ellipsoid   **FIGURE 10.38**

A *lithotripter* is a machine designed to break up kidney stones, without surgery, by producing powerful shock waves. To focus these shock waves accurately, a lithotripter uses the reflective property of ellipses. A patient is carefully positioned so that the kidney stone is at one focus, while the source of the shock waves is located at the other focus. When shock waves are emitted, they are reflected directly toward the kidney stone. As a result, it absorbs all of the energy from the shock wave and breaks up without harming the patient. This nonsurgical procedure reduces both risk and recovery time. In Exercises 93 and 94 this reflective property of ellipses is applied.

Translations of Ellipses

If the equation of an ellipse is given by either $\frac{x^2}{a^2} + \frac{y^2}{b^2} = 1$ or $\frac{x^2}{b^2} + \frac{y^2}{a^2} = 1$, then the center of the ellipse is $(0, 0)$. We can use translations of graphs to find the equation of an ellipse centered at (h, k) by replacing x with $(x - h)$ and y with $(y - k)$.

STANDARD EQUATIONS FOR ELLIPSES CENTERED AT (h, k)

An ellipse with center (h, k), and either a horizontal or vertical major axis, satisfies one of the following equations, where $a > b > 0$ and $c^2 = a^2 - b^2$ with $c \geq 0$.

$$\frac{(x - h)^2}{a^2} + \frac{(y - k)^2}{b^2} = 1$$ Major axis: horizontal; foci: $(h \pm c, k)$
                                                      Vertices: $(h \pm a, k)$

$$\frac{(x - h)^2}{b^2} + \frac{(y - k)^2}{a^2} = 1$$ Major axis: vertical; foci: $(h, k \pm c)$
                                                      Vertices: $(h, k \pm a)$

EXAMPLE 5 Translating an ellipse

Translate the ellipse with equation $\frac{x^2}{9} + \frac{y^2}{4} = 1$ so that it is centered at $(-1, 2)$. Find this equation and sketch its graph.

Algebra Review

To review translations or shifts, see Section 3.4.

SOLUTION To translate the center from $(0, 0)$ to $(-1, 2)$, replace x with $(x + 1)$ and y with $(y - 2)$. This new equation is

$$\frac{(x + 1)^2}{9} + \frac{(y - 2)^2}{4} = 1.$$

This ellipse is congruent to the given ellipse, except that it is centered at $(-1, 2)$. The given ellipse is shown in Figure 10.39, and the translated ellipse is shown in Figure 10.40.

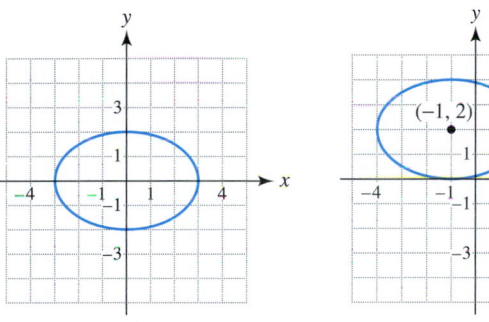

FIGURE 10.39   **FIGURE 10.40**

Now Try Exercises 31 and 33 ◆

EXAMPLE 6 Graphing an ellipse with center (h, k)

Graph the ellipse whose equation is $\frac{(x + 2)^2}{16} + \frac{(y - 2)^2}{25} = 1$. Label both the vertices and the foci.

SOLUTION The ellipse has a vertical major axis and its center is $(-2, 2)$. Since $a^2 = 25$ and $b^2 = 16$, it follows that $c^2 = a^2 - b^2 = 25 - 16 = 9$. Thus $a = 5, b = 4$, and $c = 3$. The vertices are located 5 units above and below the center of the ellipse and the foci are located 3 units above and below the center of the ellipse. That is, the vertices are $(-2, 2 \pm 5)$, or $(-2, 7)$ and $(-2, -3)$, and the foci are $(-2, 2 \pm 3)$, or $(-2, 5)$ and $(-2, -1)$. A graph of the ellipse is shown in Figure 10.41.

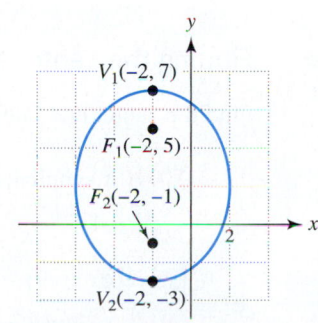

FIGURE 10.41   *Now Try Exercise 35* ◆

EXAMPLE 7 Finding the standard equation of an ellipse

Write $4x^2 - 16x + 9y^2 + 54y + 61 = 0$ in the standard form for an ellipse centered at (h, k). Identify the center and the vertices.

Algebra Review

To review completing the square, see Section 3.1. To review perfect square trinomials, see Chapter R (page R-34).

SOLUTION We can write the given equation in standard form by completing the square.

$$4x^2 - 16x + 9y^2 + 54y + 61 = 0 \qquad \text{Given equation}$$

$$4(x^2 - 4x + \underline{}) + 9(y^2 + 6y + \underline{}) = -61 \qquad \text{Distributive property}$$

$$4(x^2 - 4x + \underline{4}) + 9(y^2 + 6y + \underline{9}) = -61 + 16 + 81 \qquad \text{Complete the square.}$$

$$4(x - 2)^2 + 9(y + 3)^2 = 36 \qquad \text{Perfect square trinomials}$$

$$\frac{(x - 2)^2}{9} + \frac{(y + 3)^2}{4} = 1 \qquad \text{Divide each side by 36.}$$

 CLASS DISCUSSION

Sketch a graph of the ellipse in Example 7. Identify the foci. ◆

The center is $(2, -3)$. Because $a = 3$ and the major axis is horizontal, the vertices of the ellipse are $(2 \pm 3, -3)$ or $(5, -3)$ and $(-1, -3)$. *Now Try Exercise 45* ◆

Circles

Geometry Review

To review circles, see Chapter R (page R-2).

A circle is an ellipse where $a = b$. If a circle has radius r and center (h, k), then an equation for the circle is $\frac{(x - h)^2}{r^2} + \frac{(y - k)^2}{r^2} = 1$. Multiplying the equation by r^2 provides the following result.

> ### STANDARD EQUATION OF A CIRCLE
> The **standard equation of a circle** with center (h, k) and radius r is
> $$(x - h)^2 + (y - k)^2 = r^2.$$

EXAMPLE 8 Finding the standard equation of a circle

Find the standard equation of a circle with radius 4 and center $(5, -3)$.

SOLUTION Let $h = 5, k = -3$, and $r = 4$. The standard equation is

$$(x - 5)^2 + (y + 3)^2 = 16. \qquad \text{Now Try Exercise 55} ◆$$

EXAMPLE 9 Finding the center and radius of a circle

Find the center and radius of the circle given by $x^2 + 6x + y^2 - 2y = -6$.

SOLUTION Complete the square to write the standard equation of the circle.

$$x^2 + 6x + y^2 - 2y = -6 \qquad \text{Given equation}$$

$$(x^2 + 6x + 9) + (y^2 - 2y + 1) = -6 + 9 + 1 \qquad \text{Complete the square.}$$

$$(x + 3)^2 + (y - 1)^2 = 4 \qquad \text{Factor.}$$

The center is $(-3, 1)$, and the radius is 2. *Now Try Exercise 59* ◆

Solving Systems of Nonlinear Equations and Inequalities

In Sections 9.1 and 9.2 we discussed systems of equations and inequalities. In this subsection we revisit these topics.

EXAMPLE 10   Solving a nonlinear system of equations

Use substitution to solve the following system of equations. Give graphical support by making a sketch.

$$9x^2 + 4y^2 = 36$$
$$12x^2 + y^2 = 12$$

SOLUTION

STEP 1: Begin by solving the second equation for y^2.

$$12x^2 + y^2 = 12 \quad \text{is equivalent to} \quad y^2 = 12 - 12x^2.$$

STEP 2: Next, substitute $(12 - 12x^2)$ for y^2 in the first equation and solve for x.

$9x^2 + 4y^2 = 36$	First equation
$9x^2 + 4(12 - 12x^2) = 36$	Let $y^2 = 12 - 12x^2$.
$9x^2 + 48 - 48x^2 = 36$	Distributive property
$-39x^2 = -12$	Subtract 48; simplify.
$x^2 = \dfrac{4}{13}$	Divide by -39; reduce.
$x = \pm\sqrt{\dfrac{4}{13}}$	Square root property

STEP 3: To determine the corresponding y-values, substitute $x^2 = \frac{4}{13}$ in the equation $y = 12 - 12x^2$.

$$y^2 = 12 - 12\left(\frac{4}{13}\right) = \frac{108}{13} \quad \text{or} \quad y = \pm\sqrt{\frac{108}{13}}$$

There are four solutions: $\left(\pm\sqrt{\frac{4}{13}}, \pm\sqrt{\frac{108}{13}}\right)$.

To graph the system of equations by hand, begin by putting each equation in standard form by dividing the first equation by 36 and the second equation by 12 to obtain

$$\frac{x^2}{4} + \frac{y^2}{9} = 1 \quad \text{and} \quad x^2 + \frac{y^2}{12} = 1.$$

The graphs of these ellipses and the four solutions are shown in Figure 10.42.

Now Try Exercise 69

Note: The system of equations in Example 10 could also be solved by elimination. To do this, multiply the second equation by -4 and add.

The following formula can be used to calculate the area inside an ellipse.

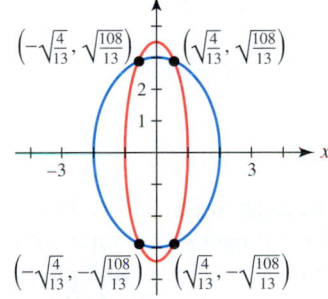

FIGURE 10.42

AREA INSIDE AN ELLIPSE

Given the standard equation of an ellipse, the area A of the region contained inside is given by $A = \pi ab$.

This formula is applied in the next example.

EXAMPLE 11

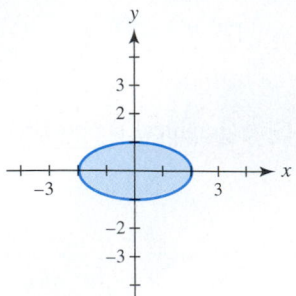

FIGURE 10.43

◆ **CLASS DISCUSSION**

Explain why the area formula for an ellipse is a generalization of the area formula for a circle. ◆

Finding the area inside an ellipse

Shade the region in the xy-plane that satisfies the inequality $x^2 + 4y^2 \leq 4$. Find the area of this region if units are in inches.

SOLUTION Begin by dividing each term in the given inequality by 4.

$$x^2 + 4y^2 \leq 4 \qquad \textit{Given inequality}$$

$$\frac{x^2}{4} + \frac{4y^2}{4} \leq \frac{4}{4} \qquad \textit{Divide by 4.}$$

$$\frac{x^2}{4} + \frac{y^2}{1} \leq 1 \qquad \textit{Simplify the fractions.}$$

The boundary of the region is the ellipse $\frac{x^2}{4} + \frac{y^2}{1} = 1$. The region *inside* the ellipse satisfies the inequality. To verify this fact, note that the test point $(0, 0)$, which is located inside the ellipse, satisfies the inequality. The solution set is shaded in Figure 10.43, and the area of this region is

$$A = \pi ab = \pi(2)(1) = 2\pi \approx 6.28 \text{ square inches.}$$

Now Try Exercise 87 ◆

EXAMPLE 12 Solving a nonlinear inequality

Shade the region in the xy-plane that satisfies the system of inequalities.

$$36x^2 + 25y^2 \leq 900$$

$$x + (y + 2)^2 \leq 4$$

SOLUTION Before sketching a graph, rewrite these two inequalities as follows.

First Inequality:
$$\frac{36}{900}x^2 + \frac{25}{900}y^2 \leq \frac{900}{900} \qquad \textit{Divide each term by 900.}$$

$$\frac{x^2}{25} + \frac{y^2}{36} \leq 1 \qquad \textit{Simplify to standard form.}$$

Second Inequality:
$$(y + 2)^2 \leq -x + 4 \qquad \textit{Subtract x.}$$

$$(y + 2)^2 \leq -(x - 4) \qquad \textit{Distributive property}$$

The first inequality represents the region inside an ellipse, as shown in Figure 10.44. The second inequality represents the region left of a parabola that opens to the left with vertex $(4, -2)$, as shown in Figure 10.45. (Note that the test point $(-2, -2)$ satisfies the second inequality since $(-2 + 2)^2 \leq -(-2 - 4)$ is a true statement. Thus we shade the region where $(-2, -2)$ is located.) The solution set for the system satisfies both inequalities and is shaded in Figure 10.46.

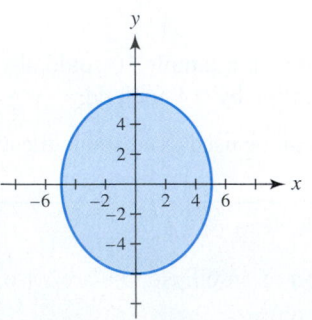

FIGURE 10.44

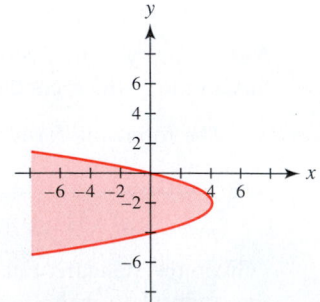

FIGURE 10.45

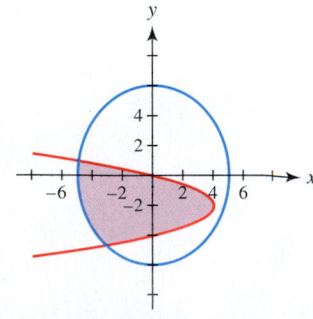

FIGURE 10.46

Now Try Exercise 83 ◆

10.2

Putting it all Together

$\mathbf{T}$he following table summarizes some important concepts about ellipses.

Concept	Equation	Example
Ellipse with center $(0, 0)$	Standard equation with $a > b > 0$ Horizontal major axis: $$\frac{x^2}{a^2} + \frac{y^2}{b^2} = 1$$ Vertical major axis: $$\frac{x^2}{b^2} + \frac{y^2}{a^2} = 1$$	$\frac{x^2}{4} + \frac{y^2}{9} = 1; a = 3, b = 2$ Center: $(0, 0)$; major axis: vertical Vertices: $(0, \pm 3)$; foci: $(0, \pm\sqrt{5})$ $(c^2 = a^2 - b^2 = 9 - 4 = 5$, so $c = \sqrt{5}$.)
Ellipse with center (h, k)	Standard equation with $a > b > 0$ Horizontal major axis: $$\frac{(x - h)^2}{a^2} + \frac{(y - k)^2}{b^2} = 1$$ Vertical major axis: $$\frac{(x - h)^2}{b^2} + \frac{(y - k)^2}{a^2} = 1$$	$\frac{(x - 1)^2}{4} + \frac{(y + 1)^2}{9} = 1; a = 3, b = 2$ Center: $(1, -1)$; major axis: vertical Vertices: $(1, -1 \pm 3)$; foci: $(1, -1 \pm \sqrt{5})$ $(c^2 = a^2 - b^2 = 9 - 4 = 5$, so $c = \sqrt{5}$.)
Circle with center (h, k) and radius r	Standard equation $$(x - h)^2 + (y - k)^2 = r^2$$ A circle is an ellipse with $a = b = r$.	$(x - 2)^2 + (y + 2)^2 = 9$ Center: $(2, -2)$; radius: $r = 3$
Area inside an ellipse	$A = \pi ab$	The area inside the ellipse given by $\frac{x^2}{49} + \frac{y^2}{9} = 1$ is $A = \pi(7)(3) = 21\pi$ square units.

10.2 Exercises

Ellipses with Center (0, 0)

Exercises 1–8: Graph the ellipse. Label the foci and the endpoints of each axis.

1. $\dfrac{x^2}{4} + \dfrac{y^2}{9} = 1$

2. $\dfrac{x^2}{9} + \dfrac{y^2}{4} = 1$

3. $\dfrac{x^2}{36} + \dfrac{y^2}{16} = 1$

4. $x^2 + \dfrac{y^2}{4} = 1$

5. $9x^2 + 5y^2 = 45$

6. $x^2 + 4y^2 = 400$

7. $25x^2 + 9y^2 = 225$

8. $5x^2 + 4y^2 = 20$

Exercises 9–12: Match the equation of the ellipse with its graph (a.–d.).

9. $\dfrac{x^2}{16} + \dfrac{y^2}{36} = 1$

10. $\dfrac{x^2}{4} + y^2 = 1$

11. $\dfrac{x^2}{16} + \dfrac{y^2}{4} = 1$

12. $\dfrac{x^2}{9} + \dfrac{y^2}{9} = 1$

a.
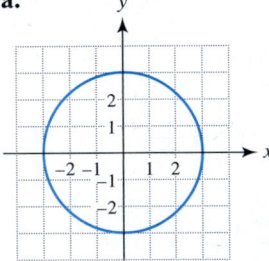

b.
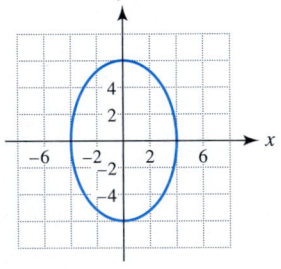

c.

d.
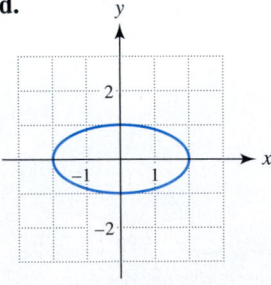

Exercises 13–16: The foci F_1 and F_2, vertices V_1 and V_2, and endpoints U_1 and U_2 of the minor axis of an ellipse are labeled in the figure. Graph the ellipse and find its standard equation. Note that the coordinates of V_1, V_2, F_1, and F_2 are integers.

13.
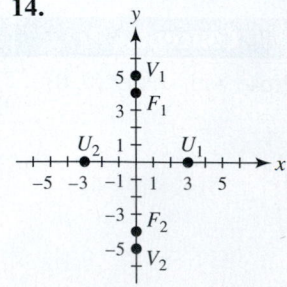

14.

15.
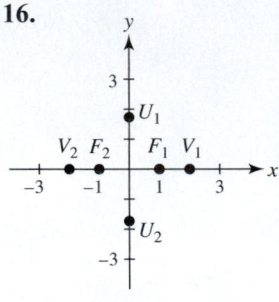

16.

Exercises 17–24: Find an equation of the ellipse, centered at the origin, satisfying the conditions. Sketch its graph.

17. Foci $(0, \pm 2)$, vertices $(0, \pm 4)$

18. Foci $(0, \pm 3)$, vertices $(0, \pm 5)$

19. Foci $(\pm 5, 0)$, vertices $(\pm 6, 0)$

20. Foci $(\pm 4, 0)$, vertices $(\pm 6, 0)$

21. Horizontal major axis of length 8, minor axis of length 6

22. Vertical major axis of length 12, minor axis of length 8

23. Eccentricity $\frac{2}{3}$, horizontal major axis of length 6

24. Eccentricity $\frac{3}{4}$, vertices $(0, \pm 8)$

Ellipses with Center (h, k)

Exercises 25–30: Sketch a graph of the ellipse.

25. $\dfrac{(x-2)^2}{4} + \dfrac{(y-1)^2}{9} = 1$

26. $\dfrac{(x+1)^2}{16} + \dfrac{(y+3)^2}{9} = 1$

27. $\dfrac{(x+1)^2}{16} + \dfrac{(y+2)^2}{25} = 1$

28. $\dfrac{(x-4)^2}{9} + \dfrac{y^2}{4} = 1$ **29.** $\dfrac{(x+2)^2}{4} + y^2 = 1$

30. $x^2 + \dfrac{(y-3)^2}{4} = 1$

Exercises 31–34: Match the equation of the ellipse with its graph (a.–d.).

31. $\dfrac{(x-2)^2}{16} + \dfrac{(y+4)^2}{36} = 1$

32. $\dfrac{(x+1)^2}{4} + \dfrac{y^2}{9} = 1$

33. $\dfrac{(x+1)^2}{9} + \dfrac{(y-1)^2}{4} = 1$

34. $\dfrac{x^2}{25} + \dfrac{(y+1)^2}{10} = 1$

a.

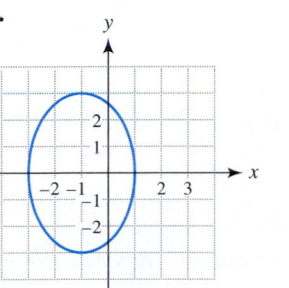

b.

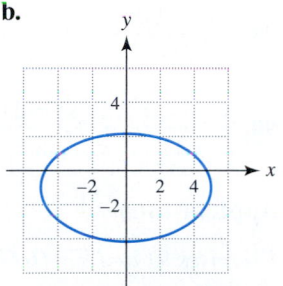

c.

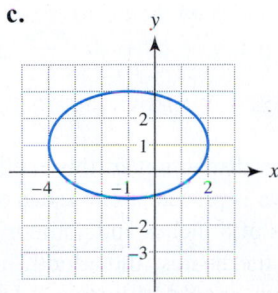

d.

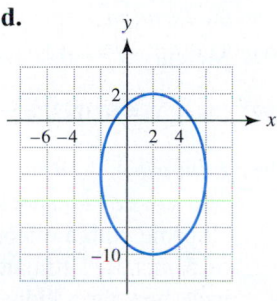

Exercises 35–38: Sketch a graph of the ellipse. Identify the foci and vertices.

35. $\dfrac{(x-1)^2}{9} + \dfrac{(y-1)^2}{25} = 1$

36. $\dfrac{(x+2)^2}{25} + \dfrac{(y+1)^2}{16} = 1$

37. $\dfrac{(x+4)^2}{16} + \dfrac{(y-2)^2}{9} = 1$

38. $\dfrac{x^2}{4} + \dfrac{(y-1)^2}{9} = 1$

Exercises 39–42: Find an equation of an ellipse that satisfies the given conditions.

39. Center $(2, 1)$, focus $(2, 3)$, and vertex $(2, 4)$

40. Center $(-3, -2)$, focus $(-1, -2)$, and vertex $(1, -2)$

41. Vertices $(\pm 3, 2)$ and foci $(\pm 2, 2)$

42. Vertices $(-1, \pm 3)$ and foci $(-1, \pm 1)$

Exercises 43 and 44: Find an (approximate) equation of the ellipse shown in the figure.

43.

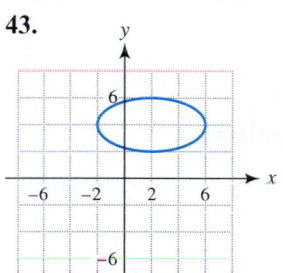

44.

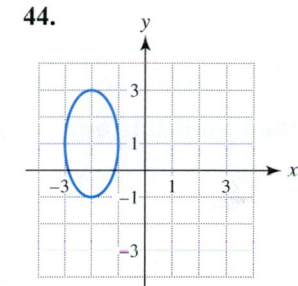

Exercises 45–52: (Refer to Example 7.) Write the equation in standard form for an ellipse centered at (h, k). Identify the center and the vertices.

45. $9x^2 + 18x + 4y^2 - 8y - 23 = 0$

46. $9x^2 - 36x + 16y^2 - 64y - 44 = 0$

47. $4x^2 + 8x + y^2 + 2y + 1 = 0$

48. $x^2 - 6x + 9y^2 = 0$

49. $4x^2 + 16x + 5y^2 - 10y + 1 = 0$

50. $2x^2 + 4x + 3y^2 - 18y + 23 = 0$

51. $16x^2 - 16x + 4y^2 + 12y = 51$

52. $16x^2 + 48x + 4y^2 - 20y + 57 = 0$

Circles

Exercises 53–58: Find the standard equation of the circle that satisfies the conditions. Graph the circle.

53. Center $(0, 0)$, radius of 4

54. Center $(1, -2)$, radius of 3

55. Center $(3, -4)$, radius of 1

56. Center $(-1, -3)$, passing through the point $(3, 0)$

57. Center $(2, 3)$ passing through $(-3, 15)$

58. Center $(-4, 5)$ passing through $(-2, 4)$

Exercises 59–62: Find the center and radius of the circle.

59. $x^2 - 4x + y^2 - 2y = 11$

60. $x^2 + 6x + y^2 - 4y = 12$

61. $x^2 + y^2 + 10y = 0$ **62.** $x^2 - 2x + y^2 + 8y = 19$

Graphing Ellipses with Technology

Exercises 63–66: Graph the ellipse.

63. $\dfrac{x^2}{15} + \dfrac{y^2}{10} = 1$ **64.** $\dfrac{(x - 1.2)^2}{7.1} + \dfrac{y^2}{3.5} = 1$

65. $4.1x^2 + 6.3y^2 = 25$ **66.** $\frac{1}{2}x^2 + \frac{1}{3}y^2 = \frac{1}{6}$

Solving Equations and Inequalities

Exercises 67–72: Solve the system of equations. Give graphical support by making a sketch.

67. $\dfrac{x^2}{4} + \dfrac{y^2}{9} = 1$ **68.** $\dfrac{x^2}{16} + \dfrac{y^2}{25} = 1$
$x + y = 3$ $-2x + y = 5$

69. $4x^2 + 16y^2 = 64$ **70.** $4x^2 + y^2 = 4$
$x^2 + y^2 = 9$ $x^2 + y^2 = 2$

71. $x^2 + y^2 = 9$ **72.** $x^2 + y^2 = 4$
$2x^2 + 3y^2 = 18$ $(x - 1)^2 + y^2 = 4$

Exercises 73–78: Solve the system of equations.

73. $\dfrac{x^2}{2} + \dfrac{y^2}{4} = 1$ **74.** $x^2 + \frac{1}{9}y^2 = 1$
$-x^2 + 2y = 4$ $x + y = 3$

75. $\dfrac{x^2}{2} + \dfrac{y^2}{4} = 1$ **76.** $\dfrac{x^2}{5} + \dfrac{y^2}{10} = 1$
$\dfrac{x^2}{4} + \dfrac{y^2}{2} = 1$ $\dfrac{x^2}{10} + \dfrac{y^2}{5} = 1$

77. $(x - 2)^2 + y^2 = 9$ **78.** $(x - 2) - y^2 = 0$
$x^2 + y^2 = 9$ $\dfrac{x^2}{4} + \dfrac{y^2}{9} = 1$

Exercises 79–86: Shade the solutions set to the system of inequalities.

79. $(x - 1)^2 + (y + 1)^2 < 4$
$(x + 1)^2 + y^2 > 1$

80. $\dfrac{x^2}{16} + \dfrac{y^2}{25} < 1$
$\dfrac{x^2}{4} + \dfrac{y^2}{9} > 1$

81. $\dfrac{x^2}{4} + \dfrac{y^2}{9} \le 1$ **82.** $\dfrac{x^2}{16} + \dfrac{y^2}{25} \le 1$
$-x + y \le 4$ $x + y \ge 2$

83. $x^2 + y^2 \le 4$ **84.** $x^2 + (y + 1)^2 \le 9$
$x^2 + (y - 2)^2 \le 4$ $(x + 1)^2 + y^2 \le 9$

85. $x^2 + y^2 \le 4$ **86.** $4x^2 + 9y^2 \le 36$
$(x + 1)^2 - y \le 0$ $x - (y - 2)^2 \ge 0$

Exercises 87–90: (Refer to Example 11.) Shade the region in the xy-plane that satisfies the given inequality. Find the area of this region if units are in feet.

87. $4x^2 + 9y^2 \le 36$

88. $9x^2 + y^2 \le 9$

89. $\dfrac{(x - 1)^2}{25} + \dfrac{(y + 2)^2}{16} \le 1$

90. $\dfrac{(x + 3)^2}{4} + \dfrac{(y - 2)^2}{8} \le 1$

Applications

Exercises 91 and 92: Orbits of Planets (Refer to Example 4.) Find an equation of the orbit for the planet. Graph its orbit and the location of the sun at a focus on the positive x-axis.

91. Mercury: $e = 0.206, a = 0.387$

92. Mars: $e = 0.093, a = 1.524$

93. *Lithotripter* (Refer to the discussion in this section.) The source of a shock wave is placed at one focus of an ellipsoid with a major axis of 8 inches and a minor axis of 5 inches. Estimate, to the nearest thousandth of an inch, how far a kidney stone should be positioned from the source.

94. *Whispering Gallery* If a large room is constructed in the shape of the upper half of an ellipsoid, it will have a unique property. Two people, standing at each of the two foci, can hear each other whispering, even though there may be considerable distance between them. Any sound emanating from one focus is reflected directly toward the other focus. See the accompanying figure. If the foci are 100 feet apart, and the maximum height of the ceiling is 40 feet, estimate the area of the floor of the room.

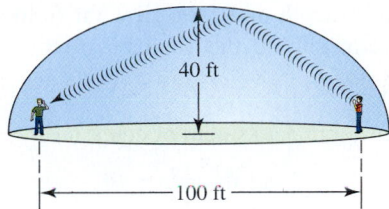

95. *Halley's Comet*   One of the most famous comets is Halley's comet. It travels in an elliptical orbit with $a = 17.95$ and $b = 4.44$ and passes by Earth roughly every 76 years. Note that each unit represents 93 million miles. The most recent date that it passed by Earth was in February 1986. (**Source:** M. Zeilik, *Introductory Astronomy and Astrophysics.*)

  (a) Write an equation for the orbit of Halley's comet, where the orbit is centered at $(0, 0)$, and the major axis lies on the x-axis.

  (b) If the sun lies (at the focus) on the positive x-axis, finds its coordinates.

  (c) Determine the maximum and minimum distances between Halley's comet and the sun.

96. *Orbit of Earth*   (Refer to Example 4.) Earth has a nearly circular orbit with $e \approx 0.0167$ and $a = 93$ million miles. Find the minimum and maximum distances between Earth and the sun. (**Source:** M. Zeilik, *Introductory Astronomy and Astrophysics.*)

97. *Arch Bridge*   An elliptical arch under a bridge is constructed so that it is 60 feet wide and has a maximum height of 25 feet, as illustrated in the accompanying figure. Find the height of the arch 15 feet from the center of the arch.

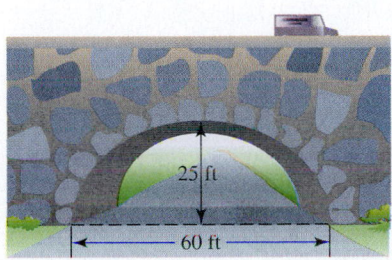

98. *Perimeter of an Ellipse*   The perimeter P of an ellipse can be approximated by

$$P \approx 2\pi\sqrt{\frac{a^2 + b^2}{2}}.$$

  (a) Approximate the distance in miles that Mercury travels in one orbit of the sun if $a = 36.0$, $b = 35.2$, and the units are in millions of miles.

  (b) If a planet has a circular orbit, does this formula give the *exact* perimeter? Explain.

99. *Satellite Orbit*   The orbit of *Explorer VII* and the outline of Earth's surface are shown in the figure. This orbit orbit can be described by the equation $\frac{x^2}{a^2} + \frac{y^2}{b^2} = 1$, where $a = 4464$ and $b = 4462$. The surface of Earth can be described by $(x - 164)^2 + y^2 = 3960^2$. Find the maximum and minimum heights of the satellite above Earth's surface if all units are in miles. (**Sources:** W. Loh; W. Thomson.)

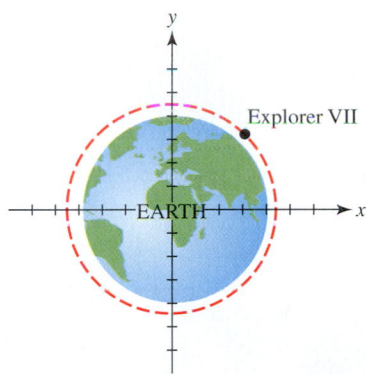

100. *Planet Velocity*   The maximum and minimum velocities in kilometers per second of a planet moving in an elliptical orbit can be calculated by

$$v_{max} = \frac{2\pi a}{P}\sqrt{\frac{1 + e}{1 - e}} \quad \text{and} \quad v_{min} = \frac{2\pi a}{P}\sqrt{\frac{1 - e}{1 + e}}.$$

In these equations, a is half the length of the major axis of the orbit in kilometers, P is its orbital period in seconds, and e is the eccentricity of the orbit. (**Source:** M. Zeilik.)

  (a) Calculate v_{max} and v_{min} for Pluto if $a = 5.913 \times 10^9$ kilometers, the period is $P = 2.86 \times 10^{12}$ seconds, and the eccentricity is $e = 0.249$.

  (b) If a planet has a circular orbit, what can be said about its orbital velocity?

Writing about Mathematics

101. Explain how the distance between the foci of an ellipse affects the shape of the ellipse.

102. Given the standard equation of an ellipse, explain how to determine the length of the major axis. How can you determine whether the major axis is vertical or horizontal?

CHECKING BASIC CONCEPTS FOR SECTIONS 10.1 AND 10.2

1. Graph the parabola defined by $x = \frac{1}{2}y^2$. Include the focus and directrix.

2. Find an equation of the parabola with focus $(-1, 0)$ and directrix $y = 3$.

3. Graph the ellipse defined by $\frac{x^2}{36} + \frac{y^2}{100} = 1$. Include the foci and label the major and minor axes.

4. Find an equation of the ellipse centered at $(3, -2)$ with a vertical major axis of length 6 and minor axis of length 4. What are the coordinates of the foci?

5. A parabolic reflector for a searchlight has a diameter of 4 feet and a depth of 1 foot. How far from the vertex of the reflector should the filament of the light bulb be located?

6. Solve the nonlinear system of equations.
$$x^2 + y^2 = 10$$
$$2x^2 + 3y^2 = 29$$

7. Write $x^2 - 4x + 4y^2 + 8y - 8 = 0$ in the standard form for an ellipse centered at (h, k). Identify the center and the vertices.

10.3 Hyperbolas

Introduction

Hyperbolas have several interesting properties. For example, if a comet passes by the sun with a high velocity, then the sun's gravity may not be strong enough to cause the comet to go into orbit. Instead, the comet will pass by the sun just once and follow a trajectory that can be described by a hyperbola. Hyperbolas also have a reflective property that is used in telescopes. In this section we learn some basic properties of hyperbolas.

- ◆ Find equations of hyperbolas
- ◆ Graph hyperbolas
- ◆ Learn the reflective property of hyperbolas
- ◆ Translate hyperbolas
- ◆ Solve systems of nonlinear equations

Equations and Graphs of Hyperbolas

The third type of conic section is a hyperbola.

HYPERBOLA

A **hyperbola** is the set of points in a plane, the difference of whose distances from two fixed points is constant. Each fixed point is called a **focus** of the hyperbola.

In Figure 10.47 a point $P(x, y)$ is shown on a hyperbola with distance d_1 from focus $F_1(c, 0)$ and distance d_2 from focus $F_2(-c, 0)$. Regardless of the location of the point $P(x, y)$, $|d_2 - d_1| = 2a$. The **transverse axis** is the line segment connecting the **vertices** $V_1(a, 0)$ and $V_2(-a, 0)$, and its length equals $2a$.

Two hyperbolas are shown in Figures 10.48 and 10.49. The coordinates of the vertices, $(\pm a, 0)$ or $(0, \pm a)$, and foci, $(\pm c, 0)$ or $(0, \pm c)$, are labeled. The two parts of the hyperbola in Figure 10.48 are the **left branch** and the **right branch**, whereas in Figure 10.49 the hyperbola has an **upper branch** and a **lower branch**. A line segment connecting the points $(0, \pm b)$ in Figure 10.48 and $(\pm b, 0)$ in Figure 10.49 is the **conjugate axis**. The lines $y = \pm \frac{b}{a}x$ and $y = \pm \frac{a}{b}x$ are **asymptotes** for each respective hyperbola. They can be used as an aid in graphing. The dashed rectangle is sometimes called the **fundamental rectangle**.

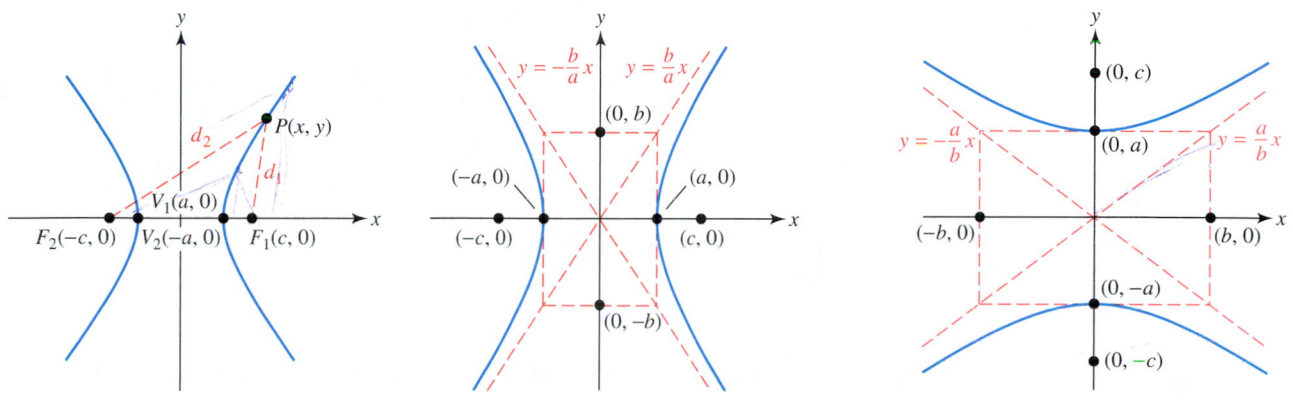

FIGURE 10.47   **FIGURE 10.48** Horizontal Transverse Axis   **FIGURE 10.49** Vertical Transverse Axis

By the vertical line test, a hyperbola cannot be represented by a function, but many can be described by the following equations. The constants *a, b,* and *c* are positive.

STANDARD EQUATIONS FOR HYPERBOLAS CENTERED AT (0, 0)

The hyperbola with center at the origin, *horizontal* transverse axis, and equation

$$\frac{x^2}{a^2} - \frac{y^2}{b^2} = 1$$

has asymptotes $y = \pm \frac{b}{a}x$, vertices $(\pm a, 0)$, and foci $(\pm c, 0)$, where $c^2 = a^2 + b^2$.

The hyperbola with center at the origin, *vertical* transverse axis, and equation

$$\frac{y^2}{a^2} - \frac{x^2}{b^2} = 1$$

has asymptotes $y = \pm \frac{a}{b}x$, vertices $(0, \pm a)$, and foci $(0, \pm c)$, where $c^2 = a^2 + b^2$.

Note: A hyperbola consists of two solid curves or branches. The asymptotes, foci, transverse axis, conjugate axis, and fundamental (dashed) rectangle are not part of the hyperbola, but are aids for sketching its graph.

One interpretation of an asymptote can be made using trajectories of comets as they approach the sun. Comets travel in parabolic, elliptic, or hyperbolic trajectories. If the speed of a comet is too slow, the gravitational pull of the sun captures the comet in an elliptical orbit.

See Figure 10.50. If the speed of the comet is too fast, the sun's gravity is too weak and the comet passes by the sun in a hyperbolic trajectory. Near the sun the gravitational pull is stronger and the comet's trajectory is curved. Farther from the sun, gravity becomes weaker and the comet eventually returns to a straight-line trajectory that is determined by the *asymptote* of the hyperbola. See Figure 10.51. Finally, if the speed is neither too slow nor too fast, the comet will travel in a parabolic path. See Figure 10.52. In all three cases, the sun is located at a focus of the conic section.

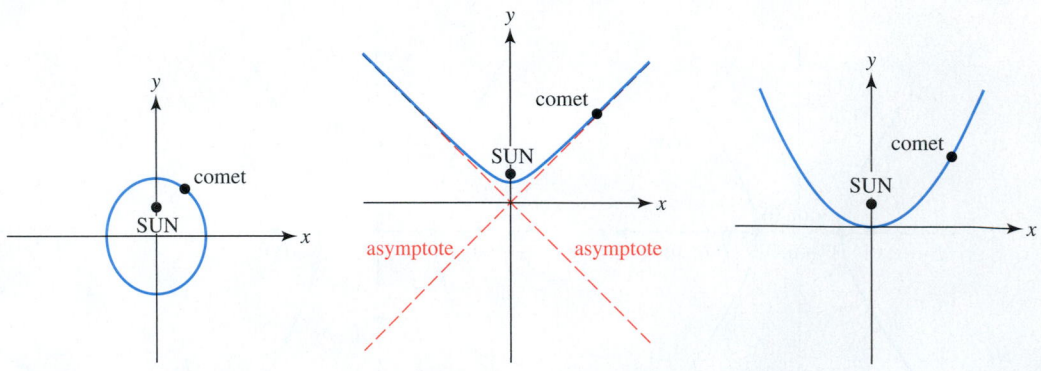

FIGURE 10.50 Elliptic Orbit   **FIGURE 10.51** Hyperbolic Path   **FIGURE 10.52** Parabolic Path

◆ **CLASS DISCUSSION**

If a comet is seen at regular intervals, what type of path must it follow? ◆

EXAMPLE 1 Sketching the graph of a hyperbola

Sketch a graph of $\frac{x^2}{4} - \frac{y^2}{9} = 1$. Label the vertices, foci, and asymptotes.

SOLUTION The equation is in standard form with $a = 2$ and $b = 3$. It has a horizontal transverse axis with vertices $(\pm 2, 0)$. The endpoints of the conjugate axis are $(0, \pm 3)$. To locate the foci find c.

$$c^2 = a^2 + b^2 = 4 + 9 = 13 \qquad \text{or}$$
$$c = \sqrt{13} \approx 3.61.$$

The foci are $(\pm\sqrt{13}, 0)$. The asymptotes are $y = \pm\frac{b}{a}x$ or $y = \pm\frac{3}{2}x$. See Figure 10.53.

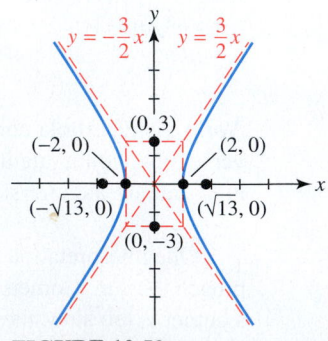

FIGURE 10.53

Now Try Exercise 1 ◆

EXAMPLE 2 Finding the equation of a hyperbola

Find the equation of the hyperbola centered at the origin with a vertical transverse axis of length 6 and focus (0, 5). Sketch a graph of the hyperbola.

SOLUTION Since the hyperbola is centered at the origin with a vertical axis, its equation is $\frac{y^2}{a^2} - \frac{x^2}{b^2} = 1$. The transverse axis has length $6 = 2a$, so $a = 3$. Since one focus is located at (0, 5), $c = 5$. We can find b by using the following equation.

$$b^2 = c^2 - a^2$$
$$b = \sqrt{c^2 - a^2}$$
$$b = \sqrt{5^2 - 3^2} = 4$$

The equation of this hyperbola is $\frac{y^2}{9} - \frac{x^2}{16} = 1$. Its asymptotes are $y = \pm\frac{a}{b}x$ or $y = \pm\frac{3}{4}x$. Its graph is shown in Figure 10.54.

Now Try Exercise 19 ◆

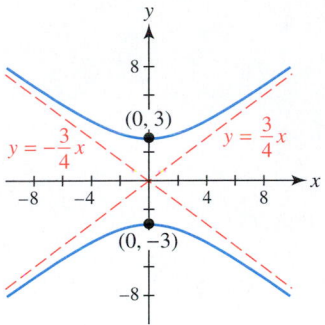

FIGURE 10.54

EXAMPLE 3 Finding the equation of a hyperbola

Find the standard equation of the hyperbola shown in Figure 10.55. Identify the vertices, foci, and asymptotes.

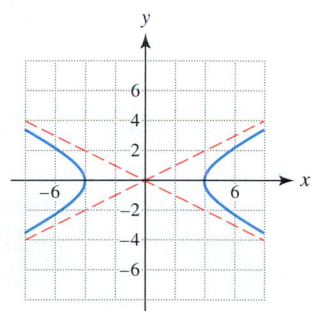

FIGURE 10.55

SOLUTION Because the hyperbola is centered at (0, 0) with a horizontal transverse axis, its equation has the form

$$\frac{x^2}{a^2} - \frac{y^2}{b^2} = 1.$$

The asymptotes are given by $y = \pm\frac{b}{a}$. In Figure 10.56 the fundamental rectangle is determined by the four points $(\pm 4, 0)$ and $(0, \pm 2)$, and its diagonals correspond to the asymptotes. It follows that $a = 4$ and $b = 2$. Thus the standard equation of the hyperbola is

$$\frac{x^2}{16} - \frac{y^2}{4} = 1.$$

The vertices are $(\pm 4, 0)$, and the asymptotes are $y = \pm\frac{1}{2}x$. To find the coordinates of the foci, find c.

$$c^2 = a^2 + b^2 = 4^2 + 2^2 = 20$$

Thus $c = \sqrt{20} \approx 4.47$, and the coordinates of the foci are $(\pm\sqrt{20}, 0)$.

Now Try Exercises 9 and 11 ◆

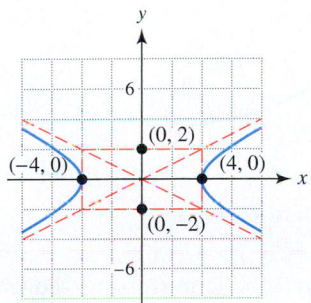

FIGURE 10.56

 **EXAMPLE 4** Graphing a hyperbola with technology

Use a graphing calculator to graph $\frac{y^2}{4.2} - \frac{x^2}{8.4} = 1$.

SOLUTION Begin by solving the given equation for y.

$$\frac{y^2}{4.2} = 1 + \frac{x^2}{8.4} \qquad \text{Add } \frac{x^2}{8.4}.$$

$$y^2 = 4.2\left(1 + \frac{x^2}{8.4}\right) \qquad \text{Multiply by 4.2.}$$

$$y = \pm\sqrt{4.2\left(1 + \frac{x^2}{8.4}\right)} \qquad \text{Square root property}$$

Graph $Y_1 = \sqrt{(4.2(1 + X^2/8.4))}$ and $Y_2 = -\sqrt{(4.2(1 + X^2/8.4))}$. See Figures 10.57 and 10.58.

$[-10, 10, 1]$ by $[-10, 10, 1]$

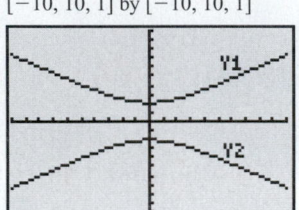

FIGURE 10.57   **FIGURE 10.58**   *Now Try Exercise 55* ◆

Reflective Property of Hyperbolas

Hyperbolas have an important reflective property. If a hyperbola is rotated about the x-axis, a **hyperboloid** is formed, as illustrated in Figure 10.59. Any beam of light that is directed toward focus F_1, will be reflected by the hyperboloid toward focus F_2. See Figure 10.60.

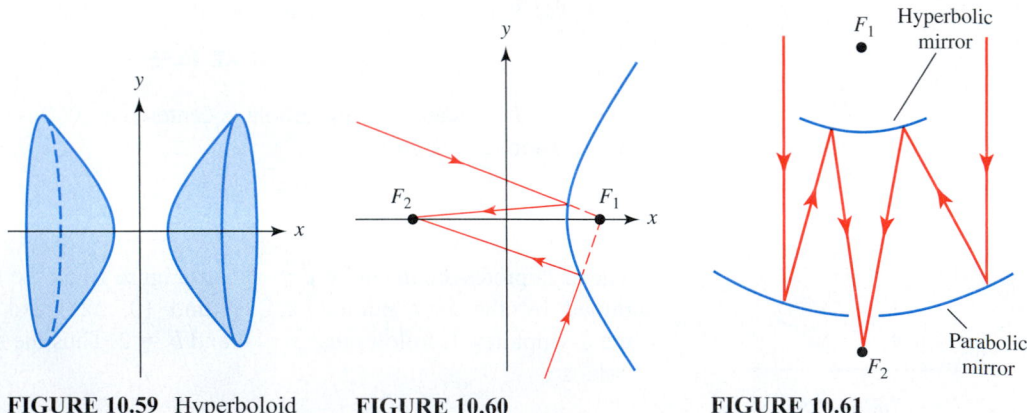

FIGURE 10.59 Hyperboloid   **FIGURE 10.60**   **FIGURE 10.61**

Telescopes sometimes make use of both parabolic and hyperbolic mirrors, as shown in Figure 10.61. When parallel rays of light from distant stars strike the large parabolic (primary) mirror, they are reflected toward its focus, F_1. A smaller hyperbolic (secondary) mirror is placed so that its focus is also located at F_1. Light rays striking the hyperbolic mirror are reflected toward its other focus, F_2, through a small hole in the parabolic mirror, and into an eye piece.

Translations of Hyperbolas

If the equation of a hyperbola is either $\frac{x^2}{a^2} - \frac{y^2}{b^2} = 1$ or $\frac{y^2}{a^2} - \frac{x^2}{b^2} = 1$, then the center of the hyperbola is (0, 0). We can use translations of graphs to find the equation of a hyperbola centered at (h, k) by replacing x with $(x - h)$ and y with $(y - k)$. The constants a, b, and c are positive.

STANDARD EQUATIONS FOR HYPERBOLAS CENTERED AT (h, k)

A hyperbola with center (h, k), and either a horizontal or vertical transverse axis, satisfies one of the following equations, where $c^2 = a^2 + b^2$.

$$\frac{(x - h)^2}{a^2} - \frac{(y - k)^2}{b^2} = 1$$

Transverse axes: horizontal
Vertices: $(h \pm a, k)$; foci: $(h \pm c, k)$
Asymptotes: $y = \pm \frac{b}{a}(x - h) + k$

$$\frac{(y - k)^2}{a^2} - \frac{(x - h)^2}{b^2} = 1$$

Transverse axes: vertical
Vertices: $(h, k \pm a)$; foci: $(h, k \pm c)$
Asymptotes: $y = \pm \frac{a}{b}(x - h) + k$

EXAMPLE 5

Graphing a hyperbola with center (h, k)

Graph the hyperbola whose equation is $\frac{(y + 2)^2}{9} - \frac{(x - 2)^2}{16} = 1$. Label the vertices, foci, and asymptotes.

SOLUTION The hyperbola has a vertical transverse axis and its center is $(2, -2)$. Since $a^2 = 9$ and $b^2 = 16$, it follows that $c^2 = a^2 + b^2 = 9 + 16 = 25$. Thus $a = 3$, $b = 4$, and $c = 5$. The vertices are located 3 units above and below the center of the hyperbola and the foci are located 5 units above and below the center of the hyperbola. That is, the vertices are $(2, -2 \pm 3)$, or $(2, 1)$ and $(2, -5)$, and the foci are $(2, -2 \pm 5)$, or $(2, 3)$ and $(2, -7)$. The asymptotes are given by

Algebra Review
To review point-slope form of a line, see Section 2.2.

$$y = \pm \frac{a}{b}(x - h) + k \quad \text{or} \quad y = \pm \frac{3}{4}(x - 2) - 2.$$

A graph of the hyperbola is shown in Figure 10.62.

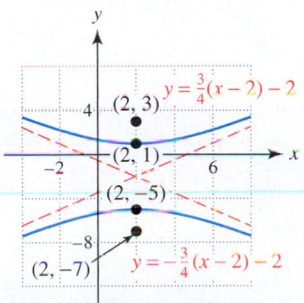

FIGURE 10.62   *Now Try Exercise 39* ◆

 **EXAMPLE 6**

Finding the standard equation of a hyperbola

Write $9x^2 - 18x - 4y^2 - 16y = 43$ in the standard form for a hyperbola centered at (h, k). Identify the center and the vertices.

Algebra Review

To review completing the square, see Section 3.1. To review perfect square trinomials, see Chapter R (page R-34).

◆ **CLASS DISCUSSION**

Sketch a graph of the hyperbola in Example 6. Identify the foci and asymptotes. ◆

SOLUTION We can write the given equation in standard form by completing the square.

$$9x^2 - 18x - 4y^2 - 16y = 43 \qquad \text{Given equation}$$

$$9(x^2 - 2x + \underline{}) - 4(y^2 + 4y + \underline{}) = 43 \qquad \text{Distributive property}$$

$$9(x^2 - 2x + \underline{1}) - 4(y^2 + 4y + \underline{4}) = 43 + 9 - 16 \qquad \text{Complete the square.}$$

$$9(x - 1)^2 - 4(y + 2)^2 = 36 \qquad \text{Perfect square trinomials}$$

$$\frac{(x - 1)^2}{4} - \frac{(y + 2)^2}{9} = 1 \qquad \text{Divide each side by 36.}$$

The center is $(1, -2)$. Because $a = 2$ and the transverse axis is horizontal, the vertices of the hyperbola are $(h \pm a, k) = (1 \pm 2, -2)$.   *Now Try Exercise 49* ◆

Solving Systems of Nonlinear Equations

If the equation of a hyperbola occurs in a system of equations, then the system of equations is nonlinear. In the next example, we determine the points where an ellipse and a hyperbola intersect.

 **EXAMPLE 7**

Solving a system of nonlinear equations

Solve the following system of equations. Give graphical support by making a sketch.

$$4x^2 - 9y^2 = 36$$
$$9x^2 + 25y^2 = 262$$

Algebra Review

To review elimination, see Section 9.2.

SOLUTION To eliminate the x-variable, multiply the first equation by -9, the second equation by 4, and then add the resulting equations.

$$-36x^2 + 81y^2 = -324 \qquad \text{Multiply first equation by } -9.$$

$$\underline{36x^2 + 100y^2 = 1048} \qquad \text{Multiply second equation by 4.}$$

$$181y^2 = 724 \qquad \text{Add.}$$

Thus $y^2 = \frac{724}{181} = 4$ or $y = \pm 2$. To determine the corresponding x-values, let $y = \pm 2$ in the given equation: $4x^2 - 9y^2 = 36$.

$$4x^2 - 9(\pm 2)^2 = 36 \qquad \text{Let } y = \pm 2.$$

$$4x^2 - 36 = 36 \qquad \text{Simplify.}$$

$$4x^2 = 72 \qquad \text{Add 36 to each side.}$$

$$x^2 = 18 \qquad \text{Divide each side by 4.}$$

$$x = \pm\sqrt{18} \qquad \text{Square root property}$$

$$x = \pm 3\sqrt{2} \qquad \sqrt{18} = \sqrt{9}\sqrt{2} = 3\sqrt{2}$$

There are four solutions to this system of nonlinear equations: $(\pm 3\sqrt{2}, \pm 2)$.

To graph these equations, rewrite them in standard form.

$$\frac{x^2}{9} - \frac{y^2}{4} = 1$$

$$\frac{x^2}{262/9} + \frac{y^2}{262/25} = 1$$

The graph of the first equation is a hyperbola centered at $(0, 0)$ with $a = 3$ and $b = 2$. The graph of the second equation is an ellipse centered $(0, 0)$ with $a = \sqrt{\frac{262}{9}} \approx 5.4$ and $b \approx \sqrt{\frac{262}{25}} \approx 3.2$. The graphs of the two equations are shown in Figure 10.63 with the four points of intersection labeled.

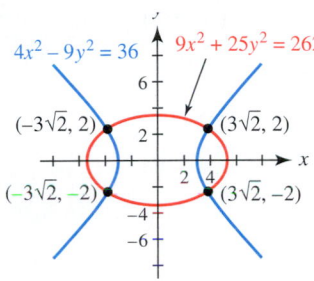

FIGURE 10.63

Now Try Exercise 63 ◆

10.3 Putting it all Together

The following table summarizes some important concepts about hyperbolas.

Concept	Equation	Example
Hyperbola with center $(0, 0)$	Standard equation Transverse axis: horizontal $$\frac{x^2}{a^2} - \frac{y^2}{b^2} = 1$$ Transverse axis: vertical $$\frac{y^2}{a^2} - \frac{x^2}{b^2} = 1$$	$\frac{y^2}{4} - \frac{x^2}{9} = 1; a = 2, b = 3$ Transverse axis: vertical Vertices $(0, \pm2)$; foci: $(0, \pm\sqrt{13})$ $(c^2 = a^2 + b^2 = 4 + 9 = 13$, so $c = \sqrt{13}$.) Asymptotes: $y = \pm\frac{2}{3}x$

Concept	Equation	Example
Hyperbola with center (h, k)	Standard equation Transverse axis: horizontal $$\frac{(x-h)^2}{a^2} - \frac{(y-k)^2}{b^2} = 1$$ Transverse axis: vertical $$\frac{(y-k)^2}{a^2} - \frac{(x-h)^2}{b^2} = 1$$	$\dfrac{(x-1)^2}{4} - \dfrac{(y+1)^2}{9} = 1$; $a = 2, b = 3$ Transverse axis: horizontal; center $(1, -1)$ Vertices $(1 \pm 2, -1)$; foci: $(1 \pm \sqrt{13}, -1)$ $(c^2 = a^2 + b^2 = 4 + 9 = 13$, so $c = \sqrt{13}$.) Asymptotes: $y = \pm\dfrac{3}{2}(x-1) - 1$

10.3 — Exercises

Hyperbolas with Center (0, 0)

Exercises 1–8: Sketch a graph of the hyperbola, including the asymptotes. Give the coordinates of the foci.

1. $\dfrac{x^2}{9} - \dfrac{y^2}{49} = 1$ **2.** $\dfrac{x^2}{16} - \dfrac{y^2}{4} = 1$

3. $\dfrac{y^2}{36} - \dfrac{x^2}{16} = 1$ **4.** $\dfrac{y^2}{4} - \dfrac{x^2}{4} = 1$

5. $x^2 - y^2 = 9$ **6.** $49y^2 - 25x^2 = 1225$

7. $9y^2 - 16x^2 = 144$ **8.** $4x^2 - 4y^2 = 100$

Exercises 9–12: Match the equation of the hyperbola with its graph (a.–d.).

9. $\dfrac{x^2}{4} - \dfrac{y^2}{6} = 1$ **10.** $\dfrac{x^2}{9} - y^2 = 1$

11. $\dfrac{y^2}{9} - \dfrac{x^2}{16} = 1$ **12.** $\dfrac{y^2}{4} - \dfrac{x^2}{4} = 1$

a.

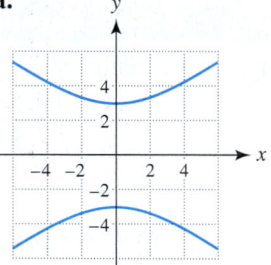

b.

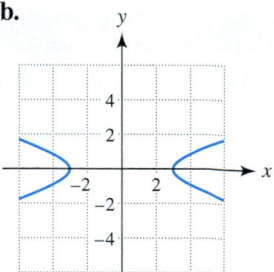

c.

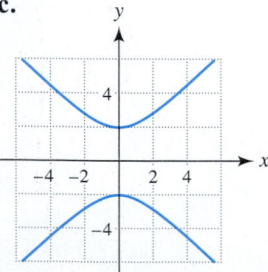

d.

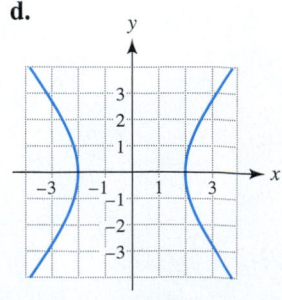

Exercises 13–16: Sketch a graph of a hyperbola, centered at the origin, with the given foci, vertices, and asymptotes shown in the figure. Find an equation of the hyperbola. Note that the coordinates of the foci and vertices are integers.

13.
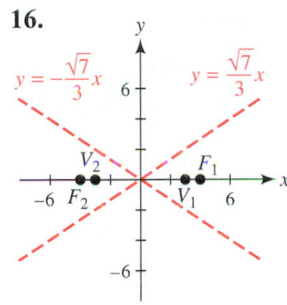

14.

15.

16.

Exercises 17–26: Determine an equation of the hyperbola, centered at the origin, satisfying the conditions. Sketch its graph, including the asymptotes.

17. Foci $(0, \pm 13)$, vertices $(0, \pm 12)$

18. Foci $(\pm 13, 0)$, vertices $(\pm 5, 0)$

19. Vertical transverse axis of length 4, foci $(0, \pm 5)$

20. Horizontal transverse axis of length 12, foci $(\pm 10, 0)$

21. Vertices $(\pm 3, 0)$, asymptotes $y = \pm \frac{2}{3}x$

22. Vertices $(0, \pm 4)$, asymptotes $y = \pm \frac{1}{2}x$

23. Endpoints of conjugate axis $(0, \pm 3)$, vertices $(\pm 4, 0)$

24. Endpoints of conjugate axis $(\pm 4, 0)$, vertices $(0, \pm 2)$

25. Vertices $(\pm \sqrt{10}, 0)$ passing through $(10, 9)$

26. Vertices $(0, \pm \sqrt{5})$ passing through $(4, 5)$

Hyperbolas with Center (h, k)

Exercises 27–32: Sketch a graph of the hyperbola.

27. $\dfrac{(x-1)^2}{16} - \dfrac{(y-2)^2}{4} = 1$

28. $\dfrac{(y+1)^2}{16} - \dfrac{(x+3)^2}{9} = 1$

29. $\dfrac{(y-2)^2}{36} - \dfrac{(x+2)^2}{4} = 1$

30. $\dfrac{(x+1)^2}{4} - \dfrac{(y-1)^2}{4} = 1$

31. $\dfrac{x^2}{4} - (y-1)^2 = 1$

32. $(y+1)^2 - \dfrac{(x-3)^2}{4} = 1$

Exercises 33–36: Match the equation of the hyperbola with its graph (a.–d.).

33. $\dfrac{(x-2)^2}{4} - \dfrac{(y+4)^2}{4} = 1$ **34.** $\dfrac{(x+1)^2}{4} - \dfrac{y^2}{9} = 1$

35. $\dfrac{(y+1)^2}{16} - \dfrac{(x-2)^2}{16} = 1$ **36.** $\dfrac{y^2}{25} - \dfrac{(x+1)^2}{9} = 1$

a.
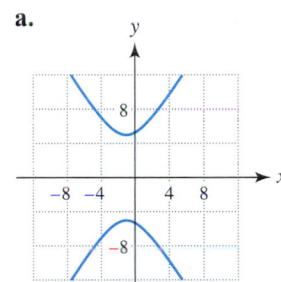

b.
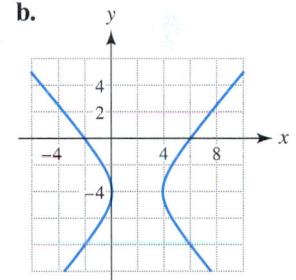

c.
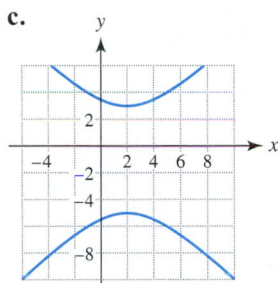

d.
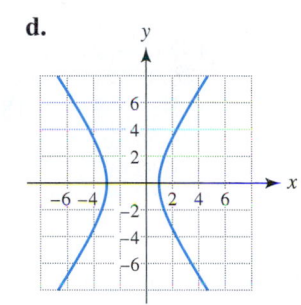

Exercises 37 and 38: Find an (approximate) equation of the hyperbola shown in the graph.

37.
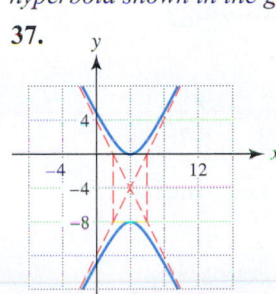

38.
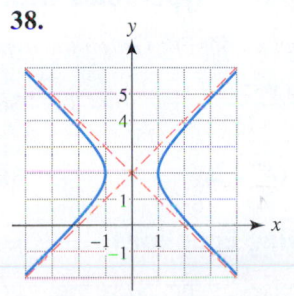

Exercises 39–42: Sketch a graph of the hyperbola including the asymptotes. Give the coordinates of the vertices and foci.

39. $\dfrac{(x-1)^2}{4} - \dfrac{(y-1)^2}{4} = 1$

40. $\dfrac{(x+2)^2}{4} - \dfrac{(y+1)^2}{16} = 1$

41. $\dfrac{(y+1)^2}{16} - \dfrac{(x-1)^2}{9} = 1$

42. $y^2 - \dfrac{(x-2)^2}{4} = 1$

Exercises 43–46: Find the standard equation of a hyperbola with center (h, k) that satisfies the given conditions.

43. Center $(2, -2)$, focus $(4, -2)$, and vertex $(3, -2)$

44. Center $(-1, 1)$, focus $(-1, 4)$, and vertex $(-1, 3)$

45. Vertices $(-1, \pm 1)$ and foci $(-1, \pm 3)$

46. Vertices $(2 \pm 1, 1)$ and foci $(2 \pm 3, 1)$

Exercises 47–54: Write the equation in standard form for a hyperbola centered at (h, k). Identify the center and the vertices.

47. $x^2 - 2x - y^2 + 2y = 4$

48. $y^2 + 4y - x^2 + 2x = 6$

49. $3y^2 + 24y - 2x^2 + 12x + 24 = 0$

50. $4x^2 + 16x - 9y^2 + 18y = 29$

51. $x^2 - 6x - 2y^2 + 7 = 0$

52. $y^2 + 8y - 3x^2 + 13 = 0$

53. $4y^2 + 32y - 5x^2 - 10x + 39 = 0$

54. $5x^2 + 10x - 7y^2 + 28y = 58$

Graphing Hyperbolas with Technology

Exercises 55–58: Graph the hyperbola.

55. $\dfrac{(y-1)^2}{11} - \dfrac{x^2}{5.9} = 1$ **56.** $\dfrac{x^2}{5.3} - \dfrac{y^2}{6.7} = 1$

57. $3y^2 - 4x^2 = 15$ **58.** $2.1x^2 - 6y^2 = 12$

Solving Equations

Exercises 59–66: Solve the system of equations. Give graphical support by making a sketch.

59. $x^2 - y^2 = 4$
   $x^2 + y^2 = 9$

60. $x^2 - 4y^2 = 16$
   $x^2 + 4y^2 = 16$

61. $\dfrac{x^2}{4} - \dfrac{y^2}{9} = 1$
   $x + y = 2$

62. $x^2 - y^2 = 4$
   $x + y = 2$

63. $8x^2 - 6y^2 = 24$
   $5x^2 + 3y^2 = 24$

64. $3y^2 - 4x^2 = 12$
   $y^2 + 2x^2 = 34$

65. $\dfrac{y^2}{3} - \dfrac{x^2}{4} = 1$
   $3x - y = 0$

66. $x^2 - 4y^2 = 16$
   $y^2 - 4x^2 = 4$

Applications

67. *Satellite Orbits*   The trajectory of a satellite near Earth can trace a hyperbola, parabola, or ellipse. If the satellite follows either a hyperbolic or parabolic path, it escapes Earth's gravitational influence after a single pass. The path that a satellite travels near Earth depends both on its velocity V in meters per second and its distance D in meters from the center of Earth. Its path is hyperbolic if $V > \dfrac{k}{\sqrt{D}}$, parabolic if $V = \dfrac{k}{\sqrt{D}}$, and elliptic if $V < \dfrac{k}{\sqrt{D}}$, where $k = 2.82 \times 10^7$ is a constant.

(**Sources:** W. Loh, *Dynamics and Thermodynamics of Planetary Entry;* W. Thomson, *Introduction to Space Dynamics.*)

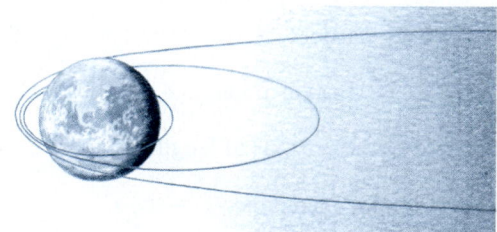

(a) When *Explorer IV* was at a maximum distance of 42.5×10^6 meters from Earth's center, it had a velocity of 2090 meters/second. Determine the shape of its trajectory.

(b) If an orbiting satellite is scheduled to escape Earth's gravity, so that it can travel to another planet, its velocity must be increased so its trajectory changes from elliptic to hyperbolic. What range of velocities would allow *Explorer IV* to leave Earth's influence when it is at a maximum distance?

(c) Explain why it is easier to change a satellite's trajectory from an ellipse to a hyperbola when D is maximum rather than minimum.

68. *Telescopes* (Refer to Figure 10.61.) Suppose that the coordinates of F_1 are (0, 5.2) and the coordinates of F_2 are (0, −5.2). If the coordinates of the vertex of the hyperbolic mirror are (0, 4.1), find the standard equation of a hyperbola whose upper branch coincides with the hyperbolic mirror.

Writing about Mathematics

69. Explain how the center and the asymptotes of a hyperbola are related to the fundamental rectangle.

70. Given the standard equation of a hyperbola, explain how to determine the length of the transverse axis. How can you determine whether the transverse axis is vertical or horizontal?

EXTENDED AND DISCOVERY EXERCISES

1. *Structure of an Atom* In 1911, Ernest Rutherford discovered the basic structure of the atom by "shooting" positively charged alpha particles with a speed of 10^7 meters per second at a piece of gold foil 6×10^{-7} meters thick. Only a small percentage of the alpha particles struck a gold nucleus head-on and were deflected directly back toward their source. The rest of the particles often followed a hyperbolic trajectory because they were repelled by positively charged gold nuclei. Thus, Rutherford proposed that the atom was composed of mostly empty space with a small and dense nucleus. The figure shows an alpha particle A initially approaching a gold nucleus N and being deflected at an angle $\theta = 90°$. N is located at a focus of the hyperbola, and the trajectory of A passes through a vertex of the hyperbola. (**Source:** Semat, H. and J. Albright, *Introduction to Atomic and Nuclear Physics*, Holt, Rinehart and Winston, 1972.)

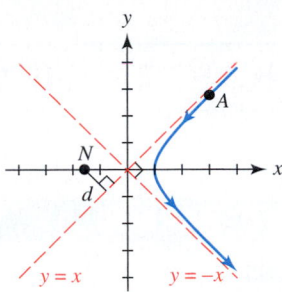

(a) Determine the equation of the trajectory of the alpha particle if $d = 5 \times 10^{-14}$ meters.

(b) What was the minimum distance between the centers of the alpha particle and the gold nucleus?

2. *Sound Detection* Microphones are placed at points (−c, 0) and (c, 0). An explosion occurs at point $P(x, y)$ having a positive x-coordinate.

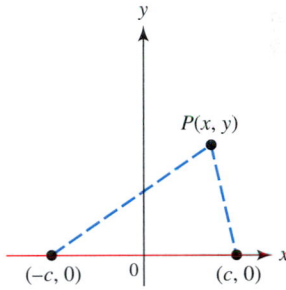

The sound is detected at the closer microphone t seconds before being detected at the farther microphone. Assume that sound travels at a speed of 330 meters per second, and show that P must be on the hyperbola

$$\frac{x^2}{330^2 t^2} - \frac{y^2}{4c^2 - 330^2 t^2} = \frac{1}{4}.$$

CHECKING BASIC CONCEPTS FOR SECTION 10.3

1. Graph the hyperbola defined by $\frac{x^2}{9} - \frac{y^2}{16} = 1$. Include the foci and asymptotes.

2. Find an equation of the hyperbola centered at (1, 3) with a horizontal transverse axis of length 6 and a conjugate axis of length 4. What are the coordinates of the foci?

3. Use the graph shown to the right to find the equation of the hyperbola.

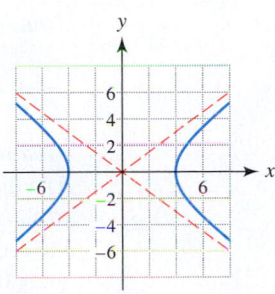

4. Write $9y^2 - 54y - 16x^2 - 32x = 79$ in the standard form for a hyperbola centered at (h, k). Identify the center and the vertices.

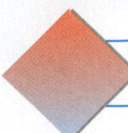

CHAPTER 10 SUMMARY

CONCEPT	EXPLANATION AND EXAMPLES

SECTION 10.1 PARABOLAS

CONIC SECTIONS

Three basic types: parabola, ellipse, and hyperbola
(Circles are examples of ellipses.)

PARABOLAS WITH VERTEX (0, 0)

Standard Forms

$x^2 = 4py$ (vertical axis)   or $y^2 = 4px$ (horizontal axis)

Meaning of p
Both the vertex-focus distance and the vertex-directrix distance are p. The sign of p determines if the parabola opens upward or downward—left or right. The focus is either $(0, p)$ or $(p, 0)$, and the directrix is either $y = -p$ or $x = -p$. Note that either $p > 0$ or $p < 0$ is possible.

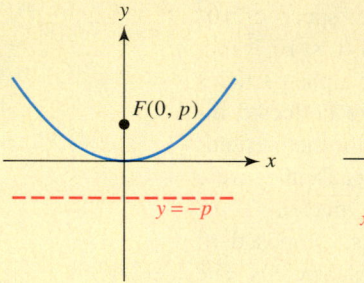

Vertical Axis: $x^2 = 4py$ Horizontal Axis: $y^2 = 4px$

PARABOLAS WITH VERTEX (h, k)

$(x - h)^2 = 4p(y - k)$ Vertical axis; focus $(h, k + p)$
                            Directrix: $y = k - p$

$(y - k)^2 = 4p(x - h)$ Horizontal axis; focus: $(h + p, k)$
                            Directrix: $x = h - p$

SECTION 10.2 ELLIPSES

ELLIPSES WITH CENTER (0, 0)

Standard Forms with a > b > 0

$\dfrac{x^2}{a^2} + \dfrac{y^2}{b^2} = 1$ (horizontal major axis)   or $\dfrac{x^2}{b^2} + \dfrac{y^2}{a^2} = 1$ (vertical major axis)

Meaning of a, b, and c
The distance from the center to a vertex is a, the distance from the center to an endpoint of the minor axis is b, and the distance from the center to a focus is c. The ratio $\frac{c}{a}$ equals the eccentricity e. The values of a, b, and c are related by $c^2 = a^2 - b^2$. The foci are either $(\pm c, 0)$ or $(0, \pm c)$, and the vertices are either $(\pm a, 0)$ or $(0, \pm a)$. If $a = b$, then the ellipse is a circle.

CONCEPT	EXPLANATION AND EXAMPLES

SECTION 10.2 ELLIPSES (CONTINUED)

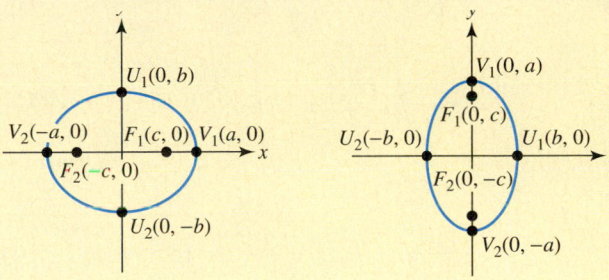

Horizontal Major Axis   Vertical Major Axis

STANDARD EQUATION OF A CIRCLE

$$(x - h)^2 + (y - k)^2 = r^2$$

Center: (h, k); radius: r

ELLIPSES WITH CENTER (h, k)

$$a > b > 0; c^2 = a^2 - b^2$$

$$\frac{(x - h)^2}{a^2} + \frac{(y - k)^2}{b^2} = 1$$

Major axis: horizontal; foci: $(h \pm c, k)$
Vertices: $(h \pm a, k)$

$$\frac{(x - h)^2}{b^2} + \frac{(y - k)^2}{a^2} = 1$$

Major axis: vertical; foci: $(h, k \pm c)$
Vertices: $(h, k \pm a)$

AREA INSIDE AN ELLIPSE

Given the standard equation of an ellipse, the area A of the region contained inside is given by $A = \pi ab$.

SECTION 10.3 HYPERBOLAS

HYPERBOLAS WITH CENTER $(0, 0)$

Standard Forms with $a, b > 0$

$$\frac{x^2}{a^2} - \frac{y^2}{b^2} = 1 \text{ (horizontal transverse axis)} \quad \text{or} \quad \frac{y^2}{a^2} - \frac{x^2}{b^2} = 1 \text{ (vertical transverse axis)}$$

Meaning of a, b, and c

The distance from the center to a vertex is a, and the distance from the center to a focus is c. The asymptotes are $y = \pm \frac{b}{a}x$ if the transverse axis is horizontal, and $y = \pm \frac{a}{b}x$ if the transverse axis is vertical. The values of a, b, and c are related by $c^2 = a^2 + b^2$. The foci are either $(\pm c, 0)$ or $(0, \pm c)$ and the vertices are either $(\pm a, 0)$ or $(0, \pm a)$.

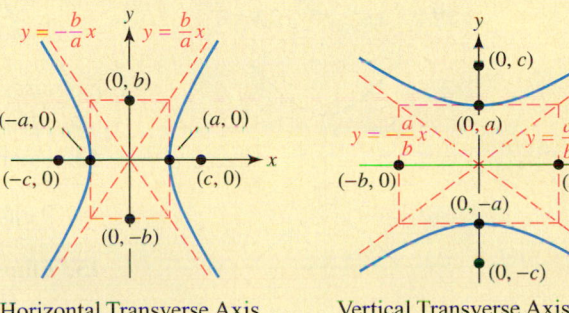

Horizontal Transverse Axis   Vertical Transverse Axis

CONCEPT	EXPLANATION AND EXAMPLES

SECTION 10.3 HYPERBOLAS (CONTINUED)

HYPERBOLAS WITH CENTER (h, k)

$a, b > 0; c^2 = a^2 + b^2$

$$\frac{(x-h)^2}{a^2} - \frac{(y-k)^2}{b^2} = 1$$

Transverse axis: horizontal
Foci: $(h \pm c, k)$; vertices: $(h \pm a, k)$
Asymptotes: $y = \pm \frac{b}{a}(x-h) + k$

$$\frac{(y-k)^2}{a^2} - \frac{(x-h)^2}{b^2} = 1$$

Transverse axis: vertical
Foci: $(h, k \pm c)$; vertices: $(h, k \pm a)$
Asymptotes: $y = \pm \frac{a}{b}(x-h) + k$

REVIEW EXERCISES

Exercises 1–6: Sketch a graph of the equation.

1. $-x^2 = y$

2. $y^2 = 2x$

3. $\dfrac{x^2}{25} + \dfrac{y^2}{49} = 1$

4. $\dfrac{y^2}{4} + \dfrac{x^2}{2} = 1$

5. $\dfrac{y^2}{4} - \dfrac{x^2}{9} = 1$

6. $x^2 - y^2 = 4$

Exercises 7–12: Match the equation with its graph (a.–f.).

7. $x^2 = 2y$

8. $y^2 = -3x$

9. $x^2 + y^2 = 4$

10. $\dfrac{x^2}{36} + \dfrac{y^2}{49} = 1$

11. $\dfrac{x^2}{4} - \dfrac{y^2}{9} = 1$

12. $\dfrac{y^2}{36} - \dfrac{x^2}{25} = 1$

a.

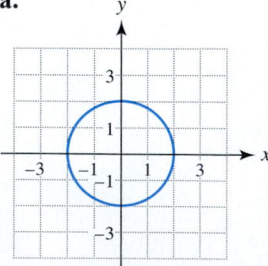

b.

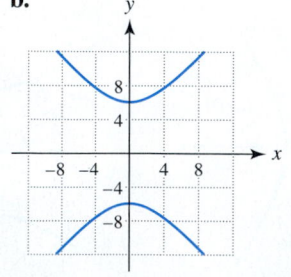

c.

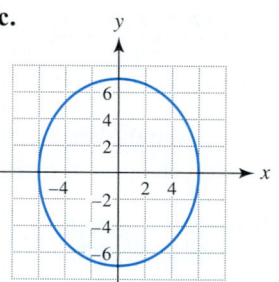

d.

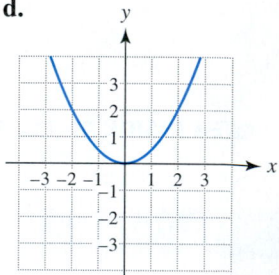

e.

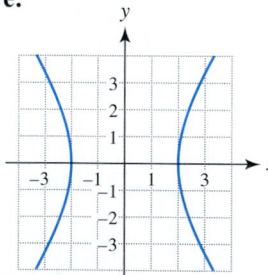

f.

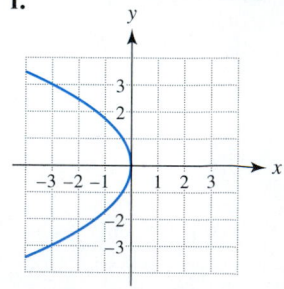

Exercises 13–18: Determine an equation of the conic section that satisfies the given conditions. Sketch its graph.

13. Parabola with focus $(2, 0)$ and vertex $(0, 0)$

14. Parabola with vertex $(5, 2)$ and focus $(5, 0)$

15. Ellipse with foci $(\pm 4, 0)$ and vertices $(\pm 5, 0)$

16. Ellipse centered at the origin with vertical major axis of length 14 and minor axis of length 8

17. Hyperbola with foci $(0, \pm 10)$ and endpoints of the conjugate axis $(\pm 6, 0)$

18. Hyperbola with vertices $(-2 \pm 3, 3)$ and foci given by $(-2 \pm 4, 3)$

Exercises 19–26: Sketch a graph of the conic section. Give the coordinates of any foci.

19. $-4y = x^2$ **20.** $y^2 = 8x$

21. $\dfrac{x^2}{25} + \dfrac{y^2}{4} = 1$ **22.** $49x^2 + 36y^2 = 1764$

23. $\dfrac{x^2}{16} - \dfrac{y^2}{9} = 1$ **24.** $\dfrac{y^2}{4} - x^2 = 1$

25. $x^2 + y^2 = 4$

26. $(x - 3)^2 + (y + 1)^2 = 9$

Exercises 27–30: Sketch a graph of the conic section. Identify the coordinates of its center when appropriate.

27. $\dfrac{(y - 2)^2}{4} + \dfrac{(x + 1)^2}{16} = 1$

28. $\dfrac{(x - 2)^2}{16} - \dfrac{y^2}{4} = 1$

29. $\dfrac{(x - 1)^2}{4} - \dfrac{(y + 1)^2}{4} = 1$

30. $(x + 2) = 4(y - 1)^2$

31. Sketch a graph of $(y - 4)^2 = -8(x - 8)$. Include the focus and the directrix.

32. Sketch a graph of $\dfrac{(x - 3)^2}{16} + \dfrac{(y + 2)^2}{9} = 1$. Include the foci.

Exercises 33–36: Graph the equation.

33. $y^2 = \tfrac{3}{4}x$ **34.** $7.1x^2 + 8.2y^2 = 60$

35. $9x^2 - 2y^2 = 17$

36. $\dfrac{(y - 1.4)^2}{7} - \dfrac{(x + 2.3)^2}{11} = 1$

Exercises 37 and 38: Write the equation in the form given by $(y - k)^2 = a(x - h)$.

37. $-2x = y^2 + 8x + 14$ **38.** $2y^2 - 12y + 16 = x$

Exercises 39–42: Write the equation in standard form for an ellipse or a hyperbola centered at (h, k). Identify the center and the vertices.

39. $4x^2 + 8x + 25y^2 - 250y = -529$

40. $5x^2 + 20x + 2y^2 - 8y = -18$

41. $x^2 + 4x - 4y^2 + 24y = 36$

42. $4y^2 + 8y - 3x^2 + 6x = 11$

43. Find the equation of a circle with center $(0, 0)$ and radius 7.

44. Find the equation of a circle with center $(-7, 6)$ and radius 20.

Exercises 45 and 46: Solve the system of equations.

45. $\dfrac{x^2}{4} + \dfrac{y^2}{4} = 1$ **46.** $x^2 - y^2 = 1$

$\dfrac{x^2}{8} + \dfrac{y^2}{2} = 1$ $x + y = 2$

Exercises 47 and 48. Shade the solution set to the system of inequalities.

47. $\dfrac{x^2}{9} + \dfrac{y^2}{4} \le 1$ **48.** $y^2 - x^2 \le 9$

$x + y \le 3$ $y - x \le 0$

Applications

49. ***Comets***   A comet travels along an elliptical orbit around the sun. Its path can be described by the equation

$$\frac{x^2}{70^2} + \frac{y^2}{500^2} = 1,$$

where units are in millions of miles.

(a) What is the comet's minimum and maximum distance from the sun?

(b) Estimate the distance that the comet travels in one orbit around the sun. (Refer to Exercise 98 from Section 10.2.)

50. ***Searchlight***   A searchlight is being constructed in the shape of a paraboloid with a depth of 7 inches and a diameter of 20 inches, as illustrated in the accompanying figure. Determine the distance d that the bulb should be from the vertex in order to have the beam of light shine straight ahead.

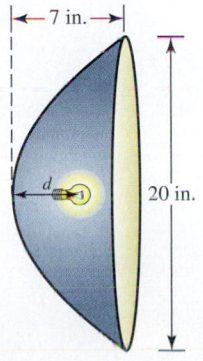

51. *Arch Bridge* An elliptical arch under a bridge is constructed so that it is 80 feet wide and has a maximum height of 30 feet, as illustrated in the accompanying figure. Find the height of the arch 10 feet from the center of the arch.

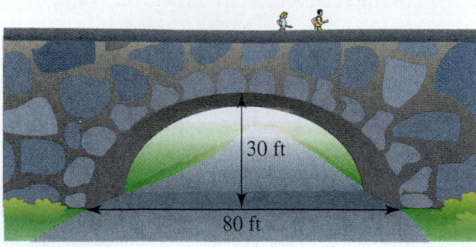

30 ft

80 ft

EXTENDED AND DISCOVERY EXERCISES

Exercises 1–5: Farthest Planet from the Sun Both Neptune and Pluto travel around the sun in elliptical orbits. For Neptune's orbit, $a = 30.10$ and for Pluto's orbit, $a = 39.44$, where the variable a represents the planet's average distance from the sun in astronomical units. (One astronomical unit equals 93 million miles.) The value of the variable a also corresponds to half the length of the major axis. Pluto has a highly eccentric orbit with $e = 0.249$, and Neptune has a nearly circular orbit with $e = 0.009$.

1. Calculate the value of c for each planet's orbit.

2. Position the sun at the *origin* of the xy-plane. Find the coordinates of the center of Neptune's orbit and the coordinates of the center of Pluto's orbit. Assume that the centers lie on the positive x-axis.

3. Find equations for Neptune's orbit and for Pluto's orbit.

4. Graph both orbits in the same xy-plane.

5. Is Pluto always the farthest planet from the sun? Explain.

11 Further Topics in Algebra

The art of asking the right questions in mathematics is more important than the art of solving them.

Georg Cantor

Mathematics permeates the fabric of modern society. At times its presence is subtle and frequently missed by the average person, but its influence is nonetheless profound. Mathematics is the language of technology—mathematics allows society to quantify its experiences.

While technology and visualization provide new ways for us to investigate mathematical problems, they are *not* replacements for mathematical understanding. The human mind is capable of mathematical insight and decision making, but does not usually perform long arithmetic calculations proficiently. On the other hand, computers and calculators are incapable of mathematical insight, but are excellent at performing arithmetic and other routine computation. In this way, technology complements the human mind.

In previous chapters we saw hundreds of examples where mathematics is used to describe physical phenom-ena and events. These included computers, CD players, cars, electricity, light, road construction, government data, telephones, medicine, ecology, business, sports, and psychology. Mathematics is diverse in its ability to adapt to new situations and solve complex problems. If any subject area is studied in enough detail, mathematics usually appears. Mathematics even describes aspects of seemingly intangible concepts such as color and photography.

This chapter introduces further topics in mathematics. It represents only a small fraction of the topics found in mathematics. Although it may be difficult to predict exactly what the twenty-first century will bring, one thing is certain—mathematics will continue to play a very important role.

Source: R. Wolff and L. Yaeger, *Visualization of Natural Phenomena.*

11.1 Sequences

- ◆ Understand basic concepts about sequences
- ◆ Learn how to represent sequences
- ◆ Identify and use arithmetic sequences
- ◆ Identify and use geometric sequences

Introduction

Sequences are a fundamental concept in mathematics and have many applications. A sequence is a function that computes an ordered list. For example, the average person in the United States uses 100 gallons of water each day. The function $f(n) = 100n$ generates the terms of the *sequence*

$$100, 200, 300, 400, 500, 600, 700, \ldots,$$

when $n = 1, 2, 3, 4, 5, 6, 7, \ldots$. This list represents the gallons of water used by the average person after n days.

Basic Concepts

A second example of a sequence involves investing money. If $100 are deposited into a savings account, paying 5% interest compounded annually, then the function defined by $g(n) = 100(1.05)^n$ calculates the account balance after n years, which is given by

$$g(1), g(2), g(3), g(4), g(5), g(6), g(7), \ldots.$$

These terms can be approximated as

$$105, 110.25, 115.76, 121.55, 127.63, 134.01, 140.71, \ldots.$$

We now define a sequence formally.

SEQUENCE

An **infinite sequence** is a function that has the set of natural numbers as its domain. A **finite sequence** is a function with domain $D = \{1, 2, 3, \ldots, n\}$, for some fixed natural number n.

Since sequences are functions, many of the concepts discussed in previous chapters apply to sequences. Instead of letting y represent the output, it is common to write $a_n = f(n)$, where n is a natural number in the domain of the sequence. The **terms** of a sequence are

$$a_1, a_2, a_3, \ldots, a_n, \ldots.$$

The first term is $a_1 = f(1)$, the second term is $a_2 = f(2)$, and so on. The ***n*th term** or **general term** of a sequence is $a_n = f(n)$.

EXAMPLE 1 Finding terms of a sequence

Write the first four terms a_1, a_2, a_3, and a_4 of each sequence, where $a_n = f(n)$.
(a) $f(n) = 2n - 5$ **(b)** $f(n) = 4(2)^{n-1}$
(c) $f(n) = (-1)^n \left(\dfrac{n}{n+1} \right)$

SOLUTION

(a) Evaluate the following.

$$a_1 = f(1) = 2(1) - 5 = -3$$

$$a_2 = f(2) = 2(2) - 5 = -1$$

In a similar manner, $a_3 = f(3) = 1$ and $a_4 = f(4) = 3$.

Algebra Review

To review exponents, see
Chapter R (page R-13).

(b) Since $f(n) = 4(2)^{n-1}$,

$$a_1 = f(1) = 4(2)^{1-1} = 4.$$

Similarly, $a_2 = 8$, $a_3 = 16$, and $a_4 = 32$.

(c) Let $f(n) = (-1)^n \left(\frac{n}{n+1} \right)$, and substitute $n = 1, 2, 3,$ and 4.

$$a_1 = f(1) = (-1)^1 \left(\frac{1}{1+1} \right) = -\frac{1}{2}$$

$$a_2 = f(2) = (-1)^2 \left(\frac{2}{2+1} \right) = \frac{2}{3}$$

$$a_3 = f(3) = (-1)^3 \left(\frac{3}{3+1} \right) = -\frac{3}{4}$$

$$a_4 = f(4) = (-1)^4 \left(\frac{4}{4+1} \right) = \frac{4}{5}$$

Note that the factor $(-1)^n$ causes the terms of the sequence to alternate signs.

Now Try Exercises 1, 3, and 9 ◆

Calculator Help

To learn how to generate a sequence,
see Appendix B (page AP-23).

Graphing calculators may be used to calculate the terms of a sequence. Figures 11.1–11.3 show the first four terms of each sequence in Example 1. The graphing calculator is in sequence mode.

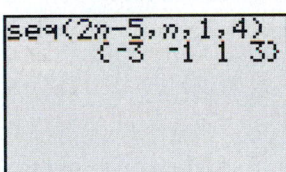

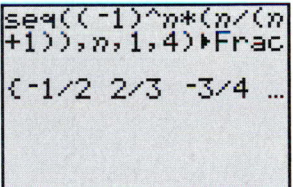

FIGURE 11.1   **FIGURE 11.2**   **FIGURE 11.3**

EXAMPLE 2 Computing terms of a sequence

The average distances of the planets from the sun display a pattern that was first described by Johann Bode in 1772. This relationship is called *Bode's law*, and was proposed even before Uranus, Neptune, and Pluto were discovered. It is a sequence defined by

$$f(1) = 0.4$$

$$f(n) = 0.3(2)^{n-2} + 0.4, \qquad \text{for } n = 2, 3, 4, \ldots, 10,$$

where $a_n = f(n)$. In this sequence, a distance of one unit corresponds to the average Earth–Sun distance. The number n represents the nth planet. The actual distances of the

TABLE 11.1

Planet	Distance
Mercury	0.39
Venus	0.72
Earth	1.00
Mars	1.52
Asteroids	2.8
Jupiter	5.20
Saturn	9.54
Uranus	19.2
Neptune	30.1
Pluto	39.5

planets from the sun, including an average distance for the asteroids from the sun, are listed in Table 11.1.   (**Source:** M. Zeilik, *Introductory Astronomy and Astrophysics*.)

(a) Find $a_4 = f(4)$, and interpret the result.

(b) Calculate the terms of the sequence defined by f. Compare them with the values in Table 11.1.

(c) If there is another planet beyond Pluto, use Bode's law to predict its distance from the sun.

SOLUTION

(a) $a_4 = f(4) = 0.3(2)^{4-2} + 0.4 = 1.6$. Bode's law predicts that the fourth planet, Mars, is located 1.6 times farther from the sun than Earth is.

(b) We will calculate $f(n)$ for $n = 1, 2, 3, \ldots, 10$. For example,

$$f(1) = 0.4 \qquad \text{and}$$
$$f(2) = 0.3(2)^{2-2} + 0.4 = 0.7.$$

Other values are found in a similar manner. The ten terms of the sequence are

$$0.4, \quad 0.7, \quad 1, \quad 1.6, \quad 2.8, \quad 5.2, \quad 10, \quad 19.6, \quad 38.8, \quad 77.2.$$

Bode's law works quite well for the seven inner planets, if we include the asteroids, but fails to predict the location of Neptune at 30.1. It appears to locate planet Pluto.

(c) Since Pluto's distance is 39.5, Bode's law predicts the next planet at 77.2.

Now Try Exercise 11 ◆

Some sequences are not defined using a general term. Instead they are defined *recursively*. With a **recursive sequence**, we must find terms a_1 through a_{n-1} before we can find a_n.

EXAMPLE 3 Finding the terms of a recursive sequence

Find the first four terms of the recursive sequence that is defined by

$$a_n = 2a_{n-1} + 1; \qquad a_1 = 3.$$

SOLUTION The sequence is defined recursively, so we must find the terms in order.

$$a_1 = 3$$
$$a_2 = 2a_1 + 1 = 2(3) + 1 = 7$$
$$a_3 = 2a_2 + 1 = 2(7) + 1 = 15$$
$$a_4 = 2a_3 + 1 = 2(15) + 1 = 31$$

The first four terms are **3**, **7**, **15**, and **31**.

Now Try Exercise 15 ◆

A population model for an insect with a life span of 1 year can be described using a recursive sequence. Suppose each adult female insect produces r female offspring that survive to reproduce the following year. Let $f(n)$ calculate the insect population during year n. Then, the number of female insects is given recursively by

$$f(n) = rf(n-1) \qquad \text{for } n > 1.$$

The number of female insects in the year n is equal to r times the number of female insects in the previous year $n - 1$. This symbolic representation of a function is fundamentally different from any other equation we have encountered thus far. The reason is that the function f is defined symbolically in terms of itself. To evaluate $f(n)$, we evaluate $f(n - 1)$. To evaluate $f(n - 1)$, we evaluate $f(n - 2)$, and so on. If we know the number of adult female insects during the first year, then we can determine the sequence. That is, if $f(1)$ is given, we can determine $f(n)$ by first computing

$$f(1), f(2), f(3), \ldots, f(n-1).$$

The next example illustrates this recursively defined sequence.   (**Source:** D. Brown and P. Rothery, *Models in Biology: Mathematics, Statistics and Computing.*)

EXAMPLE 4 Modeling insect population

Suppose that the initial density of adult female insects is 1000 per acre and $r = 1.1$. Then, the density of female insects during the year n is described by

$$f(1) = 1000$$
$$f(n) = 1.1f(n-1), \qquad n > 1.$$

(a) Rewrite this symbolic representation in terms of a_n.
(b) Find a_4 and interpret the result. Is the density of female insects increasing or decreasing?
(c) A general term for this sequence is given by $f(n) = 1000(1.1)^{n-1}$. Use this representation to find a_4.

SOLUTION

(a) Since $a_n = f(n)$ for all n, $a_{n-1} = f(n-1)$. The sequence can be expressed as

$$a_1 = 1000$$
$$a_n = 1.1a_{n-1}, \qquad n > 1.$$

(b) In order to calculate the fourth term, a_4, we must first determine a_1, a_2, and a_3.

$$a_1 = \mathbf{1000}$$
$$a_2 = 1.1a_1 = 1.1(\mathbf{1000}) = \mathbf{1100}$$
$$a_3 = 1.1a_2 = 1.1(\mathbf{1100}) = \mathbf{1210}$$
$$a_4 = 1.1a_3 = 1.1(\mathbf{1210}) = \mathbf{1331}$$

The fourth term is $a_4 = 1331$. The female population density is increasing and reaches 1331 per acre during the fourth year.

(c) Since $a_4 = f(4)$,

$$a_4 = 1000(1.1)^{4-1} = 1331.$$

It is less work to find a_n using a formula for a general term rather than a recursive formula—particularly if n is large.   *Now Try Exercise 27* ◆

◆ **CLASS DISCUSSION**
How does the value of r in Example 4 affect the population density in future years? ◆

Representations of Sequences

Sequences are functions. Therefore they have graphical, numerical, and symbolic representations.

EXAMPLE 5 Representing a sequence numerically and graphically

Let a recursive sequence be defined as follows.

$$a_1 = 3$$
$$a_n = 2a_{n-1} - 2, \qquad n > 1$$

(a) Give a numerical representation (list each term) of this sequence for $n = 1, 2, 3, 4, 5$.
(b) Graph the first five terms of this sequence.

TABLE 11.2

n	1	2	3	4	5
a_n	3	4	6	10	18

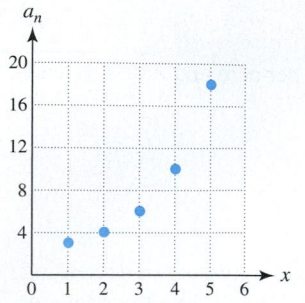

FIGURE 11.4

Calculator Help

To make a table or graph of a sequence, see Appendix B (page AP-23).

SOLUTION

(a) *Numerical Representation*   Start by calculating the first five terms of the sequence.

$$a_1 = 3$$
$$a_2 = 2a_1 - 2 = 2(3) - 2 = 4$$
$$a_3 = 2a_2 - 2 = 2(4) - 2 = 6$$
$$a_4 = 2a_3 - 2 = 2(6) - 2 = 10$$
$$a_5 = 2a_4 - 2 = 2(10) - 2 = 18$$

The first five terms are 3, 4, 6, 10, and 18. A numerical representation of the sequence is shown in Table 11.2.

(b) *Graphical Representation*   To represent these terms graphically, plot the points $(1, 3)$, $(2, 4)$, $(3, 6)$, $(4, 10)$, and $(5, 18)$, as shown in Figure 11.4. Because the domain of a sequence contains only natural numbers, the graph of a sequence is a scatterplot.

Now Try Exercises 13 and 19 ◆

A graphing calculator set in sequence mode may be used to calculate the terms of the recursive sequence in Example 5. See Figures 11.5 and 11.6.

$[0, 6, 1]$ by $[0, 20, 4]$

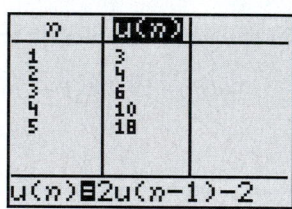

FIGURE 11.5

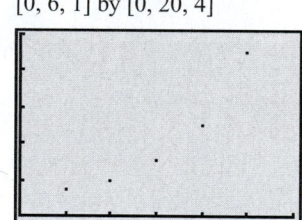

FIGURE 11.6

The next example illustrates numerical and graphical representations for a sequence involving population growth.

EXAMPLE 6   Representing sequences numerically and graphically

Frequently the population of a particular insect does not continue to grow indefinitely, as it does in Example 4. Instead, its population grows rapidly at first, and then levels off because of competition for limited resources. In one study, the behavior of the winter moth was modeled with a sequence similar to the following, where a_n represents the population density in thousands per acre at the beginning of year n.   (**Source:** G. Varley and G. Gradwell, "Population models for the winter moth.")

$$a_1 = 1$$
$$a_n = 2.85a_{n-1} - 0.19a_{n-1}^2, \qquad n \geq 2$$

(a) Give a numerical representation for $n = 1, 2, 3, \ldots, 10$. Describe what happens to the population density of the winter moth.

(b) Use the numerical representation to graph the sequence.

SOLUTION

(a) Evaluate $a_1, a_2, a_3, \ldots, a_{10}$ recursively. Since $a_1 = \mathbf{1}$,

$$a_2 = 2.85a_1 - 0.19a_1^2 = 2.85(\mathbf{1}) - 0.19(\mathbf{1})^2 = \mathbf{2.66}, \qquad \text{and}$$

$$a_3 = 2.85a_2 - 0.19a_2^2 = 2.85(\mathbf{2.66}) - 0.19(\mathbf{2.66})^2 \approx \mathbf{6.24}.$$

Approximate values for other terms are shown in Table 11.3. The figure in the margin shows the computation of the sequence using a calculator. The sequence is denoted $u(n)$ rather than a_n.

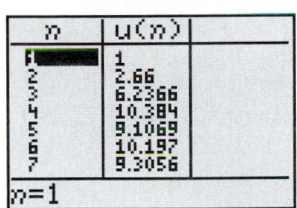

TABLE 11.3

n	1	2	3	4	5	6	7	8	9	10
a_n	1	2.66	6.24	10.4	9.11	10.2	9.31	10.1	9.43	9.98

(b) The graph of a sequence is a set of discrete points. Plot the points

$$(1, 1), (2, 2.66), (3, 6.24), \ldots, (10, 9.98),$$

as shown in Figure 11.7. At first, the insect population increases rapidly, and then oscillates about the line $y = 9.7$. (See the following Class Discussion.) The oscillations become smaller as n increases, indicating that the population density may stabilize near 9.7 thousand per acre. Some calculators can plot sequences, as shown in Figure 11.8. In this figure, the first 20 terms have been plotted.

◆ **CLASS DISCUSSION**

In Example 6, the insect population stabilizes near the value $k = 9.74$ thousand. This value of k can be found by solving the quadratic equation

$$k = 2.85k - 0.19k^2.$$

Try to explain why this is true. ◆

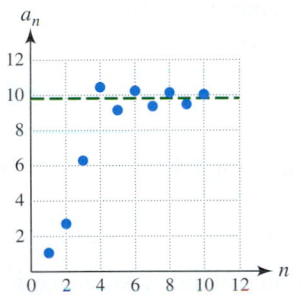

FIGURE 11.7 Insect Population

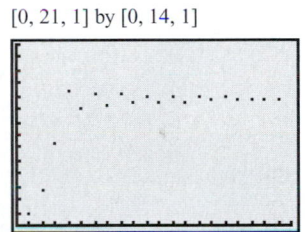

[0, 21, 1] by [0, 14, 1]

FIGURE 11.8   *Now Try Exercise 29* ◆

Arithmetic Sequences

Suppose that a person's starting salary was $20,000 per year. Thereafter this person receives a $1000 raise each year. The salary *after n years* of experience is represented by

$$f(n) = 1000n + 20,000$$

where f is a linear function. After 10 years of experience, the annual salary would be

$$f(\mathbf{10}) = 1000(\mathbf{10}) + 20,000 = \$30,000.$$

If a sequence can be defined by a linear function, it is an *arithmetic sequence.*

Algebra Review

To review linear functions, see Section 2.1.

INFINITE ARITHMETIC SEQUENCE

An **infinite arithmetic sequence** is a linear function whose domain is the set of natural numbers.

An arithmetic sequence can be defined recursively by $a_n = a_{n-1} + d$, where d is a constant. Since $d = a_n - a_{n-1}$ for each valid n, d is called the **common difference**. If $d = 0$, then the sequence is a **constant sequence**. A **finite arithmetic sequence** is similar to an infinite arithmetic sequence except its domain is $D = \{1, 2, 3, \ldots, n\}$, where n is a fixed natural number.

EXAMPLE 7 Determining arithmetic sequences

Determine if f is an arithmetic sequence for each situation.
(a) $f(n) = n^2 + 3n$
(b) A graph of f is shown in Figure 11.9.
(c) The function f is given in Table 11.4.

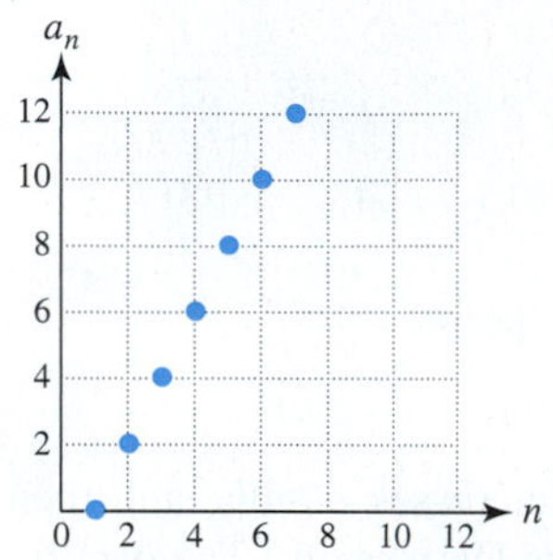

FIGURE 11.9

TABLE 11.4

n	1	2	3	4	5	6	7
$f(n)$	-1.5	0	1.5	3	4.5	6	7.5

SOLUTION
(a) This sequence is not arithmetic, because $f(n) = n^2 + 3n$ is nonlinear.
(b) The sequence in Figure 11.9 represents an arithmetic sequence because the points lie on a line. A linear function could generate these points. Notice that the slope between points is always equal to 2. This slope represents the common difference d of the sequence.
(c) The successive terms

$$-1.5, 0, 1.5, 3, 4.5, 6, 7.5$$

increase by precisely 1.5. Therefore the common difference is $d = 1.5$. Since $a_n = a_{n-1} + 1.5$ for each valid n, the sequence is arithmetic.

Now Try Exercises 63, 67, and 69 ◆

Since an arithmetic sequence is a linear function, it can always be represented by $f(n) = dn + c$, where d is the common difference and c is a constant.

EXAMPLE 8 Finding a symbolic representation for an arithmetic sequence

Find a general term $a_n = f(n)$ for each arithmetic sequence.
(a) $a_1 = 3$ and $d = -2$ (b) $a_3 = 4$ and $a_9 = 17$

SOLUTION
(a) Let $f(n) = dn + c$. Since $d = -2$, $f(n) = -2n + c$.

$$a_1 = f(1) = -2(1) + c = 3 \qquad \text{or} \qquad c = 5$$

Thus $a_n = -2n + 5$.
(b) Since $a_3 = 4$ and $a_9 = 17$, we find a linear function $f(n) = dn + c$ that satisfies the equations $f(3) = 4$ and $f(9) = 17$. The common difference is equal to the slope between the points $(3, 4)$ and $(9, 17)$.

$$d = \frac{17 - 4}{9 - 3} = \frac{13}{6}$$

It follows that $f(n) = \frac{13}{6}n + c$.

$$a_3 = f(3) = \frac{13}{6}(3) + c = 4 \qquad \text{or} \qquad c = -\frac{5}{2}$$

Thus $a_n = \frac{13}{6}n - \frac{5}{2}$.

Now Try Exercises 43 and 47 ◆

◆ **MAKING CONNECTIONS** ────────────

Linear Functions and Arithmetic Sequences   In Chapter 2 we discussed several techniques for finding symbolic representations of linear functions. These methods can be applied to finding symbolic representations of arithmetic sequences. It is important to realize that the mathematical concept of a linear function is simply being applied to the new topic of sequences.

If a_1 is the first term of an arithmetic sequence and d is the common difference, then consecutive terms of the sequence are given by

$$a_2 = a_1 + d$$
$$a_3 = a_2 + d = a_1 + d + d = a_1 + 2d$$
$$a_4 = a_3 + d = a_1 + 2d + d = a_1 + 3d$$
$$a_5 = a_4 + d = a_1 + 3d + d = a_1 + 4d$$

and, in general,

$$a_n = a_1 + (n - 1)d.$$

This result is summarized by the following.

nth TERM OF AN ARITHMETIC SEQUENCE

In an arithmetic sequence with first term a_1 and common difference d, the nth term, a_n, is given by

$$a_n = a_1 + (n - 1)d.$$

EXAMPLE 9   Finding a formula for an arithmetic sequence

Find a symbolic representation (formula) for the arithmetic sequence given by

$$9, 8.5, 8, 7.5, 7, 6.5, 6, \ldots.$$

SOLUTION   The first term is 9. Successive terms can be found by subtracting 0.5 from (or adding -0.5 to) the previous term. Therefore $a_1 = \mathbf{9}$ and $d = -\mathbf{0.5}$ and it follows that

$$a_n = a_1 + (n - 1)d \qquad \text{General formula}$$
$$= \mathbf{9} + (n - 1)(-\mathbf{0.5}) \qquad \text{Substitute.}$$
$$= -0.5n + 9.5. \qquad \text{Simplify.} \qquad \text{Now Try Exercise 31(c)} ◆$$

Geometric Sequences

Suppose that a person with a starting salary of \$20,000 per year receives a 5% raise each year. If $a_n = f(n)$ computes this salary at the *beginning* of the nth year, then

$$f(1) = 20{,}000$$
$$f(2) = 20{,}000(1.05) = 21{,}000$$
$$f(3) = 21{,}000(1.05) = 22{,}050$$
$$f(4) = 22{,}050(1.05) = 23{,}152.50$$

are salaries for the first 4 years. Each salary results from multiplying the previous salary by 1.05. A general term in the sequence can be written as

$$f(n) = 20{,}000(1.05)^{n-1}.$$

During the 10th year, the annual salary is

$$f(10) = 20{,}000(1.05)^{10-1} \approx \$31{,}027.$$

This type of sequence is a *geometric sequence* given by $f(n) = cr^{n-1}$, where c and r are constants. Geometric sequences are capable of either rapid growth or decay. The first five terms from some geometric sequences are shown in Table 11.5. The corresponding values of c and r have been included.

TABLE 11.5   Terms of Geometric Sequences

c	r	a_1, a_2, a_3, a_4, a_5
1	2	$1, 2, 4, 8, 16$
1	$\frac{1}{2}$	$1, \frac{1}{2}, \frac{1}{4}, \frac{1}{8}, \frac{1}{16}$
2	-4	$2, -8, 32, -128, 512$
3	$\frac{1}{10}$	$3, 0.3, 0.03, 0.003, 0.0003$

The terms of a geometric sequence can be found by multiplying the previous term by r. We now define a geometric sequence formally.

INFINITE GEOMETRIC SEQUENCE

An **infinite geometric sequence** is a function defined by $f(n) = cr^{n-1}$, where c and r are nonzero constants. The domain of f is the set of natural numbers.

A geometric sequence can be defined recursively by $a_n = ra_{n-1}$, where $a_n = f(n)$ and the first term is $a_1 = c$. Since $r = \frac{a_n}{a^{n-1}}$ for each valid n, r is referred to as the **common ratio**.

◆ **MAKING CONNECTIONS** ────────────

Exponential Functions and Geometric Sequences   If the domain of an exponential function, $f(x) = Ca^x$, is restricted to the natural numbers, then f represents a geometric sequence. For example, $f(x) = 3(2)^x$ generates the geometric sequence $6, 12, 24, 48, 96, \ldots$ when $x = 1, 2, 3, 4, 5, \ldots$. ◆

The next example illustrates how to recognize symbolic, graphical, and numerical representations of geometric sequences.

EXAMPLE 10 Representing geometric sequences

Decide which of the following represents a geometric sequence.
(a) $a_n = 4(0.5)^n$
(b) The sequence a_n is represented graphically in Figure 11.10.
(c) Table 11.6 is a numerical representation of a_n.

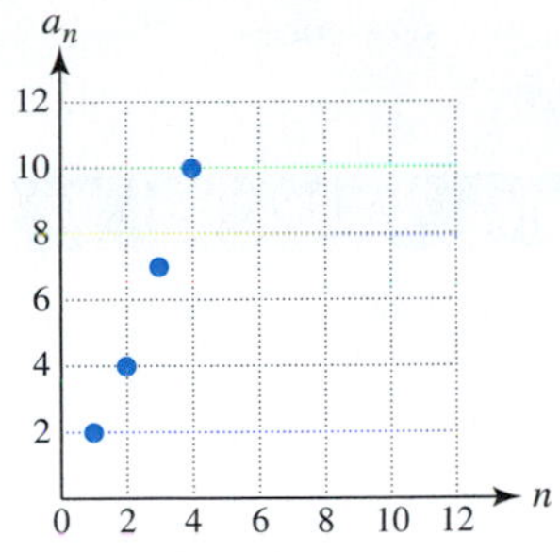

FIGURE 11.10

TABLE 11.6

n	1	2	3	4	5	6
a_n	1	−3	9	−27	81	−243

SOLUTION
(a) The formula for a_n can be written as

$$a_n = 4(0.5)^n = 4(0.5)(0.5)^{n-1} = 2(0.5)^{n-1}.$$

Thus a_n represents a geometric sequence with $c = 2$ and $r = 0.5$.
(b) The points on the graph are $(1, 2)$, $(2, 4)$, $(3, 7)$, and $(4, 10)$. Thus $a_1 = 2$, $a_2 = 4$, $a_3 = 7$, and $a_4 = 10$. Taking ratios of successive terms results in

$$\frac{a_2}{a_1} = 2, \qquad \frac{a_3}{a_2} = \frac{7}{4}, \qquad \text{and} \qquad \frac{a_4}{a_3} = \frac{10}{7}.$$

Since these ratios are not equal, there is no *common* ratio. The sequence is not geometric.
(c) Note that each term in

$$1, -3, 9, -27, 81, -243$$

results from multiplying the previous term by -3. This sequence can be written either as

$$a_n = -3a_{n-1} \qquad \text{with} \qquad a_1 = 1$$

or as

$$a_n = (-3)^{n-1}.$$

Therefore the sequence is geometric. *Now Try Exercises 71, 75, and 77* ◆

EXAMPLE 11 Finding a symbolic representation for a geometric sequence

Find a general term a_n for each geometric sequence.
(a) $a_1 = 5$ and $r = 1.12$ **(b)** $a_2 = 8$ and $a_5 = 512$

SOLUTION
(a) Since the first term is 5 and the common ratio is 1.12, $a_n = 5(1.12)^{n-1}$.
(b) We must find $a_n = cr^{n-1}$ so that $a_2 = 8$ and $a_5 = 512$. Start by determining the common ratio r. Since

$$\frac{a_5}{a_2} = \frac{cr^{5-1}}{cr^{2-1}} = \frac{r^4}{r^1} = r^3 \qquad \text{and} \qquad \frac{a_5}{a_2} = \frac{512}{8} = 64,$$

$r^3 = 64$ or $r = 4$. So, $a_n = c(4)^{n-1}$. Now,

$$a_2 = c(4)^{2-1} = 8 \qquad \text{or} \qquad c = 2.$$

Thus $a_n = 2(4)^{n-1}$. *Now Try Exercises 53 and 57* ◆

11.1

Putting it all Together

$\mathbf{T}$he following table summarizes some fundamental concepts about sequences.

Concept	Explanation	Example
Infinite sequence	A function f whose domain is the set of natural numbers; denoted $a_n = f(n)$; the terms are $a_1, a_2, a_3, \dots$.	$f(n) = n^2 - 2n$, where $a_n = f(n)$. The first three terms are $a_1 = 1^2 - 2(1) = -1$ $a_2 = 2^2 - 2(2) = 0$ $a_3 = 3^2 - 2(3) = 3$. Graphs of sequences are scatterplots.
Recursive sequence	Defined in terms of previous terms; a_1 through a_{n-1} must be calculated before a_n can be found.	$a_n = 2a_{n-1},\ a_1 = 1$ $a_1 = 1, a_2 = 2, a_3 = 4$, and $a_4 = 8$ A new term is found by multiplying the previous term by 2.
Arithmetic sequence	A *linear* function whose domain is the natural numbers; $a_n = dn + c$ or $a_n = a_{n-1} + d$, with common difference d General term is $a_n = a_1 + (n - 1)d$.	$f(n) = 2n - 1$, where $a_n = f(n)$. $a_1 = 1, a_2 = 3, a_3 = 5$, and $a_4 = 7$ Consecutive terms increase by the common difference $d = 2$. The points on the graph of this sequence lie on a line with slope 2.
Geometric sequence	$f(n) = cr^{n-1}$, where c is constant and r is the common ratio; may also be written as $a_n = ra_{n-1}$	$f(n) = 2(3)^{n-1}$, where $a_n = f(n)$. $a_1 = 2, a_2 = 6, a_3 = 18$, and $a_4 = 54$ Consecutive terms are found by multiplying the previous term by the common ratio $r = 3$. Points on the graph of a geometric sequence do not lie on a line.

11.1 — Exercises

Finding Terms of Sequences

Exercises 1–12: Find the first four terms of the sequence.

1. $a_n = 2n + 1$

2. $a_n = 3(n - 1) + 5$

3. $a_n = 4(-2)^{n-1}$

4. $a_n = 2(3)^n$

5. $a_n = \dfrac{n}{n^2 + 1}$

6. $a_n = 5 - \dfrac{1}{n^2}$

7. $a_n = (-1)^n \left(\dfrac{1}{2} \right)^n$

8. $a_n = (-1)^n \left(\dfrac{1}{n} \right)$

9. $a_n = (-1)^{n-1} \left(\dfrac{2^n}{1 + 2^n} \right)$

10. $a_n = (-1)^{n-1} \left(\dfrac{1}{3^n} \right)$

11. $a_n = 2^n + n^2$

12. $a_n = \dfrac{1}{n} + \dfrac{1}{3n}$

Exercises 13 and 14: Use the graphical representation to write the terms of the sequence.

13.

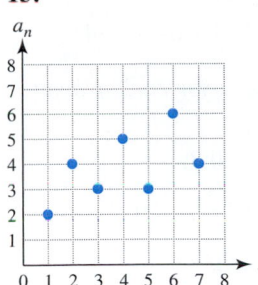

14.

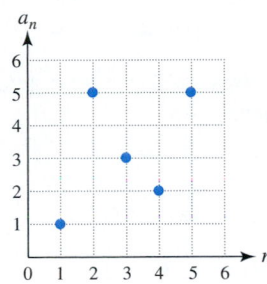

Exercises 15–24: Complete the following for the recursively defined sequence.

(a) *Find the first four terms.*

(b) *Graph these terms.*

15. $a_n = 2a_{n-1}$; $a_1 = 1$

16. $a_n = a_{n-1} + 5$; $a_1 = -4$

17. $a_n = a_{n-1} - a_{n-2}$; $a_1 = 2$, $a_2 = 5$

18. $a_n = 2a_{n-1} + a_{n-2}$; $a_1 = 0$, $a_2 = 1$

19. $a_n = a_{n-1}^2$; $a_1 = 2$

20. $a_n = \frac{1}{2}a_{n-1}^3 + 1$; $a_1 = 0$

21. $a_n = a_{n-1} + n$; $a_1 = 1$

22. $a_n = 3a_{n-1}^2$; $a_1 = 2$

23. $a_n = a_{n-1}a_{n-2}$; $a_1 = 2$, $a_2 = 3$

24. $a_n = 2a_{n-1}^2 + a_{n-2}$; $a_1 = 2$, $a_2 = 1$

Modeling Insect and Bacteria Populations

Exercises 25 and 26: Insect Population The annual population density of a species of insect after n years is modeled by a sequence. Use the graph to discuss trends in the insect population.

25.

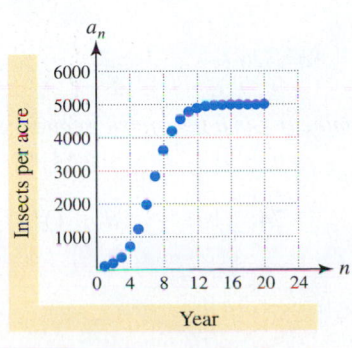

26.

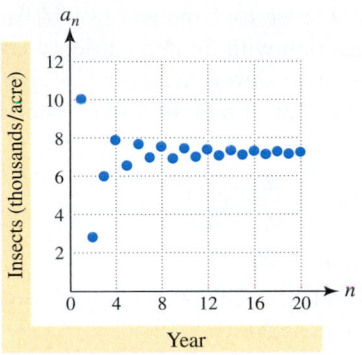

27. *Insect Population*   (Refer to Example 4.) Suppose that the density of female insects during the first year is 500 per acre with $r = 0.8$.

(a) Write a recursive sequence that describes these data, where a_n denotes the female insect density during year n.

(b) Find the terms $a_1, a_2, a_3, \ldots, a_6$. Interpret the results.

(c) Find a formula for a_n.

28. *Bacteria Growth*   Some strains of bacteria are incapable of producing an amino acid necessary for cell division, called *histidine*. If such bacteria are cultured in a medium with sufficient histidine, they can double their size and divide every 40 minutes. (**Source:** F. Hoppensteadt and C. Peskin, *Mathematics in Medicine and the Life Sciences.*)

(a) Write a recursive sequence that describes this growth where each value of n represents a 40-minute interval. Let $a_1 = 300$ represent the initial number of bacteria per milliliter. Find the first five terms.

(b) Determine the number of bacteria per milliliter after 10 hours have elapsed.

(c) Is this sequence arithmetic or geometric? Explain.

29. *Insect Population*   (Refer to Example 6.) Suppose an insect population density at the beginning of year n can be modeled by the recursively defined sequence

$$a_1 = 8$$
$$a_n = 2.9a_{n-1} - 0.2a_{n-1}^2, \qquad n > 1.$$

(a) Find the population for $n = 1, 2, 3$.

(b) Graph the sequence for $n = 1, 2, 3, \ldots, 20$. Interpret the graph.

30. *Bacteria Growth* (Refer to Exercise 28.) If bacteria are cultured in a medium with limited nutrients, competition ensues and growth slows. According to *Verhulst's model*, the number of bacteria at 40-minute intervals is given by

$$a_n = \left(\frac{2}{1 + a_{n-1}/K} \right) a_{n-1},$$

where K is a constant.

(a) Let $a_1 = 200$ and $K = 10,000$. Graph the sequence for $n = 1, 2, 3, \ldots, 20$.

(b) Describe the growth of these bacteria.

(c) Trace the graph of the sequence. Make a conjecture as to why K is called the **saturation constant**. Test your conjecture by changing the value of K.

Representations of Sequences

Exercises 31–36: The first five terms of an arithmetic sequence are given. Find

(a) *numerical,*
(b) *graphical, and*
(c) *symbolic*

representations of the sequence. Include at least eight terms of the sequence.

31. $1, 3, 5, 7, 9$

32. $4, 1, -2, -5, -8$

33. $7.5, 6, 4.5, 3, 1.5$

34. $5.1, 5.5, 5.9, 6.3, 6.7$

35. $\frac{1}{2}, 2, \frac{7}{2}, 5, \frac{13}{2}$

36. $2, 4, 6, 8, 10$

Exercises 37–42: The first five terms of a geometric sequence are given. Find

(a) *numerical,*
(b) *graphical, and*
(c) *symbolic*

representations of the sequence. Include at least eight terms of the sequence.

37. $8, 4, 2, 1, \frac{1}{2}$

38. $32, -8, 2, -\frac{1}{2}, \frac{1}{8}$

39. $\frac{3}{4}, \frac{3}{2}, 3, 6, 12$

40. $\frac{1}{27}, \frac{1}{9}, \frac{1}{3}, 1, 3$

41. $-\frac{1}{4}, -\frac{1}{2}, -1, -2, -4$

42. $9, 6, 4, \frac{8}{3}, \frac{16}{9}$

Exercises 43–52: Find a general term a_n for the arithmetic sequence.

43. $a_1 = 5, d = -2$

44. $a_1 = -3, d = 5$

45. $a_3 = 1, d = 3$

46. $a_4 = 12, d = -10$

47. $a_2 = 5, a_6 = 13$

48. $a_3 = 22, a_{17} = -20$

49. $a_1 = 8, a_4 = 17$

50. $a_1 = -2, a_5 = 8$

51. $a_5 = -4, a_8 = -2.5$

52. $a_3 = 10, a_7 = -4$

Exercises 53–62: Find a general term a_n for each geometric sequence.

53. $a_1 = 2, r = \frac{1}{2}$

54. $a_1 = 0.8, r = -3$

55. $a_3 = \frac{1}{32}, r = -\frac{1}{4}$

56. $a_4 = 3, r = 3$

57. $a_3 = 2, a_6 = \frac{1}{4}$

58. $a_2 = 6, a_4 = 24$

59. $a_1 = 10, a_2 = 2$

60. $a_1 = -5, a_3 = -125, r < 0$

61. $a_2 = -1, a_7 = -32$

62. $a_2 = \frac{9}{4}, a_4 = \frac{81}{4}, r < 0$

Identifying Types of Sequences

Exercises 63–70: Determine if f is an arithmetic sequence.

63. $f(n) = 4 - 3n^3$

64. $f(n) = 2(n - 1)$

65. $f(n) = 4n - (3 - n)$

66. $f(n) = n^2 - n + 2$

67.

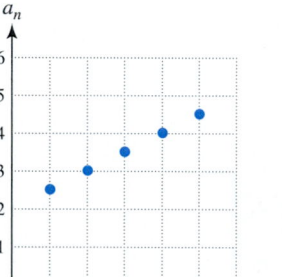

68.

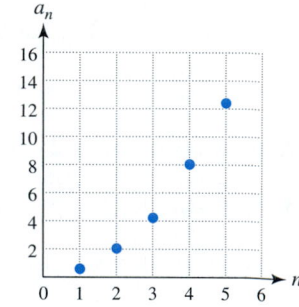

69.

n	1	2	3	4	5
$f(n)$	3	1	-1	-3	-5

70.

n	1	2	3	4	5
$f(n)$	1	4	9	16	25

Exercises 71–78: Determine if f is a geometric sequence.

71. $f(n) = 4(2)^{n-1}$

72. $f(n) = -3(0.25)^n$

73. $f(n) = -3(n)^2$

74. $f(n) = 2(n - 1)^n$

75.

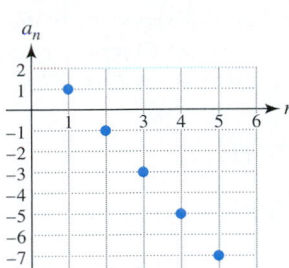

76.

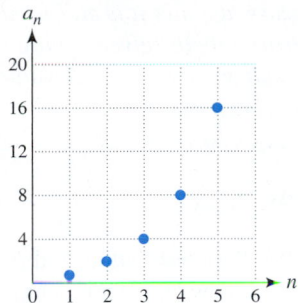

77.

n	1	2	3	4	5
$f(n)$	$\frac{1}{2}$	$\frac{3}{4}$	1	$\frac{5}{4}$	$\frac{5}{2}$

78.

n	1	2	3	4	5
$f(n)$	9	3	1	$\frac{1}{3}$	$\frac{1}{9}$

Exercises 79–84: Given the terms of a finite sequence, classify it as arithmetic, geometric, or neither.

79. $-5, 2, 9, 16, 23, 30$ **80.** $5, 2, -2, -6, -11$

81. $2, 8, 32, 128, 512$

82. $5.75, 5.5, 5.25, 5, 4.75, 4.5$

83. $100, 110, 130, 160, 200$

84. $0.7, 0.21, 0.063, 0.0189, 0.00567$

Exercises 85–88: Use the graph to determine if the sequence is arithmetic or geometric. If the sequence is arithmetic, state the sign of the common difference d and estimate its value. If the sequence is geometric, give the sign of the common ratio r and state if $|r| < 1$ or $|r| > 1$.

85.

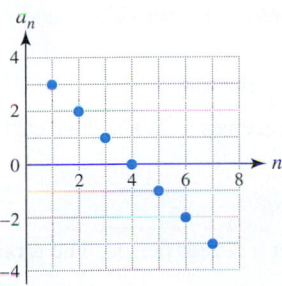

86.

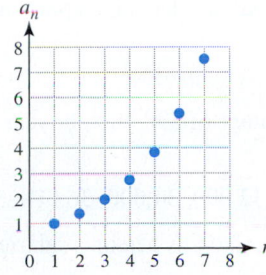

87.

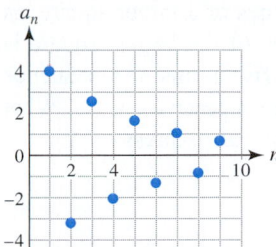

88.

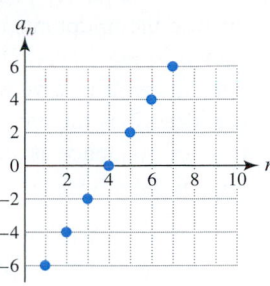

89. *Fibonacci Sequence* The *Fibonacci sequence* dates back to 1202. It is one of the most famous sequences in mathematics and can be defined recursively by

$$a_1 = 1, a_2 = 1$$

$$a_n = a_{n-1} + a_{n-2} \qquad \text{for } n > 2.$$

(a) Find the first 12 terms of this sequence.

(b) Compute $\frac{a_n}{a_{n-1}}$ when $n = 2, 3, 4, \ldots, 12$. What happens to this ratio?

(c) Show that for $n = 2, 3$, and 4 the terms of the Fibonacci sequence satisfy the equation

$$a_{n-1} \cdot a_{n+1} - a_n^2 = (-1)^n.$$

90. *Bouncing Ball* If a tennis ball is dropped, it bounces or rebounds to 80% of its initial height.

(a) Write the first five terms of a sequence that gives the maximum height attained by the tennis ball on each rebound when it is dropped from an initial height of 5 feet. Let $a_1 = 5$. What type of sequence is this?

(b) Give a graphical representation of these terms.

(c) Find a general term a_n.

91. *Salary Increases* Suppose an employee's initial salary is $30,000.

(a) If this person receives a $2000 raise for each year of experience, determine a sequence that gives the salary at the beginning of the nth year. What type of sequence is this?

(b) Suppose another employee has the same starting salary and receives a 5% raise after each year. Find a sequence that computes the salary at the beginning of the nth year. What type of sequence is this?

(c) Which salary is higher at the beginning of the 10th year, and the 20th year?

(d) Graph both sequences in the same viewing rectangle. Compare the two salaries.

92. *Area* A sequence of smaller squares is formed by connecting the midpoints of the sides of a larger square, as shown in the figure. If the area of the largest square is one square unit, give the first five terms of a sequence that describes the area of each successive square. What type of sequence is this? Write an expression for the area of the nth square.

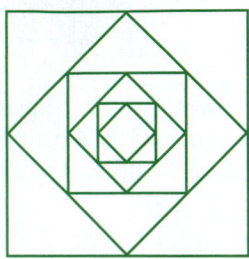

*Exercises 93–96: **Computing Square Roots*** *The following recursively defined sequence can be used to compute $\sqrt{k}$ for any positive number k.*

$$a_1 = k$$

$$a_n = \frac{1}{2}\left(a_{n-1} + \frac{k}{a_{n-1}}\right)$$

This sequence was known to Sumerian mathematicians 4000 years ago, but it is still used today. Use this sequence to approximate the given square root by finding a_6. Compare your result with the actual value. (**Source:** P. Heinz-Otto, *Chaos and Fractals*.)

93. $\sqrt{2}$ **94.** $\sqrt{11}$

95. $\sqrt{21}$ **96.** $\sqrt{41}$

97. Suppose that a_n and b_n represent arithmetic sequences. Show that their sum, $c_n = a_n + b_n$, is also an arithmetic sequence.

98. Explain why the sequence $\log 2$, $\log 4$, $\log 8$, $\log 16$, . . . , is an arithmetic sequence.

Writing about Mathematics

99. Explain how we can distinguish between an arithmetic and a geometric sequence. Give examples.

100. Compare and contrast a sequence whose nth term is given by $a_n = f(n)$, and a sequence that is defined recursively. Give examples. Which symbolic representation for defining a sequence is usually more convenient to use? Explain why.

11.2   Series

- Understand basic concepts about series
- Identify and find the sum of arithmetic series
- Identify and find the sum of geometric series
- Learn and use summation notation

Introduction

Although the terms *sequence* and *series* are sometimes used interchangeably in everyday English, they represent different mathematical concepts. In mathematics, a sequence is a function whose domain is the set of natural numbers, whereas a series is a summation of the terms in a sequence. Series have played a central role in the development of modern mathematics. Today series are often used to approximate functions that are too complicated to have simple formulas. Series are also instrumental in calculating approximations of numbers like π and e.

Basic Concepts

Suppose a person has a starting salary of \$30,000 per year and receives a \$2000 raise each year. Then,

$$30{,}000,\ 32{,}000,\ 34{,}000,\ 36{,}000,\ 38{,}000$$

are terms of the sequence that describe this person's salaries over a 5-year period. The total earned is given by the finite *series*

$$30{,}000 + 32{,}000 + 34{,}000 + 36{,}000 + 38{,}000,$$

whose sum is $170,000. Any sequence can be used to define a series. For example, the infinite sequence

$$1, \frac{1}{3}, \frac{1}{9}, \frac{1}{27}, \frac{1}{81}, \frac{1}{243}, \cdots$$

defines the terms of the infinite series

$$1 + \frac{1}{3} + \frac{1}{9} + \frac{1}{27} + \frac{1}{81} + \frac{1}{243} + \cdots.$$

We now define the concept of a series, where $a_1, a_2, a_3, \ldots, a_n, \ldots$ represent terms of a sequence.

SERIES

A **finite series** is an expression of the form

$$a_1 + a_2 + a_3 + \cdots + a_n,$$

and an **infinite series** is an expression of the form

$$a_1 + a_2 + a_3 + \cdots + a_n + \cdots.$$

An infinite series contains infinitely many terms. We must define what is meant by the sum of an *infinite* series. Let the following be a **sequence of partial sums**.

$$S_1 = a_1$$
$$S_2 = a_1 + a_2$$
$$S_3 = a_1 + a_2 + a_3$$
$$\vdots$$
$$S_n = a_1 + a_2 + a_3 + \cdots + a_n$$

If S_n approaches a real number S as $n \to \infty$, then the sum of the infinite series is S. For example, let $S_1 = 0.3$, $S_2 = 0.3 + 0.03$, $S_3 = 0.3 + 0.03 + 0.003$, and so on. Then, as $n \to \infty$, $S_n \to \frac{1}{3}$. We say that the infinite series

$$0.3 + 0.03 + 0.003 + 0.0003 + \cdots = 0.3333 \ldots = 0.\overline{3}$$

has sum $\frac{1}{3}$. Some infinite series do not have a sum S. For example, the series $1 + 2 + 3 + 4 + 5 + \cdots$ would have an unbounded or "infinite" sum.

EXAMPLE 1   Interpreting a sequence of partial sums

Suppose that a_1 represents the number of U.S. AIDS deaths reported in 1991, a_2 the number in 1992, a_3 in 1993, and so on.
(a) Write a series that represents the total number of AIDS deaths from 1995 to 2002.
(b) Interpret S_{10}.

SOLUTION
(a) The series $a_5 + a_6 + a_7 + a_8 + a_9 + a_{10} + a_{11} + a_{12}$ represents the total number of AIDS deaths reported from 1995 to 2002.
(b) Since $S_{10} = a_1 + a_2 + a_3 + \cdots + a_{10}$, it follows that S_{10} represents the total number of AIDS deaths reported from 1991 to 2000.   *Now Try Exercise 1* ◆

EXAMPLE 2   Finding partial sums

For each a_n, calculate S_4.

(a) $a_n = 2n + 1$ **(b)** $a_n = n^2$

SOLUTION

(a) Since $S_4 = a_1 + a_2 + a_3 + a_4$, start by calculating the first four terms of the sequence $a_n = 2n + 1$.

$$a_1 = 2(1) + 1 = 3; \quad a_2 = 2(2) + 1 = 5;$$
$$a_3 = 2(3) + 1 = 7; \quad a_4 = 2(4) + 1 = 9$$

Thus $S_4 = \underbrace{3 + 5 + 7 + 9}_{4 \text{ terms}} = 24$.

(b) $a_1 = 1^2 = 1; \quad a_2 = 2^2 = 4; \quad a_3 = 3^2 = 9; \quad a_4 = 4^2 = 16$

Thus $S_4 = 1 + 4 + 9 + 16 = 30$.   *Now Try Exercises 3 and 7* ◆

The techniques used to calculate π throughout history is a fascinating story. Since π is an irrational number, it cannot be represented exactly by a fraction. Its decimal expansion neither repeats nor does it have a discernible pattern. The ability to compute π was essential to the development of a society, because π appears in formulas used in construction, surveying, and geometry. In early historical records, π was given the value of 3. Later the Egyptians used a value of

$$\frac{256}{81} \approx 3.1605.$$

It was not until the discovery of series that exceedingly accurate decimal approximations of π were possible. In 2002, after 400 hours of supercomputer time, π was computed to 1.24 trillion digits. Why would anyone want to compute π to so many decimal places? One practical reason is to test electrical circuits in new computers. If a computer has a small defect in its hardware, there is a good chance that an error will appear after performing trillions of arithmetic calculations during the computation of π. (**Sources:** P. Beckmann, *A History of PI;* P. Heinz-Otto, *Chaos and Fractals.*)

EXAMPLE 3   Computing π with a series

The infinite series given by

$$\frac{\pi^4}{90} = \frac{1}{1^4} + \frac{1}{2^4} + \frac{1}{3^4} + \frac{1}{4^4} + \frac{1}{5^4} + \cdots + \frac{1}{n^4} + \cdots$$

can be used to estimate π.

(a) Approximate π by finding the sum of the first four terms.

(b) Use technology to approximate π by summing the first 50 terms. Compare the result to the actual value of π.

SOLUTION

(a) Summing the first four terms results in the following.

$$\frac{\pi^4}{90} \approx \frac{1}{1^4} + \frac{1}{2^4} + \frac{1}{3^4} + \frac{1}{4^4} \approx 1.078751929$$

This approximation can be solved for π by multiplying by 90 and then taking the fourth root. Thus

$$\pi \approx \sqrt[4]{90(1.078751929)} \approx 3.139.$$

FIGURE 11.11

Calculator Help

To find the sum of the terms of a series, see Appendix B (page AP-24).

(b) Some calculators are capable of summing the terms of a sequence, as shown in Figure 11.11. (Summing the terms of a sequence is equivalent to finding the sum of a series.) The first 50 terms of the series provides an approximation of $\pi \approx 3.141590776$. This computation matches the actual value of π for the first five decimal places.

Now Try Exercise 97 ◆

Arithmetic Series

Summing the terms of an arithmetic sequence results in an **arithmetic series**. For example, the sequence defined by $a_n = 2n - 1$ for $n = 1, 2, 3, \ldots, 7$ is the arithmetic sequence

$$1, 3, 5, 7, 9, 11, 13.$$

The corresponding arithmetic series is

$$1 + 3 + 5 + 7 + 9 + 11 + 13.$$

The following formula can be used to sum the first n terms of an arithmetic sequence. (For a proof see Exercise 3 in the Extended and Discovery Exercises at the end of this chapter.)

SUM OF THE FIRST n TERMS OF AN ARITHMETIC SEQUENCE

The **sum of the first n terms of an arithmetic sequence**, denoted S_n, is found by averaging the first and nth terms and then multiplying by n. That is,

$$S_n = a_1 + a_2 + a_3 + \cdots + a_n = n\left(\frac{a_1 + a_n}{2}\right).$$

Since $a_n = a_1 + (n - 1)d$, S_n can also be written as follows.

$$S_n = n\left(\frac{a_1 + a_n}{2}\right)$$

$$= \frac{n}{2}(a_1 + a_1 + (n - 1)d)$$

$$= \frac{n}{2}(2a_1 + (n - 1)d)$$

EXAMPLE 4 Finding the sum of an arithmetic series

Use a formula to find the sum of the arithmetic series

$$2 + 4 + 6 + 8 + \cdots + 100.$$

SOLUTION The series has $n = 50$ terms with $a_1 = 2$ and $a_{50} = 100$. We can use the formula

$$S_n = n\left(\frac{a_1 + a_n}{2}\right)$$

to find its sum.

$$S_{50} = 50\left(\frac{2 + 100}{2}\right)$$

$$= 2550$$

We can also use the formula

$$S_n = \frac{n}{2}(2a_1 + (n - 1)d)$$

with common difference $d = 2$ to find this sum.

$$S_{50} = \frac{50}{2}(2(2) + (50 - 1)2)$$

$$= 25(102)$$

$$= 2550$$

The two answers agree, as expected.　　　　　　　　　*Now Try Exercise 11* ◆

EXAMPLE 5　**Finding the sum of a finite arithmetic series**

A person has a starting annual salary of \$30,000 and receives a \$1500 raise each year.
(a) Calculate the total amount earned over 10 years.
(b) Verify this value using a calculator.

SOLUTION
(a) The arithmetic sequence describing the salary during year n is computed by

$$a_n = 30,000 + 1500(n - 1).$$

The first and tenth year's salaries are:

$$a_1 = 30,000 + 1500(1 - 1) = 30,000$$

$$a_{10} = 30,000 + 1500(10 - 1) = 43,500.$$

Thus the total amount earned during this 10-year period is

$$S_{10} = 10\left(\frac{30,000 + 43,500}{2}\right) = \$367,500.$$

This sum can also be found using $\boldsymbol{S_n = \frac{n}{2}(2a_1 + (n - 1)d)}$.

$$S_{10} = \frac{10}{2}(2 \cdot 30,000 + (10 - 1)1500) = \$367,500$$

(b) To verify this with a calculator, compute the sum

$$a_1 + a_2 + a_3 + \cdots + a_{10},$$

where $a_n = 30,000 + 1500(n - 1)$. This calculation is shown in Figure 11.12. The result of 367,500 agrees with part (a).　　　*Now Try Exercise 91* ◆

FIGURE 11.12

For tax purposes, businesses frequently depreciate equipment. Two methods of depreciation are *straight-line depreciation* and *sum-of-the-years'-digits*.

Suppose a college student buys a \$2000 computer to start a business that develops Web sites. This student estimates the life of the computer at 5 years, after which its value will be \$200. The difference between \$2000 and \$200 is \$1800, which may be deducted from his or her taxable income over a 5-year period.

In straight-line depreciation, equal portions of \$1800 are deducted over 5 years. The *sum-of-the-years'-digits* method calculates depreciation differently. A computer with a useful life of 5 years has a sum-of-the-years'-digits computed by

$$1 + 2 + 3 + 4 + 5 = 15.$$

With this method, $\frac{5}{15}$ of $1800 is deducted the first year, $\frac{4}{15}$ the second year, and so on, until $\frac{1}{15}$ is deducted the fifth year. Both depreciation methods deduct a total of $1800 over 5 years.   (**Source:** Sharp Electronics Corporation, *Conquering the Sciences.*)

EXAMPLE 6   Calculating depreciation

For each of the previous depreciation methods, complete the following.
(a) Find an arithmetic sequence that gives the amount depreciated each year.
(b) Write a series whose sum is the amount depreciated over 5 years.

SOLUTION
(a) With straight-line depreciation, $\frac{1}{5}(1800) = \$360$ is depreciated each year. The sequence describing this depreciation is

$$360, \ 360, \ 360, \ 360, \ 360.$$

For sum-of-the-years'-digits, the following depreciation schedule is computed.

$$\frac{5}{15} \text{ of } \$1800 = \$600$$

$$\frac{4}{15} \text{ of } \$1800 = \$480$$

$$\frac{3}{15} \text{ of } \$1800 = \$360$$

$$\frac{2}{15} \text{ of } \$1800 = \$240$$

$$\frac{1}{15} \text{ of } \$1800 = \$120$$

The arithmetic sequence describing these amounts is

$$600, \ 480, \ 360, \ 240, \ 120.$$

(b) The arithmetic series that describes the total amount deducted for straight-line depreciation is

$$360 + 360 + 360 + 360 + 360.$$

Although we know that the amount depreciated is equal to $1800, it can also be verified by using the following formula.

$$S_5 = 5\left(\frac{360 + 360}{2}\right) = \$1800$$

For the sum-of-the-years'-digits, the total depreciation is calculated by the series

$$600 + 480 + 360 + 240 + 120.$$

Note that this series also sums to $1800.

$$S_5 = 5\left(\frac{600 + 120}{2}\right) = \$1800$$

Now Try Exercise 95 ◆

EXAMPLE 7

Finding a term of an arithmetic series

◆ **CLASS DISCUSSION**
Explain why a formula for the sum of an infinite arithmetic series is not given. ◆

The sum of an arithmetic series with 15 terms is 285. If $a_{15} = 40$, find a_1.

SOLUTION To find a_1, we apply the sum formula

$$S_n = n\left(\frac{a_1 + a_n}{2}\right) = a_1 + a_2 + a_3 + \cdots + a_n,$$

with $n = 15$ and $a_{15} = 40$.

$$15\left(\frac{a_1 + 40}{2}\right) = 285$$

$$15(a_1 + 40) = 570 \qquad \text{Multiply by 2.}$$

$$a_1 + 40 = 38 \qquad \text{Divide by 15.}$$

$$a_1 = -2 \qquad \text{Subtract 40.} \qquad \textit{Now Try Exercise 19} ◆$$

Geometric Series

What will happen if we attempt to find the sum of the terms of an infinite geometric sequence? The mathematical concept of infinity dates back to at least the paradoxes of Zeno (450 B.C.). The following illustrates a difficulty that occurred in attempting to comprehend infinity.

Suppose that a person walks 1 mile on the first day, $\frac{1}{2}$ mile the second day, $\frac{1}{4}$ mile the third day, and so on. How far down the road would this person travel? This distance is described by the infinite series

$$1 + \frac{1}{2} + \frac{1}{4} + \frac{1}{8} + \frac{1}{16} + \frac{1}{32} + \frac{1}{64} + \cdots.$$

Does the sum of an infinite number of positive values always become infinitely large? Problems like this took centuries for mathematicians to solve.   (**Source:** H. Eves, *An Introduction to the History of Mathematics.*)

Finite Geometric Series   In a manner analogous to the way an arithmetic series was defined, a **geometric series** is the sum of the terms of a geometric sequence. In order to calculate sums of infinite geometric series, we begin by finding sums of finite geometric series.

Any finite geometric sequence can be written as

$$a_1, \ a_1r, \ a_1r^2, \ a_1r^3, \ldots, a_1r^{n-1}.$$

The summation of these n terms is a finite geometric series. Its sum S_n is expressed by

$$S_n = a_1 + a_1r + a_1r^2 + a_1r^3 + \cdots + a_1r^{n-1}.$$

To find the value of S_n, multiply this equation by r.

$$rS_n = a_1r + a_1r^2 + a_1r^3 + \cdots + a_1r^{n-1} + a_1r^n$$

Subtracting this equation from the previous equation results in

$$S_n - rS_n = a_1 - a_1r^n$$

$$S_n(1 - r) = a_1(1 - r^n)$$

$$S_n = a_1\left(\frac{1 - r^n}{1 - r}\right), \qquad \text{provided } r \neq 1.$$

This formula can be used to find the sum of the first n terms of a geometric sequence.

SUM OF THE FIRST n TERMS OF A GEOMETRIC SEQUENCE

If a geometric sequence has first term a_1 and common ratio r, then the sum of the first n terms is given by

$$S_n = a_1\left(\frac{1 - r^n}{1 - r}\right),$$

provided $r \neq 1$.

EXAMPLE 8

Finding the sum of finite geometric series

Approximate the sum for the given values of n.
(a) $1 + \frac{1}{2} + \frac{1}{4} + \cdots + \left(\frac{1}{2}\right)^{n-1}$; $n = 5$, 10, and 20
(b) $3 - 6 + 12 - 24 + 48 - \cdots + 3(-2)^{n-1}$; $n = 3$, 8, and 13

SOLUTION

(a) This geometric series has $a_1 = 1$ and $r = \frac{1}{2} = 0.5$.

$$S_5 = 1\left(\frac{1 - 0.5^5}{1 - 0.5}\right) = 1.9375$$

$$S_{10} = 1\left(\frac{1 - 0.5^{10}}{1 - 0.5}\right) \approx 1.998047$$

$$S_{20} = 1\left(\frac{1 - 0.5^{20}}{1 - 0.5}\right) \approx 1.999998$$

(b) This geometric series has $a_1 = 3$ and $r = -2$.

$$S_3 = 3\left(\frac{1 - (-2)^3}{1 - (-2)}\right) = 9$$

$$S_8 = 3\left(\frac{1 - (-2)^8}{1 - (-2)}\right) = -255$$

$$S_{13} = 3\left(\frac{1 - (-2)^{13}}{1 - (-2)}\right) = 8193$$

Now Try Exercises 37 and 39 ◆

Annuities   A sequence of deposits made at equal periods of time is called an **annuity**. Suppose A_0 dollars are deposited at the end of each year into an account that pays an annual interest rate i compounded annually. At the end of the first year the account contains A_0 dollars. At the end of the second year A_0 dollars would be deposited again. In addition, the first deposit of A_0 dollars would have received interest during the second year. Therefore the value of the annuity after 2 years is

$$A_0 + A_0(1 + i).$$

After 3 years the balance is

$$A_0 + A_0(1 + i) + A_0(1 + i)^2,$$

and after n years this amount is given by

$$A_0 + A_0(1 + i) + A_0(1 + i)^2 + \cdots + A_0(1 + i)^{n-1}.$$

This is a geometric series with first term $a_1 = A_0$ and common ratio $r = (1 + i)$. The sum of the first n terms is given by

$$S_n = A_0\left(\frac{1 - (1 + i)^n}{1 - (1 + i)}\right) = A_0\left(\frac{(1 + i)^n - 1}{i}\right).$$

EXAMPLE 9 **Finding the future value of an annuity**

Suppose that a 20-year-old worker deposits $1000 into an account at the end of each year until age 65. If the interest rate is 12%, find the future value of the annuity.

SOLUTION Let $A_0 = 1000$, $i = 0.12$, and $n = 45$. The future value of the annuity is given by

$$S_n = A_0\left(\frac{(1 + i)^n - 1}{i}\right)$$

$$= 1000\left(\frac{(1 + 0.12)^{45} - 1}{0.12}\right)$$

$$\approx \$1{,}358{,}230.$$

Now Try Exercise 49 ◆

Infinite Geometric Series In Example 8 the value of r affects the sum of a finite geometric series. If $|r| > 1$, as in part (b), then r^n becomes large in absolute value for increasing n. As a result, the sum of the series also becomes large in absolute value. On the other hand, in part (a) the common ratio satisfies $|r| < 1$. In this case the values of r^n become closer to 0 as n increases. For large values of n,

$$S_n = a_1\left(\frac{1 - r^n}{1 - r}\right) \approx a_1\left(\frac{1 - 0}{1 - r}\right) = \frac{a_1}{1 - r}.$$

This result is summarized in the following.

◆

SUM OF AN INFINITE GEOMETRIC SEQUENCE

The sum of the infinite geometric sequence with first term a_1 and common ratio r is given by

$$S = \frac{a_1}{1 - r},$$

provided $|r| < 1$. If $|r| \geq 1$, then this sum does not exist.

Infinite series can be used to describe repeating decimals. For example, the fraction $\frac{5}{9}$ can be written as the repeating decimal $0.555555\ldots$. This decimal can be expressed as an infinite series.

$$\frac{5}{9} = 0.5 + 0.05 + 0.005 + 0.0005 + \cdots + 0.5(0.1)^{n-1} + \cdots$$

In this series $a_1 = \mathbf{0.5}$ and $r = \mathbf{0.1}$. Since $|r| < 1$, the sum exists and is given by

$$S = \frac{\mathbf{0.5}}{1 - \mathbf{0.1}} = \frac{5}{9},$$

as expected.

We are now able to answer the question concerning how far a person will walk if he or she travels 1 mile on the first day, $\frac{1}{2}$ mile on the second day, $\frac{1}{4}$ mile the third day, and so on.

EXAMPLE 10 Finding the sum of an infinite geometric series

Find the sum of the infinite geometric series

$$1 + \frac{1}{2} + \frac{1}{4} + \frac{1}{8} + \cdots.$$

SOLUTION In this series, the first term is $a_1 = 1$ and the common ratio is $\frac{1}{2} = 0.5$. Its sum is

$$S = \frac{a_1}{1 - r} = \frac{1}{1 - 0.5} = 2.$$

If it were possible to walk in the prescribed manner, the total distance traveled after each day would always be less than 2 miles. *Now Try Exercise 41* ◆

Summation Notation

Summation notation is used to write series efficiently. The symbol Σ, the uppercase Greek letter *sigma*, indicates a sum.

SUMMATION NOTATION

$$\sum_{k=1}^{n} a_k = a_1 + a_2 + a_3 + \cdots + a_n.$$

The letter k is called the **index of summation**. The numbers 1 and n represent the subscripts of the first and last terms in the series. They are called the **lower limit** and **upper limit** of the summation, respectively.

EXAMPLE 11 Using summation notation

Evaluate each series.

(a) $\sum_{k=1}^{5} k^2$ (b) $\sum_{k=1}^{4} 5$ (c) $\sum_{k=3}^{6} (2k - 5)$

SOLUTION

(a) $\sum_{k=1}^{5} k^2 = 1^2 + 2^2 + 3^2 + 4^2 + 5^2 = 55$

(b) $\sum_{k=1}^{4} 5 = 5 + 5 + 5 + 5 = 20$

(c) $\sum_{k=3}^{6} (2k - 5) = \underset{k\,=\,3}{(2(3) - 5)} + \underset{k\,=\,4}{(2(4) - 5)} + \underset{k\,=\,5}{(2(5) - 5)} + \underset{k\,=\,6}{(2(6) - 5)}$

$$= 1 + 3 + 5 + 7 = 16 \qquad \text{*Now Try Exercises 63, 65, and 69* ◆}$$

EXAMPLE 12 **Writing a series in summation notation**

Write the series using summation notation.

(a) $\dfrac{1}{2^3} + \dfrac{1}{3^3} + \dfrac{1}{4^3} + \dfrac{1}{5^3} + \dfrac{1}{6^3} + \dfrac{1}{7^3} + \dfrac{1}{8^3}$

(b) $\dfrac{1}{2} + \dfrac{2}{3} + \dfrac{3}{4} + \dfrac{4}{5} + \dfrac{5}{6} + \dfrac{6}{7} + \dfrac{7}{8}$

SOLUTION

(a) The terms of the series can be written as $\dfrac{1}{k^3}$ for $k = 2, 3, 4, \ldots, 8$. Thus

$$\dfrac{1}{2^3} + \dfrac{1}{3^3} + \dfrac{1}{4^3} + \dfrac{1}{5^3} + \dfrac{1}{6^3} + \dfrac{1}{7^3} + \dfrac{1}{8^3} = \sum_{k=2}^{8} \dfrac{1}{k^3}.$$

(b) The terms of the series can be written as $\dfrac{k}{k+1}$ for $k = 1, 2, 3, \ldots, 7$. Thus

$$\dfrac{1}{2} + \dfrac{2}{3} + \dfrac{3}{4} + \dfrac{4}{5} + \dfrac{5}{6} + \dfrac{6}{7} + \dfrac{7}{8} = \sum_{k=1}^{7} \dfrac{k}{k+1}.$$

Now Try Exercises 71 and 73 ◆

Series play an essential role in mathematics and its applications, as illustrated by the next example.

EXAMPLE 13 **Expressing a series in summation notation**

Suppose that an air filter removes 90% of the impurities that enter it.

(a) Find a series that represents the amount of impurities removed by a sequence of n air filters. Express this answer in summation notation.

(b) How many air filters would be necessary to remove 99.99% of the impurities? Could 100% of the impurities be removed from the air?

SOLUTION

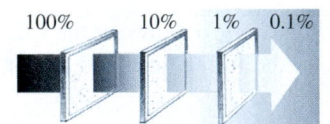

FIGURE 11.13 Percentages of Impurities Passing Through Air Filters

(a) The first filter removes 90% of the impurities so 10%, or 0.1, passes through it. Of the 0.1 that passes through the first filter, 90% is removed by the second filter, while 10% of 10%, or 0.01, passes through. Then, 10% of 0.01, or 0.001, passes through the third filter. See Figure 11.13. From this we can establish a pattern. Let 100% or 1 represent the amount of impurities entering the first air filter. The amount removed by n filters would equal

$$(0.9)(1) + (0.9)(0.1) + (0.9)(0.01) + (0.9)(0.001) + \cdots + (0.9)(0.1)^{n-1}.$$

In summation notation, this series can be written as $\sum_{k=1}^{n} 0.9(0.1)^{k-1}$.

(b) To remove 99.99% or 0.9999 of the impurities requires four air filters, since

$$\sum_{k=1}^{4} 0.9(0.1)^{k-1} = (0.9)(1) + (0.9)(0.1) + (0.9)(0.01) + (0.9)(0.001)$$

$$= 0.9 + 0.09 + 0.009 + 0.0009$$

$$= 0.9999.$$

Intuition tells us that it is impossible to remove 100% of the impurities. If k increases without a maximum value, then the fraction of impurities removed is

$$\sum_{k=1}^{\infty} (0.9)(0.1)^{k-1} = (0.9)(1) + (0.9)(0.1) + (0.9)(0.1)^2 + \cdots$$

$$= 0.99999 \ldots.$$

Notice that since the upper limit has no maximum, the symbol ∞ is used. If k is allowed to increase without bound, then the series equals the repeating decimal $0.\overline{9}$. This is an infinite geometric series with $a_1 = 0.9$ and $r = 0.1$. Its sum is

$$S = \frac{a_1}{1 - r} = \frac{0.9}{1 - 0.1} = \frac{0.9}{0.9} = 1.$$

The mathematics of this problem is telling us that it would require an infinite number of air filters to remove 100% of the impurities.   *Now Try Exercise 93* ◆

The following is a list of properties for summation notation.

PROPERTIES FOR SUMMATION NOTATION

Let $a_1, a_2, a_3, \ldots, a_n$ and $b_1, b_2, b_3, \ldots, b_n$ be sequences, and c be a constant.

1. $\displaystyle\sum_{k=1}^{n} ca_k = c \sum_{k=1}^{n} a_k$

2. $\displaystyle\sum_{k=1}^{n} (a_k + b_k) = \sum_{k=1}^{n} a_k + \sum_{k=1}^{n} b_k$

3. $\displaystyle\sum_{k=1}^{n} (a_k - b_k) = \sum_{k=1}^{n} a_k - \sum_{k=1}^{n} b_k$

4. $\displaystyle\sum_{k=1}^{n} c = nc$

5. $\displaystyle\sum_{k=1}^{n} k = \frac{n(n + 1)}{2}$

6. $\displaystyle\sum_{k=1}^{n} k^2 = \frac{n(n + 1)(2n + 1)}{6}$

These properties can be used to find sums, as illustrated in the next example.

EXAMPLE 14 Applying summation notation

Use properties for summation notation to find each sum.

(a) $\displaystyle\sum_{k=1}^{40} 5$ **(b)** $\displaystyle\sum_{k=1}^{22} 2k$ **(c)** $\displaystyle\sum_{k=1}^{14} (2k^2 - 3)$

SOLUTION

(a) $\displaystyle\sum_{k=1}^{40} 5 = 40(5) = 200$ *Property 4 with $n = 40$ and $c = 5$*

(b) $\displaystyle\sum_{k=1}^{22} 2k = 2 \sum_{k=1}^{22} k$ *Property 1 with $c = 2$ and $a_k = k$*

$\qquad\quad = 2 \cdot \dfrac{22(22 + 1)}{2}$ *Property 5 with $n = 22$*

$\qquad\quad = 506$ *Simplify.*

(c) $\displaystyle\sum_{k=1}^{14} (2k^2 - 3) = \sum_{k=1}^{14} 2k^2 - \sum_{k=1}^{14} 3$ *Property 3 with $a_k = 2k^2$ and $b_k = 3$*

$\qquad\qquad = 2 \sum_{k=1}^{14} k^2 - \sum_{k=1}^{14} 3$ *Property 1 with $c = 2$ and $a_k = k^2$*

$\qquad\qquad = 2 \cdot \dfrac{14(14 + 1)(2 \cdot 14 + 1)}{6} - 14(3)$ *Properties 6 and 4*

$\qquad\qquad = 1988$ *Simplify.*

Now Try Exercises 77, 79, and 85 ◆

11.2

Putting it all Together

$\mathbf{T}$he following table summarizes concepts related to series.

Concept	Explanation	Example
Series	A series is the summation of the terms of a sequence. A finite series always has a sum, but an infinite series may not have a sum.	$2 + 4 + 6 + 8 + \cdots + 20$ $S_4 = a_1 + a_2 + a_3 + a_4$ (partial sum) $\quad = 2 + 4 + 6 + 8$ $\quad = 20$
Arithmetic series	An arithmetic series is the summation of the terms of an arithmetic sequence. The sum of the first n terms is given by $$S_n = n\left(\frac{a_1 + a_n}{2}\right)$$ or $$S_n = \frac{n}{2}(2a_1 + (n-1)d).$$	$1 + 4 + 7 + 10 + 13 + 16 + 19$ (7 terms) $S_7 = 7\left(\dfrac{a_1 + a_7}{2}\right) = 7\left(\dfrac{1 + 19}{2}\right) = 70$ or $S_7 = \dfrac{7}{2}(2a_1 + 6d) = \dfrac{7}{2}(2(1) + 6(3)) = 70$
Geometric series	A geometric series is the summation of the terms of a geometric sequence. The sum of the first n terms is given by $$S_n = a_1\left(\frac{1 - r^n}{1 - r}\right).$$ If $|r| < 1$ then an infinite geometric series has the sum, $$S = \frac{a_1}{1 - r}.$$	$3 + 6 + 12 + 24 + 48 + 96$ (6 terms) $S_6 = 3\left(\dfrac{1 - 2^6}{1 - 2}\right) = 189$ The infinite geometric series $$1 + \frac{1}{4} + \frac{1}{16} + \frac{1}{64} + \cdots$$ has a sum $$S = \frac{1}{1 - \dfrac{1}{4}} = \frac{4}{3}.$$
Summation notation	The series $a_1 + a_2 + \cdots + a_n$ can be written as $$\sum_{k=1}^{n} a_k.$$	$1^2 + 2^2 + 3^2 + 4^2 + \cdots + 10^2$ can be written as $$\sum_{k=1}^{10} k^2.$$

11.2 — Exercises

Finding Sums of Series

1. *Prison Escapees* The table lists the number of escapees from state prisons each year. (**Source:** Bureau of Justice Statistics.)

Year	1990	1991	1992
Escapees	8518	9921	10,706

Year	1993	1994	1995
Escapees	14,035	14,307	12,249

(a) Write a series whose sum is the total number of escapees from 1990 to 1995. Find its sum.

(b) Is the series finite or infinite?

2. *Captured Prison Escapees* (Refer to the previous exercise.) The table lists the number of escapees from state prisons who were captured, including inmates who may have escaped during a previous year. (**Source:** Bureau of Justice Statistics.)

Year	1990	1991	1992
Captured	9324	9586	10,031

Year	1993	1994	1995
Captured	12,872	13,346	12,166

(a) Write a series whose sum is the total number of escapees captured from 1990 to 1995. Find its sum.

(b) Compare the number of escapees to the number captured during this time.

Exercises 3–10: For the given a_n, calculate S_5.

3. $a_n = 3n$ **4.** $a_n = n + 4$

5. $a_n = 2n - 1$ **6.** $a_n = 4n + 1$

7. $a_n = n^2 + 1$ **8.** $a_n = 2n^2$

9. $a_n = \dfrac{n}{n + 1}$ **10.** $a_n = \dfrac{1}{2n}$

Exercises 11–18: Use a formula to find the sum of the arithmetic series.

11. $3 + 5 + 7 + 9 + 11 + 13 + 15 + 17$

12. $7.5 + 6 + 4.5 + 3 + 1.5 + 0 + (-1.5)$

13. $1 + 2 + 3 + 4 + \cdots + 50$

14. $1 + 3 + 5 + 7 + \cdots + 97$

15. $-7 + (-4) + (-1) + 2 + 5 + \cdots + 98 + 101$

16. $89 + 84 + 79 + 74 + \cdots + 9 + 4$

17. The first 40 terms of the series whose terms are $a_n = 5n$

18. The first 50 terms of the series whose terms are given by $a_n = 1 - 3n$

19. The sum of an arithmetic series with 15 terms is 255. If $a_1 = 3$, find a_{15}.

20. The sum of an arithmetic series with 20 terms is 610. If $a_{20} = 59$, find a_1.

Exercises 21–30: Use a formula to find the sum of the first twenty terms for the arithmetic sequence.

21. $a_1 = 4, d = 2$ **22.** $a_1 = -3, d = \frac{2}{3}$

23. $a_1 = 10, d = -\frac{1}{2}$ **24.** $a_1 = 0, d = -4$

25. $a_1 = 4, a_{20} = 190.2$ **26.** $a_1 = -4, a_{20} = 15$

27. $a_1 = -2, a_{11} = 50$ **28.** $a_1 = 6, a_5 = -30$

29. $a_2 = 6, a_{12} = 31$ **30.** $a_8 = 4, a_{10} = 14$

Exercises 31–36: Use a formula to find the sum of the finite geometric series.

31. $1 + 2 + 4 + 8 + 16 + 32 + 64 + 128$

32. $2 + \frac{1}{2} + \frac{1}{8} + \frac{1}{32} + \frac{1}{128} + \frac{1}{512}$

33. $0.5 + 1.5 + 4.5 + 13.5 + 40.5 + 121.5 + 364.5$

34. $0.6 + 0.3 + 0.15 + 0.075 + 0.0375$

35. The first 20 terms of the series whose terms are given by $a_n = 3(2)^{n-1}$

36. The first 15 terms of the series whose terms are given by $a_n = 2\left(\frac{1}{3}\right)^n$

Exercises 37–40: Use a formula to approximate the sum for n = 4, 7, and 10.

37. $1 - \frac{1}{2} + \frac{1}{4} - \frac{1}{8} + \cdots + \left(-\frac{1}{2}\right)^{n-1}$

38. $3 - 1 + \frac{1}{3} - \frac{1}{9} + \cdots + 3\left(-\frac{1}{3}\right)^{n-1}$

39. $\frac{1}{3} + \frac{2}{3} + \frac{4}{3} + \frac{8}{3} + \cdots + \frac{1}{3}(2)^{n-1}$

40. $4 + \frac{8}{3} + \frac{16}{9} + \frac{32}{27} + \cdots + 4\left(\frac{2}{3}\right)^{n-1}$

Exercises 41–46: Find the sum of the infinite geometric series.

41. $1 + \frac{1}{3} + \frac{1}{9} + \frac{1}{27} + \frac{1}{81} + \cdots$

42. $5 + \frac{5}{2} + \frac{5}{4} + \frac{5}{8} + \frac{5}{16} + \cdots$

43. $6 - 4 + \frac{8}{3} - \frac{16}{9} + \frac{32}{27} - \frac{64}{81} + \cdots$

44. $-2 + \frac{1}{2} - \frac{1}{8} + \frac{1}{32} - \frac{1}{128} + \cdots$

45. $1 - \frac{1}{10} + \frac{1}{100} - \frac{1}{1000} + \cdots + \left(-\frac{1}{10}\right)^{n-1} + \cdots$

46. $25 - 5 + 1 - \frac{1}{5} + \cdots + 25\left(-\frac{1}{5}\right)^{n-1} + \cdots$

47. *Area* (Refer to Exercise 92, Section 11.1.) Use a geometric series to find the sum of the areas of the squares if they continue indefinitely.

48. *Perimeter* (Refer to Exercise 92, Section 11.1.) Use a geometric series to find the sum of the perimeters of the squares if they continue indefinitely.

Exercises 49–52: Annuities (Refer to Example 9.) Find the future value of each annuity.

49. $A_0 = \$2000, \quad i = 0.08, n = 20$

50. $A_0 = \$500, \quad i = 0.15, n = 10$

51. $A_0 = \$10{,}000, \quad i = 0.11, n = 5$

52. $A_0 = \$3000, \quad i = 0.19, n = 45$

Decimal Numbers and Geometric Series

Exercises 53–58: Write each rational number in the form of an infinite geometric series.

53. $\frac{2}{3}$

54. $\frac{1}{9}$

55. $\frac{9}{11}$

56. $\frac{14}{33}$

57. $\frac{1}{7}$

58. $\frac{23}{99}$

Exercises 59–62: Write the sum of each geometric series as a rational number.

59. $0.8 + 0.08 + 0.008 + 0.0008 + \cdots$

60. $0.9 + 0.09 + 0.009 + 0.0009 + \cdots$

61. $0.45 + 0.0045 + 0.000045 + \cdots$

62. $0.36 + 0.0036 + 0.000036 + \cdots$

Summation Notation

Exercises 63–70: Write out the terms of the series and then evaluate it.

63. $\sum_{k=1}^{4} (k + 1)$

64. $\sum_{k=1}^{6} (3k - 1)$

65. $\sum_{k=1}^{8} 4$

66. $\sum_{k=2}^{6} (5 - 2k)$

67. $\sum_{k=1}^{7} k^3$

68. $\sum_{k=1}^{4} 5(2)^{k-1}$

69. $\sum_{k=4}^{5} (k^2 - k)$

70. $\sum_{k=1}^{5} \log k$

Exercises 71–76: Write each series with summation notation.

71. $1^4 + 2^4 + 3^4 + 4^4 + 5^4 + 6^4$

72. $1 + \frac{1}{5} + \frac{1}{25} + \frac{1}{125} + \frac{1}{625}$

73. $1 + \frac{4}{3} + \frac{6}{4} + \frac{8}{5} + \frac{10}{6} + \frac{12}{7} + \frac{14}{8}$

74. $2 + \frac{5}{8} + \frac{10}{27} + \frac{17}{64} + \frac{26}{125} + \frac{37}{216}$

75. $1 + \frac{1}{2^2} + \frac{1}{3^2} + \frac{1}{4^2} + \frac{1}{5^2} + \cdots$

76. $1 + \frac{1}{10} + \frac{1}{100} + \frac{1}{1000} + \frac{1}{10{,}000} + \cdots$

Exercises 77–88: (Refer to Example 14.) Use properties of summation notation to find the sum.

77. $\sum_{k=1}^{60} 9$

78. $\sum_{k=1}^{43} -4$

79. $\sum_{k=1}^{15} 5k$

80. $\sum_{k=1}^{22} -2k$

81. $\sum_{k=1}^{31} (3k - 3)$

82. $\sum_{k=1}^{17} (1 - 4k)$

83. $\sum_{k=1}^{25} k^2$

84. $\sum_{k=1}^{12} 3k^2$

85. $\sum_{k=1}^{16} (k^2 - k)$

86. $\sum_{k=1}^{18} (k^2 - 4k + 3)$

87. $\sum_{k=5}^{24} k$

88. $\sum_{k=7}^{19} (k^2 + 1)$

89. Verify the formula $\sum_{k=1}^{n} k = \frac{n(n+1)}{2}$ by using the formula for the sum of the first n terms of a finite arithmetic sequence.

90. Use Exercise 89 to find the sum of the series $\sum_{k=1}^{200} k$.

91. *Stacking Logs*   A stack of logs is made in layers, with one less log in each layer. See the accompanying figure. If the top layer has 7 logs and the bottom layer has 15 logs, what is the total number of logs in the pile? Use a formula to find the sum.

92. *Stacking Logs*   (Refer to the previous exercise.) Suppose a stack of logs has 13 logs in the top layer and a total of 7 layers. How many logs are in the stack?

93. *Filter*   (Refer to Example 13.) Suppose that one filter removes half of the impurities in a water supply.
   (a) Find a series that represents the amount of impurities removed by a sequence of n filters. Express your answer in summation notation.
   (b) How many filters would be necessary to remove all of the impurities?

94. *Walking*   Suppose that a person walks 1 mile on the first day, $\frac{1}{3}$ mile on the second day, $\frac{1}{9}$ mile on the third day, and so on. Assuming that a person could walk each distance precisely, estimate how far the person would travel after a very long time?

Exercises 95 and 96: Depreciation   (Refer to Example 6.) *Let the total depreciation be T over n years.*
   (a) *Find an arithmetic sequence that gives the amount depreciated each year by using straight-line and sum-of-the-years'-digits depreciation.*
   (b) *Write a series for each method whose sum is the total amount depreciated over n years.*

95. $T = \$10,000$, $n = 4$ **96.** $T = \$42,000$, $n = 7$

Exercises 97 and 98: **The Natural Exponential Function** *The following series can be used to estimate the value of e^a for any real number a.*

$$e^a \approx 1 + a + \frac{a^2}{2!} + \frac{a^3}{3!} + \cdots + \frac{a^n}{n!},$$

where $n! = 1 \cdot 2 \cdot 3 \cdot 4 \cdots \cdot n$. Use the first eight terms of this series to approximate the given expression. Compare this estimate with the actual value.

97. e **98.** e^{-1}

Computing Partial Sums

Exercises 99–102: Use a_k and n to find $S_n = \sum_{k=1}^{n} a_k$. (Refer to Example 8.) Then, evaluate the infinite geometric series $S = \sum_{k=1}^{\infty} a_k$. Compare S to the values for S_n.

99. $a_k = \left(\frac{1}{3}\right)^{k-1}$; $n = 2, 4, 8, 16$

100. $a_k = 3\left(\frac{1}{2}\right)^{k-1}$; $n = 5, 10, 15, 20$

101. $a_k = 4\left(-\frac{1}{10}\right)^{k-1}$; $n = 1, 2, 3, 4, 5, 6$

102. $a_k = 2(-0.02)^{k-1}$; $n = 1, 2, 3, 4, 5, 6$

Writing about Mathematics

103. Discuss the difference between a sequence and a series. Give examples of each.

104. Under what circumstances can we find the sum of a geometric series? Give examples to illustrate your answer.

105. Explain how to write the series

$$\log 1 + \log 2 + \log 3 + \cdots + \log n$$

as one term.

106. Explain how to write the series

$$\log 2 - \log 4 + \log 6 - \log 8 + \cdots + (-1)^{n-1}\log 2n$$

as one term. Assume n is even.

CHECKING BASIC CONCEPTS FOR SECTIONS 11.1 AND 11.2

1. Give graphical and numerical representations of the sequence defined by $a_n = -2n + 3$, where $a_n = f(n)$. Include the first six terms.

2. Determine if the sequence is arithmetic or geometric. If it is arithmetic, state the common difference; if it is geometric, give the common ratio.
   (a) $2, -4, 8, -16, 32, -64, 128, \ldots$

   (b) $-3, 0, 3, 6, 9, 12, \ldots$

   (c) $4, 2, 1, \frac{1}{2}, \frac{1}{4}, \frac{1}{8}, \frac{1}{16}, \ldots$

3. Determine if the series is arithmetic or geometric. Use a formula to find its sum.
   (a) $1 + 5 + 9 + 13 + \cdots + 37$

   (b) $3 + 1 + \frac{1}{3} + \frac{1}{9} + \frac{1}{27} + \frac{1}{81}$

   (c) $2 + \frac{1}{2} + \frac{1}{8} + \frac{1}{32} + \cdots$

   (d) $0.9 + 0.09 + 0.009 + 0.0009 + \cdots$

4. Write each series in the previous exercise in summation notation.

5. Use properties of summation notation to find the sum.
   (a) $\displaystyle\sum_{k=1}^{15} (k + 2)$ (b) $\displaystyle\sum_{k=1}^{21} 2k^2$

6. **Bouncing Ball** A ball is dropped from a height of 6 feet. On each bounce the ball returns to $\frac{2}{3}$ of its previous height. How far does the ball travel before it comes to rest?

11.3 Counting

- ◆ Apply the fundamental counting principle
- ◆ Calculate and apply permutations
- ◆ Calculate and apply combinations

Introduction

The notion of counting in mathematics includes much more than simply counting from 1 to 100. It also includes determining the number of ways that an event can occur. For example, how many ways are there to answer a true-false quiz with ten questions? The answer involves counting the different ways that a student could answer such a quiz. Counting is an important concept that is used to calculate probabilities. Probability is discussed in Section 11.6.

Fundamental Counting Principle

Suppose that a quiz has only two questions. The first is a multiple-choice question with four choices: A, B, C, or D, and the second is a true-false (T-F) question. The tree diagram in Figure 11.14 can be used to count the ways that this quiz can be answered.

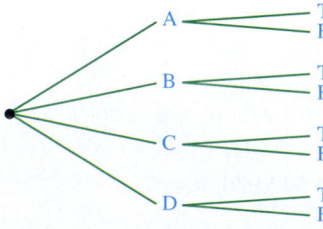

FIGURE 11.14 Different Ways to Answer a Quiz

A tree diagram is a systematic way of listing every possibility. From Figure 11.14, we can see that there are eight ways to answer the test. They are

AT, AF, BT, BF, CT, CF, DT, and DF.

For instance, CF indicates a quiz with answers of C on the first question and F on the second question.

A tree diagram is not always practical, because it can quickly become very large. For this reason mathematicians have developed more efficient ways of counting. Since the multiple-choice question has four possible answers, after which the true-false question has two possible answers, there are $4 \cdot 2 = 8$ possible ways of answering the test. This is an application of the *fundamental counting principle*, which applies to independent events. Two events are **independent** if neither event influences the outcome of the other.

FUNDAMENTAL COUNTING PRINCIPLE

Let $E_1, E_2, E_3, \ldots, E_n$ be a sequence of n independent events. If event E_k can occur m_k ways for $k = 1, 2, 3, \ldots, n$, then there are

$$m_1 \cdot m_2 \cdot m_3 \cdot \cdots \cdot m_n$$

ways for all n events to occur.

EXAMPLE 1 Counting ways to answer an exam

An exam contains four true-false questions and six multiple-choice questions. Each multiple-choice question has five possible answers. Count the number of ways that the exam can be answered.

SOLUTION Answering these ten questions can be thought of as a sequence of ten independent events. There are two ways to answer each of the first four questions, and five ways of answering each of the next six questions. The number of ways to answer the exam is

$$\underbrace{2 \cdot 2 \cdot 2 \cdot 2}_{\text{4 factors}} \cdot \underbrace{5 \cdot 5 \cdot 5 \cdot 5 \cdot 5 \cdot 5}_{\text{6 factors}} = 2^4 5^6 = 250,000.$$

Now Try Exercise 3 ◆

In the next example we count the number of different license plates possible with a given format.

EXAMPLE 2 Counting license plates

Sometimes a license plate is limited to 3 uppercase letters (A through Z) followed by 3 digits (0 through 9). For example, ABB 112 would be a valid license plate. Would this format provide enough license plates for a state with 8 million vehicles?

SOLUTION Since there are 26 letters of the alphabet, it follows that there are 26 ways to choose each of the first three letters of the license plate. Similarly, there are 10 digits, so there are 10 ways to choose each of the three digits in the license plate. By the fundamental counting principle, there are

$$26 \cdot 26 \cdot 26 \cdot 10 \cdot 10 \cdot 10 = 17,576,000$$

unique license plates that could be issued. This format for license plates could accommodate more than 8 million vehicles.

Now Try Exercise 5 ◆

One of the most important discoveries of our time was made in 1953 when the double helical structure of DNA (deoxyribonucleic acid) was identified. DNA is the material of heredity. It is composed of four bases called adenine (A), cytosine (C), guanine (G), and

thymine (T). The genetic code of a person is composed of approximately six billion of these bases stored in a long list. The nucleus of every cell contains a copy of this genetic code, tightly coiled in the shape of a double helix. The DNA of a single cell has the potential to store 30 encyclopedic volumes three times over. Any one of the four bases can occur anytime in the genetic code.

Small changes in this code can have dramatic effects. In a normal red blood cell, hemoglobin contains the genetic code

$$\text{ACTCCTG}\textbf{A}\text{GGAGGAGT,}$$

whereas a person with sickle cell anemia has the same sequence except that thymine (T) has been substituted for adenine (A):

$$\text{ACTCCTG}\textbf{T}\text{GGAGGAGT.}$$

Thus a single error in a list of six billion letters causes this serious disease. This unique list represented by the four letters A, C, G, and T defines the characteristics of an individual. Approximately 99–99.9% of this genetic code is identical for all humans. Only 0.1–1% is unique to a particular individual.   (**Source:** S. Easteal, *DNA Profiling: Principles, Pitfalls and Potential.*)

EXAMPLE 3   Counting genetic codes in DNA

Using the four letters A, C, G, and T, write a numeric expression that represents the total number of genetic codes that are possible.

SOLUTION   A human genetic code is composed of a list of six billion letters. For each letter there are four choices. By the fundamental counting principle, there are $4^{6,000,000,000}$ genetic codes of this type. (Many of these would not be valid for human beings.) This number is too large to approximate with a calculator.   *Now Try Exercise 9* ◆

EXAMPLE 4   Counting toll-free 800 telephone numbers

The number of 800 telephone numbers available for new businesses and individuals is decreasing rapidly. Count the total number of 800 numbers, if the local portion of a telephone number does not start with a 0 or 1.   (**Source:** Database Services Management.)

SOLUTION   A toll-free 800 number assumes the following form.

We can think of choosing the remaining digits for the local number as seven independent events. Since the local number cannot begin with a 0 or 1, there are eight possibilities (2 to 9) for the first digit. The remaining six digits can be any number from 0 to 9, so there are ten possibilities for each of these digits. The total is given by

$$8 \cdot 10 \cdot 10 \cdot 10 \cdot 10 \cdot 10 \cdot 10 = 8 \cdot 10^6 = 8,000,000.$$

First digit   Last 6 digits

Note: Toll-free numbers also begin with 888 and 877.   *Now Try Exercises 19 and 26* ◆

Permutations

A **permutation** is an *ordering* or *arrangement*. For example, suppose that three students are scheduled to give a speech in a class. The different arrangements of how these speeches can be ordered are called permutations. Initially, any one of the three students could give the first speech. After the first speech, there are two students remaining for the second speech. For the third speech there is only one possibility. By the fundamental counting principle, the total number of permutations is equal to

$$3 \cdot 2 \cdot 1 = 6.$$

If the students are denoted as A, B, and C, then these six permutations are ABC, ACB, BAC, BCA, CAB, and CBA. In a similar manner, if there were ten students scheduled to give speeches, the number of permutations would increase to

$$10 \cdot 9 \cdot 8 \cdot 7 \cdot 6 \cdot 5 \cdot 4 \cdot 3 \cdot 2 \cdot 1 = 3{,}628{,}800.$$

A more efficient way of writing the previous two products is to use *factorial notation.* The number $n!$ (read "*n*-factorial") is defined as follows.

---

n-FACTORIAL

For any natural number n,

$$n! = n(n-1)(n-2) \cdots (3)(2)(1)$$

and

$$0! = 1.$$

The reason for the definition of 0! will become apparent later. Factorials grow rapidly and can be computed on most calculators.

 EXAMPLE 5

Calculating factorials

Compute $n!$ for $n = 0, 1, 2, 3, 4,$ and 5 by hand. Use a calculator to find 8!, 13!, and 25!.

Calculator Help

To calculate $n!$, see Appendix B (page AP-25).

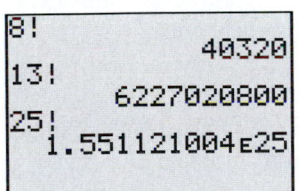

FIGURE 11.15

SOLUTION The values for $n!$ can be calculated as

$$0! = 1, \quad 1! = 1, \quad 2! = 2 \cdot 1 = 2, \quad 3! = 3 \cdot 2 \cdot 1 = 6,$$
$$4! = 4 \cdot 3 \cdot 2 \cdot 1 = 24, \quad \text{and} \quad 5! = 5 \cdot 4 \cdot 3 \cdot 2 \cdot 1 = 120.$$

Figure 11.15 shows the values of 8!, 13!, and 25!. The value for 25! is an approximation. Notice how dramatically $n!$ increases. *Now Try Exercise 27* ◆

One of the most famous unanswered questions in computing today is called the *traveling salesperson problem.* It is a relatively simple problem to state, but if someone could design a procedure to solve this problem *efficiently,* he or she would not only become famous, but would also provide a valuable method for businesses to save millions of dollars on scheduling problems, such as bus routes and truck deliveries.

One instance of the traveling salesperson problem can be stated as follows. A salesperson must begin and end at home and travel to three cities. Assuming that the salesperson can travel between any pair of cities, what route would minimize the salesperson's mileage? In Figure 11.16, the four cities are labeled A, B, C, and D. Let the salesperson

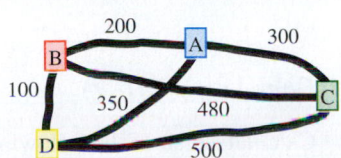

FIGURE 11.16 Traveling Salesperson Problem

live in city A. There are six routes that could be tried. They are listed in Table 11.7 with the appropriate mileage for each.

TABLE 11.7

Route	Mileage
A B C D A	200 + 480 + 500 + 350 = 1530
A B D C A	200 + 100 + 500 + 300 = 1100
A C B D A	300 + 480 + 100 + 350 = 1230
A C D B A	300 + 500 + 100 + 200 = 1100
A D B C A	350 + 100 + 480 + 300 = 1230
A D C B A	350 + 500 + 480 + 200 = 1530

The shortest route of 1100 miles occurs when the salesperson either starts at A and travels through B, D, and C, and back to A, or reverses this route. Currently, this method of listing all possible routes to find the minimum distance is the only known way to consistently find the optimal solution for any general map containing n cities. In fact, people have not been able to determine whether a *significantly* faster method even exists. (**Source:** J. Smith, *Design and Analysis of Algorithms.*)

Counting the number of routes involves the fundamental counting principle. At the first step, the salesperson can travel to any one of three cities. Once this city is selected, there are then two possible cities to choose, and then one. Finally the salesperson returns home. The total number of routes is given by

$$3! = 3 \cdot 2 \cdot 1 = 6.$$

If the salesperson must travel to 30 cities, then there are

$$30! \approx 2.7 \times 10^{32}$$

routes to check, far too many to check even with the largest supercomputers.

Next suppose that a salesperson must visit three of eight possible cities. At first there are eight cities to choose. After the first city has been visited, there are seven cities to select. Since the salesperson only travels to three of the eight cities, there are

$$8 \cdot 7 \cdot 6 = 336$$

possible routes. This number of permutations is denoted $P(8, 3)$. It represents the number of arrangements that can be made using three elements taken from a sample of eight.

PERMUTATIONS OF n ELEMENTS TAKEN r AT A TIME

If $P(n, r)$ denotes the number of permutations of n elements taken r at a time, with $r \leq n$, then

$$P(n, r) = \frac{n!}{(n - r)!} = \underbrace{n(n - 1)(n - 2) \cdots (n - r + 1)}_{r \text{ factors}}.$$

Calculator Help
To calculate $P(n, r)$, see Appendix B (page AP-25).

 **EXAMPLE 6**   Calculating $P(n, r)$

Calculate each of the following by hand. Then, support your answers by using a calculator.
(a) $P(7, 3)$ **(b)** $P(100, 2)$

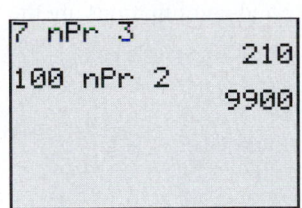

FIGURE 11.17

SOLUTION

(a) $P(7, 3) = \underbrace{7 \cdot 6 \cdot 5}_{3 \text{ factors}} = 210$

(b) $P(100, 2) = 100 \cdot 99 = 9900$. One also can compute this number as follows.

$$P(\mathbf{100}, \mathbf{2}) = \frac{\mathbf{100}!}{(\mathbf{100} - \mathbf{2})!} = \frac{100 \cdot 99 \cdot 98!}{98!} = 100 \cdot 99 = 9900$$

In this case, it is helpful to cancel 98! before performing the arithmetic. Both of these computations are performed by a calculator in Figure 11.17.

Now Try Exercises 31 and 37 ◆

 **EXAMPLE 7** Calculating permutations

If there is a class of 30 students, how many arrangements are there for 4 students to give a speech?

SOLUTION The number of permutations of 30 elements taken 4 at a time is given by

$$P(30, 4) = \underbrace{30 \cdot 29 \cdot 28 \cdot 27}_{4 \text{ factors}} = 657{,}720$$

Thus there are 657,720 ways to arrange the four speeches. *Now Try Exercise 44* ◆

In the next example we determine the number of ways that 4 people can have different birthdays. For example, if the birthdays of 4 people are February 29, May 2, June 30, and July 11, then these dates would be one way that 4 people could have different birthdays.

 **EXAMPLE 8** Counting birthdays

Count the possible ways that 4 people can have different birthdays.

SOLUTION Counting February 29, there are 366 possible birthdays. The first person could have any of the 366 possible birthdays. The second person could have any of 365 birthdays, because the first person's birthday cannot be duplicated. Similarly, the third person could have any of 364 birthdays, and the fourth person could have any of 363 birthdays. The total number of ways that 4 people could have different birthdays equals

$$P(366, 4) = 366 \cdot 365 \cdot 364 \cdot 363 \approx 1.77 \times 10^{10}.$$

Now Try Exercise 51 ◆

◆ **CLASS DISCUSSION**

Count the number of arrangements of 52 cards in a standard deck. Is it likely that there are arrangements that no one has ever shuffled at any time in the history of the world? Explain. ◆

Combinations

Unlike a permutation, a **combination** is not an ordering or arrangement, but rather a subset of a set of elements. Order is unimportant when finding combinations. For example, suppose we want to select a tennis team of two players from four people. The order in which the selection is made does not affect the final team of two players. From a set of four people, we select a subset of two players. This number of possible subsets or combinations is denoted either $C(4, 2)$ or $\binom{4}{2}$.

To calculate $C(4, 2)$, we first consider $P(4, 2)$. Denote the four players by the letters A, B, C, and D. There are $P(4, 2) = 4 \cdot 3 = 12$ permutations given by

AB, BA, AC, CA, AD, DA, BC, CB, BD, DB, CD, DC.

However, the team that comprises person A and person B is equivalent to the team with person B and person A. The sets {AB} and {BA} are equal. The valid combinations are the following two-element subsets of {A, B, C, D}.

$$\{AB\} \quad \{AC\} \quad \{AD\} \quad \{BC\} \quad \{BD\} \quad \{CD\}$$

That is, $C(4, 2) = \frac{P(4, 2)}{2!} = 6$. The relationship between $P(n, r)$ and $C(n, r)$ is given.

Calculator Help

To calculate $C(n, r)$, see Appendix B (page AP-25).

COMBINATIONS OF *n* ELEMENTS TAKEN *r* AT A TIME

If $C(n, r)$ denotes the number of combinations of *n* elements taken *r* at a time, with $r \leq n$, then

$$C(n, r) = \frac{P(n, r)}{r!} = \frac{n!}{(n - r)!\, r!}.$$

EXAMPLE **9**   Calculating $C(n, r)$

Calculate each of the following. Support your answers by using a calculator.

(a) $C(7, 3)$ **(b)** $\binom{50}{47}$

SOLUTION

(a) $C(7, 3) = \dfrac{7!}{(7 - 3)!\,3!} = \dfrac{7!}{4!\,3!} = \dfrac{7 \cdot 6 \cdot 5 \cdot 4!}{4!\,3!} = \dfrac{7 \cdot 6 \cdot 5}{3!} = \dfrac{210}{6} = 35$

(b) The notation $\binom{50}{47}$ is equivalent to $C(50, 47)$.

$$\binom{50}{47} = \frac{50!}{(50 - 47)!\,47!} = \frac{50!}{3!\,47!} = \frac{50 \cdot 49 \cdot 48 \cdot 47!}{3!\,47!}$$

$$= \frac{50 \cdot 49 \cdot 48}{3!} = \frac{117{,}600}{6} = 19{,}600$$

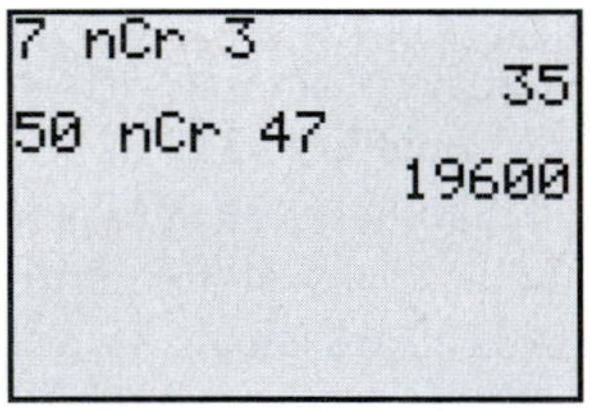

```
7 nCr 3
              35
50 nCr 47
           19600
```

FIGURE 11.18

These computations are performed by a calculator in Figure 11.18.

Now Try Exercises 57 and 63 ◆

EXAMPLE **10**   Counting combinations

A college student has 5 courses left in her major and plans to take 2 courses this semester. Assuming that this student has the prerequisites for all 5 courses, determine how many ways these 2 courses can be selected.

SOLUTION   The order in which the courses are selected is unimportant. From a set of 5 courses, the student selects a subset of 2 courses. This number of subsets is given by $C(5, 2)$.

$$C(5, 2) = \frac{5!}{(5 - 2)!\,2!} = \frac{5!}{3!\,2!} = 10$$

There are 10 ways to select 2 courses from a set of 5.

Now Try Exercise 65 ◆

 **Calculating the number of ways to play the lottery**

To win the jackpot in a lottery, a person must select 5 different numbers from 1 to 53 and then pick the powerball, which is numbered from 1 to 42. Count the ways to play the game. (**Source:** Minnesota State Lottery.)

SOLUTION From 53 numbers a player picks 5 numbers. There are $C(53, 5)$ ways of doing this. There are 42 ways to choose the powerball. Using the fundamental counting principle, the number of ways to play the game equals $C(53, 5) \cdot 42 = 120{,}526{,}770$.

Now Try Exercise 69 ◆

◆ **MAKING CONNECTIONS** —————

Permutations and Combinations A *permutation* is an arrangement (or list) of objects, where the ordering of the objects is important. For example, the expression $P(20, 9)$ would give the number of batting orders possible for 9 players selected from a team of 20 players. A *combination* is a subset of a set of objects, where the ordering of the objects is *not* important. For example, the expression $C(20, 9)$ would give the number of committees possible when 9 people are selected from a group of 20 people. The order in which committee members are selected does not affect the makeup of the committee. ◆

 **Counting committees**

How many committees of 6 people can be selected from 6 women and 3 men, if a committee must consist of at least 2 men?

SOLUTION Because there are 3 men and each committee must consist of at least 2 men, a committee can include either 2 or 3 men. The order of selection is not important. Therefore we need to consider combinations of committee members rather than permutations.

Two Men: This committee would consist of 4 women and 2 men. We can select 2 men from a group of 3 in $C(3, 2) = 3$ ways. Four women can be selected from a group of 6 in $C(6, 4) = 15$ ways. By the fundamental counting principle, there is a total of

$$C(3, 2) \cdot C(6, 4) = 3 \cdot 15 = 45$$

committees that have 2 men.

Three Men: This committee would have 3 women and 3 men. We can select 3 men from a group of 3 in $C(3, 3) = 1$ way. We can select 3 women from a group of 6 in $C(6, 3) = 20$ ways. By the fundamental counting principle, there is a total of

$$C(3, 3) \cdot C(6, 3) = 1 \cdot 20 = 20$$

committees that have 3 men.

The total number of possible committees would be $45 + 20 = 65$.

Now Try Exercise 67 ◆

11.3

Putting it all Together

The fundamental counting principle can be used to determine the number of ways a sequence of independent events can occur. Permutations are arrangements or listings, whereas combinations are subsets of a set of events.

The following table summarizes some concepts related to counting in mathematics.

Notation	Meaning	Examples
n-factorial: $n!$	$n!$ represents the product $$n(n-1)\cdots(3)(2)(1).$$	$6! = 6 \cdot 5 \cdot 4 \cdot 3 \cdot 2 \cdot 1 = 720$ $0! = 1$
$P(n,r) = \dfrac{n!}{(n-r)!}$	$P(n,r)$ represents the number of permutations of n elements taken r at a time.	The number of two-letter strings that can be formed using the four letters A, B, C, and D exactly once is given by $$P(4,2) = \frac{4!}{(4-2)!} = \frac{4!}{2!} = 12$$ or $P(4,2) = 4 \cdot 3 = 12$.
$C(n,r) = \dfrac{n!}{(n-r)!\, r!}$	$C(n,r)$ represents the number of combinations of n elements taken r at a time.	The number of committees of 3 people that can be formed from 5 people is given by $$C(5,3) = \frac{5!}{(5-3)!\,3!} = \frac{5!}{2!3!} = 10.$$

11.3 — Exercises

Counting

*Exercises 1–4: **Exam Questions**   Count the number of ways that the questions on an exam could be answered.*

1. Ten true-false questions

2. Ten multiple-choice questions with five choices each

3. Five true-false questions and ten multiple-choice questions with four choices each

4. One question involving matching ten items in one column with ten items in another column, using a one-to-one correspondence

*Exercises 5–8: **License Plates**   Count the number of possible license plates with the given constraints.*

5. Three digits followed by three letters

6. Two letters followed by four digits

7. Three letters followed by three digits or letters

8. Two letters followed by either three or four digits

*Exercises 9–12: **Counting Strings*** *Count the number of five-letter strings that can be formed with the given letters, assuming a letter can be used more than once.*

9. A, B, C   **10.** W, X, Y, Z

11. D, E, F, G, H   **12.** A, C

*Exercises 13–16: **Counting Strings*** *Count the number of strings that can be formed with the given letters, assuming each letter is used exactly once.*

13. A, B   **14.** A, B, C

15. W, X, Y, Z   **16.** V, W, X, Y, Z

17. *Combination Lock*   A briefcase has two locks. The combination to each lock consists of a three-digit number, where digits may be repeated. See the accompanying figure. How many combinations are possible? (*Hint:* The word *combination* is a misnomer. Lock combinations are permutations where the arrangement of the numbers is important.)

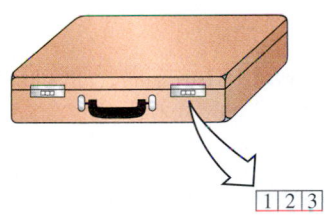

18. *Combination Lock*   A typical combination for a padlock consists of three numbers from 0 to 39. Count the number of combinations that are possible with this type of lock, if a number may be repeated.

19. *Garage Door Openers*   The code for some garage door openers consists of 12 electrical switches that can be set to either 0 or 1 by the owner. With this type of opener, how many codes are possible?   (**Source:** Promax.)

20. *Lottery*   To win the jackpot in a lottery game, a person must pick three numbers from 0 to 9 in the correct order. If a number can be repeated, how many ways are there to play the game?

21. *Radio Stations*   Call letters for a radio station usually begin with either a K or W, followed by three letters. In 1995 there were 11,834 radio stations on the air. Is there any shortage of call letters for new radio stations? (**Source:** M. Street Corporation.)

22. *Access Codes*   ATM access codes often consist of a four-digit number. How many codes are possible without giving two accounts the same access code?

23. *Computer Sale*   A computer sale offers a choice of 2 monitors, 3 printers, and 4 types of software. How many different packages can be purchased?

24. *Dice*   A red die and a blue die are thrown. How many ways are there for both dice to show an even number?

25. *Dinner Choices*   A menu offers 5 different salads, 10 different entrées, and 4 different desserts. How many ways are there to order a salad, an entrée, and a dessert?

26. *Telephone Numbers*   How many different 7-digit telephone numbers are possible if the first digit cannot be a 0 or a 1?

Exercises 27–30: Evaluate the expression.

27. 6!   **28.** 0!

29. 10!   **30.** 7!

Permutations

Exercises 31–40: Evaluate the expression.

31. $P(5, 3)$ **32.** $P(10, 2)$

33. $P(8, 1)$ **34.** $P(6, 6)$

35. $P(7, 3)$ **36.** $P(12, 3)$

37. $P(25, 2)$ **38.** $P(20, 1)$

39. $P(10, 4)$ **40.** $P(34, 2)$

41. *Standing in Line*   How many ways can 4 people stand in a line?

42. *Books on a Shelf*   How many arrangements are there of 6 different books on a shelf?

43. *Introductions*   How many ways could 5 basketball players be introduced at a game?

44. *Giving a Speech*   In how many arrangements can 3 students from a class of 15 give a speech?

45. *Traveling Salesperson* (Refer to the discussion after Example 5.) A salesperson must travel to 3 of 7 cities. Direct travel is possible between every pair of cities. How many arrangements are there for the salesperson to visit these 3 cities? Assume that traveling a route in reverse order constitutes a different route.

46. *Keys* How many distinguishable ways can 4 keys be put on a key ring?

47. *Sitting at a Round Table* How many ways can 7 people sit at a round table? Assume that a different way means that at least one person is sitting next to someone different.

48. *Batting Orders* A softball team has 10 players. How many batting orders are possible?

49. *Baseball Positions* In how many ways can 9 players be assigned to the 9 positions on a baseball team, assuming that any player can play any position?

50. *Musical Chairs Seating* In a game of musical chairs, 7 children will sit in 6 chairs arranged in a circle. One child is left out. How many (different) ways can the children sit in the chairs? Assume that a different way means that at least one child is sitting next to someone different each time the music stops.

51. *Birthdays* In how many ways can 5 people have different birthdays?

52. *Course Schedule* A scheduling committee has 1 room and 5 mathematics courses to offer to the students. In how many ways can the committee arrange the 5 courses during the day?

53. *Telephone Numbers* How many 10-digit telephone numbers are there if the first and fourth digit cannot be a 0 or a 1?

54. *Car Designs* There are 10 basic colors available for a new car along with 5 basic styles of trim. In how many ways can a person pick the color and trim?

Combinations

Exercises 55–64: Evaluate the expression.

55. $C(3, 1)$ **56.** $C(4, 3)$

57. $C(6, 3)$ **58.** $C(7, 5)$

59. $C(5, 0)$ **60.** $C(10, 2)$

61. $\binom{8}{2}$ **62.** $\binom{9}{4}$

63. $\binom{20}{18}$ **64.** $\binom{100}{2}$

65. *Lottery* To win the jackpot in a lottery, one must select 5 different numbers from 1 to 39. How many ways are there to play this game?

66. *Selecting a Committee* How many ways can a committee of 5 be selected from 8 people?

67. *Selecting a Committee* How many committees of 4 people can be selected from 5 women and 3 men, if a committee must have 2 people of each sex on it?

68. *Essay Questions* On a test involving 6 essay questions, students are asked to answer 4 questions. How many ways can the essay questions be selected?

69. *Test Questions* A test consists of two parts. In the first part a student must choose 3 of 5 essay questions, and in the second part a student must choose 4 of 5 essay questions. How many ways can the essay questions be selected?

70. *Cards* How many ways are there to draw a 5-card hand from a 52-card deck?

71. *Selecting Marbles* How many ways are there to draw 3 red marbles and 2 blue marbles from a jar that contains 10 red marbles and 12 blue marbles?

72. *Book Arrangements* A professor has 3 copies of an algebra book and 4 copies of a calculus text. How many distinguishable ways can the books be placed on a shelf?

73. *Peach Samples* How many samples of 3 peaches can be drawn from a crate of 24 peaches? (Assume that the peaches are distinguishable.)

74. *Flower Samples* A bouquet of flowers contains 3 red roses, 4 yellow roses, and 5 white roses. In how many ways can a person choose 1 flower of each type? (Assume that the flowers are distinguishable.)

75. *Permutations* Show that $P(n, n - 1) = P(n, n)$. Give an example that supports your result.

76. *Combinations* Show that $\binom{n}{r} = \binom{n}{n - r}$. Give an example that supports your result.

Writing about Mathematics

77. Explain the difference between a permutation and a combination. Give examples.

78. Explain what counting is, as presented in this section.

11.4 The Binomial Theorem

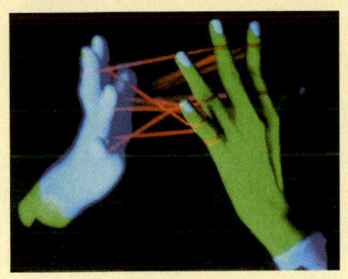

◆ Derive the binomial theorem

◆ Use the binomial theorem

◆ Apply Pascal's triangle

Introduction

In this section we discuss how to expand expressions in the form $(a + b)^n$, where n is a natural number. These expressions occur in statistics, finite mathematics, computer science, and calculus.

Derivation of the Binomial Theorem

Combinations play a central role in the development of the binomial theorem. The binomial theorem can be used to expand expressions of the form $(a + b)^n$. Before stating the binomial theorem, we begin by counting the number of strings of a given length that can be formed with only the variables a and b.

EXAMPLE 1 Calculating distinguishable strings

Count the number of distinguishable strings that can be formed with the given number of a's and b's. List these strings.
(a) Two a's, one b **(b)** Two a's, three b's   **(c)** Four a's, no b's

SOLUTION
(a) Using two a's and one b, we can form strings of length three. Once the b has been positioned, the string is determined. For example, if the b is placed in the middle position,

then the string must be *aba*. From a set of three slots, we choose one slot to place the b. This is computed by $C(3, 1) = 3$. The strings are *aab*, *aba*, and *baa*.

(b) With two a's and three b's we can form strings of length five. Once the locations of the three b's have been selected, the string is determined. For instance, if the b's are placed in the first, third, and fifth positions,

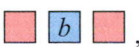

then the string becomes *babab*. From a set of five slots, we select three to place the b's. This is computed by $C(5, 3) = 10$. The ten strings are

bbbaa, bbaba, bbaab, babba, babab, baabb, abbba, abbab, ababb, and *aabbb.*

(c) There is only one string of length four that contains no b's. This is *aaaa* and is computed by

$$C(4, 0) = \frac{4!}{(4 - 0)!\, 0!} = \frac{24}{24(1)} = 1.$$

Now Try Exercises 9, 11, and 13 ◆

Next we expand $(a + b)^n$ for a few values of n, without simplifying.

$$(a + b)^1 = a + b$$
$$(a + b)^2 = (a + b)(a + b)$$
$$= aa + \mathbf{ab} + \mathbf{ba} + bb$$
$$(a + b)^3 = (a + b)(a + b)^2$$
$$= (a + b)(aa + ab + ba + bb)$$
$$= aaa + \mathbf{aab} + \mathbf{aba} + \mathbf{abb} + \mathbf{baa} + \mathbf{bab} + \mathbf{bba} + bbb$$

Notice that $(a + b)^1$ is the sum of all possible strings of length one that can be formed using a and b. The only possibilities are a and b. The expression $(a + b)^2$ is the sum of all possible strings of length two using a and b. The strings are aa, ab, ba, and bb. In a similar manner, $(a + b)^3$ is the sum of all possible strings of length three using a and b. This pattern continues for higher powers of $(a + b)$.

Strings with equal numbers of a's and equal numbers of b's can be combined into one term. For example, in $(a + b)^2$ the terms ab and ba can be combined as $2ab$. Notice that there are $C(2, 1) = 2$ distinguishable strings of length two containing one b. Similarly, in the expansion of $(a + b)^3$, the terms containing one a and two b's can be combined as

$$\mathbf{abb} + \mathbf{bab} + \mathbf{bba} = 3ab^2.$$

There are $C(3, 2) = 3$ strings of length three that contain two b's.

We can use these concepts to expand $(a + b)^4$. The expression $(a + b)^4$ consists of the sum of all strings of length four using only the letters a and b. There is $C(4, 0) = 1$ string containing no b's, $C(4, 1) = 4$ strings containing one b, and so on, until there is $C(4, 4) = 1$ string containing four b's. Thus

$$(a + b)^4 = \binom{4}{0}a^4b^0 + \binom{4}{1}a^3b^1 + \binom{4}{2}a^2b^2 + \binom{4}{3}a^1b^3 + \binom{4}{4}a^0b^4$$
$$= a^4 + 4a^3b + 6a^2b^2 + 4ab^3 + b^4.$$

These results are summarized by the binomial theorem.

BINOMIAL THEOREM

For any positive integer n and numbers a and b,

$$(a + b)^n = \binom{n}{0}a^n + \binom{n}{1}a^{n-1}b^1 + \cdots + \binom{n}{n-1}a^1b^{n-1} + \binom{n}{n}b^n.$$

We can use the binomial theorem to expand $(a + b)^n$. For example,

$$(a + b)^3 = \binom{3}{0}a^3 + \binom{3}{1}a^2b^1 + \binom{3}{2}a^1b^2 + \binom{3}{3}b^3$$
$$= 1a^3 + 3a^2b + 3ab^2 + 1b^3$$
$$= a^3 + 3a^2b + 3ab^2 + b^3.$$

Since $\binom{n}{r} = C(n, r)$, we can use the combination formula $C(n, r) = \frac{n!}{(n - r)!\, r!}$ to evaluate the binomial coefficients.

 **EXAMPLE 2** Applying the binomial theorem

Use the binomial theorem to expand the expression $(2a + 1)^5$.

SOLUTION Using the binomial theorem, we arrive at the following result.

$$(2a + 1)^5 = \binom{5}{0}(2a)^5 + \binom{5}{1}(2a)^4 1^1 + \binom{5}{2}(2a)^3 1^2 + \binom{5}{3}(2a)^2 1^3 + \binom{5}{4}(2a)^1 1^4 + \binom{5}{5} 1^5$$

$$= \frac{5!}{5!0!}(32a^5) + \frac{5!}{4!1!}(16a^4) + \frac{5!}{3!2!}(8a^3) + \frac{5!}{2!3!}(4a^2) + \frac{5!}{1!4!}(2a) + \frac{5!}{0!5!}$$

$$= 32a^5 + 80a^4 + 80a^3 + 40a^2 + 10a + 1$$

Now Try Exercise 21 ◆

Pascal's Triangle

Expanding $(a + b)^n$ for increasing values of n gives the following results.

$(a + b)^0 = $ 1

$(a + b)^1 = $ .... $1a + 1b$

$(a + b)^2 = $ .... $1a^2 + 2ab + 1b^2$

$(a + b)^3 = $ .... $1a^3 + 3a^2b + 3ab^2 + 1b^3$

$(a + b)^4 = $ .... $1a^4 + 4a^3b + 6a^2b^2 + 4ab^3 + 1b^4$

$(a + b)^5 = $ .... $1a^5 + 5a^4b + 10a^3b^2 + 10a^2b^3 + 5ab^4 + 1b^5$

Notice that $(a + b)^1$ has two terms starting with a and ending with b, $(a + b)^2$ has three terms starting with a^2 and ending with b^2, and in general $(a + b)^n$ has $n + 1$ terms starting with a^n and ending with b^n. The exponent on a decreases by 1 each successive term, and the exponent on b increases by 1 each successive term.

The triangle formed by the highlighted numbers is called **Pascal's triangle**. It can be used to efficiently compute the binomial coefficients, $C(n, r)$. The triangle consists of 1's along the sides. Each element inside the triangle is the sum of the two numbers above it. Pascal's triangle is usually written without variables as in Figure 11.19. It can be extended to include as many rows as needed.

We can use this triangle to expand powers of binomials in the form $(a + b)^n$, where n is a natural number. For example, the expression $(m + n)^4$ consists of five terms written as follows.

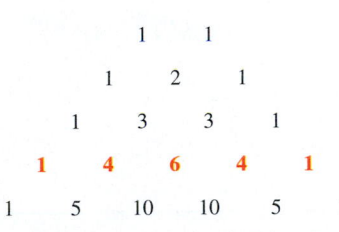

FIGURE 11.19 Pascal's Triangle

$$(m + n)^4 = \underline{} m^4 + \underline{} m^3 n^1 + \underline{} m^2 n^2 + \underline{} m^1 n^3 + \underline{} n^4$$

Since there are five terms, the coefficients can be found in the fifth row of Pascal's triangle, which is

$$1 \quad 4 \quad 6 \quad 4 \quad 1.$$

Thus,

$$(m + n)^4 = \underline{1}\, m^4 + \underline{4}\, m^3 n^1 + \underline{6}\, m^2 n^2 + \underline{4}\, m^1 n^3 + \underline{1}\, n^4.$$

$$= m^4 + 4m^3 n + 6m^2 n^2 + 4mn^3 + n^4.$$

EXAMPLE 3 Expanding expressions with the binomial theorem

Expand each of the following.
(a) $(2x + 1)^5$ **(b)** $(3x - y)^3$

SOLUTION

(a) To expand $(2x + 1)^5$, let $a = 2x$ and $b = 1$ in the binomial theorem. We can use the sixth row of Pascal's triangle to obtain the coefficients 1, 5, 10, 10, 5, and 1. Compare this solution with the solution for Example 2.

$$(2x + 1)^5 = 1(2x)^5 + 5(2x)^4(1)^1 + 10(2x)^3(1)^2 + 10(2x)^2(1)^3 + 5(2x)^1(1)^4 + 1(1)^5$$

$$= 32x^5 + 80x^4 + 80x^3 + 40x^2 + 10x + 1$$

(b) Let $a = 3x$ and $b = -y$ in the binomial theorem. Use the coefficients 1, 3, 3, and 1 from the fourth row of Pascal's triangle.

$$(3x - y)^3 = 1(3x)^3 + 3(3x)^2(-y)^1 + 3(3x)^1(-y)^2 + 1(-y)^3$$

$$= 27x^3 - 27x^2y + 9xy^2 - y^3 \qquad \textit{Now Try Exercises 33 and 35} \blacklozenge$$

The binomial theorem gives *all* of the terms of $(a + b)^n$. However, we can find any individual term by noting that the $(r + 1)$st term in the binomial expansion for $(a + b)^n$ is given by the formula $\binom{n}{r}a^{n-r}b^r$, for $0 \le r \le n$. The next example shows how to use this formula to find the $(r + 1)$st term of $(a + b)^n$.

EXAMPLE 4   Finding the *k*th term in a binomial expansion

Find the third term of $(x - y)^5$.

SOLUTION   In this example the $(r + 1)$st term is the *third* term in the expansion of $(x - y)^5$. That is, $r + 1 = 3$, or $r = \mathbf{2}$. Also, the exponent in the expression is $n = \mathbf{5}$. To get this binomial into the form $(a + b)^n$, we note that the first term in the binomial is $a = x$ and that the second term in the binomial is $b = -y$. Substituting the values for $r, n, a,$ and b in the formula for the $(r + 1)$st term yields

$$\binom{\mathbf{5}}{\mathbf{2}}(x)^{\mathbf{5}-\mathbf{2}}(-y)^{\mathbf{2}} = 10x^3y^2.$$

The third term in the binomial expansion of $(x - y)^5$ is $10x^3y^2$.   *Now Try Exercise 45* $\blacklozenge$

11.4

Putting it all Together

$\mathbf{T}$he following table summarizes topics related to the binomial theorem.

Concept	Explanation	Example
Binomial coefficient	$\binom{n}{r} = C(n, r) = \dfrac{n!}{(n - r)! \, r!}$	$\binom{5}{3} = \dfrac{5!}{(5 - 3)! \, 3!} = \dfrac{120}{2 \cdot 6} = 10$ $\binom{4}{0} = \dfrac{4!}{(4 - 0)! \, 0!} = \dfrac{24}{24 \cdot 1} = 1$

Concept	Explanation	Example
Binomial theorem	$(a + b)^n =$ $\binom{n}{0}a^n + \binom{n}{1}a^{n-1}b + \cdots + \binom{n}{n}b^n,$ for any positive integer n and real numbers a and b.	$(a + b)^3 = \binom{3}{0}a^3 + \binom{3}{1}a^2b + \binom{3}{2}ab^2 + \binom{3}{3}b^3$ $= a^3 + 3a^2b + 3ab^2 + b^3$ The binomial coefficients can also be found using the fourth row of Pascal's triangle, which is shown below.
Pascal's triangle	A triangle of numbers that can be used to find the binomial coefficients needed to expand an expression of the form $(a + b)^n$. To expand $(a + b)^n$, use row $n + 1$ of Pascal's triangle.	$$\begin{array}{ccccccccccc} & & & & & 1 & & & & & \\ & & & & 1 & & 1 & & & & \\ & & & 1 & & 2 & & 1 & & & \\ & & 1 & & 3 & & 3 & & 1 & & \\ & 1 & & 4 & & 6 & & 4 & & 1 & \\ 1 & & 5 & & 10 & & 10 & & 5 & & 1 \end{array}$$
Finding the $(r + 1)$st term of $(a + b)^n$	The $(r + 1)$st term of $(a + b)^n$ is given by $$\binom{n}{r}a^{n-r}b^r.$$ for $0 \leq r \leq n$	To find the fifth term of $(x + y)^6$, let $r + 1 = 5$ or $r = 4$, and $n = 6$. $$\binom{n}{r}a^{n-r}b^r = \binom{6}{4}x^{6-4}y^4$$ $$= 15x^2y^4.$$

11.4 Exercises

Binomial Coefficients

Exercises 1–8: Evaluate the expression.

1. $\binom{5}{4}$

2. $\binom{6}{2}$

3. $\binom{4}{0}$

4. $\binom{4}{2}$

5. $\binom{6}{5}$

6. $\binom{6}{3}$

7. $\binom{3}{3}$

8. $\binom{5}{2}$

Binomial Theorem

Exercises 9–16: (Refer to Example 1.) Calculate the number of distinguishable strings that can be formed with the given number of a's and b's.

9. Three a's, two b's

10. Five a's, three b's

11. Four a's, four b's

12. One a, five b's

13. Five a's, no b's   **14.** No a's, three b's

15. Four a's, one b **16.** Four a's, two b's

Exercises 17–30: Use the binomial theorem to expand each expression.

17. $(x + y)^2$ **18.** $(x + y)^4$

19. $(m + 2)^3$ **20.** $(m + 2n)^5$

21. $(2x - 3)^3$ **22.** $(x + y^2)^3$

23. $(p - q)^6$ **24.** $(p^2 - 3)^4$

25. $(2m + 3n)^3$ **26.** $(3a - 2b)^5$

27. $(1 - x^2)^4$ **28.** $(2 + 3x^2)^3$

29. $(2p^3 - 3)^3$ **30.** $(2r + 3t)^4$

Pascal's Triangle

Exercises 31–44: Use Pascal's triangle to help expand the expression.

31. $(x + y)^2$ **32.** $(m + n)^3$

33. $(3x + 1)^4$ **34.** $(2x - 1)^4$

35. $(2 - x)^5$ **36.** $(2a + 3b)^3$

37. $(x^2 + 2)^4$ **38.** $(5 - x^2)^3$

39. $(4x - 3y)^4$ **40.** $(3 - 2x)^5$

41. $(m + n)^6$ **42.** $(2m - n)^4$

43. $(2x^3 - y^2)^3$ **44.** $(3x^2 + y^3)^4$

Exercises 45–52: Find the specified term.

45. The fourth term of $(a + b)^9$

46. The second term of $(m - n)^9$

47. The fifth term of $(x + y)^8$

48. The third term of $(a + b)^7$

49. The fourth term of $(2x + y)^5$

50. The eighth term of $(2a - b)^9$

51. The sixth term of $(3x - 2y)^6$

52. The seventh term of $(2a + b)^9$

Writing about Mathematics

53. Explain how to find the numbers in Pascal's triangle.

54. Compare the expansion of $(a + b)^n$ with the expansion of $(a - b)^n$. Give an example.

CHECKING BASIC CONCEPTS FOR SECTIONS 11.3 AND 11.4

1. Count the ways to answer a quiz that consists of 8 true-false questions.

2. Count the number of 5-card poker hands that are possible using a standard deck of 52 cards.

3. How many distinct license plates could be made using a letter followed by 5 digits or letters?

4. Expand each expression.
     (a) $(2x + 1)^4$ **(b)** $(4 - 3x)^3$

Mathematical Induction

11.5

♦ Learn basic concepts about mathematical induction

♦ Use mathematical induction to prove statements

♦ Apply the generalized principle of mathematical induction

Introduction

The brilliant mathematician Carl Friedrich Gauss (1777–1855) proved the fundamental theorem of algebra at age 20. When he was a young child, he amazed his teacher by showing that

$$1 + 2 + 3 + 4 + \cdots + 100 = \frac{100(101)}{2}.$$

With *mathematical induction* we will be able to show, more generally, that

$$1 + 2 + 3 + 4 + \cdots + n = \frac{n(n + 1)}{2}.$$

Mathematical induction is a powerful method of proof. It is used not only in mathematics, but it is also used in computer science to prove that programs and basic concepts are correct.

Mathematical Induction

Many results in mathematics are claimed true for every positive integer. Any of these results could be checked for $n = 1$, $n = 2$, $n = 3$, and so on, but since the set of positive integers is infinite, it would be impossible to check every possible case. For example, let S_n represent the statement that the sum of the first n positive integers is $\frac{n(n + 1)}{2}$, that is,

$$S_n: 1 + 2 + 3 + \cdots + n = \frac{n(n + 1)}{2}.$$

The truth of this statement can be checked quickly for the first few values of n.

If $n = 1$, S_1 is $\qquad 1 = \frac{1(1 + 1)}{2}$, a true statement, since $1 = 1$.

If $n = 2$, S_2 is $\qquad 1 + 2 = \frac{2(2 + 1)}{2}$, a true statement, since $3 = 3$.

If $n = 3$, S_3 is $\quad 1 + 2 + 3 = \frac{3(3 + 1)}{2}$, a true statement, since $6 = 6$.

If $n = 4$, S_4 is $1 + 2 + 3 + 4 = \frac{4(4 + 1)}{2}$, a true statement, since $10 = 10$.

Since the statement is true for $n = 1, 2, 3$, and 4, and so on, can we conclude that the statement is true for all positive integers by checking a finite number of examples? The answer is no. To prove that such a statement is true for every positive integer, we use the following principle.

PRINCIPLE OF MATHEMATICAL INDUCTION

Let S_n be a statement concerning the positive integer n. Suppose that

1. S_1 is true;
2. for any positive integer k, if S_k is true, then S_{k+1} is also true.

Then, S_n is true for every positive integer n.

A proof by mathematical induction can be explained as follows. By assumption (1), the statement is true when $n = 1$. By assumption (2), the fact that the statement is true for $n = 1$ implies that it is true for $n = 1 + 1 = 2$. Using (2) again, the statement is thus true

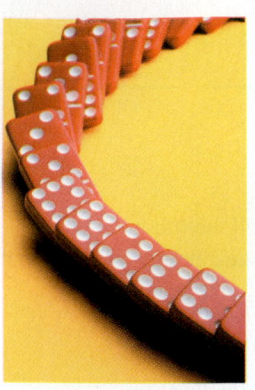

FIGURE 11.20

for $2 + 1 = 3$, for $3 + 1 = 4$, for $4 + 1 = 5$, and so on. Continuing in this way shows that the statement must be true for *every* positive integer.

The situation is similar to that of an infinite number of dominoes lined up, as suggested in Figure 11.20. If the first domino is pushed over, it pushes the next, which pushes the next, and so on, indefinitely.

Another example of the principle of mathematical induction might be an infinite ladder. Suppose the rungs are spaced so that, whenever you are on a rung, you know you can move to the next rung. Then *if* you can get to the first rung, you can go as high up the ladder as you wish.

Two separate steps are required for a proof by mathematical induction.

PROOF BY MATHEMATICAL INDUCTION

STEP 1: Prove that the statement is true for $n = 1$.

STEP 2: Show that for any positive integer k, if S_k is true, then S_{k+1} is also true.

Proving Statements

Mathematical induction is used in the next example to prove the statement S_n discussed earlier.

 **Proving an equality statement**

Let S_n represent the statement

$$1 + 2 + 3 + \cdots + n = \frac{n(n + 1)}{2}.$$

Prove that S_n is true for every positive integer n.

SOLUTION The proof by mathematical induction is as follows.

STEP 1: Show that the statement is true when $n = 1$. If $n = 1$, S_1 becomes

$$1 = \frac{1(1 + 1)}{2},$$

which is true.

STEP 2: Show that if S_k is true, then S_{k+1} is also true, where S_k is the statement

$$1 + 2 + 3 + \cdots + k = \frac{k(k + 1)}{2},$$

and S_{k+1} is the statement

$$1 + 2 + 3 + \cdots + k + (k + 1) = \frac{(k + 1)[(k + 1) + 1]}{2}.$$

Start with S_k and assume it is a true statement.

$$1 + 2 + 3 + \cdots + k = \frac{k(k + 1)}{2}$$

Add $k + 1$ to each side of this equation to obtain S_{k+1}.

$$1 + 2 + 3 + \cdots + k + (k + 1) = \frac{k(k + 1)}{2} + (k + 1)$$

Factor out $k + 1$.

$$= (k + 1)\left(\frac{k}{2} + 1\right)$$

$$= (k + 1)\left(\frac{k + 2}{2}\right)$$

$$= \frac{(k + 1)[(k + 1) + 1]}{2}.$$

This final result is the statement S_{k+1}. Therefore, if S_k is true, then S_{k+1} is also true. The two steps required for a proof by mathematical induction have been completed, so the statement S_n is true for every positive integer n.

Now Try Exercise 1 ◆

EXAMPLE 2 Proving an equality statement

Let S_n represent the statement

$$2^1 + 2^2 + 2^3 + 2^4 + \cdots + 2^n = 2^{n+1} - 2.$$

Prove that S_n is true for every positive integer n.

SOLUTION

STEP 1: Show that the statement S_1 is true, where S_1 is

$$2^1 = 2^{1+1} - 2.$$

Since $2 = 4 - 2$, S_1 is a true statement.

STEP 2: Show that if S_k is true, then S_{k+1} is also true, where S_k is

$$2^1 + 2^2 + 2^3 + \cdots + 2^k = 2^{k+1} - 2$$

and S_{k+1} is

$$2^1 + 2^2 + 2^3 + \cdots + 2^k + 2^{k+1} = 2^{(k+1)+1} - 2.$$

Start with S_k and add 2^{k+1} to each side of the equation. Then algebraically change the right side to look like the right side of S_{k+1}.

$$2^1 + 2^2 + 2^3 + \cdots + 2^k + 2^{k+1} = 2^{k+1} - 2 + 2^{k+1}$$

$$= 2 \cdot 2^{k+1} - 2$$

$$= 2^{(k+1)+1} - 2$$

The final result is the statement S_{k+1}. Therefore, if S_k is true, then S_{k+1} is also true. The two steps required for a proof by mathematical induction have been completed, so the statement S_n is true for every positive integer n.

Now Try Exercise 5 ◆

EXAMPLE 3 Proving an inequality statement

Prove that if x is a real number between 0 and 1, then for every positive integer n,

$$0 < x^n < 1.$$

SOLUTION

STEP 1: Here S_1 is the statement

$$\text{if } 0 < x < 1, \text{ then } 0 < x^1 < 1,$$

which is true.

STEP 2: S_k is the statement

$$\text{if } 0 < x < 1, \text{ then } 0 < x^k < 1.$$

To show that S_k implies that S_{k+1} is true, multiply all three parts of $0 < x^k < 1$ by x to get

$$x \cdot 0 < x \cdot x^k < x \cdot 1.$$

(Here the fact that $0 < x$ is used.) Simplify to obtain

$$0 < x^{k+1} < x.$$

Since $x < 1$,

$$0 < x^{k+1} < 1,$$

which implies that S_{k+1} is true. Therefore, if S_k is true, then S_{k+1} is true. Since both steps for a proof by mathematical induction have been completed, the given statement is true for every positive integer n.   *Now Try Exercise 15* ◆

Generalized Principle of Mathematical Induction

Some statements S_n are not true for the first few values of n, but are true for all values of n that are greater than or equal to some fixed integer j. The following slightly generalized form of the principle of mathematical induction takes care of these cases.

GENERALIZED PRINCIPLE OF MATHEMATICAL INDUCTION

Let S_n be a statement concerning the positive integer n. Let j be a fixed positive integer. Suppose that

1. S_j is true;
2. for any positive integer k, $k \geq j$, S_k implies S_{k+1}.

Then S_n is true for all positive integers n, where $n \geq j$.

 **Using the generalized principle**

Let S_n represent the statement $2^n > 2n + 1$. Show that S_n is true for all values of n such that $n \geq 3$.

SOLUTION (Check that S_n is false for $n = 1$ and $n = 2$.)

STEP 1: Show that S_n is true for $n = 3$. If $n = 3$, S_3 is

$$2^3 > 2 \cdot 3 + 1, \text{ or}$$

$$8 > 7.$$

Thus, S_3 is true.

STEP 2: Now show that S_k implies S_{k+1}, for $k \geq 3$, where

$$S_k \text{ is } 2^k > 2k + 1 \text{ and}$$
$$S_{k+1} \text{ is } 2^{k+1} > 2(k + 1) + 1.$$

Multiply each side of $2^k > 2k + 1$ by 2, obtaining

$$2 \cdot 2^k > 2(2k + 1), \text{ or}$$
$$2^{k+1} > 4k + 2.$$

Rewrite $4k + 2$ as $2(k + 1) + 2k$, giving

$$2^{k+1} > 2(k + 1) + 2k.$$

Since k is a positive integer greater than 3,

$$2k > 1.$$

It follows that

$$2^{k+1} > 2(k + 1) + 2k > 2(k + 1) + 1, \text{ or}$$
$$2^{k+1} > 2(k + 1) + 1,$$

as required. Thus, S_k implies S_{k+1}, and this, together with the fact that S_3 is true, shows that S_n is true for every positive integer n greater than or equal to 3.

Now Try Exercise 27 ◆

11.5 Putting it all Together

Some important concepts about mathematical induction are summarized in the following table.

Concept	Explanation
Principle of mathematical induction	Let S_n be a statement concerning the positive integer n. Suppose that 1. S_1 is true; 2. for any positive integer k, if S_k is true, then S_{k+1} is also true. Then, S_n is true for every positive integer n.
Proof by mathematical induction	**STEP 1:** Prove that the statement is true for $n = 1$. **STEP 2:** Show that for any positive integer k, if S_k is true, then S_{k+1} is also true.
Generalized principle of mathematical induction	Let S_n be a statement concerning the positive integer n. Let j be a fixed positive integer. Suppose that 1. S_j is true; 2. for any positive integer k, $k \geq j$, S_k implies S_{k+1}. Then S_n is true for all positive integers n, where $n \geq j$.

11.5 — Exercises

Exercises 1–14: Use mathematical induction to prove the statement. Assume that n is a positive integer.

1. $3 + 6 + 9 + \cdots + 3n = \dfrac{3n(n+1)}{2}$

2. $1 + 3 + 5 + \cdots + (2n - 1) = n^2$

3. $5 + 10 + 15 + \cdots + 5n = \dfrac{5n(n+1)}{2}$

4. $4 + 7 + 10 + \cdots + (3n + 1) = \dfrac{n(3n+5)}{2}$

5. $3 + 3^2 + 3^3 + \cdots + 3^n = \dfrac{3(3^n - 1)}{2}$

6. $1^2 + 2^2 + 3^2 + \cdots + n^2 = \dfrac{n(n+1)(2n+1)}{6}$

7. $1^3 + 2^3 + 3^3 + \cdots + n^3 = \dfrac{n^2(n+1)^2}{4}$

8. $5 \cdot 6 + 5 \cdot 6^2 + 5 \cdot 6^3 + \cdots + 5 \cdot 6^n = 6(6^n - 1)$

9. $\dfrac{1}{1 \cdot 2} + \dfrac{1}{2 \cdot 3} + \dfrac{1}{3 \cdot 4} + \cdots + \dfrac{1}{n(n+1)} = \dfrac{n}{n+1}$

10. $7 \cdot 8 + 7 \cdot 8^2 + 7 \cdot 8^3 + \cdots + 7 \cdot 8^n = 8(8^n - 1)$

11. $\dfrac{4}{5} + \dfrac{4}{5^2} + \dfrac{4}{5^3} + \cdots + \dfrac{4}{5^n} = 1 - \dfrac{1}{5^n}$

12. $\dfrac{1}{2} + \dfrac{1}{2^2} + \dfrac{1}{2^3} + \cdots + \dfrac{1}{2^n} = 1 - \dfrac{1}{2^n}$

13. $\dfrac{1}{1 \cdot 4} + \dfrac{1}{4 \cdot 7} + \cdots + \dfrac{1}{(3n-2)(3n+1)} = \dfrac{n}{3n+1}$

14. $x^{2n} + x^{2n-1}y + \cdots + xy^{2n-1} + y^{2n} = \dfrac{x^{2n+1} - y^{2n+1}}{x - y}$

Exercises 15–18: Find all positive integers n for which the given statement is not true.

15. $3^n > 6n$ **16.** $3^n > 2n + 1$

17. $2^n > n^2$ **18.** $n! > 2n$

Exercises 19–28: Prove the statement by mathematical induction.

19. $(a^m)^n = a^{mn}$ (Assume a and m are constants.)

20. $(ab)^n = a^n b^n$ (Assume a and b are constants.)

21. $2^n > 2n$, if $n \geq 3$ **22.** $3^n > 2n + 1$, if $n \geq 2$

23. If $a > 1$, then $a^n > 1$.   **24.** If $a > 1$, then $a^n > a^{n-1}$.

25. If $0 < a < 1$, then $a^n < a^{n-1}$.

26. $2^n > n^2$, for $n > 4$

27. If $n \geq 4$, then $n! > 2^n$, where
$$n! = n(n-1)(n-2) \cdots (3)(2)(1).$$

28. $4^n > n^4$, for $n \geq 5$

29. *Number of Handshakes* Suppose that each of the n ($n \geq 2$) people in a room shakes hands with everyone else, but not with himself. Show that the number of handshakes is $\dfrac{n^2 - n}{2}$.

30. *Sides of a Polygon* The series of sketches starts with an equilateral triangle having sides of length 1. In the following steps, equilateral triangles are constructed on each side of the preceding figure. The length of the sides of each new triangle is $\frac{1}{3}$ the length of the sides of the preceding triangles. Develop a formula for the number of sides of the nth figure. Use mathematical induction to prove your answer.

31. *Perimeter* Find the perimeter of the *n*th figure in Exercise 30.

32. *Area* Show that the area of the *n*th figure in Exercise 30 is

$$\sqrt{3}\left[\frac{2}{5} - \frac{3}{20}\left(\frac{4}{9}\right)^{n-1}\right].$$

33. *Tower of Hanoi* A pile of *n* rings, each ring smaller than the one below it, is on a peg. Two other pegs are attached to a board with this peg. In the game called the *Tower of Hanoi* puzzle, all the rings must be moved to a different peg, with only one ring moved at a time, and with no ring ever placed on top of a smaller ring. Find the least number of moves that would be required. Prove your result with mathematical induction.

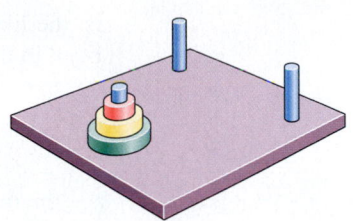

Writing about Mathematics

34. Explain the principle of mathematical induction.

35. Explain how the generalized principle of mathematical induction differs from the principle of mathematical induction.

36. When using mathematical induction, why is it important to prove that the statement holds for $n = 1$?

11.6 Probability

- ◆ Learn the basic concepts about probability
- ◆ Calculate the probability of compound events
- ◆ Calculate the probability of independent and dependent events

Introduction

Questions of chance have no doubt engaged the minds of people since antiquity. However, the mathematical treatment of probability did not occur until the fifteenth century. The birth of probability theory as a mathematical discipline began in the seventeenth century with the work of Blaise Pascal and Pierre Fermat. Today probability pervades modern society. It is used not only to determine outcomes in gambling, but also to predict weather, genetic outcomes, and the risk involved with various types of substances and behaviors.

Risk is the chance or probability that a harmful event will occur. The following activities carry an annual increased risk of death by one chance in a million: flying 1000 miles in a jet, traveling 300 miles in a car, riding 10 miles on a bicycle, smoking 1.4 cigarettes, living 2 days in New York City, having one chest X-ray, or living 2 months with a cigarette smoker. In Section 11.3 we saw that the likelihood of winning the jackpot in a lottery was one chance in 120,526,770.

Ideally, we would live in a risk-free world. However, almost every action or substance exposes people to some risk. Recognizing relative risks is important for a long and healthy life. Probability provides us with a measure of the likelihood that an event will occur. Knowledge about probability allows individuals to make informed decisions about their lives.   (**Sources:** *Historical Topics for the Mathematics Classroom, Thirty-first Yearbook*, NCTM; J. Rodricks, *Calculated Risk.*)

Definition of Probability

In the study of probability, experiments often are performed. An experiment might involve tossing a coin or measuring the cholesterol level of a patient who has had a heart attack. A result from an experiment is called an **outcome**. The set of all possible outcomes is the **sample space**. Any subset of a sample space is an **event**.

For example, if an experiment involves rolling a (fair) die, then the sample space consists of $S = \{1, 2, 3, 4, 5, 6\}$. The event $E = \{1, 6\}$ contains the outcomes of 1 or 6 showing on the die. If $n(E)$ and $n(S)$ denote the number of outcomes in E and S, then $n(E) = 2$ and $n(S) = 6$. The probability of rolling a 1 or 6 is given by $P(E) = \frac{2}{6}$. That is, the likelihood of event E occurring is two chances in six. These concepts are summarized in the following.

> ### PROBABILITY OF AN EVENT
>
> If the outcomes of a finite sample space S are equally likely, and if E is an event in S, then the **probability of E** is given by
>
> $$P(E) = \frac{n(E)}{n(S)},$$
>
> where $n(E)$ and $n(S)$ represent the number of outcomes in E and S, respectively.

Since $n(E) \leq n(S)$, the probability of an event E satisfies $0 \leq P(E) \leq 1$. If $P(E) = 1$, then event E is certain to occur. If $P(E) = 0$, then event E is impossible.

EXAMPLE 1 Drawing a card

One card is drawn at random from a standard deck of 52 cards. Find the probability that the card is an ace.

SOLUTION The sample space S consists of 52 outcomes that correspond to drawing any one of 52 cards. Each outcome is equally likely. Let E represent the event of drawing an ace. There are four aces in the deck, so event E contains four outcomes. Therefore $n(S) = 52$ and $n(E) = 4$. The probability of drawing an ace is given by

$$P(E) = \frac{n(E)}{n(S)} = \frac{4}{52} = \frac{1}{13}.$$

Now Try Exercise 15 ◆

EXAMPLE 2 Estimating probability of organ transplants

TABLE 11.8

Heart	3774
Kidney	35,025
Liver	7920
Lung	2340

Source: Coalition on Organ and Tissue Donation.

In 1997 there were 51,277 people waiting for an organ transplant. Table 11.8 lists the number of patients waiting for the most common types of transplants. Assuming none of these people need two or more transplants, approximate the probability that a transplant patient chosen at random will need
(a) a kidney or a heart; **(b)** neither a kidney nor a heart.

SOLUTION
(a) Let each patient represent an outcome in a sample space S. The event E of a transplant patient needing either a kidney or a heart contains $35{,}025 + 3774 = 38{,}799$ outcomes. The desired probability is

$$P(E) = \frac{n(E)}{n(S)} = \frac{38{,}799}{51{,}277} \approx 0.76.$$

In 1997, about 76% of transplant patients needed either a kidney or a heart.
(b) Let F be the event of a patient waiting for an organ other than a kidney or a heart. Then,

$$n(F) = n(S) - n(E) = 51{,}277 - 38{,}799 = 12{,}478.$$

The probability of F is

$$P(F) = \frac{n(F)}{n(S)} = \frac{12{,}478}{51{,}277} \approx 0.24 \quad \text{or} \quad 24\%.$$

Now Try Exercise 19 ◆

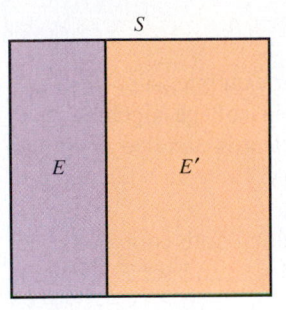

FIGURE 11.21 $E \cup E' = S$

Notice that $P(E) + P(F) = 1$ in Example 2. The events E and F are **complements** because $E \cap F = \varnothing$ and $E \cup F = S$, where $\varnothing$ denotes the *empty set*. That is, a transplant patient is either waiting for a kidney or a heart (event E), or not waiting for a kidney or a heart (event F). The complement of E may be denoted by E'.

Probability concepts can be illustrated using **Venn diagrams**. In Figure 11.21 the sample space S of an experiment is the union of the disjoint sets E and its complement E'. That is, $E \cup E' = S$ and $E \cap E' = \varnothing$.

If $P(E)$ is known, then $P(E')$ can be calculated as follows.

$$P(E') = \frac{n(E')}{n(S)} = \frac{n(S) - n(E)}{n(S)} = 1 - \frac{n(E)}{n(S)} = 1 - P(E)$$

In Example 2(b), the probability of $F = E'$ could have also been calculated by using

$$P(F) = 1 - P(E) \approx 1 - 0.76 = 0.24.$$

PROBABILITY OF A COMPLEMENT

Let E be an event and E' be its complement. If the probability of E is $P(E)$, then the probability of its complement is given by

$$P(E') = 1 - P(E).$$

EXAMPLE 3 Finding probabilities of human eye color

TABLE 11.9

	B	b
B	BB	Bb
b	bB	bb

In 1865 Gregor Mendel performed important research in genetics. His work led to a better understanding of dominant and recessive genetic traits. One example is human eye color, which is determined by a pair of genes called a *genotype*. Brown eye color B is dominant over blue eye color b. If a person has the genotype of BB, Bb, or bB, he or she will have brown eyes. A genotype of bb will result in blue eyes. A person receives one gene (B or b) from each parent. Table 11.9 shows how these two genes can be paired. (**Source:** H. Lancaster, *Quantitative Methods in Biology and Medical Sciences.*)

(a) Assuming that each genotype is equally likely, find the probability of blue eyes.
(b) What is the probability that a person has brown eyes?

SOLUTION
(a) The sample space S consists of four equally likely outcomes denoted BB, Bb, bB, and bb. The event E of blue eye color (bb) occurs once. Therefore

$$P(E) = \frac{n(E)}{n(S)} = \frac{1}{4} = 0.25.$$

(b) In this chart, brown eyes are the complement of blue eyes. The probability of brown eyes is

$$P(E') = 1 - P(E) = 1 - 0.25 = 0.75.$$

This probability also could be computed as

$$P(E') = \frac{n(E')}{n(S)} = \frac{3}{4} = 0.75,$$

since there are three genotypes that result in brown eye color.   *Now Try Exercise 40*

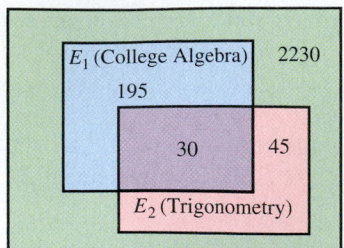

FIGURE 11.22

Compound Events

Frequently the probability of more than one event is needed. For example, suppose a college has a total of 2500 students with 225 students enrolled in college algebra, 75 in trigonometry, and 30 in both. Let E_1 denote the event that a student is enrolled in college algebra, and E_2 the event that a student is enrolled in trigonometry. Then the Venn diagram in Figure 11.22 visually describes this situation.

In this Venn diagram, it is important that the 30 students taking both courses are not counted twice. The set E_1 has a total of $195 + 30 = 225$ students, and set E_2 contains $45 + 30 = 75$ students.

EXAMPLE 4 Calculating the probability of a union

In the preceding scenario, suppose a student is selected at random. What is the probability that this student is enrolled in college algebra, trigonometry, or both?

SOLUTION We would like to find the probability $P(E_1 \text{ or } E_2)$, which is denoted $P(E_1 \cup E_2)$. Since $n(E_1 \cup E_2) = 195 + 30 + 45 = 270$ and $n(S) = 2500$,

$$P(E_1 \cup E_2) = \frac{n(E_1 \cup E_2)}{n(S)} = \frac{270}{2500} = 0.108.$$

There is a 10.8% chance that a student selected at random will be taking college algebra, trigonometry, or both.   *Now Try Exercise 35* ◆

In the previous example, it would have been incorrect to simply add the probability of a student taking algebra and the probability of a student taking trigonometry. Their sum would be

$$P(E_1) + P(E_2) = \frac{n(E_1)}{n(S)} + \frac{n(E_2)}{n(S)}$$

$$= \frac{225}{2500} + \frac{75}{2500}$$

$$= \frac{300}{2500}$$

$$= 0.12, \quad \text{or} \quad 12\%.$$

This sum is greater than $P(E_1 \cup E_2)$ because the 30 students taking both courses are counted twice in this calculation. In order to find the correct probability for $P(E_1 \cup E_2)$, we must subtract the probability of the intersection $E_1 \cap E_2$.

$$P(E_1 \cup E_2) = P(E_1) + P(E_2) - P(E_1 \cap E_2)$$

$$= \frac{n(E_1)}{n(S)} + \frac{n(E_2)}{n(S)} - \frac{n(E_1 \cap E_2)}{n(S)}$$

$$= \frac{225}{2500} + \frac{75}{2500} - \frac{30}{2500}$$

$$= \frac{270}{2500}$$

$$= 0.108, \quad \text{or} \quad 10.8\%$$

This is the same result obtained in Example 4 and suggests the following property.

> **PROBABILITY OF THE UNION OF TWO EVENTS**
>
> For any two events E_1 and E_2,
>
> $$P(E_1 \cup E_2) = P(E_1) + P(E_2) - P(E_1 \cap E_2).$$

Rolling dice

Suppose two dice are rolled. Find the probability that the dice show either a sum of eight or a pair.

SOLUTION In Table 11.10 the roll of the dice is represented by an ordered pair. For example, the ordered pair $(3, 6)$ represents the first die showing 3 and the second die 6.

TABLE 11.10   Rolling Two Dice

(1, 1)	(1, 2)	(1, 3)	(1, 4)	(1, 5)	(1, 6)
(2, 1)	**(2, 2)**	(2, 3)	(2, 4)	(2, 5)	**(2, 6)**
(3, 1)	(3, 2)	**(3, 3)**	(3, 4)	**(3, 5)**	(3, 6)
(4, 1)	(4, 2)	(4, 3)	**(4, 4)**	(4, 5)	(4, 6)
(5, 1)	(5, 2)	**(5, 3)**	(5, 4)	**(5, 5)**	(5, 6)
(6, 1)	**(6, 2)**	(6, 3)	(6, 4)	(6, 5)	**(6, 6)**

Since each die can show six different outcomes, there is a total of $6 \cdot 6 = 36$ outcomes in the sample space S. Let E_1 denote the event of rolling a sum of eight, and E_2 the event of rolling a pair. Then

$$E_1 = \{(6, 2), (5, 3), (4, 4), (3, 5), (2, 6)\} \qquad \text{and}$$
$$E_2 = \{(1, 1), (2, 2), (3, 3), (4, 4), (5, 5), (6, 6)\}.$$

The intersection of E_1 and E_2 is

$$E_1 \cap E_2 = \{(4, 4)\}.$$

Since $n(S) = 36, n(E_1) = 5, n(E_2) = 6,$ and $n(E_1 \cap E_2) = 1,$ the following can be computed.

$$P(E_1 \cup E_2) = P(E_1) + P(E_2) - P(E_1 \cap E_2)$$
$$= \frac{n(E_1)}{n(S)} + \frac{n(E_2)}{n(S)} - \frac{n(E_1 \cap E_2)}{n(S)}$$
$$= \frac{5}{36} + \frac{6}{36} - \frac{1}{36}$$
$$= \frac{10}{36} \quad \text{or} \quad \frac{5}{18}$$

This result can be verified by counting the number of boldfaced outcomes in Table 11.10. Of the 36 possible outcomes, 10 outcomes satisfy the conditions so the probability is $\frac{10}{36}$ or $\frac{5}{18}$.

Now Try Exercise 43 ◆

If $E_1 \cap E_2 = \varnothing$, then the events E_1 and E_2 are **mutually exclusive**. Mutually exclusive events have no outcomes in common so $P(E_1 \cap E_2) = 0$. In this case, $P(E_1 \cup E_2) = P(E_1) + P(E_2)$.

EXAMPLE 6 Drawing cards

Find the probability of drawing either an ace or a king from a standard deck of 52 playing cards.

SOLUTION The event E_1 of drawing an ace and the event E_2 of drawing a king are mutually exclusive. No card can be both an ace and a king. Therefore $P(E_1 \cap E_2) = 0$. The probability of drawing an ace is $P(E_1) = \frac{4}{52}$, since there are 4 aces in 52 cards. Similarly, the probability of drawing a king is $P(E_2) = \frac{4}{52}$.

$$P(E_1 \cup E_2) = P(E_1) + P(E_2)$$
$$= \frac{4}{52} + \frac{4}{52}$$
$$= \frac{8}{52} \quad \text{or} \quad \frac{2}{13}$$

Thus the probability of drawing either an ace or a king is $\frac{2}{13}$. *Now Try Exercise 45*

The next example uses concepts from both counting and probability.

EXAMPLE 7 Drawing a poker hand

A standard deck of cards contains 52 cards, consisting of four suits: hearts, diamonds, spades, and clubs. Each suit contains 13 different cards. A poker hand consists of 5 cards drawn from a standard deck of cards, and a flush occurs when the 5 cards are all the same suit. Find an expression for the probability of drawing 5 cards with the same suit in one try. Assume that the cards are not replaced.

SOLUTION Let E be the event of drawing 5 cards with the same suit. To determine $n(E)$, start by calculating the number of ways to draw a flush in a particular suit, such as hearts. From a set of 13 hearts, 5 hearts need to be drawn. There are $\binom{13}{5}$ ways to draw this hand. Because there are 4 suits, there are $4\binom{13}{5}$ ways to draw 5 cards with the same suit. Thus $n(E) = 4\binom{13}{5}$. The sample space S consists of all 5-card poker hands that can be drawn from a deck of 52 cards. There are $\binom{52}{5}$ different poker hands. Thus $n(S) = \binom{52}{5}$. The probability of a flush can now be calculated as follows.

Calculator Help

To calculate $C(n, r)$, see Appendix B (page AP-25).

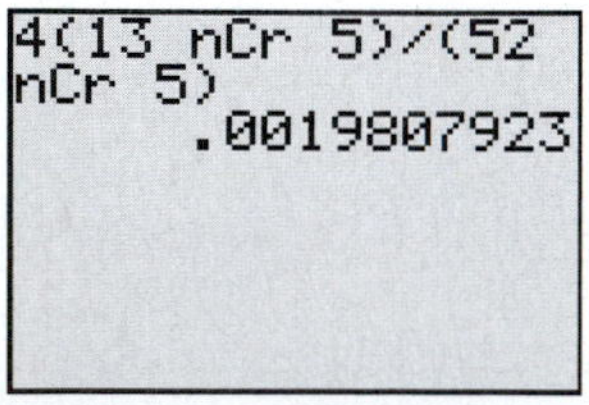

```
4(13 nCr 5)/(52
nCr 5)
      .0019807923
```

$$P(E) = \frac{n(E)}{n(S)} = \frac{4\binom{13}{5}}{\binom{52}{5}}$$

In Figure 11.23, we see that $P(E) \approx 0.00198$. Thus there is about a 0.2% chance of drawing 5 cards of the same suit (in one try) from a standard deck of 52 cards.

Now Try Exercise 31

FIGURE 11.23

Independent and Dependent Events

Two events are **independent** if they do not influence one another. Otherwise they are **dependent**. An example of independent events would be one coin being tossed twice. The result of the first toss does not affect the second toss.

PROBABILITY OF INDEPENDENT EVENTS

If E_1 and E_2 are independent events, then

$$P(E_1 \cap E_2) = P(E_1) \cdot P(E_2).$$

 **8**

Tossing a coin

Suppose a coin is tossed twice. Determine the probability that the result is two heads.

SOLUTION Let E_1 be the event of a head on the first toss, and let E_2 be the event of a head on the second toss. Then $P(E_1) = P(E_2) = \frac{1}{2}$. The two events are independent. The probability of two heads occurring is

$$P(E_1 \cap E_2) = \frac{1}{2} \cdot \frac{1}{2} = \frac{1}{4}.$$

This probability of $\frac{1}{4}$ also can be found using a tree diagram, as shown in Figure 11.24. There are four equally likely outcomes. Tosses resulting in two heads occur once.

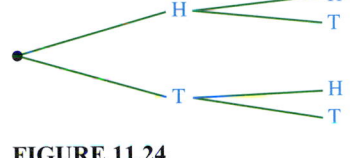

FIGURE 11.24

Now Try Exercise 21 ◆

 **9**

Rolling dice

What is the probability of rolling a sum of 12 with two dice?

SOLUTION The roll of one die does not influence the roll of the other. They are independent events. To obtain a sum of 12, both dice must show a six. Let E_1 be the event of rolling a 6 with the first die, and E_2 the event of rolling a 6 with the second die. Then $P(E_1) = P(E_2) = \frac{1}{6}$. The probability of rolling a 12 is

$$P(E_1 \cap E_2) = P(E_1) \cdot P(E_2) = \frac{1}{6} \cdot \frac{1}{6} = \frac{1}{36}.$$

◆ **CLASS DISCUSSION**

When rolling a pair of dice, what sum is most likely to appear? ◆

This can be verified by using Table 11.10. There is only 1 outcome out of 36 that results in a sum of 12.

Now Try Exercise 24 ◆

If events E_1 and E_2 influence each other, they are *dependent*. The probability of dependent events is given as follows.

PROBABILITY OF DEPENDENT EVENTS

If E_1 and E_2 are dependent events, then

$$P(E_1 \cap E_2) = P(E_1) \cdot P(E_2, \text{given that } E_1 \text{ occurred}).$$

Note: $P(E_2, \text{given that } E_1 \text{ occurred})$ is called the **conditional probability of E_2, given E_1.**

Drawing cards

Find the probability of drawing two hearts from a standard deck of 52 cards, when the first card is
(a) replaced before drawing the second card;   **(b)** not replaced.

SOLUTION

(a) Let E_1 denote the event of the first card being a heart, and E_2 the event of the second card being a heart. If the first card is replaced before the second card is drawn, the two events are independent. Since there are 13 hearts in a standard deck of 52 cards, the probability of two hearts being drawn is

$$P(E_1 \cap E_2) = P(E_1) \cdot P(E_2) = \frac{13}{52} \cdot \frac{13}{52} = \frac{1}{16}.$$

(b) If the first card is not replaced, then the outcome for the second card is influenced by the first card. Therefore the events E_1 and E_2 are dependent. The probability of drawing a second heart, given that the first card is a heart, is represented by the expression $P(E_2,$ given E_1 has occurred$) = \frac{12}{51}$. That is, if the first card drawn were a heart and removed from the deck, then there would be 12 hearts in a sample space of 51 cards.

$$P(E_1 \cap E_2) = P(E_1) \cdot P(E_2, \text{ given that } E_1 \text{ occurred}) = \frac{13}{52} \cdot \frac{12}{51} = \frac{1}{17}$$

Thus the probability of drawing two hearts is slightly less if the first card is not replaced.

Now Try Exercise 57 ◆

Calculating the probability of dependent events

Table 11.11 shows the number of students (by gender) registered for either a Spanish class or a French class. No student is taking both languages.

TABLE 11.11

	Spanish	French	Totals
Females	20	25	45
Males	40	25	65
Totals	60	50	110

If one student is selected at random, calculate each of the following.
(a) The probability that the student is female
(b) The probability that the student is taking Spanish, given that the student is female
(c) The probability that the student is female and taking Spanish

SOLUTION

(a) Let F represent the event that the student is female. Since 45 of the 110 students are female, $P(F) = \frac{45}{110} = \frac{9}{22}$.

(b) Let S be the event that the student is taking Spanish. The probability that the student is taking Spanish, given that the student is female, is $P(S,$ given $F) = \frac{20}{45} = \frac{4}{9}$, because 20 of the 45 female students are taking Spanish.

(c) The probability that the student is female and taking Spanish is calculated by

$$P(F \cap S) = P(F) \cdot P(S, \text{given } F) = \frac{45}{110} \cdot \frac{20}{45} = \frac{2}{11}.$$

Table 11.11 shows that 20 of the 110 students are female and taking Spanish. Thus the required probability is $\frac{20}{110} = \frac{2}{11}$, which is in agreement with our previous calculation.

Now Try Exercise 67 ◆

 **EXAMPLE 12** Analyzing a polygraph test

Suppose there is a 6% chance that a polygraph test will incorrectly say a person is lying when he or she is actually telling the truth. If a person tells the truth 95% of the time, what percentage of the time will the polygraph test incorrectly indicate a lie for this person?

SOLUTION Let E_1 be the event that the person is telling the truth and E_2 be the event that the polygraph test is incorrect. Then,

$$
\begin{aligned}
P(E_1 \cap E_2) &= P(E_1) \cdot P(E_2, \text{given the person is telling the truth}) \\
&= (0.95)(0.06) \\
&= 0.057 \quad \text{or} \quad 5.7\%.
\end{aligned}
$$

Now Try Exercise 63 ◆

11.6 Putting it all Together

$\mathbf{T}$he following table summarizes some concepts about probability. In this table, S denotes a sample space of equally likely outcomes, and E, E_1, and E_2 denote events in a sample space S.

Concept	Comments	Example
Probability P	A number P satisfying $0 \le P \le 1$.	$P(E) = 1$ indicates that an event E is certain to occur, $P(E) = 0$ indicates an event E is impossible, and $P(E) = 0.3$ indicates that an event E has a 30% chance of occurring.
Probability of an event	The probability of event E is $$P(E) = \frac{n(E)}{n(S)},$$ where $n(E)$ and $n(S)$ denote the number of equally likely outcomes in E and S, respectively.	The probability of rolling two dice and their sum being 3 is $\frac{2}{36}$. This is because there are $6 \cdot 6 = 36$ ways to roll two dice, and only 2 ways to roll a sum of 3. (They are 1 and 2, or 2 and 1.)
Probability of a complement	If E and E' are complementary events, then $E \cap E' = \varnothing, E \cup E' = S$, and $P(E') = 1 - P(E)$.	The probability of rolling a 6 with one die is $\frac{1}{6}$. Therefore the probability of *not* rolling a 6 is $1 - \frac{1}{6} = \frac{5}{6}$.

Concept	Comments	Example
Probability of the union of two events	The probability of E_1 and/or E_2 occurring is $P(E_1 \cup E_2) = P(E_1) + P(E_2) - P(E_1 \cap E_2)$.	If $P(E_1) = 0.5$, $P(E_2) = 0.2$, and $P(E_1 \cap E_2) = 0.1$, then $P(E_1 \cup E_2) = 0.5 + 0.2 - 0.1 = 0.6$.
Probability of independent events	Two events E_1 and E_2 are independent if they do not influence one another. $$P(E_1 \cap E_2) = P(E_1) \cdot P(E_2)$$	If E_1 represents a head on the first toss of a coin and E_2 represents a tail on the second toss, then E_1 and E_2 are independent events and $$P(E_1 \cap E_2) = \frac{1}{2} \cdot \frac{1}{2} = \frac{1}{4}.$$
Probability of dependent events	Two events E_1 and E_2 are dependent if they are not independent. $$P(E_1 \cap E_2) =$$ $$P(E_1) \cdot P(E_2, \text{given that } E_1 \text{ occurred.})$$	Drawing two hearts without replacement from a standard deck of cards is an example of dependent events. See Example 10.

11.6 — Exercises

Probability of an Event

Exercises 1–8: Determine if the number could represent a probability.

1. $\frac{11}{13}$

2. 0.995

3. 2.5

4. 1

5. 0

6. 110%

7. -0.375

8. $\frac{9}{8}$

Exercises 9–18: Find the probability of each event.

9. Tossing a head with a fair coin

10. Tossing a tail with a fair coin

11. Rolling a 2 with a fair die

12. Rolling a 5 or 6 with a fair die

13. Guessing the correct answer for a true-false question

14. Guessing the correct answer for a multiple-choice question with five choices

15. Randomly drawing a king from a standard deck of 52 cards

16. Randomly drawing a club from a standard deck of 52 cards

17. Randomly guessing a four-digit ATM access code

18. Randomly picking the winning team at a basketball game

19. The table shows some of the favorite pizza toppings. (**Source:** *USA Today.*)

Pepperoni	43%
Sausage	19%
Mushrooms	14%
Vegetables	13%

(a) If a person is selected at random, what is the probability that pepperoni is not his or her favorite topping?

(b) Find the probability that a person's favorite topping is either mushrooms or sausage.

20. (Refer to Example 2.) Find the probability that a transplant patient in 1997 was waiting for the following.
(a) A lung

(b) A lung or a liver

Probability of Compound Events

Exercises 21–30: Find the probability of the compound event.

21. Tossing a coin twice with the outcomes of two tails

22. Tossing a coin three times with the outcomes of three heads

23. Rolling a die three times and obtaining a 5 or 6 on each roll

24. Rolling a sum of 2 with two dice

25. Rolling a sum of 7 with two dice

26. Rolling a sum other than 7 with two dice

27. Rolling a die four times without obtaining a 6

28. Rolling a die four times and obtaining at least one 6

29. Drawing four consecutive aces from a standard deck of 52 cards without replacement

30. Drawing a pair (two cards with the same value) from a standard deck of 52 cards without replacement

31. *Poker Hands* (Refer to Example 7.) Calculate the probability of drawing 3 hearts and 2 diamonds in a 5-card poker hand. Assume that drawn cards are not replaced and that the 5 cards are drawn only once.

32. *Poker Hands* (Refer to Example 7.) Calculate the probability of drawing 3 kings and 2 queens in a 5-card poker hand. Assume that drawn cards are not replaced and that the 5 cards are drawn only once.

33. *Quality Control* A quality-control experiment involves selecting one string of decorative lights from a box of 20. If the string is defective, the entire box of 20 is rejected. Suppose the box contains four defective strings of lights. What is the probability of rejecting the box?

34. *Quality Control* (Refer to the previous exercise.) Suppose three strings of lights are tested. If any of the strings are defective, the entire box of 20 is rejected. What is the probability of rejecting the box if there are four defective strings of lights in a box? (*Hint:* Start by finding the probability that the box is not rejected.)

35. *Entrance Exams* A group of students is preparing for college entrance exams. It is estimated that 50% need help with mathematics, 45% with English, and 25% with both.
(a) Draw a Venn diagram representing these data.

(b) Use this diagram to find the probability that a student needs help with mathematics, English, or both.

(c) Solve part (b) symbolically by applying a probability formula.

36. *College Classes* In a college of 5500 students, 950 students are enrolled in English classes, 1220 in business classes, and 350 in both. If a student is chosen at random, find the probability that he or she is enrolled in an English class, a business class, or both.

37. *New Books Published* In 2002 a total of 119,923 new books and editions were published. The table lists the number of books published in specific areas. If a new book or edition is selected at random, find the probability that its subject area satisfies the following. (**Source:** R. R. Bowker Co.)

Art	4481
Business	4539
History	6818
Music	1614
Religion	6659
Science	7032

(a) Art or music **(b)** Neither science nor religion

38. *Death Rates* In 2000, the U.S. death rate per 100,000 people was 856. What is the probability that a person selected at random in 2000 died? (**Source:** Department of Health and Human Services.)

39. *Death Rates* In 2001, the death rate per 100,000 people between the ages of 15 and 24 was 81. What is the probability that a person selected at random from this age group died during 2001? (**Source:** Department of Health and Human Services.)

40. *AIDS* By 2001, a total of 816,147 cases of AIDS had been diagnosed. The table lists AIDS cases diagnosed in certain cities. Estimate the probability that a person diagnosed with AIDS satisfied the following conditions. (**Source:** Department of Health and Human Services.)
    (a) Resided in New York

    (b) Did not reside in New York

    (c) Resided in Los Angeles or Miami

New York	126,237
Los Angeles	43,488
San Francisco	28,438
Miami	25,357

41. *Tossing a Coin* Find the probability of tossing a coin n times and obtaining n heads. What happens to this probability as n increases? Does this agree with your intuition? Explain.

42. *Rolling Dice* Find the probability of rolling a die five times and obtaining a 6 on the first two rolls, a 5 on the third roll, and a 1, 2, 3, or 4 on the last two rolls.

43. *Rolling Dice* (Refer to Example 5.) Two dice are rolled. Find the probability that the dice show a sum of either 5 or 6.

44. *Rolling Dice* Two dice are rolled. Find the probability that the dice show a sum other than 7 or 11.

45. *Drawing Cards* (Refer to Example 6.) Find the probability of drawing a 2, 3, or 4 from a standard deck of 52 playing cards.

46. *Drawing Cards* Find the probability of drawing two cards, neither of which is an ace or a queen.

47. *Unfair Coin* Suppose a coin is not fair, but instead the probability of obtaining a head (H) is $\frac{3}{4}$ and a tail (T) is $\frac{1}{4}$. What is the probability of the following events?
    (a) HT   (b) HH

    (c) HHT   (d) THT

48. *Unfair Die* Suppose a die is not fair, but instead the probability P of each number n is listed in the table. Find the probability of each event.

n	1	2	3	4	5	6
P	0.1	0.1	0.1	0.2	0.2	0.3

(a) Rolling a number that is 4 or higher

(b) Rolling a 6 twice on consecutive rolls

49. *Dice* Suppose there are two dice, one red and one blue, having the probabilities shown in the table in the previous exercise. If both dice are rolled, find the probability of the given sum.
    (a) 12   (b) 11

50. *Garage Door Code* The code for some garage door openers consists of 12 electrical switches that can be set to either 0 or 1 by the owner. Each setting represents a different code. What is the probability of guessing someone's code at random? (**Source:** Promax.)

51. *Lottery* To win a lottery, a person must pick three numbers from 0 to 9 in the correct order. If a number may be repeated, what is the probability of winning this game with one play?

52. *Lottery* To win the jackpot in a lottery, a person must pick five numbers from 1 to 49, and then pick the powerball number from 1 to 42. If the numbers are picked at random, what is the probability of winning this game with one play?

53. *Marbles* A jar contains 22 red marbles, 18 blue marbles, and 10 green marbles. If a marble is drawn from the jar at random, find the probability that the color is the following.
    (a) Red   (b) Not red   (c) Blue or green

54. *Marbles* A jar contains 55 red marbles and 45 blue marbles. If 2 marbles are drawn from the jar at random without replacement, find the probability that the marbles satisfy the following.
    (a) Both are blue.   (b) Neither is blue.

    (c) The first marble is red and the second marble is blue.

55. *Cancer and Saccharin* Saccharin is the *least* potent carcinogen ever detected in an animal study. The dose-risk curve is shown in the figure for saccharin-induced bladder tumors in rats. Doses on the x-axis represent the percentage of the animals' diets consisting of saccharin. The associated lifetime risk R or probability of the animal developing bladder cancer is shown on the y-axis. (**Source:** J. Rodricks.)

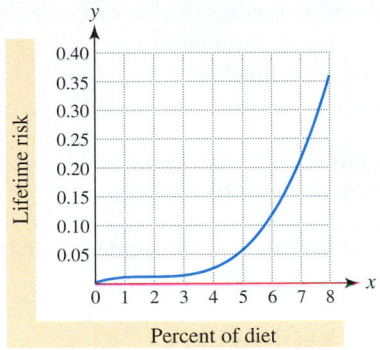

Lifetime risk — Percent of diet

(a) If the diet of a rat consists of 6% saccharin, estimate the risk of developing bladder cancer.

(b) Discuss the information shown in this graph.

56. *Horse Racing*   The favorite to win the Kentucky Derby has won the race 48 out of 122 times. If a past favorite is selected at random, what is the probability that the horse won the Kentucky Derby?   (**Source:** Churchill Downs.)

Conditional Probability and Dependent Events

57. *Drawing a Card*   Find the probability of drawing a queen from a standard deck of cards given that one card, a queen, has already been drawn and not replaced.

58. *Drawing a Card*   Find the probability of drawing a king from a standard deck of cards given that two cards, both kings, have already been drawn and not replaced.

59. *Drawing a Card*   A card is drawn from a standard deck of 52 cards. Given that the card is a face card, what is the probability that the card is a king? (*Hint:* A face card is a jack, queen, or king.)

60. *Drawing a Card*   Three cards are drawn from a deck without replacement. Find the probability that the three cards are an ace, king, and queen in that order.

61. *Drawing Marbles*   A jar initially contains 10 red marbles and 23 blue marbles. What is the probability of drawing a blue marble, given that 2 red marbles and 4 blue marbles have already been drawn?

62. *Volleyball Serve*   The probability that the first serve of a volleyball is out of bounds is 0.3, and the probability that the second serve of a volleyball is in bounds, given that the first serve was out of bounds, is 0.8. Find the probability that the first serve is out of bounds and the second serve is in bounds.

63. *Cloudy and Windy*   The probability of it being cloudy is 30% and the probability of it being cloudy and windy is 12%. Given that the day is cloudy, what is the probability that it will be windy?

64. *Rainy and Windy*   The probability of it being rainy is 80% and the probability of it being windy and rainy is 72%. Given that the day is rainy, what is the probability that it will be windy?

65. *Rolling Dice*   Two dice are rolled. If the first die is a 2, find the probability that the sum of the dice is 7 or more.

66. *Rolling Dice*   Three dice are rolled. If the first die is a 4, find the probability that the sum of the three dice is less than 12.

67. *Defective Parts*   The accompanying table shows numbers of automobile parts that are either defective or not defective.

	Type A	Type B	Totals
Defective	7	11	18
Not Defective	123	94	217
Totals	130	105	235

If one part is selected at random, calculate each of the following.
(a) The probability that the part is defective

(b) The probability that the part is type A, given that it is defective

(c) The probability that the part is type A and defective

68. *Health*   The accompanying table shows numbers of patients with two different diseases by gender. (Assume that a person cannot have both diseases.)

	Disease A	Disease B	Totals
Females	145	851	996
Males	256	355	611
Totals	401	1206	1607

If one patient is selected at random, find the probability that the patient is female and has disease B.

69. *Prime Numbers* Suppose a number from 1 to 15 is selected at random. Find the probability of each event.
(a) The number is odd.

(b) The number is even.

(c) The number is prime. (*Hint:* A natural number greater than 1 that has only itself and 1 as factors is called a **prime number**.)

(d) The number is prime and odd.

(e) The number is prime and even.

70. *Students and Classes* (Refer to Example 11.) If one student is selected at random, use Table 11.11 to calculate each of the following.
(a) The probability that the student is male

(b) The probability that the student is taking French, given that the student is male

(c) The probability that the student is male and taking French

Writing about Mathematics

71. What values are possible for a probability? Interpret different probabilities and give examples.

72. Discuss the difference between dependent and independent events. How are their probabilities calculated?

CHECKING BASIC CONCEPTS FOR SECTIONS 11.5 AND 11.6

1. Use mathematical induction to prove that
$$4 + 8 + 12 + 16 + \cdots + 4n = 2n(n + 1).$$

2. Use mathematical induction to prove that $n^2 \leq 2^n$ for $n \geq 4$.

3. Find the probability of tossing a coin four times and obtaining a head every time.

4. Find the probability of rolling a sum of 11 with two dice.

5. Find the probability of drawing four aces and a queen from a standard deck of 52 cards.

6. In 2003 there were 2.7 million high school graduates, of which 1.3 million were male. If a 2003 high school graduate is selected at random, estimate the probability that this graduate is female. (**Source:** The American College Testing Program.)

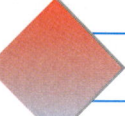

CHAPTER 11 SUMMARY

CONCEPT	EXPLANATION AND EXAMPLES

SECTION 11.1 SEQUENCES

SEQUENCES

An infinite sequence is a function whose domain is the natural numbers. Its graph is a scatterplot.

Example: $a_n = \frac{1}{2}n^2 - 2$; the first 4 terms are as follows.

$$a_1 = \frac{1}{2}(1)^2 - 2 = -\frac{3}{2}, \qquad a_2 = \frac{1}{2}(2)^2 - 2 = 0$$

$$a_3 = \frac{1}{2}(3)^2 - 2 = \frac{5}{2}, \qquad a_4 = \frac{1}{2}(4)^2 - 2 = 6$$

CONCEPT	EXPLANATION AND EXAMPLES

SECTION 11.1 SEQUENCES (CONTINUED)

A graph of the first 4 terms is shown here.

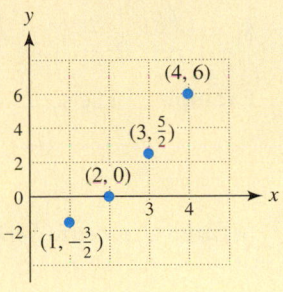

ARITHMETIC SEQUENCE

Recursive Definition:
$a_n = a_{n-1} + d$, where d is the common difference.

Function Definition:
$f(n) = dn + c$, or equivalently, $f(n) = a_1 + d(n - 1)$, where $a_n = f(n)$ and d is the common difference.

Example: $a_n = a_{n-1} + 3$, $a_1 = 4$, and $f(n) = 3n + 1$ describe the same sequence. The common difference is $d = 3$. The terms of the sequence are

$$4, 7, 10, 13, 16, 19, 22, \ldots.$$

GEOMETRIC SEQUENCE

Recursive Definition:
$a_n = ra_{n-1}$, where r is the common ratio.

Function Definition:
$f(n) = cr^{n-1}$, where $c = a_1$ and r is the common ratio.

Example: $a_n = -2a_{n-1}$, $a_1 = 3$, and $f(n) = 3(-2)^{n-1}$ describe the same sequence. The common ratio is $r = -2$. The terms of the sequence are

$$3, -6, 12, -24, 48, -96, 192, \ldots.$$

SECTION 11.2 SERIES

SERIES

A series is the summation of the terms of a sequence.

Examples: $1 + \dfrac{1}{2} + \dfrac{1}{4} + \dfrac{1}{8} + \cdots + \dfrac{1}{2^{n-1}} + \cdots$ Infinite series

$$\sum_{k=1}^{5} 2k = 2 + 4 + 6 + 8 + 10$$ Finite series

ARITHMETIC SERIES

Finite Arithmetic Series:

$$\sum_{k=1}^{n} a_k = a_1 + a_2 + a_3 + \cdots + a_n,$$

where $a_k = dk + c$ for some constants c and d.

CONCEPT	EXPLANATION AND EXAMPLES

SECTION 11.2 SERIES (CONTINUED)

ARITHMETIC SERIES
(CONTINUED)

Sum of the First n Terms:

$$S_n = n\left(\frac{a_1 + a_n}{2}\right) \quad \text{or} \quad S_n = \frac{n}{2}(2a_1 + (n-1)d).$$

Example: The series $4 + 7 + 10 + 13 + 16 + 19 + 22$ is defined by

$a_k = 3k + 1$. Its sum is $S_7 = 7\left(\frac{4 + 22}{2}\right) = 91$.

GEOMETRIC SERIES

Infinite Geometric Series:

$$\sum_{k=1}^{n} a_k = a_1 + a_2 + a_3 + \cdots + a_n + \cdots$$

where $a_k = a_1 r^{k-1}$ for some nonzero constants a_1 and r.

Sum of First n Terms:

$S_n = a_1\left(\frac{1 - r^n}{1 - r}\right)$, where a_1 is the first term and r is the common ratio.

Example: The series $3 + 6 + 12 + 24 + 48 + 96$ has $a_1 = 3$ and $r = 2$.

Its sum is $S_6 = 3\left(\frac{1 - 2^6}{1 - 2}\right) = 189$.

Sum of Infinite Geometric Series:

$S = \frac{a_1}{1 - r}$, if $|r| < 1$. S does not exist if $|r| \geq 1$.

Example: The series $4 + 1 + \frac{1}{4} + \frac{1}{16} + \frac{1}{64} + \cdots$ has sum

$$S = \frac{4}{1 - \frac{1}{4}} = \frac{16}{3}.$$

SECTION 11.3 COUNTING

**FUNDAMENTAL
COUNTING PRINCIPLE**

Let E_1 and E_2 be independent events. If event E_1 can occur m_1 ways and if event E_2 can occur m_2 ways, then there are $m_1 \cdot m_2$ ways for both events to occur.

Example: If two multiple-choice questions have 5 choices each, then there are $5 \cdot 5 = 25$ ways to answer the two questions.

FACTORIAL NOTATION

$n! = n(n - 1)(n - 2) \cdots (3)(2)(1)$

Examples: $0! = 1$; $1! = 1$; $5! = 5 \cdot 4 \cdot 3 \cdot 2 \cdot 1 = 120$

PERMUTATIONS

$P(n, r) = \frac{n!}{(n - r)!}$ represents the number of permutations or arrangements of n elements taken r at a time. Order is important when calculating a permutation.

Example: $P(5, 3) = \frac{5!}{(5 - 3)!} = \frac{120}{2} = 60$

CONCEPT	EXPLANATION AND EXAMPLES

SECTION 11.3 COUNTING (CONTINUED)

COMBINATIONS

$C(n, r) = \binom{n}{r} = \dfrac{n!}{(n - r)!\, r!}$ represents the number of combinations of n elements taken r at a time. Order is unimportant when calculating a combination.

Example: $C(5, 3) = \binom{5}{3} = \dfrac{5!}{(5 - 3)!\, 3!} = \dfrac{120}{2 \cdot 6} = 10$

SECTION 11.4 THE BINOMIAL THEOREM

BINOMIAL THEOREM

$(a + b)^n = \binom{n}{0}a^n + \binom{n}{1}a^{n-1}b + \cdots + \binom{n}{n-1}ab^{n-1} + \binom{n}{n}b^n$

Example: $(x + y)^4 = 1x^4 + 4x^3y + 6x^2y^2 + 4xy^3 + 1y^4$
The coefficients can also be found in the fifth row of Pascal's triangle.

PASCAL'S TRIANGLE

Pascal's triangle can be used to calculate the binomial coefficients when expanding the expression $(a + b)^n$.

$$
\begin{array}{ccccccccccc}
 & & & & & 1 & & & & & \\
 & & & & 1 & & 1 & & & & \\
 & & & 1 & & 2 & & 1 & & & \\
 & & 1 & & 3 & & 3 & & 1 & & \\
 & 1 & & 4 & & 6 & & 4 & & 1 & \\
1 & & 5 & & 10 & & 10 & & 5 & & 1 \\
\end{array}
$$

SECTION 11.5 MATHEMATICAL INDUCTION

PRINCIPLE OF MATHEMATICAL INDUCTION

Let S_n be a statement concerning the positive integer n. Suppose that

1. S_1 is true;
2. for any positive integer k, if S_k is true, then S_{k+1} is also true.

Then, S_n is true for every positive integer n.

SECTION 11.6 PROBABILITY

PROBABILITY

$P(E) = \dfrac{n(E)}{n(S)}$, where $n(E)$ is the number of outcomes in event E and $n(S)$ is the number of equally likely outcomes in the sample space S. Note that $0 \leq P(E) \leq 1$.

COMPOUND EVENTS

Probability of either E_1 or E_2 (or both) occurring:

$$P(E_1 \cup E_2) = P(E_1) + P(E_2) - P(E_1 \cap E_2).$$

If E_1 and E_2 are *mutually exclusive*, then $E_1 \cap E_2 = \varnothing$, and

$$P(E_1 \cup E_2) = P(E_1) + P(E_2).$$

CONCEPT	EXPLANATION AND EXAMPLES

SECTION 11.6 PROBABILITY (CONTINUED)

COMPOUND EVENTS (CONTINUED)	Probability of *both* E_1 and E_2 occurring:

If E_1 and E_2 are *independent*, then

$$P(E_1 \cap E_2) = P(E_1) \cdot P(E_2).$$

If E_1 and E_2 are *dependent*, then

$$P(E_1 \cap E_2) = P(E_1) \cdot P(E_2, \text{given that } E_1 \text{ has occurred}).$$

REVIEW EXERCISES

Exercises 1–4: Find the first four terms of the sequence.

1. $a_n = -3n + 2$

2. $a_n = n^2 + n$

3. $a_n = 2a_{n-1} + 1; a_1 = 0$

4. $a_n = a_{n-1} + 2a_{n-2}; a_1 = 1, a_2 = 4$

Exercises 5 and 6: Use the graphical representation to identify the terms of the finite sequence.

5.
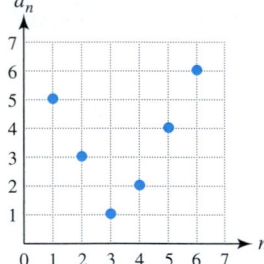

6.
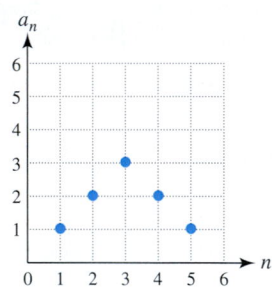

Exercises 7 and 8: Find the first five terms of the recursive sequence.

7. $a_n = 3a_{n-1} + 1; \quad a_1 = 0$

8. $a_n = a_{n-2} + 2a_{n-1}; \quad a_1 = 1, a_2 = 2$

Exercises 9–12: The first five terms of an infinite arithmetic or geometric sequence are given. Find

   (a) numerical,
   (b) graphical, and
   (c) symbolic

representations of the sequence. Include at least the first eight terms of the sequence.

9. 2, 4, 6, 8, 10

10. 3, 1, −1, −3, −5

11. 81, 27, 9, 3, 1

12. 1.5, −3, 6, −12, 24

13. Find a general term a_n for the arithmetic sequence with $a_3 = -3$ and $d = 4$.

14. Find a general term a_n for the geometric sequence with $a_1 = 2.5$ and $a_6 = -80$.

Exercises 15–18: Determine if the sequence is arithmetic, geometric, or neither.

15. $f(n) = 5 - 2n$

16. $f(n) = 3n^2$

17. $f(n) = 3(2)^n$

18. $f(n) = n + \left(\frac{1}{2}\right)^{n-1}$

19. *Height of a Ball* When a Ping-Pong ball is dropped, it rebounds to 90% of its initial height.

   (a) Write the first five terms of a sequence that gives the maximum height attained by the ball on each rebound when it is dropped from an initial height of 4 feet. Let $a_1 = 4$. What type of sequence describes these maximums?

   (b) Give a graphical representation of this sequence for the first five terms.

   (c) Find a formula for a_n.

20. *Falling Object* If air resistance is ignored, an object falls 16, 48, 80, and 112 feet during each successive 1-second interval.

   (a) What type of sequence describes these distances?

   (b) Determine how far an object falls during the sixth second.

   (c) Find a formula for the nth term of this sequence.

Exercises 21 and 22: For the given a_n, calculate S_5.

21. $a_n = 4n + 1$ **22.** $a_n = 3(4)^{n-1}$

Exercises 23–26: Use a formula to find the sum of the series.

23. $-2 + 1 + 4 + 7 + 10 + 13 + 16 + 19 + 22$

24. $2 + 4 + 6 + 8 + \cdots + 98 + 100$

25. $1 + 3 + 9 + 27 + 81 + 243 + 729 + 2187$

26. $64 + 16 + 4 + 1 + \frac{1}{4} + \frac{1}{16}$

Exercises 27–30: Find the sum of the infinite geometric series.

27. $2 + 1 + \frac{1}{2} + \frac{1}{4} + \frac{1}{8} + \frac{1}{16} + \cdots$

28. $4 - \frac{4}{3} + \frac{4}{9} - \frac{4}{27} + \frac{4}{81} - \frac{4}{243} + \cdots$

29. $0.2 + 0.02 + 0.002 + 0.0002 + 0.00002 + \cdots$

30. $0.25 + 0.0025 + 0.000025 + 0.00000025 + \cdots$

Exercises 31 and 32: Write out the terms of the series.

31. $\displaystyle\sum_{k=1}^{5} (5k + 1)$ **32.** $\displaystyle\sum_{k=1}^{4} (2 - k^2)$

Exercises 33 and 34: Write the series using summation notation.

33. $1^3 + 2^3 + 3^3 + 4^3 + 5^3 + 6^3$

34. $1 + \frac{1}{10} + \frac{1}{100} + \frac{1}{1000} + \frac{1}{10,000}$

Exercises 35 and 36: Use a formula to find the sum of the first thirty terms of the arithmetic sequence.

35. $a_1 = 5, d = -3$ **36.** $a_1 = -2, a_{10} = 16$

37. Write $\frac{2}{11}$ as an infinite geometric series.

38. Write the infinite series

$$0.23 + 0.0023 + 0.000023 + \cdots$$

as a rational number.

39. *Exam Questions* Count the ways that an exam, consisting of 20 multiple-choice questions with 4 choices each, could be answered.

40. *License Plates* Count the different license plates having 4 numeric digits followed by 2 letters.

41. *Combination Lock* A combination lock consists of 4 numbers from 0 to 49. If a number may be repeated, find the number of possible combinations.

42. *Dice* A red die and a blue die are rolled. How many ways are there for the sum to equal 4?

Exercises 43 and 44: Evaluate the expression.

43. $P(6, 3)$ **44.** $C(7, 4)$

45. *Standing in Line* In how many arrangements can 5 people stand in a line?

46. *Giving a Speech* How many arrangements are possible for 4 students to give a speech out of a class of 15?

47. *Committees* In how many ways can a committee of 3 be selected from 6 people?

48. *Committees* Find the number of committees with 3 women and 3 men that can be selected from a group of 7 women and 5 men.

49. *Test Questions* On a test involving 10 essay questions, students are asked to answer 6 questions. How many ways can the essay questions be selected?

50. *Binomial Theorem* Use the binomial theorem to expand the expression $(2x - y)^4$.

Exercises 51 and 52: Use mathematical induction to prove that the statement is true for every positive integer.

51. $1 + 3 + 5 + 7 + \cdots + (2n - 1) = n^2$

52. $2 + 2^2 + 2^3 + \cdots + 2^n = 2(2^n - 1)$

Exercises 53 and 54: Find the probability of each event.

53. A 1, 2, or 3 appears when rolling a die

54. Tossing three heads in a row using a fair coin

55. *Quality Control* A quality-control experiment involves selecting 2 batteries from a pack of 16. If either battery is defective, the entire pack of 16 is rejected. Suppose a pack contains 2 defective batteries. What is the probability that this pack will not be rejected?

56. *Venn Diagram* From a group of 82 students, 19 are enrolled in music, 22 in art, and 10 in both.
(a) Draw a Venn diagram representing these data.

(b) Use this diagram to determine the probability that a student selected at random is enrolled in music, art, or both.

(c) Solve part (b) symbolically by applying a probability formula.

57. *Marbles*   A jar contains 13 red, 27 blue, and 20 green marbles. If a marble is drawn from the jar at random, find the probability that the marble is the following.
   (a) Blue   **(b)** Not blue   **(c)** Red

58. *Cards*   Find the probability of drawing 2 diamonds from a standard deck of 52 cards without replacing the first card.

59. *Insect Populations*   The monthly density of an insect population, measured in thousands per acre, is described by the recursive sequence

$$a_1 = 100$$

$$a_n = \frac{2a_{n-1}}{1 + (a_{n-1}/4000)}, \qquad n > 1.$$

Use your calculator to graph the sequence in the window $[0, 16, 1]$ by $[0, 5000, 1000]$. Include the first 15 terms and discuss any trends illustrated by the graph.

EXTENDED AND DISCOVERY EXERCISES

*Exercises 1 and 2: **Antibiotic Resistance**   Due to the frequent use of antibiotics in society, many strains of bacteria are becoming resistant. Some types of haploid bacteria contain genetic material called plasmids. Plasmids are capable of making a strain of bacterium resistant to antibiotic drugs. Genetic engineers want to predict the resistance of various bacteria after many generations.*

1. Suppose a strain of bacterium contains two plasmids R_1 and R_2. Plasmid R_1 is resistant to the antibiotic ampicillin, whereas plasmid R_2 is resistant to the antibiotic tetracycline. When bacteria reproduce through cell division, the type of plasmids passed on to each new cell is random. For example, a daughter cell could have two plasmids of type R_1 and no plasmid of type R_2, one of each type, or no plasmid of type R_1 and two plasmids of type R_2. The probability $P_{k,j}$ that a mother cell with k plasmids of type R_1 produces a daughter cell with j plasmids of type R_1 can be calculated by the formula

$$P_{k,j} = \frac{\binom{2k}{j}\binom{4 - 2k}{2 - j}}{\binom{4}{2}}.$$

(**Source:** F. Hoppensteadt and C. Peskin, *Mathematics in Medicine and the Life Sciences.*)

(a) Compute $P_{k,j}$ for $0 \le k, j \le 2$. Assume that

$$\binom{0}{0} = 1 \text{ and } \binom{k}{j} = 0 \text{ whenever } k < j.$$

Record your results in the matrix

$$P = \begin{bmatrix} P_{00} & P_{01} & P_{02} \\ P_{10} & P_{11} & P_{12} \\ P_{20} & P_{21} & P_{22} \end{bmatrix}.$$

(b) Which elements in P are the greatest? Interpret the result.

2. (Continuation of the previous exercise.) The genetic makeup of future generations of the haploid bacterium can be modeled using matrices. Let $A = [a_1, a_2, a_3]$ be a 1×3 matrix containing three probabilities. The value of a_1 is the probability that a cell has two R_1 plasmids and no R_2 plasmid; a_2 is the probability that it has one R_1 plasmid and one R_2 plasmid; a_3 is the probability that a cell has no R_1 plasmid and two R_2 plasmids. If an entire generation of bacterium has one plasmid of each type then $A_1 = [0, 1, 0]$. In this case the bacterium is resistant to both antibiotics. The probabilities A_n for plasmids R_1 and R_2 in the nth generation of bacterium can be calculated with the matrix recurrence equation $A_n = A_{n-1}P$, where $n > 1$ and P is the 3×3 matrix determined in the previous exercise. The resulting phenomenon was not well understood until quite recently. It is now used in the genetic engineering of plasmids.

(a) If an entire strain of bacterium is resistant to both the antibiotics ampicillin and tetracycline, make a conjecture as to the drug resistance of future generations of this bacterium.

(b) Test your conjecture by repeatedly computing the matrix product $A_n = A_{n-1}P$. Let $A_1 = [0, 1, 0]$ and $n = 2, 3, \ldots, 12$. Interpret the result. (It may surprise you.)

3. The sum of the first n terms of an arithmetic series is give by $S_n = n\left(\frac{a_1 + a_n}{2}\right)$. Justify this formula using a geometric discussion. (*Hint:* Start by graphing an arithmetic sequence where n is an odd number.)

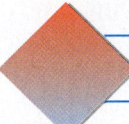

CHAPTERS 1–11 CUMULATIVE REVIEW EXERCISES

1. Find the percent change if the price of tuition and fees increases from \$8145 to \$8862. Round your answer to the nearest tenth of a percent.

2. Write 34,500 in scientific notation and 1.52×10^{-4} in standard form.

3. Evaluate $\dfrac{5 - \sqrt[3]{4}}{\pi^2 - (\sqrt{3} + 1)}$. Round your answer to the nearest hundredth.

4. Graph the relation

   $S = \{(-3, 1), (-1, 2), (0, -1), (1, 3), (-1, 1)\}$.

   Is S a function?

5. Find the exact distance between $(-4, 2)$ and $(1, -2)$.

6. Express the domain and range of the function shown in interval notation. Evaluate $f(-3)$.

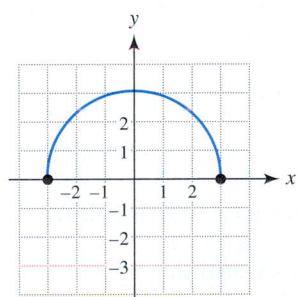

7. Graph f by hand.
   (a) $f(x) = 4x - 2$ (b) $f(x) = |2x - 1|$

   (c) $f(x) = x^2 + 2x - 3$

   (d) $f(x) = \sqrt{x - 1}$ (e) $f(x) = \dfrac{1}{x + 2}$

   (f) $f(x) = x^3 - 2$ (g) $f(x) = \sqrt[3]{x - 1}$

   (h) $f(x) = 2^x$ (i) $f(x) = 3\left(\tfrac{1}{2}\right)^x$

   (j) $f(x) = \log_2 x$

Exercises 8 and 9: Complete the following.

   (a) Evaluate $f(-3)$ and $f(a + 1)$.
   (b) Determine the domain of f.

8. $f(x) = \sqrt{1 - x}$ 9. $f(x) = \dfrac{1}{x^2 - 4}$

10. Determine if the graph represents a function.

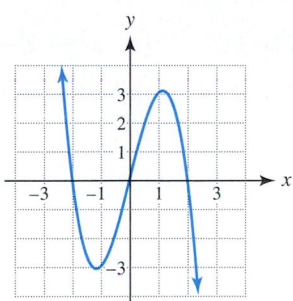

11. Find the average rate of change of $f(x) = x^3 - 4$ from $x = -2$ to $x = -1$.

12. Find the difference quotient for $f(x) = x^2 - 3x$.

13. The graph of a linear function f is shown.
    (a) Identify the slope, y-intercept, and x-intercept.

    (b) Write a formula for f.

    (c) Find any zeros of f.

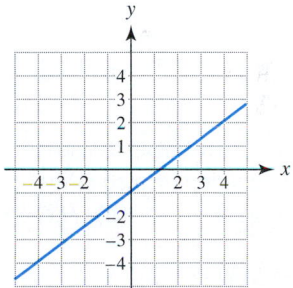

14. If $f(t) = 350 - 70t$ models the distance between a driver and home after t hours, interpret the numbers 350 and -70 in the formula for f. After how long does the driver arrive at home?

Exercises 15–17: Write an equation of a line satisfying the given conditions. Use slope-intercept form whenever possible.

15. Passing through $(2, -4)$ and $(-3, 2)$

16. Passing through the point $(-1, 3)$ and perpendicular to the line $y = -\tfrac{3}{4}x + 1$

17. Parallel to the y-axis and passing through $(-2, 4)$

18. Determine the x- and y-intercepts on the graph of $-3x + 4y = 12$. Graph the equation.

19. Solve each equation.
(a) $4(x - 2) + 1 = 3 - \frac{1}{2}(2x + 3)$

(b) $6x^2 = 13x + 5$ (c) $x^2 - x - 3 = 0$

(d) $x^3 + x^2 = 4x + 4$ (e) $x^4 - 4x^2 + 3 = 0$

(f) $\dfrac{1}{x - 3} = \dfrac{4}{x + 5}$ (g) $3e^{2x} - 5 = 23$

(h) $2\log(x + 1) - 1 = 2$

(i) $\sqrt{x + 3} + 4 = x + 1$

(j) $|3x - 1| = 5$

20. Solve each inequality. Write your answer in interval notation.
(a) $3x - 5 < x + 1$ (b) $x^2 - 4x - 5 \leq 0$

(c) $x^2 \geq 4$

(d) $(x + 1)(x - 2)(x - 3) > 0$

(e) $\dfrac{2}{x - 1} < 0$ (f) $|3x - 5| \leq 4$

(g) $|4 - x| > 0$ (h) $\dfrac{3}{4} \leq \dfrac{1 - 2x}{3} < \dfrac{5}{2}$

21. The graphs of two linear functions, f and g, are shown. Solve each equation or inequality.
(a) $f(x) = g(x)$ (b) $f(x) > g(x)$

(c) $f(x) \leq g(x)$

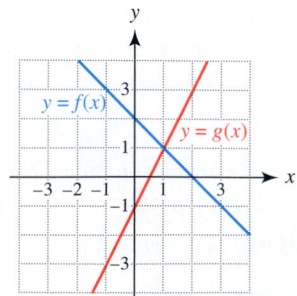

22. Graph f. Is f continuous on its domain?
$$f(x) = \begin{cases} 2x + 3 & \text{if } -3 \leq x < -1 \\ x^2 & \text{if } -1 \leq x < 1 \\ 2 - x & \text{if } 1 \leq x \leq 3 \end{cases}$$

23. Solve $-2.3x + 3.4 = \sqrt{2x^2 - 1}$ graphically. Round your answers to the nearest tenth.

24. The graph of a nonlinear function f is shown. Solve each equation or inequality. Write the solution set to each inequality in interval notation.
(a) $f(x) = 0$ (b) $f(x) > 0$ (c) $f(x) \leq 0$

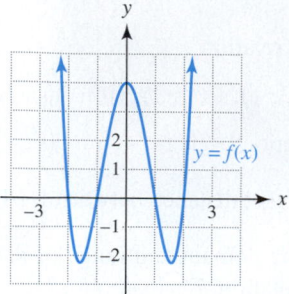

25. Write $f(x) = 3x^2 + 24x + 43$ in the vertex form given by $f(x) = a(x - h)^2 + k$.

26. Find the vertex on the graph of $f(x) = -3x^2 + 9x + 1$.

27. A graph of a quadratic function, $f(x) = ax^2 + bx + c$, is shown.
(a) State whether $a > 0$ or $a < 0$.

(b) Estimate the real solutions to $ax^2 + bx + c = 0$.

(c) Determine if the discriminant $b^2 - 4ac$ is positive, negative, or zero.

(d) Write the complete factored form of $f(x)$.

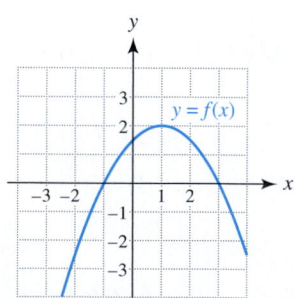

28. Use the given graph of $y = f(x)$ to sketch a graph of each expression.
 (a) $y = f(x - 1) + 2$ **(b)** $y = \frac{1}{2}f(x)$

 (c) $y = -f(x)$ **(d)** $y = f(2x)$

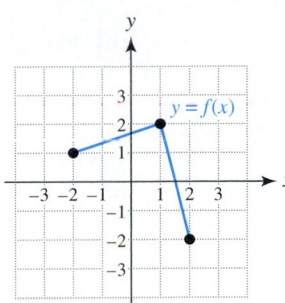

29. Use the graph of f to estimate each of the following.
 (a) Intervals where f is increasing or decreasing

 (b) The zeros of f

 (c) The coordinates of any turning points

 (d) Any local extrema

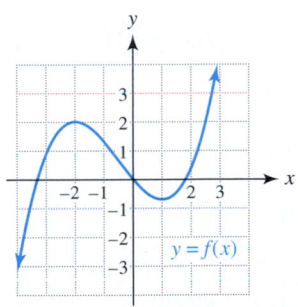

30. Determine if the function defined by $f(x) = x^3 - 5x$ is even, odd, or neither.

31. Sketch a graph of a cubic function with a negative leading coefficient, three x-intercepts, and two turning points.

32. Divide each expression.
 (a) $\dfrac{6x^4 - 2x^2 + 1}{2x^2}$ **(b)** $\dfrac{2x^4 - 3x^3 - x + 2}{x + 1}$

 (c) $\dfrac{x^3 + x^2 - 3}{x^2 + 1}$

33. A degree 4 function f has zeros $-2, -1, 1$, and 2 with leading coefficient 6. Write the complete factored form of $f(x)$.

34. A degree 6 function f has zeros $-2, 2$, and 3 with multiplicities 1, 2, and 3, respectively. If the leading coefficient is -2, write the complete factored form of $f(x)$.

35. A degree 3 function f with real coefficients has leading coefficient 3 and zeros -1 and $3i$. Write $f(x)$ in complete factored form and expanded form.

36. Write the complete factored form for the polynomial $f(x) = -2x^3 - 2x^2 + 18x + 18$.

37. Use the graph to write the complete factored form of the degree 4 polynomial $f(x)$.

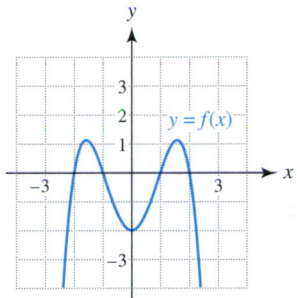

38. Write $(2 - i)(2 + 3i)$ in standard form.

39. Find all solutions, real or imaginary, to the quadratic equation $x^2 + 2x + 5 = 0$.

40. State the domain of $f(x) = \frac{2x - 5}{x + 5}$. Find any vertical or horizontal asymptotes on the graph of f.

41. Write $\sqrt[5]{(x + 1)^3}$ using rational exponents. Evaluate the expression for $x = 31$.

42. Use the tables for f and g to evaluate each expression, if possible.

x	0	1	2	3	4
$f(x)$	1	2	4	5	8

x	0	1	2	3	4
$g(x)$	4	3	2	1	0

 (a) $(f - g)(1)$ **(b)** $(f/g)(2)$

 (c) $(g \circ f)(3)$ **(d)** $(g \circ f^{-1})(5)$

43. Use the graphs of f and g to complete the following.

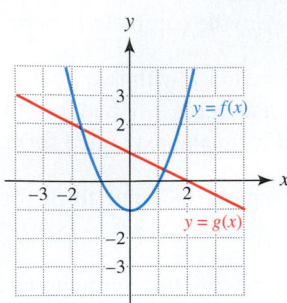

(a) $(f + g)(2)$ **(b)** $(fg)(0)$

(c) $(g \circ f)(1)$ **(d)** $(g^{-1} \circ f)(-2)$

44. Let $f(x) = \frac{1}{x + 2}$ and $g(x) = x^2 + x - 4$. Find each of the following.
(a) $(f - g)(0)$ **(b)** $(g \circ f)(-1)$

(c) $(fg)(x)$ **(d)** $(g \circ f)(x)$

45. Find $f^{-1}(x)$ if $f(x) = \frac{x}{x + 1}$.

46. Graph $f(x) = x^3 - 1$, $y = x$, and $y = f^{-1}(x)$ by hand on the same axes.

47. Find either a linear or an exponential function f that models the data in the table.

x	0	1	2	3
$f(x)$	2	6	18	54

48. There are initially 1000 bacteria in a sample and this number doubles every 2 hours.
(a) Find C and a so that $f(x) = Ca^x$ models the number of bacteria after x hours.

(b) Estimate the number of bacteria after 5.2 hours.

(c) When are there 9000 bacteria in the sample?

49. Use the graph of $y = Ca^x$ to determine values for C and a.

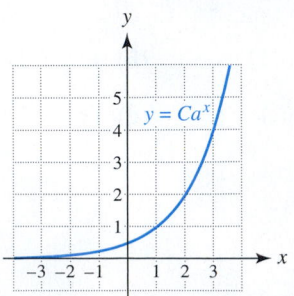

50. Five hundred dollars are deposited in an account that pays 6% annual interest compounded quarterly. Find the amount in the account after 15 years.

51. Simplify each logarithm by hand.
(a) $\log_2 \frac{1}{16}$ **(b)** $\log \sqrt{10}$

(c) $\ln e^4$ **(d)** $\log_4 2 + \log_4 32$

52. Find the domain and range of each function f.
(a) $f(x) = x^2 - 2x + 1$

(b) $f(x) = 10^x$ **(c)** $f(x) = \ln x$

(d) $f(x) = \frac{1}{x}$

53. Expand the expression $\log \sqrt{\frac{x + 1}{yz}}$.

54. Write $2 \log x + 3 \log y - \frac{1}{3} \log z$ as a logarithm of a single expression.

55. Approximate $\log_4 52$ to three decimal places.

56. Graph $y = f(x)$.
(a) $f(x) = \csc x$ **(b)** $f(x) = -2 \sin \frac{1}{2}x$

(c) $f(x) = 3 \cos \left(2\left(x - \frac{\pi}{2}\right)\right)$

(d) $f(x) = \tan (\pi x)$

57. Find the domain of $f(x) = \sec 2x$.

58. Solve each equation for all real numbers t.
(a) $\cos t = -\dfrac{\sqrt{3}}{2}$ **(b)** $\tan^2 t - 1 = 0$

(c) $\sin^2 t + \frac{1}{2} \sin t = 0$ **(d)** $\cos 2t = -1$

59. Find the complementary angle α and the supplementary angle β to $54°35'12''$.

60. Convert $5.54°$ to degrees, minutes, and seconds.

61. Convert $135°$ to radians.

62. Convert $\frac{5\pi}{4}$ radians to degrees.

63. Approximate $\cot \frac{2\pi}{9}$ to the nearest hundredth.

64. Find the values of the six trigonometric functions of an angle θ in standard position having its terminal side pass through the point $(-7, 24)$.

65. Find the exact values of the six trigonometric functions of $\theta = \frac{5\pi}{6}$.

66. Find the values of the six trigonometric functions of θ if $\cos\theta = -\frac{11}{61}$ and $\csc\theta < 0$.

67. Evaluate $\cos^{-1}\left(\frac{1}{2}\right)$.

68. Solve the right triangle shown in the figure.

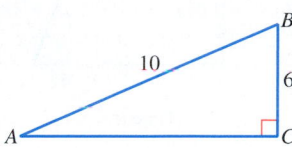

69. Simplify $(\sec t - 1)(\sec t + 1)$.

70. Verify the identity.
$$1 - \sin^2\theta + \cot^2\theta - \sin^2\theta\cot^2\theta = \cot^2\theta$$

71. Solve $2\sin 2x = \ln x$ graphically for $0 \le x < 2\pi$.

72. Solve triangle ABC. Approximate values to the nearest tenth.
   (a) $\beta = 42°, \gamma = 31°, a = 22$

   (b) $\gamma = 50°, c = 7, b = 8$

   (c) $\alpha = 44°, b = 7, c = 8$

   (d) $a = 10, b = 11, c = 12$

73. Find the area of a triangle with $\alpha = 30°, b = 15$ feet and $c = 20$ feet.

74. Let $\mathbf{a} = \langle 3, -4\rangle$ and $\mathbf{b} = \langle -5, 12\rangle$. Find the following.
   (a) $\|\mathbf{a}\|$ **(b)** $4\mathbf{b} - 2\mathbf{a}$

   (c) $\mathbf{a}\cdot\mathbf{b}$ **(d)** The angle between $\mathbf{a}$ and $\mathbf{b}$

75. Graph the parametric equations $x = 2\cos t, y = 2\sin t$ for $0 \le t \le 2\pi$.

76. Graph the polar equation $r = 1 + \sin\theta$ for $0 \le \theta < 2\pi$.

77. Find the two square roots of $-16i$.

78. Evaluate $f(2, 6)$, if $f(b, h) = \frac{1}{2}bh$.

79. Solve the system of equations.
   (a) $2x + 3y = 4$
   $\quad\ \ 2x - 5y = -12$
   (b) $-2x + \frac{1}{2}y = 1$
   $\quad\ \ 4x - y = -2$

   (c) $x^2 + y^2 = 16$
   $\quad\ \ 2x^2 - y^2 = 11$
   (d) $x + y - 2z = -6$
   $\quad\ \ 2x - y - 3z = -18$
   $\quad\ \ 3y - z = 6$

80. The variable z varies inversely as the square of x. If $z = 8$ when $x = 50$, find z when $x = 36$.

81. Graph the solution set to each inequality or system of inequalities.
   (a) $x \le 1$ **(b)** $-2x + 3y < 6$

   (c) $x + y \le 4$
   $\quad\ \ x - 2y > 6$
   (d) $x^2 + 4y^2 < 4$
   $\quad\ \ x^2 - y^2 \ge 1$

82. Find $2A + B$ and AB if
$$A = \begin{bmatrix} -1 & 0 & 2 \\ 1 & -3 & 1 \\ 0 & -3 & 4 \end{bmatrix} \quad\text{and}\quad B = \begin{bmatrix} 1 & 5 & 1 \\ -2 & 2 & 1 \\ 0 & 1 & -2 \end{bmatrix}.$$

83. Find A^{-1} if $A = \begin{bmatrix} 1 & -2 \\ -3 & 4 \end{bmatrix}$.

84. Write the linear system in the form $AX = B$ and solve by using A^{-1}.
$$\begin{aligned} x - y - 2z &= 5 \\ -x + 2y + 3z &= -7 \\ 2y + z &= -2 \end{aligned}$$

85. Calculate the determinant of each matrix.
   (a) $\begin{bmatrix} -1 & 4 \\ 2 & 3 \end{bmatrix}$ **(b)** $\begin{bmatrix} 2 & 3 & -1 \\ 3 & -1 & 5 \\ 0 & 0 & -2 \end{bmatrix}$

86. Sketch a graph of each equation. Label any foci.
   (a) $y^2 = 2x$ **(b)** $x = -\frac{1}{2}(y + 1)^2 - 2$

   (c) $\dfrac{x^2}{9} + \dfrac{y^2}{25} = 1$

   (d) $9x^2 - 18x + 4y^2 + 16y + 25 = 36$

   (e) $\dfrac{x^2}{4} - \dfrac{y^2}{9} = 1$

   (f) $\dfrac{(x + 1)^2}{16} - \dfrac{(y - 2)^2}{9} = 1$

87. Find an equation of a parabola with vertex $(0, 0)$ and focus $\left(\frac{3}{4}, 0\right)$.

88. Find an equation of an ellipse with vertices $(\pm 3, 1)$ and foci $(\pm 2, 1)$.

89. Find an equation of a hyperbola with foci $(0, \pm 13)$ and vertices $(0, \pm 5)$.

90. Find the first four terms of each sequence.

(a) $a_n = (-1)^{n-1}(3)^n$

(b) $a_n = a_{n-1}a_{n-2}$; $a_1 = 2$, $a_2 = 3$

91. Find a general term for the arithmetic sequence given that $a_1 = 4$ and $a_3 = 12$.

92. Find a general term for the geometric sequence given that $a_2 = 6$ and $r = \frac{1}{2}$.

93. Use a formula to find the sum of each sequence.

(a) $2 + 5 + 8 + 11 + \cdots + 74$

(b) $2 + 4 + 8 + 16 + \cdots + 1024$

(c) $0.2 + 0.02 + 0.002 + 0.0002 + \cdots$

94. Find the sum $\sum_{k=1}^{7}(k^2 + k)$.

95. Count the number of license plates that can be formed by three letters followed by four digits.

96. Evaluate $P(4, 2)$ and $\binom{6}{3}$.

97. Expand $(2x - 1)^4$.

98. Use mathematical induction to show that

$5 + 7 + 9 + 11 + \cdots + (2n + 3) = n(n + 4)$.

99. Find the probability of drawing a heart or an ace from a standard deck of 52 cards.

100. Find the probability of rolling a sum of 7 with two dice.

101. A number from 1 to 20 is drawn at random. Find the probability that the number is prime.

102. The probability of it being cloudy is 40% and the probability of it being cloudy *and* windy is 15%. Given that the day is cloudy, what is the probability that it will be windy?

Applications

103. *Volume of a Cone* The volume V of a cone is given by $V(r, h) = \frac{1}{3}\pi r^2 h$, where r is the radius of the cone and h is its height. If a paper cup in the shape of a cone has a volume of 10 cubic inches and a radius of 1.3 inches, find its height to the nearest tenth of an inch.

104. *Interpreting Slope* The graph shows the gallons of water in a tank after x minutes. Interpret the slope of each line segment.

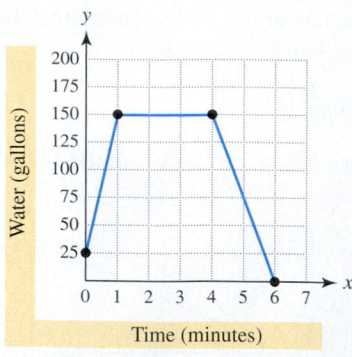

105. *Distance* At noon car A is traveling north at 50 miles per hour and is located 30 miles north of car B. Car B is traveling east at 50 miles per hour. Approximate the distance between the cars at 1:45 P.M. to the nearest tenth of a mile.

106. *Distance from Home* A driver is initially 240 miles from home, traveling toward home on a straight interstate at 60 miles per hour.

(a) Write a formula for a linear function D that models the distance between the driver and home after x hours.

(b) What is an appropriate domain for D?

(c) Graph D.

(d) Identify the x- and y-intercepts. Interpret each.

107. *Running* An athlete travels 10.5 miles in 1.3 hours, jogging at 7 miles per hour and 9 miles per hour. How long did the athlete run at each speed?

108. *Average Rate of Change* The total distance D in feet traveled by a racehorse after t seconds is given by $D(t) = 3t^2$ for $0 \le t \le 5$.

(a) Find the average rate of change of D from 0 to 1 and 3 to 4.

(b) Interpret these average rates of change.

109. *Working Together* Suppose one person can mow a lawn in 5 hours and another person can mow the same lawn in 4 hours. How long will it take to mow the lawn if they work together?

110. *Mixing Acid* Two liters of a 25% sulfuric acid solution need to be diluted to a 13% solution. How many liters of a 5% sulfuric acid solution should be mixed with the 2-liter solution?

111. *U.S. Population*   In 1980 the population of the United States was about 227 million and in 2000 it was about 280 million. (**Source:** Bureau of the Census.)
   (a) Determine a formula in the form

$$P(t) = m(t - t_1) + y_1$$

   that models these data. Let P be in millions and t be the year.

   (b) Estimate the year when the population of the United States could reach 350 million.

112. *Maximum Height*   A stone is shot upward with a slingshot. Its height s in feet after t seconds is given by $s(t) = -16t^2 + 96t + 4$. Find its maximum height.

113. *Modeling Data*   Find a quadratic function in the form $f(x) = a(x - h)^2 + k$ that models the data in the table exactly.

x	2	4	6	8
y	6	8	14	24

114. *Bicycle*   A wheel on a trailer is 21 inches in diameter and rotating at 45 radians per second. Find the speed of the trailer in feet per second.

115. *Length of a Shadow*   The angle of elevation of the sun is 42°. Find the length of the shadow cast by a building that is 105 feet tall. Round your answer to the nearest tenth of a foot.

116. *Daylight Hours*   The average number of daylight hours for each month at a location are shown in the table. Model these data by using a function of the form

$$f(t) = a \cos (b(t - c)) + d.$$

Month	1	2	3	4	5	6
Daylight Hours	9.4	10.5	12	13.5	14.6	15

Month	7	8	9	10	11	12
Daylight Hours	14.6	13.5	12	10.5	9.4	9

117. *Angle of Elevation*   A 6-foot person casts a 7-foot shadow. Estimate the angle of elevation of the sun to the nearest tenth of a degree.

118. *Surveyor*   A surveyor measures two sides of a lot in the shape of a parallelogram to be $a = 153$ feet and $c = 167$ feet. The angle between these sides is $\beta = 54°$.
   (a) Find the length of the shortest diagonal to the nearest tenth of a foot.

   (b) Estimate the area of the lot.

119. *Linear Programming*   Find the maximum value of $P = 2x + 3y$ subject to the following constraints.

$$3x + y \geq 3$$
$$3x + 2y \leq 6$$
$$x \geq 0, y \geq 0$$

120. *Airline Tickets*   Tickets for a charter flight are regularly $400, but for each additional ticket bought, the cost is reduced by $5.
   (a) Write a quadratic function C that gives the total cost of buying x tickets.

   (b) Solve $C(x) = 7000$ and interpret the result.

   (c) Find the absolute maximum for C and interpret your result. Assume that x is an integer.

121. *Dimensions of a Rectangle*   A rectangle has a perimeter of 48 inches and an area of 143 square inches. Find its dimensions.

122. *Satellite Dish*   A parabolic satellite dish has a 3-foot diameter and is 6 inches deep. How far from the vertex should the receiver be so that it is located at the focus?

123. *Bouncing Ball*   A Ping-Pong ball is dropped from 4 feet. Each time the ball bounces, it rebounds to 75% of its previous height. Use an infinite geometric series to estimate the *total* distance that the ball travels before coming to rest on the floor.

124. *Marbles*   A jar contains 15 red, 28 blue, and 34 green marbles. If a marble is drawn at random, find the probability that the marble is not blue.

Reference: Basic Concepts from Algebra and Geometry

Throughout the text there are algebra and geometry review notes that direct students to "see Chapter R." This reference chapter contains eight sections, which provide a review of important topics from algebra and geometry. Students can refer to these sections for more explanation or extra practice. Instructors can use these sections to emphasize a variety of mathematical skills.

R.1 Formulas from Geometry

◆ Use formulas for shapes in a plane
◆ Find sides of right triangles by applying the Pythagorean theorem
◆ Apply formulas to three-dimensional objects
◆ Use similar triangles to solve problems

Geometric Shapes in a Plane

In this subsection we discuss formulas related to rectangles, triangles, and circles.

Rectangles The distance around the boundary of a geometric shape in a plane is called its **perimeter**. The perimeter of a rectangle equals the sum of the lengths of its four sides. For example, the perimeter of the rectangle shown in Figure R.1 is $5 + 4 + 5 + 4 = 18$ feet. In general, the perimeter P of a rectangle with length L and width W is given by $P = 2L + 2W$.

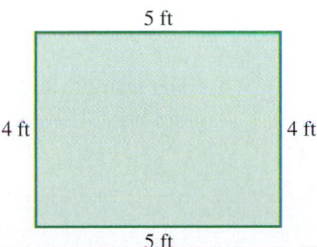

FIGURE R.1 Rectangle

The area A of a rectangle equals the product of its length and width: $A = LW$, so the rectangle in Figure R.1 has an area of $5 \cdot 4 = 20$ square feet.

Many times the perimeter or area of a rectangle is written in terms of variables. This is demonstrated in the next example.

Finding the perimeter and area of a rectangle

The length of a rectangle is three times greater than its width. If the width is x inches, write expressions that give the perimeter and area.

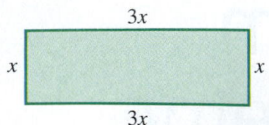

FIGURE R.2

SOLUTION The width of the rectangle is x inches, so its length is $3x$ inches. A sketch is shown in Figure R.2. The perimeter is

$$P = 2L + 2W$$
$$= 2(3x) + 2(x)$$
$$= 8x \text{ inches.}$$

The area is

$$A = LW$$
$$= 3x \cdot x$$
$$= 3x^2 \text{ square inches.}$$

Now Try Exercise 1 ◆

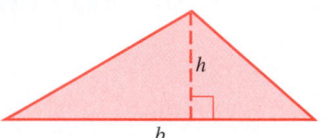

FIGURE R.3 Triangle

Triangles If the base of a triangle is b and its height is h, as illustrated in Figure R.3, then the area A of the triangle is given by

$$A = \frac{1}{2}bh.$$

Finding areas of triangles

Calculate the area of each triangle.

(a)

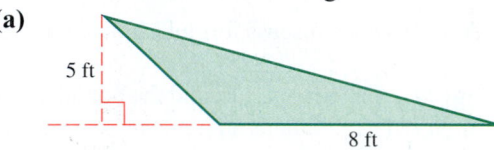

(b)

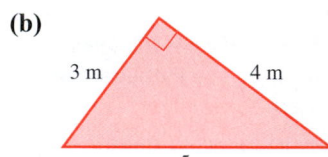

SOLUTION

(a) The triangle has a base of 8 feet and a height of 5 feet. Therefore its area is

$$A = \frac{1}{2}bh = \frac{1}{2} \cdot 8 \cdot 5 = 20 \text{ square feet.}$$

(b) The triangle shown is a right triangle. The legs of the triangle correspond to its height and base. If we let $b = 4$ meters and $h = 3$ meters, then

$$A = \frac{1}{2}bh = \frac{1}{2}(4)(3) = 6 \text{ square meters.}$$

Now Try Exercises 11 and 12 ◆

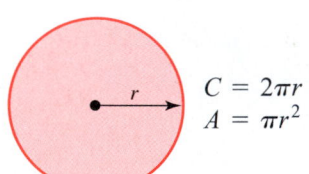

FIGURE R.4 Circle

Circles The perimeter of a circle is called its **circumference** C and is $C = 2\pi r$, where r is the radius of the circle. The area A of a circle is $A = \pi r^2$. See Figure R.4.

Finding the circumference and area of a circle

A circle has a radius of 12.5 inches. Approximate its circumference and area.

SOLUTION

Circumference: $C = 2\pi r = 2\pi(\mathbf{12.5}) = 25\pi \approx 78.5$ inches.

Area: $A = \pi r^2 = \pi(\mathbf{12.5})^2 = 156.25\pi \approx 490.9$ square inches. *Now Try Exercise 21* ◆

The Pythagorean Theorem

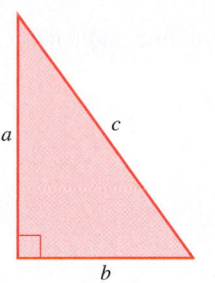

FIGURE R.5 $a^2 + b^2 = c^2$

One of the most famous theorems in mathematics is the Pythagorean theorem. It states that a triangle with legs a and b and hypotenuse c is a right triangle if and only if

$$a^2 + b^2 = c^2,$$

as illustrated in Figure R.5.

EXAMPLE 4 Applying the Pythagorean theorem

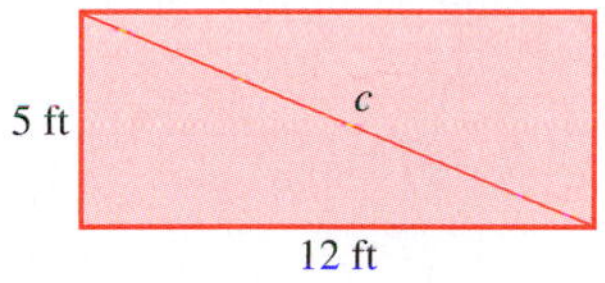

A rectangle has sides of 5 feet and 12 feet, as shown in Figure R.6. Find the diagonal c of the rectangle.

FIGURE R.6

SOLUTION The diagonal of the rectangle corresponds to the hypotenuse of a right triangle with legs of 5 feet and 12 feet. Let $a = 5$, $b = 12$, and find c.

$c^2 = \mathbf{a^2 + b^2}$	Pythagorean theorem
$c^2 = \mathbf{5^2 + 12^2}$	Substitute $a = 5$ and $b = 12$.
$c^2 = 169$	Simplify.
$c = 13$	Solve for $c > 0$.

The diagonal of the rectangle is 13 feet.   *Now Try Exercise 27* ◆

EXAMPLE 5 Finding the perimeter of a right triangle

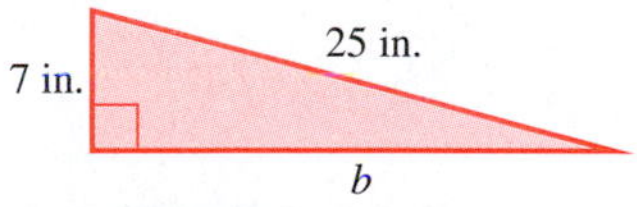

Find the perimeter of the triangle shown in Figure R.7.

FIGURE R.7

SOLUTION Given one leg and the hypotenuse of a right triangle, we can use the Pythagorean theorem to find the other leg. Let $a = 7$, $c = 25$, and find b.

$a^2 + b^2 = c^2$	Pythagorean theorem
$b^2 = c^2 - a^2$	Subtract a^2.
$b^2 = 25^2 - 7^2$	Let $a = 7$ and $c = 25$.
$b^2 = 576$	Simplify.
$b = 24$	Solve for $b > 0$.

The perimeter of the triangle is

$$a + b + c = 7 + 24 + 25 = 56 \text{ inches.}$$ *Now Try Exercise 31* ◆

Three-Dimensional Objects

Objects that occupy space have both volume and surface area. In this subsection we discuss rectangular boxes, spheres, cylinders, and cones.

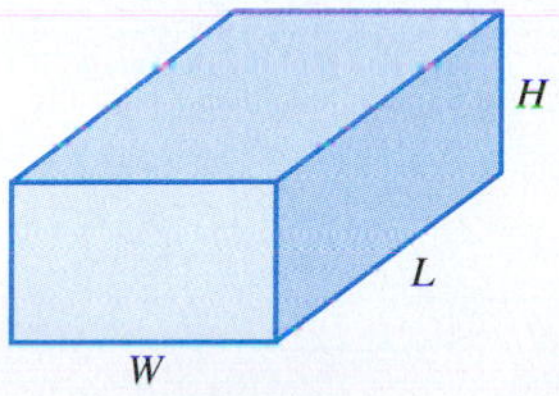

Rectangular Boxes The volume V of a rectangular box with length L, width W, and height H equals $V = LWH$. See Figure R.8. The surface area S of the box equals the sum of the areas of the six sides:

FIGURE R.8 Rectangular Box

$$S = 2LW + 2WH + 2LH.$$

EXAMPLE 6   Finding the volume and surface area of a box

The box shown in Figure R.9 has dimensions x by $2x$ by y. Find its volume and surface area.

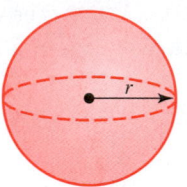

FIGURE R.9

SOLUTION

Volume: $LWH = 2x \cdot x \cdot y = 2x^2y$ cubic units

Surface Area: Base and top: $2x \cdot x + 2x \cdot x = 4x^2$
Front and back: $xy + xy = 2xy$
Left and right sides: $2xy + 2xy = 4xy$
Total surface area: $4x^2 + 2xy + 4xy = 4x^2 + 6xy$ square units

Now Try Exercise 43 ◆

Spheres   The volume V of a sphere with radius r is $V = \frac{4}{3}\pi r^3$, and its surface area S is $S = 4\pi r^2$. See Figure R.10.

FIGURE R.10 Sphere

EXAMPLE 7   Finding the volume and surface area of a sphere

Estimate, to the nearest tenth of a foot, the volume and surface area of a sphere with a radius of 5.1 feet.

SOLUTION

Volume: $V = \frac{4}{3}\pi r^3 = \frac{4}{3}\pi(\mathbf{5.1})^3 \approx 555.6$ cubic feet

Surface Area: $S = 4\pi r^2 = 4\pi(\mathbf{5.1})^2 \approx 326.9$ square feet.   *Now Try Exercise 49* ◆

Cylinders   A soup can is usually made in the shape of a cylinder. The volume of a cylinder with radius r and height h is $V = \pi r^2 h$. See Figure R.11. To find the total surface area of a cylinder, we add the area of the top and bottom to the area of the side. Figure R.12 illustrates a can cut open to determine its surface area. The top and bottom are circular with areas of πr^2 each, and the side has a surface area of $2\pi rh$. The total surface area is $S = 2\pi r^2 + 2\pi rh$.

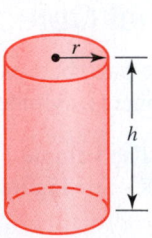

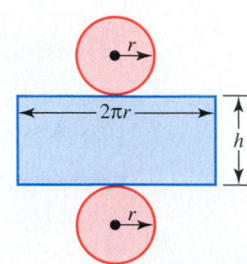

FIGURE R.11 Cylinder   **FIGURE R.12** A Cylinder Cut Open

EXAMPLE 8   Finding the volume and surface area of a cylinder

A cylinder has radius $r = 3$ inches and height $h = 2.5$ feet. Find its volume and total surface area to the nearest tenth.

SOLUTION Begin by changing 2.5 feet to 30 inches so that all units are in inches.
Volume: $V = \pi r^2 h = \pi(3)^2(30) = 270\pi \approx 848.2$ cubic inches

Total Surface Area: $S = 2\pi r^2 + 2\pi rh = 2\pi(3)^2 + 2\pi(3)(30) = 198\pi \approx 622.0$
square inches   *Now Try Exercise 57* ◆

Cones The volume of a cone with radius r and height h is $V = \frac{1}{3}\pi r^2 h$, as shown in Figure R.13. (Compare this formula with the formula for the volume of a cylinder.) Excluding the bottom of the cone, the side (or lateral) surface area is $S = \pi r \sqrt{r^2 + h^2}$. The bottom of the cone is circular and has a surface area of πr^2.

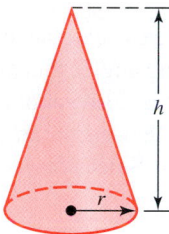

FIGURE R.13 Cone

EXAMPLE 9   Finding the volume and surface area of a cone

Approximate, to the nearest tenth, the volume and surface area (side only) of a cone with a radius of 1.45 inches and a height of 5.12 inches.

SOLUTION
Volume: $V = \frac{1}{3}\pi r^2 h = \frac{1}{3}\pi(1.45)^2(5.12) \approx 11.3$ cubic inches

Surface Area (side only): $S = \pi r \sqrt{r^2 + h^2} = \pi(1.45)\sqrt{(1.45)^2 + (5.12)^2} \approx 24.2$
square inches   *Now Try Exercise 63* ◆

Similar Triangles

The corresponding angles of **similar triangles** have equal measure, but similar triangles are not necessarily the same size. An example of two similar triangles is shown in Figure R.14. Notice that both triangles have angles of 30°, 60°, and 90°. Corresponding sides are not equal in length; however, corresponding ratios are equal. For example, in triangle ABC the ratio of the shortest leg to the hypotenuse equals $\frac{2}{4} = \frac{1}{2}$, and in triangle DEF this ratio is $\frac{3}{6} = \frac{1}{2}$.

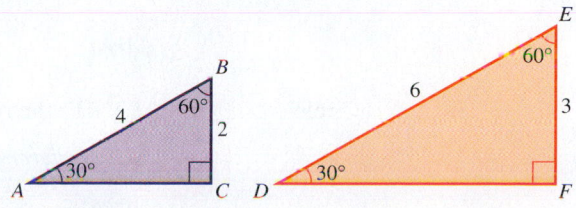

FIGURE R.14 Similar Triangles

 **Using similar triangles**

Find the length of BC in Figure R.15.

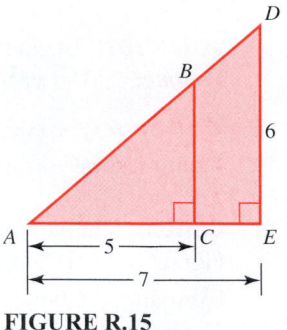

FIGURE R.15

SOLUTION   Notice that triangle ABC and triangle ADE are both right triangles. These triangles share an angle at vertex A. Therefore triangles ABC and ADE have two corresponding angles that are congruent. Because the sum of the angles in a triangle equals 180°, all three corresponding angles in these two triangles are congruent. Thus triangles ABC and ADE are similar.

Since corresponding ratios are equal, we can find BC as follows.

$$\frac{BC}{AC} = \frac{DE}{AE}$$

$$\frac{BC}{5} = \frac{6}{7}$$

Solving this equation for BC gives $BC = \frac{30}{7} \approx 4.3$.   *Now Try Exercise 69* ◆

A Summary of Geometric Formulas

The following table provides a summary of some important formulas from geometry.

Geometric Shape	Related Formulas
Rectangle	Let W be the width and L be the length. **Perimeter:** $P = 2L + 2W$ **Area:** $A = LW$
Triangle	Let s_1, s_2, and s_3 be the sides of a triangle. **Perimeter:** $P = s_1 + s_2 + s_3$ Let b be the base and h be the height. **Area:** $A = \frac{1}{2}bh$
Circle	Let r be the radius. **Circumference:** $C = 2\pi r$ **Area:** $A = \pi r^2$

Geometric Shape	Related Formulas
Rectangular box	Let L, W, and H be the length, width, and height. ***Area:*** $A = 2LW + 2WH + 2LH$ ***Volume:*** $V = LWH$
Sphere	Let r be the radius. ***Surface Area:*** $S = 4\pi r^2$ ***Volume:*** $V = \dfrac{4}{3}\pi r^3$
Cylinder	Let h be the height and r be the radius. ***Surface Area (side only):*** $S = 2\pi rh$ ***Total Surface Area:*** $S = 2\pi r^2 + 2\pi rh$ ***Volume:*** $V = \pi r^2 h$
Cone	Let h be the height and r be the radius. ***Surface Area (side only):*** $S = \pi r\sqrt{r^2 + h^2}$ ***Total Surface Area:*** $S = \pi r\sqrt{r^2 + h^2} + \pi r^2$ ***Volume:*** $V = \dfrac{1}{3}\pi r^2 h$
Units	***Perimeter (length):*** meters, inches, feet ***Surface Area (square units):*** square yards, square miles ***Volume (cubic units):*** cubic inches, cubic centimeters

R.1 — Exercises

Rectangles

Exercises 1–6: Find the area and perimeter of the rectangle with length L and width W.

1. $L = 15$ feet, $W = 7$ feet

2. $L = 16$ inches, $W = 10$ inches

3. $L = 100$ meters, $W = 35$ meters

4. $L = 80$ yards, $W = 13$ yards

5. $L = 3x$, $W = y$ 6. $L = a + 5$, $W = a$

Exercises 7–10: Find the area and perimeter of the rectangle in terms of the width W.

7. The width W is half the length.

8. Triple the width W minus 3 equals the length.

9. The length equals the width W plus 5.

10. The length is 2 less than twice the width W.

Triangles

Exercises 11 and 12: Find the area of the triangle shown in the figure.

11.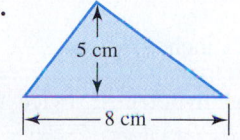

12.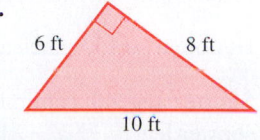

Exercises 13–20: Find the area of the triangle with base b and height h.

13. $b = 5$ inches, $h = 8$ inches

14. $b = 24$ inches, $h = 9$ feet

15. $b = 10.1$ meters, $h = 730$ meters

16. $b = 52$ yards, $h = 102$ feet

17. $b = 2x, h = 6x$ **18.** $b = x, h = x + 4$

19. $b = z, h = 5z$ **20.** $b = y + 1, h = 2y$

Circles

Exercises 21–26: Find the circumference and area of the circle. Approximate each value to the nearest tenth when appropriate.

21. $r = 4$ meters   **22.** $r = 1.5$ feet

23. $r = 19$ inches   **24.** $r = 22$ miles

25. $r = 2x$ **26.** $r = 5z$

Pythagorean Theorem

Exercises 27–34: Use the Pythagorean theorem to find the missing side of the right triangle with legs a and b, and hypotenuse c. Then calculate the perimeter. Approximate values to the nearest tenth when appropriate.

27. $a = 60$ feet, $b = 11$ feet

28. $a = 16$ inches, $b = 9$ inches

29. $a = 21$ feet, $b = 11$ yards

30. $a = 5$ centimeters, $c = 13$ centimeters

31. $a = 6$ meters, $c = 15$ meters

32. $b = 7$ millimeters, $c = 10$ millimeters

33. $b = 1.2$ miles, $c = 2$ miles

34. $a = 0.5$ kilometers, $b = 1.2$ kilometers

Exercises 35–38: Find the area of the right triangle that satisfies the conditions. Approximate values to the nearest tenth when appropriate.

35. Legs with lengths 3 feet and 6 feet

36. Hypotenuse 10 inches and leg 6 inches

37. Hypotenuse 15 inches and leg 11 inches

38. Shorter leg 40 centimeters and hypotenuse twice the shorter leg

Rectangular Boxes

Exercises 39–46: Find the volume and surface area of a rectangular box with length L, width W, and height H.

39. $L = 4$ feet, $W = 3$ feet, $H = 2$ feet

40. $L = 6$ meters, $W = 4$ meters, $H = 1.5$ meters

41. $L = 4.5$ inches, $W = 4$ inches, $H = 1$ foot

42. $L = 9.1$ yards, $W = 8$ yards, $H = 6$ feet

43. $L = 3x, W = 2x, H = x$

44. $L = 6z, W = 5z, H = 7z$

45. $L = x, W = 2y, H = 3z$

46. $L = 8x, W = y, H = z$

Exercises 47 and 48: Find the volume of the rectangular box in terms of the width W.

47. The length is twice the width W, and the height is half the width.

48. The width W is three times the height and one-third of the length.

Spheres

Exercises 49–54: Find the volume and surface area of the sphere satisfying the given condition, where r is the radius and d is the diameter. Approximate values to the nearest tenth.

49. $r = 3$ feet   **50.** $r = 4.1$ inches

51. $r = 5$ centimeters   **52.** $d = 6.4$ meters

53. $d = 10$ millimeters   **54.** $d = 16$ feet

Cylinders

Exercises 55–60: Find the volume, the surface area of the side, and the total surface area of the cylinder that satisfies the given conditions, where r is the radius and h is the height. Approximate values to the nearest tenth.

55. $r = 0.5$ feet, $h = 2$ feet

56. $r = 4.1$ centimeters, $h = 30.5$ centimeters

57. $r = 5$ inches, $h = 1.5$ feet

58. $r = \frac{1}{3}$ yard, $h = 2$ feet

59. $r = 12$ millimeters, and h is twice r

60. r is one-fourth of h, and $h = 2.1$ feet

Exercises 61 and 62: Use r to write an expression that calculates the volume of the cylinder.

61. The radius r is half the height.

62. The height is 2 units longer than the radius r.

Cones

Exercises 63–68: Approximate, to the nearest tenth, the volume and surface area (side only) of the cone satisfying the given conditions, where r is the radius and h is the height.

63. $r = 5$ centimeters, $h = 6$ centimeters

64. $r = 8$ inches, $h = 30$ inches

65. $r = 24$ inches, $h = 3$ feet

66. $r = 100$ centimeters, $h = 1.3$ meters

67. Three times r equals h, and $r = 2.4$ feet

68. Twice h equals r, and $h = 3$ centimeters

Similar Triangles

Exercises 69–74: Use the fact that triangles ABC and DEF are similar to find the value of x.

69.

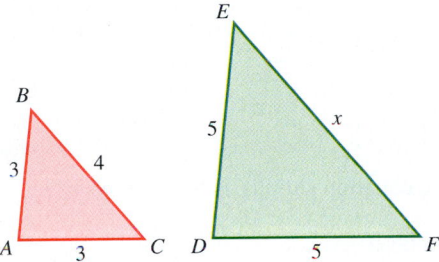

70.

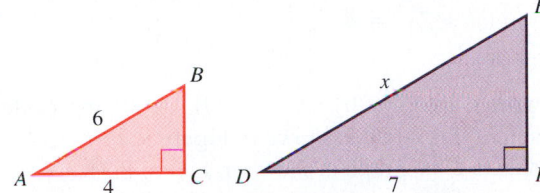

71.

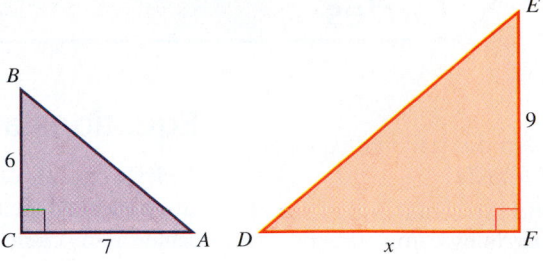

72.

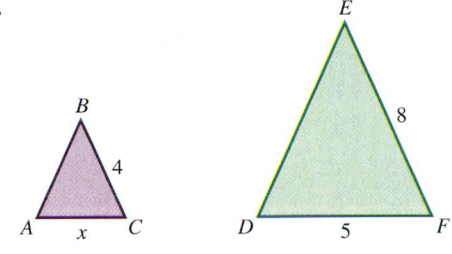

73.

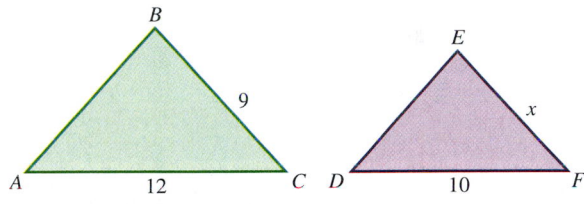

74.

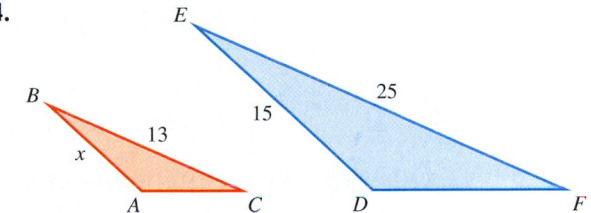

R.2   Circles

◆ Learn the standard equation of a circle

◆ Find the center and radius of a circle by completing the square

Equations and Graphs of Circles

A **circle** consists of the set of points in a plane that are equidistant from a fixed point. The distance is called the **radius** of the circle, and the fixed point is called the **center**. If we let the center of the circle be (h, k), the radius be r, and (x, y) be any point on the circle, then the distance between (x, y) and (h, k) must equal r. See Figure R.16. By the distance formula we have

$$\sqrt{(x - h)^2 + (y - k)^2} = r.$$

Squaring both sides gives

$$(x - h)^2 + (y - k)^2 = r^2.$$

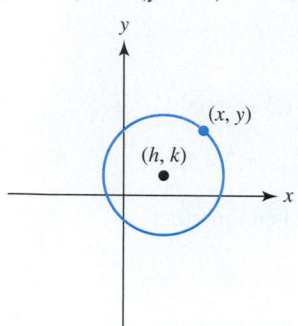

FIGURE R.16

STANDARD EQUATION OF A CIRCLE

The circle with center (h, k) and radius r has equation

$$(x - h)^2 + (y - k)^2 = r^2.$$

Note: If the center of a circle is $(0, 0)$, then the equation simplifies to $x^2 + y^2 = r^2$.

EXAMPLE 1   Finding the center and radius of a circle

Find the center and radius of the circle with the given equation. Graph each circle.
(a) $x^2 + y^2 = 9$ **(b)** $(x - 1)^2 + (y + 2)^2 = 4$

SOLUTION
(a) Because the equation can be written as $(x - 0)^2 + (y - 0)^2 = 3^2$, the center is $(0, 0)$ and its radius is 3. The graph of this circle is shown in Figure R.17.
(b) The center is $(1, -2)$ and its radius is 2. Its graph is shown in Figure R.18.

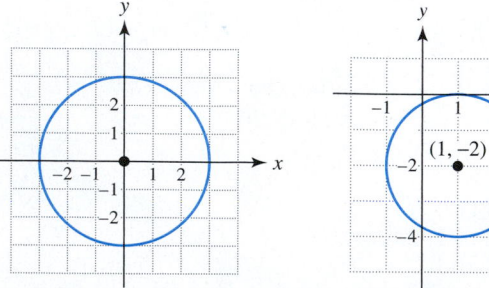

FIGURE R.17   **FIGURE R.18**

Now Try Exercises 1 and 5 ◆

 **2** Finding the equation of a circle

Find the equation of the circle that satisfies the conditions. Graph each circle.
(a) Radius 4, center $(-3, 5)$
(b) Center $(6, -3)$ with the point $(1, 2)$ on the circle.

SOLUTION
(a) Let $r = 4$ and $(h, k) = (-3, 5)$. The equation of this circle is
$$(x - (-3))^2 + (y - 5)^2 = 4^2 \quad \text{or} \quad (x + 3)^2 + (y - 5)^2 = 16$$
A graph of the circle is shown in Figure R.19.
(b) First we must find the distance between the points $(6, -3)$ and $(1, 2)$ to determine r.
$$r = \sqrt{(6 - 1)^2 + (-3 - 2)^2} = \sqrt{50} \approx 7.1$$
Since $r^2 = 50$, the equation of the circle is
$$(x - 6)^2 + (y + 3)^2 = 50.$$
Its graph is shown in Figure R.20.

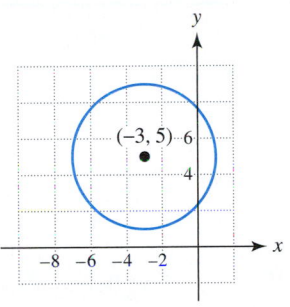

FIGURE R.19   **FIGURE R.20**

Now Try Exercises 19 and 24 ◆

Finding the Center and Radius of a Circle

When a circle's equation is not written in standard form, we can complete the square to find its center and radius. To complete the square, we use the fact that
$$x^2 + kx + \left(\frac{k}{2}\right)^2 = \left(x + \frac{k}{2}\right)^2.$$

For example, to complete the square on the expression $x^2 + 8x$, add $\left(\frac{8}{2}\right)^2 = 16$. Then,
$$x^2 + 8x + 16 = (x + 4)^2.$$

 **3** Finding the center and radius of a circle

Find the center and radius of a circle whose equation is $x^2 + 6x + y^2 - 4y - 12 = 0$. Graph the circle.

SOLUTION Begin by writing the equation as
$$(x^2 + 6x + \underline{}) + (y^2 - 4y + \underline{}) = 12.$$

To complete the square, add $\left(\frac{6}{2}\right)^2 = 9$ to the first quantity and $\left(\frac{-4}{2}\right)^2 = 4$ to the second quantity. Add the same values to the right side of the equation.
$$(x^2 + 6x + \underline{9}) + (y^2 - 4y + \underline{4}) = 12 + 9 + 4$$

Factoring gives the following result.

$$(x + 3)^2 + (y - 2)^2 = 5^2$$

The center is $(-3, 2)$ and the radius is $r = 5$. Its graph is shown in Figure R.21.

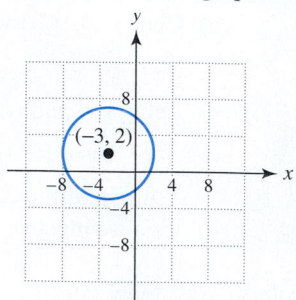

FIGURE R.21

Now Try Exercise 29 ◆

R.2 Exercises

Equations and Graphs of Circles

Exercises 1–14: Find the center and radius of the circle.

1. $x^2 + y^2 = 25$ **2.** $x^2 + y^2 = 100$

3. $x^2 + y^2 = 7$ **4.** $x^2 + y^2 = 20$

5. $(x - 2)^2 + (y + 3)^2 = 9$

6. $(x + 1)^2 + (y - 1)^2 = 16$

7. $x^2 + (y + 1)^2 = 100$ **8.** $(x - 5)^2 + y^2 = 19$

9. $5x^2 + 5y^2 = 20$ (*Hint:* Divide by 5.)

10. $3x^2 + 3y^2 = 9$ **11.** $6x^2 + 6y^2 = 30$

12. $7x^2 + 7y^2 = 49$

13. $2(x - 4)^2 + 2(y + 5)^2 = 32$

14. $4(x + 7)^2 + 4(y - 6)^2 = 9$

Exercises 15–18: Use the graph to find the standard equation of the circle.

15.

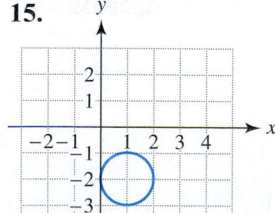

16.

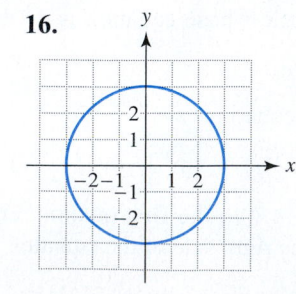

17.

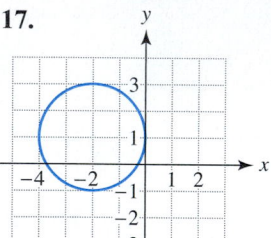

18.

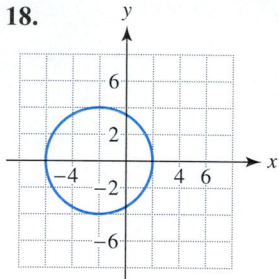

Exercises 19–26: Find the standard equation of a circle that satisfies the conditions.

19. Radius 8, center $(3, -5)$

20. Radius 5, center $(-1, 4)$

21. Radius 7, center $(3, 0)$ **22.** Radius 1, center $(0, 0)$

23. Center $(0, 0)$ with the point $(-3, -1)$ on the circle

24. Center $(3, -5)$ with the point $(4, 2)$ on the circle

25. Endpoints of a diameter $(-5, -7)$ and $(1, 1)$.

26. Endpoints of a diameter $(-3, -2)$ and $(1, -4)$.

Finding the Center and Radius of a Circle

Exercises 27–38: Find the center and radius of the circle. Graph the circle.

27. $x^2 + y^2 = 9$ **28.** $x^2 + y^2 = 1$

29. $x^2 + 2x + y^2 - 2y = 2$

30. $x^2 - 6x + y^2 - 8y = 0$

31. $x^2 + y^2 - 2y = 0$

32. $x^2 - 20x + y^2 + 64 = 0$

33. $x^2 + 18x + y^2 + 10y = -97$

34. $x^2 + 16x + y^2 - 18y + 141 = 0$

35. $2x^2 - 4x + 2y^2 + 12y + 16 = 0$ (*Hint:* Divide by 2.)

36. $3x^2 + 30x + 3y^2 - 6y + 27 = 0$

37. $4x^2 + 24x + 4y^2 + 48y + 180 = 64$

38. $2x^2 - 28x + 2y^2 - 16y + 130 = 100$

R.3 Integer Exponents

- ◆ Use bases and exponents
- ◆ Use zero and negative exponents
- ◆ Apply the product, quotient, and power rules

Bases and Positive Exponents

The area of a square that is 8 inches on a side is given by the expression

$$8 \cdot 8 = 8^2 = 64 \text{ square inches.}$$

The expression 8^2 is an exponential expression with base 8 and exponent 2. Exponential expressions occur frequently in a variety of applications. For example, suppose that an investment doubles its initial value 3 times. Then its final value is

$$2 \cdot 2 \cdot 2 = 2^3 = 8$$

times larger than its original value. Table R.1 contains examples of exponential expressions.

TABLE R.1

Expression	Base	Exponent
$2 \cdot 2 \cdot 2 = 2^3$	2	3
$6 \cdot 6 \cdot 6 \cdot 6 = 6^4$	6	4
$7 = 7^1$	7	1
$0.5 \cdot 0.5 = 0.5^2$	0.5	2
$x \cdot x \cdot x = x^3$	x	3

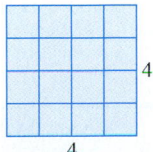

FIGURE R.22 4 Squared

We read 0.5^2 as "0.5 squared," 2^3 as "2 cubed," and 6^4 as "6 to the fourth power." The terms *squared* and *cubed* come from geometry. If the length of a side of a square is 4, then its area is

$$4 \cdot 4 = 4^2 = 16 \text{ square units,}$$

as illustrated in Figure R.22. Similarly, if the length of an edge of a cube is 4, then its volume is

$$4 \cdot 4 \cdot 4 = 4^3 = 64 \text{ cubic units,}$$

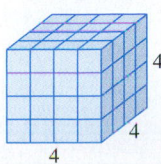

FIGURE R.23 4 Cubed

as shown in Figure R.23.

EXAMPLE  **1**   Writing numbers in exponential notation

Use the given base to write each number as an exponential expression. Check your results with a calculator.

(a) 10,000   (base 10)
(b) 27   (base 3)
(c) 32   (base 2)

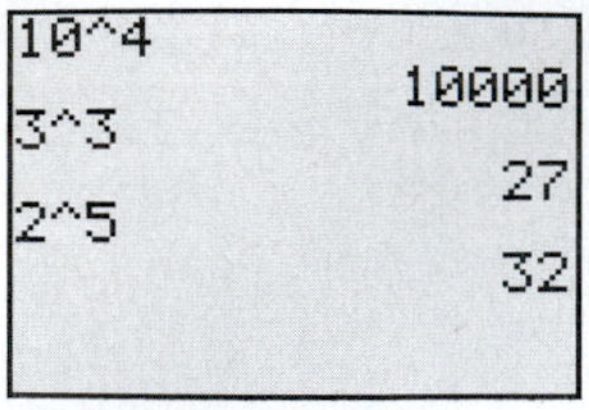

SOLUTION

(a) $10{,}000 = 10 \cdot 10 \cdot 10 \cdot 10 = 10^4$ **(b)** $27 = 3 \cdot 3 \cdot 3 = 3^3$
(c) $32 = 2 \cdot 2 \cdot 2 \cdot 2 \cdot 2 = 2^5$

These values are supported in Figure R.24, where exponential expressions are evaluated with a graphing calculator, using the ⌐^⌐ key.   *Now Try Exercise 15* ◆

```
10^4
             10000
3^3
                27
2^5
                32
```

FIGURE R.24

Zero and Negative Exponents

Exponents can be defined for any integer. If a is any nonzero real number, we define

$$a^0 = 1.$$

For example, $3^0 = 1$ and $\left(\frac{1}{7}\right)^0 = 1$. We can also define a^{-n}, where n is a positive integer, as

$$a^{-n} = \frac{1}{a^n}.$$

Thus $5^{-4} = \frac{1}{5^4}$ and $y^{-2} = \frac{1}{y^2}$.

Using these definitions, we obtain

$$\frac{1}{a^{-n}} = \frac{1}{\dfrac{1}{a^n}} = \frac{a^n}{1} = a^n.$$

Thus $\frac{1}{2^{-5}} = 2^5$ and $\frac{1}{x^{-2}} = x^2$. If a and b are nonzero numbers, then

$$\frac{a^{-n}}{b^{-m}} = \frac{\dfrac{1}{a^n}}{\dfrac{1}{b^m}} = \frac{1}{a^n} \cdot \frac{b^m}{1} = \frac{b^m}{a^n}.$$

Thus $\frac{4^{-3}}{z^{-2}} = \frac{z^2}{4^3}$. This discussion leads to the following properties for integer exponents.

INTEGER EXPONENTS

Let a and b be nonzero real numbers and m and n be positive integers. Then

1. $a^n = a \cdot a \cdot a \cdot \cdots \cdot a$ (n factors of a).
2. $a^0 = 1$.   (*Note:* 0^0 is undefined.)
3. $a^{-n} = \dfrac{1}{a^n}$ and $\dfrac{1}{a^{-n}} = a^n$.
4. $\dfrac{a^{-n}}{b^{-m}} = \dfrac{b^m}{a^n}$.
5. $\left(\dfrac{a}{b}\right)^{-n} = \left(\dfrac{b}{a}\right)^n$.

EXAMPLE 2   Evaluating expressions

Evaluate each expression.

(a) 3^{-4} (b) $\dfrac{1}{2^{-3}}$ (c) $\left(\dfrac{5}{7}\right)^{-2}$ (d) $\dfrac{1}{(xy)^{-1}}$ (e) $\dfrac{2^{-2}}{3t^{-3}}$

SOLUTION

(a) $3^{-4} = \dfrac{1}{3^4} = \dfrac{1}{3\cdot3\cdot3\cdot3} = \dfrac{1}{81}$ (b) $\dfrac{1}{2^{-3}} = 2^3 = 2\cdot2\cdot2 = 8$

(c) $\left(\dfrac{5}{7}\right)^{-2} = \left(\dfrac{7}{5}\right)^2 = \dfrac{7}{5}\cdot\dfrac{7}{5} = \dfrac{49}{25}$ (d) $\dfrac{1}{(xy)^{-1}} = (xy)^1 = xy$

Base is xy.

(e) Note that only t and not $3t$ is raised to the power of -3.

$$\frac{2^{-2}}{3t^{-3}} = \frac{t^3}{3(2^2)} = \frac{t^3}{3\cdot4} = \frac{t^3}{12}$$

Base is t.   *Now Try Exercises 31, 33, 35, 73, and 75* ◆

Powers of 10 are important because they are used in mathematics to express numbers that are either small or large in absolute value. Table R.2 may be used to simplify powers of 10. Note that, if the power decreases by 1, then the result decreases by a factor of $\frac{1}{10}$.

TABLE R.2   Powers of Ten

Power of 10	10^3	10^2	10^1	10^0	10^{-1}	10^{-2}	10^{-3}
Value	1000	100	10	1	$\frac{1}{10}$	$\frac{1}{100}$	$\frac{1}{1000}$

Product, Quotient, and Power Rules

We can calculate products and quotients of exponential expressions *provided their bases are the same*. For example,

$$3^2 \cdot 3^3 = (3\cdot3)\cdot(3\cdot3\cdot3) = 3^5.$$

This expression has a total of $2+3 = 5$ factors of 3, so the result is 3^5. To multiply exponential expressions with like bases, add exponents.

THE PRODUCT RULE

For any nonzero number a and integers m and n,

$$a^m \cdot a^n = a^{m+n}.$$

The product rule holds for negative exponents. For example,

$$10^5 \cdot 10^{-2} = 10^{5+(-2)} = 10^3.$$

EXAMPLE 3 Using the product rule

Multiply and simplify.

(a) $10^2 \cdot 10^4$ (b) $7^3 \cdot 7^{-4}$ (c) $x^3 x^{-2} x^4$ (d) $(3y^2)(2y^{-4})$

SOLUTION

(a) $10^2 \cdot 10^4 = 10^{2+4} = 10^6 = 1{,}000{,}000$ (b) $7^3 \cdot 7^{-4} = 7^{3+(-4)} = 7^{-1} = \dfrac{1}{7}$

(c) $x^3 x^{-2} x^4 = x^{3+(-2)+4} = x^5$

(d) $(3y^2)(2y^{-4}) = 3 \cdot 2 \cdot y^2 \cdot y^{-4} = 6y^{2+(-4)} = 6y^{-2} = \dfrac{6}{y^2}$

Note that 6 is not raised to the power of -2 in the expression $6y^{-2}$.

Now Try Exercises 39, 42, and 49 ◆

Consider division of exponential expressions using the following example.

$$\frac{6^5}{6^3} = \frac{6 \cdot 6 \cdot \cancel{6}^1 \cdot \cancel{6}^1 \cdot \cancel{6}^1}{\cancel{6} \cdot \cancel{6} \cdot \cancel{6}} = 6 \cdot 6 = 6^2$$

After simplifying, there are two 6s left in the numerator. The result is $6^{5-3} = 6^2 = 36$. To divide exponential expressions with like bases, subtract exponents.

> ### THE QUOTIENT RULE
>
> For any nonzero number a and integers m and n,
>
> $$\frac{a^m}{a^n} = a^{m-n}.$$

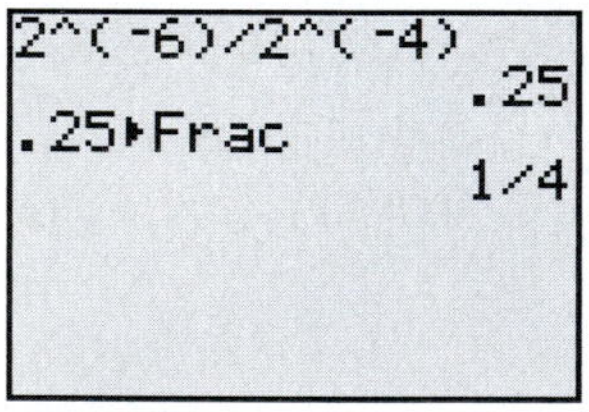

The quotient rule holds true for negative exponents. For example,

$$\frac{2^{-6}}{2^{-4}} = 2^{-6-(-4)} = 2^{-2} = \frac{1}{2^2} = \frac{1}{4}.$$

This result is supported by Figure R.25.

```
2^(-6)/2^(-4)
              .25
.25▶Frac
              1/4
```

FIGURE R.25

EXAMPLE 4 Using the quotient rule

Simplify the expression. Use positive exponents.

(a) $\dfrac{10^4}{10^6}$ (b) $\dfrac{x^5}{x^2}$ (c) $\dfrac{15x^2y^3}{5x^4y}$

SOLUTION

(a) $\dfrac{10^4}{10^6} = 10^{4-6} = 10^{-2} = \dfrac{1}{10^2} = \dfrac{1}{100}$ (b) $\dfrac{x^5}{x^2} = x^{5-2} = x^3$

(c) $\dfrac{15x^2y^3}{5x^4y} = \dfrac{15}{5} \cdot \dfrac{x^2}{x^4} \cdot \dfrac{y^3}{y^1} = 3 \cdot x^{(2-4)}y^{(3-1)} = 3x^{-2}y^2 = \dfrac{3y^2}{x^2}$

Now Try Exercises 51, 55, and 59 ◆

How should we evaluate $(4^3)^2$? To answer this question consider

$$(4^3)^2 = 4^3 \cdot 4^3 = 4^{3+3} = 4^6.$$

Similarly,

$$(x^4)^3 = x^4 \cdot x^4 \cdot x^4 = x^{4+4+4} = x^{12}.$$

These results suggest that to raise a power to a power, we must multiply the exponents.

RAISING POWERS TO POWERS

For any nonzero real number a and integers m and n,

$$(a^m)^n = a^{mn}.$$

 EXAMPLE 5   Raising powers to powers

Simplify each expression. Use positive exponents.
(a) $(5^2)^3$ **(b)** $(2^4)^{-2}$ **(c)** $(b^{-7})^5$

SOLUTION

(a) $(5^2)^3 = 5^{2 \cdot 3} = 5^6$ **(b)** $(2^4)^{-2} = 2^{4(-2)} = 2^{-8} = \dfrac{1}{2^8}$

(c) $(b^{-7})^5 = b^{-7 \cdot 5} = b^{-35} = \dfrac{1}{b^{35}}$ *Now Try Exercises 63 and 65* ◆

How can we simplify the expression $(2x)^3$? Consider the following.

$$(2x)^3 = 2x \cdot 2x \cdot 2x = (2 \cdot 2 \cdot 2) \cdot (x \cdot x \cdot x) = 2^3 x^3$$

This result suggests that to cube a product, we can cube each factor.

RAISING PRODUCTS TO POWERS

For any nonzero real numbers a and b and integer n,

$$(ab)^n = a^n b^n.$$

 EXAMPLE 6   Raising products to powers

Simplify each expression. Use positive exponents.
(a) $(6y)^2$ **(b)** $(x^2 y)^{-2}$ **(c)** $(2xy^3)^4$

SOLUTION

(a) $(6y)^2 = 6^2 y^2 = 36y^2$ **(b)** $(x^2 y)^{-2} = (x^2)^{-2} y^{-2} = x^{-4} y^{-2} = \dfrac{1}{x^4 y^2}$

(c) $(2xy^3)^4 = 2^4 x^4 (y^3)^4 = 16 x^4 y^{12}$ *Now Try Exercises 67 and 83* ◆

To simplify a power of a quotient use the following rule.

RAISING QUOTIENTS TO POWERS

For nonzero numbers a and b and any integer n,

$$\left(\frac{a}{b}\right)^n = \frac{a^n}{b^n}.$$

EXAMPLE 7 Raising quotients to powers

Simplify each expression. Use positive exponents.

(a) $\left(\dfrac{3}{x}\right)^3$ (b) $\left(\dfrac{1}{2^3}\right)^{-2}$ (c) $\left(\dfrac{3x^{-3}}{y^2}\right)^4$

SOLUTION

(a) $\left(\dfrac{3}{x}\right)^3 = \dfrac{3^3}{x^3} = \dfrac{27}{x^3}$ (b) $\left(\dfrac{1}{2^3}\right)^{-2} = \dfrac{1^{-2}}{(2^3)^{-2}} = \dfrac{1}{2^{-6}} = 2^6 = 64$

(c) $\left(\dfrac{3x^{-3}}{y^2}\right)^4 = \dfrac{3^4(x^{-3})^4}{(y^2)^4} = \dfrac{81x^{-12}}{y^8} = \dfrac{81}{x^{12}y^8}$

Now Try Exercises 69 and 71 ◆

In the next example we use several properties of exponents to simplify expressions.

EXAMPLE 8 Simplifying expressions

Write each expression using positive exponents. Simplify the result completely.

(a) $\left(\dfrac{x^2y^{-3}}{3z^{-4}}\right)^{-2}$ (b) $\dfrac{(rt^3)^{-3}}{(r^2t^3)^{-2}}$

SOLUTION

(a) $\left(\dfrac{x^2y^{-3}}{3z^{-4}}\right)^{-2} = \left(\dfrac{3z^{-4}}{x^2y^{-3}}\right)^2$ (b) $\dfrac{(rt^3)^{-3}}{(r^2t^3)^{-2}} = \dfrac{(r^2t^3)^2}{(rt^3)^3}$

$= \left(\dfrac{3y^3}{x^2z^4}\right)^2$ $= \dfrac{r^4t^6}{r^3t^9}$

$= \dfrac{9y^6}{x^4z^8}$ $= \dfrac{r}{t^3}$

Now Try Exercises 87 and 89 ◆

R.3 Exercises

Concepts

1. Identify the base and the exponent in the expression 8^3.

2. Evaluate 97^0 and 2^{-1}.

3. Write 7 cubed, using symbols.

4. Write 5 squared, using symbols.

5. Are the expressions 2^3 and 3^2 equal? Explain your answer.

6. Are the expressions -4^2 and $(-4)^2$ equal? Explain your answer.

7. $7^{-n} =$ ____

8. $6^m \cdot 6^n =$ ____

9. $\dfrac{5^m}{5^n} =$ ____

10. $(3x)^k =$ ____

11. $(2^m)^k =$ ____

12. $\left(\dfrac{x}{y}\right)^m =$ ____

13. $5 \times 10^3 =$ ____

14. $5 \times 10^{-3} =$ ____

Properties of Exponents

Exercises 15–20: (Refer to Example 1.) Write the number as an exponential expression, using the base shown. Check your result with a calculator.

15. 8 (base 2)

16. 1000 (base 10)

17. 256 (base 4)

18. $\frac{1}{64}$ (base 4)

19. 1 (base 3)

20. $\frac{1}{49}$ (base 7)

Exercises 21–38: Evaluate the expression by hand. Check your result with a calculator.

21. 5^3 22. 5^{-3}

23. -2^4 24. $(-2)^4$

25. 5^0 26. $\left(-\frac{2}{3}\right)^{-3}$

27. $\left(\frac{2}{3}\right)^3$ 28. $\frac{1}{4^{-2}}$

29. $\left(-\frac{1}{2}\right)^4$ 30. $\left(-\frac{3}{4}\right)^3$

31. 4^{-3} 32. 10^{-4}

33. $\frac{1}{2^{-4}}$ 34. $\frac{1}{3^{-2}}$

35. $\left(\frac{3}{4}\right)^{-3}$ 36. $\left(\frac{1}{2}\right)^{-2}$

37. $\frac{3^{-2}}{2^{-3}}$ 38. $\frac{10^{-4}}{4^{-3}}$

Exercises 39–50: Use the product rule to simplify the expression.

39. $6^3 \cdot 6^{-4}$ 40. $10^2 \cdot 10^5 \cdot 10^{-3}$

41. $5^{-5} \cdot 5^2$ 42. $z^2 \cdot z^3$

43. $2x^2 \cdot 3x^{-3} \cdot x^4$ 44. $3y^4 \cdot 6y^{-4} \cdot y$

45. $10^0 \cdot 10^6 \cdot 10^2$ 46. $y^3 \cdot y^{-5} \cdot y^4$

47. $5^{-2} \cdot 5^3 \cdot 2^{-4} \cdot 2^3$ 48. $2^{-3} \cdot 3^4 \cdot 3^{-2} \cdot 2^5$

49. $(2a^3)(b^2)(a^{-4})(4b^{-5})$ 50. $(3x^{-4})(2x^2)(5y^4)(y^{-3})$

Exercises 51–62: Use the quotient rule to simplify the expression. Use positive exponents to write your answer.

51. $\frac{5^4}{5^2}$ 52. $\frac{6^2}{6^{-7}}$

53. $\frac{10^{-2} \cdot 10^3}{10^{-5}}$ 54. $\frac{6^{-5} \cdot 6^2}{6}$

55. $\frac{a^{-3}}{a^2 \cdot a}$ 56. $\frac{y^0 \cdot y \cdot y^5}{y^{-2} \cdot y^{-3}}$

57. $\frac{24x^3}{6x}$ 58. $\frac{10x^5}{5x^{-3}}$

59. $\frac{12a^2b^3}{18a^4b^2}$ 60. $\frac{-6x^7y^3}{3x^2y^{-5}}$

61. $\frac{21x^{-3}y^4}{7x^4y^{-2}}$ 62. $\frac{32x^3y}{-24x^5y^{-3}}$

Exercises 63–72: Use the power rules to simplify the expression. Use positive exponents to write your answer.

63. $(5^3)^{-1}$ 64. $(-4^2)^3$

65. $(y^4)^{-2}$ 66. $(x^2)^4$

67. $(4y^2)^3$ 68. $(-2xy^3)^{-4}$

69. $\left(\frac{4}{x}\right)^3$ 70. $\left(\frac{-3}{x^3}\right)^2$

71. $\left(\frac{2x}{z^4}\right)^{-5}$ 72. $\left(\frac{2xy}{3z^5}\right)^{-1}$

Exercises 73–100: Use rules of exponents to simplify the expression. Use positive exponents to write your answer.

73. $\frac{2}{(ab)^{-1}}$ 74. $\frac{5a^2}{(xy)^{-1}}$

75. $\frac{2^{-3}}{2t^{-2}}$ 76. $\frac{t^{-3}}{2t^{-1}}$

77. $\frac{6a^2b^{-3}}{4ab^{-2}}$ 78. $\frac{20a^{-2}b}{4a^{-2}b^{-1}}$

79. $\frac{4m^2n^4}{m^{-2}n^{-3}}$ 80. $\frac{8m^{-2}n}{16m^3n^{-2}}$

81. $\frac{5r^2st^{-3}}{25rs^{-2}t^2}$ 82. $\frac{36r^{-1}(st)^2}{9(rs)^2t^{-1}}$

83. $(3x^2y^{-3})^{-2}$ 84. $(-2x^{-3}y^{-2})^3$

85. $\frac{(d^3)^{-2}}{(d^{-2})^3}$ 86. $\frac{(b^2)^{-1}}{(b^{-4})^3}$

87. $\left(\frac{3t^2}{2t^{-1}}\right)^3$ 88. $\left(\frac{-2t}{4t^{-2}}\right)^{-1}$

89. $\frac{(-m^2n^{-1})^{-2}}{(mn)^{-1}}$ 90. $\frac{(-mn^4)^{-1}}{(m^2n)^{-3}}$

91. $\left(\frac{2a^3}{6b}\right)^4$ 92. $\left(\frac{-3a^2}{9b^3}\right)^4$

93. $\frac{8x^{-3}y^{-2}}{4x^{-2}y^{-4}}$ 94. $\frac{6x^{-1}y^{-1}}{9x^{-2}y^3}$

95. $\frac{(r^2t^2)^{-2}}{(r^3t)^{-1}}$ 96. $\frac{(2rt)^2}{(rt^4)^{-2}}$

97. $\frac{4x^{-2}y^3}{(2x^{-1}y)^2}$ 98. $\frac{(ab)^3}{a^4b^{-4}}$

99. $\left(\frac{15r^2t}{3r^{-3}t^4}\right)^3$ 100. $\left(\frac{4(xy)^2}{(2xy^{-2})^3}\right)^{-2}$

R.4 Polynomial Expressions

- Perform addition and subtraction on monomials
- Perform addition and subtraction on polynomials
- Apply the distributive property
- Perform multiplication on polynomials
- Find the product of a sum and difference
- Square a binomial

Addition and Subtraction of Monomials

A term is a number, a variable, or a *product* of numbers and variables raised to powers. Examples of terms include

$$-15, \quad y, \quad x^4, \quad 3x^3z, \quad x^{-1/2}y^{-2}, \quad \text{and} \quad 6x^{-1}y^3.$$

If the variables in a term have only *nonnegative integer* exponents, the term is called a *monomial*. Examples of monomials include

$$-4, \quad 5y, \quad x^2, \quad 5x^2z^6, \quad -xy^7, \quad \text{and} \quad 6xy^3.$$

To learn how to add monomials, consider the following example. Suppose that we have 3 rectangles of the same dimension having length x and width y, as shown in Figure R.26. The total area is given by

$$xy + xy + xy.$$

This sum is equivalent to 3 times xy, which can be expressed as $3xy$. In symbols we write

$$xy + xy + xy = 3xy.$$

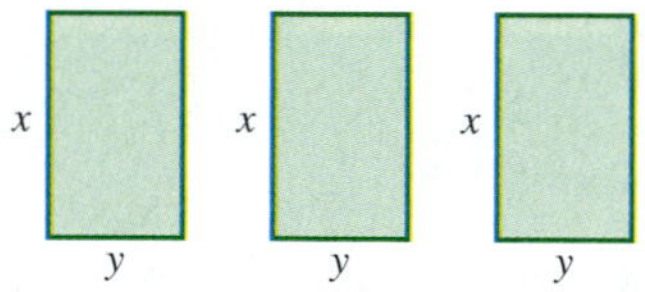

FIGURE R.26 Total Area: $3xy$

We can add these three terms because they are like terms. If two terms contain the same variables raised to the same powers, we call them like terms. We can add or subtract *like* terms, but not *unlike* terms. For example, if one cube has sides of length x, and another cube has sides of length y, their respective volumes are x^3 and y^3. The total volume of the two cubes equals

$$x^3 + y^3,$$

but we cannot combine these terms into one term because they are unlike terms. However,

$$x^3 + 3x^3 = (1 + 3)x^3 = 4x^3$$

because x^3 and $3x^3$ are like terms. To add or subtract monomials we simply combine like terms as illustrated in the next example.

EXAMPLE 1 Adding and subtracting monomials

Simplify each of the following expressions by combining like terms.
(a) $8x^2 - 4x^2 + x^3$ (b) $9x - 6xy^2 + 2xy^2 + 4x$

SOLUTION
(a) The terms $8x^2$ and $-4x^2$ are like terms, so they may be combined.

$$8x^2 - 4x^2 + x^3 = (8 - 4)x^2 + x^3 \qquad \text{Combine like terms.}$$
$$= 4x^2 + x^3 \qquad \text{Subtract.}$$

However, $4x^2$ and x^3 are unlike terms and cannot be combined.

(b) The terms $9x$ and $4x$ may be combined, as can $-6xy^2$ and $2xy^2$.

$$9x - 6xy^2 + 2xy^2 + 4x = 9x + 4x - 6xy^2 + 2xy^2 \quad \textit{Commutative property}$$
$$= (9 + 4)x + (-6 + 2)xy^2 \quad \textit{Combine like terms.}$$
$$= 13x - 4xy^2 \quad \textit{Add.}$$

Now Try Exercises 13 and 19 ◆

Addition and Subtraction of Polynomials

A polynomial is either a monomial or a sum of monomials. Examples of polynomials include

$$5x^4z^2, \qquad 9x^4 - 5, \qquad 4x^2 + 5xy - y^2, \qquad \text{and} \qquad 4 - y^2 + 5y^4 + y^5$$

1 term   2 terms   3 terms   4 terms

    Polynomials containing one variable are called polynomials of one variable. The second and fourth polynomials shown above are examples of polynomials of one variable. The leading coefficient of a polynomial of one variable is the coefficient of the monomial with highest degree. The degree of a polynomial with one variable equals the exponent of the monomial with the highest power. Table R.3 shows several polynomials of one variable along with their degrees and leading coefficients. A polynomial of degree 1 is a linear polynomial, a polynomial of degree 2 is a quadratic polynomial, and a polynomial of degree 3 is a cubic polynomial.

TABLE R.3

Polynomial	Degree	Leading Coefficient	Type
-98	0	-98	Constant
$2x - 7$	1	2	Linear
$-5z + 9z^2 + 7$	2	9	Quadratic
$-2x^3 + 4x^2 + x - 1$	3	-2	Cubic
$7 - x + 4x^2 + x^5$	5	1	Fifth Degree

    To add two polynomials, we combine like terms.

EXAMPLE 2   Adding polynomials

Simplify each expression.
(a) $(2x^2 - 3x + 7) + (3x^2 + 4x - 2)$ **(b)** $(z^3 + 4z + 8) + (4z^2 - z + 6)$

SOLUTION
(a) $(2x^2 - 3x + 7) + (3x^2 + 4x - 2) = 2x^2 + 3x^2 - 3x + 4x + 7 - 2$
$$= (2 + 3)x^2 + (-3 + 4)x + (7 - 2)$$
$$= 5x^2 + x + 5$$

(b) $(z^3 + 4z + 8) + (4z^2 - z + 6) = z^3 + 4z^2 + 4z - z + 8 + 6$
$$= z^3 + 4z^2 + (4 - 1)z + (8 + 6)$$
$$= z^3 + 4z^2 + 3z + 14$$

Now Try Exercises 29 and 35 ◆

To subtract integers we add the first integer with the *additive inverse* or *opposite* of the second integer. For example, to evaluate $3 - 5$ we perform the following operations.

$$3 - 5 = 3 + (-5) \qquad \text{Add the opposite.}$$
$$= -2 \qquad \text{Simplify.}$$

Similarly, to subtract two polynomials we add the first polynomial and the opposite of the second polynomial. To find the opposite of a polynomial, we simply negate each term. Table R.4 shows three polynomials and their opposites.

TABLE R.4

Polynomial	Opposite
$9 - x$	$-9 + x$
$5x^2 + 4x - 1$	$-5x^2 - 4x + 1$
$-x^4 + 5x^3 - x^2 + 5x - 1$	$x^4 - 5x^3 + x^2 - 5x + 1$

EXAMPLE 3   Subtracting polynomials

Simplify.
(a) $(y^5 + 3y^3) - (-y^4 + 2y^3)$ **(b)** $(5x^3 + 9x^2 - 6) - (5x^3 - 4x^2 - 7)$

SOLUTION
(a) The opposite of $(-y^4 + 2y^3)$ is $(y^4 - 2y^3)$.

$$(y^5 + 3y^3) - (-y^4 + 2y^3) = (y^5 + 3y^3) + (y^4 - 2y^3)$$
$$= y^5 + y^4 + (3 - 2)y^3$$
$$= y^5 + y^4 + y^3$$

(b) The opposite of $(5x^3 - 4x^2 - 7)$ is $(-5x^3 + 4x^2 + 7)$

$$(5x^3 + 9x^2 - 6) - (5x^3 - 4x^2 - 7) = (5x^3 + 9x^2 - 6) + (-5x^3 + 4x^2 + 7)$$
$$= (5 - 5)x^3 + (9 + 4)x^2 + (-6 + 7)$$
$$= 0x^3 + 13x^2 + 1$$
$$= 13x^2 + 1 \qquad \textit{Now Try Exercises 47 and 49} \blacklozenge$$

Distributive Properties

Distributive properties are used frequently in the multiplication of polynomials. For all real numbers a, b, and c

$$a(b + c) = ab + ac \qquad \text{and}$$
$$a(b - c) = ab - ac.$$

In the next example we use these distributive properties to multiply expressions.

EXAMPLE 4   Using distributive properties

Multiply.
(a) $4(5 + x)$ **(b)** $-3(x - 4y)$ **(c)** $(2x - 5)(6)$

SOLUTION

(a) $4(5 + x) = 4 \cdot 5 + 4 \cdot x = 20 + 4x$

(b) $-3(x - 4y) = -3 \cdot x - (-3) \cdot (4y) = -3x + 12y$

(c) $(2x - 5)(6) = 2x \cdot 6 - 5 \cdot 6 = 12x - 30$

Now Try Exercises 55, 59 and 61 ◆

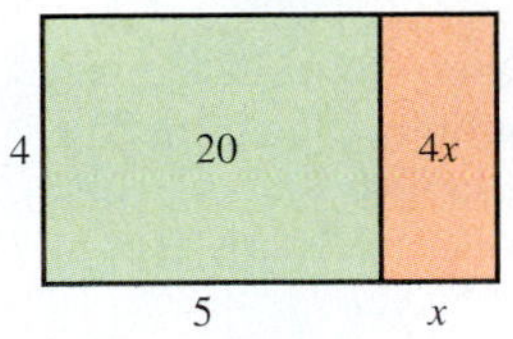

FIGURE R.27 Area: $20 + 4x$

You can visualize the solution in part (a) of the preceding example by using areas of rectangles. If a rectangle has width 4 and length $5 + x$, its area is $20 + 4x$, as shown in Figure R.27.

Multiplying Polynomials

A polynomial with two terms is a binomial and a polynomial with three terms is a trinomial. Examples are shown in Table R.5.

TABLE R.5

Monomials	$2x^2$	$-3x^4$	9
Binomials	$3x - 1$	$2x^3 - x$	$x^2 + 5$
Trinomials	$x^2 - 3x + 5$	$5x^2 - 2x + 10$	$2x^3 - x^2 - 2$

In the next example we multiply two binomials.

EXAMPLE 5 Multiplying binomials

Multiply $(x + 1)(x + 3)$.

SOLUTION To multiply $(x + 1)(x + 3)$ symbolically, we apply the distributive property.

$$\begin{aligned}(x + 1)(x + 3) &= (x + 1)(x) + (x + 1)(3) \\ &= x \cdot x + 1 \cdot x + x \cdot 3 + 1 \cdot 3 \\ &= x^2 + x + 3x + 3 \\ &= x^2 + 4x + 3\end{aligned}$$

Now Try Exercise 63 ◆

To multiply $(x + 1)$ by $(x + 3)$, we multiplied every term in $x + 1$ by every term in $x + 3$. That is,

$$\begin{aligned}(x + 1)(x + 3) &= x^2 + 3x + x + 3 \\ &= x^2 + 4x + 3.\end{aligned}$$

Note: This process of multiplying binomials is called *FOIL*. You may use it to remind yourself to multiply the first terms (F), outside terms (O), inside terms (I), and last terms (L).

Multiply the *First terms* to obtain x^2. $(x + 1)(x + 3)$

Multiply the *Outside terms* to obtain $3x$. $(x + 1)(x + 3)$

Multiply the *Inside terms* to obtain x. $(x + 1)(x + 3)$

Multiply the *Last terms* to obtain 3. $(x + 1)(x + 3)$

The following method summarizes how to multiply two polynomials in general.

> ### MULTIPLYING POLYNOMIALS
> The product of two polynomials may be found by multiplying every term in the first polynomial by every term in the second polynomial and then combining like terms.

EXAMPLE 6   Multiplying polynomials

Multiply each binomial.
(a) $(2x - 1)(x + 2)$ **(b)** $(1 - 3x)(2 - 4x)$ **(c)** $(x^2 + 1)(5x - 3)$

SOLUTION

(a) $(2x - 1)(x + 2) = 2x \cdot x + 2x \cdot 2 - 1 \cdot x - 1 \cdot 2$
$$= 2x^2 + 4x - x - 2$$
$$= 2x^2 + 3x - 2$$

(b) $(1 - 3x)(2 - 4x) = 1 \cdot 2 - 1 \cdot 4x - 3x \cdot 2 + 3x \cdot 4x$
$$= 2 - 4x - 6x + 12x^2$$
$$= 2 - 10x + 12x^2$$

(c) $(x^2 + 1)(5x - 3) = x^2 \cdot 5x - x^2 \cdot 3 + 1 \cdot 5x - 1 \cdot 3$
$$= 5x^3 - 3x^2 + 5x - 3 \qquad \textit{Now Try Exercises 63, 83, and 85} \blacklozenge$$

EXAMPLE 7   Multiplying polynomials

Multiply each expression.
(a) $3x(x^2 + 5x - 4)$ **(b)** $-x^2(x^4 - 2x + 5)$ **(c)** $(x + 2)(x^2 + 4x - 3)$

SOLUTION

(a) $3x(x^2 + 5x - 4) = 3x \cdot x^2 + 3x \cdot 5x - 3x \cdot 4$
$$= 3x^3 + 15x^2 - 12x$$

(b) $-x^2(x^4 - 2x + 5) = -x^2 \cdot x^4 + x^2 \cdot 2x - x^2 \cdot 5$
$$= -x^6 + 2x^3 - 5x^2$$

(c) $(x + 2)(x^2 + 4x - 3) = x \cdot x^2 + x \cdot 4x - x \cdot 3 + 2 \cdot x^2 + 2 \cdot 4x - 2 \cdot 3$
$$= x^3 + 4x^2 - 3x + 2x^2 + 8x - 6$$
$$= x^3 + 6x^2 + 5x - 6 \qquad \textit{Now Try Exercises 75, 77, and 81} \blacklozenge$$

Some Special Products

The following special product often occurs in mathematics.

$$(a - b)(a + b) = a \cdot a + a \cdot b - b \cdot a - b \cdot b$$
$$= a^2 + ab - ba - b^2$$
$$= a^2 - b^2$$

That is, the product of a sum and difference equals the difference of their squares.

EXAMPLE 8   Finding the product of a sum and difference

Multiply.

(a) $(x - 3)(x + 3)$ (b) $(5 + 4x^2)(5 - 4x^2)$

SOLUTION

(a) If we let $a = x$ and $b = 3$, we can apply the rule

$$(a - b)(a + b) = a^2 - b^2.$$

Thus

$$(x - 3)(x + 3) = (x)^2 - (3)^2$$
$$= x^2 - 9.$$

(b) Similarly, we can multiply as follows.

$$(5 + 4x^2)(5 - 4x^2) = (5)^2 - (4x^2)^2$$
$$= 25 - 16x^4$$

Now Try Exercises 87 and 103 ◆

Two other special products involve *squaring a binomial*:

$$(a + b)^2 = (a + b)(a + b)$$
$$= a^2 + ab + ba + b^2$$
$$= a^2 + 2ab + b^2$$

and

$$(a - b)^2 = (a - b)(a - b)$$
$$= a^2 - ab - ba + b^2$$
$$= a^2 - 2ab + b^2.$$

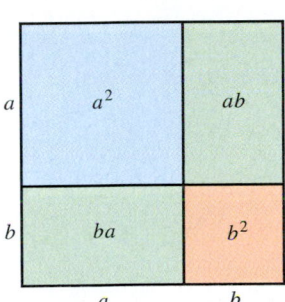

FIGURE R.28
$(a + b)^2 = a^2 + 2ab + b^2$

The first product is illustrated geometrically in Figure R.28, where each side of a square has length $(a + b)$. The area of the square is

$$(a + b)(a + b) = (a + b)^2.$$

This area can also be computed by adding the area of the four small rectangles.

$$a^2 + ab + ba + b^2 = a^2 + 2ab + b^2$$

Thus $(a + b)^2 = a^2 + 2ab + b^2$. Note that to obtain the middle term, we multiply the two terms in the binomial and double the result.

EXAMPLE 9   Squaring a binomial

Multiply.

(a) $(x + 5)^2$ (b) $(3 - 2x)^2$

SOLUTION

(a) If we let $a = x$ and $b = 5$, we can apply the formula

$$(a + b)^2 = a^2 + 2ab + b^2.$$

Thus

$$(x + 5)^2 = (x)^2 + 2(x)(5) + (5)^2$$ *To find the middle term, multiply*
$$= x^2 + 10x + 25.$$ *a and b and double the result.*

(b) Applying the formula $(a - b)^2 = a^2 - 2ab + b^2$, we find

$$(3 - 2x)^2 = (3)^2 - 2(3)(2x) + (2x)^2$$
$$= 9 - 12x + 4x^2.$$ *Now Try Exercises 93 and 99* ◆

Note: If you forget these special products, you can still use techniques learned earlier to multiply the polynomials in Examples 8 and 9. For example,

$$(3 - 2x)^2 = (3 - 2x)(3 - 2x)$$
$$= 3 \cdot 3 - 3 \cdot 2x - 2x \cdot 3 + 2x \cdot 2x$$
$$= 9 - 6x - 6x + 4x^2$$
$$= 9 - 12x + 4x^2.$$

R.4 — Exercises

Concepts

1. Give an example of a term that is a monomial.

2. What are the degree and leading coefficient of the polynomial $3x^2 - x^3 + 1$?

3. Are $-5x^3 y$ and $6xy^3$ like terms? Explain.

4. Give an example of a polynomial that has 3 terms and is degree 4.

5. Does the opposite of $x^2 + 1$ equal $-x^2 + 1$? Explain.

6. The equation $5(x - 4) = 5x - 20$ illustrates what property?

7. $(a - b)(a + b) = $ ____   8. $(a + b)^2 = $ ____

Monomials and Polynomials

Exercises 9–20: Combine like terms whenever possible.

9. $3x^3 + 5x^3$ 10. $-9z + 6z$

11. $5y^7 - 8y^7$ 12. $9x - 7x$

13. $5x^2 + 8x + x^2$ 14. $5x + 2x + 10x$

15. $9x^2 - x + 4x - 6x^2$ 16. $-y^2 - \frac{1}{2}y^2$

17. $x^2 + 9x - 2 + 4x^2 + 4x$

18. $6y + 4y^2 - 6y + y^2$ 19. $7y + 9x^2 y - 5y + x^2 y$

20. $5ab - b^2 + 7ab + 6b^2$

Exercises 21–26: Identify the degree and leading coefficient of the polynomial.

21. $5x^2 - 4x + \frac{3}{4}$ 22. $-9y^4 + y^2 + 5$

23. $5 - x + 3x^2 - \frac{2}{5}x^3$ 24. $7x + 4x^4 - \frac{4}{3}x^3$

25. $8x^4 + 3x^3 - 4x + x^5$ 26. $5x^2 - x^3 + 7x^4 + 10$

Exercises 27–36: Add the polynomials.

27. $(5x + 6) + (-2x + 6)$

28. $(5y^2 + y^3) + (12y^2 - 5y^3)$

29. $(2x^2 - x + 7) + (-2x^2 + 4x - 9)$

30. $(x^3 - 5x^2 + 6) + (5x^2 + 3x + 1)$

31. $(4x) + (1 - 4.5x)$

32. $(y^5 + y) + \left(5 - y + \frac{1}{3}y^2\right)$

33. $(x^4 - 3x^2 - 4) + \left(-8x^4 + x^2 - \frac{1}{2}\right)$

34. $(3z + z^4 + 2) + (-3z^4 - 5 + z^2)$

35. $(2z^3 + 5z - 6) + (z^2 - 3z + 2)$

36. $(z^4 - 6z^2 + 3) + (5z^3 + 3z^2 - 3)$

Exercises 37–42: Find the opposite of the polynomial.

37. $7x^3$ 38. $-3z^8$

39. $19z^5 - 5z^2 + 3z$ 40. $-x^2 - x + 6$

41. $z^4 - z^2 - 9$ 42. $1 - 8x + 6x^2 - \frac{1}{6}x^3$

Exercises 43–50: Subtract the polynomials.

43. $(5x - 3) - (2x + 4)$ 44. $(10x + 5) - (-6x - 4)$

45. $(x^2 - 3x + 1) - (-5x^2 + 2x - 4)$

46. $(-x^2 + x - 5) - (x^2 - x + 5)$

47. $(4x^4 + 2x^2 - 9) - (x^4 - 2x^2 - 5)$

48. $(8x^3 + 5x^2 - 3x + 1) - (-5x^3 + 6x - 11)$

49. $(x^4 - 1) - (4x^4 + 3x + 7)$

50. $(5x^4 - 6x^3 + x^2 + 5) - (x^3 + 11x^2 + 9x - 3)$

Exercises 51–62: Apply the distributive property.

51. $5x(x - 5)$ **52.** $3x^2(-2x + 2)$

53. $-5(3x + 1)$ **54.** $-(-3x + 1)$

55. $5(y + 2)$ **56.** $4(x - 7)$

57. $-2(5x + 9)$ **58.** $-3x(5 + x)$

59. $(y - 3)6y$ **60.** $(2x - 5)8x^3$

61. $-4(5x - y)$ **62.** $-6(3y - 2x)$

Exercises 63–74: Multiply the binomials.

63. $(y + 5)(y - 7)$ **64.** $(3x + 1)(2x + 1)$

65. $(7x - 3)(4 - 7x)$ **66.** $(3 - 2x)(3 + x)$

67. $(-2x + 3)(x - 2)$ **68.** $(z - 2)(4z + 3)$

69. $\left(x - \frac{1}{2}\right)\left(x + \frac{1}{4}\right)$ **70.** $\left(z - \frac{1}{3}\right)\left(z - \frac{1}{6}\right)$

71. $(x^2 + 1)(2x^2 - 1)$ **72.** $(x^2 - 2)(x^2 + 4)$

73. $(x + y)(x - 2y)$ **74.** $(x^2 + y^2)(x - y)$

Exercises 75–86: Multiply the polynomials.

75. $3x(2x^2 - x - 1)$ **76.** $-2x(3 - 2x + 5x^2)$

77. $-x(2x^4 - x^2 + 10)$ **78.** $-2x^2(5x^3 + x^2 - 2)$

79. $(2x^2 - 4x + 1)(3x^2)$

80. $(x - y + 5)(xy)$

81. $(x + 1)(x^2 + 2x - 3)$

82. $(2x - 1)(3x^2 - x + 6)$

83. $(2 - 3x)(5 - 2x)$

84. $(3 + z)(6 - 4z)$

85. $(x^2 + 2)(3x - 2)$

86. $(4 + x)(2x^2 - 3)$

Exercises 87–104: Multiply the expressions.

87. $(x - 7)(x + 7)$ **88.** $(x + 9)(x - 9)$

89. $(3x + 4)(3x - 4)$ **90.** $(9x - 4)(9x + 4)$

91. $(2x - 3y)(2x + 3y)$ **92.** $(x + 2y)(x - 2y)$

93. $(x + 4)^2$ **94.** $(z + 9)^2$

95. $(2x + 1)^2$ **96.** $(3x + 5)^2$

97. $(x - 1)^2$ **98.** $(x - 7)^2$

99. $(2 - 3x)^2$ **100.** $(5 - 6x)^2$

101. $3x(x + 1)(x - 1)$ **102.** $-4x(3x - 5)^2$

103. $(2 - 5x^2)(2 + 5x^2)$ **104.** $(6y - x^2)(6y + x^2)$

R.5 Factoring Polynomials

- ◆ Use common factors
- ◆ Factor by grouping
- ◆ Factor $x^2 + bx + c$
- ◆ Factor trinomials by grouping
- ◆ Factor trinomials with FOIL
- ◆ Factor the difference of two squares
- ◆ Factor perfect square trinomials
- ◆ Factor the sum and difference of two cubes

Common Factors

When factoring a polynomial, we first look for factors that are common to each term in an expression. By applying a distributive property, we can write a polynomial as two factors. For example, each term in $2x^2 + 4x$ contains a factor of $2x$.

$$2x^2 = \mathbf{2x} \cdot x$$

$$4x = \mathbf{2x} \cdot 2$$

Thus the polynomial $2x^2 + 4x$ can be factored as follows.

$$2x^2 + 4x = \mathbf{2x}(x + 2)$$

EXAMPLE 1   Finding common factors

Factor.

(a) $4x^2 + 5x$ **(b)** $6z^3 - 2z^2 + 4z$ **(c)** $4x^3y^2 + x^2y^3$

SOLUTION

(a) Both $4x^2$ and $5x$ contain a common factor of x. That is,

$$4x^2 = \textbf{\textit{x}} \cdot \textbf{4\textit{x}} \quad \text{and} \quad 5x = \textbf{\textit{x}} \cdot \textbf{5}.$$

Thus $4x^2 + 5x = \textbf{\textit{x}}(\textbf{4\textit{x} + 5})$.

(b) Each of the terms $6z^3$, $2z^2$, and $4z$ contains a common factor of $2z$. That is,

$$6z^3 = \textbf{2\textit{z}} \cdot \textbf{3\textit{z}}^2 \quad 2z^2 = \textbf{2\textit{z}} \cdot \textit{z}, \quad \text{and} \quad 4z = \textbf{2\textit{z}} \cdot \textbf{2}.$$

Thus $6z^3 - 2z^2 + 4z = \textbf{2\textit{z}}(\textbf{3\textit{z}}^2 - \textit{z} + \textbf{2})$.

(c) Both $4x^3y^2$ and x^2y^3 contain a common factor of x^2y^2. That is,

$$4x^3y^2 = \textbf{\textit{x}}^2\textbf{\textit{y}}^2 \cdot \textbf{4\textit{x}} \quad \text{and} \quad x^2y^3 = \textbf{\textit{x}}^2\textbf{\textit{y}}^2 \cdot \textbf{\textit{y}}.$$

Thus $4x^3y^2 + x^2y^3 = \textbf{\textit{x}}^2\textbf{\textit{y}}^2(\textbf{4\textit{x} + \textit{y}})$.   *Now Try Exercises 3, 5, and 15* ◆

Many times we factor out the *greatest common factor*. For example, the polynomial $15x^4 - 5x^2$ has a common factor of $5x$. We could write this polynomial as

$$15x^4 - 5x^2 = 5x(3x^3 - x).$$

However, we can also factor out $5x^2$ to obtain

$$15x^4 - 5x^2 = 5x^2(3x^2 - 1).$$

Because $5x^2$ is the common factor with the highest degree and largest coefficient, we say that $5x^2$ is the **greatest common factor** (GCF) of $15x^4 - 5x^2$.

EXAMPLE 2   Factoring greatest common factors

Factor.

(a) $6m^3n^2 - 3mn^2 + 9m$ **(b)** $-9x^3 + 6x^2 - 3x$

SOLUTION

(a) The GCF of $6m^3n^2$, $3mn^2$, and $9m$ is $3m$.

$$6m^3n^2 = \textbf{3\textit{m}} \cdot \textbf{2\textit{m}}^2\textbf{\textit{n}}^2, \qquad 3mn^2 = \textbf{3\textit{m}} \cdot \textbf{\textit{n}}^2, \quad \text{and} \quad 9m = \textbf{3\textit{m}} \cdot \textbf{3}$$

Thus $6m^3n^2 - 3mn^2 + 9m = \textbf{3\textit{m}}(\textbf{2\textit{m}}^2\textbf{\textit{n}}^2 - \textbf{\textit{n}}^2 + \textbf{3})$.

(b) Rather than factoring out $3x$, we can also factor out $-3x$ and make the leading coefficient of the remaining expression positive.

$$-9x^3 = \textbf{-3\textit{x}} \cdot \textbf{3\textit{x}}^2, \quad 6x^2 = \textbf{-3\textit{x}} \cdot \textbf{-2\textit{x}}, \quad \text{and} \quad -3x = \textbf{-3\textit{x}} \cdot \textbf{1}$$

Thus $-9x^3 + 6x^2 - 3x = \textbf{-3\textit{x}}(\textbf{3\textit{x}}^2 - \textbf{2\textit{x} + 1})$.   *Now Try Exercises 11 and 17* ◆

Factoring by Grouping

Factoring by grouping is a technique that makes use of the associative and distributive properties. The next example illustrates the first step in this factoring technique.

Consider the polynomial

$$3t^3 + 6t^2 + 2t + 4.$$

We can factor this polynomial by first grouping it into two binomials.

$$(3t^3 + 6t^2) + (2t + 4)$$ Associative property

$$3t^2(t + 2) + 2(t + 2)$$ Factor out common factors.

$$(3t^2 + 2)(t + 2)$$ $(t + 2)$

The following steps summarize factoring four terms by grouping.

FACTORING BY GROUPING

STEP 1: Use parentheses to group the terms into binomials with common factors. Begin by writing the expression with a plus sign between the binomials.

STEP 2: Factor out the common factor in each binomial.

STEP 3: Factor out the common binomial. If there is no common binomial, try a different grouping.

EXAMPLE 3   Factoring by grouping

Factor each polynomial.
(a) $12x^3 - 9x^2 - 8x + 6$ (b) $2x - 2y + ax - ay$

SOLUTION

(a) $12x^3 - 9x^2 - 8x + 6 = (12x^3 - 9x^2) + (-8x + 6)$ Write with a plus sign between binomials.

$$= 3x^2(4x - 3) - 2(4x - 3)$$ Factor out $3x^2$ and -2.

$$= (3x^2 - 2)(4x - 3)$$ Factor out $4x - 3$.

(b) $2x - 2y + ax - ay \quad = (2x - 2y) + (ax - ay)$ Group terms.

$$= 2(x - y) + a(x - y)$$ Factor out 2 and a.

$$= (2 + a)(x - y)$$ Factor out $x - y$.

Now Try Exercises 21 and 31 ◆

Factoring $x^2 + bx + c$

The product $(x + 3)(x + 4)$ can be found as follows.

$$(x + 3)(x + 4) = x^2 + 4x + 3x + 12$$
$$= x^2 + \quad 7x \quad + 12$$

The middle term $7x$ is found by calculating the sum $4x + 3x$, and the last term is found by calculating the product $3 \cdot 4 = 12$.

When we factor polynomials, we are *reversing* the process of multiplication. To factor $x^2 + 7x + 12$ we must find m and n that satisfy

$$x^2 + 7x + 12 = (x + m)(x + n).$$

Because

$$(x + m)(x + n) = x^2 + (m + n)x + mn,$$

it follows that $mn = 12$ and $m + n = 7$. To determine m and n we list factors of 12 and their sum, as shown in Table R.6.

TABLE R.6 Factor Pairs for 12

Factors	1, 12	2, 6	3, 4
Sum	13	8	7

Because $3 \cdot 4 = 12$ and $3 + 4 = 7$, we can write the factored form as

$$x^2 + 7x + 12 = (x + 3)(x + 4).$$

This result can always be checked by multiplying the two binomials.

$$(x + 3)(x + 4) = x^2 + 7x + 12$$

$$\underbrace{ \overset{3x}{} }$$

$$\underline{+4x}$$

$$\overline{7x} \quad \longleftarrow \text{The middle term checks.}$$

FACTORING $x^2 + bx + c$

To factor the trinomial $x^2 + bx + c$, find integers m and n that satisfy

$$m \cdot n = c \quad \text{and} \quad m + n = b.$$

Then $x^2 + bx + c = (x + m)(x + n)$.

EXAMPLE 4   **Factoring the form $x^2 + bx + c$**

Factor each trinomial.
(a) $x^2 + 10x + 16$ **(b)** $x^2 - 5x - 24$ **(c)** $x^2 + 7x - 30$

SOLUTION

(a) We need to find a factor pair for 16 whose sum is 10. From Table R.7 the required factor pair is $m = \mathbf{2}$ and $n = \mathbf{8}$. Thus

$$x^2 + 10x + 16 = (x + \mathbf{2})(x + \mathbf{8}).$$

TABLE R.7   Factor Pairs for 16

Factors	1, 16	**2, 8**	4, 4
Sum	17	**10**	8

(b) Factors of -24 whose sum equals -5 are $\mathbf{3}$ and $\mathbf{-8}$. Thus

$$x^2 - 5x - 24 = (x + \mathbf{3})(x - \mathbf{8}).$$

(c) Factors of -30 whose sum equals 7 are $\mathbf{-3}$ and $\mathbf{10}$. Thus

$$x^2 + 7x - 30 = (x - \mathbf{3})(x + \mathbf{10}).$$

Now Try Exercises 33 and 37 ◆

EXAMPLE 5   **Removing common factors first**

Factor completely.
(a) $3x^2 + 15x + 18$ **(b)** $5x^3 + 5x^2 - 60x$

SOLUTION

(a) If we first factor out the common factor of 3, the resulting trinomial is easier to factor.

$$3x^2 + 15x + 18 = 3(x^2 + 5x + 6)$$

Now we find m and n such that $mn = 6$ and $m + n = 5$. These numbers are 2 and 3.

$$3x^2 + 15x + 18 = 3(x^2 + 5x + 6)$$
$$= 3(x + 2)(x + 3)$$

(b) First, we factor out the common factor of $5x$. Then we factor the resulting trinomial.

$$5x^3 + 5x^2 - 60x = 5x(x^2 + x - 12)$$
$$= 5x(x - 3)(x + 4)$$

Now Try Exercises 53 and 57 ◆

Factoring Trinomials by Grouping

In this subsection we use grouping to factor trinomials in the form $ax^2 + bx + c$ with $a \neq 1$. For example, one way to factor $3x^2 + 14x + 8$ is to find two numbers m and n such that $mn = 3 \cdot 8 = 24$ and $m + n = 14$. Because $2 \cdot 12 = 24$ and $2 + 12 = 14$, $m = 2$ and $n = 12$. Using grouping, we can now factor this trinomial.

$$3x^2 + 14x + 8 = 3x^2 + 2x + 12x + 8 \qquad \text{Write } 14x \text{ as } 2x + 12x.$$
$$= (3x^2 + 2x) + (12x + 8) \qquad \text{Associative property}$$
$$= x(\mathbf{3x + 2}) + 4(\mathbf{3x + 2}) \qquad \text{Factor out } x \text{ and } 4.$$
$$= (x + 4)(\mathbf{3x + 2}) \qquad \text{Distributive property}$$

> ### FACTORING $ax^2 + bx + c$ BY GROUPING
>
> To factor $ax^2 + bx + c$ perform the following steps. (Assume that a, b, and c have no factor in common.)
>
> 1. Find numbers m and n such that $mn = ac$ and $m + n = b$. This step may require trial and error.
> 2. Write the trinomial as $ax^2 + mx + nx + c$.
> 3. Use grouping to factor this expression as two binomials.

 **EXAMPLE 6** **Factoring $ax^2 + bx + c$ by grouping**

Factor each trinomial.
(a) $12y^2 + 5y - 3$ **(b)** $6r^2 - 19r + 10$

SOLUTION

(a) In this trinomial $a = 12$, $b = 5$, and $c = -3$. Because $mn = ac$ and $m + n = b$, the numbers m and n satisfy $mn = -36$ and $m + n = 5$. Thus $m = 9$ and $n = -4$.

$$12y^2 + 5y - 3 = 12y^2 + 9y - 4y - 3 \qquad \text{Write } 5y \text{ as } 9y - 4y.$$
$$= (12y^2 + 9y) + (-4y - 3) \qquad \text{Associative property}$$
$$= 3y(4y + 3) - 1(4y + 3) \qquad \text{Factor out } 3y \text{ and } -1.$$
$$= (3y - 1)(4y + 3) \qquad \text{Distributive property}$$

(b) In this trinomial $a = 6$, $b = -19$, and $c = 10$. Because $mn = ac$ and $m + n = b$, the numbers m and n satisfy $mn = 60$ and $m + n = -19$. Thus $m = -4$ and $n = -15$.

$$
\begin{aligned}
6r^2 - 19r + 10 &= 6r^2 - 4r - 15r + 10 && \text{Write } -19r \text{ as } -4r - 15r. \\
&= (6r^2 - 4r) + (-15r + 10) && \text{Associative property} \\
&= 2r(3r - 2) - 5(3r - 2) && \text{Factor out } 2r \text{ and } -5. \\
&= (2r - 5)(3r - 2) && \text{Distributive property}
\end{aligned}
$$

Now Try Exercises 41 and 43. ◆

Factoring Trinomials with FOIL

An alternative to factoring trinomials by grouping is to use FOIL in reverse. For example, the factors of $3x^2 + 7x + 2$ are two binomials.

$$3x^2 + 7x + 2 \overset{?}{=} (\underline{\quad} + \underline{\quad})(\underline{\quad} + \underline{\quad})$$

The expressions to be placed in the four blanks are yet to be found. By the FOIL method, we know that the product of the first terms is $3x^2$. Because $3x^2 = 3x \cdot x$, we can write

$$3x^2 + 7x + 2 \overset{?}{=} (\underline{\,3x\,} + \underline{\quad})(\underline{\,x\,} + \underline{\quad}).$$

The product of the last terms in each binomial must equal 2. Because $2 = 1 \cdot 2$, we can put the 1 and 2 in the blanks, but we must be sure to place them correctly so that the product of the *outside terms* plus the product of the *inside terms* equals $7x$.

If we had interchanged the 1 and 2, we would have obtained an incorrect result.

In the next example we factor expressions of the form $ax^2 + bx + c$, where $a \neq 1$. In this situation, we may need to *guess and check* or use *trial and error* a few times before finding the correct factors.

EXAMPLE 7 Factoring the form $ax^2 + bx + c$

Factor each trinomial.
(a) $6x^2 - x - 2$ **(b)** $4x^3 - 14x^2 + 6x$

SOLUTION
(a) The factors of $6x^2$ are either $2x$ and $3x$ or $6x$ and x. The factors of -2 are either -1 and 2 or 1 and -2. To obtain a middle term of $-x$ we use the following factors.

To find the correct factorization we may need to guess and check a few times.

(b) Each term contains a common factor of $2x$, so we do the following step first.

$$4x^3 - 14x^2 + 6x = 2x(2x^2 - 7x + 3)$$

Next we factor $2x^2 - 7x + 3$. The factors of $2x^2$ are $2x$ and x. Because the middle term is negative, we use -1 and -3 for factors of 3.

$$4x^3 - 14x^2 + 6x = 2x(2x^2 - 7x + 3)$$
$$= 2x(2x - 1)(x - 3)$$

Now Try Exercises 42 and 55 ◆

Difference of Two Squares

When we factor polynomials, we are *reversing* the process of multiplying polynomials. In Section R.4 we discussed the equation

$$(a - b)(a + b) = a^2 - b^2.$$

We can use this equation to factor a difference of two squares. For example, if we want to factor $x^2 - 25$, we can substitute x for a and **5** for b in

$$a^2 - b^2 = (a - b)(a + b)$$

to get

$$x^2 - 5^2 = (x - 5)(x + 5).$$

DIFFERENCE OF TWO SQUARES

For any real numbers a and b,

$$a^2 - b^2 = (a - b)(a + b).$$

Note: The sum of two squares *cannot* be factored (using real numbers). For example, $x^2 + y^2$ cannot be factored, whereas $x^2 - y^2$ can be factored. It is important to remember that $x^2 + y^2 \neq (x + y)^2$.

EXAMPLE 8   Factoring the difference of two squares

Factor each polynomial, if possible.
(a) $9x^2 - 64$ **(b)** $4x^2 + 9y^2$ **(c)** $4a^3 - 4a$

SOLUTION
(a) Note that $9x^2 = (3x)^2$ and $64 = 8^2$.

$$9x^2 - 64 = (3x)^2 - (8)^2$$
$$= (3x - 8)(3x + 8)$$

(b) Because $4x^2 + 9y^2$ is the *sum* of two squares, it *cannot* be factored.
(c) Start by factoring out the common factor of $4a$.

$$4a^3 - 4a = 4a(a^2 - 1)$$
$$= 4a(a - 1)(a + 1)$$

Now Try Exercises 61, 65, and 69 ◆

EXAMPLE 9   Applying the difference of two squares

Factor each expression.
(a) $(n + 1)^2 - 9$ **(b)** $x^4 - y^4$ **(c)** $6r^2 - 24t^4$

SOLUTION
(a) Use $a^2 - b^2 = (a - b)(a + b)$, with $a = n + 1$ and $b = 3$.

$$(n + 1)^2 - 9 = (n + 1)^2 - 3^2 \qquad\qquad 9 = 3^2$$
$$= ((n + 1) - 3)((n + 1) + 3) \qquad \text{Difference of squares}$$
$$= (n - 2)(n + 4) \qquad \text{Combine terms.}$$

(b) Use $a^2 - b^2 = (a - b)(a + b)$, with $a = x^2$ and $b = y^2$.

$$x^4 - y^4 = (x^2)^2 - (y^2)^2 \qquad \text{Write as squares.}$$
$$= (x^2 - y^2)(x^2 + y^2) \qquad \text{Difference of squares}$$
$$= (x - y)(x + y)(x^2 + y^2) \qquad \text{Difference of squares}$$

(c) Start by factoring out the common factor of 6.

$$6r^2 - 24t^4 = 6(r^2 - 4t^4) \qquad \text{Factor out 6.}$$
$$= 6\left(r^2 - (2t^2)^2\right) \qquad \text{Write as squares.}$$
$$= 6(r - 2t^2)(r + 2t^2) \qquad \text{Difference of squares}$$

Now Try Exercises 66, 67, and 73. ◆

Perfect Square Trinomials

In Section R.4 we also showed how to expand $(a + b)^2$ and $(a - b)^2$ as follows.

$$(a + b)^2 = a^2 + 2ab + b^2 \quad \text{and} \quad (a - b)^2 = a^2 - 2ab + b^2$$

The expressions $a^2 + 2ab + b^2$ and $a^2 - 2ab + b^2$ are called **perfect square trinomials**. If we can recognize a perfect square trinomial, we can use these formulas to factor it.

PERFECT SQUARE TRINOMIALS

For any real numbers a and b,

$$a^2 + 2ab + b^2 = (a + b)^2 \quad \text{and}$$
$$a^2 - 2ab + b^2 = (a - b)^2.$$

EXAMPLE 10   Factoring perfect square trinomials

Factor.
(a) $x^2 + 6x + 9$ **(b)** $81x^2 - 72x + 16$

SOLUTION
(a) Let $a^2 = x^2$ and $b^2 = 3^2$. In a perfect square trinomial the middle term is $2ab$.

$$2ab = 2(x)(3) = 6x,$$

which equals the given middle term. Thus $a^2 + 2ab + b^2 = (a + b)^2$ implies

$$x^2 + \mathbf{6x} + 9 = (x + 3)^2.$$

(b) Let $a^2 = (9x)^2$ and $b^2 = 4^2$. In a perfect square trinomial the middle term is $2ab$.

$$2ab = 2(9x)(4) = 72x,$$

which equals the given middle term. Thus $a^2 - 2ab + b^2 = (a - b)^2$ implies

$$81x^2 - 72x + 16 = (9x - 4)^2.$$

Now Try Exercises 77 and 81 ◆

EXAMPLE 11 Factoring perfect square trinomials

Factor each expression.
(a) $9r^2 - 12rt + 4t^2$ **(b)** $25a^3 + 10a^2b + ab^2$

SOLUTION
(a) Let $a^2 = (3r)^2$ and $b^2 = (2t)^2$. To be a perfect square trinomial the middle term must equal $2ab$.

$$2ab = 2(3r)(2t) = 12rt,$$

which equals the given middle term. Thus $a^2 - 2ab + b^2 = (a - b)^2$ implies

$$9r^2 - 12rt + 4t^2 = (3r - 2t)^2.$$

(b) Start by factoring out the common factor of a. Then factor the resulting perfect square trinomial.

$$25a^3 + 10a^2b + ab^2 = a(25a^2 + 10ab + b^2)$$
$$= a(5a + b)^2$$

Now Try Exercises 87 and 89 ◆

Sum and Difference of Two Cubes

The sum or difference of two cubes may be factored. This fact is justified by the following two equations.

$$(a + b)(a^2 - ab + b^2) = a^3 + b^3 \quad \text{and}$$
$$(a - b)(a^2 + ab + b^2) = a^3 - b^3$$

These equations can be verified by multiplying the left side to obtain the right side. For example,

$$(a + b)(a^2 - ab + b^2) = a \cdot a^2 - a \cdot ab + a \cdot b^2 + b \cdot a^2 - b \cdot ab + b \cdot b^2$$
$$= a^3 - a^2b + ab^2 + a^2b - ab^2 + b^3$$
$$= a^3 + b^3.$$

SUM AND DIFFERENCE OF TWO CUBES

For any real numbers a and b,

$$a^3 + b^3 = (a + b)(a^2 - ab + b^2) \quad \text{and}$$
$$a^3 - b^3 = (a - b)(a^2 + ab + b^2).$$

EXAMPLE 12   Factoring the sum and difference of two cubes

Factor each polynomial.
(a) $x^3 + 8$ (b) $27x^3 - 64y^3$ (c) $x^6 + 8y^3$

SOLUTION
(a) Because $x^3 = (x)^3$ and $8 = 2^3$, we let $a = x$, $b = 2$, and factor. Substituting in
$$a^3 + b^3 = (a + b)(a^2 - ab + b^2)$$
gives
$$x^3 + 2^3 = (x + 2)(x^2 - x \cdot 2 + 2^2)$$
$$= (x + 2)(x^2 - 2x + 4).$$
Note that the quadratic factor does not factor further.
(b) Here, $27x^3 = (3x)^3$ and $64y^3 = (4y)^3$, so
$$27x^3 - 64y^3 = (3x)^3 - (4y)^3.$$
Substituting $a = 3x$ and $b = 4y$ in
$$a^3 - b^3 = (a - b)(a^2 + ab + b^2)$$
gives
$$(3x)^3 - (4y)^3 = (3x - 4y)((3x)^2 + 3x \cdot 4y + (4y)^2)$$
$$= (3x - 4y)(9x^2 + 12xy + 16y^2).$$
(c) Let $a^3 = (3p^3)^3$ and $b^3 = (2q^2)^3$. Then $a^3 - b^3 = (a - b)(a^2 + ab + b^2)$ implies
$$27p^9 - 8q^6 = (3p^3 - 2q^2)(9p^6 + 6p^3q^2 + 4q^4).$$

Now Try Exercises 91, 93 and 99◆

R.5 Exercises

Greatest Common Factor

Exercises 1–18: Factor out the greatest common factor.

1. $10x - 15$

2. $32 - 16x$

3. $2x^3 - 5x$

4. $3y - 9y^2$

5. $8x^3 - 4x^2 + 16x$

6. $-5x^3 + x^2 - 4x$

7. $5x^4 - 15x^3 + 15x^2$

8. $21y + 14y^3 - 7y^5$

9. $15x^3 + 10x^2 - 25x$

10. $14a^4 - 21a^2 + 35a$

11. $6r^5 - 8r^4 + 12r^3$

12. $15r^6 + 20r^4 - 10r^3$

13. $8x^2y^2 - 24x^2y^3$

14. $36xy - 24x^3y^3$

15. $18mn^2 - 12m^2n^3$

16. $24m^2n^3 + 12m^3n^2$

17. $-4a^2 - 2ab + 6ab^2$

18. $-5a^2 + 10a^2b^2 - 15ab$

Factoring by Grouping

Exercises 19–32: Use grouping to factor the polynomial.

19. $x^3 + 3x^2 + 2x + 6$

20. $4x^3 + 3x^2 + 8x + 6$

21. $6x^3 - 4x^2 + 9x - 6$

22. $x^3 - 3x^2 - 5x + 15$

23. $z^3 - 5z^2 + z - 5$

24. $y^3 - 7y^2 + 8y - 56$

25. $y^4 + 2y^3 - 5y^2 - 10y$

26. $4z^4 + 4z^3 + z^2 + z$

27. $2x^3 - 3x^2 + 2x - 3$

28. $8x^3 - 2x^2 + 12x - 3$

29. $2x^4 - x^3 + 4x - 2$ **30.** $2x^4 - 5x^3 + 10x - 25$

31. $2ax - 6bx - ay + 3by$ **32.** $ab - 3a + 2b - 6$

Factoring Trinomials

Exercises 33–58: Factor the expression completely.

33. $x^2 + 7x + 10$ **34.** $x^2 + 3x - 10$

35. $x^2 + 8x + 12$ **36.** $x^2 - 8x + 12$

37. $z^2 + z - 42$ **38.** $z^2 - 9z + 20$

39. $z^2 + 11z + 24$ **40.** $z^2 + 15z + 54$

41. $24x^2 + 14x - 3$ **42.** $25x^2 - 5x - 6$

43. $6x^2 - x - 2$ **44.** $10x^2 + 3x - 1$

45. $1 + x - 2x^2$ **46.** $3 - 5x - 2x^2$

47. $20 + 7x - 6x^2$ **48.** $4 + 13x - 12x^2$

49. $5x^3 + x^2 - 6x$ **50.** $2x^3 + 8x^2 - 24x$

51. $6x^3 + 21x^2 + 9x$ **52.** $12x^3 - 8x^2 - 20x$

53. $2x^2 - 14x + 20$ **54.** $7x^2 + 35x + 42$

55. $60t^4 + 230t^3 - 40t^2$ **56.** $24r^4 + 8r^3 - 80r^2$

57. $4m^3 + 10m^2 - 6m$ **58.** $30m^4 + 3m^3 - 9m^2$

Difference of Two Squares

Exercises 59–76: Factor the expression completely, if possible.

59. $x^2 - 25$ **60.** $z^2 - 169$

61. $4x^2 - 25$ **62.** $36 - y^2$

63. $36x^2 - 100$ **64.** $9x^2 - 4y^2$

65. $64z^2 - 25z^4$ **66.** $100x^3 - x$

67. $16x^4 - y^4$ **68.** $x^4 - 9y^2$

69. $a^2 + 4b^2$ **70.** $9r^4 + 25t^4$

71. $4 - r^2t^2$ **72.** $25 - x^4y^2$

73. $(x - 1)^2 - 16$ **74.** $(y + 2)^2 - 1$

75. $4 - (z + 3)^2$ **76.** $64 - (t - 3)^2$

Perfect Square Trinomials

Exercises 77–90: Factor the expression.

77. $x^2 + 2x + 1$ **78.** $x^2 - 6x + 9$

79. $4x^2 + 20x + 25$ **80.** $x^2 + 10x + 25$

81. $x^2 - 12x + 36$ **82.** $16z^4 - 24z^3 + 9z^2$

83. $9z^3 - 6z^2 + z$ **84.** $49y^2 + 42y + 9$

85. $9y^3 + 30y^2 + 25y$ **86.** $25y^3 - 20y^2 + 4y$

87. $4x^2 - 12xy + 9y^2$ **88.** $25a^2 + 60ab + 36b^2$

89. $9a^3b - 12a^2b + 4ab$ **90.** $16a^3 + 8a^2b + ab^2$

Sum and Difference of Two Cubes

Exercises 91–102: Factor the expression.

91. $x^3 - 1$ **92.** $x^3 + 1$

93. $y^3 + z^3$ **94.** $y^3 - z^3$

95. $8x^3 - 27$ **96.** $8 - z^3$

97. $x^4 + 125x$ **98.** $3x^4 - 81x$

99. $8r^6 - t^3$ **100.** $125r^6 + 64t^3$

101. $10m^9 - 270n^6$ **102.** $5t^6 + 40r^3$

General Factoring

Exercises 103–120: Factor the expression completely.

103. $16x^2 - 25$ **104.** $25x^2 - 30x + 9$

105. $x^3 - 64$ **106.** $1 + 8y^3$

107. $x^2 + 16x + 64$ **108.** $12x^2 + x - 6$

109. $5x^2 - 38x - 16$ **110.** $125x^3 - 1$

111. $x^4 + 8x$ **112.** $2x^3 - 12x^2 + 18x$

113. $64x^3 + 8y^3$ **114.** $54 - 16x^3$

115. $3x^2 - 5x - 8$ **116.** $15x^2 - 11x + 2$

117. $7a^3 + 20a^2 - 3a$ **118.** $b^3 - b^2 - 2b$

119. $2x^3 - x^2 + 6x - 3$ **120.** $3x^3 - 5x^2 + 3x - 5$

R.6 Rational Expressions

- ◆ Simplify rational expressions
- ◆ Perform multiplication and division on rational expressions
- ◆ Find least common factors and denominators
- ◆ Perform addition and subtraction on rational expressions
- ◆ Clear fractions from equations
- ◆ Simplify complex fractions

Simplifying Rational Expressions

When simplifying fractions, we sometimes use the **basic principle of fractions**, which states that

$$\frac{a \cdot c}{b \cdot c} = \frac{a}{b}.$$

This principle holds because $\frac{c}{c} = 1$ and $\frac{a}{b} \cdot 1 = \frac{a}{b}$. It can be used to simplify a fraction.

$$\frac{6}{44} = \frac{3 \cdot 2}{22 \cdot 2} = \frac{3}{22}$$

This same principle can also be used to simplify rational expressions. For example,

$$\frac{(z + 1)(z + 3)}{z(z + 3)} = \frac{z + 1}{z},$$

provided $z \neq -3$.

SIMPLIFYING RATIONAL EXPRESSIONS

The following principle can be used to simplify rational expressions, where A, B, and C are polynomials.

$$\frac{A \cdot C}{B \cdot C} = \frac{A}{B}, \quad B \text{ and } C \text{ are nonzero.}$$

EXAMPLE 1   Simplifying rational expressions

Simplify each expression.

(a) $\dfrac{9x}{3x^2}$ (b) $\dfrac{2z^2 - 3z - 9}{z^2 + 2z - 15}$ (c) $\dfrac{a^2 - b^2}{a + b}$

SOLUTION

(a) First factor out the greatest common factor, $3x$, in the numerator and denominator.

$$\frac{9x}{3x^2} = \frac{\mathbf{3x} \cdot 3}{\mathbf{3x} \cdot x} = \frac{3}{x}$$

(b) Start by factoring the numerator and denominator.

$$\frac{2z^2 - 3z - 9}{z^2 + 2z - 15} = \frac{(2z + 3)(\mathbf{z - 3})}{(z + 5)(\mathbf{z - 3})} = \frac{2z + 3}{z + 5}$$

(c) Start by factoring the numerator as the difference of squares.

$$\frac{a^2 - b^2}{a + b} = \frac{(a - b)(\mathbf{a + b})}{\mathbf{a + b}} = a - b$$

Now Try Exercises 1, 5, and 11 ◆

A negative sign can be placed in a fraction in a number of ways. For example,

$$-\frac{2}{3} = \frac{-2}{3} = \frac{2}{-3}$$

illustrates three fractions that are equal. This property can also be applied to rational expressions, as demonstrated in the next example.

EXAMPLE 2   Distributing a negative sign

Simplify each expression.

(a) $-\dfrac{1-z}{z-1}$ **(b)** $\dfrac{-y-2}{4y+8}$ **(c)** $\dfrac{5-x}{x-5}$

SOLUTION

(a) Start by distributing the negative sign over the numerator.

$$-\frac{1-z}{z-1} = \frac{-(1-z)}{z-1} = \frac{-1+z}{z-1} = \frac{z-1}{z-1} = 1$$

Note that the negative sign could also be distributed over the denominator.

(b) Use the distributive property to factor -1 out of the numerator and 4 out of the denominator.

$$\frac{-y-2}{4y+8} = \frac{-1(y+2)}{4(y+2)} = -\frac{1}{4}$$

(c) Start by factoring -1 out of the numerator.

$$\frac{5-x}{x-5} = \frac{-1(-5+x)}{x-5} = \frac{-1(x-5)}{x-5} = -1$$

 Now Try Exercises 4 and 9

Note: In general, $(b - a)$ equals $-1(a - b)$. As a result, if $a \neq b$, then

$$\frac{b-a}{a-b} = -1.$$

See Example 2(c).

Multiplication and Division of Rational Expressions

Multiplying and dividing rational expressions is similar to multiplying and dividing fractions.

> ### PRODUCTS AND QUOTIENTS OF RATIONAL EXPRESSIONS
>
> To multiply two rational expressions, multiply numerators and multiply denominators.
>
> $$\frac{A}{B} \cdot \frac{C}{D} = \frac{AC}{BD}, \qquad B \text{ and } D \text{ are nonzero.}$$
>
> To divide two rational expressions, multiply by the reciprocal of the divisor.
>
> $$\frac{A}{B} \div \frac{C}{D} = \frac{A}{B} \cdot \frac{D}{C}, \qquad B, C, \text{ and } D \text{ are nonzero.}$$

EXAMPLE 3   Multiplying rational expressions

Multiply.

(a) $\dfrac{1}{x} \cdot \dfrac{x+1}{2x}$ **(b)** $\dfrac{x-1}{x} \cdot \dfrac{x-1}{x+2}$

SOLUTION

(a) $\dfrac{1}{x} \cdot \dfrac{x+1}{2x} = \dfrac{1 \cdot (x+1)}{x \cdot 2x} = \dfrac{x+1}{2x^2}$

(b) $\dfrac{x-1}{x} \cdot \dfrac{x-1}{x+2} = \dfrac{(x-1)(x-1)}{x(x+2)}$

 *Now Try Exercises 15 and 19*

EXAMPLE 4   Dividing two rational expressions

Divide and simplify.

(a) $\dfrac{2}{x} \div \dfrac{2x - 1}{4x}$ (b) $\dfrac{x^2 - 1}{x^2 + x - 6} \div \dfrac{x - 1}{x + 3}$

SOLUTION

(a)
$$\dfrac{2}{x} \div \dfrac{2x - 1}{4x} = \dfrac{2}{x} \cdot \dfrac{4x}{2x - 1} \qquad \text{``Invert and multiply.''}$$

$$= \dfrac{8x}{x(2x - 1)} \qquad \text{Multiply.}$$

$$= \dfrac{8}{2x - 1} \qquad \text{Simplify.}$$

(b)
$$\dfrac{x^2 - 1}{x^2 + x - 6} \div \dfrac{x - 1}{x + 3} = \dfrac{x^2 - 1}{x^2 + x - 6} \cdot \dfrac{x + 3}{x - 1} \qquad \text{``Invert and multiply.''}$$

$$= \dfrac{(x + 1)(x - 1)}{(x - 2)(x + 3)} \cdot \dfrac{x + 3}{x - 1} \qquad \text{Factor.}$$

$$= \dfrac{(x + 1)(x - 1)(x + 3)}{(x - 2)(x - 1)(x + 3)} \qquad \text{Commutative property}$$

$$= \dfrac{x + 1}{x - 2} \qquad \text{Simplify.}$$

Now Try Exercises 17 and 29. ◆

Least Common Multiples

Two friends work part-time at a store. The first person works every fourth day, while the second person works every sixth day. If they start work on the same day, how many days pass before they both work on the same day again?

We can answer this question by listing the days that each person works.

First person: 4,   8,   **12**,   16,   20,   **24**,   28,   32,   **36**,   40
Second person: 6,   **12**,   18,   **24**,   30,   **36**,   42

After 12 days, the two friends work on the same day. The next time is after 24 days. The numbers 12 and 24 are *common multiples* of 4 and 6. (Find two more.) However, 12 is the **least common multiple** (LCM) of 4 and 6.

Another way to find the least common multiple for 4 and 6 is first to factor each number into prime numbers:

$$4 = \mathbf{2 \cdot 2} \quad \text{and} \quad 6 = \mathbf{2 \cdot 3}.$$

To find the least common multiple, list each factor the *greatest* number of times that it occurs in either factorization. Then find the product of these numbers. For our example, the factor 2 occurs two times in the factorization of 4 and only once in the factorization of 6, so list 2 two times. The factor 3 appears only once in the factorization of 6 and not at all in the factorization of 4, so list it once:

$$\mathbf{2, 2, 3}.$$

The least common multiple is their product: $\mathbf{2} \cdot \mathbf{2} \cdot \mathbf{3} = 12$.

This same procedure can also be used to find the least common multiple for two polynomials.

FINDING THE LEAST COMMON MULTIPLE

The least common multiple (LCM) of two polynomials can be found as follows.

STEP 1: Factor each polynomial completely.

STEP 2: List each factor the greatest number of times that it occurs in either factorization.

STEP 3: Find the product of this list of factors. The result is the LCM.

The next example illustrates how to use this procedure.

EXAMPLE 5 Finding least common multiples

Find the least common multiple for each pair of expressions.
(a) $4x, 5x^3$ (b) $x^2 - 2x, (x - 2)^2$ (c) $x^2 + 4x + 4, x^2 + 3x + 2$

SOLUTION
(a) **STEP 1:** Factor each polynomial completely.

$$4x = 2 \cdot 2 \cdot x \quad \text{and} \quad 5x^3 = 5 \cdot x \cdot x \cdot x$$

STEP 2: The factor 2 occurs twice, the factor 5 occurs once, and the factor x occurs at most three times. The list then is $2, 2, 5, x, x,$ and x.

STEP 3: The LCM is the product $2 \cdot 2 \cdot 5 \cdot x \cdot x \cdot x$, or $20x^3$.

(b) **STEP 1:** Factor each polynomial completely.

$$x^2 - 2x = x(x - 2) \quad \text{and} \quad (x - 2)^2 = (x - 2)(x - 2)$$

STEP 2: The factor x occurs once, and the factor $(x - 2)$ occurs at most twice. The list of factors is $x, (x - 2),$ and $(x - 2)$.

STEP 3: The LCM is the product: $x(x - 2)^2$, which is left in factored form.

(c) **STEP 1:** Factor each polynomial as follows.

$$x^2 + 4x + 4 = (x + 2)(x + 2) \quad \text{and} \quad x^2 + 3x + 2 = (x + 1)(x + 2)$$

STEP 2: The factor $(x + 1)$ occurs once and $(x + 2)$ occurs at most twice.

STEP 3: The LCM is the product $(x + 1)(x + 2)^2$, which is left in factored form.

Now Try Exercises 33, 35, and 37 ◆

Common Denominators

If each denominator is a factor of an expression, that expression is a *common denominator*. For example, $4x^2$ is a common denominator for $\frac{1}{x}, \frac{2 - x}{x^2}$, and $\frac{3}{4}$ because each denominator is a factor of $4x^2$.

$$4x^2 = x \cdot 4x \qquad x \text{ is a factor.}$$
$$4x^2 = x^2 \cdot 4 \qquad x^2 \text{ is a factor.}$$
$$4x^2 = 4 \cdot x^2 \qquad 4 \text{ is a factor.}$$

The expression $4x^3$ is also a common denominator of $\frac{1}{x}, \frac{2 - x}{x^2}$, and $\frac{3}{4}$. However, we say that $4x^2$ is the *least common denominator* (LCD) because it is the common denominator with fewest factors.

Note: A common denominator can always be found by taking the product of the denominators. However, it may not be the *least* common denominator.

EXAMPLE 6 Finding a least common denominator

Find the LCD for the given expressions.

(a) $\dfrac{1}{x}, \dfrac{1}{x-1}$ (b) $\dfrac{1}{x+2}, \dfrac{2x}{x^2-4}, \dfrac{1}{3}$

SOLUTION

(a) The LCD equals the LCM of x and $x-1$, which is their product, $x(x-1)$.

(b) A common multiple of $x+2$, x^2-4, and 3 would be their product. However, because

$$x^2 - 4 = (x+2)(x-2),$$

the LCM of the denominators is

$$(x+2)(x-2)(3)$$

which equals the LCD for the expressions.   *Now Try Exercises 39 and 43* ◆

Addition and Subtraction of Rational Expressions

Addition and subtraction of rational expressions with like denominators are performed in the following manner.

SUMS AND DIFFERENCES OF RATIONAL EXPRESSIONS

To add (or subtract) two rational expressions with like denominators, add (or subtract) their numerators. The denominator does not change.

$$\frac{A}{C} + \frac{B}{C} = \frac{A+B}{C}$$

$$\frac{A}{C} - \frac{B}{C} = \frac{A-B}{C}, \qquad C \neq 0$$

Note: If the denominators are not alike, begin by writing each rational expression, using a common denominator. The LCD equals the LCM of the denominators.

EXAMPLE 7  Adding rational expressions

Add and simplify.

(a) $\dfrac{x}{x+2} + \dfrac{3x+1}{x+2}$ (b) $\dfrac{1}{x-1} + \dfrac{2x}{x+1}$

SOLUTION

(a) The denominators are alike, so we add the numerators and keep the same denominator.

$$\frac{x}{x+2} + \frac{3x+1}{x+2} = \frac{x+3x+1}{x+2} \qquad \text{Add numerators.}$$

$$= \frac{4x+1}{x+2} \qquad \text{Combine like terms.}$$

(b) The LCM for $x - 1$ and $x + 1$ is their product, $(x - 1)(x + 1)$.

$$\frac{1}{x-1} + \frac{2x}{x+1} = \frac{1}{x-1}\cdot\frac{x+1}{x+1} + \frac{2x}{x+1}\cdot\frac{x-1}{x-1} \qquad \text{Change to a common denominator.}$$

$$= \frac{x+1}{(x-1)(x+1)} + \frac{2x(x-1)}{(x+1)(x-1)} \qquad \text{Multiply.}$$

$$= \frac{x+1+2x^2-2x}{(x-1)(x+1)} \qquad \text{Add numerators; distributive property}$$

$$= \frac{2x^2-x+1}{(x-1)(x+1)} \qquad \text{Combine like terms.}$$

Now Try Exercises 45 and 53 ◆

Subtraction of rational expressions is similar.

EXAMPLE 8 Subtracting rational expressions

Subtract and simplify.

(a) $\dfrac{3}{x^2} - \dfrac{x+3}{x^2}$ **(b)** $\dfrac{x-1}{x} - \dfrac{5}{x+5}$

SOLUTION

(a) The denominators are alike, so we subtract the numerators and keep the same denominator.

$$\frac{3}{x^2} - \frac{x+3}{x^2} = \frac{3-(x+3)}{x^2} \qquad \text{Subtract numerators.}$$

$$= \frac{3-x-3}{x^2} \qquad \text{Distributive property}$$

$$= \frac{-x}{x^2} \qquad \text{Simplify numerator.}$$

$$= -\frac{1}{x} \qquad \text{Simplify.}$$

(b) The LCD is $x(x+5)$.

$$\frac{x-1}{x} - \frac{5}{x+5} = \frac{x-1}{x}\cdot\frac{x+5}{x+5} - \frac{5}{x+5}\cdot\frac{x}{x} \qquad \text{Change to a common denominator.}$$

$$= \frac{(x-1)(x+5)}{x(x+5)} - \frac{5x}{x(x+5)} \qquad \text{Multiply.}$$

$$= \frac{(x-1)(x+5)-5x}{x(x+5)} \qquad \text{Subtract numerators.}$$

$$= \frac{x^2+4x-5-5x}{x(x+5)} \qquad \text{Multiply binomials.}$$

$$= \frac{x^2-x-5}{x(x+5)} \qquad \text{Combine like terms.}$$

Now Try Exercises 47 and 51 ◆

Clearing Fractions

To solve rational equations, it is sometimes advantageous to multiply each side by the LCD to clear fractions. For example, the LCD for the equation $\frac{1}{x+2} + \frac{1}{x-2} = 0$ is $(x + 2)(x - 2)$. Multiplying each side by the LCD results in the following.

$$(x + 2)(x - 2)\left(\frac{1}{x + 2} + \frac{1}{x - 2}\right) = 0 \qquad \text{Multiply by LCD}$$

$$\left(\frac{(x + 2)(x - 2)}{x + 2} + \frac{(x + 2)(x - 2)}{x - 2}\right) = 0 \qquad \text{Distributive property}$$

$$(x - 2) + (x + 2) = 0 \qquad \text{Simplify.}$$

$$x = 0 \qquad \text{Combine like terms and solve.}$$

This technique is applied in the next example.

EXAMPLE 9 Clearing fractions

Clear fractions from each equation and solve.

(a) $\dfrac{1}{x} - \dfrac{2}{x^2} = 0$ **(b)** $\dfrac{3}{x} + \dfrac{x}{x^2 - 1} - \dfrac{4}{x + 1} = 0$

SOLUTION

(a) The LCD is x^2.

$$x^2\left(\frac{1}{x} - \frac{2}{x^2}\right) = x^2 \cdot 0$$

$$\left(\frac{x^2}{x} - \frac{2x^2}{x^2}\right) = 0$$

$$x - 2 = 0$$

$$x = 2$$

The solution is 2.

(b) The LCD is $x(x^2 - 1) = x(x - 1)(x + 1)$.

$$x(x^2 - 1)\left(\frac{3}{x} + \frac{x}{x^2 - 1} - \frac{4}{x + 1}\right) = x(x^2 - 1) \cdot 0$$

$$\left(\frac{3x(x^2 - 1)}{x} + \frac{x(x)(x^2 - 1)}{x^2 - 1} - \frac{4x(x^2 - 1)}{x + 1}\right) = 0$$

$$\left(\frac{3x(x^2 - 1)}{x} + \frac{x^2(x^2 - 1)}{x^2 - 1} - \frac{4x(x - 1)(x + 1)}{x + 1}\right) = 0$$

$$3(x^2 - 1) + x^2 - 4x(x - 1) = 0$$

$$3x^2 - 3 + x^2 - 4x^2 + 4x = 0$$

$$4x - 3 = 0$$

$$x = \frac{3}{4}$$

The solution is $\dfrac{3}{4}$.

Now Try Exercises 63 and 71

Complex Fractions

A complex fraction is a rational expression that contains fractions in its numerator, denominator, or both. Examples of complex fractions include

$$\frac{1 + \dfrac{1}{x}}{1 - \dfrac{1}{x}}, \qquad \frac{2x}{\dfrac{4}{x} + \dfrac{3}{x}}, \qquad \text{and} \qquad \frac{\dfrac{a}{3} + \dfrac{a}{4}}{a - \dfrac{1}{a - 1}}.$$

One strategy for simplifying a complex fraction is to multiply the numerator and denominator by the LCD of the fractions in the numerator and denominator. For example, the LCD for the complex fraction

$$\frac{1 - \dfrac{1}{x}}{1 + \dfrac{1}{2x}}$$

is $2x$. To simplify, multiply the complex fraction by 1, expressed in the form $\frac{2x}{2x}$.

$$\frac{\left(1 - \dfrac{1}{x}\right) \cdot 2x}{\left(1 + \dfrac{1}{2x}\right) \cdot 2x} = \frac{2x - \dfrac{2x}{x}}{2x + \dfrac{2x}{2x}} \qquad \text{Distributive property}$$

$$= \frac{2x - 2}{2x + 1} \qquad \text{Simplify.}$$

In the next example we simplify other complex fractions.

EXAMPLE 10   Simplifying complex fractions

Simplify.

(a) $\dfrac{\dfrac{1}{x} - \dfrac{1}{y}}{x - y}$ (b) $\dfrac{\dfrac{3}{x - 1} - \dfrac{2}{x}}{\dfrac{1}{x - 1} + \dfrac{3}{x}}$

SOLUTION

(a) The LCD is the product, xy. Multiply the expression by $\frac{xy}{xy}$.

$$\frac{\left(\dfrac{1}{x} - \dfrac{1}{y}\right) \cdot xy}{(x - y) \cdot xy} = \frac{\dfrac{xy}{x} - \dfrac{xy}{y}}{xy(x - y)} \qquad \begin{array}{l}\text{Distributive and commutative} \\ \text{properties}\end{array}$$

$$= \frac{y - x}{xy(x - y)} \qquad \text{Simplify.}$$

$$= \frac{-1(x - y)}{xy(x - y)} \qquad \text{Factor out } -1.$$

$$= -\frac{1}{xy} \qquad \text{Simplify.}$$

(b) The LCD is the product, $x(x - 1)$. Multiply the expression by $\frac{x(x - 1)}{x(x - 1)}$.

$$\frac{\left(\dfrac{3}{x - 1} - \dfrac{2}{x}\right)}{\left(\dfrac{1}{x - 1} + \dfrac{3}{x}\right)} \cdot \frac{\mathbf{x(x - 1)}}{\mathbf{x(x - 1)}} = \frac{\dfrac{3x(x - 1)}{x - 1} - \dfrac{2x(x - 1)}{x}}{\dfrac{x(x - 1)}{x - 1} + \dfrac{3x(x - 1)}{x}} \qquad \text{Distributive property}$$

$$= \frac{3x - 2(x - 1)}{x + 3(x - 1)} \qquad \text{Simplify.}$$

$$= \frac{3x - 2x + 2}{x + 3x - 3} \qquad \text{Distributive property}$$

$$= \frac{x + 2}{4x - 3} \qquad \text{Combine like terms.}$$

Now Try Exercises 73 and 77 ◆

R.6   Exercises

Simplifying Rational Expressions

Exercises 1–14: Simplify the expression.

1. $\dfrac{10x^3}{5x^2}$

2. $\dfrac{24t^3}{6t^2}$

3. $\dfrac{(x - 5)(x + 5)}{x - 5}$

4. $-\dfrac{5 - a}{a - 5}$

5. $\dfrac{x^2 - 16}{x - 4}$

6. $\dfrac{(x + 5)(x - 4)}{(x + 7)(x + 5)}$

7. $\dfrac{x + 3}{2x^2 + 5x - 3}$

8. $\dfrac{2x^2 - 9x + 4}{6x^2 + 7x - 5}$

9. $-\dfrac{z + 2}{4z + 8}$

10. $\dfrac{x^2 - 25}{x^2 + 10x + 25}$

11. $\dfrac{x^2 + 2x}{x^2 + 3x + 2}$

12. $\dfrac{x^2 - 3x - 10}{x^2 - 6x + 5}$

13. $\dfrac{a^3 + b^3}{a + b}$

14. $\dfrac{a^3 - b^3}{a - b}$

Multiplication and Division of Rational Expressions

Exercises 15–30: Simplify the expression.

15. $\dfrac{1}{x^2} \cdot \dfrac{3x}{2}$

16. $\dfrac{6a}{5} \cdot \dfrac{5}{12a^2}$

17. $\dfrac{5x}{3} \div \dfrac{10x}{6}$

18. $\dfrac{2x^2 + x}{3x + 9} \div \dfrac{x}{x + 3}$

19. $\dfrac{x + 1}{2x - 5} \cdot \dfrac{x}{x + 1}$

20. $\dfrac{4x + 8}{2x} \cdot \dfrac{x^2}{x + 2}$

21. $\dfrac{(x - 5)(x + 3)}{3x - 1} \cdot \dfrac{x(3x - 1)}{(x - 5)}$

22. $\dfrac{b^2 + 1}{b^2 - 1} \cdot \dfrac{b - 1}{b + 1}$

23. $\dfrac{x^2 - 2x - 35}{2x^3 - 3x^2} \cdot \dfrac{x^3 - x^2}{2x - 14}$

24. $\dfrac{2x + 4}{x + 1} \cdot \dfrac{x^2 + 3x + 2}{4x + 2}$

25. $\dfrac{6b}{b + 2} \div \dfrac{3b^4}{2b + 4}$

26. $\dfrac{5x^5}{x - 2} \div \dfrac{10x^3}{5x - 10}$

27. $\dfrac{3a + 1}{a^7} \div \dfrac{a + 1}{3a^8}$

28. $\dfrac{x^2 - 16}{x + 3} \div \dfrac{x + 4}{x^2 - 9}$

29. $\dfrac{x + 5}{x^3 - x} \div \dfrac{x^2 - 25}{x^3}$

30. $\dfrac{x^2 + x - 12}{2x^2 - 9x - 5} \div \dfrac{x^2 + 7x + 12}{2x^2 - 7x - 4}$

Least Common Multiples

Exercises 31–38: Find the least common multiple.

31. 12, 18

32. 9, 15

33. $5a^3, 10a$

34. $6a^2, 9a^5$

35. $z^2 - 4z, (z - 4)^2$

36. $z^2 - 1, z^2 + 2z + 1$

37. $x^2 - 6x + 9, x^2 - 5x + 6$

38. $x^2 - 4, x^2 - 4x + 4$

Common Denominators

Exercises 39–44: Find the LCD for the rational expression.

39. $\dfrac{1}{x + 1}, \dfrac{1}{7}$

40. $\dfrac{1}{2x - 1}, \dfrac{1}{x + 1}$

41. $\dfrac{1}{x + 4}, \dfrac{1}{x^2 - 16}$

42. $\dfrac{4}{2x^2}, \dfrac{1}{2x + 2}$

43. $\dfrac{3}{2}, \dfrac{x}{2x + 1}, \dfrac{x}{2x - 4}$

44. $\dfrac{1}{x}, \dfrac{1}{x^2 - 4x}, \dfrac{1}{2x}$

Addition and Subtraction of Rational Expressions

Exercises 45–62: Simplify.

45. $\dfrac{4}{x + 1} + \dfrac{3}{x + 1}$

46. $\dfrac{2}{x^2} + \dfrac{5}{x^2}$

47. $\dfrac{2}{x^2 - 1} - \dfrac{x + 1}{x^2 - 1}$

48. $\dfrac{2x}{x^2 + x} - \dfrac{2x}{x + 1}$

49. $\dfrac{x}{x + 4} - \dfrac{x + 1}{x(x + 4)}$

50. $\dfrac{4x}{x + 2} + \dfrac{x - 5}{x - 2}$

51. $\dfrac{2}{x^2} - \dfrac{4x - 1}{x}$

52. $\dfrac{2x}{x - 5} - \dfrac{x}{x + 5}$

53. $\dfrac{x + 3}{x - 5} + \dfrac{5}{x - 3}$

54. $\dfrac{x}{2x - 1} + \dfrac{1 - x}{3x}$

55. $\dfrac{3}{x - 5} - \dfrac{1}{x - 3} - \dfrac{2x}{x - 5}$

56. $\dfrac{2x + 1}{x - 1} - \dfrac{3}{x + 1} + \dfrac{x}{x - 1}$

57. $\dfrac{x}{x^2 - 9} + \dfrac{5x}{x - 3}$

58. $\dfrac{a^2 + 1}{a^2 - 1} + \dfrac{a}{1 - a^2}$

59. $\dfrac{b}{2b - 4} - \dfrac{b - 1}{b - 2}$

60. $\dfrac{y^2}{2 - y} - \dfrac{y}{y^2 - 4}$

61. $\dfrac{2x}{x - 5} + \dfrac{2x - 1}{3x^2 - 16x + 5}$

62. $\dfrac{x + 3}{2x - 1} + \dfrac{3}{10x^2 - 5x}$

Clearing Fractions

Exercises 63–72: (Refer to Example 9.) Clear fractions and solve.

63. $\dfrac{1}{x} + \dfrac{3}{x^2} = 0$

64. $\dfrac{1}{x - 2} + \dfrac{3}{x + 1} = 0$

65. $\dfrac{1}{x} + \dfrac{3x}{2x - 1} = 0$

66. $\dfrac{x}{2x - 5} + \dfrac{4}{x} = 0$

67. $\dfrac{2x}{9 - x^2} + \dfrac{1}{3 - x} = 0$

68. $\dfrac{1}{1 - x^2} + \dfrac{1}{1 + x} = 0$

69. $\dfrac{1}{2x} + \dfrac{1}{2x^2} - \dfrac{1}{x^3} = 0$

70. $\dfrac{1}{x^2 - 16} + \dfrac{4}{x + 4} - \dfrac{5}{x - 4} = 0$

71. $\dfrac{1}{x} - \dfrac{2}{x + 5} + \dfrac{1}{x - 5} = 0$

72. $\dfrac{1}{x - 2} + \dfrac{1}{x - 3} - \dfrac{2}{x} = 0$

Complex Fractions

Exercises 73–84: Simplify the expression.

73. $\dfrac{1 + \dfrac{1}{x}}{1 - \dfrac{1}{x}}$

74. $\dfrac{\dfrac{1}{2} - x}{\dfrac{1}{x} - 2}$

75. $\dfrac{\dfrac{1}{x - 5}}{\dfrac{4}{x} - \dfrac{1}{x - 5}}$

76. $\dfrac{1 + \dfrac{1}{x - 3}}{\dfrac{1}{x - 3} - 1}$

77. $\dfrac{\dfrac{1}{x} + \dfrac{2 - x}{x^2}}{\dfrac{3}{x^2} - \dfrac{1}{x}}$

78. $\dfrac{\dfrac{1}{x - 1} + \dfrac{2}{x}}{2 - \dfrac{1}{x}}$

79. $\dfrac{\dfrac{1}{x + 3} + \dfrac{2}{x - 3}}{2 - \dfrac{1}{x - 3}}$

80. $\dfrac{\dfrac{1}{x} + \dfrac{2}{x}}{\dfrac{1}{x - 1} + \dfrac{x}{2}}$

81. $\dfrac{\dfrac{4}{x - 5}}{\dfrac{1}{x + 5} + \dfrac{1}{x}}$

82. $\dfrac{\dfrac{2}{x - 4}}{1 - \dfrac{1}{x + 4}}$

83. $\dfrac{\dfrac{1}{2a} - \dfrac{1}{2b}}{\dfrac{1}{a^2} - \dfrac{1}{b^2}}$

84. $\dfrac{\dfrac{1}{2x^2} - \dfrac{1}{2y^2}}{\dfrac{1}{3y^2} + \dfrac{1}{3x^2}}$

R.7   Radical Notation and Rational Exponents

◆ Use radical notation
◆ Apply rational exponents
◆ Use properties of rational exponents

Radical Notation

Recall the definition of the square root of a number a.

SQUARE ROOT

The number b is a *square root* of a if $b^2 = a$.

EXAMPLE 1   Finding square roots

Find the square roots of 100.

SOLUTION The square roots of 100 are 10 *and* -10 because $10^2 = 100$ and $(-10)^2 = 100$.
Now Try Exercise 1 ◆

Every positive number a has two square roots, one positive and one negative. Recall that the *positive* square root is called the *principal square root* and is denoted $\sqrt{a}$. The *negative square root* is denoted $-\sqrt{a}$. To identify both square roots we write $\pm\sqrt{a}$. The symbol $\pm$ is read "plus or minus." The symbol $\sqrt{\ }$ is called the **radical sign**. The expression under the radical sign is called the **radicand**, and an expression containing a radical sign is called a **radical expression**. Examples of radical expressions include

$$\sqrt{6}, \quad 5 + \sqrt{x+1}, \quad \text{and} \quad \sqrt{\frac{3x}{2x-1}}.$$

In the next example we show how to find the principal square root of an expression.

EXAMPLE 2   Finding principal square roots

Find the principal square root of each expression.

(a) 25   **(b)** 17   **(c)** 0.49   **(d)** $\dfrac{4}{9}$ **(e)** $c^2, c > 0$

FIGURE R.29

SOLUTION
(a) Because $5 \cdot 5 = 25$, the principal, or positive, square root of 25 is $\sqrt{25} = 5$.
(b) The principal square root of 17 is $\sqrt{17}$. This value is not an integer, but we can approximate it. Figure R.29 shows that $\sqrt{17} \approx 4.12$, rounded to the nearest hundredth. Note that calculators do not give exact answers when approximating many radical expressions; they give decimal approximations.
(c) Because $(0.7)(0.7) = 0.49$, the principal square root of 0.49 is $\sqrt{0.49} = 0.7$.
(d) Because $\frac{2}{3} \cdot \frac{2}{3} = \frac{4}{9}$, the principal square root of $\frac{4}{9}$ is $\sqrt{\frac{4}{9}} = \frac{2}{3}$.
(e) The principal square root of c^2 is $\sqrt{c^2} = c$, as it is given that c is positive.
Now Try Exercises 9, 11, 13, and 15 ◆

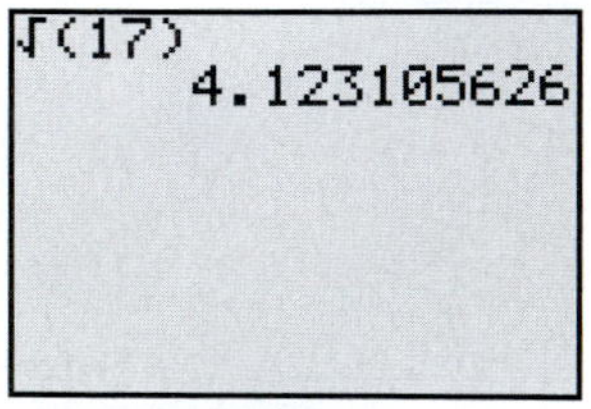

Another common radical expression is the cube root of a number a, denoted $\sqrt[3]{a}$.

CUBE ROOT

The number b is a *cube root* of a if $b^3 = a$.

Although the square root of a negative number is not a real number, the cube root of a negative number is a negative real number. *Every real number has one real cube root.*

We demonstrate how to find cube roots in the next example.

EXAMPLE 3   Finding cube roots

Find the cube root of each expression.

(a) 8   (b) −27   (c) 16   (d) $\dfrac{1}{64}$ (e) d^6

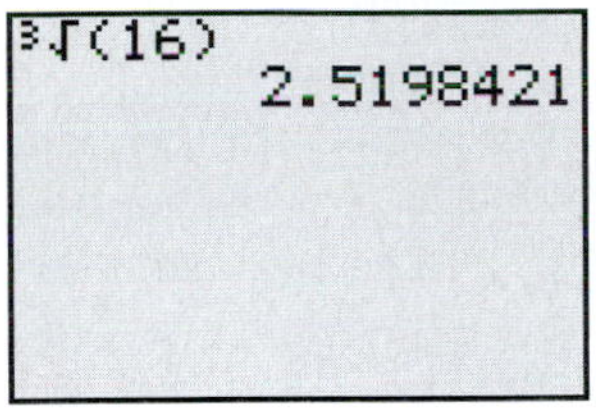

SOLUTION

(a) $\sqrt[3]{8} = 2$ because $2^3 = 2 \cdot 2 \cdot 2 = 8$.

(b) $\sqrt[3]{-27} = -3$ because $(-3)^3 = (-3)(-3)(-3) = -27$.

(c) $\sqrt[3]{16}$ is not an integer. Figure R.30 shows that $\sqrt[3]{16} \approx 2.52$.

(d) $\sqrt[3]{\dfrac{1}{64}} = \dfrac{1}{4}$ because $\left(\dfrac{1}{4}\right)^3 = \dfrac{1}{4} \cdot \dfrac{1}{4} \cdot \dfrac{1}{4} = \dfrac{1}{64}$.

(e) $\sqrt[3]{d^6} = d^2$ because $(d^2)^3 = d^2 \cdot d^2 \cdot d^2 = d^{2+2+2} = d^6$.

Now Try Exercises 17, 19, 21, 23, and 35 ◆

We can generalize square roots and cube roots to include the *n*th root of a number a. The number b is an **nth root** of a if $b^n = a$, where n is a positive integer, and the principal *n*th root is denoted $\sqrt[n]{a}$. The number n is called the **index**. For the square root the index is 2, although we usually write $\sqrt{a}$ rather than $\sqrt[2]{a}$. When n is odd, we are finding an **odd root**, and when n is even, we are finding an **even root**. The square root $\sqrt{a}$ is an example of an even root, and the cube root $\sqrt[3]{a}$ is an example of an odd root.

Note: An odd root of a negative number is a negative number, but the even root of a negative number is *not* a real number.

We find *n*th roots in the next example.

EXAMPLE 4   Finding *n*th roots

Find each root, if possible.

(a) $\sqrt[4]{16}$ (b) $\sqrt[5]{-32}$ (c) $\sqrt[4]{-81}$

SOLUTION

(a) $\sqrt[4]{16} = 2$ because $2^4 = 2 \cdot 2 \cdot 2 \cdot 2 = 16$. Note that when n is even, the principal *n*th root is positive.

(b) $\sqrt[5]{-32} = -2$ because $(-2)^5 = (-2)(-2)(-2)(-2)(-2) = -32$.

(c) The even root of a negative number is not a real number.

Now Try Exercises 33 and 41 ◆

Consider the calculations

$$\sqrt{3^2} = \sqrt{9} = 3, \quad \sqrt{(-4)^2} = \sqrt{16} = 4, \quad \text{and} \quad \sqrt{(-6)^2} = \sqrt{36} = 6.$$

In general, the expression $\sqrt{x^2}$ equals $|x|$.

```
³√(16)
          2.5198421
```

FIGURE R.30

THE EXPRESSION $\sqrt{x^2}$

For every real number x, $\sqrt{x^2} = |x|$.

EXAMPLE 5 Simplifying expressions

Write each expression in terms of an absolute value.

(a) $\sqrt{(-3)^2}$ **(b)** $\sqrt{(x+1)^2}$ **(c)** $\sqrt{z^2 - 4z + 4}$

SOLUTION

(a) $\sqrt{x^2} = |x|$, so $\sqrt{(-3)^2} = |-3| = 3$

(b) $\sqrt{(x+1)^2} = |x+1|$

(c) $\sqrt{z^2 - 4z + 4} = \sqrt{(z-2)^2} = |z-2|$ *Now Try Exercises 111, 115, and 119* ◆

Rational Exponents

When m and n are integers, the product rule states that $a^m a^n = a^{m+n}$. This rule can be extended to include exponents that are fractions. For example,

$$4^{1/2} \cdot 4^{1/2} = 4^{1/2+1/2} = 4^1 = 4.$$

That is, if we multiply $4^{1/2}$ by itself, the result is 4. Because we also know that $\sqrt{4} \cdot \sqrt{4} = 4$, this discussion suggests that $4^{1/2} = \sqrt{4}$ and motivates the following definition.

THE EXPRESSION $a^{1/n}$

If n is an integer greater than 1, then

$$a^{1/n} = \sqrt[n]{a}.$$

Note: If $a < 0$ and n is an even positive integer, then $a^{1/n}$ is not a real number.

In the next two examples, we show how to interpret rational exponents.

EXAMPLE 6 Interpreting rational exponents

Write each expression in radical notation. Then evaluate the expression to the nearest hundredth when appropriate.

(a) $36^{1/2}$ **(b)** $23^{1/5}$ **(c)** $(5x)^{1/2}$

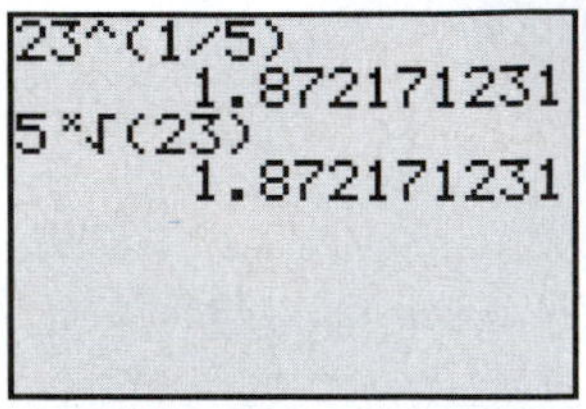

```
23^(1/5)
       1.872171231
5*√(23)
       1.872171231
```

FIGURE R.31

SOLUTION

(a) The exponent $\frac{1}{2}$ indicates a square root. Thus $36^{1/2} = \sqrt{36}$, which evaluates to 6.

(b) The exponent $\frac{1}{5}$ indicates a fifth root. Thus $23^{1/5} = \sqrt[5]{23}$, which is not an integer. Figure R.31 shows this expression approximated in both exponential and radical notation. In either case $23^{1/5} \approx 1.87$.

(c) The exponent $\frac{1}{2}$ indicates a square root, so $(5x)^{1/2} = \sqrt{5x}$.

Now Try Exercises 45, 47, and 63 ◆

Suppose that we want to define the expression $8^{2/3}$. On the one hand, using properties of exponents we have

$$8^{1/3} \cdot 8^{1/3} = 8^{1/3+1/3} = 8^{2/3}.$$

On the other hand, we have

$$8^{1/3} \cdot 8^{1/3} = \sqrt[3]{8} \cdot \sqrt[3]{8} = 2 \cdot 2 = 4.$$

Thus $8^{2/3} = 4$, and that value is obtained whether we interpret $8^{2/3}$ as either

$$8^{2/3} = (8^2)^{1/3} = \sqrt[3]{8^2} = \sqrt[3]{64} = 4.$$

or

$$8^{2/3} = (8^{1/3})^2 = (\sqrt[3]{8})^2 = 2^2 = 4$$

This result illustrates that $8^{2/3} = \sqrt[3]{8^2} = (\sqrt[3]{8})^2 = 4$ and suggests the following definition.

THE EXPRESSION $a^{m/n}$

If m and n are positive integers with $\frac{m}{n}$ in lowest terms, then

$$a^{m/n} = \sqrt[n]{a^m} = \left(\sqrt[n]{a}\right)^m.$$

Note: If $a < 0$ and n is an even integer, then $a^{m/n}$ is not a real number.

 **EXAMPLE 7**

Interpreting rational exponents

Write each expression in radical notation. Then evaluate the expression when the result is an integer.

(a) $(-27)^{2/3}$ (b) $12^{3/5}$

SOLUTION

(a) The exponent $\frac{2}{3}$ indicates that we either take the cube root of -27 and then square it or that we square -27 and then take the cube root. In either case the result will be the same. Thus

$$(-27)^{2/3} = (\sqrt[3]{-27})^2 = (-3)^2 = 9$$

or

$$(-27)^{2/3} = \sqrt[3]{(-27)^2} = \sqrt[3]{729} = 9.$$

(b) The exponent $\frac{3}{5}$ indicates that we either take the fifth root of 12 and then cube it or that we cube 12 and then take the fifth root. Thus

$$12^{3/5} = (\sqrt[5]{12})^3 \quad \text{or} \quad 12^{3/5} = \sqrt[5]{12^3}.$$

This result is not an integer.   *Now Try Exercises 51 and 65* ◆

From properties of exponents we know that $a^{-n} = \frac{1}{a^n}$, where n is a positive integer. We now define this property for negative rational exponents.

THE EXPRESSION $a^{-m/n}$

If m and n are positive integers with $\frac{m}{n}$ in lowest terms, then

$$a^{-m/n} = \frac{1}{a^{m/n}}, \qquad a \neq 0.$$

EXAMPLE 8 Interpreting negative rational exponents

Write each expression in radical notation and then evaluate.
(a) $(64)^{-1/3}$ **(b)** $(81)^{-3/4}$

SOLUTION

(a) $(64)^{-1/3} = \dfrac{1}{64^{1/3}} = \dfrac{1}{\sqrt[3]{64}} = \dfrac{1}{4}.$

(b) $(81)^{-3/4} = \dfrac{1}{81^{3/4}} = \dfrac{1}{(\sqrt[4]{81})^3} = \dfrac{1}{3^3} = \dfrac{1}{27}.$

Now Try Exercises 55 and 57 ◆

Properties of Rational Exponents

Any rational number can be written as a ratio of two integers. That is, if p is a rational number, then $p = \frac{m}{n}$, where m and n are integers. Properties for integer exponents also apply to rational exponents—with one exception. If n is even in the expression $a^{m/n}$ and $\frac{m}{n}$ is written in lowest terms, then a must be nonnegative (not negative) for the result to be a real number.

PROPERTIES OF EXPONENTS

Let p and q be rational numbers written in lowest terms. For all real numbers a and b for which the expressions are real numbers the following properties hold.

1. $a^p \cdot a^q = a^{p+q}$ Product rule for exponents

2. $a^{-p} = \dfrac{1}{a^p}, \quad \dfrac{1}{a^{-p}} = a^p$ Negative exponents

3. $\left(\dfrac{a}{b}\right)^{-p} = \left(\dfrac{b}{a}\right)^{p}$ Negative exponents for quotients

4. $\dfrac{a^p}{a^q} = a^{p-q}$ Quotient rule for exponents

5. $(a^p)^q = a^{pq}$ Power rule for exponents

6. $(ab)^p = a^p b^p$ Power rule for products

7. $\left(\dfrac{a}{b}\right)^{p} = \dfrac{a^p}{b^p}$ Power rule for quotients

In the next example, we apply these properties.

EXAMPLE 9   Applying properties of exponents

Write each expression using rational exponents and simplify. Write the answer with a positive exponent. Assume that all variables are positive numbers.

(a) $\sqrt{x} \cdot \sqrt[3]{x}$ (b) $\sqrt[3]{27x^2}$ (c) $\left(\dfrac{x^2}{81}\right)^{-1/2}$

SOLUTION

(a) $\sqrt{x} \cdot \sqrt[3]{x} = x^{1/2} \cdot x^{1/3}$ Use rational exponents.
$= x^{1/2+1/3}$ Product rule for exponents
$= x^{5/6}$ Simplify.

(b) $\sqrt[3]{27x^2} = (27x^2)^{1/3}$ Use rational exponents.
$= 27^{1/3}(x^2)^{1/3}$ Power rule for products
$= 3x^{2/3}$ Power rule for exponents

(c) $\left(\dfrac{x^2}{81}\right)^{-1/2} = \left(\dfrac{81}{x^2}\right)^{1/2}$ Negative exponents for quotients
$= \dfrac{(81)^{1/2}}{(x^2)^{1/2}}$ Power rule for quotients
$= \dfrac{9}{x}$ Power rule for exponents; simplify.

Now Try Exercises 85, 93, and 103 ◆

R.7 — Exercises

Square Roots and Cube Roots

Exercises 1–8: Find the square roots of the number. Approximate your answer to the nearest hundredth whenever appropriate.

1. 25   **2.** 49

3. 121   **4.** 36

5. $\frac{16}{25}$ **6.** $\frac{64}{81}$

7. 11   **8.** 17

Exercises 9–16: Find the principal square root of the number. Approximate your answer to the nearest hundredth whenever appropriate.

9. 144   **10.** 100

11. 23   **12.** 45

13. $\frac{4}{49}$ **14.** $\frac{16}{121}$

15. $b^2, b < 0$ **16.** $(xy)^2, xy > 0$

Exercises 17–24: Find the cube root of the number.

17. 27   **18.** 64

19. -8 **20.** -125

21. $\frac{1}{27}$ **22.** $-\frac{1}{64}$

23. b^9 **24.** $8x^6$

Radical Notation

Exercises 25–44: If possible, simplify the expression by hand. If you cannot, approximate the answer to the nearest hundredth. Variables represent any real number.

25. $\sqrt{9}$ **26.** $\sqrt{121}$

27. $-\sqrt{5}$ **28.** $\sqrt{11}$

29. $\sqrt{z^2}$ **30.** $-\sqrt{(x+2)^2}$

31. $\sqrt[3]{27}$ **32.** $\sqrt[3]{64}$

33. $\sqrt[3]{-64}$ **34.** $-\sqrt[3]{-1}$

35. $\sqrt[3]{5}$

36. $\sqrt[3]{-13}$

37. $-\sqrt[3]{x^9}$

38. $\sqrt[3]{(x+1)^6}$

39. $\sqrt[3]{(2x)^6}$

40. $\sqrt[3]{9x^3}$

41. $\sqrt[4]{81}$

42. $\sqrt[5]{-1}$

43. $\sqrt[5]{-7}$

44. $\sqrt[4]{6}$

Rational Exponents

Exercises 45–50: Write the expression in radical notation.

45. $6^{1/2}$

46. $7^{1/3}$

47. $(xy)^{1/2}$

48. $x^{2/3}y^{1/5}$

49. $y^{-1/5}$

50. $\left(\dfrac{x}{y}\right)^{-2/7}$

Exercises 51–58: Write the expression in radical notation. Then evaluate the expression when the result is an integer.

51. $27^{2/3}$

52. $8^{4/3}$

53. $(-1)^{4/3}$

54. $81^{3/4}$

55. $8^{-1/3}$

56. $16^{-3/4}$

57. $13^{-3/5}$

58. $23^{-1/2}$

Exercises 59–80: If possible, evaluate the expression by hand. If you cannot, approximate the answer to the nearest hundredth.

59. $16^{1/2}$

60. $8^{1/3}$

61. $256^{1/4}$

62. $4^{3/2}$

63. $32^{1/5}$

64. $(-32)^{1/5}$

65. $(-8)^{4/3}$

66. $(-1)^{3/5}$

67. $2^{1/2} \cdot 2^{2/3}$

68. $5^{3/5} \cdot 5^{1/10}$

69. $\left(\dfrac{4}{9}\right)^{1/2}$

70. $\left(\dfrac{27}{64}\right)^{1/3}$

71. $\dfrac{4^{2/3}}{4^{1/2}}$

72. $\dfrac{6^{1/5} \cdot 6^{3/5}}{6^{2/5}}$

73. $4^{-1/2}$

74. $9^{-3/2}$

75. $(-8)^{-1/3}$

76. $(49)^{-1/2}$

77. $\left(\dfrac{1}{16}\right)^{-1/4}$

78. $\left(\dfrac{16}{25}\right)^{-3/2}$

79. $(2^{1/2})^3$

80. $(5^{6/5})^{-1/2}$

Exercises 81–110: Simplify the expression. Assume that all variables are positive.

81. $(x^2)^{3/2}$

82. $(y^4)^{1/2}$

83. $(x^2y^8)^{1/2}$

84. $(y^{10}z^4)^{1/4}$

85. $\sqrt[3]{x^3y^6}$

86. $\sqrt{16x^4}$

87. $\sqrt{\dfrac{y^4}{x^2}}$

88. $\sqrt[3]{\dfrac{x^{12}}{z^6}}$

89. $\sqrt[3]{y^3} \cdot \sqrt[3]{y^2}$

90. $\left(\dfrac{x^6}{81}\right)^{1/4}$

91. $\left(\dfrac{x^6}{27}\right)^{2/3}$

92. $\left(\dfrac{1}{x^8}\right)^{-1/4}$

93. $\left(\dfrac{x^2}{y^6}\right)^{-1/2}$

94. $\dfrac{\sqrt{x}}{\sqrt[3]{27x^6}}$

95. $\sqrt{\sqrt{y}}$

96. $\sqrt{\sqrt[3]{(3x)^2}}$

97. $(a^{-1/2})^{4/3}$

98. $(x^{-3/2})^{2/3}$

99. $(a^3b^6)^{1/3}$

100. $(64x^3y^{18})^{1/6}$

101. $\dfrac{(k^{1/2})^{-3}}{(k^2)^{1/4}}$

102. $\dfrac{(b^{3/4})^4}{(b^{4/5})^{-5}}$

103. $\sqrt{b} \cdot \sqrt[4]{b}$

104. $\sqrt[3]{t} \cdot \sqrt[5]{t}$

105. $\sqrt{z} \cdot \sqrt[3]{z^2} \cdot \sqrt[4]{z^3}$

106. $\sqrt{b} \cdot \sqrt[3]{b} \cdot \sqrt[5]{b}$

107. $p^{1/2}(p^{3/2} + p^{1/2})$

108. $d^{3/4}(d^{1/4} - d^{-1/4})$

109. $\sqrt[3]{x}(\sqrt{x} - \sqrt[3]{x^2})$

110. $\frac{1}{2}\sqrt{x}(\sqrt{x} + \sqrt[4]{x^2})$

Exercises 111–126: Simplify the expression. Assume that all variables are real numbers.

111. $\sqrt{(-4)^2}$

112. $\sqrt{9^2}$

113. $\sqrt{y^2}$

114. $\sqrt{z^4}$

115. $\sqrt{(a+3)^2}$

116. $\sqrt{(a-b)^2}$

117. $\sqrt{(x-5)^2}$

118. $\sqrt{(2x-1)^2}$

119. $\sqrt{x^2 - 2x + 1}$

120. $\sqrt{4x^2 + 4x + 1}$

121. $\sqrt[4]{y^4}$

122. $\sqrt[4]{x^8z^4}$

123. $\sqrt[4]{x^{12}}$

124. $\sqrt[6]{x^6}$

125. $\sqrt[5]{x^5y^{10}}$

126. $\sqrt[5]{32(x+4)^5}$

R.8   Radical Expressions

- ◆ Apply the product rule
- ◆ Simplify radical expressions
- ◆ Apply the quotient rule
- ◆ Perform addition, subtraction, and multiplication on radical expressions
- ◆ Rationalize the denominator

Product Rule for Radical Expressions

The product of two (like) roots is equal to the root of their product.

PRODUCT RULE FOR RADICAL EXPRESSIONS

Let a and b be real numbers, where $\sqrt[n]{a}$ and $\sqrt[n]{b}$ are both defined. Then

$$\sqrt[n]{a} \cdot \sqrt[n]{b} = \sqrt[n]{a \cdot b}.$$

Note: The product rule works only when the radicals have the *same* index. For example, the product $\sqrt{2} \cdot \sqrt[3]{4}$ cannot be simplified because the indexes are 2 and 3. However, by using rational exponents, we can simplify this product. See Example 5(b).

We apply the product rule in the next two examples.

EXAMPLE 1 Multiplying radical expressions

Multiply each pair of radical expressions.
(a) $\sqrt{5} \cdot \sqrt{20}$ (b) $\sqrt[3]{-3} \cdot \sqrt[3]{9}$

SOLUTION
(a) $\sqrt{5} \cdot \sqrt{20} = \sqrt{5 \cdot 20} = \sqrt{100} = 10$
(b) $\sqrt[3]{-3} \cdot \sqrt[3]{9} = \sqrt[3]{-3 \cdot 9} = \sqrt[3]{-27} = -3$ *Now Try Exercises 3 and 5* ◆

EXAMPLE 2 Multiplying radical expressions containing variables

Multiply each pair of radical expressions. Assume that all variables are positive.

(a) $\sqrt{x} \cdot \sqrt{x^3}$ (b) $\sqrt[3]{2a} \cdot \sqrt[3]{5a}$ (c) $\sqrt[5]{\dfrac{2x}{y}} \cdot \sqrt[5]{\dfrac{16y}{x}}$

SOLUTION
(a) $\sqrt{x} \cdot \sqrt{x^3} = \sqrt{x \cdot x^3} = \sqrt{x^4} = x^2$
(b) $\sqrt[3]{2a} \cdot \sqrt[3]{5a} = \sqrt[3]{2a \cdot 5a} = \sqrt[3]{10a^2}$
(c) $\sqrt[5]{\dfrac{2x}{y}} \cdot \sqrt[5]{\dfrac{16y}{x}} = \sqrt[5]{\dfrac{2x}{y} \cdot \dfrac{16y}{x}}$ Product rule

$\qquad\qquad = \sqrt[5]{\dfrac{32xy}{xy}}$ Multiply fractions.

$\qquad\qquad = \sqrt[5]{32}$ Simplify.

$\qquad\qquad = 2$ $2^5 = 32$ *Now Try Exercises 27, 33, and 35* ◆

An integer a is a **perfect nth power** if there exists an integer b such that $b^n = a$. Thus 36 is a **perfect square** because $6^2 = 36$, 8 is a **perfect cube** because $2^3 = 8$, and 81 is a *perfect fourth power* because $3^4 = 81$.

The product rule for radicals can be used to simplify radical expressions. For example, because the largest perfect square factor of 50 is 25, the expression $\sqrt{50}$ can be simplified as

$$\sqrt{50} = \sqrt{25 \cdot 2} = \sqrt{25} \cdot \sqrt{2} = 5\sqrt{2}.$$

This procedure is generalized as follows.

SIMPLIFYING RADICALS (nth ROOTS)

1. Determine the largest perfect nth power factor of the radicand.
2. Use the product rule to factor out and simplify this perfect nth power.

 EXAMPLE 3   Simplifying radical expressions

Simplify each expression.

(a) $\sqrt{300}$ **(b)** $\sqrt[3]{16}$ **(c)** $\sqrt[4]{512}$

SOLUTION

(a) First note that $300 = 100 \cdot 3$ and that 100 is the largest perfect square factor of 300.

$$\sqrt{300} = \sqrt{100} \cdot \sqrt{3} = 10\sqrt{3}.$$

(b) The largest perfect cube factor of 16 is 8. Thus $\sqrt[3]{16} = \sqrt[3]{8} \cdot \sqrt[3]{2} = 2\sqrt[3]{2}$.

(c) $\sqrt[4]{512} = \sqrt[4]{256} \cdot \sqrt[4]{2} = 4\sqrt[4]{2}$ because $4^4 = 256$.

Now Try Exercises 37, 39, and 41 ◆

Note: To simplify a cube root of a negative number we factor out the negative of the largest perfect cube factor. For example, because $-16 = -8 \cdot 2$, it follows that $\sqrt[3]{-16} = \sqrt[3]{-8} \cdot \sqrt[3]{2} = -2\sqrt[3]{2}$. This procedure can be used with any negative odd root of a number.

EXAMPLE 4   Simplifying radical expressions

Simplify each expression. Assume that all variables are positive.

(a) $\sqrt{25x^4}$ **(b)** $\sqrt{32n^3}$ **(c)** $\sqrt[3]{-16x^3y^5}$ **(d)** $\sqrt[3]{2a} \cdot \sqrt[3]{4a^2b}$

SOLUTION

(a) $\sqrt{25x^4} = 5x^2$ $(5x^2)^2 = 25x^4$

(b) $\sqrt{32n^3} = \sqrt{(16n^2)2n}$ $16n^2$ is the largest perfect square factor.

$\quad = \sqrt{16n^2} \cdot \sqrt{2n}$ Product rule

$\quad = 4n\sqrt{2n}$ $(4n)^2 = 16n^2$

(c) $\sqrt[3]{-16x^3y^5} = \sqrt[3]{(-8x^3y^3)2y^2}$ $8x^3y^3$ is the largest perfect cube factor.

$\quad = \sqrt[3]{-8x^3y^3} \cdot \sqrt[3]{2y^2}$ Product rule

$\quad = -2xy\sqrt[3]{2y^2}$ $(-2xy)^3 = -8x^3y^3$

(d) $\sqrt[3]{2a} \cdot \sqrt[3]{4a^2b} = \sqrt[3]{(2a)(4a^2b)}$ Product rule

$= \sqrt[3]{(8a^3)b}$ $8a^3$ is the largest perfect cube factor.

$= \sqrt[3]{8a^3} \cdot \sqrt[3]{b}$ Product rule

$= 2a\sqrt[3]{b}$ $(2a)^3 = 8a^3$

Now Try Exercises 45, 47, and 51 ◆

The product rule for radical expressions cannot be used if the radicals do not have the same indexes. In this case we use rational exponents, as illustrated in the next example.

EXAMPLE 5   Multiplying radicals with different indexes

Simplify each expression. Write your answer in radical notation.
(a) $\sqrt{5} \cdot \sqrt[4]{5}$ **(b)** $\sqrt{2} \cdot \sqrt[3]{4}$ **(c)** $\sqrt[3]{x} \cdot \sqrt[4]{x}$

SOLUTION
(a) Because $\sqrt{5} = 5^{1/2}$ and $\sqrt[4]{5} = 5^{1/4}$,

$$\sqrt{5} \cdot \sqrt[4]{5} = 5^{1/2} \cdot 5^{1/4} = 5^{1/2+1/4} = 5^{3/4}.$$

In radical notation, $5^{3/4} = \sqrt[4]{5^3} = \sqrt[4]{125}$.
(b) First note that $\sqrt[3]{4} = \sqrt[3]{2^2} = 2^{2/3}$. Thus

$$\sqrt{2} \cdot \sqrt[3]{4} = 2^{1/2} \cdot 2^{2/3} = 2^{1/2+2/3} = 2^{7/6}.$$

In radical notation, $2^{7/6} = \sqrt[6]{2^7} = \sqrt[6]{2^6 \cdot 2^1} = \sqrt[6]{2^6} \cdot \sqrt[6]{2} = 2\sqrt[6]{2}$.
(c) $\sqrt[3]{x} \cdot \sqrt[4]{x} = x^{1/3} \cdot x^{1/4} = x^{7/12} = \sqrt[12]{x^7}$ *Now Try Exercises 55, 57, and 59* ◆

Quotient Rule for Radical Expressions

The root of a quotient is equal to the quotient of the roots.

> ### QUOTIENT RULE FOR RADICAL EXPRESSIONS
> Let a and b be real numbers, where $\sqrt[n]{a}$ and $\sqrt[n]{b}$ are both defined and $b \neq 0$. Then
> $$\sqrt[n]{\frac{a}{b}} = \frac{\sqrt[n]{a}}{\sqrt[n]{b}}.$$

EXAMPLE 6   Simplifying quotients

Simplify each radical expression. Assume that all variables are positive.
(a) $\sqrt[3]{\frac{5}{8}}$ **(b)** $\sqrt{\frac{16}{y^2}}$

SOLUTION
(a) $\sqrt[3]{\frac{5}{8}} = \frac{\sqrt[3]{5}}{\sqrt[3]{8}} = \frac{\sqrt[3]{5}}{2}$

(b) $\sqrt{\frac{16}{y^2}} = \frac{\sqrt{16}}{\sqrt{y^2}} = \frac{4}{y}$ because $y > 0$.

Now Try Exercises 7 and 21 ◆

EXAMPLE 7   Simplifying radical expressions

Simplify each radical expression. Assume that all variables are positive.

(a) $\dfrac{\sqrt{40}}{\sqrt{10}}$ (b) $\sqrt[4]{\dfrac{16x^3}{y^4}}$ (c) $\sqrt{\dfrac{5a^2}{8}} \cdot \sqrt{\dfrac{5a^3}{2}}$

SOLUTION

(a) $\dfrac{\sqrt{40}}{\sqrt{10}} = \sqrt{\dfrac{40}{10}} = \sqrt{4} = 2$

(b) $\sqrt[4]{\dfrac{16x^3}{y^4}} = \dfrac{\sqrt[4]{16x^3}}{\sqrt[4]{y^4}} = \dfrac{\sqrt[4]{16} \cdot \sqrt[4]{x^3}}{\sqrt[4]{y^4}} = \dfrac{2\sqrt[4]{x^3}}{y}$

(c) To simplify this expression, we use both the product and quotient rules.

$\sqrt{\dfrac{5a^2}{8}} \cdot \sqrt{\dfrac{5a^3}{2}} = \sqrt{\dfrac{25a^5}{16}}$ *Product rule*

$\phantom{\sqrt{\dfrac{5a^2}{8}} \cdot \sqrt{\dfrac{5a^3}{2}}} = \dfrac{\sqrt{25a^5}}{\sqrt{16}}$ *Quotient rule*

$\phantom{\sqrt{\dfrac{5a^2}{8}} \cdot \sqrt{\dfrac{5a^3}{2}}} = \dfrac{\sqrt{(25a^4)} \cdot \sqrt{a}}{\sqrt{16}}$ *Factor out largest perfect square.*

$\phantom{\sqrt{\dfrac{5a^2}{8}} \cdot \sqrt{\dfrac{5a^3}{2}}} = \dfrac{5a^2\sqrt{a}}{4}$ *$(5a^2)^2 = 25a^4$*

Now Try Exercises 13, 22 and 35 ◆

Addition and Subtraction

We can add $2x^2$ and $5x^2$ to obtain $7x^2$ because they are *like* terms. That is,

$$2x^2 + 5x^2 = (2 + 5)x^2 = 7x^2.$$

We can add and subtract **like radicals,** which have the same index and the same radicand. For example, we can add $3\sqrt{2}$ and $5\sqrt{2}$ because they are like radicals.

$$3\sqrt{2} + 5\sqrt{2} = (3 + 5)\sqrt{2} = 8\sqrt{2}$$

Sometimes two radical expressions that are not alike can be added by changing them to like radicals. For example, $\sqrt{20}$ and $\sqrt{5}$ are unlike radicals. However,

$$\sqrt{20} = \sqrt{4 \cdot 5} = \sqrt{4} \cdot \sqrt{5} = 2\sqrt{5},$$

so

$$\sqrt{20} + \sqrt{5} = 2\sqrt{5} + \sqrt{5} = 3\sqrt{5}.$$

We cannot combine $x + x^2$ because they are unlike terms. Similarly, we cannot combine $\sqrt{2} + \sqrt{5}$ because they are unlike radicals.

EXAMPLE 8   Adding radical expressions

Add the expressions and simplify.

(a) $10\sqrt{11} + 4\sqrt{11}$ (b) $5\sqrt[3]{6} + \sqrt[3]{6}$

(c) $\sqrt{12} + 7\sqrt{3}$ (d) $3\sqrt{2} + \sqrt{8} + \sqrt{18}$

SOLUTION

(a) $10\sqrt{11} + 4\sqrt{11} = (10 + 4)\sqrt{11} = 14\sqrt{11}$

(b) $5\sqrt[3]{6} + \sqrt[3]{6} = (5 + 1)\sqrt[3]{6} = 6\sqrt[3]{6}$

(c) $\sqrt{12} + 7\sqrt{3} = \sqrt{4 \cdot 3} + 7\sqrt{3}$

$\qquad\qquad = \sqrt{4} \cdot \sqrt{3} + 7\sqrt{3}$

$\qquad\qquad = 2\sqrt{3} + 7\sqrt{3}$

$\qquad\qquad = 9\sqrt{3}$

(d) $3\sqrt{2} + \sqrt{8} + \sqrt{18} = 3\sqrt{2} + \sqrt{4 \cdot 2} + \sqrt{9 \cdot 2}$

$\qquad\qquad\qquad\qquad = 3\sqrt{2} + \sqrt{4} \cdot \sqrt{2} + \sqrt{9} \cdot \sqrt{2}$

$\qquad\qquad\qquad\qquad = 3\sqrt{2} + 2\sqrt{2} + 3\sqrt{2}$

$\qquad\qquad\qquad\qquad = 8\sqrt{2}$ *Now Try Exercises 63 and 69* ◆

EXAMPLE 9   Adding radical expressions

Add the expressions and simplify. Assume that all variables are positive.

(a) $\sqrt[4]{32} + 3\sqrt[4]{2}$ (b) $-2\sqrt{4x} + \sqrt{x}$ (c) $3\sqrt{3k} + 5\sqrt{12k} + 9\sqrt{48k}$

SOLUTION

(a) Because $\sqrt[4]{32} = \sqrt[4]{16 \cdot 2} = \sqrt[4]{16} \cdot \sqrt[4]{2} = 2\sqrt[4]{2}$, we can add as follows.

$$\sqrt[4]{32} + 3\sqrt[4]{2} = 2\sqrt[4]{2} + 3\sqrt[4]{2} = 5\sqrt[4]{2}$$

(b) Note that $\sqrt{4x} = \sqrt{4} \cdot \sqrt{x} = 2\sqrt{x}$.

$$-2\sqrt{4x} + \sqrt{x} = -2(2\sqrt{x}) + \sqrt{x} = -4\sqrt{x} + \sqrt{x} = -3\sqrt{x}$$

(c) Note that $\sqrt{12k} = \sqrt{4} \cdot \sqrt{3k} = 2\sqrt{3k}$ and that $\sqrt{48k} = \sqrt{16} \cdot \sqrt{3k} = 4\sqrt{3k}$.

$3\sqrt{3k} + 5\sqrt{12k} + 9\sqrt{48k} = 3\sqrt{3k} + 5(2\sqrt{3k}) + 9(4\sqrt{3k})$

$\qquad\qquad\qquad\qquad\qquad = (3 + 10 + 36)\sqrt{3k}$

$\qquad\qquad\qquad\qquad\qquad = 49\sqrt{3k}$ *Now Try Exercises 77 and 84* ◆

Subtraction of radical expressions is similar to addition, as illustrated in the next example.

EXAMPLE 10   Subtracting radical expressions

Subtract and simplify. Assume that all variables are positive.

(a) $5\sqrt{7} - 3\sqrt{7}$ (b) $3\sqrt[3]{xy^2} - 2\sqrt[3]{xy^2}$ (c) $\sqrt{16x^3} - \sqrt{x^3}$

SOLUTION

(a) $5\sqrt{7} - 3\sqrt{7} = (5 - 3)\sqrt{7} = 2\sqrt{7}$

(b) $3\sqrt[3]{xy^2} - 2\sqrt[3]{xy^2} = (3 - 2)\sqrt[3]{xy^2} = \sqrt[3]{xy^2}$

(c) $\sqrt{16x^3} - \sqrt{x^3} = \sqrt{16} \cdot \sqrt{x^3} - \sqrt{x^3}$

$\qquad\qquad\qquad = 4\sqrt{x^3} - \sqrt{x^3}$

$\qquad\qquad\qquad = 3\sqrt{x^3}$

$\qquad\qquad\qquad = 3x\sqrt{x}$

     Now Try Exercises 67, 75, and 85 ◆

Multiplication

Some types of radical expressions can be multiplied like binomials. For example, because $(a - b)(a + b) = a^2 - b^2$ we have

$$(\sqrt{x} - 2)(\sqrt{x} + 2) = (\sqrt{x})^2 - (2)^2 = x - 4,$$

provided x is not negative. The next example demonstrates this technique.

EXAMPLE 11   Multiplying radical expressions

Multiply and simplify.

(a) $(4 + \sqrt{3})(4 - \sqrt{3})$ **(b)** $(\sqrt{b} - 4)(\sqrt{b} + 5)$

SOLUTION

(a) This expression is in the form $(a + b)(a - b)$, which equals $a^2 - b^2$.

$$(4 + \sqrt{3})(4 - \sqrt{3}) = (4)^2 - (\sqrt{3})^2$$
$$= 16 - 3$$
$$= 13$$

(b) This expression can be multiplied and then simplified.

$$(\sqrt{b} - 4)(\sqrt{b} + 5) = \sqrt{b} \cdot \sqrt{b} + 5\sqrt{b} - 4\sqrt{b} - 4 \cdot 5$$
$$= b + \sqrt{b} - 20$$

Compare this product to $(b - 4)(b + 5) = b^2 + b - 20$.

Now Try Exercises 89 and 95 ◆

Rationalizing the Denominator

Quotients containing radical expressions can appear to be different but actually be equal. For example, $\dfrac{1}{\sqrt{3}}$ and $\dfrac{\sqrt{3}}{3}$ represent the same real number even though they look like they are unequal. To show this fact, we multiply the first quotient by 1 in the form $\dfrac{\sqrt{3}}{\sqrt{3}}$:

$$\frac{1}{\sqrt{3}} \cdot \frac{\sqrt{3}}{\sqrt{3}} = \frac{1 \cdot \sqrt{3}}{\sqrt{3} \cdot \sqrt{3}} = \frac{\sqrt{3}}{3}.$$

Note: $\sqrt{b} \cdot \sqrt{b} = \sqrt{b^2} = b$ for any *positive* number b.

One way to standardize radical expressions is to remove any radical expressions from the denominator. This process is called **rationalizing the denominator**. The next example demonstrates how to rationalize the denominator of two quotients.

EXAMPLE 12   Rationalizing the denominator

Rationalize each denominator. Assume that all variables are positive.

(a) $\dfrac{3}{5\sqrt{3}}$ **(b)** $\sqrt{\dfrac{x}{24}}$

SOLUTION

(a) We multiply this expression by 1 in the form $\frac{\sqrt{3}}{\sqrt{3}}$:

$$\frac{3}{5\sqrt{3}} \cdot \frac{\sqrt{3}}{\sqrt{3}} = \frac{3\sqrt{3}}{5\sqrt{9}} = \frac{3\sqrt{3}}{5\cdot 3} = \frac{\sqrt{3}}{5}.$$

(b) Because $\sqrt{24} = \sqrt{4}\cdot\sqrt{6} = 2\sqrt{6}$, we start by simplifying the expression.

$$\sqrt{\frac{x}{24}} = \frac{\sqrt{x}}{\sqrt{24}} = \frac{\sqrt{x}}{2\sqrt{6}}$$

To rationalize the denominator we multiply this expression by 1 in the form $\frac{\sqrt{6}}{\sqrt{6}}$:

$$\frac{\sqrt{x}}{2\sqrt{6}} = \frac{\sqrt{x}}{2\sqrt{6}} \cdot \frac{\sqrt{6}}{\sqrt{6}} = \frac{\sqrt{6x}}{12}.$$

Now Try Exercises 99 and 101 ◆

If the denominator consists of two terms, at least one of which contains a radical expression, then the **conjugate** of the denominator is found by changing a + sign to a − sign or vice versa. For example, the conjugate of $\sqrt{2} + \sqrt{3}$ is $\sqrt{2} - \sqrt{3}$ and the conjugate of $\sqrt{3} - 1$ is $\sqrt{3} + 1$. In the next example, we multiply the numerator and denominator by the conjugate of the denominator to rationalize the denominator of fractions that contain radicals.

EXAMPLE 13 Rationalizing the denominator

Rationalize the denominator.

(a) $\dfrac{3 + \sqrt{5}}{2 - \sqrt{5}}$ (b) $\dfrac{\sqrt{x}}{\sqrt{x} - 2}$

SOLUTION

(a) The conjugate of the denominator is $2 + \sqrt{5}$.

$$\frac{3 + \sqrt{5}}{2 - \sqrt{5}} = \frac{(3 + \sqrt{5})}{(2 - \sqrt{5})} \cdot \frac{(2 + \sqrt{5})}{(2 + \sqrt{5})} \quad \text{Multiply by 1.}$$

$$= \frac{6 + 3\sqrt{5} + 2\sqrt{5} + (\sqrt{5})^2}{(2)^2 - (\sqrt{5})^2} \quad \text{Multiply.}$$

$$= \frac{11 + 5\sqrt{5}}{4 - 5} \quad \text{Combine terms.}$$

$$= -11 - 5\sqrt{5} \quad \text{Simplify.}$$

(b) The conjugate of the denominator is $\sqrt{x} + 2$.

$$\frac{\sqrt{x}}{\sqrt{x} - 2} = \frac{\sqrt{x}}{(\sqrt{x} - 2)} \cdot \frac{(\sqrt{x} + 2)}{(\sqrt{x} + 2)} \quad \text{Multiply by 1.}$$

$$= \frac{x + 2\sqrt{x}}{x - 4} \quad \text{Multiply.}$$

Now Try Exercises 103 and 109 ◆

R.8 Exercises

Multiplying and Dividing

Exercises 1–36: Simplify the expression. Assume that all variables are positive.

1. $\sqrt{3} \cdot \sqrt{3}$

2. $\sqrt{2} \cdot \sqrt{18}$

3. $\sqrt{2} \cdot \sqrt{50}$

4. $\sqrt[3]{-2} \cdot \sqrt[3]{-4}$

5. $\sqrt[3]{4} \cdot \sqrt[3]{16}$

6. $\sqrt[3]{x} \cdot \sqrt[3]{x^2}$

7. $\sqrt{\dfrac{9}{25}}$

8. $\sqrt[3]{\dfrac{x}{8}}$

9. $\sqrt{\dfrac{1}{2}} \cdot \sqrt{\dfrac{1}{8}}$

10. $\sqrt{\dfrac{5}{3}} \cdot \sqrt{\dfrac{1}{3}}$

11. $\sqrt{\dfrac{x}{2}} \cdot \sqrt{\dfrac{x}{8}}$

12. $\sqrt{\dfrac{4}{y}} \cdot \sqrt{\dfrac{y}{5}}$

13. $\dfrac{\sqrt{45}}{\sqrt{5}}$

14. $\dfrac{\sqrt{7}}{\sqrt{28}}$

15. $\sqrt[4]{9} \cdot \sqrt[4]{9}$

16. $\sqrt[5]{16} \cdot \sqrt[5]{-2}$

17. $\dfrac{\sqrt[5]{64}}{\sqrt[5]{-2}}$

18. $\dfrac{\sqrt[4]{324}}{\sqrt[4]{4}}$

19. $\dfrac{\sqrt{a^2 b}}{\sqrt{b}}$

20. $\dfrac{\sqrt{4xy^2}}{\sqrt{x}}$

21. $\sqrt[3]{\dfrac{x^3}{8}}$

22. $\sqrt{\dfrac{36}{z^4}}$

23. $\sqrt{4x^4}$

24. $\sqrt[3]{-8y^3}$

25. $\sqrt[4]{16x^4 y}$

26. $\sqrt[3]{8xy^3}$

27. $\sqrt{3x} \cdot \sqrt{12x}$

28. $\sqrt{6x^5} \cdot \sqrt{6x}$

29. $\sqrt[3]{8x^6 y^3 z^9}$

30. $\sqrt{16x^4 y^6}$

31. $\sqrt[4]{\dfrac{3}{4}} \cdot \sqrt[4]{\dfrac{27}{4}}$

32. $\sqrt[5]{\dfrac{4}{-9}} \cdot \sqrt[5]{\dfrac{8}{-27}}$

33. $\sqrt[4]{25z} \cdot \sqrt[4]{25z}$

34. $\sqrt[5]{3z^2} \cdot \sqrt[5]{7z}$

35. $\sqrt[5]{\dfrac{7a}{b^2}} \cdot \sqrt[5]{\dfrac{b^2}{7a^6}}$

36. $\sqrt[3]{\dfrac{8m}{n}} \cdot \sqrt[3]{\dfrac{n^4}{m^2}}$

Exercises 37–54: Simplify the radical expression by factoring out the largest perfect nth power. Assume that all variables are positive.

37. $\sqrt{200}$

38. $\sqrt{72}$

39. $\sqrt[3]{81}$

40. $\sqrt[3]{256}$

41. $\sqrt[4]{64}$

42. $\sqrt[5]{27 \cdot 81}$

43. $\sqrt[5]{-64}$

44. $\sqrt[3]{-81}$

45. $\sqrt{8n^3}$

46. $\sqrt{32a^2}$

47. $\sqrt{12a^2 b^5}$

48. $\sqrt{20a^3 b^2}$

49. $\sqrt[3]{-125x^4 y^5}$

50. $\sqrt[3]{-81a^5 b^2}$

51. $\sqrt[3]{5t} \cdot \sqrt[3]{125t}$

52. $\sqrt[4]{4bc^3} \cdot \sqrt[4]{64ab^3 c^2}$

53. $\sqrt[4]{\dfrac{9t^5}{r^8}} \cdot \sqrt[4]{\dfrac{9r}{5t}}$

54. $\sqrt[5]{\dfrac{4t^6}{r}} \cdot \sqrt[5]{\dfrac{8t}{r^6}}$

Exercises 55–62: Simplify the expression. Assume that all variables are positive and write your answer in radical notation.

55. $\sqrt{3} \cdot \sqrt[3]{3}$

56. $\sqrt{5} \cdot \sqrt[3]{5}$

57. $\sqrt[4]{8} \cdot \sqrt[3]{4}$

58. $\sqrt[5]{16} \cdot \sqrt{2}$

59. $\sqrt[4]{x^3} \cdot \sqrt[3]{x}$

60. $\sqrt[4]{x^3} \cdot \sqrt{x}$

61. $\sqrt[4]{rt} \cdot \sqrt[3]{r^2 t}$

62. $\sqrt[3]{a^3 b^2} \cdot \sqrt{a^2 b}$

Exercises 63–88: Simplify the expression. Assume that all variables are positive.

63. $2\sqrt{3} + 7\sqrt{3}$

64. $8\sqrt{7} + 2\sqrt{7}$

65. $\sqrt{x} + \sqrt{x} - \sqrt{y}$

66. $\sqrt{xy^2} - \sqrt{x}$

67. $2\sqrt[3]{6} - 7\sqrt[3]{6}$

68. $18\sqrt[3]{3} + 3\sqrt[3]{3}$

69. $3\sqrt{28} + 3\sqrt{7}$

70. $9\sqrt{18} - 2\sqrt{8}$

71. $\sqrt{44} - 4\sqrt{11}$

72. $\sqrt[4]{5} + 2\sqrt[4]{5}$

73. $2\sqrt[3]{16} + \sqrt[3]{2} - \sqrt{2}$

74. $5\sqrt[3]{x} - 3\sqrt[3]{x}$

75. $\sqrt[3]{xy} - 2\sqrt[3]{xy}$

76. $3\sqrt{x^3} - \sqrt{x}$

77. $\sqrt{4x+8} + \sqrt{x+2}$

78. $\sqrt{2a+1} + \sqrt{8a+4}$

79. $\dfrac{15\sqrt{8}}{4} - \dfrac{2\sqrt{2}}{5}$

80. $\dfrac{23\sqrt{11}}{2} - \dfrac{\sqrt{44}}{8}$

81. $2\sqrt[4]{64} - \sqrt[4]{324} + \sqrt[4]{4}$

82. $20\sqrt[3]{b^4} - 4\sqrt[3]{b}$

83. $\sqrt{64x^3} - \sqrt{x} + 3\sqrt{x}$

84. $2\sqrt{3z} + 3\sqrt{12z} + 3\sqrt{48z}$

85. $\sqrt[4]{81a^5b^5} - \sqrt[4]{ab}$

86. $\sqrt[4]{xy^5} + \sqrt[4]{x^5y}$

87. $5\sqrt[3]{\dfrac{n^4}{125}} - 2\sqrt[3]{n}$

88. $\sqrt[3]{\dfrac{8x}{27}} - \dfrac{2\sqrt[3]{x}}{3}$

Exercises 89–96: Multiply and simplify.

89. $(3+\sqrt{7})(3-\sqrt{7})$

90. $(5-\sqrt{5})(5+\sqrt{5})$

91. $(\sqrt{x}+8)(\sqrt{x}-8)$

92. $(\sqrt{ab}-3)(\sqrt{ab}+3)$

93. $(\sqrt{ab}-\sqrt{c})(\sqrt{ab}+\sqrt{c})$

94. $(\sqrt{2x}+\sqrt{3y})(\sqrt{2x}-\sqrt{3y})$

95. $(\sqrt{x}-7)(\sqrt{x}+8)$

96. $(\sqrt{ab}-1)(\sqrt{ab}-2)$

Exercises 97–112: Rationalize the denominator.

97. $\dfrac{4}{\sqrt{3}}$

98. $\dfrac{8}{\sqrt{2}}$

99. $\dfrac{5}{3\sqrt{5}}$

100. $\dfrac{6}{11\sqrt{3}}$

101. $\sqrt{\dfrac{b}{12}}$

102. $\sqrt{\dfrac{5b}{72}}$

103. $\dfrac{1}{3-\sqrt{2}}$

104. $\dfrac{1}{\sqrt{3}-2}$

105. $\dfrac{\sqrt{2}}{\sqrt{5}+2}$

106. $\dfrac{\sqrt{3}-1}{\sqrt{3}+1}$

107. $\dfrac{1}{\sqrt{7}-\sqrt{6}}$

108. $\dfrac{1}{\sqrt{8}-\sqrt{7}}$

109. $\dfrac{\sqrt{z}}{\sqrt{z}-3}$

110. $\dfrac{2\sqrt{z}}{2-\sqrt{z}}$

111. $\dfrac{\sqrt{a}+\sqrt{b}}{\sqrt{a}-\sqrt{b}}$

112. $\dfrac{1}{\sqrt{a+1}+\sqrt{a}}$

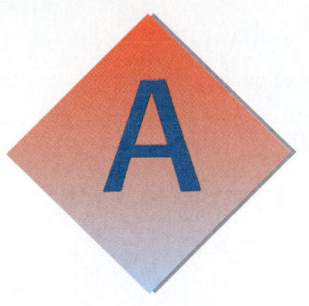

Appendix:
A Library of Functions

Basic Functions

Several important functions are used in algebra and trigonometry. The following provides symbolic, numerical, and graphical representations for some of these basic functions.

Identity Function: $f(x) = x$

x	-2	-1	0	1	2
$y = x$	-2	-1	0	1	2

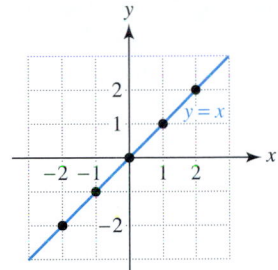

Absolute Value Function: $f(x) = |x|$

x	-2	-1	0	1	2
$y = |x|$	2	1	0	1	2

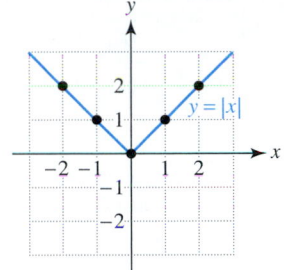

Square Function: $f(x) = x^2$

x	-2	-1	0	1	2
$y = x^2$	4	1	0	1	4

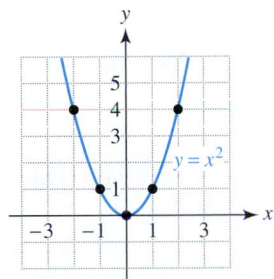

Cube Function: $f(x) = x^3$

x	-2	-1	0	1	2
$y = x^3$	-8	-1	0	1	8

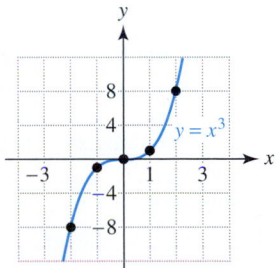

Square Root Function: $f(x) = \sqrt{x}$

x	0	1	4	9
$y = \sqrt{x}$	0	1	2	3

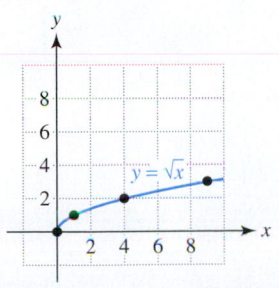

Cube Root Function: $f(x) = \sqrt[3]{x}$

x	-8	-1	0	1	8
$y = \sqrt[3]{x}$	-2	-1	0	1	2

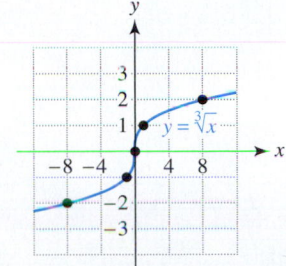

Greatest Integer Function: $f(x) = [\![x]\!]$

x	-2.5	-1.5	0	1.5	2.5
$y = [\![x]\!]$	-3	-2	0	1	2

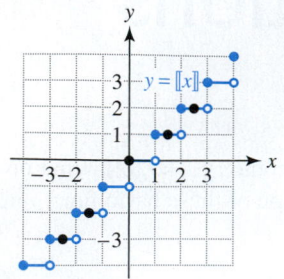

Reciprocal Function: $f(x) = \frac{1}{x}$

x	-2	-1	0	1	2
$y = \frac{1}{x}$	$-\frac{1}{2}$	-1	—	1	$\frac{1}{2}$

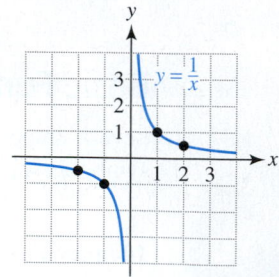

Base-2 Exponential Function: $f(x) = 2^x$

x	-2	-1	0	1	2
$y = 2^x$	$\frac{1}{4}$	$\frac{1}{2}$	1	2	4

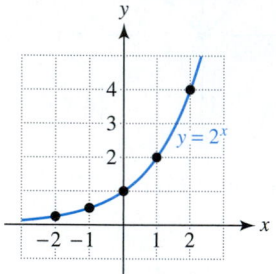

Natural Exponential Function: $f(x) = e^x$

x	-2	-1	0	1	2
$y = e^x$	e^{-2}	e^{-1}	1	e^1	e^2

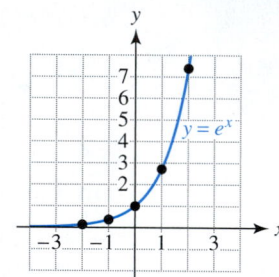

Common Logarithmic Function: $f(x) = \log x$

x	0.1	1	4	7	10
$y = \log x$	-1	0	$\log 4$	$\log 7$	1

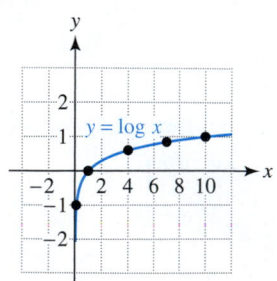

Natural Logarithmic Function: $f(x) = \ln x$

x	$\frac{1}{2}$	1	2	e	e^2
$y = \ln x$	$\ln \frac{1}{2}$	0	$\ln 2$	1	2

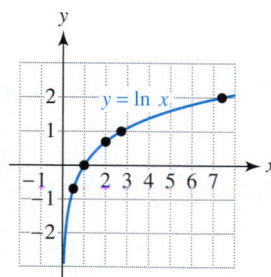

Sine Function: $f(x) = \sin x$

x	0	$\frac{\pi}{2}$	π	$\frac{3\pi}{2}$	2π
$y = \sin x$	0	1	0	-1	0

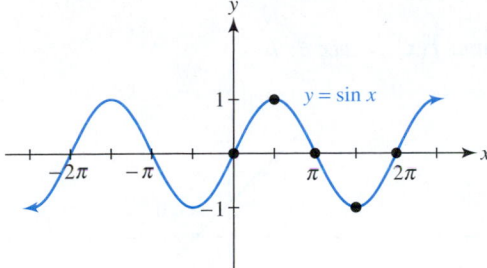

Cosine Function: $f(x) = \cos x$

x	0	$\frac{\pi}{2}$	π	$\frac{3\pi}{2}$	2π
$y = \cos x$	1	0	-1	0	1

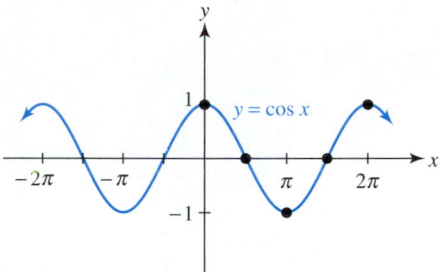

Tangent Function: $f(x) = \tan x$

x	$-\frac{\pi}{3}$	$-\frac{\pi}{4}$	0	$\frac{\pi}{4}$	$\frac{\pi}{3}$
$y = \tan x$	$-\sqrt{3}$	-1	0	1	$\sqrt{3}$

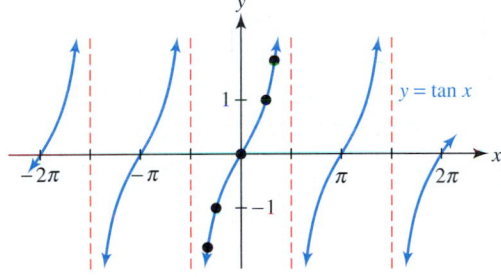

Cotangent Function: $f(x) = \cot x$

x	$\frac{\pi}{6}$	$\frac{\pi}{4}$	$\frac{\pi}{2}$	$\frac{3\pi}{4}$	$\frac{5\pi}{6}$
$y = \cot x$	$\sqrt{3}$	1	0	-1	$-\sqrt{3}$

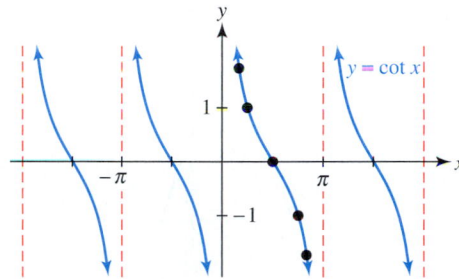

Cosecant Function: $f(x) = \csc x$

x	$\frac{\pi}{6}$	$\frac{\pi}{4}$	$\frac{\pi}{2}$	$\frac{3\pi}{4}$	$\frac{5\pi}{6}$
$y = \csc x$	2	$\sqrt{2}$	1	$\sqrt{2}$	2

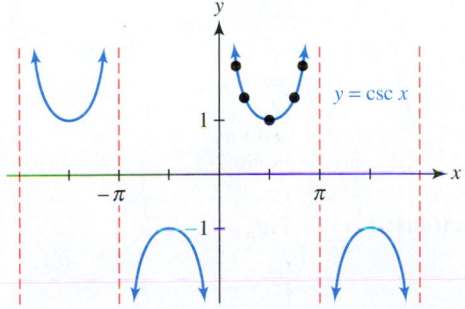

Secant Function: $f(x) = \sec x$

x	$-\frac{\pi}{3}$	$-\frac{\pi}{4}$	0	$\frac{\pi}{4}$	$\frac{\pi}{3}$
$y = \sec x$	2	$\sqrt{2}$	1	$\sqrt{2}$	2

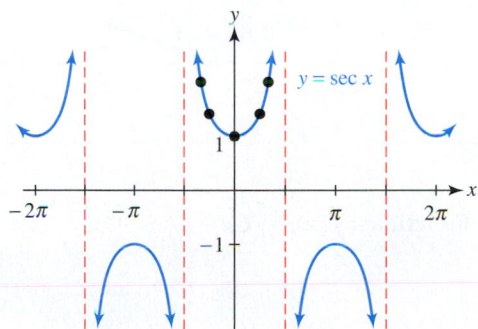

Families of Functions

Some examples of families of functions are linear, quadratic, and exponential functions. This subsection shows the formulas and graphs of some families of functions. Notice that the graphs of these functions change when their formulas are changed.

Constant Functions: $f(x) = k$

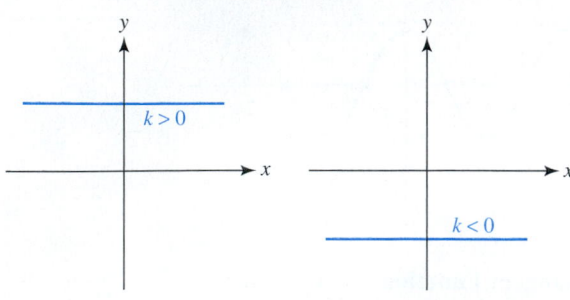

Linear Functions: $f(x) = mx + b$

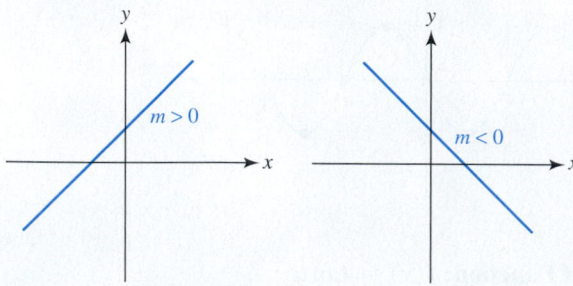

Quadratic Functions: $f(x) = ax^2 + bx + c$

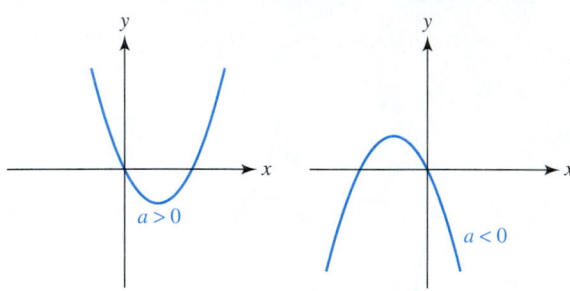

Cubic Functions: $f(x) = ax^3 + bx^2 + cx + d$

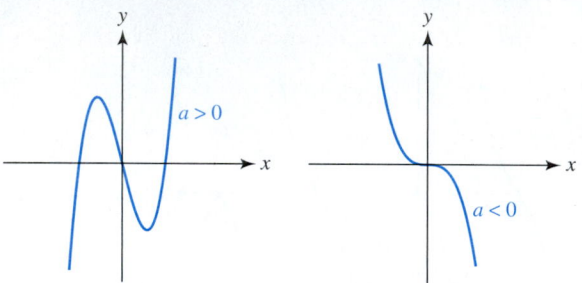

Power Functions: $f(x) = x^a, x > 0$

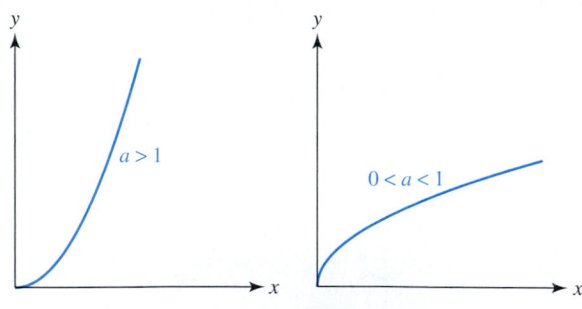

Sinusoidal Functions: $f(x) = a \sin (b(x - c)) + d$ or
$$f(x) = a \cos (b(x - c)) + d$$

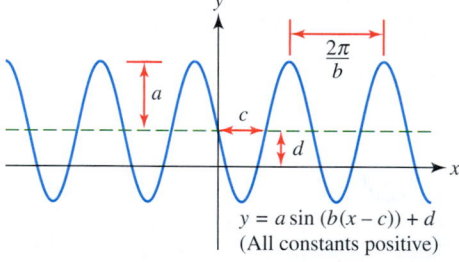

$y = a \sin (b(x - c)) + d$
(All constants positive)

Exponential Functions: $f(x) = Ca^x$

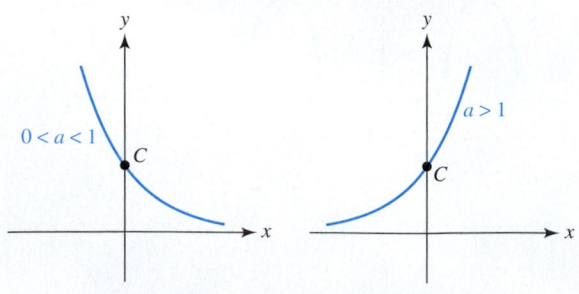

Logarithmic Functions: $f(x) = \log_a x$

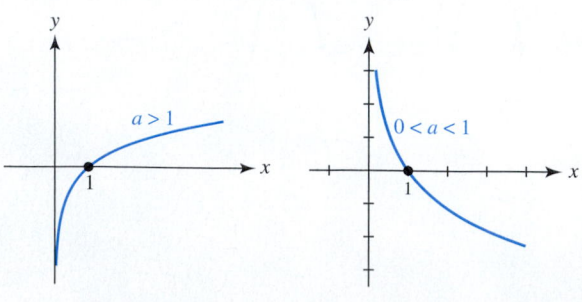

B

Appendix: Using the Graphing Calculator

Overview of the Appendix

The intent of this appendix is to provide instruction for the TI-83, TI-83 Plus, and TI-84 Plus graphing calculators that may be used in conjunction with this textbook. It includes specific keystrokes needed to work several examples from the text. Students are also advised to consult the *Graphing Calculator Guidebook* provided by the manufacturer.

Displaying Numbers in Scientific Notation

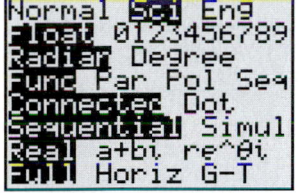

FIGURE B.1

To display numbers in scientific notation, set the graphing calculator in scientific mode (Sci), by using the following keystrokes. See Figure B.1. (These keystrokes assume that the calculator is in normal mode.)

$$\boxed{\text{MODE}} \, \boxed{\triangleright} \, \boxed{\text{ENTER}} \, \boxed{\text{2nd}} \, \boxed{\text{MODE [QUIT]}}$$

In scientific mode we can display the numbers 5432 and 0.00001234 in scientific notation, as shown in Figure B.2.

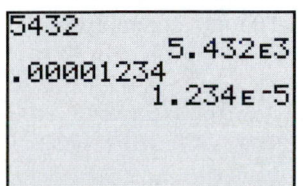

FIGURE B.2

SUMMARY: SETTING SCIENTIFIC MODE

If your calculator is in normal mode, it can be set in scientific mode by pressing

$$\boxed{\text{MODE}} \, \boxed{\triangleright} \, \boxed{\text{ENTER}} \, \boxed{\text{2nd}} \, \boxed{\text{MODE [QUIT]}}.$$

These keystrokes return the graphing calculator to the home screen.

Entering Numbers in Scientific Notation

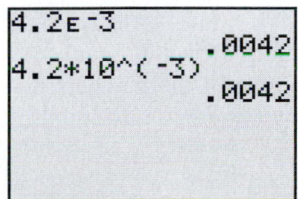

FIGURE B.3

Numbers can be entered in scientific notation. For example, to enter 4.2×10^{-3} in scientific notation, use the following keystrokes. (Be sure to use the negation key $(-)$ rather than the subtraction key.)

$$\boxed{4} \, \boxed{.} \, \boxed{2} \, \boxed{\text{2nd}} \, \boxed{,\text{[EE]}} \, \boxed{(-)} \, \boxed{3}$$

This number can also be entered using the following keystrokes. See Figure B.3.

$$\boxed{4} \, \boxed{.} \, \boxed{2} \, \boxed{\times} \, \boxed{1} \, \boxed{0} \, \boxed{\wedge} \, \boxed{(} \, \boxed{(-)} \, \boxed{3} \, \boxed{)}$$

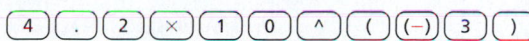

SUMMARY: ENTERING NUMBERS IN SCIENTIFIC NOTATION

One way to enter a number in scientific notation is to use the keystrokes

$$\boxed{\text{2nd}} \, \boxed{,}$$

to access an exponent (EE) of 10.

Entering Mathematical Expressions

Several expressions are evaluated in Example 7, Section 1.1. To evaluate $\sqrt[3]{131}$, use the following keystrokes from the home screen.

(MATH) (4) (1) (3) (1) ( )) (ENTER)

To calculate $\pi^3 + 1.2^2$, use the following keystrokes. (Do *not* use 3.14 for π.)

(2nd) (^[π]) (^) (3) (+) (1) (.) (2) (x^2) (ENTER)

To calculate $|\sqrt{3} - 6|$, use the following keystrokes.

(MATH) (▷) (1) (2nd) (x^2[√]) (3) ( )) (−) (6) ( )) (ENTER)

SUMMARY: ENTERING COMMON MATHEMATICAL EXPRESSIONS

To calculate a cube root, use the keystrokes (MATH) (4).

To access the number π, use the keystrokes (2nd) (^[π]).

To access the absolute value, use the keystrokes (MATH) (▷) (1).

To access the square root, use the keystrokes (2nd) (x^2[√]).

Calculating One-Variable Statistics

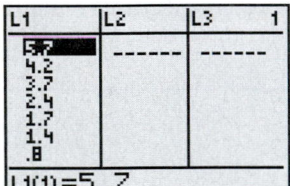

FIGURE B.4

The mean, median, minimum, and maximum for the data set in Example 2, Section 1.2 can be found with a graphing calculator. Begin by entering the data into an empty list L1 with the "STAT EDIT" menus, as shown in Figure B.4. Use the following keystrokes.

(STAT) (1) (5) (.) (7) (ENTER) (4) (.) (2) (ENTER) (3) (.) (7) (ENTER)

(2) (.) (4) (ENTER) . . . .

Continue until all 12 numbers have been entered into list L1. To calculate one-variable statistics press the following keys.

(STAT) (▷) (1) (ENTER)

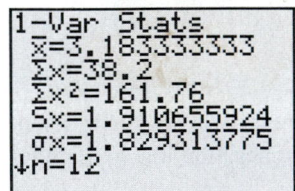

FIGURE B.5

The mean (or average) of the data is given by $\bar{x}$ = 3.183333333, as shown in Figure B.5. (If you enroll in a statistics class you will learn more about some of the numbers that appear.) In Figure B.6, the minimum is given by minX = 0.8, the maximum by maxX = 5.9, and the median by Med = 2.95.

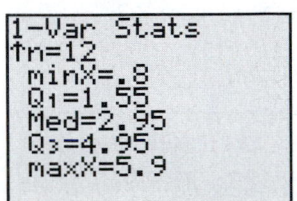

FIGURE B.6

SUMMARY: CALCULATING ONE-VARIABLE STATISTICS

Press (STAT) (1), and enter the data into list L1. Then press

(STAT) (▷) (1) (ENTER)

to calculate one-variable statistics. The mean is given by "$\bar{x}$", the minimum by "minX", the maximum by "maxX", and the median by "Med".

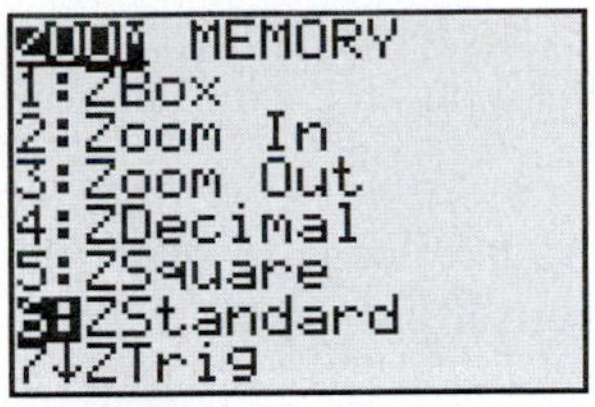

FIGURE B.7

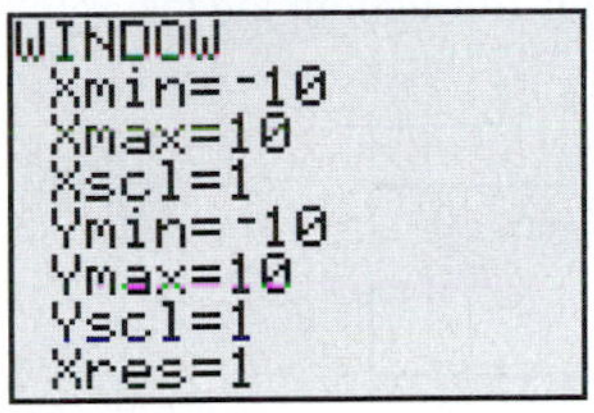

FIGURE B.8

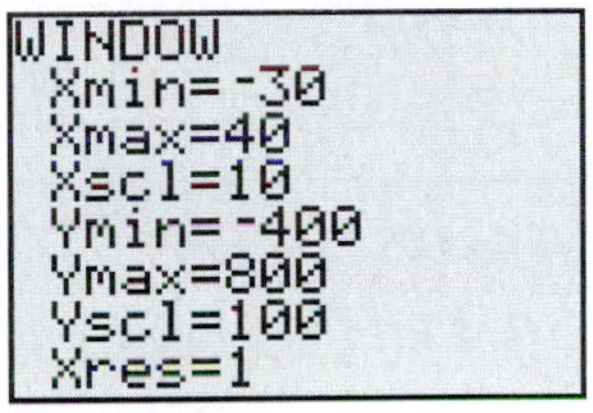

FIGURE B.9

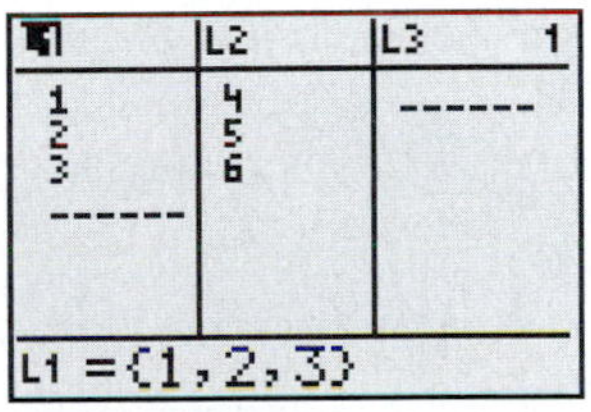

FIGURE B.10

FIGURE B.11

FIGURE B.12

Setting the Viewing Rectangle

In Example 11, Section 1.2, there are at least two ways to set the standard viewing rectangle to $[-10, 10, 1]$ by $[-10, 10, 1]$. The first involves pressing (ZOOM) followed by (6). See Figure B.7. The second method for setting the standard viewing rectangle is to press (WINDOW), and enter the following keystrokes. See Figure B.8.

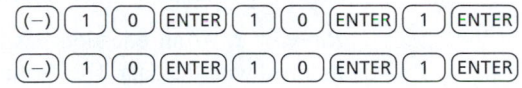

(Be sure to use the negation key $(-)$ rather than the subtraction key.) The viewing rectangle $[-30, 40, 10]$ by $[-400, 800, 100]$ can be set in a similar manner by pressing (WINDOW), as shown in Figure B.9. To see the viewing rectangle, press (GRAPH).

SUMMARY: SETTING THE VIEWING RECTANGLE

To set the standard viewing rectangle, press (ZOOM)(6). To set any viewing rectangle, press (WINDOW) and enter the necessary values. To see the viewing rectangle, press (GRAPH).

Note: You do not need to change "Xres".

Making a Scatterplot or a Line Graph

Example 12, Section 1.2 asks us to make a scatterplot with $(-5, -5)$, $(-2, 3)$, $(1, -7)$, and $(4, 8)$. To accomplish this task, begin by following these steps.

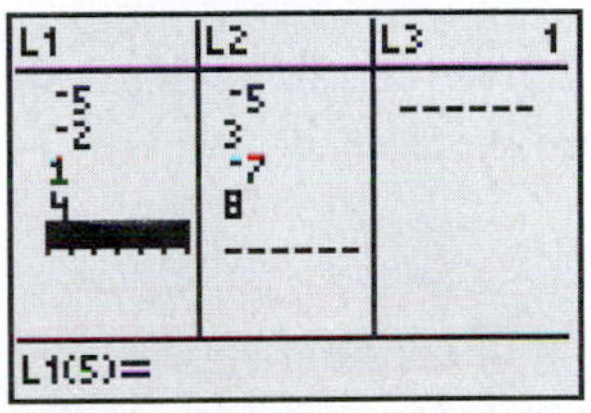

1. Press (STAT) followed by (1).
2. If list L1 is not empty, use the arrow keys to place the cursor on L1, as shown in Figure B.10. Then press (CLEAR) followed by (ENTER). This deletes all elements in the list. Similarly, if L2 is not empty, clear the list.
3. Input each x-value into list L1 followed by (ENTER). Input each y-value into list L2 followed by (ENTER). See Figure B.11.

It is essential that both lists have the same number of values—otherwise an error message appears when a scatterplot is attempted. Before these four points can be plotted, "STAT-PLOT" must be turned on. It is accessed by pressing

(2nd)(Y = [STAT PLOT]),

as shown in Figure B.12.

There are three possible "STAT PLOTS", numbered 1, 2, and 3. Any one of the three can be selected. The first plot can be selected by pressing (1). Next, place the cursor over "On" and press (ENTER) to turn "Plot1" on. There are six types of plots that can be selected. The first type is a *scatterplot* and the second type is a *line graph*, so place the cursor over the first type of plot and press (ENTER) to select a scatterplot. (To make the line graph in Example 13, Section 1.2, be sure to select the line graph.) The x-values are stored in list L1, so select L1 for "Xlist" by pressing (2nd)(1). Similarly press (2nd)(2) for the "Ylist," since the y-values are stored in list L2. Finally, there are three styles of marks that can be used to show data points in the graph. We will usually use the first, because it is largest and shows up the best. Make the screen appear as in Figure B.13. Before plotting the four data points, be sure to set an appropriate viewing rectangle. Then press (GRAPH). The data points appear as in Figure B.14.

Remark 1: A fast way to set the viewing rectangle for any scatterplot is to select the "ZoomStat" feature by pressing (ZOOM) (9). This feature automatically scales the viewing rectangle so that all data points are shown.

Remark 2: If an equation has been entered into the (Y =) menu and selected, it will be graphed with the data. This feature is used frequently to model data throughout the textbook.

$[-10, 10, 1]$ by $[-10, 10, 1]$

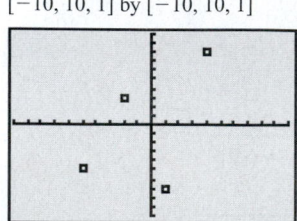

FIGURE B.13   **FIGURE B.14**

SUMMARY: MAKING A SCATTERPLOT OR A LINE GRAPH

The following are basic steps necessary to make either a scatterplot or a line graph.

1. Use (STAT) (1) to access lists L1 and L2.
2. If list L1 is not empty, place the cursor on L1 and press (CLEAR) (ENTER). Repeat for list L2, if it is not empty.
3. Enter the *x*-values into list L1 and the *y*-values into list L2.
4. Use (2nd) (Y = [STAT PLOT]) to set the appropriate parameters for the scatterplot or line graph.
5. Set an appropriate viewing rectangle. Otherwise, press (ZOOM) (9). This feature automatically sets the viewing rectangle and plots the data.

Note: (ZOOM) (9) *cannot* be used to set a viewing rectangle for the graph of a function.

Entering a Formula for a Function

To enter the formula for a function f, press (Y =). For example, use the following keystrokes after "$Y_1 = $" to enter $f(x) = 2x^2 - 3x + 7$. See Figure B.15.

(Y =) (CLEAR) (2) (X, T, θ, n) (^) (2) (−) (3) (X, T, θ, n) (+) (7)

Note that there is a built-in key to enter the variable X. If "$Y_1 = $" does not appear after pressing (Y =), press (MODE) and make sure the calculator is set in function mode denoted "Func". See Figure B.16.

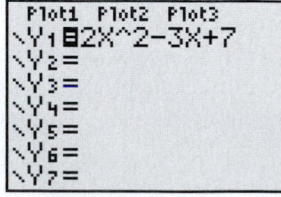

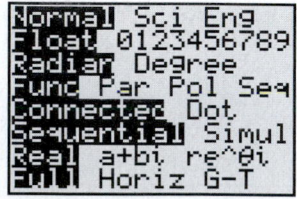

FIGURE B.15   **FIGURE B.16**

SUMMARY: ENTERING A FORMULA FOR A FUNCTION

To enter the formula for a function, press $\boxed{Y=}$. To delete an existing formula, press $\boxed{\text{CLEAR}}$. Then enter the symbolic representation for the function.

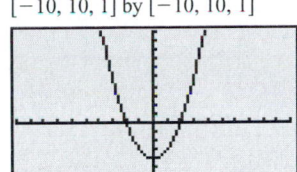

FIGURE B.17

$[-10, 10, 1]$ by $[-10, 10, 1]$

FIGURE B.18

Graphing a Function

To graph a function, such as $f(x) = x^2 - 4$, start by pressing $\boxed{Y=}$ and enter $Y_1 = X^2 - 4$. If there is an equation already entered, remove it by pressing $\boxed{\text{CLEAR}}$. The equals signs in "$Y_1 =$" should be in reverse video (a dark rectangle surrounding a white equals sign), which indicates that the equation will be graphed. If the equals sign is not in reverse video, place the cursor over it and press $\boxed{\text{ENTER}}$. Set an appropriate viewing rectangle, and then press $\boxed{\text{GRAPH}}$. The graph of f will appear in the specified viewing rectangle. See Figures B.17 and B.18.

SUMMARY: GRAPHING A FUNCTION

Use the $\boxed{Y=}$ menu to enter the formula for the function and the $\boxed{\text{WINDOW}}$ menu to set an appropriate viewing rectangle. Then press $\boxed{\text{GRAPH}}$.

Evaluating a Function Graphically

In Example 7, Section 1.3, we evaluate the function $f(x) = 0.72x + 2$ graphically at $x = 65$. Begin by entering the formula $Y_1 = .72X + 2$ into the $\boxed{Y=}$ menu. Then graph f in the appropriate viewing rectangle, as shown in Figure 1.48. To evaluate $f(65)$, use the following keystrokes, which access the CALCULATE menu.

$$\boxed{\text{2nd}}\ \boxed{\text{TRACE [CALC]}}\ \boxed{1}\ \boxed{6}\ \boxed{5}\ \boxed{\text{ENTER}}$$

See Figures 1.49 and 1.50.

SUMMARY: EVALUATING A FUNCTION GRAPHICALLY

To evaluate a function graphically, begin by graphing the function. Then use the following keystrokes to use the "value" routine in the CALCULATE menu.

$$\boxed{\text{2nd}}\ \boxed{\text{TRACE [CALC]}}\ \boxed{1}$$

Then enter the x-value where the function should be evaluated, and press $\boxed{\text{ENTER}}$.

Making a Table

In Example 7, Section 1.3, a table of values is requested. Start by pressing $\boxed{Y=}$ and then entering the formula $Y_1 = .72X + 2$, as shown in Figure B.19 on the next page. To set the table parameters, press the following keystrokes. See Figure B.20.

$$\boxed{\text{2nd}}\ \boxed{\text{WINDOW [TBLSET]}}\ \boxed{6}\ \boxed{0}\ \boxed{\text{ENTER}}\ \boxed{1}$$

These keystrokes specify a table that starts at $x = 60$ and increments the x-values by 1.

Therefore, the values of Y_1 at $x = 60, 61, 62, \ldots$ appear in the table. To create this table, press the following keys.

$$\boxed{\text{2nd}}\ \boxed{\text{GRAPH [TABLE]}}$$

One can stroll through x- and y-values by using the arrow keys. See Figure B.21. Note that there is no first or last x-value in the table.

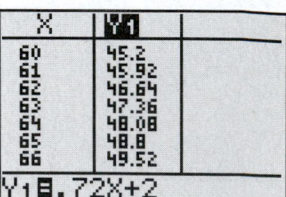

FIGURE B.19   FIGURE B.20   FIGURE B.21

SUMMARY: MAKING A TABLE OF A FUNCTION

Enter the formula for the function using $\boxed{\text{Y =}}$. Then press

$$\boxed{\text{2nd}}\ \boxed{\text{WINDOW [TBLSET]}}$$

to set the starting x-value and the increment between x-values appearing in the table. Create the table by pressing

$$\boxed{\text{2nd}}\ \boxed{\text{GRAPH [TABLE]}}.$$

Squaring a Viewing Rectangle

In a square viewing rectangle the graph of $y = x$ is a line that makes a 45° angle with the positive x-axis, a circle appears circular, and all sides of a square have the same length. An approximate square viewing rectangle can be set if the distance along the x-axis is 1.5 times the distance along the y-axis. Examples of viewing rectangles that are (approximately) square include

$$[-6, 6, 1] \text{ by } [-4, 4, 1] \quad \text{and} \quad [-9, 9, 1] \text{ by } [-6, 6, 1].$$

Square viewing rectangles can be set automatically by pressing either

$$\boxed{\text{ZOOM}}\ \boxed{4} \quad \text{or} \quad \boxed{\text{ZOOM}}\ \boxed{5}.$$

ZOOM 4 provides a decimal window, which is discussed later. See Figure B.22.

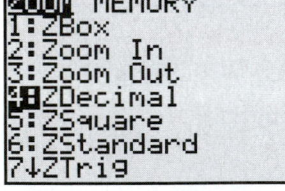

FIGURE B.22

SUMMARY: SQUARING A VIEWING RECTANGLE

Either $\boxed{\text{ZOOM}}\ \boxed{4}$ or $\boxed{\text{ZOOM}}\ \boxed{5}$ may be used to produce a square viewing rectangle. An (approximately) square viewing rectangle has the form

$$[-1.5k, 1.5k, 1] \text{ by } [-k, k, 1],$$

where k is a positive number.

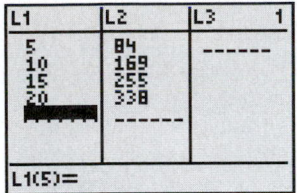

FIGURE B.23

Plotting Data and an Equation

In Example 6, Section 2.1, we plot data and graph a modeling function in the same xy-plane. (You may want to refer to the subsection on making a scatterplot and line graph in this appendix.) Start by entering the x-values into list L1 and the y-values into list L2, as shown in Figure B.23. Then press $\boxed{Y=}$ and enter the formula $Y_1 = 17X$ for $f(x)$. Make sure that "STATPLOT" is on and set an appropriate viewing rectangle. See Figures B.24 and B.25, and note that Figure B.24 shows "Plot1" in reverse video, which indicates that the scatterplot is on. Now press $\boxed{GRAPH}$ to have both the scatterplot and the graph of Y_1 appear in the same viewing rectangle, as shown in Figure B.26.

[0, 25, 5] by [0, 350, 50]

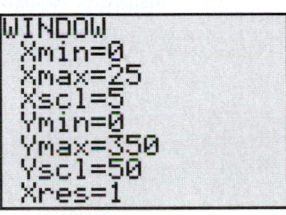

FIGURE B.24

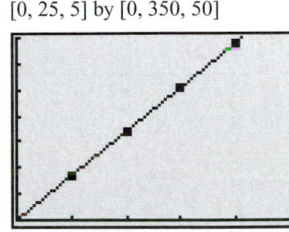

FIGURE B.25

FIGURE B.26

SUMMARY: PLOTTING DATA AND AN EQUATION

1. Enter the x-values into list L1 and the y-values into list L2 using the STAT EDIT menu. Turn on "Plot1" so that the scatterplot appears.
2. Use the $\boxed{Y=}$ menu to enter the equation to be graphed.
3. Use $\boxed{WINDOW}$ or $\boxed{ZOOM}$ to set an appropriate viewing rectangle.
4. Press $\boxed{GRAPH}$ to graph both the scatterplot and the equation in the same viewing rectangle.

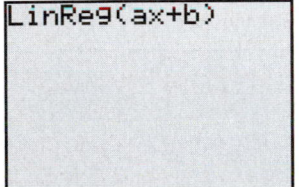

FIGURE B.27

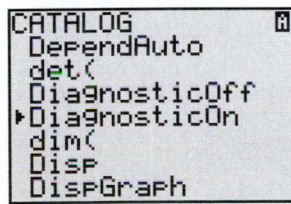

FIGURE B.28

Finding the Line of Least-Squares Fit

In Example 7, Section 2.1, the line of least-squares fit for the points (1, 1), (2, 3), and (3, 4) is found. Begin by entering the points in the same way a scatterplot is entered. See Figure 2.8, where the x-values are in list L1 and the y-values are in list L2.

After the data have been entered, perform the following keystrokes from the home screen.

$$\boxed{CLEAR}\ \boxed{STAT}\ \boxed{\triangleright}\ \boxed{4}$$

See Figure 2.9. This causes "LinReg(ax+b)" to appear on the home screen, as shown in Figure B.27. The graphing calculator assumes that the x-values are in list L1 and the y-values are in list L2. Now press $\boxed{ENTER}$. The result is shown in Figure 2.10.

If the correlation coefficient r does not appear, then enter the following keystrokes.

$$\boxed{2nd}\ \boxed{0\ [CATALOG]}$$

and scroll down until "DiagnosticsOn" is found. Press $\boxed{ENTER}$ twice. See Figures B.28 and B.29. The graph of the data and the least-squares regression line are shown in Figure 2.11.

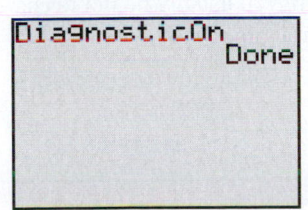

FIGURE B.29

SUMMARY: LINEAR LEAST-SQUARES FIT

1. Enter the data using (STAT)(1), as is done for a scatterplot. Input the x-values into list L1 and the y-values into list L2.
2. Press (STAT)(▷)(4) from the home screen to access the least-squares regression line. Press (ENTER) to start the computation.

FIGURE B.30

Locating a Point of Intersection

In Example 6, Section 2.3, we find the point of intersection for two lines. To find the point of intersection for the graphs of

$$f(x) = 5.91x + 13.7 \quad \text{and} \quad g(x) = -4.71x + 64.7,$$

start by entering Y_1 and Y_2, as shown in Figure B.30. Set the viewing rectangle to $[0, 12, 2]$ by $[0, 100, 10]$, and graph both equations in the same viewing rectangle, as shown in Figure 2.30. Then press the following keys to find the intersection point.

(2nd)(TRACE [CALC])(5)

See Figure B.31, where the "intersect" utility is being selected. The calculator prompts for the first curve, as shown in Figure B.32. Use the arrow keys to locate the cursor near the point of intersection and press (ENTER). Repeat these steps for the second curve, as shown in Figure B.33. Finally we are prompted for a guess. For each of the three prompts, place the free-moving cursor near the point of intersection and press (ENTER). The approximate coordinates of the point of intersection are shown in Figure 2.31.

$[0, 12, 2]$ by $[0, 100, 10]$ $[0, 12, 2]$ by $[0, 100, 10]$

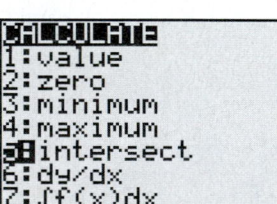   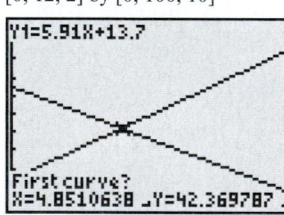   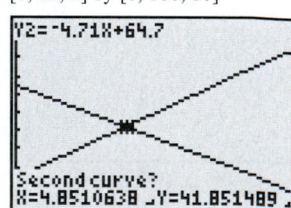

FIGURE B.31   **FIGURE B.32**   **FIGURE B.33**

SUMMARY: FINDING A POINT OF INTERSECTION

1. Graph the two functions in an appropriate viewing rectangle.
2. Press (2nd)(TRACE [CALC])(5).
3. Use the arrow keys to select an approximate location for the point of intersection. Press (ENTER) to make the three selections for "First curve?", "Second curve?", and "Guess?". (If the cursor is near the point of intersection, you usually do not need to move the cursor for each selection. Just press (ENTER) three times.)

Locating a Zero of a Function

In Example 4, Section 2.4, we locate an x-intercept or *zero* of the function f given by $f(x) = 1 - x - \left(\frac{1}{2}x - 2\right)$. Start by entering $Y_1 = 1 - X - (.5X - 2)$ into the (Y=)

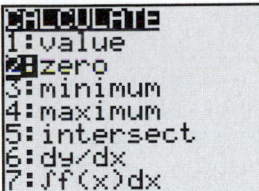

FIGURE B.34

menu. Set the viewing rectangle to $[-6, 6, 1]$ by $[-4, 4, 1]$ and graph Y_1. Afterward, press the following keys to invoke the zero finder. See Figure B.34.

$$\boxed{\text{2nd}}\ \boxed{\text{TRACE [CALC]}}\ \boxed{2}$$

The graphing calculator prompts for a left bound. Use the arrow keys to set the cursor to the left of the x-intercept and press $\boxed{\text{ENTER}}$. The graphing calculator then prompts for a right bound. Set the cursor to the right of the x-intercept and press $\boxed{\text{ENTER}}$. Finally the graphing calculator prompts for a guess. Set the cursor roughly at the x-intercept and press $\boxed{\text{ENTER}}$. See Figures B.35–B.37. The calculator then approximates the x-intercept or zero automatically, as shown in Figure 2.47.

$[-6, 6, 1]$ by $[-4, 4, 1]$ $[-6, 6, 1]$ by $[-4, 4, 1]$ $[-6, 6, 1]$ by $[-4, 4, 1]$

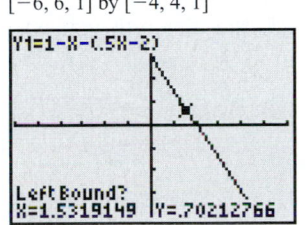 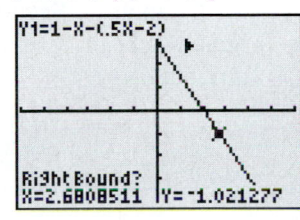 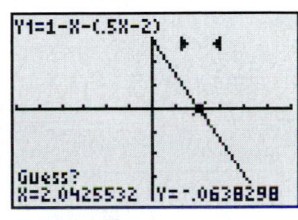

FIGURE B.35   **FIGURE B.36**   **FIGURE B.37**

SUMMARY: LOCATING A ZERO OF A FUNCTION

1. Graph the function in an appropriate viewing rectangle.
2. Press $\boxed{\text{2nd}}\ \boxed{\text{TRACE [CALC]}}\ \boxed{2}$.
3. Select the left and right bounds, followed by a guess. Press $\boxed{\text{ENTER}}$ after each selection. The calculator then approximates the zero.

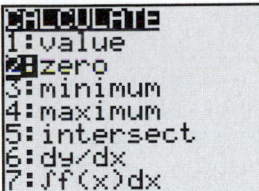

FIGURE B.38

Accessing the Greatest Integer Function

To access the greatest integer function, enter the following keystrokes from the home screen.

$$\boxed{\text{MATH}}\ \boxed{\triangleright}\ \boxed{5}$$

See Figure B.38.

SUMMARY: ACCESSING THE GREATEST INTEGER FUNCTION

1. Press $\boxed{\text{MATH}}$.
2. Position the cursor over "NUM".
3. Press $\boxed{5}$ to select the greatest integer function, which is denoted "int(".

Setting Connected and Dot Mode

In Figure 2.61 of Section 2.5 the greatest integer function is graphed in dot mode, and in Figure 2.62 it is graphed in connected mode. To set your graphing calculator in dot mode,

FIGURE B.39

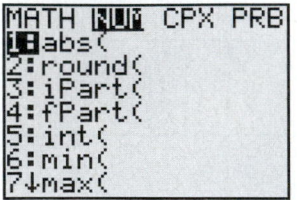

FIGURE B.40

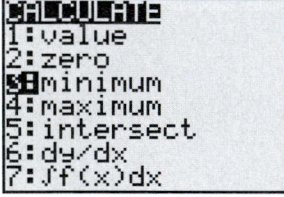

FIGURE B.41

press (MODE), position the cursor over "Dot", and press (ENTER). See Figure B.39. Graphs will now appear in dot mode rather than connected mode.

SUMMARY: SETTING CONNECTED OR DOT MODE

1. Press (MODE).
2. Position the cursor over "Connected" or "Dot". Press (ENTER).

Accessing the Absolute Value

In Example 4, Section 2.5, the absolute value is used to graph $f(x) = |x + 2|$. To graph f, begin by entering $Y_1 = abs(X + 2)$. The absolute value (abs) is accessed by pressing

(MATH) (▷) (1).

See Figure B.40.

SUMMARY: ACCESSING THE ABSOLUTE VALUE

1. Press (MATH).
2. Position the cursor over "NUM".
3. Press (1) to select the absolute value.

Finding Extrema (Minima and Maxima)

In Example 5, Section 3.1, we are asked to find a minimum point (or vertex) on the graph of $f(x) = 1.5x^2 - 6x + 4$. Start by entering $Y_1 = 1.5X^\wedge2 - 6X + 4$ from the (Y =) menu. Set the viewing rectangle to $[-4.7, 4.7, 1]$ by $[-3.1, 3.1, 1]$. Then perform the following keystrokes to find the minimum y-value.

(2nd) (TRACE [CALC]) (3)

See Figure B.41.

The calculator prompts for a left bound. Use the arrow keys to position the cursor left of the vertex and press (ENTER). Similarly, position the cursor to the right of the vertex for the right bound and press (ENTER). Finally, the graphing calculator asks for a guess between the left and right bounds. Place the cursor near the minimum point and press (ENTER). See Figures B.42–B.44. The minimum point (or vertex) is shown in Figure 3.13.

$[-4.7, 4.7, 1]$ by $[-3.1, 3.1, 1]$ $[-4.7, 4.7, 1]$ by $[-3.1, 3.1, 1]$ $[-4.7, 4.7, 1]$ by $[-3.1, 3.1, 1]$

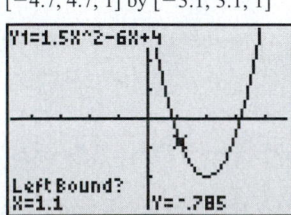   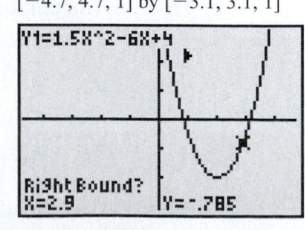   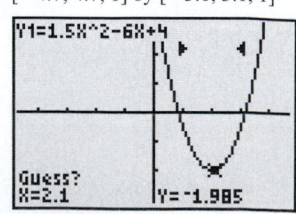

FIGURE B.42   **FIGURE B.43**   **FIGURE B.44**

A maximum of the function f on an interval can be found in a similar manner, except enter

The calculator prompts for left and right bounds, followed by a guess. Press [ENTER] after the cursor has been located appropriately for each prompt. An example of a maximum point is displayed in Figure 3.20.

SUMMARY: FINDING EXTREMA (MAXIMA AND MINIMA)

1. Graph the function in an appropriate viewing rectangle.
2. Press [2nd] [TRACE [CALC]] [3] to find a minimum point.
3. Press [2nd] [TRACE [CALC]] [4] to find a maximum point.
4. Use the arrow keys to locate the left and right x-bounds, followed by a guess. Press [ENTER] to select each position of the cursor.

Using the Ask Table Feature

In Example 10, Section 3.1, a table with x-values of 2, 3, 4, 5, 6 is created. Start by entering $Y_1 = 3(X - 4)^2 - 2$. To obtain the table shown in Figure 3.21, use the "Ask" feature rather than the "Auto" feature for the independent variable (Indpnt:). Press [2nd] [GRAPH [TABLE]]. Whenever an x-value is entered, the corresponding y-value is calculated automatically. See Figures B.45 and B.46.

FIGURE B.45

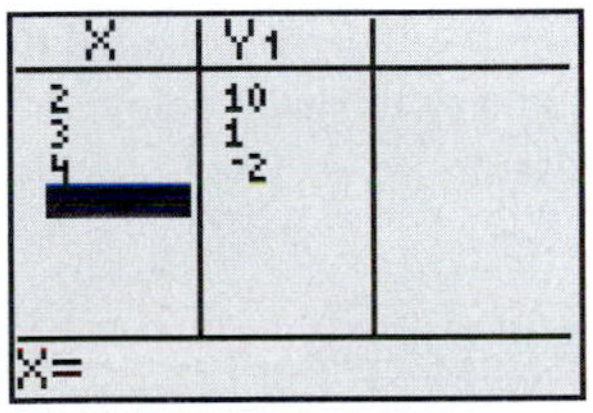

SUMMARY: USING THE ASK FEATURE FOR A TABLE

1. Enter the formula for $f(x)$ into Y_1 by using the [Y =] menu.
2. Press [2nd] [WINDOW [TBLSET]] to access "TABLE SETUP" and then select "Ask" for the independent variable (Indpnt:). "TblStart" and "ΔTbl" do not need to be set.
3. Enter x-values of your choice. The corresponding y-values will be calculated automatically.

FIGURE B.46

Finding a Nonlinear Function of Least-Squares Fit

In Example 12, Section 3.1, a quadratic function of least-squares fit is found in a manner similar to the way a linear function of least-squares fit is found. To solve Example 12, start by pressing [STAT] [1] and then enter the data points from Table 3.5, as shown in Figure 3.27. Input the x-values into list L1 and the y-values into list L2. To find the equation for a quadratic polynomial of least-squares fit, perform the following keystrokes from the home screen.

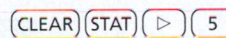

This causes the "Quadreg" to appear on the home screen. The graphing calculator assumes that the x-values are in list L1 and the y-values are in list L2, unless otherwise designated. Press [ENTER] to obtain the quadratic regression equation, as shown in Figure 3.29. A graph of the data and the regression equation are shown in Figure 3.30.

Other types of regression equations, such as cubic, quartic, power, and exponential, can be selected from the STAT CALC menu. See Figure B.47.

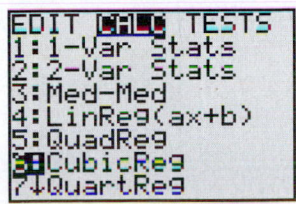

FIGURE B.47

SUMMARY: NONLINEAR LEAST-SQUARES FIT

1. Enter the data using (STAT)(1). Input the *x*-values into list L1 and the *y*-values into list L2, as is done for a scatterplot.
2. From the home screen press (STAT)(▷) and select a type of least-squares modeling function from the menu. Press (ENTER) to initiate the computation.

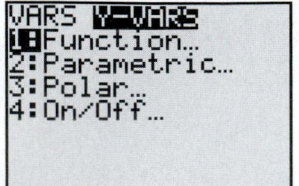

FIGURE B.48

Accessing the Variable Y₁

In Figure 3.90, Section 3.4, the expressions $-Y_1$ and $Y_1(-X)$ in the (Y=) menu are used to graph reflections. The Y_1 variable can be found by pressing the following keys. See Figures B.48 and B.49.

(VARS)(▷)(1)(1)

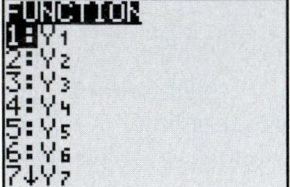

FIGURE B.49

SUMMARY: ACCESSING THE VARIABLE Y₁

1. Press (VARS).
2. Position the cursor over "Y-VARS".
3. Press (1) twice.

These keystrokes will make Y_1 appear on the screen.

Shading between Two Graphs

In Example 7, Section 3.4, the region below the graph of $f(x) = -0.4x^2 + 4$ is shaded to make it appear like a mountain, as illustrated in Figure 3.109. One way to shade below the graph of f is to begin by entering $Y_1 = -.4X^2 + 4$ after pressing (Y=). Then use the following keystrokes from the home screen.

(2nd)(PRGM [DRAW])(7)((−))(5)(,)(VARS)(▷)(1)(1)())

The expression Shade($-5, Y_1$) should appear on your home screen. See Figures B.50 and B.51. The shading utility is accessed from the DRAW menu, where this shading option requires a lower function and an upper function, respectively, separated by a comma. When (ENTER) is pressed the graphing calculator shades between the graph of the lower function and the graph of the upper function. For the lower function we have arbitrarily selected $y = -5$ because its graph lies below the graph of f and does not appear in the viewing rectangle in Figure 3.109. Instead of entering the variable Y_1, we could enter the formula $-.4X^2 + 4$ for the upper function.

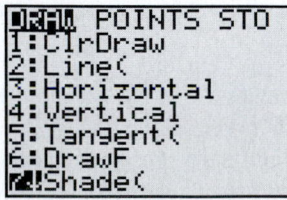

FIGURE B.50

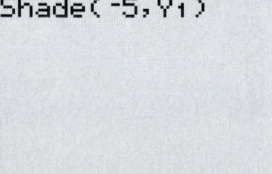

FIGURE B.51

SUMMARY: SHADING A GRAPH

1. Press (2nd) (PRGM [DRAW]) (7) from the home screen.
2. Enter a formula or a variable such as Y_1 for the lower function followed by a comma.
3. Enter a formula or a variable such as Y_2 for the upper function followed by a right parenthesis.
4. Set an appropriate viewing rectangle.
5. Press (ENTER). The region between the two graphs will be shaded.

Copying a Regression Equation into Y_1

FIGURE B.52

In Example 7, Section 4.2, we use cubic regression to model real data. The resulting formula for the cubic function shown in Figure 4.60 is quite complicated, and it is tedious to enter into "Y_1 =" by hand. A graphing calculator has the capability of copying this equation into Y_1 automatically. To do this, clear the equation for "Y_1 =". Then enter Y_1 after "CubicReg", as shown in Figure B.52. When (ENTER) is pressed, the regression equation is calculated, and then it is copied into "Y_1 =", as shown in Figure B.53. The following keystrokes may be used from the home screen. Be sure to enter the data into lists L1 and L2.

$$(STAT)\ (\triangleright)\ (6)\ (VARS)\ (\triangleright)\ (1)\ (1)\ (ENTER)$$

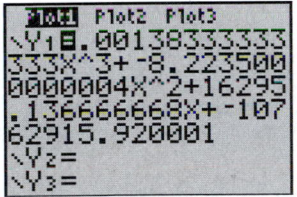

FIGURE B.53

SUMMARY: COPYING A REGRESSION EQUATION INTO Y_1

1. Clear Y_1 in the (Y =) menu, if an equation is present. Return to the home screen.
2. Select a type of regression from the STAT CALC menu.
3. Press (VARS) (▷) (1) (1) (ENTER).

Evaluating Complex Arithmetic

Complex arithmetic can be performed much like other arithmetic expressions. This is done by entering

$$(2nd)\ (.\ [i])$$

to obtain the imaginary unit i from the home screen. For example, to add the numbers $(-2 + 3i) + (4 - 6i)$, perform the following keystrokes on the home screen.

$$(\ (-) \ 2 \ + \ 3 \ (2nd)(.[i]) \ ) \ + \ (\ 4 \ - \ 6 \ (2nd)(.[i]) \ ) \ (ENTER)$$

The result is shown in the first two lines of Figure 4.76. Other complex arithmetic operations are done similarly.

SUMMARY: EVALUATING COMPLEX ARITHMETIC

Enter a complex expression in the same way as any arithmetic expression. To obtain the complex number i, use (2nd)(.[i]).

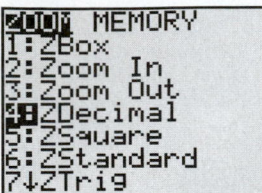

FIGURE B.54

Setting a Decimal Window

In Example 9, Section 4.5, a decimal (or friendly) window is used to trace the graph of f. With a decimal window, the cursor stops on convenient x-values. In the decimal window $[-9.4, 9.4, 1]$ by $[-9.4, 9.4, 1]$ the cursor stops on x-values that are multiples of 0.2. If we reduce the viewing rectangle to $[-4.7, 4.7, 1]$ by $[-3.1, 3.1, 1]$ the cursor stops on x-values that are multiples of 0.1. To set this smaller window automatically, press ⌊ZOOM⌋ ⌊4⌋. See Figure B.54. Decimal windows are also useful when graphing rational functions with asymptotes in connected mode.

> **SUMMARY: SETTING A DECIMAL WINDOW**
>
> 1. Press ⌊ZOOM⌋⌊4⌋ to set the viewing rectangle $[-4.7, 4.7, 1]$ by $[-3.1, 3.1, 1]$.
> 2. A larger decimal window is $[-9.4, 9.4, 1]$ by $[-6.2, 6.2, 1]$.

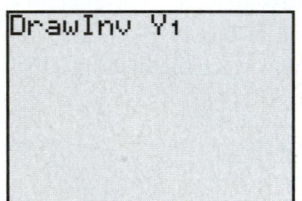

FIGURE B.55

$[-6, 6, 1]$ by $[-4, 4, 1]$

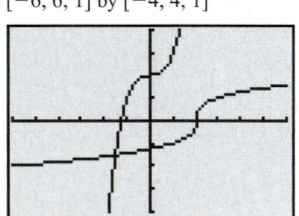

FIGURE B.56

Graphing an Inverse Function

In Example 7, Section 5.2, the inverse function of $f(x) = x^3 + 2$ is graphed. A graphing calculator can graph the inverse of a function without a formula for $f^{-1}(x)$. Begin by entering $Y_1 = X^{\wedge}3 + 2$ into the ⌊Y =⌋ menu. Then return to the home screen by pressing

⌊2nd⌋⌊MODE [QUIT]⌋.

The "DrawInv" utility may be accessed by pressing

⌊2nd⌋⌊PRGM [DRAW]⌋⌊8⌋,

followed by

⌊VARS⌋⌊▷⌋⌊1⌋⌊1⌋

to obtain the variable Y_1. See Figure B.55. Pressing ⌊ENTER⌋ causes both Y_1 and its inverse to be graphed, as shown in Figure B.56.

> **SUMMARY: GRAPHING AN INVERSE FUNCTION**
>
> 1. Enter the formula for $f(x)$ into Y_1 using the ⌊Y =⌋ menu.
> 2. Set an appropriate viewing rectangle by pressing ⌊WINDOW⌋.
> 3. Return to the home screen by pressing ⌊2nd⌋⌊MODE [QUIT]⌋.
> 4. Press ⌊2nd⌋⌊PRGM [DRAW]⌋⌊8⌋⌊VARS⌋⌊▷⌋⌊1⌋⌊1⌋⌊ENTER⌋ to create the graphs of f and f^{-1}.

Shading a System of Inequalities

In Example 6(b), Section 9.2, we are asked to shade the solution set for the system of linear inequalities $x + 3y \le 9$, $2x - y \le -1$. Begin by solving each system for y to obtain $y \le \frac{(9 - x)}{3}$ and $y \ge 2x + 1$. Then let $Y_1 = \frac{(9 - X)}{3}$ and $Y_2 = 2X + 1$, as shown in Figure 9.33. Position the cursor to left of Y_1, and press ⌊ENTER⌋ three times. The triangle that appears indicates that the calculator will shade the region below the graph of Y_1. Next locate the cursor to the left of Y_2 and press ⌊ENTER⌋ twice. This triangle indicates that the calculator will shade the region above the graph of Y_2. After setting the viewing rectangle to $[-6, 6, 1]$ by $[-6, 6, 1]$ press (GRAPH). The result is shown in Figure 9.34. The solution set could also be shaded using Shade(Y_2, Y_1) from the home screen. See Figures B.57 and B.58.

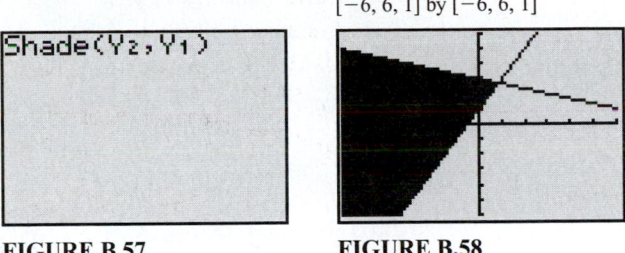

$[-6, 6, 1]$ by $[-6, 6, 1]$

FIGURE B.57   **FIGURE B.58**

SUMMARY: SHADING A SYSTEM OF EQUATIONS

1. Solve each inequality for y.
2. Enter each formula as Y_1 and Y_2 in the (Y =) menu.
3. Locate the cursor to the left of Y_1 and press (ENTER) two or three times, to shade either above or below the graph of Y_1. Repeat for Y_2.
4. Set an appropriate viewing rectangle.
5. Press (GRAPH).

Note: The "Shade" utility under the DRAW menu can also be used to shade the region *between* two graphs.

Entering the Elements of a Matrix

In Example 9, Section 9.4, the elements of a matrix are entered. The augmented matrix A is given by

$$A = \begin{bmatrix} 2 & 1 & 2 & | & 10 \\ 1 & 0 & 2 & | & 5 \\ 1 & -2 & 2 & | & 1 \end{bmatrix}.$$

Use the following keystrokes on the TI-83 Plus and TI-84 Plus to define a matrix A with dimension 3×4. (*Note*: On the TI-83 graphing calculator the matrix menu is found by pressing (MATRIX).)

(2nd) (x^{-1} [MATRIX]) (▷) (▷) (1) (3) (ENTER) (4) (ENTER)

See Figures 9.40 and 9.41.

Then input the 12 elements of the matrix A, row by row, as shown in Figure 9.41. Finish each entry by pressing (ENTER). After these elements have been entered, press

(2nd) (MODE [QUIT])

to return to the home screen. To display the matrix A, press

(2nd) (x^{-1} [MATRIX]) (1) (ENTER).

See Figure 9.42.

SUMMARY: ENTERING THE ELEMENTS OF A MATRIX *A*

1. Begin by accessing the matrix *A* by pressing (2nd)(*x*⁻¹ [MATRIX])(▷)(▷)(1).
2. Enter the dimension of *A* by pressing (*m*)(ENTER)(*n*)(ENTER), where the dimension of the matrix is *m* × *n*.
3. Input each element of the matrix, row by row. Finish each entry by pressing (ENTER). Use (2nd)(MODE [QUIT]) to return to the home screen.

Note: On the TI-83, replace the keystrokes (2nd)(*x*⁻¹ [MATRIX]) with (MATRIX).

Reduced Row-Echelon Form

In Example 9, Section 9.4, the reduced row-echelon form of a matrix is found. To find this reduced row-echelon form, use the following keystrokes from the home screen on the TI-83 Plus and TI-84 Plus.

(2nd)(*x*⁻¹ [MATRIX])(▷)(ALPHA)(APPS [B])(2nd)(*x*⁻¹ [MATRIX])(1)( ))(ENTER)

The resulting matrix is shown in Figure 9.44. On the TI-83 graphing calculator use the following keystrokes to find the reduced row-echelon form.

(MATRIX)(▷)(ALPHA)(MATRIX [B])(MATRIX)(1)( ))(ENTER)

SUMMARY: FINDING REDUCED ROW-ECHELON FORM OF A MATRIX

1. To make rref([A]) appear on the home screen, use the following keystrokes for the TI-83 Plus and TI-84 Plus graphing calculators.

(2nd)(*x*⁻¹ [MATRIX])(▷)(ALPHA)(APPS [B])(2nd)(*x*⁻¹ [MATRIX])(1)( ))(ENTER)

2. Press (ENTER) to calculate the reduced row-echelon form.
3. Use arrow keys to access elements that do not appear on the screen.

Note: On the TI-83, replace the keystrokes (2nd)(*x*⁻¹ [MATRIX]) with (MATRIX) and (APPS [B]) with (MATRIX [B]).

Performing Arithmetic Operations on Matrices

In Example 8, Section 9.5, the matrices *A* and *B* are multiplied. Begin by entering the elements for the matrices *A* and *B*. The following keystrokes can be used to define a matrix *A* with dimension 3 × 3.

(2nd)(*x*⁻¹ [MATRIX])(▷)(▷)(1)(3)(ENTER)(3)(ENTER)

Next input the 9 elements in the matrix *A*, row by row. Finish each entry by pressing (ENTER). See Figure 9.61. Repeat this process to define a matrix *B* with dimension 3 × 3.

(2nd)(*x*⁻¹ [MATRIX])(▷)(▷)(2)(3)(ENTER)(3)(ENTER)

Enter the 9 elements in *B*. See Figure 9.62. After the elements of *A* and *B* have been entered, press

$$\boxed{\text{2nd}}\ \boxed{\text{MODE [QUIT]}}$$

to return to the home screen. To multiply the expression *AB*, use the following keystrokes from the home screen.

$$\boxed{\text{2nd}}\ \boxed{x^{-1}\text{[MATRIX]}}\ \boxed{1}\ \boxed{\times}\ \boxed{\text{2nd}}\ \boxed{x^{-1}\text{[MATRIX]}}\ \boxed{2}\ \boxed{\text{ENTER}}$$

The result is shown in Figure 9.63.

SUMMARY: PERFORMING ARITHMETIC OPERATIONS ON MATRICES

1. Enter the elements of each matrix beginning with the keystrokes, $\boxed{\text{2nd}}\ \boxed{x^{-1}\text{[MATRIX]}}\ \boxed{\triangleright}\ \boxed{\triangleright}\ \boxed{k}\ \boxed{m}\ \boxed{\text{ENTER}}\ \boxed{n}\ \boxed{\text{ENTER}}$, where *k* is the menu number of the matrix and the dimension of the matrix is $m \times n$.
2. Return to the home screen by pressing $\boxed{\text{2nd}}\ \boxed{\text{MODE [QUIT]}}$.
3. Enter the matrix expression followed by $\boxed{\text{ENTER}}$. Use the keystrokes $\boxed{\text{2nd}}$ $\boxed{x^{-1}\text{[MATRIX]}}\ \boxed{k}$ to access the matrix with menu number *k* on the TI-83 Plus and TI-84 Plus.

Note: On the TI-83, replace the keystrokes $\boxed{\text{2nd}}\ \boxed{x^{-1}\text{[MATRIX]}}$ with $\boxed{\text{MATRIX}}$.

Finding the Inverse of a Matrix

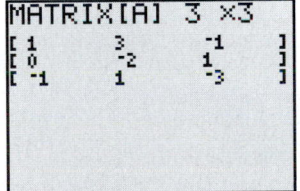

FIGURE B.59

In Example 8, Section 9.6, the inverse of *A*, denoted A^{-1}, is displayed in Figure 9.67. To calculate A^{-1}, start by entering the elements of the matrix *A*, as shown in Figure B.59. To compute A^{-1}, perform the following keystrokes from the home screen.

$$\boxed{\text{2nd}}\ \boxed{x^{-1}\text{[MATRIX]}}\ \boxed{1}\ \boxed{x^{-1}}\ \boxed{\text{ENTER}}$$

The results are shown in Figure 9.67.

SUMMARY: FINDING THE INVERSE OF A SQUARE MATRIX

1. Enter the elements of the square matrix *A*.
2. Return to the home screen by pressing

$$\boxed{\text{2nd}}\ \boxed{\text{MODE [QUIT]}}.$$

3. Perform the following keystrokes from the home screen to display A^{-1}.

$$\boxed{\text{2nd}}\ \boxed{x^{-1}\text{[MATRIX]}}\ \boxed{1}\ \boxed{x^{-1}}\ \boxed{\text{ENTER}}$$

Note: On the TI-83, replace the keystrokes $\boxed{\text{2nd}}\ \boxed{x^{-1}\text{[MATRIX]}}$ with $\boxed{\text{MATRIX}}$.

Solving a Linear System with a Matrix Inverse

In Example 8, Section 9.6, the solution to a system of equations is found. The matrix equation $AX = B$ has the solution $X = A^{-1}B$ provided A^{-1} exists, and is given by

$$AX = \begin{bmatrix} 1 & 3 & -1 \\ 0 & -2 & 1 \\ -1 & 1 & -3 \end{bmatrix} \begin{bmatrix} x \\ y \\ z \end{bmatrix} = \begin{bmatrix} 6 \\ -2 \\ 4 \end{bmatrix} = B.$$

To solve this equation, start by entering the elements of the matrices A and B. To compute the solution $A^{-1}B$, perform the following keystrokes from the home screen.

(2nd) (x^{-1} [MATRIX]) (1) (x^{-1}) ($\times$) (2nd) (x^{-1} [MATRIX]) (2) (ENTER)

The results are shown in Figure 9.68.

SUMMARY: SOLVING A LINEAR SYSTEM WITH A MATRIX INVERSE

1. Write the system of equations as $AX = B$.
2. Enter the elements of the matrices for A and B.
3. Return to the home screen by pressing

(2nd) (MODE [QUIT]).

4. Perform the following keystrokes.

(2nd) (x^{-1} [MATRIX]) (1) (x^{-1}) ($\times$) (2nd) (x^{-1} [MATRIX]) (2) (ENTER)

Note: On the TI-83, replace the keystrokes (2nd) (x^{-1} [MATRIX]) with (MATRIX).

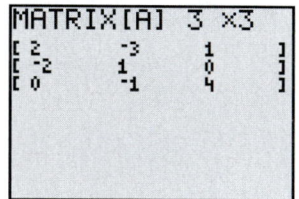

FIGURE B.60

Evaluating a Determinant

In Example 4(a), Section 9.7, a graphing calculator is used to evaluate a determinant of a matrix. Start by entering the 9 elements of the 3×3 matrix A, as shown in Figure B.60. To compute det A, perform the following keystrokes from the home screen.

(2nd) (x^{-1} [MATRIX]) ($\triangleright$) (1) (2nd) (x^{-1} [MATRIX]) (1) ( )) (ENTER)

The results are shown in Figure 9.70.

SUMMARY: EVALUATING A DETERMINANT OF A MATRIX

1. Enter the elements of the matrix A.
2. Return to the home screen by pressing

(2nd) (MODE [QUIT]).

3. Perform the following keystrokes.

(2nd) (x^{-1} [MATRIX]) ($\triangleright$) (1) (2nd) (x^{-1} [MATRIX]) (1) ( )) (ENTER)

Note: On the TI-83, replace the keystrokes (2nd) (x^{-1} [MATRIX]) with (MATRIX).

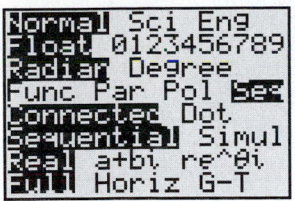

FIGURE B.61

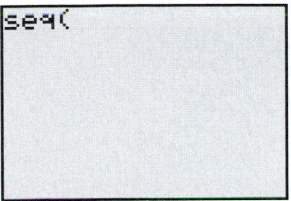

FIGURE B.62

Creating a Sequence

A graphing calculator can be used to calculate the terms of the sequence given by $f(n) = 2n - 5$ for $n = 1, 2, 3, 4$. See Example 1(a), Section 11.1. Start by setting the mode of the calculator to sequence "Seq" using the following keystrokes. See Figure B.61.

$$\boxed{\text{MODE}}\ \boxed{\triangledown}\ \boxed{\triangledown}\ \boxed{\triangledown}\ \boxed{\triangleright}\ \boxed{\triangleright}\ \boxed{\triangleright}\ \boxed{\text{ENTER}}\ \boxed{\text{2nd}}\ \boxed{\text{MODE [QUIT]}}$$

Then enter the following from the home screen.

$$\boxed{\text{2nd}}\ \boxed{\text{STAT [LIST]}}\ \boxed{\triangleright}\ \boxed{5}$$

On the home screen "seq(" will appear, as shown in Figure B.62. This sequence utility requires four things to be entered—all separated by commas. They are the formula, the variable, the subscript of the first term, and the subscript of the last term. Use the following keystrokes to obtain the first four terms (a_1, a_2, a_3, a_4) of the sequence $a_n = 2n - 5$, as shown in Figure 11.1.

$$\boxed{2}\ \boxed{\text{X, T, }\theta\text{, }n}\ \boxed{-}\ \boxed{5}\ \boxed{,}\ \boxed{\text{X, T, }\theta\text{, }n}\ \boxed{,}\ \boxed{1}\ \boxed{,}\ \boxed{4}\ \boxed{)}\ \boxed{\text{ENTER}}$$

SUMMARY: CREATING A SEQUENCE

1. To create a sequence, use the keystrokes

$$\boxed{\text{2nd}}\ \boxed{\text{STAT [LIST]}}\ \boxed{\triangleright}\ \boxed{5}.$$

2. Enter the formula, the variable, the subscript of the first term, and the subscript of the last term—all separated by commas. For example, if you want the first 10 terms of $a_n = n^2$, $(a_1, a_2, a_3, \ldots, a_{10})$, enter seq($n^2$, n, 1, 10). Be sure to set your calculator in sequence mode.

3. Pressing $\boxed{\text{ENTER}}$ causes the terms of the sequence to appear.

Entering, Tabling, and Graphing a Sequence

In Example 6, Section 11.1, a table and a graph of a sequence are created with a graphing calculator. The calculator should be set to sequence mode by entering the following keystrokes.

$$\boxed{\text{MODE}}\ \boxed{\triangledown}\ \boxed{\triangledown}\ \boxed{\triangledown}\ \boxed{\triangleright}\ \boxed{\triangleright}\ \boxed{\triangleright}\ \boxed{\text{ENTER}}$$

To enter the formula for a sequence, press $\boxed{\text{Y =}}$. See Figure B.63. Let $n\text{Min} = 1$ since the initial value of n is equal to 1. To enter $a_n = 2.85a_{n-1} - .19a_{n-1}^2$, use the following keystrokes after clearing out any old formula. (Notice that the graphing calculator uses u instead of a to denote a term of the sequence.)

$$\boxed{2}\ \boxed{.}\ \boxed{8}\ \boxed{5}\ \boxed{\text{2nd}}\ \boxed{7[u]}\ \boxed{(}\ \boxed{\text{X, T, }\theta\text{, }n}\ \boxed{-}\ \boxed{1}\ \boxed{)}\ \boxed{-}\ \boxed{.}\ \boxed{1}\ \boxed{9}$$

$$\boxed{\text{2nd}}\ \boxed{7[u]}\ \boxed{(}\ \boxed{\text{X, T, }\theta\text{, }n}\ \boxed{-}\ \boxed{1}\ \boxed{)}\ \boxed{\wedge}\ \boxed{2}$$

Since $a_1 = 1$, let $u(n\text{Min}) = \{1\}$. This can be performed as follows.

$$\boxed{\text{CLEAR}}\ \boxed{\text{2nd}}\ \boxed{(}\ \boxed{1}\ \boxed{\text{2nd}}\ \boxed{)}$$

See Figure B.63.

To create a table for this sequence, starting with a_1 and incrementing n by 1, perform the following keystrokes.

[2nd] [WINDOW [TBLSET]] [1] [ENTER] [1] [2nd] [GRAPH [TABLE]]

See Figure B.64 and the figure in the margin beside Table 11.3.

To graph the first 20 terms of this sequence, start by selecting [WINDOW]. Since we want the first 20 terms plotted, let nMin = 1, nMax = 20, PlotStart = 1, and PlotStep = 1. The window can be set as [0, 21, 1] by [0, 14, 1]. See Figure B.65. To graph the sequence, press [GRAPH]. The resulting graph is shown in Figure 11.8.

FIGURE B.63

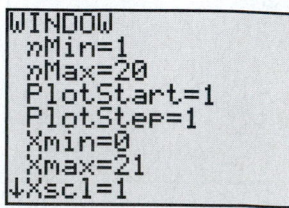

FIGURE B.64

FIGURE B.65

SUMMARY: ENTERING, TABLING, AND GRAPHING A SEQUENCE

1. Set the mode to "Seq" by using the [MODE] menu.
2. Enter the formula for the sequence by pressing [Y =].
3. To create a table of a sequence, set the start and increment values with

[2nd] [WINDOW [TBLSET]].

A table is created by pressing

[2nd] [GRAPH [TABLE]].

4. To graph a sequence, set the viewing rectangle by using [WINDOW]. Press [GRAPH] to create a graph of the sequence.

Finding the Sum of the Terms of a Series

In Example 3, Section 11.2, the sum of the series $\sum_{n=1}^{50} \left(\frac{1}{n^4} \right)$ is found by using a graphing calculator. Use the following keystrokes from the home screen.

[2nd] [STAT [LIST]] [▷] [▷] [5] [2nd] [STAT [LIST]] [▷] [5]

[1] [÷] [X, T, θ, n] [^] [4] [,] [X, T, θ, n] [,] [1] [,] [5] [0] [)] [)] [ENTER]

The results are shown in the first three lines of Figure 11.11.

SUMMARY: SUMMING A SERIES

1. Use [2nd] [STAT [LIST]] [▷] [▷] [5] to access the sum utility.
2. Use [2nd] [STAT [LIST]] [▷] [5] to access the sequence utility.
3. To use the sequence utility, see "Creating a Sequence" in this appendix.

Calculating Factorial Notation

In Example 5, Section 11.3, factorial notation is evaluated with a graphing calculator. The factorial utility is found under the MATH NUM menus. To calculate 8! use the following keystrokes from the home screen.

$$\boxed{8}\ \boxed{\text{MATH}}\ \boxed{\triangleright}\ \boxed{\triangleright}\ \boxed{\triangleright}\ \boxed{4}\ \boxed{\text{ENTER}}$$

The results are shown in the first two lines of Figure 11.15.

> ### SUMMARY: CALCULATING FACTORIAL NOTATION
>
> To calculate *n* factorial, use the following keystrokes.
>
> $$\boxed{n}\ \boxed{\text{MATH}}\ \boxed{\triangleright}\ \boxed{\triangleright}\ \boxed{\triangleright}\ \boxed{4}\ \boxed{\text{ENTER}}$$
>
> The value of *n* should be entered as a number, not a variable.

Calculating Permutations and Combinations

In Example 6(a), Section 11.3, the permutation $P(7, 3)$ is evaluated. To perform this calculation, use the following keystrokes from the home screen.

$$\boxed{7}\ \boxed{\text{MATH}}\ \boxed{\triangleright}\ \boxed{\triangleright}\ \boxed{\triangleright}\ \boxed{2}\ \boxed{3}\ \boxed{\text{ENTER}}$$

The results are shown in the first two lines of Figure 11.17.

In Example 9(a), Section 11.3, the combination $C(7, 3)$ can be calculated by using the following keystrokes.

$$\boxed{7}\ \boxed{\text{MATH}}\ \boxed{\triangleright}\ \boxed{\triangleright}\ \boxed{\triangleright}\ \boxed{3}\ \boxed{3}\ \boxed{\text{ENTER}}$$

The results are shown in the first two lines of Figure 11.18.

> ### SUMMARY: CALCULATING PERMUTATIONS AND COMBINATIONS
>
> 1. To calculate $P(n, r)$, use $\boxed{\text{MATH}}$ and select "PRB" followed by $\boxed{2}$.
> 2. To calculate $C(n, r)$, use $\boxed{\text{MATH}}$ and select "PRB" followed by $\boxed{3}$.

Graphing Parametric Equations

In Figure 8.64, Section 8.4, the parametric equations $x = t + 3, y = t^2$ for $-3 \le t \le 3$ are graphed. To set your graphing calculator in parametric mode, press $\boxed{\text{MODE}}$, position the cursor over "Par", and press $\boxed{\text{ENTER}}$. See Figure B.66. Next Press $\boxed{\text{Y} =}$ and enter the equations for x and y, as shown in Figure B.67.

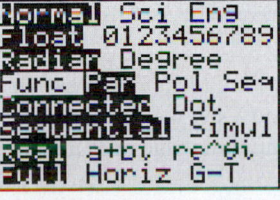

FIGURE B.66

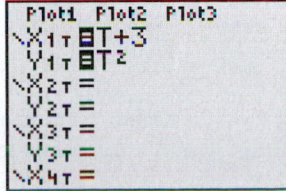

FIGURE B.67

To set a viewing rectangle, press (WINDOW). In addition to setting Xmin, Xmax, Xscl, Ymin, Ymax, and Yscl, you must set values for Tmin, Tmax, and Tstep. Tmin refers to the minimum value of t in the graph, and Tmax refers to the maximum value of t. It is given that $-3 \leq t \leq 3$, so it follows that Tmin $= -3$ and Tmax $= 3$. However, when an interval for t is not given, it may take a little experimentation to determine an appropriate interval for t. Tstep represents the increment between consecutive t-values on the graph. If Tstep is too large, the graph appears more like a line graph than a smooth curve. If Tstep is too small, the graphing calculator will take a long time to create the graph. Many times a reasonable value is Tstep $= 0.1$. See Figure 8.62. A parametric graph can be created by pressing (GRAPH).

Tables for parametric equations can be created. Press

(2nd) (WINDOW [TBLSET])

and proceed in the usual manner. Note that the variables TblStart and ΔTbl refer to t and not x. See Figure 8.63.

SUMMARY: GRAPHING PARAMETRIC EQUATIONS

1. Press (MODE), move the cursor to "Par", and press (ENTER).
2. Press (Y =) and enter the equations for x and y.
3. Press (WINDOW) and set the viewing rectangle. Be sure to set Tmin, Tmax, and Tstep. When in doubt, let Tstep $= 0.1$.
4. To make the graph appear, press (GRAPH).

Graphing in Polar Coordinates

In Figure 8.93, Section 8.5, the polar equation $r = 2 + 2 \cos \theta$ for $0° \leq \theta \leq 360°$ is graphed. To set your graphing calculator in polar coordinate mode, press (MODE), position the cursor over "Pol", and press (ENTER). See Figure B.68. Polar equations can be graphed in either degree or radian mode. To set your calculator in degree mode, position the cursor over "Degree" and press (ENTER). See Figure B.69. Next press (Y =) and enter the equation for "$r_1 =$", as shown in Figure B.70. Note that the polar equation must be solved for the variable r.

FIGURE B.68

FIGURE B.69

FIGURE B.70

To set a viewing rectangle, press (WINDOW). In addition to setting Xmin, Xmax, Xscl, Ymin, Ymax, and Yscl, you must set values for θmin, θmax, and θstep. The variable θmin refers to the minimum value of θ, and θmax refers to the maximum value of θ. Since $\cos \theta$ is periodic with 360°, the entire graph will appear if we let $0° \leq \theta \leq 360°$. Let θmin $= 0$ and θmax $= 360$. The variable θstep represents the increment between consecutive

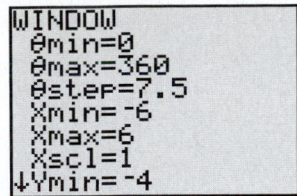

FIGURE B.71 Degree Mode

θ-values on the polar graph. If θstep is too large, the graph appears more like a line graph than a smooth curve. If θstep is too small, the graphing calculator will take a long time to create the graph. In degree mode a reasonable value for θstep is 7.5°, and in radian mode a reasonable value for θstep is 0.1 radian. See Figure B.71. A polar graph can be created by pressing (GRAPH).

Tables for polar coordinates can be created. Press

 (2nd) (WINDOW [TBLSET])

and proceed in the usual manner. Note that the variables TblStart and ΔTbl refer to θ and not x. See Figure 8.92.

SUMMARY: GRAPHING IN POLAR COORDINATES

1. Press (MODE), move the cursor to "Pol", and press (ENTER). Set the calculator to either degree or radian mode.
2. Press (Y =) and enter the polar equation.
3. Press (WINDOW) and set the viewing rectangle. Be sure to set θmin, θmax, and θstep. When in doubt, let θstep = 7.5 in degree mode and θstep = 0.1 in radian mode.
4. To make the graph appear, press (GRAPH).

Appendix C: Partial Fractions

Decomposition of Rational Expressions

The sums of rational expressions are found by combining two or more rational expressions into one rational expression. Here, the reverse process is considered: Given one rational expression, express it as the sum of two or more rational expressions. A special type of sum of rational expressions is called the **partial fraction decomposition**; each term in the sum is a **partial fraction**. The technique of decomposing a rational expression into partial fractions is useful in calculus and other areas of mathematics.

To form a partial fraction decomposition of a rational expression, follow these steps.

PARTIAL FRACTION DECOMPOSITION OF $\frac{f(x)}{g(x)}$

STEP 1: If $\frac{f(x)}{g(x)}$ is not a proper fraction (a fraction with the numerator of lower degree than the denominator), divide $f(x)$ by $g(x)$. For example,

$$\frac{x^4 - 3x^3 + x^2 + 5x}{x^2 + 3} = x^2 - 3x - 2 + \frac{14x + 6}{x^2 + 3}.$$

Then apply the following steps to the remainder, which is a proper fraction.

STEP 2: Factor $g(x)$ completely into factors of the form $(ax + b)^m$ or $(cx^2 + dx + e)^n$, where $cx^2 + dx + e$ is irreducible and m and n are positive integers.

STEP 3: **(a)** For each distinct linear factor $(ax + b)$, the decomposition must include the term

$$\frac{A}{ax + b}.$$

(b) For each repeated linear factor $(ax + b)^m$, the decomposition must include the terms

$$\frac{A_1}{ax + b} + \frac{A_2}{(ax + b)^2} + \cdots + \frac{A_m}{(ax + b)^m}.$$

STEP 4: **(a)** For each distinct quadratic factor $(cx^2 + dx + e)$, the decomposition must include the term

$$\frac{Bx + C}{cx^2 + dx + e}.$$

(b) For each repeated quadratic factor $(cx^2 + dx + e)^n$, the decomposition must include the terms

$$\frac{B_1x + C_1}{cx^2 + dx + e} + \frac{B_2x + C_2}{(cx^2 + dx + e)^2} + \cdots + \frac{B_nx + C_n}{(cx^2 + dx + e)^n}.$$

STEP 5: Use algebraic techniques to solve for the constants in the numerators of the decomposition.

To find the constants in **STEP 5**, the goal is to get a system of equations with as many equations as there are unknowns in the numerators. One method for finding these equations is to substitute values for x on each side of the rational equation formed in **STEPS 3** or **4**.

Distinct Linear Factors

EXAMPLE 1   Finding a partial fraction decomposition

Find the partial fraction decomposition of

$$\frac{2x^4 - 8x^2 + 5x - 2}{x^3 - 4x}$$

SOLUTION The given fraction is not a proper fraction; the numerator has higher degree than the denominator. Perform the division.

$$
\begin{array}{r}
2x \\
x^3 - 4x \overline{)2x^4 - 8x^2 + 5x - 2} \\
\underline{2x^4 - 8x^2 } \\
5x - 2
\end{array}
$$

The quotient is $\frac{2x^4 - 8x^2 + 5x - 2}{x^3 - 4x} = 2x + \frac{5x - 2}{x^3 - 4x}$. Now, work with the remainder fraction. Factor the denominator as $x^3 - 4x = x(x + 2)(x - 2)$. Since the factors are distinct linear factors, use **STEP 3(a)** to write the decomposition as

$$\frac{5x - 2}{x^3 - 4x} = \frac{A}{x} + \frac{B}{x + 2} + \frac{C}{x - 2}, \qquad \text{Equation 1}$$

where A, B, and C are constants that need to be found. Multiply each side of equation 1 by $x(x + 2)(x - 2)$ to get

$$5x - 2 = A(x + 2)(x - 2) + Bx(x - 2) + Cx(x + 2). \qquad \text{Equation 2}$$

Equation 1 is an identity, since both sides represent the same rational expression. Thus, equation 2 is also an identity. Equation 1 holds for all values of x except 0, -2, and 2. However, equation 2 holds for all values of x. In particular, substituting 0 for x in equation 2 gives $-2 = -4A$, so $A = \frac{1}{2}$. Similarly, choosing $x = -2$ gives $-12 = 8B$, so $B = -\frac{3}{2}$. Finally, choosing $x = 2$ gives $8 = 8C$, so $C = 1$. The remainder rational expression can be written as the following sum of partial fractions:

$$\frac{5x - 2}{x^3 - 4x} = \frac{1}{2x} + \frac{-3}{2(x + 2)} + \frac{1}{x - 2},$$

and the given rational expression can be written as

$$\frac{2x^4 - 8x^2 + 5x - 2}{x^3 - 4x} = 2x + \frac{1}{2x} + \frac{-3}{2(x + 2)} + \frac{1}{x - 2}.$$

Check the work by combining the terms on the right.   *Now Try Exercise 11* ◆

Repeated Linear Factors

EXAMPLE 2   Finding a partial fraction decomposition

Find the partial fraction decomposition of

$$\frac{2x}{(x - 1)^3}.$$

SOLUTION This is a proper fraction. The denominator is already factored with repeated linear factors. We write the decomposition as shown, by using **STEP 3(b)**.

$$\frac{2x}{(x-1)^3} = \frac{A}{x-1} + \frac{B}{(x-1)^2} + \frac{C}{(x-1)^3}$$

We clear denominators by multiplying each side of this equation by $(x-1)^3$.

$$2x = A(x-1)^2 + B(x-1) + C$$

Substituting 1 for x leads to $C = 2$, so

$$2x = A(x-1)^2 + B(x-1) + 2. \qquad \text{Equation 1}$$

We substituted the only root, and we still need to find values for A and B. However, *any* number can be substituted for x. For example, when we choose $x = -1$ (because it is easy to substitute), equation 1 becomes

$$-2 = 4A - 2B + 2$$

$$-4 = 4A - 2B$$

$$-2 = 2A - B. \qquad \text{Equation 2}$$

Substituting 0 for x in equation 1 gives

$$0 = A - B + 2$$

$$2 = -A + B. \qquad \text{Equation 3}$$

Now, we solve the system of equations 2 and 3 to get $A = 0$ and $B = 2$. The partial fraction decomposition is

$$\frac{2x}{(x-1)^3} = \frac{2}{(x-1)^2} + \frac{2}{(x-1)^3}.$$

We needed three substitutions because there were three constants to evaluate, A, B, and C. To check this result, we could combine the terms on the right.

Now Try Exercise 13 ◆

Distinct Linear and Quadratic Factors

EXAMPLE 3 Finding a partial fraction decomposition

Find the partial fraction decomposition of

$$\frac{x^2 + 3x - 1}{(x+1)(x^2+2)}.$$

SOLUTION This denominator has distinct linear and quadratic factors, where neither is repeated. Since $x^2 + 2$ cannot be factored, it is irreducible. The partial fraction decomposition is

$$\frac{x^2 + 3x - 1}{(x+1)(x^2+2)} = \frac{A}{x+1} + \frac{Bx+C}{x^2+2}.$$

Multiply each side by $(x+1)(x^2+2)$ to get

$$x^2 + 3x - 1 = A(x^2+2) + (Bx+C)(x+1). \qquad \text{Equation 1}$$

First, substitute -1 for x to get

$$(-1)^2 + 3(-1) - 1 = A[(-1)^2 + 2] + 0$$

$$-3 = 3A$$

$$A = -1.$$

Replace A with -1 in equation 1 and substitute any value for x. For instance, if $x = 0$, then

$$0^2 + 3(0) - 1 = -1(0^2 + 2) + (B \cdot 0 + C)(0 + 1)$$
$$-1 = -2 + C$$
$$C = 1.$$

Now, letting $A = -1$ and $C = 1$, substitute again in equation 1, using another number for x. For $x = 1$,

$$3 = -3 + (B + 1)(2)$$
$$6 = 2B + 2$$
$$B = 2.$$

Using $A = -1$, $B = 2$, and $C = 1$, the partial fraction decomposition is

$$\frac{x^2 + 3x - 1}{(x + 1)(x^2 + 2)} = \frac{-1}{x + 1} + \frac{2x + 1}{x^2 + 2}.$$

Again, this work can be checked by combining terms on the right.

Now Try Exercise 21 ◆

For fractions with denominators that have quadratic factors, another method is often more convenient. The system of equations is formed by equating coefficients of like terms on each side of the partial fraction decomposition. For instance, in Example 3, after each side was multiplied by the common denominator, the equation was

$$x^2 + 3x - 1 = A(x^2 + 2) + (Bx + C)(x + 1).$$

Multiplying on the right and collecting like terms, we have

$$x^2 + 3x - 1 = Ax^2 + 2A + Bx^2 + Bx + Cx + C$$
$$x^2 + 3x - 1 = (A + B)x^2 + (B + C)x + (C + 2A).$$

Now, equating the coefficients of like powers of x gives the three equations:

$$1 = A + B$$
$$3 = B + C$$
$$-1 = C + 2A.$$

Solving this system of equations for A, B, and C would give the partial fraction decomposition. The next example uses a combination of the two methods.

Repeated Quadratic Factors

 **Finding a partial fraction decomposition**

Find the partial fraction decomposition of

$$\frac{2x}{(x^2 + 1)^2(x - 1)}.$$

SOLUTION This expression has both a linear factor and a repeated quadratic factor. By **STEPS 3(a)** and **4(b)**,

$$\frac{2x}{(x^2 + 1)^2(x - 1)} = \frac{Ax + B}{x^2 + 1} + \frac{Cx + D}{(x^2 + 1)^2} + \frac{E}{x - 1}.$$

Multiplying each side by $(x^2 + 1)^2(x - 1)$ leads to

$$2x = (Ax + B)(x^2 + 1)(x - 1) + (Cx + D)(x - 1) + E(x^2 + 1)^2. \qquad \text{Equation 1}$$

If $x = 1$, equation 1 reduces to $2 = 4E$, or $E = \frac{1}{2}$. Substituting $\frac{1}{2}$ for E in equation 1 and combining terms on the right gives

$$2x = \left(A + \frac{1}{2}\right)x^4 + (-A + B)x^3 + (A - B + C + 1)x^2$$

$$+ (-A + B + D - C)x + \left(-B - D + \frac{1}{2}\right). \qquad \text{Equation 2}$$

To get additional equations involving the unknowns, equate the coefficients of like powers of x on each side of equation 2. Setting corresponding coefficients of x^4 equal, $0 = A + \frac{1}{2}$ or $A = -\frac{1}{2}$. From the corresponding coefficients of x^3, $0 = -A + B$, which means that since $A = -\frac{1}{2}$, $B = -\frac{1}{2}$. Using the coefficients of x^2, $0 = A - B + C + 1$. Since $A = -\frac{1}{2}$ and $B = -\frac{1}{2}$, $C = -1$. Finally, from the coefficients of x, $2 = -A + B + D - C$. Substituting for A, B, and C gives $D = 1$. With

$$A = -\frac{1}{2}, \quad B = -\frac{1}{2}, \quad C = -1, \quad D = 1, \quad \text{and} \quad E = \frac{1}{2},$$

the given fraction has the partial fraction decomposition

$$\frac{2x}{(x^2 + 1)^2(x - 1)} = \frac{-\frac{1}{2}x - \frac{1}{2}}{x^2 + 1} + \frac{-x + 1}{(x^2 + 1)^2} + \frac{\frac{1}{2}}{x - 1}$$

or

$$\frac{2x}{(x^2 + 1)^2(x - 1)} = \frac{-(x + 1)}{2(x^2 + 1)} + \frac{-x + 1}{(x^2 + 1)^2} + \frac{1}{2(x - 1)}.$$

Now Try Exercise 25 ◆

In summary, to solve for the constants in the numerators of a partial fraction decomposition, use either of the following methods or a combination of the two.

TECHNIQUES FOR DECOMPOSITION INTO PARTIAL FRACTIONS

Method 1 For Linear Factors

1. Multiply each side of the rational expression by the common denominator.

2. Substitute the zero of each factor in the resulting equation. For repeated linear factors, substitute as many other numbers as necessary to find all the constants in the numerators. The number of substitutions required will equal the number of constants.

Method 2 For Quadratic Factors

1. Multiply each side of the rational expression by the common denominator.

2. Collect terms on the right side of the resulting equation.

3. Equate the coefficients of like terms to get a system of equations.

4. Solve the system to find the constants in the numerators.

 **C** — **Exercises**

Exercises 1–30: Find the partial fraction decomposition for the rational expression.

1. $\dfrac{5}{3x(2x + 1)}$

2. $\dfrac{3x - 1}{x(x + 1)}$

3. $\dfrac{4x + 2}{(x + 2)(2x - 1)}$

4. $\dfrac{x + 2}{(x + 1)(x - 1)}$

5. $\dfrac{x}{x^2 + 4x - 5}$

6. $\dfrac{5x - 3}{(x + 1)(x - 3)}$

7. $\dfrac{2x}{(x + 1)(x + 2)^2}$

8. $\dfrac{2}{x^2(x + 3)}$

9. $\dfrac{4}{x(1 - x)}$

10. $\dfrac{4x^2 - 4x^3}{x^2(1 - x)}$

11. $\dfrac{4x^2 - x - 15}{x(x + 1)(x - 1)}$

12. $\dfrac{2x + 1}{(x + 2)^3}$

13. $\dfrac{x^2}{x^2 + 2x + 1}$

14. $\dfrac{3}{x^2 + 4x + 3}$

15. $\dfrac{2x^5 + 3x^4 - 3x^3 - 2x^2 + x}{2x^2 + 5x + 2}$

16. $\dfrac{6x^5 + 7x^4 - x^2 + 2x}{3x^2 + 2x - 1}$

17. $\dfrac{x^3 + 4}{9x^3 - 4x}$

18. $\dfrac{x^3 + 2}{x^3 - 3x^2 + 2x}$

19. $\dfrac{-3}{x^2(x^2 + 5)}$

20. $\dfrac{2x + 1}{(x + 1)(x^2 + 2)}$

21. $\dfrac{3x - 2}{(x + 4)(3x^2 + 1)}$

22. $\dfrac{3}{x(x + 1)(x^2 + 1)}$

23. $\dfrac{1}{x(2x + 1)(3x^2 + 4)}$

24. $\dfrac{x^4 + 1}{x(x^2 + 1)^2}$

25. $\dfrac{3x - 1}{x(2x^2 + 1)^2}$

26. $\dfrac{3x^4 + x^3 + 5x^2 - x + 4}{(x - 1)(x^2 + 1)^2}$

27. $\dfrac{-x^4 - 8x^2 + 3x - 10}{(x + 2)(x^2 + 4)^2}$

28. $\dfrac{x^2}{x^4 - 1}$

29. $\dfrac{5x^5 + 10x^4 - 15x^3 + 4x^2 + 13x - 9}{x^3 + 2x^2 - 3x}$

30. $\dfrac{3x^6 + 3x^4 + 3x}{x^4 + x^2}$

Appendix D:
Rotation of Axes

Derivation of Rotation Equations

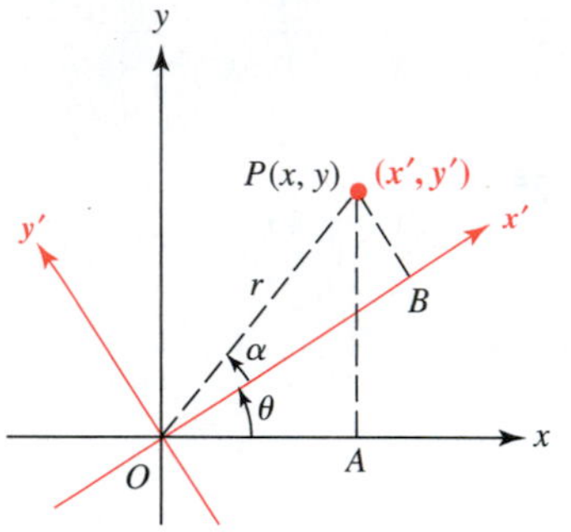

If we begin with an xy-coordinate system having origin O and rotate the axes about O through an angle θ, the new coordinate system is called a **rotation** of the xy-system. Trigonometric identities can be used to obtain equations for converting the coordinates of a point from the xy-system to the rotated $x'y'$-system. Let P be any point other than the origin, with coordinates (x, y) in the xy-system and (x', y') in the $x'y'$-system. See Figure D.1. Let $OP = r$, and let α represent the angle made by OP and the x'-axis. As shown in Figure D.1,

$$\cos(\theta + \alpha) = \frac{OA}{r} = \frac{x}{r}, \quad \sin(\theta + \alpha) = \frac{AP}{r} = \frac{y}{r},$$

$$\cos\alpha = \frac{OB}{r} = \frac{x'}{r}, \quad \sin\alpha = \frac{BP}{r} = \frac{y'}{r}.$$

These four statements can be rewritten as

$$x = r\cos(\theta + \alpha), \quad y = r\sin(\theta + \alpha), \quad x' = r\cos\alpha, \quad y' = r\sin\alpha.$$

Using the trigonometric identity for the cosine of the sum of two angles gives

$$
\begin{aligned}
x &= r\cos(\theta + \alpha) \\
&= r(\cos\theta\cos\alpha - \sin\theta\sin\alpha) \\
&= (\mathbf{r\cos\alpha})\cos\theta - (\mathbf{r\sin\alpha})\sin\theta \\
&= \mathbf{x'}\cos\theta - \mathbf{y'}\sin\theta.
\end{aligned}
$$

In the same way, by using the identity for the sine of the sum of two angles, $y = x'\sin\theta + y'\cos\theta$. This proves the following result.

ROTATION EQUATIONS

If the rectangular coordinate axes are rotated about the origin through an angle θ, and if the coordinates of a point P are (x, y) and (x', y') with respect to the xy-system and the $x'y'$-system, respectively, then the **rotation equations** are

$$x = x'\cos\theta - y'\sin\theta \quad \text{and} \quad y = x'\sin\theta + y'\cos\theta.$$

Applying a Rotation Equation

EXAMPLE 1 Finding an equation after a rotation

The equation of a curve is $x^2 + y^2 + 2xy + 2\sqrt{2}x - 2\sqrt{2}y = 0$. Find the resulting equation if the axes are rotated $45°$. Graph the equation.

FIGURE D.1

SOLUTION If $\theta = 45°$, then $\sin\theta = \frac{\sqrt{2}}{2}$ and $\cos\theta = \frac{\sqrt{2}}{2}$, and the rotation equations become

$$x = \frac{\sqrt{2}}{2}x' - \frac{\sqrt{2}}{2}y' \quad \text{and} \quad y = \frac{\sqrt{2}}{2}x' + \frac{\sqrt{2}}{2}y'.$$

Substituting these values into the given equation yields

$$x^2 + y^2 + 2xy + 2\sqrt{2}x - 2\sqrt{2}y = 0$$

$$\left[\frac{\sqrt{2}}{2}x' - \frac{\sqrt{2}}{2}y'\right]^2 + \left[\frac{\sqrt{2}}{2}x' + \frac{\sqrt{2}}{2}y'\right]^2$$

$$+ 2\left[\frac{\sqrt{2}}{2}x' - \frac{\sqrt{2}}{2}y'\right]\left[\frac{\sqrt{2}}{2}x' + \frac{\sqrt{2}}{2}y'\right]$$

$$+ 2\sqrt{2}\left[\frac{\sqrt{2}}{2}x' - \frac{\sqrt{2}}{2}y'\right] - 2\sqrt{2}\left[\frac{\sqrt{2}}{2}x' + \frac{\sqrt{2}}{2}y'\right] = 0.$$

Expanding these terms yields

$$\frac{1}{2}x'^2 - x'y' + \frac{1}{2}y'^2 + \frac{1}{2}x'^2 + x'y' + \frac{1}{2}y'^2 + x'^2 - y'^2$$

$$+ 2x' - 2y' - 2x' - 2y' = 0.$$

Collecting terms gives

$$2x'^2 - 4y' = 0$$

$$x'^2 - 2y' = 0 \qquad \text{\textit{Divide by 2.}}$$

or, finally,

$$x'^2 = 2y',$$

the equation of a parabola. The graph is shown in Figure D.2.   **Now Try Exercise 13** ◆

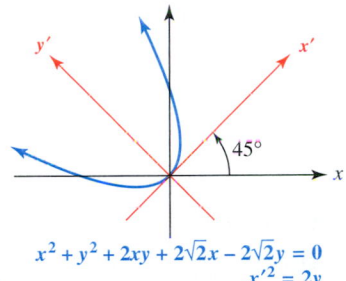

$$x^2 + y^2 + 2xy + 2\sqrt{2}x - 2\sqrt{2}y = 0$$
$$x'^2 = 2y$$

FIGURE D.2

We have graphed equations written in the form $Ax^2 + Cy^2 + Dx + Ey + F = 0$. As we saw in the preceding example, the rotation of axes eliminated the xy-term. Thus, to graph an equation that has an xy-term by hand, it is necessary to find an appropriate **angle of rotation** to eliminate the xy-term. The necessary angle of rotation can be determined by using the following result. The proof is quite lengthy and is not presented here.

ANGLE OF ROTATION

The xy-term is removed from the general equation

$$Ax^2 + Bxy + Cy^2 + Dx + Ey + F = 0$$

by a rotation of the axes through an angle θ, $0° < \theta < 90°$, where

$$\cot 2\theta = \frac{A - C}{B}.$$

This result can be used to find the appropriate angle of rotation, θ. To find the rotation equations, first find $\sin\theta$ and $\cos\theta$. The following example illustrates a way to obtain $\sin\theta$ and $\cos\theta$ from $\cot 2\theta$ without first identifying the angle θ.

EXAMPLE 2   Rotating and graphing

Rotate the axes and graph $52x^2 - 72xy + 73y^2 = 200$.

SOLUTION   Here $A = 52$, $B = -72$, and $C = 73$. By substitution,

$$\cot 2\theta = \frac{52 - 73}{-72} = \frac{-21}{-72} = \frac{7}{24}.$$

To find $\sin \theta$ and $\cos \theta$, use the trigonometric identities

$$\sin \theta = \sqrt{\frac{1 - \cos 2\theta}{2}} \quad \text{and} \quad \cos \theta = \sqrt{\frac{1 + \cos 2\theta}{2}}.$$

Sketch a right triangle and label it as in Figure D.3, to see that $\cos 2\theta = \frac{7}{25}$. (Recall that in the two quadrants for which we are concerned, cosine and cotangent have the same sign.) Then

$$\sin \theta = \sqrt{\frac{1 - \frac{7}{25}}{2}} = \sqrt{\frac{9}{25}} = \frac{3}{5} \quad \text{and} \quad \cos \theta = \sqrt{\frac{1 + \frac{7}{25}}{2}} = \sqrt{\frac{16}{25}} = \frac{4}{5}.$$

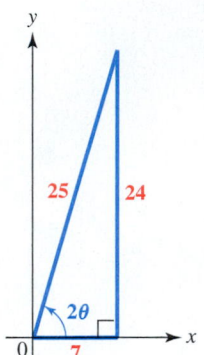

FIGURE D.3

Use these values for $\sin \theta$ and $\cos \theta$ to obtain

$$x = \frac{4}{5}x' - \frac{3}{5}y' \quad \text{and} \quad y = \frac{3}{5}x' + \frac{4}{5}y'.$$

Substituting these expressions for x and y into the original equation yields

$$52\left[\frac{4}{5}x' - \frac{3}{5}y'\right]^2 - 72\left[\frac{4}{5}x' - \frac{3}{5}y'\right]\left[\frac{3}{5}x' + \frac{4}{5}y'\right] + 73\left[\frac{3}{5}x' + \frac{4}{5}y'\right]^2 = 200.$$

This becomes

$$52\left[\frac{16}{25}x'^2 - \frac{24}{25}x'y' + \frac{9}{25}y'^2\right] - 72\left[\frac{12}{25}x'^2 + \frac{7}{25}x'y' - \frac{12}{25}y'^2\right]$$

$$+ 73\left[\frac{9}{25}x'^2 + \frac{24}{25}x'y' + \frac{16}{25}y'^2\right] = 200.$$

Combining terms gives

$$25x'^2 + 100y'^2 = 200.$$

Divide each side by 200 to get

$$\frac{x'^2}{8} + \frac{y'^2}{2} = 1,$$

an equation of an ellipse having x'-intercepts $\pm 2\sqrt{2}$ and y'-intercepts $\pm\sqrt{2}$. The graph is shown in Figure D.4. To find θ, use the fact that

$$\frac{\sin \theta}{\cos \theta} = \frac{\frac{3}{5}}{\frac{4}{5}} = \frac{3}{4} = \tan \theta,$$

from which $\theta \approx 37°$.

Now Try Exercise 17

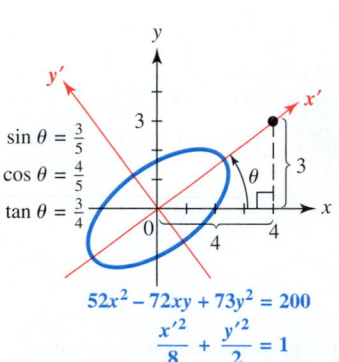

$\sin \theta = \frac{3}{5}$
$\cos \theta = \frac{4}{5}$
$\tan \theta = \frac{3}{4}$

$52x^2 - 72xy + 73y^2 = 200$
$\frac{x'^2}{8} + \frac{y'^2}{2} = 1$

FIGURE D.4

Summary of Conics with an *xy*-Term

The following summary enables us to use the general equation to decide on the type of graph to expect.

EQUATIONS OF CONICS WITH *xy*-TERM

If the general second-degree equation

$$Ax^2 + Bxy + Cy^2 + Dx + Ey + F = 0$$

has a graph, it will be one of the following:

(a) a circle or an ellipse (or a point) if $B^2 - 4AC < 0$;
(b) a parabola (or one line or two parallel lines) if $B^2 - 4AC = 0$;
(c) a hyperbola (or two intersecting lines) if $B^2 - 4AC > 0$;
(d) a straight line if $A = B = C = 0$, and $D \neq 0$ or $E \neq 0$.

 D — Exercises

Exercises 1–6: Use the summary in this section to predict the graph of the second-degree equation.

1. $4x^2 + 3y^2 + 2xy - 5x = 8$

2. $x^2 + 2xy - 3y^2 + 2y = 12$

3. $2x^2 + 3xy - 4y^2 = 0$

4. $x^2 - 2xy + y^2 + 4x - 8y = 0$

5. $4x^2 + 4xy + y^2 + 15 = 0$

6. $-x^2 + 2xy - y^2 + 16 = 0$

Exercises 7–12: Find the angle of rotation θ that will remove the xy-term in the equation.

7. $2x^2 + \sqrt{3}xy + y^2 + x = 5$

8. $4\sqrt{3}x^2 + xy + 3\sqrt{3}y^2 = 10$

9. $3x^2 + \sqrt{3}xy + 4y^2 + 2x - 3y = 12$

10. $4x^2 + 2xy + 2y^2 + x - 7 = 0$

11. $x^2 - 4xy + 5y^2 = 18$

12. $3\sqrt{3}x^2 - 2xy + \sqrt{3}y^2 = 25$

Exercises 13–16: Use the given angle of rotation to remove the xy-term and graph the equation.

13. $x^2 - xy + y^2 = 6; \theta = 45°$

14. $2x^2 - xy + 2y^2 = 25; \theta = 45°$

15. $8x^2 - 4xy + 5y^2 = 36; \sin \theta = \dfrac{2}{\sqrt{5}}$

16. $5y^2 + 12xy = 10; \sin \theta = \dfrac{3}{\sqrt{13}}$

Exercises 17–24: Remove the xy-term from the equation by performing a suitable rotation. Graph the equation.

17. $3x^2 - 2xy + 3y^2 = 8$ **18.** $x^2 + xy + y^2 = 3$

19. $x^2 - 4xy + y^2 = -5$

20. $x^2 + 2xy + y^2 + 4\sqrt{2}x - 4\sqrt{2}y = 0$

21. $7x^2 + 6\sqrt{3}xy + 13y^2 = 64$

22. $7x^2 + 2\sqrt{3}xy + 5y^2 = 24$

23. $3x^2 - 2\sqrt{3}xy + y^2 - 2x - 2\sqrt{3}y = 0$

24. $2x^2 + 2\sqrt{3}xy + 4y^2 = 5$

Exercises 25–30: In the equation, remove the xy-term by rotation. Then translate the axes and sketch the graph.

25. $x^2 + 3xy + y^2 - 5\sqrt{2}y = 15$

26. $x^2 - \sqrt{3}xy + 2\sqrt{3}x - 3y - 3 = 0$

27. $4x^2 + 4xy + y^2 - 24x + 38y - 19 = 0$

28. $12x^2 + 24xy + 19y^2 - 12x - 40y + 31 = 0$

29. $16x^2 + 24xy + 9y^2 - 130x + 90y = 0$

30. $9x^2 - 6xy + y^2 - 12\sqrt{10}x - 36\sqrt{10}y = 0$

Bibliography

Acker, A., and C. Jaschek. *Astronomical Methods and Calculations.* New York: John Wiley and Sons, 1986.

Andre-Pascal, R. *Global Energy: The Changing Outlook.* Paris, France: Organization for Economic Cooperation and Development/International Energy Agency, 1992.

Battan, L. *Weather in Your Life.* San Francisco: W. H. Freeman, 1983.

Beckmann, P. *A History of PI.* New York: Barnes and Noble, Inc., 1993.

Bell, D. *Fundamentals of Electric Circuits.* Reston, Va.: Reston Publishing Company, 1981.

Benade, A. *Fundamentals of Musical Acoustics.* New York: Oxford University Press, 1976.

Beni, G., and S. Hackwood. *Recent Advances in Robotics.* New York: John Wiley and Sons, 1985.

Beranek, L. *Noise and Vibration Control.* Washington, D.C.: Institute of Noise Control Engineering, 1988.

Brase, C. H., and C. P. Brase. *Understandable Statistics.* 5th ed. Lexington, Mass.: D. C. Heath and Company, 1995.

Brearley, J., and A. Nicholas. *This Is the Bichon Frise.* Hong Kong: TFH Publication, 1973.

Brooks, R., and J. Dieter. *Phytoarchaeology.* Portland, Or.: Dioscorides Press, 1990.

Brown, D., and P. Rothery. *Models in Biology: Mathematics, Statistics, and Computing.* West Sussex, England: John Wiley and Sons Ltd., 1993.

Brown, F., J. Hastings, and J. Palmer. *The Biological Clock.* New York: Academic Press, 1970.

Bünning, E. *The Physiological Clock.* New York: Springer-Verlag, 1967.

Burden, R., and J. Faires. *Numerical Analysis.* 5th ed. Boston: PWS-Kent Publishing Company, 1993.

Callas, D. *Snapshots of Applications in Mathematics.* Delhi, New York: State University College of Technology, 1994.

Carlson, T. "Über Geschwindigkeit und Grösse der Hefevermehrung in Würze." *Biochem. A.* 57: 313–334.

Cheney, W., and D. Kincaid. *Numerical Mathematics and Computing.* 3rd ed. Pacific Grove, Calif.: Brooks/Cole Publishing Company, 1994.

Clime, W. *The Economics of Global Warming.* Washington, D.C.: Institute for International Economics, 1992.

Cole, F. *Introduction to Meteorology.* New York: Wiley, 1980.

Conquering the Sciences. Sharp Electronics Corporation, 1986.

Cooper, J., and R. Glassow. *Kinesiology.* 2nd ed. St. Louis: The C. V. Mosby Company, 1968.

Cotton, W., and R. Pielke. *Human Impacts on Weather and Climate.* Geophysical Science Series, vol. 2. Fort Collins, Colo.: *ASTeR Press, 1992.

Craig, J. *Introduction to Robotics: Mechanics and Control.* Reading, Mass.: Addison-Wesley Publishing Company, 1989.

Crone, D. *Elementary Photogrammetry.* New York: Frederick Ungar Publishing Company, 1968.

Crownover, R. *Introduction to Fractals and Chaos.* Boston: Jones and Bartlett, 1995.

Duffet-Smith, P. *Practical Astronomy with Your Calculator.* New York: Cambridge University Press, 1988.

Easteal, S., N. McLeod, and K. Reed. *DNA Profiling: Principles, Pitfalls and Potential.* Philadelphia: Harwood Academic Publishers, 1991.

Eves, H. *An Introduction to the History of Mathematics.* 5th ed. Philadelphia: Saunders College Publishing, 1983.

Fletcher, N., and T. Rossing. *The Physics of Musical Instruments.* New York: Springer-Verlag, 1991.

Foley, J., A. van Dam, S. Feiner, J. Hughes, and R. Phillips. *Introduction to Computer Graphics.* Reading, Mass.: Addison-Wesley Publishing Company, 1994.

Foster, R., and J. Bates. "Use of mussels to monitor point source industrial discharges." *Environ. Sci. Technol.* 12: 958–962.

Freebury, H. *A History of Mathematics.* New York: Macmillan, 1961.

Freedman, B. *Environmental Ecology: The Ecological Effects of Pollution, Disturbance, and Other Stresses.* 2nd ed. San Diego: Academic Press, 1995.

Friedhoff, M., and W. Benzon. *The Second Computer Revolution: Visualization.* New York: W. H. Freeman, 1991.

Garber, N., and L. Hoel. *Traffic and Highway Engineering.* Boston, Mass.: PWS Publishing Co., 1997.

Glass, L., and M. Mackey. *From Clocks to Chaos.* Princeton, N.J.: Princeton University Press, 1988.

Glassner, A. *An Introduction to Ray Tracing.* San Diego: Academic Press, 1993.

Good, I. J. "What is the most amazing approximate integer in the universe?" *Pi Mu Epsilon Journal* 5 (1972): 314–315.

Grigg, D. *The World Food Problem.* Oxford: Blackwell Publishers, 1993.

Haber-Schaim, U., J. Cross, G. Abegg, J. Dodge, and J. Walter. *Introductory Physical Science.* Englewood Cliffs, N.J.: Prentice-Hall, Inc., 1972.

Haefner, L. *Introduction to Transportation Systems.* New York: Holt, Rinehart and Winston, 1986.

Harker, J. *The Physiology of Diurnal Rhythms.* New York: Cambridge University Press, 1964.

Harrison, F., F. Hills, J. Paterson, and R. Saunders. "The measurement of liver blood flow in conscious calves." *Quarterly Journal of Experimental Physiology* 71: 235–247.

Hartman, D. *Global Physical Climatology.* San Diego: Academic Press, 1994.

Heinz-Otto, P., H. Jürgens, and D. Saupe. *Chaos and Fractals: New Frontiers in Science.* New York: Springer-Verlag, 1993.

Hibbeler, R. *Structural Analysis.* Englewood Cliffs, N.J.: Prentice-Hall, 1995.

Hill, F. *Computer Graphics.* New York: Macmillan Publishing Company, 1990.

Hines, A., T. Ghosh, S. Loyalka, and R. Warder, Jr. *Indoor Air Quality and Control.* Englewood Cliffs, N.J.: Prentice-Hall, 1993.

Historical Topics for the Mathematics Classroom, Thirty-first Yearbook. National Council of Teachers of Mathematics, 1969.

Hoggar, S. *Mathematics for Computer Graphics.* New York: Cambridge University Press, 1993.

Hoppensteadt, F., and C. Peskin. *Mathematics in Medicine and the Life Sciences.* New York: Springer-Verlag, 1992.

Hosmer, D., and S. Lemeshow. *Applied Logistic Regression.* New York: John Wiley and Sons, 1989.

Howatson, A. *Electrical Circuits and Systems.* New York: Oxford University Press, 1996.

Howells, G. *Acid Rain and Acid Waters.* 2nd ed. New York: Ellis Horwood, 1995.

Huffman, R. *Atmospheric Ultraviolet Remote Sensing.* San Diego: Academic Press, 1992.

Huxley, J. *Problems of Relative Growth.* London: Methuen and Co. Ltd., 1932.

Jarrett, J. *Business Forecasting Methods.* Oxford: Basil Blackwell Ltd., 1991.

Karttunen, H., P. Kroger, H. Oja, M. Poutanen, K. Donner, eds. *Fundamental Astronomy.* 2nd ed. New York: Springer-Verlag, 1994.

Kerlow, I. *The Art of 3-D Computer Animation and Imaging.* New York: Van Nostrand Rienhold, 1996.

Kincaid, D., and W. Cheney. *Numerical Analysis.* Pacific Grove, Calif.: Brooks/Cole Publishing Company, 1991.

Kissam, P. *Surveying Practice.* 3rd ed. New York: McGraw-Hill, 1978.

Kline, M. *The Loss of Certainty.* New York: Oxford University Press, 1980.

Kraljic, M. *The Greenhouse Effect.* New York: The H. W. Wilson Company, 1992.

Kress, S. *Bird Life—A Guide to the Behavior and Biology of Birds.* Racine, Wisc.: Western Publishing Company, 1991.

Lack, D. *The Life of a Robin.* London: Collins, 1965.

Lancaster, H. *Quantitative Methods in Biological and Medical Sciences: A Historical Essay.* New York: Springer-Verlag, 1994.

Leder, J. *Martina Navratilova.* Mankato, Minn.: Crestwood House, 1985.

Leick, A. *GPS Satellite Surveying.* New York: Wiley, 1990.

Loh, W. *Dynamics and Thermodynamics of Planetary Entry.* Englewood Cliffs, N.J.: Prentice-Hall, 1963.

Makridakis, S., and S. Wheelwright. *Forecasting Methods for Management.* New York: John Wiley and Sons, 1989.

Mannering, F., and W. Kilareski. *Principles of Highway Engineering and Traffic Analysis.* New York: John Wiley and Sons, 1990.

Mar, J., and H. Leibowitz. *Structure Technology for Large Radio and Radar Telescope Systems.* Cambridge, Mass.: The MIT Press, 1969.

Mason, C. *Biology of Freshwater Pollution.* New York: Longman Scientific and Technical, John Wiley and Sons, 1991.

Medley, G., D. Cox, and L. Billard. "Incubation period of AIDS in patients infected via blood transfusions." *Nature* 328: 719–721.

Meeus, J. *Astronomical Algorithms.* Richmond, Va.: Willman-Bell, 1991.

Mehrotra, A. *Cellular Radio: Analog and Digital Systems.* Boston: Artech House, 1994.

Metcalf, H. *Topics in Classical Biophysics.* Englewood Cliffs, N.J.: Prentice-Hall, 1980.

Miller, A., and J. Thompson. *Elements of Meteorology.* 2nd ed. Columbus, Ohio: Charles E. Merrill Publishing Company, 1975.

Miller, A., and R. Anthes. *Meteorology.* 5th ed. Columbus, Ohio: Charles E. Merrill Publishing Company, 1985.

Moffitt, F. *Photogrammetry.* Scranton, Pa.: International Textbook Company, 1967.

Mortenson, M. *Computer Graphics: An Introduction to Mathematics and Geometry.* New York: Industrial Press Inc., 1989.

Motz, L., and J. Weaver. *The Story of Mathematics.* New York: Plenum Press, 1993.

Mueller, I., and K. Ramsayer. *Introduction to Surveying.* New York: Frederick Ungar Publishing Company, 1979.

Navarra, J. *Atmosphere, Weather and Climate.* Philadelphia: W. B. Saunders Company, 1979.

Nilsson, A. *Greenhouse Earth.* New York: John Wiley and Sons, 1992.

Paetsch, M. *Mobile Communications in the U.S. and Europe: Regulation, Technology, and Markets.* Norwood, Mass.: Artech House, Inc., 1993.

Payne, A. "Responses of the three test algae of the algal assay procedure: bottle test." *Water Res.* 9: 437–445.

Pennycuick, C. *Newton Rules Biology.* New York: Oxford University Press, 1992.

Pielou, E. *Population and Community Ecology: Principles and Methods.* New York: Gordon and Breach Science Publishers, 1974.

Pierce, J. *The Science of Musical Sound.* New York: W. H. Freeman, 1992.

Pokorny, C., and C. Gerald. *Computer Graphics: The Principles behind the Art and Science.* Irvine, Calif.: Franklin, Beedle, and Associates, 1989.

Pugh, J. *Surveying for Field Scientists.* Pittsburgh: University of Pittsburgh Press, 1975.

Raggett, G. "Modeling the Eyam plague." *The Institute of Mathematics and Its Applications* 18: 221–226.

Rayner, W., and M. Schmidt. *Elementary Surveying.* 4th ed. New York: D. Van Nostrand Company, Inc., 1963.

Resnikoff, H., and R. Wells, Jr. *Mathematics in Civilization.* New York: Dover Publications, Inc., 1984.

Rezvan, R. L. "Effectiveness of Local Ventilation in Removing Simulated Pollutants from Point Sources." *Proceedings of the Third International Conference on Indoor Air Quality and Climate,* 1984.

Riley, W., L. Sturges, and D. Morris. *Statics and Mechanics of Materials: An Integrated Approach.* New York: John Wiley and Sons, Inc., 1995.

Rist, Curtis. "The physics of foul shots." *Discover,* October 2000.

Rodricks, J. *Calculated Risk.* New York: Cambridge University Press, 1992.

Roederer, J. *Introduction to the Physics and Psychophysics of Music.* New York: Springer-Verlag, 1973.

Ronan, C. *The Natural History of the Universe.* New York: Macmillan Publishing Company, 1991.

Ryan, B., B. Joiner, and T. Ryan. *Minitab Handbook.* Boston: Duxbury Press, 1985.

Sanders, D. *Statistics: A First Course.* 5th ed. New York: McGraw-Hill, 1995.

Schlosser, W. *Challenges of Astronomy.* New York: Springer-Verlag, 1991.

Sharov, A., and I. Novikov. *Edwin Hubble, The Discoverer of the Big Bang Universe.* New York: Cambridge University Press, 1993.

Sinkov, A. *Elementary Cryptanalysis: A Mathematical Approach.* New York: Random House, 1968.

Smith, J. *Design and Analysis of Algorithms.* Boston: PWS Publishing Company, 1989.

Smith, R., and R. Dorf. *Circuits, Devices and Systems.* 5th ed. New York: John Wiley and Sons, Inc., 1992.

Speed, William. "Downloading your body." *Discover,* September 2000.

Stadler, W. *Analytical Robotics and Mechatronics.* New York: McGraw-Hill, Inc., 1995.

Stent, G. S. *Molecular Biology of Bacterial Viruses.* San Francisco: W. H. Freeman, 1963.

Taylor, W. *The Geometry of Computer Graphics.* Pacific Grove, Calif.: Wadsworth and Brooks/Cole, 1992.

Teutsch, S., and R. Churchill. *Principles and Practice of Public Health Surveillance.* New York: Oxford University Press, 1994.

Thomas, D. *Swimming Pool Operators Handbook.* National Swimming Pool Foundation of Washington, D.C., 1972.

Thomas, V. *Science and Sport.* London: Faber and Faber, 1970.

Thomson, W. *Introduction to Space Dynamics.* New York: John Wiley and Sons, 1961.

Triola, M. *Elementary Statistics.* Pearson Education, Inc., 2004.

Tucker, A., A. Bernat, W. Bradley, R. Cupper, and G. Scragg. *Fundamentals of Computing 1 Logic: Problem Solving, Programs, and Computers.* New York: McGraw-Hill, 1995.

Turner, R. K., D. Pierce, and I. Bateman. *Environmental Economics, An Elementary Approach.* Baltimore: The Johns Hopkins University Press, 1993.

Van Sickle, J. *GPS for Land Surveyors.* Chelsea, Mich.: Ann Arbor Press, 1996.

Varley, G., and G. Gradwell. "Population models for the winter moth." *Symposium of the Royal Entomological Society of London* 4: 132–142.

Walker, A. *Observation and Inference: An Introduction to the Methods of Epidemiology.* Newton Lower Falls, Mass.: Epidemiology Resources Inc., 1991.

Watt, A. *3D Computer Graphics.* Reading, Mass.: Addison-Wesley Publishing Company, 1993.

Webb, T. *Celestial Objects for Common Telescopes.* New York: Dover Publications Inc., 1962.

Weidner, R., and R. Sells. *Elementary Classical Physics,* Vol. 1, Vol. 2. Boston: Allyn and Bacon, 1965.

West, G. "Differential GPS—how accurate is it?" *Trailer Boats,* 25 (10): 72–73.

Wilcox, G., and C. Hesselberth. *Electricity for Engineering Technology.* Boston: Allyn and Bacon, 1970.

Williams, J. *The Weather Almanac 1995.* New York: Vintage Books, 1994.

Winter, C. *Solar Power Plants.* New York: Springer-Verlag, 1991.

Wolff, R., and L. Yaeger. *Visualization of Natural Phenomena.* New York: Springer-Verlag, 1993.

Wright, J. *The Universal Almanac 1997.* Kansas City: Andrews and McMeel, 1997.

Wuebbles, D., and J. Edmonds. *Primer on Greenhouse Gases.* Chelsea, Mich.: Lewis Publishers, 1991.

Zeilik, M., S. Gregory, and D. Smith. *Introductory Astronomy and Astrophysics.* 3rd ed. Philadelphia: Saunders College Publishers, 1992.

Zhao, Y. *Vehicle Location and Navigation Systems.* Boston, Mass.: Artech House, 1997.

Answers to Selected Exercises

CHAPTER 1: Introduction to Functions and Graphs

SECTION 1.1 (PP. 10–13)

1. Natural number, integer, rational number, real number
3. Integer, rational number, real number
5. Rational number, real number
7. Real number
9. Natural number, $\sqrt{9}$; integers, -3, $\sqrt{9}$; rational numbers, $-3, \frac{2}{9}, \sqrt{9}, 1.\overline{3}$; irrational numbers, $\pi, -\sqrt{2}$
11. Natural numbers, none; integer, $-\sqrt{4}$; rational numbers, $\frac{1}{3}, 5.1 \times 10^{-6}, -2.33, 0.\overline{7}, -\sqrt{4}$; irrational number, $\sqrt{13}$
13. Rational numbers   15. Rational numbers
17. Integers   19. 62.5%   21. -39.3%
23. 1.858×10^5 25. 3.892×10^{-2} 27. 2.45×10^3
29. 5.6×10^{-1} 31. -8.7×10^{-3} 33. 2.068×10^2
35. 8.54×10^5 37. 0.000001 39. $200{,}000{,}000$
41. 156.7 43. -0.568 45. $500{,}000$ 47. 4500
49. $67{,}000$ 51. 8×10^8; $800{,}000{,}000$
53. 3.5×10^{-1}; 0.35 55. 1.44×10^{-6}; 0.00000144
57. 2×10^1; 20 59. 2.1×10^{-3}; 0.0021
61. 5×10^{-3}; 0.005 63. 3.36×10^{19}
65. 8.72×10^4 67. 7.67×10^{11} 69. 5.769
71. 0.058 73. 0.419 75. -1.235 77. 15.819
79. Tuition and fees: 198.5%, CPI: 67.2%
81. $798   83. About 53,794 miles per hour
85. (a) Approximately $1820 per person   (b) Approximately $19,715 per person   (c) Estimates will vary.
87. 2.9×10^{-4} cm   89. (a) 18,466,667 ft   (b) Yes
91. 4.64%; no; yes   93. (a) $7.436\pi \approx 23.4$ in.^{3}
(b) Yes   95. (a) Increased from 1940 to 1980; decreased from 1980 to 1998   (b) 4.9×10^8 gal   (c) -20.7%

SECTION 1.2 (PP. 25–28)

1. (a)
$$\begin{array}{ccccccc} -6 & -4 & -2 & 0 & 2 & 4 & 6 & 8 \end{array}$$
(b) Max: 6; min: -2 (c) Mean: $\frac{11}{6} = 1.8\overline{3}$
3. (a)
$$\begin{array}{ccccccc} -30 & -20 & -10 & 0 & 10 & 20 & 30 & 40 \end{array}$$
(b) Max: 30; min: -20 (c) Mean: 5
5.

-30	-30	-10	5	15	25	45	55	61

(a) Max: 61; min: -30
(b) Mean: $\frac{136}{9} \approx 15.11$; median: 15; range: 91

7.

$\sqrt{15}$	4.1	$\sqrt[3]{69}$	$2^{2.3}$	2^{π}	π^2

$\sqrt{15} \approx 3.87$, $2^{2.3} \approx 4.92$, $\sqrt[3]{69} \approx 4.102$, $\pi^2 \approx 9.87$, $2^{\pi} \approx 8.82$, 4.1
(a) Max: π^2; min: $\sqrt{15}$
(b) Mean: 5.95; median: 4.51; range: $\pi^2 - \sqrt{15} \approx 6.00$
9. (a) $S = \{(-1, 5), (2, 2), (3, -1), (5, -4), (9, -5)\}$
(b) $D = \{-1, 2, 3, 5, 9\}$; $R = \{-5, -4, -1, 2, 5\}$
11. (a) $S = \{(1, 5), (4, 5), (5, 6), (4, 6), (1, 5)\}$
(b) $D = \{1, 4, 5\}$; $R = \{5, 6\}$
13. (a)
$$\begin{array}{ccccc} 100 & 300 & 500 & 700 & 900 \end{array}$$
(b) Mean: 336; median: 253.5; range: 675
The average area of the six largest islands in the world is 336,000 square miles. Half of the islands have areas of less than 253,500 square miles and half have areas that are greater. The largest difference in area between any two islands is 675,000 square miles.   (c) Greenland
15. (a)
$$\begin{array}{cccccccc} 4 & 8 & 12 & 16 & 20 & 24 & 28 & 32 \end{array}$$
(b) Mean: 19.0; median: 19.3; range: 21.7
The average of the maximum elevations for the seven continents is 19,000 feet. About half of these elevations are below 19,300 feet and about half are above. Their largest difference is 21,700 feet.   (c) Mount Everest
17. (a) $4,725,500   (b) $367,000   (c) One large salary can raise the average considerably, while the majority of the players have considerably lower salaries.
19. 16, 18, 26; no   21. 5   23. $\sqrt{29} \approx 5.39$
25. $\sqrt{133.37} \approx 11.55$ 27. 8   29. $\frac{\sqrt{17}}{4} \approx 1.03$
31. $\frac{\sqrt{2}}{2} \approx 0.71$ 33. 130   35. $\sqrt{a^2 + b^2}$
37. $\sqrt{(a - b)^2} = |a - b|$ 39. Yes
41. (a)

(b) $d = \sqrt{4100} \approx 64.0$ miles

43. 78.45 years   **45.** 543,949 inmates   **47.** 10 seconds
49. $(3, -0.5)$ **51.** $(10, 10)$ **53.** $(-2.1, -0.35)$
55. $(\sqrt{2}, 0)$ **57.** $(0, 2b)$ **59.** $(a, 0)$
61. (a) $D = \{-3, -2, 0, 7\}, R = \{-5, -3, 0, 4, 5\}$
(b) x-min: -3; x-max: 7; y-min: -5; y-max: 5

(c) & (d)

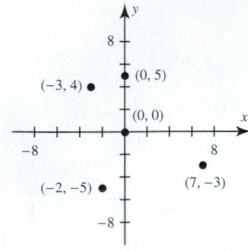

63. (a) $D = \{-4, -3, -1, 0, 2\}$,
$R = \{-2, -1, 1, 2, 3\}$
(b) x-min: -4; x-max: 2; y-min: -2; y-max: 3
(c) & (d)

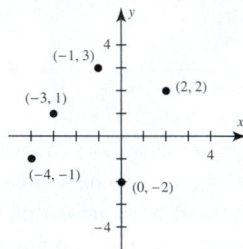

65. (a) $D = \{-35, -25, 0, 10, 75\}$,
$R = \{-55, -25, 25, 45, 50\}$
(b) x-min: -35; x-max: 75; y-min: -55; y-max: 50
(c) & (d)

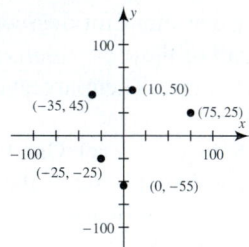

67. (a) $D = \{-0.7, 0.1, 0.5, 0.8\}$,
$R = \{-0.3, -0.1, 0, 0.4\}$
(b) x-min: -0.7; x-max: 0.8; y-min: -0.3; y-max: 0.4
(c) & (d)

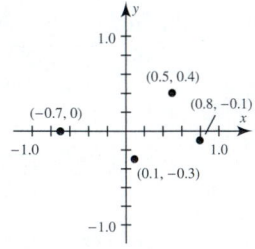

69. x-axis: 10; y-axis: 10   **71.** x-axis: 10; y-axis: 5

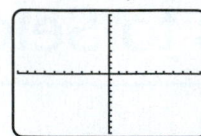   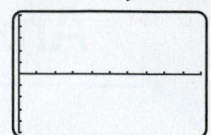

73. x-axis: 16; y-axis: 5   **75.** x-axis: 3; y-axis: 2

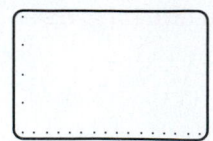   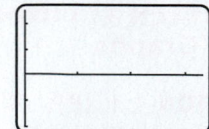

77. b.   **79.** a.
81. $[-5, 5, 1]$ by $[-5, 5, 1]$

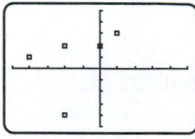

83. $[-100, 100, 10]$ by $[-100, 100, 10]$

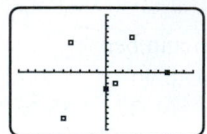

85. $[-6, 6, 2]$ by $[-12, 12, 2]$

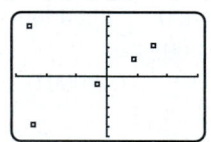

87. (a) x-min: 1979; x-max: 2001; y-min: 15; y-max: 37
(b) $[1978, 2002, 2]$ by $[10, 40, 5]$. Answers may vary.
(c)   **(d)**

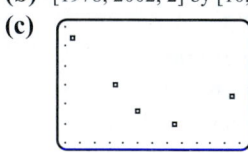

89. (a) x-min: 1996; x-max: 2004; y-min: 9.7;
y-max: 12.8
(b) $[1994, 2006, 2]$ by $[8, 14, 2]$. Answers may vary.
(c)   **(d)**

91. The data points are approximately $(1950, 6)$,
$(1960, 10)$, $(1970, 30)$, $(1980, 33)$, $(1990, 39)$, $(2000, 45)$.
Tables may vary slightly.

Year	1950	1960	1970	1980	1990	2000
Doctorates	6000	10,000	30,000	33,000	39,000	45,000

CHECKING BASIC CONCEPTS FOR SECTIONS 1.1 AND 1.2 (P. 29)

1. (a) 9.88   **(b)** 1.28   **3.** $\sqrt{72} = 6\sqrt{2} \approx 8.49$
5. Mean = 10,762.75; median = 12,941.5; range = 9262

SECTION 1.3 (PP. 42–45)

1. $(-2, 3)$ **3.** $f(7) = 8$
5.

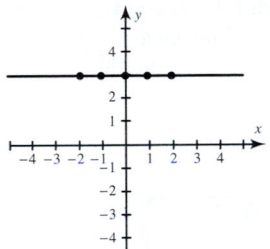

7.

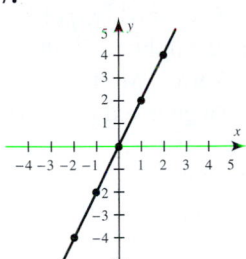

9.

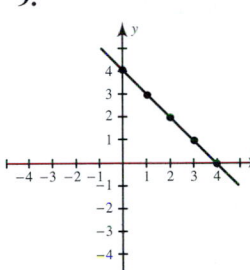

11.

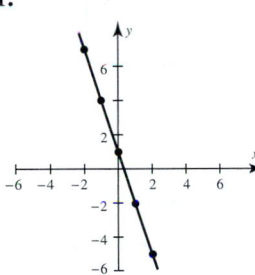

13.

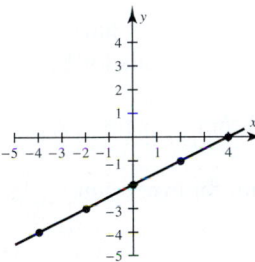

15.

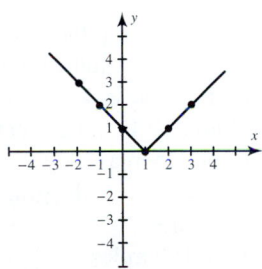

17.

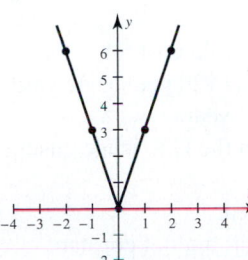

19.

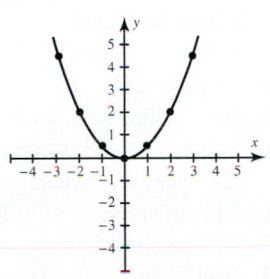

21.

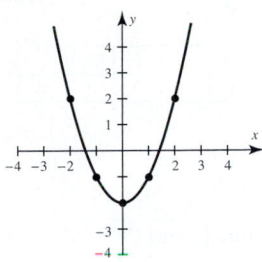

23. (a) $f(-2) = -8, f(5) = 125$ **(b)** All real numbers
25. (a) $f(-1) =$ undefined, $f(a+1) = \sqrt{a+1}$
(b) Nonnegative real numbers
27. (a) $f(-1) = -\frac{1}{2}, f(a+1) = \frac{1}{a}$ **(b)** $x \neq 1$
29. (a) $f(6) = -7, f(a-1) = -7$
(b) All real numbers
31. (a) $f(4) = \frac{1}{16}, f(-7) = \frac{1}{49}$ **(b)** $x \neq 0$
33. (a) $f(4) = \frac{1}{7}, f(a-5) = \frac{1}{a^2 - 10a + 16}$
(b) $x \neq 3, x \neq -3$
35. (a) $f(1) = 1, f(a+2) = \frac{1}{\sqrt{-a}}$ **(b)** $x < 2$
37. (a) All real numbers   **(b)** $g(-1) = -3; g(2) = 3$
(c) $g(-1) = -3; g(2) = 3$
39. (a) All real numbers   **(b)** $g(-1) = 1; g(2) = -1$
(c) $g(-1) = 1; g(2) = -1$
41. (a) All real numbers   **(b)** $g(-1) = -2; g(2) = 1$
(c) $g(-1) = -2; g(2) = 1$
43. $D = \{x | -3 \leq x \leq 3\}, R = \{y | 0 \leq y \leq 3\};$ $f(0) = 3$
45. $D = \{x | -2 \leq x \leq 4\}, R = \{y | -2 \leq y \leq 2\};$ $f(0) = -2$
47. $D =$ all real numbers, $R = \{y | y \leq 2\}; f(0) = 2$
49. $D = \{x | x \geq -1\}, R = \{y | y \leq 2\}; f(0) = 0$
51. (a) $f(2) = 7$ **(b)** $f = \{(1, 7), (2, 7), (3, 8)\}$
(c) $D = \{1, 2, 3\}, R = \{7, 8\}$
53. (a) $f(0) = -2, f(2) = 2$ **(b)** $x = 1$
55. (a) $f(0) = 0, f(2) = 4$ **(b)** $x = 0$
57. (a) $f(0) = 0, f(2) = 0$ **(b)** $x = -2, 0,$ or 2
59. (a) $[-4.7, 4.7, 1]$ by $[-3.1, 3.1, 1]$ **(b)** $f(2) = 1$
(c)

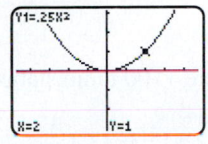

61. (a) $[-4.7, 4.7, 1]$ by $[-3.1, 3.1, 1]$ **(b)** $f(2) = 2$
(c)

63. Verbal: Square the input x.
Graphical: Numerical:
$[-10, 10, 1]$ by $[-10, 10, 1]$

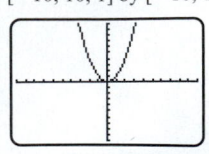

x	-2	-1	0	1	2
y	4	1	0	1	4

65. Verbal: Multiply the input by 2, add 1, and then take the absolute value.
Graphical: Numerical:
$[-6, 6, 1]$ by $[-4, 4, 1]$

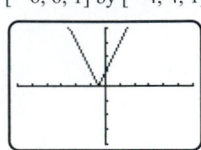

x	-2	-1	0	1	2
y	3	1	1	3	5

67. Verbal: Compute the absolute value of the input x.
Graphical: Numerical:
$[-6, 6, 1]$ by $[-4, 4, 1]$

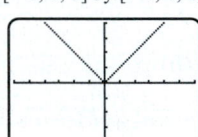

x	-2	-1	0	1	2
y	2	1	0	1	2

69. Verbal: Add 1 to the input and then take the square root of the result.
Graphical: Numerical:
$[-6, 6, 1]$ by $[-4, 4, 1]$

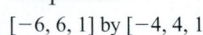
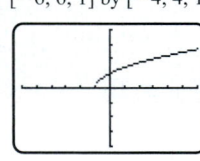

x	-2	-1	0	1	2
y	—	0	1	$\sqrt{2}$	$\sqrt{3}$

71. (a) $g = \{(-1, 2), (0, 4), (1, -3), (2, 2)\}$
(b) $D = \{-1, 0, 1, 2\}, R = \{-3, 2, 4\}$
73.

Bills (millions)	0	1	2	3	4	5	6
Counterfeit Bills	0	9	18	27	36	45	54

75. (a)

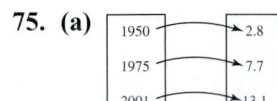

(b) $f(1975) = 7.7$; in 1975 there were 7700 radio stations on the air.
(c) $D = \{1950, 1975, 2001\}, R = \{2.8, 7.7, 13.1\}$
77. Yes. Domain and range include all real numbers.
79. No   **81.** Yes. D: $-4 \le x \le 4$; R: $0 \le y \le 4$

83. Yes. Each real number input x has exactly one cube root.
85. No. Usually, several students pass a given exam.
87. No   **89.** Yes   **91.** No   **93.** No   **95.** Yes
97. $g(x) = 12x$; $g(10) = 120$; There are 120 in. in 10 ft.
99. $g(x) = \frac{x}{4}$; $g(10) = 2.5$; There are 2.5 dollars in 10 quarters.
101. $g(x) = 86,400x$; $g(10) = 864,000$; There are 864,000 seconds in 10 days.
103. $f(x) = \frac{115}{528}x$, $f(15) \approx 3.3$; with a 15-second delay the lightning bolt was about 3.3 miles away.
105. Verbal: Multiply the input x by -5.8 to obtain the change in temperature.
Symbolic: $f(x) = -5.8x$.
Graphical: Numerical:
$[0, 3, 1]$ by $[-20, 20, 5]$

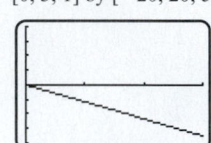
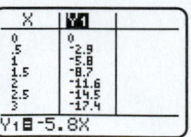

SECTION 1.4 (PP. 58–61)

1. $\frac{1}{2}$ **3.** -1 **5.** 0   **7.** -1 **9.** -8
11. $-\frac{23}{30} \approx -0.7667$ **13.** Undefined
15. $-\frac{39}{35} \approx -1.1143$
17. Slope $= 2$; the graph rises 2 units for every unit increase in x.
19. Slope $= -\frac{3}{4}$; the graph falls $\frac{3}{4}$ unit for every unit increase in x, or equivalently, the line falls 3 units for every 4-unit increase in x.
21. Slope $= 0$; the graph neither falls nor rises for every unit increase in x.
23. Slope $= -1$; the graph falls 1 unit for every unit increase in x.
25. (a) 150 miles
(b) Slope $= 75$; the car is traveling away from the rest stop at 75 miles per hour.
27. (a) Zero square yards of carpet would cost $0.
(b) Slope $= 20$ **(c)** The carpet costs $20 per square yard.
29. (a) 1980: 30.34 years; 2000: 35.2 years
(b) Slope $= 0.243$; the median age in the U.S. is increasing by 0.243 year each year.
31. Linear, slope $= 4$ **33.** Nonlinear
35. Linear, slope $= -\frac{1}{2}$ **37.** Linear, but not constant
39. Nonlinear   **41.** Linear and constant   **43.** Nonlinear
45. Nonlinear   **47.** Nonlinear   **49.** Nonlinear
51. Nonlinear

53. **(a)** 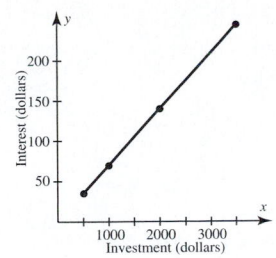   **(b)** Linear

55. **(a)** 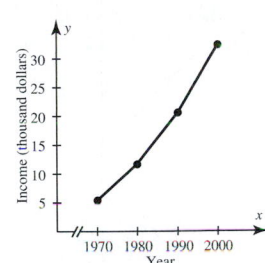   **(b)** Nonlinear

57. **(a)** No   **(b)** $f(x) = 7$
(c) $[0, 15, 3]$ by $[0, 10, 1]$

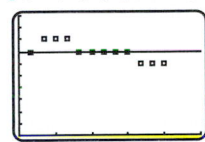

59. The velocity of the car is zero.
61. $f(x) = \dfrac{x}{16}$ **63.** $f(x) = 50x$ **65.** $f(x) = 500$
67. $f(x) = 6x + 1$

For Exercises 69–73, answers may vary.

69. 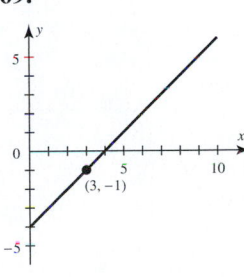   **71.**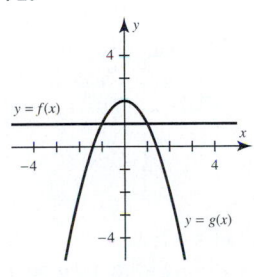

73.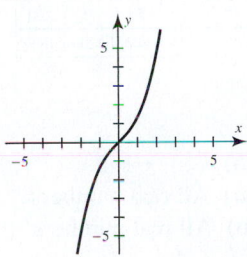

75. Average rate of change from -3 to -1 is 0.
Average rate of change from 1 to 3 is 0.
77. Average rate of change from -3 to -1 is 1.2.
Average rate of change from 1 to 3 is -1.2.
79. **(a)** 3   **(b)**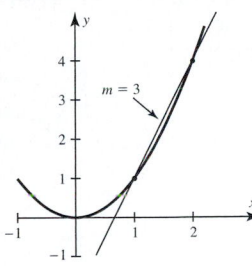

81. 7; the slope of the line passing through $(1, f(1))$
and $(4, f(4))$ is 7.
83. 26; the slope of the line passing through
$(2, f(2))$ and $(4, f(4))$ is 26.
85. 0.62; the slope of the line passing through
$(1, f(1))$ and $(3, f(3))$ is approximately 0.62.
87. **(a)** $[-10, 90, 10]$ by $[0, 80, 10]$

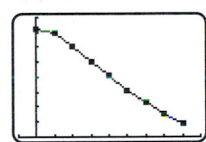

(b) $-0.24, -0.98, -0.97, -0.95, -0.93, -0.85, -0.76,$
-0.63; as a woman becomes older, her *remaining* life expectancy is reduced each year by the average rate of change.
(c) 80.1 years, 85.5 years; there is risk associated with living from 20 years old to 70 years old that does not apply to someone who is already 70 years old.
89. **(a)** 3   **(b)** 0   **91.** **(a)** $-2x - 2h$ **(b)** -2
93. **(a)** $2x + 2h + 1$ **(b)** 2
95. **(a)** $3x^2 + 6xh + 3h^2 + 1$ **(b)** $6x + 3h$
97. **(a)** $-x^2 - 2xh - h^2 + 2x + 2h$
(b) $-2x - h + 2$
99. **(a)** $2x^2 + 4xh + 2h^2 - x - h + 1$
(b) $4x + 2h - 1$
101. **(a)** $8t^2 + 16th + 8h^2$ **(b)** $16t + 8h$
(c) 64.4; the average speed of the car from 4 to 4.05 seconds is 64.4 feet per second.

1.4 EXTENDED AND DISCOVERY EXERCISES (P. 62)

1. **(a)** Yes; 2π inches per second   **(b)** No; because the area function depends on the radius squared, the area function is not a linear function and does not increase at a constant rate.

CHECKING BASIC CONCEPTS FOR SECTIONS 1.3 AND 1.4 (P. 62)

1. Symbolic: $f(x) = 5280x$

Numerical:

x	1	2	3	4	5
$f(x)$	5280	10,560	15,840	21,120	26,400

Graphical:

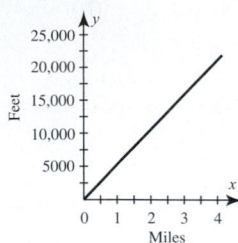

3. $\{x \mid x \geq 3\}$

5.

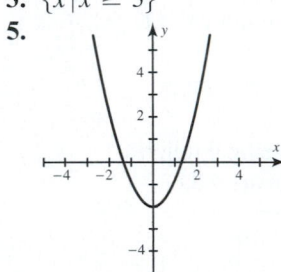

D = all real numbers, $R = \{y \mid y \geq -2\}$

7. (a) Linear   **(b)** Nonlinear
(c) Constant (and linear)

CHAPTER 1   REVIEW EXERCISES (PP. 67–71)

1. Natural number: $\sqrt{16}$; integer: $-2, 0, \sqrt{16}$;
rational number: $-2, \frac{1}{2}, 0, 1.23, \sqrt{16}$; real number:

$-2, \frac{1}{2}, 0, 1.23, \sqrt{7}, \sqrt{16}$

3. 40%   **5.** 1.891×10^6 **7.** 4.39×10^{-5}

9. 15,200

11. (a) 32.07   **(b)** 2.62   **(c)** 5.21   **(d)** 49.12

13. (a) 1.953×10^0 **(b)** 6×10^3

15.

-23	-5	8	19	24

(a) Max: 24; min: -23
(b) Mean: 4.6; median: 8; range: 47

17. (a) $S = \{(-15, -3), (-10, -1), (0, 1), (5, 3),$
$(20, 5)\}$ **(b)** $D = \{-15, -10, 0, 5, 20\}$,
$R = \{-3, -1, 1, 3, 5\}$

19. (a) $D = \{-5, -1, 0, 1, 4\}$, $R = \{-5, -2, 0, 3, 6\}$
(b) x-max: 4, x-min: -5, y-max: 6, y-min: -5
(c) & **(d)**

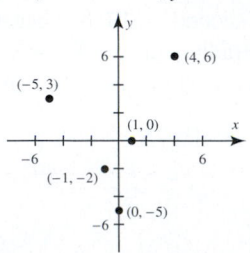

21. Not a function

$[-50, 50, 10]$ by $[-50, 50, 10]$

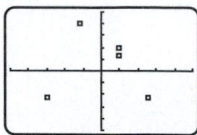

23. 10   **25.** 8   **27.** $\left(2, -\frac{3}{2}\right)$ **29.** Yes

31.   **33.**

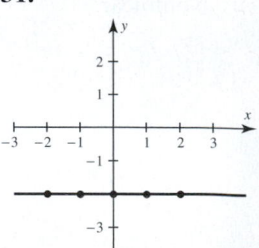

35.   **37.**

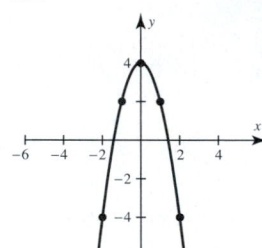 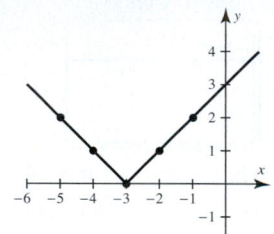

39.

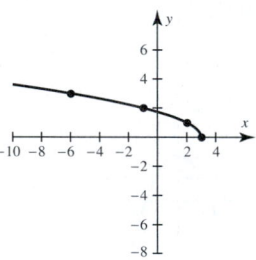

41. $f(x) = 16x$

$[0, 100, 10]$ by $[0, 1800, 300]$

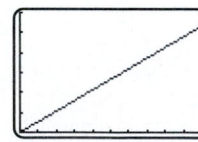

x	0	25	50	75	100
$f(x)$	0	400	800	1200	1600

43. (a) $f(-3) = 2, f(1) = -2$ **(b)** $x = -1, 3$
45. (a) $f(-8) = -2, f(1) = 1$ **(b)** All real numbers
47. (a) $f(-3) = 5, f(1.5) = 5$ **(b)** All real numbers
49. (a) $f(-10) = 97, f(a + 2) = a^2 + 4a + 1$
(b) All real numbers

51. (a) $f(-3) = \dfrac{1}{5}$, $f(a + 1) = \dfrac{1}{a^2 + 2a - 3}$

(b) $D = \{x \mid x \neq 2, x \neq -2\}$

53. No **55.** Yes **57.** Yes **59.** $f(t) = 86,400t$
61. 0 **63.** -6 **65.** $-\dfrac{3}{4}$ **67.** 0 **69.** Linear
71. Nonlinear **73.** Constant (and linear)
75.

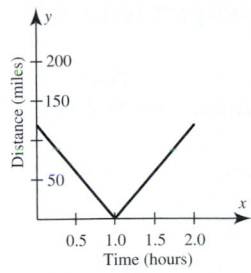

77. Linear, but not constant **79.** 5 **81.** 760 seconds
83. About 0.000796 inch **85. (a)** 140 miles **(b)** -70;
the driver is moving toward the rest stop at 70 miles per hour.
87. (a) The data decrease rapidly, indicating a very
high mortality rate during the first year.

$[-1, 5, 1]$ by $[0, 110, 10]$

(b) Yes
(c) From 0 to 1: -90; from 1 to 2: -4; from 2 to 3: -3;
from 3 to 4: -1
During the first year the population decreased, on average,
by 90 birds. The other average rates of change can be interpreted similarly.
89. (a) $[1, 5, 1]$ by $[40, 70, 5]$

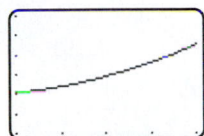

Nonlinear
(b) 2.5 **(c)** The average rate of change in outside temperature from 1 P.M. to 4 P.M. was 2.5°F per hour. The slope
of the line segment from (1, 50.5) to (4, 58) is 2.5.

$[1, 5, 1]$ by $[50, 63, 5]$

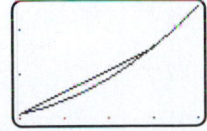

CHAPTER 1 EXTENDED AND DISCOVERY EXERCISES (PP. 70–71)

1. About 7.17 km
3. About 3.862 **5.** About 4.039

$[-3, 3, 1]$ by $[-2, 2, 1]$ $[-3, 3, 1]$ by $[-1, 3, 1]$

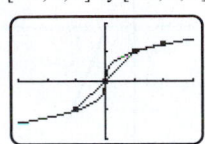 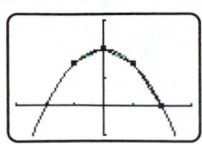

7. About 3.16; estimates will be less than the true value.
9. (a) Determine the number of square miles of Earth's
surface that are covered by the oceans. Then divide the total
volume of the water from the ice cap by the surface area of
the oceans to get the rise in sea level.
(b) About 25.7 feet **(c)** Since the average elevations of
Boston, New Orleans, and San Diego are all less than
25 feet, these cities would be under water without some type
of dike system. **(d)** About 238 feet

CHAPTER 2: Linear Functions and Equations

SECTION 2.1 (PP. 82–86)

1. Exactly **3.** Approximately
5. (a) Slope: 2; y-intercept: -1; x-intercept: 0.5
(b) $f(x) = 2x - 1$ **(c)** 0.5
7. (a) Slope: $\dfrac{3}{4}$; y-intercept: -3; x-intercept: 4
(b) $f(x) = \dfrac{3}{4}x - 3$ **(c)** 4
9. (a) Slope: 20; y-intercept: -50; x-intercept: 2.5
(b) $f(x) = 20x - 50$ **(c)** 2.5
11. Slope: 3; y-intercept: 2 **13.** Slope: $\dfrac{1}{2}$; y-intercept: -2

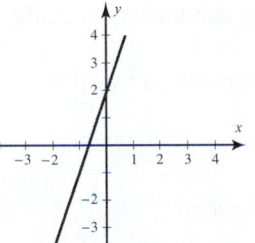 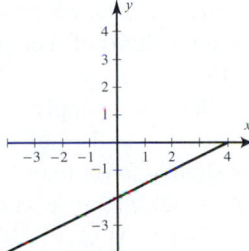

15. Slope: 0; y-intercept: -2 **17.** Slope: $-\dfrac{1}{2}$; y-intercept: 4

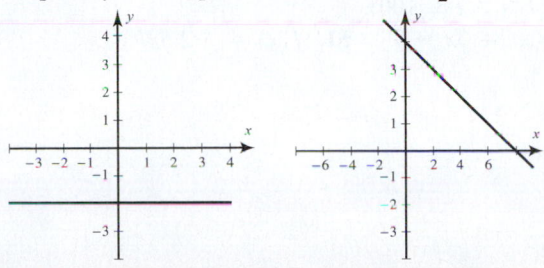

19. Slope: $\frac{1}{2}$; y-intercept: 0   **21.** Slope: -5; y-intercept: 5

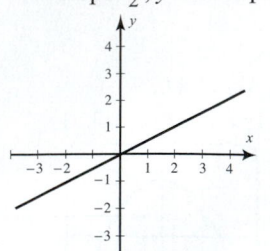 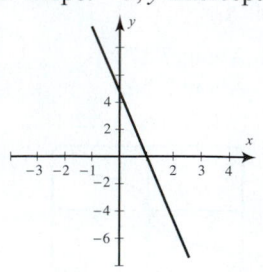

23. Slope: 20; y-intercept: -10

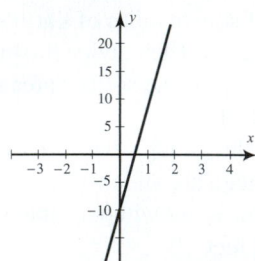

25. $f(x) = -\frac{3}{4}x + \frac{1}{3}$ **27.** $f(x) = 15x$
29. $f(x) = 0.5x + 4$ **31.** 0; 0   **33.** $-\frac{1}{4}$; $-\frac{1}{4}$
35. -3; -3 **37.** d.   **39.** c.
41. (a) $f(x) = 6x + 200$
(b)

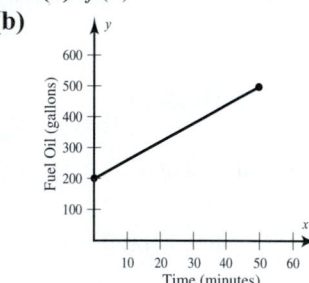

An appropriate domain is $D = \{x \mid 0 \le x \le 50\}$.
(c) y-intercept = 200, which indicates that the tank initially contains 200 gallons of fuel oil.
(d) No, the x-intercept of $-\frac{100}{3}$ corresponds to negative time, which does not apply to this problem.
43. (a) $f(x) = 16.7 - 0.26x$ **(b)** $f(12) = 13.58$, which is slightly lower than 13.9.
45. $V(t) = 32t$; t represents time in seconds; $D = \{t \mid 0 \le t \le 3\}$
47. $P(t) = 21.5 + 0.581t$; t represents years after 1900; $D = \{t \mid 0 \le t \le 100\}$
49. $f(x) = 3x - 7$ **51.** $f(x) = -2.5x$

53. (a) No   **(b)** $f(x) = 0.08x + 2.2$; answers may vary slightly.   **(c)** $[-2, 32, 4]$ by $[0, 5, 1]$

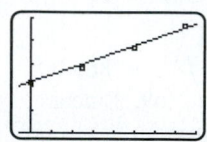

Slope = 0.08 indicates that the number of miles traveled has been increasing by about 0.08 trillion miles per year, on average.
(d) About 4.8 trillion miles; answers may vary slightly.
55. No, because with f, if the speed doubles, the braking distance only doubles instead of quadrupling.
57. (a) Positive   **(b)** $y = ax + b$, where $a \approx 3.0929$ and $b \approx -2.2143$; $r \approx 0.9989$ **(c)** $y \approx 5.209$
59. (a) Negative   **(b)** $y = ax + b$, where $a \approx -3.8857$ and $b \approx 9.3254$; $r \approx -0.9996$ **(c)** $y \approx -0.00028$; due to rounding, answers may vary slightly.
61. (a) $[-100, 1800, 100]$ by $[-1000, 28{,}000, 1000]$

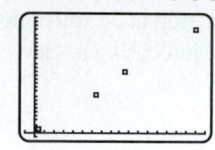

(b) $y = ax + b$, where $a \approx 14.680$ and $b \approx 277.82$
(c) 2500 light-years
63. (a) $[1973, 2000, 2]$ by $[0, 24, 2]$

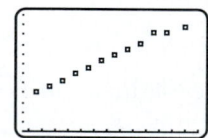

(b) $y = ax + b$, where $a \approx 0.6301$ and $b \approx -1236.4$
(c) The percentage of women in state legislatures has increased by about 0.63% per year, on average.   **(d)** 25.8%, which is higher than 22.3%

2.1 EXTENDED AND DISCOVERY EXERCISES (P. 87)

1. (a) $[-5, 125, 25]$ by $[0, 50, 10]$

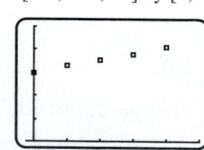

(b) $f(x) = 0.108x + 30$ **(c)** $f(65) = 37.02$ in.^{3}
3. (a) $[-0.625, 0.625, 0.1]$ by $[-0.625, 0.625, 0.1]$

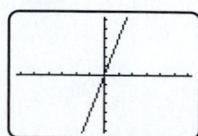

(b) A linear function would be a good approximation over a small interval.

SECTION 2.2 (PP. 99–105)

1. $y = -2(x - 1) + 2$ **3.** $y = \frac{3}{4}(x + 3) - 1$

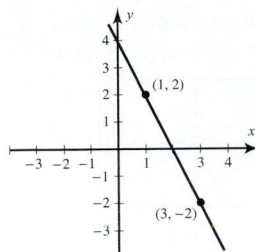
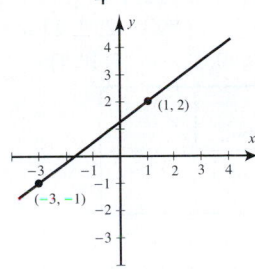

5. $y = -2.4(x - 4) + 5;\ y = -2.4x + 14.6$
7. $y = -\frac{1}{2}(x - 1) - 2;\ y = -\frac{1}{2}x - \frac{3}{2}$
9. $y = 2(x - 1980) + 5;\ y = 2x - 3955$
11. $y = \frac{2}{3}x - 1$ **13.** $y = -\frac{3}{5}x + \frac{3}{5}$
15. $y = -7.8x + 5$ **17.** $y = -\frac{1}{2}x + 45$
19. $y = 4x + 9$ **21.** $y = \frac{3}{2}x - 2960$
23. $y = -3x + 5$ **25.** $y = \frac{3}{2}x - 6$
27. $y = \frac{5}{18}x + \frac{11}{18}$ **29.** $y = \frac{2}{3}x - 2.1$
31. $y = \frac{1}{2}x + 6$ **33.** $y = x - 20$ **35.** $y = \frac{1}{2}x + \frac{9}{2}$
37. $y = -12x - 20$ **39.** $x = -5$ **41.** $y = 6$
43. $x = 4$ **45.** $x = 19$ **47.** c.   **49.** b.   **51.** e.
53. (a) $y = 1.5x - 3.2$ **(b)** When $x = -2.7$,
$y = -7.25$ (interpolation); when $x = 6.3$, $y = 6.25$
(extrapolation)
55. (a) $y = -2.1x + 105.2$ **(b)** When $x = -2.7$,
$y = 110.87$ (extrapolation); when $x = 6.3$, $y = 91.97$
(interpolation)
57. (a) $f(x) = -108(x - 1998) + 3305$ or
$f(x) = -108x + 219{,}089$; answers may vary. Approximate
(b) 3521; the estimated value is too high; extrapolation
59. x-intercept: 5; y-intercept: -4

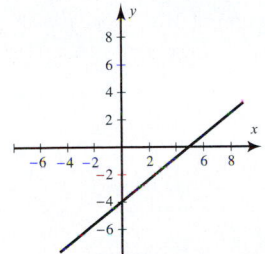

61. x-intercept: 7; y-intercept: -7

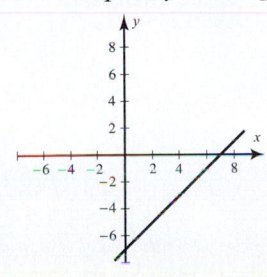

63. x-intercept: -7; y-intercept: 6

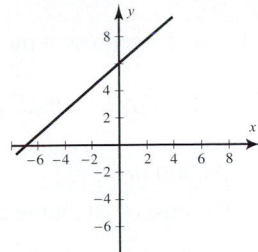

65. x-intercept: $\frac{5}{8}$; y-intercept: -5

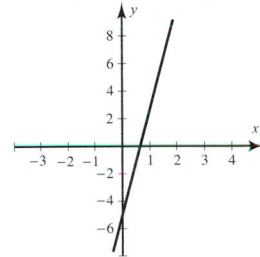

67. x-intercept: $\frac{11}{3}$; y-intercept: -11

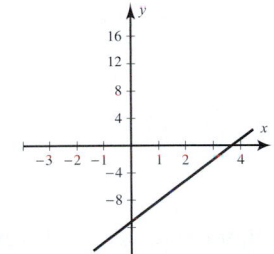

69. x-intercept: $-\frac{7}{3}$; y-intercept: 7

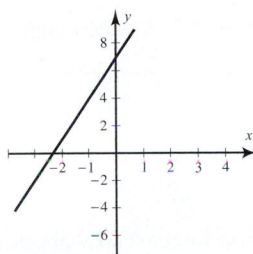

71. x-intercept: 4; y-intercept: 2

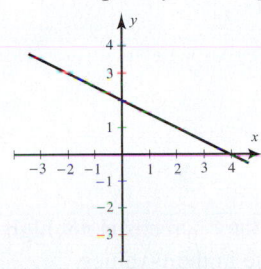

73. x-intercept: 5; y-intercept: 7; a and b represent the x- and y-intercepts respectively.

75. x-intercept: 4; y-intercept: -3; a and b represent the x- and y-intercepts respectively.

77. x-intercept: $\frac{3}{2}$; y-intercept: $\frac{5}{4}$; a and b represent the x- and y-intercepts respectively.

79. (a) $y = \frac{12,000}{7}(x - 2003) + 25,000$ or $y = \frac{12,000}{7}(x - 2010) + 37,000$; The cost of attending a private college or university is increasing by $\frac{12,000}{7} \approx \$1714$ per year, on average.   **(b)** About \$31,857

(c) $y = 1714x - 3,408,714$ (approximate)

81. (a) 11 miles per hour   **(b)** $y = 11x + 117$
(c) 117 miles   **(d)** 130.75 miles

83. (a) [1998, 2005, 1] by [0, 10, 1]

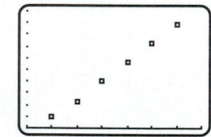

(b) Using the first and last points, $f(x) = 1.56(x - 1999) + 1$; the daily average number of worldwide spam messages increased by 1.56 billion per year, on average. Answers may vary.   **(c)** 13.5 billion

85. (a) [1997, 2003, 1] by [0, 2, 0.2]

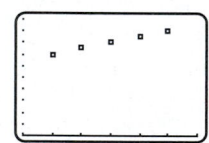

(b) $f(x) = 0.1(x - 1998) + 1.4$; the U.S. sales of Toyota vehicles has increased, by 0.1 million per year.   **(c)** Exact

87. (a) $f(x) = 0.29x + 4200$
(b) It represents the annual fixed costs.

89. (b) $f(x) = 0.38(x - 1996) + 9.7$. Answers may vary.
[1995, 2003, 1] by [9, 13, 1]

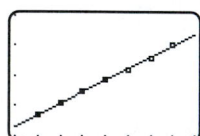

(c) The Asian-American population increased by about 0.38 million each year, on average.
(d) $f(2005) = 13.12$ million

91. (a) No, the slope is not zero.
[0, 3, 1] by [-2, 2, 1]

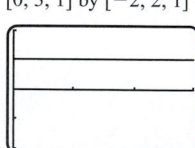

(b) The resolution of most calculator screens is not high enough to show the slight increase in the y-values.

93. (a) They do not appear to be perpendicular in the standard viewing rectangle. (Answers may vary for different calculators.)   **(b)** In [-15, 15, 1] by [-10, 10, 1] and [-3, 3, 1] by [-2, 2, 1] they appear to be perpendicular.

[-10, 10, 1] by [-10, 10, 1]   [-15, 15, 1] by [-10, 10, 1]

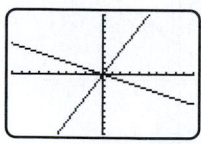

[-10, 10, 1] by [-3, 3, 1]   [-3, 3, 1] by [-2, 2, 1]

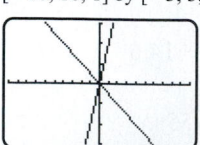

(c) The lines appear perpendicular when the distance shown along the x-axis is approximately 1.5 times longer than the distance along the y-axis.

95. $y_1 = x$, $y_2 = -x$, $y_3 = x + 2$, $y_4 = -x + 4$
97. $y_1 = x + 4$, $y_2 = x - 4$, $y_3 = -x + 4$, $y_4 = -x - 4$
99. $k = 2.5$, $y = 20$ when $x = 8$
101. $k = 0.06$, $x = \$85$ when $y = \$5.10$
103. \$1048, $k = 65.5$
105. (a) $k = 0.01$ **(b)** 1.1 mm   **(c)** No, it was a substantial thinning of the ozone layer.
107.(a) $k = \frac{8}{15}$ **(b)** $13\frac{1}{3}$ inches
109. No. Doubling the speed more than doubles the stopping distance.
111. (a) For (150, 26), $\frac{F}{x} \approx 0.173$; for (180, 31), $\frac{F}{x} \approx 0.172$; for (210, 36), $\frac{F}{x} \approx 0.171$; for (320, 54), $\frac{F}{x} \approx 0.169$; the ratios give the constant of proportionality or the coefficient of friction.
(b) $k \approx 0.17$, answers may vary.
(c) [125, 350, 25] by [0, 75, 5]

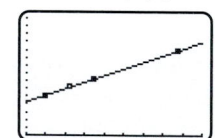

(d) 46.75 pounds

2.2 EXTENDED AND DISCOVERY EXERCISES (P. 105)

1. About 615 fish

CHECKING BASIC CONCEPTS FOR SECTIONS 2.1 AND 2.2 (P. 106)

1. Slope: -2; y-intercept: 4; x-intercept: 2

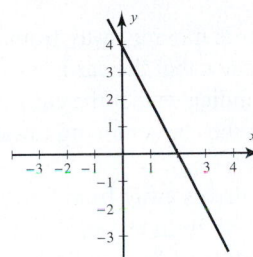

3. $f(t) = 60t + 50$, where t is in hours.

5. Horizontal: $y = 7$; vertical: $x = -4$

7. x-intercept: 6; y-intercept: -9

SECTION 2.3 (PP. 118–123)

1. Linear   **3.** Nonlinear   **5.** Linear   **7.** 4   **9.** -4

11. $\frac{32}{3}$ **13.** 4   **15.** 3   **17.** $\frac{1}{3}$ **19.** $\frac{4}{7}$ **21.** $\frac{13}{10}$

23. $\frac{17}{10}$ **25.** $-\frac{17}{8}$ **27.** $\frac{5}{17}$ **29.** $\frac{400}{7}$

31. (a) No solutions   **(b)** Contradiction

33. (a) $\frac{8}{3}$ **(b)** Conditional

35. (a) All real numbers   **(b)** Identity

37. (a) No solutions   **(b)** Contradiction

39. (a) All real numbers   **(b)** Identity

41. (a) 1990   **(b)** Conditional

43. (a) $\frac{4\sqrt{2} + 1.5}{0.5 - \sqrt{2}} \approx -7.828$ **(b)** Conditional

45. (a) No solutions   **(b)** Contradiction

47. 3   **49. (a)** 4   **(b)** 2   **(c)** -2 **51.** -1

53. 1   **55.** 4   **57.** 1.3   **59.** 0.675   **61.** 3.621

63. 2.294   **65.** -2 **67.** 3   **69.** 8.6   **71.** 3.5

73. -4 **75.** 2   **77.** 7.5   **79.** 1993

81. (a) $V(x) = 900x$ **(b)** $V(50) = 45{,}000$ cubic feet per hour   **(c)** 4.5   **(d)** $3\frac{1}{3}$ times   **83.** 1994

85. About 17.29 feet

87. (a) f is linear because the amount of oil is mixed at a constant rate.   **(b)** 0.48 pint; 0.48 pint of oil should be added to 3 gallons of gasoline to get the correct mixture.   **(c)** 12.5 gallons

89. 36.4 cubic feet   **91.** 268.6 million

93. $f(x) = 0.75x$; $42.18

95. 3.2 hours at 55 mph and 2.8 hours at 70 mph

97. (a) About 2 hours; answers may vary.

(b) $\frac{15}{8} = 1.875$ hours   **99.** About 8.33 liters

101. 1.6 pounds of $2.50 candy and 3.4 pounds of $4.00 candy   **103.** 36 inches by 54 inches

105. 25 feet by 50 feet   **107.** $1250 at 5%, $3750 at 7%

109. $-40°$F is equivalent to $-40°$C

111. (a) $C(x) = -7.5x + 15{,}090$; $L(x) = 14.75x - 29{,}500.5$ **(b)** Sales of CRT monitors decreased by 7.5 million monitors per year, on average. Sales of LCD monitors increased by 14.75 million monitors per year, on average.   **(c)** 2004   **(d)** 2004   **(e)** 2004

113. The point $(4, 3.5)$ lies on the graph of f. The graph of f is continuous, so there must be a point where it crosses the line $y = 3.5$.

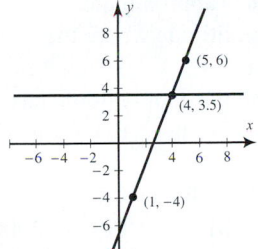

115. Because $f(2) = -1 < 0$ and $f(3) = 4 > 0$, the intermediate value property states that there exists an x-value between 2 and 3 where $f(x) = 0$.

117. Because $f(0) = -1 < 0$ and $f(1) = 1 > 0$, the intermediate value property states that there exists an x-value between 0 and 1 where $f(x) = 0$.

2.3 EXTENDED AND DISCOVERY EXERCISES (P. 123)

1. (a) Yes; since multiplication distributes over addition, doubling the lengths gives double the sum of the lengths.   **(b)** No; for example in the case of a square (a type of rectangle), the square of twice a side is four times the square of the side.

SECTION 2.4 (PP. 134–139)

1. $[5, \infty)$ **3.** $[4, 19)$ **5.** $(-\infty, -37]$

7. $[-1, \infty)$ **9.** $(-3, 5]$ **11.** $(-\infty, -2)$ **13.** $[2, \infty)$

15. $(-\infty, 10.5)$ **17.** $[13, \infty)$ **19.** $[-10, \infty)$

21. $(1, \infty)$ **23.** $\left(\frac{7}{3}, \infty\right)$ **25.** $\left[-\frac{1}{2}, 2\right]$ **27.** $[-16, 1]$

29. $\left(\frac{5}{7}, \frac{17}{7}\right]$ **31.** $(-4, 1)$ **33.** $\left[\frac{9}{2}, \frac{21}{2}\right]$ **35.** $(-\infty, -4)$

37. $\left(-\frac{1}{2}, -\frac{1}{4}\right]$ **39.** $\left(-\infty, \frac{21}{19}\right]$ **41.** $\{x \mid x \leq 2\}$

43. $\{x \mid x > 3\}$ **45.** $\{x \mid 0 \leq x \leq 2\}$

47. $\{x \mid -1 < x \leq 4\}$ **49. (a)** 2   **(b)** $(-\infty, 2)$

(c) $[2, \infty)$ **51. (a)** -2 **(b)** $(-2, \infty)$ **(c)** $(-\infty, -2]$

53. $(-\infty, 2]$ **55.** $(1, \infty)$ **57.** $\{x \mid x > 2.8\}$

59. $\{x \mid x \leq 1987.5\}$ **61.** $\{x \mid x > -1.82\}$ **63.** $[4, 6.4)$

65. $[4.6, 15.2]$ **67.** $(1, 5.5)$ **69. (a)** 8   **(b)** $x < 8$

71. $\{x \mid x < 4\}$; $\{x \mid x \geq 4\}$ **73.** $\left(-\infty, -\frac{3}{2}\right)$ **75.** $[1, 4]$

77. $\left[-\frac{1}{20}, \frac{17}{20}\right)$ **79.** $(-\infty, 25.3)$ **81.** $\left(\frac{13}{2}, \infty\right)$

83. $[0.717, \infty)$

85. (a) A loan amount of $1000 results in annual interest of $100.   **(b)** A loan amount greater than $1000 results in annual interest of more than $100.
(c) A loan amount less than $1000 results in annual interest of less than $100.   **(d)** A loan amount greater than or equal to $1000 results in annual interest of $100 or more.
87. (a) Car A is traveling faster since its graph has the greater slope.   **(b)** 2.5 hours, 225 miles   **(c)** $0 \le x < 2.5$
89. (a) $1.8 < x \le 6$, where 1.8 is approximate.
(b) The x-intercept represents the altitude where the temperature is 0°F.   **(c)** $\frac{53}{29} < x \le 6$
91. (a) The median price of a single-family home has increased, on average, by $3421 per year.   **(b)** From 1983 to 1988 (approximately)
93. (a) $T(x) = 60(x - 1997) + 100$ or $T(x) = 60(x - 2002) + 400$ **(b)** From 1998 to 2000
95. (a) $B(x) = 6(x - 2000) + 6$ or $B(x) = 6(x - 2004) + 30$ **(b)** From 2003 to 2006
97. (a) Increasing the ventilation increases the percentage of pollutants removed. A slope of 1.06 means that for each additional liter of air removed per second, the amount of pollutants decreases by 1.06%.   **(b)** From approximately 40.4 to 59.3 liters per second
99. $3.98\pi \le C \le 4.02\pi$
101. (a) $f(x) = 3x - 1.5$ **(b)** $x > 1.25$
103. (a) $f(x) = 675(x - 2002) + 47,975$ or $f(x) = 675x - 1,303,375$
(b) $x = 2002, 2003, \ldots, 2009$; from 2002 through 2009, this person's salary was less than or equal to $52,700.

2.4 EXTENDED AND DISCOVERY EXERCISES (P. 130)

1. $a < b \Rightarrow 2a < a + b < 2b \Rightarrow a < \dfrac{a + b}{2} < b$

CHECKING BASIC CONCEPTS FOR SECTIONS 2.3 AND 2.4 (P. 139)

1. (a) $[-10, 10, 1]$ by $[-10, 10, 1]$ **(b)**

(c) 2.5
3. $\left[-1, \frac{3}{2}\right]$

SECTION 2.5 (PP. 152–157)

1. (a) Max: 55 mph; min: 30 mph   **(b)** 12 miles
(c) $f(4) = 40, f(12) = 30, f(18) = 55$
(d) $x = 4, 6, 8, 12,$ and 16. The speed limit changes at each discontinuity.

3. (a) Initial: 50,000 gallons; final: 30,000 gallons
(b) $0 \le x \le 1$ or $3 \le x \le 4$ **(c)** $f(2) = 45$ thousand, $f(4) = 40$ thousand   **(d)** 5000 gallons per day
5. (a) $f(1.5) = 30; f(4) = 10$
(b) $m_1 = 20$ indicates that the car is moving away from home at 20 mph; $m_2 = -30$ indicates that the car is moving toward home at 30 mph; $m_3 = 0$ indicates that the car is not moving; $m_4 = -10$ indicates that the car is moving toward home at 10 mph.
(c) The driver starts at home and drives away from home at 20 mph for 2 hours until the car is 40 from home. The driver then travels toward home at 30 mph for 1 hour until the car is 10 miles from home. Then the car does not move for 1 hour. Finally, the driver returns home in 1 hour at 10 mph.
7. (a) $-5 \le x \le 5$
(b) $f(-2) = 2, f(0) = 3, f(3) = 6$
(c)   **(d)** f is continuous.

9. (a) $-1 \le x \le 2$
(b) $f(-2) =$ undefined, $f(0) = 0, f(3) =$ undefined
(c)   **(d)** f is not continuous.

11. (a) $-3 \le x \le 3$
(b) $f(-2) = -2, f(0) = 1, f(3) = -1$
(c)   **(d)** f is not continuous.

13.

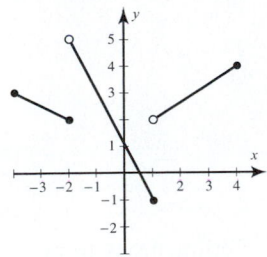

15. $f(-3) = -10, f(1) = 4, f(2) = 4, f(5) = 1$

17. f is not continuous.

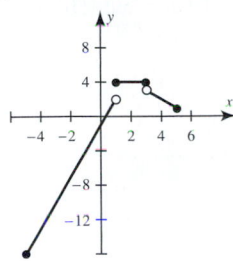

19. $g(-8) = 10, g(-2) = -2, g(2) = 2, g(8) = 5$

21. g is continuous.

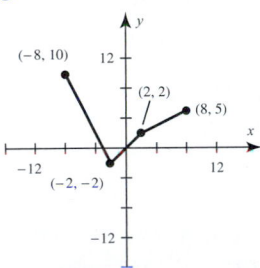

23. (a) $[-10, 10, 1]$ by $[-10, 10, 1]$

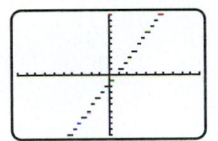

(b) $f(-3.1) = -8, f(1.7) = 2$

25. (a) $[-10, 10, 1]$ by $[-10, 10, 1]$

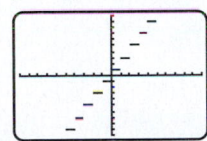

(b) $f(-3.1) = -7, f(1.7) = 3$

27. (a) $f(x) = 0.8[\![x/2]\!]$ for $6 \leq x \leq 18$

(b) $[6, 18, 1]$ by $[0, 8, 1]$

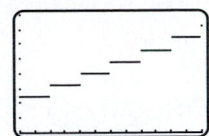

(c) $f(8.5) = \$3.20, f(15.2) = \$5.60$

29.   **31.**

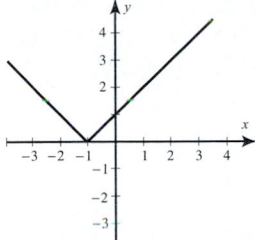   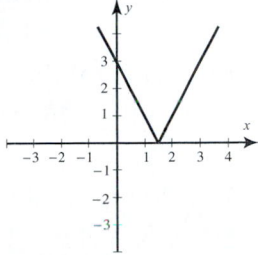

33. (a)   **(b)**

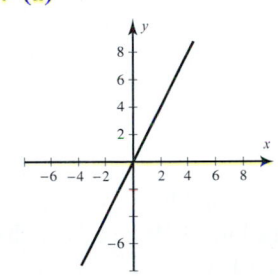   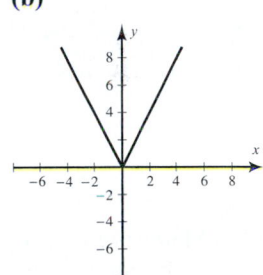

(c) $x = 0$

35. (a)   **(b)**

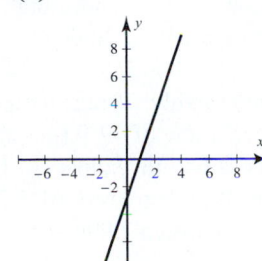   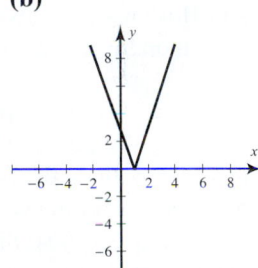

(c) $x = 1$

37. (a)   **(b)**

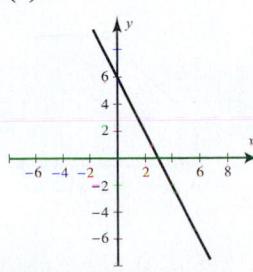   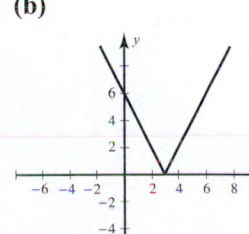

(c) $x = 3$

39. ± 2 **41.** 1 or $\frac{9}{5}$ **43.** $-\frac{1}{2}$ or 2 **45.** $-\frac{1}{3}$

47. No solutions   **49.** $-\frac{23}{12}$ or $\frac{19}{4}$ **51.** No solutions

53. -1 or $\frac{17}{5}$ **55.** -1 or 1 **57.** -3 or 2

59. (a) -1 or 7 **(b)** $-1 < x < 7$
(c) $x < -1$ or $x > 7$ **61.** -2.5 or 7.5; $-2.5 < x < 7.5$

63. $\frac{7}{3}$ or 1; $x < 1$ or $x > \frac{7}{3}$ **65.** $-\frac{17}{21}$ or $\frac{31}{21}$; $x \le -\frac{17}{21}$ or

$x \ge \frac{31}{21}$ **67.** $-\frac{1}{3}$ or $\frac{1}{3}$; $x < -\frac{1}{3}$ or $x > \frac{1}{3}$

69. There are no solutions for the equation or inequality.

71. $\left(-\frac{7}{3}, 3\right)$ **73.** $\left[-1, \frac{9}{2}\right]$ **75.** $\left(-\frac{5}{2}, \frac{11}{2}\right)$

77. $(-\infty, 1) \cup (2, \infty)$ **79.** $\left(-\infty, \frac{5}{3}\right] \cup \left[\frac{11}{3}, \infty\right)$

81. $(-\infty, -8) \cup (16, \infty)$

83.

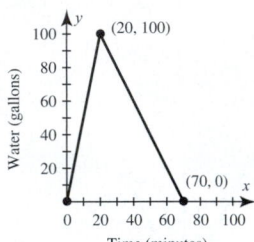

85.

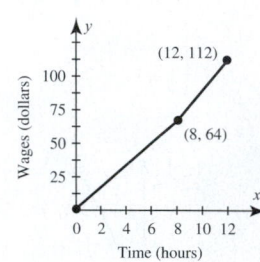

87. 6 **89.** $[-2, 4]$ **91.** $[0, \infty)$

93. (a) $19 \le T \le 67$ **(b)** The monthly average temperatures in Marquette vary between a low of 19°F and a high of 67°F. The monthly averages are always within 24 degrees of 43°F.

95. (a) $28 \le T \le 72$ **(b)** The monthly average temperatures in Boston vary between a low of 28°F and a high of 72°F. The monthly averages are always within 22 degrees of 50°F.

97. (a) $49 \le T \le 74$ **(b)** The monthly average temperatures in Buenos Aires vary between a low of 49°F (possibly in July) and a high of 74°F (possibly in January). The monthly averages are always within 12.5 degrees of 61.5°F.

99. $2.996 \le d \le 3.004$; diameters between 2.996 and 3.004 inches are acceptable.

101. $34.3 \le Q \le 35.7$

103. (a) $y = x - 1$ **(b)**

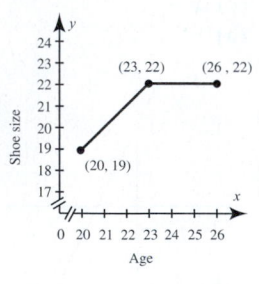

105. (a)

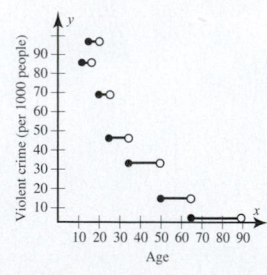

(b) The likelihood of being a victim of crime peaks from age 16 up to age 19, and then it decreases.

107. (a) Cesarean births increased at a constant rate of 6% every 5-year period between 1970 and 1985, which then leveled off at 23%.

[1965, 1995, 5] by [0, 25, 5]

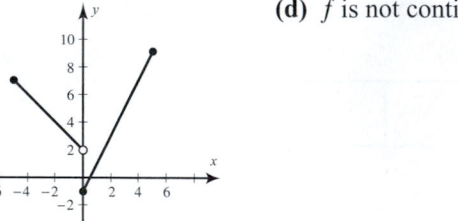

(b) $a = 1.2, b = 5, c = 23$ **(c)** $f(1978) = 14.6\%$, $f(1988) = 23\%$. In 1978, 14.6% of births were Cesarean births, and in 1988 the percentage was 23%.

2.5 EXTENDED AND DISCOVERY EXERCISES (P. 157)

1. Between approximately 23.55 feet and 24.19 feet

CHECKING BASIC CONCEPTS FOR SECTION 2.5 (P. 158)

1. (a) Domain: $[-5, 5]$
(b) $f(-2) = 4, f(0) = -1, f(3) = 5$
(c)   **(d)** f is not continuous.

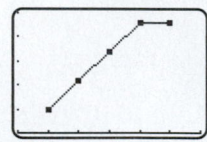

3.

5. (a) -2 or 3 **(b)** $[-2, 3]$; $(-\infty, -2) \cup (3, \infty)$

CHAPTER 2   REVIEW EXERCISES (PP. 162–166)

1. (a) Slope: -2; y-intercept: 6; x-intercept: 3
(b) $f(x) = -2x + 6$ **(c)** 3
3. $f(x) = -2.5x + 5$
5.   **7.**

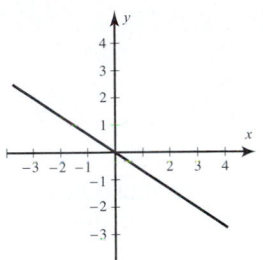

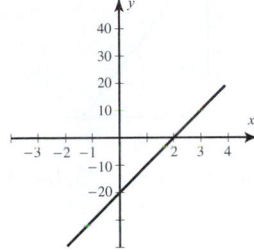

9. $f(x) = -2x - 1$ **11.** $y = 7x + 30$
13. $y = -3x + 2$ **15.** $y = -\frac{31}{57}x - \frac{368}{57}$
17. $x = 6$ **19.** $y = 3$ **21.** $y = -8$
23. x-intercept: 4;   **25.** x-intercept: 1;
    y-intercept: -5 y-intercept: 2

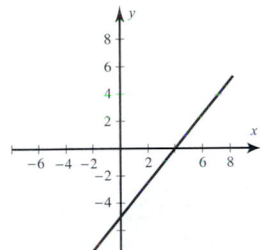

27. 6.4   **29.** $\frac{15}{7} \approx 2.143$ **31.** $\frac{1}{17}$ **33.** -2.9
35. 2   **37. (a)** All real numbers   **(b)** Identity
39. (a) -3 **(b)** Conditional   **41.** $(-3, \infty)$
43. $\left[-2, \frac{3}{4}\right)$ **45.** $(-\infty, 0) \cup (2, \infty)$ **47.** $(-\infty, 3]$
49. $(-\infty, \infty)$ **51.** $(-1, 3.5]$ **53.** $(-1, \infty)$
55. $[-2, 1]$ **57. (a)** 2   **(b)** $x > 2$ **(c)** $x < 2$
59. (a) $f(-2) = 4, f(-1) = 6, f(2) = 3, f(3) = 4$
(b) f is continuous.   **(c)** $x = -2.5$ or 2

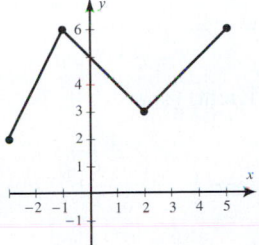

61.

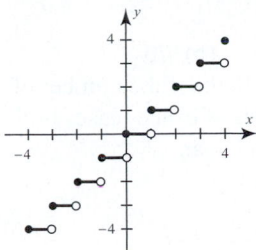

63. -2 or 7   **65.** No solutions
67. ± 3; $x < -3$ or $x > 3$
69. $\frac{17}{3}$ or -1; $x < -1$ or $x > \frac{17}{3}$ **71.** -0.5 or -0.75
73. $-3 < x < 6$ **75.** $x \le -\frac{5}{2}$ or $x \ge \frac{7}{2}$
77. (a) $x = 1990$. In 1990 the median U.S. family income
was about \$36,500.   **(b)** $x = 1990$
79. From 2001 to 2006
81. Initially, the car is at home. After traveling 30 mph for
1 hour, the car is 30 miles away from home. During the second hour the car travels 20 mph until it is 50 miles away.
During the third hour the car travels toward home at 30 mph
until it is 20 miles away. During the fourth hour the car travels away from home at 40 mph until it is 60 miles away from
home. During the last hour, the car travels 60 miles at
60 mph until it arrives home.
83.

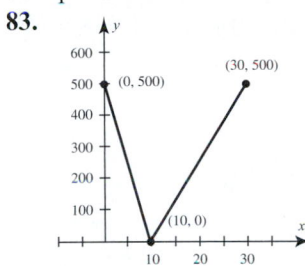

85. 155,590   **87.** 18.75 minutes
89. 0.9 hour at 7 mph and 0.9 hour at 8 mph
91. (a) $f(x) = \frac{5}{3}(x - 1994) + 504$ or $f(x) = \frac{5}{3}x - \frac{8458}{3}$
(b) 509 (answers may vary); interpolation
93. The tank initially contains 25 gallons. When $0 \le x \le 4$,
only the 5 gal/min inlet pipe is open. When $4 < x \le 8$, only
the outlet pipe is open. When $8 < x \le 12$, both inlet pipes
are open. When $12 < x \le 16$, all pipes are open. When
$16 < x \le 24$, the 2 gal/min inlet pipe and the outlet pipe are
open. When $24 < x \le 28$, all pipes are closed.
95. (a) 1972   **(b)** The function f will not continue to be
a good model far into the future. Once the water is no longer
polluted, the number of species in the river will reach its natural level.
97. (a) Positive
(b) $y = 143.45x - 280,361.2$; $r \approx 0.978$ **(c)** \$5821.55

CHAPTER 2 EXTENDED AND DISCOVERY EXERCISES (PP. 166–167)

1. (a) Approximately 5620 people **(b)** 70
(c) $0.00002 \le C \le 0.00008$ **(d)** Since if the number of people exposed is doubled, the number of cancer cases is also doubled, it is reasonable that f is linear.
3. Answers will vary **5.** 3 miles

CHAPTERS 1–2 CUMULATIVE REVIEW EXERCISES (PP. 167–170)

1. -10.6% **3.** $6,700,000; 0.000145$
5. (a) Yes **(b)** $D = \{-1, 0, 1, 2, 3\}; R = \{0, 3, 4, 6\}$
7. $\sqrt{89}$
9. (a) $D = \{x \mid -\infty < x < \infty\}; R = \{y \mid y \ge -2\}$
$f(-1) = -1$
(b) $D = \{x \mid -3 \le x \le 3\}; R = \{y \mid -3 \le y \le 2\}$
$f(-1) = -\frac{1}{2}$
11. (a) $f(2) = 7; f(a - 1) = 5a - 8$
(b) $D = \{x \mid -\infty < x < \infty\}$
13. (a) $f(2) = \sqrt{3}; f(a - 1) = \sqrt{2a - 3}$
(b) $D = \left\{x \mid x \ge \frac{1}{2}\right\}$
15. No. The graph does not pass the vertical line test.
17. 1
19. (a) $\frac{2}{3}; -2; 3$ **(b)** $f(x) = \frac{2}{3}x - 2$ **(c)** 3
21. $f(x) = -3x + \frac{4}{3}$ **23.** $-\frac{13}{2}$ **25.** $y = -\frac{11}{8}x - \frac{29}{8}$
27. $x = -1$ **29.** $y = 2x + 11$
31. x-intercept: -3; y-intercept: 2

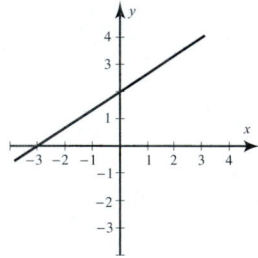

33. 1 **35.** $-\frac{24}{17}$ **37.** 3 **39.** $(-\infty, 5)$
41. $(-\infty, -2) \cup (2, \infty)$ **43.** $(0, \infty)$ **45.** $\left(-\infty, \frac{5}{8}\right]$
47. (a) 2 **(b)** $x < 2$ **(c)** $x \ge 2$ **49.** -6 or 4
51. -7 or 7 **53.** $[0, 5]$ **55.** $(-\infty, -12] \cup [28, \infty)$
57. 3.40 inches **59.** About 0.0016 inch
61. About 94.2 miles

63. (a) $D(x) = 270 - 72x$
(b) $\{x \mid 0 \le x \le 3.75\}$

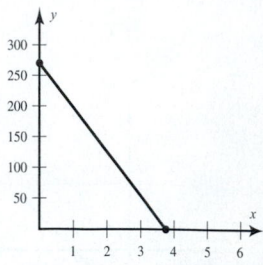

(c) x-intercept: 3.75, the driver arrives home after 3 hours and 45 minutes; y-intercept: 270, the driver is initially 270 miles from home
65. 1.25 hours at 8 mph, 0.5 hours at 10 mph
67. (a) $f(x) = 0.99(x - 2001) + 9.3$ or
$f(x) = 0.99(x - 2006) + 14.25$ **(b)** 2004
69. (a) $y = \frac{10}{3}(x - 58) + 91$
(b) $f(61) = 101$; midpoint: $(61, 101)$; answers are the same, because the midpoint lies on the graph of f.

CHAPTER 3: Quadratic Functions and Equations

SECTION 3.1 (PP. 184–188)

1. Quadratic; leading coefficient: 3; $f(-2) = 17$
3. Neither linear nor quadratic **5.** Linear
7. Quadratic; leading coefficient: -3; $f(-2) = -3$
9. Leading coefficient: positive; vertex: $(1, 0)$; axis of symmetry: $x = 1$
11. Leading coefficient: negative; vertex: $(-3, -2)$; axis of symmetry: $x = -3$
13. The graph of g is narrower than the graph of f.
15. The graph of g is wider than the graph of f and opens downward rather than upward.
17. Vertex: $(1, 2)$; leading coefficient: -3;
$f(x) = -3x^2 + 6x - 1$
19. Vertex: $(4, 5)$; leading coefficient: -2;
$f(x) = -2x^2 + 16x - 27$
21. Vertex: $\left(-5, -\frac{7}{4}\right)$; leading coefficient: $\frac{3}{4}$;
$f(x) = \frac{3}{4}x^2 + \frac{15}{2}x + 17$
23. Vertex: $\left(-\frac{3}{4}, 0\right)$; leading coefficient: $\frac{1}{2}$;
$f(x) = \frac{1}{2}x^2 + \frac{3}{4}x + \frac{9}{32}$
25. $(0, 6)$ **27.** $(3, -9)$ **29.** $(1, -1)$ **31.** $(0, 10)$
33. $\left(\frac{1}{3}, -\frac{35}{12}\right)$ **35.** $(1.9, -5.61)$ **37.** $(-0.25, 1.875)$
39. $f(x) = (x + 2)^2 - 9$; vertex: $(-2, -9)$
41. $f(x) = \left(x - \frac{3}{2}\right)^2 - \frac{9}{4}$; vertex: $\left(\frac{3}{2}, -\frac{9}{4}\right)$
43. $f(x) = 2\left(x - \frac{5}{4}\right)^2 - \frac{1}{8}$; vertex: $\left(\frac{5}{4}, -\frac{1}{8}\right)$

45. $f(x) = -\frac{1}{2}\left(x + \frac{3}{2}\right)^2 + \frac{17}{8}$; vertex: $\left(-\frac{3}{2}, \frac{17}{8}\right)$

47. $f(x) = 2(x - 2)^2 - 9$; vertex: $(2, -9)$

49. $f(x) = -3(x + 1.5)^2 + 8.75$; vertex: $(-1.5, 8.75)$

51. $f(x) = (x - 2)^2 - 2$

53. $f(x) = \frac{1}{2}(x - 2)^2 - 3$

55. $f(x) = -2(x + 1)^2 + 3$

57. $f(x) = -3(x - 2)^2 + 6$

59.
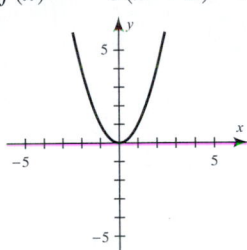

61.
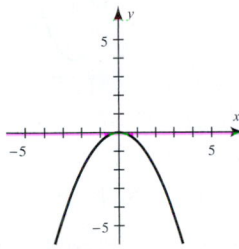

63.
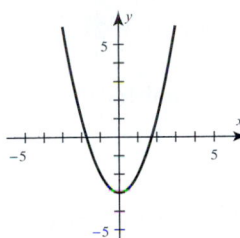

65.
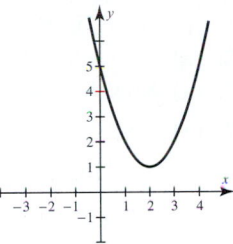

67.
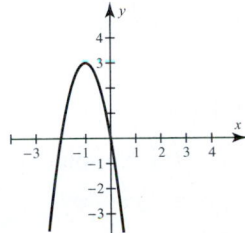

69.
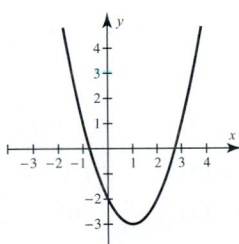

71.
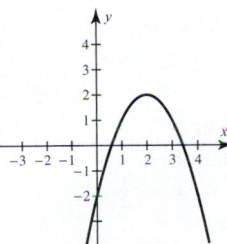

73.
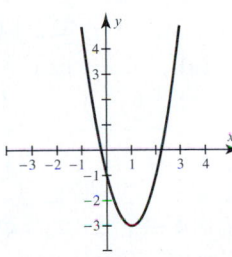

75.
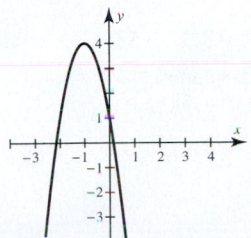

77.
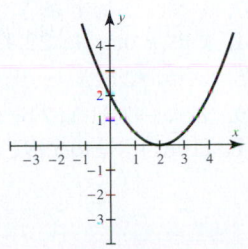

79. 250 ft by 250 ft

81. (a) $R(2) = 72$; the company receives $72,000 for producing 2000 CD players.   **(b)** 10,000   **(c)** $200,000

83. (a) 32 ft   **(b)** 34.25 ft   **85.** d.   **87.** a.

89. 146 ft after 2.75 sec   **91.** 323 ft after 6.77 sec

93. $f(x) = 2(x - 1)^2 - 3$

95. $f(x) = -2(x + 1)^2 + 4$

97. (b) $f(x) = 3100(x - 1982)^2 + 1586$. Answers may vary.

[1980, 1996, 2] by [−50000, 500000, 10000]

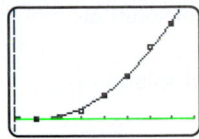

(c) $f(1998) = 795,186$ cases

99. (a) $H(t) = 2(t - 4)^2 + 90$; $D = \{0 \le t \le 4\}$

(b) $H(1.5) = 102.5$ beats per minute

101. (a) $f(x) = 0.006(x - 1900)^2 + 5.3$; answers may vary.   **(b)** 77.9 million; answers may vary.

103. $f(x) = 3.125x^2 + 2.05x - 0.9$; $f(3.5) \approx 44.56$

105. $f(x) = 1.415x^2 + 1.731x + 0.995$; $f(3.5) \approx 24.39$

107. (a) $f(x) \approx 0.000217857x^2 - 0.8680357x + 866.58$ **(b)** $f(1975) \approx 1.99$, which compares favorably with the actual value of 2.01.

3.1 EXTENDED AND DISCOVERY EXERCISES (PP. 188–189)

1. Quadratic; $y = -61$ when $x = 6$ **3.** Neither

5. (a)

x	1	2	3	4	5
$f(x)$	−2	1	6	13	22

(b) 3; 5; 7; 9   **(c)** $2x + h$; $2x + 1$

(d) 3; 5; 7; 9; the results are the same.

7. (a)

x	1	2	3	4	5
$f(x)$	0	−3	−10	−21	−36

(b) −3; −7; −11; −15   **(c)** $-4x + 3 - 2h$; $-4x + 1$

(d) −3; −7; −11; −15; the results are the same.

SECTION 3.2 (PP. 201–205)

1. $-4, 3$ **3.** $0, 2$ **5.** $-5, -\frac{3}{2}$ **7.** $-5, \frac{1}{3}$ **9.** $\frac{1}{2}, \frac{5}{6}$

11. $-3 \pm \sqrt{5}$ **13.** $\pm\frac{\sqrt{13}}{2}$

15. No real solutions   **17.** $3 \pm 2\sqrt{2}$ **19.** $\frac{-1 \pm \sqrt{13}}{3}$

21. $\frac{1}{5}$ **23.** $\frac{5 \pm \sqrt{85}}{30}$ **25.** $-\frac{5}{2}, \frac{1}{3}$ **27.** $\frac{3}{2}$ **29.** $\frac{2}{3}, 3$

31. $-2, 0$ **33.** $-2, 3$ **35.** $\pm\sqrt{3} \approx \pm 1.7$ **37.** 1.5

39. $-0.75, 0.2$ **41.** $0.7, 1.2$ **43.** $-96, 72$

45. (a) $3x^2 - 12 = 0$ **(b)** $b^2 - 4ac = 144 > 0$. There are two real solutions.   **(c)** ± 2

47. (a) $x^2 - 2x + 1 = 0$ **(b)** $b^2 - 4ac = 0$. There is one real solution.   **(c)** 1

49. (a) $x^2 - 4x + 2 = 0$ **(b)** $b^2 - 4ac = 8 > 0$.
There are two real solutions.   **(c)** $2 \pm \sqrt{2}$
51. (a) $x^2 - x + 1 = 0$
(b) $b^2 - 4ac = -3 < 0$. There are no real solutions.
53. (a) $2x^2 + 5x - 12 = 0$ **(b)** $b^2 - 4ac = 121 > 0$.
There are two real solutions.   **(c)** $-4, 1.5$
55. (a) $\frac{1}{4}x^2 + 2x + 4 = 0$ **(b)** $b^2 - 4ac = 0$.
There is one real solution.   **(c)** -4
57. (a) $\frac{1}{2}x^2 + x + \frac{13}{2} = 0$
(b) $b^2 - 4ac = -12 < 0$. There are no real solutions.
59. (a) $3x^2 + x - 1 = 0$
(b) $b^2 - 4ac = 13 > 0$. There are two real solutions.
(c) $-\frac{1}{6} \pm \frac{1}{6}\sqrt{13} \approx 0.4343, -0.7676$
61. (a) $a > 0$ **(b)** $-6, 2$ **(c)** Positive
63. (a) $a > 0$ **(b)** -4 **(c)** Zero
65. $-2 \pm \sqrt{10}$ **67.** $-\frac{5}{2} \pm \frac{1}{2}\sqrt{41}$ **69.** $1 \pm \frac{\sqrt{15}}{3}$
71. $4 \pm \sqrt{26}$ **73.** $\frac{3 \pm \sqrt{17}}{2}$ **75.** $\frac{3 \pm \sqrt{17}}{4}$
77. $\frac{7 \pm \sqrt{89}}{10}$ **79.** $\frac{-1 \pm \sqrt{97}}{12}$
81. $\{x \mid x \neq \sqrt{5}, x \neq -\sqrt{5}\}$
83. $\{t \mid t \neq -1, t \neq 2\}$ **85.** $y = \frac{-12x^2 + 1}{8}$; yes
87. $y = \pm \frac{\sqrt{12 - 3x^2}}{2}$; no   **89.** $y = \pm \frac{\sqrt{x - 50}}{5}$; no
91. $r = \pm\sqrt{\frac{3V}{\pi h}}$ **93.** $v = \pm\sqrt{\frac{2K}{m}}$
95. $t = \frac{25 \pm \sqrt{625 - 4s}}{8}$ **97.** 2.2 seconds
99. 1989   **101. (a)** R quadruples   **(b)** $a = 0.5$
(c) $x = \sqrt{1000} \approx 31.6$. The safe speed for a curve with radius 500 ft is about 31 mph or slower.
103. (a) $E(15) = 1.4$; In 2002 there were 1.4 million Wal-Mart employees.
(b) $E(x) = 0.005x^2 + 0.2$; answers may vary.
(c) $[0, 25, 5]$ by $[0, 2.6, 0.2]$

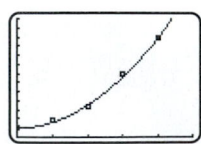

(d) About 2011; answers may vary.
105. (a) $s(t) = -16t^2 + 160t + 32$
(b) About 10.2 sec
107. 8.5 in. by 11 in.
109. 15 in. by 25 in.
111. 20 in. by 24 in.
113. About 18.23 in.

**CHECKING BASIC CONCEPTS FOR
SECTIONS 3.1 AND 3.2 (PP. 205–206)**

1. Vertex; $(1, -4)$; axis of symmetry: $x = 1$;
   x-intercepts: $-1, 3$

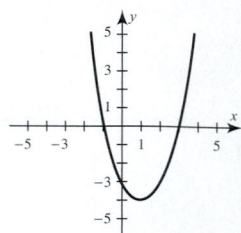

3. $f(x) = 2(x + 1)^2 + 3$
5. $f(x) = (x + 2)^2 - 7$; $(-2, -7)$; -7
7. 11 in. by 15 in.

SECTION 3.3 (PP. 213–215)

1. (a) $-3 < x < 2$ **(b)** $x \leq -3$ or $x \geq 2$
3. (a) $x = -2$ **(b)** $x \neq -2$
5. (a) No solutions   **(b)** All real numbers
7. (a) $-\frac{5}{2}, -\frac{1}{2}$ **(b)** $\left(-\frac{5}{2}, -\frac{1}{2}\right)$
(c) $\left(-\infty, -\frac{5}{2}\right) \cup \left(-\frac{1}{2}, \infty\right)$
9. (a) $-1, \frac{7}{5}$ **(b)** $(-\infty, -1) \cup \left(\frac{7}{5}, \infty\right)$ **(c)** $\left(-1, \frac{7}{5}\right)$
11. (a) $-3, 4$ **(b)** $(-3, 4)$ **(c)** $(-\infty, -3) \cup (4, \infty)$
13. (a) $\pm\sqrt{5}$ **(b)** $\left[-\sqrt{5}, \sqrt{5}\right]$
(c) $\left(-\infty, -\sqrt{5}\right] \cup \left[\sqrt{5}, \infty\right)$
15. (a) $-\frac{8}{3}, 0$ **(b)** $\left[-\frac{8}{3}, 0\right]$ **(c)** $\left(-\infty, -\frac{8}{3}\right] \cup [0, \infty)$
17. (a) $\frac{3}{2}$ **(b)** $\left(-\infty, \frac{3}{2}\right) \cup \left(\frac{3}{2}, \infty\right)$ **(c)** No solutions
19. (a) $\frac{2}{3}, \frac{5}{4}$ **(b)** $\left[\frac{2}{3}, \frac{5}{4}\right]$ **(c)** $\left(-\infty, \frac{2}{3}\right] \cup \left[\frac{5}{4}, \infty\right)$
21. (a) $-1 \pm \sqrt{2}$ **(b)** $\left(-1 - \sqrt{2}, -1 + \sqrt{2}\right)$
(c) $\left(-\infty, -1 - \sqrt{2}\right) \cup \left(-1 + \sqrt{2}, \infty\right)$
23. (a) $x < -1$ or $x > 1$ **(b)** $-1 \leq x \leq 1$
25. (a) $-6 < x < -2$ **(b)** $x \leq -6$ or $x \geq -2$
27. $-1 < x < 4$ **29.** $x < -3$ or $x > 2$
31. $-2 \leq x \leq 2$ **33.** All real numbers
35. $x \leq -2$ or $x \geq 3$ **37.** $-\frac{1}{3} < x < \frac{1}{2}$
39. $-4 \leq x \leq 10$ **41.** $-3 < x < -1$
43. $x \leq -7$ or $x \geq 5$ **45.** $x \leq 0$ or $x \geq 1$
47. $x \leq -2$ or $x \geq 3$ **49.** $-\sqrt{5} \leq x \leq \sqrt{5}$
51. $x \leq k$ or $x \geq 22.4$, where $k \approx 3.3$ **53.** $2 \leq x \leq 7$
55. $x \leq -2$ or $x \geq 5$ **57.** $-2 < x < \frac{5}{2}$
59. $x < -2 - \sqrt{7}$ or $x > -2 + \sqrt{7}$
61. From 1989 to 1992

63. (a) $163x^2 - 146x + 205 > 2000$ or
$163x^2 - 146x - 1795 > 0$, where x is positive
(b) $x > k$, where $k \approx 3.8$. This corresponds to around 1989
or later.
65. (a) The height does not change by the same amount
each 15-second interval.
(b) From 43 to 95 seconds (approximately)
(c) $[-25, 200, 25]$ by $[-5, 20, 5]$

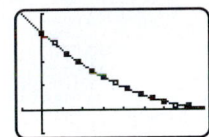

(d) From 43.8 to 93.1 seconds (approximately)

SECTION 3.4 (PP. 228–233)

1. $y = (x + 2)^2$　　**3.** $y = \sqrt{x + 3}$
5. $y = |x + 2| - 1$　　**7.** $y = (x - 2)^2 + 1$
9. $y = |x| - 3$

11. (a)　　　　　　　　**(b)**

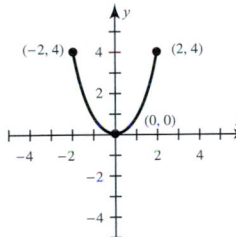
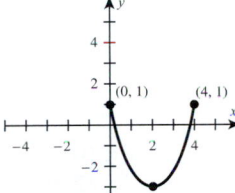

(c)

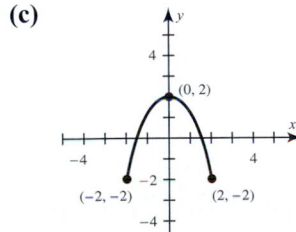

13. (a)　　　　　　　　**(b)**

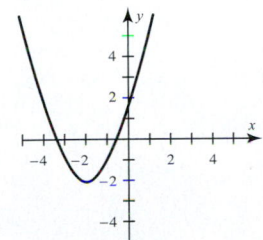
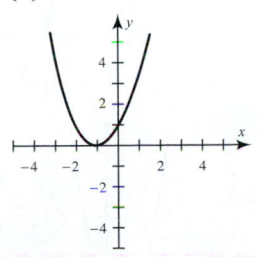

(c)

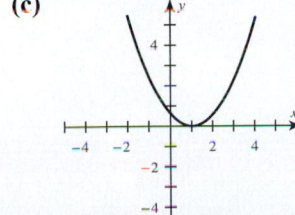

15. (a)　　　　　　　　**(b)**

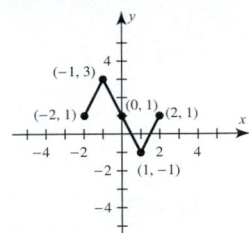
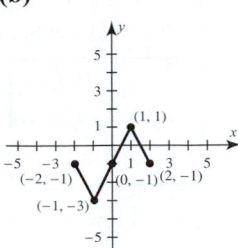

(c)

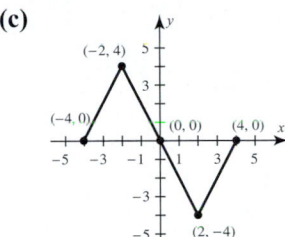

17. (a)　　　　　　　　**(b)**

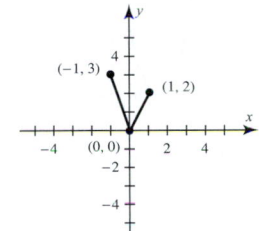
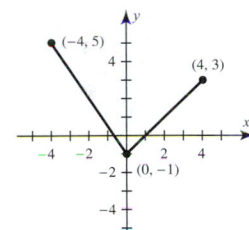

(c)

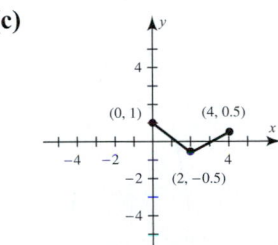

19. Shift the graph of $y = x^2$ right 3 units and upward 1 unit.
21. Shift the graph of $y = x^2$ left 1 unit and vertically
shrink it with factor $\frac{1}{4}$.
23. Reflect the graph of $y = \sqrt{x}$ across the x-axis and
shift it left 5 units.
25. Reflect the graph of $y = \sqrt{x}$ across the y-axis and vertically stretch it with factor 2.
27. Reflect the graph of $y = |x|$ across the y-axis and shift
it left 1 unit.
29. $y = (x - 2)^2 - 3$
$[-10, 10, 1]$ by $[-10, 10, 1]$

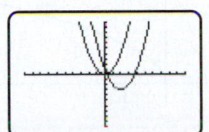

31. $y = (x + 6)^2 - 4(x + 6) + 5$
[−10, 10, 1] by [−10, 10, 1]

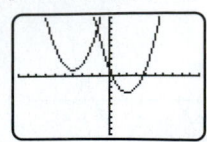

33. (a) $g(x) = 3(x + 3)^2 + 2(x + 3) - 5$
(b) $g(x) = 3x^2 + 2x - 9$
35. (a) $g(x) = -(x - 2)^3 + 2(x - 2)^2 -$
$4(x - 2) + 5$
(b) $g(x) = -(x + 8)^3 + 2(x + 8)^2 -$
$4(x + 8) - 4$
37. (a) $g(x) = -(2(x - 2)^2 - 3(x - 2) + 1)$
(b) $g(x) = 2(-(x + 4))^2 - 3(-(x + 4)) + 1$
39. (a) $g(x) = 10(x - 2000)^2 + 80$
(b) $g(x) = (-x + 20)^2$

41. 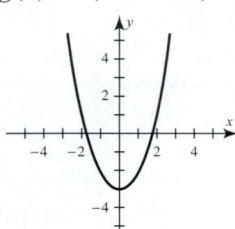   **43.**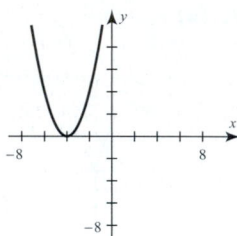

45. 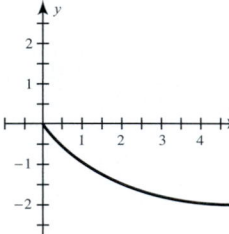   **47.**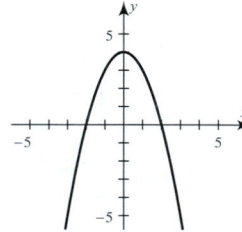

49. 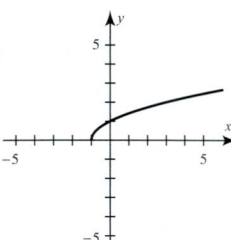   **51.**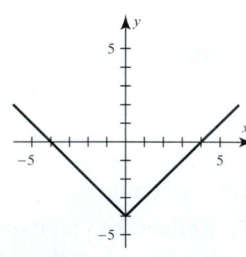

53. 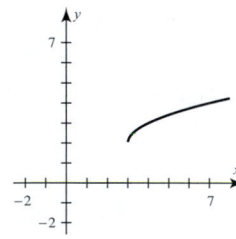   **55.**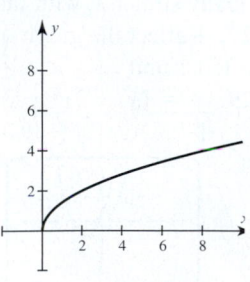

57. 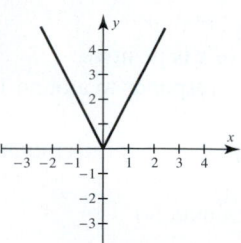   **59.**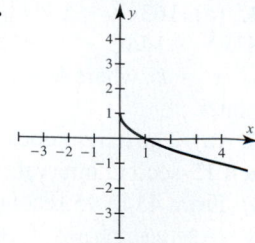

61. 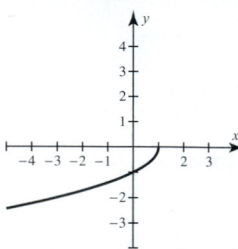   **63.**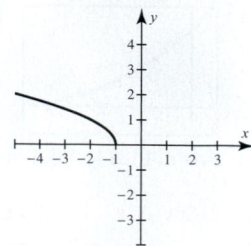

65. 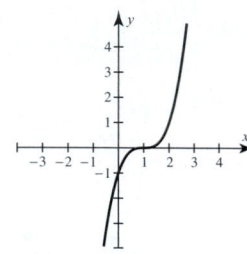   **67.**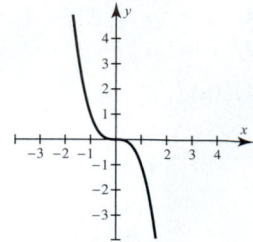

69. $(-12, 8)$, $(0, 10)$, and $(8, -2)$
71. $(-10, 7)$, $(2, 9)$, and $(10, -3)$
73. $(-12, -3)$, $(0, -4)$, and $(8, 2)$
75. $(6, 6)$, $(0, 8)$, and $(-4, -4)$
77. x-axis: $y = -(x^2 - 2x - 3)$

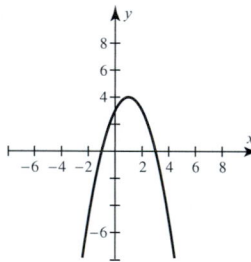

y-axis: $y = (-x)^2 - 2(-x) - 3 = x^2 + 2x - 3$

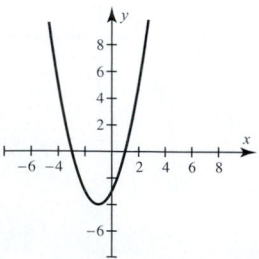

79. x-axis: $y = -(2x^2 + x - 1)$

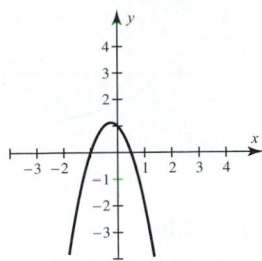

y-axis: $y = 2(-x)^2 + (-x) - 1$
$\qquad\quad = 2x^2 - x - 1$

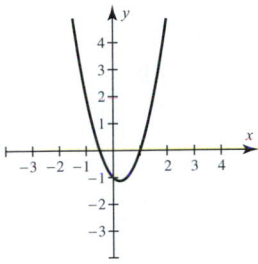

81. x-axis: $y = -(|x + 1| - 1)$

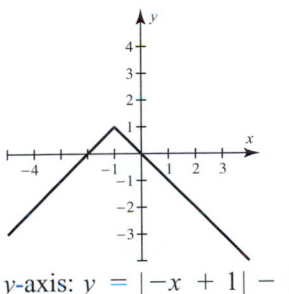

y-axis: $y = |-x + 1| - 1$

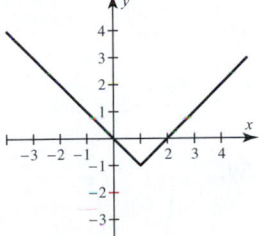

83. x-axis: y-axis:

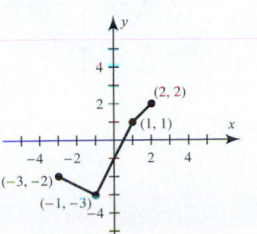 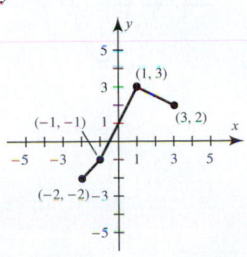

85.

x	1	2	3	4	5
$g(x)$	12	8	13	9	14

87.

x	-2	0	2	4	6
$g(x)$	5	2	-3	-5	-9

89.

x	0	1	2	3	4
$g(x)$	0	2	1	5	6

91.

x	-2	-1	0	1	2
$g(x)$	0	3	6	9	12

93. $f(x) = \frac{1}{4}(x - 1997)^2 + 3.8$
95. $g(x) = 2375(x - 1984)^2 + 5134\,(x - 1984) + 5020$
97. (a) When $x = 2$, the y-value is about 9. There are 9 hours of daylight on February 21 at 60°N latitude.
(b) Since February 21 is 2 months after the shortest day and October 21 is 2 months before, they both have approximately the same number of daylight hours. A reasonable conjecture would be 9 hours.
(c) There are approximately the same number of daylight hours either x months before or after December 21. The graph should be symmetric about the y-axis, as shown in the figure.

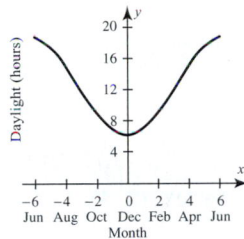

99. $y = -0.4(x - 3)^2 + 4$ (mountain)
$[-4, 4, 1]$ by $[0, 6, 1]$

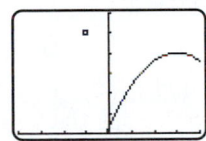

101. (a) $y = \frac{1}{20}x^2 - 1.6$
$[-15, 15, 1]$ by $[-10, 10, 1]$

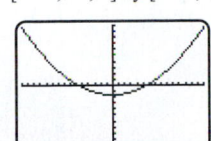

(b) $y = \frac{1}{20}(x - 2.1)^2 - 2.5$. The front reaches Columbus by midnight.
$[-15, 15, 1]$ by $[-10, 10, 1]$

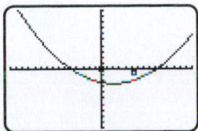

3.4 EXTENDED AND DISCOVERY EXERCISES (P. 233)

1. No, they are not commutative.

3. Yes, they are commutative.

CHECKING BASIC CONCEPTS FOR SECTIONS 3.3 AND 3.4 (P. 233)

1. (a) $[-3, 0]$; $(-\infty, -3) \cup (0, \infty)$

(b) $(-\infty, \infty)$; no solutions

3. (a) $\left(-\infty, -\sqrt{5}\right] \cup \left[\sqrt{5}, \infty\right)$

(b) All real numbers   **(c)** $\left[\frac{1-\sqrt{5}}{2}, \frac{1+\sqrt{5}}{2}\right]$

5. (a) 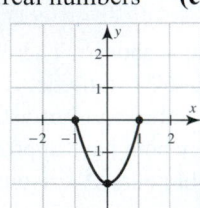   **(b)**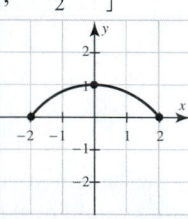

(c)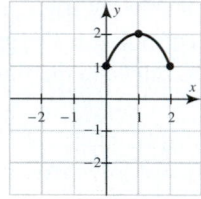

CHAPTER 3   REVIEW EXERCISES (PP. 237–240)

1. Leading coefficient: negative; vertex: $(2, 4)$; axis of symmetry: $x = 2$

3. $f(x) = -2x^2 + 20x - 49$; leading coefficient: -2

5. $f(x) = (x + 3)^2 - 10$; vertex: $(-3, -10)$

7. $\left(\frac{1}{3}, -\frac{11}{3}\right)$ **9.** $f(x) = -(x + 1)^2 + 2$

11. 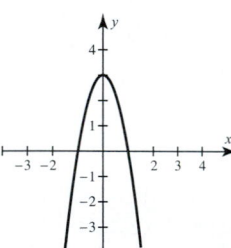   **13.**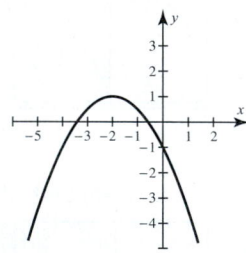

15. 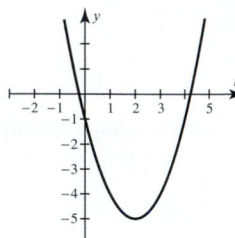   **17.**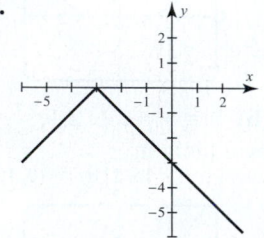

19. 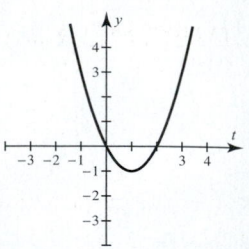   **21.**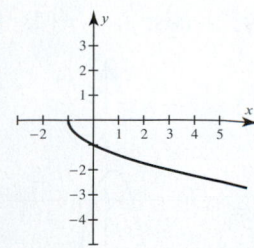

23. (a) $a > 0$ **(b)** $-3, 2$ **(c)** Positive

25. $-4, 5$ **27.** $0, 4$ **29.** $\pm\frac{\sqrt{7}}{2}$ **31.** $-\frac{7}{2}, 2$

33. $-2, 5$ **35.** $-\frac{1}{2}, \frac{5}{2}$ **37.** $-1 \pm \sqrt{6}$ **39.** $\frac{3 \pm \sqrt{11}}{2}$

41. $4 \pm \sqrt{14}$ **43.** $y = \pm\sqrt{\dfrac{2x^2 - 6}{3}}$; no

45. (a) $1, 2$ **(b)** $(1, 2)$ **(c)** $(-\infty, 1) \cup (2, \infty)$

47. $[1, 2]$ **49.** $\left(\dfrac{-19 - \sqrt{161}}{20}, \dfrac{-19 + \sqrt{161}}{20}\right)$

51. $(-\infty, -3] \cup [5, \infty)$

53. (a) $-3 < x < 2$ **(b)** $x \le -3$ or $x \ge 2$

55. $y = -f(x) = -(2x^2 - 3x + 1)$

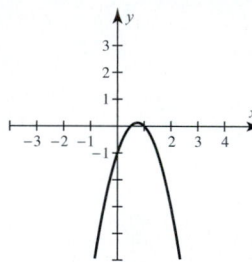

$y = f(-x) = 2(-x)^2 - 3(-x) + 1$

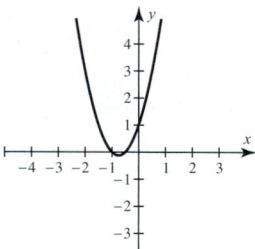

57. 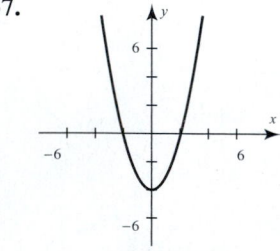   **59.**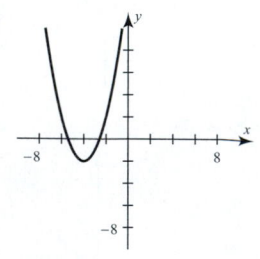

61. 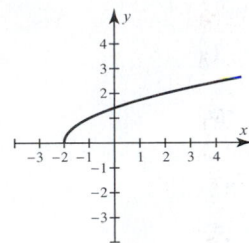 **63.**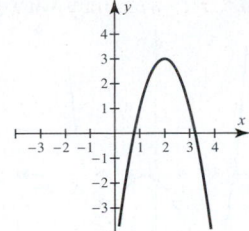

65.

x	1	2	3	4
g(x)	−1	5	6	9

67.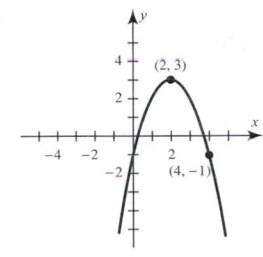

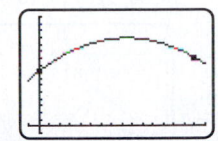

$f(x) = -(x - 2)^2 + 3$

69. 11 ft by 22 ft
71. (a) $h(0) = 5$; The stone was held 5 ft above the ground when it was released.   **(b)** 117 ft
(c) 126 ft   **(d)** After 2 seconds or 3.5 seconds
73. (a) $f(1985) \approx 4.7$. In 1985 there were about 4.7 billion people in the world.   **(b)** 6.1 billion
(c) About 2009 or late 2008
75. (a) $C(x) = x(103 - 3x)$ **(b)** $C(6) = \$510$
(c) 10 rooms (73/3 rooms cannot be rented)
(d) 17 rooms ($884)
77. (a) The change in debt is not constant each 4-year period.   **(b)** $f(x) = 1.3(x - 1980)^2 + 82$; 1990; Visa and MasterCard debt reached $212 billion in 1990.

CHAPTER 3   EXTENDED AND DISCOVERY EXERCISES (PP. 239–240)

1. (a) 23.32 ft/sec
(b) $[-1, 16, 1]$ by $[-1, 16, 1]$; Yes

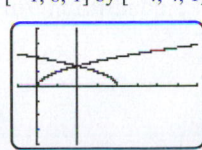

(c) 12.88 ft
3. (a) $[\,1, 8, 1]$ by $[-4, 4, 1]$

(b) The graph of $y = f(2k - x) = f(4 - x)$ is a reflection of $y = f(x)$ across the line $x = 2$.

5. (a) $[-15, 3, 1]$ by $[-3, 9, 1]$

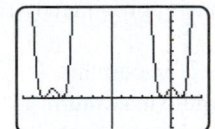

(b) The graph of $f(2k - x) = f(-12 - x)$ is a reflection of $y = f(x)$ across the line $x = -6$.
7. (a) $[0, 1200, 100]$ by $[-800, 0, 100]$

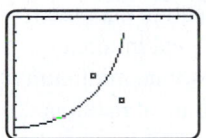

The front has reached St. Louis, but not Nashville.
(b) $g(x) = -\sqrt{750^2 - (x - 160)^2} - 110$
(c) $[0, 1200, 100]$ by $[-800, 0, 100]$

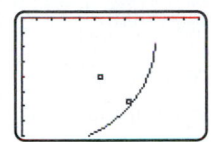

The cold front reached both cities in less than 12 hours.

CHAPTER 4: Nonlinear Functions and Equations

SECTION 4.1 (PP. 251–256)

1. Yes; degree: 3; leading coefficient: 2   **3.** No
5. Yes; degree: 4; leading coefficient: −5   **7.** No
9. Yes; degree: 0; leading coefficient: 22
11. Increasing: $[2, \infty)$; decreasing: $(-\infty, 2]$
13. Increasing: $(-\infty, -2] \cup [1, \infty)$; decreasing: $[-2, 1]$
15. Increasing: $[-8, 0] \cup [8, \infty)$; decreasing: $(-\infty, -8] \cup [0, 8]$
17. The function is neither increasing nor decreasing.
19. Increasing: $(-\infty, \infty)$; decreasing: never
21. Increasing: $[0, \infty)$; decreasing: $(-\infty, 0]$
23. Increasing: $(-\infty, 1]$; decreasing: $[1, \infty)$
25. Increasing: $[-1, \infty)$; decreasing: $(-\infty, -1]$
27. Increasing: $[1, \infty)$; decreasing: never
29. Increasing: $[-3, \infty)$; decreasing: $(-\infty, -3]$
31. Increasing: $(-\infty, \infty)$; decreasing: never
33. Increasing: $(-\infty, -2] \cup [2, \infty)$; decreasing: $[-2, 2]$
35. Increasing: $(-\infty, -2] \cup [1, \infty)$; decreasing: $[-2, 1]$
37. Increasing: $(-\infty, -1] \cup [0, 2]$; decreasing: $[-1, 0] \cup [2, \infty)$
39. (a) Local maximum: approximately 5.5; local minimum: approximately −5.5   **(b)** No absolute extrema
41. (a) Local maxima: approximately 17 and 27; local minima: approximately −10 and 24   **(b)** No absolute extrema

43. (a) Local minimum: approximately −3.2; no local maxima   **(b)** Absolute minimum: approximately −3.2; absolute maximum: 3

45. (a) Local minimum: −1; local maximum: 1
(b) Absolute minimum: −1; absolute maximum: 1

47. (a) Local minimum: −2; local maxima: 1, 2
(b) Absolute minimum: −2; absolute maximum: 2

49. (a) No local extrema   **(b)** No absolute extrema

51. (a) Local minimum: 1; no local maxima
(b) Absolute minimum: 1; no absolute maximum

53. (a) Local maximum: 4; no local minima
(b) Absolute maximum: 4; no absolute minimum

55. (a) Local minimum: $-\frac{1}{8}$; no local maxima
(b) Absolute minimum: $-\frac{1}{8}$; no absolute maximum

57. (a) Local minimum: 0; no local maxima
(b) Absolute minimum: 0; no absolute maximum

59. (a) No local extrema   **(b)** No absolute extrema

61. (a) Local minimum: −2; local maximum: 2
(b) No absolute extrema

63. (a) Local maxima: 19, −8; local minimum: −13
(b) Absolute maximum: 19; no absolute minimum

65. (a) Local minimum: −8; local maximum: 4.5
(b) Absolute minimum: −8; no absolute maximum

67. (a) Local maximum: 8; no local minima
(b) Absolute maximum: 8; no absolute minimum

69. f is odd and the point $(8, -6)$ lies on its graph.

71. Odd   **73.** Neither   **75.** Even   **77.** Even
79. Odd   **81.** Neither   **83.** Even   **85.** Even
87. Even   **89.** Neither   **91.** Odd

93. Note that $f(0)$ can be any number.

x	−3	−2	−1	0	1	2	3
$f(x)$	21	−12	−25	1	−25	−12	21

95. Answers may vary.

x	−2	−1	0	1	2
$f(x)$	5	1	2	3	4

97.   **99.** Answers may vary.

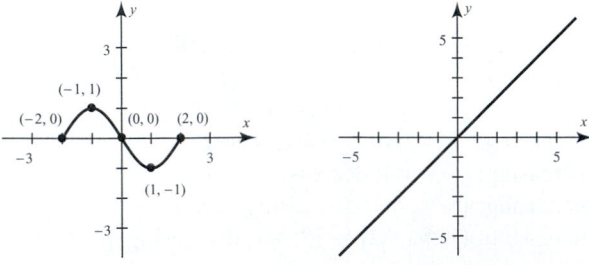

101. No. If $(2, 5)$ is on the graph of an odd function f, then so is $(-2, -5)$. Since f would pass through $(-3, -4)$ and then $(-2, -5)$, it could not always be increasing.

103. Answers may vary.

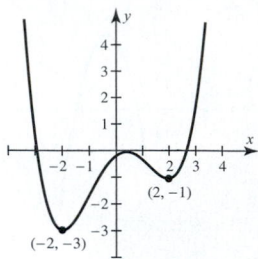

105. Answers may vary; yes but it does not have to be quadratic.

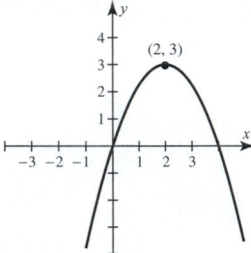

107.   **109.**

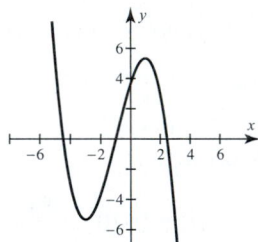   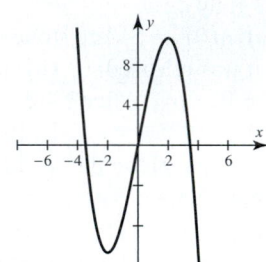

111. $(-4, 2)$ **113.** On $[0, 3]$; on $[3, 6]$

115. (a) $[0, 2.2]$; $[8.4, 14.5]$; $[20.7, 26.8]$
(b) Local maximum = 4.2, high tide is 4.2 feet; local minimum = 1.8, low tide is 1.8 feet.

117. (a) $f(5) \approx 8.32$; in 1955 U.S. consumption of energy was about 8.32 quadrillion Btu.
(b) The energy consumption increased, reached a maximum value, and then started to decrease.

[0, 30, 5] by [6, 16, 1]   [0, 30, 5] by [6, 16, 1]

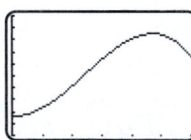   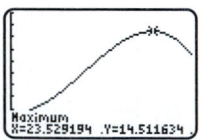

(c) Local maximum: approximately 14.5. In 1973 or 1974 maximum energy use peaked at 14.5 quadrillion Btu.

119. (a) Possible absolute maximum in January and absolute minimum in July **(b)** Absolute maximum of $140 in January and an absolute minimum of $15 in July

[1, 12, 1] by [0, 150, 10]   [1, 12, 1] by [0, 150, 10]

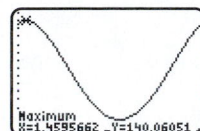

121. (a) Even **(b)** 83°F **(c)** They are equal.
(d) Monthly average temperatures are symmetric about July. July has the highest average and January the lowest. The pairs June-August, May-September, April-October, March-November, and February-December have approximately the same average temperatures.
123. (a) $h(-2) = h(2) = 336$. Two seconds before and 2 seconds after the time when the maximum height is reached, the projectile's height is 336 feet.
(b) $h(-5) = h(5) = 0$. Five seconds before and 5 seconds after the time when the maximum height is reached, the projectile is on the ground.
(c) h is an even function.

[−5, 5, 1] by [0, 500, 100]

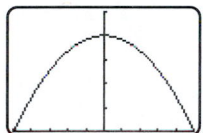

(d) $h(-x) = h(x)$ when $-5 \le x \le 5$. This means that the projectile is at the same height x seconds either before or after the time when the maximum height is attained.

4.1 EXTENDED AND DISCOVERY EXERCISES (PP. 256–257)

1. The maximum area occurs when the figure is a rectangle with length approximately 4.24 and height approximately 2.12.
3. (a) About 1 hour 54 minutes **(b)** About 2 hours 8 minutes **(c)** About 1 hour 46 minutes

SECTION 4.2 (PP. 268–273)

1. (a) The turning points are approximately (1.6, 3.6), (3, 1.2), (4.4, 3.6). **(b)** After 1.6 minutes the runner is 360 feet from the starting line. The runner turns and jogs toward the starting line. After 3 minutes the runner is 120 feet from the starting line, turns, and jogs away from the starting line. After 4.4 minutes the runner is again 360 feet from the starting line. The runner turns and jogs back to the starting line.
3. (a) 0; 0.5 **(b)** Positive **(c)** 1

5. (a) 3; −6, −1, and 6 **(b)** Negative **(c)** 4
7. (a) 4; −3, −1, 0, 1, and 2 **(b)** Positive **(c)** 5
9. (a) 2; −3 **(b)** Positive **(c)** 3
11. (a) 1; −1 and 2 **(b)** Positive **(c)** 2
13. (a) d. **(b)** (1, 0) **(c)** $x = 1$ **(d)** Local minimum: 0 **(e)** Absolute minimum: 0
15. (a) b. **(b)** (−3, 27), (1, −5) **(c)** $x \approx -4.9$, $x = 0, x \approx 1.9$ **(d)** Local minimum: −5; local maximum: 27 **(e)** Absolute minimum: none; absolute maximum: none
17. (a) a. **(b)** (−2, 16), (0, 0), (2, 16)
(c) $x \approx -2.8, x = 0, x \approx 2.8$ **(d)** Local minimum: 0; local maximum: 16 **(e)** Absolute minimum: none; absolute maximum: 16
19. (a) Degree: 1; leading coefficient: −2
(b) Up on left end, down on right end; $f(x) \to \infty$ as $x \to -\infty$, $f(x) \to -\infty$ as $x \to \infty$
21. (a) Degree: 2; leading coefficient: 1
(b) Up on both ends; $f(x) \to \infty$ as $x \to -\infty$, $f(x) \to \infty$ as $x \to \infty$
23. (a) Degree: 3; leading coefficient: −2
(b) Up on left end, down on right end; $f(x) \to \infty$ as $x \to -\infty$, $f(x) \to -\infty$ as $x \to \infty$
25. (a) Degree: 3; leading coefficient: −1
(b) Up on left end, down on right end; $f(x) \to \infty$ as $x \to -\infty$, $f(x) \to -\infty$ as $x \to \infty$
27. (a) Degree: 5; leading coefficient: 0.1
(b) Down on left end, up on right end; $f(x) \to -\infty$ as $x \to -\infty$, $f(x) \to \infty$ as $x \to \infty$
29. (a) Degree: 2; leading coefficient: $-\frac{1}{2}$
(b) Down on both ends; $f(x) \to -\infty$ as $x \to -\infty$, $f(x) \to -\infty$ as $x \to \infty$
31. (a) [−10, 10, 1] by [−10, 10, 1]

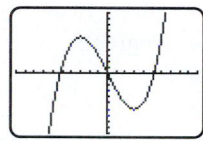

(b) (−3, 6), (3, −6) **(c)** Local minimum: −6; local maximum: 6
33. (a) [−10, 10, 1] by [−10, 10, 1]

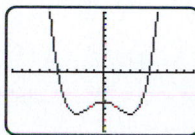

(b) There are three turning points located at (−3, −7.025), (0, −5), and (3, −7.025).
(c) Local minimum: −7.025; local maximum: −5

35. (a) $[-10, 10, 1]$ by $[-10, 10, 1]$

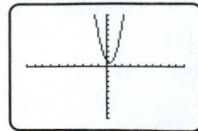

(b) $\left(\frac{1}{3}, \frac{2}{3}\right) \approx (0.333, 0.667)$

(c) Local minimum: $\frac{2}{3} \approx 0.667$

37. (a) $[-10, 10, 1]$ by $[-10, 10, 1]$

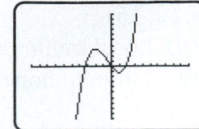

(b) $\left(-2, \frac{10}{3}\right) \approx (-2, 3.333)$, $\left(1, -\frac{7}{6}\right) \approx (1, -1.167)$

(c) Local minimum: $-\frac{7}{6} \approx -1.167$; local maximum: $\frac{10}{3} \approx 3.333$

39. As the viewing rectangle increases in size, the graphs begin to look alike. Each formula contains the term $2x^4$, which determines the end behavior of the graph for large values of $|x|$.

$[-4, 4, 1]$ by $[-4, 4, 1]$ $[-10, 10, 1]$ by $[-100, 100, 10]$

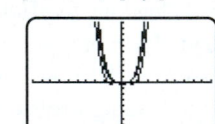

$[-100, 100, 10]$ by $[-10^6, 10^6, 10^5]$

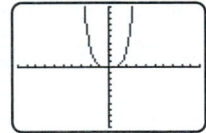

41. (a)   **(b)** Degree 1

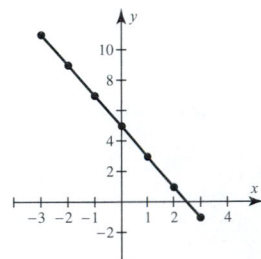

43. (a)   **(b)** Degree 2

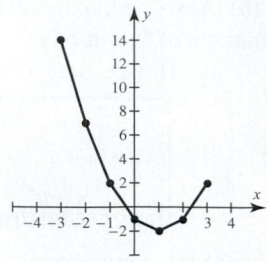

45. Answers may vary.   **47.** Answers may vary.

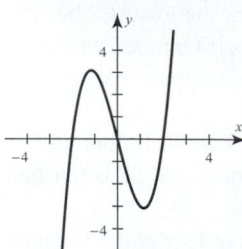   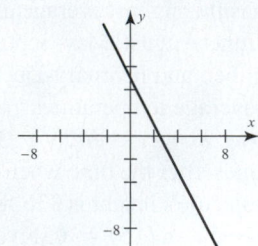

49. Not possible   **51.** Not possible
53. Answers may vary.   **55.**

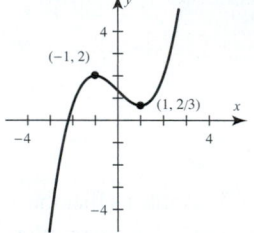   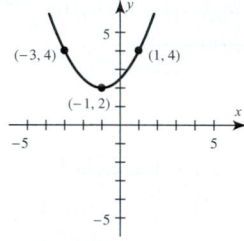

57. For f: 1; for g: 0.5; for h: 0.25. On the interval $[0, 0.5]$ the higher the degree of the function the smaller the average rate of change.

59. (a) 12.01   **(b)** 12.0001   **(c)** 12.000001
The average rate of change is approaching 12.

61. (a) $4.01\overline{6}$ **(b)** $4.00016\overline{6}$ **(c)** $4.0000016\overline{6}$
The average rate of change is approaching 4.

63. $9x^2 + 9xh + 3h^2$ **65.** $-3x^2 - 3xh - h^2 + 1$

67. $f(-2) \approx 5, f(1) \approx 0$

69. $f(-1) \approx -1, f(1) \approx 1, f(2) \approx -2$

71. $f(-3) = -63, f(1) = 3, f(4) = 10$

73. $f(-2) = 6, f(1) = 7, f(2) = 9$

75. (a)

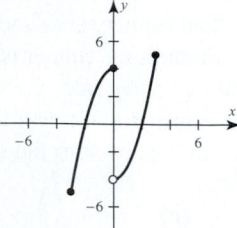

(b) f is not continuous.   **(c)** ± 2

77. (a)

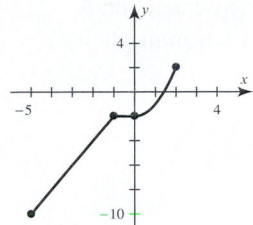

(b) f is continuous.   **(c)** $\sqrt{2}$

79. (a)

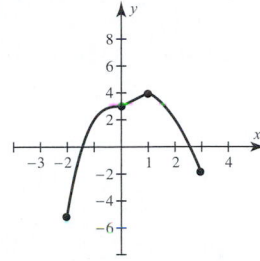

(b) f is continuous.   **(c)** $-\sqrt[3]{3}, \dfrac{\sqrt{17}+1}{2}$

81. (a) $H(-2)=0$; $H(0)=1$; $H(3.5)=1$
(b)

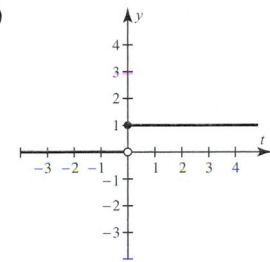

83. (a) Approximately $(1, 13)$ and $(7, 72)$
(b) The low monthly average temperature of 13°F occurs in January. The high monthly average temperature of 72°F occurs in July.
85. (a) As the speed x increases, so does the minimum sight distance y.
$[15, 75, 5]$ by $[500, 2800, 100]$

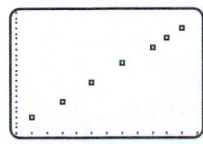

(b) Degree: 1; $D(x)=34x+130$. Answers may vary.
(c) $D(43)=1592$ ft. Answers may vary.
87. (a) $f(x) \approx 0.02352x^3 - 0.7088x^2 + 4.444x + 27.46$
(b) $f(22) \approx 32.6\%$; the estimate is too high.

89. (a) 4 seconds
$[-1, 8, 1]$ by $[-10, 170, 10]$

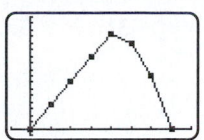

(b) From 0 to 4 sec; from 4 to 7 sec
(c) $m=36$, $a=-16$, $b=144$ **(d)** $x \approx 2.8$ or $x \approx 5.7$; the height is 100 ft at 2.8 sec and at 5.7 sec.

CHECKING BASIC CONCEPTS FOR SECTIONS 4.1 AND 4.2 (PP. 273–274)

1. (a) Increasing: $[-2, 1] \cup [3, \infty)$; decreasing: $(-\infty, -2] \cup [1, 3]$ **(b)** Local maximum: approximately 3; local minima: approximately -13 and -2
(c) Absolute minimum: approximately -13; no absolute maximum   **(d)** Approximately $-3.1, 0, 2.2,$ and 3.6; they are the same values.
3. (a) Not possible
(b) Answers may vary.

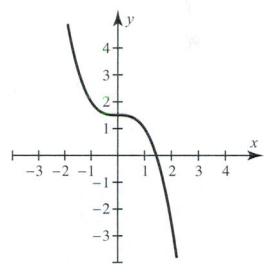

(c) Answers may vary.   **(d)** Not possible

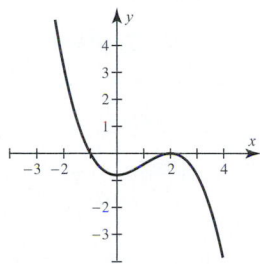

5. $f(x) \approx -1.01725x^4 + 10.319x^2 - 10$

SECTION 4.3 (PP. 288–292)

1. $\dfrac{x^3}{2} - \dfrac{3}{2x}$ **3.** $x - \dfrac{2}{3x} - \dfrac{1}{3x^3}$ **5.** $-\dfrac{1}{x^3} + \dfrac{1}{4}$
7. Quotient: $x^2 + x - 2$; remainder: 0
9. Quotient: $2x^3 - 9x^2 + 4x - 23$; remainder: 40
11. Quotient: $3x^2 + 3x - 4$; remainder: 6
13. $x^3 - 8x^2 + 15x - 6$ **15.** $x^3 - 1$
17. $4x^2 + 3x - 2 + \dfrac{4}{x-1}$ **19.** $x^2 - x + 1$
21. $3x^2 + 4x - 2 + \dfrac{2}{2x-1}$ **23.** $x^3 + 2 + \dfrac{-2}{3x-7}$

25. $5x^2 - 12 + \dfrac{30}{x^2 + 2}$ **27.** $4x + 5$

29. $x^2 - 2x + 4 + \dfrac{-1}{2x^2 + 3x + 2}$ **31.** 3   **33.** -42

35. Yes   **37.** No   **39.** $f(x) = 2\left(x - \frac{11}{2}\right)(x - 7)$

41. $f(x) = (x + 2)(x - 1)(x - 3)$

43. $f(x) = -2(x + 5)\left(x - \frac{1}{2}\right)(x - 6)$

45. $f(x) = 7(x + 3)(x - 2)$

47. $f(x) = (x + 4)(x - 2)(x - 8)$

49. $f(x) = -1(x + 8)(x + 4)(x + 2)(x - 4)$

51. $f(x) = \frac{1}{2}(x + 1)(x - 2)(x - 3)$

53. $f(x) = \frac{1}{2}(x + 1)(x - 1)(x - 2)$

55. $f(x) = -2(x + 2)(x + 1)(x - 1)(x - 2)$

57. $f(x) = 10(x + 2)\left(x - \frac{3}{10}\right)$

59. $f(x) = -3x(x - 2)(x + 3)$

61. $f(x) = x(x + 1)(x + 3)\left(x - \frac{3}{2}\right)$

63. $f(x) = (x - 1)(x - 3)(x - 5)$

65. $f(x) = -4(x + 4)\left(x - \frac{3}{4}\right)(x - 3)$

67. $f(x) = 2x(x + 2)\left(x + \frac{1}{2}\right)(x - 3)$

69. -2 (odd), 4 (even); minimum degree: 5

71. -6 (even), -1 (even), 4 (odd); minimum degree: 7

73. $f(x) = (x + 1)^2(x - 6)$

75. $f(x) = (x - 2)^3(x - 6)$

77. $f(x) = (x + 2)^2(x - 4)$

79. $f(x) = -1(x + 3)^2(x - 3)^2$

81. $f(x) = 2(x + 1)^2(x - 1)^3$

83. (a) x-intercepts: $-2, -1$; y-intercept: 4
(b) -2 has multiplicity 1; -1 has multiplicity 2
(c)

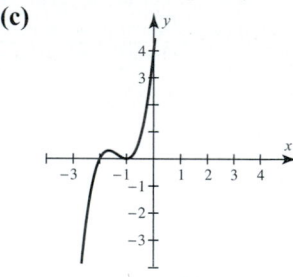

85. (a) x-intercepts: $-1, 1, 2$; y-intercept: -2
(b) -1, 1, and 2 each have multiplicity 1
(c)

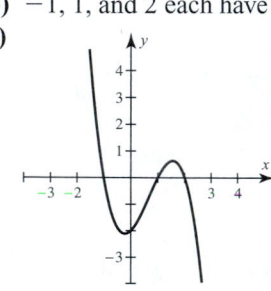

87. (a) x-intercepts: $-3, -1$; y-intercept: 3
(b) -3 has multiplicity 2; -1 has multiplicity 2
(c)

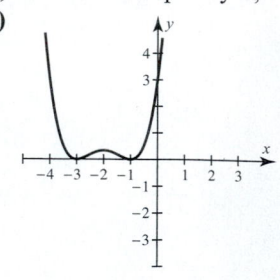

89. (a) $-3, \frac{1}{2}, 1$ **(b)** $f(x) = 2(x + 3)\left(x - \frac{1}{2}\right)(x - 1)$

91. (a) $-2, -1, 1, \frac{3}{2}$

(b) $f(x) = 2(x + 2)(x + 1)(x - 1)\left(x - \frac{3}{2}\right)$

93. (a) $\frac{1}{3}, 1, 4$ **(b)** $f(x) = 3\left(x - \frac{1}{3}\right)(x - 1)(x - 4)$

95. (a) 1   **(b)** $f(x) = (x + \sqrt{7})(x - 1)(x - \sqrt{7})$

97. $-3, 0, 2$ **99.** $-1, 1$ **101.** $-2, 0, 2$

103. $-5, 0, 5$ **105.** ± 2 **107.** $-3, 0, 6$ **109.** 0, 1

111. $-\frac{1}{4}, 0, \frac{5}{3}$ **113.** $\pm\frac{2}{3}, \pm 1$ **115.** $-1, \pm\frac{\sqrt{3}}{2}$

117. $\pm 2, \frac{1}{2}$ **119.** $\pm\frac{3}{2}, \pm\frac{\sqrt{6}}{2}$ **121.** $-2, 3$

123. $-2.01, 0.12, 2.99$ **125.** $-4.05, -0.52, 1.71$

127. $-2.69, -1.10, 0.55, 3.98$ **129.** $L = 4x + 3$; 43 ft

131. (a) As x increases, C decreases.

(b) [0, 70, 10] by [0, 22, 5]

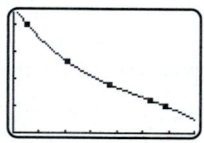

(c) $0 \leq x < 32.1$ (approximately)

133. Approximately 11.34 cm

135. (a) $f(x) \approx -0.184(x + 6.01)(x - 2.15)(x - 11.7)$

(b) The zero of -6.01 has no significance. The zeros of $2.15 \approx 2$ and $11.7 \approx 12$ indicate that during February and December the average temperature is 0°F.

137. June 2, June 22, and July 12

SECTION 4.4 (PP. 302–305)

1. $2i$ **3.** $10i$ **5.** $i\sqrt{23}$ **7.** $i\sqrt{12} = 2i\sqrt{3}$

9. $8i$ **11.** $-2 - i$ **13.** $13 - 16i$ **15.** $-1 + 6i$

17. $4 + 8i$ **19.** $5 - i$ **21.** $4 - 7i$ **23.** $-5 - 12i$

25. 4   **27.** $\frac{1}{2} - \frac{1}{2}i$ **29.** $\frac{19}{26} + \frac{9}{26}i$ **31.** $-\frac{2}{25} + \frac{4}{25}i$

33. $3i$ **35.** $\frac{1}{2} + i$ **37.** $-18.5 + 87.4i$

39. $8.7 - 6.7i$ **41.** $-117.27 + 88.11i$

43. $-0.921 - 0.236i$ **45.** $\pm i\sqrt{5}$ **47.** $\pm i\sqrt{\frac{1}{2}}$

49. $\dfrac{3}{10} \pm \dfrac{i\sqrt{11}}{10}$ **51.** $2 \pm i$ **53.** $\dfrac{3}{2} \pm \dfrac{i\sqrt{11}}{2}$

55. $-1 \pm i\sqrt{3}$ **57.** $1 \pm \dfrac{i\sqrt{2}}{2}$ **59.** $\dfrac{5}{4} \pm \dfrac{i\sqrt{7}}{4}$

61. $\dfrac{11}{8} \pm \dfrac{i\sqrt{7}}{8}$ **63. (a)** Two real zeros   **(b)** $-1, \dfrac{3}{2}$

65. (a) Two imaginary zeros   **(b)** $-\dfrac{1}{2} \pm \dfrac{i\sqrt{7}}{2}$

67. (a) Two imaginary zeros   **(b)** $-\dfrac{2}{5} \pm \dfrac{1}{5}i$

69. Two imaginary zeros

71. One real zero; two imaginary zeros

73. Two real zeros; two imaginary zeros

75. Three real zeros; two imaginary zeros

77. (a) $f(x) = (x - 6i)(x + 6i)$ **(b)** $f(x) = x^2 + 36$

79. (a) $f(x) = -1(x + 1)(x - 2i)(x + 2i)$
(b) $f(x) = -x^3 - x^2 - 4x - 4$

81. (a) $f(x) = 10(x - 1)(x + 1)(x - 3i)(x + 3i)$
(b) $f(x) = 10x^4 + 80x^2 - 90$

83. (a) $f(x) = \dfrac{1}{2}(x + i)(x - i)(x + 2i)(x - 2i)$
(b) $f(x) = \dfrac{1}{2}x^4 + \dfrac{5}{2}x^2 + 2$

85. (a) $f(x) = -2(x - (1 - i))(x - (1 + i))(x - 3)$
(b) $f(x) = -2x^3 + 10x^2 - 16x + 12$

87. $\dfrac{5}{3}, \pm 5i$ **89.** $\pm 3i, \dfrac{1}{4} \pm \dfrac{i\sqrt{7}}{4}$

91. (a) $\pm 5i$ **(b)** $f(x) = (x - 5i)(x + 5i)$

93. (a) $0, \pm i$ **(b)** $f(x) = 3(x - 0)(x - i)(x + i)$ or
$f(x) = 3x(x - i)(x + i)$

95. (a) $\pm i, \pm 2i$
(b) $f(x) = (x - i)(x + i)(x - 2i)(x + 2i)$

97. (a) $-3, 1, \pm 2i$
(b) $f(x) = (x - 1)(x + 3)(x - 2i)(x + 2i)$

99. $0, \pm i$ **101.** $2, \pm i\sqrt{7}$ **103.** $0, \pm i\sqrt{5}$

105. $0, \dfrac{1}{2} \pm \dfrac{i\sqrt{15}}{2}$ **107.** $-2, 1, \pm i\sqrt{8}$

109. $-2, \dfrac{1}{3} \pm \dfrac{i\sqrt{8}}{3}$ **111.** $Z = 10 + 6i$

113. $V = 11 + 2i$ **115.** $I = 1 + i$

4.4 EXTENDED AND DISCOVERY EXERCISE (P. 305)

1. (a) $i^1 = i, i^2 = -1, i^3 = -i, i^4 = 1, i^5 = i, i^6 = -1,$
$i^7 = -i, i^8 = 1,$ and so on.
(b) $i^n = i$ if $n = 4q + 1$
$ -1$ if $n = 4q + 2$
$ -i$ if $n = 4q + 3$
$ 1$ if $n = 4q$
where q is a whole number.

CHECKING BASIC CONCEPTS FOR SECTIONS 4.3 AND 4.4 (P. 305)

1. $x^2 - 2x + 1$
3. $f(x) = -\dfrac{1}{2}(x + 2)^2(x - 1)$;
-2 has multiplicity 2; 1 has multiplicity 1

5. $[-6, 6, 1]$ by $[-150, 150, 20]$

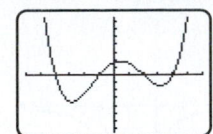

The x-intercepts are $-4, -1, 2,$ and 4.
$f(x) = (x + 4)(x + 1)(x - 2)(x - 4)$
7. Answers may vary.

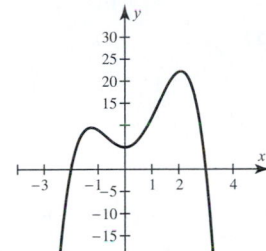

9. (a) $-\dfrac{3}{2}, 2 \pm i\sqrt{11}$ **(b)** $\dfrac{1}{6} \pm \dfrac{i\sqrt{23}}{6}$ **(c)** $\pm 2, \pm 3i$

SECTION 4.5 (PP. 319–325)

1. Yes; $D = \left\{x \mid x \neq \dfrac{5}{4}\right\}$ **3.** Yes; $D = $ all real numbers
5. No; $D = \{x \mid x \neq -1\}$ **7.** Yes; $D = $ all real numbers
9. No; $D = \{x \mid x \neq -1, x \neq 0\}$
11. Yes; $D = \{x \mid x \neq -1\}$
13. $f(x) \to \infty$ as $x \to 0^-$; $f(x) \to \infty$ as $x \to 0^+$;
$f(x) \to 0$ as $x \to -\infty$; $f(x) \to 0$ as $x \to \infty$
15. Horizontal: $y = 4$; vertical: $x = 2$; $D = \{x \mid x \neq 2\}$
17. Horizontal: $y = -4$; vertical: $x = \pm 2$;
$D = \{x \mid x \neq 2, x \neq -2\}$
19. Horizontal: $y = 0$; vertical: none; $D = $ all real numbers
21. Horizontal: $y = 2$; vertical: $x = 3$
23. Horizontal: $y = 0$; vertical: $x = \pm\sqrt{5}$
25. Horizontal: none; vertical: $x = -5$ or 2
27. Horizontal: $y = \dfrac{1}{2}$; vertical: $x = \dfrac{5}{2}$
29. Horizontal: $y = 3$; vertical: $x = 1$
31. Horizontal: none; vertical: none, since $f(x) = x - 3$
for $x \neq -3$
33. b.   **35.** d.
37. $g(x) = f(x - 3)$ **39.** $g(x) = f(x) + 2$

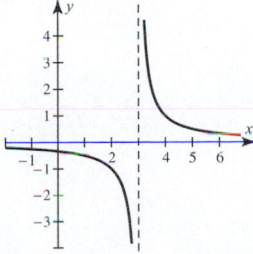

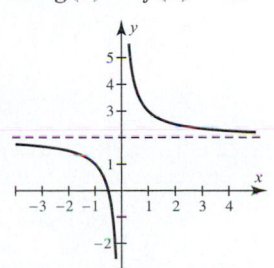

41. $g(x) = f(x + 1) - 2$ **43.** $g(x) = -2h(x - 1)$

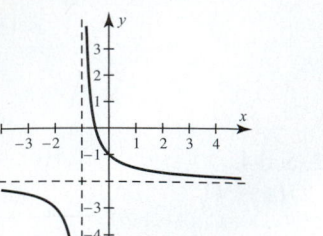   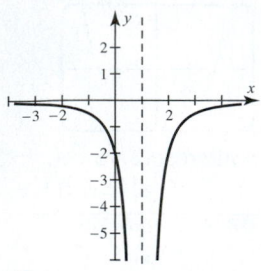

45. $g(x) = h(x + 1) - 2$ **47.**

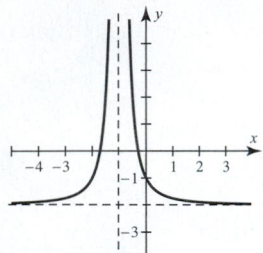   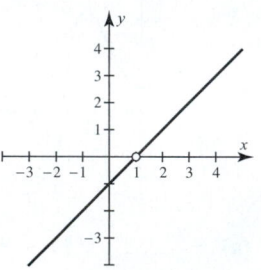

49.   **51.**

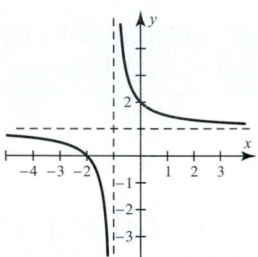   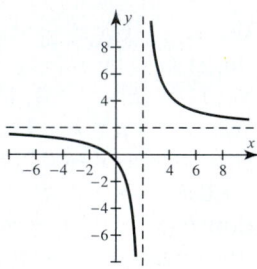

53.   **55.**

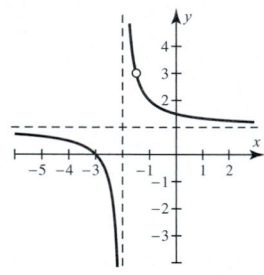   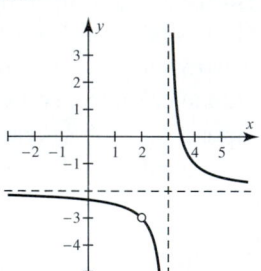

57. (a) $D = \{x | x \neq 2\}$
(b) $[-9.4, 9.4, 1]$ by $[-6.2, 6.2, 1]$

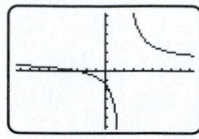

(c) Horizontal: $y = 1$; vertical: $x = 2$
(d)

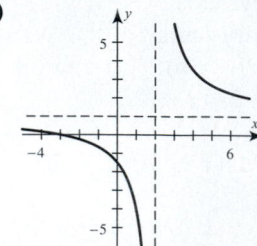

59. (a) $D = \{x | x \neq 2, x \neq -2\}$
(b) $[-9.4, 9.4, 1]$ by $[-6.2, 6.2, 1]$

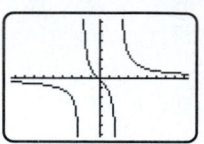

(c) Horizontal: $y = 0$; vertical: $x = \pm 2$
(d)

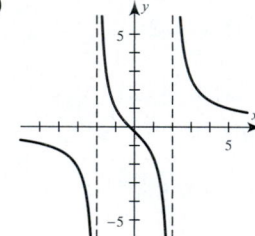

61. (a) $D = \{x | x \neq 2, x = -2\}$
(b) $[-9.4, 9.4, 1]$ by $[-9.3, 9.3, 1]$

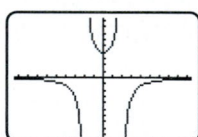

(c) Horizontal: $y = 0$; vertical: $x = \pm 2$
(d)

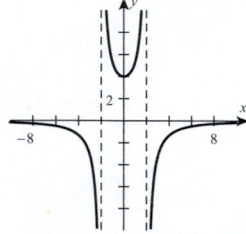

63. (a) $D = \{x \mid x \neq 2\}$
(b) $[-4.7, 4.7, 1]$ by $[-6.2, 6.2, 1]$

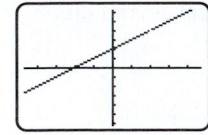

(c) Horizontal: none; vertical: none, since $f(x) = x + 2$ for $x \neq 2$
(d)

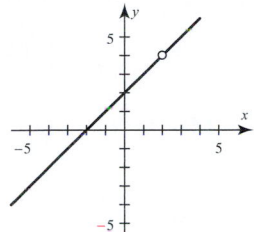

65. $y = 3$

67. $f(x) = \dfrac{x+1}{x+3}$. Answers may vary.

69. $f(x) = \dfrac{1}{x^2 - 9}$. Answers may vary.

71. (a) Slant: $y = x - 1$; vertical: $x = -1$
(b)

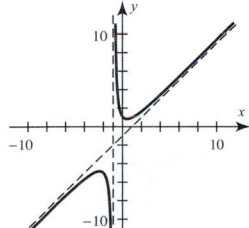

73. (a) Slant: $y = \frac{1}{2}x - 3$; vertical: $x = -2$
(b)

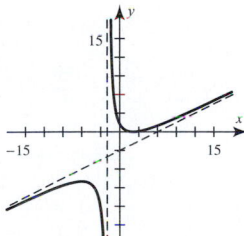

75. (a) Slant: $y = x + 3$; vertical: $x = 1$
(b)

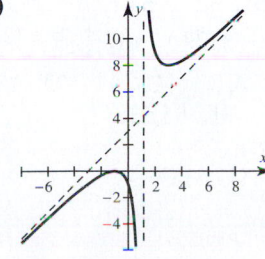

77. (a) Slant: $y = 2x + 1$; vertical: $x = \frac{1}{2}$
(b)

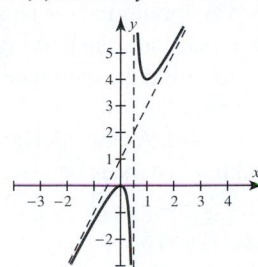

79. -3 **81.** $\frac{1}{2}, 2$ **83.** $\frac{1}{2}$ **85.** -1 **87.** $\frac{13}{8}$
89. $\pm\sqrt{2}$ **91.** No real solutions   **93.** $0, \pm 2$
95. $-\frac{5}{3}, \frac{7}{5}$ **97.** -14 **99.** No real solutions
101. (a) $T(4) = 0.25$; when vehicles leave the ramp at an average rate of 4 vehicles per minute, the wait is 0.25 minute or 15 seconds. $T(7.5) = 2$; when vehicles leave the ramp at an average rate of 7.5 vehicles per minute, the wait is 2 minutes.   **(b)** The wait increases dramatically.
103. (a) $y = 10$
$[0, 14, 1]$ by $[0, 14, 1]$

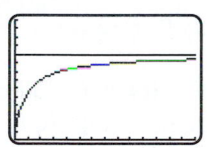

(b) When $x = 0$, there are 1 million insects.   **(c)** It starts to level off at 10 million.   **(d)** The horizontal asymptote $y = 10$ represents the limiting population after a long time.
105. (a) About 12.4 cars per minute   **(b)** 3
107. Two possible solutions: width = 7 in., length = 14 in., height = 2 in.; width $\approx$ 2.266 in., length $\approx$ 4.532 in., height $\approx$ 19.086 in.
109. (a) $A = \dfrac{x^2}{144} + \dfrac{3}{x}$ **(b)** $C = 0.1\left(\dfrac{x^2}{144} + \dfrac{3}{x}\right)$
(c) Approximately $6 \times 6 \times 3$ in.
111. (a) $f(400) = \frac{2540}{400} = 6.35$ inches. A curve designed for 60 miles per hour with a radius of 400 ft should have the outer rail elevated 6.35 in.   **(b)** As the radius x of the curve increases, the elevation of the outer rail decreases.
$[0, 600, 100]$ by $[0, 50, 5]$

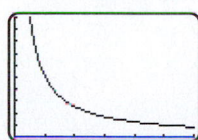

(c) The horizontal asymptote is $y = 0$. As the radius of the curve increases without bound ($x \to \infty$), the tracks become straight and no elevation or banking ($y \to 0$) is necessary.
(d) 200 ft

113. (a) $D(0.05) \approx 238$; the braking distance for a car traveling at 50 miles per hour on a 5% uphill grade is about 238 ft.   **(b)** As the uphill grade x increases, the braking distance decreases, which agrees with driving experience. **(c)** $x = \frac{13}{165} \approx 0.079$ or 7.9%
115. (a) The vertical asymptote is $x = 0$. As the coefficient of friction x decreases to 0, the braking distances become larger without a maximum.   **(b)** $\frac{25}{102}$
117. $k = 6$ **119.** $k = 8$ **121.** $T = 160$
123. $y = 2$ **125.** Becomes half as much
127. Becomes 27 times as much   **129.** $k = 0.5, n = 2$
131. $k = 3, n = 1$ **133.** 1.18 grams
135. $\sqrt{50} \approx 7$ times farther   **137.** $\frac{2}{9}$ ohm
139. F decreases by a factor of $\frac{\sqrt{2}}{2}$.

SECTION 4.6 (PP. 332–333)

1. (a) $-4, -2,$ or 2   **(b)** $(-4, -2) \cup (2, \infty)$
(c) $(-\infty, -4) \cup (-2, 2)$
3. (a) $-4, -2, 0,$ or 2   **(b)** $(-4, -2) \cup (0, 2)$
(c) $(-\infty, -4) \cup (-2, 0) \cup (2, \infty)$
5. (a) $-2, 1,$ or 2   **(b)** $(-\infty, -2) \cup (-2, 1)$
(c) $(1, 2) \cup (2, \infty)$
7. (a) 0   **(b)** $(-\infty, 0) \cup (0, \infty)$ **(c)** No solutions
9. (a) 0 or 1   **(b)** $(-\infty, 0) \cup (1, \infty)$ **(c)** $(0, 1)$
11. (a) $-2, 0,$ or 2   **(b)** $(-\infty, -2) \cup (2, \infty)$
(c) $(-2, 0) \cup (0, 2)$
13. $(-1, 0) \cup (1, \infty)$ **15.** $[-2, 0] \cup [1, \infty)$
17. $(-3, -2) \cup (2, 3)$ **19.** $(-\infty, -\sqrt{2}] \cup [\sqrt{2}, \infty)$
21. $[-2, 1] \cup [2, \infty)$ **23.** $[-3, 2]$
25. $(-\infty, 1] \cup [2, 4]$ **27.** $(-\infty, -1) \cup \left(0, \frac{4}{3}\right) \cup (2, \infty)$
29. $(-\infty, 0)$ **31.** $(-3, \infty)$ **33.** $(-2, 2)$
35. $(-\infty, 2)$ **37.** $(-\infty, -1) \cup \left(\frac{3}{2}, \infty\right)$
39. $(-\infty, -3) \cup (-1, 2)$ **41.** $(-1, 1) \cup \left[\frac{5}{2}, \infty\right)$
43. $(3, \infty)$ **45.** $(-\infty, 0) \cup \left(0, \frac{1}{2}\right] \cup [2, \infty)$
47. (a) $x \leq 36$ (approximately)   **(b)** The average line length is less than or equal to 8 cars when the average arrival rate is about 36 cars per hour or less.
49. (a) The braking distance increases.   **(b)** $0 < x \leq 0.3$
51. $\sqrt[3]{212.8} \leq x \leq \sqrt[3]{213.2}$ or (approximately) $5.97022 \leq x \leq 5.97396$ inches

CHECKING BASIC CONCEPTS FOR SECTIONS 4.5 AND 4.6 (P. 334)

1. (a) $D = \{x \mid x \neq 1\}$ **(b)** Vertical asymptote: $x = 1$; horizontal asymptote: $y = 2$
(c)

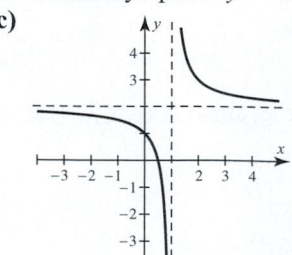

3. (a) $\frac{1}{2}$ **(b)** $-5, \frac{7}{3}$ **(c)** $\frac{23}{3}$ **(d)** No solutions
5. $(-2, -1] \cup [1, \infty)$

SECTION 4.7 (PP. 343–346)

1. 4   **3.** $\frac{1}{8}$ **5.** -9 **7.** 2   **9.** 27   **11.** 2   **13.** 9
15. $-\frac{1}{243}$ **17.** 16   **19.** $\frac{9}{4}$ **21.** $\frac{11}{6}$ **23.** $(2x)^{1/3}$
25. $z^{5/3}$ **27.** $\frac{1}{y^{3/4}}$ **29.** $x^{5/6}$ **31.** $y^{3/4}$
33. $1.2^{1.62} \approx 1.34$ **35.** $50^{3/2} - 50^{1/2} \approx 346.48$ **37.** b.
39.   **41.**

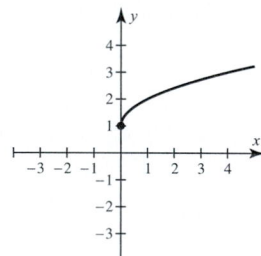   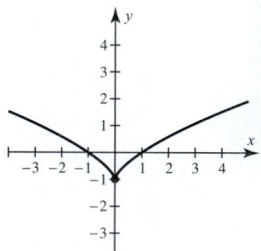

43.

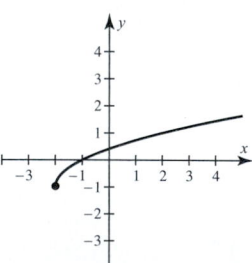

45. 2   **47.** 81   **49.** ± 32 **51.** 32   **53.** 27
55. $-1, -\frac{1}{2}$ **57.** $-\frac{1}{4}, \frac{5}{7}$ **59.** $-8, 27$ **61.** $\frac{1}{8}, \frac{64}{27}$
63. 1   **65.** 7   **67.** -1 **69.** 2, 3   **71.** 4   **73.** 15
75. $-1, 3$ **77.** -28 **79.** 2   **81.** 65,538
83. $w \approx 3.63$ lb   **85.** About 58.1 yr

87. **(a)** $a = 1960$ **(b)** $b \approx -1.2$
(c) $f(4) = 1960(4)^{-1.2} \approx 371$. If the zinc ion concentration reaches 371 milligrams per liter, a rainbow trout will survive, on average, 4 minutes.
89. **(a)** $f(2) \approx 1.06$ grams   **(b) & (c)**
Approximately 1.1 grams
91. $a \approx 3.20, b \approx 0.20$
[1, 9, 1] by [0, 6, 1]

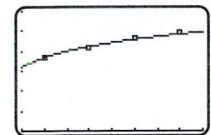

93. $a \approx 874.54, b \approx -0.49789$

4.7 EXTENDED AND DISCOVERY EXERCISES (P. 346)

1. The graph of an odd root function is always increasing; the function is negative for $x < 0$, positive for $x > 0$, and zero at $x = 0$. It is an odd function.
3. The graph of a power function in which the exponent is a negative odd integer has the y-axis as a vertical asymptote and the x-axis as a horizontal asymptote; the function is undefined at $x = 0$; the function is decreasing on $(-\infty, 0)$ and on $(0, \infty)$. The function is symmetric with respect to the origin. It is an odd function.
5. $\dfrac{1}{\sqrt{x + h} + \sqrt{x}}$

CHECKING BASIC CONCEPTS FOR SECTION 4.7 (P. 346)

1. **(a)** -8 **(b)** $\frac{1}{4}$ **(c)** 9   **3.** 8   **5.** $a = 2, b = \frac{1}{2}$

CHAPTER 4 REVIEW EXERCISES (PP. 352–357)

1. Degree: 3; leading coefficient: -7
3. **(a)**

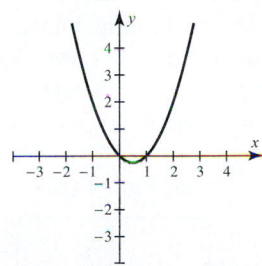

(b) Increasing: $[0.5, \infty)$; decreasing: $(-\infty, 0.5]$
5. **(a)** Local minima: $-4.5, -0.5$; local maximum: 0
(b) Absolute minimum: -4.5; absolute maximum: none

7. $[-10, 10, 1]$ by $[-100, 100, 10]$

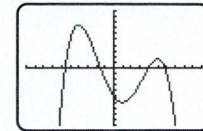

(a) Local minimum: -60.08; local maxima: 15, 75.12
(b) Absolute minimum: none; absolute maximum: 75.12
(c) Increasing: $(-\infty, -3.996] \cup [0.9995, 5.007]$; decreasing: $[-3.996, 0.9995] \cup [5.007, \infty)$, where endpoints are approximate.
 9. **(a)** One local maximum, two local minima, two x-intercepts   **(b)** Two local maxima, one local minimum, three x-intercepts   **(c)** One local maximum, two local minima, four x-intercepts
$[-10, 10, 1]$ by $[-100, 100, 10]$ $[-10, 10, 1]$ by $[-10, 10, 1]$

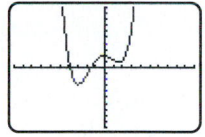   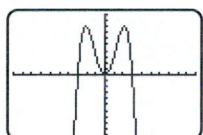

$[-10, 10, 1]$ by $[-100, 100, 10]$

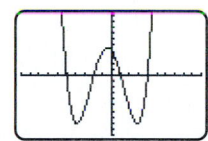

11. Even   **13.** Odd   **15.** Odd   **17.** $(1, -6)$
19. Answers may vary.   **21.** Answers may vary.

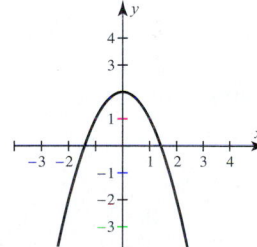

23. **(a)** Two; $-2, 0, 1$ **(b)** Negative   **(c)** 3
25. Up on left end, down on right end;
$(f(x) \to \infty$ as $x \to -\infty; f(x) \to -\infty$, as $x \to \infty)$
27. 7
29. **(a)**

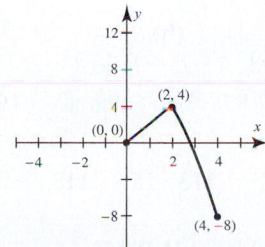

f is continuous.
(b) $f(1) = 2; f(3) = -1$ **(c)** 1, $\sqrt{6}$

31. $2x^2 - 3x - 1$ **33.** $2x^2 - 3x + 1 + \dfrac{1}{2x + 3}$

35. $f(x) = \frac{1}{2}(x - 1)(x - 2)(x - 3)$

37. $f(x) = 2(x + 4)\left(x - \frac{1}{2}\right)(x - 2)$

39. $f(x) = (x + 2)(x - 1)(x - 3)$

41. Zeros: $-3, -1, 1, 2$;
$f(x) = -(x + 3)(x + 1)(x - 1)(x - 2)$

43. (a) 1   (b) 2   (c) 3   **45.** $-3, \frac{1}{2}, 2$

47. $0, \pm\sqrt{3}$ **49.** $-2, 0, 3$ **51.** $\pm 1, \pm\sqrt{2}$

53. $-\frac{1}{2}, \pm\sqrt{3}$ **55.** $-1.88, 0.35, 1.53$ **57.** -1

59. $-10 - 11i$ **61.** $\pm\frac{3}{2}i$ **63.** $0, \pm i$

65. One real zero, two imaginary zeros

67. $f(x) = 2(x - i\sqrt{2})(x + i\sqrt{2})$

69. $f(x) = (x + i)(x - i)\left(x - \left(-\dfrac{1}{2} + \dfrac{i\sqrt{3}}{2}\right)\right)$

$\left(x - \left(-\dfrac{1}{2} - \dfrac{i\sqrt{3}}{2}\right)\right)$

71. $D = \left\{x \mid x \neq -\frac{4}{5}\right\}$; horizontal: $y = \frac{3}{5}$; vertical: $x = -\frac{4}{5}$

73. Horizontal: $y = -1$; vertical: $x = -1$ or $x = 1$;
$D = \{x \mid x \neq 1, x \neq -1\}$

75.

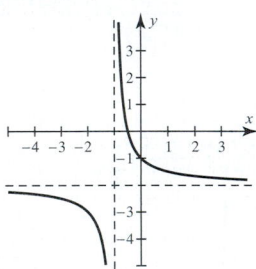

77.

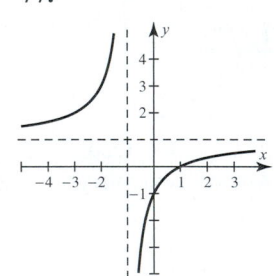

79. Answers may vary.

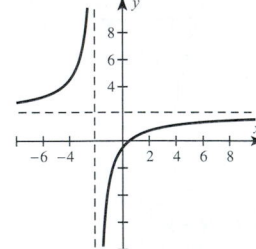

81. 4   **83.** $\frac{1}{2}$ **85.** $-\frac{1}{3}$ **87.** 5

89. (a) $(-\infty, -4) \cup (-2, 3)$ (b) $(-4, -2) \cup (3, \infty)$

91. $(-3, 0) \cup (2, \infty)$ **93.** $(-2, -1) \cup (1, 2)$

95. $(-\infty, -2) \cup \left(\frac{1}{2}, \infty\right)$ **97.** 216   **99.** 8   **101.** $x^{4/3}$

103. $y^{1/2}$ **105.** $D = \{x \mid x \geq 0\}$; $f(3) \approx 15.59$

107. 4   **109.** 6   **111.** $\frac{81}{16}$ **113.** 15   **115.** $-2, \frac{1}{3}$

117. $-1, 125$ **119.** 8

121. (a) Dog: 148; person: 69   (b) 6.4 in.

123. (a) Increasing: $[0, 0.4]$, $[1.6, 2.6]$, and $[3.4, 4]$

(b) Local maxima: approximately 18.8 and 11.8; at about 12:24 P.M. and 2:36 P.M. the wind reached relative maximums of about 18.8 mph and 11.8 mph respectively. Local minima: approximately 1.2 and 8.2; at about 1:36 P.M. and 3:24 P.M. the wind reached relative minimums of about 1.2 mph and 8.2 mph respectively.

125. (a) $81.5°F$ (b) 87.3; the ocean reaches a maximum temperature of about $87.3°F$ in late July.

127. About $10.5 \times 3.5 \times 2.6$ in.

129. (a) $a \approx 1.8342, b \approx -0.4839$
(b) 1.08 steps per second

CHAPTER 4 EXTENDED AND DISCOVERY EXERCISES (PP. 357–358)

1. (a)

$f(t) = t^2$	$t_1 = 10$ $t_2 = 11$	$t_1 = 10$ $t_2 = 10.1$	$t_1 = 10$ $t_2 = 10.01$	$t_1 = 10$ $t_2 = 10.001$
average velocity (ft/sec)	21	20.1	20.01	20.001

(b) The velocity of the bike rider is 20 ft/sec at 10 sec.

3. (a) 5   (b) 5   (c) 5, 5, 5

5. (a) $2a + 2 + h$ (b) $2a + 2$ (c) 12, 22, 32

7. (a) $3a^2 + 3ah + h^2$ (b) $3a^2$ (c) 75, 300, 675

9. (a)

f_1 in $[-10, 10, 1]$ by $[-10, 10, 1]$

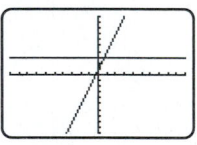

f_2 in $[-10, 10, 1]$ by $[-10, 10, 1]$

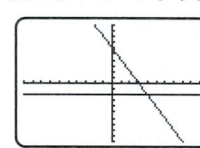

f_3 in $[-10, 10, 1]$ by $[-10, 10, 1]$

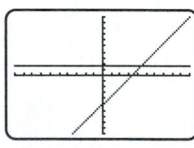

f_4 in $[-10, 10, 1]$ by $[-10, 10, 1]$

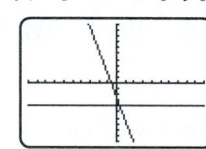

(b) Neither the graph of each linear function nor the graph of its average rate of change has any turning points.
(c) For any linear function, the graph of its (average) rate of change is a constant function whose value is equal to the slope of the graph of the linear function.

11. (a)

f_1 in $[-10, 10, 1]$ by $[-10, 10, 1]$

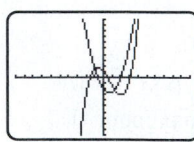

f_2 in $[-10, 10, 1]$ by $[-10, 10, 1]$

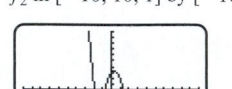

f_3 in $[-10, 10, 1]$ by $[-10, 10, 1]$

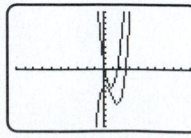

f_4 in $[-10, 10, 1]$ by $[-10, 10, 1]$

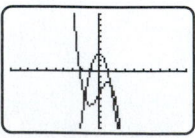

(b) The graph of a cubic function has two turning points or none; the graph of its average rate of change has one turning point. The x-coordinate of a turning point of the function corresponds to an x-intercept on the graph of the average rate of change.   **(c)** For any cubic function, the graph of its average rate of change is a quadratic function. The leading coefficients of a cubic function and its average rate of change have the same sign.

CHAPTERS 1–4 CUMULATIVE REVIEW EXERCISES (PP. 358–362)

1. 20%   **3.** -0.054 **5.** $\sqrt{185}$

7. (a)   **(b)**

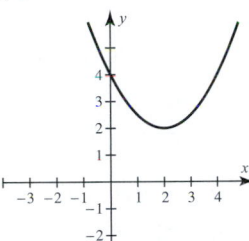   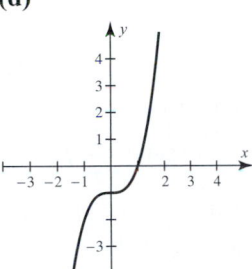

(c)   **(d)**

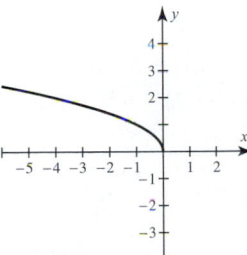   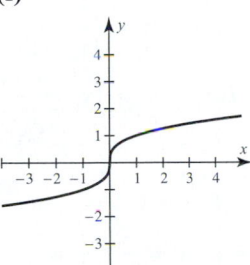

(e)   **(f)**

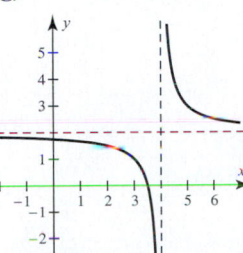   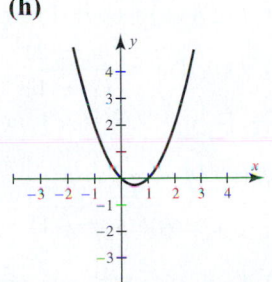

(g)   **(h)**

9. (a) $D = \{x \mid x \leq -2 \text{ or } x \geq 2\}$ **(b)** $f(2) = 0$
11. (a) $C(x) = 0.25x + 200$; $C(2000) = 700$; the cost of driving 2000 miles in one month is \$700.
13. 18   **15.** $y = -\frac{9}{5}x + \frac{7}{5}$ **17.** $y = -5$
19. Each radio costs \$15 to manufacture. The fixed cost is \$2000.   **21.** $\frac{3}{5}$ **23.** $0, \frac{8}{3}$ **25.** $-2, \frac{5}{7}$ **27.** $-3, 0, 1$
29. $\frac{1}{27}, -8$ **31.** $\frac{4}{5}$ **33.** No solutions; contradiction
35. $\left(-\infty, -\frac{1}{2}\right)$ **37.** $\left(-\infty, \frac{4}{5}\right] \cup [2, \infty)$
39. $(-\infty, -3] \cup [0, 3]$
41. (a) $-3, -1, 1, 2$
(b) $(-\infty, -3) \cup (-1, 1) \cup (2, \infty)$
(c) $[-3, -1] \cup [1, 2]$
43. $\left(3, \frac{5}{2}\right)$ **45.** $\frac{3 \pm \sqrt{13}}{2}$
47.

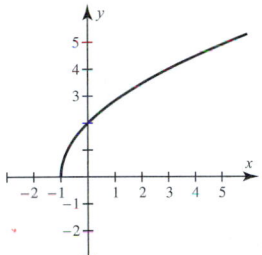

49. (a) Increasing: $(-\infty, -2] \cup [1, \infty)$; decreasing: $[-2, 1]$ **(b)** approximately $-3.3, 0,$ and 1.8
(c) $(-2, 3)$ and $(1, -1)$ **(d)** Local maximum: 3; local minimum: -1
51. Answers may vary.

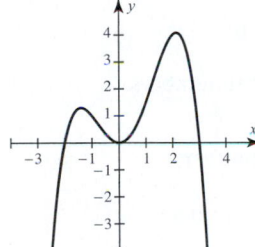

53. (a) $a - 2 + \frac{3}{a^2}$ **(b)** $2x^2 + 2x - 2 + \frac{-1}{x-1}$
(c) $x^2 + 2x - 3 + \frac{x+4}{x^2+2}$
55. $f(x) = 4(x+3)(x-1)^2(x-4)^3$
57. $f(x) = 2(x - \sqrt{3})(x + \sqrt{3})\left(x - \frac{1}{2}\right)$
59. $f(x) = -2(x+2)(x+1)(x-1)$ **61.** $-\frac{1}{2} + \frac{7}{2}i$
63. $D - \{x \mid x \neq -1, x \neq 4\}$; vertical: $x = -1, x = 4$; horizontal: $y = 0$ **65.** 3.67 in.
67. 4.0 miles   **69.** $\frac{40}{9} \approx 4.44$ hr
71. (a) $P(t) = 3(t - 1990) + 250$
(b) 2007 or late 2006   **73.** 13 by 19 in.

75. $f(x) = 2(x - 4)^2 + 6$
[0, 12, 2] by [0, 90, 10]

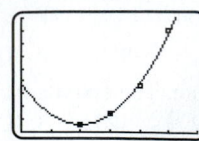

77. (a) $f(x) = 0.01(x - 1961)^2 + 13$
(b) 2013; answers may vary.

CHAPTER 5: Exponential and Logarithmic Functions

SECTION 5.1 (PP. 375–382)

1. (a) 5   **(b)** 5   **(c)** 0   **(d)** Undefined
3.

x	-2	0	2	4
$(f + g)(x)$	6	5	5	15
$(f - g)(x)$	-6	5	9	5
$(fg)(x)$	0	0	-14	50
$(f/g)(x)$	0	—	-3.5	2

5. (a) 2   **(b)** 4   **(c)** 0   **(d)** $-\frac{1}{3}$
7. (a) 2   **(b)** -3 **(c)** 2   **(d)** -2
9. (a) -5 **(b)** -5 **(c)** -3 **(d)** $-\frac{1}{3}$
11. (a) $\frac{11}{2}$ **(b)** 0   **(c)** $\frac{9}{4}$ **(d)** Undefined
13. (a) $(f + g)(x) = 2x + x^2$; all real numbers
(b) $(f - g)(x) = 2x - x^2$; all real numbers
(c) $(fg)(x) = 2x^3$; all real numbers
(d) $(f/g)(x) = \dfrac{2}{x}$; $D = \{x | x \ne 0\}$
15. (a) $(f + g)(x) = 2x^2$; all real numbers
(b) $(f - g)(x) = -2$; all real numbers
(c) $(fg)(x) = x^4 - 1$; all real numbers
(d) $(f/g)(x) = \dfrac{x^2 - 1}{x^2 + 1}$; all real numbers
17. (a) $(f + g)(x) = 2x$; $D = \{x | x \ge 1\}$
(b) $(f - g)(x) = -2\sqrt{x - 1}$; $D = \{x | x \ge 1\}$
(c) $(fg)(x) = x^2 - x + 1$; $D = \{x | x \ge 1\}$
(d) $(f/g)(x) = \dfrac{x - \sqrt{x - 1}}{x + \sqrt{x - 1}}$; $D = \{x | x \ge 1\}$
19. (a) $(f + g)(x) = 2\sqrt{x}$; $D = \{x | x \ge 0\}$
(b) $(f - g)(x) = -2$; $D = \{x | x \ge 0\}$
(c) $(fg)(x) = x - 1$; $D = \{x | x \ge 0\}$
(d) $(f/g)(x) = \dfrac{\sqrt{x} - 1}{\sqrt{x} + 1}$; $D = \{x | x \ge 0\}$

21. (a) $(f + g)(x) = \dfrac{4}{x + 1}$; $D = \{x | x \ne -1\}$
(b) $(f - g)(x) = -\dfrac{2}{x + 1}$; $D = \{x | x \ne -1\}$
(c) $(fg)(x) = \dfrac{3}{(x + 1)^2}$; $D = \{x | x \ne -1\}$
(d) $(f/g)(x) = \frac{1}{3}$; $D = \{x | x \ne -1\}$
23. (a) $(f + g)(x) = 3 - 2x$; all real numbers
(b) $(f - g)(x) = 6x - 3$; all real numbers
(c) $(fg)(x) = 6x - 8x^2$; all real numbers
(d) $(f/g)(x) = \dfrac{2x}{3 - 4x}$; $D = \left\{x \Big| x \ne \dfrac{3}{4}\right\}$
25. (a) $(f + g)(x) = \dfrac{x + 1}{2x - 4}$; $D = \{x | x \ne 2\}$
(b) $(f - g)(x) = \dfrac{1 - x}{2x - 4}$; $D = \{x | x \ne 2\}$
(c) $(fg)(x) = \dfrac{x}{(2x - 4)^2}$; $D = \{x | x \ne 2\}$
(d) $(f/g)(x) = \dfrac{1}{x}$; $D = \{x | x \ne 0, x \ne 2\}$
27. (a) $(f + g)(x) = x^2 + 3x + 2$; all real numbers
(b) $(f - g)(x) = -x^2 - x + 2$; all real numbers
(c) $(fg)(x) = (x + 2)(x^2 + 2x) = x^3 + 4x^2 + 4x$; all real numbers
(d) $(f/g)(x) = \dfrac{1}{x}$; $D = \{x | x \ne -2, x \ne 0\}$
29. (a) $(f + g)(x) = x^2 - 1 + |x + 1|$; all real numbers
(b) $(f - g)(x) = x^2 - 1 - |x + 1|$; all real numbers
(c) $(fg)(x) = (x^2 - 1)|x + 1|$; all real numbers
(d) $(f/g)(x) = \dfrac{x^2 - 1}{|x + 1|}$; $D = \{x | x \ne -1\}$
31. (a) $(f + g)(x) = \dfrac{(x - 1)(2x^2 - 2x + 5)}{(x + 1)(x - 2)}$;
$D = \{x | x \ne -1, x \ne 2\}$
(b) $(f - g)(x) = \dfrac{-3(x - 1)(2x - 1)}{(x + 1)(x - 2)}$;
$D = \{x | x \ne -1, x \ne 2\}$
(c) $(fg)(x) = (x - 1)^2$; $D = \{x | x \ne -1, x \ne 2\}$
(d) $(f/g)(x) = \dfrac{(x - 2)^2}{(x + 1)^2}$;
$D = \{x | x \ne 1, x \ne -1, \text{ and } x \ne 2\}$
33. (a) $(f + g)(x) = \dfrac{x^2 + 4x - 1}{(x - 1)^2(x + 1)}$;
$D = \{x | x \ne 1, x \ne -1\}$
(b) $(f - g)(x) = \dfrac{-x^2 - 3}{(x - 1)^2(x + 1)}$;
$D = \{x | x \ne 1, x \ne -1\}$

(c) $(fg)(x) = \dfrac{2}{(x-1)^3}$; $D = \{x \mid x \neq 1, x \neq -1\}$

(d) $(f/g)(x) = \dfrac{2(x-1)}{(x+1)^2}$; $D = \{x \mid x \neq 1, x \neq -1\}$

35. (a) $(f+g)(x) = x^{1/2}(x^2 - x + 1)$;
$D = \{x \mid x \geq 0\}$

(b) $(f-g)(x) = x^{1/2}(x^2 - x - 1)$;
$D = \{x \mid x \geq 0\}$

(c) $(fg)(x) = x^2(x-1)$; $D = \{x \mid x \geq 0\}$

(d) $(f/g)(x) = x(x-1)$; $D = \{x \mid x > 0\}$

37. (a) $Y_1 = f(x)$, $Y_2 = g(x)$, and $Y_3 = (f+g)(x)$
     $[0, 9, 1]$ by $[0, 15, 1]$

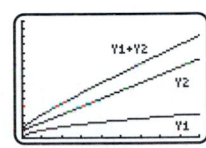

(b) To determine the graph $f + g$ add corresponding y-coordinates on the graphs of f and g.

39. (a) $g(-3) = -5$ **(b)** $g(b) = 2b + 1$
(c) $g(x^3) = 2x^3 + 1$ **(d)** $g(2x - 3) = 4x - 5$
41. (a) $g(-3) = -4$ **(b)** $g(b) = 2(b+3)^2 - 4$
(c) $g(x^3) = 2(x^3 + 3)^2 - 4$
(d) $g(2x - 3) = 8x^2 - 4$
43. (a) $g(-3) = -\frac{11}{2}$ **(b)** $g(b) = \frac{1}{2}b^2 + 3b - 1$
(c) $g(x^3) = \frac{1}{2}x^6 + 3x^3 - 1$
(d) $g(2x - 3) = 2x^2 - \frac{11}{2}$
45. (a) $g(-3) = 1$ **(b)** $g(b) = \sqrt{b + 4}$
(c) $g(x^3) = \sqrt{x^3 + 4}$
(d) $g(2x - 3) = \sqrt{2x + 1}$
47. (a) $g(-3) = 14$ **(b)** $g(b) = |3b - 1| + 4$
(c) $g(x^3) = |3x^3 - 1| + 4$
(d) $g(2x - 3) = |6x - 10| + 4$
49. (a) $g(-3)$ is undefined **(b)** $g(b) = \dfrac{4b}{b + 3}$
(c) $g(x^3) = \dfrac{4x^3}{x^3 + 3}$ **(d)** $g(2x - 3) = \dfrac{2(2x - 3)}{x}$
51. (a) $g(-3) = 4\sqrt[3]{9}$ **(b)** $g(b) = 4b^{2/3}$
(c) $g(x^3) = 4x^2$ **(d)** $g(2x - 3) = 4(2x - 3)^{2/3}$
53. (a) 5 **(b)** Undefined **(c)** 4
55. $4; 2$ **57. (a)** -4 **(b)** 2 **(c)** -4
59. (a) -3 **(b)** -2 **(c)** 0 **61. (a)** 3 **(b)** 4
63. (a) 18 **(b)** 23
65. (a) $(f \circ g)(x) = (x^2 + 3x - 1)^3$; all real numbers
(b) $(g \circ f)(x) = x^6 + 3x^3 - 1$; all real numbers
(c) $(f \circ f)(x) = x^9$; all real numbers

67. (a) $(f \circ g)(x) = 1 - x$; $D = \{x \mid x \leq 1\}$
(b) $(g \circ f)(x) = \sqrt{1 - x^2}$; $D = \{x \mid -1 \leq x \leq 1\}$
(c) $(f \circ f)(x) = x^4$; all real numbers
69. (a) $(f \circ g)(x) = 2 - 3x^3$; all real numbers
(b) $(g \circ f)(x) = (2 - 3x)^3$; all real numbers
(c) $(f \circ f)(x) = 9x - 4$; all real numbers
71. (a) $(f \circ g)(x) = 1 - x^2$; all real numbers
(b) $(g \circ f)(x) = 5 - (x - 4)^2$; all real numbers
(c) $(f \circ f)(x) = x - 8$; all real numbers
73. (a) $(f \circ g)(x) = \dfrac{1}{5x + 1}$; $D = \left\{x \mid x \neq -\dfrac{1}{5}\right\}$
(b) $(g \circ f)(x) = \dfrac{5}{x + 1}$; $D = \{x \mid x \neq -1\}$
(c) $(f \circ f)(x) = \dfrac{x + 1}{x + 2}$; $D = \{x \mid x \neq -1, x \neq -2\}$
75. (a) $(f \circ g)(x) = \sqrt{4 - x^2} + 4$;
$D = \{x \mid -2 \leq x \leq 2\}$
(b) $(g \circ f)(x) = \sqrt{4 - (x + 4)^2}$;
$D = \{x \mid -6 \leq x \leq -2\}$
(c) $(f \circ f)(x) = x + 8$; all real numbers
77. (a) $(f \circ g)(x) = \sqrt{3x - 1}$; $D = \left\{x \mid x \geq \frac{1}{3}\right\}$
(b) $(g \circ f)(x) = 3\sqrt{x - 1}$; $D = \{x \mid x \geq 1\}$
(c) $(f \circ f)(x) = \sqrt{\sqrt{x - 1} - 1}$; $D = \{x \mid x \geq 2\}$
79. (a) $(f \circ g)(x) = x$; all real numbers
(b) $(g \circ f)(x) = x$; all real numbers
(c) $(f \circ f)(x) = \dfrac{x - 9}{4}$; all real numbers
81. (a) $(f \circ g)(x) = x$; all real numbers
(b) $(g \circ f)(x) = x$; all real numbers
(c) $(f \circ f)(x) = \sqrt[3]{\sqrt[3]{x - 1} - 1}$; all real numbers
83. (a) $(f \circ g)(x) = \dfrac{x - ab + b}{3}$; all real numbers
(b) $(g \circ f)(x) = \dfrac{ax - 3ab + b}{3a}$; all real numbers
(c) $(f \circ f)(x) = \dfrac{a^2x + ab + 3b}{9}$; all real numbers

Answers may vary for Exercises 85–97.

85. $f(x) = x - 2$, $g(x) = \sqrt{x}$
87. $f(x) = x + 2$, $g(x) = 5x^2 - 4$
89. $f(x) = 2x + 1$, $g(x) = 4x^3$
91. $f(x) = x^3 - 1$, $g(x) = x^2$
93. $f(x) = x + 2$, $g(x) = -4|x| - 3$
95. $f(x) = x - 1$, $g(x) = \dfrac{1}{x^2}$
97. $f(x) = x^{1/4}$, $g(x) = x^3 - x$
99. (a) $C(x) = 1.5x + 150,000$
(b) $R(x) = 6.5x$; $R(8000) = \$52,000$
(c) $P(x) = 5x - 150,000$; $P(40,000) = \$50,000$
(d) 30,000 videos

101. (a) $I(x) = 36x$ **(b)** $C(x) = 2.54x$
(c) $F(x) = (C \circ I)(x)$ **(d)** $F(x) = 91.44x$
103. (a) $(g \circ f)(1) = 5.25$; a 1% decrease in the ozone layer could result in a 5.25% increase in skin cancer.
(b) Not possible using the given tables
105. (a) $(g \circ f)(1980) \approx 5$; in 1980 the average nighttime temperature had risen about 3.5°C since 1948, which resulted in a 5% increase in peak demand of electricity.
(b) $(f \circ g)(3)$ is meaningless because the range of g is not the same as the domain of f.
107. (a) 30,000 gallons   **(b)** $(g \circ f)(x)$ computes the gallons of water in the pool after x days.
109. (a) $r(x) = \frac{11}{6}x$ **(b)** $s(x) = \frac{11}{6}x + \frac{1}{9}x^2$
(c) $s(60) = 510$; it takes 510 feet to stop when traveling 60 mph.
111. $C = 12\pi t$ **113.** $S = \pi r^2 \sqrt{5}$
115. (a) 50   **(b)** $(C + O)(x)$ computes the total SO_2 emissions from coal and oil during year x.
(c)

x	1860	1900	1940	1970	2000
$(C+O)(x)$	2.4	12.8	26.5	50.0	78.0

117. (a)

x	1990	2000	2010	2020	2030
$h(x)$	32	35.5	39	42.5	46

(b) $h(x) = f(x) + g(x)$
119. $h(x) = 0.35x - 664.5$
121. (a) $h(x) = g(x) - f(x)$ **(b)** $h(1995) = 200$, $h(2000) = 250$ **(c)** $h(x) = 10(x - 1995) + 200$ or $h(x) = 10x - 19{,}750$
123. (a) $A(4s) = 16 \cdot \frac{\sqrt{3}}{4}s^2 = 16A(s)$; If the length of a side is quadrupled, the area increases by a factor of 16.
(b) $A(s + 2) = \frac{\sqrt{3}}{4}(s^2 + 4s + 4) = A(s) + \sqrt{3}(s + 1)$; If the length of a side increases by 2, the area increases by $\sqrt{3}(s + 1)$.
125. Let $f(x) = ax + b$ and $g(x) = cx + d$. Then $f(x) + g(x) = (ax + b) + (cx + d) = (a + c)x + (b + d)$, which is linear.
127. (a) $(f \circ g)(x) = k$; a constant function
(b) $(g \circ f)(x) = ak + b$; a constant function

5.1 EXTENDED AND DISCOVERY EXERCISES (P. 382)

1. $D = m + n$ **3.** $D = mn$

SECTION 5.2 (PP. 393–398)

1. Closing a window   **3.** Closing a book, standing up, and walking out of the classroom
5. Subtract 2 from x; $x + 2$ and $x - 2$
7. Divide x by 3 and then add 2; $3(x - 2)$ and $\frac{x}{3} + 2$
9. Subtract 1 from x and cube the result; $\sqrt[3]{x} + 1$ and $(x - 1)^3$
11. Take the reciprocal of x; $\frac{1}{x}$ and $\frac{1}{x}$

13. One-to-one   **15.** Not one-to-one
17. Not one-to-one   **19.** Not one-to-one; does not have an inverse   **21.** One-to-one; does have an inverse
23. One-to-one   **25.** Not one-to-one
27. Not one-to-one   **29.** Not one-to-one
31. Not one-to-one   **33.** Not one-to-one
35. One-to-one   **37.** No   **39.** Yes
41. $f^{-1}(x) = x^3$ **43.** $f^{-1}(x) = -\frac{1}{2}x + 5$
45. $f^{-1}(x) = \dfrac{x + 1}{3}$ **47.** $f^{-1}(x) = \sqrt[3]{\dfrac{x + 5}{2}}$
49. $f^{-1}(x) = \sqrt{x + 1}$ **51.** $f^{-1}(x) = \dfrac{1}{2x}$
53. $f^{-1}(x) = -\dfrac{2(x - 3)}{5}$ **55.** $f^{-1}(x) = -\dfrac{2x}{x - 1}$
57. $f^{-1}(x) = \dfrac{x + 1}{x - 2}$ **59.** $f^{-1}(x) = \dfrac{1}{x + 3}$
61. $f^{-1}(x) = \sqrt[3]{\dfrac{1 + x}{x}}$
63. If the domain of f is restricted to $x \geq 0$, then $f^{-1}(x) = \sqrt{4 - x}$.
65. If the domain of f is restricted to $x \geq 2$, then $f^{-1}(x) = 2 + \sqrt{x - 4}$.
67. If the domain of f is restricted to $x \geq 0$, then $f^{-1}(x) = (x - 1)^{3/2}$.
69. If the domain of f is restricted to $x \geq 0$, then $f^{-1}(x) = \sqrt{\dfrac{9 - x^2}{2}}$.
71. $f^{-1}(x) = \frac{x + 15}{5}$; D and R are all real numbers.
73. $f^{-1}(x) = x^3 + 5$; D and R are all real numbers.
75. $f^{-1}(x) = 4x + 5$; D and R are all real numbers.
77. $f^{-1}(x) = x^2 + 5$; $D = \{x \mid x \geq 0\}$ and $R = \{y \mid y \geq 5\}$
79. $f^{-1}(x) = \dfrac{1}{x} - 3$; $D = \{x \mid x \neq 0\}$ and $R = \{y \mid y \neq -3\}$
81. $f^{-1}(x) = \sqrt[3]{\dfrac{x}{2}}$; D and R are all real numbers.
83. $f^{-1}(x) = \sqrt{x}$; D and R include all nonnegative real numbers.
85.

x	5	7	9
$f^{-1}(x)$	1	2	3

For f: $D = \{1, 2, 3\}$, $R = \{5, 7, 9\}$;
for f^{-1}: $D = \{5, 7, 9\}$, $R = \{1, 2, 3\}$
87.

x	0	1	4	9	16
$f^{-1}(x)$	0	1	2	3	4

For f: $D = \{0, 1, 2, 3, 4\}$, $R = \{0, 1, 4, 9, 16\}$;
for f^{-1}: $D = \{0, 1, 4, 9, 16\}$, $R = \{0, 1, 2, 3, 4\}$

89.

x	-3	0	3	6
$f^{-1}(x)$	-8	-5	-2	1

91.

x	-8	-1	8	27
$f^{-1}(x)$	-2	-1	2	3

93. 1   **95.** 3   **97.** 5   **99.** 1

101. (a) $f(1) \approx \$110$ **(b)** $f^{-1}(110) \approx 1$ yr
(c) $f^{-1}(160) \approx 5$ years; $f^{-1}(x)$ computes the years necessary for the account to accumulate x dollars.

103. (a) 2   **(b)** 3   **(c)** 1   **(d)** 3

105. 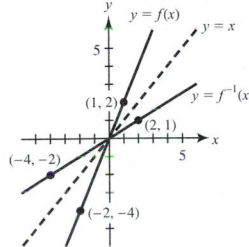   **107.**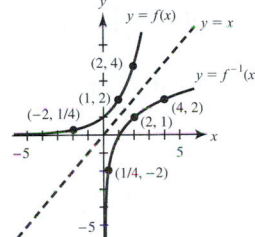

109. 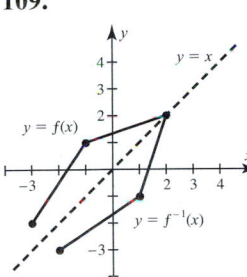   **111.**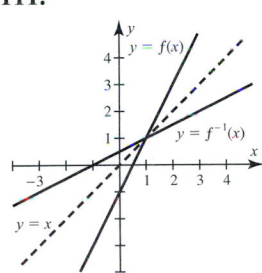

113. 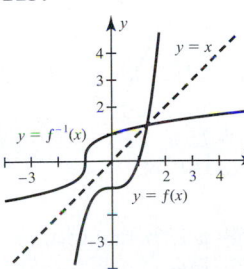   **115.**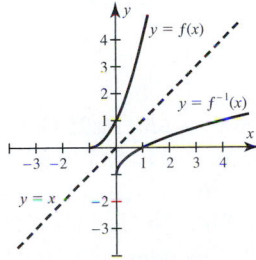

117. $Y_1 = 3X - 1, Y_2 = (X + 1)/3, Y_3 = X$
$[-4.7, 4.7, 1]$ by $[-3.1, 3.1, 1]$

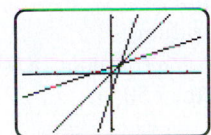

119. $Y_1 = X^3/3 - 1, Y_2 = \sqrt[3]{(3X + 3)}, Y_3 = X$
$[-4.7, 4.7, 1]$ by $[-3.1, 3.1, 1]$

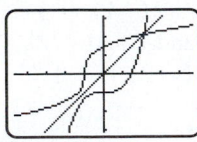

121. (a) $C(1995) = 1$ **(b)** 1995   **(c)** $C^{-1}(1) = 1995$

123. (a) Yes   **(b)** The radius r of a sphere with volume V
(c) $r = \sqrt[3]{\dfrac{3V}{4\pi}}$ **(d)** No. If V and r were interchanged, then r would represent the volume and V would represent the radius.

125. OCVJ; FUNCTIONS; if f is not one-to-one, then f^{-1} does not exist and f^{-1} is needed to decode a message.

127. (a) 135.7 pounds   **(b)** Yes
(c) $h = \dfrac{7}{25}\left(W + \dfrac{800}{7}\right) = \dfrac{7}{25}W + 32$
(d) 74; the maximum recommended height for a person weighing 150 lb is 74 in.
(e) The inverse formula computes the maximum recommended height of a person for a given weight.

129. (a) $(F \circ Y)(2) = 10{,}560$ represents the number of feet in 2 miles.   **(b)** $F^{-1}(26{,}400) = 8800$ represents the number of yd in 26,400 ft.
(c) $(Y^{-1} \circ F^{-1})(21{,}120) = 4$ represents the number of mi in 21,120 ft.

131. (a) $(Q \circ C)(96) = 1.5$ represents the number of qt in 96 tbsp.   **(b)** $Q^{-1}(2) = 8$ represents the number of cups in 2 qt.
(c) $(C^{-1} \circ Q^{-1})(1.5) = 96$ represents the number of tbsp in 1.5 qt.

133. (a) $f(1930) = 62.5, f(1980) = 65.5$; increased by 3%   **(b)** $f^{-1}(x)$ computes the year when the cloud cover was x percent.
(c) $f^{-1}(62.5) = 1930, f^{-1}(65.5) = 1980$
(d) $f^{-1}(x) = \dfrac{50}{3}(x - 62.5) + 1930$

5.2 EXTENDED AND DISCOVERY EXERCISES (PP. 398–399)

1. (a) $f^{-1}(x)$ computes the number of hours it takes to travel x miles.   **(b)** The solution to the equation $f(x) = 200$ is the number of hours it takes to travel 200 miles.   **(c)** Evaluate $f^{-1}(200)$.

3. Quadrants I and IV

CHECKING BASIC CONCEPTS FOR SECTIONS 5.1 AND 5.2 (P. 399)

1. **(a)** $(f + g)(1) = 1$ **(b)** $(f - g)(-1) = 3$
(c) $(fg)(0) = 2$ **(d)** $(f/g)(2)$ is undefined
(e) $(f \circ g)(2) = -2$ **(f)** $(g \circ f)(-2) = -1$
3. **(a)** $(f + g)(x) = x^2 + 6x - 3$
(b) $(f/g)(x) = \dfrac{x^2 + 3x - 2}{3x - 1}, x \neq \dfrac{1}{3}$
(c) $(f \circ g)(x) = 9x^2 + 3x - 4$
5. **(a)** Yes; yes; $f^{-1}(x) = x - 1$ **(b)** No; no

SECTION 5.3 (PP. 412–418)

1. $\frac{1}{8}$ **3.** 6   **5.** -18 **7.** 8   **9.** 2   **11.** e^{2x}
13. 1   **15.** 5   **17.** **(a)** 15.59   **(b)** 8.82   **(c)** -5.65
19. Linear; $f(x) = -1.25x + 2$
21. Exponential; $f(x) = 8\left(\frac{1}{2}\right)^x$
23. Exponential; $f(x) = 5(2^x)$
25. Exponential; $f(n) = 40{,}000(1.08)^{n-1}$
27. $f(x) = 2^x$ **29.** $C = 5, a = 1.5$
31. $C = 10, a = 2$ **33.** $C = 3, a = 3$
35. $C = \frac{1}{2}, a = \frac{1}{3}$
37. $C = 5000, a = 2$; x represents time in hours.
39. $C = 200{,}000, a = 0.95$; x represents the number of years after 2000.   **41.** About 11 pounds per square inch
43.   **45.**

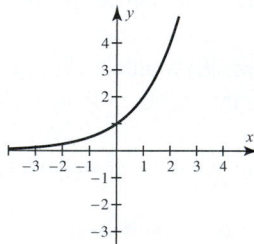   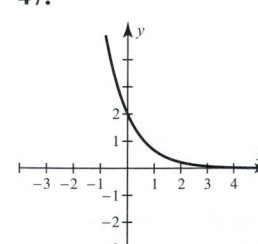

47.   **49.**

51.

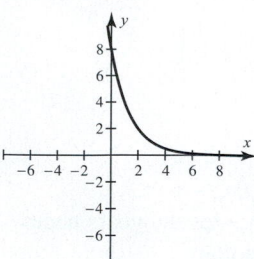

53. $C = 1, a = \frac{1}{2}$ **55.** $C = \frac{1}{2}, a = 4$
57. **(a)** $D: (-\infty, \infty); R: (0, \infty)$ **(b)** Decreasing
(c) $y = 0$ **(d)** y-intercept: 7; no x-intercept
(e) Yes; yes
59. **(i)** b.   **(ii)** d.   **(iii)** a.   **(iv)** c.
61. **(a)**   **(b)**

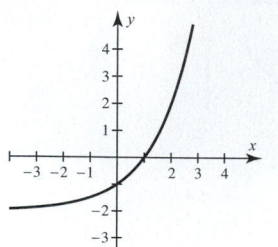   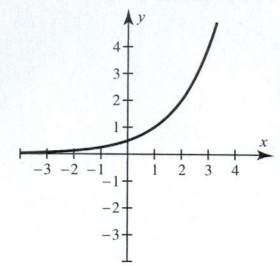

(c)   **(d)**

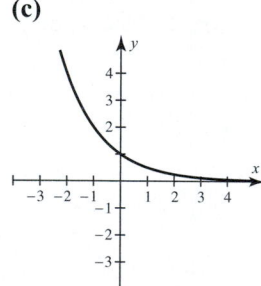   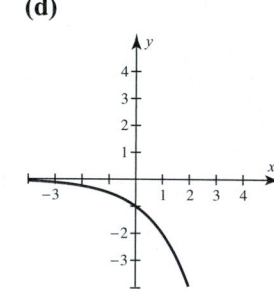

63. 22.1980   **65.** 71.2571   **67.** -0.7586
69. $841.53   **71.** $1730.97   **73.** $4451.08
75. $2072.76
77. 10%: $14,656.15; 13%: $26,553.58; a 13% rate results in considerably more interest than a 10% rate.
79. $14,326.78
81. The account with $200 will have double the money since twice the money was deposited.
83. About $1278.2 billion or $1.28 trillion
85. $19,870.65   **87.** About 27.3%   **89.** About 18,935 yr
91. **(a)** 31.7 mg, the half-life is less than 50 yr.
(b) 30.2 yr
93. **(a)** $C \approx 0.72, a \approx 1.041$; answers may vary slightly.
(b) 1.21 ppb; answers may vary slightly.
95. **(a)** About 92%   **(b)** $x \approx 0.83$ minutes
97. **(a)** $9074.47   **(b)** During the last 10 yr

99.

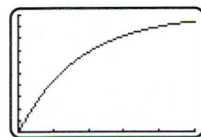

t (inches)	6	7.5	9	10.5	12
W (lb × 1000)	80	116	168	242	350

101. (a) About 6,214,000 bacteria per milliliter
(b) There will be 10 million *E. coli* bacteria after about 214 min, or about 3.6 hr.
103. (a) $P(2) \approx 0.20$ and $P(20) \approx 0.90$. There is a 20% chance that at least one tree is located within a circle having radius 2 ft and a 90% chance for a circle having radius 20 ft.
(b) The larger the circle, the more likely it is to contain a tree; Yes, $y = 1$; probability cannot be greater than 1.

[0, 25, 5] by [0, 1, 0.1]

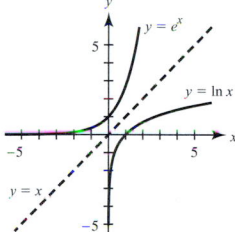

(c) $x \approx 6.1$. A circle of radius 6.1 ft has a 50–50 chance of containing at least one tree.
105. (a) $N(0) = N_0$; N_0 is the initial number of patients that are infected but have no apparent symptoms.
(b) After approximately 3.7 years, 50% of the patients who receive tainted blood transfusions develop AIDS symptoms.

5.3 EXTENDED AND DISCOVERY EXERCISES (P. 418)

1. The formulas are equivalent.   **3.** $12,914.64
5. (a) About 1.0005   **(b)** 1   **(c)** They are very similar.
7. (a) About 0.6068   **(b)** About 0.6065   **(c)** They are very similar.
9. The average rate of change near x and the value of the function at x are approximately equal.

SECTION 5.4 (PP. 430–435)

1.

x	10^0	10^4	10^8	$10^{1.2}$
log x	0	4	8	1.2

3. (a) Undefined   **(b)** -2 **(c)** $-\frac{1}{2}$ **(d)** 0
5. (a) 1   **(b)** 4   **(c)** -20 **(d)** -2
7. (a) 2   **(b)** $\frac{1}{2}$ **(c)** 3   **(d)** Undefined
9. (a) $n = 1$, $\log 79 \approx 1.898$
(b) $n = 2$, $\log 500 \approx 2.699$ **(c)** $n = 0$, $\log 5 \approx 0.6990$
(d) $n = -1$, $\log 0.5 \approx -0.3010$
11. (a) 3   **(b)** 6   **(c)** -2 **(d)** -4
13. (a) 0   **(b)** $\log 50 \approx 1.70$ **(c)** -4
(d) No real solution
15. (a) $\log 300 \approx 2.48$ **(b)** $\log 5 \approx 0.70$
(c) $\log 0.2 \approx -0.70$
17. (a) $\frac{3}{2}$ **(b)** $\frac{1}{3}$ **(c)** $\frac{3}{10}$ **(d)** -1
19. 6   **21.** $\frac{1}{2}$ **23.** 0   **25.** $\frac{1}{3}$ **27.** 2   **29.** -2
31. -1 **33.** 0   **35.** -4 **37.** k **39.** π

41. $x - 1$, for $x > 1$ **43.** $x^2 + 2$
45.

x	$\frac{1}{16}$	1	8	32
f(x)	-4	0	3	5

47.

x	6	7	21
f(x)	0	2	8

49. $\log_2 72 \approx 6.17$ **51.** $\log_5 0.25 \approx -0.86$
53. $\ln 25 \approx 3.22$ **55.** $-\ln 3 \approx -1.10$ **57.** 2
59. $\frac{2}{3}$ **61.** $\ln 23 \approx 3.14$ **63.** $\log_2 14 \approx 3.81$
65. $\ln\left(\frac{18}{5}\right) \approx 1.28$ **67.** 8   **69.** $10^{2.3} \approx 199.5262$
71. $2^{1.2} \approx 2.2974$ **73.** $e^{-2} \approx 0.1353$ **75.** 1000
77. 20   **79.** $e^{3/4} \approx 2.1170$ **81.** $e^{7/5} \approx 4.0552$
83. 16   **85.** $\dfrac{e^{6/5}}{2} \approx 1.6601$ **87.** 8
89. $a = 5, b = 2$ **91.** $(-3, \infty)$
93. $(-\infty, -1) \cup (1, \infty)$ **95.** $(-\infty, \infty)$
97. $(-\infty, 2)$ **99.** $f^{-1}(x) = \ln x$

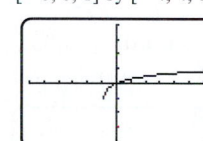

101. d.   **103.** c.
105. $D = \{x \mid x > -1\}$ **107.** $D = \{x \mid x < 0\}$
[−6, 6, 1] by [−4, 4, 1]   [−6, 6, 1] by [−4, 4, 1]

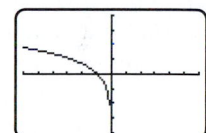

109. Decreasing

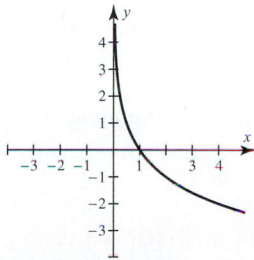

111. 105 decibels
113. $a = 7, b = 4$; about 178 km^2
115. (a) $C = 3, a = 2$ **(b)** After about 2.4 days

117. (a) Since L is increasing, heavier planes generally require longer runways.

$[0, 50, 10]$ by $[0, 6, 1]$

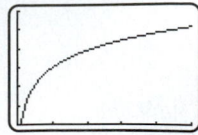

(b) No. It increases by 3000 ft.   **(c)** If the weight increases tenfold, the runway length increases by 3000 ft.
119. (a) $10^{-4.92} \approx 0.000012$ **(b)** $10^{-3.9} \approx 0.000126$
121. (a) Yugoslavia: $x = 1,000,000$;
Indonesia: $x = 100,000,000$ **(b)** 100
123. (a) $f(0) = 27$, $f(100) \approx 29.2$ in. At the eye, the barometric air pressure is 27 in; 100 mi away it is 29.2 in.
(b) The air pressure rises rapidly at first and then starts to level off.

$[0, 250, 50]$ by $[25, 30, 1]$

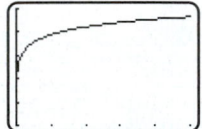

(c) About 7 miles
125. (a) About 49.4°C   **(b)** About 0.69 hr or 41.4 min
127. (a) About 0.189 or an 18.9% chance
(b) $x = -3 \ln (0.3) \approx 3.6$ min

5.4 EXTENDED AND DISCOVERY EXERCISES (P. 435)

1. (a) 1.00 **(b)** 0.50 **(c)** 0.33 **(d)** 0.25
3. (a) $T(x) = 6.5 \ln(1.3 \cdot 1.005^x)$; $T(100) \approx 4.95$
(b) $[0, 200, 50]$ by $[0, 1000, 100]$ $[0, 200, 50]$ by $[0, 10, 1]$

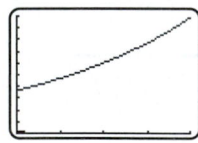   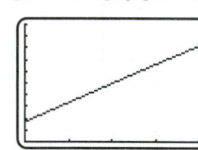

The carbon dioxide levels graph is exponential, whereas the average global temperature graph is linear.
(c) While the carbon dioxide levels increase exponentially, the average global temperature rises by a constant amount each year.

CHECKING BASIC CONCEPTS FOR SECTIONS 5.3 AND 5.4 (P. 435)

1. 1752.12; 1754.74
3. $\log_2 15$ represents the exponent k, such that $2^k = 15$. No.
5. (a) $\ln 5 \approx 1.609$ **(b)** $\log 25 \approx 1.398$
(c) $10^{1.5} \approx 31.623$
7. (a) New York: 18.2 million; Florida: 14 million
(b) Florida's population will equal New York's population in about the beginning of 2011. At this time both populations will be approximately 18.5 million.

SECTION 5.5 (PP. 442–443)

1. $\log 4 + \log 7 = \log 28 \approx 1.447$; Property 2
3. $\ln 72 - \ln 8 = \ln 9 \approx 2.197$; Property 3
5. $10 \log 2 = \log 1024 \approx 3.010$; Property 4
7. $\log_2 a + \log_2 b$ **9.** $\ln 7 + 4 \ln a$
11. $\log 6 - \log z$ **13.** $2 \log x - \log 3$
15. $\ln 2 + 7 \ln x - \ln 3 - \ln k$
17. $2 + 2 \log_2 k + 3 \log_2 x$
19. $2 + 3 \log_5 x - 4 \log_5 y$ **21.** $-1 + 3 \log_4 (x + 2)$
23. $3 \log_5 x - 4 \log_5 (x - 4)$ **25.** $\frac{1}{2} \log_2 x - 2 \log_2 z$
27. $\frac{1}{3} \ln (2x + 6) - \frac{5}{3} \ln (x + 1)$
29. $\frac{1}{3} \log_2 (x^2 - 1) - \frac{1}{2} \log_2 (1 + x^2)$ **31.** $\log 6$
33. $-\frac{3}{2} \ln 5$ **35.** $\log 2$ **37.** $\log 6$ **39.** $\log_7 5k^2$
41. $\ln x^3$ **43.** $\frac{3}{2} \log x$ **45.** $\ln \dfrac{2}{e}$ **47.** $\ln \dfrac{x^2 \sqrt{z}}{y^4}$
49. $\log (4x^{5/2})$ **51.** $\log \dfrac{(x^2 - 1)^2 (x - 2)^4}{\sqrt{y}}$
53. (a) Yes

X	Y₁	Y₂
1	.77815	.77815
2	1.3802	1.3802
3	1.7324	1.7324
4	1.9823	1.9823
5	2.1761	2.1761
6	2.3345	2.3345
7	2.4683	2.4683
X=1		

(b) By Property 2: $\log 3x + \log 2x = \log (3x \cdot 2x) = \log 6x^2$
55. (a) No

X	Y₁	Y₂
1	.40547	0
2	.40547	.69315
3	.40547	1.0986
4	.40547	1.3863
5	.40547	1.6094
6	.40547	1.7918
7	.40547	1.9459
X=1		

57. (a) Yes

X	Y₁	Y₂
1	0	0
2	1.2041	1.2041
3	1.9085	1.9085
4	2.4082	2.4082
5	2.7959	2.7959
6	3.1126	3.1126
7	3.3804	3.3804
X=1		

(b) By Property 4: $\log (x^4) = 4 \log x$
59. (a) Yes

X	Y₁	Y₂
1	0	0
2	1.3863	1.3863
3	2.1972	2.1972
4	2.7726	2.7726
5	3.2189	3.2189
6	3.5835	3.5835
7	3.8918	3.8918
X=1		

(b) By Property 4: $\ln x^4 - \ln x^2 = 4 \ln x - 2 \ln x = 2 \ln x$

61.   **63.**

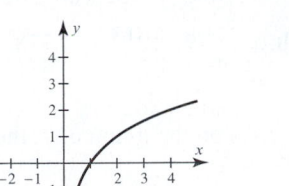   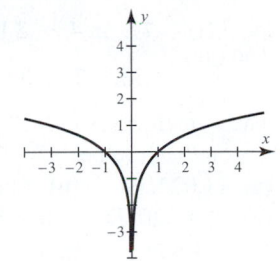

65. $\dfrac{\log 25}{\log 2} \approx 4.644$ **67.** $\dfrac{\log 130}{\log 5} \approx 3.024$

69. $\dfrac{\log 5}{\log 2} + \dfrac{\log 7}{\log 2} \approx 5.129$ **71.** $\sqrt{\dfrac{\log 46}{\log 4}} \approx 1.662$

73. $\dfrac{\log 12/\log 2}{\log 3/\log 2} = \dfrac{\log 12}{\log 3} \approx 2.262$ **75.** 4.714

77. ± 2.035 **79.** $L(x) = \dfrac{3 \ln x}{\ln 10}$; $L(50) \approx 5.097$; yes

81. 10 decibels   **83.** $I = I_0 e^{-kx}$ **85.** $t = \dfrac{\ln \frac{A}{P}}{r}$

87. $\log(1 \cdot 2^2 \cdot 3^3 \cdot 4^4 \cdot 5^5) = \log 86{,}400{,}000$

SECTION 5.6 (PP. 453–456)

1. (a) About 2   **(b)** $\ln 7.5 \approx 2.015$
3. (a) About 2   **(b)** $5 \log 2.5 \approx 1.990$
5. $\ln 1.25 \approx 0.2231$ **7.** $\log 20 \approx 1.301$
9. $-\frac{5}{6} \ln 0.4 \approx 0.7636$ **11.** $\dfrac{\ln 0.5}{\ln 0.9} \approx 6.579$
13. $1 + \dfrac{\log 4}{\log 1.1} \approx 15.55$ **15.** 1   **17.** $\frac{1}{5}$ **19.** 1, 2
21. No solutions   **23.** $\dfrac{\log 4}{\log 4 - 2 \log 3} \approx -1.710$
25. $\dfrac{3}{1 - 3 \ln 2} \approx -2.779$ **27.** $\dfrac{\log(64/3)}{\log 1.4} \approx 9.095$
29. $1980 + \dfrac{\log(8/5)}{\log 1.015} \approx 2012$ **31.** 5
33. $10^{2/3} \approx 4.642$ **35.** $\frac{1}{2}e^5 \approx 74.207$
37. $\pm\sqrt{50} \approx \pm 7.071$ **39.** 6   **41.** -39 **43.** 10^{-11}
45. $e \approx 2.718$ **47.** $2^{2.1} \approx 4.287$ **49.** $\sqrt{50} \approx 7.071$
51. $1 + e \approx 3.718$ **53.** $1 + \sqrt{2} \approx 2.414$ **55.** 2
57. 2   **59.** 4   **61.** 0.31   **63.** 1.71   **65.** 2.10
67. 1976   **69.** About 1990   **71.** About 1981
73. (a) $C = 1$, $a \approx 1.01355$
(b) $P(2010) \approx 1.14$ billion   **(c)** In about 2030
75. (a) $\dfrac{\ln 1.5}{0.09} \approx 4.505$ **(b)** \$500 invested at 9% compounded continuously results in \$750 after 4.5 years.
77. 8.25 yr, or 8 yr and 3 mo   **79.** About 8633 yr

81. (a) 0.993, or a 99.3% chance   **(b)** $x \approx 0.51$, or about half a min
83. (a) $T_0 = 32$, $D = 180$, $a \approx 0.045$ **(b)** About 139°F
(c) About 1 hr
85. (a) About 16.7°C   **(b)** About 1.09 hr
87. About 3.6 hr   **89.** About 62 days
91. $100 - 10^{1.16} \approx 85.5\%$ **93.** About 2.8 acres
95. $P = 1 - e^{-0.3214} \approx 0.275$. If a \$60 tax is placed on each ton of carbon burned, carbon dioxide emissions could decrease by 27.5%.
97. $t = \dfrac{\log(A/P)}{n \log(1 + r/n)}$

CHECKING BASIC CONCEPTS FOR SECTIONS 5.5 AND 5.6 (P. 457)

1. $2 \log x + 3 \log y - \frac{1}{3} \log z$
3. (a) $\dfrac{\log(29/5)}{\log(1.4)} \approx 5.224$ **(b)** $\frac{1}{3}$
5. (a) The temperature of the object is about 80°F.
(b) 17 min

SECTION 5.7 (PP. 462–464)

1. Logarithmic   **3.** Exponential
5. Exponential: $f(x) = 1.2(1.7)^x$
7. Logarithmic: $f(x) = 1.088 + 2.937 \ln x$
9. (a) [25, 75, 5] by [−100, 2100, 200]

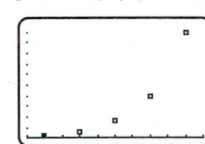

(b) $f(x) = 1.568(1.109)^x$ **(c)** About 6164 deaths per 100,000
11. (a) $a \approx 1.4734$, $b \approx 0.99986$, or
$f(x) = 1.4734(0.99986)^x$ **(b)** Approximately 0.55 kg/m^3
13. (a) The data are not linear.

[−2, 32, 5] by [0, 80, 10]

(b) $a \approx 12.42$, $b \approx 1.066$, or $f(x) = 12.42(1.066)^x$
(c) $f(39) \approx 150$. Chemical fertilizer use increased but at a slower rate than predicted by f.
15. (a) $f(x) = \dfrac{4.9955}{1 + 49.7081e^{-0.6998x}}$ **(b)** About 5 thousand per acre
17. (a) $f(25) \approx 0.065$ and $f(65) \approx 0.82$. For people age 25, 6.5% have some CHD, whereas for people age 65, 82% have some CHD.   **(b)** 48 years (approximately)

5.7 EXTENDED AND DISCOVERY EXERCISE (P. 464)

1. (a) $f(x) \approx 0.128(0.777)^x$ **(b)** After about 4.6 min

CHECKING BASIC CONCEPTS FOR SECTION 5.7 (P. 464)

1. Exponential: $f(x) \approx 0.5(1.2)^x$

3. Logistic: $f(x) \approx \dfrac{4.5}{1 + 277e^{-1.4x}}$

CHAPTER 5 REVIEW EXERCISES (PP. 468–473)

1. (a) 8   **(b)** 0   **(c)** -6 **(d)** Undefined

3. (a) 7   **(b)** 1   **(c)** 0   **(d)** $-\dfrac{9}{2}$

5. (a) 2   **(b)** 1   **7. (a)** $\sqrt{6}$ **(b)** 12

9. $(f \circ g)(x) = \left(\dfrac{1}{x}\right)^3 - \left(\dfrac{1}{x}\right)^2 + 3\left(\dfrac{1}{x}\right) - 2$;

$D = \{x \mid x \neq 0\}$

11. $(f \circ g)(x) = x$; all real numbers

13. $f(x) = x^2 + 3, g(x) = \sqrt{x}$; answers may vary.

15. Subtract 6 from x and then multiply the result by 10; $\dfrac{x}{10} + 6$ and $10(x - 6)$.

17. f is one-to-one.   **19.** f is not one-to-one; no.

21. f is not one-to-one

23. (a) $f(1) \approx \$1050$ **(b)** $f^{-1}(1200) \approx 4$ years; f^{-1} computes the number of years it takes to accumulate x dollars.

25. $f^{-1}(x) = \dfrac{x + 5}{3}$ **27.** $f^{-1}(x) = \dfrac{7x}{3 - x}$

29. $D = \{x \mid x \geq 4\}; f^{-1}(x) = \sqrt{\dfrac{x - 3}{2}} + 4, x \geq 3$
Answers may vary.

31. 1   **33.** e^{-x} **35.** $C = 3, a = 2$

37.   **39.**

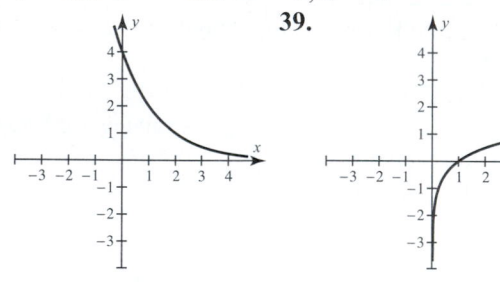

D = all real numbers $D = \{x \mid x > 0\}$

41. $C = 2, a = 2$ **43.** \$1562.71   **45.** 11.023

47. $\ln 19 \approx 2.9444$ **49.** 3   **51.** -21 **53.** 2

55. 1   **57.** $\dfrac{\log 18}{\log 3} \approx 2.631$ **59.** $\log 125 \approx 2.097$

61. $10 \ln 5.2 \approx 16.49$ **63.** $-\dfrac{1}{\log 5} \approx -1.431$

65. $10 + \dfrac{\log (29/3)}{\log (0.78)} \approx 0.869$ **67.** $f(x) = 1.5(2)^x$

69. $10^{1.5} \approx 31.62$ **71.** $e^{3.4} \approx 29.96$ **73.** 126

75. $\log 30x$ **77.** $\ln 4 - 2 \ln x$ **79.** $10^{1/4} \approx 1.778$

81. $\dfrac{100,000}{3} \approx 33,333$ **83.** 1

85. The x-intercept is b. If $(0, b)$ is on the graph of f, then $(b, 0)$ is on the graph of f^{-1}.

87. (a) $11,022/\text{ml}$ **(b)** About 1.86 hr

89. (a) $T_0 = 20, D = 80, a \approx 0.23$ **(b)** About 28.8°C
(c) About 1.4 hr

91. (a) $(g \circ f)(32) = 1$ represents the number of qt in 32 fluid oz.   **(b)** $f^{-1}(1) = 16$ represents the number of fluid oz in 1 pt.   **(c)** $(f^{-1} \circ g^{-1})(1) = 32$ represents the number of fluid oz in 1 qt.

93. (a) $W(1) \approx 19.5$ **(b)** $x \approx 2.74$. The fish weighs 50 milligrams at about 3 weeks.

(c) $x = -\dfrac{1}{0.24} \ln\left(\dfrac{1 - (50/175.6)^{1/3}}{0.66}\right) \approx 2.74$

95. After about 3.8 min

97. (a)

x	0	15	30	45	60	75	90	125
$g(x)$	7	21	57	111	136	158	164	178

(b) $g(x) = 261 - f(x)$
(c) y_1 models $g(x)$ better.
   $[-10, 140, 10]$ by $[-20, 200, 10]$

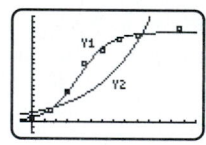

(d) $f(x) = 261 - \dfrac{171}{1 + 18.6e^{-0.0747x}}$

99. $a \approx 3.50, b \approx 0.74$, or $f(x) = 3.50(0.74)^x$
   $[0, 5, 1]$ by $[0, 3, 1]$

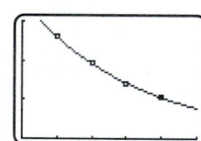

CHAPTER 5 EXTENDED AND DISCOVERY EXERCISES (PP. 472–473)

1. Answers will vary. Using power regression, the function $f(x) = 0.0002x^{1.5}$ seems to fit the data best.

3. (a) $C = 0.2; k \approx 0.0124$ **(b)** In 2018, the radiative forcing, R, could be 3 watts per square meter.

5. $\left(\dfrac{1}{\pi} \ln(640,320^3 + 744)\right)^2 \approx$

163.00000000000000000000000000000232 (there are 29 zeros after 163); it is not an integer.

CHAPTER 6: Trigonometric Functions

SECTION 6.1 (PP. 485–489)

1. (a) 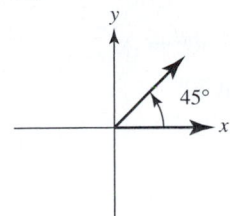   **(b)**

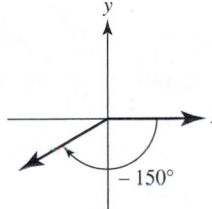

(c)   **(d)**

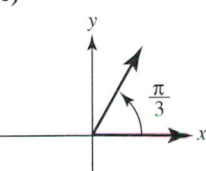   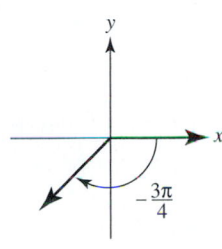

3. Answers may vary.   **5.**

 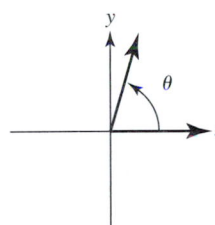   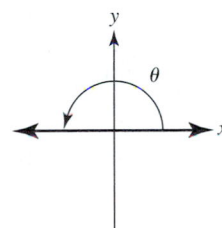

7. Answers may vary.   **9.** Answers may vary.

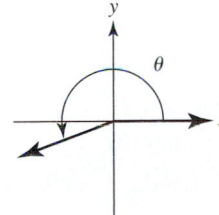   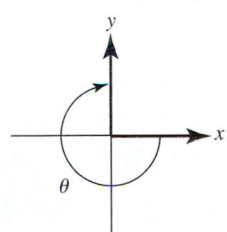

11. (a) $\frac{1}{4}$ **(b)** $\frac{1}{12}$ **(c)** $\frac{1}{6}$ **(d)** $\frac{1}{8}$

13. $510°, -210°$; answers may vary.

15. $288°, -432°$; answers may vary.

17. $\frac{5\pi}{2}, -\frac{3\pi}{2}$; answers may vary.

19. $\frac{9\pi}{5}, -\frac{11\pi}{5}$; answers may vary.

21. $\alpha = 34.1°, \beta = 124.1°$

23. $\alpha = 4°36'15'', \beta = 94°36'15''$

25. $\alpha = 66°19'25'', \beta = 156°19'25''$ **27.** $125.25°$

29. $108.76°$ **31.** $125°18'$ **33.** $51°21'36''$

35. $\theta = 2$ radians or $\theta \approx 114.6°$

37. $\theta = 1.3$ radians or $\theta \approx 74.5°$

39. (a) $\frac{\pi}{4}$ **(b)** $\frac{3\pi}{4}$ **(c)** $-\frac{2\pi}{3}$ **(d)** $-\frac{7\pi}{6}$

41. (a) $\frac{37\pi}{180}$ **(b)** 2.15 **(c)** -1.61 **(d)** 4.02

43. (a) $30°$ **(b)** $12°$ **(c)** $-300°$ **(d)** $-210°$

45. (a) $45°$ **(b)** $25.71°$ **(c)** $177.62°$ **(d)** $-143.24°$

47. $s = \frac{2\pi}{3}$ in.   **49.** $\theta = \frac{12}{5}$ radians   **51.** $r = \frac{5}{\pi}$ ft

53. $\frac{\pi}{4}$ m   **55.** π ft   **57.** $\frac{7\pi}{360}$ mi

59. 2π in., $\frac{2\pi}{15}$ in./min.   **61.** 10π in., $\frac{2\pi}{15}$ in./min

63. 16.25 ft/sec, about 11.1 mi/hr   **65.** 1.5π in.^{2}

67. 4.5π in.^{2} **69.** $\frac{17,161\pi}{3000} \approx 5.72\pi$ cm^2

71. $\frac{3\pi}{16} \approx 0.59$ ft^2 **73.** 240π in.^{2} **75.** 4292π cm^2

77. $\pm 0.036''$ **79. (a)** $1000\pi \approx 3141.6$ radians/min
(b) $15,000\pi \approx 47,123.89$ in./min or about 65.4 ft/sec
81. 810 mi   **83.** 14.5 in.

85. (a) 2.5 revolutions   **(b)** $\frac{65\pi}{6} \approx 34$ ft/sec

87. (a) $78,370$ mi/hr   **(b)** $66,630$ mi/hr
(c) $29,250$ mi/hr   **(d)** $12,160$ mi/hr
Planets farther from the sun have slower orbital velocities.
89. 137.2 m   **91.** 0.069 mi or 364 ft

93. $s = r\theta\left(\frac{\pi}{180°}\right)$, where θ is in degrees. The formula for radian measure is simpler.
95. (a) 388.8 m   **(b)** 881.8 m
97. Radius ≈ 3947 mi, circumference $\approx 24,800$ mi

SECTION 6.2 (PP. 499–503)

1.   **3**

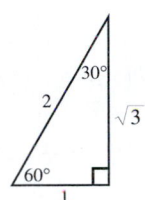   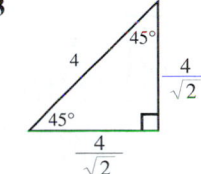

5. $\frac{\sqrt{3}}{2}$ **7.** $\frac{\sqrt{3}}{2}$ **9.** 2 **11.** 1 **13.** 1 **15.** $\frac{1}{\sqrt{2}}$

17. $\sin\theta = \frac{4}{5}, \cos\theta = \frac{3}{5}, \tan\theta = \frac{4}{3}$
$\csc\theta = \frac{5}{4}, \sec\theta = \frac{5}{3}, \cot\theta = \frac{3}{4}$

19. $\sin\theta = \frac{12}{13}, \cos\theta = \frac{5}{13}, \tan\theta = \frac{12}{5}$
$\csc\theta - \frac{13}{12}, \sec\theta - \frac{13}{5}, \cot\theta - \frac{5}{12}$

21. $a \approx 13.86, b = 8$ **23.** $b \approx 5.03, c \approx 7.83$

25. $a \approx 5.25, b \approx 6.04$ **27.** $a \approx 16.82, c \approx 23.28$

29. $a \approx 20.78$ **31.** $c \approx 168.98$

33. 3 **35.** $\frac{13}{12}$ **37.** $\frac{24}{7}$

39. $\sin 60° \approx 0.866, \cos 60° = 0.5, \tan 60° \approx 1.732,$
$\csc 60° \approx 1.155, \sec 60° = 2, \cot 60° \approx 0.577$

41. $\sin 25° \approx 0.423$, $\cos 25° \approx 0.906$, $\tan 25° \approx 0.466$,
$\csc 25° \approx 2.366$, $\sec 25° \approx 1.103$, $\cot 25° \approx 2.145$

43. $\sin 5°35' \approx 0.097$, $\cos 5°35' \approx 0.995$,
$\tan 5°35' \approx 0.098$, $\csc 5°35' \approx 10.278$,
$\sec 5°35' \approx 1.005$, $\cot 5°35' \approx 10.229$

45. $\sin 13°45'30'' \approx 0.238$, $\cos 13°45'30'' \approx 0.971$,
$\tan 13°45'30'' \approx 0.245$, $\csc 13°45'30'' \approx 4.205$,
$\sec 13°45'30'' \approx 1.030$, $\cot 13°45'30'' \approx 4.084$

47. $\sin 1.05° \approx 0.018$, $\cos 1.05° \approx 1.000$,
$\tan 1.05° \approx 0.018$, $\csc 1.05° \approx 54.570$,
$\sec 1.05° \approx 1.000$, $\cot 1.05° \approx 54.561$

49. $a = 12$, $b = 12\sqrt{3}$, $c = 12\sqrt{6}$, $d = 12\sqrt{3}$

51. $a = \dfrac{14\sqrt{3}}{3}$, $b = \dfrac{7\sqrt{3}}{3}$, $c = \dfrac{14\sqrt{3}}{3}$, $d = \dfrac{14\sqrt{6}}{3}$

53. **(a)** $\cos 20° \approx 0.9397$ **(b)** $\sin 50° \approx 0.7660$

55. **(a)** $\sec 41° \approx 1.3250$ **(b)** $\csc 27° \approx 2.2027$

57. $1500 \tan 37°30' \approx 1151$ ft

59. $100 \tan 35° \approx 70$ ft   **61.** About 7.8 ftt

63. Barnard's Star: 3.5×10^{13} mi, 5.9 light-years
Sirius: 5.1×10^{13} mi, 8.6 light-years
61 Cygni: 6.6×10^{13} mi, 11.1 light-years
Procyon: 6.7×10^{13} mi, 11.3 light-years

65. Min: 2.9×10^7 mi; max: 4.4×10^7 mi

67. 12,534 mi   **69.** 13,500 ft   **71.** $PQ \approx 194.5$ ft

73. **(a)** $d = \cot \dfrac{\theta}{2}$ **(b)** $\cot \left(\dfrac{1°45'15''}{2}\right) \approx 65.32$ m

75. **(a)** 704 ft   **(b)** 595 ft
(c) Increasing θ decreases r.

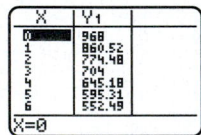

77. $d = 625\left(\dfrac{1}{\cos 54°} - 1\right) \approx 438$ ft   **79.** $A = \dfrac{\sqrt{3}}{4}s^2$

6.2 EXTENDED AND DISCOVERY EXERCISES (P. 503)

1. **(a)** $\sin \theta = \dfrac{a}{c} \Rightarrow \sin^2 \theta = \dfrac{a^2}{c^2}$

$\cos \theta = \dfrac{b}{c} \Rightarrow \cos^2 \theta = \dfrac{b^2}{c^2}$

$a^2 + b^2 = c^2$

Thus: $\sin^2 \theta + \cos^2 \theta$

$= \dfrac{a^2}{c^2} + \dfrac{b^2}{c^2} = \dfrac{a^2 + b^2}{c^2} = \dfrac{c^2}{c^2} = 1$

(b)

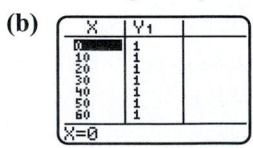

CHECKING BASIC CONCEPTS FOR SECTIONS 6.1 AND 6.2 (P. 503)

1. **(a)** $\dfrac{\pi}{4}$ **(b)** $\dfrac{5\pi}{12}$ **3.** $s = 2\pi$ in., $A = 12\pi$ in.^{2}

5. $\sin 60° = \dfrac{\sqrt{3}}{2}$, $\cos 60° = \dfrac{1}{2}$, $\tan 60° = \sqrt{3}$,
$\csc 60° = \dfrac{2}{\sqrt{3}}$, $\sec 60° = 2$, $\cot 60° = \dfrac{1}{\sqrt{3}}$

SECTION 6.3 (PP. 514–517)

1. **(a)** 13   **(b)** $\sin \theta = \dfrac{5}{13}$, $\cos \theta = \dfrac{12}{13}$

3. **(a)** 17   **(b)** $\sin \theta = \dfrac{8}{17}$, $\cos \theta = -\dfrac{15}{17}$

5. $\sin \theta = \dfrac{3}{5}$, $\cos \theta = \dfrac{4}{5}$

7. $\sin \theta = -\dfrac{2}{\sqrt{5}}$, $\cos \theta = \dfrac{1}{\sqrt{5}}$

9. $\sin \theta = -\dfrac{8}{17}$, $\cos \theta = -\dfrac{15}{17}$

11. $\sin 45° = \dfrac{1}{\sqrt{2}}$, $\cos 45° = \dfrac{1}{\sqrt{2}}$

13. $\sin(-30°) = -\dfrac{1}{2}$, $\cos(-30°) = \dfrac{\sqrt{3}}{2}$

15. $\sin \dfrac{\pi}{3} = \dfrac{\sqrt{3}}{2}$, $\cos \dfrac{\pi}{3} = \dfrac{1}{2}$

17. $\sin\left(-\dfrac{\pi}{2}\right) = -1$, $\cos\left(-\dfrac{\pi}{2}\right) = 0$

19. $\sin \dfrac{7\pi}{6} = -\dfrac{1}{2}$, $\cos \dfrac{7\pi}{6} = -\dfrac{\sqrt{3}}{2}$

21. $\sin \theta = \dfrac{2}{\sqrt{5}}$, $\cos \theta = \dfrac{1}{\sqrt{5}}$

23. $\sin \theta = -\dfrac{3}{\sqrt{10}}$, $\cos \theta = \dfrac{1}{\sqrt{10}}$

25. $\sin \theta = \dfrac{2}{\sqrt{13}}$, $\cos \theta = -\dfrac{3}{\sqrt{13}}$

27. $\sin 93.2° \approx 0.9984$, $\cos 93.2° \approx -0.0558$

29. $\sin 123°50' \approx 0.8307$, $\cos 123°50' \approx -0.5568$

31. $\sin(-4) \approx 0.7568$, $\cos(-4) \approx -0.6536$

33. $\sin \dfrac{11\pi}{7} \approx -0.9749$, $\cos \dfrac{11\pi}{7} \approx 0.2225$

35. $\sin \theta = \dfrac{3}{5}$, $\cos \theta = \dfrac{4}{5}$ **37.** $\sin \theta = -\dfrac{5}{13}$, $\cos \theta = \dfrac{12}{13}$

39. $\sin \dfrac{\pi}{2} = 1$, $\cos \dfrac{\pi}{2} = 0$

41. $\sin \dfrac{7\pi}{6} = -\dfrac{1}{2}$, $\cos \dfrac{7\pi}{6} = -\dfrac{\sqrt{3}}{2}$

43. $\sin\left(-\dfrac{3\pi}{4}\right) = -\dfrac{1}{\sqrt{2}}$, $\cos\left(-\dfrac{3\pi}{4}\right) = -\dfrac{1}{\sqrt{2}}$

45. $\sin \dfrac{5\pi}{2} = 1$, $\cos \dfrac{5\pi}{2} = 0$

47. $\sin\left(-\dfrac{\pi}{3}\right) = -\dfrac{\sqrt{3}}{2}$, $\cos\left(-\dfrac{\pi}{3}\right) = \dfrac{1}{2}$

49. $\sin 0 = 0$ **51.** $\sin \dfrac{\pi}{2} = 1$ **53.** $\sin\left(-\dfrac{\pi}{6}\right) = -\dfrac{1}{2}$

55. $\cos 0 = 1$ **57.** $\cos \dfrac{\pi}{3} = \dfrac{1}{2}$ **59.** $\cos\left(-\dfrac{3\pi}{2}\right) = 0$

61. No, their graphical and numerical representations are different.

$[-2\pi, 2\pi, \pi/2]$ by $[-4, 4, 1]$

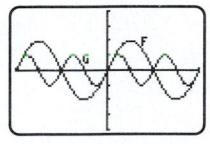 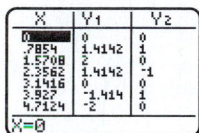

63. **(a)** $R = \{y \mid -3 \le y \le 3\}$

$[-2\pi, 2\pi, \pi/2]$ by $[-4, 4, 1]$

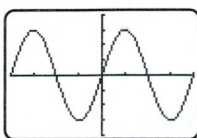

(b) $R = \{y \mid -1 \le y \le 1\}$

$[-2\pi, 2\pi, \pi/2]$ by $[-4, 4, 1]$

65. **(a)** $R = \{y \mid -1 \le y \le 3\}$

$[-2\pi, 2\pi, \pi/2]$ by $[-4, 4, 1]$

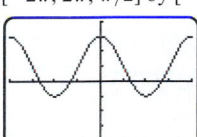

(b) $R = \{y \mid -2 \le y \le 0\}$

$[-2\pi, 2\pi, \pi/2]$ by $[-4, 4, 1]$

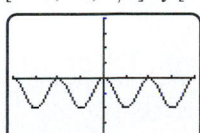

67. **(a)** 3%   **(b)** $25{,}000\left(\dfrac{3}{\sqrt{10{,}009}}\right) \approx 750$ lb

69. **(a)** V is sinusoidal and varies between -310 volts and 310 volts.
(b) $V(1/120) = 0$. After 1/120 second, the voltage is 0.
(c) $310/\sqrt{2} \approx 219$ volts

71.
(a) $[0, 13, 1]$ by $[3, 8, 1]$ **(b)** $[0, 13, 1]$ by $[3, 8, 1]$

From the close data fit shown, we may conclude that flying squirrels become active near sunset.

73. **(a)** 201 ft   **(b)** 258 ft
(c) It is easier to stop going uphill ($\theta > 0$) than downhill ($\theta < 0$).
75. **(a)** $[0, 1/100, 1/880]$ by $[-1.5, 1.5, 0.5]$

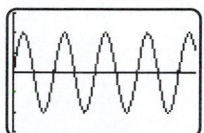

(b) $\dfrac{1}{440} \approx 0.00227$ sec   **(c)** 440 cycles per second

6.3 EXTENDED AND DISCOVERY EXERCISES (P. 517)

1. $[-2\pi, 2\pi, \pi/2]$ by $[-2, 2, 1]$

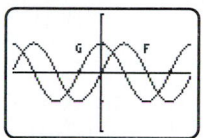

The translated graph and the graph of g are identical.

SECTION 6.4 (PP. 528–531)

1. $\sin \theta = \dfrac{12}{13}$, $\cos \theta = \dfrac{5}{13}$, $\tan \theta = \dfrac{12}{5}$
$\csc \theta = \dfrac{13}{12}$, $\sec \theta = \dfrac{13}{5}$, $\cot \theta = \dfrac{5}{12}$

3. $\sin \theta = -\dfrac{24}{25}$, $\cos \theta = \dfrac{7}{25}$, $\tan \theta = -\dfrac{24}{7}$, $\csc \theta = -\dfrac{25}{24}$, $\sec \theta = \dfrac{25}{7}$, $\cot \theta = -\dfrac{7}{24}$

5. $\sin \theta = 0$, $\cos \theta = 1$, $\tan \theta = 0$
$\csc \theta$ is undefined, $\sec \theta = 1$, $\cot \theta$ is undefined

7. $\tan \theta = \dfrac{3}{4}$, $\cot \theta = \dfrac{4}{3}$, $\csc \theta = \dfrac{5}{3}$, $\sec \theta = \dfrac{5}{4}$

9. $\sin \theta = -\dfrac{15}{17}$, $\cos \theta = -\dfrac{8}{17}$, $\tan \theta = \dfrac{15}{8}$, $\cot \theta = \dfrac{8}{15}$

11. $\sin \theta = \dfrac{5}{13}$, $\csc \theta = \dfrac{13}{5}$, $\cot \theta = \dfrac{12}{5}$, $\sec \theta = \dfrac{13}{12}$

13. $\cos \theta = \dfrac{4}{5}$, $\tan \theta = -\dfrac{3}{4}$, $\csc \theta = -\dfrac{5}{3}$, $\sec \theta = \dfrac{5}{4}$, $\cot \theta = -\dfrac{4}{3}$

15. $\sin \theta = -\dfrac{3}{5}$, $\tan \theta = \dfrac{3}{4}$, $\csc \theta = -\dfrac{5}{3}$, $\sec \theta = -\dfrac{5}{4}$, $\cot \theta = \dfrac{4}{3}$

17. -1

19. **(a)** $(0, 1)$; Answers may vary
(b) 1; Answers may vary
(c) $\tan \dfrac{\pi}{2}$ is undefined, $\cot \dfrac{\pi}{2} = 0$, $\sec \dfrac{\pi}{2}$ is undefined, $\csc \dfrac{\pi}{2} = 1$

21. **(a)** $(-1, -1)$; Answers may vary
(b) $\sqrt{2}$; Answers may vary
(c) $\tan(-135°) = 1$, $\cot(-135°) = 1$, $\sec(-135°) = -\sqrt{2}$, $\csc(-135°) = -\sqrt{2}$
23. $\sin \theta = \dfrac{1}{\sqrt{2}}$, $\cos \theta = \dfrac{1}{\sqrt{2}}$, $\tan \theta = 1$, $\csc \theta = \sqrt{2}$, $\sec \theta = \sqrt{2}$, $\cot \theta = 1$

25. $\sin \theta = -\frac{12}{13}$, $\cos \theta = \frac{5}{13}$, $\tan \theta = -\frac{12}{5}$, $\csc \theta = -\frac{13}{12}$,
$\sec \theta = \frac{13}{5}$, $\cot \theta = -\frac{5}{12}$

27. $\sin 90° = 1$, $\cos 90° = 0$, $\tan 90°$ is undefined
$\csc 90° = 1$, $\sec 90°$ is undefined, $\cot 90° = 0$

29. $\sin (-45°) = -\frac{1}{\sqrt{2}}$, $\cos (-45°) = \frac{1}{\sqrt{2}}$,
$\tan (-45°) = -1$, $\csc (-45°) = -\sqrt{2}$,
$\sec (-45°) = \sqrt{2}$, $\cot (-45°) = -1$

31. $\sin \pi = 0$, $\cos \pi = -1$, $\tan \pi = 0$, $\csc \pi$ is undefined,
$\sec \pi = -1$, $\cot \pi$ is undefined

33. $\sin \left(-\frac{\pi}{3}\right) = -\frac{\sqrt{3}}{2}$, $\cos \left(-\frac{\pi}{3}\right) = \frac{1}{2}$,
$\tan \left(-\frac{\pi}{3}\right) = -\sqrt{3}$, $\csc \left(-\frac{\pi}{3}\right) = -\frac{2}{\sqrt{3}}$,
$\sec \left(-\frac{\pi}{3}\right) = 2$, $\cot \left(-\frac{\pi}{3}\right) = -\frac{1}{\sqrt{3}}$

35. $\sin \left(-\frac{\pi}{2}\right) = -1$, $\cos \left(-\frac{\pi}{2}\right) = 0$,
$\tan \left(-\frac{\pi}{2}\right)$ is undefined, $\csc \left(-\frac{\pi}{2}\right) = -1$,
$\sec \left(-\frac{\pi}{2}\right)$ is undefined, $\cot \left(-\frac{\pi}{2}\right) = 0$

37. $\sin 360° = 0$, $\cos 360° = 1$, $\tan 360° = 0$,
$\csc 360°$ is undefined, $\sec 360° = 1$, $\cot 360°$ is undefined

39. $\sin \frac{\pi}{6} = \frac{1}{2}$, $\cos \frac{\pi}{6} = \frac{\sqrt{3}}{2}$, $\tan \frac{\pi}{6} = \frac{1}{\sqrt{3}}$, $\csc \frac{\pi}{6} = 2$,
$\sec \frac{\pi}{6} = \frac{2}{\sqrt{3}}$, $\cot \frac{\pi}{6} = \sqrt{3}$

41. $\sin \theta = \frac{4}{\sqrt{17}}$, $\cos \theta = -\frac{1}{\sqrt{17}}$, $\tan \theta = -4$
$\csc \theta = \frac{\sqrt{17}}{4}$, $\sec \theta = -\sqrt{17}$, $\cot \theta = -\frac{1}{4}$
The slope of the line equals $\tan \theta$.

43. $\sin \theta = -\frac{6}{\sqrt{37}}$, $\cos \theta = -\frac{1}{\sqrt{37}}$, $\tan \theta = 6$
$\csc \theta = -\frac{\sqrt{37}}{6}$, $\sec \theta = -\sqrt{37}$, $\cot \theta = \frac{1}{6}$
The slope of the line equals $\tan \theta$.

45. **(a)** 0.9984   **(b)** 1.0016
47. **(a)** 1.4045   **(b)** 0.7120
49. **(a)** -0.8637 **(b)** -1.1578
51. **(a)** 0.2225   **(b)** 4.4940
53. Origin symmetry   **55.** Origin symmetry

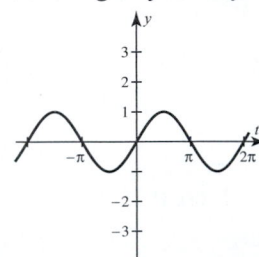   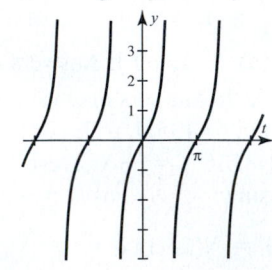

57. y-axis symmetry

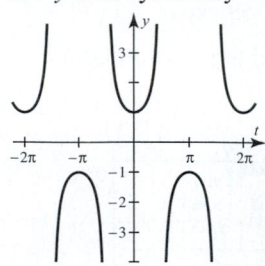

59. $D =$ all real numbers, $R = \{y \mid -1 \le y \le 1\}$,
period $= 2\pi$

61. $D = \left\{t \mid t \ne \pm\frac{\pi}{2}, \pm\frac{3\pi}{2}, \pm\frac{5\pi}{2}, \ldots\right\}$,
$R =$ all real numbers, period $= \pi$

63. $D = \left\{t \mid t \ne \pm\frac{\pi}{2}, \pm\frac{3\pi}{2}, \pm\frac{5\pi}{2}, \ldots\right\}$,
$R = \{y \mid |y| \ge 1\}$, period $= 2\pi$

65. $\tan (-90°)$ is undefined, $\tan (-45°) = -1$,
$\tan (0°) = 0$, $\tan (45°) = 1$, $\tan (90°)$ is undefined
$[-352.5°, 352.5°, 90°]$ by $[-4, 4, 1]$

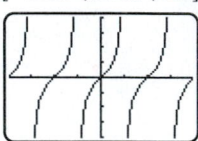

67. $\csc (-90°) = -1$, $\csc (-45°) = -\sqrt{2}$,
$\csc (0°)$ is undefined, $\csc (45°) = \sqrt{2}$, $\csc (90°) = 1$
$[-352.5°, 352.5°, 90°]$ by $[-4, 4, 1]$

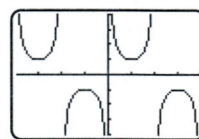

69. The zeros of $\sin t$ correspond to the asymptotes on the
graph of $\csc t$. If $\sin t = \pm 1$, then $\csc t = \pm 1$.
$[-2\pi, 2\pi, \pi/2]$ by $[-4, 4, 1]$

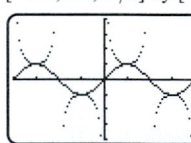

71. **(a)** $t = -\pi, 0, \pi, 2\pi$ **(b)** $t = \pm\frac{\pi}{2}, \pm\frac{3\pi}{2}$
(c) $t = -\pi, 0, \pi, 2\pi$ **(d)** No solutions
(e) No solutions   **(f)** $t = \pm\frac{\pi}{2}, \pm\frac{3\pi}{2}$

73. (a)

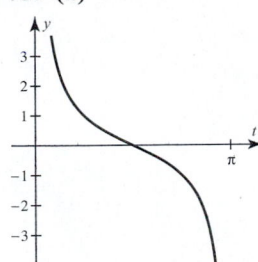

(b) Just after sunrise ($t = 0$) the shadow is very long. As the elevation of the sun increases, the shadow decreases in length until it is 0 when $t = \pi/2$. In the afternoon the shadow increases in length in the opposite direction until sunset ($t = \pi$).

75. About 17.95″   **77.** About 41%

79. [0, 20000, 5000] by [0, 4000, 1000]

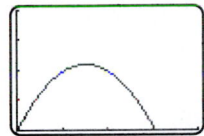

(a) Approximately (7612, 2197)   **(b)** About 15,223 ft

6.4 EXTENDED AND DISCOVERY EXERCISE (P. 531)

1. Using the small right triangle,

$$\sin \theta = \frac{\text{Opp.}}{\text{Hyp.}} = \frac{\text{Opp.}}{1} = \text{Opposite side. Similarly,}$$

$$\cos \theta = \frac{\text{Adj.}}{\text{Hyp.}} = \frac{\text{Adj.}}{1} = \text{Adjacent side.}$$

Then using the large right triangle,

$$\tan \theta = \frac{\text{Opp.}}{\text{Adj.}} = \frac{\text{Opp.}}{1} = \text{Opposite side. Similarly,}$$

$$\sec \theta = \frac{\text{Hyp.}}{\text{Adj.}} = \frac{\text{Hyp.}}{1} = \text{Hypotenuse.}$$

CHECKING BASIC CONCEPTS FOR SECTIONS 6.3 AND 6.4 (P. 531)

1. $\sin \theta = \frac{6}{\sqrt{85}}$, $\cos \theta = -\frac{7}{\sqrt{85}}$, $\tan \theta = -\frac{6}{7}$,

$\csc \theta = \frac{\sqrt{85}}{6}$, $\sec \theta = -\frac{\sqrt{85}}{7}$, $\cot \theta = -\frac{7}{6}$

3.

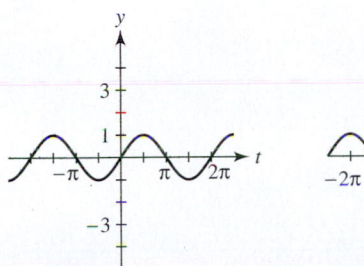
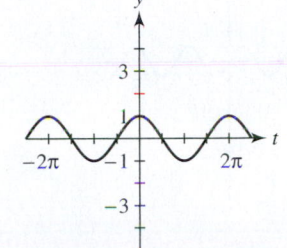

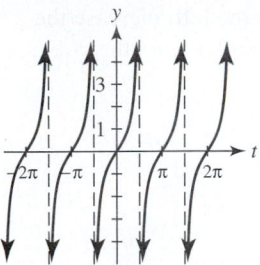

SECTION 6.5 (PP. 544–550)

1. Period: 4π; amplitude: 3

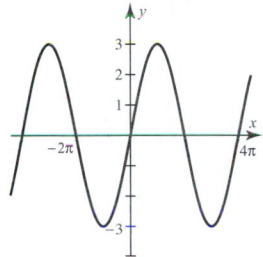

3. Period: $\frac{2\pi}{3}$; amplitude: 2

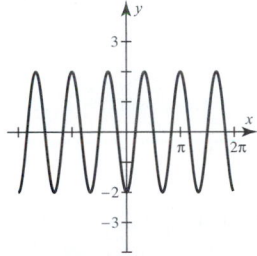

5. Period: 2; amplitude: $\frac{1}{2}$

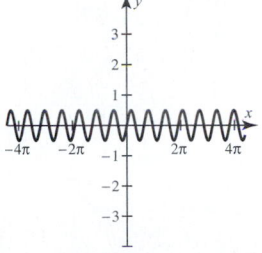

7. Shorten the period of the sine graph to π, increase the amplitude to 3, and shift the graph downward 1 unit.

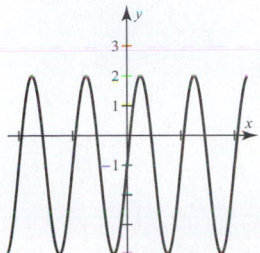

9. Shift the cosine graph $\frac{\pi}{2}$ units to the left, increase the amplitude to 2, and reflect the graph across the *x*-axis.

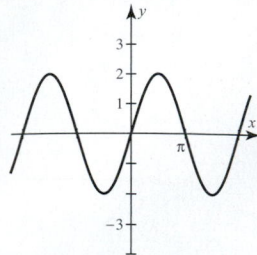

11. Shift the cosine graph 1 unit to the right, shorten the period to 2, and decrease the amplitude to $\frac{1}{2}$.

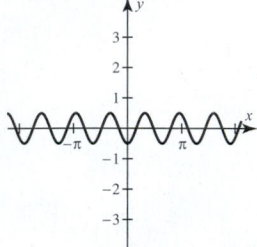

13. Amplitude: 3; period: $\frac{\pi}{2}$; phase shift: $\frac{\pi}{4}$; vertical shift: -4
15. Amplitude: 4; period: 4; phase shift: 1; vertical shift: 6
17. Amplitude: $\frac{2}{3}$; period: $\frac{\pi}{3}$; phase shift: $-\frac{\pi}{2}$; vertical shift: $-\frac{5}{2}$
19. $a = 3, b = 2$ **21.** $a = -2, b = \frac{1}{2}$
23. Amplitude: 3; period: 4π; phase shift: 0
25. Amplitude: 4; period: 2; phase shift: 1
27. c.   **29.** d.   **31.** a.   **33.** $y = 3 \sin\left(\frac{1}{2}x\right)$
35. $y = 4 \sin\left(\pi(x - 1)\right)$
37. Amplitude: 2; period: 2π; phase shift: 0

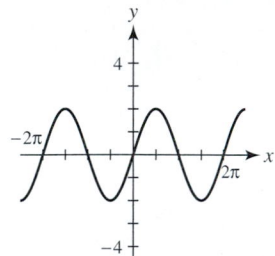

39. Amplitude: 1; period: 4π; phase shift: 0

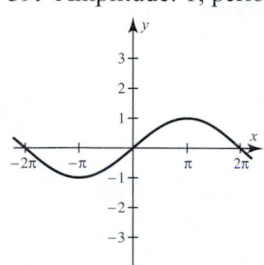

41. Amplitude: 1; period: 2π; phase shift: 0

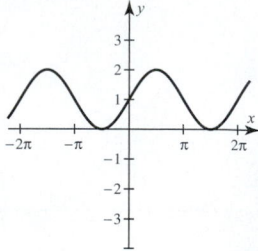

43. Amplitude: 1; period: 2; phase shift: 0

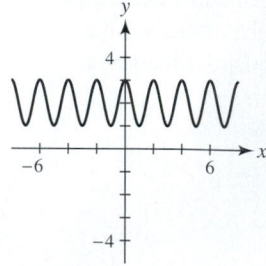

45. Amplitude: 1; period: π; phase shift: $-\pi$

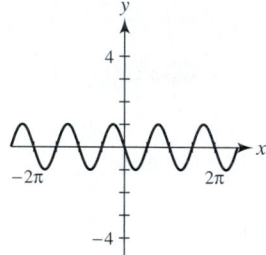

47. Amplitude: 1; period: 2π; phase shift: $\frac{\pi}{2}$

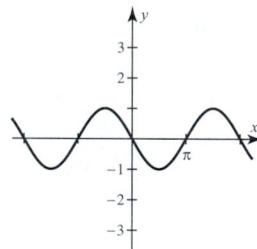

49. Amplitude: $\frac{1}{2}$; period: π; phase shift: 0

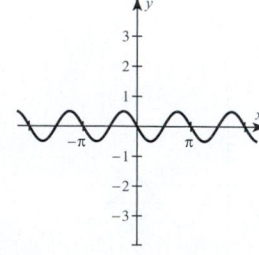

51. Amplitude: 2; period: π; phase shift: $-\frac{\pi}{4}$

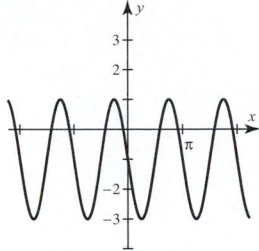

53. Amplitude: 1; period: π; phase shift: $\frac{\pi}{2}$

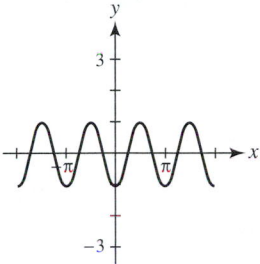

55. Amplitude $= 2$; period $= \pi$; phase shift $= 0$
$[-2\pi, 2\pi, \pi/2]$ by $[-4, 4, 1]$

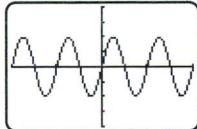

57. Amplitude $= \frac{1}{2}$; period $= \frac{2\pi}{3}$; phase shift $= -\frac{\pi}{3}$
$[-2\pi, 2\pi, \pi/2]$ by $[-4, 4, 1]$

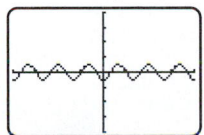

59. Amplitude $= 1$; period $= \frac{\pi}{2}$; phase shift $= 0$
$[-2\pi, 2\pi, \pi/2]$ by $[-4, 4, 1]$

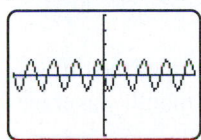

61. Amplitude $= 1$; period $= 1$; phase shift $= 0$
$[-2\pi, 2\pi, \pi/2]$ by $[-4, 4, 1]$

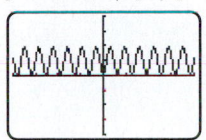

63. Amplitude $= 2$; period $= 1$; phase shift $= -\frac{1}{8}$
$[-2\pi, 2\pi, \pi/2]$ by $[-4, 4, 1]$

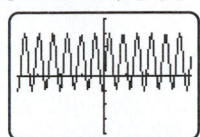

65. Period $= \frac{\pi}{2}$; phase shift $= 0$

Asymptotes: $x = \pm\frac{\pi}{4}, \pm\frac{3\pi}{4}, \pm\frac{5\pi}{4}, \pm\frac{7\pi}{4}$

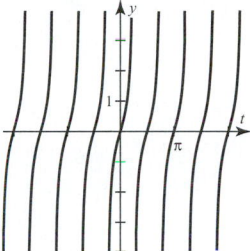

67. Period $= \pi$; phase shift $= \frac{\pi}{2}$
Asymptotes: $x = 0, \pm\pi, \pm2\pi$

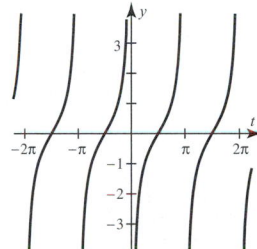

69. Period $= \frac{\pi}{2}$; phase shift $= 0$

Asymptotes: $x = 0, \pm\frac{\pi}{2}, \pm\pi, \pm\frac{3\pi}{2}, \pm2\pi$

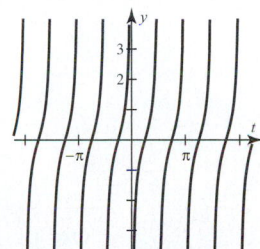

71. Period $= \dfrac{\pi}{2}$; phase shift $= \dfrac{\pi}{4}$

Asymptotes: $x = \pm\dfrac{\pi}{4}, \pm\dfrac{3\pi}{4}, \pm\dfrac{5\pi}{4}, \pm\dfrac{7\pi}{4}$

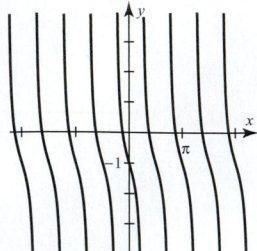

73. Period $= 4\pi$; phase shift $= 0$
Asymptotes: $x = \pm\pi$

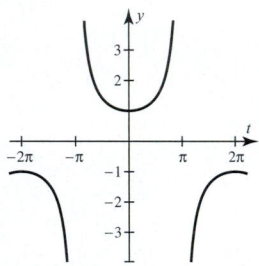

75. Period $= 2\pi$; phase shift $= \pi$
Asymptotes: $x = 0, \pm\pi, \pm 2\pi$

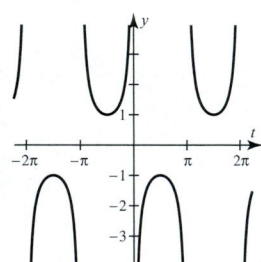

77. Period $= 6\pi$; phase shift $= \dfrac{\pi}{6}$

Asymptotes: $x = -\dfrac{4\pi}{3}, \dfrac{5\pi}{3}$

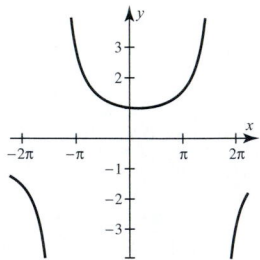

79. c.   **81.** b.
83. **(a)** $40°F, -40°F$ **(b)** Amplitude $= 40$, period $= 12$
(c) The months when the average temperature is $0°F$

85. **(a)** Amplitude $= 34$
Period $= 12$
Phase shift $= 4.3$
[0, 25, 2] by [−50, 50, 10]

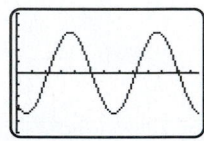

(b) $f(5) \approx 12.2°F$
$f(12) \approx -26.4°F$
(c) About $0°F$
87. **(a)** [0, 25, 2] by [0, 80, 10]

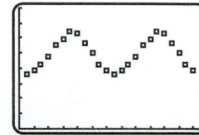

(b) $f(x) = 14\sin\left(\dfrac{\pi}{6}(x-4)\right) + 50$

(c) [0, 25, 2] by [0, 80, 10]

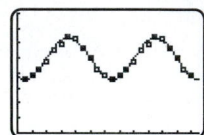

89. **(a)** $f(x) = 17\cos\left(\dfrac{\pi}{6}(x-7)\right) + 75$

(b) Yes. The period of the graph is 12 so, for example, the phase shift could be $c = 7 + 12 = 19$ or
$c = 7 - 12 = -5$.
91. About 1300 ft
93. **(a)** About 18.5 hr; June 21
(b) About 6 hr; December 21
(c) The amplitude represents half the difference in daylight between the longest and shortest day. The period represents one year. Answers may vary.
95. **(a)** Max ≈ 8 in.; min ≈ 0.5 in.
(b) Amp ≈ 3.75. The amplitude represents half the difference between the maximum and minimum monthly average precipitations.
(c) Yes. An x-intercept would correspond to no precipitation.
(d) About 51 in. (Answers may vary.)
97. **(a)** $a = 18, b = \dfrac{\pi}{6.2}, c = 9.8, d = 0$;

$f(x) = 18\cos\left(\dfrac{\pi}{6.2}(x-9.8)\right)$

(b) [0, 24, 4] by [−20, 20, 5]

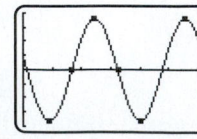

99. (a) Max ≈ 87°F in July; min ≈ 62°F in January or late December
(b) Increase

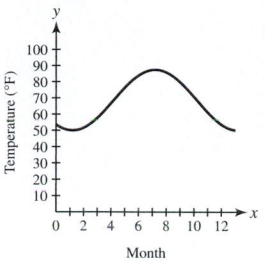

101. (a) [0, 25, 2] by [60, 90, 5]

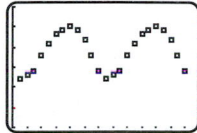

(b) $y = 6.5 \sin \left(\dfrac{\pi}{6}(x - 4) \right) + 78.5$

103. (a) The general trend in carbon dioxide levels has been increasing. During the year there are seasonal variations.
[20, 35, 5] by [320, 370, 10]

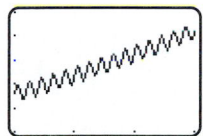

(b) The quadratic function models the general trend in carbon dioxide levels, while the sine function models the seasonal fluctuations in carbon dioxide levels.
105. (a) $s(t) = 2 \cos (4\pi t)$
(b) $s(1) = 2$. The weight is neither moving upward nor downward. At $t = 1$ the motion of the weight is changing from up to down.
107. (a) $s(t) = -3 \cos (2.5\pi t)$
(b) $s(1) = 0$. The weight is moving upward.
109. $Y_1 = 0.21 \cos (55\pi X)$
[0, 0.05, 0.01] by [−0.3, 0.3, 0.1]

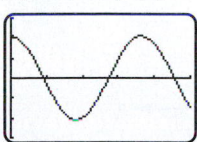

111. $Y_1 = 0.14 \cos (110\pi X)$
[0, 0.05, 0.01] by [−0.3, 0.3, 0.1]

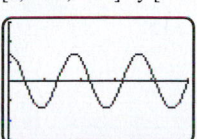

113. $y \approx 13.2 \sin (0.524x - 2.18) + 49.7$
[0, 25, 2] by [30, 80, 10]

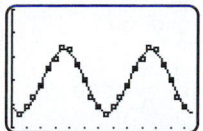

115. $y \approx 16.9 \sin (0.522x - 2.09) + 75.4$
[0, 25, 2] by [50, 100, 10]

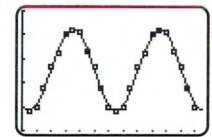

SECTION 6.6 (PP. 562–566)
1. one-to-one   **3.** π **5.** 0   **7.** $f^{-1}(x) = \dfrac{x}{3}$
9. $f^{-1}(x) = x^3$ **11.** $f^{-1}(x) = \sqrt{x} + 1$
13. *Verbal:* Subtract 6 from x.
Numerical:

x	0	1	2	3	4	5	6
$f^{-1}(x)$	−6	−5	−4	−3	−2	−1	0

Graphical:

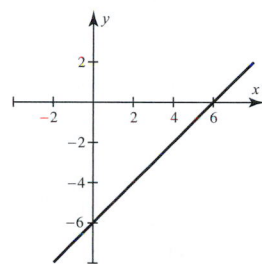

Symbolic: $f^{-1}(x) = x - 6$
15. *Verbal:* Add 1 to x and divide the result by 2.
Numerical:

x	0	1	2	3	4	5	6
$f^{-1}(x)$	0.5	1	1.5	2	2.5	3	3.5

Graphical:

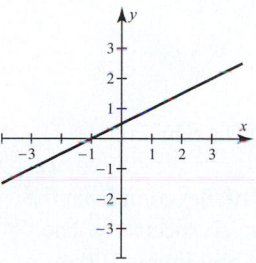

Symbolic: $f^{-1}(x) = \dfrac{x + 1}{2}$
17. [−2, 2]   **19.** $\dfrac{\pi}{2}$ **21.** $-\dfrac{\pi}{4}$ **23.** $\dfrac{2\pi}{3}$

25. (a) $\frac{\pi}{2}$ or 90° **(b)** 0 or 0° **(c)** $-\frac{\pi}{3}$ or −60°

27. (a) $\frac{\pi}{2}$ or 90° **(b)** π or 180° **(c)** $\frac{\pi}{3}$ or 60°

29. (a) $\frac{\pi}{4}$ or 45° **(b)** $-\frac{\pi}{4}$ or −45° **(c)** $\frac{\pi}{3}$ or 60°

31. (a) Undefined **(b)** 1.47 or 84.3°
(c) 2.42 or 138.6°

33. (a) β **(b)** α **(c)** β

35. 1 **37.** $\frac{3\pi}{4}$ **39.** −3

41. (a) 90° **(b)** 90° **(c)** 90°
$\sin^{-1} x + \cos^{-1} x = 90°$ whenever $-1 \le x \le 1$.

43. $\alpha = \tan^{-1} \frac{7}{24} \approx 16.3°$, $\beta = \tan^{-1} \frac{24}{7} \approx 73.7°$, $c = 25$

45. $\alpha = \sin^{-1} \frac{6}{10} \approx 36.9°$, $\beta = \cos^{-1} \frac{6}{10} \approx 53.1°$, $b = 8$

47. $\beta = 35°$, $b = \frac{5}{\tan 55°} \approx 3.5$, $c = \frac{5}{\sin 55°} \approx 6.1$

49. *Verbal:* Determine the angle (or real number) θ such that $\sin \theta = 2x$ and $-\frac{\pi}{2} \le \theta \le \frac{\pi}{2}$.

Numerical:

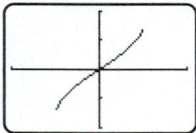

Graphical: $[-1, 1, 1]$ by $[-2, 2, 1]$

51. $\tan \theta = \frac{x}{\sqrt{1 - x^2}}$ **53.** $\cos \theta = \frac{1}{\sqrt{1 + x^2}}$

55. $\sqrt{1 - u^2}$ **57.** $\frac{\sqrt{1 - u^2}}{u}$ **59.** u **61.** 90°

63. 45° **65.** 90° **67.** 82.8° **69.** 80.5° **71.** 68.9°

73. −0.197 **75.** 1.102 **77.** −0.282

79. 1.369, 1.772 **81.** $\tan^{-1} \frac{50}{85} \approx 30.5°$

83. (a) $\tan^{-1} \frac{11}{5} \approx 65.6°$ **(b)** $\tan^{-1} \left(\frac{-3}{1}\right) \approx -71.6°$

85. $\tan^{-1} \frac{4}{10} \approx 21.8°$
87. (a) 60° **(b)** 120°
89. $\theta \approx 41.9°$
91. (a) 14.9 hr **(b)** 13.8 hr **(c)** 14.6 hr
93. $\theta_2 \approx 28.9°$
95. Let α and β represent the angles of elevation from the shrub to the shorter and taller buildings respectively. The distance from the shrub to the shorter building is $100 - x$, thus $\alpha = \arctan\left(\frac{75}{100 - x}\right)$. Similarly $\beta = \arctan\left(\frac{150}{x}\right)$. Because the angles α, θ, and β form a straight angle, $\theta = \pi - \alpha - \beta$. That is, $\theta = \pi - \arctan\left(\frac{75}{100 - x}\right) - \arctan\left(\frac{150}{x}\right)$.

6.6 EXTENDED AND DISCOVERY EXERCISE (P. 566)

1. The maximum value of θ is 38.7° when $x \approx 6.24$.
$[0, 50, 10]$ by $[0, 50, 10]$

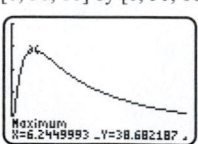

CHECKING BASIC CONCEPTS FOR SECTIONS 6.5 AND 6.6 (P. 566)

1. Amplitude = 3, period = $\frac{2\pi}{2} = \pi$, phase shift = $\frac{\pi}{4}$

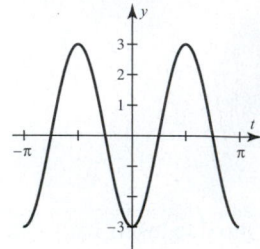

3. (a) $\sin^{-1} 0 = 0°$ **(b)** $\cos^{-1}(-1) = 180°$
(c) $\tan^{-1}(-1) = -45°$ **(d)** $\sin^{-1}\frac{1}{2} = 30°$
(e) $\tan^{-1}\sqrt{3} = 60°$
5. (a) $\sin^{-1} 0.55 \approx 0.582$ **(b)** $\cos^{-1}(-0.35) \approx 1.93$
(c) $\tan^{-1}(-2.9) \approx -1.24$

CHAPTER 6 REVIEW EXERCISES (PP. 570–574)

1. (a) **(b)**

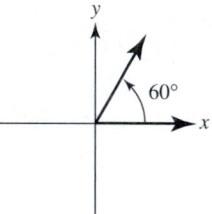 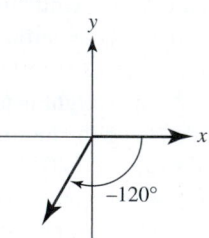

(c) **(d)**

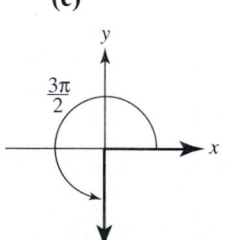 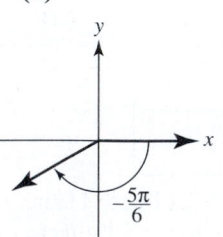

3. (a) 60° **(b)** 5° **(c)** −150° **(d)** −315°

5. 2π ft **7.** $\frac{1}{2}$ **9.** $\frac{1}{\sqrt{3}}$ **11.** $\sqrt{2}$

13. $\sin\theta = \dfrac{8}{\sqrt{145}}$, $\cos\theta = \dfrac{9}{\sqrt{145}}$, $\tan\theta = \dfrac{8}{9}$,

$\csc\theta = \dfrac{\sqrt{145}}{8}$, $\sec\theta = \dfrac{\sqrt{145}}{9}$, $\cot\theta = \dfrac{9}{8}$

15. 3

17. $\sin 25° \approx 0.423$, $\cos 25° \approx 0.906$, $\tan 25° \approx 0.466$, $\csc 25° \approx 2.366$, $\sec 25° \approx 1.103$, $\cot 25° \approx 2.145$

19. $\sin\theta = -\dfrac{2}{\sqrt{5}}$, $\cos\theta = \dfrac{1}{\sqrt{5}}$, $\tan\theta = -2$,

$\csc\theta = -\dfrac{\sqrt{5}}{2}$, $\sec\theta = \sqrt{5}$, $\cot\theta = -\dfrac{1}{2}$

21. $\sin(-45°) = -\dfrac{1}{\sqrt{2}}$, $\cos(-45°) = \dfrac{1}{\sqrt{2}}$,

$\tan(-45°) = -1$, $\csc(-45°) = -\sqrt{2}$,

$\sec(-45°) = \sqrt{2}$, $\cot(-45°) = -1$

23. $\sin\theta = \dfrac{\sqrt{3}}{2}$, $\cos\theta = -\dfrac{1}{2}$, $\tan\theta = -\sqrt{3}$,

$\csc\theta = \dfrac{2}{\sqrt{3}}$, $\sec\theta = -2$, $\cot\theta = -\dfrac{1}{\sqrt{3}}$

25. -1 **27.** 0

29. $\tan\theta = -\dfrac{4}{3}$, $\csc\theta = -\dfrac{5}{4}$, $\sec\theta = \dfrac{5}{3}$, $\cot\theta = -\dfrac{3}{4}$

31. $y = \sin x$

33. Amplitude $= 3$; period $= \pi$; phase shift $= 0$

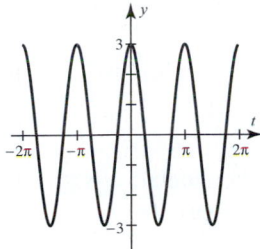

35. Amplitude $= 1$; period $= \dfrac{2\pi}{3}$; phase shift $= \dfrac{\pi}{3}$

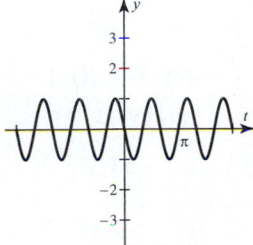

37. Amplitude $= 3$; period $= 4\pi$; phase shift $= \pi$

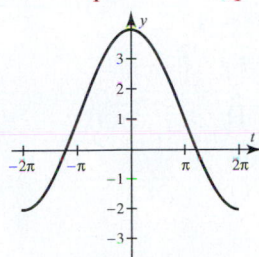

39. $a = 2$, $b = 3$

41. Period $= \dfrac{\pi}{2}$; phase shift $= 0$

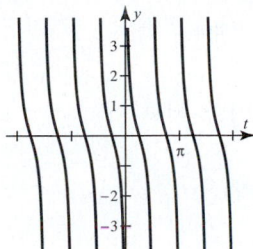

43. Period $= 2\pi$; phase shift $= -\pi$

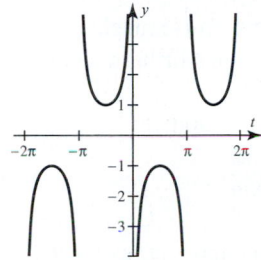

45. (a) $-\dfrac{\pi}{2}$ or $-90°$ (b) $\dfrac{\pi}{3}$ or $60°$ (c) $\dfrac{\pi}{4}$ or $45°$

47. (a) -0.64 or $-36.9°$ (b) 1.37 or $78.7°$
(c) 1.45 or $83.1°$

49. $\alpha = \tan^{-1}\dfrac{5}{3} \approx 59.0°$, $\beta = \tan^{-1}\dfrac{3}{5} \approx 31.0°$,

$c = \sqrt{34}$

51. $30°$ **53.** $30°$ **55.** $78.5°$ **57.** $71.6°$

59. -0.6435 **61.** 0.4510

63. (a) $\dfrac{\pi}{25}$ radians/sec   (b) $\pi \approx 3.14$ ft/sec

65. 89 ft

67. (a) $\sin^{-1}\dfrac{350}{6000} \approx 3.3°$ (b) $\sin^{-1}\dfrac{160}{4500} \approx 2.0°$

69. 161 mi

71. (a) $[0, 25, 2]$ by $[0, 70, 10]$

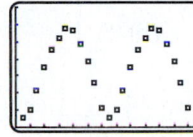

(b) $f(x) = 26\cos\left(\dfrac{\pi}{6}(x - 7)\right) + 32$

(c) $[0, 25, 2]$ by $[0, 70, 10]$

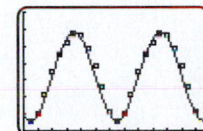

73. $y = 6\sin(x - 220°)$

CHAPTER 6 EXTENDED AND DISCOVERY EXERCISES (PP. 573–574)

1. (a) $x_Q = d \sin \theta + x_p$ and $y_Q = d \cos \theta + y_p$
(b) Approximately (233.9, 377.2)

3. (a) $f(x) = 12.5 \cos \left(\dfrac{\pi}{6} (x - 1) \right) + 61.5$

(b) [0, 25, 2] by [40, 80, 10]

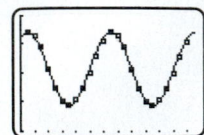

(c) The high temperatures in the Southern Hemisphere occur in January as opposed to July in the Northern Hemisphere. This affects the phase shift of f.
5. (a) For the city in the Northern Hemisphere:
$f(x) = 20 \cos \left(\frac{\pi}{6} (x - 7) \right) + 60.$
For the city in the Southern Hemisphere:
$f(x) = 20 \cos \left(\frac{\pi}{6} (x - 1) \right) + 60.$
(b) Since the maximum average high temperatures for these two cities occur 6 months apart, we may shift the graph for one city 6 units to obtain the graph for the other city.

CHAPTERS 1–6 CUMULATIVE REVIEW EXERCISES (PP. 575–580)

1. About 14.3%   **3.** 0.17
5. $D: [-3, 2]$; $R: [-2, 2]$; $f(-0.5) = -2$
7. (a) $D = \{x \mid x \le 4\}$

(b) $f(-1) = \sqrt{5}$; $f(2a) = \sqrt{4 - 2a}$

9. (a) $m = \frac{1}{2}$; y-intercept: -1; x-intercept: 2

(b) $f(x) = \frac{1}{2}x - 1$ **(c)** -2 **(d)** 2

11. $6x + 3h$ **13.** x-intercept: -10; y-intercept: 4

15. (a) $\frac{1}{5}$ **(b)** $\ln 14 \approx 2.64$ **(c)** $-\frac{4}{5}, \frac{12}{5}$ **(d)** $-1, \frac{1}{2}$
(e) 0, 1, 2   **(f)** $\pm 2, \pm \sqrt{2}$ **(g)** $-1, 2$ **(h)** 5
(i) $3 \pm \sqrt{7}$ **(j)** 65,535

17. (a) $\left(-\infty, \frac{9}{5} \right)$ **(b)** $\left(-\frac{2}{3}, \frac{7}{3} \right]$ **(c)** $\left(-\infty, -\frac{3}{2} \right] \cup [3, \infty)$

(d) $[1, 4]$ **(e)** $(-1, 0) \cup (1, \infty)$ **(f)** $\left(-2, -\frac{5}{3} \right]$

19. $f(x) = -2 \left(x - \frac{3}{2} \right)^2 + \frac{7}{2}$
21. (a) $a < 0$ **(b)** $-2, 1$ **(c)** Positive
(d) $f(x) = -2(x + 2)(x - 1)$
23. (a)   **(b)**

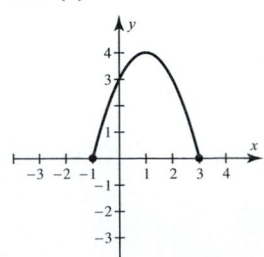   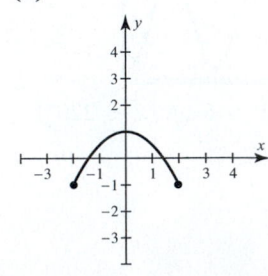

(c)   **(d)**

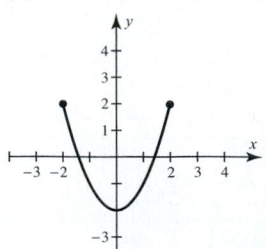   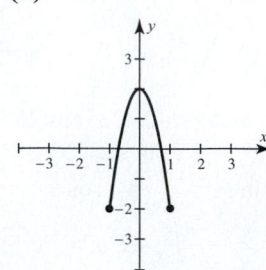

25. Answers may vary.

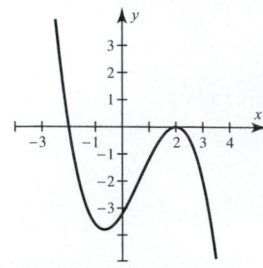

27. (a) $\dfrac{5a^2}{2} - 1 + \dfrac{2}{2a^2}$

(b) $x^3 + 3x^2 + 3x + 1 + \dfrac{1}{x - 1}$ **(c)** $x - 3 + \dfrac{4}{x^2 + 1}$

29. $f(x) = -2(x + 2)(x + 1)^2(x - 3)^2$
31. $f(x) = -x^2(x + 2)(x - 2)$
33. $D = \left\{ x \mid x \ne \frac{7}{3} \right\}$; vertical: $x = \frac{7}{3}$; horizontal: $y = \frac{2}{3}$
35. (a) 5   **(b)** Undefined   **(c)** 2   **(d)** 0
37. (a) 8   **(b)** 0   **(c)** $(f - g)(x) = x^2 + 2x$
(d) $(f \circ g)(x) = x^2 - x - 4$
39. For $(f + g)(x)$, $D = \{x \mid x \le 2 \text{ and } x \ne 0\}$; for
$(f \circ g)(x)$, $D = \{x \mid x < 2\}$
41. $f(x) = 9 \left(\frac{2}{3} \right)^x$ **43.** $C = \frac{1}{2}, a = 2$
45. (a) 2   **(b)** 4   **(c)** -2 **(d)** $\frac{1}{3}$ **(e)** 2   **(f)** 1
47. $\frac{1}{3} \log_2 (x + 2) + \frac{1}{3} \log_2 (x - 2) - \frac{1}{2} \log_2 (x^2 + 4)$
49. 4.395   **51.** $\alpha = 79°26'$; $\beta = 169°26'$ **53.** $\dfrac{5\pi}{6}$
55. $\dfrac{\pi}{4} \approx 0.79$ ft   **57.** $\dfrac{9}{\tan 30°} \approx 15.6$
59. $\sin \theta = -\dfrac{1}{\sqrt{2}}$, $\cos \theta = -\dfrac{1}{\sqrt{2}}$, $\tan \theta = 1$,
$\csc \theta = -\sqrt{2}$, $\sec \theta = -\sqrt{2}$, $\cot \theta = 1$
61. $\sin \theta = -\frac{12}{13}$, $\cos \theta = \frac{5}{13}$, $\tan \theta = -\frac{12}{5}$,
$\csc \theta = -\frac{13}{12}$, $\sec \theta = \frac{13}{5}$, $\cot \theta = -\frac{5}{12}$

63. (a)

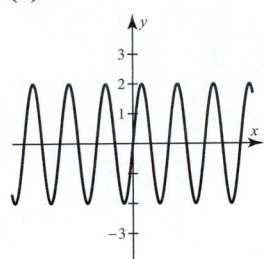

(b)

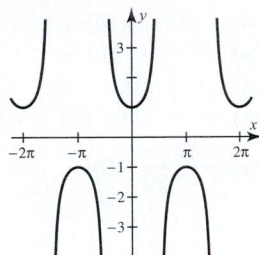

(c)

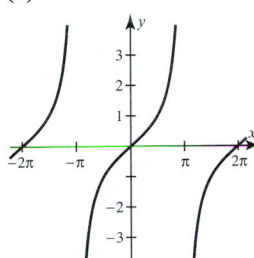

(d)

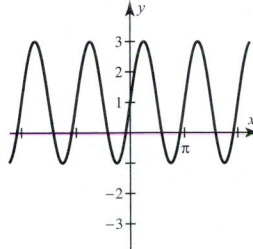

65. $\alpha \approx 33.4°$; $\beta \approx 56.6°$; $b \approx 16.7$

67. $h = \dfrac{12}{\pi} \approx 3.82$ in.

69. (a) 16 ft/sec; 48 ft/sec
(b) During the first second, the average speed is 16 ft/sec. During the next second, the average speed is 48 ft/sec. The object is speeding up.
(c) $32t + 16h$

71. 24 lb

73. (a) $C(t) = 135(t - 1980) + 800$
(b) 1996; the formula gives the correct year.

75. (a) Increasing **(b)** $V(t) = \frac{4}{3}\pi\left(\sqrt{t}\right)^3$
(c) $V(4) = \dfrac{32\pi}{3} \approx 33.5$; after 4 seconds the volume of the balloon is about 33.5 in³.

77. (a) $N(t) = 200{,}000e^{0.135t}$
(b) $N(5) \approx 392{,}807$; after 5 hours there are about 393,000 bacteria per mL in the sample.
(c) About 6.8 hr

79. $t = \dfrac{\ln\left(\frac{A}{P}\right)}{n \ln\left(1 + \frac{r}{n}\right)}$ **81.** 6.4 ft

83. $f(x) \approx 14.5 \sin\left(\dfrac{\pi}{6}(x - 4)\right) + 75.5$

CHAPTER 7: Trigonometric Identities and Equations

SECTION 7.1 (PP. 591–594)

1. $\cot \theta = 2$ **3.** $\sec \theta = \frac{7}{2}$ **5.** $\cos \theta = -\frac{1}{4}$

7. $\tan \theta = -\frac{3}{4}$, $\csc \theta = \frac{5}{3}$, $\sec \theta = -\frac{5}{4}$, $\cot \theta = -\frac{4}{3}$

9. $\cos \theta = -\frac{7}{25}$, $\tan \theta = \frac{24}{7}$, $\csc \theta = -\frac{25}{24}$, $\sec \theta = -\frac{25}{7}$

11. $\tan \theta = \frac{60}{11}$, $\cot \theta = \frac{11}{60}$, $\csc \theta = -\frac{61}{60}$, $\sec \theta = -\frac{61}{11}$

13. $\sin \theta = \frac{1}{\sqrt{2}}$, $\cos \theta = -\frac{1}{\sqrt{2}}$, $\tan \theta = -1$, $\cot \theta = -1$

15. 1 **17.** 1 **19.** 1 **21.** $\cos^2 \theta$ **23.** $\tan^2 \theta$
25. $-\tan \theta$ **27.** $\sec \theta$ **29.** $\tan^2 \theta$ **31.** $\cos x$
33. 1 **35.** Quadrant IV **37.** Quadrant III
39. Quadrant II

41. $\sin \theta = -\dfrac{\sqrt{3}}{2}$, $\tan \theta = -\sqrt{3}$, $\csc \theta = -\dfrac{2}{\sqrt{3}}$,
$\sec \theta = 2$, $\cot \theta = -\dfrac{1}{\sqrt{3}}$

43. $\sin \theta = -\frac{11}{61}$, $\cos \theta = \frac{60}{61}$, $\cot \theta = -\frac{60}{11}$,
$\csc \theta = -\frac{61}{11}$, $\sec \theta = \frac{61}{60}$

45. $\cos \theta = \frac{24}{25}$, $\tan \theta = \frac{7}{24}$, $\cot \theta = \frac{24}{7}$, $\csc \theta = \frac{25}{7}$,
$\sec \theta = \frac{25}{24}$

47. $\cos \theta = -\dfrac{\sqrt{8}}{3}$, $\tan \theta = \dfrac{1}{\sqrt{8}}$, $\cot \theta = \sqrt{8}$,
$\csc \theta = -3$, $\sec \theta = -\dfrac{3}{\sqrt{8}}$

49. $\sin \theta = -\dfrac{\sqrt{8}}{3}$, $\cos \theta = \frac{1}{3}$, $\tan \theta = -\sqrt{8}$,
$\csc \theta = -\dfrac{3}{\sqrt{8}}$, $\cot \theta = -\dfrac{1}{\sqrt{8}}$

51. $\sin \theta = \frac{3}{7}$, $\cos \theta = -\dfrac{\sqrt{40}}{7}$, $\tan \theta = -\dfrac{3}{\sqrt{40}}$,
$\sec \theta = -\dfrac{7}{\sqrt{40}}$, $\cot \theta = -\dfrac{\sqrt{40}}{3}$

53. $\sin x = \sqrt{1 - \cos^2 x}$; $\tan x = \dfrac{\sqrt{1 - \cos^2 x}}{\cos x}$

55. $\sec x = -\sqrt{1 + \tan^2 x}$; $\sin x = -\dfrac{\tan x}{\sqrt{1 + \tan^2 x}}$

57. $\cot x = -\sqrt{\csc^2 x - 1}$; $\cos x = -\dfrac{\sqrt{\csc^2 x - 1}}{\csc x}$

59. $\cos \theta = \sqrt{1 - x^2}$, 0.8586

61. $\sin \theta = -\dfrac{1}{\sqrt{1 + x^2}}$, -0.8899

63. $\tan \theta = \dfrac{x}{\sqrt{1 - x^2}}$ **65.** $-\sin 13°$ **67.** $-\tan \dfrac{\pi}{11}$

69. $\sec \dfrac{2\pi}{5}$ **71.** Yes **73.** No **75.** Yes

77. This is incorrect. The variable should be included. For example, $\cos^2(2\theta) + \sin^2(\theta) \neq 1$, but $\cos^2\theta + \sin^2\theta = 1$.

79. $d = 93{,}000{,}000\csc\theta$

81. (a) $P = 16k\cos^2(2\pi t)$

(b) Let $Y_1 = 32(\cos(2\pi X))^2$. Y_1 has a maximum value of 32 when $t = 0, 0.5, 1, 1.5, 2.0$, and Y_1 has a minimum value of 0 when $t = 0.25, 0.75, 1.25, 1.75$. The spring is either stretched or compressed the most when Y_1 is maximum.

$[0, 2, 0.5]$ by $[-1, 40, 8]$

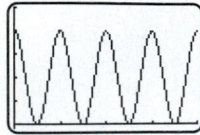

(c) $P = 16k(1 - \sin^2(2\pi t))$

83. (a) The sum of L and C equals 3.

$[0, 10^{-6}, 10^{-7}]$ by $[-1, 4, 1]$

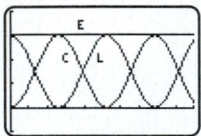

(b) Let $Y_1 = L(t)$, $Y_2 = C(t)$, and $Y_3 = E(t)$. $Y_3 = 3$ for all inputs.

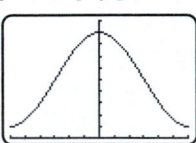

(c) $E(t) = L(t) + C(t)$

$\quad = 3\cos^2(6{,}000{,}000t) + 3\sin^2(6{,}000{,}000t)$

$\quad = 3(\cos^2(6{,}000{,}000t) + \sin^2(6{,}000{,}000t))$

$\quad = 3(1) = 3$

85. (a) The graph has y-axis symmetry. The monthly high temperatures x months before and x months after July are equal.

$[-6, 6, 1]$ by $[0, 100, 10]$

(b) f is even.

(c) $f(-x) = f(x)$

SECTION 7.2 (PP. 600–601)

1. (a) $1 - x^2$ (b) $\cos^2\theta$

3. (a) $x^2 - x$ (b) $\sec^2\theta - \sec\theta$

5. (a) $(x+1)(x+1)$ (b) $(\cos\theta + 1)(\cos\theta + 1)$

7. (a) $x(x-2)$ (b) $\sec t(\sec t - 2)$

9. (a) $\dfrac{2}{1 - x^2}$ (b) $2\csc^2\theta$

11. (a) $\dfrac{x^2 + y^2}{xy}$ (b) $\sec t\csc t$

13. (a) $y^2 + x^2$ (b) 1

15. (a) x (b) $\cos\theta$ **17.** $\sin\theta$

19. $\sin\theta - \sec\theta$ **21.** $2\tan t + \sec^2 t$ **23.** $\cos^2\theta$

25. $(1 - \tan\theta)(1 + \tan\theta)$ **27.** $(\sec t - 3)(\sec t + 2)$

29. $\sec^2\theta(\tan^2\theta + 2)$

31. $\csc^2\theta - \cot^2\theta = 1 + \cot^2\theta - \cot^2\theta = 1$

33. $(1 - \sin t)^2 = 1 - 2\sin t + \sin^2 t$ (FOIL)

35. $\dfrac{\sin t + \cos t}{\sin t} = \dfrac{\sin t}{\sin t} + \dfrac{\cos t}{\sin t} = 1 + \cot t$

37. $\sec^2\theta - 1 = (1 + \tan^2\theta) - 1$

$\qquad\qquad\quad = \tan^2\theta$

39. $\dfrac{\tan^2 t}{\sec t} = \dfrac{\sec^2 t - 1}{\sec t}$

$\qquad\quad = \sec t - \dfrac{1}{\sec t}$

$\qquad\quad = \sec t - \cos t$

41. $\dfrac{\sec t}{1 + \sec t} = \dfrac{\sec t}{1 + \sec t} \cdot \dfrac{\cos t}{\cos t}$

$\qquad\quad = \dfrac{1}{\cos t + 1}$

43. $(\sec t - 1)(\sec t + 1) = \sec^2 t - 1$

$\qquad\qquad\qquad\qquad\quad = \tan^2 t$

45. $\dfrac{1 - \sin^2\theta}{\cos\theta} = \dfrac{\cos^2\theta}{\cos\theta} = \cos\theta$

47. $\dfrac{\sec t}{\tan t} - \dfrac{\tan t}{\sec t} = \dfrac{\sec^2 t - \tan^2 t}{\sec t\tan t}$

$\qquad\qquad\qquad = \dfrac{(1 + \tan^2 t) - \tan^2 t}{\sec t\tan t}$

$\qquad\qquad\qquad = \dfrac{1}{\sec t\tan t}$

$\qquad\qquad\qquad = \cos t\cot t$

49. $\dfrac{\cot^2 t}{\csc t + 1} = \dfrac{\csc^2 t - 1}{\csc t + 1}$

$\qquad\qquad\quad = \dfrac{(\csc t + 1)(\csc t - 1)}{\csc t + 1}$

$\qquad\qquad\quad = \csc t - 1$

51.
$$\frac{\cot t}{\cot t + 1} = \frac{\cot t}{\cot t + 1} \cdot \frac{\tan t}{\tan t}$$
$$= \frac{\cot t \tan t}{\cot t \tan t + \tan t}$$
$$= \frac{1}{1 + \tan t}$$

53.
$$\frac{1}{1 - \sin t} + \frac{1}{1 + \sin t} = \frac{(1 + \sin t) + (1 - \sin t)}{(1 - \sin t)(1 + \sin t)}$$
$$= \frac{2}{1 - \sin^2 t}$$
$$= \frac{2}{\cos^2 t}$$
$$= 2 \sec^2 t$$

55.
$$\frac{\csc t + \cot t}{\csc t - \cot t} = \frac{\csc t + \cot t}{\csc t - \cot t} \cdot \frac{\csc t + \cot t}{\csc t + \cot t}$$
$$= \frac{(\csc t + \cot t)^2}{\csc^2 t - \cot^2 t}$$
$$= \frac{(\csc t + \cot t)^2}{\csc^2 t - (\csc^2 t - 1)}$$
$$= (\csc t + \cot t)^2$$

57.
$$\frac{\cos^2 t}{1 - \sin t} = \frac{\cos^2 t}{1 - \sin t} \cdot \frac{1 + \sin t}{1 + \sin t}$$
$$= \frac{\cos^2 t(1 + \sin t)}{1 - \sin^2 t}$$
$$= \frac{\cos^2 t(1 + \sin t)}{\cos^2 t}$$
$$= 1 + \sin t$$

59.
$$\frac{1}{1 + \sin \theta} = \frac{1}{1 + \sin \theta} \cdot \frac{1 - \sin \theta}{1 - \sin \theta}$$
$$= \frac{1 - \sin \theta}{1 - \sin^2 \theta}$$
$$= \frac{1 - \sin \theta}{\cos^2 \theta}$$

61.
$$\sqrt{1 - \sin^2 \theta} = \sqrt{\cos^2 \theta}$$
$$= |\cos \theta|$$
$$= \cos \theta, \text{ where } \theta \text{ is acute.}$$

63.
$$\frac{1 + 2 \sin x + \sin^2 x}{\cos^2 x} = \frac{(1 + \sin x)^2}{1 - \sin^2 x}$$
$$= \frac{(1 + \sin x)(1 + \sin x)}{(1 - \sin x)(1 + \sin x)}$$
$$= \frac{1 + \sin x}{1 - \sin x}$$

65.
$$(1 - \cos^2 x)(1 + \cos^2 x) = \sin^2 x \,(1 + (1 - \sin^2 x))$$
$$= \sin^2 x \,(2 - \sin^2 x)$$
$$= 2 \sin^2 x - \sin^4 x$$

67.
$$\cot \theta \sin \theta = \frac{\cos \theta}{\sin \theta} \cdot \sin \theta$$
$$= \cos \theta$$

69.
$$(1 - \cos^2 \theta)(1 + \tan^2 \theta) = \sin^2 \theta \sec^2 \theta$$
$$= \frac{\sin^2 \theta}{\cos^2 \theta}$$
$$= \tan^2 \theta$$

71.
$$\cos t (\tan t - \sec t) = \cos t\left(\frac{\sin t}{\cos t} - \frac{1}{\cos t}\right)$$
$$= \sin t - 1$$

73.
$$\frac{\tan(-\theta)}{\sin(-\theta)} = \frac{-\tan \theta}{-\sin \theta}$$
$$= \frac{1}{\cos \theta}$$
$$= \sec \theta$$

75. **(a)** $W(t) = 5(1 - \sin^2(120\pi t))$; $V = 25 \sin(120\pi t)$ is maximum or minimum when $\sin(120\pi t) = \pm 1$ and so $W = 0$.

(b) Whenever there is a peak or valley on the graph of V, the graph of W intersects the x-axis, which corresponds to a zero of W.

$[0, 1/15, 1/60]$ by $[-30, 30, 10]$

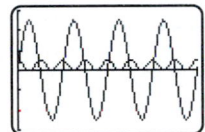

CHECKING BASIC CONCEPTS FOR SECTIONS 7.1 AND 7.2 (P. 601)

1. Quadrant III

3. **(a)** $\cos^2 \theta$ **(b)** $\tan^2 t$

5. **(a)** $(1 - \sin^2 \theta)(1 + \cot^2 \theta) = \cos^2 \theta \csc^2 \theta$
$$= \frac{\cos^2 \theta}{\sin^2 \theta} = \cot^2 \theta$$

(b)
$$\frac{\cot^2 t}{\csc t} = \frac{\csc^2 t - 1}{\csc t}$$
$$= \frac{\csc^2 t}{\csc t} - \frac{1}{\csc t}$$
$$= \csc t - \sin t$$

SECTION 7.3 (PP. 612–616)

1. $60°$ **3.** $85°$ **5.** $65°$ **7.** $\dfrac{\pi}{6}$ **9.** $\dfrac{\pi}{3}$

11. $\dfrac{\pi}{4}, \dfrac{5\pi}{4}$ **13.** $\dfrac{\pi}{3}, \dfrac{5\pi}{3}$

15. **(a)** $90°; \dfrac{\pi}{2}$ **(b)** $270°; \dfrac{3\pi}{2}$

17. **(a)** $60°, 240°; \dfrac{\pi}{3}, \dfrac{4\pi}{3}$ **(b)** $120°, 300°; \dfrac{2\pi}{3}, \dfrac{5\pi}{3}$

19. (a) $60°, 300°; \dfrac{\pi}{3}, \dfrac{5\pi}{3}$ **(b)** $120°, 240°; \dfrac{2\pi}{3}, \dfrac{4\pi}{3}$

21. (a) No solutions   **(b)** No solutions

23. (a) $\frac{1}{2}$ **(b)** $30°, 150°$

25. (a) $0, 1$ **(b)** $0°, 90°, 180°$

27. (a) $-1, 1$ **(b)** $45°, 135°, 225°, 315°$

29. (a) $-2, 1$ **(b)** $0°$

31. $\dfrac{\pi}{3}, \dfrac{2\pi}{3}, \dfrac{4\pi}{3}, \dfrac{5\pi}{3}$ **33.** No solutions   **35.** $\dfrac{\pi}{2}, \dfrac{3\pi}{2}$

37. $\dfrac{\pi}{4}, \dfrac{5\pi}{4}$ **39.** $\dfrac{\pi}{6}, \dfrac{5\pi}{6}, \dfrac{7\pi}{6}, \dfrac{11\pi}{6}$ **41.** $0, \dfrac{\pi}{2}, \pi, \dfrac{3\pi}{2}$

43. $\dfrac{\pi}{4} + \dfrac{\pi n}{2}$ **45.** No solutions   **47.** πn

49. $\dfrac{\pi}{4} + \pi n$ **51.** $\dfrac{\pi n}{2}$ **53.** $2\pi n, \dfrac{\pi}{2} + 2\pi n$

55. $\dfrac{\pi}{18} + \dfrac{2\pi}{3}n, \dfrac{5\pi}{18} + \dfrac{2\pi}{3}n$ **57.** $\dfrac{5\pi}{24} + \dfrac{\pi}{2}n, \dfrac{7\pi}{24} + \dfrac{\pi}{2}n$

59. $\dfrac{\pi}{20} + \dfrac{\pi}{5}n$ **61.** $\dfrac{7\pi}{24} + \dfrac{\pi}{2}n, \dfrac{11\pi}{24} + \dfrac{\pi}{2}n$

63. $\dfrac{3\pi}{16} + \dfrac{\pi}{2}n, \dfrac{5\pi}{16} + \dfrac{\pi}{2}n$ **65.** $\dfrac{\pi}{16} + \dfrac{\pi}{4}n$

67. $\dfrac{\pi}{16} + \dfrac{\pi}{4}n$ **69.** $0.456 + \dfrac{2\pi}{3}n, 1.638 + \dfrac{2\pi}{3}n$

71. $0.170 + \pi n, 1.401 + \pi n$ **73.** $0, 4.49$

75. $0.39, 1.96$ **77.** 3.60

79. $60°, 120°, 240°, 300°; 60° + 180° \cdot n,$
$120° + 180° \cdot n$

81. $30°, 210°; 30° + 180° \cdot n$ **83.** 1

85. $-\dfrac{\sqrt{3}}{2} \approx -0.866$ **87.** -1 **89.** 0

91. $\dfrac{1}{\sqrt{2}} \approx 0.707$

93. $180° + 360° \cdot n \; (F = 1)$
$60° + 360° \cdot n, 300° + 360° \cdot n \; (F = 0.25)$

95. $2.6, 10.7$; near February 17 and October 22

97. $2.6, 10.7$; near February 17 and October 22

99. (a) [0, 13, 1] by [200, 600, 50]

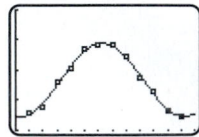

(b) $2.98 \le x \le 9.51$ (approximately). At 50°N latitude the maximum monthly hours of sunshine are greater than or equal to 350 hours roughly from March through September.

101. 0.26 **103. (a)** $0.005, 0.025, 0.045$
(b) At these times the pressure on an eardrum is maximum.

105. (a) [0, 0.01, 0.005] by [−0.005, 0.005, 0.001]

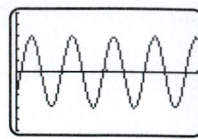

[0, 0.01, 0.005] by [−0.005, 0.005, 0.001]

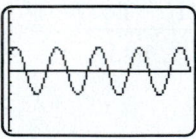

[0, 0.01, 0.005] by [−0.005, 0.005, 0.001]

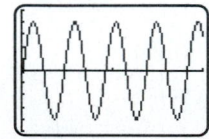

(b) Maximum: $P \approx 0.004$
(c) No. The maximum of P_1 is 0.003 and the maximum of P_2 is 0.002.

SECTION 7.4 (PP. 625–627)

1. $\dfrac{\sqrt{6} - \sqrt{2}}{4}$ **3.** $2 - \sqrt{3}$ **5.** $\dfrac{\sqrt{6} - \sqrt{2}}{4}$

7. $\dfrac{\sqrt{6} - \sqrt{2}}{4}$ **9.** $\dfrac{\sqrt{6} + \sqrt{2}}{4}$

11. (a) *Graphical:* Graphs of
$y = \sin\left(t + \dfrac{\pi}{2}\right)$ and $y = \cos t$ are the same.
Verbal: If the sine graph is translated $\dfrac{\pi}{2}$ units left, it coincides with the cosine graph.

(b) $\sin\left(t + \dfrac{\pi}{2}\right) = \sin t \cos \dfrac{\pi}{2} + \cos t \sin \dfrac{\pi}{2}$
$$= \sin t (0) + \cos t (1)$$
$$= \cos t$$

13. (a) *Graphical:* Graphs of $y = \cos(t + \pi)$ and $y = -\cos t$ are the same.
Verbal: If the cosine graph is translated π units left, it coincides with the cosine graph reflected about the x-axis.

(b) $\cos(t + \pi) = \cos t \cos \pi - \sin t \sin \pi$
$$= \cos t(-1) - \sin t(0)$$
$$= -\cos t$$

15. (a) *Graphical:* Graphs of $y = \sec\left(t - \frac{\pi}{2}\right)$ and $y = \csc t$ are the same.

Verbal: If the secant graph is translated $\frac{\pi}{2}$ units right, it coincides with the cosecant graph.

(b) $\sec\left(t - \frac{\pi}{2}\right) = \dfrac{1}{\cos(t - \pi/2)}$

$= \dfrac{1}{\cos t \cos(\pi/2) + \sin t \sin(\pi/2)}$

$= \dfrac{1}{\cos t(0) + \sin t(1)}$

$= \dfrac{1}{\sin t}$

$= \csc t$

17. $\sin\left(\dfrac{\pi}{2} - t\right) = \sin\dfrac{\pi}{2}\cos t - \cos\dfrac{\pi}{2}\sin t$

$= (1)\cos t + (0)\sin t$

$= \cos t$

19. $\sec\left(\dfrac{\pi}{2} - t\right) = \dfrac{1}{\cos\left(\dfrac{\pi}{2} - t\right)}$

$= \dfrac{1}{\cos\dfrac{\pi}{2}\cos t + \sin\dfrac{\pi}{2}\sin t}$

$= \dfrac{1}{(0)\cos t + (1)\sin t}$

$= \dfrac{1}{\sin t}$

$= \csc t$

21. (a) $\frac{56}{65}$ **(b)** $\frac{33}{65}$ **(c)** $\frac{56}{33}$ **(d)** I

23. (a) $-\frac{988}{1037}$ **(b)** $\frac{315}{1037}$ **(c)** $-\frac{988}{315}$ **(d)** IV

25. (a) $-\frac{33}{65}$ **(b)** $-\frac{56}{65}$ **(c)** $\frac{33}{56}$ **(d)** III

27. (a) $-\frac{775}{793}$ **(b)** $-\frac{168}{793}$ **(c)** $\frac{775}{168}$ **(d)** III

29. $\cos\left(t - \dfrac{\pi}{4}\right) = \cos t \cos\dfrac{\pi}{4} + \sin t \sin\dfrac{\pi}{4}$

$= \cos t\left(\dfrac{\sqrt{2}}{2}\right) + \sin t\left(\dfrac{\sqrt{2}}{2}\right)$

$= \dfrac{\sqrt{2}}{2}(\cos t + \sin t)$

31. $\tan\left(t + \dfrac{\pi}{4}\right) = \dfrac{\tan t + \tan(\pi/4)}{1 - \tan t \tan(\pi/4)}$

$= \dfrac{\tan t + 1}{1 - \tan t(1)}$

$= \dfrac{1 + \tan t}{1 - \tan t}$

33. $\dfrac{\cos(x - y)}{\cos(x + y)} = \dfrac{\cos x \cos y + \sin x \sin y}{\cos x \cos y - \sin x \sin y}$

$= \dfrac{\dfrac{\cos x \cos y}{\cos x \cos y} + \dfrac{\sin x \sin y}{\cos x \cos y}}{\dfrac{\cos x \cos y}{\cos x \cos y} - \dfrac{\sin x \sin y}{\cos x \cos y}}$

$= \dfrac{1 + \tan x \tan y}{1 - \tan x \tan y}$

35. $\dfrac{\cos(\alpha - \beta)}{\cos \alpha \sin \beta} = \dfrac{\cos \alpha \cos \beta + \sin \alpha \sin \beta}{\cos \alpha \sin \beta}$

$= \dfrac{\cos \alpha \cos \beta}{\cos \alpha \sin \beta} + \dfrac{\sin \alpha \sin \beta}{\cos \alpha \sin \beta}$

$= \dfrac{\cos \beta}{\sin \beta} + \dfrac{\sin \alpha}{\cos \alpha}$

$= \cot \beta + \tan \alpha$

$= \tan \alpha + \cot \beta$

37. $\sin 2t = \sin(t + t)$

$= \sin t \cos t + \cos t \sin t$

$= 2 \sin t \cos t$

39. $\sin(\alpha + \beta) + \sin(\alpha - \beta)$

$= \sin \alpha \cos \beta + \cos \alpha \sin \beta + \sin \alpha \cos \beta$

$\quad - \cos \alpha \sin \beta$

$= \sin \alpha \cos \beta + \sin \alpha \cos \beta$

$= 2 \sin \alpha \cos \beta$

41. $\tan(\pi - \theta) = \dfrac{\tan \pi - \tan \theta}{1 + \tan \pi \tan \theta} = \dfrac{0 - \tan \theta}{1 + (0)\tan \theta}$

$= -\tan \theta$

43. $\dfrac{\sin(x - y)}{\sin y} + \dfrac{\cos(x - y)}{\cos y}$

$= \dfrac{\sin x \cos y - \cos x \sin y}{\sin y}$

$\quad + \dfrac{\cos x \cos y + \sin x \sin y}{\cos y}$

$= \sin x\dfrac{\cos y}{\sin y} - \cos x + \cos x + \sin x\dfrac{\sin y}{\cos y}$

$= \sin x\left(\dfrac{\cos y}{\sin y} + \dfrac{\sin y}{\cos y}\right)$

$= \sin x\left(\dfrac{\cos^2 y + \sin^2 y}{\sin y \cos y}\right)$

$= \dfrac{\sin x}{\sin y \cos y}$

45. $\dfrac{\sin(x+y)}{\cos x \cos y} = \dfrac{\sin x \cos y + \cos x \sin y}{\cos x \cos y}$

$\qquad\qquad = \dfrac{\sin x \cos y}{\cos x \cos y} + \dfrac{\cos x \sin y}{\cos x \cos y}$

$\qquad\qquad = \dfrac{\sin x}{\cos x} + \dfrac{\sin y}{\cos y}$

$\qquad\qquad = \tan x + \tan y$

47. $\dfrac{7}{79}$

49. $\tan\theta = \tan(\beta - \alpha)$

$\qquad = \dfrac{\tan\beta - \tan\alpha}{1 + \tan\alpha\tan\beta}$

$\qquad = \dfrac{m_2 - m_1}{1 + m_1 m_2}$

51. $\theta = \tan^{-1}\dfrac{7}{11} \approx 32.5°$

53. **(a)** $F \approx 409$ lb   **(b)** About 0.81 or 46.2°

55. **(a)** [0, 0.02, 0.001] by [−6, 6, 1]

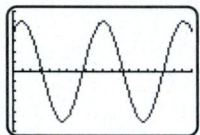

(b) $a = 5, k \approx 0.9272$

(c) $5\sin(220\,\pi t + 0.9272)$

$\qquad = 5(\sin(220\,\pi t)\cos(0.9272)$

$\qquad\quad + \cos(220\,\pi t)\sin(0.9272))$

$\qquad \approx 5(0.6\sin(220\,\pi t) + 0.8\cos(220\,\pi t))$

$\qquad = 3\sin(220\,\pi t) + 4\cos(220\,\pi t)$

7.4 EXTENDED AND DISCOVERY EXERCISES (P. 627)

1. (a) [0.15, 1.15, 0.05] by [−0.01, 0.01, 0.001]

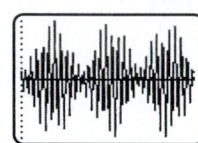

There are 3 beats in 1 second.

(b) [0.15, 1.15, 0.05] by [−0.01, 0.01, 0.001]

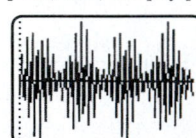

There are 4 beats in 1 second.

(c) When the frequencies are F_1 and F_2 the rate of beats per second is given by $|F_2 - F_1|$.

3. There are 3 beats in 1 second.
[0.2, 1.2, 0.05] by [−0.01, 0.01, 0.001]

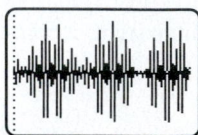

CHECKING BASIC CONCEPTS FOR SECTIONS 7.3 AND 7.4 (P. 628)

1. (a) 45°   **(b)** $\dfrac{\pi}{6}$

3. (a) $\dfrac{3\pi}{4} + \pi n$ **(b)** $\dfrac{2\pi}{3} + 2\pi n;\ \dfrac{4\pi}{3} + 2\pi n;\ 2\pi n$

5. $\sin(t - \pi) = \sin t \cos\pi - \cos t \sin\pi$

$\qquad\qquad = \sin t(-1) - \cos t(0)$

$\qquad\qquad = -\sin t$

SECTION 7.5 (PP. 641–644)

1. (a) 1   **(b)** $\dfrac{\sqrt{3}}{2}$; not equal

3. (a) 1   **(b)** $-\dfrac{1}{2}$; not equal

5. (a) 2   **(b)** Undefined; not equal

7. *Graphical:* Graphs of $y = \tan 2\theta$ and $y = 2\tan\theta$ are not the same.
Symbolic: $\tan 2\theta = \dfrac{2\tan\theta}{1 - \tan^2\theta} \neq 2\tan\theta$, unless $\tan\theta = 0$.

9. $\sin 2\theta = \dfrac{24}{25}, \cos 2\theta = \dfrac{7}{25}, \tan 2\theta = \dfrac{24}{7}$

11. $\sin 2\theta = -\dfrac{336}{625}, \cos 2\theta = -\dfrac{527}{625}, \tan 2\theta = \dfrac{336}{527}$

13. $\sin 2\theta = -\dfrac{1320}{3721}, \cos 2\theta = \dfrac{3479}{3721}, \tan 2\theta = -\dfrac{1320}{3479}$

15. $\sin 2\theta = \dfrac{336}{625}, \cos 2\theta = \dfrac{527}{625}, \tan 2\theta = \dfrac{336}{527}$

17. 0   **19.** $\dfrac{527}{625}$ **21.** $\dfrac{828}{2197}$ **23.** $1 - 2x^2$ **25.** $\dfrac{24}{25}$

27. $\sin 2\theta$ **29.** $\dfrac{1}{2}\sin 2\theta$ **31.** $\cos 4\theta$ **33.** 1

35. $\cot^2 5x$ **37.** $\dfrac{2 + \sqrt{2}}{4}$ **39.** $\dfrac{2 + \sqrt{3}}{2 - \sqrt{3}}$

41. (a) $\dfrac{\sqrt{2 + \sqrt{3}}}{2}$

(b) $-\sqrt{\dfrac{2 - \sqrt{3}}{2 + \sqrt{3}}}$ or $-\dfrac{1}{2 + \sqrt{3}}$ or $\sqrt{3} - 2$

43. (a) $\sqrt{\dfrac{2 - \sqrt{2}}{2 + \sqrt{2}}}$ or $\dfrac{\sqrt{2}}{2 + \sqrt{2}}$ or $\sqrt{2} - 1$

(b) $-\dfrac{\sqrt{2 - \sqrt{2}}}{2}$

45. $\sin 30°$ **47.** $\cos 25°$ **49.** $\tan 20°$

51. $\sin\dfrac{\theta}{2} = \dfrac{1}{\sqrt{10}}, \cos\dfrac{\theta}{2} = \dfrac{3}{\sqrt{10}}, \tan\dfrac{\theta}{2} = \dfrac{1}{3}$

53. $\sin \dfrac{\theta}{2} = -\dfrac{1}{\sqrt{26}}, \cos \dfrac{\theta}{2} = \dfrac{5}{\sqrt{26}}, \tan \dfrac{\theta}{2} = -\dfrac{1}{5}$

55. $\sin \dfrac{\theta}{2} = \dfrac{4}{5}, \cos \dfrac{\theta}{2} = \dfrac{3}{5}, \tan \dfrac{\theta}{2} = \dfrac{4}{3}$

57. $4 \sin 2x = 4(2 \sin x \cos x)$
$$= 8 \sin x \cos x$$

59. $\dfrac{2 - \sec^2 x}{\sec^2 x} = \dfrac{2}{\sec^2 x} - 1$
$$= 2 \cos^2 x - 1$$
$$= \cos 2x$$

61. $\sec 2x = \dfrac{1}{\cos 2x}$
$$= \dfrac{1}{1 - 2 \sin^2 x}$$

63. $\sin 3\theta = \sin(2\theta + \theta)$
$$= \sin 2\theta \cos \theta + \cos 2\theta \sin \theta$$
$$= (2 \sin \theta \cos \theta) \cos \theta + (1 - 2 \sin^2 \theta) \sin \theta$$
$$= 2 \sin \theta \cos^2 \theta + \sin \theta - 2 \sin^3 \theta$$
$$= 2 \sin \theta (1 - \sin^2 \theta) + \sin \theta - 2 \sin^3 \theta$$
$$= 2 \sin \theta - 2 \sin^3 \theta + \sin \theta - 2 \sin^3 \theta$$
$$= 3 \sin \theta - 4 \sin^3 \theta$$

65. $\sin 4\theta = 2 \sin 2\theta \cos 2\theta$
$$= 2(2 \sin \theta \cos \theta) \cos 2\theta$$
$$= 4 \sin \theta \cos \theta \cos 2\theta$$

67. $\dfrac{\sin 2\theta}{\sin \theta} = \dfrac{2 \sin \theta \cos \theta}{\sin \theta}$
$$= 2 \cos \theta$$

69. $2 \cos^2 \dfrac{\theta}{2} = 2\left(\dfrac{1 + \cos \theta}{2}\right)$
$$= 1 + \cos \theta$$

71.
$$\cos^4 \theta - \sin^4 \theta = (\cos^2 \theta - \sin^2 \theta)(\cos^2 \theta + \sin^2 \theta)$$
$$= (\cos 2\theta)(1)$$
$$= \cos 2\theta$$

73. $\csc 2t = \dfrac{1}{\sin 2t}$
$$= \dfrac{1}{2 \sin t \cos t}$$
$$= \dfrac{\csc t}{2 \cos t}$$

75. $\tan \dfrac{x}{2} = \dfrac{\sin(x/2)}{\cos(x/2)}$
$$= \dfrac{2 \sin(x/2) \cos(x/2)}{2 \cos(x/2) \cos(x/2)}$$
$$= \dfrac{\sin x}{2 \cos^2(x/2)}$$
$$= \dfrac{\sin x}{1 + \cos x}$$

77. (a) $\frac{1}{2}(\sin 70° - \sin 30°)$ **(b)** $\frac{1}{2}(\cos 60° + \cos 20°)$

79. (a) $\frac{1}{2}(\sin 10\theta + \sin 4\theta)$ **(b)** $\frac{1}{2}(\cos 4x - \cos 12x)$

81. (a) $2 \sin 35° \cos 5°$ **(b)** $2 \cos 40° \cos 5°$

83. (a) $2 \cos 5\theta \cos \theta$ **(b)** $2 \sin \dfrac{11x}{2} \cos \dfrac{3x}{2}$

85. $0°, 180°$ **87.** $0°, 90°, 180°, 270°$ **89.** $180°$

91. $90°, 270°$ **93.** $\pi n, \dfrac{2\pi}{3} + 2\pi n, \dfrac{4\pi}{3} + 2\pi n$

95. $\dfrac{\pi}{3} + 4\pi n, \dfrac{5\pi}{3} + 4\pi n$

97. $\dfrac{\pi}{6} + 2\pi n, \dfrac{5\pi}{6} + 2\pi n, \dfrac{3\pi}{2} + 2\pi n$

99. $\dfrac{\pi}{8} + \pi n, \dfrac{5\pi}{8} + \pi n$, or equivalently, $\dfrac{\pi}{8} + \dfrac{\pi n}{2}$

101. $\dfrac{2\pi}{3} + 4\pi n, \dfrac{10\pi}{3} + 4\pi n$

103. $\dfrac{\pi}{8} + \pi n, \dfrac{5\pi}{8} + \pi n$, or equivalently, $\dfrac{\pi}{8} + \dfrac{\pi n}{2}$

105. $\dfrac{\pi}{12} + \pi n, \dfrac{5\pi}{12} + \pi n, \dfrac{3\pi}{4} + \pi n$ **107.** $0.333, 4.379$

109. (a) $[0, 0.04, 0.01]$ by $[-500, 2500, 500]$

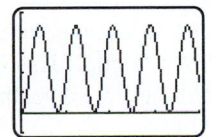

(b) $a = -1085, k = 240, d = 1085$

111. $\dfrac{1}{720} + \dfrac{n}{60}, \dfrac{5}{720} + \dfrac{n}{60}$ sec

113. (a) $d = 600\left(1 - \cos \dfrac{80°}{2}\right) \approx 140.4$ ft

(b) *Hint:* First show that $r = d + r \cos \dfrac{\beta}{2}$.

(c) No, since $\cos \dfrac{\beta}{2} \neq \frac{1}{2} \cos \beta$ in general.

115. (a) $f(t) = \cos(1394\pi t) + \cos(3266\pi t)$

(b) $f(t) = 2 \cos(2330\pi t) \cos(936\pi t)$

(c) $[0, 0.005, 0.001]$ by $[-2, 2, 1]$

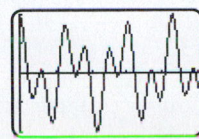

117. Let $f(t)$ model the tone for the number 3 and $g(t)$ model the tone for number 4.
(a) $f(t) = \cos(1394\pi t) + \cos(2954\pi t)$
$g(t) = \cos(1540\pi t) + \cos(2418\pi t)$
(b) $f(t) = 2\cos(2174\pi t)\cos(780\pi t)$
$g(t) = 2\cos(1979\pi t)\cos(439\pi t)$
(c) They are different. These two numbers sound different so their graphs should be different.

[0, 0.005, 0.001] by [−2, 2, 1]   [0, 0.005, 0.001] by [−2, 2, 1]

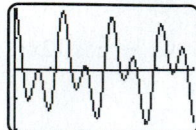   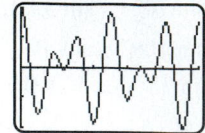

Graph of Number 3   Graph of Number 4

CHECKING BASIC CONCEPTS FOR SECTION 7.5 (P. 644)

1. $\sin 2\theta = -\dfrac{336}{625}$, $\cos 2\theta = -\dfrac{527}{625}$

3. $\sin\dfrac{\theta}{2} = \dfrac{1}{\sqrt{5}}$, $\cos\dfrac{\theta}{2} = \dfrac{2}{\sqrt{5}}$ **5.** $90°, 270°$

CHAPTER 7 REVIEW EXERCISES (PP. 648–651)

1. Quadrant II

3. $\tan\theta = -\dfrac{3}{4}$, $\csc\theta = \dfrac{5}{3}$, $\sec\theta = -\dfrac{5}{4}$, $\cot\theta = -\dfrac{4}{3}$

5. $\sin\theta = -\dfrac{7}{25}$, $\csc\theta = -\dfrac{25}{7}$, $\sec\theta = \dfrac{25}{24}$, $\cot\theta = -\dfrac{24}{7}$

7. $-\sin(13°)$ **9.** $\sec\left(\dfrac{3\pi}{7}\right)$

11. An identity is an equation that is valid for all meaningful values of the variable. An example is $\sin^2\theta + \cos^2\theta = 1$. A conditional equation is valid only for certain values of the variable. An example is $\tan\theta = 1$.
13. No. If one trigonometric function is positive, then its reciprocal must also be positive.
15. 1   **17.** 1   **19.** $\cos\theta$
21. $\sin\theta \approx 0.7776$, $\cos\theta \approx 0.6288$, $\csc\theta \approx 1.2860$, $\sec\theta \approx 1.5904$, $\cot\theta \approx 0.8086$
23. $\sin\theta \approx 0.8908$, $\tan\theta \approx -1.9604$, $\csc\theta \approx 1.1226$, $\sec\theta \approx -2.2007$, $\cot\theta \approx -0.5101$
25. $(\sin\theta + 1)(\sin\theta + 1)$
27. $(\tan\theta + 3)(\tan\theta - 3)$
29. $(\sec\theta - 1)(\sec\theta + 1) = \sec^2\theta - 1$
$= (1 + \tan^2\theta) - 1$
$= \tan^2\theta$
31. $(1 + \tan t)^2 = 1 + 2\tan t + \tan^2 t$
$= \sec^2 t + 2\tan t$

33. $\sin(x - \pi) = \sin x \cos\pi - \cos x \sin\pi$
$= \sin x(-1) - \cos x(0)$
$= -\sin x$
35. $\sin 8x = \sin(2 \cdot 4x)$
$= 2\sin 4x \cos 4x$
37. $\sec 2x = \dfrac{1}{\cos 2x}$
$= \dfrac{1}{2\cos^2 x - 1}$
39. $\cos^4 x \sin^3 x = \cos^4 x \sin^2 x \sin x$
$= \cos^4 x(1 - \cos^2 x)\sin x$
$= (\cos^4 x - \cos^6 x)\sin x$
41. $\sec^4\theta - \tan^4\theta = (\sec^2\theta - \tan^2\theta)(\sec^2\theta + \tan^2\theta)$
$= (1)(1 + \tan^2\theta + \tan^2\theta)$
$= 1 + 2\tan^2\theta$

43. $60°$ **45.** $\dfrac{2\pi}{7}$ **47.** $\dfrac{\pi}{6}, \dfrac{\pi}{2}, \dfrac{5\pi}{6}, \dfrac{3\pi}{2}$
49. (a) $60°, 240°$ **(b)** $150°, 330°$
51. $60°, 300°$ **53.** $90°$ **55.** $\dfrac{\pi}{4}, \dfrac{5\pi}{4}$
57. $\dfrac{\pi}{6} + \pi n, -\dfrac{\pi}{6} + \pi n$; $30° + 180° \cdot n$, $-30° + 180° \cdot n$
59. $\dfrac{\pi}{2} + \pi n$; $90° + 180° \cdot n$ **61.** $-\dfrac{\sqrt{2 - \sqrt{3}}}{2}$
63. $-4.43, 1.13, 4.53$
65. (a) If the sine graph is translated $\dfrac{\pi}{2}$ units right, it corresponds to the cosine graph reflected across the x-axis.
(b) $\sin\left(t - \dfrac{\pi}{2}\right) = \sin t \cos\dfrac{\pi}{2} - \cos t \sin\dfrac{\pi}{2}$
$= \sin t(0) - \cos t(1)$
$= -\cos t$
(c) Graphs of $y = \sin(t - \pi/2)$ and $y = -\cos t$ are the same.
67. (a) $\sin(\alpha + \beta) = \dfrac{63}{65}$ **(b)** $\cos(\alpha + \beta) = \dfrac{16}{65}$,
(c) $\tan(\alpha + \beta) = \dfrac{63}{16}$ **(d)** Quadrant I
69. $\sin 2\theta = -\dfrac{24}{25}$, $\cos 2\theta = -\dfrac{7}{25}$, $\tan 2\theta = \dfrac{24}{7}$
71. $\sin\dfrac{\theta}{2} = \sqrt{\dfrac{3}{8}}$, $\cos\dfrac{\theta}{2} = \sqrt{\dfrac{5}{8}}$, $\tan\dfrac{\theta}{2} = \sqrt{\dfrac{3}{5}}$
73. $\dfrac{3479}{3721}$
75. $x = 4.7, 8.7$ or about April 21 and August 22

77. (a) [0, 0.06, 0.01] by [−0.012, 0.012, 0.002]

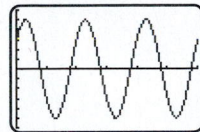

(b) $a = 0.01, k \approx 0.6435$

(c) $0.01 \sin(100\pi t + 0.6435)$

$$= 0.01 \left(\sin(100\pi t) \cos(0.6435) \right.$$
$$\left. + \cos(100\pi t) \sin(0.6435) \right)$$
$$\approx 0.01 \left(0.8 \sin(100\pi t) + 0.6 \cos(100\pi t) \right)$$
$$\approx 0.008 \sin(100\pi t) + 0.006 \cos(100\pi t)$$

79. $W(t) = 7 - 7\sin^2(240\pi t)$; When V is maximum or minimum, $W = 0$.

81. $\theta \approx 30.11°$ or 0.53 radians

CHAPTER 7 EXTENDED AND DISCOVERY EXERCISES (PP. 650–651)

1. (a) [0, 0.01, 0.002] by [−0.005, 0.005, 0.001]

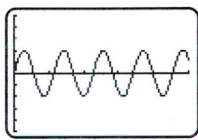

[0, 0.01, 0.002] by [−0.005, 0.005, 0.001]

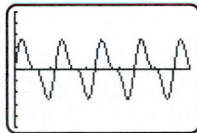

[0, 0.01, 0.002] by [−0.005, 0.005, 0.001]

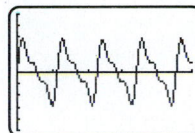

[0, 0.01, 0.002] by [−0.005, 0.005, 0.001]

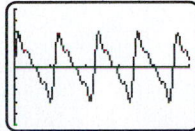

[0, 0.01, 0.002] by [−0.005, 0.005, 0.001]

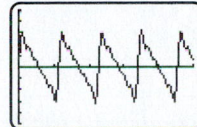

(b) The graph approximates a saw-tooth shape.

(c) The maximum pressure of P is approximately 0.00317.

(d) The pure tone is modeled by a smooth graph whereas the piano tone is modeled by a saw-tooth shape.

3. (a) [0, 0.03, 0.01] by [−1.2, 1.2, 0.5]

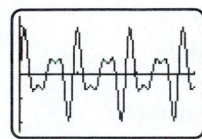

(b) $t \approx 0.0007576, 0.009850, 0.01894, 0.02803$; answers may vary slightly.

(c) 110 Hz. A person will hear the low tone 110 Hz in addition to the higher harmonic tones.

(d) A large speaker can produce the low frequency 110 Hz directly. It will be clear and contain less noise.

CHAPTER 8:   Further Topics in Trigonometry

SECTION 8.1 (PP. 661–664)

1. No   **3.** No   **5.** Yes   **7.** Yes

9. $\beta = 60°, a \approx 5.5, b \approx 4.9$

11. $\beta = 35°, a \approx 15.1, c \approx 7.4$

13. $\beta_1 \approx 57.1°, \gamma_1 \approx 76.9°, c_1 \approx 8.1$
$\beta_2 \approx 122.9°, \gamma_2 \approx 11.1°, c_2 \approx 1.6$

15. There are no solutions.

17. $\beta_1 \approx 67°13', \gamma_1 \approx 62°35', c_1 \approx 11.6$
$\beta_2 \approx 112°48', \gamma_2 \approx 17°0', c_2 \approx 3.8$

19. $\gamma = 93°, a \approx 7.8, c \approx 14.6$

21. $\beta_1 \approx 26.1°, \gamma_1 \approx 133.9°, c_1 \approx 14.7$
$\beta_2 \approx 153.9°, \gamma_2 \approx 6.1°, c_2 \approx 2.2$

23. $\gamma = 90°, \alpha = 60°, a = 10\sqrt{3} \approx 17.3$

25. $\alpha \approx 52.9°, \beta \approx 25.1°, b \approx 22.4$

27. $\beta = 10°, a \approx 92.2, c \approx 101.9$

29. There are no solutions.   **31.** There are no solutions.

33. $\alpha_1 \approx 60°56', \gamma_1 \approx 72°19', c_1 \approx 6.5$
$\alpha_2 \approx 119°4', \gamma_2 \approx 14°11', c_2 \approx 1.7$

35. $\gamma = 99°45', a \approx 84.6, b \approx 40.9$ **37.** 3629 ft

39. $d \approx 7.2$ mi   **41.** 118.0 m   **43.** $AB \approx 105.4$ ft

45. 5.7 mi   **47.** 28.8 ft   **49.** 1054 ft   **51.** 630 ft

SECTION 8.2 (PP. 671–675)

1. (a) SAS   **(b)** Law of cosines

3. (a) SSA   **(b)** Law of sines

5. (a) ASA   **(b)** Law of sines

7. (a) ASA   **(b)** Law of sines

9. $a \approx 5.4, \beta \approx 40.7°, \gamma \approx 78.3°$

11. $\alpha \approx 22.3°, \beta \approx 108.2°, \gamma \approx 49.5°$

13. $\alpha \approx 33.6°, \beta \approx 50.7°, \gamma \approx 95.7°$
15. $c \approx 28.8, \alpha \approx 116.5°, \beta \approx 28.5°$
17. $\alpha \approx 101.0°, \beta \approx 44.0°, \gamma \approx 34.9°$
Angles do not sum to 180° due to rounding.
19. $a \approx 9.0, \beta \approx 150.9°, \gamma \approx 18.6°$
21. $\alpha \approx 23.1°, \beta \approx 107.2°, \gamma \approx 49.7°$
23. $b \approx 30.7, \alpha \approx 33°26', \gamma \approx 24°24'$
25. $\alpha \approx 45.1°, \beta \approx 63.5°, \gamma \approx 71.5°$
Angles do not sum to 180° due to rounding.
27. No, since $a + b < c$.
29. No, since $89° + 112° > 180°$.
31. Yes, since we are given ASA and $\alpha + \gamma < 180°$.
33. 86.8   **35.** 5.3   **37.** 50.9   **39.** 18.3   **41.** 2.1
43. 18.3   **45.** 66   **47.** 160.4   **49.** 169 ft
51. (a) 101.3 ft   (b) $\theta \approx 0.019°$
53. 29.8 mi   **55.** 4.4 ft; 7.7 ft
57. 302 mi   **59.** 147.8 ft²
61. (a) $9\sqrt{3} \approx 15.6$ in²   (b) The results are equal.
63. $\theta \approx 40.5°$ **65.** 149,429 ft²   **67.** 21,309 ft²
69. 2000 km   **71.** 16.0 in²

CHECKING BASIC CONCEPTS FOR SECTIONS 8.1 AND 8.2 (P. 675)

1. $\beta = 74°, b \approx 16.6, c \approx 15.3$
3. (a) $b \approx 7.1, \alpha \approx 63.0°, \gamma \approx 66.0°$
(b) $\alpha \approx 110.7°, \beta \approx 37.0°, \gamma \approx 32.3°$

SECTION 8.3 (PP. 686–691)

1. (a) $a_1 \approx 3, a_2 \approx 4$ (b) $\|\mathbf{v}\| = 5$
3. (a) $a_1 \approx -5, a_2 \approx -12$ (b) $\|\mathbf{v}\| = 13$
5. (a)   (b) $\mathbf{v} = \langle 0, -20 \rangle$

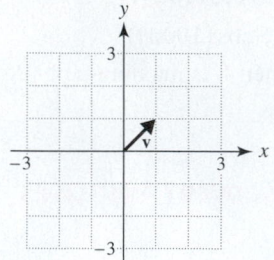

v (20 mph)

(c) $2\mathbf{v} = \langle 0, -40 \rangle$; this represents a 40-mph north wind.
$-\frac{1}{2}\mathbf{v} = \langle 0, 10 \rangle$; this represents a 10-mph south wind.
7. (a)   **v** (5 mph)
(b) $\mathbf{v} = \langle 5/\sqrt{2}, -5/\sqrt{2} \rangle$ or $\langle \frac{5}{2}\sqrt{2}, -\frac{5}{2}\sqrt{2} \rangle$
(c) $2\mathbf{v} = \langle 10/\sqrt{2}, -10/\sqrt{2} \rangle$ or $\langle 5\sqrt{2}, -5\sqrt{2} \rangle$; this represents a 10-mph northwest wind.
$-\frac{1}{2}\mathbf{v} = \langle -\frac{5}{4}\sqrt{2}, \frac{5}{4}\sqrt{2} \rangle$; this represents a 2.5-mph southeast wind.

9. (a)   (b) $\mathbf{v} = \langle 0, 30 \rangle$

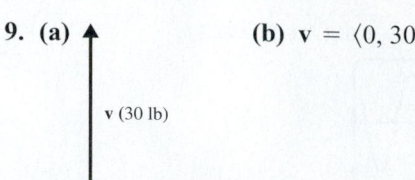

v (30 lb)

(c) $2\mathbf{v} = \langle 0, 60 \rangle$; this represents a 60-lb force upward.
$-\frac{1}{2}\mathbf{v} = \langle 0, -15 \rangle$; this represents a 15-lb force downward.
11. (a) Horizontal = 1, vertical = 1
(b) $\|\mathbf{v}\| = \sqrt{2}$; **v** is not a unit vector.
(c) $\|\mathbf{v}\|$ represents the length of **v**.

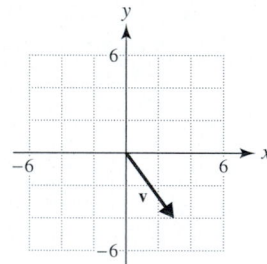

13. (a) Horizontal = 3, vertical = -4
(b) $\|\mathbf{v}\| = 5$; **v** is not a unit vector.
(c) $\|\mathbf{v}\|$ represents the length of **v**.

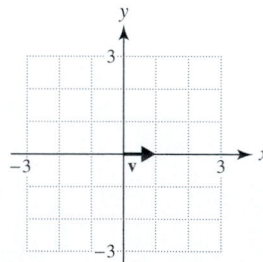

15. (a) Horizontal = 1, vertical = 0
(b) $\|\mathbf{v}\| = 1$; **v** is a unit vector.
(c) $\|\mathbf{v}\|$ represents the length of **v**.

17. (a) Horizontal = 5, vertical = 12

(b) $\|\mathbf{v}\| = 13$; $\mathbf{v}$ is not a unit vector.

(c) $\|\mathbf{v}\|$ represents the length of $\mathbf{v}$.

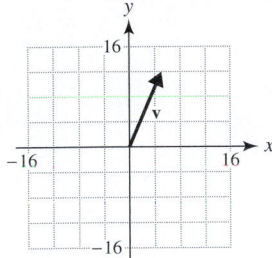

19. (a)

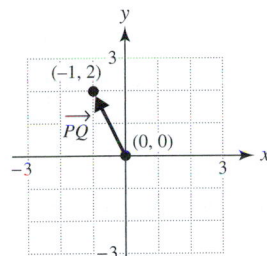

(b) $\overrightarrow{PQ} = \langle -1, 2 \rangle$ **(c)** $\|\overrightarrow{PQ}\| = \sqrt{5}$

21. (a)

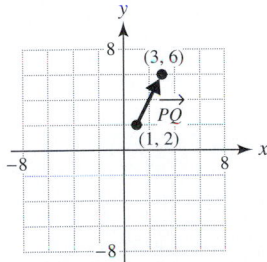

(b) $\overrightarrow{PQ} = \langle 2, 4 \rangle$ **(c)** $\|\overrightarrow{PQ}\| = \sqrt{20}$

23. (a)

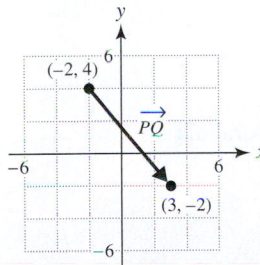

(b) $\overrightarrow{PQ} = \langle 5, -6 \rangle$ **(c)** $\|\overrightarrow{PQ}\| = \sqrt{61}$

25. (a) $\langle -4, 16 \rangle$ **(b)** $\langle -12, 0 \rangle$ **(c)** $\langle 8, -8 \rangle$

27. (a) $\langle 8, 0 \rangle$ **(b)** $\langle 0, 16 \rangle$ **(c)** $\langle -4, -8 \rangle$

29. (a) $\langle 0, 12 \rangle$ **(b)** $\langle -16, -4 \rangle$ **(c)** $\langle 8, -4 \rangle$

31. (a) $\langle 3, 2 \rangle$ **(b)** $\langle -3, 2 \rangle$

33. (a) $3\mathbf{i} - \mathbf{j}$ **(b)** $\mathbf{i} + 3\mathbf{j}$

35. (a) $\langle 0, -\frac{1}{4} \rangle$ **(b)** $\langle -2\sqrt{2}, \frac{5}{4} \rangle$

37. (a) $\frac{\sqrt{2}}{2}\mathbf{i} + \frac{\sqrt{2}+2}{2}\mathbf{j}$ **(b)** $\frac{\sqrt{2}}{2}\mathbf{i} + \frac{\sqrt{2}-2}{2}\mathbf{j}$

39. (a) 2 **(b)** $4\mathbf{i}$ **(c)** $7\mathbf{i} + 3\mathbf{j}$

41. (a) $\sqrt{5}$ **b)** $\langle -2, 4 \rangle$ **(c)** $\langle 7, 4 \rangle$

43. Horizontal = $4\cos 40° \approx 3.06$
Vertical = $4\sin 40° \approx 2.57$

45. Horizontal = $5\cos(-35°) \approx 4.10$
Vertical = $5\sin(-35°) \approx -2.87$

47. 94.2 lb   **49.** 24.4 lb

51. (a) 1 **(b)** $81.9°$ **(c)** Neither

53. (a) 0 **(b)** $90°$ **(c)** Perpendicular

55. (a) 122 **(b)** $0°$ **(c)** Parallel, same direction

57. (a) -4 **(b)** $143.1°$ **(c)** Neither

59. 150 ft-lb   **61.** 100,000 ft-lb

63. Work = 590 ft-lb, $\|\mathbf{F}\| = \sqrt{500} \approx 22.4$ lb

65. Work = 27 ft-lb, $\|\mathbf{F}\| = \sqrt{34} \approx 5.8$ lb

67. 24   **69.** 4

71. $\mathbf{v} = \langle 2, 3 \rangle$, speed = $\sqrt{13} \approx 3.6$ mph

73. $\mathbf{v} \approx \langle -364.6, -35.4 \rangle$, groundspeed ≈ 366.3 mph, bearing $\approx 264.5°$

75. Groundspeed ≈ 431.3 mph, bearing $\approx 159.1°$

77. Airspeed ≈ 149.3 mph, groundspeed ≈ 154.6 mph

79. (a) $\|\mathbf{R}\| = \sqrt{5} \approx 2.2$, $\|\mathbf{A}\| = \sqrt{1.25} \approx 1.1$.
About 2.2 inches of rain fell. The area of the opening of the rain gauge is about 1.1 square inches.
(b) $V = 1.5$; the volume of rain collected in the gauge was 1.5 cubic inches.
(c) $\mathbf{R}$ and $\mathbf{A}$ should be parallel and point in opposite directions.

81. (a) $\mathbf{c} = \mathbf{a} + \mathbf{b} = \langle 1, 4 \rangle$ **(b)** $\sqrt{17} \approx 4.1$ ft
(c) $3\mathbf{a} + \frac{1}{2}\mathbf{b} = \langle 8, 7 \rangle$

83. (a) $(2, 2)$ **(b)** $\mathbf{b} = \mathbf{a} + \mathbf{v}$

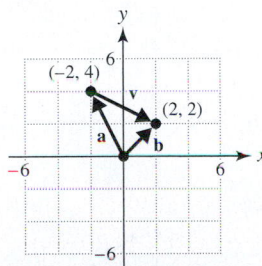

85. $W \approx 297{,}228$ ft-lb

87. Speed ≈ 180 mph, bearing $\approx 128.2°$

8.3 EXTENDED AND DISCOVERY EXERCISES (PP. 690–691)

1. Blue   **3. (a)** Blue   **(b)** White

SECTION 8.4 (PP. 699–701)

1. (a)

t	0	1	2	3
x	−1	0	1	2
y	0	2	4	6

(b)

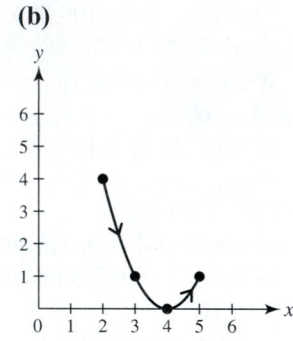

(c) Line segment

3. (a)

t	0	1	2	3
x	2	3	4	5
y	4	1	0	1

(b)

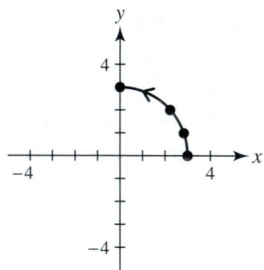

(c) Lower portion of a parabola

5. (a)

t	0	1	2	3
x	3	$\sqrt{8}$	$\sqrt{5}$	0
y	0	1	2	3

(b)

(graph)

(c) Portion of a circle with radius 3

7. (a)

t	0	1	2	3
x	0	1	2	3
y	3	$\sqrt{8}$	$\sqrt{5}$	0

(b)

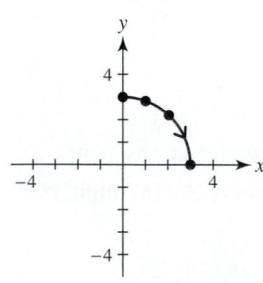

(c) Portion of a circle with radius 3

9. $y = \frac{1}{3}x - 1$; line

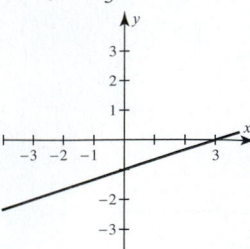

11. $x = 3(y - 1)^2$ parabola

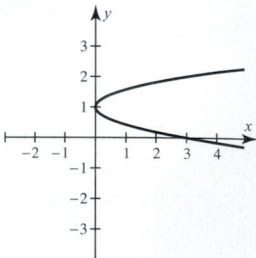

13. $x = \sqrt{1 - y^2}$; portion of a circle with radius 1

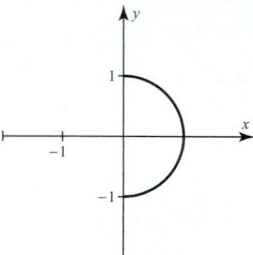

15. $y = \frac{1}{2}x^2$; portion of a parabola

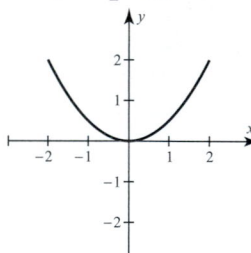

17. $y = x^2 + 4x + 5$; portion of a parabola

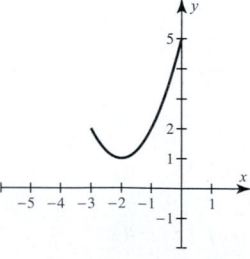

19. $x^2 + y^2 = 9$; circle with radius 3

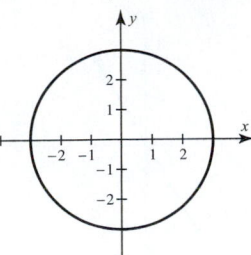

21.

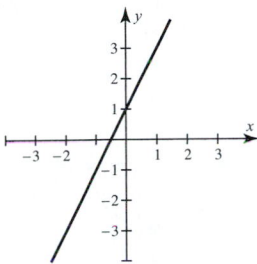

23.

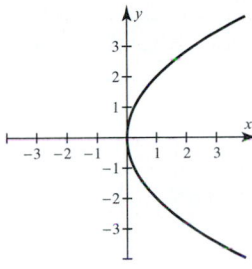

25.

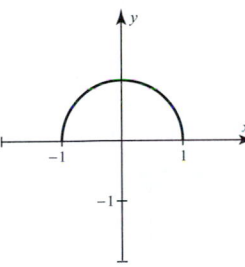

27.

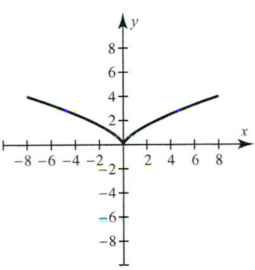

29. $[0, 6, 1]$ by $[-2, 2, 1]$

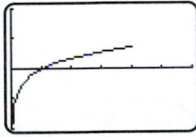

31. $[-2, 20, 2]$ by $[-1, 3, 1]$

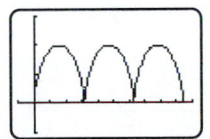

33.
$[-4.7, 4.7, 1]$ by $[-3.1, 3.1, 1]$

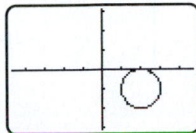

35.
$[-1.5, 1.5, 0.5]$ by $[-1, 1, 0.5]$

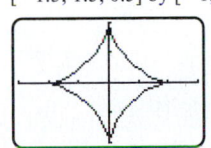

37.
$[-4.7, 4.7, 1]$ by $[-3.1, 3.1, 1]$

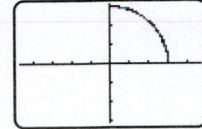

39. $x = t, y = 4 - 2t$ **41.** $x = t, y = 4 - t^2$
43. $x = t^2 + t - 3, y = t$ **45.** $x = 2 \cos t, y = 2 \sin t$
47. $x = t, y = e^{0.1t^2}$ **49.** $x = t^2 - 2t + 1, y = t$
51. (a) Traces a circle of radius 3 once
$[-4.7, 4.7, 1]$ by $[-3.1, 3.1, 1]$

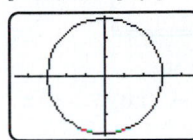

(b) Traces a circle of radius 3 twice
$[-4.7, 4.7, 1]$ by $[-3.1, 3.1, 1]$

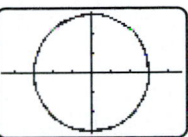

53. (a) Traces a circle of radius 3 once counterclockwise starting at $(3, 0)$
$[-4.7, 4.7, 1]$ by $[-3.1, 3.1, 1]$

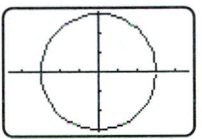

(b) Traces a circle of radius 3 once clockwise starting at $(0, 3)$
$[-4.7, 4.7, 1]$ by $[-3.1, 3.1, 1]$

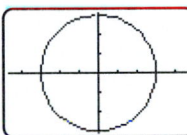

55. (a) Circle of radius 1 centered at $(-1, 2)$
$[-4.7, 4.7, 1]$ by $[-3.1, 3.1, 1]$

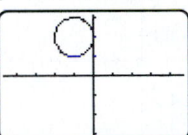

(b) Circle of radius 1 centered at $(1, 2)$
$[-4.7, 4.7, 1]$ by $[-3.1, 3.1, 1]$

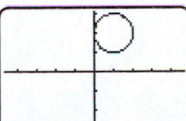

57. F
[0, 6, 1] by [0, 4, 1]

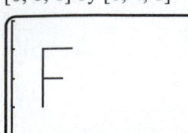

59. D
[0, 6, 1] by [0, 4, 1]

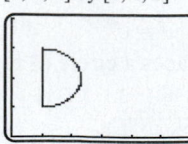

31.

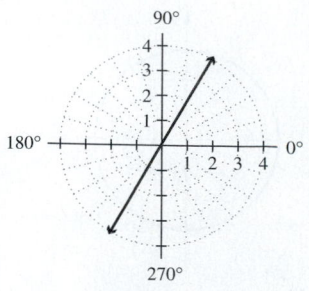

61. $x_1 = 0, y_1 = 2t; x_2 = t, y_2 = 0; 0 \le t \le 1$
63. $x_1 = \sin t, y_1 = \cos t; x_2 = 0, y_2 = t - 2; 0 \le t \le \pi.$
65. Answers may vary.
67. The ball hit at 35° travels about 128 feet. The ball hit at 50° travels about 134 feet.
69. Yes   **71.** About 285 feet
73. [−6, 6, 1] by [−4, 4, 1]

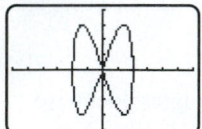

75. [−6, 6, 1] by [−4, 4, 1]

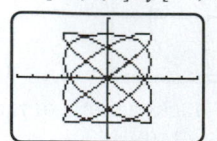

33. (a)

θ	0°	90°	180°	270°
r	2	2	2	2

(b)

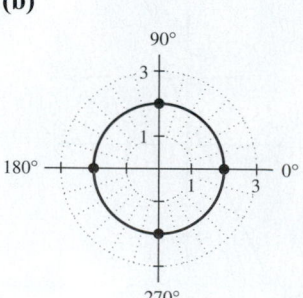

8.4 EXTENDED AND DISCOVERY EXERCISES (P. 701)

1. $F_2 = 100$ **3.** $F_2 = 300$

CHECKING BASIC CONCEPTS FOR SECTIONS 8.3 AND 8.4 (P. 701)

1. (a) $\mathbf{v} = \langle 4, 4 \rangle$ **(b)** $\|\mathbf{v}\| = 4\sqrt{2}$ **(c)** $\langle 0, 0 \rangle$
3. (a) -9 **(b)** $142.1°$

35. (a)

θ	0°	90°	180°	270°
r	0	3	0	−3

(b)

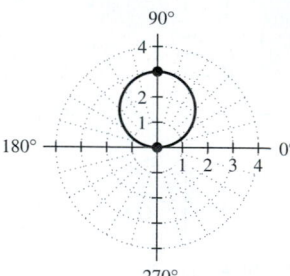

SECTION 8.5 (PP. 712−714)

1.

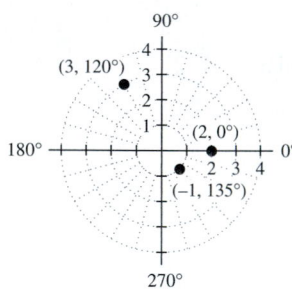

3.

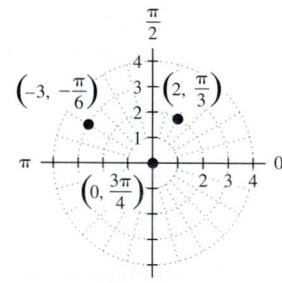

37. (a)

θ	0°	90°	180°	270°
r	2	4	2	0

(b)

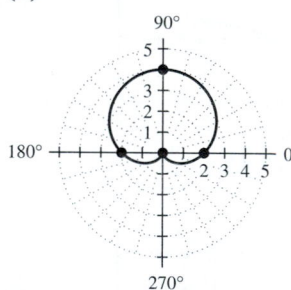

5. Yes   **7.** No   **9.** Yes
11. $\left(\frac{3}{\sqrt{2}}, \frac{3}{\sqrt{2}}\right)$ or $\left(\frac{3\sqrt{2}}{2}, \frac{3\sqrt{2}}{2}\right)$ **13.** $(0, 10)$
15. $(5, 0)$ **17.** $\left(-\frac{3}{2}, -\frac{3\sqrt{3}}{2}\right)$ **19.** $(0, -2)$
21. (a) $(3, 90°)$ **(b)** $(-3, -90°)$
23. (a) $(2, 240°)$ **(b)** $(-2, 60°)$
25. (a) $(\sqrt{18}, 315°)$ **(b)** $(-\sqrt{18}, 135°)$
27. $(25, 1.29)$ **29.** $(13, 1.97)$

39. (a)

θ	0°	90°	180°	270°
r	1	2	3	2

(b)

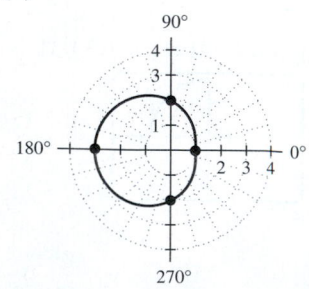

41.

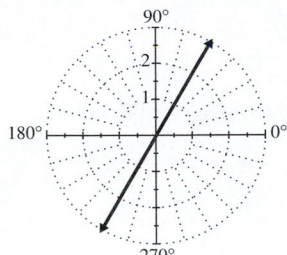

43.

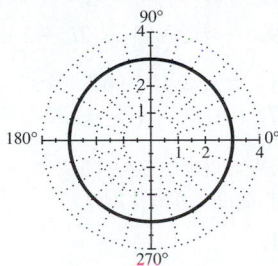

45.

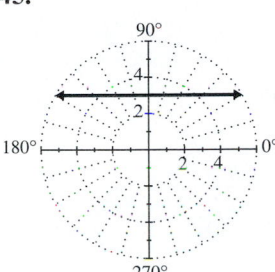

47.

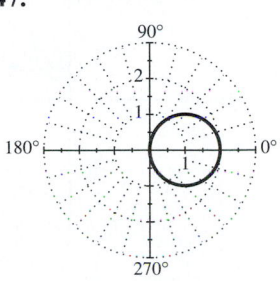

49.

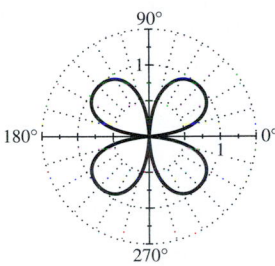

51.

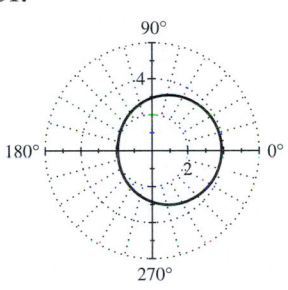

53.

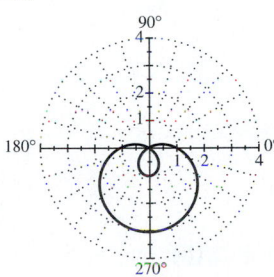

55.

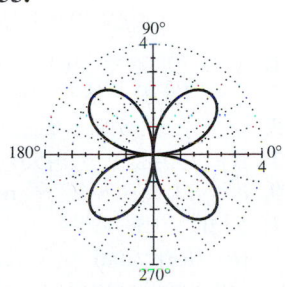

57.

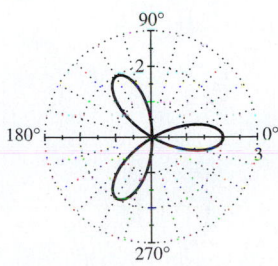

59. $r = 3 \csc \theta$ **61.** $\theta = \frac{\pi}{4}$ **63.** $r = 3$
65. $r = 2 \cos \theta$ **67.** $x^2 + y^2 = 9$ **69.** $x = 2$
71. $2x + 4y = 3$ **73.** $x^2 + y^2 = x$

75.
[−9.4, 9.4, 1] by [−6.2, 6.2, 1]

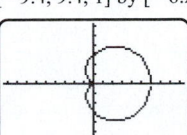

77.
[−9.4, 9.4, 1] by [−6.2, 6.2, 1]

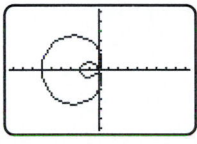

79.
[−9.4, 9.4, 1] by [−6.2, 6.2, 1]

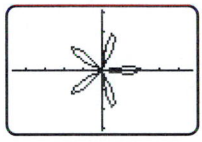

81.
[−9.4, 9.4, 1] by [−6.2, 6.2, 1]

83.
[−4.7, 4.7, 1] by [−3.1, 3.1, 1]

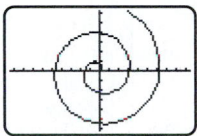

85. Use radian mode with $0 \le \theta \le \frac{9\pi}{2}$.
[−9.4, 9.4, 1] by [−6.2, 6.2, 1]

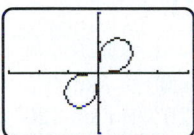

87. Let $r_1 = \sqrt{2 \sin (2\theta)}$ and $r_2 = -\sqrt{2 \sin (2\theta)}$.
[−3, 3, 1] by [−2, 2, 1]

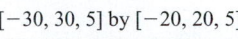

89. 30°, 150° **91.** 210°, 330° **93.** 30°, 150°
95.
[−30, 30, 5] by [−20, 20, 5]

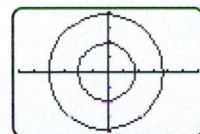

97.
[−3, 3, 1] by [−2, 2, 1]

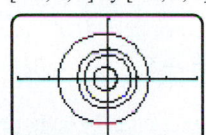

99. Inside the "figure eight" the radio signal can be received. This region is generally in an east–west direction from the two radio towers with a maximum distance of 200 miles.
[−300, 300, 100] by [−200, 200, 100]

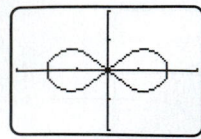

8.5 EXTENDED AND DISCOVERY EXERCISES (P. 714)

1. $y = x \tan (\ln (x^2 + y^2))$

$\dfrac{y}{x} = \tan (\ln r^2)$

$\tan \theta = \tan (\ln r^2)$

$\theta = \ln r^2$

$\theta = 2 \ln r$

$\dfrac{\theta}{2} = \ln r$

$e^{\theta/2} = r$

SECTION 8.6 (PP. 723–725)

1.

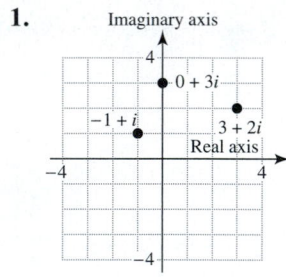

3.

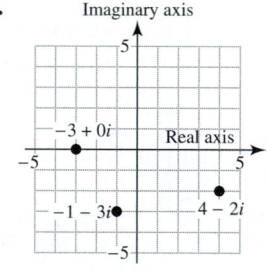

5. -5 **7.** $\sqrt{2} + i\sqrt{2}$ **9.** $-4i$ **11.** 3
13. $\sqrt{2}$ **15.** 13 **17.** 6 **19.** $\sqrt{13}$
21. $\sqrt{2}(\cos 45° + i \sin 45°)$ **23.** $5(\cos 0° + i \sin 0°)$
25. $4(\cos 90° + i \sin 90°)$ **27.** $2(\cos 120° + i \sin 120°)$
29. $2(\cos 30° + i \sin 30°)$ **31.** $2(\cos \pi + i \sin \pi)$

33. $\sqrt{8}\left(\cos \frac{3\pi}{4} + i \sin \frac{3\pi}{4}\right)$

35. $z_1 z_2 = \frac{27}{2} + \frac{27\sqrt{3}}{2}i, \quad \frac{z_1}{z_2} = \frac{3\sqrt{3}}{2} + \frac{3}{2}i$

37. $z_1 z_2 = -6, \quad \frac{z_1}{z_2} = 6i$

39. $z_1 z_2 = \frac{\sqrt{3}}{2} - \frac{1}{2}i, \quad \frac{z_1}{z_2} = \frac{1}{2} + \frac{\sqrt{3}}{2}i$

41. $8i$ **43.** 1 **45.** $-\frac{25}{2} + \frac{25\sqrt{3}}{2}i$ **47.** $-2 + 2i$
49. $-16\sqrt{3} + 16i$ **51.** $1 + i\sqrt{3}, -1 - i\sqrt{3}$
53. $-1, \frac{1}{2} + \frac{\sqrt{3}}{2}i, \frac{1}{2} - \frac{\sqrt{3}}{2}i$
55. $\frac{\sqrt{2}}{2} + \frac{\sqrt{2}}{2}i, -\frac{\sqrt{2}}{2} - \frac{\sqrt{2}}{2}i$

57. $-2, 1 + i\sqrt{3}, 1 - i\sqrt{3}$
59. $-4i, 2\sqrt{3} + 2i, -2\sqrt{3} + 2i$
61. $\pm 3, \pm 3i$ **63.** Yes   **65.** No
67. (a) $Z = 110 + 32i$ **(b)** $\sqrt{13{,}124} \approx 114.6$ ohms

CHECKING BASIC CONCEPTS FOR SECTIONS 8.5 AND 8.6 (P. 725)

1.

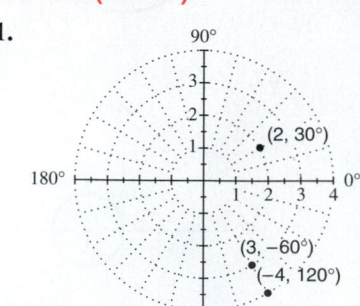

3.

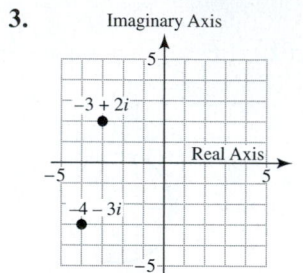

5. $\frac{\sqrt{3}}{2} + \frac{1}{2}i, -\frac{\sqrt{3}}{2} + \frac{1}{2}i, -i$

CHAPTER 8 REVIEW EXERCISES (PP. 729–733)

1. $\gamma = 70°, a = 10.1, b \approx 6.9$
3. $a \approx 5.5, \beta \approx 59.1°, \gamma \approx 78.9°$
5. $\gamma = 115°, a \approx 5.9, c \approx 16.4$
7. $\beta \approx 14.4°, \alpha \approx 145.6°, a \approx 18.2$
9. $c \approx 13.2, \beta \approx 93.6°, \alpha \approx 51.4°$
11. 53.0 **13.** 891.4
15. (a) Horizontal $= 3$, vertical $= 4$ **(b)** $\|\mathbf{v}\| = 5$
(c) $\|\mathbf{v}\|$ represents the length of $\mathbf{v}$.

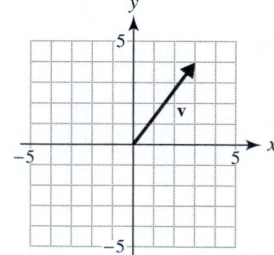

17. (a)

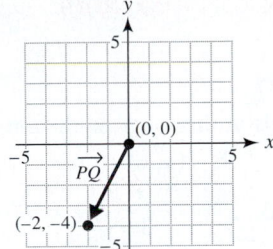

(b) $\overrightarrow{PQ} = -2\mathbf{i} - 4\mathbf{j}$ **(c)** $\|\overrightarrow{PQ}\| = \sqrt{20}$
19. (a) $2\mathbf{a} = \langle 6, -4 \rangle$ **(b)** $\mathbf{a} - 3\mathbf{b} = \langle 0, -5 \rangle$
(c) $\mathbf{a} \cdot \mathbf{b} = 1$ **(d)** $\theta \approx 78.7°$
21. (a) $2\mathbf{a} = 4\mathbf{i} + 4\mathbf{j}$ **(b)** $\mathbf{a} - 3\mathbf{b} = -\mathbf{i} - \mathbf{j}$
(c) $\mathbf{a} \cdot \mathbf{b} = 4$ **(d)** $\theta = 0°$
23. About 207.1 lb
25.

$[-4.7, 4.7, 1]$ by $[-3.1, 3.1, 1]$

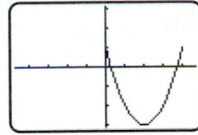

27.

$[-4.7, 4.7, 1]$ by $[-3.1, 3.1, 1]$

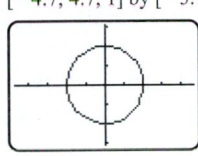

29. Answers may vary. (Graph with the axes turned off.)
(a) $x = \cos t, y = 2 \sin t; 0 \le t \le 2\pi$
(b) $x_1 = -1 + \frac{t}{\pi}, y_1 = 0; 0 \le t \le 2\pi$
$x_2 = -1, y_2 = -1 + \frac{t}{\pi}; 0 \le t \le 2\pi$
$x_3 = 1, y_3 = -1 + \frac{t}{\pi}; 0 \le t \le 2\pi$
31. (a)

θ	0°	90°	180°	270°
r	2	1	0	1

(b)

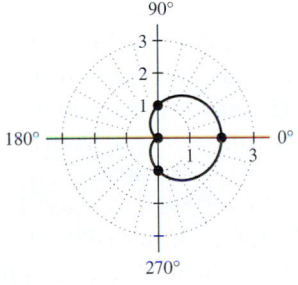

33.

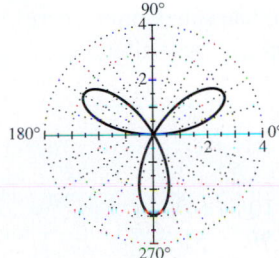

35.

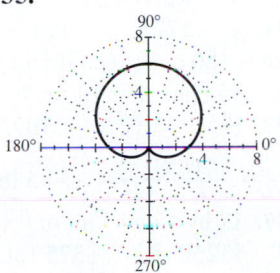

37.

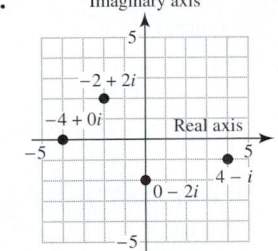

39. $z_1 z_2 = -8, \frac{z_1}{z_2} = -1 + i\sqrt{3}$
41. $\sqrt{3} + i, -\sqrt{3} - i$ **43.** 701.6 mi
45. 7204 ft **47.** 15,600 ft^2
49. (a) $\|\mathbf{I}\| = \sqrt{34,000} \approx 184.4, \|\mathbf{S}\| = \sqrt{29} \approx 5.4$
The intensity of the sun is about 184.4 watts per square
meter. The area of the solar panel is about 5.4 square meters.
(b) $W = |\mathbf{I} \cdot \mathbf{S}| = 820$. The solar panel is collecting 820
watts.
51. About 204,985 ft-lb **53.** 10,285 ft

CHAPTER 8 EXTENDED AND DISCOVERY EXERCISES (PP. 731–733)

1. (a) About 56 miles per second
(b) About 87 miles per second
3. (a) About 6.02 in. **(b)** About 7470 ft
5. (a) 40 ft **(b)** About 35.3 ft **(c)** About 100.3 ft

CHAPTERS 1–8 CUMULATIVE REVIEW EXERCISES (PP. 733–737)

1. 9.12×10^4; 0.006734 **3.** $\sqrt{65}$
5. $\{x \mid x \le 4\}; 3$ **7.** 10 **9.** $y = -\frac{7}{2}x + \frac{1}{2}$
11. 300: Initially there are 300 gallons of water in the tank.
-10: Water is being removed from the tank at a rate of 10
gallons per second.
13. $f(-1) = 1; f$ is continuous.

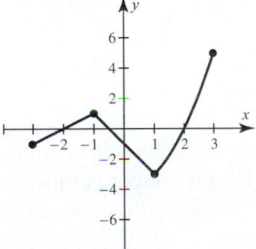

15. (a) $-2, -1, 1$ **(b)** $(-\infty, -2) \cup (-1, 1)$
(c) $[-2, -1] \cup [1, \infty)$
17. $(1, -1)$ **19.** $g(x) = x^2 + 6x + 2$
21. (a) Incr.: $[-2, 1]$; decr.: $(-\infty, -2] \cup [1, \infty)$
(b) $-3.2, -0.4, 2$ **(c)** $(-2, -15), (1, 12)$
(d) Local min: -15, local max: 12

23. Down on both ends;
$f(x) \to -\infty$ as $x \to -\infty$, $f(x) \to -\infty$ as $x \to \infty$.
25. $f(x) = 4(x + 2)(x - 3)(x - 5)$
27. $f(x) = 6(x - 1)(x + 1)(x - i)(x + i)$;
$f(x) = 6x^4 - 6$
29. $14 + 8i$
31. $D = \{x \mid x \neq \frac{2}{3}\}$; vertical: $x = \frac{2}{3}$; horizontal: $y = -1$
33. **(a)** 6 **(b)** 3 **(c)** $2x^3 + x^2 - 6x - 3$
(d) $\dfrac{1}{(2x + 1)^2 - 3}$

35. $f(x) = 2(3)^x$ **37.** \$1742.21

39. $\ln x + \frac{1}{3}\ln y - \frac{2}{3}\ln z$ **41.** $\alpha = 67°5'; \beta = 157°5'$
43. $\frac{5\pi}{4}$ **45.** $\sin\theta = \frac{5}{13}, \cos\theta = \frac{12}{13}, \tan\theta = \frac{5}{12}$
$\csc\theta = \frac{13}{5}, \sec\theta = \frac{13}{12}, \cot\theta = \frac{12}{5}$
47. $\sin\theta = -\frac{60}{61}, \cos\theta = \frac{11}{61}, \tan\theta = -\frac{60}{11}$
$\csc\theta = -\frac{61}{60}, \sec\theta = \frac{61}{11}, \cot\theta = -\frac{11}{60}$
49. $\sin\theta = -\frac{7}{25}, \cos\theta = -\frac{24}{25}, \tan\theta = \frac{7}{24}$,
$\csc\theta = -\frac{25}{7}, \sec\theta = -\frac{25}{24}, \cot\theta = \frac{24}{7}$
51. $c = 13, \theta \approx 22.6°, \beta \approx 67.4°$
53. $(\cot\theta - 1)^2$ **55.** $\frac{63}{65}$
57. **(a)** $\beta = 96, a \approx 7.8, c \approx 12.0$
(b) $\beta \approx 25.4, \gamma \approx 123.6, c \approx 9.7$
(c) $\alpha \approx 47.0, \gamma \approx 77.0, b \approx 6.8$
(d) $\alpha \approx 46.6, \beta \approx 57.9, \gamma \approx 75.5$
59. **(a)** 25 **(b)** $\langle -31, 96 \rangle$ **(c)** -323
(d) About $173.6°$
61.

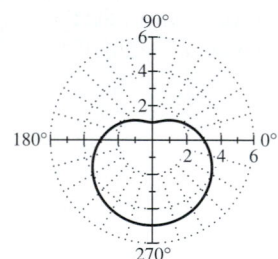

63. $m_1 = 5$: the jogger is moving away from home at 5 mph.
$m_2 = 0$: the jogger is stopped. $m_3 = 10$: the jogger is moving away from home at 10 mph. $m_4 = -6.\overline{6}$: the jogger is moving toward home at $6.\overline{6}$ mph.
65. 1.25 lb of \$3.50 candy and 3.75 lb of \$4.50 candy
67. $f(x) = -3(x - 2)^2 + 1$ **69.** 2.15 in.
71. **(a)** $C(t) = 630(t - 1980) + 3600$
(b) About 1992, which is one year off from the actual value.
73. 25 ft/sec **75.** $f(x) = 25\sin\left(\frac{\pi}{6}(x - 4)\right) + 50$
77. May and September
79. **(a)** About 247 ft **(b)** About 18,988 ft^2
81. About 249.4 ft

CHAPTER 9: Systems of Equations and Inequalities

SECTION 9.1 (PP. 750–754)

1. 20; the area of a triangle with base 5 and height 8 is 20.
3. 13 **5.** -18 **7.** $\frac{4}{5}$ **9.** $f(x, y) = y + 2x$
11. $f(x, y) = \dfrac{xy}{1 + x}$ **13.** $x = \dfrac{4y + 7}{3}; y = \dfrac{3x - 7}{4}$
15. $x = y^2 + 5; y = \pm\sqrt{x - 5}$ **17.** $x = 2y; y = \dfrac{x}{2}$
19. **(a)** $b = \dfrac{2A}{h}$ **(b)** $h = \dfrac{2A}{b}$
21. **(a)** $x = \pm\sqrt{\dfrac{z(y + 1)}{2}}$ **(b)** $y = \dfrac{2x^2}{z} - 1$
23. $(2, 1)$ **25.** $(4, -3)$ **27.** $(2, 2)$ **29.** $\left(\frac{1}{2}, -2\right)$
31. $(-2, 1)$ **33.** $(-6, 4)$ **35.** $\left(\frac{1}{2}, 2\right)$ **37.** $(6, 8)$
39. Infinitely many solutions: $\{(x, y) \mid 3x - 2y = 5\}$
41. No real solutions
43. $(4, 3)$ **45.** $(2, 1)$ **47.** $(1, -1)$ **49.** $(1, 2)$
51. $(-2, 4), (0, 0)$ **53.** $(2, 4), (4, 2)$
55. $(2, 4), (-2, -4)$ **57.** $(2, \sqrt{2})$
59. No real solutions
61. Infinitely many solutions: $\{(x, y) \mid 2x^2 - y = 5\}$
63. $(-2, 0), (2, 0)$ **65.** $(-2, -2), (0, 0), (2, 2)$
67. $(-\sqrt{8}, -\sqrt{8}), (\sqrt{8}, \sqrt{8})$ **69.** $(6, 2), (-2, -6)$
71. $l = 7, w = 5$; A rectangle with area 35 and perimeter 24 has length 7 and width 5.
73. $(-1.588, 0.239), (0.164, 1.487), (1.924, -0.351)$
75. $(1.220, 0.429)$ **77.** $(0.714, -0.169)$
79. $S(86, 185) \approx 2.1 \text{ m}^2$ **81.** $S(60, 157.48) \approx 1.6 \text{ m}^2$
83. $E(1960, \text{Male}) = 66.6$. A male born in 1960 had a life expectancy at birth of 66.6 yr.
85. **(a)** Let x represent the number of robberies in 2000 and y the number in 2001.
$x + y = 831,000$
$x - y = 15,000$
(b) & (c) $(423000, 408000)$
87. 10 hr in 2001, 11.3 hr in 2002
89. $r \approx 1.538$ in., $h \approx 6.724$ in.
91. $x \approx 177.1, y \approx 174.9$; if an athlete's maximum heart rate is 180 bpm, then it will be about 177 bpm after 5 sec and 175 bpm after 10 sec.
93. $W_1 = \dfrac{300}{1 + \sqrt{3}} \approx 109.8$ lb,
$W_2 = \dfrac{300\sqrt{3}}{\sqrt{6} + \sqrt{2}} \approx 134.5$ lb
95. 12 by 12 by 4 in. or 9.10 by 9.10 by 6.96 in.
97. \$1250 at 5%; \$3750 at 7% **99.** 0.51
101. 32.4 **103.** Approximately 10,823 watts
105. Approximately 2.54 cords
107. $S(w, h) = 0.0101(w)^{0.425}(h)^{0.725}$

9.1 EXTENDED AND DISCOVERY EXERCISES (P. 754)

1. $f(255, 0, 255)$ represents purple, since red and blue combine to make purple.

3. One possibility would be $f(0, 255, 255)$, which would be equal amounts of green and blue light.

5. 16,777,216 different colors.

SECTION 9.2 (PP. 767–772)

1. The system is consistent with solution $(2, 2)$.

3. The system is inconsistent.

5. $(14, 6)$; consistent and independent

7. $(1, 3)$; consistent and independent

9. $\{(x, y) \mid x + y = 500\}$; consistent and dependent

11. $(-2, 5)$; consistent and independent

13. Inconsistent

15. $\left(-\frac{11}{7}, \frac{12}{7}\right)$; consistent and independent

17. $(2, -4)$ **19.** $\left(\frac{1}{3}, -\frac{1}{3}\right)$

21. Infinitely many solutions of the form $\{(x, y) \mid 7x - 3y = -17\}$ **23.** No solutions

25. $\left(-2, \frac{5}{2}\right)$ **27.** $(10, 20)$

29. 10.5 in. wide, 8.5 in. high

31. 23 child tickets, 52 adult tickets   **33.** $(3, 3), (-3, 3)$

35. $(4, 3), (-4, 3), (3, 4), (-3, 4)$ **37.** $(2, 0), (-2, 0)$

39. $(1, 1)$

41.
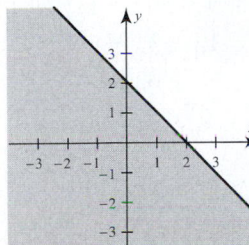

43.
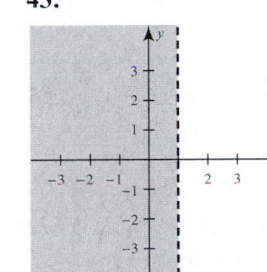

45.
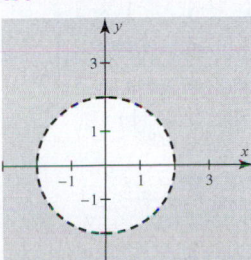

47.
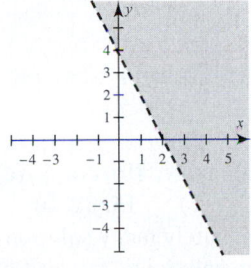

49.
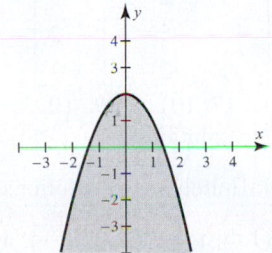

51.

53. c.; one solution is $(2, 3)$. Answers may vary.

55. d.; one solution is $(-1, -1)$. Answers may vary.

57. One solution is $(0, 2)$. Answers may vary.

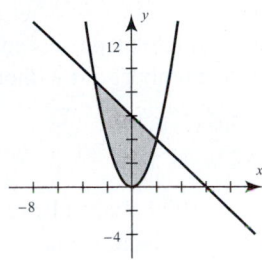

59. One solution is $(0, 0)$. Answers may vary.

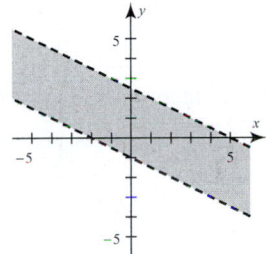

61. One solution is $(-1, 1)$. Answers may vary.

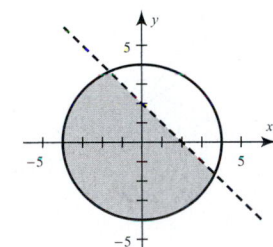

63.
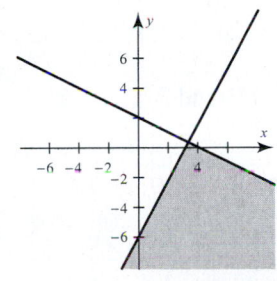

65.
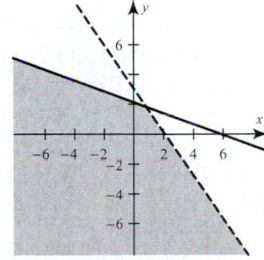

67.
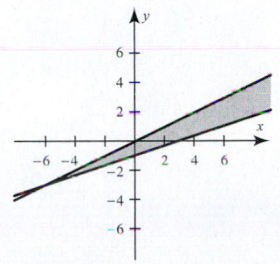

69.
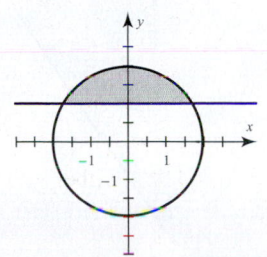

71. (a) $x + y = 3000, 0.08x + 0.10y = 264$
(b) $1800 at 8% and $1200 at 10%
73. There are no solutions. If loans totaling $3000 are at 10%, then the interest must be $300.
75. (a) Let x represent the number of television sets sold with stereo sound and y represent the number sold without stereo sound.
$$x + y = 32$$
$$\tfrac{19}{10}x - y = 0$$
(b) Approximately (11.03, 20.97); in 1999 about 11 million television sets were sold with stereo sound and about 21 million were sold without stereo sound.
77. 5.4 trillion sold in 1990; 5.5 trillion sold in 2000
79. Air speed: 520 mph; wind speed: 40 mph
81. $x = 300, y = 350$
83. This individual weighs less than recommended for his or her height.
85. $25h - 7w \le 800$
$5h - w \ge 170$
87. 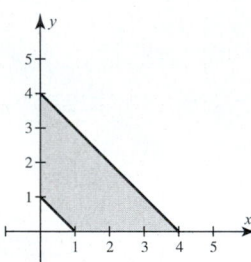   **89.**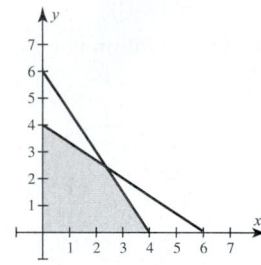

91. Maximum: 65; minimum: 8
93. Maximum: 66; minimum: 3
95. Maximum: 100; minimum: 0
97. $x + y \le 4, x \ge 0, y \ge 0$ **99.** Minimum: 6
101. Maximum: $z = 56$; minimum: $z = 24$
103. 25 radios, 30 CD players
105. 2.4 units of Brand A, 1.2 units of Brand B
107. $600   **109.** Part X: 9; part Y: 4

CHECKING BASIC CONCEPTS FOR SECTIONS 9.1 AND 9.2 (P. 772)

1. $d(13, 18) = 20$ **3.** $y = \pm\sqrt{z - x^2}$
5.

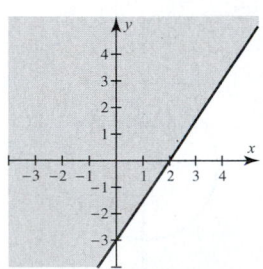

7. System: $x + y = 13{,}820$ and $x - y = 480$; the United States had 7150 nuclear warheads, and the Soviet Union had 6670 nuclear warheads.

SECTION 9.3 (PP. 778–779)

1. No   **3.** 2
5. $(0, 2, -2)$ is not, but $(-1, 3, -2)$ is a solution.
7. Both are solutions.   **9.** $(1, 2, 3)$ **11.** $(1, 0, 1)$
13. $\left(\tfrac{1}{2}, \tfrac{1}{2}, -2\right)$
15. Infinitely many solutions: $\left(\tfrac{z + 3}{2}, \tfrac{-z + 3}{2}, z\right)$
17. No solutions   **19.** $\left(-\tfrac{5}{2}, -2, 4\right)$
21. No solutions   **23.** $\left(-\tfrac{1}{2}, \tfrac{1}{2}, -\tfrac{1}{2}\right)$
25. Infinitely many solutions: $\left(\tfrac{-5z + 18}{13}, \tfrac{-6z + 19}{13}, z\right)$
27. $\left(8, -11, -\tfrac{1}{2}\right)$ **29.** $(2, 3, -4)$
31. 120 child, 280 student, and 100 adult tickets
33. No solutions; at least one student was charged incorrectly.
35. (a) $\begin{aligned} x + y + z &= 180 \\ x \phantom{{}+ y} - z &= 25 \\ -x + y + z &= 30 \end{aligned}$
(b) 75°, 55°, 50°
37. $2500 at 5%, $7500 at 7%, $10,000 at 10%
39. (a) $\begin{aligned} N + P + K &= 80 \\ N + P - K &= 8 \\ 9P - K &= 0 \end{aligned}$
(b) $(40, 4, 36)$; 40 lb of nitrogen, 4 lb of phosphorus, 36 lb of potassium

SECTION 9.4 (PP. 792–796)

1. (a) 3×1 **(b)** 2×3 **(c)** 2×2
3. Dimension: 2×3
$$\begin{bmatrix} 5 & -2 & 3 \\ -1 & 3 & -1 \end{bmatrix}$$
5. Dimension: 3×4
$$\begin{bmatrix} -3 & 2 & 1 & -4 \\ 5 & 0 & -1 & 9 \\ 1 & -3 & -6 & -9 \end{bmatrix}$$
7. $\begin{aligned} 3x + 2y &= 4 \\ y &= 5 \end{aligned}$
9. $\begin{aligned} 3x + y + 4z &= 0 \\ 5y + 8z &= -1 \\ -7z &= 1 \end{aligned}$
11. (a) Yes   **(b)** No   **(c)** Yes
13. $(5, -1)$ **15.** $(2, 0)$ **17.** $(2, 3, 1)$
19. Infinitely many solutions: $(3 - 3z, 1 + 2z, z)$
21. No solutions
23. $\begin{bmatrix} 1 & -2 & 3 & 5 \\ -3 & 5 & 3 & 2 \\ 1 & 2 & 1 & -2 \end{bmatrix}$ **25.** $\begin{bmatrix} 1 & -1 & 1 & 2 \\ 0 & 1 & -1 & 2 \\ 0 & 8 & -1 & 3 \end{bmatrix}$
27. $(-17, 10)$ **29.** $(0, 2, -1)$ **31.** $(3, 2, 1)$
33. No solutions
35. Infinitely many solutions: $\left(-1 - z, \dfrac{9 - 5z}{2}, z\right)$
37. $(1, 1, 1)$ **39.** $\left(\tfrac{1}{2}, -\tfrac{1}{2}, 4\right)$ **41.** $(-2, 5, -1)$

43. No solutions   **45.** $(-1, -2, 3)$

47. Infinitely many solutions: $\left(\dfrac{3z + 1}{7}, \dfrac{11z - 15}{7}, z\right)$

49. $(12, 3)$ **51.** $\left(-2, 4, \frac{1}{2}\right)$

53. Infinitely many solutions: $(4 - 2z, z - 3, z)$

55. No solutions   **57.** $(3, 2)$ **59.** $(-2, 1, 3)$

61. $(-2, 5, 7)$ **63.** No solutions

65. $(-9.226, -9.167, 2.440)$

67. $(5.211, 3.739, -4.655)$ **69.** $(7.993, 1.609, -0.401)$

71. (a) $F = 0.5714N + 0.4571R - 2014$ **(b)** $5700

73. Pump 1: 12 hours; pumps 2 and 3: 24 hours

75. (a)
$$\begin{aligned} x + \quad y + \quad z &= 5000 \\ x + \quad y - \quad z &= 0 \\ 0.08x + 0.11y + 0.14z &= 595 \end{aligned}$$

(b) $1000 at 8%; $1500 at 11%; $2500 at 14%

77. $(3.53, 1.62, 1.91)$

79. (a) At intersection A, incoming traffic is equal to $x + 5$. The outgoing traffic is given by $y + 7$. Therefore, $x + 5 = y + 7$. The other equations can be justified in a similar way.

(b) The three equations can be written as
$$\begin{aligned} x - y &= 2 \\ x - z &= 3 \\ y - z &= 1 \end{aligned}$$
The solution can be written as $\{(z + 3, z + 1, z) \mid z \geq 0\}$.

(c) There are infinitely many solutions since some cars could be driving around the block continually.

81. (a)
$$\begin{aligned} 1990^2 a + 1990b + c &= 11 \\ 2010^2 a + 2010b + c &= 10 \\ 2030^2 a + 2030b + c &= 6 \end{aligned}$$

(b) $f(x) = -0.00375x^2 + 14.95x - 14{,}889.125$

(c) [1985, 2035, 5] by [5, 12, 1]

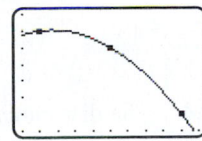

(d) Answers may vary. For example, in 2015 the predicted ratio is $f(2015) \approx 9.3$.

83. (a)
$$\begin{aligned} 3^2 a + \quad 3b + c &= 2.2 \\ 18^2 a + 18b + c &= 10.4 \\ 26^2 a + 26b + c &= 12.8 \end{aligned}$$

(b) $f(x) = -0.010725x^2 + 0.77188x - 0.019130$

(c) [0, 30, 5] by [0, 14, 1]

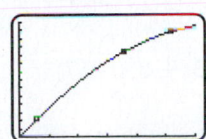

(d) Answers may vary. For example, in 1990 this percentage was $f(20) \approx 11.1$.

9.4 EXTENDED AND DISCOVERY EXERCISE (P. 796)

1. (a) $\begin{bmatrix} 1 & 19 & 57.5 & 32 & \vline & 125 \\ 1 & 26 & 65 & 42 & \vline & 316 \\ 1 & 30 & 72 & 48 & \vline & 436 \\ 1 & 30.5 & 75 & 54 & \vline & 514 \end{bmatrix}$

(b) $a \approx -552.272$, $b \approx 8.733$, $c \approx 2.859$, $d \approx 10.843$

(c) About 260 lb; a bear with a 24-inch neck, 63-inch length, and 39-inch chest weighs approximately 260 lb.

CHECKING BASIC CONCEPTS FOR SECTIONS 9.3 AND 9.4 (P. 796)

1. (a) $(3, 2, -1)$

(b) $\left\{\left(\dfrac{4 - 5z}{3}, \dfrac{5 - z}{3}, z\right) \mid z \text{ is a real number}\right\}$

(c) No solutions

3. $(2, -1, 0)$

SECTION 9.5 (PP. 807–810)

1. (a) $a_{12} = 2$, $a_{21} = 4$, a_{32} is undefined.

(b) $a_{11}a_{22} + 3a_{23} = 23$

3. (a) $a_{12} = -1$, $a_{21} = 3$, $a_{32} = 0$

(b) $a_{11}a_{22} + 3a_{23} = 13$

5. $x = 1$, $y = 1$ **7.** Not possible

9. (a) $A + B = \begin{bmatrix} 3 & 3 \\ 3 & 3 \end{bmatrix}$ **(b)** $B + A = \begin{bmatrix} 3 & 3 \\ 3 & 3 \end{bmatrix}$

(c) $A - B = \begin{bmatrix} 5 & -5 \\ -5 & 5 \end{bmatrix}$

11. (a) $A + B = \begin{bmatrix} 14 & 9 & -3 \\ 4 & -10 & 14 \\ 4 & 11 & 16 \end{bmatrix}$

(b) $B + A = \begin{bmatrix} 14 & 9 & -3 \\ 4 & -10 & 14 \\ 4 & 11 & 16 \end{bmatrix}$

(c) $A - B = \begin{bmatrix} -8 & -1 & 1 \\ -4 & 4 & -10 \\ -8 & -1 & 4 \end{bmatrix}$

13. (a) $A + B = \begin{bmatrix} 1 & -6 \\ 1 & 4 \end{bmatrix}$ **(b)** $3A = \begin{bmatrix} 6 & -18 \\ 9 & 3 \end{bmatrix}$

(c) $2A - 3B = \begin{bmatrix} 7 & -12 \\ 12 & -7 \end{bmatrix}$

15. (a) $A + B$ is undefined.

(b) $3A = \begin{bmatrix} 3 & -3 & 0 \\ 3 & 15 & 27 \\ -12 & 24 & -15 \end{bmatrix}$

(c) $2A - 3B$ is undefined.

17. (a) $A + B = \begin{bmatrix} 0 & -2 \\ -2 & 2 \\ 9 & -8 \end{bmatrix}$ (b) $3A = \begin{bmatrix} -6 & -3 \\ -15 & 3 \\ 6 & -9 \end{bmatrix}$

(c) $2A - 3B = \begin{bmatrix} -10 & 1 \\ -19 & -1 \\ -17 & 9 \end{bmatrix}$

19. $\begin{bmatrix} 0 & 2 \\ 13 & -5 \\ 0 & 1 \end{bmatrix}$ 21. $\begin{bmatrix} 2 & 6 \\ 11 & -9 \end{bmatrix}$

23. $\begin{bmatrix} 7 & 4 & 7 \\ 4 & 7 & 7 \\ 4 & 7 & 7 \end{bmatrix}$ 25. $A = \begin{bmatrix} 1 & 2 & 1 \\ 1 & 2 & 1 \\ 1 & 2 & 1 \end{bmatrix}$

27. $B = \begin{bmatrix} -1 & 1 & -1 \\ -1 & 1 & -1 \\ -1 & 1 & -1 \end{bmatrix}, A + B = \begin{bmatrix} 0 & 3 & 0 \\ 0 & 3 & 0 \\ 0 & 3 & 0 \end{bmatrix}$

29. $B = \begin{bmatrix} 3 & 3 & 3 \\ 3 & 3 & 3 \\ 3 & 3 & 3 \end{bmatrix}, B - A = \begin{bmatrix} 3 & 0 & 3 \\ 3 & 0 & 3 \\ 3 & 0 & 3 \end{bmatrix}$

31. $A = \begin{bmatrix} 3 & 3 & 3 & 3 \\ 3 & 0 & 0 & 0 \\ 3 & 3 & 3 & 0 \\ 3 & 0 & 0 & 0 \\ 3 & 0 & 0 & 0 \end{bmatrix}$

33. (a) One possibility is $A = \begin{bmatrix} 3 & 3 & 3 & 3 \\ 0 & 0 & 3 & 0 \\ 0 & 3 & 0 & 0 \\ 3 & 3 & 3 & 3 \end{bmatrix}$.

(b) $B = \begin{bmatrix} 3 & 3 & 3 & 3 \\ 3 & 3 & 3 & 3 \\ 3 & 3 & 3 & 3 \\ 3 & 3 & 3 & 3 \end{bmatrix}$

35. (a) One possibility is $A = \begin{bmatrix} 3 & 0 & 0 & 0 \\ 3 & 0 & 0 & 0 \\ 3 & 0 & 0 & 0 \\ 3 & 3 & 3 & 3 \end{bmatrix}$.

(b) $B = \begin{bmatrix} 3 & 3 & 3 & 3 \\ 3 & 3 & 3 & 3 \\ 3 & 3 & 3 & 3 \\ 3 & 3 & 3 & 3 \end{bmatrix}$

37. (a) $A = \begin{bmatrix} 12 & 4 \\ 8 & 7 \end{bmatrix}, B = \begin{bmatrix} 55 \\ 70 \end{bmatrix}$

(b) $AB = \begin{bmatrix} 940 \\ 930 \end{bmatrix}$. Tuition for Student 1 is $940, and tuition for Student 2 is $930.

39. (a) $A = \begin{bmatrix} 10 & 5 \\ 9 & 8 \\ 11 & 3 \end{bmatrix}, B = \begin{bmatrix} 60 \\ 70 \end{bmatrix}$

(b) $AB = \begin{bmatrix} 950 \\ 1100 \\ 870 \end{bmatrix}$. Tuition for Student 1 is $950, for Student 2 is $1100, and for Student 3 is $870.

41. $AB = \begin{bmatrix} -3 & 1 \\ -4 & 6 \end{bmatrix}, BA = \begin{bmatrix} 4 & 2 \\ 5 & -1 \end{bmatrix}$

43. AB and BA are undefined.

45. $AB = \begin{bmatrix} -15 & 22 & -9 \\ -2 & 5 & -3 \\ -32 & 18 & 6 \end{bmatrix}, BA = \begin{bmatrix} 5 & 14 \\ 20 & -9 \end{bmatrix}$

47. AB is undefined. $BA = \begin{bmatrix} -1 & -3 & 19 \\ -1 & 1 & -39 \end{bmatrix}$

49. AB and BA are undefined.

51. $AB = \begin{bmatrix} -1 & 15 & -2 \\ 0 & 1 & 3 \\ -1 & 14 & -5 \end{bmatrix}, BA = \begin{bmatrix} 0 & 6 & 0 \\ 6 & -5 & 9 \\ 0 & 1 & 0 \end{bmatrix}$

53. $AB = \begin{bmatrix} -1 \\ 6 \end{bmatrix}$. BA is undefined.

55. $AB = \begin{bmatrix} -7 & 8 & 7 \\ 18 & -32 & -8 \end{bmatrix}$, BA is undefined.

57. $AB = \begin{bmatrix} -5 \\ 10 \\ 17 \end{bmatrix}$, BA is undefined.

59. $AB = \begin{bmatrix} 350 \\ 230 \end{bmatrix}$; the total cost of order 1 is $350, and the total cost of order 2 is $230.

61. They both equal $\begin{bmatrix} 36 & 36 & 8 \\ -15 & -38 & -4 \\ -11 & 13 & 10 \end{bmatrix}$. The distributive property appears to hold for matrices.

63. They both equal $\begin{bmatrix} 50 & 3 & 12 \\ -6 & 55 & 8 \\ 27 & -3 & 29 \end{bmatrix}$. Matrices appear to conform to rules of algebra except that $AB \ne BA$.

9.5 EXTENDED AND DISCOVERY EXERCISES (PP. 810–811)

1. Aquamarine is represented by (0.369, 0, 0.067) in CMY.

3. $\begin{bmatrix} R \\ G \\ B \end{bmatrix} = \begin{bmatrix} 1 \\ 1 \\ 1 \end{bmatrix} - \begin{bmatrix} C \\ M \\ Y \end{bmatrix}$

SECTION 9.6 (PP. 821–824)

1. B is the inverse of A.   **3.** B is the inverse of A.

5. $k = 1$ **7.** $k = 2.5$ **9.** A **11.** A

13. **(a)** $(2, 4)$ **(b)** It will translate $(2, 4)$ to the left 2 units and downward 3 units, back to $(0, 1)$;

$$A^{-1} = \begin{bmatrix} 1 & 0 & -2 \\ 0 & 1 & -3 \\ 0 & 0 & 1 \end{bmatrix} \quad \textbf{(c)} \ I_3$$

15. $A = \begin{bmatrix} 1 & 0 & -3 \\ 0 & 1 & -5 \\ 0 & 0 & 1 \end{bmatrix}$ and $A^{-1} = \begin{bmatrix} 1 & 0 & 3 \\ 0 & 1 & 5 \\ 0 & 0 & 1 \end{bmatrix}$. A^{-1} will

translate a point 3 units to the right and 5 units upward.

17. **(a)** $BX = \begin{bmatrix} -2 \\ 0 \\ 1 \end{bmatrix} = Y$ **(b)** $B^{-1}Y = \begin{bmatrix} -\sqrt{2} \\ -\sqrt{2} \\ 1 \end{bmatrix} = X$

B^{-1} rotates the point represented by Y counterclockwise 45° about the origin.

19. **(a)** $ABX = \begin{bmatrix} 2 \\ 2 \\ 1 \end{bmatrix} = Y$ **(b)** The net result of A and B

is to translate a point 1 unit to the right and 1 unit upward.

$$AB = \begin{bmatrix} 1 & 0 & 1 \\ 0 & 1 & 1 \\ 0 & 0 & 1 \end{bmatrix}.$$

(c) Yes   **(d)** Since AB translates a point 1 unit right and 1 unit upward, the inverse of AB would translate a point 1 unit left and 1 unit downward. Therefore

$$(AB)^{-1} = \begin{bmatrix} 1 & 0 & -1 \\ 0 & 1 & -1 \\ 0 & 0 & 1 \end{bmatrix}$$

21. $A^{-1} = \begin{bmatrix} 3 & -2 \\ -1 & 1 \end{bmatrix}$ **23.** $A^{-1} = \begin{bmatrix} 5 & 2 \\ 3 & 1 \end{bmatrix}$

25. $A^{-1} = \begin{bmatrix} -\frac{1}{2} & \frac{5}{2} \\ 1 & -4 \end{bmatrix}$ **27.** $A^{-1} = \begin{bmatrix} 0 & 1 & 0 \\ 0 & 0 & 1 \\ 1 & 0 & 0 \end{bmatrix}$

29. $A^{-1} = \begin{bmatrix} -2 & 1 & -1 \\ -5 & 2 & -1 \\ 3 & -1 & 1 \end{bmatrix}$

31. $A^{-1} = \begin{bmatrix} -10 & 3 & -5 \\ 4 & -1 & 2 \\ -3 & 1 & -1 \end{bmatrix}$

33. $A^{-1} = \begin{bmatrix} -1 & -1 & -1 \\ -\frac{2}{5} & -\frac{1}{5} & -\frac{4}{5} \\ \frac{1}{5} & \frac{3}{5} & \frac{2}{5} \end{bmatrix}$ **35.** $A^{-1} = \begin{bmatrix} -10 & 30 \\ -4 & 10 \end{bmatrix}$

37. $A^{-1} = \begin{bmatrix} 0.2 & 0 & 0.4 \\ 0.4 & 0 & -0.2 \\ 1.4 & -1 & -1.2 \end{bmatrix}$

39. $A^{-1} = \begin{bmatrix} 0.5 & 0.2 & 2.1 \\ 0 & 0.2 & 1.6 \\ 0 & 0 & -1 \end{bmatrix}$

41. $A^{-1} = \begin{bmatrix} 0.5 & 0.25 & 0.25 \\ 0.25 & 0.5 & 0.25 \\ 0.25 & 0.25 & 0.5 \end{bmatrix}$

43. $A^{-1} = \begin{bmatrix} 1.2\overline{6} & 0.2\overline{6} & 0.0\overline{6} & 0.0\overline{6} \\ 0.2\overline{6} & 0.2\overline{6} & 0.0\overline{6} & 0.0\overline{6} \\ 0.0\overline{6} & 0.0\overline{6} & 0.2\overline{6} & 0.2\overline{6} \\ 0.0\overline{6} & 0.0\overline{6} & 0.2\overline{6} & 1.2\overline{6} \end{bmatrix}$

45. $AX = \begin{bmatrix} 2 & -3 \\ -3 & -4 \end{bmatrix}\begin{bmatrix} x \\ y \end{bmatrix} = \begin{bmatrix} 7 \\ 9 \end{bmatrix} = B$

47. $AX = \begin{bmatrix} \frac{1}{2} & -\frac{3}{2} \\ -1 & 2 \end{bmatrix}\begin{bmatrix} x \\ y \end{bmatrix} = \begin{bmatrix} \frac{1}{4} \\ 5 \end{bmatrix} = B$

49. $AX = \begin{bmatrix} 1 & -2 & 1 \\ 0 & 3 & -1 \\ 5 & -4 & -7 \end{bmatrix}\begin{bmatrix} x \\ y \\ z \end{bmatrix} = \begin{bmatrix} 5 \\ 6 \\ 0 \end{bmatrix} = B$

51. $AX = \begin{bmatrix} 4 & -1 & 3 \\ 1 & 2 & 5 \\ 2 & -3 & 0 \end{bmatrix}\begin{bmatrix} x \\ y \\ z \end{bmatrix} = \begin{bmatrix} -2 \\ 11 \\ -1 \end{bmatrix} = B$

53. **(a)** $AX = \begin{bmatrix} 1 & 2 \\ 1 & 3 \end{bmatrix}\begin{bmatrix} x \\ y \end{bmatrix} = \begin{bmatrix} 3 \\ 6 \end{bmatrix} = B$

(b) $X = \begin{bmatrix} -3 \\ 3 \end{bmatrix}$

55. **(a)** $AX = \begin{bmatrix} -1 & 2 \\ 3 & -5 \end{bmatrix}\begin{bmatrix} x \\ y \end{bmatrix} = \begin{bmatrix} 5 \\ -2 \end{bmatrix} = B$

(b) $X = \begin{bmatrix} 21 \\ 13 \end{bmatrix}$

57. **(a)** $AX = \begin{bmatrix} 1 & 0 & 1 \\ 2 & 1 & 3 \\ -1 & 1 & 1 \end{bmatrix}\begin{bmatrix} x \\ y \\ z \end{bmatrix} = \begin{bmatrix} -7 \\ -13 \\ -4 \end{bmatrix} = B$

(b) $X = \begin{bmatrix} 5 \\ 13 \\ -12 \end{bmatrix}$

59. **(a)** $AX = \begin{bmatrix} 1 & 2 & -1 \\ 2 & 5 & 0 \\ -1 & -1 & 2 \end{bmatrix}\begin{bmatrix} x \\ y \\ z \end{bmatrix} = \begin{bmatrix} 2 \\ -1 \\ 0 \end{bmatrix} = B$

(b) $X = \begin{bmatrix} -23 \\ 9 \\ -7 \end{bmatrix}$

61. (a) $AX = \begin{bmatrix} 1.5 & 3.7 \\ -0.4 & -2.1 \end{bmatrix}\begin{bmatrix} x \\ y \end{bmatrix} = \begin{bmatrix} 0.32 \\ 0.36 \end{bmatrix} = B$

(b) $X = \begin{bmatrix} 1.2 \\ -0.4 \end{bmatrix}$

63. (a) $AX = \begin{bmatrix} 0.08 & -0.7 \\ 1.1 & -0.05 \end{bmatrix}\begin{bmatrix} x \\ y \end{bmatrix} = \begin{bmatrix} -0.504 \\ 0.73 \end{bmatrix} = B$

(b) $X = \begin{bmatrix} 0.7 \\ 0.8 \end{bmatrix}$

65. (a) $AX = \begin{bmatrix} 3.1 & 1.9 & -1 \\ 6.3 & 0 & -9.9 \\ -1 & 1.5 & 7 \end{bmatrix}\begin{bmatrix} x \\ y \\ z \end{bmatrix} = \begin{bmatrix} 1.99 \\ -3.78 \\ 5.3 \end{bmatrix} = B$

(b) $X = \begin{bmatrix} 0.5 \\ 0.6 \\ 0.7 \end{bmatrix}$

67. (a) $AX = \begin{bmatrix} 3 & -1 & 1 \\ 5.8 & -2.1 & 0 \\ -1 & 0 & 2.9 \end{bmatrix}\begin{bmatrix} x \\ y \\ z \end{bmatrix} = \begin{bmatrix} 4.9 \\ -3.8 \\ 3.8 \end{bmatrix} = B$

(b) $X \approx \begin{bmatrix} 9.26 \\ 27.39 \\ 4.50 \end{bmatrix}$

69. Type A: \$10.99; type B: \$12.99; type C: \$14.99

71. (a) Intersection A: Incoming traffic is $x_1 + 5$ and outgoing traffic is $4 + 6$, so $x_1 + 5 = 4 + 6$.
Intersection B: Incoming traffic is $x_2 + 6$ and outgoing traffic is $x_1 + 3$, so $x_2 + 6 = x_1 + 3$.
Intersection C: Incoming traffic is $x_3 + 4$ and outgoing traffic is $x_2 + 7$, so $x_3 + 4 = x_2 + 7$.
Intersection D: Incoming traffic is $6 + 5$ and outgoing traffic is $x_3 + x_4$, so $6 + 5 = x_3 + x_4$.

(b) $AX = \begin{bmatrix} 1 & 0 & 0 & 0 \\ -1 & 1 & 0 & 0 \\ 0 & -1 & 1 & 0 \\ 0 & 0 & 1 & 1 \end{bmatrix}\begin{bmatrix} x_1 \\ x_2 \\ x_3 \\ x_4 \end{bmatrix} = \begin{bmatrix} 5 \\ -3 \\ 3 \\ 11 \end{bmatrix} = B$

The solution is $x_1 = 5$, $x_2 = 2$, $x_3 = 5$, and $x_4 = 6$.
(c) The traffic traveling west from intersection B to intersection A has a rate of $x_1 = 5$ cars per minute. The other values for x_2, x_3, and x_4 can be interpreted in a similar manner.
73. (a) $a + 1500b + 8c = 122$
$a + 2000b + 5c = 130$
$a + 2200b + 10c = 158$
(b) \$130,000

CHECKING BASIC CONCEPTS FOR SECTIONS 9.5 AND 9.6 (P. 824)

1. (a) $A + B = \begin{bmatrix} 0 & 1 & 3 \\ -1 & 5 & 3 \\ 2 & 1 & 0 \end{bmatrix}$

(b) $2A - B = \begin{bmatrix} 3 & -1 & 0 \\ -2 & -2 & 3 \\ 1 & 8 & 0 \end{bmatrix}$

(c) $AB = \begin{bmatrix} 0 & -1 & 2 \\ 3 & -1 & -1 \\ -1 & 13 & 5 \end{bmatrix}$

3. (a) $AX = \begin{bmatrix} 1 & -2 \\ 2 & 3 \end{bmatrix}\begin{bmatrix} x \\ y \end{bmatrix} = \begin{bmatrix} 13 \\ 5 \end{bmatrix} = B; X = \begin{bmatrix} 7 \\ -3 \end{bmatrix}$

(b) $AX = \begin{bmatrix} 1 & -1 & 1 \\ -1 & 1 & 1 \\ 0 & 1 & -1 \end{bmatrix}\begin{bmatrix} x \\ y \\ z \end{bmatrix} = \begin{bmatrix} 2 \\ 4 \\ -1 \end{bmatrix} = B \ X = \begin{bmatrix} 1 \\ 2 \\ 3 \end{bmatrix}$

(c) $AX = \begin{bmatrix} 3.1 & -5.3 \\ -0.1 & 1.8 \end{bmatrix}\begin{bmatrix} x \\ y \end{bmatrix} = \begin{bmatrix} -2.682 \\ 0.787 \end{bmatrix} = B;$
$X = \begin{bmatrix} -0.13 \\ 0.43 \end{bmatrix}$

SECTION 9.7 (PP. 832–833)

1. $\det A = 1 \neq 0$. A is invertible.
3. $\det A = 0$. A is not invertible.
5. $M_{12} = 10, A_{12} = -10$ **7.** $M_{22} = -15, A_{22} = -15$
9. $\det A = 3 \neq 0$. A^{-1} exists.
11. $\det A = 0$. A^{-1} does not exist.   **13.** 30   **15.** 0
17. -32 **19.** 0   **21.** 643.4   **23.** -4.484 **25.** 7
27. 6.5   **29.** $\left(-\frac{13}{9}, \frac{16}{9}\right)$ **31.** $\left(\frac{49}{2}, 19\right)$ **33.** $(5, -3)$
35. $(0.45, 0.67)$ **37.** The points are collinear.
39. The points are not collinear.

9.7 EXTENDED AND DISCOVERY EXERCISES (P. 833)

1. $(1, 3, 2)$ **3.** $(1, -1, 1)$ **5.** $(-1, 0, 4)$

CHECKING BASIC CONCEPTS FOR SECTION 9.7 (P. 833)

1. $\det A = 19$; A is invertible.

CHAPTER 9 REVIEW EXERCISES (PP. 838–841)

1. $A(3, 6) = 9$ **3.** $y = \dfrac{3x + 2}{4}$ **5.** $(1, -2)$
7. $(3, 4), (-3, -4)$ **9.** $(2, 3)$, consistentt
11. Infinitely many solutions, consistent
13. $\left(\frac{3\sqrt{2}}{2}, \frac{1}{2}\right), \left(-\frac{3\sqrt{2}}{2}, \frac{1}{2}\right)$

15. 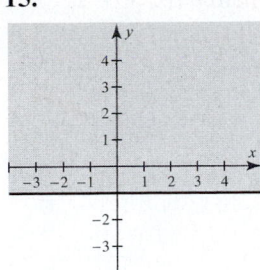 **17.**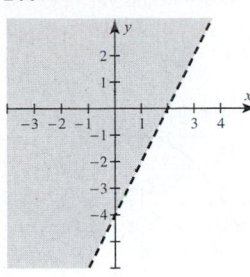

19. One solution is $(2, 2)$. Answers may vary.

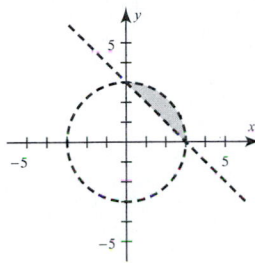

21. One solution is $(0, 2)$. Answers may vary.

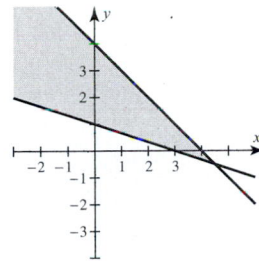

23. $(-1, 2, 1)$ **25.** No solutions   **27.** $(-9, 3)$
29. $(-2, 3, 0)$ **31.** $(-1, 3)$ **33.** $(1, -2, 3)$
35. (a) 5 **(b)** -10

37. (a) $A + 2B = \begin{bmatrix} 7 & 1 \\ -8 & 1 \end{bmatrix}$

(b) $A - B = \begin{bmatrix} -2 & -5 \\ 7 & -2 \end{bmatrix}$ **(c)** $-4A = \begin{bmatrix} -4 & 12 \\ -8 & 4 \end{bmatrix}$

39. $AB = \begin{bmatrix} -2 & -4 \\ 17 & 31 \end{bmatrix}$, $BA = \begin{bmatrix} 8 & -6 \\ -27 & 21 \end{bmatrix}$

41. $AB = \begin{bmatrix} 3 & 7 \\ -2 & 8 \end{bmatrix}$, $BA = \begin{bmatrix} 2 & -1 & 3 \\ 2 & 9 & -3 \\ 6 & 12 & 0 \end{bmatrix}$

43. B is the inverse of A.

45. (a) $AX = \begin{bmatrix} 1 & -3 \\ 2 & -1 \end{bmatrix} \begin{bmatrix} x \\ y \end{bmatrix} = \begin{bmatrix} 4 \\ 3 \end{bmatrix} = B$

(b) $X = \begin{bmatrix} 1 \\ -1 \end{bmatrix}$

47. (a) $AX = \begin{bmatrix} 11 & 31 \\ 37 & -19 \end{bmatrix} \begin{bmatrix} x \\ y \end{bmatrix} = \begin{bmatrix} -27.6 \\ 240 \end{bmatrix} = B$

(b) $X = \begin{bmatrix} 5.1 \\ -2.7 \end{bmatrix}$

49.

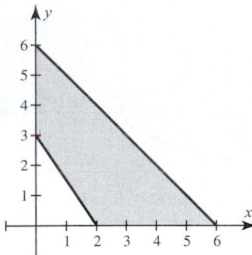

51. $A^{-1} = \begin{bmatrix} -1 & -2 \\ -1 & -1 \end{bmatrix}$ **53.** $\det A = 25$

55. $\det A = -1951 \neq 0$. A is invertible.
57. (a) $\frac{249}{99} \approx 2.5$ **(b)** They are approximately equal.
59. $l = 11, w = 7$ **61.** \$1200 at 7%, \$800 at 9%
63. Type A: \$11.49; type B: \$12.99
65. 10.5 square units
67. (a) $\pi r^2 h \geq 30, 2\pi rh \leq 45$
(b) $[0, 7, 1]$ by $[0, 7, 1]$ **(c)** $[0, 7, 1]$ by $[0, 7, 1]$

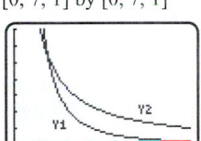   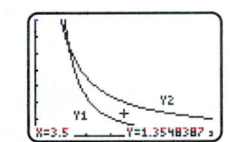

One solution is $(3.5, 1.35)$.
69. 4500   Answers may vary.

**CHAPTER 9 EXTENDED AND DISCOVERY EXERCISES
(PP. 841–843)**

1. (a) $A^{\mathrm{T}} = \begin{bmatrix} 3 & 2 & 4 \\ -3 & 6 & 2 \end{bmatrix}$ **(b)** $A^{\mathrm{T}} = \begin{bmatrix} 0 & 2 & -4 \\ 1 & 5 & 3 \\ -2 & 4 & 9 \end{bmatrix}$

(c) $A^{\mathrm{T}} = \begin{bmatrix} 5 & 1 & 6 & -9 \\ 7 & -7 & 3 & 2 \end{bmatrix}$

3. $f(x) = 2.6314x + 2.2714$ **5.** $f(x) = \dfrac{1.21}{x} + \dfrac{5.09}{x^2}$

$[-1, 6, 1]$ by $[0, 18, 2]$ $[0, 5, 1]$ by $[0, 7, 1]$

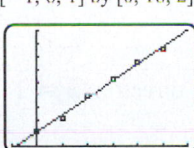   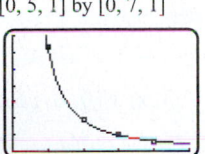

7. (a) HELP is coded as the word UBNL.
(b) LETTER is coded as the word CHHPBQ.

CHAPTER 10: Conic Sections

SECTION 10.1 (PP. 852–855)

1.

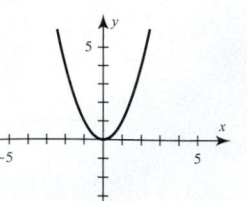

3.

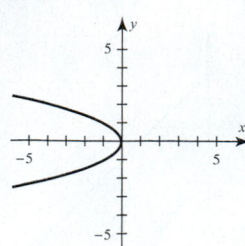

5.

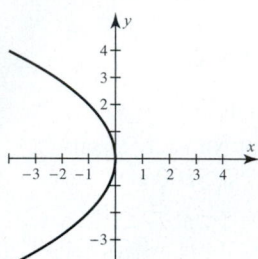

7.

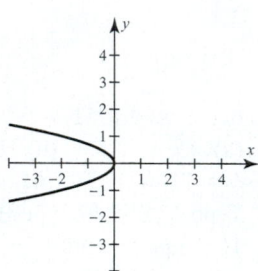

9. e.   **11.** a.   **13.** d.

15. Vertex: $V(0, 0)$; focus: $F(0, 4)$; directrix: $y = -4$

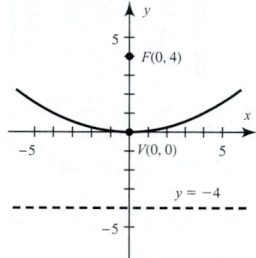

17. Vertex: $V(0, 0)$; focus: $F(2, 0)$; directrix: $x = -2$

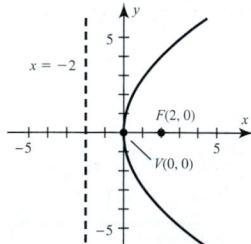

19. Vertex: $V(0, 0)$; focus: $F(-1, 0)$; directrix: $x = 1$

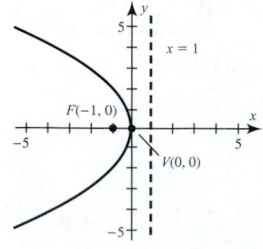

21. Vertex: $V(0, 0)$; focus: $F(0, -2)$; directrix: $y = 2$

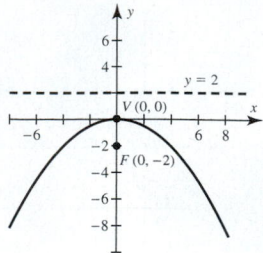

23. Vertex: $V(0, 0)$; focus: $F\left(\frac{3}{4}, 0\right)$; directrix: $x = -\frac{3}{4}$

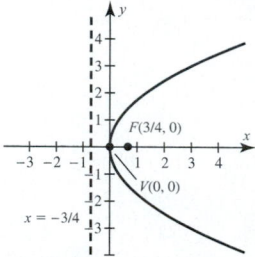

25. Vertex: $V(0, 0)$; focus: $F(-1, 0)$; directrix: $x = 1$

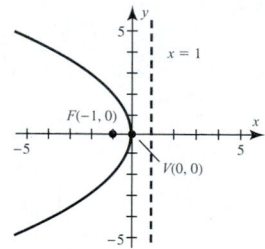

27. $x^2 = 4y$

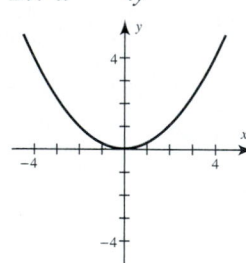

29. $y^2 = -12x$

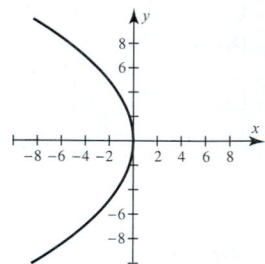

31. $x^2 = 3y$

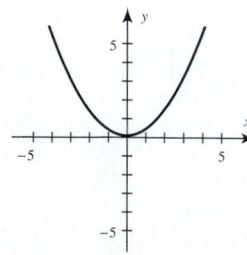

33. $y^2 = -8x$

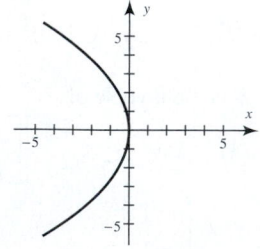

35. $y^2 = 4x$

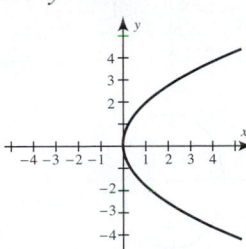

37. $y^2 = -x$

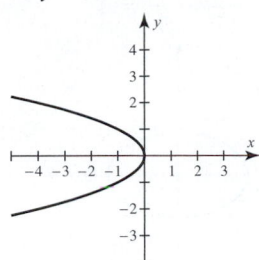

39. $y^2 = 4x$

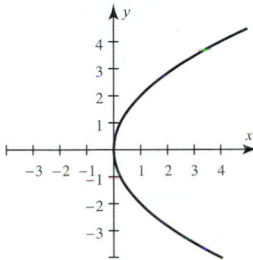

41. $x^2 = -12y$ **43.** $y^2 = -4x$

45.

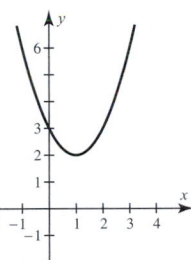

47.

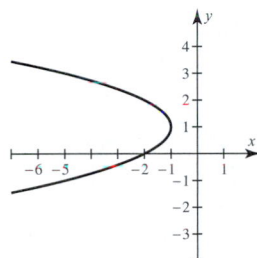

49. c.   **51.** a.

53. Vertex: $V(2, -2)$; focus: $F(2, 0)$; directrix; $y = -4$

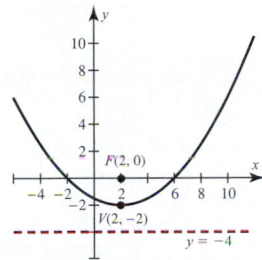

55. Vertex: $V(2, -3)$; focus: $F(1, -3)$; directrix: $x = 3$

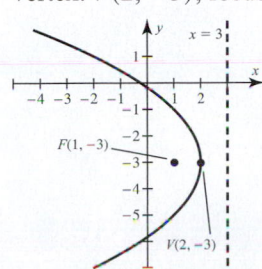

57. Vertex: $V(-2, 0)$; focus: $F(-2, -1)$; directrix: $y = 1$

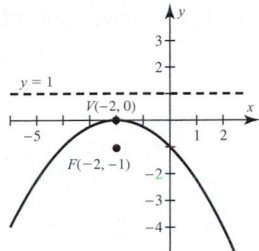

59. $x^2 = 4(y - 1)$ **61.** $y^2 = 4(x + 1)$

63. $(x + 1)^2 = -8(y - 5)$

65. $(y - 3)^2 = -\frac{9}{2}(x + 2)$

67. $(y - 0)^2 = -8\left(x + \frac{5}{4}\right)$

69. $(y + 1)^2 = \frac{1}{2}(x + 3)$ **71.** $\left(x - \frac{3}{2}\right)^2 = 2\left(y - \frac{7}{8}\right)$

73. $\left(y + \frac{1}{2}\right)^2 = \frac{5}{4}\left(x + \frac{6}{5}\right)$

75. $y = \pm\sqrt{2x}$ **77.** $y = -0.75 \pm \sqrt{-3x}$

$[-6, 6, 1]$ by $[-4, 4, 1]$ $[-6, 6, 1]$ by $[-4, 4, 1]$

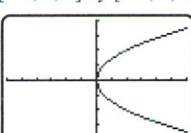   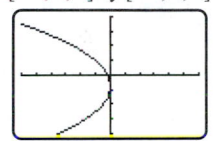

79. $y = 0.5 \pm \sqrt{3.1(x + 1.3)}$

$[-9, 9, 1]$ by $[-6, 6, 1]$

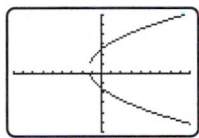

Note: If a break in the graph appears near the vertex, it should not be there. It is a result of the low resolution of the graphing calculator screen.

81. $y = -1 \pm \sqrt{\dfrac{x}{2.3}}$

$[-6, 6, 1]$ by $[-4, 4, 1]$

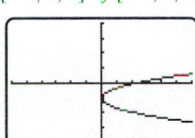

83. $p = 3$ ft   **85. (a)** $y = \frac{32}{11,025}x^2$ **(b)** About 86.1 ft

87. $\frac{25}{64}$ in.

1. Foci: $F(0, \pm\sqrt{5})$; vertices: $V(0, \pm 3)$; endpoints of the minor axis: $U(\pm 2, 0)$

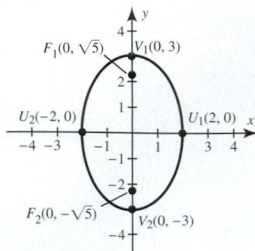

3. Foci: $F(\pm\sqrt{20}, 0)$; vertices: $V(\pm 6, 0)$; endpoints of the minor axis: $U(0, \pm 4)$

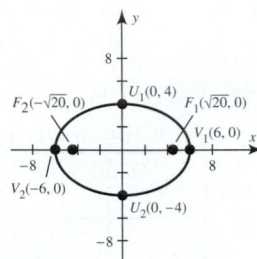

5. Foci: $F(0, \pm 2)$; vertices: $V(0, \pm 3)$; endpoints of the minor axis: $U(\pm\sqrt{5}, 0)$

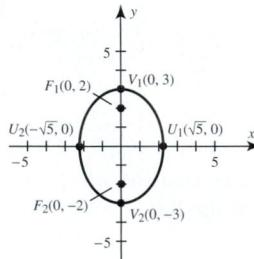

7. Foci; $F(0, \pm 4)$; vertices: $V(0, \pm 5)$; endpoints of the minor axis: $U(\pm 3, 0)$

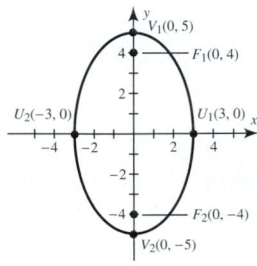

9. b.   **11.** c.

13. $\dfrac{x^2}{25} + \dfrac{y^2}{9} = 1$

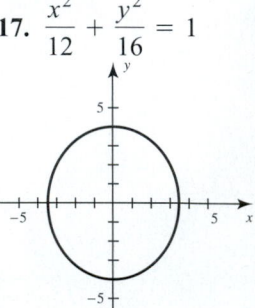

15. $\dfrac{x^2}{5} + \dfrac{y^2}{9} = 1$

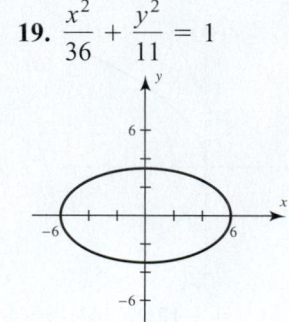

17. $\dfrac{x^2}{12} + \dfrac{y^2}{16} = 1$

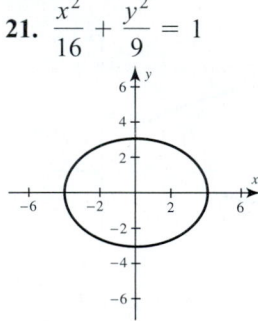

19. $\dfrac{x^2}{36} + \dfrac{y^2}{11} = 1$

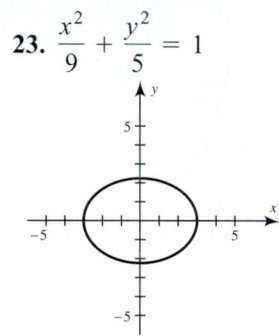

21. $\dfrac{x^2}{16} + \dfrac{y^2}{9} = 1$

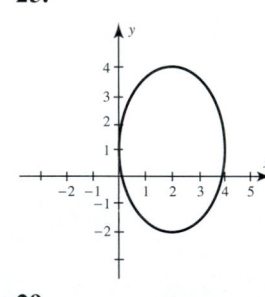

23. $\dfrac{x^2}{9} + \dfrac{y^2}{5} = 1$

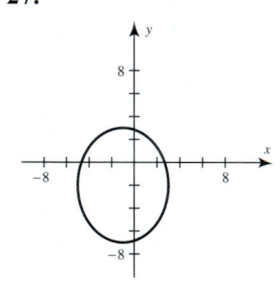

25.

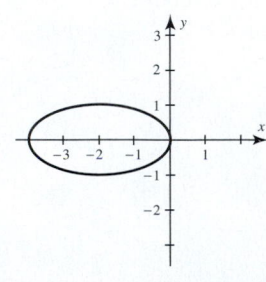

27.

29.

31. d.   **33.** c.

35. Foci: $(1, 1 \pm 4)$; vertices: $(1, 1 \pm 5)$

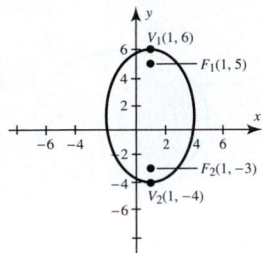

37. Foci: $(-4 \pm \sqrt{7}, 2)$; vertices: $(-4 \pm 4, 2)$

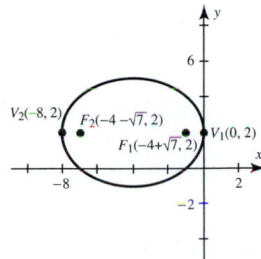

39. $\dfrac{(x-2)^2}{5} + \dfrac{(y-1)^2}{9} = 1$

41. $\dfrac{x^2}{9} + \dfrac{(y-2)^2}{5} = 1$

43. $\dfrac{(x-2)^2}{16} + \dfrac{(y-4)^2}{4} = 1$

45. $\dfrac{(x+1)^2}{4} + \dfrac{(y-1)^2}{9} = 1$; center: $(-1, 1)$;
vertices: $(-1, -2), (-1, 4)$

47. $\dfrac{(x+1)^2}{1} + \dfrac{(y+1)^2}{4} = 1$; center: $(-1, -1)$;
vertices: $(-1, -3), (-1, 1)$

49. $\dfrac{(x+2)^2}{5} + \dfrac{(y-1)^2}{4} = 1$; center: $(-2, 1)$;
vertices: $(-2 - \sqrt{5}, 1), (-2 + \sqrt{5}, 1)$

51. $\dfrac{\left(x - \frac{1}{2}\right)^2}{4} + \dfrac{\left(y + \frac{3}{2}\right)^2}{16} = 1$; center: $\left(\dfrac{1}{2}, -\dfrac{3}{2}\right)$;
vertices: $\left(\dfrac{1}{2}, \dfrac{5}{2}\right), \left(\dfrac{1}{2}, -\dfrac{11}{2}\right)$

53. $x^2 + y^2 = 16$

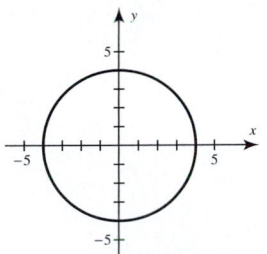

55. $(x - 3)^2 + (y + 4)^2 = 1$

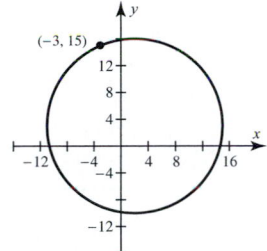

57. $(x - 2)^2 + (y - 3)^2 = 169$

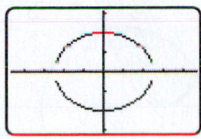

59. Center: $(2, 1)$; radius: $r = 4$
61. Center: $(0, -5)$; radius: $r = 5$

63. $y = \pm\sqrt{10\left(1 - \dfrac{x^2}{15}\right)}$

$[-6, 6, 1]$ by $[-4, 4, 1]$

65. $y = \pm\sqrt{\dfrac{25 - 4.1x^2}{6.3}}$

$[-4.7, 4.7, 1]$ by $[-3.1, 3.1, 1]$

67. $(0, 3)$, $\left(\frac{24}{13}, \frac{15}{13}\right)$

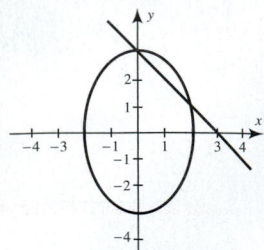

69. Four solutions: $\left(\pm\sqrt{\frac{20}{3}}, \pm\sqrt{\frac{7}{3}}\right)$

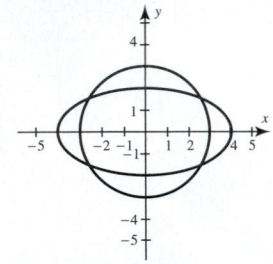

71. $(3, 0)$, $(-3, 0)$

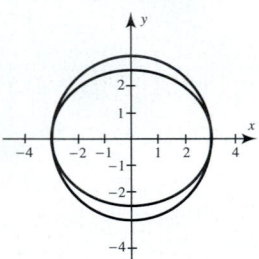

73. $(0, 2)$ **75.** Four solutions: $\left(\pm\frac{2\sqrt{3}}{3}, \pm\frac{2\sqrt{3}}{3}\right)$

77. $(1, 2\sqrt{2})$, $(1, -2\sqrt{2})$

79.

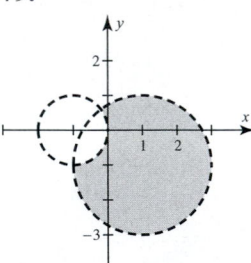

81.

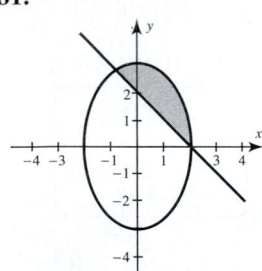

83.

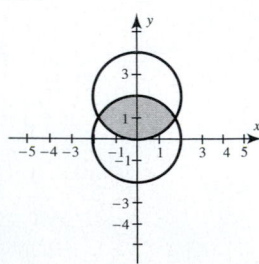

85.

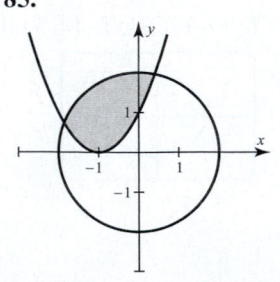

87. $A = 6\pi \approx 18.85\ \text{ft}^2$

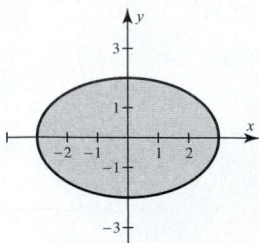

89. $A = 20\pi \approx 62.83\ \text{ft}^2$

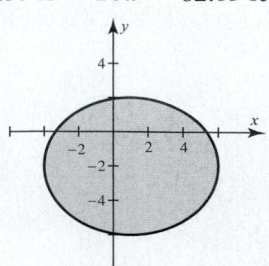

91. $\dfrac{x^2}{0.387^2} + \dfrac{y^2}{0.379^2} = 1$; sun: $(0.0797, 0)$

$[-0.6, 0.6, 0.1]$ by $[-0.4, 0.4, 0.1]$

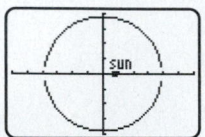

93. 6.245 in.

95. (a) $\dfrac{x^2}{17.95^2} + \dfrac{y^2}{4.44^2} = 1$ **(b)** $(17.39, 0)$

(c) Maximum: 35.34 units or about 3.3 billion mi;
minimum: 0.56 unit or about 52 million mi

97. About 21.65 ft

99. Maximum: 668 mi; minimum: 340 mi

**CHECKING BASIC CONCEPTS FOR
SECTIONS 10.1 AND 10.2 (P. 870)**

1. Vertex: $V(0, 0)$; focus: $F\left(\frac{1}{2}, 0\right)$; directrix: $x = -\frac{1}{2}$

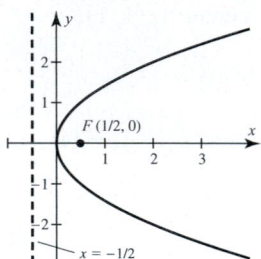

3. Foci: $F(0, \pm 8)$; vertices: $V(0, \pm 10)$; endpoints of the
minor axis: $U(\pm 6, 0)$

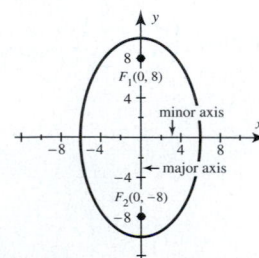

5. 1 ft   **7.** $\dfrac{(x - 2)^2}{16} + \dfrac{(y + 1)^2}{4} = 1$; center: $(2, -1)$;

vertices: $(-2, -1)$, $(6, -1)$

SECTION 10.3 (PP. 878–881)

1. Asymptotes: $y = \pm\frac{7}{3}x$; $F(\pm\sqrt{58}, 0)$

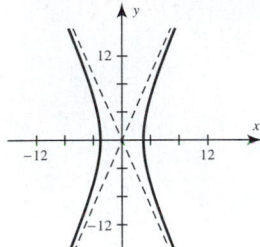

3. Asymptotes: $y = \pm\frac{3}{2}x$; $F(0, \pm\sqrt{52})$

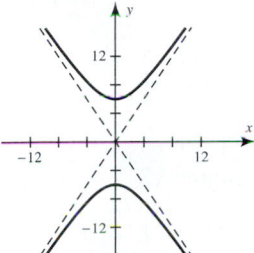

5. Asymptotes: $y = \pm x$; $F(\pm\sqrt{18}, 0)$

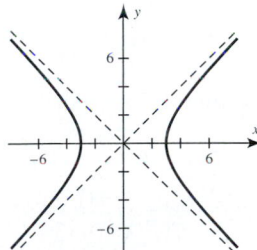

7. Asymptotes: $y = \pm\frac{4}{3}x$; $F(0, \pm 5)$

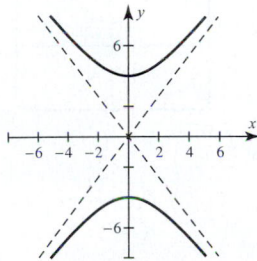

9. d.   **11.** a.

13. $\dfrac{x^2}{16} - \dfrac{y^2}{9} = 1$

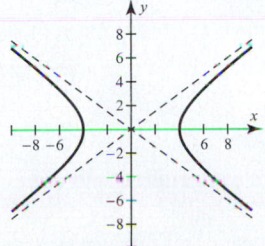

15. $\dfrac{y^2}{36} - \dfrac{x^2}{64} = 1$

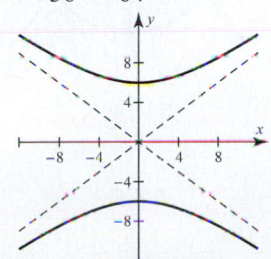

17. $\dfrac{y^2}{144} - \dfrac{x^2}{25} = 1$; asymptotes: $y = \pm\dfrac{12}{5}x$

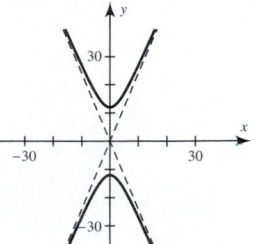

19. $\dfrac{y^2}{4} - \dfrac{x^2}{21} = 1$; asymptotes: $y = \pm\dfrac{2}{\sqrt{21}}x$

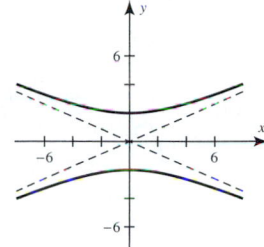

21. $\dfrac{x^2}{9} - \dfrac{y^2}{4} = 1$; asymptotes: $y = \pm\dfrac{2}{3}x$

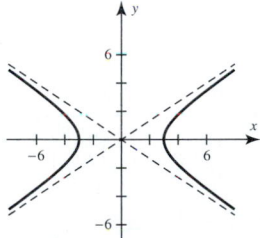

23. $\dfrac{x^2}{16} - \dfrac{y^2}{9} = 1$; asymptotes: $y = \pm\dfrac{3}{4}x$

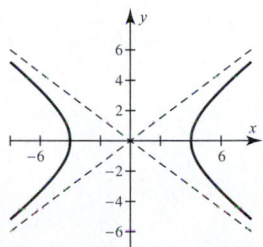

25. $\dfrac{x^2}{10} - \dfrac{y^2}{9} = 1$; asymptotes: $y = \pm\dfrac{3}{\sqrt{10}}x$

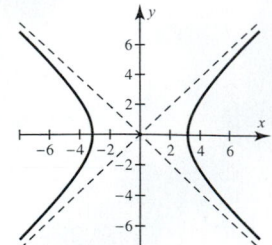

27.

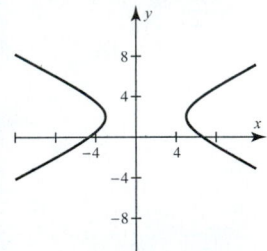

29.

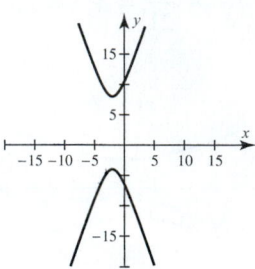

31.

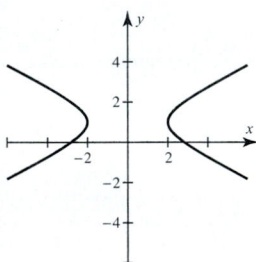

33. b.   **35.** c.   **37.** $\dfrac{(y+4)^2}{16} - \dfrac{(x-4)^2}{4} = 1$

39. Vertices: $(1 \pm 2, 1)$; foci: $(1 \pm \sqrt{8}, 1)$; asymptotes: $y = \pm(x - 1) + 1$

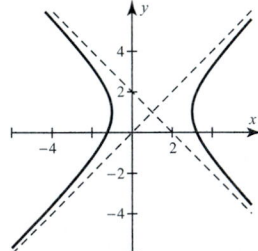

41. Vertices: $(1, -1 \pm 4)$; foci: $(1, -1 \pm 5)$; asymptotes: $y = \pm\frac{4}{3}(x - 1) - 1$

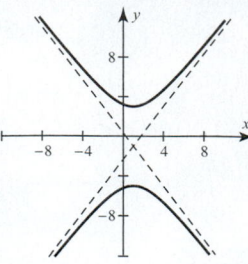

43. $(x - 2)^2 - \dfrac{(y+2)^2}{3} = 1$ **45.** $y^2 - \dfrac{(x+1)^2}{8} = 1$

47. $\dfrac{(x-1)^2}{4} - \dfrac{(y-1)^2}{4} = 1$; center: $(1, 1)$;
vertices: $(-1, 1)$, $(3, 1)$

49. $\dfrac{(y+4)^2}{2} - \dfrac{(x-3)^2}{3} = 1$; center: $(3, -4)$;
vertices: $(3, -4 - \sqrt{2})$, $(3, -4 + \sqrt{2})$

51. $\dfrac{(x-3)^2}{2} - \dfrac{(y-0)^2}{1} = 1$; center: $(3, 0)$;
vertices: $(3 - \sqrt{2}, 0)$, $(3 + \sqrt{2}, 0)$

53. $\dfrac{(y+4)^2}{5} - \dfrac{(x+1)^2}{4} = 1$; center: $(-1, -4)$;
vertices: $(-1, -4 - \sqrt{5})$, $(-1, -4 + \sqrt{5})$

55. $y = 1 \pm \sqrt{11\left(1 + \dfrac{x^2}{5.9}\right)}$ **57.** $y = \pm\sqrt{\dfrac{4x^2 + 15}{3}}$

$[-15, 15, 5]$ by $[-10, 10, 5]$ $[-9, 9, 1]$ by $[-6, 6, 1]$

59. Four solutions: $\left(\pm\sqrt{\frac{13}{2}}, \pm\sqrt{\frac{5}{2}}\right)$

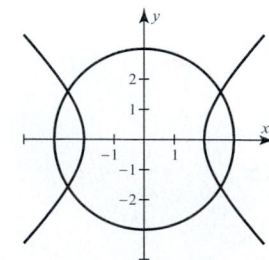

61. $(2, 0)$, $(-5.2, 7.2)$

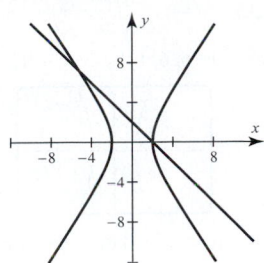

63. Four solutions: $\left(\pm 2, \pm\dfrac{2\sqrt{3}}{3}\right)$

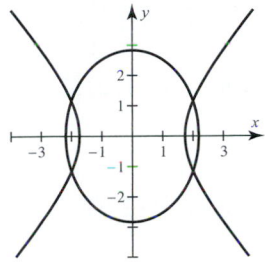

65. $\left(\dfrac{2\sqrt{11}}{11}, \dfrac{6\sqrt{11}}{11}\right)$, $\left(-\dfrac{2\sqrt{11}}{11}, -\dfrac{6\sqrt{11}}{11}\right)$

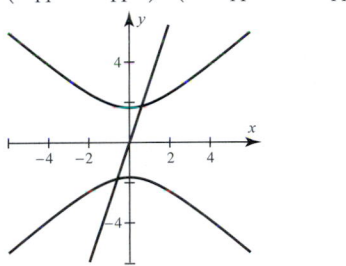

67. (a) Elliptic   **(b)** Its speed should be 4326 m/sec or greater.   **(c)** If D is larger, then $\dfrac{k}{\sqrt{D}}$ is smaller, so smaller values for V satisfy $V > \dfrac{k}{\sqrt{D}}$.

10.3 EXTENDED AND DISCOVERY EXERCISES (P. 881)

1. (a) $x = \sqrt{y^2 + 2.5 \times 10^{-27}}$; this equation represents the right half of the hyperbola.   **(b)** About 1.2×10^{-13} m

CHECKING BASIC CONCEPTS FOR SECTION 10.3 (P. 881)

1. Foci: $F(\pm 5, 0)$; asymptotes: $y = \pm\dfrac{4}{3}x$

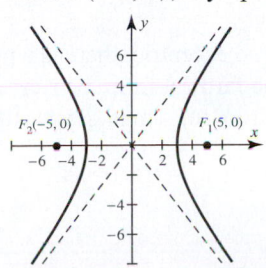

3. $\dfrac{x^2}{16} - \dfrac{y^2}{9} = 1$

CHAPTER 10 REVIEW EXERCISES (PP. 884–886)

1.

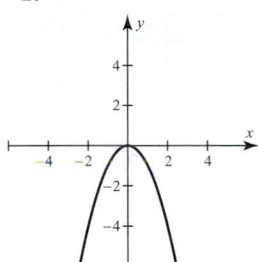

3.

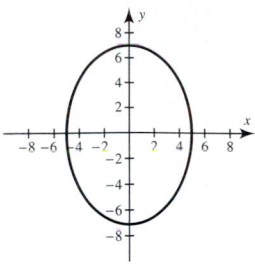

5.

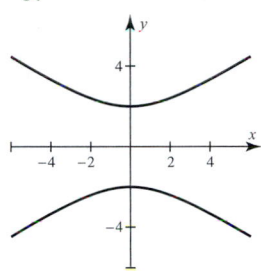

7. d.   **9.** a   **11.** e.
13. $y^2 = 8x$

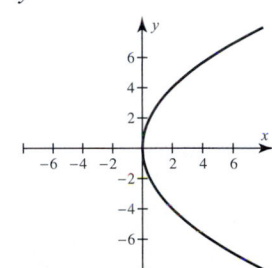

15. $\dfrac{x^2}{25} + \dfrac{y^2}{9} = 1$

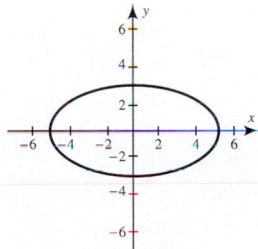

17. $\dfrac{y^2}{64} - \dfrac{x^2}{36} = 1$

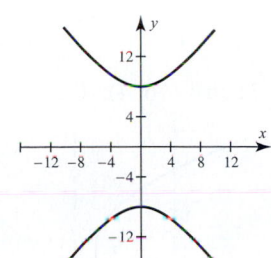

19. $F(0, -1)$

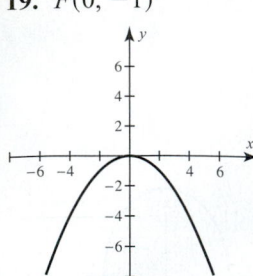

21. $F(\pm\sqrt{21}, 0)$

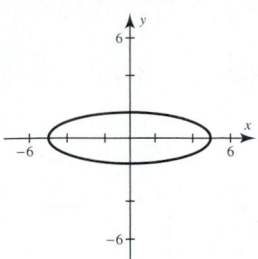

23. $F(\pm 5, 0)$

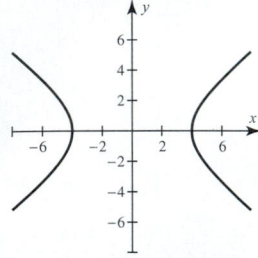

25. Both foci are located at $(0, 0)$.

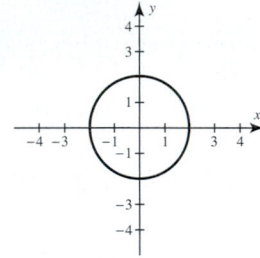

27. Center: $(-1, 2)$

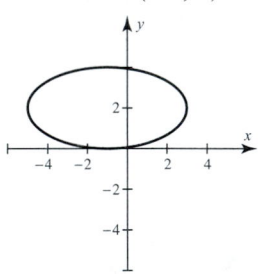

29. Center: $(1, -1)$

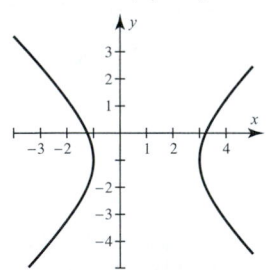

31. Focus: $F(6, 4)$; directrix: $x = 10$

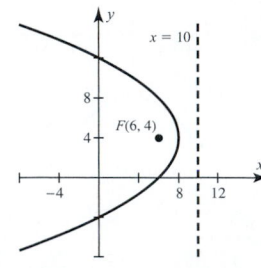

33. $y = \pm\sqrt{\frac{3}{4}}x$

$[-6, 6, 1]$ by $[-4, 4, 1]$

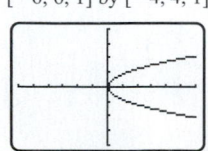

35. $y = \pm\sqrt{\dfrac{9x^2 - 17}{2}}$

$[-6, 6, 1]$ by $[-4, 4, 1]$

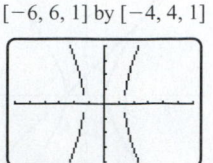

Note: If breaks in the graph appear near the vertices, they should not be there. It is a result of the low resolution of the graphing calculator screen.

37. $(y - 0)^2 = -10\left(x + \frac{7}{5}\right)$

39. $\dfrac{(x + 1)^2}{25} + \dfrac{(y - 5)^2}{4} = 1$; center: $(-1, 5)$;
vertices: $(-6, 5), (4, 5)$

41. $\dfrac{(x + 2)^2}{4} - \dfrac{(y - 3)^2}{1} = 1$; center: $(-2, 3)$;
vertices: $(-4, 3), (0, 3)$

43. $x^2 + y^2 = 49$ **45.** Four solutions: $\left(\pm\sqrt{\frac{8}{3}}, \pm\sqrt{\frac{4}{3}}\right)$

47.

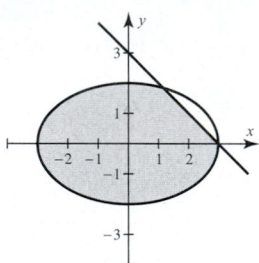

49. (a) Minimum: 4.92 million mi; maximum: 995.08 million mi

(b) $2\pi\sqrt{\dfrac{500^2 + 70^2}{2}} \approx 2243$ million mi or 2.243 billion mi

51. About 29.05 ft

CHAPTER 10 EXTENDED AND DISCOVERY EXERCISES (P. 886)

1. Neptune: 0.271; Pluto: 9.82

3. Neptune: $\dfrac{(x - 0.271)^2}{30.10^2} + \dfrac{y^2}{30.10^2} = 1$

Pluto: $\dfrac{(x - 9.82)^2}{39.44^2} + \dfrac{y^2}{38.20^2} = 1$

5. No. Because Pluto's orbit is so eccentric, there is a period of time when Pluto is not the farthest from the sun. However, its average distance a from the sun is greater than any of the other planets.

CHAPTER 11: Further Topics in Algebra

SECTION 11.1 (PP. 898–902)

1. $a_1 = 3, a_2 = 5, a_3 = 7, a_4 = 9$

3. $a_1 = 4, a_2 = -8, a_3 = 16, a_4 = -32$

5. $a_1 = \frac{1}{2}, a_2 = \frac{2}{5}, a_3 = \frac{3}{10}, a_4 = \frac{4}{17}$

7. $a_1 = -\frac{1}{2}, a_2 = \frac{1}{4}, a_3 = -\frac{1}{8}, a_4 = \frac{1}{16}$

9. $a_1 = \frac{2}{3}, a_2 = -\frac{4}{5}, a_3 = \frac{8}{9}, a_4 = -\frac{16}{17}$

11. $a_1 = 3, a_2 = 8, a_3 = 17, a_4 = 32$

13. $2, 4, 3, 5, 3, 6, 4$

15. **(a)** $a_1 = 1, a_2 = 2, a_3 = 4, a_4 = 8$
(b) $[0, 5, 1]$ by $[0, 9, 1]$

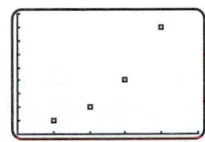

17. **(a)** $a_1 = 2, a_2 = 5, a_3 = 3, a_4 = -2$
(b) $[0, 5, 1]$ by $[-3, 6, 1]$

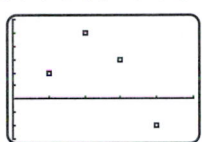

19. **(a)** $a_1 = 2, a_2 = 4, a_3 = 16, a_4 = 256$
(b) $[0, 5, 1]$ by $[0, 300, 50]$

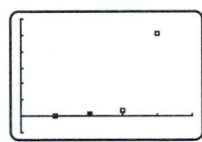

21. **(a)** $a_1 = 1, a_2 = 3, a_3 = 6, a_4 = 10$
(b) $[0, 5, 1]$ by $[0, 10, 1]$

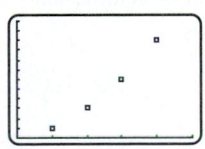

23. **(a)** $a_1 = 2, a_2 = 3, a_3 = 6, a_4 = 18$
(b) $[0, 5, 1]$ by $[0, 20, 2]$

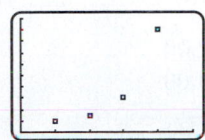

25. The insect population density increases rapidly and then levels off near 5000 per acre.

27. **(a)** $a_n = 0.8a_{n-1}, a_1 = 500$ **(b)** $a_1 = 500,$ $a_2 = 400, a_3 = 320, a_4 = 256, a_5 = 204.8,$ and $a_6 = 163.84.$ The population density decreases by 20% each year.   **(c)** $a_n = 500(0.8)^{n-1}$

29. **(a)** $a_1 = 8, a_2 = 10.4, a_3 = 8.528$
(b) $[0, 21, 1]$ by $[0, 14, 1]$

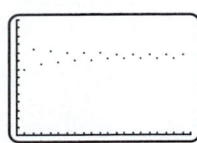

The population density oscillates above and below approximately 9.5.

31. **(a)**

n	1	2	3	4	5	6	7	8
a_n	1	3	5	7	9	11	13	15

(b) $[0, 10, 1]$ by $[0, 16, 1]$

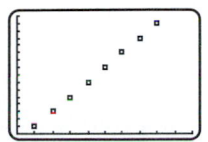

(c) $a_n = 2n - 1$

33. **(a)**

n	1	2	3	4	5	6	7	8
a_n	7.5	6	4.5	3	1.5	0	-1.5	-3

(b) $[0, 12, 1]$ by $[-4, 8, 1]$

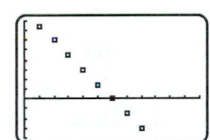

(c) $a_n = -1.5n + 9$

35. **(a)**

n	1	2	3	4	5	6	7	8
a_n	$\frac{1}{2}$	2	$\frac{7}{2}$	5	$\frac{13}{2}$	8	$\frac{19}{2}$	11

(b) $[0, 9, 1]$ by $[0, 12, 1]$

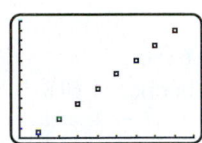

(c) $a_n = \frac{3}{2}n - 1$

37. **(a)**

n	1	2	3	4	5	6	7	8
a_n	8	4	2	1	$\frac{1}{2}$	$\frac{1}{4}$	$\frac{1}{8}$	$\frac{1}{16}$

(b) $[0, 10, 1]$ by $[-1, 9, 1]$

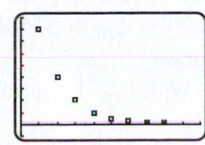

(c) $a_n = 8\left(\frac{1}{2}\right)^{n-1}$

39. (a)

n	1	2	3	4	5	6	7	8
a_n	$\frac{3}{4}$	$\frac{3}{2}$	3	6	12	24	48	96

(b) $[0, 10, 1]$ by $[-10, 110, 10]$

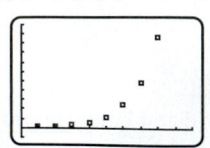

(c) $a_n = \frac{3}{4}(2)^{n-1}$

41. (a)

n	1	2	3	4	5	6	7	8
a_n	$-\frac{1}{4}$	$-\frac{1}{2}$	-1	-2	-4	-8	-16	-32

(b) $[0, 9, 1]$ by $[-36, 4, 4]$

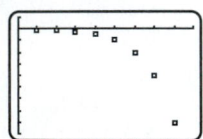

(c) $a_n = -\frac{1}{4}(2)^{n-1}$
43. $a_n = -2n + 7$ **45.** $a_n = 3n - 8$
47. $a_n = 2n + 1$ **49.** $a_n = 3n + 5$
51. $a_n = 0.5n - 6.5$ **53.** $a_n = 2\left(\frac{1}{2}\right)^{n-1}$
55. $a_n = \frac{1}{2}\left(-\frac{1}{4}\right)^{n-1}$ **57.** $a_n = 8\left(\frac{1}{2}\right)^{n-1}$
59. $a_n = 10\left(\frac{1}{5}\right)^{n-1}$ **61.** $a_n = -\frac{1}{2}(2)^{n-1}$ **63.** No
65. Yes **67.** Yes **69.** Yes **71.** Yes **73.** No
75. No **77.** No **79.** Arithmetic **81.** Geometric
83. Neither **85.** Arithmetic, $d < 0, d = -1$
87. Geometric, $r < 0, |r| < 1$
89. (a) 1, 1, 2, 3, 5, 8, 13, 21, 34, 55, 89, 144
(b) $\frac{a_2}{a_1} = 1, \frac{a_3}{a_2} = 2, \frac{a_4}{a_3} = 1.5, \frac{a_5}{a_4} = \frac{5}{3} \approx 1.6667$,
$\frac{a_6}{a_5} = \frac{8}{5} = 1.6, \frac{a_7}{a_6} = \frac{13}{8} = 1.625, \frac{a_8}{a_7} = \frac{21}{13} \approx 1.6154$,
$\frac{a_9}{a_8} = \frac{34}{21} \approx 1.6190, \frac{a_{10}}{a_9} = \frac{55}{34} \approx 1.6176$,
$\frac{a_{11}}{a_{10}} = \frac{89}{55} \approx 1.6182$, and $\frac{a_{12}}{a_{11}} = \frac{144}{89} \approx 1.6180$.
These ratios appear to approach a number near 1.618.
(c) $n = 2: a_1 \cdot a_3 - a_2^2 = (1)(2) - (1)^2 = 1 = (-1)^2$
$n = 3: a_2 \cdot a_4 - a_3^2 = (1)(3) - (2)^2 = -1 = (-1)^3$
$n = 4: a_3 \cdot a_5 - a_4^2 = (2)(5) - (3)^2 = 1 = (-1)^4$
91. (a) $a_n = 2000n + 28{,}000$ or
$a_n = 30{,}000 + 2000(n - 1)$; arithmetic
(b) $b_n = 30{,}000(1.05)^{n-1}$; geometric
(c) Since $a_{10} = \$48{,}000 > b_{10} \approx \$46{,}540$, the first salary is
higher after 10 years. Since $a_{20} = \$68{,}000 < b_{20} \approx \$75{,}809$,
the second salary is higher after 20 years.
(d) With time the geometric sequence with $r > 1$ overtakes
the arithmetic sequence.
$[0, 30, 10]$ by $[0, 150000, 50000]$

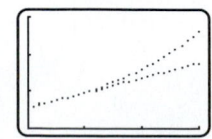

93. $a_6 \approx 1.414213562, \sqrt{2} \approx 1.414213562$
95. $a_6 = 4.582581971, \sqrt{21} \approx 4.582575695$
97. By definition,
$a_n = a_1 + (n - 1)d_1$ and $b_n = b_1 + (n - 1)d_2$.
Then $c_n = a_n + b_n$
$\qquad = [a_1 + (n - 1)d_1] + [b_1 + (n - 1)d_2]$
$\qquad = (a_1 + b_1) + [(n - 1)d_1 + (n - 1)d_2]$
$\qquad = (a_1 + b_1) + (n - 1)(d_1 + d_2)$
$\qquad = c_1 + (n - 1)d$
where $c_1 = a_1 + b_1$ and $d = d_1 + d_2$.

SECTION 11.2 (PP. 915–917)

1. (a) $8518 + 9921 + 10{,}706 + 14{,}035 +$
$14{,}307 + 12{,}249 = 69{,}736$ **(b)** Finite
3. 45 **5.** 25 **7.** 60 **9.** $\frac{71}{20}$ **11.** 80 **13.** 1275
15. 1739 **17.** 4100 **19.** 31 **21.** 460 **23.** 105
25. 1942 **27.** 948 **29.** 545 **31.** 255 **33.** 546.5
35. 3,145,725 **37.** 0.625; 0.671875; 0.666015625
39. 5; 42.333333333; 341 **41.** $\frac{3}{2}$ **43.** $\frac{18}{5}$ **45.** $\frac{10}{11}$
47. 2 **49.** \$91,523.93 **51.** \$62,278.01
53. $\frac{2}{3} = 0.6 + 0.06 + 0.006 + 0.0006 + \cdots$
55. $\frac{9}{11} = 0.81 + 0.0081 + 0.000081 + \cdots$
57. $\frac{1}{7} = 0.142857 + 0.000000142857 + \cdots$
59. $\frac{8}{9}$ **61.** $\frac{5}{11}$ **63.** $2 + 3 + 4 + 5 = 14$
65. $4 + 4 + 4 + 4 + 4 + 4 + 4 + 4 = 32$
67. $1 + 8 + 27 + 64 + 125 + 216 + 343 = 784$
69. $12 + 20 = 32$ **71.** $\sum_{k=1}^{6} k^4$ **73.** $\sum_{k=1}^{7} \left(\frac{2k}{k + 1}\right)$
75. $\sum_{k=1}^{\infty} \left(\frac{1}{k^2}\right)$ **77.** 540 **79.** 600 **81.** 1395
83. 5525 **85.** 1360 **87.** 290
89. $\sum_{k=1}^{n} k = 1 + 2 + 3 + \cdots + n$ is an arithmetic series
with $a_1 = 1$ and $a_n = n$. Its sum equals
$S_n = n\left(\frac{a_1 + a_n}{2}\right) = n\left(\frac{1 + n}{2}\right) = \frac{n(n + 1)}{2}$.
91. $S_9 = 9\left(\frac{7 + 15}{2}\right) = 99$ logs
93. (a) $\sum_{k=1}^{n} 0.5(0.5)^{k-1}$ **(b)** Infinitely many filters
95. (a) 2500, 2500, 2500, 2500
$\qquad$ 4000, 3000, 2000, 1000
$\quad$ **(b)** $2500 + 2500 + 2500 + 2500$
$\qquad$ $4000 + 3000 + 2000 + 1000$
97. $1 + 1 + \frac{1}{2} + \frac{1}{6} + \frac{1}{24} + \frac{1}{120} + \frac{1}{720} +$
$\frac{1}{5040} \approx 2.718254; e \approx 2.718282$
99. $S_2 = \frac{4}{3} \approx 1.3333, S_4 = \frac{40}{27} \approx 1.4815, S_8 \approx 1.49977$,
$S_{16} \approx 1.49999997; S = 1.5$
As n increases, the partial sums approach S.
101. $S_1 = 4, S_2 = 3.6, S_3 = 3.64, S_4 = 3.636$,
$S_5 = 3.6364, S_6 = 3.63636; S = \frac{40}{11} = 3.\overline{63}$
As n increases, the partial sums approach S.

CHECKING BASIC CONCEPTS FOR SECTIONS 11.1 AND 11.2 (P. 918)

1. $[0, 7, 1]$ by $[-10, 2, 1]$

3. (a) Arithmetic; $S_{10} = 190$

(b) Geometric; $S_6 = \frac{364}{81} \approx 4.494$

(c) Geometric; $S = \frac{8}{3} \approx 2.667$ **(d)** Geometric; $S = 1$

5. (a) 150   **(b)** 6622   **6.** 30 ft

SECTION 11.3 (PP. 926–928)

1. $2^{10} = 1024$ **3.** $2^5 4^{10} = 33,554,432$

5. $10^3 \cdot 26^3 = 17,576,000$

7. $26^3 \cdot 36^3 = 820,025,856$

9. $3^5 = 243$ **11.** $5^5 = 3125$ **13.** 2

15. 24   **17.** 1,000,000   **19.** $2^{12} = 4096$

21. No, there are 35,152 call letters possible.   **23.** 24

25. 200   **27.** 720   **29.** 3,628,800   **31.** 60   **33.** 8

35. 210   **37.** 600   **39.** 5040   **41.** 24   **43.** 120

45. 210   **47.** 360   **49.** 362,880

51. About 6.39×10^{12} **53.** 6,400,000,000   **55.** 3

57. 20   **59.** 1   **61.** 28   **63.** 190   **65.** 575,757

67. 30   **69.** 50   **71.** 7920   **73.** 2024

75. $P(n, n-1) = \frac{n!}{(n-(n-1))!} = \frac{n!}{1} = n!$ and

$P(n, n) = \frac{n!}{(n-n)!} = \frac{n!}{0!} = \frac{n!}{1} = n!$ For example,

$P(7, 6) = 5040 = P(7, 7)$.

SECTION 11.4 (PP. 933–934)

1. 5   **3.** 1   **5.** 6   **7.** 1   **9.** 10   **11.** 70   **13.** 1

15. 5   **17.** $x^2 + 2xy + y^2$

19. $m^3 + 6m^2 + 12m + 8$

21. $8x^3 - 36x^2 + 54x - 27$

23. $p^6 - 6p^5 q + 15p^4 q^2 - 20p^3 q^3 + 15p^2 q^4 - 6pq^5 + q^6$

25. $8m^3 + 36m^2 n + 54mn^2 + 27n^3$

27. $1 - 4x^2 + 6x^4 - 4x^6 + x^8$

29. $8p^9 - 36p^6 + 54p^3 - 27$

31. $x^2 + 2xy + y^2$

33. $81x^4 + 108x^3 + 54x^2 + 12x + 1$

35. $32 - 80x + 80x^2 - 40x^3 + 10x^4 - x^5$

37. $x^8 + 8x^6 + 24x^4 + 32x^2 + 16$

39. $256x^4 - 768x^3 y + 864x^2 y^2 - 432xy^3 + 81y^4$

41. $m^6 + 6m^5 n + 15m^4 n^2 + 20m^3 n^3 + 15m^2 n^4 + 6mn^5 + n^6$ **43.** $8x^9 - 12x^6 y^2 + 6x^3 y^4 - y^6$

45. $84a^6 b^3$ **47.** $70x^4 y^4$ **49.** $40x^2 y^3$ **51.** $-576xy^5$

CHECKING BASIC CONCEPTS FOR SECTIONS 11.3 AND 11.4 (P. 934)

1. $2^8 = 256$ **3.** $26 \cdot 36^5 = 1,572,120,576$

SECTION 11.5 (PP. 940–941)

1. $3 + 6 + 9 + \cdots + 3n = \frac{3n(n+1)}{2}$

(i) Show that the statement is true for $n = 1$:

$3(1) = \frac{3(1)(2)}{2}$

$3 = 3$

(ii) Assume that S_k is true:

$3 + 6 + 9 + \cdots + 3k = \frac{3k(k+1)}{2}$

Show that S_{k+1} is true:

$3 + 6 + \cdots + 3(k+1) = \frac{3(k+1)(k+2)}{2}$

Add $3(k+1)$ to each side of S_k:

$3 + 6 + 9 + \cdots + 3k + 3(k+1)$

$= \frac{3k(k+1)}{2} + 3(k+1)$

$= \frac{3k(k+1) + 6(k+1)}{2}$

$= \frac{(k+1)(3k+6)}{2}$

$= \frac{3(k+1)(k+2)}{2}$

Since S_k implies S_{k+1}, the statement is true for every positive integer n.

3–13. See the Student's Solutions Manual.

15. 1, 2   **17.** 2, 3, 4

19. $(a^m)^n = a^{mn}$

(i) Show that the statement is true for $n = 1$:

$(a^m)^1 = a^{m \cdot 1}$

$a^m = a^m$

(ii) Assume that S_k is true:

$(a^m)^k = a^{mk}$

Show that S_{k+1} is true:

$(a^m)^{k+1} = a^{m(k+1)}$

Multiply each side of S_k by a^m:

$(a^m)^k \cdot (a^m)^1 = a^{mk} \cdot a^m$

$(a^m)^{k+1} = a^{mk+m}$

$(a^m)^{k+1} = a^{m(k+1)}$

Since S_k implies S_{k+1}, the statement is true for every positive integer n.

21–29. See the Student's Solutions Manual.

31. $P = 3\left(\frac{4}{3}\right)^{n-1}$

33. See the Student's Solutions Manual.

SECTION 11.6 (PP. 950–954)

1. Yes   **3.** No   **5.** Yes   **7.** No   **9.** $\frac{1}{2}$ **11.** $\frac{1}{6}$

13. $\frac{1}{2}$ **15.** $\frac{4}{52} = \frac{1}{13}$ **17.** $\frac{1}{10,000}$ **19. (a)** 0.57 or 57%

(b) 0.33 or 33%   **21.** $\frac{1}{4}$ **23.** $\frac{1}{27}$ **25.** $\frac{6}{36} = \frac{1}{6}$

27. $\frac{625}{1296} \approx 0.482$ **29.** $\frac{1}{270,725}$

31. $\dfrac{\binom{13}{3} \cdot \binom{13}{2}}{\binom{52}{5}} \approx 0.0086$ or a 0.86% chance

33. $\frac{4}{20} = 0.2$

35. (a)

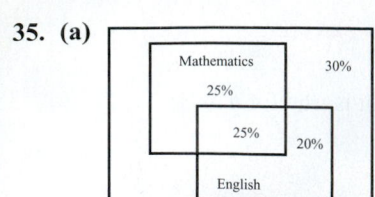

(b) 0.7 or 70%   **(c)** Let M denote the event of needing help with math and E the event of needing help with English. Then $P(M \cup E) = P(M) + P(E) - P(M \cap E) = 0.5 + 0.45 - 0.25 = 0.7$.

37. (a) $\frac{6095}{119,923} \approx 0.051$ **(b)** $\frac{106,232}{119,923} \approx 0.886$

39. $\frac{81}{100,000} = 0.00081$

41. The probability is $\left(\frac{1}{2}\right)^n$. As n increases the probability decreases. This agrees with intuition. The probability of tossing a long string of consecutive heads is small. The longer the string, the lower the chance that it will happen.

43. $\frac{9}{36} = \frac{1}{4}$ **45.** $\frac{12}{52} = \frac{3}{13}$

47. (a) $\frac{3}{16}$ **(b)** $\frac{9}{16}$ **(c)** $\frac{9}{64}$ **(d)** $\frac{3}{64}$

49. (a) 0.09   **(b)** 0.12   **51.** $\frac{1}{1000}$

53. (a) $\frac{22}{50} = 0.44$ **(b)** $\frac{28}{50} = 0.56$ **(c)** $\frac{28}{50} = 0.56$

55. (a) About 0.12 or 12%   **(b)** There is a smaller lifetime risk of bladder cancer when the percentage is under 4%. Higher percentages result in a dramatic increase in risk.

57. $\frac{3}{51} = \frac{1}{17}$ **59.** $\frac{1}{3}$ **61.** $\frac{19}{27}$ **63.** 40%   **65.** $\frac{1}{3}$

67. (a) $\frac{18}{235}$ **(b)** $\frac{7}{18}$ **(c)** $\frac{7}{235}$

69. (a) $\frac{8}{15}$ **(b)** $\frac{7}{15}$ **(c)** $\frac{2}{5}$ **(d)** $\frac{1}{3}$ **(e)** $\frac{1}{15}$

CHECKING BASIC CONCEPTS FOR SECTIONS 11.5 AND 11.6 (P. 954)

1. $4 + 8 + 12 + \cdots + 4n = 2n(n + 1)$
(i) Show that the statement is true for $n = 1$:
$$4(1) = 2(1)(1 + 1)$$
$$4 = 4$$
(ii) Assume that S_k is true:
$$4 + 8 + 12 + \cdots + 4k = 2k(k + 1)$$
Show that S_{k+1} is true:
$$4 + 8 + \cdots + 4(k + 1) = 2(k + 1)(k + 2)$$
Add $4(k + 1)$ to each side of S_k:
$$4 + 8 + 12 + \cdots + 4k + 4(k + 1)$$
$$= 2k(k + 1) + 4(k + 1)$$
$$= 2k^2 + 6k + 4$$
$$= 2(k + 1)(k + 2)$$
Since S_k implies S_{k+1}, the statement is true for every positive integer n.

3. $\frac{1}{16}$ **5.** $\dfrac{\binom{4}{4} \cdot \binom{4}{1}}{\binom{52}{5}} = \dfrac{4}{2,598,960} \approx 0.0000015$

CHAPTER 11: REVIEW EXERCISES (PP. 958–960)

1. $-1, -4, -7, -10$ **3.** $0, 1, 3, 7$ **5.** $5, 3, 1, 2, 4, 6$
7. $0, 1, 4, 13, 40$
9. (a)

n	1	2	3	4	5	6	7	8
a_n	2	4	6	8	10	12	14	16

(b) $[0, 17, 1]$ by $[0, 17, 1]$

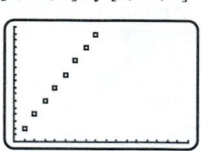

(c) $a_n = 2n$
11. (a)

n	1	2	3	4	5	6	7	8
a_n	81	27	9	3	1	$\frac{1}{3}$	$\frac{1}{9}$	$\frac{1}{27}$

(b) $[0, 10, 1]$ by $[-10, 90, 10]$

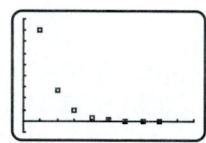

(c) $a_n = 81\left(\frac{1}{3}\right)^{n-1}$
13. $a_n = 4n - 15$ **15.** Arithmetic   **17.** Geometric
19. (a) $4, 3.6, 3.24, 2.916, 2.6244$; geometric
(b) $[0, 6, 1]$ by $[0, 6, 1]$

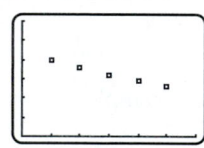

(c) $a_n = 4(0.9)^{n-1}$
21. 65   **23.** 90   **25.** 3280   **27.** 4   **29.** $\frac{2}{9}$
31. $6 + 11 + 16 + 21 + 26$ **33.** $\sum\limits_{k=1}^{6} k^3$ **35.** -1155
37. $0.18 + 0.0018 + 0.000018 + \cdots$
39. $4^{20} \approx 1.1 \times 10^{12}$ **41.** 6,250,000   **43.** 120
45. 120   **47.** 20   **49.** 210
51. $1 + 3 + 5 + \cdots + (2n - 1) = n^2$
(i) Show that the statement is true for $n = 1$:
$$2(1) - 1 = 1^2$$
$$1 = 1$$
(ii) Assume that S_k is true:
$$1 + 3 + 5 + \cdots + (2k - 1) = k^2$$
Show that S_{k+1} is true:
$$1 + 3 + \cdots + (2(k + 1) - 1) = (k + 1)^2$$
Add $2k + 1$ to each side of S_k:
$$1 + 3 + 5 + \cdots + (2k - 1) + (2k + 1)$$
$$= k^2 + 2k + 1$$
$$= (k + 1)^2$$
Since S_k implies S_{k+1}, the statement is true for every positive integer n.

53. $\frac{1}{2}$ **55.** $\frac{91}{120} \approx 0.758$

57. (a) $\frac{27}{60} = 0.45$ **(b)** $\frac{33}{60} = 0.55$ **(c)** $\frac{13}{60} \approx 0.217$

59. Initially, the population density grows slowly, then increases rapidly, and finally levels off near 4,000,000 per acre.

[0, 16, 1] by [0, 5000, 1000]

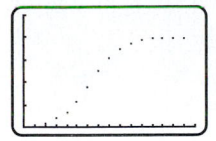

CHAPTER 11 EXTENDED AND DISCOVERY EXERCISES (P. 960)

1. (a) $P = \begin{bmatrix} 1 & 0 & 0 \\ \frac{1}{6} & \frac{2}{3} & \frac{1}{6} \\ 0 & 0 & 1 \end{bmatrix}$

(b) The greatest probabilities lie on the main diagonal: $1, \frac{2}{3}, 1$; this means that a mother cell is most likely to produce a daughter cell like itself. *Answers will vary.*

3. The quantity $\frac{a_1 + a_n}{2}$ represents not only the average of a_1 and a_n, but also the average of the terms $a_1, a_2, a_3, \ldots, a_n$. This is true whether n is odd or even. The total sum is equal to n times the average of the terms.

CHAPTERS 1–11 CUMULATIVE REVIEW EXERCISES (PP. 961–967)

1. 8.8%   **3.** 0.48   **5.** $\sqrt{41}$

7. (a)

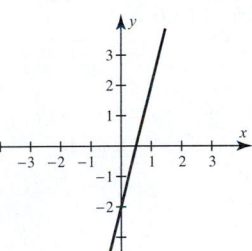

(b)

(c)

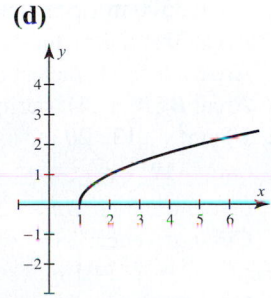

(d)

(e)

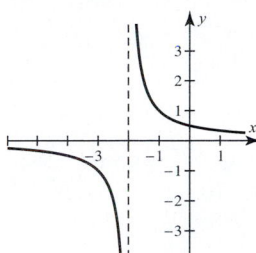

(f)

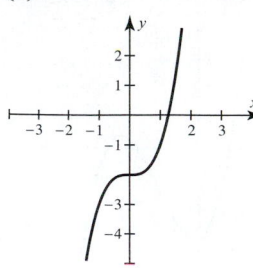

(g)

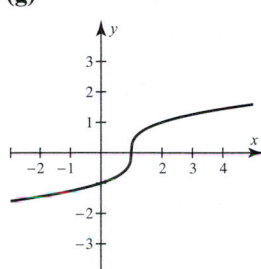

(h)

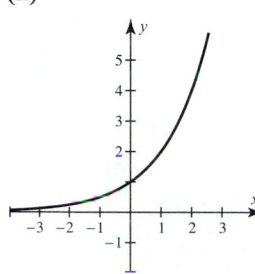

(i)

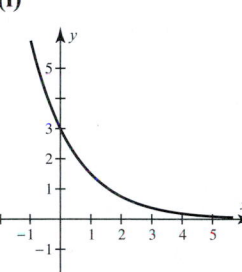

(j)

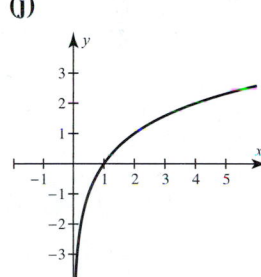

9. (a) $\frac{1}{5}; \frac{1}{a^2 + 2a - 3}$ **(b)** $D = \{x \mid x \neq -2, x \neq 2\}$

11. 7

13. (a) $\frac{3}{4}; -1; \frac{4}{3}$ **(b)** $f(x) = \frac{3}{4}x - 1$ **(c)** $\frac{4}{3}$

15. $y = -\frac{6}{5}x - \frac{8}{5}$ **17.** $x = -2$

19. (a) $\frac{17}{10}$ **(b)** $-\frac{1}{3}, \frac{5}{2}$ **(c)** $\frac{1 \pm \sqrt{13}}{2}$ **(d)** $-2, -1, 2$
(e) $\pm\sqrt{3}, \pm 1$ **(f)** $\frac{17}{3}$ **(g)** $\frac{\ln(28/3)}{2} \approx 1.117$
(h) $10^{3/2} - 1 \approx 30.623$ **(i)** 6   **(j)** $-\frac{4}{3}, 2$

21. (a) 1   **(b)** $\{x \mid x < 1\}$ **(c)** $\{x \mid x \geq 1\}$

23. $-2.8, 1.1$ **25.** $f(x) = 3(x + 4)^2 - 5$

27. (a) $a < 0$ **(b)** $-1, 3$ **(c)** Positive
(d) $f(x) = -\frac{1}{2}(x + 1)(x - 3)$

29. (a) Increasing: $(-\infty, -2] \cup [1, \infty)$; decreasing: $[-2, 1]$
(b) $-3.3, 0, 1.8$ **(c)** $(-2, 2), (1, -0.7)$
(d) Local minimum: -0.7, local maximum: 2

31. Answers may vary.

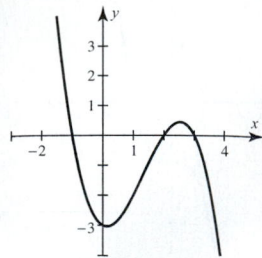

33. $f(x) = 6(x + 2)(x + 1)(x - 1)(x - 2)$
35. $f(x) = 3(x + 1)(x - 3i)(x + 3i)$;
$f(x) = 3x^3 + 3x^2 + 27x + 27$
37. $f(x) = -\frac{1}{2}(x + 2)(x + 1)(x - 1)(x - 2)$
39. $-1 \pm 2i$ **41.** $(x + 1)^{3/5}$; 8
43. **(a)** 3   **(b)** -1 **(c)** 1   **(d)** -4
45. $f^{-1}(x) = -\dfrac{x}{x - 1}$ **47.** $f(x) = 2(3)^x$
49. $C = \frac{1}{2}, a = 2$
51. **(a)** -4 **(b)** $\frac{1}{2}$ **(c)** 4   **(d)** 3
53. $\frac{1}{2}\log(x + 1) - \frac{1}{2}\log y - \frac{1}{2}\log z$ **55.** 2.850
57. $\left\{ x \mid x \neq \dfrac{\pi}{4} + \dfrac{\pi}{2}n, n \text{ is an integer} \right\}$
59. $\alpha = 35°24'48''$; $\beta = 125°24'48''$
61. $\dfrac{3\pi}{4}$ **63.** 1.19
65. $\sin\theta = \frac{1}{2}$, $\cos\theta = -\dfrac{\sqrt{3}}{2}$, $\tan\theta = -\dfrac{1}{\sqrt{3}}$,
$\csc\theta = 2$, $\sec\theta = -\dfrac{2}{\sqrt{3}}$, $\cot\theta = -\sqrt{3}$
67. $\dfrac{\pi}{3}$ **69.** $\tan^2 t$
71. Approximately 1.47, 3.48, and 4.30   **73.** 75 ft^2
75.

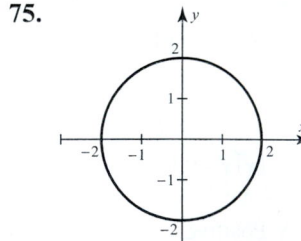

77. $-2\sqrt{2} + 2\sqrt{2}\,i$ and $2\sqrt{2} - 2\sqrt{2}\,i$
79. **(a)** $(-1, 2)$ **(b)** $\{(x, y) \mid 4x - y = -2\}$
(c) Four solutions: $(\pm 3, \pm\sqrt{7})$
(d) $\left\{ (x, y, z) \mid x = \dfrac{5z - 24}{3}, y = \dfrac{z + 6}{3}, \text{ and } z = z \right\}$

81. **(a)**

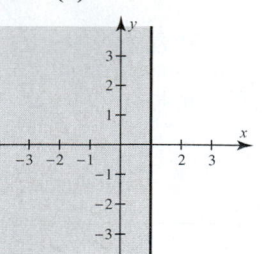

(b)

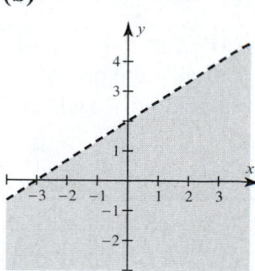

(c)

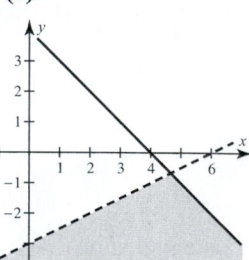

(d)

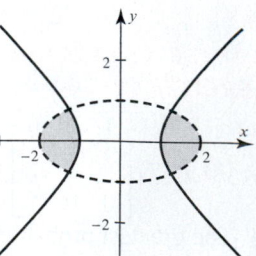

83. $\begin{bmatrix} -2 & -1 \\ -1.5 & -0.5 \end{bmatrix}$ **85.** **(a)** -11 **(b)** 22

87. $y^2 = 3x$ **89.** $\dfrac{y^2}{25} - \dfrac{x^2}{144} = 1$ **91.** $a_n = 4n$
93. **(a)** 950   **(b)** 2046   **(c)** $\frac{2}{9}$ **95.** 175,760,000
97. $16x^4 - 32x^3 + 24x^2 - 8x + 1$ **99.** $\frac{16}{52} = \frac{4}{13}$
101. $\frac{8}{20} = \frac{2}{5}$ **103.** 5.7 in.   **105.** 146.5 mi
107. 0.6 hr at 7 mph; 0.7 hr at 9 mph   **109.** $\frac{20}{9} \approx 2.22$ hr
111. **(a)** $P(t) = 2.65(t - 1980) + 227$ **(b)** 2026
113. $f(x) = 0.5(x - 2)^2 + 6$ **115.** 116.6 ft
117. 40.6°   **119.** 9   **121.** 11 by 13 in.   **123.** 28 ft

CHAPTER R: Reference: Basic Concepts from Algebra and Geometry

SECTION R.1 (PP. R-7–R-9)

1. Area: 105 ft^2; perimeter: 44 ft
3. Area: 3500 m^2; perimeter: 270 m
5. Area: $3xy$; perimeter: $6x + 2y$
7. Area: $2W^2$; perimeter: $6W$
9. Area: $W(W + 5)$; perimeter: $4W + 10$
11. 20 cm^2 **13.** 20 in.^{2} **15.** 3686.5 meters^2
17. $6x^2$ **19.** $\frac{5}{2}z^2$
21. Circumference: $8\pi \approx 25.1$ m; area: $16\pi \approx 50.3$ m^2
23. Circumference: $38\pi \approx 119.4$ in.;
area: $361\pi \approx 1134.1$ in.^{2}
25. Circumference: $4\pi x$; area: $4\pi x^2$
27. $c = 61$ ft; perimeter: 132 ft

29. $c = \sqrt{170} \approx 13.0$ yd;
perimeter: $18 + \sqrt{170} \approx 31.0$ yd
31. $b = \sqrt{189} \approx 13.7$ m;
perimeter: $21 + \sqrt{189} \approx 34.7$ m
33. $a = 1.6$ mi; perimeter: 4.8 mi   **35.** Area: 9 ft^2
37. Area: $\frac{11}{2}\sqrt{104} \approx 56.1$ in.^{2}
39. Volume: 24 ft^3; surface area: 52 ft^2
41. Volume: 216 in.^{3}; surface area: 240 in.^{2}
43. Volume: $6x^3$; surface area: $22x^2$
45. Volume: $6xyz$; surface area: $4xy + 6xz + 12yz$
47. W^3
49. Volume: $36\pi \approx 113.1$ ft^3; surface area: $36\pi \approx 113.1$ ft^2
51. Volume: $\dfrac{500\pi}{3} \approx 523.6$ cm^3;
surface area: $100\pi \approx 314.2$ cm^2
53. Volume: $\dfrac{500\pi}{3} \approx 523.6$ mm^3;
surface area: $100\pi \approx 314.2$ mm^2
55. Volume: $\frac{1}{2}\pi \approx 1.6$ ft^3; side surface area: $2\pi \approx 6.3$ ft^2;
total surface area: $\frac{5}{2}\pi \approx 7.9$ ft^2
57. Volume: $450\pi \approx 1413.7$ in.^{3}; side surface area:
$180\pi \approx 565.5$ in.^{2}; total surface area: $230\pi \approx 722.6$ in.^{2}
59. Volume: $3456\pi \approx 10{,}857.3$ mm^3;
side surface area: $576\pi \approx 1809.6$ mm^2;
total surface area: $864\pi \approx 2714.3$ mm^2
61. $2\pi r^3$
63. Volume: 157.1 cm^3; side surface area: 122.7 cm^2
65. Volume: 12.6 ft^3; side surface area: 22.7 ft^2
67. Volume: 43.4 ft^3; side surface area: 57.2 ft^2
69. $x = \frac{20}{3} \approx 6.7$ **71.** $x = \frac{21}{2} = 10.5$
73. $x = \frac{15}{2} = 7.5$

SECTION R.2 (PP. R-12–R-13)

1. Center: $(0, 0)$; radius: 5
3. Center: $(0, 0)$; radius: $\sqrt{7}$
5. Center: $(2, -3)$; radius: 3
7. Center: $(0, -1)$; radius: 10
9. Center: $(0, 0)$; radius: 2
11. Center: $(0, 0)$; radius: $\sqrt{5}$
13. Center: $(4, -5)$; radius: 4
15. $(x - 1)^2 + (y + 2)^2 = 1$
17. $(x + 2)^2 + (y - 1)^2 = 4$
19. $(x - 3)^2 + (y + 5)^2 = 64$
21. $(x - 3)^2 + y^2 = 49$
23. $x^2 + y^2 = 10$
25. $(x + 2)^2 + (y + 3)^2 = 25$

27. Center: $(0, 0)$; radius: 3

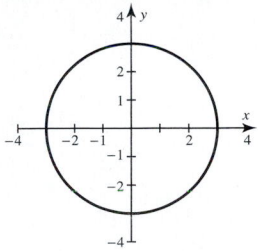

29. Center: $(-1, 1)$; radius: 2

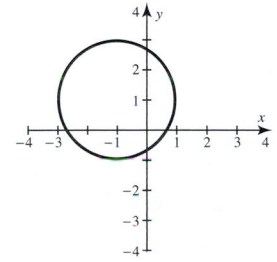

31. Center: $(0, 1)$; radius: 1

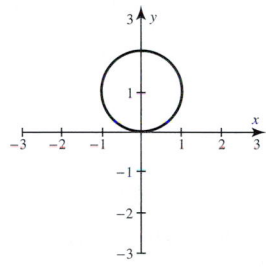

33. Center: $(-9, -5)$; radius: 3

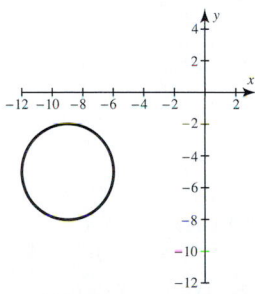

35. Center: $(1, -3)$; radius: $\sqrt{2}$

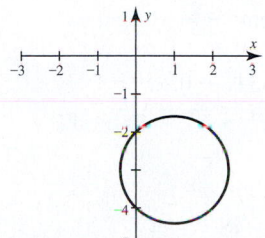

37. Center: $(-3, -6)$; radius: 4

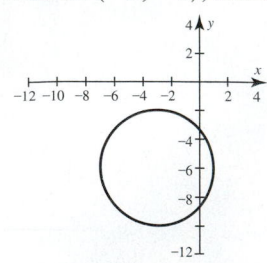

SECTION R.3 (PP. R-18–R-19)

1. Base: 8; exponent: 3   **3.** 7^3
5. No; $2^3 = 8$ and $3^2 = 9$.
7. $\dfrac{1}{7^n}$ **9.** 5^{m-n} **11.** 2^{mk} **13.** 5000   **15.** 2^3
17. 4^4 **19.** 3^0 **21.** 125   **23.** -16 **25.** 1
27. $\dfrac{8}{27}$ **29.** $\dfrac{1}{16}$ **31.** $\dfrac{1}{64}$ **33.** 16   **35.** $\dfrac{64}{27}$ **37.** $\dfrac{8}{9}$
39. $6^{-1} = \dfrac{1}{6}$ **41.** $\dfrac{1}{5^3} = \dfrac{1}{125}$ **43.** $6x^3$
45. $10^8 = 100{,}000{,}000$ **47.** $\dfrac{5}{2}$ **49.** $8a^{-1}b^{-3} = \dfrac{8}{ab^3}$
51. $5^2 = 25$ **53.** $10^6 = 1{,}000{,}000$ **55.** $\dfrac{1}{a^6}$ **57.** $4x^2$
59. $\dfrac{2b}{3a^2}$ **61.** $\dfrac{3y^6}{x^7}$ **63.** $\dfrac{1}{5^3} = \dfrac{1}{125}$ **65.** $\dfrac{1}{y^8}$
67. $4^3y^6 = 64y^6$ **69.** $\dfrac{4^3}{x^3} = \dfrac{64}{x^3}$ **71.** $\dfrac{z^{20}}{2^5 x^5} = \dfrac{z^{20}}{32x^5}$
73. $2ab$ **75.** $\dfrac{t^2}{16}$ **77.** $\dfrac{3a}{2b}$ **79.** $4m^4n^7$ **81.** $\dfrac{rs^3}{5t^5}$
83. $\dfrac{y^6}{9x^4}$ **85.** 1   **87.** $\dfrac{27t^9}{8}$ **89.** $\dfrac{n^3}{m^3}$ **91.** $\dfrac{a^{12}}{81b^4}$
93. $\dfrac{2y^2}{x}$ **95.** $\dfrac{1}{rt^3}$ **97.** y **99.** $\dfrac{125r^{15}}{t^9}$

SECTION R.4 (PP. R-26–R-27)

1. $3x^2$; answers may vary.
3. No, the powers must match for each variable.
5. No, the opposite is $-x^2 - 1$.   **7.** $a^2 - b^2$ **9.** $8x^3$
11. $-3y^7$ **13.** $6x^2 + 8x$ **15.** $3x^2 + 3x$
17. $5x^2 + 13x - 2$ **19.** $10x^2y + 2y$
21. Degree: 2; leading coefficient: 5
23. Degree: 3; leading coefficient: $-\dfrac{2}{5}$
25. Degree: 5; leading coefficient: 1
27. $3x + 12$ **29.** $3x - 2$ **31.** $-0.5x + 1$
33. $-7x^4 - 2x^2 - \dfrac{9}{2}$ **35.** $2z^3 + z^2 + 2z - 4$
37. $-7x^3$ **39.** $-19z^5 + 5z^2 - 3z$
41. $-z^4 + z^2 + 9$ **43.** $3x - 7$
45. $6x^2 - 5x + 5$ **47.** $3x^4 + 4x^2 - 4$
49. $-3x^4 - 3x - 8$ **51.** $5x^2 - 25x$ **53.** $-15x - 5$
55. $5y + 10$ **57.** $-10x - 18$ **59.** $6y^2 - 18y$

61. $-20x + 4y$ **63.** $y^2 - 2y - 35$
65. $-49x^2 + 49x - 12$ **67.** $-2x^2 + 7x - 6$
69. $x^2 - \dfrac{1}{4}x - \dfrac{1}{8}$ **71.** $2x^4 + x^2 - 1$
73. $x^2 - xy - 2y^2$ **75.** $6x^3 - 3x^2 - 3x$
77. $-2x^5 + x^3 - 10x$ **79.** $6x^4 - 12x^3 + 3x^2$
81. $x^3 + 3x^2 - x - 3$ **83.** $10 - 19x + 6x^2$
85. $3x^3 - 2x^2 + 6x - 4$ **87.** $x^2 - 49$
89. $9x^2 - 16$ **91.** $4x^2 - 9y^2$ **93.** $x^2 + 8x + 16$
95. $4x^2 + 4x + 1$ **97.** $x^2 - 2x + 1$
99. $4 - 12x + 9x^2$ **101.** $3x^3 - 3x$ **103.** $4 - 25x^4$

SECTION R.5 (PP. R-38–R-39)

1. $5(2x - 3)$ **3.** $x(2x^2 - 5)$ **5.** $4x(2x^2 - x + 4)$
7. $5x^2(x^2 - 3x + 3)$ **9.** $5x(3x^2 + 2x - 5)$
11. $2r^3(3r^2 - 4r + 6)$ **13.** $8x^2y^2(1 - 3y)$
15. $6mn^2(3 - 2mn)$ **17.** $-2a(2a + b - 3b^2)$
19. $(x + 3)(x^2 + 2)$ **21.** $(3x - 2)(2x^2 + 3)$
23. $(z - 5)(z^2 + 1)$ **25.** $y(y + 2)(y^2 - 5)$
27. $(x^2 + 1)(2x - 3)$ **29.** $(x^3 + 2)(2x - 1)$
31. $(2x - y)(a - 3b)$ **33.** $(x + 2)(x + 5)$
35. $(x + 2)(x + 6)$ **37.** $(z - 6)(z + 7)$
39. $(z + 3)(z + 8)$ **41.** $(4x + 3)(6x - 1)$
43. $(2x + 1)(3x - 2)$ **45.** $(1 - x)(1 + 2x)$
47. $(5 - 2x)(4 + 3x)$ **49.** $x(x - 1)(5x + 6)$
51. $3x(x + 3)(2x + 1)$ **53.** $2(x - 5)(x - 2)$
55. $10t^2(t + 4)(6t - 1)$ **57.** $2m(m + 3)(2m - 1)$
59. $(x - 5)(x + 5)$ **61.** $(2x - 5)(2x + 5)$
63. $4(3x - 5)(3x + 5)$ **65.** $z^2(8 - 5z)(8 + 5z)$
67. $(2x - y)(2x + y)(4x^2 + y^2)$ **69.** Does not factor
71. $(2 - rt)(2 + rt)$ **73.** $(x - 5)(x + 3)$
75. $-(z + 1)(z + 5)$ **77.** $(x + 1)^2$ **79.** $(2x + 5)^2$
81. $(x - 6)^2$ **83.** $z(3z - 1)^2$ **85.** $y(3y + 5)^2$
87. $(2x - 3y)^2$ **89.** $ab(3a - 2)^2$
91. $(x - 1)(x^2 + x + 1)$ **93.** $(y + z)(y^2 - yz + z^2)$
95. $(2x - 3)(4x^2 + 6x + 9)$
97. $x(x + 5)(x^2 - 5x + 25)$
99. $(2r^2 - t)(4r^4 + 2r^2t + t^2)$
101. $10(m^3 - 3n^2)(m^6 + 3m^3n^2 + 9n^4)$
103. $(4x - 5)(4x + 5)$
105. $(x - 4)(x^2 + 4x + 16)$ **107.** $(x + 8)^2$
109. $(x - 8)(5x + 2)$ **111.** $x(x + 2)(x^2 - 2x + 4)$
113. $8(2x + y)(4x^2 - 2xy + y^2)$
115. $(x + 1)(3x - 8)$ **117.** $a(a + 3)(7a - 1)$
119. $(x^2 + 3)(2x - 1)$

SECTION R.6 (PP. R-49–R-50)

1. $2x$ **3.** $x + 5$ **5.** $x + 4$ **7.** $\dfrac{1}{2x - 1}$
9. $-\dfrac{1}{4}$ **11.** $\dfrac{x}{x + 1}$ **13.** $a^2 - ab + b^2$ **15.** $\dfrac{3}{2x}$
17. 1   **19.** $\dfrac{x}{2x - 5}$ **21.** $x(x + 3)$

23. $\dfrac{(x-1)(x+5)}{2(2x-3)}$ **25.** $\dfrac{4}{b^3}$ **27.** $\dfrac{3a(3a+1)}{a+1}$

29. $\dfrac{x^2}{(x-5)(x^2-1)}$ **31.** 36   **33.** $10a^3$

35. $z(z-4)^2$ **37.** $(x-2)(x-3)^2$ **39.** $7(x+1)$

41. $(x+4)(x-4)$ **43.** $2(2x+1)(x-2)$

45. $\dfrac{7}{x+1}$ **47.** $-\dfrac{1}{x+1}$ **49.** $\dfrac{x^2-x-1}{x(x+4)}$

51. $\dfrac{-4x^2+x+2}{x^2}$ **53.** $\dfrac{x^2+5x-34}{(x-5)(x-3)}$

55. $\dfrac{-2(x^2-4x+2)}{(x-5)(x-3)}$ **57.** $\dfrac{x(5x+16)}{(x-3)(x+3)}$ **59.** $-\dfrac{1}{2}$

61. $\dfrac{6x^2-1}{(x-5)(3x-1)}$ **63.** -3 **65.** $-1,\dfrac{1}{3}$ **67.** -1

69. $-2,1$ **71.** $\dfrac{5}{3}$ **73.** $\dfrac{x+1}{x-1}$

75. $\dfrac{x}{3x-20}$ **77.** $\dfrac{2}{3-x}$ **79.** $\dfrac{3(x+1)}{(x+3)(2x-7)}$

81. $\dfrac{4x(x+5)}{(x-5)(2x+5)}$ **83.** $\dfrac{ab}{2(a+b)}$

25. $2x\sqrt[4]{y}$ **27.** $6x$ **29.** $2x^2yz^3$ **31.** $\dfrac{3}{2}$ **33.** $5\sqrt{z}$

35. $\dfrac{1}{a}$ **37.** $10\sqrt{2}$ **39.** $3\sqrt[3]{3}$ **41.** $2\sqrt{2}$

43. $-2\sqrt[5]{2}$ **45.** $2n\sqrt{2n}$ **47.** $2ab^2\sqrt{3b}$

49. $-5xy\sqrt[3]{xy^2}$ **51.** $5\sqrt[3]{5t^2}$ **53.** $\dfrac{3t}{r\sqrt[4]{5r^3}}$

55. $\sqrt[6]{3^5}$ **57.** $2\sqrt[12]{2^5}$ **59.** $x\sqrt[12]{x}$ **61.** $\sqrt[12]{r^{11}t^7}$

63. $9\sqrt{3}$ **65.** $2\sqrt{x}-\sqrt{y}$ **67.** $-5\sqrt[3]{6}$

69. $9\sqrt{7}$ **71.** $-2\sqrt{11}$ **73.** $5\sqrt[3]{2}-\sqrt{2}$

75. $-\sqrt[3]{xy}$ **77.** $3\sqrt{x+2}$ **79.** $\dfrac{71\sqrt{2}}{10}$

81. $2\sqrt{2}$ **83.** $(8x+2)\sqrt{x}$ or $2\sqrt{x}(4x+1)$

85. $(3ab-1)\sqrt[4]{ab}$ **87.** $(n-2)\sqrt[3]{n}$ **89.** 2

91. $x-64$ **93.** $ab-c$ **95.** $x+\sqrt{x}-56$

97. $\dfrac{4\sqrt{3}}{3}$ **99.** $\dfrac{\sqrt{5}}{3}$ **101.** $\dfrac{\sqrt{3b}}{6}$ **103.** $\dfrac{3+\sqrt{2}}{7}$

105. $10-2\sqrt{2}$ **107.** $\sqrt{7}+\sqrt{6}$ **109.** $\dfrac{z+3\sqrt{z}}{z-9}$

111. $\dfrac{a+2\sqrt{ab}+b}{a-b}$

SECTION R.7 (PP. R-56–R-58)

1. $-5,5$ **3.** $-11,11$ **5.** $-\dfrac{4}{5},\dfrac{4}{5}$ **7.** $-3.32,3.32$

9. 12   **11.** 4.80   **13.** $\dfrac{2}{7}$ **15.** $-b$ **17.** 3

19. -2 **21.** $\dfrac{1}{3}$ **23.** b^3 **25.** 3   **27.** -2.24

29. $|z|$ **31.** 3   **33.** -4 **35.** 1.71   **37.** $-x^3$

39. $4x^2$ **41.** 3   **43.** -1.48 **45.** $\sqrt{6}$

47. $\sqrt{xy}$ **49.** $\dfrac{1}{\sqrt[5]{y}}$ **51.** $\sqrt[3]{27^2}$ or $\left(\sqrt[3]{27}\right)^2$; 9

53. $\sqrt[3]{(-1)^4}$ or $\left(\sqrt[3]{-1}\right)^4$; 1   **55.** $\dfrac{1}{\sqrt[3]{8}};\dfrac{1}{2}$

57. $\dfrac{1}{\sqrt[5]{13^3}}$ or $\dfrac{1}{\left(\sqrt[5]{13}\right)^3}$ **59.** 4   **61.** 4   **63.** 2

65. 16   **67.** 2.24   **69.** $\dfrac{2}{3}$ **71.** 1.26

73. $\dfrac{1}{2}$ **75.** $-\dfrac{1}{2}$ **77.** 2   **79.** 2.83   **81.** x^3

83. xy^4 **85.** xy^2 **87.** $\dfrac{y^2}{x}$ **89.** $y^{13/6}$ **91.** $\dfrac{x^4}{9}$

93. $\dfrac{y^3}{x}$ **95.** $y^{1/4}$ **97.** $\dfrac{1}{a^{2/3}}$ **99.** ab^2 **101.** $\dfrac{1}{k^2}$

103. $b^{3/4}$ **105.** $z^{23/12}$ **107.** p^2+p **109.** $x^{5/6}-x$

111. 4   **113.** $|y|$ **115.** $|a+3|$ **117.** $|x-5|$

119. $|x-1|$ **121.** $|y|$ **123.** $|x^3|$ **125.** xy^2

SECTION R.8 (PP. R-66–R-67)

1. 3   **3.** 10   **5.** 4   **7.** $\dfrac{3}{5}$ **9.** $\dfrac{1}{4}$ **11.** $\dfrac{x}{4}$ **13.** 3

15. 3   **17.** -2 **19.** a **21.** $\dfrac{x}{2}$ **23.** $2x^2$

APPENDIX C Exercises (P. AP-33)

1. $\dfrac{5}{3x}+\dfrac{-10}{3(2x+1)}$ **3.** $\dfrac{6}{5(x+2)}+\dfrac{8}{5(2x-1)}$

5. $\dfrac{5}{6(x+5)}+\dfrac{1}{6(x-1)}$

7. $\dfrac{-2}{x+1}+\dfrac{2}{x+2}+\dfrac{4}{(x+2)^2}$ **9.** $\dfrac{4}{x}+\dfrac{4}{1-x}$

11. $\dfrac{15}{x}+\dfrac{-5}{x+1}+\dfrac{-6}{x-1}$ **13.** $1+\dfrac{-2}{x+1}+\dfrac{1}{(x+1)^2}$

15. $x^3-x^2+\dfrac{-1}{3(2x+1)}+\dfrac{2}{3(x+2)}$

17. $\dfrac{1}{9}+\dfrac{-1}{x}+\dfrac{25}{18(3x+2)}+\dfrac{29}{18(3x-2)}$

19. $\dfrac{-3}{5x^2}+\dfrac{3}{5(x^2+5)}$ **21.** $\dfrac{-2}{7(x+4)}+\dfrac{6x-3}{7(3x^2+1)}$

23. $\dfrac{1}{4x}+\dfrac{-8}{19(2x+1)}+\dfrac{-9x-24}{76(3x^2+4)}$

25. $\dfrac{-1}{x}+\dfrac{2x}{2x^2+1}+\dfrac{2x+3}{(2x^2+1)^2}$

27. $\dfrac{-1}{x+2}+\dfrac{3}{(x^2+4)^2}$

29. $5x^2+\dfrac{3}{x}+\dfrac{-1}{x+3}+\dfrac{2}{x-1}$

APPENDIX D Exercises (P. AP-37)

1. circle or ellipse or a point
3. hyperbola or two intersecting lines
5. parabola or one line or two parallel lines
7. 30°   9. 60°   11. 22.5°

13.

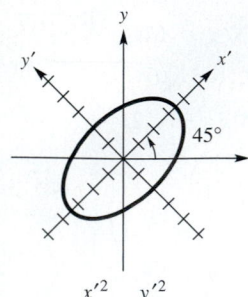

$$\frac{x'^2}{12} + \frac{y'^2}{4} = 1$$

15.

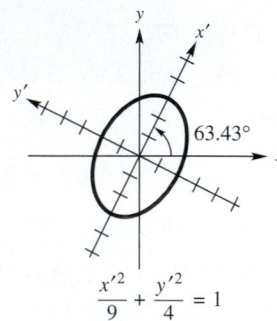

$$\frac{x'^2}{9} + \frac{y'^2}{4} = 1$$

17.

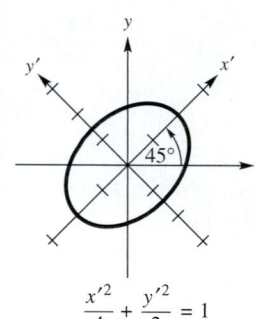

$$\frac{x'^2}{4} + \frac{y'^2}{2} = 1$$

19.

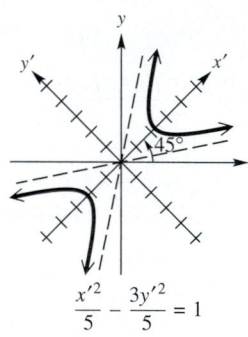

$$\frac{x'^2}{5} - \frac{3y'^2}{5} = 1$$

21.

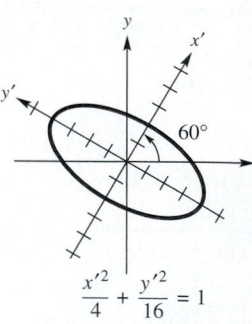

$$\frac{x'^2}{4} + \frac{y'^2}{16} = 1$$

23.

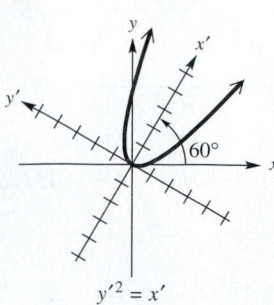

$$y'^2 = x'$$

25.

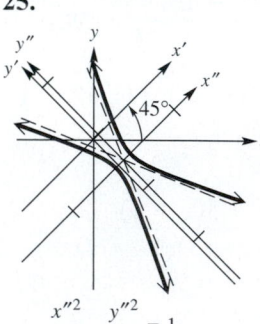

$$\frac{x''^2}{2} - \frac{y''^2}{10} = 1$$

27.

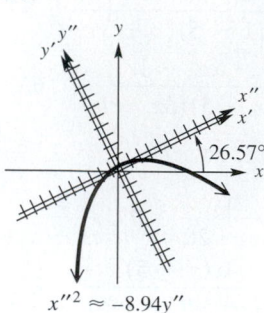

$$x''^2 \approx -8.94 y''$$

29.

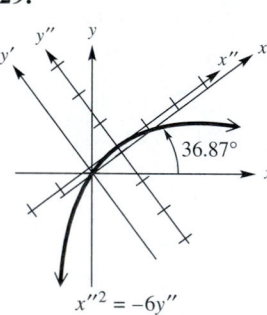

$$x''^2 = -6y''$$

Index of Applications

Index